BIOLOGY

LIFE ON EARTH

WITH PHYSIOLOGY

TENTH EDITION

Teresa Audesirk

UNIVERSITY OF COLORADO DENVER

Gerald Audesirk

UNIVERSITY OF COLORADO DENVER

Bruce E. Byers

UNIVERSITY OF MASSACHUSETTS

PEARSON

Boston Columbus Indianapolis New York San Francisco Upper Saddle River
Amsterdam Cape Town Dubai London Madrid Milan Munich Paris Montréal Toronto
Delhi Mexico City São Paulo Sydney Hong Kong Seoul Singapore Taipei Tokyo

Editor-in-Chief: Beth Wilbur
Executive Director of Development: Deborah Gale
Senior Acquisitions Editor: Star MacKenzie
Developmental Editors: Debbie Hardin & Susan Teahan
Project Editor: Nicole George-O'Brien
Assistant Editor: Frances Sink
Text Permissions Project Manager: Joseph Croscup
Text Permissions Specialist: GEX Publishing Services
Executive Managing Editor: Erin Gregg
Managing Editor: Mike Early
Production Project Manager: Camille Herrera
Production Management & Composition: S4Carlisle Publishing Services
Copyeditor: Lorretta Palagi
Proofreader: Julie Lewis
Design Manager & Cover Designer: Marilyn Perry

Interior Designer: Integra Software, Inc.
Illustrators: Imagineering
Image Lead: Donna Kalal
Photo Permissions Management: Phutu Productions
Photo Researcher: Kristin Piljay
Manufacturing Buyer: Michael Penne
Printer and Binder: Courier Kendallville
Cover Printer: Lehigh-Phoenix
Director of Editorial Content for Media: Tania Mlawer
Senior Media Producer: Laura Tommasi
Associate Media Producer: Daniel Ross
Supplement Editor: Susan Berge
Executive Marketing Manager: Lauren Harp
Cover Photo Credit: John E. Marriot/Getty Images

ABOUT THE COVER

A black bear, fattened by feasting on berries, insects, and nuts during late summer and fall, relaxes in a field of dandelions prior to hibernation. In early summer, this same field would have provided abundant dandelion flowers which bears relish after emerging—lean and hungry—from winter hibernation. Black bears are widely distributed in forested areas throughout North America. Dandelions, although they originated in Eurasia, are now distributed throughout Earth's temperate regions. Each fluffy seedhead contains 150—200 tiny dry fruits, whose parachute-like tufts allow wind to carry the enclosed seeds for long distances. Although the parent plants are dying, each may release 1,000 seeds, many of which will germinate in spring. As you look at this whimsical scene captured by Canadian photographer John E. Marriot, think about how both bear and dandelions are preparing for the harsh winter months ahead.

Library of Congress Cataloging-in-Publication Data
Audesirk, Teresa.
 Biology: life on earth with physiology/Teresa Audesirk, University of Colorado Denver, Gerry Audesirk,
 University of Colorado Denver, Bruce E. Byers, University of Massachusetts.—Tenth edition.
  pages cm
  ISBN-13: 978-0-321-79426-0
  ISBN-10: 0-321-79426-5
 1. Biology. I. Audesirk, Gerald. II. Byers, Bruce E. III. Title.
  QH308.2.A93 2014
  570—dc23
                                    2012030923

www.pearsonhighered.com

ISBN 10: 0-321-79426-5; ISBN 13: 978-0-321-79426-0 (Student edition)
ISBN 10: 0-321-84480-7; ISBN 13: 978-0-321-84480-4 (Books a la Carte)
ISBN 10: 0-321-84479-3; ISBN 13: 978-0-321-84479-8 (Instructor's Review Copy)

BOOKS | À LA CARTE EDITION

AUDESIRK AUDESIRK BYERS

TENTH EDITION

BIOLOGY

Life on Earth

WITH PHYSIOLOGY

YOUR TEXTBOOK—IN A BINDER-READY EDITION!

This unbound, three-hole punched version of your textbook lets you take only what you need to class and incorporate your own notes—all at an affordable price!

ISBN-13: 978-0-321-84480-4
ISBN-10: 0-321-84480-7

ALWAYS LEARNING

PEARSON

About the Authors

TERRY AND GERRY AUDESIRK grew up in New Jersey, where they met as undergraduates, Gerry at Rutgers University and Terry at Bucknell University. After marrying in 1970, they moved to California, where Terry earned her doctorate in marine ecology at the University of Southern

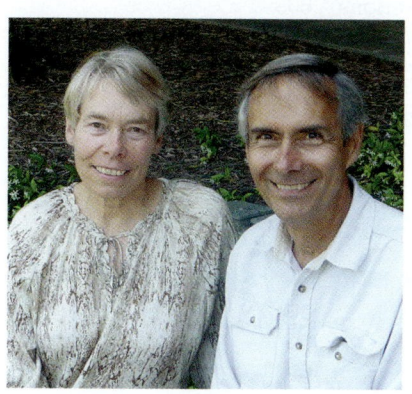

California and Gerry earned his doctorate in neurobiology at the California Institute of Technology. As postdoctoral students at the University of Washington's marine laboratories, they worked together on the neural bases of behavior, using a marine mollusk as a model system.

They are now emeritus professors of biology at the University of Colorado Denver, where they taught introductory biology and neurobiology from 1982 through 2006. In their research, funded primarily by the National Institutes of Health, they investigated the mechanisms by which neurons are harmed by low levels of environmental pollutants and protected by estrogen.

Terry and Gerry are long-time members of many conservation organizations, and share a deep appreciation of nature and of the outdoors. They enjoy hiking in the Rockies, walking and horseback riding near their home outside Steamboat Springs, and singing in the community chorus. They are delighted that their daughter Heather is pursuing solar energy research as a doctoral student at her dad's alma mater. Keeping up with the amazing and endless stream of new discoveries in biology provides them with a continuing source of fascination and stimulation.

BRUCE E. BYERS is a midwesterner transplanted to the hills of western Massachusetts, where he is a professor in the biology department at the University of Massachusetts, Amherst.

He has been a member of the faculty at UMass (where he also completed his doctoral degree) since 1993. Bruce teaches courses in evolution, ornithology, and animal behavior, and does research on the function and evolution of bird vocalizations.

With love to Jack and Lori and in loving memory of Eve and Joe

T. A. & G. A.

To Maija

B. E. B.

Brief Contents

Detailed Contents

Y chromosome

X chromosome

UNIT 5

**Animal Anatomy
and Physiology**   599

UNIT 6
Plant Anatomy and Physiology 831

Preface

THE CASE FOR SCIENTIFIC LITERACY

Climate change, biofuels vs. food and forests, bioengineering, stem cells in medicine, potential flu pandemics, the plight of polar bears and pandas, human population growth and sustainability—our students are bombarded with biology-related information and misinformation as well. The concerns that are sweeping our increasingly connected human societies are very real, urgent, and interrelated. Never have scientifically literate societies been more important to humanity's future. As educators, we feel humbled before this massive challenge. As authors, we feel hopeful that the Tenth Edition of *Biology: Life on Earth* will help lead introductory biology students along paths to understanding.

Scientific literacy requires a foundation of factual knowledge that provides a solid and accurate cognitive framework into which new information can be integrated. But more importantly, it endows people with the mental tools to apply that knowledge to their lives. Scientifically literate citizens can better understand and evaluate information from the press and Internet, distinguish fact from fiction, and make informed choices in both their personal lives and in the political arena.

This Tenth Edition of *Biology: Life on Earth* continues our tradition of:

- Helping instructors present biological information in a way that will foster scientific literacy among their students.
- Helping to inspire students with a sense of wonder about the natural world, fostering an attitude of inquiry and appreciation for the knowledge gained through science.
- Helping students to appreciate how science illuminates life, and to recognize the importance of what they are learning to their future roles in our rapidly changing world.

WHAT'S NEW IN THIS EDITION?

Each new edition gives the authors a fresh opportunity to ponder: "What can we do better?" With extensive help from reviewers, developmental editors, and our coauthors, we've answered this question with the following major changes:

- **NEW Learning Goals and Learning Outcomes** Learning Goals now begin each chapter, alerting students to the major concepts within the chapter and the expectations for student learning. Learning Outcomes at the end of each chapter restate the Learning Goals, serving as a recap of the chapter's coverage. References to the relevant Learning Outcomes are provided for each Fill-in-the-Blank and Review Question. The Learning Outcomes also correlate with those found on MasteringBiology.
- **NEW Check Your Learning self-assessment questions** Following each major chapter section, a list of questions prompts students to evaluate their understanding of the material within the section.

- **INCREASED emphasis on human health** In this edition, we have incorporated more information about health issues—which are both inherently interesting and important to students—into our nonphysiology chapters. For example, Chapter 2 now includes a discussion of radiation therapy in cancer treatment, Chapter 5 now relates membrane fluidity to "Why Your Fingers Fumble and Go Numb in the Cold," and Chapter 9 features a new Case Study, "Body, Heal Thyself."
- **CUTTING-EDGE biotechnology** The tenth edition covers important advances in this exploding and crucial area of biological research. These advances include medical applications of stem cells, the importance of epigenetics, and the use of DNA barcoding for species identification.
- **SEVEN NEW "Case Studies"** Our tenth edition features seven entirely new Case Studies that highlight topics of emerging relevance in today's world. An example is the timely "Unstable Atoms Unleashed," which explores radioactivity through the events of the 2011 tsunami that devastated Japan's Fukushima power plant.
- **MORE flexibility for users** Our former "A Closer Look" essays are now entitled "In Greater Depth" to highlight their usefulness for those instructors who wish to cover complex topics in more detail. These essays are specifically designed to be assigned by the instructor—or not—providing flexibility in the depth of coverage. The In Greater Depth topics include, for example, step-by step reactions of cellular respiration and revised descriptions of C_4 and CAM photosynthesis. This feature will support instructors who want to teach a more challenging course or cover specific topics in greater depth, and it will also satisfy students who wish to investigate a topic more thoroughly.

BIOLOGY: LIFE ON EARTH, TENK EDITION

. . . Is Organized Clearly and Uniformly

Navigational aids help students explore each chapter. An important goal of this organization is to present biology as a hierarchy of closely interrelated concepts rather than as a compendium of independent topics.

- In addition to the new Learning Goals, each chapter opens with "At a Glance" features, which allow users to see how the chapter topics are organized.
- Major sections are introduced as broad questions that stimulate thinking about the material to follow; subheadings are statements that summarize their specific content.
- A "Summary of Key Concepts" section ends each chapter, providing a concise, efficient review of the chapter's major topics.

. . . Engages and Motivates Students

Scientific literacy cannot be imposed on students—they must actively participate in acquiring the necessary information and skills. To be inspired to accomplish this, they must first recognize that biology is about their own lives. For example, we help students acquire a basic understanding and appreciation of how their own bodies function by including information about diet and weight, drugs (including tobacco and alcohol), and birth control options and how each works.

We fervently hope that students who use this text will come to see their world through keener eyes, for example, perceiving forests, fields, and ponds as vibrant and interconnected ecosystems brimming with diverse life-forms rather than as mundane features of their everyday surroundings. If we have done our job, students will also gain the interest, insight, and information they need to look at how humanity has intervened in the natural world. If they ask the question: "Is this activity sustainable?" and then use their new knowledge and critical thinking skills to seek some answers, we can be optimistic about the future.

In support of these goals, the Tenth Edition continues to offer these revised and updated features:

- **Case Studies** Each chapter opens with an attention-grabbing Case Study directly relevant to the chapter topic. Case Studies, including "Flesh-Eating Bacteria" (Chapter 36), "Bionic Ears" (Chapter 39), "Mussels Muscle In" (Chapter 27), and "Some Like it Hot—And Stinky!" (Chapter 44), are based on news events, personal interest stories, or particularly fascinating biological topics. Discussion of each Case Study continues throughout each chapter in boxed features that relate the topic to appropriate sections of the text. The "Case Study Revisited" at each chapter's end explores the topic further in light of what the chapter has covered.
- **Boxed Essays** Four categories of essays enliven this text. "Earth Watch" essays explore pressing environmental issues; "Health Watch" essays cover important or intriguing medical topics, and "Science in Action" essays explain how scientific knowledge is acquired. Although the main body of the text always provides a clear, general explanation of each topic, our "In Greater Depth" essays make this text versatile for different levels of instruction.
- **Engaging "Have You Ever Wondered" Questions and "Links to Everyday Life" Boxes** These short features explore questions that students might ask their instructors and illustrate the real-world applications of the text material to the students' lives.
- **BioEthics Questions** Many topics explored in the text are controversial and have ethical implications for human society. These include human cloning, genetically modified crops, and human impacts on other species. The BioEthics icon BioEthics alerts students and instructors to the possibility for further discussion and investigation.
- **Critical Thinking Questions** At the end of each "Case Study Revisited" section, in selected figure captions, and in the "Applying the Concepts" section at the end of each chapter, our critical thinking questions stimulate students to think about science rather than to simply memorize facts.
- **End-of-Chapter Questions** The questions that conclude each chapter allow students to review the material in different formats—fill-in-the-blank, essay, and critical thinking questions—that help them to study and test what they have learned. Additional critical thinking questions at the end of each Case Study and in selected figure captions stimulate students to think about and apply science to possible real-life situations. Answers to the fill-in-the-blank and figure caption questions are included in the back of the book. Answers or hints for the essay and critical thinking questions are included on the MasteringBiology Web site. Multiple choice type questions can be found in the reading quizzes and practice quizzes in MasteringBiology, as well as in the Instructor Test Bank.
- **Key Terms and a Complete Glossary** Boldfaced key terms are defined clearly within the text as they are introduced. These terms are also listed at the end of each chapter, providing users with a quick reference to the chapter's important vocabulary. The glossary, carefully written by the authors, provides exceptionally complete definitions for all key terms, as well as for many other important biological terms.

. . . Is a Comprehensive Learning Package

The Tenth Edition of *Biology: Life on Earth* is a complete learning package, providing updated and innovative teaching aids for instructors and learning aids for students.

Chapter-by-Chapter Summary of Important Changes

Following the revision of chapters in response to reviews by instructors and experts, the text and artwork were carefully reviewed by each of the other two authors and a developmental editor. The coauthors provided valuable insights to one another, integrating the chapters more thoroughly, improving consistency between chapters, and explaining complex concepts more clearly. Our developmental editors brought trained eyes for order and detail to our work, helping us make the writing even more student friendly. Following this intense scrutiny, each initial revision underwent a second, sometimes extensive revision. Specific changes include the following:

- **Chapter 1: An Introduction to Life on Earth** has undergone a major reorganization and now introduces the concept of life and how it evolved before delving into how scientists study life. A new "Links to Everyday Life: A Solar-Powered Big Mac?" explains how sunlight ultimately provides the calories in all the foods we eat. Another new "Links to Everyday Life: When Your Car Won't Start" applies the scientific method to a familiar problem. Many new photos illustrate this chapter. Figure 1-11 illustrating the tree of life has been revised for consistency with later chapters.

Unit 1: The Life of the Cell

- **Chapter 2: Atoms, Molecules, and Life** includes a timely new Case Study that explores radioactivity through the events of the 2011 tsunami that devastated Japan's Fukushima power plant. Radioactivity is further highlighted in the new "Science in Action: Radioactive Medicine" essay. An explanation of oxidative stress, including identification of some diseases caused by it, is now offered in the Health Watch essay.

- **Chapter 3: Biological Molecules** includes new art and text about the structure of silk protein. The methyl group was added to Table 3-1; adding this important functional group to DNA is described in Chapter 12 as a major epigenetic modification that alters gene expression. "What Makes Hair Curl?" is now revised, somewhat simplified, and presented as a "Links to Everyday Life" feature.

- **Chapter 4: Cell Structure and Function** has a Case Study Revisited section that discusses a cutting-edge product that promises to provide a revolutionary way to regrow skin on burn victims.

- **Chapter 5: Cell Membrane Structure and Function** now provides more information about how membrane fluidity influences membrane function, including a new Have You Ever Wondered feature that relates membrane fluidity to why fingers get numb in cold weather. The In Greater Depth essay now includes the relationship between phospholipid tail kinks and membrane fluidity. The section on membrane proteins has been substantially revised and reorganized. The figures illustrating cell attachment and communication have been combined and are now easier to interpret.

- **Chapter 6: Energy Flow in the Life of a Cell** includes a new Earth Watch essay, "Step on the Brakes and Recharge Your Battery," that uses principles of energy conversion to explain regenerative braking in some modern, fuel-efficient cars. Section 6.5 concerning enzyme regulation has undergone reorganization and rewriting. The Health Watch essay on lactose intolerance has been revised.

- **Chapter 7: Capturing Solar Energy: Photosynthesis** has had its Case Study featuring the asteroid impact that probably ended the reign of the dinosaurs extensively revised. The explanations of C4 and CAM photosynthesis also have been revised and placed in an In Greater Depth essay.

- **Chapter 8: Harvesting Energy: Glycolysis and Cellular Respiration** includes several figures illustrating chemical reactions, which have been changed to show all the carbon atoms. Talking boxes now improve the clarity of the two In Greater Depth essays explaining the details of glycolysis and the mitochondrial matrix reactions of cellular respiration. A new section explains how cellular respiration can obtain energy from a variety of molecules in addition to glucose.

Unit 2: Inheritance

- **Chapter 9: The Continuity of Life: Cellular Reproduction** opens with a new Case Study showing how physicians can use the body's own methods of controlling cell division to stimulate faster and more complete healing of injuries. The Health Watch describing how faulty control of cell division causes cancer has been streamlined and simplified. The discussion of cloning has been moved to Chapter 13, Biotechnology.

- **Chapter 10: Patterns of Inheritance** has been reorganized for improved clarity. A new Links to Everyday Life explores the impact of the sickle-cell allele on the health and performance of athletes who carry the allele.

- **Chapter 12: Gene Expression and Regulation** now describes epigenetic control of gene expression. Genomic imprinting and the possibility that epigenetic regulation may be inherited across generations are described in a new Health Watch.

- **Chapter 13: Biotechnology** has been updated with current information, including recently developed methods for diagnosing both inherited diseases and infections; a new gene therapy that may be able to alleviate and possibly cure AIDS; and new successes in gene therapy for severe combined immune deficiency. The discussion of cloning plants and animals has been moved from Chapter 9 to this chapter. We describe the use of DNA barcoding in species identification, and how barcoding might be used to reduce trafficking in endangered species. We have substantially revised the description of how genetically modified organisms are made.

Unit 3: Evolution and Diversity of Life

- **Chapter 14: Principles of Evolution** includes a new segment on the evolution of herbicide resistance in "superweeds." The Earth Watch on how people promote high-speed evolution includes a new example (color changes in tawny owls) of one impact of climate change on evolution.

- **Chapter 15: How Populations Evolve** includes an expanded explanation of the "gene pool," a crucial concept in evolutionary genetics. The chapter-opening Case Study now includes the increase of multidrug resistance in gonorrhea in the United States.

- **Chapter 16: The Origin of Species** presents a new "Science in Action: Seeking the Secrets of the Sea" that discusses the recently completed Census of Marine Life. A new "Earth Watch: Why Preserve Biodiversity?" explores reasons to be concerned about loss of species. New illustrated examples show how appearance can vary within a species and how different species can look alike.

- **Chapter 17: The History of Life** includes updated information on fossils found since the last edition. All dates have been updated to reflect the latest Geological Society revisions of the geological time scale. The human evolution section now includes segments on a new hypothesis on the evolution of large brains and recent findings about a near-complete skeleton of *Ardipithicus*, a full Neanderthal genome sequence, genetic evidence of a new hominin species (the Denisovans), and genomic evidence of recent rapid human evolution. The chapter's Case Study has been updated to include an emerging hypothesis about the evolutionary origin of *H. floresiensis,* and human evolutionary trees have been updated with renderings of *Ardipithicus, Sahelanthropus,* and *H. floresiensis.*

- **Chapter 18: Systematics: Seeking Order Amid Diversity** includes a new account of current views on taxonomic ranks and a new explanation of the reasons for their declining use. Included is a more explicit explanation of the rank-naming conventions that we follow in the subsequent diversity chapters. Throughout, all phylogenetic trees have been converted to a horizontal orientation, which increases readability. The eukaryote evolutionary tree figure has been revised to add previously omitted taxa and to more clearly identify the major protist clades.

- **Chapter 19: The Diversity of Prokaryotes and Viruses** contains a new Case Study on food contamination and food safety. The chapter includes substantial new discussions of lateral gene transfer and quorum sensing in bacteria. A completely new "Health Watch: Is Your Body's Ecosystem Healthy?" describes the "microbiome" (the diverse population of prokaryotes that live in and on the human body) and its effect on health. The segment on bioremediation now includes discussion of the *Deep Water Horizon* oil spill in 2010.

- **Chapter 20: The Diversity of Protists** includes a new "Links to Everyday Life: Have You Eaten Your Protists?" that describes how products derived from protists are used in both everyday foods and haute cuisine. The chapter also includes a new segment about the use of algae in biofuels.

- **Chapter 21: The Diversity of Plants** includes a new essay, "Have You Ever Wondered: Which Plants Provide Us with the Most Food?" about the relatively small number of species cultivated for human consumption. In addition, the order of topics in the chapter has been revised for improved consistency with other diversity chapters. The evolutionary tree early in the chapter has been expanded to include all of the major taxonomic groups mentioned in the text.

- **Chapter 22: The Diversity of Fungi** offers an updated discussion of fungus classification, including an accompanying evolutionary tree that reflects the latest findings, in particular, the addition of blastoclades and rumen fungi and the removal of zygomycetes as a formal group. A segment on Ug99 (a stem rust that has emerged as a threat to the human food supply) has been added to the section on human impacts. A new figure (Fig. 22-4) depicts a generalized fungal life cycle.

- **Chapter 23: Animal Diversity I: Invertebrates** now begins with the "Case Study: Physicians' Assistants," describing the use of invertebrate animals in medicine. The chapter's former Case Study on the search for the giant squid has been converted into a Science in Action feature. The section on crustaceans now includes a discussion on Antarctic krill and shrimp as a food source. The chapter features a new "Have You Ever Wondered: Why Spiders Don't Stick to Their Own Webs?" New figures show a trochophore larva, a lophophore, and how embryonic development of the coelom differs between protostomes and deuterostomes.

- **Chapter 24: Animal Diversity II: Vertebrates** offers an expanded section on marsupials that discusses Tasmanian devils and the threat to their survival posed by a transmissible cancer. The Earth Watch on amphibian extinction is updated. The chapter includes more detailed accounts of reproduction in tunicates, lancelets, and lampreys, as well as expanded coverage of placental mammals.

Unit 4: Behavior and Ecology

- **Chapter 25: Animal Behavior** includes an update in the human pheromone section to incorporate new findings that chemicals in sweat could indicate fear and stress.

- **Chapter 26: Population Growth and Regulation** has been updated with current statistics and figures related to the growth of the human population. Section 26.1 has been extensively reorganized and rewritten. A new Have You Ever Wondered? feature addresses human biotic potential. The Health Watch essay about algal blooms has been extensively revised. The In Greater Depth essay concerning logistic growth now features a specific example to increase understanding.

- **Chapter 27: Community Interactions** has had the Case Study about invasive quagga and zebra mussels revised and updated. Section 27.5, which explains keystone species, has undergone extensive revision.

- **Chapter 28: Energy Flow and Nutrient Cycling in Ecosystems** includes an expanded and updated discussion on the effects of human interference with nutrient cycling. This discussion provides more evidence and examples showing that increasing greenhouse gases cause global climate change, which in turn produces numerous impacts on Earth and its inhabitants. A new Earth Watch describes proposed geoengineering approaches for reducing climate change.

- **Chapter 29: Earth's Diverse Ecosystems** provides a reorganized discussion of how the physical properties of Earth determine the distribution of life in both terrestrial and aquatic habitats. A new "Have You Ever Wondered: If People Can Re-Create Ancient Biomes?" (à la Jurassic Park) describes attempts to re-create a "mammoth steppe" in Siberia.

- **Chapter 30: Conserving Earth's Biodiversity** includes a revised description of ecosystem services that follows the *Millennium Ecosystem Assessment*'s categories. The data on endangered species and habitat destruction have been updated to include recent findings. A new discussion on conservation easements as a way to promote biosphere reserves was added. A new "Earth Watch: Whales—The Biggest Keystones of All?" discusses how whales may help to recycle nutrients from deep waters to the surface, thereby stimulating productivity in the open oceans.

Unit 5: Animal Anatomy and Physiology

- **Chapter 31: Homeostasis and the Organization of the Animal Body** has a new "Earth Watch: Positive Feedback in the Arctic" that explains how positive feedback, originally discussed in the context of animal physiology, also applies to the entire planet, enhancing the loss of the Arctic ice cap as Earth's climate warms. The chapter includes many new and improved images of cell and tissue types.

- **Chapter 32: Circulation** features updated figures. In particular, Figure 32-1 showing open and closed circulatory systems, has been completely redrawn for better clarity. Figure 32-2b showing the three-chambered heart has been redrawn for greater accuracy. All the organs of the lymphatic system are now included in the text and in Fig. 32-18.

- **Chapter 33: Respiration** opens with an all-new Case Study that uses climbing Mt. Kilimanjaro to highlight the effects of altitude on the respiratory system. The In Greater Depth essay on gills and countercurrents has been revised and shortened to be more easily understood, and the Health Watch essay about smoking has been updated and considerably revised.

- **Chapter 34: Nutrition and Digestion** provides a Case Study that has been significantly revised to include statistics on anorexia in males and the story of a French model who died of anorexia in 2010 at the age of 28. Figure 34-9 showing adaptations of omnivore, carnivore, and herbivore teeth has been significantly upgraded and revised. A new section describes how the hormones leptin and ghrelin influence hunger. A new "Have You Ever Wondered: What's the Scoop on Bacteria in Poop?" explains the role and importance of bacteria in feces.

- **Chapter 35: The Urinary System** has been modified considerably in many places to make this complex topic more understandable and interesting. Figure 35-1 comparing invertebrate systems, Figure 35-6 showing urine formation and concentration, and Figure E35-2 showing the details of urine formation in the nephron have all been revised. The Health Watch on treatments for kidney failure has been updated to reflect new developments.

- **Chapter 36: Defenses Against Disease** now provides simplified coverage of antibody diversity in the main chapter text; a more comprehensive description of this complex process is reserved for the newly developed "In Greater Depth: How Can the Immune System Recognize So Many Different Antigens?" essay. The chapter now also includes brief descriptions of the locations and roles of the five types of antibodies. The Health Watch essay on exotic flu viruses has been updated and expanded to include discussion of 2011 research in which scientists in the United States and the Netherlands designed avian flu viruses that readily spread among mammals.

- **Chapter 37: Chemical Control of the Animal Body: The Endocrine System** begins with a new "Case Study: Life-Changing Hormones" that describes the hormonal control of development and metamorphosis in butterflies. The chapter describes the control of metamorphosis in lampreys and salmon by thyroxine. The differences between type 1 and type 2 diabetes are explained. We also describe new hypotheses about the possible roles of environmental factors in melatonin secretion and puberty in children. An updated Health Watch essay explains that in some instances solutions to endocrine disruption in aquatic ecosystems can be as straightforward as upgrading sewage treatment systems.

- **Chapter 38: The Nervous System** now integrates the general principles of nervous system organization and function into a single section. We describe the crossover of ascending and descending tracts and the functional consequences of this anatomical arrangement. An updated Science in Action essay describes potential new applications of functional magnetic resonance imaging in investigating how the brain processes images and also the possible use of this type of imaging for lie detection.

- **Chapter 39: The Senses** is now completely organized according to the stimulus that is perceived; hence, gravity and movement are covered in a separate section. The anatomy of the various senses, from free nerve endings to complex sense organs, is more clearly explained. Each receptor type is now defined within the text and in the glossary.

- **Chapter 40: Action and Support: The Muscles and Skeleton** has been reorganized. Skeletal muscle structure and function are now described first, followed by a brief comparison of skeletal, cardiac, and smooth muscles. Then we discuss how the coordinated actions of muscles and skeletons provide movement of both invertebrate and vertebrate bodies. The chapter includes a new "Have You Ever Wondered: If Exercise Affects Your Skeleton?" feature.

- **Chapter 41: Animal Reproduction** begins with a revised Case Study on breeding endangered rhinos that is integrated more closely with the chapter material, particularly with the discussions of hormonal control of ovulation and techniques of assisted reproduction. The follow-up Case Study Continued and Revisited sections ask the students to apply what they have learned about mammalian reproduction to the specific case of breeding rhinos in zoos. The complex hormonal control of the menstrual cycle has now been converted to an In Greater Depth essay; a much more simple description remains in the main text. The description and table on controlling fertility have been updated. Finally, the real answer to the question "How do porcupines mate?" is sure to appeal to students.

- **Chapter 42: Animal Development** begins with a new Case Study on the relationship between development and regeneration and the prospects for therapeutic regeneration in humans. The Health Watch essay on stem cells as treatments for disease has been updated with recent findings and applications, including successes such as tracheal transplants.

Unit 6: Plant Anatomy and Physiology

- **Chapter 43: Plant Anatomy and Nutrient Transport**
  uses a Case Study Continued box to explain the evolutionary advantage of leaf abscission in deciduous trees. The chapter includes new artwork (Fig. 43-3) illustrating the locations of various types of meristem tissue and their relationship to primary and secondary growth. Finally, we briefly explain the crucial importance of the plasma membrane of sieve-tube elements for the operation of the pressure-flow mechanism of fluid flow in phloem.
- **Chapter 44: Plant Reproduction and Development**
  includes a new introduction to plant sexual reproduction, including art depicting the essentials of the alternation of generations life cycle.
- **Chapter 45: Plant Responses to the Environment**
  features an updated Earth Watch essay on seed germination stimulated by smoke form fires that includes recent discoveries about how smoke affects hormones in plant seeds. A new Table 45-2, titled "Light and Phytochrome Activity," has been added for greater ease in understanding the relationships between the phytochromes. Changes have been made to improve precision and clarity in two figures, the first illustrating auxin-stimulated gravitropism, and the second illustrating plant communication.

ACKNOWLEDGMENTS

Biology: Life on Earth enters its 10th edition invigorated by the oversight of the excellent team at Benjamin Cummings. Beth Wilbur, our Editor-in-Chief, continues to oversee the huge enterprise with the warmth and competence that makes her such an excellent leader. Deborah Gale, Executive Director of Development, again coordinated this complex and multifaceted endeavor. Senior Acquisitions Editor Star MacKenzie did a great job of helping us form a revision plan that even further expands the text's appeal and its ability to convey fascinating information in a user-friendly manner. She listened and responded helpfully to our questions and suggestions—all while traveling extensively to share her enthusiasm for the text and its ancillary resources with educators across the country. Nicole George-OBrien has worked closely with us from the start. As Project Editor, she has done a marvelous job of keeping everything—especially the authors—on track and on schedule, not to mention helping us through the complexities of a rigorously upgraded permissions process. Her unfailingly kind and sympathetic requests and reminders made this huge project as painless as possible. As we worked with artists and developmental editors, Nicole was also a constant, behind-the-scenes presence, making sure all went smoothly. Our developmental editors, Susan Teahan and Debbie Hardin, carefully reviewed every word of the manuscripts, making sure the sometimes extensive revisions and rearrangements flowed smoothly into the existing text. Their attention to detail and numerous, thoughtful, gently worded suggestions have contributed significantly to the text's organization and clarity. Lorretta Palagi, our outstanding copyeditor, not only negotiated the intricacies of grammar and formatting, but also caught things that we had all overlooked. Susan, Debbie, and Lorretta also looked carefully at the art, checking each piece for consistency with the text and helping us with instructions to the artists. The book boasts a large number of excellent new photos, tracked down with skill and persistence by Kristin Piljay. Kristin was always cheerfully responsive to our requests for still more photos when nothing in the first batch would do.

We are grateful to Imagineering Media Services, under the direction of Project Manager Alicia Elliot, for deciphering our art instructions and patiently making new adjustments to already outstanding figures. We owe our beautifully redesigned text to Integra Software Inc. and whimsical new cover to Marilyn Perry.

The production of this text would not have been possible without the considerable efforts of Camille Herrera, Production Project Manager, and Michael Early, Managing Editor. Roxanne Klaas, Project Editor, of S4Carlisle Publishing Services, brought the art, photos, and manuscript together into a seamless and beautiful whole, graciously handling last-minute changes. We thank Lauren Harp, Executive Marketing Manager, for making sure the finished product reached your desk.

In his role as Manufacturing Buyer, Michael Penne's expertise has served us well. The ancillaries are an endeavor fully as important as the text itself. Susan Berge, Sylvia Rebert, and Frances Sink skillfully coordinated the enormous effort of producing a truly outstanding package that complements and supports the text, while David Chavez took the lead on the Instructor's Resource CD-ROM. Finally, thanks to Laura Tommasi and Daniel Ross for their efforts in developing the outstanding MasteringBiology Web site that accompanies this text.

We are extremely fortunate to be working with the Pearson Benjamin Cummings team. This Tenth Edition of **Biology: Life on Earth** reflects their exceptional abilities and dedication.

With gratitude,

TERRY AUDESIRK, GERRY AUDESIRK,
AND BRUCE BYERS

Tenth Edition Reviewers

Kammy Algiers,
  Ventura College
Erin Baumgartner,
  Western Oregon University
Bruno Borsari,
  Winona State University
Jamie Burchill,
  Troy University
Anne Casper,
  Eastern Michigan University
Mary Colavito,
  Santa Monica College
Kimberly Demnicki,
  Thomas Nelson Community College

Gerald Farr,
  Texas State University
Sandra Gibbons,
  Moraine Valley Community College
Mary Rose Grant,
  St. Louis University
Anjali Gray,
  Lourdes College
Ida Greidanus,
  Passaic Community College
Rebecca Hare,
  Cleveland County Community College

Harriette Howard-Lee Block,
Prairie View A&M University
Adam Hrincevich,
Louisiana State University
Kesmic Jackson,
Georgia State University
A. J. Karpoff,
University of Louisville
Michael Koban,
Morgan State University
Mary Lipscomb,
*Virginia Polytechnic Institute
and State University*
Fordyce Lux,
*Metropolitan State College of
Denver*
Liz Nash,
*California State University,
Long Beach*
David Rosen,
Lee College
Jack Shurley,
Idaho State University

Bill Simcik,
Lonestar College
Anna Bess Sorin,
University of Memphis
Anthony Stancampiano,
Oklahoma City University
Theresa Stanley,
Gordon College
Mary-Pat Stein,
*California State University,
Northridge*
Julienne Thomas-Hall,
Kennedy King College
Lisa Weasel,
Portland State University
Janice Webster,
Ivy Tech Community College
Marty Zahn,
*Thomas Nelson Community
College*

Previous Edition Reviewers

Mike Aaron,
*Shelton State Community
College*
W. Sylvester Allred,
Northern Arizona University
Judith Keller Amand,
*Delaware County Community
College*
William Anderson,
*Abraham Baldwin Agriculture
College*
Steve Arch,
Reed College
George C. Argyros,
Northeastern University
Kerri Lynn Armstrong,
*Community College of
Philadelphia*
Ana Arnizaut-Vilella,
*Mississippi University for
Women*
Dan Aruscavage,
*State University of New York,
Potsdam*
G. D. Aumann,
University of Houston
Vernon Avila,
San Diego State University
J. Wesley Bahorik,
*Kutztown University of
Pennsylvania*
Peter S. Baletsa,
Northwestern University
Isaac Barjis
*New York City College of
Technology*

John Barone,
Columbus State University
Bill Barstow,
University of Georgia–Athens
Mike Barton,
Centre College
Michael C. Bell,
Richland College
Colleen Belk,
*University of Minnesota,
Duluth*
Robert Benard,
*American International
College*
Heather Bennett,
Illinois College
Gerald Bergtrom,
University of Wisconsin
Arlene Billock,
*University of Southwestern
Louisiana*
Brenda C. Blackwelder,
*Central Piedmont Community
College*
Melissa Blamires,
*Salt Lake Community
College*
Karen E. Bledsoe,
Western Oregon University
Raymond Bower,
University of Arkansas
Robert Boyd,
Auburn University
Michael Boyle,
*Seattle Central Community
College*

Marilyn Brady,
*Centennial College of Applied
Arts and Technology*
David Brown,
Marietta College
Virginia Buckner,
*Johnson County Community
College*
Arthur L. Buikema, Jr.,
Virginia Polytechnic Institute
Diep Burbridge,
Long Beach City College
J. Gregory Burg,
University of Kansas
William F. Burke,
University of Hawaii
Robert Burkholter,
Louisiana State University
Matthew R. Burnham,
Jones County Junior College
Kathleen Burt-Utley,
University of New Orleans
Linda Butler,
University of Texas–Austin
W. Barkley Butler,
*Indiana University of
Pennsylvania*
Jerry Button,
Portland Community College
Bruce E. Byers,
*University of Massachusetts–
Amherst*
Sara Chambers,
Long Island University
Judy A. Chappell,
*Luzerne County Community
College*
Nora L. Chee,
Chaminade University
Joseph P. Chinnici,
*Virginia Commonwealth
University*
Dan Chiras,
University of Colorado–Denver
Nicole A. Cintas,
*Northern Virginia Community
College*
Bob Coburn,
Middlesex Community College
Joseph Coelho,
Culver Stockton College
Martin Cohen,
University of Hartford
Jay L. Comeaux,
Louisiana State University
Walter J. Conley,
*State University of New York
at Potsdam*
Mary U. Connell,
Appalachian State University
Art Conway,
Randolph-Macon College

Jerry Cook,
Sam Houston State University
Sharon A. Coolican,
Cayuga Community College
Clifton Cooper,
*Linn-Benton Community
College*
Joyce Corban,
Wright State University
Brian E. Corner,
Augsburg College
Ethel Cornforth,
San Jacinto College–South
David J. Cotter,
Georgia College
Lee Couch,
*Albuquerque Technical
Vocational Institute*
Donald C. Cox,
Miami University of Ohio
Patricia B. Cox,
University of Tennessee
Peter Crowcroft,
University of Texas–Austin
Carol Crowder,
*North Harris Montgomery
College*
Mitchell B. Cruzan,
Portland State University
Donald E. Culwell,
University of Central Arkansas
Peter Cumbie,
Winthrop University
Robert A. Cunningham,
*Erie Community College,
North*
Karen Dalton,
*Community College of
Baltimore County–Catonsville
Campus*
Lydia Daniels,
University of Pittsburgh
David H. Davis,
*Asheville-Buncombe Technical
Community College*
Jerry Davis,
*University of Wisconsin,
LaCrosse*
Douglas M. Deardon,
University of Minnesota
Lewis Deaton,
*University of Louisiana–
Lafayette*
Lewis Deaton,
*University of Southwestern
Louisiana*
Fred Delcomyn,
University of Illinois–Urbana
Joe Demasi,
Massachusetts College
David M. Demers,
University of Hartford

Lorren Denney,
Southwest Missouri State University

Katherine J. Denniston,
Towson State University

Charles F. Denny,
University of South Carolina–Sumter

Jean DeSaix,
University of North Carolina–Chapel Hill

Ed DeWalt,
Louisiana State University

Daniel F. Doak,
University of California–Santa Cruz

Christy Donmoyer,
Winthrop University

Matthew M. Douglas,
University of Kansas

Ronald J. Downey,
Ohio University

Ernest Dubrul,
University of Toledo

Michael Dufresne,
University of Windsor

Susan A. Dunford,
University of Cincinnati

Mary Durant,
North Harris College

Ronald Edwards,
University of Florida

Rosemarie Elizondo,
Reedley College

George Ellmore,
Tufts University

Joanne T. Ellzey,
University of Texas–El Paso

Wayne Elmore,
Marshall University

Thomas Emmel,
University of Florida

Carl Estrella,
Merced College

Nancy Eyster-Smith,
Bentley College

Gerald Farr,
Southwest Texas State University

Rita Farrar,
Louisiana State University

Marianne Feaver,
North Carolina State University

Susannah Feldman,
Towson University

Linnea Fletcher,
Austin Community College–Northridge

Doug Florian,
Trident Technical College

Charles V. Foltz,
Rhode Island College

Dennis Forsythe,
The Citadel

Douglas Fratianne,
Ohio State University

Scott Freeman,
University of Washington

Donald P. French,
Oklahoma State University

Harvey Friedman,
University of Missouri–St. Louis

Don Fritsch,
Virginia Commonwealth University

Teresa Lane Fulcher,
Pellissippi State Technical Community College

Michael Gaines,
University of Kansas

Cynthia Galloway,
Texas A&M University–Kingsville

Irja Galvan,
Western Oregon University

Gail E. Gasparich,
Towson University

Janet Gaston,
Troy University

Farooka Gauhari,
University of Nebraska–Omaha

John Geiser,
Western Michigan University

George W. Gilchrist,
University of Washington

David Glenn-Lewin,
Iowa State University

Elmer Gless,
Montana College of Mineral Sciences

Charles W. Good,
Ohio State University–Lima

Joan-Beth Gow,
Anna Maria College

Margaret Green,
Broward Community College

Mary Ruth Griffin,
University of Charleston

Wendy Grillo,
North Carolina Central University

David Grise,
Southwest Texas State University

Martha Groom,
University of Washington

Lonnie J. Guralnick,
Western Oregon University

Martin E. Hahn,
William Paterson College

Madeline Hall,
Cleveland State University

Georgia Ann Hammond,
Radford University

Blanche C. Haning,
North Carolina State University

Richard Hanke,
Rose State College

Helen B. Hanten,
University of Minnesota

John P. Harley,
Eastern Kentucky University

Robert Hatherill,
Del Mar College

William Hayes,
Delta State University

Kathleen Hecht,
Nassau Community College

Stephen Hedman,
University of Minnesota

Jean Helgeson,
Collins County Community College

Alexander Henderson,
Millersville University

Wiley Henderson,
Alabama A&M University

Timothy L. Henry,
University of Texas–Arlington

James Hewlett,
Finger Lakes Community College

Alison G. Hoffman,
University of Tennessee–Chattanooga

Kelly Hogan,
University of North Carolina–Chapel Hill

Leland N. Holland,
Paso-Hernando Community College

Laura Mays Hoopes,
Occidental College

Dale R. Horeth,
Tidewater Community College

Michael D. Hudgins,
Alabama State University

David Huffman,
Southwest Texas State University

Joel Humphrey,
Cayuga Community College

Donald A. Ingold,
East Texas State University

Jon W. Jacklet,
State University of New York–Albany

Rebecca M. Jessen,
Bowling Green State University

J. Kelly Johnson,
University of Kansas

James Johnson,
Central Washington University

Kristy Y. Johnson,
The Citadel

Ross Johnson,
Chicago State University

Florence Juillerat,
Indiana University–Purdue University at Indianapolis

Thomas W. Jurik,
Iowa State University

Ragupathy Kannan,
University of Arkansas, Fort Smith

Arnold Karpoff,
University of Louisville

L. Kavaljian,
California State University

Joe Keen,
Patrick Henry Community College

Jeff Kenton,
Iowa State University

Hendrick J. Ketellapper,
University of California, Davis

Jeffrey Kiggins,
Blue Ridge Community College

Aaron Krochmal,
University of Houston–Downtown

Harry Kurtz,
Sam Houston State University

Kate Lajtha,
Oregon State University

Tom Langen,
Clarkson University

Patrick Larkin,
Sante Fe College

Stephen Lebsack,
Linn-Benton Community College

Patricia Lee-Robinson,
Chaminade University of Honolulu

David E. Lemke,
Texas State University

William H. Leonard,
Clemson University

Edward Levri,
Indiana University of Pennsylvania

Graeme Lindbeck,
University of Central Florida

Jerri K. Lindsey,
Tarrant County Junior College–Northeast

Mary Lipscomb,
Virginia Polytechnic and State University

Richard W. Lo Pinto,
Fairleigh Dickinson University

Jonathan Lochamy,
Georgia Perimeter College
Jason L. Locklin,
Temple College
John Logue,
University of South Carolina–Sumter
Paul Lonquich,
California State University Northridge
William Lowen,
Suffolk Community College
Ann S. Lumsden,
Florida State University
Steele R. Lunt,
University of Nebraska–Omaha
Daniel D. Magoulick,
The University of Central Arkansas
Bernard Majdi,
Waycross College
Cindy Malone,
California State University–Northridge
Paul Mangum,
Midland College
Richard Manning,
Southwest Texas State University
Mark Manteuffel,
St. Louis Community College
Barry Markillie,
Cape Fear Community College
Ken Marr,
Green River Community College
Kathleen A. Marrs,
Indiana University–Purdue University Indianapolis
Michael Martin,
University of Michigan
Linda Martin-Morris,
University of Washington
Kenneth A. Mason,
University of Kansas
Daniel Matusiak,
St. Charles Community College
Margaret May,
Virginia Commonwealth University
D. J. McWhinnie,
De Paul University
Gary L. Meeker,
California State University, Sacramento
Thoyd Melton,
North Carolina State University
Joseph R. Mendelson III,
Utah State University

Karen E. Messley,
Rockvalley College
Timothy Metz,
Campbell University
Steven Mezik,
Herkimer County Community College
Glendon R. Miller,
Wichita State University
Hugh Miller,
East Tennessee State University
Neil Miller,
Memphis State University
Jeanne Minnerath,
St. Mary's University of Minnesota
Christine Minor,
Clemson University
Jeanne Mitchell,
Truman State University
Lee Mitchell,
Mt. Hood Community College
Jack E. Mobley,
University of Central Arkansas
John W. Moon,
Harding University
Nicole Moore,
Austin Peay University
Richard Mortenson,
Albion College
Gisele Muller-Parker,
Western Washington University
James Mulrooney,
Central Connecticut State University
Kathleen Murray,
University of Maine
Robert Neill,
University of Texas
Russell Nemecek,
Columbia College, Hancock
Harry Nickla,
Creighton University
Daniel Nickrent,
Southern Illinois University
Jane Noble-Harvey,
University of Delaware
Murad Odeh,
South Texas College
David J. O'Neill,
Community College of Baltimore County–Dundalk Campus
James T. Oris,
Miami University of Ohio
Marcy Osgood,
University of Michigan
C. O. Patterson,
Texas A&M University
Fred Peabody,
University of South Dakota

Charlotte Pedersen,
Southern Utah University
Harry Peery,
Tompkins-Cortland Community College
Luis J. Pelicot,
City University of New York, Hostos
Rhoda E. Perozzi,
Virginia Commonwealth University
Gary B. Peterson,
South Dakota State University
Bill Pfitsch,
Hamilton College
Ronald Pfohl,
Miami University of Ohio
Larry Pilgrim,
Tyler Junior College
Therese Poole,
Georgia State University
Robert Kyle Pope,
Indiana University South Bend
Bernard Possident,
Skidmore College
Ina Pour-el,
DMACC–Boone Campus
Elsa C. Price,
Wallace State Community College
Marvin Price,
Cedar Valley College
Kelli Prior,
Finger Lakes Community College
Jennifer J. Quinlan,
Drexel University
James A. Raines,
North Harris College
Paul Ramp,
Pellissippi State Technical College
Robert N. Reed,
Southern Utah University
Wenda Ribeiro,
Thomas Nelson Community College
Elizabeth Rich,
Drexel University
Mark Richter,
University of Kansas
Robert Robbins,
Michigan State University
Jennifer Roberts,
Lewis University
Frank Romano,
Jacksonville State University
Chris Romero,
Front Range Community College

Paul Rosenbloom,
Southwest Texas State University
Amanda Rosenzweig,
Delgado Community College
K. Ross,
University of Delaware
Mary Lou Rottman,
University of Colorado–Denver
Albert Ruesink,
Indiana University
Cameron Russell,
Tidewater Community College
Connie Russell,
Angelo State University
Marla Ruth,
Jones County Junior College
Christopher F. Sacchi,
Kutztown University
Eduardo Salazar,
Temple College
Doug Schelhaas,
University of Mary
Brian Schmaefsky,
Kingwood College
Alan Schoenherr,
Fullerton College
Brian W. Schwartz,
Columbus State University
Edna Seaman,
University of Massachusetts, Boston
Tim Sellers,
Keuka College
Patricia Shields,
George Mason University
Marilyn Shopper,
Johnson County Community College
Rick L. Simonson,
University of Nebraska, Kearney
Howard Singer,
New Jersey City University
Anu Singh-Cundy,
Western Washington University
Linda Simpson,
University of North Carolina–Charlotte
Steven Skarda,
Linn-Benton Community College
Russel V. Skavaril,
Ohio State University
John Smarelli,
Loyola University
Mark Smith,
Chaffey College
Dale Smoak,
Piedmont Technical College

Jay Snaric,
St. Louis Community College
Phillip J. Snider,
University of Houston
Shari Snitovsky,
Skyline College
Gary Sojka,
Bucknell University
John Sollinger,
Southern Oregon University
Sally Sommers Smith,
Boston University
Jim Sorenson,
Radford University
Mary Spratt,
University of Missouri, Kansas City
Bruce Stallsmith,
University of Alabama–Huntsville
Anthony Stancampiano,
Oklahoma City Community College
Benjamin Stark,
Illinois Institute of Technology
William Stark,
Saint Louis University
Barbara Stebbins-Boaz,
Willamette University
Kathleen M. Steinert,
Bellevue Community College
Barbara Stotler,
Southern Illinois University
Nathaniel J. Stricker,
Ohio State University
Martha Sugermeyer,
Tidewater Community College
Gerald Summers,
University of Missouri–Columbia
Marshall Sundberg,
Louisiana State University

Bill Surver,
Clemson University
Eldon Sutton,
University of Texas–Austin
Peter Svensson,
West Valley College
Dan Tallman,
Northern State University
Jose G. Tello,
Long Island University
David Thorndill,
Essex Community College
William Thwaites,
San Diego State University
Professor Tobiessen,
Union College
Richard Tolman,
Brigham Young University
Sylvia Torti,
University of Utah
Dennis Trelka,
Washington and Jefferson College
Richard C. Tsou,
Gordon College
Sharon Tucker,
University of Delaware
Gail Turner,
Virginia Commonwealth University
Glyn Turnipseed,
Arkansas Technical University
Lloyd W. Turtinen,
University of Wisconsin, Eau Claire
Robert Tyser,
University of Wisconsin, La Crosse
Robin W. Tyser,
University of Wisconsin, La Crosse
Kristin Uthus,
Virginia Commonwealth University

Rani Vajravelu,
University of Central Florida
Jim Van Brunt,
Rogue Community College
F. Daniel Vogt,
State University of New York–Plattsburgh
Nancy Wade,
Old Dominion University
Susan M. Wadkowski,
Lakeland Community College
Jyoti R. Wagle,
Houston Community College–Central
Jerry G. Walls,
Louisiana State University, Alexandria
Holly Walters,
Cape Fear Community College
Winfred Watkins,
McLennan Community College
Lisa Weasel,
Portland State University
Michael Weis,
University of Windsor
DeLoris Wenzel,
University of Georgia
Jerry Wermuth,
Purdue University–Calumet
Diana Wheat,
Linn-Benton Community College
Richard Whittington,
Pellissippi State Technical Community College
Jacob Wiebers,
Purdue University
Roger K. Wiebusch,
Columbia College
Carolyn Wilczynski,
Binghamton University
Lawrence R. Williams,
University of Houston

P. Kelly Williams,
University of Dayton
Roberta Williams,
University of Nevada–Las Vegas
Emily Willingham,
University of Texas–Austin
Sandra Winicur,
Indiana University–South Bend
Bill Wischusen,
Louisiana State University
Michelle Withers,
Louisiana State University
Chris Wolfe,
North Virginia Community College
Stacy Wolfe,
Art Institutes International
Colleen Wong,
Wilbur Wright College
Wade Worthen,
Furman University
Robin Wright,
University of Washington
Taek H. You,
Campbell University
Brenda L. Young,
Daemen College
Cal Young,
Fullerton College
Tim Young,
Mercer University
Martin Zahn,
Thomas Nelson Community College
Izanne Zorin,
Northern Virginia Community College–Alexandria
Michelle Zurawski,
Moraine Valley Community College

Case Studies place biology in a real-world context

A **Case Study** describing a true and relevant event or phenomenon runs throughout each chapter, tying biological concepts to the real world.

CHAPTER **19** The Diversity of Prokaryotes and Viruses

CASE STUDY
Unwelcome Dinner Guests

IN 2010, ANDREW LEKAS, then a student at the University of Michigan, ate a burrito at a favorite restaurant near campus. A few hours later, he began to feel sick, with symptoms that included vomiting, diarrhea, headache, and weakness. The illness ultimately became severe enough to require a hospital stay and more than a week in bed. What sickened Lekas? The lettuce in his burrito was contaminated.

As bad as Lekas's experience was, it could have been much worse. Other victims of foodborne contamination have suffered more serious consequences. For example, Stephanie Smith, a former dance instructor from Minnesota, is paralyzed from the waist down as a result of the severe illness (hemolytic uremic syndrome) she developed in 2007 after eating a tainted hamburger.

Regrettably, the ill effects of consuming contaminated food are all too common. The Centers for Disease Control and Prevention estimates that U.S. residents experience an astonishing 76 million cases of foodborne illness each year. Some of these cases are severe. For example, in 2011, at least 38 deaths were caused by a single source of contaminated cantaloupes and, in Germany, contaminated sprouts from a single supplier caused more than 850 cases of hemolytic uremic syndrome and 32 deaths. Overall, consumption of contaminated food results in about 325,000 hospitalizations and 5,200 deaths in the United States each year.

What is it that contaminates food and causes so much illness? Bacteria. The nutrients in the food you consume during meals and snacks can also provide sustenance for a wide variety of disease-causing bacteria. Some of these invisible diners may accompany your lunch to your digestive tract and take up residence there, causing unpleasant symptoms or, in many cases, serious illness.

Devising effective strategies for protecting our food supply against bacterial contamination depends in part on how well we understand the biology of bacteria. Scientific investigation of bacteria and other microbes can provide the knowledge required to detect the presence of dangerous microbes in food, develop effective treatments for foodborne illnesses, and devise safer methods for growing and processing food. Fortunately, biologists already know quite a

Hamburgers should be cook...

All chapters in the book open with a **Case Study**, a true yet extraordinary story that relates to and draws you into the science presented in the chapter. The Tenth Edition offers several new Case Studies, including Unstable Atoms Unleashed (Chapter 2 Atoms, Molecules, and Life), Unwelcome Dinner Guests (Chapter 19 The Diversity of Prokaryotes and Viruses), and Life-Changing Hormones (Chapter 37 Chemical Control of the Animal Body: The Endocrine System).

CASE STUDY continued
Unwelcome Dinner Guests

Many of the bacteria responsible for foodborne illnesses do their damage by producing toxins. For example, different populations of the bacterial species *Escherichia coli* may differ genetically, and some genetic differences can transform this normally benign inhabitant of the human digestive system into a toxin-producing pathogen. If one of these toxic strains, such as the ones designated O157:H7 and O104:H4, finds its way into a human digestive system, the bacteria attach firmly to the wall of the intestine and begin to release a toxin called shiga. Shiga toxin causes intestinal bleeding that results in painful cramping and bloody diarrhea. The toxin can damage other organs as well; victims of O157:H7 and O104:H4 often develop hemolytic uremic syndrome, a dangerous condition characterized by kidney failure and loss of red blood cells.

Every chapter contains **Case Study Continued** sections that appear when you are well into the chapter. These sections expand on the **Chapter Opening Case Studies** and connect to biological concepts you will have covered.

CASE STUDY revisited
Unwelcome Dinner Guests

How do harmful bacteria get into our food? Many foodborne illnesses result from consumption of contaminated beef. The intestinal tracts of about a third of the cattle in the United States carry bacteria that are harmful to humans, and these bacteria can be transmitted to humans when a meatpacker accidentally grinds some gut contents into hamburger. Similarly, chicken feces may splash onto eggs, setting the stage for harmful bacteria to enter the eggs through tiny cracks or when the consumer breaks the egg and its contents contact the shell. Produce such as lettuce, spinach, tomatoes, and melons can also become contaminated if farm fields are exposed to animal feces, which can be deposited by deer or wandering domestic animals or carried from nearby ranches and feedlots in dust or runoff. The warm, moist environments in which sprouts are grown provide excellent growing conditions for any harmful bacteria that may have been present on the seeds from which the sprouts were produced.

How can you protect yourself from the bacteria that share our food supply? It's easy: Clean, cook, and chill. Cleaning helps prevent the spread of pathogens. Wash your hands before preparing food, and wash all utensils and cutting boards after preparing each item. Thorough cooking is the best way to ensure that any bacteria present in food are killed. Meats, in particular, must be thoroughly cooked; food

safety experts recommend using a meat thermometer to ensure that the thickest part of cooked pork or ground beef has reached 160°F. The safe temperature for cuts of beef, veal, or lamb is 145°F; for all poultry, 165°F. The color of cooked meat can be an unreliable indicator of safety, but when a meat thermometer is unavailable, try to avoid eating meat that is still pink inside, especially ground beef. Fish should be cooked until it is opaque and flakes easily with a fork; cook eggs until both white and yolk are firm. Finally, keep stored food cold. Pathogens multiply most rapidly at temperatures between 40° and 140°F. So get your groceries home from the store and into the refrigerator or freezer as quickly as possible. Don't leave cooked leftovers unrefrigerated for more than 2 hours. Thaw frozen foods in the refrigerator or the microwave, not at room temperature. A little bit of attention to food safety can save you from unwelcome guests in your food.

Consider This
BioEthics Consumer groups contend that we can improve food safety by giving government agencies additional funding and greater authority to inspect food processing plants and order recalls of contaminated food. Opponents of such steps argue that we need not empower government agencies, because the best protection against food contamination is informed consumers, who will stop buying products from companies that have produced unsafe foods. Would you support or oppose additional government oversight of food safety?

A **Case Study Revisited** section wraps up the narrative of each chapter by connecting the biological themes described throughout the chapter with the everyday science brought out in the Case Study. The accompanying **Consider This** question allows further reflection on how the biology in the Case Study can be applied to a new situation.

NEW! Learning Goals and Learning Outcomes make it easy to track your progress

Learning Goals at the start of each chapter call attention to the biological concepts that are most important to track and understand while reading.

LEARNING GOALS

LG1 Describe the molecules and cellular structures that play central roles in the process by which information encoded in DNA is used to synthesize proteins.

LG2 Compare and contrast how information is encoded in DNA and in RNA.

LG3 Describe the functions of transcription and translation and identify the locations in the cell at which they occur.

LG4 Describe the main steps of transcription, including the molecules involved and how they interact.

LG5 Describe the main steps of translation, including the molecules involved and how they interact.

LG6 Describe the different types of mutations and explain their potential effects on protein synthesis.

LG7 Describe how gene expression is regulated.

NEW! Learning Goals at the beginning of the main narrative in each chapter state the knowledge and abilities you should master through your reading, providing you with a tool to assess your understanding of the material.

NEW! Learning Outcomes offered in the Chapter Review section, restate the Learning Goals and serve as recaps of the chapter's coverage and reminders of the concepts learned. Cross-references to Learning Outcomes in the Fill-in-the-Blank and Review Questions sections encourage you to evaluate your success in reaching each specific Learning Goal.

Learning Outcomes

In this chapter, you have learned to . . .

LO1 Describe the molecules and cellular structures that play central roles in the process by which information encoded in DNA is used to synthesize proteins.

LO2 Compare and contrast how information is encoded in DNA and in RNA.

LO3 Describe the functions of transcription and translation and identify the locations in the cell at which they occur.

LO4 Describe the main steps of transcription, including the molecules involved and how they interact.

LO5 Describe the main steps of translation, including the molecules involved and how they interact.

LO6 Describe the different types of mutations and explain their potential effects on protein synthesis.

LO7 Describe how gene expression is regulated.

Thinking Through the Concepts

Fill-in-the-Blank

1. Synthesis of RNA from the instructions in DNA is called _____. Synthesis of a protein from the instructions in mRNA is called _____. Which structure in the cell is the site of protein synthesis? _____ **LO1 LO3**

2. The three types of RNA that are essential for protein synthesis are_____, _____, and_____. Another type of RNA, which can interfere with translation, is called _____. **LO2 LO3 LO7**

3. The genetic code uses _____ (how many?) bases to code for a single amino acid. This short sequence of bases in mRNA is called a(n) _____. The complementary sequence of bases in tRNA is called a(n) _____. **LO2**

4. The enzyme _____ synthesizes RNA from the instructions in DNA. DNA has two strands, but for any given gene, only one strand, called the _____ strand, is transcribed. To begin transcribing a gene, this enzyme binds to a specific sequence of DNA bases located at the beginning of the gene. This DNA sequence is called the _____. Transcription ends when the enzyme encounters a DNA sequence at the end of the gene called the _____. **LO4**

5. Protein synthesis begins when mRNA binds to a ribosome. Translation begins with the _____ codon of mRNA and continues until a(n) _____ codon is reached. Individual amino acids are brought to the ribosome by _____. These amino acids are linked into protein by _____ bonds. **LO5**

6. There are several different types of mutations in DNA. If one nucleotide is substituted for another, this is called a(n) _____ mutation. _____ mutations occur if nucleotides are added in the middle of a gene. _____ mutations occur if nucleotides are removed from the middle of a gene. **LO6**

Review Questions

1. How does RNA differ from DNA? **LO2**

2. Name the three types of RNA that are essential to protein synthesis. What is the function of each? **LO2 LO3 LO5**

3. Define the following terms: *genetic code, codon,* and *anticodon.* What is the relationship among the bases in DNA, the codons of mRNA, and the anticodons of tRNA? **LO2 LO5**

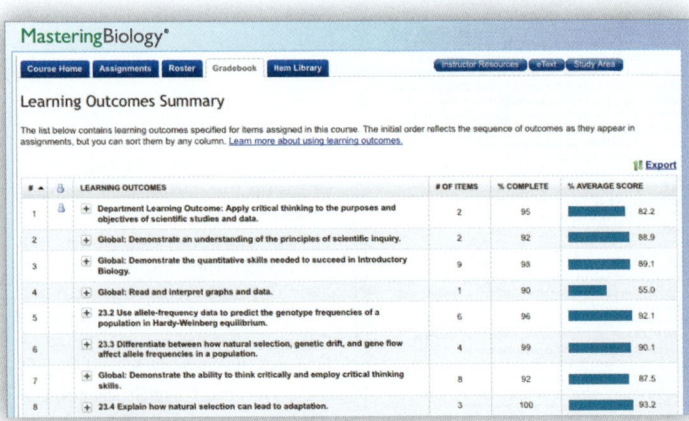

NEW! Learning Outcomes in MasteringBiology® give your instructor the tools to quantify your learning gains and to share those results with you quickly and easily.

NEW! Easily interpret evolutionary relationships using horizontal phylogenetic trees

Newly rendered, horizontal phylogenetic trees reflect the orientation used in scientific literature.

The **new phylogenetic tree orientation** allows you to read intuitively from left to right and more clearly see the relationships between evolutionary branches. Compare the new phylogenetic tree orientation (the Tenth Edition is on the left) to the old (the Ninth Edition is below).

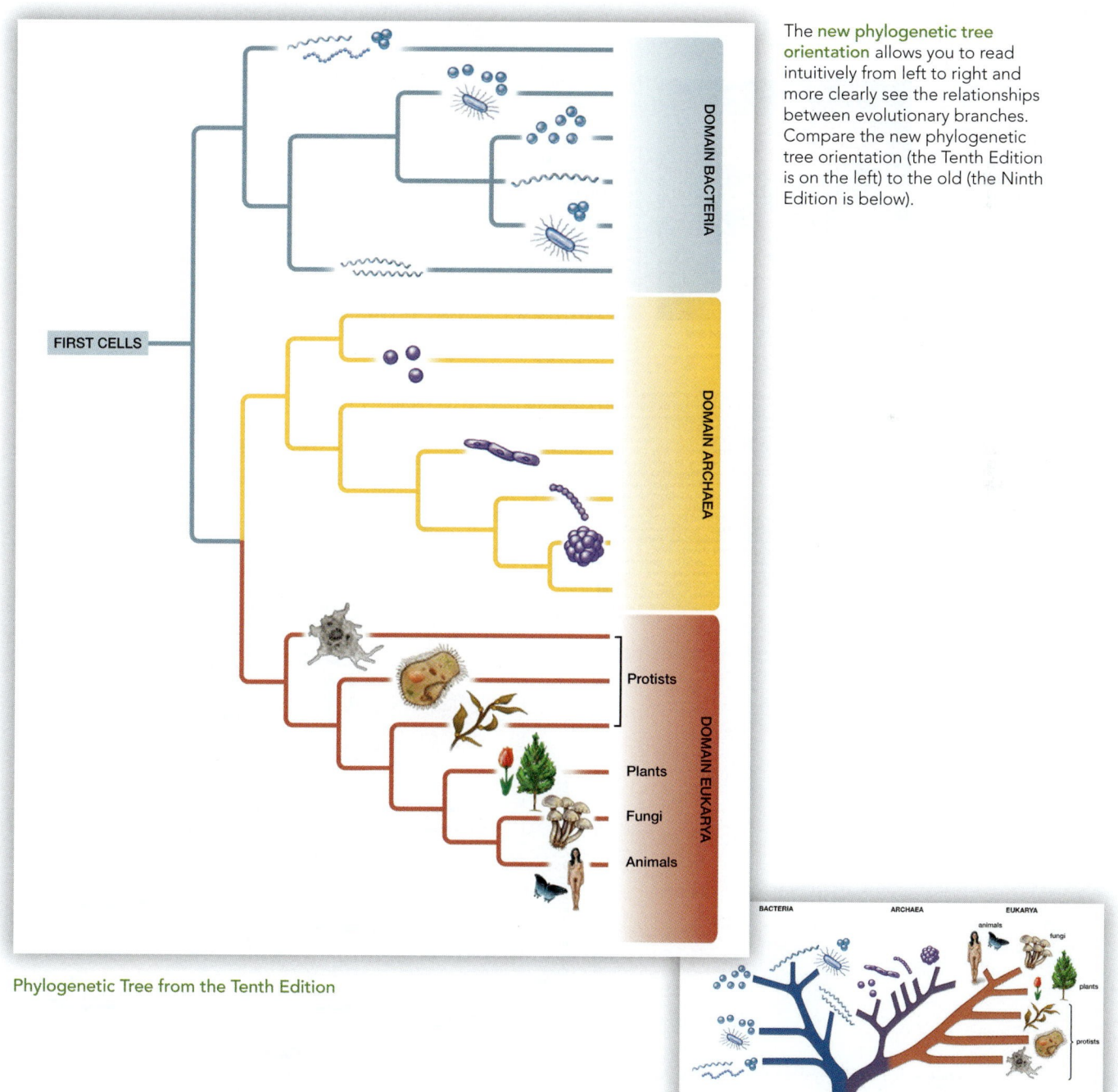

Phylogenetic Tree from the Tenth Edition

Phylogenetic Tree from the Ninth Edition

Increased coverage of health science emphasizes the relevance of biology

Health science is emphasized throughout the book with more examples of diseases, including the latest in treatments, discussions of the most current medical research, and current available data.

Health | Watch
Is Your Body's Ecosystem Healthy?

Microbiologists like to say, "You are born 100% human, but you die 90% microbial." A human fetus in the womb is more or less sterile, but as it passes down the birth canal to be born, it picks up some of the bacteria that live there, and more are transferred from the first hands to touch the new baby. From that point on, microbes accumulate steadily, acquired from the environment, from food, and from other humans. By the time a child is 3 years old, its body is home to an entire ecosystem of microorganisms. This "microbiome" includes hundreds of different species, living in huge numbers in the nose and mouth, on the scalp, in the urogenital tract, in the gut, and on almost every skin surface. All told, the microbiome of a typical person includes about 100 trillion microbial cells, about 10 microbial cells for every human one.

It is becoming increasingly apparent to physicians and researchers that the inhabitants of the microbiome are not mere hitchhikers, but instead play an important role in human health. Although some of the benefits provided by our microbial partners are well understood, such as the production by gut bacteria of nutrients essential to humans, most of the details about how the microbiome contributes to our health are more mysterious. However, much evidence suggests that its contribution is important. For example, researchers have shown that in infants with the often-fatal bowel disease necrotizing enterocolitis, the composition of the community of microbes in the gut is different from that present in healthy babies. Similarly aberrant digestive tract microbiomes have been found in adults with bowel disorders, and even in people with disorders not directly related to digestion, such as diabetes and autoimmune diseases. Such findings have led to the hypothesis that, because a person's microbiome, like a coral reef or a tropical rain forest, is a diverse ecosystem characterized by complex interactions among species, disruptions of this ecosystem compromise the microbial ecosystem's ability to perform functions essential to human health. However, scientists are not yet able to say for sure if

the disrupted microbiomes of diseased individuals are a cause or a consequence of disease.

Confirmation of the microbiome's possible role in maintaining health depends on improved knowledge of its composition and characteristics. For that reason, a major ongoing research effort is directed at identifying and characterizing all of the microbial species that compose the microbiome, and at identifying exactly how microbiomes differ among different people. The job is challenging, because the traditional way of identifying a prokaryotic species—by growing it in a culture dish in the lab—is not effective for the many species that cannot be easily cultured. Some scientists are trying the alternative approach of taking a sample of a whole microbiome and sequencing all of the DNA present in it. Some of the initial results of this approach are astonishing. For example, fecal samples from 124 people yielded a total of 1.3 million different genes, about 150 times as many as are present in the human genome. Further study of these microbial genes may reveal their functions and how they contribute to their human hosts.

In the meantime, some physicians are experimenting with treatments designed to restore ecological balance to the microbiomes of sick people. In particular, researchers have begun using fecal transplants to treat patients with certain bowel infections (Fig. E19-1). In this treatment, a sample of feces from a healthy donor is transplanted to the bowel of the sick person, with the expectation that the donor's microbial community will become established there, displacing the harmful microbes responsible for the disease. The physicians using this treatment report high success rates, much higher than those typical of treatment with antibiotics, which have been shown to destroy a healthy microbiome. These reports have encouraged regulatory agencies to overcome the yuck factor and allow more clinical trials of the fecal transplant treatment to get under way.

▲ FIGURE E19-1 **Transplanted feces could cure diseased bowels** By restoring a healthy microbial ecosystem in the recipient, fecal transplants can treat infected bowels and perhaps even diabetes, obesity, and other disorders as well.

Health Watch essays help you connect key biological concepts to the science of human health.

Anatomical and physiological figures guide you through the structures and functions of the human body. The Tenth Edition's expanded coverage of health science is illustrated throughout the chapters.

Use the book's questions to prepare for tests

Questions throughout the text provide opportunities to prepare for tests.

Major section headings pose questions, and as you read, the answers to the question are revealed, stimulating analytical thinking and modeling the importance of inquiry in scientific exploration.

13.1 WHAT IS BIOTECHNOLOGY?

13.2 HOW DOES DNA RECOMBINE IN NATURE?

13.3 HOW IS BIOTECHNOLOGY USED IN FORENSIC SCIENCE?

? HAVE **YOU** EVER WONDERED …

Why Your Fingers Fumble and Go Numb in the Cold?

Fingers are among the first parts of your body to suffer from the cold. That's because as your internal temperature begins to fall, your body conserves warmth for vital organs (such as your heart and brain) by reducing blood flow to your extremities. As a result, your hands get colder, the membranes in them become less fluid, their membrane channel proteins open and close more slowly, and the conduction of nerve impulses slows. Nerve impulses therefore do not reach your hand as fast as usual, and your fingers fumble as you try to zip your coat. When you can no longer feel your fingers, you know that their temperature has dropped so low that your sensory nerve cells are no longer transmitting signals.

Have You Ever Wondered? poses high-interest questions that students commonly ask themselves or instructors. Each question is addressed with a clear, scientifically-accurate response.

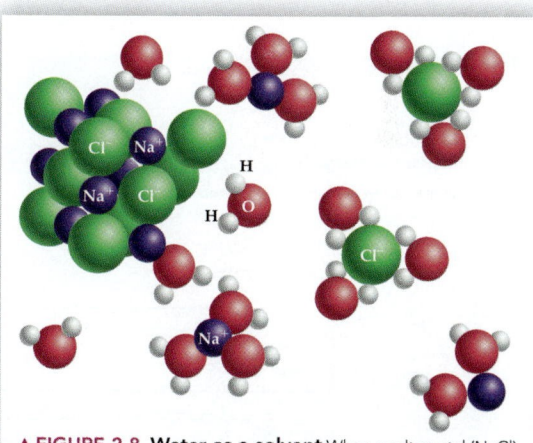

▲ FIGURE 2-8 **Water as a solvent** When a salt crystal (NaCl) is dropped into water, the water surrounds the sodium and chloride ions. The positive poles of water molecules face the Cl^-, and the negative poles face the Na^+. The ions disperse as the surrounding water molecules isolate them from one another, and the salt crystal gradually dissolves.

QUESTION If you placed a salt crystal in a nonpolar liquid (like oil), would it dissolve?

CHECK YOUR LEARNING

Can you describe how cells regulate the rate at which metabolic reactions proceed? Can you describe how poisons, drugs, and environmental conditions influence enzyme activity, and provide examples?

NEW! Check Your Learning questions are placed at the end of major sections of the chapter and encourage you to evaluate your understanding of what you have read.

Figure caption questions prompt you to think critically about the topic covered in the art and to apply what you've learned from the illustration to a new situation. (Answers to the figure questions are provided at the back of the book.)

Improve your grade—and your learning—with MasteringBiology®

Mastering is the most effective and widely used online homework and assessment system for the sciences. It delivers self-paced tutorials that focus on your course objectives, provide individualized coaching, and respond to your progress. Mastering motivates you to learn outside of class and arrive prepared for lecture or lab.

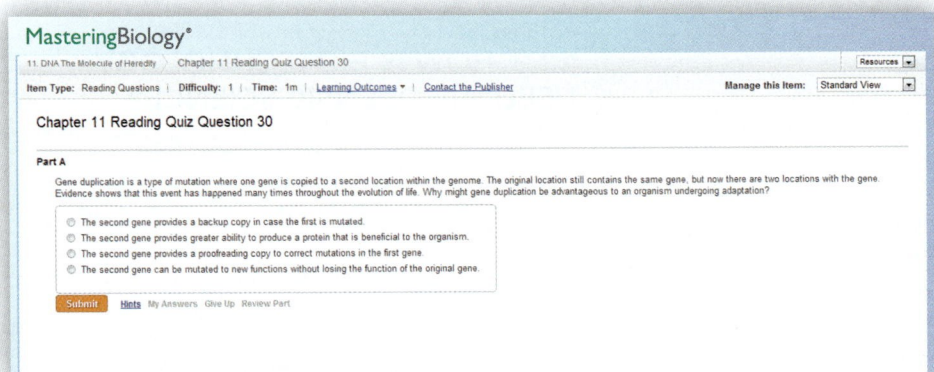

Reading Quizzes keep you on track with reading assignments. The quizzes require only 5–8 minutes for you to complete and make it possible for your instructor to understand your misconceptions before you arrive at class.

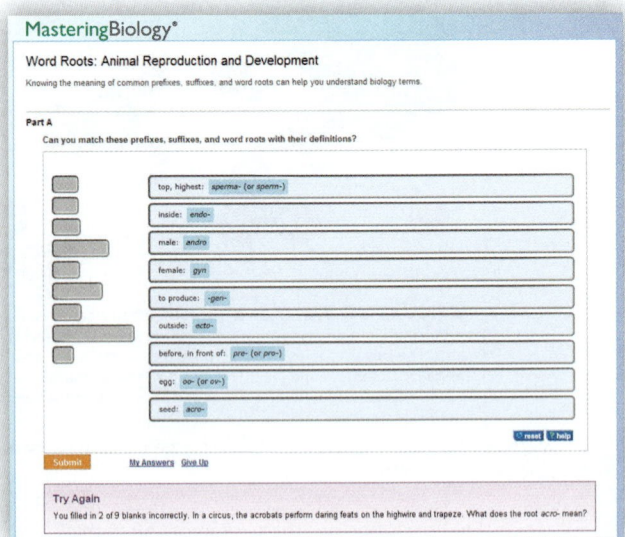

NEW! Building Vocabulary Exercises help you learn the meaning of common prefixes, suffixes, and word roots, and then ask you to apply your knowledge to learn unfamiliar biology terms.

EXPANDED! Animations help you visualize a range of biological and ecological processes. Assignable assessment questions are available for each animation.

UPDATED! Current events articles and related quizzes connect biology to everyday life and allow your instructor to monitor your scientific literacy.

Created in partnership with *ABC News*, these **video clips** cover interesting topics, including genetically altered salmon, ocean acidification, and the effect of exercise on cells.

ABC News Topics:

- Protecting the Galapagos Islands
- Henrietta Lacks's Cells
- Genetically Altered Salmon
- Exploring Evolution in the Solomon Islands
- Ocean Acidification
- The Genetics of Short Sleepers
- Modern Humans and Neanderthal DNA
- Herbicide-resistant Pigweed
- Turning Algae into Biofuel
- The Cuttlefish
- The Effect of Exercise on Cells
- Eating Behavior in Monkeys

Additional teaching and learning tools available with the Tenth Edition

Instructor Resource DVD (IR-DVD)

978-0-321-84470-5 ▪ 0-321-84470-X

This cross-platform DVD organizes all instructor and student media resources into one convenient, chapter-by-chapter resource package.

The Instructor Resource DVD includes:

- All the art and photos from the book with customizable labels
- 3-D BioFlix® animations
- Customizable PowerPoint® lecture presentations
- Clicker questions
- Digital transparencies
- Instructor guide material
- And more!

The test bank CD-ROM offers more than 4,000 questions that have been peer-reviewed, providing the questions that set the standard for quality and accuracy. The Tenth Edition includes new figure—and scenario—based questions.

Additional Resources

For Instructors

Instructor's Resources Area of MasteringBiology

Resources from the Instructor Resource DVD can also be downloaded from www.masteringbiology.com.

Instructor's Guide

978-0-321-84469-9 ▪ 0-321-84469-6

Includes revised Lecture Outlines, Lecture Activities, and Handouts to promote active learning and collaborative activities. The Instructor's Guide is also available in Microsoft Word format in the "Instructor Resources" area of MasteringBiology and on the Instructor Resource DVD.

TestGen® Test Bank (Download Only)

Adopting instructors may download the TestGen software from www.pearsonhighered.com.

Course Management Options

MasteringBiology

www.masteringbiology.com

Please contact your Pearson sales representative for more information on additional options for **online Course Management.**

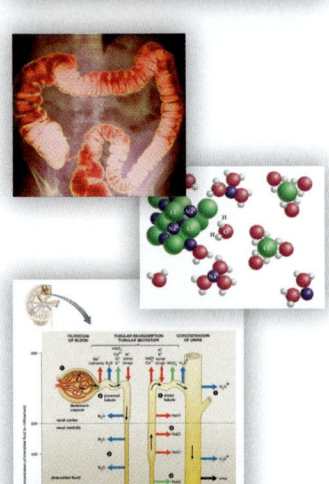

IR-DVD resources

Scientific American Current Issues in Biology

Scientific American Current Issues in Biology gives students the best of both worlds—accessible, dynamic, relevant articles from *Scientific American* magazine that present key issues in biology, paired with the authority, reliability, and clarity of Pearson's non-majors biology texts. Articles include questions to help students check their comprehension and make connections to science and society.

Scientific American is available in Volumes 1–6. This resource is available at no additional charge when packaged with a new text.

Volume 1:
978-0-8053-7507-7 ▪ 0-8053-7507-4

Volume 2:
978-0-8053-7108-6 ▪ 0-8053-7108-7

Volume 3:
978-0-8053-7527-5 ▪ 0-8053-7527-9

Volume 4:
978-0-8053-3566-8 ▪ 0-8053-3566-8

Volume 5:
978-0-321-54187-1 ▪ 0-321-54187-1

Volume 6:
978-0-321-59849-3 ▪ 0-321-59849-0

An Introduction to Life on Earth

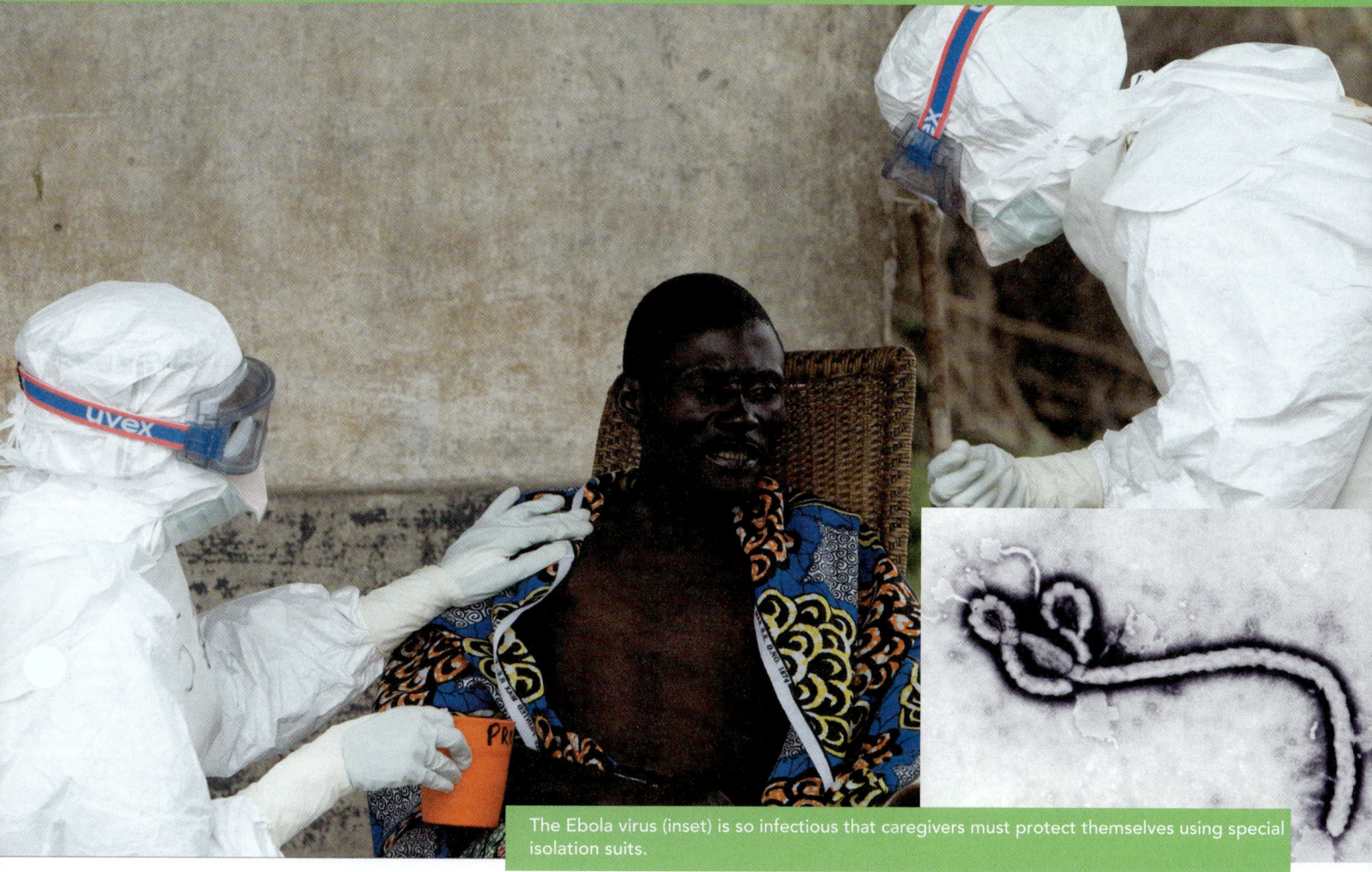

The Ebola virus (inset) is so infectious that caregivers must protect themselves using special isolation suits.

CASE STUDY

The Boundaries of Life

IN 1996 IN GABON, Africa, a hunting party discovered a newly dead chimpanzee in the bush and butchered it for food. Within a month, the hunters' village was devastated. Thirty-seven residents, including the hunters and many of their family members, had fallen ill. Before the scourge ended, more than half had died.

The villagers had fallen victim to one of Earth's most deadly pathogens—the Ebola virus (see the inset photo). The threat of Ebola hemorrhagic fever ("Ebola") strikes fear in anyone familiar with its symptoms, which begin with fever, headache, joint and muscle aches, and stomach pains, and progress to severe vomiting, bloody diarrhea, and organ failure. Internal hemorrhaging can leave victims bleeding

from every orifice. Death usually occurs within 7 to 16 days after the onset of symptoms, and there is no cure. Ebola is so contagious and deadly that caregivers wear "moonsuits" to avoid contact with any body fluids from their patients (see the chapter opener photo).

Ebola, which occurs sporadically in Africa, is one of many diseases caused by viruses. Although some viral diseases, such as smallpox and polio, have been largely eradicated, others, like the common cold and influenza, continue to make us miserable. Most alarming are viruses such as the human immunodeficiency virus (HIV), which causes acquired immunodeficiency syndrome (AIDS) and kills about 1.8 million people annually. An emerging threat is the avian influenza virus (H5N1 or "bird flu"). Although it does not spread readily from person to person,

bird flu has a very high mortality rate, and health officials are concerned that it may mutate into a far more infectious form that could cause a worldwide epidemic.

No matter how you measure it, viruses are enormously successful. Although many consist only of a small amount of genetic material surrounded by protein, viruses infect every known form of life and are the most abundant biological entity on the planet. Viruses can spread from one organism to another, rapidly increase in numbers, and mutate (undergo changes in their hereditary material). Yet in spite of these lifelike qualities, scientists disagree about whether to classify viruses as organisms or as inert parasitic biological particles. The basis for this argument may surprise you; there is no universally accepted scientific definition of life. Think about it—what is life, anyway?

AT A GLANCE

1.1 What Is Life?
1.2 What Is Evolution?

1.3 How Do Scientists Study Life?

1.4 What Is Science?

LEARNING GOALS

LG1 Describe the characteristics of living things.

LG2 Explain evolution and how three natural processes make evolution inevitable.

LG3 Explain the concept of levels of biological organization, and list the levels.

LG4 Identify the three domains of life, and describe the types of cells that make up each domain.

LG5 Explain how the three kingdoms of eukaryotic organisms are distinguished from one another.

LG6 Describe the scientific method, and provide an example of how it could be applied to an everyday problem.

LG7 Explain why experiments include control groups.

LG8 Explain how a hypothesis differs from a scientific theory.

1.1 WHAT IS LIFE?

The word **biology** comes from the Greek roots "bio" meaning "life" and "logy" meaning "the study of" (see Appendix I for more word roots). But what is life? If you look up "life" in a dictionary, you will find definitions such as "the quality that distinguishes a vital and functioning being from a dead body," but you won't discover what that "quality" is. Life is intangible and defies simple definition, even by biologists. However, most will agree that living things, or **organisms,** all share certain characteristics that, taken together, define life:

- Organisms acquire and use materials and energy.
- Organisms actively maintain organized complexity.
- Organisms perceive and respond to stimuli.
- Organisms grow.
- Organisms reproduce.
- Organisms, collectively, have the capacity to evolve.

Nonliving objects may possess some of these attributes. Crystals can grow, a desk lamp acquires energy from electricity and converts it to heat and light, but only life-forms can do it all.

Scientists in the early 1800s, observing life with simple microscopes, concluded that the **cell** is the basic unit of life (**Fig. 1-1**). A membrane separates the cell from its surroundings, enclosing a huge variety of structures and chemicals in a fluid environment. The components within the cell are precisely organized. This organization allows a staggering number of chemical reactions to occur in the proper order, at the proper location, and at the proper time so that the cell can maintain and reproduce itself.

Although the most abundant organisms on Earth exist as single cells, the qualities of life are more easily visualized in multicellular organisms such as the water flea in **Figure 1-2,** an animal smaller than this letter "o." In the sections below, we explore the characteristics of life in more depth.

Organisms Acquire and Use Materials and Energy

Organisms require both materials and energy to perform the functions of life. They obtain the materials that make up their bodies—such as minerals, water, and other simple

nucleus
cell wall
plasma membrane
organelles

▲ **FIGURE 1-1 The cell is the smallest unit of life** This artificially colored micrograph of a plant cell (a eukaryotic cell) shows a supporting cell wall (blue) that surrounds plant (but not animal) cells. Just inside the cell wall, the plasma membrane (found in all cells) has control over which substances enter and leave. Cells also contain several types of specialized organelles, including the nucleus, suspended within a fluid environment (orange).

chemical building blocks—from the air, water, soil, and from the bodies of other living things. Because life neither creates nor destroys matter, materials are continuously exchanged and recycled among organisms and their nonliving surroundings (**Fig. 1-3**).

Organisms use energy continuously to remain alive. For example, energy is required to move and to construct the complex molecules that make up an organism's body. Ultimately, the energy that sustains life comes from sunlight. Certain organisms directly capture and store solar energy using a process called **photosynthesis.** Photosynthetic organisms trap and store energy for their own use, but the energy stored in their bodies also powers all other forms of life, such as fungi and animals (ourselves included). Thus, energy flows in a one-way path from the sun through all forms of life. Eventually, it is released back to the environment as heat. The continuous flow of energy through organisms is illustrated in Figure 1-3.

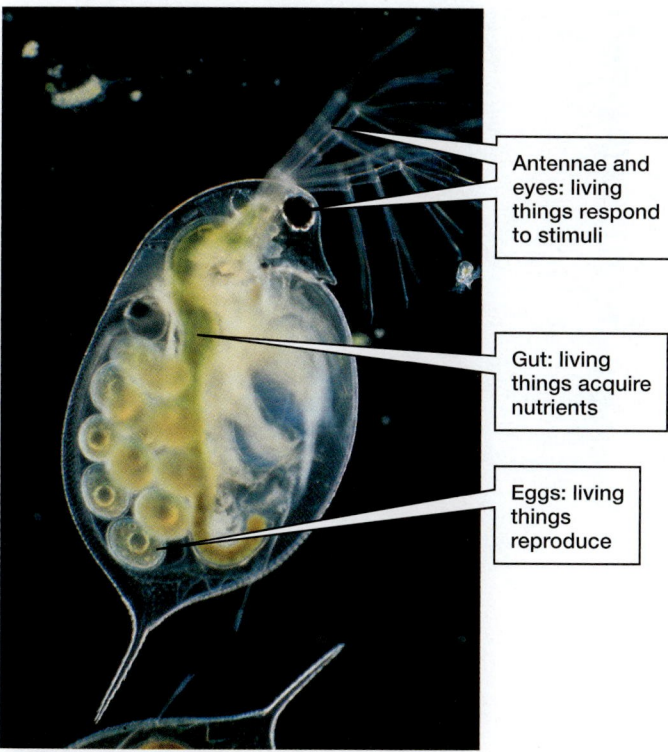

▲ FIGURE 1-2 Properties of life The water flea uses energy from photosynthetic organisms (seen as green material in its gut) to maintain its amazing complexity. Eyes and antennae respond to stimuli. This adult female is reproducing, and she herself has grown from an egg like those she now carries. All the adaptations that allow this water flea to survive, grow, and reproduce have been molded by evolution.

Antennae and eyes: living things respond to stimuli

Gut: living things acquire nutrients

Eggs: living things reproduce

⟫⟫ **LINKS** TO EVERYDAY LIFE

A Solar-Powered Big Mac?

If you've just eaten a hamburger for lunch, you may be surprised to learn that all the energy in it comes from sunlight. The bun is made from wheat, a photosynthetic organism, so it's clear that the energy in the bun comes from the sun. But how does a cow pack solar energy into its fat and muscle? To understand, think about what the cow eats. A cow's energy comes from grass and other plant material. The plants, in turn, harvested their energy via photosynthesis. Thus, the energy powering a cow is one step removed from the sunlight that energizes the plant. So, when you eat a hamburger, you consume energy that is only two steps removed from its original source: the sun.

membrane constantly pumps some chemicals in and others out. People and other mammals use energy to keep their internal organs at a warm, constant temperature that allows the proper reactions to occur in their cells. To prevent body temperature from rising too much during hot weather or vigorous exercise, we sweat and we might douse ourselves with cool water (**Fig. 1-4**). When it is cold, our bodies and those of other mammals metabolize more food than usual to generate heat. Life, then, requires very precise internal conditions that organisms must expend energy constantly to maintain.

Organisms Perceive and Respond to Stimuli

To obtain energy and nutrients and to stay alive, organisms must perceive and respond to their environments. Animals use specialized cells to detect light, temperature, sound, gravity, touch, chemicals, and many other stimuli from their external and internal surroundings. For example, when your brain detects low levels of sugar in your blood (an internal stimulus), it causes your mouth to water at the smell of food (an external stimulus). Animals have nervous and muscular systems that perceive and respond to stimuli, but plants,

Organisms Actively Maintain Organized Complexity

Whether we consider the books and papers on your desk or the components of a cell, organization tends to disintegrate into chaos unless energy is used to maintain it. (We explore this further in Chapter 6.) Living things, which are far more complex than anything people might organize, must use energy continuously to maintain themselves. For the appropriate chemical reactions to occur within a cell, for example, its

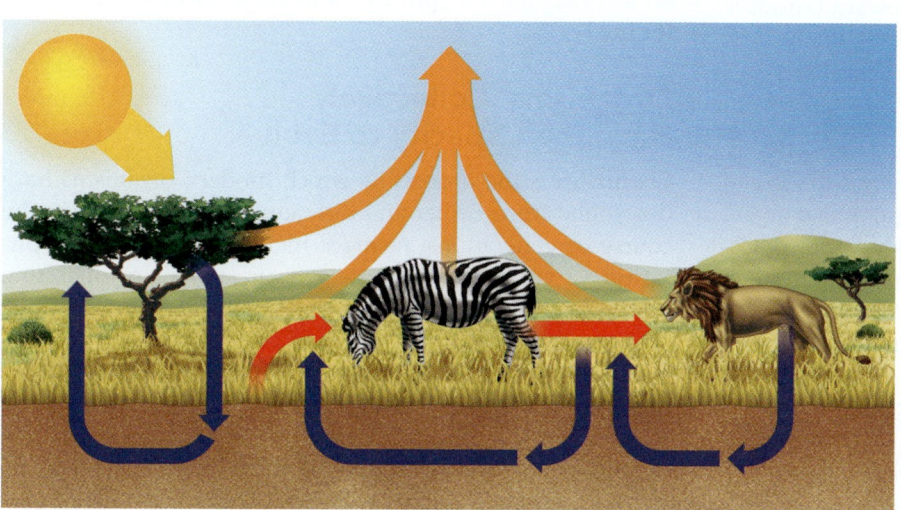

◄ FIGURE 1-3 The flow of energy and the recycling of nutrients Energy is acquired from sunlight (yellow arrow) by photosynthetic organisms, is transferred through organisms that consume other forms of life (red arrows), and is lost as heat (orange arrows) in a one-way flow. In contrast, nutrients (purple arrows) are recycled among organisms and the nonliving environment.

▲ **FIGURE 1-4 Organisms maintain relatively constant internal conditions** Evaporative cooling by water, both from sweat and from a bottle, helps this athlete maintain his body temperature during vigorous exercise.

▲ **FIGURE 1-5 Bending toward the light** Plants perceive and respond to light, which provides them with energy.

fungi, and single-celled organisms use very different mechanisms that are equally effective for their needs. Think of a plant on a windowsill sensing and then bending toward sunlight that provides energy (**Fig. 1-5**). Even many bacteria, the smallest and simplest life-forms, can move toward favorable conditions and away from harmful substances.

Organisms Grow

At some time in its life, every organism becomes larger—that is, it grows. The water flea in Figure 1-2 originally was the size of one of the eggs you see in its body. Growth requires that organisms obtain material and energy from the outside world. Single-celled organisms such as bacteria grow to about double their original size, copy their genetic material, and then divide in half to reproduce. Animals and plants use a similar process to produce more cells within their bodies, repeating the sequence until growth stops. Individual cells can also contribute to the growth of an organism by increasing in size, as occurs in muscle and fat cells in animals, and in food storage cells in plants.

Organisms Reproduce

Organisms reproduce in a variety of ways that include dividing in half, producing seeds, bearing live young (**Fig. 1-6**), and laying eggs (see Fig. 1-2). The end result is always the same: new versions of the parent organisms. Be it bacterium, slug, mushroom, or person, offspring inherit from their parents the instructions for producing and maintaining their particular forms of life. These instructions—copied in every cell and passed on to offspring—are carried in the unique structure of the hereditary molecule **deoxyribonucleic acid,** or **DNA** (**Fig. 1-7**). The complete set of DNA molecules contained in each cell of an organism can be compared to an architectural blueprint or a detailed instruction manual for life.

Organisms, Collectively, Have the Capacity to Evolve

Evolution is the process by which modern organisms have descended from earlier and different forms of life. Over the course of generations, changes in DNA within any **population** (a group of the same type of organism inhabiting the same area) are inevitable, and result in evolution. Such changes in Earth's populations of organisms, accumulating over the history of life, have resulted in the evolution of the amazing diversity of organisms that now share this planet. The next section provides a brief introduction to evolution—the most important concept in biology.

(a) Dividing *Streptococcus* bacterium

(b) Dandelion producing seeds

(c) Panda with its baby

▲ FIGURE 1-6 **Organisms reproduce**

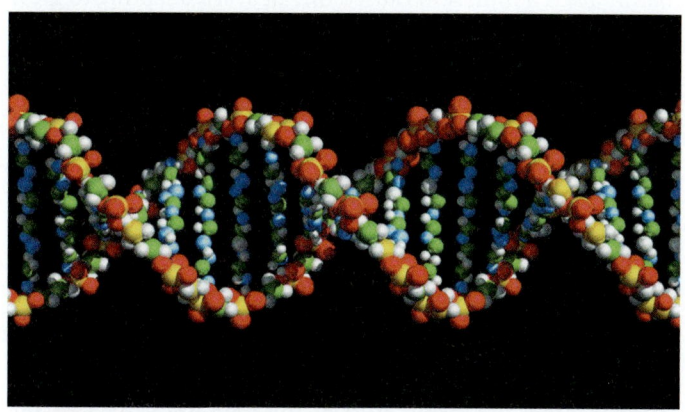

▲ FIGURE 1-7 **DNA** A model of DNA, the molecule of heredity. As James Watson, its codiscoverer, put it: "A structure this pretty just had to exist."

CASE STUDY c o n t i n u e d

The Boundaries of Life

How lifelike is a virus? Viruses release their genetic material inside cells, and hijack the infected cell's energy supplies and biochemical machinery, turning the cell into a kind of factory that churns out many copies of viral parts. These parts assemble into an army of virus particles, in a process called *replication.* The newly formed viruses then emerge from the host cell, often rupturing it in the process. Some, including HIV and the Ebola virus, acquire an outer envelope made of the host's plasma membrane. Viruses do not obtain or use their own energy or materials, they don't maintain themselves, nor do they grow. In fact, they possess only a few of the characteristics of life: Viruses respond to stimuli by binding to specific sites on the cells they attack, they evolve, and some scientists consider viral replication to be reproduction, another characteristic of life.

CHECK YOUR LEARNING

Can you describe the characteristics that define life?

1.2 WHAT IS EVOLUTION?

Evolution explains why Earth has such a huge diversity of life. At the same time, evolution accounts for the remarkable similarities among different types of organisms. For example, people have many physical features in common with chimpanzees, and our DNA differs from that of chimps by less than 5%. This is strong evidence that people and chimps descended from a common ancestor (**Fig. 1-8**), but the obvious differences reflect the differences in our evolutionary paths. How does evolution occur?

Three Natural Processes Underlie Evolution

Evolution is an automatic and inevitable outcome of three natural occurrences: (1) differences in DNA among members of a population, (2) inheritance of these differences by their offspring, and (3) *natural selection,* the increased ability of offspring that inherit certain forms and combinations of DNA molecules to survive and reproduce. Let's take a closer look at these three factors.

Mutations Are the Original Source of Differences in DNA

Look around at your classmates and notice how different they are, or observe differences among dogs in their sizes and shapes, and in the color, length, and texture of their coats. Although some of this variation (particularly among your classmates) is due to differences in environment and lifestyle, much of it is due to differences in their genes. **Genes,** which are segments of DNA, are the basic units of heredity. Before a cell divides, all of its DNA must be copied so its genes can be passed along. Just as you would make mistakes if you tried

▲ FIGURE 1-8 **Chimps and people are closely related**

to copy a blueprint by hand, cells make some errors as they copy their DNA. Changes in genes, such as those caused by these random copying errors, are called **mutations.** Mutations can also result from damage to DNA, caused for example by ultraviolet rays from sunlight, radiation released from a damaged nuclear power plant, or toxic chemicals from cigarette smoke. Just as mistakes in a blueprint will make a difference in the structure built from it, so a new cell constructed from altered DNA is likely to differ from its parent cell.

Some Mutations Are Inherited

Inherited mutations are those that occur in sperm or egg cells, thereby allowing altered DNA to pass from parent to offspring. Each cell in the offspring will carry the inherited mutation, giving the offspring traits that differ, either subtly or obviously, from its parents. Some mutations are harmful, for example, those that cause human genetic disorders such as hemophilia, sickle cell anemia, or cystic fibrosis. Many mutations have no observable effect or may change the organism in a way that is neither harmful nor beneficial, for example, by producing a different coat color on a dog. Almost all of the natural variability among traits in individuals is caused by neutral mutations that occurred in the

distant past and have been passed along through numerous generations. On rare occasions, however, an inherited mutation helps offspring to survive and reproduce more successfully than those lacking the mutation. Such infrequent events provide the raw material for evolution.

Some Inherited Mutations Help Individuals Survive and Reproduce

The most important process in evolution is natural selection, which acts on the natural variability in traits among members of a population. **Natural selection** is the process by which organisms with certain inherited traits survive and reproduce better than others in a given environment. Inherited traits that help organisms produce more healthy offspring become increasingly common in the population over many generations. Because these traits are caused by differences in genes, the genetic makeup of the population as a whole will change over time; that is, the population evolves. How might this work?

Consider a likely scenario of natural selection. Imagine that ancient beavers had short front teeth like most other mammals. If a mutation caused one beaver's offspring to grow longer front teeth, these offspring would have gnawed down trees more efficiently, built bigger dams and lodges, and eaten more bark than could beavers that lacked the mutation. Because these long-toothed beavers would obtain more food and better shelter, they would be better able to survive when times got tough, and to raise more offspring who would have inherited the genes for longer front teeth. Over time, long-toothed beavers would have become increasingly common; after many generations, all beavers would have long front teeth.

Structures, physiological processes, or behaviors that help an organism survive and reproduce in a particular environment are called **adaptations.** Most of the features that we admire so much in other life-forms, such as the long limbs of deer, the broad wings of eagles, and the mighty trunks of redwood trees, are adaptations. Adaptations help organisms escape predators, capture prey, reach the sunlight, or accomplish other feats that help ensure their survival and reproduction. The huge array of adaptations found in living things today were molded by millions of years of natural selection acting on random mutations.

What helps an organism survive today may become a liability in the future. If environments change—for example, as global climate change occurs—the genetic traits that best adapt organisms to their environments will change as well. In the case of global climate change, if a random mutation helps an organism survive and reproduce in a warmer climate, the mutation will be favored by natural selection and will increase within the population with each new generation.

Populations of the same species (organisms of the same type that can interbreed) that live in different environments will be subjected to different types of natural selection. If these differences are great enough and continue for long enough, natural selection may eventually cause the populations to become sufficiently different from one another to prevent interbreeding—a new species will have evolved.

If mutations that help an organism to adapt do not occur, a changing environment may doom a species to

▲ FIGURE 1-9 A fossil *Tyrannosaurus rex* The most widely accepted hypothesis for the extinction of dinosaurs is a massive meteorite strike that rapidly and radically altered their environment. This *Tyrannosaurus rex* (dubbed "Thomas" and displayed at the Natural History Museum of Los Angeles County) was excavated in Montana between 2003 and 2005. Weighing in at 7,000 pounds and 34 feet long, Thomas was 17 years old when he died roughly 68 million years ago.

extinction—the complete elimination of this form of life. Dinosaurs flourished for 100 million years, but they did not evolve fast enough to adapt to rapidly changing conditions, so they became extinct (**Fig. 1-9**). In recent decades, human activities such as burning fossil fuels and converting tropical forests to farmland have drastically accelerated the rate of environmental change. Mutations that better adapt organisms to their environments are quite rare, and consequently the rate of extinction has increased dramatically (described in Chapter 30).

CASE STUDY c o n t i n u e d

The Boundaries of Life

One lifelike property of viruses is their capacity to evolve. Through evolution, viruses sometimes become more infectious or more deadly, or they may gain the ability to infect new hosts. Certain types of viruses, which include Ebola, HIV, and influenza, are very sloppy in copying their genetic material and mutate about 1,000 times as often as the average animal cell. As a result, nearly every new virus carries a mutation. One consequence is that viruses such as flu can rapidly evolve into new forms, so you need a new flu shot every year. Antiviral drugs act as agents of natural selection and cause drug-resistant populations of viruses to evolve. HIV, for example, may produce 10 billion new viruses daily in an infected individual, making drug resistance inevitable. For this reason, people with AIDS are given "cocktails" of three or four different drugs, because resistance to all of them would require multiple specific mutations to occur in the same virus, an enormously unlikely event.

In the words of biologist Theodosius Dobzhansky, "Nothing in biology makes sense except in the light of evolution." The theory of evolution was formulated in the mid-1800s by two English naturalists, Charles Darwin and Alfred Russel Wallace. Since that time, it has been supported by fossils, geological studies, radioactive dating of rocks, genetics, molecular biology, biochemistry, and breeding experiments. As you will learn in the next section, people who call evolution "just a theory" profoundly misunderstand the concept of a scientific theory.

CHECK YOUR LEARNING

Can you explain how three natural processes lead inevitably to evolution?

1.3 HOW DO SCIENTISTS STUDY LIFE?

The science of biology encompasses many different areas of inquiry, each requiring different types of specialized knowledge. One way to get a sense of the scope of biology is to look at the levels of organization that occur within the living world (**Fig. 1-10**). At a large university, you may find biologists conducting research at every level, from the study of complex biological molecules (for example, how changes in DNA affect specific traits) to the biosphere (for example, how the distribution of organisms on Earth may be altered by climate change).

Life Can Be Studied at Different Levels

Figure 1-10 illustrates different levels of inquiry that make up the study of life. Starting at the bottom, each level provides the foundation for the one above it, and each higher level has new, more inclusive properties. Nearly all biological inquiry encompasses more than one of these levels, which we explore in the sections that follow.

All matter consists of **elements,** substances that cannot be broken down or converted to a simpler substance. An **atom** is the smallest particle of an element that retains all the properties of that element. For example, a diamond is a form of the element carbon. The smallest possible unit of a diamond is an individual carbon atom. Atoms may combine in specific ways to form **molecules;** for example, one oxygen atom can combine with two hydrogen atoms to form a molecule of water. Complex biological molecules containing carbon atoms—such as proteins and DNA—form the building blocks of cells, which are the basic units of life. Although many organisms exist as single cells, in multicellular organisms, cells of a similar type may combine to form **tissues,** such as the epithelial tissue that lines the stomach. Different types of tissues, in turn, unite to form functional units called **organs,** such as the entire stomach. The grouping of two or more organs that work together to perform a specific body function is called an **organ system;** for example, the stomach is part of the digestive system. Organ systems combine within complex multicellular organisms to carry out the activities of life.

Biosphere	All life on Earth and the nonliving portions of Earth that support life	
Ecosystem	A community together with its nonliving surroundings	Earth's surface snake, antelope, hawk, bushes, grass, rocks, stream
Community	Populations of different species that live in the same area and interact with one another	snake, antelope, hawk, bushes, grass
Species	All organisms that are similar enough to interbreed	
Population	All the members of a species living in the same area	herd of pronghorn antelope
Multicellular organism	An individual living thing composed of many cells	pronghorn antelope
Organ system	Two or more organs working together in the execution of a specific bodily function	the digestive system
Organ	A structure usually composed of several tissue types that form a functional unit	the stomach
Tissue	A group of similar cells that perform a specific function	epithelial tissue
Cell	The smallest unit of life	red blood cell epithelial cell nerve cell
Molecule	A combination of atoms	water glucose DNA
Atom	The smallest particle of an element that retains the properties of that element	hydrogen carbon nitrogen oxygen

▲ FIGURE 1-10 **Levels of biological organization** Each lower level provides building blocks for the one above it, which has new properties that emerge from the interplay of the levels beneath.

QUESTION What current, ongoing environmental change is likely to affect the entire biosphere?

Levels of organization also extend to groups of organisms. A population is a group of organisms of the same type (the same species) that live in a well-defined area where they can interact and interbreed with one another. A **species** consists of all organisms that are similar enough to interbreed, no matter where they are found.

A **community** is formed by populations of different species that live in the same area and interact with one another. An **ecosystem** consists of a community and the nonliving environment that surrounds it. Finally, the **biosphere** includes all life on Earth and the nonliving portions of Earth that support life.

Biologists Classify Organisms Based on Their Evolutionary Relationships

Although all forms of life share certain characteristics, evolution has produced an amazing variety of life-forms. Scientists classify organisms based on their evolutionary relatedness, placing them into three major groups, or **domains:** Bacteria, Archaea, and Eukarya (**Fig. 1-11**).

This classification reflects fundamental differences among their cell types. Members of both the Bacteria and the Archaea consist of a single, simple cell. At the molecular level, however, there are fundamental differences between them that indicate that they are only distantly related.

In contrast, members of the Eukarya have bodies composed of one or more extremely complex cells. The domain Eukarya includes a diverse collection of organisms collectively known as *protists*, and three major subdivisions called **kingdoms,** which are the fungi, plants, and animals. (You will learn far more about life's incredible diversity and how it evolved in Unit 3.)

The classification of a given organism into a domain and kingdom is based on three characteristics: the organism's cell type (simple or complex), whether the organism is **unicellular** (a single cell) or **multicellular** (composed of many cells), and how the organism acquires its energy (**Table 1-1**).

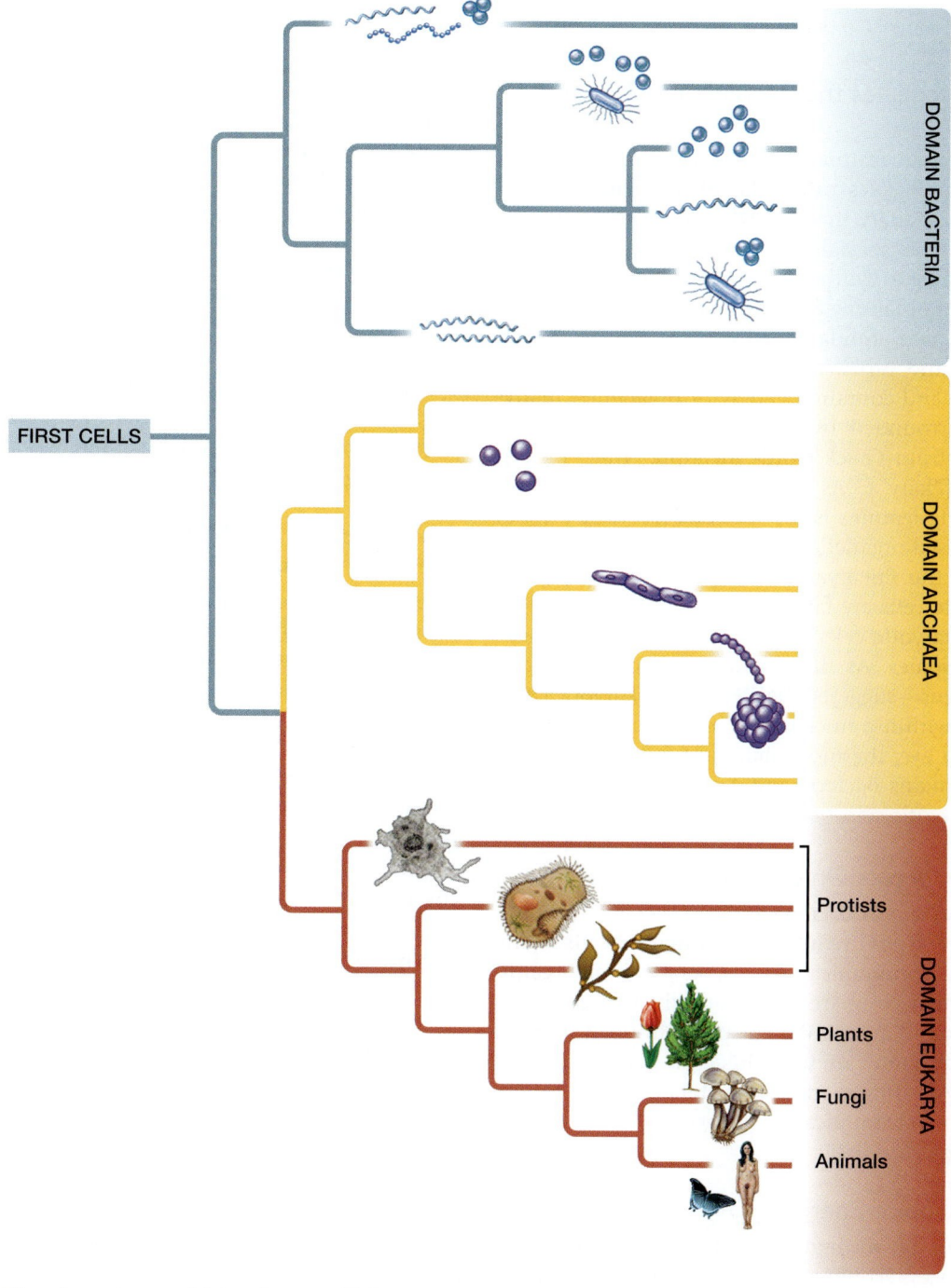

FIRST CELLS

DOMAIN BACTERIA

DOMAIN ARCHAEA

DOMAIN EUKARYA

Protists

Plants

Fungi

Animals

▲ **FIGURE 1-11 The domains and kingdoms of life**

TABLE 1-1   Some Characteristics Used in Classification of Organisms

Domain	Kingdom	Cell Type	Cell Number	Energy Acquisition
Bacteria	Not applicable	Prokaryotic	Unicellular	Autotrophic or heterotrophic
Archaea	Not applicable	Prokaryotic	Unicellular	Autotrophic or heterotrophic
Eukarya	Fungi	Eukaryotic	Multicellular	Heterotrophic, absorb food
	Plants	Eukaryotic	Multicellular	Autotrophic
	Animals	Eukaryotic	Multicellular	Heterotrophic, ingest food
	Protists*	Eukaryotic	Unicellullar or multicellular	Autotrophic or heterotrophic

*Protists are a diverse collection of organisms; their classification is under discussion in the scientific community.

Cell Type Distinguishes the Bacteria and Archaea from the Eukarya

Cells all share certain features. All cells are surrounded by a thin sheet of molecules called the **plasma membrane.** All contain the hereditary material DNA. Cells also contain **organelles,** structures specialized to carry out specific functions such as helping to synthesize large molecules, digesting food molecules, or obtaining energy (see Fig. 1-1).

There are two fundamentally different types of cells: eukaryotic and prokaryotic. **Eukaryotic** cells are extremely complex and contain a variety of organelles, many of which are surrounded by membranes. The term "eukaryotic" comes from Greek words meaning "true" ("eu") and "nucleus" ("kary"). As the name suggests, one of their most striking features is the **nucleus** (see Fig. 1-1), a membrane-enclosed organelle that contains the cell's DNA. All members of the Eukarya are composed of eukaryotic cells. **Prokaryotic** cells are far simpler and generally much smaller than eukaryotic cells, and they lack organelles enclosed by membranes. As their name—meaning "before" ("pro") the nucleus—suggests, the DNA of prokaryotic cells is not confined within a nucleus. Although they are invisible to the naked eye, the most abundant forms of life are found in the domains Bacteria and Archaea, which consist entirely of prokaryotic cells.

Multicellularity Occurs Only Among the Eukarya

Members of the domains Bacteria and Archaea are unicellular. Although a few types live in strands or mats of cells, there is relatively little communication, cooperation, or organization among them compared to multicellular organisms, which are only found among the Eukarya. Although protists are eukaryotic and many are unicellular, all plants and animals, and nearly all fungi, are multicellular; their lives depend on intimate communication and cooperation among numerous specialized cells.

The Eukarya Acquire Energy in Different Ways

One simple way of distinguishing the three kingdoms within the Eukarya is by how they obtain energy; plants are autotrophs, animals are heterotrophs and ingest (eat) their food, and fungi are heterotrophic and absorb their food (see Table 1-1). **Autotroph** means "self" ("auto") and "feeding" ("trophic"); these organisms acquire their energy through photosynthesis. Members of the plant kingdom are autotrophs (as are some archaea, some bacteria, and some protists). **Heterotroph** means "other" ("hetero") and "feeding" ("trophic"); heterotrophs cannot photosynthesize and so they rely on the bodies of other organisms for food. Some heterotrophs digest food outside their bodies and absorb the simpler molecules through their plasma membranes. These include the fungi (and many members of the domains Archaea and Bacteria). Animals (and heterotrophic protists) usually ingest chunks of food.

Biologists Use the Binomial System to Name Organisms

Biologists place organisms in a hierarchy of categories, each more inclusive than the one below it. The smallest two categories are the *genus* and *species*. A species consists of nearly identical organisms that can interbreed, and a genus includes many different species with similar characteristics. To provide a unique scientific name for each form of life, biologists use a **binomial system,** based on Latin words meaning "two" ("bi") and "names" ("nomial"), which consists of the genus and species. The genus name is capitalized, and both are italicized. The animal in Figure 1-2, for example, has the common name "water flea," but there are many types of water fleas, and people who study them need to be more precise. So *Daphnia longispina* (the water flea in Fig. 1-2) is in the genus *Daphnia* (which includes many similar species of water fleas) and the species *longispina* (referring to this particular water flea's long, tail-like spine). People are classified as *Homo sapiens*; we are the only members of this genus and species.

CHECK YOUR LEARNING

Can you describe the levels of biological organization? Can you explain how scientists name and categorize diverse forms of life?

CASE STUDY continued

The Boundaries of Life

Viruses are not included in the classification scheme of life. Viruses appear to be very ancient, although their evolutionary relationships to living things remain obscure and controversial. Some scientists hypothesize that viruses may be remnants of a precellular stage of evolution. Others suggest that viruses may have evolved from molecules of genetic material that escaped from living cells. Some scientists argue that viruses could have evolved from cells, losing virtually all cellular structures and functions. Viruses show such diversity that all of these scenarios could be possible.

1.4 WHAT IS SCIENCE?

Science can be defined as the systematic inquiry—through observation and experiment—into the origins, structure, and behavior of our living and nonliving surroundings.

Science Is Based on the Principle That All Events Have Natural Causes

Early in human history, most people operated on the premise that supernatural forces were responsible for natural events. Ancient Greeks believed that the god Zeus hurled lightning bolts from the sky, and that epileptic seizures were evidence of a visitation from the gods. It was common in the Middle Ages to believe that life arose spontaneously (*spontaneous generation*) from nonliving matter; for example, that maggots came from rotting meat.

Science, in contrast, is based on the principle that all events can be traced to natural causes that people can study objectively. Today, we realize that lightning is a massive electrical discharge, that epilepsy is a brain disorder caused by uncontrolled firing of nerve cells, and that maggots are the larval form of flies.

The Scientific Method Is an Important Tool of Scientific Inquiry

The principle that the events we observe have natural causes, even if we do not yet understand them, underlies all scientific inquiry. To learn about the world, scientists in many disciplines, including biology, use some version of the **scientific method.** This consists of six interrelated elements: *observation, question, hypothesis, prediction, experiment,* and *conclusion.* Scientific inquiry begins with an **observation** of a specific phenomenon. The observation, in turn, leads to a **question:** "What caused this?" Then, after careful study of earlier investigations, conversations with colleagues, and often after long, hard thought, a hypothesis is formed. A **hypothesis** is a proposed explanation for the phenomenon, often based on limited evidence. To be useful, the hypothesis must lead to a **prediction,** which is the expected outcome of testing if the hypothesis is correct. The prediction is tested by carefully designed additional observations or carefully controlled manipulations called **experiments.** Experiments produce results that either support or refute the hypothesis, allowing the scientist to reach a **conclusion** about whether the hypothesis is valid or not. For the conclusion to be valid, the experiment and its results must be repeatable not only by the original researcher but also by others. Like scientists, we use versions of the scientific method, although far less rigorous, in our daily lives (see "Links to Everyday Life: When Your Car Won't Start").

Biologists Test Hypotheses Using Controlled Experiments

In controlled experiments, two types of situations are established. One is a baseline or **control** situation, in which all possible variables are held constant. The other is the experimental situation, where one factor, the **variable,** is manipulated to test the hypothesis that this variable is the cause of an observation. Often the manipulation inevitably causes more than one factor to vary. For example, in "Links to Everyday Life: When Your Car Won't Start," jump starting the car might have knocked some corrosion off the battery terminal that was preventing the battery from delivering power—your battery might be fine! In real experiments, scientists must control for all the effects of any manipulation they perform, so frequently more than one control is needed. You will find a good example of this in Malte Andersson's experiments, described in "Science in Action: Controlled Experiments, Then and Now."

Valid scientific experiments must be repeatable by the researcher and by other scientists. To help ensure this, the researcher performs multiple repetitions of any experiment, often simultaneously, by setting up several replications of each control situation (*control group*) and each experimental situation (*experimental group*), all as identical as possible.

Science must also be communicated or it is useless. Good scientists publish their results, explaining their methods in detail, so others can repeat them and build on them. Francesco Redi recognized this in the 1600s when he carefully recorded the methods of his classic controlled experiment testing the hypothesis that flies caused maggots to appear on rotting meat (see "Science in Action: Controlled Experiments, Then and Now").

 LINKS TO EVERYDAY LIFE

When Your Car Won't Start

You probably use some variation of the scientific method to solve everyday problems. For example, late for an important date, you rush to your car, turn the ignition key, and make the *observation* that the car won't start. Your *question* "Why won't the car start?" leads to a *hypothesis:* The battery is dead. This leads to the *prediction* that a jump start will solve the problem. You *experiment* by attaching jumper cables from your roommate's car battery to your own. The result? Your car starts immediately, allowing the *conclusion* that your hypothesis was correct. If jump starting didn't work, what hypothesis would you test next?

SCIENCE IN ACTION   Controlled Experiments, Then and Now

A classic experiment by the Italian physician Francesco Redi (1621–1697) beautifully demonstrates the scientific method and helps to illustrate the scientific principle that events can be traced to natural causes. Redi investigated why maggots (the larva of flies) appear on spoiled meat. In Redi's time, refrigeration was unknown, and meat was stored in the open. Many people of that time believed that the appearance of maggots on meat was evidence of **spontaneous generation,** the emergence of life from nonliving matter.

Redi *observed* that flies swarm around fresh meat and that maggots appear on meat left out for a few days. He *questioned* where the maggots came from. He then formed a testable *hypothesis:* Flies produce maggots. This led to the *prediction* that keeping flies off the meat would prevent maggots from appearing. In his *experiment,* Redi wanted to test one variable—the access of flies to the meat. Therefore, he placed similar pieces of meat in each of two clean jars. He left one jar open (the control jar) and covered the other with gauze to keep out flies (the experimental jar). He did his best to keep all the other conditions the same (for example, the type of jar, the type of meat, and the temperature). After a few days, he observed maggots on the meat in the open jar but saw none on the meat in the covered jar. Redi *concluded* that his hypothesis was correct and that maggots are produced by flies, not by the nonliving meat (**Fig. E1-1**). Only through this and other controlled experiments could the age-old belief in spontaneous generation be laid to rest.

Today, more than 300 years later, the scientific method is still used. Consider the recent experiments of Malte Andersson, who investigated the mating choices of female widowbirds. Andersson *observed* that male, but not female, widowbirds have extravagantly long tails, which they display while flying across African grasslands. Andersson asked the *question:* Why do male birds have such long tails? His *hypothesis* was that females prefer to mate with long-tailed males, and so these males have more offspring, who inherit their genes for long tails.

Observation:	Flies swarm around meat left in the open; maggots appear on the meat.
Question:	Where do maggots on the meat come from?
Hypothesis:	Flies produce the maggots.
Prediction:	IF the hypothesis is correct, THEN keeping the flies away from the meat will prevent the appearance of maggots.

Experiment:

Obtain identical pieces of meat and two identical jars

Place meat in each jar

Control situation		Experimental situation
Leave the jar uncovered	**Experimental variable:** gauze prevents the entry of flies	Cover the jar with gauze
Leave exposed for several days	**Controlled variables:** time, temperature, place	Leave covered for several days
Flies swarm around and maggots appear	**Results**	Flies are kept from the meat; no maggots appear

Conclusion:	The experiment supports the hypothesis that flies are the source of maggots and that spontaneous generation of maggots does not occur.

▲ **FIGURE E1-1 The experiment of Francesco Redi illustrates the scientific method**

QUESTION Redi's experiment falsified spontaneous generation of maggots, but did his experiment convincingly demonstrate that flies produce maggots? What kind of follow-up experiment would be needed to more conclusively determine the source of maggots?

Andersson *predicted* that if his hypothesis were true, more females would build nests on the territories of males with artificially lengthened tails than on the territories of males with artificially shortened tails. To test this, he captured some males, trimmed their tails to about half their original length, and released them (*experimental* group 1). He took another group of males and glued on the tail feathers that he had removed from the first group, creating exceptionally long tails (*experimental* group 2). Finally, Andersson had two control groups. In one, the tail was cut and then glued back in place (to control for the effects of capturing the birds and manipulating their feathers). In the other control group, he simply captured and released the male birds to control for any behavioral changes caused by the stress of being caught and handled. Later, Andersson counted the number of nests that females had built on each male's territory (this indicated how many females had mated with that male). He found that males with lengthened tails had the most nests on their territories, males with shortened tails had the fewest, and control males (with normal-length tails, either untouched or cut and glued together) had an intermediate number (**Fig. E1-2**). Andersson *concluded* that his results supported the hypothesis that female widowbirds prefer to mate with long-tailed males.

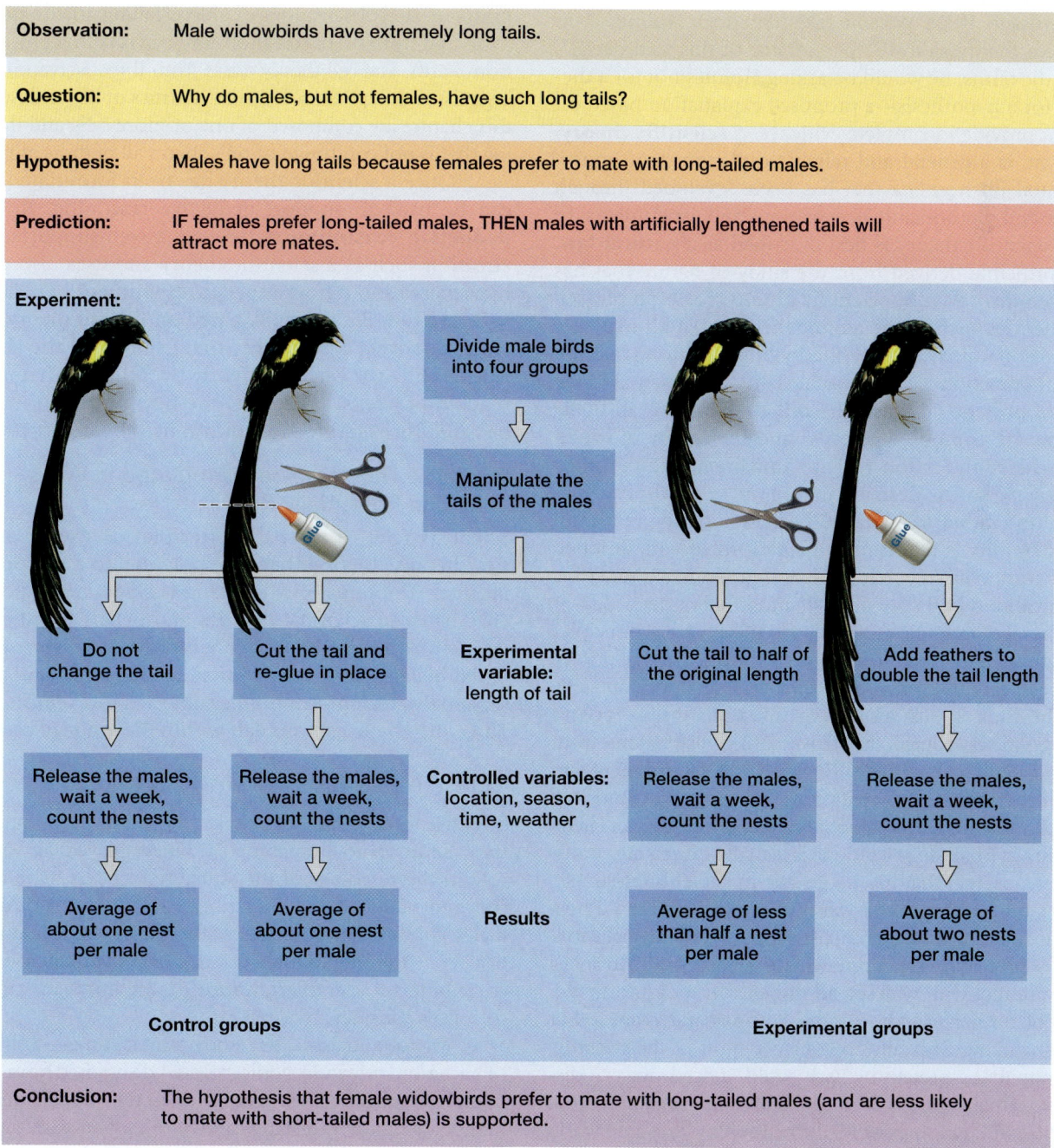

Observation: Male widowbirds have extremely long tails.

Question: Why do males, but not females, have such long tails?

Hypothesis: Males have long tails because females prefer to mate with long-tailed males.

Prediction: IF females prefer long-tailed males, THEN males with artificially lengthened tails will attract more mates.

Experiment:

Divide male birds into four groups

Manipulate the tails of the males

Do not change the tail → Release the males, wait a week, count the nests → Average of about one nest per male

Cut the tail and re-glue in place → Release the males, wait a week, count the nests → Average of about one nest per male

Experimental variable: length of tail

Controlled variables: location, season, time, weather

Results

Cut the tail to half of the original length → Release the males, wait a week, count the nests → Average of less than half a nest per male

Add feathers to double the tail length → Release the males, wait a week, count the nests → Average of about two nests per male

Control groups   **Experimental groups**

Conclusion: The hypothesis that female widowbirds prefer to mate with long-tailed males (and are less likely to mate with short-tailed males) is supported.

▲ FIGURE E1-2   **The experiment of Malte Andersson**

Experimentation using variables and controls is powerful, but it is important to recognize its limitations. In particular, scientists can seldom be sure that they have controlled for *all* possible variables, or made all the observations that could possibly refute their hypothesis. Therefore, science mandates that conclusions are always subject to revision if new experiments or observations contradict them.

Scientific Theories Have Been Thoroughly Tested

Scientists use the word "theory" in a way that differs from its everyday usage. If Dr. Watson asked Sherlock Holmes, "Do you have a theory as to the perpetrator of this foul deed?", in scientific terms, he would be asking Holmes not for a theory, but for a hypothesis—a proposed explanation based on clues that provide incomplete evidence. A **scientific theory,** in contrast, is a general and reliable explanation of important natural phenomena that has been developed through extensive and reproducible observations and experiments. In short, a scientific theory is best described as a **natural law,** a basic principle derived from the study of nature that has never been disproven by scientific inquiry. For example, scientific theories such as the atomic theory (that all matter is composed of atoms) and the theory of gravitation (that objects exert attraction for one another) are fundamental to the science of physics. Likewise, the **cell theory** (that all living organisms are composed of cells) and the theory of evolution (discussed in Section 1.2) are fundamental to the study of biology. Scientists describe these fundamental principles as "theories" rather than "facts" because a basic premise of scientific inquiry is that it must be performed with an open mind; if compelling evidence arises that renders a scientific theory invalid, that theory must be modified or discarded.

A modern example of the need to modify basic principles in the light of new scientific evidence is the discovery of prions, which are infectious proteins (see the case study for Chapter 3). Before the early 1980s, all known infectious disease agents, including viruses, copied themselves using instructions from genetic material, such as DNA. Then in 1982, neurologist Stanley Prusiner from the University of California at San Francisco published evidence that scrapie (an infectious disease of sheep that results in brain degeneration) is actually caused and transmitted by a protein that has no genetic material. Interestingly, Prusiner's original hypothesis was that scrapie was caused by a virus, but none of his experiments supported this conclusion. Infectious proteins were unknown to science, and Prusiner's results were met with widespread disbelief. It took nearly two decades of further experiments by Prusiner and many other scientists working with him to convince most of the scientific community that a protein by itself could, indeed, act as a disease agent. Finally, in 1997, Stanley Prusiner was awarded the Nobel Prize in Physiology or Medicine for his pioneering work, which introduced an entirely new biological principle.

Prions are now recognized as the agent causing mad cow disease (also called BSE for "bovine spongiform encephalitis"), which has killed not only cattle but also about 200 people who ate infected beef. Prions also cause Creutzfeldt-Jakob disease (CJD), a fatal degenerative human brain disorder. In this example, because of their willingness to modify a scientific principle in response to new data, scientists maintained the integrity of the scientific process while expanding our understanding of how diseases can occur.

Scientific Theories Involve Both Inductive and Deductive Reasoning

Scientific theories arise through **inductive reasoning,** the process of creating a broad generalization based on many observations that support it and none that contradict it. For example, the theory that Earth exerts gravitational forces on objects began from repeated observations of objects falling downward toward Earth and from no observations of objects falling upward away from Earth. The cell theory arises from the observation that all organisms that possess the characteristics of life as described in this chapter are composed of one or more cells and that nothing that is not composed of cells shares all of these attributes.

Once a scientific theory has been formulated, it then can be used to support deductive reasoning. In science, **deductive reasoning** starts with a well-supported generalization such as a scientific theory and uses it to generate hypotheses about how a specific experiment or observation will turn out. For example, based on the cell theory, if a scientist discovers a new entity that exhibits all the characteristics of life, she can confidently hypothesize that it will be made up of cells. Of course, the new organism must then be carefully examined to confirm its cellular structure.

Scientific Theories Are Formulated in Ways That Can Potentially Be Disproved

Scientists refer to the basic principles of science as "theories" because theories have the potential to be disproved, or falsified. The possibility of being falsified is a major difference between scientific theories and beliefs based on faith, which are impossible to either prove or disprove. Faith-based beliefs—for example, that each creature on Earth was separately created—cannot be subjected to scientific inquiry. Thus, such beliefs do not fall within the scope of science.

Science Is a Human Endeavor

Scientists are people, driven by the pride, fears, and ambition common to humanity. Ambition played an important role in the discovery of the structure of DNA by James Watson and Francis Crick (described in Chapter 11). Accidents, lucky guesses, controversies with competing scientists, and, of course, the intellectual curiosity of individual scientists all contribute to scientific advances. Even mistakes can play a role. Let's consider an actual case.

Microbiologists often study pure cultures—a single type of bacterium grown in sterile, covered dishes free from contamination by other bacteria and molds. At the first sign of contamination, a culture is usually thrown out, often with mutterings about sloppy technique. In the late 1920s, however, Scottish bacteriologist Alexander Fleming turned a ruined bacterial culture into one of the greatest medical advances in history.

One of Fleming's cultures became contaminated with a mold (a type of fungus) called *Penicillium*. But instead

A Petri dish contains solid growth medium

Bacteria grow in dense red colonies on the growth medium

A substance diffuses from the mold and kills the bacterial colonies, which lose their color as they die

A colony of the mold *Penicillium*

▲ FIGURE 1-12 **Penicillin kills bacteria**

of just discarding the dish, Fleming looked at it carefully and observed that no bacteria were growing near the mold (**Fig. 1-12**). He asked the question: "Why aren't bacteria growing in this region?" Fleming then formulated the hypothesis that *Penicillium* releases a substance that kills bacteria, and he predicted that a solution in which the mold had grown would contain this substance and kill bacteria. To test this hypothesis, Fleming performed an experiment. First, he grew *Penicillium* in a liquid nutrient broth. Then he filtered out the mold and poured some of the mold-free broth on a plate with a pure bacterial culture. Sure enough, something in the liquid killed the bacteria, supporting his hypothesis. This (and more experiments that confirmed his results) led to the conclusion that *Penicillium* secretes a substance that kills bacteria. Further research into these mold extracts resulted in the production of the first antibiotic—penicillin.

Fleming's experiments are a classic example of the scientific method, but they would never have happened without the combination of a mistake, an observation, and the curiosity to explore it. The outcome has saved millions of lives. As French microbiologist Louis Pasteur said, "Chance favors the prepared mind."

Knowledge of Biology Illuminates Life

Some people regard science as a dehumanizing activity, thinking that too deep an understanding of the world robs us of wonder and awe. Nothing could be further from the truth. For example, let's look at lupine flowers. Their two lower petals form a tube surrounding both male and female reproductive parts (**Fig. 1-13**). In young flowers, the weight of a bee on this tube forces pollen (carrying sperm) out of the tube onto the bee's abdomen. In older lupine flowers that are ready to be fertilized, the female part grows and emerges through the end of the tube. When a pollen-dusted bee visits, it usually deposits some pollen containing sperm on the exposed female organ. This allows the lupine to produce the seeds of its next generation.

Do these insights into lupines detract from our appreciation of them? Far from it. There is added delight in understanding the intertwined form and function of bee and flower that resulted as these organisms evolved together.

Once, as two of the authors of this text crouched beside a wild lupine, an elderly man stopped to ask what they were looking at so intently. He listened with interest as they explained what they were seeing, and immediately he went to observe another patch of lupines where bees were foraging. He, too, felt the increased sense of wonder that comes with understanding.

Throughout this text, we try to convey to you the wonder that comes with understanding, and we hope that you, too, will experience it. We also emphasize that biology is not a completed work but an ongoing exploration. We urge you to follow the journey of biological discoveries throughout your life. Don't think of biology as just another set of facts to memorize, but rather as a pathway to a new understanding of yourself and the life around you.

CHECK YOUR LEARNING

Can you outline the scientific method and explain why scientific principles are called theories?

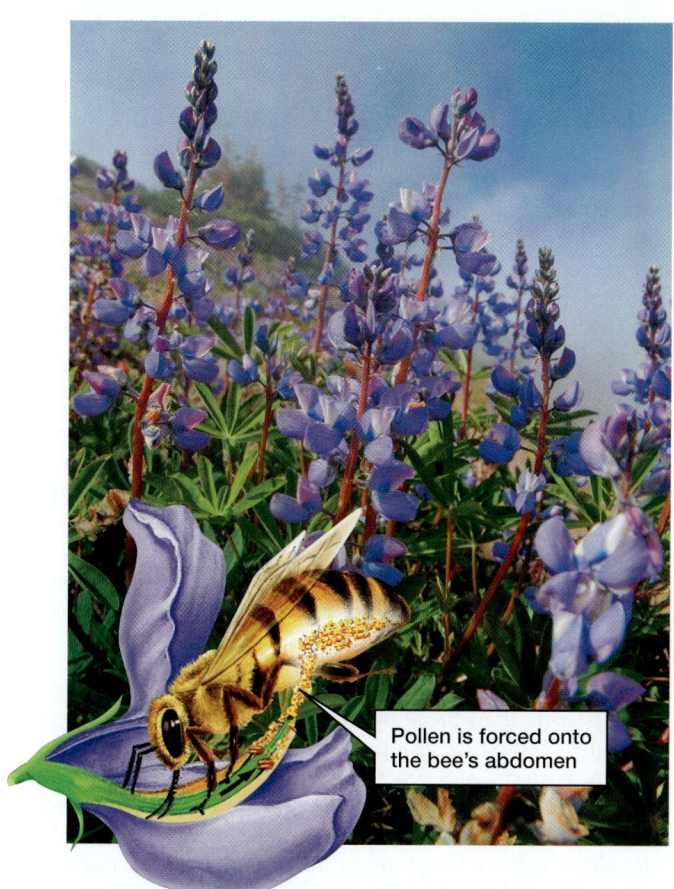

Pollen is forced onto the bee's abdomen

▲ FIGURE 1-13 **Adaptations in lupine flowers**
Understanding life helps people notice and appreciate the small wonders at their feet. (Inset) A lupine flower deposits pollen on a foraging bee's abdomen.

The Boundaries of Life

If viruses aren't forms of life, then what are they? A virus by itself is an inert particle that doesn't approach the complexity of a cell. The simplest virus, such as that causing smallpox (Fig. 1-14), consists of a protein coat that surrounds genetic material. The uncomplicated structure of viruses, coupled with amazing advances in biotechnology, have allowed researchers to synthesize viruses in the laboratory. They have accomplished this using the blueprint contained in viral genetic material and readily purchased chemicals. The first virus to be synthesized was the small, relatively simple poliovirus. This feat was accomplished in 2002 by Eckard Wimmer and coworkers at Stony Brook University, who titled their work "The Test-Tube Synthesis of a Chemical Called Poliovirus."

Protein coat

Genetic material

▲ **FIGURE 1-14 A smallpox virus**

Did these researchers create life in the laboratory?

A few scientists would say "yes," defining life by its ability to copy itself and to evolve. Wimmer himself describes viruses as entities that switch between a nonliving phase outside the cell and a living phase inside. The more stringent definition of life presented in this chapter explains why most scientists agree that viruses aren't alive. As virologist Luis Villarreal of the University of California at Irvine states, "Viruses are parasites that skirt the boundaries between life and inert matter."

Consider This

BioEthics When Wimmer and his coworkers announced that they had synthesized the poliovirus, they created considerable controversy both within and outside of the scientific community. Some people were concerned that this research jeopardized the eradication of polio, which has been nearly eliminated by widespread vaccination. They feared that deadly and highly contagious viruses might now be synthesized by bioterrorists. The researchers responded that they were merely applying current knowledge and techniques to demonstrate the principle that viruses are basically chemical entities that can be synthesized in the laboratory. Do you think scientists should synthesize viruses? What are the implications of forbidding such research?

CHAPTER REVIEW

Summary of Key Concepts

1.1 What Is Life?

Organisms acquire and use materials and energy. Materials are obtained from other organisms or the nonliving environment and are continuously recycled. Energy must be continuously captured from sunlight by photosynthetic organisms, whose bodies supply energy to all other organisms. Organisms also actively maintain organized complexity, perceive and respond to stimuli, grow, reproduce, and, collectively, evolve.

1.2 What Is Evolution?

Evolution is the scientific theory that modern organisms descended, with changes, from earlier life-forms. Evolution occurs as a consequence of (1) genetic differences, originally arising as mutations, among members of a population; (2) inheritance of these differences by offspring; and (3) natural selection of the differences that produce the best adaptations to the organisms' environment.

1.3 How Do Scientists Study Life?

Scientists identify a hierarchy of levels of organization, each more encompassing than those beneath (see Fig. 1-10). To categorize organisms, biologists group them into three major domains: Archaea, Bacteria, and Eukarya. The Archaea and Bacteria both consist of single, prokaryotic cells, but fundamental molecular differences distinguish them. Eukarya are composed of eukaryotic cells. The three kingdoms within the Eukarya are multicellular and differ in how they obtain energy. Protists are a very diverse group within the Eukarya. Some important features of the domains and kingdoms are summarized in Table 1-1. The scientific names of organisms are formed from their genus and species.

1.4 What Is Science?

Science is based on the principle that all events can be traced to natural causes that people can study objectively. Knowledge in biology is acquired through the scientific method, in which an observation leads to a question that leads to a hypothesis. The hypothesis generates a prediction that is then tested by controlled experiments. The experimental results, which must be repeatable, either support or refute the hypothesis, leading to a conclusion about the validity of the hypothesis. A scientific theory is a general explanation of natural phenomena developed through extensive and reproducible experiments and observations.

Key Terms

adaptation 6	eukaryotic 10
atom 7	evolution 4
autotroph 10	experiment 11
binomial system 10	extinction 7
biology 2	gene 5
biosphere 8	heterotroph 10
cell 2	hypothesis 11
cell theory 14	inductive reasoning 14
community 8	kingdom 9
conclusion 11	molecule 7
control 11	multicellular 9
deductive reasoning 14	mutation 6
deoxyribonucleic acid (DNA) 4	natural law 14
domain 9	natural selection 6
ecosystem 8	nucleus 10
element 7	observation 11

organ   *7*
organ system   *7*
organelle   *10*
organism   *2*
photosynthesis   *2*
plasma membrane   *10*
population   *4*
prediction   *11*
prokaryotic   *10*

question   *11*
science   *11*
scientific method   *11*
scientific theory   *14*
species   *8*
spontaneous generation   *12*
tissue   *7*
unicellular   *9*
variable   *11*

Learning Outcomes

In this chapter, you have learned to . . .

LO1 Describe the characteristics of living things.

LO2 Explain evolution and how three natural processes make evolution inevitable.

LO3 Explain the concept of levels of biological organization, and list the levels.

LO4 Identify the three domains of life, and describe the types of cells that make up each domain.

LO5 Explain how the three kingdoms of eukaryotic organisms are distinguished from one another.

LO6 Describe the scientific method, and provide an example of how it could be applied to an everyday problem.

LO7 Explain why experiments include control groups.

LO8 Explain how a hypothesis differs from a scientific theory.

Thinking Through the Concepts

Fill-in-the-Blank

1. Organisms respond to _____. Organisms acquire and use _____ and _____ from the environment. Organisms are composed of cells whose structure is both _____ and _____. Collectively, organisms _____ over time. **LO1 LO2**

2. The smallest particle of an element that retains all the properties of that element is a(n) _____. The smallest unit of life is the _____. Cells of a specific type within multicellular organisms combine to form _____. A(n) _____ consists of all of the same type of organism within a defined area. A(n) _____ consists of all the interacting populations within the same area. A(n) _____ consists of the community and its nonliving surroundings. **LO3**

3. A(n) _____ is a general explanation of natural phenomena supported by extensive, reproducible tests and observations. In contrast, a(n) _____ is a proposed explanation for observed events. To answer specific questions about life, biologists use a general process called the _____. **LO6 LO8**

4. An important scientific theory that explains why organisms are at once so similar and also so diverse is the theory of _____. This theory explains life's diversity as having originated primarily through the process of _____. **LO2**

5. The molecule that guides the construction and operation of an organism's body is called (complete term) _____, abbreviated as _____. This large molecule contains discrete segments with specific instructions; these segments are called _____. **LO1 LO2**

Review Questions

1. What properties are shared by all forms of life? **LO1**

2. Why do organisms require energy? Where does the energy come from? **LO1**

3. Define *evolution*, and explain the three natural occurrences that make evolution inevitable. **LO2**

4. What are the three domains of life? **LO4**

5. What are some differences between prokaryotic and eukaryotic cells? In which domain(s) is each found? **LO4**

6. Which kingdom(s) are heterotrophic? Which are autotrophic? **LO5**

7. What basic principle underlies all scientific inquiry? **LO6**

8. What is the difference between a scientific theory and a hypothesis? Why do scientists refer to basic scientific principles as "theories" rather than "facts"? **LO8**

9. What factors did Redi control for in his open jar of meat? What factors did Andersson control for? **LO7**

10. Explain the differences between inductive and deductive reasoning, and provide an example (real or hypothetical) of each. **LO8**

11. List the steps in the scientific method with a brief description of each step. **LO6**

Applying the Concepts

1. What properties of life does your computer possess? Which ones does it lack?

2. How would this textbook's definition of life need to be changed to allow viruses to qualify as life-forms? For prions to be considered alive?

3. Review Alexander Fleming's experiment that led to the discovery of penicillin. What would be an appropriate control for the experiment in which Fleming applied filtered medium from a *Penicillium* culture to plates of bacteria?

4. Explain an instance in which your own understanding of a phenomenon enhances your appreciation of it.

5. Provide an example (different from that in the text) of how the scientific method is used in everyday life.

Answers to Figure Caption questions and Fill-in-the-Blank questions can be found in the Answers section at the back of the book.

MB *Go to MasteringBiology for practice quizzes, activities, eText, videos, current events, and more.*

UNIT 1
The Life of the Cell

Single cells can be complex, independent organisms, such as this freshwater protist, *Dendrocometes*. A rounded attachment region anchors the cell firmly to the gills of freshwater fish or crustaceans. Tentacles, resembling microscopic antlers, snare food as water passes over them.

"Any living cell carries with it the experiences of a billion years of experimentation by its ancestors."

—Max Delbrück

The aftermath of explosions at the Fukushima nuclear power plant in Japan.

CASE STUDY
Unstable Atoms Unleashed

ON MARCH 11, 2011, an earthquake of epic magnitude—9.0 on the Richter scale—occurred off the northeast coast of Japan. It was the most violent earthquake in Japan's history, and one of the most powerful ever recorded anywhere in the world. Soon after, a tsunami caused by the quake slammed into the Fukushima Daiichi nuclear power plant on Japan's eastern coast. The waves, towering up to 46 feet, flooded the plant and knocked out its main electrical power supply and backup generators, which caused its cooling system to fail.

The cores of nuclear reactors like those in the Fukushima plant consist of large numbers of fuel rods, metal tubes filled with pellets of uranium fuel. Two thick steel containment vessels surround the nuclear core. Water is pumped continuously around the fuel rods within the inner containment vessel, absorbing the intense heat generated by the nuclear reactions within the rods. The heated water produces steam, which then expands to drive turbines that generate electricity. When loss of power shut down the plant's water pumps, operators used firefighting equipment to inject seawater into the inner containment vessel in a desperate attempt to cool it. But their efforts failed; heat and pressure cracked the inner vessel, allowing water and steam to escape. The core temperature rose to over 1,800°F (about 1,000°C), melting the tubes of the zirconium metal alloy that enclosed the fuel rods and releasing the radioactive fuel into the inner vessel.

Because of the incredibly high temperatures, the zirconium reacted with the steam to generate hydrogen gas. As the pressure of the steam and hydrogen gas increased, it threatened to rupture the outer containment vessel. To prevent this, plant operators vented the mixture—which also contained radioactive elements from the melted fuel rods—into the atmosphere. As the hot hydrogen gas encountered oxygen in the atmosphere, the two combined explosively, destroying parts of the buildings housing the containment vessels (see the chapter opener photo). Despite venting, the intense heat and the pressure it generated eventually caused the outer containment structure to leak and disgorge contaminated water into the ocean for months following the disaster. Officials evacuated tens of thousands of people living within 12 miles of the plant.

Why were people evacuated from their homes when radioactive gases were released into the atmosphere? How do the atoms of radioactive elements differ from those that are not radioactive? And what are atoms composed of?

AT A GLANCE

2.1 What Are Atoms?

2.2 How Do Atoms Interact to Form Molecules?

2.3 Why Is Water So Important to Life?

LEARNING GOALS

LG1 Describe the relationship between elements and atoms.

LG2 Describe the structure of atoms, including identification of the three subatomic particles and the characteristics of each.

LG3 Define isotopes and explain what causes some isotopes to be radioactive and how people may be helped or harmed by radioactivity.

LG4 Explain how atoms form molecules.

LG5 Describe the characteristics of the three types of chemical bonds, and provide examples of molecules in which each type of bond is found.

LG6 Describe the special properties of water and how these properties allow water to play key roles in life on Earth.

LG7 Explain the pH scale, and describe the properties of acids, bases, and buffers.

2.1 WHAT ARE ATOMS?

If you write "atom" with a pencil, you are forming the letters with graphite, a form of carbon. Now imagine cutting up the carbon into finer and finer particles. Each particle would remain graphite until the substance was eventually split into its basic subunits: individual carbon atoms. Each carbon atom is so small that 100 million of them placed in a row would span less than half an inch (1 centimeter). The structure of each atom would be identical and unique to the element carbon.

Atoms Are the Basic Structural Units of Elements

Carbon is an example of an **element**—a substance that cannot be separated into simpler substances, and cannot be converted into another substance by ordinary chemical reactions. Elements, both alone and combined with other elements, form all matter. An **atom** is the smallest unit of an element, and each atom retains all the chemical properties of that element.

Ninety-two different elements occur in nature. Each is given an abbreviation, its *atomic symbol*, based on its name (sometimes in Latin; e.g., lead is Pb, for *plumbum*). Most elements are present in only small quantities in the biosphere, and relatively few are essential to life on Earth. **Table 2-1** lists the most common elements in living things.

Atoms Are Composed of Still Smaller Particles

Atoms are composed of *subatomic particles*: **neutrons** (n), which have no charge; **protons** (p^+), which each carry a single positive charge; and **electrons** (e^-), which each carry a single negative charge. An atom as a whole is uncharged, or *neutral*, because it contains equal numbers of protons and electrons, whose positive and negative charges electrically cancel each other. Subatomic particles are so small that they are assigned their own unit of mass, measured in *atomic mass units*. As you can see in **Table 2-2**, each proton and neutron

TABLE 2-1	Common Elements in Living Organisms		
Element	Atomic Number[1]	Mass Number[2]	% by Weight in the Human Body
Oxygen (O)	8	16	65
Carbon (C)	6	12	18.5
Hydrogen (H)	1	1	9.5
Nitrogen (N)	7	14	3.0
Calcium (Ca)	20	40	1.5
Phosphorus (P)	15	31	1.0
Potassium (K)	19	39	0.35
Sulfur (S)	16	32	0.25
Sodium (Na)	11	23	0.15
Chlorine (Cl)	17	35	0.15
Magnesium (Mg)	12	24	0.05
Iron (Fe)	26	56	Trace
Fluorine (F)	9	19	Trace
Zinc (Zn)	30	65	Trace

[1]Atomic number: number of protons in the atomic nucleus.
[2]Mass number: total number of protons and neutrons.

TABLE 2-2	Mass and Charge of Subatomic Particles	
Subatomic Particle	Mass (in atomic mass units)	Charge
Neutron (n)	1	0
Proton (p^+)	1	+1
Electron (e^-)	0.00055	−1

has a mass unit of 1, while the mass of an electron is negligible compared to these larger particles. The **mass number** of an atom is the total number (which equals the total mass) of the protons and neutrons in its nucleus.

Protons and neutrons cluster together in the center of the atom, forming the **atomic nucleus.** The tiny electrons are in continuous rapid motion around the nucleus within a large, three-dimensional space, as illustrated by the two simplest atoms, hydrogen and helium, in **Figure 2-1.** *Orbital models* of atomic structure, like that shown in Figure 2-1, are extremely simplified to make atoms easy to imagine. Atoms are never drawn to scale; if they were, and if you visualize this • to be the nucleus, the electrons would be somewhere in the next room—roughly 30 feet away!

Elements Are Defined by Their Atomic Numbers

The number of protons in the nucleus—called the **atomic number**—is the feature that defines each element, making it distinct from all others. For example, every hydrogen atom has one proton, every carbon atom has six, and every oxygen atom has eight, giving these atoms atomic numbers of 1, 6, and 8, respectively. The periodic table in Appendix II organizes the elements according to their atomic numbers (rows) and their general chemical properties (columns).

Isotopes Are Atoms of the Same Element with Different Numbers of Neutrons

Although every atom of an element has the same number of protons, the atoms of that element may have different numbers of neutrons. Atoms of the same element with different numbers of neutrons are called **isotopes.** Isotopes can be distinguished from one another because each has a different mass number, which is written as a superscript preceding the atomic symbol.

Some Isotopes Are Radioactive

Most isotopes are stable, that is, their nuclei do not change spontaneously. A few, however, are **radioactive,** meaning that their nuclei spontaneously break apart, or decay. Radioactive decay emits subatomic particles (such as neutrons) that carry large amounts of energy. Earlier, we stated that one element cannot be converted into another by ordinary chemical reactions; these do not alter atomic nuclei. The decay of radioactive nuclei, in contrast, may form different elements. For example, nearly all carbon exists as stable ^{12}C. But a radioactive isotope called carbon-14 (^{14}C; 6 protons + 8 neutrons; comprising one in every trillion carbon atoms) is produced continuously by atmospheric reactions involving cosmic rays. Over thousands of years, the ^{14}C atoms spontaneously break down, forming nitrogen atoms. The ^{14}C is captured and stored in the bodies of organisms, where it exists in a fixed ratio with ^{12}C. After an organism dies, its ^{14}C disintegrates at a constant rate, reducing the ratio of ^{14}C to ^{12}C. By measuring this ratio, scientists can determine the age of formerly living artifacts up to about 50,000 years old (described in Chapter 17). This includes mummies, ancient trees and skeletons, or tools made of wood or bone.

In laboratory research, scientists often expose organisms to radioactive isotopes and trace the isotopes' movements during physiological processes. For example, experiments with radioactively labeled DNA and protein allowed scientists to conclude that DNA is the genetic material of cells (described in Chapter 11). Although they must be handled with caution, modern medicine makes extensive use of radioactive isotopes, as described in "Science in Action: Radioactive Medicine."

Some Radioactive Isotopes Damage Cells

Some radioactive isotopes release particles with enough energy that they can damage DNA, causing mutations (described in Chapter 1). People near the Fukushima power plant are being exposed to radiation in food, water, and air, increasing their risk of developing cancer. Japanese authorities have arranged to perform regular cancer screenings on hundreds of thousands of exposed children.

Nuclei and Electrons Play Complementary Roles in Atoms

Nuclei and electrons play complementary roles in atoms. Nuclei (unless they are radioactive) provide stability by resisting disturbance by outside forces; ordinary sources of energy such as heat, electricity, and light hardly affect them at all. Because its nucleus is stable, a carbon atom (^{12}C)

CASE STUDY c o n t i n u e d

Unstable Atoms Unleashed

The uranium in nuclear reactor fuel rods is mostly in the nonradioactive form (^{238}U), with about 4% radioactive uranium (^{235}U). When each ^{235}U atom splits into smaller atoms, it releases heat, which is harnessed to generate electricity, and it ejects two or three fast-moving neutrons. Neutrons hitting other ^{235}U atoms cause them to split, continuing the process. Neutrons hitting the more abundant ^{238}U atoms produce plutonium-239 (^{239}Pu). This plutonium isotope also splits apart when hit by neutrons, forming smaller atoms and releasing more heat energy that contributes to the electrical output of the power plant.

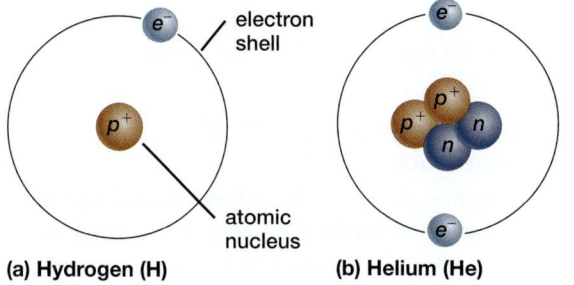

(a) Hydrogen (H)   (b) Helium (He)

▲ **FIGURE 2-1** **Atomic models** Orbital models of **(a)** hydrogen (the only atom with no neutrons) and **(b)** helium. In these simplified models, the electrons (pale blue) are represented as miniature planets, orbiting around a nucleus that contains protons (gold) and neutrons (dark blue).

SCIENCE IN ACTION Radioactive Medicine

One medically important use of radioactive isotopes is in positron emission tomography (PET) scans. For example, a solution of sugar molecules tagged with a radioactive isotope might be injected into a patient's bloodstream. More metabolically active regions of the body use more sugar for energy and accumulate larger amounts of radioactivity. The person's body is moved through a ring of detectors that respond to the energetic particles emitted as the isotope decays (Fig. E2-1a). A powerful computer uses these data to calculate precisely where the decays occur. It then generates a color-coded map of the frequency of decays within each "slice" of body passing through the detector ring. PET can be used to study the working brain, because regions activated by a specific mental task—such as a math problem—will have increased energy needs and will "light up" as they accumulate more radioactive glucose. Tumors from cancer can be detected by PET because they grow rapidly and use large amounts of glucose, showing up in PET scans as "hot spots" (Fig. E2-1b).

Radiation therapy is used to treat about half of all cancer patients. DNA is easily damaged by radiation, so rapidly dividing cancer cells (which require intact DNA to copy themselves) are particularly vulnerable. The radioactive isotope may be introduced into the bloodstream, implanted in the body near the cancer, or radiation may be directed into the tumor by an external device (Fig. E2-2). Radiation exposure slightly increases the chance that the patient will develop cancer again in the future, but most patients consider a possible cure for a present and potentially fatal cancer to be well worth the risk.

detector ring

The subject's head is placed within a ring of detectors

(a) The subject is placed in a scanner

Red indicates the highest radioactivity and blue the least; a malignant brain tumor shows clearly in red

(b) The resulting computer image

▲ FIGURE E2-1 How positron emission tomography works

▲ FIGURE E2-2 Patient receiving radiation therapy for cancer

CASE STUDY continued

Unstable Atoms Unleashed

Months after the meltdown, engineers at the Fukushima power plant discovered hot spots of radiation so intense that a person exposed for an hour would be dead within a few weeks. How could death come so fast? Extremely high doses of radiation damage DNA and other biological molecules so much that cells—particularly those that divide rapidly—can no longer function. Skin cells are destroyed. Cells lining the stomach and intestine break down, leading to nausea and vomiting. Bone marrow, where blood cells are produced, is destroyed. Lack of white blood cells allows infections to flourish, and the loss of platelets crucial for blood clotting leads to internal bleeding. Fortunately, the hot spots inside Fukushima were detected from outside the building using special cameras, so no one was exposed to such extreme levels of radioactivity.

remains carbon whether it is part of a diamond, graphite, carbon dioxide, or sugar. Electrons, in contrast, are dynamic; they can capture and release energy, and they form bonds that link atoms together, as described later.

Electrons Occupy Complex Regions Around the Nucleus

Electrons occupy complex three-dimensional regions, called **electron shells,** around the nucleus. For simplicity, we will depict these shells as increasingly large, concentric rings around the nucleus where electrons travel like planets orbiting the sun (Fig. 2-2). Each shell has a specific energy associated with it; the farther away from the nucleus, the higher the energy of the electrons occupying the shell. Imagine each shell as a step on a ladder. Climbing each step takes energy. The higher you climb, the more energy that is stored in your

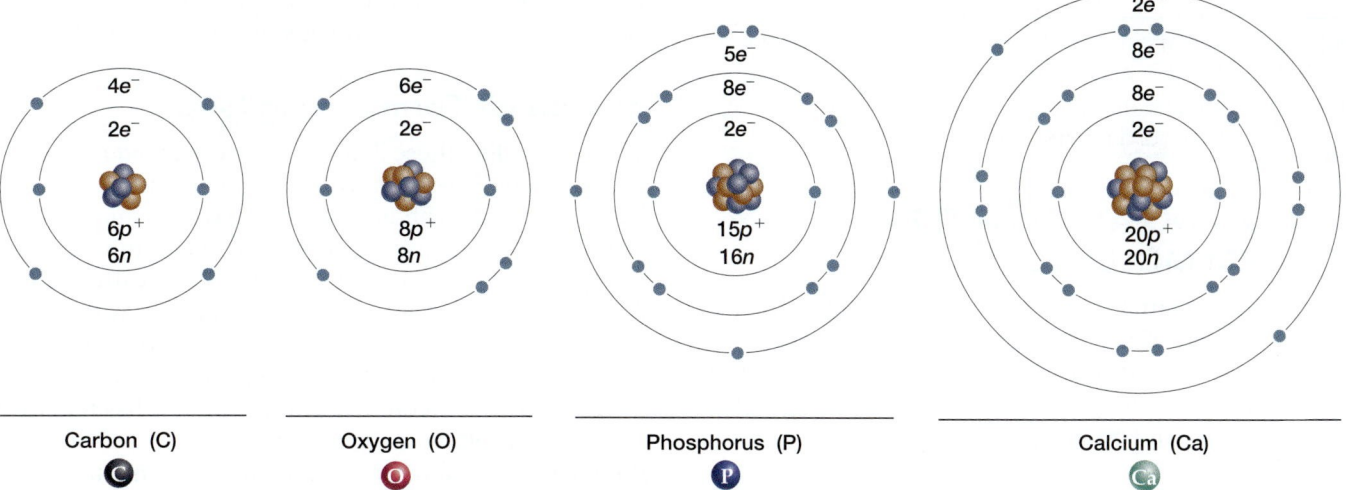

Carbon (C) Oxygen (O) Phosphorus (P) Calcium (Ca)

▲ FIGURE 2-2 **Electron shells in atoms** Most biologically important atoms have two or more shells of electrons. The first shell, the orbit closest to the nucleus, can hold two electrons; the next three shells each contain a maximum of eight electrons.

QUESTION Why do atoms that tend to react with other atoms have outer shells that are not full?

body, the more unstable you are, and the harder you would hit the ground if you fell off. Like a person on the ladder's lowest rung, electrons in the shell nearest the nucleus are in their most stable state and have the lowest energy. Electrons that occupy shells farther from the nucleus, like a person higher up a ladder, are at a higher energy state.

Electrons Can Capture and Release Energy

When an atom is excited by energy, such as heat or light, this energy can cause an electron to jump from a lower-energy electron shell to a higher-energy shell. Soon afterward, the electron spontaneously falls back into its original electron shell and releases its extra energy, often as light (**Fig. 2-3**).

We make use of this property of electrons every time we switch on a light bulb. Although fast becoming obsolete, the incandescent bulb is the easiest to understand. Electricity flows through a very thin wire, heating it to around 4,500°F (about 2,500°C) for a 100-watt bulb. The heat energy bumps some electrons in the wire into higher-energy electron shells. As the electrons drop back down into their original shells, they emit some of the energy as light. Unfortunately, about

90% of the energy absorbed by the wire of an incandescent bulb is reemitted as heat rather than light, so it's a very inefficient light source.

As Atomic Number Increases, Electrons Fill Shells Increasingly Distant from the Nucleus

Each electron shell can hold a specific number of electrons; the shell nearest the nucleus can hold only two, and the larger, more distant shells (in those atoms discussed here) hold eight. Electrons always first fill the shell nearest the nucleus and then fill the higher shells, keeping as close to the nucleus—and as stable—as possible. Elements with larger numbers of protons in their nuclei require more electrons to balance these protons, so their electrons will occupy shells at increasing distances from the nucleus.

The single electron in a hydrogen atom (H) and the two electrons in helium (He) occupy the first electron shell (see Fig. 2-1). The second shell, corresponding to a higher energy level, can hold up to eight electrons. Thus, a carbon atom (C) with six electrons will have two electrons filling its first shell and four occupying its second shell (see Fig. 2-2).

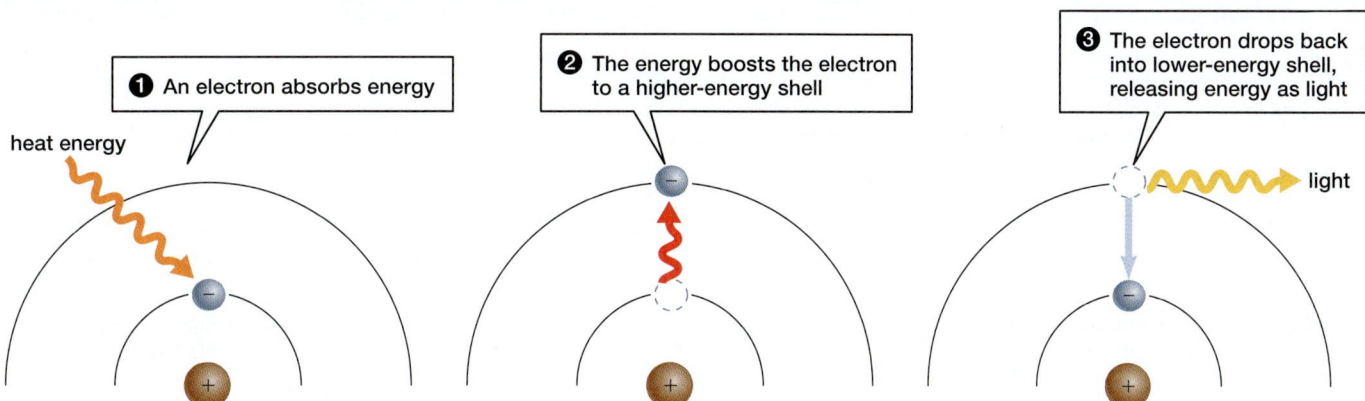

❶ An electron absorbs energy

❷ The energy boosts the electron to a higher-energy shell

❸ The electron drops back into lower-energy shell, releasing energy as light

heat energy

light

▲ FIGURE 2-3 **Energy capture and release**

CHECK YOUR LEARNING

Can you define an *element*, an *atom*, and the *subatomic particles* that make up the atom? Do you understand atomic number and mass number? Can you explain radioactivity and its dangers and benefits?

2.2 HOW DO ATOMS INTERACT TO FORM MOLECULES?

Most forms of matter that we encounter in our daily lives consist of atoms of the same or different elements linked together to form **molecules.** Simple examples are oxygen gas (O_2; two atoms of oxygen) or water (H_2O; two atoms of hydrogen and one of oxygen). How and why do molecules form?

Atoms Form Molecules to Fill Vacancies in Their Outer Electron Shells

Just as electrons are most stable in shells nearest the atomic nucleus, they are also most stable and less likely to interact with other atoms in shells that are completely full. For most elements, the number of electrons needed to balance the number of protons will fill one or more inner shells, but will not completely fill the outer shell. Atoms generally behave according to two basic principles:

- An atom will not react with other atoms when its outermost electron shell is completely full. Such an atom (e.g., helium in Fig. 2-1b) is extremely stable and is described as *inert*.

- An atom will react with other atoms if its outermost electron shell is only partially full (e.g., hydrogen in Fig. 2-1a). Such an atom is described as *reactive*.

Chemical Bonds Hold Atoms Together in Molecules

Chemical bonds are attractive forces that hold atoms together in molecules. Bonds are formed when reactive atoms gain, lose, or share electrons to gain stability and become less reactive. There are three major types of bonds: *ionic bonds, covalent bonds,* and *hydrogen bonds* (**Table 2-3**).

Ionic Bonds Form Among Ions

Atoms, including those that are reactive, have equal numbers of protons and electrons. This gives them an overall neutral charge. An atom with an almost empty outermost electron shell can become more stable by losing electrons and completely emptying its shell. An atom with a nearly full outer shell can become more stable by gaining electrons and filling its shell completely. When an atom gains an electron, it has more electrons than protons and becomes negatively charged. When an atom loses an electron, it has fewer electrons than protons, and becomes positively charged. An atom that has acquired an overall positive or negative charge is no longer called an atom; it has become an **ion.** Ions with opposite charges attract one another, and the electrical attraction between positively and negatively charged ions forms **ionic bonds.** For example, the white crystals in your salt shaker are molecules of sodium chloride linked by ionic bonds. Sodium (Na) has only one electron in its outermost electron shell, so it can become stable by losing this electron. Chlorine (Cl) has seven electrons in its outer shell, so it can become stable by gaining an electron (**Fig. 2-4a**). To form sodium chloride, then, sodium loses an electron to chlorine, forming Na^+, and chlorine picks up the electron, becoming Cl^- (**Fig. 2-4b**). The ionic bonds between sodium and chloride ions result in crystals composed of a repeating, orderly array of the two ions (**Fig. 2-4c**). As we will see later, water can break ionic bonds, as happens when water dissolves salt. Biological molecules must function in a watery environment and most are formed with covalent, rather than ionic, bonds.

Covalent Bonds Form by Sharing Electrons

Atoms with partially full outermost electron shells can become stable by sharing electrons with one another, filling both of their outer shells and forming **covalent bonds** (**Fig. 2-5**). The atoms in most biological molecules, such as proteins, sugars, and fats, are joined by covalent bonds (**Table 2-4**).

TABLE 2-3 Common Types of Bonds in Biological Molecules		
Type	**Type of Interaction**	**Example**
Ionic bond	An electron is transferred between atoms creating positive and negative ions that attract one another.	Occurs between the sodium (Na^+) and chloride (Cl^-) ions of table salt (NaCl)
Covalent bond	Electrons are shared between atoms.	
Nonpolar	Electrons are shared equally between atoms.	Occurs between the two hydrogen atoms in hydrogen gas (H_2)
Polar	Electrons are shared unequally between atoms.	Occurs between hydrogen and oxygen atoms of a water molecule (H_2O)
Hydrogen bond	Attractions occur between polar molecules in which H is bonded to O or N. The slightly positive hydrogen attracts the slightly negative O or N of a nearby polar molecule.	Occurs between water molecules, where slightly positive charges on hydrogen atoms attract slightly negative charges on oxygen atoms of nearby molecules

Sodium atom (neutral)   Chlorine atom (neutral)

11p⁺
11n

17p⁺
18n

Electron transferred

(a) Neutral atoms

Sodium ion (+)   Chloride ion (−)

11p⁺
11n

17p⁺
18n

Attraction between
opposite charges

(b) Ions

(c) An ionic compound: NaCl

▲ **FIGURE 2-4 The formation of ions and ionic bonds**
(a) Sodium has only one electron in its outer electron shell; chlorine
has seven. **(b)** Sodium can become stable by losing an electron, and
chlorine can become stable by gaining an electron. Sodium then
becomes a positively charged ion and chlorine a negatively charged
ion. **(c)** Because oppositely charged particles attract one another,
the resulting sodium and chloride ions nestle closely together in a
crystal of salt, NaCl. The arrangement of ions in salt causes it to form
cubic crystals.

Covalent Bonds May Produce Nonpolar or Polar Molecules

All atoms of the same element, and certain atoms of different elements, will share electrons equally. This creates **nonpolar covalent bonds** in which there is no charge on any
part of the molecule. For example, two hydrogen atoms
can become more stable if they share their outer electrons,
allowing each to behave almost as if it had two electrons

Same charge on
both nuclei

Electrons spend
equal time near
each nucleus

(hydrogens: uncharged)

(a) Nonpolar covalent bonding in hydrogen gas (H₂)

(oxygen: slightly negative)
(−)

Larger positive
charge

Electrons spend
more time near
the larger nucleus

8p⁺
8n

(hydrogens:
slightly positive)

Smaller positive
charge

(+)   (+)

(b) Polar covalent bonding in water (H₂O)

▲ **FIGURE 2-5 Covalent bonds involve shared electrons**
(a) In hydrogen gas, an electron from each hydrogen atom is shared
equally, forming a single nonpolar covalent bond. **(b)** Oxygen
lacks two electrons to fill its outer shell, so oxygen can form polar
covalent bonds with two hydrogen atoms, creating water. Oxygen
exerts a greater pull on the electrons than does hydrogen, so the
"oxygen end" of the molecule has a slight negative charge and the
"hydrogen end" has a slight positive charge.

TABLE 2-4   Electrons and Bonds in Atoms Common in Biological Molecules

Atom	Capacity of Outer Electron Shell	Electrons in Outer Shell	Number of Covalent Bonds Usually Formed
Hydrogen	2	1	1
Carbon	8	4	4
Nitrogen	8	5	3
Oxygen	8	6	2
Sulfur	8	6	2

Health|Watch

Watch Out for Free Radicals

Partially full outer electron shells are reactive, but this reactivity increases dramatically if the unfilled shell also contains an uneven number of electrons. Like people, electrons like to pair up. Free radicals have unpaired electrons in their outer shells. As a result they react vigorously with other molecules, capturing or releasing electrons to achieve a more stable arrangement. But in stealing or donating an electron, a free radical often leaves an unpaired electron in the molecule it attacks, creating a new free radical. This can lead to chain reactions that destroy biological molecules such as DNA that are crucial to life. Our bodies continuously produce oxygen-containing free radicals. One of these is hydrogen peroxide (H_2O_2), whose ability to break down other molecules is revealed when it bleaches hair by destroying hair pigment.

Free radicals can be formed when radiation (from sunlight, X-rays, and radioactive isotopes), chemicals in automobile exhaust and cigarette smoke, and industrial metals such as mercury and lead enter our bodies. Damage to cells caused by free radicals, called *oxidative stress*, is linked to many disorders, among them heart disease, cancer, Parkinson's disease, Alzheimer's disease, and cataracts. The gradual deterioration of the body that accompanies aging is believed by many scientists to result, at least in part, from oxidative stress accumulating over a lifetime of exposure to free radicals (**Fig. E2-3**). Fortunately our bodies produce many different molecules, collectively called **antioxidants,** that react with free radicals and render them harmless.

There is strong evidence that diets high in the natural antioxidants found in fruits and vegetables are correlated with a lower incidence of cardiovascular disease. But did you know that dark chocolate (**Fig. E2-4**) might actually be a type of health food? Recent large studies that correlate chocolate consumption and health have found that participants who regularly consumed chocolate were less likely to suffer strokes and heart attacks. Cocoa powder (the dark, bitter powder made from the seeds inside cacao pods; see Fig. E2-4, inset) contains high concentrations of flavonols, which are powerful antioxidants.

Although no controlled studies have been yet done to determine whether eating large quantities of chocolate actually reduces the risk of cancer, strokes, neurological disorders, or heart disease, there would certainly be no shortage of volunteers for this research. Before you begin an uncontrolled study on yourself, however, be aware that the most sinfully delicious chocolates are also high in fat and sugar. But—in moderation—"chocoholics" with a taste for dark chocolate have reason to relax and enjoy!

▲ **FIGURE E2-3 Free-radical damage** Ultraviolet radiation from sunlight can generate free radicals in skin, damaging molecules that give skin its elasticity and contributing to wrinkles as we age.

▲ **FIGURE E2-4 Chocolate** Cocoa is derived from cacao beans found inside pods (inset) on trees native to South America.

in its outer shell (see Fig. 2-5a). This reaction forms hydrogen gas (H_2). Because the two H nuclei are identical, their electrons spend equal time near each, and so neither end, or *pole*, of the molecule is charged. Other examples of nonpolar molecules include oxygen gas (O_2), nitrogen gas (N_2), carbon dioxide (CO_2), and biological molecules such as oils and fats (described in Chapter 3). In each of these examples, the nuclei exert an equal or very similar pull on the shared electrons.

Some atoms share electrons unequally, because one attracts the electrons more strongly than the other. Unequally shared electrons produce **polar covalent bonds** (see Fig. 2-5b). Although the molecule as a whole is electrically neutral, a polar covalent molecule has charged poles. In water (H_2O), for example, an electron is shared between each hydrogen atom and the central oxygen atom. The oxygen exerts a stronger attraction on the electrons than do the hydrogens, so the shared electrons spend more time near the oxygen atom. By attracting electrons, the oxygen pole of a water molecule becomes slightly negative, leaving each hydrogen atom slightly positive (see Fig. 2-5b).

Among atoms and molecules with unfilled outer shells, some—called **free radicals**—are so reactive that they can tear other molecules apart. Free radicals are produced in large numbers in the body by reactions that make energy available to cells. Although these reactions are essential for life, over time, the stress that free radicals place on living cells may contribute to aging and eventual death. Learn more in "Health Watch: Watch Out for Free Radicals."

Hydrogen Bonds Are Attractive Forces Between Certain Polar Molecules

Biological molecules, including sugars, proteins, and nucleic acids, often have many polar covalent bonds, between either hydrogen and oxygen or hydrogen and nitrogen. In these cases, the hydrogen is slightly positive and the oxygen or nitrogen is slightly negative. As you know, opposite charges attract. A **hydrogen bond** is the attraction between a slightly positive hydrogen and a slightly negative oxygen or nitrogen located in a nearby molecule or in another part of the same molecule. The simplest example of hydrogen bonding occurs in water. Water molecules form hydrogen bonds with one another between

their slightly positive hydrogen poles and slightly negative oxygen poles, linking the water molecules into a loosely connected network (Fig. 2-6). As you will see shortly, hydrogen bonds among water molecules give water several unusual properties. In the following chapter, we will describe how hydrogen bonding—within different parts of a single biological molecule, between nearby biological molecules, and between biological molecules and water—is crucial for life on Earth.

CHECK YOUR LEARNING

Can you explain what makes an atom reactive? Can you define *molecules* and *chemical bonds*? Are you familiar with the three different types of bonds and where they occur?

2.3 WHY IS WATER SO IMPORTANT TO LIFE?

As naturalist Loren Eiseley eloquently stated, "If there is magic on this planet, it is contained in water." Water has many special properties that all result from the polarity of its molecules and the hydrogen bonds that form among them. What makes water unique?

Water Molecules Attract One Another

Hydrogen bonds interconnect water molecules. But, like square dancers continually moving from one partner to the next—joining and releasing hands as they go—hydrogen bonds in liquid water constantly break and re-form, allowing the water to flow. In water, hydrogen bonds cause **cohesion,** the tendency for molecules of a single type to stick together. The cohesion of water plays a crucial role in the life of land plants. How does water absorbed by a plant's roots ever reach its leaves, especially if the plant is a 300-foot-tall redwood tree (Fig. 2-7a)? Water fills tiny tubes that connect the roots, stem, and leaves. Water evaporates from leaves, and just before it enters the atmosphere, each molecule pulls the one below it to the leaf's surface. Thus water's cohesiveness allows it to be pulled upward much like a chain being dragged up from the top. This works because the hydrogen bonds that link water molecules are stronger than the weight of the water in the tube, so the water chain doesn't break. Without the cohesion of water, there could be no large land plants.

O (−)
H (+)
H (+)
H (+)
O (−)
H (+)

hydrogen bonds

▲ FIGURE 2-6 **Hydrogen bonds in water** The slight charges on different parts of water molecules produce attractive forces called hydrogen bonds (dotted lines) between the oxygen and hydrogen atoms in adjacent water molecules. As water flows, these bonds are constantly breaking and re-forming.

? HAVE **YOU** EVER WONDERED . . .

Why It Hurts So Much to Do a Belly Flop?

The slap of a belly flop provides firsthand experience of the power of cohesion among water molecules. Because of the hydrogen bonds that interconnect its molecules, the water surface resists being broken. When you suddenly force water molecules apart over a large area using your belly, the result can be a bit painful. Do you think that belly-flopping into a pool of nonpolar vegetable oil would hurt as much?

(a) Cohesion helps water to reach treetops

(b) Cohesion causes surface tension

◀ FIGURE 2-7 **Cohesion among water molecules** **(a)** Cohesion caused by hydrogen bonding allows water evaporating from the leaves of plants to pull water up from the roots. **(b)** A basilisk lizard uses surface tension as it races across the water's surface to escape a predator.

Cohesion among water molecules also produces **surface tension,** the tendency for a water surface to resist being broken. Surface tension can support fallen leaves, some spiders and water insects, and even a running basilisk lizard (Fig. 2-7b).

Water Interacts with Many Other Molecules

A **solvent** is a substance that **dissolves** some other substance; in other words, a solvent completely surrounds and disperses the individual atoms or molecules of another substance. A solvent containing one or more dissolved substances is called a **solution.** Because of its polar nature, water is an excellent solvent for other polar molecules and for ions. Water dissolves many molecules that are important to life, including proteins, salts, sugars, oxygen, and carbon dioxide. Watery solutions allow them to move around freely and meet one another, allowing reactions to occur.

Let's see how table salt dissolves in water. A crystal of table salt is held together by the electrical attraction between positively charged sodium ions (Na^+) and negatively charged chloride ions (Cl^-; see Fig. 2-4c). When a salt crystal is dropped into water, the positively charged hydrogen poles of water molecules are attracted to the Cl^-, and the negatively charged oxygen poles are attracted to the Na^+. As water molecules surround the sodium and chloride ions, shielding them from interacting with each other, the ions separate from the crystal and drift away in the water—thus, the salt dissolves (Fig. 2-8).

Water dissolves polar molecules because its positive and negative poles are attracted to their oppositely charged poles. Because of their electrical attraction for water molecules, ions and polar molecules are described as **hydrophilic** (from the Greek "hydro," meaning "water," and "philic" meaning "loving"). Many biological molecules are hydrophilic and dissolve readily in water.

Gases such as oxygen and carbon dioxide are essential for biochemical reactions but are nonpolar—so how do they dissolve? These molecules are sufficiently small that they can fit into the spaces between water molecules without

▲ FIGURE 2-8 **Water as a solvent** When a salt crystal (NaCl) is dropped into water, the water surrounds the sodium and chloride ions. The positive poles of water molecules face the Cl^-, and the negative poles face the Na^+. The ions disperse as the surrounding water molecules isolate them from one another, and the salt crystal gradually dissolves.

QUESTION If you placed a salt crystal in a nonpolar liquid (like oil), would it dissolve?

disrupting the hydrogen bonds that link the water molecules. Fish swimming below the ice on a frozen lake rely on oxygen that dissolved before the ice formed, and the CO_2 they release dissolves into the water.

Larger molecules with nonpolar covalent bonds, such as fats and oils, do not dissolve in water and hence are called **hydrophobic** (Greek for "water-fearing"). Nevertheless, water has an important effect on such molecules. In sticking together, water molecules—like high school cliques—exclude oil molecules. The nonpolar oil molecules are forced into clusters, surrounded by water molecules that form hydrogen bonds with one another but not with the oil (Fig. 2-9). The tendency of oil molecules to clump together in water is

▲ FIGURE 2-9 Oil and water don't mix Yellow oil poured into water remains in discrete droplets as it rises to the surface. Oil floats because it is less dense than water.

QUESTION Why is this oil forming droplets?

Water's high specific heat protects sunbathers

▲ FIGURE 2-10 Water moderates body temperature and cools the body Because of water's high specific heat, sunbathers can absorb a great deal of heat without their body temperatures skyrocketing. Water vaporizing from perspiration also helps cool their bodies.

QUESTION Why is the skin on these sunbathers beginning to redden? Why is this dangerous?

described as a **hydrophobic interaction.** The membranes of living cells owe much of their structure to both hydrophilic and hydrophobic interactions (discussed in Chapter 5).

Water Moderates the Effects of Temperature Changes

The hydrogen bonds among polar water molecules (see Fig. 2-6) allow water to moderate temperature changes, as described in the sections that follow.

It Takes a Lot of Energy to Heat Water

The energy required to heat 1 gram of a substance by 1°C is called its **specific heat.** Water has a very high specific heat, meaning that it takes more energy to heat water than to heat most other substances. Why?

At any temperature above absolute zero (−459°F or −273°C), the atoms or molecules of a substance are in constant motion because of the heat they contain. More heat produces faster motion. Increasing the temperature of water requires the water molecules to move faster. However, because water molecules are linked by hydrogen bonds, the molecules can only move faster if the hydrogen bonds are broken more often and more rapidly. Breaking hydrogen bonds uses up a considerable amount of heat energy, which is then not available to raise the water's temperature. The same amount of heat energy could raise the temperature of nonpolar substances much more. For example, the energy needed to heat a given weight of water by 1°C would raise the temperature of an equal weight of granite rock by about

5.3°C. The high specific heat of water helps organisms, whose bodies are mostly water, to inhabit hot, sunny environments without overheating as much as they would if their bodies were mostly composed of nonpolar substances (**Fig. 2-10**). Overheating still poses a real threat, however, because the molecules in our bodies function only within a narrow range of temperatures.

It Takes a Lot of Energy to Evaporate Water

Sunbathers are cooled as water evaporates from perspiration on their skin. Water has an extremely high **heat of vaporization,** which is the amount of heat needed to cause a substance to evaporate (to change from a liquid to a vapor). Because of the polar nature of water molecules, water must absorb enough energy to break the hydrogen bonds that interconnect its molecules before the molecules can evaporate into the air. Evaporation has a cooling effect because the fastest-moving (warmest) molecules are those that vaporize, leaving cooler molecules behind.

CASE STUDY c o n t i n u e d

Unstable Atoms Unleashed

The high specific heat of water makes it an ideal coolant for nuclear power plants. Compared to nonpolar liquids like alcohol, a great deal of heat is required to raise the temperature of water. The tsunami that hit the Fukushima power plant disrupted the electrical supply running the pumps that kept water circulating over the fuel rods. Without enough water to absorb the excess heat, the zirconium alloy tubes surrounding the fuel rods melted.

Water Forms an Unusual Solid: Ice

Even solid water is unusual. Most liquids become denser when they solidify, but ice is actually less dense than liquid water. When water freezes, each molecule forms stable hydrogen bonds with four other water molecules, creating an open, hexagonal (six-sided) arrangement. This keeps the water molecules further apart than their average distance in liquid water, so ice is less dense than liquid water (**Fig. 2-11**).

When a pond or lake starts to freeze in winter, the ice floats on top, forming an insulating layer that delays the freezing of the rest of the water. This insulation allows fish and other aquatic animals to survive in the liquid water below (**Fig. 2-12**). If ice were to sink, many ponds and lakes around the world would freeze solid from the bottom up during the winter, killing plants, fish, and other underwater organisms. The ocean floor at higher latitudes would be covered with extremely thick layers of ice that would never melt.

Water-Based Solutions Can Be Acidic, Basic, or Neutral

At any given time, a tiny fraction of water molecules (H_2O) will have split into hydroxide ions (OH^-) and hydrogen ions (H^+) (**Fig. 2-13**). Pure water contains equal concentrations of OH^- and H^+. When ion-forming substances that release OH^- or H^+ are added to water, the solution may no longer have equal concentrations of OH^- and H^+ although it has a neutral charge overall. If the concentration of H^+ exceeds the concentration of OH^-, the solution is **acidic.** An **acid** is a substance that releases hydrogen ions when it dissolves in water. For example, when hydrochloric acid (HCl) is added to pure water, almost all of the HCl molecules separate into H^+ and Cl^-. Therefore, the concentration of H^+ exceeds the concentration of OH^-, and the resulting solution is acidic. Acidic substances—think lemon juice (containing citric acid) or vinegar (acetic acid)—taste sour because sour receptors on your tongue respond to excess H^+. Dental cavities are caused by bacteria that break down trapped food particles and release acid.

If the concentration of OH^- is greater than the concentration of H^+, the solution is **basic.** A **base** is a substance that combines with hydrogen ions, reducing their number. If, for instance, sodium hydroxide (NaOH) is added to water, the NaOH molecules separate into Na^+ and OH^-. Some

▲ FIGURE 2-12 **Ice floats**

OH^- ions combine with H^+ to produce H_2O, reducing the number of H^+ ions and creating a basic solution. Bases are used in many cleaning solutions. They are also in antacids like Tums™ to neutralize heartburn caused by excess hydrochloric acid in the stomach and esophagus.

▲ FIGURE 2-11 **Liquid water (left) and ice (right)**

water   hydroxide ion   hydrogen ion
(H_2O)   (OH^-)   (H^+)

▲ FIGURE 2-13 **Some water is always ionized**

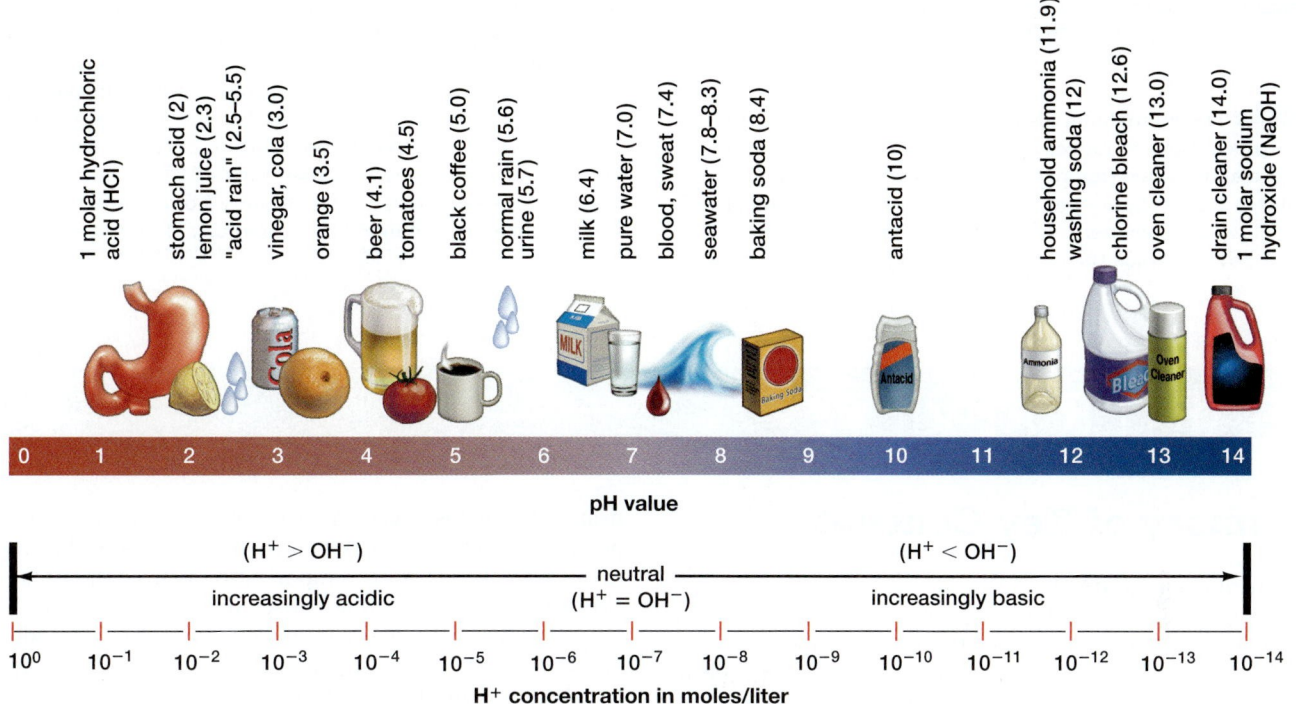

▲ FIGURE 2-14 The pH scale The pH scale reflects the concentration of hydrogen ions in a solution. Notice that pH (upper scale; 0–14) is the negative log of the H⁺ concentration (lower scale). Each unit on the scale represents a tenfold change. Lemon juice, for example, is about 10 times more acidic than orange juice, and the most severe acid rain in the northeastern United States is almost 1,000 times more acidic than normal rainfall.

The **pH scale** of 0 to 14 measures how acidic or basic a solution is (**Fig. 2-14**). Neutral pH (equal concentrations of H^+ and OH^-) is 7. Acids have a pH below 7, pure water has a pH of 7, and bases have a pH above 7. Each unit on the pH scale represents a tenfold change in the concentration of H^+. Thus, a soft drink with a pH of 3 has a concentration of H^+ that is 10,000 times that of pure water, which has a pH of 7.

A **buffer** is a type of molecule that tends to maintain a solution at a constant pH by accepting or releasing H^+ in response to small changes in H^+ concentration. In the presence of excess H^+, a buffer combines with the H^+, reducing its concentration. In the presence of excess OH^-, buffers release H^+, which combines with the OH^- to form H_2O. Mammals, including humans, maintain a pH in body fluids that is just slightly basic (pH about 7.4). If your blood became as acidic as 7.0 or as basic as 7.8, you would die because even small changes in pH cause drastic changes in both the structure and function of biological molecules. Nevertheless, living cells seethe with chemical reactions that take up or give off H^+. The pH of the body fluid remains constant because it is controlled by several different buffers.

CHECK YOUR LEARNING

Are you familiar with the unique properties of water and the importance of these properties to life? Can you explain how polar covalent and hydrogen bonds contribute to the unique properties of water? Do you understand how acids, bases, and buffers affect solutions?

CASE STUDY revisited

Unstable Atoms Unleashed

Scientists believe that the isotopes of uranium were forged in the explosion of a star and became incorporated into Earth when our solar system formed. Today the radioactive form of this rare element is mined and concentrated to help satisfy humanity's unquenchable desire for energy.

The chain reaction that generates heat in nuclear power plants begins when neutrons are released from radioactive uranium. These bombard other uranium atoms and cause them to split, in a self-sustaining chain reaction. When the tsunami struck the Fukushima plant, special neutron-absorbing rods were immediately lowered around the fuel, halting the chain reaction. But many of the atoms that result from the breakdown of uranium are also radioactive isotopes, and these continued to spontaneously decay and generate heat. This caused the unstoppable and disastrous rise in temperature and pressure that eventually burst through the containment vessels, releasing these isotopes into the environment. One isotope of particular concern is radioactive iodine (¹³¹I).

Iodine enters the body through air and food and is concentrated in the thyroid gland, where it is used to synthesize thyroid hormone. Unfortunately, the thyroid does not distinguish between radioactive and nonradioactive iodine. Children exposed to ¹³¹I are at increased risk for thyroid cancer, which may occur decades after exposure. Japanese authorities distributed iodine tablets to children near the failed reactor. This reduces the danger from ¹³¹I, because the nonradioactive iodine in the tablets competes with ¹³¹I in the thyroid and displaces it, allowing most of it to be excreted from the body. Only time will reveal the health effects of the fallout from Fukushima.

Consider This

BioEthics The Fukushima disaster led to a reassessment of safety precautions in nuclear power plants, and a worldwide dialogue about the dangers of nuclear power, which also generates waste that remains radioactive for thousands of years. How can societies evaluate and compare the safety of nuclear power versus the safety of burning fossil fuels, from which humanity gets most (about 87%) of its energy today? How can one compare the possibility of events—such as a magnitude 9 earthquake or the escape or radioactive waste—with the certainty of continued carbon dioxide emissions and the unknown health hazards of global climate change resulting from fossil fuel use? What about a third option—renewable energy, including wind and solar power? The world currently gets only a small fraction of its energy from renewable sources. Does this need to change? If so, how can change happen?

CHAPTER REVIEW

Summary of Key Concepts

2.1 What Are Atoms?

An element is a substance that can neither be broken down nor converted to different substances by ordinary chemical reactions. The smallest possible particle of an element is the atom, which is itself composed of positive protons, neutral neutrons, and negative electrons. All atoms of a given element have the same unique number of protons. Neutrons cluster with protons in atomic nuclei (except hydrogen), contributing to the mass number of the nucleus. Electrons orbit the nucleus within specific regions called electron shells. Shells at increasing distances from the nucleus contain electrons with increasing amounts of energy. Each shell can contain a fixed maximum number of electrons. An atom is most stable when its outermost shell is full. Isotopes are atoms of the same element with different numbers of neutrons. Some isotopes are radioactive; their nuclei spontaneously break down, forming new elements. This releases energy and often releases subatomic particles.

2.2 How Do Atoms Interact to Form Molecules?

Atoms gain stability by filling or emptying their outer electron shells. They do this by acquiring, losing, or sharing electrons during chemical reactions. This produces attractive forces called chemical bonds, which link atoms, forming molecules.

There are three types of bonds: ionic, covalent, and hydrogen bonds. Atoms that have lost or gained electrons are called ions. Ionic bonds are electrical attractions between negatively and positively charged ions that hold them together in the form of crystals. Covalent bonds form when atoms fill their outer electron shells by sharing electrons. In a nonpolar covalent bond, the two atoms share electrons equally. In a polar covalent bond, one atom attracts the electron more strongly than the other atom does, giving the molecule slightly positive and negative poles. Polar covalent bonds allow hydrogen bonding, the attraction between the slightly positive hydrogen on one molecule and the slightly negative regions of other polar molecules.

2.3 Why Is Water So Important to Life?

Water has many unique properties that allowed life as we know it to evolve. Water is polar and dissolves polar substances and ions. Water forces nonpolar substances, such as oil, to form clumps. Water molecules cohere to each other using hydrogen bonds, producing surface tension. Because of its high specific heat, water helps organisms to maintain a fairly stable temperature despite wide temperature fluctuations in the environment. Water's high heat of vaporization helps organisms in dry climates to cool off while conserving water. Water is almost unique in being less dense in its frozen than in its liquid state.

Pure water contains equal numbers of H^+ and OH^- (pH 7), but dissolved substances can make solutions acidic (more H^+ than OH^-) or basic (more OH^- than H^+). Buffers help maintain a constant pH.

Key Terms

acid 30	hydrophilic 28
acidic 30	hydrophobic 28
antioxidant 26	hydrophobic interaction 29
atom 20	ion 24
atomic nucleus 21	ionic bond 24
atomic number 21	isotope 21
base 30	mass number 21
basic 30	molecule 24
buffer 31	neutron 20
chemical bond 24	nonpolar covalent bond 25
cohesion 27	pH scale 31
covalent bond 24	polar covalent bond 26
dissolve 28	proton 20
electron 20	radioactive 21
electron shell 22	solution 28
element 20	solvent 28
free radical 27	specific heat 29
heat of vaporization 29	surface tension 28
hydrogen bond 27	

Learning Outcomes

In this chapter, you have learned to . . .

LO1 Describe the relationship between elements and atoms.

LO2 Describe the structure of atoms, including identification of the three subatomic particles and the characteristics of each.

LO3 Define isotopes and explain what causes some isotopes to be radioactive and how people may be helped or harmed by radioactivity.

LO4 Explain how atoms form molecules.

LO5 Describe the characteristics of the three types of chemical bonds, and provide examples of molecules in which each type of bond is found.

LO6 Describe the special properties of water and how these properties allow water to play key roles in life on Earth.

LO7 Explain the pH scale, and describe the properties of acids, bases, and buffers.

Thinking Through the Concepts

Fill-in-the-Blank

1. An atom consists of an atomic nucleus composed of positively charged _____ and uncharged _____. The number of positively charged particles in the nucleus determines the _____ of the element. Orbiting around the nucleus are _____ that occupy confined spaces called _____. **LO2**

2. An atom that has lost or gained one or more electrons is called a(n) _____. If an atom loses an electron it takes on a(n) _____ charge. If it gains an electron it takes on a(n) _____ charge. Atoms with opposite charges attract one another, forming _____ bonds. **LO5**

3. Atoms of the same element that differ in the number of neutrons in their nuclei are called _____ of one another. Some of these forms of atoms spontaneously break apart, releasing _____ particles that carry large amounts of _____. In the process, they may become different _____. Atoms that behave this way are described as being _____. **LO3**

4. An atom with an outermost electron shell that is either completely full or completely empty is described as _____. Atoms with partially full outer electron shells are _____. Covalent bonds are formed when atoms _____ electrons, filling their outer shells. **LO4 LO5**

5. Water is described as _____ because each water molecule has slightly negative and positive poles. This property allows water molecules to form _____ bonds with one another. The bonds between water molecules give water a high _____ that produces surface tension. **LO6**

Review Questions

1. Based on Table 2-1 how many neutrons are there in oxygen? In hydrogen? In nitrogen? **LO2**

2. Distinguish between atoms and molecules; between elements and compounds; and among protons, neutrons, and electrons. **LO1 LO2 LO4**

3. Define *isotopes* and describe the properties of radioactive isotopes. **LO3**

4. Compare and contrast covalent bonds and ionic bonds. **LO5**

5. Explain how polar covalent bonds allow hydrogen bonds to form, and provide an example. **LO5 LO6**

6. Why can water absorb a great amount of heat with little increase in its temperature? What is this property of water called? **LO6**

7. Describe how water dissolves a salt. **LO5 LO6**

8. Define *pH scale, acid, base,* and *buffer.* How do buffers reduce changes in pH when hydrogen ions or hydroxide ions are added to a solution? Why is this phenomenon important in organisms? **LO7**

Applying the Concepts

1. Detergents and soaps help clean by dispersing fats and oils in water so that they can be rinsed away. From your knowledge of the structure of water and the hydrophobic nature of fats, what general chemical structures (for example, polar or nonpolar parts) must a soap or detergent have, and why?

2. What do people mean when they say: "It's not the heat, it's the humidity"? What disadvantages would there be to sweating if water had a lower heat of vaporization?

Answers to Figure Caption questions and Fill-in-the-Blank questions can be found in the Answers section at the back of the book.

MB ®*Go to MasteringBiology for practice quizzes, activities, eText, videos, current events, and more.*

Friends don't eat friends. Mad cow disease emerged as a result of feeding cows chow made with protein from sheep, some of whom were infected with scrapie.

CASE STUDY

Puzzling Proteins

"YOU KNOW, LISA, I think something is wrong with me," Charlene Singh told her sister. It was 2001 when the vibrant, 22-year-old scholarship winner began to lose her memory and experience mood swings. During the next 3 years, her symptoms worsened. Singh's hands shook, she was subject to uncontrollable episodes of biting and striking people, and she became unable to walk or swallow. In June 2004, Charlene Singh became the first U.S. resident to die of the human form of mad cow disease, which she had almost certainly contracted more than 10 years earlier while living in England.

It was not until the mid-1990s that health officials recognized that mad cow disease, or bovine spongiform encephalitis (BSE), could spread to people who ate meat from infected cattle. Although millions of people may have eaten this tainted beef, fewer than 200 people worldwide have died from the human version of BSE, called variant Creutzfeldt-Jakob disease (vCJD). For those infected, however, the disease is always fatal, riddling the brains of both humans with vCJD and cows with BSE with microscopic holes, giving the brain a spongy appearance.

Where did mad cow disease come from? For centuries, it has been known that sheep can suffer from scrapie, a disease that causes similar symptoms. A leading hypothesis is that a mutated form of scrapie became capable of infecting cattle, perhaps in the early 1980s. At that time, cattle feed was often formulated with sheep parts, some of which may have harbored scrapie. BSE was first identified in British cattle in 1986, and sheep parts were banned from cattle feed in 1988. However, it was 1996 before beef exports from Britain were banned, when experts confirmed that the disease could spread to people through consumption of infected meat. As a precautionary measure at that time, more than 4.5 million cattle in Britain were slaughtered and their bodies were burned.

Why is mad cow disease particularly fascinating to scientists? In the early 1980s, Dr. Stanley Prusiner, a researcher at the University of California–San Francisco, startled the scientific world by providing evidence that a protein, which lacks genetic material (DNA or RNA), was the cause of scrapie, and that this protein could transmit the disease to experimental animals in the laboratory. He dubbed the infectious proteins "prions" (pronounced PREE-ons).

What are proteins? How do they differ from DNA and RNA? How can a protein with no hereditary material infect another organism, increase in number, and produce a fatal disease? Is vCJD still a threat? Read on to find out.

AT A GLANCE

LEARNING GOALS

LG1 Explain why carbon is important in biological molecules.

LG2 Describe dehydration synthesis and hydrolysis and explain the roles they play in forming and breaking down large biological molecules.

LG3 Describe the molecular structure, physical properties, and biological functions of the major groups of carbohydrates and include examples of each.

LG4 Describe the molecular structure, physical properties, and biological functions of the major groups of lipids and include examples of each.

LG5 Describe protein subunits and how they are joined to form proteins.

LG6 Describe the four levels of protein structure and discuss how protein structure relates to the six major functions of proteins.

LG7 Describe the molecular structure, physical properties, and biological functions of the major types of nucleotides and nucleic acids and include examples of each.

3.1 WHY IS CARBON SO IMPORTANT IN BIOLOGICAL MOLECULES?

You have probably seen fruits and vegetables grown without synthetic fertilizers or pesticides labeled as "organic." But in chemistry, the word **organic** describes molecules that have a carbon backbone bonded to hydrogen atoms. The term is derived from the ability of organisms to synthesize and use this general type of molecule. **Inorganic** molecules lack carbon atoms (examples are water and salt) or lack hydrogen atoms (carbon dioxide). Inorganic molecules are far less diverse and generally much simpler than organic molecules.

Life is characterized by an amazing variety of molecules interacting in ways that are dazzlingly complex. Interactions among molecules are governed by their structures and the chemical properties that arise from these structures. As molecules within cells interact with one another, their structures and chemical properties change. Collectively, these precisely orchestrated changes give cells the ability to acquire and use nutrients, to eliminate wastes, to move and grow, and to reproduce. As we explain below, this complexity is made possible by the versatile carbon atom.

The Unique Bonding Properties of Carbon Are Key to the Complexity of Organic Molecules

Atoms are unstable when their outermost electron shells are only partially filled (as described earlier in Chapter 2). Atoms in this state react with one another, and in so doing gain stability by sharing electrons to fill their outer shells, forming covalent bonds. Depending on the number of vacancies in their shells, two atoms can share two, four, or six electrons—forming a single, double, or triple covalent bond. The bonding patterns in the four most common types of atoms found in biological molecules

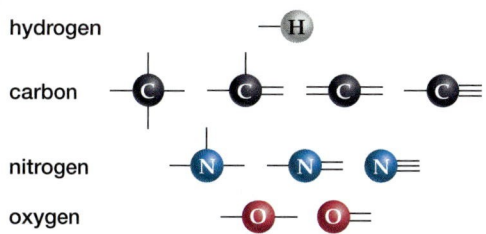

▲ **FIGURE 3-1 Bonding patterns** The bonding patterns of the four most common atoms in biological molecules. A single line indicates a single covalent bond (two shared electrons) with another atom. Two parallel lines indicate a double covalent bond (four shared electrons), and three parallel lines indicate a triple covalent bond (six shared electrons) between the atom shown and another atom.

are shown in **Figure 3-1**. The versatile carbon atom is the key to the tremendous variety of organic molecules that make life on Earth possible. The carbon atom has four electrons in its outermost shell, which can accommodate eight electrons. Therefore, a carbon atom can become stable by bonding with up to four other atoms or with fewer atoms by forming double or even triple bonds. As a result, organic molecules can assume complex shapes, including branched chains, rings, sheets, and helices.

Attached to the carbon backbone of organic molecules are **functional groups,** which are commonly occurring combinations of atoms; seven important examples are shown in **Table 3-1**. Functional groups are less stable than the carbon backbone and more likely to participate in chemical reactions, and so they are mainly responsible for the properties and chemical reactivity of organic molecules.

CHECK YOUR LEARNING

Can you define *organic* and explain why carbon is so important to life? Can you name and describe the properties of several functional groups and explain why functional groups are important in biological molecules?

TABLE 3-1   Important Functional Groups in Biological Molecules

Group	Structure	Properties	Found In
Hydroxyl		Polar; involved in dehydration and hydrolysis reactions; forms hydrogen bonds	Sugars, polysaccharides, nucleic acids, alcohols, some amino acids, steroids
Carbonyl		Polar; makes parts of molecules hydrophilic (water soluble)	Sugars (linear forms), steroid hormones, some vitamins
Carboxyl (ionized form)		Polar and acidic; the negatively charged oxygen may bond H^+, forming carboxylic acid (–COOH); involved in peptide bonds	Amino acids, fatty acids, carboxylic acids (such as acetic and citric acids)
Amino		Polar and basic; may become ionized by binding a third H^+; involved in peptide bonds	Amino acids, nucleic acids, some hormones
Sulfhydryl		Nonpolar; forms disulfide bonds in proteins	Cysteine (an amino acid); many proteins
Phosphate (ionized form)		Polar and acidic; links nucleotides in nucleic acids; forms high-energy bonds in ATP (ionized form occurs in cells)	Phospholipids, nucleotides, nucleic acids
Methyl		Nonpolar; may be attached to nucleotides in DNA (methylation), changing gene expression	Steroids, methylated nucleotides in DNA

3.2 HOW ARE ORGANIC MOLECULES SYNTHESIZED?

Although a complex molecule could be made by laboriously attaching one atom after another, the machinery of life works far more efficiently by preassembling molecular subunits and hooking them together. Just as trains are made by joining a series of train cars, small organic molecules (for example, sugars or amino acids) are joined to form longer molecules (for example, starches or proteins). The individual subunits are called **monomers** ("mono" means "one"); chains of monomers are called **polymers** ("poly" means "many").

Biological Polymers Are Formed by Removing Water and Split Apart by Adding Water

The subunits of large biological molecules are usually joined by a chemical reaction called **dehydration synthesis.** In dehydration synthesis, a hydrogen ion (H^+) is removed from one subunit and a hydroxyl ion (OH^-) is removed from a second subunit, leaving openings in the outer electron shells of atoms in the two subunits. These openings are filled when the subunits share electrons, creating a covalent bond that links them. The hydrogen ion and the hydroxyl ion combine

▲ FIGURE 3-2 **Dehydration synthesis**

to form a molecule of water (H_2O), as shown in **Figure 3-2.** This explains the literal translation of dehydration synthesis: "removing water to put together."

The reverse reaction is hydrolysis ("hydro" means "water" and "lysis" means "to break apart"). **Hydrolysis** breaks apart the molecule into its original subunits, with water donating a hydrogen ion to one subunit and a hydroxyl ion to the other (**Fig. 3-3**). Digestive enzymes use hydrolysis to

▲ FIGURE 3-3 **Hydrolysis**

TABLE 3-2 Principal Types of Biological Molecules		
Type and Structure of Molecule	**Principal Subtypes and Structures**	**Examples**
Carbohydrate: primarily carbon, oxygen, and hydrogen, in the approximate formula $(CH_2O)_{n*}$	***Monosaccharide:*** Simple sugar, often with the formula $C_6H_{12}O_6$	Glucose, fructose, galactose
	Disaccharide: Two monosaccharides bonded together	Sucrose
	Polysaccharide: Chain of monosaccharides (usually glucose)	Starch, glycogen, cellulose
Lipid: Contains a high proportion of carbon and hydrogen; most lipids are nonpolar and insoluble in water	***Triglyceride:*** three fatty acids bonded to glycerol	Oil, fat
	Wax: Variable numbers of fatty acids bonded to a long-chain alcohol	Wax in plant cuticle
	Phospholipid: Polar phosphate group and two fatty acids bonded to glycerol	Phospholipids in cell membranes
	Steroid: Four fused rings of carbon atoms with functional groups attached	Cholesterol, estrogen, testosterone
Protein: Consists of one or more chains of amino acids; may have up to four levels of structure that determine its function	***Peptide:*** Short chain of amino acids	Oxytocin
	Polypeptide: Long chain of amino acids	Hemoglobin, keratin
Nucleotide/Nucleic acid:	***Nucleotide:*** Composed of a five-carbon sugar (ribose or deoxyribose), a nitrogen-containing base, and a phosphate group	Adenosine triphosphate (ATP), cyclic adenosine monophosphate (cAMP)
	Nucleic acid: A polymer of nucleotide subunits joined by covalent bonds between their phosphate and sugar groups	Deoxyribonucleic acid (DNA), ribonucleic acid (RNA)

*n signifies the number of carbons in the molecule's backbone.

break down food. For example, the starch in a cracker consists of a series of glucose (simple sugar) molecules (see Fig. 3-8). Enzymes in our saliva and small intestines promote hydrolysis of the starch into individual sugar molecules that can be absorbed into the body.

Although there is a tremendous diversity of biological molecules, most fall into one of four general categories: carbohydrates, lipids, proteins, and nucleotides/nucleic acids (Table 3-2).

CHECK YOUR LEARNING

Can you name and describe the reactions that create biological polymers?

3.3 WHAT ARE CARBOHYDRATES?

Carbohydrate molecules are composed of carbon, hydrogen, and oxygen in the approximate ratio of 1:2:1. This ratio explains the origin of the word "carbohydrate," which literally translates to "carbon plus water." All carbohydrates are either small, water-soluble **sugars** or polymers of sugar, such as starch. If a carbohydrate consists of just one sugar molecule, it is called a **monosaccharide** ("saccharide" means "sugar"). When two monosaccharides are linked, they form a **disaccharide** ("di" means "two"). Both mono- and disaccharides are called sugars. If you've stirred sugar into coffee, you know that sugar dissolves in water; it is hydrophilic

("water loving"). This is because the hydroxyl functional groups of sugars are polar and form hydrogen bonds with polar water molecules (**Fig. 3-4**).

A polymer of many monosaccharides is called a **polysaccharide.** Most polysaccharides do not dissolve in water at body temperatures. Starches are polysaccharides that serve as energy storage molecules in cells. Other polysaccharides strengthen the cell walls of plants, fungi, and bacteria, or form a supportive armor over the bodies of insects, crabs, and their relatives.

▲ **FIGURE 3-4 Sugar dissolving in water** Glucose dissolves as the polar hydroxyl groups of each sugar molecule form hydrogen bonds with nearby water molecules.

$C_6H_{12}O_6$

(a) Chemical formula   (b) Linear, ball and stick

fructose   galactose

▲ FIGURE 3-6 **Some six-carbon monosaccharides**

(c) Ring, ball and stick   (d) Ring, simplified

▲ FIGURE 3-5 **Depictions of glucose structure**
Chemists draw the same molecule in different ways. **(a)** The chemical formula for glucose; **(b)** the linear form, which occurs when glucose is in its crystal state; **(c, d)** two depictions of the ring form of glucose, which forms when glucose dissolves in water. In (d), each unlabeled joint is a carbon atom; the carbons are numbered for easy reference. Figure 3-4 depicts the space-filling ring structure of glucose.

There Are Several Monosaccharides with Slightly Different Structures

Monosaccharides have a backbone of three to seven carbon atoms. Most of these carbon atoms have both a hydrogen (–H) and a hydroxyl group (–OH) attached to them; therefore, carbohydrates generally have the approximate chemical formula $(CH_2O)_n$, where n is the number of carbons in the backbone. When a sugar molecule is dissolved in water, such as inside a cell, its carbon backbone usually forms a ring. **Figure 3-5** and Figure 3-4 show various ways of depicting the chemical structure of the common sugar **glucose.** When you see simplified depictions later in this text, keep in mind that every unlabeled "joint" in a ring or chain is actually a carbon atom.

Glucose is the most common monosaccharide in organisms and the primary energy source of cells. Glucose has six carbons, so its chemical formula is $C_6H_{12}O_6$. Many organisms synthesize other monosaccharides that have the same chemical formula as glucose but slightly different structures. For example, some plants may store energy in *fructose* ("fruit sugar"), which we consume in fruits and juices, corn syrup, and honey (made by bees from the nectar of flowers). *Galactose* (part of lactose, or "milk sugar") is secreted by mammals in their milk as an energy source for their young (**Fig. 3-6**).

Fructose and galactose must be converted to glucose before cells can use them as a source of energy.

Other common monosaccharides, such as ribose and deoxyribose (found in the nucleic acids of RNA and DNA, respectively), have five carbons. Notice in **Figure 3-7** that deoxyribose ("deoxy" means "without oxygen") has one fewer oxygen atom than ribose because one of the hydroxyl groups in ribose is replaced by a hydrogen atom in deoxyribose.

Disaccharides Consist of Two Monosaccharides Linked by Dehydration Synthesis

Monosaccharides can be linked by dehydration synthesis to form disaccharides or polysaccharides (**Fig. 3-8**). Disaccharides are often used for short-term energy storage, especially in plants. When energy is required, the disaccharides are broken apart by hydrolysis into their monosaccharide subunits (see Fig. 3-3) and converted to glucose, which is broken down further to release energy stored in its chemical bonds. Perhaps you had toast and coffee with cream and sugar at breakfast. You stirred *sucrose* (glucose plus fructose, an energy storage molecule in sugarcane and sugar beets) into your coffee, and added cream containing *lactose* (glucose plus galactose). The disaccharide *maltose* (glucose plus glucose) is rare in nature, but it is formed when enzymes in your digestive tract hydrolyze starch, such as the starch in your toast. Other digestive enzymes then hydrolyze each maltose into two glucose molecules that cells can absorb and break down to liberate energy.

If you are on a diet, you may have used an artificial sugar substitute in your coffee. These interesting molecules are described in "Links to Everyday Life: Fake Foods."

ribose   deoxyribose

Note "missing" oxygen atom

▲ FIGURE 3-7 **Some five-carbon monosaccharides**

▲ FIGURE 3-8 Synthesis of a disaccharide Sucrose is synthesized by a dehydration reaction in which a hydrogen is removed from glucose and a hydroxyl group is removed from fructose. This forms a water molecule and leaves the two monosaccharide rings joined by single bonds to the remaining oxygen atom.

QUESTION Describe hydrolysis of this molecule.

Polysaccharides Are Chains of Monosaccharides

If you take a bite of bagel and chew on it for a few minutes, do you notice that it gradually tastes sweeter? This is because enzymes in saliva cause hydrolysis of the **starch** (a polysaccharide) in a bagel into its component glucose molecules (monosaccharides), which taste sweet. Plants often use starch (**Fig. 3-9**) as an energy-storage molecule. Animals commonly store **glycogen,** a similar polysaccharide. Both starch and glycogen are polymers of glucose molecules.

Starch, which is commonly formed in roots and seeds, consists of branched chains of up to half a million glucose subunits. Glycogen, stored in the liver and muscles of animals (including people), is also a chain of glucose subunits, but is more highly branched than starch.

Many organisms use polysaccharides as structural materials. One of the most important structural polysaccharides is **cellulose,** which makes up most of the walls of the living cells of plants, the fluffy white bolls of cotton plants, and about half the bulk of tree trunks (**Fig. 3-10**). Scientists estimate that

➤➤ LINKS TO EVERYDAY LIFE

Fake Foods

In societies blessed with an overabundance of food, obesity is a serious health problem. To address this problem, food scientists have modified biological molecules to make them noncaloric. Thus, we can now purchase foods or beverages made with artificial sweeteners or the artificial oil olestra (**Fig. E3-1**). How are these "nonbiological food molecules" made?

The sweetener sucralose is a modified sucrose molecule in which three hydroxyl groups are replaced with chlorine atoms. Sucralose tastes about 600 times as sweet as table sugar (sucrose) and provides no calories because we can't digest it. Aspartame, a combination of two amino acids (aspartic acid and phenylalanine; see Fig. 3-19a, b), tastes 200 times as sweet as sugar. Although both sweeteners have been tested extensively and found safe by the Food and Drug Administration (FDA), people with the rare genetic disorder phenylketonuria cannot break down phenylalanine and should avoid aspartame.

The artificial oil olestra tastes and feels similar to oil, and has been approved by the FDA for use in snack foods like potato chips. The olestra molecule is based on sucrose, with six to eight fatty acids attached to its carbon atoms. The fatty acid chains—which extend outward from the sucrose molecule like the arms of an octopus—prevent digestive enzymes from breaking it down. Because fat-soluble vitamins (A, D, E, and K) dissolve in olestra, foods containing olestra are supplemented with small amounts of these vitamins to compensate for those

lost as olestra moves through the digestive tract. Olestra also interferes with the absorption of carotenoids, which give fruits and vegetables their red, orange, and yellow colors and are widely believed to be part of a healthy diet. This is a possible drawback to eating olestra frequently.

▲ FIGURE E3-1 Artificial foods Artificial sweeteners and olestra are marketed to people trying to control their weight.

starch grains

(a) Potato cells

(b) A starch molecule

(c) Detail of a starch molecule

▲ FIGURE 3-9 **Starch structure and function** (a) Starch grains inside potato cells store energy that will allow the potato to generate new plants in the spring. (b) A section of a single starch molecule. Starches consist of branched chains of up to half a million glucose subunits. (c) The precise structure of the circled portion of the starch molecule in (b). Notice the linkage between the individual glucose subunits for comparison with cellulose (Fig. 3-10).

(a) Cellulose is a major component of wood

(b) A plant cell with a cell wall

(c) A close-up of cellulose fibers in a cell wall

Hydrogen bonds cross-linking cellulose molecules

bundle of cellulose molecules

cellulose fiber

Alternating bond configuration differs from starch

(d) Detail of a cellulose molecule

▲ FIGURE 3-10 **Cellulose structure and function** (a) Wood in this 3,000-year-old bristlecone pine is primarily cellulose. (b) Cellulose forms the cell wall that surrounds each plant cell. (c) Plant cell walls often consist of cellulose fibers in layers that run at right angles to each other to resist tearing in both directions. (d) Cellulose is composed of up to 10,000 glucose subunits. Compare this structure with Figure 3-9c and notice that every other glucose molecule in cellulose is "upside down."

▲ **FIGURE 3-11 Chitin structure and function** Chitin has the same glucose bonding configuration as cellulose does. In chitin, however, the glucose subunits have a nitrogen-containing functional group (green) replacing one of the hydroxyl groups. Tough, flexible chitin supports the otherwise soft bodies of arthropods (such as the spider in the photograph, insects, crabs, and their relatives) and most fungi.

plants synthesize about a trillion tons of cellulose each year, making it the most abundant organic molecule on Earth.

Cellulose, like starch, is a polymer of glucose, but in cellulose, every other glucose is "upside down" as you will see when you compare Figure 3-9c with Figure 3-10d. Although most animals easily digest starch, no vertebrates synthesize an enzyme that can attack the bonds between glucose molecules in cellulose. A few animals, such as cows and termites, harbor cellulose-digesting microbes in their digestive tracts and can benefit from the glucose subunits the microbes release. In humans, cellulose fibers pass intact through the digestive system, supplying no nutrients but providing the "roughage" that helps prevent constipation.

The supportive outer coverings (exoskeletons) of insects, crabs, and spiders are made of **chitin,** a polysaccharide in which the glucose subunits bear a nitrogen-containing functional group (**Fig. 3-11**). Chitin also stiffens the cell walls of many fungi, including mushrooms.

Carbohydrates may also form parts of larger molecules; for example, the plasma membrane that surrounds each cell is studded with proteins to which carbohydrates are attached. Nucleic acids (discussed later) also contain sugar molecules.

CHECK YOUR LEARNING

Can you describe the major classes of carbohydrates and provide examples of each type and how organisms use them?

3.4 WHAT ARE LIPIDS?

Lipids are a diverse group of molecules that contain regions composed almost entirely of hydrogen and carbon, with nonpolar carbon–carbon and carbon–hydrogen bonds. These nonpolar regions make lipids insoluble in water, or hydrophobic ("water fearing"). Some lipids are used to store energy, some form waterproof coverings on plant or animal bodies, some serve as the primary component of cellular membranes, and still others are hormones. Lipids fall into three major groups: (1) oils, fats, and waxes; (2) phospholipids; and (3) steroids.

Oils, Fats, and Waxes Are Lipids Containing Only Carbon, Hydrogen, and Oxygen

Oils, fats, and waxes are built from only three types of atoms: carbon, hydrogen, and oxygen. Each contains one or more **fatty acid** subunits, which are long chains of carbon and hydrogen with a carboxylic acid group (−COOH) at one end. Fats and oils are formed by dehydration synthesis linking three fatty acid subunits to one molecule of *glycerol*, a three-carbon molecule (**Fig. 3-12**). This structure gives fats and oils their chemical name: **triglycerides.**

▲ **FIGURE 3-12 Synthesis of a triglyceride** Dehydration synthesis links a single glycerol molecule with three fatty acids to form a triglyceride and three water molecules.

QUESTION What kind of reaction breaks this molecule apart?

(a) Fat

(b) Wax

▲ **FIGURE 3-13 Lipids in nature** (a) A fat grizzly bear ready to hibernate. If this bear stored the same amount of energy in carbohydrates instead of fat, he probably would be unable to walk. **(b)** Wax is a highly saturated lipid that remains very firm at normal outdoor temperatures. Its rigidity allows it to be used to form the strong but thin-walled hexagons of this honeycomb.

Fats and **oils** are used primarily as energy-storage molecules; they contain more than twice as many calories per gram as do carbohydrates and proteins (9 Cal/gram for fats; 4 Cal/gram for carbohydrates and proteins). For example, the bear in **Figure 3-13a** has accumulated enough fat to last him through his winter hibernation. To avoid accumulating excess fat, some people turn to foods manufactured with fat substitutes such as olestra, as described earlier in "Links to Everyday Life: Fake Foods."

Fats (such as butter and lard) are produced primarily by animals, whereas oils (such as corn, canola, and soybean oil) are found primarily in the seeds of plants. The difference between a fat, which is solid at room temperature, and an oil, which is liquid at room temperature, lies in the structure of their fatty acid subunits. In fats, the carbons of fatty acids are joined by single bonds, with hydrogen atoms at all the other bonding sites. Such fatty acids are described as **saturated,** because they contain as many hydrogen atoms as possible. Saturated fatty acid chains are straight and can pack closely together, thus forming a solid at room temperature (**Fig. 3-14a**).

If there are double bonds between some of the carbons, and consequently fewer hydrogens, the fatty acid is **unsaturated.** The double bonds in the unsaturated fatty acids of oils produce kinks in the fatty acid chains (**Fig. 3-14b**). Oils are liquid at room temperature because these kinks prevent oil molecules from packing closely together. The commercial process of hydrogenation—which breaks some of the double bonds and adds hydrogens to the carbons—can convert liquid oils to solids, but with health consequences (see "Health Watch: Cholesterol, Trans Fats, and Your Heart").

Although **waxes** are chemically similar to fats, humans and most other animals do not have the appropriate enzymes to break them down. Waxes are highly saturated and therefore solid at normal outdoor temperatures. Waxes form a waterproof coating over the leaves and stems of land

(a) A fat

(b) An oil

▲ **FIGURE 3-14 Fats and oils** (a) Fats have straight chains of carbon atoms in their fatty acid tails. **(b)** The fatty acid tails of oils have double bonds between some of their carbon atoms, creating kinks in the chains. Oils are liquid at room temperature because their kinky tails keep the molecules farther apart.

plants. Animals synthesize waxes as waterproofing for mammalian fur and insect exoskeletons. Honey bees use waxes to build elaborate honeycomb structures where they store honey and lay their eggs (**Fig. 3-13b**).

Health|Watch

Cholesterol, Trans Fats, and Your Heart

Why are so many foods advertised as "cholesterol free" or "low in cholesterol"? Although cholesterol is crucial to life, people with high levels of certain forms of cholesterol-containing particles in their blood are at increased risk for heart attacks and strokes. These particles, often called "bad cholesterol," can cause obstructions called *plaques* to form in arteries (Fig. E3-2). Plaques, in turn, may stimulate the formation of blood clots. If a clot breaks loose and blocks an artery supplying blood to the heart muscle or brain, it can cause a heart attack or a stroke. What makes cholesterol turn bad?

▲ FIGURE E3-2 **Plaque** A plaque deposit (rippled structure) partially blocks this carotid artery.

Cholesterol molecules are nonpolar, so they are transported in the blood (which is mostly water) surrounded by polar carrier molecules consisting of phospholipids and proteins. These particles are called *lipoprotein* (lipid plus protein) and they differ considerably in the percentage of cholesterol and protein they contain. Those with less cholesterol and more protein are described as *high-density lipoprotein* (HDL), because protein is denser than lipid. HDL is the good guy, because HDL transports excess cholesterol from cells to the liver, where the cholesterol is metabolized (used in bile synthesis, for example), and to organs that use it to produce steroid hormones. The particles with more lipids and less protein are *low-density lipoprotein* (LDL). These particles are carried in the blood from the liver, where most cholesterol is synthesized, to cells where cholesterol is stored or used to produce cell membranes. When there is excess LDL, it becomes the bad guy, entering cells in the artery walls, where it can build up and stimulate plaque formation.

Our diet provides only about 15% to 20% of the cholesterol in our blood, and this comes from animal-derived foods such as egg yolks, sausages, bacon, whole milk, and butter. Genetic differences cause some people to regulate their LDL cholesterol at safe levels regardless of diet and others to produce far more LDL cholesterol than they need, even on a low-cholesterol diet. Saturated fats (such as those in red meat and whole-milk products) stimulate the liver to churn out more LDL cholesterol. Diets in which unsaturated fats (found in fish, nuts, and most vegetable oils) replace saturated fats and carbohydrates are associated with a decreased risk of heart disease. Lifestyle also contributes; exercise tends to increase HDL, whereas obesity and smoking increase LDL cholesterol levels. A high ratio of LDL to HDL is correlated with an increased risk of heart disease.

The worst fats are **trans fats,** which are made artificially when hydrogen atoms are added to oils to make them solid at room temperature. This process alters the attachment points of hydrogen near the remaining double bonds, causing the kinky tails of oil to straighten. Trans fats are not metabolized in the same way as naturally occurring fats; for unknown reasons they actually decrease HDL while increasing LDL, and they are likely to place consumers at a higher risk of heart disease.

Until recently, artificially produced trans fats were abundant in commercial food products such as margarine, cookies, crackers, and French fries, because they extended the shelf life and helped retain the flavor of packaged products. Since January 2006, the FDA has required food labels to list the trans fat content, but a loophole allows up to 0.5 gram per serving to be labeled "no trans fat" in products that contain partially hydrogenated oil or vegetable shortening. Fortunately, most food manufacturers and fast-food chains are reducing or eliminating trans fats from their products.

Phospholipids Have Water-Soluble "Heads" and Water-Insoluble "Tails"

The plasma membrane that surrounds each cell contains several types of **phospholipids.** A phospholipid is similar to an oil, except that one of the three fatty acids is replaced by a phosphate group attached to any of several polar functional groups that typically contain nitrogen (Fig. 3-15). A phospholipid has two dissimilar ends. At one end are the two nonpolar fatty acid "tails," which are insoluble in water. At the other end is the phosphate–nitrogen "head" that is polar and water soluble. (These properties of phospholipids are crucial to the structure and function of cell membranes; see Chapter 5.)

Steroids Contain Four Fused Carbon Rings

All **steroids** are composed of four rings of carbon atoms. As shown in Figure 3-16, the rings share one or more sides, with various functional groups protruding from them. One steroid, *cholesterol,* is a vital component of the membranes of animal cells (Fig. 3-16a). It makes up about 2% of the human brain, where it is an important component of the lipid-rich membranes that insulate nerve cells. Cholesterol is also used by cells to synthesize other steroids, such as the female and male sex hormones *estrogen* and *testosterone* (Figs. 3-16b, c). Too much of the wrong form of cholesterol can contribute to cardiovascular disease, as explained in "Health Watch: Cholesterol, Trans Fats, and Your Heart."

variable
functional phosphate
group   group

polar head   glycerol   fatty acid tails
             backbone
(hydrophilic)   (hydrophobic)

▲ FIGURE 3-15 **Phospholipids** Phospholipids have two hydrophobic fatty acid tails attached to the glycerol backbone. The third (here, the upper) position on glycerol is occupied by a polar "head" consisting of a phosphate group to which a second (usually nitrogen-containing) variable functional group is attached. The phosphate group bears a negative charge, and the nitrogen-containing functional group bears a positive charge, making the head hydrophilic.

(a) Cholesterol   (b) Estrogen   (c) Testosterone

▲ FIGURE 3-16 **Steroids** All steroids have a similar, nonpolar molecular structure with four fused carbon rings. Differences in steroid function result from differences in functional groups attached to the rings. **(a)** Cholesterol, the molecule from which other steroids are synthesized; **(b)** the female sex hormone, estrogen (estradiol); **(c)** the male sex hormone, testosterone. Note the similarities in structure between the sex hormones.

QUESTION Why are steroid hormones able to diffuse through cell membranes to exert their effects?

CHECK YOUR LEARNING

Can you compare and contrast the structure and synthesis of fats and oils? Can you describe the functions of fats, oils, and waxes? Do you recall two reasons why cholesterol is important in the body?

3.5 WHAT ARE PROTEINS?

Proteins are molecules composed of chains of amino acids. Scientists estimate that the human body has about 100,000 different types of proteins that play important and diverse roles (Table 3-3). Most cells contain hundreds of different **enzymes,** which are proteins that promote

TABLE 3-3	Functions of Proteins
Function	**Examples**
Structural	Keratin (forms hair, nails, scales, feathers, and horns); silk (forms webs and cocoons)
Movement	Actin and myosin (found in muscle cells; allow contraction)
Defense	Antibodies (found in the bloodstream; fight disease organisms, some neutralize venoms); venoms (found in venomous animals; deter predators and disable prey)
Storage	Albumin (in egg white; provides nutrition for an embryo)
Signaling	Insulin (secreted by the pancreas; promotes glucose uptake into cells)
Catalyzing reactions	Amylase (found in saliva and the small intestine; digests carbohydrates)

specific chemical reactions. Other proteins are structural. Keratin, for example, forms hair, horns, nails, scales, and feathers (**Fig. 3-17**). Silk proteins are secreted by silk moths and spiders to make cocoons and webs. Nutritional proteins, such as albumin in egg white and casein in milk, provide amino acids to developing animals. The protein hemoglobin transports oxygen in the blood. Actin and myosin in muscle are contractile proteins that allow animal bodies to move. Some proteins are hormones (insulin and growth hormone, for example), others are antibodies (which help fight disease and infection), and a few are toxins (such as rattlesnake venom).

Proteins Are Formed from Chains of Amino Acids

Proteins are polymers of **amino acids** joined by peptide bonds. All amino acids have the same fundamental structure: a central carbon bonded to a hydrogen atom, a

(a) Hair

(b) Horn

(c) Silk

▲ FIGURE 3-17 **Structural proteins** Keratin is a common structural protein. It is the predominant protein found in **(a)** hair, **(b)** horn, and **(c)** the silk of a spider web.

variable group (R)

amino group

carboxylic acid group

hydrogen

▲ FIGURE 3-18 **Amino acid structure**

nitrogen-containing amino group ($-NH_2$), a carboxylic acid group ($-COOH$), and an "R" group that varies among different amino acids (Fig. 3-18).

Twenty amino acids are commonly found in the proteins of organisms. The R group gives each amino acid distinctive properties (Fig. 3-19). Some amino acids are hydrophilic and water soluble because their R groups are polar. Others are hydrophobic, with nonpolar R groups that are insoluble in water. The amino acid cysteine (Fig. 3-19c) has a sulfur-containing (sulfhydryl) R group that can form covalent **disulfide bonds** with the sulfur of another cysteine molecule. Disulfide bonds play important roles in proteins, as described later.

Amino Acids Are Joined by Dehydration Synthesis

Proteins are formed by dehydration synthesis, as are polysaccharides and lipids. In proteins, the nitrogen in the amino group ($-NH_2$) of one amino acid is joined to the carbon in the carboxylic acid group ($-COOH$) of another amino acid by a single covalent bond, and water is liberated (Fig. 3-20). This bond is called a **peptide bond,** and the resulting chain of two amino acids is called a **peptide,** a term used for relatively short chains of amino acids (up to 50 or so). Additional amino acids are added, one by one, until a long *polypeptide* chain is completed. Polypeptide chains in cells can range up to thousands of amino acids in length. A protein consists of one or more polypeptide chains.

A Protein Can Have As Many As Four Levels of Structure

Interactions among amino acid R groups cause twists, folds, and interconnections that give proteins their three-dimensional structure. Up to four organized levels of protein structure are possible: primary, secondary, tertiary, and quaternary (Fig. 3-21). The sequence of amino acids in a protein is described as its **primary structure** (Fig. 3-21a)

glutamic acid (glu) aspartic acid (asp) phenylalanine (phe) leucine (leu) cysteine (cys)

(a) Hydrophilic functional groups **(b) Hydrophobic functional groups** **(c) Sulfur-containing functional group**

▲ FIGURE 3-19 **Amino acid diversity** The diversity of amino acids is caused by the variable R functional group (blue backgrounds), which may be **(a)** hydrophilic or **(b)** hydrophobic. **(c)** The R group of cysteine has a sulfur atom that can form covalent bonds with the sulfur in other cysteines.

amino acid   amino acid   dehydration synthesis   peptide   water

amino group   carboxylic acid group   amino group   peptide bond

▲ FIGURE 3-20 **Protein synthesis** A dehydration reaction forms a peptide bond between the carbon of the carboxylic acid group of one amino acid and the nitrogen of the amino group of a second amino acid.

of protein. The amino acid sequence of each protein is specified by genetic instructions in a cell's DNA. The specific amino acid sequences cause polypeptides to assume simple, repeating **secondary structures:** a helix or a pleated sheet. Secondary structures are maintained by hydrogen bonds between the polar portions of amino acids. Keratin protein in hair and the subunits of the hemoglobin molecule in blood provide examples of the coiled, spring-like secondary structure called a **helix** (Fig. 3-21b).

Hydrogen bonds that form between the oxygen atoms of the carbonyl functional groups (with slightly negative charges) and the hydrogens of the amino functional groups (with slightly positive charges) hold the turns of the coils together. Other proteins, such as silk, contain polypeptide chains that repeatedly fold back upon themselves, with hydrogen bonds holding adjacent segments of the polypeptide together in a secondary **pleated sheet** arrangement (Fig. 3-22a).

(a) Primary structure: The sequence of amino acids linked by peptide bonds

(b) Secondary structure: Usually maintained by hydrogen bonds, which shape this helix

leu
val
lys
lys
gly
his
ala
lys
val
lys
pro

hydrogen bond

helix

heme group

(c) Tertiary structure: Folding of the helix results from hydrogen bonds with surrounding water molecules and disulfide bridges between cysteine amino acids

(d) Quaternary structure: Individual polypeptides are linked to one another by hydrogen bonds or disulfide bridges

▲ FIGURE 3-21 **The four levels of protein structure** Hemoglobin (illustrated) is the oxygen-carrying protein in red blood cells. The red disks represent the iron-containing heme group that binds oxygen.

QUESTION Why do most proteins, when heated excessively, lose their ability to function?

(a) Pleated sheet

stack of pleated sheets

disordered segment

strand of silk

(b) Structure of silk

▲ FIGURE 3-22 **The pleated sheet and the structure of silk protein** (a) In a pleated sheet, a single polypeptide chain is folded back upon itself repeatedly (loops of peptide connecting the ends of the chains in the sheet are not shown). Adjacent segments of the folded polypeptide are linked by hydrogen bonds (dotted lines), creating a folded (pleated) sheet-like configuration. The R groups (green) project alternately above and below the sheet. (b) Silk protein contains stacks of pleated sheets connected by disordered protein segments. Pleated sheets do not stretch, but the disordered segments do, so a strand of silk is stretchy.

CASE STUDY continued

Puzzling Proteins

Infectious prions are misfolded versions of a normal protein found throughout the body. The secondary structure of the normal prion protein is primarily helical. Infectious prions, however, fold into pleated sheets. The pleated sheets are so stable that they cannot be destroyed by high temperatures or most chemical agents, so it is very difficult to sterilize equipment contaminated by prions. More important, the altered secondary structure of prions is unaffected by the enzymes that break down normal prion protein. Because they are so stable, infectious prions accumulate destructively in the brain.

In addition to their secondary structures, many proteins are contorted into **tertiary structures (Fig. 3-21c)**. Their origami folds are determined both by the protein's secondary structure and by its environment. A protein in water, for example, one dissolved in the watery interior of a cell, folds in a way that exposes its hydrophilic amino acids to the water and hides its hydrophobic amino acids deep within the molecule. In contrast, portions of a protein embedded in the phospholipid cell membrane expose their hydrophobic functional groups to the surrounding hydrophobic tails of the phospholipids. Disulfide bonds can also contribute to tertiary structure by linking together the sulfur-containing R groups of cysteines in different regions of the polypeptide, as in keratin (see Figs. E3-3 and E3-4).

A fourth level of protein organization, called **quaternary structure,** occurs in certain proteins that contain individual polypeptides linked by hydrogen bonds, disulfide bonds, or by attractions between oppositely charged portions of different amino acids. Hemoglobin, for example, consists of four polypeptide chains held together by hydrogen bonds (**Fig. 3-21d**). Each of the four polypeptides holds an iron-containing organic molecule called a heme group (red disks in Figs. 3-21c, d), which can bind one molecule (two atoms) of oxygen.

Most proteins, such as enzymes and hormones, work in close association with other molecules. Because they have very precise roles to play, their highly organized three-dimensional structures help ensure that they interact only with other molecules of particular shapes—much like a key fits a lock. However, scientists are now discovering that many proteins are quite flexible, and only the primary structure is fixed. These *disordered proteins* or *disordered segments* of proteins can fold in a variety of ways, allowing some of them to interact with several different molecules—more like master keys that fit several locks. Some structural proteins, such as silk (**Fig. 3-22b**), have disordered segments linking sections of pleated sheets. These disordered segments allow the silk to stretch. Although the exact structure of most proteins has not yet been determined, some experts now estimate that one-third of all human proteins are disordered or include disordered segments.

The Functions of Proteins Are Related to Their Three-Dimensional Structures

Within a protein, the exact position and number of amino acids bearing specific R groups determine both the structure of the protein and its biological function. In hemoglobin, for example, amino acids bearing specific R groups must be present in precisely the right places to hold the iron-containing heme group that binds oxygen. In contrast, the polar amino acids on the outside of a hemoglobin molecule mainly keep it dissolved in the fluid of a red blood cell. Therefore, a mutation that changes the identity of one of these amino acids will not significantly alter the protein's function, as long as the substituted amino acid is also hydrophilic. A mutation that replaces a hydrophilic with a hydrophobic amino acid, however, can have catastrophic effects on the hemoglobin molecule (causing the painful and sometimes life-threatening disorder called sickle-cell anemia; see Chapters 10 and 12).

The importance of higher level protein structure becomes obvious when a protein is **denatured,** meaning that its normal three-dimensional structure is altered while leaving the primary structure intact. A denatured protein has different properties and will no longer perform its function. For example, the white of an egg consists of the protein albumin. If you fry it, the heat rips hydrogen bonds apart, destroying the protein's secondary and tertiary structure, which causes the egg white to become opaque and solid. The albumin could no longer nourish a developing chick in this denatured form. Keratin in hair is denatured by the chemicals in a permanent wave, as described in "Links to Everyday Life: What Makes Hair Curl?" Bacteria and viruses are often destroyed by denaturing their proteins using heat, ultraviolet radiation, or solutions that are highly salty or acidic. For example, canned foods are sterilized by heating them, water is sometimes sterilized with ultraviolet light, and dill pickles are preserved from bacterial attack by their salty, acidic brine.

CHECK YOUR LEARNING

Can you describe protein subunits and how proteins are synthesized? Do you understand the four levels of protein structure and why the three-dimensional structure of proteins is important? Can you list several functions of proteins and provide examples of proteins that perform each function?

3.6 WHAT ARE NUCLEOTIDES AND NUCLEIC ACIDS?

A **nucleotide** is a molecule with a three-part structure: a five-carbon sugar, a phosphate functional group, and a nitrogen-containing **base.** Bases all have carbon and nitrogen atoms linked in rings, with functional groups attached to some of the carbon atoms. Nucleotides fall into two general classes: deoxyribose nucleotides and ribose nucleotides,

▲ FIGURE 3-23 **Deoxyribose nucleotide**

depending on which type of sugar they contain (see Fig. 3-7). The bases in deoxyribose nucleotides are adenine, guanine, cytosine, and thymine. The bases in ribose nucleotides are adenine, guanine, cytosine, and uracil. Adenine and guanine have double rings, whereas cytosine, thymine, and uracil have single rings (see Fig. 3-25 and Fig. 11-3). A nucleotide with the adenine base is illustrated in **Figure 3-23.** Nucleotides may function as energy-carrier molecules, intracellular messenger molecules, or as subunits of polymers called nucleic acids, described below.

Nucleotides Act as Energy Carriers and Intracellular Messengers

Adenosine triphosphate (ATP) is a ribose nucleotide with three phosphate functional groups (**Fig. 3-24**). This molecule is formed in cells by reactions that release energy, such as the reaction that breaks down a sugar molecule. ATP stores energy in bonds between its phosphate groups, and releases energy when the bond linking the last phosphate to the ATP molecule is broken. This energy is then available to drive energy-demanding reactions such as linking amino acids to form proteins.

The ribose nucleotide cyclic adenosine monophosphate (cAMP) acts as a messenger molecule in cells. Many hormones exert their effects by stimulating cAMP to form within cells, where it initiates a series of biochemical reactions (see Chapter 37).

Other nucleotides (NAD^+ and FAD) are known as *electron carriers* because they transport energy in the form of

▲ FIGURE 3-24 **The energy-carrier molecule adenosine triphosphate (ATP)**

high-energy electrons. Their energy and electrons are used in ATP synthesis, for example, when cells break down sugar. (You will learn more about these nucleotides in Chapters 6, 7, and 8.)

DNA and RNA, the Molecules of Heredity, Are Nucleic Acids

Single nucleotides (monomers) may be strung together in long chains by dehydration synthesis, forming polymers called **nucleic acids.** In nucleic acids, an oxygen atom in the phosphate functional group of one nucleotide is covalently bonded to the sugar of the next. The polymer of deoxyribose nucleotides, called **deoxyribonucleic acid (DNA),** can contain millions of nucleotides (**Fig. 3-25**). DNA is found in the chromosomes of all cells. A DNA molecule consists of two strands of nucleotides entwined in the form of a double helix and linked by hydrogen bonds (see Fig. 3-24). Its sequence

CASE STUDY continued

Puzzling Proteins

All cells use DNA as a blueprint for producing more cells, and viruses use either DNA or RNA. Before the discovery of prions, however, no infectious agent had ever been discovered that completely lacked genetic material composed of nucleic acids. Scientists were extremely skeptical of the hypothesis that proteins could effectively "reproduce themselves" until repeated studies found no trace of genetic material associated with prions.

of nucleotides, like the letters of a biological alphabet, spells out the genetic information needed to construct the proteins of each organism. Single-stranded chains of ribose nucleotides, called **ribonucleic acid (RNA),** are copied from the DNA and direct the synthesis of proteins (see Chapters 11 and 12).

CHECK YOUR LEARNING

Can you describe the structure and provide three different functions of nucleotides? Can you explain how nucleic acids are synthesized and give two examples of nucleic acids and their functions?

CASE STUDY revisited

Puzzling Proteins

Stanley Prusiner and his associates identified the prion, a misfolded version of a protein found throughout the animal kingdom. This protein, with no trace of genetic material, is the culprit in scrapie, BSE, and vCJD. But how do prions replicate themselves and build up in the body over time? Prusiner and other researchers have strong experimental evidence supporting a radical hypothesis: The infectious prions interact with normal helical prion proteins and cause them to change their configuration into the pleated sheet configuration of the infectious form. These new "infectious prion converts" then go on to transform other normal prion proteins in an ever-expanding chain reaction. As in Charlene Singh's case, it can be a decade or more after infection before enough of the proteins have been transformed to cause disease symptoms. Fortunately, the actions taken to prevent new infections appear successful; as of mid 2012, there were no surviving vCJD victims and no new confirmed cases of vCJD; only four people in the world were known to be living with the disorder.

The emergence of vCJD has focused attention on the role of the normal prion protein. It is found in cell membranes throughout the body, so it is likely to have a variety of functions, most of which are still unknown. Normal prion protein is found in relatively high levels in the brain, a prime target of infectious prions.

Prusiner's efforts led to the recognition of a totally new disease process. A major strength of the scientific method is that hypotheses can be tested experimentally. If repeated experiments support the hypothesis, then even well-established scientific principles—for example, that

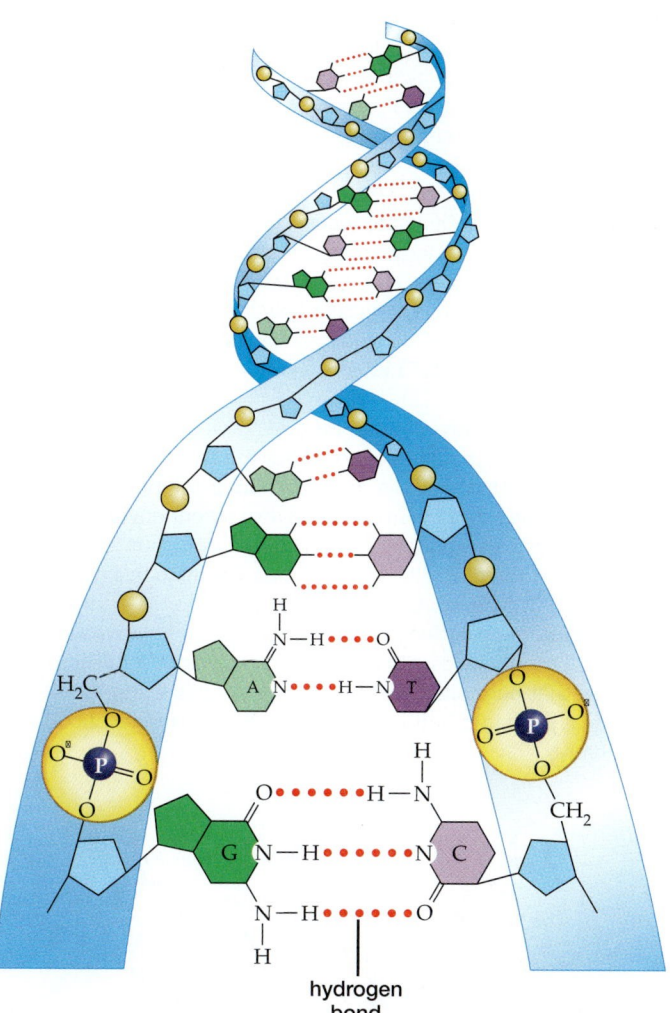

▲ **FIGURE 3-25 Deoxyribonucleic acid** Like a twisted ladder, the double helix of DNA is formed by chains of nucleotides that spiral around one another, linked by hydrogen bonds between the bases of the nucleotides in the two adjacent chains. The bases are identified by their initials: A: adenine; C: cytosine; T: thymine; G: guanine.

all infectious agents must have genetic material—must be reexamined and sometimes discarded. Prusiner's startling revision of how infectious diseases can be transmitted was recognized when he was awarded a Nobel Prize in 1997.

Consider This

A disorder called "chronic wasting disease" (CWD) of deer and elk, first identified in the late 1960s, has now been reported in at least 15 U.S. states. Like scrapie and BSE, CWD is a fatal brain disorder caused by prions. The disease spreads readily between animals in herds, apparently by contact with saliva, urine, and feces of infected animals, which all may contain prions. These infectious proteins have also been found in the muscles of deer with CWD. The U.S. Centers for Disease Control and Prevention states, "To date, no strong evidence of CWD transmission to humans has been reported." If you were a hunter in an affected region, would you eat deer or elk meat? Explain why or why not.

LINKS TO EVERYDAY LIFE

What Makes Hair Curl?

Each of the roughly 125,000 hairs on your head has a remarkable structure; **Figure E3-3** shows its assemblage of bundles within bundles, which consist mostly of keratin protein. To understand what makes hair curl, we'll focus on hydrogen and disulfide bonds. Hydrogen bonds give keratin its helical, spring-like secondary structure. When you stretch hair or get it wet, these bonds break and the hair lengthens slightly. The bonds re-form when the tension is released or the hair dries. Keratin has lots of cysteines, which can form covalent disulfide bonds with other cysteines. Whether your hair is naturally straight or curly is determined by the locations of the cysteines in your keratin, which is determined by your genes. In curly hair, disulfide bonds occur in places that contort the otherwise straight keratin molecules (**Fig. E3-4**).

Now let's curl some straight hair. We'll start by putting wet hair on curlers. The water causes keratin's hydrogen bonds to break and re-form with surrounding water molecules. As hair dries, the curlers force new hydrogen bonds to form in different places on the keratin that hold the hair in a curl. But if it's rainy or even humid outside, you can say goodbye to your curls as water breaks the new bonds and hair reverts to its natural straightness.

To permanently curl hair requires breaking and re-forming stable disulfide bonds. First the hair is soaked in a solution that breaks the disulfide bonds linking keratin molecules. The hair is then wrapped around curlers and saturated with a second solution. This solution counteracts the effects of the first, allowing disulfide bonds to re-form. The bonds form in new locations, however, because the curlers have shifted the positions of the keratin helices relative to one another. These new covalent bonds link the keratin molecules in a way that permanently maintains the curl (see Fig. E3-4). Genetically straight hair has been transformed into artificially curly hair.

▲ FIGURE E3-3 **The structure of hair** At the microscopic level, a single hair is organized into bundles within larger bundles, all consisting primarily of keratin. Individual keratin molecules are held in a helical shape by hydrogen bonds, and adjacent keratin molecules are cross-linked by disulfide bonds. These bonds give hair both elasticity and strength. Straight hair is depicted here.

▲ FIGURE E3-4 **Curly hair is contorted by disulfide bonds**

CHAPTER REVIEW

Summary of Key Concepts

3.1 Why Is Carbon So Important in Biological Molecules?

Organic molecules have a carbon backbone. They are so diverse because the carbon atom is able to form bonds with up to four other molecules. This allows organic molecules to form complex shapes, including branched chains, helices, pleated sheets, and rings. The presence of functional groups produces further diversity among biological molecules (see Table 3-1).

3.2 How Are Organic Molecules Synthesized?

Most large biological molecules are polymers synthesized by linking many smaller monomer subunits using dehydration synthesis. Hydrolysis reactions break these polymers apart. The most important organic molecules fall into four classes: carbohydrates, lipids, proteins, and nucleotides/nucleic acids (see Table 3-2).

3.3 What Are Carbohydrates?

Carbohydrates include sugars, starches, cellulose, and chitin. Sugars include monosaccharides and disaccharides. They are used for temporary energy storage and the construction of other molecules. Starches and glycogen are polysaccharides that provide longer-term energy storage in plants and animals, respectively. Cellulose forms the cell walls of plants, and chitin strengthens the exoskeletons of many invertebrates and the cell walls of fungi.

3.4 What Are Lipids?

Lipids are nonpolar, water-insoluble molecules. Oils, fats, waxes, and phospholipids all contain fatty acids, which are chains of carbon and hydrogen atoms with a carboxylic acid group at the end. Steroids all have four fused rings of carbon atoms with functional groups attached. Lipids are used for energy storage (oils and fats), as waterproofing for the outside of many plants and animals (waxes), as the principal component of cellular membranes (phospholipids and cholesterol), and as hormones (steroids).

3.5 What Are Proteins?

Proteins consist of one or more amino acid chains called polypeptides with up to four levels of structure. Primary structure is the sequence of amino acids; secondary structure consists of helices or pleated sheets. These may fold to produce tertiary structure. Proteins in which two or more polypeptides are linked have quaternary structure. Some proteins or parts of proteins are disordered and lack a stable structure. The function of a protein is determined by its shape and by how its amino acids interact with their surroundings and with each other. Proteins can be enzymes; structural molecules such as hair, silk, or horn; hormones such as insulin; or transport molecules such as hemoglobin.

3.6 What Are Nucleotides and Nucleic Acids?

A nucleotide is composed of a phosphate group, a five-carbon sugar (ribose or deoxyribose), and a nitrogen-containing base. Molecules formed from single nucleotides include energy-carrier molecules, such as ATP, and messenger molecules such as cyclic AMP. Nucleic acids are chains of nucleotides. DNA carries the hereditary blueprint, and RNA is copied from DNA and directs the synthesis of proteins.

Key Terms

adenosine triphosphate (ATP) 48	nucleic acid 49
amino acid 44	nucleotide 48
base 48	oil 42
carbohydrate 37	organic 35
cellulose 39	peptide 45
chitin 41	peptide bond 45
dehydration synthesis 36	phospholipid 43
denatured 48	pleated sheet 46
deoxyribonucleic acid (DNA) 49	polymer 36
disaccharide 37	polysaccharide 37
disulfide bond 45	primary structure 45
enzyme 44	protein 44
fat 42	quaternary structure 47
fatty acid 41	ribonucleic acid (RNA) 49
functional group 35	saturated 42
glucose 38	secondary structure 46
glycogen 39	starch 39
helix 46	steroid 43
hydrolysis 36	sugar 37
inorganic 35	tertiary structure 47
lipid 41	trans fat 43
monomer 36	triglyceride 41
monosaccharide 37	unsaturated 42
	wax 42

Learning Outcomes

In this chapter, you have learned to . . .

LO1 Explain why carbon is important in biological molecules.

LO2 Describe dehydration synthesis and hydrolysis and explain the roles they play in forming and breaking down large biological molecules.

LO3 Describe the molecular structure, physical properties, and biological functions of the major groups of carbohydrates and include examples of each.

LO4 Describe the molecular structure, physical properties, and biological functions of the major groups of lipids and include examples of each.

LO5 Describe protein subunits and how they are joined to form proteins.

LO6 Describe the four levels of protein structure and discuss how protein structure relates to the six major functions of proteins.

LO7 Describe the molecular structure, physical properties, and biological functions of the major types of nucleotides and nucleic acids and include examples of each.

Thinking Through the Concepts

Fill-in-the-Blank

1. In organic molecules made of chains of subunits, each subunit is called a(n) _____, and the chains are called _____. Carbohydrates consisting of long chains of sugars are called _____. These sugar chains can be broken down by _____reactions. Three types of carbohydrates consisting of long glucose chains are _____, _____, and _____. Three examples of disaccharides are _____, _____, and _____. **LO2 LO3**

2. Fill in the following with the appropriate type of lipid: Unsaturated, liquid at room temperature: _____; bees use to make honeycombs: _____; stores energy in animals: _____; sex hormones are synthesized from these: _____; the LDL form of this contributes to heart disease: _____; a major component of cell membranes that has polar heads: _____. **LO4**

3. Proteins are synthesized by a reaction called _____ synthesis, which releases _____. Subunits of proteins are called _____. The sequence of protein subunits is called the _____ structure of the protein. Two regular configurations of secondary protein structure are _____ and _____. Which of these secondary structures is characteristic of infectious prion proteins? _____ When a protein's secondary or higher-order structure is destroyed, the protein is said to be _____. **LO5 LO6**

4. Fill in the following with the most specific term describing the bonds involved: Create kinks in the carbon chains of fatty acids in oils: _____; maintain the helical structure of many proteins: _____; link polypeptide chains and can cause proteins to bend: _____ and _____; join the two strands of the double helix of DNA: _____; link amino acids to form the primary structure of proteins: _____. **LO4 LO5 LO6**

5. A nucleotide consists of three parts: _____, _____, and _____. A nucleotide that acts as an energy carrier is _____. The four bases found in deoxyribose nucleotides are _____, _____, _____, and _____. Two important nucleic acids are _____ and _____. The functional group that joins nucleotides in nucleic acids is _____. **LO7**

Review Questions

1. What does the term "organic" mean to a chemist? **LO1**

2. List the four principal types of biological molecules and give an example of each. **LO3 LO4 LO6 LO7**

3. What roles do nucleotides play in living organisms? **LO7**

4. How are fats and oils similar? How do they differ, and how do their differences explain whether they are solid or liquid at room temperature? **LO4**

5. Describe and compare dehydration synthesis and hydrolysis. Give an example of a substance formed by each chemical reaction, and describe the specific reaction in each instance. **LO2 LO3 LO5**

6. Distinguish among the following: monosaccharide, disaccharide, and polysaccharide. Give two examples of each. **LO3**

7. Describe the synthesis of a protein from amino acids. Then describe the primary, secondary, tertiary, and quaternary structures of a protein. **LO6**

8. Where in nature do we find cellulose? Where do we find chitin? In what way(s) are these two polymers similar? Different? **LO3**

9. Which kinds of bonds between keratin molecules are altered when hair is (a) wet and allowed to dry on curlers and (b) given a permanent wave? **LO6**

Applying the Concepts

1. In this chapter, you read that a phospholipid has a hydrophilic head and hydrophobic tails. Predict how phospholipids would organize themselves in water.

2. Fat contains twice as many calories per unit weight as carbohydrate. Compare the way fat and carbohydrates interact with water, and explain why this interaction gives fat an extra advantage for weight-efficient energy storage.

3. Some people think that consuming fat and sugar substitutes is a good way to reduce weight, whereas others object to these artificial dietary shortcuts. Explain and justify one of these two positions.

4. Saliva from infected deer can transmit chronic wasting disease, caused by misfolded prion proteins. The infectious prions can persist in soil. How might the disease pass from an infected animal to others in the herd?

5. In an alternate universe where people could digest cellulose molecules, how might this affect our way of life?

Answers to Figure Caption questions and Fill-in-the-Blank questions can be found in the Answers section at the back of the book.

(MB) *Go to MasteringBiology for practice quizzes, activities, eText, videos, current events, and more.*

CASE STUDY

Spare Parts for Human Bodies

"I DON'T THINK I've ever screamed so loud in my life." Jennifer looks back on the terrible day when boiling oil from a deep-fat fryer spilled onto her 10-month-old baby, burning more than 70% of his body. "The 911 operator said to get his clothes off, but they'd melted onto him. I pulled his socks off and the skin came right with them." A few decades ago, Zachary Jenkins' burns would have been fatal. Now, the only evidence of the burn is slightly crinkled skin. Zachary was saved by the bioengineering marvel of artificial skin.

Skin consists of several specialized cell types with complex interactions. The outer cells of skin are masters of multiplication, allowing superficial burns to heal without a trace. However, if the deeper layers of the skin are damaged, as occurs with spilled boiling oil, healing is significantly slower. Deep burns are often treated by grafting skin (including some of the deeper layer) from other sites on the body. This creates a new and painful wound where the skin for the graft is removed. In addition, for burns covering a large portion of the body, the lack of healthy skin for grafting makes this approach impossible. Although cells from a piece of the patient's skin can be grown in culture and placed over the affected area, traditional methods of growing enough cells to cover extensive burns can take weeks. Until recently, the only other option was to use skin from human cadavers or pigs. At best, these tissues serve as temporary biological bandages because the victim's body eventually rejects them. Extensive and disfiguring scars are a common legacy.

Bioengineered skin has radically changed the prognosis for burn victims. The child in this chapter's opening photograph was treated with a bioengineered artificial skin that uses skin cells obtained from the donated foreskins of infants who were circumcised at birth. This tissue, which would otherwise be discarded, provides a source of rapidly dividing cells. Cultured in the laboratory on nylon mesh, the cells secrete growth factors and protein fibers. When used for skin grafting, the foreskin cells die, but the growth factors and protein fibers remaining in the mesh stimulate the victim's own cells to grow and multiply. Therefore, new skin and blood vessels regenerate far more rapidly than with burn coverings of cadaver or pig skin.

Bioengineered artificial skin demonstrates our expanding ability to manipulate cells, the fundamental units of life. What structures make up cells? What new techniques using cells to treat burns are under development? Might medical researchers also manipulate cells into forming other replacement organs? We will explore these questions in this chapter.

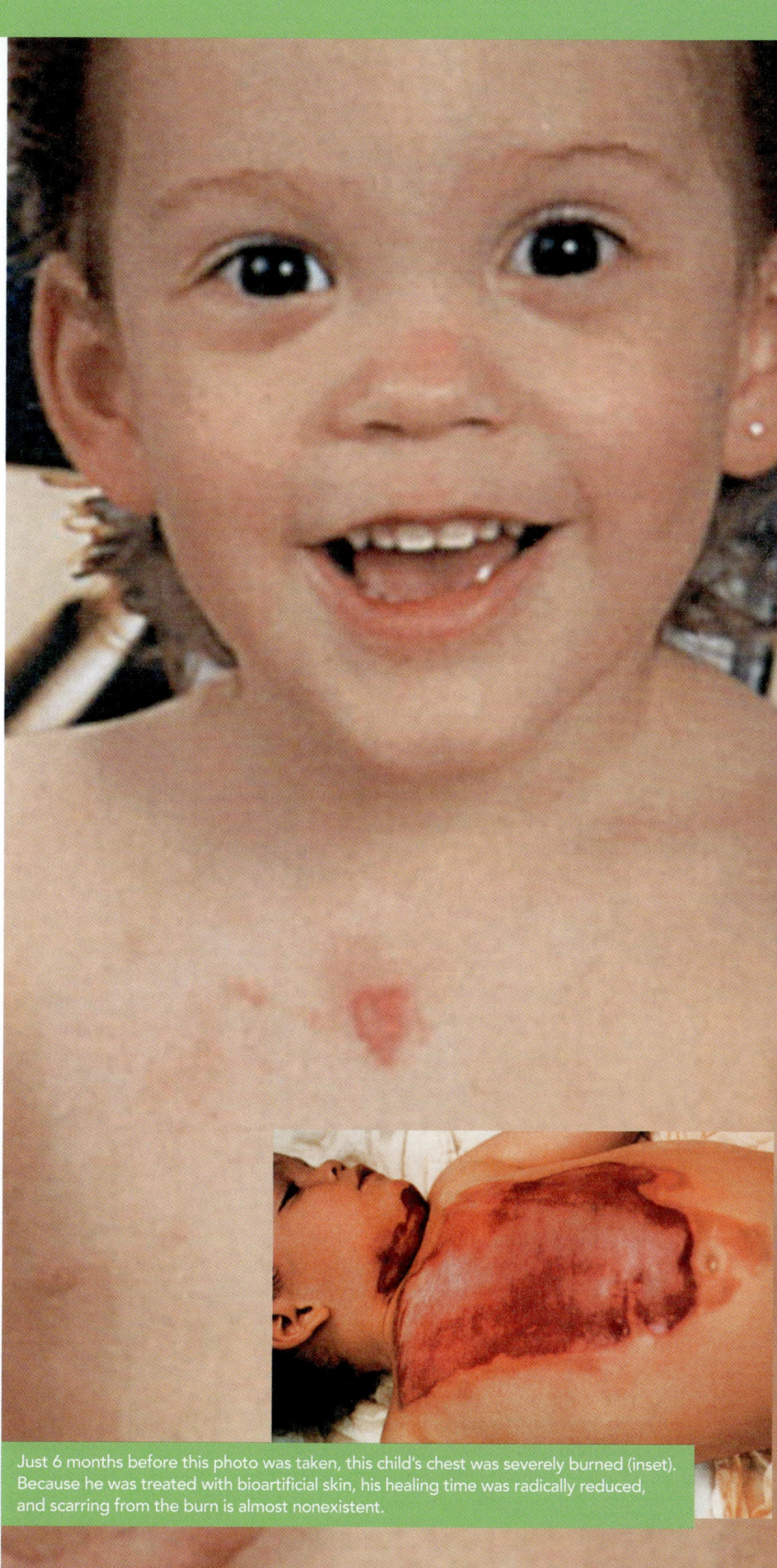

Just 6 months before this photo was taken, this child's chest was severely burned (inset). Because he was treated with bioartificial skin, his healing time was radically reduced, and scarring from the burn is almost nonexistent.

AT A GLANCE

4.1 What Is the Cell Theory?
4.2 What Are the Basic Attributes of Cells?

4.3 What Are the Major Features of Eukaryotic Cells?

4.4 What Are the Major Features of Prokaryotic Cells?

LEARNING GOALS

LG1 Summarize the cell theory.

LG2 Compare and contrast prokaryotic and eukaryotic cells.

LG3 Describe the major structures of animal cells and their functions.

LG4 Describe the structure and function of organelles that are found in plant, but not animal, cells.

LG5 Explain the structure, function, and movement of eukaryotic cilia and flagella.

LG6 Describe the structure and function of the nucleus, and compare it to the nucleoid region of prokaryotic cells.

LG7 List the parts of the endomembrane system and explain how they are interconnected.

LG8 Describe the structure and function of both the surface and the internal features of bacterial cells.

4.1 WHAT IS THE CELL THEORY?

Because cells are so small, no one had ever seen them until the first microscope was invented in the mid-1600s (see "Science in Action: The Search for the Cell"). But seeing cells was only the first step toward understanding their importance. In 1838, the German botanist Matthias Schleiden concluded that cells and substances produced by cells form the basic structure of plants, and that plant growth occurs by adding new cells. In 1839, German biologist Theodor Schwann (Schleiden's friend and collaborator) drew similar conclusions about animal cells. The work of Schleiden and Schwann provided a unifying theory of cells as the fundamental units of life. In 1855, the German physician Rudolf Virchow completed the **cell theory**—a fundamental concept of biology—by concluding that all cells come from previously existing cells.

The cell theory consists of three principles:

- Every organism is made up of one or more cells.
- The smallest organisms are single cells, and cells are the functional units of multicellular organisms.
- All cells arise from preexisting cells.

CHECK YOUR LEARNING

Can you list the three precepts of the cell theory?

4.2 WHAT ARE THE BASIC ATTRIBUTES OF CELLS?

All living things, from microscopic bacteria to the human body to a giant sequoia tree, are composed of cells. Cells perform an enormous variety of functions, including obtaining energy and nutrients, synthesizing biological molecules, eliminating wastes, interacting with other cells, and reproducing.

Most cells range in size from about 1 to 100 micrometers (μm; millionths of a meter) in diameter (**Fig. 4-1**). Why are most cells so small? The answer lies in the need for cells to exchange nutrients and wastes with their external environment through the plasma membrane. Many nutrients and wastes move into, through, and out of cells by diffusion. **Diffusion** is the process by which molecules dissolved in fluids disperse from regions where their concentration is higher to regions where concentration is lower (described in Chapter 5). Diffusion is a relatively slow process. Therefore, to meet the constant demands of cells, the interior of the cell must remain close to the external environment, keeping the cell small.

All Cells Share Common Features

All cells arose from a common ancestor at the dawn of evolution. Modern cells include the simple prokaryotic cells of bacteria and archaea, as well as the far more complex eukaryotic cells of protists, fungi, plants, and animals. All cells share important features, as described in the following sections.

The Plasma Membrane Encloses the Cell and Allows Interactions Between the Cell and Its Environment

Each cell is surrounded by an extremely thin, rather fluid membrane called the **plasma membrane** (**Fig. 4-2**). The plasma membrane and the other membranes within cells consist of a double layer (a bilayer) of phospholipids and cholesterol molecules, with many different proteins embedded within the bilayer. Important functions of the plasma membrane include:

- selectively isolating the cell's contents from the external environment;
- regulating the flow of materials into and out of the cell; and
- allowing communication with other cells and with the extracellular environment.

The phospholipid and protein components of cellular membranes have very different roles. The phospholipid bilayer helps isolate the cell from its surroundings, allowing the cell to maintain essential differences in concentrations of materials inside and outside. In contrast, proteins embedded in the plasma membrane facilitate communication between the cell and its environment. Some proteins allow passage of specific molecules or ions, while others respond to signals from molecules outside the cell and promote chemical reactions inside the cell. (The structure of the plasma membrane is covered in detail in Chapter 5.)

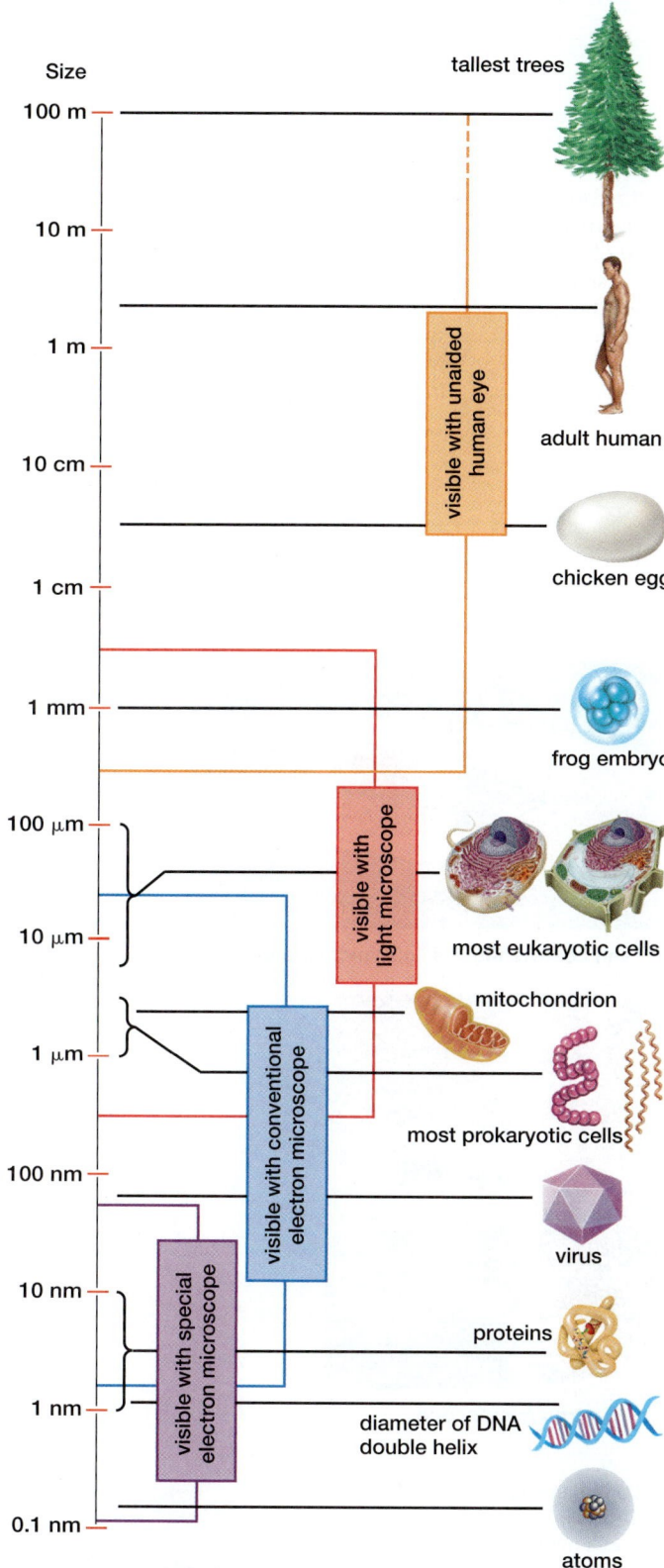

Size

- 100 m — tallest trees
- 10 m —
- 1 m — adult human
- 10 cm —
- 1 cm — chicken egg
- 1 mm — frog embryo
- 100 μm — most eukaryotic cells
- 10 μm —
- 1 μm — mitochondrion
- most prokaryotic cells
- 100 nm — virus
- 10 nm —
- proteins
- 1 nm — diameter of DNA double helix
- 0.1 nm — atoms

visible with unaided human eye

visible with light microscope

visible with conventional electron microscope

visible with special electron microscope

Units of measurement:
1 meter (m) = 39.37 inches
1 centimeter (cm) = 1/100 m
1 millimeter (mm) = 1/1,000 m
1 micrometer (μm) = 1/1,000,000 m
1 nanometer (nm) = 1/1,000,000,000 m

▲ FIGURE 4-1 **Relative sizes** Dimensions commonly encountered in biology range from about 100 meters (the height of the tallest redwood trees) through a few micrometers (the diameter of most cells) to a few nanometers (the diameter of many large molecules).

interstitial fluid (outside)

carbohydrate
glycoprotein

A phospholipid bilayer helps to isolate the cell's contents

Proteins help the cell communicate with its environment

membrane protein
channel protein
cytoskeleton
cytosol

▲ FIGURE 4-2 **The plasma membrane** The plasma membrane encloses the cell. Like all cellular membranes, it consists of a double layer of phospholipid molecules in which a variety of proteins is embedded. The membrane is supported by the cytoskeleton.

All Cells Contain Cytoplasm

The **cytoplasm** consists of all the fluid and structures that lie inside the plasma membrane but outside of the nucleus (see Figs. 4-3 and 4-4). The fluid portion of the cytoplasm in both prokaryotic and eukaryotic cells, called the **cytosol,** contains water, salts, and an assortment of organic molecules, including proteins, lipids, carbohydrates, sugars, amino acids, and nucleotides (described in Chapter 3). Most of the cell's metabolic activities—the biochemical reactions that support life—occur in the cell cytoplasm. Protein synthesis is one example. Cells must

CASE STUDY c o n t i n u e d

Spare Parts for Human Bodies

Why is skin from cadavers and pigs eventually rejected by the bodies of burn victims? The plasma membranes of all human cells bear surface molecules called glycoproteins (shown in Fig. 4-2). Each individual has unique glycoproteins that allow the person's immune system to recognize the cells as belonging to that person's own body, and not attack them as it would foreign, invading cells. But skin cells grafted from a different person or other animal bear glycoproteins that differ from those of the burn victim. This difference causes the victim's immune system to treat the skin as foreign, attack it, and eventually reject it. Fortunately, the response is gradual enough that the transplanted skin can still help the burn heal before the graft is rejected. Synthetic skin substitutes—although produced by donated human skin cells—are preferable because by the time the skin substitutes are grafted, no donor cells remain to stimulate an immune response.

SCIENCE IN ACTION The Search for the Cell

In 1665, the English scientist and inventor Robert Hooke aimed his primitive microscope at an "exceeding thin . . . piece of Cork" and saw "a great many little Boxes" (Fig. E4-1a). Hooke called the boxes "cells," because he thought they resembled the rooms occupied by monks in a monastery, which were called cells. Cork comes from the dry outer bark of the cork oak, and we now know that he was looking at the nonliving cell walls that surround all plant cells. Hooke wrote that in the living oak and other plants, "These cells [are] fill'd with juices."

In the 1670s, Dutch microscopist Anton van Leeuwenhoek constructed his own simple microscopes and observed a previously unknown world. A self-taught amateur scientist, van Leeuwenhoek's descriptions of myriad "animalcules" (mostly single-celled organisms) in rain, pond, and well water caused quite an uproar because, in those days, water was consumed without being treated. Eventually, van Leeuwenhoek made careful observations of an enormous range of microscopic specimens, including blood cells, sperm cells, and the eggs of small insects such as aphids and fleas. His discoveries struck a blow to the idea of spontaneous generation; at that time, fleas were believed to emerge spontaneously from sand or dust, and weevils were

(a) Robert Hooke's light microscope and his drawing of cork cells

(b) van Leeuwenhoek's light microscope

(c) Electron microscope

▲ FIGURE E4-1 **Microscopes yesterday and today** (a) Robert Hooke saw the walls of cork cells through his elegant light microscope, and drew them with great skill. **(b)** Hooke and van Leeuwenhoek were contemporaries. Hooke admitted that van Leeuwenhoek's microscopes produced better images, but described these extremely simple microscopes as "offensive to my eye." **(c)** This elaborate machine can perform as both a transmission electron microscope (TEM) and a scanning electron microscope (SEM).

believed to spring from grain. Although Leeuwenhoek's microscopes appear much more primitive than Hooke's, because of better lenses, they actually provided much clearer images and higher magnification (**Fig. E4-1b**).

Ever since the pioneering efforts of Hooke and van Leeuwenhoek, biologists, physicists, and engineers have collaborated in the development of a variety of advanced microscopes to view the cell and its components. *Light microscopes* use lenses made of glass or quartz to bend, focus, and transmit light rays that have passed through or been bounced off a specimen. Light microscopes provide a wide range of images, depending on how the specimen is illuminated and whether

it has been stained (**Fig. E4-2a**). The resolving power of light microscopes—that is, the smallest structure that can be distinguished under ideal conditions—is about 200 nanometers (200 nm, or 0.2 micrometers; a nanometer is a billionth of a meter). This power is sufficient to see most prokaryotic cells, and some of the internal structures of eukaryotic cells. Light microscopes can be used to observe living cells—a swimming *Paramecium*, for example.

Instead of light, electron microscopes (**Fig. E4-1c**) use beams of electrons, which are focused by magnetic fields rather than by lenses. All objects viewed using electron microscopes must be dead and sliced, dried, or coated to allow viewing.

Transmission electron microscopes pass electrons through a thin specimen and can reveal the details of interior cell structure (**Fig. E4-2c**). Some modern transmission electron microscopes can resolve structures as small as 0.05 nanometer, allowing scientists to see molecules such as DNA. *Scanning electron microscopes* bounce electrons off specimens that are dry and hard (such as shells) or that have been covered with an ultrathin coating of metal such as gold. Scanning electron microscopes can be used to view the three-dimensional surface details of structures that range in size from entire small insects down to cells and their components, with a maximum resolution of about 1.5 nm (**Figs. E4-2b, d**).

cilia   nucleus   contractile vacuole

(a) Light micrograph

(b) Scanning electron micrograph

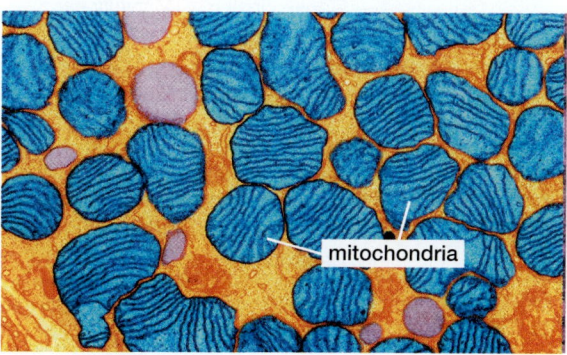

mitochondria

(c) Transmission electron micrograph

mitochondria

smooth ER

(d) Scanning electron micrograph

▲ FIGURE E4-2 **A comparison of microscope images (a)** A living *Paramecium* (a single-celled freshwater organism) photographed through a light microscope. **(b)** A scanning electron micrograph (SEM) of several *Paramecium*. **(c)** A transmission electron micrograph (TEM) showing mitochondria. **(d)** An SEM showing mitochondria and smooth endoplasmic reticulum Any colors you see in electron micrographs (SEMs or TEMs) have been added artificially.

synthesize a variety of proteins, such as those of the cytoskeleton (described later in this chapter), proteins in the plasma membrane, and all the enzymes that promote biochemical reactions.

All Cells Use DNA as a Hereditary Blueprint and RNA to Copy the Blueprint and Guide Construction of Cell Parts

The genetic material in cells is **deoxyribonucleic acid (DNA),** an inherited blueprint that stores the instructions for making all the parts of the cell and for producing new cells. This fascinating molecule (discussed in Chapters 3 and 11) contains genes consisting of precise sequences of nucleotides. During cell division, the original, or parent, cells pass exact copies of their DNA to their newly formed offspring, or daughter cells. **Ribonucleic acid (RNA),** which is chemically similar to DNA, copies the genes on DNA and helps construct proteins based on this genetic blueprint. All cells contain both DNA and RNA.

There Are Two Basic Types of Cells: Prokaryotic and Eukaryotic

All forms of life are composed of only two different types of cells. **Prokaryotic** cells ("prokaryotic" means "before the nucleus"; see Fig. 4-19) form the bodies of **bacteria** and **archaea,** the simplest forms of life. **Eukaryotic** cells ("eukaryotic" means "true nucleus"; see Figs. 4-3 and 4-4) are far more complex and make up the bodies of animals, plants, fungi, and protists. As their names suggest, one striking difference between prokaryotic and eukaryotic cells is that the genetic material of eukaryotic cells is contained within a membrane-enclosed nucleus. In contrast, the genetic material of prokaryotic cells is not enclosed within a membrane. The nucleus and other membrane-enclosed structures, collectively called **organelles,** contribute to the far greater structural complexity of eukaryotic cells. Table 4-1 summarizes the features of eukaryotic and prokaryotic cells, which we discuss further in the sections that follow.

TABLE 4-1   Functions and Distribution of Cell Structures

Structure	Function	Prokaryotes	Eukaryotes: Plants	Eukaryotes: Animals
Cell Surface				
Cell wall	Protects and supports the cell	Present	Present	Absent
Cilia	Move the cell through fluid or move fluid past the cell surface	Absent	Absent (in most)	Present
Flagella	Move the cell through fluid	Present[1]	Absent (in most)	Present
Plasma membrane	Isolates the cell contents from the environment; regulates movement of materials into and out of the cell; allows communication with other cells	Present	Present	Present
Organization of Genetic Material				
Genetic material	Encodes the information needed to construct the cell and to control cellular activity	DNA	DNA	DNA
Chromosomes	Contain and control the use of DNA	Single, circular, no proteins	Many, linear, with proteins	Many, linear, with proteins
Nucleus*	Contains chromosomes and nucleoli	Absent	Present	Present
Nuclear envelope	Encloses the nucleus; regulates movement of materials into and out of the nucleus	Absent	Present	Present
Nucleolus	Synthesizes ribosomes	Absent	Present	Present
Cytoplasmic Structures				
Ribosomes	Provide sites for protein synthesis	Present	Present	Present
Mitochondria*	Produce energy by aerobic metabolism	Absent	Present	Present
Chloroplasts*	Perform photosynthesis	Absent	Present	Absent
Endoplasmic reticulum*	Synthesizes membrane components, proteins, and lipids	Absent	Present	Present
Golgi apparatus*	Modifies, sorts, and packages proteins and lipids	Absent	Present	Present
Lysosomes*	Contain digestive enzymes; digests food and worn-out organelles	Absent	Absent (in most)	Present
Plastids*	Store food, pigments	Absent	Present	Absent
Central vacuole*	Contains water and wastes; provides turgor pressure to support the cell	Absent	Present	Absent
Other vesicles and vacuoles*	Transport secretory products; contain food obtained through phagocytosis	Absent	Present	Present
Cytoskeleton	Gives shape and support to the cell; positions and moves cell parts	Present	Present	Present
Centrioles	Produce the basal bodies of cilia and flagella	Absent	Absent (in most)	Present

[1]Some prokaryotes have structures called flagella, which lack microtubules and move in a fundamentally different way than do eukaryotic flagella.
*All of these structures are organelles, surrounded by membranes and found only in eukaryotic cells.

How Many Cells Form the Human Body?

Scientists through the years have wondered how many cells form the human body—and they don't yet agree. Estimates range from 10 trillion to 100 trillion. Interestingly, there seems to be a consensus that there are far more bacterial cells in the body than there are human cells—at least 10 times as many. Does that make us primarily prokaryotic? Hardly. Most of the bacteria are guests that reside in our digestive tracts.

CHECK YOUR LEARNING

Can you describe the structures and features shared by all cells? Can you distinguish prokaryotic from eukaryotic cells?

4.3 WHAT ARE THE MAJOR FEATURES OF EUKARYOTIC CELLS?

Eukaryotic cells make up the bodies of animals, plants, protists, and fungi; as you might imagine, these cells are extremely diverse. The single cells that make up the entire body of some

protists are sufficiently complex to perform all the necessary activities of life. Within the body of any multicellular organism there exists a huge variety of eukaryotic cells specialized to perform different functions. In this section we will focus on plant and animal cells.

The cytoplasm of all eukaryotic cells includes a variety of organelles, such as the nucleus and mitochondria, that perform specific functions within the cell; organelles are indicated by an asterisk in Table 4-1. **Figure 4-3** illustrates a generalized animal cell and **Figure 4-4** illustrates a generalized plant cell. Animal cells have structures (centrioles, lysosomes, cilia, and flagella) not found in plant cells, and plant cells have structures (central vacuoles, cell walls, and plastids including chloroplasts) not found in animal cells.

Some Eukaryotic Cells Are Supported by Cell Walls

The outer surfaces of plants, fungi, and some protists are covered with nonliving, relatively stiff coatings called **cell walls** that are secreted outside the plasma membrane. Protists may have cell walls made of cellulose, protein, or glassy silica (see Chapter 20). Plant cell walls (see Fig. 4-4) are composed primarily of cellulose, whereas fungal cell walls

▲ FIGURE 4-3 A generalized animal cell

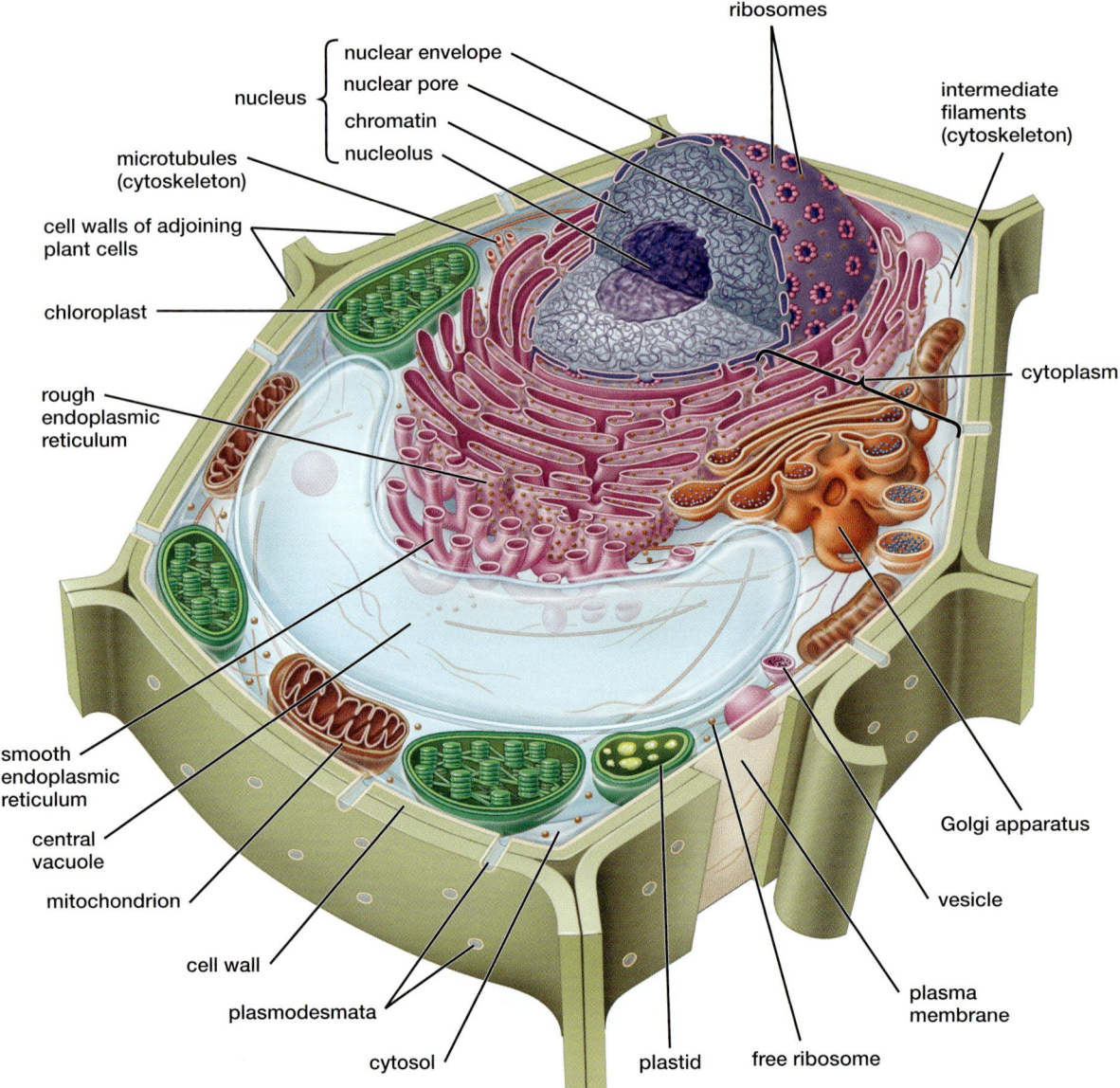

▲ FIGURE 4-4 **A generalized plant cell**

are usually made of chitin (see Chapter 3). Prokaryotic cells have cell walls made of different polysaccharides.

Cell walls protect the delicate plasma membrane and its cytoplasmic contents. Cell walls are usually porous, allowing oxygen, carbon dioxide, and water carrying dissolved molecules to flow easily through them. Plant cell walls are perforated by openings called *plasmodesmata* that connect adjacent cells (see Fig. 4-4). Plasmodesmata are lined with cell membrane and filled with cytosol.

The Cytoskeleton Provides Shape, Support, and Movement

The **cytoskeleton** is a scaffolding of protein fibers within the cytoplasm (Fig. 4-5). There are three categories of cytoskeletal proteins: thin **microfilaments** (composed of actin), medium-sized **intermediate filaments** (composed of various proteins), and thick **microtubules** (composed of tubulin) (Table 4-2).

The cytoskeleton is important in regulating the following properties of cells:

- ***Cell Shape*** In cells without cell walls, a scaffolding of intermediate filaments supports and determines the shape of the cell.
- ***Cell Movement*** Cell movement occurs as microfilaments or microtubules assemble, disassemble, or slide past one another. For example, microtubules allow movement of cilia and flagella, and microfilaments allow cells to change shape and muscle cells to contract.
- ***Organelle Movement*** Microtubules transport organelles such as vesicles and mitochondria from place to place within a cell.
- ***Cell Division*** Microtubules guide chromosome movements, and microfilaments separate the dividing cell into two daughter cells. (Cell division is covered in Chapter 9.)

Cilia and Flagella Move the Cell Through Fluid or Move Fluid Past the Cell

Both **cilia** (singular, cilium; literally meaning "eyelash") and **flagella** (singular, flagellum; literally meaning "whip")

(a) Cytoskeleton

(b) Light micrograph showing the cytoskeleton

▲ FIGURE 4-5 **The cytoskeleton (a)** The eukaryotic cytoskeleton consists of three types of proteins: microfilaments, intermediate filaments, and microtubules. **(b)** In this light micrograph, cells treated with fluorescent stains reveal microtubules and microfilaments, as well as nuclei.

are hair-like structures that propel cells through fluids or move fluids past cells. In mammals, cilia are found lining the respiratory tract and female reproductive tract; flagella are found only on sperm cells. Both are common in protists. Cilia and flagella are both covered by extensions of the plasma membrane and supported internally by microtubules of the cytoskeleton. Each cilium and flagellum contains a ring of nine fused pairs of microtubules, with an unfused pair in the center of the ring (**Fig. 4-6**).

Cilia and flagella undulate almost continuously; the energy to power this motion is supplied by mitochondria that cluster near their bases just beneath the plasma membrane. How do cilia and flagella move? In both structures, tiny sidearms of protein attach neighboring pairs of microtubules (see Fig. 4-6). These sidearms flex, using energy from adenosine triphosphate (ATP) to power their movement. Flexing of the sidearms slides the pair of microtubules past one another, causing the cilium or flagellum to bend.

In general, cilia are shorter and more numerous than flagella. Cilia produce a force on the surrounding fluid that is similar to that created by oars on the sides of a rowboat (**Fig. 4-7a**). Flagella are longer than cilia, and cells with flagella usually have only one or two of these structures.

TABLE 4-2 Major Components of the Eukaryotic Cytoskeleton			
Structure	**Appearance**	**Protein Type**	**Major Functions**
Microfilaments	Twisted double strands of protein subunits; about 7 nm in diameter	Actin	Allow muscle cell contraction; help cells change shape; separate daughter cells during cell division
Intermediate filaments	Helical subunits twisted around one another and bundled into clusters of four, which may be further twisted together; about 10 nm in diameter	Proteins differ depending on function and cell type	Provide supportive internal scaffolding that helps maintain cell shape
Microtubules	Tubes consisting of spirals of two-part protein subunits; about 25 nm in diameter	Tubulin	Transport organelles within the cell; components of cilia and flagella; guide chromosome movements during cell division

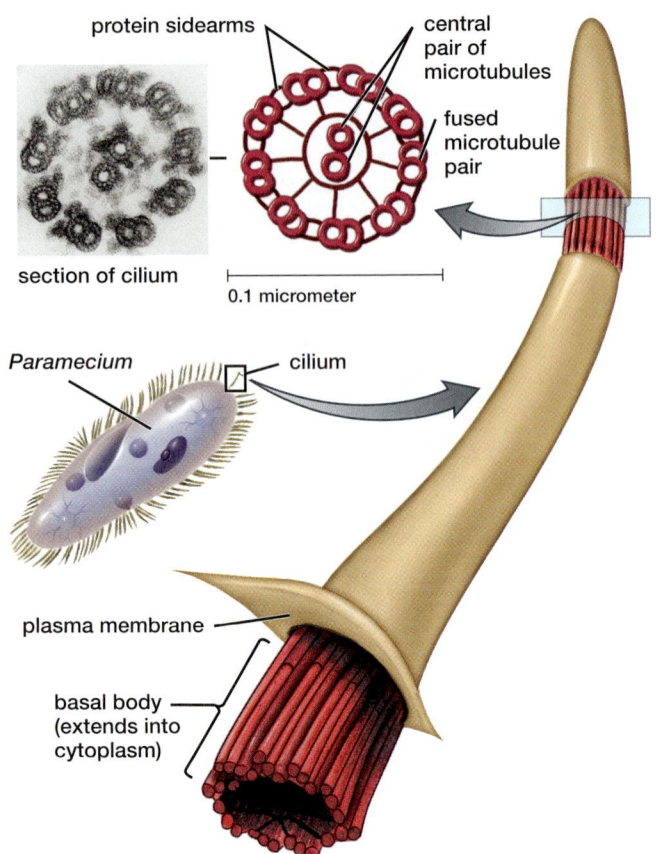

protein sidearms

central pair of microtubules

fused microtubule pair

section of cilium

0.1 micrometer

Paramecium   cilium

plasma membrane

basal body (extends into cytoplasm)

▲ **FIGURE 4-6 Cilia and flagella** Both cilia and flagella contain an outer ring of nine fused pairs of microtubules surrounding a central unfused pair. The outer pairs have sidearms made of protein that interact with adjacent pairs to provide the force for bending. Cilia and flagella arise from basal bodies, triplets of microtubules located just beneath the plasma membrane. (Only a cilium is illustrated here.)

The flagellum propels the cell through water, acting somewhat like the propeller on a motorboat (**Fig. 4-7b**).

Some unicellular organisms, such as *Paramecium* (see Figs. E4-2a, b), use cilia to swim through water; others use flagella. Flagella also propel the sperm cells of nearly all animals (see Fig. 4-7b, right). In animals, cilia usually move fluids and suspended particles past a surface. Ciliated cells line such diverse structures as the gills of oysters (over which cilia circulate water rich in food and oxygen), the female reproductive tract of vertebrates (where cilia transport the egg cell to the uterus), and the respiratory tracts of most land vertebrates (where cilia convey mucus that carries debris and microorganisms out of the air passages; see Fig. 4-7a, right).

The microtubules of each cilium and flagellum arise from a **basal body,** which consists of a ring of nine fused triplets of microtubules located just beneath the plasma membrane (see Fig. 4-6). Basal bodies are produced by **centrioles,** which are identical in structure to basal bodies. A single pair of centrioles is found near the nucleus of animal cells (see Fig. 4-3). Centrioles appear to play a role in organizing the proteins of the cytoskeleton that are active during cell division.

The Nucleus, Containing DNA, Is the Control Center of the Eukaryotic Cell

A cell's DNA stores all the information needed to construct the cell and direct the countless chemical reactions necessary for life and reproduction. Only part of the instructions in DNA is used by a cell at any given time, depending on its stage of development, its environment, and the function of the cell in a multicellular body. In eukaryotic cells, DNA is housed within the nucleus.

propulsion of fluid

cilia lining trachea

mucus-secreting cells

(a) Cilium   power stroke   plasma membrane   return stroke

direction of locomotion   propulsion of fluid

flagellum of human sperm

(b) Flagellum   continuous propulsion

▲ **FIGURE 4-7 How cilia and flagella move (a)** (Left) Cilia usually "row," providing force parallel to the plasma membrane. The "power stroke" propels the fluid past the membrane, and the "return stroke" restores the cilium to its original position; this movement is much like the arms of a swimmer doing the breast stroke. (Right) SEM of cilia lining the trachea (the tube which conducts air to the lungs); these cilia sweep out mucus and trapped particles. **(b)** (Left) Flagella move in an undulating wavelike motion, providing continuous propulsion perpendicular to the plasma membrane. Thus, a sperm's flagellum propels it forward. (Right) SEM of a human sperm cell.

QUESTION What problems would arise if the trachea were lined with flagella instead of cilia?

(a) The nucleus

(b) Nucleus of a yeast cell

▲ **FIGURE 4-8 The nucleus** **(a)** The nucleus is bounded by a double outer membrane perforated by pores. Inside the membrane are chromatin and one or more nucleoli. **(b)** SEM of the nucleus of a yeast cell. The "gatekeeper proteins" of the nuclear pore complex are colored pink. These proteins line the nuclear pores.

The **nucleus** is a large organelle with three major parts: the nuclear envelope, chromatin, and the nucleolus, shown in **Figure 4-8** (see also Fig. 4-3) and described in the following sections.

The Nuclear Envelope Allows Selective Exchange of Materials

The nucleus is isolated from the rest of the cell by a double membrane, the **nuclear envelope,** which is perforated by protein-lined nuclear pores. Water, ions, and small molecules can pass freely through the pores, but the passage of large molecules—particularly proteins, parts of ribosomes, and RNA—is regulated by specialized gatekeeper proteins called the **nuclear pore complex** (see Fig. 4-8b) that line each nuclear pore. Ribosomes stud the outer nuclear membrane, which is continuous with membranes of the rough endoplasmic reticulum, described later (see Figs. 4-3 and 4-4).

Chromatin Consists of Strands of DNA Associated with Proteins

Early observers of the nucleus noted that it was darkly colored by the stains used in light microscopy, and named the nuclear material **chromatin** (meaning "colored substance"). Biologists have since learned that chromatin consists of **chromosomes** (literally, "colored bodies") made of DNA molecules and their associated proteins. When a cell is not dividing, the chromosomes are extended into extremely long strands that are so thin that they cannot be distinguished from one another with a light microscope. During cell division, the individual chromosomes become compacted and condensed so they are easily visible with a light microscope (**Fig. 4-9**).

The genes on DNA, consisting of specific sequences of nucleotides, provide a molecular blueprint for the synthesis of a huge variety of proteins. Some of these proteins form structural components of the cell. Others regulate the movement of materials through cell membranes, and still others are enzymes that promote chemical reactions within the cell.

Proteins are synthesized in the cytoplasm, but DNA is confined to the nucleus. This means that copies of the code for proteins must be ferried out through the nuclear membrane into the cytoplasm. To accomplish this, the genetic information is copied from DNA into molecules of *messenger RNA* (mRNA). The mRNA then moves out of the nucleus and into the cytosol through the nuclear pores. In the cytosol, the sequence of nucleotides in mRNA is used to direct protein synthesis, a process that occurs on ribosomes (see Fig. 4-10). (Protein synthesis is described in Chapter 12.)

The Nucleolus Is the Site of Ribosome Assembly

Eukaryotic nuclei contain at least one **nucleolus** (plural, nucleoli; meaning "little nuclei"; see Fig. 4-8a). The nucleolus is the

▲ **FIGURE 4-9 Chromosomes** Chromosomes, seen here in a light micrograph of a dividing cell (center) in an onion root tip. The chromatin that you see in adjacent nondividing cells consists of uncondensed chromosomes.

QUESTION Why do the chromosomes in chromatin condense in dividing cells?

Spare Parts for Human Bodies

Replacing the skin cells killed by a deep burn requires a tremendous proliferation of new cells. Without special treatments, this regeneration can take up to a year and leave significant scars. But a new product may soon minimize recovery time and scars. Medical personnel remove a microscopically thin slice of healthy skin about the size of a postage stamp from the patient. Enzymes are then used to separate the skin cells from one another. These cells are suspended in liquid, sucked into a syringe-like applicator, and sprayed over the burn, covering up to 80 times the area of skin that provided the cells. This remarkably simple process takes about 20 minutes. The healthy skin cells grow and begin multiplying rapidly, spreading over the burned area to form a new layer of skin in about a week, with very little scarring. Because the patient's own skin cells are used, there is no danger of rejection. Already in use in Australia and Europe—and in clinical trials in the United States as of 2011—this product could become a standard treatment for burns and other injuries to the skin.

site of ribosome synthesis. It consists of *ribosomal RNA* (rRNA), proteins, ribosomes in various stages of synthesis, and the portions of chromosomes that carry genes that code for rRNA.

A **ribosome** is a small particle composed of a type of RNA unique to ribosomes, called ribosomal RNA, combined with proteins. Each ribosome serves as a kind of workbench for the synthesis of proteins within the cell cytoplasm. Just as a workbench can be used to construct many different objects, a ribosome can be used to synthesize a multitude of different proteins (depending on the mRNA to which it attaches). In electron micrographs of cells, ribosomes appear as dark granules, either singly or studding the membranes of the nuclear envelope and rough endoplasmic reticulum (see Fig. 4-11), or as *polyribosomes* ("many ribosomes") strung along strands of mRNA within the cytoplasm (Fig. 4-10).

▲ FIGURE 4-10 **A polyribosome** Ribosomes are strung along a messenger RNA molecule, forming a polyribosome. In the TEM (right) individual ribosomes are synthesizing multiple copies of a protein, the small strands projecting horizontally from some of the ribosomes.

Eukaryotic Cytoplasm Contains Membranes That Form the Endomembrane System

All eukaryotic cells contain internal membranes that create loosely connected compartments within the cytoplasm. These membranes, collectively called the **endomembrane system,** segregate molecules from the surrounding cytosol and ensure that biochemical processes occur in an orderly fashion. The endomembrane system creates specialized regions within the cytoplasm where an enormous variety of molecules are synthesized. The fluid property of membranes permits them to fuse with one another. In so doing they empty their contents into different compartments for various types of processing, transfer pieces of membrane among themselves, and also shuttle materials into and out of the cell.

Small temporary sacs made of membrane are called **vesicles.** One function of vesicles is to ferry biological molecules among various regions of the endomembrane system. Vesicles may also fuse with the plasma membrane, exporting their contents outside the cell (see Fig. 4-14), a process called *exocytosis* (described in Chapter 5). Conversely, the plasma membrane may extend and surround material just outside the cell, then fuse and pinch off to form a vesicle inside the cell (see Fig. 4-13), a process called *endocytosis* (see Chapter 5). How do the vesicles know where to go? Researchers have discovered that specific molecules embedded in membranes serve as "mailing labels" that provide the address for delivery of the vesicle and its payload.

The cell's endomembrane system functions to synthesize, transport, and export biological molecules, and to break down defective cellular structures. The endomembrane system includes the nuclear envelope and vesicles described earlier, as well as the endoplasmic reticulum, Golgi apparatus, and lysosomes, described in the sections that follow.

The Endoplasmic Reticulum Forms Membrane-Enclosed Channels Within the Cytoplasm

The **endoplasmic reticulum (ER)** is a series of interconnected membranes that form a labyrinth of flattened sacs and channels within the cytoplasm (reticulum translates to "network" and endoplasmic to "within the cytoplasm"; Fig. 4-11). The ER membrane typically makes up at least 50% of the total membrane of the cell. The ER plays a major role in synthesizing, modifying, and transporting biological molecules throughout the cell. For example, the ER synthesizes phospholipids and proteins for most of the membrane found within the cell. Bits of ER membrane are constantly pinched off when vesicles surround molecules such as proteins or lipids. From the ER, vesicles travel to destinations that include the Golgi apparatus, lysosomes (organelles in which digestion occurs), and the plasma membrane. As the vesicles fuse with these structures, membrane is transferred from the ER to other sites in the cell.

Rough Endoplasmic Reticulum Rough ER emerges from the ribosome-covered outer nuclear membrane (see Fig. 4-3). It is studded with ribosomes on the surface facing the cytosol, and so appears rough under the electron microscope. The

ribosomes

smooth ER

rough ER

rough ER

smooth ER

vesicles

(a) Endoplasmic reticulum may be rough or smooth

(b) Smooth and rough ER

▲ **FIGURE 4-11 Endoplasmic reticulum (a)** Ribosomes (depicted as orange dots) stud the side of the rough ER membrane that faces the cytosol. Rough ER membrane is continuous with the outer nuclear envelope. Smooth ER is less flattened and more cylindrical than rough ER, and may be continuous with rough ER. **(b)** A TEM of smooth and rough ER with vesicles.

rough ER is the most important site of protein synthesis in the cell, and it also serves to modify and transport proteins. As they are synthesized, proteins destined either for secretion outside the cell or to be used elsewhere within the cell are inserted into the interior of the ER. Here they are chemically modified and folded into their proper three-dimensional structures (see Chapter 3). Eventually the proteins accumulate in pockets of ER membrane that pinch off as vesicles, which in turn carry their protein cargo to the Golgi apparatus. Some of these proteins become membrane components such as channels through which molecules or ions can pass. Others are digestive enzymes used in the cell or exported. In cells of the pancreas, rough ER synthesizes the protein hormone insulin; in white blood cells, rough ER produces antibodies that help fight infections.

Smooth Endoplasmic Reticulum Smooth ER lacks ribosomes and is scarce in most cells. In certain cells, however, smooth ER is abundant and specialized. In some cells, smooth ER manufactures large quantities of lipids such as steroid hormones. For example, steroid sex hormones are produced by smooth ER in mammalian reproductive organs. Smooth ER is also abundant in liver cells, where it contains enzymes that detoxify harmful drugs such as alcohol and metabolic wastes such as ammonia. In muscle cells, smooth ER is enlarged and specialized to store calcium ions, the release of which allows muscle contraction.

The Golgi Apparatus Modifies, Sorts, and Packages Important Molecules

Named for the Italian physician and cell biologist Camillo Golgi, who discovered it in 1898, the **Golgi apparatus** (or simply **Golgi**) is a specialized set of membranes resembling a stack of flattened and interconnected sacs (**Fig. 4-12**). The compartments of the Golgi act like the finishing rooms of

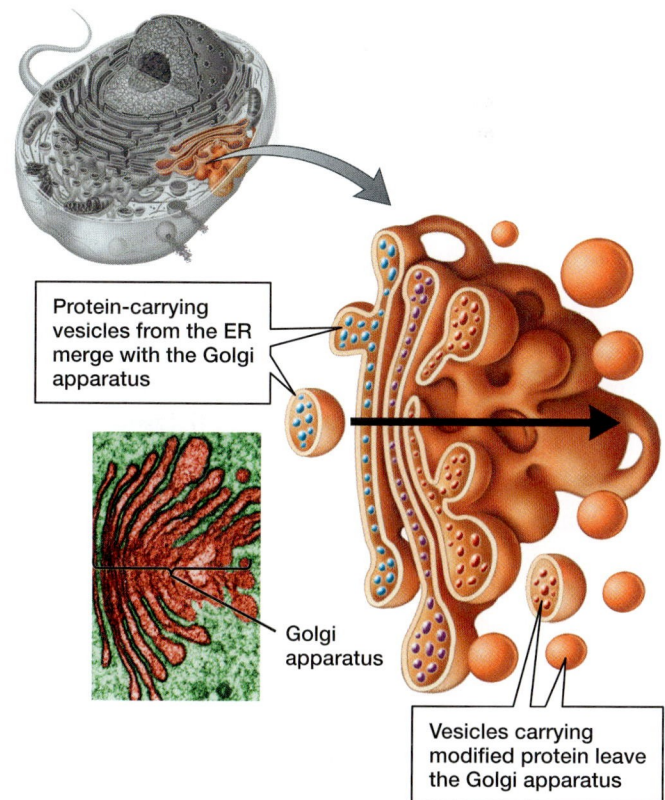

Protein-carrying vesicles from the ER merge with the Golgi apparatus

Golgi apparatus

Vesicles carrying modified protein leave the Golgi apparatus

▲ **FIGURE 4-12 The Golgi apparatus** The Golgi apparatus is a stack of flattened membranous sacs. Vesicles transport both cell membrane and the enclosed material from the ER to the Golgi. The arrow shows the direction of movement of materials through the Golgi as they are modified and sorted. Vesicles bud from the face of the Golgi opposite the ER.

a factory, where final touches are added to products to be packaged and exported. Vesicles from the rough ER fuse with the receiving side of the Golgi apparatus, adding their membranes to the Golgi and emptying their contents into the Golgi sacs. Within the Golgi compartments, some of the proteins synthesized in the rough ER are modified further. Finally, vesicles bud off from the "shipping" face of the Golgi, carrying away finished products for use in the cell or export out of the cell.

The Golgi apparatus performs the following functions:

- The Golgi modifies some molecules; an important role of the Golgi is to add carbohydrates to proteins to make glycoproteins.
- The Golgi separates various proteins received from the ER according to their destinations. For example, the Golgi apparatus separates the digestive enzymes that are bound for lysosomes from the protein hormones that the cell will secrete.
- The Golgi packages the finished molecules into vesicles that are then transported to other parts of the cell or to the plasma membrane for export.

Secreted Proteins Are Modified as They Move Through the Cell

To understand how some of the components of the endomembrane system work together, let's look at the manufacture and export of antibodies (**Fig. 4-13**). Antibodies are glycoproteins produced by white blood cells that bind to foreign invaders (such as disease-causing bacteria) and help destroy them. Antibody proteins are synthesized on ribosomes of the rough ER within white blood cells, and they are packaged into vesicles formed from ER membrane. These vesicles travel to the Golgi, where their membranes fuse with the Golgi membranes and release the antibodies into the Golgi apparatus. Within the Golgi, carbohydrates are attached to the antibodies, which are then repackaged into vesicles formed from membrane acquired from the Golgi. The vesicle containing the completed antibody travels to the plasma membrane and fuses with it, releasing the antibody outside the cell, where it will make its way into the bloodstream to help defend the body against infection.

Lysosomes Serve as the Cell's Digestive System

Lysosomes are membrane-bound sacs that digest food ranging from individual proteins to microorganisms such as bacteria. Lysosomes contain dozens of different enzymes that can break down almost any type of biological molecule. The digestive enzymes of lysosomes are manufactured in the ER and transported to the Golgi in vesicles. In the Golgi, the enzymes are sorted from other proteins and repackaged into vesicles that transport the enzymes to lysosomes (**Fig. 4-14**).

Many cells of animals and protists "eat" by endocytosis—that is, by engulfing particles just outside the cell. The plasma membrane with its enclosed food then pinches off inside the cytosol and forms a vesicle called a **food vacuole** (see Fig. 4-14). Lysosomes merge with these food vacuoles,

(interstitial fluid)

⑤ Vesicles merge with the plasma membrane and release antibodies into the interstitial fluid

(cytosol)

④ Completed glycoprotein antibodies are packaged into vesicles on the opposite side of the Golgi apparatus

vesicles

Golgi apparatus

③ Vesicles fuse with the Golgi apparatus, and carbohydrates are added as the protein passes through the compartments

② The protein is packaged into vesicles and travels to the Golgi apparatus

forming vesicle

① Antibody protein is synthesized on ribosomes and is transported into channels of the rough ER

▲ **FIGURE 4-13 A protein is manufactured and exported via the endomembrane system** The formation of an antibody is an example of the process of protein manufacture and export.

and the lysosomal enzymes digest the food into small molecules such as amino acids, monosaccharides, and fatty acids. Lysosomes also digest worn-out or defective organelles within the cell, breaking them down into their component molecules. Small biological molecules are released into the cytosol through the lysosomal membrane, where they are used in the cell's metabolic processes.

Vacuoles Serve Many Functions, Including Water Regulation, Storage, and Support

Some vacuoles are temporary large vesicles, such as food vacuoles (see Fig. 4-14). In the following sections, we describe some permanent vacuoles found in some freshwater protists and in plant cells.

Freshwater Protists Have Contractile Vacuoles

Freshwater protists such as *Paramecium* possess **contractile vacuoles,** each composed of collecting ducts, a central

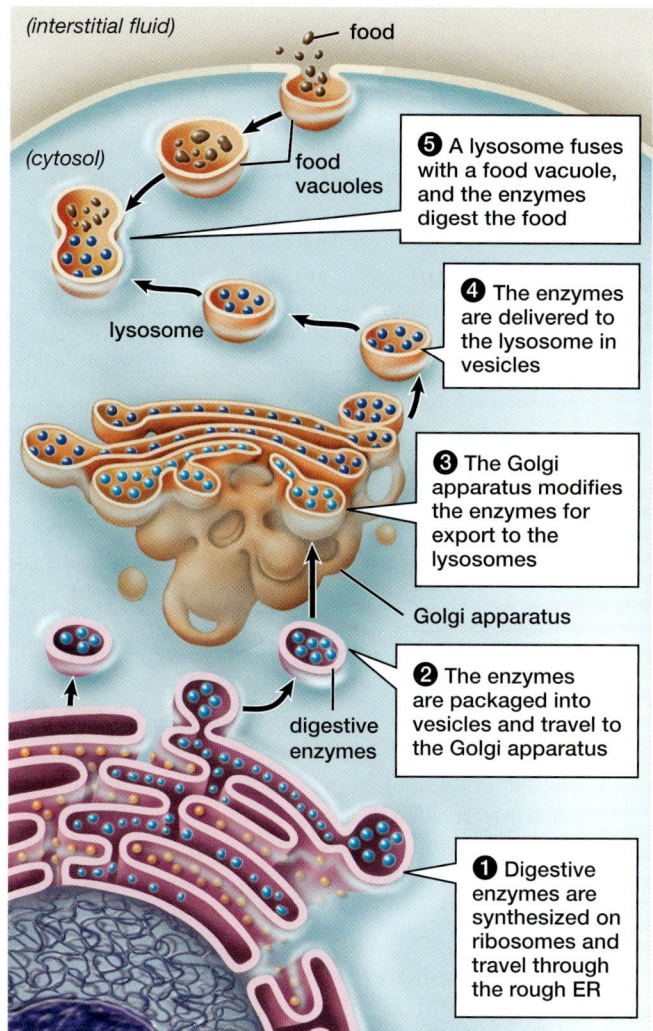

(interstitial fluid)   food

(cytosol)   food vacuoles

5 A lysosome fuses with a food vacuole, and the enzymes digest the food

lysosome

4 The enzymes are delivered to the lysosome in vesicles

3 The Golgi apparatus modifies the enzymes for export to the lysosomes

Golgi apparatus

digestive enzymes

2 The enzymes are packaged into vesicles and travel to the Golgi apparatus

1 Digestive enzymes are synthesized on ribosomes and travel through the rough ER

▲ **FIGURE 4-14 Formation and function of lysosomes and food vacuoles via the endomembrane system**

QUESTION Why is it advantageous for all membranes associated with the cell to have a fundamentally similar composition?

reservoir, and a tube leading to a pore in the plasma membrane (**Fig. 4-15**). Fresh water constantly leaks into contractile vacuoles through the protists' plasma membranes. (We describe this process, called osmosis, in Chapter 5.) The influx of water would soon burst the fragile organism if it did not have a mechanism to excrete the water. Cellular energy is used to pump salts from the cytoplasm of the protist into collecting ducts. Water follows by osmosis and drains into the central reservoir. When the reservoir of the contractile vacuole is full, it contracts, squirting the water out through a pore in the plasma membrane.

Plant Cells Have Central Vacuoles

A large **central vacuole** occupies three-quarters or more of the volume of most mature plant cells (see Fig. 4-4) and serves several functions. Its membrane helps to regulate the ion content of the cytosol, and secretes wastes and toxic

substances into the water that fills the central vacuole. Some plants store substances in central vacuoles that deter animals from munching on their otherwise tasty leaves. Vacuoles may also store sugars and amino acids not immediately needed by the cell. Blue or purple pigments stored in central vacuoles are responsible for the colors of many flowers.

Central vacuoles also provide support for plant cells. Dissolved substances in the vacuole attract water into its interior. The resulting water pressure, called *turgor pressure*, within the vacuole pushes the fluid portion of the cytoplasm up against the cell wall with considerable force. Cell walls are usually somewhat flexible, so both the overall shape and the rigidity of the cell depend on turgor pressure within the cell. Turgor pressure thus provides support for the non-woody parts of plants. (Look ahead to Fig. 5-8 to see what happens when you forget to water your houseplants.)

contractile vacuole

(a) Paramecium

Water enters the collecting ducts and fills the central reservoir

collecting ducts

central reservoir

pore

The reservoir contracts, expelling water through the pore

(b) Contractile vacuole

▲ **FIGURE 4-15 A contractile vacuole** Many freshwater protists contain contractile vacuoles. **(a)** The single-celled protist *Paramecium* lives in freshwater ponds and lakes. **(b)** An enlargement of the contractile vacuole showing its structure and activity as it collects and expels water.

Mitochondria Extract Energy from Food Molecules and Chloroplasts Capture Solar Energy

All eukaryotic cells have mitochondria that produce high-energy ATP molecules from energy-storage molecules such as sugar. The cells of plants and some protists also have chloroplasts, which can capture energy directly from sunlight and store it in sugar molecules.

Most biologists accept the **endosymbiont hypothesis** (discussed in more detail in Chapter 17), which holds that both mitochondria and chloroplasts evolved from prokaryotic bacteria. Roughly 1.7 billion years ago, these prokaryotes took up residence within other prokaryotic cells, a process called endosymbiosis (literally "living together inside"). Both mitochondria and chloroplasts are surrounded by a double membrane; the outer membrane may have come from the original host cell and the inner membrane from the guest cell. Mitochondria and chloroplasts resemble each other and both resemble prokaryotic cells in several ways. They are both the size of a typical prokaryotic cell (1 to 5 micrometers in diameter). Both have assemblies of enzymes that synthesize ATP, as would have been needed by an independent cell. Finally, both possess their own DNA and ribosomes that more closely resemble prokaryotic than eukaryotic DNA and ribosomes.

Mitochondria Use Energy Stored in Food Molecules to Produce ATP

All eukaryotic cells contain **mitochondria** (singular, mitochondrion), organelles that are sometimes called the "powerhouses" of the cell because they extract energy from food molecules and store it in the high-energy bonds of ATP. Mitochondria are found in large numbers in metabolically active cells, such as those that make up muscle, and they are less abundant in cells that are less active, such as those that make up cartilage.

Mitochondria possess a pair of membranes (**Fig. 4-16**). The outer membrane is smooth, whereas the inner membrane forms deep folds called cristae (singular, crista; meaning "crest"). The mitochondrial membranes enclose two fluid-filled spaces: the intermembrane compartment lies between the two membranes and the matrix fills the space within the inner membrane. Some of the reactions that break down high-energy molecules occur in the fluid of the matrix; the rest are conducted by a series of enzymes attached to the membranes of the cristae. (The role of mitochondria in energy production is described in detail in Chapter 8.)

Chloroplasts Are the Sites of Photosynthesis

Photosynthesis, which captures sunlight and provides the energy to power life, occurs in the chloroplasts that are found in the cells of plants and some protists. **Chloroplasts** are organelles surrounded by a double membrane (**Fig. 4-17**). The inner membrane of the chloroplast encloses a fluid called the *stroma*. Within the stroma are interconnected stacks of hollow, membranous sacs. An individual sac is called a *thylakoid*, and a stack of thylakoids is a *granum* (plural, grana).

The thylakoid membranes contain the pigment molecule **chlorophyll** (which gives plants their green color). During photosynthesis, chlorophyll captures the energy of sunlight and transfers it to other molecules in the thylakoid membranes. These molecules transfer the energy to ATP and other energy carriers. The energy carriers diffuse into the stroma, where their energy is used to drive the synthesis of sugar from carbon dioxide and water.

outer membrane

inner membrane

intermembrane space

matrix

cristae

0.1 micrometer

▲ **FIGURE 4-16 A mitochondrion** Mitochondria contain two membranes enclosing two fluid compartments: the intermembrane compartment between the outer and inner membranes and the matrix within the inner membrane. The outer membrane is smooth, but the inner membrane forms deep folds called cristae. Some structures are visible in the TEM on the right.

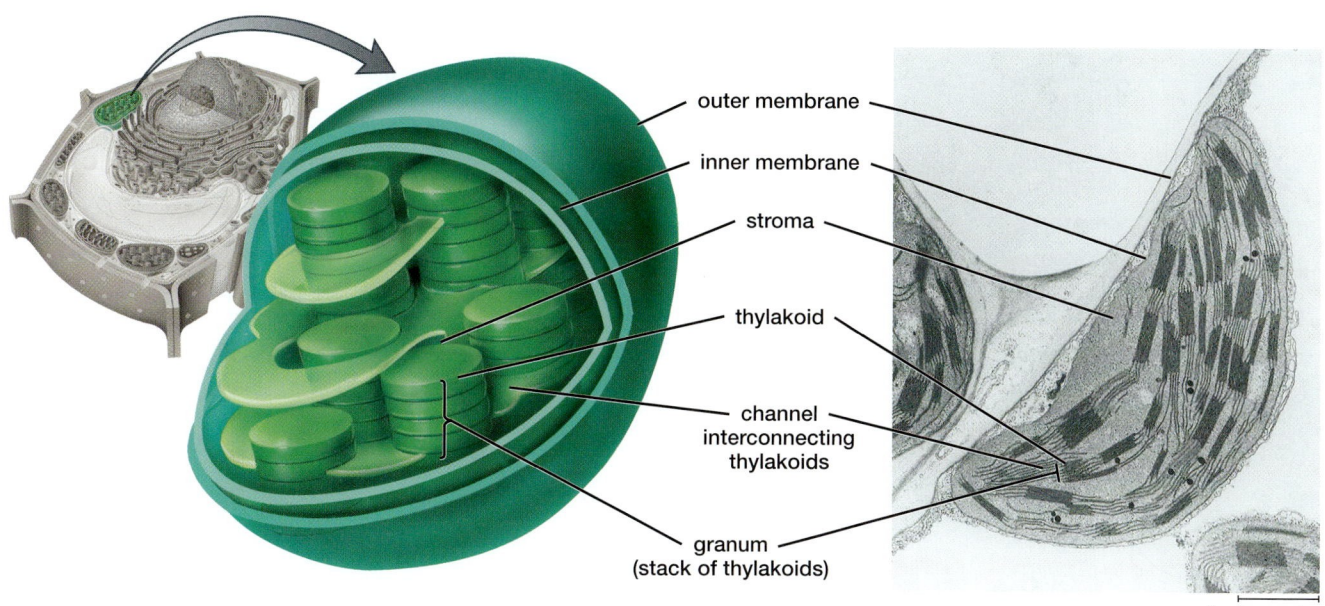

outer membrane
inner membrane
stroma
thylakoid
channel
interconnecting
thylakoids
granum
(stack of thylakoids)

1 micrometer

▲ FIGURE 4-17 **A chloroplast** Chloroplasts are surrounded by a double membrane. The fluid stroma is enclosed by the inner membrane; within the stroma are stacks of thylakoid sacs called grana. Chlorophyll is embedded in the membranes of the thylakoids.

Plants Use Some Plastids for Storage

Chloroplasts are highly specialized **plastids,** organelles found only in plants and photosynthetic protists (Fig. 4-18). Plants and photosynthetic protists use nonchloroplast types of plastids as storage containers for various molecules, including pigments that give ripe fruits their yellow, orange, or red colors. In plants that continue growing from one year to the next, plastids store photosynthetic products produced during the growing season. Most plants convert the sugars made during photosynthesis into starch, which is stored in plastids. Potatoes, for example, are composed almost entirely of cells stuffed with starch-filled plastids (see Fig. 4-18, upper right).

CHECK YOUR LEARNING

Can you define *organelles*? Do you recall which structures are found in animal but not plant cells and which are found in plant but not animal cells? Can you describe each major structure found in eukaryotic cells, including its function?

4.4 WHAT ARE THE MAJOR FEATURES OF PROKARYOTIC CELLS?

Most prokaryotic cells are small, generally less than 5 micrometers in diameter (most eukaryotic cells are from 10 to 100 micrometers in diameter) with an internal structure that is simple compared to that of eukaryotic cells (Fig. 4-19; compare this to Figs. 4-3 and 4-4). In general, prokaryotes lack the membrane-surrounded organelles that are the most prominent internal features of eukaryotic cells.

Members of the domains Archaea and Bacteria have bodies that consist of single prokaryotic cells. The Archaea superficially resemble bacteria, but recent research has revealed that they differ in many important molecular features, and they are

plastid
starch
globules

0.5 micrometer

▲ FIGURE 4-18 **A plastid** Plastids, found in the cells of plants and photosynthetic protists, are surrounded by a double outer membrane. Chloroplasts are the most familiar plastid; other types of plastids store various materials, such as the starch filling these plastids in the potato cells seen at the upper right in a TEM.

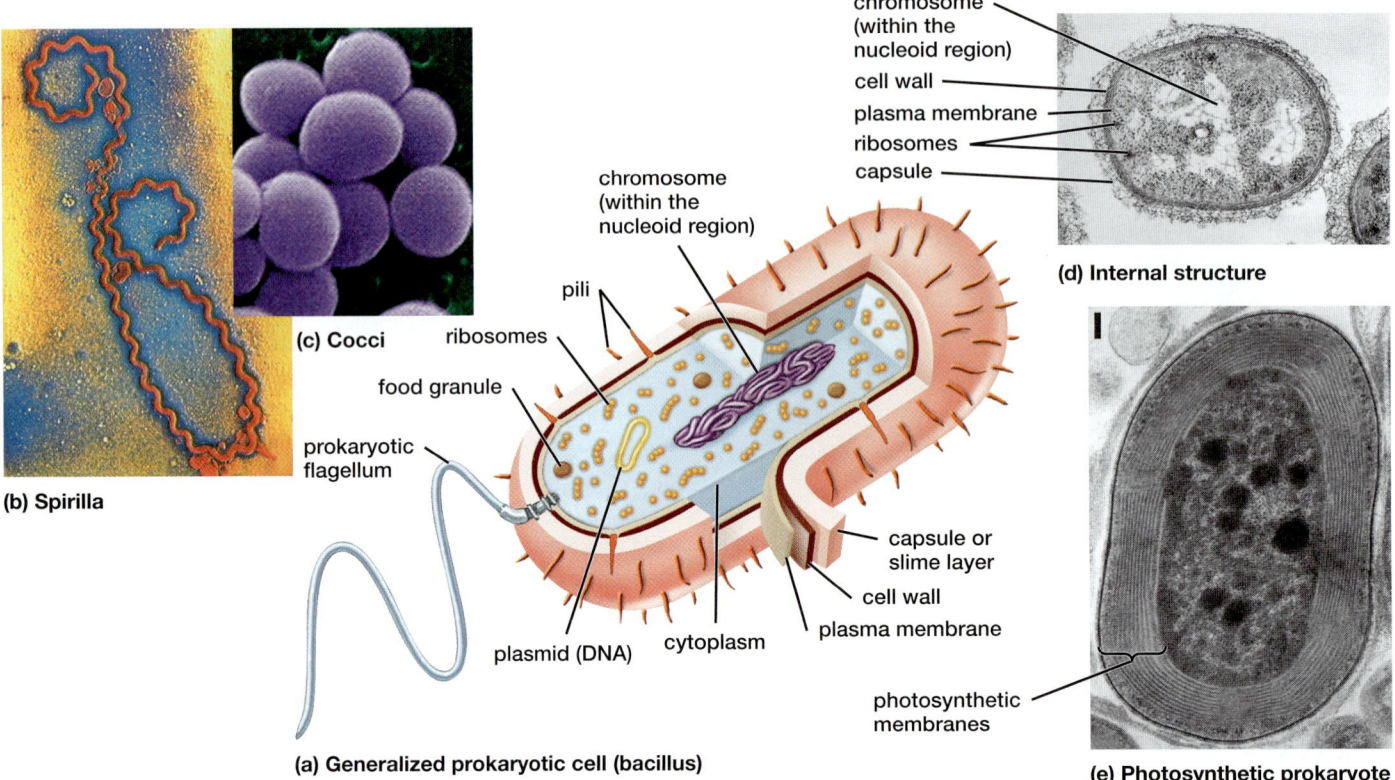

(c) Cocci

(b) Spirilla

(a) Generalized prokaryotic cell (bacillus)

(d) Internal structure

(e) Photosynthetic prokaryote

▲ **FIGURE 4-19 Prokaryotic cells are simpler than eukaryotic cells** Prokaryotes assume a variety of shapes, including **(a)** rod-shaped bacilli, **(b)** spiral-shaped spirilla, and **(c)** spherical cocci. Internal structures are revealed in the TEMs in **(d)** and **(e)**. Some photosynthetic bacteria have internal membranes where photosynthesis occurs, as shown in **(e)**.

now considered an entirely different form of life. Archaea inhabit extreme environments, such as hot springs and cow stomachs, but are increasingly being discovered in more familiar locales, such as the soil and oceans. None are known to cause disease. In this chapter we focus on the more familiar bacteria.

Prokaryotic Cells Possess Specialized Surface Features

Nearly all prokaryotic cells are surrounded by a stiff cell wall; in bacteria this is composed of *peptidoglycan* (a unique molecule consisting of short peptides linking chains of sugar molecules bearing amino groups). The cell wall protects the cell and helps the cell retain its shape. Most prokaryotes are rod-shaped *bacilli* (**Fig. 4-19a**), spiral-shaped *spirilla* (**Fig. 4-19b**), or spherical *cocci* (**Fig. 4-19c**). Several types of antibiotics, including penicillin, fight bacterial infections by interfering with cell wall synthesis, causing the bacteria to rupture. Although no prokaryotic cells possess cilia, some bacteria and archaea are propelled by flagella, the structure of which differs from that of eukaryotic flagella (see Chapter 19).

Many bacteria secrete polysaccharide coatings called *capsules* and *slime layers* outside their cell walls. Capsules and slime layers are very similar except that capsules have a more organized structure and are more difficult to remove. In bacteria such as those that cause tooth decay, diarrhea, pneumonia, or urinary tract infections, capsules and slime

layers help them adhere to specific host tissues, such as the lining of the small intestine, lungs, or bladder, or the surface of a tooth (see "Links to Everyday Life: Unwanted Guests"). Capsules and slime layers allow some bacteria to form surface films (such as may coat unwashed toilet bowls). They also protect bacteria and help prevent them from drying out.

Pili (singular, pilus; meaning "hairs") are surface proteins that project from the cell walls of many bacteria (see Fig. 4-19a). There are two types of pili: attachment pili and sex pili. *Attachment pili* are short and abundant. Like slime layers and capsules, attachment pili help bacteria adhere to structures. In some disease-causing bacteria, these pili attach to human tissues. Some types of bacteria form *sex pili*, which are few in number and quite long. A single sex pilus from one bacterium binds to a nearby bacterium of the same type and draws them together. This allows them to form a pore linking their cytoplasm through which they transfer pieces of DNA called plasmids (see Fig. 19-8). (The features of prokaryotic cells are covered in more detail in Chapter 19.)

Prokaryotic Cells Have Fewer Specialized Cytoplasmic Structures Than Do Eukaryotic Cells

The cytoplasm of a typical prokaryotic cell contains a distinctive region called the **nucleoid** (meaning "like a nucleus"; see Fig. 4-19a), consisting primarily of DNA, with some

►► LINKS TO EVERYDAY LIFE

Unwanted Guests

You may think you have "yuck mouth" in the morning, but imagine the mouths of people in the late 1600s, when dental hygiene was virtually nonexistent. It was then that the Dutch scientist Anton van Leeuwenhoek scraped white matter from his teeth. Viewing the gunk through a microscope that he had built himself, he saw swarms of cells that he called "animalcules," and which we now recognize as bacteria. Disturbed by these life-forms in his mouth, he attempted to kill them with vinegar and hot coffee—with little success. The warm, moist environment of the human mouth, particularly the crevices of the teeth and gums, is an ideal habitat for a variety of bacteria. Some produce slime layers that help them and others adhere to teeth. Thick layers of bacteria, slime, and glycoproteins from saliva make up the white substance, called plaque, which van Leeuwenhoek scraped from his teeth. Sugar in foods and beverages nourishes the bacteria, which break down the sugar into lactic acid. The acid eats away at the tooth enamel, producing tiny crevices in which the bacteria multiply, eventually producing a cavity. So, although he didn't know why, van Leeuwenhoek was right to be concerned about the "animalcules" in his mouth!

RNA and protein. The DNA within the nucleoid takes the form of a single, circular chromosome consisting of a long coiled strand of DNA that carries essential genetic information. The nucleoid is not separated from the cytoplasm by a membrane, as is the nucleus of eukaryotic cells. Most prokaryotic cells also contain small rings of DNA called **plasmids,** located outside the nucleoid. Plasmids usually carry genes that give the cell special properties; for example, some disease-causing bacteria possess plasmids that allow them to inactivate antibiotics, making the bacteria much more difficult to kill.

Bacterial cytoplasm includes ribosomes (see Fig. 4-19a). Although their role in protein synthesis is similar to that of eukaryotic ribosomes, bacterial ribosomes are smaller and contain different RNA and proteins. These ribosomes resemble those found in eukaryotic mitochondria and chloroplasts, providing support for the endosymbiont hypothesis described earlier. Prokaryotic cytoplasm also may contain food granules that store energy-rich molecules, such as glycogen, but these are not enclosed by membranes.

Although prokaryotic cells lack membrane-enclosed organelles, some bacteria use membranes to organize the enzymes responsible for a series of biochemical reactions. Such enzymes are situated in a particular sequence along the membrane to promote reactions in the proper order. For example, photosynthetic bacteria possess extensive internal membranes in which light-capturing proteins and enzymes that guide the synthesis of high-energy molecules are embedded in a specific order (**Fig. 4-19e**). Recent research has revealed that bacteria contain an extensive cytoskeleton. Although bacterial cytoskeletal proteins differ from those of the eukaryotic cytoskeleton, they play some of the same roles as in eukaryotic cells, for example, in cell division.

You might wish to go back and look at Table 4-1 at this point to review the differences between prokaryotic and eukaryotic cells.

CHECK YOUR LEARNING

Can you explain the major differences between prokaryotic and eukaryotic cells? Are you able to describe the structure and function of the major surface features of bacteria? Can you describe the internal features of bacteria, and include how some bacteria utilize internal membranes?

CASE STUDY revisited

Spare Parts for Human Bodies

Bioengineering of tissues and organs such as skin requires the coordinated efforts of biochemists, biomedical engineers, cell biologists, and physicians—experts who rarely communicated in the past. But now teams of scientists are working together to grow not only skin but also windpipes, bone, cartilage, heart valves, bladders, and breast tissue in laboratories around the world. To heal broken bones, researchers are working to develop biodegradable plastic scaffolds that incorporate protein growth factors. These molecules encourage nearby bone cells and tiny blood vessels to invade the plastic as it degrades, eventually replacing the artificial scaffold with the patient's own bone. Bioartificial blood vessels are currently being tested in animals. These are composed of a matrix of proteins deposited by cells grown on scaffolding. Replacement bladders have been created using bladder cells from patients seeded onto a bladder-shaped scaffold, incubated for about 7 weeks, and then transplanted into a patient. Some of these bladders have performed successfully in patients for years and are now in human clinical trials in preparation for the medical marketplace.

Recently the first bioartificial organ using stem cells was successfully transplanted into a man suffering from inoperable cancer that was blocking his windpipe (**Fig. 4-20**). Scientists built a Y-shaped scaffold based on three-dimensional (3-D) scans of the patient's own windpipe. They then seeded the scaffold with stem cells, which can develop into a wide range of cell types. The stem cells were taken from the patient's own bone marrow, so there was no danger of tissue rejection. The structure was bathed in special growth factors that prodded the stem cells to mature into cell types found in the windpipe.

▲ FIGURE 4-20 **A laboratory-grown windpipe**

An exciting new development in tissue engineering is bioprinting, which eliminates the need for scaffolding. Gabor Forgacs, a biophysicist at the University of Missouri–Columbia, cofounded a biotech firm that created a machine that has just reached the research market. This distant relative of the familiar ink-jet printer is designed to print organs in three dimensions. Its 3-D printer head, controlled by sophisticated computer software, ejects tiny droplets containing 10,000 or more cells onto a gel-like "paper" within which the cells can migrate and interact with one another. Layers of paper and cells are stacked and fuse together.

The bioprinting technique relies on the inherent ability of cells to communicate with one another using physical and chemical signals and to organize themselves into functional units. Forgacs and his team are currently working on printing blood vessels using droplets composed of the three types of cells that make up a blood vessel. The droplets are deposited around a central core of protein that will eventually be removed to allow blood to flow through. Over time, the three cell types spontaneously organize themselves around the core just as they do in a natural vessel. Such bioprinted blood vessels should be available in the near future. Bioprinting also offers the possibility of one day building a complete organ from a patient's own cells.

Consider This

BioEthics Forgacs foresees one of the first uses of his cell-printing technique to be the production of tissues from human cells on which to test new drugs. What do you think are the advantages of this approach compared to testing drugs on laboratory animals?

CHAPTER REVIEW

Summary of Key Concepts

4.1 What Is the Cell Theory?
The principles of the cell theory are that every living organism is made up of one or more cells, the smallest organisms are single cells, cells are the functional units of multicellular organisms, and all cells arise from preexisting cells.

4.2 What Are the Basic Attributes of Cells?
Cells are small because they must exchange materials with their surroundings by diffusion, a slow process that requires the interior of the cell to be close to the plasma membrane. All cells are surrounded by a plasma membrane that regulates the interchange of materials between the cell and its environment. All cells use DNA as a genetic blueprint and RNA to direct protein synthesis based on DNA. There are two fundamentally different types of cells: Prokaryotic cells are small and lack membrane-enclosed organelles. Eukaryotic cells have a variety of organelles, including a nucleus.

4.3 What Are the Major Features of Eukaryotic Cells?
Cells of plants, fungi, and some protists are supported by porous cell walls outside the plasma membrane. Eukaryotic cells have an internal cytoskeleton of protein filaments that shapes the cell, transports and anchors organelles, and allows cells to move. Some eukaryotic cells have cilia or flagella, extensions of the plasma membrane that contain microtubules in a characteristic pattern. These structures move fluids past the cell or move the cell through a fluid environment.

Genetic material (DNA) is contained within the nucleus, surrounded by the double membrane of the nuclear envelope. Pores in the nuclear envelope regulate the movement of molecules between nucleus and cytoplasm. The genetic material is organized into strands called chromosomes, which consist of DNA and proteins. The nucleolus consists of rRNA and ribosomal proteins, as well as the genes that code for ribosome synthesis. Ribosomes, composed of rRNA and protein, are the sites of protein synthesis.

The endomembrane system within eukaryotic cells includes the nuclear envelope, endoplasmic reticulum (ER), Golgi apparatus, and lysosomes and other vesicles. The ER forms a series of interconnected membranous compartments. The ER is a major site of new membrane production. Rough ER, continuous with the outer nuclear envelope, bears ribosomes where proteins are manufactured. These proteins are modified, folded, and transported within the ER channels. Smooth ER, which lacks ribosomes, manufactures lipids such as steroid hormones, detoxifies drugs and metabolic wastes, and stores calcium. The Golgi apparatus is a series of flattened membranous sacs. The Golgi processes and modifies materials that are synthesized in the rough ER and transported in vesicles to the Golgi. Substances modified in the Golgi are sorted and packaged into vesicles for transport elsewhere in the cell. Lysosomes are vesicles that contain digestive enzymes that digest food particles after merging with vesicles called food vacuoles. Lysosomes also digest defective organelles.

Some freshwater protists have contractile vacuoles that collect and expel excess water. Plants use central vacuoles to support the cell and may also use it to store nutrients, pigments, wastes, and toxic materials.

All eukaryotic cells contain mitochondria, which are organelles that use oxygen to complete the metabolism of food molecules, capturing some of their energy in ATP. Cells of plants and some protists contain plastids. Chloroplasts are specialized plastids that capture solar energy during photosynthesis, allowing plant cells to manufacture sugar from water and carbon dioxide. Storage plastids store pigments or starch. The endosymbiont hypothesis states that both mitochondria and chloroplasts originated from prokaryotic cells.

4.4 What Are the Major Features of Prokaryotic Cells?

All members of the domains Archaea and Bacteria consist of prokaryotic cells. Prokaryotic cells are generally smaller than eukaryotic cells and have a simpler internal structure that lacks membrane-enclosed organelles. Some bacteria have flagella. Most bacteria are surrounded by a cell wall made of peptidoglycan. Some bacteria, including many that cause disease, attach to surfaces such as human tissue using external capsules or slime layers. Others use short, numerous, hair-like protein strands called attachment pili. Sex pili, which are long and few in number, draw bacteria together to allow transfer of plasmids. Plasmids are small rings of DNA that confer special features such as antibiotic resistance. Most bacterial DNA is in a single chromosome in the nucleoid region. Bacterial cytoplasm includes ribosomes and a cytoskeleton (differing in structure from those of eukaryotic cells). Photosynthetic bacteria may have extensive internal membranes where the reactions of photosynthesis occur.

Key Terms

archaea 58	flagellum (plural,
bacteria 58	flagella) 60
basal body 62	food vacuole 66
cell theory 54	Golgi apparatus 65
cell wall 59	intermediate filament 60
central vacuole 67	lysosome 66
centriole 62	microfilament 60
chlorophyll 68	microtubule 60
chloroplast 68	mitochondrion (plural,
chromatin 63	mitochondria) 68
chromosome 63	nuclear envelope 63
cilium (plural, cilia) 60	nuclear pore complex 63
contractile vacuole 66	nucleoid 70
cytoplasm 55	nucleolus (plural,
cytoskeleton 60	nucleoli) 63
cytosol 55	nucleus 63
deoxyribonucleic acid	organelle 58
(DNA) 58	pilus (plural, pili) 70
diffusion 54	plasma membrane 54
endomembrane system 64	plasmid 71
endoplasmic reticulum	plastid 69
(ER) 64	prokaryotic 58
endosymbiont	ribonucleic acid (RNA) 58
hypothesis 68	ribosome 64
eukaryotic 58	vesicle 64

Learning Outcomes

In this chapter, you have learned to . . .

LO1 Summarize the cell theory.

LO2 Compare and contrast prokaryotic and eukaryotic cells.

LO3 Describe the major structures of animal cells and their functions.

LO4 Describe the structure and function of organelles that are found in plant, but not animal, cells.

LO5 Explain the structure, function, and movement of eukaryotic cilia and flagella.

LO6 Describe the structure and function of the nucleus, and compare it to the nucleoid region of prokaryotic cells.

LO7 List the parts of the endomembrane system and explain how they are interconnected.

LO8 Describe the structure and function of both the surface and the internal features of bacterial cells.

Thinking Through the Concepts

Fill-in-the-Blank

1. The plasma membrane is composed of two major types of molecules, _____ and _____. Which type of molecule is responsible for each of the following functions? Isolation from the surroundings: _____; interactions with other cells: _____. **LO2 LO3**

2. The _____ is composed of a network of protein fibers. The three types of cytoskeleton fibers are _____, _____, and _____. Which of these supports cilia? _____. Moves organelles? _____. Allows muscle contraction? _____. Provides a supporting framework for the cell? _____. **LO2 LO3 LO5**

3. After each description, fill in the appropriate term: "workbenches" of the cell: _____; comes in rough and smooth forms: _____; site of ribosome production: _____; a stack of flattened membranous sacs: _____; outermost layer of plant cells: _____; ferries blueprints for protein production from the nucleus to the cytoplasm: _____. **LO3 LO4**

4. Antibody proteins are synthesized on ribosomes associated with the _____. The antibody proteins are packaged into membranous sacs called _____ and are then transported to the _____. Here, what type of molecule is added to the protein? _____ After the antibody is completed, it is packaged into vesicles that fuse with the _____. **LO3 LO7**

5. After each description, fill in the appropriate structure: "powerhouses" of the cell: _____; capture solar energy: _____; outermost structure of plant cells: _____; region of prokaryotic cell containing DNA: _____; propel fluid past cells: _____; consists of the cytosol and the organelles within it: _____. **LO2 LO3 LO4 LO5 LO8**

6. Two organelles that are believed to have evolved from prokaryotic cells are _____ and _____. Evidence for this hypothesis is that both have _____ membranes, both have groups of enzymes that synthesize _____, and their _____ is similar to that of prokaryotic cells. **LO2 LO3 LO4 LO8**

7. What structure in bacterial cells is composed of peptidoglycan? _____. What structure in prokaryotic cells serves a similar function to the nucleus in eukaryotic cells? _____. Short segments of DNA that confer special features such as antibiotic resistance on bacteria are called _____. Bacterial structures called _____ pull bacterial cells together so they can transfer DNA. **LO2 LO8**

Review Questions

1. What are the three principles of the cell theory? **LO1**

2. Which cytoplasmic structures are common to both plant and animal cells, and which are found in one type but not the other? **LO3 LO4**

3. Name the proteins of the eukaryotic cytoskeleton; describe their relative sizes and major functions. **LO3**

4. Describe the nucleus and the function of each of its components, including the nuclear envelope, chromatin, chromosomes, DNA, and the nucleolus. **LO6**

5. What are the functions of mitochondria and chloroplasts? Why do scientists believe that these organelles arose from prokaryotic cells? What is this hypothesis called? **LO2 LO4 LO5 LO8**

6. What is the function of ribosomes? Where in the cell are they found? Are they limited to eukaryotic cells? **LO2 LO3**

7. Describe the structure and function of the endoplasmic reticulum (smooth and rough) and the Golgi apparatus and how they work together. **LO3 LO8**

8. How are lysosomes formed? What is their function? **LO3**

9. Diagram the structure of eukaryotic cilia and flagella. Describe how each moves and what their movement accomplishes. **LO3**

10. List the structures of bacterial cells that have the same name and function as some eukaryotic structures, but a different molecular composition. **LO2 LO8**

Applying the Concepts

1. If samples of muscle tissue were taken from the legs of a world-class marathon runner and a sedentary individual, which would you expect to have a higher density of mitochondria? Why?

2. One of the functions of the cytoskeleton in animal cells is to give shape to the cell. Plant cells have a fairly rigid cell wall surrounding the plasma membrane. Does this mean that a cytoskeleton is unnecessary for a plant cell?

3. What problems would an enormous round cell encounter? What adaptations might help a very large cell survive?

Answers to Figure Caption questions and Fill-in-the-Blank questions can be found in the Answers section at the back of the book.

(MB) *Go to MasteringBiology for practice quizzes, activities, eText, videos, current events, and more.*

CHAPTER 5

Cell Membrane Structure and Function

CASE STUDY
Vicious Venoms

THIRTEEN-YEAR-OLD JUSTIN SCHWARTZ was having a great time during his 3-week stay at a summer camp near Yosemite National Park. But that all changed after a 4.5-mile hike as Justin rested on some sunny rocks, hands hanging loosely at his sides. Suddenly, he felt a piercing pain in his left palm. A 5-foot rattlesnake—probably feeling threatened by Justin's dangling arm—had struck without warning.

His campmates stared in alarm as the snake slithered into the undergrowth, but Justin focused on his hand, where the pain was becoming agonizing and the palm was beginning to swell. He suddenly felt weak and dizzy. As counselors and campmates spent the next 4 hours carrying him down the trail, pain and discoloration spread up Justin's arm, and his hand felt as if it were going to burst. A helicopter whisked him to a hospital, where he fell unconscious. A day later, he regained consciousness at the University of California Davis Medical Center. There, Justin spent more than a month undergoing 10 surgeries. These relieved the enormous pressure from swelling in his arm, removed dead muscle tissue, and began the long process of repairing the extensive damage to his hand and arm.

Diane Kiehl's ordeal began as she dressed for an informal Memorial Day celebration with her family, pulling on blue jeans that she had tossed on the bathroom floor the previous night. Feeling a sting on her right thigh, she ripped off the jeans and watched with irritation as a long-legged spider crawled out. Living in an old house in the Kansas countryside, Diane had grown accustomed to spiders—which are often harmless—but this was an exception: a brown recluse. The two small puncture wounds seemed merely a minor annoyance until the next day, when an extensive, itchy rash appeared at the site. By the third day, intermittent pain pierced like a knife through her thigh. A physician gave her painkillers, steroids to reduce the swelling, and antibiotics to combat bacteria introduced by the spider's mouthparts. The next 10 days were a nightmare of pain from the growing sore, now covered with oozing blisters and underlain with clotting blood. It took 4 months for the lesion to heal. Even after a year, Diane sometimes felt pain in the large scar that remained.

How do rattlesnake and brown recluse spider venoms cause leaky blood vessels, disintegrating skin and tissue, and sometimes life-threatening symptoms throughout the body? What do venoms have to do with cell membranes? Find out as you read this chapter.

A rattlesnake prepares to strike. (Inset) A brown recluse spider.

AT A GLANCE

5.1 How Is the Structure of the Cell Membrane Related to Its Function?

5.2 How Do Substances Move Across Membranes?

5.3 How Do Specialized Junctions Allow Cells to Connect and Communicate?

LEARNING GOALS

LG1 Describe the components, structure, and function of cell membranes and explain how the different components contribute to membrane functions.

LG2 Explain how substances move across cell membranes.

LG3 Explain the basic concepts of diffusion and osmosis.

LG4 Describe the major types of junctions between cells and explain how these junctions function.

5.1 HOW IS THE STRUCTURE OF THE CELL MEMBRANE RELATED TO ITS FUNCTION?

All the membranes of a cell have a similar basic structure: proteins suspended in a double layer of phospholipids (**Fig. 5-1**). Phospholipids are responsible for isolating the cell's contents, whereas proteins are responsible for selectively exchanging substances and communicating with the environment, controlling biochemical reactions associated with the cell membrane, and forming connections between cells.

As you now know, all cells, as well as organelles within eukaryotic cells, are surrounded by membranes. Cell membranes perform several crucial functions:

- They selectively isolate the contents of membrane-enclosed organelles from the surrounding cytosol, and the cell's contents from the surrounding interstitial fluid.
- They regulate the exchange of essential substances between the cell and the interstitial fluid, or between membrane-enclosed organelles and the surrounding cytosol.
- They allow communication among the cells of multicellular organisms.
- They create attachments within and between cells.
- They regulate many biochemical reactions.

These are formidable tasks for a structure so thin that 10,000 membranes stacked atop one another would scarcely equal the thickness of this page. Membranes are not simply uniform sheets; they are complex structures whose different parts perform very distinct functions. Membranes vary from one tissue type to another, and their structures can also change dynamically in response to their surroundings.

Membranes Are "Fluid Mosaics" in Which Proteins Move Within Layers of Lipids

Before the 1970s, although cell biologists knew that cell membranes consist primarily of proteins and lipids, they did not know how these molecules create the structure of membranes. In 1972, S. J. Singer and G. L. Nicolson developed the **fluid mosaic model** of cell membranes, which is now known to be accurate. A **fluid** is any substance whose molecules can flow past one another; fluids include gases, liquids, and also cell membranes, whose molecules move quite freely within the thin layer. According to the fluid mosaic model, cell membranes consist of a fluid formed by a double layer (or *bilayer*) of phospholipids. A variety of different proteins form a kind of "mosaic," or patchwork, within the phospholipid bilayer. Both the phospholipids and the proteins are constantly shifting with respect to one another (see Fig. 5-1).

The Fluid Phospholipid Bilayer Helps to Isolate the Cell's Contents

A phospholipid consists of two very different parts: a "head" that is polar and hydrophilic (attracted to water) and a pair of fatty acid "tails" that are nonpolar and hydrophobic (not attracted to water). Membranes contain a variety of phospholipids with structures similar to that shown in **Figure 5-2**.

The arrangement of phospholipids in membranes arises from the fact that all cells are immersed in watery solutions. For example, single-celled organisms may live in fresh water or in the ocean, and water also saturates the cell walls that surround plant cells. The outer surfaces of animal plasma membranes are bathed in watery **interstitial fluid,** a weakly salty liquid resembling blood without its cells or large proteins. Inside the plasma membrane, the *cytosol* (the fluid portion of the cytoplasm) is mostly water. In these watery surroundings, phospholipids spontaneously arrange themselves into a double layer called a **phospholipid bilayer** (Fig. 5-3). Hydrogen bonds form between water and the hydrophilic phospholipid heads, causing the heads to face the water on either side. The phospholipid tails, being hydrophobic, cluster together within the bilayer.

The components of cell membranes are continually shifting position. The phospholipid molecules within each layer move relatively rapidly and the larger proteins move more slowly. To understand this movement, it is important to be aware that, at temperatures above absolute zero ($-459.4\,°F$ or $-273\,°C$), atoms, molecules, and ions are in constant random motion. As the temperature increases, their rate of motion increases; at temperatures that support life, these particles move rapidly indeed. Phospholipid molecules do not bond to one another, so at body temperature, their constant random motion causes them to shift

▲ **FIGURE 5-1 The plasma membrane** The plasma membrane is a bilayer of phospholipids interspersed with cholesterol molecules that form a fluid mosaic in which various proteins (blue) are embedded. Many proteins have carbohydrates attached to them, forming glycoproteins. Enzyme, recognition, connection, receptor, and transport proteins are illustrated.

about freely, although they generally maintain their orientation and only rarely flip between the two membrane layers. The flexible, fluid nature of the bilayer is very important for membrane function. As you breathe, move your eyes, and turn the pages of this book, cells in your body change shape. If plasma membranes were stiff, your cells would break open and die. This fluidity also allows

membrane-enclosed compartments to ferry substances into the cell, carry materials within the cell, and expel them to the outside, merging their membranes in the process (see Chapter 4). Most membranes maintain a consistency

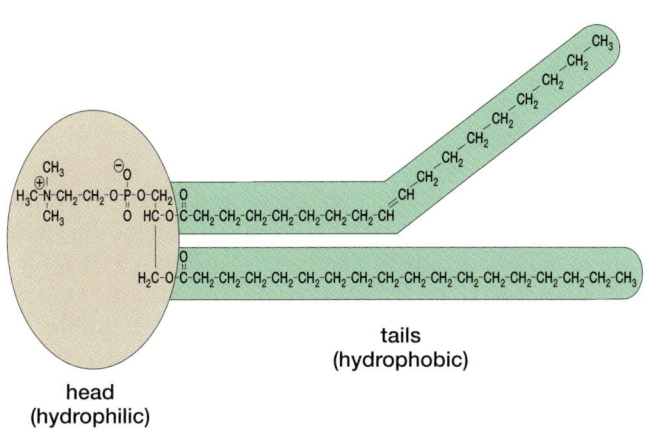

▲ **FIGURE 5-2 A phospholipid** Notice that a double bond in one of the fatty acid tails causes the tail to bend.

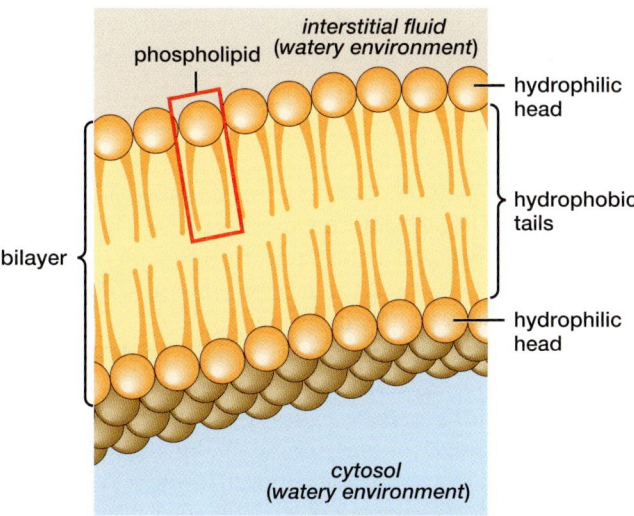

▲ **FIGURE 5-3 The phospholipid bilayer of the cell membrane**

IN GREATER DEPTH Membrane Fluidity, Function, and Phospholipids

Cell membranes must maintain an optimal fluidity to perform their many functions. Molecules are in constant random motion, and membrane phospholipids jiggle around more vigorously and maintain more distance from one another as the temperature increases. This extra space between molecules increases membrane fluidity. At cooler temperatures, membranes become less fluid as the phospholipids pack together more tightly.

Membrane fluidity is strongly influenced by the relative amounts of saturated and unsaturated fatty acid tails in membrane phospholipids. Why? Unsaturated fatty acids have one or more double bonds, causing them to kink (see Fig. 5-2). Kinked tails occupy more space and cannot pack as tightly together. Saturated fatty acid tails are straight and can pack more tightly (Fig. E5-1). But the body temperatures of many organisms—including plants, protists and bacteria, and "cold-blooded" animals such as frogs, fish, and snakes—regularly fluctuate along with environmental temperatures. To cope, these organisms actually change the composition of their membranes in response to changes in temperature, incorporating more saturated fatty acids into their membranes as the temperature rises, and replacing them with unsaturated fatty acids as the temperature falls. These molecular changes compensate for the effects of changing temperature, maintaining optimum membrane fluidity.

A few mammals have similar adaptations. For example, caribou (which live in the far north) maintain a core temperature of about 100°F (38°C), but allow the temperature of their lower legs to drop almost to freezing (32°F or 0°C),

more unsaturated   more saturated
greater fluidity   less fluidity

▲ FIGURE E5-1 **Tail kinks in phospholipids increase membrane fluidity**

▲ FIGURE E5-2 **Caribou legs are adapted to cold**

thus conserving body heat as they stand in winter snow (Fig. E5-2). In caribou legs, the membranes of cells near the chilly hooves have lots of unsaturated fatty acids, whereas those near the warmer trunk have more saturated fatty acids. This gives cell membranes the necessary fluidity throughout the leg, despite large temperature differences from thigh to hoof.

similar to room temperature olive oil. But olive oil becomes a solid if you refrigerate it, so what happens to membranes when an organism gets cold? To learn more about how organisms adapt, see "In Greater Depth: Membrane Fluidity, Function, and Phospholipids."

All animal cell membranes contain cholesterol (see Fig. 5-1), which is especially abundant in the plasma membrane. Interactions between cholesterol and phospholipids help to stabilize the membrane, making it less fluid at higher temperatures and less solid at lower temperatures. A high cholesterol content reduces the permeability of the membrane to hydrophilic substances and small molecules that would otherwise diffuse through it. This allows the

cell to exert more control over what substances enter and leave it, through the use of selective transport proteins, described later.

A few hydrophobic biological molecules can diffuse directly through the phospholipid bilayer, including steroid hormones such as estrogen and testosterone (see Chapter 3) and fat-soluble vitamins. Most molecules used by cells, including salts, amino acids, and sugars, are hydrophilic; that is, they are polar and water soluble. Such substances cannot pass through the nonpolar, hydrophobic fatty acid tails of the phospholipid bilayer. Movement of these substances in or out of the cell relies on the "mosaic" of proteins within the membrane.

CASE STUDY continued

Vicious Venoms

Some of the most devastating effects of certain snake and spider venoms occur because they contain phospholipases, enzymes that break down phospholipids. You now know that phospholipids are a major component of cell membranes, which isolate the cell's contents from its surroundings. As the phospholipids degrade, the membranes become leaky, causing the cells to die. As the venom spreads, nearby tissues can be seriously damaged—as you will see later.

A Variety of Proteins Form a Mosaic Within the Membrane

Thousands of different membrane proteins are embedded within or attached to the surfaces of the phospholipid bilayer of cell membranes. Some membrane proteins bear carbohydrate groups that project from the outer membrane surface (the branched structures in Fig. 5-1). These proteins are called **glycoproteins** ("glyco" is from the Greek, meaning "sweet," and refers to the carbohydrate portion with its sugar subunits). Membrane proteins may be grouped into five major categories based on their function: enzymes, receptor proteins, recognition proteins, connection proteins, and transport proteins, described in the sections that follow.

Enzymes

Proteins that promote chemical reactions that synthesize or break apart biological molecules are called **enzymes** (described in Chapter 6). Although many enzymes are located within the cytoplasm, some span cell membranes and others are attached to membrane surfaces. For example, some plasma membrane enzymes help synthesize the *extracellular matrix*, a web of protein and glycoprotein fibers that fills spaces between animal cells (see Fig. 5-1). Plasma membrane enzymes in cells lining the small intestine complete the digestion of carbohydrates and proteins as these nutrients are taken into the cells.

Receptor Proteins

Most cells bear dozens of types of **receptor proteins** (some of which are glycoproteins; see Fig. 5-1) that span their plasma membranes. Receptor proteins allow cells to respond to specific messenger molecules (such as hormones) carried in the bloodstream. When the appropriate messenger molecule binds to a receptor protein, the receptor is activated, often by changing shape. This activation, in turn, causes a response inside the cell (**Fig. 5-4**). The response can lead to pores in the membrane opening or closing, or enzymes becoming active or inactive. Some hormones binding to receptor proteins stimulate cell division. Receptor proteins on cells of the immune system bind to molecules on the surface of invading bacteria or viruses so they can be destroyed.

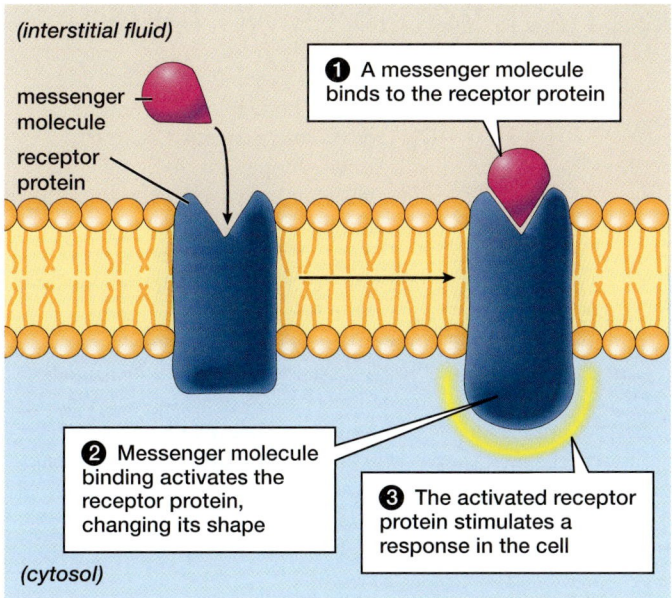

(interstitial fluid)

messenger molecule
receptor protein

❶ A messenger molecule binds to the receptor protein

❷ Messenger molecule binding activates the receptor protein, changing its shape

❸ The activated receptor protein stimulates a response in the cell

(cytosol)

▲ **FIGURE 5-4 Receptor protein activation**

Messenger molecules binding to receptor proteins often start a complex series of biochemical reactions in the cell. For example, the hormone epinephrine (adrenaline) is released in response to stress, such as you might experience if a mugger approached from a dark alley. Epinephrine binds to a specific receptor on muscle cells that activates an enzyme that causes formation of the messenger molecule cyclic adenosine monophosphate (cAMP; see Chapter 3). The cAMP then activates other enzymes, resulting in the breakdown of glycogen into glucose. The glucose is then broken down and its energy stored in adenosine triphosphate (ATP; see Chapter 3). ATP fuels muscle contraction, which you will need to escape the mugger!

Recognition Proteins

Recognition proteins are glycoproteins that serve as identification tags (see Fig. 5-1). The cells of each individual organism bear unique glycoproteins that identify the cells as "self." Immune cells ignore self-cells and attack invading cells, such as bacteria, that have different recognition proteins on their membranes. For successful organ transplants, the most important recognition glycoproteins of the donor must match those of the recipient so the organ won't be attacked by the recipient's immune system.

Connection Proteins

A diverse group of **connection proteins** anchors cell membranes in various ways. Some connection proteins help maintain cell shape by linking the plasma membrane to the cell's cytoskeleton. Other connection proteins span the plasma membrane, linking the cytoskeleton inside the cell with the extracellular matrix outside, which helps anchor the cell in place within a tissue (see Fig. 5-1). Connection proteins also form connections between adjacent cells, as described later in this chapter (see Fig. 5-16).

Transport Proteins

Transport proteins (see Fig. 5-1) span the phospholipid bilayer and regulate the movement of hydrophilic molecules across the membrane. Some transport proteins form pores (channels) that can be opened or closed to allow specific substances to pass across the membrane. Other transport proteins bind substances and conduct them through the membrane, sometimes using cellular energy. Transport proteins are described later in this chapter.

CHECK YOUR LEARNING

Can you describe the components, structure, and functions of cell membranes? Are you able to explain how the different components of cell membranes contribute to their functions? Can you describe the fluid mosaic model of cell membranes?

5.2 HOW DO SUBSTANCES MOVE ACROSS MEMBRANES?

Some substances, especially individual molecules and ions, can move across membranes by diffusing through the phospholipid bilayer or traveling through specialized transport proteins.

To provide some background, we begin our study of membrane transport with a few definitions:

- A **solute** is a substance that can be dissolved (dispersed into individual atoms, molecules, or ions) in a **solvent,** which is a fluid (usually a liquid) capable of dissolving the solute. Water, in which all biological processes occur, dissolves so many different solutes that it is sometimes called the "universal solvent."
- The **concentration** of a substance defines the amount of solute in a given amount of solvent. The concentration of sugar solution, for example, is a measure of the number of molecules of sugar contained in a given volume of solution.
- A **gradient** is a difference in certain properties—such as temperature, pressure, electrical charge, or the concentration of a solute in a fluid—between two adjacent regions. Creation of a gradient requires the expenditure of energy. Over time, gradients break down unless energy is supplied to maintain them or a perfect barrier separates them. For example, a temperature gradient breaks down as heat energy is transmitted from a higher-temperature region to a lower-temperature region. Electrical gradients can drive the movement of charged ions, which are pulled toward an oppositely charged region. Gradients of concentration or pressure cause molecules or ions to move from one region to another, equalizing the difference. Cells use energy-requiring transport proteins to generate and maintain **concentration gradients,** or differences in solute concentrations across their membranes.

Molecules in Fluids Diffuse in Response to Gradients

Recall that atoms, molecules, and ions are in constant random motion. As a result of this motion, molecules and ions in solution are continuously bombarding one another and the structures surrounding them. Over time, random movements of solutes produce a net movement from regions of high concentration to regions of low concentration, a process called **diffusion.** The greater the concentration gradient, the more rapidly diffusion occurs. If nothing opposes this diffusion (such as electrical charge, pressure differences, or physical

❶ A drop of dye is placed in water

❷ Dye molecules diffuse into the water; water molecules diffuse into the dye

❸ Both dye molecules and water molecules are evenly dispersed

dye molecules

water molecule

▶ **FIGURE 5-5 Diffusion of a dye in water**

barriers), then the random movement of molecules will continue until the substance is evenly dispersed throughout the fluid. In an analogy to gravity, molecules moving from regions of high concentration to regions of low concentration are described as moving "down" their concentration gradients.

To watch diffusion in action, place a drop of food coloring in a glass of water (Fig. 5-5). Random motion propels dye molecules both into and out of the dye droplet, but there is a net transfer of dye into the water and of water into the dye, down their concentration gradients. The net movement of dye will continue until it is uniformly dispersed in the water. If you compare the diffusion of a dye in hot and cold water, you will see that heat increases the diffusion rate because heat causes molecules to move faster.

SUMMING UP Principles of Diffusion

- Diffusion is the net movement of molecules down a gradient from high to low concentration.
- The greater the concentration gradient, the faster the rate of diffusion.
- The higher the temperature, the faster the rate of diffusion.
- If no other processes intervene, diffusion will continue until the concentrations become equal throughout the solution; that is, until the concentration gradient is eliminated.

Movement Through Membranes Occurs by Passive Transport and Energy-Requiring Transport

Gradients of ions and molecules across cell membranes are crucial for life; a cell without gradients is dead. Proteins in cell membranes must constantly use energy to create and maintain these concentration gradients because gradients underlie many crucial biochemical processes. For example, neurons depend on the flow of specific ions down their concentration gradients to produce the electrical signals that underlie sensation, movement, and thought. Plasma membranes are described as **selectively permeable** because their proteins selectively allow only certain ions or molecules to pass through, or permeate. The selective permeability of the plasma membrane creates a barrier that helps maintain the gradients that characterize all cells.

A cell's plasma membrane permits substances to move through it in two different ways: passive transport and energy-requiring transport (Table 5-1). **Passive transport** involves diffusion of substances across cell membranes down their concentration gradients, whereas **energy-requiring transport** requires that the cell expend energy to move substances across membranes for example, into or out of the cell, usually against a concentration gradient.

Passive Transport Includes Simple Diffusion, Facilitated Diffusion, and Osmosis

Diffusion can occur within a fluid or across a membrane that is permeable to the diffusing substance. Many molecules cross plasma membranes by diffusion, driven by

TABLE 5-1 Transport Across Membranes

Passive Transport	Diffusion of substances across a membrane down a gradient of concentration, pressure, or electrical charge; does not require cellular energy.
Simple diffusion	Diffusion of water, dissolved gases, or lipid-soluble molecules through the phospholipid bilayer of a membrane.
Facilitated diffusion	Diffusion of water, ions, or water-soluble molecules through a membrane via a channel or carrier protein.
Osmosis	Diffusion of water across a selectively permeable membrane from a region of higher free water concentration to a region of lower free water concentration.
Energy-Requiring Transport	Movement of substances through membranes using cellular energy, usually supplied by ATP.
Active transport	Movement of individual small molecules or ions against their concentration gradients through membrane-spanning proteins.
Endocytosis	Movement of fluids, specific molecules, or particles into a cell; occurs as the plasma membrane engulfs the substance in a membranous sac that pinches off and enters the cytosol.
Exocytosis	Movement of particles or large molecules out of a cell; occurs as a membrane within the cell encloses the material, moves to the cell surface, and fuses with the plasma membrane, allowing its contents to diffuse out.

concentration differences between the cytosol and the interstitial fluid.

Some Molecules Move Across Membranes by Simple Diffusion

Some molecules diffuse directly through the phospholipid bilayer of cell membranes, a process called **simple diffusion** (Fig. 5-6a). Very small molecules with no net charge, such as water, oxygen, and carbon dioxide, can travel across cell membranes by simple diffusion. Lipid-soluble molecules, regardless of their size, may also use this method. Such molecules include ethyl alcohol (found in alcoholic beverages), vitamins A, D, and E, and steroid hormones (described in Chapter 3). A larger concentration gradient, higher temperature, smaller molecular size, and greater solubility in lipids all increase the rate of simple diffusion.

How can water—a polar molecule—diffuse directly through the hydrophobic (literally "water-fearing") phospholipid bilayer? Because water molecules are so small and so abundant in both the cytosol and the interstitial fluid, some stray into the thicket of phospholipid tails, and the random movement of these water molecules carries them through the membrane. Because simple diffusion of water through the phospholipid bilayer is relatively slow, many cell types have specific transport proteins for water, described a bit later.

(a) Simple diffusion through the phospholipid bilayer

(b) Facilitated diffusion through carrier proteins

(c) Facilitated diffusion through channel proteins

(d) Osmosis through aquaporins or the phospholipid bilayer

▲ FIGURE 5-6 **Types of diffusion through the plasma membrane** (a) Molecules that are very small, uncharged, or lipid soluble move directly through the phospholipid bilayer by simple diffusion. Here, oxygen molecules diffuse from the interstitial fluid into the cell down their concentration gradient (red arrow). **(b)** Carrier proteins have binding sites for specific molecules (such as the glucose shown here). Binding of the transported molecule causes the carrier to change shape, shuttling the molecule across the membrane down its concentration gradient. **(c)** Facilitated diffusion through channel proteins allows ions to cross membranes. Here, chloride ions diffuse down their concentration gradient into the cell through chloride channels. **(d)** Osmosis is the diffusion of water. Water molecules can diffuse directly through the phospholipid bilayer by simple diffusion, or they can pass far more rapidly by facilitated diffusion through water channels called aquaporins.

Some Molecules Cross Membranes by Facilitated Diffusion Using Membrane Transport Proteins

The phospholipid bilayers of cell membranes are quite impermeable to larger, polar molecules (such as sugars) and to ions (for example, K^+, Na^+, Cl^-, and Ca^{2+}). The charges on these substances cause water molecules (which are polar) to cluster around them, forming an aggregation that is too large to move directly through the phospholipid bilayer. So ions and polar molecules must use specific transport proteins to move through cell membranes, a process called **facilitated diffusion.** Two types of proteins allow facilitated diffusion: carrier proteins and channel proteins. No energy is used by these transport proteins, which only facilitate diffusion down a concentration gradient either into or out of the cell.

Carrier proteins span the cell membrane and have regions that loosely bind certain ions or specific molecules such as sugars or small proteins. This binding causes carrier proteins to change shape and transfer the bound molecules across the membrane. For example, glucose carrier proteins in plasma membranes allow this sugar to diffuse from the interstitial fluid into cells, which continuously use up glucose to meet their energy needs (**Fig. 5-6b**).

Channel proteins form pores through the cell membrane. For example, the cell membranes of mitochondria and chloroplasts have pores that allow passage of a variety of water-soluble substances. In contrast, ion channel proteins (ion channels) are small and highly selective (**Fig. 5-6c**). All cells maintain gradients of several different ions across their membranes. Many ion channels help preserve these gradients by remaining closed unless they are opened by specific stimuli, such as a nerve impulse. Ion channel proteins are selective because their interior diameters limit the size of the ions that can pass through, and certain channel

protein amino acids in the pore have weak electrical charges that attract specific ions and repel others. For example, slight negative charges on some amino acids inside Na^+ channels attract Na^+ but repel Cl^-.

Many cells have specialized water channel proteins called **aquaporins** (literally, "water pores"; **Fig. 5-6d**). The narrow aquaporin channels are selective for water molecules, which are extremely small. Some amino acids within the aquaporin channel protein have slight positive charges that attract the negative poles of water but repel positive ions (see Fig. E5-4b). To learn more, see "Science in Action: The Discovery of Aquaporins."

Osmosis Is the Diffusion of Water Across Selectively Permeable Membranes

Osmosis is the diffusion of water across a selectively permeable membrane in response to gradients of concentration, pressure, or temperature. Here, we will focus on osmosis from a region of higher water concentration to a region of lower water concentration.

What do we mean when we describe a solution as having a "high water concentration" or a "low water concentration"? The answer is simple: Pure water has the highest possible water concentration. Any substance that dissolves in water (any solute) reduces the water concentration by displacing some of the water molecules in a given volume. Polar solutes or ions also attract water molecules, so the water molecules aren't free to move across a water-permeable membrane. Therefore, the higher the concentration of solute, the lower the concentration of water that is available to move across a membrane. Consequently, there will be a net movement of water molecules from the solution with more free water molecules (a lower solute concentration) into the solution with fewer free water molecules

No net flow of water

Water flows out; the sac shrinks

Water flows in; the sac expands

(a) A sac in an isotonic solution

(b) A sac in a hypertonic solution

(c) A sac in a hypotonic solution

◄FIGURE 5-7 **The effect of solute concentration on osmosis** We start with three sacs made of membrane that is selectively permeable to water (not to sugar, represented by the red balls). We place equal volumes of the same concentration of sugar water in each sac. We then immerse each in a beaker filled with a different concentration of sugar water—**(a)** isotonic, **(b)** hypertonic, or **(c)** hypotonic—to that in the sac. The figure shows the results after an hour. The blue arrows indicate relative water movement into and out of the sac.

QUESTION If the concentration of solute in the sac were doubled, how would the size of the sac compare to its size in each of the situations shown?

(a higher solute concentration). For example, water will move by osmosis from a solution with less dissolved sugar into a solution with more dissolved sugar. Because there are more free water molecules in the less-concentrated sugar solution, more water molecules will collide with—and thus move through—the water-permeable membrane on that side. The concentration of solute in water determines the solution's *osmotic strength* or its tendency to attract water across a membrane. The greater the solute concentration, the greater the osmotic strength.

Solutions with equal concentrations of solute (and thus equal concentrations of water) are described as being **isotonic** to one another ("iso" means "same"). When isotonic solutions are separated by a water-permeable membrane, there is no net movement of water between them (**Fig. 5-7a**). When a membrane that is selectively permeable to water separates solutions with different concentrations of a solute, the solution that contains a greater concentration of solute is described as being **hypertonic** ("hyper" means "greater than") to the less concentrated solution (**Fig. 5-7b**). The more dilute solution is described as **hypotonic** ("hypo" means "below"; **Fig. 5-7c**). Water tends to move through water-permeable membranes from hypotonic solutions into hypertonic solutions.

SUMMING UP Principles of Osmosis

- Osmosis is the movement of water through a selectively water-permeable membrane by simple diffusion or by facilitated diffusion through aquaporins.
- Water moves across a selectively water-permeable membrane down its concentration gradient from the side with a higher concentration of free water molecules to the side with a lower concentration of free water molecules.
- Dissolved substances, called solutes, reduce the concentration of free water molecules in a solution.
- If two solutions are separated by a membrane that is selectively permeable to water, the solution with a higher solute concentration is hypertonic and has a higher osmotic strength; the solution with the lower solute concentration is hypotonic and has a lower osmotic strength.

Osmosis Across the Plasma Membrane Plays an Important Role in the Lives of Cells

Osmosis across plasma membranes is crucial to many biological processes, including water uptake by the roots of plants, absorption of dietary water from the intestine, and the reabsorption of water in kidneys. In addition, organisms that live in fresh water must continuously expend energy to counteract osmosis because their cells are hypertonic to the surrounding water. For example, protists such as *Paramecium* use cellular energy to pump salts from the cytosol into their contractile vacuoles. Water follows by osmosis and is squirted out through a pore in the plasma membrane (see Fig. 4-15).

Nearly every living plant cell is supported by water that enters through osmosis. Most plant cells have a large central vacuole. The membrane that encloses this vacuole is rich in aquaporins. Dissolved substances stored in the vacuole make its contents hypertonic to the surrounding cytosol, which in turn is usually hypertonic to the interstitial fluid that bathes the cells. Water therefore flows into the cytosol and then into the vacuole by osmosis, creating **turgor pressure** within the cell. Turgor pressure inflates the cell, forcing the cytosol within its surrounding plasma membrane against the cell wall (**Fig. 5-8a**). If you forget to water a houseplant, the cytosol and cell vacuole lose water, causing the cell to shrink away from its cell wall. Like a leaky balloon, the plant droops as its cells lose turgor pressure (**Fig. 5-8b**). Now you know why grocery stores are always spraying their leafy produce: to keep it looking perky and fresh with full central vacuoles.

In animals, the interstitial fluid is usually isotonic to the cell cytosol. Although the concentrations of specific solutes are rarely the same both inside and outside of cells, the total concentrations of water and solutes inside and out are equal. As a result, there is no overall tendency for water to enter or leave animal cells. To demonstrate the importance of maintaining isotonic conditions between fragile cells and their surrounding interstitial fluid, we can observe cells placed in solutions with different solute concentrations. In an isotonic salt solution, red blood cells (which have an

cytoplasm   central vacuole

When water is plentiful, it fills the central vacuole, pushes the cytoplasm against the cell wall, and helps maintain the cell's shape

Water pressure supports the leaves of this impatiens plant

(a) Turgor pressure provides support

cell wall   plasma membrane

When water is scarce, the central vacuole shrinks and the cell wall is unsupported

Deprived of the support from water, the plant wilts

(b) Loss of turgor pressure causes the plant to wilt

▲ FIGURE 5-8 **Turgor pressure in plant cells** Aquaporins allow water to move rapidly in and out of the central vacuoles of plant cells. **(a)** The cell and the plant are supported by water turgor pressure. **(b)** The cell and the plant have lost turgor pressure and support due to dehydration.

QUESTION If a plant cell is placed in water containing no solutes, will the cell eventually burst? What about an animal cell? Explain.

abundance of aquaporins in their plasma membranes) retain their normal size (**Fig. 5-9a**). But if the salt solution is hypertonic to the cytosol of the blood cells, water leaves by osmosis, causing the cells to shrivel (**Fig. 5-9b**). Immersion in a hypotonic salt solution, in contrast, causes the cells to swell (and eventually burst) as water diffuses in (**Fig. 5-9c**).

Energy-Requiring Transport Includes Active Transport, Endocytosis, and Exocytosis

A multitude of cellular activities rely on the energy-expending activities of active transport, endocytosis, and exocytosis—all

necessary to sustain life. These processes are crucial to maintaining concentration gradients, acquiring food, excreting wastes, and (in multicellular organisms) communicating with other cells.

Cells Maintain Concentration Gradients Using Active Transport

By building gradients and then allowing them to run down under specific circumstances, cells regulate their biochemical reactions, respond to external stimuli, and acquire and store chemical energy. Energy stored as concentration gradients of ions powers the electrical signals of neurons, the contraction of muscles, and the generation of ATP in mitochondria and chloroplasts (see Chapters 7 and 8). But gradients do not form spontaneously—they require active transport across a membrane.

During **active transport,** membrane proteins use cellular energy to move molecules or ions across a plasma membrane *against* their concentration gradients, which means that the substances are transported from areas of lower concentration to areas of higher concentration (**Fig. 5-10**). For example, every cell must acquire some nutrients that are less concentrated in the environment than in the cell's cytoplasm. In addition, many ions, such as sodium and calcium ions, are actively transported to maintain them at much lower concentrations in the cytosol than in the interstitial fluid. Nerve cells maintain large ion concentration gradients because generating their electrical signals requires rapid, passive flow of ions when channels are opened. After these ions diffuse into or out of the cell, their concentration gradients must be restored by active transport.

Active-transport proteins span the width of the membrane and have two binding regions. One of these loosely binds with a specific molecule or ion, such as a calcium ion (Ca^{2+}), as shown in **Figure 5-10 ❶**; the second region, on the inside of the membrane, binds ATP. The ATP donates energy to the protein, causing the protein to change shape and move the calcium ion across the membrane (**Fig. 5-10 ❷**). The energy for active transport comes from breaking the high-energy bond that links the last of the three phosphate groups in ATP. As it releases its stored energy, ATP becomes ADP (adenosine diphosphate) plus a free phosphate (**Fig. 5-10 ❸**). Active-transport proteins are often called pumps because, like pumping water into an elevated storage tank, they use energy to move ions or molecules "uphill" against a concentration gradient.

Endocytosis Allows Cells to Engulf Particles or Fluids

A cell may need to acquire materials from its extracellular environment that are too large to move directly through the membrane. These materials are engulfed by the plasma membrane and are transported within the cell inside vesicles.

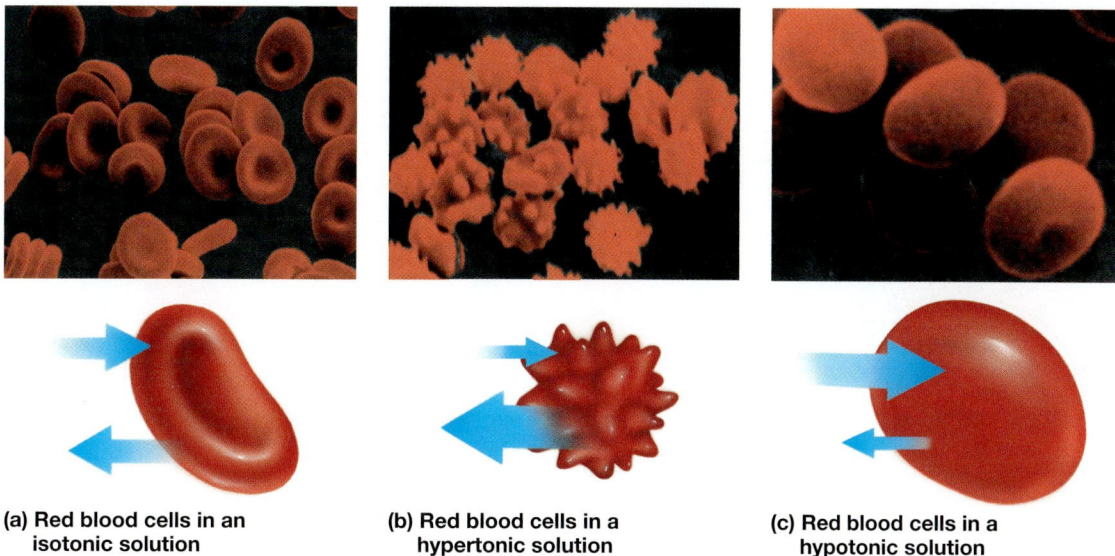

(a) Red blood cells in an
isotonic solution

(b) Red blood cells in a
hypertonic solution

(c) Red blood cells in a
hypotonic solution

▲ FIGURE 5-9 **The effects of osmosis on red blood cells** (a) The cells are immersed in an
isotonic solution and retain their normal dimpled shape. (b) The cells shrink when placed in a hypertonic
solution, as more water moves out than flows in. (c) The cells expand when placed in a hypotonic solution.

QUESTION A student pours some distilled water into a sample of blood. Returning later, she
looks at the blood under a microscope and sees no blood cells at all. What happened?

This energy-requiring process is called **endocytosis** (Greek
for "into the cell"). Here, we describe three forms of endocytosis based on the size and type of material acquired and the
method of acquisition: pinocytosis, receptor-mediated endocytosis, and phagocytosis.

In **pinocytosis** ("cell drinking"), a very small patch of
plasma membrane dimples inward as it surrounds interstitial fluid, and then the membrane buds off into the cytosol
as a tiny vesicle (**Fig. 5-11**). Pinocytosis moves a droplet of
interstitial fluid, contained within the dimpling patch of

membrane, into the cell. Therefore, the cell acquires materials in the same concentration as in the interstitial fluid.

Cells use **receptor-mediated endocytosis** to selectively take up specific molecules or complexes of molecules
(for example, packets containing protein and cholesterol)
that cannot move through channels or diffuse through the
plasma membrane (**Fig. 5-12**). Receptor-mediated endocytosis relies on specialized receptor proteins located on the
plasma membrane in thickened depressions called *coated
pits*. After the appropriate molecules bind these receptors,

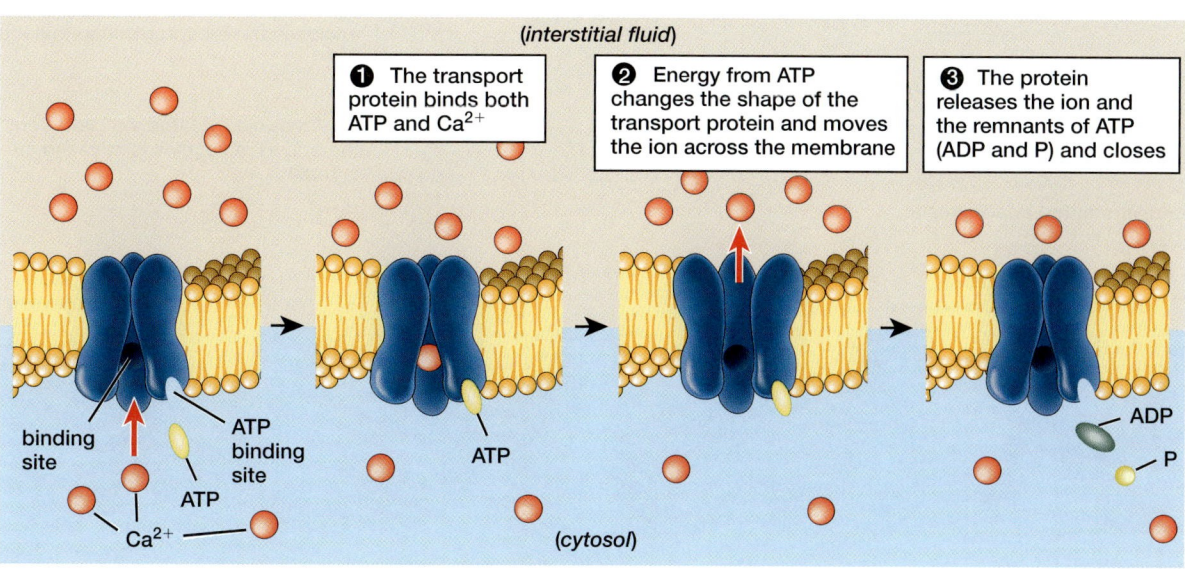

(interstitial fluid)

❶ The transport
protein binds both
ATP and Ca²⁺

❷ Energy from ATP
changes the shape of the
transport protein and moves
the ion across the membrane

❸ The protein
releases the ion and
the remnants of ATP
(ADP and P) and closes

binding
site

ATP
binding
site

ATP

Ca²⁺

ATP

ADP

P

(cytosol)

▲ FIGURE 5-10 **Active transport** Active transport uses cellular energy to move molecules across
the plasma membrane against a concentration gradient. A transport protein (blue) has an ATP binding
site and a binding site for the molecules to be transported; in this case, calcium ions (Ca^{2+}). Notice
that when ATP donates its energy, it loses its third phosphate group and becomes ADP plus a
free phosphate.

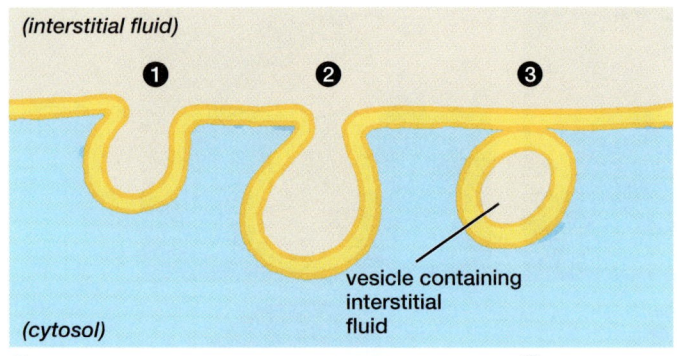

(interstitial fluid)

(cytosol)

vesicle containing interstitial fluid

❶ A dimple forms in the plasma membrane, which **❷** deepens and surrounds the interstitial fluid. **❸** The membrane encloses the interstitial fluid, forming a vesicle.

(a) Pinocytosis

interstitial fluid

cytosol

(b) TEM of pinocytosis

▲ **FIGURE 5-11 Pinocytosis** The circled numbers correspond to both **(a)** the diagram and **(b)** the transmission electron micrograph (TEM).

the coated pit deepens into a pocket that pinches off, forming a vesicle that carries the molecules into the cytosol.

Phagocytosis (from the Greek for "cell eating") moves large particles—sometimes including whole microorganisms—into the cell (**Fig. 5-13a**). When the freshwater protist *Amoeba*, for example, senses a tasty *Paramecium*, the *Amoeba* extends parts of its surface membrane. These membrane extensions are called pseudopods (Latin for "false feet"). The pseudopods fuse around the prey, thus enclosing it inside a vesicle called a **food vacuole,** for digestion (**Fig. 5-13b**). Like *Amoeba*, white blood cells use phagocytosis followed by digestion to engulf and destroy invading bacteria, in a drama that occurs constantly within your body (**Fig. 5-13c**).

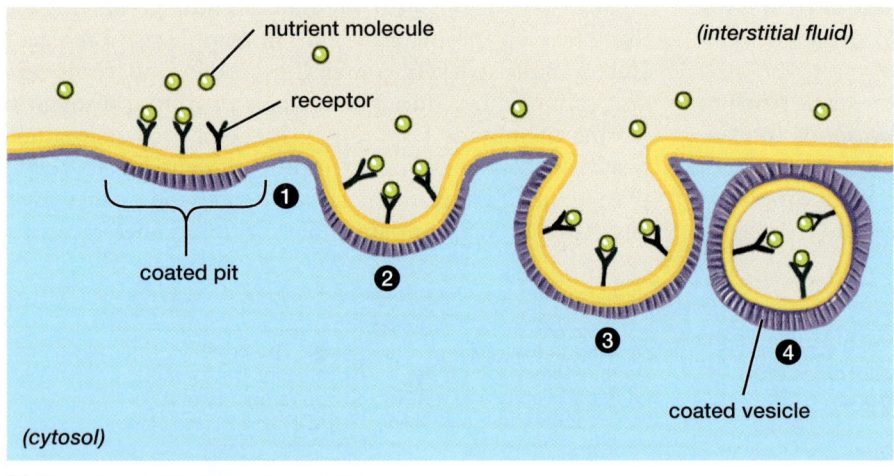

nutrient molecule

(interstitial fluid)

receptor

coated pit

❶

❷

❸

coated vesicle

❹

(cytosol)

❶ Receptor proteins for specific molecules or complexes of molecules are localized at coated pit sites.

❷ The receptors bind the molecules and the membrane dimples inward.

❸ The coated pit region of the membrane encloses the receptor-bound molecules.

❹ A vesicle ("coated vesicle") containing the bound molecules is released into the cytosol.

(a) Receptor-mediated endocytosis

extracellular particles bound to receptors

(interstitial fluid)

❶

(cytosol)

protein coating

coated pit

plasma membrane

❷

❸

coated vesicle

❹

0.1 micrometer

(b) TEM of receptor-mediated endocytosis

▲ **FIGURE 5-12 Receptor-mediated endocytosis** The numbers correspond to both **(a)** the diagram and **(b)** the transmission electron micrographs (TEM).

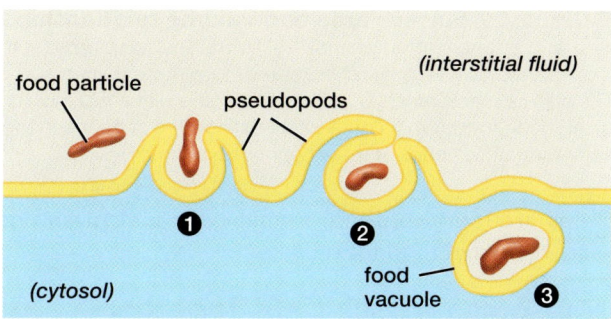

① The plasma membrane extends pseudopods toward an extracellular particle (for example, food). ② The ends of the pseudopods fuse, encircling the particle. ③ A vesicle called a food vacuole is formed containing the engulfed particle.

(a) Phagocytosis

(b) An *Amoeba* engulfs a *Paramecium*

(c) A white blood cell engulfs a disease-causing fungal cell

▲ **FIGURE 5-13 Phagocytosis (a)** The mechanism of phagocytosis. **(b)** Amoebas use phagocytosis to feed, and **(c)** white blood cells use phagocytosis to engulf disease-causing microorganisms.

Exocytosis Moves Material Out of the Cell

Cells also use energy to dispose of undigested particles of waste or to secrete substances such as hormones into the interstitial fluid, a process called **exocytosis** ("out of the cell"; **Fig. 5-14**). During exocytosis, a membrane-enclosed vesicle carrying material to be expelled moves to the cell surface, where the vesicle's membrane fuses with the cell's plasma membrane. The vesicle's contents then diffuse into the fluid outside the cell.

Exchange of Materials Across Membranes Influences Cell Size and Shape

As you learned in Chapter 4, most cells are too small to be seen with the naked eye; they range from about 1 to 100 micrometers (millionths of a meter) in diameter (see Fig. 4-1). Why are cells so small? Assuming that a cell is roughly spherical, the larger its diameter, the farther away are its innermost contents from the plasma membrane. To acquire nutrients and eliminate wastes, all parts of the cell rely on the slow process of diffusion.

In a hypothetical giant cell 8.5 inches (20 centimeters) in diameter, oxygen molecules would take more than 200 days to diffuse to the center of the cell, but by then the cell would be long dead for lack of oxygen. In addition, as a sphere enlarges, its volume increases more rapidly than does its surface area. So a large, roughly spherical cell (which would require even more nutrients and produce more wastes than a smaller cell) would have a relatively smaller area of membrane—not enough to accomplish this exchange (**Fig. 5-15**).

This constraint limits the size of most cells. However, some cells, such as nerve and muscle cells, have an extremely elongated shape that increases their membrane surface area, keeping the ratio of surface area to volume relatively high.

CHECK YOUR LEARNING

Can you explain simple diffusion, facilitated diffusion, and osmosis? Can you describe active transport, endocytosis, and exocytosis? Can you explain how the need to exchange materials across membranes influences cell size and shape?

▶ **FIGURE 5-14 Exocytosis** Exocytosis is functionally the reverse of endocytosis.

QUESTION How does exocytosis differ from diffusion of materials out of a cell?

Material is enclosed in a vesicle that fuses with the plasma membrane, allowing its contents to diffuse out

0.2 micrometer

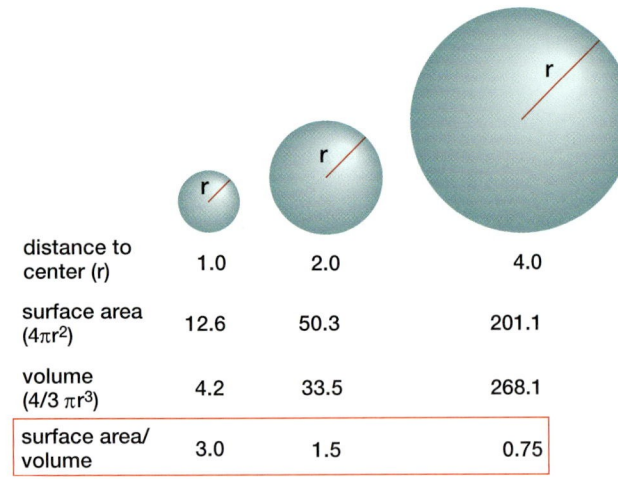

distance to center (r)	1.0	2.0	4.0
surface area ($4\pi r^2$)	12.6	50.3	201.1
volume ($4/3\,\pi r^3$)	4.2	33.5	268.1
surface area/ volume	3.0	1.5	0.75

5.3 HOW DO SPECIALIZED JUNCTIONS ALLOW CELLS TO CONNECT AND COMMUNICATE?

In multicellular organisms, some specialized structures on plasma membranes hold cells together, while others provide avenues through which cells communicate with neighboring cells. Here we discuss four major types of cell-connecting structures: desmosomes, tight junctions, gap junctions, and plasmodesmata. The first three types of junctions are found only in animal cells, and plasmodesmata are restricted to plant cells.

◀FIGURE 5-15 **Surface area and volume relationships** As the size of a sphere (or a spherical cell) increases, its volume (or the amount of cytoplasm it contains) increases much more than its surface area (its plasma membrane). This means that cells with a roughly spherical shape must remain quite small to allow the cell surface to meet the metabolic needs of the amount of cytoplasm inside.

Desmosomes Attach Cells Together

Attachment structures called **desmosomes** (**Fig. 5-16a**) join cells in tissues that are repeatedly stretched, such as the skin, intestines, and heart. These strong junctions prevent the forces on these tissues from pulling them apart. In a desmosome, anchoring proteins lie on the inner side of each membrane of adjacent cells. These anchoring proteins are attached to intermediate filaments of the cytoskeleton that extend through the cytoplasm. Linking proteins extend from each plasma membrane, spanning the narrow space between the adjacent cells and joining them firmly together.

Tight Junctions Make Cell Attachments Leakproof

Tight junctions are formed by proteins that span the plasma membranes at corresponding sites on adjacent cells. As illustrated in **Figure 5-16b**, tight junctions join cells almost as if the

(a) Desmosome   (b) Tight junctions

(c) Gap junctions   (d) Plasmodesmata

◀FIGURE 5-16 **Cell attachment structures** (a) Desmosomes form strong attachments. Anchoring proteins on adjacent plasma membranes are bound together by linking proteins. Protein filaments (intermediate filaments) attach to the cytoskeleton inside each cell, strengthening the connection. (b) Tight junction proteins of adjacent cells fuse to one another. They form a stitch-like pattern that prevents movement of substances between the plasma membranes of adjacent cells. (c) Gap junctions contain protein channels that interconnect the cytosol of adjacent cells, allowing small molecules and ions to pass between them. (d) Plasmodesmata connect the cytosol and plasma membranes of adjacent plant cells, allowing coordination of their metabolic activities.

SCIENCE IN ACTION The Discovery of Aquaporins

The French microbiologist Louis Pasteur's observation that "Chance favors the prepared mind" is as true today as it was when he first expressed it in the 1800s. Scientists have long realized that osmosis directly through the phospholipid bilayer is much too slow to account for water movement across certain cell membranes, such as those of red blood cells (see Fig. 5-9). But attempts to identify selective transport proteins for water repeatedly failed.

Then, as often happens in science, chance and prepared minds met. In the mid-1980s, Peter Agre (Fig. E5-3), then at the Johns Hopkins School of Medicine in Maryland, was attempting to determine the structure of a glycoprotein on red blood cells. The protein he isolated was contaminated, however, with large quantities of an unknown protein.

Instead of discarding the unknown protein, he and his coworkers collaborated to determine its structure. To test the hypothesis that the protein was involved with water transport, they performed an experiment using frog egg cells, whose membranes are nearly impermeable to water. Agre's team injected frog eggs with messenger RNA that coded for the mystery protein. This caused the eggs to synthesize this protein and insert it into their plasma membranes. Agre discovered that normal frog egg cells swelled only slightly when placed in a hypotonic solution, while those with the protein swelled rapidly and burst (Fig. E5-4a). Further studies revealed that no other ions or molecules could move through this channel protein, which Agre named "aquaporin." In

2000, he and other research teams reported the three-dimensional structure of aquaporin (Fig. E5-4b). Billions of water molecules can move through an aquaporin in single file every second, while larger molecules and small positively charged ions (such as hydrogen ions) cannot move through.

Many types of aquaporins have now been identified, and aquaporins have been found in all forms of life that have been

investigated. For example, the plasma membrane of the central vacuole of plant cells is rich in aquaporins, which allows the central vacuole to fill rapidly when water is available (see Fig. 5-8). In 2003, Agre shared the Nobel Prize in Chemistry for his discovery—the result of chance, careful observation, persistence, and perhaps a bit of what Agre himself describes modestly as "sheer blind luck."

(a) Frog eggs

(b) Aquaporin

▲ FIGURE E5-4 Function and structure of aquaporins
(a) The frog egg on the right has aquaporins in its plasma membrane; the egg on the left does not. They have both been immersed in a hypotonic solution for 30 seconds. The egg on the right has burst, whereas that on the left has swollen only slightly.
(b) An aquaporin (shown here in cross-section) is formed by a membrane-spanning protein. Within the channel, charged amino acids attract water molecules, promoting their movement in either direction while repelling other substances.

▲ FIGURE E5-3 Peter Agre

two adjacent membranes had been stitched together. Interlocking tight junction proteins create barriers that prevent nearly all substances from passing between the plasma membranes of the linked cells. For example, tight junctions in the bladder prevent cellular wastes in urine from leaking back into the blood. In the digestive tract, tight junctions between the cells that line it protect the rest of the body from the acids, digestive enzymes, and bacteria found in its various compartments.

Gap Junctions and Plasmodesmata Allow Direct Communication Between Cells

The cells of many tissues in the animal body are interconnected by **gap junctions** (Fig. 5-16c), clusters of channels ranging in number from a few to thousands. The channels are formed by six-sided tubes of protein called *connexons* that span the plasma membrane. Connexons line up so that their central pores link the cytosol of adjacent cells. The small size of the pore allows small water-soluble molecules—including sugars, various ions, amino acids, and small messenger molecules such as cAMP—to pass between cells, but excludes organelles and large molecules such as proteins. Gap junctions coordinate the metabolic activities of many cells. They allow electrical signals to pass extremely rapidly among certain groups of nerve cells, and they synchronize contraction of heart muscle and of smooth muscles, such as in the walls of the digestive tract, bladder, and uterus.

Plasmodesmata are channels that link nearly all adjacent plant cells. These openings are lined with plasma membrane and filled with cytosol, so the membranes and the cytosol of adjacent cells are continuous with one another at plasmodesmata (Fig. 5-16d). Many plant cells have thousands of plasmodesmata, allowing water, nutrients, and hormones to pass quite freely from one cell to another. These connections among plant cells serve a function similar to the gap junctions of animal cells in coordinating metabolic activities.

CASE STUDY revisited

Vicious Venoms

Although the "witches' brew" of rattlesnake and spider venoms differs among species, many contain phospholipases that break down membrane phospholipids, causing cells to rupture and die. Cell death destroys the tissue around a rattlesnake or brown recluse spider bite (Fig. 5-17). When phospholipases attack the membranes of capillary cells, these tiny blood vessels rupture and release blood into the tissue surrounding the wound. In extreme cases, capillary damage can lead to internal bleeding. By attacking the membranes of red blood cells, rattlesnake venoms can cause anemia (an inadequate number of oxygen-carrying red blood cells). Rattlesnake phospholipases also attack muscle cell membranes; this attack caused extensive damage in Justin Schwartz's forearm. Justin required large quantities of antivenin, which contains specialized proteins that bind and neutralize the snake venom proteins. Unfortunately, no antivenin is available for brown recluse bites, and treatment generally consists of preventing infection, controlling pain and swelling, and waiting—sometimes for months—for the wound to heal.

Although both snake and spider bites can have serious consequences, very few of the spiders and snakes found in the Americas are dangerous to people. The best defense is to learn which of these live in your area and where they prefer to hang out. If your activities bring you to such places, wear protective clothing—and always look before you reach! Knowledge can help us coexist with spiders and snakes, avoid their bites, and keep our cell membranes intact.

Consider This

Phospholipases are found in animal digestive tracts as well as in snake venom. How does the role of phospholipase in snake venom differ from its role in the snake's digestive tract?

(a) Justin's rattlesnake bite

(b) Brown recluse spider bite

▲ FIGURE 5-17 **Phospholipases in venoms can destroy cells** Both (a) Justin Schwartz's hand 36 hours after the rattlesnake bite and (b) a brown recluse spider bite show extensive tissue destruction caused by phospholipases.

Throughout this textbook, we will return many times to the concepts of membrane structure and transport described in this chapter. Understanding the diversity of membrane lipids and proteins is the key to understanding not just the isolated cell, but also entire organs, which could not perform as they do without the specialized membrane properties of their cells.

CHECK YOUR LEARNING

Can you describe the major types of junctions between cells? Can you explain how these junctions function? Are you able to explain how gap junctions and plasmodesmata allow for direct communication among cells?

CHAPTER REVIEW

Summary of Key Concepts

5.1 How Is the Structure of the Cell Membrane Related to Its Function?

Cell membrane consists of a bilayer of phospholipids in which a variety of proteins are embedded, often described as a fluid mosaic. The plasma membrane isolates the cytoplasm from the external environment, regulates the flow of materials into and out of the cell, allows communication between cells, allows attachments within and between cells, and regulates many biochemical reactions. There are five major types of membrane proteins: (1) enzymes, which promote chemical reactions; (2) receptor proteins, which bind molecules and trigger changes within the cell; (3) recognition proteins, which label the cell; (4) connection proteins, which anchor the plasma membrane to the cytoskeleton and extracellular matrix or bind cells to one another; and (5) transport proteins, which regulate the movement of most water-soluble substances through the membrane.

5.2 How Do Substances Move Across Membranes?

Diffusion is the movement of particles from regions of higher concentration to regions of lower concentration. In simple diffusion, water, dissolved gases, and lipid-soluble molecules diffuse through the phospholipid bilayer. During facilitated diffusion, carrier proteins or channel proteins allow water and water-soluble molecules to cross the membrane down their concentration gradients without expending cellular energy. Osmosis is the diffusion of water across a selectively permeable membrane down its concentration gradient through the phospholipid bilayer or through aquaporins.

Energy-requiring transport includes active transport, in which carrier proteins use cellular energy (ATP) to drive the movement of molecules across the plasma membrane against concentration gradients. Interstitial fluid, large molecules, and food particles may be acquired by endocytosis, which includes pinocytosis, receptor-mediated endocytosis, and phagocytosis. The secretion of substances, such as hormones, and the excretion of particulate cellular wastes are accomplished by exocytosis.

Cells exchange materials between the cytoplasm and the external environment by the slow process of diffusion through the plasma membrane. This requires that no part of the cell be too far from the plasma membrane, limiting the diameter of cells.

5.3 How Do Specialized Junctions Allow Cells to Connect and Communicate?

Animal cell junctions include (1) desmosomes, which attach adjacent cells and prevent tissues from tearing apart during ordinary movements; (2) tight junctions, which leak-proof the spaces between adjacent cells; and (3) gap junctions that connect the cytosol of adjacent cells. Plasmodesmata interconnect the cytosol of adjacent plant cells.

Key Terms

active transport *84*	hypertonic *83*
aquaporin *82*	hypotonic *83*
carrier protein *82*	interstitial fluid *76*
channel protein *82*	isotonic *83*
concentration *80*	osmosis *82*
concentration gradient *80*	passive transport *81*
connection protein *79*	phagocytosis *86*
desmosome *88*	phospholipid bilayer *76*
diffusion *80*	pinocytosis *85*
endocytosis *85*	plasmodesmata *90*
energy-requiring	receptor protein *79*
transport *81*	receptor-mediated
enzyme *79*	endocytosis *85*
exocytosis *87*	recognition protein *79*
facilitated diffusion *82*	selectively permeable *81*
fluid *76*	simple diffusion *81*
fluid mosaic model *76*	solute *80*
food vacuole *86*	solvent *80*
gap junction *90*	tight junction *88*
glycoprotein *79*	transport protein *80*
gradient *80*	turgor pressure *83*

Learning Outcomes

In this chapter, you have learned to . . .

LO1 Describe the components, structure, and function of cell membranes and explain how the different components contribute to membrane functions.

LO2 Explain how substances move across cell membranes.

LO3 Explain the basic concepts of diffusion and osmosis.

LO4 Describe the major types of junctions between cells and explain how these junctions function.

Thinking Through the Concepts

Fill-in-the-Blank

1. Membranes consist of a bilayer of _____. The five major categories of protein within the bilayer are: _____, _____, _____, _____, and _____ proteins. **LO1**

2. A membrane that is permeable to some substances but not to others is described as being _____. The movement of a substance through a membrane down its concentration gradient is called _____. When applied to water, this process is called _____. Channels that are specific for water are called _____. The process that moves substances through a membrane against their concentration gradient is called _____. **LO2 LO3**

3. Facilitated diffusion involves either _____ proteins or _____ proteins. Diffusion directly through the phospholipid bilayer is called _____ diffusion, and molecules that take this route must be soluble in _____ or be very small and have no net electrical charge. **LO1 LO2 LO3**

4. The three types of cell attachment structures in animal cells are: _____, _____, and _____. The structures that interconnect plant cells are called _____. **LO4**

5. After each molecule, place the two-word term that most specifically describes the *process* by which it moves through a plasma membrane: Carbon dioxide: _____; ethyl alcohol: _____; a sodium ion:_____; glucose: _____. **LO2**

6. The general process by which fluids or particles are transported into cells is called _____. Does this process require energy? _____ The specific term for engulfing fluid is _____, and the term for engulfing particles is _____. The engulfed substances are taken into the cell in membrane-enclosed sacs called _____. **LO2**

Review Questions

1. Describe and diagram the structure of a plasma membrane. What are the two principal types of molecules in plasma membranes, and what is the general function of each? **LO1**

2. Sketch the configuration that 10 phospholipid molecules would assume if placed in water. Explain why they arrange themselves this way. **LO1**

3. What are the five categories of proteins commonly found in plasma membranes, and what is the function of each one? **LO1 LO4**

4. Define *diffusion* and *osmosis*. Explain how osmosis helps plant leaves remain firm. What is the term for water pressure inside plant cells? **LO2 LO3**

5. Define *hypotonic*, *hypertonic*, and *isotonic*. What would be the fate of an animal cell immersed in each of these three types of solution? **LO2 LO3**

6. Describe the following types of transport processes in cells: simple diffusion, facilitated diffusion, active transport, pinocytosis, receptor-mediated endocytosis, phagocytosis, and exocytosis. **LO2 LO4**

7. Name the protein that allows facilitated diffusion of water. What experiment demonstrated the function of this protein? **LO1 LO2 LO3**

8. Imagine a container of glucose solution, divided into two compartments (A and B) by a membrane that is permeable to water and glucose but not to sucrose. If some sucrose is added to compartment A, how will the contents of compartment B change? Explain. **LO3**

9. Name four types of cell-to-cell junctions, and describe the function of each. Which are found in plants, and which in animals? **LO4**

Applying the Concepts

1. Different cells have somewhat different plasma membranes. The plasma membrane of a *Paramecium*, for example, is only about 1% as permeable to water as the plasma membrane of a human red blood cell. Hypothesize about why this is the case. Is *Paramecium* likely to have aquaporins in its plasma membrane? Explain your answer.

2. Predict and sketch the configuration of phospholipids that are placed in vegetable oil. Explain your prediction.

3. The fluid portion of blood, in which red blood cells are suspended, is called plasma. Is the plasma likely to be isotonic, hypertonic, or hypotonic to the red blood cells? Explain.

4. Some cells in the nervous system wrap themselves around parts of neurons, serving as insulation for the electrical signals that flow inside neurons carried by ions. Given the general roles of protein and lipids in the cell membrane, which component would you predict to be more abundant in these insulating cells? Explain.

Answers to Figure Caption questions and Fill-in-the-Blank questions can be found in the Answers section at the back of the book.

(MB) *Go to MasteringBiology for practice quizzes, activities, eText, videos, current events, and more.*

CHAPTER 6

Energy Flow in the Life of a Cell

CASE STUDY
Energy Unleashed

PICTURE THE NEW YORK MARATHON: One year, this huge throng of people from around the globe included many propelling themselves in wheelchairs using well-developed arm muscles, a woman battling cancer, a wildlife supporter running in a rhino costume, a 91-year-old man shuffling slowly, a fireman wearing full firefighting gear to honor his fallen comrades, a one-legged man using crutches, and blind people guided by the sighted. All participated in a 26-mile journey—a personal odyssey for each, and a collective testimony to human persistence, endurance, and the ability of the human body to utilize energy.

The 20,000-plus runners in the New York Marathon collectively expend more than 50 million Calories worth of energy. Once finished, they douse their overheated bodies with water and refuel themselves on snacks. Eventually, cars, buses, and airplanes—burning vast quantities of fuel and releasing enormous amounts of heat—will transport the runners back to their homes throughout the world.

What exactly is energy? Do our bodies use it according to the same principles that govern energy use in the engines of cars and airplanes? Why do our bodies generate heat, and why do we give off more heat when exercising than when watching TV?

The bodies of these runners in the New York Marathon convert stored energy to the energy of movement and heat. Their pounding footsteps shake the Verrazano Narrows Bridge.

AT A GLANCE

6.1 What Is Energy?

6.2 How Is Energy Transformed During Chemical Reactions?

6.3 How Is Energy Transported Within Cells?

6.4 How Do Enzymes Promote Biochemical Reactions?

6.5 How Are Enzymes Regulated?

LEARNING GOALS

LG1 Define the key terms used to describe energy and explain the forms that energy takes.

LG2 Summarize the first and second laws of thermodynamics.

LG3 Define *entropy* and explain how organisms create the low-entropy conditions of life.

LG4 Describe how energy is captured and released by chemical reactions.

LG5 Describe the energy-carrier molecules in cells and explain how they facilitate chemical reactions.

LG6 Explain how catalysts work and how enzymes in cells catalyze chemical reactions.

LG7 Describe how cells control their metabolic pathways by regulating enzyme activity.

6.1 WHAT IS ENERGY?

Energy is the capacity to do work. **Work,** in turn, is the transfer of energy to an object, causing the object to move. It is obvious that marathoners are working; their chests heave, their arms pump, and their legs stride, moving their bodies relentlessly forward for 26 miles. This muscular work is powered by **chemical energy,** which is energy available in the bonds of molecules. Molecules that provide this chemical energy—including sugar, glycogen, and fat—are stored in the cells of the runners' bodies. Cells utilize specialized molecules such as ATP to accept, briefly store, and transfer energy from one chemical reaction to the next.

There are two fundamental types of energy: potential energy and kinetic energy, each of which takes several forms. **Potential energy,** which is stored energy, includes the chemical energy available in biological molecules and other molecules such as fossil fuels, elastic energy stored in a wound clock spring or a drawn bow, and gravitational energy stored in water behind a dam or a roller-coaster car about to begin its downward plunge **(Fig. 6-1)**. **Kinetic energy** is the energy of movement. It includes radiant energy (such as waves of light, X-rays, and other forms of *electromagnetic radiation*), heat or thermal energy (the motion of molecules or atoms), electrical energy (also called electricity; the flow of charged particles), and any motion of larger objects, such as the plummeting roller-coaster car or running marathoners. Under the right conditions, kinetic energy can be transformed into potential energy, and vice versa. For example, the roller-coaster car converts the kinetic energy of its downward plunge into potential gravitational energy as it coasts to the top of the next rise. At a molecular level, during photosynthesis (described in Chapter 7), the kinetic energy of light is captured and transformed into the potential energy of chemical bonds. To understand energy flow and change, we need to know more about its properties and behavior.

The Laws of Thermodynamics Describe the Basic Properties of Energy

The laws of thermodynamics describe the quantity (the total amount) and the quality (the usefulness) of energy. The **first law of thermodynamics** states that energy can

▲ **FIGURE 6-1 Converting potential energy to kinetic energy** Roller coasters convert gravitational potential energy to kinetic energy as they plummet downhill.

QUESTION Could one design a roller coaster that didn't use any motors to pull the cars uphill after the cars were released from a high point?

neither be created nor destroyed by ordinary processes. (Nuclear reactions, in which matter is converted into energy, are the exception.) This means that in a hypothetical **closed system**—where energy can neither enter nor leave—the total amount of energy before and after any process will be unchanged. For this reason, the first law of thermodynamics is often called the **law of conservation of energy.** Energy can, however, be converted from one form to another.

To illustrate this law, consider a combustion engine car. Before you turn the ignition key, the energy in the car is all potential energy, stored in the chemical bonds of its fuel. As you drive, only about 20% of this potential energy is converted into the kinetic energy of motion. According to the first law of thermodynamics, however, energy is neither created nor destroyed. So what happens to the other 80% of the energy? The burning fuel is used to set the car in motion, but it also heats up the engine, the exhaust system, and the air around the car, while friction from the tires heats the road. So, as the first law dictates, the total amount of energy remains the same, although it has changed in form—much of it to waste heat.

The **second law of thermodynamics** states that when energy is converted from one form to another, the amount of useful energy decreases. In other words, all ordinary (nonnuclear) processes cause energy to be converted from more useful into less useful forms. In our combustion example, 80% of the energy stored in gasoline was converted into waste heat, which increased the random movement of molecules in the car, the air, and the road (**Fig. 6-2**). Heat is a less usable form of energy than energy stored in chemical bonds, as discussed further in "Earth Watch: Step on the Brakes and Recharge Your Battery."

Now consider the human body. Whether running or reading, your body "burns" food to release the chemical energy stored in its molecules. Your body warmth results from the heat given off, which is radiated to your surroundings. This heat is not available to power muscle contraction or to

100 units chemical energy   **80 units heat + 20 units kinetic energy energy**

▲ FIGURE 6-2 **Energy conversions result in a loss of useful energy**

help brain cells interpret written words. Thus, the second law tells us that no energy conversion process, including those that occur in the body, is 100% efficient in using energy to achieve a specific outcome.

The second law of thermodynamics also tells us something about the organization of matter. Useful energy tends to be stored in highly ordered matter, such as in the bonds of complex molecules. As a result, whenever energy is used within a closed system, there is an overall increase in the randomness and disorder of matter—be it molecules or far larger objects. We all experience this in our homes. Without energy-demanding cleaning and organizing efforts, dirty dishes accumulate, books and papers pile up, clothes collect in confusion on the floor, and the bed remains rumpled.

At the molecular level, we see the same principle at work. The chemical equation for burning gasoline is:

$$2\ C_8H_{18} + 25\ O_2 \rightarrow 16\ CO_2 + 18\ H_2O + energy$$
$$\text{(octane)} + \text{(oxygen)} \rightarrow \text{(carbon dioxide)} + \text{(water)} + \text{(energy)}$$
$$\text{(reactants)} \qquad\qquad \text{(products)}$$

Although you'll find the same number of atoms on both sides of the equation, notice that the 27 starting molecules on the left (*reactants*) are converted to 34 simple molecules on the right (*products*). The heat energy that is released imparts rapid random motion to the products and other nearby

Earth Watch

Step on the Brakes and Recharge Your Battery

As we deplete Earth's stores of fossil fuels and damage ecosystems as we dig and drill for more, we are also becoming increasingly aware of the environmental hazards of adding carbon dioxide and other pollutants to the atmosphere as we burn these fuels.

In a typical combustion engine car, whenever you step on the brake, brake pads are forced against brake discs and the resulting friction converts the kinetic energy of forward motion into waste heat. Fortunately, engineers have devised a way to capture and use some of this squandered energy. Called regenerative braking, this technology (used in trolley cars since the 1930s) is now available in hybrid or all-electric cars, which are partly or entirely driven by battery-powered electric

motors. Stepping on the brake switches the motor into a generator mode, which slows down the car and simultaneously causes the motor to generate (rather than consume) electricity. This electrical energy, derived from the kinetic energy of the car's forward motion, is stored in the chemical energy of the car's battery. Here, it will help propel the car forward when you start up again. Of course, the second law of thermodynamics tells us that each of these energy conversions will generate some waste heat, but regenerative braking wastes 30% to 50% less energy than conventional friction braking. Regenerative braking, along with other innovations to increase efficiency, should allow cars of the very near future to travel much farther on less energy.

CASE STUDY continued

Energy Unleashed

Much like a car's engine, the marathoner's muscles are only about 25% efficient in converting chemical energy into movement; much of the other 75% is lost as heat. Highly trained marathoners completing the race in about 2 hours generate enough heat to raise their body temperatures by about 1.8°F (1°C) every 5 minutes. If they were unable to dissipate the heat, they would die of heatstroke within a half hour.

molecules. This tendency toward loss of complexity, orderliness, and useful energy—and the concurrent increase in randomness, disorder, and less useful energy—is called **entropy.** To counteract entropy, energy must be infused into the system from an outside source. When the eminent Yale scientist G. Evelyn Hutchinson stated, "Disorder spreads through the universe, and life alone battles against it," he made an eloquent reference to entropy and the second law of thermodynamics.

Living Things Use the Energy of Sunlight to Create the Low-Entropy Conditions of Life

If you think about the second law of thermodynamics, you may wonder how life can exist at all. If chemical reactions, including those inside living cells, cause the amount of unusable energy to increase, and if matter tends toward increasing randomness and disorder, how can organisms accumulate the usable energy and the complex, precisely ordered molecules that characterize life? The answer is that cells, the bodies of organisms, and Earth itself are not closed systems; they receive solar energy released by nuclear reactions occurring 93 million miles away—in the sun. These reactions not only generate light and other forms of electromagnetic energy, they also produce vast increases in entropy within the sun. These reactions liberate an almost unimaginable amount of heat; the temperature of the sun's core is estimated to be 27 million °F (16 million °C).

Living things "battle against disorder" by using a continuous influx of solar light energy to synthesize complex molecules and maintain their intricate bodies. Thus, the highly ordered (and therefore low-entropy) systems that characterize life do not violate the second law of thermodynamics because they are achieved and maintained through a continuous influx of useful energy from the sun. Because of ever-increasing entropy, the solar reactions that power life on Earth will eventually cause the sun to burn out, but fortunately, not for billions of years.

CHECK YOUR LEARNING

Can you define the key terms relating to energy and thermodynamics and provide examples that illustrate these terms? Can you summarize the first and second laws of thermodynamics?

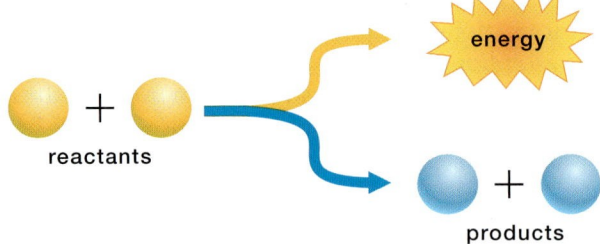

▲ **FIGURE 6-3 An exergonic reaction** The products contain less energy than the reactants.

6.2 HOW IS ENERGY TRANSFORMED DURING CHEMICAL REACTIONS?

A **chemical reaction** is a process that forms or breaks the chemical bonds that hold atoms together. Chemical reactions convert one set of chemical substances, the **reactants,** into another set, the **products.** All chemical reactions either release energy or require an overall (net) input of energy. A reaction is **exergonic** (Greek for "energy out") if it releases energy; that is, if the starting reactants contain more energy than the end products. All exergonic reactions release some of their energy as heat (**Fig. 6-3**). A reaction is **endergonic** ("energy in") if it requires a net input of energy; that is, if the products contain more energy than the reactants. Endergonic reactions require a net influx of energy from an outside source (**Fig. 6-4**).

Exergonic Reactions Release Energy

Sugar can be ignited by heat, as any cook can tell you. As it burns, sugar (glucose, for example) undergoes the same overall reaction as it does in the bodies of humans and nearly all other forms of life; sugar is combined with oxygen to produce carbon dioxide and water, generating both chemical energy and heat. The total energy in the reactant molecules (sugar and oxygen) is much higher than in the product molecules (carbon dioxide and water), so burning sugar is an exergonic reaction (**Fig. 6-5**). It may be helpful to think of exergonic reactions as running "downhill," from higher energy to lower energy.

Endergonic Reactions Require a Net Input of Energy

In contrast to what happens when sugar is burned, many reactions in living things produce products that contain more energy than do the reactants. Sugar, for example, contains far

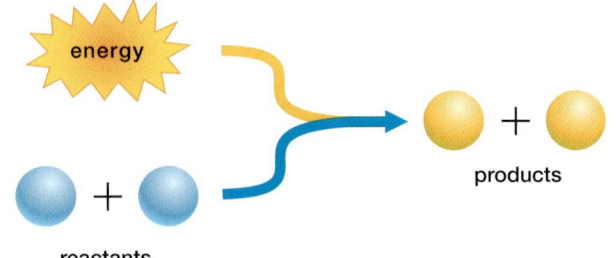

▲ **FIGURE 6-4 An endergonic reaction** The products contain more energy than the reactants.

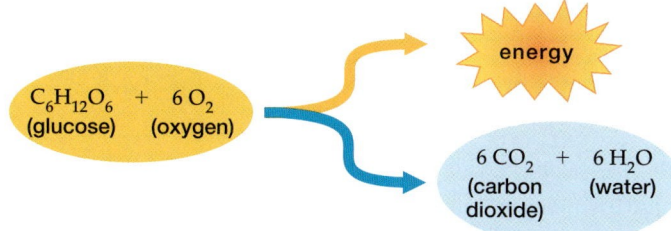

▲ FIGURE 6-5 Reactants and products of burning glucose

QUESTION Is this an endergonic or an exergonic reaction?

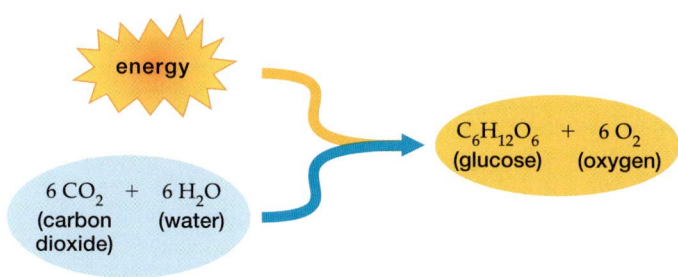

▲ FIGURE 6-6 Photosynthesis

QUESTION Is photosynthesis an endergonic or an exergonic reaction?

more energy than does the carbon dioxide and water from which it was formed. The proteins in a muscle cell contain more energy than the individual amino acids that were joined together to synthesize them. In other words, synthesizing complex biological molecules requires a net input of energy. These reactions are endergonic. Endergonic reactions will not happen spontaneously. We might call them "uphill" reactions, because the reactants contain less energy than the products do. Like pushing a rock up to the top of a hill, going from low-energy reactants to high-energy products requires a net input of energy. Where do living organisms get the energy to fuel endergonic reactions? Photosynthesis uses light energy from the sun to produce high-energy sugar from low-energy water and carbon dioxide (Fig. 6-6). Photosyn-

thetic organisms use the energy captured in sugars to synthesize proteins and all of the other complex molecules of their bodies. Virtually all other living organisms, ourselves included, use the energy captured by photosynthesis and stored in sugars and other biological molecules to power endergonic reactions, including such activities as active transport, synthesizing complex molecules, and movement.

All Chemical Reactions Require Activation Energy to Begin

All chemical reactions, even those that can continue spontaneously, require energy to get started. Think of a rock sitting at the top of a hill. It will remain there indefinitely unless a push starts it rolling down. In chemical reactions, the energy "push" is called the **activation energy** (Fig. 6-7a). Shells of negatively charged electrons surround all atoms (see Chapter 2); activation energy is the energy that is required to overcome the repulsive electrical forces between the electron shells so that they can move close enough together to react.

Activation energy can be provided by the kinetic energy of moving molecules. At any temperature higher than absolute zero (–460°F or –273°C), atoms and molecules are in constant motion. Reactive molecules moving with sufficient speed collide hard enough to force their electron

CASE STUDY c o n t i n u e d

Energy Unleashed

Marathoners rely on glycogen stored in their muscles and liver for much of the energy to power their run. Glycogen consists of chains of glucose molecules. When energy is needed, glucose molecules are cleaved from the chain and then broken down into carbon dioxide and water. This exergonic reaction generates heat and the ATP required for muscle contraction. The carbon dioxide is exhaled as the runners breathe rapidly to supply their muscles with adequate oxygen. The water generated by glucose breakdown (and a lot more that the runners drink during the race) is lost as sweat. As the sweat evaporates, it prevents the runners' bodies from overheating by carrying away the excess heat energy released by the exergonic chemical reactions in their muscles.

▶ FIGURE 6-7 Activation energy in an exergonic reaction (a) An exergonic ("downhill") reaction proceeds from high-energy reactants to lower-energy products, with a net release of energy. **(b)** Sparks from a friction gas lighter ignite gas flowing from a Bunsen burner. The sparks provide the activation energy for this exergonic reaction. Heat released by the burning gas then allows the reaction to continue spontaneously.

(a) An exergonic reaction

(b) Sparks ignite gas

shells to mingle and react. Because molecules move faster as the temperature increases, most chemical reactions occur more readily at high temperatures. For example, sparks will ignite the gas from a stove or Bunsen burner (Fig. 6-7b) and heat from a burning match can set sugar on fire. When sugar is ignited, the initial reaction combining sugar and oxygen releases enough heat to sustain the reaction, and the sugar continues to burn.

CHECK YOUR LEARNING

Can you describe how energy is captured and released by chemical reactions? Can you explain exergonic and endergonic reactions and provide examples of each? Are you able to explain activation energy?

6.3 HOW IS ENERGY TRANSPORTED WITHIN CELLS?

Most organisms are powered by the chemical energy supplied by the exergonic breakdown of glucose illustrated earlier. This energy is used to do cellular work, such as constructing complex biological molecules and causing muscles to contract. But glucose cannot be used directly to fuel these endergonic processes. Instead, the energy released by glucose breakdown is first transferred to energy-carrier molecules. **Energy-carrier molecules** are high-energy molecules that are synthesized at the site of an exergonic reaction, where they capture some of the released energy. Energy carriers behave much like rechargeable batteries: They are charged up by an exergonic reaction, and then release the energy to drive an endergonic reaction. They can then be recharged, as described later. Energy-carrier molecules only capture and transfer energy within cells; they cannot ferry energy from cell to cell, nor are they used for long-term energy storage.

ATP and Electron Carriers Transport Energy Within Cells

Many exergonic reactions in cells, such as breaking down sugars and fats, produce **adenosine triphosphate (ATP),** the most common energy-carrier molecule in the body (described in Chapter 3). ATP is a nucleotide composed of the nitrogen-containing base adenine, the sugar ribose, and three phosphate groups (see Fig. 3-23). Because ATP provides energy to drive a wide variety of endergonic reactions, it is sometimes called the "energy currency" of cells. It is produced when energy released in cells during glucose breakdown or other exergonic reactions is used to combine the lower-energy molecules **adenosine diphosphate (ADP)** and inorganic phosphate (HPO_4^{2-}, also designated P_i) (Fig. 6-8a). Because it requires an input of energy, ATP synthesis is endergonic.

ATP diffuses throughout the cell, carrying energy to sites where endergonic reactions occur. There its energy is liberated as it is broken down, regenerating ADP and P_i (Fig. 6-8b). The life span of an ATP molecule in a living cell is very short; each molecule is recycled roughly 1,400 times every day. A marathon runner may use a pound of ATP molecules every minute,

(a) ATP synthesis: Energy is stored in ATP

(b) ATP breakdown: Energy is released   ADP   phosphate

▲ FIGURE 6-8 **The interconversion of ADP and ATP**
(a) Energy is captured when a phosphate group (P_i) is added to adenosine diphosphate (ADP) to make adenosine triphosphate (ATP). (b) Energy to accomplish cellular work is released when ATP is broken down into ADP and P_i.

so if this molecule were not rapidly recycled, marathon runs would not happen. As you can see, ATP is *not* a long-term energy-storage molecule. In contrast, more stable molecules such as starch in plants and glycogen and fat in animals can store energy for hours, days, or—in the case of fat—years.

ATP is not the only energy-carrier molecule within cells. In some exergonic reactions, including both glucose breakdown and the light-capturing stage of photosynthesis, some energy is transferred to electrons. These energetic electrons, along with hydrogen ions (H^+; present in the cytosol of cells), are captured by molecules called **electron carriers.** The loaded electron carriers donate their high-energy electrons to other molecules, which are often involved in pathways that generate ATP. Common electron carriers include the nucleotide nicotinamide adenine dinucleotide (NADH) and its relative, flavin adenine dinucleotide ($FADH_2$). (You will learn more about electron carriers in Chapters 7 and 8.)

Coupled Reactions Link Exergonic with Endergonic Reactions

In a **coupled reaction,** an exergonic reaction provides the energy needed to drive an endergonic reaction (Fig. 6-9) using ATP or electron carriers as intermediaries. During photosynthesis, for example, plants use sunlight (from exergonic reactions in the sun's core) to drive the endergonic synthesis of high-energy glucose molecules from lower-energy reactants (carbon dioxide and water). Nearly all organisms use the energy released by exergonic reactions (such as the breakdown of glucose into carbon dioxide and water) to drive endergonic reactions (such as the synthesis of proteins from amino acids). Because energy is lost as heat every time it is transformed, in coupled reactions the energy released by exergonic reactions must always exceed the energy needed to drive the endergonic reactions.

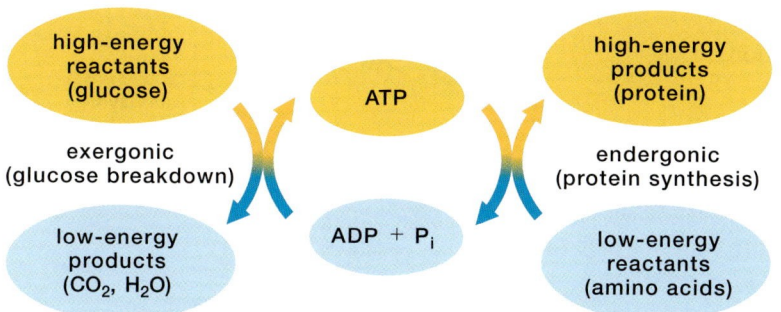

◄FIGURE 6-9 **Coupled reactions within living cells** Exergonic reactions (such as glucose breakdown) drive the endergonic reaction that synthesizes ATP from ADP and P_i. The ATP molecule carries its chemical energy to a part of the cell where energy is needed to drive an endergonic reaction (such as protein synthesis). The ADP and P_i are rejoined in an endergonic reaction that forms ATP.

The exergonic and endergonic portions of coupled reactions often occur in different places within a cell, so there must be some way to transfer the energy from the exergonic reactions that release energy to the endergonic reactions that require it. In coupled reactions, energy is transferred from place to place by energy-carrier molecules such as ATP. In its role as an intermediary in coupled reactions, ATP is constantly being synthesized to capture the energy released during exergonic reactions and then broken down to power endergonic reactions (as shown in Fig. 6-9).

CHECK YOUR LEARNING

Can you name and describe two important energy-carrier molecules in cells? Can you explain coupled reactions?

6.4 HOW DO ENZYMES PROMOTE BIOCHEMICAL REACTIONS?

Ignite sugar and it will go up in flames as it combines rapidly with oxygen, releasing carbon dioxide and water. The same overall reaction occurs in our cells, but of course they are in no danger of catching fire. Chemical reactions that released uncontrolled blasts of heat energy would not be useful to cells. On the other hand, a single molecule of sugar contains enough energy to produce dozens of molecules of ATP if the energy is released in controlled steps that are linked to ATP synthesis.

Catalysts Reduce the Energy Required to Start a Reaction

In general, the speed at which a reaction occurs is determined by its activation energy; that is, by how much energy is required to start the reaction (see Fig. 6-7a). Some reactions, such as table salt dissolving in water (see Fig. 2-8), have low activation energies and occur readily at human body temperature (approximately 98.6°F, or 37°C). In contrast, you could store sugar at body temperature for decades and it would remain virtually unchanged.

The reaction of sugar with oxygen to yield carbon dioxide and water is exergonic, but it has a high activation energy. The heat of a flame increases the rate of movement of sugar molecules and nearby oxygen molecules, causing them to collide with one another with sufficient force to overcome their activation energy and react; the sugar burns. Promoting reactions by overcoming activation energy without added heat is the job of catalysts.

Catalysts are molecules that speed up the rate of a reaction without themselves being used up or permanently altered. A catalyst speeds up a reaction by reducing the reaction's activation energy (**Fig. 6-10**). For example, consider the catalytic converters on automobile exhaust systems. Incomplete combustion of gasoline generates poisonous carbon monoxide (CO). Carbon monoxide then reacts spontaneously but slowly with oxygen in the air to form carbon dioxide:

$$2\,CO + O_2 \rightarrow 2\,CO_2 + \text{heat energy}$$

In heavy traffic, the spontaneous reaction of CO with O_2 can't keep pace with the CO emitted, and unhealthy levels of CO accumulate. Enter the catalytic converter. Catalysts such as platinum in the converter provide a surface upon which O_2 and CO combine more readily, hastening the conversion of CO to CO_2 and reducing air pollution.

All catalysts share three important properties:

- Catalysts speed up reactions by lowering the activation energy required for the reaction to begin.

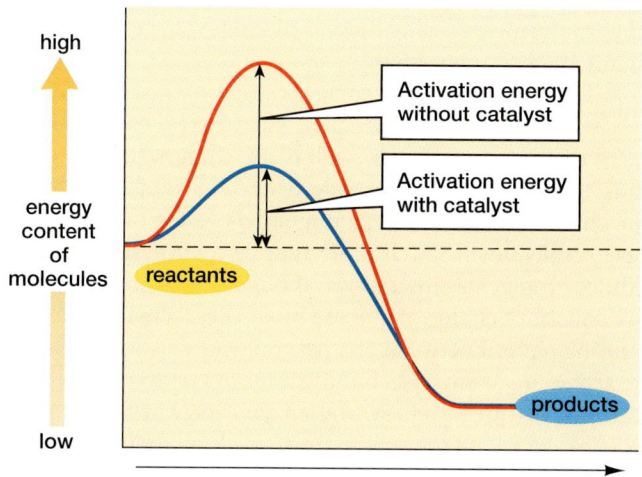

▲ FIGURE 6-10 **Catalysts such as enzymes lower activation energy** A high activation energy (red curve) means that reactant molecules must collide very forcefully to react. Catalysts lower the activation energy of a reaction (blue curve), so a much higher proportion of molecules moves fast enough to react when the molecules collide. Therefore, the reaction proceeds much more rapidly.

QUESTION Can an enzyme catalyst make an endergonic reaction occur spontaneously at body temperature?

- Catalysts can speed up both exergonic and endergonic reactions, but they cannot make an endergonic reaction occur spontaneously.
- Catalysts are not consumed or permanently changed by the reactions they promote.

Enzymes Are Biological Catalysts

Inorganic catalysts speed up a number of different chemical reactions. But indiscriminately speeding up dozens of reactions would not be useful to living organisms, and indeed would almost certainly be deadly. Instead, cells employ highly specific biological catalysts called **enzymes,** nearly all of which are proteins. A given enzyme catalyzes only a few types of chemical reactions, at most. The majority of enzymes catalyze a single reaction involving specific molecules, while leaving even very similar molecules unchanged.

Both exergonic and endergonic reactions are catalyzed by enzymes. The synthesis of ATP from ADP and P_i, for example, is catalyzed by the enzyme ATP synthase. When energy is required to drive endergonic reactions, ATP is broken down by ATPase. Enzyme names are not consistent; some add the suffix "ase" to what the enzyme does (ATP synthase), some add "ase" to the molecule upon which the enzyme acts (ATPase), and others don't use "ase" at all.

The Structure of Enzymes Allows Them to Catalyze Specific Reactions

The function of an enzyme, like the function of any protein, is determined by its structure (see Chapter 3; Fig. 3-20). Each enzyme's distinctive shape is determined by its amino acid sequence and the precise way in which the amino acid chain is twisted and folded. Enzymes use their three-dimensional structures to orient, distort, and reconfigure other molecules, causing these molecules to react, but emerging unchanged themselves.

Each enzyme has a pocket, called the **active site,** into which reactant molecules, called **substrates,** can enter. The shape of the active site, as well as the charges on the amino acids that form the active site, determine which molecules can enter. Consider the enzyme amylase, for example. Amylase breaks down starch molecules by hydrolysis, but leaves cellulose molecules intact, even though both starch and cellulose consist of chains of glucose molecules. Why? Because the bonding pattern between glucose molecules in starch fits into the active site of amylase, but the bonding pattern in cellulose does not. In the stomach, the enzyme pepsin breaks down proteins, attacking them at many sites along their amino acid chains. Certain other protein-digesting enzymes (trypsin, for example) will break bonds only between specific amino acids. Therefore, digestive systems must manufacture several different enzymes that work together to completely break down dietary protein into its individual amino acids.

How does an enzyme catalyze a reaction? You can follow the events in **Figure 6-11,** where two substrate molecules are combined into a single product. Both the shape and the charge of the active site allow substrates to enter the enzyme only in specific orientations. When appropriate

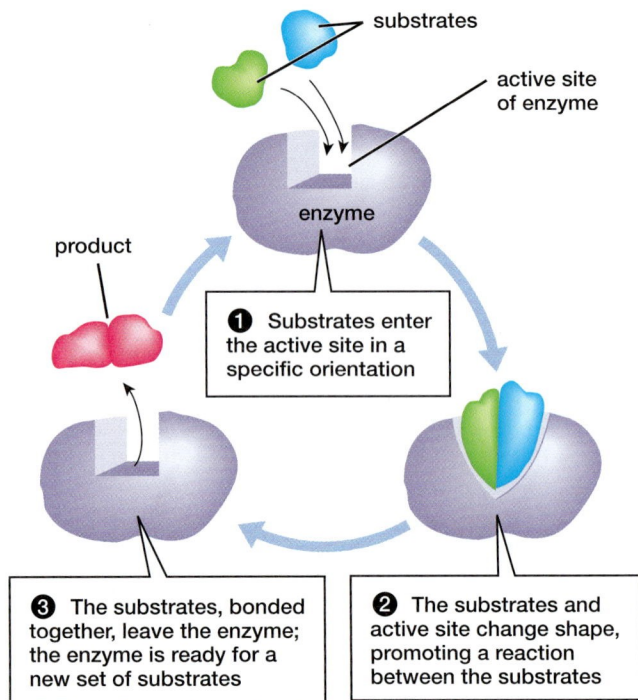

▲ FIGURE 6-11 The cycle of enzyme–substrate interactions Although this diagram shows two reactant substrate molecules combined to form a single product, enzymes can also catalyze the breakdown of a single substrate into two product molecules.

substrates enter the active site, both the substrates and active site change shape. Electrical interactions between the amino acids in the active site and the substrates may distort chemical bonds within the substrates. The combination of substrate selectivity, substrate orientation, temporary chemical bonds, and the distortion of existing bonds promotes the specific chemical reaction catalyzed by a particular enzyme. This holds true whether the enzyme is causing two molecules to react with one another, or breaking a single molecule into smaller products. When the reaction is complete, the product(s) no longer fits properly into the active site and diffuses away. The enzyme reverts to its original configuration, and it is then ready to accept another set of substrates. When substrate molecules are abundant, some fast-acting enzymes can catalyze tens of thousands of reactions per second, while others act far more slowly.

Enzymes, Like All Catalysts, Lower Activation Energy

The breakdown or synthesis of a molecule within a cell usually occurs in many small, discrete steps, each catalyzed by a different enzyme (see Fig. 6-12). Each of these enzymes lowers the activation energy for its particular reaction, allowing the reaction to occur readily at body temperature. Think of a rock climber ascending a steep cliff by finding a series of hand- and footholds that allow her to scale it one small step at a time. In a similar manner, a series of reaction steps, each requiring a small amount of activation energy and each catalyzed by an enzyme that lowers activation energy, allows the

? HAVE **YOU** EVER WONDERED …

What Makes a Firefly Glow?

The almost magical glow of fireflies comes from specialized cells in their abdomens that are bioluminescent, meaning that they produce "biological light." These cells are rich in ATP and the fluorescent chemical luciferin (Latin, meaning "light-bringer"). Luciferin and ATP serve as substrates for the enzyme luciferase. In the presence of oxygen, luciferase catalyzes a reaction that modifies luciferin and uses the energy from ATP to boost electrons briefly into a higher-energy electron shell. As they fall back into their original shell, the electrons emit their excess energy as light. The process produces so little heat that these beetles are often described as emitting "cold light."

overall reaction to surmount its high activation energy "cliff" and to proceed at body temperature.

CHECK YOUR LEARNING

Can you list the properties of catalysts and explain how they reduce activation energy? Can you explain how enzymes function as biological catalysts?

6.5 HOW ARE ENZYMES REGULATED?

The **metabolism** of a cell is the sum of all its chemical reactions. Many of these reactions, such as those that break down glucose into carbon dioxide and water, are linked in sequences called **metabolic pathways** (Fig. 6-12). In a metabolic pathway, a starting reactant molecule is converted, with the help of an enzyme, into a slightly different intermediate molecule, which is modified by yet another enzyme to form a second intermediate, and so on, until an end product is produced. Photosynthesis (Chapter 7), for example, is a metabolic pathway that results in the synthesis of high-energy molecules, including glucose. Another metabolic pathway, glycolysis, begins the breakdown of glucose (Chapter 8). Different metabolic pathways often use some of the same molecules; as a result, all the thousands of metabolic pathways within a cell are directly or indirectly interconnected.

Cells Regulate Metabolic Pathways by Controlling Enzyme Synthesis and Activity

In a test tube under constant, ideal conditions, the rate of a particular reaction will depend on how many substrate molecules diffuse into the active sites of enzyme molecules in a given time period. This, in turn, will be determined by the concentrations of the enzyme and substrate molecules. Generally, increasing the concentrations of either the enzyme or the substrate (or both) will increase the reaction rate, because it will boost the chances that the two types of molecules will meet. But living cells must precisely control the rate of reactions in their metabolic pathways, creating an environment far more complex than occurs in a test tube. Cells must keep their end products within narrow limits, even when the amounts of reactants (the substrates for enzymes) fluctuate considerably. For example, when glucose molecules flood into the bloodstream after a meal, it would not be desirable to metabolize them all at once, producing far more ATP than the cell needs. Instead, some of the glucose molecules should be stored as glycogen or fat for later use. To be effective, then, metabolic reactions within cells must be precisely regulated; they must occur at the proper times and proceed at the proper rates. Cells regulate their metabolic pathways by controlling the type, quantity, and activity levels of the enzymes they produce.

Genes That Code for Enzymes May Be Turned On or Off

A very effective way for cells to regulate enzymes is to control the quantity of enzymes they synthesize by turning the genes that code for specific enzymes on or off depending on the cell's

Initial reactant   **Intermediates**   **End products**

PATHWAY 1   A → B → C → D → E
          enzyme 1   enzyme 2   enzyme 3   enzyme 4

PATHWAY 2   F → G
                  enzyme 5   enzyme 6

▲ **FIGURE 6-12 Simplified metabolic pathways** The initial reactant molecule (A) undergoes a series of reactions, each catalyzed by a specific enzyme. The product of each reaction serves as the reactant for the next reaction in the pathway. Metabolic pathways are commonly interconnected so the product of a step in one pathway may serve as a substrate for the next enzyme in the pathway or for an enzyme in a different pathway.

Health|Watch

A Lost Enzyme Leads to Lactose Intolerance

If you enjoy ice cream and pizza, it might be hard for you to imagine life without these treats. It might even surprise you that such dairy products cannot be enjoyed by most of the world's population. Although all young children normally produce lactase (the enzyme that breaks down lactose, or "milk sugar"), about 65% of people worldwide, including 30 to 50 million people in the United States, produce less of this enzyme as they progress through childhood, a condition called *lactose intolerance*. In the worst cases of lactose intolerance, people may experience abdominal pain, flatulence, nausea, and diarrhea after consuming milk products (**Fig. E6-1**).

Why do people stop synthesizing the enzyme for this nutritious food? From an evolutionary perspective, it makes sense not to continue expending energy to produce a protein that has no function. In our early ancestors (who had not yet domesticated livestock)

▲ **FIGURE E6-1   Risky behavior?** For the majority of the world's adults, drinking milk invites unpleasant consequences.

lactase lost its function in very early childhood because, after weaning, these people no longer had access to milk—the main source of lactose. As a result, many modern adults cannot digest lactose because the gene that encodes lactase is turned off after weaning. Lactose intolerance is particularly prevalent in people of East Asian, West African, and Native American descent. Genetic studies have revealed that between 10,000 and 6,000 years ago, some people in northern Europe and the Middle East experienced mutations that allowed them to digest lactose throughout their lives. These mutations gradually spread because they provided better nutrition for members of agricultural societies, who could then obtain milk as well as meat from their livestock. Their descendants today continue to enjoy milk, ice cream, and extra-cheese pizzas.

changing needs. (You'll learn more about gene regulation in Chapter 12.) Gene regulation may cause enzymes to be synthesized in larger quantities when more of their substrate is available. Larger concentrations of enzyme make it more likely that substrate molecules will encounter the enzyme, speeding up the rate at which the reaction occurs. Consider marathon runners, who often eat high-carbohydrate meals before competing. The glucose that enters the bloodstream after a starchy meal triggers an elaborate series of metabolic adjustments. One of these is to cause cells of the pancreas to release insulin. This increase in insulin turns on the genes that code for the first enzyme in the metabolic pathway that breaks down glucose. Turning on these genes causes the cell to synthesize larger amounts of the enzyme (hexokinase), making more of it available. This enzyme adds a phosphate group to glucose, producing glucose-6-phosphate (G6P), which can be further broken down and its energy captured in ATP that will help fuel muscle cells.

Some enzymes are regulated by being synthesized only during specific stages in an organism's life. A mutation can alter this regulation, as described in "Health Watch: A Lost Enzyme Leads to Lactose Intolerance."

Some Enzymes Are Synthesized in Inactive Forms

Some enzymes are synthesized in an inactive form that is activated under the conditions found where the enzyme is needed. Examples are the protein-digesting enzymes pepsin and trypsin, mentioned earlier. Cells synthesize and release each of these enzymes in an inactive form, preventing the

enzyme from digesting and killing the cell that manufactures it. In the stomach, acid causes the inactive form of pepsin to break apart, exposing the active site of pepsin and allowing it to begin breaking down proteins from a meal. An inactive form of trypsin, which helps complete protein digestion, is released into the small intestine. The small intestine secretes an enzyme that removes the inactivating part of this protein, making functional trypsin.

Enzyme Activity May Be Controlled by Competitive or Noncompetitive Inhibition

It is not in the cell's best interest to have its enzymes churning out products all the time. Many enzymes need to be inhibited so that the cell does not use up all of its substrate or become overwhelmed by too much product. There are two general ways in which enzymes can be inhibited.

We know that for an enzyme to catalyze a reaction, its substrate must bind to the enzyme's active site (**Fig. 6-13a**). In **competitive inhibition,** a substance that is not the enzyme's normal substrate can also bind to the active site of the enzyme, competing with the substrate for the active site (**Fig. 6-13b**). Usually, a competitive inhibitor molecule has some structural similarities to the usual substrate, allowing it to occupy the active site. For example, during the metabolic pathway that produces ATP from glucose (see Chapter 8), the enzyme that acts on one intermediate molecule (oxaloacetate) to produce a second intermediate (citrate) is competitively inhibited by citrate. This is an example of a product inhibiting the enzyme that produced it. Both the product and the substrate diffuse in and out of the active site, and either can displace the other if its concentration is high enough. This example of competitive inhibition helps control the rate of glucose breakdown, because a high concentration of citrate molecules will reduce the rate of citrate production by inhibiting the enzyme that produces citrate. The best known examples of competitive inhibitors are drugs, described later.

In **noncompetitive inhibition,** a molecule binds to a site on the enzyme that is distinct from the active site. As a result, the enzyme's active site is distorted, making the enzyme less able to catalyze the reaction (**Fig. 6-13c**). Some

(a) A substrate binding to an enzyme

(b) Competitive inhibition

(c) Noncompetitive inhibition

▲ **FIGURE 6-13 Competitive and noncompetitive enzyme inhibition (a)** The normal substrate fits into the enzyme's active site when the enzyme is not being inhibited. **(b)** In competitive inhibition, a competitive inhibitor molecule that resembles the substrate enters and blocks the active site. **(c)** In noncompetitive inhibition, a molecule binds to a different site on the enzyme, distorting the active site so that the enzyme's substrate no longer fits.

noncompetitive inhibitor molecules are poisons, as described later, but many others are molecules produced by cells to regulate metabolic pathways using allosteric inhibition, as described in the following section.

Some Enzymes Are Controlled by Allosteric Regulation

An important mechanism by which cells regulate their metabolic pathways is by activating or inhibiting the activity of enzymes within the pathway. Many enzymes that participate in metabolic pathways are *allosteric enzymes*. Allosteric enzymes switch easily and spontaneously between two different configurations (allosteric literally means "other shape"). One configuration is active, while the other is inactive. Such enzymes are controlled by **allosteric regulation,** which can either activate or inhibit the enzyme. Allosteric inhibition is a form of noncompetitive inhibition in which the inhibitor molecule binds to a noncompetitive inhibition site and stabilizes the enzyme in its inactive form. Allosteric activation occurs when an activator molecule binds to a different site and stabilizes the enzyme in its active form. Activators and inhibitors bind briefly and reversibly to their allosteric regulatory sites. As a result of this temporary binding, the number of enzyme molecules being activated (or inhibited) is proportional to the numbers of activator (or inhibitor) molecules that are present at any given time.

An example of an allosteric activator molecule is ADP, which is produced by the breakdown of ATP and is then used to synthesize more ATP (see Fig. 6-8). ADP is abundant in a cell when a lot of ATP has been used up and more is needed. Under these conditions, ADP activates allosteric enzymes that function in metabolic pathways that produce ATP, stabilizing the enzymes in their active state and increasing production of ATP.

An important form of allosteric regulation is feedback inhibition. **Feedback inhibition (Fig. 6-14)** causes a metabolic pathway to stop producing its end product when the product concentration reaches an optimal level, much as a thermostat turns off a heater when a room becomes warm enough. In feedback inhibition, the activity of an enzyme near the beginning of a metabolic pathway is inhibited by the end product of that pathway; the end product acts as an allosteric inhibitor molecule.

In the metabolic pathway illustrated in Figure 6-14, a series of reactions, each catalyzed by a different enzyme, converts one amino acid to another. As the level of the end product amino acid isoleucine increases, isoleucine more frequently encounters and binds to the allosteric regulatory site on an enzyme early in the pathway, inhibiting it. Thus, when enough of the end product (isoleucine) is present, the pathway is slowed or halted. Another allosteric inhibitor is ATP, which inhibits enzymes in metabolic pathways that lead to ATP synthesis. When a cell has sufficient ATP, there is enough ATP present to block further ATP production. As ATP is used up, the pathways that produce it become active once again.

▲ FIGURE 6-14 Allosteric regulation of an enzyme by feedback inhibition
This metabolic pathway converts the amino acid threonine into the amino acid isoleucine via a series of intermediate molecules (shown here as colored shapes), each acted upon by a different enzyme (arrows). If a cell lacks isoleucine, the reactions proceed. As isoleucine builds up, it inhibits enzyme 1, blocking the pathway. When concentrations of isoleucine drop, with less inhibitor to block enzyme 1, isoleucine production resumes.

QUESTION If molecule B reversibly bound to the active site of enzyme 1, what kind of enzyme regulation would this be?

Poisons, Drugs, and Environmental Conditions Influence Enzyme Activity

Poisons and drugs that act on enzymes usually inhibit them, either competitively or noncompetitively. In addition, environmental conditions can denature enzymes, distorting the three-dimensional structure that is crucial for their function.

Some Poisons and Drugs Are Competitive or Noncompetitive Inhibitors of Enzymes

Competitive inhibitors of enzymes, including nerve gases (such as sarin) and certain insecticides (such as malathion), permanently block the active site of the enzyme acetylcholinesterase, which breaks down acetylcholine (a substance that nerve cells release to activate muscles). This allows acetylcholine to build up and overstimulate muscles, causing paralysis. Death may ensue because victims become unable to breathe. Other poisons are noncompetitive inhibitors of enzymes; these include the heavy metals arsenic, mercury, and lead. The poison potassium cyanide causes rapid death by noncompetitively inhibiting an enzyme that is crucial for the production of ATP.

Many drugs work because they act as competitive inhibitors of enzymes. For example, the antibiotic penicillin destroys bacteria by competitively inhibiting an enzyme that is crucial to synthesizing bacterial cell walls. Both aspirin and ibuprofen (Advil™) act as competitive inhibitors of an enzyme that catalyzes the synthesis of molecules that contribute to swelling, pain, and fever. Some anticancer drugs are also competitive inhibitors of enzymes. Rapidly dividing cancer cells are constantly synthesizing new strands of DNA. Some anticancer drugs resemble the subunits of DNA. These drugs compete with the normal subunits, tricking enzymes into building defective DNA, which in turn prevents the cancer cells from proliferating. Unfortunately, these anticancer drugs also interfere with the growth of other rapidly dividing cells, including those in hair follicles and lining the digestive tract. This explains why hair loss and nausea are side effects of some types of cancer chemotherapy drugs.

The Activity of Enzymes Is Influenced by Their Environment

The complex three-dimensional structures of enzymes are sensitive to environmental conditions. Recall that hydrogen bonds between polar amino acids are important in determining the three-dimensional structure of proteins (see Chapter 3). These bonds only occur within a narrow range of chemical and physical conditions, including the proper pH, temperature, and salt concentration. Thus, most enzymes have a very narrow range of conditions in which they function optimally. When conditions fall outside this range, the enzyme becomes **denatured,** meaning that it loses the exact three-dimensional structure required for it to function properly.

In humans, cellular enzymes generally work best at a pH around 7.4, the level maintained in and around our cells (**Fig. 6-15a**). For these enzymes, an acid pH alters the charges on amino acids by adding hydrogen ions to them, which in turn will change the enzyme's shape and compromise its ability to function. Enzymes that operate in the human digestive tract, however, may function outside of the pH range maintained within cells. The protein-digesting enzyme pepsin, for example, requires the acidic conditions of the stomach (pH around 2). In contrast, the protein-digesting enzyme trypsin, found in the small intestine where alkaline conditions prevail, works best at a pH close to 8 (see Fig. 6-15a).

In addition to pH, temperature affects the rate of enzyme-catalyzed reactions, which are slowed by lower temperatures and accelerated by moderately higher temperatures. This is because the rate of movement of molecules affects how likely they are to encounter the active site of an enzyme (**Fig. 6-15b**). Cooling the body can drastically

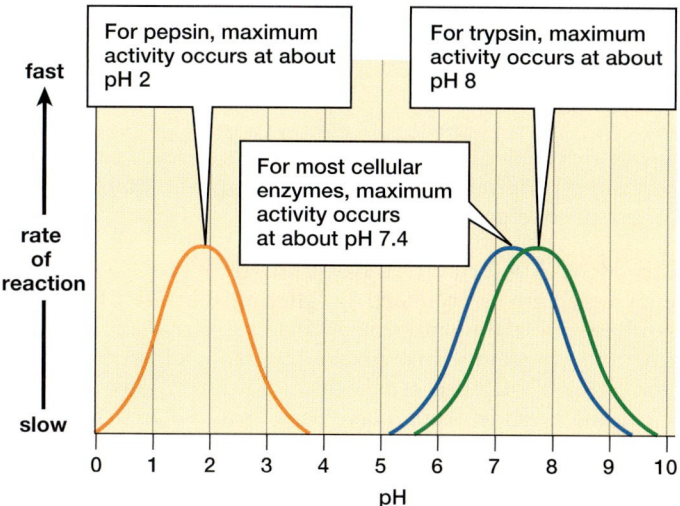

(a) Effect of pH on enzyme activity

(b) Effect of temperature on enzyme activity

▲ FIGURE 6-15 Human enzymes function best within narrow ranges of pH and temperature (a) The digestive enzyme pepsin, released into the stomach, works best at an acidic pH. Trypsin, released into the small intestine, works best at a basic pH. Most enzymes within cells work best at the pH found in the interstitial fluid and cytosol (about 7.4). (b) The maximum activity of most human enzymes occurs at human body temperature.

slow human metabolic reactions. Consider the example of a young boy who fell through the ice on a lake and was rescued and survived with no permanent disability after 20 minutes under water. At normal body temperature, the brain dies after about 4 minutes without oxygen, but the boy's body temperature and metabolic rate were lowered by the icy water, drastically reducing his need for oxygen.

In contrast, when temperatures rise too high, the hydrogen bonds that regulate protein shape may be broken apart by the excessive molecular motion, denaturing the protein. Think of the protein in egg white and how its appearance and texture are completely altered by cooking. Far lower temperatures than those required to fry an egg can still be too hot to allow enzymes to function properly. Excessive heat may be fatal; every summer, many children in the United States die from heatstroke when left unattended in overheated cars.

Food remains fresh longer in the refrigerator or freezer because cooling slows the enzyme-catalyzed reactions that allow bacteria and fungi (which can spoil food) to grow and reproduce. Before the advent of refrigeration, meat was commonly preserved by using concentrated salt solutions (think of bacon or salt pork), which kill most bacteria. Salts dissociate into ions, which form bonds with amino acids in enzyme proteins. Too much (or too little) salt interferes with the three-dimensional structure of enzymes, destroying their activity. Dill pickles are very well preserved in a vinegar-salt solution, which combines both highly salty and acidic conditions. The enzymes of organisms that live in salty environments, as you might predict, have configurations that depend on a relatively high concentration of salt ions.

CHECK YOUR LEARNING

Can you describe how cells regulate the rate at which metabolic reactions proceed? Can you describe how poisons, drugs, and environmental conditions influence enzyme activity, and provide examples?

CASE STUDY revisited

Energy Unleashed

During the course of a 26-mile race, a marathoner burns a great deal of glucose to provide enough ATP to power muscles through roughly 34,000 running steps. People store glucose molecules linked together in long branched chains of glycogen, primarily in the muscles and liver. Adults typically store about 3.5 ounces (100 grams) of glycogen in the liver, and another 9.9 ounces (280 grams) in muscles. Highly trained distance athletes may have their liver glycogen storage capacity increased by more than 50%, and the ability of their muscles to store glycogen may be more than double that of nonathletes.

Glycogen storage is crucial for marathon runners. During a marathon, a runner depletes essentially all of his or her body's stored glycogen. At this point, often about 90 minutes into the

race, the runner may experience extreme muscle fatigue, loss of motivation, and occasionally even hallucinations, because both muscles and brain are starved of energy. Runners describe this sensation as "hitting the wall" or "bonking." To store the greatest possible amount of glycogen, endurance athletes practice carbo-loading by consuming large quantities of carbohydrates (starches and sugars) during the 3 days preceding the race. By packing their livers and muscles with glycogen before the race, and by consuming energy drinks during the race, some runners manage to cross the finish line before they hit the wall.

Consider This

When a runner's body temperature begins to rise, the body activates several mechanisms, including sweating and circulating more blood to the skin. Compare this response to overheating with feedback inhibition in enzymes.

CHAPTER REVIEW

Summary of Key Concepts

6.1 What Is Energy?

Energy is the capacity to do work. Potential energy is stored energy (such as chemical energy, elastic energy, or gravitational energy). Kinetic energy is the energy of movement (including radiant energy, heat energy, and the movement of objects). Potential energy can be converted to kinetic energy, and kinetic energy can be converted to potential energy. The first law of thermodynamics states that in a closed system, the total amount of energy remains constant, although the energy may change form. The second law of thermodynamics states that any use of energy causes a decrease in the quantity of useful energy and an increase in entropy (disorder or less useful forms of energy such as heat). The highly organized, low-entropy systems that characterize life do not violate the second law of thermodynamics, because they are achieved through a continuous influx of usable energy from the sun, accompanied by an enormous increase in solar entropy.

6.2 How Is Energy Transformed During Chemical Reactions?

Chemical reactions fall into two broad categories. In exergonic reactions, the reactant molecules have more energy than do the product molecules, so the reaction releases energy. In endergonic reactions, the reactants have less energy than do the products, so the reaction requires a net input of energy. Exergonic reactions can occur spontaneously, but all reactions, including exergonic ones, require an initial input of activation energy to overcome electrical repulsions between reactant molecules. Exergonic and endergonic reactions may be coupled such that the energy liberated by an exergonic reaction drives an endergonic reaction. For example, organisms couple exergonic reactions, such as breaking down sugar, with endergonic reactions, such as synthesizing organic molecules.

6.3 How Is Energy Transported Within Cells?

Energy released by chemical reactions within a cell is captured and transported within the cell by unstable energy-carrier molecules, such as ATP and the electron carriers NADH and $FADH_2$. These molecules are the major means by which cells couple exergonic and endergonic reactions occurring at different places in the cell.

6.4 How Do Enzymes Promote Biochemical Reactions?

Cells control their metabolic reactions by regulating the synthesis and the use of enzymes, which are biological catalysts that help overcome activation energy. High activation energies slow many reactions, even exergonic ones, to an imperceptible rate under normal environmental conditions. Catalysts lower the activation energy and thereby speed up chemical reactions without being permanently changed themselves. Organisms synthesize enzyme catalysts that promote one or a few specific reactions. The reactants temporarily bind to the active site of the enzyme, causing less activation energy to be needed to produce the product. Enzymes also allow the breakdown of energy-rich molecules such as glucose in a series of small steps so that energy is released gradually and can be captured in ATP for use in endergonic reactions.

6.5 How Are Enzymes Regulated?

Enzyme action is regulated by altering the rate of enzyme synthesis, activating previously inactive enzymes, competitive and noncompetitive inhibition, and allosteric regulation, which includes feedback inhibition. Many poisons, including cyanide, some nerve gases, some insecticides, and heavy metals, act as enzyme inhibitors. Penicillin, aspirin, and cancer chemotherapy drugs also act as competitive enzyme inhibitors. Environmental conditions—including pH, salt concentration, and temperature—can promote or inhibit enzyme function by altering or preserving the enzyme's three-dimensional structure.

Key Terms

activation energy   *97*
active site   *100*
adenosine diphosphate (ADP)   *98*
adenosine triphosphate (ATP)   *98*
allosteric regulation   *103*
catalyst   *99*
chemical energy   *94*
chemical reaction   *96*
closed system   *95*
competitive inhibition   *102*
coupled reaction   *98*
denature   *104*
electron carrier   *98*
endergonic   *96*
energy   *94*
energy-carrier molecule   *98*
entropy   *96*
enzyme   *100*
exergonic   *96*
feedback inhibition   *103*
first law of thermodynamics   *94*
kinetic energy   *94*
law of conservation of energy   *95*
metabolic pathway   *101*
metabolism   *101*
noncompetitive inhibition   *103*
potential energy   *94*
product   *96*
reactant   *96*
second law of thermodynamics   *95*
substrate   *100*
work   *94*

Learning Outcomes

In this chapter, you have learned to . . .

LO1 Define the key terms used to describe energy and explain the forms that energy takes.

LO2 Summarize the first and second laws of thermodynamics.

LO3 Define *entropy* and explain how organisms create the low-entropy conditions of life.

LO4 Describe how energy is captured and released by chemical reactions.

LO5 Describe the energy-carrier molecules in cells and explain how they facilitate chemical reactions.

LO6 Explain how catalysts work and how enzymes in cells catalyze chemical reactions.

LO7 Describe how cells control their metabolic pathways by regulating enzyme activity.

Thinking Through the Concepts

Fill-in-the-Blank

1. According to the first law of thermodynamics, energy can be neither _____ nor _____. Energy occurs in two major forms: _____, the energy of movement, and _____, or stored energy. **LO1 LO2**

2. According to the second law of thermodynamics, when energy changes forms it tends to be converted from _____ useful to _____ useful forms. This leads to the conclusion that matter spontaneously tends to become less _____. This tendency is called _____. **LO2**

3. Once started, some reactions release energy and are called _____ reactions. Others require a net input of energy and are called _____ reactions. Which type of reaction will continue spontaneously once it starts? _____. Which type of reaction allows the formation of complex biological molecules from simpler molecules (for example, proteins from amino acids)? _____. When the energy released from one reaction provides the energy required by another, the two reactions are described as _____ reactions. **LO3**

4. The abbreviation ATP stands for _____. This molecule is the principal _____ molecule in living cells. The molecule is synthesized by cells from _____ and _____. This synthesis requires an input of _____, which is temporarily stored in the ATP molecule. **LO4**

5. Enzymes are nearly all (type of biological molecule) _____. Enzymes promote reactions in cells by acting as biological _____ that lower the _____. Each enzyme possesses a specialized region called a(n) _____ that binds specific biological molecules. These specific molecules are described as _____ for the enzyme. **LO5 LO6**

6. Some poisons and drugs act by _____ enzymes. When a drug is similar to the enzyme's substrate, it acts as a (two words) _____ _____. Feedback inhibition is a form of _____ regulation. **LO7**

Review Questions

1. Explain why organisms do not violate the second law of thermodynamics. What is the ultimate energy source for most forms of life on Earth? **LO2**

2. Define *potential energy* and *kinetic energy* and provide two specific examples of each. Explain how one form of energy can be converted into another. Will some energy be lost during this conversion? If so, what form will it take? **LO1 LO2**

3. Define *metabolism*, and explain how reactions can be coupled to one another. **LO3 LO4 LO7**

4. What is activation energy? How do catalysts affect activation energy? How do catalysts affect the rate of reactions? **LO3 LO5**

5. Compare breaking down glucose in a cell to setting it on fire with a match. What is the source of activation energy in each case? **LO3 LO5 LO7**

6. Compare the mechanisms of competitive and noncompetitive inhibition of enzymes. **LO7**

7. Describe the structure and function of enzymes. How is enzyme activity regulated? **LO6 LO7**

Applying the Concepts

1. One of your nerdiest friends walks in while you are vacuuming. Trying to impress her, you casually mention that you are expending electrical energy to create a lower-entropy state in your room. She comments that, ultimately, you are taking advantage of the increase in entropy of the sun to clean your room. What is she talking about?

2. The subunits of virtually all organic molecules are joined by condensation reactions and can be broken apart by hydrolysis reactions (described in Chapter 3). Why, then, does your digestive system produce separate enzymes to digest proteins, fats, and carbohydrates?

3. Suppose someone tried to refute evolution using the following argument: "According to evolutionary theory, organisms have increased in complexity through time. However, the increase in complexity contradicts the second law of thermodynamics. Therefore, evolution is impossible." Explain why their reasoning is faulty.

4. When a bear eats a salmon, can the bear use all the energy contained in the body of the fish? Why or why not? Based on your conclusion, would you predict that a natural system (such as a forest) would have more predators or more prey animals (by weight). Explain.

Answers to Figure Caption questions and Fill-in-the-Blank questions can be found in the Answers section at the back of the book.

MB *Go to MasteringBiology for practice quizzes, activities, eText, videos, current events, and more.*

CHAPTER 7 Capturing Solar Energy: Photosynthesis

A giant asteroid may have ended the reign of *Tyrannosaurus*, *Triceratops*, and the other dinosaurs of the Cretaceous period.

CASE STUDY

Did the Dinosaurs Die from Lack of Sunlight?

ABOUT 65 MILLION YEARS AGO, the Cretaceous-Tertiary (K-T) extinction event brought the Cretaceous period to a violent end, and life on Earth suffered a catastrophic blow. The fossil record indicates that a devastating mass extinction eliminated at least 50% of all forms of life known to exist at that time. The 160-million-year reign of the dinosaurs, including the massive *Triceratops* and its predator *Tyrannosaurus rex*, ended abruptly. It would be many millions of years before Earth became repopulated with a diversity of species even approaching that of the late Cretaceous.

In 1980, Luis Alvarez, a Nobel Prize–winning physicist, his geologist son Walter Alvarez, and nuclear chemists Helen Michel and Frank Asaro published a controversial hypothesis. They proposed that an invader from outer space—a massive asteroid—had brought the Cretaceous period to an abrupt and violent end. Their evidence consisted of a thin layer of clay deposited at the end of the Cretaceous period, found at sites throughout the world. Known as the "K-T boundary layer," this clay deposit contains from 30 to 160 times the iridium level typical of Earth's crust. Iridium is a silvery-white metal that is extremely rare in Earth's crust, but abundant in certain types of asteroids. How large must an iridium-rich asteroid have been to create the K-T boundary layer? The Alvarez team calculated that this iridium-enriched space rock must have been at least 6 miles (10 kilometers) in diameter. In the Alvarez scenario, as the asteroid ploughed into Earth at nearly 45,000 miles per hour, its impact released energy greater than 100 trillion tons of TNT, blasting out a plume of debris that likely extended halfway to the moon. The asteroid's fragments, plummeting back into the atmosphere in a fiery shower, ignited widespread wildfires. A shroud of dust and soot blocked the sun's rays, and cooling darkness enveloped Earth.

Although the immediate destruction from fire, earthquakes, landslides, and tsunamis must have been almost unimaginable, these paled in contrast to the prolonged effects of the collision. How could an asteroid impact have eliminated half of all forms of life? Its most damaging long-term effect would have been disruption of the most important biochemical pathway on Earth: photosynthesis. What exactly occurs during photosynthesis? What makes this process so important that interrupting it ended the age of the dinosaurs? In this chapter, we explore these questions.

AT A GLANCE

7.1 What Is Photosynthesis?

7.2 The Light Reactions: How Is Light Energy Converted to Chemical Energy?

7.3 The Calvin Cycle: How Is Chemical Energy Stored in Sugar Molecules?

LEARNING GOALS

LG1 Explain the process of photosynthesis and discuss why photosynthesis is crucial to most forms of life.

LG2 Describe the structures of plant leaves and chloroplasts and explain how these structures support the process of photosynthesis.

LG3 Identify the source of energy for photosynthesis and describe how this energy is captured and transferred during the light reactions.

LG4 Describe the inputs and products of the light reactions.

LG5 Explain how the products of the light reactions are used during the Calvin cycle and discuss how plants use the products of the Calvin cycle.

7.1 WHAT IS PHOTOSYNTHESIS?

Living things continuously expend energy to survive. For nearly all forms of life, this energy is derived from sunlight, either directly or indirectly. The only organisms capable of directly trapping solar energy are those that perform **photosynthesis,** the process by which light energy is captured and stored as chemical energy in the bonds of organic molecules such as sugar. The evolution of photosynthesis made life as we know it possible. This amazing process provides not only fuel for life but also the oxygen required to burn this fuel. In lakes and oceans, photosynthesis is performed primarily by photosynthetic protists and certain bacteria, and on land mostly by plants. Collectively, these organisms incorporate close to 100 billion tons of carbon into their bodies annually. The carbon- and energy-rich molecules of photosynthetic organisms eventually become available to feed all other forms of life. Fundamentally similar reactions occur in all photosynthetic organisms; here we will concentrate on the most familiar of these: land plants.

Leaves and Chloroplasts Are Adaptations for Photosynthesis

The leaves of plants are beautifully adapted to the demands of photosynthesis (**Fig. 7-1**). A leaf's flattened shape exposes a large surface area to the sun, and its thinness ensures that sunlight can penetrate to reach the light-trapping chloroplasts inside. Both the upper and lower surfaces of a leaf consist of a layer of transparent cells that form the **epidermis,** which protects the inner parts of the leaf while allowing light to penetrate. The outer surface of the epidermis is covered by the **cuticle,** a transparent, waxy, waterproof covering that reduces the evaporation of water from the leaf.

A leaf obtains the carbon dioxide (CO_2) necessary for photosynthesis from the air, through adjustable pores in the epidermis called stomata (singular, **stoma;** Greek for "mouth"; **Fig. 7-2**). Inside the leaf are layers of cells collectively called **mesophyll** (which translates to "middle of the leaf") where most chloroplasts are located. *Vascular bundles,* which form veins in the leaf (**Fig. 7-1b**), supply water and minerals to the leaf's cells and carry the sugars produced during photosynthesis to other parts of the plant. Cells that surround the vascular bundles are called **bundle sheath cells,** which lack chloroplasts in most plants.

Photosynthesis in plants takes place within **chloroplasts,** most of which are contained within mesophyll cells. A single mesophyll cell often contains 40 to 50 chloroplasts (**Fig. 7-1c**), which are so small (about 5 micrometers in diameter) that 2,500 of them lined up would span an average thumbnail. Chloroplasts are organelles that consist of a double outer membrane enclosing a semifluid substance, the **stroma.** Embedded in the stroma are disk-shaped, interconnected membranous sacs called **thylakoids** (**Fig. 7-1d**). Each of these sacs encloses a fluid-filled region called the thylakoid space. The chemical reactions of photosynthesis that depend on light (the light reactions) occur in and adjacent to the thylakoids. The reactions of the Calvin cycle that capture carbon from CO_2 and produce simple sugar molecules occur in the surrounding stroma.

Photosynthesis Consists of the Light Reactions and the Calvin Cycle

Starting with the simple molecules of carbon dioxide and water, photosynthesis converts the energy of sunlight into chemical energy stored in the bonds of glucose and releases oxygen as a by-product (**Fig. 7-3**). The simplest overall chemical reaction for photosynthesis is:

$$6\,CO_2 + 6\,H_2O + \text{light energy} \rightarrow C_6H_{12}O_6 \text{ (sugar)} + 6\,O_2$$

This straightforward equation obscures the fact that photosynthesis actually involves dozens of individual reactions, each catalyzed by a separate enzyme. These reactions occur in two distinct stages: the light reactions and the Calvin cycle. Each stage takes place within a different region of the

(a) Leaves

cuticle

upper epidermis

mesophyll cells

stoma

lower epidermis

chloroplasts

stoma

bundle sheath cells

vascular bundle (vein)

(b) Internal leaf structure

outer membrane

inner membrane

thylakoid

stroma

(d) Chloroplast

channel interconnecting thylakoids

(c) Mesophyll cell containing chloroplasts

▲ FIGURE 7-1 **An overview of photosynthetic structures** **(a)** In land plants, photosynthesis occurs primarily in the leaves. **(b)** A section of a leaf. **(c)** A light micrograph of a single mesophyll cell, packed with chloroplasts. **(d)** A single chloroplast, showing the stroma and thylakoids where photosynthesis occurs.

chloroplast, but the two are connected by an important link: energy-carrier molecules.

In the **light reactions,** chlorophyll and other molecules embedded in the membranes of the chloroplast thylakoids capture sunlight energy and convert some of it into chemical energy stored in the energy-carrier molecules ATP (adenosine triphosphate) and NADPH (nicotinamide adenine dinucleotide phosphate). Water is split apart, and oxygen gas is released as a by-product. In the reactions of the **Calvin cycle,** enzymes in the fluid stroma that surrounds the thylakoids use CO_2 from the atmosphere and chemical energy from the energy-carrier molecules to drive the synthesis of a three-carbon sugar that will be used to make glucose. Figure 7-3 shows the locations at which the light reactions and the Calvin

cycle occur, and illustrates the interdependence of the two processes. The "photo" part of photosynthesis refers to the capture of light energy by the light reactions in the thylakoid membranes. These reactions use light energy to charge up the energy-carrier molecules ADP (adenosine diphosphate) and $NADP^+$ (the energy-depleted form of NADPH) to form ATP and NADPH. The "synthesis" part of photosynthesis refers to the Calvin cycle, which captures carbon and uses it to synthesize sugar, powered by energy provided by ATP and NADPH. The depleted carriers—ADP and $NADP^+$—are then recharged by the light reactions into ATP and NADPH, which will fuel the synthesis of more sugar molecules.

In the following sections we will look more closely at each stage of photosynthesis.

(a) Stomata open

(b) Stomata closed

▲ FIGURE 7-2 **Stomata (a)** Open stomata allow CO_2 to diffuse in and oxygen to diffuse out. **(b)** Closed stomata reduce water loss by evaporation but prevent CO_2 from entering and oxygen from leaving.

CASE STUDY continued

Did the Dinosaurs Die from Lack of Sunlight?

More than 2 billion years ago, some bacterial (prokaryotic) cells, through chance mutations, acquired the ability to harness the energy of sunlight. Exploiting this abundant energy source, early photosynthetic cells filled the seas. As their numbers increased, oxygen began to accumulate in the atmosphere, radically altering Earth's environment. Later, plants evolved, made the transition to land, and eventually grew in luxuriant profusion. By the Cretaceous period, plants had become sufficiently abundant to provide enough food to support plant-eating giants, such as the 8-ton *Triceratops* on which huge meat-eaters, such as the 40-foot-long *Tyrannosaurus,* may have preyed or scavenged. Then, as now, energy harvested through photosynthesis sustained nearly all forms of life.

CHECK YOUR LEARNING

Can you explain why photosynthesis is important? Can you diagram the structure of leaves and chloroplasts and explain how these structures function in photosynthesis? Can you write out and explain the basic equation for photosynthesis? Are you able to summarize the main events of the light reactions and the Calvin cycle and explain the relationship between these two processes?

▲ FIGURE 7-3 **An overview of the relationship between the light reactions and the Calvin cycle** The simple molecules that provide the raw ingredients for photosynthesis (H_2O and CO_2) enter at different stages of the process and are used in different parts of the chloroplast. The O_2 liberated by photosynthesis is derived from H_2O, while the carbon used in the synthesis of sugar is obtained from CO_2. (In some later figures you will see smaller versions of this figure.)

7.2 THE LIGHT REACTIONS: HOW IS LIGHT ENERGY CONVERTED TO CHEMICAL ENERGY?

The light reactions capture the energy of sunlight, storing it as chemical energy in two different energy-carrier molecules: ATP and NADPH. The molecules that make these reactions possible, including light-capturing pigments and enzymes, are anchored in a precise array within the membranes of the thylakoid. As you read this section, notice how the thylakoid membranes and the spaces they enclose support the light reactions.

Light Is Captured by Pigments in Chloroplasts

The sun emits energy that spans a broad spectrum of electromagnetic radiation. The **electromagnetic spectrum** ranges from short-wavelength gamma rays, through ultraviolet, visible, and infrared light, to very long-wavelength radio waves (**Fig. 7-4**). Light and other electromagnetic waves are composed of individual packets of energy called **photons.** The energy of a photon corresponds to its wavelength: short-wavelength photons, such as gamma and

▲ FIGURE 7-4 Light and chloroplast pigments Visible light, a small part of the electromagnetic spectrum, consists of wavelengths that correspond to the colors of the rainbow. Chlorophyll *a* and *b* (green and blue curves, respectively) strongly absorb violet, blue, and red light, reflecting a green or yellowish-green color to our eyes. Carotenoids (orange curve) absorb blue and green wavelengths.

QUESTION Which chlorophyll would be greener and which would be more yellow-green?

X-rays, are very energetic, whereas longer-wavelength photons, such as radio waves, carry lower energies. Visible light consists of wavelengths with energies that are high enough to alter biological *pigment molecules* (light-absorbing molecules) such as chlorophyll, but not high enough to break the bonds of crucial molecules such as DNA. (It is no coincidence that these wavelengths, with just the right amount of energy, also stimulate the pigments in our eyes, allowing us to see.)

When a specific wavelength of light strikes an object such as a leaf, one of three events occurs: The light may be reflected (bounced back), transmitted (passed through), or absorbed (captured). Wavelengths of light that are reflected or transmitted can reach the eyes of an observer; these wavelengths are seen as the color of the object. Light that is absorbed can drive biological processes such as photosynthesis.

Chloroplasts contain a variety of pigment molecules that absorb different wavelengths of light. **Chlorophyll *a*,** the key light-capturing pigment molecule in chloroplasts, strongly absorbs violet, blue, and red light, but reflects green, thus giving green leaves their color (see Fig. 7-4). Chloroplasts also contain other molecules, collectively called **accessory pigments,** which absorb additional wavelengths of light energy and transfer their energy to chlorophyll *a*. Accessory pigments include chlorophyll *b*, a slightly different form of chlorophyll that absorbs some of the blue and red-orange wavelengths of light that are missed by chlorophyll *a* and reflects yellow-green light. **Carotenoids** are accessory pigments found in all chloroplasts. They absorb blue and green light and appear mostly yellow or orange because they reflect these wavelengths (see Fig. 7-4). Carotenoid accessory pigments include beta-carotene, which gives many vegetables and fruits (including carrots, squash, oranges, and cantaloupes) their orange colors. Interestingly, animals convert beta-carotene into vitamin A, which is used to synthesize the light-capturing pigment in our eyes. Thus, in a beautiful symmetry, the beta-carotene that captures light energy in plants is converted into a substance that captures light in animals as well.

Although carotenoids are present in leaves, their color is usually masked by the more abundant green chlorophyll. In temperate regions, as autumn leaves begin to die, chlorophyll breaks down before carotenoids do, revealing these bright yellow and orange pigments as fall colors (**Fig. 7-5**). (Red and purple fall leaf pigments are not involved in photosynthesis.)

The Light Reactions Occur in Association with the Thylakoid Membranes

Light energy is captured and converted into chemical energy by the light reactions that occur in and on the thylakoid membranes. These membranes contain many **photosystems,** each consisting of a cluster of chlorophyll

▲ **FIGURE 7-5 Loss of chlorophyll reveals carotenoid pigments** As winter approaches, chlorophyll in these aspen leaves breaks down, revealing yellow-to-orange carotenoid pigments.

and accessory pigment molecules surrounded by various proteins. Two photosystems—photosystem II and photosystem I—work together during the light reactions. (The photosystems are named according to the order in which they were discovered, but the light reactions use photosystem II before using photosystem I.)

Each photosystem has an electron transport chain located adjacent to it. Each **electron transport chain** consists of a series of electron-carrier molecules embedded in the thylakoid membrane. So within the thylakoid membrane, the overall path of electrons is photosystem II → electron transport chain II → photosystem I → electron transport chain I → $NADP^+$ (see Fig. 7-7).

You can think of the light reactions as a sort of pinball game: Energy is transferred from spring-driven pistons (chlorophyll molecules) to a ball (an electron). The ball is propelled upward (enters a higher-energy level). As the ball bounces back downhill, the energy it releases can be used to

? HAVE **YOU** EVER WONDERED …

What Color Might Plants Be on Other Planets?

Biologist Nancy Kiang and her colleagues at NASA have developed hypotheses about alien plant colors. M-type stars, the most abundant type in our galaxy, emit light that is redder and dimmer than that of our sun. If photosynthetic organisms happened to evolve on an Earth-like planet circling an M-type star, to glean enough energy, the plants very possibly would require pigments that would absorb all visible wavelengths of light. Such pigments would reflect almost no light back to our eyes, so these alien photosynthesizers would probably be black!

turn a wheel (generate ATP) or ring a bell (generate NADPH). With this overall scheme in mind, let's look more closely at the actual sequence of events in the light reactions.

Photosystem II Uses Light Energy to Create a Hydrogen Ion Gradient and to Split Water

As you read the descriptions below, match them to the numbered steps in **Figure 7-6.** The light reactions begin when photons of light are absorbed by pigment molecules clustered in photosystem II (**Fig. 7-6 ❶**). The energy hops from one pigment molecule to the next until it is funneled into the photosystem II reaction center (**Fig. 7-6 ❷**). The **reaction center** of each photosystem consists of a pair of specialized chlorophyll *a* molecules and a *primary electron acceptor* molecule embedded in a complex of proteins. When the energy from light reaches the reaction center, it boosts an electron from one of the reaction center chlorophylls to the primary electron acceptor, which captures the energized electron (**Fig. 7-6 ❸**).

The reaction center of photosystem II must be supplied continuously with electrons to replace those that are boosted out of it when energized by light. These replacement electrons come from water (see Fig. 7-6 ❷). Water molecules are split by an enzyme associated with photosystem II, liberating electrons that will replace those lost by the reaction center chlorophyll molecules. Splitting water also releases two hydrogen ions (H^+) that are used in forming the H^+ gradient that drives ATP synthesis (see Fig. 7-7). For every two water molecules split, one molecule of O_2 is produced.

Once the primary electron acceptor in photosystem II captures the electron, it passes the electron to the first molecule of electron transport chain II (**Fig. 7-6 ❹**). The electron then travels from one electron carrier molecule to the next, losing energy as it goes. Some of this energy is harnessed to pump H^+ across the thylakoid membrane and into the thylakoid space, where it will be used to generate ATP (**Fig. 7-6 ❺**; to be discussed shortly). Finally, the energy-depleted electron leaves electron transport chain II and enters the reaction center of photosystem I, where it replaces the electron ejected when light strikes photosystem I (**Fig. 7-6 ❻**).

Photosystem I Generates NADPH

Meanwhile, light has also been striking the pigment molecules of photosystem I. This light energy is passed to a chlorophyll *a* molecule in the reaction center (see **Fig. 7-6 ❻**). Here, it energizes an electron that is absorbed by the primary electron acceptor of photosystem I (**Fig. 7-6 ❼**). (The energized electron is immediately replaced by an energy-depleted electron from electron transport chain II.) From the primary electron acceptor of photosystem I, the energized electron is passed along electron transport chain I (**Fig. 7-6 ❽**) until it reaches $NADP^+$. When an $NADP^+$ molecule (dissolved in the fluid stroma) picks up two energetic electrons along with one hydrogen ion, the energy-carrier molecule NADPH is formed (**Fig. 7-6 ❾**).

▲ **FIGURE 7-6 Energy transfer and the light reactions of photosynthesis** Light reactions occur within or adjacent to the thylakoid membrane. The vertical axis indicates the relative energy levels of the molecules involved. ❶ Light is absorbed by photosystem II, and the energy is passed to an electron in one of the chlorophyll *a* molecules within the reaction center. ❷ The energized electron is ejected from the chlorophyll *a* molecule. ❸ The energized ejected electron is captured by a primary electron acceptor in the reaction center (purple rectangle). ❹ The high-energy electron is passed through electron transport chain II. ❺ As electron transport chain II transfers the electron along, some of the released energy is used to generate ATP. The energy-depleted electron then enters the reaction center of photosystem I, replacing an ejected electron there. ❻ Light strikes photosystem I, and the energy is passed to the electron in the reaction center chlorophyll molecules. ❼ The energized electron is ejected from the reaction center and is captured by the primary electron acceptor. ❽ The electron moves down electron transport chain I. ❾ NADPH is formed when $NADP^+$ in the stroma picks up two energized electrons, along with one H^+.

▲ FIGURE 7-7 **Events of the light reactions occur in and near the thylakoid membrane** Trace the formation of ATP and NADPH as light is captured and an energized electron travels through molecules within the thylakoid membrane. ❶ Energy released as the electron passes through electron transport chain II is harnessed to pump H^+ across the thylakoid membrane and into the thylakoid space. ❷ A high concentration gradient of H^+ is generated as energy from electrons liberated during their passage through electron transport chain II is used to pump H^+ across the membrane into the thylakoid space. ❸ During chemiosmosis, H^+ flows down its concentration gradient through ATP synthase channels, which uses energy from the gradient to generate ATP. About one ATP molecule is synthesized for every three hydrogen ions that pass through the channel.

The Hydrogen Ion Gradient Generates ATP by Chemiosmosis

Figure 7-7 shows how electrons move through the thylakoid membrane and how their energy is used to create an H^+ gradient that drives ATP synthesis through a process called **chemiosmosis.** As an energized electron travels along electron transport chain II, some of the energy the electron liberates is used to pump H^+ into the thylakoid space (**Fig. 7-7** ❶), creating a high concentration of H^+ inside the space (**Fig. 7-7** ❷) and a low concentration in the surrounding stroma. During chemiosmosis, H^+ flows back down its concentration gradient through a special type of channel that spans the thylakoid membrane. This channel, called **ATP synthase,** generates

ATP from ADP and phosphate dissolved in the stroma as the H^+ flows through the channel (**Fig. 7-7** ❸).

How is a gradient of H^+ used to synthesize ATP? Compare the H^+ gradient to water stored behind a dam at a hydroelectric plant (**Fig. 7-8**). The water behind the dam flows downward over turbines, rotating them. The turbines convert the energy of moving water into electrical energy. Like water stored behind a dam, hydrogen ions in the thylakoid space can move down their gradient. Just as the water flowing over a hydroelectric dam is channeled through turbines, the hydrogen ions can flow into the stroma only through ATP synthase channels. Like turbines generating electricity, ATP synthase captures the energy liberated by the flow of H^+ and uses it to drive ATP synthesis from ADP plus phosphate dissolved in the stroma.

1 Energy is released as water flows downhill

2 Energy is harnessed to rotate a turbine

3 The energy of the rotating turbine is used to generate electricity

▲ FIGURE 7-8 A dam allows a "water gradient" to be used to generate electricity

SUMMING UP Light Reactions

- Chlorophyll and carotenoid pigments of photosystem II absorb light that energizes and ejects an electron from a reaction center chlorophyll *a* molecule. The energized electron is captured by the primary electron acceptor molecule.

- The electron is passed from the primary electron acceptor to the adjacent electron transport chain II, where it moves from molecule to molecule, releasing energy. Some of the energy is used to create a hydrogen ion gradient across the thylakoid membrane. This gradient is used to drive ATP synthesis by chemiosmosis.

- Enzymes associated with photosystem II split water, releasing electrons that replace those ejected from the reaction center chlorophylls, and supplying H^+ that enhances the H^+ gradient for ATP production, and generating oxygen.

- Chlorophyll and carotenoid pigments in photosystem I absorb light that energizes and ejects an electron from a reaction center chlorophyll *a* molecule. The electron is replaced by an energy-depleted electron from electron transport chain II.

- The energized electron passes from the primary electron acceptor into electron transport chain I, where it moves from molecule to molecule, releasing energy.

- For every two energized electrons that exit electron transport chain I, one molecule of the energy carrier NADPH is formed from $NADP^+$ and H^+.

- The overall products of the light reactions are the energy carriers NADPH and ATP; O_2 is released as a by-product.

CHECK YOUR LEARNING

Can you list the light-capturing molecules in chloroplasts and describe their functions? Can you diagram and describe the molecules within the thylakoid membranes and explain how they capture and transfer light energy? Can you describe the products of the light reactions and the roles of these products? Can you explain how ATP is generated?

CASE STUDY continued

Did the Dinosaurs Die from Lack of Sunlight?

Scientists have calculated that the force of the asteroid impact blasted some debris so far into outer space that it took days to rain back down to Earth. Earth's rotation while the debris was aloft meant that much of the material fell on regions far from the point of impact. The plummeting chunks of rock would have made a flaming re-entry through Earth's atmosphere, setting huge fires on nearly every continent. Oxygen levels were high in the Cretaceous atmosphere, intensifying the conflagrations. A large portion of Earth's vegetation was likely consumed by fire, and many of the plants that managed to survive the fires must have succumbed during the cold, dark "global winter" that began as the planet was encompassed by soot and dust. Plant-eating animals that survived the initial blast would have soon starved, especially enormous ones like the 12-ton *Triceratops*, which needed to consume hundreds of pounds of vegetation daily. Predators such as *Tyrannosaurus*, which relied on plant-eaters for food, would have died soon afterward. During the Cretaceous, as now, interrupting the vital flow of solar energy captured by photosynthetic organisms would be catastrophic.

7.3 THE CALVIN CYCLE: HOW IS CHEMICAL ENERGY STORED IN SUGAR MOLECULES?

Our cells produce carbon dioxide as we burn sugar for energy (described in Chapter 8), but they can't form organic molecules by capturing (or *fixing*) the carbon atoms in CO_2. Although this feat can be accomplished by a few types of *chemosynthetic* bacteria that fix carbon using energy gained by breaking down inorganic molecules, nearly all carbon fixation is performed by photosynthetic organisms. The carbon is captured from atmospheric CO_2 during the Calvin cycle using energy from sunlight harnessed during the light reactions. The details of the Calvin cycle were discovered during the 1950s by chemists Melvin Calvin, Andrew Benson, and James Bassham. Using radioactive isotopes of carbon (see Chapter 2), the researchers were able to follow carbon atoms as they moved from CO_2 through the various compounds of the cycle and, finally, into sugar molecules.

The Calvin Cycle Captures Carbon Dioxide

The ATP and NADPH synthesized during the light reactions are dissolved in the fluid stroma that surrounds the thylakoids. There, these energy carriers power the synthesis of the three-carbon sugar glyceraldehyde-3-phosphate (G3P) from CO_2 during the Calvin cycle. This metabolic pathway is described as a "cycle" because it begins and ends with the same five-carbon molecule, ribulose bisphosphate (RuBP).

Keeping this in mind, for simplicity we will illustrate the cycle starting and ending with three molecules of RuBP. Each "turn" will then capture three molecules of CO_2 and produce one molecule of the simple sugar end product: G3P. The Calvin cycle is best understood if we divide it into three parts: (1) carbon fixation, (2) the synthesis of G3P, and (3) the regeneration of RuBP that allows the cycle to continue (Fig. 7-9).

During **carbon fixation,** carbon from CO_2 is incorporated into larger organic molecules. The enzyme **rubisco** combines three CO_2 molecules with three RuBP molecules to produce three unstable six-carbon molecules that immediately split in half, forming six molecules of phosphoglyceric acid (PGA, a three-carbon molecule) (**Fig. 7-9 ❶**). Because carbon fixation generates this three-carbon PGA molecule, the Calvin cycle is often referred to as the **C₃ pathway.**

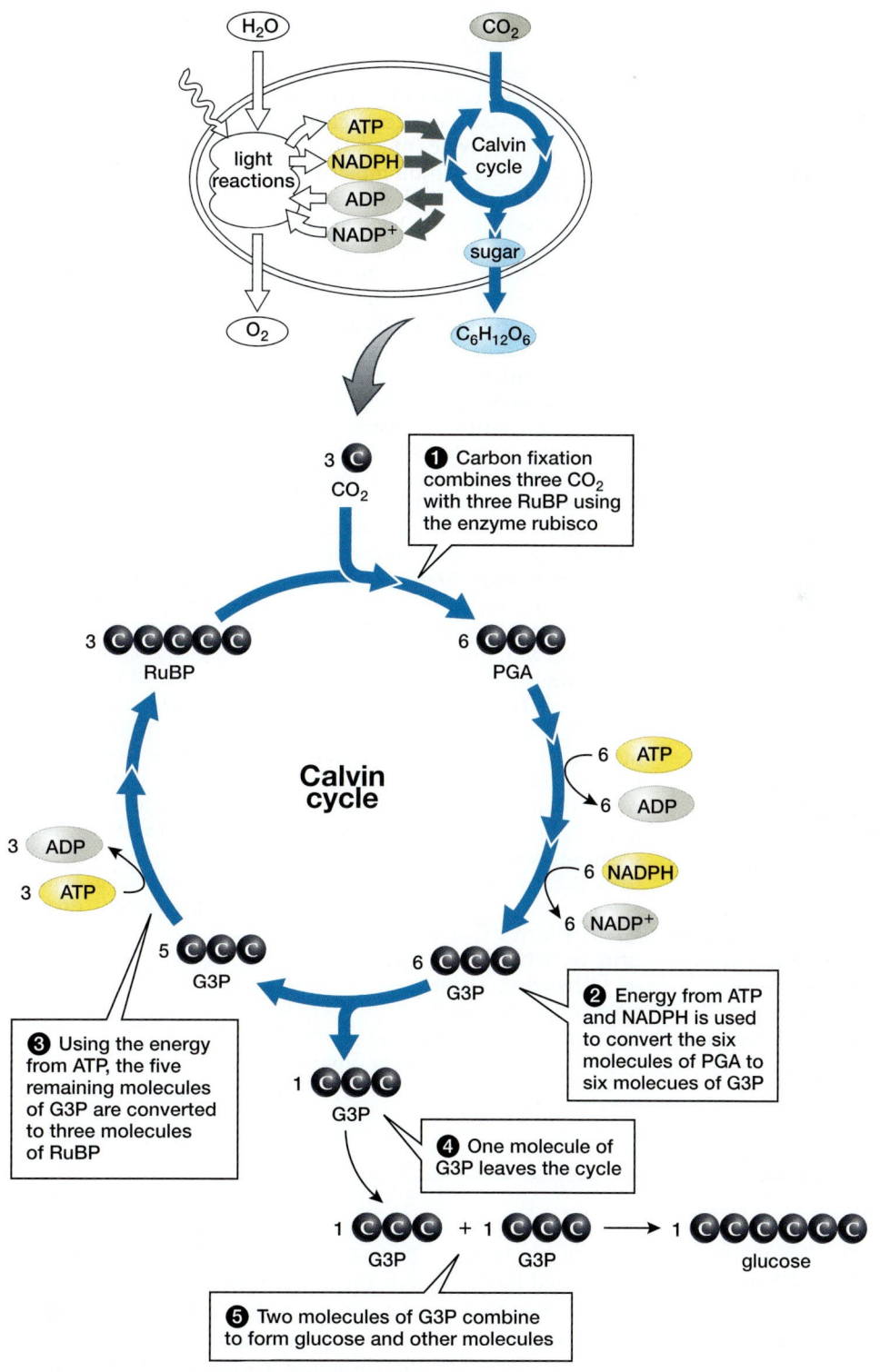

▲ FIGURE 7-9 **The Calvin cycle fixes carbon from CO_2 and produces the simple sugar G3P**

IN GREATER DEPTH Alternate Pathways Increase the Efficiency of Carbon Fixation

In hot, dry conditions, stomata remain closed much of the time to prevent water from evaporating. But this also prevents the exchange of gases, so as photosynthesis occurs, the concentration of CO_2 drops and the concentration of O_2 rises. Unfortunately O_2 as well as CO_2 can bind to the active site of the enzyme rubisco (an example of competitive inhibition; see Fig. 6-13b) and combine with RuBP, causing photorespiration. Plants, particularly fragile seedlings, may die under these circumstances because they are unable to capture enough energy to meet their metabolic needs.

Rubisco is the most abundant enzyme on Earth, and arguably one of the most important. It catalyzes the reaction by which carbon enters the biosphere, and all life is based on carbon. Why, then, is rubisco so nonselective? Apparently the necessary mutations to prevent competitive inhibition by oxygen have never occurred. Instead, over evolutionary time, flowering plants have evolved two different but related mechanisms to circumvent photorespiration: the C_4 pathway and CAM. Each involves several additional reactions and consumes more ATP than does the Calvin cycle alone, but they confer an important advantage under hot, dry conditions.

C_4 Plants Capture Carbon and Synthesize Sugar in Different Cells

In typical plants, known as C_3 *plants* (because they use the C_3 cycle, or Calvin cycle), the chloroplasts in which the Calvin cycle occurs are located primarily in mesophyll cells. No chloroplasts are found in bundle sheath cells (which surround the leaf's veins; see Fig. 7-1). In contrast, C_4 *plants* have chloroplasts in both mesophyll and bundle sheath

cells. Such plants use an initial series of reactions, called the C_4 pathway, to selectively capture carbon in their mesophyll chloroplasts. The mesophyll chloroplasts lack Calvin cycle enzymes and use the enzyme *PEP carboxylase* to fix CO_2. Unlike rubisco, PEP carboxylase is highly selective for CO_2 over O_2. PEP carboxylase causes CO_2 to react with a three-carbon molecule called phosphoenolpyruvate (PEP). This produces the four-carbon molecule oxaloacetate, from which the C_4 pathway gets its name. Oxaloacetate is rapidly converted into another four-carbon molecule, malate, which diffuses from the mesophyll cells into bundle sheath cells. The malate acts as a shuttle for CO_2.

In C_4 plants, Calvin cycle enzymes (including rubisco) are only present in the chloroplasts of the bundle sheath cells. In the bundle sheath cells, malate is broken down, forming the three-carbon molecule pyruvate and releasing CO_2. This generates a high CO_2 concentration in the bundle sheath cells (up to 10 times higher than atmospheric CO_2). The resulting high CO_2 concentration allows rubisco to fix carbon with little competition from O_2, minimizing photorespiration. The pyruvate is then actively transported back into the mesophyll cells. Here, more ATP energy is used to convert pyruvate back into PEP, allowing the cycle to continue (**Fig. E7-1a**). Plants using C_4 photosynthesis include corn, sugarcane, sorghum, some grasses, and some thistles.

CAM Plants Capture Carbon and Synthesize Sugar at Different Times

CAM plants also use the C_4 pathway, but in contrast to C_4 plants, CAM plants do not use different cell types to capture

carbon and to synthesize sugar. Instead, they perform both activities in the same mesophyll cells, but at different times; carbon fixation occurs at night, and sugar synthesis occurs during the day (**Fig. E7-1b**).

The stomata of CAM plants open at night, when less water will evaporate because temperatures are cooler and humidity is higher. Carbon dioxide diffuses into the leaf and is captured in mesophyll cells using the C_4 pathway. The malate produced by the C_4 pathway is then shuttled into the central vacuole where it is stored as malic acid until daytime. During the day, when stomata are closed to conserve water, the malic acid leaves the vacuole and re-enters the cytoplasm as malate. The malate is broken down, forming pyruvate (which will be converted to PEP) and releasing CO_2, which enters the Calvin cycle (via rubisco) to produce sugar. CAM plants include cacti, most succulents (such as the jade plant), and pineapples.

C_4 plants and CAM Pathways Are Specialized Adaptations

Because they are so much more efficient at fixing carbon, why haven't C_4 and CAM plants taken over the world by now? The trade-off is that both of these pathways consume more energy than does the Calvin cycle by itself; hence, they waste some of the solar energy that they capture. As a result, these plants only have an advantage in warm, sunny, dry environments. This explains why a lush spring lawn of Kentucky bluegrass (a C_3 plant) may be taken over by spiky crabgrass (a C_4 plant) during a hot, dry summer.

The synthesis of G3P occurs via a series of reactions using energy donated by ATP and NADPH (generated by the light reactions). During these reactions, six 3-carbon PGA molecules are rearranged to form six 3-carbon G3P molecules (**Fig. 7-9 ❷**).

Three molecules of RuBP are regenerated from five of the six G3P molecules; this regeneration is powered by ATP

generated during the light reactions (**Fig. 7-9 ❸**). The single remaining G3P molecule, the end product of photosynthesis, exits the Calvin cycle (**Fig. 7-9 ❹**).

Carbon fixation, the first step in the Calvin cycle, can be sabotaged. Unfortunately, the enzyme rubisco that allows carbon fixation is not very selective, and will cause either CO_2 or O_2 to combine with RuBP. If O_2 is fixed, the result

► **FIGURE E7-1 The C₄ pathway and the CAM pathway** Some of the reactions illustrated here occur in the cytoplasm, whereas others, including all those of the Calvin cycle, occur in chloroplasts. Both include the C₄ pathway. **(a)** The C₄ pathway occurs in plant groups including corn. **(b)** The CAM pathway occurs in plant groups including pineapples. By creating a high level of CO_2 in **(a)** the bundle sheath cells of C_4 plants and **(b)** the mesophyll cells of CAM plants (during the day, when stomata are closed), each of these pathways minimizes photorespiration.

QUESTION Why do C_3 (Calvin cycle) plants have an advantage over C_4 plants under conditions that are not hot and dry?

(a) **C₄ plants**

(b) **CAM plants**

is a wasteful process called **photorespiration.** Photorespiration prevents the Calvin cycle from synthesizing sugar, effectively derailing photosynthesis. If photorespiration could be avoided, plants could capture solar energy much more efficiently. Many researchers are working to genetically modify the enzyme rubisco to make it more selective for CO_2, with the hope of increasing the yields of crops such as wheat.

A small percentage of Earth's terrestrial plants have evolved biochemical pathways that consume a bit more energy but increase the efficiency of carbon fixation in hot, dry environments. These pathways—the **C₄ pathway** and the **crassulacean acid metabolism (CAM)** pathway—are explored in "In Greater Depth: Alternate Pathways Increase the Efficiency of Carbon Fixation."

Carbon Fixed During the Calvin Cycle Is Used to Synthesize Glucose

In reactions that occur outside of the Calvin cycle, two G3P molecules can be combined to form one six-carbon glucose molecule (**Fig. 7-9 ❺**). Glucose can then be used to synthesize sucrose (table sugar), a disaccharide storage molecule consisting of a glucose linked to a fructose. Glucose molecules are also linked together in long chains to form starch (another storage molecule) or cellulose (a major component of plant cell walls). Some glucose molecules are also broken down during cellular respiration to provide energy for the plant's cells.

Earth | Watch

Biofuels—Are Their Benefits Bogus?

When you drive your car, turn up the thermostat, or flick on your desk lamp, you are actually unleashing the energy of prehistoric sunlight trapped by prehistoric photosynthetic organisms. This is because over hundreds of millions of years, heat and pressure converted the bodies of these organisms—with their stored solar energy and carbon captured from ancient atmospheric CO_2—into coal, oil, and natural gas. Without human intervention, these fossil fuels would have remained trapped deep underground.

Global climate change is big news, and a major contributor to it is increased burning of fossil fuels by a growing human population. This combustion releases CO_2 into the atmosphere; the added carbon dioxide traps heat in the atmosphere that would otherwise radiate into space. Since we began using fossil fuels during the industrial revolution in the mid-1800s, humans have increased the CO_2 content of the atmosphere by about 38%. As a result, Earth is growing warmer, and many experts fear that a hotter future climate will place extraordinary stresses on most of Earth's inhabitants, ourselves included (see Chapter 28).

▲ **FIGURE E7-2 Cleared tropical forest** This aerial view shows the aftermath of clearing lush tropical forest, the former home of rare Sumatran tigers, elephants, leopards, orangutans, and a wealth of bird species. This will become a palm oil plantation for biofuels. (Inset) Endangered orangutans such as these are increasingly rendered homeless by deforestation, and are often killed as they are forced closer to human settlements.

To reduce CO_2 emissions and reliance on imported oil, many governments are subsidizing and promoting the use of biofuels, especially ethanol and biodiesel. Ethanol is produced by fermenting plants rich in sugars, such as sugarcane and corn, to produce alcohol (fermentation is described in Chapter 8). Biodiesel fuel is made primarily from oil derived from plants such as soybeans, canola, or palms. Because the carbon stored in biofuels was removed from the modern atmosphere by photosynthesis, burning them seems to simply restore CO_2 that was recently present in the atmosphere. Is this a solution to global climate change?

The environmental and social benefits of burning fuels derived from food crops in our gas tanks are hotly debated. Roughly 40% of the U.S. corn crop is now feeding vehicles rather than animals and people. One concern is that this added demand for such crops will drive up food prices as cars compete with animal and human consumers. Another concern is that growing corn and converting it into ethanol uses up large quantities of fossil fuel, making ethanol's "carbon footprint" little, if any, better than burning gasoline. The environmental costs of using food crops as an alternative fuel source are also enormous.

For example, Indonesia's luxurious tropical rain forests—home to orangutans, Sumatran tigers, and clouded leopards—are being destroyed to make room for oil palm plantations for biofuels (**Fig. E7-2**). In Brazil, soybean plantations for biofuels have replaced large expanses of rain forest. Ironically, clearing these forests for agriculture increases atmospheric CO_2 because rain forests trap far more carbon than the crops that replace them.

Biofuels would have far less environmental and social impact if they were not produced from food crops or by destroying Earth's dwindling rain forests. Algae show great promise as an alternative. These photosynthesizers can double their mass in a day or less. Some algae produce starch that can be fermented into ethanol; others produce oil that can become biodiesel. Researchers are also attempting to cleave cellulose into its component sugars, which would allow ethanol to be generated from corn stalks, wood chips, or grasses. This concept is being tested at several small cellulose biorefineries in the United States. Although the benefits of most biofuels in wide use today may not justify their environmental costs, there is hope that this will change as we develop better technologies to harness the energy captured by photosynthesis.

The storage products of photosynthesis are being eyed by Earth's growing and energy-hungry human population as a substitute for fossil fuels. These "biofuels" have the potential advantage of not adding additional carbon dioxide (a "greenhouse gas" that contributes to global climate change) to the atmosphere, but do they live up to their promise? We explore this question in "Earth Watch: Biofuels—Are Their Benefits Bogus?"

SUMMING UP The Calvin Cycle

The Calvin cycle can be divided into three stages:

- **Carbon fixation:** Three RuBP capture three CO_2, forming six PGA.
- **G3P synthesis:** A series of reactions, driven by energy from ATP and NADPH (from the light reactions), produces six G3P, one of which leaves the cycle and is available to form glucose.
- **RuBP regeneration:** ATP energy is used to regenerate three RuBP molecules from the remaining five G3P molecules, allowing the cycle to continue.

In a separate process outside the chloroplast, two G3P molecules produced by the Calvin cycle combine to form glucose.

CHECK YOUR LEARNING

Can you describe the function of the Calvin cycle and where it occurs? Can you list the three stages of the Calvin cycle, including the molecules that enter the cycle and those that are formed at each stage? Can you describe the fate of the simple sugar generated by the Calvin cycle?

CASE STUDY revisited

Did the Dinosaurs Die from Lack of Sunlight?

Did an asteroid end the reign of dinosaurs? The Alvarez hypothesis was initially met with skepticism. If such a cataclysmic event had occurred, where was the crater?

In 1991, scientists finally located it centered near the coastal town of Chicxulub on Mexico's Yucatán Peninsula. The crater, estimated at over 150 miles (200 to 300 km) in diameter and 10 miles (16 km) deep, was filled by debris and sedimentary rock laid down during the 65.5 million years since the impact. Ocean and dense vegetation hid most remaining traces from satellite images. The final identification of the Chicxulub crater was based on rock core samples, unusual gravitational patterns, and faint surface features. Convincing geological evidence shows that the crater was formed at the end of the Cretaceous period, the same time at which the K-T boundary layer was deposited.

Some paleontologists argue that the impact may have exacerbated more gradual changes in climate, to which the dinosaurs (with the exception of those ancestral to modern-day birds) could not adapt. Such changes might have been caused by prolonged intense volcanic activity, such as occurred at a site in India at about the time of the K-T extinction. Volcanoes spew out soot and ash, and iridium is found in higher levels in lava from Earth's molten mantle than in its crust. So furious volcanism could significantly reduce the amount of sunlight for plant growth, spew climate-changing gases into the air, and also contribute to the iridium-rich K-T boundary layer.

In 2010, alternative hypotheses to the asteroid impact were dealt a blow when an expert group of 41 researchers published a review article in the prestigious journal *Science*. This publication analyzed the previous 20 years of research by paleontologists, geochemists, geophysicists, climatologists, and sedimentation experts dealing with the K-T extinction event. The conclusion: Land and ocean ecosystems were destroyed extremely rapidly, and evidence overwhelmingly supports the asteroid impact hypothesis first proposed by the Alvarez group 30 years earlier.

Consider This

Has the publication described above changed the asteroid impact hypothesis into a scientific theory that is "set in stone"? Can the cause of the K-T extinction ever be proven? Should scientists who believed that the evidence pointed to alternative causes for the extinction now abandon their hypotheses based on this article? Does debate among scientists weaken or strengthen the role of science in society?

CHAPTER REVIEW

Summary of Key Concepts

7.1 What Is Photosynthesis?

Photosynthesis captures the energy of sunlight and uses it to convert the inorganic molecules of carbon dioxide and water into a high-energy sugar molecule, releasing oxygen as a by-product. In plants, photosynthesis takes place in the chloroplasts, using two major reaction sequences: the light reactions and the Calvin cycle.

7.2 The Light Reactions: How Is Light Energy Converted to Chemical Energy?

The light reactions occur in the thylakoids of chloroplasts. Light energizes electrons in chlorophyll molecules located in photosystems II and I. Energetic electrons jump to a primary electron acceptor, then move into adjacent electron transport chains. Energy lost as the electrons travel through electron transport chain II is used to pump hydrogen ions into the thylakoid space, creating an H^+ gradient across the thylakoid membrane.

Hydrogen ions flow down this concentration gradient through ATP synthase channels in the membrane, driving ATP synthesis. For every two energized electrons that pass through electron transport chain I, one molecule of the energy carrier NADPH is formed from $NADP^+$ and H^+. Electrons lost from photosystem II are replaced by electrons liberated by splitting water, which also generates H^+ and O_2.

7.3 The Calvin Cycle: How Is Chemical Energy Stored in Sugar Molecules?

The Calvin cycle, which occurs in the stroma of chloroplasts, uses energy from the ATP and NADPH generated during the light reactions to drive the synthesis of G3P. Two molecules of G3P may then combine to form glucose. The Calvin cycle has three parts: (1) *Carbon fixation:* Carbon dioxide combines with ribulose bisphosphate (RuBP) to form phosphoglyceric acid (PGA). (2) *Synthesis of G3P:* PGA is converted to glyceraldehyde-3-phosphate (G3P), using energy from ATP and NADPH. (3) *Regeneration of RuBP:* Five molecules of G3P are used to regenerate three molecules of RuBP, using ATP energy. One molecule of G3P exits the cycle; this G3P may be used to synthesize glucose and other molecules.

Key Terms

accessory pigment   *112*
ATP synthase   *115*
bundle sheath cells   *109*
C_3 pathway   *117*
C_4 pathway   *119*
Calvin cycle   *110*
carbon fixation   *117*
carotenoid   *112*
chemiosmosis   *115*
chlorophyll *a*   *112*
chloroplast   *109*
crassulacean acid metabolism (CAM)   *119*
cuticle   *109*
electromagnetic spectrum   *112*

electron transport chain   *113*
epidermis   *109*
light reactions   *110*
mesophyll   *109*
photon   *112*
photorespiration   *119*
photosynthesis   *109*
photosystem   *112*
reaction center   *113*
rubisco   *117*
stoma (plural, stomata)   *109*
stroma   *109*
thylakoid   *109*

Learning Outcomes

In this chapter, you have learned to . . .

LO1 Explain the process of photosynthesis and discuss why photosynthesis is crucial to most forms of life.

LO2 Describe the structures of plant leaves and chloroplasts and explain how these structures support the process of photosynthesis.

LO3 Identify the source of energy for photosynthesis and describe how this energy

is captured and transferred during the light reactions.

LO4 Describe the inputs and products of the light reactions.

LO5 Explain how the products of the light reactions are used during the Calvin cycle and discuss how plants use the products of the Calvin cycle.

Thinking Through the Concepts

Fill-in-the-Blank

1. Plant leaves contain pores called _____ that allow the plant to release _____ and take in _____. In hot dry weather, these pores are closed to prevent _____. Photosynthesis occurs in organelles called _____ that are concentrated within the _____ cells of most plant leaves. **LO2**

2. Chlorophyll *a* captures wavelengths of light that correspond to the three colors _____, _____, and _____. What color does chlorophyll reflect? _____ Accessory pigments that reflect yellow and orange are called _____. These pigments are located in clusters called _____ in the _____ membrane of the chloroplast. **LO2 LO3**

3. During the first stage of photosynthesis, light is captured and funneled into _____ chlorophyll *a* molecules. The energized electron is then passed into a _____. From here, it is transferred through a series of molecules called the _____. Energy lost during these transfers is used to create a gradient of _____. The process that uses this gradient to generate ATP is called _____. **LO3 LO4**

4. The oxygen produced as a by-product of photosynthesis is derived from _____, and the carbons used to make glucose are derived from _____. The biochemical pathway that captures atmospheric carbon is called the _____. The process of capturing carbon is called _____. **LO3 LO4 LO5**

5. In plants, the enzyme that catalyzes carbon capture is _____, which is quite nonselective and binds _____ as well as CO_2. When it binds the "wrong" molecule, this enzyme causes _____ to occur. Two pathways that reduce this process are called the _____ and the _____. **LO5**

6. Light reactions generate the energy-carrier molecules _____ and _____, which are then used in the _____ cycle. Carbon fixation combines carbon dioxide with the five-carbon molecule _____. Two molecules of _____ can be combined to produce the six-carbon sugar _____. **LO3 LO4 LO5**

Review Questions

1. Explain what would happen to life if photosynthesis ceased. Why would this occur? **LO1**

2. Write and then explain the equation for photosynthesis. **LO1**

3. Draw a simplified diagram of a leaf cross-section and label it. Explain how a leaf's structure supports photosynthesis. **LO2**

4. Draw a simplified diagram of a chloroplast, and label it. Explain how the individual parts of the chloroplast support photosynthesis. **LO2**

5. Explain photorespiration and why it occurs. Describe the two mechanisms that some groups of plants have evolved to reduce photorespiration. **LO5**

6. Trace the flow of energy in chloroplasts from sunlight to ATP, including an explanation of chemiosmosis. **LO3**

7. Summarize the events of the Calvin cycle. Where does it occur? What molecule is fixed? What is the product of the cycle? Where does the energy come from to drive the cycle? What molecule is regenerated? **LO5**

Applying the Concepts

1. Suppose an experiment is performed in which plant I is supplied with normal carbon dioxide but with water that contains radioactive oxygen atoms. Plant II is supplied with normal water but with carbon dioxide that contains radioactive oxygen atoms. Each plant is allowed to perform photosynthesis, and the oxygen gas and sugars produced are tested for radioactivity. Which plant would you expect to produce radioactive sugars, and which plant would you expect to produce radioactive oxygen gas? Explain why.

2. You continuously monitor the photosynthetic oxygen production from the leaf of a plant illuminated by white light. Explain how oxygen production will change—and why—if you place filters in front of the light source that transmit (a) only red, (b) only blue, and (c) only green light onto the leaf.

3. If you were to measure the pH in the space surrounded by the thylakoid membrane in an actively photosynthesizing plant, would you expect it to be acidic, basic, or neutral? Explain your answer.

4. Assume you want to add an accessory pigment to the photosystems of a plant chloroplast to help it photosynthesize more efficiently. What wavelengths of light would the new pigment absorb? Describe the color of your new pigment. Would this be a useful project for genetic engineers? Explain.

Answers to Figure Caption questions and Fill-in-the-Blank questions can be found in the Answers section at the back of the book.

MB *Go to MasteringBiology for practice quizzes, activities, eText, videos, current events, and more.*

Riccardo Riccò celebrates victory in Stage 6 of the 2008 Tour de France. "I had good legs today," he was quoted as saying. (Inset) Later, Riccò's team reacts to learning the outcome of his blood test.

CASE STUDY

When Athletes Boost Their Blood Counts: Do Cheaters Prosper?

AN HOUR BEFORE THE START of the 12th stage of the 2008 Tour de France, the Saunier-Duval team was in high spirits. Their star cyclist, Riccardo Riccò, had led the pack in two stages, including the grueling Stage 9, which crested two mountains over a 139-mile (224-km) course. But the team's optimism turned to consternation as they watched their top rider escorted away by police. Minutes later, the entire team withdrew from the race. Riccò's offense? Blood doping.

Blood doping packs more oxygen-carrying red blood cells into the bloodstream with the goal of enhancing athletic performance. Riccò injected the drug CERA (continuous erythropoietin receptor activator), a synthetic version of erythropoietin. The natural hormone erythropoietin (EPO) stimulates bone marrow to produce red blood cells. A healthy body produces just enough EPO to ensure that red blood cells are replaced as they age and die. If a person moves to a higher altitude where each lungful of air provides less oxygen, the body naturally compensates by increasing EPO and producing just enough more red blood cells. An injection of synthetic EPO, however, can stimulate the production of a huge number of extra red blood cells, increasing the oxygen-carrying capacity of the blood beyond normal levels and enhancing performance in sports that require endurance, such as cycling. Why is endurance improved by extra oxygen in the blood? How did this extra oxygen contribute to Riccò's "good legs"? Think about these questions as we examine the role of oxygen in the metabolic pathway that supplies energy to cells.

AT A GLANCE

8.1 How Do Cells Obtain Energy?
8.2 What Happens During Glycolysis?

8.3 What Happens During Cellular Respiration?

8.4 What Happens During Fermentation?

LEARNING GOALS

LG1 Using their overall chemical equations, explain how photosynthesis and glucose breakdown are interdependent processes.

LG2 Summarize and describe the relationship between the two stages of glucose breakdown when oxygen is available.

LG3 Explain the energy investment and energy harvesting phases of glycolysis, including their substrate and product molecules.

LG4 Summarize the relationship between glycolysis and cellular respiration and describe how the structure of the mitochondrion supports cellular respiration.

LG5 Summarize the three stages of cellular respiration, including the input and output molecules at each stage.

LG6 Explain how ATP is generated by chemiosmosis, including the role of the electron transport chain.

LG7 Describe the role of oxygen in cellular respiration.

LG8 Describe the molecules used to capture and transport energy during glycolysis and cellular respiration.

LG9 Compare the amounts of ATP generated and the mechanisms by which the ATP is produced during fermentation, glycolysis, and cellular respiration.

LG10 Explain the function of fermentation, the substrates and products of fermentation, and some ways that people put fermentation to use.

8.1 HOW DO CELLS OBTAIN ENERGY?

Cells require a continuous supply of energy to power the multitude of metabolic reactions that are essential just to stay alive. In this chapter, we describe the cellular reactions that transfer energy from energy-storage molecules, particularly glucose, to energy-carrier molecules, such as ATP.

The second law of thermodynamics tells us that every time a spontaneous reaction occurs, the amount of useful energy in a system decreases and heat is produced (see Chapter 6). Cells are relatively efficient at capturing chemical energy during glucose breakdown when oxygen is available, storing about 40% of the chemical energy from glucose in ATP molecules, and releasing the rest as heat. (If 60% waste heat sounds high, compare this to the 80% of chemical energy released as heat by conventional engines burning gasoline.)

Photosynthesis Is the Ultimate Source of Cellular Energy

The energy utilized by life on Earth comes almost entirely from sunlight, captured during photosynthesis by plants and other photosynthetic organisms, and stored in the chemical bonds of sugars and other organic molecules (see Chapter 7). Almost all organisms, including those that photosynthesize, use glycolysis and cellular respiration to break down these sugars and other organic molecules and capture some of the energy as ATP. Figure 8-1 illustrates the interrelationship between photosynthesis and the breakdown of glucose (by glycolysis and cellular respiration). Glucose ($C_6H_{12}O_6$) breakdown begins with glycolysis in the cell cytosol, liberating small quantities of ATP. Then, the end product of glycolysis is further broken down during cellular respiration in mitochondria, supplying far greater amounts of energy in ATP. In forming ATP during cellular respiration, cells use oxygen (originally released by photosynthetic organisms) and liberate both water and carbon dioxide—the raw materials for photosynthesis.

If you examine their equations on the next page, you will see that the chemical equation describing complete glucose breakdown (by glycolysis and cellular respiration) is the reverse of glucose formation by photosynthesis. The only difference is in the forms of energy involved. The light energy stored

▲ **FIGURE 8-1 Photosynthesis provides the energy released during glycolysis and cellular respiration** The products of photosynthesis are used in glycolysis and cellular respiration, the products of which are, in turn, used in photosynthesis. Photosynthesis captures sunlight energy in glucose. In photosynthetic eukaryotic cells, photosynthesis occurs in chloroplasts. Then glycolysis (in the cytosol) and cellular respiration (in the mitochondria of eukaryotic cells) liberate the chemical energy stored in glucose.

in glucose during photosynthesis is released during glucose breakdown and used to generate ATP, with some lost as heat.

Photosynthesis

$$6 CO_2 + 6 H_2O + \text{light energy} \rightarrow C_6H_{12}O_6 + 6 O_2$$

Complete Glucose Breakdown

$$C_6H_{12}O_6 + 6 O_2 \rightarrow 6 CO_2 + 6 H_2O + \text{ATP energy} + \text{heat energy}$$

Glucose Is a Key Energy-Storage Molecule

Few organisms store glucose in its simple form. Plants convert glucose to sucrose or starch for storage. Humans and many other animals store energy in molecules such as glycogen (a long chain of glucose molecules) and fat (see Chapter 3). Although most cells can use a variety of organic molecules to produce ATP, in this chapter, we focus on the breakdown of glucose, which all cells can use as an energy source. Glucose breakdown occurs in stages, starting with glycolysis and proceeding to fermentation in the absence of oxygen, and to cellular respiration if oxygen is available. Energy is captured in ATP during both glycolysis and cellular respiration, as summarized in **Figure 8-2.**

CHECK YOUR LEARNING

Can you explain how photosynthesis and glucose breakdown are related to one another using their overall chemical equations? Can you summarize glucose breakdown in the presence and absence of oxygen, explaining which reactions generate the most ATP?

8.2 WHAT HAPPENS DURING GLYCOLYSIS?

Glycolysis (from the Greek, "glyco," meaning "sweet," and "lysis," meaning "to split apart") splits the six-carbon glucose molecule into two molecules of pyruvate. Glycolysis has an *energy investment stage* and an *energy harvesting stage*, each with several steps (**Fig. 8-3**). To extract energy from glucose first requires an investment of energy from ATP. During a series of reactions, each of two ATP molecules donates a phosphate group and energy to glucose, forming an "energized" molecule of fructose bisphosphate. Fructose is a sugar very similar to glucose; "bisphosphate" ("bis" means "two") refers to the two phosphate groups acquired from the ATP molecules. Fructose bisphosphate is much more easily broken down than glucose because of the extra energy it has acquired from ATP.

▲ **FIGURE 8-2 A summary of glucose breakdown**

Next, during the energy harvesting stage, fructose bisphosphate is converted into two 3-carbon molecules of glyceraldehyde-3-phosphate, or G3P. Each G3P molecule, which retains one phosphate and some energy from ATP, then undergoes a series of reactions that convert the G3P to pyruvate. During these reactions, two ATP are generated from each G3P, for a total of four ATP per glucose molecule. Because two ATP were used up to form fructose bisphosphate, there is a net gain of only two ATP per glucose molecule during glycolysis. Additional energy is stored when two high-energy electrons and a hydrogen ion (H^+) are added to the electron-carrier **nicotinamide adenine dinucleotide (NAD⁺)** to produce **NADH.** Two molecules of NADH are produced for every glucose molecule broken down. For the details of glycolysis, see "In Greater Depth: Glycolysis."

▲ **FIGURE 8-3 The essentials of glycolysis** *Energy investment stage:* The energy of two ATP molecules is used to convert glucose into the fructose bisphosphate, which then breaks down into two molecules of G3P. *Energy harvesting stage:* The two G3P molecules undergo a series of reactions that capture energy in four ATP and two NADH molecules.

QUESTION What is the net energy yield in ATP and NADH produced?

IN GREATER DEPTH Glycolysis

Glycolysis is a series of enzyme-catalyzed reactions that break down a single molecule of glucose into two molecules of pyruvate. To help you follow the reactions in **Figure E8-1,** we show only the carbon skeletons of glucose and other molecules produced during glycolysis. Each blue arrow represents one (or occasionally more) reactions. Each reaction is catalyzed by an enzyme.

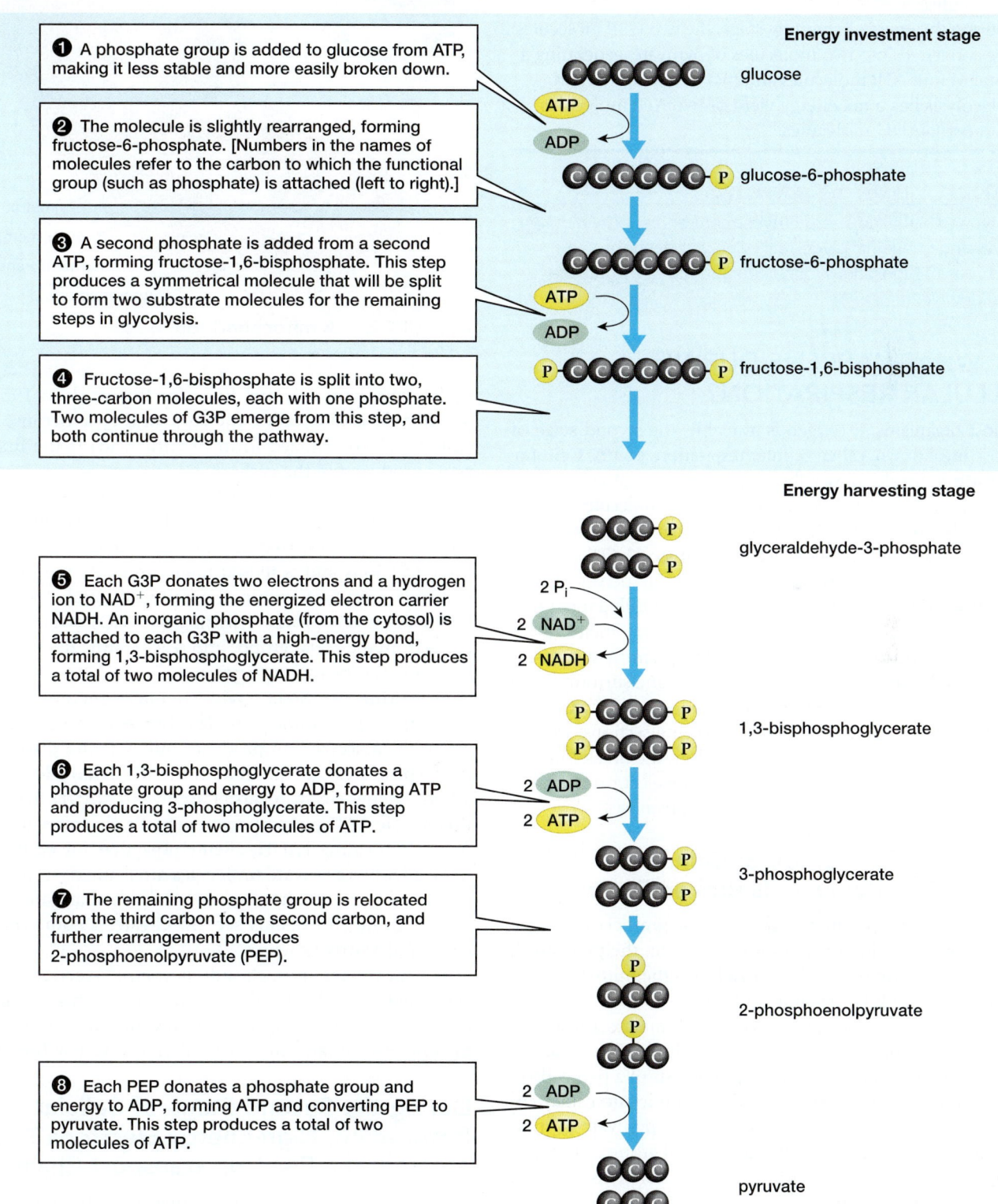

Energy investment stage

❶ A phosphate group is added to glucose from ATP, making it less stable and more easily broken down.

glucose

❷ The molecule is slightly rearranged, forming fructose-6-phosphate. [Numbers in the names of molecules refer to the carbon to which the functional group (such as phosphate) is attached (left to right).]

glucose-6-phosphate

❸ A second phosphate is added from a second ATP, forming fructose-1,6-bisphosphate. This step produces a symmetrical molecule that will be split to form two substrate molecules for the remaining steps in glycolysis.

fructose-6-phosphate

❹ Fructose-1,6-bisphosphate is split into two, three-carbon molecules, each with one phosphate. Two molecules of G3P emerge from this step, and both continue through the pathway.

fructose-1,6-bisphosphate

Energy harvesting stage

glyceraldehyde-3-phosphate

❺ Each G3P donates two electrons and a hydrogen ion to NAD$^+$, forming the energized electron carrier NADH. An inorganic phosphate (from the cytosol) is attached to each G3P with a high-energy bond, forming 1,3-bisphosphoglycerate. This step produces a total of two molecules of NADH.

2 P_i

2 NAD$^+$

2 NADH

1,3-bisphosphoglycerate

❻ Each 1,3-bisphosphoglycerate donates a phosphate group and energy to ADP, forming ATP and producing 3-phosphoglycerate. This step produces a total of two molecules of ATP.

2 ADP

2 ATP

3-phosphoglycerate

❼ The remaining phosphate group is relocated from the third carbon to the second carbon, and further rearrangement produces 2-phosphoenolpyruvate (PEP).

2-phosphoenolpyruvate

❽ Each PEP donates a phosphate group and energy to ADP, forming ATP and converting PEP to pyruvate. This step produces a total of two molecules of ATP.

2 ADP

2 ATP

pyruvate

▲ FIGURE E8-1 Glycolysis

SUMMING UP Glycolysis

- During the energy investment stage, phosphate groups and energy from each of two ATP are added to glucose to produce fructose bisphosphate.
- Fructose bisphosphate is broken down into two G3P molecules.
- During the energy harvesting stage, the two G3P molecules are converted into two molecules of pyruvate, generating a total of four ATP molecules and two NADH molecules.
- Glycolysis has a net energy yield of two ATP molecules and two NADH molecules.

CHECK YOUR LEARNING

Can you explain the energy investment and energy harvesting phases of glycolysis? Can you describe the two types of high-energy molecule produced by glucose breakdown?

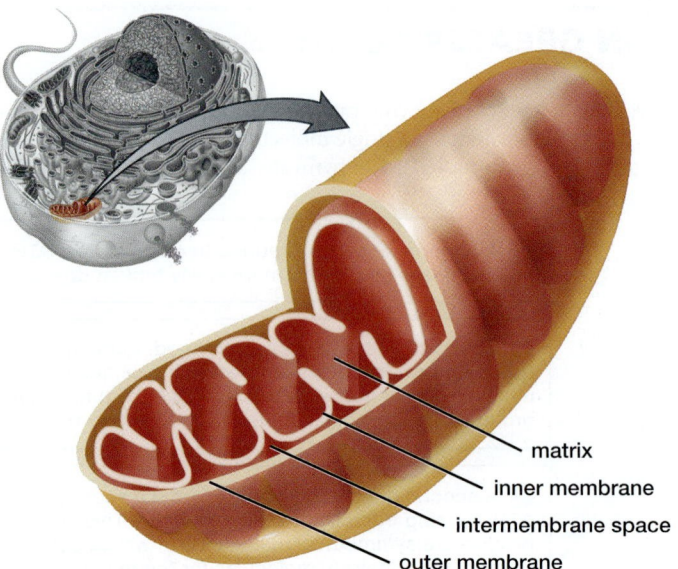

▲ FIGURE 8-4 **A mitochondrion**

matrix
inner membrane
intermembrane space
outer membrane

8.3 WHAT HAPPENS DURING CELLULAR RESPIRATION?

In most organisms, if oxygen is available, the second stage of glucose breakdown, called cellular respiration, occurs. **Cellular respiration** breaks down the two pyruvate molecules (produced by glycolysis) into six carbon dioxide molecules and six water molecules. During this process, the chemical energy from the two pyruvate molecules is used to produce 32 ATP.

In eukaryotic cells, cellular respiration occurs within **mitochondria**, organelles that are sometimes called the "powerhouses of the cell." A mitochondrion has two membranes. The inner membrane encloses a central compartment containing the fluid **matrix**, and the outer membrane surrounds the organelle, producing an **intermembrane space** between the two membranes (**Fig. 8-4**). In the following sections, we discuss the three stages of cellular respiration: the breakdown of pyruvate, the transfer of electrons along the electron transport chain, and the generation of ATP by chemiosmosis.

During the First Stage of Cellular Respiration, Pyruvate Is Broken Down

Pyruvate, the end product of glycolysis, is synthesized in the cytosol. Before cellular respiration can occur, the pyruvate is actively transported from the cytosol into the mitochondrial matrix, where cellular respiration begins.

The reactions of the mitochondrial matrix, illustrated in **Figure 8-5**, occur in two stages: the formation of acetyl CoA and the Krebs cycle. Acetyl CoA consists of a two-carbon (acetyl) functional group attached to a molecule called coenzyme A (CoA). To generate acetyl CoA, pyruvate is split, releasing CO_2 and leaving behind an acetyl group. The acetyl group reacts with CoA, forming acetyl CoA. This reaction liberates energy and stores it by transferring two high-energy electrons and a hydrogen ion to NAD^+, forming NADH.

The next set of mitochondrial matrix reactions is known as the **Krebs cycle** named after its discoverer, Hans

Krebs, who won a Nobel Prize for this work in 1953. The Krebs cycle is also called the *citric acid cycle* because citrate (the dissolved, ionized form of citric acid) is the first molecule produced in the cycle.

The Krebs cycle begins by combining acetyl CoA with a four-carbon molecule, forming a six-carbon citrate molecule and releasing CoA. CoA is not permanently altered during these reactions and is reused many times. As the Krebs cycle proceeds, enzymes within the mitochondrial matrix break down the acetyl group, releasing two CO_2 (whose carbons come from the acetyl group) and regenerating the four-carbon molecule for use in future cycles.

During the Krebs cycle, chemical energy is captured in energy-carrier molecules. The breakdown of each acetyl group from acetyl CoA produces one ATP and three NADH. It also produces one **flavin adenine dinucleotide (FADH₂)**, a high-energy electron carrier similar to NADH. During the Krebs cycle, FAD picks up two energetic electrons along with two H^+, forming $FADH_2$. Remember that for each glucose molecule, two pyruvate molecules are formed, so the energy generated per glucose molecule is twice that generated for one pyruvate (for details see "In Greater Depth: The Mitochondrial Matrix Reactions".)

During the mitochondrial matrix reactions, CO_2 is generated as a waste product. In your body, this CO_2 diffuses into your blood, which carries it to your lungs, so the air you breathe out contains more CO_2 than the air you breathe in.

During the Second Stage of Cellular Respiration, High-Energy Electrons Travel Through the Electron Transport Chain

At the end of the mitochondrial matrix reactions, the cell has gained only four ATP from the original glucose molecule (a net of two during glycolysis and two during the Krebs cycle). During glycolysis and the mitochondrial matrix reactions,

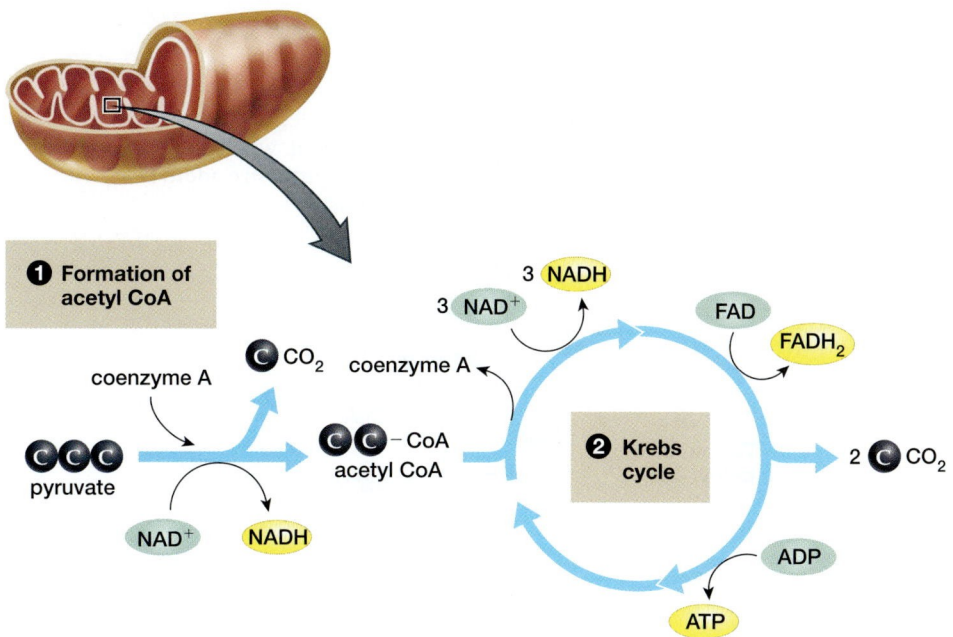

◀FIGURE 8-5 **Reactions in the mitochondrial matrix** ❶ Pyruvate reacts with CoA, forming acetyl CoA and releasing CO_2. During this reaction, two high-energy electrons and a hydrogen ion are added to NAD^+, forming NADH. ❷ When acetyl CoA enters the Krebs cycle, CoA is released for reuse. The Krebs cycle produces one ATP, three NADH, one $FADH_2$, and two CO_2 from each acetyl CoA.

however, the cell has captured many high-energy electrons in a total of 10 NADH and two $FADH_2$ for each glucose molecule. In the next stage of cellular respiration, these carriers each release two high-energy electrons into an **electron transport chain (ETC),** a series of electron transporting molecules, many copies of which are embedded in the inner mitochondrial membrane (**Fig. 8-6**). The depleted carriers are then available for recharging by glycolysis and the Krebs cycle.

The ETCs in the mitochondrial membrane serve the same function as the ETCs embedded in the thylakoid membrane of chloroplasts (see Chapter 7). High-energy electrons jump from molecule to molecule along the ETC, releasing small amounts of energy at each step. Although some energy is lost as heat, some is harnessed to pump H^+ across the inner membrane, from the matrix into the intermembrane space. This ion pumping produces a concentration gradient of H^+, which is then used to generate ATP by chemiosmosis, discussed next.

Finally, at the end of the electron transport chain, the energy-depleted electrons are transferred to oxygen, which acts as an electron acceptor. This step clears out the ETC, allowing it to accept more electrons. The energy-depleted electrons, oxygen, and hydrogen ions combine, forming water. One water molecule is produced for every two electrons that traverse the ETC.

◀FIGURE 8-6 **The electron transport chain** Many copies of the electron transport chain are embedded in the inner mitochondrial membrane. ❶ NADH and $FADH_2$ donate their energetic electrons and hydrogen ions to the electron transport chain. As the electrons pass along the chain (thick gray arrows), some of their energy is used to pump hydrogen ions from the matrix into the intermembrane space (red arrows). ❷ This creates a hydrogen ion gradient that is used to drive ATP synthesis. ❸ At the end of the electron transport chain, the energy-depleted electrons combine with oxygen and hydrogen ions in the mitochondrial matrix, forming water. ❹ The hydrogen ions flow down their concentration gradient from the intermembrane space and into the matrix through ATP synthase channels, generating ATP from ADP and phosphate.

QUESTION How would the rate of ATP production be affected by the absence of oxygen?

IN GREATER DEPTH The Mitochondrial Matrix Reactions

Mitochondrial matrix reactions occur in two stages: (1) the breakdown of pyruvate and formation of acetyl CoA and (2) the Krebs cycle (Fig. E8-2). Here we show only the carbon skeletons of the molecules. The blue arrows represent enzyme-catalyzed reactions.

First Stage: Formation of Acetyl Coenzyme A

Pyruvate is split to form CO_2 and an acetyl group. The acetyl group attaches to CoA to form acetyl CoA. Simultaneously, NAD^+ receives two high-energy electrons and a hydrogen ion to make NADH. Acetyl CoA enters the Krebs cycle.

Second Stage: The Krebs Cycle

❶ Acetyl CoA donates its acetyl group to the four-carbon molecule oxaloacetate, forming citrate. CoA is released. Water is split, donating hydrogen to CoA and oxygen to citrate.

❷ Citrate is rearranged to form isocitrate.

❸ Isocitrate forms α-ketoglutarate by releasing CO_2. Two energetic electrons and an H^+ are captured by NAD^+ to form NADH.

❹ Alpha-ketoglutarate forms succinate by releasing CO_2. Two energetic electrons and an H^+ are captured by NAD^+ to form NADH, and additional energy is captured in a molecule of ATP.

❺ Succinate is converted to fumarate. Two energetic electrons and two H^+ are captured by FAD, forming $FADH_2$.

❻ Fumarate combines with water to form malate.

❼ Malate is converted to oxaloacetate. Two energetic electrons and an H^+ ion are captured by NAD^+ to form NADH.

For each acetyl CoA, the Krebs cycle produces two CO_2, one ATP, three NADH, and one $FADH_2$. The formation of each acetyl CoA prior to the Krebs cycle also generates one CO_2 and one NADH. So for each pyruvate molecule supplied by glycolysis, the mitochondrial matrix reactions produce three CO_2, one ATP, four NADH, and one $FADH_2$. Because each glucose molecule produces two pyruvates, the number of high-energy products and CO_2 per glucose molecule will be double that from a single pyruvate. The electron-carrier molecules NADH and $FADH_2$ will deliver their high-energy electrons to the electron transport chain, where their energy will be used to synthesize additional ATP by chemiosmosis.

▲ FIGURE E8-2 The mitochondrial matrix reactions

Without oxygen to accept electrons, the electrons would be unable to move through the ETC, and H^+ would not be pumped across the inner membrane. The H^+ gradient would rapidly dissipate, and ATP synthesis by chemiosmosis would stop. Most eukaryotic cells would die within minutes without a steady supply of oxygen.

During the Third Stage of Cellular Respiration, Chemiosmosis Generates ATP

Chemiosmosis is the process by which energy is used to generate a concentration gradient of H^+ (see Chapter 7), and then some of that energy is captured in ATP as H^+ flows

When Athletes Boost Their Blood Counts: Do Cheaters Prosper?

When people and other animals exercise vigorously, they are unable to get enough air into their lungs, enough oxygen into their blood, and enough blood circulating to their muscles to allow cellular respiration to meet all their energy needs. As oxygen demand exceeds oxygen supply, muscles must rely on glycolysis (which yields far less ATP than does cellular respiration) for periods of intense exercise. This explains why some athletes, desperate for a competitive edge, may turn to illegal blood doping to increase the ability of their blood to carry oxygen.

down its gradient. As the ETC pumps H^+ across the inner membrane, it produces a high concentration of H^+ in the intermembrane space and a low concentration in the matrix (see Fig. 8-6). Expending energy to produce this nonuniform

distribution of H^+ is similar to charging a battery. This energy is released as hydrogen ions flow down their concentration gradient, a process somewhat like using the energy stored in a battery to light a bulb.

How is the released energy captured in ATP? As in the thylakoid membranes of chloroplasts, the inner membranes of mitochondria are permeable to H^+ only at ATP synthase channels. As hydrogen ions flow from the intermembrane space into the matrix through these ATP-synthesizing enzymes, the flow of ions generates ATP from ADP and inorganic phosphate dissolved in the matrix.

How does ATP escape from the mitochondrion? A carrier protein in the inner membrane uses the energy of the H^+ gradient to transport ATP out of the matrix into the intermembrane space, and simultaneously to transport ADP from the intermembrane space into the matrix. From the intermembrane space, ATP molecules diffuse through large pores in the outer mitochondrial membrane and enter the surrounding cytosol, where they provide the energy needed by the cell. The ADP released by energy-requiring activities in the cell continuously diffuses back through the mitochondrial membrane into the intermembrane space, allowing more ATP synthesis. Without this continuous recycling, life would cease. A person produces, uses, and then regenerates the equivalent of roughly his or her body weight of ATP daily.

You now know why glycolysis followed by cellular respiration generates far more ATP than glycolysis alone. **Figure 8-7** summarizes the breakdown of one glucose molecule in a eukaryotic cell with oxygen present, showing the energy produced during each stage and the general locations where the pathways occur. Chemiosmosis generates 32 ATP from high-energy electron-carrier molecules produced during glycolysis and cellular respiration.

◀ FIGURE 8-7 **The energy sources and ATP harvest from glycolysis and cellular respiration** Here we follow the energy captured from one glucose molecule. Notice that most of the ATP derived from glucose breakdown comes as a result of high-energy electrons donated by NADH and $FADH_2$ moving through the electron transport chain, which allows chemiosmosis to occur.

SUMMING UP Cellular Respiration

- In the mitochondrial matrix, each pyruvate molecule is converted into acetyl CoA, producing one NADH per pyruvate molecule and releasing one CO_2.
- As each acetyl CoA passes through the Krebs cycle, its energy is captured in one ATP, three NADH, and one $FADH_2$. The carbons of acetyl CoA are released in two CO_2 molecules.
- During the matrix reactions, the two pyruvate molecules produced from glucose during glycolysis are completely broken down, yielding two ATP and 10 high-energy electron carriers: eight NADH and two $FADH_2$. The carbon atoms from the pyruvates are released in six molecules of CO_2.
- The NADH and $FADH_2$ molecules donate their energetic electrons to the ETC embedded in the inner mitochondrial membrane.
- As they pass through the ETC, high-energy electrons release energy that is harnessed to pump H^+ into the intermembrane space.
- As they leave the ETC, the energy-depleted electrons combine with the electron acceptor oxygen and hydrogen ions to form water.
- During chemiosmosis, hydrogen ions in the intermembrane space flow down their concentration gradient through ATP synthase channels. The complete breakdown of glucose yields 32 ATP via chemiosmosis.

Cellular Respiration Can Extract Energy from a Variety of Molecules

Glucose often enters the body as starch (a long chain of glucose molecules) or sucrose (table sugar; glucose linked to fructose), but the typical human diet also provides considerable energy in the form of fat, and some from protein. This is possible because various intermediate molecules of cellular respiration can be formed by other metabolic pathways. These molecules can enter cellular respiration at appropriate stages of this metabolic pathway, and are then broken down to produce ATP. For example, the amino acids of protein can serve as an energy source. The amino acid alanine can be converted into pyruvate, which enters the very first stage of cellular respiration. Various other amino acids can be converted to intermediate molecules of the Krebs cycle, thus contributing to the ATP output of the cycle. Fats are an excellent source of energy, and serve as the major energy-storage molecule in animals. To release this energy, fatty acids (which make up most of each fat molecule; see Chapter 3) are combined with CoA and then broken down to produce acetyl CoA molecules, which enter the first stage of the Krebs cycle. This will happen if you limit your food intake and rely on stored fat for energy. But if you overeat, the fats in your meal are stored, and excess sugar and starch are also used to synthesize body fat, as described in "Health Watch: Why Can You Get Fat by Eating Sugar?"

? HAVE **YOU** EVER WONDERED . . .

Why Is Cyanide So Deadly?

Cyanide is a common weapon in old murder mysteries, where the hapless victims of the poison die almost instantly. Cyanide exerts its lethal effects by blocking the last protein in the ETC; this protein is an enzyme that combines electrons with oxygen. If these energy-depleted electrons are not carried away by oxygen, the electrons act like a plug in a pipeline. More high-energy electrons cannot travel through the ETC, so no more hydrogen can be pumped across the membrane, and ATP production by chemiosmosis comes to a screeching halt. Blocking cellular respiration with cyanide can kill a person within a few minutes.

CHECK YOUR LEARNING

Can you summarize the three stages of cellular respiration? Can you explain how ATP is generated by chemiosmosis? Are you able to describe the role of oxygen in cellular respiration?

8.4 WHAT HAPPENS DURING FERMENTATION?

Glycolysis is employed by virtually every organism on Earth, providing evidence that this is one of the most ancient of all biochemical pathways. Scientists hypothesize that the earliest forms of life appeared under the **anaerobic** (no oxygen) conditions that existed before photosynthesis evolved and enriched the air with oxygen. These pioneering life-forms relied on glycolysis for energy production, breaking down the organic molecules formed under the conditions that existed on Earth before life appeared (see Chapter 17). Many microorganisms still thrive in places where oxygen is rare or absent, such as in the stomach and intestines of animals (including humans), deep in soil, or in bogs and marshes. Most of these rely on glycolysis followed by fermentation. Some microorganisms are opportunists, using fermentation when oxygen is absent, but switching to cellular respiration when oxygen is available. Some microorganisms lack the enzymes for cellular respiration and are completely dependent on fermentation.

Fermentation Allows NAD⁺ to Be Recycled When Oxygen Is Absent

Under anaerobic conditions, the second stage of glucose breakdown is **fermentation.** During fermentation, pyruvate in the cytosol is converted either into lactate, or into ethanol and CO_2. Fermentation does not produce additional ATP, so why is it necessary? Fermentation is required to convert the NADH produced during glycolysis back to

Health | Watch

Why Can You Get Fat by Eating Sugar?

From an evolutionary perspective, feeling hungry even when we are overweight and overeating when food is abundant are highly adaptive behaviors. During times of famine, common during early human history, heavier people were more likely to survive, while lean individuals starved. It is only recently (from an evolutionary vantage point) that many people have had continuous access to high-calorie food. Craving foods laden with sugar and fat is another adaptation that helped our prehistoric ancestors survive. But today, the evolutionary drive to eat and the adaptation of storing excess food as fat lead to obesity, a growing health problem in the United States, Mexico, and many other countries throughout the world.

We don't use ATP for long-term energy storage because the large amount of energy stored in its bonds makes ATP too unstable and likely to break down. Fats, however, are stable and also store twice as much energy for their weight as do carbohydrates. Storing energy with minimum weight was important to our prehistoric ancestors who needed to move quickly to catch prey or to avoid becoming prey themselves.

Gaining fat by eating sugar or other carbohydrates is common among animals. For example, a ruby-throated hummingbird begins the summer weighing about 3 to 4 grams (a nickel weighs 5 grams). During late summer, a hummingbird feeds voraciously on the sugary nectar of flowers (Fig. E8-3),

▲ FIGURE E8-3 A ruby-throated hummingbird feeds on nectar

storing enough fat to almost double its weight. The energy in this fat allows it to make its long migration from the eastern United States, across the Gulf of Mexico and into Mexico or Central America for the winter. If the hummingbird stored sugar instead of fat, it would be too heavy to fly.

Because fat storage is so important for survival, cells have metabolic pathways that can chemically transform a variety of foods into fat. In addition to breaking down glucose, the biochemical pathways described in this chapter also participate in producing fat (a fat molecule consists of three fatty acids attached to a glycerol backbone; see Fig. 3-12). When you eat a candy bar loaded with sucrose, for example, this disaccharide is broken down into glucose and fructose. Through slightly different pathways (fructose is only metabolized in the liver), both are converted into G3P. If cells have plenty of ATP, some of this G3P is diverted from cellular respiration and used to make the glycerol backbone of fat. As the sugars continue to be broken down, acetyl CoA is formed (see Fig. 8-5). Excess acetyl CoA molecules are used as raw materials to synthesize the fatty acids that will be linked to the glycerol to form a fat molecule. Starches, such as those in potatoes or bread, are actually long chains of glucose molecules, so you can see how eating excess starch can also cause you to store fat.

NAD^+, which needs to be continuously available for glycolysis to continue.

Under **aerobic** (with oxygen) conditions, most organisms use cellular respiration, regenerating NAD^+ by donating the energetic electrons from NADH to the electron transport chain. But under anaerobic conditions, with no oxygen to allow the ETC to function, the cell must regenerate the NAD^+ using fermentation. During fermentation, the electrons from NADH (along with some hydrogen ions) are donated to pyruvate, changing the pyruvate chemically. Fermentation seems to waste the energy in NADH, because this energy is not used to generate ATP. But if the supply of NAD^+ were to be exhausted—which would happen quickly without fermentation—glycolysis would stop, energy production would cease, and the organism would soon die.

Organisms use one of two types of fermentation to regenerate NAD^+: lactic acid fermentation, which produces lactic acid from pyruvate, or alcoholic fermentation, which

generates alcohol and CO_2 from pyruvate. Because glycolysis followed by fermentation produces far less ATP than does cellular respiration, organisms that rely on fermentation must consume relatively large quantities of energy-rich molecules to meet their ATP requirements.

Some Cells Ferment Pyruvate to Form Lactate

Fermentation of pyruvate to lactate is called **lactic acid fermentation** (in the watery cytosol, lactic acid is dissolved and becomes ionized, forming lactate). When deprived of adequate oxygen, muscles do not immediately stop working. This makes sense from an evolutionary standpoint, because animals exercise most vigorously when fighting, fleeing, or pursuing prey. During such activities, the ability to continue just a little bit longer can make the difference between life and death. When muscles are sufficiently low on oxygen,

▲ **FIGURE 8-8** **Glycolysis followed by lactic acid fermentation**

QUESTION When fermentation follows glycolysis, what is the net energy yield in molecules of ATP and NADH?

they rely on glycolysis to supply its meager two ATP per glucose molecule, which may provide the energy needed for a final, brief burst of speed. Muscle cells then ferment the resulting pyruvate molecules to lactate, using electrons from NADH and hydrogen ions (**Fig. 8-8**).

For example, if you're breathing hard after sprinting to arrive at class on time, your muscles have been relying on glycolysis for part of their energy, and your lungs are working to restore the oxygen needed for cellular respiration (**Fig. 8-9a**). As oxygen is replenished, lactate is converted back into pyruvate in muscle cells, where it will be used

for cellular respiration, and also in the liver, where the pyruvate is converted back into glucose. This glucose may then be stored as glycogen or released back into the bloodstream and distributed to cells throughout the body. The burning sensation you feel as you exert your muscles to their limit is caused by the production of lactic acid, and the burn subsides as adequate oxygen becomes available to restore pyruvate.

A variety of microorganisms use lactic acid fermentation, including the bacteria that convert milk into yogurt, sour cream, and cheese. Acids, including lactic acid, taste sour and contribute to the distinctive tastes of these foods. Lactic acid also denatures milk protein, altering its three-dimensional structure. This causes the milk to thicken, giving sour cream and yogurt their semisolid textures.

CASE STUDY c o n t i n u e d

When Athletes Boost Their Blood Counts: Do Cheaters Prosper?

Why is the average speed of the 5,000-meter run in the Olympics slower than that of the 100-meter dash? During the dash, runners' leg muscles use more ATP than cellular respiration can supply. But anaerobic fermentation can only provide ATP for a short dash. Longer runs must be aerobic, and thus slower, to prevent lactic acid buildup from causing extreme fatigue, muscle pain, and cramps.

>> **LINKS** TO EVERYDAY LIFE

A Jug of Wine, a Loaf of Bread, and a Nice Bowl of Sauerkraut

Persian poet Omar Khayyam (1048–1122) described his vision of paradise on Earth as "A Jug of Wine, a Loaf of Bread—and Thou Beside Me" (**Fig. E8-4**). Historical evidence suggests that wine and beer were commercially produced at least 5,000 years ago. Yeasts are opportunists; they engage in efficient cellular respiration if oxygen is available, but switch to alcoholic fermentation if they run out of oxygen. To make champagne, fermentation is allowed to continue after the bottle is sealed, trapping carbon dioxide, which produces the tiny bubbles for which this drink is famous.

Fermentation also gives bread its airy texture. Bread contains yeast, flour, and water. After dry yeast cells are awakened from their dormant state by water, their enzymes break the starch in flour into its component glucose molecules. As the yeast cells rapidly grow and divide, they release CO_2, first during cellular respiration, and later during alcoholic fermentation when the O_2 is used up. The dough, made stretchy and resilient by kneading, traps the CO_2 gas, which expands in the heat of the oven, giving the bread its porous texture.

For thousands of years, people have relied on bacteria that utilize lactic acid fermentation to convert milk into sour cream, yogurt, and a wide variety of cheeses. In addition, lactate fermentation by salt-loving bacteria converts the sugars in cucumbers and cabbage to lactic acid. The result: pickles and sauerkraut.

▲ **FIGURE E8-4** **Some products of fermentation**

(a)   (b)

◄FIGURE 8-9 **Fermentation in action** (a) Sprinters rely on lactic acid fermentation in their leg muscle cells for their final burst of speed. (b) Bread dough rises because of the millions of tiny CO_2 bubbles released by both cellular respiration and alcoholic fermentation in yeast cells.

QUESTION Some species of bacteria use aerobic respiration and other species use fermentation. In an oxygen-rich environment, would either type be at a competitive advantage? What about in an oxygen-poor environment?

Some Cells Ferment Pyruvate to Form Alcohol and Carbon Dioxide

Many microorganisms, such as yeast (single-celled fungi), engage in **alcoholic fermentation** under anaerobic conditions. During alcoholic fermentation, pyruvate is converted into ethanol and CO_2 (rather than lactate). This releases NAD^+, which is then available to accept more high-energy electrons during glycolysis (**Fig. 8-10**). If you've ever been sufficiently bored to watch bread rise, you've seen the effects of fermentation in

action (**Fig. 8-9b**). For more about how people use fermentation, see "Links to Everyday Life: A Jug of Wine, a Loaf of Bread, and a Nice Bowl of Sauerkraut."

CHECK YOUR LEARNING

Can you explain the function of fermentation? Are you able to describe the conditions under which fermentation occurs? Can you list some examples of human uses of fermentation?

◄FIGURE 8-10 Glycolysis followed by alcoholic fermentation

CASE STUDY revisited

When Athletes Boost Their Blood Counts: Do Cheaters Prosper?

Although runners who do the 100-meter dash rely heavily on lactic acid fermentation to supply ATP, long-distance athletes, including cyclists, marathon runners, and cross-country skiers, must pace themselves. They must rely on aerobic cellular respiration for most of the race, saving the anaerobic sprint for the finish. Training for distance events focuses on increasing the capacity of the athletes' respiratory and circulatory systems to deliver enough oxygen to their muscles. Blood doping most often occurs among distance athletes seeking to increase the oxygen-carrying capacity of their blood so that cellular respiration can generate the maximum amount of ATP from glucose.

The EPO-mimicking drug CERA—that the disgraced cyclist Riccò now admits having taken—helped keep his muscles supplied with ATP by stimulating overproduction of oxygen-carrying red blood cells. In the particularly demanding mountain stages of the Tour de France, which Riccò won, his clean competitors were at a disadvantage because

their leg muscles became painfully laden with lactate from fermentation sooner than Riccò's did.

Because EPO is produced naturally in the human body, its abuse is hard to detect. CERA, developed for use by people with anemia (who have too few red blood cells), was new on the market at the time of the 2008 Tour de France, and Riccò may have assumed it would be undetectable. But CERA's manufacturer, the pharmaceutical firm Hoffman-La Roche, had provided samples of the drug to the World Anti-Doping Agency before it was marketed, allowing researchers to develop urine tests to identify users. This led to Riccò's exposure and disgrace, and his team's devastating disappointment.

Consider This

BioEthics Some athletes move to high-altitude locations to train for races run at lower altitudes, because the low oxygen levels at high altitudes stimulate increased production of red blood cells. Is this cheating? Explain your reasoning. Advances in gene therapy may one day make it possible to modify athletes' cells so that they have extra copies of the gene that produces EPO. Would this be cheating?

CHAPTER REVIEW

Summary of Key Concepts

8.1 How Do Cells Obtain Energy?

The ultimate source of energy for nearly all life is sunlight, captured during photosynthesis and stored in molecules such as glucose. Cells produce chemical energy by breaking down glucose into smaller molecules and capturing some of the released energy as ATP. During glycolysis, glucose is broken down in the cytosol, forming pyruvate and generating a small quantity of ATP and the high-energy electron-carrier NADH. If oxygen is available, the NADH from glycolysis is captured in ATP, and pyruvate is broken down through cellular respiration in the mitochondria, generating many more molecules of ATP.

8.2 What Happens During Glycolysis?

During the energy investment stage of glycolysis, glucose is energized by adding a phosphate group and some energy from each of two ATP molecules, forming fructose bisphosphate. Then, during the energy harvesting stage, a series of reactions breaks down the fructose bisphosphate into two molecules of pyruvate. This produces a net energy yield of two ATP molecules and two NADH molecules.

8.3 What Happens During Cellular Respiration?

Cellular respiration requires oxygen. Pyruvate is transported into the matrix of the mitochondria, where it reacts with coenzyme A, forming acetyl CoA, generating NADH, and releasing CO_2. The acetyl CoA enters the Krebs cycle, which releases the remaining two carbons as CO_2. One ATP, three NADH, and one $FADH_2$ are also formed for each acetyl group that goes through the cycle. In the mitochondrial matrix, each molecule of glucose that originally entered glycolysis produces a total of two ATP, eight NADH, and two $FADH_2$ (see Fig. 8-7).

The NADH and $FADH_2$ deliver their high-energy electrons to the electron transport chain (ETC) embedded in the inner mitochondrial membrane. As they pass along the ETC, energy released by the electrons is used to pump hydrogen ions across the inner membrane from the matrix to the intermembrane space, creating a hydrogen ion gradient. At the end of the ETC, the depleted electrons combine with oxygen (acting as an electron acceptor) and hydrogen ions to form water; this is the oxygen-requiring step of cellular respiration. During chemiosmosis, the energy stored in the hydrogen ion gradient is used to produce ATP as the hydrogen ions diffuse back across the inner membrane through ATP synthase channels. Chemiosmosis yields 32 ATP from the complete breakdown of a glucose molecule. These ATPs are generated by energy supplied by the two NADH formed during glycolysis, and the additional eight NADH and two $FADH_2$ produced during cellular respiration. In addition, a net of 2 ATP is formed directly during glycolysis, and two more ATPs are formed directly during the Krebs cycle. So altogether, a single molecule of glucose provides a net yield of 36 ATP when it is broken down by glycolysis followed by cellular respiration.

8.4 What Happens During Fermentation?

Glycolysis uses up NAD^+ to produce NADH as glucose is broken down into pyruvate. For these reactions to continue, NAD^+ must be continuously recycled. Under anaerobic conditions, NADH cannot release its high-energy electrons to the electron transport chain because there is no oxygen to accept them. Fermentation regenerates NAD^+ from NADH by converting pyruvate to lactate (via lactic acid fermentation) or to ethanol and CO_2 (via alcoholic fermentation), allowing glycolysis to continue.

Key Terms

aerobic *133*
alcoholic fermentation *135*
anaerobic *132*
cellular respiration *128*
chemiosmosis *130*
electron transport chain (ETC) *129*
fermentation *132*
flavin adenine dinucleotide (FAD or FADH₂) *128*

glycolysis *126*
intermembrane space *128*
Krebs cycle *128*
lactic acid fermentation *133*
matrix *128*
mitochondrion (plural, mitochondria) *128*
nicotinamide adenine dinucleotide (NAD^+ or NADH) *126*

Learning Outcomes

In this chapter, you have learned to . . .

LO1 Using their overall chemical equations, explain how photosynthesis and glucose breakdown are interdependent processes.

LO2 Summarize and describe the relationship between the two stages of glucose breakdown when oxygen is available.

LO3 Explain the energy investment and energy harvesting phases of glycolysis, including their substrate and product molecules.

LO4 Summarize the relationship between glycolysis and cellular respiration and describe how the structure of the mitochondrion supports cellular respiration.

LO5 Summarize the three stages of cellular respiration, including the input and output molecules at each stage.

LO6 Explain how ATP is generated by chemiosmosis, including the role of the electron transport chain.

LO7 Describe the role of oxygen in cellular respiration.

LO8 Describe the molecules used to capture and transport energy during glycolysis and cellular respiration.

LO9 Compare the amounts of ATP generated and the mechanisms by which the ATP is produced during fermentation, glycolysis, and cellular respiration.

LO10 Explain the function of fermentation, the substrates and products of fermentation, and some ways that people put fermentation to use.

Thinking Through the Concepts

Fill-in-the-Blank

1. The complete breakdown of glucose in the presence of oxygen occurs in two stages: _____ and _____. The first of these stages occurs in the

_____ of the cell, and the second stage occurs in organelles called _____. Which stage generates the most ATP? _____. **LO2 LO4 LO9**

2. Conditions in which oxygen is absent are described as _____. In the absence of oxygen, some microorganisms break down glucose using _____, which generates only _____ molecules of ATP. This process is followed by _____, in which no more ATP is produced, but the electron-carrier molecule _____ is regenerated so it can be used in further glucose breakdown. **LO3 LO5 LO8 LO9 LO10**

3. Yeasts in bread dough and alcoholic beverages use a type of fermentation that generates _____ and _____. Muscles pushed to their limit use _____ fermentation. Which form of fermentation is used by microorganisms that produce yogurt, sour cream, and sauerkraut? _____ **LO10**

4. The hormone _____ causes production of extra _____ cells, which increase the ability of the blood to carry _____. This gives athletes a competitive edge because their muscle cells can engage in _____ for energy production for a longer time during vigorous exercise. **LO7**

5. During cellular respiration, the electron transport chain pumps H^+ out of the mitochondrial _____ into the _____, producing a large _____ of H^+. The ATP produced by cellular respiration is generated by a process called _____. ATP is generated as H^+ travels through membrane channels linked to _____. **LO4 LO6**

6. The cyclic portion of cellular respiration is called the _____ cycle. The molecule that enters this cycle is _____. How many ATP molecules are generated by the cycle per molecule of glucose? _____ What two types of high-energy electron-carrier molecules are generated during the cycle? _____ and _____ **LO5 LO8 LO9**

Review Questions

1. Starting with glucose ($C_6H_{12}O_6$), write the overall equation for glucose breakdown in the presence of oxygen, compare this to the overall equation for photosynthesis, and explain how the energy components of the equations differ. **LO1**

2. Draw and label a mitochondrion, and explain how its structure relates to its function. **LO4**

3. What role do the following play in breaking down and harvesting energy from glucose: glycolysis, cellular respiration, chemiosmosis, fermentation, and NADH? **LO2 LO3 LO5 LO6 LO8**

4. Outline the two major stages of glycolysis. How many ATP molecules (overall) are generated per glucose molecule during glycolysis? Where in the cell does glycolysis occur? **LO3 LO8 LO9**

5. Under what conditions does fermentation occur? What are its possible products? What is its function? **LO9 LO10**

6. What three-carbon molecule is the end product of glycolysis? How are the carbons of this molecule used in the mitochondrial matrix reactions? In what form is most of the energy from the Krebs cycle captured? **LO3 LO5 LO9**

7. Describe the electron transport chain and the process of chemiosmosis. **LO6**

8. Why is oxygen necessary for cellular respiration to occur? **LO7**

9. Compare the structure of chloroplasts (described in Chapter 7) to that of mitochondria, and describe how the similarities in structure relate to similarities in function. **LO4 LO6**

Applying the Concepts

1. Some years ago a freight train overturned, spilling a load of grain. Because the grain was unusable, it was buried in the railroad embankment. Although there is no shortage of other food, the local bear population has become a nuisance by continually uncovering the grain. Yeasts are common in the soil. What do you think has happened to the grain to make the bears keep digging it up, and how is their behavior related to human cultural evolution?

2. Why can't a victim of cyanide poisoning survive by using anaerobic respiration?

3. Some species of bacteria that live at the surface of sediment on the bottom of lakes are capable of using either glycolysis plus fermentation or cellular respiration to generate ATP. There is very little circulation of water in lakes during the summer. Predict and explain what will happen to the bottommost water of lakes as the summer progresses, and describe how this situation will affect energy production by bacteria.

4. Dumping large amounts of raw sewage into rivers or lakes typically leads to massive fish kills, even if the sewage itself is not toxic to fish. What kills the fish? How might you reduce fish mortality after raw sewage is accidentally released into a small pond?

5. Why is it advantageous for different types of cells to undergo cellular respiration at different rates? How could you predict the relative rates of cellular respiration of different tissues by examining cells through a microscope?

6. Imagine a hypothetical situation in which a starving cell reaches the stage where every bit of its ATP has been depleted and converted to ADP plus phosphate. If at this point you place the cell in a solution containing glucose, will it recover and survive? Explain your answer based on what you know about glucose breakdown.

Answers to Figure Caption questions and Fill-in-the-Blank questions can be found in the Answers section at the back of the book.

MB *Go to MasteringBiology for practice quizzes, activities, eText, videos, current events, and more.*

UNIT 2

Inheritance

The capacity to blunder slightly is the real marvel of DNA. Without this special attribute, we would still be anaerobic bacteria and there would be no music.

"A structure of astounding elegance, a ladder delicately twisting into a double helix, packing into one, efficient strand all the information to create a living being."

G.Santis, Cyprus

CHAPTER 9

The Continuity of Life: Cellular Reproduction

Hines Ward of the Pittsburgh Steelers returned to action in the Super Bowl only 2 weeks after receiving platelet-rich plasma therapy for a sprained knee.

CASE STUDY

Body, Heal Thyself

ON JANUARY 18, 2009, in the first quarter of a National Football League playoff game, the Pittsburgh Steelers' star wide receiver Hines Ward sprained a ligament in his right knee. In Ward's words, "[It] was a severe injury, maybe a four- or six-week injury." But only 2 weeks later, he played in the Super Bowl, catching two passes during the Steelers victory over the Arizona Cardinals.

Healing injured ligaments is a complicated process. Ligaments consist mostly of specialized proteins, including collagen and elastin, organized in a precise, orderly arrangement that provides both strength and flexibility. A relatively small number of cells, which originally produced these proteins, are scattered throughout the ligament. When a ligament is sprained,

broken blood vessels leak blood into the site of injury. Platelets, a type of blood cell, help to form clots, preventing further bleeding, and release a number of different proteins, collectively called growth factors, into the injured ligament. The growth factors attract various types of cells to the site of injury, and stimulate cell division in both these new cells and the existing cells in the ligament. These cells produce new ligament proteins. Ideally, the injured ligament gradually develops the correct protein composition and arrangement, returning to its original size, strength, and flexibility. Unfortunately, this process is slow. To make things worse, it isn't always completely successful.

So how could Hines Ward play football in just 2 weeks? Of course, he had the best treatment that money could buy. Part of that treatment was platelet-rich plasma (PRP) therapy. In PRP therapy, blood is

taken from a patient, centrifuged, and a portion rich in platelets is injected directly into the wound site. Because the patient's own cells are used, there is no risk of rejection or introducing disease organisms from a donor's blood. Although PRP therapy for sports injuries is still considered experimental, studies in mice, rats, rabbits, horses, and humans have found that it stimulates faster and more complete healing for a wide range of damaged tissues, including bone, skin, tendons, ligaments, and muscle.

As you read this chapter, consider the following questions: How do growth factors cause cells to divide? Why don't cells in undamaged tissues divide rapidly all the time? Why are the dividing cells genetically identical to the cells they came from and, in fact, genetically identical to most of the rest of the cells of the body?

AT A GLANCE

LEARNING GOALS

LG1 Describe the fundamental components of inheritance, including chromosomes, DNA, nucleotides, and genes.

LG2 Compare and contrast the categories of cells, as distinguished by their ability to divide and differentiate.

LG3 Compare and contrast sexual reproduction and asexual reproduction.

LG4 Describe the prokaryotic cell cycle, including the mechanisms of prokaryotic fission.

LG5 Describe the structure of a eukaryotic chromosome.

LG6 Define homologous chromosome, autosome, and sex chromosome.

LG7 Distinguish between diploid and haploid cells.

LG8 Describe the eukaryotic cell cycle, including the steps and outcome of mitotic cell division.

LG9 Describe how the eukaryotic cell cycle is regulated.

LG10 Describe the steps and outcome of meiotic cell division.

LG11 Compare and contrast the three main types of eukaryotic life cycles.

LG12 Explain how meiosis and sexual reproduction produce genetic variability in populations.

9.1 WHY DO CELLS DIVIDE?

"All cells come from cells." This insight, first stated by the German physician Rudolf Virchow in the mid-1800s, captures the critical importance of cellular reproduction for all living organisms. Cells reproduce by **cell division,** in which a parent cell usually produces two **daughter cells.** In typical cell division, each daughter cell receives a complete set of hereditary information, identical to the hereditary information of the parent cell, and about half the parent cell's cytoplasm.

Cell Division Transmits Hereditary Information to Each Daughter Cell

The hereditary information of all living cells is contained in **deoxyribonucleic acid (DNA).** DNA is a polymer composed of subunits called **nucleotides** (**Fig. 9-1a;** see also Chapter 3). Each nucleotide consists of a phosphate, a sugar (deoxyribose), and one of four bases—adenine (A), thymine (T), guanine (G), or cytosine (C). A **chromosome** consists of DNA, together with proteins that organize its three-dimensional structure and regulate its use. The DNA in a chromosome consists of two long strands of nucleotides wound around each other, as a ladder would look if it were twisted into a corkscrew shape. This structure is called a double helix (**Fig. 9-1b**).

The units of inheritance, called **genes,** are segments of DNA ranging from a few hundred to many thousands of nucleotides in length. Like the letters of an alphabet, in a language with very long sentences, the specific sequences of nucleotides in genes spell out the instructions for making the proteins of a cell. (We will see in Chapters 11 and 12 how DNA encodes genetic information and how a cell regulates which genes it uses at any given time.)

For a cell to survive, it must have a complete set of genetic instructions. Therefore, when a cell divides, it cannot simply split its set of genes in half and give each daughter cell half a set. Rather, the cell must first duplicate its DNA to make two identical copies, much like making a photocopy of an instruction manual. Each daughter cell then receives a complete "DNA manual" containing all the genes.

Cell Division Is Required for Growth and Development

The familiar form of cell division in eukaryotic cells, in which each daughter cell is genetically identical to the parent cell, is called mitotic cell division (see Sections 9.4 and 9.5). Since your conception as a single fertilized egg, mitotic cell division has produced all the cells in your body, and continues every day in many organs. After cell division, the daughter cells may grow and divide again, or they may **differentiate,** becoming specialized for specific functions, such as contraction (muscle cells), fighting infections (white blood cells), or producing digestive enzymes (cells of the pancreas, stomach, and intestine). This repeating pattern of divide, grow, and (possibly) differentiate, then divide again, is called the **cell cycle** (see Sections 9.2 and 9.4).

◀FIGURE 9-1 **The structure of DNA**
(a) A nucleotide consists of a phosphate, a sugar, and one of four bases—adenine (A), thymine (T), guanine (G), or cytosine (C). A single strand of DNA consists of a long chain of nucleotides held together by bonds between the phosphate of one nucleotide and the sugar of the next. **(b)** Two DNA strands twist around one another to form a double helix.

nucleotide { phosphate, base, sugar }

(a) A single strand of DNA

(b) The double helix

Most multicellular organisms have three categories of cells, based on their abilities to divide and differentiate:

- **Stem Cells** Most of the cells formed by the first few cell divisions of a fertilized egg, and a few cells in adults, including cells in the heart, skin, intestines, fat, brain, and bone marrow, are **stem cells.** Stem cells have two important characteristics: self-renewal and the ability to differentiate into a variety of cell types. Self-renewal means that stem cells retain the capacity to divide, perhaps for the entire life of the organism. Usually, when a stem cell divides, one of its daughters remains a stem cell, maintaining the population of stem cells. The other daughter cell often undergoes several rounds of cell division, but the resulting cells eventually differentiate into specialized cell types. Some stem cells in early embryos can produce any of the specialized cell types of the entire body.
- **Other Cells Capable of Dividing** Many cells of the bodies of embryos, juveniles, and adults can also divide, but each type of cell can typically differentiate into only one or two different cell types. Dividing cells in your liver, for example, can only become more liver cells.
- **Permanently Differentiated Cells** Some cells differentiate and never divide again. For example, most of the cells in your heart and brain cannot divide.

Cell Division Is Required for Sexual and Asexual Reproduction

Organisms reproduce by either or both of two fundamentally different processes: sexual reproduction or asexual reproduction. **Sexual reproduction** in eukaryotic organisms occurs when offspring are produced by the fusion of **gametes** (sperm and eggs) generated in the gonads of adults. Cells in the adult's reproductive system undergo a specialized type of cell division called meiotic cell division, which we will describe in Section 9.8. Meiotic cell division produces daughter cells with exactly half of the genetic information of their parent cells. In animals, these daughter cells become sperm or eggs. When a sperm fertilizes an egg, the resulting offspring once again contains the full complement of genetic information.

Reproduction in which offspring are formed from a single parent, without having a sperm fertilize an egg, is called **asexual reproduction.** Asexual reproduction produces offspring that are genetically identical to the parent and to each other—they are **clones.** Bacteria (**Fig. 9-2a**) and many single-celled eukaryotic organisms, such as *Paramecium* (**Fig. 9-2b**), reproduce asexually by cell division, in which two new cells arise from each preexisting cell. Some multicellular organisms can also reproduce by asexual reproduction. For example, a *Hydra* reproduces by budding. First it grows a small replica of itself, called a bud, on its body (**Fig. 9-2c**). Eventually, the bud is able to live independently and separates from its parent, forming a new *Hydra*. Many plants and fungi can reproduce both asexually and sexually. The beautiful aspen groves of Colorado, Utah, and New Mexico, for example, develop asexually from shoots growing up from the root system of a single parent tree (**Fig. 9-2d**). Although a grove looks like a cluster of separate trees, it is often a single individual whose multiple trunks are interconnected by a common root system. Aspen can also reproduce by seeds, which result from sexual reproduction.

CHECK YOUR LEARNING

Can you describe the types of cells found in a multicellular organism, distinguished by their ability to divide and differentiate? What are the similarities and differences between sexual reproduction and asexual reproduction?

(a) Dividing bacteria

(b) Cell division in *Paramecium*

bud

(c) *Hydra* reproduces asexually by budding

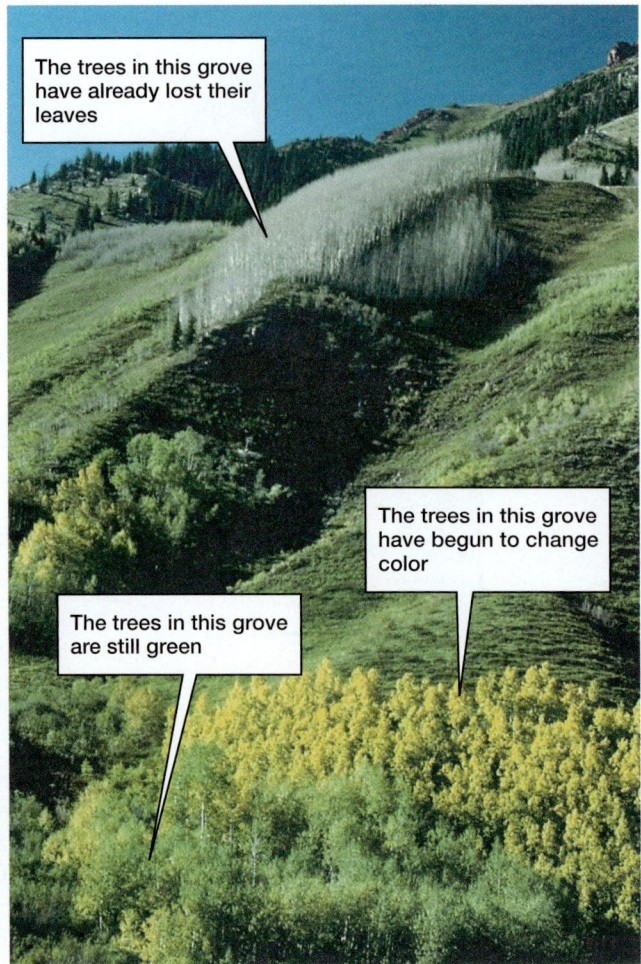

The trees in this grove have already lost their leaves

The trees in this grove have begun to change color

The trees in this grove are still green

(d) A grove of aspens often consists of genetically identical trees produced by asexual reproduction

▲ **FIGURE 9-2 Cell division enables asexual reproduction** **(a)** Bacteria reproduce asexually by dividing in two. **(b)** In single-celled eukaryotic microorganisms, such as the protist *Paramecium* that is commonly found in ponds, cell division produces two new, independent organisms. **(c)** *Hydra*, a freshwater relative of the sea anemone, grows a miniature replica of itself (a bud) on its side. When fully developed, the bud breaks off and assumes independent life. **(d)** Trees in an aspen grove are often genetically identical. Each tree grows up from the roots of a single ancestral tree. This photo shows three separate groves near Aspen, Colorado. In fall, the timing of the changing color of their leaves shows the genetic identity within a grove and the genetic difference between groves.

9.2 WHAT OCCURS DURING THE PROKARYOTIC CELL CYCLE?

The DNA of a prokaryotic cell is contained in a single, circular chromosome about a millimeter or two in circumference. Unlike eukaryotic chromosomes, prokaryotic chromosomes are not contained in a membrane-bound nucleus (see Chapter 4).

The prokaryotic cell cycle consists of a relatively long period of growth—during which the cell replicates its DNA—followed by a type of cell division called **prokaryotic fission** (**Fig. 9-3a**). (Prokaryotic fission is often called "binary fission." However, many biologists use the term binary fission to describe cell division in both prokaryotes and single-celled eukaryotes. To avoid confusion, we will use the term prokaryotic fission to describe cell division in prokaryotic cells.) The prokaryotic chromosome is usually attached at one point to the inside of the plasma membrane of the cell (**Fig. 9-3b ❶**). During the growth phase of the prokaryotic cell cycle, the DNA is replicated, producing two identical chromosomes that become attached to the plasma membrane at nearby, but separate, sites (**Fig. 9-3b ❷**). As the cell grows, new plasma membrane is added between the attachment sites of the chromosomes, pushing them apart (**Fig. 9-3b ❸**). When the cell has approximately doubled in size, the plasma membrane around the middle of the cell grows inward between the two attachment sites (**Fig. 9-3b ❹**). The plasma membrane then fuses along the equator of the cell, producing two daughter cells, each containing one of the chromosomes (**Fig. 9-3b ❺**). Because DNA replication produces two identical DNA molecules, the two daughter cells are genetically identical to one another and to the parent cell.

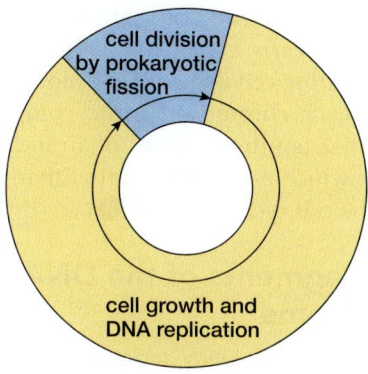

(a) The prokaryotic cell cycle

❶ The prokaryotic chromosome, a circular DNA double helix, is attached to the plasma membrane at one point.

❷ The DNA replicates and the resulting two chromosomes attach to the plasma membrane at nearby points.

❸ New plasma membrane is added between the attachment points, pushing the two chromosomes farther apart.

❹ The plasma membrane grows inward at the middle of the cell.

❺ The parent cell divides into two daughter cells.

(b) Prokaryotic fission

Under ideal conditions, the prokaryotic cell cycle proceeds rapidly. For example, the common intestinal bacterium *Escherichia coli* (usually called simply *E. coli*) can grow, replicate its DNA, and divide in about 20 minutes. Luckily, the environment in our intestines isn't ideal for bacterial growth; otherwise, the bacteria would soon outweigh the rest of our bodies!

CHECK YOUR LEARNING

Can you describe the prokaryotic cell cycle and the major events of prokaryotic fission?

9.3 HOW IS THE DNA IN EUKARYOTIC CHROMOSOMES ORGANIZED?

Eukaryotic chromosomes differ from prokaryotic chromosomes in that they are separated from the cytoplasm in a membrane-bound nucleus. Further, eukaryotic cells always have multiple chromosomes—the smallest number, 2, is found in the cells of females of a species of ant, but most animals have dozens and some ferns have more than 1,200! Finally, eukaryotic chromosomes usually contain far more DNA than do prokaryotic chromosomes. Human chromosomes, for example, contain 10 to 50 times more DNA than the typical prokaryotic chromosome. The complex events of eukaryotic cell division are largely an evolutionary solution to the problem of sorting out a large number of long chromosomes. Therefore, we will begin by taking a closer look at the structure of the eukaryotic chromosome.

The Eukaryotic Chromosome Consists of a Linear DNA Double Helix Bound to Proteins

Each human chromosome contains a single DNA double helix; depending on the chromosome, the length ranges from about 50 million to 250 million nucleotides. If their DNA were completely relaxed and extended, human chromosomes would each be about 0.6 to 3.0 inches long (15 to 75 millimeters), and a single human cell would contain about 6 feet (1.8 meters) of DNA.

Fitting this huge amount of DNA into a nucleus only a few ten-thousandths of an inch in diameter is no trivial task. For most of a cell's life, the DNA in each chromosome is wrapped around proteins called histones (**Fig. 9-4** ❶ and ❷). The DNA/histone beads further coil up like a spring or Slinky toy (**Fig. 9-4** ❸), and then these coils are bent by attaching them to "scaffolds" made of additional proteins. All of this wrapping and coiling condenses the DNA to about 1/1,000th of its original length (**Fig. 9-4** ❹). However, even this enormous degree of compaction still leaves the chromosomes much too long to be sorted out and moved into the daughter nuclei during cell division. Just as thread is easier

◀ **FIGURE 9-3 The prokaryotic cell cycle (a)** The prokaryotic cell cycle consists of growth and DNA replication, followed by prokaryotic fission. **(b)** The process of prokaryotic fission.

to organize when it is wound onto spools, sorting and transporting chromosomes is easier when they are condensed and shortened. During cell division, additional proteins fold up the DNA of each chromosome into compact structures that are less than 2 ten-thousandths of an inch long (about 4 micrometers), which is 10 times shorter than they are during the rest of the cell cycle (**Fig. 9-4 ❺**).

Genes Are Segments of the DNA of a Chromosome

Genes are sequences of DNA ranging from hundreds to thousands of nucleotides long. Each gene occupies a specific place, or **locus** (plural, loci), on a chromosome (**Fig. 9-5a**). Chromosomes vary in the number of genes they contain. Although the exact numbers remain uncertain, human chromosomes may contain as few as 70 genes (the lowest estimate for the Y chromosome) to over 3,000 genes (in chromosome 1, the largest chromosome).

In addition to genes, every chromosome has specialized regions that are crucial to its structure and function: two telomeres and one centromere (see Fig. 9-5a). **Telomeres** ("end body" in Greek) are protective caps at each end of a chromosome. Without telomeres, genes located at the ends

① DNA double helix

histone proteins

② DNA wound around histone proteins

③ Coiled DNA/histone beads

④ Loops attached to a protein scaffold; this stage of partial condensation typically occurs in a nondividing cell

protein scaffold

⑤ Folded chromosome, fully condensed in a dividing cell

▲ **FIGURE 9-4 Chromosome structure** ① A eukaryotic chromosome contains a single DNA double helix. ② The DNA is wound around proteins called histones, reducing the length by about a factor of 6. ③ Other proteins coil up the DNA/histone beads, much like a Slinky toy, reducing the length by another factor of 6 or 7. ④ These coils are attached in loops to protein "scaffolding" to complete the chromosome as it occurs during most of the life of a cell. The wrapping, coiling, and looping make the chromosome roughly 1,000 times shorter than the DNA molecule it contains. ⑤ During cell division, still other proteins fold up the chromosome, yielding about another 10-fold condensation. (Inset) The fuzzy edges visible in the electron micrograph are loops of folded chromosome.

gene loci

centromere

telomeres

(a) A eukaryotic chromosome (one DNA double helix) before DNA replication

sister chromatids

duplicated chromosome (two DNA double helices)

centromere

(b) A eukaryotic chromosome after DNA replication

independent daughter chromosomes, each with one identical DNA double helix

(c) Separated sister chromatids become independent chromosomes

▲ **FIGURE 9-5 The principal features of a eukaryotic chromosome during cell division (a)** Before DNA replication, each chromosome consists of a single DNA double helix. Genes are segments of the DNA, usually hundreds to thousands of nucleotides in length. The ends of the chromosome are protected by telomeres. **(b)** The two sister chromatids of a duplicated chromosome are held together by the centromere. **(c)** The sister chromatids separate during cell division to become two independent, genetically identical, chromosomes.

of the chromosomes would be lost during DNA replication. Telomeres also keep chromosomes from fusing with one another and forming long, unwieldy structures that probably could not be distributed properly to the daughter cells during cell division. The second specialized region of the chromosome is the **centromere.** As we will see, the centromere has two principal functions: (1) It temporarily holds two daughter DNA double helices together after DNA replication, and (2) it is the attachment site for microtubules that move the chromosomes during cell division.

Duplicated Chromosomes Separate During Cell Division

Prior to cell division, the DNA in each chromosome is replicated (see Chapter 11). At the end of DNA replication, a **duplicated chromosome** consists of two identical DNA double helices, now called sister **chromatids,** which are attached to each other at the centromere (**Fig. 9-5b**). During mitotic cell division, the two sister chromatids separate, each becoming an independent chromosome that is delivered to one of the two daughter cells (**Fig. 9-5c**).

Eukaryotic Chromosomes Usually Occur in Pairs Containing Similar Genetic Information

The chromosomes of each eukaryotic species have characteristic sizes, shapes, and patterns of bands when stained by dyes used in light microscopy. The entire set of chromosomes from a single cell is called its **karyotype** (**Fig. 9-6**). For most eukaryotic organisms, including humans, a karyotype consists of pairs of chromosomes. The two chromosomes that make up a pair are called **homologous chromosomes,** or **homologues,** from Greek words that mean "to say the same thing." Cells with pairs of homologous chromosomes are **diploid,** meaning "double." As the karyotype in Figure 9-6 shows, a typical human cell is diploid, with 23 pairs of chromosomes, for a total of 46.

Homologous chromosomes are the same length and have the same staining pattern, because both chromosomes in the pair contain the same genes arranged in the same order. These chromosomes of similar appearance and similar DNA sequences, called **autosomes,** are paired in diploid cells of both sexes. People have 22 pairs of autosomes.

In addition to autosomes, humans and most other mammals have two **sex chromosomes:** either two X chromosomes (in females) or an X and a Y chromosome (in males). Although X and Y chromosomes are quite different in size (see Fig. 9-6) and in genetic composition, a small portion of the X and Y chromosomes are homologous to each other.

Homologous Chromosomes Are Usually Not Identical

Despite their name, the two homologues in a pair don't say exactly the "same thing." Why not? A cell might make a mistake when it copies the DNA of one homologue but not the other (see Chapter 11). Or a ray of ultraviolet light from the sun might zap the DNA of one homologue but not the other. The resulting changes in the sequence of nucleotides in DNA are called **mutations.** Mutations make some

one duplicated chromosome

sister chromatids

a pair of homologous chromosomes

46 XY

1 2 3 4 5

6 7 8 9 10 11 12

13 14 15 16 17 18

sex chromosomes

19 20 21 22 X Y

◄**FIGURE 9-6 The karyotype of a human male** Staining and photographing the entire set of duplicated chromosomes from a single cell produces a karyotype. Pictures of the individual chromosomes are cut out and arranged in descending order of size. The chromosome pairs (homologues) are similar in size and have similar genetic material. Chromosomes 1 through 22 are the autosomes; the X and Y chromosomes are the sex chromosomes. Note that the Y chromosome is much smaller than the X chromosome. If this were a female karyotype, it would have two X chromosomes and no Y chromosomes.

▲ **FIGURE 9-7 Homologous chromosomes may have the same or different alleles of individual genes** The two homologues have the same genes at the same locations (loci). The homologues may have the same allele of a gene at certain loci (left) and different alleles of a gene at other loci (right).

genes on one homologue slightly different from the same genes on the other homologue. A given mutation might have happened yesterday, or it may have occurred in a sperm or egg 10,000 years ago and been inherited ever since. Such enduring mutations, inherited generation after generation, are called **alleles**—alternate forms of a given gene that may produce differences in structure or function, such as black, brown, or blond hair in humans, or different songs in birds. Thus, although homologous chromosomes have the same genes, a pair of homologues may have the same alleles of some genes and different alleles of other genes (**Fig. 9-7**). (We'll explore the consequences of having paired genes, and more than one allele of each gene, in Chapter 10.)

If we think of DNA as an instruction manual for building a cell or an organism, then mutations are like misspelled words in the manual. Although some misspellings might not matter much, others can have serious consequences. For example, single-letter misspellings in crucial genes are the alleles that can cause inherited diseases such as sickle-cell anemia and cystic fibrosis. Occasionally, however, a mutation may actually improve the DNA manual and spread throughout a species, if the organisms carrying the mutation are more likely to survive and reproduce than members of the species that lack the mutation (see Chapters 14 and 15).

Not All Cells Have Paired Chromosomes

Most cells in our bodies are diploid. However, during sexual reproduction, cells in the ovaries or testes undergo meiotic cell division (described in Section 9.8) to produce gametes (sperm or eggs). Each gamete contains only one member of each pair of autosomes and one of the two sex chromosomes. Cells that contain only one of each type of chromosome are called **haploid** (meaning "single"). In humans, a haploid cell contains one of each of the 22 autosomes, plus either an X or Y sex chromosome, for a total of 23 chromosomes.

When a sperm fertilizes an egg, fusion of the two haploid cells produces a diploid cell that once again contains pairs of homologous chromosomes. Because one homologue of each pair is inherited from the mother (in her egg), it is called a maternal chromosome. Each homologue inherited from the father (in his sperm) is called a paternal chromosome.

In biological shorthand, the number of different types of chromosomes in a species is called the haploid number and is designated n. For humans, $n = 23$ because we have 23 different types of chromosomes (22 autosomes plus one sex

chromosome). Diploid cells contain $2n$ chromosomes. Thus, the body cells of humans have 46 (2×23) chromosomes.

Not all eukaryotic organisms are diploid. The bread mold *Neurospora*, for example, has haploid cells for most of its life cycle. Many plants, on the other hand, have more than two copies of each type of chromosome, with $4n$ (tetraploid), $6n$ (hexaploid), or even more chromosomes per cell. Many common flowers, including some daylilies, orchids, lilies, and phlox, are tetraploid; most wheat is either tetraploid or hexaploid.

CHECK YOUR LEARNING

Can you describe the structure of a eukaryotic chromosome? Can you define the terms *homologous chromosome*, *autosome*, and *sex chromosome*? What is the difference between a diploid and a haploid cell?

9.4 WHAT OCCURS DURING THE EUKARYOTIC CELL CYCLE?

In the eukaryotic cell cycle, newly formed cells usually acquire nutrients from their environment, synthesize more cytoplasm and organelles, and grow larger. After a variable amount of time—depending on the organism, the type of cell, and the nutrients available—the cell may divide. Each daughter cell may then enter another cell cycle, producing more cells. Many cells, however, divide only if they receive signals, such as hormone-like molecules called growth factors, that cause them to enter another cell cycle (see Section 9.6). Other cells may differentiate and never divide again.

The Eukaryotic Cell Cycle Consists of Interphase and Mitotic Cell Division

The eukaryotic cell cycle is divided into two major phases: interphase and mitotic cell division (**Fig. 9-8**).

During Interphase, a Cell Grows in Size, Replicates Its DNA, and Often Differentiates

Most eukaryotic cells spend the majority of their time in **interphase,** the period between cell divisions. For example, some cells in human skin spend roughly 22 hours in interphase and only a couple of hours dividing. Interphase itself contains three subphases: G_1 (the first growth phase and the first gap in DNA synthesis), S (DNA synthesis), and G_2 (the second growth phase and the second gap in DNA synthesis).

A newly formed daughter cell enters the G_1 portion of interphase. During G_1, a cell carries out one or more of three activities. First, it almost always grows in size. Second, it often differentiates, developing the structures and biochemical pathways that allow it to perform a specialized function. For example, most nerve cells grow long strands, called axons, that allow them to connect with other cells, whereas liver cells produce bile, proteins that aid blood clotting, and enzymes that detoxify many poisonous materials. Third, the cell is sensitive to internal and external signals that determine whether or not it will divide. If the cell is stimulated to divide, it enters the S phase, when DNA synthesis

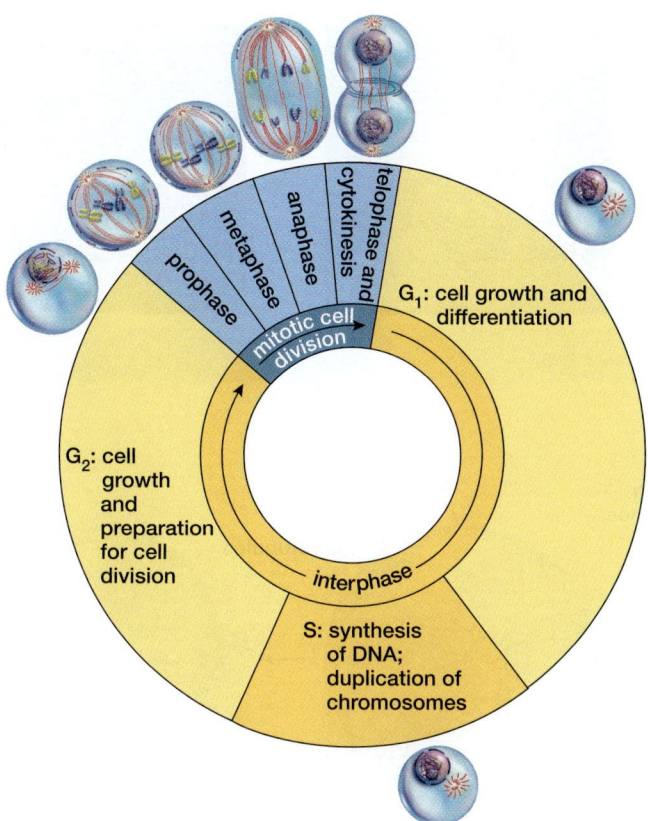

▲ FIGURE 9-8 **The eukaryotic cell cycle** The eukaryotic cell cycle consists of interphase and mitotic cell division.

(replication) occurs. The cell then proceeds to the G_2 phase, during which it may grow some more and then synthesize the proteins needed for cell division.

Many differentiated cells, such as liver cells, can be recalled from the differentiated state back into the dividing state, whereas others, such as most heart muscle and nerve cells, remain in interphase and never divide again. That is

CASE STUDY continued

Body, Heal Thyself

Like platelet-rich plasma therapy, stem cell injection is a potential treatment for wounds. For horses with many types of leg injuries, a veterinarian can take a small sample of a horse's own fat and send it to a company called Vet-Stem, which isolates stem cells from the fat. Vet-Stem sends the stem cells back to the vet, who injects them into the horse's injured leg. The stem cells divide and differentiate, producing specialized cells that repair the damaged structures.

In 2010, professional baseball pitcher Bartolo Colon received similar stem-cell therapy for a shoulder injury that never fully healed after rotator cuff surgery. In 2011, he was back pitching for the New York Yankees, with a 90+ mph fastball. Effective stem cell therapy? Or just more time to heal? No one knows for sure, but Major League Baseball seems to believe that the therapy works, and has launched an investigation into whether stem cell therapy might produce superhuman powers, and hence should be banned as a performance-enhancing drug.

one reason why heart attacks and strokes are so devastating: The dead cells usually cannot be replaced. However, the heart and brain contain a few stem cells that *can* divide. Biomedical researchers hope that, one day, these stem cells can be coaxed into dividing rapidly and repairing damaged organs.

We will investigate how the cell cycle is controlled in Section 9.6.

Mitotic Cell Division Consists of Nuclear Division and Cytoplasmic Division

Mitotic cell division, the division of one parental cell into two daughter cells, consists of two processes: mitosis and cytokinesis. **Mitosis** is the division of the nucleus. The word "mitosis" is derived from a Greek word meaning "thread," because, in its early stages, the chromosomes condense and shorten, becoming visible in a light microscope as thread-like structures. Mitosis produces two daughter nuclei, each containing one copy of each of the chromosomes that were present in the parent nucleus. **Cytokinesis** (from Greek words meaning "cell movement") is the division of the cytoplasm. Cytokinesis places about half the cytoplasm, half the organelles (such as mitochondria, ribosomes, and Golgi apparatus), and one of the newly formed nuclei into each of two daughter cells. Thus, mitotic cell division typically produces daughter cells that are physically similar and genetically identical to each other and to the parent cell.

Mitotic cell division takes place in all eukaryotic organisms. It is the mechanism of asexual reproduction in eukaryotic cells, including single-celled organisms such as yeast, amoebas, and *Paramecium*, and multicellular organisms such as *Hydra* and aspens. Mitotic cell division followed by differentiation of the daughter cells allows a fertilized egg to grow into an adult with perhaps trillions of specialized cells. Mitotic cell division also allows an organism to maintain its tissues, many of which require frequent replacement; to repair damaged parts of the body, for example, after an injury; and sometimes even to regenerate body parts. Mitotic cell division is also the mechanism by which stem cells reproduce.

CHECK YOUR LEARNING

Can you describe the events of the eukaryotic cell cycle? What is the difference between mitotic cell division and mitosis?

9.5 HOW DOES MITOTIC CELL DIVISION PRODUCE GENETICALLY IDENTICAL DAUGHTER CELLS?

Remember that the chromosomes have already been duplicated during the S phase of interphase. Therefore, when mitosis begins, each chromosome consists of two sister chromatids attached at the centromere (see Fig. 9-5b).

For convenience, biologists divide mitosis into four phases, based on the appearance and behavior of the chromosomes: prophase, metaphase, anaphase, and telophase (**Fig. 9-9**). However, these phases are not really discrete

INTERPHASE | **MITOSIS**

nuclear envelope
chromatin
nucleolus
condensing chromosomes
spindle pole
spindle microtubules
centriole pairs
beginning of spindle formation
kinetochore
spindle pole
kinetochore microtubules

(a) Late Interphase
Duplicated chromosomes are in the relaxed uncondensed state; duplicated centrioles remain clustered.

(b) Early Prophase
Chromosomes condense and shorten; spindle microtubules begin to form between separating centriole pairs.

(c) Late Prophase (also called Prometaphase)
The nucleolus disappears; the nuclear envelope breaks down; some spindle microtubules attach to the kinetochore (blue) located at the centromere of each sister chromatid.

(d) Metaphase
Kinetochore microtubules line up the chromosomes at the cell's equator.

▲ **FIGURE 9-9** **Mitotic cell division in an animal cell**

QUESTION What would the consequences be if one set of sister chromatids failed to separate at anaphase?

events; they instead form a continuum, with each phase merging into the next.

During Prophase, the Chromosomes Condense, the Spindle Microtubules Form, the Nuclear Envelope Breaks Down, and the Chromosomes Are Captured by the Spindle Microtubules

The first phase of mitosis is called **prophase** (meaning "the stage before" in Greek). During prophase, four major events occur: (1) The duplicated chromosomes condense (see Fig. 9-4), (2) the spindle microtubules form, (3) the nuclear envelope breaks down, and (4) the chromosomes are captured by the spindle microtubules (**Figs. 9-9b, c**). Chromosome condensation also causes the nucleolus to disappear. The nucleolus

consists of partially assembled ribosomes and the genes, located on several different chromosomes, that code for the RNA of the ribosomes (see Chapter 4). As the chromosomes condense, they separate from one another and ribosome synthesis ceases, so the nucleolus gradually fades away.

After the duplicated chromosomes condense, the **spindle microtubules** begin to assemble. In all eukaryotic cells, the movement of chromosomes during mitosis depends on spindle microtubules. In animal cells, the spindle microtubules originate from a region that contains a pair of microtubule-containing structures called **centrioles.** Although the cells of plants, fungi, many algae, and even some mutant fruit flies do not contain centrioles, they nevertheless form functional spindles during mitotic cell division, showing that centrioles are not required for spindle formation.

INTERPHASE

polar microtubules

chromosomes extending

nuclear envelope re-forming

microfilaments

nucleolus reappearing

(e) Anaphase
Sister chromatids separate and move to opposite poles of the cell; polar microtubules push the poles apart.

(f) Telophase
One set of chromosomes reaches each pole and begins to decondense; nuclear envelopes start to form; nucleoli begin to reappear; spindle microtubules begin to disappear; microfilaments form rings around the equator.

(g) Cytokinesis
The ring of microfilaments contracts, dividing the cell in two; each daughter cell receives one nucleus and about half of the cytoplasm.

(h) Interphase of daughter cells
Spindles disappear, intact nuclear envelopes form, and the chromosomes extend completely.

In animal cells, a new pair of centrioles forms during interphase near the previously existing pair. During prophase, the two centriole pairs migrate to opposite sides of the nucleus (see Fig. 9-9b). When the cell divides, each daughter cell will receive a pair of centrioles. Each centriole pair serves as a central point from which the spindle microtubules radiate, both inward toward the nucleus and outward toward the plasma membrane (see Fig. 9-9c). These points are called spindle poles. (Picture the cell as a globe of Earth. The spindle poles are roughly where the north and south poles would be, and the spindle microtubules correspond to the lines of longitude. As on a globe, the equator of the cell cuts across the middle, halfway between the poles.)

As the spindle microtubules form into a complete basket around the nucleus, the nuclear envelope disintegrates, releasing the duplicated chromosomes. Each sister chromatid

in a duplicated chromosome has a protein-containing structure, called a **kinetochore,** located at its centromere. The kinetochores of the two sister chromatids are arranged back-to-back, facing away from one another. The kinetochore of one sister chromatid binds to the ends of spindle microtubules leading to one pole of the cell, while the kinetochore of the other sister chromatid binds to spindle microtubules leading to the opposite pole of the cell (see Fig. 9-9c). The microtubules that bind to kinetochores are called kinetochore microtubules, to distinguish them from microtubules that do not bind to a kinetochore (discussed in the next paragraph). When the sister chromatids separate later in mitosis, the newly independent chromosomes will move along the kinetochore microtubules to opposite poles.

Other spindle microtubules, called polar microtubules, do not attach to kinetochores or to any other part of the chromosomes; rather, they have free ends that overlap along the cell's equator. As we will see, the polar microtubules will push the two spindle poles apart later in mitosis.

During Metaphase, the Chromosomes Line Up Along the Equator of the Cell

At the end of prophase, the two kinetochores of each duplicated chromosome are connected to spindle microtubules leading to opposite poles of the cell. As a result, each duplicated chromosome is connected to both spindle poles. During **metaphase** (the "middle stage"), the two kinetochores on a duplicated chromosome pull toward opposite poles of the cell. During this process, the microtubules lengthen or shorten, until each chromosome lines up along the equator of the cell, with one kinetochore facing each pole (**Fig. 9-9d**).

During Anaphase, Sister Chromatids Separate and Are Pulled to Opposite Poles of the Cell

At the beginning of **anaphase** (**Fig. 9-9e**), the sister chromatids separate, becoming independent daughter chromosomes. This separation allows each kinetochore to move its chromosome poleward, while simultaneously nibbling off the end of the attached microtubule, thereby shortening it (a mechanism appropriately called "Pac-Man" movement). One of the two daughter chromosomes derived from each parental chromosome moves to each pole of the cell. Because the daughter chromosomes are identical copies of the parental chromosomes, each cluster of chromosomes that forms at opposite poles of the cell contains one copy of every chromosome that was in the parent cell.

At about the same time, polar microtubules radiating from each pole grab one another where they overlap at the equator. These polar microtubules then simultaneously lengthen and push on one another, forcing the poles of the cell apart (see Fig. 9-9e).

During Telophase, a Nuclear Envelope Forms Around Each Group of Chromosomes

When the chromosomes reach the poles, **telophase** (the "end stage") begins (**Fig. 9-9f**). The spindle microtubules disintegrate and a nuclear envelope forms around each group of chromosomes. The chromosomes revert to their extended state, and nucleoli begin to re-form. In most cells, cytokinesis occurs during telophase, isolating each daughter nucleus in its own daughter cell (**Fig. 9-9g**). However, mitosis sometimes occurs without cytokinesis, producing cells with multiple nuclei.

During Cytokinesis, the Cytoplasm Is Divided Between Two Daughter Cells

Cytokinesis differs considerably between animal cells and plant cells. In animal cells, microfilaments attached to the plasma membrane assemble into a ring around the equator of the cell, usually late in anaphase or early in telophase (see Fig. 9-9f). The ring contracts and constricts the cell's equator, much like pulling the drawstring on a pair of sweatpants tightens the waist (see Fig. 9-9g). Eventually the "waist" of the parent cell constricts completely, dividing the cytoplasm into two new daughter cells (**Fig. 9-9h**).

Cytokinesis in plant cells is quite different, perhaps because their stiff cell walls make it impossible to divide one cell into two by pinching at the waist. Instead, carbohydrate-filled sacs called vesicles bud off the Golgi apparatus and line up along the equator of the cell between the two nuclei (**Fig. 9-10**). The vesicles fuse, producing a structure called the **cell plate,** which is shaped like a flattened sac, surrounded by membrane and filled with sticky carbohydrates. When enough vesicles have fused, the edges of the cell plate merge with the plasma membrane around the circumference of the cell. The two sides of the cell plate membrane become new plasma membranes between the two daughter cells. The carbohydrates formerly contained in the vesicles remain between the plasma membranes as the beginning of the new cell wall.

cell plate forming a new cell wall

Golgi apparatus
cell wall
plasma membrane
carbohydrate-filled vesicles

❶ Carbohydrate-filled vesicles bud off the Golgi apparatus and move to the equator of the cell.

❷ The vesicles fuse to form a new cell wall (red) and plasma membrane (yellow) between the daughter cells.

❸ Complete separation of the daughter cells.

◄ **FIGURE 9-10 Cytokinesis in a plant cell**

Body, Heal Thyself

The precision of mitotic cell division is essential for wound repair, because it ensures that the daughter cells are genetically identical to their parent cells. Imagine what might happen if DNA synthesis during interphase did not copy all of the genes accurately, or if mitotic cell division sent random numbers and types of chromosomes into the daughter cells. Some of the daughter cells might not contain all the genes needed to form the various cell types that are required to repair damaged tissues. Worse yet, some daughter cells might have genetic changes that would cause cancer, in which cells divide uncontrollably.

CHECK YOUR LEARNING

Can you describe the steps of mitotic cell division? What is the usual outcome of mitotic cell division? How does cytokinesis differ in plant and animal cells?

9.6 HOW IS THE CELL CYCLE CONTROLLED?

Cell division is regulated by a bewildering array of molecules, not all of which have been identified and studied. Nevertheless, some general principles apply to cell cycle control in most eukaryotic cells.

The Activities of Specific Proteins Drive the Cell Cycle

Normal cell cycle control works like this: During development, after an injury, or to compensate for normal wear and tear, a variety of cells in the body release hormone-like molecules called **growth factors.** Most growth factors stimulate cell division by controlling the synthesis of proteins inside the cell called cyclins, which in turn regulate the actions of enzymes called cyclin-dependent kinases. The name "cyclin" was chosen because these proteins help to govern the cell cycle. Cyclin-dependent kinases (Cdks) get their name from two features: A "kinase" is an enzyme that adds a phosphate group to another protein, stimulating or inhibiting the activity of the target protein. "Cyclin dependent" means that the kinase is active only when it binds cyclin.

Let's see how growth factors, cyclins, and Cdks stimulate cell division, for example, if you cut your skin (**Fig. 9-11**). Platelets (blood cells that are involved in clotting) accumulate at the wound site and release several growth factors, including platelet-derived growth factor and epidermal growth factor. These growth factors bind to receptors on the surfaces of cells in damaged areas of the skin (**Fig. 9-11 ❶**), stimulating the cells to synthesize cyclin proteins (**Fig. 9-11 ❷**). Cyclins bind to specific Cdks (**Fig. 9-11 ❸**), forming cyclin–Cdk complexes that promote the manufacture and activity of the proteins required for DNA synthesis (**Fig. 9-11 ❹**). The cells enter the S phase and replicate their DNA. After DNA replication is complete, other Cdks become activated during G_2 and mitosis, causing chromosome condensation, breakdown of the nuclear envelope, formation of the spindle, and attachment of the chromosomes to the spindle microtubules. Finally, still other Cdks stimulate processes that allow the sister chromatids to separate into individual chromosomes and move to opposite poles during anaphase.

(interstitial fluid)

growth factor

❶ Growth factor binds to its receptor

growth factor receptor

❷ Cyclins are synthesized

plasma membrane

cyclin

cyclin-dependent kinase (Cdk)

❹ Cyclin activates Cdk; active Cdk stimulates DNA replication

❸ Cyclin binds to Cdk

(cytosol)

◀ **FIGURE 9-11 Growth factors stimulate cell division** Progress through the cell cycle is under the overall control of cyclin and cyclin-dependent kinases (Cdks). In most cases, growth factors stimulate synthesis of cyclin proteins, which activate Cdks, starting a cascade of events that lead to DNA replication and cell division.

Why Dogs Lick Their Wounds?

The saliva of dogs, like the saliva of most mammals (including humans), contains enzymes, antibacterial compounds, epidermal growth factor (EGF), and a variety of other growth factors. When a dog licks a wound, it not only cleans out some of the dirt that may have entered, but also leaves EGF and other growth factors behind. The growth factors speed up the synthesis of cyclins, thereby stimulating the division of cells that regenerate the skin, helping to heal the wound more rapidly.

Checkpoints Regulate Progress Through the Cell Cycle

Unregulated cell division can be dangerous. If a cell contains mutations in its DNA or if daughter cells receive too many or too few chromosomes, the daughter cells may die. If they survive, they may become cancerous. To prevent this, the eukaryotic cell cycle has three major **checkpoints** (**Fig. 9-12**). At each checkpoint, proteins in the cell determine whether the cell has successfully completed a specific phase of the cycle:

- **G_1 to S** Is the cell's DNA intact and suitable for replication?
- **G_2 to Mitosis** Has the DNA been completely and accurately replicated?

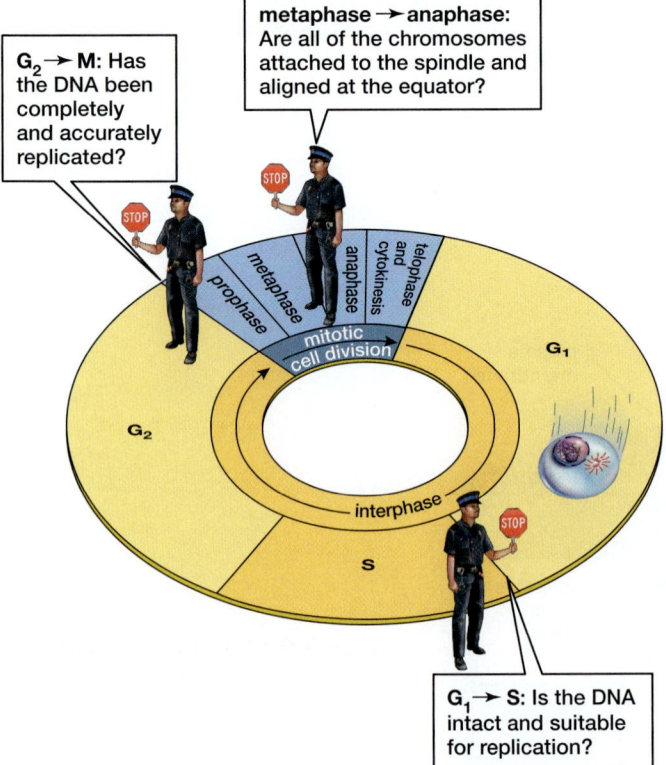

▲ **FIGURE 9-12 Control of the cell cycle** Three major checkpoints regulate a cell's transitions from one phase of the cell cycle to the next: (1) G_1 to S, (2) G_2 to mitosis (M), and (3) metaphase to anaphase.

- **Metaphase to Anaphase** Are all the chromosomes attached to the spindle and aligned properly at the equator of the cell?

The checkpoint proteins usually regulate the production of cyclins or the activity of Cdks, or both, thereby regulating progression from one phase of the cell cycle to the next. Defective checkpoint control is a primary cause of cancer, as we explore in the "Health Watch: Cancer—Running Through the Stop Signs at the Cell Cycle Checkpoints."

CHECK YOUR LEARNING

Can you describe the interactions among growth factors, cyclins, and cyclin-dependent kinases that control the eukaryotic cell cycle? How does a cell protect against producing defective daughter cells?

9.7 WHY DO SO MANY ORGANISMS REPRODUCE SEXUALLY?

Some very successful organisms routinely reproduce asexually. For example, some species of the molds *Penicillium* (which synthesizes penicillin) and *Aspergillus* (used in the manufacture of vitamin C) reproduce by "mitospores"— clouds of tiny cells produced by mitosis—and in fact have never been observed to reproduce sexually. Many of the grasses and weeds in your lawn can reproduce by sprouting whole new plants from their stems or roots. Some, like Kentucky bluegrass and dandelions, even bear flowers that can produce seeds without being fertilized! Clearly, asexual reproduction must work pretty well.

Why, then, do nearly all eukaryotic organisms reproduce sexually (even bluegrass and dandelions reproduce sexually some of the time)? As we have seen, asexual reproduction produces genetically identical offspring. In contrast, sexual reproduction shuffles genes to produce genetically unique offspring.

Sexual Reproduction May Combine Different Parental Alleles in a Single Offspring

Why is this useful? Let's consider a hypothetical situation: a prey animal avoiding being eaten by a predator. Camouflage colors can help the prey to keep from being eaten only if it stays still when it sees a predator. Camouflaged animals that constantly jump around and brightly colored animals that "freeze" when a predator appears will both probably be eaten. How might a single animal combine camouflage coloration and freezing behavior? Let's suppose that one animal has better-than-average camouflage color, while another member of the same species has more effective freezing behavior. Combining the two traits through sexual reproduction might produce some offspring that avoid predation better than either parent could. One reason that sexual reproduction is so common is that it allows useful, genetically determined traits to be combined.

How does sexual reproduction combine traits from two parents in a single offspring? As we will see shortly, meiotic cell division produces haploid cells, each containing only one homologue of each pair of chromosomes. In animals, these haploid cells usually become gametes. Consider again our hypothetical prey animals. A haploid sperm from animal A might contain alleles contributing to camouflage coloration, and a haploid egg from animal B might contain alleles that favor freezing in place at the first sign of a predator. Fusion of these gametes may combine the alleles in a single animal that has both camouflage coloration and becomes motionless when a predator approaches.

Although mutations create new alleles, and are therefore the ultimate source of the genetic variability upon which evolution is based, the rate of mutations is very low. Combining useful alleles through sexual reproduction allows for much more rapid evolution, as new combinations of alleles better adapt organisms to changing environments.

CHECK YOUR LEARNING

Can you describe an important advantage of sexual reproduction over asexual reproduction?

9.8 HOW DOES MEIOTIC CELL DIVISION PRODUCE HAPLOID CELLS?

The key to sexual reproduction in eukaryotes is **meiotic cell division,** in which a diploid cell with paired chromosomes gives rise to haploid daughter cells with unpaired chromosomes. Meiotic cell division consists of **meiosis** (the specialized nuclear division that produces haploid nuclei) followed by cytokinesis. Each daughter cell receives one homologue of each pair of chromosomes. For example, each diploid cell in your body contains 23 *pairs* of chromosomes, 46 chromosomes in all; meiotic cell division produces sperm or eggs with 23 chromosomes, one from each pair. (Fittingly, "meiosis" comes from a Greek word meaning "to diminish.")

Many of the structures and events of meiotic cell division are similar to those of mitotic cell division. However, meiotic cell division differs from mitotic cell division in several important ways. A crucial difference involves DNA replication: In mitotic cell division, the parent cell undergoes one round of DNA replication followed by one nuclear division. In meiotic cell division, the parent cell undergoes a single round of DNA replication followed by *two* nuclear divisions (**Fig. 9-13**). Thus, in meiosis, the DNA is replicated before the first division (**Fig. 9-13a**), but it is *not replicated again* between the first and second divisions.

The first division of meiosis (called *meiosis I*) separates the pairs of homologous chromosomes and sends one homologue into each of two daughter nuclei, producing two haploid nuclei. Each chromosome, however, still consists of two chromatids (**Fig. 9-13b**). The second division (called *meiosis II*) separates the chromatids into independent chromosomes and parcels one into each of two daughter nuclei. Therefore, at the end of meiosis, there are four haploid daughter nuclei, each with one copy of each homologous chromosome. Because each nucleus is usually enclosed in a different cell, meiotic cell division normally produces four haploid cells from a single diploid parent cell (**Fig. 9-13c**). Then, when a haploid sperm fuses with a haploid egg, the resulting offspring are diploid once again (**Fig. 9-14**). We'll explore the stages of meiosis in more detail in the following sections.

sister chromatids

homologous chromosomes

(a) Duplicated homologues prior to meiosis (diploid)

(b) After meiosis I (haploid)

(c) After meiosis II (haploid)

◀ FIGURE 9-13 **Meiosis halves the number of chromosomes** **(a)** Both members of a pair of homologous chromosomes are duplicated prior to meiosis. **(b)** During meiosis I, each daughter cell receives one member of each pair of homologues. **(c)** During meiosis II, sister chromatids separate into independent chromosomes, and each daughter cell receives one of these chromosomes.

Health Watch

Cancer—Running Through the Stop Signs at the Cell Cycle Checkpoints

The ultimate causes of most cancers are mutations—damage to DNA resulting from a variety of possible causes, including mistakes during replication, infection by certain viruses, exposure to ultraviolet radiation in sunlight or tanning beds, or chemicals in the environment (including pesticides, a variety of industrial products, and chemicals naturally produced by plants or fungi). In most cases, a mutation is quickly fixed by enzymes that repair DNA, or the defective cell is killed by one of the body's defenses, such as the immune system. Occasionally, however, a renegade cell survives. A single mutated cell isn't very dangerous, but if it divides repeatedly, it may form a multicellular, life-threatening tumor.

Progression through the cell cycle, including cell division, is regulated primarily by two interacting processes: (1) responses to growth factors that start or speed up the cell cycle and (2) checkpoints that stop the cell cycle if problems such as mutations in DNA or misalignment of chromosomes have occurred. Cancers develop when cells evade these controls.

Responses to Growth Factors

Most normal cells divide only when stimulated by growth factors. When a growth factor binds to its receptor on the surface of a cell, a cascade of biochemical changes results in cyclin synthesis, activation of Cdks, and progression through the cell cycle (see Fig. 9-11). Many cancerous cells have mutations, collectively called *oncogenes* (literally, "to cause cancer"), that promote uncontrolled cell division. Some oncogenes produce growth factor receptors that are permanently activated, even in the absence of growth factors (**Fig. E9-1**). Mutations in cyclin genes may cause cyclins to be synthesized at a high rate, again independently

of growth factors. Still other oncogenes cause changes that activate molecules "downstream" of growth factor receptors. The result: an abnormally large supply of activated Cdks and other molecules that stimulate cell division. When growth factors are present, cell division is stimulated even more. Like a driver who hits the accelerator instead of the brake while approaching a stop sign, a cell with these mutations is likely to barge right through the checkpoints and multiply without control.

Evading the Checkpoint Stop Signs

Cells, however, have ways of enforcing the checkpoint stop signs. All cells contain a variety of proteins collectively called *tumor suppressors*. These proteins prevent uncontrolled cell division and block the production of daughter cells that have mutated DNA. For example, the tumor suppressor protein called p53 (which means simply "a **p**rotein with a molecular mass of **53**,000") monitors the integrity of the cell's DNA. Healthy cells, with intact DNA, contain little p53. However, p53 levels rapidly increase in cells with damaged DNA. The p53 protein activates reactions in the cell that inhibit Cdks and block DNA synthesis, halting the cell cycle at the checkpoint between the G_1 and S phases. Therefore, the cell does not produce daughter cells with damaged DNA. The p53 protein also stimulates the synthesis of DNA repair enzymes. After the DNA has been repaired, p53 levels decline, Cdks become active, and the cell enters the S phase. If the DNA cannot be repaired, p53 triggers a special form of cell death called apoptosis, in which the cell cuts up its DNA and effectively commits suicide. Thus, p53 acts as a checkpoint enforcer, much

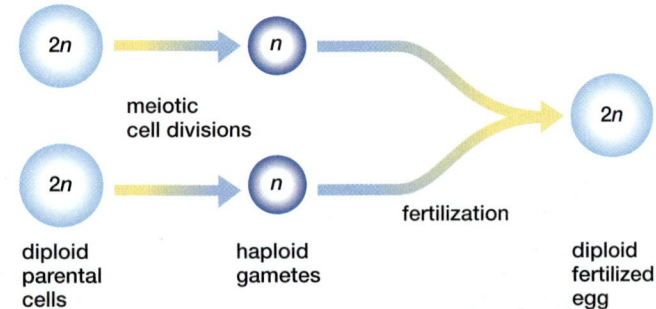

▲ FIGURE 9-14 **Meiotic cell division is essential for sexual reproduction** In sexual reproduction, specialized reproductive cells undergo meiosis to produce haploid cells. In animals, these cells become gametes (sperm or eggs). When an egg is fertilized by a sperm, the resulting fertilized egg, or zygote, is diploid.

Meiosis I Separates Homologous Chromosomes into Two Haploid Daughter Nuclei

The phases of meiosis have the same names as the similar phases in mitosis, followed by I or II to distinguish the two nuclear divisions that occur in meiosis (**Fig. 9-15**). When

meiosis I begins, the chromosomes have already been duplicated during interphase, and the sister chromatids of each chromosome are attached to one another at the centromere.

During Prophase I, Homologous Chromosomes Pair Up and Exchange DNA

During mitosis, homologous chromosomes move completely independently of each other. In contrast, during prophase I of meiosis, homologous chromosomes line up side by side and exchange segments of DNA (**Fig. 9-15a** and **Fig. 9-16**). We'll call one homologue the "maternal homologue" and the other the "paternal homologue," because one was originally inherited from the organism's mother and the other from the organism's father.

Proteins bind the maternal and paternal homologues together so that their gene loci align precisely along their entire length. Enzymes then cut through the DNA of the paired homologues at the same point and graft the cut ends together, often joining part of one of the chromatids of the maternal homologue to part of one of the chromatids of the paternal homologue, and vice versa. The binding proteins and enzymes then depart, leaving crosses, or **chiasmata** (singular, chiasma),

But what if mutation strikes the gene encoding the p53 protein? The result will often be inactive p53. Then, even if the cell's DNA is damaged, the cell skips through the G_1 to S checkpoint. Not surprisingly, about half of all cancers—including tumors of the breast, lung, brain, pancreas, bladder, stomach, and colon—have mutations in the p53 gene.

Cancer Treatments

Why is medical science, which has conquered smallpox, measles, and a host of other diseases, having such a difficult time curing cancer? One reason is that both normal and cancerous cells use the same machinery for cell division. Most treatments that slow down the multiplication of cancer cells do so by reducing growth factor responses, by inhibiting DNA replication, or by reinforcing the checkpoints in the cell cycle. Unfortunately, these treatments also inhibit the division of normal cells. Essential body parts, such as the stomach, intestine, and blood cells, require continual cell division to replace worn-out or damaged cells. The side effects of traditional chemotherapy are only too well known: hair loss, nausea, and skin disorders.

Effective and selective treatments for cancer must target cell division only, or at least mostly, in cancerous cells. There are a few treatments, some available to patients and some in clinical trials, that block growth factor actions only in cells that usually don't divide very often in adults, such as breast cells. Others inhibit certain Cdks that are not very active in adults or that are active only in certain cell types (which at least reduces the side effects). Still others target mutations that are found only in specific cancer cells and not in normal cells. Unfortunately, each of these less toxic treatments is effective only on a subset of cancers. A nontoxic, universal cure for all cancers is far off, and may not even be possible.

▲ FIGURE E9-1 A CT scan of advanced lung cancer
In women and in people who have never smoked, about 40% to 50% of lung cancer seems to be caused by too many receptors for epidermal growth factor or by mutated receptors that are active even in the absence of EGF.

like the tire-spiking strips that police sometimes use to prevent criminals from racing through roadblocks. Chances are the cell cannot plow through the G_1 to S checkpoint, so it will not continue through the cell cycle.

where the maternal and paternal chromosomes have exchanged parts (see Fig. 9-16). In human cells, each pair of homologues usually forms two or three chiasmata during prophase I. The mutual exchange of DNA between maternal and paternal chromosomes at chiasmata is called **crossing over.** If the chromosomes have different alleles, then crossing over creates genetic **recombination:** the formation of new combinations of alleles on a chromosome. Even after the exchange of DNA, the arms of the homologues remain temporarily entangled at the chiasmata. This keeps the two homologues together until they are pulled apart during anaphase I.

As in mitosis, the spindle microtubules begin to assemble outside the nucleus during prophase I. Near the end of prophase I, the nuclear envelope breaks down and spindle microtubules invade the nuclear region, capturing the chromosomes by attaching to their kinetochores.

During Metaphase I, Paired Homologous Chromosomes Line Up at the Equator of the Cell

During metaphase I, interactions between the kinetochores and the spindle microtubules move the paired homologues to the equator of the cell (**Fig. 9-15b**). Unlike in metaphase of mitosis, in which *individual* duplicated chromosomes line up along the equator, in metaphase I of meiosis, *homologous pairs* of duplicated chromosomes, held together by chiasmata, line up along the equator. Which member of a pair of homologous chromosomes faces which pole of the cell is random—the maternal homologue may face "north" for some pairs and "south" for other pairs. This randomness (also called *independent assortment*), together with genetic recombination caused by crossing over, is responsible for the genetic diversity of the haploid cells produced by meiosis.

During Anaphase I, Homologous Chromosomes Separate

Anaphase in meiosis I differs considerably from anaphase in mitosis. In anaphase of mitosis, the *sister chromatids separate* and move to opposite poles. In contrast, in anaphase I of meiosis, the sister chromatids of each duplicated homologue remain attached to each other and move to the same pole. However, the chiasmata joining the two homologues untangle, allowing the *homologues to separate* and move to opposite poles (**Fig. 9-15c**).

MEIOSIS I

(a) Prophase I
Duplicated chromosomes condense. Homologous chromosomes pair up and chiasmata occur as chromatids of homologues exchange parts by crossing over. The nuclear envelope disintegrates, and spindle microtubules form.

(b) Metaphase I
Paired homologous chromosomes line up along the equator of the cell. One homologue of each pair faces each pole of the cell and attaches to the spindle microtubules via the kinetochore (blue).

(c) Anaphase I
Homologues separate, one member of each pair going to each pole of the cell. Sister chromatids do not separate.

(d) Telophase I
Spindle microtubules disappear. Two clusters of chromosomes have formed, each containing one member of each pair of homologues. The daughter nuclei are therefore haploid. Cytokinesis commonly occurs at this stage. There is little or no interphase between meiosis I and meiosis II.

▲ FIGURE 9-15 **Meiotic cell division** In meiotic cell division, the homologous chromosomes of a diploid cell are separated, producing four haploid daughter cells. Two pairs of homologous chromosomes are shown. Yellow chromosomes are from one parent and violet chromosomes are from the other parent.

QUESTION What would be the consequences for the resulting gametes if one pair of homologues failed to separate at anaphase I?

▲ FIGURE 9-16 **Crossing over** Nonsister chromatids of different members of a homologous pair of chromosomes exchange DNA at chiasmata.

MEIOSIS II

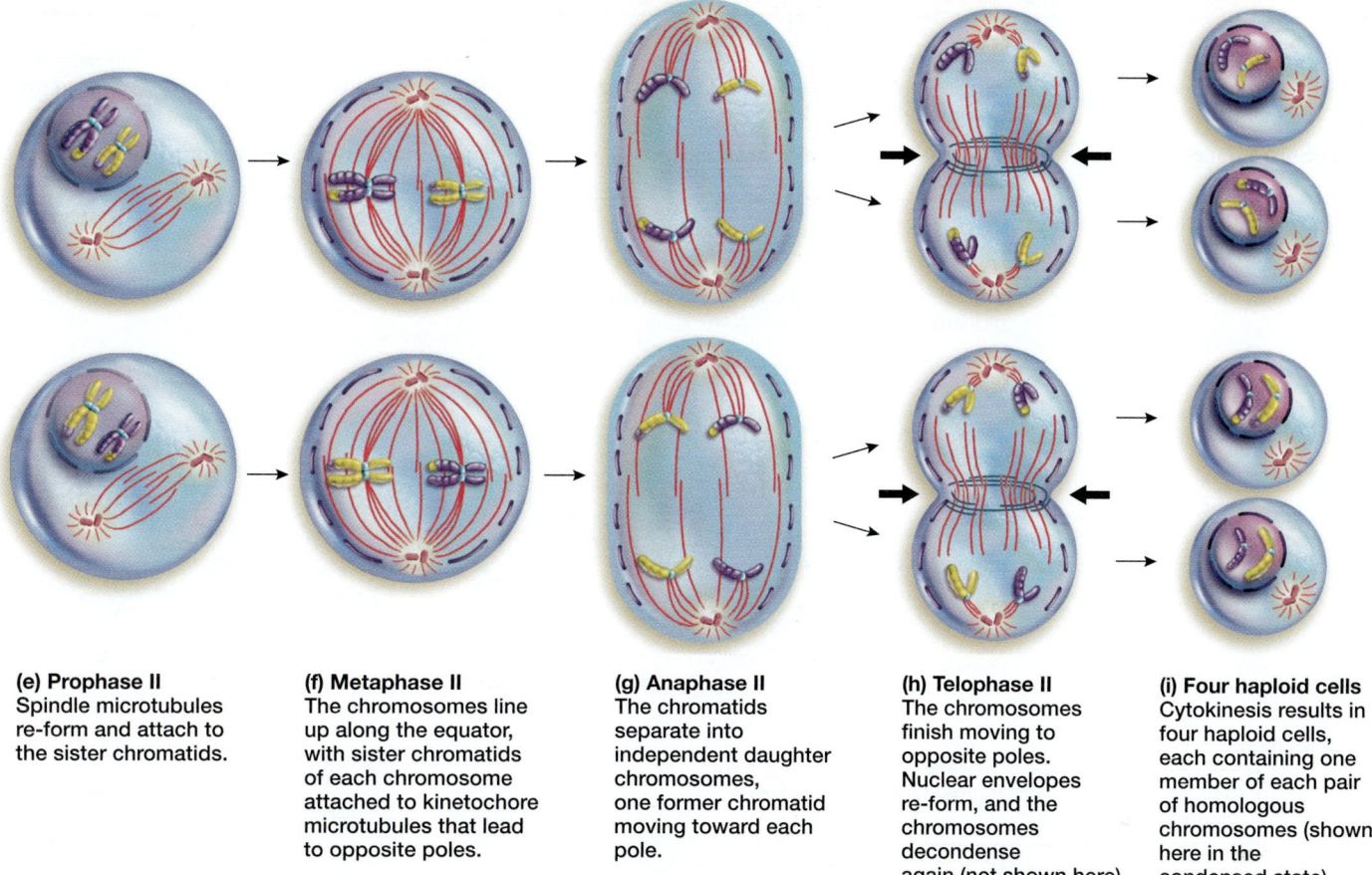

(e) Prophase II
Spindle microtubules re-form and attach to the sister chromatids.

(f) Metaphase II
The chromosomes line up along the equator, with sister chromatids of each chromosome attached to kinetochore microtubules that lead to opposite poles.

(g) Anaphase II
The chromatids separate into independent daughter chromosomes, one former chromatid moving toward each pole.

(h) Telophase II
The chromosomes finish moving to opposite poles. Nuclear envelopes re-form, and the chromosomes decondense again (not shown here).

(i) Four haploid cells
Cytokinesis results in four haploid cells, each containing one member of each pair of homologous chromosomes (shown here in the condensed state).

During Telophase I, Two Haploid Clusters of Duplicated Chromosomes Form

At the end of anaphase I, the cluster of chromosomes at each pole contains one member of each pair of homologous chromosomes. Therefore, each cluster contains the haploid number of chromosomes. In telophase I, the spindle microtubules disappear. Cytokinesis commonly occurs during telophase I (**Fig. 9-15d**). In many, but not all, organisms, nuclear envelopes re-form. Telophase I is usually followed immediately by meiosis II, with little or no intervening interphase. The chromosomes do not replicate between meiosis I and meiosis II.

Meiosis II Separates Sister Chromatids into Four Daughter Nuclei

During meiosis II, the sister chromatids of each duplicated chromosome separate in a process that is virtually identical to mitosis, though it takes place in haploid cells. During prophase II, the spindle microtubules re-form (**Fig. 9-15e**). As in mitosis, the kinetochores of the sister chromatids of each duplicated chromosome attach to spindle microtubules extending to opposite poles of the cell. During metaphase II, the duplicated chromosomes line up at the cell's equator (**Fig. 9-15f**). During anaphase II, the sister chromatids separate and move to opposite poles (**Fig. 9-15g**). Telophase II and cytokinesis conclude meiosis II as nuclear envelopes re-form, the chromosomes decondense into their extended state, and the cytoplasm divides (**Fig. 9-15h**). Both daughter cells produced in meiosis I usually undergo meiosis II, producing a total of four haploid cells from the original parental diploid cell (**Fig. 9-15i**).

Now that we have covered all of the processes in detail, examine **Table 9-1** to review and compare mitotic and meiotic cell division.

CHECK YOUR LEARNING

Can you describe the steps and outcome of meiotic cell division? What are the functions of crossing over?

TABLE 9-1 A Comparison of Mitotic and Meiotic Cell Division in Animal Cells

Feature	Mitotic Cell Division	Meiotic Cell Division
Cells in which it occurs	Body cells	Gamete-producing cells
Final chromosome number	Diploid—2n; two copies of each type of chromosome (homologous pairs)	Haploid—1n; one member of each homologous pair
Number of daughter cells	Two, identical to the parent cell and to each other	Four, containing recombined chromosomes due to crossing over
Number of cell divisions per DNA replication	One	Two
Function in animals	Development, growth, repair, and maintenance of tissues; asexual reproduction	Gamete production for sexual reproduction

MITOSIS

no stages comparable to meiosis I

interphase   prophase   metaphase   anaphase   telophase   two diploid cells

MEIOSIS

Recombination occurs

Homologues pair

Sister chromatids remain attached

interphase   prophase   metaphase   anaphase   telophase   prophase   metaphase   anaphase   telophase   four haploid cells

MEIOSIS I   MEIOSIS II

In these diagrams, comparable phases are aligned. In both mitosis and meiosis, chromosomes are duplicated during interphase. Meiosis I, with the pairing of homologous chromosomes, formation of chiasmata, exchange of chromosome parts, and separation of homologues to form haploid daughter nuclei, has no counterpart in mitosis. Meiosis II, however, is virtually identical to mitosis in a haploid cell.

9.9 WHEN DO MITOTIC AND MEIOTIC CELL DIVISION OCCUR IN THE LIFE CYCLES OF EUKARYOTES?

The life cycles of almost all eukaryotic organisms share a common overall pattern. First, two haploid cells fuse during the process of fertilization, bringing together genes from different parental organisms and endowing the resulting diploid cell with new gene combinations. Second, at some point in the life cycle, meiotic cell division occurs, re-creating haploid cells. Third, at some time in the life cycle, mitotic cell division of either haploid or diploid cells, or both, results in the growth of multicellular bodies or in asexual reproduction.

The seemingly vast differences between the life cycles of, say, ferns and humans are caused by variations in three aspects: (1) the interval between meiotic cell division and

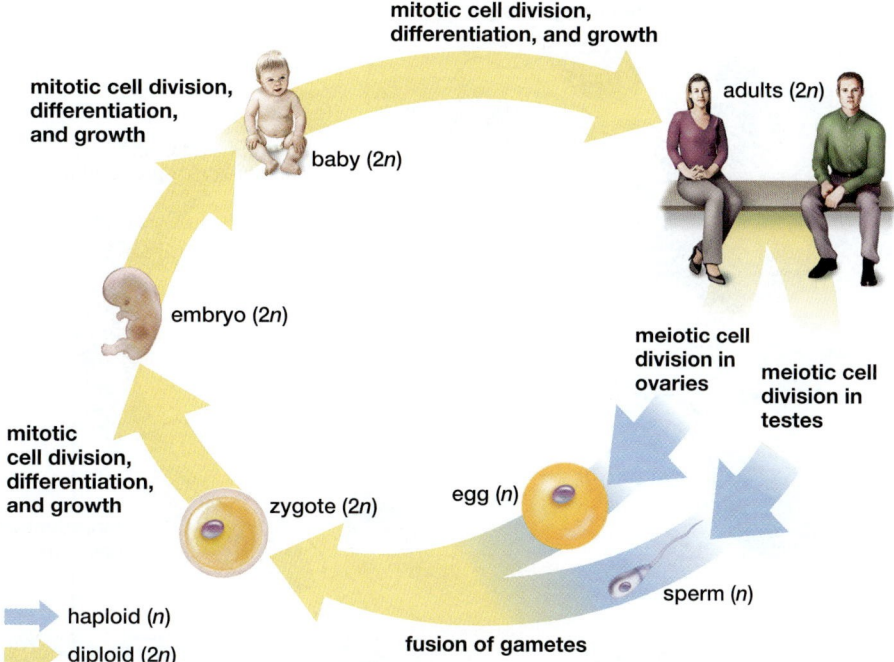

▲ **FIGURE 9-17 The human life cycle** Through meiotic cell division, the two sexes produce gametes—sperm in males and eggs in females—that fuse to form a diploid zygote. Mitotic cell division and differentiation of the daughter cells produce an embryo, child, and ultimately a sexually mature adult. The haploid stages last only a few hours to a few days; the diploid stages may survive for a century.

the fusion of haploid cells, (2) the points in the life cycle at which mitotic and meiotic cell division occur, and (3) the relative proportions of the life cycle spent in the diploid and haploid states. We can conveniently label life cycles according to the relative dominance of diploid and haploid stages.

In Diploid Life Cycles, the Majority of the Cycle Is Spent as Diploid Cells

In most animals, virtually the entire life cycle is spent in the diploid state (Fig. 9-17). Diploid adults produce short-lived haploid gametes (sperm in males, eggs in females) by meiotic cell division. These fuse to form a diploid fertilized egg, called a zygote. Growth and development of the zygote to the adult organism result from mitotic cell division and differentiation of diploid cells.

In Haploid Life Cycles, the Majority of the Cycle Is Spent as Haploid Cells

Some eukaryotes, such as many fungi and single-celled algae, spend most of their life cycles in the haploid state, in which each cell contains a single copy of each type of chromosome (Fig. 9-18). Asexual reproduction by mitotic cell division produces a population of identical, haploid cells. Under certain environmental conditions, specialized haploid reproductive cells are produced. Two of these reproductive cells fuse, forming a short-lived diploid zygote. The zygote immediately undergoes meiotic cell division, producing haploid

cells again. In organisms with haploid life cycles, mitotic cell division never occurs in diploid cells.

In Alternation of Generations Life Cycles, There Are Both Diploid and Haploid Multicellular Stages

The life cycle of plants is called *alternation of generations*, because it alternates between multicellular diploid and multicellular haploid forms. In the typical pattern (Fig. 9-19), specialized cells of a multicellular diploid adult stage (the "diploid generation") undergo meiotic cell division, producing haploid cells called spores. The spores undergo many rounds of mitotic cell division and their daughter cells differentiate, producing a multicellular haploid adult stage (the "haploid generation"). At some point, certain haploid cells differentiate into haploid gametes. Two gametes then fuse to form a diploid zygote. The zygote grows by mitotic cell division into another multicellular diploid adult stage.

In some plants, such as ferns, both the haploid and diploid stages are free-living, independent plants. Flowering plants, however, have reduced haploid stages, represented only by the pollen grain and a small cluster of cells in the ovary of the flower (see Chapters 21 and 44).

CHECK YOUR LEARNING

Can you compare and contrast the three main types of eukaryotic life cycles, and give examples of organisms that exhibit each type?

▲ **FIGURE 9-18 The life cycle of the single-celled alga *Chlamydomonas*** *Chlamydomonas* reproduces asexually by mitotic cell division of haploid cells. When nutrients are scarce, specialized haploid reproductive cells (usually from genetically different populations) fuse to form a diploid cell. Meiotic cell division then immediately produces four haploid cells, usually with different genetic compositions than either of the parental strains.

▲ **FIGURE 9-19 Alternation of generations** In plants, such as this fern, specialized cells in the multicellular diploid adult stage undergo meiotic cell division to produce haploid spores. The spores undergo mitotic cell division and differentiation of the daughter cells to produce a multicellular haploid adult stage. Sometime later, perhaps many weeks later, some of these haploid cells differentiate into sperm and eggs. These fuse to form a diploid zygote. Mitotic cell division and differentiation once again give rise to a multicellular diploid adult stage.

9.10 HOW DO MEIOSIS AND SEXUAL REPRODUCTION PRODUCE GENETIC VARIABILITY?

The individual members of most species are genetically different from one another. Further, the offspring of a single pair of parents are usually genetically different from each other and from their parents. Where do these genetic differences come from? Mutations occurring randomly over millions of years provide the original sources of genetic variability. However, mutations are rare events. Therefore, the genetic variability that occurs from one generation to the next results almost entirely from meiosis and sexual reproduction.

Shuffling of Homologues Creates Novel Combinations of Chromosomes

How does meiosis produce genetic diversity? One mechanism is the random distribution of maternal and paternal homologues to the daughter nuclei during meiosis I. Remember, at metaphase I the paired homologues line up at the cell's equator. In each pair of homologues, the maternal chromosome faces one pole and the paternal chromosome faces the opposite pole, but which homologue faces which pole is random, and is not affected by the distribution of the homologues of other chromosome pairs.

Let's consider meiosis in mosquitoes, which have three pairs of homologous chromosomes ($n = 3$, $2n = 6$). At metaphase I, the chromosomes can align in four possible configurations (**Fig. 9-20a**). Therefore, anaphase I can produce eight possible sets of chromosomes ($2^3 = 8$; **Fig. 9-20b**). A single mosquito, with three pairs of homologous chromosomes, can thus produce gametes with eight different sets of chromosomes. In people, meiosis randomly shuffles 23 pairs of homologous chromosomes, and can theoretically produce more than 8 million (2^{23}) different combinations of maternal and paternal chromosomes in their gametes.

Crossing Over Creates Chromosomes with Novel Combinations of Genes

Crossing over during meiosis produces chromosomes with combinations of alleles that differ from those of either parent. In fact, some of these new combinations may never have existed before, because homologous chromosomes cross over in new and different places at each meiotic division. In humans, 1 in 8 million gametes could have the same combination of maternal and paternal chromosomes. However, crossing over ensures that none of those chromosomes will be purely maternal or purely paternal. Even though a man produces about 100 million sperm each day, he may never produce two that carry exactly the same combinations of alleles. In all probability, every sperm and every egg is genetically unique.

Fusion of Gametes Adds Further Genetic Variability to the Offspring

At fertilization, two gametes, each containing a unique combination of alleles, fuse to form a diploid offspring. Even if we ignore crossing over, every human can theoretically produce about 8 million different gametes based solely on the random separation of the homologues. Therefore, fusion of gametes from just two people could produce 8 million × 8 million, or 64 trillion, genetically different children, which is far more people than have ever existed on Earth. Put another way, the chances that your parents could produce another child who is genetically the same as you are about $^1/_{8,000,000} \times {^1/_{8,000,000}}$, or about 1 in 64 trillion! When we factor in the almost endless variability produced by crossing over, we can confidently say that (except for identical twins) there never has been, and never will be, anyone just like you.

CHECK YOUR LEARNING

Can you explain how meiosis and sexual reproduction generate genetic variability in populations?

(a) The four possible chromosome arrangements at metaphase of meiosis I

(b) The eight possible sets of chromosomes after meiosis I

▲ **FIGURE 9-20 Random separation of homologous pairs of chromosomes produces genetic variability** For clarity, the chromosomes are depicted as large, medium, and small. The paternal chromosomes are colored yellow and the maternal chromosomes are colored violet.

CASE STUDY revisited

Body, Heal Thyself

When Hines Ward sprained the ligaments in his knee, the normal healing process started immediately. Platelets from damaged blood vessels accumulated in the injured ligament and secreted growth factors, stimulating cell division. The cells produced new ligament proteins, replacing those that were torn during the injury. Why, then, should platelet-rich plasma therapy assist healing—shouldn't platelets already present in the blood and leaking into the wound be able to do the job?

In most cases, that's exactly what happens. Nevertheless, PRP therapy can be valuable for two major reasons. First, some tissues, particularly tendons, ligaments, and some bones, have a rather sparse blood supply, so torn ligaments or tendons receive only a low dose of growth factors. As a result, cell division occurs slowly, and the injured structures often never regain their original strength. Incompletely healed tendons and ligaments may be chronically painful, especially during and after exercise, and are prone to re-injury. Injection of platelet-rich plasma can provide a larger dose of growth factors, which stimulates faster cell division and more complete healing. In a study with horses, PRP treatment resulted in better healing of tendon injuries, including more and better-organized collagen protein, and greater strength. Even tissues with a good blood supply, such as skin, may benefit from an extra dose of growth factors; some cosmetic surgeons use PRP therapy after facial surgery to improve healing.

Hines Ward's experience is an example of a second potential benefit of PRP therapy: Faster cell division may promote more rapid healing. No one, not even a weekend warrior or a couch potato, wants to be out of action longer than necessary. The extra growth factors supplied by PRP may speed up healing, even if the body would eventually have done a perfectly good job without the extra help. However, in most real-life cases, including Ward's, no one

knows for sure if healing was really faster or better because PRP therapy stimulated cell division in the injured ligament—there was, of course, no "control Ward," who was not receiving PRP for comparison.

Consider This

BioEthics   Is PRP therapy a wonder treatment that should be routinely used for most injuries? Maybe not. First, PRP apparently isn't always helpful; for example, a few studies have found little or no benefit of PRP for injuries to the Achilles tendon. Second, injecting PRP directly into a wound is quite painful. Third, PRP costs around $500 to $1,000, and is generally not covered by health insurance. In the United States and many other countries, controlling medical costs has become a major political and economic issue. Should PRP be reserved for major injuries? Or only for the wealthy? Would you want PRP for a sprained knee?

CHAPTER REVIEW

Summary of Key Concepts

9.1 Why Do Cells Divide?

Growth of multicellular eukaryotic organisms and replacement of cells that die during an organism's life occur through cell division and differentiation of the daughter cells. Asexual reproduction also occurs through cell division. A specialized type of cell division, called meiotic cell division, is required for sexual reproduction in eukaryotes.

9.2 What Occurs During the Prokaryotic Cell Cycle?

A prokaryotic cell contains a single, circular chromosome. The prokaryotic cell cycle consists of growth, replication of the DNA, and division of the cell by prokaryotic fission. The two resulting daughter cells are genetically identical to one another and to the parent cell.

9.3 How Is the DNA in Eukaryotic Chromosomes Organized?

Each chromosome in a eukaryotic cell consists of a single DNA double helix and proteins that organize the DNA. Genes are segments of DNA found at specific locations on a chromosome. During cell growth, the chromosomes are extended and accessible for use by enzymes that read their genetic instructions. During cell division, the chromosomes are duplicated and condense into short, thick structures. Eukaryotic cells typically contain pairs of chromosomes, called homologues, that carry the same genes with similar, although usually not identical, nucleotide sequences. Cells with pairs of homologous chromosomes are diploid. Cells with only one member of each chromosome pair are haploid.

9.4 What Occurs During the Eukaryotic Cell Cycle?

The eukaryotic cell cycle consists of interphase and mitotic cell division. During interphase, the cell grows and duplicates its chromosomes. Interphase is divided into G_1 (growth phase 1), S (DNA synthesis), and G_2 (growth phase 2). During G_1, many cells differentiate to perform a specific function. Some differentiated cells can reenter the dividing state; other cells remain differentiated for

the life of the organism and never divide again. Eukaryotic cells divide by mitotic cell division, which produces two daughter cells that are genetically identical to their parent cell.

9.5 How Does Mitotic Cell Division Produce Genetically Identical Daughter Cells?

The chromosomes are duplicated during interphase, prior to mitotic cell division. A duplicated chromosome consists of two identical sister chromatids that remain attached to one another at the centromere during the early stages of mitotic cell division. Mitosis (nuclear division) consists of four phases, usually accompanied by cytokinesis (cytoplasmic division) during the last phase (see Fig. 9-9):

- **Prophase** The chromosomes condense and their kinetochores attach to spindle microtubules, called kinetochore microtubules, that form at this time.
- **Metaphase** Kinetochore microtubules move the chromosomes to the equator of the cell.
- **Anaphase** The two chromatids of each duplicated chromosome separate and become independent chromosomes. The kinetochore microtubules move the chromosomes to opposite poles of the cell. Meanwhile, polar microtubules force the cell to elongate.
- **Telophase** The chromosomes decondense, and nuclear envelopes re-form around each new daughter nucleus.
- **Cytokinesis** Cytokinesis normally occurs at the end of telophase and divides the cytoplasm into approximately equal halves, each containing a nucleus. In animal cells, a ring of microfilaments pinches the plasma membrane in along the equator. In plant cells, new plasma membrane forms along the equator by the fusion of vesicles produced by the Golgi apparatus.

9.6 How Is the Cell Cycle Controlled?

Complex interactions among many proteins, particularly cyclins and cyclin-dependent protein kinases, drive the cell cycle. There are three major checkpoints where progress through the cell cycle is regulated: between G_1 and S, between G_2 and mitosis, and between metaphase and anaphase. These checkpoints ensure that the DNA is intact and replicated accurately and that the chromosomes are properly arranged for mitosis before the cell divides.

9.7 Why Do So Many Organisms Reproduce Sexually?

Genetic differences among organisms originate as mutations, which, when preserved within a species, produce alternate forms of genes, called alleles. Alleles in different individuals of a species may be combined in their offspring through sexual reproduction, creating variation among the offspring and potentially improving their likelihood of surviving and reproducing.

9.8 How Does Meiotic Cell Division Produce Haploid Cells?

Meiosis separates homologous chromosomes and produces haploid cells with only one homologue from each pair. During interphase before meiosis, chromosomes are duplicated. The cell then undergoes two specialized divisions—meiosis I and meiosis II—to produce four haploid daughter cells (see Fig. 9-15).

Meiosis I

During prophase I, homologous duplicated chromosomes, each consisting of two chromatids, pair up and exchange parts by crossing over. During metaphase I, homologues move together as pairs to the cell's equator, one member of each pair facing opposite poles of the cell. Homologous chromosomes separate during anaphase I, and two nuclei form during telophase I. Cytokinesis also usually occurs during telophase I. Each daughter nucleus receives only one member of each pair of homologues and, therefore, is haploid. The sister chromatids remain attached to each other throughout meiosis I.

Meiosis II

Meiosis II resembles mitosis in a haploid cell. The duplicated chromosomes move to the cell's equator during metaphase II. The two chromatids of each chromosome separate and move to opposite poles of the cell during anaphase II. This second division produces four haploid nuclei. Cytokinesis normally occurs during or shortly after telophase II, producing four haploid cells.

9.9 When Do Mitotic and Meiotic Cell Division Occur in the Life Cycles of Eukaryotes?

Most eukaryotic life cycles have three parts: (1) Sexual reproduction combines haploid gametes to form a diploid cell. (2) At some point in the life cycle, diploid cells undergo meiotic cell division to produce haploid cells. (3) At some point in the life cycle, mitosis of either a haploid cell, or a diploid cell, or both, results in the growth of multicellular bodies. When these stages occur, and what proportion of the life cycle is occupied by each stage, varies greatly among different species.

9.10 How Do Meiosis and Sexual Reproduction Produce Genetic Variability?

The random shuffling of homologous maternal and paternal chromosomes during meiosis I creates new chromosome combinations. Crossing over creates chromosomes with allele combinations that may never have occurred before on single chromosomes. Because of the separation of homologues and crossing over, a parent probably never produces any gametes that are completely identical. The fusion of two genetically unique gametes adds further genetic variability to the offspring.

Key Terms

allele   *146*
anaphase   *150*
asexual reproduction   *141*
autosome   *145*
cell cycle   *140*
cell division   *140*
cell plate   *150*
centriole   *148*
centromere   *145*
checkpoint   *152*
chiasma (plural, chiasmata)   *154*
chromatid   *145*
chromosome   *140*
clone   *141*
crossing over   *155*
cytokinesis   *147*
daughter cell   *140*
deoxyribonucleic acid (DNA)   *140*
differentiate   *140*
diploid   *145*
duplicated chromosome   *145*
gamete   *141*
gene   *140*

growth factor   *151*
haploid   *146*
homologous chromosome   *145*
homologue   *145*
interphase   *146*
karyotype   *145*
kinetochore   *149*
locus (plural, loci)   *144*
meiosis   *153*
meiotic cell division   *153*
metaphase   *150*
mitosis   *147*
mitotic cell division   *147*
mutation   *145*
nucleotide   *140*
prokaryotic fission   *142*
prophase   *148*
recombination   *155*
sex chromosome   *145*
sexual reproduction   *141*
spindle microtubule   *148*
stem cell   *141*
telomere   *144*
telophase   *150*

Learning Outcomes

In this chapter, you have learned to . . .

LO1 Describe the fundamental components of inheritance, including chromosomes, DNA, nucleotides, and genes.

LO2 Compare and contrast the categories of cells, as distinguished by their ability to divide and differentiate.

LO3 Compare and contrast sexual reproduction and asexual reproduction.

LO4 Describe the prokaryotic cell cycle, including the mechanisms of prokaryotic fission.

LO5 Describe the structure of a eukaryotic chromosome.

LO6 Define homologous chromosome, autosome, and sex chromosome.

LO7 Distinguish between diploid and haploid cells.

LO8 Describe the eukaryotic cell cycle, including the steps and outcome of mitotic cell division.

LO9 Describe how the eukaryotic cell cycle is regulated.

LO10 Describe the steps and outcome of meiotic cell division.

LO11 Compare and contrast the three main types of eukaryotic life cycles.

LO12 Explain how meiosis and sexual reproduction generate genetic variability in populations.

Thinking Through the Concepts

Fill-in-the-Blank

1. The genetic material of all living organisms is _____, which is contained in chromosomes. A gene is a part of a chromosome found at a specific place, or _____ . Alternate forms of a gene are called _____ . **LO1 LO5**

2. Genetically identical offspring are produced during _____ reproduction; genetically variable offspring are produced during _____ reproduction. _____ cell division is required for sexual reproduction. **LO3 LO12**

3. Prokaryotic cells divide by a process called _____. **LO4**

4. Growth and development of eukaryotic organisms occur through _____ cell division and _____ of the resulting daughter cells. _____ cells in multicellular eukaryotes remain capable of dividing throughout the life of the organism, and can differentiate into a variety of cell types. **LO2 LO8**

5. Most plants and animals have pairs of chromosomes that have similar appearance and genetic composition. These pairs are called _____. Pairs of chromosomes that are the same in both males and females are called _____. Pairs that are different in males and females are called _____. **LO6**

6. The four phases of mitosis are _____, _____, _____, and _____. Division of the cytoplasm into two cells, called _____, usually occurs during which phase? _____ **LO8**

7. Chromosomes attach to spindle microtubules at structures called _____. Some spindle microtubules, called _____ microtubules, do not bind to chromosomes, but have free ends that overlap along the equator of the cell. These microtubules push the poles of the cell apart. **LO8**

8. Meiotic cell division produces _____ haploid daughter cells from each diploid parental cell. The first haploid nuclei are produced at the end of _____. In animals, the haploid daughter cells produced by meiotic cell division are called _____. **LO7 LO10**

9. During _____ of meiosis I, homologous chromosomes intertwine, forming structures called _____. These structures are the sites of what event? _____ **LO10**

10. Three processes that promote genetic variability of offspring during sexual reproduction are _____, _____, and _____. **LO12**

Review Questions

1. Compare sexual and asexual reproduction. What type(s) of cell division are involved in each? **LO2**

2. Diagram and describe the eukaryotic cell cycle. Name the phases, and briefly describe the events that occur during each. **LO8**

3. Define *mitosis* and *cytokinesis*. What changes in cell structure would result if cytokinesis did not occur after mitosis? **LO8**

4. Diagram the stages of mitosis. How does mitosis ensure that each daughter nucleus receives a full set of chromosomes? **LO8**

5. Define the following terms: *homologous chromosome (homologue), centromere, kinetochore, chromatid, diploid,* and *haploid.* **LO6 LO8**

6. Describe and compare the process of cytokinesis in animal cells and in plant cells. **LO8**

7. How is the cell cycle controlled? Why is it important to regulate progression through the cell cycle? **LO9**

8. Diagram the events of meiosis. At which stage do homologous chromosomes separate? **LO10**

9. Describe crossing over. At which stage of meiosis does it occur? Name two functions of chiasmata. **LO10**

10. In what ways are mitosis and meiosis similar? In what ways are they different? **LO8 LO10**

11. Describe or diagram the three main types of eukaryotic life cycles. When do meiotic cell division and mitotic cell division occur in each? **LO11**

12. Describe how meiosis provides for genetic variability. If an animal had a haploid number of two (no sex chromosomes), how many genetically different gametes could it produce? (Assume no crossing over.) If it had a haploid number of five? **LO12**

Applying the Concepts

1. Most nerve cells in the adult human central nervous system, as well as heart muscle cells, do not divide. In contrast, cells lining the inside of the small intestine divide frequently. Discuss this difference in terms of why damage to the nervous system and heart muscle cells (for example, caused by a stroke or heart attack) is so dangerous. What do you think might happen to tissues such as the intestinal lining if a disorder blocked mitotic cell division in all cells of the body?

2. Cancer cells divide out of control. Side effects of chemotherapy and radiation therapy that fight cancers include loss of hair and of the intestinal lining, the latter producing severe nausea. What can you infer about the mechanisms of these treatments? What would you look for in an improved cancer therapy?

3. Some animal species can reproduce either asexually or sexually, depending on the state of the environment. Which form of reproduction do you think would have the greatest advantage in a rapidly changing environment? Which form would be most successful in a stable environment? Explain your reasoning in each case.

Answers to Figure Caption questions and Fill-in-the-Blank questions can be found in the Answers section at the back of the book.

MB *Go to MasteringBiology for practice quizzes, activities, eText, videos, current events, and more.*

CASE STUDY

Sudden Death on the Court

FLO HYMAN, 6 feet, 5 inches tall, graceful and athletic, was probably the best woman volleyball player of her time. Captain of the American women's volleyball team that won a silver medal in the 1984 Olympics, Hyman later joined a professional Japanese squad. In 1986, at the age of 31, she was taken out of a game for a short breather, and died while sitting quietly on the bench. How could this happen to someone so young and fit?

Hyman had a genetic disorder called Marfan syndrome, which affects about 1 in 5,000 to 10,000 people, probably including classical musicians and composers Sergei Rachmaninoff and Nicolo Paganini, and possibly American president Abraham Lincoln. People with Marfan syndrome are typically tall and slender, with long, flexible limbs and large hands and feet. For some people with Marfan syndrome, these characteristics lead to fame and fortune. Unfortunately, Marfan syndrome can also be deadly.

Hyman died from a ruptured aorta, the massive artery that carries blood from the heart to most of the body. Why did Hyman's aorta burst? What does a weak aorta have in common with tallness and large hands? Marfan syndrome is caused by a mutation in the gene that encodes a protein called fibrillin, an essential component of connective tissue. Many parts of your body contain connective tissue, including the tendons that attach muscles to bones, the ligaments that fasten bones to other bones in joints, and the walls of arteries. Fibrillin forms long fibers that give strength and elasticity to connective tissue. Normal fibrillin also traps certain growth factors, preventing them from stimulating excessive cell division in, for example, bone-forming cells. Defective fibrillin cannot trap these growth factors, with the result that the arms, legs, hands, and feet of people with Marfan syndrome tend to become unusually long. The combination of defective fibrillin and high concentrations of growth factors also weakens bone, cartilage, and artery walls.

Diploid organisms, including people, generally have two copies of each gene, one on each homologous chromosome. One defective copy of the fibrillin gene is enough to cause Marfan syndrome. What does this tell us about the inheritance of Marfan syndrome? Are all inherited diseases caused by a single defective copy of a gene? To find out, we must go back in time to a monastery in Moravia and visit the garden of Gregor Mendel.

Olympic volleyball silver medalist Flo Hyman was struck down by Marfan syndrome at the height of her career.

AT A GLANCE

LEARNING GOALS

LG1 Describe the relationships among chromosomes, DNA, genes, mutations, and alleles, and explain their function in inheritance.

LG2 Summarize Mendel's conclusions about the inheritance of single and multiple traits and explain how the results of his experiments support those conclusions.

LG3 Determine the genotypes and phenotypes of offspring that result from mating organisms with traits that follow simple patterns of Mendelian inheritance.

LG4 Describe the inheritance of traits characterized by incomplete dominance, codominance, polygenic inheritance, and pleiotropy.

LG5 Explain how the physical locations of genes and the events that occur during meiosis affect inheritance.

LG6 Explain the chromosomal basis of sex determination and how it influences the inheritance of sex-linked traits.

LG7 Interpret pedigrees for dominant and recessive traits.

LG8 Describe the causes and symptoms of inherited human disorders produced by defective alleles of individual genes and by abnormal numbers of chromosomes.

10.1 WHAT IS THE PHYSICAL BASIS OF INHERITANCE?

Inheritance is the process by which the traits of organisms are passed to their offspring. We will begin our exploration of inheritance with a brief overview of the structures that form its physical basis. In this chapter, we will confine our discussion to diploid organisms, including most plants and animals, that reproduce sexually by the fusion of haploid gametes.

Genes Are Sequences of Nucleotides at Specific Locations on Chromosomes

A chromosome consists of a single double helix of DNA, packaged with a variety of proteins (see Figs. 9-1 and 9-4). Segments of DNA ranging from a few hundred to many thousands of nucleotides in length are the units of inheritance—the **genes**—that encode the information needed to produce proteins, cells, and entire organisms. Therefore, genes are parts of chromosomes (**Fig. 10-1**). A gene's physical location on a chromosome is called its **locus** (plural, loci). The chromosomes of diploid organisms occur in pairs called homologues. Both members of a pair of homologues carry the same genes, located at the same loci. However, the nucleotide sequences of a given gene may differ in different members of a species, or even on the two homologues of a single individual. These different versions of a gene at a given locus are called **alleles** (see Fig. 10-1). To understand the relationship between genes and alleles, it may be helpful to think of genes as very long sentences, written in an alphabet of nucleotides instead of letters. The alleles of a gene are like slightly different spellings of words in different versions of the same nucleotide sentence.

Mutations Are the Source of Alleles

The different alleles that you have on your chromosomes were almost all inherited from your parents. But where did these alleles come from in the first place? All alleles originally arose as **mutations**—changes in the sequence of nucleotides in the DNA of a gene. If a mutation occurs in a cell that becomes a sperm or egg, it can be passed on from parent to offspring. Most of the alleles in an organism's DNA first appeared as mutations in the reproductive cells of the organism's ancestors, perhaps hundreds or even millions of years ago, and have been inherited, generation after generation, ever since. A few alleles, which we will call "new mutations," may have occurred in the reproductive cells of the organism's own parents, but this is quite rare.

An Organism's Two Alleles May Be the Same or Different

Because a diploid organism has pairs of homologous chromosomes, and both members of a pair contain the same gene loci, the organism has two copies of each gene. If both homologues have the same allele at a given gene locus, the organism is said to be **homozygous** at that locus. ("Homozygous" comes from Greek words meaning "same pair.") The chromosomes shown in Figure 10-1 are homozygous at two loci. If two homologous chromosomes have different alleles at a locus, the organism is **heterozygous** ("different pair") at that locus. The chromosomes in Figure 10-1 are heterozygous at one locus. Organisms that are heterozygous at a specific locus are often called hybrids.

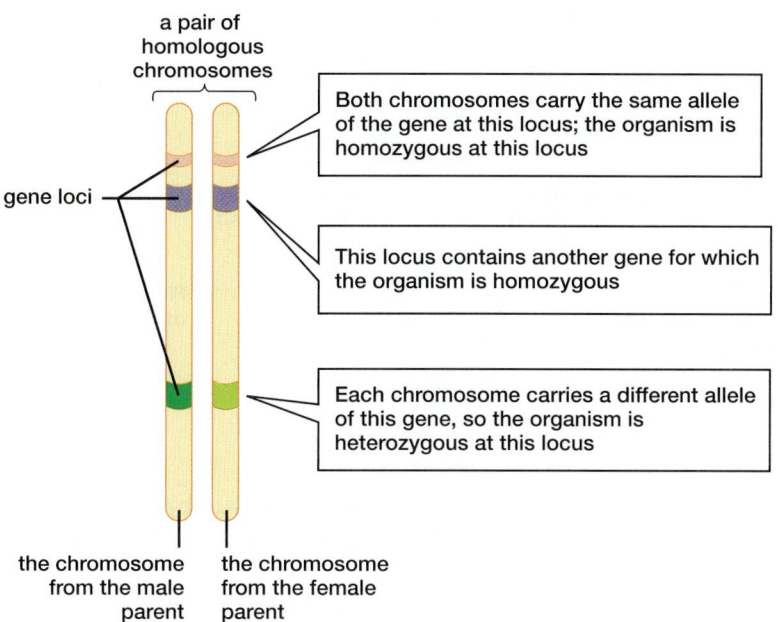

◄ FIGURE 10-1 **The relationships among genes, alleles, and chromosomes** Each homologous chromosome carries the same set of genes. Each gene is located at the same position, or locus, on its chromosome. Differences in nucleotide sequences at the same gene locus produce different alleles of the gene. Diploid organisms have two alleles of each gene, one on each homologue. The alleles on the two homologues may be the same or different.

or meiosis had been discovered, his research revealed many essential facts about genes, alleles, and the distribution of alleles in gametes and zygotes during sexual reproduction. Because his experiments are succinct, elegant examples of science in action, let's follow Mendel's paths of discovery.

Doing It Right: The Secrets of Mendel's Success

There are three key steps to any successful experiment in biology: choosing the right organism with which to work, designing and performing the experiment correctly, and analyzing the data properly. Mendel was the first geneticist to complete all three steps.

Mendel chose the edible pea as the subject for his experiments on inheritance (**Fig. 10-3**). The male reproductive structures of a flower, called stamens, produce pollen. Each pollen grain contains sperm. Pollination allows a sperm to fertilize an egg, which is located within the ovary of the flower's female reproductive structure, called the carpel. In pea flowers, the petals enclose all of the reproductive structures, preventing another flower's pollen from entering. Therefore, the eggs in a pea flower must be fertilized by sperm from the pollen of the same flower. When an organism's sperm fertilize its own eggs, the process is called **self-fertilization.**

Mendel, however, often wanted to mate two different pea plants to see what characteristics their offspring would inherit. To do this, he opened a pea flower and removed its stamens,

10.2 HOW WERE THE PRINCIPLES OF INHERITANCE DISCOVERED?

The common patterns of inheritance were discovered in the mid-1800s by an Austrian monk, Gregor Mendel (**Fig. 10-2**). Although Mendel worked long before DNA, chromosomes,

▲ FIGURE 10-2 **Gregor Mendel**

▲ FIGURE 10-3 **Flowers of the edible pea** In the intact pea flower (left), the lower petals form a container enclosing the reproductive structures—the stamens (male) and carpel (female). Pollen normally cannot enter the flower from outside, so peas usually self-pollinate and, hence, self-fertilize. If the flower is opened (right), it can be cross-pollinated by hand.

preventing self-fertilization. Then, he dusted the sticky tip of the carpel with pollen from the flower of another plant. When sperm from one organism fertilize eggs from a different organism, the process is called **cross-fertilization.**

Mendel's experimental design was simple, but brilliant. Earlier researchers had generally tried to study inheritance by simultaneously considering all of the features of entire organisms, including traits that differed only slightly among organisms. Not surprisingly, the investigators were often confused rather than enlightened. Mendel, however, chose to study traits with unmistakably different forms, such as white versus purple flowers. He also began by studying only one trait at a time.

Mendel followed the inheritance of these traits for several generations, counting the numbers of offspring with each type of trait. By analyzing these numbers, the basic patterns of inheritance became clear. Today, quantifying experimental results and applying statistical analysis are essential tools in virtually every field of biology, but in Mendel's time, numerical analysis was an innovation.

CHECK YOUR LEARNING

Can you distinguish between self-fertilization and cross-fertilization? Can you explain how Mendel used these processes in his studies of inheritance?

10.3 HOW ARE SINGLE TRAITS INHERITED?

True-breeding organisms possess a trait, such as purple flowers, that is always inherited unchanged by all offspring produced by self-fertilization. In his first set of experiments, Mendel cross-fertilized pea plants that were true-breeding for different forms of a single trait. He saved the resulting seeds and grew them the following year to determine the traits of the offspring.

In one of these experiments, Mendel cross-fertilized true-breeding, white-flowered plants with true-breeding, purple-flowered plants. This was the parental generation, denoted by the letter P. When he grew the resulting seeds, he found that all the first-generation offspring (the "first filial," or F_1 generation) produced purple flowers (**Fig. 10-4**). What had happened to the white color? The flowers of the F_1 hybrids were just as purple as their true-breeding purple parent. The white color of their true-breeding white parent seemed to have disappeared.

Mendel then allowed the F_1 flowers to self-fertilize, collected the seeds, and planted them the next spring. In the second (F_2) generation, Mendel counted 705 plants with purple flowers and 224 plants with white flowers. These numbers are approximately three-fourths purple flowers and one-fourth white flowers, or a ratio of about 3 purple to 1 white (**Fig. 10-5**). This result showed that the capacity to produce white flowers had not disappeared in the F_1 plants, but had only been hidden.

Mendel allowed the F_2 plants to self-fertilize and produce a third (F_3) generation. He found that all the white-flowered F_2 plants produced white-flowered offspring; that is, they were true-breeding. For as many generations as he

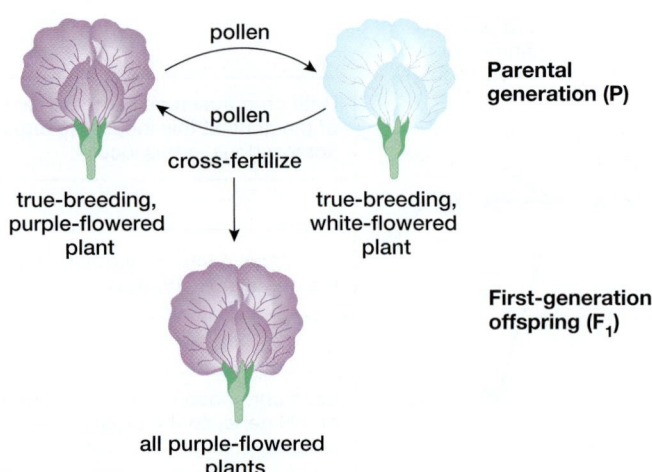

▲ **FIGURE 10-4 Cross of pea plants true-breeding for white or purple flowers** All of the offspring bear purple flowers.

had time and patience to raise, white-flowered plants always gave rise only to white-flowered offspring. In contrast, when purple-flowered F_2 plants self-fertilized, their offspring were of two types. About one-third were true-breeding for purple, but the other two-thirds were hybrids that produced both purple- and white-flowered offspring, again in the ratio of 3 purple to 1 white. Therefore, the F_2 generation included one-quarter true-breeding purple plants, one-half hybrid purple, and one-quarter true-breeding white.

The Inheritance of Dominant and Recessive Alleles on Homologous Chromosomes Can Explain the Results of Mendel's Crosses

Mendel's results, supplemented by modern knowledge of genes and chromosomes, allow us to develop a five-part hypothesis to explain the inheritance of single traits:

- Each trait is determined by pairs of discrete physical units called genes. Each organism has two alleles for each gene, one on each homologous chromosome. True-breeding, white-flowered peas have different alleles of the flower-color gene than true-breeding, purple-flowered peas do.

▲ **FIGURE 10-5 Self-fertilization of F_1 pea plants with purple flowers** Three-quarters of the offspring bear purple flowers and one-quarter bear white flowers.

- When two different alleles are present in an organism, one—the **dominant** allele—may mask the expression of the other—the **recessive** allele. The recessive allele, however, is still present. In the edible pea, the allele for purple flowers is dominant, and the allele for white flowers is recessive.

- Homologous chromosomes separate, or segregate, from each other during meiosis, thus separating the alleles they carry. This is known as Mendel's **law of segregation:** Each gamete receives only one allele of each pair of genes. When a sperm fertilizes an egg, the resulting offspring receives one allele from the father (in his sperm) and one from the mother (in her egg).

- Because homologous chromosomes separate randomly during meiosis, the distribution of alleles into the gametes is also random.

- True-breeding organisms have two copies of the same allele for a given gene, and are therefore homozygous for that gene. All of the gametes from a homozygous individual receive the same allele for that gene (**Fig. 10-6a**). Hybrid organisms have two different alleles for a given gene and so are heterozygous for that gene. Half of a heterozygote's gametes will contain one allele for that gene and half will contain the other allele (**Fig. 10-6b**).

Let's see how this hypothesis explains the results of Mendel's experiments with flower color (**Fig. 10-7**). We will use letters to represent the different alleles, assigning the uppercase letter P to the dominant allele for purple flower color and the lowercase letter p to the recessive allele for white flower color. A homozygous purple-flowered plant has two alleles for purple flower color (PP), whereas a homozygous white-flowered plant has two alleles for white flower color

(pp). Therefore, all the sperm and eggs produced by a PP plant carry the P allele, and all the sperm and eggs of a pp plant carry the p allele (**Fig. 10-7a**).

The F_1 offspring were produced when P sperm fertilized p eggs or when p sperm fertilized P eggs. In both cases, the F_1 offspring were Pp. Because P is dominant over p, all of the offspring were purple (**Fig. 10-7b**).

For the F_2 generation, Mendel allowed the heterozygous F_1 plants to self-fertilize. A heterozygous plant produces equal numbers of P and p sperm and equal numbers of P and p eggs. When a Pp plant self-fertilizes, each type of sperm has an equal chance of fertilizing each type of egg (**Fig. 10-7c**). Therefore, the F_2 generation contained three types of offspring: PP, Pp, and pp. The three types occurred in the approximate proportions of one-quarter PP (homozygous purple), one-half Pp (heterozygous purple), and one-quarter pp (homozygous white).

❓ HAVE **YOU** EVER WONDERED . . .

Why Dogs Vary So Much in Size?

All dogs evolved from wolves. Although all wolves are about the same size, dogs vary in size more than any other mammal—from huge Great Danes and Irish wolfhounds to minuscule toy breeds such as Chihuahuas and Pomeranians. Toy breeds are usually homozygous for a "small" allele of the gene that encodes insulin-like growth factor (IGF), a protein that helps to regulate body size in many mammals. Great Danes and Irish wolfhounds are usually homozygous or heterozygous for the "large" allele. The "small" allele probably arose as a mutation in dogs, because it does not exist in wolves—just imagine the fate of a Chihuahua-sized wolf in the wild!

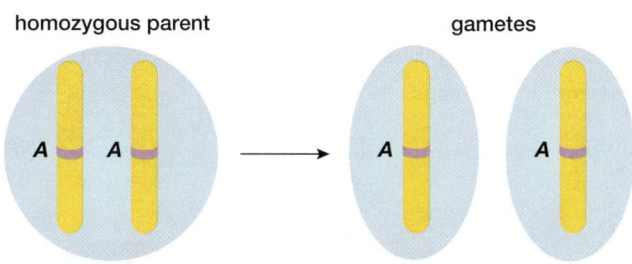

(a) Gametes produced by a homozygous parent

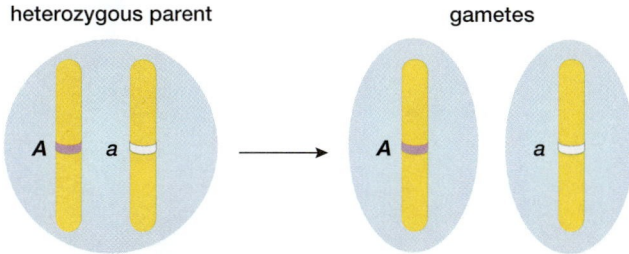

(b) Gametes produced by a heterozygous parent

▲ **FIGURE 10-6 The distribution of alleles in gametes**
(a) All of the gametes produced by homozygous organisms contain the same allele. (b) Half of the gametes produced by heterozygous organisms contain one allele, and half of the gametes contain the other allele.

Two organisms that look alike may actually have different combinations of alleles. The combination of alleles carried by an organism (for example, *PP* or *Pp*) is its **genotype.** The organism's traits, including its outward appearance, behavior, digestive enzymes, blood type, or any other observable or measurable feature, make up its **phenotype.** As we have seen, plants with either the *PP* or the *Pp* genotype have the phenotype of purple flowers. Therefore, the F₂ generation of Mendel's peas consisted of three genotypes (one-quarter *PP*, one-half *Pp*, and one-quarter *pp*), but only two phenotypes (three-quarters purple and one-quarter white).

Simple "Genetic Bookkeeping" Can Predict Genotypes and Phenotypes of Offspring

The **Punnett square method,** named after R. C. Punnett, a famous geneticist of the early 1900s, is a convenient way to predict the genotypes and phenotypes of offspring. **Figure 10-8a** shows how to use a Punnett square to determine the expected proportions of offspring that arise from the self-fertilization of a flower that is heterozygous for color (or the proportions of offspring produced by breeding any two organisms that are heterozygous for a single trait). **Figure 10-8b** shows how to calculate the proportions of offspring using the probabilities that each type of sperm will fertilize each type of egg.

As you use these genetic bookkeeping techniques, keep in mind that in a real experiment, the actual offspring will not occur in exactly the predicted proportions. Why not? Let's consider a familiar example. Each time a baby is conceived, it has a 50:50 chance of being a boy or a girl. However, many families with two children do not have one girl and one boy. The 50:50 ratio of girls to boys occurs only if we average the genders of the children in many families.

Mendel's Hypothesis Can Be Used to Predict the Outcome of New Types of Single-Trait Crosses

You have probably recognized that Mendel used the scientific method: He made an observation and used it to formulate a hypothesis. But does Mendel's hypothesis accurately predict the results of further experiments? Based on the hypothesis that heterozygous F₁ plants have one allele for purple flowers and one for white (that is, they have the *Pp* genotype), Mendel predicted the outcome of cross-fertilizing *Pp* plants with homozygous recessive white plants (*pp*): There should

▶ FIGURE 10-7 **Segregation of alleles and fusion of gametes predict the distribution of alleles and traits in the inheritance of flower color in peas** (a) The parental generation: All of the gametes of homozygous *PP* parents contain the *P* allele; all of the gametes of homozygous *pp* parents contain the *p* allele. **(b)** The F₁ generation: Fusion of gametes containing the *P* allele with gametes containing the *p* allele produces only *Pp* offspring. (Note that *Pp* is the same genotype as *pP*.) **(c)** The F₂ generation: Half of the gametes of heterozygous *Pp* parents contain the *P* allele and half contain the *p* allele. Fusion of these gametes produces *PP*, *Pp*, and *pp* offspring.

(a) Gametes produced by homozygous parents

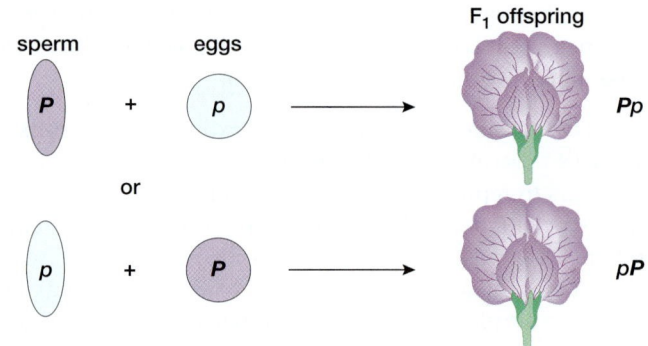

(b) Fusion of gametes produces F₁ offspring

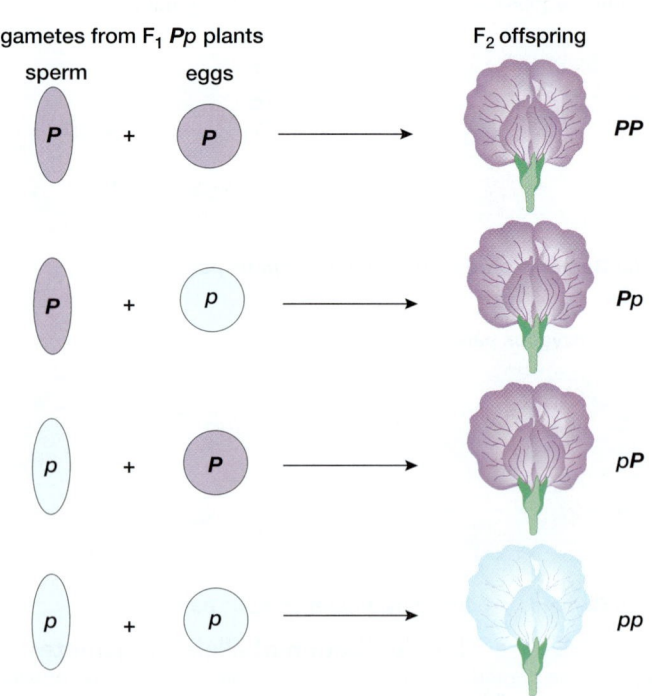

(c) Fusion of gametes from the F₁ generation produces F₂ offspring

be equal numbers of *Pp* (purple) and *pp* (white) offspring. This is indeed what he found.

This type of experiment also has practical uses. Cross-fertilization of an organism with a dominant phenotype (in this case, a purple flower) but an unknown genotype with a homozygous recessive organism (a white flower) tests whether the organism with the dominant phenotype is homozygous or heterozygous. Logically, this is called a **test cross** (Fig. 10-9). When crossed with a homozygous recessive (*pp*), a homozygous dominant (*PP*) produces all phenotypically dominant offspring, whereas a heterozygous dominant (*Pp*) yields offspring with both dominant and recessive phenotypes in a 1:1 ratio.

CHECK YOUR LEARNING

Can you describe the pattern of inheritance of a trait controlled by a single gene with two alleles, one dominant and one recessive? Can you calculate the proportions of offspring with each genotype and phenotype that would be produced by mating parents with various combinations of the two alleles?

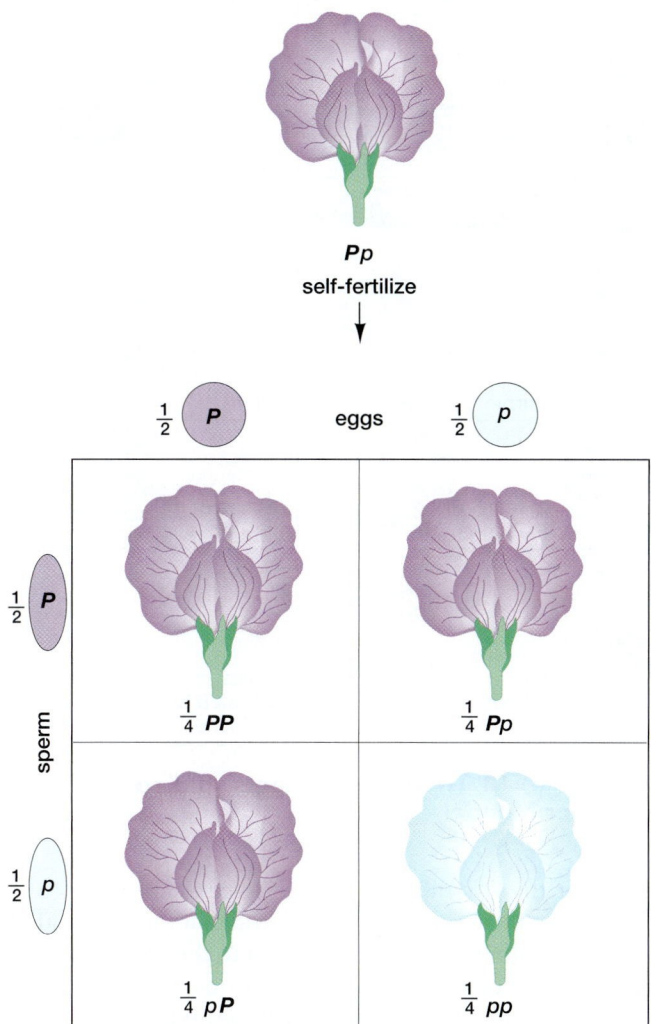

(a) Punnett square of a single-trait cross

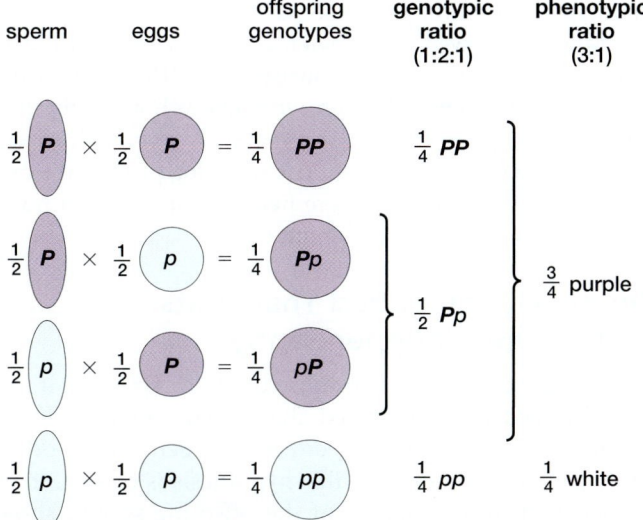

(b) Using probabilities to determine the offspring of a single-trait cross

◄ FIGURE 10-8 **Determining the outcome of a single-trait cross (a)** The Punnett square allows you to predict both genotypes and phenotypes of specific crosses; here we use it for a cross between pea plants that are heterozygous for a single trait—flower color.

(1) Assign letters to the different alleles; use uppercase for dominant alleles and lowercase for recessive alleles.

(2) Determine all the types of genetically different gametes that can be produced by the male and female parents.

(3) Draw the Punnett square, with the columns labeled with all possible genotypes of the eggs and the rows labeled with all possible genotypes of the sperm. (We also show the fractions of each genotype.)

(4) Fill in the genotype of the offspring in each box by combining the genotype of the sperm in its row with the genotype of the egg in its column. (Multiply the fraction of sperm of each type in the row headers by the fraction of eggs of each type in the column headers.)

(5) Count the number of offspring with each genotype. Note that *Pp* is the same genotype as *pP*.

(6) Convert the number of offspring of each genotype to a fraction of the total number of offspring. In this example, out of four fertilizations, only one is predicted to produce the *pp* genotype, so one-quarter of the total number of offspring produced by this cross is predicted to be white. To determine phenotypic fractions, add the fractions of genotypes that would produce a given phenotype. For example, purple flowers are produced by $\frac{1}{4}PP + \frac{1}{4}Pp + \frac{1}{4}pP$, for a total of three-quarters of the offspring.

(b) Probabilities may also be used to predict the outcome of a single-trait cross. Determine the fractions of eggs and sperm of each genotype, and multiply these fractions together to calculate the fraction of offspring of each genotype. When two genotypes produce the same phenotype (e.g., *Pp* and *pP*), add the fractions of each genotype to determine the phenotypic fraction.

QUESTION If you crossed a heterozygous *Pp* plant with a homozygous recessive *pp* plant, what would be the expected ratio of offspring? How does this differ from the offspring of a *PP* × *pp* cross? Try working this out before you read further in the text.

▲ **FIGURE 10-9 Punnett square of a test cross** An organism with a dominant phenotype may be either homozygous or heterozygous. Crossing such an organism with a homozygous recessive organism can determine whether the dominant organism is homozygous (left) or heterozygous (right).

Trait	Dominant form		Recessive form	
Seed shape	smooth		wrinkled	
Seed color	yellow		green	
Pod shape	inflated		constricted	
Pod color	green		yellow	
Flower color	purple		white	
Flower location	at leaf junctions		at tips of branches	
Plant size	tall (about 6 feet)		dwarf (about 8 to 16 inches)	

▲ **FIGURE 10-10 Traits of pea plants studied by Gregor Mendel**

10.4 HOW ARE MULTIPLE TRAITS INHERITED?

Having determined how single traits are inherited, Mendel then turned to the more complex question of the inheritance of multiple traits (**Fig. 10-10**). He began by crossbreeding plants that differed in two traits—for example, seed color (yellow or green) and seed shape (smooth or wrinkled). From other crosses of plants with these traits, Mendel already

CASE STUDY continued

Sudden Death on the Court

When a person with Marfan syndrome has children with a person without the syndrome, those children have a 50% chance of inheriting the condition. Do you think that Marfan is inherited as a dominant or a recessive allele? Why? Check your reasoning in "Case Study Revisited: Sudden Death on the Court" at the end of the chapter.

knew that the smooth allele of the seed shape gene (S) is dominant to the wrinkled allele (s), and that the yellow allele of the seed color gene (Y) is dominant to the green allele (y). He crossed a true-breeding plant with smooth, yellow seeds ($SSYY$) to a true-breeding plant with wrinkled, green seeds ($ssyy$). The $SSYY$ plant can produce only SY gametes, and the $ssyy$ plant can produce only sy gametes. Therefore, all the F_1 offspring were heterozygotes: genotypically $SsYy$ with the phenotype of smooth, yellow seeds.

Mendel allowed these heterozygous F_1 plants to self-fertilize. The F_2 generation consisted of 315 plants with smooth, yellow seeds; 101 with wrinkled, yellow seeds; 108 with smooth, green seeds; and 32 with wrinkled, green seeds—a ratio of about 9:3:3:1. The offspring produced from other crosses of plants that were heterozygous for two traits also had phenotypic ratios of about 9:3:3:1.

Mendel Hypothesized That Traits Are Inherited Independently

Mendel realized that these results could be explained if the genes for seed color and seed shape were inherited independently of each other and did not influence each other during gamete formation. If this hypothesis is correct, then for each trait, three-quarters of the offspring should show the dominant phenotype and one-quarter should show the recessive phenotype. This result is just what Mendel observed. He found 423 plants with smooth seeds (of either color) and 133 with wrinkled seeds (of either color), a ratio of about 3:1; 416 plants produced yellow seeds (of either shape) and 140 produced green seeds (of either shape), also about a 3:1 ratio. **Figure 10-11** shows how a Punnett square or probability calculation can be used to estimate the proportions of genotypes and phenotypes of the offspring of a cross between organisms that are heterozygous for two traits.

The independent inheritance of two or more traits is called the **law of independent assortment.** Multiple traits are inherited independently if the alleles of the gene controlling any given trait are distributed to gametes independently of the alleles for the genes controlling all the other traits. Independent assortment will occur when the traits being studied are controlled by genes on different pairs of homologous chromosomes. Why? During meiosis, paired homologous chromosomes line up at metaphase I. Which homologue faces which pole of the cell is random, and the orientation of one homologous pair does not influence other pairs (see Chapter 9). Therefore, when the homologues separate during anaphase I, which homologue of pair 1 moves "north" does not affect which homologue of pair 2 moves "north." The result is that the alleles of genes on different chromosomes are distributed, or assorted, independently of one another (**Fig. 10-12**).

In an Unprepared World, Genius May Go Unrecognized

In 1865, Gregor Mendel presented his results to the Brünn Society for the Study of Natural Science, and they were published the following year. His paper did not mark the beginning of genetics. In fact, it didn't make any impression at all on the study of biology during his lifetime. Apparently, very few biologists read his paper, and those who did failed to recognize its significance.

It was not until 1900 that three biologists—Carl Correns, Hugo de Vries, and Erich von Tschermak—working independently of one another and knowing nothing of Mendel's work, rediscovered the principles of inheritance. No doubt to their intense disappointment, when they searched the scientific literature before publishing their results, they found that Mendel had scooped them more than 30 years earlier. To their credit, they graciously acknowledged the important work of the Austrian monk, who had died in 1884.

CHECK YOUR LEARNING

Can you describe the pattern of inheritance of two traits, if each of the traits is controlled by a separate gene with only two alleles, one dominant and one recessive? Are you able to calculate the frequencies of the genotypes and phenotypes of the offspring that would be produced by mating organisms with various combinations of the two alleles of each gene, assuming independent assortment of the two genes?

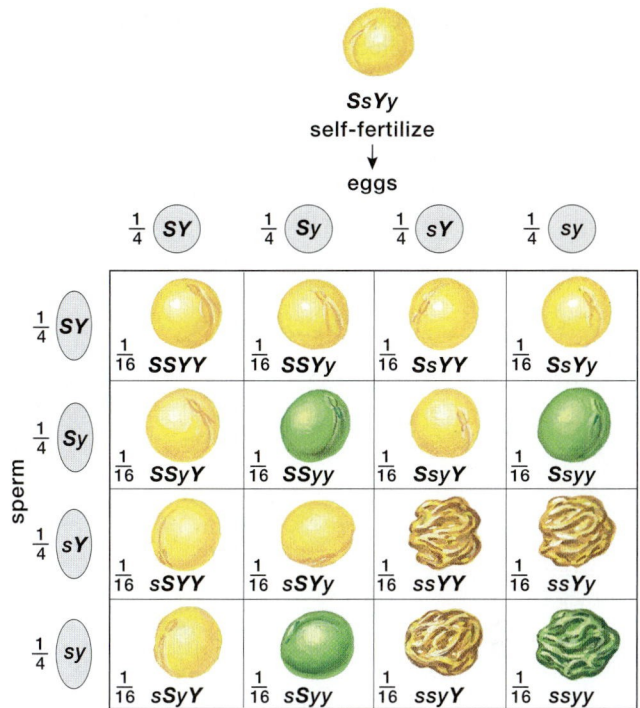

(a) Punnett square of a two-trait cross

seed shape		seed color		phenotypic ratio (9:3:3:1)
$\frac{3}{4}$ smooth	×	$\frac{3}{4}$ yellow	=	$\frac{9}{16}$ smooth yellow
$\frac{3}{4}$ smooth	×	$\frac{1}{4}$ green	=	$\frac{3}{16}$ smooth green
$\frac{1}{4}$ wrinkled	×	$\frac{3}{4}$ yellow	=	$\frac{3}{16}$ wrinkled yellow
$\frac{1}{4}$ wrinkled	×	$\frac{1}{4}$ green	=	$\frac{1}{16}$ wrinkled green

(b) Using probabilities to determine the offspring of a two-trait cross

▲ **FIGURE 10-11 Predicting genotypes and phenotypes for a cross between parents that are heterozygous for two traits** In pea seeds, yellow color (Y) is dominant to green (y), and smooth shape (S) is dominant to wrinkled (s). **(a)** Punnett square analysis. In this cross, an individual heterozygous for both traits self-fertilizes. In a cross involving two independent genes, there will be equal numbers of gametes with all of the possible combinations of alleles of the two genes—S with Y, S with y, s with Y, and s with y. Place these gamete combinations as the labels for the rows and columns in the Punnett square, and then calculate the offspring as explained in Figure 10-8. Note that the Punnett square predicts both the frequencies of combinations of traits ($\frac{9}{16}$ smooth yellow, $\frac{3}{16}$ smooth green, $\frac{3}{16}$ wrinkled yellow, and $\frac{1}{16}$ wrinkled green) and the frequencies of individual traits ($\frac{3}{4}$ yellow, $\frac{1}{4}$ green, $\frac{3}{4}$ smooth, and $\frac{1}{4}$ wrinkled). **(b)** In probability theory, the probability of two independent events is the product (multiplication) of their independent probabilities. For example, to find the probability of tossing two coins and having both come up heads, multiply the independent probabilities of each coin coming up heads ($\frac{1}{2} \times \frac{1}{2} = \frac{1}{4}$). Seed shape is independent of seed color. Therefore, multiplying the independent probabilities of the genotypes or phenotypes for each trait produces the predicted frequencies for the combined genotypes or phenotypes of the offspring. These frequencies are identical to those generated by the Punnett square.

QUESTION Can the genotype of a plant bearing smooth yellow seeds be revealed by a test cross with a plant bearing wrinkled green seeds?

▶ FIGURE 10-12 **Independent assortment of alleles** Chromosome movements during meiosis produce independent assortment of alleles, shown here for two genes. Each combination of alleles is equally likely to occur, producing gametes in the predicted proportions one-quarter *SY*, one-quarter *sy*, one-quarter *Sy*, and one-quarter *sY*.

QUESTION If the genes for seed color and seed shape were on the same chromosome rather than on different chromosomes, would their alleles assort independently? Why or why not?

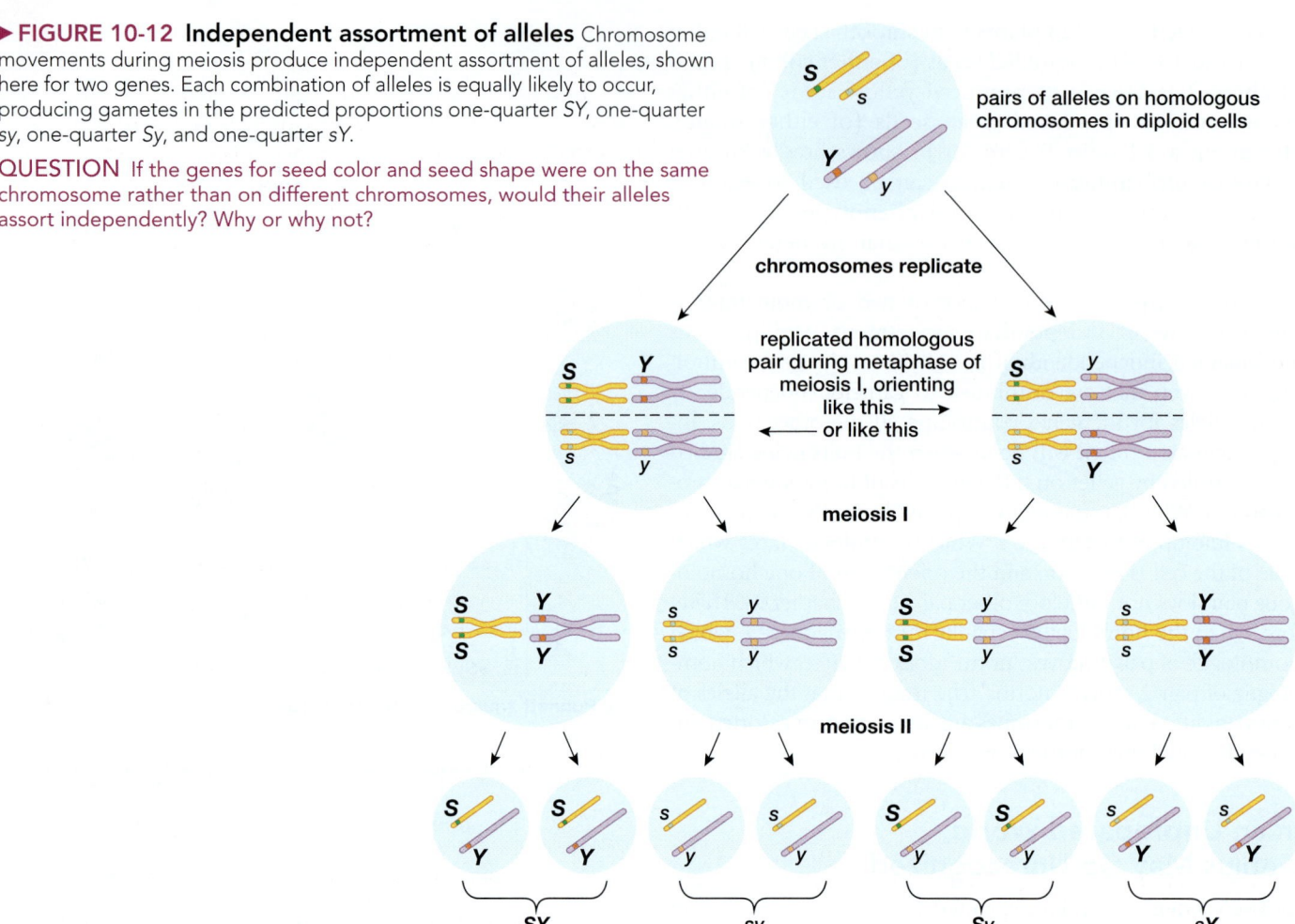

independent assortment produces four equally
likely allele combinations during meiosis

10.5 DO THE MENDELIAN RULES OF INHERITANCE APPLY TO ALL TRAITS?

In our discussion thus far, we have made some major simplifying assumptions: Each trait is completely controlled by a single gene, there are only two possible alleles of each gene, and one allele is completely dominant to the other, recessive, allele. Most traits, however, are influenced in more varied and subtle ways.

In Incomplete Dominance, the Phenotype of Heterozygotes Is Intermediate Between the Phenotypes of the Homozygotes

When one allele is completely dominant over a second allele, heterozygotes with one dominant allele have the same phenotype as homozygotes with two dominant alleles (see Figs. 10-8 and 10-9). However, in some cases the heterozygous phenotype is intermediate between the two homozygous phenotypes, a pattern of inheritance called **incomplete dominance.** In humans, hair texture is influenced by a gene with two incompletely dominant alleles, which we will call H_1 and H_2 (**Fig. 10-13**). A person with two copies of the H_1 allele

has curly hair; two copies of the H_2 allele produce straight hair. Heterozygotes, with the H_1H_2 genotype, have wavy hair. Two wavy-haired people might have children with any of the three hair types, with probabilities of one-quarter curly (H_1H_1), one-half wavy (H_1H_2), and one-quarter straight (H_2H_2).

A Single Gene May Have Multiple Alleles

Recall that all alleles originated as mutations (see Section 10.1), which may then be inherited from generation to generation. A gene may suffer many different mutations over thousands or millions of years of evolution, with the result that there are multiple alleles of many genes. Although an *individual* can have at most two different alleles (one on each of two homologous chromosomes), if we examined the genes of all the members of a *species*, we would find that many genes have dozens, even hundreds, of different alleles. Which of these alleles an offspring inherits, of course, depends on which alleles were present in its parents. There are multiple alleles for many human genetic disorders, including Marfan syndrome, Duchenne muscular dystrophy (see the "Health Watch: Muscular Dystrophy" on p. 185), and cystic fibrosis (see the case study in Chapter 12).

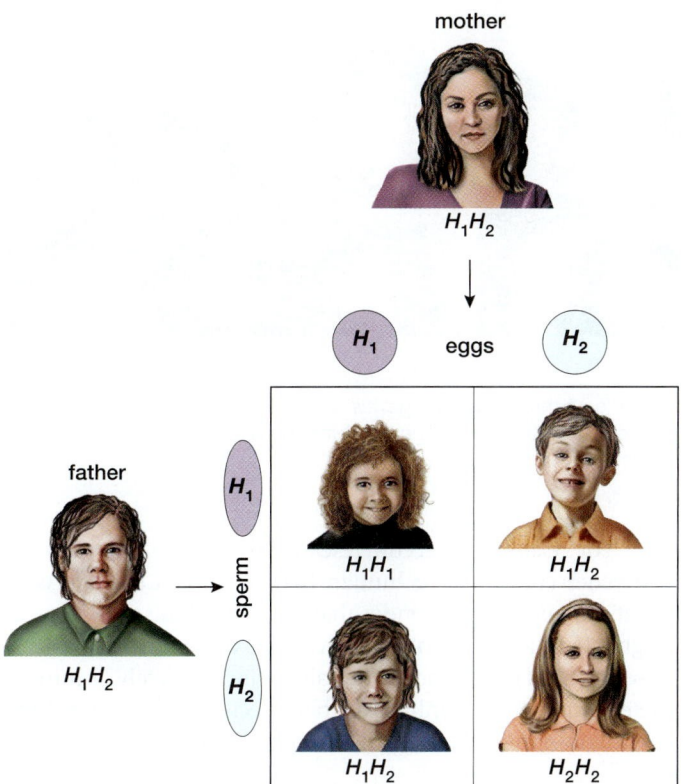

▲ FIGURE 10-13 **Incomplete dominance** The inheritance of hair texture in humans is an example of incomplete dominance. In such cases, we use capital letters for both alleles, here H_1 and H_2. Homozygotes may have curly hair (H_1H_1) or straight hair (H_2H_2). Heterozygotes (H_1H_2) have wavy hair. The children of a man and a woman, both with wavy hair, may have curly, straight, or wavy hair, in the approximate ratio of $\frac{1}{4}$ curly: $\frac{1}{2}$ wavy: $\frac{1}{4}$ straight.

Human blood types are an example of multiple alleles of a single gene. The blood types A, B, AB, and O arise as a result of three different alleles of a single gene (we will designate the alleles *A*, *B*, and *o*). This gene codes for an enzyme that adds sugar molecules to the ends of glycoproteins that protrude from the surfaces of red blood cells. Alleles *A* and *B* code for enzymes that add different sugars to the glycoproteins (we'll call the resulting molecules glycoproteins A and B, respectively). Allele *o* codes for a nonfunctional enzyme that doesn't add any sugar molecules.

A person may have one of six genotypes: *AA*, *BB*, *AB*, *Ao*, *Bo*, or *oo* (**Table 10-1**). Alleles *A* and *B* are dominant to *o*. Therefore, people with genotypes *AA* or *Ao* make only type A glycoproteins and have type A blood. Those with genotypes *BB* or *Bo* synthesize only type B glycoproteins and have type B blood. Homozygous recessive *oo* individuals lack both types of glycoproteins and have type O blood. In people with type AB blood, both enzymes are present, so their red blood cells have both A and B glycoproteins. When a heterozygote expresses the phenotypes of both of the homozygotes (in this case, both A and B glycoproteins), the pattern of inheritance is called **codominance,** and the alleles are said to be codominant to one another.

As you probably know, the fact that people may have different blood types is medically important. The human immune system produces proteins called antibodies, which bind to complex molecules that are not produced by a person's own body (if they did bind to "self" molecules, your immune system would destroy the cells of your own body). In their usual role in defending against disease, antibodies bind to molecules on the surfaces of invading bacteria or viruses, and help to destroy the invaders (see Chapter 36).

TABLE 10-1   Human Blood Group Characteristics

Blood Type	Genotype	Red Blood Cells	Has Plasma Antibodies to:	Can Receive Blood from:	Can Donate Blood to:	Frequency in U.S.
A	*AA* or *Ao*	A glycoprotein	B glycoprotein	A or O (no blood with B glycoprotein)	A or AB	42%
B	*BB* or *Bo*	B glycoprotein	A glycoprotein	B or O (no blood with A glycoprotein)	B or AB	10%
AB	*AB*	Both A and B glycoproteins	Neither A nor B glycoprotein	AB, A, B, O (universal recipient)	AB	4%
O	*oo*	Neither A nor B glycoprotein	Both A and B glycoproteins	O (no blood with A or B glycoprotein)	O, AB, A, B (universal donor)	44%

Humans show a wide range of skin tones, from almost white to very dark brown

▲ FIGURE 10-14 **Polygenic inheritance of skin color in humans** At least three separate genes, each with two incompletely dominant alleles, determine human skin color (the inheritance is actually more complex than this). The combination of complex polygenic inheritance and exposure to sunlight produces an almost continuous gradation of skin colors.

However, antibodies also complicate blood transfusions. Most antibodies are secreted by cells of the immune system and then circulate in the blood. The immune system makes antibodies that can bind to glycoproteins on red blood cells, if those glycoproteins include sugars that are different from the sugars on a person's own red blood cells. The antibodies cause red blood cells with foreign glycoproteins to clump together and rupture. The resulting clumps and fragments can clog small blood vessels and damage vital organs such as the brain, heart, lungs, or kidneys. This means that blood type must be carefully matched before a blood transfusion.

Table 10-1 summarizes human blood types and safe transfusions. Obviously, a person can donate blood to anyone with the same blood type. In addition, type O blood, lacking any sugars, can be safely transfused to all other blood types, because type O red blood cells are not attacked by antibodies in A, B, or AB blood. (The antibodies present in the transfused blood become too diluted by the much larger volume of the recipient's blood to cause problems.) People with type O blood are called "universal donors." But O blood carries antibodies to both A and B glycoproteins, so type O individuals can receive transfusions of only type O blood. Type AB blood doesn't contain antibodies against any type of red blood cells, so a person with type AB blood can receive blood from people with any other blood type; thus, they are called "universal recipients."

Many Traits Are Influenced by Several Genes

Your class probably contains people of varied heights, skin colors, and body builds—variation that cannot be divided into convenient, easily defined phenotypes. Traits such as these are influenced by interactions among two or more genes, a process called **polygenic inheritance.** As you might imagine, the more genes that contribute to a single trait, the greater the number of possible phenotypes and the smaller the differences among them. Traits that are affected by polygenic inheritance are often also strongly affected by the environment, further blurring the differences among phenotypes.

For example, recent research suggests that at least 180 genes contribute to human height. Add in the effects of nutrition, and it's no wonder that there aren't discrete steps in height. Human skin color is probably controlled by at least three different genes, each with pairs of incompletely dominant alleles. Exposure to sunlight further affects skin color, resulting in virtually continuous variation in phenotype (**Fig. 10-14**).

Single Genes Typically Have Multiple Effects on Phenotype

We have just seen that a single phenotype may result from the interaction of several genes. The reverse is also true: Single genes often have multiple phenotypic effects, a phenomenon called **pleiotropy.** For example, a mutation in a single gene in a lab mouse in 1962 produced a nude mouse (**Fig. 10-15**). Researchers rapidly discovered that nude mice are not only hairless; they lack a thymus gland, have virtually no immune response, and females do not develop functional mammary glands, so they can't nurse their pups.

▲ FIGURE 10-15 **Nude mice**

CASE STUDY continued

Sudden Death on the Court

In Marfan syndrome, a single defective fibrillin allele causes increased height, long limbs, large hands and feet, weak walls in the aorta, and often dislocated lenses in one or both eyes—a striking example of pleiotropy in humans.

The Environment Influences the Expression of Genes

An organism is not just the sum of its genes. In addition to its genotype, the environment in which an organism lives profoundly affects its phenotype. Fur color in Siamese cats vividly illustrates environmental effects on gene action. All Siamese cats are born with pale fur, but within the first few weeks, the ears, nose, paws, and tail turn dark (**Fig. 10-16**). A Siamese cat actually has the genotype for dark fur everywhere on its body. However, the enzyme that produces the dark pigment is inactive at temperatures above about 93°F (34°C). While inside their mother's uterus, unborn kittens are warm all over, so newborn Siamese kittens have pale fur on their entire bodies. After they are born, the ears, nose, paws, and tail become cooler than the rest of the body, so dark pigment is produced in those areas.

Most environmental influences are more complicated and subtle than this. Complex environmental influences are particularly common in human characteristics. We have already seen that skin color is modified by sun exposure and height is influenced by nutrition.

CHECK YOUR LEARNING

Can you describe the patterns of inheritance of traits showing incomplete dominance, codominance, and multiple alleles? Are you able to explain how polygenic inheritance and environmental influences combine to produce nearly continuous variation in many phenotypes?

▲ FIGURE 10-16 **Environmental influence on phenotype** The distribution of dark fur in the Siamese cat is an interaction between genotype and environment, producing a particular phenotype. Newborn Siamese kittens have pale fur everywhere on their bodies. In an adult Siamese, the allele for dark fur is expressed only in the cooler areas (nose, ears, paws, and tail).

10.6 HOW ARE GENES LOCATED ON THE SAME CHROMOSOME INHERITED?

Gregor Mendel knew nothing about the physical nature of genes or chromosomes. We now know that genes are parts of chromosomes, and that each chromosome contains many genes, up to several thousand in a really large chromosome. These facts have important implications for inheritance.

Genes on the Same Chromosome Tend to Be Inherited Together

Chromosomes, not individual genes, assort independently during meiosis I. Therefore, genes located on *different chromosomes* assort independently into gametes. In contrast, genes on the *same chromosome* tend to be inherited together, a phenomenon called **gene linkage.** One of the first pairs of linked genes to be discovered was found in the sweet pea, a different species from Mendel's edible pea. In sweet peas, the gene for flower color (purple vs. red) and the gene for pollen grain shape (round vs. long) are carried on the same chromosome. Thus, the alleles for these genes normally assort together into gametes during meiosis and are inherited together.

Consider a heterozygous sweet pea plant with purple flowers and long pollen (**Fig. 10-17**). The dominant purple allele of the flower-color gene and the dominant long allele of the pollen-shape gene are located on one homologous chromosome (Fig. 10-17, top). The recessive red allele of the flower-color gene and the recessive round allele of the pollen-shape gene are located on the other homologue (Fig. 10-17, bottom). Therefore, the gametes produced by this plant are likely to have either purple and long alleles or red and round alleles. This pattern of inheritance does not conform to the law of independent assortment, because the alleles for flower color and pollen shape do not segregate independently of one another, but tend to stay together during meiosis.

Crossing Over Creates New Combinations of Linked Alleles

Although they *tend* to be inherited together, genes on the same chromosome do not *always* stay together. If you cross-fertilized two sweet peas with the chromosomes shown in Figure 10-17, you might expect that all of the offspring would have either purple flowers with long pollen grains, or

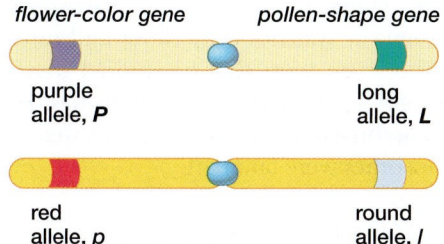

flower-color gene   pollen-shape gene

purple   long
allele, **P**   allele, **L**

red   round
allele, **p**   allele, **l**

▲ FIGURE 10-17 **Linked genes on homologous chromosomes in the sweet pea** The genes for flower color and pollen shape are on the same chromosome, so they tend to be inherited together.

red flowers with round pollen grains. (Try working this out with a Punnett square.) In reality, you would usually find a few offspring with purple flowers and round pollen, and a few with red flowers and long pollen, as if, sometimes, the genes for flower color and pollen shape became unlinked. How can this happen?

During prophase I of meiosis, homologous chromosomes sometimes exchange parts, a process called crossing over (see Chapter 9, Fig. 9-16). In most chromosomes, at least one exchange between each homologous pair occurs during each meiotic cell division. The exchange of corresponding segments of DNA during crossing over produces new allele combinations on both homologous chromosomes. Then, when homologues separate at anaphase I, the chromosomes that the haploid daughter cells receive will have different sets of alleles than the chromosomes of the parent cell had.

Crossing over during meiosis explains **genetic recombination:** new combinations of alleles of genes that are located on the same chromosome. Let's look at the sweet pea chromosomes during meiosis. During prophase I, the duplicated, homologous chromosomes pair up (**Fig. 10-18a**). Each homologue will have one or more regions where crossing over occurs. Imagine that crossing over exchanges the alleles for flower color between nonsister chromatids of the two homologues (**Fig. 10-18b**). At anaphase I, the separated homologues will each now have one chromatid bearing a piece of DNA from a chromatid of the other homologue (**Fig. 10-18c**). During meiosis II, four types of chromosomes will be distributed to the four daughter cells: two unchanged chromosomes and two recombined chromosomes (**Fig. 10-18d**).

Therefore, some gametes will be produced with each of four configurations: *PL* and *pl* (on the original parental chromosomes) and *Pl* and *pL* (on the recombined chromosomes). If a sperm with a *Pl* chromosome fertilizes an egg with a *pl* chromosome, the offspring plant will have purple flowers (*Pp*) and round pollen (*ll*). If a sperm with a *pL* chromosome fertilizes an egg with a *pl* chromosome, then the offspring will have red flowers (*pp*) and long pollen (*Ll*).

Not surprisingly, the farther apart the genes are on a chromosome, the more likely it is that crossing over will occur between them. Two genes close together on a chromosome are strongly linked and will rarely be separated by a crossover. However, if two genes are very far apart, crossing over between the genes occurs so often that they seem to be independently assorted, just as if they were on different chromosomes. When Gregor Mendel discovered independent assortment, he was not only clever and careful, he was also lucky. The seven traits that he studied were controlled by genes on only four different chromosomes. He observed independent assortment because the genes that were on the same chromosomes were far apart.

CHECK YOUR LEARNING

Can you describe how the patterns of inheritance differ between traits controlled by genes on a single chromosome and traits controlled by genes on different chromosomes?

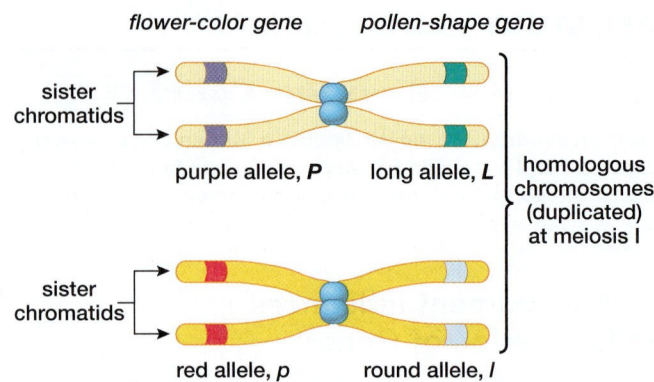

(a) Duplicated chromosomes in prophase of meiosis I

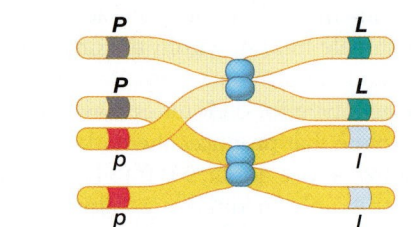

(b) Crossing over during prophase I

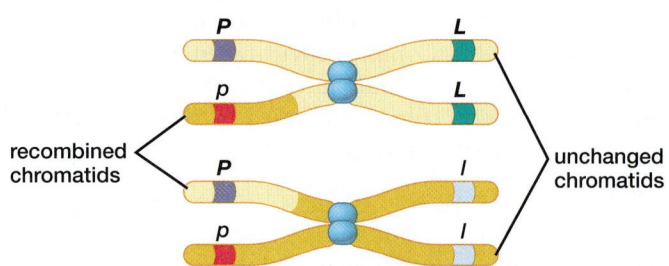

(c) Homologous chromosomes separate at anaphase I

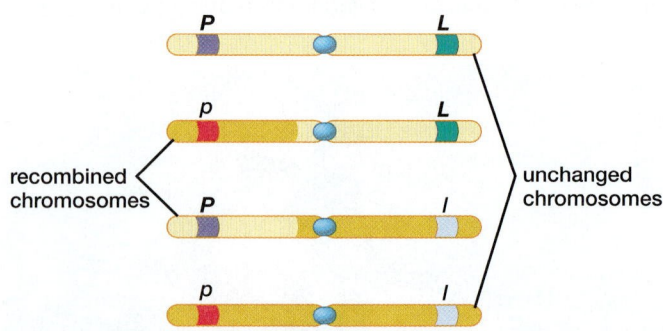

(d) Unchanged and recombined chromosomes after meiosis II

▲ **FIGURE 10-18 Crossing over recombines alleles on homologous chromosomes (a)** During prophase of meiosis I, duplicated homologous chromosomes pair up. **(b)** Nonsister chromatids of the two homologues exchange parts by crossing over. **(c)** When the homologous chromosomes separate during anaphase of meiosis I, one chromatid of each of the homologues now contains a piece of DNA from a chromatid of the other homologue. **(d)** After meiosis II, two of the haploid daughter cells receive unchanged chromosomes, and two receive recombined chromosomes. The recombined chromosomes contain allele arrangements that did not occur in the original parental chromosomes.

10.7 HOW ARE SEX AND SEX-LINKED TRAITS INHERITED?

In many animals, an individual's sex is determined by its **sex chromosomes.** In mammals, females have two identical sex chromosomes, called **X chromosomes,** whereas males have one X chromosome and one **Y chromosome** (**Fig. 10-19**). Although the Y chromosome is much smaller than the X chromosome, a small part of both sex chromosomes is homologous. As a result, the X and Y chromosomes pair up during prophase of meiosis I and separate during anaphase I. The other chromosomes, which occur in pairs that have identical appearance in both males and females, are called **autosomes.**

In Mammals, the Sex of an Offspring Is Determined by the Sex Chromosome in the Sperm

During sperm formation, the sex chromosomes segregate, and each sperm receives either the X or the Y chromosome (plus one member of each pair of autosomes). The sex chromosomes also segregate during egg formation, but because females have two X chromosomes, every egg receives one X chromosome (and one member of each pair of autosomes). Thus, a male offspring is produced if an egg is fertilized by a Y-bearing sperm, and a female offspring is produced if an egg is fertilized by an X-bearing sperm (**Fig. 10-20**).

Sex-Linked Genes Are Found Only on the X or Only on the Y Chromosome

Genes that are located only on sex chromosomes are referred to as **sex-linked.** In many animal species, the Y chromosome carries only a few genes. In humans, the Y chromosome

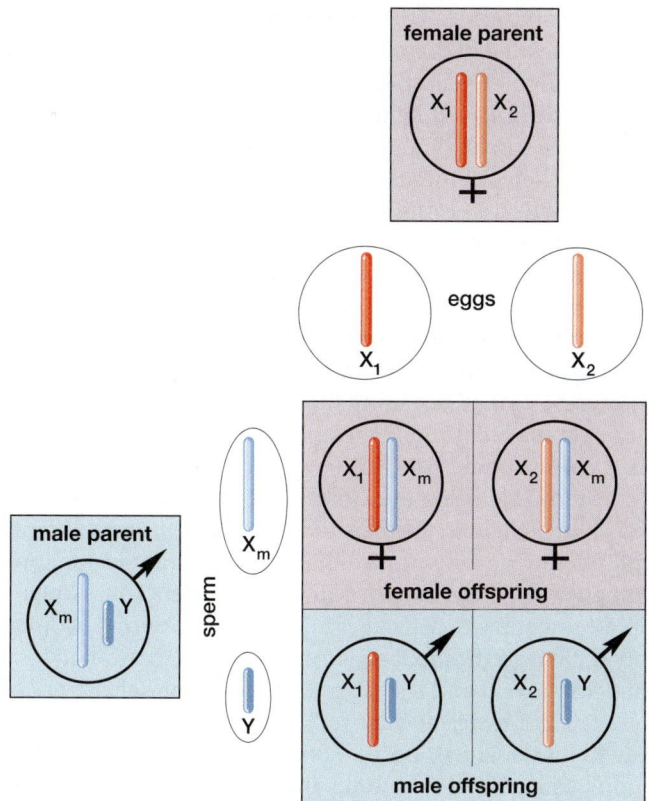

▲ **FIGURE 10-20 Sex determination in mammals** Male offspring receive their Y chromosome from their father; female offspring receive the father's X chromosome (labeled X_m). Both male and female offspring receive an X chromosome (either X_1 or X_2) from their mother.

contains a few dozen genes (probably less than 100), many of which play a role in male reproduction. The most well-known Y-linked gene is the sex-determining gene, called *SRY*. During embryonic life, the action of *SRY* sets in motion the entire male developmental pathway. Under normal conditions, *SRY* causes the male gender to be 100% linked to the Y chromosome.

In contrast to the small Y chromosome, the human X chromosome contains more than 1,000 genes, most of which have no counterpart on the Y chromosome. Few X chromosome genes have a specific role in reproduction. Most of the genes on the X chromosome determine traits that are important in both sexes, such as color vision, blood clotting, and certain structural proteins in muscles.

How does linkage of genes on the X chromosome affect inheritance? Because females have two X chromosomes, they can be either homozygous or heterozygous for genes on the X chromosome, and dominant versus recessive relationships among alleles will be expressed. Males, in contrast, fully express all the alleles they have on their single X chromosome, regardless of whether those alleles would be dominant or recessive in females.

Let's look at a familiar example: red-green color deficiency, more commonly—though usually incorrectly—called color blindness (**Fig. 10-21**). Color deficiency is caused by recessive alleles of either of two genes located on

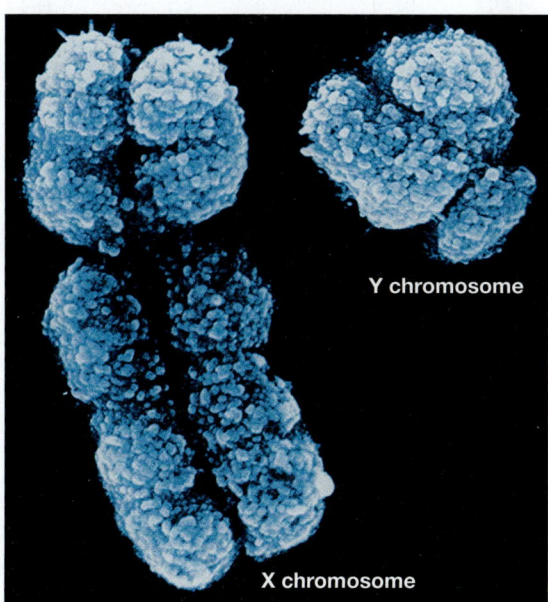

▲ **FIGURE 10-19 Human sex chromosomes** The Y chromosome (right), which carries relatively few genes, is much smaller than the X chromosome (left). Image courtesy of Indigo® Instruments: http://www.indigo.com.

the X chromosome. The normal, dominant alleles of these genes (we will call them both C) encode proteins that allow one set of color-vision cells in the eye, called cones, to be most sensitive to red light and another set to be most sensitive to green light (**Fig. 10-21a**). There are several defective recessive alleles of these genes (we will call them all c). In the most extreme cases, one of the genes will actually be missing from an X chromosome, or the defective alleles will encode proteins that make both sets of cones equally sensitive to both red and green light. Therefore, the affected person cannot distinguish red from green (**Fig. 10-21b**).

How is color deficiency inherited? A man can have the genotype CY or cY, meaning that he has a color-vision allele C or c on his X chromosome and no corresponding gene on his Y chromosome. He will have normal color vision if his X chromosome bears the C allele, or be color-deficient if it bears the c allele. A woman may be CC, Cc, or cc. Women with CC or Cc genotypes will have normal color vision; only women with cc genotypes will be color-deficient. Roughly 7% of men have defective color vision. Among women, about 93% are homozygous normal CC, 7% are heterozygous normal Cc, and less than 0.5% are color-deficient cc.

A color-deficient man (cY) will pass his defective allele only to his daughters, because only his daughters inherit his X chromosome. Usually, however, the daughters will have normal color vision, because they will also inherit a normal C allele from their mother, who is very likely homozygous normal CC.

A heterozygous woman (Cc), although she has normal color vision, has a 50% chance of passing her defective allele to her sons (**Fig. 10-21c**). Sons who receive the defective allele will be color-deficient (cY), whereas sons who inherit the functional allele will have normal color vision (CY).

CHECK YOUR LEARNING

Do you understand why sperm determine the gender of offspring in mammals? Are you able to explain why most sex-linked traits are controlled by genes on the X chromosome? Can you describe the pattern of inheritance of sex-linked traits?

10.8 HOW ARE HUMAN GENETIC DISORDERS INHERITED?

Many human diseases are influenced by genetics to a greater or lesser degree. Because experimental crosses with people are out of the question, human geneticists search medical, historical, and family records to study past crosses. Records

▶ **FIGURE 10-21 Sex-linked inheritance of red-green color deficiency** Color grids can help people with normal color vision imagine the world as seen by someone with red-green color deficiency. **(a)** Normal color vision. **(b)** To a person who is really red-green "color-blind," both grids look exactly the same. Usually, an affected person isn't completely color-blind—he sees most of the colors that normal people see, but not as well. Therefore, the grids would look very similar, but not identical. **(c)** A Punnett square showing the inheritance of color deficiency from a heterozygous woman (Cc) to her sons.

(a) Normal color vision

The individual cannot distinguish red from green

(b) Red-green color blindness

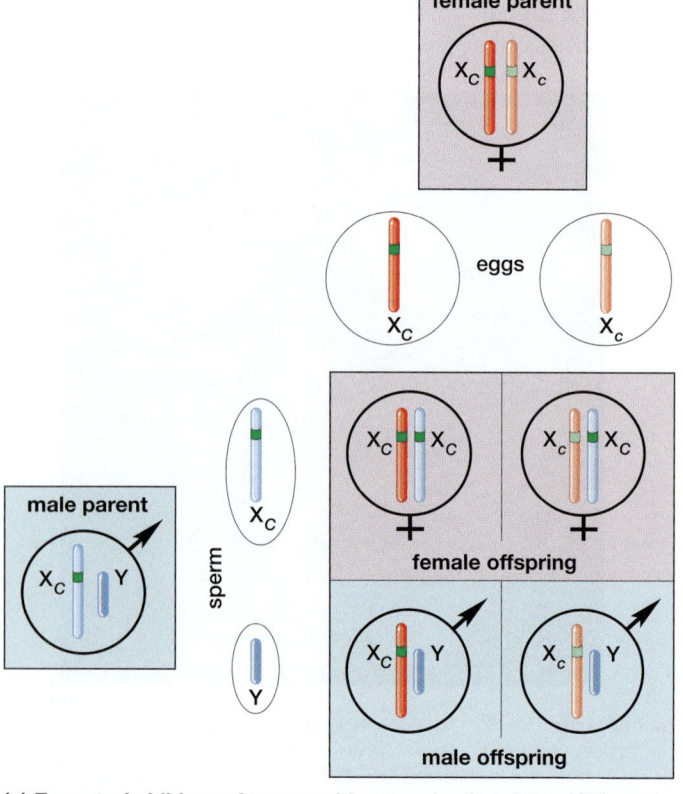

(c) Expected children of a man with normal color vision (CY), and a heterozygous woman (Cc)

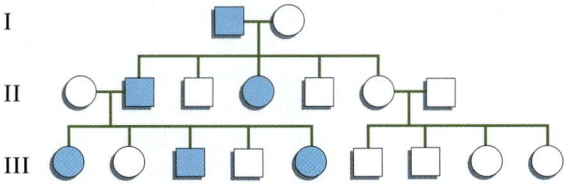

(a) A pedigree for a dominant trait

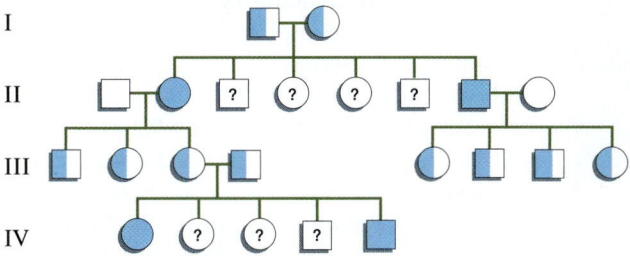

(b) A pedigree for a recessive trait

How to read pedigrees

I, II, III = generations

☐ = male   ◯ = female

☐—◯ = parents

☐ ◯ ◯ = offspring

■ or ● = shows trait

☐ or ◯ = does not show trait

▨ or ◑ = known carrier (heterozygote) for recessive trait

☐? or ◯? = cannot determine the genotype from this pedigree

▲ **FIGURE 10-22 Family pedigrees (a)** A pedigree for a dominant trait. Note that any offspring showing a dominant trait must have at least one parent with the trait (see Figs. 10-8, 10-9, and 10-11). **(b)** A pedigree for a recessive trait. Any individual showing a recessive trait must be homozygous recessive. If that person's parents did not show the trait, then both parents must be heterozygotes (carriers). Note that the genotype cannot be determined for some offspring, who may be either carriers or homozygous dominants.

extending across several generations can be arranged in the form of family **pedigrees,** diagrams that show the genetic relationships among a set of related individuals (**Fig. 10-22**).

Careful analysis of pedigrees can reveal whether a particular trait is inherited in a dominant, recessive, or sex-linked pattern. Since the mid-1960s, analysis of human pedigrees, combined with molecular genetic technology, has produced great strides in understanding human genetic diseases. For instance, geneticists now know the genes responsible for dozens of inherited diseases, including

sickle-cell anemia, hemophilia, muscular dystrophy, Marfan syndrome, and cystic fibrosis. Research in molecular genetics has increased our ability to predict genetic diseases and perhaps even to cure them (a topic we explore further in Chapter 13).

Some Human Genetic Disorders Are Controlled by Single Genes

Some common human traits, such as freckles, cleft chin, and dimples, are inherited in a simple Mendelian fashion; that is, each trait appears to be controlled by a single gene with a recessive and a dominant allele. Some human genetic disorders are also caused by defective alleles of a single gene.

Some Human Genetic Disorders Are Caused by Recessive Alleles

The human body depends on the actions of thousands of enzymes and other proteins. A mutation in an allele of the gene coding for one of these proteins can impair or destroy its function. However, the presence of one normal allele may generate enough functional protein to enable heterozygotes to be phenotypically indistinguishable from homozygotes with two normal alleles. Therefore, for many genes, a normal allele that encodes a functional protein is dominant to a mutant allele that encodes a nonfunctional protein. Put another way, a mutant allele of these genes is recessive to a normal allele. Thus, an abnormal phenotype occurs only in people who inherit two copies of the mutant allele.

A **carrier** for a genetic disorder is a person who is heterozygous, with one normal, dominant allele and one defective, recessive allele: That person is phenotypically healthy but can pass on his or her defective allele to offspring. Geneticists estimate that each of us carries recessive alleles of 5 to 15 genes that would cause serious genetic defects in homozygotes. Every time we have a child, there is a 50:50 chance that we will pass on the defective allele. This is usually harmless, because an unrelated man and woman are unlikely to possess a defective allele of the same gene, so they are unlikely to produce a child who is homozygous for a recessive genetic disease. Related couples, however (especially first cousins or closer), have inherited some of their genes from recent common ancestors, and so are more likely to carry a defective allele of the same gene. If a man and woman are both heterozygous for the same defective recessive allele, they have a 1 in 4 chance of having a child with the genetic disorder (see Fig. 10-22).

Albinism Results from a Defect in Melanin Production

An enzyme called tyrosinase is needed to produce melanin—the dark pigment in skin, hair, and the iris of the eye. Normal melanin production will occur if a person has either one or two functional tyrosinase alleles. However, if a person is homozygous for an allele that encodes defective tyrosinase, **albinism** occurs (**Fig. 10-23**). Albinism in humans and other mammals results in very pale skin and hair.

(a) Human   **(b) Wallaby**

▲ FIGURE 10-23 **Albinism (a)** Albinism occurs in most vertebrates, including people. This boy's irises are extremely pale, so his eyes are very sensitive to bright light. **(b)** The albino wallaby in the foreground is safe in a zoo, but in the wild, its bright white fur would make it very conspicuous to predators.

Sickle-Cell Anemia Is Caused by a Defective Allele for Hemoglobin Synthesis Red blood cells are packed with hemoglobin proteins, which transport oxygen and give the cells their red color. Anemia is a generic term given to a number of diseases, all characterized by a low red blood cell count or below-normal hemoglobin in the blood. **Sickle-cell anemia** is an inherited form of anemia that results from a mutation in the hemoglobin gene. A change in a single nucleotide results in an incorrect amino acid at a crucial position in the hemoglobin protein (see Section 12.4 in Chapter 12). When people with sickle-cell anemia exercise or move to high altitude, oxygen concentrations in their blood drop, and the sickle-cell hemoglobin proteins inside their red blood cells stick together. The resulting clumps of hemoglobin force red blood cells out of their usual flexible, disk shapes (**Fig. 10-24a**) into long, stiff, sickle shapes (**Fig. 10-24b**). The sickled cells are fragile and easily damaged. Anemia occurs because the sickled red blood cells are destroyed before their usual life span is completed.

The sickle shape also causes other complications. Sickle cells jam up in capillaries, causing blood clots. Tissues downstream of the clot do not receive enough oxygen. Paralyzing strokes can result if blocks occur in blood vessels in the brain.

People homozygous for the sickle-cell allele synthesize only defective hemoglobin. Consequently, many of their red blood cells become sickled, and they suffer

from sickle-cell anemia. Although heterozygotes produce about half normal and half abnormal hemoglobin, they have very few sickled red blood cells and seldom show any symptoms. Because only people who are homozygous for the sickle-cell allele usually show symptoms, sickle-cell anemia is considered to be a recessive disorder. However, during exceptionally strenuous exercise, some heterozygotes may experience life-threatening complications, as we explore in "Links to Everyday Life: The Sickle-Cell Allele and Athletics."

About 5% to 25% of sub-Saharan Africans and 8% of African Americans are heterozygous for sickle-cell anemia, but the allele is very rare in Caucasians. Why the difference? Shouldn't natural selection work to eliminate the sickle-cell allele in both African and Caucasian populations? The difference arises because heterozygotes have some resistance to the parasite that causes malaria, which is common in Africa and other places with warm

(a) Normal red blood cells

(b) Sickled red blood cells

▶ FIGURE 10-24 **Sickle-cell anemia (a)** Normal red blood cells are disk shaped with indented centers. **(b)** When blood oxygen is low, the red blood cells in a person with sickle-cell anemia become long, slender and curved, resembling a sickle.

climates, but not in colder regions such as most of Europe. This "heterozygote advantage" explains the prevalence of the sickle-cell allele in people of African origin.

Some Human Genetic Disorders Are Caused by Dominant Alleles

Some serious genetic disorders, such as Huntington disease, are caused by dominant alleles. Just as a pea plant needs only one dominant allele for purple color to bear purple flowers (see Figs. 10-7 and 10-8), so too a person needs to have only one defective dominant allele in order to suffer from these disorders. Therefore, everyone who inherits a dominant genetic disorder must have at least one parent with the disease (see Fig. 10-22a), which means that some people with dominant diseases must remain healthy enough to grow up and have children. In rare cases, a dominant allele that causes a genetic disorder may result not from an inherited allele but from a mutation in the egg or sperm of a parent who is otherwise unaffected. In this case, neither parent would have the disease.

How can a defective allele be dominant to the normal, functional allele? Some dominant alleles encode an abnormal protein that interferes with the function of the normal one. Other dominant alleles may encode proteins that carry out new, toxic reactions. Finally, dominant alleles may encode a protein that is overactive, performing its function at inappropriate times and places in the body.

Huntington Disease Is Caused by a Defective Protein That Kills Cells in Specific Brain Regions

Huntington disease is a dominant disorder that causes a slow, progressive deterioration of parts of the brain, resulting in a loss of coordination, flailing movements, personality disturbances, and eventual death. The symptoms of Huntington disease typically do not appear until 30 to 50 years of age. Therefore, before they experience their first symptoms, many Huntington victims pass the allele to their children. Geneticists isolated the Huntington gene in 1993, and a few years later, identified the gene's product, a protein they named "huntingtin." The function of normal huntingtin

---

≫ LINKS TO EVERYDAY LIFE

The Sickle-Cell Allele and Athletics

Sickle-cell anemia is considered to be a recessive trait, because only homozygous recessive people usually show any symptoms. At the molecular level, though, half the hemoglobin proteins in a heterozygote are defective. Does this really have no effect at all?

For the vast majority of heterozygotes (often described as having "sickle-cell trait"), there indeed are no health effects. However, a very small number of heterozygotes may experience serious medical problems if they participate in extreme exercise. Consider Devard and Devaughn Darling, identical twin brothers, who shared all their genes, including one copy of the sickle-cell allele (**Fig. E10-1**).

The Darling brothers starred in multiple sports in high school. Both were probable starters for the Florida State University football team when the unthinkable happened during practice on February 26, 2001: Devaughn collapsed and died. No one could prove that Devaughn's death was caused by the combination of strenuous workouts and the sickle-cell trait, but suspicions ran high. The university decided that it didn't want to risk Devard suffering the same fate, and barred Devard from playing football. Devard, however, transferred to Washington State University and played football for the Cougars for 2 years. He then played for five seasons in the National Football League.

The Darling brothers epitomize the rare, but real, dilemmas facing athletes with sickle-cell trait. Devard's football career and the accomplishments of many other heterozygotes show that having sickle-cell trait does not preclude strenuous athletics. The National Collegiate Athletic Association agrees: "Student-athletes with sickle-cell trait should not be excluded from athletics participation." However, Devaughn's tragic death underscores the need to take appropriate precautions.

▲ FIGURE E10-1 **Devard Darling runs to daylight for the Kansas City Chiefs** Devard's identical twin Devaughn died during football practice in college, probably from complications of sickle-cell trait.

Dehydration during extreme exercise, especially in hot weather, is probably the most important risk to heterozygotes, so the NCAA recommends that athletes "stay well hydrated at all times." These and other simple precautions have helped the U.S. Army to eliminate excess deaths caused by sickle-cell trait during basic training. In fact, the Army no longer even screens for sickle-cell trait. Medically appropriate and humane training procedures—realizing, for example, that failing to "tough it out" in the face of serious physical distress is not a sign of mental weakness——help all athletes, not only those with sickle-cell trait.

remains unknown. Mutant huntingtin seems to be cut up into toxic fragments inside brain cells, ultimately killing them.

Some Human Genetic Disorders Are Sex-Linked

As we described earlier, the X chromosome contains many genes that have no counterpart on the Y chromosome. Because males have only one X chromosome, they have only one allele for each of these genes. Therefore, males will show the phenotypes produced by these single alleles, even if the alleles are recessive and could be masked by dominant alleles in females.

A son receives his X chromosome from his mother and passes it only to his daughters. Thus, X-linked disorders caused by recessive alleles have a unique pattern of inheritance. Such disorders appear far more frequently in males and typically skip generations: An affected male passes the trait to a phenotypically normal, carrier daughter, who in turn bears some affected sons. The most familiar genetic defects due to recessive alleles of X-chromosome genes are red-green color deficiency (see Fig. 10-21), muscular dystrophy, and hemophilia.

Hemophilia is caused by a recessive allele on the X chromosome that results in a deficiency in one of the proteins needed for blood clotting. People with hemophilia bruise easily and may bleed extensively from minor injuries. They often have anemia due to blood loss. Nevertheless, even before modern treatment with clotting factors, some hemophiliac males survived to pass on their defective allele to their daughters, who in turn could pass it to their sons (**Fig. 10-25**). Muscular dystrophy, a fatal degeneration of the muscles in young boys, is another recessive sex-linked disorder that we describe in "Health Watch: Muscular Dystrophy" on p. **185**.

▲ **FIGURE 10-25 Hemophilia among the royal families of Europe** A famous genetic pedigree shows the transmission of sex-linked hemophilia from Queen Victoria of England (seated center front, with cane, in 1885) to her offspring and eventually to virtually every royal house in Europe, because of the extensive intermarriage of her children to the royalty of other European nations. Because Victoria's ancestors were free of hemophilia, the hemophilia allele must have arisen as a mutation either in Victoria herself or in one of her parents (or as a result of marital infidelity).

QUESTION Why is it not possible that a mutation in Victoria's husband, Albert, was the original source of hemophilia in this family pedigree?

Health|Watch

Muscular Dystrophy

When Olympic weightlifter Matthias Steiner of Germany won the gold medal in 2008 with a "clean and jerk" lift of almost 569 pounds (258 kilograms), he exerted tremendous forces on his body (Fig. E10-2). How could his muscles withstand these stresses?

Muscle cells are firmly tied together by a very long protein called dystrophin. The almost 3,700 amino acids of dystrophin form a supple yet strong rod that connects the cytoskeleton inside a muscle cell to proteins in its plasma membrane, which in turn attach to supporting proteins that surround each muscle. When a muscle contracts, the muscle cells remain intact because the forces are evenly distributed throughout each cell and to the extracellular support proteins.

Unfortunately, about 1 in 3,500 boys synthesizes seriously defective dystrophin proteins and suffers from **muscular dystrophy,** which literally means "degeneration of the muscles." The most severe form of the disease is named Duchenne muscular dystrophy; a less severe form is called Becker muscular dystrophy.

▲ FIGURE E10-2 Matthias Steiner wins the gold with a lift of almost 569 pounds.

Both forms are named after the physicians who first described the disorders. Muscular dystrophy is caused by a defective allele of the dystrophin gene (usually called the *DMD* gene because of its involvement in **D**uchenne **M**uscular **D**ystrophy). As boys with muscular dystrophy use their muscles, the lack of functioning dystrophin means that ordinary muscle contraction tears the muscle cells. The cells die and are replaced by fat and connective tissue. By the age of 7 or 8, boys with Duchenne muscular dystrophy can no longer walk. Death usually occurs in the early 20s from heart and respiratory problems.

Girls almost never have Duchenne muscular dystrophy. Why not? Because the dystrophin gene is on the X chromosome, and muscular dystrophy alleles are recessive. Therefore, a boy will suffer muscular dystrophy if he has a defective dystrophin allele on his single X chromosome, but a girl, with two X chromosomes, would need two defective copies to suffer the disorder. This virtually never happens, because a girl would have to receive one defective dystrophin allele from her mother, on one of her X chromosomes, and one from

her father, on his X chromosome. Because they suffer early disability and death, boys with Duchenne muscular dystrophy almost never have children.

This scenario makes genetic sense, but it may seem to defy the concept of evolution by natural selection. Shouldn't natural selection have almost completely eradicated defective dystrophin alleles? Actually, natural selection does rapidly eliminate defective dystrophin alleles. However, the dystrophin gene is enormous—about 2.2 *million* nucleotides long, compared to about 27 *thousand* nucleotides for the average human gene. In fact, the dystrophin gene occupies about 2% of the entire X chromosome.

Why does this matter? Remember, alleles arise as mutations in DNA. The longer the gene, the greater the chances for a mistake during DNA replication, so the mutation rate for the dystrophin gene is almost a hundred times greater than for the average gene. In fact, it is a tribute to the astounding accuracy of DNA copying that we don't all suffer from muscular dystrophy. About one-third of the boys with muscular dystrophy receive a new mutation that occurred in an X chromosome of a reproductive cell of their mother, and two-thirds inherit a preexisting allele in one of their mother's X chromosomes. The new mutations counterbalance natural selection, resulting in the steady incidence of about 1 in 3,500 boys.

Can anything be done for boys with muscular dystrophy? Right now, there are no cures, although treatments are available that slow muscle degeneration, prolong life, and make the affected boys more comfortable. However, in 2011, clinical trials showed that a novel molecular technique can trick the muscles of boys with muscular dystrophy into making partially functional dystrophin from a faulty *DMD* allele. Several research labs are trying to devise ways to colonize dystrophic muscles with stem cells that carry a completely functional copy of the *DMD* gene. If this works, the decades-old dream of curing muscular dystrophy may finally be realized.

Some Human Genetic Disorders Are Caused by Abnormal Numbers of Chromosomes

Usually, the intricate mechanisms of meiotic cell division ensure that each sperm and egg receives only one chromosome from each homologous pair (see Chapter 9). However, this elaborate dance of the chromosomes occasionally misses a step, resulting in gametes that have too many or too few chromosomes. Such errors in meiosis, called **nondisjunction,** can affect the number of sex chromosomes

or autosomes in a gamete (Fig. 10-26). Most embryos that arise from the fusion of gametes with abnormal chromosome numbers spontaneously abort, accounting for 20% to 50% of all miscarriages, but some embryos with abnormal numbers of chromosomes survive to birth or beyond.

Some Genetic Disorders Are Caused by Abnormal Numbers of Sex Chromosomes

Sperm usually carry either an X or a Y chromosome (see Fig. 10-20). Nondisjunction of sex chromosomes in males produces sperm with either no sex chromosome (often called

	Normal meiosis	Nondisjunction during meiosis I	Nondisjunction during meiosis II
Parent cell			
Meiosis I			
Meiosis II	*n* *n* *n* *n*	*n* + 1 *n* + 1 *n* − 1 *n* − 1	*n* + 1 *n* − 1 *n* *n*

▲ **FIGURE 10-26 Nondisjunction during meiosis** Nondisjunction may occur either during meiosis I or meiosis II, resulting in gametes with too many (*n* + 1) or too few (*n* − 1) chromosomes.

"O" sperm), or two sex chromosomes (XX, YY, or XY, depending on whether the nondisjunction occurred in meiosis I or II). Nondisjunction of the sex chromosomes in females produces O or XX eggs instead of eggs with one X chromosome. When normal gametes fuse with these defective sperm or eggs, the zygotes have normal numbers of autosomes but abnormal numbers of sex chromosomes (Table 10-2). The most common abnormalities are XO, XXX, XXY, and XYY. (Genes on the X chromosome are essential to survival, so any embryo without at least one X chromosome spontaneously aborts very early in development.)

Turner Syndrome (XO) About 1 in every 3,000 female babies has only one X chromosome, a condition known as **Turner syndrome** (also called monosomy X, meaning "having one X chromosome"). At puberty, hormone deficiencies prevent XO females from menstruating or developing secondary sexual characteristics, such as enlarged breasts.

Treatment with estrogen promotes physical development. However, because most women with Turner syndrome lack mature eggs, hormone treatment does not make it possible for them to bear children. Other common characteristics of women with Turner syndrome include short stature, folds of skin around the neck, and increased risk of cardiovascular disease, kidney defects, and hearing loss. Because women with Turner syndrome have only one X chromosome, they display X-linked recessive disorders, such as hemophilia and color deficiency, much more frequently than XX women do.

Trisomy X (XXX) About 1 in every 1,000 women has three X chromosomes, a condition known as **trisomy X,** or triple X. Most such women have no detectable differences from XX women, except for a tendency to be taller and to have a higher incidence of learning disabilities. Unlike women with Turner syndrome, most trisomy X women are fertile and, interestingly enough, almost always bear normal XX and XY children. Some

TABLE 10-2   Effects of Nondisjunction of the Sex Chromosomes During Meiosis

Nondisjunction in Father			
Sex Chromosomes of Defective Sperm	Sex Chromosomes of Normal Egg	Sex Chromosomes of Offspring	Phenotype
O (none)	X	XO	Female—Turner syndrome
XX	X	XXX	Female—Trisomy X
XY	X	XXY	Male—Klinefelter syndrome
YY	X	XYY	Male—Jacob syndrome

Nondisjunction in Mother			
Sex Chromosomes of Normal Sperm	Sex Chromosomes of Defective Egg	Sex Chromosomes of Offspring	Phenotype
X	O (none)	XO	Female—Turner syndrome
Y	O (none)	YO	Dies as embryo
X	XX	XXX	Female—Trisomy X
Y	XX	XXY	Male—Klinefelter syndrome

unknown mechanism must operate during meiosis to prevent an extra X chromosome from being included in their eggs.

Klinefelter Syndrome (XXY) About 1 male in every 1,000 is born with two X chromosomes and one Y chromosome. Most of these men go through life never realizing that they have an extra X chromosome. However, at puberty, some show mixed secondary sexual characteristics, including partial breast development, broadening of the hips, and small testes. These symptoms are known as **Klinefelter syndrome.** XXY men may be infertile because of low sperm count but are not impotent. They are usually diagnosed when an XXY man and his partner seek medical help because they are unable to have children.

Jacob Syndrome (XYY) **Jacob syndrome** (XYY) occurs in about 1 male in every 1,000. You might expect that an extra Y chromosome, which has few active genes, would not make very much difference, and this seems to be true in most cases. The most common effect is that XYY males tend to be taller than average. There may also be a slightly increased likelihood of learning disabilities.

Some Genetic Disorders Are Caused by Abnormal Numbers of Autosomes

Nondisjunction of the autosomes produces eggs or sperm that are missing an autosome or that have two copies of an autosome. Fusion with a normal gamete (bearing one copy of each autosome) leads to an embryo with either one or three copies of the affected autosome. Embryos that have only one copy of any of the autosomes abort so early in development that the woman never knows she was pregnant. Embryos with three copies of an autosome (trisomy) also usually spontaneously abort. However, a small fraction of embryos with three copies of chromosomes 13, 18, or 21 survive to birth. In the case of trisomy 21, the child may live into adulthood.

Trisomy 21 (Down Syndrome) An extra copy of chromosome 21, a condition called **trisomy 21,** or **Down syndrome,** occurs in about 1 of every 800 births, although this rate varies tremendously with the age of the parents (see below). Children with Down syndrome have several distinctive physical characteristics, including weak muscle tone, a small mouth held partially open because it cannot accommodate the tongue, and distinctively shaped eyelids (**Fig. 10-27**). More serious problems include heart malformations, low resistance to infectious diseases, and varying degrees of mental retardation.

The frequency of nondisjunction increases with the age of the parents, especially the mother. Only about 0.05% of children born to 20-year-old women, but more than 3% of children born to women over 45 years of age, have Down syndrome. Nondisjunction in sperm accounts for about 10% of the cases of Down syndrome, and there is only a small increase in defective sperm with increasing age of the father. Since the 1970s, it has become more common for couples to delay having children, increasing the probability of trisomy 21. Trisomy 21 can be diagnosed before birth by examining the chromosomes of fetal cells and, with less certainty, by biochemical tests and ultrasound examination of the fetus (see "Health Watch: Prenatal Genetic Screening" in Chapter 13).

CHECK YOUR LEARNING

Can you use pedigrees to determine the pattern of inheritance of a trait? Can you describe why some genetic disorders might be dominant while others are recessive, and give examples of each? Are you able to explain how nondisjunction causes offspring to have too many or too few chromosomes? Can you describe some of the human genetic disorders that are caused by nondisjunction?

(a) Karyotype showing three copies of chromosome 21

(b) Girl with Down syndrome

▲ **FIGURE 10-27 Trisomy 21, or Down syndrome (a)** This karyotype of a Down syndrome child reveals three copies of chromosome 21 (arrow). **(b)** The younger of these two sisters shows the facial features common in people with Down syndrome.

CASE STUDY revisited

Sudden Death on the Court

Medical examinations revealed that Flo Hyman's father and sister have Marfan syndrome, but her mother and brother do not. A single defective fibrillin allele is enough to cause Marfan syndrome. What can we conclude about the inheritance of Marfan syndrome from these data?

First, if even one defective fibrillin allele produces Marfan syndrome, then Hyman's mother must carry two normal alleles, because she does not have Marfan syndrome. Second, because Hyman's father has Marfan syndrome, it is very likely that Hyman inherited a defective fibrillin allele from him. The fact that her sister also has Marfan syndrome makes this virtually certain. Third, is Marfan syndrome inherited as a dominant or a recessive condition? If one defective allele is enough to cause Marfan syndrome, then this allele must be dominant and the normal allele must be recessive. Fourth, there seems to be no gender effect in Hyman's family. Marfan syndrome is not sex-linked, because the fibrillin gene is on an autosome.

Finally, if Hyman had borne children, could they have inherited Marfan syndrome from her? For a dominant disorder, any children who inherited her defective allele would develop Marfan syndrome. Therefore, her children would have had a 50% chance of inheriting Marfan syndrome. (Try working this out with a Punnett square.)

At the beginning of this chapter, we suggested that Rachmaninoff, Paganini, and Lincoln may have had Marfan syndrome. You may wonder how anyone can be sure that they actually had the disorder. Well, you're right—nobody really knows. The "diagnosis" is based on photographs and descriptions. There is considerable dispute about whether Lincoln had Marfan syndrome or some other condition. There is less controversy about Rachmaninoff, who was 6′6″ tall—extremely tall for a man born in 1873—and whose hands could span 13 white keys on a piano (try it yourself!). Most modern pianists cannot play some of Rachmaninoff's music as he originally wrote it, because the notes in some of the chords are so far apart on the keyboard.

Consider This

BioEthics If a parent has been diagnosed with Marfan syndrome, then in some instances the defective gene can be detected in an embryo. Some other genetic disorders, such as cystic fibrosis and sickle-cell anemia, can easily be detected, in adults, children, and embryos. In these recessive diseases, if two heterozygotes have children together, each of their children has a 25% chance of having the disorder. Although there is no cure for these diseases, there will probably be better treatments within a few years. If you and your spouse were both heterozygotes for a genetic disorder, would you seek prenatal diagnosis of an embryo? What would you do if your embryo were destined to be born with Marfan syndrome?

CHAPTER REVIEW

Summary of Key Concepts

10.1 What Is the Physical Basis of Inheritance?

The units of inheritance are genes, which are segments of DNA found at specific locations (loci) on chromosomes. Genes may exist in two or more slightly different, alternative forms, called alleles. When both homologous chromosomes carry the same allele at a given locus, the organism is homozygous for that particular gene. When the two homologous chromosomes have different alleles at a given locus, the organism is heterozygous for that gene.

10.2 How Were the Principles of Inheritance Discovered?

Gregor Mendel deduced many principles of inheritance in the mid-1800s, before the discovery of DNA, genes, chromosomes, or meiosis. He did this by choosing an appropriate experimental subject, designing his experiments carefully, following progeny for several generations, and analyzing his data statistically.

10.3 How Are Single Traits Inherited?

A trait is an observable or measurable feature of an organism's phenotype, such as hair texture or blood type. Traits are inherited in particular patterns that depend on the alleles that parents pass on to their offspring. Each parent provides its offspring with one allele of every gene, so the offspring inherits a pair of alleles for every gene. The combination of alleles in the offspring determines whether it displays a particular phenotype. Dominant alleles mask the expression of recessive alleles. The masking of recessive alleles can result in organisms with the same phenotype but different genotypes. Organisms with two dominant alleles (homozygous dominant) have the same phenotype as do organisms with one dominant and one recessive allele (heterozygous). Because each allele segregates randomly during meiosis, we can predict the relative proportions of offspring with a particular trait, using Punnett squares or probability.

10.4 How Are Multiple Traits Inherited?

If the genes for two traits are located on separate chromosomes, their alleles assort independently of one another into the egg or sperm; that is, the distribution of alleles of one gene into the gametes does not affect the distribution of the alleles of the other gene. Thus, breeding two organisms that are heterozygous at two loci on separate chromosomes produces offspring with nine different genotypes. For typical dominant and recessive alleles, the offspring will display only four different phenotypes (see Fig. 10-11).

10.5 Do the Mendelian Rules of Inheritance Apply to All Traits?

Not all inheritance follows the simple dominant-recessive pattern:

- In incomplete dominance, heterozygotes have a phenotype that is intermediate between the two homozygous phenotypes.

- If we examine the genes of many members of a given species, we find that many genes have more than two alleles; that is, there are multiple alleles of the gene.
- Codominance is a relationship among the alleles of a single gene in which two alleles independently contribute to the observed phenotype.
- Many traits are determined by several different genes that all contribute to the phenotype, a phenomenon called polygenic inheritance.
- Pleiotropy occurs when a single gene has multiple effects on an organism's phenotype.
- The environment influences the phenotypic expression of all traits.

10.6 How Are Genes Located on the Same Chromosome Inherited?

Genes on the same chromosome (encoded on the same DNA double helix) are linked to one another and therefore tend to be inherited together. However, crossing over will result in some recombination of alleles on each chromosome. Crossing over will occur more often the farther apart on a chromosome the genes lie.

10.7 How Are Sex and Sex-Linked Traits Inherited?

In many animals, sex is determined by sex chromosomes, often designated X and Y. In mammals, females have two X chromosomes, whereas males have one X and one Y chromosome. The rest of the chromosomes, identical in the two sexes, are called autosomes. Males include either an X or a Y chromosome in their sperm, whereas females always include an X chromosome in their eggs. Therefore, gender is determined by the sex chromosome in the sperm that fertilizes an egg.

Sex-linked genes are found on the X or Y chromosome. In mammals, the Y chromosome has many fewer genes than the X chromosome, so most sex-linked genes are found on the X chromosome. Because males have only one copy of X chromosome genes, recessive traits on the X chromosome are more likely to be phenotypically expressed in males.

10.8 How Are Human Genetic Disorders Inherited?

The genetics of humans is similar to the genetics of other animals, but is more difficult to study because experimental crosses are not feasible. Analysis of family pedigrees and, more recently, molecular genetic techniques are used to determine the mode of inheritance of human traits.

Some genetic disorders are inherited as recessive traits; therefore, only homozygous recessive persons show symptoms of the disease. Heterozygotes are called carriers; they carry the recessive allele but do not express the trait. Some other diseases are inherited as simple dominant traits. In such cases, only one copy of the dominant allele is needed to cause disease symptoms. Some human genetic disorders are sex-linked.

Errors in meiosis can result in gametes with abnormal numbers of sex chromosomes or autosomes. Many people with abnormal numbers of sex chromosomes have distinguishing physical characteristics. Abnormal numbers of autosomes typically lead to spontaneous abortion early in pregnancy. In rare instances, the fetus may survive to birth, but mental and physical deficiencies always occur, as is the case with Down syndrome (trisomy 21). The likelihood of abnormal numbers of chromosomes increases with increasing age of the mother and, to a lesser extent, the father.

Key Terms

albinism   *181*
allele   *166*
autosome   *179*
carrier   *181*
codominance *175*
cross-fertilization   *168*
dominant   *169*
Down syndrome   *187*
gene   *166*
gene linkage   *177*
genetic recombination   *178*
genotype   *170*
hemophilia   *184*
heterozygous *166*
homozygous   *166*
Huntington disease   *183*
incomplete dominance   *174*
inheritance   *166*
Jacob syndrome   *187*
Klinefelter syndrome   *187*
law of independent
    assortment   *173*

law of segregation   *169*
locus (plural, loci)   *166*
muscular dystrophy   *185*
mutation   *166*
nondisjunction   *185*
pedigree   *181*
phenotype   *170*
pleiotropy   *176*
polygenic inheritance   *176*
Punnett square method   *170*
recessive   *169*
self-fertilization   *167*
sex chromosome   *179*
sex-linked   *179*
sickle-cell anemia   *182*
test cross   *171*
trisomy 21   *187*
trisomy X   *186*
true-breeding   *168*
Turner syndrome   *186*
X chromosome   *179*
Y chromosome   *179*

Learning Outcomes

In this chapter, you have learned to . . .

LO1 Describe the relationships among chromosomes, DNA, genes, mutations, and alleles, and explain their function in inheritance.

LO2 Summarize Mendel's conclusions about the inheritance of single and multiple traits and explain how the results of his experiments support those conclusions.

LO3 Determine the genotypes and phenotypes of offspring that result from mating organisms with traits that follow simple patterns of Mendelian inheritance.

LO4 Describe the inheritance of traits characterized by

incomplete dominance, codominance, polygenic inheritance, and pleiotropy.

LO5 Explain how the physical locations of genes and the events that occur during meiosis affect inheritance.

LO6 Explain the chromosomal basis of sex determination and how it influences the inheritance of sex-linked traits.

LO7 Interpret pedigrees for dominant and recessive traits.

LO8 Describe the causes and symptoms of inherited human disorders produced by defective alleles of individual genes and by abnormal numbers of chromosomes.

Thinking Through the Concepts

Fill-in-the-Blank

1. The physical position of a gene on a chromosome is called its _____. Alternative forms of a gene are called _____. These alternative forms of genes arise as _____, which are changes in the nucleotide sequence of a gene. **LO1**

2. An organism is described as *Rr:* red. *Rr* is the organism's _____, while red color is its _____. This organism would be (homozygous/heterozygous) for this color gene. **LO2 LO3**

3. The inheritance of multiple traits depends on the locations of the genes that control the traits. If the genes are on different chromosomes, then the traits are inherited (as a group/independently). If the genes are located close together on a single chromosome, then the traits tend to be inherited (as a group/independently). Genes on the same chromosome are said to be _____. **LO3 LO5**

4. Many organisms, including mammals, have both autosomes and sex chromosomes. In mammals, males have _____ sex chromosomes and females have _____ chromosomes. The sex of offspring depends on which chromosome is present in the (sperm/egg). **LO6**

5. Genes that are present on one sex chromosome but not the other are called _____. **LO6**

6. If the phenotype of heterozygotes is intermediate between the phenotypes of the two homozygotes, this pattern of inheritance is called _____. If heterozygotes express phenotypes of both homozygotes (not intermediate, but showing both traits), this is called _____. In _____, many genes, usually with similar effects on phenotype, control the inheritance of a trait. **LO4**

Review Questions

1. Define the following terms: *gene, allele, dominant, recessive, true-breeding, homozygous, heterozygous, cross-fertilization,* and *self-fertilization.* **LO1 LO2 LO3**

2. Explain why genes located on the same chromosome are said to be linked. Why do alleles of linked genes sometimes separate during meiosis? **LO5**

3. Define *polygenic inheritance.* Why does polygenic inheritance sometimes allow parents to produce offspring that are notably different in skin color than either parent? **LO4**

4. What is sex linkage? In mammals, which sex would be most likely to show recessive sex-linked traits? **LO6**

5. What is the difference between a phenotype and a genotype? Does knowledge of an organism's phenotype always allow you to determine the genotype? What type of experiment would you perform to determine the genotype of a phenotypically dominant individual? **LO2 LO3**

6. In the pedigree of part (a) of Figure 10-22, do you think that the individuals showing the trait are homozygous or heterozygous? How can you tell from the pedigree? **LO7**

7. Define *nondisjunction,* and describe the common syndromes caused by nondisjunction of sex chromosomes and autosomes. **LO8**

Applying the Concepts

1. Sometimes the term *gene* is used rather casually. Compare the terms *allele* and *gene*.

2. In an alternate universe, all the genes in all species have only two alleles, one dominant and one recessive. Would every trait have only two phenotypes? Would all members of a species that are dominant for a given gene have exactly the same phenotype? Explain your reasoning.

3. Although American society has been described as a "melting pot," people often engage in "assortative mating," in which they marry others of similar height, socioeconomic status, race, and IQ. Discuss the consequences to society of assortative mating among humans. Would society be better off if people mated more randomly? Explain.

Genetics Problems

1. In certain cattle, hair color can be red (homozygous R_1R_1), white (homozygous R_2R_2), or roan (a mixture of red and white hairs, heterozygous R_1R_2).

   a. When a red bull is mated to a white cow, what genotypes and phenotypes of offspring could be obtained?

   b. If one of the offspring bulls in part (a) were mated to a white cow, what genotypes and phenotypes of offspring could be produced? In what proportion?

2. The palomino horse is golden in color. Unfortunately for horse fanciers, palominos do not breed true. In a series of matings between palominos, the following offspring were obtained:

   65 palominos
   32 cream-colored
   34 chestnut (reddish brown)

   What is the probable mode of inheritance of palomino coloration?

3. In the edible pea, tall (*T*) is dominant to short (*t*), and green pods (*G*) are dominant to yellow pods (*g*). List the types of gametes and offspring that would be produced in the following crosses:

   a. *TtGg* × *TtGg*
   b. *TtGg* × *TTGG*
   c. *TtGg* × *Ttgg*

4. In tomatoes, round fruit (*R*) is dominant to long fruit (*r*), and smooth skin (*S*) is dominant to fuzzy skin (*s*). A true-breeding round, smooth tomato (*RRSS*) was crossbred with a true-breeding long, fuzzy tomato (*rrss*). All the F_1 offspring were round and smooth (*RrSs*). When these F_1 plants were bred, the following F_2 generation was obtained:

   Round, smooth: 43
   Long, fuzzy: 13

   Are the genes for skin texture and fruit shape likely to be on the same chromosome or on different chromosomes? Explain your answer.

5. In the tomatoes of Problem 4, an F_1 offspring (*RrSs*) was mated with a homozygous recessive (*rrss*). The following offspring were obtained:

   Round, smooth: 583   Long, fuzzy: 602
   Round, fuzzy: 21   Long, smooth: 16

   What is the most likely explanation for this distribution of phenotypes?

6. In humans, hair color is controlled by two interacting genes. The same pigment, melanin, is present in both brown-haired and blond-haired people, but brown hair has much more of it. Brown hair (*B*) is dominant to blond (*b*). Whether any melanin can be synthesized depends on another gene. The dominant form of this second gene (*M*) allows melanin synthesis; the recessive form (*m*) prevents melanin synthesis. Homozygous recessives (*mm*) are albino. What will be the expected proportions of phenotypes in the children of the following parents?

   a. *BBMM* × *BbMm*
   b. *BbMm* × *BbMm*
   c. *BbMm* × *bbmm*

7. In humans, one of the genes determining color vision is located on the X chromosome. The dominant form (*C*) produces normal color vision; red-green color deficiency (*c*) is recessive. If a man with normal color vision marries a color-deficient woman, what is the probability of them having a color-deficient son? A color-deficient daughter?

8. In the couple described in Problem 7, the woman gives birth to a color-deficient but otherwise normal daughter. The husband sues for a divorce on the grounds of adultery. Will his case stand up in court? Explain your answer.

Answers to Figure Caption questions and Fill-in-the-Blank questions can be found in the Answers section at the back of the book.

MB *Go to MasteringBiology for practice quizzes, activities, eText, videos, current events, and more.*

CHAPTER 11

DNA: The Molecule of Heredity

Ordinary bull or incredible hulk? A tiny change in DNA makes all the difference.

CASE STUDY

Muscles, Mutations, and Myostatin

NO, THE BULL in the top photo hasn't been pumping iron or taking steroids—he's a Belgian Blue, and they always have bulging muscles. What makes a Belgian Blue look like a bodybuilder, compared to an ordinary bull, such as the Hereford in the bottom photo?

When any mammal develops, its cells divide many times, enlarge, and become specialized for a specific function. The size, shape, and cell types in any organ are precisely regulated during development, so that you don't wind up with a head the size of a basketball, or have hair growing on your liver. Muscle development is no exception. When you were very young, cells destined to form your muscles multiplied, fused together to form long, relatively thick cells with multiple nuclei, and synthesized specialized proteins that cause muscles to contract and thereby move your skeleton. A protein called myostatin, found in all mammals, puts the brakes on muscle development. "Myostatin" literally means "to make muscles stay the same," and that is exactly what it does. As muscles develop, myostatin slows down—and eventually stops—the multiplication of these pre-muscle cells. Myostatin also regulates the ultimate size of muscle cells and, therefore, their strength.

Belgian Blues have more, and larger, muscle cells than ordinary cattle do. Why? You may have already guessed—they don't produce normal myostatin. As you will learn in this chapter, proteins are synthesized from the genetic instructions contained in **deoxyribonucleic acid (DNA).** The DNA of a Belgian Blue is very slightly different from the DNA of other cattle—the Belgian Blue has a change, or mutation, in the DNA of its myostatin gene. As a result, it produces defective myostatin. Belgian Blue pre-muscle cells multiply more than normal, and the cells become extra large as they differentiate, producing remarkably buff cattle.

How does DNA contain the instructions for traits such as muscle size, flower color, or gender? How are these instructions passed, usually unchanged, from generation to generation? And why do the instructions sometimes change? To answer these questions, we must learn more about the structure and function of DNA.

AT A GLANCE

LEARNING GOALS

LG1 Describe the key experimental evidence indicating that DNA is the molecule of heredity.

LG2 Describe the structure of DNA, including nucleotides, the sugar-phosphate backbone, the double helix, and the role of hydrogen bonds.

LG3 Explain how genetic information is encoded in DNA.

LG4 Summarize the steps of DNA replication, including identification and action of the enzymes involved in each step.

LG5 Define the term *mutation*, and name and describe the different types of mutations.

11.1 HOW DID SCIENTISTS DISCOVER THAT GENES ARE MADE OF DNA?

By the late 1800s, scientists had learned that genetic information exists in discrete units that they called genes. However, they didn't know what a gene is. They knew only that genes determine many inherited traits; for example, genes determine whether roses are red, pink, yellow, or white. By the early 1900s, studies of dividing cells provided strong evidence that genes are parts of chromosomes. Soon, biochemists found that eukaryotic chromosomes are composed only of protein and DNA, so one of these must be the molecule of heredity. But which one?

Transformed Bacteria Revealed the Link Between Genes and DNA

In the late 1920s, Frederick Griffith, a British researcher, was trying to make a vaccine to prevent bacterial pneumonia, a major cause of death at that time. Some antibacterial vaccines consist of a weakened strain of the bacteria, which can't cause illness. Injecting this weakened but living strain into an animal may stimulate immunity against the disease-causing strains. Other vaccines use disease-causing (virulent) bacteria that have been killed by exposure to heat or chemicals.

Griffith experimented with two strains of the bacterium *Streptococcus pneumoniae*. One strain, R, did not cause pneumonia when injected into mice (**Fig. 11-1a**). Injecting mice with the other strain, S, caused pneumonia, killing the mice in a day or two (**Fig. 11-1b**). As expected, when the S-strain was killed and injected into mice, it did not cause disease (**Fig. 11-1c**). Unfortunately, neither the live R-strain nor the killed S-strain provided immunity against live S-strain bacteria.

Griffith also tried injecting a mixture of living R-strain bacteria and heat-killed S-strain bacteria (**Fig. 11-1d**). Because neither caused pneumonia on its own, he expected the mice to remain healthy. To his surprise, the mice sickened and died. When he autopsied the mice, he recovered living S-strain bacteria from them. Griffith hypothesized that some substance in the heat-killed S-strain changed the living, harmless R-strain bacteria into the deadly S-strain, a process he called transformation. These transformed bacteria could cause pneumonia.

Griffith never discovered an effective pneumonia vaccine, so in that sense his experiments were a failure (in fact, an effective vaccine against *Streptococcus pneumoniae* was not developed until the late 1970s). However, Griffith's experiments marked a turning point in our understanding of genetics, because other researchers suspected that the substance that causes transformation might be the long-sought molecule of heredity.

The Transforming Molecule Is DNA

In 1944, Oswald Avery, Colin MacLeod, and Maclyn McCarty discovered that the transforming molecule is DNA. They isolated DNA from S-strain bacteria, mixed it with live R-strain bacteria, and produced live S-strain bacteria. They treated some samples with protein-destroying enzymes and other samples with DNA-destroying enzymes. The protein-destroying enzymes did not prevent transformation. However, treating samples with DNA-destroying enzymes did prevent transformation. Therefore, they concluded that transformation must be caused by DNA, and not by traces of protein contaminating the DNA.

This discovery helps us to interpret the results of Griffith's experiments. Heating S-strain cells killed them but did not completely destroy their DNA. When killed S-strain bacteria were mixed with living R-strain bacteria, fragments of DNA from the dead S-strain cells entered into some of the R-strain cells and became incorporated into the chromosome of the R-strain bacteria (**Fig. 11-2**). Some of these DNA fragments contained the genes needed to cause pneumonia, transforming an R-strain cell into an S-strain cell. Thus, Avery, MacLeod, and McCarty concluded that DNA is the molecule of heredity.

Over the next decade, evidence continued to accumulate that DNA is the genetic material in many, or perhaps

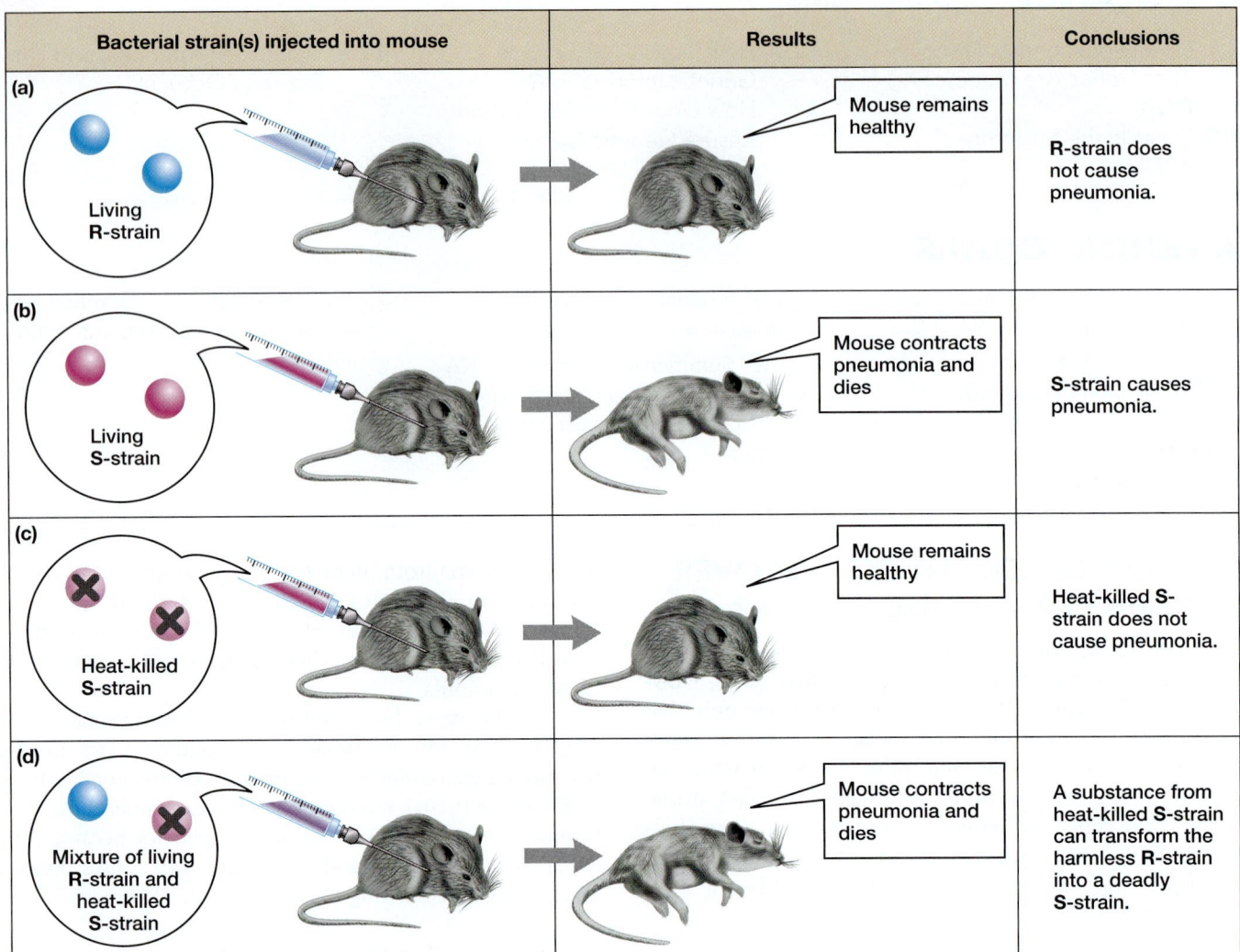

Bacterial strain(s) injected into mouse	Results	Conclusions
(a) Living **R-strain**	Mouse remains healthy	**R-strain** does not cause pneumonia.
(b) Living **S-strain**	Mouse contracts pneumonia and dies	**S-strain** causes pneumonia.
(c) Heat-killed **S-strain**	Mouse remains healthy	Heat-killed **S-strain** does not cause pneumonia.
(d) Mixture of living **R-strain** and heat-killed **S-strain**	Mouse contracts pneumonia and dies	A substance from heat-killed **S-strain** can transform the harmless **R-strain** into a deadly **S-strain**.

▲ **FIGURE 11-1 Transformation in bacteria** Griffith's discovery that bacteria can be transformed from harmless to deadly laid the groundwork for the discovery that genes are composed of DNA.

all, organisms. For example, before dividing, a eukaryotic cell duplicates its chromosomes (see Chapter 9) and exactly doubles its DNA content, but not its protein content—just what would be expected if genes are made of DNA. Virtually all of the remaining skeptics were convinced by a superb set of experiments performed in the early 1950s by Alfred Hershey and Martha Chase, which conclusively showed that DNA is the hereditary molecule of certain viruses (see "Science in Action: DNA Is the Hereditary Molecule of Bacteriophages" on pp. **196–197**).

CHECK YOUR LEARNING

Can you describe the experiments of Griffiths and those of Avery, MacLeod, and McCarty? Can you explain why these experiments showed that DNA is the hereditary molecule?

11.2 WHAT IS THE STRUCTURE OF DNA?

Knowing that genes are made of DNA does not answer critical questions about inheritance: How does DNA encode genetic information? How is DNA replicated so that a cell

can pass its hereditary information to its daughter cells? The secrets of DNA function and replication are found in the three-dimensional structure of the DNA molecule.

DNA Is Composed of Four Nucleotides

DNA consists of long chains made of subunits called **nucleotides.** Each nucleotide consists of three parts: a phosphate group, a sugar called deoxyribose, and one of four nitrogen-containing **bases.** The bases found in DNA are **adenine (A), guanine (G), thymine (T),** and **cytosine (C)** (**Fig. 11-3**). Adenine and guanine both consist of fused five- and six-member rings of carbon and nitrogen atoms, with different functional groups attached to different positions on the six-member ring. Thymine and cytosine consist of a single six-member ring of carbon and nitrogen atoms, again with different functional groups attached to different positions on the ring.

In the 1940s, biochemist Erwin Chargaff of Columbia University analyzed the amounts of the four bases in DNA from organisms as diverse as bacteria, sea urchins, fish, and humans. He found a curious consistency: Although the

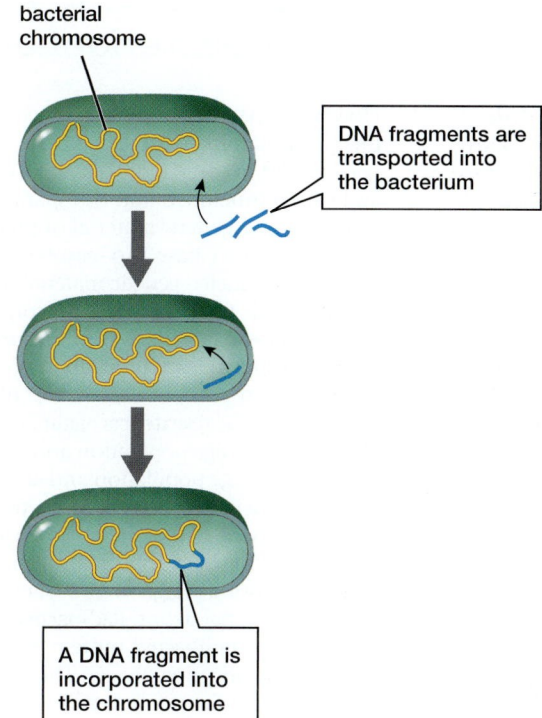

bacterial chromosome

DNA fragments are transported into the bacterium

A DNA fragment is incorporated into the chromosome

▲ FIGURE 11-2 **Molecular mechanism of transformation** Most bacteria have a single circular chromosome made of DNA. Transformation may occur when a living bacterium takes up pieces of DNA from its environment and incorporates those fragments into its chromosome.

phosphate | sugar | base = **adenine**

phosphate | sugar | base = **guanine**

phosphate | sugar | base = **thymine**

phosphate | sugar | base = **cytosine**

▲ FIGURE 11-3 **DNA nucleotides**

proportions of each base differ from species to species, for a given species, there are always equal amounts of adenine and thymine, and equal amounts of guanine and cytosine. This consistency, often called "Chargaff's rule," certainly seemed significant, but it would be almost another decade before anyone figured out what it meant about DNA structure.

DNA Is a Double Helix of Two Nucleotide Strands

Determining the structure of large biological molecules isn't easy, even today. Nevertheless, in the late 1940s, several scientists began to investigate the structure of DNA. British scientists Maurice Wilkins and Rosalind Franklin used X-ray diffraction to study the DNA molecule. They bombarded crystals of purified DNA with X-rays and recorded how the X-rays bounced off the DNA molecules (**Fig. 11-4**). As you can see, the resulting pattern does not provide a direct picture of DNA structure. However, experts like Wilkins and Franklin could extract a lot of information about DNA from the pattern. First, a molecule of DNA is long and thin, with a uniform diameter of 2 nanometers (2 billionths of a meter). Second, DNA is helical, twisted like a corkscrew or a spiral staircase. Third, DNA is a double helix; that is, two DNA strands coil around one another. Fourth, DNA consists of repeating subunits. And fifth, the phosphates are probably on the outside of the helix.

Even with the X-ray diffraction data, it was no trivial task to work out the structure of DNA. However, by combining the

diffraction data with a knowledge of how complex organic molecules bond together and an intuition that "important biological objects come in pairs," James Watson and Francis Crick deduced the structure of DNA (see "Science in Action: The Discovery of the Double Helix" on p. **199**).

Watson and Crick proposed that a single **strand** of DNA is a polymer consisting of many nucleotide subunits. The phosphate group of one nucleotide is bonded to the sugar of the next nucleotide in the strand, thus producing a **sugar-phosphate backbone** of alternating, covalently bonded sugars and phosphates (**Fig. 11-5**). The bases of the nucleotides stick out from this sugar-phosphate backbone.

All of the nucleotides in a single DNA strand are oriented in the same direction. Therefore, the two ends of a DNA strand differ; one end has a "free" or unbonded

SCIENCE IN ACTION   DNA Is the Hereditary Molecule of Bacteriophages

Certain viruses infect only bacteria and are aptly called **bacteriophages,** meaning "bacteria eaters" (**Fig. E11-1**). A bacteriophage ("phage" for short) depends on a host bacterium for every aspect of its life cycle (**Fig. E11-1b**). When a phage encounters a bacterium, it attaches to the bacterial cell wall and injects its genetic material into the bacterium. The outer coat of the phage remains outside. The bacterium cannot distinguish phage genes from its own genes, so it "reads" the phage genes and uses that information to produce more phages. Finally, one of the phage genes directs the synthesis of an enzyme that ruptures the bacterium, freeing the newly manufactured phages.

Even though many bacteriophages have intricate structures (**Fig. E11-1a**), they are chemically very simple, containing only DNA and protein. Therefore, one of these two molecules must be the phage genetic material. Alfred Hershey and Martha Chase

used this chemical simplicity to deduce that the phage genetic material is DNA.

Hershey and Chase knew that infected bacteria should contain phage genetic material, so if they could "label" phage DNA and protein, and separate the infected bacteria from the phage coats left outside, they could see which molecule entered the bacteria (**Fig. E11-2**). DNA and protein both contain carbon, oxygen, hydrogen, and nitrogen (see Chapter 3). DNA also contains phosphorus but not sulfur, whereas proteins contain sulfur but not phosphorus.

Hershey and Chase forced one population of phages to synthesize DNA using radioactive phosphorus, thereby labeling the phage DNA. Another population of phages was forced to synthesize protein using radioactive sulfur, thus labeling the phage protein. When bacteria were infected by phages containing radioactively labeled protein, the bacteria did not become radioactive.

However, when bacteria were infected by phages containing radioactive DNA, the bacteria did become radioactive. Hershey and Chase concluded that DNA, and not protein, is the genetic material of phages.

Hershey and Chase also reasoned that some of the labeled genetic material from the "parental" phages might be incorporated into the genetic material of the "offspring" phages. (You will learn more about this in Section 11.4.) In a second set of experiments, the researchers again labeled DNA in one phage population and protein in another phage population and allowed the phages to infect bacteria. After enough time had passed so that the phages had reproduced, the bacteria were broken open, and the offspring phages were separated from the bacterial debris. Radioactive DNA, but not radioactive protein, was found in the offspring phages. This second experiment confirmed the results of the first: DNA is the hereditary molecule.

(a) Structure of a bacteriophage

head / **tail** — DNA

❶ The phage attaches to a bacterium

phage — phage DNA

bacterium

bacterial chromosome

❷ The phage injects its DNA into the bacterium

❸ The phage DNA is replicated

❹ Phage parts are synthesized, using the bacterium's metabolism

❺ Complete phages are assembled

❻ The bacterial wall is destroyed; phages are released

(b) Bacteriophage life cycle

▲ **FIGURE E11-1 Bacteriophages (a)** Many bacteriophages have complex structures, including a head containing genetic material, tail fibers that attach to the surface of a bacterium, and an elaborate apparatus for injecting their genetic material into the bacterium. **(b)** A bacteriophage uses the metabolism of a bacterium to produce more phages.

Observations:
1. Bacteriophage viruses consist of only DNA and protein.
2. Bacteriophages inject their genetic material into bacteria, forcing the bacteria to synthesize more phages.
3. The outer coat of bacteriophages stays outside of the bacteria.
4. DNA contains phosphorus but not sulfur.
   - DNA can be "labeled" with radioactive phosphorus.
5. Protein contains sulfur but not phosphorus.
   - Protein can be "labeled" with radioactive sulfur.

Question: Is DNA or protein the genetic material of bacteriophages?

Hypothesis: DNA is the genetic material.

Prediction:
1. If bacteria are infected with bacteriophages containing radioactively labeled DNA, the bacteria will be radioactive.
2. If bacteria are infected with bacteriophages containing radioactively labeled protein, the bacteria will not be radioactive.

Experiment:

Radioactive phosphorus (^{32}P)   Radioactive sulfur (^{35}S)

Radioactive DNA (blue)   Radioactive protein (gold)

❶ Label the phages with ^{32}P or ^{35}S.

❷ Infect the bacteria with the labeled phages; the phages inject their genetic material into the bacteria.

❸ Whirl in a blender to break off the phage coats from the bacteria.

❹ Centrifuge to separate the phage coats (low density: stay in the liquid) from the bacteria (high density: sink to the bottom as a "pellet")

❺ Measure the radioactivity of the phages and bacteria.

Results: Bacteria are radioactive; phages are not.

Results: Phages are radioactive; bacteria are not.

Conclusion: Infected bacteria contain radioactive phosphorus but not radioactive sulfur, supporting the hypothesis that the genetic material of bacteriophages is DNA, not protein.

▲ FIGURE E11-2 **The Hershey-Chase experiment** By radioactively labeling either the DNA or the protein of bacteriophages, Hershey and Chase tested whether the genetic material of phages is DNA (left side of the experiment) or protein (right side).

Diffraction pattern of DNA

▲ **FIGURE 11-4 X-ray diffraction image of DNA** The crossing pattern of dark spots is characteristic of helical molecules such as DNA. Measurements of various aspects of the pattern indicate the dimensions of the DNA helix; for example, the distance between the dark spots corresponds to the distance between turns of the helix.

sugar, and the other end has a "free" or unbonded phosphate (**Fig. 11-5a**). Picture a long line of cars stopped on a crowded one-way street at night; the cars' headlights (free phosphates) always point forward and their taillights (free sugars) always point backward. If the cars are jammed tightly together, a pedestrian standing in front of the line of cars will see only the headlights on the first car; a pedestrian at the back of the line will see only the taillights of the last car.

Hydrogen Bonds Between Complementary Bases Hold Two DNA Strands Together in a Double Helix

Watson and Crick's crucial insight was that a complete DNA molecule in living organisms consists of two DNA strands, assembled like a twisted ladder. The sugar-phosphate backbones form the "uprights" of the DNA ladder. The "rungs" are composed of specific pairs of bases, with one member of each pair protruding from the sugar-phosphate backbone of each strand. A complete rung consists of a pair of bases held together by hydrogen bonds (see Fig. 11-5a). As the X-ray data showed, the DNA ladder isn't straight: The two strands are wound about each other to form a **double helix,** as if the ladder had been twisted lengthwise into the shape of a spiral staircase (**Fig. 11-5b**). Further, the two strands in a DNA double helix are antiparallel to one another; that is, they are oriented in opposite directions. In Figure 11-5a, note that the left-hand DNA strand has a free phosphate group at the top and a free sugar on the bottom; the ends are reversed on the right-hand DNA strand. Again imagine an evening traffic jam, this time on a crowded two-lane highway. A traffic helicopter pilot overhead would see only the headlights on cars in one lane, and only the taillights of cars in the other lane.

(a) Hydrogen bonds hold complementary base pairs together in DNA

(b) Two DNA strands form a double helix

(c) Four turns of a DNA double helix

▲ **FIGURE 11-5 The Watson-Crick model of DNA structure (a)** Hydrogen bonding between complementary base pairs holds the two strands of DNA together. Three hydrogen bonds hold guanine to cytosine, and two hydrogen bonds hold adenine to thymine. Note that each strand has a free phosphate on one end and a free sugar on the opposite end. Further, the two strands run in opposite directions. **(b)** Strands of DNA wind about each other in a double helix, like a twisted ladder, with the sugar-phosphate backbone forming the uprights and the complementary base pairs forming the rungs. **(c)** A space-filling model of DNA structure.

QUESTION Which do you think would be more difficult to break apart: an A–T base pair or a C–G base pair?

SCIENCE IN ACTION The Discovery of the Double Helix

In the early 1950s, many biologists realized that the key to understanding inheritance lay in the structure of DNA. They also knew that whoever deduced the correct structure of DNA would receive recognition, probably including a Nobel Prize. Linus Pauling of the California Institute of Technology was the leading contender. Pauling probably knew more about the chemistry of large organic molecules than any person alive. Like Rosalind Franklin and Maurice Wilkins, Pauling was an expert in X-ray diffraction techniques. In 1950, he used these techniques to show that many proteins were coiled into single-stranded helices (see Fig. 3-21b). Pauling, however, had two important handicaps. First, for years he had concentrated on protein research and, therefore, had little data about DNA. Second, he was active in the peace movement. At that time, some government officials considered such activity to be potentially subversive and a threat to national security. This latter handicap may have proved decisive.

The second most likely competitors were Wilkins and Franklin, the British scientists who had set out to determine the structure of DNA by using X-ray diffraction patterns. In fact, Franklin probably knew the most about the shape of the DNA molecule, based on her X-ray diffraction studies (see p. 195). However, the methodical approach used by Franklin and Wilkins was slow.

The door was open for the eventual discoverers of the double helix—James Watson and Francis Crick, two scientists with neither Pauling's tremendous understanding of chemical bonds nor Franklin and Wilkins' expertise in X-ray analysis. Watson and Crick did no experiments in the ordinary sense of the word. Instead, they spent their time trying to construct a model that made sense and fit the data. Because they were working in England and because Wilkins shared Franklin's data with them (probably without her knowledge), Watson and Crick were familiar with the X-ray information relating to DNA.

The X-ray data were just what Pauling lacked. Because of his presumed subversive tendencies, the U.S. State Department refused to issue Pauling a passport to leave the United States, so he could neither attend meetings at which Wilkins presented the X-ray data nor visit England to talk with Franklin and Wilkins directly. Watson and Crick knew that Pauling was working on DNA structure and were driven by the fear that he would beat them to it. In his book, *The Double Helix*, Watson recounts his belief that, if Pauling had seen the X-ray pictures, "in a week at most, Linus would have [had] the structure."

You might be thinking, "But wait just a minute! That's not fair. If the goal of science is to advance knowledge, then everyone should have access to all the data. If Pauling was the best, he should have discovered the double helix first." Perhaps. But after all, scientists are people, too. Although virtually all scientists want to see scientific progress and benefits for humanity, each individual also wants to be the one responsible for that progress and to receive the credit.

Soon after Watson and Crick proposed the double helix (**Fig. E11-3**), Watson described it in a letter to Max Delbrück, a friend and adviser at Caltech. When Delbrück told Pauling, Pauling graciously congratulated Watson and Crick on their brilliant solution. The race was over.

▲ **FIGURE E11-3** **The discovery of DNA** James Watson and Francis Crick with a model of DNA structure.

Take a closer look at the pairs of bases that form each rung of the double helix ladder (see Figs. 11-5a, b). Adenine forms hydrogen bonds only with thymine, and guanine forms hydrogen bonds only with cytosine. These A–T and G–C pairs are called **complementary base pairs.** All of the bases of the two strands of a DNA double helix are complementary to each other. For example, if one strand reads A-T-T-C-C, then the other strand must read T-A-A-G-G.

Complementary base pairs explain "Chargaff's rule"— that the DNA of a given species contains equal amounts of adenine and thymine, and equal amounts of cytosine and guanine. Because an A in one DNA strand always pairs with a T in the other strand, the amount of A always equals the amount of T. Similarly, because a G in one strand always pairs with a C in the other DNA strand, the amount of G always equals the amount of C.

Finally, look at the sizes of the bases: Because adenine and guanine consist of two fused rings, they are large, whereas thymine and cytosine, composed of only a single ring, are small. Because the double helix has only A–T and G–C pairs, all the rungs of the DNA ladder are the same width. Therefore, the double helix has a constant diameter, just as the X-ray diffraction pattern indicated.

The structure of DNA was solved. On March 7, 1953, at the Eagle Pub in Cambridge, England, Francis Crick proclaimed to the lunchtime crowd, "We have discovered the secret of life." This claim was not far from the truth. Although further data would be needed to confirm the details, within just a few years, their DNA model revolutionized much of biology, including genetics, evolutionary biology, and medicine. The revolution continues today.

CHECK YOUR LEARNING

Can you describe the four nucleotides found in DNA, how individual DNA strands are constructed, and the three-dimensional structure of DNA?

11.3 HOW DOES DNA ENCODE GENETIC INFORMATION?

Look again at the structure of DNA shown in Figure 11-5. Can you see why many scientists had trouble believing that DNA could be the carrier of genetic information? Consider the many characteristics of just one organism. How can the color of a bird's feathers, the size and shape of its beak, its ability to make a nest, and its song all be determined by a molecule made with only four different nucleotides?

Genetic Information Is Encoded in the Sequence of Nucleotides

The answer is that it's not the *number* of different nucleotides but their *sequence* that's important. Within a DNA strand, the four nucleotides can be arranged in any order, and each unique sequence of nucleotides represents a unique set of genetic instructions. An analogy might help: You don't need a lot of different letters to make up a language. English has 26 letters, but Hawaiian has only 12, and the binary language of computers uses only two "letters" (0 and 1, or "off" and "on"). Nevertheless, all three languages can spell out thousands of different words. A stretch of DNA that is just 10 nucleotides long can form more than a million different sequences of the four nucleotides. Because an organism has millions (in bacteria) to billions (in plants or animals) of nucleotides, DNA molecules can encode a staggering amount of information.

Of course, to make sense, the letters of a language must be in the correct order. Similarly, a gene must have the right nucleotides in the right sequence. Just as "friend" and "fiend" mean different things, and "fliend" doesn't mean anything, different sequences of nucleotides in DNA may encode very different pieces of information or no information at all.

In the remainder of this chapter, we will examine how DNA is replicated during cell division to ensure accurate copying of its genetic information. (In Chapter 12, we will describe how the information in DNA is used to produce the structures of living cells.)

CHECK YOUR LEARNING

Can you explain how DNA encodes hereditary information?

CASE STUDY continued

Muscles, Mutations, and Myostatin

All "normal" mammals have a DNA sequence that encodes a functional myostatin protein, which limits their muscle growth. Belgian Blue cattle have a mutation that changes a "friendly" gene to a nonsensical "fliendly" one that no longer codes for a functional protein, so they have excessive muscle development.

11.4 HOW DOES DNA REPLICATION ENSURE GENETIC CONSTANCY DURING CELL DIVISION?

In the 1850s, Austrian pathologist Rudolf Virchow realized that "all cells come from [preexisting] cells." All the trillions of cells of your body are the offspring of other cells, going all the way back to when you were a fertilized egg. Moreover, almost every cell of your body contains identical genetic information—the same genetic information that was present in that fertilized egg. When cells reproduce by mitotic cell division, each daughter cell receives a nearly perfect copy of the parent cell's genetic information. Consequently, before cell division, the parent cell must synthesize two exact copies of its DNA. A process called **DNA replication** produces these two identical DNA double helices.

DNA Replication Produces Two DNA Double Helices, Each with One Original Strand and One New Strand

How does a cell accurately copy its DNA? In their paper describing DNA structure, Watson and Crick included one of the greatest understatements in all of science: "It has not escaped our notice that the specific [base] pairing we have postulated immediately suggests a possible copying mechanism for the genetic material." In fact, base pairing is the foundation of DNA replication. Remember, the rules for base pairing are that an adenine on one strand must pair with a thymine on the other strand, and a cytosine must pair with a guanine. Therefore, the base sequence of each strand contains all the information needed to replicate the other strand.

Conceptually, DNA replication is quite simple (**Fig. 11-6**). The essential ingredients are (1) the parental DNA strands (**Fig. 11-6 ❶**), (2) **free nucleotides** that were previously synthesized in the cytoplasm and imported into the nucleus, and (3) a variety of enzymes that unwind the parental DNA double helix and synthesize new DNA strands.

First, enzymes called **DNA helicases** (meaning "enzymes that break the double helix") pull apart the parental DNA double helix, so that the bases of the two DNA strands no longer form base pairs with one another (**Fig. 11-6 ❷**). Second, enzymes called **DNA polymerases** ("enzymes that synthesize a DNA polymer") move along each separated parental DNA strand, matching bases on the parental strands with complementary free nucleotides (**Fig. 11-6 ❸**). For

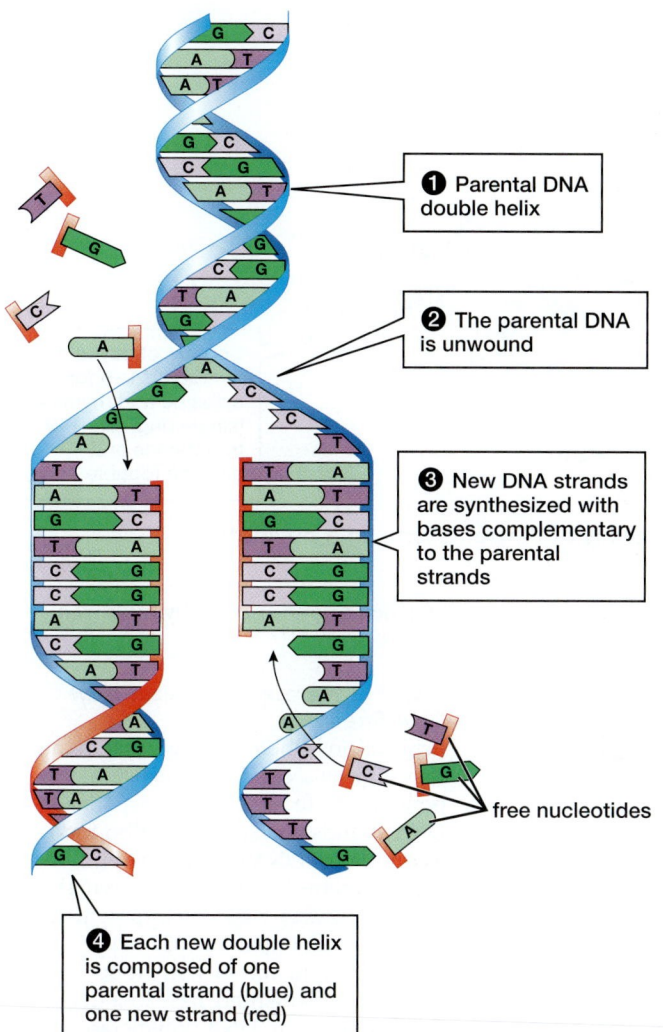

1 **Parental DNA double helix**

2 **The parental DNA is unwound**

3 **New DNA strands are synthesized with bases complementary to the parental strands**

free nucleotides

4 **Each new double helix is composed of one parental strand (blue) and one new strand (red)**

▲ FIGURE 11-6 **Basic features of DNA replication** During replication, the two strands of the parental DNA double helix separate. Free nucleotides that are complementary to those in each strand are joined to make new daughter strands. Each parental strand and its new daughter strand then form a new double helix.

example, DNA polymerase pairs an exposed adenine in the parental strand with a free thymine. DNA polymerase also connects these free nucleotides with one another to form two new DNA strands, each complementary to one of the parental DNA strands. Thus, if a parental DNA strand reads T–A–G, DNA polymerase will synthesize a new DNA strand with the complementary sequence A–T–C. For more information on how DNA is replicated, refer to "In Greater Depth: DNA Structure and Replication."

When replication is complete, each parental DNA strand and its newly synthesized, complementary daughter DNA strand wind together to form new double helixes (**Fig. 11-6** 4). Because DNA replication conserves one parental DNA strand and produces one newly synthesized strand, the process is called **semiconservative replication** (**Fig. 11-7**).

If no mistakes have been made, the base sequences of both new DNA double helices are identical to the base sequence of the parental DNA double helix and, of course, to each other.

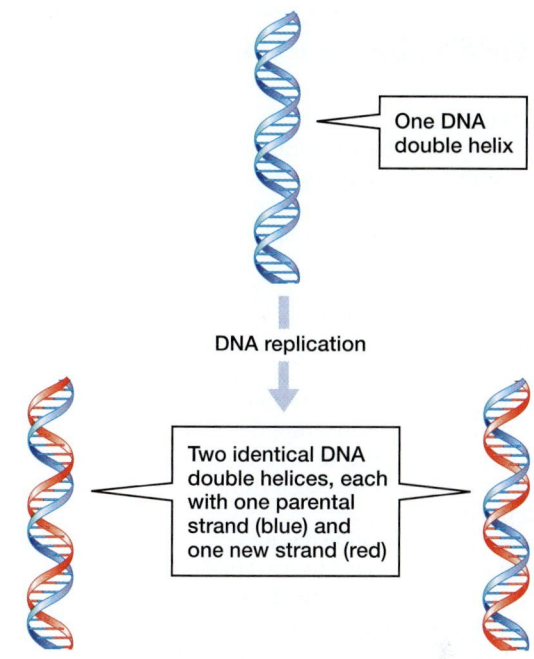

One DNA double helix

DNA replication

Two identical DNA double helices, each with one parental strand (blue) and one new strand (red)

▲ FIGURE 11-7 **Semiconservative replication of DNA**

CHECK YOUR LEARNING

Can you describe the process of DNA replication, including the enzymes involved and the actions that they perform? Can you explain why DNA replication is called "semiconservative"?

11.5 WHAT ARE MUTATIONS, AND HOW DO THEY OCCUR?

The nucleotide sequence of DNA is preserved, with great precision, from cell division to cell division, and from generation to generation. However, occasional changes in the nucleotide sequence, called **mutations,** do occur. Mutations are often harmful, much as randomly changing words in the middle of Shakespeare's *Hamlet* would probably interrupt the flow of the play. If they are really damaging, a cell or an organism inheriting such a mutation may quickly die. Some mutations, however, have no effect or, in very rare instances, are even beneficial. Mutations that are advantageous, at least in certain environments, may be favored by natural selection, and are the basis for the evolution of life on Earth (see Unit 3).

CASE STUDY c o n t i n u e d

Muscles, Mutations, and Myostatin

At the appropriate time during development, myostatin blocks the cell cycle in the G_1 phase, before DNA replication starts (see Fig. 9-8). Therefore when myostatin is present, pre-muscle cells do not enter the S phase, and do not replicate their DNA. The cells stop dividing, limiting the number of mature muscle cells. The mutated myostatin of Belgian Blue cattle does not block progression through the cell cycle. Pre-muscle cells replicate their DNA and continue to divide, producing many more muscle cells than in normal cattle.

IN GREATER DEPTH DNA Structure and Replication

DNA Structure

To fully understand DNA replication, we must return to the structure of DNA. Biochemists keep track of the atoms in a complex molecule by numbering them. In the case of a nucleotide (Fig. E11-4), the atoms that form the "corners" of the base are numbered 1 through 6 for the single rings of cytosine and thymine, or 1 through 9 for the double rings of adenine and guanine. The carbon atoms of the sugar are numbered 1' (1-prime) through 5' (5-prime). The prime symbol (') is used to distinguish atoms in the sugar from atoms in the base.

The sugar of a nucleotide has two "ends" that can be involved in synthesizing the sugar-phosphate backbone of a DNA strand: a 3' end, which has a free –OH (hydroxyl) group attached to the 3' carbon of the sugar, and a 5' end, which has a phosphate group attached to the 5' carbon. When a DNA strand is synthesized, the phosphate of one nucleotide bonds with the hydroxyl group on the sugar of the next nucleotide (Fig. E11-5).

This still leaves a free hydroxyl group on the 3' end of one nucleotide, and a free phosphate group on the 5' end of the other nucleotide. No matter how many nucleotides are joined, there is always a free hydroxyl on the 3' end of the strand and a free phosphate on the 5' end.

Because the sugar-phosphate backbones of the two strands of a double helix are antiparallel—they run in opposite directions—at one end of the double helix, one of the strands has a sugar with a free hydroxyl (the 3' end) and the other strand has a free phosphate (the 5' end). On the other end of the double helix, the free sugar and phosphate ends are reversed (Fig. E11-6).

DNA Replication

DNA replication involves three major actions (Fig. E11-7). First, the DNA double helix is unwound and the two strands are separated, allowing the nucleotide sequence to be read. Then, new DNA strands with nucleotide sequences complementary to the two original strands are synthesized. In eukaryotic cells, these new DNA strands are synthesized in short pieces, so the third step in DNA replication is to stitch the pieces together to form a continuous new strand of DNA. Each step is carried out by a distinct set of enzymes.

▲ FIGURE E11-4 Numbering of carbon atoms in a nucleotide

DNA Helicase Unwinds and Separates the Parental DNA Strands

Acting in concert with several other enzymes, DNA helicase breaks the hydrogen bonds between complementary base pairs that hold the two parental DNA strands together. This unwinds a segment of the parental double helix and separates the two strands, forming a replication bubble (Fig. E11-7 ❶ and ❷). Each replication bubble contains a replication fork at each end, where the two parental DNA strands are just beginning to be separated. Within the replication bubble, the bases of the parental DNA strands are no longer attached to one another by hydrogen bonds.

DNA Polymerase Synthesizes New DNA Strands

Replication bubbles are essential because they allow a second enzyme, DNA polymerase, to attach to the separated DNA strands. At each replication fork, a complex of DNA polymerase and other proteins binds to each parental strand (Fig. E11-7 ❸). DNA polymerase recognizes an unpaired base in the parental strand and matches it up with a complementary base in a free nucleotide. Then, DNA polymerase bonds

▲ FIGURE E11-5 Numbering of carbon atoms in a dinucleotide

▲ FIGURE E11-6 **The two strands of a DNA double helix are antiparallel**

daughter strand. Because the two strands of the parental DNA double helix are oriented in opposite directions, the DNA polymerase molecules move in opposite directions on the two parental strands (see **Fig. E11-7 ❸**).

DNA helicase and DNA polymerase work together (**Fig. E11-7 ❹**). A DNA helicase binds to the double helix and moves along, unwinding the double helix and separating the strands. Because the two DNA strands run in opposite directions, as a DNA helicase enzyme moves toward the 5′ end of one parental strand, it is simultaneously moving toward the 3′ end of the other parental strand. Now visualize two DNA polymerases landing on the two separated strands of DNA. One DNA polymerase (call it polymerase #1) can follow behind the helicase toward the 5′ end of the parental strand and can synthesize a continuous daughter DNA strand, until it runs into another replication bubble. This continuous daughter DNA strand is called the leading strand. On the other parental strand, however, DNA polymerase #2 moves *away from* the helicase: in step ❸ of **Figure E11-7**, note that the helicase moves to the left, whereas DNA polymerase #2 moves to the right. Therefore, DNA synthesis on this strand will be discontinuous: DNA polymerase #2 will synthesize a short new DNA strand, called the lagging strand, but meanwhile, the helicase continues to move to the left, unwinding more of the double helix (see **Fig. E11-7 ❹** and **❺**). Additional DNA polymerases (#3, #4, and so on) must land on this strand and synthesize more short lagging strands.

DNA Ligase Joins Segments of DNA

Multiple DNA polymerases synthesize pieces of DNA of varying lengths. Each chromosome may form hundreds of replication bubbles. Within each bubble, there will be one leading strand and dozens to thousands of lagging strands. Therefore, a cell might synthesize millions of pieces of DNA while replicating a single chromosome. How are all of these pieces sewn together? This is the job of the third major enzyme, **DNA ligase** ("an enzyme that ties DNA together"; see **Fig. E11-7 ❺**). Many DNA ligase enzymes stitch the fragments of DNA together until each daughter strand consists of one long, continuous DNA polymer.

the phosphate of the incoming free nucleotide (the 5′ end) to the sugar of the most recently added nucleotide (the 3′ end) of the growing daughter strand. In this way, DNA polymerase synthesizes the sugar-phosphate backbone of the daughter strand.

Why make replication bubbles, rather than simply starting at one end of a double helix and letting a single DNA polymerase molecule copy the DNA in one continuous piece all the way to the other end? Recall that eukaryotic chromosomes are very long: Human chromosomes range from about 50 million nucleotides in the relatively tiny Y chromosome, to about 250 million nucleotides in chromosome 1. Eukaryotic DNA is copied at a rate of about 50 nucleotides per second, so it would take about 12 to 58 days to copy a human chromosome in one continuous piece. To replicate an entire chromosome in a reasonable time, many DNA helicase enzymes open up many replication bubbles simultaneously, allowing many DNA polymerase enzymes to copy the strands in fairly small pieces all at the same time. Each individual bubble enlarges as DNA replication progresses, and the bubbles merge when they contact one another.

DNA polymerase always moves away from the 3′ end of a parental DNA strand (the end with the free hydroxyl group of the sugar) toward the 5′ end (with a free phosphate group). New nucleotides are always added to the 3′ end of the

(continued) ▶

▲ **FIGURE E11-7 DNA replication** ❶ DNA helicase enzymes separate the parental strands of a chromosome to form replication bubbles. ❷ Each replication bubble consists of two replication forks, with unwound DNA strands between the forks. ❸ DNA polymerase enzymes synthesize new pieces of DNA. ❹ DNA helicase moves along the parental DNA double helix, unwinding it and enlarging the replication bubble. DNA polymerases in the replication bubble synthesize daughter DNA strands. ❺ DNA ligase joins the small DNA segments into a single daughter strand.

QUESTION During synthesis, why doesn't DNA polymerase move away from the replication fork on both strands?

| How Much Genes Influence Athletic Prowess? | Face it—you'll never run like Usain Bolt. But how much of his fantastic abilities are genetic? In some cases, genes clearly make a huge difference. Myostatin mutations, for example, can boost speed and strength, as we explain |

in the "Case Study Revisited." However, at least 240 genes contribute to athletic performance, and the effects of most individual genes are small. Do super-athletes have all the best "athletic alleles"? Right now, no one can tell.

Accurate Replication, Proofreading, and DNA Repair Produce Almost Error-Free DNA

The specificity of hydrogen bonding between complementary base pairs makes DNA replication highly accurate. DNA polymerase incorporates incorrect bases about once in every 1,000 to 100,000 base pairs. However, completed DNA strands contain only about one mistake in every 100 million to 1 billion base pairs (in humans, usually less than one per chromosome per replication). This phenomenally low error rate is the result of DNA repair enzymes that proofread each daughter strand during and after its synthesis. For example, some forms of DNA polymerase recognize a base pairing mistake as it is made. These types of DNA polymerase pause, fix the mistake, and then continue synthesizing more DNA. Other changes in the DNA base sequence that may occur during the life of a cell are also usually fixed by DNA repair enzymes.

Toxic Chemicals, Radiation, and Occasional Errors During DNA Replication Cause Mutations

Despite this amazing accuracy, neither humans nor any other organisms have error-free DNA. A few mutations creep into the genome from the rare, unrepaired mistakes made during normal DNA replication. DNA may also be harmed by toxic chemicals (such as free radicals formed during normal cellular metabolism, some components of cigarette smoke, and toxins produced by some molds) and some types of radiation (such as ultraviolet rays in sunlight). These chemicals and radiation increase the likelihood of base pairing errors during replication. Some even damage DNA between replications. Most, but not all, of these changes in DNA sequence are fixed by repair enzymes; the changes that remain are mutations.

Mutations Range from Changes in Single Nucleotide Pairs to Movements of Large Pieces of Chromosomes

If a pair of bases is mismatched during replication, repair enzymes usually recognize the mismatch, cut out the incorrect nucleotide, and replace it with a nucleotide containing the complementary base. Sometimes, however, the enzymes replace the parental nucleotide instead of the mismatched one. The resulting base pair is complementary, but incorrect. These **nucleotide substitutions** are also called **point mutations,** because individual nucleotides in the DNA sequence are changed (**Fig. 11-8a**). An **insertion mutation** occurs when one or more nucleotide pairs are inserted into the DNA double helix (**Fig. 11-8b**). A **deletion mutation** occurs when one or more nucleotide pairs are removed from the double helix (**Fig. 11-8c**).

(a) Nucleotide substitution

original DNA sequence

substitution

nucleotide pair changed from A–T to T–A

(b) Insertion mutation

original DNA sequence

T–A nucleotide pair inserted

(c) Deletion mutation

original DNA sequence

C–G nucleotide pair deleted

▲ **FIGURE 11-8 Mutations involving single pairs of nucleotides (a)** Nucleotide substitution. **(b)** Insertion mutation. **(c)** Deletion mutation. The original DNA bases are in pale colors with black letters; mutations are in deep colors with white letters.

▲ **FIGURE 11-9 Mutations that rearrange pieces of chromosomes (a)** Inversion mutation. **(b)** Translocation of pieces of DNA between two different chromosomes. In part (a), the original DNA bases are in pale colors with black letters; mutations are in deep colors with white letters. In part (b), the DNA bases of one chromosome are in pale colors with black letters, and the DNA bases of the second chromosome are in deep colors with white letters.

Pieces of chromosomes ranging in size from a single nucleotide pair to massive pieces of DNA are occasionally rearranged. An **inversion** occurs when a piece of DNA is cut out of a chromosome, turned around, and reinserted into the gap (**Fig. 11-9a**). Finally, a **translocation** results when a chunk of DNA, often very large, is removed from one chromosome and attached to a different one (**Fig. 11-9b**).

CHECK YOUR LEARNING

Can you explain what mutations are, and how they occur? Can you describe the different types of mutations? Do you know why mutations are rare and often harmful?

CASE STUDY revisited

Muscles, Mutations, and Myostatin

Belgian Blue cattle have a deletion mutation in their myostatin gene. The result is that their cells stop synthesizing the myostatin protein about halfway through. (In Chapter 12, we'll explain why some mutations cause short, or truncated, proteins to be synthesized.) Several breeds of "double-muscled" cattle have this same deletion mutation, but other double-muscled breeds have totally different mutations. Other animals, including several breeds of dogs, such as whippets (**Fig. 11-10**), may also have myostatin mutations. The mutations are generally different than those found in any of the breeds of cattle, but produce similar phenotypic effects.

All of these mutations result in nonfunctional myostatin proteins. This fact reveals an important feature of the language of DNA: The nucleotide words must be spelled just right, or at least really close, for the resulting proteins to function. In contrast, any one of an enormous number of possible mistakes will render the proteins useless.

Humans have myostatin, too; not surprisingly, mutations can occur in the human myostatin gene. A child inherits two copies of most genes, one from each parent. About a decade ago, a child was born in Germany who inherited a point mutation in his myostatin gene from both parents. This particular point mutation results in short, inactive, myostatin proteins. At 7 months, the boy already had well-developed calf, thigh, and buttock muscles. At 4 years, he could hold a

▲ **FIGURE 11-10** **Myostatin mutation in a dog** Like Belgian Blue cattle, "bully" whippets have defective myostatin, resulting in enormous muscles.

7-pound (3.2-kilogram) dumbbell in each hand, with his arms fully extended horizontally out to his sides. (Try it—it's not that easy for many adults.)

Consider This

Mutations may be neutral, harmful, or beneficial. Into which category do myostatin mutations fall? Belgian Blue cattle are born so muscular and, consequently, so large, that they usually must be delivered by cesarean section. Myostatin-deficient mice have huge muscles, but small, brittle tendons. The result? Their oversize muscles are weakly attached to their bones. On the other hand, heterozygous whippets, with one normal and one defective myostatin allele, can run faster than ordinary whippets. Is it possible that myostatin mutations are harmful in homozygotes but beneficial in heterozygotes? How do you think that natural selection might operate in this situation?

CHAPTER REVIEW

Summary of Key Concepts

11.1 How Did Scientists Discover That Genes Are Made of DNA?

Studies by Griffith showed that genes can be transferred from one bacterial strain into another. This transfer could transform the bacterial strain from harmless to deadly. Avery, MacLeod, and McCarty showed that DNA was the molecule that could transform bacteria. Hershey and Chase found that DNA is also the hereditary material of bacteriophage viruses. Thus, genes must be made of DNA.

11.2 What Is the Structure of DNA?

DNA consists of nucleotides that are linked into long strands. Each nucleotide consists of a phosphate group, the five-carbon sugar deoxyribose, and a nitrogen-containing base. Four types of bases occur in DNA: adenine, guanine, thymine, and cytosine. The sugar of one nucleotide is linked to the phosphate of the next nucleotide, forming a sugar-phosphate backbone for each strand. The bases stick out from this backbone. Two nucleotide strands wind together to form a DNA double helix, which resembles a twisted ladder. The sugar-phosphate backbones form the sides of the ladder. The bases of each strand pair up in the middle of the helix, held together by hydrogen bonds, forming the rungs of the ladder. Only complementary base pairs can bond together in the helix: Adenine bonds with thymine, and guanine bonds with cytosine.

11.3 How Does DNA Encode Genetic Information?

Genetic information is encoded as the sequence of nucleotides in a DNA molecule. Just as a language can form thousands of words and complex sentences from a small number of letters by varying the sequence and number of letters in each word and sentence, so too DNA can encode large amounts of information with varying sequences and numbers of nucleotides in different genes. Because DNA molecules are usually millions of nucleotides long, DNA can encode huge amounts of information in its nucleotide sequence.

11.4 How Does DNA Replication Ensure Genetic Constancy During Cell Division?

When cells reproduce, they must replicate their DNA so that each daughter cell receives all the original genetic information. During DNA replication, enzymes unwind and separate part of the two parental DNA strands. Then DNA polymerase enzymes bind to each parental DNA strand. Free nucleotides form hydrogen bonds with complementary bases on the parental strands, and DNA polymerase links the free nucleotides to form new DNA strands. Therefore, the sequence of nucleotides in each newly formed strand is complementary to the sequence of a parental strand. Replication is semiconservative because both new DNA double helices consist of one parental DNA strand and one newly synthesized, complementary daughter strand. The two new DNA double helices are, therefore, duplicates of the parental DNA double helix.

11.5 What Are Mutations, and How Do They Occur?

Mutations are changes in the base sequence in DNA. DNA polymerase and other repair enzymes "proofread" the DNA, minimizing the number of mistakes during replication, but mistakes do occur. Other mutations occur as a result of radiation and damage from certain chemicals. Mutations include substitutions, insertions, deletions, inversions, and translocations. Most mutations are harmful or neutral, but a few are beneficial and may be favored by natural selection.

Key Terms

adenine (A)   *194*
bacteriophage   *196*
base   *194*
complementary base pair   *199*

cytosine (C)   *194*
deletion mutation   *205*
deoxyribonucleic acid
  (DNA)   *192*

Learning Outcomes

In this chapter, you have learned to . . .

LO1 Describe the key experimental evidence indicating that DNA is the molecule of heredity.

LO2 Describe the structure of DNA, including nucleotides, the sugar-phosphate backbone, the double helix, and the role of hydrogen bonds.

LO3 Explain how genetic information is encoded in DNA.

LO4 Summarize the steps of DNA replication, including identification and action of the enzymes involved in each step.

LO5 Define the term *mutation*, and name and describe the different types of mutations.

Thinking Through the Concepts

Fill-in-the-Blank

1. DNA consists of subunits called _____. Each subunit consists of three parts: _____, _____, and _____. **LO2**

2. The subunits of DNA are assembled by linking the _____ of one nucleotide to the _____ of the next. As it is found in chromosomes, two DNA polymers are wound together into a structure called a(n) _____. **LO2**

3. The "base pairing rule" in DNA is that adenine pairs with _____, and guanine pairs with _____. Bases that can form pairs in DNA are called _____. **LO2**

4. When DNA is replicated, two new DNA double helices are formed, each consisting of one parental strand and one new, daughter strand. For this reason, DNA replication is called _____. **LO4**

5. The DNA double helix is unwound by an enzyme called _____. Daughter DNA strands are synthesized by the enzyme _____. In eukaryotic cells, the daughter DNA strands are synthesized in pieces; these pieces are joined by the enzyme _____. **LO4**

6. Sometimes, mistakes are made during DNA replication. If uncorrected, these mistakes are called _____. When a single nucleotide is changed, this is called a(n) _____, or _____. **LO5**

Review Questions

1. Describe the experimental evidence that DNA is the hereditary material of bacteriophages. **LO1**

2. Draw the general structure of a nucleotide. Which parts are identical in all nucleotides, and which can vary? **LO2**

3. Name the four types of nitrogen-containing bases found in DNA. **LO2**

4. Describe the structure of DNA. Where are the bases, sugars, and phosphates in the structure? **LO2**

5. Which bases are complementary to one another? How are they held together in the double helix of DNA? **LO2**

6. How is information encoded in the DNA molecule? **LO3**

7. Describe the process of DNA replication. **LO4**

8. How do mutations occur? Describe the principal types of mutations. **LO5**

Applying the Concepts

1. In an alternate universe, DNA is constructed of six different nucleotides, not four as on Earth. Would you expect organisms in this universe to have more precise genetic instructions, or more different genes, than life on Earth? Would you expect the length of a typical gene to be the same, shorter, or longer than that of a typical gene on Earth?

2. Genetic information is encoded in the sequence of nucleotides in DNA. Let's suppose that the nucleotide sequence on one strand of a double helix encodes the information needed to synthesize a hemoglobin molecule. Do you think that the sequence of nucleotides on the other strand of the double helix also encodes useful information? Why? (An analogy might help. Suppose that English were a "complementary language," with letters at opposite ends of the alphabet complementary to one another; that is, A is complementary to Z, B to Y, C to X, and so on. Would a sentence composed of letters complementary to "To be or not to be?" make sense?) Finally, why do you think DNA is double-stranded?

3. **BioEthics** If you enter "myostatin blocker" into an Internet search engine, you will find sites selling substances that are claimed to block the action of myostatin, and thereby supposedly help to build larger muscles. Let's assume that they work (most probably don't). From what you have learned about myostatin in this chapter, would this be a safe, effective way to bulk up? If you were a member of the World Anti-Doping Agency, would you allow athletes to compete in the Olympics who have used myostatin blockers?

Answers to Figure Caption questions and Fill-in-the-Blank questions can be found in the Answers section at the back of the book.

MB *Go to MasteringBiology for practice quizzes, activities, eText, videos, current events, and more.*

CHAPTER 12

Gene Expression and Regulation

CASE STUDY
Cystic Fibrosis

IF ALL YOU knew was her music, you'd think Alice Martineau had it made—a young, pretty singer-songwriter under contract with a major recording label. However, like about 70,000 other people worldwide, Martineau had cystic fibrosis, a recessive genetic disorder caused by defective alleles of a gene that encodes a crucially important protein called CFTR (the CF in the name of the protein stands for "cystic fibrosis"). Cystic fibrosis occurs when a person is homozygous for these defective alleles. Before modern medical care, most people with cystic fibrosis died by age 4 or 5; even now, the average life span is only 35 to 40 years. Martineau died when she was 30.

The CFTR protein forms channels that allow chloride to move across plasma membranes down its concentration gradient. CFTR also helps sodium movement across plasma membranes. CFTR is found in many parts of the body, including the sweat glands, lungs, and intestines. In the sweat glands, CFTR helps to reclaim sodium chloride from the sweat and transport it back into the blood, so that the body doesn't lose too much salt. Probably the most crucial role of CFTR is in the cells lining the airways in the lungs. Normally, the airways are covered with a film of mucus, which traps bacteria and debris. The bacteria-laden mucus is then swept out of the lungs by cilia on the cells of the airways.

Normally, chloride moves through CFTR channels out of the airway cells into the mucus, and sodium follows. The resulting high concentration of sodium chloride causes water to move into the mucus by osmosis, resulting in a thin liquid that can be removed easily by the cilia. However, mutations in the *CFTR* gene produce defective chloride channel proteins. As a result, chloride and sodium do not move from the cells into the mucus, so water doesn't move into the mucus, either. The mucus becomes so thick that the cilia can't move it out of the lungs, leaving the airways partially clogged. Bacteria multiply in the mucus, causing chronic lung infections.

In this chapter, we examine the processes by which the instructions in genes are translated into proteins. As you will learn, changes in those instructions—mutations—alter the structure and function of proteins such as CFTR.

C is for contemplation

Alice Martineau, shown here in a portrait painted by her brother Luke, hoped that ". . . people will realize when they hear the music, I am a singer-songwriter who just happens to be ill."

AT A GLANCE

LEARNING GOALS

LG1 Describe the molecules and cellular structures that play central roles in the process by which information encoded in DNA is used to synthesize proteins.

LG2 Compare and contrast how information is encoded in DNA and in RNA.

LG3 Describe the functions of transcription and translation and identify the locations in the cell at which they occur.

LG4 Describe the main steps of transcription, including the molecules involved and how they interact.

LG5 Describe the main steps of translation, including the molecules involved and how they interact.

LG6 Describe the different types of mutations and explain their potential effects on protein synthesis.

LG7 Describe how gene expression is regulated.

12.1 HOW IS THE INFORMATION IN DNA USED IN A CELL?

Information itself doesn't do anything. For example, a blueprint may provide all the information needed to build a house, but unless that information is translated into action by construction workers, no house will be built. Likewise, although the base sequence of DNA, the molecular blueprint of every cell, contains an incredible amount of information, DNA cannot carry out any actions on its own. So how does DNA determine whether you have black, blond, or red hair or have normal lung function or cystic fibrosis?

Most Genes Contain the Information Needed to Synthesize a Protein

Long before scientists understood that genes are made of DNA, they tried to determine how genes affect the phenotype of individual cells and entire organisms. In the 1940s, biologists discovered that most genes contain the information needed to direct the synthesis of a protein (see "Science in Action: One Gene, One Protein"). Proteins, in turn, are the "molecular workers" of a cell, forming many cellular structures and the enzymes that catalyze chemical reactions within the cell. Therefore, information must flow from DNA to protein.

DNA Provides Instructions for Protein Synthesis via RNA Intermediaries

DNA cannot directly synthesize proteins. Instead, it directs protein synthesis through intermediary molecules of **ribonucleic acid,** or **RNA.** RNA is similar to DNA but differs structurally in three respects: (1) Instead of the deoxyribose sugar found in DNA, the backbone of RNA contains the sugar ribose (the "R" in RNA); (2) RNA is usually single-stranded instead of double-stranded; and (3) RNA has the base uracil instead of the base thymine (**Table 12-1**).

DNA codes for the synthesis of many types of RNA, three of which play specific roles in protein synthesis: messenger RNA, ribosomal RNA, and transfer RNA (**Fig. 12-1**). There are several other types of RNA, including RNA used as the genetic material in some viruses, such as HIV; enzymatic RNA molecules, called ribozymes, that catalyze certain chemical reactions; and "regulatory" RNA, which we will discuss later in this chapter. Here we will introduce the roles of messenger RNA, ribosomal RNA, and transfer RNA. Their interactions during protein synthesis will be explained in more detail in Section 12.3.

Messenger RNA Carries the Code for Protein Synthesis from DNA to Ribosomes

In eukaryotic cells, the DNA is located and remains stored in the nucleus, like a valuable document in a library, whereas **messenger RNA (mRNA),** like a molecular photocopy, carries the information to the cytoplasm to be used in protein synthesis (**Fig. 12-1a**). As we will see shortly, groups of three bases in mRNA, called **codons,** specify which amino acids will be incorporated into a protein.

Ribosomal RNA and Proteins Form Ribosomes

Ribosomes, the cellular structures that synthesize proteins from the instructions in mRNA, are composed of **ribosomal RNA (rRNA)** and dozens of proteins. Each ribosome consists of two subunits—one small and one large (**Fig. 12-1b**). The small subunit has binding sites for mRNA, a "start" transfer RNA, and several proteins that are essential for assembling the ribosome and starting protein synthesis. The large subunit has binding sites for two tRNA molecules and a site that catalyzes the formation of peptide bonds joining amino acids into proteins. During protein synthesis, the two subunits come together, clasping an mRNA molecule between them.

Transfer RNA Carries Amino Acids to the Ribosomes

Transfer RNA (tRNA) delivers amino acids to a ribosome, where they will be incorporated into a protein. Every cell synthesizes at least one unique type of tRNA for each amino acid. Twenty enzymes in the cytoplasm, one for each amino acid,

TABLE 12-1   A Comparison of DNA and RNA

	DNA	RNA	
Strands	Two	One	
Sugar	Deoxyribose	Ribose	
Types of bases	Adenine (A), thymine (T)	Adenine (A), uracil (U)	
	cytosine (C), guanine (G)	cytosine (C), guanine (G)	
Base pairs	DNA–DNA	RNA–DNA	RNA–RNA
	A–T	A–T	A–U
	T–A	U–A	U–A
	C–G	C–G	C–G
	G–C	G–C	G–C
Function	Contains genes; the sequence of bases in most genes determines the amino acid sequence of a protein	Messenger RNA (mRNA): carries the code for a protein-coding gene from DNA to ribosomes	
		Ribosomal RNA (rRNA): combines with proteins to form ribosomes, the structures that link amino acids to form a protein	
		Transfer RNA (tRNA): carries amino acids to the ribosomes	

recognize the different tRNA molecules and use the energy of ATP to attach the correct amino acid to one end of the tRNA molecule (**Fig. 12-1c**). These "loaded" tRNA molecules deliver their amino acids to a ribosome. A group of three bases, called an **anticodon,** protrudes from each tRNA. Complementary base pairing between codons of mRNA and anticodons of tRNA specifies which amino acids are used during protein synthesis.

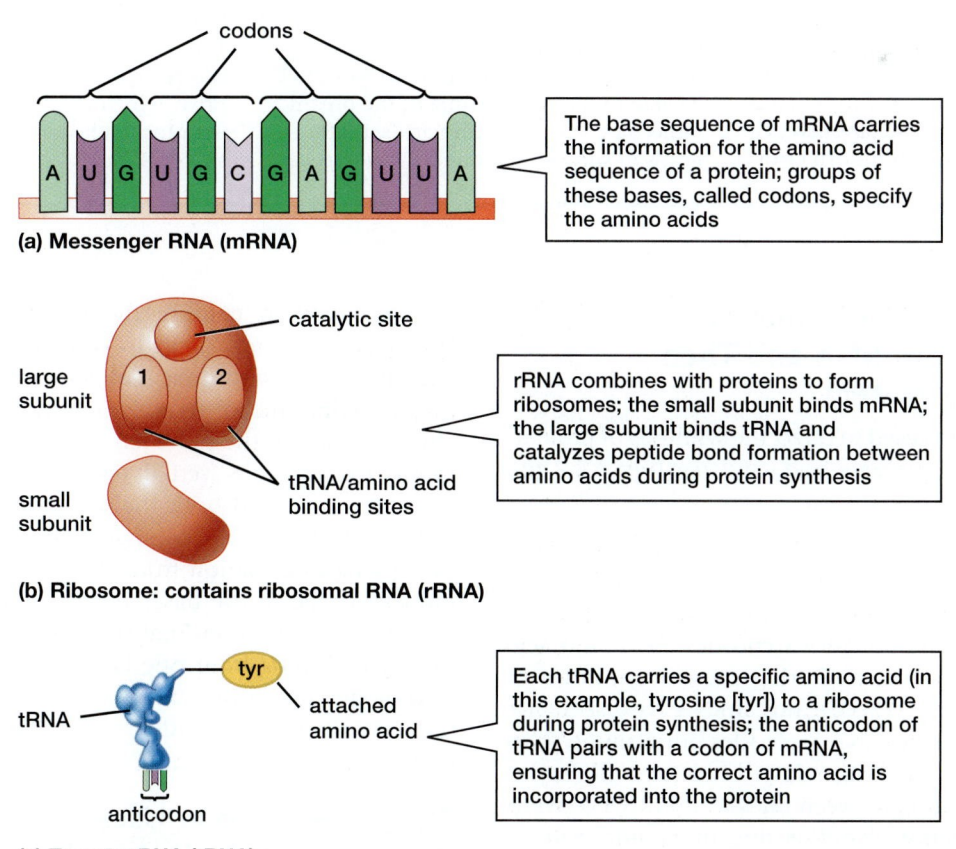

(a) Messenger RNA (mRNA)

The base sequence of mRNA carries the information for the amino acid sequence of a protein; groups of these bases, called codons, specify the amino acids

(b) Ribosome: contains ribosomal RNA (rRNA)

rRNA combines with proteins to form ribosomes; the small subunit binds mRNA; the large subunit binds tRNA and catalyzes peptide bond formation between amino acids during protein synthesis

(c) Transfer RNA (tRNA)

Each tRNA carries a specific amino acid (in this example, tyrosine [tyr]) to a ribosome during protein synthesis; the anticodon of tRNA pairs with a codon of mRNA, ensuring that the correct amino acid is incorporated into the protein

▲ FIGURE 12-1   **Cells synthesize three major types of RNA that are required for protein synthesis**

SCIENCE IN ACTION One Gene, One Protein

Ever since Gregor Mendel's work was rediscovered in the late 1800s, biologists have known that genes help to determine phenotypes as varied as hair texture, flower color, and inherited diseases such as sickle-cell anemia and cystic fibrosis. But how do individual cells and entire organisms convert their genotypes into phenotypes?

The first hint came from children born with defective metabolic pathways. Many metabolic pathways synthesize molecules in a series of steps, each catalyzed by a specific protein enzyme. The product produced by one enzyme becomes the substrate of the next enzyme, like a molecular assembly line (see Fig. 6-12). A defective pathway fails to produce its final product and may also cause intermediate molecules in the pathway to accumulate to dangerous levels.

For example, defects in the metabolism of two amino acids, phenylalanine and tyrosine, can cause albinism (no pigmentation in skin or hair; see Fig. 10-23) or several diseases with symptoms as varied as urine that turns brown when exposed to the air (alkaptonuria) or phenylalanine accumulation in the brain, leading to mental retardation (phenylketonuria). In the early 1900s, the English physician Archibald Garrod studied the inheritance of these inborn errors of metabolism. He hypothesized that (1) each inborn error

of metabolism is caused by a defective version of a specific enzyme; (2) each defective enzyme is caused by a defective allele of a single gene; and (3) therefore, at least some genes must encode the information needed for the synthesis of enzymes. In the early 1940s, geneticists George Beadle and Edward Tatum used the metabolic pathways of a common bread mold, *Neurospora crassa*, to show that Garrod's hypotheses were correct.

We usually encounter *Neurospora* growing on stale bread. However, it can survive on a much simpler diet, consisting of an energy source such as sugar, a few minerals, and vitamin B_6, because *Neurospora* can manufacture the enzymes needed to make virtually all of its own organic molecules, including amino acids. In contrast, we humans cannot synthesize many vitamins and 9 of the 20 common amino acids; we must obtain these from our food. Beadle and Tatum used *Neurospora* to test the hypothesis that many of an organism's genes encode the information needed to synthesize enzymes.

If this hypothesis was correct, a mutation in a specific gene might disrupt the synthesis of a specific enzyme, disabling one of the mold's metabolic pathways. Thus, a mutant mold wouldn't be able to synthesize some of the organic molecules, such as amino acids, that it needs to survive. The mutant *Neurospora*

could grow on a simple medium of sugar, minerals, and vitamin B_6 only if the missing organic molecules were provided.

Beadle and Tatum induced mutations by exposing *Neurospora* to X-rays, and then studied the inheritance of the metabolic pathway that synthesizes the amino acid arginine (**Fig. E12-1**). In normal molds, arginine is synthesized from citrulline, which in turn is synthesized from ornithine (**Fig. E12-1a**). Mutant A could grow only if supplemented with arginine, but not if supplemented with citrulline or ornithine (**Fig. E12-1b**). Therefore, this strain had a defect in the enzyme that converts citrulline to arginine. Mutant B could grow if supplemented with either arginine or citrulline, but not with ornithine (see Fig. E12-1b). This mutant had a defect in the enzyme that converts ornithine to citrulline. Because a mutation in a single gene affected only a single enzyme within a single metabolic pathway, Beadle and Tatum concluded that one gene encodes the information for one enzyme. Almost all enzymes are proteins, but many proteins are not enzymes (see Chapter 3). Thus, the "one gene, one enzyme" relationship was later clarified to "one gene, one protein." The importance of this observation was recognized in 1958 with a Nobel Prize, which Beadle and Tatum shared with Joshua Lederberg, one of Tatum's students.

Overview: Genetic Information Is Transcribed into RNA and Then Translated into Protein

Information in DNA is used to direct the synthesis of proteins in two steps, called *transcription* and *translation* (**Fig. 12-2** and **Table 12-2**):

1. In **transcription** (**Fig. 12-2a**), the information contained in the DNA of a gene is copied into RNA. In eukaryotic cells, transcription occurs in the nucleus.

2. The base sequence of mRNA encodes the amino acid sequence of a protein. During protein synthesis, or **translation** (**Fig. 12-2b**), this mRNA base sequence is decoded. Transfer RNA molecules bring amino acids to a ribosome. Messenger RNA binds to a ribosome, where base pairing between mRNA and tRNA converts the base sequence of mRNA into the amino acid sequence of the protein. In eukaryotic cells, ribosomes are found in the cytoplasm, so translation occurs there as well.

It's easy to confuse the terms "transcription" and "translation." It may help to compare their common English meanings with their biological meanings. In English, to "transcribe" means to make a written copy of something, almost always in the same language. In an American courtroom, for example, verbal testimony is transcribed into a written copy, and both the testimony and the transcriptions are in English. In biology, transcription is the process of copying information from DNA to RNA using the common language of the bases found in their nucleotides. In contrast, the common English meaning of "translation" is to convert words from one language to another language. In biology, translation means to convert information from the "base language" of RNA to the "amino acid language" of proteins.

The Genetic Code Uses Three Bases to Specify an Amino Acid

Before we examine transcription and translation in detail, let's see how geneticists deciphered the **genetic code**—the

(a) The metabolic pathway for synthesizing the amino acid arginine

Genotype of *Neurospora*		Supplements Added to Medium				Conclusions
		none	ornithine	citrulline	arginine	
Normal *Neurospora*						Normal *Neurospora* can synthesize arginine, citrulline, and ornithine.
Mutants with single gene defect	A					Mutant A grows only if arginine is added. It cannot synthesize arginine because it has a defect in enzyme 2; gene *A* is needed for synthesis of arginine.
	B					Mutant B grows if either arginine or citrulline is added. It cannot synthesize arginine because it has a defect in enzyme 1. Gene *B* is needed for synthesis of citrulline.

(b) Growth of normal and mutant *Neurospora* on a simple medium with different supplements

▲ **FIGURE E12-1 Beadle and Tatum's experiments with *Neurospora* mutants (a)** Each step in the metabolic pathway for arginine synthesis is catalyzed by a different enzyme. **(b)** By analyzing which supplements allowed mutant molds to grow on simple nutrient medium, Beadle and Tatum concluded that a single gene codes for the synthesis of a single enzyme.

QUESTION What result would you expect for a mutant that lacks an enzyme needed to produce ornithine?

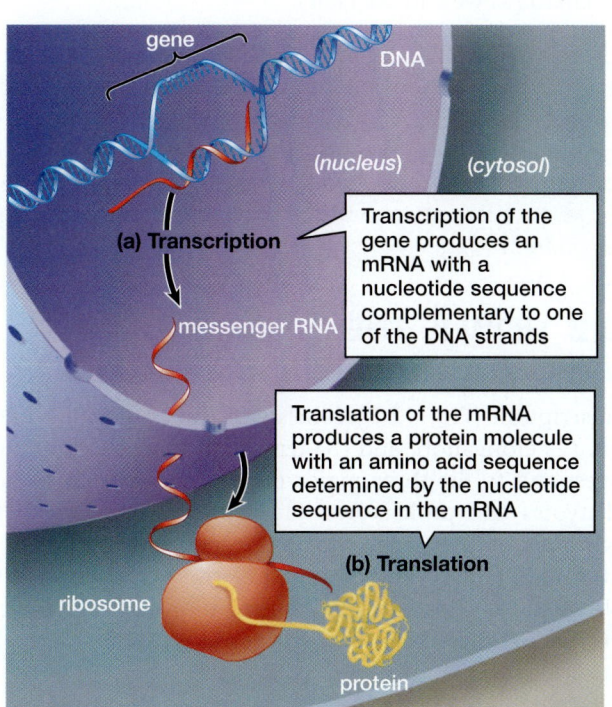

(a) Transcription

Transcription of the gene produces an mRNA with a nucleotide sequence complementary to one of the DNA strands

Translation of the mRNA produces a protein molecule with an amino acid sequence determined by the nucleotide sequence in the mRNA

(b) Translation

biological dictionary that spells out the rules for translating the base sequences in DNA and mRNA into the amino acid sequences in proteins. Both DNA and RNA contain four different bases: adenine (A); thymine (T; uracil [U] in RNA); guanine (G); and cytosine (C; see Table 12-1). However, proteins are made of 20 different amino acids, so one base cannot directly translate into one amino acid: there are not enough different bases. If a sequence of two bases codes for an amino acid, there would be 16 possible combinations (each of four possible first bases paired with each of four possible second bases, or $4 \times 4 = 16$). This still isn't enough to code for 20 amino acids. A three-base sequence

◀ **FIGURE 12-2 Genetic information flows from DNA to RNA to protein (a)** During transcription, the base sequence in a gene specifies the base sequence of a complementary RNA molecule. For protein-encoding genes, the product is an mRNA molecule that exits from the nucleus and enters the cytoplasm. **(b)** During translation, the base sequence in an mRNA molecule specifies the amino acid sequence of a protein.

TABLE 12-2 Transcription and Translation

Process	Information for the Process	Product	Major Enzyme or Structure Involved in the Process	Type of Base Pairing Required
Transcription (synthesis of RNA)	A segment of one DNA strand	One RNA molecule (e.g., mRNA, tRNA, or rRNA)	RNA polymerase	RNA with DNA: RNA bases pair with DNA bases as an RNA molecule is synthesized
Translation (synthesis of a protein)	mRNA	One protein molecule	Ribosome (also requires tRNA)	mRNA with tRNA: a codon in mRNA forms base pairs with an anticodon in tRNA

gives 64 possible combinations ($4 \times 4 \times 4 = 64$), which is more than enough. Using this reasoning, physicist George Gamow hypothesized in 1954 that sets of three bases in mRNA, called *codons*, specify each of the amino acids. In 1961, Francis Crick and three coworkers demonstrated that this hypothesis is correct.

For any language to be understood, its users must know what the words mean, where words start and stop, and where sentences begin and end. To decipher the codons, which are the "words" of the genetic code, Marshall Nirenberg and Heinrich Matthaei ground up bacteria and isolated the components needed to synthesize proteins. To this mixture, they added artificial mRNA, allowing them to control which codons were to be translated and to see which amino acids were incorporated into proteins. For example, an mRNA strand composed entirely of uracil (UUUUUUUU . . .) directed the mixture to synthesize a protein composed solely of the amino acid phenylalanine. Therefore, the triplet UUU must be the codon that translates into phenylalanine. Because the genetic code was deciphered by using these artificial mRNAs, it is usually written in terms of the base triplets in mRNA (rather than in DNA) that code for each amino acid (Table 12-3).

How does a cell recognize where codons start and stop, and where the code for an entire protein starts and stops? Translation always begins with the codon AUG, appropriately known as the **start codon.** Because AUG also codes for the amino acid methionine, all proteins originally begin with methionine, although it may be removed after the protein is synthesized. Three codons—UAG, UAA, and UGA—are **stop codons.** When the ribosome encounters any of these stop codons, it releases both the newly synthesized protein and the mRNA. Because all codons consist of three bases, and the beginning and end of a protein are specified, then "spaces" between codon "words" are unnecessary. Why? Consider what would happen if English used only three-letter words: a sentence such as THEDOGSAWTHECAT would be perfectly understandable, even without spaces between the words.

Because the genetic code has three stop codons, 61 triplets remain to specify only 20 amino acids. Therefore, several different codons may be translated into the same amino acid. For example, six codons—UUA, UUG, CUU, CUC, CUA, and CUG—all code for leucine (see Table 12-3). However, each codon specifies one, and only one, amino acid; UUA always specifies leucine, never isoleucine, glycine, or any other amino acid.

How are codons used to direct protein synthesis? Decoding the codons of mRNA is the job of tRNA and ribosomes. Remember that tRNA transports amino acids to the ribosome and that distinct tRNA molecules carry each different type of amino acid. Each of these unique tRNAs has three exposed bases, called an *anticodon*, which are complementary to the bases of a codon in mRNA. For example, the mRNA codon GUU forms base pairs with the anticodon CAA of a tRNA that has the amino acid valine attached to it. A ribosome can then incorporate valine into a growing protein chain (as we will see later).

CHECK YOUR LEARNING

Can you describe how information is encoded in DNA and RNA, and how this information flows from DNA to RNA to protein? Do you understand the difference between transcription and translation and how each process is used to convert information in DNA to the amino acid sequence of a protein?

12.2 HOW IS THE INFORMATION IN A GENE TRANSCRIBED INTO RNA?

Transcription (Fig. 12-3) consists of three steps: (1) initiation, (2) elongation, and (3) termination. These three steps correspond to the three major parts of most genes in both eukaryotes and prokaryotes: (1) a promoter region at the beginning of the gene, where transcription is started, or initiated; (2) the "body" of the gene, where elongation of the RNA strand occurs; and (3) a termination signal at the end of the gene, where RNA synthesis ceases, or terminates.

TABLE 12-3   The Genetic Code (Codons of mRNA)

First Base		Second Base: U		Second Base: C		Second Base: A		Second Base: G		Third Base
U	UUU	Phenylalanine (Phe)	UCU	Serine (Ser)	UAU	Tyrosine (Tyr)	UGU	Cysteine (Cys)	U
	UUC	Phenylalanine	UCC	Serine	UAC	Tyrosine	UGC	Cysteine	C
	UUA	Leucine (Leu)	UCA	Serine	UAA	Stop	UGA	Stop	A
	UUG	Leucine	UCG	Serine	UAG	Stop	UGG	Tryptophan (Trp)	G
C	CUU	Leucine	CCU	Proline (Pro)	CAU	Histidine (His)	CGU	Arginine (Arg)	U
	CUC	Leucine	CCC	Proline	CAC	Histidine	CGC	Arginine	C
	CUA	Leucine	CCA	Proline	CAA	Glutamine (Gln)	CGA	Arginine	A
	CUG	Leucine	CCG	Proline	CAG	Glutamine	CGG	Arginine	G
A	AUU	Isoleucine (Ile)	ACU	Threonine (Thr)	AAU	Asparagine (Asp)	AGU	Serine (Ser)	U
	AUC	Isoleucine	ACC	Threonine	AAC	Asparagine	AGC	Serine	C
	AUA	Isoleucine	ACA	Threonine	AAA	Lysine (Lys)	AGA	Arginine (Arg)	A
	AUG	Methionine (Met) Start	ACG	Threonine	AAG	Lysine	AGG	Arginine	G
G	GUU	Valine (Val)	GCU	Alanine (Ala)	GAU	Aspartic acid (Asp)	GGU	Glycine (Gly)	U
	GUC	Valine	GCC	Alanine	GAC	Aspartic acid	GGC	Glycine	C
	GUA	Valine	GCA	Alanine	GAA	Glutamic acid (Glu)	GGA	Glycine	A
	GUG	Valine	GCG	Alanine	GAG	Glutamic acid	GGG	Glycine	G

Transcription Begins When RNA Polymerase Binds to the Promoter of a Gene

The enzyme **RNA polymerase** synthesizes RNA. Near the beginning of every gene is a DNA sequence called the **promoter.** In eukaryotic cells, a promoter consists of two main parts: (1) a short sequence of bases, often TATAAA, that binds RNA polymerase; and (2) one or more other sequences, often called *transcription factor binding sites,* or *response elements.* When RNA polymerase binds to the promoter of a gene, the DNA double helix at the beginning of the gene unwinds and transcription begins (**Fig. 12-3 ❶**). Proteins called transcription factors can attach to a transcription factor binding site, enhancing or suppressing binding of RNA polymerase to the promoter and, consequently, enhancing or suppressing transcription of the gene. We will return to the important topic of gene regulation in Section 12.5.

Elongation Generates a Growing Strand of RNA

After binding to the promoter, RNA polymerase travels down one of the DNA strands, called the **template strand,** synthesizing a single strand of RNA with bases complementary to those in the DNA (**Fig. 12-3 ❷**). Like DNA polymerase, RNA polymerase always travels along the DNA template strand starting at the 3′ end of a gene and moving toward the 5′ end. Base pairing between RNA and DNA is the same as between two strands of DNA, except that uracil in RNA pairs with adenine in DNA (see Table 12-1).

After about 10 nucleotides have been added to the growing RNA chain, the first nucleotides of the RNA separate from the DNA template strand. This separation allows the two DNA strands to rewind into a double helix (**Fig. 12-3 ❸**). As transcription continues to elongate the RNA molecule, one end of the RNA drifts away from the DNA, while RNA polymerase keeps the other end attached to the template strand of the DNA (see Fig. 12-3 ❸ and **Fig. 12-4**).

Transcription Stops When RNA Polymerase Reaches the Termination Signal

RNA polymerase continues along the template strand of the gene until it reaches a sequence of DNA bases known as the *termination signal.* At this point, RNA polymerase releases the completed RNA molecule and detaches from the DNA (**Fig. 12-3 ❸** and ❹). The RNA polymerase is then available to bind to the promoter region of another gene and to synthesize another RNA molecule.

CHECK YOUR LEARNING

Can you describe the process of transcription? Do you understand how DNA, RNA, and RNA polymerase interact to produce a strand of RNA?

▶ FIGURE 12-3 **Transcription is the synthesis of RNA from instructions in DNA** A gene is a segment of a chromosome's DNA. One of the DNA strands that make up the double helix will serve as the template for the synthesis of an RNA molecule with bases complementary to the bases in the DNA strand.

QUESTION If the other DNA strand of this molecule were the template strand, in which direction would the RNA polymerase travel?

❶ **Initiation:** RNA polymerase binds to the promoter region of DNA near the beginning of a gene, separating the double helix near the promoter.

❷ **Elongation:** RNA polymerase travels along the DNA template strand (blue), unwinding the DNA double helix and synthesizing RNA by catalyzing the addition of ribose nucleotides into an RNA molecule (red). The nucleotides in the RNA are complementary to the template strand of the DNA.

❸ **Termination:** At the end of the gene, RNA polymerase encounters a DNA sequence called a termination signal. RNA polymerase detaches from the DNA and releases the RNA molecule.

❹ **Conclusion of transcription:** After termination, the DNA completely rewinds into a double helix. The RNA molecule is free to move from the nucleus to the cytoplasm for translation, and RNA polymerase may move to another gene and begin transcription once again.

12.3 HOW IS THE BASE SEQUENCE OF mRNA TRANSLATED INTO PROTEIN?

Prokaryotic and eukaryotic cells differ in the organization of their genes, how they produce a functional mRNA molecule from the instructions in their DNA, and the timing and location of translation.

Most prokaryotic genes are compact: All the nucleotides in a gene code for the amino acids in a protein. What's more, most or all of the genes for a complete metabolic pathway sit end to end on the chromosome (Fig. 12-5a). Therefore, prokaryotic cells usually transcribe a single, very long mRNA from a series of adjacent genes, each of which specifies a different protein in the pathway. Because prokaryotic cells do not have a nuclear membrane separating their DNA from the cytoplasm (see Fig. 4-19), transcription and translation are usually not separated, either in space or in time. In most cases, as an mRNA molecule begins to separate from the DNA during transcription, ribosomes immediately begin translating the mRNA into protein (Fig. 12-5b).

▲ **FIGURE 12-4 RNA transcription in action** This colorized electron micrograph shows the progress of RNA transcription in the egg of an African clawed toad. In each treelike structure, the central "trunk" is DNA and the "branches" are RNA molecules. A series of RNA polymerase molecules (too small to be seen in this micrograph) is traveling down the DNA, synthesizing RNA as they go. The beginning of the gene is on the left. The short RNA molecules on the left have just begun to be synthesized; the long RNA molecules on the right are almost finished.

QUESTION Why do you think so many mRNA molecules are being transcribed from the same gene?

In contrast, the DNA of eukaryotic cells is contained in the nucleus, whereas the ribosomes reside in the cytoplasm. Further, the genes that encode the proteins needed for a metabolic pathway in eukaryotes are not clustered together as they are in prokaryotes, but may be dispersed among several chromosomes. Finally, the RNA molecule formed during transcription is not a functional mRNA that can be immediately translated into protein.

In Eukaryotes, a Precursor RNA Is Processed to Form mRNA That Is Translated into Protein

Most eukaryotic genes consist of two or more segments of DNA with nucleotide sequences that code for a protein, interrupted by sequences that are not translated into protein. The coding segments are called **exons,** because they are *ex*pressed in protein, and the noncoding segments are called **introns,** because they are *intra*genic, meaning "within a gene" (**Fig. 12-6a**).

Transcription of a eukaryotic gene produces a very long RNA strand, often called a *precursor mRNA* or *pre-mRNA*, which starts before the first exon and ends after the last exon (**Fig. 12-6b** ❶). More nucleotides are added at the beginning and end of this pre-mRNA molecule, forming a "cap" and "tail" (**Fig. 12-6b** ❷). These nucleotides will help move the finished mRNA through pores in the nuclear envelope to the cytoplasm, bind the mRNA to a ribosome, and protect the mRNA molecule from degradation by cellular enzymes. To convert this pre-mRNA molecule into the finished mRNA, enzymes in the nucleus cut the pre-mRNA apart at the junctions between introns and exons, splice

(a) Gene organization on a prokaryotic chromosome

(b) Simultaneous transcription and translation in prokaryotes

▲ **FIGURE 12-5 Messenger RNA synthesis in prokaryotic cells (a)** In prokaryotes, many or all of the genes for a complete metabolic pathway lie side by side on the chromosome. **(b)** Transcription and translation are simultaneous in prokaryotes. In the electron micrograph, RNA polymerase (not visible at this magnification) travels from left to right on a strand of DNA. As it synthesizes an mRNA molecule, ribosomes bind to the mRNA and immediately begin synthesizing a protein (not visible). The diagram below the micrograph shows all of the key molecules involved.

together the protein-coding exons, and discard the introns (**Fig. 12-6b** ❸). The finished mRNA molecules leave the nucleus and enter the cytoplasm through pores in the nuclear envelope (**Fig. 12-6b** ❹). In the cytoplasm, the mRNA binds to ribosomes, which synthesize the protein specified by the mRNA base sequence.

Functions of Intron–Exon Gene Structure

Why do eukaryotic genes contain introns and exons? This gene structure appears to serve at least two functions. The first is to allow a cell to produce several different proteins from a single gene by splicing exons together in different ways. For example, a gene called *CT/CGRP* is transcribed in both the thyroid and the brain. In the thyroid, one splicing arrangement results in the synthesis of the hormone calcitonin,

(a) Eukaryotic gene structure

(b) RNA synthesis and processing in eukaryotes

◀ **FIGURE 12-6 Messenger RNA synthesis in eukaryotic cells** **(a)** Eukaryotic genes consist of exons (medium blue), which code for the amino acid sequence of a protein, and introns (dark blue), which do not. **(b)** Messenger RNA synthesis in eukaryotes involves several steps: ❶ transcribing the gene into a long pre-mRNA molecule; ❷ adding additional RNA nucleotides to form the cap and tail; ❸ cutting out the introns and splicing the exons together into the finished mRNA; and ❹ moving the finished mRNA out of the nucleus into the cytoplasm for translation.

which helps regulate calcium concentrations in the blood. In the brain, a different splicing arrangement results in the synthesis of a protein used as a messenger for communication between nerve cells. Alternative splicing occurs in the RNA transcribed from most eukaryotic genes. Therefore, in eukaryotes, the "one gene, one protein" rule must be reworded as "one gene, one *or more* proteins."

In the second function, fragmented genes may provide a quick and efficient way for eukaryotes to evolve new proteins with new functions. Chromosomes sometimes break apart, and their parts may reattach to different chromosomes. If the breaks occur within the introns of genes, exons may be moved intact from one chromosome to another. Most such errors would be harmful. But sometimes, the exchange of exons between genes produces new genes that enhance the survival and reproduction of the organism that carries them.

During Translation, mRNA, tRNA, and Ribosomes Cooperate to Synthesize Proteins

Like transcription, translation has three steps: (1) initiation, (2) elongation of the protein chain, and (3) termination. We will describe translation only in eukaryotic cells (**Fig. 12-7**).

Initiation: Translation Begins When tRNA and mRNA Bind to a Ribosome

A "preinitiation complex"—composed of a small ribosomal subunit, a start (methionine) tRNA, and several other proteins

(**Fig. 12-7 ❶**)—binds to the beginning of an mRNA molecule. The preinitiation complex moves along the mRNA until it finds a start (AUG) codon, which forms base pairs with the UAC anticodon of the methionine tRNA (**Fig. 12-7 ❷**). A large ribosomal subunit then attaches to the small subunit, sandwiching the mRNA between the two subunits and holding the methionine tRNA in the first tRNA binding site (**Fig. 12-7 ❸**). The ribosome is now ready to begin translation.

Elongation: Amino Acids Are Added One at a Time to the Growing Protein Chain

A ribosome holds two mRNA codons aligned with the two tRNA binding sites of the large subunit. A second tRNA, with an anticodon complementary to the second codon of the mRNA, moves into the second tRNA binding site on the large subunit (**Fig. 12-7 ❹**). The catalytic site of the large subunit breaks the bond holding the first amino acid (methionine) to its tRNA and forms a peptide bond between this amino acid and the amino acid attached to the second tRNA (**Fig. 12-7 ❺**). Interestingly, rRNA, and not one of the proteins of the large subunit, catalyzes the formation of the peptide bond. Therefore, the catalytic site of a ribosome is a ribozyme.

After the peptide bond is formed, the first tRNA is "empty," and the second tRNA carries a two-amino-acid chain. The ribosome releases the empty tRNA and shifts to the next codon on the mRNA molecule (**Fig. 12-7 ❻**). The tRNA holding the chain of amino acids also shifts, moving from the second to the first binding site of the ribosome. A new tRNA,

1 A tRNA with an attached methionine amino acid binds to a small ribosomal subunit, forming a preinitiation complex.

2 The preinitiation complex binds to an mRNA molecule. The methionine (met) tRNA anticodon (UAC) base-pairs with the start codon (AUG) of the mRNA.

3 The large ribosomal subunit binds to the small subunit. The methionine tRNA binds to the first tRNA site on the large subunit.

Elongation:

4 The second codon of mRNA (GUU) base-pairs with the anticodon (CAA) of a second tRNA carrying the amino acid valine (val). This tRNA binds to the second tRNA site on the large subunit.

5 The catalytic site on the large subunit catalyzes the formation of a peptide bond linking the amino acids methionine and valine. The two amino acids are now attached to the tRNA in the second binding site.

ribosome moves one codon to the right

6 The "empty" tRNA is released and the ribosome moves down the mRNA, one codon to the right. The tRNA that is attached to the two amino acids is now in the first tRNA binding site and the second tRNA binding site is empty.

Termination:

7 The third codon of mRNA (CAU) base-pairs with the anticodon (GUA) of a tRNA carrying the amino acid histidine (his). This tRNA enters the second tRNA binding site on the large subunit.

8 The catalytic site forms a peptide bond between valine and histidine, leaving the peptide attached to the tRNA in the second binding site. The tRNA in the first site leaves, and the ribosome moves one codon over on the mRNA.

9 This process repeats until a stop codon is reached; the mRNA and the completed peptide are released from the ribosome, and the subunits separate.

▲ **FIGURE 12-7 Translation is the process of protein synthesis** Translation decodes the base sequence of an mRNA into the amino acid sequence of a protein.

QUESTION Examine step 9. If mutations changed all of the guanine molecules visible in the mRNA sequence shown here to uracil, how would the translated peptide differ from the one shown?

with an anticodon complementary to the third codon of the mRNA, binds to the empty second site (**Fig. 12-7 ❼**). The catalytic site now joins the third amino acid to the growing protein chain (**Fig. 12-7 ❽**). The empty tRNA leaves the ribosome, the ribosome shifts to the next codon on the mRNA, and the process repeats, one codon at a time.

Termination: A Stop Codon Signals the End of Translation

When the ribosome reaches a stop codon in the mRNA, protein synthesis terminates. Stop codons do not bind to tRNA. Instead, the ribosome releases the finished protein chain and the mRNA (**Fig. 12-7 ❾**). The ribosome then disassembles into its large and small subunits.

SUMMING UP Decoding the Sequence of Bases in DNA into the Sequence of Amino Acids in Protein

Let's summarize how a cell decodes the genetic information of DNA and synthesizes a protein (**Fig. 12-8**):

a. With some exceptions, such as the genes for tRNA and rRNA, each gene codes for the amino acid sequence of a protein. The DNA of the gene consists of the template strand, which is transcribed into mRNA, and its complementary strand, which is not transcribed.

b. Transcription produces an mRNA molecule that is complementary to the template strand. Starting from the first AUG, each codon within the mRNA is a sequence of three bases that specifies an amino acid or a "stop."

c. Enzymes in the cytoplasm attach the appropriate amino acid to each tRNA, as determined by the tRNA's anticodon.

d. The mRNA moves out of the nucleus to a ribosome in the cytoplasm. Transfer RNAs carry their attached amino acids to the ribosome. There, the bases in tRNA anticodons bind to the complementary bases in mRNA codons. The ribosome catalyzes the formation of peptide bonds that join the amino acids to form a protein with the amino acid sequence specified by the sequence of bases in mRNA. When a stop codon is reached, the finished protein is released from the ribosome. This decoding chain, from DNA bases to mRNA codons to tRNA anticodons to amino acids, results in the synthesis of a protein with an amino acid sequence determined by the base sequence of a gene.

CHECK YOUR LEARNING

Can you describe the process of translation? Are you able to describe how the production of mRNA differs between prokaryotic and eukaryotic cells? Do you know how ribosomes, mRNA, and tRNA cooperate to produce a protein?

12.4 HOW DO MUTATIONS AFFECT PROTEIN STRUCTURE AND FUNCTION?

Mistakes during DNA replication, ultraviolet rays in sunlight, chemicals in cigarette smoke, and a host of other environmental factors may cause **mutations**—changes in the sequence of bases in DNA. The consequences for an organism's structure and function depend on how the mutation affects the protein encoded by the mutated gene.

CASE STUDY continued

Cystic Fibrosis

There are about 1,500 different defective alleles of the *CFTR* gene, all of which can cause cystic fibrosis. The most common defective allele is missing a single codon. The resulting lack of one crucial amino acid causes the CFTR protein to be misshapen. A normal CFTR protein is synthesized by ribosomes on rough endoplasmic reticulum (ER), enters the ER, and then is transported to the plasma membrane. The misshapen CFTR protein, however, is broken down within the ER and never reaches the plasma membrane. Four other common mutant *CFTR* alleles code for a stop codon in the middle of the protein, so translation terminates partway through. Still other mutant alleles produce proteins that are completely synthesized and inserted into the plasma membrane, but do not form functional chloride channels.

▲ **FIGURE 12-8 Complementary base pairing is required to decode genetic information** (a) The DNA of a gene contains two strands; only the template strand is used by RNA polymerase to synthesize an RNA molecule. (b) Bases in the template strand of DNA are transcribed into a complementary mRNA. Codons are sequences of three bases that specify an amino acid or a stop during protein synthesis. (c) Unless it is a stop codon, each mRNA codon forms base pairs with the anticodon of a tRNA molecule that carries a specific amino acid. (d) The amino acids borne by the tRNAs are joined to form a protein.

The Effects of Mutations Depend on How They Alter the Codons of mRNA

Mutations may be categorized as inversions, translocations, deletions, insertions, and substitutions. These different types of mutations differ greatly in how they affect DNA and, consequently, their likelihood of producing significant alterations in protein structure and function.

Inversions and Translocations

Inversions are mutations that occur when a piece of DNA is cut out of a chromosome, flipped around, and reinserted in a reversed orientation. **Translocations** are mutations that occur when a piece of DNA is removed from one chromosome and attached to another. Inversions and translocations may be relatively benign if entire genes, including their promoters, are merely moved from one place to another. In these cases, the mRNA transcribed from the gene will contain all of the original codons, and all will remain unchanged. However, if a gene is split in two, it will no longer code for a complete, functional protein. For example, almost half the cases of severe hemophilia are caused by an inversion in the gene that encodes a protein required for blood clotting.

Deletions and Insertions

In a **deletion mutation,** one or more pairs of nucleotides are removed from a gene. In an **insertion mutation,** one or more pairs of nucleotides are inserted into a gene. If one or two pairs of nucleotides are removed or added, protein function is usually completely ruined. Why? Think back to the genetic code: Three nucleotides encode a single amino acid. Therefore, deleting or inserting one or two nucleotides, or any number that isn't a multiple of three, alters all of the codons that follow the deletion or insertion. Why? Consider this sentence, composed of all three-letter words: THEDOGSAWTHECATSITANDTHEFOXRUN. Deleting or inserting a letter (deleting the first E, for example), changes all of the following words: THD OGS AWT HEC ATS ITA NDT HEF OXR UN. Most of the amino acids of a protein synthesized from an mRNA containing such a mutation will be incorrect, so the protein will be nonfunctional.

Deleting or inserting three pairs of nucleotides often has only minor effects on the protein, regardless of whether the three nucleotide pairs that were deleted or inserted make up a single codon or overlap into two codons. Returning to our model sentence, let's suppose that we delete OGS; the sentence reads: THE DAW THE CAT SIT AND THE FOX RUN, most of which still makes sense. If we add a new three-letter word, such as FAT, even in the middle of one of the original words, most of the sentence still makes sense, such as THE DOG SAF ATW THE CAT SIT AND THE FOX RUN.

Substitutions

In a **nucleotide substitution** (also called a **point mutation**), a single base pair in DNA is changed. A substitution within a protein-coding gene can produce one of four possible outcomes. To illustrate these outcomes, let's consider substitutions that occur in the gene encoding beta-globin, one of the subunits of hemoglobin, the oxygen-carrying protein in red blood cells (Table 12-4). The other type of subunit in hemoglobin is called alpha-globin; a normal hemoglobin molecule consists of two alpha and two beta subunits. In all but the last example, we will consider the results of mutations that occur in the sixth codon of the beta-globin gene (CTC in DNA, GAG in mRNA), which specifies glutamic acid—a charged, hydrophilic, water-soluble amino acid.

- **The amino acid sequence of the protein may be unchanged.** Recall that most amino acids can be encoded by several different codons. If a substitution mutation changes the beta-globin DNA base sequence from CTC to CTT, this sequence still codes for glutamic acid. Therefore, the protein synthesized from the mutated gene remains unchanged.
- **Protein function may be unchanged.** Many proteins have regions in which the exact amino acid sequence is relatively unimportant. In beta-globin, the amino acids on the outside of the protein must be hydrophilic to keep the protein dissolved in the cytoplasm of red blood cells. Exactly *which* hydrophilic amino acids are on the outside doesn't matter much. Substitutions in which the resulting amino acid is the same or functionally equivalent are

	DNA (Template Strand)	mRNA	Amino Acid	Properties of Amino Acid	Functional Effect on Protein	Disease
TABLE 12-4	**Effects of Mutations in the Hemoglobin Gene**					
Original codon 6	CTC	GAG	Glutamic acid	Hydrophilic	Normal protein function	None
Mutation 1	CTT	GAA	Glutamic acid	Hydrophilic	Neutral; normal protein function	None
Mutation 2	GTC	CAG	Glutamine	Hydrophilic	Neutral; normal protein function	None
Mutation 3	CAC	GUG	Valine	Hydrophobic	Loses water solubility; compromises protein function	Sickle-cell anemia
Original codon 17	TTC	AAG	Lysine	Hydrophilic	Normal protein function	None
Mutation 4	ATC	UAG	Stop codon	Ends translation after amino acid 16	Synthesizes only part of the protein; eliminates protein function	Beta-thalassemia

called **neutral mutations** because they do not detectably change the function of the encoded protein.

- **Protein function may be changed by an altered amino acid sequence.** A mutation from CTC to CAC replaces glutamic acid (hydrophilic) with valine (hydrophobic). This substitution is the genetic defect that causes sickle-cell anemia (see Chapter 10). The hydrophobic valines on the outside of the hemoglobin molecules cause them to clump together, distorting the shape of the red blood cells and causing serious illness.

- **Protein function may be destroyed by a premature stop codon.** A particularly catastrophic mutation occasionally occurs in the 17th codon of the beta-globin gene (TTC in DNA, AAG in mRNA). This codon specifies the amino acid lysine. A mutation from TTC to ATC (UAG in mRNA) results in a stop codon, halting translation of beta-globin mRNA before the protein is completed. People who inherit this mutant allele from both parents do not synthesize any functional beta-globin protein; they manufacture hemoglobin consisting entirely of alpha-globin subunits. This "pure alpha" hemoglobin does not bind oxygen very well. People with this condition, called beta-thalassemia, require regular blood transfusions throughout life.

CHECK YOUR LEARNING

Can you describe the different types of mutations? Can you use your knowledge of the genetic code to explain why different mutations can have such different effects on protein function?

CASE STUDY continued

Cystic Fibrosis

Cystic fibrosis is the most common lethal genetic disease among people of European descent, about 4% of whom are carriers. Why hasn't natural selection eliminated defective *CFTR* alleles? One hypothesis is that mutated *CFTR* alleles protect against cholera and typhoid. Cholera toxin activates the normal CFTR protein, causing excessive movement of chloride and sodium from intestinal cells into the cavity of the intestine. Water follows by osmosis, causing massive diarrhea that kills by dehydration and loss of salts. Defective CFTR proteins cannot be activated by cholera toxin. The CFTR protein is also the site by which typhoid bacteria enter cells, but typhoid bacteria cannot enter via mutated CFTR proteins. Obviously, such protection cannot compensate for the fatal effects of cystic fibrosis in homozygotes. However, heterozygotes, with one normal and one mutated *CFTR* allele, have nearly normal CFTR function, and might be less severely affected by cholera and typhoid. This "heterozygote advantage" may explain the high frequency of mutated *CFTR* alleles, as it does for the sickle-cell anemia allele (see Chapter 10). However, hypotheses that are biologically reasonable may not be correct: So far, epidemiological studies have failed to convincingly demonstrate a selective advantage of mutated *CFTR* alleles in human populations.

12.5 HOW IS GENE EXPRESSION REGULATED?

The complete human genome contains 20,000 to 25,000 genes. All of these genes are present in almost every body cell, but any individual cell expresses (transcribes and, if the gene product is a protein, translates) only a small fraction of them. Some genes are expressed in all cells, because they encode proteins or RNA molecules that are essential for the life of any cell. For example, all cells need to synthesize proteins, so they all transcribe the genes for tRNA, rRNA, and ribosomal proteins. Other genes are expressed exclusively in certain types of cells, at certain times in an organism's life, or under specific environmental conditions. For example, even though every cell in your body contains the gene for the milk protein casein, that gene is expressed only in women, only in certain breast cells, and only when a woman is breast-feeding.

Regulation of gene expression may occur at the level of transcription (which genes are used to make mRNA in a given cell), translation (how much protein is made from a particular type of mRNA), or protein activity (how long the protein lasts in a cell and how rapidly protein enzymes catalyze specific reactions). Although these general principles apply to both prokaryotic and eukaryotic organisms, there are some differences as well.

In Prokaryotes, Gene Expression Is Primarily Regulated at the Level of Transcription

Bacterial DNA is often organized in packages called **operons,** in which the genes for related functions lie close to one another (**Fig. 12-9a**). An operon consists of four parts: (1) a **regulatory gene,** which controls the timing or rate of transcription of other genes; (2) a promoter, which RNA polymerase recognizes as the place to start transcribing; (3) an **operator,** which governs the access of RNA polymerase to the promoter or to the (4) **structural genes,** which actually encode the related enzymes or other proteins. Whole operons are regulated as units, so that proteins that work together to perform a specific function may be synthesized simultaneously when the need arises.

Prokaryotic operons may be regulated in a variety of ways. Some operons encode enzymes that are needed by the cell just about all the time, such as the enzymes that synthesize many amino acids. Such operons are usually transcribed continuously, unless the bacterium encounters a surplus of that particular amino acid. Other operons encode enzymes that are needed only occasionally, for instance, to digest a relatively rare food. They are transcribed only when the bacterium encounters that food.

Consider the common intestinal bacterium, *Escherichia coli (E. coli).* This bacterium must live on whatever types of nutrients its host eats, and it can synthesize many different enzymes to metabolize a wide variety of foods.

regulatory gene: codes for repressor protein

operator: repressor protein binds here

promoter: RNA polymerase binds here

structural genes that code for enzymes of lactose metabolism

(a) Structure of the lactose operon

RNA polymerase

transcription blocked

a repressor protein bound to the operator site overlaps the promoter

free repressor proteins

(b) Lactose absent

RNA polymerase binds to the promoter and transcribes the structural genes

lactose bound to repressor proteins

lactose-metabolizing enzymes are synthesized

(c) Lactose present

▲ **FIGURE 12-9   Regulation of the lactose operon (a)** The lactose operon consists of a regulatory gene, a promoter, an operator, and three structural genes that code for enzymes necessary for lactose metabolism. **(b)** In the absence of lactose, repressor proteins bind to the operator of the lactose operon. RNA polymerase can bind to the promoter but cannot move past the repressor protein to transcribe the structural genes. **(c)** When lactose is present, it binds to the repressor proteins, making the repressor proteins unable to bind to the operator. RNA polymerase binds to the promoter, moves past the unoccupied operator, and transcribes the structural genes.

The genes that code for these enzymes are transcribed only when the enzymes are needed. The enzymes that metabolize lactose, the principal sugar in milk, are a case in point. The **lactose operon** contains three structural genes, each coding for an enzyme that aids in lactose metabolism (see Fig. 12-9a).

The lactose operon is shut off, or repressed, unless activated by the presence of lactose. The regulatory gene

of the lactose operon directs the synthesis of a **repressor protein** that binds to the operator site. RNA polymerase, although still able to bind to the promoter, cannot get past the repressor protein to transcribe the structural genes. Consequently, lactose-metabolizing enzymes are not synthesized (**Fig. 12-9b**).

When *E. coli* colonize the intestines of a newborn mammal, however, they find themselves bathed in lactose whenever their host nurses from its mother. Lactose molecules enter the bacteria and bind to the repressor proteins, changing their shape (**Fig. 12-9c**). The lactose–repressor complex cannot attach to the operator site. Therefore, when RNA polymerase binds to the promoter of the lactose operon, it transcribes the genes for lactose-metabolizing enzymes, allowing the bacteria to use lactose as an energy source. After the young mammal is weaned, it usually does not consume milk again. The intestinal bacteria no longer encounter lactose, the repressor proteins bind to the operator, and the genes for lactose metabolism are shut off.

In Eukaryotes, Gene Expression Is Regulated at Many Levels

In some respects, regulation of gene expression in eukaryotes is similar to regulation of expression in prokaryotes. In both, not all genes are transcribed and translated all the time. Further, controlling the rate of transcription is an important mechanism of gene regulation in both. However, eukaryotes sequester their DNA in a membrane-bound nucleus, have very different organization of both their genome and their individual genes, and cut and splice their RNA transcripts in complex ways. Further, multicellular eukaryotes have a variety of cell types, each performing distinct functions, in a single body. These differences require more sophisticated regulation of gene expression.

Gene expression in a eukaryotic cell is a multistep process, beginning with transcription of DNA and commonly ending with a protein that performs a particular function. Regulation of gene expression can occur at any of these steps, as shown in **Figure 12-10**:

❶ **Cells can control the frequency at which a gene is transcribed.** The rate of transcription of specific genes differs among organisms, among cell types in a given organism, within a given cell at different stages in the organism's life, and within a cell or organism depending on environmental conditions (see the following section).

❷ **The same gene may be used to produce different mRNAs and protein products.** The same gene may be used to produce several different protein products (as we described in Section 12.3), depending on how the pre-mRNA is spliced to form the finished mRNA that will be translated on the ribosomes.

❸ **Cells may control the stability and translation of mRNAs.** Some mRNAs are long lasting and are

◀ FIGURE 12-10 An overview of information flow in a eukaryotic cell

translated into protein many times. Others are translated only a few times before they are degraded. Recently, molecular biologists have discovered that certain small RNA molecules may block translation of some mRNAs or may target some mRNAs for destruction (see the following section).

❹ **Proteins may need to be modified before they can carry out their functions.** For instance, the protein-digesting enzymes produced by cells in your stomach wall and pancreas are initially synthesized in an inactive form, which prevents the enzymes from digesting the cells that produce them. After these inactive forms are secreted into the digestive tract, portions of the enzymes are snipped out to unveil the active site, allowing the enzymes to digest the proteins in food into smaller proteins or individual amino acids. Other modifications, such as adding or removing phosphate

groups, can temporarily activate or inactivate a protein, allowing second-to-second control of the protein's activity. Similar regulation of protein structure and function occurs in prokaryotic cells.

❺ **Cells can control the rate at which proteins are degraded.** By preventing or speeding up a protein's degradation, a cell can rapidly adjust the amount of a particular protein it contains. Protein degradation is also regulated in prokaryotic cells.

Let's examine some of these regulatory mechanisms in more detail.

Regulatory Proteins Binding to a Gene's Promoter Alter Its Rate of Transcription

The promoter regions of virtually all genes contain several different transcription factor binding sites, or response elements.

Health | Watch

A Case of Mistaken Gender

Sometime between 7 and 14 years of age, a girl usually goes through puberty: Her breasts swell, her hips widen, and she begins to menstruate. In rare instances, however, a girl may develop all of the outward signs of womanhood, but without menstruation. How can that be? If her physician performs a chromosome test, in some cases the results seem to be impossible: The girl's sex chromosomes are XY. The reason she has not begun to menstruate is that she lacks ovaries and a uterus but instead has testes inside her abdominal cavity. She has about the same concentrations of androgens (male sex hormones, such as testosterone) circulating in her blood as would be found in a boy her age. In fact, androgens have been present since early in her development. However, her cells cannot respond to them—a condition called androgen insensitivity.

The affected gene codes for a protein known as an androgen receptor. In normal males, androgens bind to the receptor proteins, stimulating the transcription of genes that help to produce many male features, including the formation of a penis and the descent of the testes into sacs outside the body cavity. Androgen insensitivity is caused by defective androgen receptors. There are more than 200 mutant alleles of the androgen receptor gene. The most serious are mutations that create a premature stop codon. As you know, these mutations are likely to have catastrophic effects on protein structure and function.

Because the androgen receptor gene is on the X chromosome, a person who is genetically male (XY) inherits a single allele for the androgen receptor. If this allele is seriously defective, the person will not synthesize functional androgen receptor proteins. The person's cells will be unable to respond to testosterone, and male characteristics will not develop. In many respects, female development is the "default" option in humans, and without

functional androgen receptors, the affected person's body will develop female characteristics. Thus, a mutation that changes the nucleotide sequence of a single gene, causing a single type of defective protein to be produced, can cause a person who is genetically male to look like and perceive herself to be female (**Fig. E12-2**).

▲ **FIGURE E12-2 Androgen insensitivity leads to female features** This person has an X and a Y chromosome. She has testes that produce testosterone, but a mutation in her androgen receptor genes makes her cells unable to respond to testosterone, resulting in her female appearance.

Therefore, whether these genes are transcribed depends on which transcription factors are synthesized by the cell and whether those transcription factors are active. For example, when cells are exposed to free radicals (see Chapter 2), a transcription factor binds to antioxidant response elements in the promoters of several genes. As a result, the cell produces enzymes that break down free radicals to harmless substances.

Many transcription factors require activation before they can affect gene transcription. One of the best-known examples is the role that the sex hormone, estrogen, plays in controlling egg production in birds. The gene for albumin, the major protein in egg white, is not transcribed in winter when birds are not breeding and estrogen levels are low. During the breeding season, the ovaries of female birds release estrogen, which enters cells in the oviduct and binds to a transcription factor (usually called an estrogen receptor). The complex of estrogen and its receptor then attaches to an estrogen response element in the promoter of the

albumin gene, making it easier for RNA polymerase to bind to the promoter and start transcribing mRNA. The mRNA is translated into large amounts of albumin. Similar activation of gene transcription by steroid hormones occurs in other animals, including humans. The importance of hormonal regulation of transcription during development is illustrated by genetic defects in which receptors for sex hormones are nonfunctional (see "Health Watch: A Case of Mistaken Gender").

Epigenetic Controls Alter Gene Transcription and Translation

Epigenetics (which means "in addition to genetics") is the study of how cells and organisms change gene expression and function without changing the base sequence of their DNA. There is disagreement about what processes should be considered to be epigenetic. In general, however, epigenetic control works in three ways: (1) modification of DNA; (2) modification of chromosomal proteins; and (3) changing

? HAVE **YOU** EVER WONDERED ...

Why Bruises Turn Colors?

Bruises typically progress from purple to green to yellow. This sequence is visual evidence of the control of gene expression. If you bang your shin on a chair, blood vessels break and release red blood cells, which burst and spill their hemoglobin. Hemoglobin and its iron-containing heme group are dark bluish-purple in the deoxygenated state, so fresh bruises are purple. Heme, which is toxic to the liver, kidneys, brain, and blood vessels, stimulates transcription of the heme oxygenase gene. Heme oxygenase is an enzyme that converts heme to biliverdin, which is green. A second enzyme, which is always present because its gene is always expressed, converts biliverdin to bilirubin, which is yellow. The bruise finally disappears as bilirubin moves to the liver, which secretes it into the bile. You can follow the detoxification of heme by watching your bruise change color.

transcription and translation through the actions of several types of RNA collectively called noncoding, or regulatory, RNA. Many types of epigenetic controls can be inherited from parent to daughter cell during mitotic cell division. Some may even be inherited from one generation to the next, as we explore in "Health Watch: The Strange World of Epigenetics."

Epigenetic Modification of DNA May Suppress Transcription Certain enzymes in a cell add methyl groups ($-CH_3$) to cytosine bases in specific locations in the cell's DNA, a process called *methylation*. If a gene or its promoter has lots of methyl groups, the gene usually will not be copied into mRNA, and so its instructions will not be used to make proteins. The number and location of methyl groups on DNA are important in normal development and in some diseases, such as cancer. In cancer cells, growth factor genes (see Chapter 9) often have too few methyl groups. This can cause the genes to be transcribed at a very high level, producing high concentrations of growth factors that inappropriately stimulate cell division. If tumor suppressor genes have too many methyl groups, shutting down their transcription, the body is robbed of one of its most effective weapons against cancer. Defective epigenetic control has also been implicated in disorders as varied as heart disease, obesity, infertility, and type 2 (adult-onset) diabetes.

Epigenetic Modification of Histones May Enhance Transcription In eukaryotic chromosomes, DNA is wound around "spools" made of proteins called *histones* (see Chapter 9). When the DNA is tightly wound, RNA polymerase can't get to the promoters of genes, so transcription occurs

slowly, if at all. However, when acetyl groups ($-COCH_3$) are added to histones, the DNA partially unwinds and RNA polymerase has better access to the promoters, making gene transcription easier.

Noncoding RNA May Alter Transcription or Translation

Protein-coding genes make up only a small percentage of human DNA. Does that mean that the rest of our DNA is pointless? Far from it. Recently, molecular biologists have found that some of this DNA is transcribed into *noncoding RNA*, which may exert major effects on gene expression.

MicroRNA and RNA Interference As you know, mRNA is transcribed from DNA and translated into protein, which then performs cellular functions such as catalyzing biochemical reactions. How much protein is synthesized depends both on how much mRNA is made and on how rapidly and for how long mRNA is translated. Enter RNA interference. The DNA of organisms as diverse as plants, roundworms, and people contains hundreds of genes that do not code for proteins, but that are nevertheless transcribed into RNA. In some cases, RNA polymerase transcribes a long piece of noncoding RNA. Cellular enzymes cut the RNA transcript into very short RNA strands appropriately named **microRNA.** Each microRNA is complementary to part of a specific mRNA. These microRNA molecules interfere with translation of the mRNA (hence, the name "RNA interference"). In some cases, these small RNA strands base-pair with the complementary mRNA, forming a little section of double-stranded RNA that cannot be translated. In other cases, the short RNA strands combine with enzymes to cut up complementary mRNA, which also prevents translation.

Why should a cell interfere with the translation of its own mRNA? RNA interference is important for the development of eukaryotic organisms. For example, in mammals microRNAs influence the development of the heart and brain, the secretion of insulin by the pancreas, and even learning and memory. Unfortunately, defects in microRNA production—either too much or too little of certain microRNAs—can lead to cancer or heart disease.

Some organisms use microRNA to defend against disease. Many plants produce microRNA that is complementary to the nucleic acids (usually RNA) of plant viruses. When the microRNA finds complementary viral RNA molecules, it directs enzymes to cut up the viral RNA, thereby preventing viral reproduction.

Altering Transcription with Noncoding RNA

Other types of noncoding RNA can affect gene transcription. Some noncoding RNAs inhibit the binding of RNA polymerase to specific gene promoters, thereby blocking transcription. Other noncoding RNAs stimulate or inhibit epigenetic changes to DNA or histones in specific locations on specific chromosomes. These noncoding RNAs may enhance

Health Watch

The Strange World of Epigenetics

We have described many ways in which cells control gene expression. Most of these work for times ranging from a few seconds to a few days and then fade away. Epigenetic controls, however, often work for the life of an organism. Some may even be passed down from parent to offspring.

Epigenetic controls are important in ensuring proper control of gene transcription and translation. For example, adding methyl groups to the promoter of the insulin gene turns off transcription. All of the cells of early embryos have methylated, silenced insulin genes. Later in development, methyl groups on the insulin gene are selectively removed in cells destined to become insulin-secreting cells of the pancreas. Every other cell of the body has methylated, silenced insulin genes. Some cells, such as those in the intestinal lining, divide every day or two—thousands of divisions during a lifetime. Throughout all of these divisions, the insulin genes remain methylated. How? Recall that DNA replication is semiconservative (see Chapter 11). When an intestinal cell divides, both daughter cells inherit one parental DNA strand with methyls on the insulin gene and one new DNA strand without methyls on the gene. However, an enzyme in the daughter cells adds the parental methyl pattern to the daughter DNA strand. The result is that all intestinal cells have silenced insulin genes.

In the vast majority of cases, methyl patterns on DNA are erased during meiotic cell division or gamete development, so epigenetic changes do not pass from generation to generation. However, there are some exceptions. Methyl groups may be added to certain clusters of genes in either the sperm or the egg, resulting in genomic imprinting, in which a given gene will be expressed only if it is inherited from either the father or the mother, respectively. For example, the large, meaty buttocks of sheep called callipyge (Greek for "beautiful butt") develop only if the callipyge mutation is inherited from the father. In humans, Angelman syndrome, a rare genetic disease characterized by seizures, speech defects, and motor disabilities, is the result of a deletion mutation in chromosome 15. Angelman syndrome occurs only when the mutation has been inherited from the mother. The normal, functional genes on the father's chromosome are silenced by methyl groups, and cannot compensate for the mother's mutation.

Epigenetic changes that last for many generations have been documented in bacteria, protists, fungi, plants, and animals. This leads to a provocative question: In people, can epigenetic changes caused by parents' life experiences or environment become a long-lasting, even permanent, part of the inheritance of their offspring?

For now, the answer seems to be "Maybe." No one can perform controlled, multigenerational experiments on people, so good data are hard to obtain. Therefore, evidence for multigenerational epigenetic inheritance in humans will most likely be a "natural experiment" in which some major event affected a fairly large number of people. One such natural experiment has already occurred in a remote northern area of Sweden called Norrbotten. Until fairly recent times, Norrbotten was extremely isolated. Little food entered or left the region. If crop harvests were good, people stuffed themselves during the following winter; if harvests were bad, people starved. Researchers tracked birth and death records, and correlated those with harvests during the 1800s. They found that the grandsons of boys who lived during the years of abundant harvests, and therefore probably overate, lived remarkably shorter lives—from 6 to 32 years shorter, depending on how the data were analyzed—than the grandsons of boys who suffered through winters of near starvation. Similar effects were found in girls. The results are intriguing, but no one knows what genes were involved or whether epigenetic methyl groups on DNA caused the difference.

---

or reduce transcription, depending on the exact nature of the epigenetic controls that are affected.

Perhaps the best-known noncoding RNA silences transcription in mammalian X chromosomes. As you know, male mammals have an X and a Y chromosome (XY), and females have two X chromosomes (XX). As a consequence, females have the capacity to synthesize mRNA from genes on their two X chromosomes, whereas males, with only one X chromosome, may produce only half as much. In 1961, Mary Lyon, an English geneticist, hypothesized that one of the two X chromosomes in females is inactivated in some way, so that its genes are not expressed. Subsequent research showed that she was correct. In female mammals, one of the X chromosomes is inactivated, and about 85% of its genes are not transcribed. Early in embryonic development (about the 16th day in humans), one X chromosome in each of a female's cells begins to produce large amounts of a noncoding RNA molecule called Xist. Xist RNA coats most of the X chromosome, condenses it into a tight mass, and prevents transcription. The condensed X chromosome, called a **Barr body** after its discoverer, Murray Barr, appears as a discrete spot in the nuclei of the cells of female mammals (Fig. 12-11).

Usually, large clusters of cells (with each cluster descended from a single cell of the early embryo) have the same X chromosome inactivated. As a result, the bodies of female mammals are composed of patches of cells in which one of the X chromosomes is fully active and patches of cells in which the other X chromosome is active. The results of this phenomenon are easily observed in calico cats (Fig. 12-12). The X chromosome of a cat contains a gene encoding an enzyme that produces fur pigment. There are two common alleles of this gene: One produces orange fur

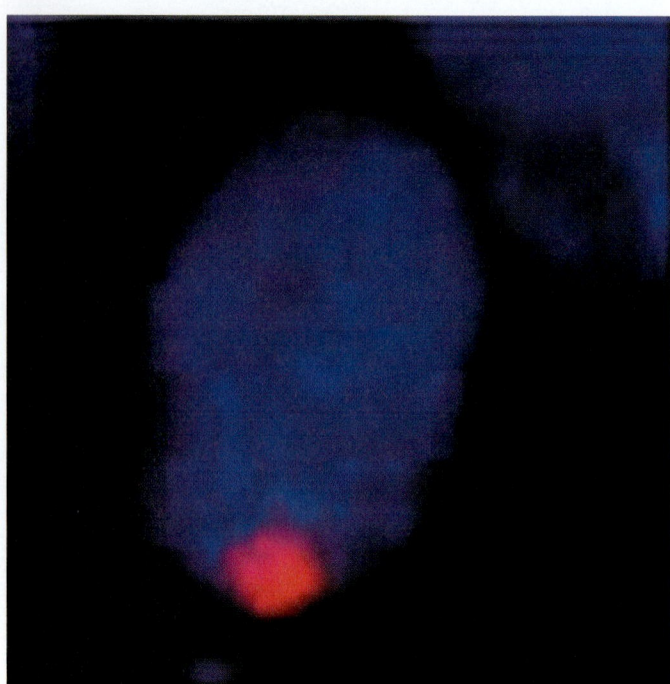

▲ **FIGURE 12-11 A Barr body** The red spot at the bottom of the nucleus is an inactivated X chromosome called a Barr body. In this fluorescence micrograph, the Barr body is stained with a dye that binds to the Xist RNA coating the inactivated X chromosome.

and the other produces black fur. If one X chromosome in a female cat has the orange allele and the other X chromosome has the black allele, the cat will have patches of orange and black fur. These patches represent areas of skin that developed from cells in the early embryo in which different X chromosomes were inactivated. Calico coloring is almost exclusively found in female cats. Because male cats

▲ **FIGURE 12-12 Inactivation of the X chromosome regulates gene expression** This female calico kitten carries a gene for orange fur on one X chromosome and a gene for black fur on her other X chromosome. Inactivation of different X chromosomes produces the black and orange patches. The white color is due to an entirely different gene that prevents pigment formation altogether.

normally have only one X chromosome, which is active in all of their cells, male cats usually have black fur or orange fur, but not both.

CHECK YOUR LEARNING

Can you describe the ways in which information flow from DNA to RNA to protein synthesis to protein function can be regulated? Do you understand which controls over gene expression are likely to be very brief, which may be long lasting, and why they differ?

CASE STUDY revisited

Cystic Fibrosis

All of the defective alleles of the *CFTR* gene are recessive to the functional *CFTR* allele. People who are heterozygous, with one normal *CFTR* allele and one copy of any of the defective alleles, produce enough functional CFTR protein for adequate chloride transport. Therefore, they produce normal, watery secretions in their lungs and do not develop cystic fibrosis. Someone with two defective alleles will produce only proteins that don't work properly and will develop the disease.

Genetic diseases such as cystic fibrosis can't be "cured" in the way that an infection can be cured by killing offending bacteria or viruses. Typically, genetic diseases are treated by replacing the lost function, such as by giving insulin to diabetics, or by relieving the symptoms. In cystic fibrosis, the most common treatments relieve some of the symptoms. These treatments include antibiotics, medicines that open the airways, and physical therapy to drain the lungs.

What ultimately happens to a person with cystic fibrosis depends on how defective the mutant alleles are. Canadian triathlete Lisa Bentley, for example, has a relatively mild case of cystic fibrosis (**Fig. 12-13**). However, during a

▲ **FIGURE 12-13 Lisa Bentley, sometimes called the Iron Queen, wins another triathlon.**

9-hour triathlon, she produces copious amounts of very salty sweat. It is a constant challenge for Bentley to keep her body supplied with salt during a race. Nevertheless, Bentley has won 11 triathlons, including the Australian Ironman Triathlon five straight years, from 2002 through 2006. Bentley carefully controls her diet, especially her salt intake. Vigorous exercise helps to clear out her lungs. She also scrupulously avoids situations where she might be exposed to contagious diseases. Her cystic fibrosis hasn't kept her from becoming one of the finest athletes in the world.

Consider This

BioEthics Most biomedical research in the United States is funded by government agencies, such as the National Institutes of Health (NIH). The NIH funds research on

genetic disorders such as cystic fibrosis and muscular dystrophy; infectious diseases such as AIDS and tuberculosis; and a host of other conditions, including heart disease and cancer. The NIH also supports research on diseases, such as malaria, that are uncommon in the United States but claim hundreds of thousands of lives each year in impoverished countries. How do you think that NIH funds should be spent—in proportion to disease severity and incidence in the United States, which might mean that almost all NIH funds should be spent on heart disease and cancer; or according to the prospective life span of the victims, so that diseases of the young, such as cystic fibrosis, receive much more? Does the United States have a responsibility to develop treatments for people in other countries suffering from diseases that are rare in America?

CHAPTER REVIEW

Summary of Key Concepts

12.1 How Is the Information in DNA Used in a Cell?

Genes are segments of DNA that can be transcribed into RNA and, for most genes, translated into protein. Transcription produces the three types of RNA needed for translation: mRNA, tRNA, and rRNA. Messenger RNA carries the genetic information of a gene from the nucleus to the cytoplasm, where ribosomes use the information to synthesize a protein. Ribosomes are composed of rRNA and proteins, organized into large and small subunits. There are many different tRNAs. Each tRNA binds a specific amino acid and carries it to a ribosome for incorporation into a protein. The genetic code consists of codons, sequences of three bases in mRNA that specify the start of translation (start codon), the amino acids in the protein chain, or the end of protein synthesis (stop codons).

12.2 How Is the Information in a Gene Transcribed into RNA?

Within an individual cell, only certain genes are transcribed. When the cell requires the product of a gene, RNA polymerase binds to the promoter region of the gene and synthesizes a single strand of RNA. This RNA is complementary to the template strand in the gene's DNA double helix. Cellular proteins called transcription factors may bind to DNA near the promoter and enhance or suppress transcription of a given gene.

12.3 How Is the Base Sequence of mRNA Translated into Protein?

In prokaryotic cells, all of the nucleotides of a protein-coding gene code for amino acids; therefore, the RNA transcribed from the gene is the mRNA that will be translated on a ribosome. In eukaryotic cells, protein-coding genes consist of two parts: exons, which code for amino acids in a protein, and introns, which do not. The entire gene, including both introns and

exons, is transcribed into a pre-mRNA molecule. The introns of the pre-mRNA are cut out and the exons are spliced together to produce a finished mRNA.

In eukaryotes, mRNA carries the genetic information from the nucleus to the cytoplasm, where ribosomes use this information to synthesize a protein. The two ribosomal subunits come together at the start codon of the mRNA molecule to form the complete protein-synthesizing assembly. Transfer RNAs deliver the appropriate amino acids to the ribosome for incorporation into the growing protein. Which tRNA binds and, consequently, which amino acid is delivered depend on base pairing between the anticodon of the tRNA and the codon of the mRNA. Two tRNAs, each carrying an amino acid, bind simultaneously to the ribosome; the large subunit catalyzes the formation of peptide bonds between the amino acids. As each new amino acid is attached, one tRNA detaches, and the ribosome moves over one codon, binding to another tRNA that carries the next amino acid specified by mRNA. Addition of amino acids to the growing protein continues until a stop codon is reached, causing the ribosome to disassemble and to release both the mRNA and the newly formed protein.

12.4 How Do Mutations Affect Protein Structure and Function?

A mutation is a change in the nucleotide sequence of a gene. Mutations can be caused by mistakes in base pairing during replication, by chemical agents, and by environmental factors such as radiation. Common types of mutations include inversions, translocations, deletions, insertions, and substitutions (point mutations). Mutations may be neutral or harmful, but in rare cases a mutation will promote better adaptation to the environment and thus will be favored by natural selection.

12.5 How Is Gene Expression Regulated?

The expression of a gene requires that it be transcribed and translated and that the resulting protein perform some action within the cell. Which genes are expressed in a cell at any given time is regulated by the function of the cell, the developmental stage of the organism, and the environment. Control of gene regulation can occur at many steps. The amount of mRNA synthesized from a particular gene can be regulated by increasing or decreasing the rate of its transcription, as well as by changing the stability of the mRNA itself. Rates of translation of mRNAs can also be regulated. Regulation of transcription and translation affects how many protein molecules are produced from a particular gene. Even after they are synthesized, many proteins must be modified before they can function. Proteins also vary in how rapidly they are degraded in a cell. In epigenetic regulation, adding methyl groups to DNA often suppresses gene transcription, whereas adding acetyl groups to histones increases gene transcription. Noncoding RNA may suppress transcription, speed up mRNA degradation, or inhibit translation of mRNA.

Key Terms

anticodon 211	operon 222
Barr body 227	point mutation 221
codon 210	promoter 215
deletion mutation 221	regulatory gene 222
epigenetics 225	repressor protein 223
exon 217	ribonucleic acid (RNA) 210
genetic code 212	ribosomal RNA (rRNA) 210
insertion mutation 221	ribosome 210
intron 217	RNA polymerase 215
inversion 221	start codon 214
lactose operon 223	stop codon 214
messenger RNA (mRNA) 210	structural gene 222
microRNA 226	template strand 215
mutation 220	transcription 212
neutral mutation 222	transfer RNA (tRNA) 210
nucleotide substitution 221	translation 212
operator 222	translocation 221

Learning Outcomes

In this chapter, you have learned to . . .

LO1 Describe the molecules and cellular structures that play central roles in the process by which information encoded in DNA is used to synthesize proteins.

LO2 Compare and contrast how information is encoded in DNA and in RNA.

LO3 Describe the functions of transcription and translation and identify the locations in the cell at which they occur.

LO4 Describe the main steps of transcription, including the molecules involved and how they interact.

LO5 Describe the main steps of translation, including the molecules involved and how they interact.

LO6 Describe the different types of mutations and explain their potential effects on protein synthesis.

LO7 Describe how gene expression is regulated.

Thinking Through the Concepts

Fill-in-the-Blank

1. Synthesis of RNA from the instructions in DNA is called _____. Synthesis of a protein from the instructions in mRNA is called _____. Which structure in the cell is the site of protein synthesis? _____ **LO1 LO3**

2. The three types of RNA that are essential for protein synthesis are_____, _____, and_____. Another type of RNA, which can interfere with translation, is called _____. **LO2 LO3 LO7**

3. The genetic code uses _____ (how many?) bases to code for a single amino acid. This short sequence of bases in mRNA is called a(n) _____. The complementary sequence of bases in tRNA is called a(n) _____. **LO2**

4. The enzyme _____ synthesizes RNA from the instructions in DNA. DNA has two strands, but for any given gene, only one strand, called the _____ strand, is transcribed. To begin transcribing a gene, this enzyme binds to a specific sequence of DNA bases located at the beginning of the gene. This DNA sequence is called the _____. Transcription ends when the enzyme encounters a DNA sequence at the end of the gene called the _____. **LO4**

5. Protein synthesis begins when mRNA binds to a ribosome. Translation begins with the _____ codon of mRNA and continues until a(n) _____ codon is reached. Individual amino acids are brought to the ribosome by _____. These amino acids are linked into protein by _____ bonds. **LO5**

6. There are several different types of mutations in DNA. If one nucleotide is substituted for another, this is called a(n) _____ mutation. _____ mutations occur if nucleotides are added in the middle of a gene. _____ mutations occur if nucleotides are removed from the middle of a gene. **LO6**

Review Questions

1. How does RNA differ from DNA? **LO2**

2. Name the three types of RNA that are essential to protein synthesis. What is the function of each? **LO2 LO3 LO5**

3. Define the following terms: *genetic code, codon,* and *anticodon.* What is the relationship among the bases in DNA, the codons of mRNA, and the anticodons of tRNA? **LO2 LO5**

4. How is mRNA formed from a eukaryotic gene? **LO4**

5. Diagram and describe protein synthesis. **LO5**

6. Explain how complementary base pairing is involved in both transcription and translation. **LO4 LO5**

7. Describe the principal mechanisms of regulating gene expression. **LO7**

8. Define *mutation.* Are most mutations likely to be beneficial or harmful? Explain your answer. **LO6**

Applying the Concepts

1. **BioEthics** As you have learned in this chapter, many factors influence gene expression, including hormones. The use of anabolic steroids and growth hormones among athletes has created controversy in recent years. Hormones certainly affect gene expression, but, in the broadest sense, so do vitamins and foods. What do you think are appropriate guidelines for the use of hormones? Should athletes take steroids or growth hormones? Should children at risk of being unusually short be given growth hormones? Should parents be allowed to request growth hormones for a child of normal height in the hope of producing a future basketball player?

2. Many years ago, some researchers reported that they could transfer learning from one animal (a flatworm) to another by feeding trained animals to untrained animals. Further, they claimed that RNA was the active molecule of learning. Given your knowledge of the roles of RNA and protein in cells, do you think that a *specific* memory (for example, remembering the base sequences of codons of the genetic code) could be encoded by a *specific* molecule of RNA and that this RNA molecule could transfer that memory to another person? In other words, in the future, could you learn biology by popping an RNA pill? If so, how would this work? If not, can you propose a reasonable hypothesis for the results with flatworms? How would you test your hypothesis?

3. Androgen insensitivity is inherited as a simple recessive trait, because one copy of the normal androgen receptor allele produces sufficient amounts of androgen receptors. Given this information and your knowledge of the chromosomal basis of inheritance, can androgen insensitivity be inherited, or must it arise as a new mutation each time it occurs? If it can be inherited, would inheritance be through the mother or through the father? Why?

Answers to Figure Caption questions and Fill-in-the-Blank questions can be found in the Answers section at the back of the book.

MB *Go to MasteringBiology for practice quizzes, activities, eText, videos, current events, and more.*

DNA profiling proved that Dennis Maher, shown here with attorney Aliza Kaplan, was innocent of the crimes for which he spent 19 years in prison.

CASE STUDY
Guilty or Innocent?

"IT'S OK TO CRY," Innocence Project attorney Aliza Kaplan told Dennis Maher, on his way to court in 2003. Maher seemed calm as District Attorney Martha Coakley asked the judge to dismiss the charges for which Maher had been imprisoned for over 19 years. The judge ordered Maher's immediate release. Maher and his family hugged and wept in the hall outside.

In 1984, Maher had been sentenced to life imprisonment for two counts of rape and one of attempted rape. As it turned out, his only crimes were living in the vicinity of the rapes, wearing a red sweatshirt, and looking like the real assailant. All three victims picked Maher out of lineups. How can three people all identify the wrong man? It was dark, the assaults were swift, and, obviously, the victims were tremendously stressed. Contrary to popular belief, eyewitness testimony is very unreliable. Studies have found that eyewitness identification is wrong about 35% to 80% of the time, depending on the conditions used in the experiments.

You have probably already guessed what led to Maher's eventual exoneration—DNA evidence. In 1993, while watching the *Phil Donahue Show* in prison, Maher learned about the Innocence Project, which seeks to exonerate people who have been wrongly convicted of serious crimes. The Innocence Project agreed to help, but hit a brick wall—no biological evidence was available for any of the cases.

Finally, 7 years later, an Innocence Project law student found the semen-stained underwear of one of the rape victims, lost in a box in a courthouse storage room. A few months later, a semen specimen from the second rape turned up as well. DNA profiling proved that Maher was not the assailant in either case.

In this chapter, we'll investigate the techniques of biotechnology that now pervade much of modern life. How do crime scene investigators decide that two DNA samples match? How can biotechnology diagnose inherited disorders? Should biotechnology be used to change the genetic makeup of crops, livestock, or even people?

AT A GLANCE

LEARNING GOALS

LG1 Describe natural processes that recombine DNA.

LG2 Define polymerase chain reaction, short tandem repeats, gel electrophoresis, and DNA probes, and describe how electrophoresis and the polymerase chain reaction work.

LG3 Describe how DNA profiles are generated, and explain how they can be used to identify individual people.

LG4 List the purposes for genetically modifying organisms, and describe the steps involved in creating a genetically modified organism.

LG5 Describe the potential drawbacks of genetically modified organisms.

LG6 Summarize the findings that resulted from the Human Genome Project, and describe the potential benefits of those findings.

LG7 Explain how biotechnology can be used to diagnose and treat inherited disorders.

13.1 WHAT IS BIOTECHNOLOGY?

Biotechnology is the use, and especially the alteration, of organisms, cells, or biological molecules to produce food, drugs, or other goods. Some aspects of biotechnology are ancient. For example, people have used yeast cells to produce bread, beer, and wine for the past 10,000 years. Prehistoric art and archaeological discoveries indicate that many plants and animals, including wheat, grapes, dogs, pigs, and cattle, were domesticated and selectively bred for desirable traits 6,000 to 15,000 years ago. For instance, selective breeding for desirable inherited traits rapidly transformed relatively slim wild boars, with long tusks and fierce temperaments, into much heavier, more placid domestic pigs. Selective breeding is still an important tool for improving livestock and crops.

In addition to selective breeding, modern biotechnology frequently uses **genetic engineering** to isolate and manipulate the genes that control inherited characteristics. Genetically engineered cells or organisms have genes deleted, added, or changed. Genetic engineering can be used to learn more about how cells and genes work; to develop better treatments for diseases; to produce valuable biological molecules, including hormones and vaccines; and to improve plants and animals for agriculture. A relatively familiar application of modern biotechnology is **cloning**—making identical copies of individual genes or even entire organisms. We describe how genes are cloned in Section 13.4. Natural and technological methods of cloning whole organisms are explored in "Science in Action: Carbon Copies—Cloning in Nature and the Lab."

A key tool in modern biotechnology is **recombinant DNA,** which is DNA that has been altered to contain genes or parts of genes from different organisms. Large amounts of recombinant DNA can be produced in bacteria, viruses, or yeast, and then transferred into other species. Plants and animals that contain DNA that has been modified or derived from other species are called **transgenic** or **genetically modified organisms (GMOs).**

Modern biotechnology includes many methods of analyzing and manipulating DNA, whether or not the DNA is subsequently put into a cell or an organism. For example, determining the nucleotide sequence of DNA is crucial for fields as diverse as forensic science, medicine, and evolutionary biology.

In this chapter, we will provide an overview of the applications of biotechnology and their impacts on society. We will also briefly describe some of the important methods used in those applications. We will organize our discussion around five major themes: (1) recombinant DNA mechanisms found in nature; (2) biotechnology in criminal forensics; (3) biotechnology in agriculture, specifically transgenic plants and animals; (4) analyzing the entire genomes of humans and other organisms; and (5) biotechnology in medicine, focusing on the diagnosis and treatment of inherited disorders.

CHECK YOUR LEARNING

Can you describe applications of genetic engineering and recombinant DNA?

13.2 HOW DOES DNA RECOMBINE IN NATURE?

The process of recombining DNA is not unique to modern laboratories. Many natural processes can transfer DNA from one organism to another, sometimes even to organisms of different species.

SCIENCE IN ACTION   Carbon Copies—Cloning in Nature and the Lab

The word "cloning" may evoke images of Dolly the sheep or even *Star Wars: Attack of the Clones*, but nature has been quietly cloning for hundreds of millions of years. How are clones produced, either in nature or in the lab?

Cloning in Nature

Sexual reproduction relies on meiotic cell division, the production of gametes, and fertilization, and usually produces genetically unique offspring. In contrast, asexual reproduction (see Fig. 9-2) relies on mitotic cell division. Because mitotic cell division creates daughter cells that are genetically identical to the parent cell, offspring produced by asexual reproduction are genetically identical to their parents—they are clones. Many clusters of plants, including strawberries, grasses, and aspen groves (see Fig. 9-2) are clones.

Cloning Plants: A Familiar Application in Agriculture

Humans have been in the cloning business a lot longer than you might think. For example, consider navel oranges, which don't produce seeds. How do they reproduce? Navel orange trees are propagated by cutting a piece of stem from an adult navel tree and grafting it onto the top of the root of a seedling of another type of orange tree. The cells of the aboveground, fruit-bearing parts of the grafted tree are clones of the original navel orange stem. All navel oranges originated from a single mutant bud of an orange tree discovered in Brazil in the early 1800s, and propagated asexually ever since. Two navel orange trees were brought from Brazil to Riverside, California, in the 1870s—one of them is still there! All American navel orange trees are clones of these two trees.

Cloning Mammals

But, of course, the popular perception of the word "cloning" isn't about growing strawberries or making navel orange trees: It's about making whole new animals, usually by implanting a nucleus from one animal into a fertilized egg taken from another animal. In the 1950s, John Gurdon and his colleagues did this in frogs. They destroyed the nuclei of frog eggs, and then inserted new nuclei, taken from embryonic frog cells. Some of the resulting cells developed into complete frogs. By the 1990s, several labs had cloned mammals using nuclei from embryos, but it wasn't until 1996 that Ian Wilmut of the Roslin Institute in Edinburgh, Scotland, cloned a mammal using a nucleus taken from an adult. The resulting cloned sheep was the famous Dolly (Fig. E13-1).

Why bother to clone mammals? Suppose that a pharmaceutical company genetically engineered a cow that secreted a valuable molecule, such as an antibiotic, in its milk. Genetic engineering techniques are extremely expensive and somewhat hit or miss, so the company might successfully produce only one profitable cow. This cow could then be cloned, creating a whole herd of antibiotic-producing cows. Cloned cows that produce more milk or meat, and cloned pigs tailored to be organ donors for humans, already exist.

Cloning might also help rescue critically endangered species, many of which don't reproduce well in zoos. As Richard Adams of Texas A&M University put it, "You could repopulate the world [with an endangered species] in a matter of a couple of years. Cloning is not a trivial pursuit."

Cloning: An Imperfect Technology

Unfortunately, cloning mammals is beset with difficulties. An egg is subjected to severe trauma when its nucleus is sucked out or destroyed and a new nucleus is inserted. Often, the egg may simply die. Even if the egg survives and starts dividing, it may not develop properly. If the eggs do develop into viable embryos, the embryos must be implanted into the uterus of a surrogate mother. Many clones die or are aborted during gestation, often with serious or fatal consequences for the surrogate mother. Even if the clone survives gestation and birth, it may have defects, commonly a deformed head, lungs, or heart. Not surprisingly, these problems translate into a phenomenally high failure rate. It took 277 tries to produce Dolly. The first cloned dog, produced in 2005, was the sole survivor of 1,095 embryos implanted into 123 surrogate mothers, resulting in only three pregnancies, with two live puppies, one of which died within a month.

The Future of Cloning

Modern cloning technology has now successfully cloned dogs, cows, cats, sheep, horses, and a variety of other animals. As the process becomes more routine, it also brings ethical questions. While hardly anyone objects to cloning navel oranges, and few would refuse antibiotics or other medicinal products from cloned livestock, many people think that cloning pets is a frivolous luxury—especially when you consider that almost 10 million unwanted dogs and cats are euthanized in the United States every year. For the foreseeable future, it seems likely that mammalian cloning will be restricted to demonstrations of its feasibility (that is, to answer such questions as "Can we clone water buffalos and camels?" The answer is yes, we can) and to replicate particularly valuable animals, such as endangered species or champion horses.

What about cloning people? A horror story? Or a modern-day version of Shakespeare's "brave new world that has such people in it"? In principle, humans can probably be cloned. However, in the words of Ian Wilmut, who cloned Dolly: ". . . the possibility for harm far outweighs any currently conceivable benefits." Some people believe that human cloning is inevitable, eventually. Maybe so, but with huge technical and ethical hurdles to overcome, not anytime soon.

Sexual Reproduction Recombines DNA

Homologous chromosomes exchange DNA by crossing over during meiosis I (see Chapter 9). As a result, each chromosome in a gamete contains a mixture of alleles from the two parental chromosomes. Thus, every egg and sperm contains recombinant DNA, derived from the two parents. When a sperm fertilizes an egg, the resulting offspring also contains recombinant DNA.

Transformation May Combine DNA from Different Bacterial Species

In **transformation,** bacteria pick up pieces of DNA from the environment (Fig. 13-1). The DNA may be part of the chromosome from another bacterium (Fig. 13-1a), even from another species, or it may be tiny circular DNA molecules called **plasmids** (Fig. 13-1b). Many types of bacteria

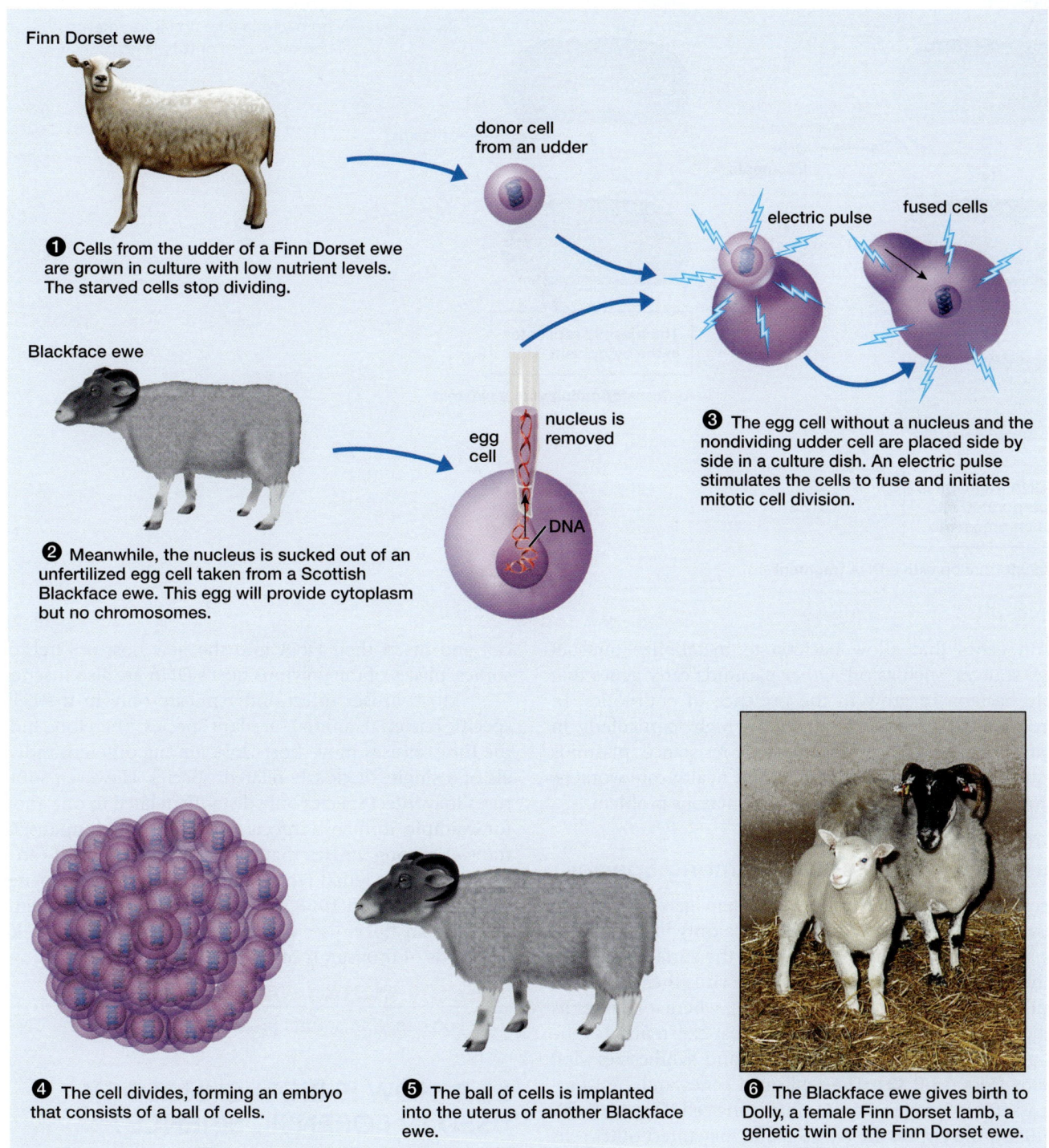

Finn Dorset ewe

❶ Cells from the udder of a Finn Dorset ewe are grown in culture with low nutrient levels. The starved cells stop dividing.

donor cell from an udder

electric pulse   fused cells

Blackface ewe

egg cell   nucleus is removed

DNA

❷ Meanwhile, the nucleus is sucked out of an unfertilized egg cell taken from a Scottish Blackface ewe. This egg will provide cytoplasm but no chromosomes.

❸ The egg cell without a nucleus and the nondividing udder cell are placed side by side in a culture dish. An electric pulse stimulates the cells to fuse and initiates mitotic cell division.

❹ The cell divides, forming an embryo that consists of a ball of cells.

❺ The ball of cells is implanted into the uterus of another Blackface ewe.

❻ The Blackface ewe gives birth to Dolly, a female Finn Dorset lamb, a genetic twin of the Finn Dorset ewe.

▲ FIGURE E13-1 The making of Dolly

contain plasmids, which range in length from about 1,000 to 100,000 nucleotides. (For comparison, the *E. coli* chromosome is around 4,600,000 nucleotides long.) A single bacterium may contain dozens or even hundreds of copies of a plasmid. When the bacterium dies, its plasmids are released into the environment, where they may be picked up by other bacteria of the same or different species. In addition, living bacteria can often pass plasmids directly to other living bacteria. Plasmids may also move from bacteria to yeast, transferring genes from a prokaryotic cell to a eukaryotic cell.

What use are plasmids? A bacterium's chromosome contains all the genes the cell normally needs for basic survival. However, genes carried by plasmids may permit the bacteria to thrive in novel environments. Some plasmids

bacterial chromosome

DNA fragments

A DNA fragment is incorporated into the chromosome

(a) Transformation with a DNA fragment

bacterial chromosome

plasmid

The plasmid replicates in the cytoplasm

(b) Transformation with a plasmid

◄ **FIGURE 13-1 Transformation in bacteria** Bacterial transformation occurs when living bacteria take up **(a)** fragments of chromosomes or **(b)** plasmids.

contain genes that allow bacteria to metabolize unusual energy sources, such as oil. Other plasmids carry genes that enable bacteria to grow in the presence of antibiotics. In environments where antibiotic use is high, particularly in hospitals, bacteria carrying antibiotic-resistance plasmids can quickly spread among patients and health care workers, making antibiotic-resistant infections a serious problem.

Viruses May Transfer DNA Among Species

Viruses, which are often little more than genetic material encased in a protein coat, can reproduce only inside cells. A virus attaches to specific molecules on the surface of a suitable host cell (**Fig. 13-2 ❶**). Usually the virus then enters the cytoplasm of the host cell (**Fig. 13-2 ❷**), where it releases its genetic material (**Fig. 13-2 ❸**). The host cell replicates the viral genetic material (DNA or RNA) and synthesizes viral proteins (**Fig. 13-2 ❹**). The replicated genes and viral proteins assemble into new viruses inside the cell (**Fig. 13-2 ❺**). Eventually, the viruses are released and may infect other cells (**Fig. 13-2 ❻**).

Some viruses can transfer genes from one organism to another. In these instances, the viral DNA is inserted into one of the host cell's chromosomes (see Fig. 13-2 ❸). The viral DNA may remain there for days, months, or even years. Every time the cell divides, it replicates the viral DNA along with its own DNA. (Researchers believe that about 3% of the human genome consists of "fossil" viral genes, inserted into our DNA thousands to millions of years ago.) When new viruses are finally produced, some of the host cell's genes may be attached to the viral DNA. If such recombinant viruses infect another

cell and insert their DNA into the new host cell's chromosomes, pieces of the previous host's DNA are also inserted.

Most viruses infect and replicate only in the cells of specific bacterial, animal, or plant species. Therefore, most of the time, viruses move host DNA among different individuals of a single, or closely related, species. However, some viruses may infect species only distantly related to one another; for example, influenza infects birds, pigs, and humans. Gene transfer among viruses that infect multiple species can produce extremely lethal recombined viruses. This happened in 1957 and again in 1968, when recombination between bird and human flu viruses caused global epidemics that killed hundreds of thousands of people.

CHECK YOUR LEARNING

Can you describe natural processes that recombine DNA?

13.3 HOW IS BIOTECHNOLOGY USED IN FORENSIC SCIENCE?

Applications of DNA biotechnology vary, depending on the goals of those who use it. Forensic scientists need to identify victims and criminals; biotechnology firms need to identify specific genes and insert them into organisms such as bacteria, animals, or plants; and biomedical firms and physicians need to detect defective alleles and, ideally, devise ways to fix them or to insert normally functioning alleles into patients. We will begin by describing a few common methods of manipulating DNA and then discuss their application in forensic DNA analysis. Later, we will investigate how biotechnology is used in agriculture and medicine.

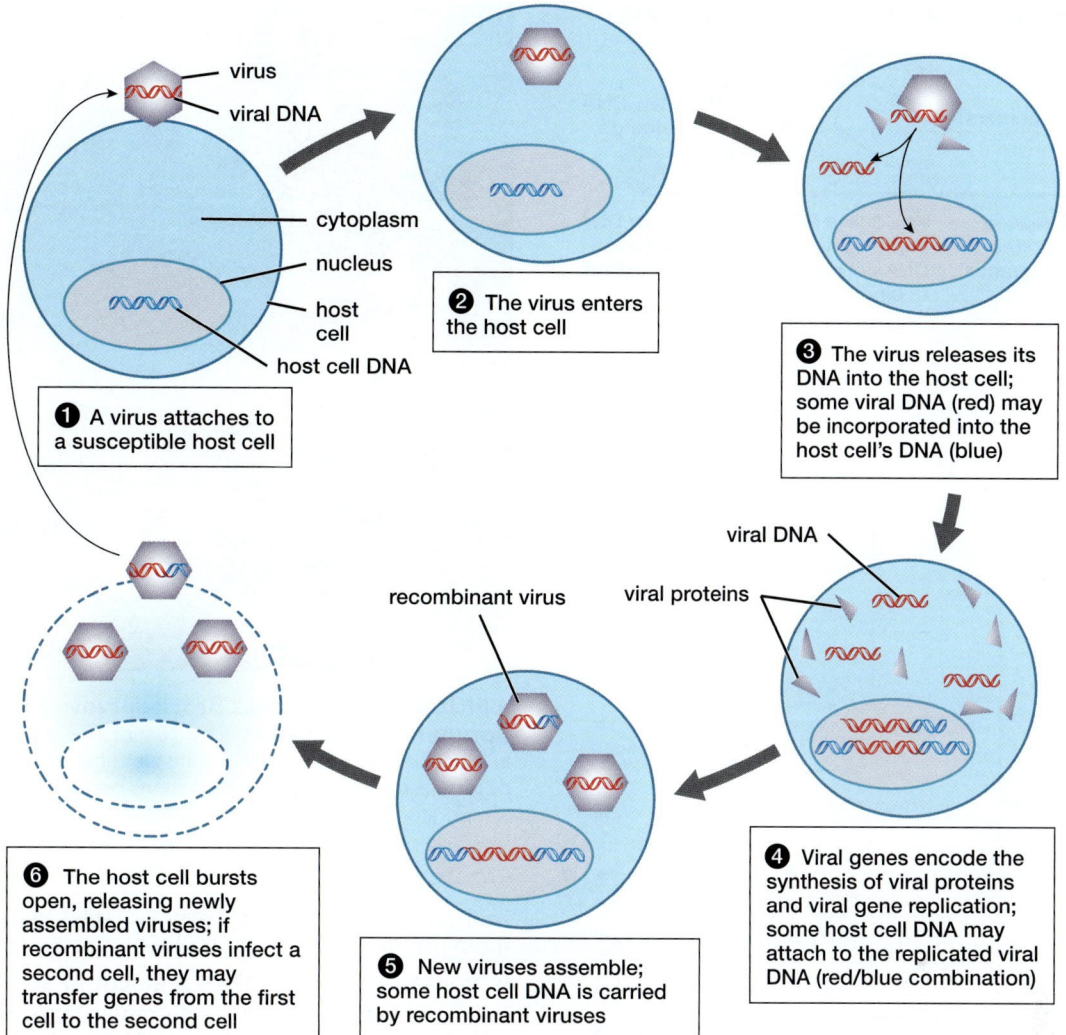

▲ FIGURE 13-2 The life cycle of a typical virus In some cases, viral infections may transfer DNA from one host cell to another.

The Polymerase Chain Reaction Amplifies DNA

Developed by Kary Mullis in 1986, the **polymerase chain reaction (PCR)** can be used to make billions, even trillions, of copies of selected pieces of DNA. PCR is so crucial to molecular biology that it earned Mullis a share in the Nobel Prize for Chemistry in 1993. PCR involves two major steps: (1) marking the DNA segment to be copied and (2) running repetitive reactions to make multiple copies. First, the desired DNA segment is bracketed with two short pieces of DNA, called *primers*. Primers are usually manufactured in a machine called a DNA synthesizer, which can be programmed to make short pieces of DNA with any desired sequence of nucleotides. The nucleotide sequence of one primer is complementary to the beginning of the target DNA segment on one strand of the double helix, and the sequence of the other primer is complementary to the beginning of the target DNA on the other strand. DNA polymerase recognizes the primers as the place where DNA replication should begin.

Second, the marked DNA is copied. DNA is mixed with primers, free nucleotides, and DNA polymerase in a small test tube. The primers bind to the desired DNA segment. The reaction mixture is then cycled through a series of temperature changes (**Fig. 13-3a**):

1. The test tube is heated to 194° to 203°F (90° to 95°C) (**Fig. 13-3a ❶**). High temperatures break the hydrogen bonds between complementary bases, separating the DNA into single strands.
2. The temperature is lowered to about 122°F (50°C) (**Fig. 13-3b ❷**), which allows the two primers to form complementary base pairs with the target DNA.
3. The temperature is raised to 158° to 162°F (70° to 72°C) (**Fig. 13-3a ❸**). DNA polymerase uses the free nucleotides to make copies of the DNA segment bounded by the primers. Most DNA polymerases do not function at temperatures much higher than 105°F (40°C). However, PCR uses a special DNA polymerase isolated from bacteria that live in hot springs

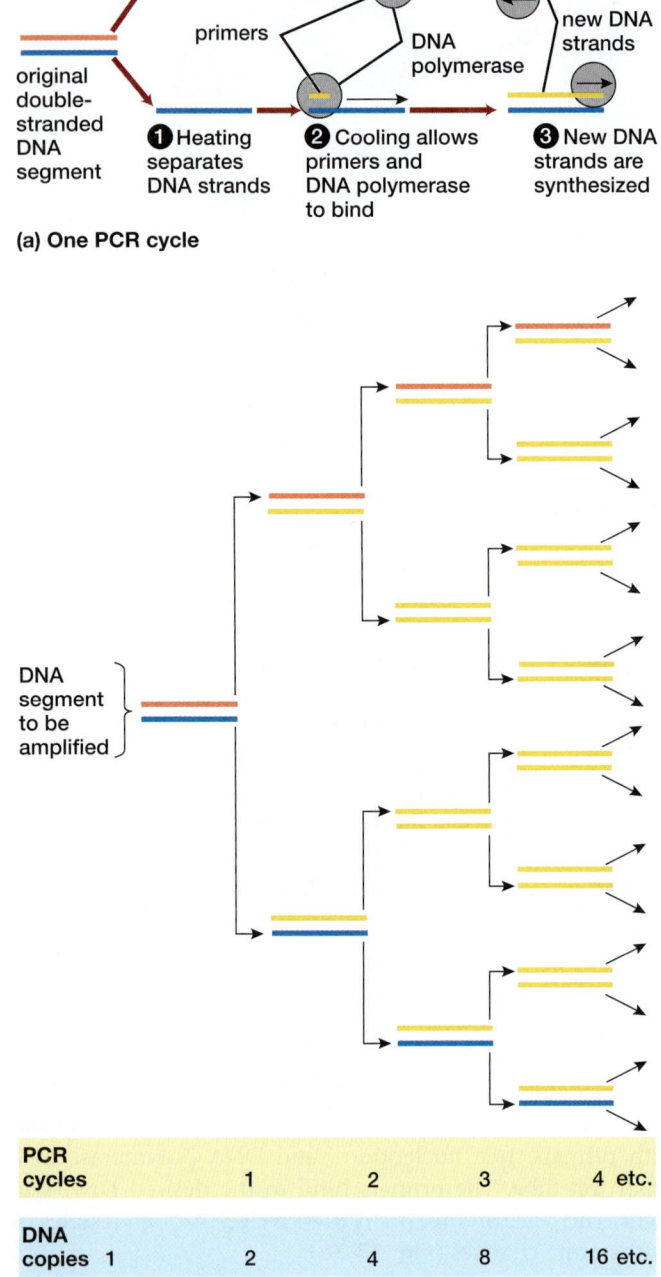

194°F (90°C) 122°F (50°C) 158°F (70°C)

primers DNA polymerase new DNA strands

original double-stranded DNA segment

1 Heating separates DNA strands

2 Cooling allows primers and DNA polymerase to bind

3 New DNA strands are synthesized

(a) One PCR cycle

DNA segment to be amplified

PCR cycles		1	2	3	4 etc.
DNA copies	1	2	4	8	16 etc.

(b) Each PCR cycle doubles the number of copies of the DNA

▲ **FIGURE 13-3 PCR copies a specific DNA sequence**
(a) The polymerase chain reaction consists of a cycle of heating, cooling, and warming that is typically repeated 30 to 40 times. **(b)** Each cycle doubles the amount of target DNA. After a little more than 30 cycles, a billion copies of the target DNA have been synthesized.

(**Fig. 13-4**), which actually works best at these high temperatures.

4. This cycle is repeated, usually 30 to 40 times, until the free nucleotides have been used up.

In PCR, the amount of DNA doubles with every temperature cycle (**Fig. 13-3b**). Twenty PCR cycles make about a

▲ **FIGURE 13-4 Thomas Brock surveys Mushroom Spring** Brock discovered the bacterium *Thermus aquaticus* in Mushroom Spring in Yellowstone National Park. The DNA polymerase from *T. aquaticus* functions best at the high temperatures required by PCR.

million copies, and a little over 30 cycles make a billion copies. Each cycle takes only a few minutes, so PCR can produce billions of copies of a DNA segment in an afternoon, starting, if necessary, from a single molecule of DNA. The DNA is then available for forensics, cloning, making transgenic organisms, or many other purposes.

Differences in Short Tandem Repeats Are Used to Identify Individuals by Their DNA

In many criminal investigations, PCR is used to amplify the DNA so that there is enough to compare the DNA left at a crime scene with a suspect's DNA. How do crime labs compare DNA? After years of painstaking work, forensics experts have found that specific segments of DNA, called **short tandem repeats (STRs)**, can be used to identify people with astonishing accuracy. Think of STRs as very small, stuttering genes (**Fig. 13-5**). STRs are *short* (about 20 to 250 nucleotides), *repeating* (consisting of the same sequence of 2 to 5 nucleotides repeated up to 50 times), and *tandem* (having all of the repetitions right alongside one another). As with any gene, there may be alternative forms, or alleles, of an STR. In any given STR, the different alleles simply have different numbers of repeats of the same short nucleotide sequence. To identify individuals from DNA samples, the U.S. Department of Justice established a standard set of 13 STRs that have highly variable numbers of repeats in different people. Most crime labs also examine a gene that shows whether the DNA sample came from a man or a woman.

Eight side-by-side (tandem) repeats
of the same four-nucleotide sequence

▲ FIGURE 13-5 **Short tandem repeats are common in noncoding regions of DNA** This STR contains the sequence AGAT, repeated from 7 to 15 times in different individuals.

CASE STUDY continued

Guilty or Innocent?

When a law student finally located semen samples in the Dennis Maher case, the Innocence Project team needed to find out if the samples collected from the rape victims came from Maher. Fortunately, DNA doesn't degrade very fast, so even old DNA samples, such as those in the Maher case, usually have STRs that are mostly intact. First, lab technicians amplified the DNA with PCR so that they had enough material to analyze. Then, they determined whether the DNA from the semen samples matched Maher's DNA.

Forensics labs use PCR primers that amplify only the STRs and the DNA immediately surrounding them. Because STR alleles vary in how many repeats they contain, they vary in size: An STR allele with more repeats is larger than one with fewer repeats. Therefore, a forensic lab needs to identify each STR in a DNA sample and then determine its size to find out which alleles occur in the sample.

Modern forensics labs use sophisticated and expensive machines to analyze STRs. Most of these machines, however, are based on two methods that are used in molecular biology labs around the world: first, separating DNA segments by size, and second, labeling specific DNA segments of interest.

Gel Electrophoresis Separates DNA Segments

A mixture of DNA pieces can be separated by a technique called **gel electrophoresis** (Fig. 13-6). First, a laboratory technician loads the DNA mixture into shallow grooves, or wells, in a slab of agarose, a carbohydrate purified from certain types of seaweed (Fig. 13-6 ❶). Agarose forms a gel, which is simply a meshwork of fibers with holes of various sizes between the fibers. The gel is put into a chamber with electrodes connected to each end. One electrode is made positive and the other negative; therefore, current will flow between the electrodes through the gel. How does this process separate pieces of DNA? Remember, the phosphate groups in the backbones of DNA are negatively charged. When electrical current flows through the gel, the negatively charged DNA fragments move toward the positively charged electrode. Because smaller fragments slip through the holes in the gel more easily than larger fragments do, they move

more rapidly toward the positively charged electrode. Eventually the DNA fragments are separated by size, forming distinct bands on the gel (Fig. 13-6 ❷).

DNA Probes Are Used to Label Specific Nucleotide Sequences

Unfortunately, the DNA bands are invisible. There are several dyes that stain DNA, but these are often not very useful in either forensics or medicine. Why not? Because there may be many DNA fragments of approximately the same size. For example, five or six different STRs with the same numbers of repeats might be mixed together in the same band. How can a technician identify a *specific* STR? Well, how does *nature* identify sequences of DNA? Right—by base pairing!

When the gel is finished running, the technician treats it with chemicals that break apart the double helices into single DNA strands. These DNA strands are transferred out of the gel onto a piece of paper made of nylon (Fig. 13-6 ❸). Because the DNA samples are now single-stranded, pieces of synthetic DNA, called **DNA probes,** can base-pair with specific DNA fragments in the sample. DNA probes are short pieces of single-stranded DNA that are complementary to the nucleotide sequence of a given STR (or any other DNA of interest in the gel). The DNA probes are labeled, either by radioactivity or by attaching colored molecules to them. Therefore, a given DNA probe will label certain DNA sequences, but not others (Fig. 13-7).

To locate a specific STR, the paper is bathed in a solution containing a DNA probe that will base-pair with, and therefore bind to, only that particular STR (Fig. 13-6 ❹). Any extra DNA probe is then washed off the paper. The result: The DNA probe shows where that specific STR ran in the gel (Fig. 13-6 ❺). (Visualizing the DNA fragments with radioactive or colored DNA probes is standard procedure in many research applications. In forensics labs, the STRs are directly labeled with colored molecules during PCR and are immediately visible in the gel, so DNA probes are not necessary.)

Unrelated People Almost Never Have Identical DNA Profiles

DNA samples run on STR gels produce a pattern, called a **DNA profile** (Fig. 13-8). The positions of the bands on the gel are determined by the numbers of repeats of the four-nucleotide sequence of each STR allele. If the same STRs are analyzed, then all samples of a person's DNA produce the same profile every time.

What does a DNA profile tell us? As with any gene, every person has two alleles of each STR, one on each homologous chromosome. The two alleles of a given STR might have the same number of repeats (the person would be homozygous for that STR) or a different number of repeats (the person would be heterozygous). For example, in the D16 STR samples shown on the right side of Figure 13-8, the first person's DNA has a single band at 12 repeats (this person is homozygous for the D16 STR), but the second person's DNA has two bands—at 13 and 12 repeats (this

1 DNA samples are pipetted into wells (shallow slots) in the gel. Electrical current is sent through the gel (negative at the end with the wells and positive at the opposite end).

power supply

pipetter

gel

wells

2 Electrical current moves the DNA segments through the gel. Smaller pieces of DNA move farther toward the positive electrode.

DNA "bands" (not yet visible)

3 The gel is placed on special nylon "paper." Electrical current drives the DNA out of the gel onto the nylon.

gel

nylon paper

4 The nylon paper with the DNA bound to it is bathed in a solution of labeled DNA probes (red) that are complementary to specific DNA segments in the original DNA sample.

solution of DNA probes (red)

nylon paper

5 Complementary DNA segments are labeled by the probes (red bands).

▲ **FIGURE 13-6 Gel electrophoresis and labeling with DNA probes separates and identifies segments of DNA.**

person is heterozygous for the D16 STR). If you look closely at all of the DNA samples in Figure 13-8, you will see that, although the DNA from some people had the same repeats for one of the STRs (e.g., the second, fourth, and fifth samples for D16), no one's DNA had the same repeats for all four STRs.

Are 13 STRs enough to uniquely identify people, given the huge human population? Worldwide, different people may have as few as 5 to as many 38 repeats in a given STR. Although there are some complicating factors that forensics labs need to take into account, let's take a simple case: Assume that a crime lab analyzes five STRs, each with 10 possible numbers of repeats (for example, all people have either 6, 7, 8, 9, 10, 11, 12, 13, 14, or 15 repeats). Let's also assume that all of the numbers of repeats occur with equal probability in the human population, that is, one in ten, or 1/10. Finally, STRs are independently assorted (see Chapter 10), so the probability of two unrelated people sharing the same number of repeats of all five STRs is simply the product (multiplication) of the separate probabilities, or $1/10 \times 1/10 \times 1/10 \times 1/10 \times 1/10 = 1$ chance in 100,000.

With 13 STRs, containing up to 38 repeats each, the chances of a random match are incredibly small. A perfect match of both alleles for all 13 STRs used in the United States means that there is far less than one chance in a trillion that the two DNA samples matched purely by random chance. For complicated statistical reasons, there are probably a few unrelated people in the world who have the same DNA profile, but the odds of anyone who would be a likely suspect in a criminal case being misidentified are extremely low. Finally, a *mismatch* in DNA profiles is absolute proof that two samples did not come from the same source. In the Maher case, when DNA profiling showed that the STR alleles in the semen did not match Maher's, that ruled him out as the perpetrator.

STR #1: The probe base-pairs and binds to the DNA

STR #2: The probe cannot base-pair with the DNA, so it does not bind

▲ **FIGURE 13-7 DNA probes base-pair with complementary DNA sequences** A short, single-stranded piece of DNA is labeled with a colored molecule (red ball). This labeled DNA will base-pair with a target strand of DNA with a complementary base sequence (top), but not with a noncomplementary strand (bottom).

In the United States, anyone convicted of certain crimes (assault, burglary, attempted murder, etc.) must give a blood sample. Crime lab technicians then determine the criminal's DNA profile, and code the results as the number of repeats in each STR. The profile is stored in computer files at a state agency, at the FBI, or both. (On *CSI* and other TV crime shows, when the actors refer to "CODIS," that acronym stands for "Combined DNA Index System," a DNA profile database kept on FBI computers.) Because all U.S. forensic labs use the same 13 STRs, computers can easily determine if DNA left behind at another crime scene matches one of the profiles stored in the CODIS database. If the STRs match, then the odds are overwhelming that the crime scene DNA was left by the person with the matching profile. If there aren't any matches, the crime scene DNA profile will remain on file. Sometimes, years later, a new DNA profile will match an archived crime scene profile, and a "cold case" will be solved (see "Case Study Revisited" at the end of this chapter).

People aren't the only organisms that can be identified by the DNA sequences, of course. An international group of government and private organizations is putting together the "Barcode of Life" to enable rapid DNA identification of all of the species of life on Earth. DNA barcodes have both serious and entertaining applications, as we explain in "Earth Watch: What's Really in That Sushi?"

CHECK YOUR LEARNING

Can you explain the uses of electrophoresis, DNA probes, and the polymerase chain reaction, and how they work? Do you understand how DNA profiles are produced, and why a DNA profile is usually unique to each individual person?

▲ **FIGURE 13-8 DNA profiling** The lengths of short tandem repeats of DNA form characteristic patterns on a gel. This gel displays four different STRs (Penta D, CSF, etc.). The columns of evenly spaced yellow bands on the far left and far right sides of the gel show the number of repeats in the different STR alleles. DNA samples from 13 different people were run between these standards, resulting in one or two yellow bands in each vertical lane. The position of each band corresponds to the number of repeats in that STR allele (more repeats means more nucleotides, which in turn means that the allele is larger). (Photo courtesy of Dr. Margaret Kline, National Institute of Standards and Technology.)

QUESTION For any single person, a given STR always has either one or two bands. Why? Further, single bands are always about twice as bright as each band of a pair. For example, in the D16 STR on the right, the single bands of the first and third DNA samples are twice as bright as the pairs of bands of the second, fourth, and fifth samples. Why?

Earth|Watch

What's Really in That Sushi?

Walk into a Japanese restaurant or a sushi bar and chances are that the most expensive item on the menu is some type of tuna sushi. But is it really tuna? In 2008, two New York teenagers, Kate Stoeckle and Louisa Strauss, decided to find out (Fig. E13-2). Sounds difficult—after all, the fish are beheaded, cleaned, and skinned, and only a chunk of meat is presented to the diner—but biotechnology makes it simple, using DNA barcoding.

DNA barcoding sequences a small fragment of DNA from a gene found in the mitochondria of virtually all eukaryotic organisms—a fragment only 650 nucleotides long (Fig. E13-3). So far, no two species have been found to have the same nucleotide sequence in this particular piece of DNA. Thus, DNA barcoding is a simple, inexpensive way to identify species with absolute certainty.

Kate and Louisa visited restaurants and grocery stores, and brought home samples of raw fish. They cut off little pieces from each sample, preserved them in alcohol, and sent them off to a lab at the University of Guelph for barcoding. Surprise! About a quarter of the sushi samples were imposters. One specimen sold as red snapper was actually Acadian redfish, an endangered species. One "tuna" sushi turned out to be tilapia, a freshwater species often raised in fish farms. It turns out that some restaurants mislabel half their sushi.

DNA barcoding is useful for more than just checking up on your local sushi bar. Commercial laboratories will now barcode a sample for just a few dollars, with a one-day turnaround. Because of its high precision and low cost, barcoding is rapidly gaining acceptance as a way to identify agricultural pests such as fruit flies and public health threats such as disease-carrying mosquitoes. The U.S. Environmental Protection Agency is barcoding insect larvae to assess the health of streams—different species of insects live in clean vs. polluted streams. The Federal Aviation Administration barcodes feathers to find out what kinds of birds crash into planes.

In the future, DNA barcoding might help to stop illegal trafficking in endangered species, which is extremely lucrative (thought to be second only to illegal narcotics). Meat, skin, feathers, and many other animal parts are difficult to identify visually. And how many customs officials can tell the difference between greenhouse orchids and rare orchids plucked from a rain forest? Humans might be deceived, but barcoding can't be fooled. The hope is that, within a few years, small devices will be created that analyze suspect samples in an hour or so, perhaps even in minutes, and compare the barcode with publicly available databases. The day may come when barcoding not only verifies your sushi but puts a stop to the exploitation of endangered species.

▲ FIGURE E13-2 Kate Stoeckle and Louisa Strauss with their research subjects

▲ FIGURE E13-3 **DNA barcoding** The different colors in the barcodes represent different bases in the DNA sequence of a fragment of a mitochondrial gene. Closely related organisms have more similar barcodes than distantly related organisms do, but every species has a unique barcode.

CASE STUDY continued

Guilty or Innocent?

Different countries use different STRs, and different numbers of STRs, for DNA profiling. Recently, the countries of the European Union pooled their databases, so that a DNA profile from a crime in, say, Italy, could be compared against profiles in Great Britain or France. Unfortunately, because they use different STRs, there may be as few as six STRs in common among the DNA profiles of member countries. As a result, comparing crime scene DNA with the European Union database can result in lots of false positives—matches that occur purely by chance. Fortunately, mismatches will still eliminate innocent suspects.

13.4 HOW IS BIOTECHNOLOGY USED TO MAKE GENETICALLY MODIFIED ORGANISMS?

The techniques of PCR, gel electrophoresis, and identifying DNA sequences with specific probes have applications far beyond forensic science. In combination with other procedures that we will discuss shortly, biotechnology can also be used to identify, isolate, and modify genes; combine genes from different organisms; and move genes from one species to another. Let's see how some of these techniques can be used to make genetically modified organisms (GMOs).

There are three major steps to making a GMO: (1) Obtain the desired gene, (2) clone the gene, and (3) insert the gene into the cells of the host organism. Various technologies can be used for each step, each involving complex procedures. We will provide only a brief overview of the general processes.

The Desired Gene Is Isolated or Synthesized

Two common methods are used to obtain a gene. For a long time, the only practical method was to isolate the target gene from the organism that possessed it. Chromosomes can be isolated from cells of the gene donor, cut up with enzymes (see below), and DNA fragments containing the desired gene can be separated from the rest of the DNA by gel electrophoresis (see Fig. 13-6). Today, biotechnologists can often synthesize the gene—or a modified version of it—in the lab, using DNA synthesizers.

The Gene Is Cloned

Once the gene has been obtained, it can be used to make transgenic organisms, shared with other scientists around the world, or used for medical treatments. Therefore, it is useful, or even essential, to have a huge number of copies of the gene. The simplest way to generate lots of copies of the gene is to let living organisms do it, by **DNA cloning.** In DNA cloning, the gene is usually inserted into single-celled organisms, such as bacteria or yeasts, that multiply very rapidly, manufacturing copies of the gene as they do.

The most common method of DNA cloning is to insert the gene into a bacterial plasmid (see Fig. 13-1), which will be replicated when bacteria containing the plasmid multiply. Inserting the gene into a plasmid, rather than the bacterial chromosome, also allows it to be easily separated from the majority of the bacterial DNA. The target gene may be further purified from the plasmid, or the whole plasmid may be used to make transgenic organisms, including plants, animals, or other bacteria.

Genes are inserted into plasmids using **restriction enzymes,** each of which cuts DNA at a specific nucleotide sequence. There are hundreds of different restriction enzymes. Many cut straight across the double helix of DNA. Others make a staggered cut, snipping the DNA in a different location on each of the two strands, so that single-stranded sections hang off the ends of the DNA. These single-stranded regions are commonly called "sticky ends," because they can base-pair with, and thus stick to, other single-stranded pieces of DNA with complementary bases (**Fig. 13-9**). Restriction enzymes that make a staggered cut are used in DNA cloning.

To insert a gene into a plasmid, the same restriction enzyme is used to cut the DNA on both sides of the gene and to split open the circle of plasmid DNA (**Fig. 13-10 ❶**). As a result, the ends of the DNA containing the gene and the opened-up plasmid both have complementary nucleotides in their sticky ends, and can base-pair with each other. When the cut genes and plasmids are mixed together, some of the genes will be temporarily inserted between the cut ends of the plasmids, held together by their complementary sticky ends. Adding DNA ligase (see Chapter 11) permanently bonds the genes into the plasmids (**Fig. 13-10 ❷**).

Bacteria are then transformed with these recombinant plasmids. Under the right conditions, when the bacteria multiply, they replicate the plasmids, too. Huge vats of bacteria can be easily grown, producing as many copies of the gene as anyone could ever need.

▲ FIGURE 13-9 Restriction enzymes cut DNA at specific nucleotide sequences

QUESTION Restriction enzymes are isolated from bacteria. Why would bacteria synthesize enzymes that cut up DNA? (*Hint:* Bacteria can be infected by viruses called bacteriophages; see Chapter 11.) Why wouldn't a bacterium's restriction enzymes destroy the DNA of its own chromosome?

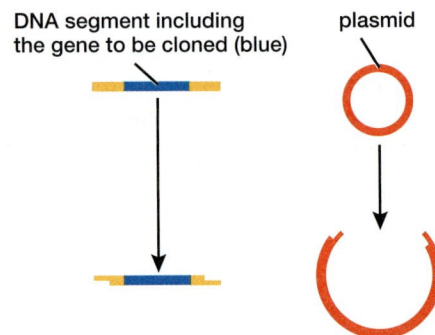

① The plasmid and the DNA segment containing the desired gene are cut with the same restriction enzyme.

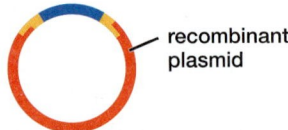

② The plasmids and the DNA segment containing the gene, both with the same complementary sticky ends, are mixed together; DNA ligase bonds the genes into the plasmids.

▲ FIGURE 13-10 **Inserting a gene into a plasmid for DNA cloning**

The Gene Is Inserted into a Host Organism

Now comes the hard part—**transfecting** the host organism. The gene must be inserted into the host and expressed in the appropriate cells, at the appropriate times, and at the desired level. Host organisms may be transfected by one of several different methods. In some cases, the recombinant plasmids or genes purified from them are inserted into harmless bacteria or viruses, called *vectors*, and then the host organism is infected with them. In the ideal case, the bacteria or viruses insert the new gene into the chromosomes of the host organism's cells, where it becomes a permanent part of the host's genome, and is replicated whenever the host's DNA is replicated. This is how some plants were transfected with genes for herbicide resistance and insect resistance (see Section 13.5).

A simpler method is to use a "gene gun." Microscopically small pellets of gold or tungsten are coated with DNA (either plasmids or purified genes) and then shot at cells or organisms. This process is literally "hit or miss," but is often quite effective for plants, cells in culture, and sometimes even whole animals (usually small ones, such as roundworms or fruit flies). Gene guns are often used when host organisms are easily available in large numbers, so that a low success rate doesn't really matter.

Finally, plasmids or purified genes can be directly injected into animal cells, usually fertilized eggs (**Fig. 13-11**). Tiny glass pipettes are loaded with a suitable solution containing the DNA. The pipettes have tips that are sharp enough to impale a cell without damaging it. Pressure applied to the back of the pipette pushes some of the DNA into the cell.

Sharp, thin pipette to inject DNA into cell

Smooth, blunt pipette to hold cell in place

▲ FIGURE 13-11 **Transfecting a fertilized egg by injecting foreign DNA** The large pipette on the right holds the egg stationary during the procedure. The small, sharp pipette on the left penetrates the egg and injects DNA.

CHECK YOUR LEARNING

Can you explain how genes are inserted into a plasmid, and why that is useful in making a genetically modified organism? Can you describe the procedures used to transfect an organism with a foreign gene?

13.5 HOW IS BIOTECHNOLOGY USED IN AGRICULTURE?

The main goals of agriculture are to grow as much food as possible, as cheaply as possible, with minimal loss from pests such as insects and weeds. Many commercial farmers and seed suppliers have turned to biotechnology to achieve these goals. Many people, however, feel that the risks of genetically modified food to human health or the environment are not worth the benefits. We will explore this controversy in Section 13.8.

Many Crops Are Genetically Modified

According to the U.S. Department of Agriculture (USDA), in 2011, 88% of the corn, 90% of the cotton, and 94% of the soybeans grown in the United States were transgenic; that is, they contained genes from other species (**Table 13-1**). Globally, 15.4 million farmers planted more than 365 million acres of land with transgenic crops in 2010.

Crops are most commonly modified to improve their resistance to insects, herbicides, or both. Herbicide-resistant crops allow farmers to kill weeds without harming their crops. Less competition from weeds means more water, nutrients, and light for the crops, and hence, larger harvests. Many herbicides kill plants by inhibiting an enzyme that is used by plants, fungi, and some bacteria—but not animals—to synthesize specific amino acids. Without these amino acids, the plants die because they cannot synthesize proteins.

TABLE 13-1   Genetically Engineered Crops with USDA Approval

Genetically Engineered Trait	Potential Advantage	Examples
Resistance to herbicide	Application of herbicide kills weeds but not crop plants, producing higher crop yields	Beet, canola, corn, cotton, flax, potato, rice, soybean, tomato
Resistance to pests	Crop plants suffer less damage from insects, producing higher crop yields	Corn, cotton, potato, rice, soybean
Resistance to disease	Plants are less prone to infection by viruses, bacteria, or fungi, producing higher crop yields	Papaya, potato, squash
Sterile	Transgenic plants cannot cross with wild varieties, making them safer for the environment and more economically productive for the seed companies that produce them	Chicory, corn
Altered oil content	Oils can be made healthier for human consumption or can be made similar to more expensive oils (such as palm or coconut)	Canola, soybean

Many herbicide-resistant transgenic crops have been given a bacterial gene encoding an enzyme that functions even in the presence of these herbicides, so the transgenic plants continue to synthesize normal amounts of amino acids and proteins.

The insect resistance of many crops has been enhanced by giving them a gene, called *Bt*, from the bacterium *Bacillus thuringiensis*. The protein encoded by the *Bt* gene damages the digestive tract of insects, but not mammals. Transgenic *Bt* crops often suffer far less damage from insects than do regular crops, so farmers can apply less pesticide to their fields (Fig. 13-12).

Genetically Modified Plants May Be Used to Produce Medicines

The tools of biotechnology can also be used to insert medically useful genes into plants, producing medicines down on the "pharm." For example, a plant can be engineered to produce harmless proteins that are normally found in disease-causing bacteria or viruses. If these proteins resist digestion in the stomach and intestine, simply eating such plants could act as a vaccination against the disease-causing organisms. Several years ago, such "edible vaccines" were touted as a great way to provide vaccinations—no need to produce purified vaccines, no refrigeration needed, and, of course, no needles. Recently, however, many biomedical researchers have warned that edible vaccine plants may not really be a good idea, because there is no simple way to control the dose: Too little and the user doesn't develop decent immunity; too much, and the vaccine proteins might be harmful. Nevertheless, producing vaccine proteins in plants is still worthwhile. To make typical vaccines, pharmaceutical companies would just have to extract and purify the proteins before use. Plant-produced vaccines against hepatitis B, measles, rabies, tooth decay, flu, infant diarrhea, and many other diseases are in animal or clinical trials.

Molecular biologists can also engineer plants to produce human antibodies to combat various diseases. When a disease-causing microbe invades your body, it takes several days for your immune system to respond and produce

▲ **FIGURE 13-12   *Bt* plants resist insect attack** Transgenic cotton plants expressing the *Bt* gene (right) resist attack by bollworms, which eat cotton seeds. The transgenic plants therefore produce far more cotton than nontransgenic plants do (left).

? HAVE YOU EVER WONDERED . . .

If the Food You Eat Has Been Genetically Modified?

In addition to obvious things—tortilla chips, soy sauce, and margarine—corn and soy products are found in an amazing variety of foods. For example, corn syrup is an ingredient in foods as diverse as soda, ketchup, and bran flakes; soybean oil or protein is an important ingredient in cookies, cake mixes, and veggie burgers. Almost all of the corn and soy grown in the United States is genetically modified, with the result that about 80% of the packaged foods in American supermarkets contain substances made from GM plants. Many countries, including the European Union, require labeling of GM foods, but the U.S. Food and Drug Administration does not, so in America you can't tell if a food contains GM ingredients just by looking at the label. Therefore, unless you are extremely motivated and careful, you probably eat GM foods.

enough antibodies to overcome the infection. Meanwhile, you feel terrible, and might even die if the disease is serious enough. A direct injection of large quantities of the right antibodies might be able to cure the disease very quickly. Although none have yet entered medical practice, plant-derived antibodies against the bacteria that cause tooth decay and the virus that causes rabies have been developed.

Genetically Modified Animals May Be Useful in Agriculture and Medicine

Producing transgenic animals usually involves injecting the desired DNA, often incorporated into a disabled virus that cannot produce disease, into a fertilized egg. The egg is allowed to divide a few times in culture before being implanted into a surrogate mother. If the offspring are healthy and express the foreign gene, they are then bred with one another to produce homozygous transgenic animals. So far, it has proven difficult to create commercially valuable transgenic livestock, but several companies are working on it.

For example, biotechnology companies have made genetically modified sheep that produce more wool; cattle that produce more protein in their milk; and pigs that produce omega-3 fatty acids in their meat (omega-3 fatty acids are thought to provide a variety of health benefits). In 2010 researchers successfully produced transgenic chickens that cannot spread the H5N1 influenza virus, which causes avian flu. Worldwide, many millions of chickens have been killed to stop outbreaks of avian flu, so flu-resistant chickens could potentially be very valuable birds. There are even goats that secrete spider silk proteins in their milk. Spider silk is far stronger than steel or Kevlar®, the fiber usually used in bulletproof vests, so the hope is that lightweight, impenetrable vests could be made using silk protein from these goats.

Biotechnologists are also developing animals that produce medicines, such as human antibodies or other essential proteins. For example, there are genetically engineered sheep whose milk contains a protein, alpha-1-antitrypsin, that may prove valuable in treating cystic fibrosis and emphysema. Other GM sheep produce human clotting factors, which could be used to treat hemophilia. Other livestock have been engineered so that their milk contains erythropoietin (a hormone that stimulates red blood cell synthesis) or clot-busting proteins (to treat heart attacks caused by blood clots in the coronary arteries).

Finally, biomedical researchers have made a large number of transgenic animals, primarily mice, which carry genes associated with human diseases, such as Alzheimer's disease, Marfan syndrome, and cystic fibrosis. These animals are used to investigate the causes of disease and to develop possible treatments.

CHECK YOUR LEARNING

Can you describe the advantages of genetically modified crops and animals in agriculture, and provide some examples? Can you list some examples of how genetically modified animals might be useful in medicine?

13.6 HOW IS BIOTECHNOLOGY USED TO LEARN ABOUT THE GENOMES OF HUMANS AND OTHER ORGANISMS?

Genes influence virtually all the traits of human beings, including gender, size, hair color, intelligence, and susceptibility to disease organisms and toxic substances in the environment. To begin to understand how our genes influence our lives, the Human Genome Project was launched in 1990, with the goal of determining the nucleotide sequence of all the DNA in our entire set of genes, called the human genome.

By 2003, this joint project of molecular biologists in several countries had sequenced the human genome with an accuracy of about 99.99%. The human genome contains between 20,000 and 25,000 genes, comprising approximately 2% of the DNA. Some of the other 98% consists of promoters and regions that regulate how often individual genes are transcribed, but it's not really known what most of our DNA does.

Why do scientists want to sequence the genomes of people and other organisms? First, many genes were discovered whose functions are unknown. Using the genetic code to translate the DNA sequences of novel genes, biologists can predict the amino acid sequences of the proteins they encode. Comparing these proteins to familiar proteins whose functions are already known will enable us to find out what many of these newly discovered genes do.

Second, knowing the nucleotide sequences of human genes will have an enormous impact on medical practice. Over 2,000 genes are known to be associated with human inherited diseases. Defective alleles of these genes may cause or predispose people to develop medical conditions such as sickle-cell anemia, cystic fibrosis, breast cancer, alcoholism, schizophrenia, heart disease, Alzheimer's disease, and many others. Ongoing human genome research involves sequencing the DNA of many different people, looking for small differences in the DNA sequences of alleles that might increase a person's susceptibility to infectious diseases, toxic pollutants, or chemicals in tobacco smoke, or alter a person's responses to drugs. Someday, this knowledge will allow custom-tailored treatments for many diseases. Knowing the genomes of infectious viruses and bacteria should help in developing vaccines or treatments for the diseases they cause.

Third, the Human Genome Project, along with companion projects that have sequenced the genomes of organisms as diverse as bacteria, fungi, mice, and chimpanzees, helps us to appreciate our place in the evolution of life on Earth. For example, the DNA of humans and chimps differs by less than 5%. Comparing the similarities and differences may help biologists to understand what genetic differences help to make us human, and why we are susceptible to certain diseases that chimps are not. Recently, researchers have deciphered partial genomes for Neanderthals and recently discovered ancient hominids, called Denisovans. Modern human populations, depending on their origins, may have up to a few percent Neanderthal or Denisovan genes.

CHECK YOUR LEARNING

Can you describe the purposes and findings of the Human Genome Project? Do you understand why it is useful to know the DNA sequences of the genomes of other organisms?

13.7 HOW IS BIOTECHNOLOGY USED FOR MEDICAL DIAGNOSIS AND TREATMENT?

For over two decades, biotechnology has been used to diagnose some inherited disorders, even in fetuses (see "Health Watch: Prenatal Genetic Screening," page 250). More recently, medical researchers have begun using biotechnology in an attempt to cure, or at least treat, genetic diseases.

DNA Technology Can Be Used to Diagnose Inherited Disorders

A person inherits a genetic disease because he or she inherits one or more defective alleles, which differ from normal, functional alleles because they have different nucleotide sequences. Most methods of diagnosing genetic disorders begin with PCR, to make multiple copies of specific genes, and sometimes specific alleles. Restriction enzymes or DNA probes may then be used to identify defective alleles.

Using PCR to Obtain Disease-Specific Alleles

Recall that PCR uses specific DNA primers to determine which DNA sequences to amplify. Thanks to many years of research by hundreds of researchers, the DNA sequences of the genes responsible for many genetic disorders are now known. PCR can be used to isolate and amplify disease-specific genes for various types of diagnostic procedures (see below). In some cases, medical testing companies have designed primers that amplify only the defective alleles that cause a given disorder and not the normal alleles that do not cause disease, so PCR itself can be a diagnostic tool.

Restriction Enzymes May Cut Different Alleles of a Gene at Different Locations

Sickle-cell anemia is an inherited form of anemia—not having enough red blood cells—caused by a point mutation in which thymine replaces adenine near the beginning of the globin gene (see Chapters 10 and 12). A common diagnostic test for sickle-cell anemia relies on the fact that restriction enzymes cut DNA only at specific nucleotide sequences. To diagnose the presence of the sickle-cell allele, DNA is extracted from cells of a patient, a parent who might be a carrier of the allele, or even a fetus (see "Health Watch: Prenatal Genetic Screening," page 250). PCR is used to amplify a section of DNA that includes the mutation site. A restriction enzyme called MstII can cut the normal sequence (CCTGAGGAG), but not the sickle-cell sequence (CCTGTGGAG). The result is that MstII cuts the normal globin allele in half, but the sickle-cell allele remains intact. Gel electrophoresis easily separates the intact sickle-cell allele from the pieces of the normal allele.

Different Alleles Bind to Different DNA Probes

Cystic fibrosis is a disease caused by a defect in a protein, called CFTR, that normally helps to move chloride ions across the plasma membranes of many cells, including those in the lungs, sweat glands, and intestines (see the case study in Chapter 12). In the lungs, lack of adequate chloride transport results in thick mucus lining the airways, frequent bacterial infections, and eventual death. There are more than 1,500 different *CFTR* alleles, all at the same locus, each encoding a different, defective CFTR protein. People with either one or two normal alleles synthesize enough functioning CFTR proteins that they do not develop cystic fibrosis. People with two defective alleles (they may be the same or different alleles) do not synthesize functional transport proteins, and therefore they develop cystic fibrosis.

How can anyone hope to diagnose a disorder that may be caused by any of 1,500 different alleles? Fortunately, 32 alleles account for about 90% of the cases of cystic fibrosis—the rest of the alleles are extremely rare. Nevertheless, it would probably be hopeless to try to find unique restriction enzyme cut sites for even 32 alleles, so the methodology used to diagnose sickle-cell anemia cannot be used for cystic fibrosis.

However, each defective allele has a different nucleotide sequence. As you know, a DNA strand will form perfect base pairs only with a perfectly complementary strand. Several companies now produce cystic fibrosis "arrays," which are pieces of specialized filter paper to which single-stranded DNA probes are bound (Fig. 13-13). Each probe is complementary to one strand of a different *CFTR* allele (Fig. 13-13a). A person's DNA is tested for cystic fibrosis by cutting it into small pieces, separating the pieces into single strands, and labeling the strands with a colored molecule (Fig. 13-13b). The array is then bathed in the resulting solution of labeled DNA fragments. Under the right conditions, only a perfect complementary strand of the person's DNA will bind to any given probe on the array, thereby showing which *CFTR* alleles the person possesses (Fig. 13-13c). A similar test using DNA probes is also available for the diagnosis of sickle-cell anemia.

In the near future, routine medical diagnosis might use greatly expanded versions of this type of analysis, called DNA microarrays, to determine which bacteria or viruses are causing an infection. In one application, called the Virochip, thousands of small DNA probes, each complementary to a part of a specific viral gene, are placed on a small chip. Nucleic acids are isolated from the patient and labeled with a fluorescent dye. The chip is bathed in the labeled nucleic acid solution. Different spots light up on the chip, depending on what virus has infected the patient. Similar arrays are under development for the diagnosis of bacterial infections.

DNA microarrays could also help to provide more effective, customized medical care. Different people have different alleles of hundreds of genes; these may cause them to be more or less susceptible to many diseases, or to respond more or less well to various treatments. Someday, physicians might be able to use a DNA microarray containing thousands of probes for disease-related alleles, to determine which susceptibility alleles each patient carries, and tailor medical care

DNA probe for normal *CFTR* allele

DNA probes for 10 different mutant *CFTR* alleles

(a) Linear array of probes for cystic fibrosis

colored molecule

piece of patient's DNA

A T C A T C T T T G G T G

(b) *CFTR* **allele labeled with a colored molecule**

#1

Homozygous for normal *CFTR* alleles—the person is phenotypically normal

#2

One normal and one defective *CFTR* allele—the person is phenotypically normal

#3

Two different defective *CFTR* alleles—the person develops cystic fibrosis

(c) Linear arrays with labeled DNA samples from three different people

▲ FIGURE 13-13 **A cystic fibrosis diagnostic array**
(a) A typical diagnostic array for cystic fibrosis consists of special paper to which DNA probes complementary to the normal *CFTR* allele (far left spot) and several of the most common defective *CFTR* alleles (the other 10 spots) are attached. **(b)** DNA from a patient is cut into small pieces, separated into single strands, and the *CFTR* alleles are labeled with colored molecules. **(c)** The array is bathed in a solution of the patient's labeled DNA. The labeled DNA binds to different spots on the array, depending on which *CFTR* alleles the patient possesses.

accordingly. Several companies already produce small arrays tailored to investigate gene activity in specific diseases, such as cancers of the breast, prostate, and immune system. Some hospitals use them to provide patients with treatments that are most likely to succeed with their specific cancer.

Finally, several companies offer individualized DNA screening to the public. These tests use DNA microarrays to look for alleles that might predispose an individual to develop heart disease, breast cancer, arthritis, or other diseases. The companies even offer health advice, presumably based on the individual DNA analysis, although most of the advice—to exercise, keep blood pressure and cholesterol levels down, keep weight under control, and refrain from smoking—would be a good idea for anyone, regardless of what alleles they carry.

DNA Technology Can Help to Treat Disease

There are two principal applications of DNA technology for treating disease: (1) producing medicines, mostly in bacteria, using recombinant DNA techniques and (2) **gene therapy,** which seeks to cure diseases by inserting, deleting, or altering genes in a patient's cells.

Using Biotechnology to Produce Medicines

Thanks to recombinant DNA technology, several medically important proteins are now routinely made in bacteria. The first human protein made by recombinant DNA technology was insulin. Prior to 1982, when recombinant human insulin was first licensed for use, the insulin needed by people with diabetes was extracted from the pancreases of cattle or pigs slaughtered for meat. Although the insulin from these animals is very similar to human insulin, the slight differences caused an allergic reaction in about 5% of people with diabetes. Recombinant human insulin does not cause allergic reactions.

Other human proteins, such as growth hormone and clotting factors, can also be produced in transgenic bacteria. Some of these proteins were formerly obtained from either human blood or human cadavers; these sources are expensive and sometimes dangerous. As you probably know, blood can be contaminated by the human immunodeficiency virus (HIV), which causes acquired immune deficiency syndrome (AIDS). Cadavers may also contain several hard-to-diagnose infectious diseases, such as Creutzfeldt-Jakob syndrome, in which an abnormal protein can be passed from the tissues of an infected cadaver to a patient and cause fatal brain degeneration (see the Chapter 3 case study). Engineered proteins grown in bacteria or other cultured cells avoid these dangers. Some of the types of human proteins produced by recombinant DNA technology are listed in **Table 13-2.**

Treating AIDS by Gene Therapy

The human immunodeficiency virus enters several kinds of immune cells (mostly white blood cells), including a type called *helper T cells* that play a crucial role in responses to infection. HIV kills helper T cells. When the body's supply of helper T cells becomes too low, the immune response falters, ordinarily trivial infections become life threatening, and full-blown AIDS develops. Untreated, AIDS is almost always fatal within a few years.

Why isn't AIDS always fatal? HIV binds to a receptor protein, called CCR5, found on the surface of susceptible immune cells. HIV then moves into the cells and begins its deadly infectious cycle. But a tiny number of people carry a mutation and don't make CCR5 receptors, so they can't be infected with the usual strains of HIV.

Biotechnology offers the possibility of eliminating the CCR5 receptor in patients with AIDS and curing, or at least greatly alleviating, their disease. Molecular biologists can manufacture specialized enzymes tailor-made to cut up specific genes, such as the one that encodes the CCR5 receptors. The treatment would work like this: Immune cells are removed from a patient, and the *CCR5* gene is damaged with the enzyme. Although the cells try to repair the damaged

TABLE 13-2   Examples of Medical Products Produced by Recombinant DNA Methods

Type of Protein	Purpose	Example	Method of Production
Human hormones	Used in the treatment of diabetes and growth deficiency	Humulin™ (human insulin)	Human gene inserted into bacteria
Human cytokines (regulate immune system function)	Used in bone marrow transplants and to treat cancers and viral infections, including hepatitis and genital warts	Leukine™ (granulocyte-macrophage colony stimulating factor)	Human gene inserted into yeast
Antibodies (immune system proteins)	Used to fight infections, cancers, diabetes, organ rejection, and multiple sclerosis	Herceptin™ (antibodies to a protein expressed in some breast cancer cells)	Recombinant antibody genes inserted into cultured hamster cells
Viral proteins	Used to generate vaccines against viral diseases and for diagnosing viral infections	Engerix-B™ (hepatitis B vaccine)	Viral gene inserted into yeast
Enzymes	Used in the treatment of heart attacks, cystic fibrosis, and other diseases, and in the production of cheeses and detergents	Activase™ (tissue plasminogen activator)	Human gene inserted into cultured hamster cells

DNA, about a quarter of them fail, and can never make CCR5 receptors again. These CCR5-deleted cells are transfused back into the patient. In two small clinical trials, the numbers of functioning immune cells were greatly increased in most of the patients receiving this treatment.

Mature immune cells don't live forever, so this treatment would probably have to be repeated at intervals for the rest of the patient's life. However, all the cells of the immune system originate from stem cells in the bone marrow. Stem cells are capable of dividing throughout a person's life, continuing to produce one or more (often many) different types of daughter cells (see Chapter 9 for more information about stem cells). Ideally, stem cells could be removed from a patient, genetically "repaired," and replaced into the patient. In the case of HIV, all of the helper T cells that derive from the CCR5-deleted stem cells would also lack CCR5 and resist HIV infection. In addition, natural selection should operate in the patient's body: HIV would continue to kill off the unrepaired cells, but not the CCR5-deleted cells. Eventually, the patient's entire population of helper T cells should consist of CCR5-deleted cells, and the cure should be permanent.

Treating Severe Combined Immune Deficiency by Gene Therapy

Severe combined immune deficiency (SCID) is a rare disorder in which a child fails to develop an immune system. About 1 in 80,000 children is born with some form of SCID. Infections that would be trivial in a normal child become life threatening. In some cases, if the child has an unaffected relative with a similar genetic makeup, a bone marrow transplant from the healthy relative can give the child functioning stem cells, so that he or she can develop a working immune system. Most children with SCID, however, die before their first birthday.

Most forms of SCID are caused by defective recessive alleles of one of several genes. In one type of SCID, affected children are homozygous recessive for a defective allele that normally codes for an enzyme called adenosine deaminase (this condition is often called ADA-SCID). In 1990, gene therapy was performed

on 4-year-old Ashanti DeSilva, who suffered from ADA-SCID. Some of her white blood cells were removed, genetically altered with a virus containing a functional version of her defective allele, and then returned to her bloodstream. The treatment was a partial success, but not a complete cure. Ashanti, now a healthy adult, continues to receive regular injections of a form of adenosine deaminase to boost her immune system. More recent clinical trials have used a somewhat different gene therapy to cure ADA-SCID. The researchers removed bone marrow stem cells from children with ADA-SCID, inserted a functional copy of the adenosine deaminase gene into the cells, and returned the repaired cells into the children. Because bone marrow stem cells continue to produce new white blood cells throughout life, the hope is that these children might be permanently cured. As of late 2011, all 27 children treated in this way are alive and healthy, and so far, 19 have needed no additional treatments to enhance their immune responses.

A second type of SCID, called X-linked SCID, is caused by a defective recessive allele of a gene located on the X chromosome. Twenty children have been given gene therapy to insert a functional copy of this gene into their bone marrow stem cells. Eighteen appear to be cured, some for as long as 10 years after treatment. However, gene therapy for X-linked SCID is not without risks: Several children developed leukemia, apparently because the gene insertion turned on an oncogene (see Chapter 9). More recent methods seem to have reduced, and perhaps eliminated, this danger.

Although the parents knew there were dangers associated with gene therapy, as one mother put it, "We didn't really have a choice." Now, instead of dying as infants, or living encapsulated within a sterile bubble, isolated from human contact, these children can live normal lives—going to school, playing soccer, and riding horses.

CHECK YOUR LEARNING

Can you explain how biotechnology is used to diagnose both inherited and infectious diseases? Can you describe the procedures and advantages of gene therapy to cure inherited diseases?

Health Watch

Prenatal Genetic Screening

Prenatal diagnosis of a variety of genetic disorders, including cystic fibrosis, sickle-cell anemia, muscular dystrophy, and Down syndrome, requires samples of fetal cells or chemicals produced by the fetus. Presently, three techniques are commonly used to obtain samples for prenatal diagnosis: amniocentesis, chorionic villus sampling, and maternal blood collection.

Amniocentesis

The human fetus, like all animal embryos, develops in a watery environment. A waterproof membrane called the amnion surrounds the fetus and holds the fluid. As the fetus develops, it releases various chemicals (often in its urine) and sheds some of its cells into the amniotic fluid. When a fetus is 15 weeks or older, amniotic fluid can be collected safely by a procedure called **amniocentesis.**

First, the physician determines the position of the fetus by ultrasound scanning. How does that work? You probably already know how bats find moths at night: They shriek extremely high-frequency sound (far above the limits of human hearing) and listen for echoes bouncing off the moth's body. To locate a fetus with ultrasound, high-frequency sound is broadcast into a pregnant woman's abdomen, and sophisticated instruments convert the resulting echoes into a real-time image of the fetus (**Fig. E13-4**). Using the ultrasound image as a guide, the physician carefully inserts a sterilized needle through the abdominal wall, the uterus, and the amnion (being sure to avoid the fetus and placenta), and withdraws 10 to 20 milliliters of amniotic fluid (**Fig. E13-5**). Amniocentesis carries a slight risk of miscarriage, about 0.5% or less.

Chorionic Villus Sampling

The chorion is a membrane that is produced by the fetus and becomes part of the placenta. The chorion produces many small projections, called villi. In **chorionic villus sampling (CVS),** a physician inserts a small tube into the uterus through the mother's vagina and suctions off a few villi for analysis (see Fig. E13-5). The loss of a few villi does not harm the fetus. CVS has two major advantages over amniocentesis. First, it can be done much earlier in pregnancy—as early as the 8th week, but usually between the 10th and 12th weeks. This is especially important if the woman is contemplating a therapeutic abortion if her fetus has a major defect. Second, the sample contains far more fetal cells than can be obtained by amniocentesis. However, CVS appears to have a slightly greater risk of causing a miscarriage than amniocentesis does. Also, because the chorion is outside of the amniotic sac, CVS does not obtain a sample of the amniotic fluid. Finally, in some cases chorionic cells have chromosomal abnormalities that are in fact not present in the fetus, which complicates karyotyping. For these reasons, CVS is less commonly performed than amniocentesis.

amniotic fluid

head

neck

torso

▲ **FIGURE E13-4   A human fetus imaged with ultrasound**

Maternal Blood Collection

A tiny number of fetal cells cross the placenta and enter the mother's bloodstream as early as the sixth week of pregnancy. Separating fetal cells (perhaps as few as one per milliliter of blood) from the huge numbers of maternal cells is challenging, but it can be done. Surprisingly, there is also fetal DNA floating free in the mother's blood. In addition, a variety of proteins and other chemicals produced by the fetus may enter the mother's bloodstream.

Analyzing the Samples

Both amniotic fluid and maternal blood are briefly centrifuged to separate the cells from the fluids (amniotic fluid or plasma, respectively). Biochemical analysis may be performed to measure the concentrations of specific hormones, enzymes, or other proteins in the fluids. For example, if the amniotic fluid contains high concentrations of an embryonic protein called alpha-fetoprotein, this indicates that the fetus may have nervous system disorders, such as spina bifida, in which the spinal cord is incomplete, or anencephaly, in which major portions of the brain fail to develop. Specific combinations of alpha-fetoprotein, estrogen, and other chemicals in maternal plasma (detected by triple screen, quad screen, or penta screen tests, depending on whether three, four, or five chemicals are analyzed) indicate the likelihood of Down syndrome, spina bifida, or certain other disorders. However, these biochemical assays do not provide definitive diagnoses. Therefore, if they indicate that a disorder is present, then other tests, such as karyotyping, DNA analysis of fetal cells, or highly detailed

13.8 WHAT ARE THE MAJOR ETHICAL ISSUES OF MODERN BIOTECHNOLOGY?

Modern biotechnology offers the promise—some would say the threat—of greatly changing our lives, and the lives of many other organisms on Earth. Is humanity capable of handling the responsibility of biotechnology? Here we will explore two important issues: the use of genetically modified organisms in agriculture and the prospects for genetically modifying human beings.

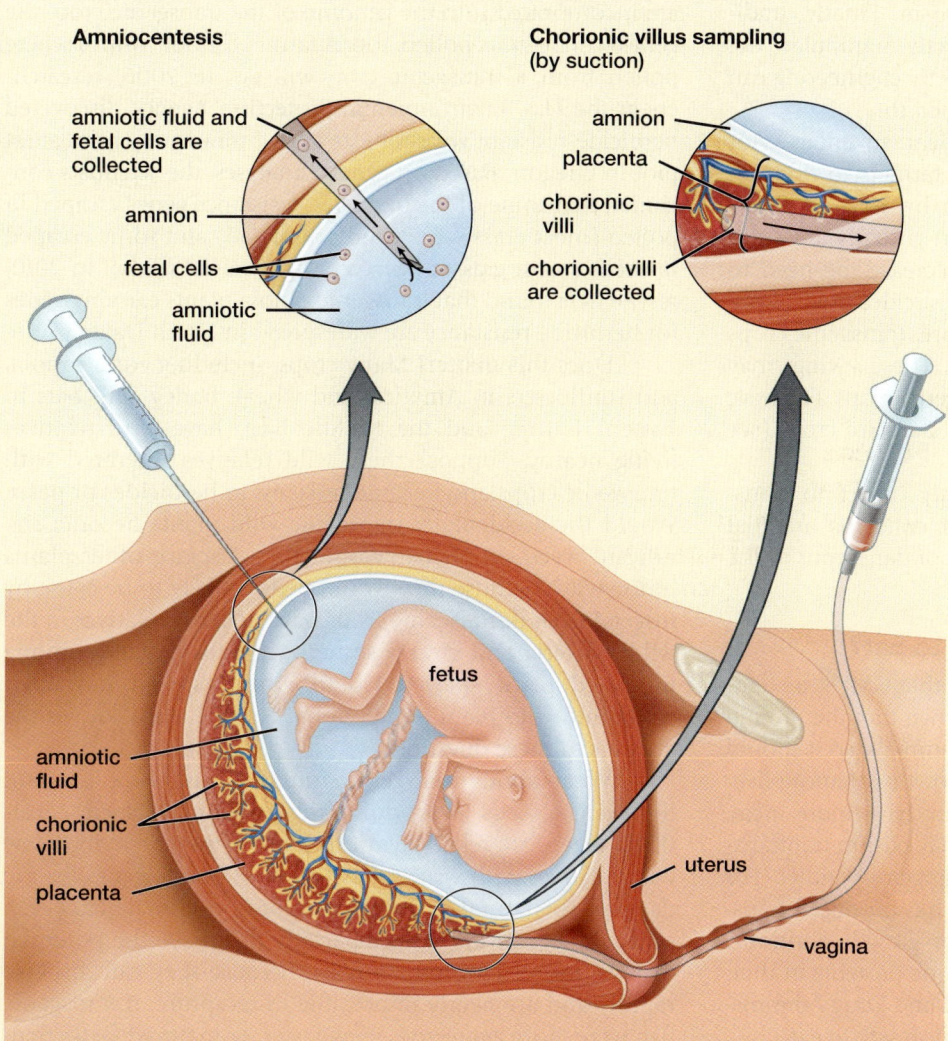

Amniocentesis

amniotic fluid and fetal cells are collected

amnion

fetal cells

amniotic fluid

Chorionic villus sampling (by suction)

amnion

placenta

chorionic villi

chorionic villi are collected

fetus

amniotic fluid

chorionic villi

placenta

uterus

vagina

◀ **FIGURE E13-5 Prenatal sampling techniques** The two most common ways of obtaining samples for prenatal diagnosis are amniocentesis and chorionic villus sampling. (In reality, CVS is usually performed when the fetus is much younger than the one depicted in this illustration.)

Biotechnology techniques can be used to analyze fetal DNA for many defective alleles, such as those that cause sickle-cell anemia or cystic fibrosis. Modified PCR techniques can also be used to diagnose Down syndrome, eliminating the need to culture fetal cells obtained by amniocentesis. In Great Britain, PCR is often used as a stand-alone method for diagnosing Down syndrome. In the United States, karyotyping remains the gold standard.

Prenatal diagnosis using fetal DNA obtained from maternal blood samples is advancing rapidly. Researchers in Hong Kong recently developed methods based on PCR and DNA sequencing to diagnose Down syndrome using fetal DNA found in maternal blood. Several companies now offer paternity testing based on fetal cells or fetal DNA isolated from maternal blood.

You may think that prenatal diagnosis is useful mainly when the parents are contemplating an abortion if the fetus has a serious genetic disorder. However, prenatal diagnosis

ultrasound examination of the fetus, are employed to find out whether or not the fetus actually has one of these conditions.

Fetal cells, or at least fetal DNA, are required for karyotyping or DNA analysis. Amniotic fluid contains very small numbers of fetal cells, so to obtain enough cells for karyotyping or DNA analysis, the usual procedure is to grow the cells in culture for a week or two. The large number of fetal cells obtained by CVS means that karyotyping and DNA analyses can usually be performed without culturing the cells first (although some labs do culture the cells for a few days). Karyotyping the fetal cells can show if there are too many or too few copies of the chromosomes, and if any chromosomes show structural abnormalities. Down syndrome, for example, results from the presence of three copies of chromosome 21 (see Chapter 10).

can also be used to provide better care for an affected infant. For example, if the fetus is homozygous for the sickle-cell allele, some therapeutic measures can be taken beginning at birth. In particular, regular doses of penicillin greatly reduce bacterial infections that otherwise kill about 15% of homozygous children. Further, knowing that a child has the disorder ensures correct diagnosis and rapid treatment during a "sickling crisis," when malformed red blood cells clump and block blood flow. Fetuses with spina bifida are often delivered by planned cesarean section, both to reduce damage to the spinal cord during birth and so that a surgical team can be standing by in case immediate surgery on the infant is necessary. Comparable medical or behavioral interventions can be planned for several other disorders as well.

Should Genetically Modified Organisms Be Permitted in Agriculture?

The aims of traditional and modern agricultural biotechnology are the same: to modify the genetic makeup of living organisms to make them more useful. However, there are

three principal differences. First, traditional biotechnology is slow; many generations of selective breeding are usually necessary to produce useful new strains of plants or animals. Genetic engineering, in contrast, can potentially introduce massive genetic changes in a single generation. Second, traditional biotechnology almost always recombines genetic material from the same, or at least very closely related, species, whereas genetic engineering can recombine DNA from very different species in a single organism. Finally, traditional biotechnology has no way to directly manipulate the DNA sequence of genes themselves. Genetic engineering can produce new genes never before seen on Earth.

The best transgenic crops have clear advantages for farmers. Herbicide-resistant crops allow farmers to rid their fields of weeds, which reduce harvests by 10% or more, through the use of powerful herbicides at virtually any stage of crop growth. Insect-resistant crops decrease the need to apply pesticides, saving the cost of the pesticides themselves, as well as tractor fuel and labor. Therefore, transgenic crops may produce larger harvests at less cost. These savings may be passed along to the consumer. Transgenic crops also have the potential to be more nutritious than standard crops (see "Health Watch: Golden Rice").

However, many people strenuously object to transgenic crops or livestock. The principal concerns are that GMOs may be harmful to human health or dangerous to the environment.

Are Foods from GMOs Dangerous to Eat?

In most cases, there is no reason to think that GMOs are dangerous to eat. For example, tests have shown that the protein encoded by the *Bt* gene is not toxic to mammals, so it should not be dangerous to human health. If growth-enhanced livestock are ever marketed, they will simply have more meat, composed of exactly the same proteins that exist in nontransgenic animals, so they shouldn't be dangerous either. For example, a company called AquaBounty has produced transgenic salmon containing extra genes for growth hormone. The fish grow faster, but will have the same proteins in their flesh as any other salmon. The U.S. Food and Drug Administration has declared the AquaBounty salmon safe to eat.

On the other hand, some people might be allergic to genetically modified plants. In the 1990s, a gene from Brazil nuts was inserted into soybeans in an attempt to improve the balance of amino acids in soybean protein. It was soon discovered that people allergic to Brazil nuts would probably also be allergic to the transgenic soybeans. These transgenic soybean plants never made it to the farm. The U.S. Food and Drug Administration now requires all new transgenic crop plants to be tested for allergenic potential.

Are GMOs Hazardous to the Environment?

The environmental effects of GMOs are much more debatable. One clear positive effect of *Bt* crops is that farmers apply less pesticide to their fields. This should translate into less pollution of the environment and less harm to the farmers. For example, in 2002 and 2003, Chinese farmers planting *Bt* rice reduced pesticide use by 80% compared to farmers planting conventional rice. Further, they suffered no instances of pesticide poisoning, compared to about 5% of farmers planting conventional rice. A 10-year study in Arizona showed that *Bt* cotton allowed farmers to use less pesticide while obtaining the same yields of cotton.

On the other hand, *Bt* or herbicide-resistance genes might spread outside a farmer's fields. Because these genes are incorporated into the genome of the transgenic crop, the genes will be in its pollen, too. A farmer cannot control where pollen from a transgenic crop will go. In 2006, researchers at the U.S. Environmental Protection Agency discovered herbicide-resistant grasses more than 2 miles away from a test plot in Oregon. Based on genetic analyses, the scientists concluded that some of the herbicide-resistance genes escaped in pollen (most grasses are wind pollinated) and some escaped in seeds (most grasses have very lightweight seeds). In 2010, researchers found that transgenic canola plants carrying genes for herbicide resistance are widespread in North Dakota.

Does this matter? Many crops, including corn, canola, and sunflowers in America, and wheat, barley, and oats in Eastern Europe and the Middle East, have wild relatives living nearby. Suppose these wild relatives interbred with transgenic crops and became resistant to herbicides or pests. Would the accidentally transgenic wild plants become significant weed problems? Would they displace other plants in the wild, because they would be less likely to be eaten by insects? Even if transgenic crops have no close relatives in the wild, bacteria and viruses sometimes transfer genes among unrelated plant species. Could viruses spread unwanted genes into wild plant populations? No one really knows the answers to these questions.

What about transgenic animals? Unlike pollen and lightweight seeds, most domesticated animals, such as cattle or sheep, are relatively immobile. Further, most have few wild relatives with which they might exchange genes, so the dangers to natural ecosystems appear minimal. However, some transgenic animals, especially fish, have the potential to pose more significant threats, because they can disperse rapidly and are nearly impossible to recapture. If transgenic fish were more aggressive, grew faster, or matured faster than wild fish, they might replace native populations. One possible way out of this dilemma, pioneered by AquaBounty, is to raise only sterile transgenic fish, so that any escapees would die without reproducing and thus have minimal impact on natural ecosystems. Skeptics worry that the sterilization process may not be 100% effective, leaving open the possibility of fertile transgenic fish escaping into the wild.

Should the Genome of Humans Be Changed by Biotechnology?

Many of the ethical implications of human applications of biotechnology are fundamentally the same as those connected with other medical procedures. For example, long before biotechnology enabled prenatal testing for cystic fibrosis or sickle-cell anemia, trisomy 21 (Down syndrome) could be diagnosed in embryos by simply counting the chromosomes in cells taken from amniotic fluid (see "Health Watch: Prenatal Genetic Screening"). Whether parents should use such information as a basis for choosing an abortion or to prepare to care for the affected child generates enormous controversy. Other ethical concerns, however, have arisen purely as a result of advances

Health|Watch

Golden Rice

Rice is the principal food for about two-thirds of the people on Earth. Rice provides carbohydrates and some protein, but is a poor source of many vitamins, including vitamin A. Unless people eat enough fruits and vegetables, they often lack sufficient vitamin A, and may suffer from poor vision, immune system defects, and damage to their respiratory, digestive, and urinary tracts. According to the World Health Organization, more than 100 million children suffer from vitamin A deficiency. As a result, each year 250,000 to 500,000 children become blind, principally in Asia, Africa, and Latin America; half of those children die. Vitamin A deficiency typically strikes the poor, because rice may be all they can afford to eat. In 1999, biotechnology provided a possible remedy: rice genetically engineered to contain beta-carotene, a pigment that makes daffodils bright yellow and that the human body easily converts into vitamin A.

Creating a rice strain with high levels of beta-carotene wasn't simple. However, funding from the Rockefeller Institute, the European Community Biotech Program, and the Swiss Federal Office for Education and Science enabled molecular biologists Ingo Potrykus and Peter Beyer to tackle the task. They inserted three genes into the rice genome, two from daffodils and one from a bacterium. As a result, "Golden Rice" grains synthesize beta-carotene.

The trouble was, the original Golden Rice didn't make very much beta-carotene, so people would have had to eat enormous amounts to get enough vitamin A. The Golden Rice community didn't give up. It turns out that daffodils aren't the best source for genes that direct beta-carotene synthesis. Golden Rice 2, with genes from corn, produces over 20 times more beta-carotene than the original Golden Rice does, and consequently is bright yellow (**Fig. E13-6**). About 2 cups of cooked Golden Rice 2 should provide enough beta-carotene to equal the full recommended daily amount of vitamin A. Golden Rice 2 was given, free, to the Humanitarian Rice Board for experiments and planting in Southeast Asia.

However, Golden Rice faces other hurdles. Many people strongly resist large-scale planting of Golden Rice (or other transgenic crops). Getting Golden Rice genes into the popular Asian strains required years of traditional genetic crosses. By 2007, the International Rice Research Institute succeeded in incorporating the carotene-synthesizing genes of Golden Rice into Asian rice strains, and the first field trials of these Golden

▲ **FIGURE E13-6 Golden rice** The high beta-carotene content of Golden Rice 2 gives it a bright yellow color. Normal rice lacks beta-carotene and is off-white.

Rice varieties began in the Philippines in April 2008. Locally adapted strains of Golden Rice 2 are expected to be available for farmers in the Philippines to plant in 2013.

Is Golden Rice the best way, or the only way, to solve the problems of malnutrition in poor people? Perhaps not. For one thing, many poor people's diets are deficient in many nutrients, not just vitamin A. To help solve that problem, the Bill and Melinda Gates Foundation is funding research to increase the levels of vitamin E, iron, and zinc in rice. Further, not all poor people eat mostly rice. In parts of Africa, sweet potatoes are the main source of calories. Eating orange, instead of white, sweet potatoes, has dramatically increased vitamin A intake for many of these people. Finally, in many parts of the world, governments and humanitarian organizations have started vitamin A supplementation programs. In some parts of Africa and Asia, as many as 80% of the children receive large doses of vitamin A a few times when they are very young. Some day, the combination of these efforts may result in a world in which no children suffer blindness from the lack of a simple nutrient in their diets.

---

in biotechnology. For instance, should people be allowed to select, or even change, the genomes of their offspring?

On July 4, 1994, a girl in Colorado was born with Fanconi anemia, an inherited disorder that is fatal without a transplant of stem cells from the bone marrow of a genetically compatible donor. Her parents wanted another child—a very special child. They wanted one without Fanconi anemia, of course, but they also wanted a child who could serve as a stem cell donor for their daughter. They went to Yury Verlinsky of the Reproductive Genetics Institute for help. Verlinsky used the parents' sperm and eggs to create dozens of embryos in culture. The embryos were tested both for the genetic defect and for tissue compatibility with the couple's daughter. Verlinsky chose an embryo with the desired genotype and implanted it into the mother's uterus. Nine months later, a son was born. Blood from his umbilical cord provided stem cells to transplant into his sister's bone marrow. Today, the girl's bone marrow failure has been cured, although she will always have anemia and many accompanying symptoms. Were these appropriate uses of genetic screening? Should dozens of embryos be created, knowing that the vast majority will be discarded? Is this ethical if it is the only way to save the life of another child?

The same technologies used to insert genes into stem cells to cure SCID could be used to insert or change the genes

parents with a genetic disease

fertilized egg with a defective gene

embryo with a genetic defect

cell removed and cultured

treated culture

therapeutic gene

viral vector

genetically corrected cell from culture

egg cell without a nucleus

genetically corrected egg cell

genetically corrected clone of the original embryo

healthy baby

◄ **FIGURE 13-14 Using biotechnology to correct genetic defects in human embryos** In this hypothetical example, human embryos would be derived from eggs fertilized by sperm in culture, with one or both of the parents having a genetic disorder. When an embryo containing a defective gene grows into a small cluster of cells, a single cell would be removed from the embryo, and the defective gene in the cell replaced using an appropriate vector, usually a disabled virus. The nucleus of another egg cell (taken from the same mother) would be removed. The genetically repaired cell could then be injected into the egg whose nucleus had been removed. The now repaired egg cell would then be implanted in the woman's uterus for fetal development.

of fertilized eggs (Fig. 13-14). Suppose it were possible to insert functional *CFTR* alleles into human eggs, thereby preventing cystic fibrosis. Would this be an ethical change to the human genome? What about making bigger football players or more beautiful supermodels? If and when the technology is developed to cure diseases, it will be difficult to prevent it from being used for nonmedical purposes. Who will determine which uses are appropriate and which are trivial vanity?

CHECK YOUR LEARNING

Can you explain why people might be opposed to the use of genetically modified organisms in agriculture? Can you envision circumstances in which it would be ethical to modify the genome of a human fertilized egg?

CASE STUDY r e v i s i t e d

Guilty or Innocent?

DNA profiling is often considered the gold standard of evidence, because of its extremely low error rate. Thanks to the power of DNA profiling, Dennis Maher is now a free man. And he's not alone. DNA evidence has proved the innocence of more than 300 people in the United States who had been wrongfully convicted of major crimes—and 17 of those people had spent time on death row.

Many people exonerated after years in prison have a hard time adjusting to life "outside." Maher, however, is a real success story. Soon after his release, he got a job servicing trucks for Waste Management, a trash pickup and recycling company. He met his wife, Melissa, through an online dating service, using the name DNADennis. They have two children, one named Aliza after his Innocence Project attorney. Does Maher harbor any resentment against the system that cost him 19 years of his life? Sure. As he puts it, "I lost what I can never get back." But he also says, "I don't have time to be angry. If I'm an angry person, I won't have the things I have in my life."

If you're a fan of TV crime shows such as *CSI*, you know that DNA profiling also helps police to solve cases. Most police departments in large cities now have a "cold case unit" that takes a fresh look at stale evidence, sometimes leading to convictions many years after a perpetrator seemed to have gotten away with a crime. In Pasadena, California, for example, a 72-year-old woman was murdered in her home in 1987, before DNA profiling had been invented. Although police thought they knew the assailant, they didn't have enough evidence to prove it, so the man was freed. In 2007, however, Pasadena set up an Unsolved Homicide Case Unit. Their first success was to use DNA profiling to identify, arrest, and convict the murderer, now serving 15 years to life in prison.

Consider This

BioEthics Who are the heroes in these stories? There are the obvious ones, of course—Innocence Project attorney Aliza Kaplan, the law students who spent many hours slogging through poorly labeled boxes searching for evidence, the members of the Pasadena Police Department, and, of course, Dennis Maher. But what about molecular biologist Kary Mullis, who discovered PCR? Or Thomas Brock, whose discovery of *Thermus aquaticus* in Yellowstone hot springs provided the source of heat-stable DNA polymerase that is so essential to PCR (see Fig. 13-4)? Or the hundreds of biologists, chemists, and mathematicians who developed procedures for gel electrophoresis, DNA labeling, and statistical analysis of sample matching?

Scientists often say that science is worthwhile for its own sake, and that it is difficult or impossible to predict which discoveries will lead to the greatest benefits for humanity. Nonscientists, when asked to pay the costs of scientific projects, are sometimes skeptical of such claims. How do you think that public support of science should be allocated? Fifty years ago, would you have voted to give Thomas Brock public funds to see what types of organisms lived in hot springs?

CHAPTER REVIEW

Summary of Key Concepts

13.1 What Is Biotechnology?

Biotechnology is the use, and especially the alteration, of organisms, cells, or biological molecules to produce food, drugs, or other goods. Modern biotechnology generates altered genetic material by genetic engineering. Genetic engineering frequently involves the production of recombinant DNA by combining DNA from different organisms. When DNA is transferred from one organism to another, the recipients are called transgenic or genetically modified organisms (GMOs). Major applications of modern biotechnology include increasing our understanding of gene function, treating disease, improving agriculture, and solving crimes.

13.2 How Does DNA Recombine in Nature?

DNA recombination occurs naturally through processes such as sexual reproduction; bacterial transformation, in which bacteria acquire DNA from plasmids or other bacteria; and viral infection, in which viruses incorporate fragments of DNA from their hosts and transfer the fragments to members of the same or other species.

13.3 How Is Biotechnology Used in Forensic Science?

Specific regions of very small quantities of DNA, such as might be obtained at a crime scene, can be amplified by the polymerase chain reaction. The most common regions used in forensics are short tandem repeats (STRs). The STRs are separated by gel electrophoresis and made visible with DNA probes. The pattern of STRs, called a DNA profile, can be used to match DNA found at a crime scene with DNA from suspects with extremely high accuracy.

13.4 How Is Biotechnology Used to Make Genetically Modified Organisms?

There are three steps to making a genetically modified organism. First, the desired gene is obtained from another organism, or, less commonly, synthesized. Second, the gene is cloned, often into a bacterial plasmid, to provide multiple copies of the gene. Third, the gene is inserted into a host organism, often through the action of bacteria or viruses, with gene guns, or by injection into cells (especially fertilized eggs).

13.5 How Is Biotechnology Used in Agriculture?

Many crop plants have been modified by the addition of genes that promote herbicide resistance or insect resistance. Plants may also be modified to produce human proteins, vaccines, or antibodies. Transgenic animals may be produced as well, with properties such as faster growth, increased production of valuable products such as milk, or the ability to produce human proteins, vaccines, or antibodies.

13.6 How Is Biotechnology Used to Learn About the Genomes of Humans and Other Organisms?

Techniques of biotechnology were used to discover the complete nucleotide sequence of the human genome. This knowledge is being used to learn the identities and functions of new genes, to discover medically important genes, to explore genetic variability among individuals, and to better understand the evolutionary relationships between humans and other organisms.

13.7 How Is Biotechnology Used for Medical Diagnosis and Treatment?

Biotechnology may be used to diagnose genetic disorders such as sickle-cell anemia or cystic fibrosis. For example, in the diagnosis of sickle-cell anemia, specific restriction enzymes can cut the normal globin allele but not the sickle-cell allele. The resulting DNA fragments of different lengths may then be separated and identified by gel electrophoresis. In the diagnosis of cystic fibrosis, DNA probes complementary to various cystic fibrosis alleles are placed on a DNA array. Base-pairing of a patient's DNA to specific probes on the array identifies which alleles are present in the patient.

Inherited diseases are caused by defective alleles of crucial genes. Genetic engineering may be used to insert functional alleles of these genes into normal cells, stem cells, or even into eggs to correct the genetic disorder.

13.8 What Are the Major Ethical Issues of Modern Biotechnology?

The use of genetically modified organisms in agriculture is controversial for two major reasons: food safety and potentially harmful effects on the environment. In general, GMOs contain proteins that are harmless to mammals, are readily digested, or are already found in other foods. The transfer of potentially allergenic proteins to normally nonallergenic foods can be avoided by thorough testing. Environmental effects of GMOs are more difficult to predict. It is possible that foreign genes, such as those for pest or herbicide resistance, might be transferred to wild plants, with resulting damage to agriculture and/or disruption of ecosystems. If they escape, highly mobile transgenic animals might displace their wild relatives.

Genetically selecting or modifying human embryos is highly controversial. As technologies improve, society may be faced with decisions about the extent to which parents should be allowed to correct or enhance the genomes of their children.

Key Terms

amniocentesis *250*
biotechnology *233*
chorionic villus
   sampling (CVS) *250*
cloning *233*
DNA cloning *243*
DNA probe *239*
DNA profile *239*
gel electrophoresis *239*
gene therapy *248*
genetic engineering *233*

genetically modified
   organism (GMO) *233*
plasmid *234*
polymerase chain
   reaction (PCR) *237*
recombinant DNA *233*
restriction enzyme *243*
short tandem repeat (STR) *238*
transfect *244*
transformation *234*
transgenic *233*

Learning Outcomes

In this chapter, you have learned to . . .

LO1 Describe natural processes that recombine DNA.

LO2 Define polymerase chain reaction, short tandem repeats, gel electrophoresis, and DNA probes, and describe how electrophoresis and the polymerase chain reaction work.

LO3 Describe how DNA profiles are generated, and explain how they can be used to identify individual people.

LO4 List the purposes for genetically modifying

organisms, and describe the steps involved in creating a genetically modified organism.

LO5 Describe the potential drawbacks of genetically modified organisms.

LO6 Summarize the findings that resulted from the Human Genome Project, and describe the potential benefits of those findings.

LO7 Explain how biotechnology can be used to diagnose and treat inherited disorders.

Thinking Through the Concepts

Fill-in-the-Blank

1. _____ are organisms that contain DNA that has been modified (usually through use of recombinant DNA technology) or derived from other species. **LO4**

2. _____ is the process whereby bacteria pick up DNA from their environment. This DNA may be part of a chromosome or it may be tiny circles of DNA called _____. **LO1**

3. The _____ is a technique for multiplying DNA in the laboratory. **LO2**

4. Matching DNA samples in forensics uses a specific set of small "genes" called _____. The alleles of these genes in different people vary in the _____ of the allele. The pattern of these alleles that a given person possesses is called his or her _____. **LO3**

5. Pieces of DNA can be separated according to size by a process known as _____. The identity of a

specific sample of DNA is usually determined by binding a synthetic piece of DNA called a(n) _____, which binds to the sample DNA by _____. **LO2**

Review Questions

1. Describe three natural forms of genetic recombination, and discuss the similarities and differences between recombinant DNA technology and these natural forms of genetic recombination. **LO1**

2. What is a plasmid? How are plasmids involved in bacterial transformation? **LO1**

3. What is a restriction enzyme? How can restriction enzymes be used to splice a piece of human DNA into a plasmid? **LO4**

4. Describe the polymerase chain reaction. **LO2**

5. What is a short tandem repeat? How are short tandem repeats used in forensics? **LO3**

6. How does gel electrophoresis separate pieces of DNA? **LO2**

7. How are DNA probes used to identify specific nucleotide sequences of DNA? How are they used in the diagnosis of genetic disorders? **LO2 LO7**

8. Describe several uses of genetic engineering in agriculture. **LO5**

9. Describe several uses of genetic engineering in human medicine. **LO7**

10. Describe amniocentesis and chorionic villus sampling, including the advantages and disadvantages of each. What are their medical uses? **LO7**

Applying the Concepts

1. **BioEthics** In a 2004 Web survey conducted by the Canadian Museum of Nature, 84% of the people polled said they would eat a genetically modified banana that contained a vaccine against an infectious disease, whereas only 47% said they would eat a GM banana with extra vitamin C produced by the action of a rat gene. Do you think that this difference in acceptance of GMOs is scientifically valid? Why or why not?

2. **BioEthics** As you may know, many insects have evolved resistance to common pesticides. Do you think that insects might evolve resistance to *Bt* crops? If this is a risk, do you think that *Bt* crops should be planted anyway? Why or why not?

3. **BioEthics** Discuss the ethical issues that surround the release of genetically modified organisms (plants, animals, or bacteria) into the environment. What could go wrong? What precautions might prevent the problems you listed from occurring? What benefits do you think would justify the risks?

4. All children born with X-linked SCID are boys. Can you explain why?

Answers to Figure Caption questions and Fill-in-the-Blank questions can be found in the Answers section at the back of the book.

MB *Go to MasteringBiology for practice quizzes, activities, eText, videos, current events, and more.*

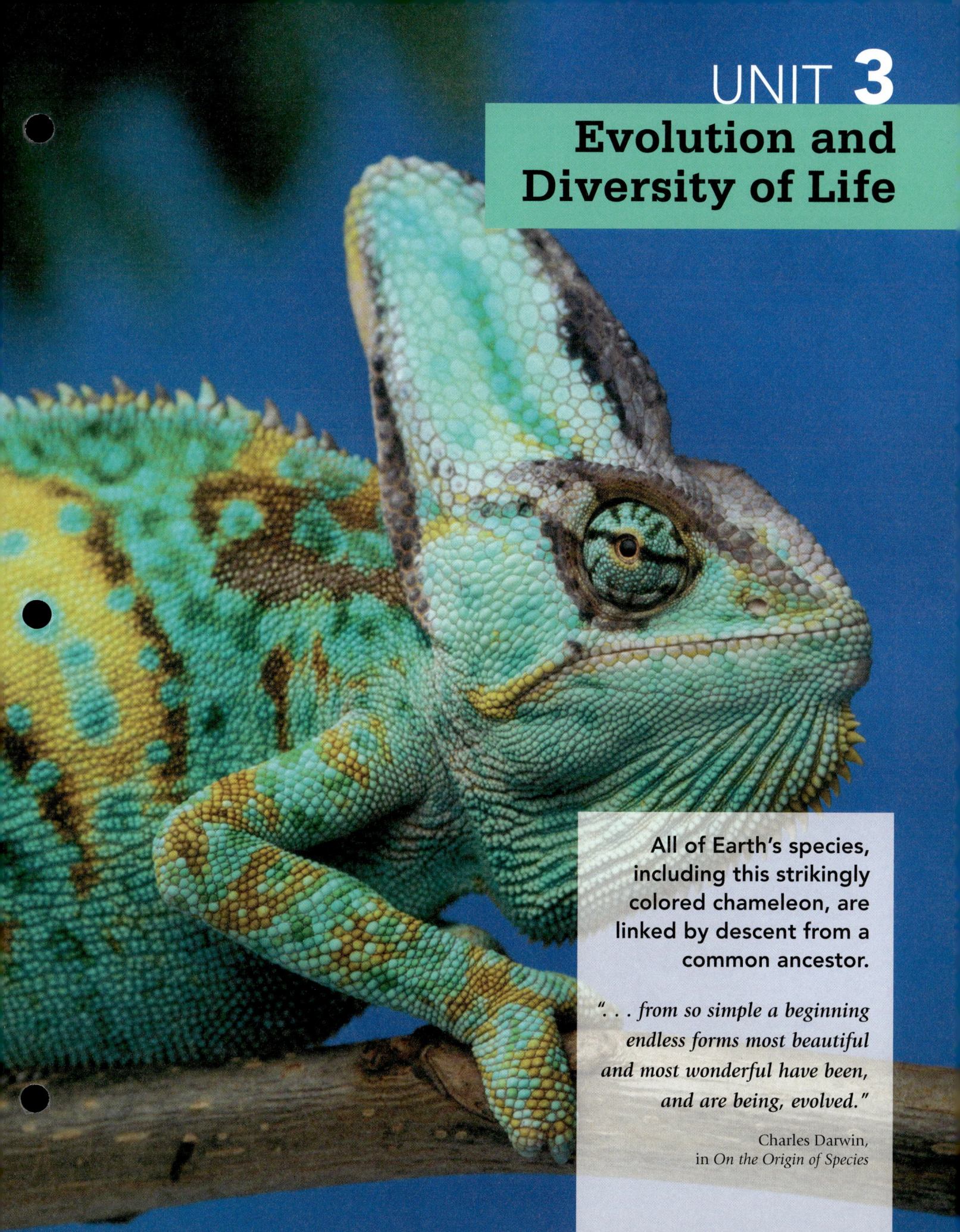

UNIT 3
Evolution and Diversity of Life

All of Earth's species, including this strikingly colored chameleon, are linked by descent from a common ancestor.

"... from so simple a beginning endless forms most beautiful and most wonderful have been, and are being, evolved."

Charles Darwin,
in *On the Origin of Species*

CASE STUDY

What Good Are Wisdom Teeth and Ostrich Wings?

HAVE YOU HAD YOUR WISDOM TEETH REMOVED YET? If not, it's probably only a matter of time. Almost all of us will visit an oral surgeon to have our wisdom teeth extracted. There's just not enough room in our jaws for these rearmost molars, and removing them is the best way to prevent the pain, infections, and gum disease that can accompany the development of wisdom teeth. Removal is harmless because we don't really need wisdom teeth.

If you've already suffered through a wisdom tooth extraction, you may have found yourself wondering why we even have these extra molars. Biologists hypothesize that we have them because our apelike ancestors had them and we inherited them, even though we don't need them. The presence in a living species of structures that have no current essential function, but that *are* useful in other living species, demonstrates shared ancestry among these species.

The connection between evolutionary ancestry and traits that have no essential function is illustrated by flightless birds. Consider the ostrich, a bird that can grow to 8 feet tall and weigh 300 pounds (see the photo at left). These massive creatures cannot fly. Nonetheless, they have wings, just as sparrows and ducks do. Why do ostriches have wings? Because the common ancestor of sparrows, ducks, and ostriches had wings, as do all of its descendants, even those that cannot fly. The bodies of today's organisms may contain now-useless hand-me-downs from their ancestors.

AT A GLANCE

14.1 How Did Evolutionary Thought Develop?

14.2 How Does Natural Selection Work?

14.3 How Do We Know That Evolution Has Occurred?

14.4 What Is the Evidence That Populations Evolve by Natural Selection?

LEARNING GOALS

LG1 Define *evolution*.

LG2 Describe pre-Darwin hypotheses and ideas that set the stage for the later emergence of the theory of evolution by natural selection.

LG3 Explain the mechanism of natural selection and the logic by which Darwin deduced it, and describe how natural selection affects populations.

LG4 Describe the main types of evidence that evolution has occurred, and provide an example of each type.

LG5 Compare and contrast similarities due to homology and similarities attributable to convergent evolution.

LG6 Describe the main lines of evidence that populations evolve by natural selection and provide examples of such evidence.

14.1 HOW DID EVOLUTIONARY THOUGHT DEVELOP?

When you began studying biology, you may not have seen a connection between your wisdom teeth and an ostrich's wings. But the connection is there, provided by the concept that unites all of biology: **evolution,** or change over time in the characteristics of a population. (A **population** consists of all the individuals of one species in a particular area.)

Modern biology is based on our understanding that life has evolved, but early scientists did not recognize this fundamental principle. The main ideas of evolutionary biology became widely accepted only after the publication of Charles Darwin's work in the nineteenth century. Nonetheless, the intellectual foundation on which these ideas rest developed gradually over the centuries before Darwin's time.

Early Biological Thought Did Not Include the Concept of Evolution

Pre-Darwinian science, heavily influenced by theology, held that all organisms were created simultaneously by God and that each distinct life-form remained fixed and unchanging from the moment of its creation. This explanation of how life's diversity arose was elegantly expressed by the ancient Greek philosophers, especially Plato and Aristotle. Plato (427–347 B.C.) proposed that each object on Earth is merely a temporary reflection of its divinely inspired "ideal form." Plato's student Aristotle (384–322 B.C.) categorized all organisms into a linear hierarchy that he called the "Ladder of Nature" (**Fig. 14-1**).

These ideas formed the basis of the view that the form of each type of organism is permanently fixed. This view reigned unchallenged for nearly 2,000 years. By the eighteenth century, however, several lines of newly emerging evidence began to undermine this static view of creation.

Exploration of New Lands Revealed a Staggering Diversity of Life

The Europeans who explored and colonized Africa, Asia, and the Americas were often accompanied by naturalists who observed and collected the plants and animals of these previously unknown (to Europeans) lands. By the 1700s, the accumulated observations and collections of the naturalists had begun to reveal the true scope of life's variety. The number of species, or different types of organisms, was much greater than anyone had suspected.

Stimulated by the new evidence of life's incredible diversity, some eighteenth-century naturalists began to take note of some fascinating patterns. They noticed, for example, that each geographic area had its own distinctive set of species. In addition, the naturalists saw that some of the species in a given location closely resembled one another, yet differed in some characteristics. To some scientists of the day, the differences between the species of different geographic areas and the existence of clusters of similar species within areas seemed inconsistent with the idea that species were fixed and unchanging. (You may wish to refer to the timeline in **Fig. 14-2** as you read the following historical account.)

A Few Scientists Speculated That Life Had Evolved

A few eighteenth-century scientists went so far as to speculate that species had, in fact, changed over time. For example, the French naturalist Georges Louis LeClerc (1707–1788), known by the title Comte de Buffon, suggested that the original creation provided a relatively small number of founding species, after which some might have "improved" or "degenerated," perhaps after moving to new geographic areas. That is, Buffon suggested that species had changed over time through natural processes.

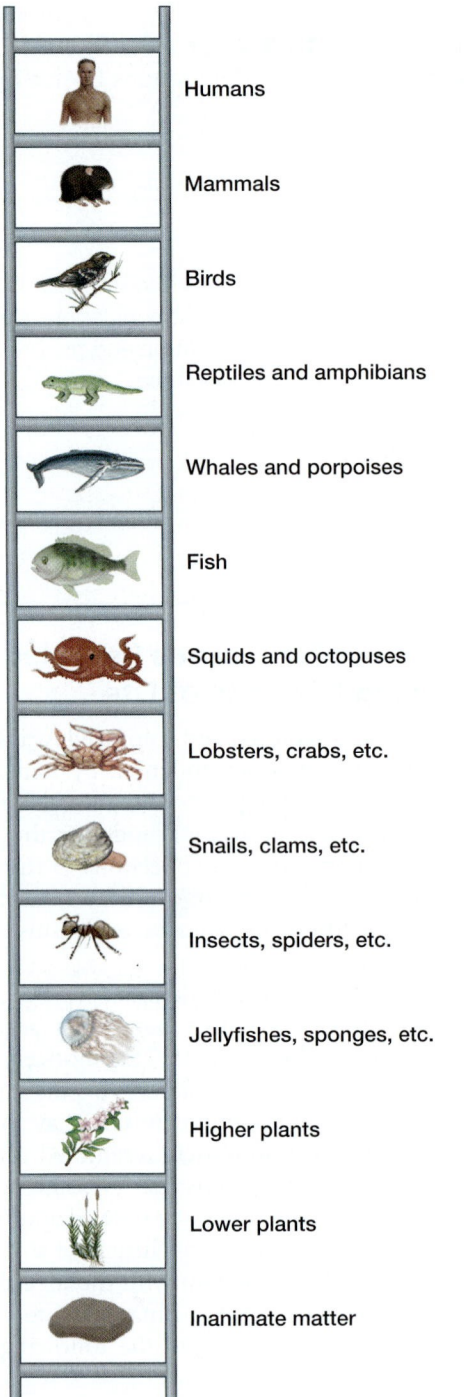

Humans

Mammals

Birds

Reptiles and amphibians

Whales and porpoises

Fish

Squids and octopuses

Lobsters, crabs, etc.

Snails, clams, etc.

Insects, spiders, etc.

Jellyfishes, sponges, etc.

Higher plants

Lower plants

Inanimate matter

▲ **FIGURE 14-1 Aristotle's Ladder of Nature** In Aristotle's view, fixed, unchanging species can be arranged in order of increasing closeness to perfection, with inferior types at the bottom and superior types above.

Fossil Discoveries Showed That Life Has Changed over Time

As Buffon and his contemporaries pondered the implications of new biological discoveries, developments in geology cast further doubt on the idea of permanently fixed species. Especially important was the discovery, during excavations for roads, mines, and canals, of rock fragments that resembled parts of living organisms. People had known of such objects since the fifteenth century, but most thought they were ordinary rocks that wind, water, or people had worked into lifelike forms. As more and more organism-shaped rocks were discovered, however, it became obvious that they were **fossils,** the preserved remains or traces of organisms that had died long ago (**Fig. 14-3**). Many fossils are bones, wood, shells, or their impressions in mud that have been petrified, or converted to stone. Fossils also include other kinds of preserved traces, such as tracks, burrows, pollen grains, eggs, and feces.

By the beginning of the nineteenth century, some pioneering investigators realized that the distribution of fossils in rock was also significant. Many rocks occur in layers, with newer layers positioned over older layers. The British surveyor William Smith (1769–1839), who studied rock layers and the fossils embedded in them, recognized that certain fossils were always found in the same layers of rock. Further, the organization of fossils and rock layers was consistent across different areas: Fossil type A could always be found in a rock layer resting beneath a younger layer containing fossil type B, which in turn rested beneath a still-younger layer containing fossil type C, and so on.

Scientists of the period also discovered that fossil remains showed a remarkable progression. Most fossils found in the oldest layers were very different from modern organisms, and the resemblance to modern organisms gradually increased in progressively younger rocks (**Fig. 14-4**). Many of the fossils were from plant or animal species that had gone *extinct*; that is, no members of the species still lived on Earth.

Putting all of these facts together, some scientists came to an inescapable conclusion: Different types of organisms had lived at different times in the past.

Some Scientists Devised Nonevolutionary Explanations for Fossils

Despite the growing fossil evidence, many scientists of the period did not accept the proposition that species changed and new ones arose over time. To account for extinct species while preserving the notion of a single creation by God, Georges Cuvier (1769–1832) advanced the idea of *catastrophism*. Cuvier, a French anatomist and paleontologist, hypothesized that a vast supply of species was created initially. Successive catastrophes (such as the Great Flood described in the Bible) produced layers of rock and destroyed many species, fossilizing some of their remains in the process. The organisms of the modern world, he speculated, are the species that survived the catastrophes.

Geology Provided Evidence That Earth Is Exceedingly Old

Cuvier's hypothesis of a world shaped by successive catastrophes was challenged by the work of the geologist Charles Lyell (1797–1875). Lyell, building on the earlier thinking of James Hutton (1726–1797), considered the forces of wind,

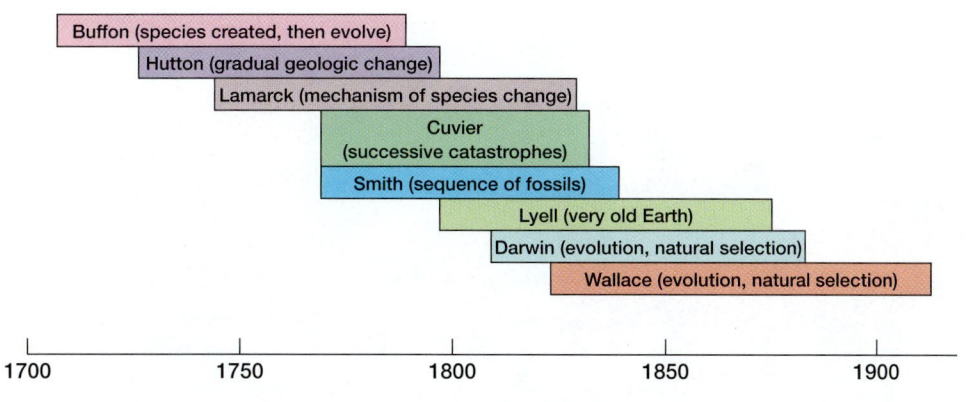

water, and volcanoes and concluded that there was no need to invoke catastrophes to explain the findings of geology. Don't flooding rivers lay down layers of sediment? Don't lava flows produce layers of basalt? Shouldn't we conclude, then, that layers of rock are evidence of ordinary natural processes, occurring repeatedly over long periods of time? This concept, that Earth's present landscape was produced by past action of the same gradual geological processes that we observe today, is called *uniformitarianism*. Acceptance of uniformitarianism by scientists of the time had a profound impact, because the idea implies that Earth is very old.

Before the 1830 publication of Lyell's evidence in support of uniformitarianism, few scientists suspected that Earth could be more than a few thousand years old. Counting generations in the Old Testament, for example, yields a maximum age of 4,000 to 6,000 years. An Earth this young poses problems for the idea that life has evolved. For example, ancient writers such as Aristotle described wolves, deer, lions, and other organisms that were identical to those present in Europe more than 2,000 years later. If organisms had changed so little over that time, how could whole new

species possibly have arisen if Earth was created only a couple of thousand years before Aristotle's time?

But if, as Lyell suggested, rock layers thousands of feet thick were produced by slow, natural processes, then Earth must be old indeed, many millions of years old. Lyell, in fact, concluded that Earth was eternal. Modern geologists estimate that Earth is about 4.5 billion years old (see "Science in Action: How Do We Know How Old a Fossil Is?" on p. 317).

Lyell (and his intellectual predecessor Hutton) showed that there was enough time for evolution to occur. But what was the mechanism? What process could cause evolution?

Some Pre-Darwin Biologists Proposed Mechanisms for Evolution

One of the first scientists to propose a mechanism for evolution was the French biologist Jean Baptiste Lamarck (1744–1829). Lamarck was impressed by the sequences of organisms in rock layers. He observed that older fossils tend to be less like existing organisms than are more recent

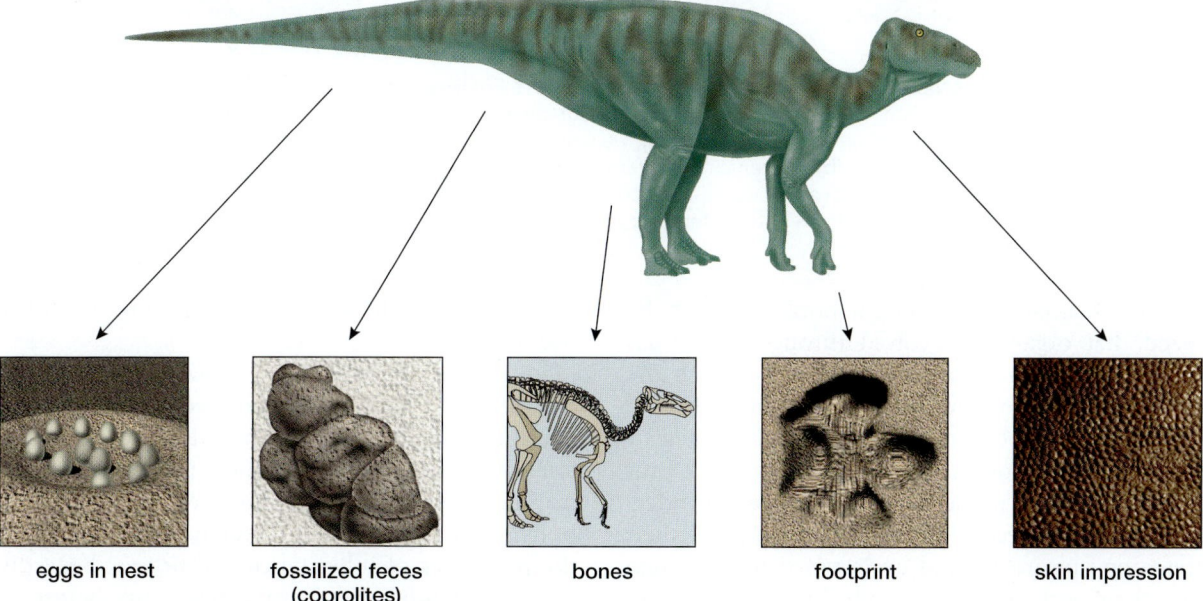

| eggs in nest | fossilized feces (coprolites) | bones | footprint | skin impression |

▲ FIGURE 14-3 **Types of fossils** Any preserved part or trace of an organism is a fossil.

(a) Trilobite (b) Seed ferns (c) *Allosaurus*

▲ FIGURE 14-4 **Different fossils are found in different rock layers** Fossils provide strong support for the idea that today's organisms were not created all at once but arose over time by the process of evolution. If all species had been created simultaneously, we would not expect **(a)** the earliest trilobites to be found in older rock layers than **(b)** the earliest seed ferns, which in turn would not be expected in older layers than **(c)** dinosaurs, such as *Allosaurus*. Trilobites first appeared about 520 million years ago, seed ferns (which were not actually ferns but had fern-like foliage) about 380 million years ago, and dinosaurs about 230 million years ago.

fossils. In 1809, Lamarck published a book in which he hypothesized that organisms evolved through the inheritance of acquired characteristics, a process in which the bodies of living organisms are modified through the use or disuse of parts, and these modifications are inherited by offspring. Why would bodies be modified? Lamarck proposed that all organisms possess an innate drive for perfection. For example, if ancestral giraffes tried to improve their lot by stretching upward to feed on leaves that grow high up in trees, their necks became slightly longer as a result. Their offspring would inherit these longer necks and then stretch even farther to reach still higher leaves. Eventually, this process would produce modern giraffes with very long necks indeed.

Today, we understand how inheritance works and can see that Lamarck's proposed evolutionary process could not work as he described it. Acquired characteristics are not inherited. The fact that a prospective father pumps iron doesn't mean that his child will look like a champion bodybuilder. Remember, though, that in Lamarck's time the principles of inheritance had not yet been discovered. (Gregor Mendel was born a few years before Lamarck's death, and his work with inheritance in pea plants was not widely recognized until 1900—see Chapter 10.) In any case, Lamarck's insight that inheritance plays an important role in evolution had an important influence on the later biologists who discovered the key mechanism of evolution.

Darwin and Wallace Proposed a Mechanism of Evolution

By the mid-1800s, a growing number of biologists had concluded that present-day species had evolved from earlier ones. But how? In 1858, Charles Darwin (1809–1882) and Alfred Russel Wallace (1823–1913), working separately, provided convincing evidence that evolution was driven by a simple yet powerful process.

Although their social and educational backgrounds were very different, Darwin and Wallace were quite similar in some respects. Both had traveled extensively in the tropics and had studied the plants and animals living there. Both found that some species differed in only a few features (Fig. 14-5). Both were familiar with the fossils that had been discovered, many of which showed a trend through time of increasing similarity to modern organisms. Finally, both were aware of the studies of Hutton and Lyell, who had proposed that Earth is extremely ancient. These facts suggested to both Darwin and Wallace that species change over time. Both men sought a mechanism that might cause such evolutionary change.

Of the two, Darwin was the first to propose a mechanism for evolution, which he sketched out in 1842 and described more fully in an essay in 1844. He sent the essay to a few colleagues, but did not submit it for publication, perhaps because he was fearful of the controversy that publication would cause. Some historians wonder if Darwin would ever have published his ideas had he not received, some 16 years after his initial draft, a paper by Wallace that outlined ideas remarkably similar to Darwin's own. Darwin realized that he could delay no longer.

In separate but similar papers that were presented to the Linnaean Society in London in 1858, Darwin and Wallace each described the same mechanism for evolution. Initially, their papers had little impact. The secretary of the society, in fact, wrote in his annual report that nothing very interesting happened that year. Fortunately, the next year, Darwin published his monumental book, *On the Origin of Species by Means of Natural Selection*, which attracted a great deal of attention to the new ideas about how species evolve. (To learn more about Darwin's life, see "Science in Action: Charles Darwin—Nature Was His Laboratory.")

CHECK YOUR LEARNING

Can you identify some of the thinkers whose ideas set the stage for the development of the theory of evolution? Can you describe the key ideas of those thinkers? Can you define *evolution*?

(a) Large ground finch, beak suited to large seeds

(b) Small ground finch, beak suited to small seeds

(c) Warbler finch, beak suited to insects

(d) Vegetarian tree finch, beak suited to leaves

◀ FIGURE 14-5 Darwin's finches, residents of the Galápagos Islands Darwin studied a group of closely related species of finches on the Galápagos Islands. Each species specializes in eating a different type of food and has a beak of characteristic size and shape because past individuals whose beaks were best suited to exploit each local food source produced more offspring than did individuals with less effective beaks.

SCIENCE IN ACTION Charles Darwin—Nature Was His Laboratory

In 1831, when Charles Darwin was 22 years old (**Fig. E14-1**), he secured a position as "gentleman companion" to Captain Robert Fitzroy of the HMS *Beagle*. The *Beagle* soon embarked on a 5-year surveying expedition along the coastline of South America and then around the world.

Darwin's voyage on the *Beagle* sowed the seeds for his theory of evolution. In addition to his duties as companion to the captain, Darwin served as the expedition's official naturalist, whose task was to observe and collect geological and biological specimens. The *Beagle* sailed to South America and made many stops along its coast. There Darwin observed the plants and animals of the tropics and was stunned by the huge diversity of species compared with the number found in Europe.

Although he had boarded the *Beagle* convinced of the permanence of species, Darwin's experiences soon led him to doubt it. He discovered a snake with rudimentary hind limbs, calling it "the passage by which Nature joins the lizards to the snakes" (**Fig. E14-2**). Darwin also noticed that penguins used their wings to paddle through the water rather than fly through the air, and he observed a snake that had no rattles but vibrated its tail like a rattlesnake. If a creator had individually created each animal in its present form, to suit its present environment, what could be the purpose behind these makeshift arrangements?

Perhaps the most significant stopover of the voyage was the month spent on the Galápagos Islands off the northwestern coast of South America. There, Darwin found huge tortoises. Different islands were home to distinctively different types of tortoises. Darwin also found several types of mockingbirds and, as with the tortoises, different islands had different mockingbird species. Could the differences in these organisms have arisen after they became isolated from one another on separate islands?

From his experience as a naturalist, Darwin realized that members of a species typically compete with one another for survival. He also recognized that the competition's winners and losers were determined not by chance, but by the characteristics and abilities of the individual competitors. Darwin understood that if these characteristics were inherited by their possessors' offspring, species would evolve by natural selection.

When Darwin finally published *On the Origin of Species* in 1859, his evidence for evolution by natural selection had become truly overwhelming. Although its full implications would not be realized for decades, Darwin's theory has become a unifying concept for virtually all of biology.

▲ FIGURE E14-1 **A painting of Charles Darwin as a young man**

▲ FIGURE E14-2 **The vestigial remnants of hind legs in a snake** Some snakes have small "spurs" where their distant ancestors had rear legs. In some species, these vestigial structures even retain claws.

14.2 HOW DOES NATURAL SELECTION WORK?

Darwin and Wallace proposed that life's huge variety arose by a process of descent with modification, in which individuals in each generation differ slightly from the members of the preceding generation. Over long stretches of time, these small differences accumulate to produce major transformations.

Darwin and Wallace's Theory Rests on Four Postulates

The chain of logic that led Darwin and Wallace to their proposed process of evolution turns out to be surprisingly simple and straightforward. It is based on four postulates about populations:

Postulate 1 Individual members of a population differ from one another in many respects.

Postulate 2 At least some of the differences among members of a population are due to characteristics that may be passed from parent to offspring.

Postulate 3 In each generation, some individuals in a population survive and reproduce successfully but others do not.

Postulate 4 The fate of individuals is not determined entirely by chance or luck. Instead, an individual's likelihood of survival and reproduction depends on its characteristics. Individuals with advantageous traits survive longest and leave the most offspring, a process known as **natural selection.**

Darwin and Wallace understood that if all four postulates were true, populations would inevitably change over time. If members of a population have different traits, and if the individuals that are best suited to their environment leave more offspring, and if those individuals pass their favorable traits to their offspring, then the favorable traits will become more common in subsequent generations. The characteristics of the population will change slightly with each generation. This process is evolution by natural selection.

Are the four postulates true? Darwin thought so, and devoted much of *On the Origin of Species* to describing supporting evidence. Let's briefly examine each postulate, in some cases with the advantage of knowledge that had not yet come to light during the lifetimes of Darwin and Wallace.

Postulate 1: Individuals in a Population Vary

The accuracy of postulate 1 is apparent to anyone who has glanced around a crowded room. People differ in size, eye color, skin color, and many other physical features. Similar variability is present in populations of other organisms, although it may be less obvious to the casual observer (**Fig. 14-6**).

Postulate 2: Traits Are Passed from Parent to Offspring

The principles of genetics had not yet been discovered when Darwin published *On the Origin of Species*. Therefore, although observation of people, pets, and farm animals

▲ **FIGURE 14-6 Variation in a population of snails** Although these snails are all members of the same population, no two are exactly alike.

QUESTION Is sexual reproduction required to generate the variability in structures and behaviors that is necessary for natural selection?

seemed to show that offspring generally resemble their parents, Darwin and Wallace did not have scientific evidence in support of postulate 2. Mendel's later work, however, demonstrated conclusively that particular traits can be passed to offspring. Since Mendel's time, genetics researchers have produced a detailed picture of how inheritance works.

Postulate 3: Some Individuals Fail to Survive and Reproduce

Darwin's formulation of postulate 3 was heavily influenced by Thomas Malthus's *Essay on the Principle of Population* (1798), which described the perils of unchecked growth of human populations. Darwin was keenly aware that organisms can produce far more offspring than are required merely to replace the parents. He calculated, for example, that a single pair of elephants would multiply to a population of 19 million in 750 years if each descendant had six offspring that lived to reproduce.

But we aren't overrun with elephants. The number of elephants, like the number of individuals in most natural populations, tends to remain relatively constant. Therefore, more organisms must be born than survive long enough to reproduce. In each generation, many individuals must die young. Even among those that survive, many must fail to reproduce, produce few offspring, or produce less-vigorous offspring that, in turn, fail to survive and reproduce. As you might expect, whenever biologists have measured reproduction in a population, they have found that some individuals have more offspring than others.

Postulate 4: Survival and Reproduction Are Not Determined by Chance

If unequal reproduction is the norm in populations, what determines which individuals leave the most offspring? A large amount of scientific evidence has shown that reproductive success depends on an individual's characteristics. For example, scientists found that larger male elephant seals in a California population had more offspring than smaller males (because females were more likely to mate with large males). In a Colorado population of snapdragons, plants with white flowers had more offspring than plants with yellow flowers (because pollinators found white flowers more attractive). These results, and hundreds of other similar ones, show that in the competition to survive and reproduce, winners are for the most part determined not by chance but by the traits they possess.

Natural Selection Modifies Populations over Time

Observation and experiment suggest that the four postulates of Darwin and Wallace are sound. Logic suggests that the resulting consequence ought to be change over time in the characteristics of populations. In *On the Origin of Species*, Darwin proposed the following example: "Let us take the case of a wolf, which preys on various animals, securing [them] by . . . fleetness. . . . The swiftest and slimmest wolves would have the best chance of surviving, and so be preserved or selected. . . . Now if any slight innate change of habit or structure benefited an individual wolf, it would have the best chance of surviving and of leaving offspring. Some of its young would probably inherit the same habits or structure, and by the repetition of this process, a new variety might be formed." The same logic applies to the wolf's prey; the fastest or most alert or best camouflaged would be most likely to avoid predation and would pass these traits to its offspring.

Notice that natural selection acts on individuals within a population. The influence of natural selection on the fates of individuals eventually has consequences for the population as a whole. Over generations, the population changes as the percentage of individuals inheriting favorable traits increases. An individual cannot evolve, but a population can.

Although it is easier to understand how natural selection would cause changes within a species, under the right circumstances, the process might produce entirely *new* species. We will discuss the circumstances that might give rise to new species in Chapter 16.

CHECK YOUR LEARNING

Can you explain how natural selection works and how it affects populations? Can you describe the logic, based on four postulates, by which Darwin deduced that populations must evolve by natural selection?

14.3 HOW DO WE KNOW THAT EVOLUTION HAS OCCURRED?

Today, evolution is an accepted scientific theory. A scientific theory is a general explanation of important natural phenomena, developed through extensive, reproducible observations (see Chapter 1). An overwhelming body of evidence supports the conclusion that evolution has occurred. The key lines of evidence come from fossils, comparative anatomy (the study of how body structures differ among species), embryology (the study of developing organisms in the period from fertilization to birth or hatching), biochemistry, and genetics.

Fossils Provide Evidence of Evolutionary Change over Time

If it is true that many fossils are the remains of species ancestral to modern species, we might expect to find fossils in a progressive series that starts with an ancient organism, progresses through several intermediate stages, and culminates in a modern species. Such series have indeed been found. For example, fossils of the ancestors of modern whales illustrate stages in the evolution of an aquatic species from land-dwelling ancestors (Fig. 14-7). Series of fossil giraffes, elephants, horses, and mollusks also show the evolution of body structures over time. These fossil series suggest that new species evolved from, and replaced, previous species.

Comparative Anatomy Gives Evidence of Descent with Modification

Fossils provide snapshots of the past that allow biologists to trace evolutionary changes, but careful examination of today's organisms can also uncover evidence of evolution. Comparing the bodies of organisms of different species can reveal similarities that can be explained only by shared ancestry and differences that could result only from evolutionary change during descent from a common ancestor. In this way, the study of comparative anatomy has supplied strong evidence that different species are linked by a common evolutionary heritage.

Homologous Structures Provide Evidence of Common Ancestry

A body structure may be modified by evolution to serve different functions in different species. The forelimbs of birds and mammals, for example, are variously used for flying, swimming, running over several types of terrain, and grasping objects, such as branches and tools. Despite this enormous diversity of function, the internal anatomy of all bird and mammal forelimbs is remarkably similar (Fig. 14-8). It seems inconceivable that the same bone arrangements would be used to serve such diverse functions if each animal had been created separately. Such similarity is exactly what we would expect, however, if bird and mammal forelimbs were derived from a common ancestor. Through natural selection, the ancestral forelimb has undergone different modifications

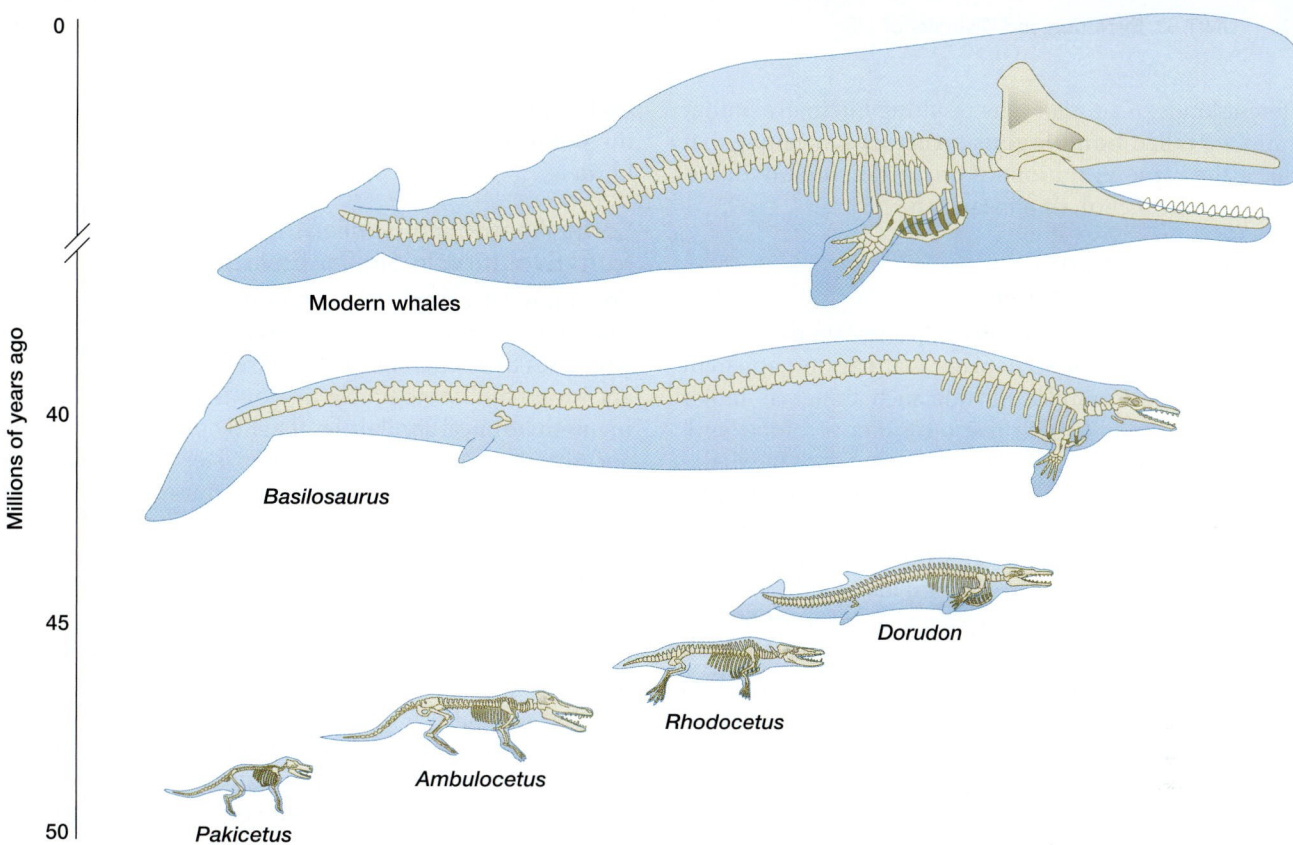

▲ FIGURE 14-7 The evolution of the whale During the past 50 million years, whales have evolved from four-legged land dwellers, to semi-aquatic paddlers, to fully aquatic swimmers with shrunken hind legs, to today's sleek ocean dwellers.

QUESTION The fossil history of some kinds of modern organisms, such as sharks and crocodiles, shows that their structure and appearance have changed very little over hundreds of millions of years. Is this lack of change evidence that such organisms have not evolved during that time?

▲ FIGURE 14-8 Homologous structures Despite wide differences in function, the forelimbs of all of these animals contain the same set of bones, inherited from a common ancestor. The different colors of the bones highlight the correspondences among the various species.

in different kinds of animals. The resulting internally similar structures are called **homologous structures,** meaning that they have the same evolutionary origin despite any differences in current function or appearance.

Functionless Structures Are Inherited from Ancestors

Evolution by natural selection also helps explain the curious circumstance of **vestigial structures** that have no apparent essential function. Examples include such things as molar teeth in vampire bats (which live on a diet of blood and, therefore, don't chew their food) and pelvic bones in whales and certain snakes (**Fig. 14-9**). Both of these vestigial structures are clearly homologous to structures that are found in—and used by—other vertebrates (animals with a backbone). The continued existence in animals of structures for which they have no use is best explained as a sort of "evolutionary baggage." For example, the ancestral mammals from which whales evolved had four legs and a well-developed set of pelvic bones (see Fig. 14-7). Whales do not have hind legs, yet they have small pelvic and leg bones embedded in their sides. During whale evolution, losing the hind legs provided an advantage, better streamlining the body for movement through water. The result is the modern whale with small, useless pelvic bones that persist because they have shrunk to the point that they no longer constitute a survival-reducing burden.

Some Anatomical Similarities Result from Evolution in Similar Environments

The study of comparative anatomy has demonstrated the shared ancestry of life by identifying a host of homologous structures that different species have inherited from common ancestors, but comparative anatomists have also identified many anatomical similarities that do not stem from common ancestry. Instead, these similarities stem from **convergent evolution,** in which natural selection causes non-homologous structures that serve similar functions to resemble one another. For example, both birds and insects have wings, but this similarity did not arise from evolutionary modification of a structure that both birds and insects inherited from a common ancestor. Instead, the similarity arose from parallel modification of two different, non-homologous structures. Because natural selection favored flight in both birds and insects, the two groups evolved wings of roughly similar appearance, but the similarity is superficial. Such outwardly similar but non-homologous structures are called **analogous structures** (**Fig. 14-10**). Analogous structures are typically

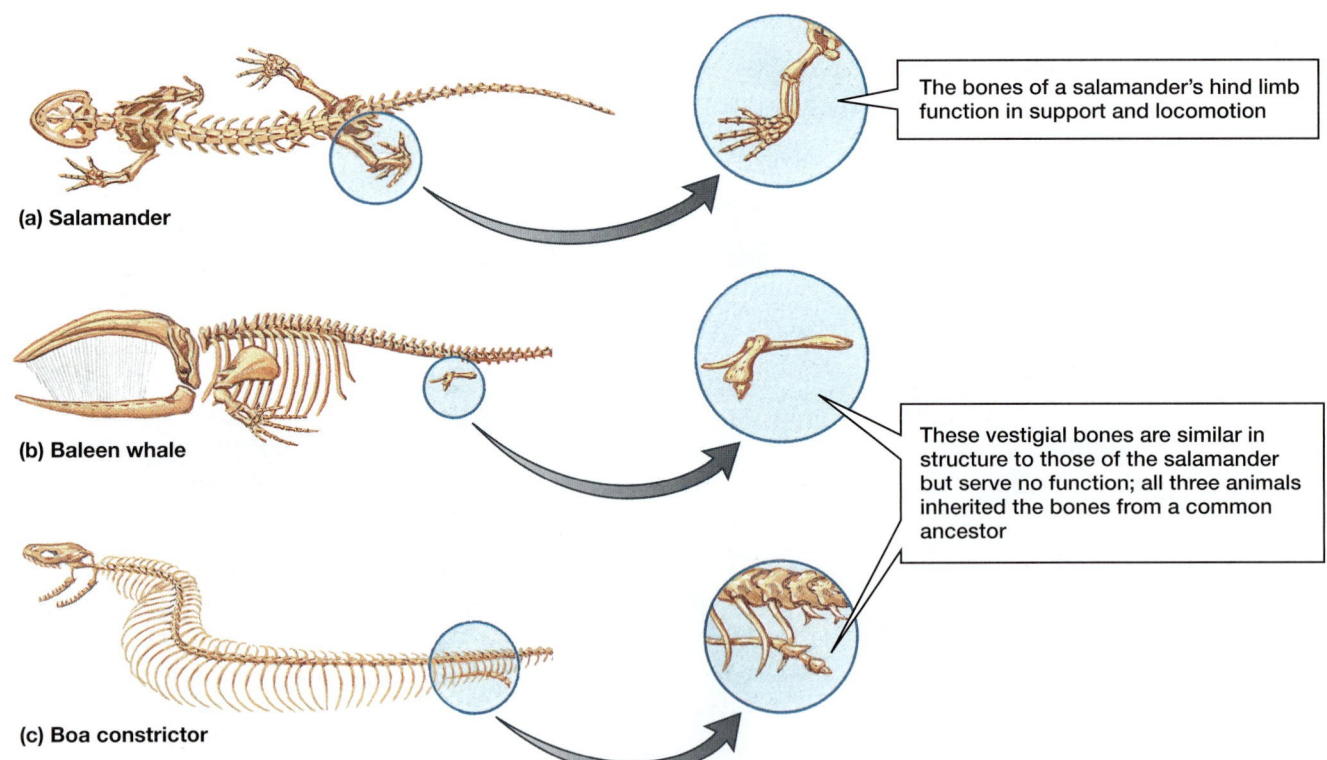

(a) Salamander

The bones of a salamander's hind limb function in support and locomotion

(b) Baleen whale

(c) Boa constrictor

These vestigial bones are similar in structure to those of the salamander but serve no function; all three animals inherited the bones from a common ancestor

▲ FIGURE 14-9 **Vestigial structures** Many organisms have vestigial structures that serve no apparent function. The **(a)** salamander, **(b)** baleen whale, and **(c)** boa constrictor all inherited hind limb bones from a common ancestor. These bones remain functional in the salamander but are vestigial in the whale and snake.

QUESTION Compile a list of human vestigial structures. For each structure, name the corresponding homologous structure in a nonhuman species.

❓ HAVE **YOU** EVER WONDERED ...

Why Backaches Are So Common?

Between 70% and 85% of people will experience lower back pain at some point in life and, for many people, the condition is chronic. This state of affairs is an unfortunately painful consequence of the evolutionary process. We walk upright on two legs, but our distant ancestors walked on all fours. Thus, natural selection formed our vertically oriented spine by remodeling one whose normal orientation was parallel to the ground. Our spinal anatomy evolved some modifications in response to its new posture but, as is often the case with evolution, the changes involved some trade-offs. The arrangements of bone and muscle that permit our smooth, bipedal gait also generate vertical compression of the spine, and the resulting pressure can, and frequently does, cause painful damage to muscle and nerve tissues.

(a) Damselfly

(b) Swallow

▲ **FIGURE 14-10 Analogous structures** Convergent evolution can produce outwardly similar structures that differ anatomically, such as the wings of **(a)** insects and **(b)** birds.

QUESTION Are a peacock's tail and a dog's tail homologous structures or analogous structures?

very different in internal anatomy, because the parts are not derived from common ancestral structures.

Embryological Similarity Suggests Common Ancestry

In the early 1800s, German embryologist Karl von Baer noted that all vertebrate embryos look quite similar to one another early in their development (**Fig. 14-11**). In their early embryonic stages, fish, turtles, chickens, mice, and humans all develop tails and gill slits (also called gill grooves). But among this group of animals, only fish retain gills as adults, and only fish, turtles, and mice retain substantial tails.

Why do vertebrates that are so different have similar developmental stages? The only plausible explanation is that ancestral vertebrates possessed genes that directed the development of gills and tails. All of their descendants still have those genes. In fish, these genes are active throughout development, resulting in adults with fully developed tails and gills. In humans and chickens, these genes are active only during early developmental stages, and the structures are lost or inconspicuous in adults.

Modern Biochemical and Genetic Analyses Reveal Relatedness Among Diverse Organisms

Biologists have been aware of anatomical and embryological similarities among organisms for centuries, but it took the emergence of modern technology to reveal similarities at the molecular level. Biochemical similarities among organisms provide perhaps the most striking evidence of their evolutionary relatedness. Just as relatedness is revealed by homologous anatomical structures, it is also revealed by homologous molecules.

Today's scientists have access to a powerful tool for revealing molecular homologies: DNA sequencing. It is now possible to quickly determine the sequence of nucleotides in a DNA molecule and to compare the DNA of different organisms. For example, consider the gene that encodes the protein cytochrome *c* (see Chapters 11 and 12 for information on DNA and how it encodes proteins). Cytochrome *c* is present in all plants and animals (and many single-celled organisms) and performs the same function in all of them. The sequence of nucleotides in the gene for cytochrome *c* is similar in these diverse species (**Fig. 14-12**). The widespread presence of the same complex protein, encoded by the same gene and performing the same function, is evidence that the common ancestor of plants and animals had cytochrome *c* in its cells. At the same time, though, the sequence of the cytochrome *c* gene differs slightly in different species,

(a) Lemur   (b) Pig   (c) Human

▲ FIGURE 14-11 **Embryological stages reveal evolutionary relationships** Early embryonic stages of a **(a)** lemur, **(b)** pig, and **(c)** human, showing strikingly similar anatomical features.

showing that variations arose during the independent evolution of Earth's multitude of plant and animal species.

Some biochemical similarities are so fundamental that they extend to all living cells. For example:

- All cells have DNA as the carrier of genetic information.
- All cells use RNA, ribosomes, and approximately the same genetic code to translate that genetic information into proteins.
- All cells use roughly the same set of 20 amino acids to build proteins.
- All cells use ATP as a cellular energy carrier.

The most plausible explanation for such widespread sharing of complex and specific biochemical traits is that the traits are homologies. That is, they arose only once, in the common ancestor of all living things, from which all of today's organisms inherited them.

CHECK YOUR LEARNING

Can you describe the evidence that evolution has occurred? Can you explain the difference between similarity due to homology and similarity due to convergent evolution?

◀ FIGURE 14-12 **Molecular similarity shows evolutionary relationships** The DNA sequences of the genes that code for cytochrome c in a human and a mouse. Of the 315 nucleotides in the gene, only 30 (shaded blue) differ between the two species.

CASE STUDY continued

What Good Are Wisdom Teeth and Ostrich Wings?

Just as anatomical homology can lead to vestigial structures such as human wisdom teeth and the wings of flightless birds, genetic homology can lead to vestigial DNA sequences. For example, most mammal species produce an enzyme, L-gulonolactone oxidase, that catalyzes the last step in the production of vitamin C. The species that produce the enzyme are able to do so because they all inherited the gene that encodes it from a common ancestor. Humans, however, do not produce L-gulonolactone oxidase, so we can't produce vitamin C ourselves and must consume it in our diets. But even though we don't produce the enzyme, our cells do contain a stretch of DNA with a sequence very similar to that of the enzyme-producing gene present in rats and most other mammals. The human version, though, does not encode the enzyme (or any protein). We inherited this stretch of DNA from an ancestor that we share with other mammal species, but in us, the sequence has undergone a change that rendered it nonfunctional. (The change probably did not confer a strong disadvantage, because our ancestors got sufficient vitamin C in their diets.) The nonfunctional sequence remains as a vestigial trait, evidence of our shared ancestry.

14.4 WHAT IS THE EVIDENCE THAT POPULATIONS EVOLVE BY NATURAL SELECTION?

We have seen that evidence of evolution comes from many sources. But what is the evidence that evolution occurs by the process of natural selection?

Controlled Breeding Modifies Organisms

One line of evidence supporting evolution by natural selection is **artificial selection,** the breeding of domestic plants and animals to produce specific desirable features. The various dog breeds provide a striking example of artificial selection (Fig. 14-13). Dogs descended from wolves, and even today, the two will readily crossbreed. With few exceptions, however, modern dogs do not closely resemble wolves. Some breeds are so different from one another that they would be considered separate species if they were found in the wild. Humans produced these radically different dogs in a few thousand years by doing nothing more than repeatedly selecting individuals with desirable traits for breeding. Therefore, it is quite plausible that natural selection could, by an analogous process acting over hundreds of millions of years, produce the spectrum of living organisms. Darwin was so impressed by the connection between artificial selection and natural selection that he devoted a chapter of *On the Origin of Species* to the topic.

Evolution by Natural Selection Occurs Today

Additional evidence of natural selection comes from scientific observation and experimentation. The logic of natural selection gives us no reason to believe that evolutionary change is limited to the past. After all, inherited variation and

(a) Gray wolf

(b) Diverse dogs

▲ FIGURE 14-13 Dog diversity illustrates artificial selection A comparison of (a) the ancestral dog (the gray wolf, *Canis lupus*) and (b) various breeds of dog. Artificial selection by humans has caused great divergence in the size and shape of dogs in only a few thousand years.

competition for access to resources are certainly not limited to the past. If Darwin and Wallace were correct that those conditions lead inevitably to evolution by natural selection, then researchers ought to be able to detect evolutionary change as it occurs. And they have. Next, we will consider some examples that give us a glimpse of natural selection at work.

Brighter Coloration Can Evolve When Fewer Predators Are Present

On the island of Trinidad, guppies live in streams that are also inhabited by several species of larger, predatory fish that frequently dine on guppies (Fig. 14-14). In upstream portions of these streams, however, the water is too shallow for the predators, and guppies are free of danger from predators.

▲ FIGURE 14-14 **Guppies evolve to become more colorful in predator-free environments** Male guppies (top) are more brightly colored than females (bottom). Some male guppies are more colorful than others. In environments without predators, brighter males are selected; in environments with predators, duller males are selected.

When scientists compared male guppies in an upstream area with ones in a downstream area, they found that the upstream guppies were much more brightly colored than the downstream guppies. The scientists knew that the source of the upstream population was guppies that had found their way up into the shallower waters many generations earlier.

The explanation for the difference in coloration between the two populations stems from the sexual preferences of female guppies. The females prefer to mate with the most brightly colored males, so the brightest males have a large advantage when it comes to reproduction. In predator-free areas, male guppies with the bright colors that females prefer have more offspring than duller males. Bright color, however, makes guppies more conspicuous to predators and, therefore, more likely to be eaten. Thus, where predators are common, they act as agents of natural selection by eliminating the bright-colored males before they can reproduce. In these areas, the duller males have the advantage and produce more offspring. The color difference between the upstream and downstream guppy populations is a direct result of natural selection.

Natural Selection Can Lead to Herbicide and Pesticide Resistance

Natural selection is also evident in weed species that have evolved resistance to the herbicides with which we try to control them. Successful agriculture depends on farmers' ability to kill weeds that compete with crop plants, but many members of weed species can no longer be killed by a formerly lethal dose of glyphosate, the active ingredient in Roundup, the world's most widely used herbicide. How did these glyphosate-resistant "superweeds" arise? They arose because the herbicide has acted as an agent of natural selection. Consider, for example, giant ragweed, one of the highly destructive weed species that are now resistant to glyphosate in some

places. When a field is sprayed with Roundup, almost all of the giant ragweed plants there are killed, because glyphosate inactivates an enzyme that is essential to plants' survival. A few ragweed plants, however, survive, and researchers have discovered that some of these survivors carry a mutation that causes them to produce a tremendous amount of the enzyme that glyphosate attacks, more enzyme than the usual dose of glyphosate can destroy. In the face of repeated applications of Roundup, the formerly rare protective mutation has become common in many giant pigweed populations. (For additional examples of how humans influence evolution, see "Earth Watch: People Promote High-Speed Evolution.")

The evolution of glyphosate-resistant superweeds was a direct result of changes in agricultural practice. In the 1990s, the biotechnology company Monsanto began selling seeds that had been genetically engineered to produce crops that are not harmed by glyphosate. These "Roundup Ready" crops, which now account for the vast majority of soybean, corn, and cotton plantings in the United States and other countries, allow farmers to freely apply glyphosate to their fields without fear of harming crop plants. As a result, use of glyphosate has skyrocketed. Today, large-scale agriculture is extremely dependent on this single herbicide, even as its effectiveness declines steadily due to the evolution of resistant weeds.

Just as weeds have evolved to resist herbicides, many of the insects that attack crops have also evolved resistance to the pesticides that farmers use to control them. Such resistance has been documented in more than 500 species of crop-damaging insects, and virtually every pesticide has fostered the evolution of resistance in at least one insect species. We pay a heavy price for this evolutionary phenomenon. The additional pesticides that farmers apply in their attempts to control resistant insects cost almost $2 billion each year in the United States alone and add millions of tons of poisons to Earth's soil and water.

Experiments Can Demonstrate Natural Selection

In addition to observing natural selection in the wild, scientists have also devised numerous experiments that confirm the action of natural selection. For example, one group of evolutionary biologists released small groups of *Anolis sagrei* lizards onto 14 small Bahamian islands that were previously uninhabited by lizards (**Fig. 14-15**). The original lizards came from a population on Staniel Cay, an island with tall vegetation, including plenty of trees. In contrast, the islands to which the small colonial groups were introduced had few or no trees and were covered mainly with small shrubs and other low-growing plants.

The biologists returned to those islands 14 years after releasing the colonists and found that the original small groups of lizards had given rise to thriving populations of hundreds of individuals. On all 14 of the experimental islands, lizards had legs that were shorter and thinner than those of lizards from the original source population on Staniel Cay. In just over a decade, it appeared, the lizard populations had changed in response to new environments.

Why had the new lizard populations evolved shorter, thinner legs? Shorter, thinner legs allow for more agility and

▲ FIGURE 14-15 **Anole leg size evolves in response to a changed environment**

maneuverability on narrow surfaces, whereas longer, thicker legs permit greater running speed. On Staniel Cay, with its thick-branched trees, speed is more important than maneuverability for avoiding predators, so natural selection favored fast-running lizards whose legs were as long and thick as possible. But when the lizards were moved to an environment with only thin-branched bushes, individuals with long, thick legs were at a disadvantage. In the new environment, more agile individuals with shorter, thinner legs were better able to escape predators and survive to produce a greater number of offspring. Thus, members of subsequent generations had shorter, thinner legs on average.

CASE STUDY revisited

What Good Are Wisdom Teeth and Ostrich Wings?

Wisdom teeth are but one of many human anatomical structures that appear to no longer serve an important function (Fig. 14-16). Darwin himself noted many of these "useless, or nearly useless" traits in the very first chapter of *On the Origin of Species* and declared them to be prime evidence that humans had evolved from earlier species.

Body hair is another vestigial human trait. It seems to be an evolutionary relic of the fur that kept our distant ancestors warm (and that still warms our closest evolutionary relatives, the great apes). Not only do we retain useless body hair, we also still have arrector pili, the muscle fibers that allow other mammals to puff up their fur for better insulation. In humans, these vestigial structures just give us goose bumps.

Though humans don't have and don't need a tail, we nonetheless have a tailbone. The tailbone consists of a few tiny vertebrae fused into a small structure at the base of the backbone, where a tail would be if we had one. People born without a tailbone or who have theirs surgically removed suffer no ill effects.

Consider This

BioEthics Advocates of the view that all organisms were created simultaneously by God argue that there

Selection Acts on Random Variation to Favor Traits That Work Best in Particular Environments

Two important points underlie the evolutionary changes just described:

- **The variations on which natural selection works are produced by chance mutations.** Bright coloration in Trinidadian guppies, extra enzyme in giant ragweed plants, and shorter legs in Bahamian lizards were not *produced* by the female mating preferences, Roundup herbicide, or thinner branches. The mutations that produced each of these beneficial traits arose spontaneously.
- **Natural selection favors organisms that are best adapted to a particular environment.** Natural selection is not a process for producing ever-greater degrees of perfection. Natural selection does not select for the "best" in any absolute sense, but only for what is best in the context of a particular environment, which varies from place to place and which may change over time. A trait that is advantageous under one set of conditions may become disadvantageous if conditions change. For example, in the presence of glyphosate herbicide, producing copious amounts of an essential enzyme yields an advantage to a giant ragweed plant, but under natural conditions producing unnecessary extra enzyme would be a huge waste of precious energy and nutrients.

CHECK YOUR LEARNING

Can you describe some observations and experiments that demonstrate that populations evolve by natural selection?

are no vestigial organs because if any function at all can be attributed to a structure, it cannot be considered functionless, even if its removal has no effect. Thus, according to this view, ostrich wings are not evidence of evolution, because they *can* be used to brush off biting insects. Is this a valid argument?

▲ FIGURE 14-16 **Wisdom teeth** Squeezed into a jaw that is too short to contain them, wisdom teeth often become impacted—unable to erupt through the surface of the gum. The leftmost upper and lower teeth in this X-ray image are impacted wisdom teeth.

Earth | Watch

People Promote High-Speed Evolution

You probably don't think of yourself as a major engine of evolution. Nonetheless, as you go about the routines of your daily life, you are contributing to what is perhaps today's most significant cause of rapid evolutionary change. Human activity has changed Earth's environments tremendously, and the biological logic of natural selection, spelled out so clearly by Darwin, tells us that environmental change leads inevitably to evolutionary change. Thus, by changing the environment, humans have become a major agent of natural selection.

Unfortunately, many of the evolutionary changes we have caused have turned out to be bad news for us. Our liberal use of pesticides has selected for resistant pests that frustrate efforts to protect our food supply. By overmedicating ourselves with antibiotics and other drugs, we have selected for resistant "supergerms" and diseases that are ever more difficult to treat (see Chapter 15). Heavy fishing in the world's oceans has favored smaller fish that can slip through nets more easily, thereby selecting for slow-growing fish that remain small even as mature adults. As a result, fish of many commercially important species are now so small that our ability to extract food from the sea is compromised.

Our use of pesticides, antibiotics, and fishing technology has caused evolutionary changes that threaten our health and welfare, but the scope of these changes may be dwarfed by those that will arise from human-caused modification of Earth's climate. Human activities, especially activities that use energy derived from fossil fuels, modify the climate by contributing to global warming. In coming years, species' evolution will be influenced by environmental changes associated with a warming climate, such as reduced ice and snow, longer growing seasons, and shifts in the life cycles of other species that provide food or shelter.

There is growing evidence that global warming is already causing evolutionary change. Warming-related evolution has been found in a number of plant and animal populations. For example, red squirrels in a Canadian colony closely monitored by scientists now produce litters 18 days earlier, on average, than they did 10 years ago. The change is tied to a warming climate, because spring now arrives earlier and spruce trees produce earlier crops of seeds, the squirrels' only food. Squirrels that breed earlier gain a competitive edge by better exploiting the warmer weather and more abundant food. And, because the time at which a squirrel gives birth is influenced by the animal's genetic makeup, early-breeding squirrels pass to their offspring the genes that confer this advantage. As a result, the genetic makeup of the squirrel population is changing, and early-breeding squirrels are becoming more common than later-breeding ones.

In Finland, tawny owls have changed color in response to a warmer climate (Fig. E14-3). Owls of this species come in two varieties, gray or brown. In the past, most owls were gray, and

▲ **FIGURE E14-3 Tawny owl populations have evolved in response to global climate change** The proportion of brown owls is increasing, because brown owls survive better than gray ones over winters with less snow.

the brown variety was rare. Today, however, about 40% of Finnish owls are brown, and researchers have shown that the increase in brown owls was probably caused by climate change. In particular, the researchers found that in snowy winters, gray owls survive much better than brown owls, perhaps because they are better camouflaged against the snow and therefore suffer less predation by eagles. In winters with less snow, brown owls are better camouflaged and more likely to survive. As temperatures have risen in recent decades, snowy winters have become increasingly rare and the resulting reduced snow cover has favored the survival of brown owls. Thus, a brown owl is more likely than a gray one to survive the winter and, because feather color is inherited, produce brown offspring. The proportion of brown owls in the population has grown in response to natural selection associated with warming temperatures.

The available evidence suggests that global climate change will have an enormous evolutionary impact, potentially affecting the evolution of almost every species. How will these evolutionary changes affect us and the ecosystems on which we depend? This question is not readily answerable, because the path of evolution is not predictable. We can hope, however, that careful monitoring of evolving species and increased understanding of evolutionary processes will help us take appropriate steps to safeguard our health and well-being as Earth warms.

CHAPTER REVIEW

Summary of Key Concepts

14.1 How Did Evolutionary Thought Develop?

Historically, the most common explanation for the origin of species was the divine creation of each species in its present form, and species were believed to remain unchanged after their creation. This view was challenged by evidence from fossils, geology, and biological exploration, most notably in the tropics. Since the middle of the nineteenth century, scientists have realized that species originate and evolve by the operation of natural processes that change the genetic makeup of populations.

14.2 How Does Natural Selection Work?

Charles Darwin and Alfred Russel Wallace independently proposed the theory of evolution by natural selection. Their theory expresses the logical consequences of four postulates about populations. If (1) populations are variable, (2) the variable traits can be inherited, (3) there is unequal reproduction, and (4) differences in reproductive success depend on the traits of individuals, then the characteristics of successful individuals will be "naturally selected" and become more common over time.

14.3 How Do We Know That Evolution Has Occurred?

Many lines of evidence indicate that evolution has occurred, including the following:

- Fossils of ancient species tend to be simpler in form than modern species. Sequences of fossils have been discovered that show a graded series of changes in form. Both of these observations would be expected if modern species evolved from older species.
- Species thought to be related through evolution from a common ancestor possess many similar anatomical structures. For example, amphibians, reptiles, birds, and mammals have very similar forearm bones.
- Stages in early embryological development are quite similar among very different types of vertebrates.
- Similarities in biochemical traits such as the use of DNA as the carrier of genetic information support the notion of descent of related species through evolution from common ancestors.

14.4 What Is the Evidence That Populations Evolve by Natural Selection?

Similarly, many lines of evidence indicate that natural selection is the chief mechanism driving changes in the characteristics of species over time, including the following:

- Inheritable traits have been changed rapidly in populations of domestic animals and plants by selectively breeding organisms with desired features (artificial selection). The immense variations in species produced in a few thousand years by artificial selection makes it almost inevitable that much larger changes would be wrought by hundreds of millions of years of natural selection.
- Evolution can be observed today. Both natural and human activities can drastically change the environment over short periods of time. Inherited characteristics of species have been observed to change significantly in response to such environmental changes.

Key Terms

analogous structure	268	homologous structure	268
artificial selection	271	natural selection	265
convergent evolution	268	population	259
evolution	259	vestigial structure	268
fossil	260		

Learning Outcomes

In this chapter, you have learned to . . .

LO1 Define *evolution*.

LO2 Describe pre-Darwin hypotheses and ideas that set the stage for the later emergence of the theory of evolution by natural selection.

LO3 Explain the mechanism of natural selection and the logic by which Darwin deduced it, and describe how natural selection affects populations.

LO4 Describe the main types of evidence that evolution has occurred, and provide an example of each type.

LO5 Compare and contrast similarities due to homology and similarities attributable to convergent evolution.

LO6 Describe the main lines of evidence that populations evolve by natural selection and provide examples of such evidence.

Thinking Through the Concepts

Fill-in-the-Blank

1. The flipper of a seal is homologous with the _____ of a bird, and both of these are homologous with the _____ of a human. The wing of a bird and the wing of a butterfly are described as _____ structures that arose as a result of _____ evolution. Remnants of structures in animals that have no use for them, such as the small hind leg bones of whales, are described as _____ structures. **LO4 LO5**

2. The finding that all organisms share the same genetic code provides evidence that all descended from a(n) _____. Further evidence is provided by the fact that all cells use roughly the same set of _____ to build proteins, and all cells use the molecule _____ as an energy carrier. **LO4**

3. Georges Cuvier espoused a concept called _____ to explain layers of rock with embedded fossils. Charles

Lyell, building on the work of James Hutton, proposed an alternative explanation called _____, which states that layers of rock and many other geological features can be explained by gradual processes that occurred in the past just as they do in the present. This concept provided important support for evolution because it required that Earth be extremely _____. **LO2**

4. The process by which inherited characteristics of populations change over time is called _____. Variability among individuals is the result of chance changes called _____ that occur in the hereditary molecule _____. **LO1 LO2**

5. The process by which individuals with traits that provide an advantage in their natural habitats are more successful at reproducing is called _____. People who breed animals or plants can produce large changes in their characteristics in a relatively short time, a process called _____. **LO3 LO6**

6. Darwin's postulate 2 states that _____. The work of _____ provided the first experimental evidence for this postulate. **LO3**

Review Questions

1. Selection acts on individuals, but only populations evolve. Explain why this statement is true. **LO3**

2. Distinguish between catastrophism and uniformitarianism. How did these hypotheses contribute to the development of evolutionary theory? **LO2**

3. Describe Lamarck's theory of inheritance of acquired characteristics. Why is it invalid? **LO2**

4. What is natural selection? Describe how natural selection might have caused unequal reproduction among the ancestors of a fast-swimming predatory fish, such as the barracuda. **LO3**

5. Describe how evolution occurs. In your description, include discussion of the reproductive potential of species, the stability of natural population sizes, variation among individuals of a species, inheritance, and natural selection. **LO3**

6. What is convergent evolution? Give an example. **LO5**

7. How do biochemistry and molecular genetics contribute to the evidence that evolution has occurred? **LO4**

Applying the Concepts

1. In discussions of untapped human potential, it is commonly said that the average person uses only 10% of his or her brain. Is this conclusion likely to be correct? Explain your answer in terms of natural selection.

2. Both the theory of evolution by natural selection and the theory of special creation (which states that all species were simultaneously created by God) have had an impact on evolutionary thought. Discuss why one is considered to be a scientific theory and the other is not.

3. Does evolution through natural selection produce "better" organisms in an absolute sense? Are we climbing the "Ladder of Nature"? Defend your answer.

4. In what sense are humans currently acting as agents of selection on other species? Name some traits that are *favored* by the environmental changes humans cause.

5. Darwin and Wallace's discovery of natural selection is one of the great revolutions in scientific thought. Some scientific revolutions spill over and affect the development of philosophy and religion. Is this true of evolution? Does (or should) the idea of evolution by natural selection affect the way humans view their place in the world?

6. In an alternate universe in which the climate of a planet changes much more rapidly than has been the case on Earth, what differences would you expect to find in the planet's fossil record?

Answers to Figure Caption questions and Fill-in-the-Blank questions can be found in the Answers section at the back of the book.

(**MB**) ®*Go to MasteringBiology for practice quizzes, activities, eText, videos, current events, and more.*

CHAPTER 15

How Populations Evolve

Staphylococcus aureus, a common source of human infections, is among the many bacterial species that have evolved resistance to antibiotics.

CASE STUDY

Evolution of a Menace

ON A FEBRUARY DAY IN 2008, a 20-year-old student arrived at the health center at Western Washington University. He had been bothered by a lingering cough for a couple of weeks and, when his symptoms worsened to include a fever and vomiting, he sought medical attention. The health center staff quickly determined that the student had pneumonia and began treatment. His condition deteriorated, however, and he was transferred to the local hospital. A few days later, he died.

Why couldn't doctors save a previously healthy young man from a normally curable disease? Because the victim's pneumonia was caused by methicillin-resistant *Staphylococcus aureus* (MRSA) bacteria. *Staphylococcus aureus*, sometimes referred to as "staph," is a common bacterium that

can infect the skin, blood, or respiratory system. Many staph infections can be successfully treated with antibiotics, but MRSA bacteria are antibiotic resistant and cannot be killed by many of the most commonly used antibiotics. Until recently, MRSA infections occurred almost exclusively in hospitals. Today, however, resistant staph is widespread, and more than 10% of MRSA infections occur outside of hospitals, in homes, schools, and workplaces.

In the United States, MRSA infections kill about 19,000 people each year. And, unfortunately, *Staphylococcus* is by no means the only disease-causing bacterium that is becoming less vulnerable to antibiotics. For example, antibiotic resistance has also appeared in the bacteria that cause tuberculosis, a disease that kills almost 2 million people each year. In an increasing number of tuberculosis cases, the disease does not respond to any of the drugs commonly used to treat it. Multidrug

resistance is also becoming more prevalent in the bacteria responsible for the widespread sexually transmitted disease gonorrhea. In addition, drug resistance is common in the bacteria that cause food poisoning, blood poisoning, dysentery, pneumonia, meningitis, and urinary tract infections. We are experiencing a global onslaught of resistant "supergerms," and are facing the specter of diseases that cannot be cured.

Many physicians and scientists believe that the most effective way to combat the rise of resistant diseases is to reduce the use of antibiotics. Why might such a strategy be effective? Because the upsurge of antibiotic resistance is a consequence of evolutionary change in populations of bacteria, and the agent of this change is natural selection imposed by antibiotic drugs. To understand how the crisis of antibiotic resistance arose and to devise a strategy to resolve it, we must have a clear understanding of the mechanisms by which populations evolve.

AT A GLANCE

LEARNING GOALS

LG1 Define an *equilibrium population* and describe the conditions under which a population is expected to remain at evolutionary equilibrium.

LG2 Use the vocabulary of population genetics to describe the main causes of

evolution and explain how each of these causes affects evolution.

LG3 Describe how variation in mating success among the individuals of a species can influence the species' evolution.

LG4 Compare and contrast directional selection, stabilizing selection, and disruptive selection.

15.1 HOW ARE POPULATIONS, GENES, AND EVOLUTION RELATED?

If you live in an area with a seasonal climate and you own a dog or cat, you have probably noticed that your pet's fur gets thicker and heavier as winter approaches. Has the animal evolved? No. The changes that we see in an individual organism over the course of its lifetime are not evolutionary changes. Instead, evolutionary changes occur from generation to generation, causing descendants to be different from their ancestors.

Furthermore, we can't detect evolutionary change across generations by looking at a single set of parents and offspring. For example, if you observed that a 6-foot-tall man had an adult son who stood 5 feet tall, could you conclude that humans were evolving to become shorter? Obviously not. Rather, if you wanted to learn about evolutionary change in human height, you would begin by measuring many humans over many generations to see if the average height is changing with time. Evolution is a property not of individuals but of populations. A **population** is a group that includes all the members of a species living in a given area.

Recognizing that evolution is a population-level phenomenon was one of Darwin's key insights. But populations are composed of individuals, and the actions and fates of individuals determine which characteristics will be passed to descendant populations. In this fashion, inheritance provides the link between the lives of individual organisms and the evolution of populations. We will therefore begin our discussion of the processes of evolution by reviewing some principles of genetics as they apply to individuals. We will then extend those principles to the genetics of populations.

Genes and the Environment Interact to Determine Traits

Each cell of every organism contains genetic information encoded in the DNA of its chromosomes. A *gene* is a segment of DNA located at a particular place on a chromosome (see Chapter 10). The sequence of nucleotides in a gene encodes the sequence of amino acids in a protein, usually an enzyme that catalyzes a particular reaction in the cell. At a given

gene's location, different members of a species may have slightly different nucleotide sequences, called *alleles*. Different alleles generate different forms of the same enzyme. In this way, various alleles of the genes that influence eye color in humans, for example, help produce eyes that are brown, or blue, or green, and so on.

In any population of organisms, there are usually two or more alleles of each gene. An individual of a diploid species whose alleles of a particular gene are both the same is *homozygous* for that gene, and an individual with different alleles for that gene is *heterozygous*. The specific alleles borne on an organism's chromosomes (its *genotype*) interact with the environment to influence the development of its physical and behavioral traits (its *phenotype*).

Let's illustrate these principles with an example. A black hamster's coat is colored black because a chemical reaction in its hair follicles produces a black pigment. When we say that a hamster has the allele for a black coat, we mean that a particular stretch of DNA on one of the hamster's chromosomes contains a sequence of nucleotides that codes for the enzyme that catalyzes a pigment-producing reaction that results in a black coat. A hamster with the allele for a brown coat has a different sequence of nucleotides at the corresponding chromosomal position. That different sequence codes for an enzyme that cannot produce black pigment. If a hamster is homozygous for the black allele (two black alleles) or is heterozygous (one black allele and one brown allele), its fur contains the pigment and is black. But if a hamster is homozygous for the brown allele, its hair follicles produce no black pigment and its coat is brown (**Fig. 15-1**). Because the hamster's coat is black even when only one copy of the black allele is present, the black allele is considered *dominant* and the brown allele *recessive*.

The Gene Pool Comprises All of the Alleles in a Population

In studying evolution, looking at the process in terms of its effects on genes has proven to be an enormously useful tool. In particular, evolutionary biologists have made excellent use of the tools of a branch of genetics, called population genetics, that deals with the frequency, distribution, and inheritance of alleles in populations. To take advantage of this

Each chromosome has one allele of the coat-color gene

Coat-color allele *B* is dominant, so heterozygous hamsters have black coats

phenotype

genotype   *BB*   *Bb*   *bb*

chromosomes

           B   *B*   *B*   *b*   *b*   *b*

          homozygous   heterozygous   homozygous

◄FIGURE 15-1 **Alleles, genotype, and phenotype in individuals**
An individual's particular combination of alleles is its genotype. The word "genotype" can refer to the alleles of a single gene (as shown here), to a set of genes, or to all of an organism's genes. An individual's phenotype is determined by its genotype and environment. "Phenotype" can refer to a single trait, a set of traits, or all of an organism's traits.

powerful aid to understanding evolution, you will need to learn a few of the basic concepts of population genetics.

Population genetics defines a **gene pool** as a set that contains all of the alleles of all of the genes from all of the individuals in a population. You can think of a gene pool as the contents of an imaginary bucket into which each member of a population has tossed one copy of its genotype. Thus, the number of copies of each allele in the gene pool is equal to the sum of (1) the number of individuals that carry a only single copy of the allele, and (2) twice the number of individuals that carry two copies. A gene pool is not an actual physical entity but is instead a mental construct that can help us understand the process of evolution.

In addition, each particular gene can be considered to have its own gene pool, which comprises all of the alleles of that specific gene in a population (Fig. 15-2). If we counted the number of copies of each allele present in the genotype of each individual in the population, and then added up the counts, we could determine the relative proportion of each allele in the gene pool. An allele's proportion in the gene pool is its **allele frequency.** For example, the population of 25 hamsters portrayed in Figure 15-2 contains 50 alleles of the gene that controls coat color (because hamsters are diploid and each hamster thus has two copies of each gene). Twenty of those 50 alleles are of the type that codes for black coats, so the frequency of that allele in the population is 20/50 = 0.40 (or 40%).

Evolution Is the Change of Allele Frequencies Within a Population

A casual observer might define evolution on the basis of changes in the outward appearance or behaviors of the members of a population. A population geneticist, however, looks at a population and sees a gene pool that just happens to be divided into the packages that we call individual organisms. So, many of the outward changes that

we observe in the individuals that make up the population can also be viewed as the visible expression of underlying changes to the gene pool. A population geneticist, therefore, defines evolution as changes over time in the allele frequencies of a gene pool. Evolution is change in the genetic makeup of populations over generations.

The Equilibrium Population Is a Hypothetical Population in Which Evolution Does Not Occur

It is easier to understand what causes populations to evolve if we first consider the characteristics of a population that would *not* evolve. In 1908, English mathematician Godfrey H. Hardy and German physician Wilhelm Weinberg independently developed a simple mathematical model of a non-evolving population. This model, now known as the **Hardy–Weinberg principle,** showed that under certain conditions, allele frequencies and genotype frequencies in a population will remain constant no matter how many generations pass. (For more information about the model, see "In Greater Depth: The Hardy–Weinberg Principle.") In other words, this population will not evolve. Population geneticists use the term **equilibrium population** for this hypothetical non-evolving population in which allele frequencies do not change as long as the following conditions are met:

- There must be no mutation.
- There must be no **gene flow.** That is, there must be no movement of alleles into or out of the population (as would be caused, for example, by the movement of organisms into or out of the population).
- The population must be very large.
- All mating must be random, with no tendency for certain genotypes to mate with specific other genotypes.
- There must be no natural selection. That is, all genotypes must reproduce with equal success.

Population: 25 individuals

The gene pool of the coat-color gene contains 20 copies of allele *B* and 30 copies for allele *b*

Gene pool: 50 alleles

◄**FIGURE 15-2 A gene pool** In diploid organisms, each individual in a population contributes two alleles of each gene to the gene pool.

Under these conditions, allele frequencies in a population will remain the same indefinitely. If one or more of these conditions is violated, then allele frequencies may change: The population will evolve.

As you might expect, few natural populations are truly in equilibrium. What, then, is the importance of the Hardy–Weinberg principle? The Hardy–Weinberg conditions are useful starting points for studying the mechanisms of evolution. In the following sections, we will examine some of these conditions, show that natural populations typically fail to meet them, and illustrate the consequences of such failures. In this way, we can better understand both the inevitability of evolution and the processes that drive evolutionary change.

CHECK YOUR LEARNING

Can you define *evolution* in terms of concepts from population genetics? Can you define *equilibrium population* and describe the conditions under which a population is expected to remain at equilibrium?

15.2 WHAT CAUSES EVOLUTION?

Population genetics theory predicts that the Hardy–Weinberg equilibrium can be disturbed by deviations from any of its five conditions. Therefore, we can predict five major causes of evolutionary change: mutation, gene flow, small population size, nonrandom mating, and natural selection.

Mutations Are the Original Source of Genetic Variability

A population remains in evolutionary equilibrium only if there are no **mutations** (changes in DNA sequence). Most mutations occur during cell division, when a cell makes a copy of its DNA. Sometimes, errors occur during the copying process and the copied DNA does not match the original. Most such errors are quickly corrected by cellular systems that identify and repair DNA copying mistakes, but some changes in nucleotide sequence slip past the repair systems. An unrepaired mutation in a cell that gives rise to gametes (eggs or sperm) may be passed to offspring and enter the gene pool of a population.

Inherited Mutations Are Rare But Important

How significant is mutation in changing the gene pool of a population? For any given gene, only a tiny proportion of a population inherits a new mutation from the previous generation. For example, a new mutant version of a typical human gene will appear in only about 1 out of every 100,000 gametes produced and, because new individuals are formed by the fusion of two gametes, in only about 1 of every 50,000 newborns. Therefore, mutation by itself generally causes only very small changes in the frequency of any particular allele.

Despite the rarity of inherited mutations of any particular gene, the cumulative effect of mutations is essential to evolution. Most organisms have a large number of different genes, so even if the rate of mutation is low for any one gene, the sheer number of possibilities means that each new generation of a population is likely to include some mutations. For example, geneticists estimate that humans have between 20,000 and 25,000 different genes and, because each of us has two copies of each gene, a person carries between 40,000 and 50,000 alleles. Thus, even if each allele has, on average, only a 1 in 100,000 chance of mutation, most newborns will probably carry one or two mutations. These mutations are new alleles—new variations on which other evolutionary processes can work. As such, they are the foundation of evolutionary change. Without mutations there would be no evolution.

Mutations Are Not Goal Directed

A mutation does not arise as a result of, or in anticipation of, the needs of an organism. A mutation simply happens and may produce a change in a structure or function of the organism. Whether that change is helpful or harmful or neutral, now or in the future, depends on environmental conditions over which the organism has little or no control (**Fig. 15-3**). The mutation merely provides a potential for evolutionary change. Other processes, especially natural selection, may act to spread the mutation through the population or to eliminate it.

❶ Start with bacterial colonies that have never been exposed to antibiotics

❷ Use velvet to transfer colonies to identical positions in three dishes containing the antibiotic streptomycin

❸ Incubate the dishes

❹ Only streptomycin-resistant colonies grow; the few colonies are in the exact same positions in each dish

◄ **FIGURE 15-3 Mutations occur spontaneously** This experiment demonstrates that mutations occur spontaneously and not in response to environmental conditions. When bacterial colonies that have never been exposed to antibiotics are exposed to the antibiotic streptomycin, only a few colonies grow. The observation that these surviving colonies grow in the exact same positions in all dishes shows that the mutations for resistance to streptomycin were present in the original dish before exposure to streptomycin.

QUESTION If it were true that mutations do occur in response to the presence of antibiotics, how would the result of this experiment have differed from the actual result?

Evolution of a Menace

If mutations are rare and random, why do mutant alleles that confer antibiotic resistance arise so commonly in bacterial populations? The seemingly inevitable emergence of resistance results in large measure from the huge populations and very short generation times that are typical among bacteria. A single drop of human saliva contains about 150 million bacteria, and an entire human body contains hundreds of trillions. With so many bacteria present, even an extremely rare resistance mutation that occurs in only a tiny percentage of the population will be present in some individuals. Also, because many mutations occur during cell division, rapid reproduction creates many opportunities for mutations to arise. Bacteria reproduce very rapidly, as often as every 15 minutes in some species. This rapid reproduction, taking place in huge populations, results in a high likelihood that mutations leading to antibiotic resistance will occur in bacterial populations.

Gene Flow Between Populations Changes Allele Frequencies

The movement of alleles between populations, known as *gene flow*, changes how alleles are distributed among populations. When individuals move from one population to another and interbreed at the new location, alleles are transferred from one gene pool to another. For example, baboons live in social groupings called troops, and some individuals—usually juvenile males—routinely leave their troop and move to new populations. If the departing baboons are fortunate, they join another troop and achieve sufficient social status to breed. In this way, the male offspring of one troop can carry alleles to the gene pools of other troops.

In some kinds of organisms, alleles move between populations only at certain stages of the life cycle. In flowering plants, for example, most gene flow is due to the movement of seeds and pollen (**Fig. 15-4**). Pollen, which contains sperm cells, may be carried long distances by wind or by animal pollinators. If the pollen ultimately reaches the flowers of a different population of its species, it may fertilize eggs and add its collection of alleles to the local gene pool. Similarly, seeds may be borne by wind, water, or animals to distant locations where they can germinate to become part of a population far from their place of origin.

The main evolutionary effect of gene flow is to increase the genetic similarity of different populations of a species. To help understand why, picture two glasses, one containing fresh water and the other salty seawater. If a few spoonfuls are transferred from the glass containing seawater to the glass containing fresh water, the water in the freshwater glass becomes saltier. In the same way, movement of alleles from one population to another tends to change the gene pool of the destination population so that it is more similar to the source population.

If alleles move continually back and forth between different populations, the gene pools of the different populations will, in effect, mix. This mixing prevents the development of large differences in the genetic compositions of the populations. But if gene flow between populations of a species is

▲ FIGURE 15-4 **Pollen can be an agent of gene flow** Pollen, drifting on the wind, can carry alleles from one population to another.

blocked, the resulting genetic differences may grow so large that one of the populations becomes a new species. (We will discuss this process in Chapter 16.)

Allele Frequencies May Change by Chance in Small Populations

Allele frequencies in populations can be changed by chance events other than mutations. For example, if bad luck prevents some members of a population from reproducing, their alleles will ultimately be removed from the gene pool, altering its makeup. What kinds of bad-luck events can randomly prevent some individuals from reproducing? Seeds can fall into a pond or a parking lot and thus never sprout; flowers can be destroyed by a hailstorm or a wildfire before they are pollinated; any organism can be killed by a flood or a volcanic eruption before reproducing. Any event that arbitrarily cuts lives short or otherwise allows only a random subset of a population to reproduce can cause random changes in allele frequencies. The process by which chance events change allele frequencies is called **genetic drift.**

To see how genetic drift works, imagine a population of 20 hamsters in which the frequency of the black coat color allele B is 0.50 and the frequency of the brown coat color allele b is 0.50 (**Fig. 15-5**, top). If all of the hamsters in the population were to interbreed to yield another population of 20 animals, the frequencies of the two alleles would not change in

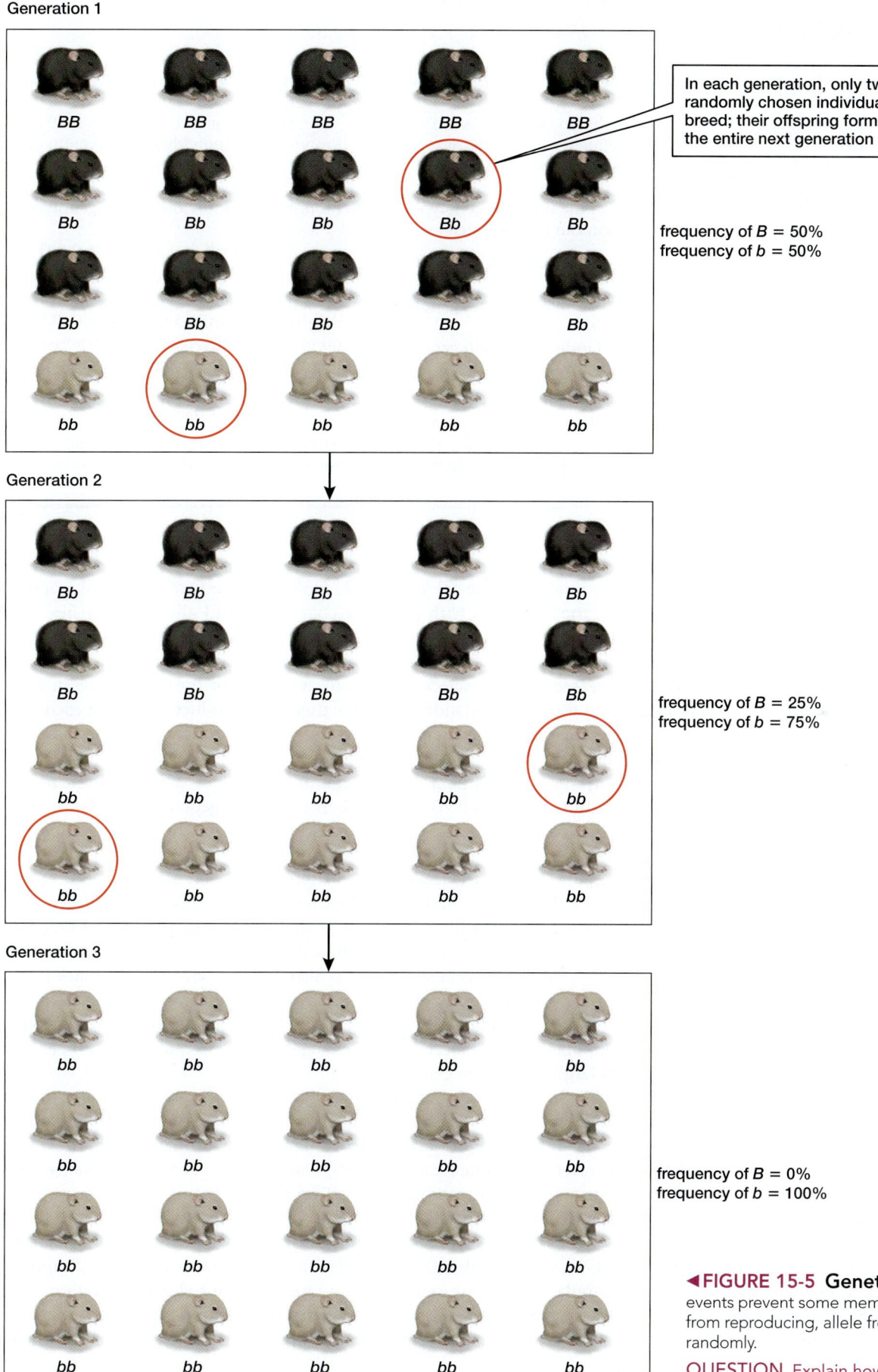

Generation 1

In each generation, only two randomly chosen individuals breed; their offspring form the entire next generation

frequency of B = 50%
frequency of b = 50%

Generation 2

frequency of B = 25%
frequency of b = 75%

Generation 3

frequency of B = 0%
frequency of b = 100%

◀FIGURE 15-5 **Genetic drift** If chance events prevent some members of a population from reproducing, allele frequencies can change randomly.

QUESTION Explain how the distribution of genotypes in generation 2 is calculated.

IN GREATER DEPTH The Hardy–Weinberg Principle

The Hardy–Weinberg principle states that allele frequencies will remain constant over time in the gene pool of a large population in which there is random mating but no mutation, no gene flow, and no natural selection. In addition, Hardy and Weinberg showed that if allele frequencies do not change in an equilibrium population, the proportion of individuals with a particular genotype will also remain constant.

To better understand the relationship between allele frequencies and the occurrence of genotypes, picture an equilibrium population whose members carry a gene that has two alleles, A_1 and A_2. Note that each individual in this population must carry one of three

possible diploid genotypes (combinations of alleles): A_1A_1, A_1A_2, or A_2A_2.

Suppose that in our population's gene pool, the frequency of allele A_1 is p, and the frequency of allele A_2 is q. Hardy and Weinberg demonstrated that the proportions of the different genotypes in the population can be calculated as:

Proportion of individuals with genotype
$A_1A_1 = p^2$
Proportion of individuals with genotype
$A_1A_2 = 2pq$
Proportion of individuals with genotype
$A_2A_2 = q^2$

For example, if, in our population's gene pool, 70% of the alleles of a gene are A_1 and 30% are A_2 (that is, $p = 0.7$ and $q = 0.3$), then genotype proportions would be:

Proportion of individuals with genotype
$A_1A_1 = 49\%$
(because $p^2 = 0.7 \times 0.7 = 0.49$)
Proportion of individuals with genotype
$A_1A_2 = 42\%$
(because $2pq = 2 \times 0.7 \times 0.3 = 0.42$)
Proportion of individuals with genotype
$A_2A_2 = 9\%$
(because $q^2 = 0.3 \times 0.3 = 0.09$)

Because every member of the population must possess one of the three genotypes, the three proportions must always add up to one. For this reason, the expression that relates allele frequency to genotype proportions can be written as:

$$p^2 + 2pq + q^2 = 1$$

where the three terms on the left side of the equation represent the three genotypes.

the next generation. But if we instead allow only two, randomly chosen hamsters (the ones circled in Fig. 15-5, top) to breed and become the parents of the next generation of 20 animals, allele frequencies might be quite different in generation 2 (**Fig. 15-5**, center; the frequency of B has decreased and the frequency of b has increased). And if breeding in the second generation were again restricted to two randomly chosen hamsters (circled in Fig. 15-5, center), allele frequencies might change again in generation 3 (**Fig. 15-5**, bottom). Allele frequencies will continue to change in random fashion for as long as reproduction is restricted to a random subset of the population. Note that the changes caused by genetic drift can include the disappearance of an allele from the population, as illustrated by the disappearance of the B allele (and therefore the black coat phenotype) in generation 3 of the example shown in Figure 15-5.

Population Size Matters

Genetic drift occurs to some extent in all populations, but it occurs more rapidly and has a greater effect in small populations than in large ones. If a population is sufficiently large, chance events are unlikely to significantly alter its genetic composition, because random removal of a few individuals' alleles won't have a big impact on allele frequencies in the population as a whole. In a small population, however, a particular allele may be carried by only a few organisms. Chance events could eliminate most or all copies of such an allele from the population.

To see how population size affects genetic drift, let's return to generation 1 of our hypothetical hamster population (see Fig. 15-5, top). Three-quarters of the hamsters are black and one-quarter are brown; the frequencies of alleles B

and b are each 50%. Now imagine two additional populations with the same coat colors and allele frequencies, one with only eight individuals and one with 20,000.

Now let's picture reproduction in our two populations. Let's select, at random, a quarter of the individuals in each population and allow them to reproduce, such that each pair of hamsters in both populations produces eight offspring and then dies. In the large population, 5,000 hamsters reproduce, yielding a new generation of 20,000. What are the chances that all 20,000 members of the new generation will be brown? Just about nil; that would happen only if all 5,000 randomly chosen breeders just happened to be brown (bb) hamsters. In fact, it would be extremely unlikely for even 15,000 offspring hamsters to be brown, or for 12,000 to be brown. The most likely outcome is that about a quarter of the breeders will be brown and three-quarters black, which would yield a new generation that is also 25% brown and 75% black (with allele frequencies of B = 50% and b = 50%), just as in the original population. In the large population, then, we would not expect a major change in allele frequencies from generation to generation.

In the small population of eight, the situation is different. If we again choose a random quarter of the population to reproduce, only two hamsters will breed. Given the allele frequencies in the population, there is a 12.5% chance that both reproducers will be brown. If this occurs—a not terribly unlikely outcome—the next generation of eight hamsters will consist entirely of brown individuals. It is therefore possible that the allele for black coat color could disappear from the smaller population within a single generation.

One way to test these predictions about genetic drift in large versus small populations is to write a computer program that simulates how the frequencies of the alleles could change over many generations in which only a random subset of the population breeds. **Figure 15-6a** shows the results from four runs of such a simulation of our large population. The initial frequency of each allele is set at 50%. Notice that the frequency of allele B remains close to its initial frequency.

Figure 15-6b shows the fate of allele B in four runs of a simulation of our small population. In one of the four runs (red line), allele B reaches a frequency of 100% in the second generation, meaning that all the hamsters in the second and following generations are black. In another run, the frequency of B falls to zero in the third generation (blue line), and the population subsequently is all brown. One of the two hamster phenotypes disappeared in half of the simulations, and the outcome of a given simulation was much less predictable in the small population than it was in the large one.

A Population Bottleneck Can Cause Genetic Drift

Two causes of genetic drift, the population bottleneck and the founder effect, further illustrate the impact that small population size may have on the allele frequencies of a species. In a **population bottleneck,** a population is drastically reduced by, for example, a natural catastrophe or overhunting. After such a bottleneck, only a few individuals are available to contribute genes to the next generation. Population bottlenecks can rapidly change allele frequencies and can reduce genetic variability by eliminating alleles (**Fig. 15-7a**).

Even if the population later increases, the genetic effects of the bottleneck may remain for hundreds or thousands of generations.

Loss of genetic variability due to bottlenecks has been documented in numerous species, including the northern elephant seal (**Fig. 15-7b**). The elephant seal was hunted almost to extinction in the 1800s; by the 1890s, only about 20 individuals survived. Dominant male elephant seals typically monopolize breeding, so a single male may have fathered all the offspring at this extreme bottleneck point. Since the late nineteenth century, elephant seals have increased in number to about 30,000 individuals, but biochemical analysis shows that all northern elephant seals are genetically almost identical. With so little genetic variation, the elephant seal has little potential to evolve in response to environmental changes (see "Earth Watch: The Perils of Shrinking Gene Pools"). Because of the limited genetic variation in this species, the species remains vulnerable to extinction regardless of how many elephant seals there are.

Isolated Founding Populations May Produce Bottlenecks

The **founder effect** occurs when isolated colonies are founded by a small number of organisms. A flock of birds, for instance, that becomes lost during migration or is blown off course by a storm may settle on an isolated island. This small founder group may, by chance, have allele frequencies that are very different from the frequencies of the parent population. If this is the case, the gene pool of the future

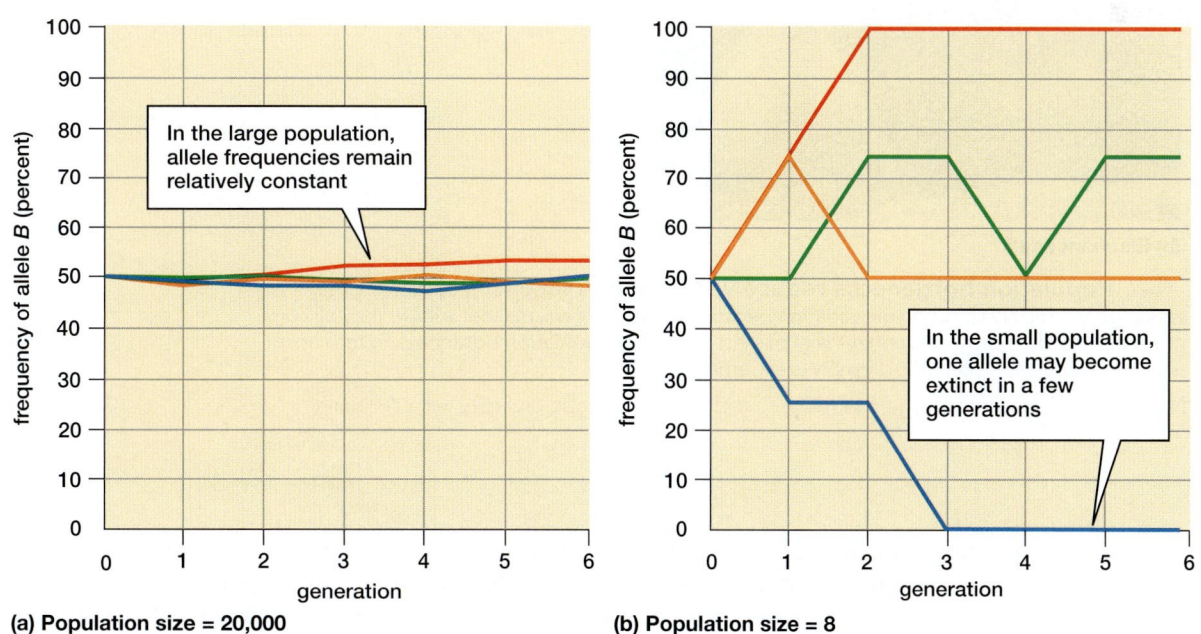

(a) Population size = 20,000

(b) Population size = 8

▲ FIGURE 15-6 **The effect of population size on genetic drift** Each colored line represents one computer simulation of the change over time in the frequency of allele B in **(a)** a large population and **(b)** a small population. Half of the alleles in each starting population were B (50%) and, in each generation, randomly chosen individuals reproduced.

EXERCISE Sketch a graph that shows the result you would predict if the simulation were run four times with a population size of 20.

(a) Simulation of a population bottleneck

The gene pool of a population contains equal numbers of red, blue, yellow, and green alleles

A bottleneck event drastically reduces the size of the population

By chance, the gene pool of the reduced population contains mostly blue and a few yellow alleles

After the population grows and returns to its original size, blue alleles predominate; red and green alleles have disappeared

(b) Elephant seals

▲ FIGURE 15-7 **Population bottlenecks reduce variation** (a) A population bottleneck may drastically reduce genotypic and phenotypic variation because the few organisms that survive may all carry similar sets of alleles. (b) The northern elephant seal passed through a population bottleneck in the recent past. As a result, the population's genetic diversity is extremely low.

QUESTION If a population grows large again after a bottleneck, genetic diversity will eventually increase. Why?

population in the new location will be quite unlike that of the larger population from which it sprang. Consider, for example, the Amish inhabitants of Lancaster County, Pennsylvania, who are descended from only 200 or so eighteenth-century immigrants. Among today's Lancaster County Amish, a set of genetic defects known as Ellis–van Creveld syndrome is far more common than it is among the general population (Fig. 15-8). The prevalence of the syndrome among the Amish stems from a single couple among

the original immigrants who carried the Ellis–van Creveld allele. Because the founder population was so small, this single occurrence meant that the allele was carried by a comparatively high proportion of the population (1 or 2 carriers out of 200 versus about 1 in 1,000 in the general population). This high initial allele frequency, a result of the founder effect, combined with subsequent genetic drift, has led to extraordinarily high levels of Ellis–van Creveld syndrome among this Amish group.

(b) A six-fingered hand

◀FIGURE 15-8 **A human example of the founder effect (a)** An Amish woman with her child, who suffers from a set of genetic defects known as Ellis–van Creveld syndrome. Symptoms of the syndrome include short arms and legs, **(b)** extra fingers, and, in some cases, heart defects. The founder effect accounts for the prevalence of Ellis–van Creveld syndrome among the Amish residents of Lancaster County, Pennsylvania.

(a) A child with Ellis–van Creveld syndrome

Mating Within a Population Is Almost Never Random

The effects of *nonrandom mating* can play a significant role in evolution, because organisms seldom mate strictly randomly. For example, many organisms have limited mobility and tend to remain near their place of birth, hatching, or germination. In such species, most of the offspring of a given parent live in the same area and thus, when they reproduce, there is a good chance that they will be related to their reproductive partners. Such sexual reproduction between relatives is called *inbreeding*.

Because relatives are genetically similar, inbreeding tends to increase the number of individuals that inherit the same alleles from both parents and are therefore homozygous for many genes. This increase in homozygotes can have harmful effects, such as increased occurrence of genetic diseases or defects. Many gene pools include harmful recessive alleles that persist in the population because their negative effects are masked in heterozygous carriers (which have only a single copy of the harmful allele). Inbreeding, however, increases the odds of producing homozygous offspring with two copies of the harmful allele.

In animals, nonrandom mating can also arise if individuals have preferences or biases that influence their choice of mates. The snow goose is a case in point. Individuals of this species come in two "color phases"; some snow geese are white, and others are blue-gray (**Fig. 15-9**). Although both white and blue-gray geese belong to the same species, mate choice is not random with respect to color. The birds exhibit a strong tendency to mate with a partner of the same color. This preference for mates that are similar is known as *assortative mating*.

Neither inbreeding nor assortative mating by themselves will alter allele frequencies in a population. Nonetheless, they can have large effects on the distribution of different genotypes, and thus on the distribution of phenotypes, in the population.

All Genotypes Are Not Equally Beneficial

In a hypothetical equilibrium population, individuals of all genotypes survive and reproduce equally well; no genotype has any advantage over the others. This condition, however, is probably met only rarely, if ever, in real populations. Even though many alleles are neutral, in the sense that organisms possessing any of several alleles are equally likely to survive and reproduce, some alleles confer an advantage on their possessor. Any time an allele provides, in Alfred Russel Wallace's words, "some little superiority," the individuals

▲ FIGURE 15-9 **Nonrandom mating among snow geese** Snow geese, which have either white plumage or blue-gray plumage, are most likely to mate with other birds of the same color.

Earth Watch

The Perils of Shrinking Gene Pools

Many of Earth's species are in danger. According to the World Conservation Union, more than 20,000 species of plants and animals are currently threatened with extinction. For most of these endangered species, the main threat is habitat destruction. When a species' habitat shrinks, its population size almost invariably follows suit.

Many people, organizations, and governments are concerned about the plight of endangered species and are working to protect them and their habitats. The hope is that these efforts will not only prevent the extinction of endangered species but will also restore their numbers so that they no longer need protection. Unfortunately, a population that has already become small enough to warrant endangered status is likely to undergo evolutionary changes that increase its chances of going extinct. The principles of evolutionary genetics that we've explored in this chapter can help us understand these changes.

One problem is that, in small populations, mating choices are limited and a high proportion of matings may be between close relatives. This inbreeding increases the odds that offspring will be homozygous for harmful recessive alleles. These less-fit individuals may die before reproducing, further reducing the size of the population.

The greatest threat to small populations, however, stems from their inevitable loss of genetic diversity (**Fig. E15-1**). From our discussion of population bottlenecks, it is apparent that, when populations shrink to very small sizes, many of the alleles that were present in the original population will not be represented in the gene pool of the remnant population. Furthermore, we have seen that genetic drift in small populations will cause many of the surviving alleles to subsequently disappear permanently from the population (see Fig. 15-6b). Because genetic drift is a random process, many of the lost alleles will be advantageous ones that were previously favored by natural selection. Inevitably, the number of different alleles in the population grows ever smaller. As ecologist Thomas Foose aptly put it, "Gene pools are being converted into gene puddles." Even if the size of an endangered population eventually begins to grow, the damage has already been done; lost genetic diversity is regained only very slowly, over hundreds of generations.

Why does it matter if a population's genetic diversity is low? Low diversity creates two main risks. First, the fitness of the population as a whole is reduced by the loss of advantageous

▲ **FIGURE E15-1 Only a few hundred Sumatran rhinoceroses remain**

alleles that underlie adaptive traits. A less-fit population is unlikely to thrive. Second, a genetically impoverished population lacks the variation that will allow it to adapt when environmental conditions change. When the environment changes, as it inevitably will, a genetically uniform species is less likely to contain individuals well suited to survive and reproduce under the new conditions. A species unable to adapt to changing conditions is at very high risk of extinction.

BioEthics   What can be done to preserve the genetic diversity of endangered species? The best solution, of course, is to preserve plenty of diverse types of habitat so that species never become endangered in the first place. The human population, however, has grown so large and has thus appropriated so large a share of Earth's resources that this solution is impossible in many places. For many species, the only solution is to ensure that areas of preserved habitat are large enough to hold populations of sufficient size to contain most of a threatened species' total genetic diversity. If, however, circumstances dictate that preserved areas must be small, it is important that the small areas be linked by corridors of the appropriate habitat, so that gene flow among populations in the small preserves can increase the spread of new and beneficial alleles.

? HAVE **YOU** EVER WONDERED . . .

Why Antibiotics Can't Cure a Cold?

Antibiotics kill or inhibit the reproduction of bacteria, usually by blocking a metabolic pathway that the bacteria use to replicate DNA, synthesize proteins, or build a cell wall. Colds and flu, however, are caused by viruses. Viruses do not have metabolism and are not affected by antibiotics. So, taking antibiotics when you have a cold or the flu won't help you get better.

who carry it are favored by **natural selection,** the process in which individuals with traits that help them survive and reproduce leave more offspring than do individuals that lack those traits. We examine the impact of natural selection in greater depth in the next section.

Table 15-1 summarizes the different causes of evolution.

CHECK YOUR LEARNING

Can you describe how mutation, gene flow, genetic drift, nonrandom mating, and natural selection affect evolution?

CASE STUDY continued

Evolution of a Menace

Antibiotic resistance evolves by natural selection. To see how, imagine a hospital patient with an infected wound. A doctor decides to treat the infection with an intravenous drip of penicillin. As the antibiotic courses through the patient's blood vessels, millions of bacteria die before they can reproduce. A few bacteria, however, carry a rare allele that codes for an enzyme that destroys penicillin. The bacteria carrying this rare allele are able to survive and reproduce, and their offspring inherit the penicillin-destroying allele. After a few generations, the frequency of the penicillin-destroying allele has soared to nearly 100%, and the frequency of the normal allele has declined to near zero. As a result of natural selection imposed by the antibiotic's killing power, the population of bacteria within the patient's body has evolved. The gene pool of the population has changed, and natural selection, in the form of bacterial destruction by penicillin, has caused the change.

TABLE 15-1 Causes of Evolution

Process	Consequence
Mutation	Creates new alleles; increases variability
Gene flow	Increases similarity of different populations
Genetic drift	Causes random change of allele frequencies; can eliminate alleles
Nonrandom mating	Changes genotype frequencies, but not allele frequencies
Natural and sexual selection	Increases frequency of favored alleles; produces adaptations

15.3 HOW DOES NATURAL SELECTION WORK?

Unlike the other causes of evolution that we have discussed, natural selection shapes the evolution of populations as they adapt to their changing environments. Studying the adaptive evolution that results from natural selection has been a major focus of evolutionary biology.

Natural Selection Stems from Unequal Reproduction

The British economist Herbert Spencer, writing in 1864, coined the phrase "survival of the fittest" to summarize the process that Darwin had named natural selection. But this formulation is not quite accurate: Natural selection favors traits that increase their possessors' survival only to the extent that improved survival leads to improved reproduction. A trait that improves survival may, for example, increase the likelihood that an individual survives long enough to reproduce, or might increase an organism's life span and, therefore, its number of opportunities to reproduce. But ultimately, it is reproductive success that determines the future of an individual's alleles, and the prevalence in the next generation of the traits associated with those alleles. Thus, the main driver of natural selection is differences in reproduction: Individuals bearing certain alleles leave more offspring (who inherit those alleles) than do other individuals with different alleles. In the terminology of evolutionary biology, individuals with greater lifetime reproductive success are said to have greater **fitness** than do individuals with lower reproductive success.

Natural Selection Acts on Phenotypes

Although we have defined evolution as changes in the genetic composition of a population, it is important to recognize that natural selection does not act directly on the genotypes of individual organisms. Rather, natural selection acts on phenotypes, the structures and behaviors displayed by the members of a population. This selection of phenotypes, however, inevitably affects the genotypes present in a population, because phenotypes and genotypes are closely tied. For example, we know that a pea plant's height is strongly influenced by the plant's alleles of certain genes. If a population of pea plants were to encounter environmental conditions that favored taller plants, then taller plants would leave more offspring. These offspring would carry the alleles that contributed to their parents' height. Thus, if natural selection favors a particular phenotype, it will necessarily also favor the underlying genotype.

Some Phenotypes Reproduce More Successfully Than Others

As we have seen, natural selection simply means that some phenotypes reproduce more successfully than others do. This simple process is such a powerful agent of change because only the fittest phenotypes pass traits to subsequent generations. But what makes a phenotype fit? Successful phenotypes are those that have the best **adaptations**—characteristics that help an individual survive and reproduce in a particular environment.

An Environment Has Nonliving and Living Components

Individual organisms must cope with an environment that includes not only nonliving physical factors but also the other living organisms with which the individual interacts. The nonliving component of the environment includes such factors as climate, availability of water, and availability of soil nutrients. These nonliving factors play a large role in determining the traits that help an organism to survive and reproduce. However, adaptations also arise because of interactions with the living component of the environment, namely, other organisms. A simple example illustrates this concept.

Consider a buffalo grass plant growing in a small patch of soil in the eastern Wyoming plains. The plant's roots must be able to take up enough water and minerals for growth and reproduction, and to that extent, it must be adapted to its nonliving environment. But even in the dry prairies of Wyoming, this requirement is relatively trivial, provided that the plant is alone and protected in its square yard of soil. In reality, however, many other plants—other buffalo grass plants as well as other grasses, sagebrush bushes, and annual wildflowers—also sprout in that same patch of soil. If our buffalo grass is to survive, it must compete with the other plants for resources. Its long, deep roots and efficient methods of mineral uptake have evolved not because the plains are dry but because the buffalo grass must share the dry prairies with other plants. Further, buffalo grass must also coexist with animals that wish to eat it, such as the cattle and other plant-eating animals that graze the prairie. So over time, tougher, harder-to-eat buffalo grass plants survived better and reproduced more on the prairie than did less-tough buffalo grass plants. As a result, buffalo grass leaves are quite tough; embedded silica compounds reinforce them.

Competition Acts As an Agent of Selection

As the example of buffalo grass shows, one of the major sources of natural selection is **competition** with other organisms for scarce resources. Competition for resources is most intense among members of the same species because, as Darwin wrote in *On the Origin of Species*, "they frequent the same districts, require the same food, and are exposed to the same dangers." In other words, no two competing organisms have such similar requirements for survival as do two members of the same species. Although different species may also compete for the same resources, they generally do so to a lesser extent than do individuals within a species.

Both Predator and Prey Act As Agents of Selection

When two species interact extensively, each exerts strong selection on the other. When one evolves a new feature or modifies an old one, the other typically evolves new adaptations in response. This process in which species mutually affect one another's evolution is called **coevolution.** Perhaps the most familiar form of coevolution is found in predator–prey relationships.

Predation describes any interaction in which one organism consumes another. In some instances, coevolution between predators (those that do the consuming) and prey (those that are consumed) is a sort of biological arms race, with each side evolving new adaptations in response to escalations by the other. Darwin used the example of wolves and deer: Wolf predation selects against slow or careless deer, thus leaving faster, more-alert deer to reproduce and pass on these traits. The resulting alert, swift deer select in turn against slow, clumsy wolves, because such predators cannot acquire enough food.

Antibiotic Resistance Illustrates Key Points About Natural Selection

The example of antibiotic resistance highlights some important features of natural selection.

Natural selection does not cause genetic changes in individuals. Alleles for antibiotic resistance arise spontaneously in some bacteria, long before the bacteria encounter an antibiotic. Antibiotics do not cause resistance to appear; their presence merely favors the survival of bacteria with antibiotic-destroying alleles over that of bacteria without such alleles.

Natural selection acts on individuals, but it is populations that are changed by evolution. The agent of natural selection—in this example, antibiotics—acts on individual bacteria. As a result, some individuals reproduce and some do not. However, it is the population as a whole that evolves as its allele frequencies change.

Evolution by natural selection is not progressive; it does not make organisms "better." The traits favored by natural selection change as the environment changes. Resistant bacteria are favored only when antibiotics are present. At a later time, when the environment no longer contains antibiotics, resistant bacteria may be at a disadvantage relative to other bacteria.

Sexual Selection Favors Traits That Help an Organism Mate

In many animal species, males have conspicuous features such as bright colors, long feathers or fins, or elaborate antlers. Males may also exhibit elaborate courtship behaviors or sing loud, complex songs. Although these extravagant features typically play a role in mating, they also seem to be at odds with efficient survival and reproduction. Exaggerated ornaments and displays may help males gain access to females, but they also make the males more conspicuous and thus vulnerable to predators. Darwin was intrigued by this apparent contradiction. He coined the term **sexual selection** to describe the special kind of selection that acts on traits that help an animal acquire a mate.

Darwin recognized that sexual selection could be driven either by sexual contests among males or by female preference for particular male phenotypes. Male–male competition for access to females can favor the evolution of features that provide an advantage in fights or ritual displays of aggression (**Fig. 15-10**). Female mate choice provides a second source of sexual selection. In animal species in which females actively choose their mates, females often seem to prefer males with the most elaborate ornaments or most extravagant displays (**Fig. 15-11**). Why?

One hypothesis is that male structures, colors, and displays that do not enhance survival might instead provide a female with an outward sign of a male's condition. Only a

▲ **FIGURE 15-10 Competition between males favors the evolution, through sexual selection, of structures for ritual combat** Two male bighorn sheep spar during the fall mating season. In many species, the losers of such contests are unlikely to mate, while winners enjoy tremendous reproductive success.

QUESTION If we studied a population of bighorn sheep and were able to identify the father and mother of each lamb born, would you predict that the difference in number of offspring between the most reproductively successful adult and the least successful adult would be greater for males or for females?

vigorous, energetic male can survive when burdened with conspicuous coloration or a large tail that might make him more vulnerable to predators. Conversely, males that are sick or under parasitic attack are dull and frumpy compared with healthy males. A female that chooses the brightest, most ornamented male is also choosing the healthiest, most vigorous male. By doing so, she gains fitness if, for example, the most vigorous male provides superior parental care to offspring or if he carries alleles for disease resistance that will be inherited by offspring and help ensure their survival. Females thus gain a reproductive advantage by choosing the most highly ornamented males, and the traits (including the exaggerated ornament) of these flashy males will be passed to subsequent generations.

▲ FIGURE 15-11 **The peacock's showy tail has evolved through sexual selection** The ancestors of today's peahens were apparently picky when deciding on a male with which to mate, favoring males with longer and more colorful tails.

Selection Can Influence Populations in Three Ways

Natural selection and sexual selection can lead to various patterns of evolutionary change. Evolutionary biologists group these patterns into three categories (**Fig. 15-12**):

- **Directional selection** favors individuals with an extreme value of a trait and selects against both average individuals

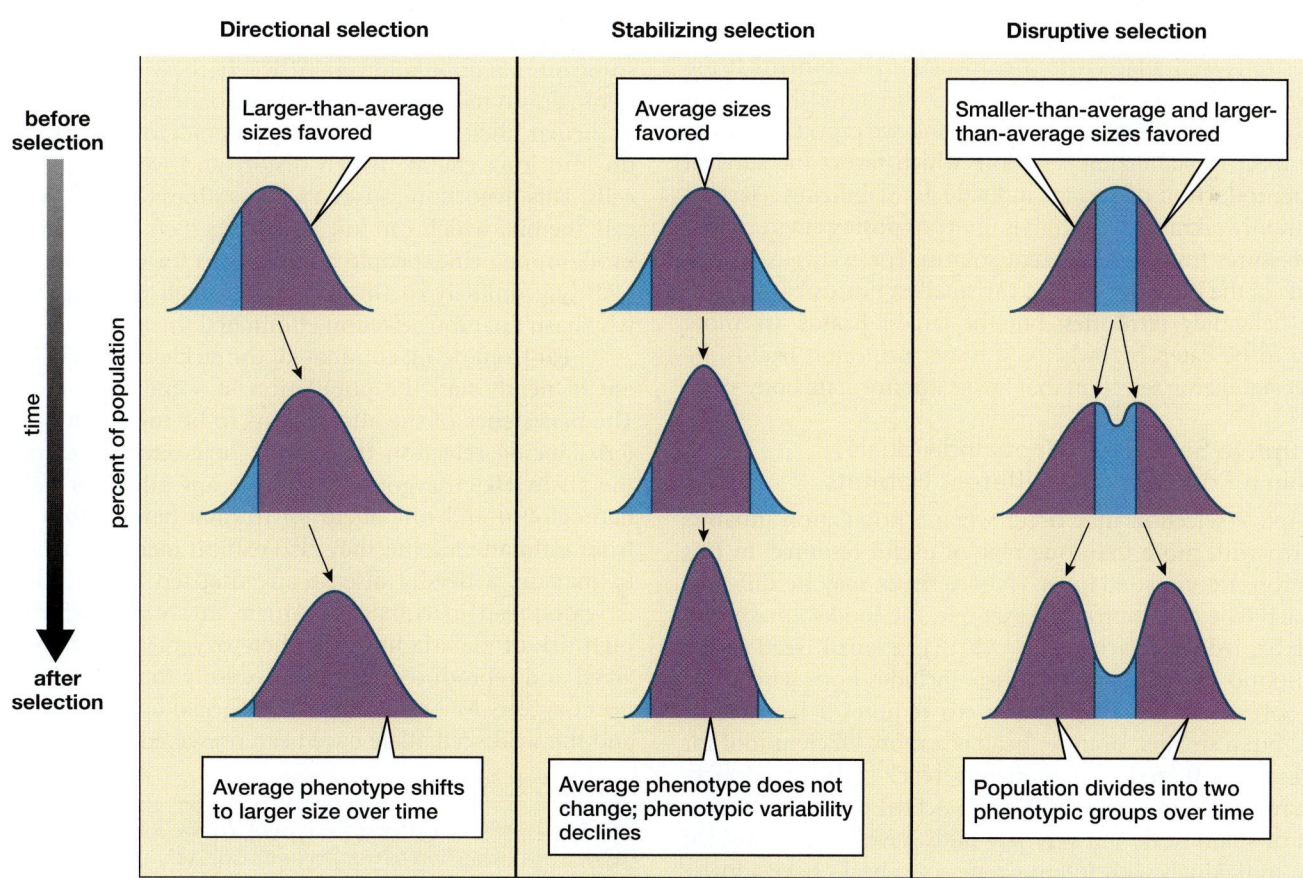

▲ FIGURE 15-12 **Three ways that selection affects a population over time** A graphical illustration of three ways natural and/or sexual selection, acting on a normal distribution of phenotypes, can affect a population over time. In all graphs, the blue areas represent individuals that are selected against—that is, the individuals that do not reproduce as successfully as do the individuals in the purple range.

QUESTION When selection is directional, is there any limit to how extreme the trait under selection will become? Why or why not?

and individuals at the opposite extreme. For example, directional selection might favor small size and select against both average and large individuals in a population.

- **Stabilizing selection** favors individuals with the average value of a trait (for example, intermediate body size) and selects against individuals with extreme values.
- **Disruptive selection** favors individuals at both extremes of a trait (for example, both large and small body sizes) and selects against individuals with intermediate values.

Directional Selection Shifts Character Traits in a Specific Direction

If environmental conditions change in a consistent way, a species may respond by evolving in a consistent direction. For example, during past "Ice Age" periods in which Earth's climate cooled considerably, many mammal species evolved thicker fur. The evolution of antibiotic resistance in bacteria is another example of directional selection: When antibiotics are present in a bacterial species' environment, individuals with greater resistance reproduce more prolifically than do individuals with less resistance.

Stabilizing Selection Acts Against Individuals Who Deviate Too Far from the Average

Directional selection can't go on forever. What happens once a species is well adapted to a particular environment? If the environment is unchanging, most new variations that appear will be harmful. Under these conditions, we expect species to be subject to stabilizing selection, which favors the survival and reproduction of average individuals. Stabilizing selection commonly occurs when a trait is under opposing environmental pressures from two different sources. For example, among lizards of the genus *Aristelliger*, the smallest lizards have a hard time defending territories, but the largest lizards are more likely to be eaten by owls. As a result, *Aristelliger* lizards are under stabilizing selection that favors intermediate body size.

Disruptive Selection Adapts Individuals Within a Population to Different Habitats

Disruptive selection may occur when a population inhabits an area with more than one type of useful resource. In this situation, the most adaptive characteristics may be different for each type of resource. For example, the food source of the black-bellied seedcracker (**Fig. 15-13**), a small, seed-eating bird found in the forests of Africa, includes both hard seeds and soft seeds. Cracking hard seeds requires a large, stout beak, but a smaller, pointier beak is a more efficient tool for processing soft seeds. Consequently, black-bellied seedcrackers have beaks in one of two sizes. A bird may have a large beak or small beak, but very few birds have a medium-sized beak; individuals with intermediate-sized beaks have a lower survival rate than individuals with either large or small beaks. Disruptive selection in black-bellied seedcrackers thus favors birds with large beaks and birds with small beaks, but not those with medium-sized beaks.

▲ **FIGURE 15-13 Black-bellied seedcrackers** As a result of disruptive selection, each black-bellied seedcracker has either a large (left) or small (right) beak.

Black-bellied seedcrackers represent an example of *balanced polymorphism*, in which two or more phenotypes are maintained in a population. In many cases of balanced polymorphism, multiple phenotypes persist because each is favored by a separate environmental factor. For example, consider two different forms of hemoglobin that are present in some human populations in Africa. In these populations, the hemoglobin molecules of people who are homozygous for a particular allele produce defective hemoglobin that clumps up into long chains, which distort and weaken red blood cells. This distortion causes a serious illness known as sickle-cell anemia, which can kill its victims. Before the advent of modern medicine, people homozygous for the sickle-cell allele were unlikely to survive long enough to reproduce. So why hasn't natural selection eliminated the allele?

Far from being eliminated, the sickle-cell allele is present in nearly half the population in some areas of Africa. The persistence of the allele seems to be the result of counterbalancing selection that favors heterozygous carriers of the allele. Heterozygotes, who have one allele for defective hemoglobin and one allele for normal hemoglobin, suffer from mild anemia but they also exhibit increased resistance to malaria, a deadly disease affecting red blood cells that is widespread in equatorial Africa. In areas of Africa with high risk of malaria infection, heterozygotes must have survived and reproduced more successfully than either type of homozygote. As a result, both the normal hemoglobin allele and the sickle-cell allele have been preserved.

CHECK YOUR LEARNING

Can you describe why selection of phenotypes can affect the evolution of genotypes? Can you explain how competition and predation influence evolution? Can you explain how sexual selection works and describe examples of its outcome? Can you compare and contrast directional selection, stabilizing selection, and disruptive selection?

CASE STUDY revisited

Evolution of a Menace

The evolution of antibiotic resistance in populations of bacteria—such as the MRSA bacteria that cause so many deadly infections—is a direct consequence of natural selection by antibiotics. When a population of disease-causing bacteria begins to grow in a human body, physicians try to halt population growth by introducing an antibiotic drug to the bacteria's environment. Although many bacteria are killed, some survive because they have genomes with a mutant allele that confers resistance. Bacteria carrying the resistance allele produce a disproportionately large share of offspring, which inherit the allele. Soon, resistant bacteria predominate in the population.

By introducing massive quantities of antibiotics into bacterial environments, humans have accelerated the pace of the evolution of antibiotic resistance. Each year, U.S. physicians write more than 100 million prescriptions for antibiotics; the Centers for Disease Control and Prevention estimates that about half of these prescriptions are unnecessary.

Although medical use and misuse of antibiotics is the most important source of natural selection for antibiotic resistance, antibiotics also pervade the environment outside our bodies. Our food supply, especially meat, contains a portion of the 29 million pounds of antibiotics that are fed to farm animals in the United States each year. In addition, Earth's soils and water are laced with antibiotics that enter the environment through human and livestock wastes, and from the antibacterial soaps and cleansers that are now routinely used in many households and workplaces. As a result of this massive alteration of the environment, resistant bacteria are now found not only in hospitals and the bodies of sick people but also in our food, water, and soil. Susceptible bacteria are under constant attack, and resistant strains have little competition. In our fight against disease, we have rashly overlooked some basic principles of evolutionary biology and are now paying a heavy price.

Consider This

Microbiologists have discovered that alleles associated with antibiotic resistance are present in bacteria that live in soil, even in environments that are comparatively free of antibiotic pollution from human activities. Why are such alleles present (albeit at low levels) in bacterial populations? Conversely, if resistance alleles are beneficial, why are they rare in natural populations of bacteria?

CHAPTER REVIEW

Summary of Key Concepts

15.1 How Are Populations, Genes, and Evolution Related?

Evolution is change in the frequencies of alleles in a population's gene pool. Allele frequencies in a population will remain constant over generations only if the following conditions are met: (1) There is no mutation, (2) there is no gene flow, (3) the population is very large, (4) all mating is random, and (5) all genotypes reproduce equally well (that is, there is no natural selection). Understanding what happens when these conditions are not met helps reveal the mechanisms of evolution.

15.2 What Causes Evolution?

Evolutionary change is caused by mutation, gene flow, small population size, nonrandom mating, and natural selection.

- Mutations are random, undirected changes in DNA composition. Although most mutations are neutral or harmful to the organism, some prove advantageous in certain environments. Mutations are rare and do not by themselves change allele frequencies very much, but they provide the raw material for evolution by other processes.
- Gene flow is the movement of alleles between different populations of a species. Gene flow tends to reduce differences in the genetic composition of different populations.
- If a population is small, chance events may reduce the survival and reproduction of a disproportionate number of individuals that bear a particular allele, thereby greatly changing the allele's frequency in the population; this is genetic drift.

- Nonrandom mating, such as assortative mating and inbreeding, can change the distribution of genotypes in a population, in particular by increasing the proportion of homozygotes.
- The survival and reproduction of organisms are influenced by their phenotypes. Because phenotype depends at least partly on genotype, natural selection tends to favor the persistence of certain alleles at the expense of others.

15.3 How Does Natural Selection Work?

Natural selection is driven by differences in reproductive success among different genotypes. Natural selection stems from the interactions of organisms with both the living and non-living parts of their environments. When two species interact intensively, both of them may evolve in response. Such coevolution can result from any type of interaction between organisms, including competition and predation. Phenotypes that help organisms mate can evolve by sexual selection.

Key Terms

adaptation 289	gene pool 279
allele frequency 279	genetic drift 282
coevolution 290	Hardy–Weinberg principle 279
competition 290	mutation 281
directional selection 291	natural selection 288
disruptive selection 292	population 278
equilibrium population 279	population bottleneck 285
fitness 289	predation 290
founder effect 285	sexual selection 290
gene flow 279	stabilizing selection 292

Learning Outcomes

In this chapter, you have learned to . . .

LO1 Define an *equilibrium population* and describe the conditions under which a population is expected to remain at evolutionary equilibrium.

LO2 Use the vocabulary of population genetics to describe the main causes of evolution and explain how each of these causes affects evolution.

LO3 Describe how variation in mating success among the individuals of a species can influence the species' evolution.

LO4 Compare and contrast directional selection, stabilizing selection, and disruptive selection.

Thinking Through the Concepts

Fill-in-the-Blank

1. The _____ provides a simple mathematical model for a non-evolving population, also called a(n) _____ population, in which _____ frequencies do not change over time. Are such populations likely to be found in nature? _____ **LO1**

2. Different versions of the same gene are called _____. These versions arise as a result of changes in the sequence of _____ that form the gene. These changes are caused by _____. An individual with two identical copies of a given gene is described as being _____ for that gene, while an individual with two different versions of that gene is described as _____. **LO2**

3. An organism's _____ refers to the specific alleles found within its chromosomes, while the traits that these alleles produce are called its _____. Which of these does natural selection act on? _____ **LO2**

4. A random form of evolution is called _____. This form of evolution will only occur in populations that are _____. Two important causes of this form of evolution are _____ and _____. Which of these would apply to a population started by a breeding pair that was stranded on an island? _____ **LO2**

5. Competition is most intense between members of a(n) _____. Predators and their prey act as agents of _____ on one another, resulting in a form of evolution called _____. This results in the evolution of characteristics called _____ that help both predators and their prey survive and reproduce. **LO2**

6. The evolutionary fitness of an organism is measured by its success at _____. The fitness of an organism can change if its _____ changes. **LO2**

Review Questions

1. What is a gene pool? How would you determine the allele frequencies in a gene pool? **LO2**

2. Define *equilibrium population*. Outline the conditions that must be met for a population to stay in genetic equilibrium. **LO1**

3. How does population size affect the likelihood of changes in allele frequencies by chance alone? Can significant changes in allele frequencies (that is, evolution) occur as a result of genetic drift? **LO2**

4. If you measured the allele frequencies of a gene and found large differences from those predicted by the Hardy–Weinberg principle, would that prove that natural selection is occurring in the population you are studying? Review the conditions that lead to an equilibrium population, and explain your answer. **LO2**

5. People like to say that "you can't prove a negative." Study the experiment in Figure 15-3 again, and comment on what it demonstrates. **LO2**

6. Describe the three ways in which natural selection can affect a population over time. Which way(s) is (are) most likely to occur in stable environments, and which way(s) might occur in rapidly changing environments? **LO4**

7. What is sexual selection? How is sexual selection similar to and different from other forms of natural selection? **LO3**

Applying the Concepts

1. In North America, the average height of adult humans has been increasing steadily for decades. Is directional selection occurring? What data would justify your answer?

2. Malaria is rare in North America. In populations of African Americans, what would you predict is happening to the frequency of the hemoglobin allele that leads to sickling in red blood cells? How would you go about determining whether your prediction is true?

3. By the 1940s, the whooping crane population had been reduced to fewer than 50 individuals. Thanks to conservation measures, its numbers are now increasing. What special evolutionary problems do whooping cranes face now that they have passed through a population bottleneck?

4. In many countries, conservationists are trying to design national park systems so that "islands" of natural area (the big parks) are connected by thin "corridors" of undisturbed habitat. The idea is that this arrangement will allow animals and plants to migrate between refuges. Why would such migration be important?

5. A preview question for Chapter 16: A *species* is all the populations of organisms that potentially interbreed with one another but that are reproductively isolated from (cannot interbreed with) other populations. Using the five conditions of the Hardy–Weinberg principle as a starting point, what factors do you think would be important in the splitting of a single ancestral species into two modern species?

Answers to Figure Caption questions and Fill-in-the-Blank questions can be found in the Answers section at the back of the book.

MB *Go to MasteringBiology for practice quizzes, activities, eText, videos, current events, and more.*

The saola, unknown to science until 1992, is one of a number of previously undiscovered species found in the mountains of Vietnam.

CASE STUDY
Lost World

THE STEEP, RAIN-DRENCHED SLOPES of Vietnam's Annamite Mountains are remote and forbidding, cloaked in tropical mists that lend an air of mystery and concealment to the forested peaks. As it turns out, this remote refuge conceals a most astonishing biological surprise: the saola, a hoofed, horned antelope that was unknown to science until the early 1990s. The discovery of a new species of large mammal at this late date was a complete shock. After centuries of human exploration and exploitation of every corner of the world's forests, deserts, and savannas, scientists were certain that no large mammal species could have escaped

detection. As long ago as 1812, French naturalist Georges Cuvier wrote that "there is little hope of discovering new species of large quadrupeds." And yet, the saola—3 feet high at the shoulder, weighing up to 200 pounds and sporting 20-inch black horns—remained hidden in Annamite Mountain forests, outside the realm of scientific knowledge.

It is surprising that the saola stayed hidden from scientists for so long because people tend to notice large animals. But as the discovery of the saola showed, even relatively conspicuous organisms may remain unknown if they live in a sufficiently remote area. The Annamite Mountains, isolated by inhospitable terrain and the wars fought in Vietnam during the last century, concealed the

saola and other undiscovered plant, mammal, and reptile species. Similarly, the remote Foja Mountains of New Guinea hid two previously unknown bird species, several new mammal species, and dozens of new species of butterflies, frogs, and flowering plants, which were not discovered until scientific expeditions to the region in 2006 and 2008. A recent survey of unexplored underwater habitats in Indonesia revealed more than 50 new species of fish, coral, and shrimp. A curious biologist might wonder why these distinctive species are concentrated in particular geographic areas. But before we can fully consider that question, we will need to explore the evolutionary process by which new species arise.

AT A GLANCE

LEARNING GOALS

LG1 Define *species* and explain why it is difficult to develop a universally applicable criterion for distinguishing species.

LG2 Explain the biological species concept and its limitations.

LG3 Describe the factors that can make it difficult to tell different species apart.

LG4 Describe the isolating mechanisms that restrict gene flow between different species.

LG5 Explain how new species arise.

LG6 Explain the process of adaptive radiation and describe examples.

LG7 Identify the major factors that cause extinction.

LG8 Interpret an evolutionary tree diagram.

16.1 WHAT IS A SPECIES?

Although Darwin brilliantly explained how evolution shapes complex organisms, his ideas did not fully explain life's diversity. In particular, the process of natural selection cannot by itself explain how living things came to be divided into groups, with each group distinctly different from all other groups. When we look at big cats, we don't see a continuous array of different tiger phenotypes that gradually grades into a lion phenotype. We see lions and tigers as separate, distinct types with no overlap. Each distinct type is known as a species.

In everyday life, most of us make unthinking use of an informal, nonscientific conception of species. We perceive sparrows as clearly different from eagles, which are obviously different from ducks. But we sometimes run into trouble when we try to make finer distinctions; it is not easy to distinguish among different species of sparrows, especially if we don't have a precise idea of what constitutes a species. How, then, do scientists make these finer distinctions?

Each Species Evolves Independently

Today, biologists define a **species** as a group of populations that evolves independently. Each species follows a separate evolutionary path because alleles rarely move between the gene pools of different species. This definition, however, does not clearly state the standard by which such evolutionary independence is judged. The most widely used standard defines species as "groups of actually or potentially interbreeding natural populations, which are reproductively isolated from other such groups." This definition, known as the *biological species concept*, is based on the observation that **reproductive isolation** (inability to successfully breed outside the group) ensures evolutionary independence.

The biological species concept has two major limitations. First, because the definition is based on patterns of sexual reproduction, it does not help us determine species boundaries among asexually reproducing organisms. Second, it is not always practical or even possible to directly observe whether members of two different groups interbreed. Thus, a biologist who wishes to determine if a group

of organisms is a separate species must often make the determination without knowing for sure if group members breed with organisms outside the group.

Despite the limitations of the biological species concept, most biologists accept it for identifying species of sexually reproducing organisms. However, alternative definitions are required by scientists who study bacteria and other organisms that mainly reproduce asexually. Even some biologists who study sexually reproducing organisms prefer species definitions that do not depend on a property (reproductive isolation) that can be difficult to measure. Several such alternatives to the biological species concept have been proposed. (One that has gained many adherents is described in Chapter 18.)

Appearance Can Be Misleading

Biologists have found that some organisms with very similar appearances belong to different species. For example, the cordilleran flycatcher and the Pacific-slope flycatcher are so similar that even experienced birdwatchers cannot tell them apart (**Fig. 16-1**). Likewise, there are virtually no visible differences between the Asian mosquito species *Anopheles dirus* and *Anopheles harrisoni*. This similarity in appearance, however, disguises a crucial difference that is important to people. *A. dirus* spreads the often deadly disease malaria from person to person, but *A. harrisoni* does not generally spread malaria.

Superficial similarity can sometimes hide multiple species. Researchers recently discovered that the species previously known as the two-barred flasher butterfly is actually a group of at least 10 different species. The caterpillars of the different species do differ in appearance, but the adult butterflies are all so similar that their species identities went undetected for more than two centuries after the butterfly was first described and named.

Conversely, differences in appearance do not always mean that two populations belong to different species (**Fig. 16-2**). For example, if you encounter a Northwestern garter snake, it may be brown, black, gray, green, or some

(a) Cordilleran flycatcher

(b) Pacific-slope flycatcher

▲ FIGURE 16-1 **Members of different species may be similar in appearance**
The **(a)** cordilleran flycatcher and **(b)** Pacific-slope flycatcher are different species.

shade in between, and it may be striped or unstriped. If it is striped, the stripes may be broad or narrow, and could be any of a variety of colors. Despite their diversity of appearance, though, all Northwestern garter snakes are members of the same species.

CHECK YOUR LEARNING

Can you describe how biologists define *species* and explain why it is difficult to develop a criterion for distinguishing species? Are you able to describe the biological species concept and discuss its limitations? Can you list some of the reasons why it is hard to tell different species apart?

16.2 HOW IS REPRODUCTIVE ISOLATION BETWEEN SPECIES MAINTAINED?

What prevents different sexually reproducing species from interbreeding? The traits that prevent interbreeding and maintain reproductive isolation are called **isolating mechanisms** (Table 16-1). Isolating mechanisms provide a clear benefit to individuals. An individual that breeds with a member of another species will probably produce no offspring (or offspring that are unfit or sterile), thereby wasting its reproductive effort and failing to contribute to future

(a) A green-striped Northwestern garter snake

(b) A red-striped Northwestern garter snake

▲ FIGURE 16-2 **Members of a species may differ in appearance (a)** This green-striped Northwestern garter snake and **(b)** this red-striped Northwestern garter snake are members of the same species.

TABLE 16-1 Mechanisms of Reproductive Isolation

Premating Isolating Mechanisms: factors that prevent organisms of two species from mating
- **Geographic isolation:** The species do not interbreed because a physical barrier separates them.
- **Ecological isolation:** The species do not interbreed even if they are within the same area because they occupy different habitats.
- **Temporal isolation:** The species do not interbreed because they breed at different times.
- **Behavioral isolation:** The species do not interbreed because they have different courtship and mating rituals.
- **Mechanical incompatibility:** The species do not interbreed because their reproductive structures are incompatible.

Postmating Isolating Mechanisms: factors that prevent organisms of two species from producing vigorous, fertile offspring after mating
- **Gametic incompatibility:** Sperm from one species cannot fertilize eggs of another species.
- **Hybrid inviability:** Hybrid offspring fail to survive.
- **Hybrid infertility:** Hybrid offspring are sterile or have low fertility.

generations. Thus, natural selection favors traits that prevent reproduction across species boundaries.

Premating Isolating Mechanisms Prevent Mating Between Species

Reproductive isolation can be maintained by a variety of mechanisms, but those that prevent mating attempts are especially effective. The mechanisms that prevent mating between species are collectively called **premating isolating mechanisms.** We next describe the most important types of such premating isolating mechanisms.

Members of Different Species May Be Prevented from Meeting

Members of different species cannot mate if they never get near one another. **Geographic isolation** prevents interbreeding between populations that do not come into contact because they live in different, physically separated places (**Fig. 16-3**). However, we cannot determine if geographically separated populations are actually distinct species. Should the physical barrier separating the two populations disappear (a new channel might connect two previously isolated

lakes, for example), the reunited populations might interbreed freely and not be separate species after all. Geographic isolation, therefore, is usually not considered to be a mechanism that maintains reproductive isolation between species. Instead, it is a mechanism that allows new species to form. If populations cannot interbreed after geographic barriers have been eliminated, then other premating isolating mechanisms must have developed.

Different Species May Occupy Different Habitats

Two populations that use different resources may spend time in different habitats within the same general area and thus exhibit **ecological isolation.** White-crowned sparrows and white-throated sparrows, for example, have extensively overlapping geographic ranges. The white-throated sparrow, however, frequents dense thickets, whereas the white-crowned sparrow inhabits fields and meadows, seldom penetrating far into dense growth. The two species may coexist within a few hundred yards of one another and yet seldom meet during the breeding season. A more dramatic example is provided by the more than 300 species of fig wasp (**Fig. 16-4**). In most cases, fig wasps of a given species breed in (and pollinate) the fruits of one particular species of fig, and each fig

(a) Kaibab squirrel

(b) Abert squirrel

◀ **FIGURE 16-3 Geographic isolation** To determine if these two squirrels are members of different species, we must know if they are "actually or potentially interbreeding." Unfortunately, it is hard to tell, because **(a)** the Kaibab squirrel lives only on the north rim of the Grand Canyon and **(b)** the Abert squirrel lives on the south rim. The two populations are geographically isolated but still quite similar. Have they diverged enough since their separation to become reproductively isolated? Because they remain geographically isolated, we cannot say for sure.

▲ FIGURE 16-4 Ecological isolation This female fig wasp is carrying fertilized eggs from a mating that took place within a fig. She will find another fig of the same species, enter it through a pore, lay eggs, and die. Her offspring will hatch, develop, and mate within the fig. Because each species of fig wasp reproduces only in its own particular fig species, each wasp species is reproductively isolated.

species hosts only one or two species of pollinating wasp. Thus, fig wasps of different species only rarely encounter one another during breeding, and pollen from one fig species is not ordinarily carried to flowers of a different species.

Different Species May Breed at Different Times

Even if two species occupy similar habitats, they cannot mate if they have different breeding seasons, a phenomenon called **temporal isolation** (time-based isolation). For example, the spring field cricket and the fall field cricket both occur in many areas of North America but, as their names suggest, the former species breeds in spring and the latter in autumn. As a result, the two species do not interbreed.

In plants, the reproductive structures of different species may mature at different times. For example, Bishop pines and Monterey pines grow together near Monterey on the California coast (Fig. 16-5), but the two species release their sperm-containing pollen (and have eggs ready to receive the

pollen) at different times: The Monterey pine releases pollen in early spring, the Bishop pine in summer. For this reason, the two species do not interbreed under natural conditions.

Different Species May Have Different Courtship Signals

Among animals, elaborate courtship colors and behaviors can prevent mating with members of other species. Signals and behaviors that differ from species to species create **behavioral isolation.** For example, the extravagant plumes and arresting pose of a courting male Raggiana bird of paradise are conspicuous indicators of his species, and there is little chance that females of another species will mate with him by mistake (Fig. 16-6). Among frogs, males are often impressively indiscriminate, jumping on every female in sight, regardless of the species, when the spirit moves them. Females, however, approach only male frogs that utter the call appropriate to their species. If they do find themselves in an unwanted embrace, they make the "release call," which causes the male to let go. As a result, few **hybrids**—offspring of parents of different species—are produced.

Differing Sexual Organs May Foil Mating Attempts

In rare instances, a male and a female of different species attempt to mate. Their attempt is likely to fail. Among animal species with internal fertilization (in which the sperm is deposited inside the female's reproductive tract), the male's and female's sexual organs simply may not fit together. Incompatible body shapes may also make copulation between species impossible. For example, snails of species whose shells have left-handed spirals may be unable to successfully copulate with closely related snails whose shells have right-handed spirals, because the shell mismatch is accompanied by a body orientation mismatch that prevents the genitals of the two species from lining up properly during attempted copulations. Among plants, differences in flower size or structure may prevent pollen transfer between species because the differing flowers may attract different pollinators. Isolating mechanisms of this type are called **mechanical incompatibilities.**

(a) Bishop pine

(b) Monterey pine

◀ FIGURE 16-5 Temporal isolation (a) Bishop pines and (b) Monterey pines coexist in nature. In the laboratory they produce fertile hybrids. In the wild, however, they do not interbreed, because they release pollen at different times of the year.

▲ FIGURE 16-6 **Behavioral isolation** The mate-attraction display of a male Raggiana bird of paradise includes distinctive posture, movements, plumage, and vocalizations that do not resemble those of other bird-of-paradise species.

Postmating Isolating Mechanisms Limit Hybrid Offspring

When premating isolating mechanisms fail or have not yet evolved, members of different species may mate. If, however, all resulting hybrid offspring die during development, then the two species are still reproductively isolated from one another. Even if hybrid offspring are able to survive, if these hybrids are less fit than their parents or are themselves infertile, the two species may still remain separate, with little or no gene flow between them. Mechanisms that prevent the formation of vigorous, fertile hybrids between species are called **postmating isolating mechanisms.**

One Species' Sperm May Fail to Fertilize Another Species' Eggs

Even if a male inseminates a female of a different species, his sperm may not be able to fertilize her eggs, an isolating mechanism called **gametic incompatibility.** For example, in animals with internal fertilization, fluids in the female reproductive tract may weaken or kill sperm of other species.

Gametic incompatibility may be an especially important isolating mechanism in species, such as marine invertebrate animals and wind-pollinated plants, that reproduce by scattering gametes in the water or in the air. For example, sea urchin sperm cells contain a protein that allows them to bind to eggs. The structure of the protein differs among species so that sperm of one sea urchin species cannot bind to the eggs of another species. In abalones (a type of mollusk), eggs are surrounded by a membrane that can be penetrated only by sperm containing a particular enzyme. Each abalone species has a distinctive version of the enzyme, so hybrids are rare, even though several species of abalones coexist in the same waters and spawn during the same period. Among plants, similar chemical incompatibility may prevent the germination of pollen from one species that lands on the stigma (pollen-catching structure) of the flower of another species.

Hybrid Offspring May Fail to Survive or Reproduce

If cross-species fertilization does occur, the resulting hybrid may be unable to survive, a situation called **hybrid inviability.** The genetic instructions directing development of the two species may be so different that hybrids abort early in development. For example, captive leopard frogs can be induced to mate with wood frogs, and the matings generally yield fertilized eggs. The resulting embryos, however, inevitably fail to survive more than a few days.

In other animal species, a hybrid might survive to adulthood but fail to reproduce because it exhibits ineffective breeding behavior. Hybrids between certain species of lovebirds, for example, have great difficulty building nests. Members of each parental species inherit a particular innate behavior for carrying nest material; one species tucks the material under its rump feathers and the other carries it in its beak. Hybrids, however, use a confused mixture of the two behaviors. They repeatedly attempt to tuck nest material under their feathers, but are unable to do so because they don't release the material from their beaks. Hybrids with such ineffective nest-building behavior probably could not reproduce in the wild.

Hybrid Offspring May Be Infertile

Most animal hybrids, such as the mule (a cross between a horse and a donkey) and the liger (a zoo-based cross between a lion and a tiger), are sterile (**Fig. 16-7**). This **hybrid infertility** prevents hybrids from passing on their genetic material to offspring, thus blocking gene flow between the two parent populations. A common reason for hybrid infertility is the failure of chromosomes to pair properly during meiosis, so that eggs and sperm fail to develop.

CHECK YOUR LEARNING

Can you describe the main types of premating and postmating reproductive isolating mechanisms? Can you provide an example or two of each type of mechanism?

▲ FIGURE 16-7 **Hybrid infertility** This liger, the hybrid offspring of a lion and a tiger, is sterile. The gene pools of its parent species remain separate.

16.3 HOW DO NEW SPECIES FORM?

Despite his exhaustive exploration of the process of natural selection, Charles Darwin did not propose a complete mechanism of **speciation,** the process by which new species form. Today, however, biologists recognize that speciation depends on two processes: isolation and genetic divergence.

- **Isolation of populations:** If individuals move freely between two populations, interbreeding and the resulting gene flow will cause changes in one population to soon become widespread in the other as well. Thus, two populations cannot grow increasingly different unless something happens to block interbreeding between them. Speciation depends on isolation.
- **Genetic divergence of populations:** It is not sufficient for two populations simply to be isolated. They will become separate species only if, during the period of isolation, they evolve sufficiently large differences. The differences must be large enough that, if the isolated populations are reunited, they can no longer interbreed and produce vigorous, fertile offspring. That is, speciation is complete only if divergence results in evolution of an isolating mechanism. Such differences can arise by chance (genetic drift), especially if at least one of the isolated populations is small (see Chapter 15). Large genetic differences can also arise through natural selection, if the isolated populations experience different environmental conditions.

Speciation always requires isolation followed by divergence, but these steps can take place in several different ways. Evolutionary biologists group the different pathways to speciation into two broad categories: **allopatric speciation,** in which two populations are geographically separated from one another, and **sympatric speciation,** in which two populations share the same geographic area. (To learn more about how scientists study the outcome of speciation, see "Science in Action: Seeking the Secrets of the Sea.")

Geographic Separation of a Population Can Lead to Allopatric Speciation

New species can arise by allopatric speciation when an impassible barrier physically separates different parts of a population.

Organisms May Colonize Isolated Habitats

A small population can become isolated if it moves to a new location (**Fig. 16-8**). For example, some members of a population of land-dwelling organisms might colonize an oceanic island. The colonists might be birds, flying insects, fungal spores, or wind-borne seeds blown by a storm. More earthbound organisms might reach the island on a drifting raft of vegetation torn from the mainland coast. Whatever the means, we know that such colonization occurs

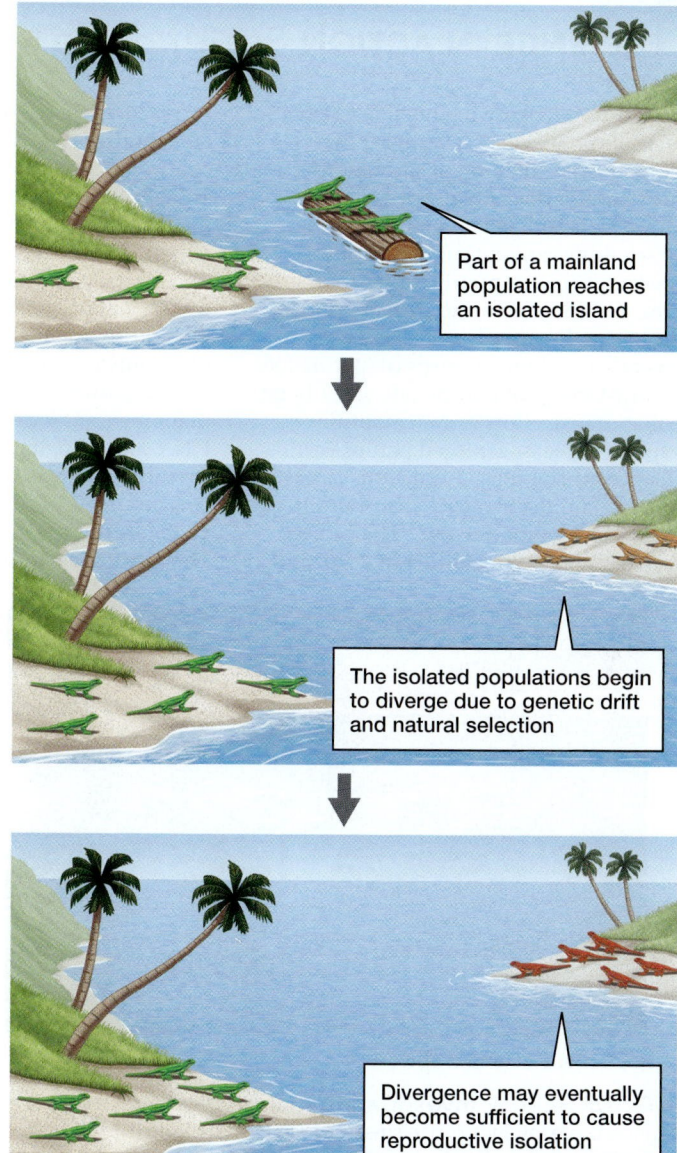

Part of a mainland population reaches an isolated island

The isolated populations begin to diverge due to genetic drift and natural selection

Divergence may eventually become sufficient to cause reproductive isolation

▲ **FIGURE 16-8 Allopatric isolation and divergence**
In allopatric speciation, some event causes a population to be divided by an impassable geographic barrier. One way the division can occur is by colonization of an isolated island. The two now-separated populations may diverge genetically. If the genetic differences between the two populations become large enough to prevent interbreeding, then the two populations constitute separate species.

QUESTION Make a list of events or processes that could cause geographic subdivision of a population. Are the items on your list sufficient to account for formation of the millions of species that have inhabited Earth?

regularly, given the presence of living things on even the remotest islands.

Isolation by colonization need not be limited to islands. For example, different coral reefs may be separated by miles of open ocean, so any reef-dwelling sponges, fishes, or algae that were carried by ocean currents to a distant reef

SCIENCE IN ACTION Seeking the Secrets of the Sea

In addition to studying how new species arise, biologists also investigate the outcome of millennia of speciation: life's current diversity of species. However, even after several centuries of scientific exploration, much of this diversity remains poorly understood. One ambitious effort to increase our understanding of life's diversity is the recently concluded Census of Marine Life.

The Census of Marine Life was a huge collaborative effort to systematically explore the least explored part of Earth, the ocean. The census aimed to "assess and explain the diversity, distribution, and abundance of marine life." The project lasted 10 years, concluding in 2010, and

involved 2,700 scientists from more than 80 countries (**Fig. E16-1a**). The scientists conducted 540 expeditions at a cost of $650 million. The expeditions spanned the globe from the tropics to polar regions, explored coastal waters and the open ocean, and examined life from the water's surface down to its deepest depths. The census scientists studied a huge range of organisms—from microbes to whales—counting them, tracking their movements, mapping their locations, analyzing their DNA, and bringing many back to the lab for further study.

The efforts of the census scientists yielded a massive amount of information. The researchers found living organisms

in every habitat that they explored, including ocean depths that lack oxygen. They documented previously unknown animal migration routes, and compiled millions of records of organisms' locations and abundance that can serve as a baseline for tracking the effects of human activities on marine organisms. The census also identified "hot spots" in the ocean where life is especially abundant, and discovered more than 6,000 new species (**Fig E16-1b**). Overall, the Census of Marine Life dramatically increased our knowledge of life in the oceans, and demonstrated the value of large-scale, coordinated scientific investigation of biodiversity.

(a) Census scientist at work

(b) Yeti crab

▲ **FIGURE E16-1 Investigating marine biodiversity (a)** A researcher with the Census of Marine Life examines organisms in a lighted aquarium. **(b)** The yeti crab, which lives on the ocean floor in 7,500-foot-deep Pacific waters, is among the species discovered by the Census of Marine Life.

would be effectively isolated from their original populations. Any bounded habitat, such as a lake or a mountaintop, can isolate arriving colonists.

Geological and Climate Changes May Divide Populations

Isolation can also result from landscape changes that divide a population. For example, rising sea levels might transform a coastal hilltop into an island, isolating the residents. New rock from a volcanic eruption can divide a previously continuous sea or lake, splitting populations. A river that changes course can also divide populations, as can a newly formed mountain range. Climate shifts, such as those that happened in past ice ages, can change the distribution of vegetation and strand portions of populations in isolated patches of suitable habitat. You can probably imagine many

other scenarios that could lead to the geographic subdivision of a population.

Over the history of Earth, many populations have been divided by continental drift. Earth's continents float on molten rock and slowly move about the surface of the planet. On a number of occasions during Earth's long history, continental landmasses have broken into pieces that subsequently moved apart (see Fig. 17-11). Each of these breakups must have split a multitude of populations.

Natural Selection and Genetic Drift May Cause Isolated Populations to Diverge

If two populations become geographically isolated for any reason, there will be no gene flow between them. If the environments of the locations differ, then natural selection may favor different traits in the different locations, and the populations may accumulate genetic differences. Alternatively, genetic differences may arise if one or more of the separated populations is small enough that substantial genetic drift occurs, which may be especially likely in the aftermath of a founder event in which a few individuals become isolated from the main body of the species. In either case, genetic differences between the separated populations may eventually become large enough to make interbreeding impossible. At that point, the two populations will have become separate species. Most evolutionary biologists believe that geographic isolation followed by allopatric speciation has been the most common source of new species, especially among animals.

Genetic Isolation Without Geographic Separation Can Lead to Sympatric Speciation

Genetic isolation—limited gene flow—is required for speciation, but populations can become genetically isolated without geographic separation. Thus, new species can arise by sympatric speciation.

Ecological Isolation Can Reduce Gene Flow

If a geographic area contains two distinct types of habitats (each with distinct food sources, places to raise young, and so on), different members of a single species may begin to specialize in one habitat or the other. If conditions are right, natural selection in the two different habitats may lead to the evolution of different traits in the two groups. Eventually, these differences may become large enough to prevent members of the two groups from interbreeding, and the formerly single species will have split into two species. Such a split seems to be taking place right before biologists' eyes, so to speak, in the case of the fruit fly *Rhagoletis pomonella*.

Rhagoletis is a parasite of the American hawthorn tree. This fly lays its eggs in the hawthorn's fruit; when the maggots hatch, they eat the fruit. About 150 years ago, scientists noticed that *Rhagoletis* had begun to infest apple trees, which were introduced into North America from Europe. Today, it appears that *Rhagoletis* is splitting into two species—one that breeds on apples, and one that breeds on hawthorns (**Fig. 16-9**). The two groups have evolved substantial genetic differences, some of which—such as those that affect the time of year at which adult flies emerge and begin to mate—are important for survival on a particular host plant.

The two kinds of flies will become two species only if they maintain reproductive separation. Apple trees and hawthorns typically grow in the same areas, and flies, after all, can fly. So why don't apple flies and hawthorn flies interbreed and cancel out any genetic differences between them? First, female flies usually lay their eggs in the same type of fruit in which they developed. Males also tend to prefer the same type of fruit in which they developed. Therefore, apple-liking males will encounter and mate with apple-liking females. Second, apples mature 2 or 3 weeks later than do hawthorn fruits, and the two types of flies emerge with

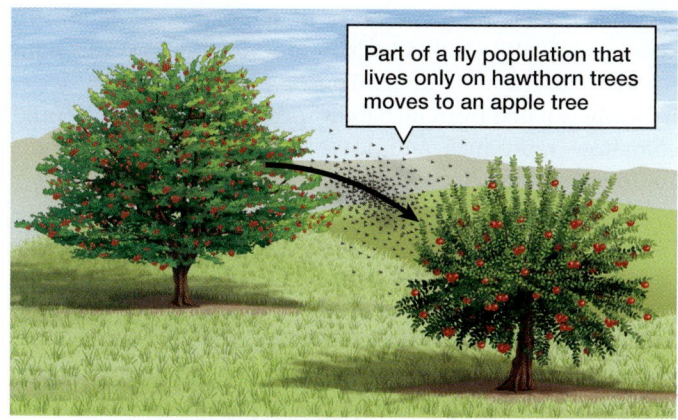

Part of a fly population that lives only on hawthorn trees moves to an apple tree

The flies living on the apple tree do not encounter the flies living on the hawthorn tree, so the populations diverge

▲ **FIGURE 16-9 Sympatric isolation and divergence** In sympatric speciation, some event blocks gene flow between two parts of a population that remains in a single geographic area. One way in which this genetic isolation can occur is if a portion of a population begins to use a previously unexploited resource, such as when some members of an insect population shift to a new host plant species (as has occurred in the fruit fly species *Rhagoletis pomonella*). The two now-isolated populations may diverge genetically. If the genetic differences between the two populations become large enough to prevent interbreeding, then the two populations constitute separate species.

QUESTION How might future scientists test whether the current *R. pomonella* has become two species?

timing appropriate for their chosen host fruit. Thus, the two varieties of flies have very little chance of meeting. Although some interbreeding between the two types of flies occurs, it seems they are well on their way to speciation. Will they make it? Entomologist Guy Bush suggests, "Check back with me in a few thousand years."

Mutations Can Lead to Genetic Isolation

In some instances, new species can arise nearly instantaneously as a result of mutations that change the number of chromosomes in an organism's cells. The acquisition of multiple copies of each chromosome is known as **polyploidy** and has been a frequent cause of sympatric speciation. In general, polyploid individuals cannot mate successfully with normal diploid individuals. Thus, a polyploid mutant is genetically isolated from its parent species. If, however, it could somehow reproduce and leave offspring, its descendants could form a new, reproductively isolated species.

Polyploid plants are more likely than polyploid animals to be able to reproduce, so speciation by polyploidy is more common in plants than in animals. Many plants can either self-fertilize or reproduce asexually, or both. Most animals, however, cannot self-fertilize or reproduce asexually. Therefore, a polyploid plant is much more likely than a polyploid animal to become the founding member of a new, polyploid species.

Under Some Conditions, Many New Species May Arise

In the same way that the history of your family can be represented by a family tree, the history of life can be represented by an *evolutionary tree*. The base of the evolutionary tree of life represents Earth's earliest organisms, and each of the endmost branches represents one of today's living species. Each fork in the branches represents a speciation event, when one species split into two. Hypotheses and discoveries about the evolutionary relationships among species are often communicated by depictions of a portion of life's evolutionary tree (**Fig. 16-10a**).

In some cases, many new species have arisen in a relatively short time (**Fig. 16-10b**). This process, called **adaptive radiation,** can occur when populations of one species invade a variety of new habitats and evolve in response to the differing environmental pressures in those habitats.

Adaptive radiation has occurred many times and in many groups of organisms, typically when species encounter a wide variety of unoccupied habitats. For example, episodes of adaptive radiation took place when some wayward finches colonized the Galápagos Islands, when a

CASE STUDY c o n t i n u e d

Lost World

It is not surprising that the forests of New Guinea's Foja Mountains would be home to a variety of distinctive species. New Guinea is, after all, an island. It is likely that, in the past, populations colonized the island and became genetically isolated from mainland populations, thereby initiating the process of speciation. Similarly, the distinctive species recently discovered on Indonesian coral reefs arose because reefs form in shallow seas, and they are typically surrounded by deep, open-ocean waters that many reef-dwelling organisms cannot easily cross. But what about the saola and the other unique species of the Ammanite Mountains? How might populations inhabiting mainland forests in Vietnam have become isolated from other populations? Try to think of a possible answer to this question. We will return to this topic in the "Case Study Revisited" section at the end of this chapter.

population of cichlid fish reached isolated Lake Malawi in Africa, and when an ancestral silversword plant species arrived at the Hawaiian Islands (**Fig. 16-11**). These events gave rise to adaptive radiations of 13 species of Darwin's finches in the Galápagos, more than 300 species of cichlids in Lake Malawi, and 30 species of silversword plants in Hawaii. In these examples, the invading species faced no competitors except other members of their own species, and all the available habitats were rapidly exploited by new species that evolved from the original invaders.

CHECK YOUR LEARNING

Can you describe the two general steps that are required for a new species to arise? Do you understand the difference between allopatric and sympatric speciation, and are you able to describe each process? Can you explain adaptive radiation and describe the process by which it might arise? Are you able to interpret an evolutionary tree diagram?

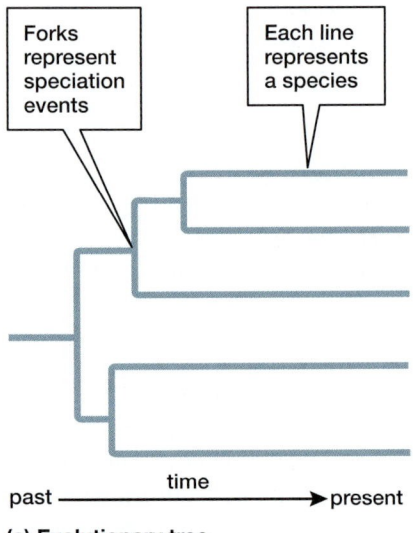

Forks represent speciation events

Each line represents a species

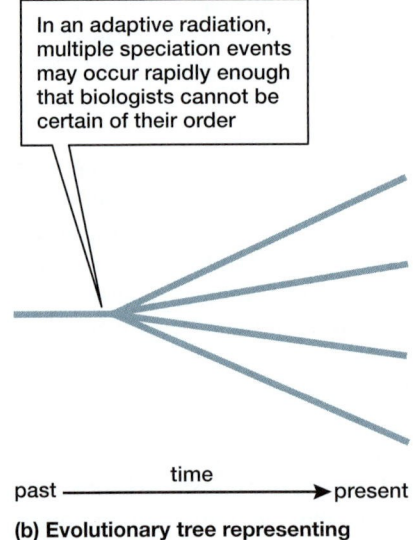

In an adaptive radiation, multiple speciation events may occur rapidly enough that biologists cannot be certain of their order

past ——— time ——→ present

(a) Evolutionary tree

past ——— time ——→ present

(b) Evolutionary tree representing adaptive radiation

◄ **FIGURE 16-10 Interpreting evolutionary trees** Evolutionary history is often represented by **(a)** an evolutionary tree, a graph in which the horizontal axis plots time. In **(b)**, an evolutionary tree representing an adaptive radiation, many lines may branch from a single point. This pattern reflects biologists' uncertainty about the order in which the multiple speciation events of the radiation took place. With more research, it may be possible to replace the "starburst" pattern with a more informative tree.

(a) Ahinahina

(b) Waialeale dubautia

(c) Kupaoa

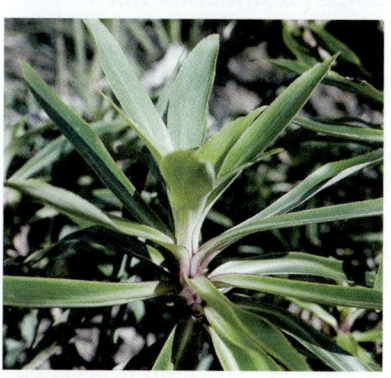

(d) Na'ena'e 'ula

▲ **FIGURE 16-11 Adaptive radiation** About 30 species of silversword plants inhabit the Hawaiian Islands. These species are found nowhere else, and all of them descended from a single ancestral population within a few million years. This adaptive radiation has led to a collection of closely related species of diverse form and appearance, with an array of adaptations for exploiting the many different habitats in Hawaii, from warm, moist rain forests to cool, barren volcanic mountaintops.

? HAVE **YOU** EVER WONDERED …

How Many Species Inhabit the Planet?

One way to determine the number of species on Earth might be to simply count them. You could comb the scientific literature to find all the species that scientists have discovered and named, and then tally up the total number. One attempt to do just that, the Catalogue of Life project, has compiled an online searchable database that listed 1,368,009 species as of 2011. But even the Catalogue of Life can't tell you how many species are on Earth.

Why doesn't counting work? Because most of the planet's species remain undiscovered. Relatively few scientists are engaged in the search for new species, and many undiscovered species are small and inconspicuous, or live in poorly explored habitats such as the floor of the ocean or the topmost branches of tropical rain forests. So, no one knows the actual number of species on Earth. But biologists agree that the number must be much higher than the number of named species. Estimates range from 2 million to 100 million or more. One recent study used a new method to estimate that approximately 8.7 million species inhabit Earth.

16.4 WHAT CAUSES EXTINCTION?

Every living organism must eventually die, and the same is true of species. Just like individuals, species are "born" (through the process of speciation), persist for some period of time, and then perish. The ultimate fate of any species is **extinction,** the death of the last of its members. In fact, at least 99.9% of all the species that have ever existed are now extinct. The natural course of evolution, as revealed by fossils, is continual turnover of species as new ones arise and old ones become extinct.

The immediate cause of extinction is probably always environmental change, in either the nonliving or the living parts of the environment. Environmental changes that can lead to extinction include habitat destruction and increased competition among species. In the face of such changes, species with small geographic ranges or highly specialized adaptations are especially susceptible to extinction.

Localized Distribution Makes Species Vulnerable

Species vary widely in their range of distribution and, hence, in their vulnerability to extinction. Some species, such as herring gulls, white-tailed deer, and humans, inhabit entire

▲ FIGURE 16-12 Very localized distribution can endanger a species The Devil's Hole pupfish is found in only one spring-fed water hole in the Nevada desert. This and other isolated small populations are at high risk of extinction.

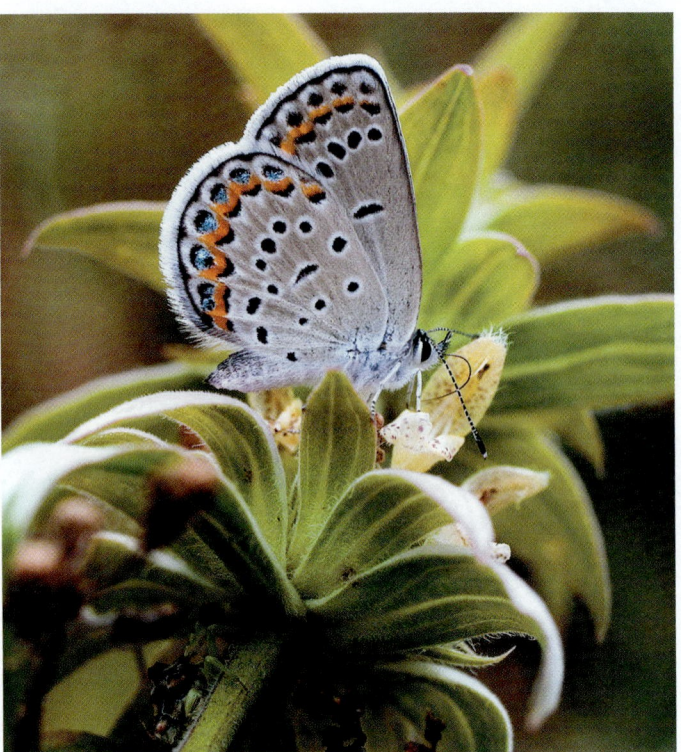

▲ FIGURE 16-13 Extreme specialization places species at risk The Karner blue butterfly feeds exclusively on the blue lupine, found in dry forests and clearings in the northeastern United States. Such behavioral specialization renders the butterfly extremely vulnerable to any environmental change that may exterminate its single host plant species.

QUESTION If specialization puts a species at risk for extinction, how could this hazardous trait have evolved?

continents or even the whole Earth; others, such as the Devil's Hole pupfish (**Fig. 16-12**), have extremely limited ranges. Obviously, if a species inhabits only a very small area, any disturbance of that area could easily result in extinction. If Devil's Hole dries up due to a drought or well drilling nearby, its pupfish will immediately vanish. Conversely, wide-ranging species will not succumb to local environmental catastrophes.

Overspecialization Increases the Risk of Extinction

Another factor that may make a species vulnerable to extinction is overspecialization. Each species evolves adaptations that help it survive and reproduce in its environment. In some cases, these adaptations include specializations that favor survival in a particular and limited set of environmental conditions. The Karner blue butterfly, for example, feeds only on the blue lupine plant (**Fig. 16-13**). The butterfly is therefore found only where the plant thrives. But the blue lupine has become quite rare because farms and development have largely replaced its habitat of sandy, open woods and clearings in northeast North America. If the lupine disappears, the Karner blue butterfly will surely become extinct along with it.

Interactions with Other Species May Drive a Species to Extinction

Interactions such as competition and predation serve as agents of natural selection (see Chapter 15). In some cases, these same interactions can lead to extinction rather than to adaptation.

Organisms compete for limited resources in all environments. If a species' competitors evolve superior adaptations and the species doesn't evolve fast enough to keep up, it may become extinct. A particularly striking example of extinction through competition occurred in South America, beginning about 3 million years ago. At that time, the isthmus of Panama rose above sea level and formed a land

bridge between North America and South America. After the previously separated continents were connected, the mammal species that had evolved in isolation on each continent were able to mix. Many species did indeed expand their ranges, as North American mammals moved southward and South American mammals moved northward. As they moved, each species encountered resident species that occupied the same kinds of habitats and exploited the same kinds of resources. The ultimate result of the ensuing competition was that North American species that moved south diversified and underwent an adaptive radiation that displaced the vast majority of the South American species, many of which went extinct. Clearly, evolution had bestowed on the North American species some (as yet unknown) set of adaptations that enabled their descendants to exploit resources more efficiently and effectively than their South American counterparts could.

Habitat Change and Destruction Are the Leading Causes of Extinction

Habitat change, both contemporary and prehistoric, is the single greatest cause of extinctions. Present-day habitat destruction due to human activities is proceeding at a rapid

Earth|Watch

Why Preserve Biodiversity?

Today, most extinctions occur in the tropics, where the vast majority of species live. The main cause of these extinctions is environmental change, especially habitat destruction. Unfortunately, tropical habitats are being rapidly destroyed and disrupted by human activities (**Fig. E16-2**). For example, a recent analysis of thousands of satellite photos concluded that worldwide tropical rain-forest cover has decreased by about 23,000 square miles per year for the past decade. Most of the lost forest was destroyed by logging or to clear land for agriculture. Similarly, a worldwide survey of coral reefs revealed that about 20% of Earth's reef area has already been destroyed and an additional 25% is severely damaged, again mostly as the result of human influences such as pollution.

The rapid destruction of habitats in the tropics is causing many species to go extinct, as their homes disappear. Recent ecological research suggests that the current rate of extinction is extremely high, perhaps higher than ever before in the history of life on Earth. Does it matter? Is there any reason for us to try to slow the loss of biodiversity?

One possible reason to protect Earth's biodiversity is an ethical one. As the only species on the planet with the power to destroy other species, perhaps we have an ethical obligation to protect them from extinction. But even if you do not agree with this ethical argument, there may be compelling practical reasons to save other species from destruction. Our ecological self-interest may be at stake.

For example, Earth's species form communities, highly complex webs of interdependent life-forms whose interactions sustain one another. These communities play a crucial role in processes that purify the air we breathe and the water we drink, build the rich topsoil in which we grow our crops, provide the bounty of food that we harvest from the oceans, and decompose and detoxify our waste. We depend entirely on these "ecosystem services." When our activities cause species to disappear from communities, we take a big risk. If we remove too many species, or remove some especially crucial species, we may disrupt the finely tuned processes of the community and undermine its ability to sustain us.

▲ **FIGURE E16-2 Biodiversity threatened** Destruction of tropical rain forests by indiscriminant logging threatens Earth's greatest storehouse of biological diversity.

pace (see "Earth Watch: Why Preserve Biodiversity?"). Many biologists believe that we are presently in the midst of the fastest-paced and most widespread episode of species extinction in the history of life. Loss of tropical forests is especially devastating to species diversity. As many as half the species presently on Earth may be lost during the next 50 years as the tropical forests that contain them are cut for timber or to clear land for cattle and crops. (In Chapter 17, we will discuss extinctions due to prehistoric habitat change.)

CHECK YOUR LEARNING

Can you describe the main causes of extinction? Can you describe some examples of living species that are at risk of extinction?

CASE STUDY revisited

Lost World

One possible explanation for the distinctive collection of species found in the Annamite Mountains of Vietnam lies in the geological history of the region. During the ice ages that have occurred repeatedly during the past million years or so, the area covered by tropical forests must have shrunk dramatically. Organisms that depended on the forests for survival would have been restricted to any remaining "islands" of forest, isolated from their fellows in other, distant patches of forest. What is now the Annamite Mountain region may well have been an isolated forest during periods of glacial advance. As we learned in this chapter, this kind of

(continued)

isolation can set the stage for allopatric speciation and may have created the conditions that gave rise to the saola and other unique denizens of Vietnamese forests.

Ironically, we have discovered the lost world of Vietnamese animals at a moment when that world is in danger of disappearing. Economic development in Vietnam has brought logging and mining to ever more remote regions of the country, and Annamite Mountain forests are being cleared at an unprecedented rate. The increasing local human population means that animals are hunted heavily; most of our knowledge of the saola comes from carcasses found in local markets. In recent years, saola sightings have become increasingly rare. Despite intensive searching by biologists, there has been only one verified observation of a live saola since an unattended wildlife camera snapped a photo of a passing saola in 1999. In late 2010, scientists received word that a saola had been captured by some residents of a remote mountain village. By the time the researchers reached the village, though, the captive animal was very weak; it died a couple of days later.

All of the newly discovered mammals of Vietnam are quite rare, seen only infrequently even by local hunters. Fortunately, the Vietnamese government has established a number of national parks and nature preserves in key areas. Only time will tell if these measures are sufficient to ensure the survival of the mysterious mammals of the Annamites.

Consider This

BioEthics For those who believe in the inherent value of protecting and preserving other species, the precarious state of rare species around the world poses profound ethical dilemmas. In many cases, the habitat destruction that endangers other species also helps make space for the farmland, housing, and workplaces needed by our growing human population. How can we reconcile the conflict between valid human needs and the needs of endangered species? Furthermore, it is becoming clear that, even with the best of intentions, we cannot save all of the species currently threatened with extinction. The resources available to preserve and manage protected habitats are limited, and we must make choices that will allow some species to survive and others to perish. If all species are precious, how can we make such choices? Who should decide which species will live and which will die, and what criteria should be used?

CHAPTER REVIEW

Summary of Key Concepts

16.1 What Is a Species?

According to the biological species concept, a species consists of all the populations of organisms that are potentially capable of interbreeding under natural conditions and that are reproductively isolated from other populations.

16.2 How Is Reproductive Isolation Between Species Maintained?

Reproductive isolation between species may be maintained by one or more of several mechanisms, collectively known as premating isolating mechanisms and postmating isolating mechanisms. Premating isolating mechanisms include geographic isolation, ecological isolation, temporal isolation, behavioral isolation, and mechanical incompatibility. Postmating isolating mechanisms include gametic incompatibility, hybrid inviability, and hybrid infertility.

16.3 How Do New Species Form?

Speciation, the formation of new species, takes place when gene flow between two populations is reduced or eliminated and the populations diverge genetically. Most commonly, speciation is allopatric—gene flow is restricted by geographic isolation. However, speciation can also be sympatric—gene flow is restricted by ecological isolation or by mutations that cause polyploidy. Whether genetic isolation initially arises allopatrically or sympatrically, speciation is completed by subsequent genetic divergence of the separated populations through genetic drift or natural selection.

16.4 What Causes Extinction?

Factors that cause extinctions include competition among species and habitat destruction. Localized distribution and overspecialization increase a species' vulnerability to extinction.

Key Terms

adaptive radiation   *304*
allopatric speciation   *301*
behavioral isolation   *299*
ecological isolation   *298*
extinction   *305*
gametic
   incompatibility   *300*
geographic isolation   *298*
hybrid   *299*
hybrid infertility   *300*
hybrid inviability   *300*
isolating mechanism   *297*

mechanical
   incompatibility   *299*
polyploidy   *303*
postmating isolating
   mechanism   *300*
premating isolating
   mechanism   *298*
reproductive isolation   *296*
speciation   *301*
species   *296*
sympatric speciation   *301*
temporal isolation   *299*

Learning Outcomes

In this chapter, you have learned to . . .

LO1 Define *species* and explain why it is difficult to develop a universally applicable criterion for distinguishing species.

LO2 Explain the biological species concept and its limitations.

LO3 Describe the factors that can make it difficult to tell different species apart.

LO4 Describe the isolating mechanisms that restrict gene flow between different species.

LO5 Explain how new species arise.

LO6 Explain the process of adaptive radiation and describe examples.

LO7 Identify the major factors that cause extinction.

LO8 Interpret an evolutionary tree diagram.

Thinking Through the Concepts

Fill-in-the-Blank

1. A species is a group of _____ that evolves _____. The biological species concept identifies species on the basis of their _____. The biological species concept cannot be applied to species that reproduce _____. **LO1 LO2**

2. Fill in the following with the appropriate isolating mechanism: Occurs when members of two populations have different courtship behaviors: _____; occurs when hybrid offspring fail to survive to reproduce: _____; occurs when members of two populations have different breeding seasons: _____; occurs when sperm from one species fails to fertilize the eggs of another species: _____; occurs when the sexual organs of two species are incompatible: _____. **LO4**

3. Formation of a new species occurs when two populations of an existing species first become _____ and then _____. The process in which geographic separation of parts of a population leads to the formation of new species is called _____. Isolated populations may diverge through the action of _____ or _____. **LO5**

4. The process by which many new species arise in a relatively short period of time is known as _____. This process often occurs when a species arrives in a previously unoccupied _____. **LO6**

5. A species may be at higher risk of extinction if its geographic range includes a(n) _____ area, or if its food or habitat requirements are _____. The leading direct cause of extinction is _____. **LO7**

Review Questions

1. Define the following terms: *species, speciation, allopatric speciation,* and *sympatric speciation*. Explain how allopatric and sympatric speciation might work, and give a hypothetical example of each. **LO1 LO2 LO5**

2. Many of the oak tree species in central and eastern North America hybridize (interbreed). Are they "true species"? **LO2**

3. Review the material on the possibility of sympatric speciation in *Rhagoletis* flies. What types of genotypic, phenotypic, or behavioral data would convince you that the two forms have become separate species? **LO2 LO3 LO4**

4. A drug called colchicine prevents cell division after the chromosomes have doubled at the start of meiosis. Describe how you would use colchicine to produce a new polyploid plant species. **LO5**

5. What are the two major types of reproductive isolating mechanisms? Give examples of each type, and describe how they work. **LO4**

Applying the Concepts

1. Why do you suppose there are so many *endemic* species— that is, species found nowhere else—on islands? Why have the overwhelming majority of recent extinctions occurred on islands?

2. A biologist you've met claims that the fact that humans are pushing other species into small, isolated populations is good for biodiversity because these are the conditions that lead to new speciation events. Comment.

3. Southern Wisconsin is home to several populations of gray squirrels (*Sciurus carolinensis*) with black fur. Design a study to determine if the squirrels with black fur are actually a separate species.

4. It is difficult to gather data on speciation events in the past or to perform interesting experiments about the process of speciation. Does this difficulty make the study of speciation "unscientific"? Should we abandon the study of speciation?

Answers to Figure Caption questions and Fill-in-the-Blank questions can be found in the Answers section at the back of the book.

MB *Go to MasteringBiology for practice quizzes, activities, eText, videos, current events, and more.*

The skull of *Homo floresiensis*, a recently discovered diminutive human relative, is dwarfed by the skull of a modern *Homo sapiens*.

CASE STUDY
Little People, Big Story

MUCH OF OUR KNOWLEDGE of the evolutionary history of life comes from the work of paleontologists, the scientists who study fossils. Not long ago, a group of paleontologists, digging beneath the floor of a cave on the Indonesian island of Flores, discovered what they at first believed to be the fossil skeleton of a human child. Closer examination of the skeleton, however, revealed that it instead belonged to a fully grown adult, but one that was no more than 3 feet tall. The researchers gave this extraordinary creature the nickname "Hobbit."

Unlike today's small humans, such as pygmies or pituitary dwarves, Hobbit had a very small brain, smaller even than the brain of a typical chimpanzee. Furthermore, the shapes and arrangements of the bones in Hobbit's wrist, shoulder, and other parts of its skeleton were unlike those of anatomically modern humans. On the basis of these findings, many researchers have concluded that Hobbit was not simply a small *Homo sapiens* but was instead a different, related species. The new species was dubbed *Homo floresiensis*.

Some paleontologists are skeptical of the conclusion that the specimen represents a new human species. The skeptics contend that Hobbit might simply be a small-bodied *H. sapiens* who was victimized by a skeleton-deforming disease. Hobbit proponents, however, maintain that the available evidence supports the claim that *H. floresiensis* is a distinct species.

If Hobbit is indeed *H. floresiensis*, we shared Earth with close relatives until much more recently than was previously thought. Hobbit's bones are about 18,000 years old. Until they were unearthed, scientists believed that we had long been the only surviving member of the human family tree. Hobbit's discovery, however, raises the possibility that, not too long ago, in the forests of Flores, representatives of our species encountered members of another, tiny human species.

Although the tale of *H. floresiensis* is especially significant in our human-centered view of the world, it is but one thread among the millions that together make up the story of life's evolution. We therefore turn our attention from our hobbit-like cousin to a brief tour of some of the highlights of life's history.

AT A GLANCE

LEARNING GOALS

LG1 Describe a scenario for the origin of life and summarize evidence that supports the scenario.

LG2 Describe the major evolutionary events and innovations that occurred during the period in which all organisms were single celled.

LG3 Describe the earliest multicellular organisms and the factors associated with the origin of multicellularity.

LG4 Describe the adaptations associated with the evolution of increased animal diversity in the oceans.

LG5 Describe the adaptations associated with the transition to terrestrial life.

LG6 Describe the transitions and innovations associated with the origin and evolution of the major groups of land plants.

LG7 Describe the transitions and innovations associated with the origin and evolution of the major groups of vertebrates.

LG8 Describe the evolutionary history of humans and the factors that may have fostered humans' distinctive adaptations.

LG9 Explain the possible causes and evolutionary impact of mass extinctions.

17.1 HOW DID LIFE BEGIN?

Pre-Darwinian thought held that all species were simultaneously created by God a few thousand years ago. Further, until the nineteenth century most people thought that new members of species sprang up all the time, through **spontaneous generation** from both nonliving matter and other, unrelated forms of life. In 1609, a French botanist wrote, "There is a tree . . . frequently observed in Scotland. From this tree leaves are falling; upon one side they strike the water and slowly turn into fishes, upon the other they strike the land and turn into birds." Medieval writings abound with similar observations. Microorganisms were thought to arise spontaneously from broth, maggots from meat, and mice from mixtures of sweaty shirts and wheat.

In 1668, the Italian physician Francesco Redi disproved the maggots-from-meat hypothesis simply by keeping flies (whose eggs hatch into maggots) away from uncontaminated meat (see "Science in Action: Controlled Experiments, Then and Now," on pp. **12–13**). In the mid-1800s, Louis Pasteur in France and John Tyndall in England disproved the broth-to-microorganism idea by showing that microorganisms did not appear in sterile broth unless the broth was first exposed to existing microorganisms in the surrounding environment (**Fig. 17-1**). Although Pasteur and Tyndall's work effectively demolished the notion of spontaneous generation, it did not address the question of how life on Earth originated in the first place. Or, as the biochemist Stanley Miller put it, "Pasteur never proved it didn't happen once, he only showed that it doesn't happen all the time."

The First Living Things Arose from Nonliving Ones

Modern scientific ideas about the origin of life began to emerge in the 1920s, when Alexander Oparin in Russia and John B. S. Haldane in England noted that today's oxygen-rich atmosphere would not have permitted the spontaneous

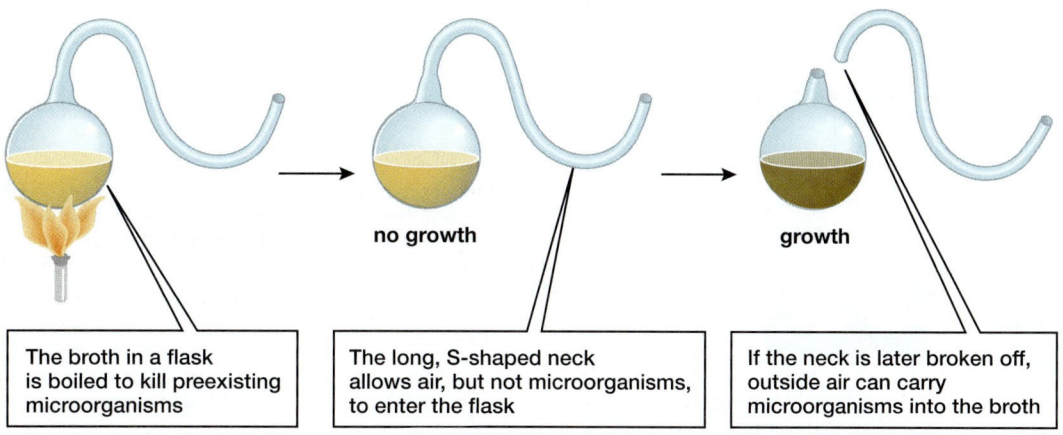

no growth

growth

The broth in a flask is boiled to kill preexisting microorganisms

The long, S-shaped neck allows air, but not microorganisms, to enter the flask

If the neck is later broken off, outside air can carry microorganisms into the broth

▲ FIGURE 17-1 **Spontaneous generation refuted** Louis Pasteur's experiment disproving the spontaneous generation of microorganisms in broth.

formation of the complex organic molecules necessary for life. Oxygen reacts readily with other molecules, disrupting chemical bonds. Thus, an oxygen-rich environment tends to keep molecules simple.

Oparin and Haldane speculated that the atmosphere of the young Earth must have contained very little oxygen and that, under such atmospheric conditions, complex organic molecules could have arisen through ordinary chemical reactions. Some kinds of molecules could persist in the lifeless environment of early Earth better than others and would therefore become more common over time. This chemical version of the "survival of the fittest" is called *prebiotic* (meaning "before life") evolution. In the scenario envisioned by Oparin and Haldane, prebiotic chemical evolution gave rise to progressively more complex molecules and eventually to living organisms.

Organic Molecules Can Form Spontaneously Under Prebiotic Conditions

Inspired by the ideas of Oparin and Haldane, Stanley Miller and Harold Urey set out in 1953 to simulate prebiotic evolution in the laboratory. They knew that, on the basis of the chemical composition of the rocks that formed early in Earth's history, geochemists had concluded that the early atmosphere probably contained virtually no oxygen gas, but did contain other substances, including methane (CH_4),

ammonia (NH_3), hydrogen (H_2), and water vapor (H_2O). Miller and Urey simulated the oxygen-free atmosphere of early Earth by mixing these components in a flask. An electrical spark mimicked the intense energy of early Earth's lightning storms. In this experimental microcosm, the researchers found that simple organic molecules appeared after just a few days (**Fig. 17-2**). The experiment showed that small molecules likely present in the early atmosphere can combine to form larger organic molecules if electrical energy is present. (Recall from Chapter 6 that reactions that synthesize biological molecules from smaller ones are endergonic— they consume energy.) Similar experiments by Miller and others have produced amino acids, peptides, nucleotides, adenosine triphosphate (ATP), and other molecules characteristic of living things.

In recent years, new evidence has convinced most geochemists that the actual composition of Earth's early atmosphere probably differed from the mixture of gases used in the pioneering Miller–Urey experiment. This improved understanding of the early atmosphere, however, has not undermined the basic finding of the Miller–Urey experiment. Additional experiments with more realistic (but still oxygen-free) simulated atmospheres or simulated early oceans have also yielded organic molecules. In addition, these experiments have shown that electricity is not the only suitable energy source. Other energy sources that

◀ FIGURE 17-2 The experimental apparatus of Stanley Miller and Harold Urey Life's very earliest stages left no fossils, so evolutionary scientists have pursued a strategy of re-creating in the laboratory the conditions that may have prevailed on early Earth. The mixture of gases in this spark chamber simulates Earth's early atmosphere.

QUESTION How would the experiment's result change if oxygen (O_2) were included in the spark chamber?

were available on early Earth, such as heat or ultraviolet (UV) light, have also been shown to drive the formation of organic molecules in experimental simulations of prebiotic conditions. Thus, even though we may never know exactly what the earliest atmosphere was like, we can be confident that organic molecules formed on early Earth.

Additional organic molecules probably arrived from space when meteorites and comets crashed into Earth's surface. Analysis of present-day meteorites recovered from impact craters on Earth has revealed that some meteorites contain relatively high concentrations of amino acids and other simple organic molecules. Laboratory experiments suggest that these molecules could have formed in interstellar space before plummeting to Earth. When small molecules known to be present in space were placed under space-like conditions of very low temperature and pressure and bombarded with UV light, larger organic molecules were produced.

Organic Molecules Can Accumulate Under Prebiotic Conditions

Prebiotic synthesis was neither very efficient nor very fast. Nonetheless, in a few hundred million years, large quantities of organic molecules accumulated in the early Earth's oceans. Today, most organic molecules have a short life because they are either digested by living organisms or they react with atmospheric oxygen. Early Earth, however, lacked both life and free oxygen, so organic molecules would not have been exposed to these threats.

Still, the prebiotic molecules must have been threatened by the sun's high-energy UV radiation, because early Earth lacked an ozone layer. The *ozone layer* is a region high in today's atmosphere that is enriched with ozone molecules, which form when incoming solar energy splits some O_2 molecules in the outer atmosphere into individual oxygen atoms (O) that then react with O_2 to form O_3 (ozone). The resulting high-altitude layer of ozone molecules absorbs some of the sun's UV light before it reaches Earth's surface. Early Earth, however, had no ozone layer, because there was little or no oxygen gas in the atmosphere and therefore no ozone formation.

Before the ozone layer formed, UV bombardment must have been fierce. UV radiation, as we have seen, can provide energy for the formation of organic molecules but it can also break them apart. Some sites, however, such as those beneath rock ledges or at the bottoms of even fairly shallow seas, would have been protected from UV radiation. In these locations, organic molecules may have accumulated.

Clay May Have Catalyzed the Formation of Larger Organic Molecules

In the next stage of prebiotic evolution, simple molecules combined to form larger molecules. The chemical reactions that formed the larger molecules required that the reacting molecules be packed closely together. Scientists have proposed several processes by which the required high concentrations might have been achieved on early Earth. One possibility is that small molecules accumulated on the surfaces of clay particles, which may have a small electrical charge that attracts dissolved molecules with the opposite charge. Clustered on such a clay particle, small molecules would have been sufficiently close together to allow chemical reactions between them. Researchers have demonstrated the plausibility of this scenario with experiments in which adding clay to solutions of dissolved small organic molecules catalyzed the formation of larger, more complex molecules, including RNA. Such molecules might have formed on clay at the bottom of early Earth's oceans or lakes and gone on to become the building blocks of the first living organisms.

RNA May Have Been the First Self-Reproducing Molecule

Although all modern organisms use DNA to encode and store genetic information, it is unlikely that DNA was the earliest informational molecule. DNA can reproduce itself only with the help of large, complex protein enzymes, but the instructions for building these enzymes are encoded in DNA itself. For this reason, the origin of DNA's role as life's information storage molecule poses a "chicken and egg" puzzle: DNA requires proteins, but those proteins require DNA. It is thus difficult to construct a plausible scenario for the origin of self-replicating DNA. It is therefore likely that the current DNA-based system of information storage evolved from an earlier system.

RNA Can Act as a Catalyst

A prime candidate for the first self-replicating informational molecule is RNA. In the 1980s, Thomas Cech and Sidney Altman, working with the single-celled organism *Tetrahymena*, discovered a cellular reaction that was catalyzed not by a protein, but by a small RNA molecule. Because this special RNA molecule performed a function previously thought to be performed only by protein enzymes, Cech and Altman decided to give their catalytic RNA molecule the name **ribozyme** (Fig. 17-3).

In the years since their discovery, researchers have found dozens of naturally occurring ribozymes that catalyze a variety of different reactions, including cutting other RNA molecules and splicing RNA fragments together. Ribozymes have also been found in ribosomes, where they catalyze the attachment of amino acid molecules to growing proteins. In addition, researchers have been able to synthesize various ribozymes in the laboratory, including some that can catalyze the replication of small RNA molecules. The most effective replication ribozyme so far synthesized can copy RNA sequences up to 95 nucleotides long.

Earth May Once Have Been an RNA World

The discovery that RNA molecules can act as catalysts for diverse reactions, including RNA replication, provides support for the hypothesis that life arose in an "RNA world." According to this view, the current era of DNA-based life was preceded by one in which RNA served as both the information-carrying genetic molecule and the catalyst for

▲ FIGURE 17-3 A computer-generated model of a ribozyme This RNA molecule, isolated from the single-celled organism *Tetrahymena*, acts like an enzyme, catalyzing metabolic reactions.

its own replication. This RNA world may have emerged after hundreds of millions of years of prebiotic chemical synthesis, during which RNA nucleotides would have been among the molecules synthesized. (A recent experiment showed that RNA nucleotides can assemble spontaneously from simpler molecules.) After reaching a sufficiently high concentration, perhaps on clay particles, the nucleotides probably bonded together to form short RNA chains.

Let's suppose that, purely by chance, one of these RNA chains was a ribozyme that could catalyze the production of copies of itself. This first self-reproducing ribozyme probably wasn't very good at its job and likely produced copies with lots of errors. These mistakes were the first mutations. Like modern mutations, most undoubtedly ruined the catalytic abilities of the "daughter molecules," but a few may have been improvements. Such improvements set the stage for natural selection among RNA molecules, as variant ribozymes with increased speed and accuracy of replication copied themselves more rapidly than did less efficient RNA molecules, and thereby became increasingly common. Molecular evolution in the RNA world proceeded until, by some still unknown chain of events, RNA gradually receded into its present role as an intermediary between DNA and protein enzymes.

Membrane-Like Vesicles May Have Enclosed Ribozymes

Self-replicating molecules on their own do not constitute life; in all living cells such molecules are contained within some kind of enclosing membrane. The precursors of the earliest biological membranes may have been simple structures that formed spontaneously from purely physical, mechanical processes. For example, chemists have shown that if water containing proteins and lipids is agitated to simulate waves beating against ancient shores, the proteins and lipids combine to form hollow structures called *vesicles*. These hollow balls resemble living cells in several respects. They have a well-defined outer boundary that separates their internal contents from the external solution. If the composition of the vesicle is right, a "membrane" forms that is remarkably similar in appearance to a real cell membrane. Under certain conditions, vesicles can absorb material from the external solution, grow, and even divide.

If a vesicle happened to surround the right ribozymes, it would form something resembling a living cell. We could call it a **protocell,** structurally similar to a cell but not alive. In the protocell, ribozymes and any other enclosed molecules would have been protected from degradation by free-roaming reactive molecules in the primordial soup. Nucleotides and other small molecules might have diffused across the membrane and been used to synthesize new ribozymes and other complex molecules. After sufficient growth, the vesicle may have divided, with a few copies of the ribozymes becoming incorporated into each daughter vesicle. If this process occurred, the evolution of the first cells would be nearly complete.

Was there a particular moment when a nonliving protocell gave rise to a living organism? Probably not. Like most evolutionary transitions, the change from protocell to living cell was a continuous process, with no sharp boundary between one state and the next.

But Did All This Really Happen?

The scenario just described, although plausible and consistent with many research findings, is by no means certain. One of the most striking aspects of origin-of-life research is a great diversity of assumptions, experiments, and contradictory hypotheses. Researchers disagree about whether life arose in quiet terrestrial pools, in the sea, in hot deep-sea vents, or in polar ice. A few researchers even argue that life arrived on Earth from space. Can we draw any firm conclusions from the research conducted so far? No, but we can make a few reasonable deductions.

First, the experiments of Miller and others show that amino acids, nucleotides, and other organic molecules, along with simple membrane-like structures, are likely to have formed in abundance on early Earth. Second, chemical evolution had long periods of time and huge areas of the Earth available to it. Given sufficient time and a sufficiently large pool of reactant molecules, even extremely rare events can occur many times. And given the vast expanses of time and space available, each small step on the path from primordial soup to living cell had ample opportunity to take place.

Most biologists accept that the origin of life was probably an inevitable consequence of natural laws. We should emphasize, however, that this proposition cannot be tested definitively. The origin of life left no record, and researchers exploring this mystery can proceed only by developing a hypothetical scenario and then conducting laboratory investigations to determine if the scenario's steps are chemically and biologically possible and plausible.

CHECK YOUR LEARNING

Can you describe a likely scenario for the origin of life? For each step in the scenario, can you describe some evidence that suggests the step is plausible?

17.2 WHAT WERE THE EARLIEST ORGANISMS LIKE?

When Earth first formed about 4.5 billion years ago, it was quite hot (Fig. 17-4). A multitude of meteorites smashed into the forming planet, and the kinetic energy of these extraterrestrial rocks was converted into heat on impact. Still more heat was released by the decay of radioactive atoms. The rock composing Earth melted, and heavier elements such as iron and nickel sank to the center of the planet, where they remain molten even today. Nonetheless, geological evidence suggests that Earth was cool enough for water to exist in liquid form by 4.3 billion years ago. Once liquid water was available, the prebiotic evolution that ultimately led to the first living organisms could begin.

The oldest fossil organisms found so far are in rocks that are about 3.4 billion years old. (Their age was determined using radiometric dating techniques; see "Science in Action: How Do We Know How Old a Fossil Is?") Chemical traces in older rocks have led some paleontologists to believe that life is even older, perhaps as old as 3.9 billion years.

The period in which life began is known as the Precambrian era. This interval was designated by geologists and paleontologists, who have devised a hierarchical naming system of eras, periods, and epochs to delineate the immense span of geological time (Table 17-1).

The First Organisms Were Anaerobic Prokaryotes

The first cells to arise in Earth's oceans were **prokaryotes,** cells whose genetic material was not contained within a nucleus. These cells probably obtained nutrients and energy by absorbing organic molecules from their environment. There was no oxygen gas in the atmosphere, so the cells must have metabolized the organic molecules anaerobically. (You may recall from Chapter 8 that anaerobic metabolism yields only small amounts of energy.)

Thus, the earliest cells were primitive anaerobic bacteria. As these bacteria multiplied, they must have eventually used up the organic molecules produced by prebiotic chemical reactions. Simpler molecules, such as carbon dioxide and water, would still have been very abundant, as was energy in the form of sunlight. What was lacking, then, was not materials or energy but energetic molecules—molecules in which energy is stored in chemical bonds.

▲ FIGURE 17-4 **Early Earth** Life began on a planet characterized by abundant volcanic activity, frequent electrical storms, repeated meteorite strikes, and an atmosphere that lacked oxygen gas.

TABLE 17-1   The History of Life on Earth

Era	Period	Epoch	Millions of Years Ago	Major Events
Cenozoic	Quaternary	Recent	0.01–present	Evolution of genus *Homo*
		Pleistocene	2.6–0.01	
	Tertiary	Pliocene	5–2.6	Widespread flourishing of birds, mammals, insects, and flowering plants
		Miocene	23–5	
		Oligocene	34–23	
		Eocene	56–34	
		Paleocene	65–56	
Mesozoic	Cretaceous		146–65	Flowering plants appear and become dominant Mass extinction of marine and terrestrial life, including dinosaurs
	Jurassic		202–146	Dominance of dinosaurs and conifers First birds
	Triassic		251–202	First mammals and dinosaurs Forests of gymnosperms and tree ferns
Paleozoic	Permian		299–251	Massive marine extinctions, including trilobites Flourishing of reptiles and the decline of amphibians
	Carboniferous		359–299	Forests of tree ferns and club mosses Dominance of amphibians and insects First reptiles and conifers
	Devonian		416–359	Fishes and trilobites flourish First amphibians, insects, seeds, and pollen
	Silurian		444–416	Many fishes, trilobites, and mollusks First vascular plants
	Ordovician		488–444	Dominance of arthropods and mollusks in the ocean Invasion of land by plants and arthropods First fungi
	Cambrian		542–488	Marine algae flourish Origin of most marine invertebrate phyla First fishes
Precambrian			About 1,000	First animals (soft-bodied marine invertebrates)
			1,200	First multicellular organisms
			2,000	First eukaryotes
			2,200	Accumulation of free oxygen in the atmosphere
			3,500	Origin of photosynthesis (in cyanobacteria)
			3,900–3,500	First living cells (prokaryotes)
			4,000–3,900	Appearance of the first rocks on Earth
			4,600	Origin of the solar system and Earth

SCIENCE IN ACTION   How Do We Know How Old a Fossil Is?

Until the twentieth century, geologists could date rock layers and their accompanying fossils only in a *relative* way: Fossils found in deeper layers of rock were generally older than those found in shallower layers. But a few decades after the discovery of radioactivity in 1896, it became possible to determine absolute dates, within limits. The nuclei of radioactive elements spontaneously break down, or decay, into other elements. For example, carbon-14 (usually written ^{14}C) decays by emitting an electron to become nitrogen-14 (^{14}N). Each radioactive element decays at a rate that is independent of temperature, pressure, or the chemical compound of which the element is a part. The time it takes for half of a radioactive element's nuclei to decay at this characteristic rate is called its *half-life*. The half-life of ^{14}C, for example, is 5,730 years.

How are radioactive elements used in determining the age of rocks? If we know the rate of decay and measure the proportion of decayed nuclei to undecayed nuclei, we can estimate how much time has passed since these radioactive elements became trapped in the rock. This process is called *radiometric dating*.

A particularly straightforward radiometric dating technique measures the decay of potassium-40 (^{40}K), which has a half-life of about 1.25 billion years, into argon-40 (^{40}Ar). Potassium-40 is commonly found in volcanic rocks such as granite and basalt, and the ^{40}Ar it decays into is a gas. Suppose that a volcano erupts with a massive lava flow, covering the countryside. All the ^{40}Ar, being a gas, will bubble out of the molten lava, so when the lava first cools and solidifies into rock, it will not contain any ^{40}Ar (**Fig. E17-1**). Over time, however, any ^{40}K present in the hardened lava will decay into ^{40}Ar, with half of the ^{40}K decaying every 1.25 billion years. This ^{40}Ar gas will be trapped in the rock. A geologist could take a sample of the rock and measure the ratio of ^{40}K to ^{40}Ar to determine the rock's age. For example, if the analysis finds equal amounts of the two elements, the geologist will conclude that the lava hardened 1.25 billion years ago. Such age estimates are quite reliable. If a fossil is found beneath a lava flow dated at, say, 500 million years, then we know that the fossil is at least that old.

▲ **FIGURE E17-1 The relationship between time and the decay of radioactive ^{40}K to ^{40}Ar**

QUESTION Uranium-235, with a half-life of 713 million years, decays to lead-207. If you analyze a rock and find that it contains uranium-235 and lead-207 in a ratio of 1:1, how old is the rock?

Some Organisms Evolved the Ability to Capture the Sun's Energy

Eventually, some cells evolved the ability to use the energy of sunlight to drive the synthesis of complex, high-energy molecules from simpler molecules; in other words, photosynthesis appeared. Photosynthesis requires a source of hydrogen, and the very earliest photosynthetic bacteria probably used hydrogen sulfide gas dissolved in water for this purpose (much as today's purple photosynthetic bacteria do). Eventually, however, Earth's supply of hydrogen sulfide (which is produced mainly by volcanoes) must have run low. The shortage of hydrogen sulfide set the stage for the evolution of photosynthetic bacteria that were able to use the planet's most abundant source of hydrogen—water (H_2O).

Water-based photosynthesis converts water and carbon dioxide to energetic molecules of sugar, releasing oxygen as a by-product. The emergence of this new method for capturing energy introduced significant amounts of free oxygen to the atmosphere for the first time. At first, the newly liberated oxygen was quickly consumed by reactions with other molecules in the atmosphere and in Earth's crust. One especially common reactive atom in the crust was iron, and much of the new oxygen combined with iron atoms to form huge deposits of iron oxide (rust). As a result, iron oxide is abundant in rocks formed during this period.

After all the accessible iron had turned to rust, the concentration of oxygen gas in the atmosphere began to increase. Chemical analysis of rocks suggests that significant amounts of oxygen first appeared in the atmosphere about 2.3 billion years ago, produced by bacteria that were probably very similar to modern cyanobacteria. (Because Earth's supply of oxygen molecules is continually recycled, you will undoubtedly breathe in some oxygen molecules today that were expelled 2 billion years ago by one of these early bacteria.)

Aerobic Metabolism Arose in Response to Dangers Posed by Oxygen

Oxygen is potentially very dangerous to living things, because it can react with organic molecules, breaking them down. Many of today's anaerobic bacteria perish when exposed to what is for them a deadly poison—oxygen. The accumulation of oxygen in the atmosphere of early Earth probably exterminated many organisms and fostered the evolution of cellular mechanisms for detoxifying oxygen. This crisis for evolving life also provided the environmental pressure for the next great advance: the ability to use oxygen in metabolism. This ability not only provides a defense against the chemical action of oxygen, but actually channels oxygen's destructive power through aerobic respiration to generate useful energy for the cell. (See Chapter 8 for more information on aerobic respiration.) Because the amount of energy available to a cell is vastly increased when oxygen is used to metabolize food molecules, aerobic cells had a significant selective advantage.

Some Organisms Acquired Membrane-Enclosed Organelles

Hordes of bacteria would offer a rich food supply to any organism that could eat them. Paleobiologists speculate that, once this potential prey population appeared, predation would have evolved quickly. These early predators were probably prokaryotes that were larger than typical bacteria. In addition, they must have lost the rigid cell wall that surrounds most bacterial cells, so that their flexible plasma membrane was in contact with the surrounding environment. Thus, the predatory cells were able to envelop smaller bacteria in an infolded pouch of membrane and in this fashion engulf whole bacteria as prey.

These early predators were probably capable of neither photosynthesis nor aerobic metabolism. Although they could ingest smaller bacteria, they metabolized them inefficiently. By about 1.7 billion years ago, however, one predator probably gave rise to the first eukaryotic cell. Eukaryotic cells differ from prokaryotic cells in that they have an elaborate system of internal membranes, many of which enclose organelles such as a nucleus that contains the cell's genetic material. Organisms composed of one or more eukaryotic cells are known as **eukaryotes.**

The Internal Membranes of Eukaryotes May Have Arisen Through Infolding of the Plasma Membrane

The internal membranes of eukaryotic cells may have originally arisen through inward folding of the cell membrane of a single-celled predator. If, as in most of today's bacteria, the DNA of the eukaryotes' ancestor was attached to the inside of its cell membrane, an infolding of the membrane near the point of DNA attachment may have pinched off and become the precursor of the cell nucleus.

In addition to the nucleus, other key eukaryotic structures include the organelles used for energy metabolism: mitochondria (in all eukaryotes) and chloroplasts (in plants and algae). How did these organelles evolve?

Mitochondria and Chloroplasts May Have Arisen from Engulfed Bacteria

The **endosymbiont hypothesis** proposes that early eukaryotic cells acquired the precursors of mitochondria and chloroplasts by engulfing certain types of bacteria. These cells and the bacteria trapped inside them (*endo* means "within") gradually entered into a *symbiotic* relationship, a close association between different types of organisms over an extended time. How might this have happened?

Let's suppose that an anaerobic predatory cell captured an aerobic bacterium for food, as it often did, but for some reason failed to digest this particular prey (**Fig. 17-5 ❶**). The aerobic bacterium remained alive and well, protected from other predatory cells. In fact, it was better off than ever, because the cytoplasm of its predator-host was chock-full

❶ An anaerobic, predatory prokaryotic cell engulfs an aerobic bacterium

aerobic bacterium

❷ Descendants of the engulfed bacterium evolve into mitochondria

photosynthetic bacterium

❸ The mitochondria-containing cell engulfs a photosynthetic bacterium

❹ Descendants of the photosynthetic bacterium evolve into chloroplasts

▲ FIGURE 17-5 **The probable origin of mitochondria and chloroplasts in eukaryotic cells**

QUESTION Scientists have identified a living bacterium believed to be descended from the endosymbiont that gave rise to mitochondria. Would you expect the DNA sequence of this modern bacterium to be most similar to the sequence of DNA from a plant chloroplast, an animal cell nucleus, or a plant mitochondrion?

of half-digested food molecules, the remnants of anaerobic metabolism. The aerobe absorbed these molecules and used oxygen to metabolize them, thereby gaining enormous amounts of energy. So abundant were the aerobes' food resources, and so bountiful their energy production, that the aerobes must have leaked energy, probably as ATP or similar molecules, back into their host's cytoplasm. The anaerobic predatory cell with its symbiotic bacteria could now metabolize food aerobically, gaining a great advantage over other anaerobic cells and leaving a greater number of offspring. Eventually, the endosymbiotic bacterium lost its ability to live independently of its host, and the mitochondrion was born (**Fig. 17-5 ❷**).

One of these successful new cellular partnerships managed a second feat: It captured a photosynthetic bacterium and again failed to digest its prey (**Fig. 17-5 ❸**). The bacterium flourished in its new host and gradually evolved into the first chloroplast (**Fig. 17-5 ❹**). Other eukaryotic organelles may have also originated through endosymbiosis. Many biologists believe that cilia, flagella, centrioles, and microtubules may all have evolved from a symbiosis between a spirilla-like bacterium (a form of bacterium with an elongated corkscrew shape) and an early eukaryotic cell.

Evidence for the Endosymbiont Hypothesis Is Strong

Evidence that supports the endosymbiont hypothesis includes the many distinctive biochemical features shared by eukaryotic organelles and living bacteria. In addition, mitochondria and chloroplasts each contain their own minute supply of DNA, which many researchers interpret as remnants of the DNA originally contained within the engulfed bacteria.

Another kind of support comes from *living intermediates*, organisms alive today that are similar to hypothetical ancestors and thus help show that a proposed evolutionary pathway is plausible. For example, the amoeba *Pelomyxa palustris* lacks mitochondria but hosts a permanent population of aerobic bacteria that carry out much the same role. Similarly, a variety of corals, some clams, a few snails, and at least one species of *Paramecium* harbor a collection of photosynthetic algae in their cells (**Fig. 17-6**). These examples of modern cells that host bacterial endosymbionts suggest that similar symbiotic associations could have occurred almost 2 billion years ago and led to the first eukaryotic cells.

CHECK YOUR LEARNING

Can you describe scenarios for the major evolutionary events and innovations that occurred during the period in which all organisms were single celled, such as the origins of photosynthesis, atmospheric oxygen, aerobic respiration, and eukaryotic organelles? Can you state the order in which these events occurred, and list evidence that supports the scenarios you described?

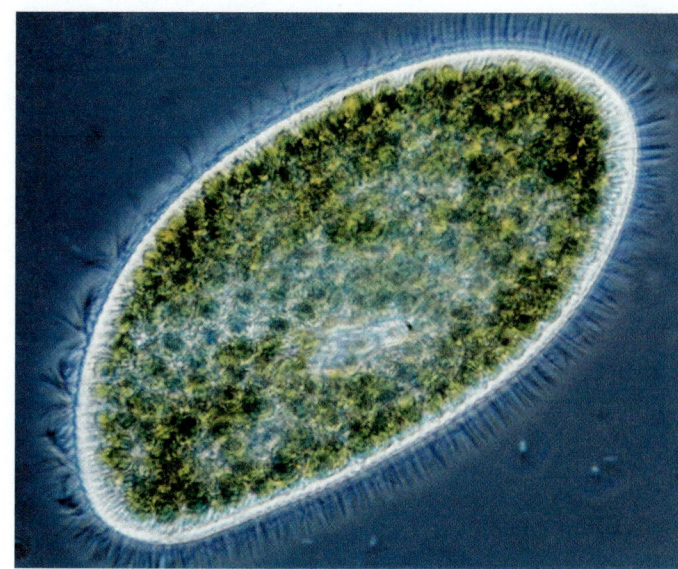

▲ **FIGURE 17-6 Symbiosis within a modern cell** The ancestors of the chloroplasts in today's plant cells may have resembled *Chlorella*, the green, photosynthetic, single-celled algae living symbiotically within the cytoplasm of the *Paramecium* pictured here.

17.3 WHAT WERE THE EARLIEST MULTICELLULAR ORGANISMS LIKE?

Once predation had evolved, increased size became an advantage. In the marine environments to which life was restricted, a larger cell could easily engulf a smaller cell and would also be difficult for other predatory cells to ingest. But enormous single cells have problems. The larger a cell becomes, the less surface membrane is available per unit volume of cytoplasm (see Fig. 5-16). Thus, as a cell grows larger, the process of diffusion across its plasma membrane becomes progressively less able to accommodate the oxygen and nutrients that must move into the cell and the waste products that must move out.

There are only two ways in which an organism larger than a millimeter or so in diameter can survive. First, it can have a low metabolic rate so that it doesn't need much oxygen or produce much carbon dioxide. This strategy seems to work for certain very large single-celled algae. Alternatively, an organism can be multicellular; that is, it can consist of many small cells packaged into a larger, unified body.

Some Algae Became Multicellular

The oldest fossils of multicellular organisms are about 1.2 billion years old. They consist of impressions of multicellular algae that arose from single-celled eukaryotic organisms containing chloroplasts. Multicellularity would have provided at least two advantages for these organisms. First, large, many-celled algae would have been difficult for single-celled predators to engulf. Second, specialization of cells would have provided the potential for staying in one place in the brightly lit waters of the shoreline, as rootlike structures burrowed in sand or clutched onto rocks, while leaflike structures floated above in the sunlight. The green, brown, and red algae lining our shores today—some, such as the brown kelp, which can be more than 200 feet long—are the descendants of these early multicellular algae.

Animal Diversity Arose in the Precambrian Era

The oldest known unequivocal traces of animal include fossil embryos found in Precambrian deposits that are 630 million years old. Older fossils that may demonstrate the earlier existence of animals include 850-million-year-old mesh-like structures that look a lot like the collagen protein web that is left behind when a modern sponge dies and decays. Fossils of apparently adult animal bodies first appear in rocks laid down between 610 million and 544 million years ago. Some of these ancient invertebrate animals (animals lacking a backbone) are quite different in appearance from any animals that appear in later fossil layers and may represent types of animals that left no descendants. Other fossils in these rock layers, however, appear to be ancestors of today's animals. Ancestral sponges and jellyfish appear in the oldest layers, followed later by ancestors of worms, mollusks, and arthropods.

The full range of modern invertebrate animals, however, does not appear in the fossil record until the Cambrian period, marking the beginning of the Paleozoic era, about 542 million years ago. (The phrase "fossil record" is a shorthand reference to the entire collection of all fossil evidence that has been found to date.) These Cambrian fossils reveal an adaptive radiation (see p. 304) that had already yielded a diverse array of complex body plans. Almost all of the major groups of animals on Earth today were already present in the early Cambrian. The apparently sudden appearance of so many different kinds of animals suggests that these groups actually arose earlier, but their early evolutionary history is not preserved in the fossil record.

Predation Favored the Evolution of Improved Mobility and Senses

The early diversification of animals was probably driven in part by the emergence of predatory lifestyles. For example, coevolution of predator and prey favored animals that were more mobile than their predecessors. Mobile predators gained an advantage from an ability to travel over wide areas in search of suitable prey; mobile prey benefited if they were able to make a speedy escape. The evolution of efficient movement was often associated with the evolution of greater sensory capabilities and more complex nervous systems. Senses for detecting touch, chemicals, and light became highly developed, along with nervous systems capable of handling the sensory information and directing appropriate behaviors.

By the Silurian period (444 million to 416 million years ago), life in Earth's seas included an array of anatomically complex animals, including armored trilobites, shelled

ammonites, and the chambered nautilus (Fig. 17-7). The nautilus survives today in almost unchanged form in deep Pacific waters.

Skeletons Improved Mobility and Protection

In many Paleozoic animal species, mobility was enhanced in part by the origin of hard external body coverings known as **exoskeletons.** Exoskeletons improved mobility by providing hard surfaces to which muscles attached. These attachments made it possible for animals to use their muscles to move appendages used to swim through the water or crawl over the seafloor. Exoskeletons also provided support for animals' bodies and protection from predators.

About 530 million years ago, one group of animals— the fishes—developed a new form of body support and muscle attachment: an internal skeleton. These early fishes were inconspicuous members of the ocean community, but by 400 million years ago, fishes were a diverse and prominent group. By and large, fishes proved to be faster than the invertebrates, with more acute senses and larger brains. Eventually, they became the dominant predators of the open seas.

CHECK YOUR LEARNING

Can you describe fossil evidence of the earliest multicellular organisms and the earliest animals? Can you describe the advantages that fostered the origin of multicellularity and the adaptations associated with the later increase in animal diversity?

17.4 HOW DID LIFE INVADE THE LAND?

A compelling subplot in the long tale of life's history is the story of life's invasion of land. In moving to solid ground after more than 3 billion years of a strictly watery existence, organisms had many obstacles to overcome. Life in the sea provides buoyant support against gravity, but on land an organism must bear its weight against the crushing force of gravity. The sea provides ready access to life-sustaining water, but adequate water may not be easily available to a terrestrial organism. Sea-dwelling plants and animals can reproduce by means of mobile sperm or mobile eggs (or both), which swim or drift to each other through the water. The gametes of land-dwellers, however, must be protected from drying out.

Despite the obstacles to life on land, the vast empty spaces of the Paleozoic landmass represented a tremendous evolutionary opportunity. The potential rewards of terrestrial life were especially great for plants. Water strongly absorbs light, so even in the clearest water, photosynthesis is limited to the upper few hundred meters of depth, and usually much less. Out of the water, the dazzling brightness of the sun permits rapid photosynthesis. Furthermore, terrestrial soils are rich storehouses of nutrients, whereas seawater tends to be low in some nutrients, particularly nitrogen and phosphorus. Finally, the Paleozoic sea swarmed with plant-eating

(a) Silurian scene

(b) Trilobite

(c) Ammonite

(d) *Nautilus*

▲ **FIGURE 17-7 Diversity of ocean life during the Silurian period (a)** Life characteristic of the oceans during the Silurian period, 444 million to 416 million years ago. Among the most common fossils from that time are **(b)** the trilobites and their predators, the nautiloids, and **(c)** the ammonites. **(d)** This living *Nautilus* is very similar in structure to the Silurian nautiloids, showing that a successful body plan may exist virtually unchanged for hundreds of millions of years.

animals, but the land was devoid of animal life. Thus, the plants that first colonized the land would have had ample sunlight, untouched nutrient sources, and no predators.

Some Plants Became Adapted to Life on Dry Land

In moist soils at the water's edge, a few small green algae began to grow, taking advantage of the sunlight and nutrients. These algae didn't have large bodies to support them against the force of gravity, and, living right in the film of water on the soil, they could easily obtain water. About 475 million years ago, some of these algae gave rise to the first multicellular land plants. Initially simple, low-growing forms, land plants eventually evolved solutions to two of the main difficulties of plant life on land: obtaining and conserving water and staying upright despite gravity and winds. Water-resistant coatings on aboveground parts reduced water loss by evaporation, and rootlike structures delved into the soil, mining water and minerals. Specialized cells formed tissues (called vascular tissues) that contained tubes to conduct water from roots to leaves. Extra-thick walls surrounding certain cells enabled stems to stand erect.

Primitive Land Plants Retained Swimming Sperm and Required Water to Reproduce

Reproduction out of water presented challenges. Plants produce sperm and eggs, as animals do, and these gametes must meet to produce the next generation. The first land plants had swimming sperm, presumably much like those of today's mosses and ferns. Consequently, the earliest plants were restricted to swamps and marshes or to areas with abundant rainfall, where the ground would occasionally be covered with water. Here, the sperm and eggs could be released into the water, and sperm could swim to reach an egg. Later plants with swimming sperm prospered during periods in which the climate was warm and moist. For example, the Carboniferous period (359 million to 299 million years ago) was characterized by vast forests of giant tree ferns, club mosses, and horsetails (**Fig. 17-8**).

Seed Plants Encased Sperm in Pollen Grains

Meanwhile, some plants inhabiting drier regions had evolved a means of reproduction that no longer depended on water. The eggs of these plants were retained on the parent plant,

◄ **FIGURE 17-8 The swamp forest of the Carboniferous period** The treelike plants in this artist's reconstruction are giant horsetails, most species of which are now extinct.

QUESTION Why are today's ferns and club mosses so small in comparison to their giant ancestors?

and the sperm were encased in drought-resistant pollen grains that were carried by the wind from plant to plant. When the pollen grains landed near an egg, they released sperm cells directly into living tissue, eliminating the need for a surface film of water. The fertilized egg remained on the parent plant, where it developed inside a seed, which provided protection and nutrients for the developing embryo.

The earliest seed-bearing plants appeared in the late Devonian period (375 million years ago) and produced their seeds along branches, without any specialized structures to hold them. By the middle of the Carboniferous period, however, a new kind of seed-bearing plant had arisen. These plants, called **conifers,** protected their developing seeds inside cones. Conifers, which as wind-pollinated plants did not depend on water for reproduction, flourished and spread during the Permian period (299 to 251 million years ago), when mountains rose, swamps drained, and the climate became much drier. The conifers' good fortune, however, was not shared by the tree ferns and giant club mosses, which, with their swimming sperm, largely went extinct.

Flowering Plants Enticed Animals to Carry Pollen

About 140 million years ago, during the Cretaceous period, the flowering plants appeared, having evolved from a group of conifer-like plants. Many flowering plants are pollinated by insects and other animals, and this mode of pollination seems to have conferred an evolutionary advantage. Flower pollination by animals can be far more efficient than pollination by wind. Wind-pollinated plants must produce an enormous amount of pollen because the vast majority of pollen grains fail to reach their target. Today, flowering plants dominate the land, except in cold northern regions, where conifers still prevail.

Some Animals Became Adapted to Life on Dry Land

After land plants evolved, providing potential food sources for other organisms, animals emerged from the sea. The earliest evidence of land animals comes from fossils that are about 430 million years old. The first animals to move onto land were **arthropods** (the group that today includes insects, spiders, scorpions, centipedes, and crabs). Why arthropods? The answer seems to be that they already possessed certain structures that, purely by chance, were suited to life on land. Foremost among these structures was an exoskeleton, such as the shell of a lobster or crab. Exoskeletons are both waterproof and strong enough to support a small animal against the force of gravity.

For millions of years, arthropods had the land and its plants to themselves, and for tens of millions of years more, they were the dominant land animals. Dragonflies with a wingspan of 28 inches (70 centimeters) flew among the Carboniferous tree ferns, while millipedes 6.5 feet (2 meters) long munched their way across the swampy forest floor. Eventually, however, the arthropods' splendid isolation came to an end.

Amphibians Evolved from Lobefin Fishes

About 400 million years ago, a group of Silurian fishes called the lobefins appeared, probably in fresh water. **Lobefins** had two important features that would later enable their descendants to colonize land: (1) stout, fleshy fins with which they crawled about on the bottoms of shallow, quiet waters, and (2) an outpouching of the digestive tract that could be filled with air, like a primitive lung. One group of lobefins inhabited very shallow ponds and streams, which shrank during droughts and often became oxygen poor. By taking air into

▲ FIGURE 17-9 **A fish that walks on land** Some modern fishes, such as this mudskipper, walk on land. As did the ancient lobefin fishes that gave rise to amphibians, mudskippers use their strong pectoral fins to move across dry areas in their swampy habitats.

QUESTION Does the mudskipper's ability to walk on land constitute evidence that lobefin fishes were the ancestors of amphibians?

their lungs, these lobefins could obtain oxygen anyway. Some began to use their fins to crawl from pond to pond in search of prey or water, as some fish do today (**Fig. 17-9**).

The benefits of feeding on land and moving from pool to pool favored the evolution of a group of animals that could stay out of water for longer periods and that could move about more effectively on land. With improvements in lungs and legs, **amphibians** evolved from lobefins, first appearing in the fossil record about 370 million years ago. To an amphibian, the Carboniferous swamp forests were a kind of paradise: no predators to speak of, abundant prey, and a warm, moist climate. As had the insects and millipedes, some amphibians evolved gigantic size, including salamanders more than 10 feet (3 meters) long.

Despite their success, the early amphibians were not fully adapted to life on land. Their lungs were simple sacs without very much surface area, so they had to obtain some of their oxygen through their skin. Therefore, their skin had to be kept moist, a requirement that restricted them to swampy habitats. Further, amphibian sperm and eggs could not survive in dry surroundings and had to be deposited in water. So, although amphibians could move about on land, they could not stray too far from the water's edge. Along with the tree ferns and club mosses, amphibians declined when the climate turned dry at the beginning of the Permian period about 299 million years ago.

Reptiles Evolved from Amphibians

As the conifers were evolving on the fringes of the swamp forests, a group of amphibians was also evolving adaptations to drier conditions. These amphibians ultimately gave rise to the **reptiles,** which had three major adaptations to life on land. First, reptiles evolved shelled, waterproof eggs that enclosed a supply of water for the developing embryo. Thus, reptiles could lay eggs on land and avoid the dangerous swamps full of fish and amphibian predators. Second, ancestral reptiles evolved scaly, water-resistant skin that reduced the loss of body water to the dry air. Finally, reptiles evolved improved lungs that were able to provide the entire oxygen supply of an active animal. As the climate dried during the Permian period, reptiles became the dominant land vertebrates, relegating amphibians to the swampy backwaters where most remain today.

A few tens of millions of years later, the climate became wetter again. This period saw the evolution of some very large reptiles, in particular, the dinosaurs (**Fig. 17-10**). The variety of dinosaur forms was enormous—large and

◀ FIGURE 17-10 **A reconstruction of a Cretaceous forest** By the Cretaceous period, flowering plants dominated terrestrial vegetation. Dinosaurs, such as the predatory pack of 6-foot-long *Velociraptors* shown here, were the preeminent land animals. Although small by dinosaur standards, *Velociraptors* were formidable predators with great running speed, sharp teeth, and deadly, sickle-like claws on their hind feet.

small, fleet-footed and ponderous, predators and plant eaters. Dinosaurs were among the most successful animals ever, if we consider persistence as a measure of success. They flourished for more than 100 million years, until about 65 million years ago, when the last dinosaurs went extinct. No one is certain why they died out, but the aftereffects of a gigantic meteorite's impact with Earth seem to have been the final blow (as discussed in Section 17.5).

Even during the age of dinosaurs, many reptiles remained quite small. One major difficulty faced by small reptiles is maintaining a high body temperature. A warm body is advantageous for an active animal, because warmer nerves and muscles work more efficiently. But a warm body loses heat to the environment unless the air is also warm. Heat loss is a big problem for small animals, which have a larger surface area per unit of volume than do larger animals. Many species of small reptiles have evolved slow metabolisms and cope with the heat-loss problem by confining activity to times when the air is sufficiently warm. One group of reptiles, however, followed a different evolutionary pathway. Members of this group, the birds, evolved insulation, in the form of feathers. (Birds were formerly placed in their own taxonomic group, separate from reptiles. For more information on why birds are now understood to be a type of reptile, see "In Greater Depth: Phylogenetic Trees" on pp. 342–343.)

In ancestral birds, feathers, which are modified scales, helped retain body heat. Consequently, these animals could be active in cool habitats and during the night, when their scaly relatives became sluggish. Later, some ancestral birds evolved longer, stronger feathers on their forelimbs, perhaps under selection for better ability to glide from trees or to jump after insect prey. Ultimately, feathers evolved into structures capable of supporting powered flight. Fully developed, flight-capable feathers are present in 150-million-year-old fossils, so the earlier insulating structures that eventually developed into flight feathers must have been present well before that time.

Reptiles Gave Rise to Mammals

Unlike the egg-laying reptiles, **mammals** evolved live birth and the ability to feed their young with secretions of the mammary (milk-producing) glands. Ancestral mammals also developed hair, which provided insulation. Because soft tissues like the uterus and mammary glands do not generally fossilize, we may never know when these structures first appeared or what their intermediate forms looked like. Hair, however, is sometimes preserved in fossils, albeit rarely. The oldest known hair belongs to a small aquatic mammal that was fossilized about 160 million years ago, so mammals have presumably had hair for at least that long.

The earliest fossil mammal unearthed thus far is almost 200 million years old. Early mammals, which can be distinguished by distinctive skeletal features, thus coexisted with the dinosaurs. They were mostly small creatures. The largest known mammal from the dinosaur era was about the size of a modern raccoon, but most early mammal species were far smaller than that. When the dinosaurs went extinct, however, mammals colonized the habitats left empty by the

extinctions. Mammals prospered, diversifying into the array of forms that we see today.

CHECK YOUR LEARNING

Can you describe the transitions and innovations associated with the origin and evolution of the major groups of land plants and the major groups of vertebrates? Can you describe the advantages gained by the first plants and animals to colonize land?

17.5 WHAT ROLE HAS EXTINCTION PLAYED IN THE HISTORY OF LIFE?

If there is a lesson in the great tale of life's history, it is that nothing lasts forever. The story of life can be read as a long series of evolutionary dynasties, with each new dominant group rising, ruling the land or the seas for a time and, inevitably, falling into decline and extinction. Dinosaurs are the most famous of these fallen dynasties, but the list of extinct groups known only from fossils is impressively long. Despite the inevitability of extinction, however, the overall trend has been for species to arise at a faster rate than they disappear, so the number of species on Earth has tended to increase over time.

Evolutionary History Has Been Marked by Periodic Mass Extinctions

Over much of life's history, the origin and disappearance of species has proceeded in a steady, relentless manner. This slow and steady turnover of species, however, has been interrupted by episodes of **mass extinction.** These mass extinctions are characterized by the relatively sudden disappearance of a wide variety of species over a large part of Earth. The most dramatic episode of all, which occurred 251 million years ago at the end of the Permian period, wiped out more than 90% of the world's species, destroying entire ecosystems. Life came perilously close to disappearing altogether.

Climate Change Contributed to Mass Extinctions

Mass extinctions have had a profound impact on the course of life's history, repeatedly redrawing the picture of life's diversity. What could have caused such dramatic changes in the fortunes of so many species? Many evolutionary biologists believe that changes in climate must have played an important role. When the climate changes, as it has done many times over the course of Earth's history, organisms that are adapted for survival in one climate may be unable to survive in a drastically different climate. In particular, at times when warm climates gave way to drier, colder climates with more variable temperatures, species may have gone extinct after failing to adapt to the harsh new conditions.

One cause of climate change is the shifting positions of continents. These movements are sometimes called *continental drift.* Continental drift is caused by **plate tectonics,** in which the Earth's surface, including the continents and the seafloor, is divided into plates that rest atop a viscous but fluid layer and move slowly about. As the plates wander, their positions may change in latitude (**Fig. 17-11**). For example, 340 million years ago, much of North America was located at or near the equator, an area characterized by consistently warm and wet tropical weather. But as time passed, plate tectonics carried the continent up into temperate and arctic regions. As a result, the once tropical climate was replaced by a regime of seasonal changes, cooler temperatures, and less rainfall. Plate tectonics continues today; the Atlantic Ocean, for example, widens by a few centimeters each year.

Catastrophic Events May Have Caused the Biggest Mass Extinctions

Geological data indicate that most mass extinction events coincided with periods of climatic change. To many scientists, however, the rapidity of mass extinctions suggests that the slow process of climate change could not, by itself, be responsible for such large-scale disappearances of species. Perhaps more sudden events also play a role. For example, catastrophic geological events, such as massive volcanic eruptions, could rapidly kill all organisms. Geologists have found evidence of past volcanic eruptions so huge that they make the 1980 Mount St. Helens explosion look like a firecracker by comparison.

The search for the causes of mass extinctions took a fascinating turn in the early 1980s when Luis and Walter Alvarez proposed that the extinction event of 65 million years ago, which wiped out the dinosaurs and many other species, was caused by the impact of a huge meteorite. The Alvarezes' idea was met with great skepticism when it was first introduced, but geological research since that time has generated a great deal of evidence that a massive impact did indeed occur 65 million years ago. In fact, researchers have identified the Chicxulub crater, a 100-mile-wide crater buried beneath the Yucatan Peninsula of Mexico, as the impact site of a giant meteorite—6 miles (10 kilometers) in diameter—that collided with Earth around the time that dinosaurs disappeared.

Could this immense meteorite strike have caused the mass extinction that coincided with it? No one knows for sure, but scientists suggest that such a massive impact would have thrown so much debris into the atmosphere that the entire planet would have been plunged into darkness for a period of years. With little light reaching the planet, temperatures would have dropped precipitously and the photosynthetic capture of energy (on which all life ultimately depends) would have declined drastically. The worldwide "impact winter" would have spelled doom for the dinosaurs and a host of other species.

▶ **FIGURE 17-11 Continental drift from plate tectonics**
The continents are passengers on plates moving on Earth's surface as a result of plate tectonics. **(a)** About 340 million years ago, much of what is now North America was positioned at the equator. **(b)** All the plates eventually fused together into one gigantic landmass, which geologists call Pangaea. **(c)** Gradually Pangaea broke up into Laurasia and Gondwanaland, which itself eventually broke up into West and East Gondwana. **(d)** Further plate motion eventually resulted in the current positions of the modern-day continents.

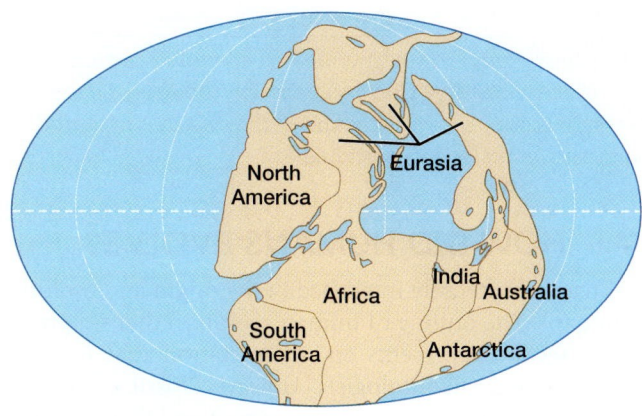

(a) 340 million years ago

(b) 225 million years ago

(c) 135 million years ago

(d) Present

CHECK YOUR LEARNING

Can you explain how extinction has affected the course of evolutionary history? Can you describe the likely causes of mass extinctions in general and of the one that occurred 65 million years ago in particular?

17.6 HOW DID HUMANS EVOLVE?

Scientists are intensely interested in the origin and evolution of humans. The outline of human evolution that we present in this section represents an interpretation that is widely shared among paleontologists. However, fossil evidence of human evolution is comparatively scarce and therefore open to a variety of interpretations. Thus, some paleontologists would disagree with some aspects of the scenario we present.

Humans Inherited Some Early Primate Adaptations for Life in Trees

Humans are members of a mammal group known as **primates,** which also includes lemurs, monkeys, and apes. The oldest primate fossils are 55 million years old, but because primate fossils are relatively rare compared with those of many other animals, the first primates probably arose considerably earlier but left no fossil record. Early primates probably fed on fruits and leaves, and were adapted for life in the trees. Many modern primates retain the tree-dwelling lifestyle of their ancestors (**Fig. 17-12**). The common heritage of humans and other primates is reflected in a set of physical characteristics that was present in the earliest primates and that persists in many modern primates, including humans.

? HAVE **YOU** EVER WONDERED …

If Extinct Species Can Be Revived by Cloning?

Scientists have cloned a number of animal species, including mice, dogs, cats, horses, and cows. Could the technology of cloning be used to bring back extinct species? In principle, yes, provided that perfectly preserved DNA of the extinct species is available. Such DNA could be transferred to an egg from a closely related, living species, and the egg implanted in a surrogate mother of that species.

For example, researchers have suggested that it might be possible to clone a woolly mammoth, using an elephant surrogate mother and DNA extracted from 20,000-year-old mammoths found frozen beneath the Siberian tundra. Most scientists, however, believe that any DNA recovered from a fossil mammoth would be far too degraded for use in cloning, and synthesizing an entire mammoth genome (its sequence is now almost fully known) is beyond the capabilities of current technology. The odds of success might be greater for another proposed project, which would use DNA from a preserved museum specimen to revive the Tasmanian tiger, an Australian mammal that has been extinct for only 70 years. If cloning recently extinct species proves to be possible, do you think it would be a good idea?

Binocular Vision Provided Early Primates with Accurate Depth Perception

One of the earliest primate adaptations seems to have been large, forward-facing eyes (see Fig. 17-12). Jumping from branch to branch is risky business unless an animal can

(a) Tarsier

(b) Lemur

(c) Macaque

◀ **FIGURE 17-12 Representative primates** The **(a)** tarsier, **(b)** lemur, and **(c)** lion-tail macaque monkey all have a relatively flat face, with forward-looking eyes providing binocular vision. All also have color vision and grasping hands. These features, retained from the earliest primates, are shared by humans.

accurately judge where the next branch is located. Accurate depth perception was made possible by binocular vision, provided by forward-facing eyes with overlapping fields of view. Another key adaptation was color vision. We cannot, of course, tell if a fossil animal had color vision, but since modern primates have excellent color vision, it seems reasonable to assume that earlier primates did as well. Many primates feed on fruit, and color vision helps to distinguish ripe fruit from green leaves.

Early Primates Had Grasping Hands

Early primates had long, grasping fingers that could wrap around and hold onto tree limbs. This adaptation to tree dwelling was the basis for later evolution of human hands that could perform both a *precision grip* (used for delicate maneuvers such as picking up small objects and sewing) and a *power grip* (used for powerful actions, such as thrusting with a spear or swinging a hammer).

A Large Brain Facilitated Hand–Eye Coordination and Complex Social Interactions

Primates have brains that are larger, relative to their body size, than the brains of almost all other animals. No one really knows for certain which environmental factors favored the evolution of large brains. It seems reasonable, however, that controlling and coordinating rapid locomotion through trees, the dexterous movements of the hands in manipulating objects, and binocular, color vision would be facilitated by increased brain power. Most primates also have fairly complex social systems, which require relatively high intelligence. If sociality promoted increased survival and reproduction, the benefits to individuals of successful social interaction might have favored the evolution of a larger brain.

The Oldest Hominin Fossils Are from Africa

On the basis of comparisons of DNA from modern chimps, gorillas, and humans, researchers estimate that the **hominin** line (humans and their fossil relatives) diverged from the ape lineage sometime between 5 million and 8 million years ago. The fossil record, however, suggests that the split must have occurred at the early end of that range. Paleontologists working in the African country of Chad discovered fossils of a hominin, *Sahelanthropus tchadensis*, that lived more than 6 million years ago (**Fig. 17-13**). *Sahelanthropus* is clearly a hominin because it shares several anatomical features with later members of the group. But because this oldest known member of our family also exhibits other features that are more characteristic of apes, it may represent a point on our family tree that is close to the split between apes and hominins.

In addition to *Sahelanthropus*, two other hominin species—*Orrorin tugenensis* and *Ardipithecus ramidus*—are known from African fossils appearing in rocks that are between 4 million and 6 million years old. Most of our knowledge of these species is based on fossil finds that include only small portions of skeletons. But one specimen, a fairly complete 4.4-million-year-old *Ardipithecus* skeleton, has revealed

▲ **FIGURE 17-13 The earliest hominin** This nearly complete skull of *Sahelanthropus tchadensis*, which is more than 6 million years old, is the oldest hominin fossil yet found.

some intriguing features of this early hominin. The structure of its legs, feet, hands, and pelvis suggest that *Ardipithecus* could walk upright, though it probably also climbed trees in its forest habitat. Its canine teeth were small, like those of modern humans and unlike the large, fang-like canines of today's apes.

A more extensive record of early hominin evolution begins about 4 million years ago. That date marks the beginning of the fossil record of the genus *Australopithecus* (**Fig. 17-14**), a group of African hominin species with brains larger than those of their forebears but still much smaller than those of modern humans.

Early Hominins Could Stand and Walk Upright

It is possible that even the earliest hominins walked upright. The discoverers of *Sahelanthropus* and *Orrorin* argue that the leg and foot bones of these earliest hominins have characteristics that indicate bipedal locomotion, but this conclusion will remain speculative until more complete skeletons of these species are found. However, the *Ardipithecus* skeleton shows that hominins capable of upright posture had arisen by 4.4 million years ago, and the earliest australopithecines (as the various species of *Australopithecus* are collectively known) had knee joints that allowed them to straighten their legs fully, permitting efficient bipedal (upright, two-legged) locomotion. Footprints almost 4 million years old, discovered in Tanzania by anthropologist Mary Leakey, show that the earliest australopithecines could, and at least sometimes did, walk upright.

The reasons for the evolution of bipedal locomotion among the early hominins remain poorly understood. Perhaps hominins that could stand upright gained an advantage in gathering or carrying food. Whatever its cause, the early evolution of upright posture was extremely important in

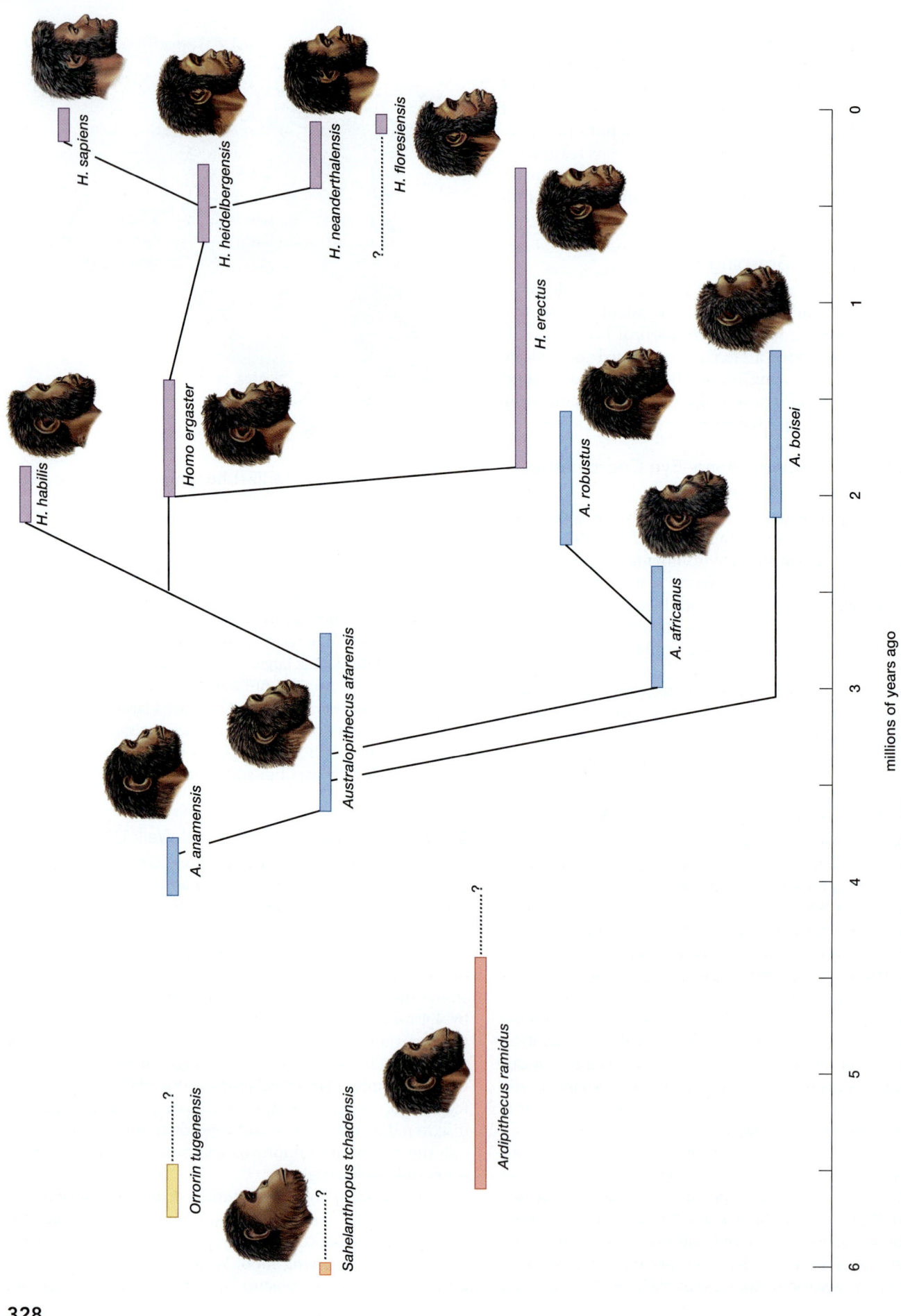

▲ **FIGURE 17-14 A possible evolutionary tree for humans** This hypothetical family tree shows facial reconstructions of representative specimens. Although many paleontologists consider this to be the most likely human family tree, there are several alternative interpretations of the known hominin fossils. Fossils of the earliest hominins are scarce and fragmentary, so the relationship of these species to later hominins remains unknown.

the evolutionary history of hominins, because it freed their hands from use in walking. Later hominins were thus able to carry weapons, manipulate tools, and eventually achieve the cultural revolutions produced by modern *Homo sapiens*.

Several Species of *Australopithecus* Emerged in Africa

The oldest australopithecine species, represented by fossilized teeth, skull fragments, and arm bones, was unearthed near an ancient lake bed in Kenya from sediments that were dated as being between 3.9 million and 4.1 million years old. The species was named *Australopithecus anamensis* by its discoverers (*anam* means "lake" in the local Ethiopian language). The second most ancient australopithecine, called *Australopithecus afarensis*, was discovered in the Afar region of Ethiopia. Fossil remains of this species as old as 3.9 million years have been unearthed. The *A. afarensis* line apparently gave rise to at least two distinct forms: small, omnivorous species such as *A. africanus* (which was similar to *A. afarensis* in size and eating habits), and larger, herbivorous species such as *A. robustus* and *A. boisei*. All of the australopithecine species had gone extinct by 1.2 million years ago. Before disappearing, however, one of these species gave rise to a new branch of the hominin family tree, the genus *Homo* (see Fig. 17-14).

The Genus *Homo* Diverged from the Australopithecines 2.5 Million Years Ago

Hominins that are sufficiently similar to modern humans to be placed in the genus *Homo* first appear in African fossils that are about 2.5 million years old. Among the earliest African *Homo* fossils are *H. habilis* (see Fig. 17-14), a species whose body and brain were larger than those of the australopithecines but that retained the apelike long arms and short legs of their australopithecine ancestors. In contrast, the skeletal anatomy of *H. ergaster*, a species whose fossils first appear 2 million years ago, has limb proportions more like those of modern humans. This species is believed by many paleoanthropologists (scientists who study human origins) to be on the evolutionary branch that led ultimately to our own species, *H. sapiens*. In this view, *H. ergaster* was the common ancestor of two distinct branches of hominins. The first branch led to *H. erectus*, which was the first hominin species to leave Africa. The second branch from *H. ergaster* ultimately led to *H. heidelbergensis*, some of which migrated to Europe and gave rise to the Neanderthals, *H. neanderthalensis*. Meanwhile, back in Africa, another branch split off from the *H. heidelbergensis* lineage. This branch became *H. sapiens*—modern humans.

The Evolution of *Homo* Was Accompanied by Advances in Tool Technology

Hominin evolution is closely tied to the development of tools, a hallmark of hominin behavior. The oldest tools discovered so far were found in 2.5-million-year-old East African rocks, concurrent with the early emergence of the genus *Homo*. Early *Homo*, whose molar teeth (the rearmost teeth in the jaw) were

much smaller than those of the genus's australopithecine ancestors, might first have used stone tools to break and crush tough foods that were hard to chew. Hominins constructed their earliest tools by striking one rock with another to chip off fragments. During the next several hundred thousand years, toolmaking techniques in Africa gradually became more advanced. By 1.7 million years ago, tools had become more sophisticated. Flakes were chipped symmetrically from both sides of a rock to form double-edged tools ranging from hand axes, used for cutting and chopping, to points, probably used on spears (**Figs. 17-15a, b**). *Homo ergaster* and other bearers

(a) *Homo habilis*

(b) *Homo ergaster*

(c) *Homo neanderthalensis*

▲ **FIGURE 17-15 Representative hominin tools (a)** *Homo habilis* produced only fairly crude chopping tools called hand axes, usually unchipped on one end to hold in the hand. **(b)** *Homo ergaster* manufactured much finer tools. The tools were typically sharp all the way around the stone; at least some of these blades were probably tied to spears rather than held in the hand. **(c)** Neanderthal tools were works of art, with extremely sharp edges made by flaking off tiny bits of stone. In comparing these weapons, note the progressive increase in the number of flakes taken off the blades and the corresponding decrease in flake size. Smaller, more numerous flakes produce a sharper blade and suggest more insight into toolmaking and finer control of hand movements.

of these weapons presumably ate meat, probably acquired from both hunting and scavenging for the remains of prey killed by other predators. Double-edged tools were carried to Europe at least 600,000 years ago by migrating populations of *H. heidelbergensis*, and the Neanderthal descendants of these emigrants took stone-tool construction to new heights of skill and delicacy (Fig. 17-15c).

Neanderthals Had Large Brains and Excellent Tools

Neanderthals first appeared in the European fossil record about 150,000 years ago. By about 70,000 years ago, they had spread throughout Europe and western Asia. By 30,000 years ago, however, the species was extinct.

Contrary to the popular image of a hulking, stoop-shouldered "caveman," Neanderthals were quite similar to modern humans in many ways. Although more heavily muscled, Neanderthals walked fully erect, were dexterous enough to manufacture finely crafted stone tools, and had brains that, on average, were slightly larger than those of modern humans. Many European Neanderthal fossils show heavy brow ridges and a broad, flat skull, but others, particularly from areas around the eastern shores of the Mediterranean Sea, are somewhat more physically similar to *H. sapiens*.

Despite the physical and technological similarities between Neanderthals and *H. sapiens*, there is no solid archaeological evidence that Neanderthals ever developed an advanced culture that included such characteristically human endeavors as art, music, and rituals. Some anthropologists argue that, because their skeletal anatomy shows that they were physically capable of making the sounds required for speech, Neanderthals might have acquired language. This interpretation of Neanderthal anatomy, however, is not unanimously accepted. In general, the available evidence of the Neanderthal way of life is limited and open to different interpretations, and anthropologists are engaged in a sometimes heated debate about how advanced Neanderthal culture became.

Neanderthals and *Homo sapiens* May Have Interbred

Though some anthropologists argue that Neanderthals were simply a variety of *H. sapiens*, most agree that Neanderthals were a separate species, *H. neanderthalensis*. Dramatic evidence in support of this hypothesis has come from researchers who have isolated DNA from Neanderthal skeletons. This endeavor recently culminated with the extraction and isolation of an entire Neanderthal genome from 38,000-year-old bones found in a cave in Croatia. After successfully isolating the Neanderthal DNA, the researchers determined the nucleotide sequence of much of the genome and compared it to whole-genome sequences from several modern humans. From these comparisons, the researchers deduced that the evolutionary branch leading to Neanderthals diverged from the ancestral human line between 270,000 and 440,000 years ago, thousands of years before the emergence of modern *H. sapiens*. However, the sequence comparison also revealed that up to 4% of a modern non–African human's DNA is similar to distinctively Neanderthal sequences. This finding suggests that our ancestors interbred with Neanderthals, probably about 60,000 years ago. Because modern Africans do not carry the Neanderthal sequences but all other people do, interbreeding with Neanderthals must have occurred after *H. sapiens* had left Africa but before modern humans spread around the world.

Scientists' ability to obtain DNA from ancient bones has also revealed a previously unknown hominin. The new hominin's existence was discovered by sequencing DNA extracted from a single finger bone found in deposits laid down between 30,000 and 50,00 years ago in Denisova Cave in Siberia. Analysis of the sequence showed that the bone came from a hominin that is evolutionarily distinct from both *H. neanderthalensis* and *H. sapiens*. Though the Denisovan hominin is so far known only by its DNA, one bone, and a tooth, paleoanthropologists suspect that it's only a matter of time until skeletons turn up. Thus, it appears that modern humans at one time shared the planet (or at least some parts of it) not only with Neanderthals, but also with Denisovans and the diminutive hominin species *H. floresiensis* (see this chapter's case study).

Modern Humans Emerged Less Than 200,000 Years Ago

The fossil record shows that anatomically modern humans appeared in Africa at least 160,000 years ago and possibly as long as 195,000 years ago. The location of these fossils suggests that *Homo sapiens* originated in Africa, but most of our knowledge about our own early history comes from European and Middle Eastern fossils of *H. sapiens*, collectively known as Cro-Magnons (after the district in France in which their remains were first discovered). Cro-Magnons appeared about 90,000 years ago. They had domed heads, smooth brows, and prominent chins (just like us). Their tools were precision instruments similar to the stone tools that were still used in a few cultures as recently as the 1960s.

Behaviorally, Cro-Magnons seem to have been similar to, but more sophisticated than, Neanderthals. Artifacts from 30,000-year-old Cro-Magnon archaeological sites include elegant bone flutes, graceful carved ivory sculptures, and evidence of elaborate burial ceremonies (Fig. 17-16). Perhaps the most remarkable accomplishment of Cro-Magnons is the magnificent art left in caves in places such as Altamira in Spain and Lascaux and Chauvet in France (Fig. 17-17). The oldest cave paintings so far found are more than 30,000 years old, and even the oldest ones make use of sophisticated artistic techniques. No one knows exactly why these paintings were made, but they attest to minds as capable as our own.

Cro-Magnons and Neanderthals Lived Side by Side

Cro-Magnons coexisted with Neanderthals in Europe and the Middle East for perhaps as many as 50,000 years before the Neanderthals disappeared. The genetic analyses described geographical areas for such a long time. The persistence in one area of two similar but distinct groups for tens of thousands of years seems inconsistent with both interbreeding and direct competition. Perhaps the competition between *H. neanderthalensis* and *H. sapiens* was indirect, so that the two species were able to coexist for a time in the same habitat, until the superior ability of *H. sapiens* to exploit the available resources slowly drove Neanderthals to extinction.

Several Waves of Hominins Emigrated from Africa

The human family tree is rooted in Africa, but hominins found their way out of Africa on numerous occasions. For example, *H. erectus* reached tropical Asia almost 2 million years ago and apparently thrived there, eventually spreading across Asia (**Fig. 17-18a**). Similarly, *H. heidelbergensis* made it to Europe at least 780,000 years ago. It is increasingly clear that the genus *Homo* made repeated long-distance emigrations. What is less clear is how all this wandering is related to the origin of modern *H. sapiens*. According to the "African replacement" hypothesis (the basis of the scenario outlined earlier), *H. sapiens* emerged in Africa and dispersed less than 150,000 years ago, spreading into the Near East, Europe, and Asia and replacing all other hominins (see Fig. 17-18a). But some paleoanthropologists believe that populations of *H. sapiens* evolved simultaneously in many regions from the already widespread populations of *H. erectus*. According to this "multiregional origin" hypothesis, continued migrations and interbreeding among *H. erectus* populations in different regions of the world maintained them as a single species as they gradually evolved into *H. sapiens* (**Fig. 17-18b**). Although an increasing number of studies of modern human DNA support the African replacement model of the origin of our species, both hypotheses are consistent with the fossil record. Therefore, the question remains unsettled.

▲ **FIGURE 17-16   Paleolithic burial** This 24,000-year-old grave shows evidence that Cro-Magnon people ritualistically buried their dead. The body was covered with a dye known as red ocher, then buried with a headdress made of snail shells and a flint tool in its hand.

earlier show that Cro-Magnons interbred with Neanderthals, so some researchers hypothesize that Neanderthals were essentially absorbed into the human genetic mainstream. Other scientists disagree, noting that the DNA evidence reveals only relatively limited interbreeding, and suggest that the later-arriving Cro-Magnons simply overran and displaced the less-well-adapted Neanderthals.

Neither hypothesis does a good job of explaining how the two kinds of hominins managed to occupy the same

CASE STUDY c o n t i n u e d

Little People, Big Story

What is the ancestry of *H. floresiensis*, the Hobbit of Flores Island in Indonesia? Some clues point toward an intriguing scenario. First, the only other evidence of early hominin habitation on Flores consists of stone tools found at an 840,000-year-old site, which suggests that the ancestors of *H. floresiensis* arrived on Flores at least that long ago. The only hominin known to have been present in Asia at that time is *H. erectus*, so researchers initially hypothesized that *H. floresiensis* descended from a population of *H. erectus* that became isolated on Flores and subsequently evolved very small size. However, recent analyses of *H. floresiensis* skeletons point to a different conclusion. Although the relatively modern structure of *H. floresiensis*'s small skull places the hobbit firmly in the genus *Homo*, the skeleton below the neck is more primitive, with a structure similar to that of the ancient African hominins of the genus *Australopithecus*. The *Australopithecus*-like features of Hobbit's below-the-neck skeleton suggest that the ancestor of *H. floresiensis* was a small australopithecine that left Africa and migrated to Asia more than a million years ago.

FIGURE 17-17   The sophistication of Cro-Magnon people Cave paintings by Cro-Magnons have been remarkably preserved by the relatively constant underground conditions of the Chauvet-Pont-d'Arc cave in France.

→ *Homo erectus*
→ *Homo sapiens*

(a) African replacement hypothesis

(b) Multiregional hypothesis

▲ FIGURE 17-18 **Competing hypotheses for the evolution of *Homo sapiens* (a)** The "African replacement" hypothesis suggests that *H. sapiens* evolved in Africa, then migrated throughout the Near East, Europe, and Asia, displacing the other hominin species that were present in those regions. **(b)** The "multiregional" hypothesis suggests that populations of *H. sapiens* evolved in many regions simultaneously from the already widespread populations of *H. erectus*.

QUESTION Paleontologists recently discovered fossil hominins with features characteristic of modern humans in 160,000-year-old sediments in Africa. Which hypothesis does this new evidence support?

The Evolutionary Origin of Large Brains May Be Related to Meat Consumption and Cooking

The main physical features that distinguish us from our closest relatives, the apes, are our upright posture and large, highly developed brains. As described earlier, upright posture arose very early in hominin evolution, and hominins walked upright for several million years before large-brained *Homo* species arose. What circumstances might have caused the evolution of increased brain size? Many explanations have been proposed, but little direct evidence is available; hypotheses about the evolutionary origins of large brains are necessarily speculative.

One proposed explanation for the origin of large brains suggests that they evolved in response to increasingly complex social interactions. In particular, fossil evidence suggests that, beginning about 2 million years ago, hominin social life began to include a new type of activity—the cooperative hunting of large game. The resulting access to significant amounts of meat must have fostered a need to develop methods for distributing this valuable, limited resource among group members. Some anthropologists hypothesize that the individuals best able to manage this social interaction would have been more successful at gaining a large share of meat and using their share advantageously. Perhaps this social management was best accomplished by individuals with larger, more powerful brains, and natural selection therefore favored such individuals. Observations of chimpanzee societies have shown that the distribution of group-hunted meat often involves intricate social interactions in which meat is used to form alliances, repay favors, gain access to sexual partners, placate rivals, and so on. Perhaps the mental skill required to plan, assess, and remember such interactions was the driving force behind the evolution of our large, clever brains.

Whatever the nature of the advantages that favored individuals with larger brains, such brains could not have evolved without some mechanism to provide the large amount of energy necessary to grow and maintain a large volume of brain tissue. Some researchers speculate that cooking was the breakthrough that freed up the required extra energy. Cooked food is more digestible than raw food and requires far less chewing, so cooked food provides more nutrients with less effort expended. Thus, cooking by early hominins might have removed the limit that had previously restricted brain size. However, larger brains first arose in *H. erectus* at least 2 million years ago, and the earliest direct archaeological evidence of controlled fires is only 790,000 years old. Proponents of the cooking hypothesis suggest that cooking actually did arise 2 million years ago, and the lack of evidence of cooking fires that old is simply due to the incompleteness of the hominin fossil record.

Sophisticated Culture Arose Relatively Recently

Even after the evolution of comparatively large brains in species such as *H. erectus*, more than a million years passed before the origin of modern humans and their extremely large brains. And even after the first appearance of modern *H. sapiens*, more than 100,000 years passed before the appearance of any archaeological evidence of the distinctively human characteristics that were made possible by a large brain: language, abstract thought, and advanced culture. The evolutionary origin of these human traits is another unresolved question, in part because direct evidence of the transition to advanced culture may never be found. Early humans capable of language and symbolic thought would not necessarily have created artifacts that indicated these capabilities. We can uncover some clues by studying our ape relatives, which possess less-complex versions of many human behaviors and mental processes. Their behavior might resemble that of ancestral hominins. Nonetheless, the late, seemingly rapid origin of advanced human culture remains a puzzle.

Biological Evolution Continues in Humans

Until recently, most evolutionary biologists agreed that the evolution of human bodies by natural selection had slowed or halted after we began to live in advanced societies and were thereby shielded from the daily struggle for existence that dominates the lives of other organisms. Today, however, our ever-growing ability to rapidly sequence DNA has made it possible for researchers to analyze sequences of a large and growing number of human genomes, and these analyses have led to a surprising conclusion: People have evolved rapidly since the advent of music, art, language, and the other hallmarks of advanced culture, and they continue to evolve today. Many of our genes show the telltale signs of evolution by natural selection in recent millennia. In many cases, the exact functions of these genes remain unknown, but researchers have determined the functions of some recent evolutionary changes. For example, the alleles required to digest milk have arisen and become fixed in some populations within the past 7,000 years. Similarly, researchers have identified 20 genes that have evolved rapidly and recently in Tibetans; all of the genes are involved in the physiological response to high-altitude, low-oxygen conditions. The alleles that became fixed in Tibetan populations all evolved within the 3,000 years that have passed since a founding population colonized the Tibetan plateau, one of the highest-altitude inhabited places on Earth.

Culture Also Evolves

Human evolution in recent millennia has also included a great deal of cultural evolution, the evolution of information and behaviors that are transmitted from generation to generation by learning. Our recent evolutionary success, for example, was engendered not so much by new physical adaptations as by a series of cultural and technological revolutions. The first such revolution was the development of tools, which began with the early hominins. Tools increased the efficiency with which food and shelter could be acquired and thus increased the number of individuals that could survive within a given ecosystem. About 10,000 years ago, human culture underwent a second revolution as people discovered how to grow crops and domesticate animals. This agricultural revolution dramatically increased the amount of food that could be extracted from the environment, and the human population surged, increasing from about 5 million at the dawn of agriculture to around 750 million by 1750. The subsequent Industrial Revolution gave rise to the modern economy and its attendant improvements in public health. Longer lives and lower infant mortality led to truly explosive population growth, and today Earth's population is 7 billion and still growing.

Human cultural evolution and the accompanying increases in human population have had profound effects on the continuing biological evolution of other life-forms. Our agile hands and minds have transformed much of Earth's terrestrial and aquatic habitats. Humans have become the most powerful agent of natural selection. In the words of the late evolutionary biologist Stephen Jay Gould, "We have become, by the power of a glorious evolutionary accident called intelligence, the stewards of life's continuity on Earth. We did not ask for this role, but we cannot renounce it. We may not be suited for it, but here we are."

CHECK YOUR LEARNING

Can you describe the evolutionary history of humans and the factors that may have fostered humans' distinctive adaptations? Are you able to name and describe some characteristics of the hominin species that played key roles in humans' evolutionary history? Can you describe the key features of the most recent phase of human evolution?

CASE STUDY revisited

Little People, Big Story

The discovery of the hobbit-sized *Homo floresiensis* was exciting to many people in part because it suggested that our species may have more close relatives than we previously suspected and that at least some of them lived tantalizingly close to the present. Plus, the idea of a society of tiny humans seems to have an inherent appeal. But the discovery also raises a host of fascinating evolutionary questions.

For example, if *H. floresiensis* is in fact descended from a small australopithecine that left Africa 1 or 2 million years ago, then the current consensus narrative of the migrations associated with human evolution would be upended. That narrative, based on fossil evidence, holds that *H. erectus*, with its large body and relatively large brain, was the earliest hominin to leave Africa and the only one to reach Asia until the arrival of *H. sapiens* more than a million years later. But if *H. floresiensis* has an australopithecine ancestor, then hominin emigration from Africa began very early (perhaps before the origin of *Homo*) and must have included more instances of departure than are currently revealed in the fossil record.

If, alternatively, *H. floresiensis* descended from *H. erectus*, how did Hobbits come to be so small? Large animal species that are isolated on islands sometimes evolve smaller body size. For example, the now-extinct elephants of Flores were only about 4 feet tall. Biologists suggest that the absence of large predators on most islands eliminates much of the benefit of large size, conferring an advantage on smaller individuals that require less food. This kind of dynamic could have fostered the evolution of small stature in *H. floresiensis*.

Regardless of the identity of *H. floresiensis*'s ancestor, how did it get to the island of Flores? Unlike some islands,

Flores was never connected to the mainland. Archaeologists generally agree that hominins did not construct boats until 60,000 years ago at the earliest. So how did Hobbit's ancestor end up on Flores at least 800,000 years before the invention of boats? Perhaps they drifted there on clumps of floating vegetation, but we may never ever know for sure.

Consider This

BioEthics *Homo floresiensis* was found on an island. If you were searching for evidence of other undiscovered recent hominin species, would you concentrate your search on islands? Why or why not?

CHAPTER REVIEW

Summary of Key Concepts

17.1 How Did Life Begin?

Before life arose, energy from lightning, ultraviolet light, and heat formed organic molecules from water and the components of primordial Earth's atmosphere—methane, ammonia, hydrogen, and water vapor. The organic molecules formed probably included nucleic acids, amino acids, short proteins, and lipids. By chance, some molecules of RNA may have had enzymatic properties, catalyzing the assembly of copies of themselves from nucleotides in Earth's waters. These may have been the forerunners of life. Protein-lipid vesicles enclosing these ribozymes may have formed protocells.

17.2 What Were the Earliest Organisms Like?

The oldest fossils, about 3.5 billion years old, are of prokaryotic cells that fed by absorbing organic molecules that had been synthesized in the environment. Because there was no free oxygen in the atmosphere, their energy metabolism must have been anaerobic. As the cells multiplied, they depleted the organic molecules that had been formed by prebiotic synthesis. Some cells developed the ability to synthesize their own food molecules by using simple inorganic molecules and the energy of sunlight. These earliest photosynthetic cells were probably ancestors of today's cyanobacteria.

Photosynthesis releases oxygen as a by-product, and by about 2.3 billion years ago significant amounts of free oxygen were accumulating in the atmosphere. Aerobic metabolism, which generates more cellular energy than does anaerobic metabolism, probably arose about this time.

Eukaryotic cells had evolved by about 1.7 billion years ago. The first eukaryotic cells probably arose as symbiotic associations between predatory prokaryotic cells and other bacteria. Mitochondria may have evolved from aerobic bacteria engulfed by predatory cells. Chloroplasts may have evolved from photosynthetic cyanobacteria by a similar process.

17.3 What Were the Earliest Multicellular Organisms Like?

Multicellular organisms evolved from eukaryotic cells and first appeared in the seas about 1.2 billion years ago. Multicellularity offers several advantages, including greater size. In plants, increased size offered some protection from predation. Specialization of cells allowed plants to anchor themselves in the nutrient-rich, well-lit waters near the shore. For animals, multicellularity allowed more efficient predation and more effective escape from predators. These in turn provided environmental pressures for faster locomotion, improved senses, and greater intelligence. Diverse animals forms appear in the fossil record beginning about 600 million years ago; fish were the predominant marine animals by about 400 million years ago.

17.4 How Did Life Invade the Land?

The first land organisms were probably algae. The first multicellular land plants appeared about 475 million years ago. Life on land required special adaptations for support of the body, reproduction, and the acquisition, distribution, and retention of water, but the land also offered abundant sunlight and protection from aquatic herbivores. Soon after land plants evolved, arthropods invaded the land.

The earliest land vertebrates evolved from lobefin fishes, which had leglike fins and a primitive lung. A group of lobefins evolved into the amphibians about 370 million years ago. Reptiles evolved from amphibians, with several further adaptations for life on land. One reptile group, the birds, evolved feathers that provided insulation and facilitated flight. Mammals, whose bodies are insulated by hair, descended from a reptile group.

17.5 What Role Has Extinction Played in the History of Life?

The history of life has been characterized by constant turnover of species as some go extinct and are replaced by new ones. Mass extinctions, in which large numbers of species disappear within a relatively short time, have occurred periodically. Mass extinctions were probably caused by some combination of climate change and catastrophic events, such as volcanic eruptions and meteorite impacts.

17.6 How Did Humans Evolve?

One group of mammals evolved into the tree-dwelling primates. Some primates descended from the trees, and these were the ancestors of apes and humans. The oldest known hominin fossils are between 6 million and 7 million years old and were found in Africa. The australopithecines arose in Africa about 4 million years ago. These hominins walked erect, had larger brains than did their forebears, and fashioned primitive tools. One group of australopithecines gave rise to a line of hominins in the genus *Homo*. *Homo* arose in Africa, but populations of several *Homo* species migrated from Africa and spread to other geographic areas. In the last of these migrations, *Homo sapiens*, characterized by a large brain and advanced tool technology, dispersed from Africa to Asia and Europe.

Key Terms

amphibian *323*
arthropod *322*
conifer *322*
endosymbiont
    hypothesis *318*
eukaryote *318*
exoskeleton *320*
hominin *327*
lobefin *322*

mammal *324*
mass extinction *324*
plate tectonics *324*
primate *326*
prokaryote *315*
protocell *314*
reptile *323*
ribozyme *313*
spontaneous generation *311*

Learning Outcomes

In this chapter, you have learned to . . .

LO1 Describe a scenario for the origin of life and summarize evidence that supports the scenario.

LO2 Describe the major evolutionary events and innovations that occurred during the period in which all organisms were single celled.

LO3 Describe the earliest multicellular organisms and the factors associated with the origin of multicellularity.

LO4 Describe the adaptations associated with the evolution of increased animal diversity in the oceans.

LO5 Describe the adaptations associated with the transition to terrestrial life.

LO6 Describe the transitions and innovations associated with the origin and evolution of the major groups of land plants.

LO7 Describe the transitions and innovations associated with the origin and evolution of the major groups of vertebrates.

LO8 Describe the evolutionary history of humans and the factors that may have fostered humans' distinctive adaptations.

LO9 Explain the possible causes and evolutionary impact of mass extinctions.

Thinking Through the Concepts

Fill-in-the-Blank

1. Because there was no oxygen in the earliest atmosphere, the first cells must have derived energy by _____ metabolism of organic molecules. Oxygen was introduced into the atmosphere when some microbes developed the ability to _____ and released oxygen as a by-product. Oxygen was _____ to many of the earliest cells, but some cells evolved the ability to use oxygen in _____ respiration, which provided far more _____. **LO2**

2. The molecule _____ became a candidate for the first self-replicating information-carrying molecule when Tom Cech and Sidney Altman discovered that some of these molecules can act as _____, which they called _____. **LO1**

3. Complex cells that contain a nucleus and other organelles are called _____ cells. A compelling explanation for the origin of these complex cells is the _____ hypothesis. One observation that supports this hypothesis is that mitochondria have their own _____ and _____. **LO2**

4. The sperm of early land plants had to reach the egg by _____, limiting them to _____ environments. An important adaptation of plants to dry land was the evolution of _____, which enclosed sperm in a drought-resistant coat. **LO5**

5. Early plants that protected their seeds within cones are called _____. These relied on _____ to carry their pollen. Later, some plants evolved _____, which attracted animals, particularly _____ that carried their pollen. Animal pollination is much more _____ than wind pollination. **LO6**

6. The first animals to live on land were _____ because their external skeletons, also called _____, supported the animals' weight, while protecting their bodies from _____. **LO5**

7. Amphibians gave rise to _____, which had three important adaptations to life on dry land: shelled, waterproof _____; scaly, water-resistant _____; and more efficient _____. **LO5 LO7**

Review Questions

1. What is the evidence that life might have originated from nonliving matter on early Earth? What kind of evidence would you like to see before you would accept this hypothesis? **LO1**

2. Explain the endosymbiont hypothesis for the origin of chloroplasts and mitochondria. **LO2**

3. Name two advantages of multicellularity for plants and two for animals. **LO3**

4. What advantages and disadvantages would terrestrial existence have had for the first plants to invade the land? For the first land animals? **LO5**

5. Outline the major adaptations that emerged during the evolution of vertebrates from fish to amphibians to reptiles to birds and mammals. Explain how these adaptations increased the fitness of the various groups for life on land. **LO7**

6. Outline the evolution of humans from early primates. Include in your discussion such features as binocular vision, grasping hands, bipedal locomotion, toolmaking, and brain expansion. **LO8**

Applying the Concepts

1. What is cultural evolution? Is cultural evolution more or less rapid than biological evolution? Why?

2. Do you think that studying our ancestors can shed light on the behavior of modern humans? Why or why not?

3. A biologist might answer the age-old question "What is life?" by saying, "The ability to self-replicate." Do you agree with this definition? If so, why? If not, how would you define life in biological terms?

4. **BioEthics** Extinctions have occurred throughout the history of life on Earth. Why should we care if humans are causing a mass extinction event now?

5. The "African replacement" and "multiregional origin" hypotheses of the evolution of *Homo sapiens* make contrasting predictions about the extent and nature of genetic divergence among human races. One predicts that races are old and highly diverged genetically; the other predicts that races are young and little diverged genetically. What data would help you determine which hypothesis is closer to the truth?

6. In biological terms, what do you think was the most significant event in the history of life? Explain your answer.

Answers to Figure Caption questions and Fill-in-the-Blank questions can be found in the Answers section at the back of the book.

MB ®*Go to MasteringBiology for practice quizzes, activities, eText, videos, current events, and more.*

CHAPTER 18

Systematics: Seeking Order Amid Diversity

CASE STUDY
Origin of a Killer

ONE OF THE WORLD'S most frightening diseases is also one of its most mysterious. Acquired immune deficiency syndrome (AIDS) appeared seemingly out of nowhere, and when it was first recognized in the early 1980s, no one knew what caused it or where it came from. Scientists raced to solve the mystery and, within a few years, had identified the infectious agent that causes AIDS: human immunodeficiency virus (HIV). Once HIV had been identified, researchers turned their attention to the question of its origin.

Finding the source of HIV required an evolutionary approach. To ask "Where did HIV come from?" is really to ask "What kind of virus was the ancestor of HIV?" To answer this question, researchers began by identifying the closest relatives of HIV; when a biologist concludes that two viruses are closely related, it means that they share a recent common ancestor from which both evolved. Thus, comparing HIV with its closest relatives allowed researchers to infer the characteristics of their common ancestor.

The researchers who explored the ancestry of HIV discovered that its closest relatives are found not among other viruses that infect humans, but among those that infect monkeys and apes. In fact, the latest research on HIV's evolutionary history has concluded that the closest relative of HIV-1 (the type of HIV that is most responsible for the worldwide AIDS epidemic) is a virus strain that infects a particular chimp subspecies that inhabits a limited range in West Africa. Therefore, the ancestor of the virus that we now know as HIV-1 did not evolve from a preexisting human virus. Instead, a chimpanzee virus must have acquired mutations that allowed it to infect humans and cause a deadly disease.

Biologists studying the evolutionary history of type 1 human immunodeficiency virus (HIV-1) discovered that the virus, which causes AIDS, probably originated in chimpanzees.

AT A GLANCE

18.1 How Are Organisms Named and Classified?

18.2 What Are the Domains of Life?
18.3 Why Do Classifications Change?

18.4 How Many Species Exist?

LEARNING GOALS

LG1 Describe how organisms are named and classified.

LG2 Describe the traits that systematists use for reconstructing phylogeny and explain why some similarities can be used for such reconstruction and some cannot.

LG3 Describe life's three domains and explain the evolutionary relationships among them.

LG4 Explain why classifications of organisms can change.

LG5 Compare and contrast the biological species concept and the phylogenetic species concept and describe the benefits and limitations of each.

LG6 Report current estimates of the number of species on Earth and explain why it is difficult to determine this number with certainty.

18.1 HOW ARE ORGANISMS NAMED AND CLASSIFIED?

To study and discuss organisms, biologists must name them. The branch of biology that is concerned with naming and classifying organisms is known as **taxonomy.** (A *taxon*—plural, *taxa*—is a named species or a named group of species). The basis of modern taxonomy was established by the Swedish naturalist Carl von Linné (1707–1778), who called himself Carolus Linnaeus, a Latinized version of his name. One of Linnaeus's most enduring achievements was the introduction of the two-part scientific name.

Each Species Has a Unique, Two-Part Name

The **scientific name** of an organism is a two-part Latin name that designates its genus and species. A **genus** is a group that includes a number of very closely related species; each

species within a genus includes populations of organisms that can potentially interbreed under natural conditions. For example, the genus *Sialia* (bluebirds) includes three species: the eastern bluebird (*Sialia sialis*), the western bluebird (*Sialia mexicana*), and the mountain bluebird (*Sialia currucoides*) (**Fig. 18-1**). Although the three species are similar, bluebirds normally breed only with members of their own species.

In a scientific name, the genus name is presented first, followed by the species name. By convention, scientific names are always underlined or *italicized*. The first letter of the genus name is always capitalized, and the first letter of the species name is always lowercase. The species name is never used alone but is always paired with its genus name.

Each two-part scientific name is unique, so referring to an organism by its scientific name rules out any chance of ambiguity or confusion. For example, the bird *Gavia immer* is commonly known in North America as the common loon, in

  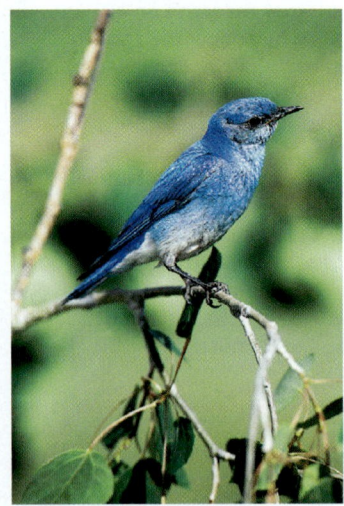

(a) Eastern bluebird　　**(b) Western bluebird**　　**(c) Mountain bluebird**

▲ FIGURE 18-1 **Three species of bluebird** Despite their obvious similarity, these three species of bluebird—(a) the eastern bluebird (*Sialia sialis*), (b) the western bluebird (*Sialia mexicana*), and (c) the mountain bluebird (*Sialia currucoides*)—evolve independently because they do not interbreed.

Great Britain as the northern diver, and by still other names in non–English-speaking countries. But the Latin scientific name *Gavia immer* is recognized by biologists worldwide, overcoming language barriers and allowing precise communication.

Modern Classification Emphasizes Patterns of Evolutionary Descent

In addition to naming species, biologists also classify them. Prior to the 1859 publication of Darwin's *On the Origin of Species*, classification served mainly to facilitate the study and discussion of organisms, much as a library's online catalog facilitates our ability to find a book. But after Darwin demonstrated that all organisms are linked by common ancestry, biologists began to recognize that classification ought to reflect and describe the pattern of evolutionary relatedness among organisms. Today, the process of classification focuses almost exclusively on reconstructing **phylogeny,** or evolutionary history. The science of reconstructing phylogeny is known as **systematics.** Systematists communicate their hypotheses about phylogeny by constructing evolutionary trees (see Fig. 16-10).

Systematists Identify Features That Reveal Evolutionary Relationships

As systematists seek to reconstruct the tree of life, they must do so without much direct knowledge of evolutionary history. Because systematists can't see into the past, they must infer it as best they can, on the basis of similarities among living organisms. Not all similarities are useful for constructing phylogenetic trees, however. Some observed similarities stem from convergent evolution (see pp. 267–269) in organisms that are not closely related, and such similarities are not useful for inferring evolutionary history. Instead, systematists use similarities that exist because two kinds of organisms both inherited a characteristic from a common ancestor. Therefore, the scientists who devise classifications must distinguish informative similarities caused by common ancestry from uninformative similarities that result from convergent evolution. In the search for informative similarities, biologists look at many kinds of characteristics.

Historically, the most important and useful distinguishing characteristics have been anatomical. Systematists look carefully at similarities in both external body structure and internal structures, such as skeletons and muscles. For example, homologous structures such as the finger bones of dolphins, bats, seals, and humans provide evidence of a common ancestor (see Fig. 14-8). To detect relationships among more closely related species, biologists may use microscopes to discern finer details, such as the external structure of the pollen grains of a flowering plant (**Fig. 18-2**).

Modern Systematics Relies on Molecular Similarities to Reconstruct Phylogeny

Recent advances in the techniques of molecular genetics have revolutionized studies of evolutionary relationships by allowing scientists to determine genetic similarities among

Pollen grains

▲ **FIGURE 18-2 Microscopic structures may be used to classify organisms** The shape and surface features of pollen grains are among the finely detailed structures that can be useful in classification. Such structures can reveal similarities and differences between species that are not apparent in larger and more easily visible structures.

organisms. Today's systematists rely mainly on the nucleotide sequences of DNA (that is, organisms' genotypes) to investigate relatedness among different types of organisms.

The logic underlying such molecular systematics is straightforward. It is based on the observation that when a single species divides into two species, the gene pool of each resulting species begins to accumulate mutations. The particular mutations present in each species' gene pool, however, will differ because the species are now evolving independently, with no gene flow between them. As time passes, more and more genetic differences accumulate. So, a systematist who has obtained DNA sequences from representatives of both species can compare the two species' nucleotide sequences at any given location in the genome. Fewer differences indicate more closely related organisms (species with a relatively recent common ancestor). The process by which systematists use genetic (and anatomical) similarities to reconstruct evolutionary history is discussed in "In Greater Depth: Phylogenetic Trees" on pp. 342–343.

CASE STUDY continued

Origin of a Killer

Analysis of nucleotide sequences was absolutely required to construct the phylogeny of viruses that revealed the ancestor of HIV. The closely related viruses included in the phylogeny are all but indistinguishable on the basis of appearance and structure; the differences between them are revealed only by their nucleotide sequences. Thus, scientific sleuthing about the origin of HIV would have been impossible before the modern era of routine DNA sequencing.

▲ **FIGURE 18-4 Clades form a nested hierarchy** Any group that includes all the descendants of a common ancestor is a clade. Some of the clades represented on this evolutionary tree are shaded in different colors. Note that smaller clades nest within larger clades.

▲ **FIGURE 18-3 Human and chimp chromosomes are similar** Chromosomes from different species can be compared by means of banding patterns that are revealed by staining. The comparison illustrated here, between human chromosomes (left member of each pair; H) and chimpanzee chromosomes (C), reveals that the two species are genetically very similar. In fact, the entire genomes of both species have been sequenced and are about 96% identical. The numbering system shown is that used for human chromosomes; note that human chromosome 2 corresponds to a combination of two chimp chromosomes. Yunis, J. J. et al. 1980. Science 208:1145–1148.

In some cases, similarity of DNA sequences will be reflected in the structure of chromosomes. For example, both the DNA sequences and the chromosomes of chimpanzees and humans are extremely similar, showing that these two species shared a common ancestor in the not too distant past (**Fig. 18-3**). (Some findings about genetic variation in humans are described in "Links to Everyday Life: Small World.")

Systematists Name Groups of Related Species

Although systematists communicate their findings about phylogeny mainly by presenting evolutionary trees, they also name groups of species. In keeping with their emphasis on reconstructing evolutionary history, systematists give formal names only to groups that include all the organisms descended from a common ancestor. Such groups are known as **clades.** If you examine an evolutionary tree, you will see that clades can be arranged in a hierarchy, with smaller clades nested within larger ones (**Fig. 18-4**).

When systematists name a clade, the name itself does not convey much information about the clade. For example, the name does not reveal much about the clade's size or breadth. Is the clade a relatively large, broadly inclusive one, such as the one that includes all mammals? Or is it a smaller, narrower clade, perhaps one that includes only the three species of zebra? The clade's size and breadth would be obvious if we could view its evolutionary tree, but if we don't have access to the tree, we will need clues other than the clade's name in order to understand its scope.

One possible way to signal the relative size and inclusiveness of named clades is to place them into categories called *taxonomic ranks.* This approach has a long history; Linnaeus placed each species into a series of ranked categories on the basis of its resemblance to other species. The Linnaean classification system eventually came to include eight major ranks: *domain, kingdom, phylum, class, order, family, genus,* and *species.* These ranks form a nested hierarchy in which each level includes all of the other levels below it; each domain contains a number of kingdoms; each kingdom contains a number of phyla; each phylum includes a number of classes; each class includes a number of orders; and so on. As we move down the hierarchy, smaller and smaller groups are included. Thus, if you knew that a certain named clade was a

» LINKS TO EVERYDAY LIFE

Small World

In light of humankind's intense curiosity about the origin of our species, it is not surprising that systematists have devoted a great deal of attention to uncovering the evolutionary history of *Homo sapiens*. Though much of this inquiry has focused on revealing the evolutionary connections between modern humans and the species to which we are most closely related, the techniques and methods of systematics have also been used to assess the evolutionary relationships among different populations within our species. Biologists have compared DNA sequences from human populations in many different parts of the world. Different investigators have compared different portions of the human genome, and the scientists collaborating on the ambitious 1000 Genomes Project have made substantial progress toward their goal of sequencing the whole genomes of 2,500 people from around the world. As a result, a large amount of data has been gathered, and some interesting findings have emerged.

First, genetic divergence among human populations is very low compared to that of other animal species. For example, the range of genetic differences among all of Earth's humans is only one-tenth the size of the differences among the deer mice of North America (and many other species have even more genetic variability than deer mice). Clearly, all humans are genetically very similar, and the differences among different human populations are tiny.

It is also increasingly apparent that most of the genetic variability that does exist among humans can be found in African populations. The range of genetic differences found within sub-Saharan African populations is greater than the range between African populations and any non-African population. For many genes, all known variants are found in Africa, and no non-African population contains any distinctive variants; rather, non-African populations contain subsets of the African set of variants. This finding strongly suggests that *H. sapiens* originated in Africa, and that we have not lived anywhere else long enough to evolve many differences from our African ancestors.

---

phylum, and a second named clade was a genus within the phylum, you would have some understanding of relative sizes and scopes of the two clades.

The system of taxonomic ranks, however, comes with its own problems. For example, traditional taxonomic ranks can lead to false perceptions of equivalence between different groups of equivalent rank. For example, Felidae (cats) and Orchidae (orchids) are both families. You might reasonably think that, since the two groups have the same rank, they are in some sense evolutionarily equivalent. This conclusion, however, would be mistaken. For example, the common ancestor of the Felidae lived about 30 million years ago, whereas the common ancestor of the Orchidae lived more than 100 million years ago. Moreover, Felidae contains about 35 species, but Orchidae contains more than 20,000 species. The two groups have very different sizes and evolutionary histories and, to many systematists, a ranking system that implies that they are equivalent is not very useful.

Another problem with taxonomic ranks arises from new discoveries about evolutionary history that can require seemingly absurd assignments of taxonomic rank. Consider, for example, the discovery that modern birds are a surviving group of dinosaurs, and therefore part of the dinosaur clade. Traditionally, Dinosauria (the dinosaurs) is an order within the Class Reptilia (the reptiles). But if dinosaurs form an order, and birds are a subset of dinosaurs, then Aves (all birds) must be a family (the next rank down from order). But how can that be, when classical taxonomists have already classified birds into 29 orders containing about 235 families? A traditional bird family comprises a group no broader than, say, the 120 species of wood-warblers, but the logic of ranking seems to require that we now place all 10,000 bird species, with representatives as evolutionarily distant as flamingos and hummingbirds, into a single family. In the case of birds and many other organisms, it seems impossible to reconcile traditional, rank-based classifications with emerging knowledge of evolutionary history.

Use of Taxonomic Ranks Is Declining

Because taxonomic ranks do not accurately communicate information about evolutionary history, today's systematists have de-emphasized the Linnaean classification system. Many systematists do not assign taxonomic ranks to the clades they name and instead concentrate on using data to construct accurate evolutionary trees, rather than on subjective evaluations of whether a given clade should be called a kingdom, a phylum, a class, an order, or a family. As a result, use of Linnaean taxonomic ranks is declining.

In the chapters of this text that describe the diversity of life (Chapters 19–24), our use of taxonomic ranks will vary according to the practices of the biologists who study the different organisms we discuss. In most chapters, we will follow the emerging convention of avoiding ranks, instead using the term "taxonomic group" (as a synonym for clade) to describe a collection of related species, relying on context and explanation to make clear the relative scope and breadth of a named clade. In some chapters, however, we will make selective use of a few Linnaean ranks. For example, we will follow the tradition of using "kingdom" to refer to the three clades that contain, respectively, all animals, all plants, and all fungi. Similarly, in the two chapters about animals, we will designate certain clades as phyla, in keeping with tradition in animal systematics. And we will refer to the three broadest, most inclusive of life's clades as **domains.**

IN GREATER DEPTH Phylogenetic Trees

Systematists strive to develop a system of classification that reflects the phylogeny of organisms. Thus, the main job of systematists is to reconstruct phylogeny. Reconstructing the evolutionary history of all of Earth's organisms is, of course, a huge task, so each systematist typically chooses to work on some particular portion of the history.

The result of a phylogenetic reconstruction is usually represented by a diagram. These diagrams can take a number of different forms, but all of them show the sequence of branching events in which ancestral species split to give rise to descendant species. For this reason, diagrams of phylogeny are generally tree-like (though the tree may be oriented in any of a variety of directions; in this text we usually orient the tree horizontally, with the branch tips on the right.)

These trees can present the phylogeny of any specified set of taxa. Thus, phylogenetic trees can show evolutionary history at different scales. Systematists might reconstruct, for example, a tree of 10 species in a particular genus of clams, or a tree of 25 clades of animals, or a tree of the three domains of life.

After selecting the taxa to include, a systematist is ready to begin building a tree. Most systematists use the *cladistic* approach to reconstruct phylogenetic trees. Under the cladistic approach,

relationships among taxa are revealed by the occurrence of similarities known as *synapomorphies*. A synapomorphy is a trait that is similar in two or more taxa because these taxa inherited a "derived" version of the trait that had changed from its original state in a common ancestor. For example, the presence of feathers is a synapomorphy that links all living birds and distinguishes them from other vertebrates. The common ancestor of birds and crocodiles (their closest living relatives) had scales, which were converted into feathers—the derived state—in the lineage leading to birds but not in the lineage leading to crocodiles. The formation of synapomorphies is illustrated in Figure E18-1.

In the imaginary scenario illustrated in Figure E18-1, we can easily identify synapomorphies because we know the ancestral state of the trait and the subsequent changes that took place. In real life, however, systematists would not have direct knowledge of the ancestor, which lived in the distant past and whose identity is unknown. Without this direct knowledge, a systematist observing a similarity between two taxa is faced with a challenge. Is the observed similarity a synapomorphy, or does it have some other cause, such as convergent evolution? The cladistic approach provides methods for identifying synapomorphies, but

it remains possible to mistakenly use a shared trait that is not in fact a synapomorphy. To guard against such errors, systematists use numerous traits to build a tree, thereby minimizing the influence of any single trait.

In the last phase of the tree-building process, the systematist compares different possible trees. For example, three taxa can be arranged in three different branching patterns (Fig. E18-2). Each branching pattern represents a different hypothesis about the evolutionary history of taxa A, B, and C. Which hypothesis is most likely to represent the true history of the three taxa? The one in which the taxa on adjacent branches share synapomorphies. For example, imagine that a systematist identified a number of synapomorphies that are shared by taxa A and B but are not found in C, but has found no synapomorphies that link taxa B and C or taxa A and C. In this case, tree 1 in Figure E18-2 is the best-supported hypothesis.

With large numbers of taxa, the number of possible trees grows dramatically. Similarly, a large number of traits also complicates the problem of identifying the tree best supported by the data. Fortunately, however, systematists have developed sophisticated computer programs to help cope with these complications.

Under the cladistic approach, phylogenetic trees play a key role in

① This short DNA sequence is present in an ancestral species

CGT AGA TAC

② After the ancestral population splits into two descendant species, evolutionary change in one species replaces the G in the second position with C; in the other descendant species, T replaces A in the sixth position

③ Each descendant species splits into two species but there are no subsequent sequence changes in any of the descendant species

CGT AGT TAC

CCT AGA TAC

A
CGT AGT TAC

B
CGT AGT TAC

Species A and Species B share a derived trait (T in the sixth position)

C
CCT AGA TAC

D
CCT AGA TAC

Species C and Species D share a derived trait (C in the second position)

◀ FIGURE E18-1 Related taxa are linked by shared derived traits (synapomorphies) A derived trait is one that has been modified from the ancestral version of the trait. When two or more taxa share a derived trait, the shared trait is said to be a synapomorphy. The hypothetical scenario illustrated here shows how synapomorphies arise.

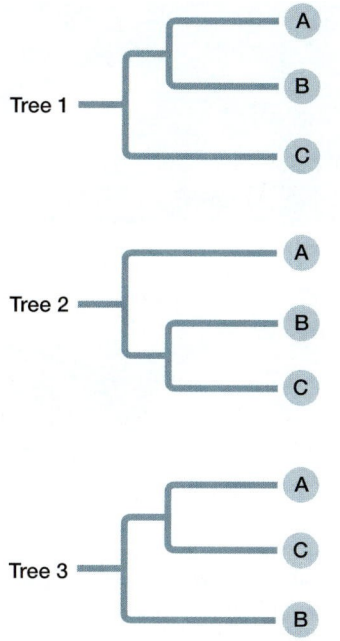

▲ FIGURE E18-2 **The three possible trees for three taxa**

classification. Each named group should contain only organisms that are more closely related to one another than to any organisms outside the group. So, for example, the members of the clade Canidae (which includes dogs, wolves, foxes, and coyotes) are more closely related to each other than to any member of any other clade. Another way to state this principle is to say that each designated group should contain *all* of the living descendants of a common ancestor (**Fig. E18-3a**). In the terminology of cladistic systematics, such groups are said to be *monophyletic*. (Note that "monophyletic group" is therefore synonymous with "clade.")

Some names, especially names that predate the cladistic approach, designate groups that contain some, but not all, of the descendants of a common ancestor. Such groups are *paraphyletic*. For example, the taxon historically known as the reptiles—snakes, lizards, turtles, and crocodilians—is paraphyletic. To see why,

examine the tree in **Figure E18-3b**. Find the branch that represents the common ancestor of crocodilians, snakes, lizards, and turtles (it is at the base of the tree). Then examine the tree again and make a list of all of the descendants of that common ancestor. Your list, if you performed this mental exercise correctly, includes the birds. That is, birds are part of the monophyletic group that includes all living descendants of the common ancestor that gave rise to crocodilians, snakes, lizards, and turtles. Therefore, the reptiles (Reptilia) constitute a monophyletic clade only if birds are included in the group. If we omit the birds, the taxon Reptilia is paraphyletic and, according to cladistic principles, is not a valid group name. Nonetheless, you will probably continue to encounter the word "reptiles" used in its older, technically incorrect sense, because so many people are accustomed to using it that way.

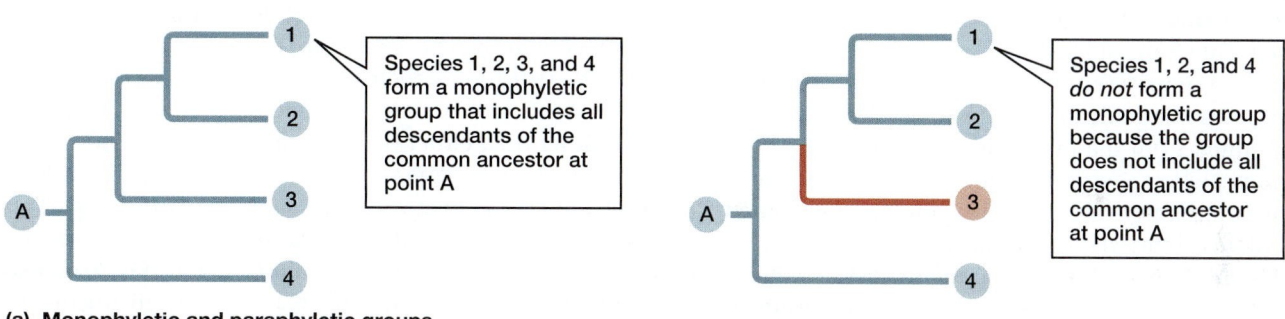

(a) **Monophyletic and paraphyletic groups**

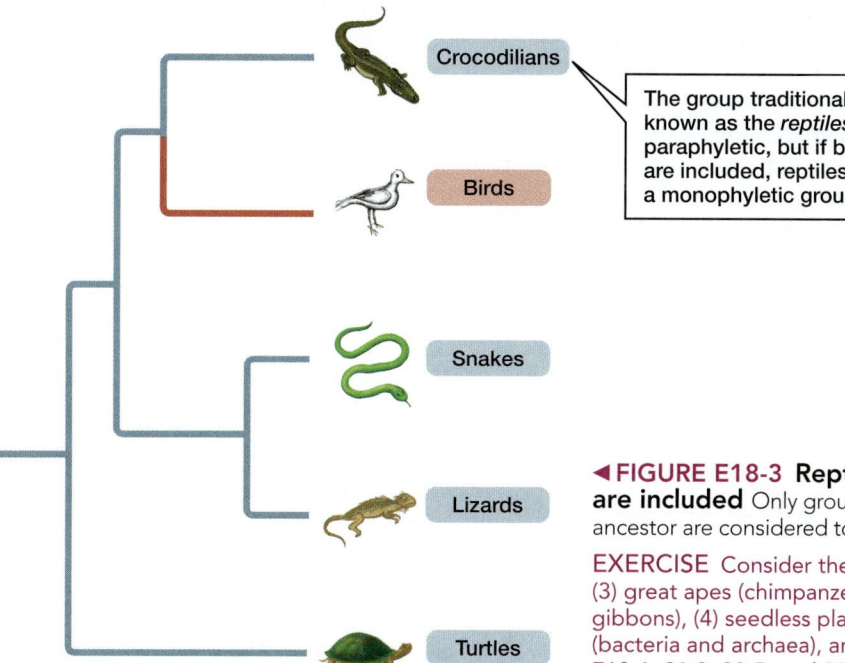

Crocodilians

Birds

The group traditionally known as the *reptiles* is paraphyletic, but if birds are included, reptiles form a monophyletic group

Snakes

Lizards

Turtles

(b) **Reptiles**

◀ FIGURE E18-3 **Reptiles are a monophyletic group only if birds are included** Only groups that contain all of the descendants of a common ancestor are considered to be monophyletic groups.

EXERCISE Consider the following list of groups: (1) protists, (2) fungi, (3) great apes (chimpanzees, pygmy chimpanzees, gorillas, orangutans, and gibbons), (4) seedless plants (ferns, mosses, and liverworts), (5) prokaryotes (bacteria and archaea), and (6) animals. Using Figures 18-6, 18-7, E18-1, 21-3, 22-5, and 23-1 for reference, identify the monophyletic groups on the list.

CHECK YOUR LEARNING

Can you explain why scientific names are necessary? Can you describe the type of similarities that systematists use to reconstruct phylogeny and provide some examples of traits that systematists might examine to find such similarities? Are you able to describe the system of Linnaean taxonomic ranks and discuss the pros and cons of using it?

18.2 WHAT ARE THE DOMAINS OF LIFE?

If we picture the common ancestor of all living things as the trunk at the very base of the tree of life, we might ask: Which clades arose from the earliest branching of the trunk? Each of the earliest branches must have given rise to a huge clade of descendant species. These early-branching clades will be the largest ones that systematists can distinguish.

By the 1970s, most systematists had concluded from the available evidence that early splits in the tree of life divided all species into five kingdoms. The five-kingdom system placed all prokaryotic organisms into a single kingdom and divided the eukaryotes into four kingdoms. Among the eukaryotes, the five-kingdom system recognized three kingdoms of multicellular organisms (plants, animals, and fungi) and placed all of the remaining, mostly single-celled eukaryotes in a single kingdom.

As new data accumulated and understanding of phylogeny grew, however, scientific assessment of life's fundamental categories was gradually revised. A key element of this revision stemmed from the pioneering work of microbiologist Carl Woese, who showed that biologists had overlooked a key event in the early history of life, one that demanded a new and more evolutionarily accurate classification of life.

Woese and other biologists interested in the evolutionary history of microorganisms studied the biochemistry of prokaryotic organisms. The researchers, by studying nucleotide sequences of the RNA that is found in organisms' ribosomes, discovered that prokaryotes fall into two large groups, each with its own distinctive version of ribosomal RNA. Woese dubbed these two groups the Bacteria and the Archaea (Fig. 18-5). The very large number of differences between the ribosomal RNA sequences of Bacteria and Archaea indicate that their common ancestor lived a long, long time ago.

Despite superficial similarities in their appearance under the microscope, Bacteria and Archaea differ at the

(a) A bacterium

(b) An archaean

▲ **FIGURE 18-5 Two domains of prokaryotic organisms** Although similar in appearance, **(a)** *Pseudomonas aeruginosa* and **(b)** *Methanococcus jannaschii* are less closely related than a mushroom and an elephant. *Pseudomonas* is in the domain Bacteria, and *Methanococcus* is in Archaea.

most fundamental molecular level; for example, they differ dramatically in the chemical composition of their cell walls and in the structure of their RNA polymerase molecules.

Bacteria and Archaea are no more closely related to one another than either one is to any eukaryote. The tree of life split into three parts very early in the history of life, long before the appearance of plants, animals, and fungi. This early split is reflected in a modern classification scheme that divides life into three domains: **Bacteria, Archaea,** and **Eukarya** (Fig. 18-6). Eukarya includes all organisms with eukaryotic cells, including plants, fungi, animals, and an array of clades of mostly singled-celled organisms collectively known as *protists*. Figure 18-7 shows the evolutionary relationships among some members of the domain Eukarya.

CHECK YOUR LEARNING

Can you name and briefly describe the three domains of life? Can you explain how scientists discovered that prokaryotes fall into two domains?

BACTERIA   ARCHAEA   EUKARYA
                                    animals   fungi
                                              plants
                                              protists

◄ **FIGURE 18-6 The tree of life** The three domains of life represent the three main "branches" on the tree of life.

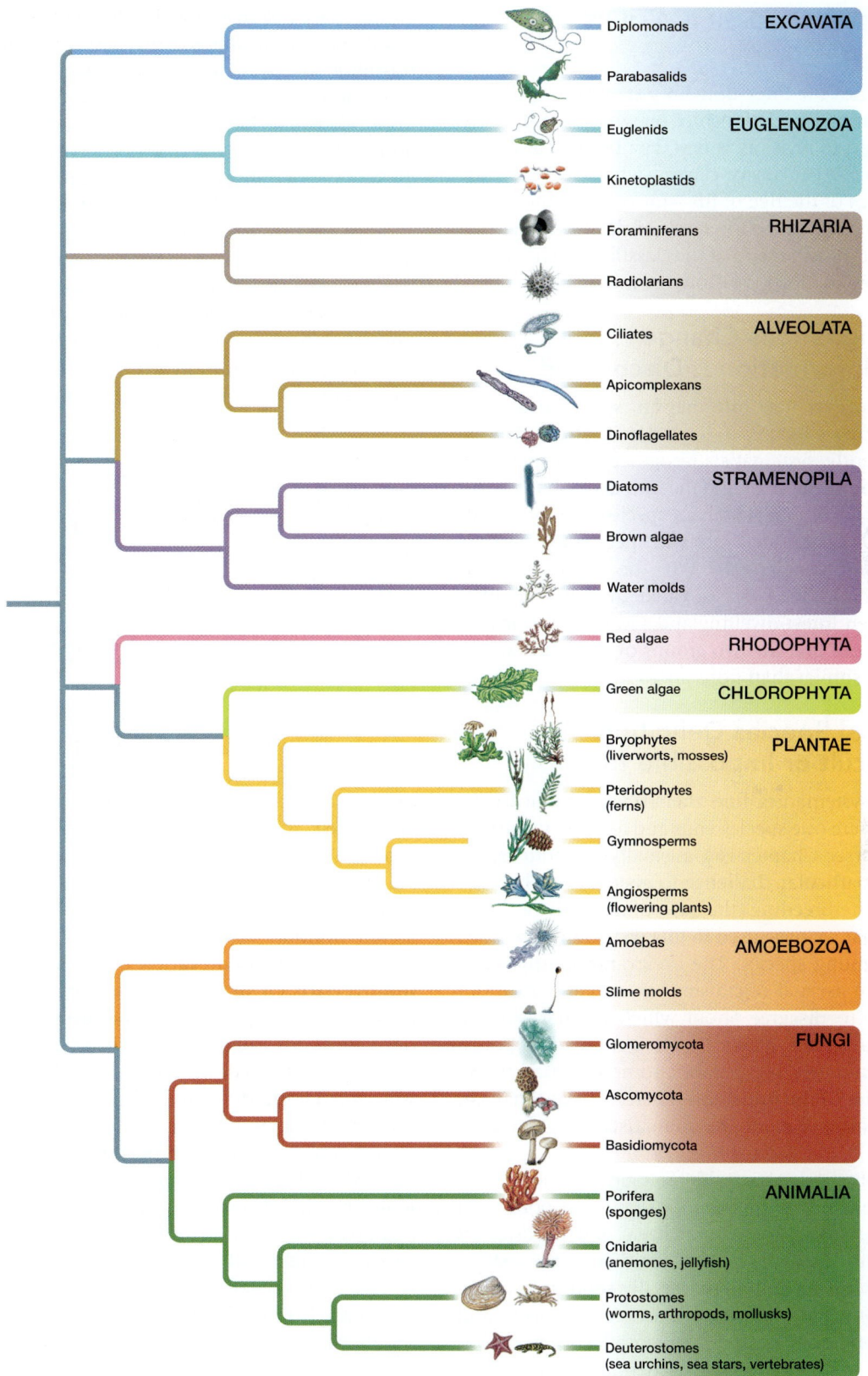

▲ FIGURE 18-7 An evolutionary tree of eukaryotes Some of the major evolutionary lineages within the domain Eukarya are shown. The term "protist" refers to the many eukaryotes that are not plants, animals, or fungi.

18.3 WHY DO CLASSIFICATIONS CHANGE?

As the emergence of the three-domain system shows, the hypotheses of evolutionary relationships on which classification is based are subject to revision as new data become available. Even the largest, most inclusive clades—which represent the earliest branchings of the tree of life—must sometimes be rearranged. Such changes at the top levels of classification occur only rarely, but at the other end of the classification hierarchy, among species designations, revisions are more frequent.

Species Designations Change When New Information Is Discovered

As researchers uncover new information, systematists regularly propose changes in species-level classifications. For example, until recently, systematists recognized two species of elephant, the African elephant and the Indian elephant. Now, however, we recognize three elephant species; the former African elephant is now divided into two species, the savanna elephant and the forest elephant. Why the change? Genetic analysis of elephants in Africa revealed that there is little gene flow between forest-dwelling and savanna-dwelling elephants. It turns out that the two groups of elephants are no more genetically similar than lions are to tigers.

The Biological Species Definition Can Be Difficult or Impossible to Apply

In some cases, systematists find themselves unable to say with certainty where one species ends and another begins. As discussed earlier (see Chapter 16), asexually reproducing organisms pose a particular challenge to systematists, because the criterion of interbreeding (the basis of the biological species definition that we have used in this text) cannot be used to distinguish among species. The irrelevance of this criterion in studies of asexual organisms leaves plenty of room for investigators to disagree about which asexual populations constitute a species, especially when comparing groups with similar phenotypes. For instance, some systematists recognize 200 species of the British blackberry (a plant that can reproduce asexually), but other systematists recognize only 20 species.

The difficulty of applying the biological species definition to asexual organisms applies to a significant portion of Earth's organisms. Most bacteria, archaea, and protists, for example, reproduce asexually most of the time. Some systematists argue that we need a more universally applicable definition of species, one that won't exclude asexual organisms and that doesn't depend on the criterion of reproductive isolation.

A number of alternative species definitions have been proposed, but none has been sufficiently compelling to displace completely the biological species definition. One alternative definition, however, has been gaining adherents in recent years. The *phylogenetic species concept* defines a species as the smallest diagnosable group that contains all the descendants of a single common ancestor. In other words, if we draw an evolutionary tree that describes the pattern of ancestry among a collection of organisms, each distinctive branch on the tree constitutes a separate species, regardless of whether the individuals represented by that branch can interbreed with individuals from other branches. As you might suspect, rigorous application of the phylogenetic species concept would vastly increase the number of different species recognized by systematists.

Proponents and opponents of the phylogenetic species concept are currently engaged in a vigorous debate about the merits of this definition of species. Perhaps one day the phylogenetic species concept will replace the biological species concept as the "textbook definition" of species. In the meantime, classifications will continue to be debated and revised as systematists learn more about evolutionary relationships, particularly with the application of techniques used in molecular genetics.

CHECK YOUR LEARNING

Can you explain why phylogenetic classifications sometimes change? Can you describe the limitations of the biological species concept and explain how it differs from the phylogenetic species concept?

18.4 HOW MANY SPECIES EXIST?

The challenge of reconstructing the evolutionary history of Earth's species is complicated by the fact that most species remain undiscovered and undescribed. Scientists do not know even within an order of magnitude how many species share our world. Each year, between 7,000 and 10,000 new species are named, most of them insects, many from tropical rain forests. The total number of named species is currently about 1.5 million. However, many scientists believe that 7 million to 10 million species may exist, and estimates range as high as 100 million.

The number and variety of Earth's species constitute its **biodiversity.** Of all the species that have been identified thus far, about 5% are prokaryotes and protists. An additional 20% or so are plants and fungi, and the rest are animals. This distribution has little to do with the actual diversity of these organisms and a lot to do with the size of the organisms, how easy they are to classify, how accessible they are, and the number of scientists studying them. Historically, systematists have chiefly focused on large or conspicuous organisms in temperate regions, but biodiversity is greatest among small, inconspicuous organisms in the Tropics. In addition to the overlooked species on land and in shallow waters, an entire "continent" of species lies largely unexplored on the deep-sea floor. From the relatively limited samples available, scientists estimate that hundreds of thousands of unknown species may reside there.

Although about 5,000 species of prokaryotes have been described and named, most prokaryotic diversity remains undiscovered. Consider a study by Norwegian scientists, who analyzed DNA to count the number of different bacterial species present in a small sample of forest soil. To distinguish among species, the researchers arbitrarily defined bacterial DNA as coming from separate species if it differed by at least 30% from that of any other bacterial DNA in the sample. Using this criterion, they reported more than 4,000 species of bacteria in their soil sample and an equal number of species in a sample of shallow marine sediment.

Our ignorance of the full extent of life's diversity adds a new dimension to the tragedy of the destruction of tropical rain forests. Although these forests cover only about 6% of Earth's land area, they are believed to be home to two-thirds of the world's existing species, most of which have never been studied or named. Because these forests are being destroyed so rapidly, Earth is losing many species that we will never even know existed! Consider this: In 1990, a new species of primate, the black-faced lion tamarin, was discovered in a small patch of dense rain forest on an island just off the east coast of Brazil (Fig. 18-8). Had the patch of forest been cut before this squirrel-sized monkey was discovered, its existence would have remained undocumented. At current rates of deforestation, most of the tropical rain forests, with their undescribed wealth of life, will be gone within the next century.

CHECK YOUR LEARNING

Can you explain why the number of described species is much lower than the actual number? Can you explain why it is difficult to accurately estimate how many species are on Earth?

▲ FIGURE 18-8 **The black-faced lion tamarin** Researchers estimate that no more than 400 individuals remain in the wild; captive breeding may be the black-faced lion tamarin's only hope for survival.

CASE STUDY revisited

Origin of a Killer

What evidence has persuaded evolutionary biologists that HIV originated in apes and monkeys? To understand the evolutionary thinking behind this conclusion, examine the evolutionary tree shown in Figure 18-9. This tree illustrates the phylogeny of HIV and its close relatives, the simian immunodeficiency viruses (SIVs), as revealed by a comparison of RNA sequences among different viruses.

Notice the positions on the tree of the four human viruses (two strains of HIV-1 and two of HIV-2; a strain is a genetically distinct subgroup of a particular type of virus). The branch leading to strain 1 of HIV-1 is directly adjacent to the branch leading to strain 1 of chimpanzee SIV. The adjacent branches indicate that strain 1 of HIV-1 is more closely related to a chimpanzee virus than to strain 2 of HIV-1. Similarly, strain 1 of HIV-2 is more closely related to pig-tailed macaque SIV than to strain 2 of HIV-2. Both HIV-1 and HIV-2 are more closely related to ape or monkey viruses than to one another.

The only way for the evolutionary history shown in the tree to have emerged is if viruses jumped between host species. If HIV had evolved strictly within human hosts, the human

viruses would be each other's closest relatives. Because the human viruses do not cluster together on the phylogenetic tree, we can infer that cross-species infection occurred, probably on multiple occasions. The most likely means of transmission is human consumption of monkeys (HIV-2)

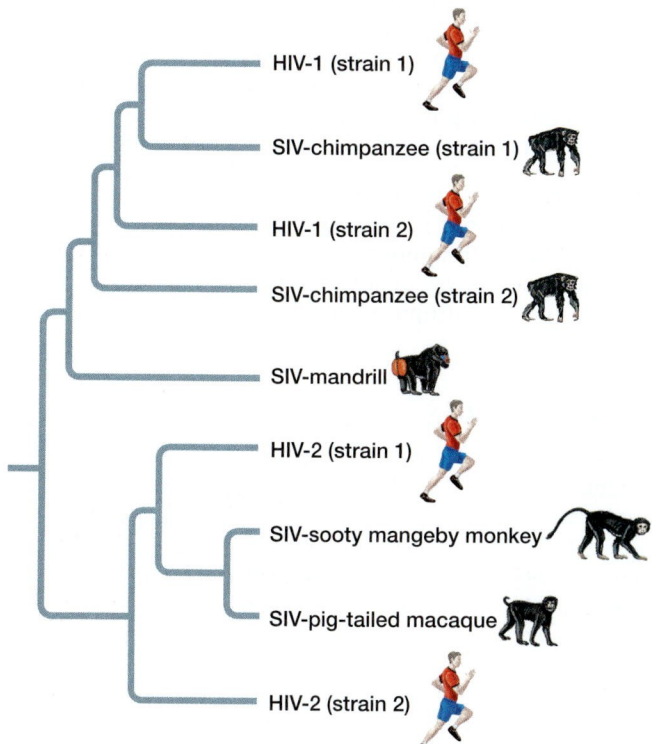

▶ FIGURE 18-9 **Evolutionary analysis helps reveal the origin of HIV** In this phylogeny of some immunodeficiency viruses, the viruses with human hosts do not cluster together. This lack of congruence between the evolutionary histories of the viruses and their host species suggests that the viruses must have jumped between host species.

and chimpanzees (HIV-1). Recently, a strain of SIV that is especially closely related to HIV-1 was found in members of a population of chimps that inhabit forests in the southeastern corner of the Central African country of Cameroon. It is likely that viruses from this chimp population made the chimp-to-human jump that started the HIV epidemic.

Consider This

BioEthics Can understanding the evolutionary origin of HIV help researchers devise better ways to treat AIDS and control its spread? More generally, how can evolutionary thinking help advance medical research?

CHAPTER REVIEW

Summary of Key Concepts

18.1 How Are Organisms Named and Classified?

The scientific name of an organism is composed of its genus name and species name. Systematists use anatomical and molecular similarities among organisms to reconstruct the evolutionary relationships among species, and depict the results of their reconstructions in tree diagrams. On the basis of these evolutionary trees, systematists name clades (groups that include the species descended from a common ancestor). Clades nest within larger clades to form a nested hierarchy of categories. In Linnaean classification, the different clades in a hierarchy are assigned taxonomic ranks. The eight major ranks, in order of decreasing inclusiveness, are domain, kingdom, phylum, class, order, family, genus, and species.

18.2 What Are the Domains of Life?

The three domains of life, each representing one of three main branches of the tree of life, are Bacteria, Archaea, and Eukarya. Plants, fungi, and animals are among the clades within the domain Eukarya.

18.3 Why Do Classifications Change?

Classifications are subject to revision as new information is discovered. Species boundaries may be hard to define, particularly in the case of asexually reproducing species. However, systematics is essential for precise communication and contributes to our understanding of the evolutionary history of life.

18.4 How Many Species Exist?

Although only about 1.5 million species have been named, estimates of the total number of species range up to 100 million. New species are being identified at the rate of 7,000 to 10,000 annually, mostly in tropical rain forests.

Key Terms

Archaea	344	genus	338
Bacteria	344	phylogeny	339
biodiversity	346	scientific name	338
clade	340	species	338
domain	341	systematics	339
Eukarya	344	taxonomy	338

Learning Outcomes

In this chapter, you have learned to . . .

LO1 Describe how organisms are named and classified.

LO2 Describe the traits that systematists use for reconstructing phylogeny and explain why some similarities can be used for such reconstruction and some cannot.

LO3 Describe life's three domains and explain the evolutionary relationships among them.

LO4 Explain why classifications of organisms can change.

LO5 Compare and contrast the biological species concept and the phylogenetic species concept and describe the benefits and limitations of each.

LO6 Report current estimates of the number of species on Earth and explain why it is difficult to determine this number with certainty.

Thinking Through the Concepts

Fill-in-the-Blank

1. The science of naming and classifying organisms is called _____. The related science of reconstructing and depicting evolutionary history is called _____. A group consisting of all organisms descended from a particular common ancestor is a(n) _____. **LO1**

2. A scientific name consists of a(n) _____ name followed by a(n) _____ name. Both parts of a scientific name are in _____ (a language). The first letter of the first word in a scientific name is always _____, and both parts of the name are printed in _____ letters. **LO1**

3. In Linnaean classification, the eight major taxonomic ranks, in descending order of inclusiveness, are _____, _____, _____, _____, _____, _____, and _____. The three domains of life are _____, _____, and _____. **LO1 LO3**

4. Systematists determine the evolutionary relationships among species mainly on the basis of similarities in _____ and _____. **LO2**

5. Species designations may change when researchers _____. The biological species definition is difficult to apply to organisms that _____. An alternative species definition, known as the _____, instead designates species on the basis of _____. **LO4 LO5**

6. The number of named species is about_____, but the actual number of species on Earth is estimated to be between _____ and _____, with estimates as high as _____. **LO6**

Review Questions

1. What contributions did Linnaeus and Darwin make to modern taxonomy? **LO1**

2. What features would you study to determine whether a dolphin is more closely related to a fish or to a bear? **LO2**

3. What techniques might you use to determine whether the extinct cave bear is more closely related to a grizzly bear or to a black bear? **LO2**

4. Only a small fraction of the total number of species on Earth has been scientifically described. Why? **LO6**

5. In England, "daddy longlegs" refers to a long-legged fly, but the same name refers to a spider-like animal in the United States. How do scientists attempt to avoid such confusion? **LO1**

6. Why are species designations of asexually reproducing organisms more likely to differ among different systematists than are the species designations of sexually reproducing organisms? **LO4**

Applying the Concepts

1. There are many areas of disagreement about the classification of organisms. For example, there is no consensus about whether the red wolf is a distinct species or about how many kingdoms are within the domain Bacteria. What difference does it make whether biologists consider the red wolf a species or into which kingdom a bacterial species falls? As Shakespeare put it, "What's in a name?"

2. BioEthics The pressures created by human population growth and economic expansion place storehouses of biological diversity such as the Tropics in peril. The seriousness of the situation is clear when we consider that probably only 1 out of every 20 tropical species is known to science at present. What arguments can you make for preserving biological diversity in poor and developing countries, such as those in many areas of the Tropics? Does such preservation require that these countries sacrifice economic development? Suggest some solutions to the conflict between the growing demand for resources and the importance of conserving biodiversity.

3. During major floods, only the topmost branches of submerged trees may be visible above the water. If you were asked to sketch the branches below the surface of the water solely on the basis of the positions of the exposed tips, you would be attempting a reconstruction somewhat similar to the "family tree" by which taxonomists link various organisms according to their common ancestors (analogous to branching points). What sources of error do both exercises share? What advantages do modern taxonomists have?

4. The Florida panther, found only in the Florida Everglades, is currently classified as an endangered species, protecting it from human activities that could lead to its extinction. It has long been considered a subspecies of cougar (mountain lion), but recent mitochondrial DNA studies have shown that the Florida panther may actually be a hybrid between North American and South American cougars. Should the Florida panther be protected by the Endangered Species Act?

Answers to Figure Caption questions and Fill-in-the-Blank questions can be found in the Answers section at the back of the book.

MB *Go to MasteringBiology for practice quizzes, activities, eText, videos, current events, and more.*

CHAPTER 19

The Diversity of Prokaryotes and Viruses

Hamburgers should be cooked thoroughly to eliminate dangerous bacteria.

CASE STUDY

Unwelcome Dinner Guests

IN 2010, ANDREW LEKAS, then a student at the University of Michigan, ate a burrito at a favorite restaurant near campus. A few hours later, he began to feel sick, with symptoms that included vomiting, diarrhea, headache, and weakness. The illness ultimately became severe enough to require a hospital stay and more than a week in bed. What sickened Lekas? The lettuce in his burrito was contaminated.

As bad as Lekas's experience was, it could have been much worse. Other victims of foodborne contamination have suffered more serious consequences. For example, Stephanie Smith, a former dance instructor from Minnesota, is paralyzed from the waist down as a result of the severe illness (hemolytic uremic syndrome) she developed in 2007 after eating a tainted hamburger.

Regrettably, the ill effects of consuming contaminated food are all too common. The Centers for Disease Control and Prevention estimates that U.S. residents experience an astonishing 76 million cases of foodborne illness each year. Some of these cases are severe. For example, in 2011, at least 38 deaths were caused by a single source of contaminated cantaloupes and, in Germany, contaminated sprouts from a single supplier caused more than 850 cases of hemolytic uremic syndrome and 32 deaths. Overall, consumption of contaminated food results in about 325,000 hospitalizations and 5,200 deaths in the United States each year.

What is it that contaminates food and causes so much illness? Bacteria. The nutrients in the food you consume during meals and snacks can also provide sustenance for a wide variety of disease-causing bacteria. Some of these invisible diners may accompany your lunch to your digestive tract and take up residence there, causing unpleasant symptoms or, in many cases, serious illness.

Devising effective strategies for protecting our food supply against bacterial contamination depends in part on how well we understand the biology of bacteria. Scientific investigation of bacteria and other microbes can provide the knowledge required to detect the presence of dangerous microbes in food, develop effective treatments for foodborne illnesses, and devise safer methods for growing and processing f Fortunately, biologists already know quite a bit about microorganisms. In this chapter, we'll explore some of that knowledge.

AT A GLANCE

19.1 Which Organisms Are Members of the Domains Archaea and Bacteria?

19.2 How Do Prokaryotes Survive and Reproduce?

19.3 How Do Prokaryotes Affect Humans and Other Organisms?

19.4 What Are Viruses, Viroids, and Prions?

LEARNING GOALS

LG1 Compare and contrast archaea and bacteria.

LG2 Describe the variety of sizes, shapes, and other physical characteristics among prokaryotes.

LG3 Describe the range of environments in which prokaryotes live.

LG4 Describe the variety of methods by which prokaryotes acquire energy.

LG5 Describe how prokaryotes reproduce and exchange genetic material.

LG6 Describe adaptations that help protect prokaryotes from environmental threats.

LG7 Explain how prokaryotes affect eukaryotes.

LG8 Describe the structure and characteristics of viruses, viroids, and prions and the effects they can have on host organisms.

19.1 WHICH ORGANISMS ARE MEMBERS OF THE DOMAINS ARCHAEA AND BACTERIA?

Earth's first organisms were prokaryotes, single-celled organisms that lacked organelles such as the nucleus, chloroplasts, and mitochondria. (See Chapter 4 for a comparison of prokaryotic and eukaryotic cells.) For the first 1.5 billion years or more of life's history, all life was prokaryotic. Even today, prokaryotes are extraordinarily abundant. A drop of seawater contains hundreds of thousands of prokaryotic organisms, and a spoonful of soil contains billions. The average human body is home to trillions of prokaryotes, which live on the skin, in the mouth, and in the stomach and intestines. In terms of abundance, prokaryotes are Earth's predominant form of life.

Both bacteria and archaea are usually very small, ranging from about 0.2 to 10 micrometers in diameter. In comparison, the diameters of eukaryotic cells range from about 10 to 100 micrometers. About 250,000 average-sized bacteria or archaea could congregate on the period at the end of this sentence, though a few species of bacteria are larger. The largest known bacterium (*Thiomargarita namibiensis*) is as much as 700 micrometers in diameter, as big as the tip of a ballpoint pen and visible to the naked eye.

The cell walls produced by prokaryotic cells give characteristic shapes to different types of bacteria and archaea. The most common shapes are spherical, rod-shaped, and corkscrew-shaped (**Fig. 19-1**).

Bacteria and Archaea Are Fundamentally Different

Two of life's three domains, **Bacteria** and **Archaea,** consist entirely of prokaryotes. Bacteria and archaea are superficially similar in appearance under the microscope, but have striking structural and biochemical differences that reveal the ancient evolutionary separation between the two groups. For example, the cell walls of bacteria are strengthened by molecules of *peptidoglycan*, a polysaccharide that also incorporates some amino acids. Peptidoglycan is unique to bacteria, and the cell walls of archaea do not contain it. Bacteria and archaea also differ in the structure and composition of their plasma membranes, ribosomes, and enzymes involved in RNA synthesis, as well as in the mechanics of basic processes such as transcribing the instructions encoded in DNA and synthesizing proteins.

Classification of Prokaryotes Within Each Domain Is Difficult

The sharp differences between archaea and bacteria make distinguishing the two domains a straightforward matter, but classification within each domain is challenging. Historically, the main challenge arose because prokaryotes are very small and structurally simple; they do not exhibit the huge array of anatomical differences that can be used to infer the evolutionary history of plants, animals, and other eukaryotes. Consequently, prokaryotes have historically been classified on the basis of such features as shape, means of locomotion, pigments, nutrient requirements, the appearance of colonies (clusters of individuals that descended from a single cell), and staining properties. For example, the Gram stain, a staining technique, distinguishes two types of cell-wall construction in bacteria. Depending on the results of the stain, bacteria are classified as either *gram-positive* or *gram-negative*.

In recent years, DNA sequence comparisons have revealed that many of the observable similarities and differences that informed traditional prokaryotic classification do not accurately reflect evolutionary history. New DNA sequence data have helped systematists identify newly recognized *clades* (groups of species united by descent from a common ancestor) within Bacteria and Archaea. Sequence comparisons have also helped suggest which of the traditional groupings are most likely to represent true clades.

But DNA sequence data have also revealed that prokaryote classification is even more challenging than previously

(a) Spherical

(b) Rod-shaped

(c) Corkscrew-shaped

▲ **FIGURE 19-1 Three common prokaryote shapes**
(a) Spherical bacteria of the genus *Staphylococcus*, **(b)** rod-shaped bacteria of the genus *Escherichia*, and **(c)** corkscrew-shaped bacteria of the genus *Borrelia*.

thought. In particular, it is now apparent that prokaryote evolutionary history has included a large amount of *lateral gene transfer*, the movement of genes from one species to another. Past (and ongoing) genetic mixing among even distantly related prokaryotes has yielded a very complex evolutionary history that is extremely difficult to reconstruct. A systematist who uses standard methods to construct an evolutionary tree of a set of prokaryotes based on sequences of a particular set of genes may get a result different from that of a systematist who uses a different set of genes to classify the same set of species. Thus, prokaryote evolutionary history

is perhaps best represented by a diagram portraying a web of interconnected strands (representing both descent from ancestors and lateral gene transfer), rather than by the kind of tree-like diagram typically used to represent phylogeny. Given the continual exchange of genes among prokaryotes, some systematists wonder if it is even conceptually accurate to say that the prokaryotic domains contain independently evolving units equivalent to eukaryotic species. The conceptual and practical challenges of prokaryote systematics mean that classification of prokaryotes remains a work in progress.

CHECK YOUR LEARNING

Can you describe some differences between bacteria and archaea? Can you describe the typical sizes and shapes of prokaryotes?

19.2 HOW DO PROKARYOTES SURVIVE AND REPRODUCE?

The abundance of prokaryotes is due in large part to adaptations that allow members of the two prokaryotic domains to inhabit and exploit a wide range of environments. In this section, we discuss some of the traits that help prokaryotes survive and thrive.

Some Prokaryotes Are Motile

Many bacteria and archaea adhere to a surface or drift passively in liquid surroundings, but some are motile—they can move about. Many of these motile prokaryotes have **flagella** (singular, flagellum), hair-like extensions that can rotate rapidly to propel the organism through its liquid environment. Prokaryotic flagella may appear singly at one end of a cell, in pairs (one at each end of the cell), as a tuft at one end of the cell (**Fig. 19-2a**), or scattered over the entire cell surface. The use of flagella to move allows prokaryotes to disperse into new habitats, migrate toward nutrients, and leave unfavorable environments.

The structure of prokaryote flagella is different from the structure of eukaryotic flagella (see pp. **60–61** for a description of the eukaryotic flagellum). In bacterial flagella, a unique wheel-like structure embedded in the bacterial membrane and cell wall allows the flagellum to rotate (**Fig. 19-2b**). Archaeal flagella are thinner than bacterial flagella and are constructed of different proteins. The structure of the archaeal flagellum, however, is not yet as well understood as that of the bacterial flagellum.

Many Bacteria Form Protective Films on Surfaces

Many prokaryotes continually synthesize signaling molecules and secrete them to the surrounding environment. If a large number of prokaryotes gathers in one place, the signaling molecules become concentrated enough that the molecules begin to move into neighboring cells across their plasma membranes and combine with receptors inside the cells. The activated receptors trigger cellular processes that would

(a) A flagellated archaean

cell wall

outer membrane

peptidoglycan layer

"wheel-and-axle" base

plasma membrane

(b) The structure of the bacterial flagellum

▲ FIGURE 19-2 **The prokaryote flagellum (a)** A flagellated archaean of the genus *Aquifex* uses its flagella to move toward favorable environments. **(b)** In bacteria, a unique "wheel-and-axle" arrangement anchors the flagellum within the cell wall and plasma membrane, enabling the flagellum to rotate rapidly. This diagram represents the flagellum of a gram-negative bacterium; the flagella of gram-positive bacteria lack the outermost "wheels."

not otherwise be activated. Thus, the behavior of prokaryotes may change when population density grows sufficiently high, a process known as *quorum sensing*.

Among the most common changes induced by quorum sensing is formation of biofilms. In a **biofilm,** one or more species of prokaryote aggregate to form a community that is typically surrounded by sticky protective slime. The slime, composed of polysaccharide or protein, is secreted by the prokaryotes and both protects them and helps them adhere to surfaces. One familiar biofilm is dental plaque, which is formed by the bacteria that inhabit the mouth (**Fig. 19-3**).

The protection afforded by biofilms helps defend the embedded bacteria against a variety of attacks, including those launched by antibiotics and disinfectants. As a result, biofilms formed by bacteria harmful to humans can be very difficult to eradicate. Many infections of the human body

▲ FIGURE 19-3 **The cause of tooth decay** Bacteria in the human mouth form a slimy biofilm that helps them cling to tooth enamel and protects them from threats in the environment. In this micrograph, individual bacteria (colored green or yellow) are visible, embedded in the brown biofilm. The bacteria-laden biofilm can cause tooth decay.

take the form of biofilms, including those responsible for tooth decay, gum disease, and ear infections. Biofilms also cause many of the hospital-acquired infections that affect 2 million Americans each year and kill 90,000 of them. Such biofilms can form in wounds and surgical incisions, as well as on implanted medical devices such as catheters, pacemakers, and artificial hips and knees.

Protective Endospores Allow Some Bacteria to Withstand Adverse Conditions

When environmental conditions become inhospitable, many rod-shaped bacteria form protective structures called **endospores.** An endospore, which forms inside a bacterium, consists of the bacterium's genetic material and a few enzymes encased within a thick protective coat (**Fig. 19-4**). After an endospore forms, the bacterial cell that contains it breaks open, and the spore is released to the environment. Metabolic activity ceases until the spore encounters favorable conditions, at which time metabolism resumes and the spore develops into an active bacterium.

Endospores are resistant even to extreme environmental conditions. Some can withstand boiling for an hour or more. Endospores are also able to survive for extraordinarily long periods. In the most astonishing example of such longevity, scientists recently discovered endospores that had been sealed inside rock for 250 million years. After being carefully extracted from their rocky tomb, the spores were incubated in test tubes. Amazingly, live bacteria developed from the ancient spores, which were older than the oldest dinosaur fossils.

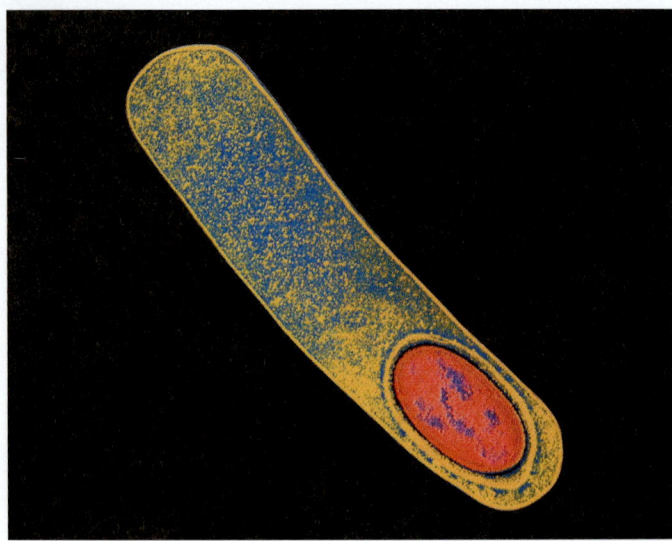

▲ FIGURE 19-4 **Spores protect some bacteria** A resistant endospore (red oval) has formed inside a bacterium of the genus *Clostridium*.

QUESTION What might explain the observation that most species of endospore-forming bacteria live in soil?

▲ FIGURE 19-5 **Some prokaryotes thrive in extreme conditions** Hot springs harbor bacteria and archaea that are both heat and mineral tolerant. Several species of cyanobacteria paint these hot springs in Yellowstone National Park with vivid colors; each species is confined to a specific area determined by temperature range.

QUESTION Some of the enzymes that have important uses in molecular biology procedures are extracted from prokaryotes that live in hot springs. Can you guess why?

CASE STUDY c o n t i n u e d

Unwelcome Dinner Guests

A few of the bacteria that commonly cause foodborne illness form endospores. For example, *Bacillus cereus*, a bacterial species that includes strains that cause vomiting or diarrhea in people unlucky enough to consume them, forms endospores that are widespread in soil and dust. If some spores find their way into warm, moist food, they can develop and give rise to a thriving population of bacteria. *B. cereus* spores are somewhat resistant to heat but, fortunately for us, they can be destroyed by thorough cooking.

Prokaryotes Are Specialized for Specific Habitats

Prokaryotes occupy virtually every habitat, including those where extreme conditions keep out other forms of life. For example, some bacteria thrive in near-boiling environments, such as the hot springs of Yellowstone National Park (**Fig. 19-5**). Many archaea live in even hotter environments, including deep-sea vents, where superheated water is spewed through cracks in Earth's crust at temperatures of up to 230°F (110°C). Prokaryotes can also survive at the extremely high pressures found deep beneath Earth's surface and in very cold environments, such as in Antarctic sea ice.

Extreme chemical conditions do not prevent colonization by prokaryotes, either. Thriving colonies of bacteria and archaea live in the Dead Sea, for example, where a salt concentration seven times that of the oceans precludes all other life, and in waters that are as acidic as vinegar or as alkaline as household ammonia. Given their ability to survive such extreme environments, it is not surprising that prokaryote communities also reside in a full range of more moderate habitats, including in and on the healthy human body.

No single species of prokaryote, however, is as versatile as these examples may suggest. In fact, most prokaryotes are specialists. One species of archaea that inhabits deep-sea vents, for example, grows optimally at 223°F (106°C) and stops growing altogether at temperatures below 194°F (90°C). Bacteria that live on the human body are also specialized; different species colonize the skin, the mouth, the respiratory tract, the large intestine, and the urogenital tract. (To learn more about how the bacteria in our bodies might affect our health, see "Health Watch: Is Your Body's Ecosystem Healthy?")

❓ HAVE **YOU** EVER WONDERED …

What Causes Bad Breath?

Unpleasant breath odors are caused mainly by prokaryotes that live in the mouth. The warm, moist human mouth cavity hosts a diverse microbial community that includes more than 600 prokaryote species. Many of these species acquire energy and nutrients by breaking down mucus, food particles, and dead cells. The by-products of this breakdown can include foul-smelling gases, some of which are also emitted by feces or decaying bodies.

The highest concentration of bad-breath prokaryotes is found at the base of the tongue. This location may be especially hospitable to microbes owing to the accumulation of mucus that drains down into the back of the throat from the nose. So, if you gargle with mouthwash to control your bad breath, thrust your tongue forward so the mouthwash can reach the base of your tongue.

Health Watch

Is Your Body's Ecosystem Healthy?

Microbiologists like to say, "You are born 100% human, but you die 90% microbial." A human fetus in the womb is more or less sterile, but as it passes down the birth canal to be born, it picks up some of the bacteria that live there, and more are transferred from the first hands to touch the new baby. From that point on, microbes accumulate steadily, acquired from the environment, from food, and from other humans. By the time a child is 3 years old, its body is home to an entire ecosystem of microorganisms. This "microbiome" includes hundreds of different species, living in huge numbers in the nose and mouth, on the scalp, in the urogenital tract, in the gut, and on almost every skin surface. All told, the microbiome of a typical person includes about 100 trillion microbial cells, about 10 microbial cells for every human one.

▲ FIGURE E19-1 **Transplanted feces could cure diseased bowels** By restoring a healthy microbial ecosystem in the recipient, fecal transplants can treat infected bowels and perhaps even diabetes, obesity, and other disorders as well.

It is becoming increasingly apparent to physicians and researchers that the inhabitants of the microbiome are not mere hitchhikers, but instead play an important role in human health. Although some of the benefits provided by our microbial partners are well understood, such as the production by gut bacteria of nutrients essential to humans, most of the details about how the microbiome contributes to our health are more mysterious. However, much evidence suggests that its contribution is important. For example, researchers have shown that in infants with the often-fatal bowel disease necrotizing enterocolitis, the composition of the community of microbes in the gut is different from that present in healthy babies. Similarly aberrant digestive tract microbiomes have been found in adults with bowel disorders, and even in people with disorders not directly related to digestion, such as diabetes and autoimmune diseases. Such findings have led to the hypothesis that, because a person's microbiome, like a coral reef or a tropical rain forest, is a diverse ecosystem characterized by complex interactions among species, disruptions of this ecosystem compromise the microbial ecosystem's ability to perform functions essential to human health. However, scientists are not yet able to say for sure if

the disrupted microbiomes of diseased individuals are a cause or a consequence of disease.

Confirmation of the microbiome's possible role in maintaining health depends on improved knowledge of its composition and characteristics. For that reason, a major ongoing research effort is directed at identifying and characterizing all of the microbial species that compose the microbiome, and at identifying exactly how microbiomes differ among different people. The job is challenging, because the traditional way of identifying a prokaryotic species—by growing it in a culture dish in the lab—is not effective for the many species that cannot be easily cultured. Some scientists are trying the alternative approach of taking a sample of a whole microbiome and sequencing all of the DNA present in it. Some of the initial results of this approach are astonishing. For example, fecal samples from 124 people yielded a total of 1.3 million different genes, about 150 times as many as are present in the human genome. Further study of these microbial genes may reveal their functions and how they contribute to their human hosts.

In the meantime, some physicians are pushing ahead with treatments designed to restore ecological health to the microbiomes of sick people. In particular, a few doctors have begun using fecal transplants to treat patients with severe bowel infections (Fig. E19-1). In this treatment, a small amount of feces from a healthy donor is transplanted into the bowel of the sick person, with the expectation that the transplanted microbial community will become established and spread, displacing the harmful microbes responsible for the infection. The physicians using this treatment report very high success rates, much higher than those typical of treatment with antibiotics, which have been shown to decimate the gut microbiome. These reports have encouraged funding agencies to overcome the yuck factor and fund large-scale clinical trials of the fecal transplant treatment, which are now under way.

Prokaryotes Have Diverse Metabolisms

Prokaryotes are able to colonize diverse habitats partly because they have evolved diverse methods of acquiring energy and nutrients from the environment. For example, unlike eukaryotes, many prokaryotes are **anaerobes;** their metabolisms do not require oxygen. Their ability to inhabit oxygen-free environments allows prokaryotes to exploit habitats in which eukaryotes could not survive. Some anaerobes, such as many of the archaea found in hot springs and the bacterium that causes tetanus, are actually poisoned by oxygen. Others are opportunists, engaging in anaerobic respiration when oxygen is lacking and switching to aerobic respiration (a more efficient process; see pp. **128–132**) when oxygen becomes available. Many prokaryotes are strictly aerobic and require oxygen at all times.

Whether aerobic or anaerobic, different prokaryote species can extract energy from an amazing array of substances. Prokaryotes subsist not only on the sugars, carbohydrates, fats, and proteins that we usually think of as foods, but also on compounds that are inedible or even poisonous to humans, including petroleum, methane (the main component of natural gas), and solvents such as benzene and toluene. Some prokaryotes can even metabolize inorganic molecules, including hydrogen, sulfur, ammonia, and iron. The process of metabolizing inorganic molecules sometimes yields by-products that are useful to other organisms. For example, certain bacteria release sulfates or nitrates, crucial plant nutrients, into the soil.

Some species of bacteria, such as the cyanobacteria (**Fig. 19-6**), use photosynthesis to capture energy directly from sunlight. Like green plants, photosynthetic bacteria possess chlorophyll. Most species produce oxygen as a by-product of photosynthesis, but some, known as the sulfur bacteria, use hydrogen sulfide (H_2S) instead of water (H_2O) in photosynthesis, releasing sulfur instead of oxygen. No photosynthetic archaea are known.

Prokaryotes Reproduce by Fission

Most prokaryotes reproduce asexually by **prokaryotic fission** (also called binary fission), a form of cell division that is much simpler than mitotic cell division (see pp. **142–143**). Prokaryotic fission produces genetically identical copies of the original cell (**Fig. 19-7**). Under ideal conditions, some prokaryotic cells can divide about once every 20 minutes, potentially giving rise to sextillions (10^{21}) of offspring in a single day. This extraordinarily rapid reproduction allows prokaryotes to exploit temporary habitats, such as a mud puddle or warm pudding.

Rapid reproduction also allows bacterial populations to evolve quickly. Recall that many mutations, the source of genetic variability, are the result of mistakes in DNA

▲ **FIGURE 19-7   Reproduction in prokaryotes** Prokaryotic cells reproduce by prokaryotic fission. In this color-enhanced electron micrograph, an *Escherichia coli*, a normal inhabitant of the human intestine, is dividing.

QUESTION   What is the main advantage of prokaryotic fission, compared to sexual reproduction?

replication during cell division (see pp. **201, 205–206**). Thus, the rapid, repeated cell division of prokaryotes provides ample opportunity for new mutations to arise and also allows mutations that enhance survival to spread quickly.

Prokaryotes May Exchange Genetic Material Without Reproducing

Although prokaryotic reproduction is generally asexual and does not involve genetic recombination, some bacteria and archaea nonetheless exchange genetic material. In these species, DNA is transferred from a donor to a recipient in a process called **conjugation.** The plasma membranes of two conjugating prokaryotes fuse temporarily to form a cytoplasmic bridge across which DNA travels. In bacteria, donor cells may use specialized extensions called *sex pili* that attach to a recipient cell, drawing it closer to allow conjugation (**Fig. 19-8**). Conjugation produces new genetic combinations that may allow the resulting bacteria to survive under a greater variety of conditions. In some cases, genetic material may be exchanged even between individuals of different species.

Much of the DNA transferred during bacterial conjugation is contained within a structure called a **plasmid,** a small, circular DNA molecule that is separate from the single bacterial chromosome. Plasmids may carry genes for antibiotic resistance or genes that are also found on the main bacterial chromosome.

CHECK YOUR LEARNING

Can you describe the range of environments inhabited by prokaryotes and the variety of methods by which they acquire energy? Can you describe adaptations that help protect prokaryotes from environmental threats? Can you explain how prokaryotes reproduce and exchange genetic material?

▲ **FIGURE 19-6   Cyanobacteria** Micrograph of cyanobacterial filaments.

▲ **FIGURE 19-8 Conjugation: prokaryotic "mating"** During conjugation, one prokaryote acts as a donor, transferring DNA to the recipient. In this micrograph, two *Neisseria gonorrhoeae* are connected by a long sex pilus. The sex pilus will retract, drawing the recipient bacterium (at left) to the donor bacterium.

19.3 HOW DO PROKARYOTES AFFECT HUMANS AND OTHER ORGANISMS?

Although they are largely invisible to us, prokaryotes play a crucial role in life on Earth. Plants and animals (including humans) are utterly dependent on prokaryotes. Prokaryotes help plants and animals obtain vital nutrients and help break down and recycle wastes and dead organisms. We could not survive without prokaryotes, but their impact on us is not always beneficial. Some of humanity's most deadly diseases stem from microbes.

Prokaryotes Play Important Roles in Animal Nutrition

Many eukaryotic organisms depend on close associations with prokaryotes. For example, most animals that eat leaves—including cattle, rabbits, koalas, and deer—can't themselves digest cellulose, the principal component of plant cell walls. Instead, these animals depend on certain bacteria that have the ability to break down cellulose. These bacteria live in the animals' digestive tracts, where they liberate nutrients from plant tissue that the animals are unable to break down themselves. Without such bacteria, leaf-eating animals could not survive.

Prokaryotes also have important effects on human nutrition. Many foods, including cheese, yogurt, and sauerkraut, are produced by the action of bacteria. Bacteria also inhabit your intestines. These bacteria feed on undigested food and synthesize such nutrients as vitamin K and vitamin B_{12}, which the human body absorbs.

Prokaryotes Capture the Nitrogen Needed by Plants

Humans could not live without plants, and plants are entirely dependent on bacteria. In particular, plants are unable to capture nitrogen from that element's most abundant reservoir, the atmosphere. Plants need nitrogen to grow. To acquire it, they depend on **nitrogen-fixing bacteria,** which live both in soil and in specialized nodules, small, rounded lumps on the roots of certain plants (legumes, which include alfalfa, soybeans, lupines, and clover; **Fig. 19-9**). The nitrogen-fixing bacteria capture nitrogen gas (N_2) from air trapped in the soil and combine it with hydrogen to produce ammonium (NH_4^+), a nitrogen-containing nutrient that plants can use directly.

(a) Nodules on roots

(b) Nitrogen-fixing bacteria within nodules

▲ FIGURE 19-9 **Nitrogen-fixing bacteria in root nodules (a)** Special chambers called nodules on the roots of a legume provide a protected environment for nitrogen-fixing bacteria. **(b)** This scanning electron micrograph shows the nitrogen-fixing bacteria inside cells within the nodules.

QUESTION If all of Earth's nitrogen-fixing prokaryotes were to die suddenly, what would happen to the concentration of nitrogen gas in the atmosphere?

Prokaryotes Are Nature's Recyclers

Prokaryotes play a crucial role in recycling waste. Most prokaryotes obtain energy by breaking down complex organic molecules (molecules that contain carbon and hydrogen). Such prokaryotes find a plentiful source of organic molecules in the waste products and dead bodies of plants and animals. By consuming and thereby decomposing these wastes, prokaryotes prevent wastes from accumulating in the environment. In addition, decomposition by prokaryotes releases the nutrients contained in wastes. Once released, the nutrients become available for reuse by living organisms.

Prokaryotes perform their recycling service wherever organic matter is found. They are important decomposers in lakes and rivers, in the oceans, and in the soil and groundwater of forests, grasslands, deserts, and other terrestrial environments. The recycling of nutrients by prokaryotes and other decomposers provides the raw materials needed for continued life on Earth.

Prokaryotes Can Clean Up Pollution

Many of the pollutants that are produced as by-products of human activity are organic compounds. As such, these pollutants can potentially serve as food for archaea and bacteria. Many of them are, in fact, consumed; the range of compounds consumed by prokaryotes is staggering. Nearly anything that human beings can synthesize—including detergents, many toxic pesticides, and harmful industrial chemicals such as benzene and toluene—can be broken down by some prokaryote.

Even oil can be broken down by prokaryotes. Soon after the tanker *Exxon Valdez* spilled 11 million gallons of crude oil into Prince William Sound, Alaska, in 1989, researchers sprayed oil-soaked beaches with a fertilizer that encouraged the growth of natural populations of oil-eating bacteria. Within 15 days, the oil deposits on these beaches were noticeably reduced in comparison with unsprayed areas. However, oil-eating bacteria are having a much slower effect on the 200 million gallons of oil released into the Gulf of Mexico during the *Deep Water Horizon* well blowout in 2010. The oil from the blowout was released, and largely remains, deep underwater where the temperature is cold and prokaryote metabolism is slow. In addition, it is not practical to encourage bacterial growth by fertilizing the vast area over which the oil has spread.

The practice of manipulating conditions to stimulate breakdown of pollutants by living organisms is known as **bioremediation.** Improved methods of bioremediation could dramatically increase our ability to clean up toxic waste sites and polluted groundwater. A great deal of current research is therefore devoted to identifying prokaryote species that are especially effective for bioremediation and discovering practical methods for manipulating these organisms to improve their usefulness.

Some Bacteria Pose a Threat to Human Health

Despite the benefits some bacteria provide, the feeding habits of certain bacteria threaten our health and well-being. These **pathogenic** (disease-producing) bacteria synthesize toxic substances that cause disease symptoms. So far, no pathogenic archaea have been identified.

Some Anaerobic Bacteria Produce Dangerous Poisons

Some bacteria produce toxins that attack the nervous system. One such toxin is produced by *Clostridium tetani*, the bacterium that causes tetanus, a sometimes fatal disease whose symptoms include painful, uncontrolled contraction of muscles throughout the body. *C. tetani* are anaerobic bacteria that survive as spores until introduced into a favorable, oxygen-free environment. A deep puncture wound may allow tetanus bacteria to penetrate a human body and reach a place where they will be protected from contact with oxygen. As they multiply, the bacteria release their toxin into the body's bloodstream.

Another anaerobic *Clostridium* species that produces a dangerous neurotoxin is *Clostridium botulinum*. *C. botulinum* occurs naturally in soil, but may also thrive in a sealed container of canned food that has been improperly sterilized. Such foodborne *C. botulinum* is dangerous because botulinum toxin is among the most toxic substances known; a single gram is enough to kill 15 million people.

Humans Battle Bacterial Diseases Old and New

Bacterial diseases have had a significant impact on human history. Perhaps the most infamous example is bubonic plague, or "Black Death," which killed 100 million people during the mid-fourteenth century. In many parts of the world, one-third or more of the population died. Plague is caused by a highly infectious bacterium that is spread by fleas that feed on infected rats and then move to human hosts. Although bubonic plague has not reemerged as a large-scale epidemic, about 2,000 to 3,000 people worldwide are still diagnosed with the disease each year.

Some bacterial pathogens seem to emerge suddenly. Lyme disease, for example, was unknown until 1975. This disease, named after the town of Old Lyme, Connecticut, where it was first described, is caused by the spiral-shaped bacterium *Borrelia burgdorferi*. The bacterium is carried by deer ticks, which

CASE STUDY continued

Unwelcome Dinner Guests

Many of the bacteria responsible for foodborne illnesses do their damage by producing toxins. For example, different populations of the bacterial species *Escherichia coli* may differ genetically, and some genetic differences can transform this normally benign inhabitant of the human digestive system into a toxin-producing pathogen. If one of these toxic strains, such as the ones designated O157:H7 and O104:H4, finds its way into a human digestive system, the bacteria attach firmly to the wall of the intestine and begin to release a toxin called shiga. Shiga toxin causes intestinal bleeding that results in painful cramping and bloody diarrhea. The toxin can damage other organs as well; victims of O157:H7 and O104:H4 often develop hemolytic uremic syndrome, a dangerous condition characterized by kidney failure and loss of red blood cells.

transmit it to the humans they bite. At first, the symptoms resemble flu, with chills, fever, and body aches. If untreated, weeks or months later the victim may experience rashes, bouts of arthritis, and in some cases abnormalities of the heart and nervous system. Both physicians and the general public are becoming more familiar with the disease, so more victims are receiving treatment before serious symptoms develop.

Perhaps the most frustrating pathogens are those that come back to haunt us long after we believed that we had them under control. Tuberculosis, a bacterial disease once almost vanquished in developed countries, is again on the rise in the United States and elsewhere. Two sexually transmitted bacterial diseases, gonorrhea and syphilis, have reached epidemic proportions around the globe. Cholera, a water-transmitted bacterial disease that flourishes when raw sewage contaminates drinking water or fishing areas, is under control in developed countries but remains a major killer in poorer parts of the world.

Common Bacterial Species Can Be Harmful

Some pathogenic bacteria are so widespread and common that we cannot expect to ever be totally free of their damaging effects. For example, different species of the abundant streptococcus bacterium produce several ailments. One streptococcus causes tooth decay. Another causes pneumonia by stimulating an immune response that clogs the lungs with fluid. Yet another streptococcus has gained fame as the "flesh-eating bacterium" (see "Case Study: Flesh-Eating Bacteria" in Chapter 36). About 500 to 1,000 Americans each year develop necrotizing fasciitis (as the "flesh-eating" infection is more properly known), and about 15% of these victims die. The streptococci enter through broken skin and produce toxins that either destroy flesh directly or stimulate an overwhelming and misdirected attack by the immune system against the body's own cells. A limb can be destroyed in hours, and in some cases, only amputation can halt the rapid tissue destruction. In other cases, these rare strep infections sweep through the body, causing death within a matter of days.

CHECK YOUR LEARNING

Can you explain how prokaryotes affect animal and plant nutrition? Are you able to explain prokaryotes' role in nutrient recycling? Can you describe how prokaryotes help clean up pollution? Can you describe some of the pathogenic bacteria that threaten human health?

19.4 WHAT ARE VIRUSES, VIROIDS, AND PRIONS?

Although **viruses** are generally found in close association with living organisms, most biologists do not consider viruses to be alive because they lack many of the traits that characterize life. For example, they are not cells—nor are they composed of cells. Further, they cannot, on their own, accomplish the basic tasks that living cells perform. Viruses have no ribosomes on which to make proteins, no cytoplasm, no ability to synthesize organic molecules, and no capacity to extract and use the energy stored in such molecules. They possess no membranes of their own and cannot grow or reproduce on their own. The simplicity of viruses seems to place them outside the realm of living things.

A Virus Consists of a Molecule of DNA or RNA Surrounded by a Protein Coat

Viruses are tiny; most are much smaller than even the smallest prokaryotic cell (Fig. 19-10). Virus particles are so small (0.05–0.2 micrometer in diameter) that they can be seen

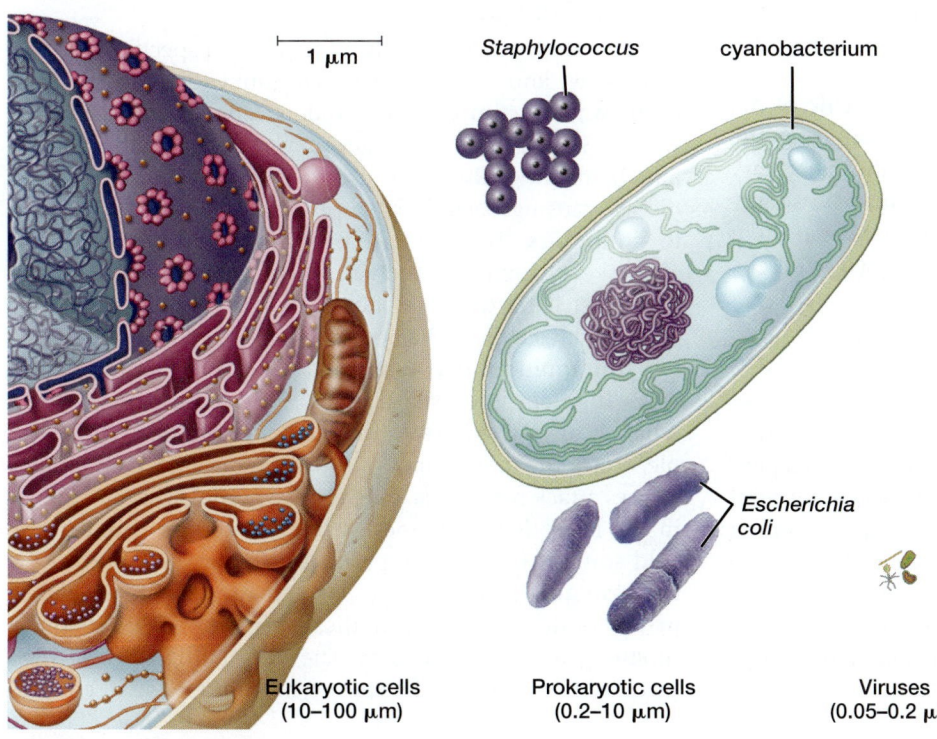

1 μm

Staphylococcus

cyanobacterium

Escherichia coli

Eukaryotic cells
(10–100 μm)

Prokaryotic cells
(0.2–10 μm)

Viruses
(0.05–0.2 μm)

◄ FIGURE 19-10 **The sizes of microorganisms** The relative sizes of eukaryotic cells, prokaryotic cells, and viruses (1 μm = 1/1,000 millimeter).

▶ FIGURE 19-11 **Viruses come in a variety of shapes** Viral shape is determined by the nature of the virus's protein coat.

(a) Rabies virus

(b) Bacteriophage

(c) Tobacco mosaic viruses

(d) Influenza viruses

only under the enormous magnification of an electron microscope. Under such magnification, one can see that viruses assume a great variety of shapes (Fig. 19-11).

Viruses consist of two major parts: a molecule of hereditary material and a coat of protein surrounding the molecule. Depending on the type of virus, the hereditary molecule may be either DNA or RNA and may be single-stranded or double-stranded, linear or circular. The protein coat may be surrounded by an envelope formed from the plasma membrane of the host cell (Fig. 19-12).

Viruses Require a Host to Reproduce

A virus can reproduce only inside a **host** cell—the cell that the virus infects. Viral reproduction begins when a virus penetrates a host cell. After the virus enters the host cell, the viral genetic material takes command. The hijacked host cell then uses the instructions encoded in the viral genes to produce the components of new viruses. The pieces are rapidly assembled, and an army of new viruses bursts forth to invade and conquer neighboring cells (see "In Greater Depth: Virus Replication" on p. 362).

Viruses Are Host Specific

Each type of virus is specialized to attack a specific host cell. As far as we know, no organism is immune to all viruses. Even bacteria fall victim to viral invaders called **bacteriophages** (Fig. 19-13). Bacteriophages may soon become important in treating diseases caused by bacteria, because many disease-causing bacteria have become increasingly resistant to antibiotics. Treatments based on bacteriophages could also take advantage of the viruses' specificity, attacking only the targeted bacteria and not the many other harmless or beneficial bacteria in the body.

In multicellular organisms such as plants and animals, different viruses specialize in attacking particular cell types. Viruses responsible for the common cold, for example, attack the membranes of the respiratory tract, and rabies viruses attack nerve cells. One type of herpes virus specializes in the mucous membranes of the mouth and lips, causing cold sores; a second type produces similar sores on or near the genitals. Herpes viruses take up permanent residence in the body, erupting periodically (typically during times of stress) as infectious sores. The devastating disease AIDS (acquired immune deficiency syndrome), which cripples

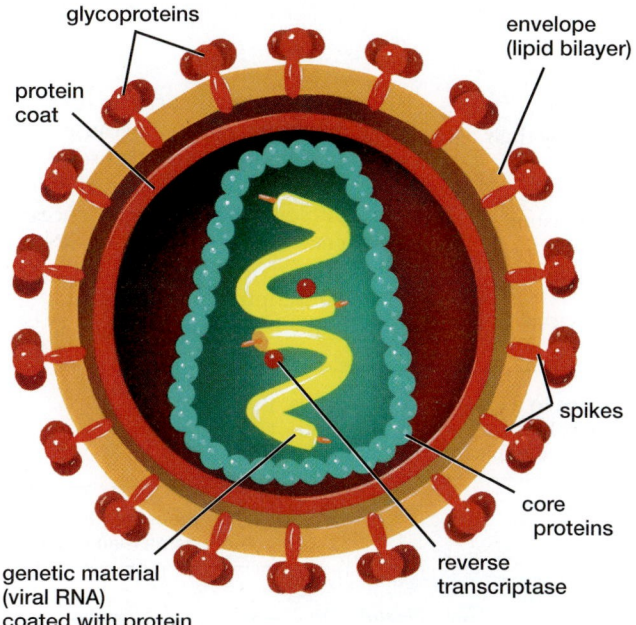

glycoproteins

protein coat

envelope (lipid bilayer)

spikes

core proteins

reverse transcriptase

genetic material (viral RNA) coated with protein

▲ FIGURE 19-12 **Viral structure and replication** A cross-section of HIV, the virus that causes AIDS. Inside, a protein coat surrounds genetic material and molecules of reverse transcriptase, an enzyme that catalyzes the transcription of DNA from the viral RNA template after the virus enters a host cell. This virus is among those that also have an outer envelope formed from the host cell's plasma membrane. Spikes made of glycoprotein (protein and carbohydrate) project from the envelope and help the virus attach to its host cell.

QUESTION Why are viruses unable to replicate outside of a host cell?

the body's immune system, is caused by a virus that attacks a specific type of white blood cell that controls the body's immune response. Viruses also cause some types of cancer, such as T-cell leukemia (a cancer of the white blood cells), liver cancer, and cervical cancer.

▲ FIGURE 19-13 **Some viruses infect bacteria** In this electron micrograph, bacteriophages are seen attacking a bacterium. They have injected their genetic material inside, leaving their protein coats clinging to the bacterial cell wall.

QUESTION In biotechnology, viruses are often used to transfer genes from the cells of one species to the cells of another. Which properties of viruses make them useful for this purpose?

Viral Infections Are Difficult to Treat

Because viruses depend on the cellular machinery of their hosts, the illnesses they cause are difficult to treat. The antibiotics that are often effective against bacterial infections are useless against viruses, and antiviral agents may destroy host cells as well as viruses. Despite the difficulty of attacking viruses as they "hide" within cells, a number of antiviral drugs have been developed. Many of these drugs destroy or block the function of enzymes that the targeted virus requires for replication.

Unfortunately, the benefits of most antiviral drugs are limited because many viruses quickly evolve resistance to the drugs. Mutation rates can be very high in viruses, in part because many viruses lack mechanisms for correcting errors that occur during replication of genetic material. It is thus common that when a population of viruses is under attack by an antiviral drug, a mutation will arise that confers resistance to the drug. The resistant viruses prosper and replicate in great numbers, eventually spreading to new human hosts. Ultimately, resistant viruses predominate, and a formerly helpful antiviral drug is rendered ineffective.

Some Infectious Agents Are Even Simpler Than Viruses

Viroids are infectious particles that lack a protein coat and consist of nothing more than short, circular strands of RNA. Despite their simplicity, viroids are able to enter the nucleus of a host cell and direct the synthesis of new viroids. About a dozen crop diseases, including cucumber pale fruit disease, avocado sunblotch, and potato spindle tuber disease, are caused by viroids. No viroid is known to infect animals.

Simple infectious agents known as *prions* attack mammalian nervous systems. In the 1950s, physicians studying the Fore, a primitive tribe in New Guinea, were puzzled to observe numerous cases of a fatal degenerative disease of the nervous system, which the Fore called kuru. The symptoms of kuru—loss of coordination, dementia, and ultimately death—were similar to those of the rare but more

CASE STUDY continued

Unwelcome Dinner Guests

Although most foodborne illnesses are caused by bacteria, a few are caused by viruses. For example, hepatitis A is caused by a virus with a single-stranded RNA genome. The virus is often transmitted in food, usually when the food has been handled by an infected person who has been lax about hand washing. Some people infected with hepatitis A do not exhibit symptoms, but many experience flu-like symptoms accompanied by jaundice (yellowish skin). Although additional complications can arise, most victims recover within a few months. Hepatitis A is one of the few foodborne diseases for which a vaccination exists.

IN GREATER DEPTH Virus Replication

Viruses multiply, or replicate, using their own genetic material, which—depending on the virus—consists of single-stranded or double-stranded RNA or DNA. This material serves as a blueprint for the viral proteins and genetic material required to make new viruses. Viral enzymes may participate in replication as well, but the overall process depends on the biochemical machinery of the host cell. Viruses cannot replicate outside of living cells.

The process of viral replication varies considerably among the different types of viruses, but most modes of replication are variations of a general sequence of events:

Penetration To replicate, a virus must enter a host cell. Some viruses are engulfed by a host cell (endocytosis) after binding to receptors on the cell's plasma membrane that stimulate endocytosis. Other viruses are coated with an envelope that can fuse with the host's membrane. The viral genetic material is then released into the cytoplasm.

Synthesis Viruses redirect the host cell's protein synthesis machinery to produce many copies of the viral proteins, and the viral genetic material is replicated many times. Transcription of the viral genome to messenger RNA uses nucleotides from the host cell, and protein synthesis uses the host cell's ribosomes, transfer RNAs, and amino acids.

Assembly The viral genetic material and enzymes are surrounded by their protein coat.

Release Viruses emerge from the host cell by "budding" from the plasma membrane or by bursting the cell.

Figure E19-2 shows the life cycle of the human immunodeficiency virus

(HIV), the *retrovirus* that causes AIDS. Retroviruses are so named because a key step in their replication uses single-stranded RNA as a template to make double-stranded DNA, a process that reverses the normal DNA-to-RNA pathway. Retroviruses accomplish this reverse transcription by using a viral enzyme called *reverse transcriptase*.

❶ A virus attaches to a receptor on the host's plasma membrane; its core disintegrates, and viral RNA enters the cytoplasm

envelope
coat
core
RNA
reverse transcriptase

(cytoplasm)

❻ Viruses bud from the plasma membrane

(nucleus)

DNA

RNA

mRNA

❷ Viral reverse transcriptase produces DNA, using viral RNA as a template

❸ DNA enters the nucleus and is incorporated into the host chromosomes; it is transcribed into mRNA and more viral RNA, which move to the cytoplasm

❹ Viral proteins are synthesized, using mRNA

❺ Viral proteins and RNA are assembled

HIV virus, a retrovirus, invades a white blood cell

▲ FIGURE E19-2 How viruses replicate

widespread Creutzfeldt-Jakob disease in humans and of scrapie and bovine spongiform encephalopathy, diseases of domestic livestock (see "Case Study: Puzzling Proteins" in Chapter 3). Each of these diseases typically results in brain tissue that is spongy—riddled with holes. The researchers in New Guinea eventually determined that kuru was transmitted by ritual cannibalism; members of the Fore tribe honored their dead by consuming their brains. This practice has since stopped, and kuru has virtually disappeared. Clearly, kuru was caused by an infectious agent transmitted by infected brain tissue—but what was that agent?

In 1982, neurologist Stanley Prusiner published evidence that scrapie (and, by extension, kuru, Creutzfeldt-Jakob disease, and a number of other, similar afflictions) is caused by an infectious agent that consists only of protein.

This idea seemed preposterous at the time, because most scientists believed that infectious agents must contain genetic material such as DNA or RNA to replicate. But Prusiner and his colleagues were able to isolate the infectious agent from scrapie-infected hamsters and demonstrate that it contained no nucleic acids. The researchers called these infectious protein particles **prions** (Fig. 19-14).

How can a protein replicate itself and be infectious? Research over the decades since Prusiner's discovery has shown that prions are misfolded versions of a common protein called PrP. PrP is found in the membranes of neurons and is required for normal neuron function, although its precise role remains unknown. Sometimes, copies of the PrP molecule become folded into the wrong shape and are thus transformed into infectious prions. If misfolded prion

▲ FIGURE 19-14 **Prions: puzzling proteins** A section from the brain of a cow infected with bovine spongiform encephalopathy contains fibrous clusters of prion proteins.

protein molecules are introduced into a healthy mammal, they can induce other, normal copies of the PrP molecule to become transformed into prions, which in turn induce still other conversions of normal PrP to the prion version. Eventually, this chain reaction leads to a concentration of prions high enough to cause nerve cell damage and degeneration. Why would a slight alteration to a normally benign protein turn it into a dangerous cell killer? No one knows.

No One Is Certain How These Infectious Particles Originated

The origin of viruses, viroids, and prions is uncertain. Some scientists believe that the huge variety of mechanisms for self-replication among viruses reflects their status as evolutionary remnants of the very early history of life, before the more familiar large, double-stranded DNA molecules evolved. Another possibility is that viruses and viroids may be the descendants of parasitic cells that lost their capacity for independent metabolism. These ancient parasites may have been so successful at exploiting their hosts that they eventually lost the ability to synthesize all of the molecules required for survival on their own and became dependent on the host's biochemical machinery. Whatever the origin of these infectious particles, their success poses a continuing challenge to living things.

CHECK YOUR LEARNING

Can you describe the structure and characteristics of viruses, viroids, and prions? Can you describe the effects they can have on host organisms?

CASE STUDY revisited

Unwelcome Dinner Guests

How do harmful bacteria get into our food? Many foodborne illnesses result from consumption of contaminated beef. The intestinal tracts of about a third of the cattle in the United States carry bacteria that are harmful to humans, and these bacteria can be transmitted to humans when a meatpacker accidentally grinds some gut contents into hamburger. Similarly, chicken feces may splash onto eggs, setting the stage for harmful bacteria to enter the eggs through tiny cracks or when the consumer breaks the egg and its contents contact the shell. Produce such as lettuce, spinach, tomatoes, and melons can also become contaminated if farm fields are exposed to animal feces, which can be deposited by deer or wandering domestic animals or carried from nearby ranches and feedlots in dust or runoff. The warm, moist environments in which sprouts are grown provide excellent growing conditions for any harmful bacteria that may have been present on the seeds from which the sprouts were produced.

How can you protect yourself from the bacteria that share our food supply? It's easy: Clean, cook, and chill. Cleaning helps prevent the spread of pathogens. Wash your hands before preparing food, and wash all utensils and cutting boards after preparing each item. Thorough cooking is the best way to ensure that any bacteria present in food are killed. Meats, in particular, must be thoroughly cooked; food

safety experts recommend using a meat thermometer to ensure that the thickest part of cooked pork or ground beef has reached 160°F. The safe temperature for cuts of beef, veal, or lamb is 145°F; for all poultry, 165°F. The color of cooked meat can be an unreliable indicator of safety, but when a meat thermometer is unavailable, try to avoid eating meat that is still pink inside, especially ground beef. Fish should be cooked until it is opaque and flakes easily with a fork; cook eggs until both white and yolk are firm. Finally, keep stored food cold. Pathogens multiply most rapidly at temperatures between 40° and 140°F. So get your groceries home from the store and into the refrigerator or freezer as quickly as possible. Don't leave cooked leftovers unrefrigerated for more than 2 hours. Thaw frozen foods in the refrigerator or the microwave, not at room temperature. A little bit of attention to food safety can save you from unwelcome guests in your food.

Consider This

BioEthics Consumer groups contend that we can improve food safety by giving government agencies additional funding and greater authority to inspect food processing plants and order recalls of contaminated food. Opponents of such steps argue that we need not empower government agencies, because the best protection against food contamination is informed consumers, who will stop buying products from companies that have produced unsafe foods. Would you support or oppose additional government oversight of food safety?

CHAPTER REVIEW

Summary of Key Concepts

19.1 Which Organisms Are Members of the Domains Archaea and Bacteria?

Archaea and bacteria are unicellular and prokaryotic. Although archaea and bacteria are morphologically similar, they are not closely related and differ in several fundamental features, including cell-wall composition, ribosomal RNA sequence, and membrane lipid structure. A cell wall determines the characteristic shapes of prokaryotes: spherical, rod-shaped, or corkscrew-shaped.

19.2 How Do Prokaryotes Survive and Reproduce?

Certain types of bacteria can move about using flagella; others form spores that disperse widely and withstand inhospitable environmental conditions. Bacteria and archaea have colonized nearly every habitat on Earth, including hot, acidic, very salty, and anaerobic environments.

Prokaryotes obtain energy in a variety of ways. Some, including the cyanobacteria, rely on photosynthesis. Others break down inorganic or organic molecules to obtain energy. Many are anaerobic, able to obtain energy when oxygen is not available. Prokaryotes reproduce by prokaryotic fission and may exchange genetic material by conjugation, in which DNA is transferred from a donor to a recipient.

19.3 How Do Prokaryotes Affect Humans and Other Organisms?

Some bacteria are pathogenic, causing disorders such as pneumonia, tetanus, botulism, and the sexually transmitted infections gonorrhea and syphilis. Most prokaryotes, however, are harmless to humans and play important roles in natural ecosystems. Some live in the digestive tracts of animals that eat leaves, where the prokaryotes break down cellulose. Nitrogen-fixing bacteria enrich the soil and aid in plant growth. Many other bacteria live off the dead bodies and wastes of other organisms, liberating nutrients for reuse.

19.4 What Are Viruses, Viroids, and Prions?

Viruses are parasites consisting of a protein coat that surrounds genetic material. They are noncellular and unable to move, grow, or reproduce outside a living cell. They invade cells of a specific host and use the host cell's energy, enzymes, and ribosomes to produce more virus particles, which are liberated when the cell ruptures. Many viruses are pathogenic to humans, including those causing colds and flu, herpes, AIDS, and certain forms of cancer.

Viroids are short strands of RNA that can invade a host cell's nucleus and direct the synthesis of new viroids. To date, viroids are known to cause only certain diseases of plants.

Prions have been implicated in diseases of the nervous system, such as kuru, Creutzfeldt-Jakob disease, scrapie, and bovine spongiform encephalopathy. Prions are unique in that they lack genetic material. They are composed solely of mutated prion protein, which may catalyze the formation of more prions from normal prion protein.

Key Terms

anaerobe   355
Archaea   351
Bacteria   351
bacteriophage   360
biofilm   353
bioremediation   358
conjugation   356
endospore   353
flagellum (plural, flagella)   352
host   360
nitrogen-fixing bacterium   357
pathogenic   358
plasmid   356
prion   362
prokaryotic fission   356
viroid   361
virus   359

Learning Outcomes

In this chapter, you have learned to . . .

LO1 Compare and contrast archaea and bacteria.

LO2 Describe the variety of sizes, shapes, and other physical characteristics among prokaryotes.

LO3 Describe the range of environments in which prokaryotes live.

LO4 Describe the variety of methods by which prokaryotes acquire energy.

LO5 Describe how prokaryotes reproduce and exchange genetic material.

LO6 Describe adaptations that help protect prokaryotes from environmental threats.

LO7 Describe how prokaryotes affect eukaryotes.

LO8 Describe the structure and characteristics of viruses, viroids, and prions and the effects they can have on host organisms.

Thinking Through the Concepts

Fill-in-the-Blank

1. _____ have peptidoglycan in their _____, but _____ do not. **LO1**

2. The size of prokaryotic cells is _____ than the size of eukaryotic cells. The most common shapes of prokaryotes are _____, _____, and _____. **LO2**

3. Many prokaryotes use _____ to move about. Some prokaryotes secrete slime that protects them when they aggregate in communities called _____. Other prokaryotes can survive long periods and extreme conditions by producing protective structures called _____. **LO6**

4. _____ bacteria inhabit environments that lack oxygen. _____ bacteria capture energy from sunlight. **LO3 LO4**

5. Prokaryotes reproduce by _____, and may sometimes exchange genetic material through the process of _____. **LO5**

6. The plant nutrient ammonium is produced by _____ bacteria in the soil. Prokaryotes

that live in the digestive tracts of cows and rabbits break down _____ in the leaves that those mammals eat. **LO7**

7. Diseases caused by pathogenic bacteria include _____, _____, and _____. Harmful strains of *E. coli* can be transmitted to humans by consumption of _____, _____, or _____. **LO7**

8. A virus consists of a molecule of _____ or _____ surrounded by a(n) _____ coat. A virus cannot reproduce unless it enters a(n) _____ cell. A virus that infects bacteria is known as a(n) _____. **LO8**

Review Questions

1. Describe some of the ways in which prokaryotes obtain energy and nutrients. **LO4**

2. What are nitrogen-fixing bacteria, and what role do they play in ecosystems? **LO7**

3. Describe some of the extreme environments in which prokaryotes are found. What parts of the human body are inhabited by prokaryotes? **LO3**

4. What is an endospore? What is its function? **LO6**

5. What is conjugation? What role do plasmids play in conjugation? **LO5**

6. Why are prokaryotes especially useful in bioremediation? **LO7**

7. Describe the structure of a typical virus. How do viruses replicate? **LO8**

8. Describe some examples of how prokaryotes are helpful to humans and some examples of how they are harmful to humans. **LO7**

9. How do archaea and bacteria differ? How do prokaryotes and viruses differ? **LO1 LO8**

Applying the Concepts

1. In some developing countries, antibiotics can be purchased without a prescription. Why do you think this is done? What biological consequences would you predict?

2. Before the discovery of prions, many (perhaps most) biologists would have agreed with the statement "It is a fact that no infectious organism or particle can exist that lacks nucleic acid (such as DNA or RNA)." What lessons do prions teach us about nature, science, and scientific inquiry? (You may wish to review Chapter 1 to help answer this question.)

3. Argue for and against the statement "Viruses are alive."

Answers to Figure Caption questions and Fill-in-the-Blank questions can be found in the Answers section at the back of the book.

MB *Go to MasteringBiology for practice quizzes, activities, eText, videos, current events, and more.*

The photosynthetic protist *Caulerpa taxifolia* is an unwanted invader in temperate seas.

CASE STUDY
Green Monster

IN CALIFORNIA, it is a crime to possess, transport, or sell *Caulerpa*. Is *Caulerpa* an illicit drug or some kind of weapon? No—it is merely a small green seaweed. Why, then, do California's lawmakers want to ban it from their state?

The story of *Caulerpa*'s rise to public enemy status begins in the early 1980s at the Wilhelmina Zoo in Stuttgart, Germany. There, the keepers of a saltwater aquarium found that the tropical seaweed *Caulerpa taxifolia* was an attractive companion and background for the tropical fish on display. Even better, years of captive breeding at the zoo had yielded a strain of the seaweed that was well suited to life in an aquarium. The new strain was particularly hardy and could survive in waters considerably cooler than the tropical waters in which wild *Caulerpa* is found. The zoo staff was happy to send cuttings to other institutions that wished to use it in aquarium displays.

One institution that received a cutting was the Oceanographic Museum of Monaco, located on the shore of the Mediterranean Sea. In 1984, a visiting marine biologist discovered a small patch of *Caulerpa* growing in the waters just below the museum. Presumably, someone cleaning an aquarium had dumped the water into the sea and thereby inadvertently introduced *Caulerpa* to the Mediterranean.

By 1989, the *Caulerpa* patch had grown to cover a few acres. It grew as a continuous mat that seemed to exclude most of the other organisms that normally inhabit the Mediterranean Sea floor. The local herbivores, such as sea urchins and fish, did not feed on *Caulerpa*.

It soon became apparent that *Caulerpa* spreads rapidly, is not controlled by predation, and displaces native species. By the mid-1990s, biologists were alarmed to find *Caulerpa* all along the Mediterranean coast from Spain to Italy. Today, it grows in extensive beds throughout the Mediterranean and covers an ever-expanding area on the seafloor.

Despite the threat it poses to ecosystems, *Caulerpa* is a fascinating creature. We will return to *Caulerpa* and its biology after an overview of the protists, a group that includes green algae such as *Caulerpa*, along with a host of other organisms.

AT A GLANCE

LEARNING GOALS

LG1 Define *protist* and describe the variety of morphologies, modes of nutrition, and modes of reproduction among protists.

LG2 Describe key aspects of the evolutionary history of protists.

LG3 Use examples to describe the effects of protists on other organisms, including humans.

LG4 Describe the major protist taxonomic groups and representative members of each group.

20.1 WHAT ARE PROTISTS?

Two of life's domains, Bacteria and Archaea, contain only prokaryotes. The third domain, Eukarya, includes all eukaryotic organisms. The most conspicuous Eukarya are plants, fungi, and animals (which we will discuss in Chapters 21 through 24). The remaining eukaryotes constitute a diverse array of organisms collectively known as **protists.** Protists do not form a clade—a group consisting of all the descendants of a particular common ancestor—so systematists do not use the term "protist" as a formal group name. Instead, protist is a term of convenience that refers to any eukaryote that is not a plant, animal, or fungus.

Most protists are single celled and are invisible to us as we go about our daily lives. If we could somehow shrink to their microscopic scale, we might be more impressed with their beautiful forms, their varied lifestyles, their astonishingly diverse modes of reproduction, and the structural and physiological complexity that is possible within the limits of a single cell.

Protists Use Diverse Modes of Nutrition

Three major modes of nutrition are represented among protists. Protists may ingest their food, absorb nutrients from their surroundings, or capture solar energy directly by photosynthesis. Protists that ingest their food are generally predators. Predatory single-celled protists may have flexible cell membranes that can change shape to surround and engulf food such as bacteria. Protists that feed in this manner typically use finger-like extensions called **pseudopods** to engulf prey. Other predatory protists create tiny currents that sweep food particles into mouth-like openings in the cell. Whatever the means by which food is ingested, once it is inside the protist cell, it is typically packaged into a membrane-surrounded *food vacuole* for digestion.

Protists that absorb nutrients directly from the surrounding environment may be free living or may live inside the bodies of other organisms. The free-living types live in soil and other environments that contain dead organic matter, where they act as decomposers. Most absorptive feeders, however, live inside other organisms. In most cases, these protists are parasites whose feeding activity harms the host species.

Photosynthetic protists are abundant in oceans, lakes, and ponds. Most float suspended in the water, but some live in close association with other organisms, such as corals or clams. These associations appear to be mutually beneficial; some of the solar energy captured by the photosynthetic protists is used by the host organism, which provides shelter and protection for the protists.

Protist photosynthesis takes place in chloroplasts. Chloroplasts are the descendants of ancient photosynthetic bacteria that took up residence inside a larger cell in a process known as *endosymbiosis* (see Chapter 17). In addition to the original instance of endosymbiosis that created the first protist chloroplast, there have been several later occurrences of *secondary endosymbiosis* in which a nonphotosynthetic protist engulfed a photosynthetic, chloroplast-containing protist. Ultimately, most components of the engulfed species disappeared, leaving only a chloroplast surrounded by four membranes. Two of these membranes are from the original, bacteria-derived chloroplast; one is from the engulfed protist; and one is from the food vacuole that originally contained the engulfed protist. Multiple occurrences of secondary endosymbiosis account for the presence of photosynthetic species in a number of different, unrelated protist groups.

Protists Use Diverse Modes of Reproduction

Most protists reproduce asexually; an individual divides by mitotic cell division to yield two individuals that are genetically identical to the parent cell (**Fig. 20-1a**). Many protists, however, are also capable of sexual reproduction, in which two individuals contribute genetic material to an offspring that is genetically different from either parent. Nonreproductive processes that combine the genetic material of different individuals are also common among protists (**Fig. 20-1b**).

In many protist species that are capable of sexual reproduction, most reproduction is nonetheless asexual. Sexual reproduction occurs only infrequently, at a particular time of year or under certain circumstances, such as a crowded environment or a shortage of food. The details of sexual reproduction and the resulting life cycles vary tremendously among different types of protists.

Protists Affect Humans and Other Organisms

Protists have important impacts, both positive and negative, on human lives. The primary positive impact actually benefits all living things and stems from the ecological role of photosynthetic marine protists. Just as plants do on land,

(a) Reproducing by cell division

(b) Exchanging genetic material

▲ **FIGURE 20-1 Protistan reproduction and gene exchange (a)** *Micrasterias*, a green alga, reproduces asexually by cell division. **(b)** Two *Euplotes* ciliates exchange genetic material across a cytoplasmic bridge.

QUESTION What do biologists mean when they say that sex and reproduction are uncoupled in most protists?

photosynthetic protists in the oceans capture solar energy and make it available to the other organisms in the ecosystem. Thus, the marine ecosystems on which humans depend for food in turn depend on protists. Further, in the process of using photosynthesis to capture energy, the protists release oxygen gas that helps replenish the oxygen removed from the atmosphere by respiration (recall from Chapter 8 that cellular respiration consumes oxygen).

On the negative side of the ledger, many human diseases are caused by parasitic protists. The diseases caused by protists include some of humanity's most prevalent ailments and some of its deadliest afflictions. Protists also cause a number of plant diseases, some of which attack crops that are important to humans. In addition to causing diseases, some marine protists release toxins that can accumulate to harmful levels in coastal areas.

CHECK YOUR LEARNING

Can you define *protist* and describe the various ways in which protists acquire nutrients and reproduce? Can you describe a scenario for the evolutionary origin of protist chloroplasts? Are you able to describe the major effects of protists on people and other organisms?

20.2 WHAT ARE THE MAJOR GROUPS OF PROTISTS?

Genetic comparisons are helping systematists gain a better understanding of the evolutionary history of protist groups. Because systematists strive to devise classification systems that reflect evolutionary history, the new genetic information has fostered a revision of protist classification. Some protist species that had been previously grouped together on the basis of physical similarity actually belong to independent evolutionary lineages that diverged from one another very early in the history of eukaryotes. Conversely, some protist groups bearing little physical resemblance to one another have been revealed to share a common ancestor, and have therefore been classified together. The process of revising protist classification, however, is far from complete. Thus, our understanding of the eukaryotic tree of life is still "under construction"; many of the tree's branches are in place, but others await new information that will allow systematists to place them alongside their closest evolutionary relatives. (One hypothesis about the phylogeny of the major eukaryote groups is depicted in Fig. 18-7.)

Past classifications of protists grouped species according to their mode of nutrition, but the old categories do not accurately reflect our current understanding of phylogeny. Nonetheless, biologists still use terminology that refers to groups of protists that share particular characteristics but are not necessarily related. For example, photosynthetic protists are collectively known as **algae** (singular, alga), and single-celled, nonphotosynthetic protists are collectively known as **protozoa** (singular, protozoan).

In the following sections, we will explore a brief sampling of protist diversity. Some key features of the protist groups we describe are summarized in **Table 20-1.**

Excavates Lack Mitochondria

Excavates are named for a feeding groove that gives the appearance of having been "excavated" from the surface of the cell. Excavates are anaerobes (can live and grow without oxygen) and lack mitochondria. It is likely that the excavates' ancestors did possess mitochondria, but these organelles were lost early in the evolutionary history of the group. The two largest groups of excavates are the diplomonads and the parabasalids.

Diplomonads Have Two Nuclei

The single cells of **diplomonads** have two nuclei and move about by means of multiple flagella. A parasitic diplomonad, *Giardia*, poses a health problem in the United States, particularly to hikers who drink from what appear to be pure

TABLE 20-1   The Major Groups of Protists

Group		Subgroup	Locomotion	Nutrition	Representative Features	Representative Genus
Excavates		Diplomonads	Swim with flagella	Heterotrophic (i.e., consume other organisms)	Lack mitochondria; inhabit soil or water, or may be parasitic	*Giardia* (intestinal parasite of mammals)
		Parabasalids	Swim with flagella	Heterotrophic	Lack mitochondria; parasites or commensal symbionts	*Trichomonas* (causes the sexually transmitted infection trichomoniasis)
Euglenozoans		Euglenids	Swim with one flagellum	Photosynthetic	Have an eyespot; all fresh water	*Euglena* (common pond-dweller)
		Kinetoplastids	Swim with flagella	Heterotrophic	Inhabit soil or water, or may be parasitic	*Trypanosoma* (causes African sleeping sickness)
Stramenopiles (chromists)		Water molds	Swim with flagella (gametes)	Heterotrophic	Filamentous	*Plasmopara* (causes downy mildew)
		Diatoms	Glide along surfaces	Photosynthetic	Have silica shells; most marine	*Navicula* (glides toward light)
		Brown algae	Nonmotile	Photosynthetic	Seaweeds of temperate oceans	*Macrocystis* (forms kelp forests)
Alveolates		Dinoflagellates	Swim with two flagella	Photosynthetic	Many bioluminescent; often have cellulose walls	*Gonyaulax* (causes red tide)
		Apicomplexans	Nonmotile	Heterotrophic	All parasitic; form infectious spores	*Plasmodium* (causes malaria)
		Ciliates	Swim with cilia	Heterotrophic	Include the most complex single cells	*Paramecium* (fast-moving pond-dweller)
Rhizarians		Foraminiferans	Extend thin pseudopods	Heterotrophic	Have calcium carbonate shells	*Globigerina*
		Radiolarians	Extend thin pseudopods	Heterotrophic	Have silica shells	*Actinomma*
Amoebozoans		Amoebas	Extend thick pseudopods	Heterotrophic	Have no shells	*Amoeba* (common pond-dweller)
		Acellular slime molds	Slug-like mass oozes over surfaces	Heterotrophic	Form multinucleate plasmodium	*Physarum* (forms a large, bright orange mass)
		Cellular slime molds	Amoeboid cells extend pseudopods; slug-like mass crawls over surfaces	Heterotrophic	Form pseudoplasmodium with individual amoeboid cells	*Dictyostelium* (often used in laboratory studies)
Red algae			Nonmotile	Photosynthetic	Some deposit calcium carbonate; mostly marine	*Porphyra* (used to make sushi wrappers)
Green algae			Swim with flagella (some species)	Photosynthetic	Closest relatives of land plants	*Ulva* (sea lettuce)

mountain streams. *Cysts* (tough structures that enclose the organism during one phase of its life cycle) of this diplomonad are released in the feces of infected humans, dogs, or other animals; a single gram of feces may contain 300 million cysts. Once outside the animal's body, the cysts may enter freshwater streams and even community reservoirs. If a mammal drinks infected water, the cysts develop into the adult form in the small intestine of their mammalian host (**Fig. 20-2**). In humans, infections can cause severe diarrhea, dehydration, nausea, vomiting, and cramps. Fortunately, these infections can be cured with drugs, and deaths from *Giardia* infections are uncommon.

Parabasalids Include Mutualists and Parasites

Parabasalids are anaerobic, flagellated protists named for the presence in their cells of a distinctive structure called the parabasal body, which consists of densely packed Golgi vesicles (see Chapter 4). All known parabasalids live inside animals. For example, this group includes several species that inhabit the digestive systems of some species of wood-eating termites. The termites cannot themselves digest the cellulose in wood, but the parabasalids can. Thus, the insect and the protist are in a mutually beneficial relationship. The termite delivers food to the parabasalids in its gut; as the parabasalids digest the food, some of the nutrients released become available for use by the termite.

In other cases, the host animal does not benefit from a parabasalid's presence but is instead harmed. For example, in humans, the parabasalid *Trichomonas vaginalis* causes the sexually transmitted infection trichomoniasis (**Fig. 20-3**). *Trichomonas* inhabits the mucous layers of the urinary and reproductive tracts, and uses its flagella to move along them. When conditions are favorable, a *Trichomonas* population can grow rapidly. Infected women may experience unpleasant symptoms, including vaginal itching and discharge. Infected men do not usually display symptoms, but they can transmit the infection to sexual partners.

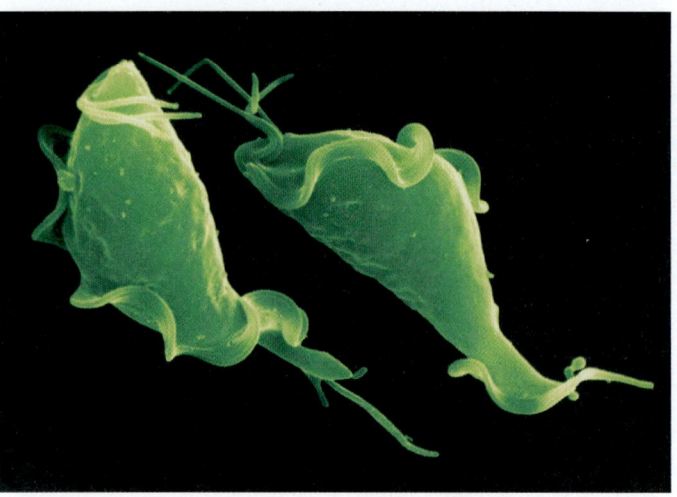

▲ FIGURE 20-3 **Trichomonas causes a sexually transmitted infection** The parabasalid *Trichomonas vaginalis* infects the urinary and reproductive tracts of men and women. Women, however, are far more likely to experience unpleasant symptoms.

Euglenozoans Have Distinctive Mitochondria

In most **euglenozoans,** the folds of the inner membrane of the cell's mitochondria have a distinctive shape that appears, under the microscope, as a stack of disks. Two major groups of euglenozoans are the euglenids and the kinetoplastids.

Euglenids Lack a Rigid Covering and Swim by Means of Flagella

Euglenids are single-celled protists that live mostly in fresh water and are named after the group's best-known representative, *Euglena*, a complex single cell that moves about by whipping its flagellum through water (**Fig. 20-4**). Euglenids lack a rigid outer covering, so some can move by wriggling as well

▲ FIGURE 20-2 *Giardia:* **The curse of campers**
A diplomonad (genus *Giardia*) that may infect water—causing gastrointestinal disorders for the people who drink it—is shown here in a human small intestine.

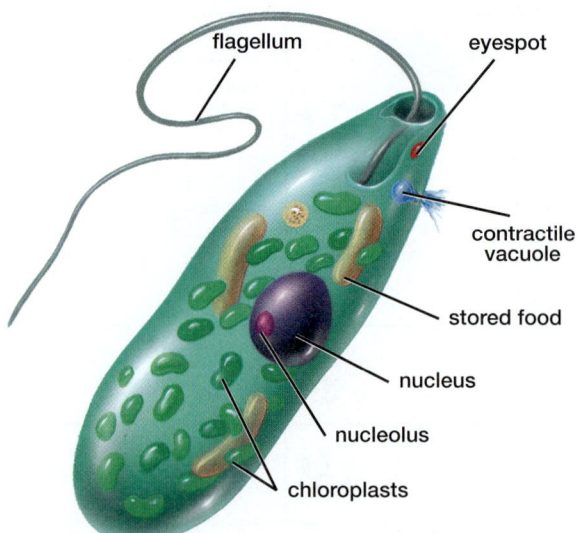

▲ FIGURE 20-4 *Euglena,* **a representative euglenid**
Euglena's elaborate single cell is packed with green chloroplasts, which will disappear if the protist is kept in darkness.

as by whipping their flagella. Many euglenids are photosynthetic, but some species instead absorb or engulf food. Some euglenids possess simple light-sensing organelles consisting of a photoreceptor, called an *eyespot*, and an adjacent patch of pigment. The pigment shades the photoreceptor only when light strikes from certain directions, enabling the organism to determine the direction of the light source. Using information from the photoreceptor, the flagellum propels the protist toward light levels appropriate for photosynthesis.

Some Kinetoplastids Cause Human Diseases

The DNA in the mitochondria of **kinetoplastids** is arranged in complex assemblies called kinetoplasts, in which many copies of the circular mitochondrial genome are interlinked to form distinctive disk-shaped structures. Most kinetoplastids possess at least one flagellum, which may propel the organism, sense the environment, or ensnare food. Some kinetoplastids are free living, inhabiting soil and water; others live inside other organisms in a relationship that may be mutually beneficial or parasitic. A dangerous parasitic kinetoplastid in the genus *Trypanosoma* is responsible for African sleeping sickness, a potentially fatal disease (**Fig. 20-5**). Like many parasites, this organism has a complex life cycle, part of which is spent in the tsetse fly. While feeding on the blood of a mammal, an infected fly can transmit saliva containing the trypanosome to the mammal. The parasite then develops in the new host (which may be a human) and enters the bloodstream. The trypanosome may then be ingested by another tsetse fly that bites the host, thus beginning a new cycle of infection.

Stramenopiles Have Distinctive Flagella

The **stramenopiles** (also known as *chromists*) form a group whose shared ancestry was discovered through genetic comparison. All members of the group have fine, hair-like projections on their flagella (though in many stramenopiles, flagella are present only at certain stages of the life cycle). Despite their shared evolutionary history, however, stramenopiles display a wide range of forms. Some are photosynthetic and some are not; most are single celled, but some are multicellular. Three major stramenopile groups are the water molds, the diatoms, and the brown algae.

Water Molds Have Had Important Impacts on Humans

The **water molds** (also known as *oomycetes*) form a small group of protists, many of which are shaped as long filaments that aggregate to form cottony tufts. These tufts are superficially similar to structures produced by some fungi, but this resemblance is due to convergent evolution (see pp. 268–269), not shared ancestry. Many water molds are decomposers that live in water and damp soil. Some species have profound economic impacts on humans. For example, a water mold causes a disease of grapes, known as *downy mildew*. Its inadvertent introduction into France from the United States in the late 1870s nearly destroyed the French wine industry. A water mold is also responsible for *late blight*, a devastating disease of potatoes. When this protist was accidentally introduced into Ireland in about 1845, it destroyed nearly the entire potato crop, causing a devastating potato famine during which as many as 1 million people in Ireland starved and many more immigrated to the United States to avoid the famine.

Diatoms Are Encased Within Glassy Walls

The **diatoms,** photosynthetic stramenopiles found in both fresh and salt water, produce protective shells of *silica* (glass), some of exceptional beauty (**Fig. 20-6**). These shells consist of top and bottom halves that fit together like a pillbox or petri dish. Accumulations of diatoms' glassy walls over millions of years have

▲ **FIGURE 20-5 A disease-causing kinetoplastid** This photomicrograph shows human blood that is heavily infested with the corkscrew-shaped, parasitic kinetoplastid *Trypanosoma*, which causes African sleeping sickness.

▲ **FIGURE 20-6 Some representative diatoms** This photomicrograph illustrates the intricate, microscopic beauty and variety of the glassy walls of diatoms.

produced fossil deposits of "diatomaceous earth" that may be hundreds of meters thick. This slightly abrasive substance is widely used in products such as toothpaste and metal polish.

Diatoms form part of the **phytoplankton,** the single-celled photosynthesizers that float passively in the upper layers of Earth's lakes and oceans. Phytoplankton play an immensely important ecological role. Marine phytoplankton account for about 50% of all photosynthetic activity on Earth, absorbing carbon dioxide, recharging the atmosphere with oxygen, and supporting the complex web of aquatic life.

Brown Algae Are Multicellular

Though most photosynthetic protists, such as diatoms, are single celled, some form multicellular aggregations that are commonly known as seaweeds. Although some seaweeds seem to resemble plants, they lack many of the distinctive features of plants. For example, no seaweed has true roots or shoots.

The stramenopiles include one group of seaweeds, the brown algae, which are named for the brownish-yellow pigments that (in combination with green chlorophyll) increase the seaweed's light-gathering ability. Almost all brown algae are marine. The group includes the dominant seaweed species that dwell along rocky shores in the temperate (cooler) oceans of the world, including the eastern and western coasts of the United States. Brown algae live in habitats ranging from nearshore waters, where they cling to rocks that are exposed at low tide, to far offshore. Several species use gas-filled floats to support their bodies (**Fig. 20-7a**). Some of the giant kelp found along the Pacific coast reach heights of 175 feet (53 meters), and may grow more than 6 inches (15 centimeters) in a single day. With their dense growth and towering height, kelp form undersea forests that provide food, shelter, and breeding areas for marine animals (**Fig. 20-7b**).

Alveolates Include Parasites, Predators, and Phytoplankton

The **alveolates** are single-celled organisms that have distinctive, small cavities beneath the surface of their cells. As with the stramenopiles, the evolutionary link among the alveolates was long obscured by the variety of structures and ways of life among the group members, but was revealed by molecular comparisons. Some alveolates are photosynthetic, some are parasitic, and some are predatory. The major alveolate groups are the dinoflagellates, apicomplexans, and ciliates.

Dinoflagellates Swim by Means of Two Whip-Like Flagella

Though most **dinoflagellates** are photosynthetic, there are also some nonphotosynthetic species. Dinoflagellates are named for the two whip-like flagella that propel them (**Fig. 20-8**). One flagellum encircles the cell, and the second projects behind it. Some dinoflagellates are enclosed only by a plasma membrane; others have cellulose walls that resemble armor plates. Although some species live in fresh water, dinoflagellates are especially abundant in the ocean, where they are an important component of the phytoplankton and

(a) Fucus

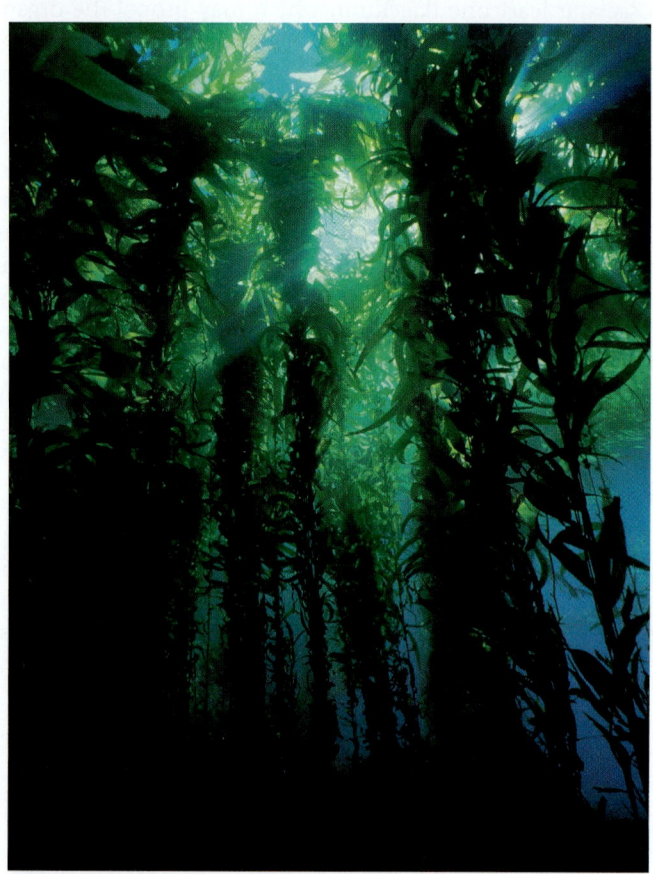

(b) Kelp forest

▲ FIGURE 20-7 **Brown algae, multicellular protists**
(a) *Fucus*, a genus found near shores, is shown here exposed at low tide. Notice the gas-filled floats, which provide buoyancy in water. **(b)** The giant kelp *Macrocystis* forms underwater forests off southern California.

a food source for larger organisms. Many dinoflagellates are bioluminescent, producing a brilliant blue-green light when disturbed by motion in the water.

Warm water that is rich in nutrients may bring on a dinoflagellate population explosion. Dinoflagellates can become so numerous that the water is dyed red by the color of their bodies, causing a "red tide" (**Fig. 20-9**). During red tides, fish

▲ FIGURE 20-8 **A dinoflagellate** This dinoflagellate has two flagella: a longer one that extends from a slot, and a shorter one that lies in a groove that encircles the cell.

often die by the thousands, suffocated by clogged gills or by the oxygen depletion that results from the decay of billions of dinoflagellates. But dinoflagellate explosions can benefit oysters, mussels, and clams, which have a feast, filtering millions of the protists from the water and consuming them. In the process, however, the mollusks' bodies accumulate concentrations of a nerve toxin produced by the dinoflagellates. Dolphins, seals, sea otters, and humans who eat affected mollusks may be stricken with potentially lethal paralytic shellfish poisoning.

▲ FIGURE 20-9 **A red tide** The explosive reproductive rate of certain dinoflagellates under the right environmental conditions can produce concentrations so great that their microscopic bodies dye the seawater red or brown.

CASE STUDY c o n t i n u e d

Green Monster

Red tides have become increasingly common in recent years. One reason for this increased incidence is that dinoflagellate species that can cause red tides have been inadvertently spread around the world by humans. The dinoflagellates travel mainly in seawater that is pumped into the ballast tanks of cargo ships and then discharged at distant ports. When released into new waters, populations of toxin-producing dinoflagellates may grow explosively, just as the invasive *Caulerpa* seaweed often spreads uncontrollably when introduced to environments free of its normal predators and parasites.

Apicomplexans Are Parasitic and Have No Means of Locomotion

All **apicomplexans** (sometimes known as *sporozoans*) are parasitic, living inside the bodies and sometimes inside the individual cells of their hosts. They form infectious spores—resistant structures transmitted from one host to another through food, water, or the bite of an infected insect. As adults, apicomplexans have no means of locomotion. Many have complex life cycles, a common feature of parasites. A well-known example is the malaria parasite *Plasmodium* (**Fig. 20-10**). Parts of its life cycle are spent in the body of a female *Anopheles* mosquito. The mosquito is not harmed by the presence of *Plasmodium*, and may eventually bite a human and pass the protist to the unfortunate victim. The protist develops in the victim's liver, then enters the blood, where it reproduces rapidly in red blood cells. When the blood cells rupture, they release large quantities of spores, which cause the recurrent fever of malaria. Uninfected mosquitoes may acquire the parasite by feeding on the blood of a malaria victim, spreading the parasite when they bite another person.

Ciliates Are the Most Complex of the Alveolates

Ciliates, which inhabit fresh and salt water, represent the peak of unicellular complexity. They possess many specialized organelles, including cilia, the short hair-like outgrowths for which they are named. In the well-known freshwater genus *Paramecium*, rows of cilia cover the organism's entire body surface (**Fig. 20-11**). The coordinated beating of the cilia propels the cell through the water at a rate of 1 millimeter per second—a protistan speed record. Although only a single cell, *Paramecium* responds to its environment as if it had a well-developed nervous system. Confronted with a noxious chemical or a physical barrier, the cell immediately backs up by reversing the beating of its cilia and then proceeds in a new direction. Some ciliates, such as *Didinium*, are accomplished predators (**Fig. 20-12**).

Rhizarians Have Thin Pseudopods

Protists in a number of different groups possess flexible plasma membranes that they can extend in any direction to form finger-like projections called pseudopods, which they use in locomotion and for engulfing food. The pseudopods

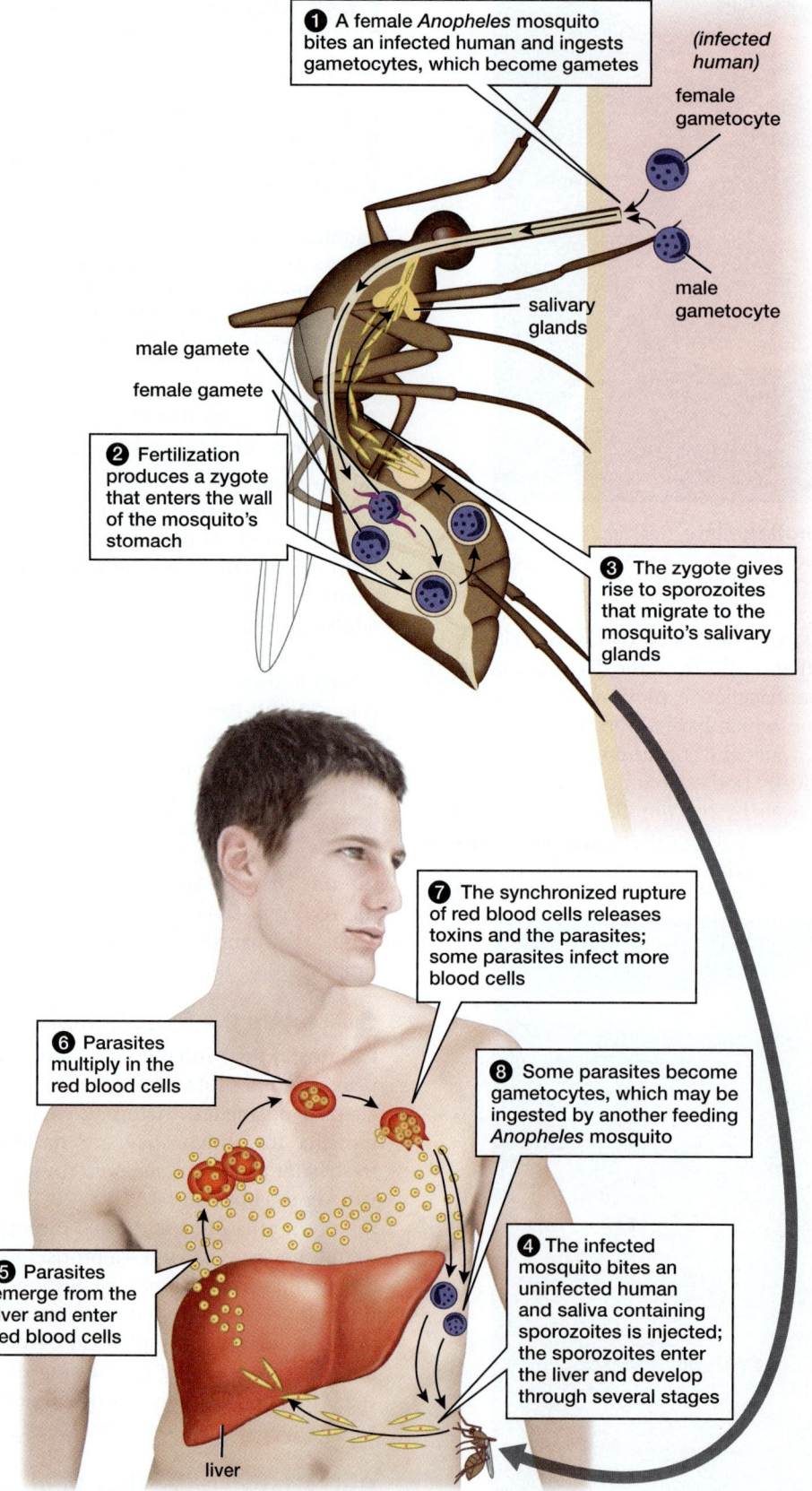

❶ A female *Anopheles* mosquito bites an infected human and ingests gametocytes, which become gametes

(infected human)

female gametocyte

male gametocyte

salivary glands

male gamete

female gamete

❷ Fertilization produces a zygote that enters the wall of the mosquito's stomach

❸ The zygote gives rise to sporozoites that migrate to the mosquito's salivary glands

❼ The synchronized rupture of red blood cells releases toxins and the parasites; some parasites infect more blood cells

❻ Parasites multiply in the red blood cells

❽ Some parasites become gametocytes, which may be ingested by another feeding *Anopheles* mosquito

❺ Parasites emerge from the liver and enter red blood cells

❹ The infected mosquito bites an uninfected human and saliva containing sporozoites is injected; the sporozoites enter the liver and develop through several stages

liver

▲ FIGURE 20-10 The life cycle of the malaria parasite

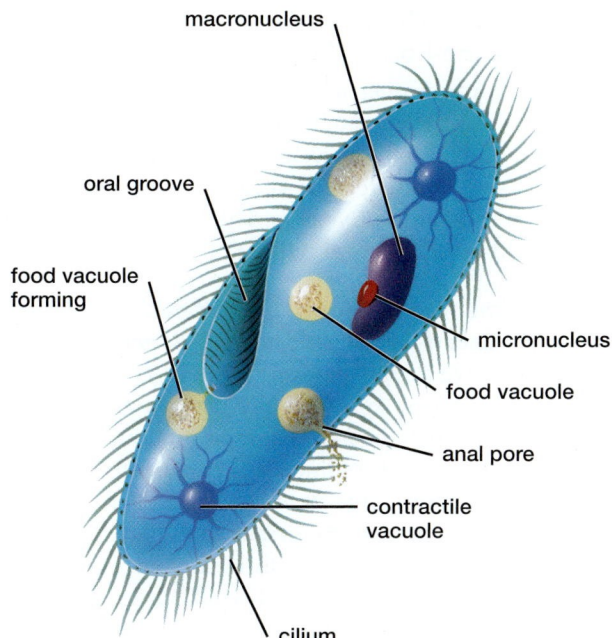

FIGURE 20-11 **The complexity of ciliates** The ciliate *Paramecium* illustrates some important ciliate organelles. The oral groove acts as a mouth, food vacuoles—miniature digestive systems—form at its apex, and waste is expelled by exocytosis through an anal pore. Contractile vacuoles regulate water balance.

of **rhizarians** are thin and thread-like. In many species in this group, the pseudopods extend through hard shells. Rhizarians include the foraminiferans and the radiolarians.

Fossil Foraminiferan Shells Form Chalk

The **foraminiferans** are primarily marine protists that produce beautiful shells. Their shells are constructed mostly

FIGURE 20-12 **A microscopic predator** In this scanning electron micrograph, the predatory ciliate *Didinium* attacks a *Paramecium*. Notice that the cilia of *Didinium* are confined to two bands, whereas *Paramecium* has cilia over its entire body. Ultimately, the predator will engulf and consume its prey. This microscopic drama could take place on a pinpoint with room to spare.

of calcium carbonate (chalk; **Fig. 20-13a**). These elaborate shells are pierced by myriad openings through which pseudopods extend. The chalky shells of dead foraminiferans, sinking to the ocean bottom and accumulating over millions of years, have resulted in immense deposits of limestone, such as those that form the famous White Cliffs of Dover, England.

Radiolarians Have Glassy Shells

Like foraminiferans, **radiolarians** have thin pseudopods that extend through hard shells. The shells of radiolarians, however, are made of glass-like silica (**Fig. 20-13b**). In some areas of the ocean, radiolarian shells raining down over vast stretches of time have accumulated to form thick layers of sediment.

(a) A foraminiferan

(b) A radiolarian

FIGURE 20-13 **Foraminiferans and radiolarians (a)** The chalky shell of a foraminiferan, and **(b)** the delicate, glassy shell of a radiolarian. In life, thin pseudopods, which sense the environment and capture food, would extend out through the openings in the shells.

▲ **FIGURE 20-14 An amoeba** Lobose amoebas are active predators that move through water to engulf food with thick, blunt pseudopods.

Amoebozoans Have Pseudopods and No Shells

Amoebozoans move by extending finger-shaped pseudopods, which may also be used for feeding. Amoebozoans generally do not have shells. The major groups of amoebozoans are the amoebas and the slime molds.

Amoebas Have Thick Pseudopods

Amoebas, sometimes known as *lobose amoebas* to distinguish them from other protists that have pseudopods, are common in freshwater lakes and ponds (**Fig. 20-14**). Many amoebas are predators that stalk and engulf prey, but some species are parasites. One parasitic form causes amoebic dysentery, a disease that is prevalent in warm climates. The dysentery-causing amoeba multiplies in the intestinal wall, triggering severe diarrhea.

Slime Molds Are Decomposers That Inhabit the Forest Floor

The physical form of *slime molds* seems to blur the boundary between a colony of separate individuals and a single, multicellular individual. The life cycle of the slime mold consists of two phases: a mobile feeding stage and a stationary reproductive stage called a *fruiting body*. There are two main types of slime molds: acellular and cellular.

Acellular Slime Molds Form a Multinucleate Mass of Cytoplasm Called a Plasmodium The **acellular slime molds,** also known as *plasmodial* slime molds, consist of a mass of cytoplasm that may spread thinly over an area of several square yards. Although the mass contains thousands of diploid nuclei, the nuclei are not confined in separate cells surrounded by plasma membranes. This structure, called a **plasmodium,** explains why these protists are described as "acellular" (without cells). The plasmodium oozes through decaying leaves and rotting logs, engulfing food such as bacteria and particles of organic material. The mass may be bright yellow or orange; a large plasmodium can be rather startling (**Fig. 20-15a**). Dry conditions or starvation stimulate the plasmodium to form a fruiting body, on which haploid spores are produced (**Fig. 20-15b**). The spores are dispersed and germinate under favorable conditions, eventually giving rise to a new plasmodium.

Cellular Slime Molds Live as Independent Cells But Aggregate into a Pseudoplasmodium When Food Is Scarce The **cellular slime molds,** also known as *social amoebas,* live in soil as independent haploid cells that move and feed by producing pseudopods. In the best-studied genus, *Dictyostelium,* individual cells release a chemical signal when food becomes scarce. This signal attracts nearby cells into a dense aggregation that forms a slug-like mass called a **pseudoplasmodium** ("false plasmodium") because, unlike a true plasmodium, it consists of individual cells (**Fig. 20-16**). A pseudoplasmodium can be viewed as a colony of individuals, because the cells that compose it are not all genetically

(a) Plasmodium

(b) Fruiting bodies

▲ **FIGURE 20-15 An acellular slime mold (a)** A plasmodium oozes over a stone on the damp forest floor. **(b)** When food becomes scarce, the mass differentiates into fruiting bodies in which spores are formed.

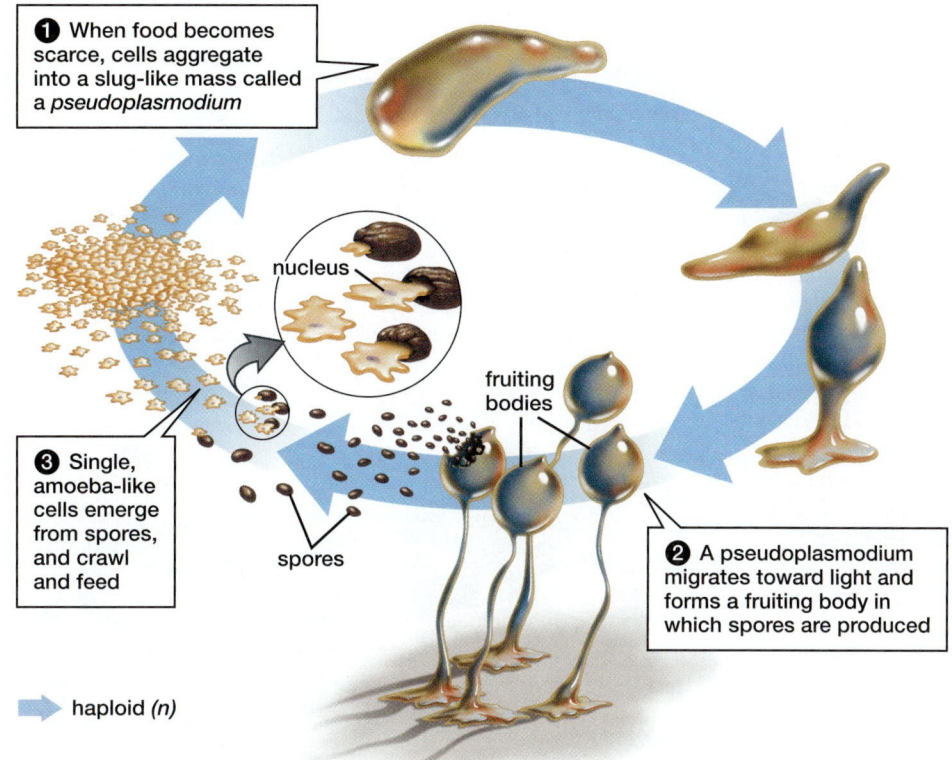

❶ When food becomes scarce, cells aggregate into a slug-like mass called a *pseudoplasmodium*

nucleus

fruiting bodies

❸ Single, amoeba-like cells emerge from spores, and crawl and feed

spores

❷ A pseudoplasmodium migrates toward light and forms a fruiting body in which spores are produced

haploid *(n)*

◄ FIGURE 20-16 **The life cycle of a cellular slime mold**

identical. In some ways, however, a pseudoplasmodium is more like a multicellular organism, because its cells differentiate into different cell types, with different types serving different functions. A pseudoplasmodium moves about in slug-like fashion, migrating toward an aboveground spot suitable for spore dispersal, where its cells differentiate to convert the structure to a fruiting body. Haploid spores formed within the fruiting body are dispersed by wind and germinate directly into new single-celled individuals.

Red Algae Contain Red Photosynthetic Pigments

The red algae are multicellular, photosynthetic seaweeds (Fig. 20-17). These protists range in color from bright red to nearly black; they derive their hue from red pigments that mask their green chlorophyll. Red algae are found almost exclusively in marine environments. They dominate in deep, clear tropical waters, where their red pigments absorb the deeply penetrating blue-green light and transfer this light energy to chlorophyll, where it is used in photosynthesis.

Some species of red algae deposit calcium carbonate (which forms limestone) in their tissues, and contribute to the formation of reefs. Red algae also contain gelatinous substances with commercial uses in products such as paints, cosmetics, and food (see "Links to Everyday Life: Have You Eaten Your Protists Today?"). However, the major importance of these and other algae lies in their photosynthetic

▲ FIGURE 20-17 **Red algae** Red coralline algae from the Mediterranean Sea. Coralline algae, which deposit calcium carbonate within their bodies, contribute to coral reefs in tropical waters.

ability; the energy they capture helps support nonphotosynthetic organisms in marine ecosystems.

Green Algae Are Closely Related to Land Plants

Green algae, a large and diverse group of photosynthetic protists, include both multicellular and unicellular species. Most live in freshwater ponds and lakes, but some live in the seas. Some green algae, such as *Spirogyra*, form thin filaments

⟫ LINKS TO EVERYDAY LIFE

Have You Eaten Your Protists Today?

There is a good chance that you ate some algae today. Or to be more precise, there is a good chance that you ate a substance extracted from one of these photosynthetic protists. Consider, for example, carrageenan, the name for a family of polysaccharides extracted from the cell walls of various species of red algae. Carrageenan melts at a relatively low temperature and, after cooling, forms a gel that remains stable at room temperature. These properties have proved useful to food processors, and carrageenan is widely used as a thickener and stabilizer in commercially produced foods including ice cream, yogurt, chocolate milk, soymilk, jellies, soups, salad dressings, and lunchmeats. Agar, a combination of polysaccharides also extracted from red algae, is another industrial food additive that is also used as a gelling and thickening agent in traditional Asian cuisine. Alginic acid, extracted from brown algae such as giant kelp, is likewise found in many of the products that fill supermarket shelves. Solutions of salts derived from alginic acid, such as sodium alginate, gel when mixed with calcium salts, a property that olive processors, for example, use to their advantage. To produce the stuffed olives beloved by martini-drinkers, manufacturers inject a slurry of ground pimento and sodium alginate into the space formerly occupied by the olive pit, and the slurry sets into solid form when the olive is dunked in a calcium salt solution.

In addition to their uses in traditional Asian cooking and industrial food production, carrageenan, agar, and alginic acid have assumed prominent roles in the trendy restaurants that specialize in what has come to be called "molecular gastronomy." Some chefs enjoy using cutting-edge food science and technology to prepare innovative dishes. So if you were to splurge on an expensive meal at a fancy restaurant, you might find yourself eating agar spaghetti infused with arugula (**Fig. E20-1**), spherical pearls of alginate with liquid mango juice centers, or soft-boiled carrageenan "eggs" that taste like bacon.

▲ **FIGURE E20-1   Algae in haute cuisine** This arugula-flavored "spaghetti" consists largely of agar, a gelatinous extract of red algae. The balsamic vinegar pearls that garnish the dish are also made of agar. (The tomato slices and mozzarella are genuine.)

HAVE **YOU** EVER WONDERED …

What Are Sushi Wrappers Made of?

If you like sushi, you've probably eaten a sushi roll, in which rice and other items are surrounded by a tasty, blackish-green wrapper. The wrapper is made from the dried bodies of a multicellular protist, the red alga *Porphyra*. *Porphyra* is grown commercially, often in large coastal "farms" where the seaweed grows attached to vast nets suspended from the ocean's surface. After harvest, the seaweeds are shredded, pulped, pressed into sheets, and dried, in a process very similar to papermaking.

from long chains of cells (**Fig. 20-18a**). Other species of green algae form colonies containing clusters of cells that are somewhat interdependent and form a structure intermediate between unicellular and multicellular forms. These colonies range from a few cells to a few thousand cells, as in species of *Volvox*. Most green algae are small, but some marine species are large. For example, the green alga *Ulva*, or sea lettuce, is similar in size to the leaves of its namesake (**Fig. 20-18b**).

Some species of green algae are currently under intensive cultivation by companies that hope to use them for commercial production of biofuels (**Fig. 20-18c**). Fuels based on algae could in principle replace dwindling fossil fuels with a renewable fuel whose production and use releases less carbon dioxide into the atmosphere. However, efforts to develop an efficient, economically viable process for converting algae to fuel have not yet been successful.

The green algae are of special interest because, unlike other groups that contain multicellular, photosynthetic protists, green algae are closely related to plants. Plants share a common ancestor with some types of green algae, and many researchers believe that the very earliest plants were similar to today's multicellular green algae.

CHECK YOUR LEARNING

Can you list the major protist taxonomic groups and the key characteristics of each group? Can you describe some examples of how members of each group affect humans?

(a) *Spirogyra*

(c) Growing green algae for biofuel

(b) *Ulva*

▲ **FIGURE 20-18 Green algae (a)** *Spirogyra* is a filamentous green alga composed of strands only one cell thick. **(b)** *Ulva* is a multicellular green alga that assumes a leaf-like shape. **(c)** Biofuels produced from algae grown at a facility like the one pictured may one day fill a significant portion of our energy needs, if technical obstacles can be overcome.

CASE STUDY revisited

Green Monster

Caulerpa taxifolia, the invasive seaweed that threatens to overrun the Mediterranean, is a green alga. This species and other members of its genus have very unusual bodies. Outwardly, they appear plantlike, with rootlike structures that attach to the seafloor and with other structures that look like stems and leaves, rising to a height of several inches. Despite its seeming similarity to a plant, however, a *Caulerpa* body consists of a single, extremely large cell. The entire body is surrounded by a single, continuous cell membrane. The interior consists of cytoplasm that contains numerous cell nuclei but is not subdivided. That a single cell can take such a complex shape is extraordinary.

A potential problem with *Caulerpa*'s single-celled organization might arise when its body is damaged, perhaps by wave action or when a predator takes a bite out of it. When the cell membrane is breached, there is nothing to prevent all of the cytoplasm from leaking out, an event that would be fatal. But *Caulerpa* has evolved a defense against this potential calamity. Shortly after the cell membrane breaks, it is quickly filled with a "wound plug" that closes the gap. After the plug is established, the cell begins to grow and regenerates any lost portion of the body.

This ability to regenerate is a key component of the ability of *Caulerpa taxifolia* to spread rapidly in new environments. If part of a *Caulerpa* body breaks off and drifts to a new location, it can regenerate a whole new body. The regenerated individual becomes the founder of a new, quickly growing colony—and these quickly growing colonies might appear anywhere in the world. Authorities in many countries worry that the aquarium strain of *Caulerpa* could invade their coastal waters, unwittingly transported by ships from the Mediterranean or released by careless aquarists. In fact, invasive *Caulerpa* is no longer restricted to the Mediterranean. It has been found in two locations on the California coast and in at least eight bodies of water in Australia. Local authorities in both countries mobilized to control the invading algae. Officials in California believe that, after several years of effort, their program of poisoning the invader has eradicated it for now. Australia has not been so fortunate, and *Caulerpa taxifolia* continues to spread there.

Consider This

BioEthics Is it important to stop the spread of *Caulerpa*? Governments invest substantial resources to combat introduced species and prevent their populations from increasing and dispersing. Why might this be a wise use of funds? Can you think of some arguments against spending time and money for this purpose?

CHAPTER REVIEW

Summary of Key Concepts

20.1 What Are Protists?

"Protist" is a term of convenience that refers to any eukaryote that is not a plant, animal, or fungus. Most protists are single, highly complex eukaryotic cells, but some form colonies, and some, such as seaweeds, are multicellular. Protists exhibit diverse modes of nutrition, reproduction, and locomotion. Photosynthetic protists form much of the phytoplankton, which plays a key ecological role. Some protists cause human diseases; others are crop pests.

20.2 What Are the Major Groups of Protists?

Protist groups include excavates (diplomonads and parabasalids), euglenozoans (euglenids and kinetoplastids), stramenopiles (water molds, diatoms, and brown algae), alveolates (dinoflagellates, apicomplexans, and ciliates), rhizarians (foraminiferans and radiolarians), amoebozoans (amoebas and slime molds), red algae, and green algae (the closest relatives of plants).

Key Terms

acellular slime mold	376	foraminiferan	375
alga (plural, algae)	368	kinetoplastid	371
alveolate	372	parabasalid	370
amoeba	376	phytoplankton	372
amoebozoan	376	plasmodium	376
apicomplexan	373	protist	367
cellular slime mold	376	protozoan (plural,	
ciliate	373	protozoa)	368
diatom	371	pseudoplasmodium	376
dinoflagellate	372	pseudopod	367
diplomonad	368	radiolarian	375
euglenid	370	rhizarian	375
euglenozoan	370	stramenopile	371
excavate	368	water mold	371

Learning Outcomes

In this chapter, you have learned to . . .

LO1 Define *protist* and describe the variety of morphologies, modes of nutrition, and modes of reproduction among protists.

LO2 Describe key aspects of the evolutionary history of protists.

LO3 Use examples to describe the effects of protists on other organisms, including humans.

LO4 Describe the major protist taxonomic groups and representative members of each group.

Thinking Through the Concepts

Fill-in-the-Blank

1. Many predatory protists engulf prey with finger-shaped extensions called _____. Protists that absorb nutrients from their surroundings may act as _____ of dead organic matter or as harmful _____ of larger living organisms. **LO1**

2. Photosynthetic protists are collectively known as _____; nonphotosynthetic, single-celled protists are collectively known as _____. **LO1**

3. Protist chloroplasts surrounded by four-layer membranes arose evolutionarily through _____, in which an ancestral nonphotosynthetic protist engulfed but did not digest a(n) _____. **LO2**

4. The disease-causing parasite *Giardia* is a member of the _____ group; the protist that causes malaria is a member of the _____ group; and the protist that causes sleeping sickness is a member of the _____ group. **LO3 LO4**

5. The plant diseases downy mildew and late blight are caused by protists in the _____ group. Slime molds are members of the _____ group. **LO3 LO4**

6. Protists that make up a large proportion of Earth's phytoplankton include _____ and _____. The protist group most closely related to land plants is _____. **LO2 LO3 LO4**

Review Questions

1. List the major differences between prokaryotes and protists. **LO1**

2. What is secondary endosymbiosis? **LO2**

3. What is the importance of dinoflagellates in marine ecosystems? What can happen when they reproduce rapidly? **LO3**

4. What is the major ecological role played by single-celled algae? **LO3**

5. Which protist group consists entirely of parasitic forms? **LO4**

6. Which protist groups include seaweeds? **LO4**

7. Which protist groups include species that use pseudopods? **LO4**

Applying the Concepts

1. Recent research shows that ocean water off southern California has become 2° to 3°F (1° to 1.5°C) warmer during the past four decades, possibly due to the greenhouse effect. This warming has indirectly led to a depletion of nutrients in the water and thus a decline in photosynthetic protists such as diatoms. What effects is this warming likely to have on life in the oceans?

2. The internal structure of many protists is much more complex than that of cells of multicellular organisms. Does this mean that the protist is engaged in more complex activities than the multicellular organism is? If not, why are protistan cells more complicated?

3. What are some important benefits and services provided by protists to Earth's other organisms?

Answers to Figure Caption questions and Fill-in-the-Blank questions can be found in the Answers section at the back of the book.

MB *Go to MasteringBiology for practice quizzes, activities, eText, videos, current events, and more.*

CASE STUDY

Queen of the Parasites

THE FLOWER OF THE STINKING CORPSE LILY makes a strong impression. For one thing, it's huge; a single flower may be 3 feet across. It also has a rather strange appearance, consisting largely of fleshy lobes that are almost fungus-like. But as its name implies, the thing that makes a stinking corpse lily almost impossible to ignore is its aroma, which has been described as "a penetrating smell more repulsive than any buffalo carcass in an advanced stage of decomposition."

Unlike most plants, a stinking corpse lily has no visible leaves, roots, or stems. In fact, it is a parasite, and its body is completely embedded in the tissue of its host, a vine in the grape family. Because it has no leaves, the stinking corpse lily cannot produce any food of its own, but instead draws all of its nutrition from its host. The parasite becomes visible outside the body of its host only when one of its cabbage-shaped flower buds pushes through the surface of the host's stem and its gigantic, stinking flower opens for a week or so before shriveling and falling off. If a male and a female flower happen to be open simultaneously and close together, the female flower may be fertilized and produce seeds. A seed that is dispersed in the droppings of an animal that has consumed it, and that happens to land on a stem of the host species, may germinate and penetrate a new host.

When you think of plants, you might first think of their most obvious feature: green leaves that capture solar energy by photosynthesis. It may seem odd, then, that this chapter about plants begins with a peculiar plant that does not photosynthesize. Oddities such as the stinking corpse lily, however, serve as reminders that evolution does not always follow a predictable pathway, and that even an adaptation as seemingly valuable as the ability to live on sunlight can be lost.

The huge, foul-smelling flower of the stinking corpse lily is a treat for visitors to Asian rain forests.

AT A GLANCE

LEARNING GOALS

LG1 Describe the key features of plants.

LG2 Describe the evolutionary history of plants and the adaptations that equip plants for life on land.

LG3 Compare and contrast vascular plants and nonvascular plants.

LG4 Describe the major plant taxonomic groups and list representative members of each group.

LG5 Describe the key steps in the life cycles of mosses, ferns, gymnosperms, and flowering plants.

LG6 Describe the effects that plants have on other organisms, including humans.

21.1 WHAT ARE THE KEY FEATURES OF PLANTS?

Plants are the most conspicuous living things in almost every landscape on Earth. Unless you are in a polar region, an especially dry desert, or a densely populated urban area, you live surrounded by plants. The plants that dominate Earth's forests, grasslands, parks, lawns, orchards, and farm fields are such familiar parts of the backdrop to our daily lives that we tend to take them for granted. But if we take some time to look more closely at our green companions, we might gain a greater appreciation for the adaptations that have made them so successful and for the properties that make them essential to our own survival.

What distinguishes plants from other organisms? Plants exhibit three characteristic traits: photosynthesis, multicellular embryos, and alternation of generations, as explained below. Each of these traits also occurs in some other kinds of organisms, but only plants combine all three.

Plants Are Photosynthetic

Perhaps the most noticeable feature of plants is their green color. The color comes from the presence of the pigment chlorophyll in many plant tissues. Chlorophyll plays a crucial role in photosynthesis, the process by which plants use energy from sunlight to convert water and carbon dioxide to sugar (see Chapter 7). Chlorophyll and photosynthesis, however, are not unique to plants; they are also present in many types of protists and prokaryotes.

Plants Have Multicellular, Dependent Embryos

Plants are distinguished from other photosynthetic organisms by their characteristic embryos. A plant embryo is multicellular and is attached to and dependent on its parent. As it grows and develops, the embryo receives nutrients from the tissues of the parent plant. Such multicellular, dependent embryos are not found among photosynthetic protists.

Plants Have Alternating Multicellular Haploid and Diploid Generations

Plant reproduction is characterized by a type of life cycle called **alternation of generations** (Fig. 21-1). In organisms with alternation of generations, separate diploid and haploid generations alternate with one another. (Recall that a diploid organism has paired chromosomes; a haploid organism has unpaired chromosomes.) In the diploid ($2n$) generation, the body consists of diploid cells and is known as the **sporophyte.** (The multicellular embryo described in the previous section is part of the diploid sporophyte generation.) Certain cells of sporophytes undergo meiosis to produce haploid (n) reproductive cells called *spores*. The haploid spores develop into multicellular, haploid bodies called **gametophytes.**

A gametophyte ultimately produces male and female haploid *gametes* (sperm and eggs) by mitosis. Gametes, like spores, are reproductive cells but, unlike spores, an individual gamete by itself cannot develop into a new individual. Instead, two gametes of opposite sexes must meet and fuse to form a new diploid individual. In plants, gametes produced by gametophytes fuse to form a diploid *zygote* (a fertilized egg), which develops into a diploid embryo. The embryo develops into a mature sporophyte, and the cycle begins again.

CHECK YOUR LEARNING

Can you describe the features that distinguish plants from other kinds of organisms?

21.2 HOW HAVE PLANTS EVOLVED?

The modern green algae known as stoneworts (Fig. 21-2) are plants' closest living relatives. The evolutionary relationship between plants and stoneworts has been revealed by DNA comparisons, and is reflected in other similarities between plants and green algae. For example, green algae and plants use the same type of chlorophyll and accessory pigments in photosynthesis. In addition, both plants and green algae store food as starch and have cell walls made of cellulose. In contrast, the photosynthetic pigments, food-storage molecules,

▲ **FIGURE 21-1 Alternation of generations in plants** As shown in this generalized depiction of a plant life cycle, a diploid sporophyte generation produces haploid spores through meiotic cell division. The spores develop into a haploid gametophyte generation that produces haploid gametes by mitotic cell division. The fusion of these gametes results in a diploid zygote that develops into the sporophyte plant.

▲ **FIGURE 21-2** *Chara*, **a stonewort** The green algae known as stoneworts are plants' closest living relatives.

and cell walls of other photosynthetic protists, such as the red algae and the brown algae, differ from those of plants.

The Ancestors of Plants Lived in Water

The ancestors of plants were photosynthetic protists, perhaps similar to stoneworts. Like modern stoneworts, the protists that gave rise to plants presumably lacked true roots, stems, leaves, and complex reproductive structures such as flowers or cones, features that appeared only later in the evolutionary history of plants. In addition, the ancestors of plants were confined to watery habitats.

For these ancestors of plants, life in water had many advantages. For example, in water, a body is bathed in a nutrient-rich solution, is supported by buoyancy, and is not likely to dry out. In addition, life in water facilitates reproduction, because gametes and zygotes can be carried by water currents or propelled by flagella.

Early Plants Invaded Land

Despite the benefits of aquatic environments, early plants invaded habitats on land. Today, most plants live on land. The move to land brought its own advantages, including access to sunlight unimpeded by water that might block its rays, and access to nutrients contained in surface rocks. However, the move to land also imposed some challenges; plants could no longer rely on watery surroundings to provide support, moisture, nutrients, and transportation for gametes and zygotes. As a result, life on land has favored the evolution in plants of traits that help meet these environmental challenges: structures that support the body and conserve water, conducting cells that transport water and nutrients to all parts of the plant, and processes that disperse gametes and zygotes by methods that are independent of water.

Plant Bodies Evolved to Resist Gravity and Drying

Some of the key adaptations to life on land arose early in plant evolution, and they are now found in virtually all land plants. These early adaptations include:

- Roots or rootlike structures that anchor the plant and absorb water and nutrients from the soil.
- A waxy **cuticle** that covers the surfaces of leaves and stems and that limits the evaporation of water (see Fig. 7-1).
- Pores called **stomata** (singular, stoma) in the leaves and stems that open to allow gas exchange but close when water is scarce, reducing the amount of water lost to evaporation (see Fig. 7-2).

Other key adaptations occurred somewhat later in the transition to terrestrial life, and are now widespread but not universal among plants (most nonvascular plants, a group described later, lack these traits):

- Conducting cells that transport water and minerals upward from the roots and that move photosynthetic products from the leaves to the rest of the plant body.
- The stiffening substance **lignin,** a rigid polymer that impregnates the conducting cells and supports the plant body against the force of gravity.

Plants Evolved Protection for Their Embryos and Sex Cells That Disperse Without Water

The most widespread groups of plants, collectively known as seed plants, are characterized by especially well-protected and well-provisioned embryos and by sex cells that do not rely on water for dispersal. The key adaptations of these plant groups are seeds, pollen, and, in the flowering plants, flowers and fruits.

Early seed plants gained an advantage over their competitors by producing seeds, which provided protection and nourishment for developing embryos and the potential for more effective dispersal. Early seed plants also produced dry, microscopic pollen grains that allowed wind, instead of water, to carry the male gametes. Later came the evolution of flowers, which enticed animal pollinators that were able to deliver pollen more precisely than did wind. Fruits also attracted animal foragers, which consumed the fruit and dispersed its seeds in their feces.

More Recently Evolved Plants Have Smaller Gametophytes

The evolutionary history of plants has been marked by a tendency for the sporophyte generation to become increasingly prominent, and for the longevity and size of the gametophyte generation to shrink (see Table 21-1). Thus, the earliest plants are believed to have been similar to today's nonvascular plants, which have a sporophyte that is smaller than the gametophyte

CASE STUDY c o n t i n u e d

Queen of the Parasites

The stinking corpse lily, with its huge, 3-foot-wide flowers, apparently evolved from an ancestor with tiny flowers. A recent analysis of DNA sequences revealed that the plant group most closely related to the group that includes the stinking corpse lily is the spurges, plants with mostly tiny flowers. The analysis also showed that the common ancestor of spurges and corpse lilies probably had flowers that were about $\frac{1}{80}$th the size of modern stinking corpse lily flowers.

and remains attached to it. In contrast, plants that originated somewhat later, such as ferns and the other seedless vascular plants, feature a life cycle in which the sporophyte is dominant, and the gametophyte is a much smaller, independent plant. Finally, in the most recently evolved group of plants, the seed plants, gametophytes are microscopic and barely recognizable as an alternate generation. These tiny gametophytes, however, still produce the eggs and sperm that unite to form the zygote that develops into the diploid sporophyte.

CHECK YOUR LEARNING

Can you describe the probable ancestor of plants? Can you identify the closest living relatives of plants and explain their similarities to and differences from plants? Can you describe the adaptations that equip plants for life on land?

21.3 WHAT ARE THE MAJOR GROUPS OF PLANTS?

Two major groups of land plants arose from ancient algal ancestors (**Fig. 21-3** and **Table 21-1**). Members of one group, the **nonvascular plants** (also called *bryophytes*), require a moist environment to reproduce and thus straddle the boundary between aquatic and terrestrial life, much like the amphibians of the animal kingdom. The other group, the

vascular plants (also called *tracheophytes*), has been able to colonize drier habitats.

Nonvascular Plants Lack Conducting Structures

Nonvascular plants retain some characteristics of their algal ancestors. Their gametes are dispersed by water, and they lack true roots, leaves, and stems. They do possess rootlike anchoring structures called *rhizoids* that bring water and nutrients into the plant body, but nonvascular plants lack well-developed structures for conducting water and nutrients. They must instead rely on slow diffusion or poorly developed conducting tissues to distribute water and other nutrients. As a result, their body size is limited. Size is also limited by the absence of the stiffening agent lignin in their bodies. Without lignin, nonvascular plants cannot grow upward very far. Most nonvascular plants are less than 1 inch (2.5 centimeters) tall.

Nonvascular Plants Include the Liverworts, Hornworts, and Mosses

The nonvascular plants include three groups: liverworts, hornworts, and mosses. Liverworts and hornworts are named for their shapes. The gametophytes of certain liverwort species have a lobed form reminiscent of the shape of a liver

TABLE 21-1	Features of the Major Plant Groups					
Group	Subgroup	Relationship of Sporophyte and Gametophyte	Transfer of Reproductive Cells	Early Embryonic Development	Dispersal	Water and Nutrient Transport Structures
Nonvascular plants	Liverworts Hornworts Mosses	The gametophyte is dominant—the sporophyte develops from a zygote retained on a gametophyte	Motile sperm swim to a stationary egg retained on a gametophyte	Occurs within the archegonium of a gametophyte	Haploid spores are carried by wind	Absent
Vascular plants	Club mosses Horsetails and ferns	The sporophyte is dominant—it develops from a zygote retained on a gametophyte	Motile sperm swim to a stationary egg retained on a gametophyte	Occurs within the archegonium of a gametophyte	Haploid spores are carried by wind	Present
	Gymnosperms	The sporophyte is dominant—the microscopic gametophyte develops within a sporophyte	Wind-dispersed pollen carries sperm to a stationary egg in a cone	Occurs within a protective seed containing a food supply	Seeds containing a diploid sporophyte embryo are dispersed by wind or animals	Present
	Angiosperms	The sporophyte is dominant—the microscopic gametophyte develops within a sporophyte	Pollen, dispersed by wind or animals, carries sperm to a stationary egg within a flower	Occurs within a protective seed containing a food supply; the seed is encased within fruit	Fruit, carrying seeds, is dispersed by animals, wind, or water	Present

▲ FIGURE 21-3 **Evolutionary tree of some major plant groups**

(a) Liverwort

(b) Hornwort

(c) Moss

(d) *Sphagnum* bog

▲ FIGURE 21-4 **Nonvascular plants** The plants shown here are less than a half-inch (about 1 centimeter) in height. **(a)** Liverworts grow in moist, shaded areas. The palmlike structures on the female plants shown here hold eggs. Male plants produce sperm that swim through a film of water to reach and fertilize the eggs. **(b)** The hornlike sporophytes of hornworts grow upward from the gametophyte body. **(c)** Moss plants, showing the stalks that carry spore-bearing capsules. **(d)** Mats of *Sphagnum* moss cover moist bogs in northern regions.

QUESTION Why are all nonvascular plants short?

(**Fig. 21-4a**). Hornwort sporophytes generally have a spiky shape that appears hornlike to some observers (**Fig. 21-4b**). Liverworts and hornworts are most abundant in areas where moisture is plentiful, such as in moist forests and near the banks of streams and ponds.

Mosses are the most diverse and abundant of the nonvascular plants (**Fig. 21-4c**). Like liverworts and hornworts, mosses are most likely to be found in moist habitats. Some mosses, however, have a waterproof covering that retains moisture, preventing water loss. Many of these mosses are also able to survive the loss of much of the water in their bodies; they dehydrate and become dormant during dry periods but absorb water and resume growth when moisture returns. Such mosses can survive in deserts, on bare rock, and in far northern and southern latitudes where humidity is low and liquid water is scarce for much of the year.

Mosses of the genus *Sphagnum* are especially widespread, living in moist habitats in northern regions around the world.

In many of these wet northern habitats, *Sphagnum* is the most abundant plant, forming extensive mats (**Fig. 21-4d**). Because decomposition is slow in cold climates and because *Sphagnum* contains compounds that inhibit bacterial growth, dead *Sphagnum* may decay very slowly. As a result, partially decayed moss tissue can accumulate in deposits that can, over thousands of years, become hundreds of feet thick. These deposits are known as peat. Peat has long been harvested for use as fuel, a practice that continues today in Ireland, Finland, Russia, and other northern countries. Now, however, peat is more often harvested for use in horticulture. Dried peat can absorb many times its own weight in water, making it useful as a soil conditioner and as a packing material for transporting live plants.

The Reproductive Structures of Nonvascular Plants Are Protected

Nonvascular plants require moisture to reproduce, but they have evolved some traits that facilitate reproduction on land

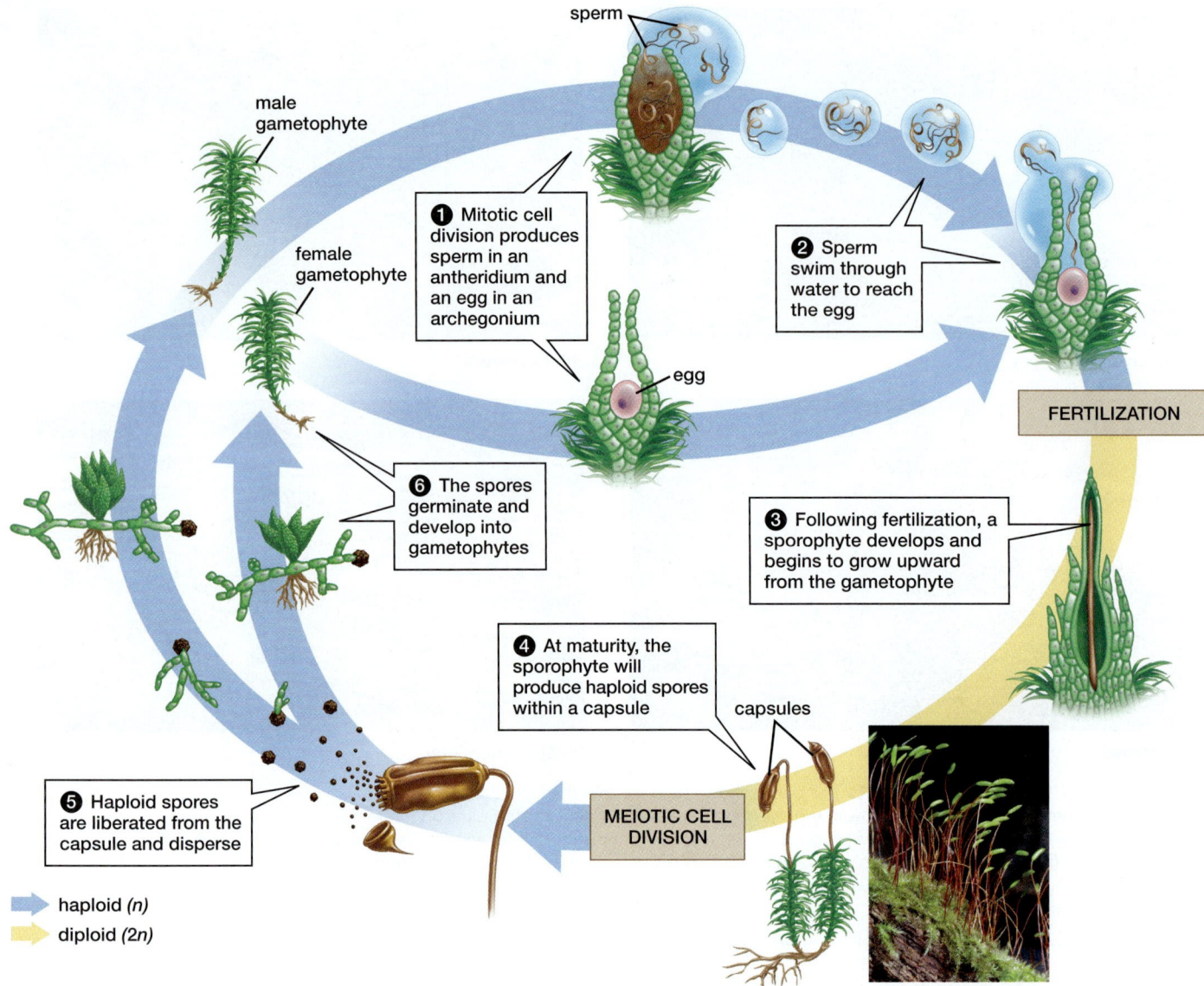

sperm

male
gametophyte

female
gametophyte

1 Mitotic cell division produces sperm in an antheridium and an egg in an archegonium

2 Sperm swim through water to reach the egg

egg

FERTILIZATION

6 The spores germinate and develop into gametophytes

3 Following fertilization, a sporophyte develops and begins to grow upward from the gametophyte

4 At maturity, the sporophyte will produce haploid spores within a capsule

capsules

5 Haploid spores are liberated from the capsule and disperse

MEIOTIC CELL DIVISION

➤ haploid *(n)*
➤ diploid *(2n)*

▲ **FIGURE 21-5 Life cycle of a moss** The photo shows moss plants; the short, leafy green plants are haploid gametophytes; the reddish brown stalks are diploid sporophytes.

(Fig. 21-5). For example, the reproductive structures of non-vascular plants are enclosed, which prevents the gametes from drying out. There are two types of reproductive structures: **archegonia** (singular, archegonium), in which eggs develop, and **antheridia** (singular, antheridium), where sperm are formed (Fig. 21-5 **1**). In some nonvascular plant species, both archegonia and antheridia are located on the same plant; in other species, each individual plant is either male or female.

In all nonvascular plants, the sperm must swim to the egg through a film of water (Fig. 21-5 **2**). (Nonvascular plants that live in drier areas can reproduce only when it rains.) After fertilization, the zygote is retained in the archegonium, where the embryo grows and matures into a small diploid sporophyte that remains attached to the parent gametophyte plant (Fig. 21-5 **3**). At maturity, the sporophyte produces reproductive capsules. Within each capsule, haploid spores are produced by meiotic cell division (Fig. 21-5 **4**).

When the capsule is opened, spores are released and dispersed by the wind (Fig. 21-5 **5**). If a spore lands in a suitable environment, it may develop into another haploid gametophyte plant (Fig. 21-5 **6**).

Vascular Plants Have Conducting Cells That Also Provide Support

Vascular plants are distinguished by specialized groups of tube-shaped conducting cells. These cells are impregnated with the stiffening substance lignin and serve both supportive and conducting functions. They allow vascular plants to grow taller than nonvascular plants, both because of the extra support provided by lignin and because the conducting cells allow water and nutrients absorbed by the roots to move to the upper portions of the plant. Another difference between vascular plants and nonvascular plants is that in vascular plants, the diploid sporophyte is the larger, more

(a) Club moss

(c) Fern

(b) Horsetail

(d) Tree fern

▲ **FIGURE 21-6 Some seedless vascular plants** Seedless vascular plants are found in moist woodland habitats. **(a)** The club mosses (sometimes called ground pines) grow in temperate forests. **(b)** The giant horsetail extends long, narrow branches in a series of rosettes at regular intervals along the stem. Its leaves are insignificant scales. At right is a cone-shaped spore-forming structure. **(c)** The leaves of this deer fern are emerging from coiled, immature leaves called fiddleheads. **(d)** Although most fern species are small, some, such as this tree fern, retain the large size that was common among ferns of the Carboniferous period.

QUESTION In each of these photos, is the pictured structure a sporophyte or a gametophyte?

conspicuous generation; in nonvascular plants, the haploid gametophyte is more evident.

The vascular plants can be divided into two groups: the seedless vascular plants and the seed plants.

The Seedless Vascular Plants Include the Club Mosses, Horsetails, and Ferns

Like the nonvascular plants, seedless vascular plants have swimming sperm and require water for reproduction. As their name implies, they do not produce seeds but rather propagate by spores. Present-day seedless vascular plants—the club mosses, horsetails, and ferns—are much smaller than their ancestors, which dominated the landscape in the Carboniferous period (359 million to 299 million years ago; see Fig. 17-8). (Today, seed plants are more prominent.)

Club Mosses and Horsetails Are Small and Inconspicuous

The club mosses, which despite their common name are not actually mosses, are now limited to representatives a few inches in height (**Fig. 21-6a**). Their leaves are small and scalelike, resembling the leaflike structures of mosses. Club mosses of the genus *Lycopodium*, commonly known as ground pine, form a beautiful ground cover in some temperate coniferous and deciduous forests.

Modern horsetails belong to a single genus, *Equisetum*, that contains only 15 species, most less than 3 feet tall (**Fig. 21-6b**). The bushy branches of some species lend them the common name horsetails; the leaves are reduced to tiny scales on the branches. They are also called "scouring rushes" because all species of *Equisetum* deposit large

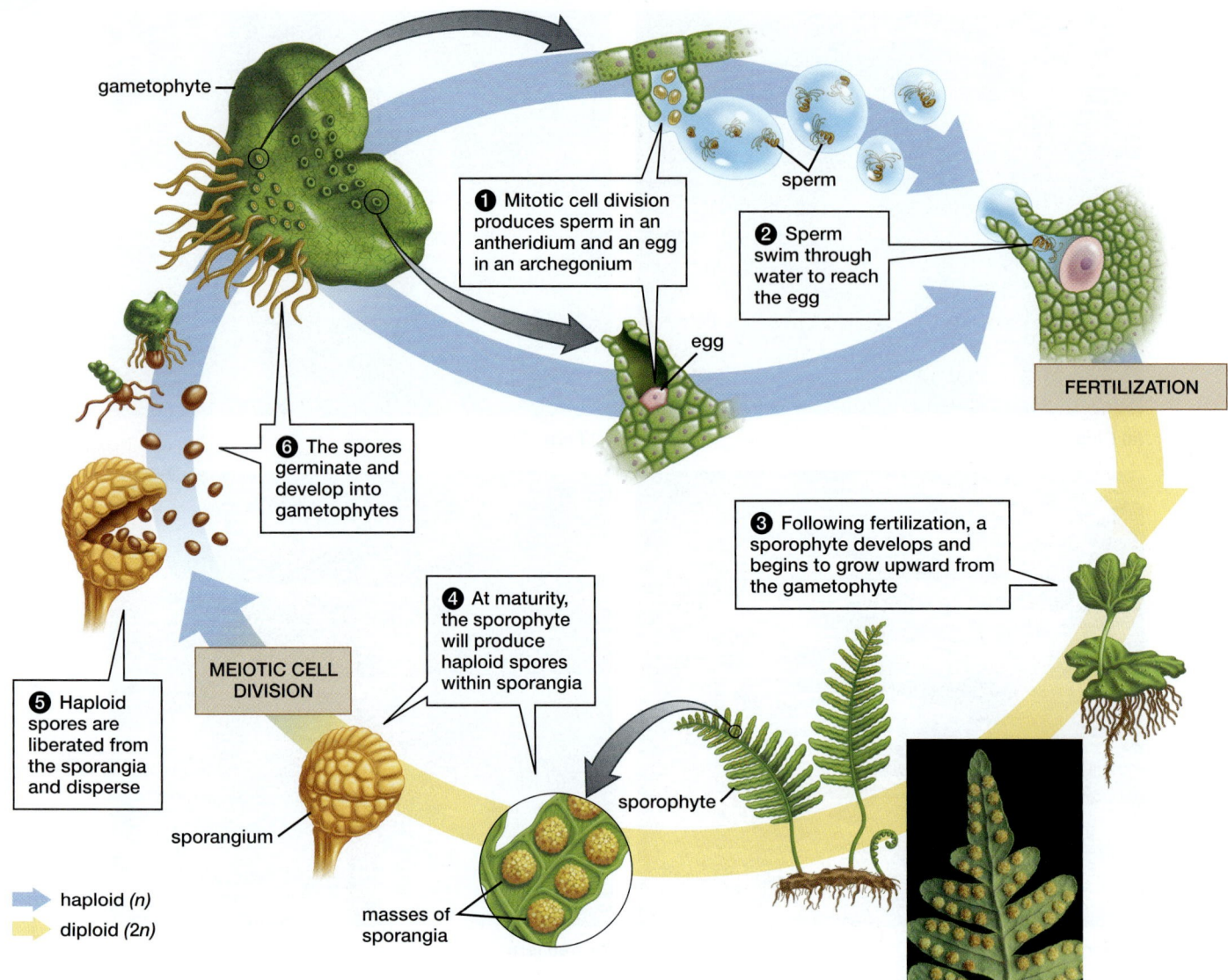

gametophyte

① Mitotic cell division produces sperm in an antheridium and an egg in an archegonium

sperm

② Sperm swim through water to reach the egg

egg

FERTILIZATION

⑥ The spores germinate and develop into gametophytes

③ Following fertilization, a sporophyte develops and begins to grow upward from the gametophyte

④ At maturity, the sporophyte will produce haploid spores within sporangia

MEIOTIC CELL DIVISION

⑤ Haploid spores are liberated from the sporangia and disperse

sporangium

sporophyte

masses of sporangia

haploid *(n)*

diploid *(2n)*

▲ FIGURE 21-7 **Life cycle of a fern**

amounts of silica (glass) in their outer layer of cells, giving them an abrasive texture. Early European settlers of North America used horsetails to scour pots and floors.

Ferns Are Broad-Leaved and Diverse

The ferns, with 12,000 species, are the most diverse of the seedless vascular plants (**Fig. 21-6c**). In the tropics, tree ferns still reach heights reminiscent of their ancestors from the Carboniferous period (**Fig. 21-6d**). Ferns are the only seedless vascular plants that have broad leaves.

In fern reproduction, gametes are produced in archegonia and antheridia on the tiny fern gametophyte (**Fig. 21-7 ❶**). Sperm are released into water and swim to reach an egg in an archegonium (**Fig. 21-7 ❷**). If fertilization occurs, the resulting zygote develops into a sporophyte plant, which grows upward from its parent, the gametophyte (**Fig. 21-7 ❸**). On a mature sporophyte fern plant, which is much larger than the gametophyte, haploid spores are produced in structures

called *sporangia* that form on special leaves of the sporophyte (**Fig. 21-7 ❹**). The sporangia open to release the spores, which are dispersed by the wind (**Fig. 21-7 ❺**). If a spore lands in a spot with suitable conditions, it germinates and develops into a gametophyte plant (**Fig. 21-7 ❻**).

The windborne spores of ferns make them especially effective at colonizing locations that lack abundant plant life. For example, just 2 years after a massive volcanic eruption that destroyed most life on the island of Krakatau in 1883, visitors reported that ferns blanketed the previously denuded landscape. Similarly, fern abundance increased dramatically following the catastrophic asteroid impact that caused the extinction of dinosaurs and many other species about 65 million years ago. Spores of fossil ferns are extremely abundant in 65-million-year-old rocks at many locations around the world; these "spore spikes" are interpreted as evidence that massive fires followed the asteroid impact, burning up most vegetation and creating an opening for widespread colonization by ferns.

The Seed Plants Are Aided by Two Important Adaptations: Pollen and Seeds

The seed plants are distinguished from nonvascular plants and seedless vascular plants by their production of pollen and seeds. In seed plants, gametophytes (which produce the sex cells) are tiny. The female gametophyte is a small group of haploid cells that produces the egg. The male gametophyte is the **pollen** grain. Pollen grains are dispersed by wind or by animal pollinators such as bees. In this way, sperm move through the air to fertilize egg cells. This airborne transport means that the distribution of seed plants is not limited by the need for water through which sperm can swim to the egg.

Analogous to the eggs of birds and reptiles, **seeds** consist of an embryonic sporophyte plant, a supply of food for the embryo, and a protective outer coat (**Fig. 21-8**). The *seed coat* maintains the embryo in a state of suspended animation or dormancy until conditions are suitable for growth. The stored food helps sustain the emerging plant until it develops roots and leaves and can make its own food by photosynthesis.

Seed plants are grouped into two general types: gymnosperms, which lack flowers, and angiosperms, the flowering plants.

Gymnosperms Are Nonflowering Seed Plants

Gymnosperms evolved earlier than the flowering plants. Early gymnosperms coexisted with the forests of seedless vascular plants that prevailed during the Carboniferous period. During the subsequent Permian period (299 million to 251 million years ago), however, gymnosperms became the predominant plant group and remained so until the rise of the flowering plants more than 100 million years later. Most of these early gymnosperms are now extinct. Today, only four groups of gymnosperms survive: ginkgos, cycads, gnetophytes, and conifers.

Only One Ginkgo Species Survives

Ginkgos have a long evolutionary history. They were widespread during the Jurassic period, which began 202 million years ago. Today, however, they are represented by the single species *Ginkgo biloba*, the maidenhair tree (**Fig. 21-9a**). Ginkgo trees are either male or female; female trees bear foul-smelling, fleshy seeds the size of cherries. Because they are more resistant to pollution than are most other trees, ginkgos (usually the male trees) have been extensively planted in U.S. cities. In the past few decades, the leaves of the ginkgo have gained attention as an herbal supplement that purportedly improves memory.

Cycads Are Restricted to Warm Climates

Like ginkgos, cycads were diverse and abundant in the Jurassic period but have since dwindled. Today approximately 160 species survive, most of which dwell in tropical or subtropical climates. Cycads have large, finely divided leaves and bear a superficial resemblance to palms or large ferns (**Fig. 21-9b**).

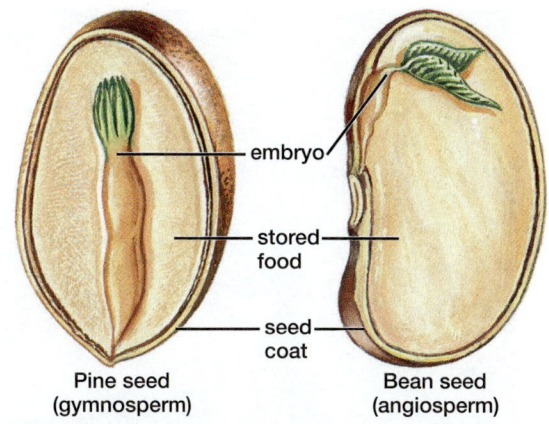

(a) Seeds

Pine seed (gymnosperm) Bean seed (angiosperm)

embryo — stored food — seed coat

(b) Dandelion

(c) Coconut

▲ **FIGURE 21-8 Seeds** **(a)** Seeds from a gymnosperm (left) and an angiosperm (right). Both consist of an embryonic plant and stored food confined within a seed coat. **(b)** The tiny seeds of the dandelion are dispersed by the wind, held aloft by parachute-like tufts that are part of the fruit. **(c)** The massive, armored seeds (protected inside the fruit) of the coconut palm, which can survive prolonged immersion in seawater as they traverse oceans.

QUESTION Can you think of some adaptations that help protect seeds from destruction by animal consumption?

(a) Gingko

(b) Cycad

(c) Gnetophyte

(d) Conifer

▲ **FIGURE 21-9 Gymnosperms (a)** The ginkgo, or maidenhair tree, is widely cultivated as a shade or ornamental tree. **(b)** A cycad. Common in the age of dinosaurs, these are now limited to about 160 species. Like ginkgos, cycads have separate sexes. **(c)** The leaves of the gnetophyte *Welwitschia* can live to be hundreds of years old. **(d)** The needle-shaped leaves of conifers are protected by a waxy surface layer.

Most cycads are about 3 feet (1 meter) in height, although some species can reach 65 feet (20 meters).

The tissues of cycads contain potent toxins. Despite the presence of these toxins, people in some parts of the world use cycad seeds, stems, and roots for food. Careful preparation and processing removes the toxins before the plants are consumed. Nonetheless, cycad toxins are the suspected cause of neurological problems that occur in societies, such as the Chamorro people of the Mariana Islands, that use cycads for food. Cycad toxins can also harm grazing livestock.

About half of all cycad species are classified as threatened or endangered. The main threats to cycads are habitat destruction, competition from introduced species, and harvesting for the horticultural trade. A large specimen of a rare cycad highly prized by collectors can sell for thousands of dollars. Because cycads grow slowly, recovery of endangered populations is uncertain.

Gnetophytes Include the Odd *Welwitschia*

The gnetophytes include about 70 species of shrubs, vines, and small trees. Leaves of gnetophyte species in the genus *Ephedra* contain alkaloid compounds that act in humans as stimulants and appetite suppressants. For this reason,

Ephedra is widely used as an energy booster and weight-loss aid. However, following reports of sudden deaths of *Ephedra* users and publication of several studies linking *Ephedra* consumption to increased risk of heart problems, the U.S. Food and Drug Administration banned the sale of products containing this gnetophyte.

The gnetophyte *Welwitschia mirabilis* is among the most distinctive of plants (**Fig. 21-9c**). Found only in the extremely dry deserts of southwest Africa, *Welwitschia* has a deep taproot that can extend as far as 100 feet (30 meters) down into the soil. Above the surface, the plant has a fibrous stem. Two (and only two) leaves grow from the stem. The leaves are never shed and remain on the plant for its entire life, which can be very long. The oldest *Welwitschia* are more than 2,000 years old, and a typical life span is about 1,000 years. The strap-like leaves continue to grow for that entire period, spreading over the ground. The older portions of the leaves, whipped by the wind for centuries, may shred or split, giving the plant its characteristic gnarled and tattered appearance.

Conifers Are Adapted to Cool Climates

Though other gymnosperm groups like gingkos and cycads are drastically reduced from their former prominence, the

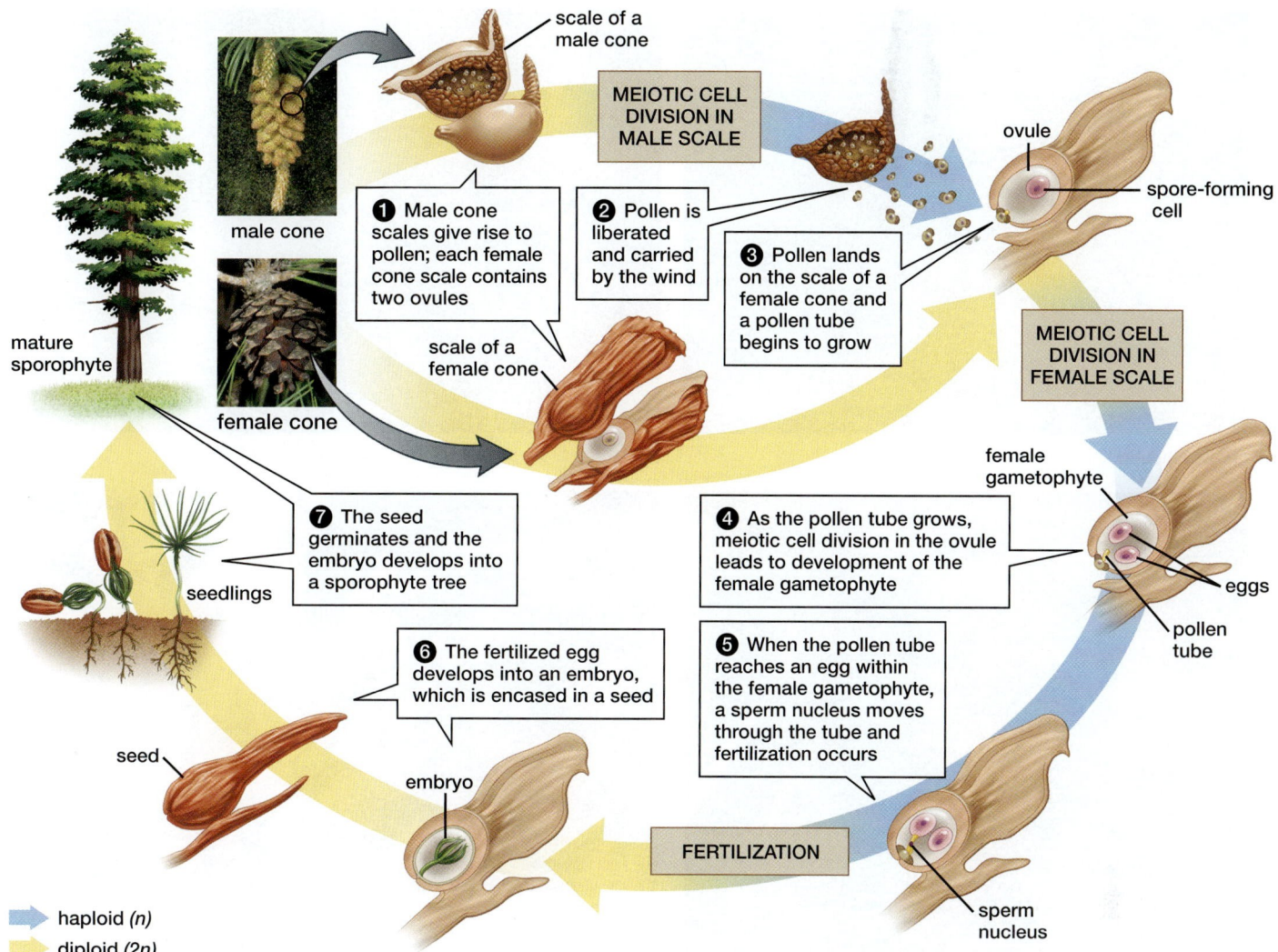

scale of a male cone

MEIOTIC CELL DIVISION IN MALE SCALE

1 Male cone scales give rise to pollen; each female cone scale contains two ovules

2 Pollen is liberated and carried by the wind

3 Pollen lands on the scale of a female cone and a pollen tube begins to grow

ovule

spore-forming cell

MEIOTIC CELL DIVISION IN FEMALE SCALE

male cone

female cone

scale of a female cone

mature sporophyte

7 The seed germinates and the embryo develops into a sporophyte tree

seedlings

4 As the pollen tube grows, meiotic cell division in the ovule leads to development of the female gametophyte

female gametophyte

eggs

pollen tube

6 The fertilized egg develops into an embryo, which is encased in a seed

5 When the pollen tube reaches an egg within the female gametophyte, a sperm nucleus moves through the tube and fertilization occurs

seed

embryo

FERTILIZATION

sperm nucleus

haploid (n)

diploid (2n)

▲ FIGURE 21-10 **Life cycle of the pine**

conifers still dominate large areas of our planet. Conifers, whose 500 species include pines, firs, spruce, hemlocks, and cypresses, are most abundant in the cold latitudes of the far north and at high elevations where conditions are dry. Not only is rainfall limited in these areas, but water in the soil remains frozen and unavailable during the long winters.

Conifers are adapted to these dry, cold conditions in three ways. First, most conifers retain green leaves throughout the year, enabling these plants to continue photosynthesizing and growing slowly during times when most other plants become dormant. For this reason, conifers are often called evergreens. Second, conifer leaves are thin needles covered with a thick, waterproof surface that minimizes evaporation (**Fig. 21-9d**). Finally, conifers produce an "antifreeze" in their sap that enables them to continue transporting nutrients in below-freezing temperatures. This substance gives them their fragrant piney scent.

Reproduction is similar in all conifers, so let's examine the reproductive cycle of a pine tree (**Fig. 21-10**). The tree itself is the diploid sporophyte, and it produces both male and female cones (**Fig. 21-10 ❶**). Male cones are relatively small (typically about ¾ inch long), delicate structures consisting

of scales in which pollen (the male gametophyte) develops. Each female cone consists of a series of woody scales arranged in a spiral around a central axis. At the base of each scale are two **ovules** (unfertilized seeds), within which diploid spore-forming cells arise.

Male cones release pollen during the reproductive season and then disintegrate (**Fig. 21-10 ❷**). The amount of pollen released is immense; inevitably, some pollen grains land by chance on female cone scales (**Fig. 21-10 ❸**). In the aftermath of such a pollination event, a pollen grain sends out a pollen tube that slowly burrows into an ovule. As the pollen tube grows, the diploid spore-forming cell in the ovule undergoes meiosis to produce haploid spores, one of which gives rise to a haploid female gametophyte, within which egg cells develop (**Fig. 21-10 ❹**). After nearly 14 months, the tube finally reaches the egg cell and releases the sperm that fertilize it (**Fig. 21-10 ❺**). The resulting zygote becomes enclosed in a seed as it develops into an embryo—a tiny embryonic sporophyte plant (**Fig. 21-10 ❻**). The seed is liberated when the cone matures and its scales separate. If it lands in a suitable patch of soil, it may germinate and grow into a sporophyte tree (**Fig. 21-10 ❼**).

(a) Duckweed

(c) Grass

(b) Eucalyptus

(d) Butterfly weed

▲ **FIGURE 21-11 Angiosperms (a)** The smallest angiosperm is the duckweed, found floating on ponds. These specimens are about ⅛ inch (3 millimeters) in diameter. **(b)** The largest angiosperms are eucalyptus trees, which can reach 325 feet (100 meters) in height. **(c)** Grasses (and many trees) have inconspicuous flowers and rely on wind for pollination. More conspicuous flowers, such as those on **(d)** this butterfly weed and on a eucalyptus tree (**b**, inset), entice insects and other animals that carry pollen between individual plants.

QUESTION What are the advantages and disadvantages of wind pollination? What are the advantages and disadvantages of pollination by animals? Why do both types of pollination persist among the angiosperms?

Angiosperms Are Flowering Seed Plants

Flowering plants, or **angiosperms,** have been Earth's predominant plants for more than 100 million years. The group is incredibly diverse, with more than 230,000 species. Angiosperms range in size from the diminutive duckweed (**Fig. 21-11a**) to the towering eucalyptus tree (**Fig. 21-11b**). From desert cactus to tropical orchids to grasses to parasitic stinking corpse lilies, angiosperms rule over the plant kingdom. Their enormous success is due in part to three major adaptations: flowers, fruits, and broad leaves.

Flowers Attract Pollinators

Flowers, the structures in which both male and female gametes are formed, may have evolved when gymnosperm ancestors formed an association with animals (most likely insects) that carried their pollen from plant to plant. According to this scenario, the relationship between these ancient gymnosperms and their animal pollinators was so beneficial that natural selection favored the evolution of showy flowers that advertised the presence of pollen to insects and other animals (**Figs. 21-11b, d**). The animals benefited by eating

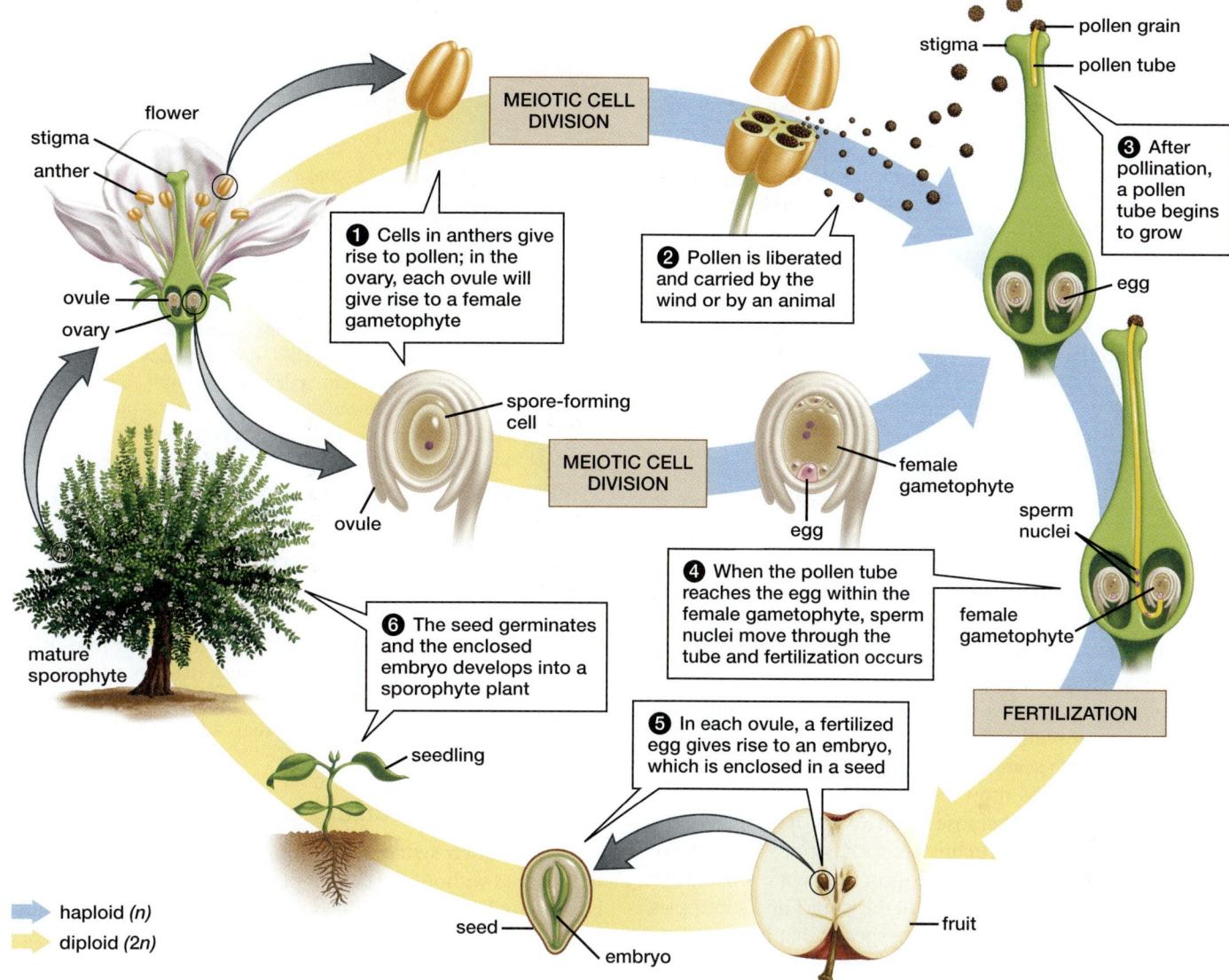

stigma — **pollen grain**
pollen tube

❸ After pollination, a pollen tube begins to grow

egg

MEIOTIC CELL DIVISION

flower

stigma
anther

❶ Cells in anthers give rise to pollen; in the ovary, each ovule will give rise to a female gametophyte

❷ Pollen is liberated and carried by the wind or by an animal

ovule
ovary

spore-forming cell

MEIOTIC CELL DIVISION

ovule

female gametophyte

egg

sperm nuclei

female gametophyte

❹ When the pollen tube reaches the egg within the female gametophyte, sperm nuclei move through the tube and fertilization occurs

mature sporophyte

❻ The seed germinates and the enclosed embryo develops into a sporophyte plant

❺ In each ovule, a fertilized egg gives rise to an embryo, which is enclosed in a seed

FERTILIZATION

seedling

➡ **haploid (n)**
➡ **diploid (2n)**

seed
embryo

fruit

▲ FIGURE 21-12 **Life cycle of a flowering plant**

some of the protein-rich pollen, whereas the plant benefited from the animals' unwitting transportation of pollen from plant to plant. With this animal assistance, many flowering plants no longer needed to produce prodigious quantities of pollen and send it flying on the fickle winds to ensure fertilization. But there are nonetheless many wind-pollinated angiosperms (**Fig. 21-11c**).

In the angiosperm life cycle (**Fig. 21-12**), flowers develop on the dominant sporophyte plant. In the flower, female gametophytes develop from ovules within a structure called the *ovary*; male gametophytes (pollen) are formed inside a structure called the *anther* (**Fig. 21-12 ❶**). During the reproductive season, pollen is released from the anthers and carried away on the wind or by animal pollinators (**Fig. 21-12 ❷**). If a pollen grain lands on a *stigma*, a sticky pollen-catching structure of the flower, a pollen tube begins to grow from the pollen grain (**Fig. 21-12 ❸**). The tube bores through the stigma and extends toward the female gametophyte, within which an egg cell has developed. Fertilization occurs when the pollen tube reaches the egg cell (**Fig. 21-12 ❹**). The

resulting zygote develops into an embryo enclosed in a seed formed from the ovule (**Fig. 21-12 ❺**). After it is dispersed, the seed may germinate and give rise to a sporophyte plant (**Fig. 21-12 ❻**).

Fruits Encourage Seed Dispersal

The ovary surrounding the seeds of an angiosperm matures into a **fruit**, the second adaptation that has contributed to the success of angiosperms. Just as flowers encourage animals to transport pollen, so, too, many fruits entice animals to disperse seeds. If an animal eats a fruit, many of the enclosed seeds may pass through the animal's digestive tract unharmed, perhaps to fall at a suitable location for germination. Not all fruits, however, depend on edibility for dispersal. Dog owners are well aware, for example, that some fruits (called burs) disperse by clinging to animal fur. Other fruits, such as those of maples, form wings that carry the seed through the air. The variety of dispersal mechanisms made possible by fruits has helped the angiosperms invade nearly all terrestrial habitats.

Queen of the Parasites

Why does the flower of a stinking corpse lily smell like rotting meat? Though the smell is utterly revolting to humans, it is attractive to blowflies and other insects that normally feed on and lay their eggs in decaying flesh. When such insects visit a male stinking corpse lily, they may carry away pollen that can fertilize a nearby female flower.

In many angiosperm species, flowers contain nectar that provides food for animal pollinators. But no such nectar reward awaits a fly that enters the flower of a stinking corpse lily. Instead, a fly attracted by the flower's stench searches in vain for putrefying meat, its movement guided toward the flower's cache of sticky pollen by grooves and hairs inside the flower. Eventually, the fly departs, coated in pollen. In essence, the fly has been tricked by the plant into providing a service for no reward. Thus, the stinking corpse lily is a master exploiter: It takes advantage of both the host vines that provide its food and the flies that facilitate its reproduction.

Broad Leaves Capture More Sunlight

The third feature that gives angiosperms an advantage in warmer, wetter climates is broad leaves. When water is plentiful, as it is during the warm growing season of temperate and tropical climates, broad leaves provide an advantage by collecting more sunlight for photosynthesis than the slender needles of conifers can. In regions with seasonal variations in growing conditions, many trees and shrubs drop their leaves during periods when water is in short supply, because being leafless reduces evaporative water loss. In temperate climates, such periods occur during the fall and winter, at which time most temperate angiosperm trees and shrubs drop their leaves. In the tropics and subtropics, most angiosperms are evergreen, but species that inhabit tropical areas where periods of drought are common may drop their leaves to conserve water during the dry season.

The advantages of broad leaves are offset by some evolutionary costs. In particular, broad, tender leaves are much more appealing to herbivores than are the tough, waxy needles of conifers. As a result, angiosperms have developed a range of defenses against mammalian and insect herbivores. These adaptations include physical defenses such as thorns, spines, and resins that toughen the leaves. The evolutionary struggle for survival has also led to a host of chemical defenses—compounds that make plant tissue poisonous or distasteful to potential predators. Many of the compounds responsible for chemical defense have properties that humans have exploited for medicinal and culinary uses. Medicines such as aspirin and codeine, stimulants such as nicotine and caffeine, and spices such as mustard and pepper are all derived from angiosperm plants.

CHECK YOUR LEARNING

Can you explain how vascular plants and nonvascular plants differ? Can you describe the major plant taxonomic groups and representative members of each group? Are you able to describe the key steps in the life cycles of mosses, ferns, gymnosperms, and flowering plants?

Which Plants Provide Us with the Most Food?

Although thousands of plant species have edible parts, people exploit only a small proportion of them for food. In fact, the vast majority of the plant-derived food consumed by humans comes from only 20 species. The fruits (grains) of just three grass species—corn (maize), wheat, and rice—provide about half the calories consumed by people worldwide. (The average person in the United States consumes about 200 pounds of these grains each year.) In terms of annual production, the big three are followed, in order, by soybeans, barley, sorghum, millet, and peanuts. Looking further down the list, the world's most abundantly produced foods that are *not* grains or legumes are potatoes and cassava (a root that is a staple in parts of Africa and South America).

21.4 HOW DO PLANTS AFFECT OTHER ORGANISMS?

As plants survive, grow, and reproduce, they alter and influence Earth's landscape and atmosphere in ways that are tremendously beneficial to the rest of the planet's inhabitants, including humans. Humans also reap additional benefits by actively exploiting plants.

Plants Play a Crucial Ecological Role

The complex ecosystems that host terrestrial life could not be maintained without the help of plants. Plants make vital contributions to the food, air, soil, and water that sustain life on land.

Plants Capture Energy That Other Organisms Use

Plants provide food, directly or indirectly, for all of the animals, fungi, and nonphotosynthetic microbes on land. Plants use photosynthesis to capture solar energy, and they convert part of the captured energy into leaves, shoots, seeds, and fruits that are eaten by other organisms. Many of these consumers of plant tissue are themselves eaten by still other organisms. Plants are the main providers of energy and nutrients to terrestrial ecosystems, and life on land depends on plants' ability to manufacture food from sunlight.

Plants Help Maintain the Atmosphere

In addition to their role as food suppliers, plants make essential contributions to the atmosphere. For example, plants produce oxygen gas as a by-product of photosynthesis, and by doing so they continually replenish oxygen in the atmosphere. Without plants' contribution, atmospheric oxygen would be rapidly depleted by the oxygen-consuming respiration of Earth's multitude of organisms.

Plants Build and Protect Soil

Plants also help create and maintain soil. When a plant dies, its stems, leaves, and roots become food for fungi, prokaryotes, and other decomposers. Decomposition breaks the plant tissue into tiny particles of organic matter that become part of the soil. Organic matter improves the ability of soil to hold water and nutrients, thereby making the soil more fertile and better able to support the growth of living plants. The roots of those living plants help hold the soil together and keep it in place. Soils from which vegetation has been removed are susceptible to erosion by wind and water (Fig. 21-13).

Plants Help Keep Ecosystems Moist

Plants take up water from the soil and retain some of it in their tissues. By doing so, plants slow the rate at which water escapes from terrestrial ecosystems, and increase the amount of water available to meet the needs of the ecosystems' inhabitants. By reducing the amount of water runoff, plants also reduce the chances of destructive flooding. Thus, floods can be more frequent in areas in which forests, grasslands, or marshes have been destroyed by human activities.

Plants Provide Humans with Necessities and Luxuries

It would be difficult to exaggerate the degree to which human populations depend on plants. Neither our explosive population growth nor our rapid technological advance would have been possible without plants.

Plants Provide Shelter, Fuel, and Medicine

Plants are the source of the wood that is used to construct housing for a large portion of Earth's human population. For much of human history, wood was also the main fuel for warming dwellings and for cooking. Wood is still the most important fuel in many parts of the world. Coal, another important fuel, is composed of the remains of ancient plants that have been transformed by geological processes.

Plants have also supplied many of the medicines on which modern health care depends. Important drugs that were originally found in and extracted from plants include aspirin, the heart medication digoxin, the cancer treatments Taxol® and vinblastine, the malaria drug quinine, the painkillers codeine and morphine, and many more.

In addition to harvesting useful material from wild plants, humans have domesticated a host of useful plant species. Through generations of selective breeding, people have modified the seeds, stems, roots, flowers, and fruits of favored plant species to provide themselves with food and fiber. It is difficult to imagine life without corn, rice, potatoes, apples, tomatoes, cooking oil, cotton, and the myriad other staples that domestic plants provide.

Plants Provide Pleasure

Despite the obvious contributions of plants to human well-being, our relationship with plants seems to be based on something more profound than their ability to help us meet our material needs. Though we appreciate the practical value of wheat and wood, our most emotionally powerful connections with plants are purely sensual. Many of life's pleasures

▲ FIGURE 21-13 **Plants protect soil** Damage to natural vegetation, such as the deforestation of this mountainside, leaves the underlying soil vulnerable to erosion.

come to us courtesy of our plant partners. We delight in the beauty and fragrance of flowers, and present them to others as symbols of our most sublime and inexpressible emotions. Quite a few of us spend hours of our leisure time tending gardens and lawns, for no reward other than the pleasure and satisfaction we derive from observing the fruits of our labor. In our homes, we reserve space not only for members of our families, but also for our houseplant companions. We feel compelled to line our streets with trees, and we seek refuge from the stress of daily life in parks with abundant plant life. Clearly, plants help fill our desires as well as our needs.

CHECK YOUR LEARNING

Can you describe some of the effects that plants have on other organisms, including humans?

CASE STUDY revisited

Queen of the Parasites

The 17 or so parasitic plant species of the genus *Rafflesia*, which includes the stinking corpse lily, are found in the moist forests of Southeast Asia, a habitat that is disappearing rapidly as forests are cleared for agriculture and development. The geographic range of the stinking corpse lily is limited to the dwindling forests of the Malaysian peninsula and the Indonesian islands of Borneo and Sumatra; the species is rare and endangered. The government of Indonesia has established parks and reserves in an effort to help protect the stinking corpse lily, but—as is often the case in developing countries—a forest that is protected on paper may still be vulnerable in reality.

Perhaps the best hope for the continued survival of the largest *Rafflesia* is the growing realization among the rural residents of Sumatra and Borneo that the spectacular, putrid-smelling flowers of the stinking corpse lily might lure interested tourists to their countries. Under an innovative conservation program that seeks to take advantage of this potential for ecotourism, people who live in the vicinity of the stinking corpse lily can become caretakers of the plants. These assigned caretakers watch over the plants and, in return, may charge a small fee to curious visitors. Local inhabitants have been given an economic incentive to protect this rare parasitic plant.

Consider This

A parasitic lifestyle is unusual among plants, but it is not really rare. Many plant species are parasitic, and systematists estimate that parasitism has evolved at least nine different times over the evolutionary history of plants. Given the obvious benefits of photosynthesis, why has parasitism (which is often accompanied by loss of photosynthetic capability) evolved repeatedly in photosynthetic plants?

CHAPTER REVIEW

Summary of Key Concepts

21.1 What Are the Key Features of Plants?
Plants are photosynthetic, multicellular organisms that exhibit alternation of generations, in which a haploid gametophyte generation alternates with a diploid sporophyte generation. Unlike their green algae relatives, plants have multicellular, dependent embryos.

21.2 How Have Plants Evolved?
Photosynthetic protists, probably aquatic green algae, gave rise to the first plants. Ancestral plants were probably similar to modern multicellular algae such as stoneworts, which are plants' closest living relatives.

Early plants invaded terrestrial habitats, and modern plants exhibit a number of key adaptations for terrestrial existence: rootlike structures for anchorage and for absorption of water and nutrients; a waxy cuticle that slows the loss of water through evaporation; stomata that can open, allowing gas exchange, or close, preventing water loss; conducting cells that transport water and nutrients throughout the plant; and a stiffening substance, called lignin, that impregnates the conducting cells and helps support the plant body.

Plant reproductive structures suitable for life on land include a smaller male gametophyte (pollen) that allows wind to replace water in carrying sperm to eggs; seeds that nourish, protect, and help disperse developing embryos; flowers that attract animals, which carry pollen more precisely and efficiently than wind; and fruits that entice animals to disperse seeds.

There has been a general evolutionary trend toward a reduction in size of the haploid gametophyte, which is dominant in nonvascular plants but microscopic in seed plants.

21.3 What Are the Major Groups of Plants?
Two major groups of plants, nonvascular plants and vascular plants, arose from their ancient algal ancestors. Nonvascular plants, including the hornworts, liverworts, and mosses, are small, simple land plants that lack conducting cells. Although some have adapted to dry areas, most live in moist habitats. Nonvascular plant reproduction requires water through which the sperm swim to the egg.

In vascular plants, a system of conducting cells—stiffened by lignin—conducts water and nutrients absorbed by the roots into the upper portions of the plant and supports the body as well.

Thanks to this support system, seedless vascular plants, including the club mosses, horsetails, and ferns, can grow larger than nonvascular plants. As in nonvascular plants, the sperm of seedless vascular plants must swim to the egg for sexual reproduction to occur.

Vascular plants with seeds have two major additional adaptive features: pollen and seeds. Seed plants are often classified into two categories: gymnosperms and angiosperms. Gymnosperms include ginkgos, cycads, gnetophytes, and the highly successful conifers. These plants were the first fully terrestrial plants to evolve. Their success on dry land is partially due to the evolution of the male gametophyte into the pollen grain. Pollen protects and transports the sperm, eliminating the need for them to swim to the egg. The seed, a protective resting structure containing an embryo and a supply of food, is a second important adaptation contributing to the success of seed plants.

Angiosperms, the flowering plants, dominate much of the land today. In addition to pollen and seeds, angiosperms also produce flowers and fruits. The flower allows angiosperms to use animals as pollinators. In contrast to wind, animals can in some cases carry pollen farther and with greater accuracy and less waste. Fruits may attract animal consumers, which incidentally disperse the seeds in their feces.

21.4 How Do Plants Affect Other Organisms?
Plants play a key ecological role, capturing energy for use by inhabitants of terrestrial ecosystems, replenishing atmospheric oxygen, creating and stabilizing soils, and slowing the loss of water from ecosystems. Plants are also exploited by humans to provide food, fuel, building materials, medicines, and aesthetic pleasure.

Key Terms

alternation of
  generations   *382*
angiosperm   *394*
antheridium (plural,
  antheridia)   *388*
archegonium (plural,
  archegonia)   *388*
conifer   *393*
cuticle   *384*
flower   *394*
fruit   *395*

gametophyte   *382*
gymnosperm   *391*
lignin   *384*
nonvascular plant   *385*
ovule   *393*
pollen   *391*
seed   *391*
sporophyte   *382*
stoma (plural, stomata)   *384*
vascular plant   *385*

Learning Outcomes

In this chapter, you have learned to . . .

LO1 Describe the key features of plants.

LO2 Describe the evolutionary history of plants and the adaptations that equip plants for life on land.

LO3 Compare and contrast vascular plants and nonvascular plants.

LO4 Describe the major plant taxonomic groups and list representative members of each group.

LO5 Describe the key steps in the life cycles of mosses, ferns, gymnosperms, and flowering plants.

LO6 Describe the effects that plants have on other organisms, including humans.

Thinking Through the Concepts

Fill-in-the-Blank

1. Scientists hypothesize that the ancestors of plants were _____. There are two major types of plants; those that lack conducting cells are called _____ and those with conducting cells are called _____. All plants produce multicellular _____ and exhibit a complex life cycle called _____. **LO1 LO2 LO3**

2. Plant adaptations to life on land include a(n) _____, which reduces evaporation of water, and _____, which open to allow gas exchange but close when _____ is scarce. In addition, the bodies of vascular plants gain increased support from _____ impregnated with the polymer _____; these structures also help _____ and _____ to move within the plant body. **LO2**

3. Seedless vascular plants must reproduce when conditions are wet because their sperm must _____. Two adaptations that allow seed plants to reproduce more efficiently on dry land are _____ and _____. The seed plants fall into two major categories: the nonflowering _____ and the flowering _____. Flowers were favored by natural selection because they _____. Fruits were favored by natural selection because they _____. **LO2 LO3**

4. Three groups of nonvascular plants are _____, _____, and _____. Three groups of seedless vascular plants are _____, _____, and _____. Today, the most diverse group of plants is the _____. **LO4**

Review Questions

1. What is meant by "alternation of generations"? What two generations are involved? How does each reproduce? **LO1**

2. Explain the evolutionary changes in plant reproduction that adapted plants to increasingly dry environments. **LO2 LO5**

3. Describe evolutionary trends in the life cycles of plants. Emphasize the relative sizes of the gametophyte and sporophyte. **LO2 LO3 LO5**

4. From which algal group did green plants probably arise? Explain the evidence that supports this hypothesis. **LO2**

5. List the structural adaptations necessary for the invasion of dry land by plants. Which of these adaptations are possessed by nonvascular plants? By ferns? By gymnosperms and angiosperms? **LO2 LO4**

6. The number of species of flowering plants is greater than the number of species in the rest of the plant kingdom combined. What feature(s) are responsible for the enormous success of angiosperms? Explain why. **LO2 LO4**

7. List the adaptations of gymnosperms that have helped them become the dominant trees in dry, cold climates. **LO2 LO4**

8. What is a pollen grain? What role has it played in helping plants colonize dry land? **LO2 LO4 LO5**

9. The majority of all plants are seed plants. What is the advantage of a seed? How do plants that lack seeds meet the needs served by seeds? **LO2 LO4 LO5**

Applying the Concepts

1. You are a geneticist working for a firm that specializes in plant biotechnology. Explain what *specific* parts (fruit, seeds, stems, roots, etc.) of the following plants you would try to alter by genetic engineering, what changes you would try to make, and why: (a) corn, (b) tomatoes, (c) wheat, and (d) avocados.

2. **BioEthics** Prior to the development of synthetic drugs, more than 80% of all medicines were of plant origin. Even today, indigenous tribes in remote Amazonian rain forests can provide a plant product to treat virtually any ailment. Herbal medicine is also widely and successfully practiced in China. Most of these drugs are unknown to the Western world. But the forests from which much of this plant material is obtained are being converted to agriculture. We are in danger of losing many of these potential drugs before they can be evaluated by Western medicine. What steps can you suggest to preserve these natural resources while also allowing nations to direct their own economic development?

3. Only a few hundred of the more than 200,000 species of plants have been domesticated for human use. One example is the almond. The domestic almond is nutritious and harmless, but its wild precursor can cause cyanide poisoning. The oak makes potentially nutritious seeds (acorns) that contain very bitter-tasting tannins. If we could breed the tannin out of acorns, they might become a delicacy. Why do you suppose we have failed to domesticate oaks?

Answers to Figure Caption questions and Fill-in-the-Blank questions can be found in the Answers section at the back of the book.

MB *Go to MasteringBiology for practice quizzes, activities, eText, videos, current events, and more.*

These honey mushrooms are part of the visible portion of the largest organism on Earth.

CASE STUDY

Humongous Fungus

WHAT IS THE LARGEST organism on Earth? A reasonable guess might be the world's largest animal, the blue whale, which can be 100 feet long and weigh 400,000 pounds. But the blue whale is dwarfed by the General Sherman tree, a giant sequoia that is 275 feet high and whose weight is estimated at 6,200 *tons*. Even these two behemoths, however, can't match the real record-holder, the fungus *Armillaria ostoyae*, also known as the honey mushroom.

The largest known *Armillaria* is a specimen in Oregon that spreads over almost 2,400 acres (about 3.4 square miles) and probably weighs even more than the General Sherman tree. Despite its huge size, no one has actually seen the monster fungus, because it is largely underground. Its only aboveground parts are brown mushrooms that sprout occasionally from the creature's gigantic body. Just beneath the surface, however, the fungus spreads through the soil by means of long, string-like structures that extend until they encounter the tree roots on which *Armillaria* subsists.

How can researchers be sure that the Oregon fungus is truly one single individual and not many intertwined individuals? The strongest evidence is genetic. Researchers gathered *Armillaria* tissue samples from throughout the area thought to be inhabited by a single individual and compared DNA extracted from the samples. All were genetically identical, demonstrating that they came from the same individual.

It may seem strange that the world's largest organism went unnoticed until very recently, but the lives of fungi typically take place outside of our view. Nonetheless, fungi play a fascinating role in human affairs. Read on to find out more about the inconspicuous but often influential fungi.

AT A GLANCE

LEARNING GOALS

LG1 Describe the key features of fungi.

LG2 List the major fungus taxonomic groups and describe their key characteristics.

LG3 Summarize the fungal life cycle and describe the key steps in the life cycles of basidiomycetes, ascomycetes, and bread molds.

LG4 Describe effects that fungi have on other organisms, including humans.

LG5 Describe ecologically important symbiotic associations between fungi and other organisms.

22.1 WHAT ARE THE KEY FEATURES OF FUNGI?

When you think of a fungus, you probably picture a mushroom. Most fungi, however, do not produce mushrooms. And even in those that do, the mushrooms are just temporary reproductive structures. The main body is typically concealed beneath the soil or inside a piece of decaying wood. So, to fully appreciate fungi, we must look beyond the conspicuous structures we encounter on the forest floor, at the edges of our lawns, or on top of a pizza. A closer look at fungi reveals a group of eukaryotic, mostly multicellular organisms that play a key role in the web of life and whose lifestyle differs in fascinating ways from that of plants or animals.

Fungal Bodies Consist of Slender Threads

The body of almost every fungus is a **mycelium** (plural, mycelia; **Fig. 22-1a**), which is an interwoven mass of one-cell-thick, threadlike filaments called **hyphae** (singular, hypha; **Figs. 21-1b, c**). In some species, hyphae consist of single elongated cells with numerous nuclei; in other species, hyphae are subdivided—by partitions called **septa** (singular, septum)—into many cells, each containing one or more nuclei. Pores in the septa allow cytoplasm to stream between cells, distributing nutrients. Like plant cells, fungal cells are surrounded by cell walls. Unlike plant cells, however, fungal cell walls are strengthened by *chitin*, the same substance found in the hard outer surface (exoskeleton) of insects, crabs, and their relatives.

Fungi cannot move. They compensate for this lack of mobility with hyphae that can grow rapidly in any direction within a suitable environment. In this way, the fungal mycelium can quickly spread into aging bread or cheese, beneath the bark of decaying logs, or into the soil. Periodically, the hyphae differentiate into reproductive structures that project above the surface beneath which the mycelium grows. These

(a) Mycelium **(b) Hyphae** **(c) Hypha cross-section**

▲ FIGURE 22-1 **The filamentous body of a fungus (a)** A fungal mycelium spreads over decaying vegetation. The mycelium is composed of **(b)** a tangle of microscopic hyphae, only one cell thick, portrayed in cross-section **(c)** to show their internal organization.

QUESTION Which features of a fungus's body structure are adaptations related to its method of acquiring nutrients?

structures, including mushrooms, puffballs, and the powdery molds on spoiled food, represent only a fraction of the complete fungal body, but are typically the only part of the fungus that we can easily see.

Fungi Obtain Their Nutrients from Other Organisms

Like animals, fungi survive by breaking down nutrients stored in the bodies or wastes of other organisms. Some fungi digest the bodies of dead organisms. Others are parasitic, feeding on living organisms and causing disease. Some live in close, mutually beneficial relationships with other organisms that provide food. There are even a few predatory fungi, which attack tiny worms in soil (**Fig. 22-2**).

Unlike most animals, fungi do not ingest food. Instead, they secrete enzymes that digest complex molecules outside their bodies, breaking down the molecules into smaller subunits that can be absorbed. Fungal hyphae can penetrate deeply into a source of nutrients and, because the hyphae are only one cell thick, each cell in a fungal body is in position to absorb nutrients directly from the surrounding environment. Their mode of securing nutrition serves fungi well. Almost every biological material can be consumed by at least one fungal species, so nutritional support for fungi is likely to be present in nearly every habitat.

Fungi Can Reproduce Both Asexually and Sexually

Fungi develop from **spores**—haploid (contain only a single copy of each chromosome) cells that can give rise to a new individual. Fungal spores are tiny and extraordinarily mobile, even though most lack a means for self-propulsion. They are distributed far and wide as hitchhikers on the outside of animal bodies, as passengers inside the digestive systems of animals that have eaten them, or as airborne drifters, cast aloft by chance or shot into the atmosphere by elaborate reproductive structures (**Fig. 22-3**). Spores are often produced in great numbers; a single giant puffball may contain 5 trillion spores.

(a) Earthstar

(b) *Pilobolus*

▲ FIGURE 22-3 **Some fungi can eject spores** (a) A ripe earthstar mushroom, struck by a drop of water, releases a cloud of spores that will be dispersed by air currents. (b) The delicate, translucent reproductive structures of *Pilobolus*, which inhabits horse manure, literally blow their tops when ripe, dispersing the black, spore-containing caps up to 3 feet away. Spores that adhere to grass remain there until consumed by a grazing herbivore, perhaps a horse. Later (likely some distance away), the horse will deposit a fresh pile of manure containing *Pilobolus* spores that have passed unharmed through its digestive tract.

▲ FIGURE 22-2 **Nemesis of nematodes** The fungus *Arthrobotrys*, also known as the nematode (roundworm) strangler, traps its prey in a noose-like modified hypha. When a nematode wanders into the noose, its presence stimulates the noose cells to swell with water. In a fraction of a second, the noose constricts, trapping the worm. Fungal hyphae then penetrate and feast on their prey.

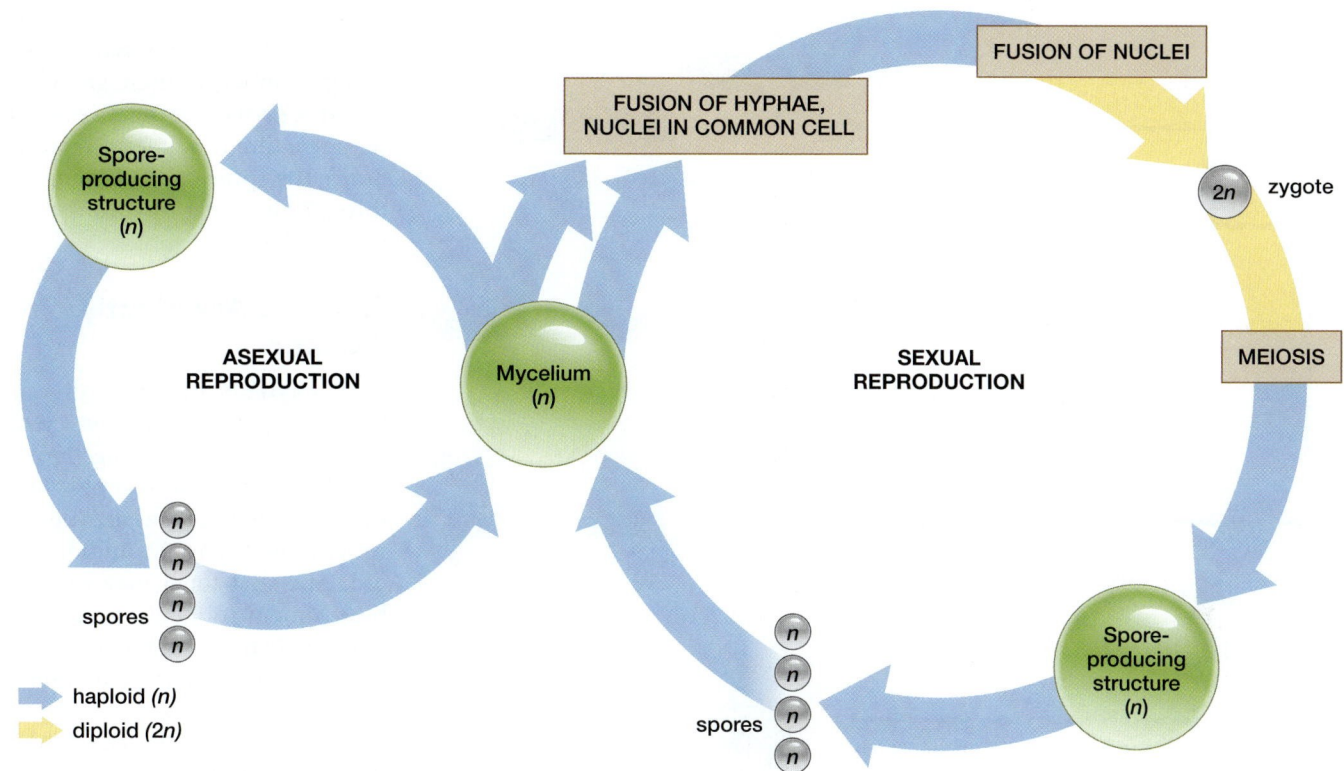

▲ **FIGURE 22-4 Generalized life cycle of fungi** In asexual reproduction, haploid hyphae in a mycelium give rise to structures that produce haploid spores by mitotic cell division. In sexual reproduction, haploid hyphae of different, compatible mating types fuse, resulting in cells that contain nuclei from both parents. These nuclei subsequently fuse, generating a diploid zygote that undergoes meiosis to yield haploid spores.

In general, fungi are capable of both asexual and sexual reproduction (**Fig. 22-4**). For the most part, fungi reproduce asexually under stable conditions, with sexual reproduction occurring mainly under conditions of environmental change or stress. Both asexual and sexual reproduction ordinarily involve the production of spores within special fruiting bodies that project above the mycelium.

Asexual Reproduction Produces Haploid Spores by Mitosis

The mycelia and spores of fungi are haploid. A haploid mycelium produces haploid asexual spores by mitosis. If an asexual spore is deposited in a favorable location, it will begin mitotic divisions and develop into a new mycelium. This simple reproductive cycle results in the rapid production of a genetically identical clone of the original mycelium.

Sexual Reproduction Produces Haploid Spores by Meiosis

Diploid structures form only during a brief period of the sexual portion of the fungal life cycle. Sexual reproduction begins when a filament of one mycelium comes into contact with a filament from a second mycelium that is of a different, but compatible, mating type. (The different mating types of fungi are analogous to the different sexes of animals, except that in fungi there are often more than two mating types.) If conditions are suitable, the two hyphae may fuse,

so that nuclei from the two different hyphae share a common cell. This merger of hyphae is followed by fusion of the two different haploid nuclei to form a diploid zygote. The zygote then undergoes meiosis to form haploid sexual spores. These spores are dispersed, germinate, and divide by mitosis to form new haploid mycelia. Unlike the cloned offspring produced by asexual spores, these sexually produced fungal bodies are genetically distinct from either parent.

CHECK YOUR LEARNING

Can you describe the structure of a typical fungus? Are you able to explain how fungi obtain energy and nutrients and how they reproduce?

22.2 WHAT ARE THE MAJOR GROUPS OF FUNGI?

Nearly 100,000 species of fungi have been described, but this number represents only a fraction of the true diversity of these organisms. Many new species are discovered and described each year, and mycologists (scientists who study fungi) estimate that the number of undiscovered species of fungus is well over a million. Fungus species are classified into six main taxonomic groups: Chytridiomycota (chytrids), Neocallimastigomycota (rumen fungi), Blastocladiomycota (blastoclades), Glomeromycota (glomeromycetes), Basidiomycota (basidiomycetes), and Ascomycota (ascomycetes)

Chytridiomycota

Neocallimastigomycota

Blastocladiomycota

Glomeromycota

Basidiomycota

Ascomycota

▲ **FIGURE 22-5 Evolutionary tree of the major groups of fungi**

by their swimming spores, which require water for dispersal. (Though many members of these groups live in water, even those that live on land require a film of water for reproduction.) The spores propel themselves through the water by means of one or more flagella.

Chytrids Are Mostly Aquatic

Most **chytrids** live in fresh water, but a few species are marine. Chytrid spores have a single flagellum on one end. The oldest known fossil fungi are chytrids that are found in rocks more than 600 million years old. Ancestral fungi may well have been similar to today's aquatic and marine chytrids, so fungi probably originated in a watery environment before colonizing land.

Most chytrid species feed on dead aquatic plants or other debris in watery environments, but some species are parasites of plants or animals. One such parasitic chytrid is believed to be a major cause of the current worldwide die-off of frogs, which threatens many species and has apparently already caused the extinction of several. (For more on the decline of frogs, see "Earth Watch: Frogs in Peril" on p. 454.)

Rumen Fungi Live in Animal Digestive Tracts

The **rumen fungi** are anaerobic (they do not require oxygen) and reside mainly in the digestive tracts of plant-eating animals such as cows, sheep, kangaroos, elephants, and iguana lizards. These animals are not able to digest cellulose (a major component of plant tissue) themselves but instead rely on symbiotic organisms that inhabit their guts. The rumen fungi are among these organisms; they produce enzymes that digest cellulose and the resulting breakdown product nourishes both the fungi and their animal hosts. The spores of most rumen fungi have multiple flagella, which may form a tuft at one end of the spore.

Blastoclades Have a Nuclear Cap

Blastoclades (**Fig. 22-6**) are distinguished by some characteristic features, such as a distinctive structure called the *nuclear cap* that is found near the nucleus of blastoclade spores. The nuclear cap is formed by the spore's ribosomes.

(**Fig. 22-5** and **Table 22-1**). Some fungi, however, are not members of any of these six groups. Most of these unclassified fungal species were historically placed in the taxonomic group Zygomycota (zygomycetes), but recent analysis of DNA sequences has revealed that zygomycetes do not constitute a clade. (A clade is a group consisting of all the descendants of a particular common ancestor.) Because systematists prefer to give formal names only to clades, fungal classification may soon be revised to place the species previously classified as zygomycetes into several new taxonomic groups.

Chytrids, Blastoclades, and Rumen Fungi Produce Swimming Spores

The members of three taxonomic groups of fungi—the chytrids, rumen fungi, and blastoclades—are distinguished

TABLE 22-1 The Major Taxonomic Groups of Fungi

Common Name (Latin Name)	Reproductive Structures	Cellular Characteristics	Economic and Health Impacts	Representative Genera
Chytrids (Chytridiomycota)	Form haploid or diploid flagellated spores	Septa are absent	Contribute to the decline of frog populations	*Batrachochytrium* (frog pathogen)
Rumen fungi (Neocallimastigomycota)	Form haploid or diploid flagellated spores	Septa are absent	Help enable cattle, horses, sheep to subsist on plants	*Neocallimastix* (lives in herbivore digestive systems)
Blastoclades (Blastocladiomycota)	Form haploid or diploid flagellated spores	Septa are absent	None known	*Allomyces* (aquatic decomposer)
Glomeromycetes (Glomeromycota)	Form haploid asexual spores, often in clusters	Septa are absent	Form mycorrhizae (mutualistic, symbiotic associations with plant roots)	*Glomus* (widespread mycorrhizal partner)
Basidiomycetes (Basidiomycota)	Sexual reproduction involves formation of haploid basidiospores on club-shaped basidia	Septa are present	Cause smuts and rusts on crops; include some edible mushrooms	*Amanita* (poisonous mushroom); *Polyporus* (shelf fungus)
Ascomycetes (Ascomycota)	Form haploid sexual ascospores in saclike ascus	Septa are present	Cause molds on fruit; can damage textiles; cause Dutch elm disease and chestnut blight; include yeasts and morels	*Saccharomyces* (yeast); *Ophiostoma* (causes Dutch elm disease)
"Zygomycetes" (not a formally designated taxonomic group)	Form diploid sexual zygospores	Septa are absent	Cause soft fruit rot and black bread mold	*Rhizopus* (causes black bread mold); *Pilobolus* (dung fungus)

▲ **FIGURE 22-6 Blastoclade filaments** These filaments of the blastoclade fungus *Allomyces* are in the midst of sexual reproduction. The orange structures visible on many of the filaments will release male gametes; the clear swollen structures will release female gametes. Blastoclade gametes are flagellated, and these swimming reproductive structures aid dispersal of members of this mostly aquatic group.

Blastoclades live in fresh water or in soil, and some are parasites of plants or aquatic invertebrates such as water fleas or mosquito larvae. Their spores have a single flagellum.

Glomeromycetes Associate with Plant Roots

Almost all **glomeromycetes** live in intimate contact with the roots of plants. In fact, the hyphae of glomeromycetes actually penetrate the cells of the roots, and form microscopic branching structures inside the cells (**Fig. 22-7**). This

▲ **FIGURE 22-7 Glomeromycete in a plant cell** Glomeromycete hyphae penetrate the cells of plants with which the fungus forms mutually beneficial associations. Inside the host plant's root cells, the fungus develops characteristic branching structures.

invasion of the plant's cells does not appear to harm the plant. To the contrary, glomeromycetes provide benefits to the plants they inhabit. This type of beneficial association between fungi and plant roots is known as a *mycorrhiza* and is described in more detail later in this chapter.

Glomeromycete reproduction is not fully understood; sexual reproduction by a member of the group is yet to be observed. During asexual reproduction, glomeromycetes produce clusters of spores by mitotic cell division. The spores form at the tips of hyphae that typically remain outside the host plant cell. When the spores germinate, hyphae grow into the surrounding soil, but the new fungus survives only if its germinating hyphae reach a plant root.

Basidiomycetes Produce Club-Shaped Reproductive Structures

Basidiomycetes are called the **club fungi** because they produce club-shaped reproductive structures. Members of this phylum typically reproduce sexually (**Fig. 22-8**). Hyphae of different mating types (designated "+" and "−") fuse (**Fig. 22-8 ❶**) to form hyphae in which each cell contains two nuclei, one from each parent (**Fig. 22-8 ❷**). These hyphae grow into an underground mycelium that, in response to appropriate environmental conditions, gives rise to an aboveground fruiting body that consists of densely aggregated hyphae (**Fig. 22-8 ❸**). Some of the hyphae in the fruiting body develop into club-shaped reproductive cells called **basidia** (singular, basidium) that, like their precursor cells, contain two haploid nuclei (**Fig. 22-8 ❹**). In each basidium, the two nuclei fuse to yield a diploid nucleus (**Fig. 22-8 ❺**). The diploid nucleus divides by meiosis and gives rise to four haploid *basidiospores* (**Fig. 22-8 ❻**). If it falls on fertile ground, a basidiospore may germinate and form haploid hyphae (**Fig. 22-8 ❼**).

CASE STUDY continued

Humongous Fungus

Because the underground bodies of basidiomycetes such as *Armillaria* grow at a relatively steady rate, the age of a mycelium can be estimated by measuring the area over which its aboveground reproductive structures spread. On the basis of such measurements, it is apparent that basidiomycetes can live for hundreds of years. Some are even older than that. For example, the researchers who discovered the gigantic *Armillaria* in Oregon estimate that it took at least 2,400 years to grow to its current size.

Basidiomycete fruiting bodies are familiar to most of us as mushrooms, puffballs, shelf fungi, and stinkhorns (**Fig. 22-9**). On the undersides of mushrooms are leaflike gills on which basidia are produced. Basidiospores are released by the billions through openings in the tops of puffballs or from the gills of mushrooms and are dispersed by wind and water.

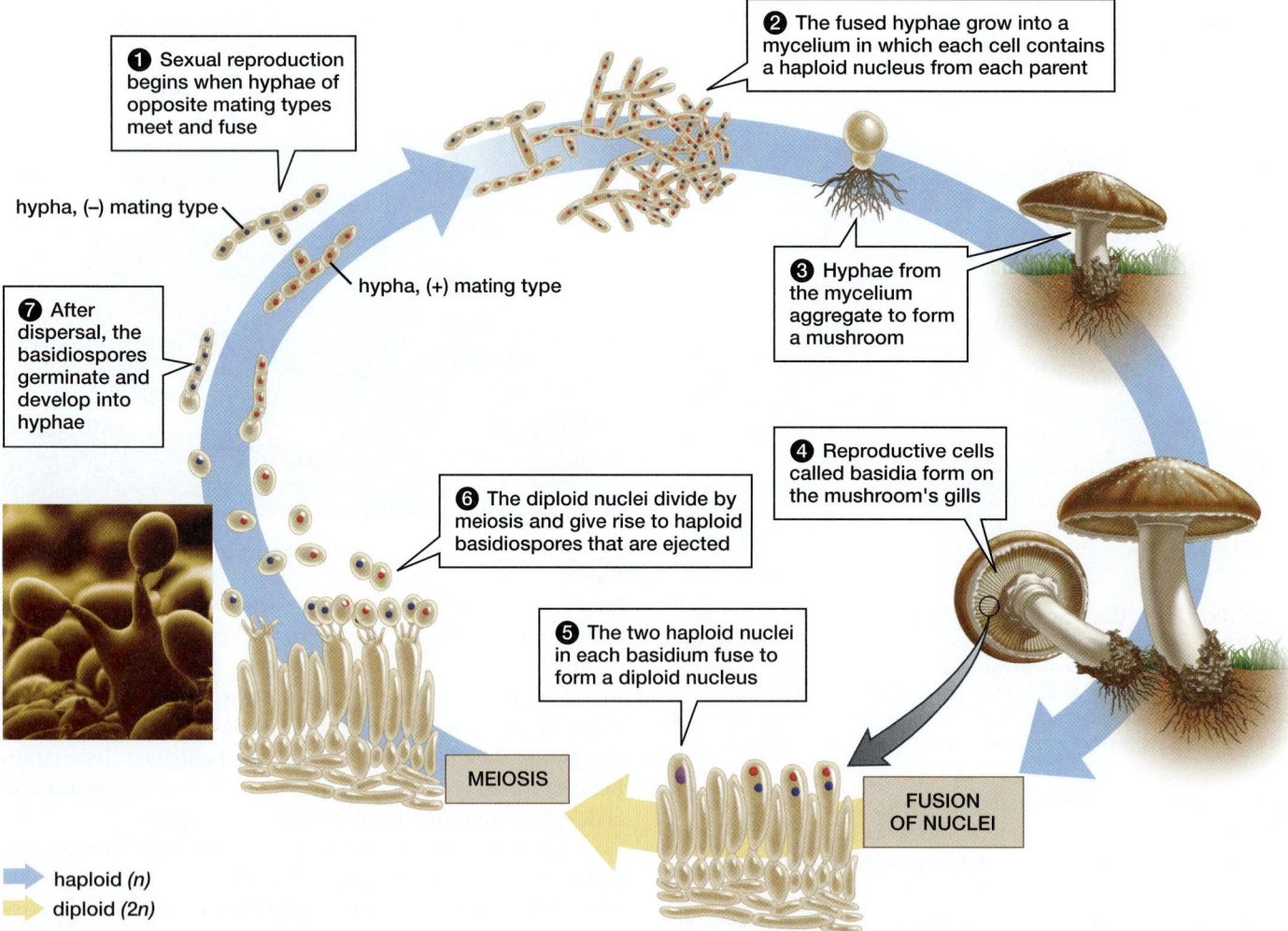

❶ Sexual reproduction begins when hyphae of opposite mating types meet and fuse

❷ The fused hyphae grow into a mycelium in which each cell contains a haploid nucleus from each parent

hypha, (−) mating type

hypha, (+) mating type

❼ After dispersal, the basidiospores germinate and develop into hyphae

❸ Hyphae from the mycelium aggregate to form a mushroom

❻ The diploid nuclei divide by meiosis and give rise to haploid basidiospores that are ejected

❹ Reproductive cells called basidia form on the mushroom's gills

❺ The two haploid nuclei in each basidium fuse to form a diploid nucleus

MEIOSIS

FUSION OF NUCLEI

→ haploid *(n)*
→ diploid *(2n)*

▲ **FIGURE 22-8 The life cycle of a typical basidiomycete** The photo shows two basidiospores attached to a basidium.

QUESTION If two spores from the same basidium each germinate and the resulting hyphae come into contact, can sexual reproduction follow?

(a) Puffball

(b) Shelf fungus

(c) Stinkhorn

▲ **FIGURE 22-9 Diverse basidiomycetes (a)** The giant puffball *Lycoperdon giganteum* may produce up to 5 trillion spores. **(b)** Shelf fungi, some the size of dessert plates, are conspicuous on trees. **(c)** The spores of stinkhorns are carried on the outside of a slimy cap that smells terrible to humans, but appeals to flies. The flies lay their eggs on the stinkhorn, and inadvertently disperse the spores that stick to their bodies.

QUESTION Are the structures shown in these photos haploid or diploid?

▲ **FIGURE 22-10 A mushroom fairy ring** Mushrooms emerge in a fairy ring from an underground fungal mycelium, growing outward from a central point where a single spore germinated, perhaps centuries ago.

In many cases, spores give rise to hyphae that grow outward from the original spore in a roughly circular pattern as the older hyphae in the center die. The subterranean body periodically sends up numerous mushrooms, which emerge in a ringlike pattern called a fairy ring (**Fig. 22-10**).

Ascomycetes Form Spores in a Saclike Case

The **ascomycetes,** or **sac fungi,** reproduce both asexually and sexually (**Fig. 22-11**). In asexual reproduction, spores are produced at the tips of specialized hyphae and, after dispersal, develop into new hyphae (**Fig. 22-11 ❶**). During sexual reproduction, spores are produced by a more complex sequence of events that begins, in a typical ascomycete, when hyphae of different mating types (+ and −) come into contact (**Fig. 22-11 ❷**). The two hyphae form reproductive structures that become linked by a connecting bridge. Haploid nuclei move across the bridge from the (−) reproductive structure to the (+) one, so that the (+) structure contains multiple nuclei from both parents (**Fig. 22-11 ❸**). The (+) structures that now contain the pooled nuclei develop into hyphae that are incorporated into a fruiting body (**Fig. 22-11 ❹**). At the tips of some of these hyphae, a saclike case called an **ascus** (plural, asci) forms (**Fig. 22-11 ❺**). At this stage, each ascus contains two haploid nuclei. These nuclei fuse to yield a single diploid nucleus (**Fig. 22-11 ❻**), which then divides by meiosis to yield four haploid nuclei (**Fig. 22-11 ❼**). These four nuclei divide by mitosis and develop into eight haploid spores known as *ascospores* (**Fig. 22-11 ❽**). Eventually, the ascus ruptures, liberating its ascospores. If the spores land in an appropriate location, they may germinate and develop into hyphae (**Fig. 22-11 ❾**).

Some ascomycetes live in decaying forest vegetation and form either beautiful cup-shaped reproductive structures (**Fig. 22-12a**) or corrugated, mushroom-like fruiting bodies called *morels* (**Fig. 22-12b**). The ascomycetes also include

❶ In asexual reproduction, haploid spores develop at the tips of hyphae, disperse, and germinate to form new hyphae

❷ Sexual reproduction begins when hyphae of opposite mating types meet and form connected reproductive structures

fruiting body

spores

ASEXUAL REPRODUCTION

❸ Haploid nuclei move from the (-) to the (+) structure

❹ The structures containing the pooled (+) and (-) nuclei develop into hyphae that are incorporated into a fruiting body

❺ The tips of some hyphae in the fruiting body form asci that contain two haploid nuclei

ascus

hypha, (–) mating type

hypha, (+) mating type

❾ The ascus bursts, dispersing spores that germinate and grow into hyphae

SEXUAL REPRODUCTION

hyphae

❽ The haploid nuclei divide by mitosis and give rise to eight ascospores

❼ The diploid nucleus divides by meiosis to produce four haploid nuclei

FUSION OF NUCLEI

❻ The haploid nuclei in an ascus fuse to form a diploid nucleus

MEIOSIS

haploid (n)

diploid (2n)

▲ **FIGURE 22-11** **The life cycle of a typical ascomycete** Some asci rising from hyphae are shown in the photo.

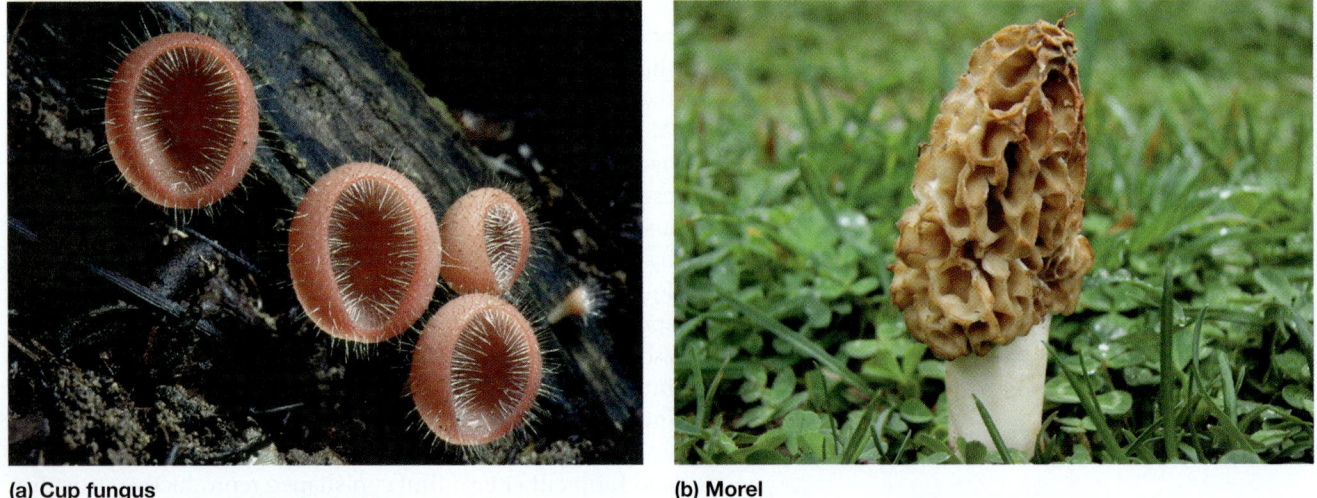

(a) Cup fungus

(b) Morel

▲ **FIGURE 22-12** **Diverse ascomycetes** **(a)** The cup-shaped fruiting body of the scarlet cup fungus. **(b)** The morel, an edible delicacy. (Consult an expert before sampling any wild fungus—some are deadly!)

many of the colorful molds that attack stored food and destroy fruit and grain crops and other plants, as well as the species that produces penicillin, the first antibiotic. Yeasts, some of the few unicellular fungi, are also ascomycetes.

Bread Molds Are Among the Fungi That Can Reproduce by Forming Diploid Spores

Many of the species formerly assigned to the **zygomycetes** live in soil or on decaying plant or animal material. These species include those belonging to the genus *Rhizopus*, which cause the familiar annoyances of soft fruit rot and

black bread mold. The life cycle of the black bread mold, which reproduces both asexually and sexually, is depicted in **Figure 22-13**. Asexual reproduction is initiated by the formation of haploid spores in black spore cases called **sporangia** (singular, sporangium; **Fig. 22-13 ❶**). These spores disperse through the air and, if they land on a suitable substrate (such as a piece of bread), germinate to form new haploid hyphae.

If extensions from two hyphae of different mating types (+ and −) come into contact, sexual reproduction may ensue (**Fig. 22-13 ❷**). The two hyphae fuse to form a *zygosporangium* that contains multiple haploid nuclei from the two parents (**Fig. 22-13 ❸**). As the zygosporangium develops, it

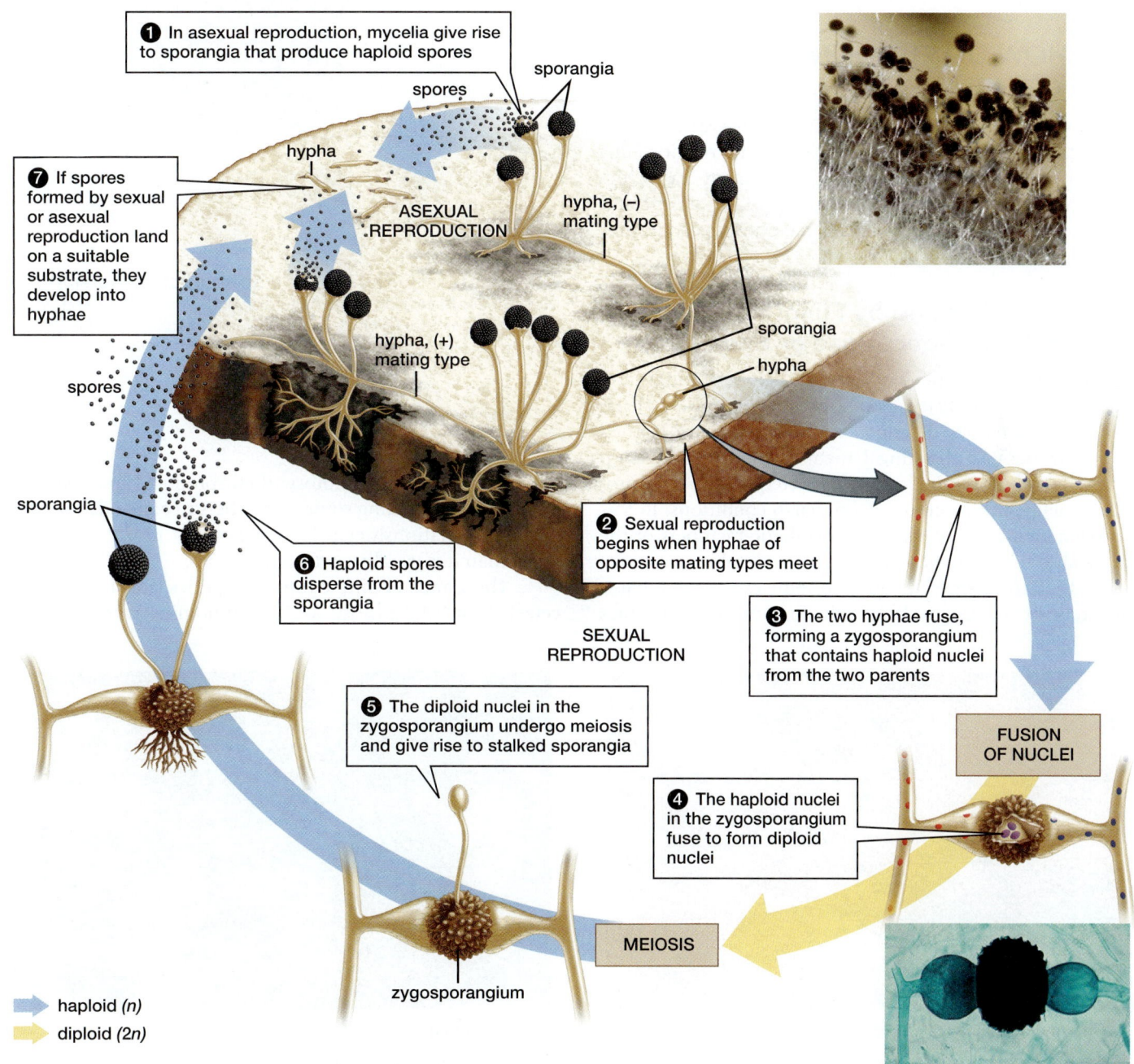

❶ In asexual reproduction, mycelia give rise to sporangia that produce haploid spores

spores
sporangia
hypha

ASEXUAL REPRODUCTION

hypha, (−) mating type

sporangia

hypha

❼ If spores formed by sexual or asexual reproduction land on a suitable substrate, they develop into hyphae

hypha, (+) mating type

spores

sporangia

❻ Haploid spores disperse from the sporangia

❷ Sexual reproduction begins when hyphae of opposite mating types meet

SEXUAL REPRODUCTION

❸ The two hyphae fuse, forming a zygosporangium that contains haploid nuclei from the two parents

FUSION OF NUCLEI

❺ The diploid nuclei in the zygosporangium undergo meiosis and give rise to stalked sporangia

❹ The haploid nuclei in the zygosporangium fuse to form diploid nuclei

MEIOSIS

zygosporangium

→ haploid *(n)*
→ diploid *(2n)*

▲ **FIGURE 22-13 The life cycle of a bread mold**

becomes tough and resistant, and can remain dormant for long periods until environmental conditions are favorable for growth. Inside the zygosporangium, the haploid nuclei fuse to produce diploid nuclei (**Fig. 22-13 ❹**). When conditions are favorable, the diploid nuclei undergo meiosis and give rise to stalked sporangia (**Fig. 22-13 ❺**). The sporangia produce haploid spores that disperse (**Fig. 22-13 ❻**), germinate, and develop into new haploid hyphae (**Fig. 22-13 ❼**).

CHECK YOUR LEARNING

Can you list and describe the six main taxonomic groups of fungi and explain why some fungal species are not assigned to any of these groups? Are you able to describe the life cycles of a typical basidiomycete, ascomycete, and bread mold?

22.3 HOW DO FUNGI INTERACT WITH OTHER SPECIES?

Many fungi live in direct contact with another species for a prolonged period. Such intimate, long-term relationships are known as *symbiotic* relationships. In many cases, the fungal member of a symbiotic relationship is parasitic and harms its host. But some symbiotic relationships are mutually beneficial.

Lichens Are Formed by Fungi That Live with Photosynthetic Algae or Bacteria

Lichens are symbiotic associations between fungi and single-celled green algae or cyanobacteria (**Fig. 22-14**). Lichens are sometimes described as fungi that have learned to garden, because the fungal member of the partnership "tends" the photosynthetic algal or bacterial partner by providing shelter and protection from harsh conditions. In this protected environment, the photosynthetic members of the partnership use sunlight to manufacture simple sugars, producing food for themselves but also some excess food that is consumed by the fungus. In fact, the fungus often consumes

algal layer

fungal hyphae

attachment structure

▲ **FIGURE 22-14 The lichen: a symbiotic partnership**
Most lichens have a layered structure bounded on the top and bottom by an outer layer formed from fungal hyphae. The fungal hyphae emerge from the lower layer, forming attachments that anchor the lichen to a surface, such as a rock or a tree. An algal layer in which the alga and fungus grow in close association lies beneath the upper layer of hyphae.

the lion's share of the photosynthetic product (up to 90% in some species), leading some researchers to conclude that the symbiotic relationship in lichens is really much more one sided than it is usually portrayed.

Thousands of different fungal species (mostly ascomycetes) form lichens (**Fig. 22-15**), combining with one of a

(a) Encrusting lichen

(b) Leafy lichen

▲ **FIGURE 22-15 Diverse lichens** (a) A colorful encrusting lichen, growing on dry rock, illustrates the tough independence of this symbiotic combination of fungus and algae. Pigments produced by the fungal partner are responsible for the bright orange color. (b) A leafy lichen grows on a rock.

much smaller number of algal or bacterial species. Together, these organisms form a unit so tough and self-sufficient that lichens are among the first living things to colonize newly formed volcanic islands, because many lichens can grow on bare rock. Brightly colored lichens also invade other inhospitable habitats ranging from deserts to the Arctic. Understandably, lichens in extreme environments grow very slowly; arctic colonies, for example, may expand as slowly as 1 to 2 inches per 1,000 years. Despite their slow growth, lichens can persist for long periods of time; some arctic lichens are more than 4,000 years old.

Mycorrhizae Are Fungi Associated with Plant Roots

Mycorrhizae (singular, mycorrhiza) are important symbiotic associations between fungi and plant roots. More than 5,000 species of mycorrhizal fungi grow in intimate association with plant roots, including those of most tree species. The hyphae of mycorrhizal fungi surround the plant root and invade the root cells (**Fig. 22-16**).

Mycorrhizae Help Feed Plants

The association between plants and mycorrhizae benefits both the fungi and their plant partners. The mycorrhizal fungi receive energy-rich sugar molecules that are produced photosynthetically by plants and passed from their roots to the fungi. In return, the fungi absorb minerals and organic nutrients from the soil, passing some of them directly into the root cells. Phosphorus and nitrogen, key nutrients that are crucial for plant growth, are among the molecules that mycorrhizae transport from soil to roots. Mycorrhizal fungi also absorb water and pass it to the plant—an advantage for plants in dry, sandy soils.

The partnership between mycorrhizae and plants makes a crucial contribution to the health of Earth's plants. Plants that are deprived of mycorrhizal fungi tend to be smaller and less vigorous than are plants with mycorrhizal partners. Thus, the presence of mycorrhizae increases the overall productivity of Earth's plant communities, enhancing their ability to support the animals and other organisms that depend on them.

Endophytes Are Fungi That Live Inside Plant Stems and Leaves

The intimate association between fungi and plants is not limited to root mycorrhizae. Fungi have also been found living inside the aboveground tissues of virtually every plant species that has been tested for their presence. Some of these *endophytes* (organisms that live inside other organisms) are parasites that cause plant diseases, but many, perhaps most, are beneficial to the host plant. The best-studied examples of beneficial fungal endophytes are the ascomycete species that live inside the leaf cells of many species of grass. These fungi produce substances that are distasteful or toxic to insects and grazing mammals and thus help protect the grass plants from those predators.

The antipredator protection provided by fungal endophytes is sufficiently effective that agricultural scientists are working hard to discover a way to grow grasses free of endophytes and therefore more palatable to grazing animals. Horses, cows, and other agriculturally important grazers tend to avoid eating grasses that contain endophytes. When the only available food is endophyte-containing grass, animals that eat it experience poor health and slow growth.

Some Fungi Are Important Decomposers

Some fungi, acting as mycorrhizae and endophytes, play a major role in the growth and preservation of plant tissue. Other fungi, however, play a similarly major role in its destruction, by acting as decomposers. Many fungal species can digest lignin or cellulose, the molecules that make up wood; some species can digest both molecules. Thus, when a tree or other woody plant dies, fungi can completely decompose its remains.

Fungi consume not only dead wood but the dead of all kingdoms. The fungi that are *saprophytes* (feeding on dead organisms) return the dead tissues' component substances to the ecosystems from which they came. The extracellular digestive activities of saprophytic fungi liberate nutrients that can be used by plants. If fungi and bacteria were suddenly to disappear, the consequences would be disastrous. Nutrients would remain locked in the bodies of dead plants and animals, the recycling of nutrients would grind to a halt, soil fertility would rapidly decline, and waste and organic debris would accumulate. In short, ecosystems would collapse.

▲ **FIGURE 22-16 Mycorrhizae enhance plant growth** Hyphae of mycorrhizae that have penetrated the tissue of a plant root. Plants grow significantly better in a symbiotic association with these fungi, which help make nutrients and water available to the roots.

CHECK YOUR LEARNING

Can you describe and explain the significance of some symbiotic associations involving fungi, including lichens, mycorrhizae, and endophytes? Can you explain how fungi help recycle nutrients?

22.4 HOW DO FUNGI AFFECT HUMANS?

The average person gives little thought to fungi, except perhaps for an occasional, momentary appreciation for the mushrooms on a pizza. Nonetheless, fungi affect our lives in more ways than you might imagine.

Fungi Attack Plants That Are Important to People

Fungi cause the majority of plant diseases, and some of the plants that they infect are important to humans. For example, fungal pathogens have a devastating effect on the world's food supply. Especially damaging are the basidiomycete plant pests descriptively called *rusts* and *smuts*, which cause billions of dollars' worth of damage to grain crops annually (Fig. 22-17). For example, Ug99 (so named because it was discovered in Uganda in 1999), an especially virulent strain of a wheat disease called black stem rust, is currently a major threat to the world's supply of wheat. (Wheat is one of the world's most important crops, in terms of total calories contributed.) None of the wheat varieties that feed much of the world's population is resistant to Ug99, whose spores can travel long distances by wind. As a result, wheat crops have been decimated over a large and growing area of Africa, Central Asia, and the Middle East. If Ug99 spreads to the massive wheat crops of China and India, serious food shortages might occur.

Fungal diseases also affect the appearance of our landscape. The American elm and the American chestnut—two tree species that were once prominent in many of America's parks, yards, and forests—were destroyed on a massive scale by the ascomycetes that cause Dutch elm disease and chestnut blight. Today, few people can recall the graceful forms of large elms and chestnuts, which are now almost entirely absent from the landscape.

Fungi continue to attack plant tissues long after they have been harvested for human use. To the dismay of homeowners, a host of different fungal species attack wood, causing it to rot. Fungi can also cause significant damage to cotton and wool textiles, especially in warm, humid climates where molds flourish.

The fungal impact on agriculture and forestry is not entirely negative, however. Fungal parasites that attack insects and other arthropod pests can be an important ally in pest control (Fig. 22-18a). Farmers who wish to reduce their dependence on toxic and expensive chemical pesticides are increasingly turning to biological methods of pest control, including the application of "fungal pesticides." Fungal pathogens are currently used to control termites, rice weevils, tent caterpillars, aphids, citrus mites, and other pests. In addition, biologists have discovered that certain fungi attack and kill the mosquito species that transmit malaria (Fig. 22-18b). Plans are under way to enlist these fungi in the fight against malaria, one of the world's deadliest diseases.

Fungi Cause Human Diseases

The fungi include parasitic species that attack humans directly. Some of the most familiar fungal diseases are those caused by ascomycetes that attack the skin, resulting in athlete's foot, jock itch, and ringworm. These diseases, though unpleasant, are not life threatening and can usually be treated with antifungal

CASE STUDY continued

Humongous Fungus

The *Armillaria* fungus species that grew to massive size in Oregon harms trees in the forests it inhabits. As the fungus feeds on roots, it causes "root rot" that weakens or kills trees. This root rot provides aboveground evidence of *Armillaria*'s existence; the giant Oregon specimen was first identified by examining aerial photos to find forested areas with many dead trees.

(a) Corn smut

(b) Black stem rust

▲ FIGURE 22-17 Smuts and rusts (a) Corn smut is a basidiomycete pathogen that destroys millions of dollars' worth of corn each year. Even a pest like corn smut has its admirers, though. In Mexico this fungus is known as *huitlacoche* and is considered to be a great delicacy. (b) Another basidiomycete, black stem rust, currently threatens millions of acres of wheat in Africa and Asia.

(a) Pest-killing fungus

▲ FIGURE 22-18 **Helpful fungal parasites (a)** Fungi such as *Cordyceps* are used by farmers to control insect pests. **(b)** A healthy malaria-carrying mosquito (top) infected by *Beauveria* is transformed to a fungus-encrusted corpse in less than 2 weeks.

(b) Disease-fighting fungus

ointments. Prompt treatment can also usually control another common fungal disease, vaginal infections caused by the yeast *Candida albicans* (**Fig. 22-19**). Fungi can also infect the lungs if victims inhale spores of disease-causing fungal species such as those that cause valley fever and histoplasmosis. Like other fungal infections, these diseases can, if promptly diagnosed, be controlled with antifungal drugs. If untreated, however, they can develop into serious, systemic infections. Singer Bob Dylan, for instance, became gravely ill with histoplasmosis when a fungus infected the membrane surrounding his heart.

Fungi Can Produce Toxins

In addition to their role as agents of infectious disease, some fungi produce toxins that are dangerous to humans. Of particular concern are toxins produced by fungi that grow on grains and other foodstuffs that have been stored in moist conditions. For example, molds of the genus *Aspergillus*

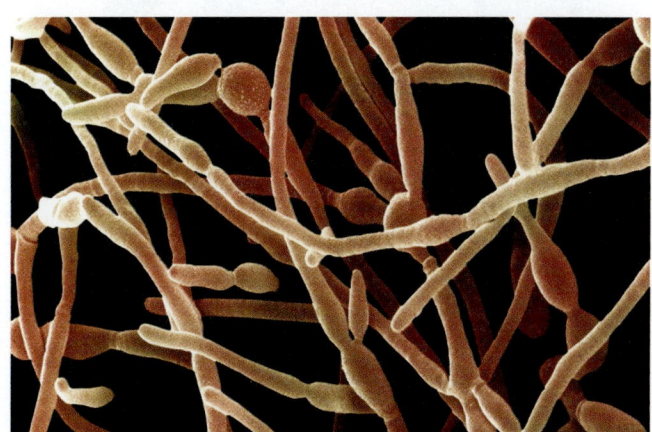

▲ FIGURE 22-19 **The unusual yeast** Most yeasts are single-celled, but some, such as *Candida* (shown here), may form multicellular filaments. *Candida* is a common cause of vaginal infections.

produce highly toxic, carcinogenic compounds known as aflatoxins. Some foods, such as peanuts, seem especially susceptible to attack by *Aspergillus*. Since aflatoxins were discovered in the 1960s, food growers and processors have developed methods for reducing the growth of *Aspergillus* in stored crops, so aflatoxins have been largely eliminated from the nation's peanut butter supply.

One infamous toxin-producing fungus is the ascomycete *Claviceps purpurea*, which infects rye plants and causes a disease known as ergot. This fungus produces several toxins, which can affect humans if infected rye is ground into flour and consumed. This happened frequently in northern Europe in the Middle Ages, with devastating effects. At that time, ergot poisoning was typically fatal, and victims experienced terrible symptoms before dying. One ergot toxin constricts blood vessels and reduces blood flow. The effect can be so extreme that gangrene develops and limbs shrivel and fall off. Other ergot toxins cause symptoms that include a burning sensation, vomiting, convulsive twitching, and vivid hallucinations. Today, new agricultural techniques have effectively eliminated ergot poisoning, but the hallucinogenic drug LSD, which is derived from a component of the ergot toxins, remains as a legacy of this disease.

Many Antibiotics Are Derived from Fungi

Fungi have also had positive impacts on human health. The modern era of lifesaving antibiotic medicines was ushered in by the discovery of penicillin, which is produced by an ascomycete mold (**Fig. 22-20**; see Fig. 1-12). Penicillin is still used, along with other fungi-derived antibiotics such as oleandomycin and cephalosporin, to combat bacterial diseases. Other important drugs are also derived from fungi, including cyclosporin, which is used to suppress the immune response after an organ transplant, so that the body is less likely to reject the transplanted organs.

▲ **FIGURE 22-20** *Penicillium* *Penicillium* growing on an orange. Reproductive structures, which coat the fruit's surface, are visible, while hyphae, beneath, draw nourishment from inside. The antibiotic penicillin was first isolated from this fungus.

QUESTION Why do some fungi produce antibiotic chemicals?

Fungi Make Important Contributions to Gastronomy

Fungi make important contributions to the human diet. We consume some fungi directly, including wild and cultivated basidiomycete mushrooms and ascomycetes such as morels and the rare and prized truffle (**Fig. 22-21**). (Use caution if you gather your own wild fungi; see "Links to Everyday Life: Collect Carefully.")

▲ **FIGURE 22-21** **The truffle** Truffles are the underground, spore-containing structures of an ascomycete that forms a mycorrhizal association with the roots of oak trees.

 LINKS TO EVERYDAY LIFE

Collect Carefully

In the early 1980s, doctors at a California hospital noticed a curious trend. Over a period of a few months, the number of patients admitted because of food poisoning had risen dramatically. Many of those admitted had died. What caused this sudden outbreak of poisoning? Further investigation revealed that, in almost all cases, the victims were recent immigrants from Laos or Cambodia. Struggling to adjust to their new country, they had been thrilled to find that California's forests contained mushrooms that looked just like the ones they had collected for food back in Asia. Unfortunately, the similarity was superficial; the California mushrooms were, in fact, poisonous species. The immigrants' nostalgic pursuit of "comfort food" had tragic consequences.

In general, immigrants from countries where mushroom collecting is common have proved to be especially likely to be poisoned by toxic mushrooms. But immigrants are not the only victims. Each year, a number of small children, inexperienced collectors, and unlucky guests at gourmet dinners make unexpected trips to the hospital after eating poisonous wild mushrooms.

It can be fun and rewarding to collect wild mushrooms, which offer some of the richest and most complex flavors a human can experience. But if you decide to go collecting, be careful, because some of the deadliest poisons known to humankind are found in mushrooms. Especially noted for their poisons are certain species in the genus *Amanita*, which have suggestive common names such as death cap and destroying angel (**Fig. E22-1**). These names are apt, because

even a single bite of one of these mushrooms can be lethal. Damage from *Amanita* toxins is most severe in the liver, where the toxins tend to accumulate. Often, a victim of *Amanita* poisoning can be saved only by undergoing a liver transplant. So be sure to protect your health by inviting an expert to join your mushroom-hunting expeditions.

▲ **FIGURE E22-1** **The destroying angel** Mushrooms produced by the basidiomycete *Amanita virosa* can be deadly.

? HAVE **YOU** EVER WONDERED …

Why Truffles Are So Expensive?

Although many fungi are prized as food, none is as avidly sought as the truffle. The finest Italian truffles may sell for as much as $1,500 per pound, and unusually large specimens can fetch spectacular prices. A 3.3-pound Italian white truffle recently sold at auction for $330,000!

Why such high prices? Truffles develop underground, and it takes some work to find one. In fact, humans can't do it alone and need help from other species. Some animals, especially pigs, are attracted to the aroma of a mature truffle. If a pig follows the smell to a truffle, it will dig the fungus up and devour it. That's why, traditionally, truffle collectors have used muzzled pigs to hunt their quarry.

Today, trained dogs are the most common assistants to truffle hunters. Dogs are necessary even on the farms where much of today's truffle crop is laboriously grown. The difficulty of cultivating and harvesting truffles accounts in part for their high price. And no one has figured out how to cultivate the prized white truffle. The only way to acquire one is to follow the nose of a truffle-hunting dog or pig.

The role of fungi in cuisine also has less-visible manifestations. For example, some of the world's most famous cheeses, including Roquefort, Camembert, Stilton, and Gorgonzola, gain their distinctive flavors from ascomycete molds that grow on them as they ripen. Perhaps the most important and pervasive fungal contributors to our food supply, however, are the single-celled ascomycetes (and a few species of basidiomycetes) known as *yeasts*.

Wine and Beer Are Made Using Yeasts

The discovery that yeasts could be harnessed to enliven our culinary experience is surely a key event in human history. Among the many foods and beverages that depend on yeasts for their production are bread, wine, and beer, which are consumed so widely that it is difficult to imagine a world without them. All derive their special qualities from fermentation by yeasts. Fermentation occurs when yeasts extract energy from sugar and, as by-products of the metabolic process, emit carbon dioxide and ethyl alcohol (see "Links to Everyday Life: A Jug of Wine, a Loaf of Bread, and a Nice Bowl of Sauerkraut" in Chapter 8).

As yeasts consume the fruit sugars in grape juice, the sugars are converted to alcohol, and wine is the result. Eventually, the increasing concentration of alcohol kills the yeasts, ending fermentation. If the yeasts in fermenting wine die before all available grape sugar is consumed the wine will be sweet; if the sugar is exhausted, the wine will be dry.

Beer is brewed from grain (usually barley), but yeasts cannot effectively consume the carbohydrates in grain. For the yeasts to do their work, the barley grains must have sprouted (recall that grains are actually seeds). Germination converts the grains' carbohydrates to sugar, so the sprouted barley provides an excellent food source for the yeasts. As with wine, fermentation converts sugars to alcohol, but beer brewers capture the carbon dioxide by-product as well, giving the beer its characteristic bubbly carbonation.

Yeasts Make Bread Rise

In bread making, carbon dioxide is the crucial fermentation product. The yeasts added to bread dough do produce alcohol as well as carbon dioxide, but the alcohol evaporates during baking. In contrast, the carbon dioxide is trapped in the dough, where it forms the bubbles that give bread its light, airy texture (and saves us from a life of eating sandwiches made of crackers).

So the next time you're enjoying a slice of French bread with Camembert cheese and a nice glass of chardonnay, or a slice of pizza and a cold bottle of your favorite brew, you might want to quietly give thanks to the yeasts. Our diets would certainly be a lot duller without the help we get from fungal assistants.

CHECK YOUR LEARNING

Can you explain how fungi affect agriculture? Are you able to describe examples of how fungi affect human health? Can you describe the role of fungi in the production of cheese, wine, beer, and bread?

CASE STUDY revisited

Humongous Fungus

Why do *Armillaria* fungi grow so large? Their size is due in part to their ability to form structures called rhizomorphs, which consist of hyphae bundled together inside a protective rind. The rhizomorphs can extend long distances through nutrient-poor areas to reach new sources of food. The *Armillaria* fungus can thus grow beyond the boundaries of a particular food-rich area.

Another factor that may contribute to the gigantic size of the Oregon *Armillaria* is the climate in which it was found. In the dry climate of eastern Oregon, fungal fruiting bodies form only rarely, so the colossal *Armillaria* rarely produces spores. In the absence of spores that might grow into new individuals, the existing individual faces little competition for resources and is free to grow and fill an increasingly large area.

The discovery of the Oregon specimen is merely the latest chapter in a long-running, good-natured "fungus war" that began in 1992 with the discovery of the first humongous fungus, a 37-acre *Armillaria gallica* growing in Michigan. Since that initial landmark discovery, research groups in Michigan, Oregon, and elsewhere have engaged in a friendly competition to find the largest fungus. Will the current record ever be topped? Stay tuned.

Consider This

Because the entire Oregon *Armillaria* grew from a single spore, all of its cells are genetically identical. However, it is unlikely that any substances are transported through the entire 3.4-square-mile mycelium. And there is no continuous skin or bark or membrane that covers the entire mycelium and separates it from the environment as a unit. Is the fungus's genetic unity sufficient evidence for it to be considered a single individual, or is greater physiological integration required? Do you think that the claim of "world's largest organism" is valid?

CHAPTER REVIEW

Summary of Key Concepts

22.1 What Are the Key Features of Fungi?

Fungal bodies generally consist of filamentous hyphae, which are either multicellular or multinucleated and form large, intertwined networks called *mycelia*. Fungal nuclei are generally haploid. A cell wall of chitin surrounds fungal cells. All fungi feed by secreting digestive enzymes outside their bodies and absorbing the liberated nutrients.

Fungal reproduction is varied and complex. Asexual reproduction can occur through mitotic production of haploid spores. Sexual reproduction occurs when compatible haploid nuclei fuse to form a diploid zygote, which undergoes meiosis to form haploid spores. Both asexual and sexual spores produce haploid mycelia through mitosis.

22.2 What Are the Major Groups of Fungi?

The major taxonomic groups of fungi are Chytridiomycota (chytrids), Neocallimastigomycota (rumen fungi), Blastocladiomycota (blastoclades), Glomeromycota (glomeromycetes), Basidiomycota (basidiomycetes), and Ascomycota (ascomycetes). Their characteristics are summarized in Table 22-1.

22.3 How Do Fungi Interact with Other Species?

A lichen is a symbiotic association between a fungus and algae or cyanobacteria. The fungal partner provides shelter for the algae or cyanobacteria, which convert sunlight to energy-rich molecules that nourish the fungus. Mycorrhizae are associations between fungi and the roots of most vascular plants. The fungus derives photosynthetic nutrients from the plant roots and, in return, carries water and nutrients into the root from the surrounding soil. Endophytes are fungi that grow inside the leaves or stems of plants and that may help protect the plants that harbor them. Saprophytic fungi are extremely important decomposers in ecosystems. Their filamentous bodies penetrate soil and decaying organic material, liberating nutrients through extracellular digestion.

22.4 How Do Fungi Affect Humans?

Many plant diseases are caused by parasitic fungi. Some parasitic fungi can help control insect crop pests. Others can cause human diseases, including ringworm, athlete's foot, and common vaginal infections. Some fungi produce toxins that can harm humans. Nonetheless, fungi add variety to the human food supply, and fermentation by fungi helps make wine, beer, and bread.

Key Terms

ascomycete   *407*
ascus (plural, asci)   *407*
basidiomycete   *406*
basidium (plural, basidia)   *406*
blastoclade   *404*
chytrid   *404*
club fungus   *406*
glomeromycete   *405*
hypha (plural, hyphae)   *401*
lichen   *410*
mycelium (plural, mycelia)   *401*
mycorrhiza (plural, mycorrhizae)   *411*
rumen fungus   *404*
sac fungus   *407*
septum (plural, septa)   *401*
sporangium (plural, sporangia)   *409*
spore   *402*
zygomycete   *409*

Learning Outcomes

In this chapter, you have learned to . . .

LO1 Describe the key features of fungi.

LO2 List the major fungus taxonomic groups and describe their key characteristics.

LO3 Summarize the fungal life cycle and describe the key steps in the life cycles of basidiomycetes, ascomycetes, and bread molds.

LO4 Describe effects that fungi have on other organisms, including humans.

LO5 Describe ecologically important symbiotic associations between fungi and other organisms.

Thinking Through the Concepts

Fill-in-the-Blank

1. The portions of a fungus that are visible to the naked eye are often structures specialized for _____. These structures release tiny _____, which are dispersed to produce new fungi. **LO1 LO3**

2. The fungal body is a(n) _____, and is composed of microscopic threads called _____ that may be subdivided into many cells by _____. The cell walls of fungi are strengthened by _____. **LO1**

3. In fungi, asexual spores are produced by _____ cell division and have _____ set(s) of chromosomes. Sexual spores are produced by _____ cell division in a(n) _____ and have _____ set(s) of chromosomes. **LO1 LO3**

4. Fill in the blanks in the following sentences with the common names of fungal taxonomic groups. Almost all _____ live in intimate association with plant roots. Flagellated, swimming spores are produced by _____. Mushrooms and puffballs are reproductive structures of _____. **LO2**

5. _____ are symbiotic associations of fungi and green algae. _____ are symbiotic, mutually beneficial associations of fungi and plant roots. Some fungi are _____ that live inside the aboveground tissues of plants. **LO5**

6. Fungi are the only decomposers that can digest _____. The fungi that cause breads to rise and wines to ferment are _____. Human ailments caused by fungi include _____ and _____. **LO4**

Review Questions

1. Describe the structure of the fungal body. How do fungal cells differ from most plant and animal cells? **LO1**

2. What portion of the fungal body is represented by mushrooms, puffballs, and similar structures? Why are these structures elevated above the ground? **LO1 LO3**

3. What two plant diseases, caused by parasitic fungi, have had an enormous impact on forests in the United States? In which taxonomic group are these fungi found? **LO2 LO4**

4. List some fungi that attack crops. To which taxonomic group do they belong? **LO2 LO4**

5. Describe asexual reproduction in fungi. **LO1 LO3**

6. List the major taxonomic groups of fungi, describe some key features of each group, and give an example of a fungus in each group. **LO2**

7. Describe how a fairy ring of mushrooms is produced. Why is the diameter related to its age? **LO3**

8. Describe two symbiotic relationships between a fungus and another organism. In each case, explain how each partner in these associations is affected. **LO4 LO5**

Applying the Concepts

1. Dutch elm disease in the United States is caused by an *exotic*—that is, an organism (in this case, a fungus) introduced from another part of the world. What damage has this introduction done? What other fungal pests fall into this category? Why are parasitic fungi particularly likely to be transported out of their natural habitat? What can governments do to limit this importation?

2. The discovery of penicillin revolutionized the treatment of bacterial diseases. However, penicillin is now rarely prescribed. Why is this?

3. The discovery of penicillin was the result of a chance observation by an observant microbiologist, Alexander Fleming. How would you search systematically for new antibiotics produced by fungi? Where would you look for these fungi?

4. Fossil evidence indicates that mycorrhizal associations between fungi and plant roots existed in the late Paleozoic era, when the invasion of land by plants began. This evidence suggests an important link between mycorrhizae and the successful invasion of land by plants. Why might mycorrhizae have been important fungi in the colonization of terrestrial habitats by plants?

5. What ecological consequences would occur if humans, using a new and deadly fungicide, destroyed all fungi on Earth?

Answers to Figure Caption questions and Fill-in-the-Blank questions can be found in the Answers section at the back of the book.

(MB) *Go to MasteringBiology for practice quizzes, activities, eText, videos, current events, and more.*

CHAPTER 23

Animal Diversity I: Invertebrates

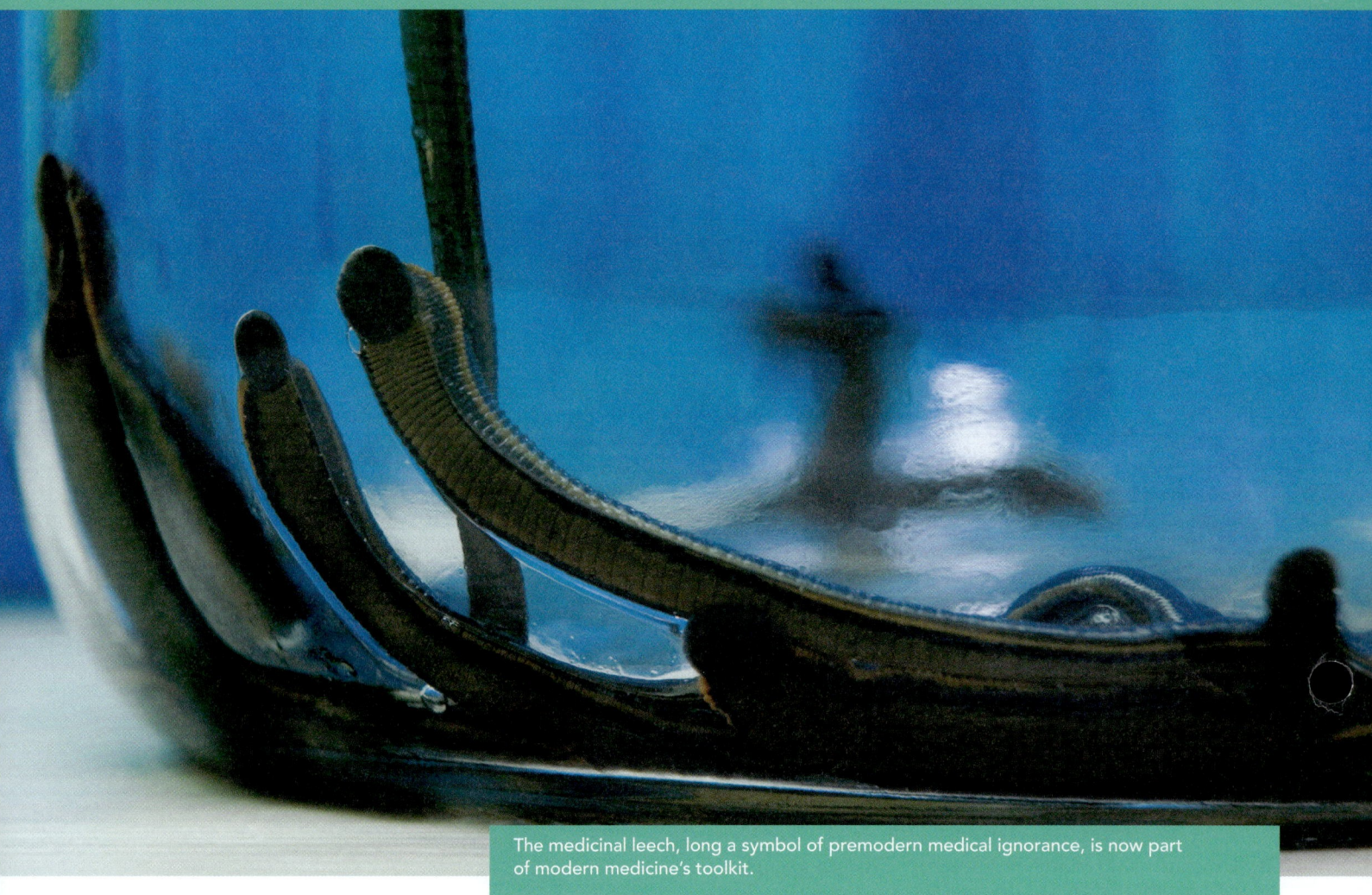

The medicinal leech, long a symbol of premodern medical ignorance, is now part of modern medicine's toolkit.

CASE STUDY
Physicians' Assistants

MODERN MEDICINE is a high-tech enterprise, dependent on multimillion-dollar machines, sophisticated medical devices, and drugs developed via cutting-edge chemistry and biotechnology. But despite the prevalence of advanced technology in medicine, physicians also get low-tech assistance from an unlikely source: *invertebrates* (animals without a backbone). Consider, for example, the medicinal leech. For more than 2,000 years, healers enlisted these parasitic segmented worms for treatment of a wide range of illnesses. For much of human medical history, treatment with leeches was based on the hope that the creatures would suck out the "tainted" blood that was believed to be the primary cause of disease.

As the actual causes of disease were discovered, however, medical use of leeches declined. By the beginning of the twentieth century, leeches no longer had a place in the toolkit of modern medicine and had become a symbol of the ignorance of an earlier age. Today, however, medicinal leeches are making a surprising comeback.

Currently, leeches are used to treat a surgical complication known as venous insufficiency. This complication is especially common in reconstructive surgery, such as surgery to reattach a severed finger or repair a disfigured face. In such cases, surgeons are often unable to reconnect all of the veins that would normally carry blood away from tissues. Eventually, new veins will grow, but in the meantime blood may accumulate in the repaired tissue. Unless the excess blood is removed, it will coagulate,

causing clots that can deprive the tissue of the oxygen and nutrients it needs to live. Fortunately, leeches can help. Applied to the affected area, the leeches get right to work, making small, painless incisions and sucking blood into their stomachs. To aid them in their blood-removal task, the leeches' saliva contains a mixture of chemicals that dilate blood vessels and prevent blood from clotting. Although the chemical brew in the saliva is an adaptation that helps leeches consume blood more efficiently, it also helps the patient by promoting blood flow in the damaged tissue. In this way, leeches provide a painless, effective treatment for venous insufficiency.

Although relatively few of them have medical uses, invertebrate animals account for a large share of Earth's known biodiversity. In this chapter, we explore and describe the diversity of invertebrates.

AT A GLANCE

23.1 What Are the Key Features of Animals?

23.2 Which Anatomical Features Mark Branch Points on the Animal Evolutionary Tree?

23.3 What Are the Major Animal Phyla?

LEARNING GOALS

LG1 Describe the key features of animals.

LG2 Describe the anatomical features that mark major branching points on the animal evolutionary tree, and explain the functions and significance of these features.

LG3 Describe the major animal taxonomic groups and representative members of each group.

LG4 Describe the effects that invertebrate animals have on humans.

23.1 WHAT ARE THE KEY FEATURES OF ANIMALS?

It is difficult to devise a concise definition of the term "animal." No single feature uniquely defines animals, so the group is defined by a list of characteristics. None of these characteristics is unique to animals, but together they distinguish animals from members of other taxonomic groups:

- Animals are eukaryotes.
- Animals are multicellular.
- Animal cells lack a cell wall.
- Animals obtain their energy by consuming other organisms.
- Animals typically reproduce sexually.
- Animals are motile (able to move about) during some stage of their lives.
- Most animals are able to respond rapidly to external stimuli.

CHECK YOUR LEARNING

Can you list the characteristics that collectively distinguish animals from other kinds of organisms?

23.2 WHICH ANATOMICAL FEATURES MARK BRANCH POINTS ON THE ANIMAL EVOLUTIONARY TREE?

By the Cambrian period, which began 542 million years ago, most of the animal phyla that currently populate Earth were already present. Unfortunately, the Precambrian fossil record is very sparse, and does not reveal much about the early evolutionary history of animals. Therefore, systematists have looked to features of animal anatomy, embryological development, and DNA sequences for clues about animal history. These investigations have shown that certain features mark major branching points on the animal evolutionary tree and represent milestones in the evolution of the different body plans of modern animals (**Fig. 23-1**). In the following sections, we will explore these evolutionary milestones and their legacies in the bodies of modern animals.

Lack of Tissues Separates Sponges from All Other Animals

One of the earliest major innovations in animal evolution was the appearance of **tissues**—groups of similar cells integrated into a functional unit. Today, the bodies of almost all animals include tissues; the only animals that have retained the ancestral lack of tissues are the sponges. In sponges, individual cells may have specialized functions, but they act more or less independently and are not organized into tissues. This unique feature of sponges suggests that the split between sponges and the evolutionary branch leading to all other animal phyla must have occurred very early in the history of animals.

Animals with Tissues Exhibit Either Radial or Bilateral Symmetry

The evolutionary advent of tissues coincided with the first appearance of body symmetry; all animals with true tissues also have symmetrical bodies. An animal is said to be symmetrical if it can be bisected along at least one plane such that the resulting halves are mirror images of one another.

The symmetrical, tissue-bearing animals can be divided into two groups, one containing animals that exhibit **radial symmetry** (**Fig. 23-2a**) and one with animals that exhibit **bilateral symmetry** (**Fig. 23-2b**). In radial symmetry, any plane through a central axis divides the object into roughly equal halves. In contrast, a bilaterally symmetrical animal can be divided into roughly mirror-image halves only along one particular plane through the central axis.

The difference between radially and bilaterally symmetrical animals reflects another major branching point in the animal evolutionary tree. This split separated the ancestors of the radially symmetrical cnidarians (sea jellies, corals, and anemones) and ctenophores (comb jellies) from the ancestors of the remaining animal phyla, all of which are bilaterally symmetrical.

► **FIGURE 23-1 An evolutionary tree of some major animal phyla** Based on Casey W. Dunn et al., Broad phylogenomic sampling improves resolution of the animal tree of life, © 2008 Nature Publishing Group.

No tissues

Tissues

Radial symmetry, two tissue layers

Bilateral symmetry, cephalization, three tissue layers

Cuticle molted

Protostome development

Deuterostome development

Porifera (sponges)

Cnidaria (sea jellies, corals, anemones)

Ctenophora (comb jellies)

Nematoda (roundworms)

Arthropoda (insects, arachnids, crustaceans)

Platyhelminthes (flatworms)

Annelida (segmented worms)

Mollusca (clams, snails, octopuses, squid)

Echinodermata (sea stars, sea urchins, sea cucumbers)

Chordata (lancelets, sea squirts, vertebrates)

Ecdysozoans

Lophotrochozoans

Protostomes

Deuterostomes

Bilaterally symmetrical

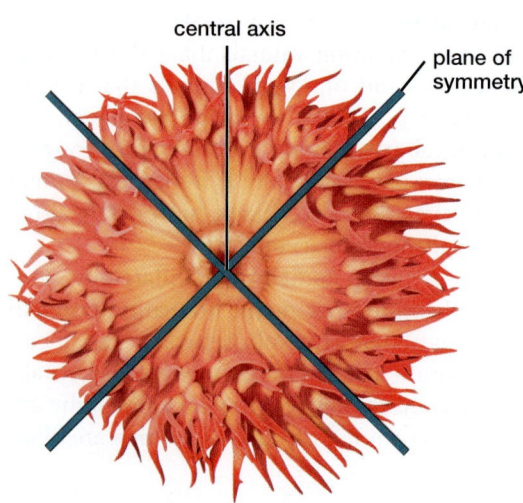

(a) Radial symmetry

central axis

plane of symmetry

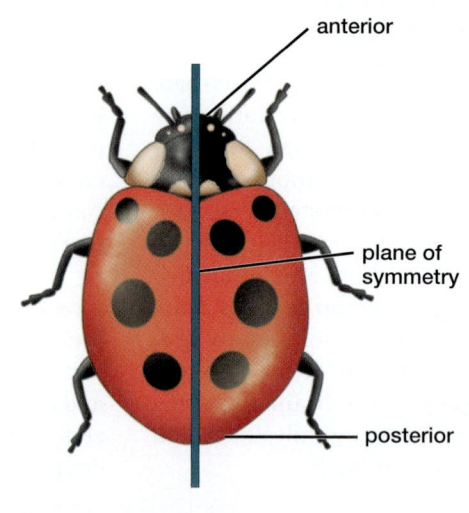

(b) Bilateral symmetry

anterior

plane of symmetry

posterior

◄ **FIGURE 23-2 Body symmetry and cephalization** **(a)** Animals with radial symmetry, such as this sea anemone, lack a well-defined head. Any plane that passes through the central axis divides the body into mirror-image halves. **(b)** Animals with bilateral symmetry, such as this beetle, have an anterior head end and a posterior tail end. The body can be split into two mirror-image halves only along a particular plane that runs down the midline.

Radially Symmetrical Animals Have Two Embryonic Tissue Layers; Bilaterally Symmetrical Animals Have Three

The distinction between radial and bilateral symmetry in animals is closely tied to a corresponding difference in the number of tissue layers, called *germ layers*, that arise during embryonic development. Embryos of animals with radial symmetry have two germ layers: an inner layer of **endoderm** (which gives rise to the tissues that line the gut cavity) and an outer layer of **ectoderm** (which gives rise to the tissues that cover the outside of the body.) Embryos of bilaterally symmetrical animals add a third germ layer, **mesoderm,** which lies between the endoderm and the ectoderm. In bilaterally symmetrical animals, endoderm differentiates to form the tissues that line most hollow organs, respiratory surfaces, and the gut; ectoderm forms nerve tissue and the tissues on the outer surface of the body; and mesoderm forms muscle and, when present, the circulatory and skeletal systems.

The parallel evolution of symmetry type and number of germ layers helps us make sense of the potentially puzzling case of the echinoderms (sea stars, sea urchins, and sea cucumbers). Adult echinoderms are radially symmetrical, yet our evolutionary tree places them squarely within the bilaterally symmetrical group. It turns out that echinoderms have three germ layers, as well as several other characteristics (some described later) that unite them with the bilaterally symmetrical animals. So, the immediate ancestors of echinoderms must have been bilaterally symmetrical, and the group subsequently evolved radial symmetry (a case of convergent evolution). To this day, larval echinoderms retain bilateral symmetry.

Bilaterally Symmetrical Animals Have Heads

Radially symmetrical animals tend either to be *sessile* (fixed to one spot, like sea anemones) or to drift around on currents (like sea jellies). Because such animals do not actively propel themselves in a particular direction, all parts of their bodies are more or less equally likely to encounter food. In contrast, most bilaterally symmetrical animals are *motile* (move under their own power), and resources such as food are most likely to be encountered by the part of the animal that is closest to the direction of movement. The evolution of bilateral symmetry was therefore accompanied by **cephalization,** the concentration of sensory organs and a brain in a defined head region. Cephalization produces an *anterior* (head) end, where sensory cells, sensory organs, clusters of nerve cells, and organs for ingesting food are concentrated. The other end of a cephalized animal is designated *posterior* and may feature a tail (see Fig. 23-2b).

Most Bilateral Animals Have Body Cavities

The members of many bilateral animal phyla have a fluid-filled cavity between the digestive tube (or gut, where food is digested and absorbed) and the outer body wall. In an animal with a body cavity, the gut and body wall are separated by a fluid-filled space, creating a "tube-within-a-tube" body plan. Body cavities are absent in radially symmetrical animals, so it is likely that this feature arose sometime after the split between radially and bilaterally symmetrical animals.

A body cavity can serve a variety of functions. In an earthworm it acts as a kind of skeleton, providing support for the body and a framework against which muscles can act. In other animals, internal organs are suspended within the fluid-filled cavity, which serves as a protective buffer between the organs and the outside world.

Body Cavity Structure Varies Among Phyla

The most widespread type of body cavity is a **coelom,** a fluid-filled cavity that is completely lined with a thin layer of tissue that develops from mesoderm (**Fig. 23-3a**). Phyla whose members have a coelom are called *coelomates*. The annelids (segmented worms), arthropods (insects, spiders,

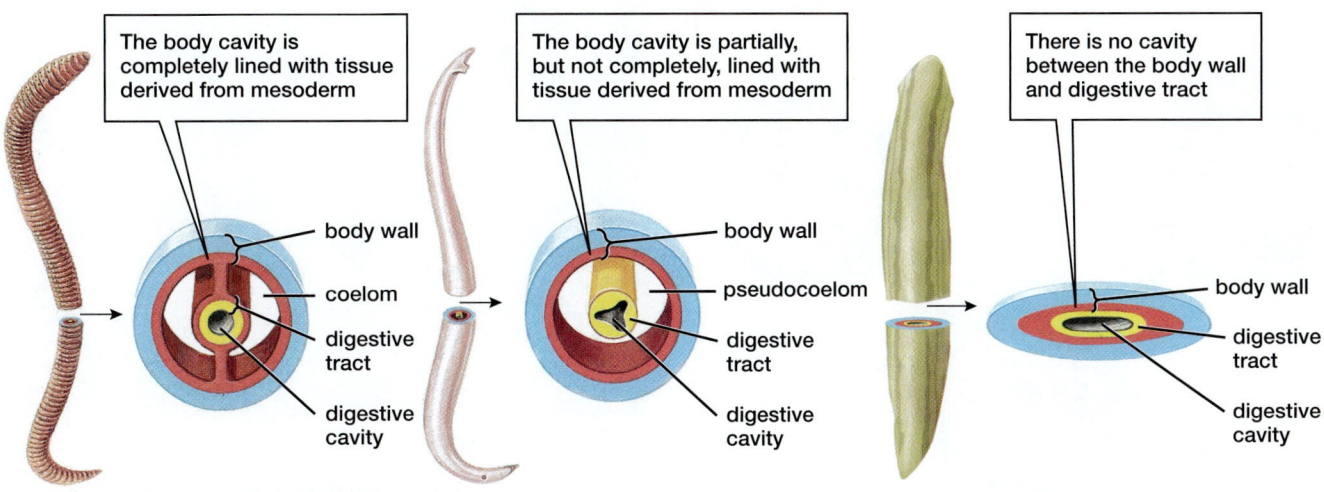

The body cavity is completely lined with tissue derived from mesoderm

body wall
coelom
digestive tract
digestive cavity

(a) "True" coelom (annelids, chordates)

The body cavity is partially, but not completely, lined with tissue derived from mesoderm

body wall
pseudocoelom
digestive tract
digestive cavity

(b) "False" or pseudocoelom (roundworms)

There is no cavity between the body wall and digestive tract

body wall
digestive tract
digestive cavity

(c) No coelom (flatworms)

▲ **FIGURE 23-3 Body cavities (a)** Annelids have a true coelom. **(b)** Roundworms are pseudocoelomates. **(c)** Flatworms have no cavity between the body wall and digestive tract. (Tissues shown in blue are derived from ectoderm, those in red from mesoderm, and those in yellow from endoderm.)

crustaceans), mollusks (clams and snails), echinoderms, and chordates (which include humans) are coelomates.

Some animals have a body cavity that is *not* completely surrounded by mesoderm-derived tissue. This type of cavity is known as a **pseudocoelom.** Phyla whose members have a pseudocoelom are collectively known as *pseudocoelomates* (**Fig. 23-3b**). The roundworms (nematodes) are the largest pseudocoelomate group.

Some phyla of bilateral animals have no body cavity at all and are known as *acoelomates*. For example, flatworms have no cavity between their gut and body wall; instead, the space is filled with solid tissue (**Fig. 23-3c**).

Bilateral Organisms Develop in One of Two Ways

Among the bilateral animal phyla, embryological development follows a variety of pathways. These diverse developmental pathways, however, can be grouped into two categories, known as **protostome** development and **deuterostome** development (**Fig. 23-4**). In protostome development, the coelom (when present) forms within the space between the body wall and the digestive cavity. In deuterostome development, the coelom forms as an outgrowth of the digestive cavity. The two types of development also differ in the pattern of cell division immediately after fertilization and in the method by which the mouth and anus are formed. Protostomes and deuterostomes represent distinct evolutionary branches within the bilateral animals. Annelids, arthropods, flatworms, roundworms, and mollusks exhibit protostome development; echinoderms and chordates are deuterostomes.

Protostomes Include Two Distinct Evolutionary Lines

The protostome animal phyla fall into two groups, which correspond to two different lineages that diverged early in the evolutionary history of protostomes. One group, the *ecdysozoans*, includes phyla such as the arthropods and roundworms, whose members have bodies covered by an outer layer that is periodically shed. The other group is known as the *lophotrochozoans* and includes phyla whose members have a special feeding structure called a lophophore (**Fig. 23-5a**), as well as phyla whose members pass through a particular type of developmental stage called a trochophore larva (**Fig. 23-5b**). The flatworms, annelids, and mollusks are lophotrochozoan phyla.

(a) Lophophore

Protostome development

> Solid mass of mesoderm splits to form coelom

Deuterostome development

> Mesoderm pockets pinch off of digestive cavity to form coelom

mesoderm

digestive cavity

coelom

▲ FIGURE 23-4 **Formation of the coelom in protostomes and deuterostomes** The difference depicted here is one of several between protostome and deuterostome embryonic development.

(b) Trochophore larva

▲ FIGURE 23-5 **Lophotrochozoan characteristics** Members of the lophotrochozoan phyla, which include flatworms, segmented worms, and mollusks, exhibit either **(a)** a feeding structure known as a lophophore or **(b)** a distinctive swimming larval form called a trochophore.

Can you describe how the body organization of sponges differs from that of all other animals? Are you able to describe the different types of body symmetry, embryonic tissue layer arrangements, body cavities, and embryonic development present among tissue-bearing animals? Can you give examples of animal groups with each type of body symmetry, tissue layer arrangement, body cavity, and embryonic development? Can you name and describe the two main lineages within the protostome animals?

23.3 WHAT ARE THE MAJOR ANIMAL PHYLA?

For convenience, biologists often place animals in one of two major categories: **vertebrates,** those with a backbone (or vertebral column), and **invertebrates,** those lacking a backbone. The vertebrates—fish, amphibians, reptiles, and mammals (see Chapter 24)—are perhaps the most conspicuous animals from a human point of view, but less than 3% of all known animal species are vertebrates. The vast majority of animals are invertebrates. Biologists recognize about 27 phyla of animals; some key phyla are summarized in Table 23-1.

The earliest animals probably originated from colonies of single-celled protists whose members had become specialized to perform distinct roles within the colony. In our survey of the invertebrate animals, we will begin with the sponges, whose body plan most closely resembles the probable ancestral protist colonies.

Sponges Are Simple, Sessile Animals

Sponges (Porifera) are found in most aquatic environments. Most of Earth's 5,000 sponge species live in salt water, where they inhabit ocean waters warm and cold, deep and shallow. In addition, some sponges live in freshwater habitats such as lakes and rivers. Adult sponges live attached to rocks or other underwater surfaces. They are generally sessile, though researchers have demonstrated that some species, at least in aquaria, are able to move about (very slowly—a few millimeters per day). Sponges come in a variety of shapes and sizes. Some species have a well-defined shape, but others grow free-form over underwater rocks (**Fig. 23-6**). The largest sponges can grow to more than 3 feet (1 meter) in height.

Most sponges are **hermaphroditic**—that is, they possess both male and female sexual organs. They typically reproduce sexually, a process that begins when sperm are released into the water. A sperm cell may enter the body of another sponge and be transported to an egg (eggs are retained in the sponge's body). Fertilized eggs develop inside the adult into active larvae that escape through the openings in the sponge body. Water currents disperse the larvae to new areas, where they settle and develop into adult sponges.

(a) Encrusting sponge

(b) Tubular sponge

(c) Vase-shaped sponge

▲ FIGURE 23-6 **The diversity of sponges** Sponges come in a wide variety of sizes, shapes, and colors. Some, such as **(a)** this encrusting sponge, grow in a free-form pattern over undersea rocks. Others may be **(b)** tubular or **(c)** vase shaped.

QUESTION Sponges are often described as the most "primitive" of animals. How can such a primitive organism have become so diverse and abundant?

TABLE 23-1   Comparison of the Major Animal Phyla

Common Name (Phylum)		Sponges (Porifera)	Sea Jellies, Corals Anemones (Cnidaria)	Flatworms (Platyhelminthes)
Body plan	Level of organization	Cellular—lack tissues and organs	Tissue—lack organs	Organ system
	Germ layers	Absent	Two	Three
	Symmetry	Absent	Radial	Bilateral
	Cephalization	Absent	Absent	Present
	Body cavity	Absent	Absent	Absent
	Segmentation	Absent	Absent	Absent
Internal systems	Digestive system	Intracellular	Gastrovascular cavity; some intracellular	Gastrovascular cavity
	Circulatory system	Absent	Absent	Absent
	Respiratory system	Absent	Absent	Absent
	Excretory system (fluid regulation)	Absent	Absent	Canals with ciliated cells
	Nervous system	Absent	Nerve net	Head ganglia with longitudinal nerve cords
	Reproduction	Sexual; asexual (budding)	Sexual; asexual (budding)	Sexual (some hermaphroditic); asexual (body splits)
	Support	Endoskeleton of spicules	Hydrostatic skeleton	Hydrostatic skeleton
	Number of known species	5,000	9,000	20,000

Sponges Lack Tissues

Sponge bodies do not contain tissues; in some ways, a sponge resembles a colony of single-celled organisms. The colony-like properties of sponges were revealed in an experiment performed by embryologist H. V. Wilson in 1907. Wilson mashed a sponge through a piece of silk, thereby breaking it apart into single cells and cell clusters. He then placed these tiny bits of sponge into seawater and waited for 3 weeks. By the end of the experiment, the cells had reaggregated into a functional sponge, demonstrating that individual sponge cells had been able to survive and function independently.

A sponge's body is perforated by numerous tiny pores, through which water enters, and by fewer, large openings, through which water is expelled (**Fig. 23-7**). Within the sponge, water travels through canals. As water passes through the sponge, oxygen is extracted, microorganisms are filtered out and taken into individual cells where they are digested, and wastes are released.

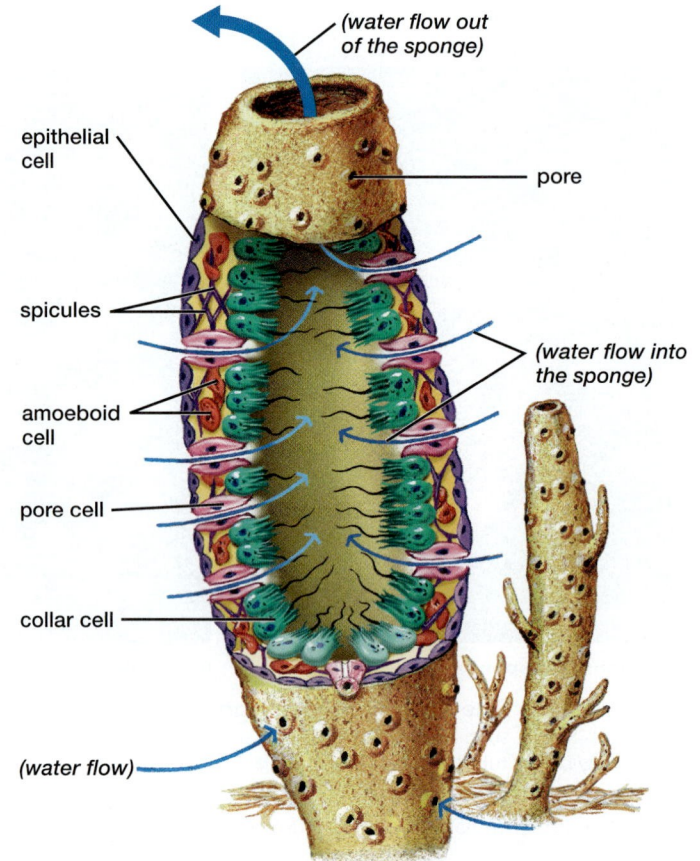

(water flow out of the sponge)

epithelial cell

pore

spicules

(water flow into the sponge)

amoeboid cell

pore cell

collar cell

(water flow)

▶ **FIGURE 23-7   The body plan of sponges** Water enters through numerous tiny pores in the sponge body and exits through a larger opening. Microscopic food particles are filtered from the water.

Segmented Worms (Annelida)	Clams, Snails, Octopuses, Squid (Mollusca)	Insects, Arachnids, Crustaceans (Arthropoda)	Roundworms (Nematoda)	Sea Stars, Sea Urchins, Sea Cucumbers (Echinodermata)
Organ system	Organ system	Organ system	Organ system	Organ system
Three	Three	Three	Three	Three
Bilateral	Bilateral	Bilateral	Bilateral	Bilateral larvae, radial adults
Present	Present	Present	Present	Absent
Coelom	Coelom	Coelom	Pseudocoel	Coelom
Present	Absent	Present	Absent	Absent
Separate mouth and anus	Separate mouth and anus	Separate mouth and anus	Separate mouth and anus	Separate mouth and anus (usually)
Closed	Open	Open	Absent	Absent
Absent	Gills, lungs	Tracheae, gills, or lungs	Absent	Tube feet, skin gills
Nephridia	Nephridia	Excretory glands resembling nephridia	Excretory gland	Absent
Head ganglia with paired ventral cords; ganglia in each segment	Well-developed brain in some cephalopods; several paired ganglia, most in the head; nerve network in the body wall	Head ganglia with paired ventral nerve cords; ganglia in segments, some fused	Head ganglia with dorsal and ventral nerve cords	Head ganglia absent; nerve ring and radial nerves; nerve network in the skin
Sexual (some hermaphroditic)	Sexual (some hermaphroditic)	Usually sexual	Sexual (some hermaphroditic)	Sexual (some hermaphroditic); asexual by regeneration (rare)
Hydrostatic skeleton	Hydrostatic skeleton	Exoskeleton	Hydrostatic skeleton	Endoskeleton of plates beneath outer skin
9,000	50,000	1,000,000	12,000	6,500

Sponge Cells Are Specialized for Different Functions

Sponges have three major cell types, each with a specialized role (see Fig. 23-7). Flattened *epithelial cells* cover the animal's outer body surfaces. Some epithelial cells are modified into *pore cells*, which surround pores, controlling their size and thereby regulating the entry of water. The pores close when harmful substances are present. *Collar cells* bear flagella that extend into the inner cavity. The beating flagella maintain a flow of water through the sponge. The collars that surround the flagella act as fine sieves, filtering out microorganisms that are then ingested by the cell. Some of the food is passed to the *amoeboid cells*. These cells roam freely between the epithelial and collar cells, digesting and distributing nutrients, producing reproductive cells, and secreting small skeletal projections called *spicules*. Spicules, which may be composed of calcium carbonate (chalk), silica (glass), or protein, form an internal skeleton that provides support for the sponge's body (see Fig. 23-7).

Some Sponges Contain Chemicals Useful to Humans

Because sponges remain in one spot and have no protective shell, they are vulnerable to predators such as fish, turtles, and sea slugs. Many sponges, however, contain chemicals that are toxic or distasteful to potential predators. Fortunately for people, a number of these chemicals have proved to be valuable medicines. For example, the drug spongistatin, a compound first isolated from a sponge, is an emerging treatment for the fungal infections that frequently sicken people with AIDS. Other medicines derived from sponges include some promising new anticancer drugs. The discovery of these medicines has raised hopes that, as more species are screened by researchers, sponges will become a major source of new drugs.

Cnidarians Are Well-Armed Predators

The *cnidarians* (Cnidaria) include sea jellies (also known as jellyfish), corals, sea anemones, and hydrozoans. The roughly 9,000 known species of cnidarians are confined to watery habitats; most are marine. Most species are small, from a few millimeters to a few inches in diameter, but the largest sea jellies can be 8 feet across and have tentacles 150 feet long. All cnidarians are predators.

Cnidarians Have Tissues and Two Body Types

The cells of cnidarians are organized into distinct tissues, including contractile tissue that acts like muscle. Cnidarian nerve cells are organized into tissue called a *nerve net*, which branches through the body and controls the contractile

(a) Anemone

(b) Sea jelly

(c) Coral

(d) Sea wasp

◄ **FIGURE 23-8 Cnidarian diversity (a)** A red-spotted anemone spreads its tentacles to capture prey. **(b)** A sea jelly drifts in the ocean, its tentacles hanging down. **(c)** A close-up of coral reveals the extended tentacles of the polyps. **(d)** A sea wasp, a type of sea jelly whose stinging cells contain one of the most toxic of all known venoms. The venom quickly kills prey, such as the shrimp shown here, that brush against a sea wasp's tentacles.

QUESTION In each of these photos, is the pictured organism a polyp or a medusa?

tissue to bring about movement and feeding behavior. However, most cnidarians lack organs and have no brain.

Cnidarians come in a variety of forms (**Fig. 23-8**), all of which are variations on two basic body plans: the *polyp* (**Fig. 23-9a**) and the *medusa* (**Fig. 23-9b**). The tubular polyp is adapted to a life spent quietly attached to rocks. The polyp has tentacles that reach upward to grasp and immobilize prey. The medusa floats in the water and is carried by currents, its bell-shaped body trailing tentacles like multiple fishing lines.

Many cnidarian life cycles include both polyp and medusa stages, though some species live only as polyps and others only as medusae. Both polyps and medusae develop from just two germ layers—the interior endoderm and the

(a) Polyp

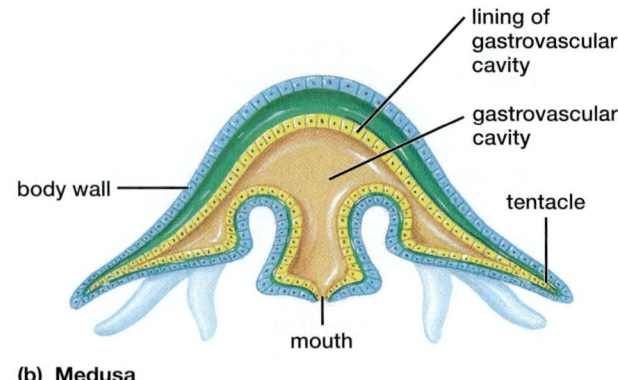

(b) Medusa

▲ **FIGURE 23-9 Polyp and medusa (a)** The polyp form is seen in sea anemones (see Fig. 23-8a) and the individual polyps within a coral (see Fig. 23-8c). **(b)** The medusa form, seen in the sea jelly (see Fig. 23-8b), resembles an inverted polyp. (Tissues shown in blue are derived from ectoderm, those in yellow from endoderm.)

exterior ectoderm; between those layers is a jelly-like substance. Polyps and medusae are radially symmetrical, with body parts arranged in a circle around the mouth and digestive cavity (see Fig. 23-2a).

Reproduction varies considerably among different types of cnidarians, but one pattern is fairly common in species with both polyp and medusa stages. In such species, polyps typically reproduce by asexual **budding** in which the polyp produces miniature versions of itself that drop off and assume an independent existence. Under certain conditions, however, budding may instead give rise to medusae. After a medusa grows to maturity, it may release gametes (sperm or eggs) into the water. If a sperm and egg meet, they may fuse to form a zygote that develops into a free-swimming, ciliated larva. The larva may eventually settle on a hard surface, where it develops into a polyp.

Cnidarians Have Stinging Cells

Cnidarian tentacles are armed with *cnidocytes*, cells containing structures that, when touched, explosively inject poisonous or sticky filaments into prey (**Fig. 23-10**). These stinging cells, found only in cnidarians, are used to capture prey. Cnidarians do not actively hunt. Instead, they wait for their victims to blunder, by chance, into the grasp of their enveloping tentacles. Stung and firmly grasped, the prey is forced through an expandable mouth into a digestive sac, the *gastrovascular cavity* (see Fig. 23-9). Digestive enzymes secreted into this cavity break down some of the food, and further digestion occurs within the cells lining the cavity. Because the gastrovascular cavity has only a single opening, undigested material is expelled through the mouth when digestion is completed. Although this two-way traffic prevents continuous feeding, it is adequate to support the low energy demands of these animals.

The venom of some cnidarians can cause painful stings in humans; the stings of a few sea jelly species can even be life threatening. The most deadly of these species is the sea wasp, *Chironex fleckeri* (see Fig. 23-8d), which is found in the waters off northern Australia and Southeast Asia, and can grow to be about 12 inches (30 centimeters) in diameter. The amount of venom in a single sea wasp could kill up to 60 people, and the victim of a serious sting may die within minutes.

Many Corals Secrete Hard Skeletons

One group of cnidarians, the corals, is of particular ecological importance (see Fig. 23-8c). In many coral species, polyps form colonies, and each member of the colony secretes a hard skeleton of calcium carbonate. The skeletons persist long after the organisms die, serving as a base to which other individuals attach themselves. The cycle continues until, after thousands of years, massive coral reefs are formed.

Coral reefs are found in both cold and warm oceans. Cold-water reefs form in deep waters and, though widely distributed, have only recently attracted the attention of researchers and are not yet well studied. The more familiar warm-water coral reefs are restricted to clear, shallow waters in the tropics. Here, coral reefs form undersea habitats that are the basis of an ecosystem of stunning diversity and unparalleled beauty.

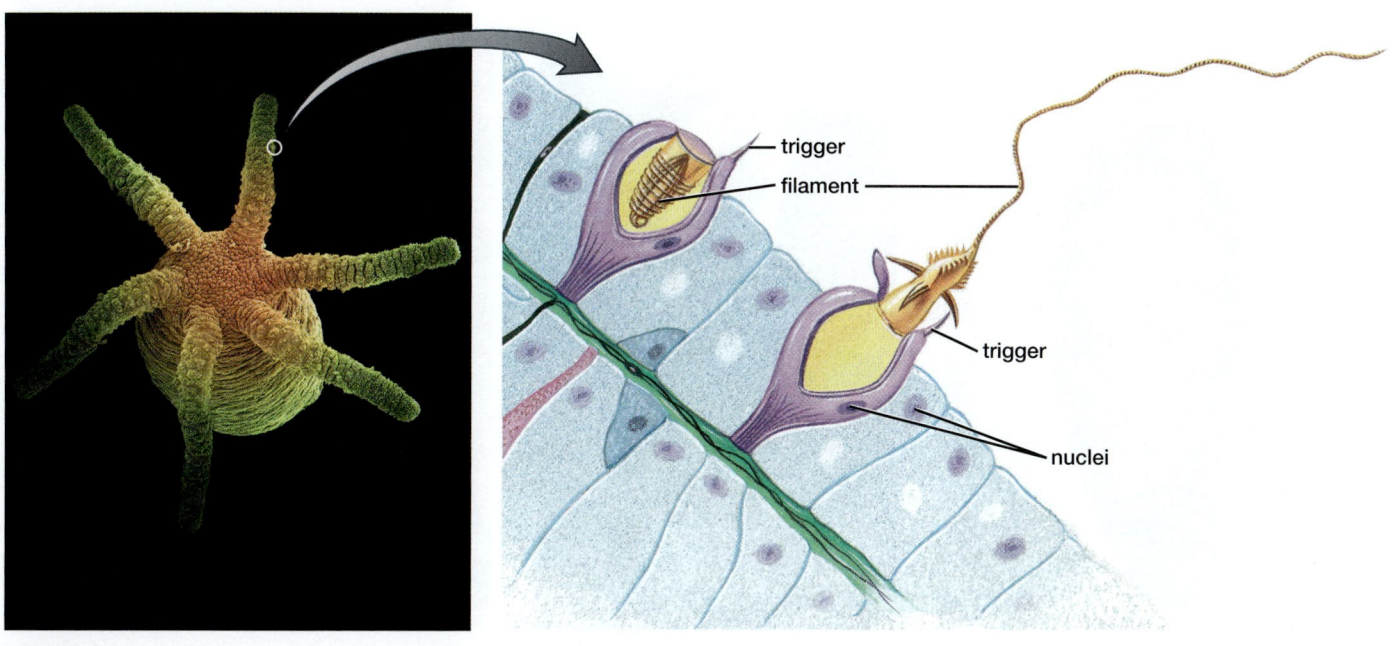

(a) Hydra   **(b) Cnidocytes**

▲ **FIGURE 23-10 Cnidarian weaponry: the cnidocyte (a)** Cnidarian tentacles, such as those of the hydra, contain cnidocyte cells. **(b)** At the slightest touch to its trigger, a structure within a cnidocyte cell violently expels a poisoned filament, which may penetrate prey.

Comb Jellies Use Cilia to Move

The roughly 150 species of radially symmetrical *comb jellies* (Ctenophora) are superficially similar in appearance to some cnidarians, but form a distinct evolutionary lineage. Most comb jellies are less than an inch (2.5 cm) in diameter, but a few species can grow to more than 3 feet (1 m) across. Comb jellies move by means of cilia, which are arranged in eight rows known as combs. Although most comb jellies are colorless and transparent or translucent, light scattered by the beating cilia of the combs can appear as eight ever-changing rainbows of color on the comb jelly's body (**Fig. 23-11**).

All comb jellies are carnivores. Most species inhabit coastal or oceanic waters and eat tiny invertebrate animals (in some cases including other, smaller comb jellies), which are captured with sticky tentacles. (Comb jellies lack the stinging cells that characterize cnidarians.) Almost all comb jellies are hermaphroditic. Each individual releases both sperm and eggs to the surrounding water. Fertilized eggs float freely in the water until larvae emerge and gradually develop into adult comb jellies.

▲ **FIGURE 23-11 A comb jelly** This comb jelly's body is unpigmented, but rows of cilia refract light to produce iridescent colors.

Flatworms May Be Parasitic or Free Living

The *flatworms* (Platyhelminthes) are aptly named; they have a flat, ribbon-like shape. They are bilaterally symmetrical (see Fig. 23-2b). Many of the approximately 20,000 flatworm species are parasites (**Fig. 23-12a**). (**Parasites** are organisms that live in or on the body of another organism, called a *host*, which is harmed as a result of the relationship.) Nonparasitic, free-living flatworms inhabit freshwater, marine, and moist terrestrial habitats. They tend to be small and inconspicuous (**Fig. 23-12b**), but some are brightly colored, spectacularly patterned residents of tropical coral reefs (**Fig. 23-12c**).

Flatworms can reproduce both sexually and asexually. Free-living species may reproduce by cinching themselves around the middle until they separate into two halves, each of which regenerates its missing parts. All flatworm species can reproduce sexually; most are hermaphroditic. This trait allows a flatworm to reproduce through self-fertilization, a great advantage to a parasitic worm that may be the only one of its kind present in its host.

(a) Fluke

(b) Freshwater flatworm

(c) Marine flatworm

▲ **FIGURE 23-12 Flatworm diversity (a)** This fluke is an example of a parasitic flatworm. **(b)** Eyespots are clearly visible in the head of this free-living, freshwater flatworm. **(c)** Many of the flatworms that inhabit tropical coral reefs are brightly colored.

Flatworms Have Organs But Lack Respiratory and Circulatory Systems

Unlike cnidarians, flatworms have organs, in which tissues are grouped into functional units. For example, most free-living flatworms have sense organs, including eyespots (see Fig. 23-12b) that detect light and dark, and cells that respond to chemical and tactile stimuli. To process information, a flatworm has clusters of nerve cells called **ganglia** (singular, ganglion) in its head, forming a simple brain. Paired neural structures called **nerve cords** conduct nervous signals to and from the ganglia.

Flatworms lack respiratory and circulatory systems. In the absence of a respiratory system, gases are exchanged by direct diffusion between body cells and the environment. This mode of respiration is possible because the small size and flat shape of flatworm bodies ensure that no body cell is very far from the surrounding environment. In the absence of a circulatory system to carry them, nutrients move directly from the digestive tract to body cells. The digestive cavity has a branching structure that reaches all parts of the body and allows digested nutrients to diffuse into nearby cells. The digestive cavity has only one opening to the environment, so wastes pass out through the mouth.

Some Flatworms Are Harmful to Humans

Some parasitic flatworms can infect humans. For example, tapeworms can infect people who eat improperly cooked beef, pork, or fish that has been infected by the worms. Tapeworm larvae form encapsulated resting structures, called *cysts*, in the muscles of these animals. The cysts hatch in the human digestive tract, where the young tapeworms attach themselves to the lining of the intestine. There they may grow to a length of more than 20 feet (7 meters), absorbing digested nutrients directly through their outer surface and eventually releasing packets of eggs that are shed in the host's feces. If pigs, cows, or fish eat food contaminated with infected human feces, the eggs hatch in the animal's digestive tract, releasing larvae that burrow into its muscles and form cysts, thereby continuing the infective cycle (**Fig. 23-13**).

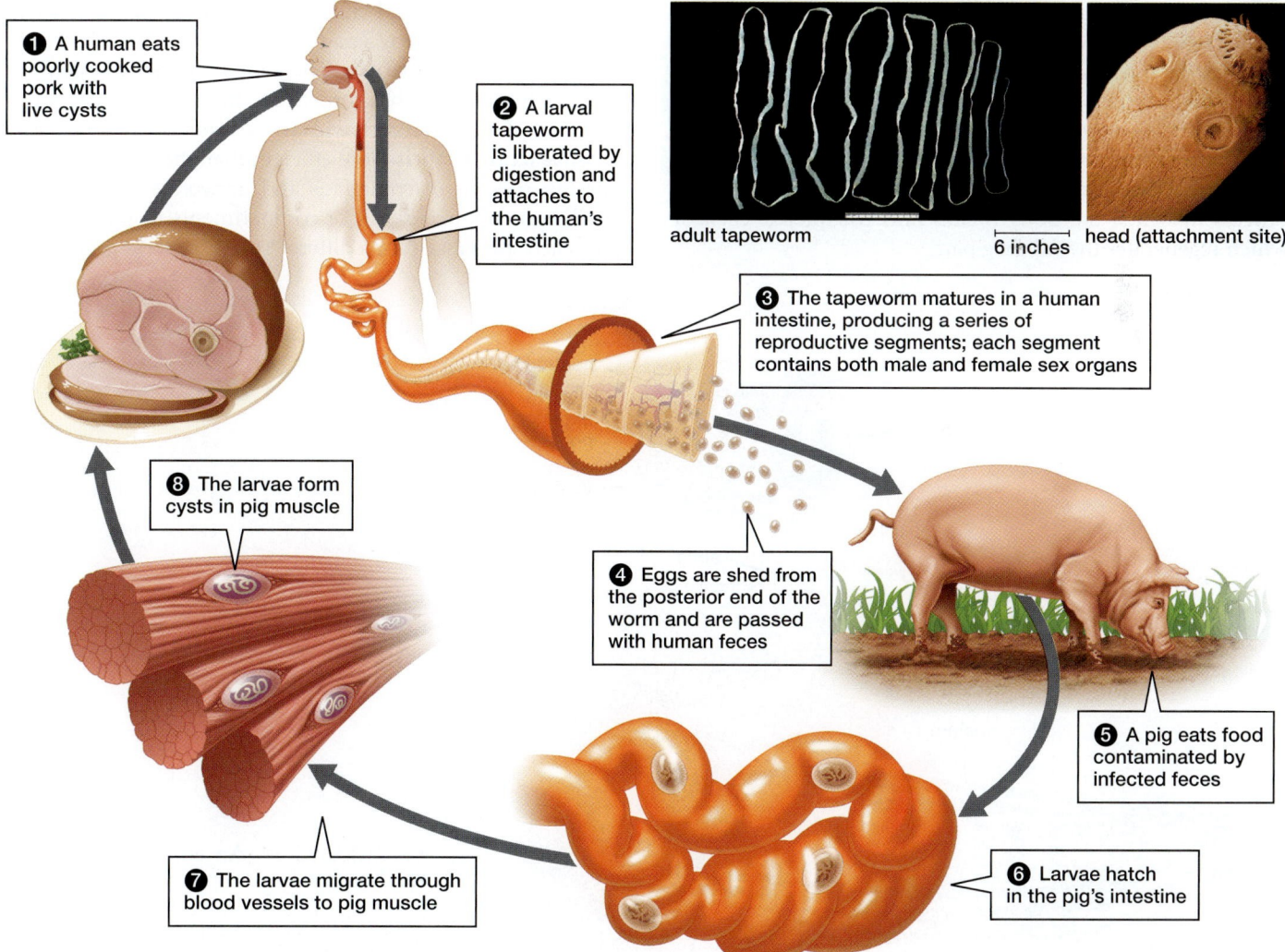

❶ A human eats poorly cooked pork with live cysts

❷ A larval tapeworm is liberated by digestion and attaches to the human's intestine

adult tapeworm 6 inches head (attachment site)

❸ The tapeworm matures in a human intestine, producing a series of reproductive segments; each segment contains both male and female sex organs

❽ The larvae form cysts in pig muscle

❹ Eggs are shed from the posterior end of the worm and are passed with human feces

❺ A pig eats food contaminated by infected feces

❼ The larvae migrate through blood vessels to pig muscle

❻ Larvae hatch in the pig's intestine

▲ FIGURE 23-13 **The life cycle of the human pork tapeworm**
QUESTION Why have tapeworms evolved a long, flat shape?

Another group of parasitic flatworms includes the *flukes* (see Fig. 23-12a). Of these, the most devastating are liver flukes (common in Asia) and blood flukes, such as those of the genus *Schistosoma*, which cause the disease schistosomiasis. Like most parasites, flukes have a complex life cycle that includes an intermediate host (a snail, in the case of *Schistosoma*). Prevalent in Africa and parts of South America, schistosomiasis affects an estimated 200 million people worldwide. Its symptoms include diarrhea, anemia, and possible brain damage.

Annelids Are Segmented Worms

The bodies of *annelids* (Annelida) are divided into a series of similar repeating segments. Externally, this **segmentation** appears as a series of ringlike depressions on the surface. Internally, most of the segments contain identical copies of nerves, excretory structures, and muscles.

Sexual reproduction is common among annelids. Some species are hermaphroditic; others have separate sexes. Fertilization may be external or internal. External fertilization, in which sperm and eggs are released into the surrounding environment, is found mainly in species that live in water. In internal fertilization, two individuals copulate and sperm are transferred directly from one to the other. In hermaphroditic species, sperm transfer may be mutual, with each partner both donating and receiving sperm. In addition, some annelids can reproduce asexually, typically by a process in which the body breaks into two pieces, each of which regenerates the missing part.

Annelids Are Coelomates and Have Organ Systems

Annelids have a fluid-filled true coelom between the body wall and the digestive tract (see Fig. 23-3a). The incompressible fluid in the coelom of many annelids is confined by the partitions between the segments and serves as a **hydrostatic skeleton,** a supportive framework against which muscles can act. A hydrostatic skeleton allows earthworms to burrow through soil.

Annelids have well-developed organ systems. For example, annelids have a **closed circulatory system** that distributes gases and nutrients throughout the body. In closed circulatory systems (including yours), blood remains confined to the heart and blood vessels. In the earthworm, for example, blood with oxygen-carrying hemoglobin is pumped through well-developed vessels by five pairs of "hearts" (**Fig. 23-14**). These hearts are actually short segments of specialized blood vessels that contract rhythmically. The blood is filtered and wastes are removed by excretory organs called *nephridia* (singular, nephridium) and excreted to the environment through small pores. Nephridia resemble the individual tubules of the vertebrate kidney. The annelid nervous system consists of a simple brain in the head and a series of repeating paired segmental ganglia joined by a pair of ventral nerve cords that pass along the length of the body. The annelid digestive system includes a tubular gut that runs from the mouth to the anus. This kind of digestive tract, with two openings and a one-way digestive path, is much more efficient than the single-opening digestive systems of cnidarians and flatworms. Digestion in annelids occurs in a series of compartments in the digestive tract, each specialized for a different phase of food processing.

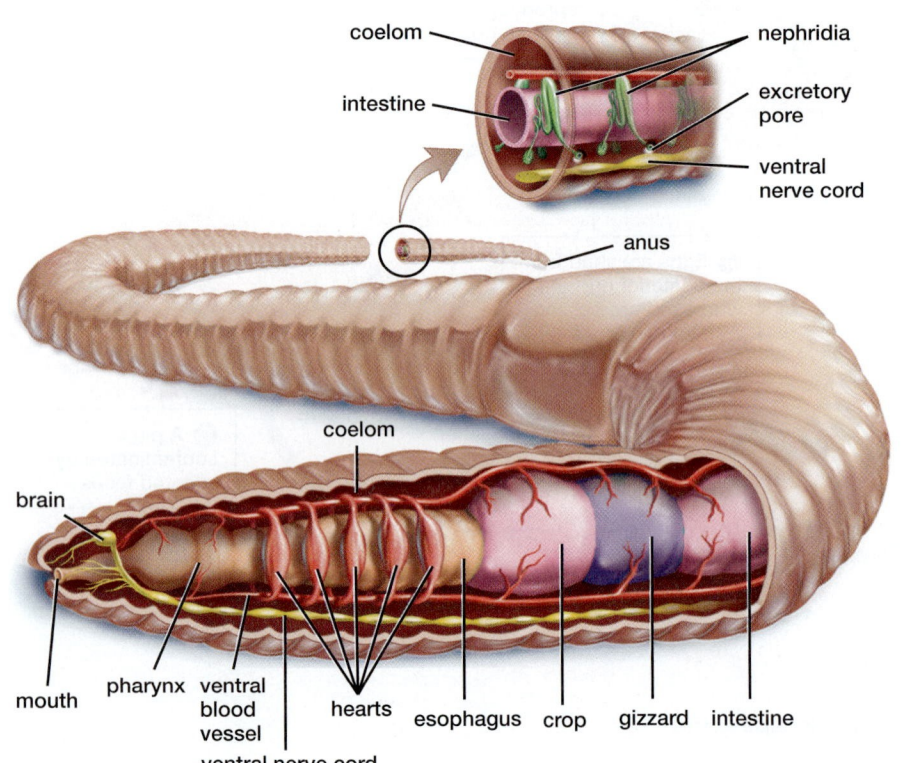

◄FIGURE 23-14 **An annelid, the earthworm** This diagram shows an enlargement of segments, many of which are repeating similar units separated by partitions.

QUESTION What advantage does a digestive system with two openings have relative to digestive systems with only a single opening (like that of the flatworms)?

Annelids Include Oligochaetes, Polychaetes, and Leeches

The 9,000 species of annelids fall into three main subgroups: the *oligochaetes*, the *polychaetes*, and the *leeches*. The oligochaetes include the familiar earthworm and its relatives. Charles Darwin, arguably the greatest of all biologists, devoted substantial time to the study of earthworms. He was especially impressed by earthworms' role in improving soil fertility. More than a million earthworms may live in an acre of land, tunneling through soil, consuming and excreting soil particles and organic matter. These actions help ensure that air and water can move easily through the soil and that organic matter is continually mixed into it, creating conditions that are favorable for plant growth. In Darwin's view, the activity of earthworms has had such a significant impact on agriculture that "it may be doubted whether there are many other animals which have played so important a part in the history of the world." The impact of earthworms, however, can also be negative. In some areas of North America, non-native, invasive earthworms disrupt the normal structure of forest soils, harming native forest plants.

Polychaetes live primarily in the ocean. Some polychaetes have paired fleshy paddles on most of their segments that are used in locomotion. Others live in tubes from which they project feathery gills that both exchange gases and filter the water for microscopic food (**Figs. 23-15a, b**).

Leeches (**Fig. 23-15c**) live in freshwater or moist terrestrial habitats and are either carnivorous or parasitic. Many carnivorous leeches prey on smaller invertebrates; some suck the blood of larger animals.

Most Mollusks Have Shells

If you have ever enjoyed a bowl of clam chowder, a plate of oysters on the half shell, or some sautéed scallops, you are indebted to *mollusks* (Mollusca). Mollusks are very diverse; in terms of number of known species, the 50,000 mollusks

CASE STUDY continued

Physicians' Assistants

The medicinal leech's role in medical treatment is reflected in its formal approval as a "medical device" by the U.S. Food and Drug Administration (FDA). Only one supplier, the French company Ricarimpex, has been approved to provide leeches for medical use in the United States. At Ricarimpex, leeches begin life in artificial ponds that simulate their natural habitat. Shortly after they hatch, young leeches are moved to a laboratory, where they spend 18 months feeding on chicken blood served in sheep-gut condoms. Once the leeches reach adult size (about 2½ inches long), they are packed into moist cotton bags and shipped out.

are second (albeit a distant second) only to the arthropods. These diverse species exhibit a range of lifestyles, from sessile forms such as mussels that spend their adult lives in one spot, filtering microorganisms from the water, to active, voracious predators of the ocean depths, such as the giant squid. Most mollusks are protected by hard shells of calcium carbonate. Others, however, lack shells. Some unshelled mollusks pursue prey and escape predators by moving swiftly; others are slow moving, but are protected by bodies that incorporate toxic or distasteful chemicals. Mollusks, with the exception of some snails and slugs, are aquatic.

The circulatory systems of most mollusks include a feature not seen in annelids: the **hemocoel,** or blood cavity. Blood empties into the hemocoel, where it bathes the internal organs directly. This arrangement is known as an **open circulatory system.** Mollusks also have a *mantle,* an extension of the body wall that forms a chamber for the gills and, in shelled species, secretes the shell (**Fig. 23-16**). The mollusk nervous system, like that of annelids, consists of ganglia connected by nerves, but in many mollusks more of the ganglia are concentrated in the brain. Reproduction

(a) Polychaete gills   **(b) Deep-sea polychaete**   **(c) Leech**

▲ **FIGURE 23-15 Diverse annelids (a)** The brightly colored gills of a polychaete annelid. The rest of the worm's body is hidden inside a tube embedded in the coral that is visible in the background. **(b)** This polychaete lives near deep-sea vents where the water temperature may reach 175°F (80°C). **(c)** This leech, a freshwater annelid, shows numerous segments. The sucker encircles its mouth, allowing it to attach to its prey.

QUESTION Why does pouring salt on a leech harm it?

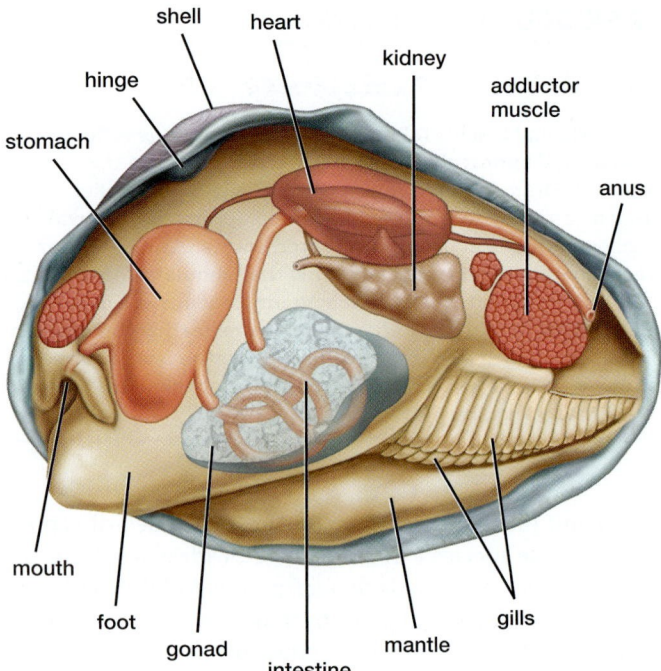

▲ **FIGURE 23-16 A bivalve mollusk** The body plan of a clam, showing the mantle, foot, gills, shell, and other features that are seen in most (but not all) mollusk species. The clam uses its adductor muscles to close its shell.

is sexual; some species have separate sexes, and others are hermaphroditic.

Among the many subgroups of mollusks, we will discuss three in more detail: gastropods, bivalves, and cephalopods.

Gastropods Are One-Footed Crawlers

Snails and slugs—collectively known as *gastropods*—crawl on a muscular *foot*, and many are protected by shells that vary widely in form and color (**Fig. 23-17a**). Not all gastropods are shelled, however. Sea slugs, for example, lack shells, but

their brilliant colors warn potential predators that they are poisonous or foul tasting (**Fig. 23-17b**).

Gastropods feed with a *radula*, a flexible ribbon of tissue studded with spines that is used to scrape algae from rocks or to grasp larger plants or prey. Most snails use gills, typically enclosed in a cavity beneath the shell, for respiration. Gases can also diffuse readily through the skin of most gastropods. The few gastropod species that live in terrestrial habitats (including the destructive garden snails and slugs) use a simple lung for breathing.

Bivalves Are Filter Feeders

Included among the *bivalves* are scallops, oysters, mussels, and clams (**Fig. 23-18**). Bivalves possess two shells connected by a flexible hinge. A strong muscle clamps the shells closed in response to danger; this muscle is what you are served when you order scallops in a restaurant.

Clams use a muscular foot for burrowing in sand or mud. In mussels, which live attached to rocks, the foot is smaller and helps secrete threads that anchor the animal to the rocks. Scallops lack a foot and move by a sort of jet propulsion achieved by flapping their shells together.

Bivalves are filter feeders, using their gills as both respiratory and feeding structures. Water circulates over the gills, which are covered with a thin layer of mucus that traps microscopic food particles. Food is conveyed to the mouth by the beating of cilia on the gills.

Cephalopods Are Marine Predators

The *cephalopods* include octopuses, nautiluses, cuttlefish, and squids (**Fig. 23-19**). The largest invertebrates, the giant squid and the colossal squid, belong to this group (see "Science in Action: The Search for a Sea Monster"). All cephalopods are carnivores, and all are marine. In these mollusks, the foot has evolved into tentacles with well-developed sensory abilities for detecting prey. Prey are

(a) Snail

(b) Sea slug

▲ **FIGURE 23-17 The diversity of gastropod mollusks (a)** A Florida tree snail displays a brightly striped shell and eyes at the tip of stalks that retract instantly if touched. **(b)** The brilliant colors of many sea slugs warn potential predators that they are distasteful.

(a) Scallop

(b) Mussels

▲ FIGURE 23-18 **The diversity of bivalve mollusks**
(a) This scallop parts its hinged shells to allow the intake of water from which food will be filtered. The blue spots visible along the mantle just inside the upper and lower shells are simple eyes. **(b)** Mussels attach to rocks in dense aggregations exposed at low tide. White barnacles (which are arthropods) are attached to the mussel shells and surrounding rock.

(a) Octopus

(b) Squid

(c) Nautilus

▲ FIGURE 23-19 **The diversity of cephalopod mollusks**
(a) In emergencies, an octopus can jet backward by vigorously contracting its mantle. Octopuses and squid can emit clouds of dark purple ink to confuse pursuing predators. **(b)** A squid can move by contracting its mantle to generate jet propulsion, which pushes the animal backward through the water. **(c)** The chambered nautilus secretes a shell with internal, gas-filled chambers that provide buoyancy. Note the well-developed eyes and the tentacles used to capture prey.

grasped by suction disks on the tentacles and may be immobilized by a paralyzing venom in the saliva before being torn apart by beaklike jaws.

Cephalopods move rapidly by jet propulsion, which is generated by the forceful expulsion of water from the mantle cavity. Octopuses may also travel along the seafloor by using their tentacles like multiple undulating legs. The rapid movements and active lifestyles of cephalopods are made possible in part by their closed circulatory systems. Cephalopods are the only mollusks with a closed circulation system, which transports oxygen and nutrients more efficiently than does an open circulatory system.

Cephalopods have highly developed brains and sensory systems. The cephalopod eye rivals our own in complexity. The cephalopod brain, especially that of the octopus, is exceptionally large and complex. It is enclosed in a skull-like case of cartilage and endows the octopus with highly developed capabilities to learn and remember. In the laboratory, octopuses can rapidly learn to solve a maze, associate symbols with food, or open a screw-cap jar to obtain food. In the wild, some octopuses use tools; veined octopuses clean mud from buried coconut shells, stack them up for transport, and turn them into a protective shelter.

SCIENCE IN ACTION   The Search for a Sea Monster

The giant squid is one of the world's largest invertebrate animals, reaching lengths of 40 feet (13 meters) or more. Each of its huge eyes can be as large as a human head. The squid's ten tentacles, two of which are longer than the others, are covered with powerful suckers. The suckers contain sharp, clawlike hooks to better grasp prey, which is then pulled toward the squid's mouth, where a heavily muscled beak tears the food. Clearly, the giant squid is one of the most imposing organisms on Earth, yet we know almost nothing about its habits and lifestyle. A single brief video recording is our only observation of an adult giant squid alive in its natural habitat very deep in the ocean.

Clyde Roper, a biologist at the Smithsonian Institution, has devoted much of his professional life to a quest to view and study live giant squids. Roper's search for the giant squid led him to organize three major expeditions. The first of these searched waters near the Azores islands in the Atlantic Ocean. Because sperm whales are known to prey upon giant squid, Roper believed that the whales might lead him to the squids. To test this idea, he and his team affixed video cameras to sperm whales, thus allowing the scientists to see what the whales were seeing. These "crittercams" revealed a great deal of new information about sperm whale behavior but, alas, no footage of giant squids.

The next Roper-led expedition took place in the Kaikoura Canyon, an area of very deep water (3,300 feet, or 1,000 meters) off the coast of New Zealand. Cameras were again deployed on sperm whales, but this time the mobile cameras were supplemented by a stationary, baited camera and an unmanned, remote-controlled submarine. Again, however, a large investment of time, money, and equipment yielded no squid sightings.

A few years later, Roper assembled a team of scientists for a return to Kaikoura Canyon. This time, the group was able to use Deep Rover, a one-person submarine that could carry an observer to depths of 2,200 feet. The scientists used Deep Rover to explore the canyon, following sperm whales in hopes that the huge mammals would

lead them to giant squid. Unfortunately, the team again failed to find a squid.

Roper has pursued his search for the giant squid with extraordinary persistence, but he is not alone in seeking a glimpse of the creature. Other research teams have also been on the lookout for giant squid, and it was one of these groups that finally secured the first (and so far only) visual

record of living giant squid. Working off the coast of Japan, the researchers placed a video camera on a long, baited fishing line. Long hours of dragging the line through the water at a depth of 3,000 feet were eventually rewarded with images of a giant squid that attacked the bait (**Fig. E23-1a**). Later, a squid attacking the bait was lured to the surface (**Fig. E23-1b**).

(a) Giant squid at depth

(b) Giant squid at surface

▲ FIGURE E23-1 **Giant squid** **(a)** Among the first-ever images of a giant squid was this shot of the animal approaching a baited line. **(b)** A giant squid drawn to the ocean's surface.

Arthropods Are the Most Diverse and Abundant Animals

In terms of both number of individuals and number of species, no other animal phylum comes close to the *arthropods* (Arthropoda), which include insects, arachnids, myriapods, and crustaceans. About 1 million arthropod species have been discovered, and scientists estimate that millions more remain undescribed.

Arthropods Have Appendages and an External Skeleton

All arthropods have paired, jointed appendages and an **exoskeleton,** an external skeleton that encloses the arthropod body like a suit of armor. Secreted by the *epidermis* (the outer layer of skin), the exoskeleton is composed chiefly of protein and a polysaccharide called *chitin* (see Fig. 3-11). The external skeleton protects against predators and is responsible for arthropods' greatly increased agility relative to their wormlike ancestors. The exoskeleton provides rigid attachment sites for muscles, but also becomes thin and flexible at joints, thereby increasing the range of movement of the appendages. This combination of rigidity and flexibility makes possible the flight of the bumblebee and the intricate, delicate manipulations of the spider as it weaves its web (**Fig. 23-20**). The exoskeleton also contributed enormously to the arthropod invasion of the land by providing a watertight covering for delicate, moist tissues such as those used for gas exchange. (Arthropods were the earliest land-dwelling animals; see p. 322.)

Like a suit of armor, however, the arthropod exoskeleton poses some problems. First, because it cannot expand as the animal grows, the exoskeleton must periodically be shed, or **molted,** and replaced with a larger one (**Fig. 23-21**). Molting uses energy and leaves the animal temporarily vulnerable to predators until the new skeleton hardens. The exoskeleton is also heavy, and its weight increases exponentially as an animal grows. It is no coincidence that the largest arthropods are crustaceans (crabs and lobsters), whose watery habitat supports much of their weight.

▲ FIGURE 23-21 **The exoskeleton must be molted periodically** A cicada emerges from its outgrown exoskeleton (left).

Arthropods Have Specialized Segments and Adaptations for Active Lifestyles

Arthropods are segmented, but their segments tend to be few and specialized for different functions such as sensing the environment, feeding, and movement (**Fig. 23-22**). For example, in insects, sensory and feeding structures are concentrated on the front segment, known as the *head*, and digestive structures are largely confined to the *abdomen*, which is the rear segment. Between the head and the abdomen is the *thorax*, the segment to which structures used in locomotion, such as wings and walking legs, are attached.

Efficient gas exchange is required to supply adequate oxygen to the muscles, which allows the rapid flying, swimming, or running displayed by many arthropods. In aquatic arthropods such as crustaceans, gas exchange is accomplished by gills. In terrestrial arthropods, gas exchange is performed either by lungs (in arachnids) or by *tracheae* (singular, trachea), a network of narrow, branching respiratory tubes that open to the surrounding environment and that penetrate all parts of the body. Most arthropods have open circulatory systems, like those of mollusks, in which blood directly bathes the organs in a hemocoel.

▲ FIGURE 23-20 **The exoskeleton allows precise movements** A garden orb spider begins to wrap a captured wasp in silk. Such dexterous manipulations are made possible by the exoskeleton and jointed appendages characteristic of arthropods.

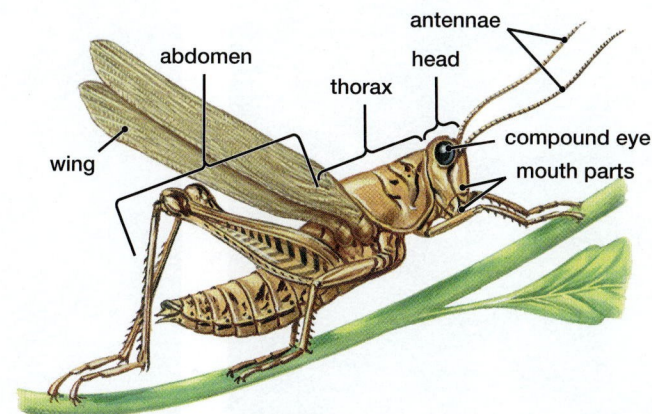

▲ FIGURE 23-22 **Segments are fused and specialized in insects** Insects, such as this grasshopper, show fusion and specialization of body segments into a distinct head, thorax, and abdomen. Segments are visible on the abdomen.

Most arthropods possess well-developed sensory and nervous systems. Arthropod sensory systems often include **compound eyes,** which have multiple light detectors (Fig. 23-23), and acute chemical and tactile senses. The arthropod nervous system consists of a brain composed of fused ganglia, and a series of additional ganglia along the length of the body that are linked by a ventral nerve cord. This well-developed nervous system, combined with sophisticated sensory abilities, has permitted the evolution of complex behaviors in many arthropods.

Insects Are the Only Flying Invertebrates

The number of described *insect* species is about 850,000, roughly three times the total number of known species in all other groups of animals combined (Fig. 23-24). Insects have a single pair of antennae and three pairs of legs, usually supplemented by two pairs of wings. Insects' capacity for flight distinguishes them from all other invertebrates and has contributed to their enormous success. As anyone who has pursued a fly can testify, flight helps in escaping from predators. It also allows the insect to find widely dispersed food. Swarms of locusts can travel 200 miles a day in search of food; researchers tracked one swarm on a journey that totaled almost 3,000 miles. Flight requires rapid and efficient gas exchange, which insects accomplish by means of tracheae.

▲ FIGURE 23-23 **Arthropods possess compound eyes** This scanning electron micrograph shows the compound eye of a horse fly. Compound eyes consist of an array of similar light-gathering and sensing elements whose orientation gives the arthropod a wide field of view. Insects have reasonably good image-forming ability and good color discrimination.

(a) Aphid

(b) Ant

(c) Beetle flying

(d) Moth larva

◀ FIGURE 23-24 **The diversity of insects (a)** Aphids suck sugar-rich juice from plants. **(b)** The bullet ant can inflict an extremely painful sting. **(c)** A June beetle displays its two pairs of wings as it comes in for a landing. The outer wings protect the abdomen and the inner wings, which are relatively thin and fragile. **(d)** Caterpillars are larval forms of moths or butterflies. This caterpillar larva of a silk moth can produce clicking sounds with its mouthparts. The sounds may serve to warn predators that the caterpillar is distasteful.

During their development, insects undergo **metamorphosis,** a radical change from a juvenile body form to an adult body form. In insects with complete metamorphosis, the immature stage, called a **larva** (plural, larvae), is worm shaped (for example, the maggot of a blowfly or the caterpillar of a moth or butterfly). The larva hatches from an egg, grows by eating voraciously and shedding its exoskeleton several times, and then forms a nonfeeding stage called a **pupa** (plural, pupae). Encased in an outer covering, the pupa body undergoes a drastic modification, emerging in its adult form. The adults mate and lay eggs, continuing the cycle. Metamorphosis may include a change in diet as well as in shape, thereby eliminating competition for food between adults and juveniles. In some cases, the change in diet allows the different stages to exploit different foods when they are most available. For instance, a caterpillar that feeds on new green shoots in springtime metamorphoses into a butterfly that drinks nectar from the summer's blooming flowers. Some insects, such as grasshoppers and crickets, undergo a more gradual metamorphosis (called incomplete metamorphosis), hatching as young that bear some resemblance to the adult, then gradually acquiring more adult features as they grow and molt.

Biologists classify the amazing diversity of insects into several dozen groups. We will describe three of the largest ones here.

Butterflies and Moths The butterflies and moths make up what is perhaps the most conspicuous and best-studied group of insects. The brightly colored, often iridescent wing patterns of many butterfly and moth species arise from pigments and light-refracting structures in the scales that cover the wings of all members of this group. (The scales are the powdery substance that rubs off onto your hand when you handle a butterfly or moth.) Butterflies fly mainly during the day, moths at night (though there are exceptions to this general rule, such as the hummingbird-like hawk moths that are often seen feeding by day in flower gardens).

The evolution of butterflies and moths has been closely tied to the evolution of flowering plants. Butterflies and moths feed almost exclusively on flowering plants, both as caterpillars and as adults. Many species of flowering plants depend in turn on butterflies and moths for pollination.

Bees, Ants, and Wasps The bees, ants, and wasps are known to many people by their painful stings. Many species in this group are equipped with a stinger that extends from the abdomen and can be used to inject venom into the victim of a sting. The venom may be extremely toxic but, fortunately for human stinging victims, each insect carries only a tiny amount. Nonetheless, the amount is often sufficient to cause considerable pain. Only females have stingers, which are used by many stinging species to help defend a nest from attack by a potential predator. Defense, however, is not the only use for stingers. Many wasps, for example, act as parasites when reproducing: These wasps lays their eggs inside the body of another species, typically a moth or butterfly caterpillar, which becomes food for the wasp larva after it

hatches. Before laying its egg, the wasp may sting the caterpillar, paralyzing it.

The social behavior of some ant and bee species is extraordinarily intricate. They may form huge colonies with complex organization in which individuals specialize in particular tasks such as foraging, defense, reproduction, or rearing larvae. The organization and division of labor in these insect societies require sophisticated communication and learning. Social insects accomplish remarkable tasks. For example, honeybees manufacture and store food (honey), and some ant species "farm" by cultivating fungi in underground chambers or "milking" aphids by inducing them to secrete a nutritious liquid.

Beetles Roughly one-third of all known insect species are beetles. Beetles exhibit a huge variety of shapes, sizes, and lifestyles. All beetles, however, have hard, protective coverings over their wings. Many destructive agricultural pests are beetles, such as the Colorado potato beetle, the grain weevil, and the Japanese beetle. However, others, such as lady beetles, are predators that help control insect pests.

One of this group's most impressive adaptations is found in the bombardier beetle. This species defends itself against ants and other enemies by emitting a toxic spray from a nozzle-like structure at the end of its abdomen. The beetle is able to precisely aim the spray, which emerges with explosive force and at temperatures higher than 200°F (93°C). To avoid harming itself, the beetle produces this hot, toxic brew only when needed, by combining two harmless substances.

Most Arachnids Are Predatory Meat Eaters

The *arachnids* include spiders, mites, ticks, and scorpions (**Fig. 23-25**). All arachnids have eight walking legs, and most are carnivorous. Many subsist on a liquid diet of blood or predigested prey. For example, spiders, the most numerous arachnids, first immobilize their prey with a paralyzing venom. They then inject digestive enzymes into the helpless victim (typically an insect) and suck in the resulting soup. Arachnids breathe using either tracheae, lungs, or both.

In contrast to the compound eyes of insects and crustaceans, arachnids have simple eyes, each with a single lens. Most spiders have eight eyes placed in such a way as to give them a panoramic view of predators and prey. The eyes are sensitive to movement, and in some spider species— especially those that hunt actively and have no webs—the eyes are thought to form images. Most spider perception, however, is not through their eyes but through sensory hairs found over much of the body. Some of a spider's hairs are touch sensitive and help the animal perceive prey, mates, and surroundings. Other hairs are sensitive to chemicals and function as organs of smell and taste. Hairs also respond to vibrations in the air, ground, or web, allowing spiders to detect nearby movement by predators, prey, or other spiders.

Among the distinctive features of spiders is their production of protein threads known as silk. Spiders manufacture silk in special glands in their abdomens and use it to perform a variety of functions, such as building webs that capture prey, wrapping up and immobilizing captured prey

(a) Spider

(b) Scorpion

(c) Ticks

▲ **FIGURE 23-25 The diversity of arachnids (a)** The tarantula is one of the largest spiders but is relatively harmless. **(b)** Scorpions, found in warm climates such as the deserts of the southwestern United States, paralyze their prey with venom from a stinger at the tip of the abdomen. A few species can harm humans. **(c)** Ticks before (left) and after feeding on blood. The uninflated exoskeleton is flexible and folded, allowing the animal to become grotesquely bloated while feeding.

(see Fig. 23-20), constructing protective shelters for them-selves, making cocoons to surround their eggs, and making "draglines" that connect a spider to a web or other surface and support its weight if it drops from its perch. Spider silk is

? HAVE **YOU** EVER WONDERED …

Why Spiders Don't Stick to Their Own Webs?

If you have ever accidentally walked into the web of a spider that uses a sticky web to capture insect prey, then you know that the strands of the web can be very sticky indeed. And the small insects one sees trapped on such webs seem to be quite firmly snared. So how does the spider that built the web scuttle so easily across it without getting stuck? One key factor is that the web is not entirely sticky. The web's scaffolding is made from non-sticky strands, with sticky ones restricted to the capture area. So a moving spider can often simply stay on the non-sticky strands and avoid the sticky ones. But the spider's distinctive claws also help. These claws, in conjunction with specialized hairs on the tips of its legs, allow a spider leg to grip a single web strand and then release it. A spider moving in this fashion, grasping strands with only the very tips of its legs, will have only a tiny surface area in contact with the web at any one time. Thus, even if the spider grips a sticky strand with one of its legs, it can easily pull loose, just as you can if you step on a wad of chewing gum. In contrast, an unsuspecting fly that crashes into the web will contact a number of strands with many parts of its body simultaneously, and be stuck fast.

an amazingly light, strong, and elastic fiber. It can be stronger than a steel wire of the same size, yet is as elastic as rubber. Human engineers have long sought to develop a fiber with this combination of strength and elasticity. Despite careful study of the structure of spider silk, no comparable man-made substance has been successfully manufactured.

Myriapods Have Many Legs

The *myriapods* include the centipedes and millipedes, whose most prominent feature is an abundance of legs (**Fig. 23-26**). Most millipede species have between 100 and 300 legs; the species with the largest number of legs can have up to 750. Centipedes are not quite as leggy; most have around 70 legs. Both centipedes and millipedes have one pair of antennae. The legs and antennae of centipedes are longer and more delicate than those of millipedes. Myriapods have very simple eyes that detect light and dark but do not form images. In some species, the number of eyes can be high—up to 200—but other species lack eyes altogether. Myriapods respire by means of tracheae.

Myriapods inhabit terrestrial environments exclusively, living mostly in soil or leaf litter or under logs and rocks. Centipedes are generally carnivorous, capturing prey (mostly other arthropods) with their frontmost legs, which are modi-fied into sharp claws that inject poison into prey. Bites from large centipedes can be painful to humans. In contrast, most millipedes are not predators but instead feed on de-caying vegetation and other debris. When attacked, many millipedes defend themselves by secreting a foul-smelling, distasteful liquid.

(a) Centipede

(b) Millipede

▲ FIGURE 23-26 **The diversity of myriapods (a)** Centipedes and **(b)** millipedes are common nocturnal arthropods. Each segment of a centipede's body holds one pair of legs, while each millipede segment has two pairs.

Most Crustaceans Are Aquatic

The *crustaceans*, including crabs, crayfish, lobsters, shrimp, and barnacles, are the only arthropods that live primarily in the water (**Fig. 23-27**). Crustaceans range in size from microscopic species that live in the spaces between grains of sand to the largest of all arthropods, the Japanese spider crab, with legs spanning nearly 12 feet (4 meters). Crustaceans have two pairs of sensory antennae, but the rest of their appendages are highly variable in form and number, depending on the habitat and lifestyle of the species. Most crustaceans have compound eyes similar to those of insects, and nearly all respire using gills.

Crustaceans are an important food source for larger animals. For example, small crustaceans called krill are abundant in the Southern Ocean and are the main food of whales, seals, seabirds, and other animals. People eat a lot of crustaceans, too. Shrimp, for example, are the most consumed seafood in the United States, by far. Per-capita consumption of shrimp has doubled during the past two decades. Today, most of the shrimp we eat are farm raised, mainly in coastal areas of Asia and South America. Unfortunately, widespread shrimp farming has had adverse ecological consequences, mainly because large areas of ecologically important mangrove forests have been cleared for shrimp farming.

▶ FIGURE 23-27 **The diversity of crustaceans (a)** The microscopic water flea is common in freshwater ponds. Notice the eggs developing within the body. **(b)** The sowbug, found in dark, moist places such as under rocks, leaves, and decaying logs, is one of the few crustaceans to invade the land successfully. **(c)** The hermit crab protects its soft abdomen by inhabiting an abandoned snail shell. **(d)** The gooseneck barnacle uses a tough, flexible stalk to anchor itself to rocks, boats, or even animals such as whales. Other types of barnacles attach with shells that resemble miniature volcanoes (see Fig. 23-18b). Early naturalists thought barnacles were mollusks until they observed barnacles' jointed legs (seen here extending into the water).

(a) Water flea

(b) Sowbug

(c) Hermit crab

(d) Barnacles

Roundworms Are Abundant and Mostly Tiny

Although you may be blissfully unaware of their presence, *roundworms* (Nematoda) are nearly everywhere. Roundworms, also called *nematodes*, have colonized nearly every habitat on Earth, and they play an important role in breaking down organic matter. They are extremely numerous; a single rotting apple may contain 100,000 roundworms. Billions thrive in each acre of topsoil. In addition, almost every plant and animal species hosts several parasitic nematode species.

In addition to being abundant and ubiquitous, roundworms are diverse. Although only about 12,000 roundworm species have been named, there may be as many as 500,000. Most, such as the one shown in **Figure 23-28**, are microscopic, but some parasitic forms reach a meter in length.

Roundworms Are Pseudocoelomates with a Simplified Body Plan

Roundworms have a rather simple body plan, featuring a tubular gut and a fluid-filled pseudocoelom that surrounds the organs and forms a hydrostatic skeleton (see Fig. 23-3b). A tough, flexible, nonliving cuticle encloses and protects the thin, elongated body and is periodically molted. Sensory organs in the roundworm head transmit information to a simple "brain," composed of a nerve ring.

Nematodes lack circulatory and respiratory systems. Because most nematodes are extremely thin and have low energy requirements, diffusion suffices for gas exchange and distribution of nutrients. Most nematodes reproduce sexually, and the sexes are separate; the male (usually smaller than the female) fertilizes the female by placing sperm inside her body.

A Few Roundworm Species Are Harmful to Humans

During your life, you may become host to one of the 50 species of roundworms that infect humans. Most such worms are relatively harmless, but there are important exceptions. For example, hookworm larvae (found in soil in some tropical regions) can bore into human feet, enter the bloodstream, and travel to the intestine, where they cause continuous bleeding. Another dangerous roundworm parasite, *Trichinella*, causes

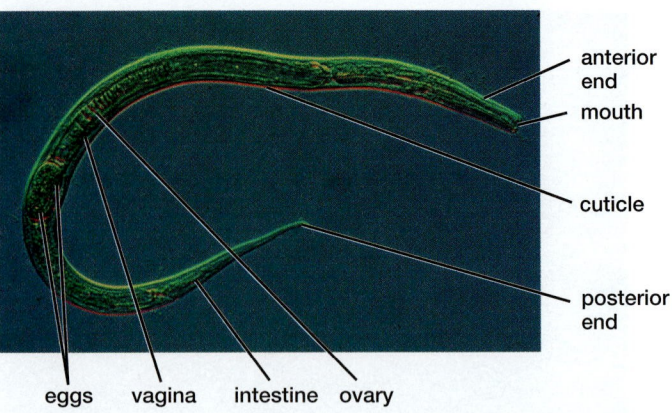

▲ FIGURE 23-28 A freshwater nematode Eggs can be seen inside this female freshwater nematode.

the disease trichinosis. *Trichinella* worms can infect people who eat improperly cooked infected pork, which can contain up to 15,000 larval cysts per gram (**Fig. 23-29a**). The cysts hatch in the human digestive tract and invade blood vessels and muscles, causing bleeding and muscle damage.

Parasitic roundworms can also endanger domestic animals. Dogs, for example, are susceptible to heartworm (**Fig. 23-29b**), which is transmitted by mosquitoes. In the southern United States, and increasingly in other parts of the country, heartworm poses a severe threat to the health of unprotected pets.

Echinoderms Have a Calcium Carbonate Skeleton

Echinoderms (Echinodermata) are found only in marine environments, and their common names tend to evoke their saltwater habitats: sand dollars, sea urchins, sea stars (or starfish), sea cucumbers, and sea lilies (**Fig. 23-30**). The name "echinoderm" (Greek, "hedgehog skin") stems from the bumps or spines that extend from the skin of most echinoderms. These spines are especially well developed in sea urchins and much reduced in sea stars and sea cucumbers. Echinoderm bumps and spines are actually extensions of an

(a) *Trichinella*

(b) Heartworms

▲ FIGURE 23-29 Some parasitic nematodes (a) Encysted larva of the *Trichinella* worm in the muscle tissue of a pig, where it may live for up to 20 years. **(b)** Adult heartworms in the heart of a dog. The juveniles are released into the bloodstream, where they may be ingested by mosquitoes and passed to another dog by the bite of an infected mosquito.

(a) Sea cucumber

(b) Sea urchin

(c) Sea star

▲ **FIGURE 23-30** **The diversity of echinoderms** (a) A sea cucumber feeds on debris in the sand. (b) A sea urchin's spines are actually projections of the internal skeleton. (c) A sea star typically has five arms.

CASE STUDY continued

Physicians' Assistants

Some people have used hookworms and other parasitic roundworms to treat autoimmune diseases or severe allergies, in which the immune system inappropriately attack's the body's own tissues. Parasitic roundworms have evolved the ability to survive longer in their host by suppressing the host's immune system, and this immune suppression seems to ease the symptoms of people with immune system disorders who have infected themselves with parasitic worms. Unlike medical treatment with leeches, however, parasitic roundworms do not have the FDA's stamp of approval. Worm infections, after all, can make a person sick, and immune suppression associated with worm infections can increase susceptibility to other infectious diseases. Despite the risks, however, patients desperate for relief from debilitating, chronic disorders may infect themselves. Clinical researchers are working hard to discover methods that would make treatment with parasitic worms safe and effective.

endoskeleton (internal skeleton) composed of plates of calcium carbonate that lie beneath the outer skin.

Echinoderms Are Bilaterally Symmetrical as Larvae and Radially Symmetrical as Adults

Echinoderms exhibit deuterostome development and are linked by common ancestry with the other deuterostome phyla, including the chordates (described in Chapter 24). Deuterostomes form a group of branches on the larger evolutionary tree of bilaterally symmetrical animals, but in echinoderms bilateral symmetry is expressed only in embryos and free-swimming larvae. An adult echinoderm, in contrast, is radially symmetrical and lacks a head. This absence of cephalization is consistent with the sluggish existence of echinoderms. Most echinoderms move very slowly as they feed on algae or small particles sifted from sand or water. Some echinoderms are predators. Sea stars, for example, slowly pursue even slower-moving prey, such as snails or clams.

Echinoderms Have a Water-Vascular System

Echinoderms move on numerous tiny *tube feet*, delicate cylindrical projections that extend from the lower surface of the body and terminate in a suction cup. Tube feet are part of a unique echinoderm feature, the *water-vascular system*, which functions in locomotion, respiration, and food capture (**Fig. 23-31**). Seawater enters through an opening (the *sieve plate*) on the animal's upper surface and is conducted through a circular central canal, from which branch a number of radial canals. These canals conduct water to the tube feet, each of which is controlled by a muscular squeeze bulb known as an *ampulla*. Contraction of the bulb forces water into the tube foot, causing it to extend. The suction cup may be pressed against the seafloor or a food object, to which it adheres tightly until its internal pressure is released.

Some Echinoderm Organ Systems Are Simplified

Echinoderms have a relatively simple nervous system with no distinct brain. Movements are loosely coordinated by a system consisting of a nerve ring that encircles the esophagus, radial nerves to the rest of the body, and a nerve network through the epidermis. In sea stars, simple receptors for light and chemicals are concentrated on the arm tips, and sensory cells are scattered over the skin. In some brittle star species, light receptors are associated with tiny lenses, smaller than the width of a human hair, that gather light and focus it on receptors. These "microlenses" are formed from crystals of calcite (calcium carbonate) and their optical quality is excellent, far superior to that of any human-created lens of comparable size. Researchers hypothesize that each of the thousands of lenses on a brittle star forms a tiny image, and that the animal combines the resulting multitude of images to form a view of the surrounding environment.

Echinoderms lack a circulatory system, although movement of the fluid in their well-developed coelom serves this function. Gas exchange occurs through the tube feet and, in some forms, through numerous tiny "skin gills" that project through the epidermis. Most species have separate sexes and reproduce by shedding sperm and eggs into the water, where fertilization occurs.

Many echinoderms are able to regenerate lost body parts, and these regenerative powers are especially potent in sea stars. In fact, a single arm of a sea star is capable of developing into a whole animal, provided that part of the central body is attached to it. Before this ability was widely

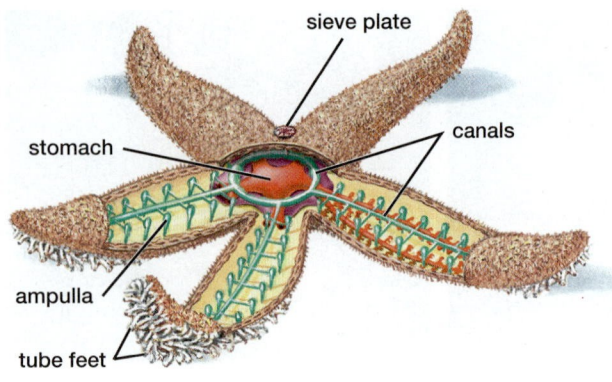

sieve plate

stomach

canals

ampulla

tube feet

(a) Sea star body plan

(b) Sea star consuming a mussel

▲ FIGURE 23-31 **The water-vascular system of echinoderms** (a) Changing pressure inside the seawater-filled water-vascular system extends or retracts the tube feet. **(b)** The sea star often feeds on mollusks such as this mussel. A feeding sea star attaches numerous tube feet to the mussel's shells, exerting a relentless pull. Then, the sea star turns the delicate tissue of its stomach inside out, extending it through its centrally located ventral mouth. The stomach can fit through an opening in the bivalve shells that is as narrow as 1 millimeter wide. Once pushed between the shells, the stomach tissue secretes digestive enzymes that weaken the mollusk, causing it to open further. Partially digested food is transported to the upper portion of the stomach, where digestion is completed.

appreciated, mussel fishermen often tried to rid mussel beds of predatory sea stars by hacking them into pieces and throwing the pieces back. Needless to say, the strategy backfired.

The Chordates Include the Vertebrates

The *chordates* (Chordata) include the vertebrate animals and also a few groups of invertebrates, such as the sea squirts and the lancelets. (We will discuss these invertebrate chordates and their vertebrate relatives in Chapter 24.)

CHECK YOUR LEARNING

Can you describe the basic body plans of sponges, cnidarians, comb jellies, flatworms, annelids, mollusks, arthropods, roundworms, and echinoderms? For each of these groups, are you able to list some member organisms? For each group, can you describe its member organisms' nervous and sensory systems and methods of circulation, gas exchange, digestion, and reproduction? Can you give examples of the effects invertebrate animals have on humans?

CASE STUDY r e v i s i t e d

Physicians' Assistants

Another invertebrate animal that has been recruited for medical duty is the blowfly or, more precisely, blowfly larvae, commonly known as maggots (Fig. 23-32). Blowfly maggots have proved to be effective at ridding wounds and ulcers of dead and dying tissue. If such tissue is not removed, it can interfere with healing or lead to infection. Traditionally, dead tissue in wounds is removed by a physician wielding a scalpel, but maggots offer an increasingly common alternative treatment. In this treatment, a bandage containing day-old, sterile maggots is applied to the wound. The maggots consume dead or dying tissue, secreting digestive enzymes that do not harm healthy skin or bone. After a few days, the maggots have grown to the size of rice kernels and are removed. The treatment is repeated until the wound is clean.

Consider This

BioEthics Medical treatment with invertebrate animals is typically much less expensive than drugs, surgery, and other options. Nonetheless, very little research funding is directed to promising but challenging medical uses

▲ FIGURE 23-32 **Blowfly maggots can clean wounds**

of animals, such as treatment of autoimmune diseases with parasitic roundworms. Would it be a good idea to increase research on such treatments, or are they too inherently risky?

CHAPTER REVIEW

Summary of Key Concepts

23.1 What Are the Key Features of Animals?

Animals are multicellular, sexually reproducing organisms that acquire energy by consuming other organisms. Most animals can perceive and react rapidly to environmental stimuli and are motile at some stage in their lives. Their cells lack cell walls.

23.2 Which Anatomical Features Mark Branch Points on the Animal Evolutionary Tree?

The earliest animals had no tissues, a feature retained by modern sponges. All other modern animals have tissues. Animals with tissues can be divided into radially symmetrical and bilaterally symmetrical groups. During embryonic development, radially symmetrical animals have two germ layers; bilaterally symmetrical animals have three. Bilaterally symmetrical animals also tend to have sense organs and clusters of neurons concentrated in the head, a process called cephalization. Bilateral phyla can be divided into two main groups, one of which undergoes protostome development, the other of which undergoes deuterostome development. Protostome phyla can in turn be divided into ecdysozoans and lophotrochozoans. Some phyla of bilaterally symmetrical animals lack body cavities, but most have either pseudocoeloms or true coeloms.

23.3 What Are the Major Animal Phyla?

The bodies of sponges (Porifera) are typically free-form in shape and are sessile. Sponges have relatively few types of cells. Despite the division of labor among the cell types, there is little coordination of activity. Digestion occurs exclusively within the individual cells.

The sea jellies, corals, anemones, and hydrozoans (Cnidaria) have tissues. A simple network of nerve cells directs the activity of contractile cells, allowing loosely coordinated movements. Digestion is extracellular, occurring in a central gastrovascular cavity with a single opening. Cnidarians exhibit radial symmetry, an adaptation to both the free-floating lifestyle of the medusa and the sedentary existence of the polyp.

Comb jellies (Ctenophora) are superficially similar to sea jellies but form a separate taxonomic group. Comb jellies are radially symmetrical, mostly small carnivores that move using eight rows of cilia.

Flatworms (Platyhelminthes) have a distinct head with sensory organs and a simple brain. A system of canals that form a network throughout the body aids in excretion. They lack a body cavity.

The segmented worms (Annelida) are the most complex of the worms, with a well-developed closed circulatory system and excretory organs that resemble the basic unit of the vertebrate kidney. The segmented worms have a compartmentalized digestive system, like that of vertebrates, which processes food in a sequence. Annelids also have a true coelom, a fluid-filled space between the body wall and the internal organs.

The clams, snails, octopuses, and squid (Mollusca) lack a skeleton; some forms protect the soft, moist, muscular body with a single shell (many gastropods and a few cephalopods) or a pair of hinged shells (the bivalves). The lack of a waterproof external covering limits this phylum to aquatic and moist terrestrial habitats. Although the body plan of gastropods and bivalves limits the complexity of their behavior, the cephalopod's tentacles are capable of precisely controlled movements. The octopus has the most complex brain and the best-developed learning capacity of any invertebrate.

Arthropods (Arthropoda), the insects, arachnids, millipedes and centipedes, and crustaceans, are the most diverse and abundant animals on Earth. They have invaded nearly every available terrestrial and aquatic habitat. Jointed appendages and well-developed nervous systems make possible complex, finely coordinated behavior. The exoskeleton (which conserves water and provides support) and specialized respiratory structures (which remain moist and protected) enable the insects and arachnids to inhabit dry land. The diversification of insects has been enhanced by their ability to fly. Crustaceans, which include the largest arthropods, are restricted to moist, usually aquatic habitats and respire using gills.

The pseudocoelomate roundworms (Nematoda) possess a separate mouth and anus and a cuticle layer that is molted.

The sea stars, sea urchins, and sea cucumbers (Echinodermata) are an exclusively marine group. Like other complex invertebrates and chordates, echinoderm larvae are bilaterally symmetrical; however, the adults show radial symmetry. This, in addition to a primitive nervous system that lacks a definite brain, adapts them to a relatively sedentary existence. Echinoderm bodies are supported by a nonliving internal skeleton that sends projections through the skin. The water-vascular system, which functions in locomotion, feeding, and respiration, is a unique echinoderm feature.

The chordates (Chordata) include two invertebrate groups, the lancelets and sea squirts, as well as the vertebrates.

Key Terms

bilateral symmetry   *419*
budding   *427*
cephalization   *421*
closed circulatory system   *430*
coelom   *421*
compound eye   *436*
deuterostome   *422*
ectoderm   *421*
endoderm   *421*
endoskeleton   *441*
exoskeleton   *435*
ganglion (plural, ganglia)   *429*
hemocoel   *431*
hermaphroditic   *423*

hydrostatic skeleton   *430*
invertebrate   *423*
larva (plural, larvae)   *437*
mesoderm   *421*
metamorphosis   *437*
molt   *435*
nerve cord   *429*
open circulatory system   *431*
parasite   *428*
protostome   *422*
pseudocoelom   *422*
pupa (plural, pupae)   *437*
radial symmetry   *419*
segmentation   *430*
tissue   *419*
vertebrate   *423*

Learning Outcomes

In this chapter, you have learned to . . .

LO1 Describe the key features of animals.

LO2 Describe the anatomical features that mark major branching points on the animal evolutionary tree, and explain the functions and significance of these features.

LO3 Describe the major animal taxonomic groups and representative members of each group.

LO4 Describe the effects that invertebrate animals have on humans.

Thinking Through the Concepts

Fill-in-the-Blank

1. Animals obtain energy by _____; they generally reproduce _____, and their cells lack _____. **LO1**

2. Bilaterally symmetrical animals have _____ embryonic tissue layers, known as _____, _____, and _____. Radially symmetrical animals have _____ tissue layers; they lack the _____ layer. **LO2**

3. Animals that have an anterior and posterior end are said to be _____. The anterior end of such animals often contains structures used for _____ and _____. Coelomate animals have a body _____ that is completely lined with tissue derived from _____. **LO2**

4. Lophotrochozoans and ecdysozoans (molting animals) are two large clades of animals that exhibit _____ development. The other major type of development found in bilateral animals is known as _____ development, and is present in _____ and _____. **LO2**

5. Animals that lack a backbone are described as _____; those that possess a backbone are _____. The vast majority of all animals fall into which of these two groups? _____ The only animals that lack tissues are _____, whose bodies resemble a colony of _____. Sea anemones and corals are _____. Earthworms and leeches are _____. **LO2 LO3**

6. Three major groups within the mollusks are the two-shelled clams and scallops called _____; the foot-crawling snails and slugs called _____; and the tentacled squid and octopuses called _____. Members of the largest animal phylum are called _____. Three important groups within this phylum are the six-legged, often flying _____; the eight-legged spiders and mites called _____; and the mostly aquatic _____. **LO3**

7. Phyla that contain animals with segmented bodies include _____ and _____. In a(n) _____ system, blood is confined to blood vessels. In a(n) _____ system, blood bathes internal organs within a cavity called the _____. **LO3**

8. For each of the following distinctive structures, name the animal group in which it is found: mantle:_____; cnidocyte: _____; water-vascular system: _____; paired, jointed appendages: _____. **LO3**

Review Questions

1. List the characteristics that, taken together, distinguish animals from other kinds of organisms. **LO1**

2. List the distinguishing characteristics of each phylum discussed in this chapter, and give an example of a member of each phylum. **LO3**

3. Briefly describe each of the following adaptations, and explain its adaptive significance: bilateral symmetry, cephalization, closed circulatory system, coelom, radial symmetry, segmentation. **LO2**

4. Describe and compare the respiratory systems of the four major arthropod groups. **LO3**

5. Describe the advantages and disadvantages of the arthropod exoskeleton. **LO3**

6. In which of the three major mollusk groups is each of the following characteristics found? **LO3**

   a. two hinged shells
   b. a radula
   c. tentacles
   d. some sessile members
   e. the best-developed brains
   f. numerous eyes

7. Give three functions of the water-vascular system of echinoderms. **LO3**

8. To what lifestyle is radial symmetry an adaptation? Bilateral symmetry? **LO2 LO3**

9. List some invertebrate animals that can harm humans and, for each animal that you list, name its taxonomic group. **LO3 LO4**

Applying the Concepts

1. **BioEthics** Insects are the largest group of animals on Earth. Insect diversity is greatest in the Tropics, where habitat destruction and species extinction are occurring at an alarming rate. What biological, economic, and ethical arguments can you advance to persuade people and governments to preserve this biological diversity?

2. Discuss at least three ways in which the ability to fly has contributed to the success and diversity of insects.

3. Discuss and defend the attributes you would use to define biological success among animals. Are humans a biological success by these standards? Why?

Answers to Figure Caption questions and Fill-in-the-Blank questions can be found in the Answers section at the back of the book.

MB *Go to MasteringBiology for practice quizzes, activities, eText, videos, current events, and more.*

CHAPTER 24

Animal Diversity II: Vertebrates

Would you be shocked to learn that *Tyrannosaurus* still walked the Earth? The discovery of modern coelacanth fishes was no less surprising.

CASE STUDY

Fish Story

ON DECEMBER 22, 1938, Marjorie Courtenay-Latimer received a phone call that would lead to one of the most spectacular discoveries in biological history. The call was from a local fisherman whom Courtenay-Latimer, the curator of a small museum in South Africa, had asked to collect some fish specimens for the museum. His boat had returned from its most recent voyage and was waiting at the town dock. Dutifully, Courtenay-Latimer went to the boat and began sorting through the fish that were strewn across the deck. Later, she wrote, "I noticed a blue fin sticking up from beneath the pile. I uncovered the specimen, and, behold,

there appeared the most beautiful fish I had ever seen." In addition to its beauty, the fish had some odd features, including fins that were stumpy and lobed, unlike the fins of any other living species.

Courtenay-Latimer did not recognize the strange fish, but she knew it was unusual. She tried to find a place to refrigerate it, but in her small town she was unable to find a cold storage facility willing to store a fish. In the end, she was able to save only the skin. Undaunted, she made some drawings of the fish and used them to attempt an identification. To her amazement, the creature did not resemble any species known to inhabit the waters off South Africa, but did seem similar to members of a family of fishes known

as coelacanths. The only problem with this assessment was that coelacanths were known only from fossils. The earliest coelacanth fossils were found in 400-million-year-old rocks and, as far as anyone knew, the group had been extinct for 80 million years!

Perplexed, Courtenay-Latimer sent her drawings to J. L. B. Smith, a fish expert at Rhodes University. Smith was astounded when he saw the sketch, later writing that "a bomb seemed to burst in my brain." Although bitterly disappointed that the specimen's bones and internal organs had been lost, Smith arranged to view the preserved skin. Ultimately, he confirmed the astonishing news that coelacanths still swam in Earth's waters.

AT A GLANCE

24.1 What Are the Key Features of Chordates?

24.2 Which Animals Are Chordates?

24.3 What Are the Major Groups of Vertebrates?

LEARNING GOALS

LG1 Describe the key features of chordates.

LG2 Describe the adaptations that mark major branching points on the chordate evolutionary tree, and explain the functions and significance of these features.

LG3 Describe the major chordate taxonomic groups and representative members of each group.

24.1 WHAT ARE THE KEY FEATURES OF CHORDATES?

Humans are members of a taxonomic group known as the chordates (Chordata). Chordates (**Fig. 24-1**) include not only other bony animals like the birds and apes, but also the tunicates (sea squirts) and small fishlike creatures called lancelets. What characteristics do we share with these animals that seem so different from us?

All Chordates Share Four Distinctive Structures

All chordates have deuterostome development (which is also characteristic of echinoderms; see pp. 420–421) and are further united by four features that all possess at some stage of their lives: a dorsal nerve chord, a notochord, pharyngeal gill slits, and a post-anal tail.

Dorsal Nerve Cord

The **nerve cord** of chordates lies above the digestive tract, running lengthwise along the dorsal (upper) portion of the body. In contrast, the nerve cords of other animals lie in a ventral position, below the digestive tract (see Fig. 23-14). A chordate's nerve cord is hollow—its center is filled with fluid, unlike the nerve cords of other animals, which are not hollow and have solid nerve tissue throughout. During embryonic development in chordates, the nerve cord develops a thickening at its anterior end that becomes a brain.

Notochord

The **notochord** is a stiff but flexible rod that extends along the length of the body, between the digestive tract and the nerve cord. It provides support for the body and an attachment site for muscles. In many chordates, the notochord is present only during early stages of development and disappears as a skeleton develops.

Pharyngeal Gill Slits

Pharyngeal gill slits are located in the pharynx (the cavity behind the mouth). In some chordates the slits form functional openings for gills (organs for gas exchange in water); in others they appear only as grooves during an early stage of development.

Post-Anal Tail

The **post-anal tail** is a posterior extension of the chordate body that extends past the anus and contains muscle tissue and the rearmost portion of the nerve cord. Other animals lack this kind of tail.

This list of distinctive chordate structures may seem puzzling because, although humans are chordates, at first glance we seem to lack every feature except the nerve cord. But evolutionary relationships are sometimes seen most clearly during early stages of development, and it is during our embryonic phase that we develop, and subsequently lose, our notochord, our gill slits, and our tails (**Fig. 24-2**).

CHECK YOUR LEARNING

Can you list and explain the features that distinguish chordates from other animals?

24.2 WHICH ANIMALS ARE CHORDATES?

Chordates include three clades (groups that include all of the descendants of a common ancestor): the tunicates, the lancelets, and the craniates.

Tunicates Include Sea Squirts and Salps

The tunicates (Urochordata) are a group of about 1,600 species of marine invertebrate chordates. Tunicates are small, with lengths ranging from a few millimeters to 1 foot (30 centimeters). The group includes immobile, filter-feeding, vase-shaped animals known as sea squirts (**Fig. 24-3a**). Much of a sea squirt's body is occupied by its pharynx, which is like a basket perforated by gill slits and lined with mucus. Water enters the sea squirt's body through an *incurrent siphon*, passes into the pharynx at its top, moves through the gill slits, and exits the body through an *excurrent siphon*. Food particles are trapped in the basket's mucous lining.

Adult sea squirts are sessile—they live firmly attached to a surface. Their ability to move is limited to forceful contractions of their saclike bodies, which can send a jet of seawater into the face of anyone who plucks one from its undersea home; hence, the name sea squirt. Although adult sea squirts are immobile, their larvae swim actively and possess the four chordate features (see Fig. 24-3a, left). Some

Urochordata
(tunicates)

Cephalochordata
(lancelets)

Myxini
(hagfishes)

Petromyzontiformes
(lampreys)

Chondrichthyes
(sharks, rays)

Actinopterygii
(ray-finned fishes)

Actinistia
(coelacanths)

Dipnoi
(lungfishes)

Amphibia
(frogs,
salamanders)

Reptilia
(turtles, snakes
crocodiles,
birds)

Mammalia
(mammals)

Craniates

Vertebrates

Tetrapods

Dorsal nerve cord,
notochord,
pharyngeal gill slits,
post-anal tail

Skull

Vertebral column

Jaws

Lungs

Lobed fins

Limbs

Amniotic
egg

Hair,
milk

▶ FIGURE 24-1 An evolutionary tree of
the chordates The talking boxes indicate the
points at which key traits first appeared.

eye   heart   liver

gill slit

tail

limb bud
(future leg)

limb bud
(future arm)

◀ FIGURE 24-2 Chordate features
in the human embryo This 5-week-old
human embryo is about 1 centimeter long
and clearly shows a tail and external gill slits
(more properly called grooves, since they do
not penetrate the body wall). Although the
tail will disappear completely, the gill grooves
contribute to the formation of the lower jaw.

▲ **FIGURE 24-3 Invertebrate chordates (a)** The sea squirt larva (left) exhibits all the diagnostic features of chordates. The adult sea squirt (a type of tunicate, middle) has lost its tail and notochord and has assumed a sedentary life, as shown in the photo (right). **(b)** A lancelet, a fishlike invertebrate chordate. The adult organism exhibits all the chordate features.

tunicates remain mobile throughout their lives. For example, barrel-shaped tunicates known as salps live in the open ocean and move by contracting an encircling band of muscle, which forces a jet of water out of the back of the animal, propelling it forward.

Most tunicates are *hermaphroditic* (each individual possesses both male and female sex organs). They may reproduce asexually or sexually. In asexual reproduction, miniature versions of an adult grow from its body and then drop off. In sexual reproduction, sperm are broadcast into the surrounding water, and fertilize eggs that (depending on species) are either also broadcast into the water or retained inside the tunicate's body. In species that retain eggs inside the body, swimming sperm must enter the body to fertilize the eggs, and the resulting larvae must swim out.

Lancelets Are Marine Filter-Feeders

The 30 or so species of lancelets (Cephalochordata) form another group of invertebrate chordates. Lancelets are small (2 inches, or about 5 centimeters, long), fishlike animals

that retain all four chordate features as adults (**Fig. 24-3b**). An adult lancelet spends most of its time half-buried in the sandy sea bottom, with only the anterior end of its body exposed. The motion of cilia in the pharynx draws seawater into the lancelet's mouth. As the water passes through the pharyngeal gill slits, a film of mucus filters tiny food particles from the water. The captured food particles are transported to the lancelet's digestive tract.

Lancelets have separate sexes and reproduce sexually. At particular times of year, most of the males and females in an area simultaneously release gametes (eggs and sperm) into the surrounding water. Fertilized eggs develop into microscopic larvae, which swim slowly and drift about during several weeks of continuing growth and development before dropping to the seabed and completing their transformation to the adult form.

Craniates Have a Skull

The **craniates** include all chordates that have a skull that encloses the brain. The skull may be composed of bone or cartilage, a tissue that resembles bone but is less brittle and more flexible. The earliest known craniates, whose fossils were found in 530-million-year-old rocks, resembled lancelets but had brains, skulls, and eyes. However, the mouths of the earliest craniates lacked jaws.

Today, craniates include two subgroups: the hagfishes and the *vertebrates*. Table 24-1 summarizes some characteristics of the craniate groups described in the rest of this chapter.

Hagfishes Are Slimy Residents of the Ocean Floor

As did ancestral craniates, hagfishes (Myxini) lack jaws. Instead, they use a tonguelike, tooth-bearing structure to grind and tear food. A hagfish body is stiffened by a notochord, but its skeleton is limited to a few small cartilaginous elements, one of which forms a rudimentary skull. Because hagfishes lack skeletal elements that surround the nerve cord to form a vertebral column, most systematists do not consider them to be vertebrates. Instead, hagfishes represent the craniate group that is most closely related to the vertebrates.

The 75 or so species of hagfishes are exclusively marine (Fig. 24-4). They respire using gills, have a two-chambered heart, and are ectothermic—that is, they depend on heat from the external environment to regulate their body temperature. (Gills, two-chambered hearts, and ectothermy are also found in all vertebrate fishes.) Hagfishes live near the ocean floor, often burrowing in the mud, and feed primarily on worms. They will, however, eagerly attack dead and dying fish, using their teeth to burrow into a fish's body and consume its soft internal organs.

Hagfishes secrete massive quantities of slime as a defense against predators. Despite their well-deserved reputation as "slimeballs of the sea," hagfishes are avidly pursued by many commercial fishermen, because the leather industry in some parts of the world provides a market for hagfish skin. Most leather items that purport to be "eel skin" are in fact made from tanned hagfish skin.

▲ **FIGURE 24-4 Hagfishes** Hagfishes live in communal burrows in mud, feeding on worms.

Vertebrates Have a Backbone

Vertebrates are animals in which the embryonic notochord is replaced during development by a backbone, or **vertebral column,** composed of bone or cartilage. The vertebral column of a vertebrate supports its body, provides attachment sites for muscles, and protects the delicate nerve cord and brain. It is also part of a living internal skeleton that can grow and repair itself.

The early history of vertebrates was characterized by an array of strange, now-extinct jawless fishes, many of which were protected by bony armor plates. About 425 million years ago, jawless fishes gave rise to a group of fish that possessed an important new structure: jaws. Jaws allowed fish to grasp, tear, or crush their food, permitting them to exploit a much wider range of food sources than could jawless fish. Today, most (but not all) vertebrates have jaws.

TABLE 24-1 Comparison of Craniate Groups				
Group	**Fertilization**	**Respiration**	**Heart Chambers**	**Body Temperature Regulation**
Hagfishes (Myxini)	External	Gills	Two	Ectothermic
Lampreys (Petromyzontiformes)	External	Gills	Two	Ectothermic
Cartilaginous fishes (Chondrichthyes)	Internal	Gills	Two	Ectothermic
Ray-finned fishes (Actinopterygii)	External[1]	Gills	Two	Ectothermic
Coelacanths (Actinistia)	Internal	Gills	Two	Ectothermic
Lungfishes (Dipnoi)	External	Gills and lungs	Two	Ectothermic
Amphibians (Amphibia)	External or internal[2]	Skin, gills, and lungs	Three	Ectothermic
Reptiles (Reptilia)	Internal	Lungs	Three[3]	Ectothermic[4]
Mammals (Mammalia)	Internal	Lungs	Four	Endothermic

[1]A relatively small number of ray-finned fish have internal fertilization.
[2]External in most frogs and toads; internal in caecilians and most salamanders.
[3]Except for birds and crocodilians, which have four chambers.
[4]Except for birds, which are endothermic.

CASE STUDY continued

Fish Story

In the years since Courtenay-Latimer's discovery that coelacanths are not extinct, scientists have had the opportunity to investigate the creature's anatomy. The body of the living coelacanth has some unusual features. For example, adult coelacanths retain a notochord, the body-stiffening rod that most other vertebrates lose during embryonic development. In addition, a coelacanth's brain is very small relative to its body size. The brain of a 90-pound (40-kilogram) coelacanth weighs only 1 or 2 grams (less than a tenth of an ounce). The tiny brain occupies less than 2% of the space in the cranial cavity; the rest is filled with fat.

Vertebrates have other adaptations that have contributed to their successful invasion of most habitats. One such adaptation is paired appendages. These first appeared as fins in fish and served as stabilizers for swimming. Over millions of years, some fins were modified by natural selection into legs that allowed animals to crawl onto dry land, and later into wings that allowed some to take to the air. Another adaptation that has contributed to the success of vertebrates is an increase in the size and complexity of their brains and sensory structures, which allow vertebrates to perceive their environment in detail and to respond to it in a great variety of ways.

CHECK YOUR LEARNING

Can you name and describe the chordates that are not craniates? Can you name and describe the craniates that are not vertebrates? Are you able to describe the key adaptations of vertebrates?

24.3 WHAT ARE THE MAJOR GROUPS OF VERTEBRATES?

Today, vertebrates include lampreys, cartilaginous fishes, ray-finned fishes, coelacanths, lungfishes, amphibians, reptiles, and mammals.

Some Lampreys Parasitize Fish

Like hagfishes, the roughly 35 species of lampreys (Petromyzontiformes) are jawless. A lamprey is recognizable by the large, rounded sucker that surrounds its mouth and by the single nostril on the top of its head. The nerve cord of a lamprey is protected by segments of cartilage, so lampreys are considered to be true vertebrates. They live in both fresh and salt waters, but the marine forms must return to fresh water to spawn. Lampreys migrate up shallow streams to spawn; eggs are deposited and fertilized in depressions that groups of lampreys excavate in the gravel streambed. The adults die a short time after spawning. After the young hatch, they spend several years in the stream as larvae, eating algae, before maturing and moving downstream to their adult habitat in an ocean, lake, or river.

▲ FIGURE 24-5 **Lampreys** Some adult lampreys are parasitic, attaching to fish with sucker-like mouths lined with rasping teeth (inset).

Adult lampreys of some species are parasitic. A parasitic lamprey uses its tooth-lined mouth to attach itself to a larger fish (Fig. 24-5). Using rasping teeth on its tongue, the lamprey excavates a hole in the host's body wall, through which it sucks blood and body fluids. Beginning in the 1920s, parasitic lampreys spread into the Great Lakes. There, in the absence of effective predators, they have multiplied prodigiously and greatly reduced commercial fish populations, including the lake trout. Vigorous measures to control the lamprey population have allowed some recovery of the other fish populations of the Great Lakes.

Cartilaginous Fishes Are Marine Predators

The cartilaginous fishes (Chondrichthyes) include 625 marine species, among them the sharks, skates, and rays (Fig. 24-6). Unlike hagfishes and lampreys (but like all other vertebrates), cartilaginous fishes have jaws. They are graceful predators whose skeleton is formed entirely of cartilage. Their bodies are protected by leathery skin roughened by tiny scales. Although some must swim to circulate water through their gills, most can pump water across their gills. In contrast to the external fertilization that characterizes reproduction in almost all other fish, cartilaginous fish have internal fertilization, in which a male deposits sperm directly into a female's reproductive tract. Some cartilaginous fishes are very large. A whale shark, for example, can grow to more than 45 feet (14 meters) in length, and a manta ray may be more than 20 feet (6 meters) wide.

Although some sharks feed by filtering plankton (tiny animals and protists) from the water, most are predators of larger prey such as other fishes, marine mammals, sea turtles, crabs, or squid. Many sharks attack their prey with strong jaws that contain several rows of razor-sharp teeth; the back rows move forward as the front teeth are lost to age and use (see Fig. 24-6a).

Most sharks avoid humans, but large sharks of some species can be dangerous to swimmers and divers. However, shark attacks on people are rare. A U.S. resident is 30 times more likely to die from a lightning strike than from a shark

(a) Shark

(b) Ray

▲ FIGURE 24-6 **Cartilaginous fishes** (a) A shark displaying several rows of teeth. As the frontmost teeth are lost, they are replaced by the new ones behind them. Both sharks and rays lack a swim bladder and tend to sink toward the bottom when they stop swimming. **(b)** The tropical blue-spotted stingray swims by graceful undulations of lateral extensions of its body.

attack, and a beachgoer is far more likely to drown than to be bitten by a shark. Nonetheless, shark attacks do occur. During 2011, for example, there were 75 documented attacks in the world, 12 of them fatal.

Skates and rays are mostly bottom dwellers with flattened bodies, wing-shaped fins, and thin tails (see Fig. 24-6b). Rays are generally larger than skates, but the most notable difference between the two groups is that rays give birth to live young, whereas skates lay eggs. Most skates and rays eat invertebrates. Some ray species defend themselves with a spine near their tail that can inflict dangerous wounds, and others produce a powerful electrical shock that can stun their prey.

Ray-Finned Fishes Are the Most Diverse Vertebrates

The vertebrate diversity crown belongs to the ray-finned fishes (Actinopterygii). About 24,000 species have been identified, and scientists estimate that perhaps twice this number exist, with many undiscovered species inhabiting deep waters and remote areas. Ray-finned fishes are found in nearly every watery habitat, both freshwater and marine.

Ray-finned fishes are distinguished by the structure of their fins, which are formed by webs of skin supported by bony spines. In addition, ray-finned fishes have skeletons made of bone, a trait they share with the lobe-finned fishes and limbed vertebrates discussed later in this chapter. The skin of ray-finned fishes is covered with interlocking scales that provide protection while allowing for flexibility. Most ray-finned fishes have a swim bladder, a sort of internal balloon that allows a fish to float effortlessly at any level in the water. The swim bladder evolved from lungs, which were present (along with gills) in the ancestors of modern ray-finned fishes.

The ray-finned fishes include not only a large number of species but also a huge variety of different forms and lifestyles (**Fig. 24-7**). These range from snakelike eels to flattened flounders; from sluggish bottom feeders that probe the seafloor to speedy, streamlined predators that range in open water; from brightly colored reef dwellers to translucent, luminescent deep-sea dwellers; from the massive 3,000-pound (1,350-kilogram) mola to the tiny stout infantfish, which weighs in at about 0.00003 ounce (1 milligram).

Ray-finned fishes are an extremely important source of food for humans. Unfortunately, however, our appetite for ray-finned fishes, combined with increasingly effective high-tech methods for finding and catching them, has had a devastating impact on fish populations. Populations of almost all economically important ray-finned fish species have declined drastically. If overfishing continues, fish stocks are likely to collapse.

Coelacanths and Lungfishes Have Lobed Fins

Although almost all fish with bony skeletons belong to the ray-finned group, some bony fishes are members of two other groups, the coelacanths (Actinistia) and the lungfishes (Dipnoi). Coelacanths are described in this chapter's case study (see the chapter-opening photo). The six species of lungfishes are found in freshwater habitats in Africa, South America, and Australia (**Fig. 24-8**). Lungfishes have both gills and lungs. They tend to live in stagnant waters that may be low in oxygen, and their lungs allow them to supplement their supply of oxygen by breathing air directly. Lungfishes of several species are able to survive even if the pools they inhabit dry up completely. These fish burrow into mud and seal themselves in mucus-lined chambers. There, they breathe through their lungs as their metabolic rate declines drastically. When the rains return and the pools refill, the lungfishes leave their burrows and resume their underwater way of life.

Lungfishes and coelacanths are sometimes called *lobe-fins*, because members of both groups have fleshy fins that contain rod-shaped bones surrounded by a thick layer of muscle. This trait is indicative of the groups' shared ancestry,

▼ **FIGURE 24-7 The diversity of ray-finned fishes** Ray-finned fishes have colonized nearly every aquatic habitat. **(a)** This female deep-sea anglerfish attracts prey with a living lure that projects just above her mouth. The fish is ghostly white; at the 6,000-foot (1,800-meter) depth where anglers live, no light penetrates and thus colors are unnecessary. Male deep-sea anglerfish are extremely small and remain attached to the female as permanent parasites, always available to fertilize her eggs. Two parasitic males can be seen attached to this female. **(b)** This tropical green moray eel lives in rocky crevices. The small fish (a banded cleaner goby) on its lower jaw eats parasites that cling to the moray's skin. **(c)** A sea horse may anchor itself with its prehensile tail (adapted for grasping) while feeding on small crustaceans.

QUESTION With regard to water regulation (maintaining the proper amount of water in the body), how does the challenge faced by a freshwater fish differ from that faced by a saltwater fish?

(a) Anglerfish   **(b) Moray eel**   **(c) Sea horse**

though the two lineages have been evolving separately for hundreds of millions of years.

In addition to the coelacanths and lungfishes, several other lineages of lobefins arose early in the evolutionary history of jawed fish. Members of one of these other lineages evolved modified fleshy fins that, in an emergency, could be used as legs, allowing the fish to drag itself from a drying puddle to a deeper pool. This lineage left descendants that survive today. These survivors are the **tetrapods** (from the Greek for "four feet"), which instead of fins have limbs that can support their weight on land and digits (fingers or toes) on the ends of those limbs. The tetrapods include amphibians, reptiles, and mammals.

▲ **FIGURE 24-8 Lungfishes are lobe-finned fish** Among the fishes, lungfishes are the group most closely related to land-dwelling vertebrates.

Amphibians Live a Double Life

The early tetrapods that made the first vertebrate invasion of land were amphibians. Today, the 6,300 species of amphibians (Amphibia) straddle the boundary between aquatic and terrestrial existence (Fig. 24-9). The limbs of amphibians show varying degrees of adaptation to movement on land, from the belly-dragging crawl of salamanders to the long leaps of frogs. A three-chambered heart (in contrast to the two-chambered heart of fishes) circulates blood more efficiently, and lungs replace gills in most adult forms. Amphibian lungs, however, are relatively inefficient and must be supplemented by the skin, which serves as an additional respiratory organ. This respiratory function requires that the skin remain moist, a constraint that greatly restricts the range of amphibian habitats on land.

Many amphibians are also tied to moist habitats by their breeding behavior, which requires water. For example, as in most fishes, fertilization in frogs and toads is generally external and takes place in water, where the sperm can swim to the eggs. The eggs must remain moist, because they are protected only by a jelly-like coating that leaves them vulnerable to water loss by evaporation. Different amphibian species keep their eggs moist in different ways, but many species simply lay their eggs in water. In some amphibian species, fertilized eggs develop into aquatic larvae such as the tadpoles of some frogs and toads. These aquatic larvae undergo a dramatic transformation into semiterrestrial adults, a metamorphosis that gives the amphibians their name, which means "double life." Their double life and their thin, permeable skin have made amphibians particularly vulnerable to pollutants and environmental degradation, as described in "Earth Watch: Frogs in Peril."

Frogs and Toads Are Adapted for Jumping

The frogs and toads, with 5,600 species, are the most diverse group of amphibians. Adult frogs and toads move about by hopping and leaping, and their bodies are well adapted for this mode of locomotion. Their hind legs are long relative to their body size (much longer than their forelegs), and they do not have a tail. The names "frog" and "toad" do not describe distinct evolutionary groups, but are instead used informally to distinguish two combinations of characteristics that are common among members of this branch of the amphibians. In general, frogs have smooth, moist skin, live in or near water, and have long hind limbs suitable for leaping; toads have bumpy, drier skin, live on land, and have shorter hind limbs suitable for hopping. Many frogs and toads (and

other amphibians) contain toxic substances that make them distasteful to predators. In a few species, such as the golden poison dart frog of South America, the protective chemical is extremely toxic. The toxin from a single golden poison dart frog could kill several adult humans.

Most Salamanders Have Tails

Most salamanders have a lizard-like body: slender, with four legs of roughly equal size and a long tail (**Fig. 24-9c**). Some salamanders, however, have only very small legs; members of these species may have an eel-like appearance. Most of the roughly 550 species of salamanders live on land, often in moist, protected places, such as beneath rocks or logs on a forest floor. But members of some species are fully aquatic and spend their entire lives in the water. Even members of land-dwelling species generally move to ponds or streams to breed. In almost all salamander species, eggs hatch into aquatic larvae that use external

(a) Tadpole

(b) Frog

(c) Salamander

(d) Caecilian

▲ **FIGURE 24-9** **"Amphibian" means "double life"** The double life of amphibians is illustrated by the bullfrog's transition from **(a)** a completely aquatic larval tadpole to **(b)** an adult leading a semiterrestrial life. **(c)** The red salamander is restricted to moist habitats in the eastern United States. **(d)** Caecilians are legless, mostly burrowing amphibians.

QUESTION What advantages might amphibians gain from their "double life"?

Earth Watch

Frogs in Peril

Frogs and toads have lived in Earth's ponds and swamps for nearly 150 million years, somehow surviving the Cretaceous catastrophe that extinguished the dinosaurs and so many other species about 65 million years ago. Their evolutionary longevity, however, doesn't protect them from the environmental changes wrought by human activities. During the past three decades, herpetologists (biologists who study reptiles and amphibians) from around the world have documented an alarming decline in amphibian populations. Thousands of species of frogs, toads, and salamanders are dramatically decreasing in number, and many have gone extinct.

This is a worldwide phenomenon; population crashes have been reported from every part of the globe (Fig. E24-1). Yosemite toads and yellow-legged frogs are disappearing from the mountains of California. Leopard frogs, formerly abundant throughout North America, are becoming rare in the United States. Of nearly 100 species of harlequin frog known from Central and South America, only 10 can still be found. Even amphibians in protected areas are dying. In the Cape Peninsula National Park in South Africa, the only remaining population of Rose's ghost frog has shrunk dramatically and the species is now critically endangered. The gastric brooding frog of Australia—which fascinated biologists by swallowing its eggs, brooding them in its stomach and later regurgitating fully formed offspring—was once abundant and seemed safe in a national park. In 1980, however, the gastric brooding frog disappeared and hasn't been seen since.

▲ FIGURE E24-1 **Amphibians in danger** The corroboree toad is rapidly declining in its native Australia. The thin water-permeable and gas-permeable skin of frogs and toads (and the jelly-like coating around their eggs) make them vulnerable to both air and water pollutants.

The causes of the worldwide decline in amphibian diversity are not fully understood, but researchers have discovered that frogs and toads in many places are succumbing to infection by a pathogenic fungus. The fungus has been found in the skin of dead and dying frogs in widespread locations, including Australia, Central America, and the western United States. In those places, discovery of the fungus has coincided with massive frog and toad die-offs, and most herpetologists agree that the fungus is causing the deaths.

It seems unlikely, however, that the fungus alone is responsible for the worldwide decline of amphibians. For one thing, die-offs have occurred in many places where the fungus has not been found. In addition, many herpetologists believe that the fungal epidemic would not have arisen if the frogs and toads had not first been weakened by other stressors. So, if the fungus is not doing all of the damage on its own, what are the other possible causes of amphibian decline? All of the most likely causes stem from human modification of the biosphere—the portion of Earth that sustains life.

Habitat destruction, especially the draining of wetlands that are hospitable to amphibian life, is one major cause of the decline. Amphibians are also vulnerable to toxic substances in the environment. For example, researchers found that frogs exposed to trace amounts of atrazine, a widely used herbicide that is found in virtually all fresh water in the United States, suffer severe damage to their reproductive tissues. The unique biology of amphibians makes them especially susceptible to poisons in the environment. Amphibian bodies at all stages of life are protected only by a thin, permeable skin that pollutants can easily penetrate.

Amphibian eggs can also be damaged by ultraviolet (UV) light. Researchers have demonstrated that the eggs of some species of frogs in the Pacific Northwest are sensitive to damage from UV light and that the most sensitive species are experiencing the most drastic declines. Unfortunately, many parts of Earth are subject to intense UV radiation levels, because atmospheric pollutants have caused a thinning of the protective ozone layer.

Many scientists believe that the troubles of amphibians signal an overall deterioration of Earth's ability to support life. According to this line of reasoning, the highly sensitive amphibians are providing an early warning of environmental degradation that will eventually affect more resistant organisms as well. Equally worrisome is the observation that amphibians are not just sensitive indicators of the health of the biosphere but also crucial components of many ecosystems. They may keep insect populations in check, and serve as food for larger carnivores. Their decline will further disrupt the balance of these delicate communities.

gills to breathe. In some species, the larvae do not metamorphose, but instead retain the larval form even as adults.

Alone among vertebrates, salamanders can regenerate lost limbs. This ability has attracted the attention of researchers interested in regenerative medicine, which seeks treatments that enable human bodies to repair or regenerate damaged tissues and organs. The researchers hope that increased understanding of the mechanisms by which salamanders regenerate limbs will lead to effective treatments for humans.

Caecilians Are Limbless, Burrowing Amphibians

The caecilians form a small (175 species) group of legless amphibians that live in tropical regions. At first glance, a caecilian's appearance is reminiscent of an earthworm, though the larger species, which can be up to 5 feet (1.5 meters) long, might be mistaken for a snake (Fig. 24-9d). Most caecilians are burrowing animals that live underground, though a few species are aquatic. Caecilians eyes are very small and often covered by skin. As a result, caecilian vision is probably limited to detecting light.

Reptiles Are Adapted for Life on Land

The reptiles (Reptilia) include lizards, snakes, alligators, crocodiles, turtles, and birds (Fig. 24-10). Reptiles evolved from an amphibian ancestor about 250 million years ago.

Reptiles Haves Scales and Shelled Eggs

Some reptiles, particularly desert dwellers such as many tortoises and lizards, are completely independent of their aquatic origins. They achieved this independence through a series of adaptations, three of which are especially notable: (1) Reptiles evolved a tough, scaly skin that resists water loss and protects the body. (2) Reptiles evolved internal fertilization, in which the male deposits sperm within the female's body. (3) Reptiles evolved a shelled **amniotic egg,** which can be buried in sand or dirt, far from water. The shell prevents the egg from drying out on land. An internal membrane, the **amnion,** encloses the embryo in the watery environment that all developing animals require (Fig. 24-11).

In addition to these features, reptiles have more efficient lungs than do amphibians, and do not use their skin as a respiratory organ. Reptile circulatory systems include three-chambered or (in birds, alligators, and crocodiles) four-chambered hearts that segregate oxygenated and deoxygenated blood more effectively than do amphibian hearts.

Lizards and Snakes Share a Common Evolutionary Heritage

Lizards and snakes together form a distinct lineage containing about 6,800 species. The common ancestor of snakes and lizards had limbs, which are retained by most lizards but have been lost in snakes. The limbed ancestry of snakes is revealed by remnants of hind limb bones found in some snake species.

▲ FIGURE 24-11 **The amniotic egg** A crocodile struggles free of its egg. The amniotic egg encapsulates the developing embryo in a fluid-filled membrane (the amnion), ensuring that development occurs in a watery environment, even if the egg is far from water.

(a) Snake

(b) Alligator

(c) Tortoise

▲ FIGURE 24-10 **The diversity of reptiles (other than birds) (a)** This scarlet king snake has a color pattern very similar to that of the poisonous coral snake, which potential predators avoid. This mimicry helps the harmless king snake elude predation. **(b)** The outward appearance of the American alligator, found in swampy areas of the South, is almost identical to that of 150-million-year-old fossil alligators. **(c)** The tortoises of the Galápagos Islands, Ecuador, may live to be more than 100 years old.

Most lizards are small predators that eat insects or other small invertebrates, but a few lizard species are quite large. The Komodo dragon, for example, can reach 10 feet (3 meters) in length and weigh more than 200 pounds (90 kilograms). These giant lizards live in Indonesia and have powerful jaws and inch-long teeth that enable them to prey on large animals including deer, goats, and pigs. The Komodo dragon, however, does not rely on its teeth alone to kill its prey. It also produces a potent venom that flows from a gland in its jaw into the wound of a bitten victim. If an animal bitten by a Komodo dragon is not immediately killed, the venom ensures that it will likely die soon after the attack. The lizard simply waits patiently until its wounded, poisoned prey dies.

Most snakes are active, predatory carnivores and have a variety of adaptations that help them acquire food. For example, many snakes have special sense organs that help track prey by sensing small temperature differences between a prey's body and its surroundings. Some snake species immobilize prey with venom that is delivered through hollow teeth. Snakes also have a distinctive jaw joint that allows the jaws to distend so that the snake can swallow prey much larger than its head.

Alligators and Crocodiles Are Adapted for Life in Water

Crocodilians, as the 21 species of alligators and crocodiles are collectively known, are found in coastal and inland waters of the warmer regions of Earth. They are well adapted to an aquatic lifestyle, with eyes and nostrils located high on their heads so that they are able to remain submerged for long periods with only the uppermost portion of the head above the water's surface. Crocodilians have strong jaws and conical teeth that they use to crush and kill the fish, birds, mammals, turtles, and amphibians that they eat.

Parental care is extensive in crocodilians, which bury their eggs in mud nests. Parents guard the nest until the young hatch and then carry their newly hatched young in their mouths, moving them to safety in the water. Young crocodilians may remain with their mother for several years.

Turtles Have Protective Shells

The 240 species of turtles occupy a variety of habitats, including deserts, streams and ponds, and the ocean. This variety of habitats has fostered a variety of adaptations, but all turtles are protected by a hard, boxlike shell that is fused to the vertebrae, ribs, and collarbone. Turtles have no teeth but have instead evolved a horny beak. The beak is used to eat a variety of foods; some turtles are carnivores, some are herbivores, and some are scavengers. The largest turtle, the leatherback, is an ocean dweller that can grow to 6 feet (2 meters) or more in length and feeds largely on sea jellies. Leatherbacks and other marine turtles must return to land to breed, and often undertake extraordinary long-distance migrations to reach the beaches on which they bury their eggs.

Birds Are Feathered Reptiles

One very distinctive group of reptiles is the birds (Fig. 24-12). Although the 9,600 species of birds have traditionally been classified as a group separate from reptiles, biologists have shown that birds are really a subset of the reptiles (see pp. 342–343 for a more complete explanation). The first birds appear in the fossil record roughly 150 million years ago (Fig. 24-13) and are distinguished from other reptiles by feathers, which are essentially a highly specialized version of reptilian scales. Modern birds retain scales on their legs—evidence of the ancestry they share with the rest of the reptiles.

Bird anatomy and physiology are dominated by adaptations that help them fly. In particular, birds are exceptionally light for their size. Lightweight bones reduce the weight

(a) Hummingbird
(b) Frigate bird
(c) Ostrich

▲ FIGURE 24-12 **The diversity of birds** (a) The delicate hummingbird beats its wings about 60 times per second and weighs about 0.15 ounce (4 grams). (b) This young frigate bird, a fish-eater from the Galápagos Islands, has nearly outgrown its nest. (c) The ostrich, the largest of all birds, weighs more than 300 pounds (135 kilograms); its eggs weigh more than 3 pounds (1,500 grams).

QUESTION Although the ancestor of all birds could fly, many bird species—such as the ostrich—cannot. Why do you suppose flightlessness has evolved repeatedly among birds?

▲ **FIGURE 24-13 Archaeopteryx, the earliest-known bird** *Archaeopteryx* is preserved in 150-million-year-old limestone. Feathers, a feature unique to birds, are clearly visible, but traits characteristic of birds' ancestors are also apparent: Unlike a modern bird, *Archaeopteryx* had teeth, a bony tail, and claws on its forelimbs.

? HAVE YOU EVER WONDERED …

Which Vertebrates Have Gone into Space?

Humans have walked on the moon and now travel almost routinely into orbit around Earth. But before a person was first launched into space, other animals were sent there, to make sure that a vertebrate body could survive such journeys. In the 1940s and 1950s, rhesus monkeys, mice, and dogs were launched into space as passengers aboard missiles. The first living thing to orbit Earth was a dog named Laika, who was aboard the Soviet satellite *Sputnik 2* during its historic journey in November 1957. In the decades since, the roster of space-faring vertebrates has grown to include dozens of dogs, monkeys, and mice, as well as cats, rats, chimpanzees, and various species of tortoises, frogs, fish, and salamanders. Many of these animal travelers, especially in the early days of space travel, did not survive the experience. Without their sacrifices, however, human space travel would not have become possible.

of the bird skeleton, and many bones present in other reptiles have been lost in the course of evolution or fused with other bones. Bird reproductive organs shrink considerably during nonbreeding periods, and female birds possess only a single ovary, further minimizing weight. Feathers serve as lightweight extensions of the wing and the tail surfaces that provide the lift and control required for flight; feathers also provide lightweight protection and insulation for the body.

Birds are also able to maintain body temperatures high enough to allow their muscles and metabolic processes to operate at peak efficiency, supplying the power to fly regardless of the temperature of the external environment. This physiological ability to maintain an internal temperature that is usually higher than that of the surrounding environment is characteristic of both birds and mammals, which are therefore sometimes described as warm blooded or endothermic. In contrast, the body temperature of ectothermic (cold-blooded) animals—invertebrates, fish, amphibians, and reptiles other than birds—varies with the temperature of their environment, though these animals may exert some control of their body temperature by their behavior (such as basking in the sun or seeking shade).

Endothermic animals such as birds have a high metabolic rate, which increases their demand for energy and requires efficient oxygenation of tissues. Therefore, birds must eat frequently, and they possess circulatory and respiratory adaptations that help meet the need for efficiency. A bird's four-chambered heart prevents mixing of oxygenated and deoxygenated blood. The respiratory system of birds is supplemented by air sacs that provide a continuous supply of oxygenated air to the lungs, even while the bird exhales.

Mammals Provide Milk to Their Offspring

One branch of the tetrapod evolutionary tree gave rise to a group that evolved hair and diverged to form the mammals (Mammalia). The mammals first appeared approximately 250 million years ago but did not diversify and become prominent on land until after the dinosaurs went extinct roughly 65 million years ago. In most mammals, fur protects and insulates the warm body. Like birds, alligators, and crocodiles, mammals have four-chambered hearts that increase the amount of oxygen delivered to the tissues. Many mammals are fast and agile, and have legs adapted for running rather than crawling.

Mammals are named for the milk-producing **mammary glands** used by all female mammals to suckle their young. In addition to these unique glands, the mammalian body has sweat, scent, and sebaceous (oil-producing) glands, none of which is found in other vertebrates. Mammals have brains that are more highly developed than the brains of any other vertebrate group, giving mammals unparalleled curiosity and learning ability. Relatively long periods of parental care after birth allow some mammals to learn extensively under parental guidance. Humans and other primates are exceptional examples. In fact, the large brains of humans have been the major factor leading to human domination of Earth.

The 4,600 species of mammals include three main lineages: monotremes, marsupials, and placental mammals.

Monotremes Are Egg-Laying Mammals

Unlike other mammals, **monotremes** lay eggs rather than giving birth to live young. This group includes only three species: the platypus and two species of spiny anteaters, also known as echidnas (**Fig. 24-14**). Monotremes are found only in Australia (the platypus and short-beaked echidna) and New Guinea (the long-beaked echidna).

Echidnas are terrestrial and eat insects or earthworms that they dig out of the ground. Platypuses forage for food in the water, diving below the surface to capture small

(a) Platypus

(b) Spiny anteater

▲ **FIGURE 24-14 Monotremes (a)** Monotremes, such as this platypus, lay leathery eggs resembling those of reptiles. Platypuses live in burrows that they dig in the banks of rivers, lakes, or streams. **(b)** The short limbs and heavy claws of spiny anteaters (also known as echidnas) help them unearth insects and earthworms to eat. The stiff spines that cover a spiny anteater's body are modified hairs.

vertebrate and invertebrate animals. Platypus bodies are well adapted for this aquatic lifestyle, with a streamlined shape, webbed feet, a broad tail, and a fleshy bill.

Monotreme eggs have leathery shells and are incubated for 10 to 12 days by the mother. Echidnas have a special pouch for incubating eggs, but platypus eggs are held for incubation between the mother's tail and belly. Newly hatched monotremes are tiny and helpless and feed on milk secreted by the mother. Monotremes, however, lack nipples. Milk from the mammary glands oozes through ducts on the mother's abdomen and soaks the fur around the ducts. The young then suck the milk from the fur.

Marsupial Diversity Reaches Its Peak in Australia

In all mammals except the monotremes, embryos develop in the uterus, a muscular organ in the female reproductive tract. The lining of the uterus combines with membranes derived from the embryo to form the **placenta,** a structure that allows gases, nutrients, and wastes to be exchanged between the circulatory systems of the mother and embryo.

In **marsupials,** embryos develop in the uterus for only a short period. Marsupial young are born at a very immature stage of development. Immediately after birth, a marsupial crawls to a nipple, firmly grasps it, and, nourished by milk, completes its development. In most but not all marsupial species, this postbirth development takes place in a protective pouch.

Only one marsupial species, the Virginia opossum, is native to North America. The majority of the 275 species of marsupials are found in Australia, where marsupials such as kangaroos have come to be seen as emblematic of the island continent. Kangaroos are the largest and most conspicuous of Australia's marsupials; the largest species, the red kangaroo, may be 7 feet tall (about 2 meters) and can make 30-foot (9-meter) leaps when moving at top speed. Though kangaroos are perhaps the most familiar marsupials, the group encompasses species with a range of sizes, shapes, and lifestyles, including koalas, wombats, and the Tasmanian devil (**Fig. 24-15**).

The Tasmanian devil, a carnivorous predator the size of a small dog, is among the marsupial species at risk of extinction. Tasmanian devil populations were decimated by hunting until the species was protected by law and began to recover in the 1940s. Today, however, the recovery is threatened by a new form of cancer that appeared suddenly around 1996. Unlike most cancers, the one that afflicts Tasmanian devils is transmissible—it spreads from animal to animal. Tumors grow on the faces of affected animals, which (like all Tasmanian devils) often bite other animals on the face during fighting or sex. Tumor cells may enter a bite wound. The resulting facial tumors usually kill infected animals within a few months. Researchers estimate that the population of Tasmanian devils has decreased by 60% to 80% since the cancer epidemic began.

Placental Mammals Inhabit Land, Air, and Sea

Most mammal species are **placental** mammals (**Fig. 24-16**), so named because their placentas are far more complex than those of marsupials. Compared to marsupials, placental mammals retain their young in the uterus for a much longer period, so that offspring complete their embryonic development before being born.

The largest groups of placental mammals, in terms of number of species, are the rodents and the bats. Rodents account for almost 40% of all mammal species. Most rodent species are rats or mice, but the group also includes squirrels, hamsters, guinea pigs, porcupines, beavers, woodchucks, chipmunks, and voles. The largest rodent, the capybara, is found in South America and can weigh up to 110 pounds (50 kilograms).

About 20% of mammal species are bats, the only mammals to have evolved wings and powered flight. Bats are nocturnal and spend the daylight hours roosting in caves, rock crevices, trees, or people's houses. Most bat species have evolved adaptations for feeding on a particular kind of food. Some bats eat fruit; others feed on nectar from night-blooming flowers. Most bats are predators, including species that hunt frogs, fish, or even other bats. A few

(a) Wallaby **(b) Wombat** **(c) Tasmanian devil**

▲ **FIGURE 24-15 Marsupials (a)** Marsupials, such as the wallaby, give birth to extremely immature young who develop within the mother's protective pouch. **(b)** The wombat is a burrowing marsupial whose pouch opens toward the rear of its body to prevent dirt and debris from entering the pouch during tunnel digging. One of the wombat's predators is **(c)** the Tasmanian devil, the largest carnivorous marsupial.

(a) Capybara **(b) Whale** **(c) Bat**

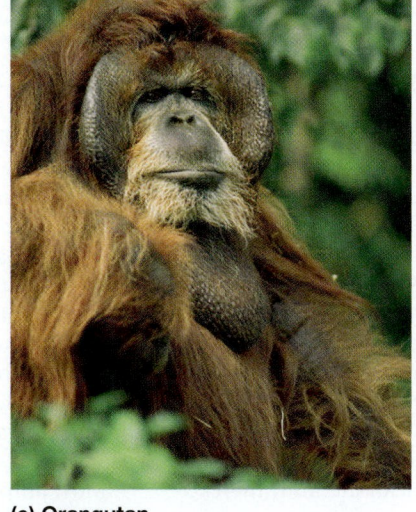

(d) Cheetah **(e) Orangutan**

◄ **FIGURE 24-16 The diversity of placental mammals (a)** The South American capybara is the world's largest rodent. **(b)** A humpback whale may migrate 15,000 or more miles each year. **(c)** A bat, the only type of mammal capable of true flight, navigates at night by using a kind of sonar. Large ears help the animal detect echoes as its high-pitched cries bounce off nearby objects. **(d)** Mammals are named after the mammary glands with which females nurse their young, as illustrated by this mother cheetah. **(e)** Orangutans are gentle, intelligent apes that occupy swamp forests in limited areas of the Tropics but are endangered by hunting and habitat destruction.

species (vampire bats) subsist entirely on blood that they lap up from incisions they make in the skin of sleeping mammals or birds. Most predatory bats, however, feed on flying insects, which they detect by echolocation. To echolocate, a bat emits short pulses of high-pitched sound (too high for humans to hear). The sounds bounce off objects in the surrounding environment to produce echoes, which the bat hears and uses to identify and locate insect prey.

Although the majority of placental mammal species are rodents or bats, the other placental mammals are diverse in form and include many species that loom large in the human imagination. For example, many people are fascinated by the sometimes human-like social behavior of our closest relatives, the chimpanzees, gorillas, and other great apes. Many of us are awed by the grace and power of large carnivores such as lions, cheetahs, tigers, and wolves, even as populations of these keystone predators (see Chapter 27) have declined due to hunting and habitat destruction. And many are fascinated by the 70 species of whales, placental mammals that evolved from terrestrial ancestors and recolonized the ocean. The largest whale species, the blue whale, can grow to more than 100 feet long (more than 30 meters) and is the largest animal known to have existed in the history of Earth. Due to heavy hunting in the past, the blue whale is now a threatened species, with a worldwide population of no more than 8,000 to 9,000 individuals.

CHECK YOUR LEARNING

Can you describe the key features of lampreys, cartilaginous fishes, ray-finned fishes, coelacanths, lungfishes, amphibians, reptiles, and mammals? Can you name and describe the main subgroups included within each of these groups?

CASE STUDY revisited

Fish Story

After Marjorie Courtenay-Latimer's discovery of the coelacanth, J. L. B. Smith dedicated himself to searching for more coelacanth specimens in the waters off South Africa. He didn't find one until 1952, when fishermen from the Comoro Islands, having seen leaflets that offered a reward for a coelacanth, contacted Smith with the news that they had one in their possession. Smith immediately booked a flight to the Comoros, and reportedly wept for joy upon holding the 88-pound coelacanth awaiting him.

In the years since Smith's trip, about 200 additional coelacanths have been caught by fishermen, mostly in waters around the Comoros but also around nearby Madagascar and off the coasts of Mozambique and South Africa. Scientists thought that the fish's range was restricted to this relatively small area in the western Indian Ocean, and it was therefore something of a shock when a few specimens were discovered in Indonesia, more than 6,000 miles away. DNA tests showed that these Indonesian coelacanths were members of a second species.

The known populations of coelacanths are small, consisting of a few hundred individuals, and appear to be declining. Part of this decline is due to fishing, though coelacanths are mostly caught accidentally by fishermen searching for more commercially desirable species. Conservation efforts in South Africa and the Comoros thus focus largely on introducing fishing methods that will reduce the chances of accidentally snaring a coelacanth.

Consider This

BioEthics Is it worth spending money to try to protect coelacanths from being accidentally killed by fishermen, or should scarce resources be instead devoted to preserving more ecologically important species and habitats?

CHAPTER REVIEW

Summary of Key Concepts

24.1 What Are the Key Features of Chordates?
All chordates possess a notochord; a dorsal, hollow nerve cord; pharyngeal gill slits; and a post-anal tail at some stage in their development.

24.2 Which Animals Are Chordates?
The chordates include three taxonomic groups: the tunicates, the lancelets, and the craniates. Tunicates are invertebrate filter-feeders that include the sessile sea squirts and the motile salps. The lancelets are also invertebrate filter-feeders, and live partially buried in sandy seafloors. Craniates include all animals with skulls: the hagfishes and the vertebrates. Hagfishes are jawless, eel-shaped craniates that lack a backbone and, therefore, are not vertebrates.

24.3 What Are the Major Groups of Vertebrates?
Lampreys are jawless vertebrates; the best-known lamprey species are parasites of fish. Cartilaginous fishes have skeletons made entirely of cartilage and bodies protected by leathery skin. They breathe with gills and reproduce using internal fertilization. Ray-finned fishes have bony skeletons and fins consisting of webs of skin supported by bony spines. Their skin in protected by interlocking scales, and they breathe with gills.

Coelacanths and lungfishes are collectively known as lobefins, because of their fleshy, bone-containing fins. Coelacanths were thought to be extinct until the discovery of living ones in 1938. They breathe with gills; lungfishes have both gills and lungs and can survive out of the water during the dry season.

Most amphibians have simple lungs for breathing air. Most are confined to relatively damp terrestrial habitats because of their need to keep their skin moist, their need for water to facilitate external fertilization, and their aquatic larvae.

Reptiles—with their well-developed lungs, dry skin covered with relatively waterproof scales, internal fertilization, and an amniotic egg with its own water supply—are well adapted to the driest terrestrial habitats. One group of reptiles, the birds, has additional adaptations, such as an elevated body temperature, that allow the muscles to respond rapidly regardless of the temperature of the environment. The bird body is adapted for flight, with feathers, lightweight bones, efficient circulatory and respiratory systems, and well-developed eyes.

Mammals have insulating hair and (except for monotreme mammals) give birth to live young that are nourished with milk. The mammalian nervous system is the most complex in the animal kingdom, providing mammals with enhanced learning ability that helps them adapt to changing environments.

Key Terms

amnion 455	pharyngeal gill slit 446
amniotic egg 455	placenta 458
craniate 448	placental 458
mammary gland 457	post-anal tail 446
marsupial 458	tetrapod 452
monotreme 457	vertebral column 449
nerve cord 446	vertebrate 449
notochord 446	

Learning Outcomes

In this chapter, you have learned to . . .

LO1 Describe the key features of chordates.

and explain the functions and significance of these features.

LO2 Describe the adaptations that mark major branching points on the chordate evolutionary tree,

LO3 Describe the major chordate taxonomic groups and representative members of each group.

Thinking Through the Concepts

Fill-in-the-Blank

1. In chordates, the nerve cord is _____ and runs along the _____ side of the body. During at least one stage of a chordate's life, it has a tail that extends past its _____ and its body is stiffened by a(n) _____ that runs along its length. **LO1**

2. Animals that are chordates but not vertebrates include _____, _____, and _____. Craniates are animals that have a(n) _____. Animals that are craniates but not vertebrates include _____. **LO2 LO3**

3. Both gills and lungs are present in adult _____. Sharks and rays have internal skeletons composed of _____. The vertebrate group with the largest number of species is _____. Lampreys have teeth but lack _____. **LO2 LO3**

4. Among tetrapod groups, hair is found in _____; the skin is a respiratory organ in _____; shelled, amniotic eggs are found in _____; aquatic larvae with gills are found in _____. **LO2 LO3**

5. The only mammals that lay eggs are _____. The only vertebrates that regenerate lost limbs are _____. The only mammals with powered flight are _____. **LO3**

Review Questions

1. Briefly describe each of the following adaptations, and explain the adaptive significance of each: vertebral column, jaws, limbs, amniotic egg, feathers, placenta. **LO2**

2. List the vertebrate groups that have each of the following: **LO2 LO3**

   a. a skeleton of cartilage
   b. a two-chambered heart
   c. amniotic egg
   d. endothermy
   e. a four-chambered heart
   f. a placenta
   g. lungs supplemented by air sacs

3. List four distinguishing features of chordates. **LO1**

4. Describe the ways in which amphibians are adapted to life on land. In what ways are amphibians still restricted to a watery or moist environment? **LO2 LO3**

5. List the adaptations that distinguish reptiles from amphibians and help reptiles adapt to life in dry terrestrial environments. **LO2 LO3**

6. List the adaptations of birds that contribute to their ability to fly. **LO2 LO3**

7. How do mammals differ from birds, and what adaptations do they share? **LO2 LO3**

8. How has the mammalian nervous system contributed to the success of mammals? **LO2 LO3**

Applying the Concepts

1. Are hagfishes vertebrates or invertebrates? On which characteristics did you base your answer? Is it important to be able to place them in one category or the other? Why?

2. Is the decline of amphibian populations of concern to humans? Why is it important to understand the causes of these phenomena?

Answers to Figure Caption questions and Fill-in-the-Blank questions can be found in the Answers section at the back of the book.

MB *Go to MasteringBiology for practice quizzes, activities, eText, videos, current events, and more.*

UNIT 4

Human observers
are captivated by
the bright colors and
ethereal beauty of coral
reefs, which are among
the most diverse,
productive, and fragile
of Earth's ecosystems.

*"When we try to pick out
anything by itself, we find
it hitched to everything
else in the Universe."*

—John Muir,
in *My First Summer in the Sierra* (1911)

Behavior and Ecology

CASE STUDY
Sex and Symmetry

WHAT MAKES A MAN SEXY? According to a growing body of research, it's his symmetry. Female sexual preference for symmetrical males was first documented in insects. For example, biologist Randy Thornhill found that symmetry accurately predicts the mating success of male Japanese scorpionflies (see the inset photo). In Thornhill's experiments and observations, the most successful males were those whose left and right wings were equal or nearly equal in length. Males with one wing longer than the other were less likely to copulate; the greater the difference between the two wings, the lower the likelihood of mating success.

Thornhill's work with scorpionflies led him to wonder if the effects of male symmetry also extend to humans. To test the hypothesis that female humans find symmetrical males more attractive, Thornhill and colleagues began by measuring symmetry in some young adult males. Each man's degree of symmetry was assessed by measurements of his ear length and the width of his foot, ankle, hand, wrist, elbow, and ear. From these measurements, the researchers derived an index that summarized the degree to which the size of these features differed between the right and left sides of the body.

The researchers next gathered a panel of heterosexual female observers who were unaware of the nature of the study and showed them photos of the faces of the measured males. As predicted by the researchers' hypothesis, men judged by the panel to be most attractive were also the most symmetrical. Apparently, a man's attractiveness to women is correlated with his body symmetry.

Why might females prefer symmetrical males? Consider this question as you read about animal behavior.

Both this male scorpionfly and this male human are exceptionally attractive to females of their species. The secret to their sex appeal may be that both have highly symmetrical bodies.

AT A GLANCE

LEARNING GOALS

LG1 Compare and contrast innate behavior and learned behavior, and explain why no behavior can be accurately described as fully innate or fully learned.

LG2 Compare and contrast habituation, trial-and-error learning, insight learning, and imprinting.

LG3 Describe how animals use different sensory systems to communicate, and summarize the characteristics, advantages, and disadvantages of communicating with visual, acoustic, and chemical signals.

LG4 Describe animal behaviors that have evolved as a result of competition for resources, and explain the functions of these behaviors.

LG5 Describe signals and behaviors animals use to attract mates, and explain the functions and evolutionary basis of these signals and behaviors.

LG6 Summarize hypotheses about why animals play, and describe examples of such play.

LG7 Describe animal societies and social behaviors among animals, and summarize the main causes, advantages, and disadvantages of social behavior and of living in a group.

LG8 Summarize biological explanations of human behaviors, and describe the methods and findings that form the basis of such explanations.

25.1 HOW DO INNATE AND LEARNED BEHAVIORS DIFFER?

Behavior is any observable activity of a living animal. For example, a moth flies toward a bright light, a honeybee flies toward a cup of sugar-water, and a housefly flies toward a piece of rotting meat. Bluebirds sing, wolves howl, and frogs croak. Mountain goats butt heads in ritual combat, chimpanzees groom one another, ants attack a termite that approaches an anthill. Humans dance, play sports, and wage wars. Even the most casual observer sees many fascinating examples of animal behavior each day.

Innate Behaviors Can Be Performed Without Prior Experience

Innate behaviors are performed in reasonably complete form the first time an animal of the right age and motivational state encounters a particular stimulus. (The proper motivational state for feeding, for example, is hunger.) Scientists can demonstrate that a behavior is innate by depriving an animal of the opportunity to learn it. For example, red squirrels, which in the wild bury nuts in the fall for retrieval during the winter, can be raised from birth in a bare cage on a liquid diet, providing them with no experience of nuts, digging, or burying. Nonetheless, such a squirrel will, when presented with nuts for the first time, carry one to the corner of its cage, and then make covering and patting motions with its forefeet, demonstrating that nut burying is an innate behavior.

Some innate behaviors can be recognized by their occurrence immediately after birth, before any opportunity for learning presents itself. Consider, for example, the common cuckoo, a bird species in which females lay eggs in the nests of other bird species, to be raised by the unwitting adoptive parents. Soon after a cuckoo egg hatches, the cuckoo chick performs the innate behavior of shoving the nest owner's eggs (or baby birds) out of the nest, eliminating its competitors for food (**Fig. 25-1**).

Learned Behaviors Require Experience

Natural selection may favor innate behaviors in many circumstances. For instance, a gull chick pecks at its parent's bill very soon after hatching, which is to the chick's advantage because pecking stimulates the parent to feed it. But in other circumstances, rigidly fixed behavior patterns may be less useful. For example, a male red-winged blackbird presented with a stuffed female blackbird will often attempt to copulate with the stuffed bird, a behavior that obviously will produce no offspring. In many situations, a degree of behavioral flexibility is advantageous.

The capacity to make changes in behavior on the basis of experience is called **learning.** This deceptively simple definition encompasses a vast array of phenomena. A toad learns to avoid distasteful insects; a baby shrew learns which adult is its mother; a human learns to speak a language; a sparrow learns to use the stars for navigation. Each of the many examples of animal learning represents the outcome of a unique evolutionary history, so learning is as diverse as animals themselves. Nonetheless, it can be useful to categorize types of learning, keeping in mind that the categories are only rough guides and that many examples of learning will not fit neatly into any category.

Habituation Is a Decline in Response to a Repeated Stimulus

A common form of simple learning is **habituation,** defined as a decline in response to a repeated stimulus. The ability to habituate prevents an animal from wasting its energy

(a) A cuckoo chick ejects an egg

(b) A foster parent feeds a cuckoo

▲ FIGURE 25-1 **Innate behavior (a)** The cuckoo chick, just hours after it hatches and before its eyes have opened, evicts the eggs of its foster parents from the nest. **(b)** The parents, responding to the stimulus of the cuckoo chick's wide-gaping mouth, feed the chick, unaware that it is not related to them.

QUESTION The cuckoo chick benefits from its innate behavior, but the foster parent harms itself with its innate response to the cuckoo chick's begging. Why hasn't natural selection eliminated this disadvantageous innate behavior?

and attention on irrelevant stimuli. This form of learning is displayed by even the simplest animals. For example, a sea anemone will retract its tentacles when first touched, but gradually stops retracting if touching is repeated frequently (Fig. 25-2).

The ability to habituate is generally adaptive. If a sea anemone withdrew every time it was brushed by a strand of waving seaweed, the animal would waste a great deal of energy, and its retracted posture would prevent it from snaring food. Humans habituate to many stimuli; city dwellers

habituate to nighttime traffic sounds, as country dwellers do to choruses of crickets and tree frogs. Each may initially find the other's habitat unbearably noisy at first, but eventually stops responding to the novel sounds.

Conditioning Is a Learned Association Between a Stimulus and a Response

A more complex form of learning is **trial-and-error learning,** in which animals acquire new and appropriate responses to stimuli through experience. Many animals are

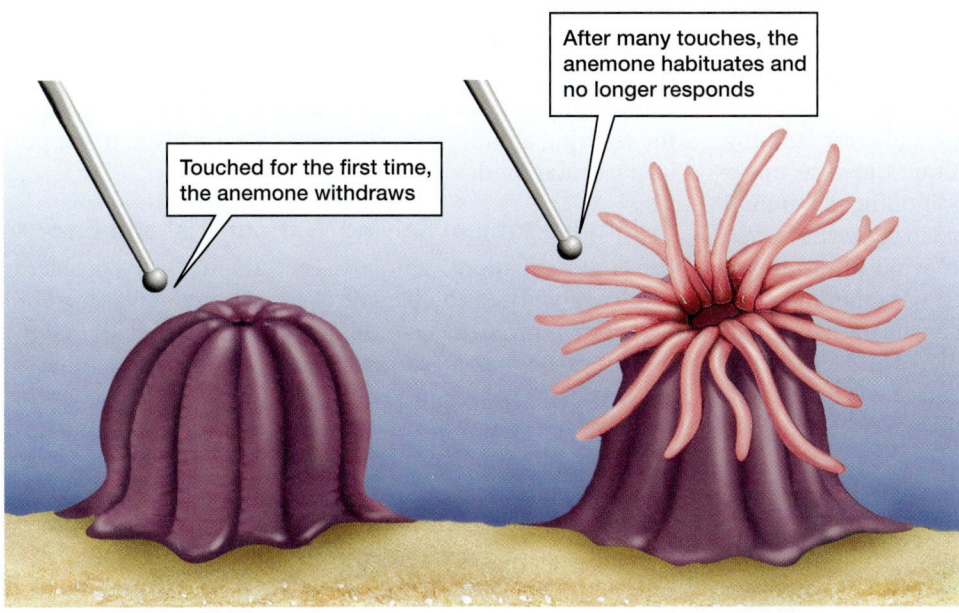

After many touches, the anemone habituates and no longer responds

Touched for the first time, the anemone withdraws

◀ FIGURE 25-2 **Habituation in a sea anemone**

1 A naive toad is presented with a bee.

2 While trying to eat the bee, the toad is stung painfully on the tongue.

3 Presented with a harmless robber fly, which resembles a bee, the toad cringes.

4 The toad is presented with a dragonfly.

5 The toad immediately eats the dragonfly, demonstrating that the learned aversion is specific to bees.

▲ **FIGURE 25-3 Trial-and-error learning in a toad**

faced with naturally occurring rewards and punishments and can learn to modify their behavior in response to them. For example, a hungry toad that captures a bee quickly learns to avoid future encounters with bees (**Fig. 25-3**). After only one experience with a stung tongue, a toad ignores bees and even other insects that resemble them.

Trial-and-error learning is an important factor in the behavioral development of many animal species and often occurs during play and exploratory behavior. This type of learning also plays a key role in human behavior—allowing, for example, a child to learn which foods taste good or bad, that a stove can be hot, and not to pull a cat's tail.

Some interesting properties of trial-and-error learning have been revealed by a laboratory technique known as **operant conditioning.** During operant conditioning, an animal learns to perform a behavior (such as pushing a lever or pecking a button) to receive a reward or to avoid punishment. This technique is most closely associated with the American comparative psychologist B. F. Skinner, who designed the "Skinner box," in which an animal is isolated and allowed to train itself. The box might contain a lever that, when pressed, ejects a food pellet. If the animal accidentally bumps the lever, a food reward appears. After a few such occurrences, the animal learns the connection between pressing the lever and receiving food and begins to press the lever repeatedly.

Operant conditioning has been used to train animals to perform tasks far more complex than pressing a lever, and

has revealed that species differ in their propensity to learn associations, more easily learning those that are relevant to their own needs. For example, if a rat is given a distinctively flavored food containing a substance that makes the rat sick, the animal learns to avoid eating that food in the future. In contrast, it is very difficult to train a rat to rear up on its hind legs in response to a particular sound or visual cue. The difference can be explained by asking which learning task is more likely to benefit a wild rat. Clearly, avoiding sickening foods is beneficial to animals such as rats that eat a wide variety of foods. However, the animal gains no obvious benefit from learning to stand up in response to a noise. In general, the learning abilities of each species have evolved to support its particular mode of life.

Insight Is Problem Solving Without Trial and Error

In certain situations, animals seem able to solve problems suddenly, without the benefit of prior experience. This kind of sudden problem solving is sometimes called **insight learning,** because it seems at least superficially similar to the process by which humans mentally manipulate concepts to arrive at a solution. We cannot, of course, know for sure if non-human animals experience similar mental states when they solve problems.

In 1917, the animal behaviorist Wolfgang Kohler showed that a hungry chimpanzee, without any training, could stack boxes to reach a banana suspended from the ceiling. This type of problem solving was once believed to be limited to very intelligent types of animals such as chimpanzees, but similar abilities may also be present in species that we tend to view as less intelligent. For example, Robert Epstein and colleagues performed an experiment that showed that pigeons may be capable of insight learning. In the experiment, caged pigeons (whose wings had been clipped to prevent flight) were first trained to perform two unrelated tasks in return for food rewards. The tasks were to push a small box around the cage and to peck at a small plastic banana. Later, the trained birds were presented with a novel situation: a plastic banana that hung above their reach in a cage that also contained a small box. Many of the pigeons pushed the box to a position beneath the plastic banana and climbed atop the box to peck the faux fruit. A pigeon trained to execute the necessary physical movements can also solve the suspended banana problem.

There Is No Sharp Distinction Between Innate and Learned Behaviors

Although the terms "innate" and "learned" can help us describe and understand behaviors, these words can also lull us into an oversimplified view of animal behavior. In practice, no behavior is totally innate or totally learned. Instead, all behaviors are mixtures of the two.

Seemingly Innate Behavior Can Be Modified by Experience

Behaviors that seem to be performed correctly on the first attempt without prior experience can later be modified by experience. For example, a newly hatched gull chick is able to peck at a red spot on its parent's beak (Fig. 25-4), an innate behavior that causes the parent to regurgitate food for the chick to eat. Biologist Niko Tinbergen studied this pecking behavior and found that the pecking response of very young chicks was triggered by the long, thin shape and red color of the parent's bill. In fact, when Tinbergen offered newly hatched chicks a thin, red rod with white stripes painted on it, they pecked at it more often than at a real beak. Within a few days, however, the chicks learned enough about the appearance of their parents that they began pecking more frequently at models more closely resembling the parents. After 1 week, the young gulls recognized their parents' appearance enough to prefer models of their own species to models of a closely related species. Eventually, the young birds learned to beg only from their own parents.

Habituation can also fine-tune an organism's innate responses to environmental stimuli. For example, young birds crouch down when a hawk flies over but ignore harmless birds such as geese. Early scientific observers hypothesized that only the very specific shape of predatory birds provoked crouching. Using an ingenious model, Niko Tinbergen and Konrad Lorenz (two of the founding fathers of **ethology,** the study of animal behavior) tested and confirmed this hypothesis. When moved in one direction, the model resembled a goose and chicks ignored it. When its movement was

▲ FIGURE 25-4 **Innate behaviors can be modified by experience** A gull chick pecks at the red spot on its mother's bill, causing her to regurgitate food.

reversed, the model resembled a hawk and elicited crouching behavior from the chicks. Further research, however, revealed that newborn chicks instinctively crouch when *any* object moves over their heads. Over time, their response habituates to things that soar by harmlessly and frequently, such as leaves, songbirds, and geese. Predators are much less common, and the novel shape of a hawk continues to elicit instinctive crouching. Thus, learning modifies the innate response, making it more advantageous.

Learning May Be Governed by Innate Constraints

Learning always occurs within boundaries that help increase the chances that only the appropriate behavior is acquired. For example, even though young robins hear the singing of sparrows, warblers, finches, and other birds, the young birds do not imitate the songs of these other species. Instead, young robins learn only the songs of adult robins. The robin's ability to learn songs is limited to those of its own species, and the songs of other species are excluded from the learning process.

The innate constraints on learning are perhaps most strikingly illustrated by **imprinting,** a form of learning in which an animal's nervous system is rigidly programmed to learn a certain thing only during a certain period of development. The information learned during this *sensitive period* is incorporated into behaviors that are not easily altered by further experience.

Imprinting is best known in birds such as geese, ducks, and chickens. These birds learn to follow the animal or object that they most frequently encounter during an early sensitive period. In nature, a mother bird is likely to be nearby during the sensitive period, so her offspring imprint on her. In the laboratory, however, these birds may imprint on a toy train or other moving object (**Fig. 25-5**). If given a choice, however, they select an adult of their own species.

All Behavior Arises Out of Interactions Between Genes and the Environment

Many early ethologists saw innate behaviors as rigidly controlled by genetic factors and viewed learned behaviors as determined exclusively by an animal's environment. Today, however, ethologists realize that, just as no behavior is wholly innate or wholly learned, no behavior can be caused strictly by genes or strictly by the environment. Instead, all behavior develops out of an interaction between genes and the environment. The relative contributions of heredity and learning vary among animal species and among behaviors within an individual.

A great deal of evidence demonstrates the existence of both genetic and environmental components in the

development of behaviors. For example, consider bird migration. It is well known that migratory birds must learn by experience how to navigate with celestial cues; young, developing birds deprived of the opportunity to observe the apparent rotation of the night sky are not, as adults, able to orient correctly at night. However, this kind of learning is not the only factor involved; bird migration behavior also has an inherited component.

At the close of the summer, many birds leave their breeding habitats and head for their winter territory, which may be hundreds or even thousands of miles away. Many of these migrating birds are traveling for the first time, because they hatched only a few months earlier. Amazingly, these naive birds depart at the proper time, head in the proper direction, and locate the proper wintering location, all without following more experienced birds (which typically depart a few weeks in advance of the first-year birds). Somehow, these young birds execute a very difficult task the first time they try it. Thus, it seems that birds must be born with the ability to migrate; it must be "in their genes." Indeed, birds hatched and raised in isolation indoors still orient in the proper migratory direction when autumn comes, apparently without the need for any learning or experience.

The conclusion that birds must have a genetically controlled ability to migrate in the right direction has been further supported by hybridization experiments with blackcap warblers. This species breeds in Europe and migrates to Africa, but populations from different areas travel by different routes. Blackcaps from western Europe travel in a southwesterly direction to reach Africa, whereas birds from eastern Europe travel to the southeast (**Fig. 25-6**). If birds from the two populations are crossbred in captivity, however, the hybrid offspring exhibit migratory orientation due south, which is intermediate between the orientations of the two parents. This result suggests that parental genes—of which offspring inherit a mixture—influence migratory direction.

CHECK YOUR LEARNING

Can you explain the difference between innate behavior and learned behavior? Are you able to describe habituation, trial-and-error learning, insight learning, and imprinting? Can you explain why no behavior can be accurately described as fully innate or fully learned?

25.2 HOW DO ANIMALS COMMUNICATE?

Animals frequently broadcast information. The sounds uttered, movements made, and chemicals emitted by animals can reveal their location, level of aggression, readiness to mate, and so on. If this information evokes a response from other individuals, and if that response tends to benefit the sender and the receiver, then a communication channel can form. **Communication** is the production of a signal by one organism that causes another organism to change its behavior in a way beneficial to both.

▲ FIGURE 25-5 **Konrad Lorenz and imprinting** Konrad Lorenz, known as the "father of ethology," is followed by goslings that imprinted on him shortly after they hatched. They follow him as they would their mother.

▲ FIGURE 25-6 **Genes influence migratory behavior** Blackcap warblers from western Europe begin their fall migration by flying in a southwesterly direction, but those from eastern Europe fly to the southeast when they begin migrating.

QUESTION If young blackcap warblers from a wild population in western Europe were transported to eastern Europe and reared to adulthood in a normal environment, in which direction would you expect them to orient?

Although animals of different species may communicate (picture a cat, its tail erect and bushy, hissing at a dog), most animals communicate primarily with members of their own species. Potential mates may communicate, as may parents and offspring. Communication is also often used to help resolve the conflicts that arise when members of a species compete directly with one another for food, space, and mates.

The ways in which animals communicate are astonishingly diverse and use all of the senses. In the following sections, we will look at communication by visual displays, sound, chemicals, and touch.

Visual Communication Is Most Effective over Short Distances

Animals with well-developed eyes use visual signals to communicate. Visual signals can be *active*, in which a specific movement (such as baring fangs) or posture (such as lowering the head) conveys a message (**Fig. 25-7**). Alternatively, visual signals may be *passive*, in which case the size, shape, or color of the animal conveys important information, commonly about its sex and reproductive state. For example, when female mandrills become sexually receptive, they develop a large, brightly colored swelling on their buttocks (**Fig. 25-8**). Active and passive signals can be combined, as illustrated by the lizard in **Figure 25-9**.

▲ FIGURE 25-7 **Active visual signals** The wolf signals aggression by lowering its head, ruffling the fur on its neck and along its back, facing its opponent with a direct stare, and exposing its fangs. These signals can vary in intensity, communicating different levels of aggression.

Like all forms of communication, visual signals have both advantages and disadvantages. On the plus side, visual signals communicate instantaneously, and active visual signals can be rapidly changed to convey a variety of messages in a short period. Visual communication is quiet and unlikely to alert distant predators, although the signaler does make itself conspicuous to those nearby. On the negative side, visual signals are generally ineffective in dense vegetation or in darkness, and are limited to close-range communication.

▲ FIGURE 25-8 **A passive visual signal** The female mandrill's colorfully swollen buttocks serve as a passive visual signal that she is fertile and ready to mate.

▲ FIGURE 25-9 **Active and passive visual signals combined** A South American anole lizard raises his head high (an active visual signal), revealing a colored throat pouch (a passive visual signal) that warns others to keep their distance.

Communication by Sound Is Effective over Longer Distances

The use of sound overcomes many of the shortcomings of visual displays. Like visual displays, acoustic signals (signals consisting of sound) reach receivers almost instantaneously. But unlike visual signals, acoustic signals can be transmitted through darkness, dense forests, and murky water. Acoustic signals can also be effective over longer distances than visual signals. For example, the low, rumbling calls of an African elephant can be heard by elephants several miles away, and the songs of humpback whales are audible for hundreds of miles. Likewise, the howls of a wolf pack carry for miles on a still night. Even the small kangaroo rat produces a sound (by striking the desert floor with its hind feet) that is audible 150 feet (45 meters) away. The advantages of long-distance transmission, however, are offset by an important disadvantage: Predators and other unwanted receivers can also detect an acoustic signal from a distance and can use that signal to locate the signaler.

? HAVE **YOU** EVER WONDERED …

Why Fireflies Flash? In many parts of the world, warm nights are enlivened by the myriad flashing lights of fireflies. Humans are entranced and entertained by the pulsing luminescence of these flying beetles, but what does the display do for the fireflies? It's a mating signal. Each of the many species of firefly has its own distinctive pattern of flashes, which helps ensure that a nocturnal firefly searching for a mate can find a partner of its own species. Firefly evolution has produced a method of visual signaling that works in the dark.

Like visual displays, acoustic signals can be varied to convey rapidly changing messages. An individual can convey different messages by varying the pattern, volume, or pitch of a sound. In a study of vervet monkeys in Kenya in the 1960s, ethologist Thomas Struhsaker found that the monkeys produced different calls in response to threats from each of their major predators: snakes, leopards, and eagles. Later, other researchers reported that the response of vervet monkeys to each of these calls is appropriate to the particular predator. For example, the "bark" that warns of a leopard or other four-legged carnivore causes monkeys on the ground to take to trees and those in trees to climb higher. The "rraup" call, which advertises the presence of an eagle or other hunting bird, causes monkeys on the ground to look upward and take cover, whereas monkeys already in trees drop to the shelter of lower, denser branches. The "chutter" call that indicates the presence of a snake causes the monkeys to stand up and search the ground for the predator.

The use of sound is by no means limited to birds and mammals. Male crickets produce species-specific songs that attract female crickets of the same species. The high-pitched whine of the female mosquito as she prepares to bite alerts nearby males that she may soon have the blood meal necessary for laying eggs. Male water striders vibrate their legs, sending species-specific patterns of vibrations through the water, attracting mates and repelling other males (**Fig. 25-10**). Many species of fish produce croaks, grunts, or other sounds.

Chemical Messages Persist Longer But Are Hard to Vary

Chemical substances that are produced by individuals and that influence the behavior of other members of the species are called **pheromones.** Pheromones can carry messages over long distances and take little energy to produce. Unlike visual or sound signals that may attract predators, pheromones are typically not detectable by other species. In addition, a pheromone can act as a kind of signpost, persisting over time and conveying a message long after the signaling animal has departed. Wolf packs, hunting over areas of nearly 400 square miles (about 1,000 square kilometers), warn other packs of their presence by marking the boundaries of

▲ **FIGURE 25-10 Communication by vibration** The light-footed water strider relies on the surface tension of water to support its weight. By vibrating its legs, the water strider sends signals that radiate out over the surface of the water. These vibrations advertise the strider's species and sex to others nearby.

their travels with urine that contains pheromones. As anyone who has walked a dog can attest, the domesticated dog reveals its wolf ancestry by staking out its neighborhood with urine that carries the chemical message "I live in this area." (For an example of how humans take advantage of animals' ability to detect chemical messages, see "Links to Everyday Life: Mine Finders.")

Chemical communication requires animals to synthesize a different substance for each message. As a result, chemical signaling systems communicate fewer and simpler messages than do sight- or sound-based systems. In addition, pheromone signals cannot easily convey rapidly changing messages. Nonetheless, chemicals can effectively convey critical information.

Many pheromones cause an immediate change in the behavior of the animal that detects them. For example, foraging termites that discover food lay a trail of pheromones from the food to the nest, and other termites follow the trail (**Fig. 25-11**). Pheromones can also stimulate physiological changes in the animal that detects them. For example, the queen honeybee produces a pheromone called *queen substance*, which prevents other females in the hive from becoming sexually mature. Similarly, mature males of some mouse species produce urine containing a pheromone that influences female reproductive physiology. The pheromone stimulates newly mature females to become fertile and sexually receptive. It will also cause a female mouse that is newly pregnant by another male to abort her litter and become sexually receptive to the new male.

Humans have harnessed the power of pheromones to combat insect pests. The sex attractant pheromones of some agricultural pests, such as the Japanese beetle and the gypsy moth, have been successfully synthesized. These synthetic pheromones can be used to disrupt mating or to lure these insects into traps. Controlling pests with pheromones has major environmental advantages over conventional pesticides, which kill beneficial as well as harmful insects and foster the

≫ **LINKS** TO EVERYDAY LIFE

Mine Finders

Although people detect their surroundings mainly by sight and sound, animals of many other species have a highly developed sense of smell, with odor-detection abilities that far exceed ours. In some cases, people have taken advantage of animal olfactory abilities to solve human problems. Consider, for example, the problem of unexploded land mines. More than 100 million of these explosive devices remain buried in countries around the world, where they were planted during past wars and forgotten. They pose a major threat to the safety of millions of mostly poor, rural people. Unfortunately, the process of removing mines is slow, expensive, and very dangerous. Animals can help; for example, dogs can smell the explosive in a land mine. Dogs, however, are heavy enough to detonate a mine, so dogs trained to find mines are at risk of getting themselves blown up.

Recently, however, a new, better, animal assistant has been drafted to help find mines. In Mozambique, home to as many as 11 million mines, rats find mines (**Fig. E25-1**). In particular, Gambian giant pouched rats have been trained to sniff out mines and, in return for a banana or peanut reward, scratch the ground vigorously when they find a mine. The rats are very good at detecting mines and are too light to detonate the ones they find. They also work very quickly; in 1 hour,

▲ **FIGURE E25-1 A Gambian giant pouched rat at work, detecting land mines**

two rats can search an area that would consume 2 weeks' time for a trained human with a metal detector. Rats have proved to be an unlikely ally in the battle to solve one of humanity's most dangerous self-inflicted problems.

▲ FIGURE 25-11 **Communication by chemical messages** A trail of pheromones, secreted by termites from their own colony, orients foraging termites toward a source of food.

evolution of pesticide-resistant insects. In contrast, each pheromone is specific to a single species and does not promote the spread of resistance, because insects resistant to the attraction of their own pheromones do not reproduce successfully.

Communication by Touch Helps Establish Social Bonds

Communication by physical contact often serves to establish and maintain social bonds among group members. This function is especially apparent in humans and other primates, who have many gestures—including kissing, nuzzling, patting, petting, and grooming—that serve important social functions (Fig. 25-12a).

Communication by touch is not limited to primates, however. For example, in many other mammal species, close physical contact helps cement the bond between parent and offspring. In addition, species in which sexual activity is

preceded or accompanied by physical contact can be found throughout the animal kingdom (Fig. 25-12b).

CHECK YOUR LEARNING

Can you compare the advantages and disadvantages of visual, acoustic, and chemical signals? Are you able to describe some functions of animal communication? Can you describe examples of animal communication that uses visual, acoustic, chemical, and tactile (touch) signals?

25.3 HOW DO ANIMALS COMPETE FOR RESOURCES?

The contest to survive and reproduce stems from the scarcity of resources relative to the reproductive potential of populations. The resulting competition underlies many of the most frequent types of interactions between animals.

Aggressive Behavior Helps Secure Resources

One of the most obvious manifestations of competition for resources such as food, space, or mates is **aggression,** or antagonistic behavior, between members of the same species. Aggressive behavior includes physical combat between rivals. A fight, however, can injure its participants; even the victorious animal might not survive to pass on its genes. As a result, natural selection has favored the evolution of displays or rituals for resolving conflicts. Aggressive displays allow competitors to assess each other and determine a winner on the basis of size, strength, and motivation, rather than on the basis of wounds inflicted. Thanks to communication, most aggressive encounters end without physical damage to the participants.

During aggressive displays, animals may exhibit weapons, such as claws and fangs (Fig. 25-13a), and often make themselves appear larger (Fig. 25-13b). Competitors often stand upright and erect their fur, feathers, ears, or fins (see Fig. 25-7). These visual displays are typically accompanied by vocal signals

(a) Baboons

(b) Land snails

▲ FIGURE 25-12 **Communication by touch** (a) An adult olive baboon grooms a juvenile. Grooming both reinforces social relationships and removes debris and parasites from the fur. (b) Touch is also important in sexual communication. These land snails engage in courtship behavior that will culminate in mating.

(a) A male baboon

(b) Sarcastic fringeheads

▲ FIGURE 25-13 **Aggressive displays** (a) Threat display of the male baboon. Despite the potentially lethal fangs so prominently displayed, aggressive encounters between baboons rarely cause injury. **(b)** The aggressive display of many male fish, such as these sarcastic fringeheads, includes elevating the fins and flaring the gill covers, thus making the body appear larger.

such as growls, croaks, roars, or chirps. Fighting tends to be a last resort when displays fail to resolve a dispute.

In addition to aggressive visual and vocal displays, many animal species engage in ritualized combat. Deadly weapons may clash harmlessly (**Fig. 25-14**) or may not be used at all. In many cases, these encounters involve shoving rather than

slashing. The ritual thus allows contestants to assess the strength and the motivation of their rivals, and the loser slinks away in a submissive posture that minimizes the size of its body.

Dominance Hierarchies Help Manage Aggressive Interactions

Aggressive interactions use a lot of energy, can cause injury, and can disrupt other important tasks, such as finding food, watching for predators, or raising young. Thus, there are advantages to resolving conflicts with minimal aggression. In a **dominance hierarchy,** each animal establishes a rank that determines its access to resources. Although aggressive encounters occur frequently while the dominance hierarchy is being established, once each animal learns its place in the hierarchy, disputes are infrequent, and the dominant individuals obtain the most access to the resources needed for reproduction, including food, space, and mates. For example, domestic chickens, after some initial squabbling, sort themselves into a reasonably stable "pecking order." Thereafter, all birds in the group defer to the dominant bird, all but the dominant bird give way to the second most dominant, and so on. In wolf packs, one member of each sex is the dominant, or "alpha," individual to whom all others of that sex are subordinate. Among male bighorn sheep, dominance is reflected in horn size (**Fig. 25-15**).

Animals May Defend Territories That Contain Resources

In many animal species, competition for resources takes the form of **territoriality,** the defense of an area where important resources are located. The defended area may include places to mate, raise young, feed, or store food. Territorial animals generally restrict most or all of their activities to the

▲ FIGURE 25-14 **Displays of strength** Ritualized combat of fiddler crabs. Oversized claws, which could severely injure another animal, grasp harmlessly. Eventually one crab, sensing greater vigor in his opponent, retreats unharmed.

▲ **FIGURE 25-15 A dominance hierarchy** The dominance hierarchy of the male bighorn sheep is signaled by the size of the horns. The backward-curving horns, which have clearly not evolved to inflict injury, are used in ritualized combat.

▲ **FIGURE 25-16 A feeding territory** Acorn woodpeckers live in communal groups that excavate acorn-sized holes in dead trees, and stuff the holes with green acorns for dining during the lean winter months. The group defends the trees vigorously against other groups of acorn woodpeckers and against acorn-eating birds of other species, such as jays.

defended area and advertise their presence there. Territories may be defended by males, females, a mated pair, or entire social groups. However, territorial behavior is most commonly seen in adult males, and territories are usually defended against members of the same species, who compete most directly for the resources being protected.

Territories are as diverse as the animals defending them. For example, a territory can be a tree where a woodpecker stores acorns (**Fig. 25-16**), a small depression in a lake floor used as a nesting site by a cichlid fish, a hole in the sand that is home to a crab, or an area of forest that provides food for a squirrel.

Territoriality Reduces Aggression

Acquiring and defending a territory requires considerable time and energy, yet territoriality is seen in animals as diverse as worms, arthropods, fish, birds, and mammals. The fact

that organisms as distantly related as worms and humans independently evolved similar behavior suggests that territoriality provides some important advantages. Although the particular benefits depend on the species and the type of territory it defends, we can make some broad generalizations. First, as with dominance hierarchies, once a territory is established through aggressive interactions, relative peace prevails as boundaries are recognized and respected. One reason for this stability is that an animal is highly motivated to defend its territory and will often defeat even larger, stronger animals that attempt to invade it. Conversely, an animal outside its territory is much less secure and more easily defeated. This principle was demonstrated by Niko Tinbergen in an experiment using stickleback fish (**Fig. 25-17**).

▲ **FIGURE 25-17 Territory ownership and aggression** Niko Tinbergen's experiment demonstrating the effect of territory ownership on aggressive motivation.

Competition for Mates May Be Based on Territories

For males of many species, successful territorial defense has a direct impact on reproductive success. In these species, males defend territories, and females are attracted to high-quality territories, which might have features such as large size, abundant food, and secure nesting areas. Males who successfully defend the best territories have the greatest chance of mating. For example, experiments have shown that male stickleback fish that defend large territories are more successful in attracting mates than are males that defend small territories. Females that select males with the best territories increase their own reproductive success.

Animals Advertise Their Occupancy

Territories are advertised through sight, sound, and smell. If a territory is small enough, its owner's mere presence, reinforced by aggressive displays toward intruders, can provide sufficient defense. A mammal that has a territory but cannot always be present may use pheromones to scent-mark the boundaries of its territory. For example, male rabbits use pheromones secreted by anal and chin glands to mark their territories. Hamsters rub the areas around their dens with secretions from glands in their flanks.

Vocal displays are a common form of territorial advertisement. Male sea lions defend a strip of beach by swimming up and down in front of it, calling continuously. Male crickets produce a specific pattern of chirps to warn other males away from their burrows. Birdsong is a striking example of territorial defense. The husky trill of the male seaside sparrow is part of an aggressive display, warning other males to steer clear of his territory (**Fig. 25-18**). In fact, male sparrows that are unable to sing are unable to defend territories. The importance of singing to seaside sparrows' territorial defense was elegantly demonstrated by ornithologist M. Victoria McDonald, who captured territorial males and performed an operation that left them

▲ **FIGURE 25-18 Defense of a territory by song** A male seaside sparrow announces ownership of his territory.

temporarily unable to sing but still able to utter the other, shorter and quieter signals in their vocal repertoires. The songless males were unable to defend territories or attract mates, but regained their lost territories when they recovered their singing ability.

CHECK YOUR LEARNING

Can you describe the function of aggressive behavior? Are you able to explain why natural selection has favored the evolution of aggressive signals and displays, territoriality, and dominance hierarchies? Can you describe how animals advertise their presence on a territory?

25.4 HOW DO ANIMALS FIND MATES?

In many sexually reproducing animal species, mating involves copulation or other close contact between males and females. Before animals can successfully mate, however, they must identify one another as members of the same species, as members of the opposite sex, and as being sexually receptive. In many species, finding an appropriate potential partner is only the first step. Often the male must demonstrate his quality before the female will accept him as a mate. The need to fulfill all of these requirements has resulted in the evolution of a diverse and fascinating array of courtship behaviors.

Signals Encode Sex, Species, and Individual Quality

Individuals that waste energy and gametes by mating with members of the wrong sex or wrong species are at a disadvantage in the contest to reproduce. Thus, natural selection favors behaviors by which animals communicate their sex and species to potential mates.

Many Mating Signals Are Acoustic

Animals often use sounds to advertise their sex and species. Consider the raucous nighttime chorus of male tree frogs, each singing a species-specific song. Male grasshoppers and crickets also advertise their sex and species by their calls, as does a male fruit fly with a buzz he produces by vibrating one wing.

Signals that advertise sex and species may also be used by potential mates in comparisons among rival suitors. For example, the male bellbird uses its deafening song to defend large territories and to attract females from great distances. A female flies from one territory to another, alighting near each male in his tree. The male, beak gaping, leans directly over the flinching female and utters an earsplitting note. The female apparently endures this noise to compare the songs of the various males, perhaps choosing the loudest as a mate.

Visual Mating Signals Are Also Common

Many species use visual displays for courting. Male fence lizards, for example, bob their heads in a species-specific rhythm, and females prefer the rhythm of their own species.

▶ **FIGURE 25-19 Sexual displays**
(a) During courtship, a male gardener bowerbird builds a bower out of twigs and decorates it with colorful items that he gathers. **(b)** A male frigate bird inflates his scarlet throat pouch to attract passing females.

QUESTION The male bowerbird provides no protection, food, or other resources to his mate or offspring. Why, then, do females carefully compare the bowers of different males before choosing a mate?

(a) A bowerbird bower

(b) A male frigate bird

The elaborate construction projects of the male gardener bowerbird and the scarlet throat of the male frigate bird serve as flashy advertisements of sex, species, and male quality (**Fig. 25-19**). Sending these extravagant signals must be risky, because they make it much easier for predators to locate the sender. For males, the added risk is an evolutionary necessity, because females won't mate with males that lack the appropriate signal. Females, in contrast, typically do not need to attract males or assume the risk associated with a conspicuous signal, so in many species females are drab in comparison to males (**Fig. 25-20**).

The intertwined functions of sex recognition and species recognition, advertisement of individual quality, and synchronization of reproductive behavior commonly require a complex series of signals, both active and passive, by both sexes. Such signals are beautifully illustrated by the complex

▲ **FIGURE 25-20 Sex differences in guppies** As in many animal species, the male guppy (top) is brighter and more colorful than the female.

CASE STUDY c o n t i n u e d

Sex and Symmetry

If females are attracted to symmetrical males, we might expect females to assess the symmetry of male visual signals that function in mate attraction. For example, each male house finch has a patch of bright red feathers on the crown of his head, and researchers have shown that males whose crown patches are brightly colored are more likely to attract a mate than are males with dull patches. But males with patches that are both colorful *and* highly symmetrical have the highest mating success of all.

underwater "ballet" executed by the male and female three-spined stickleback fish (**Fig. 25-21**).

Chemical Signals Can Bring Mates Together

Pheromones can also play an important role in reproductive behavior. A sexually receptive female silk moth, for example, sits quietly and releases a chemical message that can be detected by males up to 3 miles (5 kilometers) away. The exquisitely sensitive and selective receptors on the antennae of the male silk moth respond to just a few molecules of the substance, allowing him to travel upwind along a concentration gradient to find the female (**Fig. 25-22a**).

Water is an excellent medium for dispersing chemical signals, and fish commonly use a combination of pheromones and elaborate courtship movements to ensure the synchronous release of gametes. Mammals, with their highly developed sense of smell, often rely on pheromones released by the female during her fertile periods to attract males (**Fig. 25-22b**).

CHECK YOUR LEARNING

Can you describe how signals function in courtship and mating? Can you describe some examples of mating-related communication with acoustic, visual, and chemical signals?

1 A male, inconspicuously colored, leaves the school of males and females to establish a breeding territory.

2 As his belly takes on the red color of the breeding male, he displays aggressively at other red-bellied males, exposing his red underside.

3 Having established a territory, the male begins nest construction by digging a shallow pit that he will fill with bits of algae cemented together by a sticky secretion from his kidneys.

4 After he tunnels through the nest to make a hole, his back begins to take on the blue courting color that makes him attractive to females.

5 An egg-carrying female displays her enlarged belly to him by assuming a head-up posture. Her swollen belly and his courting colors are passive visual displays.

6 Using a zigzag dance, he leads her to the nest.

7 After she enters, he stimulates her to release eggs by prodding at the base of her tail.

8 He enters the nest as she leaves and deposits sperm, which fertilize the eggs.

▲ FIGURE 25-21 **Courtship of the three-spined stickleback**

(a) Antennae detect pheromones (b) Noses detect pheromones

▲ FIGURE 25-22 **Pheromone detectors (a)** Male moths find females not by sight but by following airborne pheromones released by females. These odors are sensed by receptors on the male's huge antennae, whose enormous surface area maximizes the chances of detecting the female scent. **(b)** When dogs meet, they typically sniff each other near the base of the tail. Scent glands there broadcast information about the bearer's sex and interest in mating.

QUESTION Female dogs use a pheromone to signal readiness to mate, but female mandrills (see Fig. 25-8) signal mating readiness with a visual signal. What differences would you predict between the two species' methods of searching for food?

25.5 WHY DO ANIMALS PLAY?

Many animals play. Pygmy hippopotamuses push one another, shake and toss their heads, splash in the water, and pirouette on their hind legs. Otters delight in elaborate acrobatics. Bottlenose dolphins balance fish on their snouts, throw objects, and carry them in their mouths while swimming. Baby vampire bats chase, wrestle, and slap each other with their wings. Pigface, a giant African softshell turtle who lived at the National Zoo in Washington, D.C., for more than 50 years, would spend hours each day batting a ball around his enclosure. Even octopuses have been seen playing a game: pushing objects away from themselves and into a current, then waiting for the objects to drift back, only to push them back into the current to start the cycle over again.

Animals Play Alone or with Other Animals

Play can be solitary, as when a single animal manipulates an object, such as a cat with a ball of yarn, or the dolphin with its fish, or a macaque monkey making and playing with a snowball. Play can also be social. Often, young animals of the same species play together, but parents may join them. Social play typically includes chasing, fleeing, wrestling, kicking, and gentle biting (**Fig. 25-23**).

Play seems to lack any clear immediate function and is abandoned in favor of feeding, courtship, and escaping from danger. Young animals play more frequently than do adults. Play typically borrows movements from other behaviors (attacking, fleeing, stalking, and so on) and uses considerable energy. Also, play is potentially dangerous. Many young humans and other animals are injured, and some are killed, during play. In addition, play can distract an animal from the presence of danger while making it conspicuous to predators. So, why do animals play?

Play Aids Behavioral Development

It is likely that play has survival value and that natural selection has favored those individuals who engage in playful activities. One of the best explanations for the survival value of play is the *practice hypothesis*. It suggests that play allows young animals to gain experience in behaviors that they will use as adults. By performing these acts repeatedly in play, an animal practices skills that will later be important in hunting, fleeing, and social interactions.

Play is most intense early in life when the brain develops and crucial neural connections form. John Byers, a biologist at the University of Idaho, has observed that species with large brains tend to be more playful than species with small brains. Because larger brains are generally linked to greater learning ability, this relationship supports the hypothesis that adult skills are learned during juvenile play. Watch children roughhousing or playing tag, and you will see how play might foster strength and coordination and develop skills that might have helped our hunting ancestors survive.

CHECK YOUR LEARNING

Can you describe the characteristics of animal play and some examples of it? Are you able to describe a hypothesis about the function of play?

25.6 WHAT KINDS OF SOCIETIES DO ANIMALS FORM?

Sociality, the tendency to associate with others and form groups, is a widespread feature of animal life. Most animals interact at least a little with other members of their species. Many spend the bulk of their lives in the company of others, and a few species have developed complex, highly structured societies.

(a) Chimpanzees

(b) Polar bears

(c) Red foxes

▲ FIGURE 25-23 **Young animals at play**

Group Living Has Advantages and Disadvantages

Living in a group has both costs and benefits, and a species will not evolve social behavior unless the benefits of doing so outweigh the costs. Benefits to social animals include:

- Increased abilities to detect, repel, and confuse predators.
- Increased hunting efficiency or increased ability to spot localized food resources.
- Advantages resulting from the potential for division of labor within the group.
- Increased likelihood of finding mates.

On the negative side, social animals may encounter:

- Increased competition for limited resources.
- Increased risk of infection from contagious diseases.
- Increased risk that offspring will be killed by other members of the group.
- Increased risk of being spotted by predators.

Sociality Varies Among Species

The degree to which animals of the same species cooperate varies from one species to the next. Some types of animals, such as the mountain lion, are basically solitary; interactions between adults consist of brief aggressive encounters and mating. Other types of animals cooperate on the basis of changing needs. For example, the coyote is solitary when food is abundant, but hunts in packs when food becomes scarce.

Many animals form loose social groups, such as pods of dolphins, schools of fish, flocks of birds, and herds of musk oxen (**Fig. 25-24**). Participation in such groups can provide benefits. For example, the characteristic spacing of fish in schools or the V-pattern of geese in flight provides a hydrodynamic or aerodynamic advantage for each individual in the group, reducing the energy required for swimming or flying. Some biologists hypothesize that herds of antelope or schools of fish confuse predators; their myriad bodies make it difficult for the predator to focus on and pursue a single individual.

A small number of species, most of which are insects or mammals, form highly integrated cooperative societies. As you read the following section, you may notice that some cooperative societies are based on behavior that seems to sacrifice the individual for the good of the group. There are many examples: Young, mature Florida scrub jays may remain at their parents' nest and help them raise subsequent broods instead of breeding; worker ants often die in defense of their nest; ground squirrels may sacrifice their own lives to warn the rest of their group about an approaching predator. These behaviors are examples of **altruism**—behavior that decreases the reproductive success of one individual to benefit another.

Forming Groups with Relatives Fosters the Evolution of Altruism

How might altruistic behavior evolve? When individuals perform self-sacrificing deeds, why aren't the alleles that contribute to this behavior eliminated from the gene pool? One possibility

▶ **FIGURE 25-24 Cooperation in loosely organized social groups** A herd of musk oxen functions as a unit when threatened by predators such as wolves. Males form a circle, horns pointed outward, around the females and young.

is that other members of the group are close relatives of the altruistic individual. Because close relatives share alleles, the altruistic individual may promote the survival of its own alleles through behaviors that maximize the survival of its close relatives. This concept is called **kin selection.** Kin selection helps explain the self-sacrificing behaviors that contribute to the success of cooperative societies. Cooperative behavior is illustrated in the following sections, which describe two examples of complex societies, one in an insect and one in a mammal.

Honeybees Live Together in Rigidly Structured Societies

Perhaps the most puzzling of all animal societies are those of the bees, ants, and termites. Scientists have long struggled to explain the evolution of a social structure in which most individuals never breed, but instead labor intensively to feed and protect the offspring of a different individual. Whatever its evolutionary explanation, the intricate organization of a social insect colony is fascinating. In these communities, the individual is a mere cog in an intricate, smoothly running machine and could not survive by itself.

Individual social insects are born into one of several castes within the society. These castes are groups of similar individuals that perform a specific function. For example, honeybees emerge from their larval stage into one of three major preordained roles. One role is that of *queen.* Only one queen is tolerated in a hive at any time. Her functions are to produce eggs (up to 1,000 per day for a lifetime of 5 to 10 years) and to regulate the lives of the workers. Male bees, called *drones,* serve merely as mates for the queen. Soon after the queen hatches, drones lured by her sex pheromones swarm around her, and she mates with as many as 15 of them. This relatively brief "orgy" supplies her with sperm that will last a lifetime, enough to fertilize more than 3 million eggs. Their sexual chore accomplished, the drones become superfluous and are eventually driven out of the hive or killed.

The hive is run by the third class of bees, sterile female *workers.* A worker's tasks are determined by her age and by conditions in the colony. A newly emerged worker starts life as a "waitress," carrying food such as honey and pollen to the queen, to other workers, and to developing larvae. As she matures, special glands begin to produce wax, and she becomes a builder, constructing perfectly hexagonal cells of wax in which the queen deposits her eggs and the larvae develop. She also takes shifts as a "maid," cleaning the hive and removing the dead, and as a guard, protecting the hive against intruders. Her final role in life is that of a forager, gathering pollen and nectar, food for the hive. She spends nearly half of her 2-month life in this role. At times during her foraging phase, she may act as a scout, seeking new and rich sources of nectar. If she finds one, she returns to the hive and communicates its location to other foragers. She communicates by means of the **waggle dance,** an elegant form of symbolic expression (**Fig. 25-25**).

Pheromones play a major role in regulating the lives of social insects. Honeybee drones are drawn irresistibly to the queen's sex pheromone (*queen substance*), which she releases during her mating flights. Back at the hive, she uses the same substance to maintain her position as the only fertile female. The queen substance is licked off her body and passed among the workers, rendering them sterile. The queen's presence and health are signaled by her continuing production of queen substance; a decrease in production alters the behavior of the workers. Almost immediately they begin building extra-large royal cells. The workers feed the larvae that develop in these cells a special glandular secretion known as royal jelly. This unique food alters the development of the growing larvae so that, instead of a worker, a new queen emerges from the royal cell. The old queen then leaves the hive, taking a swarm of workers with her to establish residence elsewhere. If more than one new queen emerges, a battle to the death ensues. The victorious queen takes over the hive.

Naked Mole Rats Form a Complex Vertebrate Society

The nervous systems of vertebrates are far more complex than those of insects, and we might therefore expect vertebrate societies to be proportionately more complex. With

If the dance is performed on a vertical wall inside the hive, the angle (from vertical) of the waggle run represents the angle between the sun and the food source

up

40°

If the dance is performed on a horizontal surface outside, the waggle run is aimed at the food source

The rate of circling communicates the distance to the food source

▲ **FIGURE 25-25 Bee language: the waggle dance** A forager, returning from a rich source of nectar, performs a waggle dance that communicates the distance and direction of the food source as other foragers crowd around her, touching her with their antennae. The bee moves in a straight line while shaking her abdomen back and forth ("waggling") and buzzing her wings. She repeats this dance over and over in the same location, circling back in alternating directions.

the exception of human society, however, they are not. Perhaps the most unusual society among non-human mammals is that of the naked mole rat (**Fig. 25-26**). These nearly blind, nearly hairless relatives of guinea pigs live in large underground colonies in southern Africa and have a form of social organization much like that of honeybees. The colony is dominated by the queen, a single reproducing female to whom all other members are subordinate.

The queen is the largest individual in the colony and maintains her status by aggressive behavior, particularly shoving. She prods and shoves lazy workers, stimulating them to become more active. As in honeybee hives, there is a division of labor among the workers, in this case based on size. Small, young rats clean the tunnels, gather food, and dig more tunnels. Tunnelers line up head to tail and pass excavated dirt

along the completed tunnel to an opening. Just below the opening, a larger mole rat flings the dirt into the air, adding it to a cone-shaped mound. Biologists observing this behavior from the surface dubbed it "volcanoing." In addition to volcanoing, large mole rats defend the colony against predators and members of other colonies.

If another female begins to become fertile, the queen apparently senses changes in the estrogen levels of the subordinate female's urine. The queen then selectively shoves the would-be breeder, causing stress that prevents her rival from ovulating. When the queen dies, a few of the females gain weight and begin shoving one another. The aggression may escalate until a rival is killed. Ultimately, a single female becomes dominant. Her body lengthens, and she assumes the queenship and begins to breed. Litters averaging 14 pups are produced about four times a year. During the first month, the queen nurses her pups, and the workers feed the queen. Then the workers begin feeding the pups solid food.

CHECK YOUR LEARNING

Can you list the advantages and disadvantages of living in a group and describe the different degrees of sociality that occur among mammals? Are you able to explain the conditions under which altruism is most likely to evolve? Can you describe the societies of honeybees and naked mole rats?

▲ **FIGURE 25-26 Naked mole rat workers**

25.7 CAN BIOLOGY EXPLAIN HUMAN BEHAVIOR?

The behaviors of humans, like those of all other animals, have an evolutionary history. Thus, the techniques and concepts of ethology can help us understand and explain human behavior. Human ethology, however, will remain a less rigorous science than animal ethology. We cannot treat people as laboratory animals, devising experiments that control and manipulate the factors that influence their attitudes and actions. Nevertheless, many scientists have taken an ethological, evolutionary approach to human behavior, and their work has had a major impact on our view of ourselves.

The Behavior of Newborn Infants Has a Large Innate Component

Because newborn infants have not had time to learn, we can assume that much of their behavior is innate. The rhythmic movement of an infant's head in search of its mother's breast is an innate behavior that is expressed in the first days after birth. Sucking, which can be observed even in a human fetus, is also innate (**Fig. 25-27**). Other behaviors seen in newborns and even premature infants include grasping with the hands and feet and making walking movements when the body is held upright and supported.

Another example is smiling, which can occur soon after birth. Initially, smiling can be induced by almost any object looming over the newborn. This initial indiscriminate response, however, is soon modified by experience. Infants up to 2 months old will smile in response to a stimulus consisting of two dark, eye-sized spots on a light background, which at that stage of development is a more potent stimulus for smiling than is an accurate representation of a human face. But as the child's development continues, learning and further development of the nervous system interact to limit the response to more correct representations of a face.

Newborns in their first 3 days of life can be conditioned to produce certain rhythms of sucking when their mother's voice is used as reinforcement. In experiments, infants preferred their own mothers' voices to other female voices, as indicated by their responses (**Fig. 25-28**). The infant's ability to learn his or her mother's voice and respond positively to it within days of birth has strong parallels to imprinting and may help initiate bonding with the mother.

Young Humans Acquire Language Easily

One of the most important insights from studies of animal learning is that animals tend to have an inborn predilection for specific types of learning that are important to their species' mode of life. In humans, one such inborn predilection is for the acquisition of language. Young children are able to acquire language rapidly and nearly effortlessly; they typically acquire a vocabulary of 28,000 words before the age of 8. Research suggests that we are born with a brain that is already primed for this early facility with language. For example, a human fetus begins responding to sounds during the third trimester of pregnancy, and researchers have demonstrated that infants are able to distinguish among consonant sounds by 6 weeks after birth. In one experiment, infants sucked on a pacifier that contained a force transducer to record the sucking rate. The infants were conditioned to suck at a higher rate in response to playback of adult voices making various consonant sounds. When one

▲ **FIGURE 25-28 Newborns prefer their mother's voice** Using a nipple connected to a computer that plays audio tapes, researcher William Fifer demonstrated that newborns can be conditioned to suck at specific rates in order to listen to their own mothers' voices through headphones. For example, if the infant sucks faster than normal, her mother's voice is played; if she sucks more slowly, another woman's voice is played. Researchers found that infants easily learned and were willing to work hard at this task just to listen to their own mothers' voices, presumably because they had become used to her voice in the womb.

▲ **FIGURE 25-27 A human instinct** Thumb sucking is a difficult habit to discourage in young children, because sucking on appropriately sized objects is an instinctive, food-seeking behavior. This fetus sucks its thumb at about 4 months of development.

sound (such as "ba") was presented repeatedly, the infants became habituated and decreased their sucking rate. But when a new sound (such as "pa") was presented, sucking rate increased, revealing that the infants perceived the new sound as different.

Behaviors Shared by Diverse Cultures May Be Innate

Another way to study the innate bases of human behavior is to compare simple acts performed by people from diverse cultures. This comparative approach has revealed several gestures that seem to form a universal, and therefore probably innate, human signaling system. Such gestures include facial expressions for pleasure, rage, and disdain, and greeting movements such as an upraised hand or the "eye flash" (in which the eyes are widely opened and the eyebrows rapidly elevated). The evolution of the neural pathways underlying these gestures presumably depended on the advantages that accrued to both senders and receivers from sharing information about the emotional state and intentions of the sender. A species-wide method of communication was perhaps especially important before the advent of language and later remained useful during encounters between people who shared no common language.

Certain complex social behaviors are widespread among diverse cultures. For example, the incest taboo (avoidance of mating with close relatives) seems to be universal across human cultures (and even across many species of non-human primates). It seems unlikely, however, that a shared belief could be encoded in our genes. Some biologists have suggested that the taboo is instead a cultural expression of an evolved, adaptive behavior. According to this hypothesis, close contact among family members early in life suppresses sexual desire, and this response arose because of the negative consequences of inbreeding (such as a higher incidence of genetic diseases). The hypothesis does not require us to assume an innate social belief, but rather proposes that we inherit a learning program that causes us to undergo a kind of imprinting early in life.

Humans May Respond to Pheromones

Although the main channels of human communication are through the eyes and ears, humans also seem to respond to chemical messages. The possible existence of human pheromones was hinted at in the early 1970s, when biologist Martha McClintock found that the menstrual cycles of roommates and close friends tended to become synchronized. McClintock suggested that the synchrony resulted from some chemical signal between the women, but almost 30 years passed before she and her colleagues uncovered more conclusive evidence that a pheromone was at work.

In 1998, McClintock's research group asked nine female volunteers to wear cotton pads in their armpits for 8 hours each day during their menstrual cycles. The pads were then disinfected with alcohol and swabbed above the upper lips of another set of 20 female subjects (who reported that they could detect no odors other than alcohol on the pads). The subjects were exposed to the pads in this way each day for 2 months, with half the group sniffing secretions from women in the early (preovulation) part of the menstrual cycle, while the other half was exposed to secretions from later in the cycle (postovulation). Women exposed to early-cycle secretions had shorter-than-usual menstrual cycles, and women exposed to late-cycle secretions had delayed menstruation. It appears that women release different pheromones, with different effects on receivers, at different points in the menstrual cycle.

Other research suggests that people may also be able to detect chemical indicators of fear or stress. For example, in 2009 Lilianne Mujica-Parodi and her colleagues performed an experiment in which they collected sweat samples from 144 people. Half the people were sampled during a first-time skydive with a 1-minute freefall, the other half (the controls) were sampled during a bout of exercise. Test subjects then smelled one of the two types of samples while undergoing brain imaging. A brain region called the amygdala, which is associated with strong emotions such as fear and rage, was active in subjects who smelled the "stress sweat" from the skydivers, but not in the subjects who smelled the exercise sweat. A chemical present in the sweat of emotionally stressed people apparently triggered a similar emotion, or at least a similar brain response, in people exposed to the chemical.

Although McClintock's and Mujica-Parodi's experiments offer strong evidence for the existence of human pheromones, little else is known about chemical communication in humans. The actual molecules that caused the effects documented by these and other experiments remain unknown. Receptors for chemical messages have not yet been found in humans, and we don't know if "menstrual pheromones" and "stress pheromones" are examples of an important communication system or merely isolated cases of a vestigial ability. Despite the hopeful advertisements for "sex attraction pheromones" on late-night television, chemical communication in humans is a scientific mystery awaiting a solution.

Studies of Twins Reveal Genetic Components of Behavior

Twins present an opportunity to examine the hypothesis that differences in human behavior are related to genetic differences. If a particular behavior is heavily influenced by genetic factors, we would expect to find that *identical twins* (which arise from a single fertilized egg and have identical genes) are more likely to share the behavior than are *fraternal twins*

CASE STUDY continued

Sex and Symmetry

Does symmetry have a scent? In one study, researchers measured the body symmetry of 80 men and then issued a clean T-shirt to each one. Each subject wore his shirt to bed for two consecutive nights. A panel of 82 women sniffed the shirts and rated their scents for "pleasantness" and "sexiness." Which shirts had the sexiest, most pleasant scents? The ones worn by the most symmetrical men. The researchers concluded that women can identify symmetrical men by their scent.

(which arise from two individual eggs fertilized by different sperm and are no more similar genetically than are other siblings). Data from twin studies, and from other within-family investigations, have tended to confirm the heritability of many human behavioral traits. These studies have documented a significant genetic component for traits such as activity level, alcoholism, sociability, anxiety, intelligence, dominance, and even political attitudes. On the basis of tests designed to measure many aspects of personality, identical twins are about two times more similar in personality than are fraternal twins.

The most fascinating twin findings come from observations of identical twins separated soon after birth, reared in different environments, and reunited for the first time as adults. Identical twins reared apart have been found to be as similar in personality as those reared together, indicating that the differences in their environments had little influence on their personality development. They have been found to share nearly identical taste in jewelry, clothing, humor, food, and names for children and pets. In some cases these separated twins share personal idiosyncrasies such as giggling, nail biting, drinking patterns, hypochondria, and mild phobias.

Biological Investigation of Human Behavior Is Controversial

The field of human behavioral genetics is controversial, especially among nonscientists, because it challenges the long-held belief that environment is the most important determinant of human behavior. As discussed earlier in this chapter, we now recognize that all behavior has some genetic basis and that complex behavior in non-human animals typically combines elements of both innate and learned behaviors. Thus, it seems certain that our own behavior is influenced by both our evolutionary history and our cultural heritage. The debate over the relative importance of heredity and environment in determining human behavior continues and is unlikely ever to be fully resolved. Human ethology is not yet recognized as a rigorous science, and it will always be hampered because we can neither view ourselves with detached objectivity nor experiment with people as if they were laboratory rats. Despite these limitations, there is much to be learned about the interaction of learning and innate tendencies in humans.

CASE STUDY revisited

Sex and Symmetry

In the experiment described at the beginning of this chapter, women found males with the most symmetrical bodies to have the most attractive faces. But how did the women know which males were most symmetrical? After all, the researchers' measurement of male symmetry was based on small differences in the sizes of body parts that the female judges did not even see during the test.

Perhaps male body symmetry is reflected in facial symmetry, and females prefer symmetrical faces. To test this hypothesis, a group of researchers used computers to alter photos of male faces, either increasing or decreasing their symmetry (**Fig. 25-29**). Then heterosexual female observers rated each face for attractiveness. The observers had a strong preference for more symmetrical faces.

Why would females prefer to mate with symmetrical males? The most likely explanation is that symmetry indicates good physical condition. Disruptions of normal embryological development can cause bodies to be asymmetrical, so a highly symmetrical body indicates healthy, normal development. Females that mate with individuals whose health and vitality are announced by their symmetrical bodies are likely to have offspring that are similarly healthy and vital.

Consider This

Is our perception of human beauty determined by cultural standards, or is it part of our biological makeup, the product of our evolutionary heritage? What evidence would persuade you that beauty is a biological phenomenon or that it is a cultural one?

▲ **FIGURE 25-29 Faces of varying symmetry** Researchers used sophisticated software to modify facial symmetry. From left: a face modified to be less symmetrical; the original, unmodified face; a face modified to be more symmetrical; a perfectly symmetrical face.

CHAPTER REVIEW

Summary of Key Concepts

25.1 How Do Innate and Learned Behaviors Differ?

Although all animal behavior is influenced by both genetic and environmental factors, biologists distinguish between innate behaviors, whose development is not highly dependent on external factors, and learned behaviors, which require more extensive experiences with environmental stimuli in order to develop. Innate behaviors can be performed properly the first time an animal encounters the appropriate stimulus, whereas learned behavior changes in response to the animal's social and physical environment.

The distinction between innate and learned behavior is often blurred in naturally occurring behaviors. Learning allows animals to modify innate responses so that they occur only with appropriate stimuli. Imprinting, a form of learning with innate constraints, is possible only at a certain time in an animal's development.

25.2 How Do Animals Communicate?

Communication allows animals of the same species to interact effectively in their quest for mates, food, shelter, and other resources. Animals communicate through visual signals, sound, chemicals (pheromones), and touch. Visual communication is quiet and can convey rapidly changing information. Visual signals are active (body movements) or passive (body shape and color). Sound communication can also convey rapidly changing information, and it is effective when vision is impossible. Pheromones can be detected after the sender has departed, conveying simple messages over time. Physical contact reinforces social bonds and is a part of many premating rituals.

25.3 How Do Animals Compete for Resources?

Although many competitive interactions are resolved through aggression, serious injuries are rare. Most aggressive encounters are settled by displays that communicate the motivation, size, and strength of the combatants.

Some species establish dominance hierarchies that minimize aggression. On the basis of initial aggressive encounters, each animal acquires a status in which it defers to more dominant individuals and, in turn, dominates subordinates. When resources are limited, dominant animals obtain the largest share and are most likely to reproduce.

Territoriality, a behavior in which animals defend areas where important resources are located, also minimizes aggressive encounters. In general, territorial boundaries are respected, and the best-adapted individuals defend the richest territories and produce the most offspring.

25.4 How Do Animals Find Mates?

Successful reproduction requires that animals recognize the species, sex, and sexual receptivity of potential mates. In many species, animals also assess the quality of potential mates. These requirements have contributed to the evolution of sexual displays that use all forms of communication.

25.5 Why Do Animals Play?

Animals of many species engage in seemingly wasteful (and sometimes dangerous) play behavior. Play behavior in young animals has been favored by natural selection, probably because it provides opportunities to practice and perfect behaviors that will later be crucial for survival and reproduction.

25.6 What Kinds of Societies Do Animals Form?

Social living has both advantages and disadvantages, and species vary in the degree to which their members cooperate. Some species form cooperative societies. The most rigid and highly organized are those of the social insects such as the honeybee, in which the members follow rigidly defined roles throughout life. These roles are maintained through both genetic programming and the influence of certain pheromones. Naked mole rats exhibit the most complex and rigid vertebrate social interactions, resembling those of social insects.

25.7 Can Biology Explain Human Behavior?

Researchers are increasingly investigating whether human behavior is influenced by evolved, genetically inherited factors. This emerging field is controversial. Because we cannot freely experiment on humans, and because learning plays a major role in nearly all human behavior, investigators must rely on studies of newborn infants, comparative cultural studies, and studies of identical and fraternal twins. Evidence is mounting that our genetic heritage plays a role in personality, intelligence, simple universal gestures, our responses to certain stimuli, and our tendency to learn specific things such as language at particular stages of development.

Key Terms

aggression 472	insight learning 466
altruism 479	kin selection 480
behavior 464	learning 464
communication 468	operant conditioning 466
dominance hierarchy 473	pheromone 470
ethology 467	territoriality 473
habituation 464	trial-and-error
imprinting 467	learning 465
innate 464	waggle dance 480

Learning Outcomes

In this chapter, you have learned to . . .

LO1 Compare and contrast innate behavior and learned behavior, and explain why no behavior can be accurately described as fully innate or fully learned.

LO2 Compare and contrast habituation, trial-and-error learning, insight learning, and imprinting.

LO3 Describe how animals use different sensory systems to communicate, and, summarize the

characteristics, advantages, and disadvantages of communicating with visual, acoustic, and chemical signals.

LO4 Describe animal behaviors that have evolved as a result of competition for resources, and explain the functions of these behaviors.

LO5 Describe signals and behaviors animals use to attract mates, and explain the functions and evolutionary basis of these signals and behaviors.

LO6 Summarize hypotheses about why animals play, and describe examples of such play.

LO7 Describe animal societies and social behaviors among animals, and summarize the main causes, advantages, and disadvantages of social behavior and of living in a group.

LO8 Summarize biological explanations of human behaviors, and describe the methods and findings that form the basis of such explanations.

Thinking Through the Concepts

Fill-in-the-Blank

1. In general, animal behaviors arise from an interaction between the animal's _____ and its _____. Some behaviors are performed correctly the first time an animal encounters the proper _____. Such behaviors are described as _____. **LO1**

2. Play is almost certainly an adaptive behavior because it uses considerable _____, it can distract the animal from watching for _____, and it may cause _____. The most likely explanation for why animals play is the "_____ hypothesis," which states that play teaches the young animal _____ that will be useful as a(n) _____. This hypothesis is supported by the observation that animals that have larger _____ and are most capable of _____ are more likely to play. **LO6**

3. One of the simplest forms of learning is _____, defined as a decline in response to a(n) _____, harmless stimulus. A different type of learning in which an animal's nervous system is rigidly programmed to learn a certain behavior during a certain period in its life is called _____. The time frame during which such learning occurs is called the _____. **LO2**

4. Animals often deal with competition for resources through _____ behavior. Such conflicts are often resolved through _____, which allow the competing animals to assess each other without _____ each other. Animals that resolve conflicts this way often display their _____, and make their bodies appear _____. **LO4**

5. The defense of an area where important resources are located is called _____. Examples of important resources that may be defended include places to _____, _____, _____, and _____. Such resources are most commonly defended by which sex? _____ Are these spaces usually defended against members of the same species or against members of different species? _____ **LO4**

6. After each form of communication, list a major advantage in the first blank, and a major disadvantage in the second blank: Pheromones: _____; _____. Visual displays: _____; _____. **LO3**

7. To attract a mate, a male animal may produce signals that advertise the male's _____, _____, and/or _____. **LO5**

Review Questions

1. Explain why neither the term "innate" nor "learned" adequately describes the behavior of any given organism. **LO1**

2. Explain why animals play. Include the features of play in your answer. **LO6**

3. List four senses through which animals communicate, and give one example of each form of communication. For each sense listed, present both advantages and disadvantages of that form of communication. **LO3**

4. A bird will ignore a squirrel in its territory, but will act aggressively toward a member of its own species. Explain why. **LO4**

5. Why are most aggressive encounters among members of the same species relatively harmless? **LO4**

6. Discuss the advantages and disadvantages of group living. **LO7**

7. In what ways do naked mole rat societies resemble those of the honeybee? **LO7**

8. What kinds of evidence might indicate that a particular human behavior is innate or that variation in a human behavior is influenced by genetic differences? **LO8**

Applying the Concepts

1. Male mosquitoes orient toward the high-pitched whine of the female, and female mosquitoes, the only sex that sucks blood, are attracted to the warmth, humidity, and carbon dioxide exuded by their prey. Using this information, design a mosquito trap or killer that exploits a mosquito's innate behaviors. Then, design one for moths.

2. You raise honeybees but are new at the job. Trying to increase honey production, you introduce several queens into the hive. What is the likely outcome? What different things could you do to increase production?

3. Describe and give an example of a dominance hierarchy. What role does it play in social behavior? Give a human parallel, and describe its role in human society. Are the two roles similar? Why or why not? Repeat this exercise for territorial behavior in humans and in another animal.

4. You are manager of an airport. Planes are being endangered by large numbers of flying birds, which can be sucked into and disable the engines. Without harming the birds, what might you do to discourage them from nesting and flying near the airport and its planes?

Answers to Figure Caption questions and Fill-in-the-Blank questions can be found in the Answers section at the back of the book.

MB *Go to MasteringBiology for practice quizzes, activities, eText, videos, current events, and more.*

CHAPTER 26

Population Growth and Regulation

Bemused-looking heads survey the deforested landscape of Easter Island. If they could speak, would these statues tell of a population that outgrew the environment's capacity to sustain it?

CASE STUDY

The Mystery of Easter Island

WHY DO CIVILIZATIONS VANISH? Among those who have pondered this question were the first Europeans to reach Easter Island, a remote island in the South Pacific. When the European explorers arrived in the 1700s, they were mystified by the enormous stone statues partially embedded in the island's grassy landscape. The island's few inhabitants at the time had no record or memory of the statues' creators, nor could the current inhabitants have transported or erected the massive structures themselves. Moving statues that weigh up to 80 tons over the 6 miles from the nearest stone quarry, and then maneuvering them into an upright position, would have required long ropes and strong timbers. Europeans found Easter Island devoid of anything that could have furnished the wood to make rollers or fibers to make rope. There were almost no trees at the time, and none of the island's shrubs grew higher than 10 feet.

An important clue to the mystery of Easter Island was revealed by scientists who studied the pollen grains from layers of ancient sediments. Because the age of each sediment layer can be determined by radioactive carbon dating, and because each plant species can be identified by the unique appearance of its pollen, pollen analysis can show how vegetation has changed over time. The researchers discovered that the island was once covered by a diverse forest, including toromiro trees that make excellent firewood; hauhau trees that could supply fiber for rope; and Easter Island palm trees, with long, straight trunks that would have made good rollers for moving statues.

The island's first settlers probably arrived sometime between the years 800 and 1200 A.D., and by about 1400, almost all of Easter Island's trees were gone. Most researchers agree that the demise of the forest began with the arrival of humans, who cleared land for agriculture and used the trees for firewood and construction materials. Apparently, the culture responsible for the Easter Island statues disappeared along with the forest. Could there have been a connection between these two disappearances? How do populations change over time? What natural processes control population size? How have people circumvented these controls? You'll find out as you read this chapter.

AT A GLANCE

26.1 How Does Population Size Change?

26.2 How Is Population Growth Regulated?

26.3 How Are Populations Distributed in Space and Age?

26.4 How Is the Human Population Changing?

LEARNING GOALS

LG1 Describe the factors that cause population size to change, including the interactions among population growth, growth rate, and population size.

LG2 Describe and graph exponential and logistic growth and, using examples, explain the conditions under which exponential and logistic growth occur.

LG3 List five factors that influence biotic potential, and explain why natural selection has favored a high biotic potential.

LG4 List examples of density-dependent and density-independent environmental resistance and describe how each controls population size.

LG5 Graph and then explain the possible fates of a population that overshoots the carrying capacity of its environment.

LG6 Explain three types of spatial distribution and three types of survivorship curves and discuss the factors that cause each.

LG7 Describe the advances that have allowed continued exponential human population growth since prehistoric times.

LG8 Graph and then explain the stages of the demographic transition, noting where developed and developing countries fall along the graph.

LG9 Draw and then explain general age structure diagrams of growing, stable, and shrinking populations and discuss how these diagrams predict future growth.

26.1 HOW DOES POPULATION SIZE CHANGE?

A **population** consists of all the members of a particular species that live within an **ecosystem,** which includes all the living and nonliving components of a defined geographical area. Your campus, for example, probably hosts populations of birds, squirrels, grass, trees, shrubs, and students in campus dorms. Each population forms an integral part of a larger **community,** defined as a group of interacting populations. Communities, in turn, exist within ecosystems. A natural ecosystem can be as small as a pond or as large as an ocean; it can be a field, a forest, or an island. The enormous ecosystem that encompasses all of Earth's habitable surface is called the **biosphere. Ecology** (from the Greek "oikos," meaning "a place to live") is the study of the interrelationships of organisms with each other and with their nonliving environment. We now begin our exploration of ecology with an overview of populations.

Changes in Population Size Result from Natural Increases and Net Migration

In natural ecosystems, some populations remain relatively stable in size over time, some undergo yearly cycles, and still others change sporadically in response to complex environmental variables. As individuals invade new ecosystems, for example, their population might grow rapidly and then either stabilize or plummet.

Population size changes through births, deaths, and net migration. The **natural increase** of a population is the difference between births and deaths. Although it may sound strange, natural "increase" can be negative (a decrease) if deaths exceed births. The net migration of a population is the difference between **immigration** (migration into the population) and **emigration** (migration out). A population thus grows when the sum of natural increase and net migration is positive and declines when this sum is negative. A simple equation for the change in population size within a given time span is:

$$\text{change in population size} = \text{natural increase} + \text{net migration}$$
$$\text{(births} - \text{deaths)} \quad \text{(immigration} - \text{emigration)}$$

In the wild, most migration is emigration of young animals out of the population. Young locusts, for example, may emigrate in huge swarms in response to overcrowding. And the term "lone wolf" comes from young wolves that emigrate from their home packs and immigrate into new areas where they may join an existing pack or (if they find a mate) start a pack of their own. As the members of a population reproduce, emigration in some species helps to keep the original population at a size that is relatively stable and consistent with the resources available to support it. Although migration can be significant in some natural populations, for simplicity, we will ignore migration and use only birth and death rates in our calculations of population growth.

Populations Grow Based on the Birth Rate, the Death Rate, and the Population Size

The size of most natural populations of organisms fluctuates over the course of a year because reproduction tends to be seasonal and many offspring die early in life. Scientists studying such populations must always do their counts at the same point in the reproductive cycle of the species, which usually means at the same time each year.

Growing populations add individuals in proportion to the population's size, much like a bank account accumulates compound interest. If conditions remain the same, a population will grow at a constant percentage of its size over a given time interval. The **growth rate** (r) of a population can be expressed as the percentage change in the population size per unit time (we will use 1 year). The population growth rate is its **birth rate** (b) minus its **death rate** (d). (Here, because we are ignoring migration, the population's growth rate is the same as its rate of natural increase.)

$$\underset{\text{(birth rate)}}{b} \quad - \quad \underset{\text{(death rate)}}{d} \quad = \quad \underset{\text{(growth rate)}}{r}$$

If the birth rate exceeds the death rate, the population growth rate will be positive and population size will increase. If the death rate exceeds the birth rate, the growth rate will be negative and the population will decrease. Now, let's calculate the growth rate of a deer herd, using rates expressed as percentages of the population. Assume that in a year, the females in a deer herd of 100 give birth to 30 fawns, but 10 fawns and 10 adult deer die. This birth rate is 30% (30 out of 100) and the death rate is 20%, for an annual growth rate of 10%:

$$\underset{\text{(birth rate in \%)}}{30} \quad - \quad \underset{\text{(death rate in \%)}}{20} \quad = \quad \underset{\text{(growth rate in \%)}}{10}$$

To calculate population growth (G), the number of individuals added to a population in a given time interval, we multiply the growth rate (r) by the population size (N) at the beginning of the time interval:

$$\underset{\substack{\text{(population growth} \\ \text{per unit time)}}}{G} \quad = \quad \underset{\text{(growth rate)}}{r} \quad \times \quad \underset{\substack{\text{(population} \\ \text{size)}}}{N}$$

If you think about this equation, you will realize that, at a constant growth rate (r), the value of G will increase with each successive time interval. To calculate the growth in numbers of deer during a specific time interval, we express percentages as decimal fractions, so that 10%, for example, becomes 0.10. The growth (G) of our deer herd (N) equals $0.10 \times 100 = 10$ during the first year. If the growth rate remains constant, then the herd begins its second year with 110 deer (N), and then adds another 10% ($110 \times 0.10 = 11$ deer). The herd begins

its third year with 121 deer, and so on. Of course we won't see exactly 12.1 deer added during the fourth year. For real populations, this equation only provides estimates, because the growth rate of a population predicts its future size *only if environmental conditions remain constant* so that births and deaths continue to occur at constant rates. But does the environment ever remain constant? We explore this question in later sections.

If Births Exceed Deaths, Exponential Growth Occurs

A constant growth rate (r) produces exponential growth. During **exponential growth,** an ever-larger number is added to the population during each succeeding time period. This pattern of growth will occur in any population in which each individual, on average over the course of its life span, produces more than one offspring that survives to reproduce. If the size of an exponentially growing population is graphed against time, a characteristic shape called a **J-curve** will be produced. To illustrate the J-curve, let's look at an example of golden eagles. We assume that golden eagles live for 30 years and that each pair produces two offspring annually after reaching sexual maturity. **Figure 26-1** illustrates the growth of two eagle populations (each founded by a single breeding pair) over their life span. The red line assumes breeding begins at age 4, and the blue line at age 6. In each case, exponential growth occurs. But after 24 years, the population whose members begin reproducing at 4 years of age is more than six times as large as the population whose members begin reproducing at age 6 (see the table in Fig. 26-1). By postponing reproduction, people can also significantly slow population growth. For example, if each woman bears three children in her early teens, the population will grow far more rapidly than if each woman bears five children starting after age 30.

The death rate also has a major impact on population size. **Figure 26-2** compares three hypothetical bacterial populations experiencing different death rates. Notice that the three curves have the same J-shape. As long as births exceed deaths,

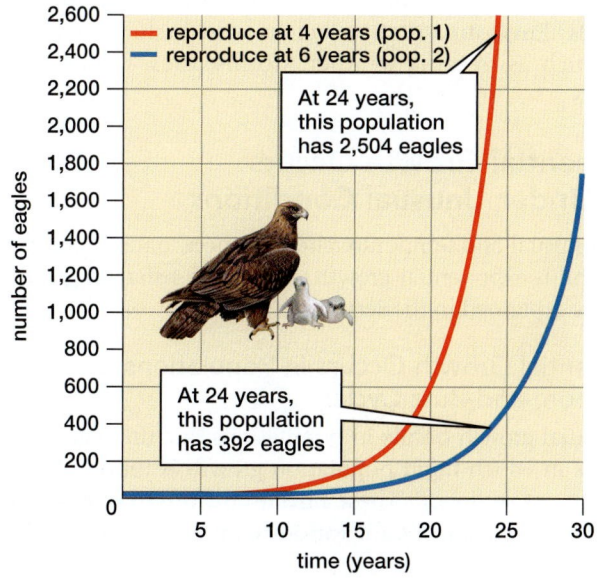

Time (years)	Number of eagles (pop. 1)	Number of eagles (pop. 2)
0	2	2
6	8	4
12	52	18
18	362	86
24	2,504	392
30	17,314	1,764

◄ **FIGURE 26-1 Exponential growth curves are J-shaped** The graph shows the growth of two hypothetical populations of eagles, each starting with a single pair, but differing in the age at which breeding begins. During the 30-year lifetime of the founding pairs, the population that begins reproducing at age 4 reaches 10 times the size of the population that begins reproducing at age 6 (see the table).

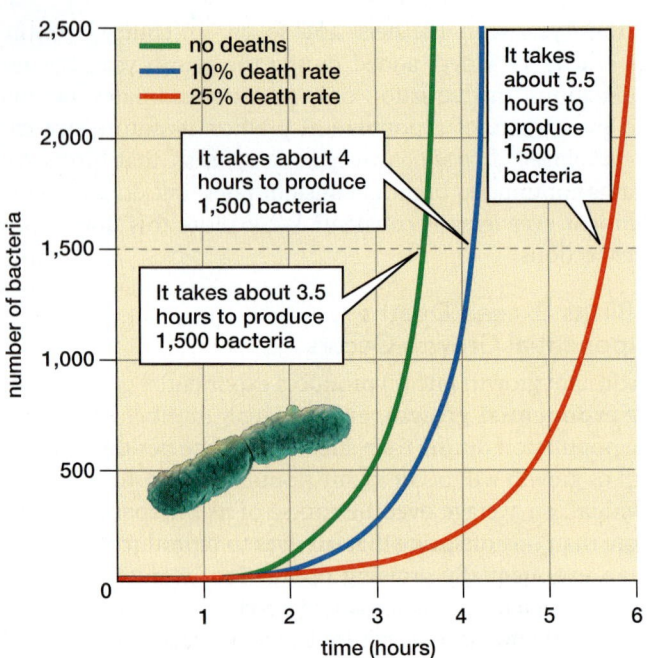

▲ **FIGURE 26-2 The effect of death rates on population growth** This graph assumes that a bacterial population doubles every 20 minutes. Notice that all show the characteristic J-shape of exponential growth curves, although it takes the population with the higher death rate longer to achieve any given size.

the population eventually becomes enormous, but as the death rate increases, it takes longer to reach any given population size.

Biotic Potential Determines the Maximum Rate at Which a Population Can Grow

The ability to produce many offspring is an inherited trait. Natural selection has favored organisms whose attributes adapt them to their environments and that pass these adaptations on to as many healthy offspring as possible. **Biotic potential** refers to the maximum rate at which a particular population could increase. Calculations of biotic potential assume ideal conditions (unlimited resources and no predators) that allow a maximum birth rate and a minimum death rate. Although the average number of offspring an individual produces each year varies from millions (for an oyster) to one or fewer, each healthy organism has the potential to replace itself many times during its reproductive lifetime.

In a stable population, on the average only one offspring per individual survives to have its own young. Because life in the wild is fraught with danger, a high biotic potential helps ensure that some offspring live long enough to reproduce. Biotic potential differs among species. Factors that influence biotic potential include:

- The age at which the organism first reproduces.
- The frequency of reproduction.
- The average number of offspring produced each time the organism reproduces.
- The length of the organism's reproductive life span.
- The death rate of individuals under ideal conditions.

HAVE **YOU** EVER WONDERED ...

How Many Children Can One Woman Bear?

The often-cited example for the greatest amount of childbearing is the wife of Feodor Vassilyev, a Russian peasant, who reportedly had 69 children (16 sets of twins, 7 triplets, and 4 quadruplets) between 1725 and 1765, but no reliable record confirms this. Former reality TV star Michelle Duggar, of Arkansas, probably comes close to demonstrating the realistic biotic potential of humans. Duggar was born in 1966, and between 1988 and 2009, she gave birth to 19 children (including two sets of twins), all of whom remained healthy as of 2012.

CHECK YOUR LEARNING

Can you explain how migration and natural increase cause population size to change? Can you describe how population growth is calculated and how the growth rate interacts with population size to determine how populations grow per unit time? Can you define *biotic potential* and list the factors that influence it?

26.2 HOW IS POPULATION GROWTH REGULATED?

In 1859, Charles Darwin wrote: "There is no exception to the rule that every organic being naturally increases at so high a rate that, if not destroyed, the Earth would soon be covered by the progeny of a single pair." Clearly, population growth cannot continue indefinitely. In the following sections we discuss how population size results from the interaction between biotic potential and **environmental resistance,** or all the curbs on population growth imposed by the living and nonliving environment. Examples of environmental resistance include interactions among organisms such as predation and competition for limited resources. Environmental resistance also includes natural events such as freezing weather, storms, fires, floods, and droughts.

Exponential Growth Occurs Only Under Unusual Conditions

Under unusual and temporary circumstances, natural populations exhibit exponential growth, producing J-shaped growth curves, as explained in the sections that follow.

Exponential Growth Occurs in Populations with Boom-and-Bust Cycles

Exponential growth occurs in populations that undergo regular cycles in which rapid population growth is followed by a sudden, massive die-off. These **boom-and-bust cycles** occur in a variety of organisms for various complex reasons. Many short-lived, rapidly reproducing species—from photosynthetic microorganisms to insects—have seasonal population cycles that are linked to changes in rainfall, temperature, or nutrient availability, as is shown for a population of photosynthetic bacteria in **Figure 26-3a.** Boom-and-bust cycles in these and other aquatic microorganisms can impact human health, as described in "Health Watch: Boom-and-Bust Cycles Can Be Bad News."

In temperate climates, insect populations grow rapidly during the spring and summer, then crash with the killing hard frosts of winter. Female houseflies, for example, lay about 120 eggs at a time. Because they hatch and mature within 2 weeks, seven generations can occur during a spring and summer breeding season. The housefly's biotic potential is so enormous that (without environmental resistance) the seventh generation would contain roughly 6 trillion

flies—all descended from a single pregnant female. More complex factors can produce roughly 4-year cycles for small rodents such as voles and lemmings (**Fig. 26-3b**), and much longer population cycles for hares, muskrats, and grouse. Lemming populations, for example, may grow until they overgraze their fragile arctic tundra ecosystem. Lack of food, increasing populations of predators, and social stress caused by crowding may all contribute to a sudden high mortality. Many deaths occur as waves of lemmings emigrate from regions of high population density. During these dramatic mass movements, lemmings are easy targets for predators, while many others drown as they encounter bodies of water and attempt to swim across. The reduced lemming population eventually contributes to a decline in predator numbers, as well as a recovery of the plant community on which the lemmings feed. These responses, in turn, set the stage for the next round of exponential growth in the lemming population (see Fig. 26-3b).

Exponential Growth Occurs Temporarily When Environmental Resistance Is Reduced

In populations that do not experience boom-and-bust cycles, exponential growth may occur temporarily under special circumstances—for example, if food supply or habitat is increased or if predation is reduced. For example, the whooping crane was nearly driven to extinction by habitat loss and hunting. In 1940, with only 15 birds remaining, human predation was banned. Crane habitats are being preserved and restored. With environmental resistance from people reduced, the crane population is now exhibiting the J-curve characteristic of exponential growth (**Fig. 26-4**). It remains among the world's rarest birds, so continued population growth will be necessary to help ensure the whooping crane's long-term survival.

Exponential growth can also occur when individuals invade a new habitat with favorable conditions and little competition. **Invasive species** (see Chapter 27) are organisms with a high biotic potential that are introduced (deliberately or accidentally) into ecosystems where they did not evolve and where they encounter little environmental resistance. Invasive species often show explosive population growth. For example, people introduced 100 cane toads into Australia in 1935 to control beetles that were destroying sugar cane. Cane toad females lay 7,000 to 35,000 eggs at a time, and in their new environment, the cane toads encountered few predators. Spreading outward from their release point, cane toads now inhabit an area in Australia of nearly 400,000 square miles and are migrating rapidly into new habitats, threatening native species by eating or displacing them. The cane toad population, now estimated at well over 200 million, continues to grow exponentially. An invasive species may also have contributed to the demise of Easter Island's forests, as described in "Case Study Continued: The Mystery of Easter Island."

As you will learn in the next section, all populations that exhibit exponential growth must eventually either stabilize or drop, sometimes precipitously in a population crash.

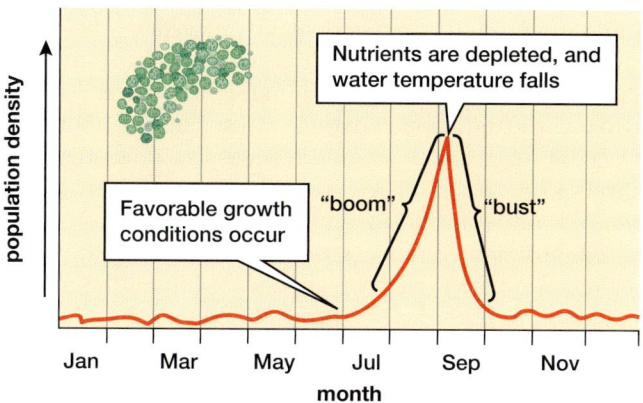

(a) A boom-and-bust cycle in photosynthetic bacteria

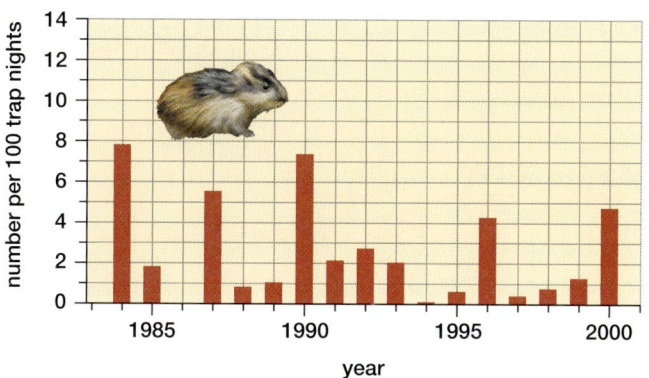

(b) Boom-and-bust cycles in a lemming population in the Canadian Arctic.

▲ **FIGURE 26-3 Boom-and-bust population cycles** (a) The density of a hypothetical population of photosynthetic bacteria in a lake. These microorganisms persist at a low level until early July. Then conditions become favorable for growth, and exponential growth occurs until early September, when the population plummets. (b) This lemming population from the Canadian Arctic follows a roughly 3- to 4-year cycle of boom and bust. The data are based on live trapping.

QUESTION What factors might make the data in the lemming graph somewhat erratic and irregular?

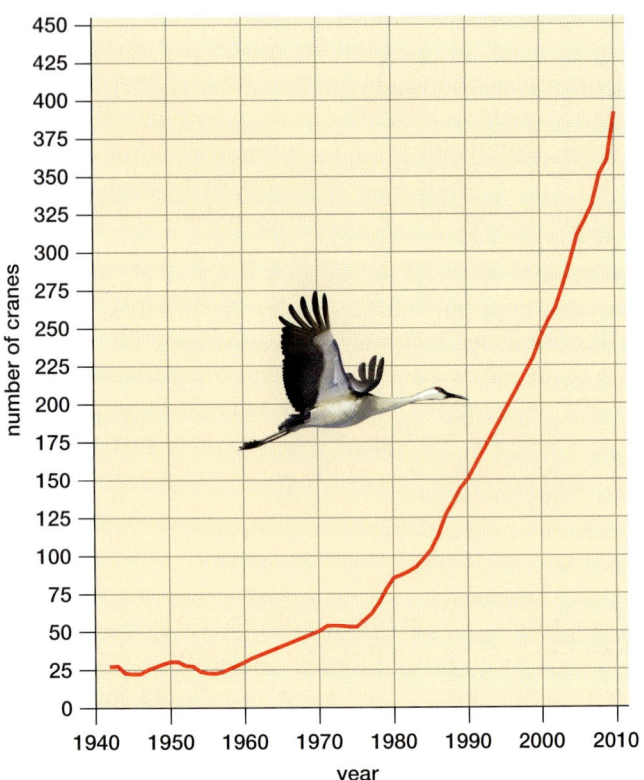

▲ **FIGURE 26-4 Exponential growth of wild whooping cranes** Hunting and habitat destruction had reduced the world's whooping crane population to 21 individuals before they were protected in 1940. By 2005, their wild population had grown to 340 individuals, and by 2010 it stood at 383. Note the J-curve characteristic of exponential growth. Data from Canadian Wildlife Service and U.S. Fish and Wildlife Service. 2007. *International recovery plan for the whooping crane.* Ottawa: Recovery of Nationally Endangered Wildlife (RENEW); Albuquerque, NM: U.S. Fish and Wildlife Service.

Environmental Resistance Limits Population Growth

Environmental resistance eventually stops exponential growth and ideally produces a balance between the size of a population and the resources available to support it. For example, imagine a sterile culture dish in which nutrients are constantly replenished and wastes removed. If a small number of living skin cells were added, they would attach to the bottom and begin reproducing by mitotic cell division. If you counted the cells daily under a microscope and graphed their numbers, for a while, your graph would resemble the J-curve characteristic of exponential growth. But as the cells began to occupy all the available space on the dish, their growth rate would slow and eventually drop to zero; the cell population size would be determined by the resource of space.

Logistic Growth Occurs When New Populations Stabilize as a Result of Environmental Resistance

At a certain point, your graph of skin cell numbers will resemble the one shown in **Figure 26-5a.** This growth pattern, described as **logistic population growth,** is characteristic

of populations that increase up to the maximum number that their environment can sustain, and then stabilize. The maximum population size that can be sustained by an ecosystem for an extended period of time without damage to the ecosystem is called the ecosystem's **carrying capacity (K).** The curve that results when logistic growth is graphed is called an **S-curve,** after its general shape. The formula for logistic growth (which is based on the assumption that the population size cannot overshoot *K*) is explained in "In Greater Depth: A Look at Logistic Population Growth."

In nature, an increase in population size (*N*) above carrying capacity (*K*) can be sustained for a short time. This is dangerous, however, because a population above carrying capacity is living at the expense of resources that cannot regenerate as fast as they are being depleted. A small overshoot above *K* is likely to be followed by a decrease in both *K* and *N*, until the resources recover and the original *K* is restored.

If a population far exceeds the carrying capacity of its environment, the consequences are more severe, because in this situation, the excess demands placed on the ecosystem are likely to destroy essential resources (such as Easter Island's forests), which may be unable to recover. This can permanently and severely reduce *K*, causing the population to decline to a fraction of its former size or to disappear entirely (**Fig. 26-5b**). For example, when reindeer were introduced onto an island with no large predators, their population increased rapidly, seriously overgrazing the lichen they relied on for food. Starvation then caused the reindeer population to plummet, as shown in **Figure 26-6.**

Logistic population growth can occur in nature when a species moves into a new habitat, as ecologist John Connell demonstrated for barnacle populations colonizing bare rock along a rocky ocean shoreline (**Fig. 26-7**). Initially, the new settlers find ideal conditions that allow their population to grow exponentially. As population density increases, however, individuals increasingly compete with one another, particularly for space, energy, and nutrients. Laboratory

CASE STUDY c o n t i n u e d

The Mystery of Easter Island

Rats accompanied Polynesian colonizers to Easter Island, possibly as stowaways in the islanders' boats, or they may have been brought intentionally as a source of food. On Easter Island, these rats encountered little predation and plenty of food and nesting sites among the original forests. With little environmental resistance, the population of this invasive species flourished, possibly growing into the millions over several years. The exponentially growing rat population would have devoured Easter Island's ground-nesting birds. They also ate palm nuts, as revealed by the high percentage of rat-gnawed nuts discovered in some Easter Island caves and in Easter Island sediments, which have also yielded an abundance of rat bones. Rats eating seeds undoubtedly contributed to the extinction of the Easter Island palm trees and the elimination of the island's other forest plants along with the communities of animals they sheltered and supported.

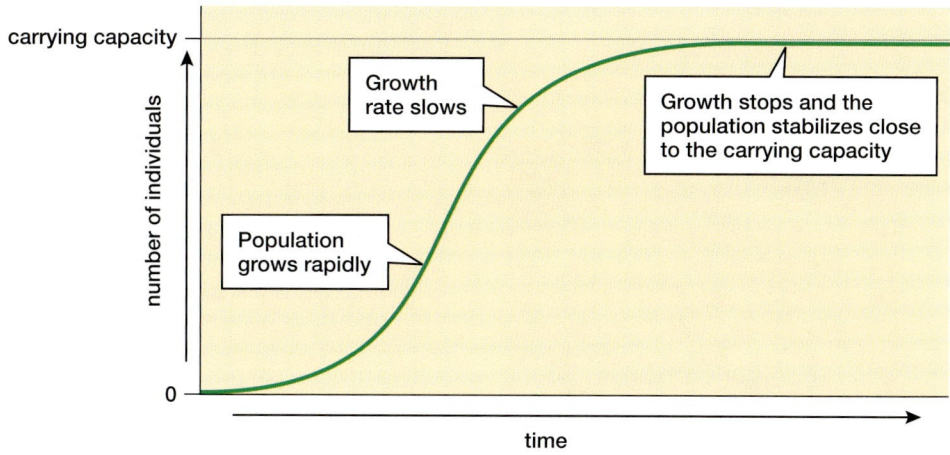

carrying capacity

number of individuals

Growth
rate slows

Growth stops and the
population stabilizes close
to the carrying capacity

Population
grows rapidly

0

time

(a) An S-shaped growth curve stabilizes at carrying capacity

◄FIGURE 26-5 **The S-curve
of logistic population growth**
(a) During logistic growth, a population
will remain small for a time before
its numbers rise dramatically.
Eventually, the growth rate slows as
the population encounters increasing
density-dependent environmental
resistance. Population growth finally
ceases at or near carrying capacity
(K). The result is a curve shaped like a
"lazy S." **(b)** Populations can overshoot
carrying capacity (K), but only for a
limited time. Three possible results are
illustrated.

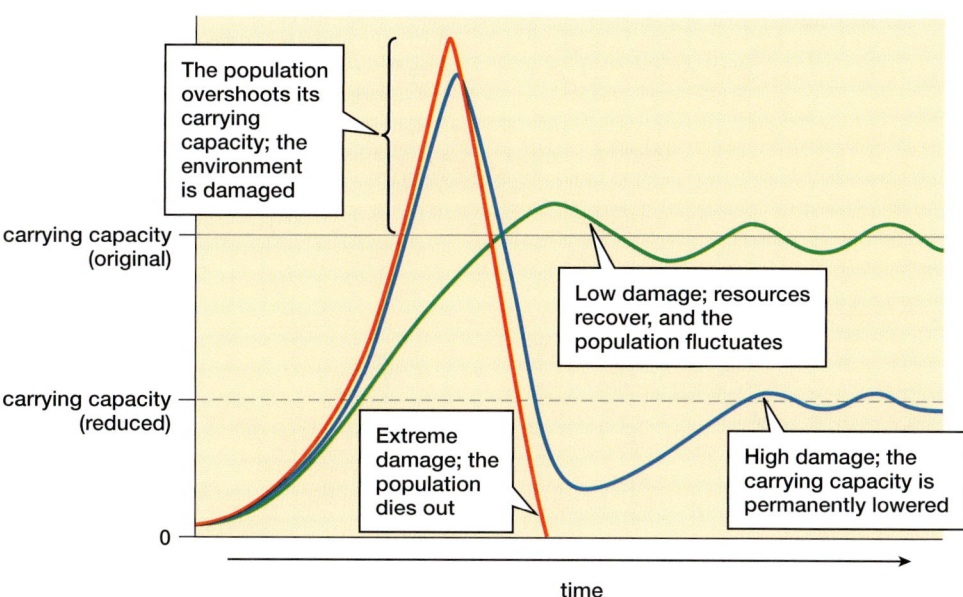

The population
overshoots its
carrying
capacity; the
environment
is damaged

carrying capacity
(original)

Low damage; resources
recover, and the
population fluctuates

carrying capacity
(reduced)

Extreme
damage; the
population
dies out

High damage; the
carrying capacity is
permanently lowered

0

time

(b) Consequences of exceeding carrying capacity

experiments using fruit flies have shown that competition for resources can control population size by reducing both the birth rate and the average life span of the competing flies. During logistic population growth, as environmental resistance increases, population growth slows and eventually stops at approximately the carrying capacity of the environment. In nature, conditions are never completely stable, so carrying capacity and population sizes within a community will vary somewhat from year to year.

Two forms of environmental resistance usually maintain populations at or below the carrying capacity of their environment. **Density-independent** factors limit population size regardless of the population density (the number of individuals per unit of area). **Density-dependent** factors, in contrast, increase in effectiveness as the population density increases. Nutrients, energy, and space, the primary determinants of carrying capacity, are all density-dependent regulators of population size. In the following

sections, we will look more closely at how these factors limit population growth.

Density-Independent Factors Limit Populations Regardless of Density

The most important natural density-independent factors are climate and weather, which are responsible for most boom-and-bust population cycles. Many insects and annual plant populations are limited in size by the number of individuals that can be produced before the first hard freeze. Such populations are controlled by the climate because they typically do not reach carrying capacity before winter sets in. Weather can also significantly alter natural populations from year to year. Hurricanes, droughts, floods, and fire can have profound effects on local populations, regardless of density.

Human activities limit the growth of natural populations in density-independent ways. Pesticides and pollutants can cause dramatic declines in natural populations. Before

Health Watch

Boom-and-Bust Cycles Can Be Bad News

Although documented since Biblical times, during the past three decades, population explosions of sometimes toxic, single-celled photosynthetic protists and bacteria have occurred with increasing frequency throughout the world. Collectively called *harmful algal blooms (HABs)*, these population booms of toxic microorganisms kill fish and sicken people. They also cause major economic losses to the shellfish industry because clams, mussels, and scallops feed on these organisms and concentrate the poisons in their bodies, posing a hazard to consumers. Some of the most damaging poisons are neurotoxins produced by dinoflagellates (protists), such as *Karenia brevis* (Fig. E26-1), which can reach densities of 20 million per liter of water. This and other protist species can cause red tides (see Fig. 20-11) that result in massive fish kills, often in late summer.

Many of the bacteria and protists that produce harmful algal blooms are natural residents of lakes and coastal waters. What causes these populations to "bloom and boom"? The

▲ FIGURE E26-1 **One cause of red tides** The dinoflagellate *Karenia brevis* (seen in this artificially colored SEM) causes red tides in coastal waters of Florida and in the Gulf of Mexico.

reasons are complex and vary with the species, but warm water temperatures and adequate nutrients, such as phosphorus and nitrogen, are required. Runoff of these nutrients from human agricultural activities is increasing the frequency and intensity of HABs throughout the world. Global climate change may be contributing to the problem because warmer waters foster more rapid growth of these protists and also extend their growing season. The booming populations go "bust" when the enormous populations of cells deplete the local water of nutrients, and falling water temperatures in the autumn and winter further decrease their reproductive rate.

Government agencies are using satellite images to track changes in water color caused by HABs and are funding research into the causes of these explosive blooms of microorganisms, with the goal of accurately forecasting these events and safely suppressing them.

*Data not taken for 1943–1946

▲ FIGURE 26-6 **The effects of exceeding carrying capacity** In 1911, the U.S. government introduced 25 reindeer onto St. Paul Island off the Alaskan coast to provide a continuing meat supply for the island's residents. With abundant food and no predators, the herd grew exponentially (initial J-curve). In 1938, the herd had increased to 2,046 reindeer, roughly three times the island's estimated carrying capacity. The lichen that provide reindeer food during the winter months were seriously overgrazed and unable to recover under the conditions of intense browsing imposed by the large reindeer population. By 1950, only 8 reindeer remained.

▲ FIGURE 26-7 **A logistic curve in nature** Barnacles are crustaceans whose larvae are carried in ocean currents to rocky seashores where they settle, attach permanently to rock, and grow into the shelled adult form. On a bare rock, the number of settling larvae and newly metamorphosed juveniles produce a logistic growth curve as competition for space limits their population density.

it was banned in the 1970s, the pesticide DDT drastically reduced populations of predatory birds, including eagles, ospreys, and pelicans; a variety of pollutants continue to adversely affect wildlife (as you will learn in Chapter 28). Although well-regulated hunting can help maintain animal populations in a healthy balance with available resources, overhunting has driven some animal species to extinction. Examples from the United States are the once-abundant

IN GREATER DEPTH A Look at Logistic Population Growth

To understand the S-curve of logistic growth, first review the formula on p. 489 for exponential growth ($G = r \times N$). The logistic growth equation starts with this formula but multiplies it by the factor $(K - N)/K$ that models environmental resistance. This new factor increasingly limits growth as the population size approaches the carrying capacity of its local environment.

The equation for logistic population growth is:

$$G = rN\frac{(K - N)}{K}$$

Let's start with the value $(K - N)$. When we subtract the current population (N) from the carrying capacity (K), we get the number of individuals that can still be added to the current population before it reaches carrying capacity. Now, if we divide this new number by K, we get the fraction of the carrying capacity that can still be added to the current population before it reaches carrying capacity and stops growing (when $G = 0$). For example, consider a population of 10 foxes, with an annual growth rate of 1%, introduced onto an island whose carrying capacity for foxes is 1,000. The initial population ($N = 10$) is small, so the limiting factor $(K - N)/K$ is $(1,000 - 10)/1,000$ or 0.99. Because 0.99 is extremely close to 1.0, this limiting factor has almost no effect

on the exponential growth formula, and will continue to have little impact on population growth for many years. But as N increases over time, $K - N$ will eventually approach zero. (So when there are 900 foxes, $(1,000 - 900)/1,000 = 0.1$, which will greatly reduce annual growth). As G becomes increasingly small, the curve begins to level off. When N equals K, then $(K - N) = 0$, and population growth will cease because $G = rN$ multiplied by $(0)/K = 0$, as illustrated by the final horizontal portion of the logistic curve (see Fig. 26-5a). So when our fox population reaches the island's carrying capacity of 1,000, it will stop increasing and stabilize at about this number.

CASE STUDY continued

The Mystery of Easter Island

There are no witnesses to the events on Easter Island, but if an expanding human population exceeded the island's carrying capacity, the increasing demand for wood and cropland would have contributed to the destruction of the island's forests. This would have reduced or eliminated habitats for native species, which would also have been increasingly hunted for food, driving many species to extinction. Deforestation is also likely to change the local climate, making it hotter and drier and less able to support life. The barren landscape that replaced the lush forests of Easter Island is a dramatic example of what may happen when overpopulation permanently and dramatically reduces the capacity of a region to sustain people and other forms of life. Islands are particularly vulnerable to such drastic events, partly because their populations cannot emigrate. For Earth's expanding human population, however, the entire planet is an island.

passenger pigeon and the colorful Carolina parakeet, neither of which will ever be seen again. In addition, habitat destruction by humans, a density-independent factor, is the single greatest threat to wildlife worldwide; in the United States, this has almost certainly driven the magnificent ivory-billed woodpecker to extinction.

Density-Dependent Factors Become More Effective as Population Density Increases

Populations of organisms with a life span of more than a year have evolved adaptations that allow them to survive density-independent controls imposed by seasonal changes, such as cold and lack of food during winter. Many mammals, for example, develop thick coats and store fat for the winter; some also hibernate. Migration is another coping mechanism; many birds migrate long distances to find food and a hospitable climate. Most trees and bushes survive the

rigors of winter by entering a period of dormancy, dropping their leaves and drastically slowing their metabolic activities.

For long-lived species in undisturbed habitats, the most important elements of environmental resistance are density dependent. Density-dependent factors exert a negative feedback effect on population size, because they become increasingly effective as the population density increases.

Predators Exert Density-Dependent Controls on Populations Predators are organisms that eat other organisms, called their **prey.** Often, prey are killed directly and eaten (**Fig. 26-8a**), but not always. When deer browse on the buds of maple trees, for example, or when gypsy moth larvae feed on the leaves of oaks, the trees are harmed but not killed.

Predation becomes increasingly influential as prey populations grow, because most predators eat a variety of prey, depending on what is most abundant and easiest to find. Coyotes might eat more field mice when the mouse population is high, but switch to eating more ground squirrels as the mouse population declines. In this way, predators often exert density-dependent population control over more than one prey population.

Predator populations often grow as their prey becomes more abundant, which make them even more effective as control agents. For predators such as the arctic fox and snowy owl, which rely heavily on lemmings for food, the number of offspring produced is determined by the abundance of prey. Snowy owls (**Fig. 26-8b**) hatch up to 12 chicks when lemmings are abundant, but may not reproduce at all in years when the lemming population has crashed.

In some cases, an increase in predators might cause a dramatic decline in the prey population, which in turn may result in an eventual decline in the predator population. This pattern can result in out-of-phase **population cycles** of predators and prey. In natural ecosystems, both predators and prey are subjected to a variety of other influences, so

(a) Predators often kill weakened prey

(b) Predator populations often increase when prey are abundant

▲ FIGURE 26-8 **Predators help control prey populations**
(a) A pack of grey wolves has brought down an elk that may have been weakened by age or parasites. **(b)** The snowy owl produces more chicks when prey (such as lemmings) are abundant.

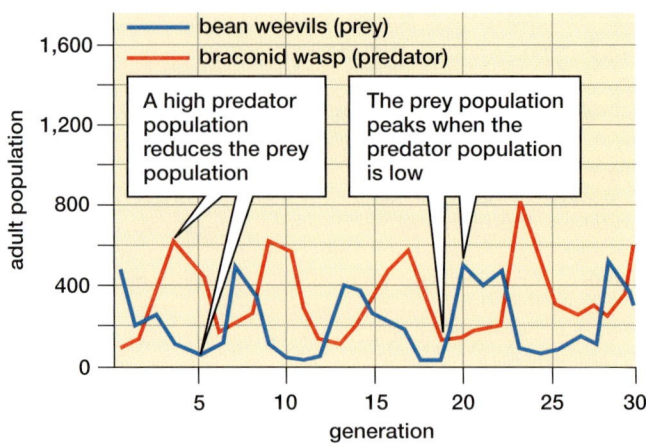

▲ FIGURE 26-9 **Experimental predator–prey cycles**
Tiny braconid wasps lay their eggs on bean weevil larvae, which provide food for the newly hatched wasp larvae. A large weevil population ensures a high survival rate for wasp offspring, increasing the predator population. Then, under intense predation, the weevil population plummets, reducing the food available to the next generation of wasps, whose population declines as a result. Reduced predation then allows the weevil population to increase rapidly, and so on.

clear-cut examples of such cycles in nature are rare. However, out-of-phase population cycles of predators and their prey have been demonstrated under controlled laboratory conditions (Fig. 26-9).

Predators may contribute to the overall health of prey populations by culling those individuals that are poorly adapted, weakened by age, or unable to find adequate food and shelter. In this way, predation may maintain healthy prey populations near a density that can be sustained by the resources of the ecosystem.

Parasites Spread More Rapidly Among Dense Populations A **parasite** feeds on a larger organism, its **host,** harming it. Although some kill their hosts, many parasites benefit by having their host remain alive. Parasites include tapeworms that live in the intestines of mammals, ticks that cling to the host's skin, and disease-causing microorganisms. Most parasites cannot travel long distances, so they spread more readily among hosts in dense populations. For example, plant diseases spread readily through acres of densely

planted crops, and childhood diseases spread rapidly through schools and day-care centers. Even when parasites do not kill their hosts directly, they influence population size by weakening their hosts and making them more susceptible to death from other causes, such as harsh weather or predators. Organisms weakened by parasites are also less likely to reproduce. Parasites, like predators, more often contribute to the death of less-fit individuals, producing a balance in which the host population is regulated but not eliminated.

This balance can be destroyed if parasites or predators are introduced into regions where local prey species have had no opportunity to evolve defenses against them. With little environmental resistance, these invasive species reproduce explosively to the detriment of their hosts or prey. The smallpox virus, inadvertently carried by traveling Europeans during colonial times, caused heavy losses of life among native inhabitants of North America, Hawaii, South America, and Australia. The chestnut blight fungus, introduced from Asia, has almost eliminated wild chestnut trees from U.S. forests. Introduced rats almost certainly contributed to the loss of most native birds on Easter Island, and introduced rats and mongooses have exterminated several of Hawaii's native bird populations.

Competition for Resources Helps Control Populations
The resources that determine carrying capacity—space, energy, and nutrients—may be inadequate to support all the organisms that need them. **Competition,** the interaction among individuals who attempt to use the same limited resource, limits population size in a density-dependent manner. There are two major forms of competition: **interspecific competition** (competition among individuals of different species) and **intraspecific competition** (competition

(a) Gypsy moths laying eggs

(b) Gypsy moth caterpillars

◀ FIGURE 26-10 **Scramble competition (a)** Gypsy moths gather on tree trunks to lay egg masses. **(b)** Hundreds of caterpillars will hatch from each egg mass and compete with one another.

among individuals of the same species). Because the needs of members of the same species for water and nutrients, shelter, breeding sites, light, and other resources are almost identical, intraspecific competition is an important density-dependent mechanism of population control.

Organisms have evolved several ways to deal with intraspecific competition. Some organisms, including most plants and many insects, engage in **scramble competition,** a kind of free-for-all with resources as the prize. For example, gypsy moth females each lay a mass of up to 1,000 eggs on tree trunks in eastern North America. As the eggs hatch, armies of caterpillars crawl up the tree (Fig. 26-10). Huge outbreaks of this invasive species can completely strip large trees of their leaves in a few days. Under these conditions, competition for food may be so great that most of the caterpillars die before they can metamorphose into egg-laying moths. Another example of scramble competition is seen in plants, whose seeds may sprout in dense clusters. As they grow, those that sprouted first begin to shade the smaller ones, and their larger root systems absorb much of the available water from the soil, causing the younger ones to wither and die.

Many animals (and even a few plants) have evolved **contest competition,** in which social or chemical interactions determine access to important resources. Territorial species, including wolves, many fish, rabbits, and songbirds, defend an area that contains important resources such as food or places to raise their young. When the population exceeds the size that can be sustained by available resources, only the best-adapted individuals are able to defend territories that supply adequate food and shelter. Those without territories may not reproduce (reducing the future population), or they may fail to obtain adequate food or shelter and become easy prey for predators.

As population densities increase and competition becomes more intense, some types of animals (including lemmings and locusts) react by emigrating. Emigrating swarms of locusts periodically plague parts of Africa, consuming all vegetation in their path (Fig. 26-11).

Density-Independent and Density-Dependent Factors Interact to Regulate Population Size

The size of a population at any given time is the result of complex interactions between density-independent and density-dependent forms of environmental resistance. For example, a stand of pines weakened by drought (a density-independent

▲ FIGURE 26-11 **Emigration** In response to overcrowding and lack of food, locusts emigrate in swarms, devouring nearly all vegetation (and even feeding on each other) as they go.

QUESTION What benefits does mass emigration give to animals such as locusts or lemmings? What parallels can you find to recent human emigrations?

factor) may more readily fall victim to the pine beetle (density dependent). Likewise, a caribou weakened by hunger (density dependent) and attacked by parasites (density dependent) is more likely to be killed by an exceptionally cold winter (density independent).

Human activities are increasingly imposing density-independent limitations on natural populations. Examples include bulldozing grasslands and their prairie dog towns to build shopping malls and housing tracts or felling rain forests to replace them with croplands. These activities also reduce the carrying capacity of the environment, which in turn exerts density-dependent limits on future population sizes.

CHECK YOUR LEARNING

Are you able to describe exponential growth, the conditions under which it occurs, and what shape of curve it produces when graphed? Can you explain the stages of the logistic growth curve and describe the possible outcomes when populations overshoot carrying capacity? Can you explain the two major forms of environmental resistance and provide examples of each?

26.3 HOW ARE POPULATIONS DISTRIBUTED IN SPACE AND AGE?

Populations of different types of organisms show characteristic spacing of their members, determined by their behavioral characteristics and their environments. In addition, each population exhibits patterns of reproduction and survival that are characteristic of its species.

Populations Exhibit Different Spatial Distributions

Spatial distribution describes how individuals within a population are distributed within a given area. Spatial distribution may vary with time, changing with the breeding season, for example. Ecologists recognize three major types of spatial distribution: clumped, uniform, and random (**Fig. 26-12**).

Populations whose members live in groups exhibit a **clumped distribution.** Examples include family or social groupings, such as elephant herds, wolf packs, prides of lions, flocks of birds, and schools of fish (**Fig. 26-12a**). What are the advantages of clumping? Birds in flocks benefit from many eyes to spot food, such as a tree full of fruit. Schooling fish and flocks of birds also may confuse predators with their sheer numbers. Predators, in turn, sometimes hunt in groups, cooperating to bring down larger prey (see Fig. 26-8a). Some species form temporary groups to mate and care for their young. Other plant or animal populations cluster because resources are localized. In grasslands, for example, stands of cottonwood trees cluster along the banks of streams.

Organisms with a **uniform distribution** maintain a relatively constant distance between individuals. This is most common among animals that exhibit territorial behaviors that evolved to maintain their access to limited resources. Territorial behavior is more common among animals during their breeding seasons. Seabirds may space

(a) Clumped distribution

(b) Uniform distribution

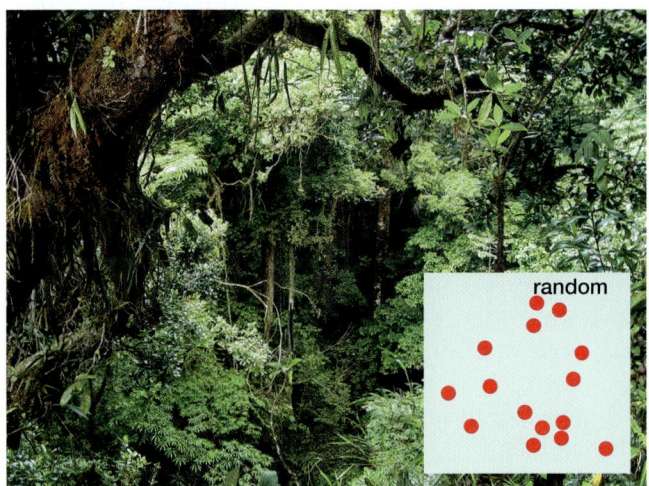

(c) Random distribution

▲ **FIGURE 26-12 Spatial distribution in populations**
(a) A school of fish may confuse predators with their numbers.
(b) These gannets occupy evenly spaced nests along the ocean shore. **(c)** Because of uniformly good growing conditions, many rain-forest plants grow wherever their seeds fall.

their nests evenly along the shore, just out of reach of one another (**Fig. 26-12b**). Among plants, mature desert creosote bushes are often spaced very evenly. Research has shown that this spacing results from competition among their root systems, which occupy a roughly circular area around each plant. The roots efficiently absorb water and other nutrients from the soil, preventing survival of bushes that germinate close by.

Organisms with a **random distribution** are relatively rare. Such individuals do not form social groups. The resources they need are more or less equally available throughout the area they inhabit, and those resources are not scarce enough to require territorial spacing. Trees and other plants in rain forests come close to being randomly distributed (**Fig. 26-12c**). There are probably no vertebrate species that maintain a random distribution throughout the year; most interact socially, at least during the breeding season.

Populations Exhibit Different Age Distributions

Animals of different species differ considerably in their chances of dying at any given phase of their life cycle. Some species produce many offspring that are provided with very few resources; most of these offspring die before they can reproduce. Others produce few offspring, which are each given far more resources and often survive to reproduce. To determine the pattern of survivorship, researchers construct **survivorship tables,** which track groups of organisms (born at the same time) throughout their lives, recording how many survive in each succeeding year (or other unit of time; **Fig. 26-13a**). If these numbers are graphed, they reveal the **survivorship curves** characteristic of the species in the particular environment where the data were collected. Three types of survivorship curves—described as late loss, constant loss, and early loss, according to the part of the life cycle during which most deaths occur—are shown in **Figure 26-13b**.

Late-loss populations produce convex survivorship curves. Such populations have relatively low juvenile death rates, and most individuals survive to old age. Late-loss survivorship curves are characteristic of humans and other large and long-lived animals such as elephants and mountain sheep. These species produce relatively few offspring, which are protected and nourished by their parents during early life.

Constant-loss populations produce straight-line survivorship curves and consist of individuals that have an equal chance of dying at any time during their life span. This pattern is seen in some birds such as gulls and the American robin, in some species of turtles, and in laboratory populations of organisms that reproduce asexually, such as hydra and bacteria.

Early loss populations exhibit concave survivorship curves. These curves are characteristic of organisms that produce large numbers of offspring that receive little or no care or resources from the parents after they hatch or germinate. Many of these species engage in scramble competition for resources early in life. The death rate is very high among the young, but those individuals that reach adulthood have a reasonable chance of surviving to old age. Most invertebrates, many fish and amphibians, and most plants exhibit early loss survivorship curves. A female oyster, for example, may release millions of eggs into the ocean each year, a cane toad may lay up to 35,000 eggs, and a single cattail stalk may send 250,000 seeds to the winds. None of these offspring receive parental care and nearly all die.

CHECK YOUR LEARNING

Can you explain the three basic types of distribution within populations? Can you graph and describe the three general types of survivorship curves?

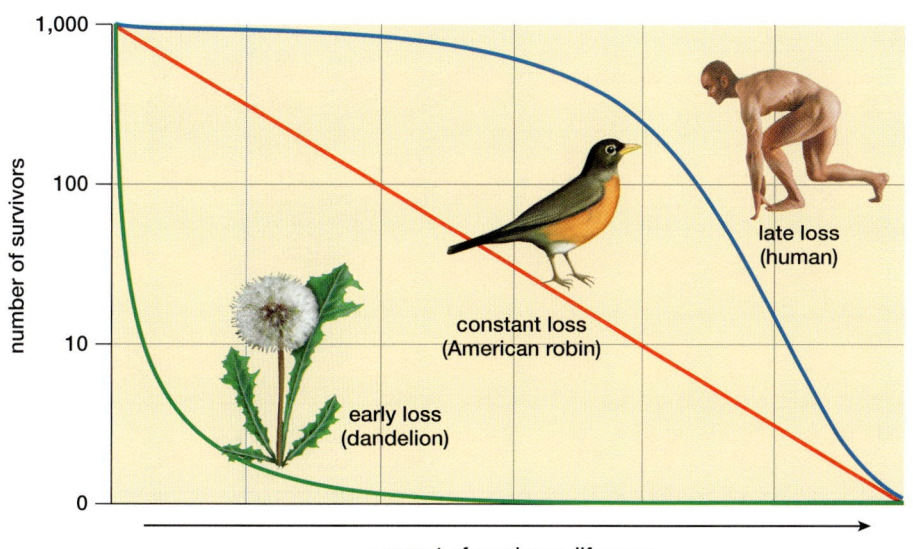

Age	Number of survivors
0 (birth)	100,000
10	99,124
20	98,713
30	97,754
40	96,489
50	93,698
60	87,967
70	76,241
80	54,117
90	22,312
100	2,523

(a) A survivorship table

(b) Survivorship curves

▲ FIGURE 26-13 **Survivorship tables and survivorship curves** (a) A survivorship table for the U.S. population in 2004, showing how many people are expected to remain alive at increasing ages for each 100,000 people born. Plotting these data will produce a curve similar to the blue curve in part (b). (b) Three types of survivorship curve are shown. Data in part (a) from Arias, E. 2007. United States life tables, 2004. *National Vital Statistics Reports 56*(9). Hyattsville, MD: National Center for Health Statistics.

26.4 HOW IS THE HUMAN POPULATION CHANGING?

No force on Earth rivals that exerted by humans. We possess enormous brainpower and dexterous hands that can shape the environment to our demands. As our species was evolving, natural selection favored those with the ability and the drive to bear and nurture many offspring, which helped ensure that a few would survive. Ironically, this characteristic now threatens us and the biosphere on which we depend.

The Human Population Continues to Grow Rapidly

Compare the graph of human population growth in **Figure 26-14** with the exponential growth curves in Figures 26-1 and 26-2. The time spans are different, but each has the J-curve characteristic of exponential growth. The human population initially grew slowly; it took roughly 200,000 years to reach 1 billion. In the table within Figure 26-14, note the decreasing amount of time required to add billions of people, a characteristic of exponential growth. But also notice that we have been adding billions at a relatively constant rate since the 1970s. This suggests that, although the human population continues to grow rapidly, it may no longer be growing exponentially. Are humans starting to enter the final bend of the

S-shaped logistic growth curve (see Fig. 26-5a) that will eventually lead to a stable population? Time will tell. However, despite the fact that our annual growth rate (rate of natural increase) has declined from 1.8% in 1960 to 1.2% in 2011, Earth's human population is adding people faster than ever in history. Having reached 7 billion in late 2011, our numbers now grow by about 83 million each year; this is more than 227,000 every day. Why hasn't environmental resistance put an end to our continued population growth?

In contrast to non-human populations, people have responded to environmental resistance by devising ways to overcome it. To accommodate our growing numbers, we have altered the face of the globe. Is there an ultimate limit to Earth's carrying capacity? Have we already reached or possibly even exceeded it? We explore these questions in "Earth Watch: Have We Exceeded Earth's Carrying Capacity?"

A Series of Advances Has Increased Earth's Capacity to Support People

Human population growth has been spurred by a series of advances, each of which circumvented some type of environmental resistance, increasing Earth's carrying capacity for people. Early humans discovered fire, invented tools and weapons, built shelters, and designed protective clothing, a series of *technical advances* that increased carrying capacity.

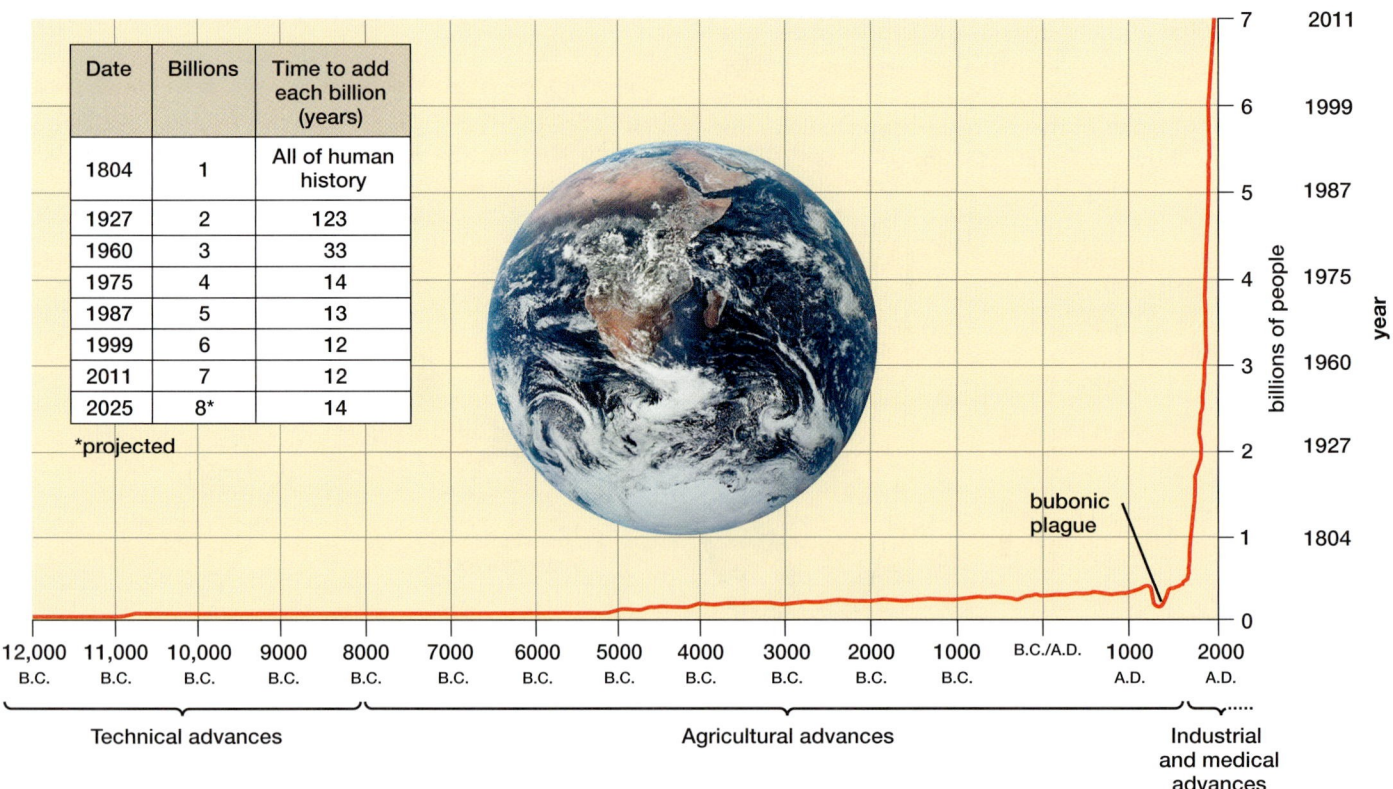

Date	Billions	Time to add each billion (years)
1804	1	All of human history
1927	2	123
1960	3	33
1975	4	14
1987	5	13
1999	6	12
2011	7	12
2025	8*	14

*projected

Technical advances | Agricultural advances | Industrial and medical advances

▲ **FIGURE 26-14 Human population growth** The human population from the Stone Age to the present has shown continued exponential growth as various advances overcame environmental resistance. Note the time intervals over which additional billions were added. (Inset) Earth is an island of life in a sea of emptiness; its space and resources are limited. Photo courtesy of NASA.

QUESTION What do you think this curve will look like when we reach the year 2200? 2500? Explain.

Earth|Watch

Have We Exceeded Earth's Carrying Capacity?

In Côte d'Ivoire, a country in western Africa, the government is waging a battle to protect some of its rapidly dwindling tropical rain forest from illegal hunters, farmers, and loggers. Officials destroy the shelters of the squatters, who immediately return and rebuild. One such squatter, Sep Djekoule, explained: "I have ten children and we must eat. The forest is where I can provide for my family, and everybody has that right." His words exemplify the conflict between population growth and environmental protection, between the human drive to have many children and Earth's finite resources. How many people can Earth sustain?

The Global Footprint Network, consisting of an international group of scientists and professionals from many fields, is attempting to assess humanity's *ecological footprint* (see Chapter 30). This project compares human demand for resources to Earth's capacity to supply these resources in a sustainable manner. "Sustainable" means that the resources could be renewed indefinitely, and the ability of the biosphere to supply them would not be diminished over time. Their question boils down to: "Is humanity living on the 'interest' produced by our global endowment, or are we eating into the 'principal' and so reducing its ability to support us?" The group concluded that, in 2007, humanity was already consuming 150% of the resources that were sustainably available. In other words, to avoid damaging Earth's resources (thus reducing Earth's carrying capacity), our population in 2007 required 1.5 Earths. But by 2012, we had added well over 300 million more people. A population that exceeds carrying capacity damages its ecosystem, reducing its future ability to sustain that population. Because people have used their technological prowess to overcome environmental resistance, our collective ecological footprint now tramples Earth's sustainable resource base, reducing Earth's future capacity to support us. For example, each year, overgrazing and deforestation decrease the productivity of land, especially in *developing countries* where average living standards, educational levels, and life expectancy are relatively low. The human population now uses more than 37% of Earth's total land area to grow crops and livestock. Despite this, the United Nations estimates that more than 900 million people are undernourished, including an estimated 30% of the population of sub-Saharan Africa (**Fig. E26-2**, inset). Erosion reduces the ability of land to support both crops and grazing livestock (**Fig. E26-2**). The quest for farmland drives people to clear-cut forests in places where the soil is poorly suited for agriculture. The demand for wood also causes large areas to be deforested annually, causing the runoff of much-needed fresh water, the erosion of precious topsoil, the pollution of rivers and oceans, and an overall reduction in the ability of the land and water to support future crops and livestock, and to support the fish and other wild animals that people harvest for food. Our consumption of food, wood, and—more recently—biofuels (crops that provide fuel) drives the destruction of tens of millions of acres of rain forest annually, causing the extinction of species on an unprecedented scale (see Chapter 30).

The United Nations estimates that more than 30% of commercial ocean fish populations are being harvested unsustainably, and more than 50% more are being harvested at their maximum sustainable yield. In parts of India, China,

▲ **FIGURE E26-2 Overgrazing can lead to the loss of productive land** Human activities, including overgrazing, deforestation, and poor agricultural practices reduce the productivity of the land. (Inset) An expanding human population, coupled with a loss of productive land, can lead to tragedy.

Africa, and the United States, underground water stores are being depleted to irrigate cropland far faster than they are being refilled by natural processes. Because irrigated land supplies about 40% of human food crops, water shortages can rapidly lead to food shortages.

Our present population, at its present level of technology, is clearly "overgrazing" the biosphere. As the 5.7 billion people in less-developed countries strive to increase their standard of living, the damage to Earth's ecosystems accelerates. We all want to enjoy luxuries far beyond bare survival, but unfortunately, the resources currently demanded to support the high standard of living in developed countries are unattainable for most of Earth's inhabitants. For example, supporting the world population sustainably at the average standard of living in the United States would require 4.7 Earths. Technology can help us improve agricultural efficiency, conserve energy and water, reduce pollutants, and recycle far more of what we use. But technological advances are expensive to develop and difficult for many developing countries to implement. In the long run, no amount of technological innovation will compensate for our biotic potential, which we must restrain if we expect Earth to continue to support us.

Inevitably, the human population will stop growing. Either we must voluntarily reduce our birth rate or various forms of environmental resistance, including disease and starvation, will dramatically increase human death rates. Hope for the future lies in recognizing the signs of human overgrazing and responding by reducing our population before we cause further damage to the biosphere, diminishing its ability to support people and the other precious and irreplaceable forms of life on Earth.

Tools and weapons allowed them to hunt more effectively and obtain additional high-quality food, while shelter and clothing expanded the habitable areas of the globe.

Domesticated crops and animals had supplanted hunting and gathering in many parts of the world by about 8000 B.C. These *agricultural advances* provided people with a larger, more dependable food supply, further increasing Earth's carrying capacity for humans. An increased food supply resulted in a longer life span and more childbearing years, but a high death rate from disease continued to restrain population growth. Human population growth continued slowly for thousands of years until major *industrial and medical advances* permitted a population explosion. These advances began in England in the mid-eighteenth century, and spread throughout Europe and North America during the nineteenth and into the twentieth century. Medical progress dramatically decreased the death rate by reducing environmental resistance caused by disease. The discovery of bacteria and their role in infection resulted in better control of bacterial diseases through improved sanitation and, later, antibiotics. Vaccines for diseases such as smallpox reduced deaths from viral infections.

The Demographic Transition Explains Trends in Population Size

Today, countries are often described as either developed or developing. People in **developed countries**—including Australia, New Zealand, Japan, and countries in North America and Europe—benefit from a relatively high standard of living, with access to modern technology and medical care, including readily available contraception. Average income is relatively high, education and employment opportunities are available to both sexes, and death rates from infectious diseases are low. But fewer than 20% of the world's people live in developed countries. In the **developing countries** of Central and South America, Africa, and much of Asia—home to more than 80% of humanity—the average person lacks these advantages.

The historical rate of population growth in developed countries has changed over time in reasonably predictable stages, producing a pattern called the **demographic transition** (Fig. 26-15). Before major industrial and medical advances occurred, today's developed countries were in the *pre-industrial stage*, with relatively small and stable populations whose high birth rates were balanced by high death rates. This was followed by the *transitional stage*, in which food production increased and health care improved. These advances caused death rates to fall, while birth rates remained high, leading to an explosive rate of natural increase. During the *industrial stage*, birth rates fell as more people moved from small farms to cities (where children were less important as a source of labor), contraceptives became more readily available, and opportunities for women to work outside the home increased. Most developed countries are now in the *post-industrial stage* of the demographic transition, and, with the exception of the United States (which we discuss later), their populations are relatively stable, with low birth and death rates.

A population's *fertility rate* reflects the average number of children that each woman bears. If immigration and emigration rates are balanced, a population will eventually stabilize if parents, on average, have just the number of children required to replace themselves; this is called **replacement level fertility (RLF).** Replacement level fertility is 2.1 children per woman (rather than exactly 2) because not all children survive to maturity.

World Population Growth Is Unevenly Distributed

In developing countries, such as most countries in Central and South America, Asia (excluding China and Japan), and Africa, medical advances have decreased death rates and increased life span, but birth rates remain relatively high. Although China is also a developing country, years ago, as its population approached 1 billion, the Chinese government recognized the

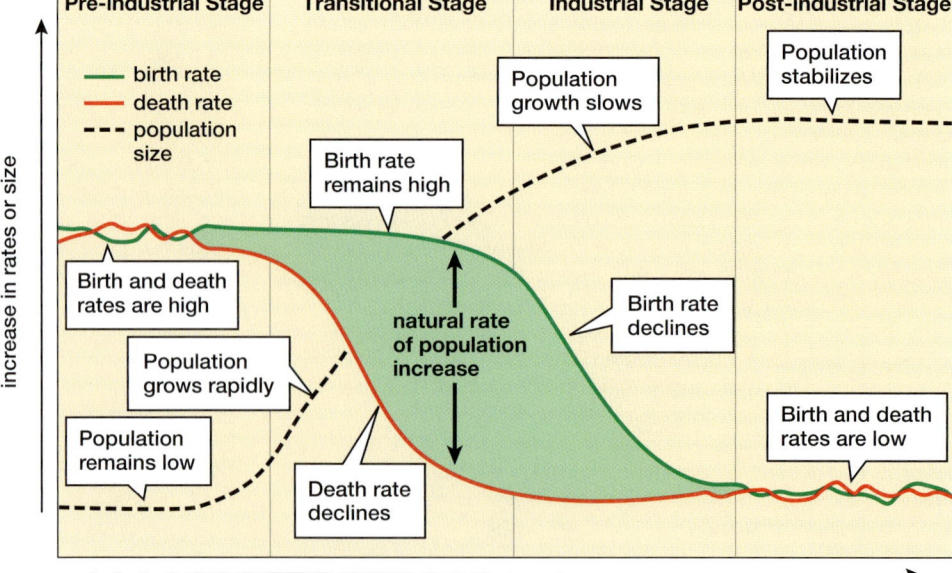

▶ FIGURE 26-15 **The demographic transition** A demographic transition typically begins with a relatively stable and small size with high birth and death rates. Death rates decline first, causing a population increase. Then birth rates decline, causing the population to stabilize at a higher number with relatively low birth and death rates.

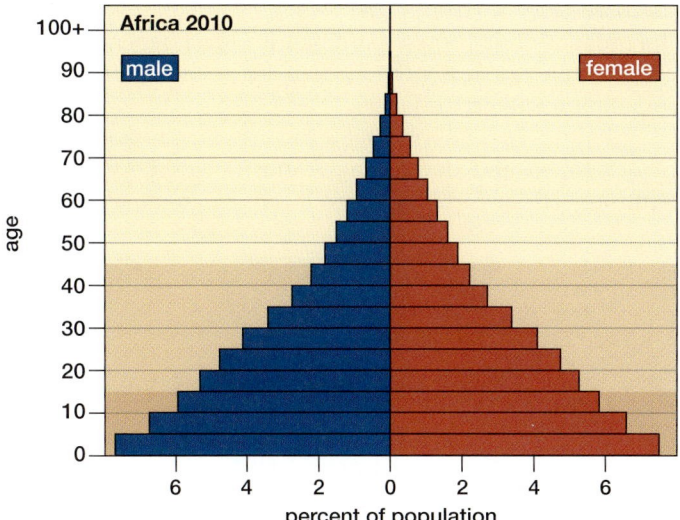

(a) Africa: A rapidly growing population

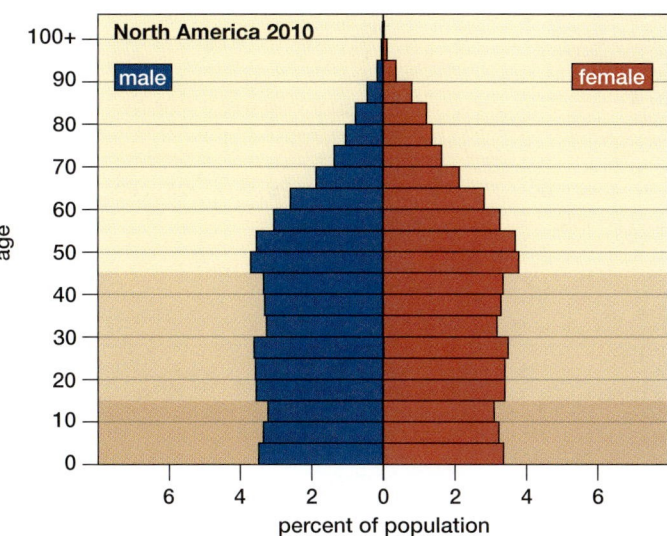

(b) North America: A slowly growing population

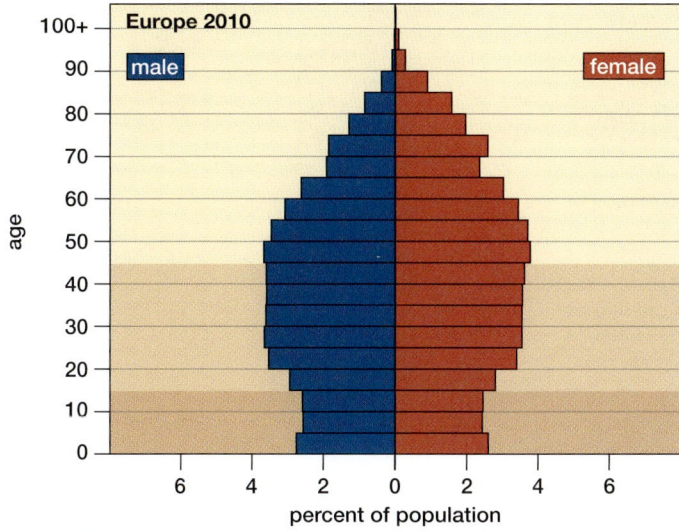

(c) Europe: A slowly declining population

negative impacts of continued population growth and instituted social reforms (many of them quite punitive) that brought China's fertility rate below replacement level.

Most other developing countries are within the late-transitional or the industrial stage of the demographic transition. In many of these nations, adult children provide financial security for aging parents. Young children may also contribute significantly to the family income by working on farms or sometimes in factories. Social factors drive population growth in countries where children confer prestige and where religious beliefs promote large families. Also, in developing countries, many individuals who would like to limit their family size lack access to contraceptives. For example, in the West African nation of Nigeria, only 10% of couples use modern contraceptive methods, and the average woman bears 5.7 children. Nigeria is suffering from soil erosion, water pollution, and the loss of forests and wildlife, suggesting that its carrying capacity is already compromised. With 43% of its 162 million people under the age of 15, however, continued population growth is inevitable, for reasons described later.

As in Nigeria, population growth is highest in the countries that can least afford it, as a result of a type of positive feedback. As more people compete for the same limited resources, poverty increases. Poverty diverts children away from schools and into activities that help support their families. A lack of education and lack of access to contraceptives then contributes to continued high birth rates. Of the 7 billion people on Earth in 2011, about 5.8 billion resided in developing countries. Although fertility rates in some developing countries, such as Brazil, have declined because of social changes and increased access to contraceptives, the prospects for world population stabilization in the near future are almost nonexistent.

The Age Structure of a Population Predicts Its Future Growth

Age structure diagrams show age groups on the vertical axis and the numbers (or percentages) of individuals in each age group on the horizontal axis, with males and females shown on opposite sides. Age structure diagrams all rise to a peak that reflects the maximum human life span, but the shape of the rest of the diagram reveals whether the population is expanding, stable, or shrinking. If adults in the reproductive age group (15 to 44 years) are having more children (the 0- to 14-year age group) than are needed to replace themselves, the population is above RLF and is expanding; its age structure will be roughly triangular (**Fig. 26-16a**). If the

◀ **FIGURE 26-16 Age structure diagrams (a)** The age structure for Africa illustrates a rapidly growing population that is projected to at least double by 2050. **(b)** North America represents a more slowly growing population that is still projected to add 100 million by 2050. **(c)** Europe's age structure is of a slowly shrinking population, projected to decline by 19 million by 2050. Background colors from bottom to top indicate age groups: prereproductive children (0 to 14 years), reproductive age adults (15 to 44 years), and postreproductive adults (45 to 100 years). Data from United Nations, Department of Economic and Social Affairs of the United Nations Secretariat, Population Division. *World Population Prospects: The 2010 Revision; medium-level projection.*

adults of reproductive age have just the number of children needed to replace themselves, the population is at RLF. A population that has been at RLF for many years will have an age structure diagram with relatively straight sides (**Fig. 26-16b**). In shrinking populations, the reproducing adults have fewer children than are required to replace themselves, causing the age structure diagram to narrow at the base (**Fig. 26-16c**). The *median age* (age at which half the population is younger and half is older) depends on the age structure; the lower the median age, the more rapidly the population will expand.

Figure 26-17 shows age structures for the populations of developed and developing countries for the year 2010, with projections for 2050. Even if rapidly growing countries

(a) Developed countries

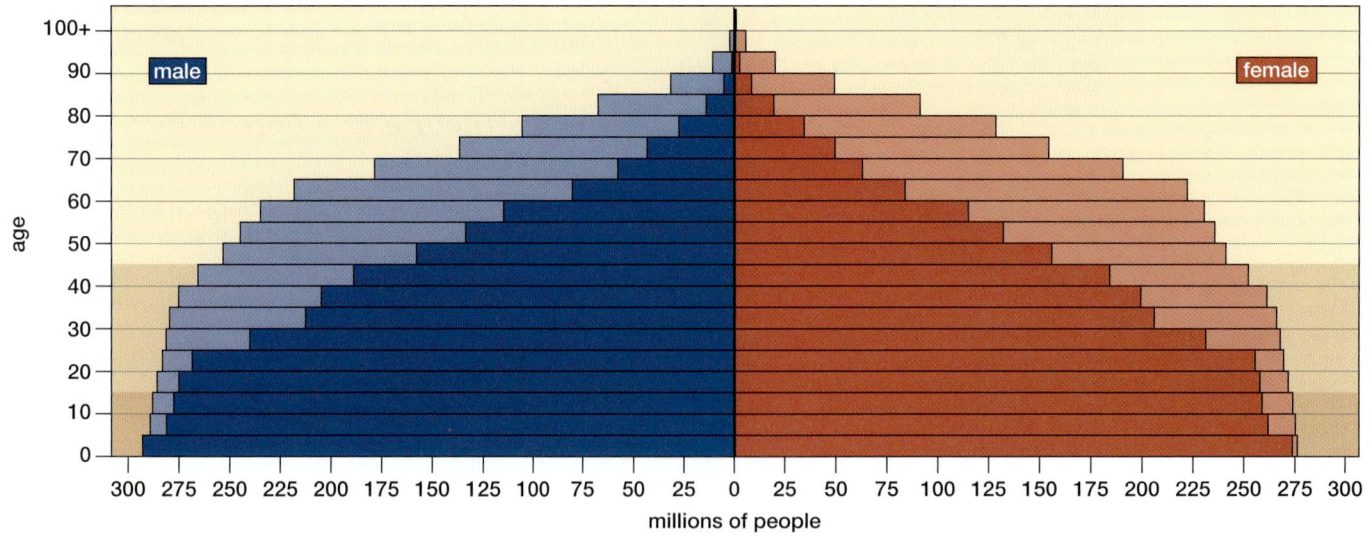

(b) Developing countries

▲ **FIGURE 26-17 Age structure diagrams of developed and developing countries**
Note that the predicted difference in the number of children compared to the number of adults in developing countries is smaller in 2050, as these populations approach RLF. The large numbers of young people who will be entering their childbearing years will cause continued growth. Data from United Nations, Department of Economic and Social Affairs of the United Nations Secretariat, Population Division. *World Population Prospects: The 2010 Revision; medium-level projection.*

QUESTION How does a fertility rate above RLF produce positive feedback (in which a change creates a situation that amplifies itself) for population growth?

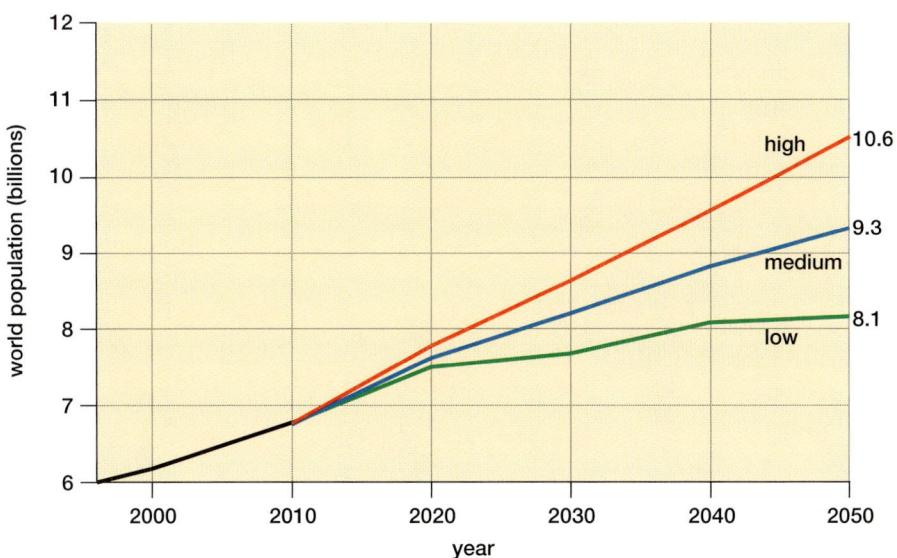

◀ FIGURE 26-18 United Nations
world population projections
Notice that only the lowest projection is for
population stabilization by 2050. Data from
United Nations, Department of Economic and Social
Affairs of the United Nations Secretariat, Population
Division. *World Population Prospects: The 2010 Revision;
medium-level projection*

QUESTION What factors could make the
projected values for 2050 inaccurate? If
inaccurate, do you think they are more likely
to be high or low? Explain.

were to achieve RLF immediately, their populations would
continue to increase for decades. Why? When the number of
children exceeds the number of reproducing adults, this creates momentum for future growth as these children mature
and enter their reproductive years. For example, when China
reached RLF in the early 1990s, about 27% of its population
was under age 15, and its median age was 26. Because of this
momentum, China has since grown by roughly 176 million
people and continues to grow. China is now closer to stabilization, however, because its median age is currently 34 and
only 19.5% are younger than 15. In a stable human population, fewer than 20% of individuals will fall into this age
group. In many African nations, children make up well over
40% of the population.

The United Nations has developed high, medium,
and low projections for future growth based on assumptions about fertility rates (Fig. 26-18). For the year 2050,
the medium projection is that Earth's population will have
increased by about 33% to over 9.3 billion, with 8 billion
people living in the developing nations. And the population will still be growing, although far more slowly than at
present.

Fertility in Some Nations
Is Below Replacement Level

Table 26-1 provides growth rates for various world regions.
In Europe, the average annual change in population is 0%,
with an average fertility rate of 1.6—substantially below
RLF—as many women delay or forgo having children.
This situation raises concerns about the availability of future workers and taxpayers to support the resulting temporary increase in the percentage of elderly people. Several
European countries are offering or considering incentives
(such as large tax breaks) for couples to have children at
an earlier age, which shortens the generation time and

increases the population. Japan's government is also concerned about the country's low fertility rate (1.4) and provides subsidies to encourage larger families—even though
Japan, which is about the size of the U.S. state of Montana,
is already home to 128 million people (equivalent to 41%
of the entire U.S. population).

Although a reduced population will ultimately offer
tremendous benefits for both the world's people and the
biosphere that sustains them, current economic structures in
countries throughout the world are based on growing populations. The difficult adjustments necessitated as populations
decline—or even merely stabilize—motivate governments to
adopt policies that encourage more childbearing and continued growth.

TABLE 26-1 Average Population Statistics by World Region: 2011		
Region	**Fertility Rate**	**Rate of Natural Increase (%)**
World	2.5	1.2
Developing countries	2.6	1.4
Africa	4.7	2.4
Latin America/ Caribbean	2.2	1.2
Asia*	2.6	1.4
China	1.5	0.5
Developed countries	1.7	0.2
Europe	1.6	0.0
North America	1.9	0.5

*Excluding China.

The U.S. Population Is Growing Rapidly

The United States has a population of more than 313 million and a growth rate of about 0.7% per year (adding one person every 15 seconds), and is the fastest-growing developed country in the world (**Fig. 26-19**). Continued immigration, which accounts for about 30% of the population increase, will ensure growth for the indefinite future, unless the U.S. fertility rate (2.0 in 2011) drops sufficiently below RLF to compensate for the influx of people.

The rapid growth of the U.S. population has major environmental implications both for local ecosystems and for the planet; for example, the average U.S. resident uses nearly four times as much energy as the average person worldwide.

CHECK YOUR LEARNING

Can you explain the advances that have allowed exponential growth of the human population? Are you able to explain why this growth continues today? Can you explain the demographic transition? Can you explain age structure diagrams and how their shape predicts future changes in population size?

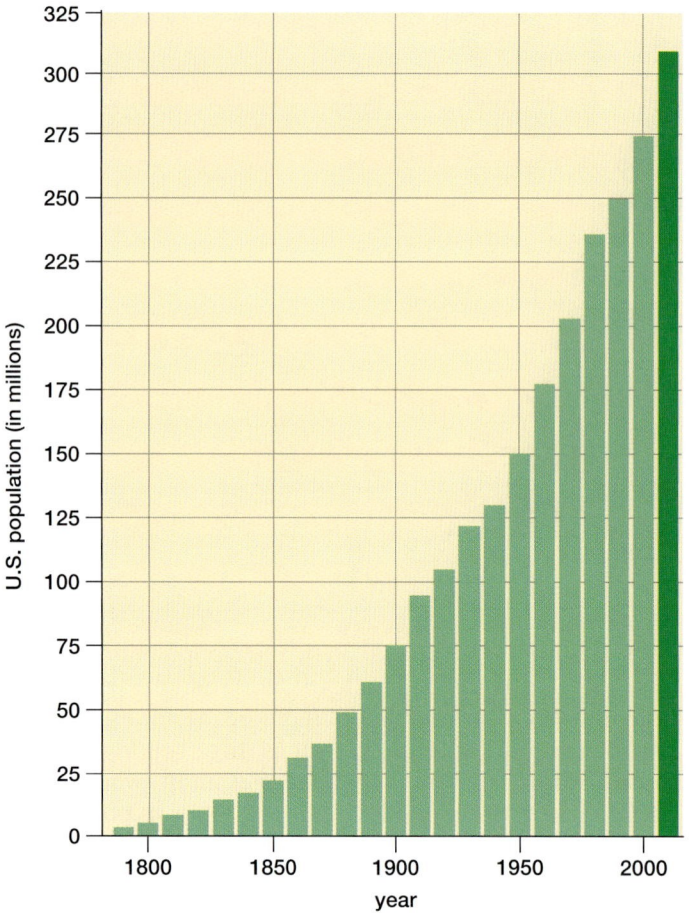

▶ **FIGURE 26-19 United States population growth**
Since 1790, U.S. population growth has produced a J-shaped curve typical of exponential growth. Each bar represents the endpoint of a 10-year interval; the dark green bar is the most recent. The U.S. Census Bureau predicts that the U.S. population will reach 341 million by 2020. Data from the U.S. Census Bureau, Population Division.

QUESTION At what stage of the S-curve is the U.S. population? What factors do you think will cause it to stabilize, and when? Might the recent recession reduce population growth?

CASE STUDY revisited

The Mystery of Easter Island

Radioactive carbon dating of sediments, examination of pollen, and archeological excavations suggest that when Polynesians arrived on Easter Island, they found a forested paradise. Ancient bones provide evidence that 6 species of land-dwelling birds and 25 species of seabirds nested on the island. The now-extinct Easter Island palm tree provided edible nuts and sap that could be made into syrup and wine. Its trunks could be made into large canoes from which the islanders harpooned porpoises, whose bones were abundant in refuse heaps dated between 900 and 1300.

The changing composition of refuse heaps over time attests to the changing diet of the natives as their island ecosystem became degraded and their choices increasingly limited. Palms disappeared around 1400, and porpoise bones disappeared from refuse heaps at around 1500, probably as the last remaining canoes became unusable. All of Easter Island's land-dwelling birds disappeared long ago, and only

one species of seabird still nests there. Considerable evidence supports the hypothesis that the population grew until it exceeded the environment's capacity to support it. Like the reindeer on St. Paul Island (see Fig. 26-6), Easter Island's human population is likely to have damaged the ecosystem on which it depended.

What can we learn from these reconstructed events? Easter Island provides a sobering example of what may become of our home planet in the not-too distant future. We must stop our burgeoning numbers from depleting Earth's limited resources. All of humanity inhabits an island in the vast sea of outer space; emigration is not an option.

Consider This

BioEthics With increasing evidence that the still-growing human population may have already exceeded Earth's carrying capacity, should it be considered a basic human right for couples to bear as many children as they choose? Explain your position. If you answer "no," what steps do you think society should take to limit population growth?

CHAPTER REVIEW

Summary of Key Concepts

26.1 How Does Population Size Change?

Individuals join populations through births and deaths, which result in a natural increase, and through immigration and emigration, which produce net migration. Ignoring migration, the population growth rate (r) is its birth rate (b) minus its death rate (d). Population growth (G), the increase during a given time interval, equals the growth rate (r) multiplied by the population size (N). All organisms have the biotic potential to more than replace themselves over their lifetimes, resulting in population growth. A constant growth rate produces exponential growth.

26.2 How Is Population Growth Regulated?

Carrying capacity (K) is the maximum size at which a population may be sustained indefinitely by an ecosystem. K is determined by limited resources such as space, nutrients, and energy. Environmental resistance generally maintains populations at or below the carrying capacity. Above K, populations deplete their resource base, leading to (1) the population stabilizing near K; (2) reduction of K and a permanently reduced population; or (3) the population being eliminated from the area. Population growth is restrained by density-independent forms of environmental resistance (weather and climate) and density-dependent forms of resistance (competition, predation, and parasitism).

26.3 How Are Populations Distributed in Space and Age?

Clumped distribution may occur for social reasons or around limited resources. Uniform distribution is often the result of territorial spacing. Random distribution is rare, occurring when individuals do not interact socially and when resources are abundant and evenly distributed.

Late-loss (convex) curves are characteristic of long-lived species with few offspring, which receive parental care, most of which survive to old age. Species with constant-loss curves have an equal chance of dying at any age. Early loss (concave) curves are typical of organisms that produce numerous offspring, most of which die before reaching maturity.

26.4 How Is the Human Population Changing?

The human population has exhibited exponential growth for an unprecedented time, the result of a combination of high birth rates and technological, agricultural, industrial, and medical advances that have overcome several types of environmental resistance and increased Earth's carrying capacity for humans. Age structure diagrams depict numbers of males and females in groups of increasing age. Expanding populations have pyramidal age structures; stable populations show rather straight-sided age structures; and shrinking populations have age structures that are constricted at the base.

Most people live in developing countries with growing populations. Momentum from birth rates above replacement level fertility ensures decades of continued growth. The United States is the fastest-growing developed country, owing both to relatively high birth rates and immigration.

Key Terms

age structure diagram 503
biosphere 488
biotic potential 490
birth rate 489
boom-and-bust cycle 490
carrying capacity (K) 492
clumped distribution 498
community 488
competition 496
constant-loss population 499
contest competition 497
death rate 489
demographic transition 502
density-dependent 493
density-independent 493
developed country 502
developing country 502
early loss population 499
ecology 488
ecosystem 488
emigration 488
environmental resistance 490
exponential growth 489
growth rate 489

host 496
immigration 488
interspecific competition 496
intraspecific competition 496
invasive species 491
J-curve 489
late-loss population 499
logistic population growth 492
natural increase 488
parasite 496
population 488
population cycle 495
predator 495
prey 495
random distribution 499
replacement level fertility (RLF) 502
scramble competition 497
S-curve 492
survivorship curve 499
survivorship table 499
uniform distribution 498

Learning Outcomes

In this chapter, you have learned to . . .

LO1 Describe the factors that cause population size to change, including the interactions among population growth, growth rate, and population size.

LO2 Describe and graph exponential and logistic growth and, using examples, explain the conditions under which exponential and logistic growth occur.

LO3 List five factors that influence biotic potential, and explain why natural selection has favored a high biotic potential.

LO4 List examples of density-dependent and density-independent environmental resistance and describe how each controls population size.

LO5 Graph and then explain the possible fates of a population that overshoots the carrying capacity of its environment.

LO6 Explain three types of spatial distribution and three types of survivorship curves and discuss the factors that cause each.

LO7 Describe the advances that have allowed continued exponential human population growth since prehistoric times.

LO8 Graph and then explain the stages of the demographic transition, noting where developed and developing countries fall along the graph.

LO9 Draw and then explain general age structure diagrams of growing, stable, and shrinking populations and discuss how these diagrams predict future growth.

Thinking Through the Concepts

Fill-in-the-Blank

1. Two examples of density-independent forms of environmental resistance are _____ and _____. Two important types of density-dependent environmental resistance are _____ and _____. **LO4**

2. Graphs that plot how the numbers of individuals born at the same time change over time are called _____. The specific type of curve that applies to a dandelion that releases 300 seeds, most of which never germinate, is called _____. The curve for humans is an example of _____. **LO6**

3. The type of growth that occurs in a population that grows by 0.1% per year is _____. Does this form of growth add the same number of individuals each year? _____ What shape of curve is generated if this type of growth is graphed? _____ Can this type of growth be sustained indefinitely? _____ **LO2 LO5**

4. The maximum population size that can be sustained indefinitely without damaging the environment is called the _____. A growth curve in which a population first grows logarithmically and then levels off at (or below) this maximum sustainable size is called a(n) _____ curve, or a(n) _____ curve. **LO2 LO5**

5. The type of spatial distribution likely to occur when resources are localized is _____. The type of spatial distribution that results when pairs of animals defend breeding territories is _____. The least common form of distribution is _____. **LO6**

6. An expanding population has an age structure diagram shaped like a(n) _____. If the sides of an age structure diagram are roughly vertical, the population is _____. The shape of the age structure diagram for developing countries collectively is _____. **LO9**

7. A population grows whenever the number of _____ plus _____ exceeds the number of _____ plus _____. The growth rate of a population increases whenever the age at which the organism first reproduces _____, when the frequency of reproduction _____, and when the length of the organism's reproductive life span _____. **LO1 LO3**

Review Questions

1. Define *biotic potential*, list the factors that influence it, and explain why natural selection has favored it. **LO1 LO3**

2. Write and describe the meaning of the equation for population growth using the variables G, r, and N. **LO1**

3. Draw, name, and describe the properties of the growth curve of a population without environmental resistance. **LO2**

4. Define *environmental resistance* and distinguish between density-independent and density-dependent forms of environmental resistance. List and explain three examples of each. **LO1 LO4**

5. What is logistic population growth? What is K? **LO2 LO5**

6. Describe three different possible consequences of exceeding carrying capacity. Sketch these scenarios on a graph. Explain your answer. **LO5**

7. Distinguish between populations showing concave and convex survivorship curves. **LO6**

8. Explain why environmental resistance has not prevented exponential human population growth since prehistoric times and provide examples. Can this continue? Explain why or why not. **LO5 LO7**

9. Draw the general shape of age structure diagrams characteristic of (a) expanding, (b) stable, and (c) shrinking populations. Label all the axes. Explain why you can predict the next several decades of growth by the current age structure of populations. **LO9**

10. Sketch and label the graph showing the general stages of the demographic transition and explain the changes that influence the growth and size of the population over time. **LO8**

Applying the Concepts

1. **BioEthics** The United States has historically high immigration levels. Discuss the pros and cons of allowing high levels of immigration into the United States.

2. Research a developing country (such as Nigeria or Bangladesh or Pakistan) with rapid population growth and find out what factors sustain that growth and why. Explain these factors and assess the likelihood that the fertility rate of this country will drop in the near future.

3. Why is the concept of carrying capacity difficult to apply to human populations?

4. **BioEthics** Why are some countries with shrinking populations trying to increase their birth rates? Do you think this is wise? Explain why or why not.

Answers to Figure Caption questions and Fill-in-the-Blank questions can be found in the Answers section at the back of the book.

MB *Go to MasteringBiology for practice quizzes, activities, eText, videos, current events, and more.*

CASE STUDY
Mussels Muscle In

HOW DOES A TOWN on the shore of Lake Erie find itself without water? The answer: when the town's water intake pipes are clogged with hundreds of millions of zebra mussels. Monroe, Michigan, was nearly brought to a standstill by these newcomers that had arrived only a few years earlier as hitchhikers in the ballast water of a cargo ship. The ballast water, which was discharged from the ship near Lake Erie, had been acquired in the Black Sea of southeastern Europe and contained millions of microscopic mussel larvae. In the Great Lakes, zebra mussels and quagga mussels—close relatives that arrived in a similar manner slightly later— found an ideal habitat with plenty of food and no major predators or competitors. Populations of zebra and quagga mussels have exploded, spreading to all of the Great Lakes, the Ohio and Mississippi Rivers and their tributaries, and to lakes in at least 15 states.

Zebra and quagga mussels belong to a large group of mollusks that includes clams and scallops. Both average about the size of a fingernail. They can reach enormous densities; in parts of northern Lake Michigan, quagga mussels coat the lake bed at densities up to 19,000 per square yard.

Mussels feed by filtering water and sifting out phytoplankton, the photosynthetic microorganisms that harvest the solar energy that sustains most freshwater communities. In just over a decade, invading quagga mussels have massively altered the ecosystems of two of Earth's largest bodies of fresh water: Lake Huron and Lake Michigan. Their continuous filtering of lake water has dramatically reduced phytoplankton populations. As phytoplankton become scarce, populations of microscopic animals (zooplankton) that feed on phytoplankton also decline. Then populations of small fish, which rely on zooplankton for food, plummet and no longer provide adequate food for larger fish, which form the basis of important fisheries.

Think about zebra and quagga mussels as you read about the community interactions that characterize healthy ecosystems. Why have these unwelcome imports been so enormously successful? What impact, if any, do they have on one another? Can anything control them?

Quagga mussels cover a sandal found at Lake Mead in Nevada, far from the Great Lakes where these mussels were first introduced.

AT A GLANCE

LEARNING GOALS

LG1 Explain the four basic types of community interaction and discuss why each is important to the community.

LG2 Define the *ecological niche* concept and use it to explain why there is more competition within a species than there is among species.

LG3 Explain how interspecific competition causes coevolution and results in resource partitioning.

LG4 Describe predator–prey relationships and provide examples of the coevolution of predators and their prey.

LG5 Compare parasite–host with predator–prey relationships and provide some examples of each.

LG6 Explain mutualism and provide examples of symbiotic mutualistic relationships as well as other types of mutualism that are less intimate and extended.

LG7 Define the term *keystone species* and provide examples of some keystone predators, including a description of how they influence the structure of their communities.

LG8 Define the term *succession* and compare primary succession with secondary succession.

27.1 WHY ARE COMMUNITY INTERACTIONS IMPORTANT?

An ecological **community** consists of all the populations that interact with one another within a defined area. Community interactions such as competition, predation, and parasitism can limit the size of populations (as discussed in Chapter 26), maintaining a balance between resources and the numbers of individuals consuming them.

Community interactions exert strong evolutionary forces on the species involved. For example, in killing prey that is easiest to catch, predators spare those individuals with better defenses against predation. These better-adapted individuals then produce offspring, and over time, their inherited characteristics increase within the prey population. Thus, as community interactions limit population size, they simultaneously shape the bodies and behaviors of the interacting populations. The process by which two interacting species act on one another as agents of natural selection is called **coevolution.**

The major community interactions are competition, predation, parasitism, and mutualism. They can be classified according to whether each of the species is harmed or helped by the interaction, as shown in Table 27-1.

CHECK YOUR LEARNING

Can you define a *community* and explain why community interactions are important? Can you list four major types of community interaction?

27.2 HOW DOES THE ECOLOGICAL NICHE INFLUENCE COMPETITION?

Each species occupies a unique **ecological niche** that encompasses all aspects of its way of life. The concept of the

TABLE 27-1 Interactions Among Species

Type of Interaction	Effect on Species A	Effect on Species B
Competition between A and B	Harms	Harms
Predation by A on B	Benefits	Harms
Parasitism by A on B	Benefits	Harms
Mutualism between A and B	Benefits	Benefits

ecological niche is important for understanding how competition within and among species selects for adaptations in body form and behavior. Although the word *niche* may call to mind a small cubbyhole, in ecology, it means much more. One important aspect of the ecological niche is the organism's physical home, or habitat. The primary habitat of a white-tailed deer in the United States, for example, is the eastern deciduous forest. In addition, an ecological niche includes all the physical environmental conditions necessary for the survival and reproduction of a given species. These can include nesting or denning sites, climate, the type of nutrients the species requires, its optimal temperature range, the amount of water it needs, the pH and salinity of the water or soil it may inhabit, and (for plants) the degree of sun or shade it can tolerate. Finally, the ecological niche also encompasses the entire "role" that a given species performs within an ecosystem, including what it eats (or whether it obtains energy from photosynthesis) and the other species with which it competes. Although different species share many aspects of their ecological niche with others, no two species ever occupy exactly the same ecological niche within the same natural community.

Competition Occurs Whenever Two Organisms Attempt to Use the Same, Limited Resources

Competition is an interaction in which individuals of the same or different species attempt to use the same, limited resources, particularly energy, nutrients, or space. **Interspecific competition** (*inter* means "between") refers to competitive interactions among members of different species, as can occur if they feed on the same things or require similar breeding areas. For example, interspecific competition occurs between zebra mussels and quagga mussels, who both consume phytoplankton. The greater the overlap between the ecological niches of two species, the greater the amount of competition between them. Interspecific competition is detrimental to all of the species involved because it reduces their access to resources that are in limited supply.

Adaptations Reduce the Overlap of Ecological Niches Among Coexisting Species

Just as no two organisms can occupy exactly the same physical space at the same time, no two species can inhabit exactly the same ecological niche simultaneously and continuously. This important concept, called the **competitive exclusion principle,** was formulated in 1934 by Russian biologist G. F. Gause. This principle leads to the hypothesis that if a researcher forces two species with very similar niches to compete for the same limited resource, inevitably, one will outcompete the other, and the species that is less well adapted to the experimental conditions will die out.

To test this hypothesis, Gause used two species of the protist *Paramecium* (*P. aurelia* and *P. caudatum*). Grown separately on the same food (bacteria), both species thrived, feeding on bacteria suspended in the water of their test tubes (**Fig. 27-1a**). But when Gause placed the two species together, one (*P. aurelia*) grew more rapidly and always eliminated, or "competitively excluded," the other (*P. caudatum*; **Fig. 27-1b**). Gause then repeated the experiment, pairing *P. caudatum* with a different species, *P. bursaria*, which fed mostly on bacteria that had settled to the bottom of the test tube. These two species of *Paramecium* were able to coexist indefinitely because they preferred feeding in different places, and thus occupied slightly different niches.

Ecologist Robert MacArthur further explored the competitive exclusion principle, this time under natural conditions, by carefully observing five species of North American warbler. These birds all hunt for insects and nest in spruce trees. Although the niches of these birds appear to overlap considerably, MacArthur found that each species concentrates its search for food in specific regions within spruce trees, employs different hunting tactics, and nests at a slightly different time. The five species of warblers have evolved behaviors that reduce the overlap of their niches, which reduces interspecific competition (**Fig. 27-2**).

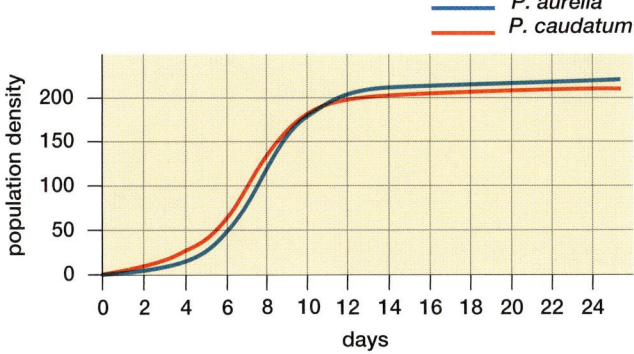

(a) Grown in separate flasks

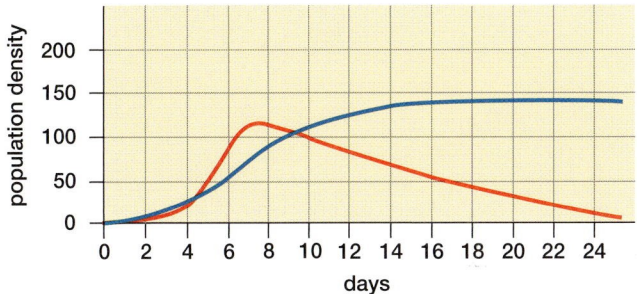

(b) Grown in the same flask

▲ **FIGURE 27-1 Competitive exclusion (a)** Raised separately with a constant food supply, both *Paramecium aurelia* and *P. caudatum* show the S-curve typical of a population that initially grows rapidly and then stabilizes. **(b)** Raised together and forced to occupy the same niche, *P. aurelia* always eventually outcompetes *P. caudatum* and causes that population to die off. Data from Gause, G. F. 1934. *The Struggle for Existence.* Baltimore: Williams & Wilkins.

QUESTION Explain how competitive exclusion could contribute to the threat posed by an invasive species.

This phenomenon of dividing up resources, called **resource partitioning,** is the outcome of the coevolution of different species with extensive (but not total) niche overlap. A famous example of resource partitioning was discovered by Charles Darwin among related species of finches on the Galápagos Islands. Different finch species that shared the same island had evolved different bill sizes and shapes and different feeding behaviors that reduced the competition among them (see Chapter 14).

Competition Among Species May Reduce the Population Size and Distribution of Each

Although natural selection reduces niche overlap among different species, those with similar niches still compete for limited resources, restricting the size and distribution of both populations. A classic study of this effect of interspecific competition was performed by ecologist Joseph Connell using barnacles, which are shelled crustaceans that attach permanently to rocks and other surfaces.

Barnacles of the genus *Chthamalus* share rocky ocean shores with barnacles of the genus *Balanus*, and their niches overlap considerably. Both live in the *intertidal zone*, an area

| Yellow-rumped warbler | Bay-breasted warbler | Cape May warbler | Black-throated green warbler | Blackburnian warbler |

▲ FIGURE 27-2 **Resource partitioning** Each of these five insect-eating species of North American warblers searches for food in slightly different parts of spruce trees. They reduce competition by occupying very similar, but not identical, niches. Adapted from MacArthur, R. H. 1958. Population ecology of some warblers of Northeastern coniferous forest. *Ecology* 39:599–619.

of the shore that is alternately covered and exposed by the tides. Connell found that *Chthamalus* dominates the upper intertidal zone and *Balanus* dominates the middle intertidal zone, even though the middle habitat is suitable for both species. When he scraped off *Balanus*, the *Chthamalus* population increased, spreading downward into the middle intertidal region where its competitor had been scraped off. This demonstrated that *Chthamalus* would occupy the middle intertidal zone if it didn't have to cope with competition from *Balanus*, a larger, faster-growing barnacle. But *Chthamalus* tolerates drier conditions than does *Balanus*, so it takes over in the upper intertidal, where only the highest tides submerge the barnacles. As this example illustrates, interspecific competition can limit both the size and the distribution of competing populations.

Competition Within a Species Is a Major Factor Controlling Population Size

Individuals of the same species have the same requirements for resources and thus occupy the same ecological niche. For this reason, **intraspecific competition** is the most intense form of competition because all of the members of the species compete for all of the same resources. Intraspecific competition (*intra* means "within") exerts strong density-dependent environmental resistance, limiting population size (see Chapter 26). Intraspecific competition is one of the main factors driving evolution by natural selection, in which individuals that are better equipped to obtain scarce resources are more likely to reproduce successfully, passing their heritable traits to their offspring.

CHECK YOUR LEARNING

Can you compare interspecific and intraspecific competition and use the ecological niche concept to explain which is more intense? Can you explain how competitive exclusion leads to resource partitioning?

CASE STUDY c o n t i n u e d

Mussels Muscle In

The ecological niches of zebra and quagga mussels overlap with those of native North American freshwater mussels and clams, with which they compete for both space and food. Because they reproduce far more rapidly, the imported species have outcompeted the natives, often literally covering them; thousands of zebra mussels have been found attached to a single large native mussel. As a result, several species of native freshwater clams have been nearly eliminated from parts of the Great Lakes. But quagga mussels also compete directly with zebra mussels. Quaggas use food more efficiently, they can thrive in colder and deeper water, and quaggas occupy both lake-bottom sediments and hard surfaces, whereas zebra mussels only colonize hard surfaces. As a result, quagga mussels have rapidly displaced zebra mussels in the Great Lakes. Because they can colonize more of the lakebed, quagga mussels pose an even greater threat to the lake communities than do zebra mussels.

27.3 HOW DO PREDATOR–PREY INTERACTIONS SHAPE EVOLUTIONARY ADAPTATIONS?

Predators eat other organisms. Although we generally think of predators as being *carnivores* (animals that eat other animals), ecologists sometimes include *herbivores* (animals that eat plants) in this general category. In this discussion, we define predation in this more general way, to include the quagga mussel that filters microscopic phytoplankton from water, the grass-eating pika (**Fig. 27-3a**), the bat homing in on a moth (**Fig. 27-3b**), and the more familiar example of an owl eating a mouse (**Fig. 27-3c**). Predators are generally less abundant than their prey (as explained in Chapter 28). To survive, predators must feed and prey must avoid becoming food. Therefore, predator and prey exert intense selective pressure on one another, resulting in coevolution.

(a) Pika

(b) Long-eared bat

(c) Eagle owl

▲ **FIGURE 27-3 Forms of predation (a)** A pika, whose preferred food is grass, is a small relative of the rabbit and lives in the Rocky Mountains. **(b)** A long-eared bat uses a sophisticated echolocation system to hunt moths, which in turn have evolved special sound detectors and behaviors to avoid capture. **(c)** An eagle owl feasts on a mouse.

QUESTION Describe some examples of coevolution of predators and prey.

As prey become more difficult to catch, predators must become more adept at hunting. Coevolution has endowed the mountain lion with tearing teeth and claws, and has given the hunted fawn dappled spots that serve as camouflage, as well as the behavior of lying perfectly still when its mother is away. Coevolution has produced the keen eyesight of the hawk and owl, which is countered by the earthy colors of their mouse and ground-squirrel prey. Evolution in response to predation has also produced the milkweed's toxic chemicals, the skunk's noxious spray, and the coral snake's venom (see Figs. 27-4, 27-8, and 27-10b). In "Earth Watch: Invasive Species Disrupt Community Interactions," we describe what happens when natural checks and balances are circumvented by transporting predatory or competing organisms into ecological communities whose members are not adapted to deal with them.

? HAVE YOU EVER WONDERED …

Why Deer Freeze in Headlights?

Some prey animals have evolved not only camouflaged coloring but also behaviors that enhance their survival. In response to a predator, they will often remain motionless (making the camouflage more effective) and only bolt at the last minute, when it becomes obvious that the predator has spotted or smelled them. Your car, resembling a very large predator, is likely to evoke both of these instinctive behaviors. Although they work well in natural situations, these responses increase the animal's chances of becoming roadkill.

Some Predators and Prey Have Evolved Counteracting Adaptations

Plants have evolved a variety of chemical adaptations that deter their herbivorous predators. Grasses embed tough silicon (glassy) substances in their blades that make them difficult to chew. This has created selection pressure for grazing animals with longer, harder teeth. Over evolutionary time, as grasses evolved tougher blades that discouraged predation, horses evolved longer teeth with thicker enamel coatings that resist wear and abrasion from the tough grasses.

The adaptations of echolocating bats and their moth prey (see Fig. 27-3b) provide excellent examples of how both body structures and behaviors are molded by coevolution. Most bats are nighttime hunters that emit pulses of sound that are so high pitched that people can't hear them. Using the echoes that occur as their sounds bounce back from nearby objects, bats perceive a sonar image of their surroundings, which allows them to navigate and detect prey.

Under selection pressure from this unusual prey-locating system, some moths (a favorite prey of bats) have evolved ears that are particularly sensitive to the pitches used by echolocating bats. When they hear a bat, these moths take evasive action, flying erratically or dropping to the ground. The bats, in turn, have evolved the ability to counter this defense by switching the frequency of their sound pulses away from the moth's sensitivity range. Some moths interfere with the bats' echolocation by producing their own high-frequency clicks. In still another coevolutionary adaptation, a bat that is hunting a clicking moth may turn off its own sound pulses temporarily (thus avoiding detection) and follow the moth's clicks to capture it.

Earth | Watch

Invasive Species Disrupt Community Interactions

Invasive species are species that are introduced into an ecosystem in which they did not evolve and that are harmful to human health, the environment, or the economy of a region. Invasive species often spread widely because they find few effective forms of environmental resistance, such as strong competitors, predators, or parasites, in their new environment. Their unchecked population growth may seriously damage the ecosystem as they outcompete or prey on native species.

Not all non-native species become pests; the ones that do typically reproduce rapidly, disperse widely, and thrive under a relatively wide range of environmental conditions. Invasive plants often spread by sprouting from roots as well as seeds, and some aquatic forms generate new plants from fragments. Invasive animals often eat a wide variety of foods. By evading the checks and balances imposed by millennia of coevolution, invasive species are wreaking havoc on natural ecosystems throughout the world. Some examples follow.

English house sparrows were introduced into the United States on several occasions, starting in the 1850s, in the hope that they would control caterpillars feeding on shade trees. In 1890, European starlings were released into Central Park in New York City by a group attempting to introduce all the birds mentioned in the works of Shakespeare. Both bird species have spread throughout the continental United States. Their success has reduced the populations of some native songbirds, such as bluebirds and purple martins, with which they compete for nesting sites. Red fire ants from South America were accidentally introduced into Alabama on shiploads of lumber in the 1930s and have since spread throughout the southern United States. Fire ants kill native ants, birds, and young reptiles. Their mounds can ruin farm fields, and their fiery stings and aggressive temperament can make backyards inhospitable. Imported cane toads have proven unstoppable as they spread through northeastern Australia, outcompeting native frogs and killing potential predators with their toxic secretions (**Fig. E27-1a**).

Invasive plants also threaten natural communities. In the 1930s and 1940s, the Japanese vine kudzu was planted extensively in the southern United States to control soil erosion. Today, kudzu is a major pest, blocking life-giving sunlight from trees, and even engulfing small, abandoned houses (**Fig. E27-1b**). Both water hyacinth and purple loosestrife were introduced into the United States as ornamental plants. Water hyacinth now clogs waterways in southern states, slowing boat traffic and displacing natural vegetation. Purple loosestrife aggressively invades wetlands, where it outcompetes native plants and reduces food and habitat for native animals (**Fig. E27-1c**).

Invasive species rank second only to habitat destruction in pushing endangered species toward extinction. Recently, wildlife officials have made cautious attempts to reestablish biological checks and balances by importing predators or parasites (called *biocontrols*) to attack invasive species. This practice is fraught with danger, because new imports can have unpredicted and possibly disastrous effects on native wildlife. The cane toad, for example, was introduced into Australia from South America in the 1930s to control introduced beetles that threatened the sugarcane crop.

Despite the risks of imported biocontrols, there are often few realistic alternatives, because poisons kill native

(a) Cane toad

(b) Kudzu

(c) Purple loosestrife

▲ **FIGURE E27-1 Invasive species (a)** The cane toad (native to central and South America) was imported to Australia, where it outcompetes native toads and frogs. **(b)** The Japanese vine kudzu, imported to the southern United States from Asia, can rapidly cover entire trees and small buildings. **(c)** Purple loosestrife, native to Eurasia, now displaces native plants and reduces food and habitat for native animals in wetlands throughout the United States.

and non-native organisms indiscriminately. Biologists now carefully screen proposed biocontrols to make sure they are specific for the intended invasive species, and there have been a number of successful introductions. For example, imported beetles are now among the most important ways of controlling purple loosestrife in North America. Professionals and volunteers work together to raise and release these beetles by the millions, helping to restrain this invasive weed.

Predators and Prey May Engage in Chemical Warfare

The evolution of counteracting defenses may give rise to a kind of "chemical warfare" between predators and prey. Many plants, including milkweeds, synthesize toxic and distasteful chemicals. As plants evolved these defensive toxins, certain insects evolved increasingly efficient ways to detoxify or even use these substances. The result is that nearly every toxic plant is eaten by at least one type of insect. For example, monarch butterflies lay their eggs on milkweed; when their larvae hatch, they consume this poisonous plant (Fig. 27-4). The caterpillars not only tolerate the milkweed poison but store it in their tissues as a defense against their own predators. The stored toxin is retained in the metamorphosed monarch butterfly (see Fig. 27-9a). Viceroy butterflies (see Fig. 27-9b) use a similar strategy, storing a bitter compound from willows (eaten by the larvae) in the tissues of the adult.

Toxins can be used both to attack and to defend. The venom of spiders and snakes, such as the coral snake (see Fig. 27-10b), both paralyzes their prey and deters predators. Other chemicals are purely defensive. These include the clouds of ink that certain mollusks (including squid, octopuses, and some sea slugs) emit when a predator attacks. These colorful chemical "smoke screens" confuse predators and mask the prey's escape. Another dramatic example of chemical defense is seen in the bombardier beetle. In response to the bite of an ant, the beetle releases secretions from special glands into an abdominal chamber. There, enzymes catalyze an explosive chemical reaction that shoots a toxic, boiling-hot spray onto the attacker.

▲ FIGURE 27-4 **Chemical warfare** A monarch caterpillar feeds on milkweed that contains a powerful toxin.

QUESTION Why is the caterpillar colored with conspicuous stripes?

Looks Can Be Deceiving for Both Predators and Prey

An old saying goes that the best hiding place may be in plain sight. Both predators and prey have evolved colors, patterns, and shapes that resemble their surroundings. Such disguises, called **camouflage,** render plants and animals inconspicuous, even when they are in full view (Fig. 27-5).

Some animals closely resemble specific objects such as leaves, twigs, seaweed, thorns, or even bird droppings (Figs. 27-6a–c). Camouflaged animals tend to remain motionless; a crawling bird dropping would ruin the disguise. Whereas many camouflaged animals resemble parts of plants, a few succulent desert plants have evolved to resemble small rocks, hiding them from animals seeking the water stored in the plants' bodies (Fig. 27-6d).

(a) Sand dab fish adjust camouflage to different backgrounds   **(b) A camouflaged horned lizard**

▲ FIGURE 27-5 **Camouflage by blending in (a)** Sand dabs (flounder fish) are flat, bottom-dwelling ocean fish whose mottled colors closely resemble the sand on which they rest. Both their colors and patterns can be modified by nervous signals to better blend with their backgrounds.
(b) This horned lizard, found in California from the coast to mountain foothills, helps protect itself from predation by snakes and hawks by resembling its surroundings of leaf litter.

(a) Citrus swallowtail larva

(b) Leafy sea dragon

(c) Thorn treehoppers

(d) Living rock cacti

▲ FIGURE 27-6 **Camouflage by resembling specific objects** (a) A citrus swallowtail butterfly caterpillar, whose color and shape resemble a bird dropping, sits motionless on a leaf. **(b)** The leafy sea dragon (an Australian "seahorse" fish) has evolved extensions of its body that mimic the algae in which it often hides. **(c)** These Florida thorn treehopper insects avoid detection by resembling thorns on a branch. **(d)** These cacti of the American Southwest are appropriately called "living rock cacti."

QUESTION In general, how might such camouflage have evolved?

Predators that ambush prey are also aided by camouflage. For example, the spotted snow leopard becomes inconspicuous on a mountainside as it watches for prey (**Fig. 27-7a**). The frogfish closely resembles the sponges and algae-covered rocks where it lurks motionless on the ocean floor, awaiting smaller fish to swallow (**Fig. 27-7b**).

Some prey animals have evolved very differently, exhibiting bright **warning coloration.** These animals may taste bad, inflict a venomous sting or bite (as bees or coral snakes do), or produce a big stink when bothered (**Fig. 27-8**). The eye-catching colors seem to declare "Attack at your own risk!"

Mimicry refers to a situation in which members of one species have evolved to resemble another species. By sharing a similar warning-color pattern, several poisonous species may all benefit. Mimicry among different distasteful species is called *Müllerian mimicry*. For example, toxic monarch butterflies have wing patterns strikingly similar to those of equally distasteful viceroy butterflies (**Fig. 27-9**). Birds that become ill from consuming one species are likely to avoid the other as well. A toad that is stung while attempting to eat a bee is likely to avoid not only bees but other black and yellow striped insects (such as yellow jacket wasps) without ever tasting one.

Once warning coloration evolved, there arose a selective advantage for harmless animals to resemble venomous ones, an adaptation called *Batesian mimicry*. Through Batesian mimicry, the harmless hoverfly avoids predation by resembling a bee (**Fig. 27-10a**), and the nonvenomous scarlet king snake is protected by brilliant warning

(a) A camouflaged snow leopard

(b) A camouflaged frogfish

▲ **FIGURE 27-7 Camouflage assists predators** **(a)** As it waits for prey, which includes gazelles, wild sheep, and deer, the rare snow leopard from the mountains of Mongolia avoids detection with its camouflage coloration. **(b)** Combining camouflage and aggressive mimicry, a frogfish waits in ambush, its lumpy, yellow body matching the sponge-encrusted rock on which it rests. Above its mouth dangles a tiny lure that closely resembles a small fish. The lure attracts small predators, which will suddenly find themselves to be prey instead. The frogfish can expand its mouth by a factor of 12 in a few thousandths of a second, instantly sucking in its prey.

▲ **FIGURE 27-8 Warning coloration** The vivid stripe and the tail display behavior of the skunk advertise its ability to make any attacker miserable.

▼ **FIGURE 27-9 Müllerian mimicry** Nearly identical warning coloration protects both **(a)** the distasteful monarch and **(b)** the equally distasteful viceroy butterfly.

(a) Monarch (distasteful)

(b) Viceroy (distasteful)

(a) Bee (venomous)

Hoverfly (nonvenomous)

(b) Coral snake (venomous)

Scarlet king snake (nonvenomous)

▲ **FIGURE 27-10 Batesian mimicry (a)** A bee, which is capable of stinging (left), is mimicked by the stingless hoverfly (right). **(b)** The warning coloration of the venomous coral snake (left) is mimicked by the harmless scarlet king snake (right).

coloration that closely resembles that of the highly venomous coral snake (**Fig. 27-10b**).

Certain prey species use another form of mimicry: **startle coloration.** Several insects and even some vertebrates (such as the false-eyed frog) have evolved patterns of color that closely resemble the eyes of a much larger, and possibly dangerous, animal (**Fig. 27-11**). If a predator gets close, the prey suddenly reveals its eyespots, startling the predator and sometimes allowing the prey to escape.

A sophisticated variation of startle coloration is seen in snowberry flies, which are hunted by territorial jumping spiders (**Fig. 27-12**). When a spider approaches, the fly spreads its wings and moves them in a jerky dance. Seeing this display, the spider is likely to flee the fly. Why? The fly's wing pattern looks like the legs of the jumping spider, and the wings' jerky movements mimic the behavior of a jumping spider driving another spider from its territory.

Some predators have evolved **aggressive mimicry,** in which they entice their prey to come close by resembling something attractive to the prey. For example, by using a rhythm of flashes that is unique to each species, female fireflies attract males to mate. But in one species, the females sometimes mimic the flashing pattern of a different species, attracting males that they kill and eat. The frogfish (see Fig. 27-7b) is not only camouflaged but exhibits aggressive mimicry by dangling a wriggling lure that resembles a small fish just above its mouth. Fish attracted by the lure are engulfed in a split second if they get too close.

CHECK YOUR LEARNING

Can you define *predation* and provide examples of how predators and their prey have coevolved? Can you describe some examples of chemical warfare between predators and prey? Can you define and provide examples of *camouflage, warning coloration,* Batesian and Müllerian *mimicry, startle coloration,* and *aggressive mimicry*?

(a) False-eyed frog

(b) Peacock moth

(c) Swallowtail caterpillar

▲ FIGURE 27-11 **Startle coloration** (a) When threatened, the false-eyed frog of South America raises its rump, which resembles the eyes of a large predator. (b) The peacock moth from Trinidad is well camouflaged, but should a predator approach, it suddenly opens its wings to reveal spots resembling large eyes. (c) Predators of this caterpillar larva of the Eastern tiger swallowtail butterfly are deterred by its resemblance to a snake. The caterpillar's head is the "snake's" nose, and it bears two sets of eyespots.

(a) Jumping spider (predator)

(b) Snowberry fly (prey)

▲ FIGURE 27-12 **A prey mimics its predator** (a) When a jumping spider approaches, (b) the snowberry fly spreads its wings, revealing a pattern that resembles spider legs. The fly enhances the effect by performing a jerky, side-to-side dance that resembles the leg-waving display of a jumping spider defending its territory.

27.4 WHAT ARE PARASITISM AND MUTUALISM?

Parasites live in or on their prey, which are called **hosts,** usually harming or weakening them but not immediately killing them. Some parasite–host relationships are *symbiotic;* that is, the relationships involve a close, long-term physical association between the participating species. Parasites are generally much smaller and more numerous than their hosts. Familiar parasites include tapeworms, fleas, ticks, and the numerous types of disease-causing protists, bacteria, and viruses. Many parasites, such as the protist that causes malaria, have complex life cycles involving two or more hosts (see Fig. 20-12). The roundworm parasite highlighted in "Science in Action: A Parasite Makes Ants Berry Appealing to Birds" appears to use both ant and bird hosts during its life

cycle. There are very few parasitic vertebrates, but the lamprey (see Fig. 24-5), which attaches itself to a host fish and sucks its blood, is one example.

Parasites and Their Hosts Act as Agents of Natural Selection on One Another

The variety of infectious bacteria and viruses and the precision of the immune system that counters their attacks are evidence of the powerful forces of coevolution between parasitic microorganisms and their hosts. A specific example is the malaria parasite, which spends part of its life cycle in red blood cells and has selected for a human allele that causes red blood cells to become distorted and resistant to infection. Although individuals who inherit two copies of this mutated allele will have sickle-cell anemia, in some regions

SCIENCE IN ACTION A Parasite Makes Ants Berry Appealing to Birds

On a field trip to the rain forest of Panama in May 2005, ecologist Steve Yanoviak and fellow researchers were examining a colony of tree-dwelling black ants and found a few with strikingly red abdomens (Fig. E27-2). When sliced open, each abdomen disgorged hundreds of eggs in which tiny parasitic roundworms were developing. Both the roundworm species and its effect on the ants were new to science, so the team eagerly investigated their chance observation.

To test the hypothesis that only adult ants with red abdomens harbor roundworms, the researchers dissected 300 all-black adults and found none to be infected, while all those with red abdomens were infected. Noticing that the infected abdomens broke off readily, they measured the force required to pluck abdomens from infected ants versus normal ants, and discovered that only infected abdomens could be pulled off without dislodging the ant from its twig. Although ants of this species usually bite and emit foul-smelling, distasteful chemicals when handled, the infected ants did not exhibit these behaviors, allowing birds to easily dine on them. Further, their red abdomens strongly resembled, in both color and shape, berries from trees on which both the ants and birds foraged (see Fig. E27-2).

Based on these observations, Yanoviak and coworkers hypothesized that the roundworms trick berry-eating birds into dispersing their eggs by causing the ants' abdomens to resemble berries. To investigate this, the team dissected ants at all stages of development, and found that the parasites infect larval ants and develop inside as the ant matures. When the roundworms produce eggs that are ready to be released, the infected ants' abdomens become enlarged and red.

To test whether the eggs could survive passing through a bird's digestive tract, and thus be dispersed in its droppings, the researchers fed an infected ant abdomen to a chicken, and later, found hundreds of healthy roundworm eggs in its feces. But how did the larval ants become infected? The researchers hypothesized that ant larvae might eat bird droppings containing roundworm eggs. The team collected food particles that adult ants were carrying to their colony to feed their larvae, and discovered that, indeed, many of these were bits of bird droppings.

The researchers hypothesized that the roundworms spend most of their life cycle inside the abdomens of ants, but then rely on fruit-eating birds to disperse their eggs. Fruit-eating birds visit many trees and leave droppings that are collected by ants from many colonies. So, natural selection would favor roundworms whose larvae caused their host ants' abdomens to swell and mimic ripe fruit, causing birds to disperse them widely.

As always in science, new findings trigger new questions; for example, how does the roundworm infection cause the ant abdomen to turn red? How does infection change the ants' defensive behaviors? We can be sure that answers to these questions will raise still others.

▲ FIGURE E27-2 **An ant resembles berries** The abdomen of a parasitized ant closely resembles berries on trees in which the ants live. Birds eating these berries may pluck off the ant's abdomen. Photo courtesy of Steven Yanoviak, PhD., University of Arkansas at Little Rock.

of sub-Saharan Africa, 5% to 25% of the human population carry the mutated sickle-cell allele because of the protection it confers against malaria.

Both Species Benefit from Mutualistic Interactions

Mutualism refers to interactions between species in which both benefit. Many mutualistic relationships are symbiotic. If you see colored patches on rocks, they are probably lichens, a symbiotic mutualistic association between an alga and a fungus (Fig. 27-13a). The fungus provides support and protection while obtaining food from the photosynthetic alga, the bright colors of which are actually light-trapping pigments. Mutualistic associations also occur in the digestive tracts of cows and termites, where protists and bacteria find food and shelter. The microorganisms break down cellulose, making its component sugar molecules available both to themselves and to the animals that harbor them. In our own intestines, mutualistic bacteria synthesize vitamins, such as vitamin K, which we absorb and use. Plants called legumes benefit by providing chambers in their roots that house nitrogen-fixing bacteria (see Fig. 19-9). These bacteria are among the few organisms that can acquire nitrogen gas from the air and chemically modify it into a form that plants can use as a nutrient.

(a) Lichen

(b) Clownfish

▲ **FIGURE 27-13** **Mutualism (a)** This brightly colored lichen growing on bare rock is a mutualistic relationship between an alga and a fungus. **(b)** The clownfish snuggles unharmed among the stinging tentacles of an anemone.

The clownfish of the southern Pacific Ocean is coated with a layer of protective mucus, allowing it to take shelter among the venomous tentacles of certain species of anemones (**Fig. 27-13b**). In this mutualistic symbiotic association, the anemone provides the clownfish with protection from predators, while the clownfish cleans its anemone, defends it from predators, and brings it bits of food.

Many mutualistic relationships are not intimate and extended, and so are not symbiotic. Consider, for example, the relationship between plants and the insects that pollinate them. The insects fertilize the plants by carrying plant sperm (found in pollen grains), and benefit by sipping nectar and sometimes eating pollen. Both the bee and hoverfly in Figure 27-10a are important plant pollinators.

CHECK YOUR LEARNING

Can you explain parasitic and mutualistic relationships? Can you provide examples of each?

27.5 HOW DO KEYSTONE SPECIES INFLUENCE COMMUNITY STRUCTURE?

In some communities, a particular species, called a **keystone species,** plays a major role in determining community structure—a role that is out of proportion to the species' abundance in the community. If the keystone species is removed from its community, normal community interactions are significantly altered and the relative abundance of other species changes dramatically.

In the African savanna, the African elephant is a keystone predator. By grazing on small trees and bushes (**Fig. 27-14a**), elephants prevent the encroachment of forests and help maintain the grassland community, along with its diverse population of grazing mammals and their predators.

Although elephants alter plant communities directly, large predators, such as wolves and cougars (**Fig. 27-14b;** see also Fig. 26-8a), may have a surprising impact on terrestrial forests and riverbank vegetation. By keeping populations of deer and elk in check, these large carnivores can help restore the health of forests and stream banks that would be overgrazed if the deer and elk populations were left unchecked. Because all of this vegetation provides food, nesting sites,

(a) African elephant

(b) Cougar

(c) Northern sea otter

▲ **FIGURE 27-14** **Keystone species (a)** The elephant is a keystone species on the African savanna. **(b)** The cougar, found in isolated habitats throughout North and South America, helps to control grazing animals such as deer. **(c)** The northern sea otter rests in a kelp bed while cradling a sea urchin, one of its favorite foods.

and shelter for diverse communities of smaller animals, the entire community structure depends on the presence of such predators.

Identifying a keystone species can be a difficult task. Many have been recognized only after their loss has had dramatic, unforeseen consequences. The intricate and tangled web of community interactions is beautifully illustrated by the plight of the northern sea otter in the Aleutian Islands of southwestern Alaska (**Fig. 27-14c**). This sea otter population had rebounded after hunting for their pelts was banned in 1911. Kelp forests, sometimes described as the "rain forests of the ocean," flourished around islands in waters where the otters were abundant. But starting in the mid-1990s, otter numbers mysteriously plummeted, with some populations declining by 90%. As a result, the numbers of sea urchins, a favored food of otters, skyrocketed. Sea urchins are a major predator of kelp, and they rapidly deforested the seabed, eliminating the diverse community of fish, mollusks, and crustaceans that the kelp forests once fed and sheltered. Clearly, sea otters are a keystone species near the Aleutian Islands. But what is killing them? Killer whales, which had coexisted with sea otters in the past, were increasingly observed dining on otters. Why? Populations of seals and stellar sea lions—the whales' preferred prey—have declined drastically, forcing the whales to dine on smaller prey, such as otters. Wildlife biologists hypothesize that seal and sea lion populations have plummeted, at least in part, because overfishing in the North Pacific has severely depleted their food supply.

CHECK YOUR LEARNING

Can you explain the concept of a keystone species? Can you provide some examples of the effects that removing keystone species can have on their communities?

CASE STUDY continued

Mussels Muscle In

Many scientists are convinced that the voracious appetites and staggering numbers of zebra and quagga mussels have decimated populations of a Great Lakes keystone species, the shrimplike crustacean *Diporeia*. How? By competing with the shrimp for phytoplankton—and winning. As the mussels have invaded southern Lake Michigan, spring blooms of phytoplankton, particularly diatoms, have declined by about 70%. *Diporeia*, a bottom-feeder about 0.5 inch (1.25 cm) in length, relies heavily on the bodies of diatoms that drift down after their typical spring blooms. Since the mid-1990s *Diporeia* populations have declined by roughly 95% throughout Lake Michigan, providing strong support for the hypothesis that the invasive mussels are responsible. The loss of *Diporeia* has profound implications for Great Lakes communities because this crustacean is a major food source for a variety of small fish that serve as food for larger commercial fish species, including salmon, trout, and walleye.

27.6 HOW DO COMMUNITY INTERACTIONS CAUSE CHANGE OVER TIME?

In a mature terrestrial ecosystem, the populations that make up the community interact with one another and with their nonliving environment in intricate ways. But this tangled web of life did not spring fully formed from bare rock or naked soil; rather, it emerged in stages over a long period. These stages are called succession. **Succession** is a gradual change in a community and its nonliving environment in which assemblages of plants and animals replace one another in a sequence that is reasonably predictable. During succession, (1) early organisms modify the environment in ways that favor later organisms; (2) end-stage organisms suppress earlier organisms but tolerate one another, producing a stable community; and (3) there is a general trend toward more species and longer-lived species.

Succession begins with an ecological disturbance, an event that disrupts the ecosystem by altering its community, or its *abiotic* (nonliving) environment, or both. The precise changes that occur during succession are as diverse as the environments in which succession occurs, but we can recognize certain general stages. Succession starts with a few hardy lichens or plants called **pioneers.** The pioneers alter the environment in ways that favor competing plants, which gradually displace the pioneers. If allowed to continue, succession progresses to a diverse and relatively stable **climax community.** Alternatively, recurring disturbances can maintain a community in an earlier, or **subclimax,** stage of succession. Our discussion of succession will focus on plant communities, which dominate the landscape and provide both food and habitat for animals and microorganisms.

There Are Two Major Forms of Succession: Primary and Secondary

Succession takes one of two major forms: primary and secondary. During **primary succession,** a community gradually forms in a location where there are no remnants of a previous community, and often no trace of life at all. The disturbance that sets the stage for primary succession may be a glacier scouring the landscape down to bare rock, or it may be a volcano producing a new island in the ocean or creating a layer of newly hardened lava on land (**Fig. 27-15a**). Building a community from scratch through primary succession typically requires thousands or even tens of thousands of years.

During **secondary succession,** a new community develops after an existing ecosystem is disturbed in a way that leaves significant remnants of the previous community behind, such as soil and seeds. For example, beavers, landslides, or people may dam streams, causing marshes, ponds, or lakes to form. A landslide or avalanche may strip a swath of trees from a mountainside. When Mount St. Helens in Washington State erupted in 1980, it left patchy remnants of

(a) Kilauea, in Hawaii (primary succession)

(b) Mt. St. Helens, Washington State (secondary succession)

(c) Yellowstone National Park, Wyoming (secondary succession)

▲ FIGURE 27-15 **Succession in progress** (a) Primary succession. Left: The Hawaiian volcano Kilauea has erupted repeatedly since 1983, sending rivers of lava over the surrounding countryside. Right: A pioneer fern takes root in a crack in hardened lava. **(b)** Secondary succession. Left: On May 18, 1980, the explosion of Mount St. Helens in Washington State devastated the surrounding pine forest ecosystem. Right: This photo from 2009 shows the rebound of life over three decades. Because traces of the former ecosystem remained after the explosion, this is an example of secondary succession. **(c)** Secondary succession. Left: In the summer of 1988, extensive fires swept through the forests of Yellowstone National Park in Wyoming. Right: Trees and flowering plants are thriving in the sunlight, and wildlife populations are rebounding as secondary succession occurs.

QUESTION People have suppressed fires for decades. What are the implications of fire suppression for forest ecosystems and succession?

forest and a thick layer of nutrient-rich ash that encouraged a rapid proliferation of new life (**Fig. 27-15b**). Fire is another common disturbance that leads to secondary succession. The residues of burnt trees and shrubs are high in plant nutrients. Fires also spare some trees and many healthy roots. Some plants produce seeds that can withstand fire, or may even require it in order to sprout. Cones of lodgepole pines (**Fig. 27-15c**) are opened to release their seeds by the heat of fire. Thus, fires may promote rapid regeneration of forests and other communities.

Primary Succession May Begin on Bare Rock

Figure 27-16 illustrates primary succession on Isle Royale, Michigan, an island in northern Lake Superior that was scraped down to bare rock by glaciers that retreated roughly 10,000 years ago. Weathering produces cycles of freezing and thawing that cause rocks to crack, producing fissures, and erode rock surface layers, producing small particles. Rainwater dissolves some of the minerals from rock particles, making them available to pioneer organisms.

Weathered rock provides an attachment site for pioneering lichens, which obtain energy through photosynthesis and acquire some of their mineral nutrients by dissolving rock with an acid that they secrete. As the lichens spread over the rock surface, certain drought-tolerant mosses begin growing in the rock cracks. Fortified by nutrients liberated by the lichens, the mosses form a dense mat that traps dust, tiny rock particles, and bits of organic debris. The mosses will eventually cover and kill many of the lichens that made their growth possible.

As some mosses die and decompose each year, their bodies add nutrients to a thin layer of new soil, while the living moss mat acts like a sponge, absorbing and trapping moisture. Within the moss, seeds of larger plants, such as bluebell and yarrow, germinate. Then as these plants die, their decomposing bodies contribute to the growing layer of soil.

As woody shrubs such as blueberry and juniper take advantage of the newly formed soil, the mosses and remaining lichens may be shaded out and buried by decaying leaves and vegetation. Eventually, trees such as jack pine, black spruce, and aspen take root in the deeper crevices, and the sun-loving shrubs are shaded out. Within the forest, shade-tolerant seedlings of taller or faster-growing trees thrive, including balsam fir, paper birch, and white spruce. In time, these trees tower over and replace the original trees, which are intolerant of shade. After a thousand years or more, a tall climax forest thrives on what was once bare rock.

An Abandoned Farm Will Undergo Secondary Succession

Figure 27-17 illustrates secondary succession on an abandoned farm in the southeastern United States. The pioneer species are sun-loving, fast-growing annual plants such as

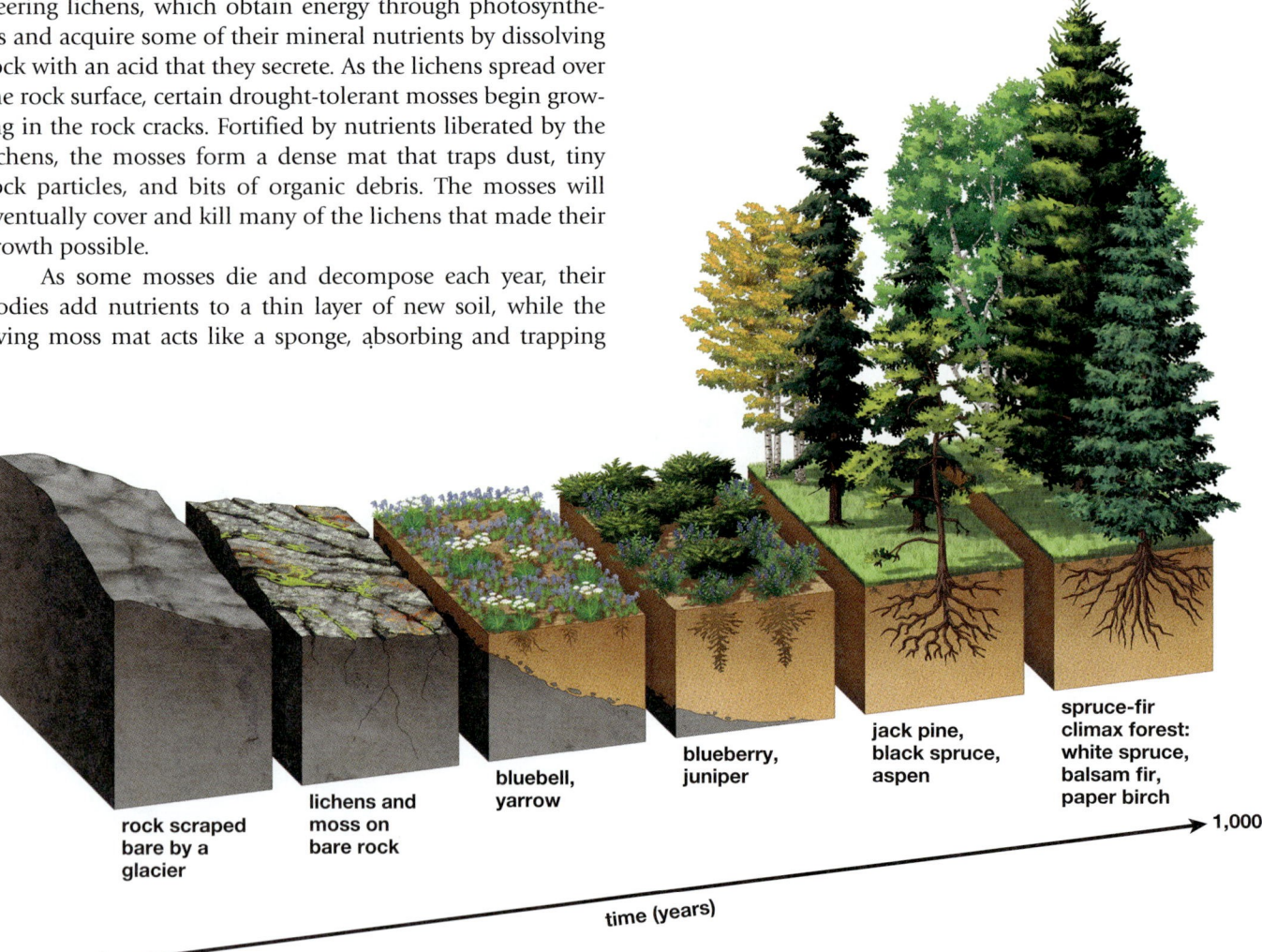

rock scraped bare by a glacier

lichens and moss on bare rock

bluebell, yarrow

blueberry, juniper

jack pine, black spruce, aspen

spruce-fir climax forest: white spruce, balsam fir, paper birch

1,000

time (years)

0

▲ **FIGURE 27-16 Primary succession** Primary succession as it occurs on bare rock exposed as glaciers retreated from Isle Royale in Lake Superior in upper Michigan. Notice that the soil deepens over time, gradually burying the bedrock and allowing trees to take root.

ragweed, crabgrass, and Johnson grass, which take root in the rich soil. Such species generally produce large numbers of easily dispersed seeds that help them colonize open spaces, but they don't compete well against longer-lived (perennial) species that grow larger over the years and shade out the pioneers.

After a few years, perennial plants such as asters, goldenrod, Queen Anne's lace, and perennial grasses move in, followed by woody shrubs such as blackberry and smooth sumac. These plants reproduce rapidly and dominate for many years as pine and cedar trees become established. Then, roughly two decades after a field is abandoned, an evergreen forest dominated by pines takes over and persists for several more decades.

As is common during succession, the new forest alters conditions in ways that favor its competitors. The shade of the pine forest inhibits the growth of its own seedlings, while favoring the growth of hardwood trees, whose seedlings are shade tolerant. Slow-growing hardwoods such as oak and hickory, which take root beneath the pines, begin to replace the aging pine trees after about 70 years. Roughly a century after the field was abandoned, the region is covered by relatively stable climax forest dominated by oak and hickory.

Succession Also Occurs in Ponds and Lakes

In freshwater ponds or lakes, succession occurs not only through changes within the pond or lake but also through an influx of nutrients from outside the ecosystem. Sediments and nutrients carried in by runoff from the surrounding land have a particularly large impact on small freshwater lakes,

ponds, and bogs, which gradually undergo succession to dry land (Fig. 27-18). In forests, meadows may be produced by lakes undergoing succession. As the lake fills in from the edges, grasses colonize the newly formed soil. As the lake shrinks and the area of meadow expands, trees will encroach around the meadow's edges.

Succession Culminates in a Climax Community

Succession ends with a reasonably stable climax community, which perpetuates itself if it is not disturbed by external forces (such as fire, parasites, invasive species, or human activities). The populations within a climax community have ecological niches that allow them to coexist without supplanting one another. In general, climax communities have more species and more types of community interactions than do early stages of succession. The plant species that dominate climax communities generally live longer and tend to be larger than pioneer species, particularly in climax forests. Extensive areas of characteristic climax plant

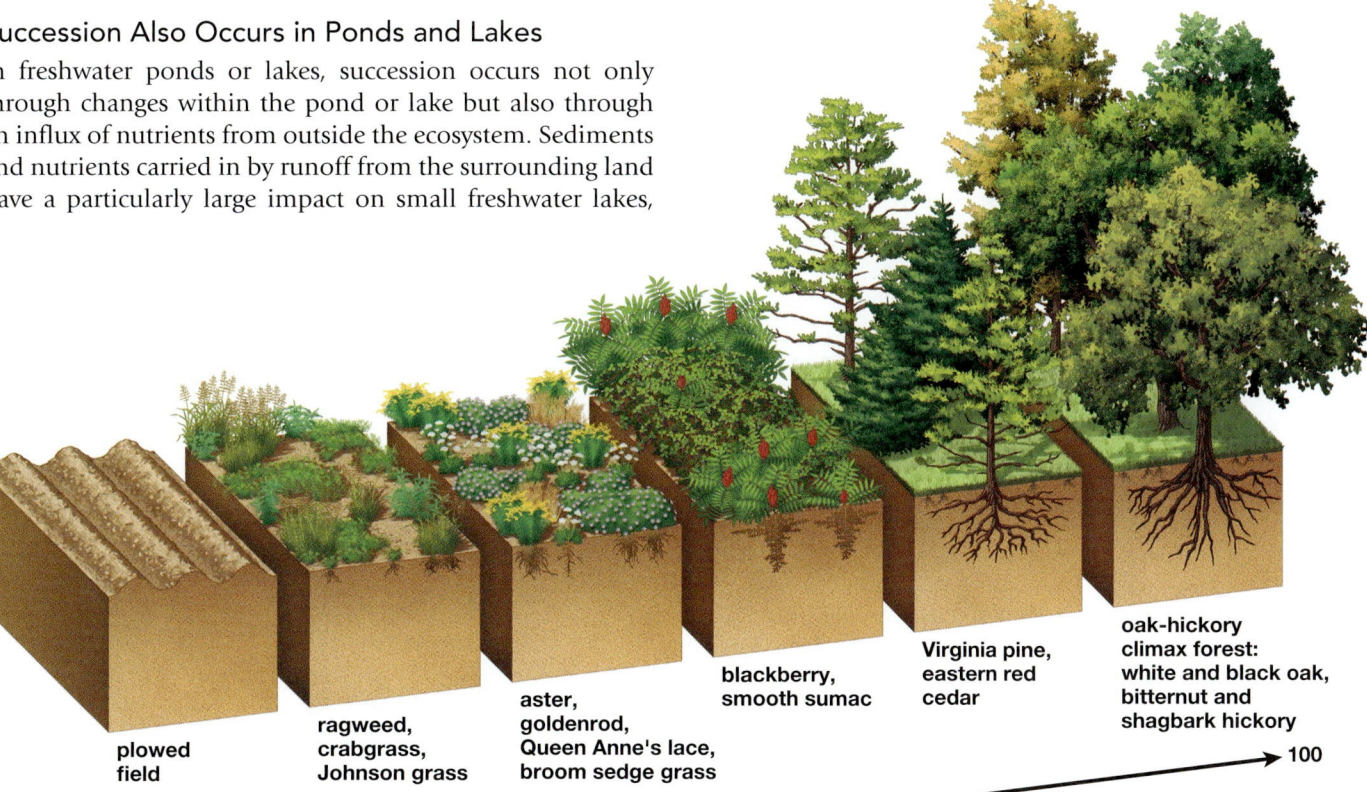

plowed field

ragweed, crabgrass, Johnson grass

aster, goldenrod, Queen Anne's lace, broom sedge grass

blackberry, smooth sumac

Virginia pine, eastern red cedar

oak-hickory climax forest: white and black oak, bitternut and shagbark hickory

0 — time (years) → 100

▲ FIGURE 27-17 **Secondary succession** Secondary succession as it might occur on a plowed, abandoned farm field in North Carolina, in the southeastern United States. Notice that a thick layer of soil is present from the beginning, which greatly speeds up the process compared to primary succession.

▲ **FIGURE 27-18 Succession in a small freshwater pond** In small ponds, succession is speeded by an influx of materials from the surroundings. Over time, the decaying bodies of aquatic plants and marsh-loving cattails and grasses build up sediment and eventually soil. This provides anchorage for more terrestrial plants. Eventually, the pond will fill in and become dry land.

communities are called **biomes,** and include deserts, grasslands, and several types of forests (described in Chapter 28).

If you have traveled, you may have noticed that the type of climax community varies dramatically from one area to the next. For example, if you drive through Colorado or Wyoming, you will see a shortgrass prairie climax community on the eastern plains (in those rare areas where it has not been replaced by farms), pine-spruce forests in the mountains, tundra on the mountain summits, and sagebrush-dominated communities in the western valleys. The exact nature of the climax community is determined by numerous geological and climatic variables, including temperature, rainfall, elevation, latitude, type of rock (which determines the type of nutrients available), and exposure to sun and wind. Natural events such as windstorms, avalanches, and fires started by lightning may destroy sections of climax forest, reinitiating secondary succession and producing a patchwork of various successional stages within an ecosystem.

In many forests throughout the United States, rangers now allow fires set by lightning to run their course, recognizing that this natural process is important for the maintenance of the entire ecosystem. Fires liberate nutrients and kill some (but usually not all) of the trees. As a result, more nutrients and sunlight reach the forest floor, encouraging the growth of subclimax plants. The combination of climax and subclimax regions within the ecosystem provides habitats for a far larger number of species than would either climax or subclimax vegetation alone.

Some Ecosystems Are Maintained in Subclimax Stages

Frequent disturbances can maintain some ecosystems in subclimax stages for very long periods of time. The tallgrass prairie that once covered northern Missouri and Illinois is a subclimax stage of an ecosystem whose climax community is deciduous forest. The subclimax prairie was maintained by periodic fires, some set by lightning and others deliberately set by Native Americans centuries ago to increase grazing land for bison. Forest now encroaches, but some tallgrass prairie preserves are being sustained by carefully managed burning.

Farms, gardens, and lawns are subclimax communities maintained by frequent, intentional disturbance. Grains are specialized grasses characteristic of the early stages of succession, and farmers spend a great deal of time, energy, and herbicides to prevent competitors, such as other grasses, wildflowers, and woody shrubs, from taking over. The suburban lawn is a subclimax ecosystem maintained by regular mowing, which destroys woody colonizers, and the application of herbicides, which selectively kill pioneer species such as crabgrass and dandelions.

CHECK YOUR LEARNING

Can you explain the process of succession and its general stages? Can you define *primary succession*, *secondary succession*, *subclimax ecosystem*, and *climax ecosystem*?

CASE STUDY revisited

Mussels Muscle In

Zebra and quagga mussels have all the characteristics that make invasive species successful in their introduced homes. They have phenomenal biotic potential—an adult female of either species releases up to 1 million eggs annually. Their microscopic larvae can be carried for miles in water currents, allowing them to rapidly colonize the length of a river. Because these mussels can survive out of water for days, those that cling to the hulls of boats can be easily transferred to other lakes and rivers, where they rapidly establish new colonies.

As the zebra and quagga mussels continue to spread, some researchers suggest that lack of phytoplankton, their primary food, may eventually slow their proliferation in the Great Lakes. Lake Michigan and Lake Erie, for example, have become strikingly clearer since the mussels' invasion because phytoplankton populations have plummeted. But food depletion is not an appealing solution to the mussel problem, because phytoplankton directly or indirectly support most of the life within the Great Lakes.

Ironically, zebra and quagga mussels have been implicated in Lake Erie's worst phytoplankton bloom in decades. Unusually heavy runoff from snow and rain in early 2011 carried large amounts of fertilizer from farm fields into the lake. By October, a thick green scum of the photosynthetic bacteria *Microcystis* was visible from space (**Fig. 27-19**). *Microcystis* blooms have occurred with increasing frequency since the mussels invaded. Why? Zebra and quagga mussels filter nearly all the phytoplankton from the water that flows over their gills. They then consume the diatoms but reject the *Microcystis*, releasing them back into the lake. With diatom populations severely depleted, the bacteria have far less competition for nutrients in the water, which allows their populations to skyrocket.

▲ **FIGURE 27-19 Phytoplankton bloom** Green plumes formed by filaments of the microscopic photosynthetic bacteria *Microcystis* are visible from space near the shore of Lake Erie in 2011.

Consider This

BioEthics Mussel larvae settle rapidly on boat hulls. Thus, to prevent the spread of the larvae, a boat that has been in the water for 24 hours must be washed with a hot-water pressure washer or dried for at least 5 days before being launched in another body of water. Will recreational boaters comply voluntarily? Do you think there is any practical way to restrict the spread of these mussels to still more lakes and reservoirs? What measures would be required to accomplish this?

CHAPTER REVIEW

Summary of Key Concepts

27.1 Why Are Community Interactions Important?
Ecological communities consist of all the interacting populations within an ecosystem. Community interactions influence population size, and the interacting populations within communities act on one another as agents of natural selection. Thus, community interactions shape the bodies and behaviors of members of the interacting populations.

27.2 How Does the Ecological Niche Influence Competition?
The ecological niche defines all aspects of a species' habitat and interactions with its living and nonliving environments. Each species occupies a unique ecological niche. Interspecific competition occurs whenever the niches of two species within a community overlap. When two species with the same niche are forced (under laboratory conditions) to occupy the same ecological niche, one species always outcompetes the other. Species within natural communities have evolved in ways that avoid excessive niche overlap, with behavioral and physical adaptations that allow resource partitioning. Interspecific competition limits both the population size and the distribution of competing species. Intraspecific competition also limits populations because individuals of the same species occupy the same ecological niche and compete with one another for all of their needs.

27.3 How Do Predator–Prey Interactions Shape Evolutionary Adaptations?
Predators eat other organisms and are generally larger and less abundant than their prey. Predators and prey act as strong agents of natural selection on one another. Both predators and prey have evolved a variety of toxic chemicals for attack and defense. Plants have evolved defenses ranging from poisons to overall toughness. These defenses, in turn, have selected for predators that can detoxify poisons and grind down tough tissues. Prey animals have evolved a variety of protective colorations that render them either inconspicuous (camouflage) or startling (startle coloration) to their predators. Some prey are poisonous, distasteful, or venomous, and exhibit warning coloration by which they are readily recognized and avoided by predators, whereas others have evolved to resemble other, more distasteful organisms through mimicry.

27.4 What Are Parasitism and Mutualism?

Parasites feed on larger, less abundant host organisms, harming them but usually not killing them immediately. Parasites include disease-causing viruses and microorganisms as well as animals such as fleas and tapeworms. Mutualism benefits both interacting species. Some mutualistic interactions are symbiotic, such as those of cows and their cellulose-digesting microorganisms. Other mutualistic interactions are more temporary, such as those that occur between plants and the animals that pollinate them.

27.5 How Do Keystone Species Influence Community Structure?

Keystone species have a greater influence on community structure than can be predicted by their numbers. If a keystone species is removed from a community, the structure of the community is significantly altered.

27.6 How Do Community Interactions Cause Change over Time?

Succession is a change in a community and its nonliving environment over time. During succession, plants alter the environment in ways that favor their competitors, thus producing a somewhat predictable progression of dominant species. Primary succession, which may take thousands of years, occurs where no remnant of a previous community existed, such as on bare rock. Secondary succession occurs much more rapidly because it builds on the remains of a disrupted community, such as an abandoned field or the remnants of a forest after a fire. Uninterrupted succession ends with a climax community, which tends to be self-perpetuating unless acted on by outside forces, such as fire or human activities. Some ecosystems, including tallgrass prairie and farm fields, are maintained in relatively early, subclimax stages of succession by periodic disruptions.

Key Terms

aggressive mimicry	518	keystone species	521
biome	526	mimicry	516
camouflage	515	mutualism	520
climax community	522	parasite	519
coevolution	510	pioneer	522
community	510	predator	512
competition	511	primary succession	522
competitive exclusion principle	511	resource partitioning	511
		secondary succession	522
ecological niche	510	startle coloration	518
host	519	subclimax	522
interspecific competition	511	succession	522
intraspecific competition	512	warning coloration	516

Learning Outcomes

In this chapter, you have learned to . . .

LO1 Explain the four basic types of community interaction and discuss why each is important to the community.

LO2 Define the *ecological niche* concept and use it to explain why there is more competition within a species than there is among species.

LO3 Explain how interspecific competition causes coevolution and results in resource partitioning.

LO4 Describe predator–prey relationships and provide examples of the coevolution of predators and their prey.

LO5 Compare parasite–host with predator–prey relationships and provide some examples of each.

LO6 Explain mutualism and provide examples of symbiotic mutualistic relationships as well as other types of mutualism that are less intimate and extended.

LO7 Define the term *keystone species* and provide examples of some keystone predators, including a description of how they influence the structure of their communities.

LO8 Define the term *succession* and compare primary succession with secondary succession.

Thinking Through the Concepts

Fill-in-the-Blank

1. Organisms that interact serve as agents of _____ on one another. This results in _____, which is the process by which species evolve adaptations to one another. Four types of community interactions described in this chapter are _____, _____, _____, and _____. **LO1 LO3**

2. Predators may be meat eaters, called _____, or plant eaters, called _____. Both predators and their prey may blend into their surroundings by using _____. Predators are generally _____ and less _____ than their prey. **LO4**

3. Competition occurs whenever two different populations within a community have overlapping _____. The concept that no two species with identical niches can coexist indefinitely is called the _____ principle. **LO2 LO3**

4. Fill in the types of coloration or mimicry: Used by a prey to signal that it is distasteful: _____; used by a moth with large eyespots on its wings:_____; mimicry of a poisonous animal by a nonpoisonous animal: _____; mimicry used by a predator to attract its prey: _____. **LO4**

5. Fill in the appropriate type of community interaction: Bacteria, living in the human gut, that synthesize vitamin K: _____; bacteria that cause illness: _____; a deer eating grass: _____; a tick sucking blood: _____; a bee pollinating a flower:_____; kudzu covering trees:_____. **LO1 LO5 LO6**

6. A somewhat predictable change in community structure over time is called_____. This process takes two forms. Which of these forms would start with bare rock? _____ Which would occur after a forest fire? _____ A relatively stable community that is the end product of this process is called a _____ community. A mowed lawn in suburbia is an example of a _____ community. **LO8 LO9**

Review Questions

1. Define an *ecological community*, and explain the four important types of community interactions. **LO1**

2. Describe several key ways in which specific plants and animals protect themselves from being eaten. For each, think of and describe an adaptation that might evolve in predators of these species that would overcome their defenses. **LO4**

3. Explain how resource partitioning is a logical outcome of the competitive exclusion principle. **LO3**

4. Define *parasitism* and *mutualism* and provide an example of each. **LO5 LO6**

5. Define *succession*. Which type of succession would occur on a clear-cut forest (where all trees have been logged) and why? **LO8**

6. Provide examples of two climax and two subclimax communities. How do they differ? **LO9**

7. What is an invasive species? Why are they destructive? What general adaptations do invasive species possess? **LO2 LO4 LO7**

8. What is a keystone species? How can a keystone species within a community be identified? **LO7**

Applying the Concepts

1. Herbivorous animals that eat seeds are considered by some ecologists to be predators of plants, and herbivorous animals that eat leaves are considered to be parasites of plants. Discuss the validity of this classification scheme.

2. An ecologist visiting an island finds two very closely related species of birds, one of which has a slightly larger bill than the other. Interpret this finding with respect to the competitive exclusion principle and the ecological niche, and explain both concepts.

3. As a frogfish sits camouflaged on the ocean floor, wiggling its lure; a small fish approaches the lure and is eaten, while a very large predatory fish fails to notice the frogfish. Describe all the possible types of community interactions and adaptations that these organisms exhibit.

4. Design an experiment to determine whether the kangaroo is a keystone species in the Australian outback.

5. Why is it difficult to study succession? Suggest some ways you would approach this challenge for a few different ecosystems.

Answers to Figure Caption questions and Fill-in-the-Blank questions can be found in the Answers section at the back of the book.

MB *Go to MasteringBiology for practice quizzes, activities, eText, videos, current events, and more.*

CHAPTER 28 Energy Flow and Nutrient Cycling in Ecosystems

CASE STUDY

Dying Fish Feed an Ecosystem

THE SOCKEYE SALMON in Katmai National Park in Alaska have a remarkable life cycle. After hatching in shallow depressions in the gravel bed of a swiftly flowing stream, the juvenile salmon spend 1 to 3 years in fresh water, often in a nearby lake. Then the young fish, only 4 to 6 inches long, begin the changes in physiology that will prepare them for life in the sea. They head downstream in rivers that eventually flow into estuaries, where the river's fresh water and the ocean's salt water mix. In the estuary, the salmon complete their transformation, and then head out to sea.

In the open ocean, the young salmon grow rapidly, feeding on small fish and crustaceans. A few years later, when they reach sexual maturity, an instinctive drive compels them to return to fresh water to spawn. However, not just any stream or river will do. The salmon swim along the coast until the unique scent of their home stream entices them to swim inland. They usually spend a few days in the estuary where the home stream joins the sea, as their bodies undergo the changes that allow them to survive in fresh water. Then, battling swift currents and leaping up small waterfalls, the salmon that are lucky enough to avoid the jaws of brown bears and the talons of eagles carry their precious payload of sperm and eggs upstream to renew the cycle of life. The salmon die soon after spawning.

The fishes' journey back to their birthplace is remarkable in another way. Nutrients almost always flow downstream, washed from the land into the ocean. But salmon, filled with muscle and fat acquired from feeding in the ocean, bring ocean nutrients back to the land. From bears to eagles to spruce trees, this flow of nutrients supports the extraordinarily rich ecosystems of the Alaskan coast. How is the movement of nutrients tied to the flow of energy from the sun that supports all life on Earth? Consider this question as we explore ecosystem function in this chapter.

A brown bear intercepts a salmon on its journey back to its home stream to spawn.

AT A GLANCE

LEARNING GOALS

LG1 Compare and contrast biotic and abiotic components of ecosystems and describe how nutrients and energy move between these two components.

LG2 Describe the trophic levels in an ecosystem and explain how energy flows through trophic levels.

LG3 Explain why detritivores and decomposers are essential to ecosystem function.

LG4 Explain why energy transfer between trophic levels is inefficient and discuss how this inefficiency determines the relative abundance of organisms at each trophic level.

LG5 Explain the general principles that govern nutrient cycling in an ecosystem.

LG6 Describe the hydrologic, nitrogen, carbon, and phosphorus cycles.

LG7 Describe how human interference with nutrient cycling causes acid deposition, damages aquatic ecosystems, and causes global climate change.

LG8 Describe the evidence that climate change is occurring and discuss key impacts of climate change on Earth's ecosystems.

28.1 HOW DO NUTRIENTS AND ENERGY MOVE THROUGH ECOSYSTEMS?

All **ecosystems** consist of two components. The **biotic** component of an ecosystem is the community of living organisms—bacteria, fungi, protists, plants, and animals—in a given area. The **abiotic** component of an ecosystem consists of all the nonliving physical or chemical aspects of the environment, such as climate, light, temperature, the availability of water, and minerals in the soil. Interactions within biological communities and between communities and their abiotic environment determine the movement of energy and nutrients through ecosystems. Two basic principles underlie this movement: Nutrients recycle within and among ecosystems, whereas energy flows through ecosystems (Fig. 28-1).

Nutrients are atoms and molecules that organisms obtain from their environment. The same nutrient atoms have been sustaining life on Earth for about 3.5 billion years. Your body undoubtedly includes oxygen, carbon, hydrogen, and nitrogen atoms that were once part of a dinosaur or a wooly mammoth. Nutrients are transported around the planet and converted to different molecular forms, but they do not leave Earth.

Energy, in contrast, takes a one-way journey through ecosystems. Solar energy is captured by photosynthetic bacteria, algae, and plants, and then flows from organism to organism. Eventually, however, all of life's energy is converted to heat that is given off to the environment and cannot be used to drive the chemical reactions of living organisms. Therefore, life requires a continuous input of energy.

CHECK YOUR LEARNING

Can you explain why nutrients cycle within and among ecosystems, whereas energy flows through ecosystems?

28.2 HOW DOES ENERGY FLOW THROUGH ECOSYSTEMS?

Ninety-three million miles from Earth, thermonuclear reactions in the sun convert hydrogen into helium, transforming a relatively small amount of matter into enormous quantities of energy. A tiny fraction of this energy reaches Earth in the form of electromagnetic waves, including heat (infrared light), visible light, and ultraviolet light. About half the energy that reaches Earth is visible light. Much of the solar energy is reflected back into space by the atmosphere, clouds, and Earth's surface. Some is absorbed by Earth and its atmosphere, warming the planet. Less than 0.03% of the energy reaching our planet from the sun is captured by photosynthetic organisms, and supports life on Earth.

Energy Enters Ecosystems Through Photosynthesis

Plants, algae, and photosynthetic bacteria acquire nutrients such as carbon, nitrogen, oxygen, and phosphorus from the abiotic portions of ecosystems. Photosynthetic organisms capture the energy of sunlight and use it to join these nutrient atoms into sugars, starches, proteins, nucleic acids, and all the other biological molecules of their bodies, storing some of the captured energy in the chemical bonds of these molecules. As we shall see shortly, the nutrients and energy contained in biological molecules move from photosynthetic organisms to nonphotosynthetic organisms, such as animals, fungi, and most bacteria. Thus, photosynthesizers bring both energy and nutrients into ecosystems.

Energy Is Passed from One Trophic Level to the Next

Energy flow through ecosystems begins with photosynthetic organisms and passes through several levels of nonphotosynthetic organisms that feed on the photosynthesizers or on each other.

▶ **FIGURE 28-1 Energy flow, nutrient cycling, and feeding relationships in ecosystems** The energy of sunlight (yellow arrow) enters an ecosystem during photosynthesis by organisms collectively called producers, which store some of the energy in the biological molecules of their bodies. This biologically available energy (red arrows) is then passed to nonphotosynthetic organisms collectively called consumers. Energy in wastes and dead bodies supports detritivores and decomposers. Every organism loses some energy as heat (orange arrows), so useful energy gradually becomes unavailable to living organisms. Therefore, ecosystems require a continuous input of energy. In contrast, nutrients (purple arrows) are continually recycled.

QUESTION Why is some energy in any process always lost as heat?

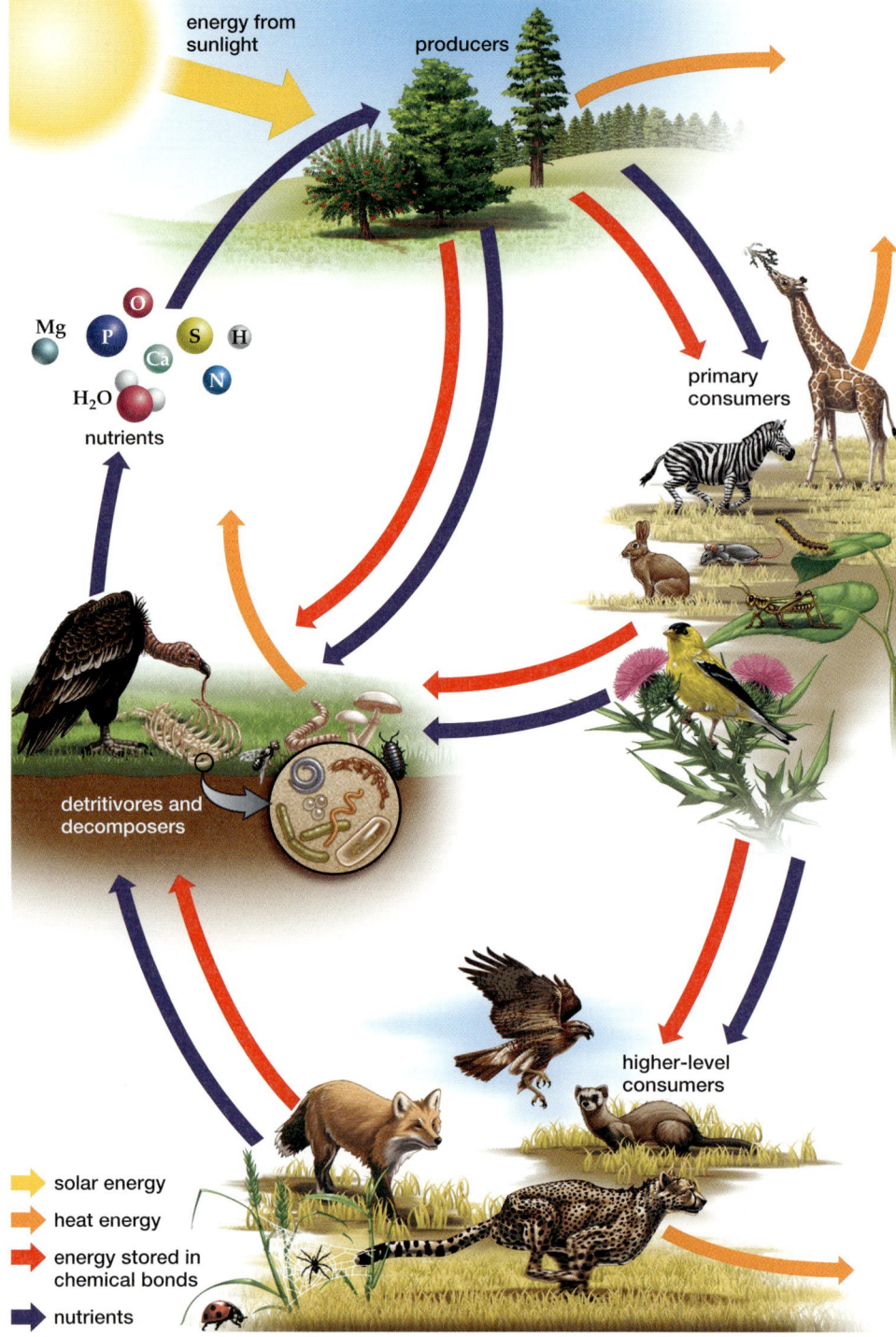

Each category of organisms is called a **trophic level** (literally, "feeding level"). Photosynthetic organisms, from oak trees in a forest to single-celled algae in the ocean, form the first trophic level. These organisms are called **producers,** or **autotrophs** (Greek, meaning "self-feeders"), because they produce food for themselves using inorganic nutrients and solar energy. In doing so, they directly or indirectly produce food for nearly all other forms of life as well. Organisms that cannot photosynthesize, called **consumers,** or **heterotrophs** ("other-feeders"), must acquire energy and most of their nutrients prepackaged in the molecules that are part of the bodies of other organisms.

There are several levels of consumers. **Primary consumers** feed directly on producers. These **herbivores** (from Latin words meaning "plant eaters"), which include animals such as grasshoppers, mice, and zebras, form the second trophic level. **Carnivores** ("meat eaters"), such as spiders, hawks, and salmon, make up the higher-level consumers. Carnivores act as **secondary consumers** when they prey on herbivores. Some carnivores at least occasionally eat other carnivores; when doing so, they occupy the fourth trophic level, and are called **tertiary consumers.** In some instances, particularly in the oceans, there are even higher trophic levels.

Net Primary Production Is a Measure of the Energy Stored in Producers

The amount of life that an ecosystem can support is determined by the amount of energy captured by the producers in that ecosystem. The energy that photosynthetic organisms in a given area store in their bodies over a given period of time (for example, calories per square meter per year) is called **net primary production.** Mass is much easier to determine than energy. **Biomass,** or dry biological material, is usually a good measure of the energy stored in organisms' bodies. Therefore, net primary production is usually presented as grams of biomass per square meter per year (**Fig. 28-2**).

The net primary production of an ecosystem is influenced by many factors, including the amount of sunlight reaching the producers, the availability of water and nutrients, and the temperature. In the desert, for example, lack of water limits production. In the open ocean, light is a limiting factor in deep waters, and lack of nutrients limits production in most surface waters. In ecosystems where all resources are abundant, such as tropical rain forests, production is high.

An ecosystem's contribution to Earth's total production is determined both by the ecosystem's productivity and by the portion of Earth that the ecosystem covers. The oceans have low net primary production, but they cover about 70% of Earth's surface, so they contribute about 25% of Earth's total production. This is about the same overall contribution as tropical rain forests, which have a high net primary production, but which cover less than 5% of Earth's surface.

Food Chains and Food Webs Describe Feeding Relationships Within Communities

A **food chain** is a linear feeding relationship that includes a single species in each trophic level that is fed upon by a single species in the trophic level just above it (**Fig. 28-3**). Different ecosystems support radically different food chains. Plants are the dominant producers in land-based (terrestrial) ecosystems (**Fig. 28-3a**). Plants support plant-eating insects, reptiles, birds, and mammals, each of which may be preyed on by other animals. In contrast, microscopic photosynthetic protists and bacteria collectively called **phytoplankton** (from Greek words meaning "plant drifters") are the dominant producers in most aquatic food chains, such as those found in lakes and oceans (**Fig. 28-3b**). Phytoplankton support a diverse group of consumers called **zooplankton** ("animal drifters"), which consist mainly of protists and small shrimp-like crustaceans. These are eaten primarily by fish, which in turn are eaten by larger fish.

Animals in natural communities often do not fit neatly into the categories of primary, secondary, and tertiary consumers depicted in simple food chains. A **food web** shows many interconnected food chains, and more accurately describes the actual feeding relationships within a community (**Fig. 28-4**). Some animals, such as raccoons, bears, rats, and humans, are **omnivores** ("everything eaters"), and they act as primary, secondary, and occasionally tertiary consumers. A hawk, for instance, is a secondary consumer when it eats a mouse (an herbivore) and a tertiary consumer when it eats a meadowlark that feeds on insects. A carnivorous plant such as the Venus flytrap can tangle the food web further by acting as both a photosynthesizing producer and a spider-trapping tertiary consumer.

Detritivores and Decomposers Release Nutrients for Reuse

Among the most important strands in a food web are the detritivores and decomposers. **Detritivores** ("debris eaters") are an army of mostly small and often unnoticed organisms, including nematode worms, earthworms, millipedes, dung beetles, slugs, and the larvae of some flies. They eat the refuse of life, such as fallen leaves and the wastes and dead bodies of other organisms A few large vertebrates such as vultures are also detritivores.

Decomposers are primarily fungi and bacteria. They feed on the same material as detritivores, but they do not ingest chunks of organic matter, as detritivores do. Instead, they secrete digestive enzymes outside their bodies, where the enzymes break down nearby organic material. The decomposers

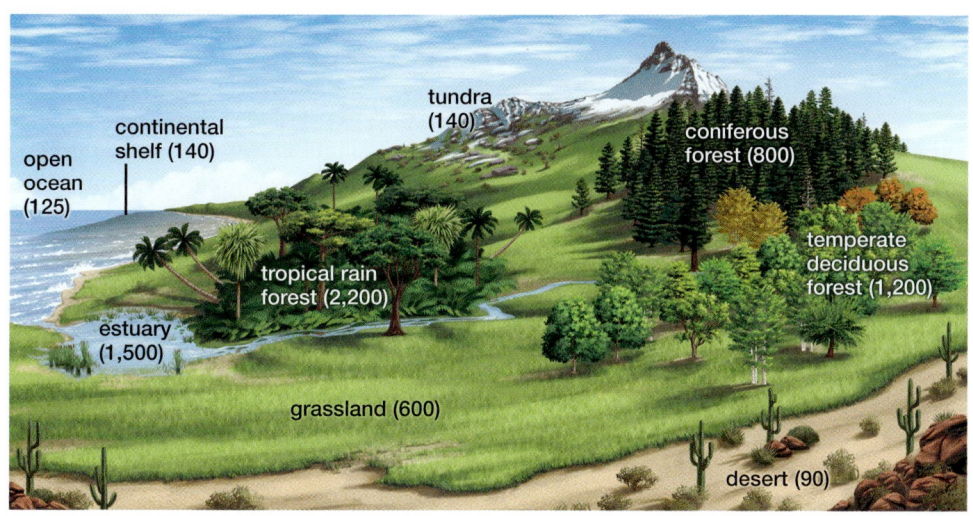

▲ FIGURE 28-2 **Net primary production in ecosystems** The average net primary production of some terrestrial and aquatic ecosystems is shown, measured in grams of biological material produced per square meter per year.

QUESTION What factors contribute to the differences in productivity among ecosystems?

(a) A simple terrestrial food chain

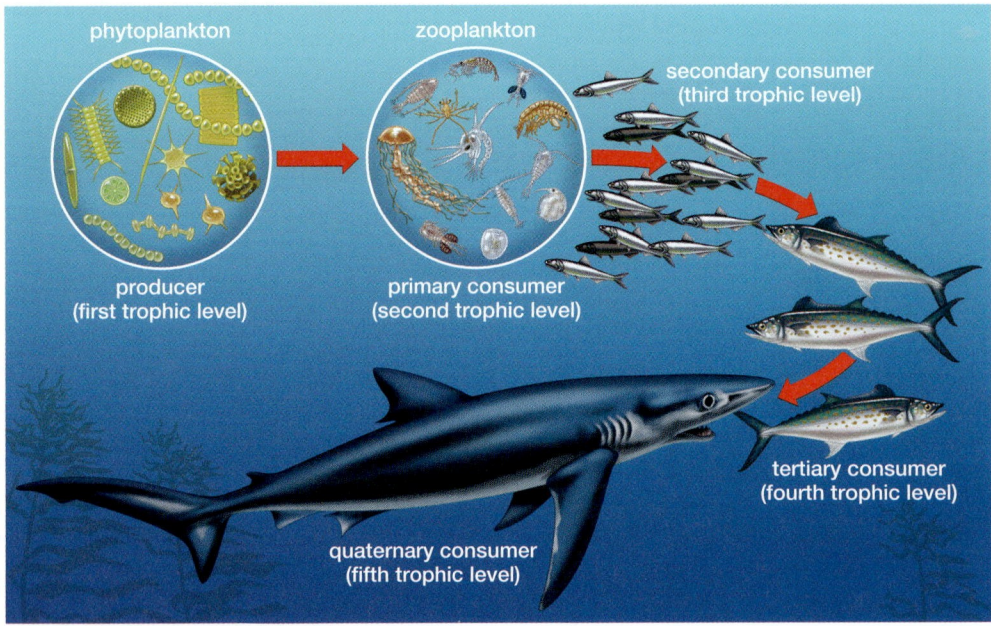

(b) A simple marine food chain

▲ FIGURE 28-3 **Food chains on land and sea**

absorb some of the resulting nutrient molecules, but many of the nutrients remain in the environment. Mushrooms in a lawn or the blue-gray fuzz you may notice on old bread are fungal decomposers hard at work, and the smelly slime on spoiled meat signals the presence of bacterial decomposers.

Detritivores and decomposers are absolutely essential to life on Earth. They reduce the bodies and wastes of other organisms to simple molecules—such as carbon dioxide, ammonia, and minerals—that return to the atmosphere, soil, and water. Without detritivores and decomposers, ecosystems would gradually be buried by accumulated wastes and dead bodies, whose nutrients would be unavailable to enrich the

soil and water. Eventually, plants and other photosynthetic organisms would be unable to obtain enough nutrients. With the producers eliminated, both energy and nutrients would cease to enter the ecosystem, and organisms at higher trophic levels, including people, would disappear as well.

Energy Transfer Between Trophic Levels Is Inefficient

A fundamental principle of the branch of physics called thermodynamics is that energy use is never completely efficient (see Chapter 6). For example, as your car burns

◀ **FIGURE 28-4 A simplified grassland food web** The animals pictured in the foreground include a vulture (a detritus feeder), a bull snake, a ground squirrel, a burrowing owl, a badger, a mouse, and a shrew (which looks like a small mouse but is carnivorous). In the middle distance you'll see a grouse, a meadowlark, a grasshopper, and a jackrabbit. In the far distance, look for pronghorn antelope, a hawk, a wolf, and bison.

QUESTION Using this figure as a guide, draw a food chain with four trophic levels, and identify each level.

gasoline, only about 20% of the resulting energy is used to move the car; the other 80% is lost as heat. Inefficiency is also the rule in living systems: Waste heat is produced by all the biochemical reactions that keep cells alive. For example, splitting the chemical bonds of adenosine triphosphate (ATP) to power muscular contraction releases heat energy; this is why shivering or walking briskly on a cold day warms your body.

Energy transfer from one trophic level to the next is quite inefficient. When a grasshopper (a primary consumer) eats grass (a producer), only some of the solar energy trapped by the grass is available to the insect. Some was converted into the chemical bonds of cellulose, which a grasshopper cannot digest. In addition, although grasses don't feel warm to the touch, virtually every chemical reaction in the grass's leaves, stems, and roots gives off some energy as low-level heat, which thus isn't available to any other living organism. Therefore, only a fraction of the energy captured by the producers of the first trophic level can be used by organisms in the second trophic level. If the grasshopper is eaten by a robin (the third trophic level), the bird will not obtain all the energy that the insect acquired from the plants. Some of the energy will have been used up to power hopping, flying, and eating. Some energy will be found in the grasshopper's indigestible exoskeleton. Much of it will have been lost as heat. Likewise, most of the energy in a robin's body will be unavailable to a hawk that may consume it.

Although energy transfer between trophic levels varies significantly among communities, the average net transfer of energy from one trophic level to the next is roughly 10%. This means that, in general, the amount of energy stored in primary consumers is only about 10% of the energy stored in the bodies of producers. In turn, the bodies of secondary consumers contain roughly 10% of the energy stored in

primary consumers. This inefficient energy transfer between trophic levels is called the "10% law." An **energy pyramid** illustrates the energy relationships between trophic levels—widest at the base, and progressively narrowing in higher trophic levels (**Fig. 28-5**). A biomass pyramid for a given community typically has the same general shape as the community's energy pyramid.

As a result of the inefficiency of energy transfer between trophic levels, the predominant organisms in a community are almost always plants, because they have the most

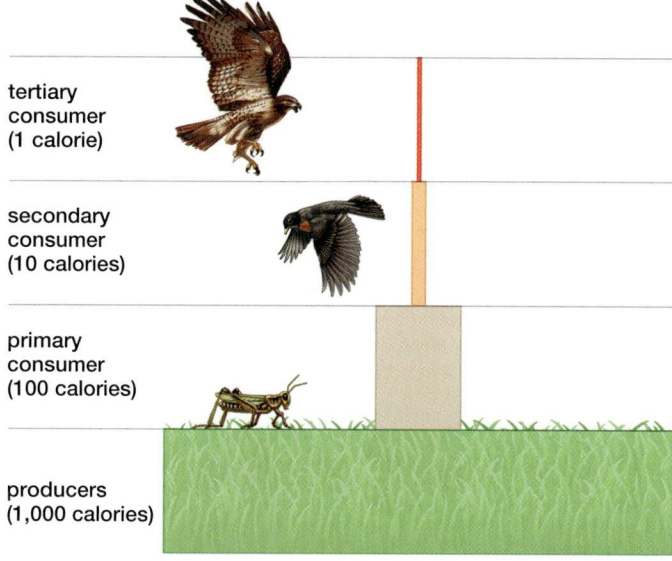

tertiary consumer (1 calorie)

secondary consumer (10 calories)

primary consumer (100 calories)

producers (1,000 calories)

▲ **FIGURE 28-5 An energy pyramid for a grassland ecosystem** The width of each rectangle is proportional to the energy stored at that trophic level. Representative organisms for the first four trophic levels in a U.S. grassland ecosystem illustrated here are grass, a grasshopper, a robin, and a red-tailed hawk.

Health | Watch

Biological Magnification of Toxic Substances

During the 1950s and 1960s, wildlife biologists in the United States witnessed an alarming decline in populations of several predatory birds, especially fish-eaters such as cormorants, ospreys, brown pelicans, and bald eagles. Some, including the brown pelican and the bald eagle, came perilously close to extinction. What had gone wrong? They were being poisoned by a pesticide called DDT. To control insects, the aquatic ecosystems supporting these birds had been sprayed with relatively low amounts of DDT. However, the birds were found to contain concentrations of DDT up to a million times greater than the concentration in the water where their fish prey lived. Low concentrations of DDT probably had little or no harmful effect in the fish, but the enormously higher concentrations in the birds who ate the fish caused them to lay thin-shelled eggs that broke under the parents' weight during incubation.

The birds were victims of *biological magnification*, the process by which toxic substances accumulate in increasingly high concentrations in animals occupying higher trophic levels. Substances that undergo biological magnification (such as DDT) share two properties that make them dangerous. First, neither animals nor decomposers can readily break them down into harmless substances—that is, they are not easily *biodegradable*. Second, such substances tend to be stored in living tissue, particularly in fat.

How does biological magnification work? Some toxic chemicals are readily absorbed by an animal's digestive tract, often as it absorbs fats from its foods. The chemicals are then stored in the animal's body. Because energy transfer between trophic levels is inefficient, an herbivore must eat large quantities of producers just to stay alive, let alone to grow or reproduce. By repeatedly eating food with low levels of toxic chemicals and storing them in its own body, an herbivore may gradually accumulate dangerous concentrations of the chemicals. This process—eat, absorb, and store—continues up the trophic levels. By the third or fourth trophic level, a carnivore may be eating prey that already has fairly high

concentrations of the toxic chemicals: A bald eagle may eat a salmon that ate smaller fish that ate small crustaceans that ate phytoplankton, so the magnification can become enormous (**Fig. E28-1**). Animals exposed to high levels of pesticides and other persistent pollutants are at increased risk for infertility, some types of cancer, heart disease, suppressed immune function, and neurological damage.

Mercury, a potent neurotoxin, is a particular cause for concern. Because it is an element, mercury can never be broken down. Further, mercury accumulates in muscle as well as fat. High mercury levels in a few types of long-lived, predatory fish—such as swordfish, shark, and albacore tuna—have prompted the U.S. Food and Drug Administration to advise young children and women of childbearing age to limit their consumption of these fish. In the United States, coal-fired power plants are the largest single source of mercury contamination. Atmospheric mercury can be wafted thousands of miles from these plants, so nowhere on Earth is really free of mercury contamination.

Fortunately, there is good news. Organic pollutants do eventually break down in the environment, and populations of predatory birds have recovered significantly since DDT was banned in the United States in 1973. Nesting pairs of bald eagles in the lower 48 states of the United States have increased from fewer than 500 in the 1960s to 10,000 today. Another 10,000 to 12,000 nesting pairs live in southern Alaska. In 2004, 158 countries worldwide agreed to restrict or ban the production and use of a dozen "persistent organic pollutants," including DDT and several other highly toxic pesticides (DDT may still be used for malaria control). Additional countries have since agreed to the restrictions, bringing the total to 176. And at the end of 2011, the U.S. Environmental Protection Agency issued new rules that require reductions in emissions of many toxic substances from power plants, including mercury, arsenic, nickel, and selenium.

▲ **FIGURE E28-1 Biological magnification** Because they eat carnivorous fish, bald eagles feed at a high trophic level, which causes fat-soluble toxic substances to accumulate in their bodies.

---

energy available to them, as sunlight. The most abundant animals are herbivores. Carnivores are relatively scarce, because there is far less energy available to support them.

Energy losses within and between trophic levels mean that long-lived animals at higher trophic levels eat many times their body weight in food that they obtain from lower trophic levels—consider, for example, how much food you eat in a year, even if you stay the same weight. If the food contains certain types of toxic substances, they may be stored, and therefore become more concentrated, in the bodies of animals in higher trophic levels. This **biological magnification** can lead to harmful, even fatal, effects, as we

explore in the "Health Watch: Biological Magnification of Toxic Substances."

CHECK YOUR LEARNING

Can you name the trophic levels in a community and give examples of organisms found in each trophic level? Can you explain how energy flows through an ecosystem? Are you able to explain why detritivores and decomposers are essential to ecosystem function? Can you explain how the inefficiency of energy transfer between trophic levels determines the relative abundances of organisms in the different trophic levels?

CASE STUDY continued

Dying Fish Feed an Ecosystem

When a sockeye salmon eats a smaller fish, the salmon is typically a tertiary consumer (the fourth trophic level), because the small fish was a secondary consumer (third trophic level), feeding on zooplankton (primary consumers in the second trophic level) that themselves fed on phytoplankton (first trophic level). When an Alaskan brown bear eats a salmon, it is therefore on the fifth trophic level. If we consider a simplified food chain with only one type of organism at each trophic level, application of the 10% law means that a food chain containing a single 1,000-pound bear will also contain 10,000 pounds of salmon, 100,000 pounds of smaller fish, a million pounds of zooplankton, and 10 million pounds of phytoplankton. Ecologist Paul Colinvaux captured this essential truth in the title of his book, *Why Big Fierce Animals Are Rare*. Brown bears, tigers, and sperm whales have always been, and will always be, uncommon—and special—inhabitants of our planet.

28.3 HOW DO NUTRIENTS CYCLE WITHIN AND AMONG ECOSYSTEMS?

As noted earlier, nutrients are elements and small molecules that form the chemical building blocks of life. Some, called **macronutrients,** are required by organisms in large quantities. These include water, carbon, hydrogen, oxygen, nitrogen, phosphorus, sulfur, and calcium. **Micronutrients,** including zinc, molybdenum, iron, selenium, and iodine, are required only in trace quantities. **Nutrient cycles,** also called *biogeochemical cycles,* describe the pathways that macronutrients and micronutrients follow as they move from their major sources in the abiotic parts of ecosystems, called **reservoirs,** through living communities and back again. In the following sections, we describe the cycles of water, carbon, nitrogen, and phosphorus.

The Hydrologic Cycle Has Its Major Reservoir in the Oceans

The **hydrologic cycle** (Fig. 28-6) is the pathway that water takes as it travels from its major reservoir—the oceans—through the atmosphere, to smaller reservoirs in freshwater lakes, rivers, and groundwater, and then back again to the oceans. The hydrologic cycle differs from most other nutrient cycles in that the biotic portion of ecosystems plays only a small role—in other words, the hydrologic cycle would continue even if life on Earth disappeared.

The hydrologic cycle is driven by solar heat energy, which evaporates water from oceans, lakes, and streams. When water vapor condenses in the atmosphere, the water falls back to Earth as rain or snow. The water then flows downhill in rivers that eventually empty into the oceans. Oceans cover about 70% of Earth's surface and contain more than 97% of Earth's water. Another 2% of the total water is trapped in ice, leaving only 1% as liquid fresh water. Because oceans cover so much of Earth's surface, most evaporation occurs from them and most precipitation falls back onto them.

Of the water that falls on land, some is absorbed by the roots of plants; much of this water is returned to the atmosphere by evaporation from plant leaves. Most of the rest of the water that falls on land evaporates from the soil, lakes, and streams; a portion runs back to the oceans; an extremely minuscule fraction is stored in the bodies of living organisms; and some enters natural underground reservoirs called **aquifers.** Aquifers are composed of water-permeable rock, such as sandstone, or sediments such as sand or gravel, which are saturated with water. They are often tapped to supply water for household use and for irrigating crops. Water movement from the surface into aquifers

reservoirs

processes

water vapor in the atmosphere

precipitation over land

precipitation over the ocean

evaporation from the land and from the leaves of plants

evaporation from the ocean

evaporation from lakes and rivers

lakes and rivers

runoff from rivers and land

water in the ocean

seepage through soil into groundwater

extraction for agriculture

groundwater, including aquifers

◄ FIGURE 28-6 The hydrologic cycle

is usually slow, and in many areas of the world—including China, India, Northern Africa, and the Midwestern United States—water is being pumped out of aquifers faster than it is being replenished. Eventually, water shortages will force significant changes in agriculture as the aquifers are depleted.

The hydrologic cycle is crucial for terrestrial communities because it continually restores the fresh water needed for land-based life. As you study the nutrient cycles that follow, keep in mind that nutrients in the soil must be dissolved in soil water to be taken up by the roots of plants or to be absorbed by bacteria. Plant leaves can only take up carbon dioxide gas after it has dissolved in a thin layer of water coating the cells inside the leaf. The hydrologic cycle doesn't depend on terrestrial organisms, but they would rapidly disappear without it.

The Carbon Cycle Has Major Reservoirs in the Atmosphere and Oceans

Carbon atoms form the framework of all organic molecules. The **carbon cycle** (Fig. 28-7) is the pathway that carbon takes from its major short-term reservoirs in the atmosphere and oceans, through producers and into the bodies of consumers, detritivores, and decomposers, and then back again to its reservoirs. Carbon enters a community when producers capture carbon dioxide (CO_2) during photosynthesis. On land, photosynthetic organisms acquire CO_2 from the atmosphere, where CO_2 makes up about 0.039% (390 parts per million)

of all atmospheric gases. CO_2 dissolved in water provides aquatic producers such as phytoplankton with the CO_2 they need for photosynthesis.

The carbon taken up by photosynthesizers is "fixed" in biological molecules such as sugars and proteins. Producers return some of this carbon to the atmosphere or water as CO_2 generated by cellular respiration, but much of the carbon remains stored in the biological molecules of their bodies. When primary consumers eat the producers, they acquire this stored carbon. Primary consumers and the organisms in higher trophic levels release CO_2 during respiration, excrete carbon compounds in their feces, and store the rest of the carbon in their bodies. All living things eventually die, and their bodies are broken down by detritivores and decomposers, whose cellular respiration returns CO_2 to the atmosphere and oceans. The complementary processes of uptake by photosynthesis and release by cellular respiration continually recycle carbon from the abiotic to the biotic portions of an ecosystem and back again.

Some carbon, however, cycles much more slowly. Much of Earth's carbon is bound up in limestone rock, formed from calcium carbonate ($CaCO_3$) deposited on the ocean floor in the shells of prehistoric phytoplankton. The movement of carbon from this source to the atmosphere and back again requires millions of years, so this process makes very little contribution to the carbon cycling that supports ecosystems. **Fossil fuels,** which include coal, oil, and natural

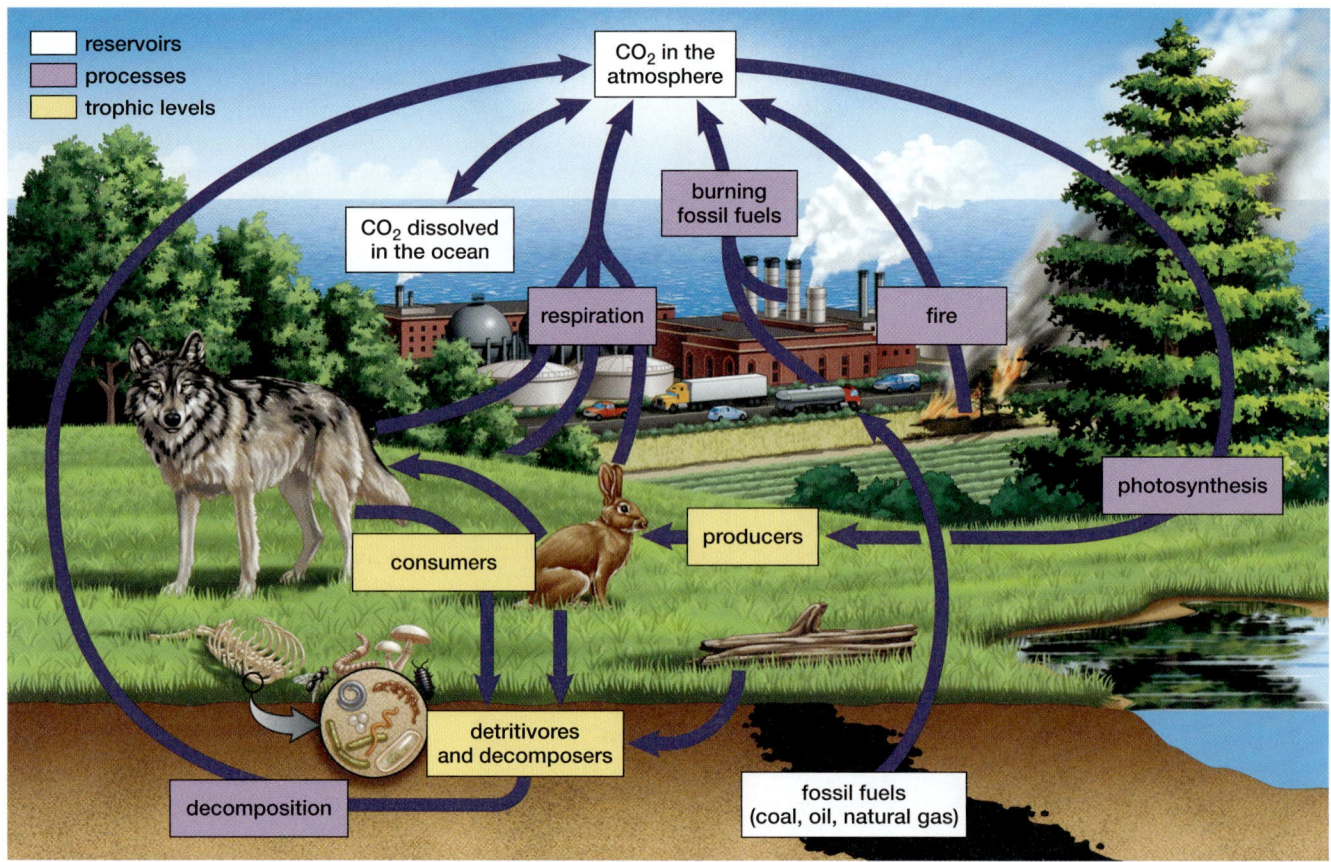

▲ FIGURE 28-7 The carbon cycle

gas, are additional long-term reservoirs for carbon. These substances were produced over millions of years from the remains of prehistoric organisms buried deep underground and subjected to high temperature and pressure. In addition to carbon, the energy of prehistoric sunlight (originally captured by photosynthetic organisms) is trapped in these deposits. When humans burn fossil fuels to tap this stored energy, CO_2 is released into the atmosphere, with potentially serious consequences, as we describe in Section 28.4.

The Nitrogen Cycle Has Its Major Reservoir in the Atmosphere

Nitrogen is a crucial component of proteins, many vitamins, nucleotides (such as ATP), and nucleic acids (such as DNA). The **nitrogen cycle** (Fig. 28-8) is the pathway taken by nitrogen from its primary reservoir—nitrogen gas (N_2) in the atmosphere—to much smaller reservoirs of ammonia and nitrate in soil and water, through producers, consumers, detritivores and decomposers, and back to its reservoirs.

The atmosphere contains about 78% nitrogen gas, but plants and most other producers cannot use nitrogen in this form—they require either ammonia (NH_3) or nitrate (NO_3^-). A few types of bacteria that live in soil or water can convert N_2 into ammonia, in a process called **nitrogen fixation.** Some nitrogen-fixing bacteria enter into a mutually beneficial relationship with certain plants, called *legumes,* in which the bacteria live in swellings on the plants' roots (see Chapter 19). Legumes such as alfalfa, soybeans, clover, and peas are extensively planted on farms, in part because they release excess ammonia produced by the bacteria, thus fertilizing the soil. Other bacteria in soil and water convert ammonia to nitrate. A small amount of nitrate is also produced during electrical storms, when the energy of lightning combines nitrogen and oxygen gases to form nitrogen oxide compounds. These nitrogen oxides fall to the ground dissolved in rain and are eventually converted to nitrate.

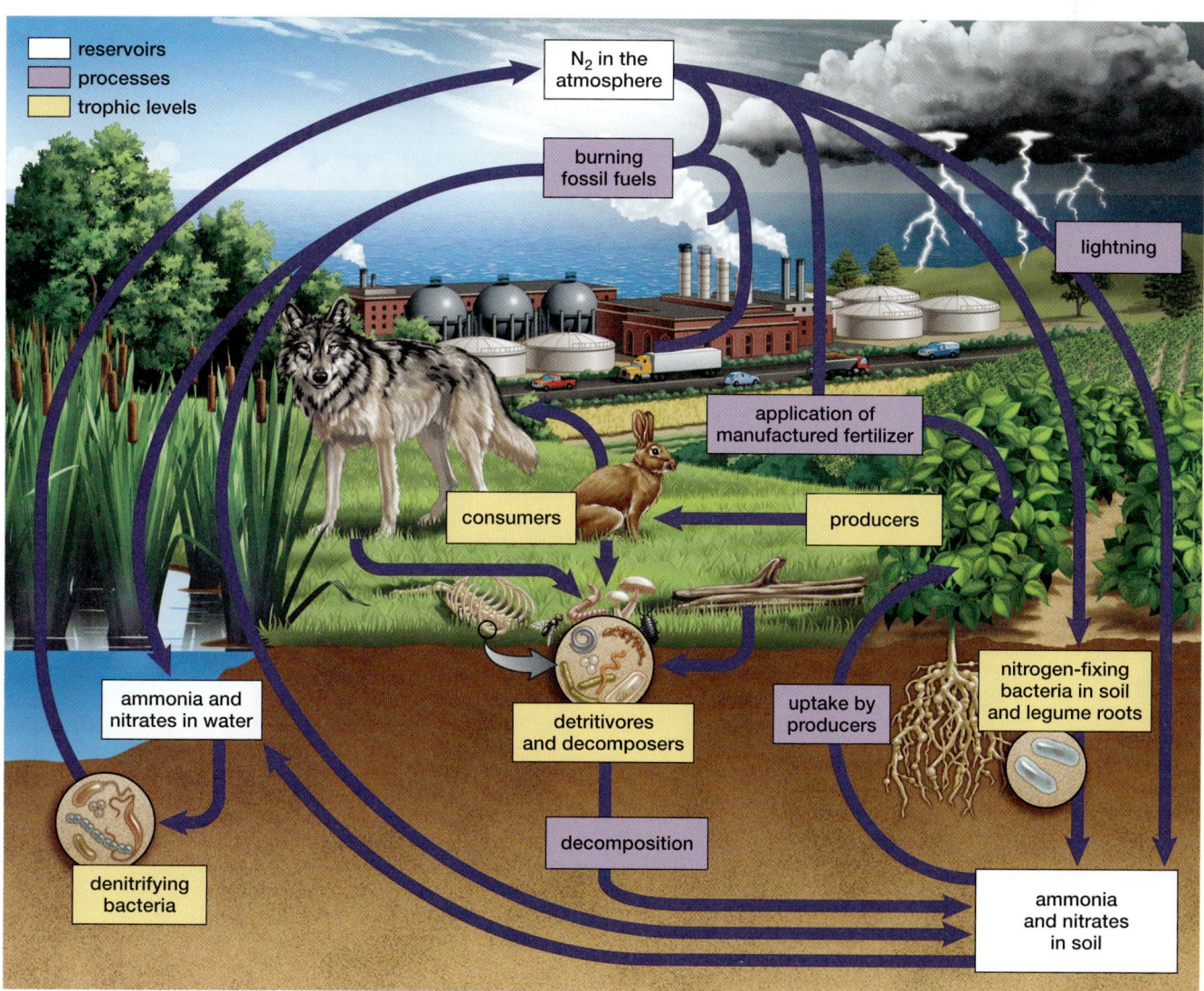

▲ **FIGURE 28-8 The nitrogen cycle**

QUESTION What incentives cause humans to capture nitrogen from the air and pump it into the nitrogen cycle? What are some consequences of human augmentation of the nitrogen cycle?

Ammonia and nitrate are absorbed by producers and are incorporated into biological molecules such as proteins and nucleic acids. These are passed through successively higher trophic levels as primary consumers eat the producers and are themselves eaten. At each trophic level, bodies and wastes are broken down by decomposers, which liberate ammonia back into the soil and water. The nitrogen cycle is completed by **denitrifying bacteria.** These residents of wet soil, swamps, and estuaries break down nitrate, releasing nitrogen gas back into the atmosphere (see Fig. 28-8).

People significantly manipulate the nitrogen cycle, both deliberately and unintentionally. As noted earlier, farmers plant legumes to fertilize their fields. Fertilizer factories combine N_2 from the atmosphere with hydrogen generated from natural gas, fixing the nitrogen as ammonia, which is then often converted to nitrate or urea (an organic nitrogen compound). About 150 million tons of nitrogen-based fertilizer are applied to farms each year. In addition, the heat produced by burning fossil fuels combines atmospheric N_2 and O_2, generating nitrogen oxides that form nitrates. These human activities now dominate the nitrogen cycle.

The Phosphorus Cycle Has Its Major Reservoir in Rock

Phosphorus is found in biological molecules such as nucleic acids and the phospholipids of cell membranes. It also forms a major component of vertebrate teeth and bones. The **phosphorus cycle** (Fig. 28-9) is the pathway taken by phosphorus from its primary reservoir in rocks to much smaller reservoirs in soil and water, through producers and into consumers, detritivores and decomposers, and then back to its reservoirs.

Throughout its cycle, almost all phosphorus is bound to oxygen, forming phosphate (PO_4^{3-}). There are no gaseous forms of phosphate, so there is no atmospheric reservoir in the phosphorus cycle. As phosphate-rich rocks are exposed by geological processes, some of the phosphate is dissolved by rain and flowing water, which carries it into soil, lakes, and the ocean, forming the smaller reservoirs of phosphorus that are directly available to ecological communities. Dissolved phosphate is absorbed by producers, which incorporate it into biological molecules. From producers, phosphate is passed through food webs; at each level, excess phosphate is excreted. Ultimately, detritivores and decomposers return the phosphate to the soil and water, where it may then be reabsorbed by producers or may become bound to sediments and eventually re-formed into rock.

Some of the phosphate dissolved in fresh water is carried to the oceans. Although much of this phosphate ends up in marine sediments, some is absorbed by marine producers and is eventually incorporated into the bodies of invertebrates and fish. Some of these, in turn, are consumed by seabirds, which excrete large quantities of phosphate back onto the land. At one time, seabird excrement (called guano) deposited along the western coast of South America was the

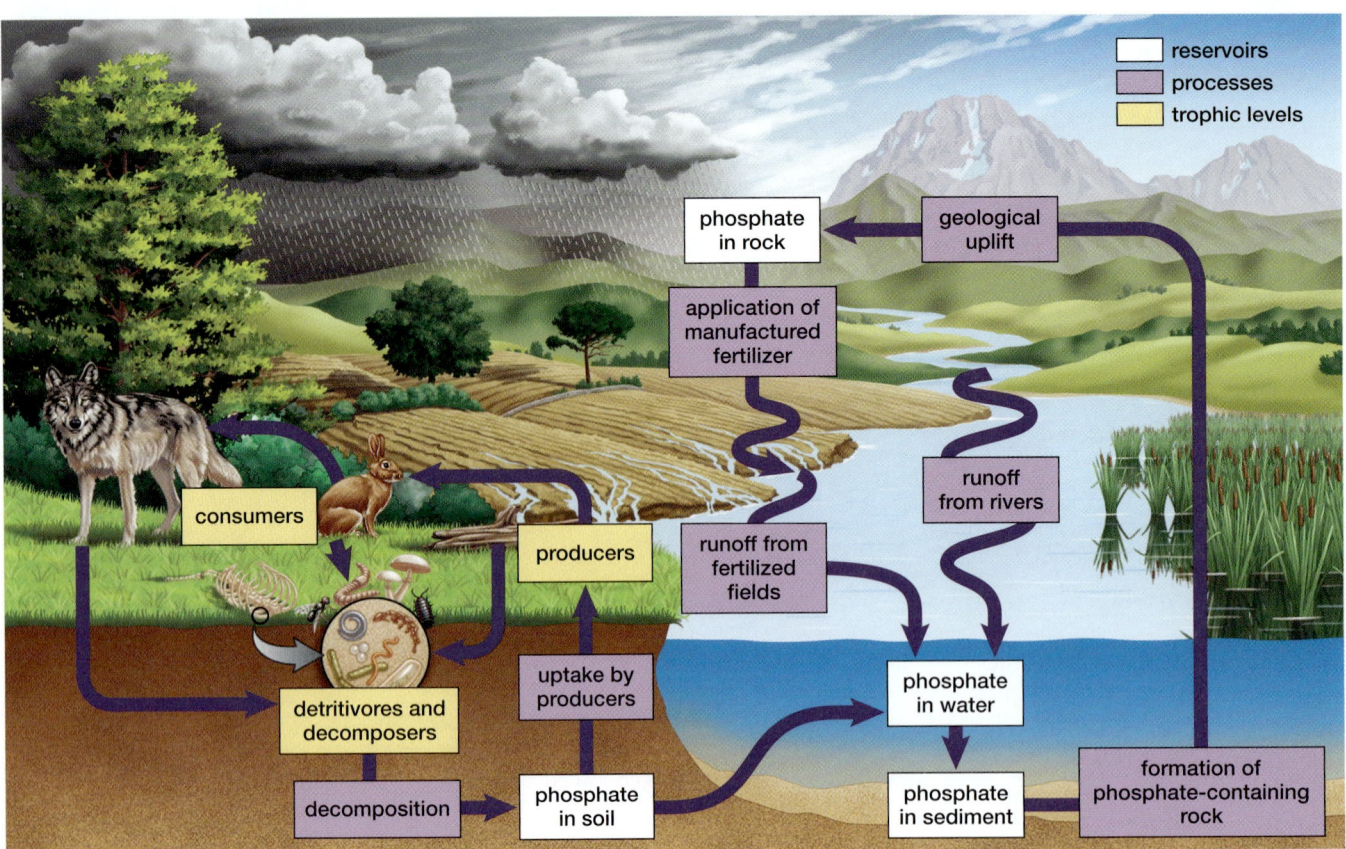

▲ FIGURE 28-9 The phosphorus cycle

major source of the world's phosphate fertilizer. Demand for phosphate is now so great that most phosphate fertilizer is extracted from phosphate-rich rock.

CHECK YOUR LEARNING

Can you explain why nutrients cycle within and among ecosystems? Are you able to describe the hydrologic, nitrogen, carbon, and phosphorus cycles?

28.4 WHAT HAPPENS WHEN HUMANS DISRUPT NUTRIENT CYCLES?

Ancient peoples, with small populations and limited technology, had relatively little impact on nutrient cycles. However, as the human population grew and technology increased, people began to act more independently of natural ecosystem processes. The Industrial Revolution, which began in earnest in the mid-nineteenth century, resulted in a tremendous increase in our reliance on energy stored in fossil fuels for heat, light, transportation, industry, and agriculture. Fertilizer use on commercial farms grew exponentially. Today, human use of fossil fuels and chemical fertilizers has significantly disrupted the global nutrient cycles of nitrogen, phosphorus, sulfur, and carbon.

Overloading the Nitrogen and Phosphorus Cycles Damages Aquatic Ecosystems

Each year, roughly 40 million tons of phosphate fertilizer are produced from mined phosphate rock, and about 150 million tons of nitrogen-based fertilizer are manufactured from atmospheric nitrogen. These fertilizers are applied to farm fields to help satisfy the agricultural demands of a growing human population.

When water washes over the land from rainfall or irrigation, it dissolves and carries away some of the phosphate and nitrogen-based fertilizer. As the water drains into lakes, rivers, and ultimately the oceans, these fertilizers can disrupt the delicate balance of food webs by overstimulating the growth of phytoplankton. The resulting phytoplankton "blooms" can turn clear water into an opaque green soup. As the phytoplankton die, their bodies sink into deeper water, where they provide a feast for decomposer bacteria. Cellular respiration by these decomposers uses up most of the available dissolved oxygen. Deprived of oxygen, aquatic invertebrates and fish either leave the area or die and decompose (making the problem worse). A dramatic example occurs in the Gulf of Mexico off the coast of Louisiana. In the American Midwest, spring rains and melting snows wash enormous quantities of nitrates and phosphates off fertilized farm fields into streams that eventually flow into the Mississippi River. The Mississippi then empties the fertilizers into the Gulf of Mexico. In summer, when sunlight becomes stronger and the Gulf warms, the fertilizers create an algal bloom (Fig. 28-10). As the algae die and their bodies decompose, a "dead zone" develops, where oxygen levels are so low that few organisms can survive. Hurricanes and

▲ **FIGURE 28-10 An algal bloom in the Gulf of Mexico** Nutrients, especially nitrate and phosphate, wash off farmland in the Midwest and move down the Mississippi River, which flows from upper left to the center of the image, where it ends in the Mississippi Delta. When these nutrients enter the Gulf, they fertilize an explosive growth of algae, visible in this satellite photo as hazy green swirls near the coast.

tropical storms break up the dead zone each autumn, but it reappears the following summer. The Gulf of Mexico dead zone now typically covers 6,000 to 8,500 square miles each summer. Worldwide, dead zones are increasing in both size and number as agricultural activities intensify.

Overloading the Sulfur and Nitrogen Cycles Causes Acid Deposition

Natural processes put both nitrogen oxides and sulfur oxides into the atmosphere. Fires and lightning produce several types of nitrogen oxides, including nitrate; volcanoes, hot springs, and decomposers release sulfur dioxide (SO_2). However, combustion of fossil fuels now produces most of the nitrogen and sulfur oxides entering the atmosphere. For example, burning of sulfur-containing fossil fuels, primarily coal, accounts for about 75% of all sulfur dioxide emissions worldwide. When combined with water vapor in the atmosphere, nitrogen oxides and sulfur dioxide are converted to nitric acid and sulfuric acid. Days later and often hundreds of miles from the source, these acids fall to Earth in rain or snow. This "acid rain"—more accurately called **acid deposition**—was first recognized in New Hampshire, where a sample of rain collected in 1963 had a pH of 3.7—about the same as that of orange juice, and 20 to 200 times more acidic than unpolluted rain, which usually has a pH between 5 and 6.

Acid deposition damages forests, can render lakes lifeless, and even eats away at buildings and statues (Fig. 28-11). In the United States, acid deposition is most damaging to natural ecosystems in New England, the mid-Atlantic states, the upper Midwest, Western mountains, and the state of Florida, where most of the rocks and soils have little buffering capacity to neutralize acids. Upstate New York and New England are doubly vulnerable, because the

▲ **FIGURE 28-11 Acid deposition is corrosive** These two identical building decorations in Brooklyn, New York, show the effects of acid deposition. On the left, the decoration has been restored to its original state; on the right, an unrestored decoration has eroded almost completely away.

▲ **FIGURE 28-12 Acid deposition can destroy forests** The Camel's Hump in Vermont shows the damage caused by acid deposition. Note that virtually all of the older, mature trees are dead and bare. Since the 1990s, lower sulfur and nitrogen emissions from power plants have reduced acid rain in New England, and young trees are beginning to recolonize the forest.

prevailing westerly winds that sweep across North America carry sulfates and nitrates from coal-burning power plants and industries in the Midwest directly over these states. In the Adirondack and Catskill Mountains of New York and the White and Green Mountains of New England, acid rain has rendered many lakes and ponds too acidic to support fish and the food webs that sustain them.

How does acid deposition harm ecosystems? Acid deposition increases the exposure of organisms to toxic metals—such as aluminum, mercury, lead, and cadmium—all of which are far more soluble in acidified water than in water of neutral pH. Aluminum, for example, is a component of the most common minerals in soil. Acid conditions dissolve aluminum out of the soil into soil water and lakes, where it inhibits plant growth and kills fish. Calcium and magnesium, which are essential nutrients for all plants, are leached out of the soil by acid precipitation.

Plants in acidified soil become weak and more vulnerable to infection and damage by insects. Populations of both red spruce and sugar maples are declining in much of the Northeast. The decline has been shown to be caused by acid conditions, coupled with drought, insect attack, and climate change. About half of the red spruce and one-third of the sugar maples in the Green Mountains of Vermont have been killed (**Fig. 28-12**).

Since 1990, government regulations have resulted in substantial reductions in emissions of both sulfur dioxide and nitrogen oxides from U.S. power plants—sulfur dioxide emissions are down about 40%, and nitrogen oxide levels have been reduced by more than 50%. Air quality has improved, and rain has become less acidic, although large areas of the northeastern United States still receive rain with a pH below 5.0.

Damaged ecosystems recover slowly. Adirondack lakes are gradually becoming less acidic, but many are still much more acidic than before acid deposition began. If acid deposition is completely eliminated, eventually the lakes will return to their normal pH. Most aquatic life should then recover in 3 to 10 years, depending on the species, although the lakes may need to be restocked with trout. Forests will take much longer to recover, because the life span of trees is so long and because soil chemistry changes more slowly.

Interfering with the Carbon Cycle Is Changing Earth's Climate

The fate of sunlight entering Earth's atmosphere is shown in **Figure 28-13.** Some of the energy from sunlight (**Fig. 28-13** ❶) is reflected back into space by the atmosphere (particularly clouds), and by Earth's surface, especially by areas covered with snow or ice (**Fig. 28-13** ❷). Most sunlight, however, strikes relatively dark areas of the surface (land, vegetation, and open water) and is converted into heat (**Fig. 28-13** ❸) that is radiated into the atmosphere (**Fig. 28-13** ❹). Although most of this heat continues on into space (**Fig. 28-13** ❺), water vapor, CO_2 and several other **greenhouse gases** trap some of the heat in the atmosphere (**Fig. 28-13** ❻). This is a natural process called the **greenhouse effect,** which keeps our atmosphere relatively warm and allows life on Earth as we know it.

For Earth's temperature to remain constant, the total amount of energy entering and leaving Earth's atmosphere must be equal. If atmospheric concentrations of greenhouse gases increase, more heat is retained than is radiated into space, causing Earth to warm. Greenhouse gases are in fact increasing, largely because people burn fossil fuels, releasing CO_2. Other important greenhouse gases include methane (CH_4), released by agricultural activities, landfills, and coal mining, and nitrous oxide (N_2O), released by agricultural activities and burning fossil fuels. CO_2, however, contributes by far the largest share of the greenhouse effect caused by human activities, so we will focus our discussion on this molecule.

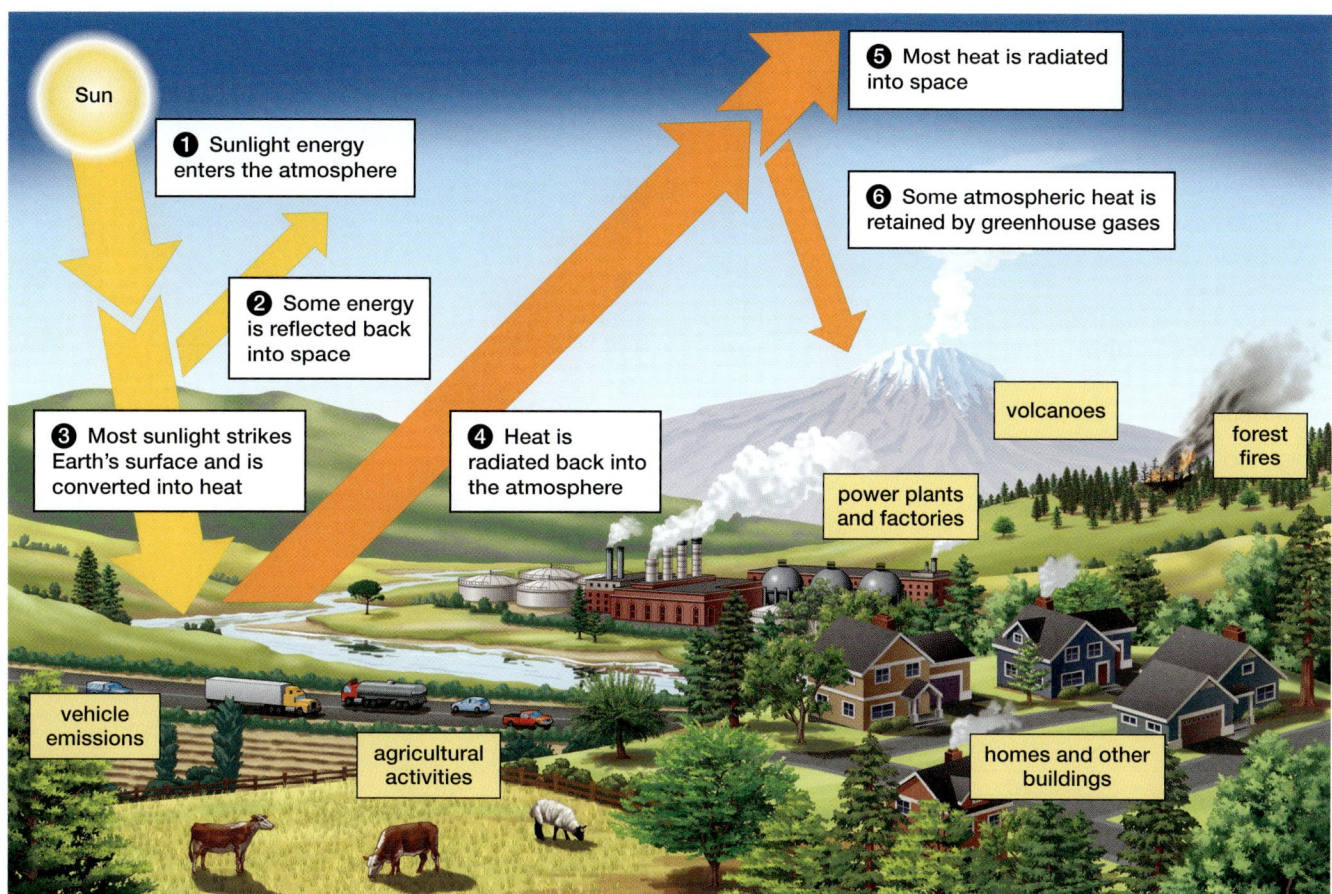

▲ FIGURE 28-13 **The greenhouse effect** Incoming sunlight warms Earth's surface and is radiated back to the atmosphere. Greenhouse gases, released by natural processes and substantially augmented by human activities (both shown in yellow rectangles), absorb increasing amounts of this heat, raising global temperatures.

Burning Fossil Fuels Is Causing Climate Change

Since the mid-1800s, human societies have increasingly relied on energy from fossil fuels. As we burn fossil fuels in our power plants, factories, and cars, we harvest the energy of ancient sunlight and release CO_2 into the atmosphere. Burning fossil fuels accounts for about 80% to 85% of the CO_2 that human activities release into the atmosphere each year.

A second source of added atmospheric CO_2 is **deforestation,** which destroys tens of millions of forested acres annually and accounts for about 15% to 20% of humanity's CO_2 emissions. Deforestation is occurring principally in the Tropics, where rain forests are rapidly being converted to agricultural land to feed growing populations and to supply the world's demand for biofuels, such as ethanol and biodiesel. The carbon stored in the trees returns to the atmosphere when they are cut down and burned. A third, very minor source of CO_2 is volcanic activity. Only about 1% as much CO_2 enters the atmosphere from volcanoes as from human activities.

Collectively, human activities release about 35 to 40 billion tons of CO_2 into the atmosphere each year. About half of this carbon is absorbed by oceans, plants, and soil, with the rest remaining in the atmosphere. As a result, since 1850—when people began burning large quantities of

fossil fuels during the Industrial Revolution—the CO_2 content of the atmosphere has increased by about 40%—from 280 parts per million (ppm) to 392 ppm—and is growing by about 2 ppm annually (**Fig. 28-14a**). Based on analyses of gas bubbles trapped in ancient Antarctic ice, scientists have determined that the atmospheric CO_2 content is now higher than at any time during the past 650,000 years.

A large and growing body of evidence indicates that human release of CO_2 and other greenhouse gases has amplified the natural greenhouse effect and thereby altered the global climate. Surface temperature data, recorded from thousands of weather stations around the world and from satellites that measure temperatures over the oceans, show that Earth has warmed by about 1°F (0.6°C) since 1970 (**Fig. 28-14b**). The decade from 2001 to 2010 was the warmest ever recorded up to that time; in fact, all but 1 of the 10 warmest years on record occurred within that decade.

The overall impact of increased greenhouse gases is now usually called **climate change,** which includes both global warming and many other effects on our climate and Earth's ecosystems. Although a 1°F increase may not sound like much, our warming climate already has had widespread effects. Spring snow cover in the Northern Hemisphere is declining. Glaciers

(a) Atmospheric CO₂

(b) Global surface temperature

▲ FIGURE 28-14 **Global temperature increases parallel atmospheric CO_2 increases** **(a)** Yearly average CO_2 concentrations in parts per million. These measurements were recorded at 11,155 feet (3,400 meters) above sea level, near the summit of Mauna Loa, Hawaii. **(b)** Global surface temperatures. Because global temperature varies considerably from year to year, this temperature graph shows trends by averaging each year with the 4 years preceding it. Data for both graphs from the National Oceanic and Atmospheric Administration.

(a) Muir Glacier, 1941

(b) Muir Glacier, 2004

▲ FIGURE 28-15 **Glaciers are melting** Photos taken from the same vantage point in **(a)** 1941 and **(b)** 2004 document the retreat of the Muir Glacier in Glacier Bay National Park, Alaska.

are retreating worldwide; the World Glacier Monitoring Service reports that about 90% of the world's mountain glaciers are shrinking, and that this trend seems to be accelerating (Fig. 28-15). Glacier National Park, Montana, named for its spectacular abundance of glaciers, had 150 glaciers in 1910; now, only 25 remain—and the remaining glaciers are significantly smaller than they were in the recent past. The oceans are warming, which causes their water to expand and occupy more volume; as a result, sea levels are rising. During the past 30 years, the Arctic ice cap has become almost 50% thinner and 35% smaller in area, and is shrinking by about 10% per decade. Finally, in 2011, scientists compiled the results of 53 studies that examined changes in the distribution of more than 1,000 species of terrestrial plants and animals. The species' ranges are moving toward the poles at an average rate of about 10.5 miles (17 kilometers) per decade—just what would be expected if they are moving in response to a warming planet.

Climate scientists predict that a warming atmosphere will cause more severe storms, including stronger hurricanes; greater amounts of rain or snow in single storms (a phenomenon already observed in the northeastern United States

during the past half-century); and more frequent and more prolonged droughts. Increased CO_2 also makes the oceans more acidic, which disturbs many natural processes, including the ability of many marine animals, such as snails and corals, to make their shells and skeletons.

Continued Climate Change Will Disrupt Ecosystems and Endanger Many Species

What does the future hold? Predictions of continued climate change are based on sophisticated computer models developed and run independently by climate scientists around the world. As the models continue to improve, they match past climate with ever-greater accuracy, providing increasing confidence in their predictions for the future. The models also provide evidence that natural causes, such as changes in the output of the sun, cannot account for the recent warming. The models match the data only when human carbon emissions are included in the calculations. The Intergovernmental Panel on Climate Change (IPCC) is a consortium of hundreds of climate scientists and other experts from 130 nations. In their 2007 report, the IPCC predicted that even under the best-case scenario in which a concerted worldwide effort is made to reduce greenhouse gas emissions, the average global temperature will rise by at least 3.2°F (1.8°C)

? HAVE **YOU** EVER WONDERED ...

How Big Your Carbon Footprint Is?

Each of us affects Earth through the choices we make. A *carbon footprint* is a measure of the impact that human activities have on climate, based on the quantity of greenhouse gases they emit. Our personal carbon footprints give us a sense of our individual impacts. For example, each gallon of gasoline burned releases 19.6 pounds (8.9 kg) of CO_2 into the air. So, if your car gets 20 miles to the gallon, then each mile that you drive will add about a pound of CO_2 to the atmosphere.

The Web sites of the U.S. Environmental Protection Agency and several environmental organizations provide household emissions calculators that allow you to estimate your carbon footprint, and provide advice on how you can reduce it. To get started, type "carbon footprint" into an Internet search engine.

by the year 2100. The IPCC's high-level emissions scenario projects an increase of 7.2°F (4.0°C) (**Fig. 28-16**). These changes in climate will be difficult to stop, let alone reverse, as we explore in "Earth Watch: Geoengineering—A Solution to Climate Change?"

Even if the more optimistic predictions are correct, the consequences for natural ecosystems will be profound. Thousands of species will change their ranges, moving away from the Equator toward the poles or higher up mountainsides. Some plants and animals will find it easier to move than others will, either because they are intrinsically more mobile (such as birds) or because they can move great distances while reproducing (such as some plants that produce lightweight, wind-borne seeds). It is highly unlikely that entire communities of organisms can just pack up and move, intact. Some will move farther poleward faster than others, with potentially serious consequences for certain species. For example, if a predator's main prey moves poleward faster than the predator can, then the predator may be left behind without a food supply.

Some species, particularly those on mountains or in the Arctic and Antarctic, will have nowhere to go. For example, the loss of summer sea ice is bad news for polar bears and other marine mammals that rely on ice floes as nurseries for their young and as staging platforms for hunting fish or seals. As summer ice diminishes, both walrus and polar bear populations are moving onto land to give birth, putting the adults farther away from their prime hunting grounds. As the adults crowd together onto small beaches, instead of being spread out over sea ice, they sometimes endanger their own young; in 2009, for example, 131 walrus calves were trampled to death on a beach when the adults panicked and stampeded. In 2008, in response to its declining numbers and the projected continued loss of its habitat, the U.S. Fish and Wildlife Service designated the polar bear as a threatened species—the first species to be listed primarily as a result of global climate change. Complete loss of sea ice, which climate models predict could occur within the next century, might cause the extinction of polar bears in the wild. Penguins in the Antarctic may face similar dangers. Many species of penguins walk for miles across ice sheets to their breeding grounds. As the ice melts and breaks up, their journeys become much more difficult. Ten of the 17 species of penguins are already on the Red List of threatened species developed by the International Union for Conservation of Nature.

Some of the movement of species may have direct impacts on human health. Many diseases, especially those carried by mosquitoes and ticks, are currently restricted to tropical or subtropical parts of the planet. Like other animals, these disease vectors will probably spread poleward as a result of warming temperatures, bringing their diseases, such as malaria, dengue fever, yellow fever, and Rift Valley fever, with them. On the other hand, it may become so hot and dry in parts of the tropics that mosquitoes and some other insects may have shortened life spans, thus reducing vector-borne diseases in these regions. Although computer models can predict temperature changes, no one can confidently predict the resulting overall effects on human health.

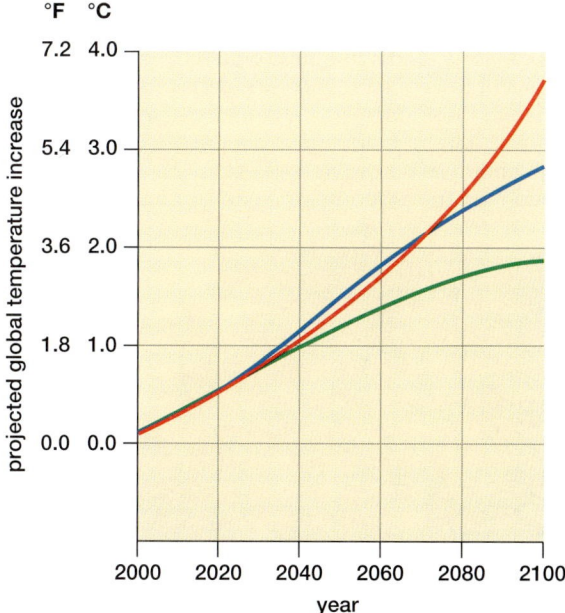

▲ **FIGURE 28-16 Projected range of temperature increases** The IPCC projections for the twenty-first century are based on three different scenarios of greenhouse gas emissions. The red, blue, and green data lines are projections based on high, moderate, and substantially reduced growth of greenhouse gas emissions, respectively. Even under the most optimistic assumptions, a continued increase in global temperatures is predicted. Global temperature change is expressed relative to the average temperature from 1980 to 1999. Source: Data from IPCC, 2007, *Fourth Assessment Report: Summary for Policymakers.*

CHECK YOUR LEARNING

Can you explain how human activities have disrupted nutrient cycles? Can you describe how human interference with nutrient cycles causes acid deposition, damages aquatic ecosystems, enhances the greenhouse effect, and causes climate change? Can you describe some of the evidence that Earth is warming and some of the impacts of climate change on Earth's ecosystems?

Earth Watch

Geoengineering—A Solution to Climate Change?

What should be done about climate change? The obvious solution is to reduce our emissions of CO_2 and other greenhouse gases.

But that isn't as simple as it sounds. Modern human societies depend on the energy of fossil fuels, and cannot just stop burning them overnight. In addition, CO_2 lasts a long time in the atmosphere, so the CO_2 emitted today will still be warming the globe years from now. As a result, some scientists and policymakers are pondering a temporary fix while we make the transition to carbon-neutral, renewable energy sources. Perhaps we can slow down climate change with *geoengineering*—altering certain fundamental characteristics of our planet to counteract the warming effect of greenhouse gases. Although geoengineering is a complex subject, there are two main approaches: shading the planet and removing some CO_2 from the atmosphere.

Shading the Planet

Although CO_2 traps heat and warms Earth, some other molecules, such as a few sulfur compounds, reflect sunlight back into space, cooling the planet. Whenever a massive volcano erupts, it spews millions of tons of sulfur dioxide miles high into the atmosphere. In 1991, when Mt. Pinatubo erupted in the Philippines (Fig. E28-2a), it blasted about 20 million tons of sulfur dioxide as high as 22 miles skyward. For the next couple of years, global temperatures were slightly cooler, because more sunlight was reflected back into space by sulfur high in the atmosphere. In the early 2000s, global emissions of sulfur from power plants, especially in China, rose by about 25% (Fig. E28-2b). At the same time, there appeared to be a pause in the upward trend of planetary temperatures. Climate models suggest that the pause in global warming could be quite precisely accounted for by the shading effect of the added sulfur in the atmosphere. Based on data such as these, some have suggested that governments could slow global warming by sending planes loaded with sulfur compounds high into the atmosphere, where they would deliberately release the sulfur.

Removing CO_2

Both engineering and biological approaches to removing CO_2 from the atmosphere have been proposed. The most commonly proposed engineering method is often called scrubbing: In this scheme, towers would suck in air and remove the CO_2 through one of several possible chemical reactions that produce solid carbonate compounds. The cost of scrubbing has been estimated to be between about $100 and $600 per ton of carbon. Current carbon emissions are about 10 billion tons per year (35 to 40 billion tons of CO_2), so scrubbing would be an expensive proposition.

A possible biological approach to removing CO_2 is to fertilize the oceans. In many parts of the open ocean, the nutrient that limits phytoplankton growth is iron, which is an essential part of some of the enzymes involved in ATP production in both mitochondria and chloroplasts. The proposal is to spread powdered iron on open ocean waters, triggering phytoplankton blooms. The phytoplankton would then take up CO_2 during photosynthesis and store some of the carbon in their bodies. When the phytoplankton die, they would sink, carrying the carbon to the ocean depths, where it should remain for many years.

Will Geoengineering Work? Is It Worth the Risks?

Many people question the feasibility and desirability of geoengineering. Atmospheric sulfur causes respiratory damage in people and acid deposition that damages ecosystems. In addition, simply shading the planet would do nothing to reduce the increasing acidification of the oceans. No one really knows if ocean fertilization would work, or what other effects it might have on ocean ecosystems. Scrubbing is very expensive, and something would have to be done with the billions of tons of carbon compounds produced every year. Despite these problems, the risks of climate change are so great that more and more people are thinking about geoengineering as a last-ditch way to buy humanity enough time to convert to carbon-neutral energy sources.

(a) The eruption of Mt. Pinatubo.

(b) Smoke and sulfur dioxide billow from a Chinese power plant.

▲ FIGURE E28-2 Sulfur emissions both cool and pollute (a) The 1991 eruption of Mt. Pinatubo in the Philippines injected millions of tons of sulfur dioxide miles up in the atmosphere. The sulfur compounds reflected sunlight and cooled the planet for a few years. (b) Increasing emissions of sulfur dioxide from Chinese power plants during the early 2000s both cooled the planet and caused significant damage to people's health.

CASE STUDY revisited

Dying Fish Feed an Ecosystem

The sockeye salmon's return to an Alaskan stream is unforgettable (Fig 28-17). Even after the fish have run the gauntlet of brown bears and bald eagles, hundreds remain, their brilliant red bodies writhing in water so shallow that it barely covers them. A female beats her tail, excavating a shallow depression in the gravel where she releases her eggs; a male then showers them with sperm. But sperm and eggs aren't the only cargo the salmon carry upstream from the ocean. About 95% of a sockeye's body mass was acquired during its years of feeding in the ocean, so it carries enormous amounts of energy and nutrients upstream with it. Sunlight energy, originally captured by phytoplankton, then transferred to zooplankton and smaller fish, is now stored in the bodies of

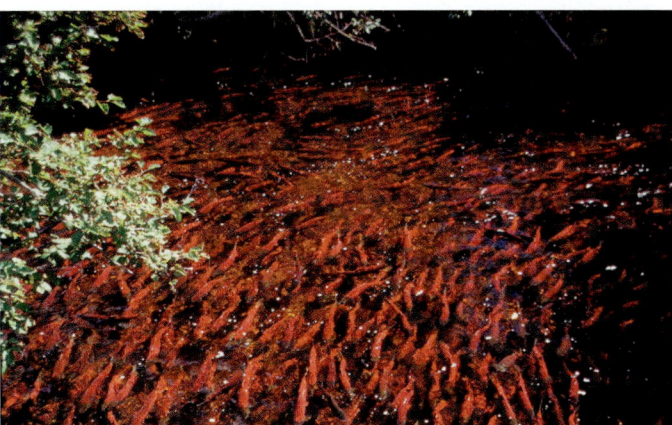

▲ FIGURE 28-17 **Spawning sockeye salmon**

the salmon, where it becomes available to terrestrial predators and scavengers. During a summer gorging on salmon, a brown bear may put on as much as 400 pounds of fat, which serves as its vital energy supply while it hibernates during the long Alaskan winter. The local mink populations also profit. Females nurse their young during the salmon runs, when a virtually inexhaustible food supply, mostly in the form of half-eaten salmon carcasses, is available.

The salmon bring back nutrients from the oceans, too, especially nitrogen. Nitrogen from ocean sources can be distinguished from terrestrial nitrogen by the ratio of two isotopes of nitrogen, ^{14}N and ^{15}N. Ecologists have found that 50% to 70% of the nitrogen near some streams in Alaska originated in the ocean and was brought upstream in the bodies of salmon. In some places, more than half of this salmon-derived nitrogen was carried some distance away from the streams by bears, who usually eat only the choicest parts of the salmon and leave the rest. Is salmon-derived nitrogen important? It is to Sitka spruce, which can grow three times faster near salmon streams than near similar streams without salmon runs. It's also important to the next generation of salmon. Because salmon carcasses fertilize nearby lakes, stimulating the growth of phytoplankton and consequently the zooplankton upon which newly hatched salmon feed, the adult salmon indirectly feed their young.

Consider This

BioEthics Dams, river pollution, and overfishing have depleted many salmon populations—some are listed as endangered or threatened under the Endangered Species Act. Some people argue that because these salmon are also raised commercially in fish farms, the depletion of wild populations doesn't really matter. Based on what you have learned in this chapter, do you think that wild populations of salmon should be protected and efforts made to increase their numbers?

CHAPTER REVIEW

Summary of Key Concepts

28.1 How Do Nutrients and Energy Move Through Ecosystems?

Ecosystems are sustained by a continuous input of energy from sunlight and the recycling of nutrients. Energy enters the biotic portion of ecosystems through photosynthesis and then flows through the ecosystem from organism to organism. Nutrients are obtained by organisms from their living and nonliving environment and are recycled within and among ecosystems.

28.2 How Does Energy Flow Through Ecosystems?

Photosynthetic organisms act as conduits of both energy and nutrients into biological communities. The energy of sunlight is captured by photosynthetic organisms, also called producers. Producers make up the first trophic (feeding) level in an ecosystem. Herbivores (plant eaters, also called primary consumers) form the second trophic level. Carnivores (meat eaters) are secondary consumers when they prey on herbivores and tertiary or

higher-level consumers when they eat other carnivores. Omnivores, which consume both plants and other animals, occupy multiple trophic levels. Food chains represent feeding relationships in which each trophic level is represented by one organism. In natural ecosystems, feeding relationships are complex and are described as food webs. Detritivores and decomposers feed on dead bodies and wastes. Decomposers (mostly bacteria and fungi) liberate nutrients as simple molecules that reenter nutrient cycles.

The higher the trophic level, the less energy available to sustain it. In general, only about 10% of the energy captured by organisms at one trophic level is available to organisms in the next higher level.

28.3 How Do Nutrients Cycle Within and Among Ecosystems?

A nutrient cycle depicts the movement of a particular nutrient from its reservoir, usually in the abiotic portion of the ecosystem, through the biotic portion of the ecosystem, and back to its reservoir.

In the hydrologic cycle, the major reservoir of water is the oceans. Solar energy evaporates water, which returns to Earth as precipitation. Water flows into lakes and underground aquifers, and is carried by rivers to the oceans. In the carbon cycle, the short-term reservoirs are CO_2 in the oceans and the atmosphere. Carbon enters producers via photosynthesis. From producers, carbon is passed through the food web and released to the atmosphere as CO_2 during cellular respiration. Burning fossil fuels also releases CO_2 into the atmosphere. In the nitrogen cycle, the major reservoir consists of nitrogen gas in the atmosphere. Nitrogen gas is captured by nitrogen-fixing bacteria, which release ammonia. Other bacteria convert ammonia to nitrate. Plants get their nitrogen from nitrates and ammonia. Nitrogen passes from producers to consumers and is returned to the environment through excretion and the activities of detritivores and decomposers. Nitrogen gas is returned to the air by denitrifying bacteria. In the phosphorus cycle, the principal reservoir consists of phosphate in rocks. Phosphate dissolves in water, is absorbed by photosynthetic organisms, and is passed through food webs. Some phosphate is excreted, and the rest is returned to the soil and water by decomposers. Some is carried to the oceans, where it is deposited in marine sediments.

28.4 What Happens When Humans Disrupt Nutrient Cycles?

Human activities often produce and release more nutrients than nutrient cycles can efficiently process. For example, the use of enormous quantities of fertilizer by agricultural activities has disrupted many aquatic ecosystems. By burning fossil fuels, humans have overloaded the natural cycles for sulfur, nitrogen, and carbon. In the atmosphere, sulfur dioxide and nitrogen oxide are converted to sulfuric acid and nitric acid, which fall to Earth as acid deposition, which has harmful effects on freshwater lakes and forests.

Burning fossil fuels has substantially increased atmospheric carbon dioxide. Climate scientists have concluded that increased CO_2 causes increased global temperatures. Increased temperatures cause climate change that is manifested in more extreme weather, melting glaciers, thinning Arctic ice caps, warming and acidifying oceans, rising sea levels, and changing distributions and seasonal activities of wildlife.

Key Terms

abiotic 531	energy pyramid 535
acid deposition 541	food chain 533
aquifer 537	food web 533
autotroph 532	fossil fuel 538
biological	greenhouse effect 542
magnification 536	greenhouse gas 542
biomass 533	herbivore 532
biotic 531	heterotroph 532
carbon cycle 538	hydrologic cycle 537
carnivore 532	macronutrient 537
climate change 543	micronutrient 537
consumer 532	net primary
decomposer 533	production 533
deforestation 543	nitrogen cycle 539
denitrifying bacteria 540	nitrogen fixation 539
detritivore 533	nutrient 531
ecosystem 531	nutrient cycle 537

omnivore 533	reservoir 537
phosphorus cycle 540	secondary consumer 532
phytoplankton 533	tertiary consumer 532
primary consumer 532	trophic level 532
producer 532	zooplankton 533

Learning Outcomes

In this chapter, you have learned to . . .

LO1 Compare and contrast biotic and abiotic components of ecosystems and describe how nutrients and energy move between these two components.

LO2 Describe the trophic levels in an ecosystem and explain how energy flows through trophic levels.

LO3 Explain why detritivores and decomposers are essential to ecosystem function.

LO4 Explain why energy transfer between trophic levels is inefficient and discuss how this inefficiency determines the relative abundance of organisms at each trophic level.

LO5 Explain the general principles that govern nutrient cycling in an ecosystem.

LO6 Describe the hydrologic, nitrogen, carbon, and phosphorus cycles.

LO7 Describe how human interference with nutrient cycling causes acid deposition, damages aquatic ecosystems, and causes global climate change.

LO8 Describe the evidence that climate change is occurring and discuss key impacts of climate change on Earth's ecosystems.

Thinking Through the Concepts

Fill-in-the-Blank

1. Nearly all life gets its energy from _____, which is captured by the process of _____. In contrast, _____ are constantly recycled during processes called _____. **LO1 LO2 LO5**

2. Photosynthetic organisms are called either _____ or _____. The energy that these store and make available to other organisms is called _____. **LO2**

3. Feeding levels within ecosystems are also called _____. An illustration of these levels with only one organism at each level is called a(n) _____. Feeding relationships are most accurately depicted as _____. **LO2**

4. In general, only about _____ percent of the energy available in one trophic level is captured by the level above it. **LO4**

5. Photosynthetic organisms make up the first trophic level. Organisms in higher trophic levels are collectively called _____ or _____. Photosynthetic organisms are consumed by organisms collectively called _____ or _____. Animals that feed on other animals are called _____ or _____. Organisms that feed on wastes and dead bodies are called _____ and _____. **LO2 LO3**

6. During the nitrogen cycle, nitrogen gas is captured from its atmospheric reservoir by _____ in the soil,

and is then returned to this reservoir by _____.
The two forms of nitrogen that are used by plants are
_____ and _____. **LO5 LO6**

7. Two relatively short-term reservoirs for carbon are the
_____ and _____. Carbon in these
reservoirs is in the form of _____. Two long-
term reservoirs for carbon are _____ and
_____. **LO5 LO6**

Review Questions

1. What makes the movement of energy through ecosystems
fundamentally different from the movement of
nutrients? **LO1 LO2 LO5**

2. What is a producer? What trophic level does it occupy, and
what is its importance in ecosystems? **LO2**

3. Define *net primary production*. Would you predict higher
productivity in a farm pond or an alpine lake? Explain
your answer. **LO2**

4. List the first three trophic levels. Among the consumers,
which are most abundant? Use the "10% law" to explain
why you would predict that there will be a greater biomass
of plants than herbivores in any ecosystem. **LO2 LO4**

5. How do food chains and food webs differ? Which is
the more accurate representation of actual feeding
relationships in ecosystems? **LO2**

6. Define *detritivore* and *decomposer* and explain their
importance in ecosystems. **LO3**

7. Trace the movement of carbon from one of its reservoirs
through the biotic community and back to the reservoir.
How have human activities altered the carbon cycle,
and what are the implications for Earth's future
climate? **LO5 LO6 LO7 LO8**

8. Explain how nitrogen gets from the atmosphere into a
plant's body. **LO6**

9. Trace a pathway of a phosphorus molecule from a
phosphate-rich rock into a carnivore. What makes the
phosphorus cycle fundamentally different from the
carbon and nitrogen cycles? **LO6**

10. Trace the movement of a water molecule from the ocean,
through a plant body, and back to the ocean, describing
all the intermediate stages and processes. **LO6**

Applying the Concepts

1. BioEthics   Humans are omnivores who can feed on
several trophic levels. Discuss how the inefficiency of
energy transfer between trophic levels might apply to
how many humans can be fed, with what environmental
impacts, by people eating fundamentally different diets.

2. Define and give an example of *biological magnification*.
What qualities are present in materials that undergo
biological magnification? In which trophic level are the
problems worst, and why?

3. BioEthics   Discuss the contribution of human population
growth to (a) acid rain and (b) global climate change.

4. Describe what would happen to a population of deer if
all predators were removed and hunting banned. Include
effects on vegetation as well as on the deer population itself.

*Answers to Figure Caption questions and Fill-in-the-Blank questions
can be found in the Answers section at the back of the book.*

MB ®*Go to MasteringBiology for practice quizzes, activities,
eText, videos, current events, and more.*

CASE STUDY
Food of the Gods

WHAT DO CHOCOLATE AND COFFEE have in common? Some would say they are both necessities of life. Every year, the average person in the United States eats about 11 pounds of chocolate and drinks about 1,100 cups of coffee. And Americans aren't the most voracious consumers of either chocolate or coffee: Scandinavians and Germans devour about twice as much chocolate and coffee as Americans do. Coffee and chocolate have been considered delicacies for hundreds of years. In fact, there must have been chocoholics in Sweden at least 300 years ago: Carolus Linnaeus, the Swedish scientist who invented scientific taxonomy, named the cacao tree *Theobroma cacao*—in Greek, "theobroma" means "food of the gods."

Coffee and cocoa both come from seeds of rain-forest plants. Coffee "beans" are the seeds of the coffee plant, which originated in Ethiopia, Africa. Cocoa is produced from the seeds of the cacao tree native to rain forests in South and Central America. Cacao and coffee plants are now widely cultivated in the Tropics, including South and Central America, Africa, and Southeast Asia. Worldwide, about 4 million tons of cocoa and 8 million tons of coffee are produced each year.

The original varieties of both cacao and coffee grew in the shade of taller trees—in fact, full sunlight killed the plants, especially seedlings. In Central and South America, cacao and coffee were often grown as an understory plant in the rain forest. These plantations provided a multilevel, diverse habitat that supported monkeys, frogs, flowers, and about 200 species of birds. The forest vegetation absorbed water and protected the soil from erosion. The shade discouraged weed growth, and detritivores and decomposers recycled fallen leaves into plant nutrients.

In the 1960s and 1970s, however, new varieties of cacao and coffee plants were developed that thrived in full sun and yielded more seeds. As world demand for chocolate and coffee increased, more and more cacao and coffee plants were grown as monocultures—growing only one crop and eradicating competing plants—in full-sun plantations. What is lost when rain forests are converted to monoculture plantations? Or when wetlands are filled in for housing developments? To answer these questions, we must first understand the properties of the communities that make up life on Earth.

Harvesting a cacao pod. Many South American farmers are switching back to traditional sustainable methods of cacao farming. (Inset) An opened cacao pod reveals the seeds embedded in a sweet white pulp.

AT A GLANCE

29.1 What Determines the Distribution of Life on Earth?

29.2 What Factors Influence Earth's Climate?

29.3 What Are the Principal Terrestrial Biomes?

29.4 What Are the Principal Aquatic Biomes?

LEARNING GOALS

LG1 List the four requirements for life on Earth, and explain which are most important in determining the nature and distribution of terrestrial and aquatic biomes.

LG2 Describe how Earth's curvature, tilt on its axis, and orbit about the Sun affect climate and the distribution of precipitation and temperature at different latitudes.

LG3 Explain how oceans, continents, and mountains affect climate.

LG4 Describe the principal terrestrial biomes and the impacts of human activities on them.

LG5 Describe the principal aquatic biomes and the impacts of human activities on them.

29.1 WHAT DETERMINES THE DISTRIBUTION OF LIFE ON EARTH?

Of all the planets in our solar system, only Earth teems with life. However, both the type of living organisms—cacti or palm trees, lobsters or jaguars—and their abundance vary enormously over the surface of our planet. This variability arises from an uneven distribution of four requirements for life on Earth:

- Nutrients from which to construct living tissue,
- Energy to power metabolic activities,
- Liquid water to serve as a medium in which metabolic activities occur, and
- Temperatures that keep water in its liquid state and that allow metabolic activities to occur.

With extremely rare exceptions, energy enters an ecosystem as sunlight, which is captured by plants and other photosynthetic organisms and then transferred through food webs to nonphotosynthetic organisms (see Chapter 28). Therefore, the distribution of life on Earth is largely determined by the requirements of the photosynthesizers. Although there are exceptions, we can make two major generalizations about how the location and abundance of the four requirements of life affect photosynthetic organisms.

First, in the oceans, and even in many lakes and rivers, liquid water is almost always available, even if it might be beneath a surface layer of ice. However, only the top 650 feet (200 meters) of a body of water, and usually much less, receives enough sunlight for photosynthesis. Even here, photosynthesis may be limited, because surface waters usually contain low levels of many nutrients. Temperature is also important, because many aquatic organisms, such as corals, thrive only within a fairly narrow temperature range. Therefore, sunlight energy, nutrients, and temperature are usually the factors that determine the distribution of life in aquatic ecosystems.

Second, energy from sunlight and nutrients in the soil are both relatively plentiful on land. Terrestrial plants require liquid water in the soil, at least during part of the year, because plants require water for their metabolic activities and to replace water that evaporates from their leaves. Different types of plants, such as cacti, redwoods, and grasses, have very different requirements for soil moisture, both in overall amount and in its availability at different times of the year. Soil moisture depends on precipitation and temperature. Generally speaking, the more precipitation, the more moisture the soil contains. In addition, high temperatures evaporate water from the soil, drying it out, while prolonged freezing temperatures turn soil water to ice, making it unavailable to plants. Therefore, temperature and precipitation largely determine the distribution of life on land.

These requirements for life on Earth occur in specific patterns and locations, resulting in characteristic communities of living organisms in regions that extend over thousands, sometimes (in the oceans) even millions, of square miles. These large-scale communities are called **biomes,** often named after their principal types of vegetation, such as deciduous forests or grasslands.

You probably recognize temperature and precipitation, measured over hours or days, as two of the principal components of weather. Weather patterns that prevail for years or centuries in a particular region make up its **climate.** Because the types of vegetation that dominate in terrestrial biomes are determined by long-term patterns of temperature and precipitation, the climate in any particular region is a good predictor of its biome (**Fig. 29-1**). Therefore, before we begin our survey of terrestrial biomes, we will explore how the physical features of Earth and its movement in the solar system affect climate.

CHECK YOUR LEARNING

Can you name the four requirements for life on Earth? Can you explain which of these requirements are most important in determining the nature and distribution of terrestrial and aquatic biomes?

▲ FIGURE 29-1 Rainfall and temperature influence the distribution of terrestrial ecosystems On land, rainfall and temperature determine the amount of soil moisture available to support plant growth.

29.2 WHAT FACTORS INFLUENCE EARTH'S CLIMATE?

Weather and climate are driven by a great thermonuclear engine: the sun. Solar energy reaches Earth in a range of wavelengths—from short, high-energy ultraviolet (UV) rays, through visible light, to the long infrared wavelengths that we experience as heat (see Chapter 7). Before it reaches Earth's surface, however, sunlight is modified by the atmosphere. Some of the sunlight is reflected back into space and some is absorbed by molecules in the atmosphere, but most solar energy reaches Earth's surface, heating the planet. Fortunately, most UV radiation, which can damage biological molecules, including DNA, does not reach the surface. UV radiation is absorbed by an **ozone layer** in the middle atmosphere, or stratosphere. During the twentieth century, humans produced a number of chemicals that began to deplete the ozone layer. In a remarkable demonstration of international cooperation, almost all the nations of the world agreed to limit the production of

ozone-depleting chemicals, as we explore in "Earth Watch: Plugging the Ozone Hole."

The climate of different locations on Earth varies tremendously. These variations arise from the physical properties of our planet. Among the most important are the Earth's curvature, its tilted axis, and the fact that it orbits the sun rather than staying put in one place. As we will see, these factors cause uneven heating of Earth's surface. Uneven heating, in conjunction with Earth's rotation on its axis, generates air and ocean currents, which are in turn modified by the presence, location, and topography of the continents.

Earth's Curvature and Tilt on Its Axis Determine the Angle at Which Sunlight Strikes the Surface

Average yearly temperatures are determined by the amount of sunlight that reaches a given region, which in turn

Earth | Watch

Plugging the Ozone Hole

UV light is so energetic that it can damage biological molecules. UV causes sunburn, premature aging of the skin, and skin cancer. Fortunately, more than 97% of UV radiation is filtered out by an ozone-enriched region of the stratosphere called the ozone layer, which begins about 6 miles (10 kilometers) above Earth's surface and extends out to about 30 miles (50 kilometers) above the surface. UV light striking ozone and oxygen gas causes reactions that both break down and regenerate ozone. In the process, the UV radiation is converted to heat, and the overall level of ozone remains reasonably constant—or it did before humans intervened.

In 1985, British atmospheric scientists published the startling news that springtime levels of stratospheric ozone over Antarctica had declined by more than 30% since 1979. By the mid-1990s, the **ozone hole** over Antarctica had worsened, with springtime ozone only about 50% of its original levels. Ozone levels remain about the same today (**Fig. E29-1**). Photosynthesis by phytoplankton—the producers for marine ecosystems, and the basis for food webs that support penguins, seals, and whales—is reduced under the ozone hole. Although ozone layer depletion is most severe over Antarctica, the ozone layer is somewhat reduced over most of the world, including the continental United States.

The ozone hole is caused primarily by human production and release of chlorofluorocarbons (CFCs). These chemicals were once widely used in the production of foam plastic, as coolants in refrigerators and air conditioners, as aerosol spray propellants, and as cleansers for electronic parts. CFCs are very stable, and were considered safe. Their stability, however, proved to be a major problem because they remain chemically unchanged as they slowly rise into the stratosphere. UV light in the stratosphere causes them to break down and release chlorine atoms, which in turn catalyze the breakdown of ozone.

Fortunately, major steps have been taken toward plugging the ozone hole. The 1987 Montreal Protocol and its subsequent amendments set limits and established phase-out periods for several ozone-depleting chemicals. In a remarkable worldwide effort, 197 countries have signed the treaty. Since 2000, the ozone layer has begun to show some small signs of recovery. However, because CFCs persist in the atmosphere for many years, full recovery is not expected until 2060, possibly even later.

low   **ozone levels**   high

(a) Antarctic ozone hole, September 1979   **(b) Antarctic ozone hole, September 2011**

▲ **FIGURE E29-1 The Antarctic ozone hole** Satellite images show ozone concentrations above the Antarctic in September 1979, before significant ozone depletion occurred, and in September 2011. Low concentrations of ozone are shown in blue and purple. There are fluctuations from year to year, but the area of the hole (about 9.5 million square miles; 25 million square kilometers) and the lowest concentration of ozone in the hole (about 50% of the concentration before depletion began) are currently both about the same as they were in the mid-1990s. Images courtesy of NASA.

depends on latitude. Latitude is a measure of the distance north or south from the equator, expressed in degrees. The equator is defined as 0° latitude, and the poles are at 90° north and south latitudes. Sunlight hits the equator relatively directly (perpendicular to the surface) throughout the year. However, because Earth is a sphere, the farther away from the equator, the more slanted the sunlight, so a given amount of sunlight is spread out over a larger area. In addition, slanting sunlight at high latitudes must travel through more of Earth's atmosphere than vertical sunlight

▶ FIGURE 29-2 **Earth's curvature and tilt produce seasons and climate** Temperatures are highest at the equator and lowest at the poles. Sunlight strikes the Equator nearly vertically year-round. Further toward the poles, sunlight hits Earth's surface at an angle that spreads a given amount of sunlight over a much larger land surface. The tilt of Earth on its axis causes seasonal variations in how directly sunlight strikes different parts of Earth's surface.

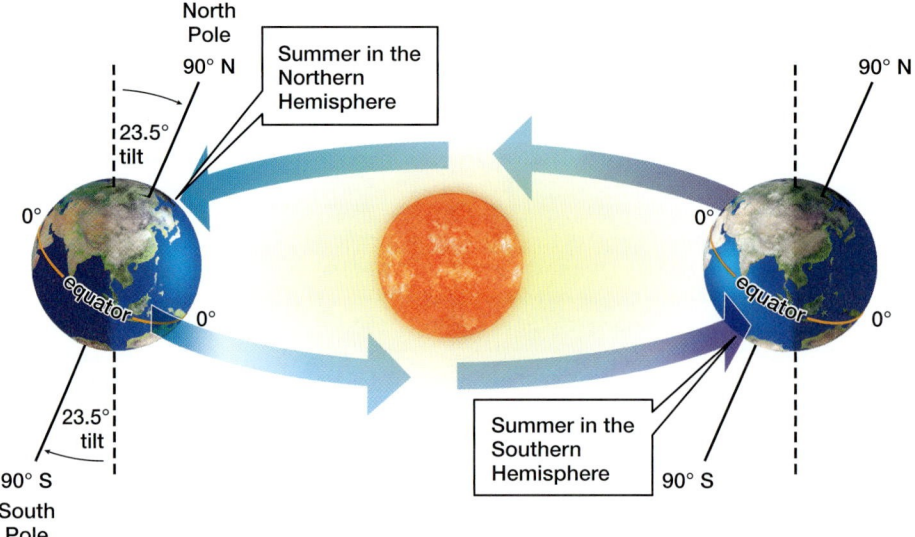

at the equator does, further reducing the amount of solar energy that reaches the surface.

Earth is also tilted on its axis, about 23.5° relative to a line perpendicular to the plane of its orbit around the sun (**Fig. 29-2**). During the course of a year, the tilted axis causes latitudes north and south of the equator to experience regular, significant changes in the angle and duration of sunlight, resulting in pronounced seasons. When Earth's position in its orbit causes the Northern Hemisphere to be tilted toward the sun, this hemisphere receives more direct sunlight (the sun is said to be "high in the sky") and experiences summer (Fig. 29-2, left). Simultaneously, the Southern Hemisphere is tilted away from the sun, receives more slanted sunlight,

and thus experiences winter. Six months later, conditions are reversed (Fig. 29-2, right): It is summer in the Southern Hemisphere and winter in the Northern Hemisphere. Because sunlight hits the equator fairly directly throughout the year, the Tropics remain warm year-round.

Air Currents Produce Large-Scale Climatic Zones That Differ in Temperature and Precipitation

The angle at which sunlight strikes Earth's surface produces climatic zones with markedly different temperatures and precipitation (**Fig. 29-3**). Averaged over the course of a

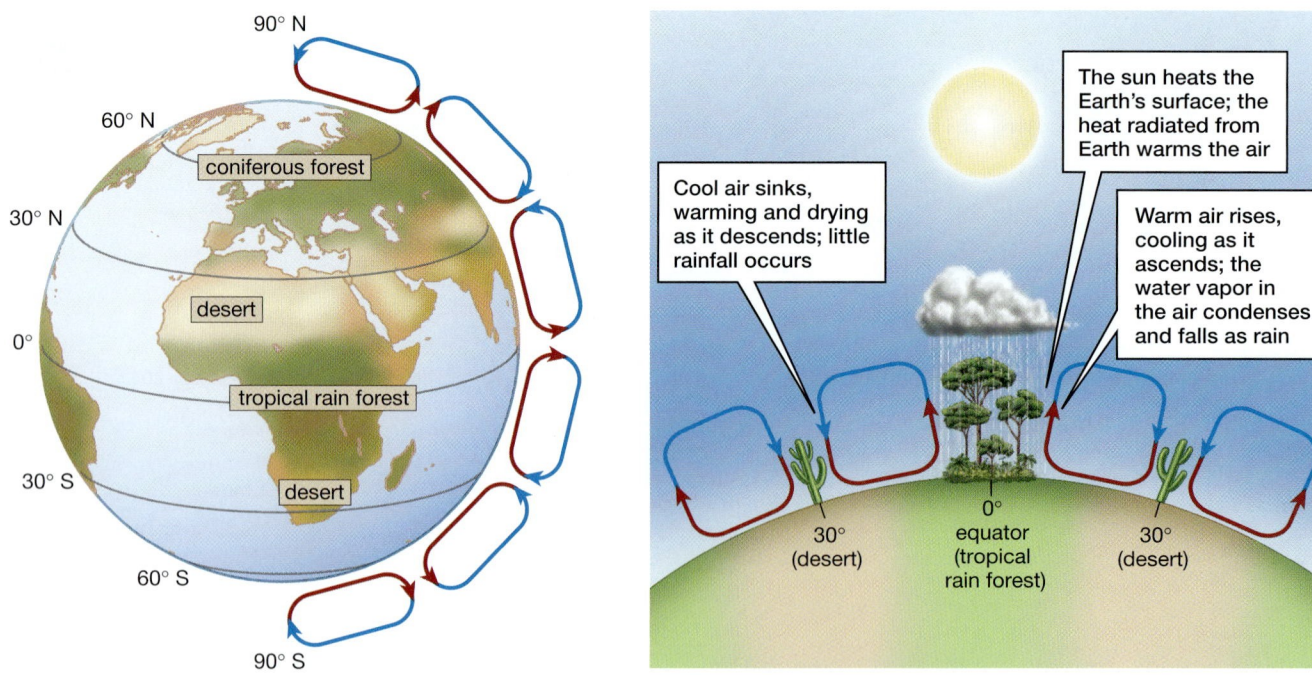

(a) Global air circulation patterns

(b) Air circulation affects climate

▲ FIGURE 29-3 **Air currents and climatic zones (a)** Warm air (red) rises at about 0° and 60° latitudes, and cool air (blue) falls at about 30° and 90° latitudes. **(b)** Air circulation patterns produce broad climatic zones.

year, sunlight travels through the smallest possible thickness of atmosphere—and therefore has its maximum warming effect—at the equator. Warm air is less dense than cool air, so air heated by the warm surface rises at the equator. Warm air can also hold more moisture than cool air can. Therefore, the warm air at the equator is also laden with water evaporated by solar heat. As the water-saturated air rises, it cools. Water condenses out of the air and falls as rain. The direct rays of the sun and abundant rainfall at the equator create a warm, wet climate, where rain forests flourish.

After the moisture has fallen from the rising equatorial air, cooler, drier air remains. The continuing upward flow of air from the equatorial region pushes this air north and south. At about 30° N and 30° S latitudes, the air has cooled enough to sink. As it sinks, the air is warmed by heat radiated from Earth's surface. By the time it reaches the surface, the air is both warm and very dry. The major deserts of the world are found near these latitudes. After reaching the desert surface, the warm, dry air flows north and south, some moving back toward the equator and some moving

toward the poles. This general circulation pattern is repeated farther north and south, with rising air producing relatively abundant precipitation, but cooler temperatures, at around 60° N and 60° S. These climatic conditions favor the growth of coniferous and deciduous forests. The poles receive very little sunlight, and so they are extremely cold. Sinking air also causes the poles to be extremely dry.

These patterns of air circulation predict that climate zones should occur in bands corresponding to latitude. The actual locations of Earth's biomes agree fairly well with this overall pattern, but the fit isn't perfect by any means (Fig. 29-4). The differences are largely caused by three factors: the rotation of Earth on its axis, the existence of the continents, and the presence of large mountain ranges on the continents.

Climate Variability Is Affected by Proximity to Oceans

The interaction between Earth's rotation and air flowing north and south from 30° N and 30° S latitudes determines average

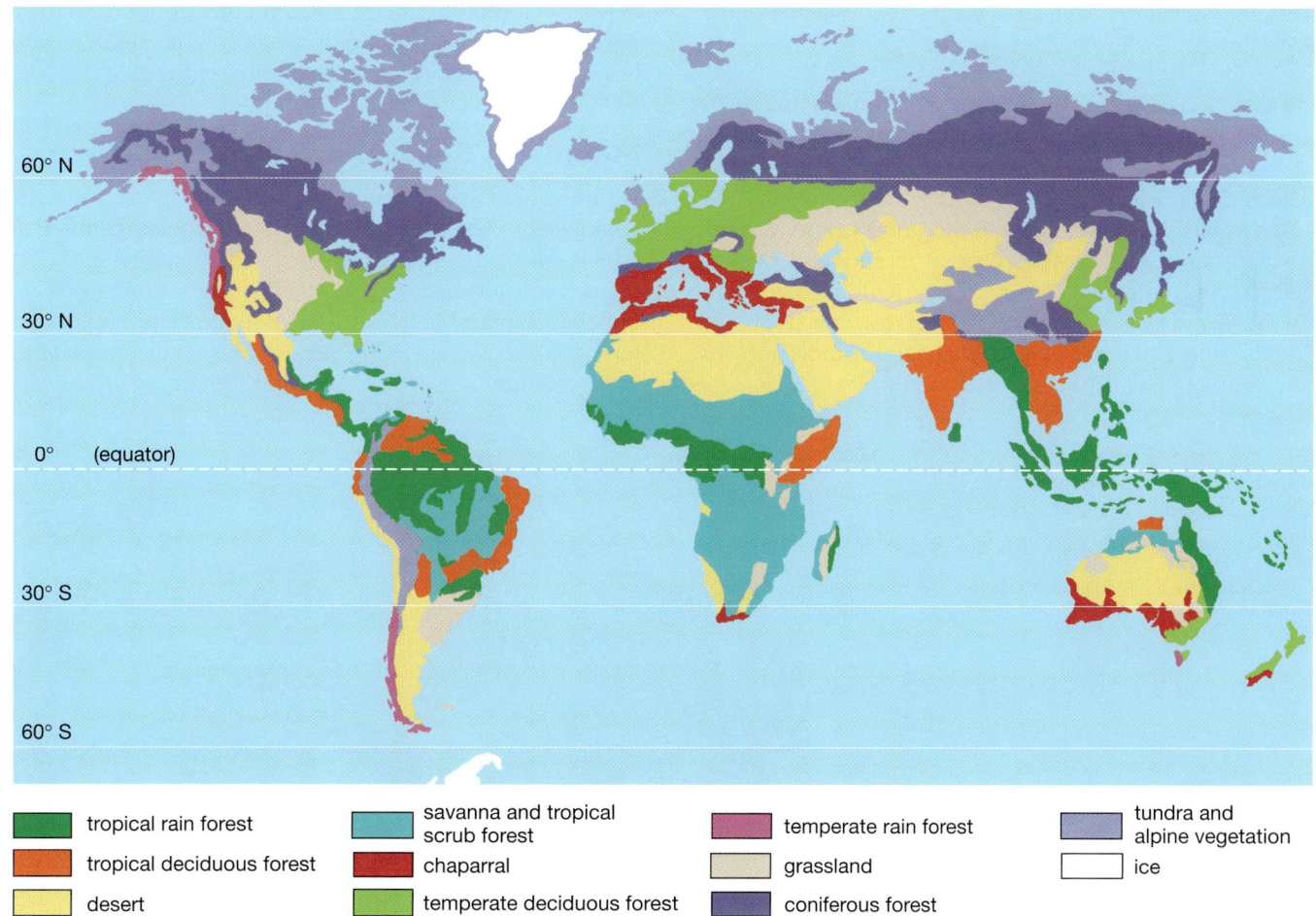

■ tropical rain forest	■ savanna and tropical scrub forest	■ temperate rain forest	■ tundra and alpine vegetation
■ tropical deciduous forest	■ chaparral	■ grassland	□ ice
■ desert	■ temperate deciduous forest	■ coniferous forest	

▲ FIGURE 29-4 **The world's terrestrial biomes** Although mountain ranges and the immense size of continents complicate the distribution of biomes, fairly consistent patterns remain. Tundra and coniferous forests are in the northernmost parts of the Northern Hemisphere, whereas the deserts of Mexico, the Sahara, Saudi Arabia, South Africa, and Australia are located around 30° N and 30° S latitudes. Tropical rain forests are found near the equator. Note that there is little land in the Southern Hemisphere between 45° S and the Antarctic continent; therefore, coniferous forest and tundra biomes are rare.

▲ **FIGURE 29-5 Ocean circulation patterns** Gyres flow clockwise in the Northern Hemisphere and counterclockwise in the Southern Hemisphere. Some ocean currents, such as the Gulf Stream, carry warm water from the Tropics toward the poles. Others, such as the California Current, carry cold water from polar regions toward the equator.

wind directions: east to west between the equator and 30° N and 30° S latitudes, and west to east north of 30° N and south of 30° S latitudes. Friction between winds and the ocean surface produces ocean currents. If there were no continents, then ocean currents would flow around the globe, east to west near the equator, and west to east north of 30° N and south of 30° S. However, the continents interrupt the currents, breaking them into roughly circular patterns called **gyres,** which circulate clockwise in the Northern Hemisphere and counterclockwise in the Southern Hemisphere (**Fig. 29-5**).

Interactions between prevailing winds, ocean currents, and the sheer sizes of the continents profoundly affect terrestrial climates. Because water both heats and cools more slowly than land or air does, the interiors of continents have much more extreme temperatures than coastal regions do. For example, the average high temperature in San Francisco on the coast of central California ranges from about 58°F (14°C) in winter to 71°F (22°C) in summer. Sacramento, only about 80 miles inland, has average high temperatures of 54°F (12°C) in winter and 92°F (33°C) in summer. In St. Louis, Missouri, about 1,700 miles east of San Francisco and 600 miles from the nearest ocean, high temperatures average about 38°F (3°C) in winter and 90°F (32°C) in summer.

Ocean gyres further modify some coastal climates. Some gyres carry warm water from the Tropics to coastal regions located relatively far from the equator. This creates warmer, moister climates than would be expected at these latitudes. For example, the Gulf Stream, carrying warm water from the Caribbean up the coast of North America and across the Atlantic (see Fig. 29-5), is responsible for the mild, infamously moist climate of the British Isles. Other ocean currents, such as the California current, move cold water from near the poles down toward the equator, causing cooler climates than would be expected in regions adjacent to these currents.

Mountains Complicate Climate Patterns

Variations in elevation within continents significantly affect climate. As elevation increases, air becomes thinner and cooler. The temperature drops approximately 3.5°F (2°C) for every 1,000 feet (305 meters) in elevation, so increasing elevation and increasing latitude have similar effects on terrestrial ecosystems (**Fig. 29-6**). Even near the equator, lofty mountains, such as Mount Kilimanjaro in Tanzania

▶ FIGURE 29-6 **Effects of elevation on temperature** Climbing a mountain in the Northern Hemisphere is like heading north; in both cases, increasingly cool temperatures produce a similar series of biomes.

Water vapor is carried from the ocean by the prevailing winds

Water falls as rain or snow as the air rises and cools

Cool, dry air sinks, warms, and absorbs water from the land

moist climate

dry climate in the rain shadow

◀FIGURE 29-7 **Mountains create rain shadows**

(19,341 ft) and Chimborazo in Ecuador (20,565 ft), may be snowcapped much of the year.

Mountains also modify patterns of precipitation. When water-laden air is forced to rise as it meets a mountain, it cools. Because cooling reduces the air's ability to hold water, the water condenses as rain or snow on the windward side of the mountain. The air warms again as it travels down the far (lee) side of the mountain, so it absorbs water from the land, creating a local dry area called a **rain shadow** (Fig. 29-7). For example, the Sierra Nevada range of California wrings moisture from westerly winds blowing off the Pacific Ocean. On the western side of the mountains, heavy winter snows provide moisture for forests of pine, fir, and massive sequoias. In the rain shadow on the east side of the Sierra Nevada, the Mojave Desert receives only about 5 inches of rain a year, and supports mostly cacti and drought-resistant bushes.

CHECK YOUR LEARNING

Can you distinguish between weather and climate? Can you explain how Earth's curvature, tilt on its axis, and orbit around the Sun affect climate? Can you explain how temperature and precipitation interact to determine soil moisture and the distribution of terrestrial biomes? Do you understand how winds, ocean currents, continents, and mountains affect climate and the distribution of biomes?

29.3 WHAT ARE THE PRINCIPAL TERRESTRIAL BIOMES?

In the following sections, we discuss the major terrestrial biomes, beginning at the equator and working our way poleward. We also discuss some of the impacts of human activities on these biomes. (You will learn more about human impacts on Earth in Chapter 30.)

Tropical Rain Forests

Near the equator, the average temperature is between 77° and 86°F (25° and 30°C), with little variation during the year. Rainfall ranges from 100 to 160 inches (250 to 400 centimeters) annually. These evenly warm, moist conditions create the most productive biome on Earth, the **tropical rain forest,** dominated by broadleaf evergreen trees (Fig. 29-8). Extensive rain forests are found in Central and South America, Africa, and Southeast Asia.

Rain forests have the highest **biodiversity,** or total number of species, of any biome on Earth. Although rain forests cover less than 5% of Earth's total land area, ecologists estimate that they contain 5 to 8 million species, representing half to two-thirds of the world's biodiversity. For example, in a 3-square-mile (about 5-square-kilometer) tract of rain forest in Peru, scientists counted more than 1,300 butterfly species and 600 bird species. For comparison, the entire United States is home to only about 600 butterfly species and 700 bird species.

Tropical rain forests typically have several layers of vegetation. The tallest trees reach 150 feet (50 meters), towering above the rest of the forest. Below is a fairly continuous canopy of treetops at about 90 to 120 feet (30 to 40 meters). Another layer of shorter trees typically stands below the canopy. Woody vines grow up the trees. Collectively, these plants capture most of the sunlight. Shorter plants often have enormous, dark-green leaves, an adaptation that allows them to carry out photosynthesis and grow in the dim light that reaches the forest floor.

In tropical rain forests, edible plant material close to the ground is relatively scarce, so most of the animals—including birds, monkeys, and insects—inhabit the trees. Competition for the nutrients that do reach the ground is intense among both plants and animals. For example, when a monkey defecates high up in the canopy, hundreds of dung beetles converge on the droppings within minutes after the waste hits the ground. Plants absorb nutrients almost as soon as soil decomposers release them from wastes or dead plants and animals. This rapid recycling means that almost all the nutrients in a rain forest are stored in the vegetation, leaving the soil relatively infertile.

Human Impacts Because of infertile soil and heavy rains, agriculture in rain forests is risky and destructive. If the trees are cut and carried away for lumber, few nutrients remain to support crops. If the trees are burned, releasing nutrients into the soil, the heavy year-round rainfall quickly dissolves

▲ FIGURE 29-8 **The tropical rain-forest biome** Towering trees reach for the light in the dense tropical rain forest. Amid their branches dwells the most diverse assortment of life on Earth, including (clockwise from upper left) a tree-dwelling orchid, a red-eyed tree frog, and a fruit-eating toucan.

QUESTION How does a biome with such poor soil support the highest plant productivity and the greatest animal diversity on Earth?

the nutrients and carries them away, leaving the soil depleted after only a few seasons of cultivation.

Nevertheless, rain forests are being felled for lumber or burned for ranching or farming at an alarming rate (Fig. 29-9). Conservative estimates indicate that about 13 to 25 million acres of tropical rain forest are lost each year (some estimates are much higher)—the area of a football field every 1.5 to 3 seconds. About half of the world's total rain forests have now been lost, taking with them a substantial portion of Earth's biodiversity, which can never be replaced. In addition, like all forests, rain forests absorb carbon dioxide and release oxygen. About 10% to 20% of the CO_2 released into the atmosphere by human activities comes from cutting and burning tropical rain forests, intensifying the greenhouse effect and accelerating climate change. Fortunately, some areas have been set aside as protected preserves, and some reforestation efforts are under way, with local residents becoming more involved in conservation efforts.

▲ FIGURE 29-9 **Burning the Amazon rain forest** The burned area will be converted to ranching or agriculture, but both are doomed to fail eventually due to poor soil quality. Fires and the smoke they produce also threaten adjacent forests and their myriad inhabitants, and increase atmospheric CO_2.

CASE STUDY continued

Food of the Gods

Cocoa farming is one of the major sources of income in many West African nations, where 2 million farmers produce about 70% of the world's cocoa. In the past few years, several groups, including the World Cocoa Foundation, the Bill and Melinda Gates Foundation, the Rainforest Alliance, and industry giants such as Hershey and Mars, have pledged significant resources to develop sustainable cocoa farming in West Africa. The efforts include providing modern technology to cocoa farmers, helping to increase their annual production and income; developing disease- and drought-resistant strains of cacao trees; reducing the use of child labor; preventing environmental degradation; and preserving rain-forest biodiversity.

Tropical Deciduous Forests

Slightly farther from the equator, although annual rainfall is still high, there are pronounced wet and dry seasons. In these areas, which include much of India as well as parts of Southeast Asia, South America, and Central America, **tropical deciduous forests** grow. During the dry season, the trees cannot get enough water from the soil to compensate for evaporation from their leaves. Many plants shed their leaves during the dry season ("deciduous" literally means "falling off"), which minimizes water loss. If the rains fail to return on schedule, the trees do not leaf out until the drought ends.

Human Impacts Human activities impact tropical deciduous forests in much the same ways that they affect tropical rain forests. Logging, burning to clear land for agriculture, and cutting for firewood all contribute to deforestation of tropical deciduous forests. Fortunately, many tropical deciduous trees "stump-sprout" after logging. Therefore, if disturbances are not too severe and not too frequent, tropical deciduous forests often recover fairly quickly, with nearly the same species as were present before the disturbances occurred.

Tropical Scrub Forests and Savannas

Along the edges of the tropical deciduous forest, reduced rainfall produces the **tropical scrub forest** biome, dominated by deciduous trees that are shorter and more widely spaced than in tropical deciduous forests. Between the scattered trees, sunlight penetrates to ground level, which allows grass to grow. Still farther from the equator, the climate grows drier, and grasses become the dominant vegetation, with only scattered trees; this biome is the **savanna** (Fig. 29-10).

Rainfall in savannas ranges from about 12 to 40 inches (30 to 100 centimeters) a year, almost all falling during a

▲ FIGURE 29-10 **The African savanna** Giraffes feed on savanna trees and share this biome with (clockwise from left) lions, herds of zebra, and rare black rhinos.

▶ FIGURE 29-11 **The desert biome (a)** Under the most extreme conditions of heat and drought, deserts can be almost devoid of life, such as these sand dunes of the Sahara Desert in Africa. **(b)** Throughout much of Utah and Nevada, the Great Basin Desert presents a landscape of widely spaced shrubs, such as sagebrush and greasewood.

(a) Sahara dunes

(b) Utah desert

rainy season lasting 3 or 4 months. When the dry season arrives, rain might not fall for months, and the soil becomes hard, dry, and dusty. Grasses are well adapted to this type of climate, growing very rapidly during the rainy season and dying back to drought-resistant roots during the dry season. Only a few specialized trees, such as the thorny acacia or the water-storing baobab, can survive the devastating savanna dry seasons.

The African savanna supports the most diverse and impressive array of large mammals on Earth. These mammals include numerous herbivores, such as antelope, wildebeest, water buffalo, elephants, and giraffes; and carnivores such as lions, leopards, hyenas, and wild dogs.

Human Impacts Africa's rapidly expanding human population threatens the wildlife of the savanna. For example, poaching for rhino horns has driven the black rhinoceros to the brink of extinction. Poaching also endangers the African elephant, a keystone species in this ecosystem. The abundant grasses that make the savanna a suitable habitat for so much wildlife also make it suitable for grazing domestic cattle. Fences erected to contain cattle disrupt the migration of the herds of wild herbivores as they search for food and water.

Deserts

Even drought-resistant grasses need at least 10 to 20 inches (25 to 50 centimeters) of rain a year, depending on the temperature and the precipitation's seasonal distribution. Biomes where annual rainfall is 10 inches or less are called **deserts.** Although we tend to think of them as hot, deserts are defined by their lack of rain rather than by their temperatures. In the Gobi Desert of Asia, for example, although the summers are very hot, average temperatures are below freezing for half the year. Desert biomes are found on every continent, typically at around 30° N and 30° S latitudes, and also in the rain shadows of major mountain ranges.

Deserts vary in just how dry they are. At one extreme are the Atacama Desert in Chile and parts of the Sahara Desert in Africa, where it almost never rains and no vegetation grows (**Fig. 29-11a**). More commonly, deserts are characterized by widely spaced vegetation and large areas of bare ground (**Fig. 29-11b**).

Only highly specialized plants can grow in deserts. Although unrelated, both cacti (mostly in the Western Hemisphere) and euphorbs (mostly in the Eastern Hemisphere; **Fig. 29-12**) have shallow, spreading roots that rapidly absorb rainwater before it evaporates. Their thick stems store water when it is available. Spines protect the plants from herbivores that would otherwise eat the stems for both nutrition and water. Evaporation is minimized because the leaves, if any, are very small (the spines of cacti are highly modified leaves); typically, photosynthesis occurs in the green, fleshy stem. A heavy wax coating on the stems further reduces water loss.

Some deserts have a very brief rainy season, in which a whole year's rain falls in just a few storms. Specialized annual wildflowers take advantage of the brief period of moisture to sprout from seed, grow, flower, and produce seeds of their own in a month or less (**Fig. 29-13**).

Desert animals are also adapted to survive heat and drought. Few animals are active during the hot summer days.

(a) Cactus

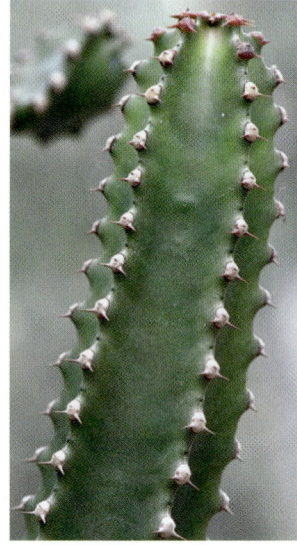

(b) Euphorb

▲ FIGURE 29-12 **Environmental demands shape physical characteristics** Evolution in response to similar desert conditions has molded the bodies of **(a)** cacti and **(b)** euphorbs into nearly identical shapes, although they are not closely related species.

QUESTION Describe the similar environmental demands that operate on these two different types of plants.

▲ FIGURE 29-13 **Desert wildflowers** After a relatively wet spring, this Arizona desert is carpeted with wildflowers. Through much of the year—and sometimes for several years—annual wildflower seeds lie dormant, waiting for adequate spring rains to fall.

Many desert denizens take refuge from the heat in underground burrows that stay relatively cool and moist. In North American deserts, nocturnal (night-active) animals include jackrabbits, bats, kangaroo rats, and burrowing owls (Fig. 29-14). Reptiles such as snakes, turtles, and lizards adjust their activity cycles depending on the temperature. In summer, they may be active only around dawn and dusk. Kangaroo rats and many other small desert animals survive without ever drinking. They acquire water from their food and from water produced as a by-product of cellular respiration (see Chapter 8). Larger animals, such as desert bighorn sheep, depend on permanent water holes during the driest times of the year.

Human Impacts Desert ecosystems are fragile. In the Mojave Desert of California, tread marks left by tanks during the 1940s are still visible today. Desert soil is stabilized by strands of bacteria that intertwine among sand grains. The tanks, and more recently off-road vehicles, have destroyed this crucial network, causing soil erosion and reducing nutrients available to the desert's slow-growing plants. Desert soil may require hundreds of years to fully recover from heavy vehicle use.

Human activities also contribute to **desertification,** the process by which relatively dry regions are converted to desert as a result of drought coupled with misuse of the land. When people overharvest bushes and trees for firewood, graze too many livestock, and deplete both surface and groundwater to grow crops, the native vegetation becomes extremely vulnerable to drought. Loss of vegetation in turn allows the soil to erode, further decreasing the land's productivity. Desertification has severely impacted the Sahel region in Africa just south of the Sahara Desert (Fig. 29-15). In 2011, eleven African countries, in cooperation with the Global Environment Facility, proposed building a "Great Green Wall" of trees and bushes, about 9 miles (15 kilometers) wide, clear across the continent, in an effort to revegetate the degraded environment and stop desertification in the Sahel.

(a) A kangaroo rat

(b) A burrowing owl

▲ FIGURE 29-14 **Desert dwellers (a)** Kangaroo rats and **(b)** burrowing owls spend the hottest part of the day in burrows, emerging at night to feed.

Chaparral

Many coastal regions that border on deserts, such as those in southern California and much of the Mediterranean, support the **chaparral** biome (Fig. 29-16). The annual rainfall is up to 30 inches (76 centimeters), nearly all of which falls during cool, wet winters. Summers are hot and dry. Chaparral plants consist mainly of drought-resistant shrubs and small trees. Their leaves are usually small and are often coated with tiny

▲ FIGURE 29-15 **Desertification** A rapidly growing human population coupled with drought and poor land use has reduced the ability of many dry regions, such as the Sahel in Africa, to support life.

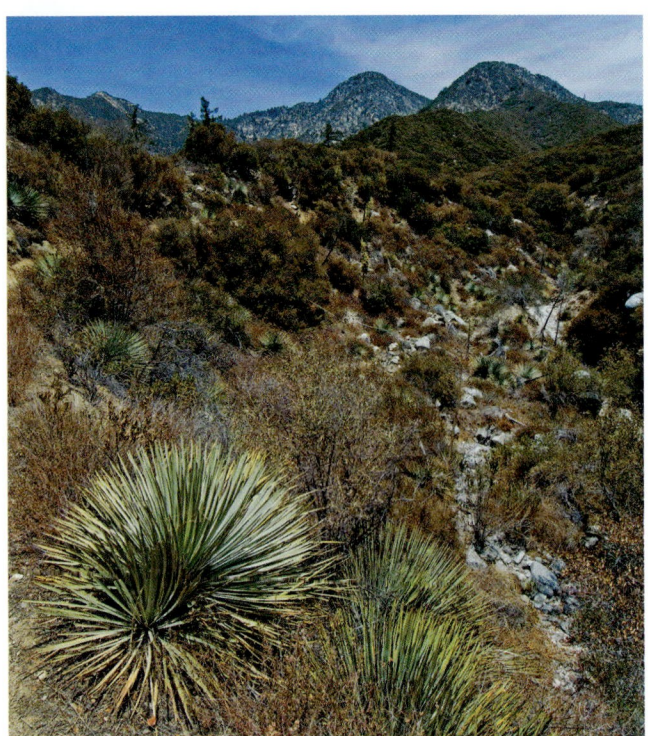

▲ **FIGURE 29-16** **The chaparral biome** Limited to warm, dry coastal regions and maintained by fires caused by lightning, this hardy biome is characterized by drought-resistant shrubs and small trees, such as these seen in the western foothills of the San Gabriel mountains in Southern California.

▲ **FIGURE 29-17** **Tallgrass prairie** In the central United States, moisture-bearing winds out of the Gulf of Mexico produce summer rains, encouraging the lush growth of tall grasses and abundant wildflowers. Periodic fires, now carefully managed, prevent the encroachment of forest and maintain this subclimax biome.

QUESTION Why is tallgrass prairie one of the most endangered biomes in the world?

hairs or waxy layers that reduce evaporation during the dry summer months. Chaparral is adapted to fire. Many shrubs regrow from their roots after fires. Others have seeds that are stimulated to germinate by chemicals found in smoke.

Human Impacts People enjoy living in warm, dry climates adjacent to oceans, so development for housing is a major threat to chaparral biomes. In more rugged terrain, especially in southern Europe, chaparral has been cleared for grazing, olive groves, and other types of agriculture.

Grasslands

Grassland, or *prairie*, biomes are typically located in the centers of continents, such as North America and Eurasia, and receive 10 to 30 inches (25 to 75 centimeters) of rain annually. In general, these biomes have a continuous cover of grass and virtually no trees, except along rivers. In tallgrass prairie—in North America, originally found from Texas to southern Canada—grasses reach up to 6 feet in height. An acre of natural tallgrass prairie may support 200 to 400 different species of native plants (**Fig. 29-17**). Areas further west, which receive less rainfall, support midgrass and shortgrass prairies (**Fig. 29-18**). In these grasslands, prairie dogs and ground squirrels provide food for eagles, foxes, coyotes, and bobcats. Pronghorns browse in western grasslands, and bison survive in preserves.

▲ **FIGURE 29-18** **Shortgrass prairie** Shortgrass prairie is characterized by low-growing grasses. In addition to many wildflowers, life in shortgrass prairies includes (clockwise from left) bison (in preserves), pronghorns, and prairie dogs.

Why do grasslands lack trees? Water and fire are the crucial factors in the competition between grasses and trees. The hot, dry summers and frequent droughts of the midgrass and shortgrass prairies can be tolerated by grass but are fatal to trees. In tallgrass prairies, forests are the climax ecosystems. Historically, however, tree growth was suppressed by a combination of occasional severe drought and frequent fires caused by lightning or set by Native Americans. Although fire kills trees, the root systems of grasses survive.

Human Impacts Grasses growing and decomposing for thousands of years produced the most fertile soil in the world. In the early nineteenth century, North American grasslands supported an estimated 60 million bison. Today, the midwestern U.S. grasslands have been largely converted to farm and range land, and cattle have replaced bison.

Prairie dog colonies and the eagles and ferrets that hunted them have become a rare sight as their habitat has been displaced by ranching and, more recently, by suburban sprawl. Wolves, once common, have been eliminated from the prairies. In some regions, overgrazing has destroyed the native grasses, allowing woody sagebrush to flourish (**Fig. 29-19**). Undisturbed grasslands are now largely confined to protected areas. Tallgrass prairie is one of the most endangered ecosystems in the world. Only about 1% remains, in tiny remnants restored by planting native species and maintained by controlled burning.

Temperate Deciduous Forests

At their eastern edge, the North American grasslands merge into the **temperate deciduous forest** biome (**Fig. 29-20**). Temperature deciduous forests are also found in much of Europe and eastern Asia. More precipitation occurs in temperate deciduous forests than in grasslands (30 to 60 inches,

▲ **FIGURE 29-20 The temperate deciduous forest biome** Temperate deciduous forests of the eastern United States are inhabited by (clockwise from left) white-tailed deer (this biome's largest herbivore) and birds such as this blue jay. In spring, a profusion of woodland wildflowers (such as these hepaticas) blooms briefly before the trees produce leaves that shade the forest floor.

or 75 to 150 centimeters). The soil retains enough moisture for trees to grow, shading out most grasses.

Winters in temperate deciduous forests often have long periods of below-freezing weather, when liquid water is not available. The trees drop their leaves in the fall and remain dormant through the winter, conserving scarce water. During the brief time in spring when the ground has thawed but the emerging leaves have not yet blocked off the sunlight, abundant wildflowers grace the forest floor.

Insects and other arthropods are numerous and conspicuous in deciduous forests. The decaying leaf litter on the forest floor also provides food and suitable habitat for bacteria, earthworms, fungi, and small plants. A variety of vertebrates—including mice, shrews, squirrels, raccoons, deer, bears, and many species of birds—dwell in deciduous forests.

Human Impacts Large predatory mammals such as black bears, wolves, bobcats, and mountain lions were formerly

▲ FIGURE 29-19 **Sagebrush desert or shortgrass prairie?** Biomes are influenced by human activities as well as by temperature, rainfall, and soil. The shortgrass prairie field on the right has been overgrazed by cattle, causing the grasses to be replaced by sagebrush.

abundant in the eastern United States, but hunting and habitat loss have severely reduced their numbers. Deer thrive due to a lack of natural predators. Clearing for lumber, agriculture, and housing dramatically reduced deciduous forests in the United States. Virgin (uncut) deciduous forests are almost nonexistent, but the last half-century has seen extensive regrowth of deciduous forests on abandoned farms and formerly logged land.

Temperate Rain Forests

On the U.S. Pacific Coast, from the lowlands of the Olympic Peninsula in Washington State to southeast Alaska, lies a **temperate rain forest** (Fig. 29-21). Temperate rain forests are also located along the southeastern coast of Australia, the southwestern coast of New Zealand, and parts of Chile and Argentina. As in the tropical rain forest, temperate rain forests experience a tremendous amount of rain. In North America, these biomes typically receive more than 55 inches (140 centimeters) of rain annually—and as much as 12 feet a year in some areas. The nearby ocean keeps the temperature moderate.

Most of the trees in the temperate rain forest are conifers, such as spruce, Douglas fir, and hemlock. The forest floor and tree trunks are typically covered with mosses and ferns. Fungi thrive in the moisture and enrich the soil. As in tropical rain forests, so little light reaches the forest floor that tree seedlings usually cannot become established. Whenever one of the forest giants falls, however, it opens up a patch of light, and new seedlings quickly sprout, often right atop the fallen log.

Human Impacts Tall, straight trees are extremely valuable for lumber, and consequently many temperate rain forests have been logged. In the mild, wet climate, the forests regrow quickly. However, some animals, such as the spotted owl, dwell mainly in old-growth forests that are hundreds of years old, and have become rare as a result of extensive logging.

Northern Coniferous Forests

North of the grasslands and temperate forests stretches the **northern coniferous forest** (also called the *taiga;* Fig. 29-22). The northern coniferous forest, which is the largest terrestrial biome on Earth, stretches across North America, Scandinavia, and Siberia, nearly encircling the globe. It includes parts of Alaska and the northern United States, and much of southern Canada. Very similar forests occur in many mountain ranges, including the Cascades, the Sierra Nevadas, and the Rocky Mountains.

Conditions in the northern coniferous forest are much harsher than in temperate deciduous forests, with long, cold winters and short growing seasons. About 16 to 40 inches (40 to 100 centimeters) of precipitation occur annually, much of it as snow. The conical shape and narrow, stiff needles of conifers allow them to shed snow efficiently. Their small, wax-covered needles minimize water loss during the long winters, when water remains frozen. By retaining their leaves, these evergreens conserve the energy that deciduous

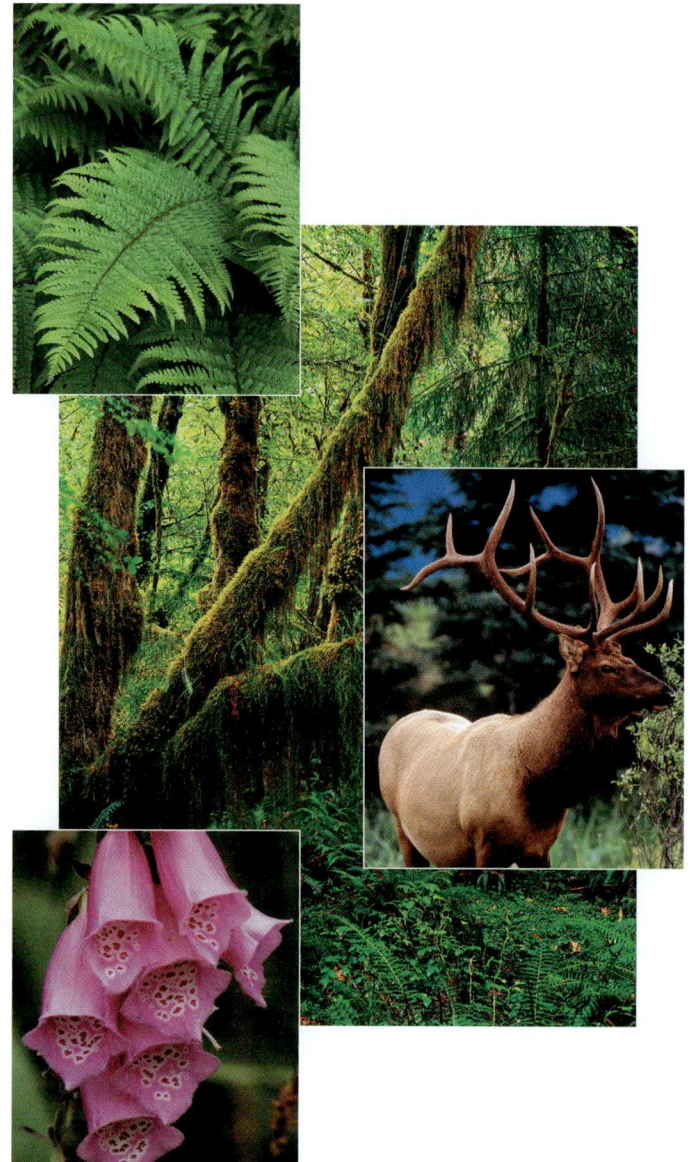

▲ **FIGURE 29-21 The temperate rain-forest biome** The Hoh River temperate rain forest in Olympic National Park receives about 12 feet of rain annually. Ferns, mosses, and wildflowers grow in the pale green light of the forest floor. Denizens of this rain forest include (clockwise from upper left) ferns, such as this lady fern, elk, and flowering foxglove.

trees must expend to grow new leaves, and they are ready to take advantage of good growing conditions when spring arrives. Large mammals—including black bears, moose, deer, and wolves—still roam the northern coniferous forest, as do wolverines, lynxes, foxes, bobcats, and snowshoe hares. These forests also serve as breeding grounds for many of North America's bird species.

Human Impacts Clear-cutting for papermaking and lumber has leveled huge expanses of northern coniferous forest in both Canada and the U.S. Pacific Northwest

▲ FIGURE 29-22 **The northern coniferous forest biome** The small needles and conical shape of conifers allow them to shed heavy snows. (Lower left) A Canada lynx catches a snowshoe hare. (Upper right) A great horned owl waits for nightfall, when it will begin to hunt.

? HAVE **YOU** EVER WONDERED …

If People Can Re-Create Ancient Biomes?

Remember *Jurassic Park*? OK, maybe resurrecting *T. rex* and *Velociraptor* is a little far fetched. But how about a Pleistocene Park? Russian scientist Sergey Zimov thinks that the mosses and shrubs, and possibly even the coniferous forests, of northern Siberia are artificial landscapes, created because humans wiped out mammoths, bison, wooly rhinos, and most other large herbivores about 10,000 years ago. Zimov wants to re-create what he calls the "mammoth steppe"—a vast grassland supporting herds of herbivores and the carnivores that prey on them. So far, he and his team have introduced Yakutian horses, musk ox, bison, and elk to Pleistocene Park, a huge protected area in Siberia. Reindeer and moose were already there, along with gray wolves, lynx, brown bears, and foxes. Horses seem to be crucial to restoring the grasslands (**Fig. E29-2**). Wherever there are enough horses, grasslands are returning. As for mammoths, that's a bit more like *Jurassic Park*, although some people think that genetic engineering might possibly succeed someday, using DNA from frozen mammoth carcasses, elephants as surrogate mothers, and a lot of luck. But even without mammoths, the "re-wilding" of Pleistocene Park is well under way.

▲ FIGURE E29-2 **Yakutian horses seem to be an essential component of Pleistocene Park**

▲ **FIGURE 29-23 Clear-cutting** Coniferous forests are vulnerable to clear-cutting, as seen in this forest in Alberta, Canada. Clear-cutting is a relatively simple and inexpensive means of logging compared to selective harvesting of trees, but its environmental costs are high. Erosion diminishes the fertility of the soil, slowing new growth. Further, the dense stands of similarly aged trees that typically regrow are more vulnerable to attack by parasites than a natural stand of trees of various ages would be.

(Fig. 29-23). Demand is also increasing to extract natural gas and oil, often from unconventional sources such as oil sands. Nevertheless, much of Canada's coniferous forest remains intact. Encouragingly, in 2008, the provincial governments of Ontario and Quebec pledged to protect half of the publicly owned coniferous forest and to manage the remainder sustainably.

Tundra

The biome furthest north is the arctic **tundra,** a vast treeless region bordering the Arctic Ocean (Fig. 29-24). Conditions in the tundra are severe. Winter temperatures often reach −40°F (−55°C) or below, with howling winds. Precipitation averages 10 inches (25 centimeters) or less each year, making this region a freezing desert. Even during the summer, frosts are frequent, and the growing season may last only a few weeks. A similar climate and quite similar tundra are found at high elevations on mountains worldwide.

The cold climate of the arctic tundra results in **permafrost,** a permanently frozen layer of soil. Soil above the permafrost thaws each summer, usually to a depth of 2 feet (60 centimeters) or more. When the summer thaws arrive, the underlying permafrost limits the ability of soil to absorb the water from melting snow and ice, and so the tundra becomes a marsh.

Because of the extreme cold, the brief growing season, and the permafrost, which limits the depth of roots, trees do not grow in the tundra. Nevertheless, the ground is carpeted with small perennial flowers, dwarf willows, and large lichens called "reindeer moss," a favorite food of caribou. The summer marshes also provide superb mosquito habitat. These and other insects feed about 100 different species of birds, most of which migrate here to nest and raise their young during the brief summer feast. The tundra vegetation also supports arctic hares and lemmings (small rodents) that are eaten by wolves, owls, and arctic foxes.

▲ **FIGURE 29-24 The tundra biome** Life on the tundra is seen here in Denali National Park, Alaska, turning color in autumn. (Clockwise from left) Tundra animals, such as caribou and arctic fox, can regulate blood flow in their legs, keeping them just warm enough to prevent frostbite while preserving precious body heat for the brain and other vital organs. Perennial plants such as this frost-covered bearberry grow low to the ground, avoiding the chilling tundra wind.

Human Impacts The tundra is among the most fragile of all terrestrial biomes because of its short growing season. A willow 4 inches (10 centimeters) high may be 50 years old. Alpine tundra is readily damaged by off-road vehicles and even hikers. Fortunately for the inhabitants of the arctic tundra, the impacts of civilization are localized around oil-drilling sites, pipelines, mines, and scattered military bases. The most significant threat to the tundra is climate change. Shrubs and trees are replacing tundra along its southern margins. Climate models suggest that more than a third of Earth's tundra may be lost by the end of this century.

CHECK YOUR LEARNING

Can you describe the principal terrestrial biomes and discuss how temperature and precipitation interact to determine their characteristic plant life? Can you describe some human impacts on terrestrial biomes?

29.4 WHAT ARE THE PRINCIPAL AQUATIC BIOMES?

Of the four requirements for life, aquatic ecosystems typically provide abundant water and appropriate temperatures. However, sunlight in aquatic ecosystems decreases with depth, as it is absorbed by water and blocked by suspended particles. In addition, nutrients in aquatic ecosystems tend to be concentrated in sediments on the bottom, so where nutrients are high, light levels tend to be low.

Freshwater Lakes

Freshwater lakes form when natural depressions fill with water from groundwater seepage, streams, or runoff from rain or melting snow. Large lakes in temperate climates have distinct zones of life (**Fig. 29-25**). Near the lake shore is a shallow **littoral zone,** where plants find both abundant sunlight and nutrients. Littoral zone communities are the most diverse regions of lakes. They support algae; plants such as cattails, bullrushes, and water lilies, anchored in the bottom near the shore; and submerged plants that flourish in slightly deeper waters.

The greatest diversity of animal life is also found in the littoral zone, although many of the animals, especially fish, spend time in more than one zone. Littoral vertebrates include frogs, aquatic snakes, turtles, and fish such as pike, bluegill, and perch; invertebrates include insect larvae, snails, flatworms, and crustaceans such as crayfish. Littoral waters are also home to small organisms collectively called **plankton** (from a Greek word meaning "drifters"). Photosynthetic

▲ FIGURE 29-25 **Lake life zones** A typical large lake has three life zones: a nearshore littoral zone with rooted plants, an open-water limnetic zone, and a deep, dark profundal zone.

protists and bacteria are called **phytoplankton.** Nonphotosynthetic protists and tiny crustaceans that feed on phytoplankton make up the **zooplankton.**

As the water increases in depth, plants are unable to anchor themselves to the bottom and still receive enough sunlight for photosynthesis. This open-water region is divided into an upper **limnetic zone,** in which enough light penetrates to support photosynthesis by plankton, and a lower **profundal zone,** in which light is too weak for photosynthesis to occur (see Fig. 29-25). Plankton and fish dominate in the limnetic zone. Organisms that live in the profundal zone are nourished by organic matter that drifts down from the littoral and limnetic zones and by sediment washed in from the land. Inhabitants of the profundal zone include catfish, which mainly feed on the bottom, and detritivores and decomposers such as crayfish, aquatic worms, clams, leeches, and bacteria.

Freshwater Lakes Are Classified According to Their Nutrient Content

Freshwater lakes may be described as *oligotrophic* (Greek, "poorly fed"), *eutrophic* ("well fed"), or *mesotrophic* (between these two extremes, or "middle fed"). Here, we describe the characteristics of oligotrophic and eutrophic lakes.

Oligotrophic lakes contain few nutrients and support relatively little life. Many oligotrophic lakes were formed by glaciers that scraped depressions in bare rock. The lakes are now fed by mountain streams and snowmelt. Because there is little sediment or microscopic life to cloud the water, oligotrophic lakes are clear, and light penetrates deeply. With few bacteria to compete for oxygen, fish that require well-oxygenated water, such as trout, thrive in oligotrophic lakes.

Eutrophic lakes receive relatively large inputs of sediments, organic material, and inorganic nutrients (such as phosphates and nitrates) from their surroundings, allowing them to support dense plant communities (**Fig. 29-26**). They are murky from suspended sediment and dense phytoplankton populations, so the limnetic zone is shallow. The

▲ **FIGURE 29-26 A eutrophic lake** Rich in dissolved nutrients carried from the land, eutrophic lakes support dense growths of algae, phytoplankton, and both floating and rooted plants.

dead bodies of limnetic zone inhabitants sink into the profundal zone, where they feed decomposer organisms. The metabolic activities of these decomposers use up oxygen, so the profundal zone of eutrophic lakes is often very low in oxygen and supports little life.

Gradually, as nutrient-rich sediment accumulates, oligotrophic lakes tend to become eutrophic, a process called eutrophication. Although large lakes may last for millions of years, eutrophication may eventually cause lakes to undergo succession to dry land (see Chapter 27).

Human Impacts Nutrients carried into lakes from farms, feedlots, sewage, and even fertilized suburban lawns accelerate eutrophication. Overfertilized lakes sometimes experience massive algal blooms, followed by die-offs and decomposition that deplete the water of oxygen and kill most of the fish. Phosphate-free detergents, more effective sewage treatment, careful application of fertilizers, and proper location and operation of feedlots lessen the danger of eutrophication.

Streams and Rivers

Streams often originate in mountains, the *source region* shown in **Figure 29-27,** where runoff from rain and melting snow cascades over impervious rock. Little sediment reaches the streams, phytoplankton is sparse, and the water is clear and cold. Algae adhere to rocks in the streambed, where insect larvae find food and shelter. Turbulence keeps mountain streams well oxygenated, providing a home for trout that feed on insect larvae and smaller fish.

At lower elevations, in the *transition zone,* small streams merge, forming wider, slower-moving streams and small rivers. The water warms slightly, and more sediment is carried in, providing nutrients that allow aquatic plants, algae, and phytoplankton to proliferate. Fish such as bass, bluegills, and yellow perch (all of which require less oxygen than trout do) are found in such waterways.

As the land becomes lower and flatter, the river warms, widens, and slows, meandering back and forth. The water becomes murky with sediment and dense populations of phytoplankton. Decomposer bacteria deplete the oxygen in deeper water, but carp and catfish can still thrive even though oxygen levels are fairly low. When precipitation or snowmelt is high, the river may flood the surrounding flat land, called a *floodplain,* depositing a rich layer of sediment over the adjoining terrestrial ecosystem.

Rivers drain into lakes or into other rivers that ultimately lead to the oceans. Close to sea level, most rivers move slowly, depositing their sediment. The sediment interrupts the river's flow, breaking it into small winding channels that form an *estuary* (described below) where the river empties into the ocean.

Human Impacts Rivers are sometimes channelized (deepened and straightened) to facilitate boat traffic, to prevent flooding, and to allow farming along their banks. Channelization increases erosion, because water flows more rapidly

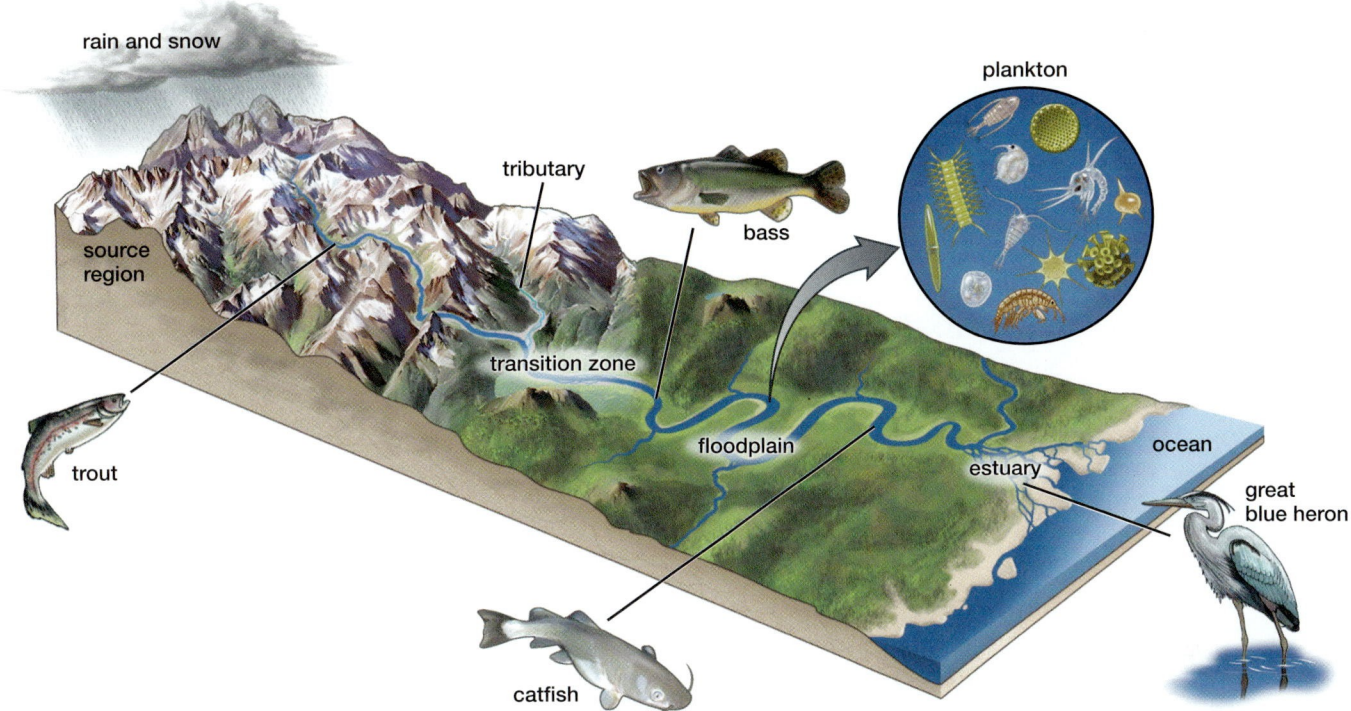

▲ **FIGURE 29-27** **From stream to river to sea** At high elevations, precipitation feeds fast-flowing, clear streams, which grow and slow as they are joined by tributaries at lower elevations. Many become slow-moving rivers that weave through a floodplain, dropping nutrient-rich sediment. Many rivers create estuaries where they meet the ocean. The communities of freshwater organisms change as the water flows from mountains to the ocean.

in straightened rivers. In addition, where natural flooding has been prevented, floodplain soil no longer receives the nutrients formerly deposited by floodwaters.

In the United States, both Pacific and Atlantic salmon populations have been greatly reduced by hydroelectric dams, water diversion for agriculture, erosion from logging operations, and overfishing. On both coasts of the United States, federal, state, and local groups are working to restore clean, free-flowing rivers that support salmon and rich wildlife communities. Some dams in Washington State and Maine have been removed to allow salmon once again to migrate upstream to spawn, in some cases for the first time in more than 150 years.

Freshwater Wetlands

Freshwater **wetlands,** also called marshes, swamps, or bogs, are regions where soil is covered or saturated with water. Wetlands support dense growths of algae and phytoplankton, as well as both floating and rooted plants, including water-tolerant grasses and some trees, such as bald cypress. Wetlands provide breeding grounds, food, and shelter for a great variety of birds (cranes, grebes, herons, kingfishers, and ducks), mammals (beavers, muskrats, and otters), freshwater fish, and invertebrates such as crayfish and dragonflies.

Freshwater wetlands are among the most productive ecosystems in North America. Many occur around the margins of lakes or in the floodplains of rivers. Wetlands act as giant

sponges, absorbing water and then gradually releasing it into rivers, making wetlands important safeguards against flooding and erosion. Wetlands also serve as natural water filters and purifiers. As water flows slowly through wetlands, suspended particles fall to the bottom. Wetland plants and phytoplankton absorb nutrients such as nitrates and phosphates that have washed from the land. Toxic substances, including pesticides and heavy metals (such as lead and mercury), may be absorbed by wetland plants and sediments. Soil-dwelling bacteria break down some pesticides, rendering them harmless.

Human Impacts About half of the freshwater wetlands in the United States (outside of Alaska) have been lost as a result of being drained and filled for agriculture, housing, and commercial uses. Destruction of wetlands makes nearby water more susceptible to pollutants, reduces wildlife habitat, and may increase the severity of floods.

Fortunately, many communities in the United States have recognized the water-purifying capability of wetlands and have constructed small wetlands to help clean wastewater. Additionally, local, state, and federal agencies have cooperated to protect existing wetlands and restore some that have been degraded. One of the largest ecosystem restorations ever attempted, the Comprehensive Everglades Restoration Plan, is currently under way in Florida (see Chapter 30). These actions have combined to slow wetland loss in the United States, although a recent survey found that the lower 48 states still lose about 14,000 acres of wetland each year.

Marine Biomes

The oceans can be divided into life zones characterized by the amount of light they receive and by their proximity to the shore (Fig. 29-28). The **photic zone** consists of relatively shallow waters (to a depth of about 650 feet, or 200 meters) where the light is strong enough to support photosynthesis. Below the photic zone lies the **aphotic zone,** which extends to the ocean floor, with a maximum depth of about 36,000 feet (11,000 meters). Light in the aphotic zone is inadequate for photosynthesis. In its upper regions, a murky twilight prevails, but deeper waters are completely dark. In the

▲ **FIGURE 29-28 Ocean life zones** Photosynthesis can occur only in the sunlit photic zone, which includes the intertidal zone, nearshore zone, and the upper waters of the open ocean. Approximate depths of various regions are shown, although these vary considerably depending on the clarity of the water; note that the depths are not drawn to scale. Nearly all the organisms that spend their lives in the aphotic zone rely on organic material that drifts down from the photic zone. The average depth of the ocean floor is about 13,000 feet (4,000 meters), but it reaches about 36,000 feet (11,000 meters) in the Mariana trench, located off Japan in the Pacific Ocean.

aphotic zone, nearly all the energy to support life must be extracted from the excrement and bodies of organisms that sink down from the photic zone above.

Because their water levels rise and fall with the tides, oceans do not have a defined shoreline. Instead, the **intertidal zone,** where the land meets the ocean, is alternately covered and exposed by the tides. The **nearshore zone** extends out to sea from the low-tide line, with gradually increasing depth as the continental shelf slopes downward. The nearshore zone is usually considered to end, and the **open ocean** to begin, where the water is deep enough that wave action no longer affects the bottom, even during strong storms.

Shallow Water Marine Biomes

As in freshwater lakes, the major concentrations of life in the oceans are found in shallow waters where nutrients and light are both abundant. Such locations include estuaries, the intertidal zone, and kelp forests and coral reefs, which are mostly located in the nearshore zone.

Estuaries Wetlands that form where rivers meet oceans are called **estuaries (Fig. 29-29a).** The waters of estuaries vary in salinity. High tides, for example, bring an influx of seawater, while heavy rains bring a pulse of fresh water down the river. Estuaries support enormous biological productivity and

(a) An estuary

(b) A tide pool

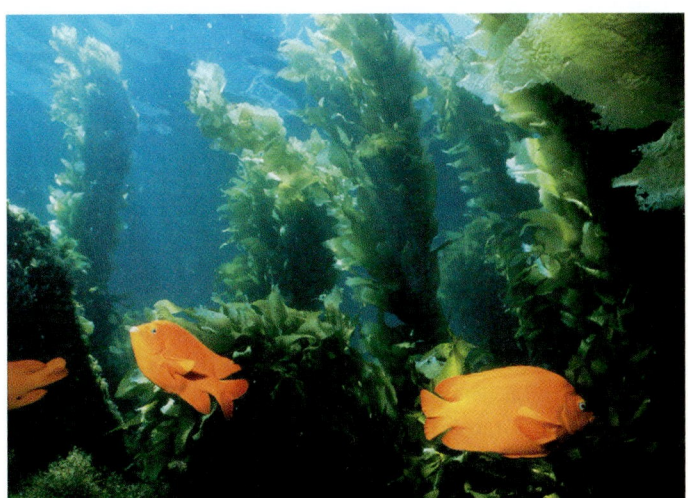

(c) An underwater kelp forest

(d) A tropical coral reef

▲ FIGURE 29-29 **Shallow water marine biomes (a)** Life flourishes in estuaries, where fresh river water mixes with seawater. Salt marsh grasses thrive here, providing shelter for fish and invertebrates that are eaten by egrets (shown here) and many other birds. **(b)** Although pounded by waves and baked by the sun, tide pools in the intertidal zone harbor a brilliant diversity of invertebrates. **(c)** Kelp forests are home to a stunning array of invertebrates and fish, such as these bright orange Garibaldis. **(d)** Coral reefs provide habitat for an extremely diverse community of fish and invertebrates.

QUESTION Why do estuaries and other coastal ecosystems have higher productivity than the open ocean?

diversity. Many commercially important species, including shrimp, oysters, clams, crabs, and a variety of fish, spend part of their lives in estuaries. Dozens of species of birds, including ducks, swans, and shorebirds, feed and nest in estuaries.

Intertidal Zones In the intertidal zone, organisms must be adapted to survive both in the air and in seawater as the tides rise and fall. During heavy rains, tide pools and mudflats may also experience significant dilution of the seawater. On rocky shores, barnacles (shelled crustaceans) and mussels (mollusks) filter phytoplankton from the water at high tide and close their shells at low tide to resist drying. Farther down, sea stars pry open mussels to eat, sea urchins feast on algae coating the rocks, and anemones spread their tentacles to catch passing crustaceans and small fish (**Fig. 29-29b**). The intertidal zone of sandy shores and mudflats typically has less diversity but still contains life, including sand crabs and burrowing worms.

Kelp Forests Kelp are enormous brown algae that can grow as tall as 160 feet (50 meters; see Chapter 20). Kelp often occur in dense stands called **kelp forests,** found throughout the world in cool waters of the nearshore zone (**Fig. 29-29c**). Under ideal conditions, kelp can grow nearly 2 feet (about 50 centimeters) in a single day. Kelp forests provide food and shelter for an amazing variety of animals, including annelid worms, sea anemones, sea urchins, snails, sea stars, lobsters, crabs, fish, seals, and otters.

Coral Reefs Corals are relatives of anemones and sea jellies. Some corals build skeletons of calcium carbonate. These skeletons accumulate over hundreds or thousands of years, building **coral reefs** (**Fig. 29-29d**). Coral reefs are most abundant in tropical waters of the Pacific and Indian Oceans, the Caribbean, and the Gulf of Mexico as far north as southern Florida, where water temperatures typically range between 68° and 86°F (20° and 30°C). Coral reefs provide anchorage, shelter, and food for an extremely diverse community of algae, fish, and invertebrates (such as shrimp, sponges, and octopuses). The reefs are home to more than 90,000 known species, with possibly a million or more yet to be discovered.

Most reef-building corals harbor unicellular photosynthetic protists, called dinoflagellates, in their bodies. The relationship is mutually beneficial: The dinoflagellates benefit from high nutrient and carbon dioxide levels within the corals. In return, the dinoflagellates provide the corals with food derived from photosynthesis. Because their dinoflagellates require sunlight for photosynthesis, reef-building corals can thrive only within the photic zone, usually at depths of less than 130 feet (40 meters). Dinoflagellates give many corals their brilliant colors.

Human Impacts Human population growth is increasing the conflict between preserving coastal ecosystems as wildlife habitat and developing these areas for energy extraction, housing, harbors, and marinas. Estuaries are threatened by runoff from farming operations, which often provide a glut of nutrients from fertilizer and livestock excrement. This fosters excessive growth of algae and photosynthetic bacteria. When these organisms die, they provide nutrients that stimulate the growth of decomposers, whose metabolism depletes the water of oxygen, killing both fish and invertebrates.

Coral reefs face multiple threats. Anything that diminishes the water's clarity harms the coral's photosynthetic partners and hinders coral growth. Runoff from farming, agriculture, logging, and construction carries silt and excess nutrients. Mollusks, turtles, fish, crustaceans, and the corals themselves are harvested from reefs faster than they can reproduce. These are used for food or are sold to tourists, shell enthusiasts, and aquarium owners. Removing predatory fish and invertebrates from reefs sometimes leads to an explosion of algae that smother the reefs, or to increased populations of sea urchins and sea stars that eat the coral.

Even though they require warm water, coral reefs are vulnerable to global warming. When waters become too warm, corals expel their colorful photosynthetic dinoflagellates and appear to be bleached (see Chapter 30). The dinoflagellates return if the water cools, but when water temperatures remain too high for too long, the corals eventually starve.

There is some good news, however. Many countries now recognize the enormous benefits of coral reefs, including economic benefits from tourism, and are working to protect their reefs. Australia's Great Barrier Reef Marine Park and the Papahānaumokuākea Marine National Monument in the Hawaiian Islands protect enormous reef systems. Collectively, about 20,000 species thrive in these two hot spots of biodiversity.

The Open Ocean

Beyond the coastal regions lie vast areas of the ocean in which the bottom is too deep to allow plants to anchor and still receive enough light to grow. Therefore, most life in the open ocean depends on photosynthesis by phytoplankton drifting in the photic zone. Phytoplankton are consumed by zooplankton, such as tiny shrimp-like crustaceans, which in turn are eaten by larger invertebrates, small fish, and even some marine mammals, such as humpback and blue whales (**Fig. 29-30**).

The amount of life in the open ocean varies tremendously from place to place, largely caused by differences in nutrient availability. Nutrients in the photic zone are constantly incorporated into the bodies of organisms and carried into the ocean depths when the creatures die and sink. These nutrients are replenished from two major sources: runoff from the land and upwelling from the ocean depths. **Upwelling** brings cold, nutrient-laden water from the ocean depths to the surface. Major areas of upwelling occur around Antarctica and along western coastlines, including those of California, Peru, and West Africa. The blue clarity of tropical waters is due to a lack of nutrients, which limits the concentration of phytoplankton in the water. Nutrient-rich waters that support a large phytoplankton community are greenish and relatively murky.

are increasingly being established throughout the world, causing substantial improvements in the diversity, number, and size of marine animals. Nearby areas also benefit because the reserves act as nurseries, helping to restore populations outside the reserves.

The Ocean Floor

Because the amount of light in the aphotic zone is inadequate for photosynthesis, most of the food on the ocean floor comes from the excrement and dead bodies that drift down from above. Nevertheless, life is found on the ocean floor in amazing quantity and variety, including worms, sea cucumbers, sea stars, mollusks, squid, and fish of bizarre shapes (**Fig. 29-31**). Some of these animals generate their own light, a phenomenon known as bioluminescence. Some fish maintain colonies of bioluminescent bacteria in special visible chambers on their bodies. Bioluminescence may help bottom dwellers to attract prey or mates. Little is known of the behavior and ecology of these exotic creatures, which almost never survive being brought to the surface.

Recently, entire communities, including species new to science, have been found feeding on the dead bodies of whales, each of which contains an average of 40 tons of food. When a whale carcass reaches the ocean floor, fish, crabs, worms, and snails swarm over it, extracting nutrients from its flesh and bones. Bone-eating "zombie worms," first described in 2005, consist mostly of rootlike structures that tunnel into the bone and absorb nutrients. Anaerobic bacteria continue bone breakdown, and clams, worms, mussels, and crustaceans move in to feed on the bacteria.

Hydrothermal Vent Communities In 1977, geologists exploring the Galápagos Rift (an area of the Pacific floor where the plates that form Earth's crust are separating) found vents they called "black smokers" emitting superheated water blackened with sulfides and other minerals. Surrounding these vents was a rich **hydrothermal vent community** of pink fish, blind white crabs, enormous mussels, white clams, sea anemones, giant tube worms, and a species of snail sporting iron-laden armor plates (**Fig. 29-32**). Hundreds of new species have been found in these specialized habitats, which have now been discovered in many deep-sea regions where separating tectonic plates allow material from Earth's interior to spew out.

In this unique ecosystem, sulfur bacteria serve as the producers. They harvest energy from a source that is deadly to most other forms of life—hydrogen sulfide, discharged from cracks in Earth's crust. Like photosynthesis, hydrogen sulfide **chemosynthesis** manufactures organic molecules from carbon dioxide. However, chemosynthesis uses hydrogen sulfide, not sunlight, as an energy source. Many vent animals consume the sulfur bacteria directly, whereas others, such as the giant tube worm, harbor chemosynthetic bacteria within their bodies and live off the by-products of bacterial metabolism. The worm derives its red color from a unique form of hemoglobin that transports hydrogen sulfide to its symbiotic bacteria.

The bacteria and archaea that inhabit the vent communities can survive at remarkably high temperatures; some

▲ **FIGURE 29-30 The open ocean** The open ocean supports fairly abundant life in the photic zone, including (clockwise from left) a small sea jelly, which is a member of the zooplankton; phytoplankton, the ocean's photosynthetic producers; and marine mammals such as this humpback whale and its calf.

Human Impacts Two major threats to the open ocean are pollution and overfishing. For example, plastic refuse, blown off the land or deliberately dumped at sea, is often mistaken for food by sea turtles, gulls, porpoises, seals, and whales. Animals that consume this refuse may die from clogged digestive tracts. Oil contaminates the open ocean from oil-tanker spills, runoff from improper disposal on land, and leakage from offshore oil wells. Some components of oil cause lethal developmental defects in a variety of marine organisms.

Many fish populations are harvested unsustainably, as a result of increased demand for fish and highly efficient fishing technologies (see Chapter 30). For example, the once abundant cod populations off eastern Canada collapsed in 1992, prompting a fishing moratorium that still continues. Populations of haddock, swordfish, tuna, and many other types of seafood have also declined dramatically because of overfishing. Some sharks, which are keystone predators in ocean food webs, are now endangered due to overfishing.

Efforts are now being made to prevent overfishing. Many countries have established quotas on fish whose populations are threatened. Marine reserves, where fishing is prohibited,

▶ **FIGURE 29-31 Denizens of the deep** The skeleton of a whale provides an undersea nutrient bonanza. A "zombie worm" (bottom left) can insert its rootlike lower body deep into the bones of the decomposing whale carcass. Other denizens of the deep include (upper left) a nearly transparent deep ocean squid with short tentacles below bulging eyes, and (lower right) a viperfish, whose huge jaws and sharp teeth allow it to grasp and swallow its prey whole.

▶ **FIGURE 29-32 Hydrothermal vent communities** "Black smokers" spew superheated water rich in minerals that provide both energy and nutrients to the vent community. Giant red tube worms may reach 9 feet (nearly 3 meters) in length and live up to 250 years. (Left) The foot of this snail is protected by scales coated with iron sulfide.

can live at 248°F (106°C; the tremendous pressure in the deep ocean prevents water from boiling at temperatures well above its sea-level boiling point). Scientists are investigating how the enzymes and other proteins of these heat-loving microbes can continue to function at temperatures that would destroy the proteins in our bodies.

CASE STUDY revisited

Food of the Gods

World demand for both coffee and chocolate will probably continue to soar. Can coffee and cacao farmers meet this demand, enjoy a reasonable income, and simultaneously help to preserve rain forests? Traditional cultivation techniques—such as coffee grown in "rustic plantations" and cacao plants grown using a method called "cabruca farming"—provide hope.

Full-sun plantations, in which the original vegetation has been completely removed and replaced with monocultures of coffee or cacao, tend to have the least biodiversity. Shady plantations provide more biodiversity, but some shady plantations are better than others. For example, most coffee from Mexico is grown in shady plantations ("shade-grown coffee"), but often under only a few types of trees that do not produce a sufficiently diverse habitat for rain-forest birds and other species. Coffee grown on rustic plantations that promote diversity and sustainable production can be certified by the Smithsonian Migratory Bird Center and the Rainforest Alliance. To achieve such certification, plantations must include a diversity of tree species. In many cases, the canopy trees serve as an additional source of food or income for the farmers, providing citrus fruits, bananas, guavas, and lumber. Some rustic plantations provide a home for more than 150 different species of birds, which feast on the fruit, or on insects that thrive in the trees and moist soil.

Deforestation has claimed 93% of the roughly 300 million acres of rain forest that once towered over Brazil's Atlantic coast. In the remaining forest, conservation programs are now helping farmers reintroduce the traditional cabruca method of growing cacao trees, which involves thinning the rain forest and planting amid its natural diversity. Joao Tavares, whose family has been using this technique for four generations, has planted many native trees on his farm, where orchids abound and the forest resounds with birdsong. "We understand that we have to preserve the cabruca," he explains, "even if you have less production." He adds that on farms where the forest has been cleared to plant dense stands of cacao trees, "you have more production, but you have lots of problems. You have more disease, more insects, so we decide to preserve."

Consider This

Sustainably grown coffee and cocoa-based products typically cost more than similar foods grown in full-sun plantations. The higher price reflects the value of protecting the rain forests and the ecosystem services that they provide (see Chapter 30). In the long run, which do you think would be economically more efficient: full-sun plantations with the rain forest removed or rustic plantations with the rain forest at least partially preserved?

CHAPTER REVIEW

Summary of Key Concepts

29.1 What Determines the Distribution of Life on Earth?

The requirements for life on Earth include nutrients, energy, liquid water, and an appropriate temperature range. On land, sunlight energy and nutrients are usually plentiful; the distribution of life is largely determined by soil moisture, which in turn is determined by precipitation and temperature. In aquatic ecosystems, liquid water is readily available; sunlight, nutrients, and temperature determine the distribution and abundance of life. The requirements for life occur in specific patterns and locations on Earth, resulting in characteristic large-scale communities called biomes.

29.2 What Factors Influence Earth's Climate?

Because of Earth's curvature, the sun's rays are nearly vertical and pass through the least amount of atmosphere at the equator; toward the poles, the rays are more slanted and must penetrate more atmosphere. Thus, the equator is uniformly warm, whereas higher latitudes have lower overall temperatures. Earth's tilt on its axis causes seasonal variations in climate at northern and southern latitudes as Earth orbits the sun. Rising warm air and sinking cool air in regular patterns from north to south produce areas of low and high moisture. These patterns are modified by the presence and topography of continents and by proximity to oceans.

29.3 What Are the Principal Terrestrial Biomes?

The limiting factors for life on land are temperature and soil moisture. Large regions with similar climates have similar vegetation, determined by the interaction of temperature and precipitation on the availability of soil water.

The tropical rain-forest biome, located near the equator, is warm and wet year-round. Tropical rain forests are dominated by broadleaf evergreen trees. Most nutrients are found in the vegetation. Most animals live in the trees. Rain forests have the highest productivity and biodiversity on Earth. Slightly farther from the equator, wet seasons alternate with dry seasons during which trees shed their leaves, producing tropical deciduous forests. Scrub forests and savannas receive less rain than tropical deciduous forests and have extended dry seasons. Savannas are characterized by widely spaced trees with grass growing beneath. The African savanna is home to the world's most extensive herds of large mammals. Most deserts, which receive less than 10 inches of rain annually, are located around 30° N and 30° S latitudes, or in the rain shadows of mountain ranges. In deserts, plants are widely spaced and have adaptations to conserve water. Animals have both behavioral and physiological mechanisms to avoid excessive heat and to conserve water. Chaparral exists in desert-like conditions that are moderated by their proximity to a coastline, allowing drought-resistant bushes and small trees to thrive. Grasslands occur in the centers of continents. These biomes have a continuous grass cover and few trees. Relatively low precipitation, fires, and severe droughts prevent the growth of trees. Grasslands have the world's richest soils and have largely been converted to agriculture. Temperate deciduous forests, whose broadleaf trees drop their leaves in winter, dominate the eastern half of the United States, and are also found in Europe and eastern Asia. Moderate precipitation and lack of severe droughts allow the growth of deciduous trees, which shade the forest floor, preventing the growth of grasses. Temperate rain forests, dominated by conifers, occur in coastal regions with both high rainfall and moderate temperatures. They are found along the northern Pacific coast of the United States, the southeastern coast of Australia, the southwestern coast of New Zealand, and parts of Chile and Argentina. The northern coniferous forest nearly encircles Earth below the arctic region. It is dominated by conifers whose small, waxy needles are adapted to conserve water and take advantage of the short growing season. The tundra is a frozen desert where permafrost prevents the growth of trees and the bushes remain stunted. Tundra is found both in the Arctic and on mountain peaks.

29.4 What Are the Principal Aquatic Biomes?

Sunlight energy and nutrients are the two major limiting factors in the distribution and abundance of life in aquatic biomes. Sunlight is strong enough for photosynthesis only in shallow waters. Nutrients are found in bottom sediments, washed in from surrounding land or provided by upwelling in nearshore ocean waters. Temperature also plays a role in determining the distribution of life in aquatic biomes.

In freshwater lakes, the littoral zone receives both sunlight and nutrients, and supports the most life. The limnetic zone is the well-lit region of open water where photosynthetic protists thrive. In the deep profundal zone of large lakes, light is inadequate for photosynthesis, and most energy is provided by detritus. Oligotrophic lakes are clear, low in nutrients, and support sparse communities. Eutrophic lakes are rich in nutrients and support dense communities. During succession, lakes shift from an oligotrophic to a eutrophic condition.

Streams begin at a source region, often in mountains, where water is provided by rain and snow. Source water is generally clear, high in oxygen, and low in nutrients. In the transition zone, streams join to form rivers that carry sediment from land and support a larger community. On their way to lakes or oceans, rivers enter relatively flat floodplains, where they deposit nutrients, take a meandering path, and spill over the land during floods.

Most life in the oceans is found in shallow water, where sunlight can penetrate, and is concentrated near the continents, particularly in areas of upwelling, where nutrients are most plentiful. Estuaries are highly productive areas where rivers meet the ocean. The intertidal zone, alternately covered and exposed by tides, harbors organisms that can withstand waves and exposure to air. Kelp forests grow in cool, nutrient-rich coastal areas, and provide food and shelter for many fish and invertebrates, as well as seals and otters. Coral reefs, formed by the skeletons of corals, are primarily found in shallow water in warm tropical seas. This complex habitat supports an extremely diverse undersea ecosystem. In the open ocean, most life is found in the photic zone, where light supports photosynthesis by phytoplankton. In the lower aphotic zone, life is supported by nutrients that drift down from the photic zone. The deep ocean floor lies within the aphotic zone. Whale carcasses provide a nutrient bonanza that supports a succession of unique communities over a span of many decades. Specialized vent communities, supported by chemosynthetic bacteria, thrive at great depths in superheated waters where Earth's crustal plates are separating.

Key Terms

aphotic zone 570	oligotrophic lake 568
biodiversity 557	open ocean 571
biome 551	ozone hole 553
chaparral 561	ozone layer 552
chemosynthesis 573	permafrost 566
climate 551	photic zone 570
coral reef 572	phytoplankton 568
desert 560	plankton 567
desertification 561	profundal zone 568
estuary 571	rain shadow 557
eutrophic lake 568	savanna 559
grassland 562	temperate deciduous
gyre 556	forest 563
hydrothermal vent	temperate rain forest 564
community 573	tropical deciduous
kelp forest 572	forest 559
intertidal zone 571	tropical rain forest 557
limnetic zone 568	tropical scrub forest 559
littoral zone 567	tundra 566
nearshore zone 571	upwelling 572
northern coniferous	wetlands 569
forest 564	zooplankton 568

Learning Outcomes

In this chapter, you have learned to . . .

LO1 List the four requirements for life on Earth, and explain which are most important in determining the nature and distribution of terrestrial and aquatic biomes.

LO2 Describe how Earth's curvature, tilt on its axis, and orbit about the Sun affect climate and the distribution of precipitation and temperature at different latitudes.

LO3 Explain how oceans, continents, and mountains affect climate.

LO4 Describe the principal terrestrial biomes and the impacts of human activities on them.

LO5 Describe the principal aquatic biomes and the impacts of human activities on them.

Thinking Through the Concepts

Fill-in-the-Blank

1. The tilt of Earth on its axis produces _____. Coastal climates are more moderate because of _____. A dry region on the side of a mountain range that faces away from the direction of prevailing winds is called a(n) _____. **LO2 LO3**

2. Of the four major requirement of life, which are the most important in determining the nature and distribution of terrestrial biomes? _____, _____ Which are most important for aquatic ecosystems? _____, _____, _____ **LO1**

3. The most biologically diverse terrestrial ecosystems are _____. The most biologically diverse aquatic ecosystems are _____. **LO4 LO5**

4. In the _____ biome, most of the nutrients are found in the bodies of plants, rather than in the _____. In the _____ biome, there are pronounced seasons, and trees drop their leaves in winter. In the _____ biome, there is too little rain to support trees, but the soil is so rich that most of the biome has been converted to farmland. The _____ biome is characterized by less than 10 inches of rainfall annually. A biome in which grasses are the dominant vegetation and trees are widely spaced, with a huge diversity of large mammals, is the _____. Stunted vegetation grows in the freezing desert biome known as the _____. **LO4**

5. The shallow portion of a large freshwater lake is called the _____. Photosynthetic plankton is called _____; nonphotosynthetic plankton is called _____. The open-water portion of a lake is divided into two zones, the upper _____ and the lower _____. Lakes that are low in nutrients are described as _____. Lakes high in nutrients are described as _____. The most diverse freshwater ecosystems are _____. **LO5**

6. The primary producers of the open ocean are mainly _____. In the deep ocean, many fish produce light through _____. Hydrothermal vent communities are supported by bacteria that obtain energy from the process of _____, using the compound _____ as an energy source. Water may be at temperatures above the surface boiling point near hydrothermal vents, but it does not boil because of the _____. **LO5**

Review Questions

1. Explain how air currents contribute to the formation of rain forests and large deserts. **LO2**

2. What are large, roughly circular ocean currents called? What effect do they have on climate, and where is that effect strongest? **LO3**

3. Explain why traveling up a mountain in the Northern Hemisphere takes you through biomes similar to those you would encounter by traveling north for a long distance. **LO3**

4. Where are the nutrients of the tropical rain-forest biome concentrated? Why is life in the tropical rain forest concentrated high above the ground? **LO4**

5. Explain two undesirable effects of agriculture in the tropical rain-forest biome. **LO4**

6. List some adaptations of desert cactus plants and desert animals to heat and drought. **LO4**

7. What is desertification? **LO4**

8. How are trees of the northern coniferous forest adapted to a lack of water and a short growing season? **LO4**

9. How do deciduous and coniferous biomes differ? **LO4**

10. What environmental factor best explains why the natural biome is shortgrass prairie in eastern Colorado, tallgrass prairie in Illinois, and deciduous forest in Ohio? **LO4**

11. Where is life in the oceans most abundant, and why? **LO5**

12. Distinguish among the littoral, limnetic, and profundal zones of lakes in terms of their location and the communities they support. **LO5**

13. Distinguish between oligotrophic and eutrophic lakes. Describe a natural scenario and a human-created scenario under which an oligotrophic lake might be converted to a eutrophic lake. **LO5**

14. Compare the source, transition, and floodplain zones of streams and rivers. **LO5**

15. Distinguish between the photic and aphotic zones. How do organisms in the photic zone obtain nutrients? How are nutrients obtained in the aphotic zone? **LO5**

Applying the Concepts

1. In which terrestrial biome is your college or university located? Discuss similarities and differences between your location and the general description of that biome in the text. In the city or town where your campus is located, how has human domination modified community interactions?

2. Fairbanks, Alaska, the plains of eastern Montana, and Tucson, Arizona, all have about the same annual precipitation. Explain why these locations contain very different vegetation.

3. Using Figures 29-3 and 29-4 as starting points, explain why terrestrial biomes are not evenly distributed in bands of latitude across Earth's surface. Explain how your proposed mechanisms apply to two specific locations.

Answers to Figure Caption questions and Fill-in-the-Blank questions can be found in the Answers section at the back of the book.

MB *Go to MasteringBiology for practice quizzes, activities, eText, videos, current events, and more.*

CHAPTER **30** Conserving Earth's Biodiversity

So many monarch butterflies cluster on their roosts in the mountains of central Mexico that their weight bends the branches of the trees.

CASE STUDY

The Migration of the Monarchs

IT MAY WELL BE the greatest show on Earth. Each fall, the monarch butterflies of eastern North America—hundreds of millions of them—migrate south to spend the winter in the mountains of central Mexico. Imagine insects weighing about half a gram each, spread out over millions of square miles during the summer months, flying for several weeks, and ending up in just a few groves of trees. And none of these monarchs has ever been to Mexico before! Their great-great grandparents, now long dead, are the closest relatives to have ever seen Mexico. How the migrating monarchs find these groves is a continuing puzzle for researchers.

Without these wintering sites in Mexico, the entire monarch population east of the Rocky Mountains would vanish. Conditions in these groves of fir and pine trees are just right for overwintering monarchs. A thick canopy of needles protects them from snow and rain. The groves are cool enough to slow down the butterflies' metabolism so they don't starve to death but are not so cold that they freeze. These sites are so essential that Mexico and the United Nations have included most of the groves in a Monarch Butterfly Biosphere Reserve.

However, the monarch reserves are owned not by the Mexican government but by the local people, most of whom are poor farmers. The trees that are so essential to monarch survival are also an important economic resource for the farmers, providing firewood and lumber. To complicate matters even further, rogue loggers, sometimes armed with automatic rifles, also covet the trees.

The conflicting claims of monarch butterflies, farmers, and loggers illustrate a crucial question facing people worldwide: How can human needs be met without destroying the natural environment? In this chapter, we will explore the discipline of conservation biology, which attempts to devise viable solutions for the preservation and sustainable development of our planet.

AT A GLANCE

LEARNING GOALS

LG1 Summarize the key concepts and goals of conservation biology.

LG2 Describe different ecosystem services, and explain how biodiversity helps provide and maintain those services.

LG3 Explain the goals of ecological economics, and describe some examples of its application.

LG4 Describe how human activities threaten biodiversity and ecosystem services, and identify the effects of these activities.

LG5 Describe strategies, actions, practices, and policies that can help preserve biodiversity and ecosystem services.

LG6 Describe the principles of sustainable development.

30.1 WHAT IS CONSERVATION BIOLOGY?

Conservation biology is the scientific discipline devoted to understanding and preserving Earth's **biodiversity**—the amazing variety of living organisms that inhabit Earth. Conservation biologists study and seek to conserve biodiversity at different levels:

- **Genetic diversity** The success and survival of a species depend on the variety and relative frequencies of different alleles in its gene pool. Genetic diversity is crucial for a species to adapt to changing environments.
- **Species diversity** The variety and relative abundance of the different species that make up a community are important for the functioning and sometimes even the survival of the community.
- **Ecosystem diversity** Ecosystem diversity includes the variety of both communities and the nonliving environment on which the communities depend. Diverse communities protect ecosystems by providing services such as slowing the runoff of rain and melting snow, degrading wastes, and generating oxygen.

CHECK YOUR LEARNING

Can you describe the goals of conservation biology? Can you explain the importance of the three levels of biodiversity that conservation biologists study and seek to protect?

30.2 WHY IS BIODIVERSITY IMPORTANT?

The vast majority of people live in cities or suburbs. Most of our food comes in packages from a supermarket. We may spend weeks without glimpsing an ecosystem in its natural state. So why should we care about preserving biodiversity? Many people would say that species and ecosystems are worth preserving for their own sake. But even if you disagree with that statement, a very practical reason for preserving biodiversity is simple self-interest: Ecosystems are essential for human well-being—indeed, for our very survival.

Ecosystem Services Are Practical Uses for Biodiversity

Ecosystem services are the benefits that people obtain from ecosystems (**Fig. 30-1**). Ecosystem services can be grouped into two general categories: (1) natural substances and other benefits that people directly use and (2) processes, sometimes called supporting services, that are required to sustain the direct benefits.

Some Ecosystem Services Provide Direct Benefits to People

People use a wide variety of goods and services that are found in, or depend on, healthy ecosystems.

- **Oxygen** Plants on land and photosynthetic microorganisms in the oceans produce almost all of the oxygen that people and other animals breathe.
- **Water** Natural ecosystems, including forests, grasslands, and wetlands, purify water by removing sediments and pollutants.
- **Food** People throughout the world eat wild-grown food. For example, the United Nations Food and Agriculture Organization estimates that, in 2008, about 20 pounds (9 kilograms) of wild fish and other seafood were caught per person, worldwide. In parts of Africa, Asia, and South America, wild animals provide an important source of protein for an often poorly nourished population. Even in developed countries, hunting for food and sport is important to the economy of many rural areas.
- **Wood** In many less-developed countries, rural residents rely on wood from local forests for heat and cooking. Sustainable harvesting practices allow rain forests to provide valuable hardwoods such as teak and ipe ("Brazilian

MOUNTAIN AND POLAR
Food
Fiber
Fresh water
Erosion control
Climate regulation
Recreation and ecotourism
Aesthetic values
Spiritual values

INLAND WATER
Rivers and other wetlands
Fresh water
Food
Pollution control
Flood regulation
Sediment retention
and transport
Disease regulation
Nutrient cycling
Recreation and
ecotourism
Aesthetic values

CULTIVATED
Food
Fiber
Fresh water
Dyes
Timber
Pest regulation
Biofuels
Medicines
Nutrient cycling
Aesthetic values
Cultural heritage

COASTAL
Food
Fiber
Timber
Fuel
Climate regulation
Waste processing
Nutrient cycling
Storm and wave protection
Recreation and ecotourism
Aesthetic values

FOREST AND
WOODLANDS
Food
Timber
Fresh water
Fuelwood
Flood regulation
Disease regulation
Carbon sequestration
Local climate regulation
Medicines
Recreation
Aesthetic values
Spiritual values

DRYLANDS
Food
Fiber
Fuelwood
Local climate regulation
Cultural heritage
Recreation and ecotourism
Spiritual values

URBAN
Parks and gardens
Air quality regulation
Water regulation
Local climate regulation
Cultural heritage
Recreation
Education

MARINE
Food
Climate regulation
Nutrient cycling
Recreation

ISLAND
Food
Fresh water
Recreation
and ecotourism

Source: Millennium Ecosystem Assessment

▲ FIGURE 30-1 **Ecosystem services**

walnut"). These woods are used locally or exported to developed countries, where they are valued for furniture, flooring, and other products.

- **Medicines** In many less-developed countries, the majority of people rely on traditional medicines that are derived primarily from wild plants. More than three-quarters of the medicines commonly prescribed in the United States contain active ingredients that are—or were originally—extracted from natural sources, mostly plants.

- **Recreation** Many, perhaps most, people take pleasure in "returning to nature." In the United States, more than 450 million visitors flock to national parks and national forests each year. Hundreds of millions more go to wildlife refuges and state parks. In many rural areas, the local economy depends on money spent by visitors who come to hike, camp, hunt, fish, or photograph nature. Worldwide, the economic value of outdoor recreation is estimated to be $3 trillion annually.

Ecotourism, in which people travel to observe unique biological communities, is a rapidly growing recreational industry. Examples of ecotourism destinations include tropical coral reefs and rain forests, the Galápagos Islands, the African savanna, and even Antarctica (**Fig. 30-2**). Ecotourism helps to support the Monarch Butterfly

Biosphere Reserve, which is visited by tens of thousands of people each year.

Some Ecosystem Services Sustain the Direct Benefits

Supporting services are required for ecosystems to continue to provide direct benefits, such as oxygen, food, and clean water, to people.

- **Soil Formation** It can take hundreds of years to build up a single inch of soil. The rich soils of the Midwestern United States accumulated under natural grasslands over thousands of years. Farmers have converted these grasslands into one of the most productive agricultural regions in the world.

Soil, with its diverse community of decomposers and detritivores (including bacteria, fungi, worms, and insects), plays a major role in breaking down wastes and recycling their nutrients. People rely on soils to decompose waste products from industry, sewage, agriculture, and forestry. Thus, soil serves some of the same functions as a water purification plant. Soil communities are also crucial to nearly every nutrient cycle. For example, nitrogen-fixing bacteria in soil convert atmospheric nitrogen into a form that plants can use.

(a) Scuba diving in a coral reef in the Red Sea

(b) Viewing wildlife in Africa

(c) Spotting penguins in Antarctica

▲ FIGURE 30-2 **Ecotourism** Carefully managed ecotourism represents a sustainable use of natural ecosystems, generating revenue and providing an incentive to preserve wildlife habitat.

- **Erosion and Flood Control** Plants block wind that blows away loose soil. Their roots further stabilize the soil and increase its ability to hold water, reducing both erosion and flooding. Deforestation is thought to have contributed to massive flooding triggered by heavy rains in the Sierra Madre mountains of Mexico in 1998 and 2005, in Haiti in 2008, and in Pakistan and Australia in 2010 and 2011 (**Fig. 30-3**).
- **Climate Regulation** By providing shade, reducing temperatures, and serving as windbreaks, plant communities have a major impact on local climates. Forests dramatically influence the water cycle, as water evaporating from leaves returns to the atmosphere. In the Amazon rain forest, one-third to one-half of the rain is water that evaporated from leaves. Extensive clear-cutting of rain forests can cause the local climate to become hotter and drier, making it harder for the logged rain forest to regenerate and damaging nearby intact forests as well.

▲ FIGURE 30-3 **Loss of flood control services** Although triggered by heavy monsoon rains, massive deforestation contributed to catastrophic flooding in Pakistan in 2010 and 2011.

Forests also affect the global climate. They absorb carbon dioxide from the atmosphere, storing the carbon in trunks, roots, and branches. As the trees decompose or are burned, they release CO_2, which contributes to global climate change. About 10% to 20% of the carbon dioxide produced by human activities results from deforestation.
- **Genetic Resources** The wealth of genes found in wild plants is an often-overlooked ecosystem service. According to the UN Food and Agriculture Organization, most of our food is supplied by only 12 crop plants, such as rice, wheat, and corn. Researchers have identified genes in wild relatives of these domesticated plants that might be transferred into crops to increase their productivity or provide greater resistance to disease, drought, and salt accumulation in irrigated soil. For example, some wild relatives of wheat have genes that confer considerable salt tolerance. Researchers are working to transfer these genes into domestic wheat. In addition, wild plants might be developed into food crops that are more nutritious or better suited to local growing conditions.

Ecological Economics Attempts to Measure the Monetary Value of Ecosystem Services

Historically, people have assumed that ecosystem services are free and unlimited. Therefore, the value of ecosystem services has seldom been taken into account when making decisions about land use, farming practices, power generation, and a host of other human activities. Fortunately, that is beginning to change. The modern discipline of **ecological economics** attempts to determine the monetary value of ecosystem services and to assess the trade-offs that occur when natural ecosystems are damaged to make way for human activities.

For example, a farmer planning to divert water from a wetland to irrigate a crop would traditionally weigh the benefit of increased crop production against the cost of the project's labor and materials. On the other hand, the wetland provides many services, such as neutralizing pollutants,

▲ FIGURE 30-4 **Ashokan Reservoir** Reservoirs in the Catskill Mountains supply New York City with extraordinarily clean water.

controlling floods, and providing breeding grounds for fish, birds, and many other animals. If the loss of ecosystem services were factored into the cost–benefit analysis, the intact wetland might well be more valuable than the crop. The reality, however, is that the economic benefits from projects that damage ecosystems often go to individuals, whereas the costs are usually borne by society as a whole. Thus, in a market economy, it is difficult to apply the principles of ecological economics except for projects designed and funded by government agencies.

New York City offers an excellent example of government planning to preserve ecosystem services. The city obtains most of its water from the Catskill Mountains, 120 miles away in upstate New York (Fig. 30-4). In 1997, realizing that its water was being polluted by sewage and agricultural runoff as the Catskills were developed, city officials calculated that it would cost $6 billion to $8 billion to build a water filtration plant, plus an additional $300 million annually to run it. Recognizing that the same water purification service was provided by the ecosystems of the Catskill Mountains, city officials decided to invest in protecting them, purchasing large tracts of land and keeping them in a reasonably natural state. The U.S. Environmental Protection Agency certified that a filtration plant will not be needed until 2017, at the earliest, if the city continues to preserve land in the Catskills.

Sometimes, people realize that the best way to reap the benefits of ecosystem services is to restore degraded ecosystems to health. In "Earth Watch: Restoring the Everglades," we describe just such a project designed to undo human manipulation of the largest wetland ecosystem in the United States.

What Is the Monetary Value of Earth's Ecosystem Services?

In 1997, an international team of ecologists, economists, and geographers calculated that ecosystem services provide about *$33 trillion* in benefits to humanity every year, almost twice the world's annual gross domestic product (an estimate of the market value of all goods and services produced everywhere in the world). A study published in 2002 estimated that ecosystem services provide benefits that are more than four times the world's gross domestic product. Whichever figure is used, it is clear that ecosystem services are incredibly valuable from an economic standpoint. However, in 2005, the *Millennium Ecosystem Assessment*—a report resulting from 4 years of effort by more than 1,300 scientists in 95 countries—concluded that 60% of Earth's ecosystem services were being degraded. Although we rely on Earth's ecosystems for services of enormous value, we are exploiting those ecosystems in a way that cannot be sustained.

Biodiversity Supports Ecosystem Function

Many people work to preserve biodiversity for its own sake, or because they consider it to be the right thing to do. There are also pragmatic reasons: Biodiversity is often crucial to the ability of ecosystems to provide many services. A recent study concluded that areas with the highest biodiversity also tended to be areas providing the greatest ecosystem services. Why is biodiversity important to ecosystem function?

One way in which biodiversity might protect ecosystems, sometimes called the "redundancy hypothesis," is that several species in a community may have functionally equivalent roles. For example, several species of bees in an ecosystem may pollinate flowers. If human activities exterminate a few of these species, the remaining ones may increase their population size and pollinate most or all of the flowers, as long as the ecosystem operates under typical conditions. If, however, the ecosystem is stressed—by drought, for example—the bee species that happened to survive human disturbances may not survive the drought, resulting in significantly less pollination and, hence, less plant reproduction.

The "rivet hypothesis" postulates that superficially similar species may have somewhat different positions in the web of ecosystem stability. In an airplane wing, the loss of a couple of rivets may not be catastrophic, but lose rivets in strategic places and the entire wing falls apart. Similarly, in an ecosystem, the loss of a few critical species may cause collapse. Returning to our bee example, some species of bees specialize in pollinating specific species of flowers. Eliminating one of these bee species may mean that some species of plants no longer reproduce. Any animals that specialize in feeding on those plants will die along with them. If just a few critical bee species disappear, then many plant and animal species will also die off, and the whole ecosystem may collapse.

Some species, called **keystone species,** are neither redundant nor one of many rivets, but are fundamentally essential to the function of the ecosystem. Think of the analogy that inspired the phrase: A keystone sits at the top of a stone arch and holds all the other pieces in place. Remove the keystone, and the whole arch collapses.

Earth | Watch

Restoring the Everglades

In 1948, the U.S. Congress authorized the Central and Southern Florida Project. Canals, levees, and other structures were built to control flooding, irrigate farms, and provide drinking water for new developments surrounding the Everglades, the massive wetland in southern Florida. In the 1960s, just north of the Everglades, the meandering, 103-mile-long Kissimmee River was straightened into a 56-mile-long canal, eliminating most of the surrounding wetlands (**Fig. E30-1**). As the Everglades and other wetlands diminished, the native plants and animals that depended on them dwindled. In their place, invasive species flourished. The natural water-purifying functions of the wetlands were also lost. During the next 50 years, people became aware that a serious mistake had been made.

With half of the Everglades' original area converted to agriculture, housing, and other development, the state of Florida and the U.S. government launched the Comprehensive Everglades Restoration Plan in 2000 (**Fig. E30-2**). One of the largest ecosystem restorations ever attempted, the plan will remove 240 miles of canals and levees, reestablish natural river flow, restore wetlands, and recycle wastewater. More than 400,000 acres of land will be purchased and restored; as of 2010, more than 200,000 acres had already been purchased and 36,000 acres of wetlands rebuilt. The Tamiami Trail, a major road that prevents water from northern Florida from moving down into Everglades National Park, is being rebuilt with a series of bridges (one will be more than a mile long) that will allow reasonably natural water flow once again. Twenty-four miles of the Kissimmee River have been restored, with another 15 to 20 miles to go. Bird populations have already rebounded along restored portions of the river, and water quality has improved.

The restoration plan envisions another 20 years of work, but the end result should be a revitalized Everglades that provides lasting benefits for both people and wildlife.

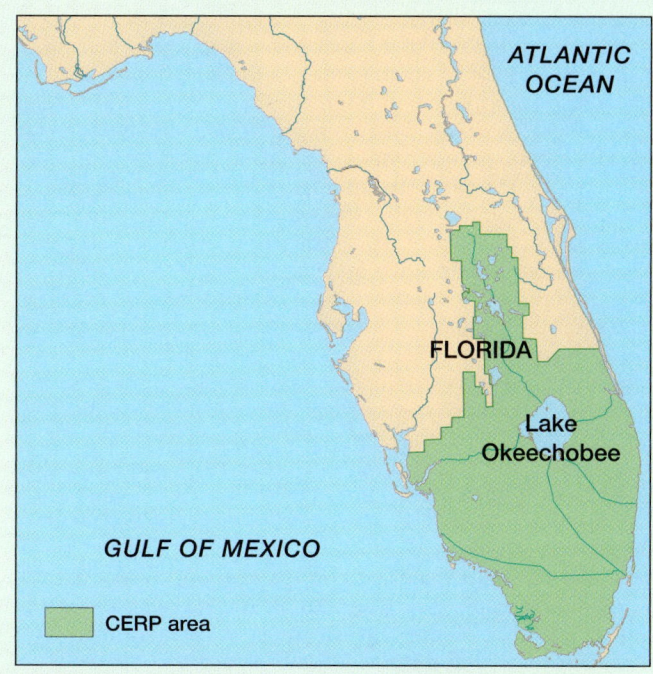

▲ **FIGURE E30-2 Restoring the Everglades** The area of Florida impacted by the Comprehensive Everglades Restoration Plan (CERP).

(a) Natural Kissimmee River before channelization

(b) Channelized Kissimmee River

▲ **FIGURE E30-1 Florida's Kissimmee River**

Earth | Watch

Whales—The Biggest Keystones of All?

The oceans are a lot emptier than they used to be. No one knows for sure how many whales originally roamed the seas, but it is believed that commercial whaling during the past 400 years has reduced most whale populations by at least 90%. Because whales swim at the top of the food chain, you would think that whale prey, from giant squid to shrimp-like krill, should be experiencing a population boom, right?

Actually, they're not. There are probably a variety of reasons why whale prey are not prospering, including ocean warming and overharvesting of krill, but Steve Nicol of the Australian Antarctic Division thinks that the loss of whales contributes to the decline of many species, all the way down to the single-celled photosynthetic algae that support the entire ocean ecosystem. How can that be? Nicol believes that the problem is quite simple: Sunlight is found only in the top layers of the ocean, while most nutrients are found at the bottom.

Most living things are denser than seawater. For example, single-celled algae sink a few yards each day. Both feces and dead animals sink much faster, some as much as a half-mile a day. Ocean currents and winter storms bring some of these nutrients up to the surface, but not all.

Enter the great whales. Many whales feed at substantial depths, from a few hundred feet to as much as a half mile below the surface. Whales, of course, return to the surface to breathe—and to poop (Fig. E30-3). Whales release huge plumes of buoyant feces that effectively bring nutrients from the depths back to the surface. For example, a critical nutrient for photosynthetic algae is iron, which is very scarce in most surface waters. As much as a quarter of all the iron in surface waters is locked away in the bodies of krill, and therefore is not available to the algae. Further, people harvest krill for health supplements and chicken feed, removing this iron from the oceans. Whales, however, recycle krill iron by eating and digesting the krill, and then defecating—whale feces have

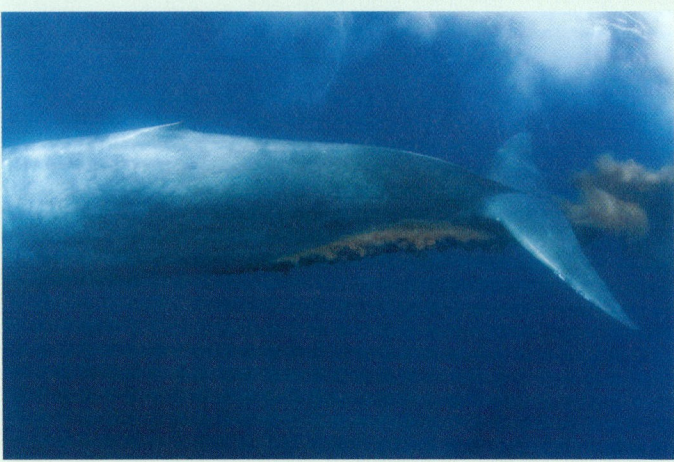

▲ FIGURE E30-3 **Whale feces fertilize the oceans** Whales, such as the blue whale shown here, release huge plumes of semi-liquid feces that drift in surface waters, providing essential nutrients for photosynthetic algae.

about 10 million times as much iron as seawater does. By fertilizing the growth of single-celled algae upon which the krill feed, blue and other baleen whales may actually increase krill populations.

If Nicol is correct, then whales are truly the keystone species of the oceans, with incredible ecosystem value. Most whale species are endangered. But many people would argue that even whale species that are not endangered are far too important to hunt, at least not until their numbers have increased tremendously. By continuing to hunt the great whales, we are prying out the keystone, and the arch of ocean productivity is teetering on the brink of collapse.

Similarly, in a biological community, a keystone species is one that plays an essential role, a role that is usually more important than would be predicted by the size of its population or by a superficial glance at its position in the food web. For example, the great whales may be keystone species in the oceans, as we explore in "Earth Watch: Whales—The Biggest Keystones of All?"

The bottom line is that species diversity is important to ecosystem function, whether the individual species are redundant, rivets, or keystones. In truth, we often don't understand ecosystem function well enough to tell which species play which roles, so protecting all species is important.

CHECK YOUR LEARNING

Can you describe the various ecosystem services provided to humanity by intact, functioning ecosystems? Can you explain why biodiversity helps to maintain functioning ecosystems?

30.3 IS EARTH'S BIODIVERSITY DIMINISHING?

No species lasts forever. Over the course of evolutionary time, species arise, flourish for various periods of time, and go extinct. If all species are fated to eventual extinction, why should we worry about modern extinctions? Because the rate of extinction during modern times has become extraordinarily high.

Extinction Is a Natural Process, But Rates Have Risen Dramatically in Recent Years

The fossil record indicates that, in the absence of cataclysmic events, extinctions occur naturally at a very low rate. According to the *Millennium Ecosystem Assessment*, this background extinction rate ranges from about 0.1 to 1 extinction per 1,000 species per 1,000 years. However, the fossil record also provides evidence of five major **mass extinctions,** during which many species were eradicated in a relatively short period of time (see Chapter 17). The most recent major extinction happened roughly 65 million years ago, abruptly ending the age of dinosaurs. Sudden changes in the environment, such as might be caused by enormous meteor impacts or rapid climate change, are the most likely explanations for mass extinctions.

The *Millennium Ecosystem Assessment* estimates that human activities are producing extinction rates of about 50 to 100 extinctions per 1,000 species per 1,000 years, or about 50 to 1,000 times the background rate that would occur in the absence of human activities. Although not all biologists agree that these rates are high enough to substantially reduce overall biodiversity, most have concluded that people are causing a sixth mass extinction. The lack of complete consensus reflects the difficulty of measuring extinction rates. Because biologists have identified only a fraction of Earth's species, biologists are unsure of the proportion of species that have already or may soon become extinct.

Extinctions of birds and mammals are best documented, although these represent only about 0.1% of the world's species. Since the 1500s, we have lost about 2% of all mammal species and 1.3% of all bird species. The background extinction rate for birds is thought to be about one species every 400 years. However, in the past 500 years, the extinction rate has been about one species per year—losses attributable almost entirely to human activities.

Each year, a "Red List" of at-risk species is published by the International Union for Conservation of Nature (IUCN)—the world's largest conservation network, consisting of member organizations in 140 countries, including 200 government agencies, more than 800 nongovernmental organizations, and about 11,000 scientists and other experts. Species are described as **vulnerable, endangered,** or **critically endangered,** depending on how likely they are to become extinct in the near future. Species that fall into any of these categories are described as **threatened.** In 2011, the Red List contained 19,265 threatened species, including 12% of all birds, 21% of mammals, and 28% of amphibians. In 2011, the U.S. Fish and Wildlife Service listed almost 1,400 threatened and endangered species in the United States alone. Why are so many species in danger of extinction?

CHECK YOUR LEARNING

Can you define *mass extinction*? Can you explain why biologists fear that a mass extinction is occurring as a result of human activities?

30.4 WHAT ARE THE MAJOR THREATS TO BIODIVERSITY?

The worldwide decline in biodiversity has two principal causes: (1) the increasingly large fraction of the Earth's resources used to support human beings, and (2) the direct impacts of human activities, including habitat destruction, overexploitation of wild populations, introduction of invasive species, pollution, and global climate change.

Humanity's Ecological Footprint Exceeds Earth's Resources

The human **ecological footprint** is an estimate of the area of Earth's surface required to produce the resources we use and absorb the wastes we generate. A complementary concept, **biocapacity,** is an estimate of the sustainable resources and waste-absorbing capacity actually available on Earth. Although related to the concept of carrying capacity (explained in Chapter 26), both the footprint and biocapacity calculations are subject to change as new technologies influence the way people use resources. The calculations are intended to be conservative and to avoid overstating human impacts. The calculations also assume that humans can use the entire planet, without reserving any of it for the rest of life on Earth.

In 2007, the biocapacity available for each of the 6.7 billion people then living on Earth was 4.5 acres (1.8 hectares), but the average human footprint was 6.7 acres (2.7 hectares). In other words, we exceeded biocapacity by about 50%: In the long run, we would need about 1.5 Earths to support humanity at 2007 consumption and population levels (**Fig. 30-5**). Countries vary enormously in their ecological footprints, from about 12 to 24 acres per person for wealthy countries such as most of Europe, Canada, Australia, New Zealand, and the United States, to as little as 1 to 2 acres per person for poor countries such as most of those in Africa. Since these estimates were made, the human

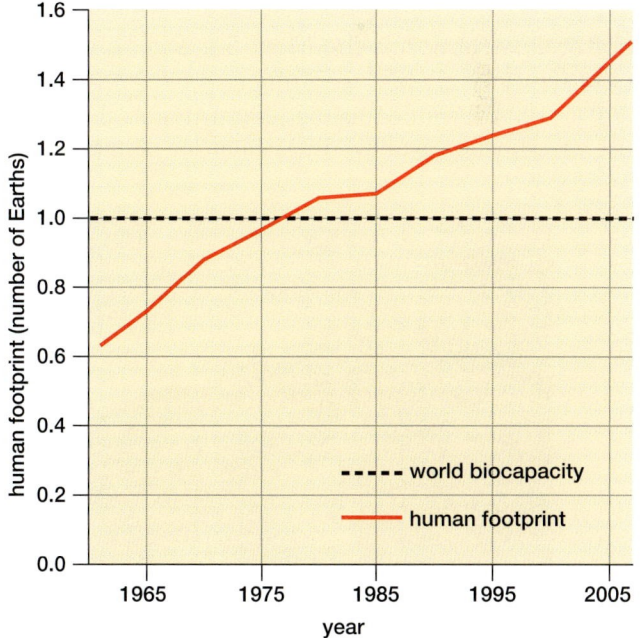

▲ FIGURE 30-5 **Human demand exceeds Earth's estimated biocapacity** Humanity's ecological footprint from 1961 to 2007, expressed as a fraction of Earth's total sustainable biocapacity (dashed line at 1.0). In 1961, we were using a little more than half of Earth's biocapacity. By 2007, we would have needed about 1.5 Earths to support us, at current rates of consumption, in a sustainable manner. (Because of the difficulty of obtaining and analyzing the necessary data, footprint calculations for 2007 were first published in 2010.) Data from the World Wildlife Fund, the Zoological Society of London, and the Global Footprint Network (2010), *The Living Planet Report.*

population has grown by more than 300 million, while Earth's total biocapacity has not significantly increased.

Running such an ecological deficit is possible only on a temporary basis. Imagine a savings account that must support you for the rest of your life. If you preserve the capital and live on the interest, the account will sustain you indefinitely. But if you withdraw the capital to support an extravagant lifestyle or a growing family, you will soon run out of money. By degrading Earth's ecosystems, humanity is drawing down Earth's ecological capital. As the human population grows and countries such as India and China raise their living standards, the strain on Earth's resources will increase.

Human Activities Directly Threaten Biodiversity

Habitat destruction, overexploitation, invasive species, pollution, and global climate change pose the greatest dangers to biodiversity. Imperiled species often face several of these threats simultaneously. For example, the major decline in frog populations worldwide has resulted from a combination of habitat destruction, invasive species, pollution, and a virulent fungal infection that many experts believe is linked to global climate change (see "Earth Watch: Frogs in Peril" in Chapter 24). Coral reefs, home to about one-third of marine fish species, suffer from a combination of overharvesting, pollution, and global warming. The IUCN estimates that as many as one-third of all reef-building corals are threatened with extinction.

Habitat Destruction Is the Most Serious Threat to Biodiversity

The IUCN has identified habitat destruction as the leading cause of extinction worldwide. Habitat loss imperils more

CASE STUDY continued

The Migration of the Monarchs

Monarch summer habitat is shrinking in the United States and Canada. Monarch caterpillars eat only milkweeds, which grow mainly in disturbed areas such as roadsides and pastures. Herbicides and mowing have greatly reduced the wild milkweed population. If you have a garden, you can help monarch populations by planting milkweeds (*Asclepias* species). Most have beautiful pink or orange flowers, with lots of nectar for other butterfly species and hummingbirds as well.

than 85% of all endangered mammals, birds, and amphibians. Although decreasing quality and area of all types of habitat imperil biodiversity, the most serious threat is the loss of tropical rain forests, home to about half of Earth's plant and animal species. Conservative estimates suggest that about 13 to 25 million acres of tropical rain forest are lost each year (some estimates are much higher). In more familiar terms, that's 20,000 to 40,000 square miles per year (the state of Kentucky is about 40,000 square miles), or the area of a football field every 1.5 to 3 seconds. The primary cause of destruction of tropical rain forests is the conversion of land to agriculture, to create both small subsistence farms and huge plantations and ranches that supply beef, coffee, soybeans, palm oil, sugarcane, and biofuels, mostly to developed countries (**Fig. 30-6**; see also "Earth Watch: Biofuels—Are Their Benefits Bogus?" in Chapter 7).

Even when a natural ecosystem is not completely destroyed, it may become broken into small pieces (**Fig. 30-7**). This **habitat fragmentation** is a serious threat to wildlife. Some species of U.S. songbirds, such as the ovenbird and Acadian flycatcher, may need up to 600 acres of continuous

(a) A clear-cut in a tropical rain forest

(b) Plantations seen from space

▲ **FIGURE 30-6 Habitat destruction** The loss of habitat due to human activities is the greatest single threat to biodiversity worldwide. **(a)** Clear-cuts in a tropical rain forests may take decades to regrow. **(b)** This image of soybean plantations in a Bolivian rain forest was photographed by astronauts on the International Space Station.

▲ FIGURE 30-7 **Habitat fragmentation** Fields isolate forest patches in Paraguay.

QUESTION Which types of species do you think are most likely to disappear from small patches of forest?

forest to find food, mates, and breeding sites. Big cats are also threatened by habitat fragmentation. Mountain lions in Florida and southern California are often killed while attempting to cross highways that divide up their habitat. In the 1970s, India established a series of forest reserves intended to protect the endangered Bengal tiger. Originally interconnected, the reserves have now become islands in a sea of development, forcing the estimated 1,400 remaining tigers into isolated patches of woodland.

To be truly functional, a preserve must support a **minimum viable population (MVP),** the smallest isolated population that can persist in spite of natural events, including inbreeding, disease, fires, and floods. The MVP for any species is influenced by many factors, including the quality of the environment, the species' average life span, its fertility, and the number of young that typically reach maturity. Most wildlife experts think that an MVP of Bengal tigers must include at least 50 females—more than are found in most of India's tiger reserves.

However, the news is not all bad. Many countries are working to preserve critical habitat. One of the largest protected habitats is the Papahānaumokuākea Marine National Monument in the Hawaiian Islands, designated in 2006. This national monument covers 84 million acres of the Pacific Ocean and is home to about 7,000 species of birds, fish, and marine mammals. Some species depend on such huge reserves; for others, critical habitat may be a few patches of sandy beach. In "Earth Watch: Saving Sea Turtles," you will learn about an innovative sea turtle conservation program in Brazil, which not only preserves turtle nesting sites on some of Brazil's many beaches but helps local communities to prosper, too.

Overexploitation Threatens Many Species

Overexploitation is the hunting or harvesting of natural populations at a rate that exceeds their ability to replenish their numbers. Overexploitation of many species has increased as a growing demand for wild animals and plants has been coupled with technological advances that have greatly increased our efficiency at harvesting them. For example, overfishing is the single greatest threat to marine life, causing dramatic declines of many species, including cod, many sharks, red snapper, five species of tuna, and swordfish. The UN Food and Agriculture Organization estimates that about 32% of global fish populations are overexploited, and another 53% are being fished to their maximum sustainable yield. Other experts conclude that as much as 70% of the world's fisheries are overexploited.

Paradoxically, both poverty and wealth can contribute to overexploitation, particularly of endangered species. Rapidly growing populations in less-developed countries increase the demand for animal products, as hunger and poverty drive people to harvest all that can be sold or eaten, legally or illegally, without regard to its rarity. As Callum Rankine of the World Wildlife Fund explains, "It's extremely difficult to get people to live sustainably. Often they are just concerned with trying to live."

Rich consumers may also fuel the exploitation of endangered species, by paying high prices for illegal products such as elephant-tusk ivory, rare orchids, and exotic birds. Although good data about black market activities are difficult to come by (for obvious reasons), the sale of endangered species, or products derived from them, is extremely lucrative, thought to be second only to the sale of narcotics.

Invasive Species Displace Native Wildlife and Disrupt Community Interactions

Humans have transported a multitude of species around the world—everything from thistles to camels. In many cases, the introduced species cause no great harm. Sometimes, however, non-native species become invasive: They increase in number at the expense of native species, competing with them for food or habitat, or preying on them directly (see Chapter 27). Although scientists differ about exactly what constitutes an "invasive" species (for example, how large the invasive population must be, and how seriously it must impact a native species), the Center for Invasive Species and Ecosystem Health lists almost 2,800 invasive species in the United States, mostly plants and insects. About half of all threatened U.S. species suffer from competition with or predation by invasive species.

Island ecosystems are particularly vulnerable to invasive species. On islands, populations of native plants and animals are usually low; the native species are often found nowhere else in the world; and, if they can't compete with the invaders, the natives cannot easily leave to find new homes. For example, the Hawaiian Islands have lost about 1,000 species of native plants and animals since their settlement by humans, mostly caused by either overexploitation

Earth Watch

Saving Sea Turtles

Six of the seven species of sea turtles are threatened with extinction. Why? Most sea turtles don't begin to breed until they are 20 to 50 years old. Then, when they reach reproductive age, the females must swim hundreds, even thousands, of miles to reach their nesting grounds, often on the same beaches where they hatched. The turtles drag themselves ashore, excavate a hole in the sand, deposit their eggs, and return to the sea (Fig. E30-4a). The eggs may be eaten by domestic dogs, foxes, wild pigs, raccoons, and a host of other predators and parasites. After about 2 months, baby turtles emerge from the surviving eggs and begin their difficult journey to adulthood. Seabirds and crabs attack them as they crawl to the ocean (Fig. E30-4b). Once there, the hatchlings are a tasty morsel for fish.

As if these dangers weren't enough, people pose many additional threats to sea turtles. The nesting beaches attract poachers who find females and their eggs easy prey. Turtle meat and eggs are a delicacy in many cultures, turtle shells make beautiful jewelry, and turtle skin makes fancy leather. Turtles are also caught, both deliberately and accidentally, in fish lines and nets. The beautiful beaches attract tourists who may frighten nesting females. Hatchlings find the sea by crawling toward the brightest area in sight—but the brightest place on the beach may now be, not the ocean, but the lights of a resort.

Since 1980, the conservation organization TAMAR (from the Portuguese *tar*tarugas *mar*inhas, or "sea turtle") has reduced these threats for the five species of sea turtles that nest along the Brazilian coast. TAMAR founders realized that for sea turtle conservation to succeed, fishermen and local villagers had to participate. Today, most of the 400 TAMAR employees are fishermen. Instead of hunting sea turtles, they free turtles caught in nets and patrol the beaches during nesting season. TAMAR biologists tag females and trace their travels. The fishermen fend off the (now rare) turtle poachers, identify nests in risky locations, and relocate the eggs to better beach sites or to a nearby hatchery. As of 2010, TAMAR had helped more than10 million hatchlings reach the sea.

TAMAR has been successful because the project organizers have engaged local communities as partners in turtle protection. Money flows into the local economies as ecotourists come to see baby turtles, visit turtle museums, and buy souvenirs made by local residents. TAMAR also sponsors communal gardens, day-care centers, and environmental education activities. Recognizing that the economic benefits derived from preserving turtles far outweigh the money that can be made from hunting them, local residents eagerly

(a) A green turtle excavating a nest

(b) A turtle hatchling heads for the sea

▲ FIGURE E30-4 **Endangered sea turtles (a)** A female green turtle scoops sand with powerful flippers, creating a cavity where she will bury about 100 eggs. **(b)** After incubating in the sand for about 2 months, the eggs hatch. Here a hatchling heads for the sea, where (if it survives) it will spend 20 to 50 years before reaching sexual maturity.

participate in turtle conservation. The success of TAMAR not only underscores the need for community support for the sustainable use of any natural resource, but highlights how successful such efforts can be.

---

or competition and predation by invasive species. Most of the native wildlife of Hawaii remains in danger: As of 2010, the U.S. Fish and Wildlife Service found that 142 animal species and 354 plant species in Hawaii are endangered, by far the largest number in any state.

Many invasive species are transported unintentionally, but some are deliberately introduced. Pigs and goats, released by early Polynesian settlers to provide food, have devastated native plants in Hawaii and other Pacific islands. Mongooses,

small cat-sized carnivores native to Asia and Africa, were deliberately imported to Hawaii in the 1800s to control accidentally introduced rats. Now both mongooses and rats pose major threats to Hawaii's native ground-nesting birds.

Lakes are also especially vulnerable to invasive species. The Great Lakes of the United States and Canada host dozens of invasive species, including zebra mussels, which disrupt entire food webs by feeding on photosynthetic phytoplankton (see Chapter 27) and lampreys, which attach to

(a) Blue rock hunter cichlid

(b) Nile perch

▲ **FIGURE 30-8 Invasive species endanger native wildlife (a)** Lake Victoria was home to hundreds of species of stunningly colored cichlid fish, such as the blue rock hunter cichlid pictured here. **(b)** The Nile perch, introduced into Lake Victoria for fishermen, has proven to be a disaster for native fish.

fish such as lake trout and kill them by draining their body fluids. Lake Victoria in Africa was once home to about 400 to 500 different species of cichlid fish that were found nowhere else on Earth (**Fig. 30-8a**). Enormous predatory Nile perch (**Fig. 30-8b**) and much smaller plankton-feeding tilapia were introduced into Lake Victoria in the mid-1900s. The combination of predation by Nile perch, competition from tilapia, pollution, and algal blooms (brought on by nutrients from surrounding farms draining into the lake) has caused a mass extinction of cichlids; only about 200 species remain.

Pollution Is a Multifaceted Threat to Biodiversity

Pollution takes many forms, including synthetic chemicals such as plasticizers, flame retardants, and pesticides; toxic metals such as mercury, lead, and cadmium; and high levels of nutrients, usually from sewage or agricultural runoff.

Because synthetic chemicals are often lipid soluble, even small amounts in the environment may accumulate to toxic levels in the fatty tissue of animals (see Chapter 28). In the mid-twentieth century, for example, the insecticide DDT accumulated in many predatory birds, causing them to lay eggs with shells so thin that they cracked when the parents sat on them during incubation. Recently, an enormous controversy has arisen over a chemical called bisphenol A, used in many plastics. Bisphenol A appears to mimic the actions of estrogen and disrupt reproduction in animals and people, although researchers disagree about whether current human exposures are high enough to cause harm (see Chapter 37).

Many heavy metals are naturally bound up in rocks, and thus rendered harmless. However, mining, industrial processes, and burning fossil fuels release heavy metals into the environment. Even extremely low levels of certain heavy metals, such as mercury and lead, are toxic to virtually all organisms.

Finally, nutrients in excessive amounts become pollutants. For example, burning of fossil fuels releases nitrogen and sulfur compounds, disrupting their natural biogeochemical cycles and causing acid rain that threatens forests and lakes (see Chapter 28).

Global Climate Change Is an Emerging Threat to Biodiversity

Burning of fossil fuels, coupled with deforestation, has substantially increased atmospheric carbon dioxide levels. As predicted by climatologists, this increase has been accompanied by an overall increase in global temperatures (see Chapter 28). In response to this warming trend, some species are shifting their ranges further toward the poles and many plants and animals are beginning springtime activities earlier in the year.

The rapid pace of human-induced climate change challenges the ability of species to adapt. Scientists at the Convention on Biological Diversity, an international organization with more than 150 member countries, have concluded that warmer conditions have already contributed to some extinctions and are likely to cause many more. Although it is difficult to predict all the impacts of global climate change, they likely include the following:

- Deserts may become hotter and drier, making survival more difficult for their inhabitants.
- Warmer conditions will probably force some species to retreat toward the poles or up mountains to stay within the climate zones in which they can survive and reproduce. Relatively immobile species, especially plants, may be unable to retreat fast enough to stay within a suitable temperature range, because they typically only "move" as far and as fast as wind or animals disperse their seeds.

▶FIGURE 30-9 **Global climate change threatens biodiversity (a)** Pikas live at high altitudes in the Rocky Mountains; as the climate warms, suitable pika habitat may disappear right off the tops of the mountains. **(b)** Pine bark beetles have killed many of the lodgepole pines on this hillside. Reddish-brown trees bear dead needles; in a year or two, the needles will fall off the lifeless trees. **(c)** Living corals usually contain photosynthetic algae that provide nourishment for the coral. When the water warms too much, corals lose their algae and become strikingly white; without the algae to help feed them, they often die.

(a) A pika gathers plants for the winter

- Cool habitat may disappear completely from mountaintops. Animals that live at high altitudes, such as pikas in the Rocky Mountains (**Fig. 30-9a**), face shrinking habitat as the mountains warm. Some local populations on isolated mountains have already vanished.
- Insect pests that were previously killed by frost or sustained freezes may spread and thrive. For example, in the northern and central Rocky Mountains, pine bark beetles were formerly controlled partly by sustained, extremely cold weather in the winter. In the past 20 years, however, these beetles have reached epidemic levels, so that most of the mature lodgepole pines in the Rockies are expected to die within the next decade (**Fig. 30-9b**).
- Coral reefs require warm water, but too much warming causes bleaching and coral death (**Fig. 30-9c**). Coral reefs have already suffered massive damage in the Seychelles Islands, American Samoa, Sri Lanka, the coasts of Tanzania and Kenya, and parts of the Australian Great Barrier Reef.

CHECK YOUR LEARNING

Can you explain the concepts of ecological footprints and biocapacity, and how these are interrelated? Can you describe how habitat destruction, overexploitation, invasive species, pollution, and global climate change threaten biodiversity?

(b) Pine bark beetles have killed these lodgepole pines

30.5 HOW CAN CONSERVATION BIOLOGY HELP TO PRESERVE BIODIVERSITY?

Research in conservation biology can help to devise strategies for conserving biodiversity. Four important goals of conservation biology are:

- To understand the impact of human activities on species, populations, communities, and ecosystems.
- To preserve and restore natural communities.
- To stop further loss of Earth's biodiversity.
- To foster sustainable use of Earth's resources.

Within the life sciences, conservation biologists enlist the help of ecologists, wildlife managers, geneticists, botanists, and zoologists. Effective conservation depends on expertise and support from people outside of biology as well. These include government leaders at all levels, who establish environmental policy and laws; environmental lawyers, who help enforce laws protecting species and their habitats; and

(c) Bleached corals (white) are usually dead or dying

ecological economists, who help place a value on ecosystem services. Social scientists provide insight into the ways that people in different cultures use their environments. Educators help students understand how ecosystems function, how they support human life, and how people can either disrupt or preserve them. Conservation organizations identify areas of concern, provide educational materials, and organize grassroots support by individuals. Finally, individual choices and actions help to determine whether conservation efforts succeed.

Habitat Preservation Is Essential to Preserving Biodiversity

Because habitat destruction and fragmentation are key factors threatening biodiversity, habitat preservation is essential. Protected reserves, connected by wildlife corridors, are vital to conserving natural ecosystems.

Core Reserves Preserve All Levels of Biodiversity

Core reserves are natural areas protected from most human uses except low-impact recreation. Ideally, a core reserve encompasses enough space to preserve ecosystems with all their biodiversity. Core reserves should also be large enough to withstand storms, fires, and floods without losing species.

To establish effective core reserves, ecologists must estimate the smallest areas required to sustain MVPs of the species that require the most space. The sizes of these *minimum critical areas* vary significantly among species, and also depend on the availability of food, water, and shelter. In general, large predators in arid environments need a larger minimum critical area than small herbivores in lush environments. However, it is difficult to make precise estimates of minimum critical areas for many species.

Corridors Connect Critical Animal Habitats

One fact stands out in estimating minimum critical areas, especially for reserves that include large predators: In today's crowded world, an individual core reserve is seldom large enough to maintain biodiversity and complex community interactions by itself. **Wildlife corridors,** which are strips of protected land linking core reserves, allow animals to move relatively freely and safely between habitats that would otherwise be isolated (**Fig. 30-10**). Corridors effectively increase the size of smaller reserves by connecting them. Ideally, both core reserves and corridors are surrounded by buffer zones supporting human activities that are compatible with wildlife. Buffer zones prevent high-impact uses such as clear-cutting, mining, and housing from impacting wildlife in the core region.

Sometimes, an effective wildlife corridor can be as narrow as an underpass beneath a highway. In densely populated Southern California, plans for a development of more than 1,000 new houses in the hills south of Los Angeles were abandoned and freeway exits were closed after wildlife biologists found that mountain lions used the Coal Canyon underpass to move between habitat in the Chino Hills north of the freeway and the Santa Ana Mountains to the south. The underpass and its surroundings have been restored to a more natural state, encouraging mountain lions and other wild animals to cross safely

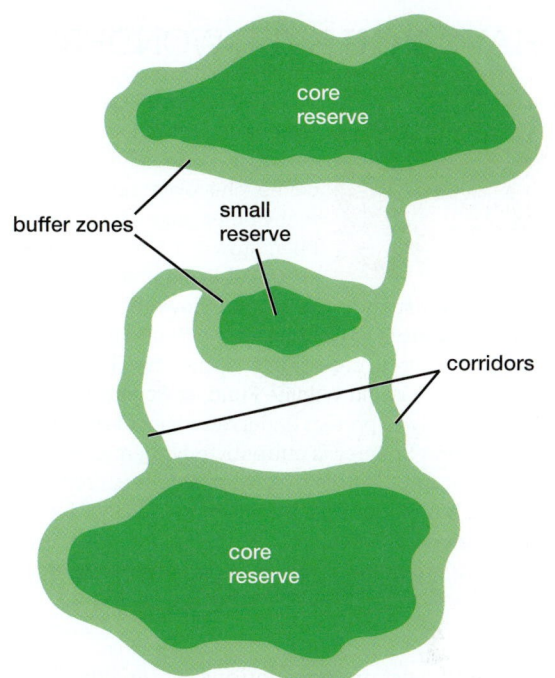

▲ **FIGURE 30-10 Wildlife corridors connect reserves**
QUESTION How do wildlife corridors reduce species extinction in small reserves?

▲ **FIGURE 30-11 Wildlife corridors connect habitats**
Asphalt has been removed and traffic barred from the underpass at Coal Canyon beneath the Riverside Freeway south of Los Angeles, allowing mountain lions to move safely between habitats on either side.

beneath the freeway (**Fig. 30-11**). People, too, use the corridor—in fact, the wildlife corridor is now featured in hiking guidebooks.

In the northern Rocky Mountains, a coalition of conservation groups and scientists has proposed a series of wildlife corridors linking existing core reserves, such as Yellowstone, Grand Teton, and Glacier National Parks, with one another and with nearby ecosystems. These interconnected habitats would sustain populations of grizzly bears, elk, wolves, and mountain lions.

CHECK YOUR LEARNING

Can you describe some strategies and policies that can preserve natural ecosystems and their associated biodiversity? Can you define the terms *core reserve* and *wildlife corridor*, and explain the relationship between them?

30.6 WHY IS SUSTAINABILITY ESSENTIAL FOR A HEALTHY FUTURE?

Natural ecosystems share certain features that allow them to persist and flourish. Among the most important characteristics of sustainable ecosystems are:

- Diverse communities,
- Relatively stable populations that remain within the carrying capacity of the environment,
- Recycling and efficient use of raw materials, and
- Reliance on renewable sources of energy.

Environments that have been modified by human development often do not possess these characteristics. As a result, many human-modified ecosystems may not be sustainable in the long run. How can we meet our needs in ways that sustain the ecosystems on which we depend?

Sustainable Development Promotes Long-Term Ecological and Human Well-Being

In *Caring for the Earth: A Strategy for Sustainable Living*, the IUCN stated that **sustainable development** "meets the needs of the present without compromising the ability of future generations to meet their own needs." It explains that "Humanity must take no more from nature than nature can replenish. This in turn means adopting lifestyles and development paths that respect and work within nature's limits. It

can be done without rejecting the many benefits that modern technology has brought, provided that technology also works within those limits."

Unfortunately, in modern human society, "sustainable development" is almost an oxymoron, because "development" so often means replacing natural ecosystems with human infrastructure such as houses, factories, and shopping centers. Most people in developed countries have achieved a high quality of life, but they have done so by exploiting, in an unsustainable manner, the direct and indirect services provided by ecosystems and by using large quantities of nonrenewable energy.

However, evidence from all parts of the world shows that such activities are unraveling the tapestry of natural communities and undermining Earth's ability to support life. As individuals and governments recognize the need to change, an increasing number of projects are being developed that are intended to meet human needs sustainably. We describe two of these in the following sections.

Biosphere Reserves Provide Models for Conservation and Sustainable Development

A world network of **Biosphere Reserves** has been designated by the United Nations. The goal of Biosphere Reserves is to maintain biodiversity and evaluate techniques for sustainable human development while preserving local cultural values. Biosphere Reserves typically consist of three regions. In the central core reserve, only research, tourism, and some traditional cultural uses are allowed. In surrounding buffer zones, people may engage in low-impact activities, such as recreation, research, environmental education, and carefully regulated forestry and grazing. Outside the buffer zone is a transition area that supports settlements, tourism, fishing, and agriculture, all (ideally) operated sustainably (**Fig. 30-12**). The first Biosphere Reserves were designated in 1976. There are now more than 580 sites in 114 countries, including the United States.

National governments nominate sites in their countries for reserve designation and continue to own and manage

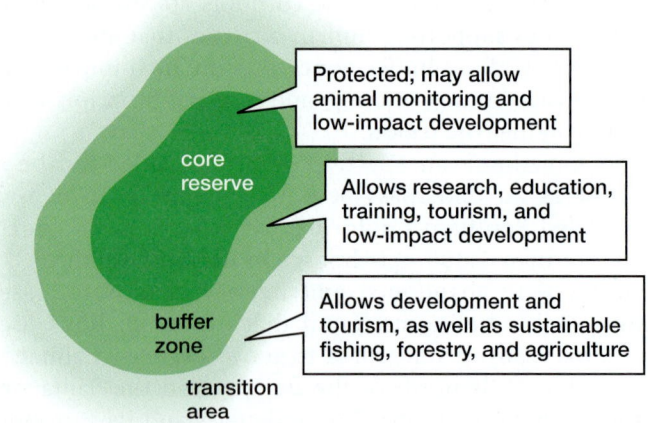

Protected; may allow animal monitoring and low-impact development

Allows research, education, training, tourism, and low-impact development

Allows development and tourism, as well as sustainable fishing, forestry, and agriculture

core reserve

buffer zone

transition area

▲ **FIGURE 30-12 The design of an ideal Biosphere Reserve**

CASE STUDY continued

The Migration of the Monarchs

"Monarch Butterfly Biosphere Reserve" is a nice name, but often a designation like that isn't accompanied by any mechanism to make it a reality. However, the World Wildlife Fund, with a grant from the Packard Foundation, has established a $5,000,000 trust fund to help farmers find alternate sources of income rather than log their land, thus helping to keep an intact buffer zone around the core monarch groves. The fund is also used to establish forested corridors connecting the groves where the monarchs overwinter.

them. This has greatly reduced opposition to the reserves, but as a result of their voluntary nature, few completely adhere to the ideal Biosphere Reserve model. In the United States, most of the 47 Biosphere Reserves are national parks and national forests. Much of the land in buffer and transition zones is privately owned, and the landowners may be unaware of its designation. Often, funding is inadequate to compensate landowners for restricting development or for promoting and coordinating sustainable development.

However, many states, counties, and even cities offer conservation easements, whereby a landowner gives up the right to develop property, usually in return for some sort of tax credit. Conservation easements can be powerful tools for preserving natural habitat at low cost, whether or not they are part of Biosphere Reserves. In Virginia, for example, more than 500,000 acres of woods, farmland, and wildlife habitat have been preserved through conservation easements.

Sustainable Agriculture Preserves Productivity with Reduced Impact on Natural Communities

The greatest loss of natural habitat occurs when people convert natural ecosystems to farms. For example, in the midwestern United States, tens of millions of acres of grasslands have been converted to farmland, principally for growing corn, wheat, and soybeans. Because farms typically grow only one or a few crops, and most of the plants are harvested for human consumption, both plant and animal diversity on farms is very low, compared to the natural habitat before the farms were established.

Farming is necessary to feed humanity. Further, to make a reasonable living, farmers must produce large amounts of food at low cost. This often leads to unsustainable practices that interfere with ecosystem services. For example, allowing fields to remain bare after harvesting often increases soil erosion, as wind and rain remove the exposed soil. Applying insecticides often indiscriminately kills pest insects, their natural predators, and pollinators. In many parts of the world, irrigation relies on underground water supplies that are being depleted faster than they can be replenished by rain and snow. Further, because both underground and surface water supplies contain some salt, evaporation of irrigation water often leaves enough salt behind to reduce soil fertility.

Fortunately, many farmers recognize the benefits of sustainable agriculture (Table 30-1). The **no-till** cropping

TABLE 30-1 Agricultural Practices Affect Sustainability

	Unsustainable Agriculture	Sustainable Agriculture
Soil erosion	Allows soil to erode far faster than it can be replenished because the remains of crops are plowed under, leaving the soil exposed until new crops grow.	No-till agriculture greatly reduces soil erosion. Planting strips of trees as windbreaks reduces wind erosion.
Pest control	Uses large amounts of pesticides to control crop pests.	Trees and shrubs near fields provide habitat for insect-eating birds and predatory insects. Reducing insecticide use helps to protect birds and insect predators.
Fertilizer use	Uses large amounts of synthetic fertilizer.	No-till agriculture retains nutrient-rich soil. Animal wastes are used as fertilizer. Legumes that replenish soil nitrogen (such as soybeans and alfalfa) are alternated with crops that deplete soil nitrogen (such as corn and wheat).
Water quality	Allows runoff from bare soil to contaminate water with pesticides and fertilizers. Allows excessive amounts of animal wastes to drain from feedlots.	Animal wastes are used to fertilize fields. Plant cover left by no-till agriculture reduces nutrient runoff.
Irrigation	May excessively irrigate crops, using groundwater pumped from natural underground storage at a rate faster than the water is replenished by rain or snow.	Modern irrigation technology reduces evaporation and delivers water only when and where it is needed. No-till agriculture reduces evaporation.
Crop diversity	Relies on a small number of high-profit crops, which encourages outbreaks of insects or plant diseases and leads to reliance on large quantities of pesticides.	Alternating crops and planting a wider variety of crops reduces the likelihood of major outbreaks of insects and diseases.
Fossil fuel use	Uses large amounts of nonrenewable fossil fuels to run farm equipment, produce fertilizer, and apply fertilizers and pesticides.	No-till agriculture reduces the need for plowing and fertilizing.

(a) Cotton seedlings emerge in a no-till field in North Carolina   **(b) The same field one month later**

▲ **FIGURE 30-13 No-till agriculture (a)** A cover crop of wheat has been killed with an herbicide. Cotton seedlings thrive amid the dead wheat, which anchors soil and reduces evaporation. **(b)** Later in the season, the same field shows a healthy cotton crop mulched by the dead wheat. Photos courtesy of Dr. George Naderman, Former Extension Soil Specialist (retired), College of Agriculture and Life Sciences, North Carolina State University, Raleigh, NC.

technique, which leaves the residue of harvested crops in the fields to form mulch for the next year's crops, represents one possible component of sustainable agriculture (**Fig. 30-13**). In 2009, no-till methods were used on about 88 million acres in the United States (about one-third of all croplands). No-till farming requires less plowing, saving 3 to 14 gallons of diesel fuel per acre.

On the other hand, most no-till farmers spray herbicides to kill weeds. Some of the herbicide may blow off the fields and damage nearby natural habitats. In addition, some herbicides may harm animals. Although controversial, some studies indicate that atrazine, an herbicide commonly used in no-till agriculture, damages the reproductive systems of amphibians and perhaps other animals. The residues of last year's crops may contain pathogens such as fungi, which would be reduced by plowing in conventional agriculture but may require the use of pesticides in a no-till field.

Organic farmers typically do not use synthetic herbicides, insecticides, or fertilizers. Some organic farmers use no-till methods, but most plow their fields at least every other year to help kill weeds. Organic farming relies on natural predators to control pests and on soil microorganisms to degrade animal and crop wastes, thereby recycling their nutrients. Diverse crops reduce outbreaks of pests and diseases that attack a single type of plant. There is an ongoing debate about the relative productivity of organic versus conventional farming, and whether organic farming with plowing, or no-till farming with herbicides, is better for the soil and the natural environment.

In the best-case scenario, farmers would grow a variety of crops, use agricultural practices that retain soil

fertility, and use as little energy and as few potentially toxic chemicals as possible. Insect pests would be controlled by predators such as birds and predatory insects and by crop rotation, so that pests that specialize on particular crop plants would not find a feast laid out for them year after year. Fields would be relatively small, separated by strips of natural habitat for native plants and animals. In practice, there is substantial disagreement among both farmers and agricultural experts about whether all of these goals can be met while still yielding large harvests and keeping costs manageable.

Because the loss of ecosystem services is not factored into the costs of unsustainable farming practices, food produced unsustainably tends to be cheaper, at least in the short term. This is why organic fruits are usually more expensive in your local supermarket. In the long run, if typical commercial farming results in salty soils, outbreaks of crop diseases and pests, or loss of topsoil, then it will be more expensive than sustainable agriculture is. Many projects, such as the University of California's Sustainable Agriculture Research and Education Program based in Davis, California, support research and educate farmers about the advantages of sustainable agriculture and how to practice it.

The Future of Earth Is in Your Hands

How should we manage our planet so that it provides a healthy, satisfying life for the current generation of people, while simultaneously retaining biodiversity and the resources needed for future generations? No one can give a single, simple answer. However, three interrelated questions must be considered: (1) What should human lifestyles look like? (2) What technologies can support those lifestyles in a sustainable way? (3) How many people can Earth support and in what lifestyle?

Changes in Lifestyle and Use of Appropriate Technologies Are Essential

The billions of people on Earth will never all agree on exactly what is needed for a happy, fulfilling life. Nearly everyone would agree, however, that a minimal lifestyle must include adequate food and clothing, clean air and water, good health care and working conditions, educational and career opportunities, and access to natural environments. Most of Earth's people live in less-developed countries and lack at least some of these necessities.

Without a sustainable approach to development, there can be no long-term improvement in the quality of human life—in fact, it might even decline. We must make choices about which technologies are sustainable and how to make the transition from the realities of today to a hoped-for tomorrow. For example, in the long run, unless energy sources such as nuclear fusion become a reality, sustainable living must rely on renewable energy sources—solar, wind, geothermal, and wave energy—that don't produce highly toxic wastes or more carbon dioxide than the planet can recycle.

Human Population Growth Is Unsustainable

The root causes of environmental degradation are simple: too many people using too many resources and generating too much waste. As the IUCN eloquently stated in *Who Will Care for the Earth?* ". . . the central issue [is] how to bring human populations into balance with the natural ecosystems that sustain them."

In the long run, that balance cannot be achieved if the human population continues to grow. Given the lifestyle to which the vast majority of people on Earth aspire, many are convinced that the balance cannot be maintained even with our current population, and yet we add 75 million to 80 million people each year. No matter how simple our diets, how efficient our housing, how low-impact our farming techniques, or how much we reuse and recycle, continued population growth will eventually overwhelm our best efforts.

Let's return to our comparison of Earth's biocapacity and the human ecological footprint (**Fig. 30-14**). As you can see, the rapid increase in humanity's ecological footprint between 1961 and 2007 (red line) is roughly paralleled by our rapid population increase (blue line). The ecological footprint *per person* (green line) has been nearly constant for almost 40 years—in other words, the average person was using about the same amount of Earth's biocapacity in 2007 as in 1970. If the human population had not increased, the total human ecological footprint would

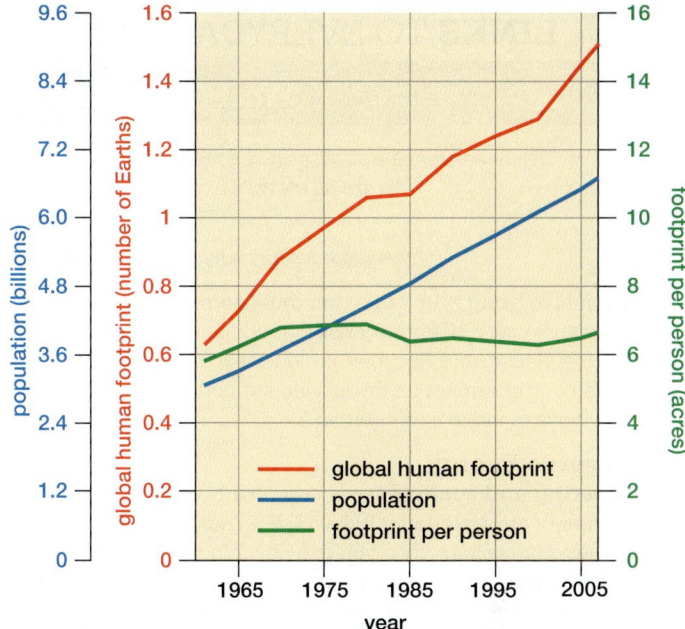

▲ FIGURE 30-14 **Human population growth threatens sustainability** Between 1961 and 2007, human population growth (blue line) increased at approximately the same rate as the global human ecological footprint (red line). The footprint per person (green line) has remained almost the same since 1970, meaning that the increase in our global footprint has resulted almost entirely from population growth. Data from the World Wildlife Fund, the Zoological Society of London, and the Global Footprint Network (2010), *The Living Planet Report.*

still be well below Earth's biocapacity, but because there are so many more of us today, the total human footprint has climbed far above Earth's biocapacity. Eliminating, and probably reversing, population growth is essential if we wish to improve the quality of life for the 7 billion of us already here, provide the potential for a similar quality of life for our descendants, and save what is left of Earth's biodiversity for future generations.

The Choices Are Yours

This chapter has provided some examples of human activities that will help protect biodiversity, save species from extinction, and advance sustainable development. Look around your campus and community—what is being done sustainably? What isn't? What would it take to make the necessary changes? In "Links to Everyday Life: What Can Individuals Do?" we suggest some ways that individuals can live more sustainably and help protect life on Earth.

CHECK YOUR LEARNING

Can you describe the principles of sustainable development? Can you discuss how population, technology, and lifestyle choices interact to affect sustainability?

LINKS TO EVERYDAY LIFE

What Can Individuals Do?

*"There are no passengers on spaceship earth.
We are all crew."*

—Marshall McLuhan
(twentieth-century educator and philosopher)

Sustainable living is an ethic that must permeate all levels of human society, beginning with individuals. The adage "Reduce, Reuse, and Recycle" provides excellent advice to minimize your impact on Earth's life-support systems. Here are some ways to make a difference:

Conserve Energy

- **Heating and cooling** Don't heat your house over 68°F in winter or air condition it below 78°F in summer. Reduce the heating or cooling while you're away. When you purchase or remodel a home, consider energy-efficient features such as passive solar heating, good insulation, an attic fan, double-glazed windows (with "low-E" coating to reduce heat transfer), and tight weather stripping. Plant deciduous trees on the south side of your home for shade in summer (when the trees are covered with leaves) and sun in winter (after the leaves have dropped off). If possible, purchase renewable energy, usually wind power, from your energy provider.
- **Hot water** Take shorter showers and switch to low-flow shower heads. Wash only full loads in your washing machine and dishwasher; use cold water to wash clothes; don't prewash your dishes. Insulate and turn down the temperature on your water heater.
- **Appliances** When you choose a major appliance, look for the most energy-efficient models. Don't use your dryer in the summer—put up a clothesline. Turn off unused lights and appliances. Replace incandescent light bulbs with fluorescent or LED bulbs wherever possible.
- **Transportation** Choose the most fuel-efficient car that meets your needs, and use it efficiently by combining errands. Use public transit, carpool, walk, bicycle, or telecommute when possible.

Conserve Materials

- **Recycle** Look into recycling options in your community, and recycle everything that is accepted. Explore composting (there are excellent informational sites on the Internet). Support and encourage campus and community recycling efforts.
- **Buy recycled material** Purchase recycled paper products. You can also buy decking and carpet made from recycled plastic bottles.

- **Reuse** Reuse anything possible, such as manila envelopes, file folders, and both sides of paper. Refill your water bottle. Reuse your grocery bags. Give away—rather than throw away—serviceable clothing, toys, and furniture. Make rags out of unusable old clothes and use them instead of disposable cleaning materials.
- **Conserve water** If you live in a dry area, plant drought-resistant vegetation around your home to reduce water usage. Turn off the faucet while you brush your teeth. Consider buying a low-flush toilet.

Support Sustainable Practices

- **Food choices** Buy locally grown produce that does not require long-distance shipping. Look for coffee with the "Bird-Friendly™" or "Rainforest Alliance Certified" seal of approval, and request it at your local coffee shop. Reduce your meat consumption. Search the Internet to find out which fish at your local supermarket have been harvested sustainably.
- **Harmful chemicals** Limit the use of harsh cleaners, insecticides, and herbicides that may contaminate water and soil.

Magnify Your Efforts

- **Support organized conservation efforts** Join conservation groups and donate money for conservation efforts. Find these on the Internet, and sign up for e-mail alerts that educate you about environmental legislation and make it easy for you to contact your government representatives and express your views. Join the Campus Climate Challenge and reduce energy use on your campus.
- **Volunteer** Join grassroots efforts to change the world. Volunteer for local campus and community projects that improve the environment.
- **Make your vote count** Investigate candidates' stands and voting records on environmental issues, and consider this information when choosing which candidate to support.
- **Educate** Through your words and actions, share your concern for sustainability with your family, friends, and community. Write letters to the editor of your school or local newspaper, to local businesses, and to elected officials. Recruit other concerned students and lobby for change.
- **Reduce population growth** Consider the consequences of the enormous and expanding human population when you plan your family. Adoption, for example, allows people to raise large families while simultaneously contributing to the welfare of humanity and the environment.

CASE STUDY revisited

The Migration of the Monarchs

Sure, a monarch butterfly is beautiful, and a hundred million of them are spectacular. But can they help people to build a house or feed their families? Indeed they can, because it is not a case of monarchs and trees versus the local people. The reality is that the monarchs and trees can help the farmers to prosper.

Aerial photographs show continued deforestation in the Monarch Butterfly Biosphere Reserve, mostly by

illegal loggers. But do these loggers replant the forest? No. Although they make large profits for a few years, the loggers leave the farmers with bare slopes that cannot provide the ecosystem services typical of a healthy forest, such as preventing soil erosion, retaining and gradually releasing clean water, and providing wildlife habitat. Fortunately, environmental and social organizations, in collaboration with the Mexican government, have planted more than 5 million trees in the past decade—almost 700,000 in 2010 alone. In an ideal world, helping the farmers would be motive enough to replant the forests; in the real world, it probably wouldn't happen if the monarchs weren't there.

But how can the farmers make a living in the Monarch Butterfly Biosphere Reserve? Several organizations are helping Mexican agricultural experts to train the farmers in sustainable agriculture. One of the most profitable "crops" in the reserve is, ironically, trees. The soil and climate provide ideal conditions for rapid tree growth. Some conifers mature in less than 20 years. Planting seedlings today allows some harvesting in only 5 years, for firewood and Christmas trees. After 15 years, the trees are large enough for commercial lumber. If, meanwhile, the trees are continually replanted, the cycle can continue indefinitely, and the old-growth groves needed by the monarchs can be left alone.

Another source of income for the farmers is ecotourism. Tourists flock to the reserve each year to see the butterflies. Although ecotourism is not without its own problems, if properly regulated, it can both preserve the forest and provide significant income opportunities for the local people, who serve as guides to the monarch groves, and offer food, accommodations, and souvenirs for the tourists.

Consider This

BioEthics How do you think that society should deal with environmental problems that lack a "poster species" like the monarch butterfly? Are the ecosystem services provided by forests in, say, Guatemala or Vietnam less important because they don't happen to be home to a few hundred million spectacular butterflies?

CHAPTER REVIEW

Summary of Key Concepts

30.1 What Is Conservation Biology?
Conservation biology is the scientific discipline devoted to understanding and preserving Earth's biodiversity, including diversity at the genetic, species, and ecosystem levels.

30.2 Why Is Biodiversity Important?
Biodiversity is a source of goods, such as food, fuel, building materials, and medicines. Biodiversity provides ecosystem services such as forming soil, purifying water, controlling floods, moderating climate, and providing genetic reserves and recreational opportunities. The emerging discipline of ecological economics attempts to measure the contribution of ecosystem goods and services to the economy, and estimates the costs of losing them to unsustainable development.

30.3 Is Earth's Biodiversity Diminishing?
Natural communities have a low background extinction rate. Many biologists believe that human activities are currently causing a mass extinction, increasing extinction rates by a factor of 50 to 1,000. According to the IUCN Red List, in 2011, more than 19,000 plants and animals were threatened with extinction.

30.4 What Are the Major Threats to Biodiversity?
The ecological footprint estimates the area of Earth required to support the human population at any given level of consumption and waste production. Biocapacity estimates the resources and waste-absorbing capacity actually available. The human footprint is already exceeding Earth's biocapacity, leaving less and less to support other forms of life. Major threats to biodiversity include habitat destruction and fragmentation as ecosystems are converted to human uses; overexploitation as populations of wild animals and plants are harvested beyond their ability to regenerate; invasive species; pollution; and global climate change.

30.5 How Can Conservation Biology Help to Preserve Biodiversity?
Conservation biology seeks to identify the diversity of life, explore the impact of human activities on natural ecosystems, and apply this knowledge to conserve species and foster the survival of healthy, self-sustaining communities. Conservation efforts include conserving wild ecosystems by establishing wildlife reserves connected by wildlife corridors, with the goal of preserving functional communities and self-sustaining populations.

30.6 Why Is Sustainability Essential for a Healthy Future?
Sustainable development meets present needs without compromising the future. Such development requires that people maintain biodiversity, recycle raw materials, and rely on renewable resources. Biosphere Reserves promote conservation and sustainable development. A shift to sustainable farming is crucial for conserving soil and water, reducing pollution and energy use, and preserving biodiversity. Human population growth is unsustainable and is driving consumption of resources beyond nature's ability to replenish them. We must bring our population into line with Earth's ability to support us, leaving room and resources for all forms of life.

Key Terms

biocapacity *585*
biodiversity *579*
Biosphere Reserve *592*
conservation biology *579*
core reserve *591*
critically endangered species *585*
ecological economics *581*
ecological footprint *585*
ecosystem services *579*
endangered species *585*
habitat fragmentation *586*

keystone species *582*
mass extinction *584*
minimum viable population (MVP) *587*
no-till *593*
overexploitation *587*
sustainable development *592*
threatened species *585*
vulnerable species *585*
wildlife corridor *591*

Learning Outcomes

In this chapter, you have learned to . . .

LO1 Summarize the key concepts and goals of conservation biology.

LO2 Describe different ecosystem services, and explain how biodiversity helps provide and maintain those services.

LO3 Explain the goals of ecological economics, and describe some examples of its application.

LO4 Describe how human activities threaten biodiversity and ecosystem services, and identify the effects of these activities.

LO5 Describe strategies, actions, practices, and policies that can help preserve biodiversity and ecosystem services.

LO6 Describe the principles of sustainable development.

Thinking Through the Concepts

Fill-in-the-Blank

1. Three levels of biodiversity are _____, _____, and _____. If the population of a species becomes too small, it is likely to have lost much of its _____ diversity. **LO1**

2. Products or processes by which functioning ecosystems benefit humans are collectively called _____. Four important examples of these benefits include _____, _____, _____, and _____. **LO2**

3. Many of the benefits that humans derive from functioning ecosystems, such as purifying water, have traditionally been considered to be free. The discipline of _____ tries to quantify the monetary value of these benefits. **LO3**

4. The major threats to biodiversity include _____, _____, _____, _____, and _____. For most endangered species, _____ is probably the major threat. **LO4**

5. The smallest population of a species that is likely to be able to survive in the long term is called the _____. When suitable habitat for a given species is split up into areas that are too small to support a large enough population, this is called _____. One way in which conservation biologists seek to maintain large enough populations is to set up core reserves of suitable habitat, connected by _____. **LO5**

6. A Native American saying tells us that "We do not inherit the Earth from our ancestors, we borrow it from our children." If this principle guided our activities, we would practice _____ development. **LO6**

Review Questions

1. Define *conservation biology*. What are some of the disciplines it draws on, and how does each discipline contribute to it? **LO1 LO3**

2. What are the three different levels of biodiversity, and why is each one important? **LO1**

3. What is ecological economics? Why is it important? **LO3**

4. List the types of goods and services that natural ecosystems provide. **LO2**

5. What five specific threats to biodiversity are described in this chapter? Provide an example of each. **LO4**

6. Why is the TAMAR turtle project a good model for conservation and sustainable development? **LO5 LO6**

Applying the Concepts

1. **BioEthics** What are the ethical foundations of conservation biology? Do you agree with them? Why or why not?

2. List some reasons that the ecological footprints of U.S. residents are by far the largest in the world. Looking at your own life, how could you reduce the size of your footprint? How does the ecological footprint of U.S. residents extend into the Tropics?

3. Search for and describe some examples of habitat destruction, pollution, and invasive species in the region around your home or campus. Predict how each of these might affect specific local populations of native animals and plants.

4. Identify a dense suburban development near your home or school. Redesign it to make it into a sustainable development. (This would make a good group project.)

5. What economic arguments would conventional farmers be likely to raise against switching to organic farming techniques and other sustainable agricultural methods? What would be the advantages to farmers? How does this affect consumers?

Answers to Figure Caption questions and Fill-in-the-Blank questions can be found in the Answers section at the back of the book.

MB *Go to MasteringBiology for practice quizzes, activities, eText, videos, current events, and more.*

Animal Anatomy and Physiology

The animal body is an exquisite expression of the elegance with which evolution has linked form to function.

"Know thyself"

—Inscribed above the entrance to the Temple of Apollo, home of the Oracle of Delphi

Three-hundred pound defensive ends couldn't defeat Korey Stringer, but heatstroke did.

CASE STUDY
Overheated

WHAT DID ATHLETES KOREY STRINGER, All-Pro offensive lineman for the Minnesota Vikings, and Luke Roach, a marathon runner, have in common? They both died of heatstroke while exercising. How can that happen to young, healthy men?

Body temperatures in healthy people usually range from about 97° to 99°F (36° to 37°C), whether asleep or exercising, in air temperatures from below freezing to well over 100°F (over 37°C). This narrow range of body temperature is an example of homeostasis—the ability to maintain a fairly constant internal environment in the face of varying external conditions. But sometimes temperature homeostasis fails, and the body temperature rises significantly, a condition called hyperthermia. Extreme hyperthermia had devastating results for Stringer and Roach.

Korey Stringer was fit, but huge— 6 feet, 4 inches tall, weighing over 330 pounds. In August 2001, the Vikings held intense workouts, in full pads, under conditions of high heat and humidity. Stringer's strenuous exercise generated enormous amounts of heat that his body couldn't get rid of. By the time Stringer was taken to a hospital, his body temperature was 108.8°F (over 42.5°C). He suffered multiple organ failure and died early the next morning.

Smaller athletes, like Luke Roach, are usually at less risk of heatstroke, but may still succumb during extremely strenuous, prolonged exercise. Roach died near the finish line of the Chicago Marathon, even though the temperature was below 50°F (10°C), so you might think that Roach shouldn't have been in danger. But running produces lots of heat, and marathoners run for 2 or 3 hours without stopping. Sometimes a runner will just push too hard for too long, and overheat without even knowing it. One of the early symptoms of heatstroke is confusion, so it's easy for even a trained athlete to miss his body's warning signs.

As you read this chapter, consider the many aspects of your physiology, in addition to temperature, that maintain homeostasis. How does the body recognize not only its optimal temperature, but the proper concentrations of sugar, salts, and oxygen in the blood? What kinds of actions does the body take to correct imbalances?

AT A GLANCE

31.1 Homeostasis: How Do Animals Regulate Their Internal Environment?

31.2 How Is the Animal Body Organized?

LEARNING GOALS

LG1 Define *homeostasis* and explain how homeostasis is maintained in animal bodies.

LG2 Define the terms *tissue*, *organ*, and *organ system*.

LG3 Describe the main categories of tissues, and explain the characteristics and functions of the major types within each category.

LG4 Describe the structure and functions of skin.

LG5 Describe the main components and functions of human organ systems.

31.1 HOMEOSTASIS: HOW DO ANIMALS REGULATE THEIR INTERNAL ENVIRONMENT?

Whether you are sitting in your room, sweltering in a jungle, or shivering in a blizzard, most of the cells of your body—for instance, in your heart, brain, and muscles—maintain an almost constant temperature. Further, whether you are standing in parched desert air or swimming in a pool, the ocean, or the Great Salt Lake, your cells are bathed in a liquid, called interstitial fluid, that has an almost constant composition despite enormous differences in the water and salt content of the outside environment.

Humans, of course, are not the only animals to maintain benign, stable internal conditions in the face of harsh, often rapidly changing external conditions. Salmon, for example, live most of their lives in the ocean, but mate and spawn in fresh water. The salt concentration in a salmon's cells, blood, and interstitial fluid is much lower than that of sea water but much higher than that of fresh water. Therefore, when a salmon is in the ocean, it will tend to lose water by osmosis; in fresh water, osmosis will drive water into its body (see Chapter 5 for more information about osmosis). Nevertheless, salmon neither shrivel up in the ocean nor burst open in fresh water. A complete list of organisms with such adaptations would be almost endless.

The "internal constancy" of animal bodies was first recognized by French physiologist Claude Bernard in the mid-nineteenth century. In the 1920s, Walter Cannon coined the term **homeostasis** to describe the ability of an organism to maintain its internal environment within the narrow limits that allow optimal cell functioning. Although the word "homeostasis" (meaning "to stay the same") implies a static, unchanging state, the internal environment actually seethes with activity as the body continuously adjusts to varying internal and external conditions.

The Internal Environment Is Maintained in a State of Dynamic Constancy

The internal state of an animal body can be described as a dynamic constancy. Many physical and chemical changes do occur (the dynamic aspect), but the net result of all this activity is that physical and chemical conditions are kept within the range that cells require to function (the constant aspect). In addition to temperature and concentrations of water and salt in body fluids, homeostatic mechanisms regulate a host of other conditions, including glucose concentrations, pH (acid-base balance), hormone secretion, and concentrations of oxygen and carbon dioxide.

Why are cells so particular about their surroundings? Animal cells generate and use large quantities of ATP to sustain their life processes (see Chapter 6). Continuous supplies of high-energy molecules (primarily glucose) and oxygen are required to carry out the reactions that generate most of this ATP. Thus, energy production helps explain the importance of keeping glucose and oxygen levels stable.

Almost every biochemical reaction in a cell is catalyzed by a specific protein whose ability to function depends on its three-dimensional structure, maintained, in part, by hydrogen bonds (see Chapters 2 and 3). These crucial but vulnerable bonds can be disrupted by an environment that is too hot, too salty, too acidic, or too basic. Thus, the need to maintain hydrogen bonds and the protein function that depends on them helps explain the requirement for a narrow range of temperature, salt, and pH. Reactions between carbon dioxide and water help to control the pH of the blood and interstitial fluid (see Chapter 33), so animals must regulate their CO_2 concentrations.

Animals Vary in Their Homeostatic Abilities

Birds and mammals are extremely proficient at maintaining homeostasis for virtually all of their internal conditions. Some animals, however, have reduced homeostasis for one or more aspects of their internal environment. For example, many marine invertebrates, including snails, crabs, and worms that live in tide pools, cannot regulate the overall concentration of their body fluids. If hot, dry air evaporates some of the water in a tide pool, increasing its salt concentration, the body fluids of these animals become more salty. If heavy rains dilute the seawater in the pool, the animals' body fluids become less salty.

Probably the most common and best studied variability in homeostasis among animals is a dramatic difference in their ability to regulate body temperature.

Animals Vary in How—and How Well— They Regulate Body Temperature

You have probably seen mammals and birds described as "warm-blooded," and reptiles, amphibians, fish, and invertebrates as "cold-blooded." However, these terms are often misleading. For example, the bodies of desert pupfish (**Fig. 31-1a**) may reach over 100°F (37.8°C) when the summer sun heats the springs and small ponds in which they live, so sometimes pupfish are quite warm-blooded. Could we classify animals by how constant their body temperatures are? Well, hummingbirds have body temperatures as high as 105°F (40.5°C) while foraging during the day (**Fig. 31-1b**), but may cool down to 55°F (13°C) during the night.

To avoid such confusion, scientists usually classify animals according to their major source of body warmth. **Ectotherms** (Greek for "outside heat") derive most of their heat from the environment. Reptiles, amphibians, and most fish and invertebrates are ectotherms. In the simplest cases—an earthworm in its burrow or a fish in a stream—an ectotherm's body temperature will be the same as the temperature of its environment. **Endotherms** (Greek for "inside heat") produce most of their heat by metabolic reactions. Birds and mammals are the principal endotherms on Earth, although some other types of animals occasionally generate significant amounts of heat metabolically.

The body temperatures of most ectotherms vary quite a bit as the external temperature changes over the course of hours, days, or weeks. However, through behavior or by occupying a very constant environment, some ectotherms can maintain quite stable body temperatures. For example, insects and lizards often warm up by basking in the sun (**Fig. 31-1c**). The pupfish mentioned earlier can tolerate water temperatures ranging from 36° to 113°F (2.2° to 45°C), but can breed only within a narrow range of temperatures. During breeding season, a pupfish regulates its temperature quite precisely by swimming to different areas of its pool as the temperature changes. In the deep ocean, the temperature is so constant (typically from about 32° to 37.5°F, or 0° to 3°C) that ectothermic deep-sea fish experience little variation in body temperature throughout their lives.

Most endotherms maintain a fairly constant body temperature between about 95°F and 106°F (about 35° to 41°C). There are both benefits and costs to keeping this warm. A major benefit is that a warm body usually can sense its environment better, respond more quickly, and move faster than a cold body. The major cost, of course, is the energy required to maintain a high body temperature. Many ectotherms, such as butterflies, bees, and most lizards, and even a few endotherms, such as hummingbirds, cool down and become inactive at night, thereby conserving energy, but warm up and become active during the day. At night, butterflies and bees become so cool that they cannot fly; lizards are often too sluggish to hunt or escape predators. At daybreak, bees shiver and butterflies beat their wings to generate metabolic heat, while lizards seek a warm, sunlit stone, providing the heat they need to resume their active lifestyles. Hummingbirds simply increase their metabolism to warm up at dawn.

How does an animal sense the conditions within its body and adjust them when necessary? The internal environment is maintained by mechanisms collectively known as feedback systems.

Feedback Systems Regulate Internal Conditions

There are two types of feedback systems: (1) negative feedback systems, which counteract the effects of changes in the internal environment and are principally responsible for maintaining homeostasis; and (2) positive feedback systems, which drive rapid, self-limiting changes, such as those that occur during childbirth.

(a) Desert pupfish

(b) Ruby-throated hummingbird

(c) Iguana

▲ **FIGURE 31-1 Warm-blooded or cold-blooded? (a)** Because "cold-blooded" fish such as this desert pupfish may be quite warm, and **(b)** "warm-blooded" animals like this hummingbird can become quite cool, biologists classify animals as endothermic or ectothermic, depending on the source of body warmth. **(c)** This iguana basking in the sun illustrates a behavioral mechanism that many ectotherms use to regulate body temperature.

Negative Feedback Reverses the Effects of Changes

The most important mechanism governing homeostasis is **negative feedback,** in which a change causes responses that counteract the change. The overall result of negative feedback is to return the system to its original condition.

Negative feedback is a common feature of both living and nonliving systems. Negative feedback regulates almost every aspect of a living organism's physiology, including body temperature; levels of glucose, hormones, water, salts, and oxygen in the blood; and even standing upright. Mechanical devices often incorporate negative feedback, for example, to keep a constant temperature in your house or hot-water heater, fill the tank of your toilet, or keep your car battery charged while you drive.

All negative feedback systems contain three principal components: a *sensor*, a *control center*, and an *effector*. The sensor detects the current condition, the control center compares that condition to a desired state called the set point, and the effector produces an output that restores the desired condition. Let's see how negative feedback systems work, first in the familiar example of heating your home and then in the control of body temperature.

In the negative feedback system that controls the temperature of your home on a cold day, the sensor is a thermometer, the control center is a thermostat, and the effector is a heater (**Fig. 31-2a**). The thermometer detects the room temperature, and sends that information to the thermostat, where the actual temperature is compared to the set point of the desired temperature. If the room temperature is below the set point, then the thermostat signals the heater to turn on and generate heat. The heater warms the room, restoring the temperature to the set point, which causes the thermostat to turn off the heater.

Because none of these components is perfect or responds instantaneously, the room temperature actually gets a bit cooler than the set point before the thermostat turns on the heater, and a bit warmer than the set point before the heater stops. The temperature oscillates above and below the set point, illustrating the dynamic constancy characteristic of homeostasis: The temperature does change over time (it is dynamic), but it always stays within a narrow range (it is relatively constant).

Endothermic animals use negative feedback systems to maintain their internal temperature despite fluctuations in the temperature around them (**Fig. 31-2b**). In humans and other mammals, the temperature control center is located in a part of the brain called the hypothalamus. Nerve endings in the abdomen, skin, large veins, and the hypothalamus itself act as temperature sensors and transmit this information

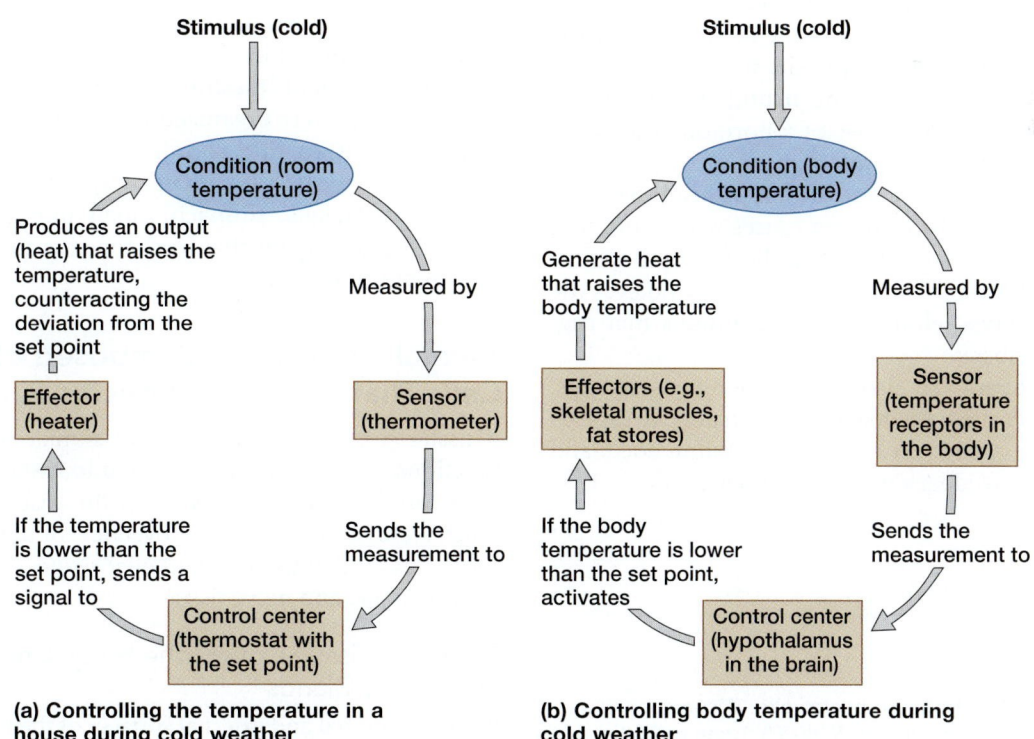

(a) Controlling the temperature in a house during cold weather

(b) Controlling body temperature during cold weather

▲ **FIGURE 31-2 Negative feedback maintains homeostasis** In negative feedback, responses to a stimulus (at the top of each diagram) counteract the effects of the stimulus. **(a)** Negative feedback regulates a specific condition (here, the temperature in your house) within narrow limits. **(b)** Most aspects of animal physiology, such as the control of body temperature, are regulated by negative feedback.

QUESTION What would happen if a cold, shivering mammal ingested a poison that destroyed all of its body's nerve endings that detect heat?

to the hypothalamus. If the body temperature falls below the set point (the normal body temperature), the hypothalamus activates mechanisms that generate and conserve heat, such as burning fats, shivering (small, rapid muscular contractions), or moving to a warmer place. The blood vessels supplying nonvital areas of the body (such as the face, hands, feet, and skin) are constricted, reducing heat loss and diverting warm blood to the body core (including the brain, heart, and other internal organs). When normal body temperature is restored, the sensors signal the hypothalamus to switch off the actions that generate and conserve heat.

The temperature control system can also act to reduce body temperature. If body temperature rises over the set point, the hypothalamus sends out signals that cause the blood vessels leading to the skin to open up, allowing warm blood to flow to the skin, where the heat can be radiated out to the surrounding air. Sweat glands produce a watery secretion that cools the body as the water evaporates from the skin. The fatigue and discomfort caused by elevated body temperature usually cause people to rest, so the body generates less heat.

Positive Feedback Enhances the Effects of Changes

In **positive feedback,** a change produces a response that amplifies the change. Positive feedback is relatively rare in biological systems, but does occur, for example, during childbirth. The early contractions of labor push the baby's head against the cervix at the base of the uterus, causing the cervix to stretch and open. Nerve cells in the cervix react to the stretching by signaling the hypothalamus, which responds by triggering the release of a hormone called oxytocin that stimulates more and stronger uterine contractions. Stronger contractions cause the baby's head to stretch the cervix even more, which in turn causes release of more oxytocin. In biology, positive feedback always stops, eventually. During childbirth, the delivery of the baby relieves the pressure on the cervix, eliminating the stimulus that triggered positive feedback.

Positive feedback can also occur in nonliving systems. In "Earth Watch: Positive Feedback in the Arctic," we explore how the shrinking Arctic ice cap may cause unintended, and unwanted, positive feedback in Earth's climate.

CASE STUDY continued

Overheated

Positive feedback contributes to heatstroke. When the body temperature rises above 104°F (40°C), brain cells die. The negative feedback mechanisms that usually bring the body's temperature back to normal malfunction, and positive feedback takes over: A heatstroke victim often doesn't sweat, even as his body temperature skyrockets. His metabolic rate may increase, generating even more heat, and his body temperature may soar as high as 108°F. All too often, this runaway positive feedback is terminated only by death.

CHECK YOUR LEARNING

Are you able to explain the difference between ectotherms and endotherms and give examples of each? Can you define *homeostasis* and explain why negative feedback is crucial in achieving homeostasis? Do you understand why positive feedback is rare in living organisms?

31.2 HOW IS THE ANIMAL BODY ORGANIZED?

Animals maintain homeostasis by performing dozens, perhaps hundreds, of functions simultaneously, coordinating the activity of thousands to trillions of cells. For example, we have seen that the hypothalamus in the brain, temperature sensors in multiple parts of the body, and sweat glands and blood vessels in the skin are all crucial to controlling human body temperature—and this list by no means includes all of the structures that are involved. This coordination of complex body systems is based on a simple hierarchy of structures:

$$cells \rightarrow tissues \rightarrow organs \rightarrow organ\ systems$$

An example of this hierarchy is illustrated in **Figure 31-3**. Cells are the fundamental units of all living organisms (see Chapter 1). In an animal body, a **tissue** is composed of dozens to billions of structurally similar cells that work together to carry out a particular task. Tissues are the building blocks of **organs,** which are discrete structures that perform complex functions. Examples of organs include the stomach, small intestine, kidneys, and urinary bladder. Organs, in turn, are organized into **organ systems,** groups of organs that act in a coordinated manner. For example, the digestive system is an organ system consisting of the mouth, esophagus, stomach, small intestine, large intestine, and other organs that work together to digest food and absorb nutrients from it.

Animal Tissues Are Composed of Similar Cells That Perform a Specific Function

A tissue consists of cells that are similar in structure and functions. A tissue may also include extracellular components produced by its cells, as in the case of cartilage and bone. Here we present a brief overview of the four major categories of animal tissues: epithelial tissue, connective tissue, muscle tissue, and nerve tissue.

Epithelial Tissue Covers the Body, Lines Its Cavities, and Forms Glands

Both the internal and external surfaces of the body—including the skin and the insides of the circulatory, digestive, respiratory, reproductive, and urinary systems—are covered by epithelial tissue (**Fig. 31-4**). **Epithelial tissue** consists of sheets of cells firmly attached to one another by connections such as desmosomes and tight junctions (see Chapter 5). Sheets of epithelial tissue are also attached to an underlying layer of fibrous proteins, called the *basement membrane.*

▲ FIGURE 31-3 Cells, tissues, organs, and organ systems The animal body is composed of cells, which make up tissues, which combine to form organs that work together as organ systems.

Labels in figure 31-3:
- **Tissues:** epithelial, connective
- **Cells:** epithelial cells
- arteriole
- venule
- lymphatic vessel
- muscle
- **Organ:** stomach
- pancreas
- stomach
- large intestine
- mouth
- pharynx
- esophagus
- small intestine
- gallbladder
- liver
- **Organ system:** digestive system

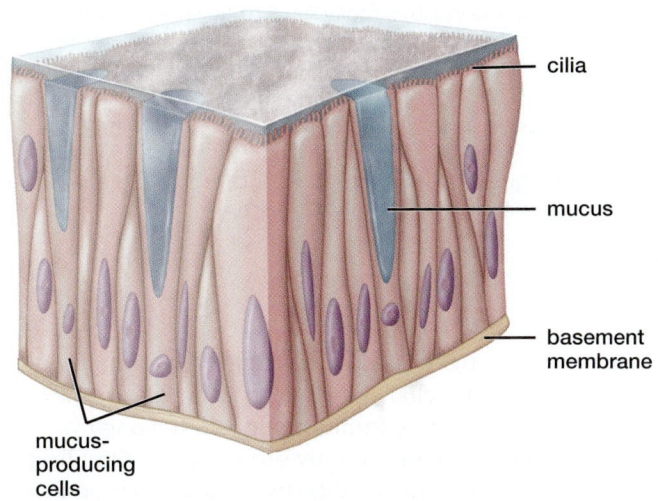

(a) Lining of the lungs (simple epithelium)
- flattened cells
- basement membrane

(b) Lining of the trachea (simple epithelium)
- cilia
- mucus
- basement membrane
- mucus-producing cells

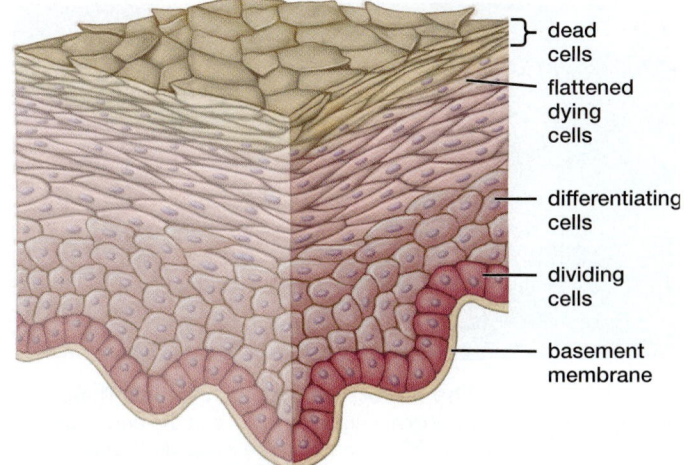

(c) Skin epidermis (stratified epithelium)
- dead cells
- flattened dying cells
- differentiating cells
- dividing cells
- basement membrane

▲ FIGURE 31-4 Epithelial tissue (a) Thin, flattened cells in a single layer form the epithelial tissue that lines the lungs, where exchange of gases by diffusion is important. **(b)** The trachea is lined with elongated epithelial cells bearing cilia, interspersed with cells that secrete mucus. **(c)** The epidermis of the skin consists of multilayered, stratified epithelial tissue covered by a protective layer of dead cells. Beneath all epithelial tissue is a thin layer of fibrous proteins, called the basement membrane.

Earth | Watch

Positive Feedback in the Arctic

In positive feedback, changing conditions cause responses that enhance the change. Climate scientists fear that global climate change has triggered positive feedback in the Arctic, and that the result may be the planetary equivalent of hyperthermia.

The North Pole lies in the middle of the Arctic Ocean, about 400 miles from the nearest land. People have walked to the Pole, because the ocean used to be covered with ice 6 to 15 feet thick (2 to 5 meters). However, as the planet heats up, its effects are felt most strongly in the Arctic, which is warming about twice as fast as the global average. As a result, during the past 30 years the Arctic ice cap has become almost 50% thinner and 35% smaller in area (**Fig. E31-1**). The cap is currently shrinking by about 10% per decade.

▲ **FIGURE E31-2 Positive feedback accelerates the loss of the Arctic ice cap** Most of the sunlight hitting sea ice is reflected, making it appear white. Most of the sunlight hitting the open ocean is absorbed, so it is darker and heats up much faster.

▲ **FIGURE E31-1 The Arctic ice cap is shrinking** These satellite images show the minimum extent of ice at the end of the summer in 1979 (top) and 2011 (bottom). The ice cover fluctuates from year to year, but overall it shrank by about 35% during this time.

What does this have to do with positive feedback? The problem lies in how much sunlight is reflected by ice compared to open ocean waters (**Fig. E31-2**). Sea ice typically reflects 50% to 70% of incoming sunlight back into space; the rest is absorbed, warming the ice. Snow cover on top of the ice may reflect as much as 80% of the light. In contrast, seawater reflects only 5% to 10% of the light, so almost all of the energy in sunlight goes into warming the water. Classic positive feedback: As the Arctic ice sheet melts, more water is exposed. The exposed water absorbs more sunlight than the ice did, and becomes warmer, which melts more ice, exposing still more water.

The global consequences of melting the Arctic ice cap are difficult to predict. The planet as a whole will warm, of course. Weather systems in the Northern Hemisphere will undoubtedly change. The circulation patterns in the oceans (see Chapter 29) may shift, stop, or even reverse, with potentially serious consequences for our climate.

When an athlete pushes really hard to finish a marathon in hot weather, he is running an experiment with his body: Can he keep his temperature under control? What are the consequences if he cannot? By pumping carbon dioxide and other greenhouse gases into the atmosphere, humanity is running an experiment with our planet. The melting ice cap in the Arctic suggests that the experiment is producing a positive feedback cycle that may be hard to stop.

Some of these protein fibers are secreted by the epithelial cells themselves; others are secreted by connective tissue cells (see the next section). The basement membrane provides support, flexibility, and strength to the epithelial tissue, which is obviously important in organs that are subject to stretch and strain, such as the skin. In the gallbladder and

urinary bladder, tight junctions seal off the extracellular spaces between epithelial cells, which is crucial to prevent bile or urine, respectively, from leaking into the body.

A **simple epithelium** is only one cell thick. Simple epithelium lines the respiratory, digestive, urinary, reproductive, and circulatory systems. Simple epithelia vary in structure, depending on the organ and its function. For example, the simple epithelium that lines the lungs consists of a layer of thin, flattened cells that allows rapid diffusion of gases between the lungs and the bloodstream (**Fig. 31-4a**). The epithelium that lines the trachea (windpipe) is more complex, consisting of both short and tall cells bearing cilia, interspersed with glandular cells that secrete mucus (**Fig. 31-4b**). The mucus traps inhaled debris, and the cilia sweep it out of the trachea.

Stratified epithelium, which is several cells thick and can usually withstand considerable wear and tear, is mostly found in the skin and just inside body openings that are continuous with the skin, such as the mouth and anus (**Fig. 31-4c**). Cells at the base of stratified epithelium divide rapidly, giving rise to daughter cells that form the surface layers. In the skin, the daughter cells become filled with strong proteins (mostly keratin) as they differentiate. Eventually these cells die and form the tough, flexible, waterproof outer surface.

An important property of both simple and stratified epithelia is that their cells are continuously lost and replaced. For example, consider the abuse suffered by the epithelia of your mouth and skin. Scalded by coffee and scraped by corn chips, the epithelium of the mouth would be destroyed within a few days if it did not replace itself continuously. The surface cells of the skin epithelium are worn away by rubbing on clothing, bath towels, and furniture, and are replaced by new cells produced in the deeper layers, resulting in complete renewal about every 2 weeks.

Glands are cells or groups of cells specialized to secrete substances such as hormones or digestive enzymes outside the cell. Most glands are composed of specialized epithelial cells. (Two glands, the posterior pituitary and adrenal medulla, are nervous, not epithelial, tissue.) Some glands are single cells, such as the mucus-producing epithelial cells of the trachea (see Fig. 31-4b). Most glands are multicellular, including those that secrete saliva, sweat, milk, and most hormones.

Glands are classified into two categories: exocrine glands and endocrine glands. **Exocrine glands** secrete substances into a body cavity or onto the body surface, usually through a narrow tube or duct. Examples of exocrine glands are sweat glands (see Fig. 31-11 later in this chapter), mammary glands (which secrete milk), and several glands that release digestive enzymes and other substances into the mouth, stomach, or small intestine. **Endocrine glands,** such as the ovaries, testes, thyroid, and pituitary, do not have ducts. They secrete hormones into the interstitial fluid, from which the hormones diffuse into nearby capillaries. **Hormones** are chemicals that are produced in small quantities and are transported in the bloodstream to distant parts of the body, where they regulate the activity of other cells (see Chapter 37).

Connective Tissues Have Diverse Structures and Functions

The defining characteristic of all **connective tissue** is that a large proportion of the connective tissue consists of an extracellular matrix—which may be a fluid almost as thin as water or as thick as molasses, or a solid as hard as rock—with a relatively small number of cells embedded within it. The matrix contains a high concentration of proteins, secreted by these connective tissue cells. The most abundant protein in connective tissue is usually collagen. Most types of connective tissue support and strengthen other tissues, and help to bind the cells of those other tissues together in coherent structures, such as skin or muscles. Connective tissues fall into three main categories: loose connective tissue, dense connective tissue, and specialized connective tissue (**Fig. 31-5**).

Loose Connective Tissue Loose connective tissue is the most abundant form of connective tissue, consisting of a thick fluid containing scattered cells that secrete protein fibers (**Fig. 31-5a**). This flexible tissue connects, supports, and

(a) Loose connective tissue   **(b) Dense connective tissue**   **(c) Specialized connective tissue**

▲ **FIGURE 31-5 Connective tissue (a)** Loose connective tissue underlies epithelial tissue and supports almost every organ in the body. **(b)** The dense connective tissue in tendons and ligaments is tremendously strong in the direction in which its protein fibers are aligned, but can be easily disrupted by forces acting perpendicular to the aligned fibers. **(c)** Cartilage is a specialized connective tissue consisting of scattered cells surrounded by the collagen and glycoproteins that they secrete.

surrounds other tissue types, and it forms an internal framework for organs such as the liver. Loose connective tissue combines with epithelial tissue to form membranes. (In biology, the term *membrane* means a sheet-like structure. This term is used for both the phospholipid bilayers of cells [see Chapters 4 and 5] and thin multicellular structures composed of epithelial and connective tissues, such as the mucous membranes that line the digestive and respiratory tracts.)

Dense Connective Tissue Dense connective tissue is packed with collagen fibers (**Fig. 31-5b**). In **tendons** (which connect muscles to bones) and **ligaments** (which connect bones to bones), the collagen fibers are arranged parallel to one another. In other types of dense connective tissue, such as the lower layers of the dermis of the skin and the capsules that surround many internal organs and muscles, the collagen fibers form an irregular meshwork. Both designs provide flexibility and strength. Tendons and ligaments have tremendous strength, but only in the direction in which the collagen fibers are oriented (which is why twisting a knee in the wrong direction can rupture a ligament). The irregular mesh of collagen in skin and muscle capsules resists tearing in all directions, but isn't as strong as the parallel arrangement in tendons and ligaments.

Specialized Connective Tissues This diverse group includes cartilage, bone, fat, blood, and lymph. **Cartilage** consists of widely spaced cells surrounded by a thick, nonliving matrix composed of collagen secreted by the cartilage cells (**Fig. 31-5c**). Cartilage covers the ends of bones at joints, provides the supporting framework for the respiratory passages, supports the nose and ears, and forms shock-absorbing pads between the vertebrae. Cartilage is fairly flexible, but can snap if bent too far. **Bone** resembles cartilage, but its matrix is hardened by deposits of calcium phosphate. Bone forms in concentric circles around a central canal, which contains blood vessels (**Fig. 31-6**). (We will discuss cartilage and bone in Chapter 40.)

▲ **FIGURE 31-6 Bone** Concentric circles of bone (a specialized connective tissue), deposited around a central canal that contains blood vessels, are clearly visible in this micrograph. Individual bone cells appear as dark spots trapped in small chambers within the hard matrix that the cells deposit.

▲ **FIGURE 31-7 Adipose tissue** Adipose tissue is specialized connective tissue made up almost exclusively of fat cells. Stored fat occupies most of the volume of the cell.

Fat cells, collectively called **adipose tissue**, are modified for long-term energy storage (**Fig. 31-7**). Adipose tissue is especially important for animals adapted to cold environments because it both stores energy and insulates the body.

Although they are liquids, **blood** and **lymph** are considered specialized forms of connective tissues because they are composed mostly of an extracellular matrix (in these tissues, a watery liquid) in which proteins and individual cells are suspended. The cellular portion of blood consists of red blood cells (which transport oxygen), white blood cells (which fight infection), and cell fragments called platelets (which aid in blood clotting) (**Fig. 31-8**). These are all suspended in a liquid called plasma. Lymph consists largely of liquid that has leaked out of blood capillaries (the smallest of the blood vessels), plus white blood cells. (You will learn more about blood and lymph in Chapter 32.)

Muscle Tissue Has the Ability to Contract

The long, thin cells of muscle tissue contract (shorten) when stimulated and then relax passively when the stimulation stops. There are three types of muscle tissue: skeletal, cardiac, and smooth (**Fig. 31-9**). Contraction of **skeletal muscle** (**Fig. 31-9a**) is stimulated by the nervous system and is generally under voluntary, or conscious, control. As its name implies, its main function is to move the skeleton, as occurs when you walk or turn the pages of this text.

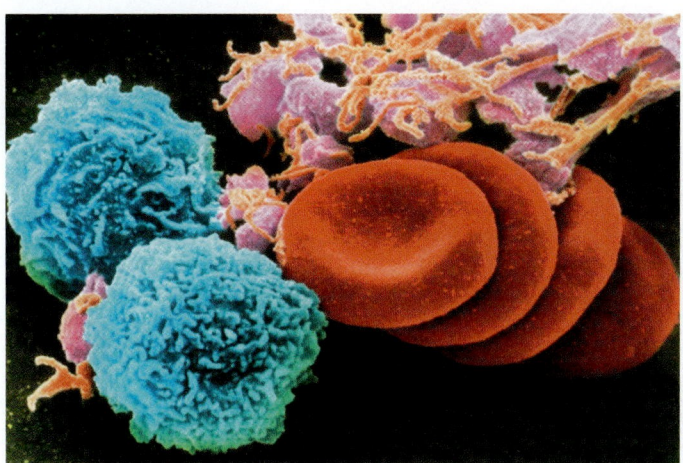

▲ FIGURE 31-8 **Blood cells** This colorized scanning electron micrograph shows red blood cells (dimpled red disks), white blood cells (colored blue in this image), and cell fragments called platelets (colored purple). The cells are suspended in plasma, the extracellular matrix of this specialized connective tissue. In this image, the platelets are enmeshed in protein strands that help to form a clot.

QUESTION Why is blood considered a form of connective tissue?

CASE STUDY c o n t i n u e d

Overheated

During strenuous exercise, your skeletal muscles generate 10 to 15 times more heat than they do when you're simply resting—enough to raise your body temperature about 2°F every 5 minutes or so, if you can't get rid of the extra heat. If it is humid and your sweat doesn't evaporate fast enough, or if you're wearing heavy clothing (let alone football pads), it won't take you long to enter the first stages of heatstroke. Visit the website of the Korey Stringer Institute at the University of Connecticut to find out how to recognize the symptoms of exertional heatstroke, and learn what steps you can take to prevent it.

Cardiac muscle is located only in the heart (Fig. 31-9b). Unlike skeletal muscle, it is spontaneously active and involuntary (not under conscious control). Cardiac muscle cells are interconnected by gap junctions, through which electrical signals spread rapidly throughout the heart, stimulating the cardiac muscle cells to contract in a coordinated fashion.

Smooth muscle (Fig. 31-9c) is found throughout the body, embedded in the walls of the digestive and respiratory tracts, uterus, bladder, larger blood vessels, skin, and even in the iris of the eye. Smooth muscle produces slow, sustained contractions that may be stimulated by the nervous system, by stretching, or by hormones and certain other chemicals. Smooth muscle contraction is typically involuntary. (You will learn more about muscles in Chapter 40.)

Nerve Tissue Is Specialized to Produce and Conduct Electrical Signals

You owe your ability to sense and respond to the world to **nerve tissue,** which makes up the brain, the spinal cord, and the nerves in all parts of the body. Nerve tissue is composed of two types of cells: nerve cells, also called neurons, and glial cells (Fig. 31-10).

Neurons (Fig. 31-10a) are specialized to generate electrical signals and to conduct these signals to other neurons, muscles, or glands. A typical nerve cell consists

(a) Skeletal muscle

(b) Cardiac muscle

(c) Smooth muscle

▲ FIGURE 31-9 **Muscle tissue** Fibrous proteins inside muscle cells cause them to shorten, or contract, when they are stimulated. **(a)** In skeletal muscle, the regular arrangement of these proteins makes this type of muscle tissue appear striped, or "striated," when viewed under a microscope. **(b)** In cardiac muscle, the contractile proteins are also aligned in stripes. Clusters of gap junctions (dark bands between cells) interconnect the cytoplasm of adjacent cardiac muscle cells. **(c)** Smooth muscle fibers are spindle-shaped cells in which the contractile proteins are not consistently aligned; hence, they appear "smooth."

(a) Neuron

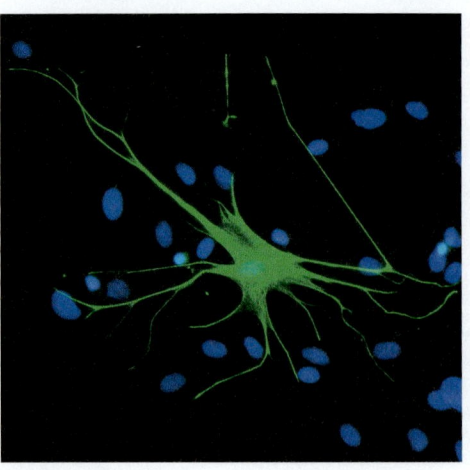

(b) Glial cell

◄**FIGURE 31-10 Nerve tissue**
Nerve tissue consists of neurons and glial cells. **(a)** Neurons are specialized to receive and transmit signals. The long strand emerging from the cell body is the axon, which sends signals to other cells; the short yellow spikes on the cell body are dendrites, which receive signals from other cells. **(b)** The brain has several different types of glial cells; this one, called an astrocyte, helps to nourish neurons and protect them from damage.

of *dendrites*, which are specialized to receive signals from other neurons or from the environment; a *cell body*, which contains the nucleus and carries out most of the neuron's metabolism; an *axon*, which carries the neuron's electrical signal to a muscle, gland, or other neuron; and the *terminals* on the ends of an axon, which transmit information to the other cells.

Glial cells (Fig. 31-10b) surround, support, electrically insulate, and protect neurons. Glial cells also regulate the composition of the interstitial fluid in the nervous system. (We discuss nerve tissue in Chapter 38.)

Organs Include Two or More Interacting Tissue Types

Organs are formed from at least two tissue types that function together. Most organs, however, consist of all four tissue types, with varying types and proportions of epithelial, connective, muscle, and nerve tissues. Most organs function as part of organ systems. In the following section, we describe the skin as a representative organ that includes all four of the tissue types.

The Skin Illustrates the Properties of Organs

The skin consists of an outer layer of epithelial tissue underlain by connective tissue. Coursing through the connective tissue are blood vessels, nerves, muscles, and glands (Fig. 31-11). Far more than a simple covering for the body, the skin is essential for survival. For example, it is such an important barrier against the entry of disease-causing microorganisms and the evaporation of precious body water that the large-scale destruction of skin—as occurs with extensive burns—can prove fatal.

The **epidermis,** or outer layer of the skin, is a stratified epithelial tissue (see Fig. 31-11). It is covered by a protective layer of dead cells produced by the underlying living epidermal cells. These dead cells are packed with the protein keratin, which helps make the skin elastic, tough, and fairly waterproof.

Immediately beneath the epidermis lies a layer of connective tissue, the **dermis.** Embedded in the dermis are many structures composed of epithelial, muscle, and nervous tissues. Arterioles (small arteries) snake throughout the dermis, carrying blood pumped from the heart into a meshwork of capillaries that nourish both the dermal and epidermal tissue. The capillaries empty into venules (small veins) in the dermis. Loss of heat through the skin is regulated by neurons controlling the diameter of the arterioles. To conserve body heat, smooth muscles in the arterioles contract, reducing the diameter of the arterioles and restricting blood flow through them. To cool the body, these smooth muscles relax, allowing the arterioles to open up and flood the capillary beds with blood, thus releasing excess heat. Lymph vessels collect and carry off interstitial fluid within the dermis. Various sensory nerve endings that respond to temperature, touch, pressure, vibration, and pain are scattered throughout the dermis and epidermis and provide information to the nervous system.

Specialized epithelial cells dip down from the epidermis into the dermis, forming **hair follicles.** Cells in the base of the follicles divide rapidly, and their daughter cells fill with keratin. As new cells form in the base of the follicle, older, dying cells are pushed upward to the surface of the skin, emerging as hairs.

The dermis also contains glands derived from epithelial tissue. Sweat glands produce watery secretions that cool the skin and excrete substances such as salts and urea. Sebaceous glands secrete an oily substance (sebum) that lubricates the epithelium.

Skin also contains muscle tissue. All skin has muscle cells within the walls of the arterioles that supply blood to the skin. In addition, hairy skin has tiny muscles attached to the hair follicles that can cause the hairs of the skin to "stand on end" in response to signals from motor neurons. Most mammals can increase the height of their insulating fur in cold weather by erecting their hairs. This reaction is useless for retaining heat in people, whose hair, flat or erect, is far too sparse to provide much insulation. We merely experience "goose bumps" when these muscles contract.

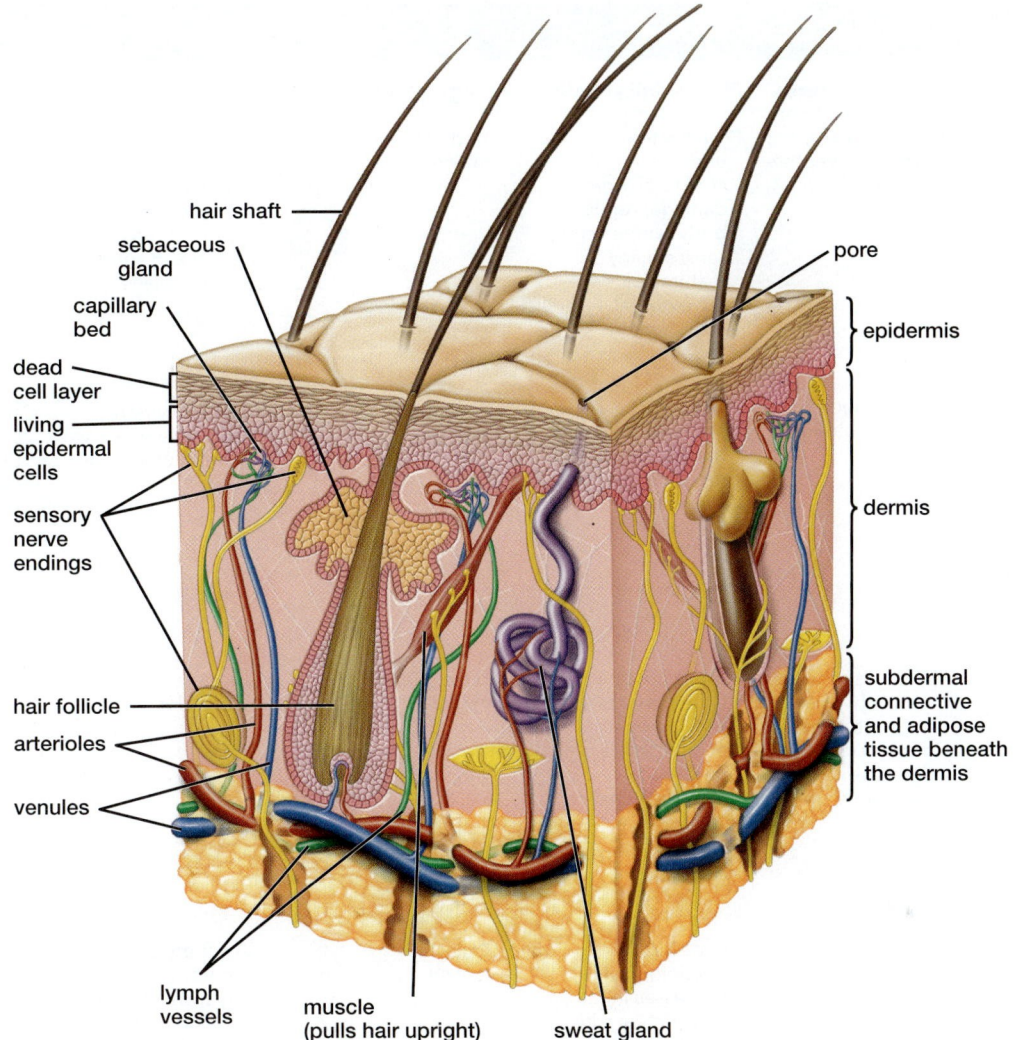

▲ FIGURE 31-11 Skin is an organ Mammalian skin, a representative organ, is shown in cross-section. The skin contains epithelial, connective, muscle, and nervous tissue. A basement membrane of fibrous proteins holds the epidermis to the dermis, but is too thin to be seen at this scale.

? HAVE **YOU** EVER WONDERED . . .

Why You Get Goose Bumps When You Are Frightened?

No matter what the temperature, you may get goose bumps when you're scared. This is a relic of our hairier evolutionary past. In threatening situations, many mammals erect their fur, which makes them appear larger. Large animals are likely to scare off potential opponents, so erect fur may prevent unnecessary fighting. People, of course, don't have enough hair on their bodies for this trick to work.

Finally, below the dermis (and thus technically not part of the skin) is a layer of adipose tissue interspersed with fibrous proteins produced by connective tissue cells. The fat helps to insulate the body, preserving heat and making it easier to maintain a constant body temperature in cool weather. The fat also serves as an energy storage depot.

Organ Systems Consist of Two or More Interacting Organs

Organ systems consist of two or more individual organs (in some cases, located in different parts of the body) that work together to perform a common function. The major organ systems of the vertebrate body and their representative organs and functions are described in **Table 31-1**.

CHECK YOUR LEARNING

Can you explain the differences between tissues, organs, and organ systems, and describe their relationships to one another? Are you able to describe the four types of tissues? Do you understand how the four tissue types interact in an organ such as the skin? Can you name and describe the major human organ systems?

TABLE 31-1 Major Vertebrate Organ Systems

Organ System	Major Structures	Physiological Role	Organ System	Major Structures	Physiological Role
Integumentary system	Skin, hair, nails, sensory receptors, various glands	Protects the underlying structures from damage; regulates body temperature; senses many features of the external environment	Respiratory system	Nose, pharynx, trachea, lungs (mammals, birds, reptiles, amphibians), gills (fish and some amphibians)	Provides a large area for gas exchange between the blood and the environment; allows oxygen acquisition and carbon dioxide elimination
Circulatory system	Heart, blood vessels, blood	Transports nutrients, gases, hormones, metabolic wastes; also assists in temperature control	Lymphatic/immune system	Lymph, lymph nodes and vessels, white blood cells	Carries fat and excess fluids to the blood; destroys invading microbes
Digestive system	Mouth, esophagus, stomach, small and large intestines, glands producing digestive secretions	Supplies the body with nutrients that provide energy and materials for growth and maintenance	Urinary system	Kidneys, ureters, bladder, urethra	Maintains homeostatic conditions within the bloodstream; filters out cellular wastes, certain toxins, and excess water and nutrients
Nervous system	Brain, spinal cord, peripheral nerves	Controls physiological processes in conjunction with the endocrine system; senses the environment, directs behavior	Endocrine system	A variety of hormone-secreting glands and organs, including the hypothalamus, pituitary, thyroid, pancreas, adrenals, ovaries, and testes Male Female	Controls physiological processes, typically in conjunction with the nervous system
Skeletal system	Bones, cartilage, tendons, ligaments	Provides support for the body, attachment sites for muscles, and protection for internal organs	Muscular system	Skeletal muscle Smooth muscle Cardiac muscle	Moves the skeleton Controls movement of substances through hollow organs (digestive tract, large blood vessels) Initiates and implements heart contractions
Male reproductive system	Testes, seminal vesicles, prostate gland, penis	Produces sperm and sex hormones, inseminates female	Female reproductive system	Ovaries, oviducts, uterus, vagina, mammary glands	Produces egg cells and sex hormones, nurtures developing offspring

CASE STUDY revisited

Overheated

Heatstroke is not confined to athletes. In fact, when the air temperature becomes extremely high, especially coupled with high humidity, people can die of heatstroke sitting in their living rooms. Although athletes typically receive more coverage in the news, death from heatstroke is far more likely to occur among the elderly during summer heat waves. Farm workers, toiling under the hot sun, are also at considerable risk.

Children are especially vulnerable to heatstroke. In very young children, temperature homeostasis is not fully developed. Children produce more heat than adults would under similar conditions, and they do not sweat as much as adults do. In addition, children playing outdoors in hot weather may become so excited that they are unaware that they may be overheating.

Perhaps the biggest danger of heatstroke in children is being left in a car on a sunny day, which kills 35 to 40 children in the United States each year. Even if the outside temperature isn't blazing hot, the inside of a closed car can easily reach 100°F (about 38°C) in just a few minutes. When it's over 90°F outside (32°C), the temperature in a closed car can skyrocket to 140°F (60°C) in less than an hour. Leaving a window rolled down a couple of inches helps somewhat, but often not enough. In warm weather, no one should *ever* leave a child in a car, even for a few minutes—no errand, no matter how important or how brief you think it might be, can possibly justify the risk.

Consider This

BioEthics Computer models of global climate change predict not only higher overall temperatures, but also increased heat waves, with peak temperatures far exceeding the average warming. During the devastating European heat wave in the summer of 2003, between 35,000 and 52,000 "excess deaths" occurred, probably almost all from the heat. The vast majority of climate scientists are convinced that the buildup of carbon dioxide in the atmosphere is causing global climate change. Before taking potentially expensive actions to reduce carbon dioxide emissions, economists and politicians try to estimate the costs. Should excess deaths be included as a cost of inaction? If so, what value should we place on a life?

CHAPTER REVIEW

Summary of Key Concepts

31.1 Homeostasis: How Do Animals Regulate Their Internal Environment?

Homeostasis refers to the dynamic equilibrium within the animal body in which internal conditions including temperature, salt, oxygen, glucose, pH, and water levels are maintained within a range in which proteins can function and energy can be made available. Animals differ in temperature regulation. Ectotherms derive most of their body warmth from the environment. Endotherms derive most of their heat from metabolic activities and tend to regulate their body temperatures within a narrow range.

Homeostatic conditions are maintained through negative feedback, in which a change triggers a response that counteracts the change and restores the original conditions. There are a few instances of positive feedback, in which a change initiates events that intensify the change (such as uterine contractions leading to childbirth), but these situations are all self-limiting.

31.2 How Is the Animal Body Organized?

The animal body is composed of organ systems, each consisting of two or more organs. Organs, in turn, are made up of tissues. A tissue is a group of cells and extracellular material that form a structural and functional unit specialized for a specific task. Animal tissues include epithelial, connective, muscle, and nerve tissue.

Epithelial tissue forms coverings over internal and external body surfaces and also gives rise to glands. Connective tissue usually contains considerable extracellular material in which cells and proteins are embedded, and includes the dermis of the skin, bone, cartilage, tendons, ligaments, fat, and blood. Muscle tissue is specialized to produce movement by contraction. There are three types of muscle tissue: skeletal, cardiac, and smooth. Nerve tissue, composed of neurons and glial cells, generates and conducts electrical signals.

An organ contains at least two tissue types. Mammalian skin is a representative organ. The epidermis, an epithelial tissue, covers and protects the dermis beneath it. The dermis consists of connective tissue that contains blood and lymph vessels, sweat and sebaceous glands, hair follicles, muscles that erect the hairs, and a variety of sensory nerve endings. Animal organ systems include the integumentary, respiratory, circulatory, lymphatic/immune, digestive, urinary, nervous, endocrine, skeletal, muscular, and reproductive systems, summarized in Table 31-1.

Key Terms

adipose tissue *608*
blood *608*
bone *608*
cardiac muscle *609*
cartilage *608*
connective tissue *607*
dermis *610*
ectotherm *602*
endocrine gland *607*
endotherm *602*
epidermis *610*
epithelial tissue *604*
exocrine gland *607*
gland *607*
glial cell *610*
hair follicle *610*

homeostasis *601*
hormone *607*
ligament *608*
lymph *608*
negative feedback *603*
nerve tissue *609*
neuron *609*
organ *604*
organ system *604*
positive feedback *604*
simple epithelium *606*
skeletal muscle *608*
smooth muscle *609*
stratified epithelium *607*
tendon *608*
tissue *604*

Learning Outcomes

In this chapter, you have learned to . . .

LO1 Define *homeostasis* and explain how homeostasis is maintained in animal bodies.

LO2 Define the terms *tissue*, *organ*, and *organ system*.

LO3 Describe the main categories of tissues, and explain the characteristics and functions of the major types within each category.

LO4 Describe the structure and functions of skin.

LO5 Describe the main components and functions of human organ systems.

Thinking Through the Concepts

Fill-in-the-Blank

1. The ability of the body to maintain its internal conditions within the narrow range required by cells to function is called _____. The most important mechanism governing this ability is _____. **LO1**

2. The four levels of organization of the animal body, from the smallest to the most inclusive, are _____, _____, _____, and _____. **LO2**

3. Fill in the appropriate tissue type (some tissue types may be appropriate for more than one blank): Supports and strengthens other tissues: _____; forms glands: _____; includes the blood: _____; includes the dermis of the skin: _____; covers the body and lines its cavities: _____; can shorten when stimulated: _____; includes glial cells: _____; includes adipose tissue: _____. **LO3**

4. Glands with ducts connecting them to the epithelium are called _____ glands. Glands with no ducts are called _____ glands; most of these secrete _____ (a general term for a type of messenger molecule). **LO3**

5. Fill in the appropriate type(s) of muscle (there may be more than one answer for some blanks): Contracts rhythmically and spontaneously: _____; is controlled voluntarily: _____; contains orderly arrangements of fibrous proteins, giving a striped appearance: _____; is not under voluntary control: _____; is found in the walls of the digestive tract: _____; moves the skeleton: _____. **LO3**

Review Questions

1. Define *homeostasis*, and explain how negative feedback helps to maintain it. Describe one example of homeostasis in the human body. **LO1**

2. Define and compare *ectotherms* and *endotherms*. Provide an example of each. Are "cold-blooded" and "warm-blooded" accurate ways to describe them? Explain. **LO1**

3. Explain positive feedback, and provide one physiological example. Explain why this type of feedback is relatively rare in physiological processes. **LO1**

4. Explain what goes on in your body to restore temperature homeostasis when you become overheated by exercising on a hot, humid day. **LO1**

5. Describe the structure and functions of epithelial tissue. **LO3**

6. What property distinguishes connective tissue from all other tissue types? List three general types of connective tissue, and briefly describe the function of each type. **LO3**

7. Describe the skin as an organ. Include the various tissues that compose it, and briefly describe the role of each tissue. **LO4**

8. Name the human organ systems, and briefly describe the components and functions of each. **LO5**

Applying the Concepts

1. Why does life on land present particular difficulties in maintaining homeostasis?

2. Advertisements for a sports drink say that "water is not enough" for people who are exercising, and that it is important to consume a liquid that contains flavoring, carbohydrates, potassium, calcium, and sodium. Do you think that consuming such a sports drink during exercise can help maintain homeostasis? Explain your answer.

3. Third-degree burns are usually painless. Skin regenerates only from the edges of these wounds. Second-degree burns are often very painful. Skin regenerates from cells located at the burn edges, in hair follicles, and in sweat glands. First-degree burns are painful but heal rapidly from undamaged epidermal cells. Using this information, draw the depth of first-, second-, and third-degree burns on Figure 31-11.

4. Imagine you are a health care professional teaching a prenatal class for fathers. Create a design for a machine with sensors, electrical currents, motors, and so forth that would illustrate the feedback relationships involved in labor and childbirth that a layperson could understand.

Answers to Figure Caption questions and Fill-in-the-Blank questions can be found in the Answers section at the back of the book.

MB *Go to MasteringBiology for practice quizzes, activities, eText, videos, current events, and more.*

CASE STUDY
Running with Heart

DOES JUST THINKING about running a marathon make you feel tired? Donald Arthur, a 66-year-old retired Harlem bookkeeper, has completed 46 marathons in 33 states and is training for his next. He won't accomplish his goal until he completes another 17 marathons, one in every state of the United States. Amazing? It gets better. In 1996 Arthur would become exhausted just chewing his food. Walking a block was a feat that could take an hour. His heart was failing from dilated cardiomyopathy.

Dilated cardiomyopathy (based on Greek, meaning "expanded cardiac muscle disease") causes the heart muscle to become stretched and weakened and renders it unable to adequately pump blood through the body, a condition called *heart failure*. This deprives the entire body of oxygen. But the situation is worsened because heart failure also causes fluid to accumulate in the lungs, making oxygen less available to the blood. Although cardiomyopathy strikes both sexes and people of all ages, its most common victims are middle-aged men. Some people inherit a tendency to develop this disorder, which is also associated with excessive alcohol consumption, the use of drugs such as cocaine and methamphetamine, and infections that attack heart valves or heart muscle. In serious cases such as Arthur's, lack of oxygen renders the victim short of breath even at rest, and makes any sort of exertion nearly impossible. Without dramatic intervention, the situation gradually worsens and is often fatal. Donald Arthur's doctors had given him about 3 months to live when a donated heart became available.

Just 15 months after his transplant, Donald Arthur—a former self-proclaimed "confirmed couch potato"—completed the first of his 10 New York marathons and began a new life. He helped to start Transplant Speakers International, an organization that spreads the word about organ and tissue donation.

How does heart failure differ from a heart attack? Why are hearts likely to fail? How do circulatory systems supply the body with oxygen and nutrients? How have circulatory systems changed over evolutionary time? Read on to find out.

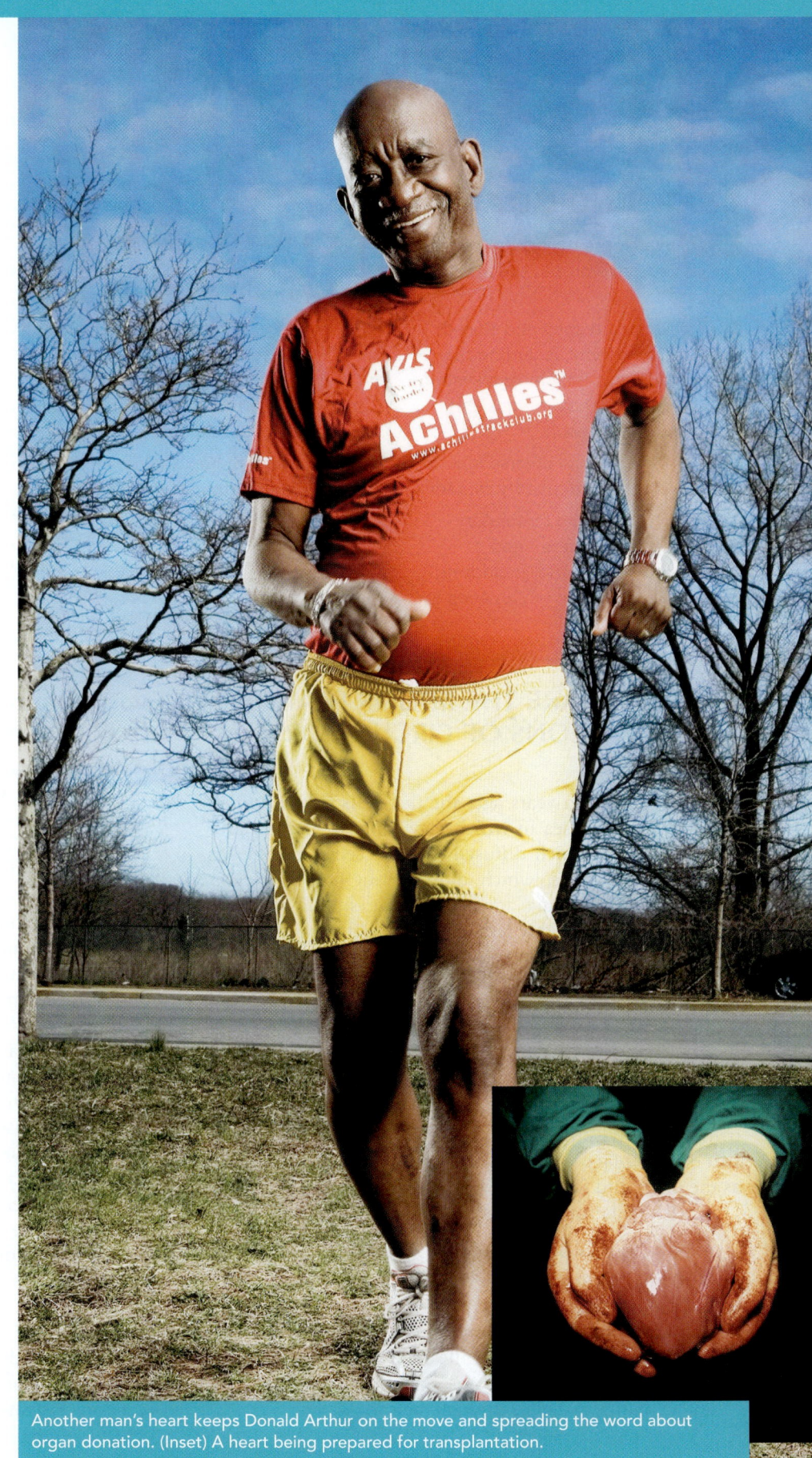

Another man's heart keeps Donald Arthur on the move and spreading the word about organ donation. (Inset) A heart being prepared for transplantation.

AT A GLANCE

LEARNING GOALS

LG1 Name the major features of all circulatory systems and the function of each feature.

LG2 Compare open and closed circulatory systems, including the animal groups in which each system is found, and the advantages and disadvantages of each.

LG3 Describe the diverse functions of the vertebrate circulatory system.

LG4 Describe the circulatory patterns of two-, three-, and four-chambered hearts, including the animal groups in which each is found.

LG5 Describe the structure of cardiac muscle and how the heart's pacemaker and its connections cause coordinated contraction of the atria and ventricles.

LG6 Describe the cardiac cycle, including the role of the heart's valves in maintaining proper blood flow.

LG7 Explain how the nervous system and hormones influence heart rate, the distribution of blood within the body, and red blood cell production.

LG8 Describe blood plasma and the three cell-based components of blood, explaining the function of each and including the events that occur during blood clotting.

LG9 Describe the structure and function of each of the five types of blood vessels.

LG10 List the organs and vessels of the lymphatic system and explain the functions of each, including where and how lymph is produced.

32.1 WHAT ARE THE MAJOR FEATURES AND FUNCTIONS OF CIRCULATORY SYSTEMS?

Almost four billion years ago, the first cells were nurtured by the primordial sea in which they evolved. The sea provided nutrients, which diffused into the cells, and the sea washed away wastes, which diffused out from the cells. Today, microorganisms, including some simple multicellular animals, still rely almost exclusively on diffusion to exchange nutrients and wastes with the environment. Sponges, for example, circulate seawater through pores in their bodies, bringing the environment close to each cell (see Figs. 23-6 and 33-1a). In more complex animals, individual cells are farther from the outside world, but they still require short diffusion distances, so that adequate nutrients reach the cells and the animals aren't poisoned by their own wastes. With the evolution of the circulatory system, a sort of "internal sea" arose that performs a similar function as the ocean did for the first cells.

All circulatory systems have three major components:

- A pump, the **heart,** that keeps the blood circulating.
- A liquid, **blood,** that serves as a medium of transport for gases, nutrients, and cellular wastes.
- A system of tubes, **blood vessels,** consisting of *arteries* that carry blood away from the heart, *veins* that carry blood toward the heart, and *capillaries* that link arteries and veins and exchange materials through their walls.

Two Types of Circulatory Systems Are Found in Animals

The circulatory systems of animals take two different forms: open and closed. **Open circulatory systems** are present in many invertebrates, including all arthropods (such as crustaceans, spiders, and insects) and most mollusks (such as snails and clams). An animal with an open circulatory system has one or more simple hearts, some blood vessels, and a series of interconnected chambers within the body, collectively called a **hemocoel** (**Fig. 32-1a**). Within the hemocoel, which occupies 20% to 40% of the body volume, tissues and internal organs are directly bathed in **hemolymph,** a fluid that functions both as blood and as the interstitial fluid that surrounds each cell within multicellular organisms. In insects, a single large dorsal blood vessel is modified into a series of contracting chambers in the abdomen. These chambers form the insect heart (see Fig. 32-1a). Each chamber contains a slit-like opening called an *ostium* (plural, ostia) that functions as a one-way valve. The ostia are forced shut as the heart chambers contract, propelling the hemolymph forward through the dorsal vessel. The hemolymph is pumped into hemocoel compartments of the head and then flows back through the body into the abdominal portions of the hemocoel. As the heart chambers relax, they expand, drawing hemolymph back in through the ostia.

Animals with open circulatory systems expend less energy on circulation than do animals with closed systems. But open circulatory systems maintain relatively low blood pressures dorsal vessel | heart portion of dorsal vessel | ostia (valve-like)

hemocoel

(a) Open circulatory system

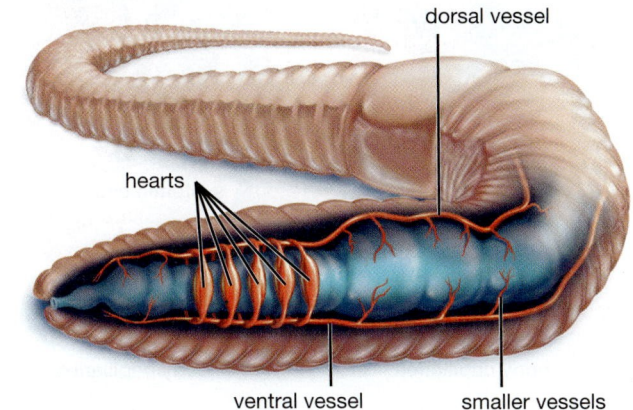

dorsal vessel

hearts

ventral vessel   smaller vessels

(b) Closed circulatory system

▲ **FIGURE 32-1 Open and closed circulatory systems**
(a) In the open circulatory systems of most invertebrates, one or more hearts pump hemolymph through vessels into the hemocoel, where blood directly bathes the internal organs. The hemolymph of insects, such as this grasshopper, is transparent and sometimes pale green. **(b)** In a closed circulatory system, blood remains confined within the heart(s) and the blood vessels. In the earthworm, five contractile vessels serve as hearts that pump blood through major ventral and dorsal vessels, from which smaller interconnecting vessels branch. Earthworm blood contains red hemoglobin.

QUESTION Why doesn't insect hemolymph need oxygen-carrying molecules, such as hemoglobin?

and are less efficient at supplying oxygen and nutrients to tissues than are closed systems. As a result, open circulatory systems are generally found in relatively sedentary organisms. How can an active insect obtain enough oxygen with an open circulatory system? An insect's blood does not carry oxygen to its tissues. Insects have outsourced this gas-exchange function to a system of gas-filled tubes (called tracheae; see **Fig. 33-4**) that have openings to the air and branch extensively throughout their tissues, carrying oxygen close to each cell.

In contrast to an open circulatory system, a **closed circulatory system** confines the blood to the heart and to vessels that carry blood throughout the body. Blood pressure and flow rates are higher than is possible in an open system. A closed circulatory system is better able to direct blood to specific tissues as needed. Thus, closed circulatory systems are extremely efficient at delivering oxygen and nutrients to the tissues of active animals with high metabolic rates. Closed circulatory systems are present in all vertebrates (such

as fishes, reptiles, and mammals) and in a few invertebrates, including very active mollusks (squid and octopuses) and, perhaps surprisingly, earthworms, where five contractile vessels serve as hearts (**Fig. 32-1b**). Earthworms benefit from this efficient system because they must burrow through dense soil where little oxygen is available. The closed circulatory systems of vertebrates are described in the sections that follow.

The Vertebrate Circulatory System Has Diverse Functions

The circulatory system supports all the other organ systems in the body. In vertebrates, the circulatory system performs the following functions:

- Transports oxygen from the gills or lungs to the tissues, and transports carbon dioxide from the tissues to the gills or lungs.
- Distributes nutrients from the digestive system to all body cells.
- Transports toxic substances to the liver for detoxification, and transports cellular wastes to the kidneys where they are filtered from the blood and excreted.
- Distributes hormones from the glands and organs that produce them to the tissues upon which they act.
- Helps regulate body temperature by adjusting blood flow.
- Helps stop bleeding and heal wounds by producing blood clots.
- Protects the body from diseases by circulating white blood cells and antibodies.

In the following sections we examine the three parts of the circulatory system: the heart, the blood, and the blood vessels, with emphasis on the human system. Finally, we describe the lymphatic system, which works closely with the circulatory system.

CHECK YOUR LEARNING

Can you explain the major features of all circulatory systems? Are you able to compare open and closed circulatory systems? Can you list the functions of the vertebrate circulatory system?

32.2 HOW DOES THE VERTEBRATE HEART WORK?

The vertebrate heart consists of muscular chambers capable of strong contractions. Chambers called **atria** (singular, atrium) collect blood. Contractions of the atria send blood into the **ventricles,** chambers whose contractions circulate blood through the lungs and to the rest of the body.

The Two-Chambered Heart of Fishes Was the First Vertebrate Heart to Evolve

Fish hearts consist of two main contractile chambers: a single atrium that empties into a single ventricle (**Fig. 32-2a**). Blood pumped from the ventricle passes first through the

▶ FIGURE 32-2 **Evolution of the vertebrate heart** (a) The earliest vertebrate heart to evolve was the two-chambered heart of fishes. (b) Amphibians and most reptiles have three-chambered hearts in which two atria empty into a single ventricle. (c) The hearts of crocodilians, birds, and mammals combine two separate pumps that maintain very different pressures and prevent mixing of oxygenated and oxygen-poor blood. (Because these hearts face the reader, left and right appear reversed. In art throughout this text, oxygenated blood is red and oxygen-poor blood is blue.)

■ oxygen-poor blood
■ oxygenated blood

gill capillaries   lung capillaries   lung capillaries

ventricle

atrium

pulmonary circuit   pulmonary circuit

atria   ventricle

systemic circuit   atria   ventricles   systemic circuit

body capillaries   body capillaries   body capillaries

(a) Two-chambered heart (fishes)   **(b) Three-chambered heart (amphibians and some reptiles)**   **(c) Four-chambered heart (crocodiles, birds, and mammals)**

gills, where the blood picks up oxygen and releases carbon dioxide. The blood travels directly from the gills through the rest of the body, delivering oxygen to the tissues and picking up carbon dioxide. Blood from the body then returns to the single atrium. The bodies of fishes are supported by water, so their hearts do not need to pump blood against gravity. Their blood pressure is low compared to most terrestrial vertebrates and drops considerably as the blood travels through the microscopic gill capillaries and enters the veins. Undulations of the tail and body during swimming help to drive blood back toward the heart.

Increasingly Complex and Efficient Hearts Evolved in Terrestrial Vertebrates

Over the course of evolution, vertebrates emerged from the sea. As fishes gave rise to amphibians (such as salamanders and frogs), a three-chambered heart evolved, with two atria and one ventricle (**Fig. 32-2b**). Reptiles evolved from amphibians, and most reptiles (including snakes, turtles, and lizards) also have three-chambered hearts.

An important circulatory adaptation in these terrestrial vertebrates is **double circulation,** which creates two separate circuits of blood. The **pulmonary circuit** ("pulmonary" refers to lungs) directs blood from the heart through the lungs, where carbon dioxide is exchanged for oxygen, and back to the heart. The **systemic circuit** carries blood between the heart and the rest of the body, where oxygen is exchanged for carbon dioxide. In the three-chambered heart,

blood from the systemic circulation enters the right atrium, blood from the pulmonary circuit enters the left atrium, and both atria empty into the single ventricle. Amphibian and reptile hearts have internal features that direct most of the oxygen-poor blood into the right portion of the ventricle where it is pumped to the lungs, and direct most of the oxygenated blood into the left portion of the ventricle, which pumps it to the rest of the body.

Four-Chambered Hearts Consist of Two Separate Pumps

A few groups of reptiles, such are birds (now classified as reptiles) and crocodiles, as well as all mammals, have four-chambered hearts (**Fig. 32-2c**). The four-chambered heart—with its right atrium and right ventricle completely isolated from its left atrium and left ventricle—acts like two hearts beating as one (**Fig. 32-3**). The "right heart" deals with oxygen-poor blood. The right atrium receives oxygen-depleted blood from the body through the two largest **veins** (vessels that carry blood toward the heart), the superior vena cava and the inferior vena cava. After filling with blood, the right atrium contracts, forcing the blood into the right ventricle. Contraction of the right ventricle then sends the oxygen-poor blood to the lungs through the pulmonary **arteries** (vessels that carry blood away from the heart). The "left heart" deals with oxygenated blood. Oxygen-rich blood from the lungs enters the left atrium through the pulmonary veins and is then squeezed into the left ventricle.

aorta

superior
vena cava

pulmonary artery
(to left lung)

left atrium

pulmonary artery
(to right lung)

pulmonary veins
(from left lung)

pulmonary veins
(from right lung)

atrioventricular valve

right
atrium

semilunar valves

atrioventricular valve

left ventricle

thicker muscle
of left ventricle

inferior
vena cava

right
ventricle

descending aorta
(to lower body)

▲ **FIGURE 32-3 The human heart and its valves and vessels** The muscular wall of the left ventricle is thicker than the right because it must pump blood throughout the body. One-way semilunar valves separate the aorta from the left ventricle, and the pulmonary artery from the right ventricle. Atrioventricular valves separate each atrium from its corresponding ventricle.

A strong contraction of the left ventricle, the heart's most muscular chamber, sends the oxygenated blood coursing out through the largest artery, the aorta, and then to the rest of the body.

Valves Maintain the Direction of Blood Flow

The directionality of blood flow is maintained by one-way valves (see Figs. 32-3 and 32-5). Pressure in one direction opens them easily, but reverse pressure forces them closed. **Atrioventricular valves** allow blood to flow from the atria into the ventricles, but prevent the blood from flowing back into the atria when the ventricles contract. **Semilunar valves** (from Latin, meaning "half-moon" valves, referring to their shape) allow blood to enter the pulmonary artery and the aorta when the ventricles contract, but prevent blood from returning as the ventricles relax.

Cardiac Muscle Is Present Only in the Heart

Most of the heart consists of a specialized type of muscle, **cardiac muscle,** found nowhere else in the body. Cardiac muscle cells are small, branched, and packed with an orderly array of protein strands that give them a striped appearance (**Fig. 32-4**). Cardiac muscle cells are linked to one another by **intercalated discs,** which appear as bands between the

cells. Here, adjacent cell membranes are attached to one another by junctions called desmosomes (see Fig. 5-16a), which prevent the strong heart contractions from pulling the muscle cells apart. Intercalated discs also contain gap junctions, which allow the electrical signals that trigger contractions to spread directly and rapidly from one muscle cell to adjacent ones. This causes the interconnected regions of cardiac muscle to contract almost synchronously.

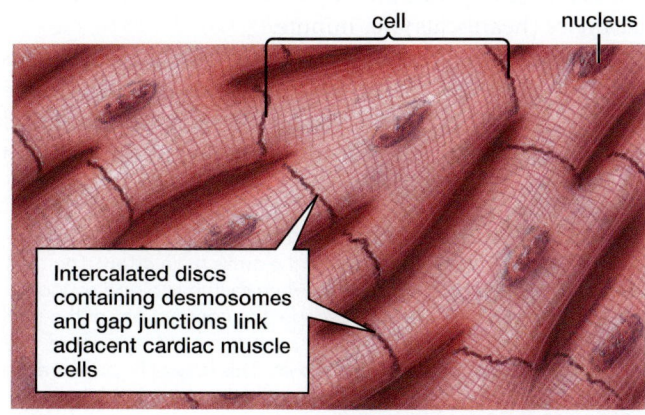

cell

nucleus

Intercalated discs containing desmosomes and gap junctions link adjacent cardiac muscle cells

▲ FIGURE 32-4 **The structure of cardiac muscle**

QUESTION If a muscle is exercised regularly, it increases in size. Why is the resting heart rate of a well-conditioned athlete slower than the heart rate of a less active person?

Oxygenated blood is pumped to the body through the systemic circuit

Oxygen-poor blood is pumped to the lungs through the pulmonary circuit

Oxygen-poor blood from the body enters the right ventricle

Oxygenated blood from the lungs enters the left ventricle

Blood fills both atria and begins to flow passively into the ventricles

❶ **Atrial systole** (0.1 second): Both atria contract, forcing blood into the ventricles

❷ **Ventricular systole** (0.3 second): Both ventricles contract, forcing blood through the pulmonary and systemic circuits. Systolic pressure is measured here

❸ **Diastole** (0.4 second): The heart relaxes, ending the cycle. Diastolic pressure is measured here

▲ FIGURE 32-5 **The cardiac cycle**

The Coordinated Contractions of Atria and Ventricles Produce the Cardiac Cycle

The human heart beats about 100,000 times each day. Each heartbeat is actually a series of coordinated events, called the **cardiac cycle** (Fig. 32-5). During each cycle, the two atria first contract in synchrony, emptying their contents into the ventricles, a process called *atrial systole*. A fraction of a second later, during *ventricular systole*, the two ventricles contract simultaneously, forcing blood into arteries that exit the heart. Then, during *diastole*, both atria and both ventricles relax briefly before the cardiac cycle repeats. In a typical resting person, the complete cycle occurs in just under 1 second, or about 70 times per minute. **Heart rate** is the number of cardiac cycles (heartbeats) per minute.

Your blood pressure consists of two measurements, described in Figure 32-6. **Systolic pressure** (the higher of the two) is generated in the arteries by the muscular left ventricle as it pumps blood through the systemic circuit.

▶ FIGURE 32-6 **Measuring blood pressure** First, the cuff is inflated until its pressure closes off the arm's main artery. The pressure is then gradually reduced. When rhythmic blood sounds are first heard through the stethoscope, the contractions of the left ventricle have just overcome the pressure in the cuff, and some blood gets through with each heartbeat. This is systolic pressure. Cuff pressure is then further reduced until no pulse sounds are audible, indicating that blood is flowing continuously through the artery, and the pressure *between* ventricular contractions is just overcoming the cuff pressure. This is diastolic pressure. The numbers are in millimeters of mercury, a standard measure of pressure, also used in barometers.

Diastolic pressure (the lower of the two) is the pressure in the arteries as the heart rests between contractions.

A blood pressure reading of less than 120/80 mmHg (millimeters of mercury) is considered healthy, and a pressure of 140/90 mmHg or higher is defined as high blood pressure. High blood pressure, or **hypertension,** is caused

pressure gauge

cuff

A stethoscope detects pulse sounds

The cuff is inflated, putting pressure on the artery

CASE STUDY continued

Running with Heart

Heart failure is a frequent and potentially fatal result of dilated cardiomyopathy, the condition from which Donald Arthur suffered. Often, the stretched and weakened left ventricle can't pump out all the oxygenated blood that it receives from the lungs. As a result, blood backs up into the pulmonary veins, increasing the blood pressure in the lungs. The blood pressure in the pulmonary circuit is normally only about one-eighth of the pressure in the systemic circuit. Increased pulmonary blood pressure, as in cardiomyopathy, forces fluid from the blood out through the thin walls of the lung capillaries and into the air spaces of the lungs. As happens when a person drowns, fluid in the lungs prevents adequate oxygen from reaching the blood as it circulates through the lungs.

by the constriction of small arteries, which causes resistance to blood flow and strain on the heart. Some people have a genetic tendency toward hypertension, but it is also associated with smoking, obesity, lack of exercise, high alcohol consumption, stress, and aging. High blood pressure contributes to "hardening of the arteries," described in "Health Watch: Repairing Broken Hearts" on pp. 628–629.

Electrical Impulses Coordinate the Sequence of Heart Chamber Contractions

The contraction of the heart is initiated and coordinated by a *pacemaker*, a cluster of specialized heart muscle cells that produces spontaneous electrical signals at a regular rate. The heart's pacemaker is the **sinoatrial (SA) node,** located in the upper wall of the right atrium (**Fig. 32-7**). Electrical signals from the SA node pass freely and rapidly through gap junctions into the connecting cardiac muscle cells and then throughout the atria.

During the cardiac cycle, the atria contract first and empty their contents into the ventricles, then refill while the ventricles contract. This requires a slight delay between the atrial and the ventricular contractions. How is this accomplished? First, the SA node initiates a wave of contraction (**Fig. 32-7 ❶**) that sweeps through the right and left atria, which contract in synchrony (**Fig. 32-7 ❷**). The signal then reaches a barrier of tissue between the atria and the ventricles that cannot conduct electrical signals. Here, the excitation is channeled through the **atrioventricular (AV) node,** a small mass of specialized muscle cells located in the floor of the right atrium (**Fig. 32-7 ❸**). The impulse is conducted slowly at the AV node, briefly delaying conduction into fibers that stimulate ventricular contraction. This delay gives the atria

❶ An electrical signal from the sinoatrial (SA) node starts atrial contraction

❷ The electrical signal spreads through the atria, causing them to contract

❸ The signal enters the atrioventricular (AV) node, which transmits it to the AV bundle with a slight delay

❹ The signal travels through the AV bundle branches to the base of the ventricles

❺ Purkinje fibers transmit the signal to ventricular cardiac muscle cells, causing contraction from the base upward

SA node

AV node

AV bundle

Inexcitable tissue separates the atria and ventricles

AV bundle branches

Purkinje fibers

▲ FIGURE 32-7 **The heart's pacemaker and its connections**

time to complete the transfer of blood into the ventricles before ventricular contraction begins.

From the AV node, the signal to contract spreads along specialized tracts of rapidly conducting muscle fibers. These tracts start with the thick cluster of fibers called the *atrioventricular bundle* (*AV bundle*), which sends branches to the lower portion of both ventricles (**Fig. 32-7 ❹**). Here, the AV bundle branches separate and give rise to *Purkinje fibers*, which transmit the electrical signal to the surrounding cardiac muscle cells, sending a wave of contraction from the base of the ventricles upward within the ventricular walls (**Fig. 32-7 ❺**). This ventricular contraction forces blood up into the pulmonary artery and the aorta.

A variety of disorders can interfere with the complex series of events that produce the normal cardiac cycle. When the pacemaker fails—or if other areas of the heart become more excitable and usurp the pacemaker's role—rapid, uncoordinated, weak contractions called **fibrillation** may occur. Fibrillation of the ventricles is soon fatal, because blood is not pumped by the quivering muscle. A defibrillator applies a jolt of electricity to the heart, synchronizing the contraction of the cardiac muscle, and (if successful) allowing the pacemaker to resume its normal coordinating function.

The Nervous System and Hormones Influence Heart Rate

Your heart rate is finely tuned to your body's activity level, whether you are running to class or basking in the sun. On its own, the SA node pacemaker would maintain a steady rhythm of about 100 beats per minute. However, nerve impulses and hormones significantly alter the heart rate. In a resting individual, the parasympathetic nervous system, which regulates body systems during periods of rest, slows the heart rate to roughly 70 beats per minute. When exercise or stress creates a demand for greater blood flow to the muscles, the sympathetic nervous system, which prepares the body for emergency action, accelerates the heart rate and increases the force of cardiac muscle contractions. The adrenal glands simultaneously

release the hormone epinephrine (adrenaline) into the blood, which reinforces these effects when it reaches the heart.

CHECK YOUR LEARNING

Can you describe the three types of vertebrate hearts and the structure of cardiac muscle? Are you able to trace the flow of blood through the four-chambered heart, naming the structures through which the blood passes and explaining the function of each? Can you explain the cardiac cycle and how electrical impulses are conducted through the human heart?

32.3 WHAT IS BLOOD?

Blood, sometimes called the "river of life," has two major components: (1) a liquid called **plasma,** which constitutes 55% to 60% of the blood volume, and (2) a cell-based component consisting of *red blood cells*, *white blood cells*, and *platelets* suspended in the plasma (**Fig. 32-8**). The average person has roughly 5.3 quarts (about 5 liters) of blood, so if

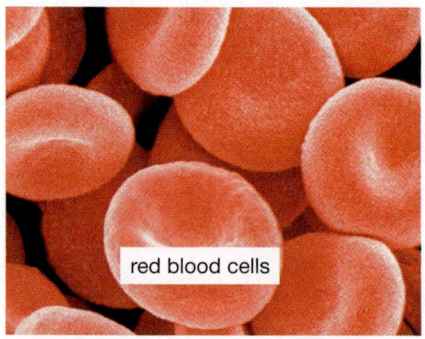
(a) Erythrocytes (red blood cells)

(b) Leukocyte (white blood cell)

(c) Megakaryocyte forming platelets

▲ **FIGURE 32-8 Types of blood cells (a)** This SEM shows the biconcave shape of erythrocytes. **(b)** A leukocyte called a neutrophil is shown surrounded by much smaller red blood cells. **(c)** Platelets are membrane-enclosed fragments of megakaryocytes.

QUESTION Why does dietary iron deficiency cause anemia (too few red blood cells to provide adequate oxygen)?

TABLE 32-1 Blood Components and Their Functions

Plasma Components (about 55% of blood volume)	Functions
Water	Dissolves other components; gives blood its fluidity
Major Proteins	
Albumin	Maintains blood osmotic strength; binds and transports some hormones and fatty acids
Globulins	Serve as antibodies that fight infection; bind and transport some hormones, ions, and other molecules
Fibrinogen	Gives rise to fibrin, which promotes clotting
Ions (sodium, potassium, calcium, magnesium, chloride, bicarbonate, hydrogen)	Maintain pH; allow neuronal activity; allow muscle contraction; facilitate enzyme activity
Nutrients (simple sugars, amino acids, lipids, vitamins, oxygen)	Provide materials for cellular metabolism
Wastes (urea, carbon dioxide, ammonia)	By-products of cellular metabolism that are transported in blood to sites of elimination
Hormones	Signaling molecules that are transported in blood to their target cells

Cell-Based Components (about 45% of blood volume)	Functions
Erythrocytes (5,000,000 per mm^3)	Transport oxygen
Leukocytes (5,000–10,000 per mm^3)	All fight infection and disease
Neutrophils	Engulf and destroy bacteria
Eosinophils	Kill parasites
Basophils	Produce inflammation
Lymphocytes	Mount an immune response
Monocytes	Mature into macrophages, which engulf debris, foreign cells, and foreign molecules
Platelets (250,000 per mm^3)	Essential for blood clotting

you donate a pint of blood, you are giving only about 10% of your blood. The components of blood are summarized in Table 32-1.

Plasma Is Primarily Water in Which Proteins, Salts, Nutrients, and Wastes Are Dissolved

Although plasma is about 90% water, this clear, pale-yellow fluid has more than 100 different types of molecules dissolved in it. The plasma transports hormones, nutrients, and cellular wastes. It also contains a variety of ions; some of these maintain blood pH, while others are crucial for the functioning of nerve and muscle cells.

Proteins make up the largest component of dissolved molecules by weight. The three most common plasma proteins are *albumin*, *globulins*, and *fibrinogen*. Albumin helps to maintain the blood's osmotic strength, thus preventing too much fluid from diffusing out of the plasma through capillary walls. Some globulins are antibodies that play an important role in the immune response (see Chapter 36).

Fibrinogen is important in blood clotting, described later in this chapter.

The Cell-Based Components of Blood Are Formed in Bone Marrow

Of the three cell-based components in blood—red blood cells, white blood cells, and platelets—only the white blood cells are complete, functional cells. The mature red blood cells of mammals are technically not cells because they lose their nuclei during development. Platelets are actually small fragments of cells. All three cell-based components originate from cells that reside in **bone marrow**, a tissue within the cavities of large bones where blood cells form (see Fig. 40-10).

Red Blood Cells Carry Oxygen from the Lungs to the Tissues

About 99% of all blood cells are the oxygen-carrying red blood cells, also called **erythrocytes.** A red blood cell is shaped like a ball of clay squeezed between a thumb and forefinger

▲ FIGURE 32-9 **Hemoglobin** Each of four peptide chains (two each of two types) surrounds an iron-containing heme molecule (red disk) which binds oxygen.

▲ FIGURE 32-10 **Red blood cell regulation by negative feedback**

QUESTION Some endurance athletes cheat by injecting large doses of erythropoietin. How does this provide a competitive advantage?

(Fig. 32-8a). The red color of erythrocytes is produced by the large, iron-containing protein **hemoglobin** (Fig. 32-9), which transports almost all of the oxygen carried in the blood. Each hemoglobin molecule can bind and carry four molecules of oxygen, one on each heme group (the red discs in Fig. 32-9). Hemoglobin takes on a bright cherry-red color when bound to oxygen, and becomes a deeper maroon-red color after it releases oxygen, appearing bluish in veins seen through the skin. Hemoglobin binds loosely to oxygen, picking it up in the capillaries of the lungs, where the oxygen concentration is high, and releasing it in other tissues of the body where the oxygen concentration is lower (see Chapter 33).

Erythrocytes have an average life span of about 4 months. Every second, more than 2 million red blood cells (100 billion daily) die and are replaced by new ones formed in the bone marrow. Dead erythrocytes are broken down in the spleen and liver. Iron from hemoglobin is returned to the bone marrow where it is reused to synthesize hemoglobin that will be packaged in newly forming red blood cells. Although this recycling process is efficient, some iron is lost during bleeding from injury or from menstruation, and small amounts are excreted daily in feces, so some iron must be provided in the diet.

Negative Feedback Regulates Red Blood Cell Numbers

The number of erythrocytes determines how much oxygen the blood can carry. The red blood cell number is maintained by a negative feedback system that involves the hormone **erythropoietin.** Erythropoietin, produced by the kidneys and released into the blood in response to low oxygen levels, stimulates the bone marrow to increase production of red blood cells (Fig. 32-10). Low oxygen may be caused by a loss of blood, insufficient production of hemoglobin, high altitude (where less oxygen is available), or conditions that interfere with gas exchange in the lungs, such as lung disease and heart failure. When a healthy oxygen level is restored, erythropoietin production declines, and the rate of red blood cell production returns to normal. People with advanced kidney disease often suffer from *anemia*—an inadequate number of erythrocytes—because their failing kidneys do not produce adequate erythropoietin.

CASE STUDY c o n t i n u e d

Running with Heart

Victims of heart failure often have elevated levels of erythropoietin, produced as their bodies attempt to compensate for the inadequate oxygen reaching their tissues. The amount by which this hormone increases is correlated with the severity of their heart failure. Despite elevated erythropoietin, many victims of heart failure suffer from anemia, although no one knows why. Anemia forces the weakened left ventricle to circulate blood with less oxygen to fuel its efforts. In some cases, giving additional erythropoietin to heart failure patients with anemia may improve their condition.

White Blood Cells Defend the Body Against Disease

White blood cells, called **leukocytes,** are larger than red blood cells, but they can crawl, change shape, and ooze through far narrower spaces than erythrocytes can, including through capillary walls. There are five types of leukocytes: *neutrophils, eosinophils, basophils, lymphocytes,* and *monocytes* (Fig. 32-8b; see also Table 32-1). Their life spans range from hours to years, and altogether, they make up less than 1% of the cellular portion of the blood. Leukocytes protect the body against disease (see Chapter 36). Monocytes, for example, enter tissues and transform into macrophages (literally, "big eaters"). Macrophages engulf bacteria (Fig. 32-11) and cellular debris; in the spleen and liver they also break down dead red blood cells.

Platelets Are Cell Fragments That Aid in Blood Clotting

Platelets are pieces of large cells called *megakaryocytes.* Megakaryocytes remain in the bone marrow, where they pinch off membrane-enclosed chunks of cytoplasm that

▲ **FIGURE 32-11 A white blood cell attacks bacteria** The large pale blue cell (a macrophage) has formed cytoplasmic extensions that are engulfing a cluster of rod-shaped tuberculosis bacteria (pink).

become platelets (**Fig. 32-8c**). The platelets, which survive for about 10 days, enter the blood and play a central role in blood clotting.

Blood Clotting Plugs Damaged Blood Vessels

Blood clotting is a complex process that protects animals from losing excessive amounts of blood, not only from trauma, but also from the minor wear and tear that occurs with normal activities. Clotting begins when blood comes in contact with injured tissue, for example, a break in a blood vessel wall (**Fig. 32-12**). Platelets adhere to collagen proteins

exposed on the ruptured wall and form a *platelet plug* that partially blocks the opening. Both the adhering platelets and the ruptured blood vessel cells release a variety of substances that initiate complex reactions among circulating plasma proteins. These reactions cause the plasma protein *prothrombin* to be converted into the active enzyme **thrombin.** Thrombin cleaves the plasma protein *fibrinogen*, causing it to form insoluble protein strands called **fibrin.** Fibrin strands adhere to one another, forming a fibrous network around the aggregated platelets.

This protein web traps more platelets and blood cells, primarily erythrocytes (**Fig. 32-13**), increasing the density of the clot. Platelets adhering to the fibrous mass send out sticky projections that grip one another. The cross-linked platelets eventually contract, squeezing the fibrin web and forcing fluid out. This creates a denser, stronger clot (on the skin it is called a scab). The contraction also pulls the damaged surfaces of the wound closer together, which promotes healing.

Hemophilia is a genetic disorder whose victims lack a plasma protein called clotting factor. Clotting factor causes platelets to adhere to one another and plug damaged vessels. Because they lack this factor, people with hemophilia may experience internal bleeding from everyday bumps and normal joint movements, as well as excessive bleeding from wounds.

CHECK YOUR LEARNING

Can you describe each component of blood and explain its function? Do you understand how the number of red blood cells in the body is regulated? Are you able to explain the sequence of events during blood clotting?

❶ Damaged cells expose collagen, which activates platelets, causing them to stick and form a plug

❷ Both damaged cells and activated platelets release chemicals that convert prothrombin into the enzyme thrombin

❸ Thrombin catalyzes the conversion of fibrinogen into protein fibers called fibrin, which forms a meshwork around the platelets and traps red blood cells

collagen fibers

platelets

red blood cells

blood vessel

platelet plug

prothrombin   thrombin

fibrin

thrombin

fibrinogen

▲ **FIGURE 32-12 Blood clotting** Injured tissue and adhering platelets cause a series of biochemical reactions among blood proteins that lead to clot formation. A simplified sequence is shown here.

▲ FIGURE 32-13 **A blood clot** Threadlike fibrin protein strands produce a tangled, sticky mass that traps blood cells and eventually forms a clot.

32.4 WHAT ARE THE TYPES AND FUNCTIONS OF BLOOD VESSELS?

Blood circulates throughout the body within a network of blood vessels; some of the major blood vessels of the human circulatory system are illustrated in **Figure 32-14**. Blood leaving the heart travels from arteries to arterioles to capillaries, then into venules, and finally, to veins, which return it to the heart (**Fig. 32-15**). With the exception of capillaries, blood vessels all have a fundamentally similar structure, with three cellular layers. Capillary walls are made of a single layer of *endothelial cells* (a type of epithelial cell; see Chapter 31). The larger vessels are lined with endothelial cells, but have two additional cell layers: a middle layer of smooth muscle cells and an outer layer of connective tissue (see Fig. 32-15). What happens when blood vessels rupture, are narrowed by deposits of cholesterol, or are blocked by clots, damming the river of life? We explore these questions in "Health Watch: Repairing Broken Hearts" on pp. 628–629.

Arteries and Arterioles Carry Blood Away from the Heart

Arteries carry blood away from the heart. The walls of these vessels are thicker and far more elastic than are those of veins (see Fig. 32-15). With each surge of blood from the ventricles, the arteries expand slightly, like thick-walled balloons. As their elastic walls recoil between heartbeats, the arteries actually help pump the blood and keep it flowing steadily into the smaller vessels. Arteries branch into smaller

▶ FIGURE 32-14 **The human circulatory system** Arteries (emphasized on the person's right side) carry blood from the heart, and veins (emphasized on the person's left side) carry blood toward the heart. All arteries except the pulmonary artery carry oxygenated blood, and all veins except the pulmonary veins carry oxygen-poor blood. The microscopic lung capillaries are shown greatly enlarged.

jugular vein
aorta
superior vena cava
coronary artery
liver
inferior vena cava
intestine
femoral artery

carotid artery
pulmonary artery
lung capillaries
heart
kidney
femoral vein

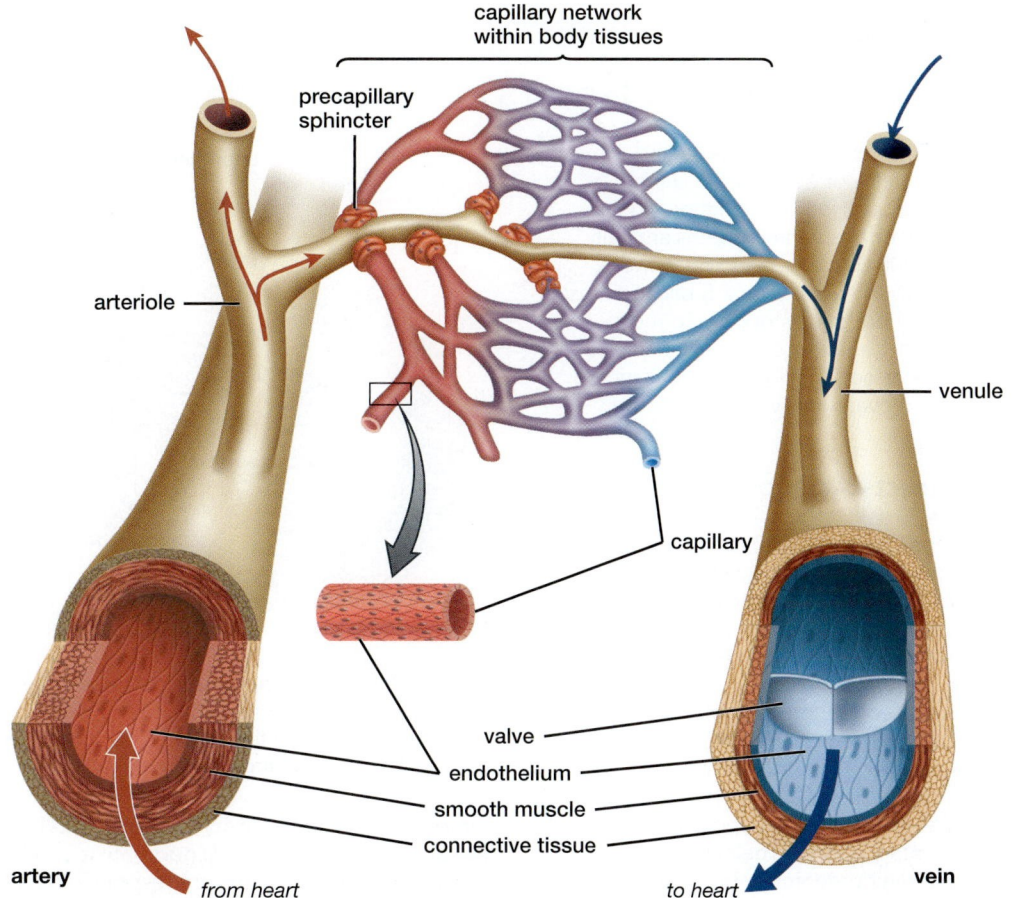

▲ FIGURE 32-15 Structures and interconnections of blood vessels Arteries and arterioles are more muscular than veins and venules. Oxygenated blood moves from arteries to arterioles to capillaries. Capillary walls are only one endothelial cell thick, allowing them to exchange gases and nutrients with their surroundings.

QUESTION Could the vessels as they are color-coded here be part of the circulation to the lungs? Why or why not?

diameter vessels called **arterioles,** which play a major role in determining how blood is distributed within the body, as described later.

Capillaries Allow Exchange of Nutrients and Wastes

Arterioles conduct blood into elaborate networks of tiny **capillaries** (see Fig. 32-15, middle), which are microscopically narrow vessels. An important function of the circulatory system is to allow individual body cells to exchange nutrients and wastes with the blood by diffusion. Capillaries, with their thin walls, are the only vessels in which this exchange can occur; larger vessels are adapted to conduct blood rather than exchange substances, and their walls are relatively impermeable. Capillaries are so narrow (about 10 micrometers) that red blood cells must pass through them single file (**Fig. 32-16**). Capillaries are also so numerous that most body cells are no more than 100 micrometers (slightly less than the thickness of four pages of this book) from a capillary. This is close enough for diffusion to work effectively. To supply all of the body's trillions of cells requires roughly 50,000 miles (about

Red blood cells must pass through capillaries in single file

Capillary walls are thin and permeable to gases, nutrients, and cellular wastes

▲ FIGURE 32-16 Red blood cells travel single file through a capillary

QUESTION Why does oxygen move out of capillaries into body cells, while carbon dioxide moves into capillaries from body cells?

Health | Watch

Repairing Broken Hearts

Cardiovascular disease (CVD; disorders of the heart and blood vessels) is the leading cause of death in the United States. According to the American Heart Association, CVD is responsible for about one out of every three deaths annually in the United States—about 800,000 lives—and no wonder. The heart must contract vigorously more than 2.5 billion times during an average lifetime without once stopping to rest, forcing blood through a lengthy network of vessels. Because the heart may weaken or the vessels may become constricted, blocked, or ruptured, the cardiovascular system is a prime candidate for malfunction.

Atherosclerosis Obstructs Arteries

Atherosclerosis (from the Greek "athero," meaning "gruel" or "paste"; and "scleros," meaning "hard") is caused by deposits, called **plaques,** within the artery walls. These deposits cause the walls of the arteries to thicken and lose their elasticity. They tend to form in people with high levels of "bad" cholesterol and low levels of "good" cholesterol. What is the difference between the two?

Cholesterol is transported through the bloodstream in two types of packets called *low-density lipoprotein* (LDL; often called "bad" cholesterol) and *high-density lipoprotein* (HDL; often called "good" cholesterol). Both LDL and HDL consist of cholesterol surrounded by a shell of proteins and phospholipids that make it soluble in the watery plasma. Their cholesterol molecules are the same, but their predominant proteins differ, and LDL has less total protein in its shells than does HDL. LDL carries cholesterol from the liver to the body cells, including the cells of the artery walls, where it may contribute to plaques. HDL transports cholesterol to the liver, thus removing it from the bloodstream and reducing overall blood cholesterol levels.

Plaque formation is often initiated by minor damage to the endothelium lining an artery. The damaged endothelium attracts white blood cells, which burrow beneath it and ingest large quantities of LDL cholesterol and other lipids (**Fig. E32-1**). The bloated bodies of these macrophages contribute to a growing fatty core of plaque. Meanwhile, smooth muscle cells from below the endothelium migrate into the core, absorb more fat and cholesterol, and add to the plaque. They also produce proteins that form a fibrous cap that covers the fatty core.

As the plaque grows, its fibrous cap may rupture, exposing clot-promoting factors within the plaque. A blood clot then forms, which further obstructs the artery. The clot may

▲ **FIGURE E32-1 Plaque clogs an artery** Compare this artery to that shown in Figure 32-15. The yellow deposits inside this artery form a plaque. If the fibrous cap ruptures (as illustrated here), a blood clot forms, obstructing the artery.

completely block the artery (see Fig. E32-1) or it may break free and be carried in the blood until it blocks a narrower part of the artery. Arterial clots are responsible for the most serious consequences of atherosclerosis: heart attacks and strokes.

A **heart attack** occurs if a *coronary artery*, which supplies the heart muscle, is blocked. This deprives the hard-working cardiac muscle of blood and the oxygen it carries, leading to the death of some heart muscle cells. If a sufficiently large area of cardiac muscle dies, the heart stops. Although heart attacks are the major cause of death from atherosclerosis, this disease also causes plaques and clots to form in arteries other than the coronary artery. A **stroke,** sometimes called a "brain attack," is the death of brain cells from lack of oxygen when their blood supply is interrupted by a clot or the rupture of a vessel. Depending on the extent and location of brain damage, strokes can be fatal, and stroke survivors may suffer neurological problems, including partial paralysis; difficulty remembering, communicating, or learning; mood swings; or personality changes.

When a heart or brain attack strikes, rapid treatment can minimize the damage and significantly increase the victim's

80,500 kilometers) of capillaries—enough to encircle the globe twice. The rate of blood flow drops considerably as blood moves through this narrow, lengthy capillary network, allowing more time for diffusion to occur.

Interstitial fluid acts as an intermediary between body cells and capillary blood, supplying the cells with nutrients and accepting their wastes and other secretions. Substances take various routes across the thin capillary walls. Gases, water, lipid-soluble hormones, and fatty acids can diffuse directly through the endothelial cell membranes. Some small proteins

are ferried across the endothelial cell membranes in vesicles, but large proteins such as albumin remain behind in the blood. Small, water-soluble nutrients, such as salts, glucose, and amino acids, enter the interstitial fluid through narrow spaces between adjacent capillary cells. White blood cells can also ooze through these crevices to engulf foreign particles, but erythrocytes and platelets remain inside the capillaries.

Relatively high pressure within capillaries that branch directly from arterioles causes large quantities of fluid to leak through the capillary walls. Pressure within the capillaries then

The callouts in the figure read:

(a) A tiny drill grinds away the plaque

(b) A balloon is inflated, compressing the plaque

(c) A wire mesh stent is placed in the opened artery

▲ **FIGURE E32-2 Angioplasty unclogs arteries** A narrowed artery may be opened **(a)** by drilling out the plaque or **(b)** by inflating a tiny balloon inside. **(c)** Following angioplasty, a metal mesh stent is often inserted to maintain the opening.

chances of survival. Blood clots can be dissolved by injecting a special clot-busting protein; this treatment works best if administered within a few hours after the attack occurs.

Treatment of Atherosclerosis

Atherosclerosis is promoted by high blood pressure, cigarette smoking, obesity, diabetes, lack of exercise, genetic predisposition, and high LDL-cholesterol levels in blood. Treatment of atherosclerosis includes changes in diet and lifestyle; but if this fails, drugs may be prescribed to lower cholesterol. If a person has had a heart attack or suffers from **angina,** which is chest pain caused by insufficient blood flow to the heart, he or she may be a candidate for surgery to widen or bypass the obstructed artery.

Angioplasty refers to techniques that widen obstructed coronary arteries (**Fig. E32-2**). A physician threads a flexible tube through an artery in the upper leg or arm and guides it into the clogged artery. The tube may be equipped with a tiny drill bit, which shears off the plaque in microscopic pieces that are carried away in the blood (**Fig. E32-2a**), or it may have a small balloon at its tip, which is inflated to compress the plaque (**Fig. E32-2b**). After the procedure, physicians may insert a wire mesh tube, called a stent, into the artery to help keep it open (**Fig. E32-2c**). In more severe cases, coronary bypass surgery may be performed. This operation bypasses obstructed coronary arteries with segments of vein, usually taken from the patient's leg (**Fig. E32-3**) or with segments of artery from the patient's forearm.

The callouts in the figure read:

❸ The grafted vein bypasses the obstruction

❶ A coronary artery is blocked here

❷ A segment of vein is removed from the leg

▲ **FIGURE E32-3 Coronary bypass surgery**

drops as blood travels toward the venules (see Fig. 32-15). The high osmotic pressure of the blood inside the capillaries (due largely to albumin protein) draws water back into the vessels by osmosis as blood approaches the venous end of the capillaries. As water moves into the capillaries, diluting the blood, dissolved substances in the interstitial fluid diffuse back into the capillaries along their concentration gradients. Thus, about 85% of the fluid that leaks out of capillary networks branching from arterioles is restored to the bloodstream on the venous side of each capillary network. As you will learn later in

this chapter, the lymphatic system returns the remaining excess interstitial fluid to the blood (see Fig. 32-18).

Veins and Venules Carry Blood Back to the Heart

After picking up carbon dioxide and other wastes from cells, capillary blood drains into larger vessels called **venules,** which empty into still larger veins (see Fig. 32-15, right). Veins provide a low-resistance pathway that conducts blood back toward the heart. The walls of veins are thinner, less muscular, and more expandable than those of arteries. When veins are compressed, as in muscle contraction, one-way valves keep blood flowing toward the heart (**Fig. 32-17**).

Blood pressure in the veins is too low to move blood back to the heart from lower body parts such as the feet and legs. However, the pressure changes that occur during breathing, as well as the contractions of skeletal muscle during exercise, help return blood to the heart by squeezing the veins and forcing blood through them. Prolonged sitting or standing still can cause swollen ankles because, without muscle contractions to compress the veins, venous blood tends to pool in the lower legs. This increases blood pressure in the leg capillaries, which then absorb less interstitial fluid. Regular extended periods of sitting or standing can contribute to the formation of varicose veins in the lower legs. In this condition, veins just below the skin become permanently swollen with blood because their valves have been stretched and weakened by pooling blood.

If blood pressure should fall—for instance, after extensive bleeding—veins can help restore it. In such cases, the sympathetic nervous system (which prepares the body for emergency action) stimulates contraction of the smooth muscles in both the vein and artery walls. This decreases the internal volume of the vessels and raises blood pressure.

Arterioles Control the Distribution of Blood Flow

Arterioles carry blood to capillaries. Their muscular walls are influenced by nerves, hormones, and chemicals produced by nearby tissues. This allows arterioles to contract and relax in response to the needs of the tissues and organs they supply. In a suspense novel, you might read: "Her face went pale as she gazed at the bloodstained floor." Skin becomes pale when the arterioles supplying the skin capillaries constrict because the sympathetic nervous system and epinephrine from the adrenal glands has stimulated the smooth muscle in their walls to contract. This redirects blood away from the skin and into the muscles, where it may be needed for vigorous action.

In extremely cold weather, fingers and toes can become frostbitten because the sympathetic nervous system causes arterioles that supply blood to the extremities to constrict. This shunts blood to vital organs, such as the heart and brain, which cannot function properly if their temperature drops. By minimizing blood flow to heat-radiating extremities, the body conserves heat. On a hot summer day, in contrast, you become flushed as arterioles in the skin expand and deliver more blood to the skin capillaries. This enables the body to dissipate excess heat to the air outside, helping to reduce the body temperature.

The flow of blood in capillaries is also regulated by tiny rings of smooth muscle called **precapillary sphincters**, which surround the junctions between arterioles and capillaries (see Fig. 32-15). These open and close in response to local chemical changes. For example, the accumulation of carbon dioxide, lactic acid, or other cellular wastes signals the need for increased blood flow to bring oxygen to the tissues. These signals cause the precapillary sphincters and the muscles in the walls of nearby arterioles to relax, allowing more blood to flow through the capillaries.

CHECK YOUR LEARNING

Can you describe and compare the structures of arteries, arterioles, capillaries, venules, and veins? Can you explain the functions of each of these types of blood vessels? Do you understand how the distribution of blood flow is controlled?

32.5 HOW DOES THE LYMPHATIC SYSTEM WORK WITH THE CIRCULATORY SYSTEM?

The **lymphatic system** includes the lymphatic organs, as well as an extensive system of lymphatic vessels, which eventually feed into the circulatory system (**Fig. 32-18**). This organ system performs the following functions:

- Returns excess interstitial fluid to the bloodstream.
- Transports fats from the small intestine to the bloodstream.
- Filters aged blood cells and other debris from the blood.
- Defends the body by exposing bacteria and viruses to lymphocytes and macrophages.

Valve is open, allowing blood flow upward

closed valve

Muscle contraction compresses vein

Valve is closed, blocking blood flow downward

(a) Muscle relaxed   **(b) Muscle contracted**

▲ **FIGURE 32-17 Valves direct blood flow in veins**

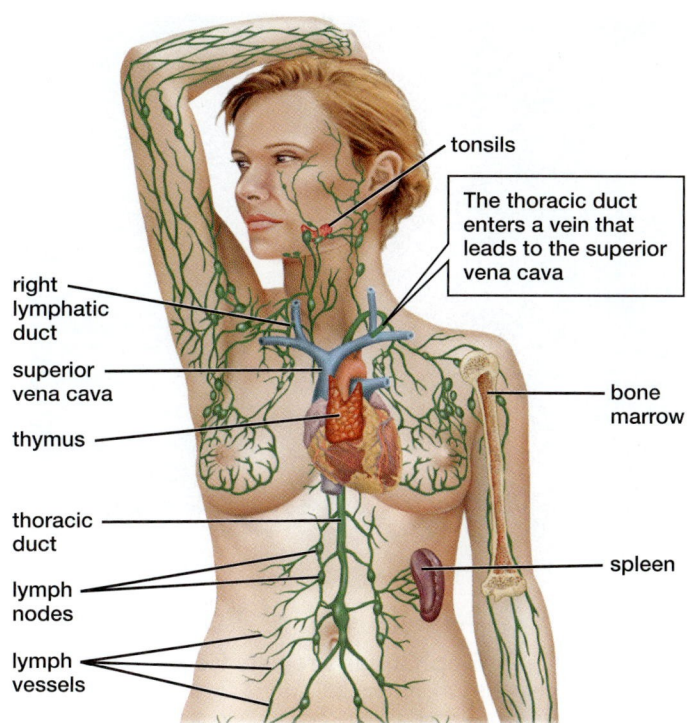

◄FIGURE 32-18 **The human lymphatic system** Lymphatic vessels and lymphatic organs consisting of bone marrow, thymus, tonsils, and the spleen are illustrated. Lymph is returned to blood veins via the right lymphatic duct and via the larger thoracic duct.

In the following sections, we emphasize the first three functions, in which the lymphatic system works intimately with the circulatory system (the role of the lymphatic system in defense of the body will be covered in Chapter 36).

Lymphatic Vessels Resemble the Capillaries and Veins of the Circulatory System

The smallest lymphatic vessels, called **lymphatic capillaries,** resemble blood capillaries in that they branch extensively throughout the body and their walls are only one cell thick. Lymphatic capillaries are far more permeable than blood capillaries, however, and are absent from bone and the central nervous system. Unlike blood capillaries, which form a continuous interconnected network, lymphatic capillaries "dead-end" in the interstitial fluid surrounding body cells (**Fig. 32-19**). Interstitial fluid flows into the lymphatic capillaries and becomes lymph.

▲ FIGURE 32-19 **Lymph capillary structure** Lymph capillaries end blindly in the body tissues. Here, pressure from the accumulation of interstitial fluid leaking from capillaries forces the fluid into the lymph capillaries as well as back into the venous side of the capillary network.

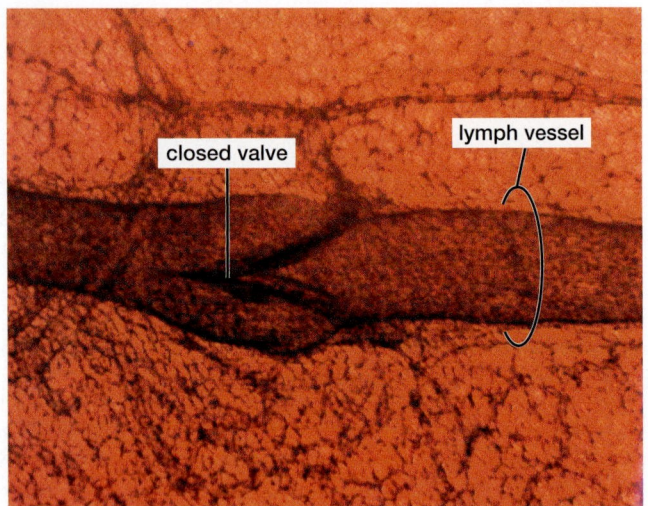

▲ FIGURE 32-20 **A valve in a lymph vessel** Lymphatic vessels (like blood veins) have internal one-way valves that direct the flow of lymph toward the large lymphatic ducts into which they empty.

QUESTION Which direction is the lymph flowing in this vessel?

▲ FIGURE 32-21 **Elephantiasis results from blocked lymphatic vessels** When a parasitic worm scars and blocks lymphatic vessels, preventing fluid from returning to the bloodstream, the affected area can become massively swollen.

From the lymphatic capillaries, lymph is channeled into increasingly large lymphatic vessels, which resemble the veins of the circulatory system in that both have similar walls, and both possess one-way valves that control the direction of fluid movement (**Fig. 32-20**). As the larger lymphatic vessels fill, stretch stimulates contractions of smooth muscles in their walls, pumping the lymph toward the heart. As with veins, further impetus for lymph flow through lymphatic vessels comes from internal pressure changes caused by breathing and the contraction of nearby skeletal muscles during exercise.

The Lymphatic System Returns Interstitial Fluid to the Blood

As described earlier, dissolved substances are exchanged between the capillaries and body cells via interstitial fluid. This fluid is pressure filtered out of blood plasma through capillary walls. In an average person, each day, blood capillaries leak out about 3 or 4 more quarts (roughly 3 to 4 liters) of interstitial fluid than they reabsorb. One function of the lymphatic system is to return this excess fluid to the blood.

As interstitial fluid accumulates around body cells, increasing pressure forces it through flap-like openings between the cells of the lymphatic capillary walls. Acting like one-way doors, these valves allow substances to enter, but not leave, the lymphatic capillaries. The lymphatic system transports this interstitial fluid—called **lymph** after it has entered lymphatic vessels—back to the circulatory system. Lymph vessels empty into the thoracic duct or the right lymphatic duct (see Fig. 32-18). These ducts discharge the lymph into large veins near the base of the neck, which merge into the superior vena cava, which enters the heart. The importance of the lymphatic system in returning fluid to the bloodstream is illustrated by elephantiasis

(**Fig. 32-21**). This disfiguring condition is caused by a parasitic roundworm that infects, scars, and blocks lymphatic vessels, preventing them from transporting interstitial fluid back to the bloodstream.

The Lymphatic System Transports Fats from the Small Intestine to the Blood

After a fatty meal, fat-transporting particles may make up 1% of the lymph, giving it a milky white color. How does this occur? The small intestine is richly supplied with lymph capillaries called *lacteals* (see Fig. 34-16). After absorbing digested fats, intestinal cells release fat-transporting particles into the interstitial fluid. These fatty particles are too large to diffuse into blood capillaries but can easily move into lymphatic capillaries. The lymphatic system then delivers the fat to the venous blood.

Lymphatic Organs Help Defend the Body

Organs of the lymphatic system (see Fig. 32-18) include bone marrow (where lymphocytes are produced) and the **thymus** (where lymphocytes mature during childhood). The remaining lymphatic organs—tonsils, lymph nodes, and the spleen—house macrophages and lymphocytes that proliferate in readiness to mount an immune response (described in Chapter 36). The **tonsils** are patches of lymphatic tissue that stand guard within the pharynx. Along the lymphatic vessels, hundreds of small swellings called **lymph nodes** filter the lymph past white blood cells. The **spleen,** a fist-sized lymphatic organ between the stomach and diaphragm, filters blood past macrophages and lymphocytes much as the lymph nodes filter lymph. The spongy interior of the spleen also breaks down red blood cells, and stores blood, which can be released if bleeding causes blood volume to fall.

CHECK YOUR LEARNING

Can you explain how the lymphatic system returns interstitial fluid to the blood, and why this is important? Do you understand the role of the lymphatic system in fat absorption from the intestines? Can you describe how lymphatic organs help defend the body from infection?

CASE STUDY revisited

Running with Heart

The phone call that saved and then transformed Donald Arthur's life came on a midsummer evening. Fitzgerald "Poochie" Gittens, a 25-year-old father of two, had been shot to death in a case of mistaken identity. His heart was a match for Arthur. Gittens' grieving mother Margaret made the difficult decision to donate his heart so that her son's untimely death could preserve the life of a complete stranger.

After crossing the finish line in New Jersey—his 33rd marathon in his 23rd state—Donald Arthur reached for his cell phone and called Margaret, whom he has now adopted as if she were his own mother. "Hey, Mom. Finished," he said. "Yeah, Poochie and I did another one."

Each year, about 400,000 people in the United States are diagnosed with cardiomyopathy, which causes nearly 25,000 deaths. Roughly 2,200 heart transplants are performed annually in the United States, but there are more than 3,100 individuals awaiting a replacement heart. About 70% of heart recipients survive for 5 years, and about half remain alive at the 10-year mark. Donald Arthur has beaten the odds.

Arthur's most memorable race was in 1999, when Poochie Gittens' brother, Mack, accompanied him in the New York Marathon. As they crossed the Verrazano Narrows Bridge, Donald grabbed Mack's hand and placed it on his chest, so that Mack could feel the rapidly beating heart. "This is your brother," he said. Looking back, Arthur recalls, "Maybe you could only see two people, but there were three of us running that day." Gittens' mother was waiting at the finish line to place their finisher's medals around their necks. Arthur, who does his marathons at a rapid walk, says, "It's not the race, but the opportunity to spread the word and inspire people about organ transplants. You can't imagine the rewards."

Consider This

BioEthics Gittens' mother, in shock from her son's death, found it extremely difficult to allow his organs to be donated, but she has never regretted the decision. If more organs were donated, more people like Donald Arthur would be alive today. What arguments do potential donors use *against* organ donations? Would you want to donate your own organs or allow a relative's organs to be donated after death?

CHAPTER REVIEW

Summary of Key Concepts

32.1 What Are the Major Features and Functions of Circulatory Systems?

Circulatory systems transport blood rich in dissolved nutrients and oxygen close to each cell, releasing nutrients and absorbing wastes. All circulatory systems have three major parts: one or more hearts that pump blood, the blood itself, and a system of blood vessels. Invertebrates have open or closed circulatory systems. In open systems, found in most invertebrates, hemolymph is pumped by the heart into a hemocoel, where the hemolymph directly bathes the internal organs. A few invertebrates and all vertebrates have closed systems, in which blood is confined to the heart and blood vessels. Vertebrate circulatory systems transport gases, hormones, nutrients, and wastes; help regulate body temperature; and help defend the body against disease.

32.2 How Does the Vertebrate Heart Work?

The vertebrate heart evolved from two chambers in fishes, to three in amphibians and some reptiles, to four in birds, crocodiles, and mammals. In the four-chambered heart, blood is pumped separately to the lungs at low pressure and to the rest of the body at much greater pressure. The heart also maintains complete separation of oxygenated and oxygen-poor blood. Oxygen-poor blood is collected from the body in the right atrium and then passed to the right ventricle, which pumps it to the lungs. Oxygenated blood from the lungs enters the left atrium, passes to the left ventricle, and is pumped to the rest of the body.

The cardiac cycle consists of atrial contraction followed by ventricular contraction. The direction of blood flow is maintained by valves within the heart. The contractions of the heart are initiated by the sinoatrial node, the heart's pacemaker, and coordinated by the atrioventricular node, the AV bundle, and Purkinje fibers. The heart rate can be modified by the nervous system and by hormones such as epinephrine.

32.3 What Is Blood?

Blood is composed of fluid plasma and cell-derived components. Plasma consists of water that contains proteins, hormones, nutrients, gases, and wastes. Red blood cells, or erythrocytes, are packed with hemoglobin, which carries oxygen. Their numbers are regulated by the hormone erythropoietin. There are five types of white blood cells, or leukocytes, which fight infection. Platelets, which are fragments of megakaryocytes, are important for blood clotting.

32.4 What Are the Types and Functions of Blood Vessels?

Blood leaving the heart travels through arteries, arterioles, capillaries, venules, veins, and then back to the heart. Each vessel is specialized for its role. Elastic, muscular arteries conduct blood from the heart, leading to smaller arterioles that empty into capillaries. The microscopically narrow, thin-walled capillaries allow exchange of materials between the body cells and the blood. Venules and veins provide a path of low resistance back to the heart, with one-way valves that maintain the direction of blood flow. The distribution of blood is regulated by the constriction and dilation of arterioles under the influence of the sympathetic nervous system, and by local factors, such as the amount of carbon dioxide in the tissues. Local factors also regulate precapillary sphincters, which control blood flow to the capillaries.

32.5 How Does the Lymphatic System Work with the Circulatory System?

The human lymphatic system consists of the lymphatic vessels and the lymphatic organs, which are the bone marrow, thymus, tonsils, lymph nodes, and spleen. The lymphatic system removes excess interstitial fluid that leaks through blood capillary walls and delivers it back to the circulatory system. It transports fats to the bloodstream from the small intestine, and fights infection by providing sites where lymphocytes can mature, proliferate, and mount an immune response.

Key Terms

angina   629
arteriole   627
artery   618
atherosclerosis   628
atrioventricular (AV) node   621
atrioventricular valve   619
atrium (plural, atria)   617
blood   616
blood clotting   625
blood vessel   616
bone marrow   623
capillary   627
cardiac cycle   620
cardiac muscle   619
closed circulatory system   617
diastolic pressure   620
double circulation   618
erythrocyte   623
erythropoietin   624
fibrillation   622
fibrin   625
heart   616
heart attack   628
heart rate   620
hemocoel   616
hemoglobin   624

hemolymph   616
hypertension   620
intercalated disc   619
interstitial fluid   628
leukocyte   624
lymph   632
lymph node   632
lymphatic capillary   631
lymphatic system   630
open circulatory system   616
plaque   628
plasma   622
platelet   624
precapillary sphincter   630
pulmonary circuit   618
semilunar valve   619
sinoatrial (SA) node   621
spleen   632
stroke   628
systemic circuit   618
systolic pressure   620
thrombin   625
thymus   632
tonsil   632
vein   618
ventricle   617
venule   629

Learning Outcomes

In this chapter, you have learned to . . .

LO1 Name the major features of all circulatory systems and the function of each feature.

LO2 Compare open and closed circulatory systems, including the animal groups in which each system is found, and the advantages and disadvantages of each.

LO3 Describe the diverse functions of the vertebrate circulatory system.

LO4 Describe the circulatory patterns of two-, three-, and four-chambered hearts, including the animal groups in which each system is found.

LO5 Describe the structure of cardiac muscle and how the heart's pacemaker and its connections cause coordinated contraction of the atria and ventricles.

LO6 Describe the cardiac cycle, including the role of the heart's valves in maintaining proper blood flow.

LO7 Explain how the nervous system and hormones influence heart rate, the distribution of blood within the body, and red blood cell production.

LO8 Describe blood plasma and the three cell-based components of blood, explaining the function of each and including the events that occur during blood clotting.

LO9 Describe the structure and function of each of the five types of blood vessels.

LO10 List the organs and vessels of the lymphatic system and explain the functions of each, including where and how lymph is produced.

Thinking Through the Concepts

Fill-in-the-Blank

1. Animals with open circulatory systems utilize a fluid called _____ that serves the function of both blood and _____ fluid. This fluid is passed among chambers in the body that are collectively called the _____. **LO2**

2. Fill in the following with the appropriate heart chamber, including the side it is on: Receives blood from the body: _____; has the thickest wall: _____; contains the SA node: _____; pumps blood to the lungs: _____; pumps blood into the aorta: _____. **LO3 LO4 LO5**

3. The heart's pacemaker is called the (complete term) _____. The pacemaker is composed of specialized _____. The pacemaker first sends impulses that stimulate contraction through both _____. The (complete term) _____ introduces a delay in the transmission of pacemaker signals into the ventricles. Signals are conducted from this node directly into fibers called the _____. If the pacemaker loses control of heart contractions, ineffective quivering of the heart muscle, called _____, may occur. **LO4 LO5 LO6**

4. Fill in the following with the appropriate type of blood vessel: Allow exchange of wastes and nutrients between blood and body cells: _____; have the thickest walls: _____; transport blood toward the heart: _____; receive blood from the capillaries: _____; deliver blood into capillaries:_____; carry blood away from the heart: _____. **LO9**

5. During clot formation, damaged cells activate cell fragments called _____. These cell fragments release chemicals that cause the enzyme _____ to be produced. This enzyme catalyzes the conversion of a blood protein called _____ into strands of _____, which create a framework for the clot. **LO8**

6. Fill in the following with the appropriate cell-based blood component, using the scientific term: Formed by megakaryocytes: _____; there are five types of these: _____; carry oxygen: _____; include macrophages: _____; lack a nucleus: _____; fight infection and disease: _____; contain hemoglobin: _____. **LO7**

7. Lymph is _____ that has entered lymphatic vessels. The lymphatic system returns excess fluid that has leaked out of blood _____ to the circulatory system. Another function of the lymphatic system is to transport _____ from the small intestine. Lymphatic vessels have _____ that keep lymph flowing in the proper direction. The lymphatic organ that filters blood is the _____. **LO9**

Review Questions

1. List the major structures of all circulatory systems, and the specific functions of vertebrate circulatory systems. **LO1**

2. Describe and compare the features of open and closed circulatory systems; include the animal groups in which they are found and some advantages and disadvantages of each. **LO2**

3. Explain how two- and three-chambered vertebrate hearts work and in which animal groups each is found. **LO3**

4. Explain the structure of cardiac muscle and how it helps produce coordinated contraction within in each heart chamber. **LO4**

5. List the three cell-based components of blood and describe the structure and principal functions of each. **LO7**

6. Explain the sequence of events that causes the mammalian heart to beat, including the names of the structures involved. **LO4**

7. Describe veins, capillaries, and arteries, noting their similarities and differences. **LO8**

8. Distinguish among blood plasma, interstitial fluid, and lymph. **LO7 LO9**

9. Describe the cardiac cycle, and relate the contractions of the atria and ventricles to the two readings taken during the measurement of blood pressure. **LO5**

10. List the components of the lymphatic system, and describe three important functions of this system. **LO9 LO10**

11. Explain how the sympathetic nervous system influences blood distribution and blood pressure. **LO6**

12. Describe how the number of red blood cells is regulated by a negative feedback system. **LO6**

13. In what way do veins and lymphatic vessels resemble one another? How are they different? **LO8 LO9**

Applying the Concepts

1. In addition to completely separating oxygenated and oxygen-poor blood, what major advantage is there to the four-chambered heart with its two separate pumps?

2. **BioEthics** Considering the prevalence of cardiovascular disease and the high, increasing costs of treating it, certain treatments may not be available to all who might need them. What factors would you take into account if you had to ration cardiovascular procedures, such as heart transplants or bypass surgery?

3. Joe, an overweight 45-year-old executive of a major corporation who works 60-hour weeks, has been feeling chest pain when he tries to play basketball with his son on weekends. What is Joe experiencing, what are his lifestyle change or treatment options, and how does each work?

Answers to Figure Caption questions and Fill-in-the-Blank questions can be found in the Answers section at the back of the book.

MB *Go to MasteringBiology for practice quizzes, activities, eText, videos, current events, and more.*

Climbing Mt. Kilimanjaro is a breathtaking experience. (Inset) A proud Alistair Cook who summited Kilimanjaro in celebration of his upcoming 70th birthday. Tragically, he died of a heart attack just moments after his triumph.

CASE STUDY

The High Life Can Lead to Death

AN EXCERPT FROM THE DIARY of Jeanette Johnson, who reached the summit of Mt. Kilimanjaro on New Year's Day, 2001, reads: "It was a perfect night, and perfectly quiet, except for the sounds of deep coughing. They were coming from the tent of a team member named Craig—a super-fit cross-country skier from upstate New York and the strongest guy on our team." Her team was a diverse group of tourists brought together by the desire to test themselves at heights few would ever reach. Jeanette was writing from their camp at 12,800 feet, which they reached after the second day of their ascent: "Craig

hacked all night, and by morning he was coughing up blood."

Mt. Kilimanjaro, a dormant volcano in Tanzania, is Africa's highest peak. The 19,340-foot summit is often billed by tour companies as "a walk," requiring no previous climbing experience or skills and accessible to most people in good physical condition. The peak has been conquered by Martha Stewart, a group of British celebrities, a 92-year-old man, and Kyle Maynard—born with stumps for arms and legs—who did a fast crawl to the top. At the summit, because atmospheric pressure is reduced and the air is thinner, each breath takes in 50% less oxygen than at sea level.

Kilimanjaro is the most climbed mountain in the world. Every year, over 20,000 people start the climb, about 8,000 of these reach the summit, and about 10 people die in the attempt. Death is usually from heart attacks or from a condition in which fluid accumulates in the lungs, called high-altitude pulmonary edema (HAPE).

Why might the fittest member in a climbing group develop life-threatening respiratory symptoms? How does the human body adjust to a rapid increase in altitude and a resulting decrease in oxygen availability? How do respiratory systems function intimately with circulatory systems to supply oxygen continuously to every cell in the body?

AT A GLANCE

LEARNING GOALS

LG1 Explain how organismal respiration supports cellular respiration.

LG2 Describe how respiratory and circulatory systems work together to allow gas exchange between the body cells and the environment.

LG3 List and explain three adaptations for diffusion found in all respiratory systems.

LG4 Explain and provide examples of how respiration may occur without specialized respiratory organs.

LG5 Describe the four stages of gas exchange in organisms with well-developed respiratory and circulatory systems, including where these stages occur in fish and mammals.

LG6 Compare the respiratory systems of insects, amphibians, snakes, and birds.

LG7 Describe the two different types of gills, and explain how fish gills extract oxygen from water.

LG8 Describe the conducting portion of the human respiratory system, including the function of each part.

LG9 Describe the mechanics and control of human breathing.

LG10 Explain how the structure of alveoli allows gas exchange between air and capillary blood.

LG11 Compare the mechanisms by which oxygen and carbon dioxide are transported in the bloodstream.

33.1 WHY EXCHANGE GASES AND WHAT ARE THE REQUIREMENTS FOR GAS EXCHANGE?

Late again! Sprinting up two flights of stairs to your classroom, your calves are burning. (Remembering Chapter 8, you think, "Ah ha! That's lactic acid building up—my muscle cells are using fermentation because they can't get enough oxygen for cellular respiration.") As you slip into your seat, quietly panting and feeling your heart pounding, the discomfort eases. Less exertion coupled with rapid breathing ensures that adequate oxygen (O_2) is now available. You are experiencing firsthand the connection between breathing and generating ATP through cellular respiration.

The Exchange of Gases Supports Cellular Respiration

As cellular respiration converts the energy in nutrients (such as sugar) into the ATP that supplies cellular energy, the process requires a steady supply of O_2 and generates carbon dioxide (CO_2) as a waste product. Organismal respiration, which we will simply call **respiration,** is the process by which organisms exchange gases with the environment, taking in oxygen and releasing carbon dioxide (CO_2), in support of cellular respiration. The rapid beating of your heart as you relax after your sprint reminds you that your circulatory system works in close harmony with your respiratory system to deliver O_2 and remove CO_2 from each cell of your body.

Gas Exchange Through Cells and Tissues Relies on Diffusion

All organisms that obtain energy through cellular respiration rely on the process of diffusion to acquire O_2 and to eliminate CO_2. Although animal respiratory structures are quite diverse, they all facilitate diffusion through three adaptations:

- Respiratory surfaces—even those in contact with air—remain moist, because cell membranes are always moist, and only gases dissolved in water can diffuse into or out of cells.
- Respiratory surfaces are very thin to minimize diffusion distances.
- Respiratory surfaces have a sufficiently large surface area in contact with the environment to allow adequate gas exchange by diffusion to meet the needs of the organism.

During diffusion, individual molecules move from areas of higher concentration to areas of lower concentration (see Chapter 5), so gas exchange by diffusion requires concentration gradients of gases. To maintain these gradients, the air or water in which an animal lives must flow past the respiratory surface, continuously supplying O_2 and carrying away CO_2. Such movement is called **bulk flow** because the masses of molecules that form the flowing substance, such as air or water, move together (or "in bulk") through relatively large spaces. The transfer of gases between the environment and the body cells during respiration requires both bulk flow and diffusion.

CHECK YOUR LEARNING

Can you explain how organismal respiration supports cellular respiration? Are you able to explain the adaptations of respiratory surfaces for diffusion? Do you understand how bulk flow increases diffusion rates?

33.2 HOW DO RESPIRATORY ADAPTATIONS MINIMIZE DIFFUSION DISTANCES?

Because of the time required for gases to move by diffusion, this process is only adequate to support cellular respiration over very short distances. Evolution has provided a variety of ways to keep diffusion distances short—usually less than a millimeter—as you will see in the examples that follow.

Relatively Inactive Animals May Lack Specialized Respiratory Organs

For animals that are relatively sluggish and don't require large amounts of ATP, diffusion may provide adequate oxygen. Sponges (**Fig. 33-1a**), Earth's simplest animals, live in aquatic—mostly marine—environments. Sponges use ciliated cells to create a current of water, which moves by bulk flow through pores in their bodies into a central chamber. The water then flows out through one or more larger openings. This circulation, aided by ocean currents, brings a continuous supply of water that is relatively high in O_2 and low in CO_2

to within diffusing distance of each cell of the sponge. Sea jellies, corals, and anemones (cnidarians; **Fig. 33-1b**) have an extremely thin outer skin. This allows gases to move readily by diffusion into and out of their outer cell layers as seawater flows over their bodies. Cnidarians use muscle cells to generate additional bulk flow of water into and out of a central chamber called the *gastrovascular cavity* (also used in digestion, see Fig. 34-7). The gastrovascular cavity brings water close enough to the internal cells that diffusion meets their needs for gas exchange. Flatworms also possess a gastrovascular cavity; in some species this is used for gas exchange as well as for digestion. In all flatworms, gas exchange by diffusion occurs through an extensive gas-permeable skin surface. Their flattened shapes, which gives them the name "flatworms," ensures that all their cells are close to the skin (**Fig. 33-1c**). Bulk flow of water over the bodies of these aquatic animals maintains the necessary diffusion gradients.

Some more complex animals, such as earthworms, combine a large skin surface for diffusion, produced by the worms' elongated shape, with a well-developed circulatory system (see Fig. 32-1b) that transports gases in the blood by bulk flow. As oxygen diffuses through the skin and into skin capillaries that carry it away, this process maintains the concentration gradient, favoring the continued inward diffusion of O_2. Meanwhile, CO_2 diffuses from cells into the blood and is carried in capillaries to the skin, where it diffuses out. If an earthworm's skin dries out, it will suffocate, because diffusion of gases into and out of cells requires a moist surface.

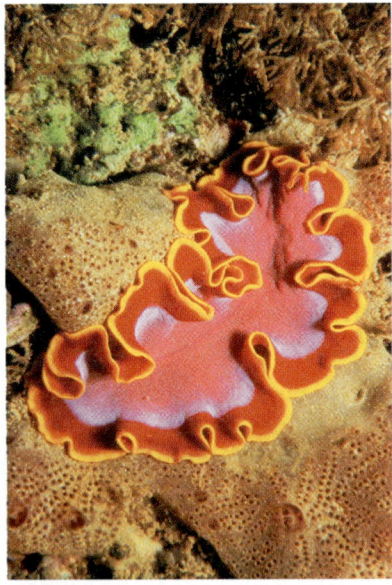

(a) Sponge   **(b) Sea jelly**   **(c) Flatworm**

▲ **FIGURE 33-1 Some animals lack specialized respiratory structures** Animals without respiratory systems have relatively low metabolic demands and large, moist body surfaces. **(a)** Flagellated cells draw currents of oxygen-rich water through numerous pores in the body of the sponge and expel oxygen-poor water through one or more larger openings. **(b)** Cells in the bell-shaped body of a sea jelly have a low metabolic rate; ocean currents and swimming movements provide adequate gas exchange. **(c)** The large outer surface of this marine flatworm exchanges gases with the water.

Respiratory Systems and Circulatory Systems Often Work Together to Facilitate Gas Exchange

Most relatively large, active animals have **respiratory systems** that consist of respiratory organs that work as a unit to facilitate gas exchange between the animal and its environment. Respiratory systems nearly always work closely with circulatory systems that carry gases close to all body cells. Major organs of gas exchange include gills in many aquatic vertebrates and invertebrates, tracheae in insects, and lungs in terrestrial vertebrates. For animals with interacting circulatory and respiratory systems, gas exchange occurs in the following stages, illustrated for mammals in **Figure 33-2**:

❶ Bulk flow carries air or water, relatively high in O_2 and low in CO_2, past a respiratory surface. The flow is usually propelled by muscular movements.

❷ O_2 diffuses from the air or water through the respiratory surface and into the capillaries of the circulatory system, while CO_2 diffuses out of the capillaries and through the respiratory surface to the air or water.

❸ Bulk flow of blood transports gases between the respiratory system and the tissues. Blood is pumped throughout the body by the heart.

❹ Diffusion transfers O_2 out of the capillaries and into nearby tissues, and transfers CO_2 from the tissues into capillaries.

Gills Facilitate Gas Exchange in Aquatic Environments

Gills are the respiratory structures of many aquatic animals. The simplest type of gill, found in some amphibians (see Fig. 33-5a) and in nudibranch (literally, "naked gill") mollusks, consists of many thin projections of the body surface that protrude into the surrounding water (**Fig. 33-3**).

In general, gills are elaborately branched or folded, increasing their surface area. Just beneath their delicate outer membranes, gills have a dense profusion of capillaries that carry blood close to the gill surface, where gas exchange occurs. Most types of fish protect their delicate gill membranes beneath a bony flap, the *operculum*. Fish create a continuous current over their gills by pumping water into their mouths and ejecting it out through the opercular opening (see Fig. E33-1). Some fishes, such as mackerel, tuna, and some sharks, swim with their mouths open to help create a water current over their gills.

Fish face a challenge in extracting enough O_2 from water to support their active lifestyles. There are only about 3% as many oxygen molecules in a given volume of fresh water as there are in the same volume of air (ocean water contains even less). Because water is about 800 times as dense as air, pumping enough water over gills to obtain adequate oxygen uses up far more energy than breathing air does. In response to these challenges, fish have evolved a very efficient method for exchanging gases with water, called **countercurrent exchange**. Within the gill, water and blood flow in opposite directions, maintaining a concentration gradient

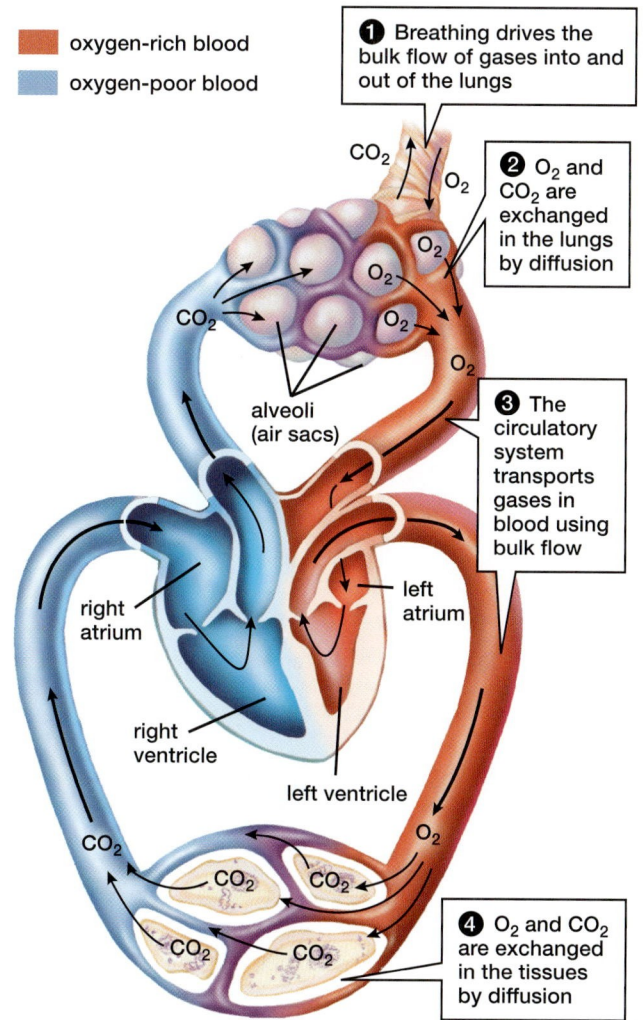

- oxygen-rich blood
- oxygen-poor blood

❶ Breathing drives the bulk flow of gases into and out of the lungs

CO_2 O_2

❷ O_2 and CO_2 are exchanged in the lungs by diffusion

CO_2 O_2 O_2 O_2 O_2

alveoli (air sacs)

❸ The circulatory system transports gases in blood using bulk flow

right atrium

left atrium

right ventricle

left ventricle

CO_2 O_2

CO_2 CO_2 CO_2 CO_2

❹ O_2 and CO_2 are exchanged in the tissues by diffusion

▲ FIGURE 33-2 **An overview of gas exchange in mammals**

▲ FIGURE 33-3 **External gills on a mollusk** The feathery projections from the back of this nudibranch mollusk are gills used for gas exchange.

through the entire length of the gill, which can cause most of the oxygen to diffuse into the gill from the water. This is described in "In Greater Depth: Gills and Gases—Countercurrent Exchange."

Terrestrial Animals Have Internal Respiratory Structures

Gills are useless in air because they collapse and dry out, so new adaptations were needed to allow animals to make the transition from water to dry land over evolutionary time. Respiratory structures evolved that protected, supported, and retained the moisture of delicate cell membranes through which gas exchange must occur. These structures include tracheae in insects and lungs in vertebrates.

Insects Respire Using Tracheae

The respiratory system of an insect consists of **tracheae** (singular, trachea), a system of branching, air-filled tubes. Tracheae conduct air in and out of the body through openings called **spiracles,** arrayed in rows along each side of the body surface (**Figs. 33-4a, b**). Tracheae, which are reinforced with chitin (a principal component of the insect's external skeleton), penetrate the body tissues and branch into microscopic, open-ended tubes called *tracheoles* (**Fig. 33-4c**).

Tracheoles deliver air close to each body cell, minimizing diffusion distances for O_2 and CO_2. Some large insects use pumping movements of their abdomens to increase air movement in and out of the tracheae.

In most animals, the respiratory system moves oxygen into the circulatory system, which distributes oxygen to the body cells. However, the open circulatory systems of insects (see Fig. 32-1) cannot move blood fast enough to supply sufficient oxygen to support strenuous activity, such as flying. Tracheae, by supplying oxygen close to the individual cells, compensate for the insect's relatively inefficient circulatory system.

Terrestrial Vertebrates Respire Using Lungs

Lungs are chambers containing moist respiratory surfaces that are protected within the body, where water loss is minimized and the body wall provides support. The first vertebrate lung probably appeared in a freshwater fish and consisted of an outpocketing of the digestive tract. This simple lung supplemented the gills, helping the fish survive in stagnant water where O_2 is scarce.

Amphibians, which evolved from fish, straddle the boundary between aquatic and terrestrial life. Amphibians, such as frogs, use gills during their aquatic larval (tadpole) stage, but most lose their gills and develop simple, sac-like lungs as they metamorphose into more terrrestrial adults

(a) Insect respiratory system

(b) Spiracle and tracheae

(c) Gas exchange pathway

▲ **FIGURE 33-4 Insects breathe using tracheae (a)** The tracheae of insects, such as this beetle, branch intricately throughout the body; air moves in and out through spiracles in the body wall. **(b)** This light microscope image shows the tracheae branching outward from a spiracle. **(c)** Tracheae branch into microscopic, open-ended tracheoles that conduct air to body cells for gas exchange.

IN GREATER DEPTH Gills and Gases—Countercurrent Exchange

To meet the challenge of extracting oxygen from water, which has an oxygen concentration that is always far lower than that of air, fish pump water into their mouths and over their gills (**Fig. E33-1a**). Gills of bony fish contain several supporting gill arches, each bearing a series of paired gill filaments of tissue through which gas exchange occurs (**Fig. E33-1b**). Each filament is served by two blood vessels; one carries oxygen-poor blood from the body, and the other carries oxygen-rich blood back to the body. The filaments are composed of thin leaflets of membrane called lamellae (singular, lamella). Each lamella spans the narrow space between the incoming and outgoing vessels and is filled by a capillary bed that absorbs oxygen from the surrounding water. Oxygen-rich water entering the gills flows past the lamellar capillary beds in the direction opposite to the flow of blood, a process called *countercurrent flow* (**Fig. E33-1c**). Fluids flowing past one another in opposite directions (countercurrents) maximize the rate and efficiency of diffusion of solutes (such as oxygen) between them

along their concentration gradients. This transfer of solute is called *countercurrent exchange*.

As illustrated in **Figure E33-2**, countercurrent exchange maintains a gradient of solute so that diffusion occurs—in this case, O_2 from water to blood—along the entire length of the gill capillaries. In the figure, notice that the oxygen-poor fluid in the upper tube

(representing gill capillary blood) acquires oxygen continuously as it flows from right to left past fluid in the lower tube (representing water flowing in the opposite direction over the gills). Countercurrent exchange is so efficient that some fish can extract 85% of the O_2 from the water flowing over their gills; humans, in contrast, extract only about 25% of the oxygen from the air we breathe.

▲ **FIGURE E33-2 Countercurrent exchange** During countercurrent exchange, small but continuing concentration gradients are maintained, and most of the oxygen from the water is eventually transferred to the capillary blood. The numbers and corresponding colors represent hypothetical percentages of oxygen saturation. Red arrows indicate the direction of diffusion.

(a) Water flows over gills   **(b) Gill structure**   **(c) Lamella**

▲ **FIGURE E33-1 Gills exchange gases with water (a)** Fish pump water in through their mouths and out over their gills. **(b)** Water flows past a dense array of paired gill filaments. **(c)** Each filament consists of a series of lamellae. Water flows over the lamellae in the opposite direction from the blood flowing through the lamellar capillaries.

(Figs. 33-5a, b). Most amphibians also rely to a considerable extent on diffusion of gases through their thin, moist skin, which is rich in capillaries.

In reptiles (snakes, lizards, turtles, and birds) and in mammals, relatively waterproof skin is covered with scales (Fig. 33-5c), feathers, or fur. This reduces water loss in dry environments, but also prevents the skin from serving as a respiratory organ. To compensate, the lungs of reptiles and mammals have a far larger surface area for gas exchange than do those of amphibians.

Bird respiratory systems differ substantially from those of other terrestrial vertebrates, including other reptiles. The bird lung has adaptations that allow exceptionally efficient gas exchange, providing adequate O_2 to support the enormous energy demands of flight. In the bird respiratory system, the lungs are connected to seven to nine inflatable air sacs, which serve as reservoirs for air. In Figure 33-6a, these have been grouped into anterior air sacs (just in front of the lungs) and posterior air sacs (just behind the lungs).

The lungs of birds are rigid and filled with perforated tubes called *parabronchi* (Fig. 33-6c), which are barely visible to the naked eye. Parabronchi are open at both ends, allowing air to flow completely through the lungs. The ends of the parabronchi connect to the air sacs. Parabronchi are surrounded by tissue that is riddled with interconnected microscopic air spaces. The air spaces are surrounded by a dense capillary network where gas exchange occurs.

The unique organization of air sacs and lungs allows a one-way flow of fresh, oxygen-rich air through the lungs. Air moves in through the nostrils, travels back to the posterior air sacs via a route that bypasses the lungs, then forward through

(a) Tadpole

(b) Adult bullfrog

(c) Snake

▲ FIGURE 33-5 **Amphibians and reptiles have different respiratory adaptations (a)** The bullfrog, an amphibian, begins life as a fully aquatic tadpole with feathery external gills. **(b)** During metamorphosis into an air-breathing adult, the frog's gills are lost and replaced by simple sac-like lungs. In both tadpole and adult, gas exchange also occurs by diffusion through moist skin. **(c)** Scaled reptiles, such as this snake, cannot respire through their skin; their lungs compensate with a larger surface area for gas exchange.

QUESTION How do the respiratory adaptations of amphibians influence the range of habitats in which amphibians are found?

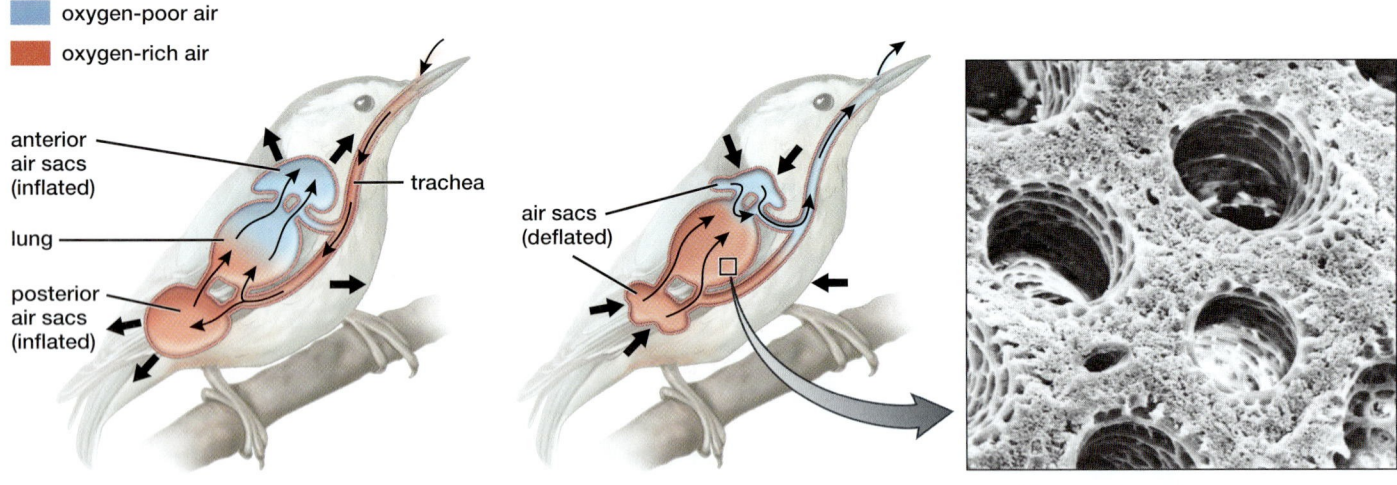

- oxygen-poor air
- oxygen-rich air

anterior air sacs (inflated)

trachea

lung

posterior air sacs (inflated)

air sacs (deflated)

(a) Breathing in

(b) Breathing out

(c) SEM of parabronchi in lung tissue

▲ FIGURE 33-6 **The bird respiratory system** In addition to rather rigid lungs, birds have anterior and posterior groups of flexible air sacs where no gas exchange occurs. Bold arrows indicate expansion and contraction of the chest and air sacs; thin arrows indicate air flow. **(a)** Breathing in expands the air sacs, drawing fresh, oxygen-rich air (red) past the lungs and into the posterior air sacs, sending some of this air into the lungs, and moving "used" oxygen-poor air (blue) out of the lungs and into the anterior air sacs. **(b)** Breathing out deflates the chest, compressing the air sacs and forcing used, oxygen-poor air from the anterior sacs out through the nostrils, and fresh air out of the posterior sacs to fill the lungs. **(c)** Open-ended, tubular parabronchi conduct air through the lungs as the air flows from posterior to anterior air sacs. The porous regions visible between the parabronchi are filled with capillaries and air spaces where gas exchange occurs.

the lungs, into the anterior air sacs, and back out of the body. As inhalation (breathing in) draws fresh air into the posterior sacs (Fig. 33-6a), the incoming air displaces the fresh air stored in the posterior sacs, forcing it forward through the lungs, where O_2 is extracted. Simultaneously, inhalation draws "used" oxygen-poor air out of the lungs and into the anterior sacs. Then, as the bird exhales (breathes out), all of the air sacs are compressed, forcing the oxygen-poor air from the anterior sacs out through the bird's nostrils, and fresh air from the posterior sacs into the lungs (**Fig. 33-6b**). Thus, bird lungs receive fresh air both as the bird inhales and as it exhales.

CHECK YOUR LEARNING

Can you explain the four stages of gas exchange in animals with circulatory systems and lungs? Can you explain the respiratory adaptations of fish? Can you describe the general respiratory adaptations for life on land, and the specific adaptations of amphibians and reptiles including birds?

33.3 HOW IS AIR CONDUCTED THROUGH THE HUMAN RESPIRATORY SYSTEM?

The respiratory system in humans and other mammals can be divided into two parts: the **conducting portion** and the **gas-exchange portion.** The conducting portion consists of a series of passageways that carry air into and out of the gas-exchange portion within the lungs, where oxygen and carbon dioxide are exchanged through the blood.

The Conducting Portion of the Respiratory System Carries Air to the Lungs

Air enters through the nose or the mouth, passes through the nasal cavity or oral cavity into a chamber, the **pharynx** (which is shared with the digestive tract), and then travels through the **larynx,** where sounds are produced (**Fig. 33-7**).

(a) Human respiratory system

(b) Alveoli with capillaries

▲ **FIGURE 33-7 The human respiratory system** (a) The major structures of the human respiratory system are illustrated here. The pulmonary artery carries oxygen-poor blood (blue) to the lungs; the pulmonary vein carries oxygen-rich blood (red) back to the heart. **(b)** Close-up of alveoli (interiors shown in cutaway section), their arterioles, and their surrounding network of capillaries.

QUESTION Why is oxygen-rich blood depicted as red and oxygen-poor blood as blue?

The opening to the larynx is guarded by the **epiglottis,** a flap of tissue supported by cartilage. During normal breathing, the epiglottis is tilted upward, allowing air to flow freely into the larynx. During swallowing, the epiglottis folds downward and covers the larynx, directing substances into the esophagus (see Fig. 34-13). If an individual attempts to inhale and swallow at the same time, this reflex may fail and food can become lodged in the larynx, blocking air from entering the lungs. What should you do if you see this happen? Use the **Heimlich maneuver,** described in **Figure 33-8,** which is easy to perform and has saved countless lives.

Within the larynx, or "voice box," are the **vocal cords,** bands of elastic tissue controlled by muscles. Muscular contractions can cause the vocal cords to partially obstruct the air passage through the larynx. Exhaled air then causes the vocal cords to vibrate, producing the tones of speech or song. Stretching the cords changes the pitch of the tones, which can be articulated into words by movements of the tongue and lips.

Inhaled air travels past the larynx into the **trachea,** a flexible tube whose walls are reinforced with semicircular bands of stiff cartilage. Within the chest, the trachea splits into two large branches called **bronchi** (singular, bronchus), one leading to each lung. Inside the lung, each bronchus branches repeatedly into ever-smaller tubes. Finally, these divide into **bronchioles** about 1 millimeter in diameter. The walls of bronchi and bronchioles are encased in smooth muscle, which regulates their diameter. During activities that require extra oxygen, such as exercise, the smooth muscle relaxes, allowing more air to enter. Bronchioles narrow and lead to microscopic **alveoli** (singular, alveolus), tiny air sacs where gas exchange occurs (see Fig. 33-7).

During its passage through the conducting system, air is warmed and moistened. Much of the dust and bacteria it carries is trapped in mucus secreted by cells that line the respiratory passages. The mucus, with its trapped debris, is continuously swept upward toward the pharynx by cilia that line the bronchioles, bronchi, and trachea. Upon reaching the pharynx, the mucus is coughed up or swallowed. Smoking interferes with this cleansing process by paralyzing the cilia (see "Health Watch: Smoking—A Life and Breath Decision").

Air Is Inhaled Actively and Exhaled Passively

Breathing occurs as inhalation and exhalation cause bulk flow of air into and out of the lungs. **Inhalation** occurs when the chest cavity is enlarged. The lower boundary of the chest cavity is formed by a sheet of muscle, the **diaphragm,** which domes upward when relaxed. During inhalation, the diaphragm is contracted, which pulls it downward. The rib muscles also contract during inhalation, lifting the ribs up and outward. Both of these muscular movements enlarge the chest cavity (**Fig. 33-9a**). When the chest cavity is expanded, the lungs inflate within it, because fluid fills the narrow space between the lungs and the inner chest wall, forming an air-tight seal. As the lungs expand, their increased volume reduces the pressure inside them, drawing air in. (A puncture wound to the chest is dangerous in part because it can allow air to penetrate between the chest wall and the lungs, breaking the seal and preventing the lungs from inflating when the chest cavity expands.)

Exhalation occurs spontaneously when the muscles that cause inhalation are relaxed. As the diaphragm relaxes, it domes upward; at the same time, the ribs move downward and inward. Both of these movements decrease the size of the chest cavity, forcing air out of the lungs (**Fig. 33-9b**). Additional air can be forced out by contracting the abdominal muscles. After exhalation, the lungs still contain some air, which helps prevent the thin alveoli from collapsing and fills the spaces within the conducting portion of the respiratory system. A typical breath in an average-sized adult moves about a pint (roughly 500 ml) of air through the respiratory system. Because air must fill the conducting portion, only about three quarters of the inhaled and exhaled air reaches the alveoli. During exercise, deeper breathing can move four times this volume of air per breath.

Breathing Rate Is Controlled by the Respiratory Center of the Brain

Imagine having to think about every breath. Fortunately, breathing occurs rhythmically and automatically without conscious thought. But, unlike the heart muscle, the diaphragm and rib muscles used in breathing are not self-activating; each contraction causing inhalation is stimulated

object ejected

lungs compressed

diaphragm pushed upward

❶ Grasp the hands between the navel and breastbone

❷ Quickly and forcefully pull upward and toward your body

▲ **FIGURE 33-8 The Heimlich maneuver can save lives** Pressing forcefully and suddenly upward and inward just below the diaphragm forces air from the lungs, which can expel food blocking the trachea. Repeat the maneuver if necessary.

(a) Inhalation   **(b) Exhalation**

▲ **FIGURE 33-9** **The mechanics of breathing** **(a)** During inhalation, rhythmic nerve impulses from the brain stimulate the diaphragm and the muscles surrounding the ribs to contract. This increases the size of the chest cavity, causing air to rush in. **(b)** During exhalation, these muscles relax, forcing air out of the lungs.

QUESTION Why does contracting the diaphragm muscle (which decreases its size) increase the volume of the chest cavity?

by impulses from nerve cells. These impulses originate in the **respiratory center,** which is located in the medulla, a portion of the brain just above the spinal cord (see Fig. 38-12). Nerve cells in the respiratory center generate cyclic bursts of electrical signals (action potentials) that cause contraction.

CASE STUDY continued

The High Life Can Lead to Death

Above 10,000 feet (3,048 meters), more than 70% of people climbing Kilimanjaro suffer from acute mountain sickness, which causes many climbers to turn around before reaching the summit. Few reach Kilimanjaro's summit without symptoms such as headache, nausea and vomiting, dizziness, dehydration, insomnia, and shortness of breath—even during rest. The cause? Climbing too high, too fast, without giving the body adequate time to adjust to the decreased availability of oxygen.

Adjustments to high altitude include breathing faster and deeper, which circulates more air through the lungs. Bone marrow produces more red blood cells to carry oxygen. Blood pressure and heart rate increase. Reduced oxygen causes the smooth muscles in artery and vein walls to relax, allowing more blood to reach the tissues. To allow this acclimation to occur, experts recommend ascending only 1,000 feet a day above 10,000 feet, and also acclimating for a day after each 3,000-foot gain. But many tours lead people from Kilimanjaro's base at about 6,000 feet to its peak at 19,340 in just 4 or 5 days.

During the intervals between bursts, the muscles relax, causing exhalation.

The respiratory center receives input from several sources and adjusts the breathing rate and volume to meet the body's changing needs. The respiratory rate is primarily regulated by CO_2 receptors, also located in the medulla. These adjust the breathing rate to maintain an extremely constant, low level of CO_2 in the blood, while also ensuring that O_2 levels remain adequate. For example, if you run to class, your muscles need extra ATP. This requires an increase in cellular respiration, which generates more CO_2. Receptors in the medulla detect the increased CO_2 and cause you to breathe faster and more deeply, eliminating more CO_2 and taking in more O_2. These receptors are extremely sensitive; a 16% increase in CO_2 will cause a person to double the amount of air moving through the lungs. There are also O_2 and CO_2 receptors in the aorta and carotid arteries that stimulate the respiratory center. These receptors are also primarily sensitive to increased CO_2 in the blood, but a major decline in blood O_2 will also activate them.

CHECK YOUR LEARNING

Can you describe the structures and functions of the conducting portion of the human respiratory system? Can you describe the mechanics of breathing and the control of breathing rate?

Health | Watch

Smoking—A Life and Breath Decision

Each year in the United States, smoking-related diseases kill about 443,000 people prematurely, accounting for one out of every five deaths. These preventable afflictions include lung cancer, several other forms of cancer, heart disease, emphysema, chronic bronchitis, and stroke. Despite decades of warning about smoking's dangers, about 19% of U.S. adults continue to smoke.

The nicotine in tobacco is a powerfully addictive drug—as addictive as cocaine or heroin. Like cocaine and heroin, nicotine activates the brain's reward center more intensely than do natural stimuli. The brain adapts by becoming less sensitive to these drugs, requiring larger quantities to experience the same rewarding effect. When a person tries to quit smoking, withdrawal symptoms may include nicotine craving, depression, anxiety, irritability, headaches, difficulty concentrating, and disturbed sleep.

As smoke is inhaled, toxic substances paralyze the cilia lining the trachea and bronchi; a single cigarette can inactivate them for a full hour. Because cilia remove inhaled particles, smoking inhibits them just when they are needed most. The visible portion of cigarette smoke consists of billions of microscopic carbon particles. Adhering to these particles are about 250 different toxic substances, at least 69 of which can cause cancer. With the cilia incapacitated, the particles enter the lungs.

Cigarette smoke also impairs the white blood cells that defend the bronchioles and alveoli by engulfing foreign particles and bacteria, rendering the lungs more vulnerable to these invaders. Consequently, still more bacteria, dust, and smoke particles enter the lungs. In response to the irritation of cigarette smoke, the respiratory system produces more mucus that helps snare foreign particles. But without the cilia sweeping it along, the mucus builds up and can obstruct the airways, producing the familiar "smoker's cough." Like a car air

filter that's never been changed, the lungs of a heavy smoker are literally blackened; compare the color of the normal lung in **Figure E33-3a** with the smoker's lung in **Figure E33-3b**. The longer the delicate tissues of the lungs are exposed to the carcinogens on the trapped particles, the greater the chance that cancer will develop (**Fig. E33-3c**).

Some smokers will develop chronic bronchitis, a persistent lung infection characterized by coughing, swelling of the lining of the respiratory tract, an increase in the production of mucus, and a reduction in the number and activity of cilia. The result is decreased air flow to the alveoli, causing some victims to require supplemental oxygen.

Smoking also promotes *atherosclerosis*, thickening of the arterial walls by fatty deposits that can lead to heart attacks (see Chapter 32). Carbon monoxide, present in high levels in cigarette smoke, reduces the blood's oxygen-carrying capacity and increases the workload on the heart. Chronic bronchitis and emphysema compound this problem. As a result, smokers are far more likely than nonsmokers to die of heart disease.

The carbon monoxide in cigarette smoke also contributes to reproductive problems—which are more common in smokers than in nonsmokers—because it deprives the developing fetus of oxygen. These complications include an increased incidence of miscarriages, lower birth weight babies, and—later in life—more learning and behavioral problems in their children. In spite of this, about 13% of women in the United States smoke during pregnancy.

Passive smoking, or breathing secondhand smoke, poses health hazards to nonsmoking children and adults. Children whose parents smoke have decreased lung capacity and are more likely to contract bronchitis, pneumonia, ear infections, coughs, and colds. These children are also more likely to develop asthma and to have more frequent and severe asthma attacks when they breathe secondhand smoke. Nonsmoking

33.4 HOW DOES GAS EXCHANGE OCCUR IN THE HUMAN RESPIRATORY SYSTEM?

As air moves by bulk flow into and out of the lungs, gases enter and leave the blood by diffusion within the lungs (see Fig. 33-2). Like all mammalian respiratory systems, the human lung provides a moist, well-protected air cavity where exquisitely fragile membranes separate atmospheric air from the bloodstream.

Gas Exchange Occurs in the Alveoli

Inside each lung, a dense, tree-like branching system of bronchioles conducts air to the alveoli, which cluster around the end of each bronchiole like a bunch of grapes. In an average adult, the two lungs combined have approximately 300 million alveoli. These microscopic chambers (0.2 millimeter in diameter) give magnified lung tissue the appearance of a

pink sponge. The inside of the alveoli are coated with a thin layer of watery fluid containing a *surfactant* (a detergent-like substance composed of proteins and lipids), which prevents the fragile alveolar surfaces from sticking together and collapsing when air is exhaled. Gases dissolve in this fluid as they pass in or out of the alveolus.

A network of capillaries covers most of the alveolar surface (see Fig. 33-7b). The walls of the alveoli consist of a single layer of epithelial cells. The **respiratory membrane,** through which gases diffuse, consists of the alveolar wall and the single layer of endothelial cells that form the capillary wall, held together by protein fibers that these cells secrete (**Fig. 33-10**). The respiratory membrane creates a microscopically small diffusion distance (0.5 to 1 micrometer) between the air in the lungs and the blood in capillaries, allowing for very efficient transfer of gases.

The lungs provide an enormous moist surface for gas exchange. In an average adult, the total surface area of the alveoli is about 1,500 square feet (roughly 145 square meters;

(a) A normal lung   **(b) A smoker's lung**   **(c) Lung cancer**

▲ **FIGURE E33-3 Smoking damages the lungs** **(a)** A normal lung. **(b)** A lung of a smoker who died of emphysema is both blackened and collapsed because its alveoli have ruptured. **(c)** The lung cancer is visible as a large pale mass; the surrounding lung is blackened by trapped smoke particles.

spouses of smokers face a 30% higher risk of both heart attack and lung cancer than do spouses of nonsmokers. The Centers for Disease Control and Prevention estimate that secondhand smoke is responsible for about 50,000 deaths annually in the United States, including 46,000 from heart disease and 3,400 from lung cancer.

A smoker who quits will begin to feel better almost immediately, as her blood level of CO declines, and her heart rate and blood pressure begin to return to normal. In a few weeks, her smoker's cough will diminish and blood circulation will improve. After several months, the ability of her lungs to exchange gases will increase, her sense of smell will improve, and food will taste better. People who quit at any age reduce their chances of dying of smoking-related diseases, but those who quit by age 30 are 90% less likely to die of these causes than if they had continued.

about 80 times the skin surface area of a human adult, and about one-third the area of a basketball court).

Oxygen and Carbon Dioxide Are Transported in Blood Using Different Mechanisms

Blood picks up O_2 from the air in the lungs and supplies it to the body tissues, simultaneously absorbing CO_2 from the tissues and releasing it into the lungs. These exchanges occur because the diffusion gradients favor them; in the lungs, O_2 is high and CO_2 is low, whereas in body cells, CO_2 is high and O_2 is low (see Fig. 33-10).

▶ **FIGURE 33-10 Gas exchange between alveoli and capillaries** The respiratory membrane consists of alveolar and capillary walls that are each only one cell thick. Gases diffuse through this membrane between the lungs and the circulatory system. The inner alveolus is coated in a slippery surfactant fluid that prevents the alveolar membranes from adhering to one another.

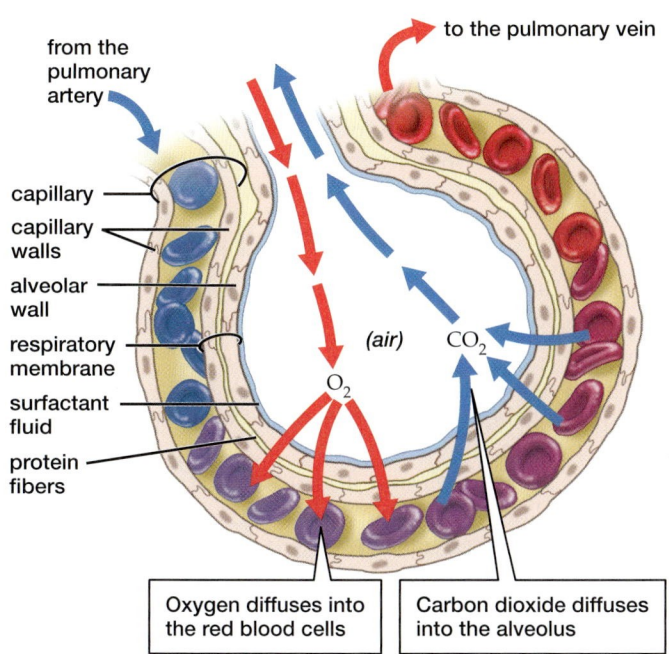

to the pulmonary vein

from the pulmonary artery

capillary

capillary walls

alveolar wall

respiratory membrane

surfactant fluid

protein fibers

(air)

O_2

CO_2

Oxygen diffuses into the red blood cells

Carbon dioxide diffuses into the alveolus

CASE STUDY continued

The High Life Can Lead to Death

Some people who ascend rapidly to high altitudes, such as Jeanette's team member Craig, are stricken with high-altitude pulmonary edema. HAPE strikes susceptible individuals regardless of age, sex, or fitness. Craig exhibited classic symptoms of HAPE; gurgling sounds could be heard in his chest, and he coughed up frothy fluid tinged pink with blood. Like a drowning person, many of his lung alveoli were filled with fluid instead of air, preventing gas exchange. His skin had the bluish tinge of oxygen-poor blood. Oxygen-starved brains function poorly, and HAPE victims often lose touch with reality, believing that they can continue safely. Some even become incoherent. At the end of day 3, his words slurring, Craig insisted that he'd be fine by morning. But when experts from other hiking teams saw him, they realized that Craig was on the verge of death; the fluid in his lungs was suffocating him.

Nearly all (about 98%) of the O_2 carried by blood is bound to **hemoglobin** (Fig. 33-11a), a large protein that gives red blood cells their color (see Fig. 32-9). Each hemoglobin molecule can carry up to four O_2 molecules. By removing O_2 from solution in the plasma, hemoglobin enhances the gradient that causes O_2 to move from the lungs into the plasma. This process allows blood to carry far more O_2 than if the O_2 were simply dissolved in the plasma. As O_2 binds hemoglobin, the protein changes its shape, which alters its color. Oxygenated or "oxygen-rich" blood is a bright cherry-red, and deoxygenated or "oxygen-poor" blood is maroon-red, which appears bluish through the skin. For this reason, we depict oxygen-rich blood as red, and oxygen-poor blood as blue.

Carbon dioxide produced by cellular respiration in body cells diffuses through the interstitial fluid into nearby capillaries and then is carried in the bloodstream to the respiratory membranes of the alveoli. Blood transports CO_2 in three different ways; roughly 10% is dissolved in the plasma (Fig. 33-11b ❶), about 20% is bound to hemoglobin (Fig. 33-11b ❷), and most of it (about 70%) combines with water to form bicarbonate ions (HCO_3^-; Fig. 33-11b ❸) in the following reaction:

$$H_2O + CO_2 \rightarrow H^+ + HCO_3^-$$

The HCO_3^- then diffuses into the plasma (Fig. 33-11b ❹). Most of the H^+ remains bound to hemoglobin in the red blood cells; this helps prevent the blood plasma from becoming too acidic (a serious condition that can lead to coma and death). These reactions keep the amount of CO_2 dissolved in plasma low, which increases the gradient for CO_2 to diffuse into the plasma from the body cells where CO_2 levels are relatively high.

The CO_2 gradient is reversed when the blood reaches the lung capillaries. The dissolved CO_2 diffuses from the

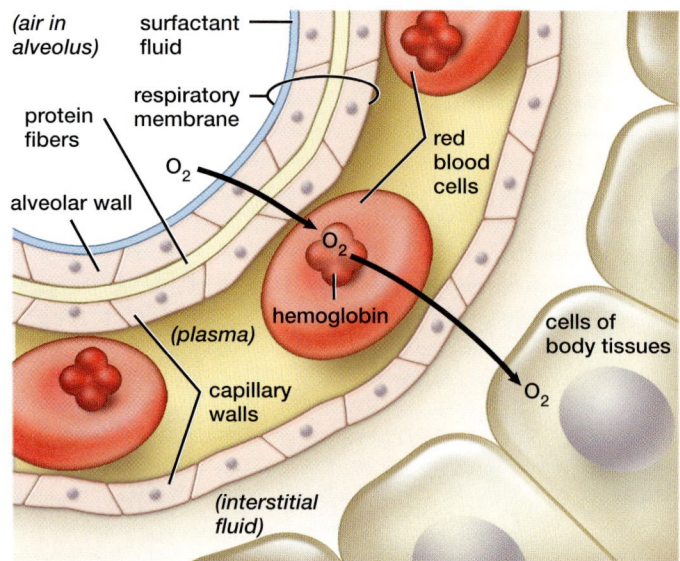

(a) O_2 transport from the lungs to the tissues

(b) CO_2 transport from the tissues to the lungs

▲ FIGURE 33-11 **Oxygen and carbon dioxide transport**
(a) The high O_2 content of the air in the alveoli favors diffusion of O_2 through the respiratory membrane into the alveolar capillaries. Here, O_2 loosely binds to hemoglobin and is transported to the body cells. At the body cells where O_2 is lower, O_2 diffuses off the hemoglobin, out of the capillaries, through the interstitial fluid, and into the cells. **(b)** CO_2 diffuses from body cells through interstitial fluid and into capillaries. ❶ A small amount of CO_2 is carried dissolved in plasma. ❷ Some CO_2 binds to hemoglobin for transport. ❸ Most CO_2 from tissues is combined with H_2O by an enzyme in red blood cells, forming HCO_3^- and H^+. The released H^+ binds to hemoglobin and ❹ the HCO_3^- diffuses into the plasma for transport. ❺ In the alveolar capillaries, as CO_2 levels drop, HCO_3^- diffuses back into red blood cells, where the enzyme recombines it with H^+, forming H_2O and CO_2. ❻ The CO_2 diffuses into the plasma and then into the alveolus.

QUESTION Why is it important for the hydrogen ions (generated when bicarbonate is formed) to remain bound to hemoglobin?

plasma into the alveoli; reduced CO_2 in the plasma causes bicarbonate to regenerate CO_2 and H_2O, in the reverse reaction (**Fig. 33-11b ❺**):

$$H^+ + HCO_3^- \rightarrow H_2O + CO_2$$

The resulting CO_2 and H_2O leave the red blood cells and enter the plasma. This increases plasma CO_2, which promotes diffusion of CO_2 through the respiratory membrane and into the alveoli, from which it is expelled during exhalation (**Fig. 33-11b ❻**).

CHECK YOUR LEARNING

Can you explain how the structure of the alveoli facilitates gas exchange? Are you able to explain how O_2 and CO_2 are transported and exchanged between the lungs and body cells?

? HAVE **YOU** EVER WONDERED …

Why People Can Die from Breathing Carbon Monoxide?

Carbon monoxide (CO) is a toxic gas produced when fuel is not completely burned. Some CO is released by engines, furnaces, charcoal grills, and cigarettes. This gas can be deadly because it binds to the same sites on hemoglobin that bind O_2, but adheres more than 200 times as tightly. As a result, carbon monoxide remains bound to hemoglobin for several hours, preventing the hemoglobin from transporting O_2. People can die of asphyxiation from breathing air with as little as 0.1% CO. Hemoglobin bound to CO appears bright red, just as if it were bound to O_2. Although victims of asphyxiation from other causes have bluish lips and nail beds because their hemoglobin is deoxygenated, the lips and nail beds of victims of CO poisoning are a healthy pink color.

CASE STUDY revisited

The High Life Can Lead to Death

Summit fever—the intense drive to reach the top of a mountain after investing so much and getting so close—can endanger climbers. With judgment impaired from lack of oxygen, some people continue forging onward and upward, even while drowning in their own body fluids. Fortunately, fellow climbers immediately helped Craig descend rapidly, which saved his life. Jeanette met him later, recovered but chagrined, at the mountain's base.

Not all HAPE victims are so fortunate. On New Year's morning, as Jeanette's group approached Uhuru peak—Kilimanjaro's highest point—a woman on her way down suddenly collapsed and stopped breathing. Earlier, though obviously suffering, she had convinced her tour leader that she was OK, and had been left to descend slowly with her husband and a porter. As Jeanette and her team tried desperately to save her using CPR, the 53-year-old woman succumbed to high-altitude pulmonary edema. The porter led her grieving husband down the mountain, alone. The victim had been extremely fit, had regularly led hikes near her home in California, where she had made over 250 mountain ascents.

HAPE seems to randomly strike unacclimated people as they exert themselves while ascending rapidly to high altitudes. For unknown reasons, HAPE victims fail to acclimate adequately to diminished oxygen. Compared to the physiological adjustments of typical climbers, the blood vessels of HAPE victims do not expand as much, nor does their breathing rate increase as much to compensate for the lower oxygen levels at high altitude. As a result, these unfortunate climbers experience lower blood oxygen than their better-adjusted companions. But it gets worse. The vessels that carry blood to and from the lungs of HAPE sufferers respond abnormally to low blood oxygen, greatly increasing the blood pressure in lung capillaries. This added pressure causes fluid from the blood to leak through the capillary walls and into the alveoli. In severe cases such as Craig's, many capillaries rupture under the increased pressure, causing blood to leak into the alveoli. Immediate rapid descent is the only reliable treatment. Hopefully researchers will someday devise a simple test to determine who is susceptible to HAPE, allowing people to reach high summits with less risk to their lives.

Consider This

If you planned to climb Mt. Kilimanjaro, what activities could you engage in beforehand to help protect you from HAPE? When would be the best time—with respect to your climb of Kilimanjaro—to undertake these activities?

CHAPTER REVIEW

Summary of Key Concepts

33.1 Why Exchange Gases and What Are the Requirements for Gas Exchange?

The respiratory system of animals supports cellular respiration. Oxygen-rich air is inhaled and supplies O_2 to the blood, which carries it to cells throughout the body for use in reactions that generate ATP. Blood also picks up CO_2 (a waste product of cellular respiration) from body cells and transports it to the lungs, where it is exhaled into the atmosphere.

The exchange of O_2 and CO_2 between the body and the environment occurs by diffusion across a moist surface. Obtaining adequate oxygen by diffusion requires respiratory surfaces that are moist, thin, and sufficiently large to provide adequate

O_2 for cellular respiration. The transfer of gases between the environment and the body cells during respiration requires both diffusion down concentration gradients and bulk flow to maintain these gradients. Air or water moves by bulk flow past the respiratory surface and gases are carried in blood by bulk flow. Gases move by diffusion across membranes between the respiratory system and the capillaries and between the capillaries and body cells.

33.2 How Do Respiratory Adaptations Minimize Diffusion Distances?

In watery environments, animals with very small or flattened bodies and low metabolic demands may lack specialized respiratory organs and rely exclusively on diffusion through the body surface. Larger, more active animals have evolved specialized respiratory systems. Animals in aquatic environments such as fish and larval amphibians, often possess gills. On land, respiratory surfaces must be protected, supported, and kept moist internally. These requirements have led to the evolution of tracheae in insects and lungs in terrestrial vertebrates. Amphibians, nonbird reptiles, and birds have increasingly specialized respiratory systems that support increasing demands for gas exchange.

33.3 How Is Air Conducted Through the Human Respiratory System?

The human respiratory system includes a conducting portion that consists of the nose and mouth, pharynx, larynx, trachea, bronchi, and bronchioles, which lead to the gas-exchanging alveoli. Inhalation actively draws air into the lungs by contracting the diaphragm and the rib muscles, which expands the chest cavity. Exhalation occurs passively when the diaphragm and rib muscles relax, reducing the volume of the chest cavity and expelling the air. Respiration is controlled by the medulla's respiratory center. The respiratory rate is controlled by receptors that monitor CO_2 and O_2 levels in the blood.

33.4 How Does Gas Exchange Occur in the Human Respiratory System?

Bronchioles conduct air to microscopic air sacs called alveoli where gas exchange with capillary blood occurs. Gases diffuse through the respiratory membrane, which consists of a single layer each of alveolar and capillary cells. Blood in the dense capillary network surrounding the alveoli absorbs O_2 from the alveolar air and releases CO_2 into it.

Oxygen entering lung capillaries is picked up by hemoglobin within red blood cells. Blood transports the O_2 to the body tissues, where it diffuses out. Carbon dioxide diffuses from the tissues into the blood. Most is transported as bicarbonate ions (HCO_3), some is bound to hemoglobin, and a small amount is carried dissolved in blood plasma. Upon reaching the lung capillaries, CO_2 diffuses out into the alveoli to be exhaled.

Key Terms

alveolus (plural, alveoli) *644*
bronchiole *644*
bronchus (plural, bronchi) *644*
bulk flow *637*
conducting portion *643*
countercurrent exchange *639*
diaphragm *644*
epiglottis *644*
exhalation *644*
gas-exchange portion *643*
gill *639*
Heimlich maneuver *644*
hemoglobin *648*
inhalation *644*
larynx *643*
lung *640*
pharynx *643*
respiration *637*
respiratory center *645*
respiratory membrane *646*
respiratory system *639*
spiracle *640*
trachea *644*
tracheae *640*
vocal cords *644*

Learning Outcomes

In this chapter, you have learned to . . .

LO1 Explain how organismal respiration supports cellular respiration.

LO2 Describe how respiratory and circulatory systems work together to allow gas exchange between the body cells and the environment.

LO3 List and explain three adaptations for diffusion found in all respiratory systems.

LO4 Explain and provide examples of how respiration may occur without specialized respiratory organs.

LO5 Describe the four stages of gas exchange in organisms with well-developed respiratory and circulatory systems, including where these stages occur in fish and mammals.

LO6 Compare the respiratory systems of insects, amphibians, snakes, and birds.

LO7 Describe the two different types of gills, and explain how fish gills extract oxygen from water.

LO8 Describe the conducting portion of the human respiratory system, including the function of each part.

LO9 Describe the mechanics and control of human breathing.

LO10 Explain how the structure of alveoli allows gas exchange between air and capillary blood.

LO11 Compare the mechanisms by which oxygen and carbon dioxide are transported in the bloodstream.

Thinking Through the Concepts

Fill-in-the-Blank

1. Three types of organisms that lack specialized respiratory systems are _____, _____, and _____. These animals rely entirely on the process of _____ for gas exchange. **LO3 LO4**

2. Air enters insect bodies through openings called _____. Insect _____ are tubes that conduct air close to each body cell; this is necessary because the _____ of insects cannot adequately distribute gases. **LO6**

3. Which part of the conducting portion of the respiratory system is shared with the digestive tract? _____ What structure normally keeps food from entering the larynx? _____ After passing through the larynx, air travels in sequence through the _____, _____, and _____, after which it enters the gas-exchange portion of the respiratory system, which consists of the _____. **LO8 LO10**

4. In the lungs, oxygen diffuses into blood vessels called _____. Most oxygen in the blood is carried by a protein called _____. Most carbon dioxide in the blood is transported in the form of _____ ions. Cells require oxygen and release carbon dioxide because they are generating energy using the process of _____. **LO1 LO10 LO11**

5. The respiratory membrane is found in the _____. This membrane consists of the wall of a(n)_____ and the wall of a(n) _____ joined by protein strands. The respiratory membrane is (how many?) _____ cells thick. The inside of the respiratory membrane is coated with a watery fluid containing _____. **LO10**

6. Air is inhaled when the _____ and the rib muscles contract, making the chest cavity _____. In contrast, exhalation is a(n) _____ process, caused by allowing these muscles to _____. Breathing is driven by neurons of the _____ located in the _____ of the brain. The respiratory rate is increased when receptors detect an excess of _____ in the blood. **LO9**

Review Questions

1. Describe how fish gills work, including the basic concept of countercurrent exchange. Why is this process important to allow fish to extract oxygen from water? **LO5 LO7**

2. How does the respiratory system of a frog change when it undergoes metamorphosis? Why are these changes necessary? **LO6 LO7**

3. Describe the respiratory system of birds and how it allows oxygen-rich air to enter the lungs during both inhalation and exhalation. **LO6**

4. Trace the route taken by air in the mammalian respiratory system, listing the structures through which air flows and the structure in which gas exchange occurs. **LO8 LO10**

5. Explain some characteristics of animals in moist environments that make specialized respiratory systems unnecessary. **LO3 LO4**

6. How is breathing initiated? How are breathing rate and depth adjusted, and which blood gas is most tightly regulated? **LO9**

7. What events occur during human inhalation? During exhalation? Which of these is always an active process? **LO9**

8. Compare bulk flow and diffusion. Explain how bulk flow and diffusion interact to promote gas exchange between air and blood and between blood and tissues. **LO2 LO5**

9. Compare CO_2 and O_2 transport in the blood. Include the source and destination of each, and what physiological process makes gas exchange necessary. **LO1 LO2 LO11**

10. Explain how the structure and arrangement of alveoli make them well suited for their role in gas exchange. **LO2 LO3 LO10**

Applying the Concepts

1. BioEthics Heart–lung transplants are performed in some cases in which both organs have been damaged by cigarette smoking. Given a scarcity of organ donors, what criteria would you use in selecting a recipient for such a transplant?

2. Nicotine is a drug in tobacco that is responsible for several of the effects that smokers crave. Discuss the advantages and disadvantages of low-nicotine cigarettes.

3. Explain the effects of smoking on the developing fetus and young children. If a pregnant friend continued to smoke, how would you deal with the situation?

4. Discuss why a brief exposure to carbon monoxide is much more dangerous than a brief exposure to carbon dioxide.

5. Mary, a strong-willed 3-year-old, threatens to hold her breath until she dies if she doesn't get her way. Can she carry out her threat? Explain.

Answers to Figure Caption questions and Fill-in-the-Blank questions can be found in the Answers section at the back of the book.

MB *Go to MasteringBiology for practice quizzes, activities, eText, videos, current events, and more.*

Supermodel Chloe Memisevic. (Inset) Recovered anorexic Carré Otis, a former supermodel, is now the model of health.

Dying to Be Thin

REFLECTING ON HER CAREER, former supermodel Carré Otis (inset) stated: "The sacrifices I made were life-threatening. I had entered a world that seemed to support a 'whatever it takes' mentality to maintain abnormal thinness." For many models, performers, and others in the public eye, meeting expectations for thinness is a continuing battle that sometimes leads to tragedy. Having conquered anorexia, Otis is now a spokesperson for the National Eating Disorders Association. She hopes to help others avoid the damage her body suffered. "It was common for the young girls I worked with to have a heart attack; if an eating disorder is not treated, it can be a fatal disease."

Although eating disorders occur most frequently in women, about 10% of anorexia sufferers are men. One 17-year-old male with anorexia explained, "I would wake up a few minutes early, . . . then pump out about a hundred push-ups, do some crunches and then get in the shower, get dressed, come downstairs, hide [my breakfast], then flush it when I was going up to brush my teeth. . . . And then I'd pump out some more push-ups."

Eating disorders include two particularly debilitating conditions: anorexia and bulimia. Celebrities who have battled eating disorders include Jane Fonda, the late Princess Diana, Paula Abdul, Elton John, as well as Isabella Caro, one of several fashion models who have died from complications of anorexia. People with anorexia experience an intense fear of gaining weight; although their bodies become skeletal, they see themselves as fat. In response, they eat very little food and often exercise compulsively, burning off essentially all their body fat and breaking down muscle tissue to supply their energy needs. In addition to damaging muscle tissue, anorexia disrupts digestive, reproductive, endocrine, and cardiac functions.

People with bulimia—some of whom are anorexic, while others maintain a normal weight—engage in binge eating, consuming enormous amounts of food in a short period of time. To purge the food from their bodies, they then induce vomiting or overdose with laxatives and may also engage in excessive exercise.

What are some treatments for under- and overeating? What nutrients do our bodies require, and how do our bodies extract nutrients from the food we eat? Read on to find out.

AT A GLANCE

34.1 What Nutrients Do Animals Need?

34.2 How Does Digestion Occur?

34.3 How Do Humans Digest Food?

LEARNING GOALS

LG1 List the six major categories of nutrients and the role each plays in maintaining the body.

LG2 Describe the role of energy in the diet, including how food energy is measured, which nutrients provide the most energy, and how the body stores energy.

LG3 List and explain the five tasks that all digestive systems must accomplish.

LG4 Describe three different modes of digestion in invertebrates, including the names of the structures involved.

LG5 Describe the digestive specializations of herbivorous, omnivorous, and carnivorous vertebrates, and the distinctive features of bird and cow digestive systems.

LG6 List the structures, in order, through which food passes in the human digestive system, describe the function of each structure, and explain how food is moved through it.

LG7 Describe the physical and chemical breakdown of food starting with the mouth and ending with the small intestine; for each digestive compartment, include the sources and functions of the digestive secretions that act there.

LG8 Describe nutrient absorption in the small intestine, including how different nutrients are absorbed and how the structure of the small intestine facilitates absorption.

LG9 Explain where and how feces are produced and eliminated and what feces are composed of.

LG10 Explain how the nervous system and hormones regulate digestion and food intake.

34.1 WHAT NUTRIENTS DO ANIMALS NEED?

Whether you eat cauliflower or a candy bar, your food contains important nutrients. **Nutrients** are substances obtained from food and water that organisms need to maintain health. Water is essential, and often contains dissolved minerals. Animal nutrients fall into six major categories: carbohydrates, lipids, proteins, minerals, vitamins, and water. These substances provide energy and raw materials to synthesize the molecules of life.

Most Energy Is Provided by Carbohydrates and Lipids

Cells rely on a continuous supply of energy to maintain their incredible complexity and wide range of activities. Deprived of this energy, cells begin to die within minutes. Nutrients that provide energy are carbohydrates (sugars and starches), lipids (fats and oils), and proteins. Although people's diets vary considerably, in a typical U.S. diet, carbohydrates provide about 50% of the total energy, fats and oils about 35%, and proteins about 15% (well above the world average for protein). These molecules are broken down by digestion, which frees their subunits to be used to generate ATP during cellular respiration (see Chapter 8).

Energy from Nutrients Is Measured in Calories

A **calorie** is the amount of energy required to raise the temperature of 1 gram of water by 1°C. In fact, scientists measure the calories stored in food by burning it completely and measuring the heat produced. The energy content of foods is expressed in units of 1,000 calories (kilocalories), or **Calories** (with a capital "C"). About 60% of the caloric energy in the food we eat is released as heat. The remaining 40% is captured in ATP, which is continuously utilized during metabolic reactions including those that power muscle contraction, active transport across membranes, and the synthesis of biological molecules.

The average human body at rest burns roughly 70 Calories per hour, but this value is influenced by several factors. For example, muscle burns more Calories than fat does, so a muscular individual, even at rest, consumes more calories than a person carrying the same weight in fat. People also differ in **metabolic rate,** the speed at which cellular reactions occur. Exercise significantly boosts Caloric requirements; well-trained athletes can temporarily burn nearly 20 Calories per minute during vigorous exercise. The Calorie content of some foods and the time required to burn off these Calories during various types of exercise is provided in Table 34-1.

Carbohydrates Are a Source of Quick Energy

Carbohydrates include sugars such as glucose, from which cells derive most of their energy; sucrose (table sugar); and polysaccharides, long chains of sugar molecules (see Chapter 3). Cellulose, starch, and glycogen are all polysaccharides composed of chains of glucose. Cellulose, the major structural component of plant cell walls, is the most abundant carbohydrate on the planet, but only a few types of animals are able to digest it, as described later. Starch is the

TABLE 34-1 Approximate Energy Expended by a 150-Pound Person for Different Activities					
		Time to "Work Off"			
Activity	Calories/Hour	500 Calories (Cheeseburger)	340 Calories (Ice cream cone)	70 Calories (Apple)	40 Calories (Broccoli, 1 cup)
Running (6 mph)	700	43 min	26 min	6 min	3 min
Cross-country skiing (moderate)	560	54 min	32 min	7.5 min	4 min
Roller skating	490	1 hr 1 min	37 min	8.6 min	5 min
Bicycling (11 mph)	420	1 hr 11 min	43 min	10 min	6 min
Walking (3 mph)	250	2 hr	1 hr 12 min	17 min	10 min
Frisbee® playing	210	2 hr 23 min	1 hr 26 min	20 min	11 min
Studying	100	5 hr	3 hr	42 min	24 min

principal energy-storage material of plants, and a major energy source for humans and many other animals.

Glycogen is used by animals for short-term energy storage. Animals, including humans, store glycogen in the liver and muscles. Athletes sometimes "carbo-load" before competing by eating meals rich in carbohydrates, such as potatoes and pasta, to store as much glycogen as possible. Although humans can accumulate hundreds of pounds of fat, most can store less than a pound of glycogen. During exercise such as running, the body uses glycogen as a source of quick energy; one pound of glycogen can power about 18 miles of running.

Fats and Oils Are the Most Concentrated Energy Source

Fats and oils are the most concentrated sources of energy, containing over twice as many Calories per unit weight as carbohydrates or proteins do (about 9 Calories per gram for fats compared to about 4 Calories per gram for proteins and carbohydrates).

When an animal's diet provides more energy than it expends, most of the excess energy is stored as body fat. In addition to its high caloric content, fat is hydrophobic, so it neither attracts water nor dissolves in water, as carbohydrates and proteins do. For this reason, fat deposits do not cause extra water to accumulate in the body, so fat stores more Calories with less weight than other molecules do. Minimizing weight allows an animal to move faster (important for escaping predators and hunting prey) and to use less energy when it moves (important when food supplies are limited).

In addition to storing energy, fat deposits may provide insulation. Fat, which conducts heat at only one-third the rate of other body tissues, is often stored in a layer directly beneath the skin. Birds and mammals that live in polar climates or in cold ocean waters—such as penguins, seals, whales, and walruses—are particularly dependent on this insulating layer, which reduces the amount of energy they must expend to keep warm (**Fig. 34-1**).

▲ FIGURE 34-1 **Fat provides insulation** This harp seal pup can withstand the icy waters of the arctic seas because a thick layer of fat insulates it from the cold. The fat is transferred from the mother who nurses the baby seal with milk that contains up to 48% fat. During the 12-day nursing period, the pup gains nearly 5 pounds each day.

An Evolutionary Tendency to Store Fat Can Lead to Obesity When Food Is Abundant

Because humans evolved under the same food constraints as other animals did, we have inherited a strong tendency to eat when food is available, often eating more than we need immediately, because we may require the energy later, when food is scarce. Some modern societies now have access to almost unlimited high-calorie food. In this environment, our natural tendency to overeat can become a liability, and some of us need to exert considerable willpower to avoid storing excessive amounts of fat and becoming overweight. The **body mass index (BMI)** is based on height and weight, and is a common tool for estimating a healthy weight. For people with average amounts of muscle, a BMI between 18.5 and 24.9 is considered healthy. In the United States, about 33% of all U.S. adults are overweight (BMI between 25 and 29.9) and an additional 33% are obese (BMI of 30 or more); see "Health Watch: Eating to Death?" later in this chapter.

CASE STUDY continued

Dying to Be Thin

A BMI below 18.5 is considered underweight. Carrè Otis, who once carried less than 100 pounds on her 5-foot, 10-inch frame, then had a BMI of 14.2. In response to the death of several models from complications of anorexia, the fashion trade show "Madrid Fashion Week" in Spain has banned any model with a BMI under 18 from its shows, and Israel now requires models to have a BMI of at least 18.5. Unfortunately, the cult of thinness persists and ultrathin women such as Chloe Memisevic (in our chapter-opening photo), whose estimated BMI is less than 15, still model for major fashion houses.

Essential Nutrients Provide the Raw Materials for Health

To maintain health, our diets must provide our bodies with a variety of raw materials. Our cells can synthesize most of the molecules our bodies require, but they cannot synthesize certain raw materials, called **essential nutrients,** which must be supplied in the diet. For example, although carbohydrates are an important source of energy and part of a balanced diet, they are not considered essential nutrients because our bodies can synthesize those we need.

Essential nutrients differ for different animals. For example, vitamin C (ascorbic acid) is an essential nutrient for people, but not for most other animals because they can synthesize it. Essential nutrients for humans include certain fatty acids and amino acids, a variety of minerals and vitamins, and water.

Certain Fatty Acids Are Essential in the Human Diet

Fats and oils are more than just a source of energy—some provide **essential fatty acids.** Essential fatty acids serve as raw materials used to synthesize molecules involved in a wide range of physiological activities: They help us to absorb fat-soluble vitamins (described later) and are important in cell division, fetal development, and the immune response. Important sources of essential fatty acids include fish oils, canola oil, soybean oil, flaxseed, and walnuts.

Amino Acids Form the Building Blocks of Protein

Proteins form muscle, connective tissue, nails, and hair. They also act as enzymes, receptors on cell membranes, and antibodies. The human body cannot synthesize 9 (adults) or 10 (infants) of the 20 different amino acids used in proteins. These **essential amino acids** must be obtained from protein-rich foods such as meat, milk, eggs, corn, beans, and soybeans. Because many plant proteins are deficient in some of the essential amino acids, vegetarians must consume a variety of plants (for example, legumes, grains, and corn) whose proteins collectively provide all of them. Protein deficiency can cause a debilitating condition called kwashiorkor (**Fig. 34-2**), which occurs most frequently in poverty-stricken countries.

▲ **FIGURE 34-2 Kwashiorkor is caused by protein deficiency** In a child with kwashiorkor, the abdomen swells because low levels of blood albumin protein decrease the blood's osmotic strength, allowing fluids to leak from blood capillaries into the interstitial fluid of the abdominal area. Muscles atrophy due to lack of protein.

Minerals Are Elements Required by the Body

Minerals are elements that play many crucial roles in animal nutrition. Minerals such as calcium, magnesium, and phosphorus are major constituents of bones and teeth. Sodium, calcium, and potassium are essential for muscle contraction and the conduction of nerve impulses. Iron is a central component of each hemoglobin molecule in blood, and iodine is found in hormones produced by the thyroid gland. We also require trace amounts of several other minerals listed in **Table 34-2.** Because no organism can manufacture elements, all essential minerals must be obtained from food or dissolved in drinking water.

Vitamins Play Many Roles in Metabolism

"Take your vitamins!" is a familiar refrain in many households with children. **Vitamins** are a diverse group of organic molecules that animals cannot synthesize, but that are necessary for cell function, growth, and development. The vitamins essential in human nutrition are listed in **Table 34-3.** Human vitamins are grouped into two categories: water soluble and fat soluble.

Water-Soluble Vitamins Water-soluble vitamins include vitamin C and the eight compounds that make up the B-vitamin complex. Because these substances dissolve in the watery blood plasma and are filtered out by the kidneys, they are not stored in the body in appreciable amounts. Therefore, these vitamins must be constantly provided in the diet.

TABLE 34-2 Important Minerals for Humans

Mineral	Dietary Sources	Important Roles in the Body	Deficiency Symptoms
Calcium	Milk, cheese, leafy vegetables	Helps in bone and tooth formation and maintenance; aids in blood clotting; contributes to nerve impulse transmission and muscle contraction	Stunted growth, rickets, osteoporosis
Phosphorus	Milk, cheese, meat, poultry, grains	Helps maintain pH of body fluids; contributes to bone and tooth formation; component of ATP and of phospholipids in cell membranes	Muscular weakness, weakening of bone
Potassium	Meats, milk, fruits	Helps maintain pH and osmotic strength of body fluids; important in nervous system activity	Nausea, muscular weakness, paralysis
Chlorine	Table salt	Helps maintain pH and osmotic strength of body fluids; component of HCl produced by gastric glands; important in nervous system activity	Muscle cramps, apathy, reduced appetite
Sodium	Table salt	Helps maintain pH and osmotic strength of body fluids; important in nervous system activity	Muscle cramps, nausea
Magnesium	Whole grains, leafy vegetables, dairy products, legumes, nuts	Helps to activate many enzymes	Tremors, muscle spasms, and weakness, irregular heartbeat, hypertension
Iron	Meats, legumes, nuts, whole grains, leafy vegetables	Component of hemoglobin and many enzymes	Iron-deficiency anemia (weakness, reduced resistance to infection)
Fluorine	Fluoridated water, seafood	Component of teeth and bones	Increased tooth decay; may increase risk of osteoporosis
Zinc	Seafood, meat, cereals, nuts, legumes	Constituent of several enzymes; component of proteins required for normal growth, smell, and taste	Retarded growth, learning impairment, depressed immunity
Iodine	Iodized salt, seafood, dairy products, many vegetables	Component of thyroid hormones	Goiter (enlarged thyroid gland)
Chromium	Meats, whole grains	Helps maintain normal blood glucose levels	Elevated insulin in blood; increased risk of adult-onset diabetes

Most water-soluble vitamins act as *coenzymes*; that is, they work in conjunction with enzymes to promote chemical reactions that supply energy or synthesize biological molecules. Because each vitamin participates in several metabolic processes, a deficiency of a single vitamin can have wide-ranging effects (see Table 34-3). For example, deficiency of the B vitamin niacin causes the swollen tongue and skin lesions of pellagra (Fig. 34-3), as well as some nervous disorders. Folic acid, another B vitamin, is required to synthesize thymine, a component of DNA; folic acid deficiency impairs cell division throughout the body. As you might predict, it is particularly important for pregnant women to get enough folic acid to supply the rapidly growing fetus. Folic acid deficiency can also lead to a reduction in red blood cells, resulting in anemia. For folic acid to function properly, trace amounts of vitamin B_{12} are required. In the human diet, vitamin B_{12} is obtained only by eating meat and dairy products or vitamin B_{12}–supplemented foods.

Fat-Soluble Vitamins Fat-soluble vitamins can be stored in body fat and may accumulate in the body over time. The fat-soluble vitamins A, D, E, and K have a variety of functions (see Table 34-3). Vitamin A is used to synthesize the

▲ FIGURE 34-3 **Pellagra is caused by niacin deficiency** Symptoms of pellagra include the scaly, reddish-brown skin lesions visible on this individual's legs.

TABLE 34-3 Important Vitamins for Humans

Vitamin	Dietary Sources	Functions in Body	Deficiency Symptoms
Water Soluble			
B Complex			
Vitamin B$_1$ (thiamin)	Meat, cereals, peas and soybeans, fish	Coenzyme in glucose metabolism	Beriberi (muscle weakness, peripheral nerve changes, heart failure)
Vitamin B$_2$ (riboflavin)	Shellfish, dairy products, eggs	Component of coenzymes involved in energy metabolism	Reddened lips, cracks at the corners of the mouth, light sensitivity, blurred vision
Niacin	Meats, fish, leafy vegetables	Component of coenzymes involved in energy metabolism	Pellagra (skin and gastrointestinal lesions, and nervous and mental disorders)
Vitamin B$_5$ (pantothenic acid)	Meats, whole grains, legumes, eggs; product of intestinal bacteria	Component of coenzyme A, with a role in cellular respiration	Fatigue, sleep disturbances, impaired coordination
Vitamin B$_6$ (pyridoxine)	Meats, whole grains, tomatoes, potatoes	Coenzyme in amino acid metabolism	Irritability, convulsions, skin disorders, increased risk of heart disease
Biotin	Legumes, vegetables, meats; product of intestinal bacteria	Coenzyme in amino acid metabolism and cellular respiration	Fatigue, depression, nausea, skin disorders, muscular pain
Folic acid	Meats, green and leafy vegetables, whole grains, eggs; product of intestinal bacteria	Coenzyme in nucleic and amino acid metabolism	Anemia, gastrointestinal disturbances, diarrhea, nervous system defects in fetus and low birth weight
Vitamin B$_{12}$	Meats, eggs, dairy products	Coenzyme in nucleic acid metabolism and other metabolic pathways	Anemia, neurological disturbances
Vitamin C (ascorbic acid)	Citrus and other fruits, tomatoes, leafy vegetables	Helps maintain cartilage, bone, and dentin (hard tissue of teeth); helps in collagen synthesis	Scurvy (degeneration of skin, teeth, gums, and blood vessels, and epithelial hemorrhages)
Fat Soluble			
Vitamin A	Green, yellow, and red vegetables; liver; fortified dairy products	Component of visual pigment; helps maintain skin and other epithelial cells; promotes normal development of teeth and bones	Night blindness, permanent blindness, increased susceptibility to infections
Vitamin D	Cod liver oil, eggs, fortified dairy products	Promotes bone growth and mineralization; increases calcium absorption	Rickets (bone deformities) in children, skeletal deterioration
Vitamin E (tocopherol)	Nuts, whole grains, leafy vegetables, vegetable oils	Antioxidant; may reduce cellular damage from free radicals	Neurological damage
Vitamin K	Leafy vegetables; product of intestinal bacteria	Important in blood clotting	Bleeding, internal hemorrhages

light-capturing molecule in the retina of the eye; a deficiency can lead to impaired night vision. Vitamin D is required for normal bone formation; a deficiency can lead to bone deformities, such as occur in rickets (Fig. 34-4). Sunlight stimulates vitamin D synthesis in the skin; however, people who get little exposure to sunlight may not synthesize enough. Researchers have discovered that many adult women in the United States, particularly those with dark skin (which reduces the penetration of sunlight), have inadequate levels of vitamin D. Breast-fed children born to vitamin D–deficient mothers are at particular risk for rickets. Vitamin E is an antioxidant, neutralizing free radicals that form in the body (see Chapter 2). Vitamin K helps to regulate blood clotting.

The overwhelming variety of tempting processed foods in a typical U.S. supermarket makes it easy to make poor nutritional choices. To help people make better-informed choices, the U.S. government has placed nutritional guidelines, called "ChooseMyPlate," on an interactive Web site. Commercially packaged foods are also required to provide complete information about Calorie, fiber, fat, sugar, and vitamin content (Fig. 34-5). Fast-food restaurants often make this information available on wall charts, handouts, and their Web sites.

The Human Body Is About Sixty Percent Water

A person can survive far longer without food than without water, which makes up roughly 60% of total body weight. All metabolic reactions occur in a watery solution, and water participates directly in hydrolysis reactions (see Chapter 3) that break down proteins, carbohydrates, and fats into simpler molecules. Water is the principal component of saliva, blood, lymph, interstitial fluid, and the cytosol within each cell. By

▲ **FIGURE 34-4 Rickets is caused by vitamin D deficiency**
Lack of vitamin D in early childhood prevents bones from absorbing sufficient calcium, making them soft and unable to bear weight without bending. This can result in permanent bone deformities.

Nutrition Facts

Serving Size 1/2 Cup (140g)
Servings About 3

Amount Por Serving		
Calories 70		Calories from Fat 0
		% Daily Value*
Total Fat 0g		0%
Saturated Fat 0g		0%
Trans Fat 0g		
Cholesteroi 0mg		0%
Sodium 10mg		0%
Total Carbohydrate 17g		6%
Dietary Fiber 1g		2 %
Sugars 17g		
Protein 1g		

Vitamin A	6%	• Vitamin C	35%
Calcium	2%	• Iron	4%

*Percent Daily Values are based on a 2,000 calorie diet. Your Daily Values may be higher or lower depending on Your calorie needs:

		Calories:	2,000	2,500
Total Fat	Less than		65g	80g
Sat Fat	Less than		20g	25g
Cholesterol	Less than		300mg	300mg
Sodium	Less than		2,400mg	2,400mg
Total Carbohydrate			300g	375g
Dietary Fiber			25g	30g

▲ **FIGURE 34-5 Food labeling** The U.S. government requires complete nutritional labeling of foods, as illustrated by this example. The weight (in grams) of various nutrients—such as fat, cholesterol, and sodium—is shown as a percentage of the recommended daily intake (% Daily Value), assuming a 2,000-Calorie diet.

QUESTION Does this item look like a healthy food choice? Explain.

sweating, people use the evaporation of water to keep from overheating. Urine, which is mostly water, is necessary to eliminate cellular waste products from the body (see Chapter 35).

CHECK YOUR LEARNING

Can you list the six major types of nutrients and the role of each in the body? Do you understand how food energy is measured and how energy is obtained, used, and stored by the body? Can you define *essential nutrients* and describe five different categories of these?

34.2 HOW DOES DIGESTION OCCUR?

Digestion is the process that physically grinds up food and then chemically breaks it down. The animal **digestive system** consists of a series of compartments in which food is processed, as well as organs that produce secretions that aid in digestion. The digestive system takes in food, breaks down complex food molecules into simpler molecules, absorbs nutrients, and then expels leftover wastes from the body.

Animals eat the bodies of other organisms, but some body parts are difficult to digest. Each plant cell, for example, is supported by a wall of indigestible cellulose. Animal bodies may be covered with indigestible chitin, fur, scales, or feathers. In addition, the complex lipids, proteins, and carbohydrates in food do not occur in a form that can be used directly. These nutrients must be broken down before they can be absorbed and distributed to the cells of the animal that has consumed them, where they are recombined in various ways. The digestive tracts of animals are diverse, having evolved to meet the challenges posed by a variety of diets. All digestive systems, however, must accomplish five tasks:

1. **Ingestion** Food is brought into the digestive tract through an opening, usually called a **mouth.**
2. **Mechanical digestion** The food is physically broken down into smaller pieces with a larger surface area for attack by digestive enzymes.
3. **Chemical digestion** Particles of food are exposed to enzymes and other digestive secretions that break down large molecules into smaller subunits.
4. **Absorption** The small subunits are transported out of the digestive tract and into the body through cells lining the digestive tract. Although the digestive tract surrounds and acts on food, nutrients do not actually enter the body until they are absorbed.
5. **Elimination** Indigestible materials are expelled from the body.

In the following sections, we explore a few of the diverse mechanisms by which animal digestive systems accomplish these functions.

In Sponges, Digestion Occurs Within Single Cells

Sponges are the only animals that lack a digestive system and rely exclusively on **intracellular digestion,** in which all digestion occurs within individual cells. As you might

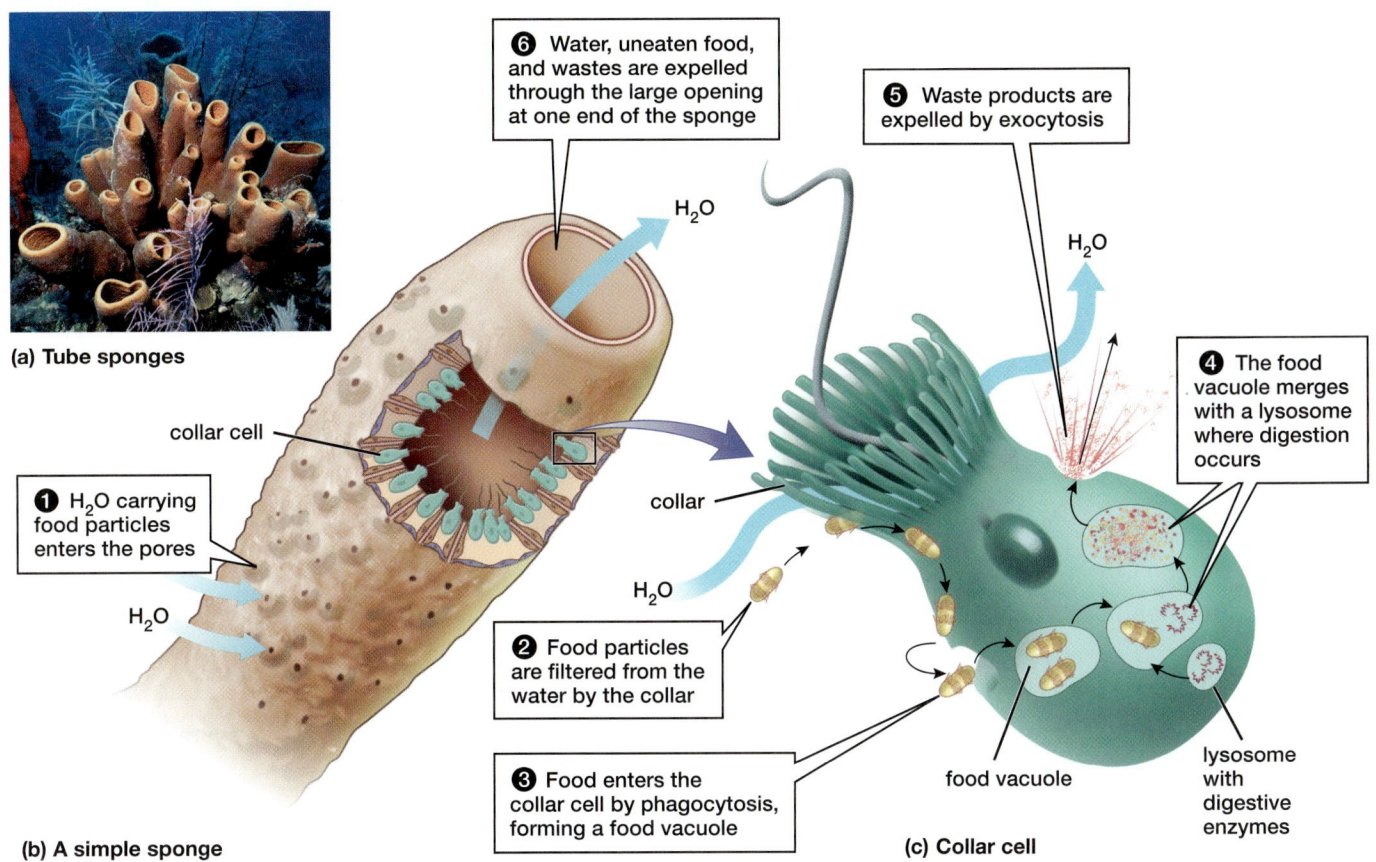

6 Water, uneaten food, and wastes are expelled through the large opening at one end of the sponge

5 Waste products are expelled by exocytosis

(a) Tube sponges

collar cell

1 H_2O carrying food particles enters the pores

4 The food vacuole merges with a lysosome where digestion occurs

collar

H_2O

H_2O

H_2O

2 Food particles are filtered from the water by the collar

3 Food enters the collar cell by phagocytosis, forming a food vacuole

food vacuole

lysosome with digestive enzymes

(b) A simple sponge

(c) Collar cell

▲ **FIGURE 34-6 Intracellular digestion in a sponge (a)** Tube sponges photographed in the Virgin Islands. **(b)** The anatomy of a simple sponge, showing the direction of water flow and the location of the collar cells. **(c)** Here, a single collar cell is enlarged to show the intracellular digestion of single-celled organisms, which are filtered from the water, trapped on the outside of the collar, engulfed, and digested.

suspect, this limits their food to microscopic particles. Sponges circulate seawater through pores in their bodies (**Fig. 34-6 ❶**). Specialized collar cells inside the sponge filter microscopic organisms from the water (**Fig. 34-6 ❷**) and ingest them using the process of phagocytosis ("cell eating"; see Chapter 5). Once ingested by a cell, the food is enclosed in a temporary sac called a **food vacuole** (**Fig. 34-6 ❸**). The vacuole fuses with a **lysosome,** a membrane-enclosed packet of digestive enzymes within the cell (**Fig. 34-6 ❹**). Food is broken down within the vacuole into smaller molecules that can be absorbed into the cell cytoplasm. Undigested remnants are expelled by exocytosis (**Fig. 34-6 ❺**) and exit the sponge through a large opening in its body wall (**Fig. 34-6 ❻**).

The Simplest Digestive System Is a Chamber with One Opening

All animals except sponges have evolved a chamber within the body where chunks of food are broken down by enzymes that act outside the cells, a process called **extracellular digestion.** One of the simplest of these chambers is found in cnidarians, such as sea anemones, *Hydra,* and sea jellies. The cnidarian digestive chamber is called a **gastrovascular cavity,** and has a single opening through which food is

ingested and wastes are ejected. The animal's stinging tentacles capture smaller animals and push their prey through the mouth into the gastrovascular cavity (**Fig. 34-7 ❶**). Gland cells lining the cavity secrete enzymes that begin digesting the prey (**Fig. 34-7 ❷**). Nutritive cells lining the cavity then absorb the nutrients and also engulf partly digested food particles by phagocytosis. Further digestion is intracellular, within food vacuoles in the nutritive cells (**Fig. 34-7 ❸**). Because undigested wastes are expelled back through the mouth, only one meal can be processed at a time.

Most Animals Have Tubular Digestive Systems with Specialized Compartments

A saclike digestive system is unsuitable for animals that must eat frequently. Most animals, including invertebrates such as mollusks, arthropods, echinoderms, and earthworms, as well as all vertebrates, have digestive systems that are basically one-way tubes that begin with a mouth and end with an anus. Specialized regions within the tube process the food in an orderly sequence, taking it in through the mouth, physically grinding it up, enzymatically breaking it down, absorbing the nutrients, and finally, expelling the wastes through the anus.

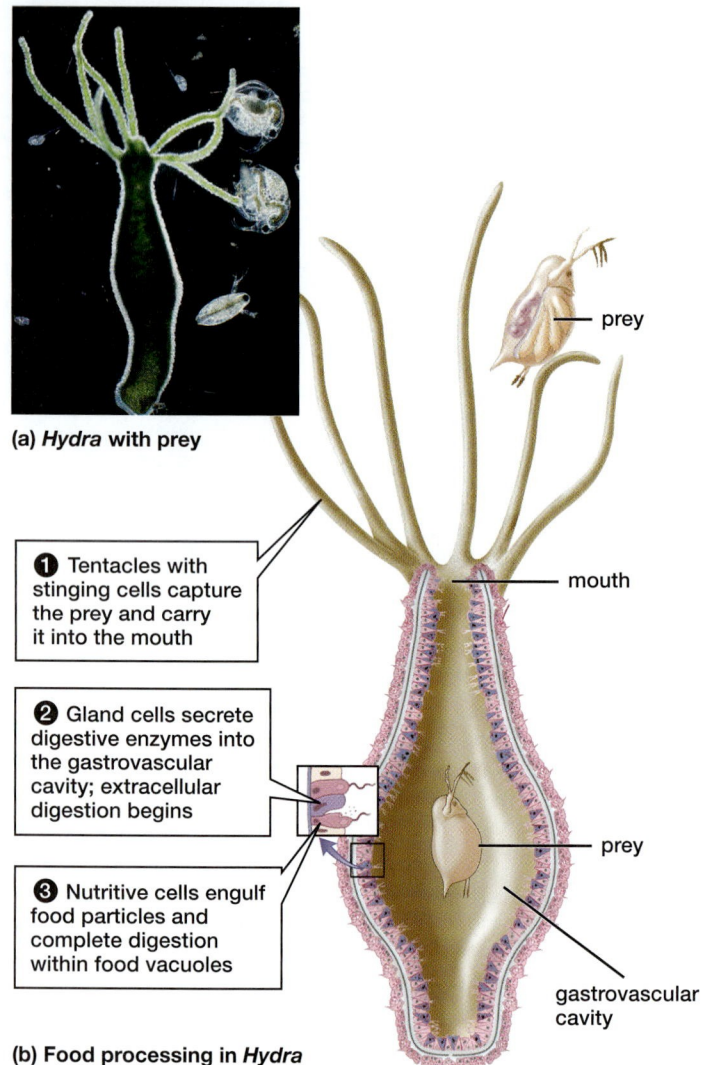

(a) *Hydra* with prey

❶ Tentacles with stinging cells capture the prey and carry it into the mouth

❷ Gland cells secrete digestive enzymes into the gastrovascular cavity; extracellular digestion begins

❸ Nutritive cells engulf food particles and complete digestion within food vacuoles

mouth

prey

gastrovascular cavity

(b) Food processing in *Hydra*

▲ **FIGURE 34-7** **Digestion in a sac** **(a)** A *Hydra* has captured and ingested a waterflea (*Daphnia*, a tiny crustacean) into its gastrovascular cavity. **(b)** After nutrients are absorbed, undigested waste will be expelled out through the mouth.

The earthworm provides a good example (**Fig. 34-8**). As it burrows, the worm ingests soil and bits of plant material that pass through the **esophagus,** a muscular tube leading from the mouth to the *crop,* an expandable sac where food is stored. The material is then gradually released into the *gizzard,* where ingested sand grains and muscular contractions grind it into smaller particles. In the intestine, enzymes attack the food particles, and the resulting small molecules are absorbed into the worm's body. The remaining soil containing undigested organic material is expelled through the anus.

Vertebrate Digestive Systems Are Specialized According to Their Diets

Different types of animals have radically different diets. **Carnivores,** such as wolves, cats, seals, and predatory birds, eat other animals. **Herbivores,** which eat only plants, include seed-eating birds; grazing animals, such as deer, camels, and cows; and many rodents, such as mice. **Omnivores,** such as

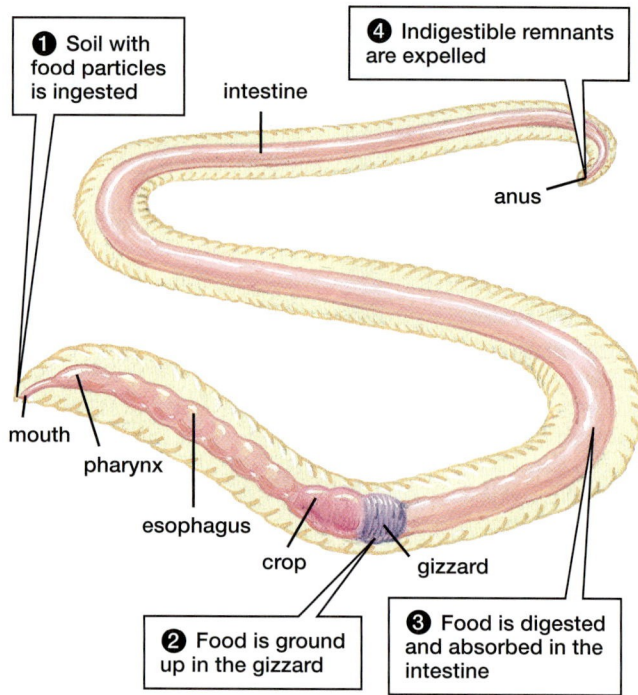

❶ Soil with food particles is ingested

❹ Indigestible remnants are expelled

intestine

anus

mouth

pharynx

esophagus

crop

gizzard

❷ Food is ground up in the gizzard

❸ Food is digested and absorbed in the intestine

▲ **FIGURE 34-8** **A tubular digestive system** The earthworm has a one-way digestive system that passes food through a series of compartments, each specialized to play a specific role in breaking down food and absorbing it.

humans, bears, and raccoons, consume and are adapted to digest both animal and plant sources of food. Specialized digestive systems allow animals with different diets to extract the maximum amount of nutrients from their foods. The major organs of the vertebrate digestive system are the mouth, esophagus, stomach, small intestine, and large intestine—but some vertebrates have additional specialized chambers.

Teeth Accommodate Different Diets

Teeth are adapted to diet. The varied, omnivorous diet of humans has selected for our distinctive set of teeth. We have thin, flat incisors for shearing off food. Because we don't catch prey with our mouths, our canines are small, and in some people they are not even sharp, resembling incisors. Our premolars and molars have relatively large, irregular surfaces for crushing and grinding (**Fig. 34-9a**).

If you have a dog, look carefully in its mouth. Carnivores have very small incisors, but greatly enlarged canines for stabbing and tearing flesh. Their molars and premolars have sharp edges for shearing through tendon and bone. None of their teeth are adapted for grinding or chewing; they tend to swallow their food in chunks (**Fig. 34-9b**).

Herbivores—such as horses, rabbits, deer, and cows—generally lack canine teeth. In cows and other ruminants (described below), the upper incisors are replaced by a tough, leathery *dental pad.* Ruminant herbivores pull vegetation into their mouths with large flexible tongues and bite it off between their lower incisors and dental pads. Horses and rabbits (which are not ruminants) have well-developed incisors for snipping plants. The premolars and molars of herbivores have large flat surfaces for grinding up tough plant material (**Fig. 34-9c**).

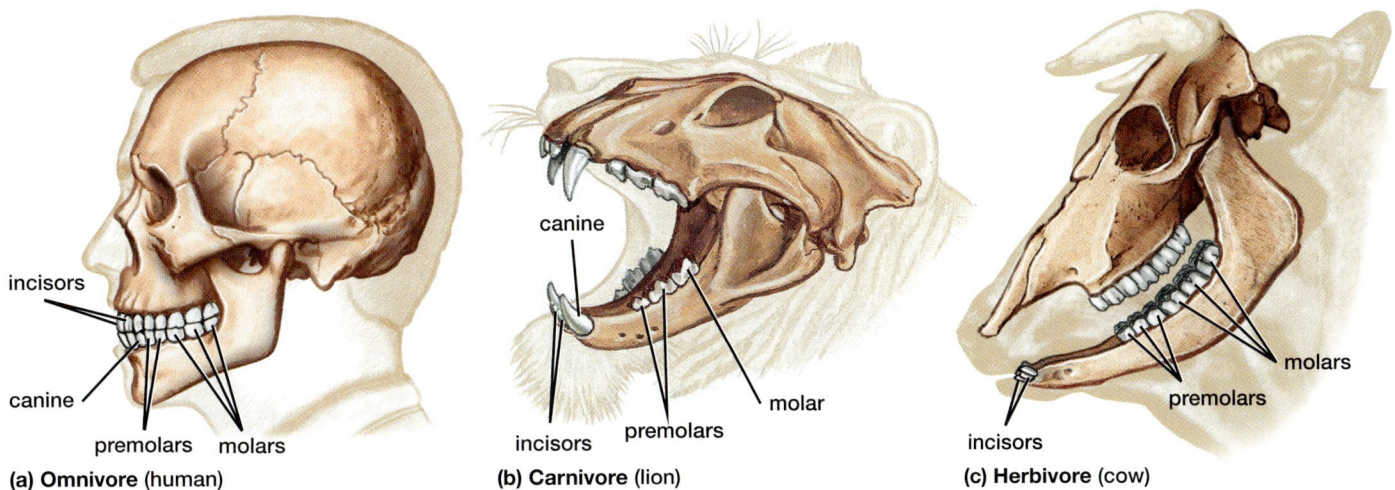

canine

incisors

canine

premolars   molars

(a) Omnivore (human)

incisors   premolars

molar

(b) Carnivore (lion)

molars

premolars

incisors

(c) Herbivore (cow)

▲ **FIGURE 34-9 Teeth have evolved to suit different diets (a)** Humans have cutting incisors, reduced canines, and flattened premolars and molars for grinding up plant and animal food. **(b)** Carnivores have large canines for grasping and killing prey, reduced incisors, and premolars and molars adapted for cutting rather than grinding. **(c)** Herbivores have flattened premolars and molars for grinding up tough plant fibers, and they usually lack canine teeth. Cows and other ruminants also lack upper incisors.

Birds' Stomachs Grind Food

Birds lack teeth and swallow their food whole. In many birds, food is stored in an expandable crop (**Fig. 34-10**). The food then passes gradually into two stomach chambers. The tube-like first chamber secretes protein-digesting enzymes that begin protein breakdown, while the second, the gizzard, is a thick-walled, muscular, grinding chamber lined with ridges or plates made of the protein keratin (which also forms the bird's beak). The gizzard crushes and grinds food using muscular contractions. Many birds swallow sand or small stones that lodge in the gizzard and aid in the grinding process. From the gizzard, pulverized food particles are released into the small intestine, where they are further digested and their nutrients are absorbed.

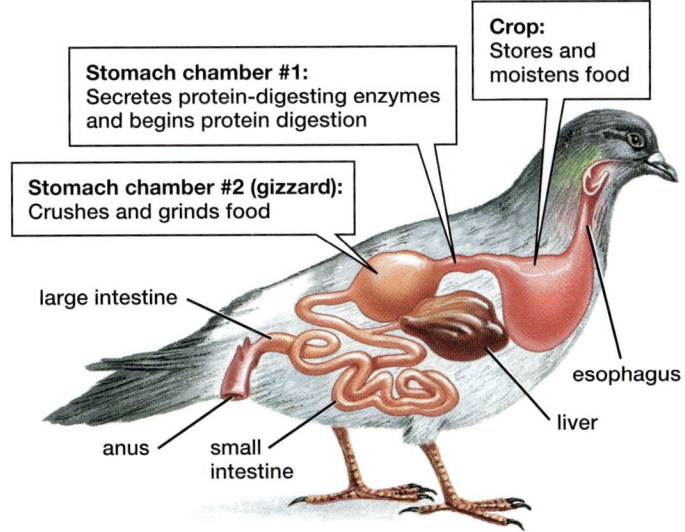

Crop: Stores and moistens food

Stomach chamber #1: Secretes protein-digesting enzymes and begins protein digestion

Stomach chamber #2 (gizzard): Crushes and grinds food

large intestine

anus   small intestine

esophagus

liver

▲ **FIGURE 34-10 Bird digestive adaptations**

QUESTION How can birds grind their food without teeth?

Specialized Stomach Chambers Allow Ruminants to Digest Cellulose

The cellulose surrounding each plant cell is potentially one of the most abundant food energy sources on Earth; nevertheless, if humans were restricted to a cow's diet of grass, we would soon starve. Although cellulose, like starch, consists of long chains of glucose molecules, cellulose resists the attack of animal digestive enzymes because of the way bonds link its glucose molecules together (see Chapter 3).

Ruminant animals, including cows, sheep, goats, camels, and hippos, can obtain energy from cellulose. Ruminants have multiple stomach chambers (**Fig. 34-11**). The first chamber is the *rumen*; a cow rumen can hold nearly 40 gallons (about 150 liters). This chamber is home to a variety of microorganisms that produce enzymes that break down cellulose and other carbohydrates and then ferment the resulting sugars. These microbial populations continuously renew themselves using this energy source. These microorganisms release small organic molecules that supply at least half of the cow's energy needs; most of these are absorbed through the rumen wall.

After partial digestion in the rumen, the plant material enters the reticulum, where it is formed into masses called cud. The cud is regurgitated, chewed, and then swallowed back into the rumen. (Ruminant animals can often be seen placidly *ruminating*, or chewing their cud.) The extra chewing exposes more of the cellulose and cell contents to the rumen's microorganisms, which digest it further.

Gradually, the partially digested plant material and many microorganisms are released into the *omasum*, where water, salts, and the remaining small organic molecules released by the microorganisms are absorbed. The remaining food then enters the *abomasum*, where acid and protein-digesting enzymes are secreted and protein digestion begins. Here, the cow digests not only plant proteins, but also the microorganisms that accompany the partially digested food

Rumen: Houses microorganisms that convert cellulose to small organic molecules that the rumen absorbs

small intestine

large intestine

anus

esophagus

Reticulum: Forms cud, which is regurgitated and re-chewed

Omasum: Absorbs water, salts, and small organic molecules released by the microorganisms

Abomasum: Produces acid and protein-digesting enzymes that begin protein digestion

◄ **FIGURE 34-11 Ruminants have a multichambered stomach**

QUESTION In addition to the ability to digest cellulose, what other nutritional benefits might ruminants gain by having microorganisms in their guts?

from the rumen. The cow then absorbs most of the products of digestion through the walls of its small intestine.

Small Intestine Length Is Correlated with Diet

For the majority of vertebrates, most chemical digestion and absorption of nutrients occurs in the small intestine. Herbivores require a relatively long small intestine, which provides more opportunity to extract nutrients from their bulky diet of plants. Carnivores, which consume a high-protein diet, have shorter small intestines than herbivores because proteins are relatively easy to digest, and protein digestion begins in the stomach, as described in more detail later.

This link between diet and small intestine length is strikingly illustrated during frog development. A juvenile tadpole is an algae-eating herbivore with an elongated small intestine. As it metamorphoses into a carnivorous (usually insect-eating) adult frog, its small intestine shortens to about one-third of its length in the tadpole.

CHECK YOUR LEARNING

Can you list and describe the five basic tasks of all digestive systems? Are you able to compare the various ways that invertebrates digest food? Can you explain how vertebrate digestive systems are specialized for different diets?

34.3 HOW DO HUMANS DIGEST FOOD?

The human digestive system (**Fig. 34-12**) is adapted for processing the wide variety of foods in our omnivorous diet. Each compartment has particular digestive secretions associated with it, as summarized in **Table 34-4**.

TABLE 34-4 Digestive Secretions in Humans

Site of Production	Secreted Substance	Source of Secretion	Role in Digestion
Mouth	Salivary amylase	Salivary glands	Breaks down starch into disaccharide sugars
	Mucus, water	Salivary glands	Lubricates and dissolves food
Stomach lining	Pepsin	Gastric glands in stomach	Breaks down proteins into peptides
	Hydrochloric acid	Gastric glands in stomach	Allows pepsin to work; kills some bacteria; aids in mineral absorption
	Mucus	Cells lining the stomach	Protects the stomach from digesting itself
Small intestine	Proteases	Pancreas	Break down peptides into smaller peptides and amino acids
	Bile	Liver	Emulsifies lipids
	Lipase	Pancreas	Breaks down lipids into fatty acids and glycerol
	Sodium bicarbonate	Pancreas	Neutralizes acidic chyme from the stomach
	Pancreatic amylase	Pancreas	Breaks down starch into disaccharides
	Peptidases	Small intestine	Breaks down small peptides into amino acids
	Disaccharidases	Small intestine	Split disaccharides into monosaccharides
	Mucus	Small intestine	Protects the intestine from digestive secretions

▶ FIGURE 34-12 **The human digestive tract**

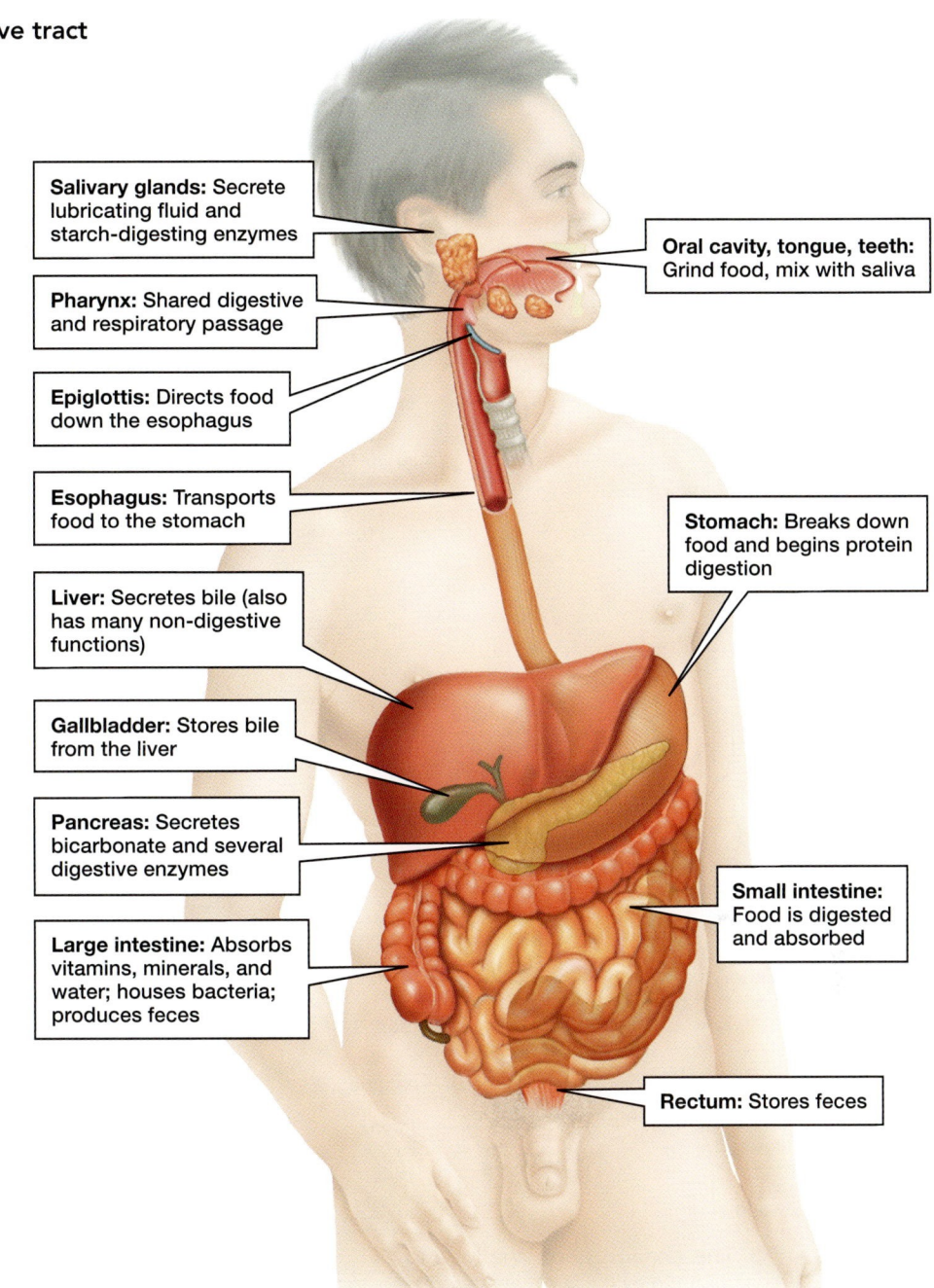

Salivary glands: Secrete lubricating fluid and starch-digesting enzymes

Oral cavity, tongue, teeth: Grind food, mix with saliva

Pharynx: Shared digestive and respiratory passage

Epiglottis: Directs food down the esophagus

Esophagus: Transports food to the stomach

Stomach: Breaks down food and begins protein digestion

Liver: Secretes bile (also has many non-digestive functions)

Gallbladder: Stores bile from the liver

Pancreas: Secretes bicarbonate and several digestive enzymes

Small intestine: Food is digested and absorbed

Large intestine: Absorbs vitamins, minerals, and water; houses bacteria; produces feces

Rectum: Stores feces

Mechanical and Chemical Digestion Begin in the Mouth

As you take a bite of food, your mouth waters and you begin chewing. These actions begin both the mechanical and the chemical digestion of food. While the teeth pulverize the food, the first phase of chemical digestion occurs as three pairs of salivary glands pour out saliva. Human salivary glands produce about 1 to 1.5 quarts (about 1.0 to 1.5 liters) of saliva daily.

Saliva has many functions. Water and mucus in saliva lubricate food to ease swallowing. Antibacterial agents in saliva help to guard against infection. Saliva also contains the digestive enzyme **amylase,** which begins breaking starches

into disaccharides (sugars with two glucose subunits; see Chapter 3). The water in saliva dissolves some molecules, such as acids and sugars, exposing them to taste receptor cells, located in clusters called *taste buds,* on the tongue. Taste buds help to identify the type and the quality of the food.

The muscular tongue manipulates chewed food into a mass and presses it back into the **pharynx,** a cavity that connects the mouth with the esophagus (**Fig. 34-13a**). The pharynx is shared with the respiratory system, in which it conducts air from the nose and mouth to the larynx, which leads to the trachea. This arrangement occasionally causes problems, as anyone who has ever choked on a piece of food well knows. Normally, however, the swallowing reflex

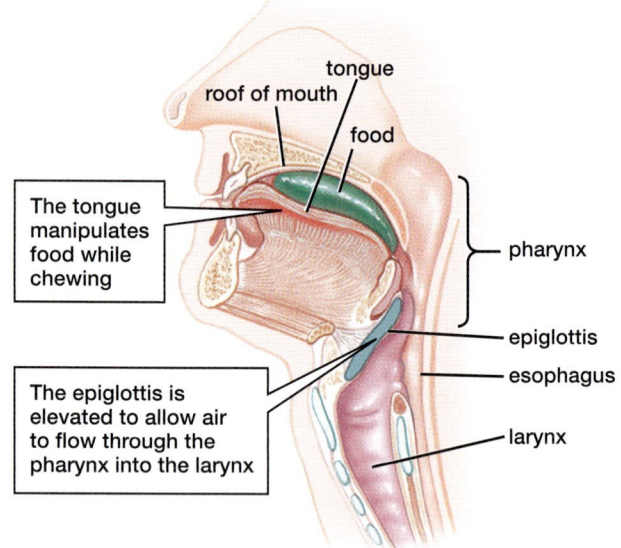

The tongue manipulates food while chewing

tongue
roof of mouth
food

pharynx

epiglottis
esophagus

The epiglottis is elevated to allow air to flow through the pharynx into the larynx

larynx

(a) Before swallowing

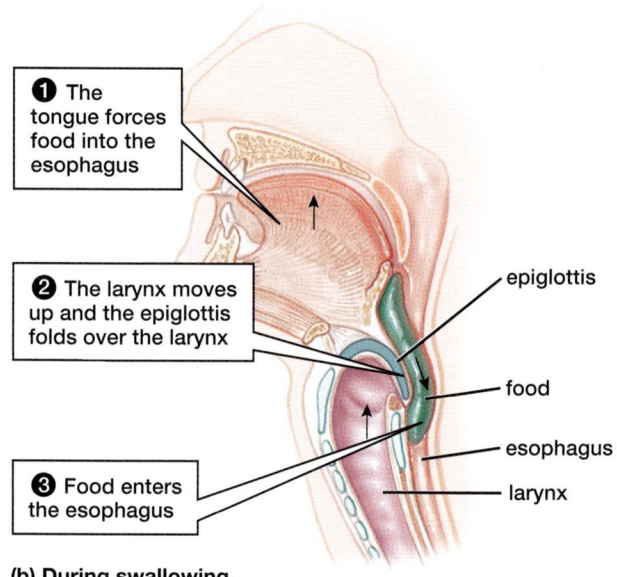

❶ The tongue forces food into the esophagus

❷ The larynx moves up and the epiglottis folds over the larynx

epiglottis

food

❸ Food enters the esophagus

esophagus

larynx

(b) During swallowing

▲ **FIGURE 34-13 The challenge of swallowing**
(a) Swallowing is complicated by the fact that the esophagus (part of the digestive system) and the larynx (part of the respiratory system) both open into the pharynx. **(b)** During swallowing, the larynx moves upward beneath the epiglottis. The epiglottis folds down over the larynx, sealing off the opening to the respiratory system and directing food down the esophagus instead.

elevates the larynx so it meets the **epiglottis,** a flap of tissue containing flexible cartilage that blocks off the respiratory passages, directing food into the esophagus (**Fig. 34-13b**).

The Esophagus Conducts Food to the Stomach, Where Mechanical and Chemical Digestion Continue

Swallowing forces food into the esophagus, a muscular tube that propels food from the mouth to the stomach. Mucus

lower esophageal sphincter

muscle layers

pyloric sphincter

small intestine

stomach lining

connective tissue covering

▲ **FIGURE 34-14 The stomach**

secreted by cells lining the esophagus protects it from abrasion and also lubricates the food during its passage. Muscles surrounding the esophagus produce a wave of contraction that begins just above the swallowed mass and progresses down the esophagus, forcing the food toward the stomach. This muscular action, called **peristalsis,** occurs throughout the digestive tract, pushing food products through the esophagus, stomach, intestines, and finally, out through the anus. Peristalsis is so effective that a person can actually swallow when upside down.

The **stomach** in humans is a muscular sac with a folded inner lining that allows it to expand so that we can eat rather large, infrequent meals (**Fig. 34-14**). Some carnivores take this ability to an extreme. A lion, for instance, may consume 40 pounds (18 kilograms) of meat at one meal, and then spend the next few days digesting it. In adult humans, the stomach can comfortably hold about a quart (1 liter), although this varies with body size. Food is retained in the stomach by two rings of circular muscles, called **sphincter muscles.** The sphincter at the top, called the *lower esophageal sphincter*, keeps food and stomach acid from sloshing up into the esophagus while the stomach churns. It opens briefly just after swallowing, allowing food to enter the stomach. A second sphincter, the *pyloric sphincter*, separates the lower portion of the stomach from the upper small intestine. This muscle regulates the passage of food into the small intestine.

The stomach has four distinct functions. First, the stomach stores and releases food gradually into the small intestine at a rate suitable to allow the small intestine to completely digest the food and absorb its nutrients. Second, the muscular stomach walls produce a variety of churning contractions that disrupt chunks of food, breaking them up into much smaller pieces that are more readily attacked by digestive enzymes. Third, the stomach begins protein breakdown using secretions from the gastric glands (see Table 34-4). Fourth, the stomach's gastric glands secrete hormones to regulate digestive activity.

The **gastric glands** are clusters of specialized epithelial cells that line millions of microscopic pits in the stomach lining. Gastric gland secretions include mucus, hydrochloric acid (HCl), and the protein pepsinogen. The hydrochloric acid

SCIENCE IN ACTION Do Bacteria Cause Ulcers?

Ulcers occur when localized areas of the tissue layers that line the stomach or duodenum (the portion of the small intestine nearest the stomach) become eroded (Fig. E34-1). Ulcer victims can experience burning pain, nausea, and, in severe cases, bleeding. Before the 1990s, doctors believed that most ulcers were caused mainly by overproduction of stomach acid and treated their patients with antacids, a bland diet, and stress-reduction programs. The ulcers were eased by the reduction in stomach acid, but commonly recurred when the treatment stopped. Now, we know that most ulcers are caused by a type of bacterium, and antibiotics have become the standard ulcer treatment. How did researchers make the link between bacteria and ulcers?

In the 1980s, J. Robin Warren, a pathologist at the Royal Perth Hospital in Australia, noticed that samples of inflamed stomach tissue were commonly infected with a spiral-shaped bacterium. He discussed his work with Barry Marshall, then a trainee in internal medicine at the same hospital, and they collaborated to test the hypothesis that the bacterium, later named *Helicobacter pylori*, caused stomach inflammation and ulcers.

The medical community was skeptical because these bacteria are found in the stomachs of many people without ulcers. To demonstrate conclusively that a particular bacterium caused the disease, Warren and Marshall followed a protocol that researchers often use to find disease-causing microbes: First, confirm the presence of these bacteria in all animals infected with the disease; second, grow the bacteria in culture; third, infect experimental animals with the cultured bacteria and demonstrate that they develop the disease; and fourth, re-isolate and culture the identical type of bacterium from the diseased animals.

Marshall and Warren tried to grow *H. pylori* from the stomachs of ulcer patients by incubating the bacteria at body temperature in culture dishes on a nutrient medium, with no success. But the lab technicians were discarding the cultures after 2 days if no growth was visible. As is common in research, chance provided an opportunity for scientific insight when a technician accidentally left the dishes in an incubator over a holiday break, and the resulting 5-day-old cultures showed colonies of the slow-growing bacteria.

Next, Marshall and another coworker tried to infect piglets (whose digestive systems resemble those of humans) with the cultured bacteria but the piglets did not develop ulcers from *H. pylori*. Frustrated, Marshall took an unusual and dangerous approach to the problem; he experimented on himself. After undergoing an exam with an endoscope (a tiny camera threaded down his esophagus) that showed his own stomach to be free of inflammation, Marshall swallowed a culture of about a billion *H. pylori*. During the following week, he began feeling ill, and samples of his stomach tissue showed that the mucus-producing stomach lining was damaged, thinned, and heavily infected with the bacteria. This led to the hypothesis that taking antibiotics to kill the bacteria would relieve his symptoms, which is exactly what happened. Although this experiment was dangerous, had a sample size of one, and was not going to be repeated by others, it bolstered Marshall's hypothesis and set the stage for further research. After Marshall and Warren as well as other independent researchers performed different types of studies using large sample sizes and adequate controls, the hypothesis was accepted.

Scientists now know that *H. pylori* colonize the protective mucus that coats the lining of the stomach and duodenum. In the process, these bacteria weaken the mucous layer and increase stomach acid production, making the stomach and duodenum more susceptible to attack by stomach acid and protein-digesting enzymes. The body's immune response to the infection further contributes to the tissue destruction. In the United States, *H. pylori* causes about 90% of ulcers; most will be cured by a 2-week treatment with antibiotics.

For their findings, based on careful observations, chance, and the scientific method, Warren and Marshall were awarded the Nobel Prize for Physiology or Medicine in 2005.

▲ **FIGURE E34-1 An ulcer** This ulcer was photographed through an endoscope.

gives the stomach fluid a very acidic pH of 1 to 3, about the same as lemon juice. This destroys many microbes, such as bacteria and viruses, that are inevitably swallowed along with food. Pepsinogen is the inactive form of *pepsin*, a type of **protease**—a protein-digesting enzyme that breaks proteins into shorter chains of amino acids called peptides. The stomach's acidity converts pepsinogen into pepsin (which works best in this acidic environment). Why not secrete pepsin in the first place? Gastric glands secrete inactive pepsinogen because pepsin would digest the very cells that synthesize it. Mucus, secreted by gastric gland cells and by epithelial cells throughout the stomach, coats the stomach lining and serves as a barrier to self-digestion. The protection, however, is not perfect, so cells of the stomach epithelium are replaced every few days. The stomach produces all of the ingredients necessary to digest itself if its protective mucous barriers are breached. Indeed, this is what happens when a person develops ulcers, as described in "Science in Action: Do Bacteria Cause Ulcers?"

Food in the stomach is gradually converted to a thick, acidic liquid called **chyme,** which consists of digestive secretions and partially digested food. Peristaltic waves (about

CASE STUDY continued

Dying to Be Thin

Stomach acid can be very destructive to the digestive tracts of people with bulimia, many of whom vomit several times daily. The strong acid in the stomach contents dissolves the protective enamel of teeth, making them extremely prone to decay. Stomach acid also damages tissues of the gums, throat, and esophagus. In extreme cases, the explosive pressure of repeated vomiting can tear or rupture the esophagus—a true medical emergency. Frequent vomiting weakens the stomach lining, allowing acid to attack the stomach wall, which may produce ulcers, and can even cause the stomach to rupture, a situation that can be fatal.

three per minute) then propel the chyme toward the small intestine, forcing about a teaspoon of liquid chyme through the pyloric sphincter with each wave. Small chunks of food cannot pass through the sphincter and remain in the stomach for further breakdown. Depending on the amount and type of food eaten, it takes about 4 to 6 hours to empty the stomach after a meal. Churning movements of an empty stomach may be heard as rumblings and felt as hunger pangs.

Only a few substances, including alcohol and certain drugs, can enter the bloodstream through the stomach wall. Food in the stomach slows alcohol absorption; hence, the advice "never drink on an empty stomach" is sound.

Most Chemical Digestion and Nutrient Absorption Occur in the Small Intestine

The **small intestine** is a long, muscular tube that receives chyme from the stomach, completes the chemical digestion of food molecules in the chyme, and absorbs these smaller nutrient molecules into the body. The small intestine consists of three segments: the duodenum, the jejunum, and the ileum. The short *duodenum* receives chyme from the stomach, receives digestive secretions from the gallbladder and the pancreas, secretes hormones that help control digestion, and begins the absorption of nutrients. In the *jejunum* and finally the *ileum* (which empties into the large intestine), digestion and nutrient absorption are completed. Like the stomach, the small intestine is protected from digesting itself by mucus secreted by specialized cells in its lining.

A Variety of Digestive Substances Are Found in the Small Intestine

After the stomach releases chyme into the small intestine, chemical digestion is accomplished with the aid of other digestive secretions from the liver, the pancreas, and cells lining the small intestine itself (**Fig. 34-15**).

The Liver Produces Bile The **liver** is perhaps the most versatile organ in the body. The liver stores fats and carbohydrates for energy, regulates blood glucose levels, synthesizes blood proteins, stores iron and certain vitamins, converts toxic ammonia

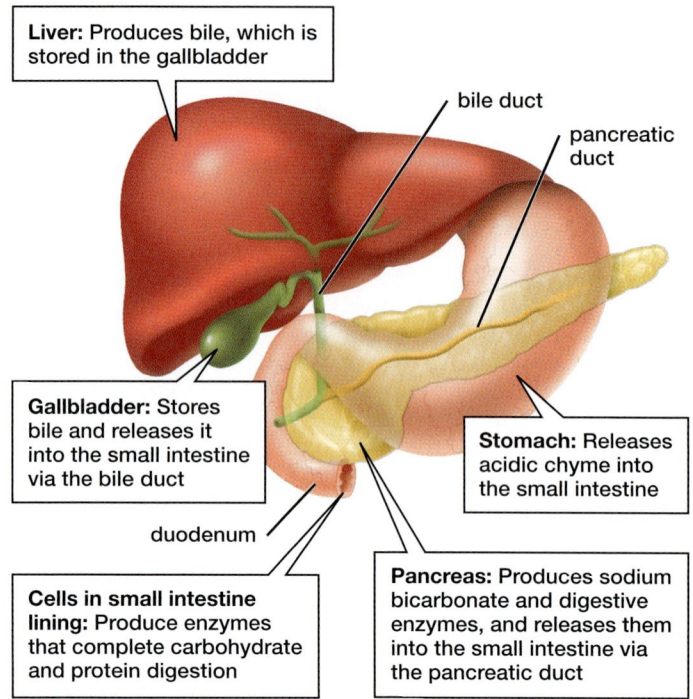

Liver: Produces bile, which is stored in the gallbladder

bile duct

pancreatic duct

Gallbladder: Stores bile and releases it into the small intestine via the bile duct

Stomach: Releases acidic chyme into the small intestine

duodenum

Cells in small intestine lining: Produce enzymes that complete carbohydrate and protein digestion

Pancreas: Produces sodium bicarbonate and digestive enzymes, and releases them into the small intestine via the pancreatic duct

▲ **FIGURE 34-15 Sources of digestive secretions utilized in the small intestine**

(released when amino acids are broken down) into urea, and detoxifies harmful substances such as nicotine and alcohol.

The role of the liver in digestion is to produce **bile,** a greenish liquid consisting primarily of water and bile salts that breaks down ingested fats. *Bile salts*, synthesized from cholesterol, have a hydrophilic end that is attracted to water and a hydrophobic end that interacts with fats. They disperse lipids into microscopic particles in the watery chyme, much as dish detergent disperses fat or oil from a fry pan. The particles expose a large surface area to attack by **lipases,** lipid-digesting enzymes. Bile synthesized in the liver is stored and concentrated in the **gallbladder.** Hormonal signals from the duodenum in response to the entry of chyme cause the gallbladder to contract and expel bile through the *bile duct*, which empties into the duodenum (see Fig. 34-15).

The Pancreas Supplies Pancreatic Juice The **pancreas** lies in the loop between the stomach and small intestine (see Fig. 34-15). In addition to releasing hormones that help regulate blood sugar (see Chapter 37), the pancreas produces a digestive secretion called **pancreatic juice.** In response to hormones secreted by the duodenum as chyme enters, the pancreas sends pancreatic juice into the duodenum through the *pancreatic duct* (see Fig. 34-15). Pancreatic juice is a mixture of water, sodium bicarbonate (which neutralizes the acidic chyme), and several digestive enzymes, including pancreatic amylase, lipase, and proteases (see Table 34-4). Pancreatic digestive enzymes work best in the slightly alkaline (basic) environment created by sodium bicarbonate in the pancreatic juice. Pancreatic amylase breaks down carbohydrates, lipase attacks lipids, and proteases split proteins and peptides, whose breakdown began in the stomach.

Enzymes in the Intestinal Wall Complete Digestion The epithelium of the small intestine consists primarily of cells whose plasma membranes form fringes of microscopic projections called **microvilli** ("tiny hairs"; singular, microvillus), which provide a large surface area for absorption. These microvilli also bear enzymes called *peptidases* that break down peptides into amino acids and *disaccharidases* that split disaccharide sugars into monosaccharides. An example is the enzyme lactase, which splits the disaccharide lactose (milk sugar) into the monosaccharides glucose and galactose. Although everyone makes lactase during infancy, many people stop synthesizing lactase in adulthood, and become lactose intolerant (see Chapter 6).

Most Absorption Occurs in the Small Intestine

In addition to being the principal site of chemical digestion, the small intestine is also the major site of nutrient absorption into the body.

The Intestinal Lining Provides a Huge Surface Area for Absorption In a living human adult, the small intestine is about 1 inch (2.5 centimeters) in diameter and about 8 to 10 feet (2.5 to 3 meters) long. (Measurements of 20 feet or more are from cadavers, where all muscle tone is lost.) In addition to being quite long, the small intestine has numerous folds and projections, giving it an internal surface area that is roughly 600 times that of a smooth tube of the same length (**Fig. 34-16a**). Tiny protrusions called **villi** (singular, villus; from Latin, meaning "hair") cover the entire folded surface of the intestinal wall (**Figs. 34-16b, c**). Villi, which are about 1/25th of an inch (about 1 millimeter) long, make the intestinal lining appear velvety to the naked eye. The plasma membranes of the epithelial cells covering the villi

are folded into microvilli (**Fig. 34-16d**). Taken together, the specializations of the lining of the adult small intestine give it a surface area of about 2,700 square feet (about 250 square meters), almost the size of a doubles tennis court.

Unsynchronized contractions of the circular muscles of the small intestine, called *segmentation movements*, slosh the chyme back and forth, bringing nutrients into contact with the enormous absorptive surface of the small intestine. When absorption is complete, coordinated peristaltic waves conduct the leftovers into the large intestine.

Nutrients Are Absorbed by Various Means Nutrients absorbed by the small intestine include water, monosaccharides, amino acids and short peptides, fatty acids, vitamins, and minerals. Some nutrients enter the cells lining the small intestine by diffusion and others by active transport. Water follows by osmosis. The water and most other nutrients then enter the network of blood capillaries located within each of the villi.

Fatty acids released by the digestion of fats and oils take a distinctive pathway. Clustered together with cholesterol and fat-soluble vitamins, they diffuse directly through intestinal epithelial cell membranes. Inside the epithelial cells, these substances are assembled and coated with proteins to form particles called **chylomicrons,** which are then released inside the villus. The chylomicrons are too large to enter blood capillaries, and instead diffuse through the porous wall of the **lacteal,** a lymph capillary that ends blindly within each villus (see Fig. 34-16c). Chylomicrons are then transported in lymph by the lymphatic system, which eventually empties into a large vein near the heart (see Fig. 32-18). Excess fat can accumulate to health-threatening levels, and some obese people choose surgery to lose weight, as described in "Health Watch: Eating to Death?"

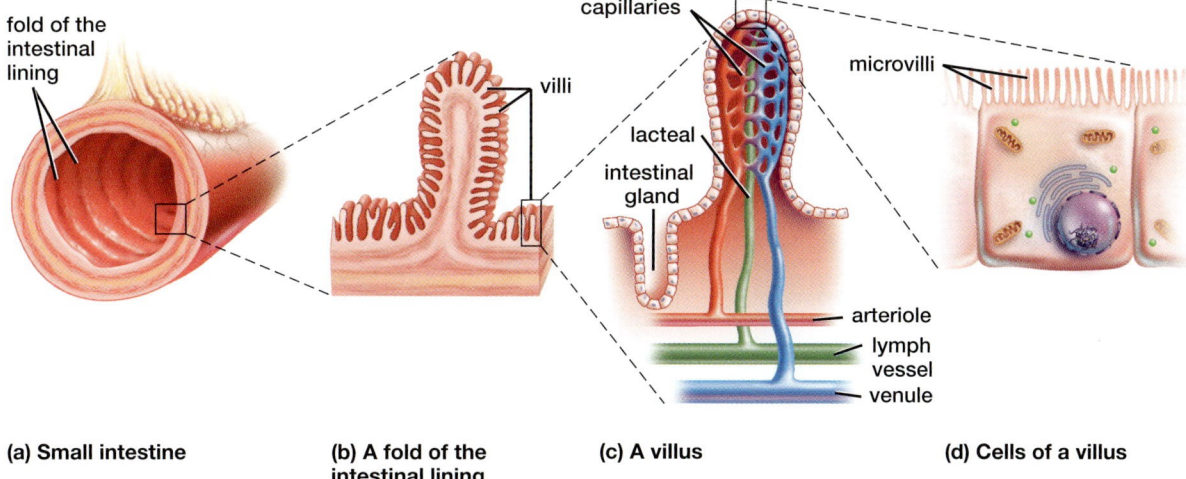

(a) Small intestine   (b) A fold of the intestinal lining   (c) A villus   (d) Cells of a villus

▲ FIGURE 34-16 **The structure of the small intestine** (a) Visible folds in the intestinal lining are carpeted with (b) tiny projections called villi, which extend from the folded intestinal lining. (c) Each villus contains a network of capillaries and a central lymph capillary called a lacteal. Most digested nutrients enter the capillaries, but fats enter the lacteal. (d) The plasma membranes of the epithelial cells covering each villus bear microvilli.

QUESTION What might the anatomy of the digestive system be like if the internal folds, villi, and microvilli of the small intestine had not evolved?

Health Watch

Eating to Death?

Patrick Deuel's doctor gave him an ultimatum: be hospitalized and lose weight, or die. At 1,072 pounds, he was imprisoned by his size; a wall had to be removed to get him out of his bedroom for medical treatment. Although Deuel's case is extreme, obesity is a growing epidemic in many countries. Roughly two-thirds of U.S. adults are now overweight or obese.

Are some people simply born to be fat? If so, Deuel is one of them; his doctor described him as obese when he was only 3 months old. Researchers have found scores of different genes that influence weight, and it is clear that some people have far more difficulty remaining slim than others. But our genes haven't changed appreciably in the past 30 years, so the current epidemic of obesity is almost certainly caused by changes in our environment, which now offers an abundance of convenient, delicious, and inexpensive high-calorie foods, and by changes in our behavior, which has become more sedentary.

Obese individuals have a higher risk of liver disease, gallstones, sleep apnea (interruptions of breathing during sleep), diabetes, some cancers, arthritis, hypertension, heart disease, and stroke. People are described as *morbidly obese* when they have a BMI above 40 or are 100 pounds or more over their ideal weight. Nearly 6% of people in the United States are morbidly obese, and their weight poses a severe risk to their health and longevity. Although the healthiest way to lose weight is to reduce caloric intake and increase exercise, many people who are obese have repeatedly failed in their attempts. This failure has spurred the development of surgical fixes that act directly on the digestive tract. Although surgical approaches to weight loss are somewhat risky, doctors may recommend surgery for those whose obesity seems to pose an even more serious health risk than the surgery.

After losing over 400 pounds on a supervised diet and exercise program, Deuel underwent *gastric bypass surgery*, which accounts for more than half of the weight-loss surgeries performed annually in the United States (**Fig. E34-2a**). This procedure blocks off all but a small pouch at the top of the stomach, and connects this pouch directly to the region of the small intestine below the duodenum. As a result, only very small amounts of solid food can be consumed at a sitting, and fewer calories are absorbed from food because the absorptive area of the small intestine is reduced. Although this surgery has proven effective in reducing weight, it restricts absorption of calcium, iron, and vitamins D and B_{12}, so vitamin and mineral supplementation is recommended to compensate for these losses.

The second most common form of weight-loss surgery is *gastric band surgery*. As shown in **Figure E34-2b,** this operation limits food intake by placing an inflatable band around the upper portion of the stomach, converting the food-collecting portion of the stomach into a pouch with a small opening through which digested food can pass. The band can be inflated or deflated to adjust the size of the opening from the pouch into the lower portion of the stomach. Although the gastric band, on average, produces less dramatic results than gastric bypass surgery, it has fewer complications, it is reversible, and it is less likely to lead to nutrient deficiencies because it does not bypass any of the small intestine.

Deuel reached a low of 370 pounds 2 years after gastric bypass surgery, but 5 years after the surgery, he was back up to 560 pounds; not the man he used to be, but still morbidly obese. Although there are many weight-loss success stories, there are, unfortunately, no fail-safe cures.

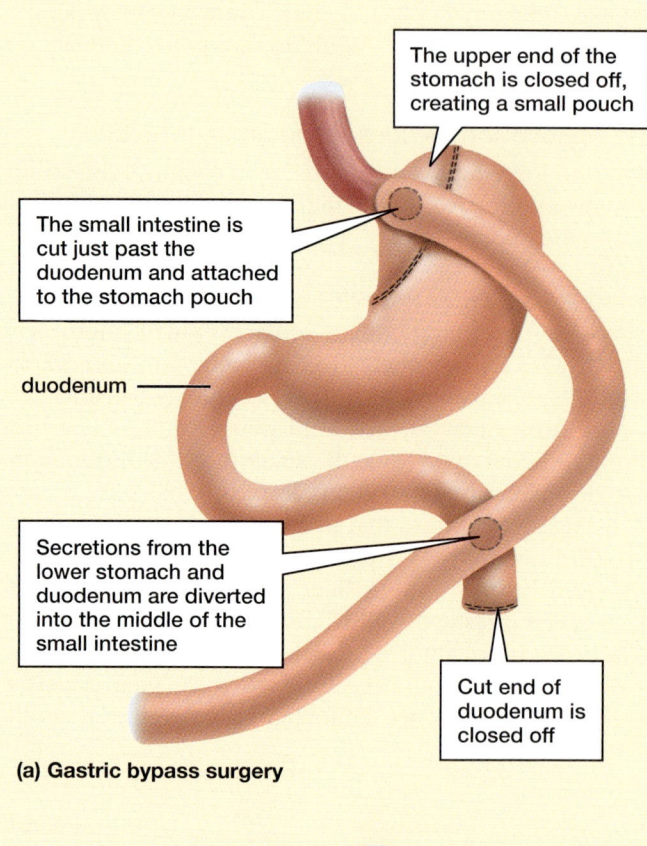

The upper end of the stomach is closed off, creating a small pouch

The small intestine is cut just past the duodenum and attached to the stomach pouch

duodenum

Secretions from the lower stomach and duodenum are diverted into the middle of the small intestine

Cut end of duodenum is closed off

(a) Gastric bypass surgery

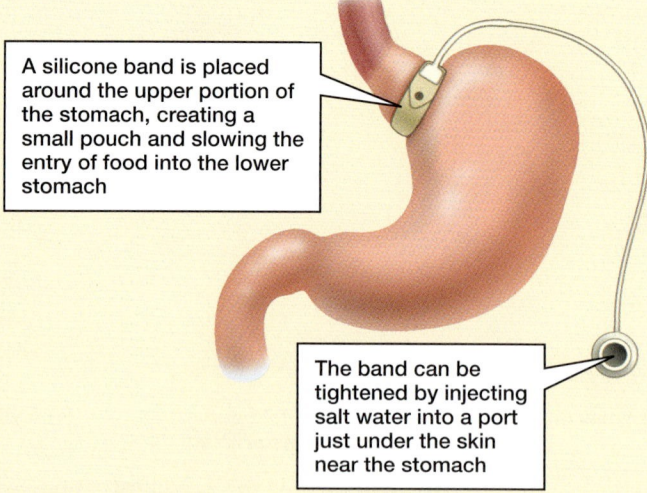

A silicone band is placed around the upper portion of the stomach, creating a small pouch and slowing the entry of food into the lower stomach

The band can be tightened by injecting salt water into a port just under the skin near the stomach

(b) Gastric band surgery

▲ **FIGURE E34-2 Weight-loss surgeries**

Water Is Absorbed and Feces Are Formed in the Large Intestine

The **large intestine** in a living adult human is about 2.5 inches (6.5 centimeters) in diameter and about 5 feet (1.5 meters) long, wider but far shorter than the small intestine (see Fig. 34-12). Most of the large intestine is called the **colon;** its final 6-inch chamber is the **rectum.** The leftovers of digestion—indigestible fiber, small amounts of unabsorbed nutrients, and water—flow into the large intestine. In the large intestine, these wastes support a flourishing population of bacteria. Some of the bacteria in the large intestine earn their keep by synthesizing vitamin B_{12}, thiamin, riboflavin, and vitamin K (a typical human diet would be deficient in vitamin K without these bacteria). Epithelial cells of the large intestine absorb the vitamins as well as most leftover water and salts.

After absorption is complete, any remaining material is compacted into semisolid **feces.** Feces consist primarily of water, indigestible wastes, mucus, some unabsorbed nutrients, and a host of bacteria. Feces are transported by peristaltic movements until they reach the rectum. Expansion of this chamber stimulates the urge to defecate. The anal opening is controlled by two sphincter muscles—an inner one that is under involuntary control, and an outer muscle that can be consciously controlled. Although defecation is a reflex, it comes under voluntary control in early childhood.

Digestion Is Controlled by the Nervous System and Hormones

As you sit down to lunch and hungrily begin eating, your body is coordinating a complex series of events that will convert the meal into nutrients circulating in your blood. Not surprisingly, the secretions and muscular activity of the digestive tract are coordinated by both nerve signals and hormones.

Food Triggers Nervous System Responses

Especially if you are hungry, the sight, smell, taste, and even the thought of food stimulates the hypothalamus, which controls many involuntary responses of the nervous system (see Chapter 38). The hypothalamus stimulates nervous pathways that prepare the digestive system to process food. For example, salivation increases, and the stomach produces more acid and protective mucus. As food moves through the digestive tract, its bulk activates stretch receptors. The stretch receptors stimulate local nervous reflexes that cause peristalsis and segmentation movements.

Hormones Help Regulate Digestive Activity

Hormones secreted by the digestive system enter the bloodstream and circulate through the body, acting on specific receptors within the digestive tract or the brain. Like most hormones, they are regulated by negative feedback. For example, nutrients in chyme, particularly amino acids and peptides from protein digestion, stimulate gastric glands in the stomach lining to release the hormone **gastrin** into the bloodstream. Gastrin travels back to the stomach and stimulates further acid secretion, which promotes protein digestion. When the pH of the stomach reaches a low level (high acidity), this inhibits gastrin secretion, which in turn inhibits further acid production.

Two additional hormones, **secretin** and **cholecystokinin,** are released into the bloodstream by cells of the duodenum in response to the acidity of chyme and the nutrients—particularly peptides and fats—in chyme. These two hormones help regulate the chemical environment within the small intestine and the rate at which the chyme enters the small intestine, promoting optimal digestion and absorption of nutrients. Together, secretin and cholecystokinin reduce stomach acid production, preventing excess acidity. They also slow peristaltic stomach contractions, reducing the rate at which chyme is forced into the small intestine and allowing more time for digestion and absorption to occur. They increase bile production by the liver and bile release from the gallbladder. Finally, secretin and cholecystokinin increase the production and release of pancreatic juice into the small intestine.

Hormones Regulate Hunger

Scientists are discovering an ever-increasing array of hormones related to hunger and fat storage; some of these act on the brain, and many contribute in complex ways to a person's weight. Two appetite-regulating hormones discovered in the 1990s are leptin and ghrelin.

Leptin is a peptide secreted by fat cells. When calorie intake is restricted, leptin in the bloodstream is reduced, stimulating the hypothalamus to trigger hunger. Although an increase in circulating leptin seems to contribute to a sense of fullness, injections of leptin do not help obese people lose weight. Researchers are now combining leptin

? HAVE **YOU** EVER WONDERED . . .

What's the Scoop on Bacteria in Poop?

You've learned that feces are the leftovers of digestion, but about one-third of fecal dry weight actually consists of bacteria. The cells that make up your body are outnumbered 10:1 by the bacteria in your large intestine. We each host a gut community that includes many hundreds of different bacterial species; as one specialist puts it: "Poo is a zoo." Human bacterial assemblages differ somewhat based on sex, genes, environment, and diet. The bacterial menagerie in your gut helps control your immune responses, prevents some disease-causing bacteria from establishing themselves, and releases both cancer-fighting and cancer-promoting substances from the foods you eat. In addition, these symbiotic bacteria convert some indigestible food fiber into small fatty acids that contribute 4% to 10% of your dietary calories. Different proportions of bacterial types in the human gut affect how much energy different people extract from the same diet. Weight-loss diets can cause a shift toward a bacterial community that produces fewer fatty acids. This suggests that the miniature zoos in our guts may actually influence whether or not we become overweight.

with other appetite-regulating substances in the quest for an antiobesity drug.

Ghrelin is a peptide secreted by gastric gland cells in the stomach lining. An increase in circulating ghrelin occurs prior to mealtime, stimulating hunger via the hypothalamus. Ghrelin decreases after a meal, reducing hunger. Researchers compared ghrelin blood levels in people who lost weight through diet with those who underwent gastric bypass surgery. Ghrelin in dieters was elevated, which would tend to make them feel hungry and so regain weight. In contrast, gastric bypass surgery often caused large reductions in circulating ghrelin, so people not only ate less but also felt less hungry. Like most hormones, ghrelin has multiple roles in the body—many of them beneficial. Not surprisingly, pharmaceutical researchers are seeking safe ways to block ghrelin's ability to stimulate hunger.

CHECK YOUR LEARNING

Can you trace food through the human digestive tract, describing the structures and functions of each digestive compartment? Can you explain how the small intestine absorbs nutrients, and how its structure facilitates nutrient absorption? Can you describe feces and how they are produced and eliminated? Can you explain how the nervous system and hormones regulate digestion and food intake?

CASE STUDY revisited

Dying to Be Thin

Anorexia and bulimia can cause severe malnutrition and—in extreme cases—death from starvation. These disorders also reduce the levels of certain crucial ions in the blood, interfering with heart muscle contractions and increasing the risk of death from cardiac arrest. Eating disorders also play havoc with the digestive system because of frequent vomiting.

The causes of eating disorders are complex and poorly understood. Genes apparently play a role; people with a close relative who has an eating disorder are about five times as likely to develop such a disorder themselves. Mental problems (such as anxiety and depression) and personality traits (such as perfectionism, low self-esteem, and a high need for acceptance and achievement) seem to predispose people to eating disorders. Cases often begin during adolescence, when bodies and brains are undergoing rapid changes. Magazines, TV, and Internet ads bombard susceptible young people with the message that being skinny is a route to acceptance, beauty, and riches.

Isabelle Caro (**Fig. 34-17**), a French model who died at age 28, had suffered from anorexia since the age of 13. She is best known for her efforts to help raise awareness and help other victims of this disorder by posing nude for billboards in Italy featuring the words: "No—Anorexia." The year before Caro went public with her battle in 2007, the weight of her 5-foot, 4-inch body had dipped to 55 pounds, causing her to lapse into a coma and spend months in intensive care. She emerged determined to fight the demons of anorexia, struggling to eat tiny portions and even rediscovering the delicious flavor of ice cream. Sadly, her efforts came too late for her ravaged body, and she died in 2010.

Anorexia and bulimia are difficult to treat and to overcome. Victims are given nutritional therapy to help them recover from malnutrition. Psychotherapy is usually necessary, and antidepressant drugs may be helpful. Because many victims hide or deny their problems, and because treatment is expensive, the majority of sufferers are inadequately treated; fewer than half experience complete recovery. For Carré Otis, the wake-up call came when, at age 30, she required surgery to repair her heart, damaged from years of malnutrition. "I needed to make a change, or clearly my body would not

▲ **FIGURE 34-17 Anorexia** Isabelle Caro in 2008.

hold up. At that moment, I finally realized how out of control I was, and knew that I was not ready to die. I was ready to embark on the road to recovery."

Consider This

BioEthics Media and fashion houses that glamorize slimness have been blamed for the prevalence of eating disorders. Why do you think extreme thinness is made so appealing? Are there appropriate measures that a free society can take to reverse or limit this message? Do you support the idea of a minimum BMI for models?

CHAPTER REVIEW

Summary of Key Concepts

34.1 What Nutrients Do Animals Need?

All animals require nutrients that provide energy; for humans, these are primarily carbohydrates and lipids, with a small percent derived from protein. Food energy is measured in Calories. Excess energy from food is stored in body fat, the most concentrated energy source. Each type of animal has specific requirements for essential nutrients that it cannot synthesize, but that are required for cellular function. For humans, these include essential fatty acids, essential amino acids, minerals, vitamins, and water.

34.2 How Does Digestion Occur?

Digestion is the mechanical and chemical breakdown of food that converts complex molecules into simpler molecules that can be absorbed and used by the organism. In some animals, digestion is entirely intracellular, as occurs within individual cells of sponges, which lack digestive systems. The simplest digestive system is the saclike gastrovascular cavity in organisms such as *Hydra*. Most digestive systems consist of a one-way tube along which specialized compartments process food in an orderly sequence. Specialized digestive systems allow different animals to utilize a wide variety of foods. Digestive systems must accomplish five tasks: ingestion of food, mechanical digestion, chemical digestion, absorption of nutrients, and elimination of wastes.

34.3 How Do Humans Digest Food?

In humans, digestion begins in the mouth, where food is physically broken down by chewing and chemical digestion of starches is initiated by saliva. Food is then conducted to the stomach by peristaltic waves of the esophagus. In the acidic environment of the stomach, food is churned into smaller particles, and protein digestion begins. Gradually, acidic chyme is released into the small intestine, where it is neutralized by sodium bicarbonate from the pancreas. In the small intestine, secretions from the pancreas and liver, and from the cells of the intestinal epithelium, complete the chemical digestion of proteins, fats, and carbohydrates. In the small intestine, the simple molecular products of digestion are absorbed into the bloodstream, either directly via capillaries or indirectly through the lymphatic system, for distribution to the body cells. The colon portion of the large intestine absorbs most of the remaining water and salts, as well as vitamins manufactured by intestinal bacteria. The remaining wastes become feces that are passed to the rectum, whose distention triggers the urge to defecate.

Digestion is regulated by the nervous system and hormones. The smell and taste of food and the action of chewing trigger the secretion of saliva into the mouth. Nutrients, particularly peptides, stimulate the stomach to produce gastrin, which causes acid production and peristalsis. As chyme enters the small intestine, it stimulates the release of secretin and cholecystokinin by cells of the duodenum. These hormones increase the release of digestive secretions into the small intestine and slow stomach acid production and peristalsis. Important hunger-regulating hormones include leptin (produced by fat cells) and ghrelin (produced by the stomach). Low leptin and high ghrelin levels in the blood trigger hunger via the hypothalamus.

Key Terms

absorption   658
amylase   663
bile   666
body mass index (BMI)   654
calorie   653
Calorie   653
carnivore   660
chemical digestion   658
cholecystokinin   669
chylomicron   667
chyme   665
colon   669
digestion   658
digestive system   658
elimination   658
epiglottis   664
esophagus   660
essential amino acid   655
essential fatty acid   655
essential nutrient   655
extracellular digestion   659
feces   669
food vacuole   659
gallbladder   666
gastric gland   664
gastrin   669
gastrovascular cavity   659
ghrelin   670
herbivore   660

ingestion   658
intracellular digestion   658
lacteal   667
large intestine   669
leptin   669
lipase   666
liver   666
lysosome   659
mechanical digestion   658
metabolic rate   653
microvillus (plural, microvilli)   667
mineral   655
mouth   658
nutrient   653
omnivore   660
pancreas   666
pancreatic juice   666
peristalsis   664
pharynx   663
protease   665
rectum   669
ruminant   661
secretin   669
small intestine   666
sphincter muscle   664
stomach   664
villus (plural, villi)   667
vitamin   655

Learning Outcomes

In this chapter, you have learned to . . .

LO1 List the six major categories of nutrients and the role each plays in maintaining the body.

LO2 Describe the role of energy in the diet, including how food energy is measured, which nutrients provide the most energy, and how the body stores energy.

LO3 List and explain the five tasks that all digestive systems must accomplish.

LO4 Describe three different modes of digestion in invertebrates, including the names of the structures involved.

LO5 Describe the digestive specializations of herbivorous, omnivorous, and carnivorous vertebrates, and the distinctive features of bird and cow digestive systems.

LO6 List the structures, in order, through which food passes in the human digestive system, describe the function of each structure, and explain how food is moved through it.

LO7 Describe the physical and chemical breakdown of food starting with the mouth and ending with the small intestine; for each digestive compartment, include the sources and functions of the digestive secretions that act there.

LO8 Describe nutrient absorption in the small intestine, including how different nutrients are absorbed and how the structure of the small intestine facilitates absorption.

LO9 Explain where and how feces are produced and eliminated and what feces are composed of.

LO10 Explain how the nervous system and hormones regulate digestion and food intake.

Thinking Through the Concepts

Fill-in-the-Blank

1. Two general roles for nutrients are to provide _____ and _____. Nutrients required in small amounts that often assist in enzyme function are called _____. Nutrients that are elements are called _____. Nutrients that cannot be synthesized by the body are called _____ nutrients. **LO1**

2. Sponges rely exclusively on _____ digestion. Cnidarians such as *Hydra* digest food in a(n) _____. Earthworms have a(n) _____ digestive system. Ruminants can break down _____ only because of microorganisms in their stomachs. **LO4 LO5**

3. The enzyme called _____ is present in saliva and begins breaking down _____. Digestion of _____ begins in the stomach. Stomach acid converts the inactive substance _____ into the active enzyme _____. Nearly all fat digestion occurs in the _____. **LO6 LO7**

4. The five major processes carried out by the digestive systems are _____, _____, _____, _____, and _____. **LO3**

5. In humans, a cavity called the _____ is shared by the _____ and _____ systems. A flap called the _____ prevents food from entering the trachea during the act of _____. Food is pushed through the digestive system by muscular contractions called _____. Circular muscles that control movement into and out of organs such as the stomach are called _____. **LO6**

6. Fats are dispersed by a secretion called _____. This secretion is produced by the _____ and is stored in the _____. Products of fat digestion are formed into chylomicrons inside intestinal _____ cells. The chylomicrons are released inside the villus and diffuse through the wall of the _____, which is part of the _____ system. **LO8**

7. Fill in the hormones. Release is stimulated by amino acids and peptides in the stomach: _____; increased secretion causes hunger: _____; released by fat cells: _____; reduce stomach acid production: _____ and _____; produced by cells in stomach gastric glands: _____ and _____; stimulate pancreatic juice secretion: _____ and _____. **LO10**

Review Questions

1. List the six general types of nutrients. Which two provide the most energy for humans? Which of these stores the most energy in the body, and why? **LO1 LO2**

2. Give an example of a vertebrate and an invertebrate that both use gizzards. Describe the general structure and function of the gizzard and how it works. **LO4 LO5**

3. Vertebrates can be grouped into three categories based on their diets; list and briefly define these categories, giving one example of each. Which group has the shortest small intestine, and why? **LO5**

4. Why is the stomach both muscular and expandable? **LO6**

5. List and describe the functions of the secretions of the gastric glands. **LO7 LO10**

6. List the digestive substances secreted into the small intestine, and describe the origin and function of each. **LO7 LO8**

7. Name and describe the muscular movements of the human digestive system, where they occur, and their functions. **LO6**

8. Vitamin C is an essential vitamin for humans but not for dogs. Certain amino acids are essential for humans but not for plants. Explain. **LO1**

9. Name four structural or functional adaptations of the human small intestine that contribute to effective digestion and absorption. **LO8**

10. Describe the processing of a piece of steak, starting in the mouth and ending with the absorption of individual amino acids. **LO6 LO7 LO8**

11. Explain the composition of feces and how the large intestine forms and expels them. **LO9**

12. Describe how and where the three important digestive system hormones described in this chapter coordinate digestion. **LO7 LO10**

Applying the Concepts

1. Wanting to lose weight, Jason decides to use artificial sweetener in the cookies he's baking. Since the sweetener is hundreds of times as sweet as sugar, the amount he uses to replace the 2 cups of sugar in his cookie recipe is only 1 teaspoon. His friend smirks and uses sugar to make the same recipe. The other ingredients are the same: 1.5 cups of butter and 3 cups of flour. If Jason and his friend's cookies are each the same size, how do Jason's cookies compare in calories? Explain.

2. Small birds have high metabolic rates, efficient digestive tracts, and high-calorie diets. Some birds consume roughly one-third of their body weight in food daily. Why do you think birds rarely consume leaves or grass?

3. Symbiotic protozoa in the digestive tracts of termites produce cellulose-digesting enzymes that allow the termites to utilize wood as a food source. In return, termites provide protozoa with food and shelter. Imagine that the human digestive tract is gradually invaded, over many generations, by symbiotic protozoa capable of digesting cellulose. What adaptive changes in human body structure and function might be selected for as a result?

4. Discuss why people get fat and how modern types and availability of food are contributing to an epidemic of obesity. In a typical day, what stimuli do you encounter that tend to make you (or your friends) overeat?

Answers to Figure Caption questions and Fill-in-the-Blank questions can be found in the Answers section at the back of the book.

MB *Go to MasteringBiology for practice quizzes, activities, eText, videos, current events, and more.*

CASE STUDY
Paying It Forward

ANTHONY DEGIULIO HAD A DREAM— he wanted to save someone's life. At first, it was a nebulous ambition, but it started to take shape as he watched a segment of the TV show *60 Minutes* that highlighted living kidney donation. DeGiulio, who hadn't realized that a live person could donate a kidney, immediately saw this as a way to achieve his ambition. "I want to do this," he told his wife.

DeGiulio called New York-Presbyterian Hospital, and set in motion a series of events described as an "altruistic chain" that gave new life not to one person, but to four. The chain of events was possible because three people in the area needed kidneys, and each had a family member eager to donate, but were prevented from doing so because of incompatibility between their tissue types. Barbara Asofsky, a nursery school teacher, had known for 5 years that she would need a kidney transplant. When DeGiulio's kidney was found to be a good match for Asofsky, her husband, Douglas, was happy to donate the kidney he had hoped to give to his wife—but couldn't because of incompatibility—to a stranger instead. The fortunate stranger was Alina Binder, a student at Brooklyn College. Alina's father, Michael, was a good match for Andrew Novak, a telecommunications technician. Finally, Andrew's sister, Laura Nicholson, donated her kidney to Luther Johnson, a hotel kitchen steward.

Do kidney transplant recipients lead completely normal lives? How is a donor kidney removed, and what are the risks to the donor? For people with kidney failure, are there alternatives to a kidney transplant? Why are kidneys so important, and how do they work? Read on to find out.

Barbara Asofsky hugs Good Samaritan Anthony DeGiulio, whose kidney restored her health and started a chain of kidney donations.

AT A GLANCE

35.1 What Are the Major Functions of Urinary Systems?

35.2 What Are Some Examples of Invertebrate Urinary Systems?

35.3 What Are the Structures of the Mammalian Urinary System?

35.4 How Is Urine Formed?

35.5 How Do Vertebrate Urinary Systems Help Maintain Homeostasis?

LEARNING GOALS

LG1 Explain the two major functions of urinary systems and the general processes that accomplish these functions.

LG2 Describe how urea is produced from protein.

LG3 Describe and compare the urinary systems of freshwater flatworms, earthworms, and insects.

LG4 Draw, label, and explain the functions of each major part of the human urinary system, including the structures within the human kidney and those involved in the blood supply to the kidney.

LG5 Diagram and label a nephron and collecting duct, and explain each stage of urine production within these structures.

LG6 Explain the role of antidiuretic hormone, the distal tubule, the loop of Henle, and the collecting duct in controlling blood osmolarity in mammals and how mammalian nephrons are adapted for different environments.

LG7 Describe how the kidneys help regulate blood ion levels, pH, oxygen levels, and blood pressure.

LG8 Describe osmoregulation in freshwater and saltwater fish.

35.1 WHAT ARE THE MAJOR FUNCTIONS OF URINARY SYSTEMS?

Urinary systems are organ systems that produce and eliminate urine. **Urine** is a watery fluid that contains a variety of substances that have been filtered from the blood or (in many invertebrates) from the interstitial fluid that bathes all cells. Urine contains waste products of cellular metabolism, foreign substances, and some ions and nutrients in excess of the body's needs. By producing and eliminating urine, urinary systems play two major roles in most animals: They excrete cellular wastes and they maintain homeostasis.

Excretion Eliminates Wastes from the Body

Excretion is the general process by which wastes are eliminated from the body. Although excretion is an important role of urinary systems, this process also occurs through the respiratory system, where carbon dioxide is released into the air, and through the digestive tract, where undigested material is excreted in feces.

Urinary systems excrete cellular wastes. The primary cellular wastes in urine are the nitrogen-containing compounds: ammonia, urea, and uric acid. These *nitrogenous wastes* are formed when cells break down the amino acids liberated by protein digestion (**Fig. 35-1**). Within cells, some amino acids are used to synthesize new proteins. Those amino acids that aren't required for protein synthesis have their amino groups removed, allowing the rest of the molecule to be used to synthesize other molecules or to provide a source of cellular energy. The liberated amino groups ($-NH_2$) are released by the cells as **ammonia** (NH_3), which is toxic. Ammonia is the primary nitrogenous waste of freshwater organisms, which avoid harm by releasing the

ammonia continually into their watery environments as it is formed, often through skin or gills. Terrestrial animals that collect and store their urine convert ammonia into less toxic **urea** (in mammals: see Fig. 35-1) or uric acid (in birds and other reptiles, and insects).

Homeostasis Maintains Relatively Constant Internal Conditions

The second function of the urinary system is to help maintain **homeostasis,** the relatively constant internal environment that cells need in order to function properly (see Chapter 31). Urinary systems play a crucial role in homeostasis by filtering the blood or the interstitial fluid to adjust water content, pH, and the concentrations of a variety of small molecules and ions in body fluids.

CHECK YOUR LEARNING

Can you describe the two major functions of urinary systems?

35.2 WHAT ARE SOME EXAMPLES OF INVERTEBRATE URINARY SYSTEMS?

Sponges (see Fig. 34-6) and cnidarians (such as *Hydra;* see Fig. 34-7), lack urinary systems. For these simple animals, diffusion and active transport through cell membranes into the surrounding water is adequate to excrete cellular wastes and maintain homeostasis. The urinary systems of flatworms act primarily to maintain water balance, and those in more complex invertebrates act as filters that eliminate wastes and regulate the composition of the body fluids, as described in the following sections.

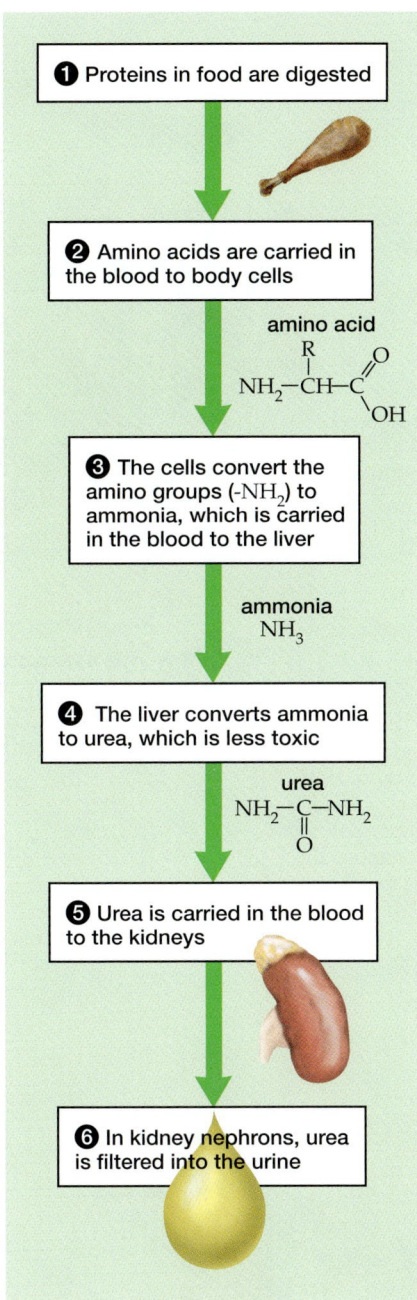

❶ Proteins in food are digested

❷ Amino acids are carried in the blood to body cells

amino acid

$$NH_2-CH-C\underset{OH}{\overset{O}{\diagdown}}$$
$$| R$$

❸ The cells convert the amino groups (-NH₂) to ammonia, which is carried in the blood to the liver

ammonia
NH_3

❹ The liver converts ammonia to urea, which is less toxic

urea
$$NH_2-\underset{O}{\overset{||}{C}}-NH_2$$

❺ Urea is carried in the blood to the kidneys

❻ In kidney nephrons, urea is filtered into the urine

▲ FIGURE 35-1 **Urea formation and excretion**

Protonephridia Filter Interstitial Fluid in Flatworms

The urinary systems of freshwater flatworms consist of **protonephridia** (singular, protonephridium). Protonephridia include tubules that branch throughout the body within the interstitial fluid that surrounds the flatworm's tissues. At intervals along the tubules are ciliated "flame cells," whose beating cilia resemble a flickering flame (**Fig. 35-2a**). Flame cells produce a current that draws interstitial fluid into the tubules through slit-like openings. Most dissolved substances are reabsorbed as the fluid flows through the tubule, and watery urine is excreted through the excretory pores. In

planarian flatworms, which inhabit lakes and streams, water continually enters the body by osmosis. Their simple urinary systems primarily collect and excrete this excess water. The large body surface of flatworms allows most cellular wastes to diffuse directly into the environment without passing through the urinary system.

Malpighian Tubules Filter the Hemolymph of Insects

Insects have open circulatory systems, in which *hemolymph* (a fluid that serves as both blood and interstitial fluid) fills the *hemocoel* (the body cavity) and bathes the internal tissues and organs directly. Insect urinary systems consist of **Malpighian tubules,** small tubes that extend outward from the intestine and end blindly within the hemolymph (**Fig. 35-2b**). Wastes and dissolved nutrients move from the hemolymph into the Malpighian tubules both by diffusion and by active transport, and water follows by osmosis. The urine flows through the Malpighian tubules into the intestine, where active transport returns important dissolved substances to the insect's hemolymph, and water follows by osmosis. The remaining urine is excreted from the digestive tract along with feces.

Nephridia Filter Interstitial Fluid in Annelid Worms and Mollusks

In some invertebrates, such as earthworms (and other annelids) and mollusks, the urinary system consists of tubular structures called **nephridia** (singular, nephridium). In the earthworm, the **coelom** (the body cavity enclosing the internal organs) is filled with interstitial fluid into which wastes and nutrients from the blood diffuse. Each nephridium begins with a funnel-like opening, the *nephrostome*, ringed with cilia that conduct interstitial fluid into a narrow, twisted tubule surrounded by a network of capillaries (**Fig. 35-2c**). As the fluid traverses the tubule, salts and other nutrients are reabsorbed back into the capillary blood, leaving wastes and excess water behind. The resulting urine is then excreted through an opening in the body wall called the *nephridiopore*. Most of the earthworm's body segments contain a pair of nephridia. As you study the kidney tubules of vertebrates, notice their similarities to nephridia.

CHECK YOUR LEARNING

Can you describe and compare the urinary systems of freshwater flatworms, insects, and earthworms?

35.3 WHAT ARE THE STRUCTURES OF THE MAMMALIAN URINARY SYSTEM?

The mammalian urinary system consists of the kidneys, ureters, bladder, and urethra. These organs filter small molecules and ions out of the blood, return essential ions and nutrients, and collect and excrete excess substances and cellular wastes. In the sections that follow, we discuss the mammalian urinary system, using the human system as an example.

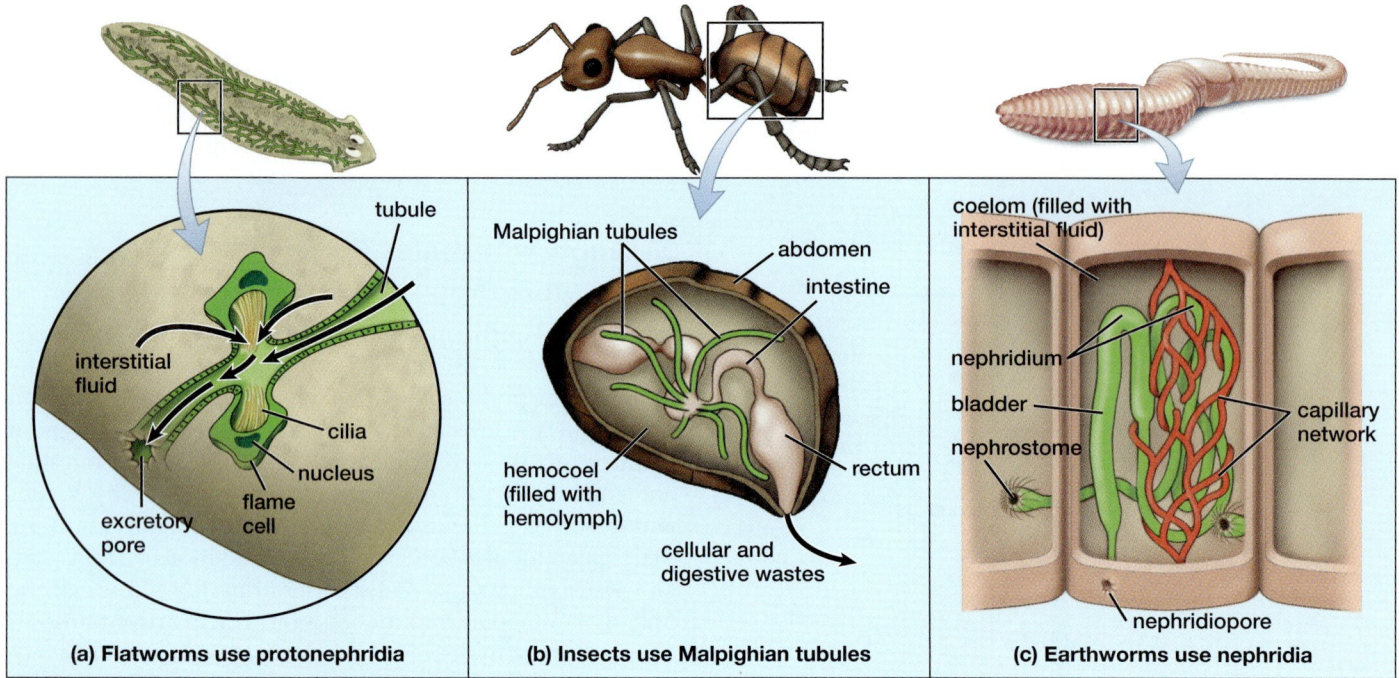

tubule

interstitial fluid

cilia

nucleus

flame cell

excretory pore

(a) Flatworms use protonephridia

Malpighian tubules

abdomen

intestine

hemocoel (filled with hemolymph)

rectum

cellular and digestive wastes

(b) Insects use Malpighian tubules

coelom (filled with interstitial fluid)

nephridium

bladder

nephrostome

capillary network

nephridiopore

(c) Earthworms use nephridia

▲ **FIGURE 35-2 Some invertebrate urinary systems (a)** The protonephridia of freshwater flatworms such as *Planaria* consist of tubules that collect and excrete urine (primarily excess water) from the interstitial fluid. The water is propelled along the tubules to the excretory pores by ciliated flame cells. **(b)** Malpighian tubules of insects extend from the intestine, filter the hemolymph that fills the hemocoel, and produce concentrated urine, which is excreted with the feces. **(c)** Paired nephridia in most segments of the earthworm filter interstitial fluid from the coelom. Interstitial fluid enters the nephrostome, and is carried by the tubule into the adjacent segment where nutrients are reclaimed into surrounding capillaries. Urine is collected in a bladder and exits the nephridiopore.

Different Structures of the Human Urinary System Produce, Store, and Excrete Urine

Human **kidneys** are paired organs, resembling a kidney bean in both shape and color, located at about waist level on either side of the spinal column (**Fig. 35-3**). The outermost layer of each kidney where urine is formed is the **renal cortex** (renal is from the Latin, meaning "kidney" and "cortex" means "bark"). Beneath the renal cortex lies the **renal medulla** ("kidney marrow"), which allows the kidney to produce concentrated urine that conserves water. The renal medulla surrounds a branched, funnel-like chamber called the **renal pelvis** ("kidney bucket"), which collects urine and conducts it into the ureter (**Fig. 35-4**). The **ureter** is a narrow, muscular tube that contracts rhythmically to propel the urine to the **bladder,** a hollow, muscular chamber that collects and stores urine.

The average adult bladder can hold about a pint (500 milliliters) of urine, but the desire to urinate is triggered by smaller amounts. As accumulating urine expands the bladder wall, this eventually activates stretch receptors that trigger reflexive contractions. Urine is retained in the bladder by two circular sphincter muscles. The internal sphincter, located at the junction of the bladder and the urethra, opens automatically during these contractions. The external sphincter, slightly below the internal sphincter, is under voluntary control, allowing the brain to suppress urination unless the

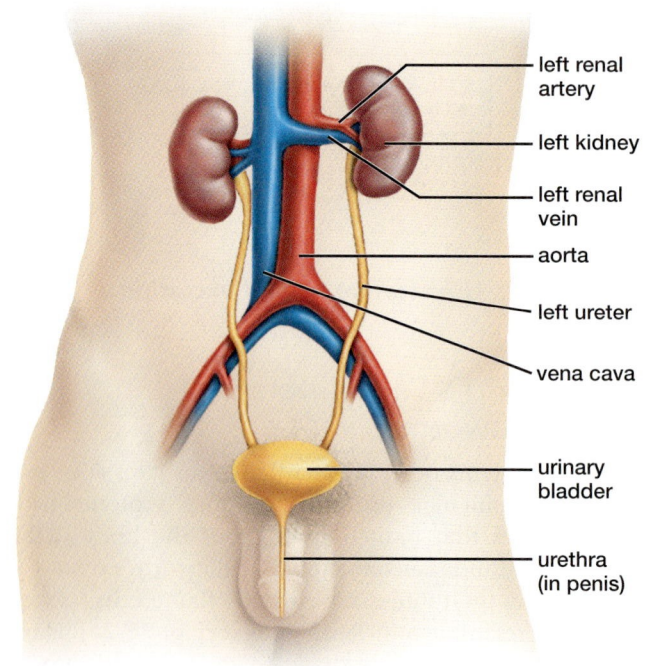

left renal artery

left kidney

left renal vein

aorta

left ureter

vena cava

urinary bladder

urethra (in penis)

▲ **FIGURE 35-3 The human urinary system and its blood supply**

▲ FIGURE 35-4 The internal structure and blood supply of the human kidney A single nephron is drawn considerably larger than normal (and further enlarged in the inset) to show its location in the kidney and its relationship to the collecting duct. Collecting ducts empty the urine into the renal pelvis, which funnels urine into the ureter. Yellow arrows show the path of urine flow.

bladder becomes overly full. Urine exits the body through the **urethra,** a single narrow tube about 1.5 inches long in the female and about 8 inches long in the male (where it extends through the penis).

Large Quantities of Blood Flow Through the Kidneys

To accomplish their functions of excreting cellular wastes and maintaining homeostasis, the kidneys filter large quantities of blood. Blood enters each kidney through a **renal artery** and exits through a **renal vein** (see Figs. 35-3 and 35-4). Nearly one-quarter of the volume of blood pumped by each heartbeat travels through the kidneys, which together receive more than 1 quart of blood (about 1 liter) every minute. All of the blood plasma is filtered through the kidneys about 60 times daily, allowing these organs to maintain the composition of the blood within narrow limits.

The immense filtering capacity of the kidneys explains why kidney donors and kidney recipients can each survive with only one kidney; half of the usual capacity is fully adequate to maintain homeostasis under normal conditions. Should both kidneys fail, however, death would occur very

CASE STUDY continued

Paying It Forward

When a person's kidneys fail, wastes accumulate in the blood, and imbalances in ion concentrations occur. Most people awaiting kidney transplants are kept alive using *hemodialysis*, a treatment in which wastes and excess water are filtered from the blood by an elaborate machine. Although people may remain on hemodialysis for many years, the treatment is far from ideal. Whereas healthy kidneys work nonstop, hemodialysis treatments are performed only about three times weekly. Patients must carefully monitor their diets between sessions, and those whose kidneys cannot produce urine must also carefully limit fluid intake to minimize the accumulation of water in their blood.

soon without treatment, as described in "Health Watch: When the Kidneys Collapse."

Nephrons in the Kidneys Filter Blood and Produce Urine

Each human kidney contains roughly 1 million microscopic filters, called **nephrons** (**Fig. 35-5;** see also Fig. 35-4, inset). Nephrons are packed together in the renal cortex, with a portion of each nephron extending into the renal medulla.

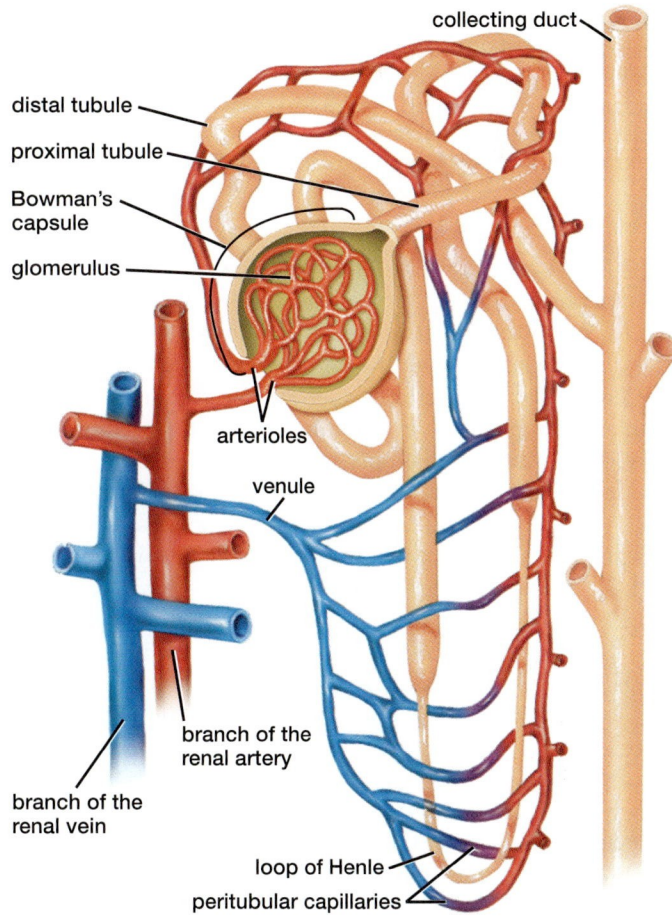

▲ FIGURE 35-5 An individual nephron and its blood supply

Health | Watch

When the Kidneys Collapse

Each year in the United States, about 90,000 people die of kidney failure, also called end-stage renal disease (ESRD). The most common causes of ESRD are diabetes and high blood pressure, both of which damage the glomerular capillaries, but the kidneys can also be damaged by infection or overdoses of some painkilling medicines. Kidney failure is typically treated by hemodialysis or by kidney transplantation from a living or newly deceased donor.

During hemodialysis, the patient's blood is diverted from the body and pumped through a machine with tubes made of a membrane suspended in dialyzing fluid. Like glomerular capillaries, the membrane has pores that are too small to permit the passage of blood cells and large proteins but large enough to pass small molecules. Dialyzing fluid has normal blood levels of salts and nutrients. Thus, only molecules or ions whose concentrations are higher than normal in the patient's blood (such as urea and H^+) diffuse into the dialyzing fluid. A typical hemodialysis session takes at least 4 hours, usually three times a week (**Fig. E35-1**). The 400,000 people in the United States who rely on hemodialysis can survive for many years, but their blood composition fluctuates, and toxic substances accumulate between sessions.

A less common form of dialysis is peritoneal dialysis, which can be done at home. Dialyzing fluid is pumped through a tube implanted directly into the abdominal cavity, which is lined with a natural membrane called the peritoneum. Waste products from blood circulating in peritoneal capillaries gradually diffuse into the dialysis fluid, which is replaced about 4 times daily, or circulated by a pump that operates overnight.

An automated wearable artificial kidney, which would be a tremendous advance over conventional dialysis, may be

▲ FIGURE E35-1 **A patient on hemodialysis**

marketed soon. This peritoneal dialysis machine uses far less expensive dialysis fluid than does a traditional peritoneal dialysis machine. The fluid is pumped through disposable filters instead of being discarded. The machine, the size of a typical woman's purse, will allow patients far greater mobility plus the benefits of continuous filtering.

Also in the works is a bioartificial kidney, which would be wearable or implanted in the body. The device uses a cartridge filled with cultured kidney tubule cells, which would perform a far greater range of homeostatic functions than does dialysis alone. People with ESRD have reason to be optimistic.

Blood carried to the kidney by the renal artery enters each nephron through a microscopically narrow arteriole. Within each nephron the arteriole branches further, forming a dense capillary meshwork called the **glomerulus,** the first portion of each nephron (see Fig. 35-5). The glomerular capillaries empty into an outgoing arteriole (in contrast to most capillaries, which empty into venules; see Chapter 32). This arteriole gives rise to **peritubular capillaries** ("peritubular" means "around the tubule") which form a network surrounding the nephron. The peritubular capillaries conduct the blood into a venule that joins the renal vein (see Fig. 35-5).

Each nephron has two major parts. In the first portion, the glomerulus, fluid is filtered out of the blood through the porous capillary walls. The second portion consists of a long, twisted **tubule** ("little tube"). Urine formation occurs in the tubule, which has four major sections. The tubule begins with a cuplike chamber called **Bowman's capsule,** which surrounds the glomerulus and collects fluid filtered out of the blood from the glomerular capillaries. The remaining sections of the tubule return water as well as essential molecules and ions to the blood, while retaining and concentrating wastes. From Bowman's capsule, fluid is conducted into

the **proximal tubule,** then into the **loop of Henle.** In most nephrons, the loop of Henle extends down from the renal cortex into the uppermost portion of the renal medulla. However, in about 15% of human nephrons, long loops of Henle extend deep into the renal medulla (see Fig. 35-4, inset). The loop of Henle produces and maintains a high salt concentration in the interstitial fluid of the renal medulla, which helps the kidney to retain water by producing concentrated urine. The filtrate is converted into urine in the **distal tubule.**

In the renal cortex, the distal tubule empties urine into a **collecting duct,** a larger tube that is not part of the nephron. Each of the thousands of collecting ducts within the kidney receives fluid from many nephrons. Collecting ducts conduct urine from the renal cortex, through the renal medulla, and into the renal pelvis (see Fig. 35-4). As urine flows through the collecting duct, additional water may leave the urine as required to maintain homeostasis.

CHECK YOUR LEARNING

Can you list and describe the major structures of the human urinary system? Can you diagram and describe the structures within the kidney? Can you describe the blood supply to each kidney? Can you draw and label a nephron?

35.4 HOW IS URINE FORMED?

Urine is produced in the nephrons of the kidneys during three processes: **filtration,** which filters water and most small dissolved molecules out of the blood; **tubular reabsorption,** during which water and nutrients filtered out of the blood by the nephron are restored, leaving wastes such as urea behind; and **tubular secretion,** during which wastes and excess ions remaining in the blood are secreted into the nephron. All movement of dissolved substances between the nephron and nearby capillaries occurs through the interstitial fluid that surrounds these structures.

Glucose, amino acids, fatty acids, and vitamins are important nutrients in blood. Other nutrients include ions, such as calcium (Ca^{2+}), sodium (Na^+), chloride (Cl^-), and potassium (K^+). Some ions are important in regulating blood pH; these include hydrogen (H^+) and bicarbonate (HCO_3^-). The urine retains wastes, such as urea, and excesses of water, ions, and some vitamins. The urine may also contain foreign substances, such as drugs and molecules produced when drugs are metabolized.

Filtration Removes Small Molecules and Ions from the Blood

Filtration, the first step in urine formation, occurs when water carrying small dissolved molecules and ions is forced through the walls of the nephron's glomerular capillaries (**Fig. 35-6 ❶**). The blood pressure within the glomerular capillaries is far higher than in most body capillaries, and the glomerular capillary walls are far more porous than most capillary walls. Therefore, as blood flows through the glomerular capillaries, about 20% of the blood's fluid is forced out through the capillary walls. Blood cells and large proteins are too large to penetrate the capillary walls, however, and remain in the blood. The fluid filtered out of the glomerular capillaries, called **filtrate,** is collected in Bowman's capsule and then continues through the tubule.

Tubular Reabsorption Returns Important Substances to the Blood

During tubular reabsorption, nearly all the water, essential ions, and organic nutrients such as glucose and amino acids that were removed during filtration are returned to the blood (**Fig. 35-6 ❷**; see also Fig. E35-2). These important substances are conveyed by diffusion or active transport through the wall of the nephron and into the peritubular capillaries, which return them to the bloodstream. Filtered waste products such as urea remain in the tubule. The reabsorption of water and inorganic nutrient ions such as Na^+, Cl^-, and Ca^{2+} is precisely regulated by circulating hormones, so that excess water and ions remain in the tubule for excretion. Although most tubular reabsorption takes place in the proximal tubule, some also occurs in the distal tubule.

❶ **Filtration:** Water, nutrients, and wastes are filtered from the glomerular capillaries into the Bowman's capsule of the nephron

❷ **Tubular reabsorption and tubular secretion:** In the proximal and distal portions of the tubule, most water and nutrients are reabsorbed into the blood, and additional wastes are secreted by active transport into the tubule from the blood

blood leaving the glomerulus

blood entering the glomerulus

glomerulus

Bowman's capsule

proximal tubule

loop of Henle

collecting duct

distal tubule

← tubular reabsorption
← tubular secretion

❹ **Concentration of urine:** In the collecting duct, urine may become more concentrated than the blood as water leaves by osmosis

❸ **Generation of high solute concentration:** The loop of Henle and the collecting duct both help to produce a high solute concentration in the interstitial fluid

▲ **FIGURE 35-6 Urine formation and concentration** The major functions of each part of the nephron and the collecting duct are shown in this simplified diagram.

Tubular Secretion Actively Transports Substances into the Tubule for Excretion

Tubular secretion actively transports wastes and excess ions from the blood into the tubule (see Fig. 35-6 ❷). Secreted substances include excess K^+ and H^+, small quantities of ammonia, some water-soluble medicinal and recreational drugs (including penicillin, aspirin, nicotine, cannabis, cocaine, and many more), some food additives, certain pesticides, and some poisons such as arsenic. Tubular secretion occurs primarily in the proximal tubule, but some also occurs in the distal tubule. The secretion of ions (K^+ and H^+) in the distal tubule is regulated by circulating hormones to compensate for changing conditions within the body. Filtrate is called urine after it exits the distal tubule.

CHECK YOUR LEARNING

Can you describe the three stages of urine formation, including the nutrients and wastes that are involved in each process? Can you describe where in the nephron each process occurs?

35.5 HOW DO VERTEBRATE URINARY SYSTEMS HELP MAINTAIN HOMEOSTASIS?

Vertebrate urinary systems help maintain homeostasis in several ways, including:

- Retaining small organic nutrients (such as glucose and amino acids) and inorganic nutrient ions (such as Na^+, K^+, Cl^-, and Ca^{2+}) within the blood and interstitial fluid as described above;
- Regulating the water and ion content of the blood to maintain the proper blood osmolarity;
- Maintaining the proper pH of the blood by regulating hydrogen and bicarbonate ion concentrations; and
- Secreting substances that help regulate blood pressure and blood oxygen content.

The Kidneys Regulate the Water and Ion Content of the Blood

One important function of the kidney is to regulate blood **osmolarity,** a measure of its concentration of ions and other solutes. Human kidneys filter out about half a cup (4 ounces or 125 milliliters) of water and solutes from the blood each minute. If the kidneys were unable to return this water to the blood, we would need to drink nearly 50 gallons of water a day to replace the water lost in our urine! Consequently, the urinary system needs to restore nearly all (about 99%) of the water filtered out through the glomeruli.

Reabsorption of Water Occurs Passively Through the Tubule and Collecting Duct

Reabsorption of water into the blood occurs passively by osmosis as the filtrate travels through the tubule and collecting duct. When the filtrate enters the distal tubule, most (75% to 80%) of its water has already been reabsorbed, but the filtrate is still considerably more dilute than the surrounding interstitial fluid in the renal cortex. From this point on, additional reabsorption of water is precisely regulated to maintain the blood's osmolarity within narrow limits. If fluid intake has been excessive, water will be left behind in the filtrate, and watery urine will be produced until the normal blood volume is restored. This requires that the membranes of the distal tubule and collecting duct cells remain quite impermeable to water, retaining water in the urine. If fluid intake has been low, tubular reabsorption will remove more water from the filtrate. Additional water will be reabsorbed as the urine flows through the collecting duct, and the urine will become concentrated.

How is water permeability regulated? The interstitial fluid within the renal medulla contains high concentrations of salt and urea. Salt enters the fluid from the loop of Henle, and urea enters from the collecting duct (**Fig. 35-6 ❸**). As urine travels through the collecting duct to the renal pelvis, it passes through this interstitial fluid, which becomes increasingly concentrated in salt and urea from the upper to the lower region of the renal medulla. As the difference in osmolarity between the urine in the collecting duct and the surrounding interstitial fluid increases, water leaves the urine by osmosis and is carried off by the peritubular capillaries. The urine in the collecting duct can become as concentrated as the surrounding interstitial fluid—up to four times the osmolarity of normal blood (**Fig. 35-6 ❹**).

Water Permeability of the Distal Tubule and Collecting Duct Is Regulated by ADH

The amount of water reabsorbed in the distal tubule and collecting duct depends on their permeability to water. Water permeability throughout the tubule and collecting duct is controlled by the numbers of **aquaporins,** which are water channel proteins in the cell membranes of those structures. The membranes of the proximal tubule and the descending portion of the loop of Henle have large numbers of aquaporins at all times, so these membranes remain highly permeable to water. In contrast, aquaporin numbers in the distal tubule and the collecting duct membranes are adjusted as the body's needs change. These numbers are controlled by **antidiuretic hormone (ADH;** "diuretic" means "increasing urine production," so an "*anti*diuretic" decreases urine production).

ADH is secreted by the posterior pituitary gland, is carried in the bloodstream, and stimulates cells of the distal tubule and collecting duct to insert aquaporin proteins into their membranes. In the absence of ADH, these membranes are nearly impermeable to water. The more ADH, the more aquaporins are inserted, and the greater the permeability of the distal tubule and collecting duct membranes to water. In this way, ADH levels control the osmolarity of the blood by producing urine that contains varying amounts of water.

How are ADH levels controlled? Receptors in the hypothalamus monitor blood osmolarity. For example, when your body becomes dehydrated—as might occur if you were dripping sweat while exercising in the hot sun—your blood osmolarity rises. When blood osmolarity exceeds an optimal level, the hypothalamus stimulates the pituitary gland to release ADH. In response to ADH, cells of the distal tubule and collecting duct insert additional aquaporins into their membranes (Fig. 35-7). As urine flows through the distal tubule and collecting duct, the more concentrated interstitial fluid draws water out by osmosis. The water enters peritubular capillaries and is restored to the bloodstream.

But reducing water loss will not, by itself, restore blood to its normal osmolarity. To compensate for lost water, receptors in the hypothalamus simultaneously activate a thirst center, also in the hypothalamus, that stimulates the desire to drink and restore water to the blood.

Blood osmolarity is controlled by negative feedback. When blood osmolarity returns to normal, the receptors in the hypothalamus signal the pituitary to reduce ADH release down to a baseline level. If you drink too much water, ADH secretion will be temporarily blocked, causing you to excrete large amounts of very dilute urine (with only about one-third the osmolarity of the blood). When normal blood osmolarity is restored, ADH secretion is increased to baseline levels.

Mammalian Nephrons Are Adapted to the Availability of Water

Only mammals and birds have loops of Henle that allow them to produce urine that is more concentrated than their blood. We have described human nephrons with loops of Henle that extend deep into the renal medulla, but in fact

1 Heat causes water loss and dehydration through sweating

2 Receptors in the hypothalamus detect the increased blood osmolarity and signal the pituitary gland

3 The pituitary gland releases ADH into the bloodstream

4 ADH increases the permeability of the distal tubule and the collecting duct, allowing more water to be reabsorbed into the blood

5 Water is retained in the body and concentrated urine is produced

▲ FIGURE 35-7 Dehydration stimulates ADH release and water retention

QUESTION Describe the feedback process that would occur if you drank far more water than your body needed.

only about 15% of human nephrons have these long loops. The rest have short loops that barely penetrate the renal medulla. Why do we have two types of nephrons? Active transport of salt out of long loops of Henle allows maximal reabsorption of water. However, active transport also uses a lot of ATP energy, which could instead be used for feeding, growth, or reproduction. Therefore, the availability of water has been a force of natural selection driving the evolution of mammalian kidneys that contain different percentages of long-looped and short-looped nephrons.

Mammals adapted to dry climates generally have kidneys containing a high percentage of nephrons with long loops of Henle (Fig. 35-8). This causes a higher concentration of salt to be produced in the interstitial fluid of the kidney medulla, thus allowing more water to be reclaimed from the urine as it travels through the collecting ducts. The masters of urine concentration are desert rodents such as kangaroo rats (see Fig. 29-14a), which can produce urine with 14 times the osmolarity of their blood. Not surprisingly, kangaroo rats have only very long-looped nephrons. With their extraordinary ability to conserve water, kangaroo rats do not need to drink; they rely entirely on water contained in their food and on metabolic reactions that produce water.

▲ FIGURE 35-8 **A well-adapted desert dweller** Camels, which are native to deserts, have extremely long loops of Henle, and can produce urine with over nine times their blood osmolarity.

IN GREATER DEPTH How the Nephron Forms Urine

The complex structure of the nephron is finely adapted to its function of maintaining homeostasis. **Figure E35-2** illustrates the processes occurring in each portion. These processes produce filtrate from the blood and convert it into urine. The osmolarity of the interstitial fluid (measured in milliosmols) is shown along the left vertical axis. The normal osmolarity of blood and interstitial fluid (with the exception of that in the renal medulla) is regulated throughout the body at about 300 milliosmols. Within the renal medulla, however, the osmolarity of the interstitial fluid reaches about 1,200 milliosmols. If excess water has been consumed, the urine will have an osmolarity lower than that of normal blood and interstitial fluid—as low as 100 milliosmols. If too little water has been consumed for the body's needs, the urine may have an osmolarity of up to 1,200 milliosmols, four times that of normal blood and interstitial fluid.

The following descriptions of how urine concentration is regulated refer to the circled numbers in the figure.

❶ **Glomerular capillaries and Bowman's capsule** Filtration occurs here. During filtration, water and small dissolved nutrients and wastes are forced out of the glomerular capillaries and into Bowman's capsule. This produces filtrate resembling blood plasma without its large proteins, with the same osmolarity as the blood and interstitial fluid. The filtrate enters the proximal tubule.

❷ **Proximal tubule** Most tubular reabsorption and tubular secretion occur here. Tubular reabsorption moves water and most dissolved nutrients from the filtrate back through the walls of the tubule and into the interstitial fluid, where they are reabsorbed into the peritubular capillaries. Dissolved nutrients include amino acids, glucose, and various ions, such as sodium (Na^+), chloride (Cl^-), potassium (K^+), calcium (Ca^{2+}), and bicarbonate (HCO_3^-). Some of these are actively transported from the filtrate and

others diffuse out. Water (H_2O) follows these solutes by osmosis. Because both H_2O and its solutes have been reabsorbed, the osmolarity of the filtrate in the proximal tubule remains about the same as the blood. Excess hydrogen ions (H^+) and some foreign substances (such as drugs) are actively transported by tubular secretion from the capillary blood into the proximal tubule.

❸ **Descending portion of the loop of Henle** A salt concentration gradient exists in the surrounding interstitial fluid. The descending portion of the loop is permeable to H_2O, but not NaCl. Water leaves the filtrate by osmosis as the osmolarity of the interstitial fluid surrounding it increases. Loss of H_2O increases the concentration of the filtrate within the tubule. At the bottom of the loop of Henle, the filtrate reaches the same osmolarity as the surrounding interstitial fluid (about 1,200 milliosmols).

❹ **Thin segment of the ascending portion of the loop of Henle** Salt diffuses out here. This segment is permeable only to NaCl (not to H_2O). As the filtrate flows upward through the decreasing salt concentration in the interstitial fluid, NaCl diffuses out because NaCl is more concentrated in the filtrate than in the surrounding interstitial fluid. Because H_2O cannot follow the NaCl, this process contributes to the high salt concentration of the interstitial fluid. As it flows upward through the thin segment, the filtrate becomes less concentrated than the surrounding interstitial fluid, because it has lost NaCl but retained H_2O.

❺ **Thick segment of the ascending portion of the loop of Henle** Here, NaCl is actively pumped out of the filtrate, contributing further to the high osmolarity of the surrounding interstitial fluid. Like the thin segment, the thick segment of the loop of Henle is impermeable to H_2O. Actively transporting NaCl out

of the filtrate further reduces the osmolarity of the filtrate to about 100 milliosmols, about one-third the osmolarity of blood and interstitial fluid within the renal cortex.

❻ **Distal tubule** The distal tubule receives the relatively dilute filtrate from the loop of Henle. Some tubular secretion and some tubular reabsorption occur in the distal tubule, fine-tuning blood composition in response to the changing needs of the body. These processes in the distal tubule are primarily controlled by hormones (such as ADH) released in response to changes in blood plasma composition. The distal tubule secretes excess K^+, H^+, and some drugs from the blood into the filtrate for excretion. The distal tubule also reabsorbs Ca^{2+} and NaCl, restoring them to the blood. Once the filtrate leaves the distal tubule, it is called urine.

❼ **Collecting duct** Urine can become quite concentrated here. The collecting duct carries urine through the increasing solute concentration of the interstitial fluid within the renal medulla. The high interstitial fluid solute levels are generated by the loop of Henle (as described earlier) and by the lower collecting duct (as described next). Water loss by osmosis is controlled by ADH. At the highest ADH levels, the collecting duct becomes so permeable to water that the osmolarity of the urine equilibrates with the osmolarity of the surrounding interstitial fluid, reaching about 1,200 milliosmols in humans.

❽ **Lower collecting duct** Urea is added to the interstitial fluid here, as the duct passes through the renal medulla (black arrow). Urea and NaCl contribute about equally to the total osmolarity in the renal medulla, making them equally important to producing concentrated urine. Considerable urea also remains as a waste product in the urine, where it is typically 50 times as concentrated as in blood plasma.

▼ FIGURE E35-2 Details of urine formation A single nephron, showing the movement of materials through different regions. The peritubular capillaries that surround the nephron (see Fig. 35-5) have been omitted for simplicity. The concentration of dissolved substances in the filtrate within the nephron is indicated by the intensity of yellow color; black arrows within the tubule indicate the direction of filtrate flow. Outside the nephron, darker shades of beige represent higher concentrations of salt and urea in the surrounding interstitial fluid. The dashed line marks the boundary between the renal cortex and the renal medulla.

FILTRATION
OF BLOOD

TUBULAR REABSORPTION
TUBULAR SECRETION

CONCENTRATION
OF URINE

osmotic concentration of interstitial fluid (in milliosmols)

300

300

300

600

900

1,200

HCO_3^-
Ca^{2+}
Cl^- H^+
Na^+ some
nutrients H_2O K^+ drugs

H^+
K^+
NaCl some
Ca^{2+} drugs HCO_3^- H_2O*

❶

❷ proximal
tubule

Bowman's
capsule

renal cortex

renal medulla

H_2O

❻ distal
tubule

NaCl

H_2O*

❼

H_2O

❺ NaCl

H_2O

❸

NaCl

H_2O*

(interstitial fluid)

❹ NaCl

urea

❽

H_2O

H_2O*

mechanism
under study

osmosis

active transport

diffusion

* controlled by ADH

loop of Henle

collecting duct

In contrast, mammals adapted to habitats with an abundance of fresh water typically have short loops of Henle. For example, beavers, which live along streams, have only short-looped nephrons and can only concentrate their urine to about twice their blood osmolarity.

The Kidneys Help Maintain Blood pH

The maintenance of proper pH (from 7.38 to 7.42) is crucial to a multitude of cellular metabolic processes, including the functioning of enzymes and neurons. However, normal cellular metabolic activities—such as breaking down proteins and fats or synthesizing ATP—produce an excess of H^+, which makes solutions acidic. The most important defense against pH change is provided by buffers in the blood, particularly bicarbonate (HCO_3^-), which can combine with H^+ to form H_2CO_3. The kidneys help maintain blood pH by secreting excess H^+ into the tubule for excretion and by reabsorbing HCO_3^- into the blood. These processes occur in both the proximal and distal tubules of the nephrons.

The Kidneys Help Regulate Blood Pressure and Oxygen Levels

When blood pressure falls, as can occur with excessive blood loss, the kidneys release the enzyme **renin** into the bloodstream. Renin catalyzes the formation of the hormone **angiotensin** from a protein circulating in the blood. Angiotensin helps combat low blood pressure in three major ways: (1) It stimulates release of another hormone (aldosterone; see Chapter 37) that causes the proximal tubules of the nephrons to reabsorb more Na^+ into the blood. This causes more water to follow by osmosis and helps to maintain blood volume. (2) It stimulates ADH release, causing more water to be reclaimed from the distal tubule and collecting duct. (3) It causes arterioles throughout the body to constrict, increasing blood pressure.

The kidneys also help maintain blood oxygen at levels that support the body's needs. The availability of oxygen is reduced if blood is lost, if lung disease reduces oxygen uptake, or at high altitudes, where each lungful of air supplies less oxygen. In response to reduced oxygen, the kidneys release the hormone **erythropoietin** (see Chapter 33).

CASE STUDY continued

Paying It Forward

Kidney failure almost always causes anemia, because the kidneys can no longer produce enough erythropoietin to stimulate adequate red blood cell production. Fortunately, human erythropoietin can be manufactured by genetically engineered cells grown in culture and administered to patients who have anemia caused by kidney failure. However, a functioning kidney will respond to the changing needs of the body, producing the proper amounts of this hormone at the appropriate times—a definite advantage for transplant recipients.

Erythropoietin is carried in the blood to the bone marrow and stimulates the bone marrow to produce more oxygen-transporting red blood cells.

Fish Face Special Challenges in Their Aquatic Environments

Animals that are constantly immersed in a solution that has either a lower (hypotonic) or a higher (hypertonic) osmolarity than their body fluids have evolved special mechanisms to maintain a homeostatic balance of water and salt within their bodies, a process called **osmoregulation.**

Freshwater fish, which live in a hypotonic environment, maintain a plasma osmolarity nearly 100 times the osmolarity of their freshwater surroundings. As these fish circulate water over their gills to exchange gases, some water continuously leaks into their bodies by osmosis, and salts diffuse out (**Fig. 35-9a**). Freshwater fish acquire salt from their food, and also through their gills, which use active transport to pump salt into their bodies from the surrounding water.

(a) Freshwater fish

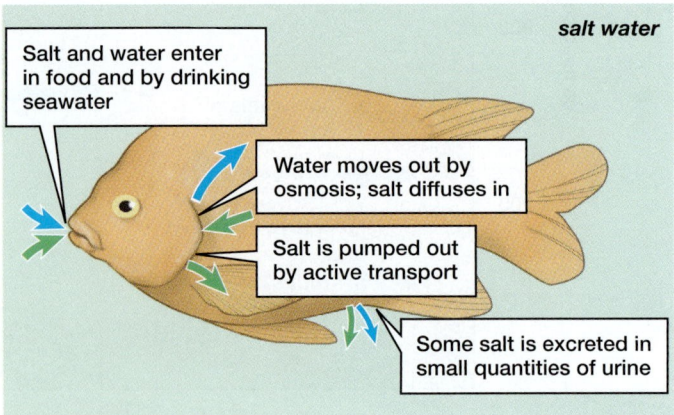

(b) Saltwater fish

▲ **FIGURE 35-9 Osmoregulation in fish (a)** Urinary systems of freshwater fish, such as this bluegill, must contend with large amounts of water entering their bodies by osmosis. Some water enters during feeding as well. **(b)** Urinary systems of saltwater fish, such as this Garibaldi, must conserve water, which constantly diffuses out of their bodies into the more salty surrounding seawater.

To compensate for the entry of excess water, freshwater fish never drink (although they take in some water as they feed), and their kidneys retain salt and excrete large quantities of extremely dilute urine.

Saltwater fish live in a hypertonic environment; seawater has a solute concentration two to three times that of their body fluids. As a result, water continuously leaves their tissues by osmosis, and salt continuously diffuses in and is also acquired with food. Most saltwater fish drink to restore their lost water, and excrete the excess salt by active transport through their gills (**Fig. 35-9b**). Fish nephrons completely lack loops of Henle, so fish cannot produce urine that is more concentrated than their blood. To conserve water, the kidneys of most saltwater fish excrete only very small quantities of urine, containing some salts that their gills did not eliminate. Cartilaginous fishes such as sharks and rays have evolved a different solution for conserving water. These saltwater fish store urea in their tissues at a concentration so high that it would kill most other vertebrates. The stored urea gives the tissues in sharks and rays approximately the same osmolarity as the surrounding seawater, so they avoid losing water by osmosis.

CHECK YOUR LEARNING

Can you explain how the nephron and collecting duct help control blood osmolarity? Do you understand the role of ADH in water reabsorption? Can you explain how kidneys help control blood pH, blood pressure, and blood oxygen content? Can you describe how mammalian kidneys are adapted to wet and dry environments? Do you understand how and why the urinary systems of freshwater and saltwater fish differ?

CASE STUDY revisited

Paying It Forward

Since the 1950s, when living kidney donation was first recognized as a viable alternative to cadaver organ donors, family and friends have come forward to offer a kidney to a victim of kidney failure. Ideally, in addition to blood type, several important glycoproteins of the donor such as major histocompatibility complex (MHC) proteins, which identify cells as belonging to a particular individual (see Chapter 36), should match those of the recipient. This reduces the chances that the recipient's immune system will attack the donated kidney as if it were a foreign invader. But, with the exception of identical twins, no two people have perfectly matching tissues. This means that people with kidney transplants must take immune-suppressing drugs for the remainder of their lives, making them vulnerable to infections and some types of cancer. Despite this drawback, a transplanted kidney is by far the best option for those lucky enough to receive one.

To remove a kidney from the donor, surgeons generally use a technique called laparoscopic surgery, where they make three or four incisions about one-half inch long through which they insert surgical tools, including a tiny video camera. The kidney is removed through an incision about 2½ inches long, put on ice, and rushed to its recipient. The operation takes 3 to 4 hours; donors remain in the hospital for about 3 days and return to work in about 3 weeks. Donating a kidney requires major surgery, with its associated risks. Kidney donors will also lack a back-up kidney in the unlikely event that their remaining kidney fails. But a recent study of deaths among 80,000 kidney donors during a 15-year period found no greater mortality among this group (once they had recovered from their surgery) than among non-donors.

Since 2008, when DeGiulio's donated kidney started a chain that saved four lives (**Fig. 35-10**), such "domino donation chains" have become longer and more frequent. For example, during a 4-month period in 2011, 17 hospitals in 11 states from California to New Jersey matched 30 people—who might otherwise have died—with kidneys from 30 donors they had never met. This heroic enterprise was started by Good Samaritan Rick Ruzzamenti, who got the idea from the desk clerk at his yoga studio, who had mentioned to him that

▲ **FIGURE 35-10 The domino chain of kidney donations**

she had donated a kidney to a friend. Of the 13,000 to 14,000 kidney transplants that are performed annually in the United States, about 6,000 are from living donors who are usually relatives or close friends of the recipients. The approximately 60,000 eligible individuals awaiting a kidney transplant who haven't found compatible donors ardently hope that domino donation chains continue to be forged.

When DeGiulio shared his plans to donate a kidney with his friends and family members, some of them told him he was "nuts." DeGiulio disagrees. "It's easily the greatest thing I've ever done in my life, and also the easiest decision I ever made," said the 32-year-old donor. Ruzzamenti had a similar reaction. "People think it's so odd that I'm donating a kidney," he told the transplant coordinator at his hospital, but "I think it's so odd that they think it's so odd. . . . It causes a shift in the world."

Consider This

BioEthics Would you donate a kidney to a friend or family member whose kidneys were failing? Would you consider starting a domino donation chain by donating a kidney to a stranger?

CHAPTER REVIEW

Summary of Key Concepts

35.1 What Are the Major Functions of Urinary Systems?

Urinary systems produce and eliminate urine, which contains waste products of cellular metabolism, foreign substances, and some ions and nutrients in excess of the body's needs. In producing urine, urinary systems perform two basic functions: excretion of cellular wastes and maintenance of homeostasis. Urinary systems maintain homeostasis by adjusting the water content, pH, and the concentrations of small molecules and ions in body fluids.

35.2 What Are Some Examples of Invertebrate Urinary Systems?

The flatworm's simple urinary system consists of protonephridia, a network of tubules that branch throughout the body, collecting wastes and excess water. Flame cells create a current that forces the urine out of the body through urinary pores. Insects use Malpighian tubules that filter hemolymph within the hemocoel of their open circulatory systems. Malpighian tubules release urine into the intestine for elimination. Earthworms use nephridia to filter the interstitial fluid that fills the body cavity. Nutrients are reabsorbed into the interstitial fluid, and the urine is released through nephridiopores.

35.3 What Are the Structures of the Mammalian Urinary System?

The human urinary system consists of paired kidneys and ureters, a bladder, and a urethra. Kidneys produce urine, which is conducted by the ureters to the bladder, where it is stored. Distention of the bladder wall triggers urination, during which urine passes out of the body through the urethra. Kidneys have an outer renal cortex packed with filtering units called nephrons. Some nephrons extend into the underlying renal medulla, where urine is concentrated. The innermost kidney chamber is the renal pelvis, which funnels urine into the ureter. Large quantities of blood enter each kidney through a renal artery and exit through a renal vein. The renal cortex of the kidney is packed with nephrons, which filter the blood. Each nephron includes the capillaries of the glomerulus and a tubule consisting of Bowman's capsule, the proximal tubule, the loop of Henle and the distal tubule. The distal tubule empties into the collecting duct, which carries urine through the renal medulla and releases it into the renal pelvis. Peritubular capillaries surround the tubule of the nephron and reabsorb important substances into the blood while secreting additional wastes into the urine.

35.4 How Is Urine Formed?

Urine formation in the nephron begins with filtration through the glomerular capillaries. Here water and most small dissolved molecules are filtered out of the blood in a filtrate that is collected in Bowman's capsule and enters the tubule of the nephron. Tubular reabsorption in the proximal and distal portions of the tubule restores most water, nutrients, and important ions to the blood, leaving wastes such as urea behind. Tubular secretion in the proximal and distal tubule actively transports wastes and excess ions that still remain in the blood into the filtrate. Peritubular capillaries surrounding the nephron allow exchange of substances between the blood and the filtrate by way of the interstitial fluid. The filtrate is called urine after it exits the nephron.

35.5 How Do Vertebrate Urinary Systems Help Maintain Homeostasis?

Vertebrate urinary systems regulate the water and ion content of the blood, maintain the proper blood pH, retain nutrients, and secrete substances that help regulate blood pressure and blood oxygen content. Water is returned to blood from the filtrate by reabsorption in the tubule and collecting duct. As urine flows through the collecting duct to the renal pelvis, it passes through the interstitial fluid in the renal medulla. This interstitial fluid becomes increasingly concentrated in salt (added by the loop of Henle) and urea (added by the collecting duct).

Dehydration increases blood osmolarity, which causes the posterior pituitary to secrete antidiuretic hormone (ADH) into the blood. ADH causes the aquaporins to be inserted into the distal tubule and collecting duct membranes, increasing their water permeability. High water permeability allows water to be reabsorbed into the blood from the distal tubule and from the collecting duct as it passes through the increasing concentration of salt and urea in the renal medulla.

The kidneys help regulate blood pressure by secreting the enzyme renin when blood pressure falls. Renin catalyzes the formation of angiotensin from a blood protein. Angiotensin increases Na^+ reabsorption, stimulates ADH release, and constricts arterioles, elevating blood pressure. Erythropoietin, released from the kidneys when the oxygen content of the blood is low, stimulates bone marrow to produce red blood cells.

Mammals that live where water is abundant tend to have short loops of Henle and produce dilute urine, whereas those that live where water is scarce have long loops of Henle and produce concentrated urine. Freshwater fish generate large quantities of dilute urine and actively transport salt in through their gill tissues. Saltwater fish drink seawater, actively transport salt out of their gill tissues, and produce very little urine.

Key Terms

ammonia   674
angiotensin   684
antidiuretic hormone
   (ADH)   680
aquaporin   680
bladder   676
Bowman's capsule   678
coelom   675
collecting duct   678
distal tubule   678
erythropoietin   684
excretion   674
filtrate   679
filtration   679
glomerulus   678
homeostasis   674
kidney   676
loop of Henle   678
Malpighian tubule   675
nephridium (plural,
   nephridia)   675

nephron   677
osmolarity   680
osmoregulation   684
peritubular capillaries   678
protonephridium (plural,
   protonephridia)   675
proximal tubule   678
renal artery   677
renal cortex   676
renal medulla   676
renal pelvis   676
renal vein   677
renin   684
tubular reabsorption   679
tubular secretion   679
tubule   678
urea   674
ureter   676
urethra   677
urinary system   674
urine   674

Learning Outcomes

In this chapter, you have learned to . . .

LO1 Explain the two major functions of urinary systems and the general processes that accomplish these functions.

LO2 Describe how urea is produced from protein.

LO3 Describe and compare the urinary systems of freshwater flatworms, earthworms, and insects.

LO4 Draw, label, and explain the functions of each major part of the human urinary system, including the structures within the human kidney and those involved in the blood supply to the kidney.

LO5 Diagram and label a nephron and collecting duct,

and explain each stage of urine production within these structures.

LO6 Explain the role of antidiuretic hormone, the distal tubule, the loop of Henle, and the collecting duct in controlling blood osmolarity in mammals and how mammalian nephrons are adapted for different environments.

LO7 Describe how the kidneys help regulate blood ion levels, pH, oxygen levels, and blood pressure.

LO8 Describe osmoregulation in freshwater and saltwater fish.

Thinking Through the Concepts

Fill-in-the-Blank

1. Fill in the urinary organ of the following animals: earthworm: _____; insects: _____; flatworms: _____. Which of these organs most closely resembles the nephrons of vertebrates?_____. **LO3**

2. The human kidney consists of an outer layer called the _____, an underlying layer called the _____, and a funnel-like, central chamber

called the _____. Urine is funneled from the central chamber into the _____, which leads to a storage organ, the _____, which empties through the _____. **LO4**

3. The five parts of the nephron, in the order in which filtrate passes through them, are: _____, _____, _____, _____, _____. **LO5**

4. Fill in with the appropriate term: Most tubular reabsorption occurs here: _____; final concentration of urine occurs here: _____; creates a salt concentration gradient in the interstitial fluid: _____; location of the concentration gradient within the kidney: _____. **LO5 LO6**

5. Fill in the following substances: produced from ammonia and excreted in urine: _____; secreted into the tubule when blood pH is too low: _____; actively transported out of the ascending portion of the loop of Henle: _____; leaves the tubule by osmosis: _____. **LO2 LO4 LO6 LO7**

6. The kidney helps maintain an internal balance called _____. To help regulate blood pressure, the kidney secretes the enzyme _____, which converts a substance circulating in the bloodstream into the hormone _____. To control oxygen levels, the kidneys secrete the hormone _____, which acts on the bone marrow to increase production of _____. **LO1 LO7**

7. If blood osmolarity increases, receptors in the _____ detect this and signal the _____ gland to increase release of _____. This hormone acts on the walls of the _____ and the _____, causing them to insert more water pores called _____ into their membranes and, thus, cause the urine to become more _____. **LO6**

Review Questions

1. Explain the two major functions of urinary systems, and list the processes that accomplish these functions. **LO1 LO5**

2. Trace a urea molecule through the mammalian body, starting in the bloodstream and ending outside the body. **LO4 LO5**

3. What is the function of the loop of Henle? The collecting duct? Antidiuretic hormone? **LO6**

4. Discuss the differences in function of the two major capillary beds in the kidneys: the glomerular capillaries and the peritubular capillaries. **LO4 LO5**

5. Describe the processes of filtration, tubular reabsorption, and tubular secretion. **LO5**

6. Describe the role of the kidneys as organs of homeostasis. **LO6 LO7**

7. Briefly describe and compare the urinary systems of flatworms, insects, and earthworms. **LO3**

8. Explain and contrast osmoregulation in freshwater and saltwater fish. **LO8**

Applying the Concepts

1. Would you expect the restrooms in a restaurant with a bar to be used more than those in restaurants that don't serve liquor? Explain.

2. Some weight-loss diets are high in protein and very low in carbohydrates. Why is it advisable to drink more water than usual on such diets?

3. Imagine yourself lost in a desert in the searing noonday sun, with an empty water bottle. How does your urinary system respond?

4. In his poem "The Rime of the Ancient Mariner," Samuel Taylor Coleridge wrote: "Water, water, everywhere, nor any drop to drink." Seawater has more than four times the osmolarity of blood. Why can't a person avoid dying of thirst by drinking seawater?

5. Some people advocate drinking one's own urine, claiming significant health benefits. What problems do you see with this "alternative therapy"? Are there conditions under which this might be advisable?

Answers to Figure Caption questions and Fill-in-the-Blank questions can be found in the Answers section at the back of the book.

MB *Go to MasteringBiology for practice quizzes, activities, eText, videos, current events, and more.*

Jim Henson, creator of the Muppets, was killed by a fast-acting, massive infection with *Streptococcus pyogenes*.

CASE STUDY

Flesh-Eating Bacteria

ON MAY 4, 1990, JIM HENSON, creator of the Muppets and the original voice of Kermit the Frog, felt unusually tired and had a sore throat. By May 13, he felt considerably worse, and consulted a physician, who didn't find any signs of pneumonia or other serious illness. However, on May 15, Henson began coughing up blood. Three hours later, at New York-Presbyterian Hospital, he couldn't breathe, and was placed on a respirator. Despite the best care and multiple antibiotics, Jim Henson died just 20 hours later.

Henson was killed by *Streptococcus pyogenes*. Chances are, you've been infected by this type of bacterium—it causes strep throat. However, some strains of *S. pyogenes* are nastier than others, and some people are more susceptible than others. In these instances, *S. pyogenes* may cause one of two deadly streptococcal diseases: streptococcal toxic shock syndrome or necrotizing soft tissue infection. Henson died of streptococcal toxic shock syndrome when *S. pyogenes* invaded his lungs, causing massive inflammation and multiple organ failure.

In necrotizing soft tissue infection, *S. pyogenes* enter a cut or open sore, and rapidly destroy huge amounts of tissue, leading to the common name "flesh-eating bacteria." Although this sounds overly dramatic, it's pretty accurate. Consider Monica Jorge, who gave birth to a healthy baby girl by planned cesarean section in 2007. No one knows how the bacteria entered Jorge's body, but within hours, *S. pyogenes* had destroyed most of her abdominal wall. To eliminate the infection, surgeons had to remove her uterus, ovaries, part of her large intestine, and large parts of both arms and both legs.

Despite occasional devastating failures such as these, the body's defenses against infection by bacteria and other microbes are usually highly effective. As you learn about these defenses, consider a few questions: How does your body distinguish "self" from "invader"? How does the immune system kill invaders? How can microbes such as *S. pyogenes* evade the immune system and cause such rapid and deadly disease?

AT A GLANCE

LEARNING GOALS

LG1 Name the body's external barriers against disease, and describe the processes by which those barriers protect the body against infection.

LG2 Describe the processes by which the innate immune response protects against infection, including the roles of white blood cells, the inflammatory response, and fever.

LG3 Name the organs, tissues, cells, and molecules that make up the adaptive immune system, and describe their structures and functions.

LG4 Describe how the immune system distinguishes self from non-self.

LG5 Describe the processes by which the adaptive immune system protects against infection, including humoral immunity, cell-mediated immunity, and the role of helper T cells in the immune response.

LG6 Explain the process of acquired immunity.

LG7 Explain how antibiotics and vaccines supplement the human immune system.

LG8 Define *allergy*, *autoimmune disease*, and *immune deficiency disease*, describe how each occurs, and give examples of each.

LG9 Explain how the immune system combats cancer, and describe current medical treatments against cancer.

36.1 HOW DOES THE BODY DEFEND ITSELF AGAINST DISEASE?

The environment teems with **microbes,** which include microscopic living organisms such as bacteria, protists, and many fungi, and viruses, which are usually not considered to be alive. Most microbes live in water or in the soil. Of those that live in animal bodies, most are harmless, and some are even beneficial. For example, cattle do not produce enzymes that break down cellulose; without cellulose-digesting bacteria in their digestive tracts to do the job for them, they would starve in the midst of a pasture full of grass.

Some microbes, however, are **pathogens,** a term derived from Greek words meaning "to produce disease." Pathogens, of course, aren't trying to make us miserable. They're merely reproducing and seeking new hosts. In some cases, this process is mostly an inconvenience—when you have a cold, sneezing broadcasts viruses to everyone around you. Other diseases are far more dangerous. Cholera, for example, causes diarrhea. In places with inadequate sanitation, diarrhea may allow cholera bacteria to enter the water supply and spread to new hosts. Meanwhile, the diarrhea may be so devastating that the original host dies of dehydration.

Most microbial diseases, such as cholera, measles, tetanus, and chicken pox, have infected people for hundreds or even thousands of years. In recent decades, new pathogens, and more deadly strains of familiar pathogens, have arisen. Many are viruses that you've probably read about in the news: HIV, Ebola virus, West Nile virus, swine (H1N1) flu, and avian (H5N1) flu. We are also endangered by deadly, often new, strains of bacteria. For example, the intestinal bacterium *Escherichia coli* is usually harmless—indeed beneficial, because it produces vitamin K in our large intestines. However, one strain, called O157:H7, usually acquired by eating undercooked ground beef, causes food poisoning, sometimes with fatal consequences. *Staphylococcus aureus* bacteria ("staph") occur frequently (and usually harmlessly) on the skin and in the nasal passages, but some strains cause fatal toxic shock syndrome or prolonged infections if they penetrate through the skin or mucous membranes. And, as described in the case study, *Streptococcus pyogenes* can destroy much of a person's body in just a few hours. To make things worse, some bacteria, including certain strains of *Staphylococcus aureus* and *Mycobacterium tuberculosis* (which causes TB), have become resistant to most antibiotics.

Given the number and diversity of disease-causing organisms, you might wonder, "Why don't we get sick more often?" Over evolutionary time, animals and their pathogens have engaged in a constantly escalating battle. As animals evolved more sophisticated defense systems, pathogens in turn evolved more effective tactics for penetrating those defenses. This "arms race" has honed our defenses into a stunningly complex system that resists most attacks by microbes.

Vertebrate Animals Have Three Major Lines of Defense: Nonspecific External Barriers, Nonspecific Internal Defenses, and Specific Internal Defenses

Vertebrates (animals with backbones) have evolved three major forms of protection against disease (**Fig. 36-1**):

- **Nonspecific external barriers** These barriers prevent most disease-causing microbes from entering the body. They are primarily anatomical structures, such as skin and cilia, and secretions, such as tears, saliva, and mucus. These barriers cover the external surfaces of the body and line body cavities that open to the external environment, including the respiratory, digestive, and urogenital tracts.
- **Nonspecific internal defenses** If the external barriers are breached, a variety of nonspecific internal defenses, collectively called the **innate immune response,** swing into action. Some white blood cells engulf foreign particles or destroy infected cells. Chemicals released by damaged body cells and proteins released by white blood cells trigger inflammation and fever (in animals that regulate their body temperature). Both the external barriers and the nonspecific internal defenses operate regardless of the exact nature of the invader.
- **Specific internal defenses** The final line of defense is the **adaptive immune response,** in which immune cells selectively destroy the specific invading toxin or microbe and then "remember" the invader, allowing a faster response if it reappears in the future.

The major similarities and differences between innate immunity and adaptive immunity are summarized in **Table 36-1.** The internal defenses involve a large number of different cell types, briefly described in **Table 36-2.** Most of these cells and their roles in defending the body will be unfamiliar to you now, but Table 36-2 will be a useful guide as you read the rest of the chapter.

Invertebrate Animals Possess the Nonspecific Lines of Defense

Invertebrates are protected only by nonspecific external barriers and nonspecific internal defenses. Different invertebrates possess an enormous range of external barriers, ranging from exoskeletons to slimy secretions. Internally, invertebrates have white blood cells that attack pathogens and secrete proteins that neutralize invaders or the toxins they release.

Interestingly, the amino acid sequences of some defensive proteins are similar in vertebrates (including humans, birds, and fish) and invertebrates (including mollusks, worms, and insects). This similarity suggests that the genes coding for these proteins have been passed down from a common ancestor that lived hundreds of millions of years ago.

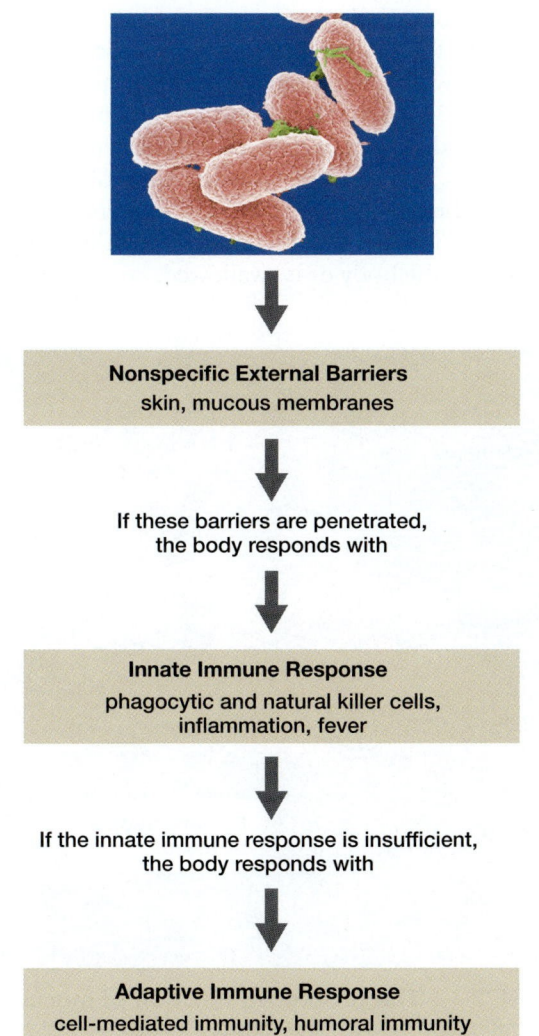

Nonspecific External Barriers
skin, mucous membranes

↓

If these barriers are penetrated, the body responds with

↓

Innate Immune Response
phagocytic and natural killer cells, inflammation, fever

↓

If the innate immune response is insufficient, the body responds with

↓

Adaptive Immune Response
cell-mediated immunity, humoral immunity

▲ FIGURE 36-1 **Levels of defense against infection**

TABLE 36-1 Innate and Adaptive Immune Responses to Invasion		
Characteristic	**Innate Immune Response**	**Adaptive Immune Response**
Specificity of response to invasion	Nonspecific	Specific for particular microbes
Speed of response to the first invasion by a given microbe	Immediate	Delay of several days to 2 weeks
Magnitude of the response to a second invasion by the same microbe	Identical to the first response	Enormously enhanced response; usually no disease symptoms occur (the animal has become immune to that specific microbe)
Distribution in the animal kingdom	Invertebrates and vertebrates	Vertebrates only

TABLE 36-2 The Body's Cellular Arsenal Against Disease

Type of Cell	Function
Neutrophils	White blood cells that engulf invading microbes
Dendritic cells	White blood cells that engulf invading microbes and present antigens to lymphocytes
Macrophages	White blood cells that engulf invading microbes and present antigens to lymphocytes
Natural killer cells	White blood cells that destroy infected or cancerous cells
Mast cells	Connective tissue cells that release histamine; important in the inflammatory response
B cells	Lymphocytes (a type of white blood cell) that produce antibodies
Memory B cells	Offspring of B cells that provide future immunity against invasion by the same antigen
Plasma cells	Offspring of B cells that secrete antibodies into the bloodstream
T cells	Lymphocytes (a type of white blood cell) that regulate the immune response or kill infected cells or cancerous cells
Cytotoxic T cells	T cells that destroy infected body cells or cancerous cells
Helper T cells	T cells that stimulate immune responses by both B cells and cytotoxic T cells
Memory T cells	Offspring of cytotoxic or helper T cells that provide future immunity against invasion by the same antigen
Regulatory T cells	T cells that suppress immune attack against the body's own cells and molecules; important in preventing autoimmune diseases

CHECK YOUR LEARNING

Can you explain the relationship between the terms "microbe" and "pathogen"? Can you name and describe the defenses that are characteristic of vertebrates and those that are characteristic of invertebrates?

36.2 HOW DO NONSPECIFIC DEFENSES FUNCTION?

The ideal defenses are barriers that prevent invaders from entering the body in the first place. However, if these barriers are breached, the body has several nonspecific methods of killing any of a wide variety of invading microbes.

The Skin and Mucous Membranes Form Nonspecific External Barriers to Invasion

In animal bodies, the first line of defense consists of the surfaces with direct exposure to the environment: the skin and the mucous membranes of the digestive, respiratory, and urogenital tracts.

The Skin and Its Secretions Block Entry by Microbes and Provide an Inhospitable Environment for Microbial Growth

The outer surface of human skin consists of dry, dead cells filled with proteins similar to those in hair and nails. Many microbes that land on the skin cannot obtain the water and nutrients they need to survive. The skin is also protected by secretions from sweat glands and sebaceous (oil) glands. These secretions contain natural antibiotics, such as lactic acid, that inhibit the growth of many bacteria and fungi. Despite these defenses, the skin is a veritable zoo of microbes, including both exotic types that are specialized for growth in the relative desert of the skin and the more familiar, and more dangerous, *Streptococcus* and *Staphylococcus* bacteria. However, if you practice reasonable personal hygiene, their populations are usually fairly low, and they seldom penetrate through unbroken skin into the tissues below.

Mucus, Antibacterial Proteins, and Ciliary Action Defend the Mucous Membranes Against Microbes

Mucous membranes deploy several defenses against infection. First, as their name implies, mucous membranes secrete mucus, which traps microbes that enter the nose or mouth (**Fig. 36-2**). Further, mucus contains antibacterial proteins, including *lysozyme*, which kills bacteria by digesting their cell walls, and *defensin*, which makes holes in bacterial plasma membranes. Finally, cilia on the membranes sweep up the mucus, microbes and all, until it is either coughed or sneezed out of the body or is swallowed.

Bacteria trapped by mucus

▲ **FIGURE 36-2 The protective function of mucus** Mucus traps microbes and debris in the respiratory tract. In this colored micrograph, bacteria are caught in mucus atop the cilia. The cilia then sweep both the mucus and bacteria out of the body.

If microbes are swallowed, they enter the stomach, where they encounter protein-digesting enzymes and extreme acidity, both of which are often lethal. Further along, the intestines contain bacteria that are harmless to people but that secrete substances that destroy invading bacteria or fungi. In the urinary tract, the slight acidity of urine inhibits bacterial growth. In females, acidic secretions and mucus help protect the vagina. Finally, fluids released by the body, including tears, urine, diarrhea, and vomit, help to expel invaders.

The Innate Immune Response Nonspecifically Combats Invading Microbes

Despite these defenses, many disease-causing microbes enter the body through the mucous membranes or through cuts in the skin. However, pathogens that get through the external barriers encounter three types of nonspecific innate immune responses:

- **Protection by white blood cells** The body has a standing army of white blood cells, or **leukocytes,** many of which are specialized to attack and destroy invading cells or the body's own cells if they have been infected by viruses.
- **The inflammatory response** A wound, with its combination of tissue damage and invading microbes, provokes an inflammatory response, which recruits leukocytes to the site of injury and walls off the injured area, isolating the infected tissue from the rest of the body.
- **Fever** If a population of microbes succeeds in establishing a major infection, the body may produce a fever, which both slows down microbial reproduction and enhances the body's own fighting abilities.

Phagocytes and Natural Killer Cells Destroy Invading Microbes

One group of leukocytes, collectively called **phagocytes,** ingests invaders by *phagocytosis* (see Chapter 5). Three important types of phagocytes are **macrophages** (literally, "big eaters"), **neutrophils,** and **dendritic cells.** These cells travel within the bloodstream, ooze out through capillary walls, and patrol the body's tissues (Fig. 36-3a), where they consume bacteria and foreign substances that have penetrated the skin or mucous membranes (Fig. 36-3b). As described later, dendritic cells and macrophages also play a critical role in the adaptive immune response.

Nonspecific defense against viruses is the job of another type of leukocyte, called the **natural killer cell.** Viruses enter body cells and use the cells' own metabolism to manufacture more viruses (see Fig. 13-2 and "In Greater Depth: Virus Replication" in Chapter 19). Natural killer cells destroy virus-infected cells. Killing infected cells before the viruses have had enough time to reproduce and spread to other cells can stop viral infections before they do very much damage to the body as a whole. How do natural killer cells distinguish virus-infected cells, which should be killed, from healthy cells, which should be spared?

The surfaces of normal body cells display a set of proteins, collectively called the **major histocompatibility complex (MHC).** MHC proteins differ from species to species, and, in fact, differ from individual to individual within a species. Therefore, MHC proteins are unique to each person (except identical twins, who have the same MHC proteins), and identify the body's own cells as "self." Natural killer cells patrol the body, killing any "non-self" cells that they encounter, while sparing self cells. Virus-infected cells often lack some MHC proteins, so they look like non-self to natural killer cells. Natural killer cells are not phagocytic; instead, they release proteins that bore holes in the membranes of infected or cancerous cells and secrete enzymes through the holes. Shot full of holes and chewed on by enzymes, the attacked cells soon die.

Most cancerous cells have missing or altered MHC proteins. Therefore, they are also identified as non-self by natural killer cells, and are destroyed in the same way as virus-infected cells are.

The Inflammatory Response Attracts Phagocytes to Injured or Infected Tissue

Whether you catch the flu or get a splinter in your finger, you will experience inflammation. The **inflammatory response**

(a) A macrophage leaves a capillary and enters a wound

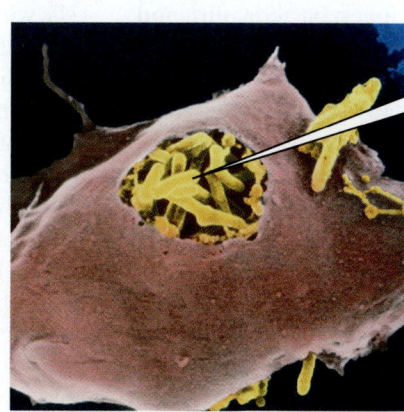

Bacteria visible through a hole in the macrophage's plasma membrane

(b) A macrophage stuffed with bacteria that it has ingested

▲ **FIGURE 36-3 The attack of the macrophages**

dead cell layer

epidermis

dermis

① Tissue damage carries bacteria into the wound

② Wounded cells release chemicals (red) that stimulate mast cells

③ Mast cells release histamine (blue)

④ Histamine increases capillary blood flow and permeability

⑤ Phagocytes leave the capillaries and ingest bacteria and dead cells

▲ **FIGURE 36-4 The inflammatory response**

QUESTION How does the inflammatory response help white blood cells to leave the capillaries?

causes injured tissues to become warm, red, swollen, and painful (*inflame* literally means "to set on fire").

The inflammatory response begins when damaged cells release chemicals that cause certain cells in the connective tissue, called **mast cells,** to release **histamine** and other substances (**Fig. 36-4**). Histamine relaxes the smooth muscle surrounding arterioles, thereby increasing blood flow, and makes capillary walls leaky. Extra blood flowing through leaky capillaries drives fluid from the blood into the wounded area. Thus, histamine accounts for the redness, warmth, and swelling of the inflammatory response. Other chemicals released by wounded cells and mast cells, and some substances produced by the microbes themselves, attract macrophages and other types of white blood cells. These cells squeeze out through the leaky capillary walls and ingest bacteria, dirt, and cellular debris (see Figs. 36-3 and 36-4). In some cases, pus, a thick whitish mixture of dead bacteria, debris, and living and dead white blood cells, may accumulate. (Flesh-eating bacteria cause large amounts of pus to form in a wound; in fact, the name *pyogenes* means "pus-producer.")

Other chemicals released by injured cells initiate blood clotting (see Chapter 32). Clots plug up damaged blood vessels, reducing blood loss and preventing microbes from entering the bloodstream. Swelling and some of the chemicals released by the injured tissue cause pain, which usually leads to protective behaviors that reduce the likelihood of further injury.

Fever Combats Large-Scale Infections

If invaders breach these defenses and cause a serious infection, they may trigger a **fever.** The onset of fever is controlled by the hypothalamus, the part of the brain housing temperature-sensing nerve cells that serve as the body's thermostat. In humans, the thermostat is set at about 97° to 99°F (36° to 37°C). Certain types of bacteria, as well as the phagocytic cells that respond to an infection, produce chemicals called *pyrogens* (literally, "fire-producers"). Pyrogens travel in the bloodstream to the hypothalamus and raise the thermostat's set point. The body responds with increased fat metabolism (which generates more heat), constriction of blood vessels supplying the skin (which reduces heat loss through the skin), and behaviors such as shivering and bundling under blankets. Pyrogens also cause other cells to reduce the concentration of iron in the blood.

How does fever combat infection? An elevated body temperature increases the activity of phagocytic white blood cells while simultaneously slowing bacterial reproduction. The iron deficiency accompanying a fever also hampers bacterial multiplication. Meanwhile, fever causes the cells of the adaptive immune system to multiply more rapidly, hastening the onset of an effective adaptive immune response. In addition, fever stimulates cells infected by viruses to produce a protein called interferon, which travels to other cells and increases their resistance to viral attack. Interferon also stimulates the natural killer cells that destroy virus-infected body cells.

On the other hand, high fevers are certainly uncomfortable, and may be dangerous. For example, extremely high fevers may cause brain damage, although this is rare. Because fevers have the potential to be either beneficial or harmful, you should consult your physician about whether you should take fever-reducing medicines such as aspirin, acetaminophen, or ibuprofen.

CASE STUDY continued

Flesh-Eating Bacteria

The intact skin is an effective barrier against flesh-eating bacteria. However, if the skin is cut, *S. pyogenes* may enter the body. *S. pyogenes* bacteria are surrounded by capsules made of a polysaccharide called hyaluronic acid, which is also a major component of mammalian connective tissue. Hence, phagocytic cells often do not recognize the capsules as foreign and, consequently, do not attack *S. pyogenes*.

CHECK YOUR LEARNING

Are you able to describe the external barriers to infection, and explain how they function? Can you name the major components of the innate immune response, and describe how each of them combats invasion by microbes? Do you understand why external barriers and the innate immune response are nonspecific?

36.3 WHAT ARE THE KEY COMPONENTS OF THE ADAPTIVE IMMUNE SYSTEM?

The nonspecific defenses are not impregnable. If they are breached, the body mounts a highly specific and coordinated adaptive immune response directed against the particular organism that has successfully colonized the body.

The essential features of the adaptive immune response were recognized more than 2,000 years ago by the Greek historian Thucydides. He observed that sometimes a person would contract a disease, recover, and never catch that particular disease again—the person had become immune. With rare exceptions, immunity to one disease confers no protection against other diseases. Thus, the adaptive immune response attacks one specific type of microbe, overcomes it, and provides future protection against that microbe but no others.

The Adaptive Immune System Consists of Cells and Molecules Dispersed Throughout the Body

As befits a system that must patrol the entire body for invading microbes, the **adaptive immune system** (often called simply the immune system) is distributed throughout the body, with concentrations of cells in certain locations (**Fig. 36-5**). The adaptive immune system consists of three major components:

- **Immune cells** The adaptive immune response is produced by interactions among several types of white blood cells, including macrophages, dendritic cells, and lymphocytes. Some, such as macrophages and dendritic cells, play a role in both the innate and adaptive immune responses (see Table 36-2). Unique to the adaptive immune response are specialized white blood cells called **lymphocytes.** There are two types of lymphocytes: **B cells**

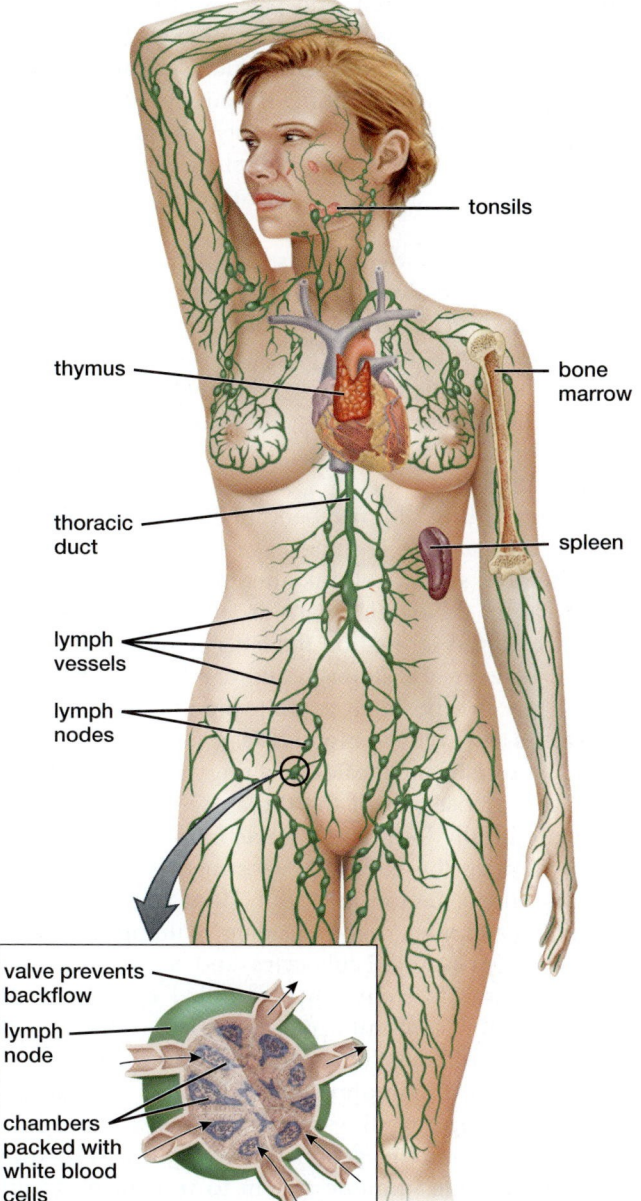

▲ **FIGURE 36-5 The lymphatic system contains much of the immune system** The cells of the immune system are formed in the bone marrow and mature either there or in the thymus. Most mature immune cells reside in the lymph nodes or spleen. As lymph travels through the lymph nodes, most microbial invaders are attacked and killed.

and **T cells.** Both B cells and T cells arise from dividing stem cells in the bone marrow. Some of the resulting daughter cells complete their development in the bone marrow, becoming B (for *bone*) cells. Other precursor cells migrate from the marrow, through the circulatory system, to the thymus (see below), where they develop into T (for *thymus*) cells. We will describe the functions of B and T cells shortly.

- **Tissues and organs** The cells of the immune system reside in many locations in the body, including bone marrow, the vessels of the lymphatic system, the lymph nodes, the thymus, the spleen, and patches of specialized connective tissue, such as the tonsils. Some types of immune cells are widely, if thinly, distributed in mucous membranes, especially those lining the airways, digestive tract, and vagina. Although the circulatory system is not usually considered part of the immune system, it carries immune cells and the proteins they secrete (see below) throughout the body.

The human body contains approximately 500 **lymph nodes** scattered along the lymph vessels (see Fig. 36-5). Lymph nodes contain masses of macrophages and lymphocytes lining narrow passages through which the lymph flows. When you have a disease that causes "swollen glands," these are actually lymph nodes swollen with white blood cells, bacteria, debris from dead cells, and fluid.

The **thymus** is located beneath the breastbone slightly above the heart. It is quite large in infants and young children but starts to shrink after puberty. The thymus is essential for the development of T cells. The **spleen** is a fist-sized organ located on the left side of the abdominal cavity, between the stomach and diaphragm. The spleen filters blood, exposing it to white blood cells that destroy microbes and aged red blood cells.

The **tonsils** are located in a ring around the pharynx (the uppermost part of the throat; see Fig. 33-7). Here, they are ideally located to sample microbes entering the body through the mouth. Macrophages and other white blood cells in the tonsils directly destroy many invading microbes and often trigger an adaptive immune response.

- **Secreted proteins** Leukocytes and some other cells secrete many different proteins, called **cytokines,** that are used for communication between cells. A large number of different types of proteins in the blood, collectively called **complement,** assist the immune system in killing invading microbes (see Section 36.5). **Antibodies,** proteins produced by B cells and their offspring, help the immune system to recognize invading microbes and destroy them (see Sections 36.4 and 36.5). Although antibodies are exclusively part of the adaptive immune response, some cytokines and complement proteins are involved in both the innate and adaptive responses. For example, some cytokines stimulate innate immune responses such as inflammation, activation of natural killer cells, and fever.

All adaptive immune responses include the same three steps. First, lymphocytes recognize an invading microbe, identifying it as non-self; second, they launch an attack; and third, they retain a memory of the invader that allows them to ward off future infections by the same type of microbe.

CHECK YOUR LEARNING

Can you describe the adaptive immune response, and explain how it differs from the innate immune response? Are you able to list the major components of the adaptive immune system, and describe how each component functions?

36.4 HOW DOES THE ADAPTIVE IMMUNE SYSTEM RECOGNIZE INVADERS?

To understand how the adaptive immune system recognizes invaders and initiates a response, we must answer three related questions: How do lymphocytes recognize foreign cells and molecules? How can lymphocytes produce specific responses to so many different types of cells and molecules? How do they avoid mistaking the body's own cells and molecules for invaders?

The Adaptive Immune System Recognizes Invaders' Complex Molecules

From the perspective of the immune system, a bacterium and a human differ from one another because each contains specific, complex molecules that the other doesn't. Your immune cells distinguish your body's cells and molecules from those of all the other organisms on Earth, including other people, because some of your complex molecules, particularly your MHC proteins, are unique to you (unless you have an identical twin), and some of the complex molecules of all other organisms are unique to them. These complex molecules—usually proteins, polysaccharides, or glycoproteins—are called **antigens,** because they are "*anti*body *gen*erating" molecules; that is, they can provoke an immune response, including the production of antibodies.

Antigens are often located on the surfaces of invading microbes. Many viral antigens become incorporated into the plasma membranes of infected body cells. Viral or bacterial antigens are also "displayed" on the plasma membranes of dendritic cells and macrophages that engulf them. Other antigens, such as toxins released by bacteria, may be dissolved in blood plasma, lymph, or interstitial fluid. Lymphocytes produce two types of proteins that recognize, bind, and help destroy foreign cells and molecules: antibodies and T-cell receptors.

Antibodies are manufactured only by B cells and their offspring. Antibodies are Y-shaped proteins composed of two pairs of peptide chains: one pair of identical large (heavy) chains and one pair of identical small (light) chains (**Fig. 36-6**). Both heavy and light chains consist of a **variable region,** which differs among antibodies, and a **constant region,** which is the same in all antibodies of a given type (see below).

The light and heavy chains combine to form the two functional parts of an antibody: the "arms" of the Y and the "stem" of the Y. The variable regions, located at the tips of the arms, form sites that bind antigens (see Fig. 36-6). Each binding site has a particular size, shape, and electrical charge, so only certain antigen molecules can fit in and bind to the antibody. The sites are so specific that each antibody can bind only a few, very similar, types of antigens. Every B cell produces its own unique antibody, with variable regions that differ from those of the antibodies produced by all other B cells.

The stem of the Y is composed of the constant region of the heavy chain, and is the same for any antibody of a

▲ FIGURE 36-6 Antibody structure Antibodies are Y-shaped proteins composed of two pairs of peptide chains (light chains and heavy chains). Constant regions on both chains form the stem of the Y. The variable regions on the two chains form a specific binding site at the end of each arm of the Y. Different antibodies have different variable regions, forming unique binding sites.

QUESTION Why do antibody molecules have both constant and variable regions?

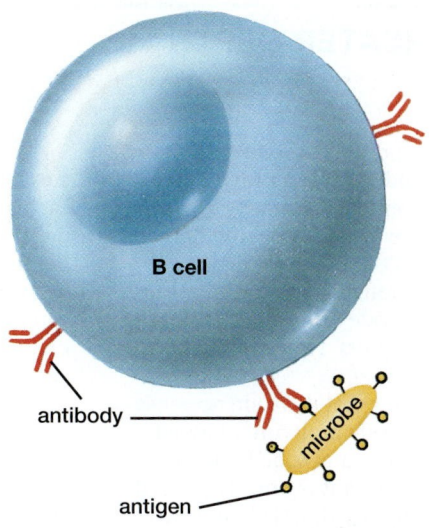

▲ FIGURE 36-7 Antibodies on the surface of B cells bind to antigens on invading microbes

particular type. There are five major types of antibodies, given the abbreviations IgD, IgM, IgG, IgA, and IgE. These differ in location and function. IgM and IgD antibodies are attached to the surfaces of B cells. When an IgM or IgD antibody on the surface of a B cell binds an antigen, this triggers multiplication of the B cell (see below). IgG antibodies are the most common type found in blood, lymph, and interstitial fluid; these are the major antibodies that actually help to destroy invading microbes. IgG antibodies also cross the placenta and defend the developing fetus against disease. IgA antibodies are secreted onto the surfaces of the digestive and respiratory tracts, and into saliva and breast milk; they help to combat invaders entering via the mouth and nasal passages, and provide temporary defenses for nursing infants whose immune systems have not yet fully developed. IgE antibodies are responsible for allergic reactions. The constant region of an antibody determines where the antibody is located (for example, on the surface of a B cell or secreted into breast milk) and what role it plays in the immune response (see Sections 36.5 and 36.8).

Antibodies serve two functions in the adaptive immune response: (1) recognizing foreign antigens and triggering the response against the invaders, and (2) helping to destroy the invading cells or molecules that bear the antigens. In an antibody's recognition role, the stem of an antibody anchors it in the plasma membrane of the B cell that produced it. The antibody's two arms stick out from the B cell, sampling the blood and lymph for antigen molecules (**Fig. 36-7**). When an arm of the antibody encounters an antigen with a compatible chemical structure, it binds to it. Antigen–antibody binding stimulates the B cell to divide, as we will describe shortly.

T-cell receptors are found only on the surfaces of T cells. Every T cell produces T-cell receptors that differ from those of all other T cells. Variable regions, which form highly specific binding sites for antigens, protrude from the cell surface. When an antigen binds to a T-cell receptor, the T cell becomes activated in ways that depend on the type of T cell (see below).

Both antibodies and T-cell receptors play crucial roles in helping to destroy invading microbes, as we will describe in Section 36.5.

The Adaptive Immune System Can Recognize Millions of Different Antigens

During your lifetime, your body will be challenged by a multitude of invaders. Your classmates may sneeze cold and flu viruses into the air you breathe. Your food may contain bacteria or molds. A mosquito carrying West Nile virus may bite you. Fortunately, the adaptive immune system can recognize and respond to millions of potentially harmful antigens. How can it accomplish this remarkable feat?

B and T cells cannot design and build antibodies and T-cell receptors that fit particular invading antigens. Rather, B and T cells randomly synthesize millions of different antibodies and T-cell receptors. At any given time, the human body contains perhaps 100 million different antibodies and even more T-cell receptors. This vast array is simply there, waiting, much like clothes hanging on racks in a department store. Given enough clothing from which to choose, most of us will find something suitable that fits. Similarly, invading antigens almost always encounter at least a few antibodies and T-cell receptors that will bind them. "In Greater Depth: How Can the Immune System Recognize So Many Different Antigens?" explains how your body synthesizes many millions of antibodies and T-cell receptors from the instructions in just a few hundred genes.

IN GREATER DEPTH How Can the Immune System Recognize So Many Different Antigens?

B and T cells produce millions of different antibodies and T-cell receptors. Antibodies and T-cell receptors are proteins, which are encoded by genes. However, humans have only 20,000 to 25,000 genes, so there aren't enough genes in the entire genome to encode millions of antibodies and T-cell receptors. How can the body produce so many?

The answer is that there are no genes for entire antibody molecules. Instead, B cells have genes that code for *parts* of antibodies—constant regions (C), variable regions (V), and joining (J) or diversity (D) regions that connect the two (Fig. E36-1). Humans have about 40 genes for the variable region of the heavy chain; 27 and 6 genes, respectively, for the diversity and joining regions; and one gene for the constant region of each type of antibody (IgM, IgE, etc.). Humans have about 70 genes for the variable region of the light chain, 5 genes for the joining region, and one constant region gene (Fig. E36-1a).

As each B cell develops, it randomly cuts out and discards all but one gene for each antibody part, and assembles two unique composite antibody genes from the individual genes that it keeps—a heavy-chain gene consisting of one variable, one diversity, one joining, and one constant region gene; and a light-chain gene consisting of one variable, one joining, and one constant region gene (Fig. E36-1b). Antibodies are produced from the instructions in these composite genes (Fig. E36-1c), theoretically yielding more than 3 million unique combinations.

Much more diversity arises because the process of cutting the DNA and splicing the individual genes together is imprecise. A few nucleotides are often added to or cut out of the genes before they are spliced together. The result: Immunologists estimate that perhaps 15 to 20 *billion* unique antibodies are possible. Each B cell probably produces an antibody that is different from the one produced by every other B cell (except its own daughter cells).

How can gene splicing produce so many different composite genes? An analogy might help: dealing cards for 5-card poker. Although a single deck contains only 52 cards, they can be dealt out in 2,598,960 unique 5-card hands. Similarly, each B cell is "dealt" an "antibody hand" consisting of a few genes randomly selected from hundreds of genes that encode the antibody components (including the differences created by imprecise splicing), resulting in billions of possible unique antibodies.

T-cell receptors are encoded by a different set of genes, but the process that results in the diversity of antibodies is similar for T-cell receptors. In fact, there are more component genes available for constructing the composite genes coding for complete T-cell receptors, so there may be as many as a quadrillion different possible T-cell receptors!

▶ **FIGURE E36-1**
Recombination produces antibody genes (a) The precursor cells that give rise to B cells have many genes that code for parts of antibodies, illustrated here as variable (V), diversity (D), joining (J), and constant (C) region genes. Each colored band represents an individual gene for an antibody part. There are five genes for the constant regions of the heavy chain and one gene for the constant region of the light chain. Only a few of the many possible variable, diversity, and joining genes are shown here. **(b)** Genes for antibody parts are spliced together to form composite genes that encode complete antibodies. For the heavy chain, a composite antibody gene contains one gene segment each for the variable, diversity, joining, and constant regions; for the light chain, one each for the variable, joining, and constant regions. In this illustration, the complete antibody genes assembled in cells 1, 2, and 3 are composed of different genes for the antibody parts, as is signified by the different numbers assigned to each gene type. **(c)** As a result, the variable regions for the antibodies produced by cells 1, 2, and 3 differ from one another.

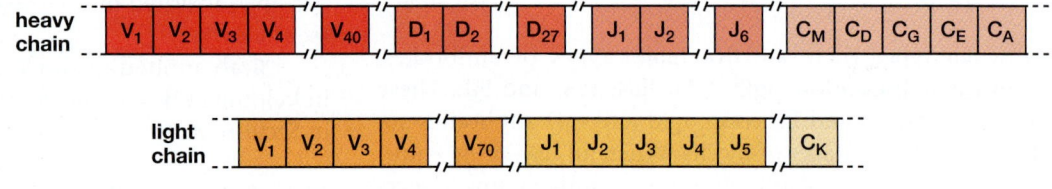

(a) Genes for parts of the heavy chain (top) and light chain (bottom) of antibodies

(b) Composite antibody genes in three different B cells

(c) Antibodies synthesized by these three B cells

The Adaptive Immune System Distinguishes Self from Non-Self

Your body manufactures thousands of different proteins, including MHC proteins that are unique to you. These proteins can stimulate powerful immune responses in other people's bodies; that is, they can act as antigens. This, of course, is why transplants are often rejected, unless the transplant recipient takes drugs to suppress the immune response. Why don't antigens produced by your own body arouse your immune system? Two important mechanisms normally prevent self-immunity: the continuous presence of the body's own antigens during immune cell development, and modulation of the immune response by regulatory T cells.

Because the production of antibodies and T-cell receptors is a random process, some newly formed immune cells do produce antibodies or T-cell receptors that can bind to the body's own proteins and polysaccharides, treating them as antigens. However, if these immature immune cells contact antigens that bind to their antibodies or T-cell receptors, they undergo *apoptosis*, or programmed cell death, in which they essentially commit cellular suicide. Thus, one way that the immune system distinguishes self from non-self is by destroying immune cells that would attack the body's own molecules.

Not all self-reactive B and T cells are eliminated in this way. One of the functions of **regulatory T cells** is to prevent any remaining self-reactive lymphocytes from attacking the body and causing an autoimmune disease (see Section 36.8).

CHECK YOUR LEARNING

Do you understand the mechanisms by which the adaptive immune system recognizes invading microbes? Can you describe the structure of antibodies and how antibodies form specific binding sites for antigens? Can you explain why the immune system does not treat the body's own cells and molecules as foreign?

36.5 HOW DOES THE ADAPTIVE IMMUNE SYSTEM ATTACK INVADERS?

An invading microbe usually includes a handful of different antigens, each capable of binding to only a few antibodies or T-cell receptors. Your immune system contains millions of B and T cells, each bearing antibodies or T-cell receptors that differ from the antibodies or T-cell receptors of every other immune cell in your body. The benefit to having millions of unique immune cells is that almost any invader will provoke an adaptive immune response. The drawback to having very small numbers of cells that can recognize any given invader is that a handful of cells isn't enough to kill the invaders immediately. It usually takes 1 or 2 weeks to mount a good immune response to the first exposure to an invading microbe, as the responding cells multiply and differentiate. In the meantime, you become ill, and may even die, if the development of the immune response loses its race with the multiplying microbes and the damage they cause to your body.

The adaptive immune system simultaneously launches two types of attack against microbial invaders: humoral immunity and cell-mediated immunity. Although humoral and cell-mediated immunity are not completely independent, we consider them separately to make them easier to understand.

Humoral Immunity Is Produced by Antibodies Dissolved in the Blood

Humoral immunity is provided by B cells and the antibodies that they secrete into the bloodstream. Each of the millions of different B cells in your body bears its own unique type of antibody on its surface (**Fig. 36-8**). When a microbe enters the body, the antibodies on a few of these B cells can bind to antigens on the invader. Antigen–antibody binding causes these B cells, but no others, to multiply rapidly. This process is called **clonal selection**, because the antigens "select" which B cells will multiply, and the resulting daughter cells are clones—cells that are genetically identical to the selected B cells.

The daughter cells differentiate into two cell types: *memory B cells* and **plasma cells.** Memory B cells do not release antibodies, but they play an important role in future immunity to the invader that stimulated their production, as we will see later. Plasma cells become enlarged and packed with rough endoplasmic reticulum, which synthesizes huge quantities of antibodies (see Fig. 36-8). These antibodies are released into the bloodstream (hence the name "humoral" immunity; to the ancient Greeks, blood was one of the four "humors," or body fluids). The secreted antibodies bear the same antigen binding site that was found in the antibodies located on the surface of the original parent B cell.

Clonal selection, multiplication of activated B cells, differentiation of daughter cells into plasma cells, and antibody secretion by plasma cells all take time. This is why it may take a couple of weeks to completely recover from an infection.

Humoral Antibodies Have Multiple Modes of Action Against Invaders

Antibodies in the blood combat invading molecules or microbes in three principal ways. First, the circulating antibodies may bind to a foreign molecule, virus, or cell, and render it harmless, a process called neutralization. For example, if the active site of a toxic enzyme in snake venom is covered with antibodies, it cannot harm your body (**Fig. 36-9a**). Many viruses, such as those that cause rabies, gain entry into your body's cells when a protein on the virus binds to a molecule on the surface of a cell. If antibodies cover up this viral protein, the neutralized virus cannot enter a cell.

Second, antibodies may coat the surface of invading molecules, viruses, or cells, and make it easier for macrophages and other phagocytes to destroy them (**Fig. 36-9b**). Remember, the variable regions on the "arms" of an antibody bind to antigens on invaders, so the constant regions that make up the "stems" stick out into the blood or interstitial fluid. Macrophages bind to the antibody stems, engulf the antibody-coated invaders, and digest them.

► **FIGURE 36-8 Clonal selection of B cells by invading antigens**
❶ The immune system contains millions of B cells, each with a unique antibody. ❷ An invading antigen binds to the antibody on only one B cell, stimulating the B cell to divide. ❸ Clones of the B cell result. ❹ The B cells' daughter cells differentiate into plasma cells and memory B cells. Ordinary B cells are small, with little endoplasmic reticulum (top). ❺ As they differentiate into plasma cells, they become much larger, almost entirely filled with rough endoplasmic reticulum that synthesizes antibodies.

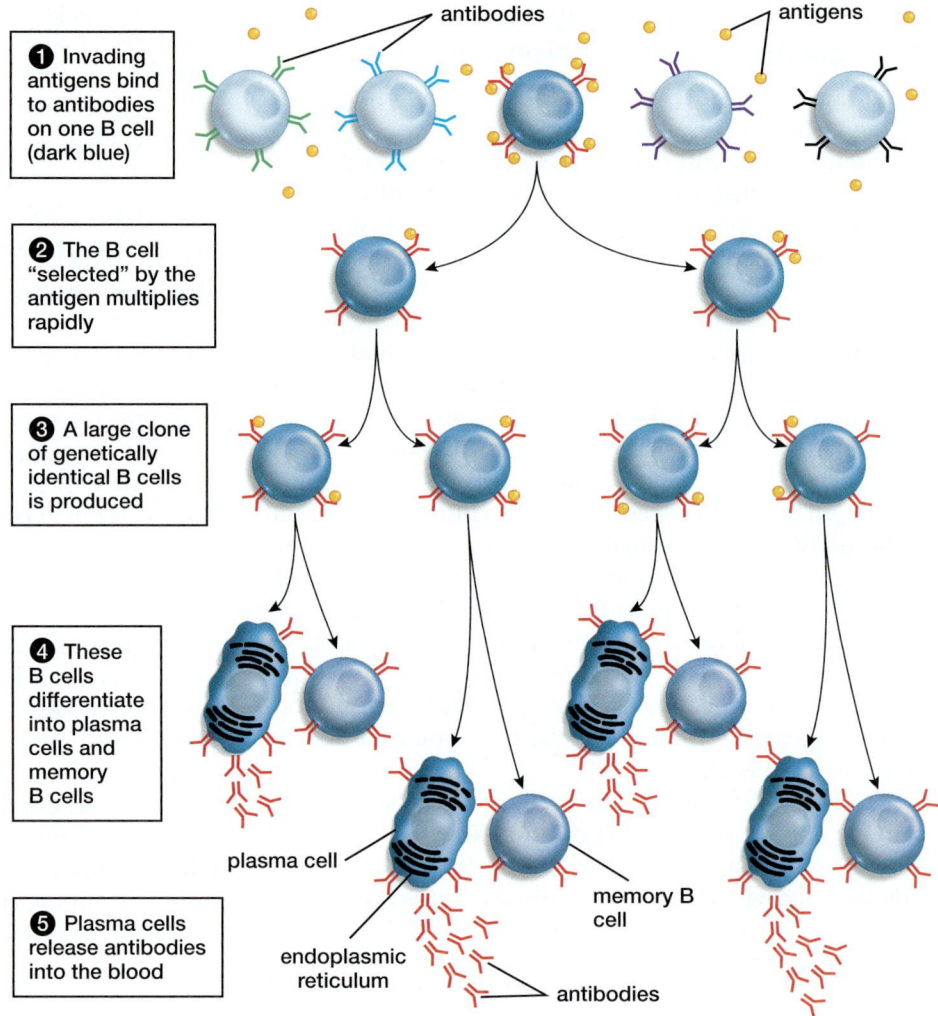

❶ Invading antigens bind to antibodies on one B cell (dark blue)

antibodies

antigens

❷ The B cell "selected" by the antigen multiplies rapidly

❸ A large clone of genetically identical B cells is produced

❹ These B cells differentiate into plasma cells and memory B cells

plasma cell

memory B cell

❺ Plasma cells release antibodies into the blood

endoplasmic reticulum

antibodies

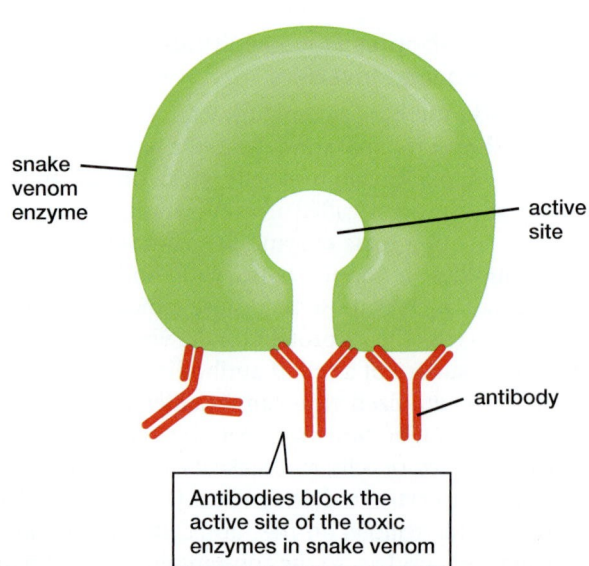

snake venom enzyme

active site

antibody

Antibodies block the active site of the toxic enzymes in snake venom

(a) Antibodies neutralize toxic molecules by covering up their active sites

macrophage

antibody

antigen

microbe

microbe

(b) Antibodies bind to antigens on a microbe and promote phagocytosis by macrophages

▲ **FIGURE 36-9 Antibody action**

Third, when antibodies bind to antigens on the surface of a microbe, the antibodies interact with complement proteins that are always present in the blood. Some complement proteins punch holes in the plasma membrane of the microbe, killing it. Other complement proteins make it easier for phagocytes to ingest the invaders.

CASE STUDY c o n t i n u e d

Flesh-Eating Bacteria

Some complement proteins enhance phagocytosis of invading bacteria. However, the flesh-eating *S. pyogenes* bacterium produces a protein on its surface that inhibits the binding of complement proteins, helping to protect the bacterium from complement-stimulated phagocytosis.

Humoral Immunity Fights Invaders That Are Outside Cells

Antibodies are large proteins that cannot readily cross plasma membranes, so they usually do not enter cells. Therefore, a humoral response is effective against microbes or their toxins only when they are outside cells, generally in the blood, lymph, or interstitial fluid. Bacteria, bacterial toxins, and some fungi and protists are usually vulnerable to the humoral immune response. Viruses are exposed to the humoral response when they are outside of the body's cells—for example, when they are spreading from cell to cell in the interstitial fluid—but are safe from antibody attack when they are inside a cell. Fighting viral infections, therefore, requires help from the cell-mediated immune response.

Cell-Mediated Immunity Is Produced by Cytotoxic T Cells

Cell-mediated immunity is produced by a type of T cell called the **cytotoxic T cell,** which attacks virus-infected body cells and cells that have become cancerous. Although the process is complex, in essence, it works like this: When a cell is infected by a virus, some pieces of viral proteins are brought to the surface of the infected cell and "displayed" on the outside of the plasma membrane. Cytotoxic T cells, each bearing its own unique type of T-cell receptor, drift about, occasionally bumping into the displayed viral antigens. When a cytotoxic T cell with an appropriate matching T-cell receptor binds to a viral antigen, the cytotoxic T cell squirts proteins onto the surface of the infected cell. These proteins form pores in the infected cell's plasma membrane. Enzymes, also secreted by the cytotoxic T cell, pass through the pores, killing the infected cell. If the infected cell is killed before the virus has finished multiplying, then no new viruses are produced, and the viral infection cannot spread to other cells. Cancer cells often display unusual proteins on their surfaces that cytotoxic T cells recognize as foreign, and are killed by the same mechanism (**Fig. 36-10**).

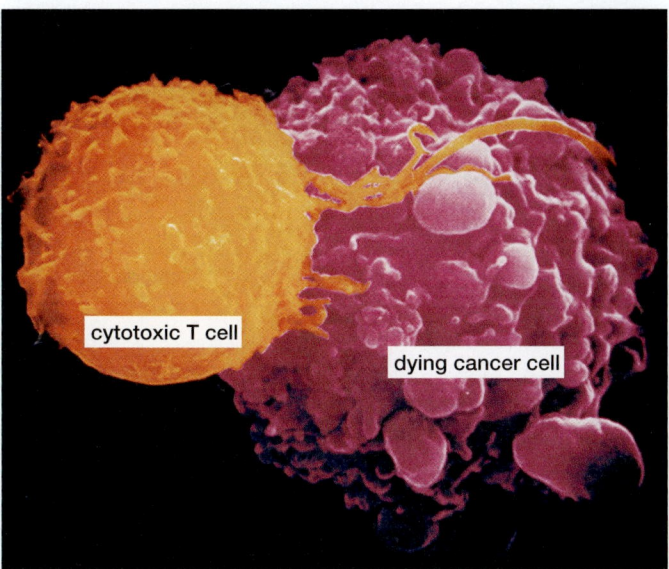

▲ **FIGURE 36-10 Cell-mediated immunity in action** This scanning electron micrograph captures a cytotoxic T cell in the act of attacking a cancer cell.

Helper T Cells Enhance Both Humoral and Cell-Mediated Immune Responses

B cells and cytotoxic T cells cannot fight microbial invasions by themselves; they require the assistance of **helper T cells.** Helper T cells bear T-cell receptors that bind to antigens displayed on the surfaces of dendritic cells or macrophages that have engulfed and digested invading microbes. Only helper T cells bearing matching T-cell receptors can bind to any particular antigen. When its receptor binds an antigen, a helper T cell multiplies rapidly. Its daughter cells differentiate and release cytokines that stimulate cell division and differentiation in both B cells and cytotoxic T cells. In fact, B cells and cytotoxic T cells usually make a significant contribution to defense against disease only if they simultaneously bind to an antigen *and* receive stimulation by cytokines from helper T cells.

Figure 36-11 compares the humoral immune response with the cell-mediated immune response, and shows the role of helper T cells in both.

CHECK YOUR LEARNING

Can you describe the processes by which the adaptive immune system works, including humoral immunity and cell-mediated immunity? Do you understand why helper T cells are so important in the adaptive immune response?

36.6 HOW DOES THE ADAPTIVE IMMUNE SYSTEM REMEMBER ITS PAST VICTORIES?

After recovering from a disease, you remain immune to that particular microbe for many years, perhaps a lifetime. How does the adaptive immune system accomplish this feat? The plasma cells and cytotoxic T cells that conquered the disease generally live only a few days. However, some of the

HUMORAL IMMUNITY

Targets invaders outside cells (e.g., viruses, bacteria, fungi, protists, and toxins)

HELPER T CELLS

Stimulate both humoral and cell-mediated immunity by releasing cytokines

CELL-MEDIATED IMMUNITY

Targets defective body cells (e.g., infected cells and cancer cells), transplants

virus

viral antigen

Viral antigens presented on the surfaces of dendritic cells or macrophages, and infected cells

dendritic cell or macrophage

infected cell

B-cell antibodies bind to viral antigens and stimulate the B cells to divide and differentiate

antibody

T-cell receptors bind to viral antigens

B cell

cytokines

helper T cell

cytotoxic T cell

Cytokines released by helper T cells stimulate B cells and cytotoxic T cells

plasma cell

memory B cell

memory helper T cell

memory cytotoxic T cell

cytotoxic T cell

infected cell

Plasma cells secrete antibodies into the blood and interstitial fluid

Memory cells confer future immunity to this specific virus but not to any other microbes

Cytotoxic T cells release pore-forming proteins that destroy infected cells

▲ **FIGURE 36-11 A summary of humoral and cell-mediated immune responses**

daughter cells of the original B cells, cytotoxic T cells, and helper T cells that responded to the infection differentiate into **memory B cells** and **memory T cells** that survive for many years. Memory cells bear the same antibodies or T-cell receptors as the original parent cells did. If the body is reinvaded by this same type of microbe, memory cells recognize the invader and mount an immune response. Memory B cells rapidly produce a clone of plasma cells, secreting antibodies that combat this second invasion. Memory T cells produce clones of helper T cells and cytotoxic T cells that are also specific for the "remembered" invader. There are far more

memory cells than original B, cytotoxic T, or helper T cells that responded to the first infection. Further, memory cells respond faster than the original parent cells could. Therefore, memory cells usually produce an immune response to a second infection that is so fast and so large that the body fends off the attack before you suffer any disease symptoms: You have become immune (**Fig. 36-12**).

Acquired immunity confers long-lasting protection against many diseases—smallpox, measles, mumps, and chicken pox, for example. Acquired immunity may fail, however, if disease organisms mutate rapidly and produce

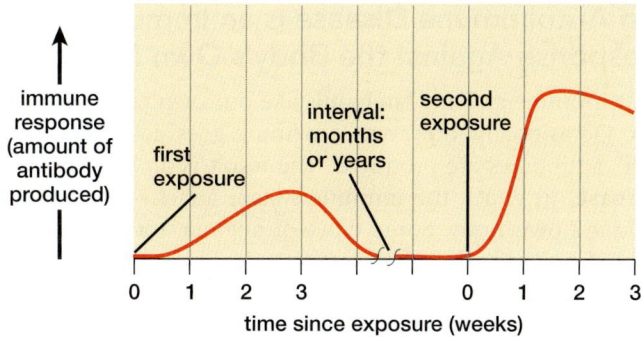

▲ **FIGURE 36-12 Acquired immunity** The immune system responds sluggishly to the first exposure to a disease-causing organism as B and T cells are selected and multiply. A second exposure activates memory cells formed during the first response, making the second response both faster and larger.

new antigens that the memory cells do not recognize (see "Health Watch: Exotic Flu Viruses").

CHECK YOUR LEARNING

Do you understand why the immune response to the first exposure to a microbe is relatively slow, and why responses to future infections by the same microbe are much faster? Can you explain the role of memory cells in acquired immunity?

36.7 HOW DOES MEDICAL CARE ASSIST THE IMMUNE RESPONSE?

For most of human history, the battle against disease was fought by the immune response alone. Now, however, the immune response has a powerful assistant: medical treatment. Here we describe two of the most important medical tools: antibiotics and vaccinations.

Antibiotics Slow Down Microbial Reproduction

Antibiotics are chemicals that help to combat infection by slowing down the multiplication of bacteria, fungi, or protists. Although antibiotics usually do not destroy every single disease-causing microbe in the body, they may kill enough of them to give the immune system time to finish the job. However, antibiotics are potent agents of natural selection, favoring the survival and reproduction of microbes that can withstand their effects (see the case study "Evolution of a Menace" in Chapter 15). The occasional mutant microbe that is resistant to an antibiotic will pass on the genes for resistance to its offspring. The result: Resistant microbes thrive while susceptible microbes die off. Eventually, many antibiotics become ineffective in treating diseases.

Antibiotics are not useful against viruses, because they target metabolic processes that viruses do not possess. Until recently, little could be done to help people suffering from viral infections except to treat the symptoms and hope that the immune system would triumph. Now, drugs are available that target different stages of the viral cycle of infection, including attachment to a host cell, replication of viral parts, assembly of new viruses within the host cell, and the release of these

viruses to infect more of the body's cells (see Chapter 19). Antiviral drugs are available to treat HIV, herpes virus (cold and genital sores), and the flu virus.

Vaccinations Produce Immunity Against Disease

A **vaccine** stimulates an immune response by exposing a person to antigens produced by a pathogen. Vaccines often consist of weakened or killed microbes (that cannot cause disease) or some of the pathogen's antigens, usually synthesized using genetic engineering techniques. When the body is exposed to a weakened pathogen or its antigens, it produces an army of memory cells that confer immunity against living, dangerous microbes of the same type. Like immunity acquired by recovering from a real illness, immunity stimulated by an effective vaccine produces such a rapid and large immune response to a subsequent invasion by the living pathogen that you don't experience any symptoms at all. Learn more about vaccines in "Science in Action: The Discovery of Vaccines."

CHECK YOUR LEARNING

Can you explain how antibiotics and vaccines supplement the human immune response? Do you understand how antibiotic resistance arises?

36.8 WHAT HAPPENS WHEN THE IMMUNE SYSTEM MALFUNCTIONS?

Occasionally, the immune system launches inappropriate attacks, undermining health instead of promoting it. In addition, the immune system may suffer from disorders that decrease its effectiveness.

Allergies Are Misdirected Immune Responses

More than 50 million people in the United States suffer from **allergies,** which are immune reactions to harmless substances. Common allergies include those to pollen, mold spores, bee or wasp venoms, and some foods, such as milk, eggs, fish, wheat, tree nuts, or peanuts.

An allergic reaction begins when allergy-causing antigens, called *allergens*, enter the body and bind to "allergy antibodies" (IgE antibodies) on a special type of B cell. This B cell proliferates, producing plasma cells that pour out allergy antibodies into the plasma (**Fig. 36-13**). The antibodies attach to mast cells, mostly in the respiratory and digestive tracts. If allergens later bind to these attached antibodies, they cause the mast cells to release histamine, which causes leaky capillaries and other symptoms of inflammation (see Fig. 36-4). In the respiratory tract, histamine also increases the secretion of mucus. Thus, airborne substances such as pollen grains, which typically enter the nose and throat, may trigger the runny nose, sneezing, and congestion typical of "hay fever." Food allergies usually cause intestinal cramps and diarrhea. In some cases, such as severe allergic reactions to peanuts or bee stings, the inflammatory response in the airways is so strong that the airways may completely close, causing death by suffocation.

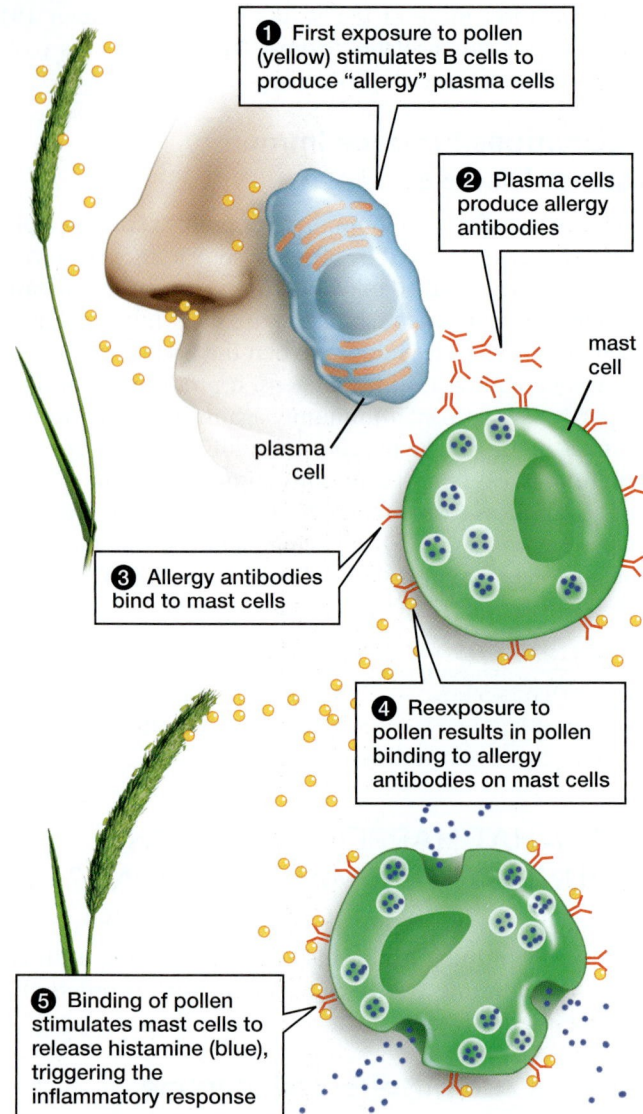

① First exposure to pollen (yellow) stimulates B cells to produce "allergy" plasma cells

② Plasma cells produce allergy antibodies

mast cell

plasma cell

③ Allergy antibodies bind to mast cells

④ Reexposure to pollen results in pollen binding to allergy antibodies on mast cells

⑤ Binding of pollen stimulates mast cells to release histamine (blue), triggering the inflammatory response

▲ FIGURE 36-13 An allergic reaction to pollen
QUESTION What might be the evolutionary advantage of allergic reactions? (*Hint:* Are there harmful substances or organisms that might provoke allergic reactions?)

 HAVE **YOU** EVER WONDERED …

Why You Get Colds So Often? If the immune system is so good at remembering past assaults and developing long-lasting immunity, why do many people get several colds every year? The problem is that there are at least 200 different viruses that cause cold symptoms, and immunity to one usually doesn't confer immunity to the others. Cold viruses also mutate rapidly, and some of the mutations allow an "old" virus to look "new" to the immune system. Thus, a large number of constantly mutating cold viruses keeps one step ahead of the human immune system.

An Autoimmune Disease Is an Immune Response Against the Body's Own Molecules

Our immune systems rarely mistake our own cells for invaders. Occasionally, however, something goes awry, and "anti-self" antibodies are produced. The result is an **autoimmune disease,** in which the immune system attacks a component of one's own body. Some types of anemia, for example, are caused by antibodies that destroy a person's red blood cells. Many cases of type 1 diabetes begin when the immune system attacks the insulin-secreting cells of the pancreas. Other autoimmune diseases include rheumatoid arthritis (which affects the joints), myasthenia gravis (skeletal muscle), multiple sclerosis (central nervous system), and systemic lupus (almost any part of the body).

Unfortunately, at present, there are no cures for autoimmune diseases. For some, replacement therapy can alleviate the symptoms—for instance, by giving insulin to diabetics. Alternatively, autoimmune disorders can be reduced with drugs that suppress the immune response. Immune suppression, however, also reduces responses to the assaults of disease microbes, so this therapy has major drawbacks. In the future, it might be possible to stimulate the activity of specific types of regulatory T cells that can damp down autoimmune responses; however, treatments based on regulatory T cells appear to be many years away from the doctor's office.

Immune Deficiency Diseases Occur When the Body Cannot Mount an Effective Immune Response

There are two very different types of disorders in which the immune system cannot combat routine infections. About 1 in 500 people has some type of hereditary immune deficiency, with a genetic defect that impairs immune responses to a greater or lesser extent. About 1 in 80,000 children is born with *severe combined immune deficiency*, a group of genetic defects in which few or no immune cells are formed, and which is invariably fatal without treatment. Much more frequent are acquired immune deficiencies that arise during a person's lifetime. There are many causes of acquired immune deficiencies, including radiation, chemotherapy, immune suppressive drugs, diabetes, leukemia, and pathogenic microbes. We will limit our discussions to severe combined immune deficiency and acquired immune deficiency syndrome (AIDS).

Severe Combined Immune Deficiency Is an Inherited Disorder

A child with **severe combined immune deficiency (SCID)** may survive for a few months after birth, protected by IgG and IgA antibodies acquired from the mother during pregnancy or in her milk. Once these antibodies are lost, however, common infections can prove fatal, because the child cannot generate an effective immune response. One form of therapy is to transplant bone marrow (from which immune cells arise) from a healthy donor into the child. Bone

Health | Watch

Exotic Flu Viruses

Almost everyone has heard of avian flu, also called bird flu, or H5N1 (Fig. E36-2). In the early 2000s, news reports suggested that we faced the real possibility of a worldwide epidemic that might kill millions of people. In 2009, swine flu, or H1N1, reared its head, infecting tens of millions of people by the end of that year. Are avian flu and swine flu especially infectious or dangerous? How do they differ from regular human flu?

Every winter, a wave of "ordinary" human flu sweeps across the world. Hundreds of thousands of the elderly, newborn, or infirm die. Most healthy adults who catch the flu—millions each year—do not die, but they suffer respiratory distress, fever, and muscle aches. They survive because their adaptive immune systems produce antibodies and cytotoxic T cells that eventually overcome the flu virus. Their immune systems also produce memory cells that lie in wait for the next flu season.

So why can you get the flu again and again? Flu viruses mutate very rapidly. Therefore, this year's flu antigens are similar to, but not exactly the same as, last year's flu antigens. Memory cells from the previous year do respond, but not as effectively as if the mutations had not occurred. Therefore, you get sick, although your partial immunity reduces the severity and duration of your illness.

In some years, however, the flu virus antigens are fundamentally different. These new strains arise when mutations in flu viruses that normally infect other animals allow these viruses to infect people, or when genes from human and bird or pig viruses combine in a single virus. Avian and swine flu viruses have very different antigens than human flu viruses do. With no useful memory from previous years, the adaptive immune system takes a week or two to do anything really effective against the virus. By then, depending on how deadly the virus is, it might be too late. In 1918, an avian flu virus mutated and killed more than 50 million people worldwide.

Avian Flu

Is today's avian flu likely to cause a similar catastrophe? Modern avian flu first appeared in 1997 in Hong Kong, infecting 18 people, 6 of whom died. Health authorities concluded that the flu came from infected chickens. To prevent further human illness, the government ordered every chicken in Hong Kong—about 1.5 million birds—to be killed. Nevertheless, avian flu spread across the world, with deaths reported in countries as far apart as Turkey, Indonesia, and Egypt. As of August 2012, about 60% of the 608 confirmed human cases of avian flu have been fatal.

Today's avian flu isn't very efficient at passing from birds to people, and seems almost totally incapable of being transmitted from person to person, so it is unlikely to cause a worldwide epidemic. However, additional mutations might make it more infectious. In 2011, in an effort to learn what genetic changes would allow avian flu to spread from person to person, scientists in the United States and the Netherlands designed avian flu viruses that infect, and spread easily between, ferrets (a domesticated weasel with an immune

▲ FIGURE E36-2 The avian flu virus—beautiful, but deadly

system very similar to a human's). At first, the U.S. National Science Advisory Board for Biosecurity asked the researchers not to publish the details of their experimental methods, for fear that terrorists might use the techniques to develop a deadly strain of avian flu that would be highly contagious in people. However, some scientists argued that naturally occurring mutations might someday produce a similarly infectious avian flu virus, and that the knowledge gained by studying the artificially synthesized virus might help to prevent an epidemic. In March 2012, the advisory board and the World Health Organization both decided that the benefits of sharing the results among legitimate researchers outweighed the risks, and recommended full publication.

Swine Flu

The H1N1 swine flu virus is a combination of genes from human, avian, and pig viruses. Compared to avian flu, H1N1 flu is much more infectious to people, but it is also much less deadly. In the 2009–2010 flu season, the U.S. Centers for Disease Control and Prevention estimated that H1N1 flu had infected 40 to 90 million Americans (no one knows for sure), causing over 12,000 deaths, a fatality rate below 0.1%. At least half of the people who died had other risk factors; obesity, diabetes, and pregnancy appear to make people more susceptible to complications of H1N1 flu. Because H1N1 flu is so widespread, the annual flu vaccination that many people get each year now includes protection against the H1N1 virus.

SCIENCE IN ACTION The Discovery of Vaccines

Today, it's difficult to imagine the terror that smallpox created just a few hundred years ago, and the tactics that people would use to avoid it. This highly contagious disease killed about 30% of its victims and left most of the survivors disfigured with pitted skin. The pits were caused by blisters filled with pus. Almost 1,000 years ago, in an attempt to prevent smallpox, inhabitants of China and India inoculated themselves by grinding up dried blisters from people with mild cases of smallpox and inhaling the resulting powder. In another early form of inoculation, a sharp knife was dipped in fresh pus, again from someone with a mild case, and used to cut the skin of an uninfected person. Most recipients developed relatively minor symptoms and could resist later exposure to severe forms of the disease.

In the early 1700s, Lady Mary Wortley Montague, wife of the English ambassador to Turkey, observed smallpox inoculation firsthand. Upon returning to England, she convinced some aristocrats, including members of the British royal family, to inoculate their children. Deliberately exposing people to smallpox, however mild, was a frightening prospect. Inoculated patients would be sick for a few days to a couple of weeks, and 1% to 2% died—one of the sons of King George III of England died from the procedure. Nevertheless, smallpox inoculation became increasingly widespread. For example, in 1777, George Washington ordered soldiers of the Continental Army to be inoculated.

Meanwhile, people noticed that milkmaids, who often contracted cowpox while milking cows, seldom caught smallpox. In 1796, the English surgeon and experimental biologist Edward Jenner obtained fluid from cowpox blisters on milkmaid Sarah Nelmes and inoculated an 8-year-old boy, James Phipps, with the fluid (Fig. E36-3). A few months later, Jenner inoculated Phipps with smallpox, and the boy remained healthy. After repeating the procedure several times with other children, and getting the same results, Jenner published his findings. Jenner's cowpox inoculation was rapidly adopted in Europe and eventually worldwide.

Nearly a century passed before vaccination was applied to other infectious diseases. The French microbiologist Louis Pasteur, among the first to recognize the role of microbes in causing disease, experimented with chicken cholera in the late 1800s. He grew cholera bacteria in culture and found that they caused the fatal disease when injected into chickens. The story goes that Pasteur went on vacation during the summer, leaving a cholera culture behind. When he came back, he resumed his experiments, injecting chickens with bacteria from the old culture. The chickens became ill but didn't die. Assuming that he had done something wrong, he grew a fresh culture. As luck would have it, he inoculated some of the same chickens that had survived the earlier injection with the spoiled culture. To his surprise, these chickens remained healthy, while "naïve" chickens died. Pasteur hypothesized that weakened bacteria in the spoiled culture protected against later infection by healthy bacteria. He coined the term "vaccine" (from the Latin "vacca," meaning "cow"), in recognition of Jenner's pioneering work with cowpox. Pasteur later applied the technique to anthrax in sheep and to rabies in humans.

Vaccines are perhaps the greatest triumph of modern medicine, preventing deadly diseases before they occur. Vaccines have eradicated smallpox from the planet. Completely eliminating polio seems close at hand; only about 600 cases occurred in 2011, confined to a handful of countries in Asia and Africa. Measles may be next. These diseases, which formerly killed millions of people each year, may soon become only bad memories, thanks to the efforts of Jenner, Pasteur, and thousands of biomedical scientists worldwide.

◄ FIGURE E36-3
Edward Jenner inoculates a child with cowpox to protect against smallpox

marrow transplants sometimes result in enough immune cell production to confer normal immune responses. Gene therapy, an experimental treatment in which genetic engineering is used to insert functional genes into a child's own cells, has been used to create a working immune system in a few dozen children with SCID (see Chapter 13).

AIDS Is an Acquired Immune Deficiency Disease

The most common immune deficiency disease is **acquired immune deficiency syndrome,** or **AIDS.** The United Nations Program on HIV/AIDS estimates that in 2010, about 1.8 million people worldwide died of AIDS and nearly 2.6 million more became infected, bringing the total infected population to 34 million. AIDS is caused by **human immunodeficiency viruses (HIV).** These viruses undermine the immune system by infecting and destroying helper T cells, which stimulate both the cell-mediated and humoral immune responses (see Fig. 36-11). AIDS does not kill people directly, but its victims become increasingly susceptible to other diseases as their helper T-cell populations decline.

Because HIV cannot survive for long outside the body, it can be transmitted only through direct contact with an infected person's broken skin, mucous membranes, or virus-laden body fluids, including blood, semen, vaginal secretions, and breast milk. HIV infection can be spread by sexual activity, by sharing a needle with an intravenous drug user, or through a blood transfusion (this is now rare in developed countries because all donated blood is screened for HIV). A woman infected with HIV can transmit the virus to her child during pregnancy, childbirth, or breast-feeding.

HIV enters a helper T cell and hijacks the cell's metabolic machinery, forcing it to make more viruses, which then emerge, taking an outer coating of T-cell membrane with them (**Fig. 36-14**). Early in the infection, as the immune system fights the virus, the victim may develop a fever, rash, muscle aches, headaches, and enlarged lymph nodes. After several months, the rate of viral replication slows. Enough helper T cells remain that infected individuals are able to resist disease, and they generally feel quite well. This condition often persists for several years. However, helper T-cell levels continue to decline, and eventually the immune response becomes too weak to overcome routine infections. At this point the person is considered to have AIDS. The life expectancy for untreated AIDS victims is 1 to 2 years.

Several drugs can slow down the replication of HIV and thereby slow the progress of AIDS. Combinations of drugs targeting different stages of viral replication have been particularly effective, and a complete AIDS treatment has now been combined into a once-a-day pill. HIV-positive people who receive the best medical care might now live a normal life span, although nobody knows for sure, because the most effective drugs have been available for only about 15 years.

The best solution would be a vaccine against HIV. This is a major challenge, partly because HIV disables the immune response that a vaccine depends on. Further, HIV has an incredibly high mutation rate, perhaps a thousand times faster than that of flu viruses, which themselves have a high mutation rate. Single infected individuals may harbor different strains of HIV because of mutations that occurred within their bodies after they were first infected. Despite years of effort, as of 2012, there is still no effective vaccine against HIV.

CHECK YOUR LEARNING

Can you describe allergies, autoimmune diseases, and immune deficiency diseases, including how each occurs?

36.9 HOW DOES THE IMMUNE SYSTEM COMBAT CANCER?

Cancer, the uncontrolled replication of the body's own cells, is one of the most dreaded diseases in the world, and with good reason. More than 550,000 people in the United States will die of cancer this year, a fatality rate second only to heart disease. Worldwide, cancer will kill more than 7.5 million people. Cancers may be triggered by many causes, including environmental factors (for example, UV radiation or smoking), faulty genes, mistakes during cell division, and viruses. All of these triggers produce cancer by sabotaging the mechanisms that normally control the growth of the body's cells (see Chapter 9).

The Immune System Recognizes Most Cancerous Cells as Foreign

Cancer cells form in our bodies every day. Fortunately, the immune system destroys nearly all of them before they have a chance to proliferate and spread. How are cancer cells weeded out? The surfaces of most cancer cells either lack some of the usual proteins found on normal body cells (such as some MHC proteins) or bear proteins that are unique to cancers, especially cancers caused by viruses. Natural killer cells and cytotoxic T cells recognize these protein differences as markers of non-self cells, and destroy the cancer cells (see Fig. 36-10). However, some cancer cells evade detection because they do not bear antigens that allow the immune system to recognize them as foreign.

Vaccination Can Prevent Some Cancers

Viruses cause some cancers of the liver, mouth, throat, and penis, some types of leukemia, and probably all cases of cervical cancer. In the United States, preventive vaccines have been approved that protect against the virus that causes hepatitis B, which triggers some cases of liver cancer, and against the two types of human papilloma viruses (HPV) that cause most cases of cervical cancer.

Vaccines May Someday Help to Cure Cancer

Researchers at the U.S. National Cancer Institute, universities, and pharmaceutical companies are developing "treatment vaccines" to cure cancers after they occur. Some of these vaccines would provide a patient with antigens commonly found on cells of the type of cancer that the patient has. The antigens would be enhanced in various ways to boost the patient's immune response against them. Other prospective treatment

▲ **FIGURE 36-14 HIV causes AIDS** In this electron micrograph, human immunodeficiency viruses are seen emerging from a helper T cell, acquiring an outer envelope of plasma membrane (green) in the process. The plasma membrane coating will help them infect new cells.

vaccines consist of antigens from a patient's own tumor cells. Still another approach is to take antigen-presenting dendritic cells from a patient, expose them to antigens from cancer cells, and force them to multiply rapidly in culture. The resulting daughter cells would then be injected back into the patient. In principle, this large number of activated dendritic cells should stimulate the patient's own anticancer immune response.

Most Medical Treatments for Cancer Depend on Selectively Killing Cancerous Cells

In the meantime, medical care for most cancers still relies on early detection and the traditional treatments of surgery, radiation, and chemotherapy. Surgically removing the tumor is the first step in treating many cancers, but it can be difficult to remove every bit of cancerous tissue. Tumors can be bombarded with radiation, which can destroy even microscopic clusters of cancer cells by disrupting their DNA, thus preventing cell division. Unfortunately, neither surgery nor radiation is effective against cancer that has spread throughout the body. Chemotherapy is commonly used to supplement surgery or radiation, or to combat cancers that cannot be treated any other way. Chemotherapy drugs attack the machinery of cell division, and so they are somewhat selective for cancer cells, which divide more frequently than normal cells do. Unfortunately, chemotherapy inevitably also kills some healthy, dividing cells. Damage to dividing cells in patients' hair follicles and intestinal lining by chemotherapy produces its well-known side effects of hair loss, nausea, and vomiting.

CHECK YOUR LEARNING

Can you explain how the immune system attacks cancerous cells, even though they are your own body's cells? Are you able to describe current medical treatments against cancer?

CASE STUDY revisited

Flesh-Eating Bacteria

Film-maker Woody Allen once said, "Just because you're paranoid doesn't mean they aren't out to get you." This certainly seems to apply to *S. pyogenes*, which avoids the immune response and damages the body in so many different ways. In addition to the tactics described earlier, *S. pyogenes* releases several toxins, some that kill ordinary body cells and some that specifically kill phagocytic cells that might otherwise destroy the bacteria. *S. pyogenes* also produces enzymes that trigger a cascade of reactions that dissolve blood clots, thereby making it easier for the bacteria to enter the bloodstream and spread throughout the body.

Perhaps most devastating of all, *S. pyogenes* produces proteins that are "superantigens," which massively activate the immune system—causing a response 10,000 to 100,000 times stronger than the response to infection by most other bacteria. This enormous immune response releases huge quantities of cytokines, causing the overwhelming inflammation of streptococcal toxic shock syndrome. Remember, inflammation causes leaky capillaries with increased blood flow, so affected tissues swell up with the extra fluid. If massive inflammation occurs in the lungs, then the victim cannot breathe, and may die. Inflammation caused by superantigens is probably what killed Jim Henson.

In Monica Jorge's case, drastic surgery saved her life, but at the cost of large parts of her arms and legs. Jorge's story came to the attention of TV talk-show host Oprah Winfrey, who dubbed her the "Warrior Mom" for her courage in the face of such adversity. In 2010–2011, Oprah and Nate Berkus, a design expert with his own TV show, teamed up to build the Jorge family a home customized to accommodate her disabilities.

Consider This

BioEthics *S. pyogenes* isn't the only flesh-eating bacterium; several other bacteria can also rapidly destroy tissue. Even so, such infections are quite rare—perhaps 500 to 1,000 cases occur each year in the United States. However, 25% to 70% of the victims die. A physician who suspects a flesh-eating infection begins antibiotic treatment immediately.

Antibiotic treatment for anyone with an infected cut, even if there is no evidence of flesh-eating bacteria, would certainly reduce suffering, disfigurement, and mortality. On the other hand, frequent use of antibiotics is a potent agent of natural selection favoring antibiotic-resistant bacteria. How do you think that the medical community should balance the benefits of antibiotic treatment for individuals against the overall risk to society of increased antibiotic resistance among bacteria?

CHAPTER REVIEW

Summary of Key Concepts

36.1 How Does the Body Defend Itself Against Disease?

Nonspecific external barriers, including the skin and mucous membranes, prevent disease-causing organisms from easily entering the body. Nonspecific internal defenses, collectively called the innate immune response, consist of white blood cells, inflammation, and fever. These defenses destroy microbes, toxins, and cancerous and infected body cells. The adaptive immune response selectively destroys a specific toxin or microbe and "remembers" the invader, allowing a faster response if the invader reappears.

36.2 How Do Nonspecific Defenses Function?

The skin and its secretions physically block the entry of microbes into the body and inhibit their growth. The mucous membranes of the respiratory and digestive tracts secrete antibiotic substances and mucus that traps microbes. If microbes do enter the body, they are engulfed by phagocytic white blood cells. Natural killer cells secrete proteins that kill infected or cancerous cells. Injuries stimulate the inflammatory response, in which chemicals are released that attract phagocytic white blood cells, increase blood flow, and make capillaries leaky. Later, blood clots wall off the injury site. Fever is caused by chemicals produced by bacteria or released by white blood cells in response to infection. High temperatures inhibit bacterial growth and accelerate the immune response.

36.3 What Are the Key Components of the Adaptive Immune System?

Most of the cells of the adaptive immune system are located in bone marrow, thymus, spleen, and lymphatic vessels and nodes. These cells include macrophages, dendritic cells, and two types of lymphocytes: B cells and T cells. The cells secrete cytokine proteins that allow communication between cells. B cells also secrete antibodies that combat infection.

36.4 How Does the Adaptive Immune System Recognize Invaders?

Cells of the adaptive immune system recognize large, complex molecules, called antigens, produced by invading microbes and cancerous cells. Antibodies (on B cells) and T-cell receptors (on T cells) bind to specific antigens and trigger responses in their respective cell types. Antibodies are Y-shaped proteins composed of a constant region and a variable region. Each antibody has specific sites that bind only one or a few types of antigen. Each B cell synthesizes only one type of antibody, unique to that particular cell and its offspring. The body contains millions of different antibodies and T-cell receptors. Any antigen entering the body is likely to bind to some of these antibodies and T-cell receptors and trigger an immune response.

Both foreign invaders and the body's own cells have antigens that can potentially bind antibodies and T-cell receptors. However, immature immune cells die if they bind antigen. Because the body's own proteins are continuously present during immune cell differentiation, self-reactive immune cells are usually destroyed. Therefore, only foreign antigens normally provoke an immune response.

36.5 How Does the Adaptive Immune System Attack Invaders?

Only those B and T cells that are activated by antigen binding multiply and produce a specific immune response to an invading microbe, a process called clonal selection. B cells give rise to plasma cells, which secrete antibodies into the bloodstream, causing humoral immunity. Antibodies destroy microbes or their toxins while they are outside of body cells. Cytotoxic T cells destroy some microbes, cancer cells, and virus-infected cells on contact, causing cell-mediated immunity. Helper T cells stimulate both the humoral and cell-mediated immune responses.

36.6 How Does the Adaptive Immune System Remember Its Past Victories?

Some progeny cells of both B and T cells are long-lived memory cells. If the same antigen reappears in the bloodstream, these memory cells are immediately activated, dividing rapidly and causing an immune response that is much faster and more effective than the original response.

36.7 How Does Medical Care Assist the Immune Response?

Antibiotics kill microbes or slow down their reproduction, allowing the body's defenses more time to respond and exterminate the invaders. Vaccinations are injections of antigens from disease organisms, in some cases the weakened or dead microbes themselves. An immune response is evoked by the antigens, providing memory and a rapid response should a real infection occur later.

36.8 What Happens When the Immune System Malfunctions?

Allergies are immune responses to normally harmless foreign substances. Certain B cells treat these as antigens and produce "allergy antibodies" that bind to mast cells. When exposed to the antigen, the mast cells release histamine, causing a local inflammatory response. Autoimmune diseases arise when the immune system mistakes the body's own cells for foreign invaders and destroys them. Immune deficiency diseases occur when the immune system cannot respond strongly enough to ward off usually minor diseases. Immune deficiency diseases may be innate, such as severe combined immune deficiency (SCID), or acquired during a person's lifetime, such as acquired immune deficiency syndrome (AIDS).

36.9 How Does the Immune System Combat Cancer?

Cancer is a population of the body's cells that multiplies without control. Cancerous cells are often recognized as non-self by the immune system and destroyed by natural killer cells and cytotoxic T cells. Cancer may be triggered by genetic factors, environmental factors, mistakes during cell division, or viruses. Vaccines can help to prevent certain virus-caused cancers. Other vaccines are under development to try to cure cancer once it occurs.

Key Terms

Learning Outcomes

In this chapter, you have learned to . . .

LO1 Name the body's external barriers against disease, and describe the processes by which those barriers protect the body against infection.

LO2 Describe the processes by which the innate immune response protects against infection, including the roles of white blood cells, the inflammatory response, and fever.

LO3 Name the organs, tissues, cells, and molecules that make up the adaptive immune system, and describe their structures and functions.

LO4 Describe how the immune system distinguishes self from non-self.

LO5 Describe the processes by which the adaptive immune system protects against infection, including humoral immunity, cell-mediated immunity, and the role of helper T cells in the immune response.

LO6 Explain the process of acquired immunity.

LO7 Explain how antibiotics and vaccines supplement the human immune system.

LO8 Define *allergy, autoimmune disease,* and *immune deficiency disease,* describe how each occurs, and give examples of each.

LO9 Explain how the immune system combats cancer, and describe current medical treatments against cancer.

Thinking Through the Concepts

Fill-in-the-Blank

1. External defenses against microbial invasion include the _____, and the mucous membranes that line the _____, _____, and _____ tracts. **LO1**

2. Nonspecific internal defenses against disease include _____, which engulf and digest microbes; _____, which destroy cells that have been infected by viruses; the _____, provoked by injury; and _____, an elevation of body temperature that slows microbial reproduction and enhances the body's defenses. **LO2**

3. The adaptive immune response is stimulated when the body is invaded by complex proteins or polysaccharides collectively called _____. These molecules bind to one of two types of protein receptors of the immune system: _____ or _____. **LO3 LO5**

4. An antibody consists of four protein chains, two _____ chains and two _____ chains. Each is composed of a(n) _____ region and a(n) _____ region. The _____ regions form the binding site for antigen. **LO3**

5. _____ immunity is provided by B cells and their daughter cells, called _____, which secrete antibodies into the blood plasma. _____ immunity is provided by T cells. _____ T cells kill body cells that have been infected by viruses. _____ T cells produce cytokines that stimulate immune responses in both B cells and T cells. Protection against future invasions by microbes bearing the same antigens is provided by _____ cells of both B and T types. **LO5 LO6**

6. In medical practice, a(n) _____ provides antigens that stimulate an immune response, to protect against infection without actually causing disease. **LO7**

7. A(n) _____ occurs when the immune system produces a response to a harmless substance such as pollen. In a(n) _____, the immune system cannot mount an effective response, even to dangerous infections; these may be inborn or acquired. When the immune system attacks a person's own body, this is called a(n) _____. **LO8**

Review Questions

1. What are the human body's three lines of defense against invading microbes? Which are nonspecific (that is, act against all types of invaders) and which are specific (act only against a particular type of invader)? **LO1 LO2 LO3**

2. How do natural killer cells and cytotoxic T cells destroy their targets? **LO2 LO5**

3. Describe humoral immunity and cell-mediated immunity. Include in your answer the types of immune cells involved in each, the location of antibodies and receptors that bind foreign antigens, and the mechanisms by which invading cells are destroyed. **LO5**

4. Diagram the structure of an antibody. What parts bind to antigens? Why does each antibody bind only to a specific antigen? **LO3**

5. How does the immune system combat cancer? **LO9**

6. How does the body distinguish "self" from "non-self"? **LO4**

7. What are memory cells? How do they contribute to long-lasting immunity to specific diseases? **LO6**

8. What is a vaccination? How does it confer immunity to a disease? **LO7**

9. How does the inflammatory response help the body resist disease? What allergy symptoms does it cause? **LO2 LO8**

10. Distinguish between autoimmune diseases and immune deficiency diseases, and give one example of each. **LO8**

11. Describe the causes and eventual outcome of AIDS. How is HIV spread? **LO8**

Applying the Concepts

1. Why is it essential that antibodies and T-cell receptors bind only large molecules (such as proteins) and not small molecules (such as amino acids)?

2. The essay "Health Watch: Exotic Flu Viruses" states that the flu virus is different each year. If that is true, what good does it do to get a "flu shot"?

3. Organ transplant patients often receive the drug cyclosporine. This drug inhibits the production of a cytokine that stimulates helper T cells to proliferate. How does cyclosporine prevent rejection of transplanted organs? Some patients who received successful transplants many years ago are now developing various kinds of cancers. Propose a hypothesis to explain this phenomenon.

Answers to Figure Caption questions and Fill-in-the-Blank questions can be found in the Answers section at the back of the book.

MB *Go to MasteringBiology for practice quizzes, activities, eText, videos, current events, and more.*

Hormones control the metamorphosis of a caterpillar into a butterfly, such as this Asian swallowtail.

CASE STUDY

Life-Changing Hormones

SO YOU THINK that the hormone-induced changes—physical and emotional—that girls and boys experience during puberty are amazing? Well, of course, they are. But metamorphosis in Asian swallowtail butterflies is truly astounding.

As you probably know, a butterfly egg hatches into a caterpillar. There are several stages of caterpillar growth, after which the caterpillar forms a pupa. The body inside the pupa undergoes a remarkable transformation, with some parts completely disappearing and new structures such as wings developing. Eventually, the pupal case splits open, and the adult butterfly emerges.

A caterpillar is covered by a rigid coating, called the cuticle. The cuticle

is divided into segments by flexible grooves, so the caterpillar can move, but it can't get much larger. Therefore, a growing caterpillar faces two major problems: how to get out of its too-tight suit of armor, and what it should be when it escapes—a larger caterpillar or a pupa. The solutions lie in interactions between two hormones: molting hormone and juvenile hormone.

A growing caterpillar stretches the grooves connecting its segments. Sensory cells in the grooves detect the stretch and signal the caterpillar's brain to release molting hormone. Molting hormone causes the caterpillar to detach from its old cuticle and secrete a soft new cuticle underneath it. The caterpillar then pumps itself full of air, which splits open the old, stiff cuticle and stretches out the new, soft one to accommodate future growth.

What emerges from the old cuticle depends on the concentration of juvenile hormone. Juvenile hormone levels are very high in newly hatched caterpillars and slowly decline over time. Each of the first four molts produces another, larger caterpillar. At the end of the fifth caterpillar stage, juvenile hormone levels plummet, and the caterpillar forms a pupa. The combination of very low juvenile hormone levels and high molting hormone levels transforms the caterpillar into a winged butterfly.

How can hormones cause such enormous differences in appearance and behavior, changing boys and girls into men and women, or caterpillars into butterflies? As we will see in this chapter, hormones influence virtually every aspect of an animal's life, from regulating metabolism and growth to triggering puberty or metamorphosis.

AT A GLANCE

37.1 How Do Animal Cells Communicate?

37.2 How Do Endocrine Hormones Produce Their Effects?

37.3 What Are the Structures and Functions of the Mammalian Endocrine System?

LEARNING GOALS

LG1 List and describe the four major mechanisms of cell-to-cell communication in animals.

LG2 Describe the pathway and mechanisms of action of an endocrine hormone, beginning with its secretion from an endocrine cell and ending with its effects on a target cell.

LG3 Describe the physiological effects of peptide hormones, amino acid-derived hormones, and steroid hormones, including the responses in target cells.

LG4 Explain how hormone release is controlled.

LG5 Describe the mammalian endocrine system, including the names and locations of the principal endocrine glands, the hormones that the endocrine glands release, and the principal responses that the hormones produce in the body.

37.1 HOW DO ANIMAL CELLS COMMUNICATE?

The cells of an animal's body must communicate with one another to ensure the proper functioning of the whole organism. Should leg muscles just hold up a standing body or should they make the legs run? Should blood flow be directed to the intestines for digestion or to the muscles for movement? These and hundreds of other messages are sent throughout the body every minute. Methods of communication among cells fall into four broad categories: direct, synaptic, paracrine, and endocrine (Table 37-1).

In direct communication, some tissues, such as heart muscle, have gap junctions that link the cytoplasm of adjacent cells, allowing ions and electrical signals to flow between them (see Chapters 5 and 32). This type of communication is very fast but requires the cells to be in intimate contact with one another.

In the other three types of communication—synaptic, paracrine, and endocrine—the "sending" cells release

TABLE 37-1 How Cells Communicate

Communication	Chemical Messengers	Mechanism of Transmission	Examples
Direct	Ions, small molecules	Direct movement through gap junctions linking the cytoplasm of adjacent cells	Ions flowing between cardiac muscle cells
Synaptic	Neurotransmitters	Diffusion from a neuron across a narrow space (synaptic cleft) to an adjacent cell	Acetylcholine, dopamine
Paracrine	Local hormones	Diffusion through the interstitial fluid to nearby cells	Prostaglandins, histamine
Endocrine	Endocrine hormones	Carried in the bloodstream to nearby or distant cells	Insulin, estrogen, growth hormone

chemical messengers. The messengers move to "receiving" cells and alter the receiving cells' physiology by binding to **receptors,** which are specialized proteins located either on the surface of, or inside, the receiving cells. When a messenger binds to a receptor, the receiving cell responds in a way that is determined by the messenger, the receptor, and the type of cell. These responses can be as varied as muscle contraction, secretion of milk in lactating women, or active transport of salt by cells in the kidney.

Every cell has dozens of receptor proteins, each capable of binding a specific chemical messenger and stimulating a particular response. Cells with receptors that bind a messenger and respond to it are **target cells** for that messenger. Cells without the correct receptors cannot respond to the messenger and are not target cells.

The chemical messengers released during synaptic, paracrine, and endocrine communication are given a variety of names. Even biologists don't always agree on the names and their definitions. We define a **hormone** as a chemical that is secreted into the interstitial fluid and carried by the interstitial fluid or bloodstream to other cells, where it causes specific effects in target cells. As described below, **local hormones** may be produced by small glands, clusters of cells, or even single cells. They diffuse through the interstitial fluid and influence cells in the immediate vicinity of the sending cells. **Endocrine hormones** are produced by endocrine glands. They travel in the bloodstream and may influence cells either close to the gland or in distant parts of the body. Sometimes the same molecule may be a local or endocrine hormone, depending on where it is produced and acts. Testosterone, for example, is synthesized in the testes. As an endocrine hormone, it is carried throughout the body and exerts its well-known effects on the development of beards, muscles, and behavior. In the testes itself, testosterone acts as a local hormone, stimulating sperm development. A **neurotransmitter** is a chemical messenger that is specifically produced by a nerve cell and acts at extremely close range—usually about 20 nanometers (billionths of a meter)—to influence another cell.

Synaptic communication is used in the nervous system. Electrical signals within individual nerve cells can send information throughout the brain and to the farthest reaches of the body in a fraction of a second. Then, the nerve cell communicates with other cells at junctions called *synapses*. At a synapse, a nerve cell elicits responses from a target cell by releasing neurotransmitters that cross the tiny space between the end of a nerve cell and its target. These responses may be very brief, such as reflexes, or very long lasting, such as learning. (We will explore the nervous system in detail in Chapter 38.)

In this chapter, we will focus on paracrine and endocrine communication.

Paracrine Communication Acts Locally

In **paracrine communication,** cells release local hormones that diffuse through the interstitial fluid to nearby cells ("para" means "alongside" in Greek). Because the sending and receiving cells are close to one another, paracrine communication tends to be very fast. Local hormones have mostly short-range actions, either because they are degraded soon after release, or because cells close by take them up so fast that they cannot get very far from the cells that secrete them.

Local hormones include histamine, which is released as part of the allergic and inflammatory responses, and many of the cytokines by which cells of the immune system communicate with one another (see Chapter 36). Nitric oxide, a gas produced by cells lining many blood vessels, acts as a local hormone. It diffuses rapidly into muscle cells surrounding the vessels, causing them to relax. This causes the blood vessel to expand, which increases blood flow. Nitric oxide is also a neurotransmitter in the brain and elsewhere in the body.

An important group of local hormones is the **prostaglandins,** which are modified fatty acids secreted by cells throughout the body. Prostaglandins have diverse roles, depending on the type of prostaglandin and the target cell. For example, during childbirth, prostaglandins both cause the cervix to dilate and help to stimulate uterine contractions. Prostaglandins contribute to pain and inflammation (such as occurs in arthritic joints). Drugs such as aspirin, acetaminophen (Tylenol), and ibuprofen provide relief from these symptoms by inhibiting the enzymes that synthesize prostaglandins.

Endocrine Communication Uses the Circulatory System to Carry Hormones to Target Cells Throughout the Body

Endocrine communication begins with the secretion of hormones by **endocrine glands** (Fig. 37-1). An endocrine gland may be a well-defined cluster of cells whose principal function is hormone secretion, as is the case with the thyroid and pituitary glands. Alternatively, an endocrine gland may consist of clusters of cells, or even scattered individual cells, embedded in organs that have multiple functions, such as the pancreas, ovary, or testes. In all cases, the secretory cells of an endocrine gland are embedded within a network of capillaries. Endocrine cells secrete their hormones into the interstitial fluid surrounding the capillaries. The hormones diffuse into the capillaries and are carried in the blood throughout the body in only a few seconds.

Although an endocrine hormone may reach nearly all of the body's cells, only target cells, with receptors capable of binding that specific hormone, can respond. The hormone *oxytocin*, for example, stimulates the contraction of uterine muscles during childbirth, because these muscle cells have receptors that bind oxytocin. However, oxytocin does not cause most of the other muscles of the body to contract, because their cells do not have the necessary receptors. A woman's uterine muscles, therefore, contain target cells for oxytocin, whereas her biceps do not.

The changes induced by hormones may be prolonged and irreversible, as in the onset of puberty or the transformation of a caterpillar into a butterfly. More typically, the

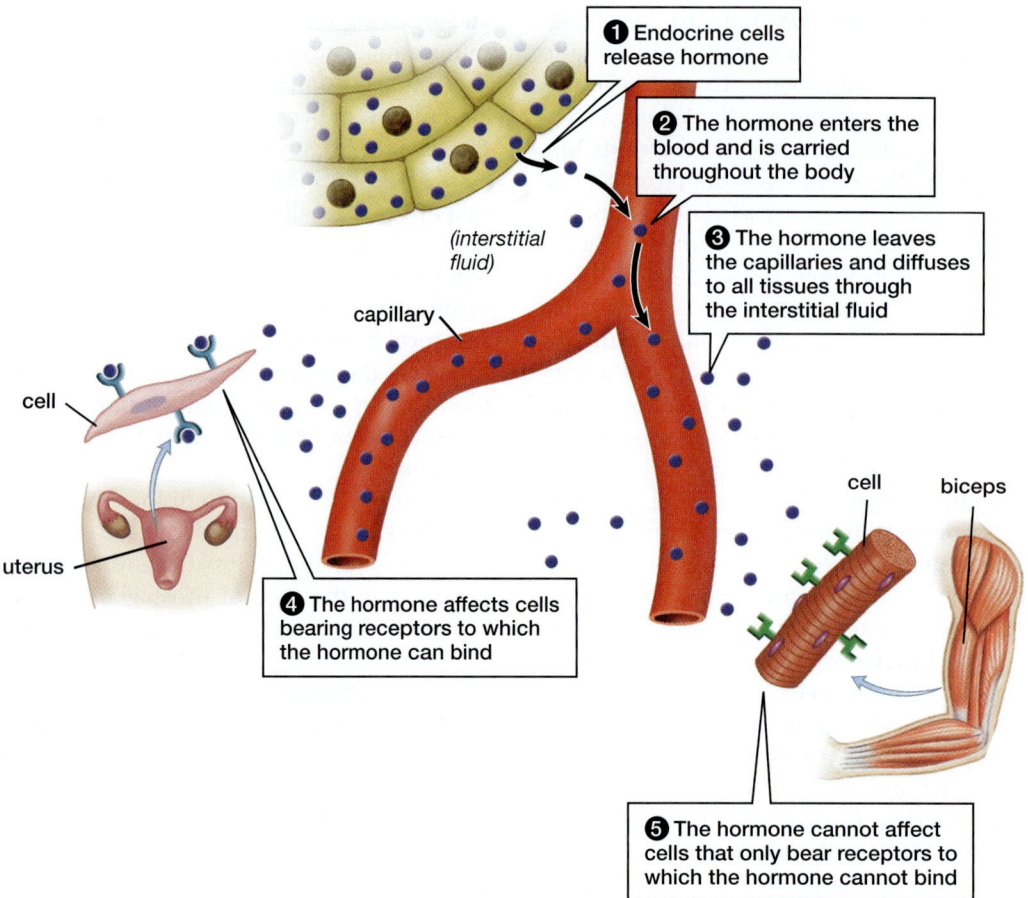

① Endocrine cells release hormone

② The hormone enters the blood and is carried throughout the body

③ The hormone leaves the capillaries and diffuses to all tissues through the interstitial fluid

(interstitial fluid)

capillary

cell

uterus

④ The hormone affects cells bearing receptors to which the hormone can bind

cell   biceps

⑤ The hormone cannot affect cells that only bear receptors to which the hormone cannot bind

▲ **FIGURE 37-1 Hormone release, distribution, and reception**

changes are temporary and reversible, and help to regulate the physiological systems of the animal body with a time course of seconds to hours.

CHECK YOUR LEARNING

Can you describe the four methods of communication between cells in an animal body? Are you able to explain why some cells respond to chemical messengers and others do not? Can you describe the pathway by which endocrine hormones move from secreting cells to target cells, often in far-distant parts of the body?

37.2 HOW DO ENDOCRINE HORMONES PRODUCE THEIR EFFECTS?

There are three classes of vertebrate endocrine hormones: **peptide hormones,** which are chains of amino acids; **amino acid-derived hormones,** which are composed of one or two modified amino acids; and **steroid hormones,** which are synthesized from cholesterol. These hormones bind to receptors located on the surface of, or inside, a target cell. Hormone-receptor binding then causes one or both of two major effects: (1) stimulating changes in the metabolism of the cell, usually by activating or inhibiting enzymes, or (2) affecting gene transcription, thereby changing the amounts or types of proteins that the cell synthesizes.

Peptide Hormones and Amino Acid-Derived Hormones Usually Bind to Receptors on the Surfaces of Target Cells

Most peptide hormones and amino acid-derived hormones are soluble in water but not in lipids, so these hormones cannot diffuse through the phospholipid bilayer of the plasma membrane. Therefore, peptide hormones and amino acid-derived hormones bind to receptors on the surface of a target cell (**Fig. 37-2 ①**). Hormone–receptor binding activates an enzyme that synthesizes a molecule, called a **second messenger,** inside the cell (**Fig. 37-2 ②**). A common second messenger is **cyclic adenosine monophosphate,** or **cyclic AMP** (see Chapter 3). The second messenger transfers the signal from the first messenger—the hormone—to other molecules within the cell, often activating specific intracellular enzymes (**Fig. 37-2 ③**), which then initiate a chain of biochemical reactions (**Fig. 37-2 ④**).

The reactions vary depending on the hormone, the second messenger, and the target cell. For example, epinephrine (an amino acid-derived hormone also called adrenaline) prepares the body to deal with emergency situations. It stimulates the synthesis of cyclic AMP in both heart muscle cells and liver cells. Cyclic AMP acts differently, however, in the two cell types. Cyclic AMP causes heart muscle cells to contract more strongly, which increases blood flow. In liver cells, cyclic

▲ **FIGURE 37-2 Actions of peptide hormones and amino acid-derived hormones on target cells** Peptide hormones and amino acid-derived hormones usually stimulate target cells by binding to receptors on the plasma membrane, which causes the cell to synthesize a second messenger molecule that sets off a cascade of intracellular biochemical reactions.

AMP activates enzymes that break down glycogen (a starch-like polysaccharide) to glucose. The glucose is released into the bloodstream, providing the entire body with a source of quick energy.

Steroid Hormones Usually Bind to Receptors Inside Target Cells

Steroid hormones are lipid soluble. Although they sometimes bind to receptors on the surface of a target cell, most steroid hormones bind intracellular receptors. Steroid hormone action begins with the hormone diffusing through the plasma membrane into the cytoplasm of a cell (**Fig. 37-3 ❶**). Steroid hormones diffuse into every cell they encounter, but only cells with appropriate receptors can respond and be targets for a particular hormone. Once inside a target cell, steroid hormones attach to receptors located either in the cytoplasm or in the nucleus, forming a hormone–receptor complex (**Fig. 37-3 ❷**). If the receptors are in the cytoplasm, the hormone–receptor complex moves into the nucleus. The complex then binds to specific genes (**Fig. 37-3 ❸**), and stimulates the transcription of messenger RNA (**Fig. 37-3 ❹**). The messenger RNA travels to the cytoplasm and directs protein synthesis (**Fig. 37-3 ❺**). The steroid hormone testosterone, for example, increases the synthesis of proteins involved

in the development of the testes, muscles, beard, pubic hair, larynx (voice box), and many other parts of the body. The increased muscle mass, and therefore strength, produced by testosterone and chemically similar *anabolic steroids* has tempted many athletes to try to boost their competitiveness by taking these "performance-enhancing" drugs, as we explore in "Health Watch: Anabolic Steroids—Fool's Gold?".

Although it is not a steroid, thyroid hormone has a similar mechanism of action. Thyroid hormone is transported across the membrane by carrier proteins. Once inside cells, thyroid hormone binds to intracellular receptors and activates transcription of specific genes.

CASE STUDY continued

Life-Changing Hormones

Insect molting hormone is a steroid. Like mammalian steroid hormones, molting hormone penetrates through the plasma membranes of cells and attaches to a receptor in the nucleus. The hormone–receptor complex binds to specific genes. Molting hormone stimulates transcription of dozens of genes, including ones involved in cell division and the development of specific structures in the caterpillar or butterfly.

Health | Watch

Anabolic Steroids—Fool's Gold?

Many athletes, at levels from high school through the Olympics and professional leagues, take anabolic steroid hormones to become stronger and faster—some even admit it. Top athletes who have been penalized for taking steroids include former world record sprinters Marion Jones (Fig. E37-1) and Tim Montgomery, shot putter Kevin Toth, and runners Antonio Pettigrew, Regina Jacobs, and Michelle Collins. Since 2005, Major League Baseball has suspended more than 30 players for using performance-enhancing drugs, mostly steroids. And when you see a Hollywood actor who has bulked up by 20 or 30 pounds in just a few months, you have to wonder how he got that way.

How do anabolic steroids work? They mimic the muscle-building actions of testosterone, only more so. Anabolic steroids and testosterone bind to the same intracellular receptors. However, many steroids activate these receptors at lower concentrations than testosterone can. Further, if you inject testosterone into your body, it will be completely metabolized within a few hours, whereas some anabolic steroids remain in your system for weeks. Therefore, the strength and duration of hormone–receptor action in target cells may be much greater for anabolic steroids than for natural testosterone.

Anabolic steroids have legitimate medical uses, for example, in men with low natural testosterone levels, in boys with delayed puberty, or (rarely) to combat osteoporosis in the elderly. However, steroids are likely to cause unwanted side effects, some possibly serious. Obviously, no one does clinical trials on secret, illegal drugs, so the potentially harmful effects of anabolic steroids are often uncertain. However, the National Institute on Drug Abuse and the U.S. Anti-Doping Agency warn against anabolic steroid use, for many reasons.

In both men and women, testosterone affects reproductive physiology, cholesterol metabolism, bone and hair growth, mood, and behavior. Anabolic steroids can have similar, although often greater, effects. Steroids may cause acne and depress the immune system. Anabolic steroids have been linked to increases in blood pressure and decreases in the "good" form of cholesterol (HDL)—both are risk factors for heart attacks and strokes. Anabolic steroids can cause bone growth to shut down prematurely, so young steroid users may never reach their full potential height.

In males, anabolic steroids create a negative feedback effect that reduces the natural production of testosterone. Tricked by these testosterone mimics, the anterior pituitary releases smaller amounts of hormones that are required for testes development and sperm production, so the testes often shrink and sperm count drops. Finally, men produce enzymes that convert some anabolic steroids into estrogen, which may cause partial breast development.

In females, anabolic steroids promote male-like body changes, including deepening of the voice, increased facial hair, and even pattern baldness. Testosterone-like hormones also interfere with egg development and ovulation.

Are bulging muscles worth the risks? Some athletes say they willingly risk long-term damage to their bodies to win Olympic gold. And the payday in professional sports can be astronomical—the highest-paid players in the National Basketball Association and Major League Baseball make many millions of dollars each season. For the more typical man or woman, whose principal benefits are admiring glances from potential dating partners and competitors, well. . . . Before *you* decide, be sure to find reliable information from authoritative sources such as the National Institutes of Health or the Mayo Clinic—not from body-building websites or online steroid retailers.

▲ **FIGURE E37-1 Tarnished gold** Marion Jones lost five Olympic medals because of her steroid abuse.

steroid hormone   (interstitial fluid)

❷ The hormone binds to a receptor in the nucleus or to a receptor in the cytoplasm that carries it into the nucleus

❸ The hormone–receptor complex binds to DNA and causes RNA polymerase to bind to a promoter site for a specific gene

❶ A steroid hormone diffuses through the plasma membrane

DNA

plasma membrane

ribosome

hormone receptor

RNA polymerase

❺ The mRNA leaves the nucleus, then attaches to a ribosome and directs the synthesis of a specific protein product

mRNA

❹ RNA polymerase catalyzes the transcription of DNA into messenger RNA (mRNA)

gene

new protein

nuclear envelope

(cytoplasm)   *(nucleus)*

▲ **FIGURE 37-3 Steroid hormone action on target cells** Steroid hormones often stimulate target cells by binding to intracellular receptors, which creates a hormone–receptor complex that activates gene transcription and ultimately results in the synthesis of new, or increased amounts of, specific proteins.

Hormone Release Is Regulated by Feedback Mechanisms

The release of most hormones is controlled by **negative feedback:** a response to a change that tends to counteract the change and restore the system to its original condition (see Chapter 31). For example, suppose you have jogged a few miles on a hot, sunny day and have lost a quart of water through perspiration. In response to the loss of water from your bloodstream, your pituitary gland releases antidiuretic hormone (ADH), which causes your kidneys to reabsorb water and to produce only small amounts of very concentrated urine (see Chapter 35). But let's say that you arrive home and drink 2 quarts of water, replacing twice as much water as you lost in sweat. If your body retained this extra water, it might raise your blood pressure and damage your heart. Negative feedback, however, acts to restore the original condition. When enough water has entered your blood to bring its volume back to normal, ADH secretion is turned off. Your kidneys start producing lots of watery urine, ridding your body of the extra quart of water. Look for other examples of negative feedback throughout this chapter.

In a few cases, hormone release is temporarily controlled by **positive feedback,** in which the response to a change enhances the change. For example, contractions of the uterus early in childbirth push the baby's head against the cervix (a ring of connective tissue between the uterus and the vagina), which causes the cervix to stretch. Stretching the cervix sends nervous signals to the mother's brain, causing the release of oxytocin. Oxytocin stimulates further contractions of the uterine muscles,

pushing the baby harder against the cervix, which stretches further, causing still more oxytocin to be released. However, positive feedback cannot continue indefinitely. In the case of childbirth, the positive feedback between oxytocin release and uterine contractions ends when the infant is born. After delivery, the cervix is no longer stretched, so oxytocin release stops.

CHECK YOUR LEARNING

Can you name the three classes of endocrine hormones? Can you describe how hormones affect target cells? Do you understand the processes of negative and positive feedback, and can you provide an example of how each is used in regulating hormone release?

37.3 WHAT ARE THE STRUCTURES AND FUNCTIONS OF THE MAMMALIAN ENDOCRINE SYSTEM?

The mammalian **endocrine system** consists of the endocrine glands and the hormones they produce (**Fig. 37-4** and **Table 37-2**). In the following sections, we will focus on the hypothalamus–pituitary complex, the thyroid gland, the pancreas, the sex organs, and the adrenal glands.

Hormones of the Hypothalamus and Pituitary Gland Regulate Many Functions Throughout the Body

The **hypothalamus** is a part of the brain that contains clusters of specialized nerve cells called **neurosecretory cells,**

TABLE 37-2 Major Mammalian Endocrine Glands and Their Hormones

Endocrine Gland	Hormone	Type of Chemical	Principal Function
Hypothalamus (to anterior pituitary)	Releasing and inhibiting hormones	Peptides	Releasing hormones stimulate the release of hormones from the anterior pituitary; inhibiting hormones inhibit the release of hormones from the anterior pituitary
Anterior pituitary	Follicle-stimulating hormone (FSH)	Peptide	Females: stimulates the growth of follicles in the ovary, secretion of estrogen, and perhaps ovulation
			Males: stimulates sperm development
	Luteinizing hormone (LH)	Peptide	Females: stimulates ovulation, growth of the corpus luteum, and the secretion of estrogen and progesterone
			Males: stimulates the secretion of testosterone
	Thyroid-stimulating hormone (TSH)	Peptide	Stimulates the thyroid to release thyroxine
	Adrenocorticotropic hormone (ACTH)	Peptide	Stimulates the adrenal cortex to release hormones, especially glucocorticoids such as cortisol
	Prolactin (PRL)	Peptide	Stimulates milk synthesis and secretion by the mammary glands
	Growth hormone (GH)	Peptide	Stimulates growth, protein synthesis, and fat metabolism; inhibits sugar metabolism
Hypothalamus (via posterior pituitary)	Antidiuretic hormone (ADH)	Peptide	Promotes reabsorption of water from the kidneys; constricts arterioles
	Oxytocin	Peptide	Females: stimulates contraction of uterine muscles during childbirth, milk ejection, and maternal behaviors
			Males: may facilitate ejaculation of sperm
Thyroid	Thyroxine	Amino acid derivative	Increases the metabolic rate of most body cells; increases body temperature; regulates growth and development
	Calcitonin	Peptide	Inhibits the release of calcium from bones; decreases the blood calcium concentration
Parathyroid	Parathyroid hormone	Peptide	Increases blood calcium by stimulating calcium release from bones, absorption by the intestines, and reabsorption by the kidneys
Pancreas	Insulin	Peptide	Decreases blood glucose by increasing uptake of glucose into cells and conversion of glucose to glycogen, especially in the liver; regulates fat metabolism
	Glucagon	Peptide	Converts glycogen to glucose, raising blood glucose levels
Testes[1]	Testosterone	Steroid	Stimulates the development of genitalia and male secondary sexual characteristics; stimulates sperm development
Ovaries[1]	Estrogen	Steroid	Causes the development of female secondary sexual characteristics and the maturation of eggs; promotes the development of the uterine lining
	Progesterone	Steroid	Stimulates the development of the uterine lining and the formation of the placenta
Adrenal cortex	Glucocorticoids (e.g., cortisol)	Steroid	Increase blood sugar; regulate sugar, lipid, and fat metabolism; have anti-inflammatory effects
	Mineralocorticoids (aldosterone)	Steroid	Increase reabsorption of salt in the kidney
	Testosterone	Steroid	Causes masculinization of body features, growth
Adrenal medulla	Epinephrine (adrenaline) and norepinephrine (noradrenaline)	Amino acid derivatives	Increase levels of sugar and fatty acids in the blood; increase metabolic rate; increase the rate and force of contractions of the heart; constrict some blood vessels

Other Sources of Hormones

Endocrine Gland	Hormone	Type of Chemical	Principal Function
Pineal gland	Melatonin	Amino acid derivative	Regulates seasonal reproductive cycles and sleep–wake cycles; may regulate onset of puberty
Thymus	Thymosin	Peptide	Stimulates maturation of T cells of the immune system
Kidney	Erythropoietin	Peptide	Stimulates red blood cell synthesis in bone marrow
Digestive tract[2]	Secretin, gastrin, ghrelin, cholecystokinin, and others	Peptides	Control secretion of mucus, enzymes, and salts in the digestive tract; regulate peristalsis; regulate appetite
Heart	Atrial natriuretic peptide (ANP)	Peptide	Increases salt and water excretion by the kidneys; lowers blood pressure
Fat cells	Leptin	Peptide	Regulates appetite; stimulates immune function; promotes blood vessel growth; required for the onset of puberty

[1]See Chapters 41 and 42.
[2]See Chapter 34.

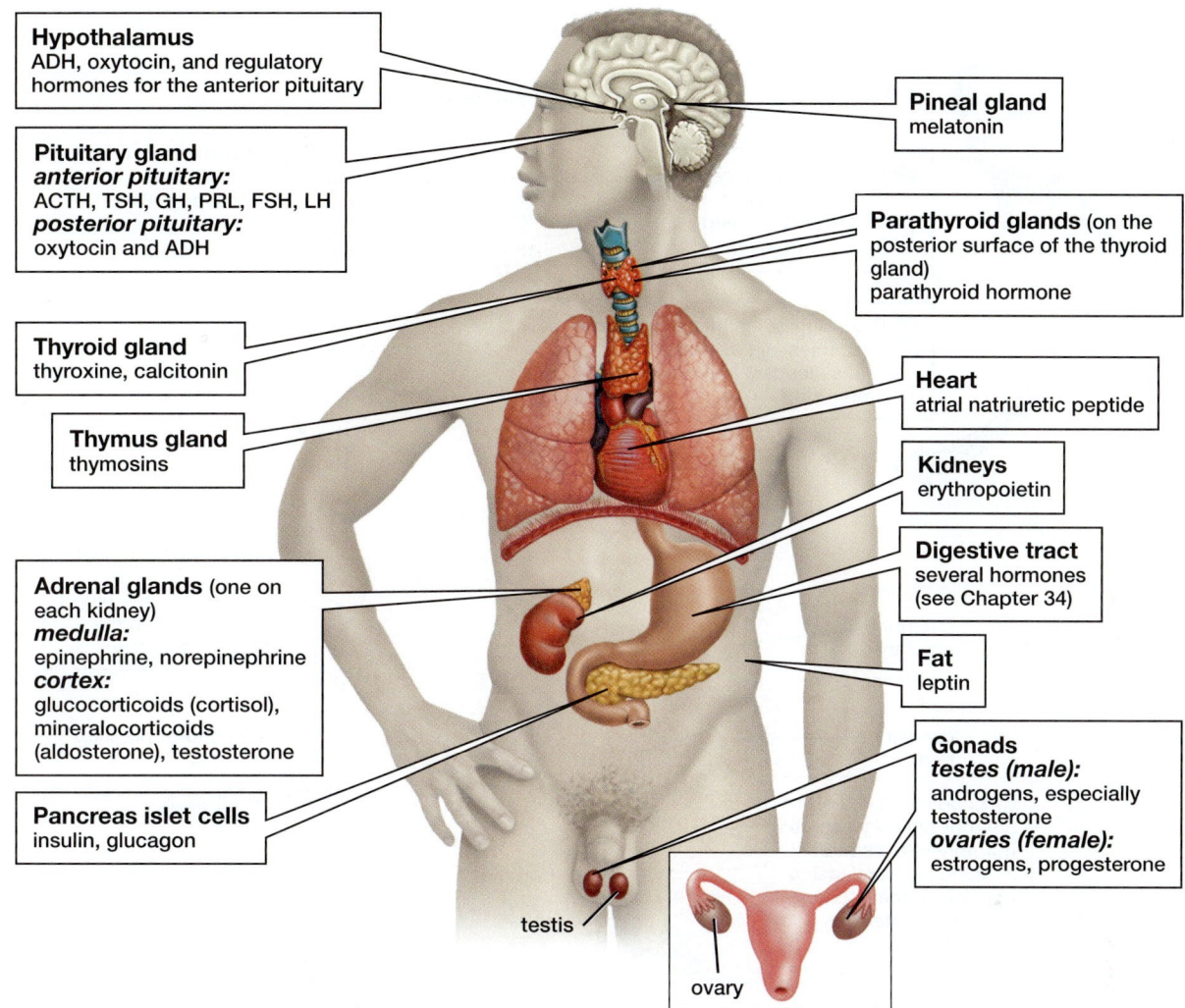

Hypothalamus
ADH, oxytocin, and regulatory hormones for the anterior pituitary

Pituitary gland
anterior pituitary:
ACTH, TSH, GH, PRL, FSH, LH
posterior pituitary:
oxytocin and ADH

Thyroid gland
thyroxine, calcitonin

Thymus gland
thymosins

Adrenal glands (one on each kidney)
medulla:
epinephrine, norepinephrine
cortex:
glucocorticoids (cortisol), mineralocorticoids (aldosterone), testosterone

Pancreas islet cells
insulin, glucagon

Pineal gland
melatonin

Parathyroid glands (on the posterior surface of the thyroid gland)
parathyroid hormone

Heart
atrial natriuretic peptide

Kidneys
erythropoietin

Digestive tract
several hormones
(see Chapter 34)

Fat
leptin

Gonads
testes (male):
androgens, especially testosterone
ovaries (female):
estrogens, progesterone

testis

ovary

▲ **FIGURE 37-4 The major mammalian endocrine glands and their hormones**

which synthesize peptide hormones, store them, and release them when stimulated (**Fig. 37-5,** top). The **pituitary gland** is a pea-sized gland connected to the hypothalamus by a stalk (**Fig. 37-5,** bottom). The pituitary consists of two distinct parts: the **anterior pituitary** and the **posterior pituitary.**

The Anterior Pituitary Produces and Releases Multiple Hormones

Four of the hormones released by the anterior pituitary regulate hormone production in other endocrine glands. **Follicle-stimulating hormone (FSH)** and **luteinizing hormone (LH)** stimulate the production of sperm and testosterone in the testes of males and the production of eggs, estrogen, and progesterone in the ovaries of females. (We will discuss the roles of FSH and LH in Chapter 41.) **Thyroid-stimulating hormone (TSH)** stimulates the thyroid gland to release its hormones, and **adrenocorticotropic hormone (ACTH;** "hormone that stimulates the adrenal cortex") causes the release of the hormone *cortisol* from the adrenal cortex. We will examine the effects of thyroid and adrenal cortical hormones later in this chapter.

The remaining hormones of the anterior pituitary do not act on other endocrine glands. **Prolactin,** in conjunction with other hormones, stimulates the development of milk-producing mammary glands in the breasts during pregnancy. **Growth hormone** acts on nearly all the body's cells by increasing protein synthesis, promoting the use of fats for energy, and regulating carbohydrate metabolism. During childhood, growth hormone stimulates bone growth. Much of the normal variation in human height is due to differences in the secretion of, or responses to, growth hormone. Too little growth hormone—or defective receptors for it—causes some cases of dwarfism; too much can cause gigantism (**Fig. 37-6**).

A major advance in the treatment of pituitary dwarfism occurred in 1981 when molecular biologists successfully inserted the gene for human growth hormone into bacteria, which then churned out large quantities of the hormone. Previously, the main commercial source of growth hormone was human cadavers, from which tiny amounts were extracted at great cost. Thanks to the new, cheaper source, children with underactive pituitary glands, who would previously have been extremely short, can now achieve normal height. Growth hormone supplements are sometimes used by athletes or action-movie actors to increase strength and muscular appearance. The International Olympic Committee and

hypothalamus

1 Neurosecretory cells of the hypothalamus produce releasing and inhibiting hormones

1 Neurosecretory cells of the hypothalamus produce oxytocin and ADH

2 Releasing or inhibiting hormones (green circles) are secreted into capillaries leading into the anterior pituitary

2 Oxytocin and ADH (blue triangles) are secreted into the blood via capillaries in the posterior pituitary

blood flow

anterior pituitary

endocrine cell

capillary bed

3 Endocrine cells of the anterior pituitary secrete hormones (red squares) in response to releasing hormones; the pituitary hormones enter the bloodstream

pituitary

posterior pituitary

capillary bed

blood flow

▲ **FIGURE 37-5 The hypothalamus–pituitary system** The left side of the diagram (green circled numbers) shows the relationship between the hypothalamus and the anterior pituitary, and the right side (blue circled numbers) shows the relationship between the hypothalamus and the posterior pituitary. Releasing hormones are shown as green circles, hormones from the anterior pituitary as red squares, and hormones from the hypothalamus/posterior pituitary as blue triangles.

most professional sports ban growth hormone supplements as performance-enhancing drugs.

Hypothalamic Hormones Control the Release of Hormones in the Anterior Pituitary Neurosecretory cells of the hypothalamus produce six hormones that regulate the release of these hormones from the anterior pituitary (**Fig. 37-5** green step ❶). The hypothalamic regulatory hormones are called **releasing hormones** or **inhibiting hormones,** depending on whether they stimulate or inhibit the release of a particular hormone in the anterior pituitary.

The releasing and inhibiting hormones are secreted into a capillary bed in the stalk connecting the hypothalamus to the pituitary (**Fig. 37-5** green step ❷), and travel a short distance through blood vessels to a second capillary bed that surrounds the endocrine cells of the anterior pituitary. There, the releasing and inhibiting hormones diffuse out of the capillaries and

bind to receptors on the surfaces of the endocrine cells, regulating the release of their hormones (**Fig. 37-5** green step ❸).

The Posterior Pituitary Releases Hormones Synthesized by Cells in the Hypothalamus

The hypothalamus contains two types of neurosecretory cells that grow thin fibers, called axons, into the posterior pituitary (**Fig. 37-5** blue step ❶). The axons end in a capillary bed into which they release hormones that are then carried by the bloodstream to the rest of the body (**Fig. 37-5** blue step ❷). These neurosecretory cells synthesize and release either antidiuretic hormone (ADH) or oxytocin.

Antidiuretic hormone (ADH; literally, "a hormone that reduces urination") helps prevent dehydration by causing the kidneys to absorb water and return it to the bloodstream. However, wastes must still be excreted from the body, so with less water to dilute salts and waste products, the urine

▲ **FIGURE 37-6 When the anterior pituitary malfunctions**
An improperly functioning anterior pituitary may produce too much or too little growth hormone. Too little causes dwarfism, and too much causes gigantism.

QUESTION Why is it easier to treat dwarfism than gigantism?

is highly concentrated and often deep yellow (see Chapter 35). Alcohol inhibits the release of ADH, thereby increasing urination. People who drink strong alcoholic beverages may lose more water in urine than they gain from drinking, and become dehydrated. Dehydration contributes to the headache and generally miserable sensations known as a hangover.

Oxytocin, as we described earlier, causes contractions of the muscles of the uterus during childbirth. It also triggers the "milk letdown reflex" in nursing mothers by causing muscle tissue within the mammary glands of the breasts to contract in response to stimulation by the suckling infant. This contraction ejects milk from the saclike milk glands into the nipples (**Fig. 37-7**).

Oxytocin also acts directly in the brain, causing behavioral effects. In rats, for example, injecting oxytocin into the brain causes virgin females to exhibit maternal behaviors, such as building a nest and retrieving pups that have strayed. In humans (both men and women), oxytocin may play a role in emotions, including trust and both romantic and maternal love (see the case study "How Do I Love Thee?" in Chapter 38).

The Thyroid and Parathyroid Glands Influence Metabolism and Calcium Levels

The **thyroid gland** lies at the front of the neck, nestled just below the larynx (**Fig. 37-8**). The thyroid produces two hormones: thyroxine and calcitonin. The **parathyroid gland**

consists of two pairs of small disks of endocrine cells, one pair on each side of the back of the thyroid gland. These cells release parathyroid hormone.

Thyroxine Influences Energy Metabolism

Thyroxine, or thyroid hormone, is an iodine-containing amino acid derivative. By stimulating the synthesis of enzymes that break down glucose and provide energy, and perhaps by directly stimulating the mitochondria, thyroxine increases metabolic rate in most of the cells of the body. Low thyroxine makes people feel mentally and physically sluggish. They may lose appetite but still gain weight, and they become less tolerant of cold (because the body generates less heat when its metabolic rate is low). Excess thyroxine leads to restlessness and irritability, increased appetite, and intolerance to heat.

In juvenile vertebrates, including humans, thyroxine helps regulate growth by stimulating both metabolic rate and nervous system development. In people, undersecretion of thyroid hormone early in life can cause cretinism, a condition characterized by reduced mental and physical development. Fortunately, early diagnosis and thyroxine supplementation can reverse this condition.

An iodine-deficient diet can reduce the production of thyroxine and trigger a feedback that attempts to restore normal hormone levels by increasing the number of thyroxine-producing cells. The thyroid gland becomes enlarged and may form a bulge in the neck, a condition called **goiter** (**Fig. 37-9**). Even minor iodine deficiency, far smaller than the deficit required to produce goiter, reduces thyroxine production. Hundreds of millions of people, almost entirely in less-developed countries, have iodine-deficient diets. Iodine deficiency in pregnant women and young children is the leading preventable cause of mental retardation. Iodized salt is a simple, cheap solution to iodine deficiency, at a cost of less than $2 per ton of salt.

Thyroxine Release Is Controlled by the Hypothalamus and Anterior Pituitary

Levels of thyroxine in the bloodstream are fine-tuned by negative feedback (**Fig. 37-10**). Thyroid stimulating hormone–releasing hormone (TSH-releasing hormone) produced by neurosecretory cells in the hypothalamus (**Fig. 37-10 ❶**) travels to the anterior pituitary and causes the release of TSH (**Fig. 37-10 ❷**). TSH travels in the bloodstream to the thyroid and stimulates the release of thyroxine (**Fig. 37-10 ❸**). Secretion of TSH-releasing hormone and TSH are regulated by negative feedback. Adequate levels of thyroxine circulating in the bloodstream inhibit the secretion of both TSH-releasing hormone from the hypothalamus and TSH from the anterior pituitary, thus inhibiting further release of thyroxine from the thyroid gland (**Fig. 37-10 ❹**).

Thyroxine Has Varied Effects in Different Vertebrates

Thyroxine governs many functions in non-human vertebrates. For example, in amphibians and many fish, thyroxine stimulates metamorphosis from a larval to an adult body form.

► **FIGURE 37-7 Hormones and breast-feeding** Feedback between an infant and its mother regulates the control of milk letdown by oxytocin during breast-feeding. The cycle begins with the infant's suckling and continues until the infant is full and stops suckling. When the nipple is no longer stimulated, oxytocin release stops, the muscles relax, and milk flow ceases.

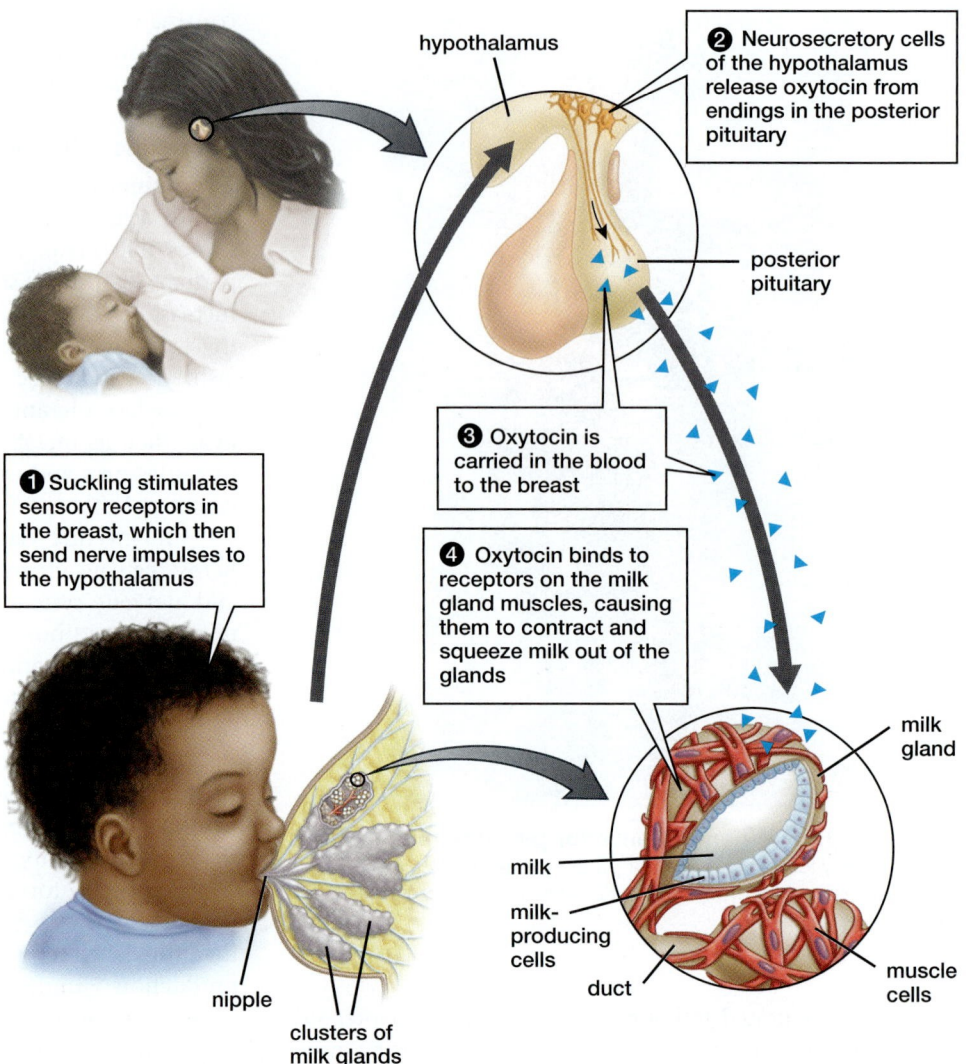

hypothalamus

❷ Neurosecretory cells of the hypothalamus release oxytocin from endings in the posterior pituitary

posterior pituitary

❸ Oxytocin is carried in the blood to the breast

❶ Suckling stimulates sensory receptors in the breast, which then send nerve impulses to the hypothalamus

❹ Oxytocin binds to receptors on the milk gland muscles, causing them to contract and squeeze milk out of the glands

milk gland

milk

milk-producing cells

duct

muscle cells

nipple

clusters of milk glands

A frog hatches out of its egg as an aquatic tadpole, which looks a bit like a fat-headed fish, with gills, a large finned tail, and no legs. A few weeks later, increasing thyroxine levels stimulate the tadpole to sprout legs and resorb its tail. Another few weeks, and the transformation into a small frog is complete. In lampreys and salmon, thyroxine helps trigger the bodily transformations that allow the young fish, which are born in fresh water, to thrive in seawater, where they grow up and mature. Thyroxine also regulates the seasonal molting of most vertebrates. From snakes to birds to your family dog, surges of thyroxine stimulate the shedding of skin, feathers, or hair.

Parathyroid Hormone and Calcitonin Regulate Calcium Metabolism

The proper concentration of calcium is essential to nerve and muscle function. **Parathyroid hormone** from the parathyroid gland and **calcitonin** from the thyroid work together to maintain nearly constant calcium levels in the blood and body fluids.

The skeleton serves as a "bank" into which calcium can be deposited or withdrawn as necessary. If blood calcium levels drop, parathyroid hormone causes the bones to

larynx

The thyroid gland produces thyroxine and calcitonin

thyroid gland

esophagus

parathyroid glands

The parathyroid glands produce parathyroid hormone

trachea

▲ **FIGURE 37-8 The thyroid and parathyroid glands** The thyroid gland wraps around the front of the larynx in the neck. The tiny parathyroid glands sit in the back of the thyroid.

▲ FIGURE 37-9 **Goiter** An iodine-deficient diet often causes enlargement of the thyroid gland. Seldom seen in developed countries, goiter is all too common in less-developed countries where people lack iodized salt in their diets.

release calcium. It also causes the kidneys to reabsorb more calcium during urine production, and to return the calcium to the blood. The increased blood calcium then inhibits further release of parathyroid hormone in a negative feedback loop. If blood calcium gets too high, the thyroid releases calcitonin, which inhibits the release of calcium from bone. In most vertebrates, calcitonin is important in regulating calcium concentrations in the blood, and may even promote bone growth. In humans, the actions of calcitonin appear to be minor compared to those of parathyroid hormone.

The Pancreas Has Both Digestive and Endocrine Functions

In its digestive role, the **pancreas** produces bicarbonate and several enzymes that are released into the small intestine, promoting the breakdown of food (see Chapter 34). The endocrine portion of the pancreas consists of clusters of **islet cells.** Each islet cell produces one of two peptide hormones: **insulin** or **glucagon.**

Insulin and Glucagon Control Glucose Levels in the Blood

Insulin and glucagon have opposing effects on carbohydrate and fat metabolism. Insulin reduces blood glucose, whereas glucagon increases it (**Fig. 37-11**). Together, the two hormones help keep blood glucose nearly constant. When blood glucose rises (for example, after you have eaten; **Fig. 37-11 ❶**), the pancreas releases insulin (**Fig. 37-11 ❷**). Insulin causes body cells to take up glucose (**Fig. 37-11 ❸**) and either metabolize it

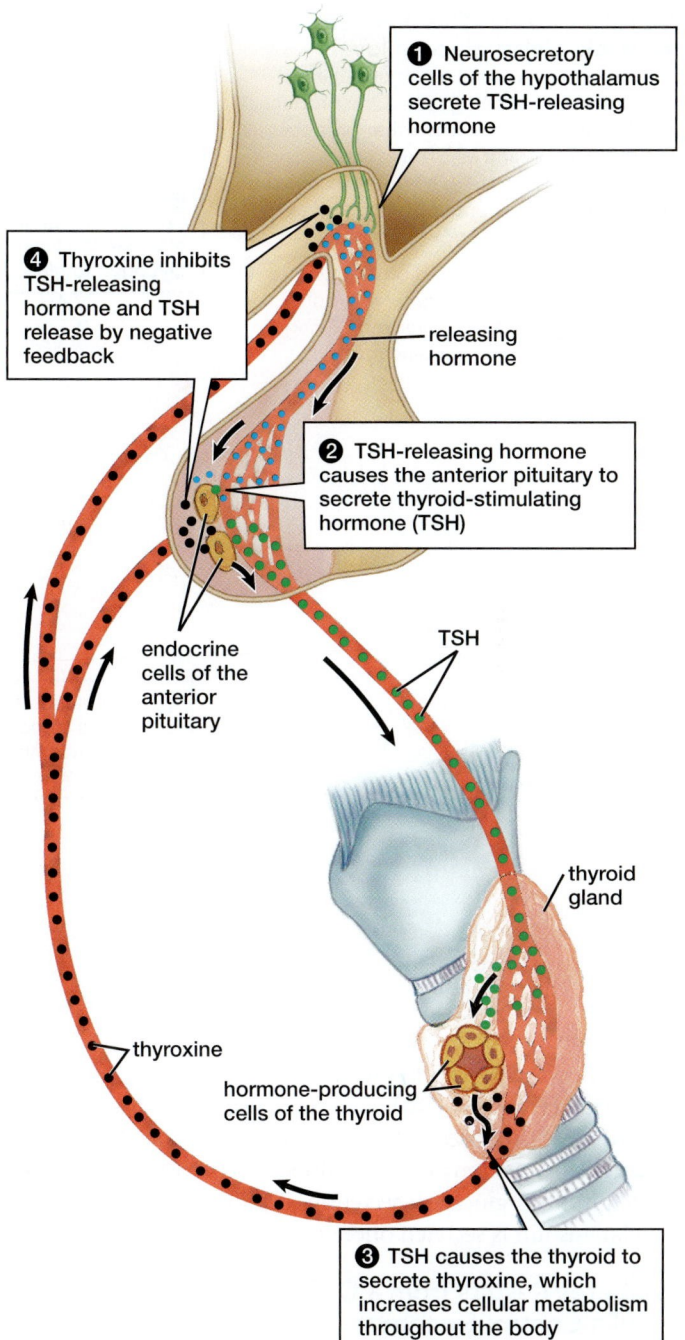

❶ Neurosecretory cells of the hypothalamus secrete TSH-releasing hormone

❹ Thyroxine inhibits TSH-releasing hormone and TSH release by negative feedback

releasing hormone

❷ TSH-releasing hormone causes the anterior pituitary to secrete thyroid-stimulating hormone (TSH)

endocrine cells of the anterior pituitary

TSH

thyroid gland

thyroxine

hormone-producing cells of the thyroid

❸ TSH causes the thyroid to secrete thyroxine, which increases cellular metabolism throughout the body

▲ FIGURE 37-10 **Negative feedback in thyroid gland function** The concentration of thyroxine in the bloodstream regulates the secretion of TSH-releasing hormone and TSH by negative feedback.

QUESTION A common test of thyroid gland function is to measure the amount of thyroid-stimulating hormone circulating in the blood. What would you hypothesize is wrong in a person who has an abnormally high level of TSH?

for energy or convert it to fat or glycogen (a polymer of glucose that is stored in the liver and skeletal muscles). When blood glucose levels drop (for example, after you've skipped breakfast or run a 10-kilometer race; **Fig. 37-11 ❹** and **❺**), insulin secretion is inhibited and glucagon secretion is stimulated (**Fig. 37-11 ❻**). Glucagon activates an enzyme in the liver that

▶ **FIGURE 37-11 The pancreas controls blood glucose levels**
Pancreatic islets contain two populations of hormone-producing cells: One produces insulin (green arrows); the other produces glucagon (blue arrows). These two hormones cooperate in a two-part negative feedback loop to control blood glucose concentrations (red arrows).

QUESTION How would blood glucose be affected in a person who was born with a mutation that prevented glucagon receptors from binding glucagon?

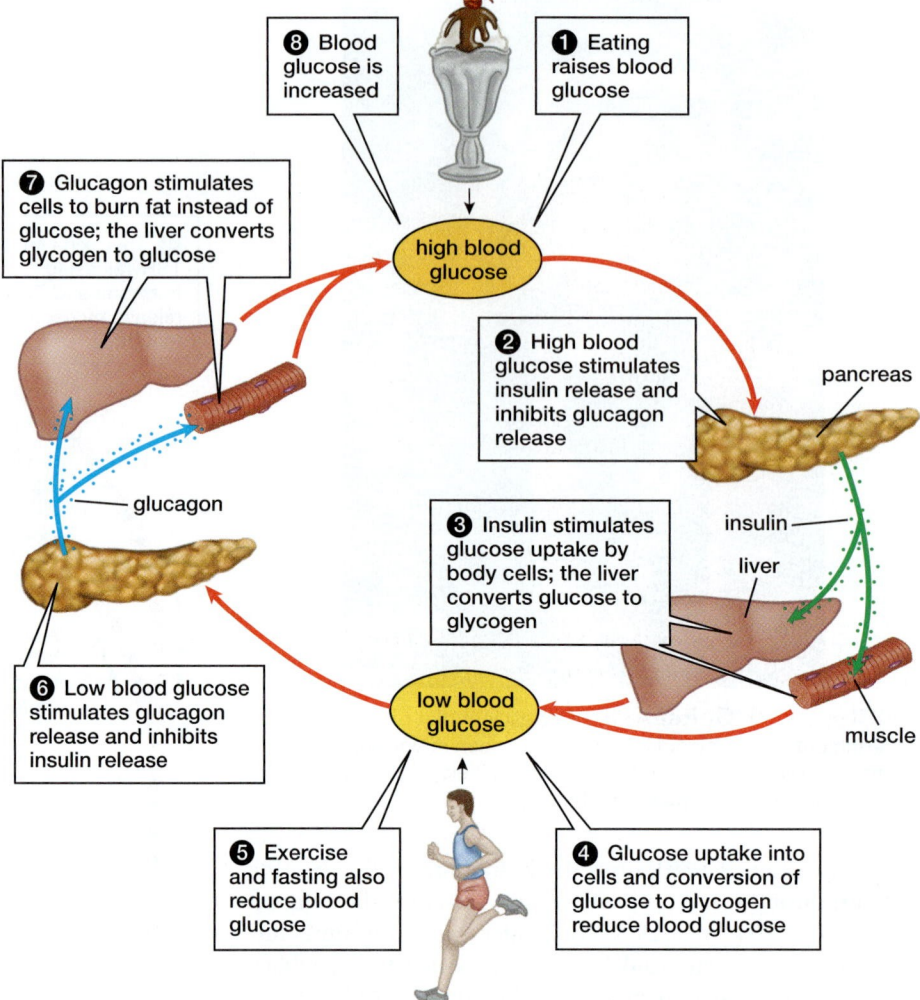

❽ Blood glucose is increased

❶ Eating raises blood glucose

❼ Glucagon stimulates cells to burn fat instead of glucose; the liver converts glycogen to glucose

high blood glucose

❷ High blood glucose stimulates insulin release and inhibits glucagon release

pancreas

glucagon

❸ Insulin stimulates glucose uptake by body cells; the liver converts glucose to glycogen

insulin
liver

❻ Low blood glucose stimulates glucagon release and inhibits insulin release

low blood glucose

muscle

❺ Exercise and fasting also reduce blood glucose

❹ Glucose uptake into cells and conversion of glucose to glycogen reduce blood glucose

breaks down glycogen to glucose, which is then released into the blood (Fig. 37-11 ❼). Glucagon also promotes fat breakdown, which releases fatty acids that can be metabolized for energy. These actions increase blood glucose (Fig. 37-11 ❽), which inhibits glucagon secretion. If blood glucose levels rise too far, insulin is secreted once again.

Diabetes Results from a Defective Insulin Control System

Lack of insulin production or the failure of target cells to respond to insulin results in **diabetes mellitus.** In either case, blood glucose levels are high and fluctuate wildly with food intake. Because many body cells can't take up glucose unless they are stimulated by insulin, the cells of people with diabetes rely heavily on fats as an energy source instead. Providing lots of fats to the cells requires extremely high levels of lipids—including cholesterol—in the blood. Some of this fat and cholesterol settles in the blood vessels, forming plaques that partially clog them. As a result, about 65% of people with diabetes die of cardiovascular disease or strokes. Fat deposits in small blood vessels in the kidneys and the retina of the eye may cause kidney failure and blindness.

There are two forms of diabetes, type 1 and type 2. Type 1 diabetes usually begins in childhood, and is caused by a malfunctioning immune system that destroys the insulin-producing islet cells. Insulin replacement therapy profoundly improves the health of people with type 1 diabetes, but usually requires frequent blood testing and insulin injections, and can't fully mimic natural control of energy metabolism. Insulin pumps—about the size of a cell phone, usually worn attached to a belt—can deliver insulin virtually continuously, eliminating the need for insulin injections. Some insulin pumps measure blood glucose levels in real time and deliver appropriate insulin doses accordingly.

Type 2 diabetes usually begins in adulthood and is most commonly associated with obesity and lack of exercise. In the early stages, people with type 2 diabetes produce adequate amounts of insulin. However, their bodies are *insulin resistant*—their responses to insulin aren't strong enough to control blood glucose. Fat tissue releases specific types of harmful fatty acids that make body cells less responsive to insulin. The more fat in the body, the higher the concentration of these fatty acids in the blood. Therefore, weight loss and exercise are the usual first-line treatments for type 2 diabetes. If those methods fail, a drug called metformin is usually prescribed. Metformin reduces glucose production by the liver and increases insulin responsiveness by many cells, resulting in increased glucose uptake. These actions help to reduce blood glucose levels.

The Sex Organs Produce Both Gametes and Sex Hormones

The **testes** in males and the **ovaries** in females are important endocrine organs (see Fig. 37-4). The testes secrete several steroid hormones, collectively called **androgens;** the most important of these is **testosterone.** The ovaries secrete two types of steroid hormones: **estrogen** and **progesterone.** Sex hormones influence development in both genders, and affect brain function and behavior throughout life. (The roles of sex hormones in sperm and egg production, the menstrual cycle, and pregnancy are discussed in Chapter 41.)

Sex Hormone Levels Increase During Puberty

Sex hormones play a key role in puberty, the phase of life during which the reproductive systems of both sexes become mature and functional. Puberty begins when, for reasons not fully understood, the hypothalamus starts to secrete increasing amounts of releasing hormones, which in turn stimulate the anterior pituitary to secrete more luteinizing hormone (LH) and follicle-stimulating hormone (FSH). LH and FSH stimulate target cells in the testes or ovaries to produce higher levels of sex hormones.

The resulting increase in sex hormones affects target cells throughout the body. Both sexes develop pubic and underarm hair. Testosterone, secreted by the testes, stimulates the development of male secondary sexual characteristics, including body and facial hair, a deep voice, and muscle growth. Testosterone also promotes sperm cell production. In females, estrogen secreted by the ovaries stimulates breast development and the maturation of the female reproductive system, including egg production. Progesterone prepares the female reproductive tract to receive and nourish the fertilized egg.

Studies in a wide variety of animals, as well as a few studies in humans, have revealed that environmental pollutants from agricultural and industrial activities can disrupt hormone signaling. These "endocrine disruptors" frequently mimic or block the actions of sex hormones—sometimes with devastating effects—as described in "Earth Watch: Endocrine Deception."

The Adrenal Glands Secrete Hormones That Regulate Metabolism and Responses to Stress

The **adrenal glands** (Latin for "on the kidney") consist of two very different parts: the adrenal cortex and the adrenal medulla (**Fig. 37-12**).

The Adrenal Cortex Produces Steroid Hormones

The outer layer of each adrenal gland is the **adrenal cortex** ("cortex" is Latin for "bark"). The cortex secretes three types of steroid hormones: glucocorticoids, mineralocorticoids, and small amounts of testosterone. As their names imply,

The adrenal medulla secretes epinephrine and norepinephrine

The adrenal cortex secretes glucocorticoids, mineralocorticoids, and testosterone

kidney

▲ FIGURE 37-12 **The adrenal glands** Atop each kidney sits an adrenal gland, composed of an outer cortex and an inner medulla.

glucocorticoids help control glucose metabolism, while **mineralocorticoids** regulate salt metabolism.

Glucocorticoid release is stimulated by adrenocorticotropic hormone (ACTH) from the anterior pituitary, which in turn is stimulated by a releasing hormone produced by the hypothalamus. Glucocorticoids are released in response to stimuli such as stress, trauma, or temperature extremes. **Cortisol** is by far the most abundant glucocorticoid. Cortisol increases blood glucose levels by stimulating glucose production, inhibiting the uptake of glucose by muscle cells, and promoting the use of fats for energy.

Mineralocorticoid hormones regulate the mineral (salt) content of the blood. The most important mineralocorticoid is **aldosterone,** which helps to control sodium concentrations. Sodium ions are the most abundant positive ions in blood and interstitial fluid. The sodium ion concentration gradient across plasma membranes (high in the interstitial fluid, low in the cytoplasm) is crucial to many cellular events, including the production of electrical signals by nerve cells. If blood sodium falls, the adrenal cortex releases aldosterone, which causes the kidneys and sweat glands to

❓ HAVE **YOU** EVER WONDERED …

Why You Often Get Sick When You're Stressed?

In addition to its effects on glucose metabolism, cortisol also inhibits the immune response. If you sometimes get sick during final exam week, stress-induced cortisol release may be responsible. Why should the body ever suppress its immune response? Although finals are not usually life threatening, many other stresses, such as finding enough food or encountering predators, can be. Because activation of the immune system requires a lot of energy and makes people and other mammals feel sluggish, suppressing the immune response in favor of dealing with immediate stresses may save an animal's life.

Earth|Watch

Endocrine Deception

Some synthetic organic compounds that enter the environment can mimic or block the actions of certain hormones, most commonly estrogen, testosterone, or thyroxine. These **endocrine disruptors** include some pesticides, herbicides, ingredients in plastics, flame retardants, detergents, sunscreens, and polychlorinated biphenyls (used in sealants and paints and as insulating fluids in power transformers). Probably the most potent endocrine disruptor in the environment is ethinylestradiol, a synthetic estrogen commonly found in birth control pills.

In the most common form of endocrine disruption, a synthetic chemical enters cells and binds to estrogen receptors. The results, including overactivation of estrogen receptors, inappropriate timing of receptor activation, or activation in males, may alter development. Endocrine disruptors exert harmful effects in a wide variety of vertebrates, including feminization in males, masculinization in females, reproductive cancers, malformed sex organs, altered blood hormone levels, and reduced fertility. Feminization of males—such as abnormal testes, and sometimes even egg production—is the most common effect of estrogenic endocrine disruptors.

In one of the best-known cases, agricultural runoff and a pesticide spill near Lake Apopka in Florida polluted the lake's water with large quantities of several endocrine disruptors, including the pesticide DDT and its breakdown products. Wildlife biologists noted an alarming decline in the alligator population of the lake. Many eggs never hatched. Males had high estrogen, low testosterone, small penises, and abnormal testes. Females typically had exceptionally high estrogen levels and abnormal ovaries.

Although alarming, the Lake Apopka results were the result of massive exposure to endocrine disruptors. Do smaller doses, likely to be found in fairly clean water, also have harmful effects? Indeed they may. In 2003 and 2004, researchers from the University of Colorado sampled a common minnow in Boulder Creek, which runs through Boulder, Colorado (**Fig. E37-2**). They found that upstream of the city's sewage outfall, about half the fish were males and half were females, but downstream of the outfall, more than 80% of the fish were females. In lab studies, male fish exposed to wastewater from the treatment plant were rapidly feminized. In 2007, the city of Boulder upgraded its sewage treatment plant, converting to a system that uses microbes to more thoroughly digest organic molecules, including many potential endocrine disruptors. It worked—a follow-up study in 2009 and 2010 found that the wastewater no longer caused feminization of fish.

Are endocrine disruptors harmful to people? Some, such as polychlorinated biphenyls, certainly are. Others probably are harmful at high enough concentrations. The debates among

▲ **FIGURE E37-2 Sampling fish in Boulder Creek** Although Boulder Creek does not look polluted, it formerly contained chemicals that disrupted the reproductive systems of fish.

toxicologists, reproductive biologists, industry officials, and government regulators focus on several questions: What are the minimum concentrations that affect people? Do those concentrations actually occur in people? Might mixtures of several endocrine disruptors, all at very low concentrations, add up to harmful effects? How could we definitively prove either safety or harm, given that human experiments cannot be performed?

Some known endocrine disruptors, such as DDT and polychlorinated biphenyls, have been banned in many countries. For suspected endocrine disruptors, the situation is more unsettled. In 2008, Canada banned the use of a chemical called bisphenol A (BPA) in plastic baby bottles; the European Union followed suit in 2011. In 2010, the U.S. Food and Drug Administration, although not banning BPA, recommended that parents should limit their use of baby products containing BPA. Meanwhile, companies such as Disney First Years, Gerber, and Evenflow stopped using BPA in baby bottles, and Walmart and Toys 'R' Us stopped selling baby bottles and sippy cups containing BPA.

People disagree about whether these actions by governments, bottle manufacturers, and retailers were based on sound scientific evidence, public opinion, or the "precautionary principle" that when harm is suspected, the burden of proof falls on those who claim that a product is safe. These debates play out in the media, so you can follow the arguments yourself, in newspapers, online, or in magazines such as *Scientific American*, *Science News*, or *New Scientist*.

---

retain sodium and return it to the blood. Meanwhile, almost all foods contain sodium, so eating helps to replenish the body's sodium stores. When the combination of dietary sodium intake and aldosterone-induced sodium conservation returns blood sodium concentrations back to normal levels, aldosterone secretion is shut off—another example of negative feedback.

In both women and men, the adrenal cortex also produces the male sex hormone testosterone, although in much smaller amounts than are produced by the testes. Tumors of the adrenal cortex can lead to excessive testosterone release, causing masculinization of women. Many of the "bearded ladies" who once appeared in circus sideshows probably had this condition.

The Adrenal Medulla Produces Amino Acid-Derived Hormones

The **adrenal medulla** is located in the center of each adrenal gland. It produces two hormones in response to stress or exercise—**epinephrine** and a small quantity of **norepinephrine** (also called adrenaline and noradrenaline, respectively). These hormones prepare the body for emergency action. They increase heart and respiratory rates, increase blood pressure, cause blood glucose levels to rise, and direct blood flow away from the digestive tract and toward the brain and muscles. They also cause the air passages to the lungs to expand, allowing larger volumes of air to enter and leave the lungs. For this reason, epinephrine is often administered to anyone whose airways become constricted, such as during a severe allergic reaction or an asthma attack.

Hormones Are Also Produced by the Pineal Gland, Thymus, Kidneys, Digestive Tract, Heart, and Fat Cells

The **pineal gland** is located between the two hemispheres of the brain (see Fig. 37-4). The pineal produces the hormone **melatonin,** an amino acid derivative. Melatonin is secreted in a daily cycle, very little during the day and much more during the night. In mammals, the cycle is driven by light perceived by the eyes. In some vertebrates, such as the frog, the pineal itself contains photoreceptive cells. The skull above it is thin, so the pineal can detect sunlight and thus day length. By responding to day lengths characteristic of different seasons, the pineal helps to regulate the reproductive cycles of many animals.

Despite years of research, the function of the pineal gland and melatonin in humans is still not well understood. Melatonin secretion may influence sleep–wake cycles. Darkness increases melatonin production and bright light inhibits it, and there is some evidence that secretion of melatonin at night promotes sleep. Consequently, melatonin is sometimes used as a sleeping aid or to overcome jet lag, although most research suggests that the effects are fairly small.

Children produce much more melatonin than adults. Melatonin secretion drops sharply after puberty, leading some to hypothesize that melatonin may suppress the onset of puberty, although no one has yet demonstrated a convincing mechanism. In developed countries, the average age of puberty has dropped markedly since the 1800s; in girls, the average onset of puberty has dropped from about 16 or 17 in the mid-1800s to 10 or 11 today. Some researchers think that artificial light—including indoor lighting and TV—has effectively made our days longer, thereby reducing melatonin release. Other factors, including better nutrition and changes in social interactions, might also play a role.

The **thymus** is located in the chest cavity behind the breastbone (see Fig. 37-4). The thymus produces the hormone **thymosin,** which stimulates the development of specialized white blood cells (T cells) that play crucial roles in the immune response (see Chapter 36). The thymus is large in children but, under the influence of sex hormones,

gradually decreases in size after puberty. As a result, the elderly produce fewer new T cells than younger people do and, hence, are more susceptible to new diseases.

The kidneys release **erythropoietin** when the oxygen content of the blood is low. Erythropoietin stimulates the bone marrow to increase red blood cell production (see Chapter 32). The kidneys also produce an enzyme called *renin* in response to low blood pressure, for example, after profuse bleeding from a wound. Renin catalyzes the production of the hormone **angiotensin** from proteins in the blood. Angiotensin raises blood pressure by constricting arterioles. It also stimulates the release of aldosterone by the adrenal cortex, causing the kidneys to return sodium to the blood. The resulting high salt concentration attracts and retains water, increasing blood volume and pressure.

The stomach and small intestine produce a bewildering variety of peptide hormones, including gastrin, ghrelin, secretin, cholecystokinin, and several others. These hormones regulate several aspects of nutrition, such as the secretion of digestive hormones, movement of food within the stomach and small intestine, and the sensations of hunger and fullness (see Chapter 34).

Even the heart releases a hormone. If the blood volume is too high—for example, if you drink too much water—the atria become overfilled, which stretches their walls and stimulates the release of **atrial natriuretic peptide (ANP).** ANP inhibits the release of ADH and aldosterone and increases the excretion of sodium. By reducing reabsorption of water and salt by the kidneys, ANP helps to lower blood volume.

Finally, who would have thought that fat could be an endocrine organ? Fat cells release the peptide hormone **leptin.** Mice genetically engineered to lack the gene for leptin become obese (**Fig. 37-13**), and leptin injections cause them to lose weight. It is thought that the amount of leptin released from fat informs the body how much fat it has stored and therefore how much to eat. Therefore, if mice already have lots of stored fat, high leptin levels cause them to eat less. Unfortunately, leptin doesn't work well as a human weight-loss aid. Many obese people have high levels of leptin but

▲ **FIGURE 37-13 Leptin helps regulate body fat** The mouse on the left has been genetically engineered to lack the gene for the hormone leptin.

seem to be relatively insensitive to it. However, researchers have discovered other functions of leptin. It appears to stimulate the growth of new capillaries, speed wound healing, and stimulate the immune system. Leptin is also required for the onset of puberty and the development of secondary sexual characteristics.

CASE STUDY revisited

Life-Changing Hormones

Many hormones have multiple actions. In butterflies, juvenile hormone and molting hormone control whether a caterpillar molts into another, larger caterpillar, a pupa, or a winged adult. In addition, these same hormones help determine the appearance of the caterpillars, by stimulating or inhibiting the expression of genes that control the colors and patterns in the caterpillars' cuticles. This is dramatically illustrated by changes in the colors of Asian swallowtail caterpillars.

When an Asian swallowtail egg hatches, the tiny caterpillar has a very strange color pattern—it looks almost exactly like a bird dropping (**Fig. 37-14**). The caterpillars that emerge after the first three molts also look like bird droppings. This color pattern is caused by high concentrations of juvenile hormone. However, juvenile hormone concentrations decline over time, and something remarkable happens at the fourth molt. By this time, juvenile hormone levels have dropped below a critical concentration, and the fifth-stage caterpillar emerges bright green, with brown and white bands and two prominent "eyespots" on one end.

A likely hypothesis is that these color changes evolved because they help the caterpillars avoid being eaten by birds. The bird-dropping pattern is very conspicuous, but not very appetizing. The caterpillars usually remain motionless during the day—as a bird dropping should—and feed at night. However, as the caterpillars become larger with each successive molt, they may become larger than most bird droppings are. The final, green caterpillar stage blends in with the leaves it eats. If the camouflage fails and a bird comes too close, the caterpillar rears its head, showing the eyespots. The caterpillar now looks like a snake, which may scare away the bird.

CHECK YOUR LEARNING

Can you name the major mammalian endocrine glands, their locations in the body, the major hormones they produce, and give examples of at least one effect of each hormone? Do you understand how negative feedback regulates the secretion and actions of thyroxine, oxytocin, and insulin?

▲ **FIGURE 37-14 Asian swallowtail caterpillars** The first four caterpillar stages mimic bird droppings (left); the final caterpillar stage is camouflaged by its green coloration (right).

Consider This

Some insects are serious pests, eating crops or carrying diseases. Most insecticides target the insect nervous system. Unfortunately, these insecticides can also affect the nervous systems of other animals if the insecticide isn't applied carefully, in the right places and at the right concentrations. Synthetic chemicals that mimic or antagonize the effects of insect hormones may be attractive alternatives for insect control. How do you think that chemicals mimicking or antagonizing juvenile hormone or molting hormone might act as insecticides? What are the potential advantages and disadvantages?

CHAPTER REVIEW

Summary of Key Concepts

37.1 How Do Animal Cells Communicate?

In multicellular organisms, communication among cells occurs through gap junctions directly linking cells, by diffusion of chemicals to nearby cells (neurotransmitters secreted by nerve cells or local hormones such as prostaglandins), or by transport of chemicals within the bloodstream (endocrine hormones). A chemical messenger acts only on target cells bearing receptors that can bind the chemical. Vertebrate endocrine hormones are produced by glands embedded in capillary beds. The hormones are secreted into the interstitial fluid, diffuse into the capillaries, and are transported in the bloodstream to other parts of the body.

37.2 How Do Endocrine Hormones Produce Their Effects?

Vertebrate endocrine hormones fall into one of three classes: peptides, amino acid derivatives, and steroids. Most hormones act on their target cells in one of two ways: (1) Peptide hormones and amino acid-derived hormones bind to receptors on the surface of a target cell and activate intracellular second messengers, such as cyclic AMP, which then alter the cell's metabolism or change the rate of transcription of specific genes. (2) Steroid hormones usually diffuse through the plasma membrane of a target cell and bind to receptors inside the cell. The hormone–receptor complex changes the transcription of specific genes. Thyroid hormones are transported across the plasma membrane into a cell, where they bind to receptors and change gene transcription.

Hormone action is usually regulated by negative feedback, a process in which a hormone causes changes that inhibit further secretion of that hormone. In rare instances, such as childbirth, hormone release may be temporarily controlled by positive feedback.

37.3 What Are the Structures and Functions of the Mammalian Endocrine System?

The major endocrine glands of the human body are the hypothalamus–pituitary complex, the thyroid and parathyroid glands, the pancreas, the sex organs, and the adrenal glands. The hormones released by these glands and their actions are summarized in Table 37-2. Other structures that produce hormones include the pineal gland, thymus, kidneys, digestive tract, heart, and fat cells.

Key Terms

adrenal cortex   725
adrenal gland   725
adrenal medulla   727
adrenocorticotropic hormone (ACTH)   719
aldosterone   725
amino acid-derived hormone   714
androgen   725
angiotensin   727
anterior pituitary   719
antidiuretic hormone (ADH)   720
atrial natriuretic peptide (ANP)   727
calcitonin   722
cortisol   725
cyclic adenosine monophosphate (cyclic AMP)   714
diabetes mellitus   724
endocrine communication   713
endocrine disrupter   726
endocrine gland   713
endocrine hormone   713
endocrine system   717
epinephrine   727
erythropoietin   727
estrogen   725
follicle-stimulating hormone (FSH)   719
glucagon   723
glucocorticoid   725
goiter   721
growth hormone   719
hormone   713
hypothalamus   717
inhibiting hormone   720
insulin   723

islet cell   723
leptin   727
local hormone   713
luteinizing hormone (LH)   719
melatonin   727
mineralocorticoid   725
negative feedback   717
neurosecretory cell   717
neurotransmitter   713
norepinephrine   727
ovary   725
oxytocin   721
pancreas   723
paracrine communication   713
parathyroid gland   721
parathyroid hormone   722
peptide hormone   714
pineal gland   727
pituitary gland   719
positive feedback   717
posterior pituitary   719
progesterone   725
prolactin   719
prostaglandin   713
receptor   713
releasing hormone   720
second messenger   714
steroid hormone   714
synaptic communication   713
target cell   713
testis (plural, testes)   725
testosterone   725
thymosin   727
thymus   727
thyroid gland   721
thyroid-stimulating hormone (TSH)   719
thyroxine   721

Learning Outcomes

In this chapter, you have learned to . . .

LO1   List and describe the four major mechanisms of cell-to-cell communication in animals.

LO2   Describe the pathway and mechanisms of action of an endocrine hormone, beginning with its secretion from an endocrine cell and ending with its effects on a target cell.

LO3   Describe the physiological effects of peptide hormones, amino acid-derived hormones, and

steroid hormones, including the responses in target cells.

LO4   Explain how hormone release is controlled.

LO5   Describe the mammalian endocrine system, including the names and locations of the principal endocrine glands, the hormones that the endocrine glands release, and the principal responses that the hormones produce in the body.

Thinking Through the Concepts

Fill-in-the-Blank

1. Endocrine hormones are molecules released by cells that are parts of the _____. These cells are embedded in capillary beds, so the hormones enter the bloodstream and move throughout the body. Only specific cells of the body, called _____, can respond to any given hormone, because only these cells bear proteins, called _____, that can bind the hormone. **LO1 LO2**

2. Most endocrine hormones fall into three chemical classes: _____, _____, and _____. _____ and _____ are mostly water soluble and bind to receptors on the surfaces of cells. These typically stimulate the synthesis of intracellular molecules called _____, which activate enzymes and change the cell's metabolism. _____ are lipid soluble and bind to receptors in the cytoplasm or nucleus. The hormone–receptor complex typically binds to DNA and causes _____. **LO3**

3. A part of the brain called the _____ controls the activity of the pituitary gland. Specialized nerve cells in this brain area, called _____, release the hormones _____ or _____ from the endings of their axons in the posterior pituitary. **LO5**

4. The major hormones produced by the anterior pituitary gland are (in any order): _____, _____, _____, _____, _____, and _____. **LO5**

5. The pancreas releases the hormone _____ when blood glucose levels become too high; it causes many cells of the body to take up glucose. When the pancreas produces too little of this hormone, or body cells cannot respond to it, a disorder called _____ results. _____ is released when blood glucose levels become too low; it causes the liver to break down the starch-like storage molecule, _____, and release glucose into the blood. **LO5**

6. The male sex organs, called the _____, release the sex hormone _____. The female sex organs,

called the _____, release two hormones, _____ and _____. **LO5**

7. The adrenal cortex releases three major types of steroid hormones: _____, _____, and _____. The adrenal medulla releases the amino acid-derived hormones _____ and _____. **LO5**

Review Questions

1. Which chemical class of hormones usually attaches to membrane receptors on target cells? What cellular events usually follow? **LO3**

2. Which chemical class of hormones usually binds to receptors inside target cells? What cellular events usually follow? **LO3**

3. Diagram the process of negative feedback, and give an example of negative feedback in the control of hormone action. **LO4**

4. What are the major endocrine glands in the human body, and where are they located? **LO5**

5. Describe the structure and function of the hypothalamus–pituitary complex. Describe how releasing hormones regulate the secretion of hormones by cells of the anterior pituitary. Name the hormones of the anterior pituitary, and give one function of each. **LO5**

6. Describe how the hormones of the pancreas act together to regulate the concentration of glucose in the blood. **LO5**

7. Compare the adrenal cortex and adrenal medulla by answering the following questions: Where are they located within the adrenal gland? Which hormones do they produce? Which organs do their hormones target? **LO5**

Applying the Concepts

1. A student decides to do a science project on the effect of the thyroid gland on frog metamorphosis. She sets up three aquaria with tadpoles. She adds thyroxine to the water of one, the drug thiouracil to a second, and nothing to the third. Thiouracil destroys thyroxine. Assuming that the student uses appropriate concentrations, predict what will happen.

2. Suggest a hypothesis about the endocrine system to explain why many birds lay their eggs in the spring. Why do poultry farmers who produce eggs keep the lights on at night in their chicken coops?

3. BioEthics Some parents who are interested in college sports scholarships for their children are asking physicians to prescribe growth hormone treatments, even though their children are in the normal height range. What biological and ethical dilemmas does this raise for the parents, children, physicians, coaches, and college scholarship boards?

4. Why do you think that leptin might be required for the onset of puberty in girls?

Answers to Figure Caption questions and Fill-in-the-Blank questions can be found in the Answers section at the back of the book.

MB *Go to MasteringBiology for practice quizzes, activities, eText, videos, current events, and more.*

Love: "A fire sparkling in lovers' eyes . . . a madness most discreet," or just the right mixture of chemicals deep within lovers' brains? (Inset) Research on prairie voles provides insights into the neurochemical basis of love.

CASE STUDY

How Do I Love Thee?

"But soft! What light through yonder window breaks?

It is the east, and Juliet is the sun."

—Romeo and Juliet,
Act II, scene II

SHAKESPEARE'S *ROMEO AND JULIET* dramatically portrays the power of romantic love, for which two teenage lovers defy their families, risk their fortunes and their futures, and ultimately give up their lives. Romance is, of course, not the only manifestation of love. A mother's love for her child is just as strong. People have also defined their lives and willingly died for love of God and love of country. Just what *is* love? Are all these types of love distinct, or are they related? What happens in the brain when two lovers meet, or a mother cradles her infant? No one knows for sure—not in people, anyway. Surprisingly, neuroscientists know a lot about love—or at least about pair bonding and sex—in a small rodent called the prairie vole.

If Juliet had been a prairie vole, her first encounter with Romeo would have released a flood of oxytocin, the same hormone that causes uterine contractions during childbirth. The oxytocin would have bound to receptors in a few small areas of her brain, causing nerve cells to release dopamine, often called the reward chemical of the brain. She would have felt wonderful. What's more, she would have linked that euphoric feeling to Romeo. In a Romeo vole, some of the molecules and brain regions would have differed, but the end result would have been similar: a flood of dopamine, giving him the ultimate high and making him believe that he could attain that feeling again only with Juliet. So, the two voles would mate and bond for life. They would build a nest, live together, and raise their young.

How do humans and other animals perceive and react to their world? For example, how do they become calm or excited, fearful or smitten? How do animals respond with appropriate behaviors such as resting, eating, or mating? The answers to these questions lie in the nervous system.

AT A GLANCE

LEARNING GOALS

LG1 Identify the structures of a typical neuron, and explain how each structure functions.

LG2 Describe resting potentials, action potentials, inhibitory postsynaptic potentials, and excitatory postsynaptic potentials, and explain the relationships among these potentials.

LG3 Explain the process of synaptic transmission.

LG4 Describe the cellular components of a typical neural pathway and how each component functions during information processing.

LG5 Explain how the nervous system codes for the nature and intensity of a stimulus.

LG6 Describe and distinguish among the central, peripheral, somatic, and autonomic nervous systems.

LG7 Diagram the human brain and spinal cord, labeling the major structures within each, and describe the principal functions of each of those structures.

LG8 Explain how the left and right cerebral hemispheres of the human brain differ in function.

LG9 Compare and contrast short-term memory and long-term memory.

38.1 WHAT ARE THE STRUCTURES AND FUNCTIONS OF NERVE CELLS?

The nervous system contains two principal cell types: **neurons,** often called nerve cells, and **glia.** As we will see, neurons are the cells that carry out the principal jobs of all nervous systems: They receive, process, and transmit information, and control movements of the body. Glia assist neuronal function by providing nutrients, regulating the composition of the interstitial fluid that bathes the neurons, modulating communication among neurons, and speeding up the movement of electrical signals within neurons. Although glia are very important—the nervous system could not function without them—in this chapter we will focus on the structure and function of neurons.

The Functions of a Neuron Are Localized in Separate Parts of the Cell

A neuron is a highly specialized cell. In both vertebrates and invertebrates, a typical neuron includes four major structures: dendrites, a cell body, an axon, and synaptic terminals (**Fig. 38-1**). These four structures perform the four main functions of a neuron:

1. Receive information from the internal or external environment or from other neurons.
2. Process this information, often along with information from other sources, and produce an electrical signal.
3. Conduct the electrical signal, sometimes for a considerable distance, to a junction where it meets another cell.
4. Transmit information to other cells, such as other neurons or the cells of muscles or glands.

Dendrites Respond to Stimuli

Dendrites, branched tendrils protruding from the cell body, perform the "receive information" function (**Fig. 38-1 ❷** and ❼). Dendrites provide a large surface area for receiving signals, either from the environment or from other neurons. Dendrites of sensory neurons produce electrical signals in response to specific stimuli from the external environment—such as pressure, odor, or light—or the internal environment, such as body temperature or blood pH. Dendrites of neurons in the brain and spinal cord usually respond to chemicals, called **neurotransmitters,** which are released by other neurons. **Table 38-1** lists a few neurotransmitters and some of their functions.

The Cell Body Processes Signals from the Dendrites

Electrical signals travel down the dendrites and converge on the neuron's **cell body,** which performs the "process information" function (**Fig. 38-1 ❸**). The cell body adds up the electrical signals that it receives from the dendrites. As we will see shortly, some of these signals are positive and some are negative. If their sum is sufficiently positive, the neuron will produce a large, rapid electrical signal called an **action potential** (**Fig. 38-1 ❹**). The neuron's cell body also contains organelles, such as the nucleus, mitochondria, endoplasmic reticulum, and Golgi apparatus, and performs typical cellular activities, such as synthesizing complex molecules and coordinating the cell's metabolism.

The Axon Conducts Action Potentials Long Distances

In a typical neuron, a long, thin strand called an **axon** extends outward from the cell body. The axon conducts action potentials (**Fig. 38-1 ❺**) from the cell body to the axon's end, where it contacts other cells (**Fig. 38-1 ❶** and ❻). The axons of certain neurons stretch from your spinal cord to your toes, a distance of about 3 feet (approximately a meter), making these neurons the longest cells in the body. **Nerves** are axons bundled together much like wires in an electrical cable. In vertebrates, nerves emerge from the brain and spinal cord and extend to all regions of the body.

① Synaptic terminals: Transmit signals from other neurons

② Dendrites: Receive signals from other neurons

③ Cell body: Integrates signals; coordinates the neuron's metabolic activities

④ An action potential starts here

⑤ Axon: Conducts the action potential

⑥ Synaptic terminals: Transmit signals to other neurons

⑦ Dendrites (of other neurons): Receive signals

neurotransmitters

dendrite

receptors

synaptic terminal

synapse

◄FIGURE 38-1 **A neuron, its specialized parts, and their functions** The red arrows indicate action potentials moving from the cell body down the axon to the synaptic terminals.

TABLE 38-1 Some Important Neurotransmitters

Neurotransmitter	Location in the Nervous System	Some Major Functions
Acetylcholine	Motor neuron-to-muscle synapses; autonomic nervous system, many areas of the brain	Activates skeletal muscles; activates target organs of the parasympathetic nervous system
Dopamine	Midbrain	Important in emotion, rewards, and the control of movement
Norepinephrine (noradrenaline)	Sympathetic nervous system	Activates target organs of the sympathetic nervous system
Serotonin	Midbrain, pons, and medulla	Influences mood and sleep
Glutamate	Many areas of the brain and spinal cord	Major excitatory neurotransmitter in the CNS
Glycine	Spinal cord	Major inhibitory neurotransmitter in the spinal cord
GABA (gamma amino butyric acid)	Many areas of the brain and spinal cord	Major inhibitory neurotransmitter in the brain
Endorphins	Many areas of the brain and spinal cord	Influence mood, reduce pain sensations
Nitric oxide	Many areas of the brain	Important in forming memories

At Synapses, Signals Are Transmitted from One Cell to Another

A neuron communicates with another cell, or *innervates* it, at a site called a **synapse.** A typical synapse consists of (1) the **synaptic terminal,** which is a swelling at the end of an axon of the "sending" neuron; (2) a "receiving" neuron (usually its dendrites), a muscle cell, or gland cell; and (3) a small gap separating the two (Fig. 38-1 ❻). Most synaptic terminals contain neurotransmitters that are released in response to an action potential reaching the terminal. The neurotransmitters diffuse across the gap, bind to receptors on the plasma membrane of the receiving neuron, and stimulate a response in this cell. Therefore, at a synapse, the output of the first cell becomes the input to the second cell.

CHECK YOUR LEARNING

Can you describe the structure of a typical neuron? Are you able to explain the functions of each part of a neuron?

38.2 HOW DO NEURONS PRODUCE AND TRANSMIT INFORMATION?

As a general rule, information is carried *within* a neuron by electrical signals, and information is transmitted *between* neurons by neurotransmitters released from one neuron and received by a second neuron.

Information Within a Single Neuron Is Carried by Electrical Signals

An unstimulated, inactive neuron maintains a constant electrical voltage difference, or potential, across its plasma membrane, similar to the voltage across the poles of a battery. This voltage, called the **resting potential,** is always negative inside the cell and ranges from about –40 to –90 millivolts (mV; thousandths of a volt).

If a neuron is stimulated, the voltage inside it may become either more negative or less negative (Fig. 38-2). If the potential becomes sufficiently less negative, it reaches a level called **threshold,** and triggers an action potential. During an action potential, the neuron's voltage rises rapidly to about +50 mV. An action potential lasts a few milliseconds (thousandths of a second) before the cell's negative resting potential is restored. The plasma membranes of axons are specialized to conduct action potentials from a neuron's cell body to the axon's synaptic terminals. Unlike electrical voltages in metal wires, which become smaller the farther they travel, action potentials are conducted from cell body to axon terminal with no change in voltage. "In Greater Depth: Electrical Signaling in Neurons" examines the cellular mechanisms of resting and action potentials.

Myelin Speeds Up the Conduction of Action Potentials

How fast an action potential travels varies tremendously among axons. In general, the thicker the axon, the faster the action potential moves. A much more effective way to speed

▲ **FIGURE 38-2 Electrical events during an action potential** ❶ A neuron maintains a voltage across its plasma membrane, called the resting potential, of about –60 mV with respect to the outside. ❷ Stimulation from the environment or other cells may make the neuron more negative (downward deflection) or less negative (upward deflection). ❸ If the voltage becomes 10 to 20 mV less negative than the resting potential, the neuron reaches threshold, and ❹ produces a brief positive potential called an action potential. ❺ After a millisecond or two, the voltage across the neuron's plasma membrane returns to the resting potential.

up conduction is to cover the axon with a fatty insulation called **myelin** (Fig. 38-3). Myelin is formed by glial cells that wrap themselves around the axon. Each myelin wrapping covers about 0.2 to 2 mm of axon, leaving short segments of naked axon, called nodes, in between. In an axon without a myelin covering, action potentials travel continuously but fairly slowly, typically about 3 to 6 feet (1 to 2 meters) per second. In contrast, action potentials in myelinated axons "jump" rapidly from node to node. In some myelinated axons, action potentials move as fast as 330 feet (100 meters) per second.

Neurons Use Chemicals to Communicate with One Another at Synapses

Think of an action potential as a packet of information moving down an axon. Once it reaches the synaptic terminal, this information must be transmitted to another cell, either another neuron or a cell in a muscle or gland. At what are called *electrical synapses,* electrical activity can pass directly from neuron to neuron through gap junctions connecting the insides of the cells (see Chapter 5). Although electrical synapses occur in many places in the mammalian nervous system, neurons far more frequently use chemicals to communicate with one another or with cells of muscles or glands. We will confine our discussion to these *chemical synapses.*

In ordinary English, the word "transmit" means "to send something," and that is exactly what happens at a synapse (Fig. 38-4). Two neurons do not actually touch at a synapse: a tiny gap, the **synaptic cleft,** separates the sending neuron, or **presynaptic neuron,** from the receiving neuron, or **postsynaptic neuron.** The presynaptic neuron

An action potential jumps from node to node, greatly speeding up conduction down the axon

myelin-producing glial cell

node

myelin

axon

myelin sheath

axon

◄FIGURE 38-3 A myelinated axon Many vertebrate axons are covered with myelin, which consists of insulated wrappings of the plasma membranes of specialized glial cells. Action potentials occur only at the nodes between each myelin wrapping, jumping from node to node (red arrows), with almost no time spent traveling beneath the myelin.

presynaptic neuron

synapse

synaptic vesicle

synaptic terminal of presynaptic neuron

neurotransmitters

❶ An action potential is initiated

❷ The action potential reaches the synaptic terminal of the presynaptic neuron

❸ The positive charge of the action potential causes the synaptic vesicles to release neurotransmitters

❹ Neurotransmitters bind to receptors on the postsynaptic neuron

dendrite of postsynaptic neuron

synaptic cleft

❻ Neurotransmitters are taken back into the synaptic terminal, are degraded, or diffuse out of the synaptic cleft

❺ Neurotransmitter binding causes ion channels to open, and ions flow in or out

neurotransmitter

ions

receptor

postsynaptic neuron

◄FIGURE 38-4 The structure and function of a synapse

QUESTION Imagine an experiment in which the neurons pictured here are bathed in a solution containing a nerve poison. The presynaptic neuron is stimulated and produces an action potential, but this does not result in a PSP in the postsynaptic neuron. When the experimenter adds some neurotransmitter to the synapse, the postsynaptic neuron still produces no PSP. How does the poison act to disrupt nerve function?

IN GREATER DEPTH Electrical Signaling in Neurons

Potassium Permeability Produces the Resting Potential

The resting potential is based on a balance between chemical and electrical gradients across a membrane that is selectively permeable to specific ions. The cytoplasm of a neuron contains a high concentration of positively charged potassium ions (K^+) and large, negatively charged organic molecules such as ATP and proteins (Fig. E38-1a, bottom). The interstitial fluid outside the neuron contains high concentrations of positively charged sodium ions (Na^+) and negatively charged chloride ions (Cl^-). The concentration gradients of Na^+ and K^+ are maintained by an active transport protein in the plasma membrane called the **sodium-potassium (Na^+-K^+) pump,** which pumps K^+ into and Na^+ out of the cell.

In an unstimulated neuron, only K^+ can cross the plasma membrane, by traveling through specific membrane proteins called resting K^+ channels (see Fig. E38-1a, bottom). There are also voltage-gated Na^+ and K^+ channels in the membrane; as their name suggests, they have "gates" on their channels that are opened or closed by the voltage across the plasma membrane. In an unstimulated neuron, these voltage-gated channels are closed. We will describe their function in a moment.

Because the K^+ concentration is higher inside the cell than outside, K^+ diffuses out of the cell through the resting K^+ channels, leaving the negatively charged organic ions behind (see Fig. E38-1a, bottom). As positively charged K^+ leaves, the inside of the cell becomes more and more negative. K^+ is, of course, electrically attracted by this negative voltage. Eventually, the negative voltage inside the cell becomes large enough so that the rate of K^+ diffusing out, driven by its concentration gradient, is exactly counterbalanced by the rate of K^+ being pulled back in by electrical attraction. The resulting negative voltage is the neuron's resting potential.

Changes in Permeability to Sodium and Potassium Produce an Action Potential

Action potentials occur when the resting potential is changed, becoming less negative and reaching the threshold voltage (about 10 to 20 mV less negative than the resting potential). At threshold, voltage-gated Na^+ channels open. Because there is a much higher concentration of Na^+ outside the neuron than inside, Na^+ rapidly diffuses into the neuron, making the inside positive (Fig. E38-1b).

Voltage-gated Na^+ channels stay open a very short time and then close. Meanwhile, voltage-gated K^+ channels open, allowing K^+ to flow out of the cell, which restores the negative resting potential (Fig. E38-1c).

Action Potentials Are Conducted Down Axons Without Changing in Size

Neurobiologists often say that a neuron "fires" action potentials. The analogy is apt: To shoot a gun, you pull the trigger. If you don't pull the trigger hard enough, the gun doesn't fire. But jerking the trigger really hard doesn't make the bullet come out of the barrel any faster. Similarly, action potentials are all-or-none: If the neuron does not reach threshold, there will be no action potential. If threshold is reached, a full-size action potential will occur and travel the entire length of the axon, but the action potential will be the same voltage and travel at the same speed, whether the neuron just barely reaches threshold or far exceeds it.

An action potential usually starts where the axon emerges from a neuron's cell body. As Na^+ enters the axon here, its positive charge repels other positively charged ions in the axon's cytoplasm. Think of a pool table, with a dozen balls lined up in a row, touching each other. If you hit the ball on one end with the cue ball, the ball on the opposite end instantly shoots off, and the balls in the middle remain where they were. Similarly, when Na^+ enters during an action potential, its positive charge repels other positively charged ions farther along inside the axon, almost instantaneously changing the resting potential of nearby areas, so that the potential becomes more positive and exceeds threshold. This causes voltage-gated Na^+ channels in these nearby areas to open, beginning a new action potential. Na^+ enters at these new locations slightly farther along the axon (see Fig. E38-1c, top), starting the whole process over. Because a new, full-size action potential is produced again and again all the way down the axon, the action potential travels to the end of the axon without losing voltage. As the wave of positive charge passes a given point along the axon, the resting potential is restored as voltage-gated K^+ channels open and K^+ flows out (see Fig. E38-1c).

Note that the Na^+-K^+ pump plays no role in an individual action potential. Only a tiny fraction of the total K^+ and Na^+ in and around a neuron is exchanged during each action potential, so the concentration gradients of K^+ and Na^+

---

sends neurotransmitter molecules across the gap to the postsynaptic neuron.

Communication between neurons begins with an action potential in a presynaptic neuron, usually beginning near the cell body (Fig. 38-4 ❶) and traveling down the axon until it reaches a synaptic terminal (Fig. 38-4 ❷). The synaptic terminal contains scores of tiny sacs, called vesicles, each full of neurotransmitter molecules. When the action potential invades the synaptic terminal, the inside of the terminal becomes positively charged, triggering a cascade of changes that causes some of the vesicles to release neurotransmitters into the synaptic cleft (Fig. 38-4 ❸). The neurotransmitter molecules diffuse across the cleft and bind to receptor proteins on the outer surface of the plasma membrane of the postsynaptic neuron (Fig. 38-4 ❹).

Synapses Produce Inhibitory or Excitatory Postsynaptic Potentials

At most synapses, the receptor proteins on the postsynaptic neuron control ion channels that span the neuron's plasma membrane. When neurotransmitter molecules bind to these receptor proteins, they open these channels. Depending on which channels are associated with a particular receptor, ions such as sodium (Na^+), potassium (K^+), calcium (Ca^{2+}), or chloride (Cl^-) may move through the channels (Fig. 38-4 ❺), causing a small, brief change in voltage, called a **postsynaptic potential (PSP).** If the postsynaptic neuron becomes more negative (the downward deflection in Fig. 38-2 ❷), its resting potential moves farther away from threshold. This change in voltage is called an **inhibitory postsynaptic potential (IPSP),** because it inhibits the postsynaptic neuron,

(a) The resting potential

(b) The action potential

(c) The resting potential is restored

▲ FIGURE E38-1   **The mechanisms underlying resting and action potentials** The upper illustrations in each part of the figure show a section of an axon, with the important ion movements across the boxed portion of the plasma membrane during **(a)** the resting potential, **(b)** the rising phase of an action potential, and **(c)** the falling phase of an action potential. In the upper illustration of part (c), a new action potential has begun, farther along in the axon (Na$^+$ enters the axon). The lower illustrations show the distribution of ions inside and outside the axon, the important ion channels controlling the resting and action potentials, and the movements of ions through the channels in the boxed portion of the axon.

scarcely change. Therefore, Na$^+$-K$^+$ pump activity is not needed to restore the resting potential, which, as we have seen, results from opening voltage-gated K$^+$ channels at the end of the action potential. In the long run (minutes to hours), however, the activity of the Na$^+$-K$^+$ pump is crucial, because it maintains the concentration gradients of both K$^+$ and Na$^+$.

making it less likely to fire an action potential. If the postsynaptic neuron becomes less negative (the upward deflection in Fig. 38-2 ❷), then its resting potential will move closer to threshold. Consequently, this voltage change is called an **excitatory postsynaptic potential (EPSP),** because it excites the postsynaptic neuron, making it more likely to fire an action potential. "In Greater Depth: Synaptic Transmission" explains the mechanisms by which neurotransmitters binding to receptors cause PSPs.

Neurotransmitter Action Is Usually Brief

Consider what would happen if transmitters from a presynaptic neuron stimulated a postsynaptic cell, and the postsynaptic cell never stopped responding. You might, for example, contract your biceps muscle, flex your arm, and have it stay

flexed forever! Fortunately, the nervous system has several ways of ending neurotransmitter action. Some neurotransmitters—notably acetylcholine, the transmitter that stimulates skeletal muscle cells—are rapidly broken down by enzymes in the synaptic cleft. Many other neurotransmitters are transported back into the presynaptic neuron (**Fig. 38-4** ❻). Neurotransmitters also diffuse out of the synaptic cleft.

Integration of Postsynaptic Potentials Determines the Activity of a Neuron

The dendrites of a single neuron often receive EPSPs and IPSPs from the synaptic terminals of hundreds or even thousands of presynaptic neurons. Most of these PSPs are small, rapidly fading signals, but they travel far enough to reach the cell body of the postsynaptic neuron. The voltages

IN GREATER DEPTH Synaptic Transmission

When an action potential reaches a presynaptic terminal, the positive charge inside the terminal opens voltage-gated ion channels that are selectively permeable to calcium (Ca^{2+}). The concentration of Ca^{2+} outside the terminal is about 10,000 times higher than the concentration inside. Therefore, Ca^{2+} diffuses into the terminal, setting off a cascade of events that causes vesicles to release their neurotransmitters into the synaptic cleft.

The transmitters diffuse across the cleft and bind to receptor proteins on the postsynaptic cell. This typically has one of two effects. In some synapses, the result of transmitter–receptor binding is similar to what happens when a peptide hormone binds to its receptor (see Fig. 37-2): Intracellular messengers are synthesized and the metabolism of the postsynaptic cell changes. At most synapses, however, the receptor proteins are linked to ion channels, and neurotransmitter binding opens the channels, causing postsynaptic potentials (PSPs; **Fig. E38-2**).

If the channels are permeable to Na^+ (**Fig. E38-2a**), then Na^+ diffuses down its concentration gradient into the postsynaptic neuron. Na^+ entry makes the inside of the cell less negative. Such voltage changes are called excitatory postsynaptic potentials (EPSPs), because, if the postsynaptic neuron becomes sufficiently less negative, it will reach threshold and produce an action potential.

If the channels are permeable to K^+ (**Fig. E38-2b**), then K^+ diffuses out of the cell, making it more negative. Making the cell more negative inhibits the production of action potentials in the postsynaptic cell, so this voltage change is called an inhibitory postsynaptic potential (IPSP).

(a) Excitatory postsynaptic potential (EPSP)

Neurotransmitter binding opens an ion channel permeable to Na^+; Na^+ enters the postsynaptic neuron, eliciting an EPSP

Neurotransmitter binding opens an ion channel permeable to K^+; K^+ leaves the postsynaptic neuron, eliciting an IPSP

(b) Inhibitory postsynaptic potential (IPSP)

▶ **FIGURE E38-2 Neurotransmitter binding to receptor proteins opens ion channels** (a) The ionic mechanism of an EPSP. (b) The ionic mechanism of an IPSP.

of all the PSPs that reach the cell body at about the same time are added up, a process called **integration.** If the excitatory and inhibitory postsynaptic potentials, when added together, raise the electrical potential inside the neuron above threshold, the postsynaptic cell will produce an action potential.

CHECK YOUR LEARNING

Can you describe resting and action potentials? Can you explain how an action potential in a presynaptic neuron causes a response in a postsynaptic neuron? Do you understand the difference between inhibitory postsynaptic potentials and excitatory postsynaptic potentials?

CASE STUDY continued

How Do I Love Thee?

Different animal species, and even different individuals within a species, have different numbers and types of receptors for neurotransmitters. For example, prairie voles, which are monogamous, have a very high density of receptors for dopamine in certain parts of their brains. In contrast, the promiscuous montane vole has much lower concentrations of dopamine receptors in these same brain regions. In humans, people with social phobias, who avoid many social situations for fear of embarrassment, have fewer dopamine receptors in reward areas of the brain than do people who are good at social interactions.

38.3 HOW DOES THE NERVOUS SYSTEM PROCESS INFORMATION AND CONTROL BEHAVIOR?

The nervous system performs marvelous feats of computation, stores prodigious amounts of information, and directs complex behaviors. These accomplishments arise from three interacting properties: specialization of individual neurons; huge, yet orderly, networks of connections between neurons; and outputs from the nervous system to specific muscles and glands that actually perform the behaviors dictated by the nervous system.

Most behaviors are controlled by pathways composed of four elements:

1. **Sensory neurons** respond to a stimulus, either internal or external to the body.
2. **Interneurons** receive signals from sensory neurons, hormones, neurons that store memories, and many other sources.
3. **Motor neurons** receive instructions from sensory neurons or interneurons and activate muscles or glands.
4. **Effectors,** usually muscles or glands, perform the response directed by the nervous system.

These four elements, when properly connected, carry out the basic operations required of any nervous system:

1. Determine the type of stimulus (sensory neurons).
2. Determine and signal the intensity of a stimulus (sensory neurons and interneurons).
3. Integrate information from many sources (interneurons).
4. Direct appropriate behaviors (interneurons, motor neurons, and effectors).

The Nature of a Stimulus Is Encoded by Specialization of Sensory Neurons and Their Connections to Specific Parts of the Brain

The senses inform the brain about the properties of its environment, both outside the body (such as images, sounds, or odors) and inside the body (such as the body temperature or the concentration of salts and glucose in the blood). Sensory neurons are specialized to respond to specific stimuli. For example, some sensory neurons respond to touch, but not temperature, light, or chemicals; others respond to specific chemicals but not to touch or light, or even to other chemicals (see Chapter 39). Information gathered by the senses is always converted to action potentials, either directly in the sensory neuron or in interneurons, such as in the retina of the eye or in the inner ear. The sensory information is then sent to the brain in the form of these action potentials. Given that all action potentials are fundamentally the same, how can the brain determine *what* a stimulus is?

The nervous system encodes the nature of a stimulus— touch or temperature, for example— by which sensory neurons respond to the stimulus and to which parts of the brain the axons of those neurons connect. For example, your brain interprets action potentials that occur in the axons of your optic nerves (which connect the eye to the brain) as the sensation of light. Long ago, a German physiologist sat in a dark room and poked himself in the eye, which caused neurons in his eye to produce action potentials that traveled to his brain. (As they say on TV, do *not* try this yourself!) The result? He "saw stars" because his brain interpreted action potentials in his optic nerve as light. Thus, you distinguish hot from cold temperatures, or the bitterness of coffee from the sweetness of sugar, because these diverse stimuli result in action potentials in different sensory neurons, with axons that connect to different areas of your brain.

The Intensity of a Stimulus Is Encoded by the Frequency of Action Potentials

Because all action potentials are roughly the same size and duration, no information about the strength, or intensity, of a stimulus (for example, the loudness of a sound) can be encoded in a single action potential. Intensity is coded in two ways (Fig. 38-5). First, intensity can be signaled by the frequency of action potentials in a single neuron. The more intense the stimulus, the faster the neuron fires action potentials. Second, many neurons may respond to the same type of stimulus. Stronger stimuli excite more of these neurons, whereas weaker stimuli excite fewer. Thus, intensity can also be signaled by the number of similar neurons firing at the same time. A gentle touch may cause a single touch receptor in your skin to fire action potentials very slowly (Fig. 38-5a); a hard poke may cause several touch receptors to fire, some very rapidly (Fig. 38-5b).

The Nervous System Processes Information from Many Sources

Your brain is continuously bombarded by sensory stimuli from both inside and outside the body. The brain must evaluate these inputs, determine which ones are important, and decide how to respond. In many instances, multiple sensory neurons converge onto a smaller number of interneurons in the brain. Some of these interneurons add up the postsynaptic potentials that result from the activity of the sensory neurons and initiate action accordingly. Depending on the relative strengths of the different sensory inputs, these interneurons may produce action potentials and stimulate motor neurons to produce appropriate behaviors.

The Nervous System Produces Outputs to Muscles and Glands

Motor neurons in the brain, the spinal cord, or in the sympathetic and parasympathetic nervous systems (described later) stimulate activity in effectors—the muscles or glands that actually produce behaviors. The same principles of connectivity and intensity coding that we just described for sensory inputs are used for the brain's outputs to effectors. Which muscles or glands are activated is determined by their connections to the brain or spinal cord. For example, different motor neurons activate your biceps muscles and

▶ **FIGURE 38-5 Signaling stimulus intensity** The intensity of a stimulus is signaled by the rate at which individual sensory neurons produce action potentials and by the number of sensory neurons activated. **(a)** In this example, a gentle touch activates only the closest sensory neuron, which fires action potentials at a low rate. **(b)** A hard poke activates multiple sensory neurons, causing the closest one to fire rapidly and more distant ones to fire more slowly.

QUESTION On the basis of this information, how do you think skin areas that are especially sensitive to touch differ from less sensitive areas?

(a) Gentle touch

sensory neuron 1

sensory neuron 2

+40 / −70 sensory neuron 1

+40 / −70 sensory neuron 2

Sensory neuron 1 fires slowly; sensory neuron 2 is silent

(b) Hard poke

sensory neuron 1

sensory neuron 2

+40 / −70 sensory neuron 1

+40 / −70 sensory neuron 2

time

Sensory neurons 1 and 2 both fire

the muscles in your face. How strongly a muscle contracts is determined by how many motor neurons connect to it, and how fast those motor neurons fire action potentials.

Behaviors Are Controlled by Networks of Neurons in the Nervous System

Simple behaviors, such as reflexes (see Section 38.5), may be controlled by activity in as few as two or three neurons—a sensory neuron, a motor neuron, and perhaps an interneuron in between—ultimately stimulating a single muscle. In humans, simple reflexes, such as the familiar knee-jerk or pain-withdrawal reflexes, are produced by neurons in the spinal cord.

More complex behaviors are organized by interconnected neural pathways, in which several types of sensory input (along with memories, hormones, and other factors) converge on a set of interneurons. The interneurons integrate PSPs from these multiple sources and stimulate motor neurons that direct activity in the appropriate muscles and glands. Thousands or even millions of neurons, mostly in the brain, may be required to perform complex actions such as playing the piano.

CHECK YOUR LEARNING

Are you able to list the four components of a neural pathway and explain how these components carry out the functions of a nervous system? Can you explain how the brain interprets the type and intensity of a sensory stimulus? Are you able to explain how the brain determines which muscles to contract and how strongly to contract them?

38.4 HOW ARE NERVOUS SYSTEMS ORGANIZED?

All animals have one of two basic types of nervous systems: a diffuse nervous system, such as that of cnidarians (*Hydra*, jelly-fish, anemones, and their relatives; **Fig. 38-6a**), or a centralized

nervous system, found in more complex organisms. Nervous system architecture is highly correlated with an animal's body plan and lifestyle.

Cnidarians are radially symmetrical (see Chapter 23). Because they have no "front end," there has been no evolutionary pressure to concentrate the senses in one place. For example, a *Hydra* sits anchored to a rock at the bottom of a pond, so prey or predators are equally likely to come from any direction. Cnidarian nervous systems are composed of a network of neurons, often called a **nerve net**, woven through the animal's tissues. Here and there we find a cluster of neurons, called a **ganglion** (plural, ganglia), but nothing resembling a real brain.

Almost all other animals are bilaterally symmetrical, with definite head and tail ends. Because the head is usually the first part of the body to encounter food, danger, and potential mates, it is advantageous to have sense organs concentrated there. Sizable ganglia evolved that integrate the information gathered by the senses and direct appropriate actions. Over evolutionary time, the major sense organs became localized in the head and the ganglia became centralized into a brain. This trend is clearly seen in the invertebrates (**Figs. 38-6b, c**), but reaches its peak in the vertebrates, in which nearly all the cell bodies of the nervous system reside in the brain or spinal cord.

CHECK YOUR LEARNING

Can you describe the anatomy of diffuse and centralized nervous systems and provide examples of animals with each type?

38.5 WHAT ARE THE STRUCTURES AND FUNCTIONS OF THE HUMAN NERVOUS SYSTEM?

The nervous system of all mammals, including humans, can be divided into two parts: central and peripheral. Each of these has further subdivisions (**Fig. 38-7**). The **central**

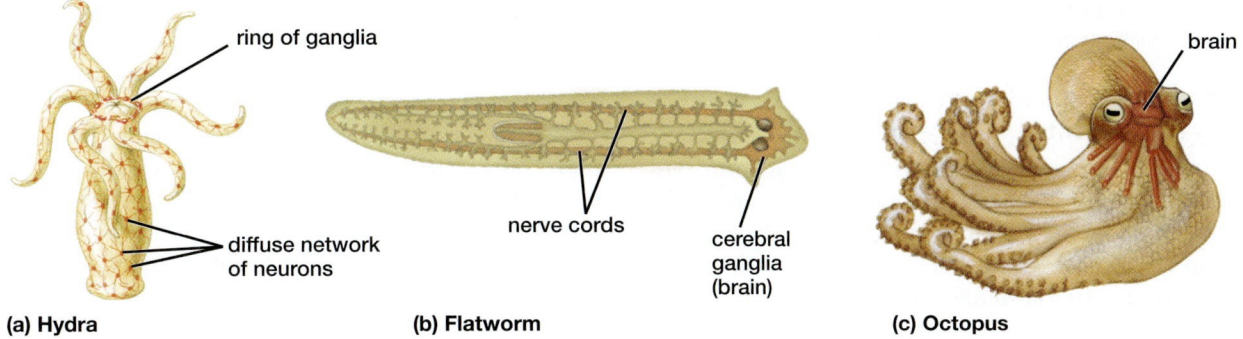

ring of ganglia

brain

diffuse network
of neurons

nerve cords

cerebral
ganglia
(brain)

(a) Hydra

(b) Flatworm

(c) Octopus

▲ **FIGURE 38-6 Nervous system organization (a)** The diffuse nervous system of *Hydra* contains a few concentrations of neurons at the bases of tentacles, but no brain. Neural signals are conducted in all directions throughout the body. **(b)** The flatworm has a nervous system that is less diffuse, with a cluster of ganglia in the head. **(c)** An octopus has a large, complex brain and learning capabilities rivaling those of some mammals.

nervous system (CNS) consists of the **brain** and **spinal cord.** The **peripheral nervous system (PNS)** consists of neurons and axons that lie outside the CNS.

The Peripheral Nervous System Links the Central Nervous System with the Rest of the Body

The cell bodies of some sensory neurons, including many that detect touch, pressure, temperature, stretch, and pain, are located in the PNS, often in clusters called dorsal root ganglia

that are found alongside the spinal cord (see Fig. 38-10 later in this chapter). The cell bodies of neurons that control involuntary movements and activities, such as dilation or constriction of the pupils and airways, secretion of saliva, and sexual arousal, are also mostly, but not entirely, found in the PNS.

The peripheral nerves consist of three major categories of axons: (1) axons of sensory neurons that carry sensory information from all parts of the body to the CNS; (2) axons of motor neurons that innervate the skeletal muscles, providing control of voluntary movements; and (3) axons of neurons that control involuntary movements and activities.

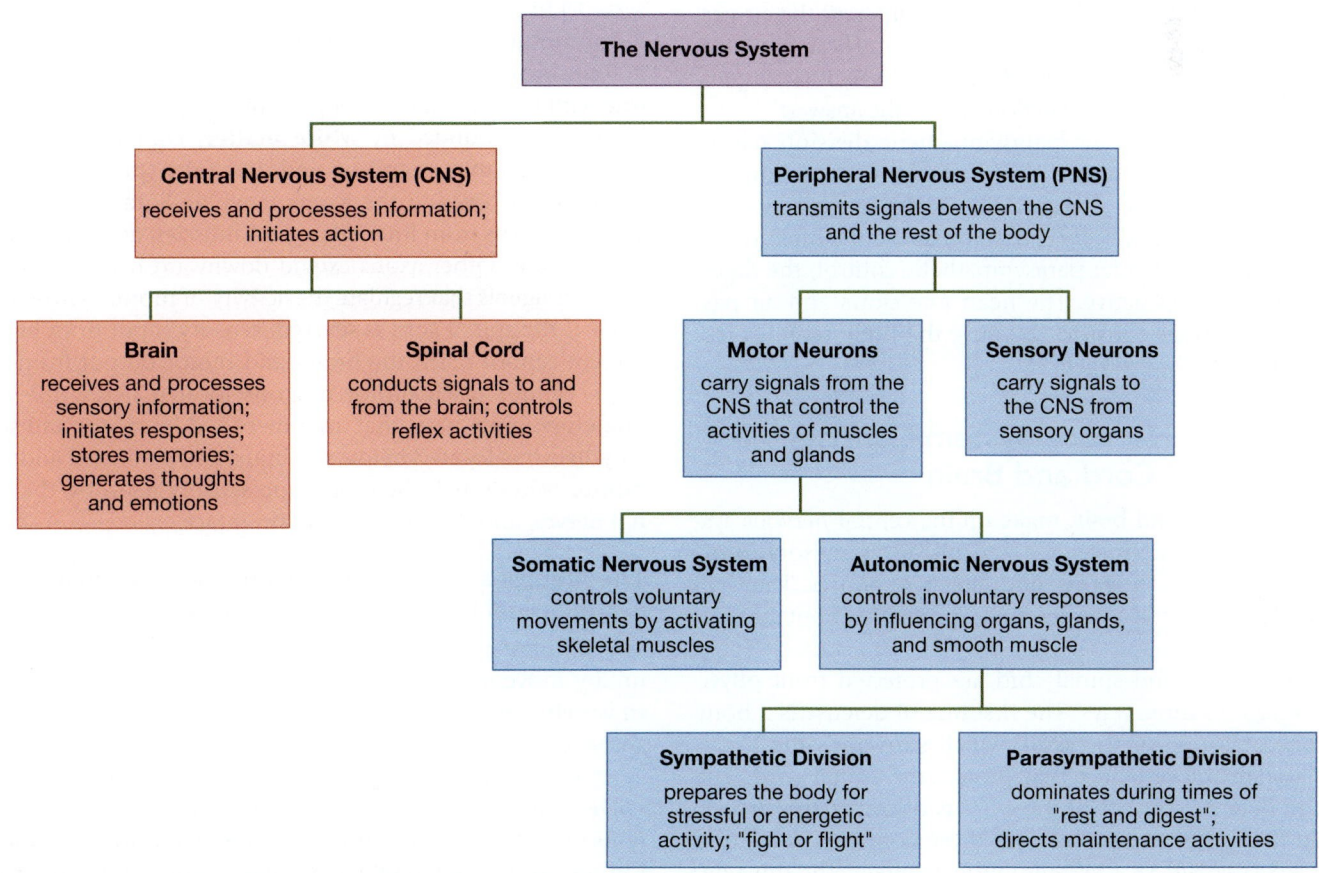

The Nervous System

Central Nervous System (CNS)
receives and processes information; initiates action

Peripheral Nervous System (PNS)
transmits signals between the CNS and the rest of the body

Brain
receives and processes sensory information; initiates responses; stores memories; generates thoughts and emotions

Spinal Cord
conducts signals to and from the brain; controls reflex activities

Motor Neurons
carry signals from the CNS that control the activities of muscles and glands

Sensory Neurons
carry signals to the CNS from sensory organs

Somatic Nervous System
controls voluntary movements by activating skeletal muscles

Autonomic Nervous System
controls involuntary responses by influencing organs, glands, and smooth muscle

Sympathetic Division
prepares the body for stressful or energetic activity; "fight or flight"

Parasympathetic Division
dominates during times of "rest and digest"; directs maintenance activities

▲ **FIGURE 38-7 Organization and functions of the vertebrate nervous system**

Motor activities of the PNS are controlled by either the somatic nervous system or the autonomic nervous system.

The Somatic Nervous System Controls Voluntary Movement

Motor neurons of the **somatic nervous system** form synapses with skeletal muscles and control voluntary movement. As you take notes or lift a coffee cup, your somatic nervous system is in charge. The cell bodies of most somatic motor neurons are located in the spinal cord (see Fig. 38-10). Their axons go directly to the muscles they control.

The Autonomic Nervous System Controls Involuntary Actions

Neurons of the **autonomic nervous system** innervate many parts of the body, including the heart, smooth muscles in the respiratory tract and blood vessels, and many glands, and produce mostly involuntary actions (**Fig. 38-8**). The autonomic nervous system consists of two parts: the **sympathetic division** and the **parasympathetic division.** In general, these two divisions innervate the same organs, but produce opposite effects.

The neurons of the sympathetic division release the neurotransmitter norepinephrine onto their target organs, preparing the body for stressful or energetic activity, such as fighting, escaping, or taking an exam. During such "fight-or-flight" activities, the sympathetic nervous system reduces activity in the digestive tract, redirecting some of its blood supply to the muscles of the arms and legs. The heart rate accelerates. The pupils of the eyes open wider, admitting more light, and the air passages in the lungs expand, letting in more air. These responses may also happen if you are suddenly asked to answer a question in class—especially if you don't know the answer!

The neurons of the parasympathetic division release the neurotransmitter acetylcholine onto their target organs. The parasympathetic division dominates during maintenance activities that can be carried on at leisure, often called "rest and digest." Under parasympathetic control, the digestive tract becomes active. The heart rate slows and air passages in the lungs constrict, because the body requires less blood flow and less oxygen.

The Central Nervous System Consists of the Spinal Cord and Brain

The spinal cord and brain make up the central nervous system (CNS). The CNS receives and processes sensory information, generates thoughts, and directs responses. The CNS consists primarily of interneurons—probably about 100 billion of them.

The brain and spinal cord are protected from physical damage in three ways. The first line of defense is a bony armor, consisting of the skull, which surrounds the brain, and a chain of vertebrae that protect the spinal cord. The second defense consists of three layers of connective tissues, called meninges, that lie beneath these bones and surround the CNS (see Fig. 38-12a later in the chapter). The third de-

fense, lying between the layers of the meninges, is the cerebrospinal fluid, a clear liquid similar to blood plasma. Cerebrospinal fluid cushions the brain and spinal cord and nourishes the CNS.

The brain is protected from damaging chemicals in the bloodstream because the walls of brain capillaries are far less permeable than are the capillaries in the rest of the body. This **blood–brain barrier** selectively transports needed materials into the brain, while keeping many dangerous substances out.

The Spinal Cord Controls Many Reflexes and Conducts Information to and from the Brain

The spinal cord (**Fig. 38-9**) extends from the base of the brain to the lower back. Nerves carrying axons of sensory neurons emerge from the dorsal (back) part of the spinal cord, and nerves carrying axons of motor neurons emerge from the ventral (front) part. These nerves merge to form the spinal nerves that innervate most of the body. Because of their resemblance to the roots of a tree that merge into a single trunk, these branches are called the dorsal and ventral roots of the spinal nerves, respectively. Swellings on each dorsal root, called the **dorsal root ganglia,** contain the cell bodies of sensory neurons.

In the center of the spinal cord is a butterfly-shaped area of **gray matter.** ("Gray matter" is really a misnomer, because most nervous tissue is pinkish-tan. It turns gray when it is preserved.) In the spinal cord, gray matter consists of the cell bodies of motor neurons that control voluntary muscles and the autonomic nervous system, plus interneurons that communicate with the brain and other parts of the spinal cord. The gray matter is surrounded by **white matter,** containing myelin-coated axons that extend up and down the spinal cord—the fatty myelin covering the axons is white. Some axons carry sensory signals from internal organs, muscles, and the skin up to the brain. Other axons extend downward from the brain, carrying signals that regulate the activity of motor neurons.

If the spinal cord is severed, sensory input from below the cut cannot reach the brain, and motor output from the brain cannot reach motor neurons located below the cut. Therefore, body parts that are innervated by motor and sensory neurons located below the injury are paralyzed and feel numb, even though the motor and sensory neurons, the spinal nerves, and the muscles remain intact.

The Neurons That Control Many Reflexes Reside in the Spinal Cord and Peripheral Nervous System

The simplest type of behavior is the **reflex,** a largely involuntary movement of a body part in response to a stimulus. In vertebrates, many reflexes are produced by neurons in the spinal cord and the peripheral nervous system.

Let's examine the pain-withdrawal reflex, which involves neurons of both the central and peripheral nervous systems (**Fig. 38-10**). If you bring your hand too close to a flame, the resulting tissue damage activates pain sensory neurons

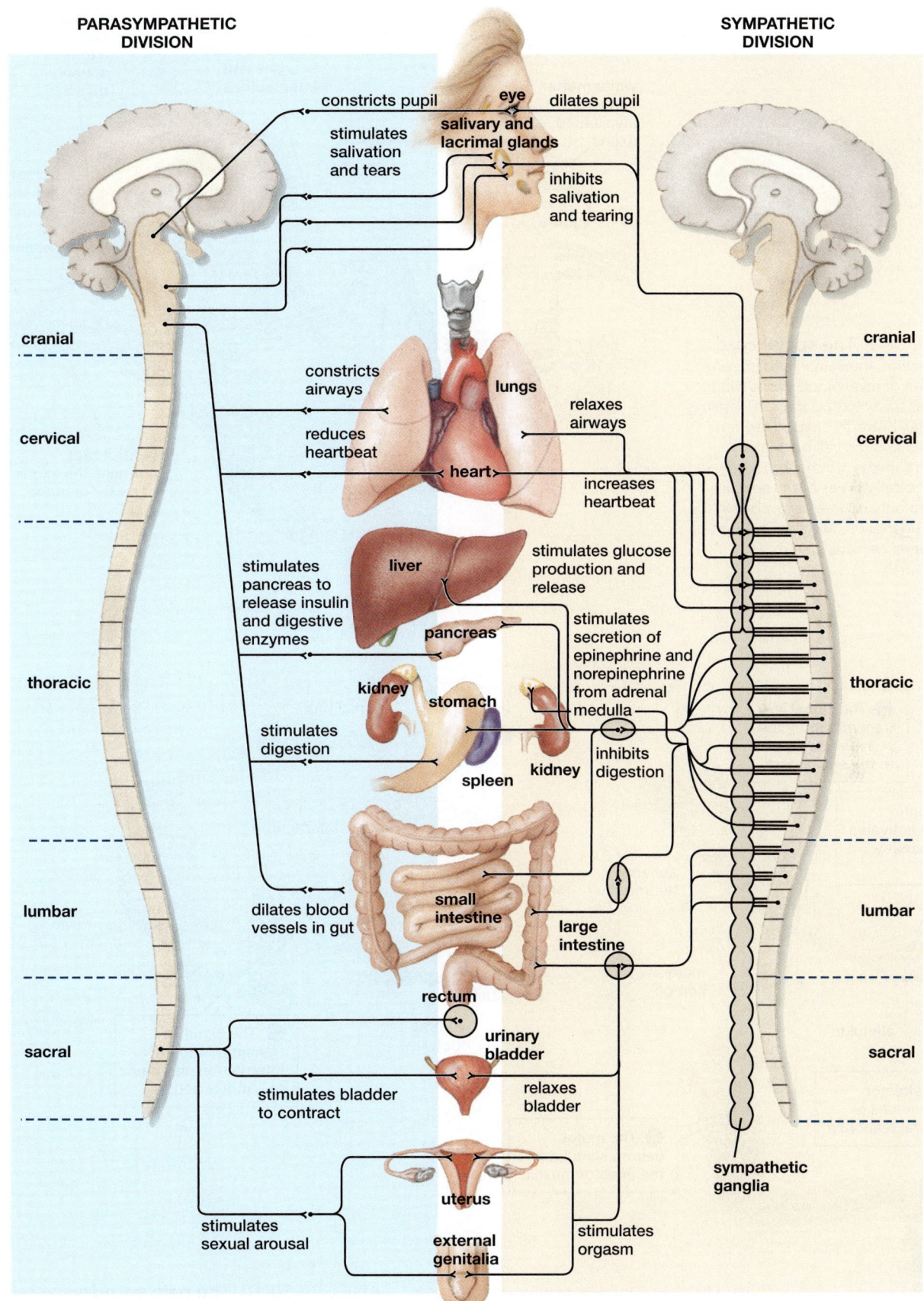

PARASYMPATHETIC DIVISION

SYMPATHETIC DIVISION

constricts pupil

eye

dilates pupil

stimulates salivation and tears

salivary and lacrimal glands

inhibits salivation and tearing

cranial

cranial

constricts airways

lungs

relaxes airways

cervical

reduces heartbeat

heart

increases heartbeat

cervical

stimulates pancreas to release insulin and digestive enzymes

liver

stimulates glucose production and release

pancreas

stimulates secretion of epinephrine and norepinephrine from adrenal medulla

thoracic

kidney

stomach

thoracic

stimulates digestion

spleen

kidney

inhibits digestion

dilates blood vessels in gut

small intestine

large intestine

lumbar

lumbar

rectum

urinary bladder

sacral

stimulates bladder to contract

relaxes bladder

sacral

sympathetic ganglia

stimulates sexual arousal

uterus

external genitalia

stimulates orgasm

▲ **FIGURE 38-8 The autonomic nervous system** The autonomic nervous system has two divisions, sympathetic and parasympathetic, which supply nerves to many of the same organs but generally produce opposite effects. Activation of the autonomic nervous system is involuntarily commanded by signals from the hypothalamus.

gray matter
contains the cell
bodies of motor
neurons and
interneurons

dorsal root
contains the
axons of
sensory
neurons

white matter
contains
myelinated
axons

**dorsal root
ganglion**
contains the
cell bodies of
sensory neurons

**spinal
nerve**

ventral root
contains the axons
of motor neurons

▶ FIGURE 38-9 **The spinal cord**
In cross-section, the spinal cord has an
outer region of myelinated axons (white
matter) that travel to and from the brain
and an inner, butterfly-shaped region
of dendrites and the cell bodies of
interneurons and motor neurons (gray
matter). The cell bodies of the sensory
neurons are outside the spinal cord in the
dorsal root ganglion, and thus are part of
the peripheral nervous system.

❷ The signal is
transmitted by the
pain sensory neuron
to the spinal cord

dorsal root

❶ A painful
stimulus activates
a pain sensory
neuron

interneuron

**sensory
neuron**

**spinal
cord**

❸ The signal is
transmitted to an
interneuron and then
to a motor neuron

**motor
neuron**

stimulus

**ventral
root**

❺ The effector
muscle causes a
withdrawal response

❹ The motor
neuron stimulates
the effector muscle

◀ FIGURE 38-10 **The pain-withdrawal reflex**

QUESTION Why might a paralyzed victim of a spinal
cord injury, when pricked with a pin on a paralyzed part
of his or her body, have a normal pain-withdrawal reflex
but feel no pain?

(Fig. 38-10 **❶**). Action potentials in the axons of these neurons travel up a spinal nerve and enter the spinal cord through a dorsal root (Fig. 38-10 **❷**). Within the gray matter of the cord, the pain sensory neuron stimulates an interneuron, which stimulates a motor neuron (Fig. 38-10 **❸**). Action potentials in the axon of the motor neuron leave the spinal cord through a ventral root and travel in a spinal nerve to a muscle. Stimulating the muscle (Fig. 38-10 **❹**) causes it to contract, withdrawing your hand away from the flame (Fig. 38-10 **❺**).

Many spinal cord interneurons also have axons that extend up to the brain. Action potentials in these axons inform the brain about burnt hands and may trigger more complex behaviors, such as shrieks and learning about the dangers of open flames. The brain in turn sends action potentials down axons that connect to interneurons and motor neurons in the spinal cord. These signals from the brain can modify spinal reflexes. With enough training or motivation, you can suppress the pain-withdrawal reflex. To rescue a child from a burning crib, for example, you could reach into the flames.

Some Complex Actions Are Coordinated Within the Spinal Cord

The wiring for some fairly complex activities also resides in the spinal cord. The neurons and interconnections needed for the basic movements of walking and running, for example, are contained in the spinal cord. The advantage of this arrangement is probably an increase in speed and coordination, because messages do not have to travel all the way up to the brain and back down again merely to swing a leg forward while walking. The brain's role in these behaviors is to initiate, guide, and modify the firing of spinal motor neurons, based on conscious decisions (Where are you going? How fast should you walk?). To maintain balance, the brain also uses sensory input from the muscles to regulate motor neuron activity and adjust the way the muscles move.

The Brain Consists of Many Parts That Perform Specific Functions

All vertebrate brains consist of three major parts: the **hindbrain, midbrain,** and **forebrain** (Fig. 38-11a). Scientists believe that in the earliest vertebrates, these three anatomical divisions were also functional divisions: The hindbrain governed automatic behaviors such as breathing and heart rate, the midbrain controlled vision, and the forebrain dealt largely with the sense of smell. In nonmammalian vertebrates, the three divisions remain prominent (Figs. 38-11b, c). However, in mammals—particularly in humans—the midbrain has been significantly reduced in size, while the forebrain has been greatly enlarged (Figs. 38-11d, e). The major structures of the human brain are shown in Figure 38-12.

The Hindbrain Consists of the Medulla, Pons, and Cerebellum

In humans and other mammals, the hindbrain is composed of the medulla, the pons, and the cerebellum (Fig. 38-12a).

(a) Embryonic vertebrate brain

(b) Shark brain

(c) Goose brain

(d) Horse brain

(e) Human brain

▲ **FIGURE 38-11 A comparison of vertebrate brains**
(a) The brains of modern vertebrate embryos, thought to be similar to the brains of the distant ancestors of today's vertebrates, consist of three distinct regions: the forebrain, midbrain, and hindbrain. **(b)** The brain of an adult shark maintains this basic organization. **(c)** In the goose, the midbrain is reduced, and the cerebrum and cerebellum are larger. **(d, e)** In mammals, especially humans, the cerebrum is very large compared to the other brain regions.

▶ FIGURE 38-12 The human brain
Structures of the human brain, seen
as if the brain were sliced **(a)** through
the midline between the cerebral
hemispheres, showing the hindbrain,
midbrain, and forebrain, and **(b)** from ear
to ear, showing mostly the cerebrum.
Not all brain structures are visible in
these sections.

(a) A lateral section of the human brain

(b) A cross-section of the brain

The **medulla** resembles an enlarged extension of the spinal cord, with neuron cell bodies at its center, surrounded by a layer of myelin-covered axons. The medulla controls automatic functions such as breathing, heart rate, blood pressure, and swallowing. Certain neurons in the **pons,** located just above the medulla, appear to influence transitions between sleep and wakefulness and between stages of sleep. Other neurons in the pons regulate the rate and pattern of breathing.

The **cerebellum** is crucial for coordinating movements. It receives information both from command centers in the forebrain that control movement and from position sensors in muscles and joints. By comparing information from these two sources, the cerebellum guides smooth, accurate motions and body position. The cerebellum is also involved in *motor learning.* As you learn to write or play the guitar, your forebrain directs each separate movement. After you have become skilled, your forebrain still determines what behavior to perform (for example, what words to write), but your cerebellum becomes largely responsible for ensuring that the actions are carried out appropriately. Not surprisingly, the cerebellum is especially large in animals, such as bats and birds, that perform aerial maneuvers (see Fig. 38-11c).

Health|Watch

Drugs, Neurotransmitters, and Addiction

Chances are you know someone who is addicted. How can substances such as cocaine, alcohol, and nicotine so profoundly alter people's lives? The answer lies in how these drugs affect neurotransmitter action and in how the nervous system adapts to those effects.

Many addictive drugs, including cocaine, meth (methamphetamine), and ecstasy (MDMA), target synapses in the brain's reward circuitry that use the transmitters dopamine or serotonin. Normally, after releasing a transmitter, the presynaptic neurons in these synapses immediately pump the transmitter back in, thus limiting its effects. Cocaine, meth, and ecstasy block the pumps, so the transmitter concentration increases around the postsynaptic receptors, enhancing synaptic transmission. Because serotonin and especially dopamine make people feel good, cocaine and meth are highly rewarding and users want to repeat the experience. Meanwhile, the brain, in response to excessive dopamine stimulation, reduces the number of dopamine receptors in the synapses and increases the number of dopamine pumps. Therefore, when not taking the drugs, the user feels "down," because he has less dopamine available (because of the increased number of pumps clearing dopamine out of the synaptic cleft) and fewer receptors to respond to what little dopamine remains. Over time, he needs more and more cocaine or meth just to feel OK, let alone to get high: He has become an addict (Fig. E38-3).

By blocking serotonin pumps, ecstasy causes a temporary but massive increase in serotonin. Users report feelings of pleasure, increased energy, heightened sensory awareness, and improved rapport with other people. Ecstasy users may incur long-term damage to serotonin-producing neurons, and may suffer deficits in learning and memory. Ecstasy may also damage dopamine-producing neurons.

Alcohol stimulates receptors for the neurotransmitter gamma aminobutyric acid (GABA), thereby enhancing inhibitory neuronal signals, and blocks receptors for glutamate, which reduces excitatory signals. Together, these changes produce alcohol's well-known relaxing effects. However, when a person drinks frequently, the brain compensates by decreasing GABA receptors and increasing glutamate receptors. Without alcohol, an alcoholic feels jittery and nervous—in short, overstimulated. Alcohol also initially increases dopamine neurotransmission, so the drinker feels good. Later, alcohol suppresses dopamine release.

▲ **FIGURE E38-3 Addiction** An addict experiences extreme physical and emotional distress without his drug, because his nervous system has adapted to its presence.

These effects combine to produce addiction—an alcoholic needs alcohol just to feel normal, and more and more of it to feel good.

Nicotine in cigarette smoke stimulates receptors that normally respond to acetylcholine. Overstimulation of these receptors activates other neurons that increase dopamine release, thereby contributing to smoking's pleasurable and addictive properties.

To overcome addiction, drug users must undergo the misery caused by a nervous system that is deprived of a drug to which it has adapted. Although transmitter release and receptor concentrations eventually return to normal, drug cravings often recur periodically, suggesting that the brains of addicts have been permanently altered, in poorly understood but important ways.

In some respects, love strongly resembles addiction—producing feelings of pleasure, trust, and euphoria. But is love really like addiction? More likely, it's the other way around. Reward circuits in the brain probably evolved because they promote sex, pair bonding, infant care, positive social interactions, and a host of other useful feelings and behaviors. By activating these same circuits, cocaine, alcohol, and nicotine hijack essential brain functions but with destructive results.

The Midbrain Contains Clusters of Neurons That Contribute to Movement, Arousal, and Emotion

The midbrain is quite small in humans (see Figs. 38-11e and 38-12a). It contains an auditory relay center and clusters of neurons that control reflex movements of the eyes. For example, if you're sitting in class and someone races through the door, centers in your midbrain are alerted and direct your gaze to the new, potentially interesting or threatening, visual stimulus. The midbrain also contains neurons that produce the transmitter dopamine. One of these clusters of neurons, called the *substantia nigra* (Fig. 38-12b), helps to control movement (see the description of the basal ganglia, below). Another cluster is an essential part of the "reward

circuit" that is responsible for pleasurable sensations and, unfortunately, addiction, as we explore in "Health Watch: Drugs, Neurotransmitters, and Addiction."

The midbrain also contains a portion of the **reticular formation.** The reticular formation consists of dozens of interconnected groups of neurons in the medulla, pons, and midbrain that send axons to the forebrain. These neurons receive input from virtually every sense, from every part of the body, and from many areas of the brain as well. The reticular formation plays a role in sleep and wakefulness, emotion, and some movements and reflexes. It filters sensory inputs before they reach the conscious regions of the brain. Activities of the reticular formation allow you to read and concentrate in the presence of a variety of distracting stimuli, such as music or the smell of coffee. The fact that a mother wakens upon hearing the faint cry of her infant but sleeps through loud traffic noise outside her window testifies to the effectiveness of the reticular formation in screening sensory inputs to the conscious parts of the brain.

The Forebrain Includes the Thalamus, Hypothalamus, and Cerebrum

The **thalamus** channels sensory information from all parts of the body to the cerebral cortex (see Figs. 38-12a, b). In fact, input from all of the senses, except olfaction, passes through the thalamus on its way to the cerebral cortex. Signals traveling from the spinal cord, cerebellum, medulla, pons, and reticular formation also pass through the thalamus.

The **hypothalamus** (literally, "under the thalamus") contains many clusters of neurons. Some release hormones into the blood or control the release of hormones from the pituitary gland (see Chapter 37). Other parts of the hypothalamus direct the activities of the autonomic nervous system. The hypothalamus maintains homeostasis by influencing body temperature, food intake, water balance, heart rate, blood pressure, the menstrual cycle, and circadian rhythms.

The **cerebrum** consists of two **cerebral hemispheres.** Each hemisphere is composed of an outer **cerebral cortex,** several clusters of neurons beneath the cortex near the thalamus, and bundles of axons, some that interconnect the two hemispheres and others that link the hemispheres with the midbrain and hindbrain (see Figs. 38-12a, b).

Structures in the Interior of the Cerebrum Clusters of neurons in the **amygdala** (see Fig. 38-13 later) produce sensations of pleasure, fear, or sexual arousal when stimulated. In people, electrically stimulating the amygdala causes feelings of rage or fear. Damage to the amygdala early in life eliminates the ability both to feel fear and to recognize fearful facial expressions in other people.

The **hippocampus** (see Fig. 38-12b) plays an important role in the formation of long-term memory, particularly of places, and is thus required for learning, as discussed in more detail later in this chapter. The hippocampus is evolutionarily very ancient—all vertebrates have a homologous part of the brain. What's more, the hippocampus-like region appears to be involved in "place learning" in most, perhaps all, vertebrates. For example, blue jays, Steller's jays, and nutcrackers store seeds for the winter and must remember where their caches are. These birds have a larger hippocampus than most other birds do. In humans, London taxi drivers, who must memorize that city's maze of streets, have larger-than-average hippocampi; repeated, intensive place learning apparently causes the hippocampus to become larger.

The **basal ganglia** consist of structures deep within the cerebrum (see Fig. 38-12b), as well as the substantia nigra in the midbrain. These structures are important in the overall control of movement. The motor part of the cerebral cortex directs specific movements, such as which motor neurons to fire to activate just the right muscles to pick up a pen. The

basal ganglia appear to be essential to the decision to initiate a particular movement and suppress other movements, for example, to pick up the pen but not simultaneously attempt to comb your hair with the same hand. This function is most clearly illustrated by two major disorders of the basal ganglia. In Parkinson's disease, the substantia nigra degenerates, and affected people have a hard time starting a movement—the "freezing" symptom. In Huntington's disease, basal ganglia in the cerebrum degenerate, and affected people make involuntary, undirected movements.

The **limbic system** is a diverse group of forebrain structures, including the hypothalamus, amygdala, and hippocampus, as well as nearby regions of the cerebral cortex, located in a ring between the thalamus and cerebral cortex (**Fig. 38-13**). These structures help to produce emotions and emotional behaviors, including fear, rage, calm, hunger, thirst, and sexual responses. Other brain regions are also involved in emotions,

▲ **FIGURE 38-13 The limbic system** The limbic system includes several structures in the forebrain involved in producing emotions and responding to emotions perceived in others.

including other parts of the cerebral cortex, the midbrain, the hindbrain, and possibly even the spinal cord.

The Cerebral Cortex The cerebral cortex is the thin outer layer of each cerebral hemisphere, with billions of neurons packed in a highly organized way into a sheet just a few millimeters thick. The cortex is folded into *convolutions*, which are raised, wrinkled ridges that increase its surface area to over two square yards—about the area of a single bed! Neurons in the cortex receive sensory information, process it, direct voluntary movements, create memories, and allow us to be creative and even envision the future. The cortexes in each hemisphere communicate with each other through a large band of axons, the **corpus callosum** (see Fig. 38-12).

Each hemisphere of the cerebral cortex is divided into four anatomical regions: the frontal, parietal, occipital, and temporal lobes (**Fig. 38-14**). The cortex can also be divided into functional regions. Primary sensory areas are regions of the parietal, temporal, and occipital lobes that receive input from senses such as the ears, eyes, temperature receptors in the skin, and so on. In the primary sensory areas, these inputs are converted into subjective impressions, such as sound, light, heat, or cold. Nearby association areas interpret the stimuli. Sounds, for example, may be interpreted as speech, music, dogs barking, car horns honking, and so forth. Primary motor areas in the frontal lobe command movements by stimulating motor neurons in the spinal cord that activate muscles, allowing you to walk to class or play a video game. An adjacent area, called the premotor area, receives inputs from sensory association areas and other parts of the cortex, and directs the motor area to produce movements. The prefrontal cortex, the part of the frontal lobe situated directly behind the bones of the forehead, is involved in complex brain functions such as short-term memory, decision making, predicting the consequences of actions, controlling aggression, planning for the future, and working for delayed rewards.

Neurons in the motor areas of each hemisphere of the cerebral cortex send axons down through the hindbrain into the spinal cord. The axons cross the midline and end up on the opposite side of the spinal cord, where they stimulate motor neurons that innervate the skeletal muscles (see Figs. 38-9 and 38-10). Similarly, sensory pathways from most of the body enter the spinal cord and cross over to the opposite side on their way to the two hemispheres. Because both sensory and motor pathways cross the midline on their way to and from the cerebral cortex, the left hemisphere of the cortex receives sensation from the right side of the body and controls movement of the right side of the body. The right hemisphere senses and controls the left side of the body.

Damage to the cortex from trauma, stroke, or a tumor results in specific deficits, such as problems with speech, difficulty reading, or the inability to sense or move specific parts of

CASE STUDY continued

How Do I Love Thee?

Dopamine-containing neurons in the midbrain send axons to several parts of the forebrain. When people fall in love, oxytocin stimulates dopamine release in these areas, contributing to emotional attachment, lack of fear and critical judgment of the loved one, and fond memories of times spent together.

◄ **FIGURE 38-14 The cerebral cortex** Different regions of the human left cerebral cortex are specialized to perform specific functions. A map of the right cerebral cortex would be similar, except that speech and language areas would not be as well developed.

the body, depending on where the damage occurs. Most brain cells of adults cannot be replaced, so if a brain region is destroyed, these deficits may be permanent. Fortunately, however, training can sometimes allow undamaged regions of the cortex to take control over and restore some of the lost functions.

How Do Neuroscientists Learn About the Functions of Brain Regions?

As we described the functions of different parts of the brain, you may have wondered, "How do they know *that*?" Historically, the functions of different parts of the brain were discovered by examining the behaviors and abilities of people who suffered brain injuries, often in accidents or wars.

Consider the case of Phineas Gage (Fig. 38-15). In 1848, Gage was setting an explosive charge to clear rocks from a railroad line under construction, when the gunpowder triggered prematurely. The blast blew a 13-pound steel rod through his skull, severely damaging both of his frontal lobes. Amazingly, Gage survived his wounds. However, his personality changed radically. Before the accident, Gage was conscientious, industrious, and well liked. After his recovery, he became impetuous, profane, and incapable of working toward a goal.

Studies of people with brain injuries, such as Phineas Gage, have revealed that many parts of the brain are highly specialized. One patient with very localized damage to the left frontal lobe is unable to name fruits and vegetables, although he can name everything else. Some other victims of localized brain damage are unable to recognize faces.

Although evaluating the behavioral effects of brain damage has played an important role in discovering the functions of the various parts of the brain, modern neuroscientists have powerful, often noninvasive, techniques to visualize activity in the intact brain. Positron emission tomography and magnetic resonance imaging allows researchers to watch the brain in real time while a person performs a task, as we explore in "Science in Action: Neuroimaging—Observing the Brain in Action" on p. 752.

The Left and Right Sides of the Brain Are Specialized for Different Functions

Although the two cerebral hemispheres are similar in appearance, this symmetry does not extend to function. We have already seen that the two hemispheres receive sensory information from and control movements of opposite sides of the body. There are major differences in the control of intellectual functions as well.

Beginning in the 1950s, Roger Sperry of the California Institute of Technology studied people whose hemispheres had been separated by cutting the corpus callosum to prevent the spread of epilepsy from one hemisphere to the other. Severing the corpus callosum prevented the two hemispheres from communicating with one another. Sperry made use of the fact that the optic nerves (which are not severed by the surgery) follow a pathway that causes the left half of each visual field to be "seen" by the right hemisphere and the right half to be seen by the left hemisphere (Fig. 38-16). Under everyday conditions,

▲ FIGURE 38-15 A revealing accident Studies of the skull of Phineas Gage have enabled scientists to re-create the path taken by the steel rod that was blown through his head.

even in someone with a severed corpus callosum, rapid eye movements inform both hemispheres about both the left and right visual fields. However, Sperry used an ingenious device that projected different images onto the left and right visual fields and thus sent different signals to each hemisphere.

When Sperry projected an image of a nude figure onto only the left visual field, the subjects would smile but claim to have seen nothing. The same figure projected onto the right visual field was readily described verbally. These and later experiments revealed that the left hemisphere is usually dominant in speech, reading, writing, language comprehension, mathematical ability, and logical problem solving. The right hemisphere is superior to the left in musical skills, artistic ability, recognizing faces, spatial visualization, and the ability to recognize and express emotions.

However, the left–right dichotomy is not complete and unchanging. A person who has suffered a stroke that damaged the left hemisphere typically shows symptoms such as loss of speaking ability. These deficits can often be partially overcome through training, even though the left hemisphere itself has not recovered. This observation suggests that the right hemisphere has some language capabilities.

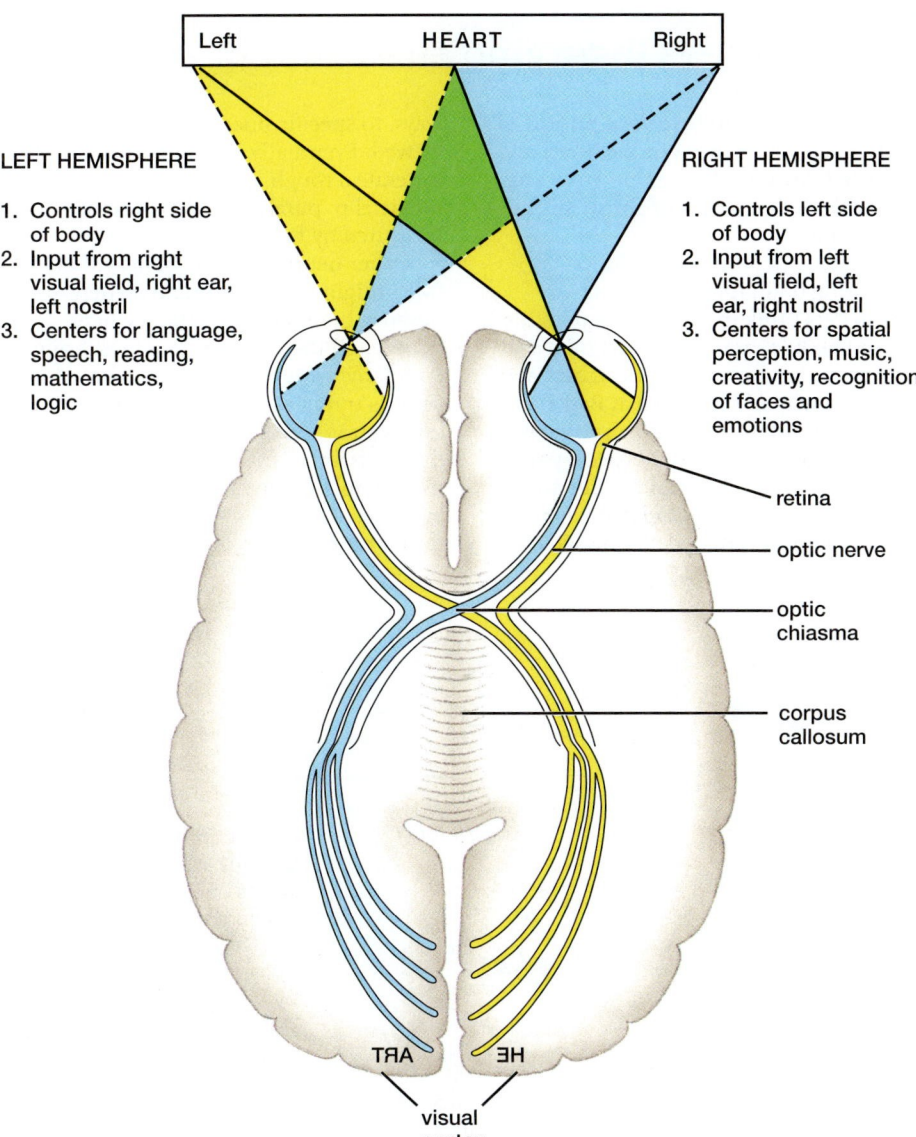

◄ FIGURE 38-16 Specialization of the cerebral hemispheres Each half of the retina of each eye "sees" the opposite visual field. The axons from the half-retinas that see the left visual field send information to the right hemisphere, and vice versa. Therefore, with a quick glance at the word "heart" (before you had a chance to move your eyes), the right hemisphere would perceive "he" and the left hemisphere would perceive "art." In addition to receiving inputs from different parts of the visual field, the two hemispheres typically control the opposite sides of the body and are specialized for a variety of functions.

Interestingly, women have a slightly larger corpus callosum than men do, suggesting a gender difference in the extent of interconnections between the two hemispheres. Imaging neural activity in the brains of normal subjects as they perform various mental tasks has provided further evidence of this difference: When subjects were asked to compare two word lists to find rhyming words, a specific region of the left cortex of male subjects became active, but in women, similar areas in both the left and right hemispheres were activated.

Learning and Memory Involve Biochemical and Structural Changes in Specific Parts of the Brain

Most neurobiologists and psychologists agree that learning has two phases: **short-term memory** and **long-term memory.** For example, if you look up a number in the phone book, you will probably remember the number long enough to dial it, but will then forget it. This is short-term memory. But if you call the number frequently, eventually you will remember the number more or less permanently—it has been stored in long-term memory.

The frontal and parietal lobes of the cerebral cortex, and some of the basal ganglia deep in the cerebrum, are the primary sites of short-term memory. Most short-term memory probably requires the repeated activity of a particular neural circuit in the brain. As long as the circuit is active, the memory stays. In other cases, short-term memory may rely on short-lived biochemical changes that temporarily strengthen synapses between specific neurons.

In contrast, long-term memory seems to be structural. It may require the formation of new, long-lasting synaptic connections between neurons or the long-term strengthening of existing but weak synapses (for example, by increasing neurotransmitter release or increasing the number of receptors for the neurotransmitter). For many memories, including memories of places, facts, and specific events, converting short-term memory into long-term memory seems to involve the hippocampus, which is believed to process new memories and transfer them to the cerebral cortex for permanent

SCIENCE IN ACTION Neuroimaging—Observing the Brain in Action

For most of human history, the brain has been a mystery. Now, however, imaging techniques provide exciting insights into brain function. These techniques include positron emission tomography (PET) and functional magnetic resonance imaging (fMRI). How do these work? Both rely on the fact that regions of the brain that are most active need the most energy; hence, they use more glucose and attract a greater flow of oxygenated blood than less-active areas do.

In typical PET scans, scientists inject the subject with a radioactive form of glucose and monitor levels of radioactivity, which reflect differences in glucose metabolism and hence in brain activity. These differences are translated by a computer into colors on images of the brain. For example, PET scans have confirmed that specific aspects of language occur in distinct areas of the cerebral cortex (Fig. E38-4).

Functional MRI detects differences in the way oxygenated and deoxygenated blood respond to a powerful magnetic field. Active brain regions can be distinguished with fMRI without using radioactivity and over much shorter time spans than those required by PET. Consequently, fMRI has become the favored tool for localizing specific brain functions. Functional MRI scans have been used to determine the parts of the brain that are most active during various emotional states. For example, when a person is frightened, the amygdala lights up. When people in love view photos of their lovers, other areas in the brain are activated. Most of these same areas are also activated by drugs of abuse, such as cocaine.

What are the limits of fMRI in evaluating brain function? Some researchers think that it may be possible to read people's minds, at least on a superficial level. In 2011, researchers at the University of California at Berkeley collected fMRI data while people viewed short movie clips. Computers compared brain activity to the actual movies and built up models that correlated which brain regions responded, and in what ways, to specific images that the subjects viewed. Eventually, the computer could generate a rough approximation of the movie clip, purely from brain activity measured by fMRI scans.

Other neuroscientists believe that fMRI might be the ultimate lie detector, because brain activity differs when people tell the truth or lie. At least one company, with the catchy name No Lie MRI, is trying to commercialize this application. As of 2011, U.S. courts have ruled that fMRI lie detection is not reliable enough to be admitted as evidence. But with a few more years of development, who knows?

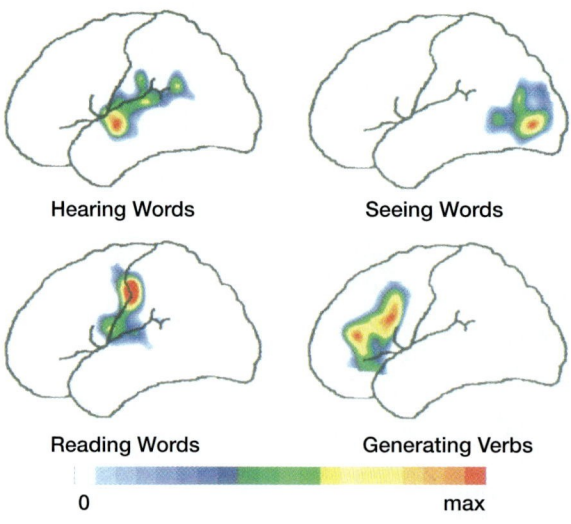

Hearing Words   Seeing Words

Reading Words   Generating Verbs

0   max

◄FIGURE E38-4
Localization of language tasks PET scans reveal the different cortical regions involved in language-related tasks, based on research by Marcus Raichle of the Washington University School of Medicine in St. Louis. The scale ranges from white (lowest brain activity) to red (highest).

storage. Although long-term memory probably resides in many areas of the cerebrum, some research suggests that the temporal and frontal lobes are particularly important. The cerebellum and basal ganglia are crucial for learning and storing habits and physical skills (motor learning).

CHECK YOUR LEARNING

Can you distinguish between the central and peripheral nervous systems, and the somatic and autonomic nervous systems? Are you able to describe the autonomic nervous system and its divisions, and provide some examples of activities controlled by each division? Can you label diagrams of the human brain and spinal cord, naming the principal structures and describing their functions? Are you able to describe the different functions that are usually controlled by the left and right halves of the human cerebral hemispheres? Can you distinguish between short-term and long-term memory?

CASE STUDY revisited

How Do I Love Thee?

"... heaven is here,
Where Juliet lives."

—Romeo and Juliet,
Act III, scene III

What happens in the human brain when we fall in love? Although people aren't just big prairie voles, you may be surprised—and perhaps disappointed—to learn that people and prairie voles show striking similarities in both brain function and hormones during emotional encounters. For example, magnetic resonance imaging has shown that areas of the human brain that contain oxytocin and dopamine are activated when people are shown pictures of their lovers or children, but not when they are shown images of equally attractive and familiar people to whom they feel no emotional bond. Some of these areas are the same ones that are activated in prairie voles and that appear to be important in motivation and reward. In humans, as in voles, oxytocin probably plays an important role in attraction and commitment. Oxytocin also reduces stress and inhibits the amygdala, a part of the brain involved in feeling fear. Oxytocin enhances trust, even between complete strangers. Finally, oxytocin levels increase in women and men during sexual encounters.

What about different kinds of love? Brain scans reveal that some of the same brain areas are activated by photos of lovers and of children. Other areas are activated by one or the other, but not both. Still other brain areas, particularly those involved in critical decision making and social judgments, are turned off, at least while seeing the lover or child, who almost always seems to be better than he or she really is.

Consider This

BioEthics You can buy oxytocin online, in the form of a cologne, perfume, or nasal spray. The claim is that it will make other people trust you more, thus aiding both business and sexual conquests. Scientific research indeed shows that oxytocin enhances trust when sprayed into the nose, which provides direct access to olfactory neurons and the brain. Assuming that wearing oxytocin cologne produces a high enough concentration of oxytocin in the air to affect anyone's brain, would you buy it? Why or why not?

CHAPTER REVIEW

Summary of Key Concepts

38.1 What Are the Structures and Functions of Nerve Cells?

Nervous systems are composed of cells called neurons. A neuron has four major functions, which are reflected in its structure: (1) dendrites receive information from the environment or from other neurons; (2) the cell body adds together electrical signals from synapses on the dendrites and on the cell body itself, with the resulting electrical potential determining whether or not the neuron produces an action potential; (3) the axon conducts the action potential to its synaptic terminal; and (4) synaptic terminals release neurotransmitters that transmit the signal to glands, muscles, or other neurons.

38.2 How Do Neurons Produce and Transmit Information?

An unstimulated neuron maintains a negative resting potential inside the cell. Signals received from other neurons are small, rapidly fading changes in potential called postsynaptic potentials. Inhibitory and excitatory postsynaptic potentials (IPSPs and EPSPs) make the neuron less likely or more likely, respectively, to produce an action potential. If postsynaptic potentials, added together within the cell body, bring the neuron to threshold, an action potential will be triggered. An action potential is a wave of positive charge that travels, undiminished in magnitude, along an axon to its synaptic terminals.

A synapse consists of the synaptic terminal of the presynaptic neuron, a specialized region of the postsynaptic neuron, and the gap between the neurons, called the synaptic cleft. Neurotransmitters from the presynaptic neuron, released in response to an action potential, diffuse across the synaptic cleft, bind to receptors in the postsynaptic cell's plasma membrane, and produce either an EPSP or an IPSP.

38.3 How Does the Nervous System Process Information and Control Behavior?

Neural pathways usually have four elements: (1) sensory neurons, (2) interneurons, (3) motor neurons, and (4) effectors (muscles or glands). These elements carry out four operations: (1) determining the type of stimulus, (2) determining and signaling the intensity of the stimulus, (3) integrating information from many sources, and (4) directing appropriate behaviors.

38.4 How Are Nervous Systems Organized?

Nervous systems may be diffuse (distributed throughout the body, usually in radially symmetrical animals) or centralized (with most of the senses and nervous system in the head, in bilaterally symmetrical animals). Complex nervous systems are centralized.

38.5 What Are the Structures and Functions of the Human Nervous System?

The nervous system of humans and other mammals consists of the central nervous system and the peripheral nervous system. The peripheral nervous system is divided into sensory and motor portions. The motor portion consists of the somatic nervous system (which controls voluntary movement) and the autonomic nervous system (which directs involuntary responses). The autonomic nervous system is further subdivided into the sympathetic and parasympathetic divisions.

The central nervous system consists of the brain and spinal cord. The spinal cord contains neurons that control voluntary muscles and the autonomic nervous system; neurons that communicate with the brain and other parts of the spinal cord; axons leading to and from the brain; and neural pathways for reflexes and certain simple behaviors.

The brain consists of three parts: the hindbrain, midbrain, and forebrain. Each is further subdivided into distinct regions. The hindbrain in humans consists of the medulla and pons, which control involuntary functions (such as breathing), and the cerebellum, which coordinates complex motor activities (such as typing). In humans, the small midbrain contains clusters of neurons that help to control movement, arousal, and emotion. The midbrain and hindbrain contain the reticular formation, which is a filter and relay for sensory stimuli. The forebrain includes the thalamus, a sensory relay station that shuttles information to and from conscious centers in the forebrain; the hypothalamus, which is responsible for maintaining homeostasis and directs much of the activity of the autonomic nervous system; and the cerebrum, the center for information processing, memory, and initiation of voluntary actions. The cerebral cortex includes primary sensory and motor areas, together with association areas that analyze sensory information and plan movements. A group of forebrain structures called the limbic system, together with many other parts of the brain, controls the perception and expression of emotions.

The cerebral hemispheres are specialized. In general, the left hemisphere controls the right side of the body and dominates speech, reading, writing, language comprehension, mathematical ability, and logical problem solving. The right hemisphere controls the left side of the body and specializes in recognizing faces and spatial relationships, producing artistic and musical abilities, and recognizing and expressing emotions.

Memory takes two forms: Short-term memory is electrical or chemical, whereas long-term memory probably involves structural changes that increase the effectiveness or number of synapses. The hippocampus is an important site for the transfer of information from short-term into long-term memory.

Key Terms

action potential 732	long-term memory 751
amygdala 748	medulla 746
autonomic nervous system 742	midbrain 745
	motor neuron 739
axon 732	myelin 734
basal ganglion 748	nerve 732
blood–brain barrier 742	nerve net 740
brain 741	neuron 732
cell body 732	neurotransmitter 732
central nervous system (CNS) 740	parasympathetic division 742
cerebellum 746	peripheral nervous system (PNS) 741
cerebral cortex 748	pons 746
cerebral hemisphere 748	postsynaptic neuron 734
cerebrum 748	postsynaptic potential (PSP) 736
corpus callosum 749	presynaptic neuron 734
dendrite 732	reflex 742
dorsal root ganglion 742	resting potential 734
effector 739	reticular formation 747
excitatory postsynaptic potential (EPSP) 737	sensory neuron 739
forebrain 745	short-term memory 751
ganglion (plural, ganglia) 740	sodium-potassium (Na^+-K^+) pump 736
glia 732	somatic nervous system 742
gray matter 742	spinal cord 741
hindbrain 745	sympathetic division 742
hippocampus 748	synapse 734
hypothalamus 748	synaptic cleft 734
inhibitory postsynaptic potential (IPSP) 736	synaptic terminal 734
integration 738	thalamus 748
interneuron 739	threshold 734
limbic system 748	white matter 742

Learning Outcomes

In this chapter, you have learned to . . .

LO1 Identify the structures of a typical neuron, and explain how each structure functions.

LO2 Describe resting potentials, action potentials, inhibitory postsynaptic potentials, and excitatory postsynaptic potentials, and explain the relationships among these potentials.

LO3 Explain the process of synaptic transmission.

LO4 Describe the cellular components of a typical neural pathway and how each component functions during information processing.

LO5 Explain how the nervous system codes for the nature and intensity of a stimulus.

LO6 Describe and distinguish among the central, peripheral, somatic, and autonomic nervous systems.

LO7 Diagram the human brain and spinal cord, labeling the major structures within each, and describe the principal functions of each of those structures.

LO8 Explain how the left and right cerebral hemispheres of the human brain differ in function.

LO9 Compare and contrast short-term memory and long-term memory.

Thinking Through the Concepts

Fill-in-the-Blank

1. An individual nerve cell, also called a(n) _____, includes structures specialized to perform different functions. The "input" end of a nerve cell, called a(n) _____, receives information from the environment or from other nerve cells. The _____ contains the nucleus and other typical organelles of a eukaryotic cell. Electrical signals are sent down the _____, a long, thin strand that leads to the _____, where the nerve cell sends out its signal to other cells. **LO1**

2. When they are not being stimulated, neurons have an electrical charge across their membranes called the resting potential. This potential is _____ charged inside. When a neuron receives a sufficiently large stimulus, and reaches a potential called the _____, it produces an action potential. This causes the neuron to become _____ charged inside. **LO2**

3. When an action potential reaches a synaptic terminal, it causes the release of a chemical called a(n) _____. This chemical binds to protein _____ on the postsynaptic cell, causing a change in potential. If the postsynaptic cell becomes less negative, this change in potential is called a(n) _____. If the postsynaptic cell becomes more negative, it is called a(n) _____. **LO3**

4. The _____ is part of the peripheral nervous system that innervates skeletal muscles. Smooth muscles and glands are innervated by the _____, which has two divisions: the _____, active during "fight or flight," and the _____, active during "rest and digest." **LO6**

5. The human hindbrain consists of three parts: the _____, the _____, and the _____. One of these, the _____, is important is coordinating complex movements such as typing. **LO7**

6. The cerebral cortex consists of four lobes: the _____, _____, _____, and _____. The visual centers are located in the _____ lobe. The primary motor areas are in the _____ lobe. **LO7**

Review Questions

1. Diagram a synapse. How are signals transmitted from one neuron to another at a synapse? **LO3**

2. How does the brain perceive the intensity of a stimulus? The type of stimulus? **LO5**

3. What are the four elements of a neuron-to-muscle pathway? Describe how these elements function in the human pain-withdrawal reflex. **LO4**

4. Draw a cross-section of the spinal cord. What types of neurons are located in the spinal cord? Explain why severing the spinal cord paralyzes the body below where the cut occurs. **LO7**

5. Describe the functions of the following parts of the human brain: medulla, cerebellum, reticular formation, thalamus, amygdala, and cerebral cortex. **LO7**

6. What structure connects the two cerebral hemispheres? Describe the functions of the two hemispheres in most right-handed people. **LO8**

7. Explain the differences between short-term memory and long-term memory. **LO9**

Applying the Concepts

1. BioEthics Parkinson's disease is caused by the degeneration of dopamine-producing cells in the substantia nigra, a part of the midbrain important in controlling movement. Some physicians have reported improvement after injecting cells taken from the same general brain region of an aborted fetus into appropriate parts of the brain of a Parkinson's patient. Discuss this type of surgery from as many viewpoints as possible: ethical, financial, practical, and so on. Based on your responses, is fetal transplant surgery the answer to curing Parkinson's disease?

2. If the axons of human spinal cord neurons were unmyelinated, would you expect the spinal cord to be larger or smaller? Would you move faster or slower? Explain your answer.

3. What is the adaptive value of reflexes? Why couldn't all behaviors be controlled by reflexes?

Answers to Figure Caption questions and Fill-in-the-Blank questions can be found in the Answers section at the back of the book.

(MB)® *Go to MasteringBiology for practice quizzes, activities, eText, videos, current events, and more.*

For many deaf people, cochlear implants reveal a new world of sound.

CASE STUDY
Bionic Ears

SAMANTHA BRILLING was born hearing impaired, although not totally deaf. She wore hearing aids for many years, but they didn't provide normal hearing. Sounds that most of us take for granted were missing for Samantha, including music and conversation with groups of her friends. Today, she can hear almost normally, thanks to a cochlear implant. As you will learn in this chapter, the cochlea is the part of the ear that converts sound to electrical signals, which then travel in the auditory nerve to the brain. A cochlear implant is a microprocessor-controlled device that partially replaces cochlear function in a deaf or hearing-impaired person by converting sound to electrical pulses that stimulate the auditory nerve.

Although she doesn't have perfect hearing, Samantha can now enjoy music, carry on a conversation in a crowded room, and know when someone out of sight is calling to her. She can hear her professors in class. In fact, Samantha can hear so well that she has held several jobs that require accurate communication with the general public, including working as a receptionist at her college alumni center and volunteering at a local hospital.

How do your ears produce the sensation of sound? How does a "bionic ear" replace a biological cochlea? How can electrical impulses, whether generated biologically or bionically, be understood by the brain as speech, bird calls, or music? And what about your other senses—how do you perceive the odor of pine trees, the colors of blooming flowers, the silky feel of a baby's skin, or the "burn" of hot peppers?

AT A GLANCE

LEARNING GOALS

LG1 Describe the important characteristics of sensory receptors, and summarize how nervous systems perceive environmental stimuli.

LG2 Define the five major types of sensory receptors and provide examples of each.

LG3 Describe the structures and functions of the mammalian ear, including how the ear converts sound waves into electrical activity, how it signals the intensity and pitch of sound, and how it signals the direction of gravity and movement of the head.

LG4 Describe the structures and functions of the mammalian eye, including how light is focused, how photoreceptors convert light into electrical activity, and how the retina signals the color and intensity of light.

LG5 Compare and contrast the senses of taste and olfaction.

LG6 Explain how pain is perceived.

39.1 HOW DO ANIMALS SENSE THEIR ENVIRONMENT?

You're probably used to the notion that animals have five senses—touch, hearing, vision, smell, and taste. These are, of course, crucial senses that animals use to perceive stimuli in their external environment. However, animals detect many other external stimuli as well, such as temperature and the direction of gravity. Some can detect magnetic or electrical fields, and use this information to move around in murky water, to migrate, or to find prey. In addition, animals need to evaluate their internal environment, including conditions such as blood pH, oxygen, and glucose levels; body temperature; the positions of their limbs; and how full their bladders are. Detecting stimuli from both the external and internal environments is the function of the senses.

All sensory perception begins with a **receptor:** a molecule, cell, or multicellular structure that produces a response when it is acted on by a stimulus. Every cell has many types of receptor molecules, such as membrane proteins that bind hormones and trigger responses in the cell. A **sensory receptor** is a specialized cell (often a neuron) that produces an electrical signal in response to an environmental stimulus—that is, it translates an environmental stimulus into the language of the nervous system. Sensory receptors can be grouped into five major categories, according to the stimuli to which they respond (Table 39-1).

Many sensory receptors are nerve cells with specialized dendrites, often called *free nerve endings*, that are located in the skin, the digestive and respiratory tracts, the bladder, and many other body parts. The plasma membranes of these dendrites have receptor proteins that respond to stimuli such as heat, cold, touch, or pain. Other sensory receptors are housed in a **sense organ**—a structure that includes both the sensory receptors and accessory structures that play essential roles in detecting specific stimuli. The most familiar sense

TABLE 39-1 Principal Categories of Vertebrate Sensory Receptors

Category of Receptor	Stimuli	Sensory Cell Type	Location
Thermoreceptor	Heat, cold	Free nerve ending	Skin, brain
Mechanoreceptor	Vibration, motion, gravity	Hair cell	Inner ear
	Vibration, pressure, touch	Free nerve endings and endings surrounded by accessory structures	Skin
	Stretch	Specialized nerve endings in muscles or joints	Muscles, tendons
Photoreceptor	Light	Rod, cone	Retina of the eye
Chemoreceptor	Odor (airborne molecules)	Olfactory receptor	Nasal cavity
	Taste (waterborne molecules)	Taste receptor	Tongue and oral cavity
Pain receptor	Chemicals released by tissue injury; extreme heat or cold; excessive stretch; acid	Free nerve ending	Widespread in the body

organs are the eyes and ears, which contain accessory structures such as the lens and the eardrum. However, sensory receptors for vibration, pressure, odors, and tastes are also located within sense organs. Perhaps the most striking example of the importance of the accessory structures for sensory reception occurs in the mammalian inner ear. As we will describe in Section 39.3, virtually identical sensory receptor cells respond to the pull of gravity, motion of the head, and musical notes ranging from rumbling bass to screechingly high treble, depending on the accessory structures that enclose them.

The Senses Inform the Brain About the Nature and Intensity of Environmental Stimuli

For sensory information about the environment to be useful, the brain must determine the nature of the stimulus—light, sound, or touch, for example—and the strength, or intensity, of the stimulus—bright or dim light, loud or soft sounds, and so on. All sensory receptors produce electrical signals in response to environmental stimuli. Further, no matter how much complex processing may occur within a sense organ, all sensory information ultimately reaches the brain as action potentials traveling in axons that connect the sense to the brain (see Chapter 38). Given that all action potentials are fundamentally the same, how can the brain recognize the nature and intensity of a stimulus?

Determining the nature of a stimulus begins with sensory receptor cells, each of which contains receptor molecules that respond to certain stimuli and not others. Each sensory receptor is linked to a specific set of axons that connect to particular locations in the brain or spinal cord. Electrical activity in specific regions of the brain is then interpreted as a specific form of sensory perception. In mammals, for example, neurons that detect odors send axons to a part of the brain called the olfactory bulb. The activity of specific sets of neurons in the olfactory bulb results in the perception of specific odors, such as coffee or roses.

The linkage of stimulus type to sensory receptor to axons that lead to a particular brain region produces the first principle of sensory perception: The type of stimulus is encoded by which neurons in the brain are activated.

Determining the **intensity** of a stimulus—loud versus soft sounds, for instance—also begins with sensory receptor cells. When a sensory receptor is stimulated, it produces an electrical signal called a **receptor potential** (Fig. 39-1). In many sensory receptor neurons, the receptor potential may bring the neuron above threshold and trigger action potentials. The axons of these sensory receptor neurons often connect directly with the central nervous system (CNS).

Unlike action potentials, which are always the same size (see Chapter 38), receptor potentials vary in size with the intensity of a stimulus—the stronger the stimulus, the larger the receptor potential (Fig. 39-2). A small receptor potential may barely reach threshold and, hence, will produce only a few action potentials; a large receptor potential will go far above threshold, causing a high frequency of action potentials.

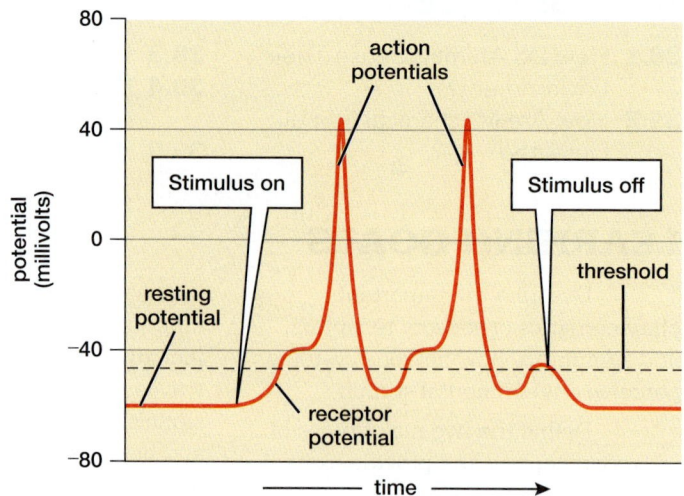

▲ FIGURE 39-1 **Converting an environmental stimulus to action potentials** In most sensory receptor neurons, an environmental stimulus generates a receptor potential that makes the resting potential less negative. If the receptor potential is large enough, it reaches threshold and triggers action potentials.

Some sensory receptors, such as those in the inner ear, do not have axons. Most of these sensory receptors form synapses with neurons that do have axons connecting to the CNS (see Fig. 38-4 for a description of the anatomy and function of a synapse). Receptor potentials in this type of sensory receptor cause neurotransmitters to be released onto a neuron, ultimately stimulating action potentials that move down the neuron's axon to the CNS. In such cases, the stronger the stimulus, the larger the receptor potential, which causes the release of larger amounts of transmitter onto the neurons, producing a higher frequency of action potentials traveling to the CNS.

The linkage of stimulus intensity to receptor potential size to action potential frequency produces the second principle of sensory perception: The intensity of a stimulus is encoded by the frequency of action potentials reaching the brain.

CHECK YOUR LEARNING

Can you list and describe the five major types of sensory receptors, and give an example of each? Can you describe how the nervous system codes for the type of stimulus and the intensity of a stimulus? Can you define the term *receptor potential* and explain how the size of a receptor potential helps to code for stimulus intensity?

CASE STUDY continued

Bionic Ears

Cochlear implants work because the brain interprets action potentials in axons of the auditory nerve as sound, regardless of the actual stimulus triggering them. Alessandro Volta, who invented the battery in 1800, discovered this principle a few years later. He stuck a metal rod in his ear and connected it to a battery. He felt a jolt and heard a sound like boiling water. Like Volta's rod, a modern cochlear implant stimulates action potentials in axons of the auditory nerve, although with much more sensitivity and precision.

(a) Weak stimulus

A weak stimulus produces a small receptor potential that reaches just above threshold and triggers only a few action potentials

threshold

▲FIGURE 39-2 Stimulus intensity is encoded by the frequency of action potentials (a) Weak stimuli cause small receptor potentials that barely reach threshold (dashed line) and produce only a few action potentials. (b) Strong stimuli cause large receptor potentials that reach far above threshold, producing many action potentials.

(b) Strong stimulus

A strong stimulus produces a large receptor potential that reaches far above threshold and triggers many action potentials

threshold

39.2 HOW ARE MECHANICAL STIMULI DETECTED?

Mechanoreceptors, found throughout the human body, respond to physical deformation such as the stretching, dimpling, or bending of the various parts of the body. Mechanoreceptors include receptors in the skin that respond to touch, vibration, or pressure; stretch receptors in many internal organs, including the intestines, stomach, urinary bladder, and muscles; and receptors in the inner ear that respond to sound, gravity, or movement (see Sections 39.3 and 39.4).

Embedded in the human skin are the endings of several types of mechanoreceptor neurons, which produce receptor potentials when their membranes are stretched or dimpled (Fig. 39-3). The free nerve endings of some touch receptors produce sensations of touch, itching, or tickling. The endings of other receptors are enclosed in accessory structures— for example, *Pacinian corpuscles*, which respond to changes in pressure such as rapid vibrations or a sharp poke; *Meissner's corpuscles*, which respond to light touch or slow vibrations; and *Ruffini corpuscles*, which respond to steady pressure. The density of mechanoreceptors in the skin varies tremendously over the surface of the body. For example, each square inch

of fingertip has hundreds of touch receptors, but on the back, there may be less than one per square inch.

Mechanoreceptors in many hollow organs, such as the stomach and urinary bladder, signal fullness by responding to stretch. Mechanoreceptors in the joints, also responding to stretch, let us know whether the joints are straight or bent, and by how much. Stretch receptors in muscles tell the brain if the muscle is contracted or stretched, and by how much. Joint and muscle receptors combine to inform the brain of the position, angle, and force of movement of arms, legs, hands, and feet—you don't have to look at them or consciously think about what they're doing. Imagine how annoying it would be if you had to watch your fork on its way to your mouth to avoid stabbing yourself in the face!

Mechanoreceptors are important to all animals. Spiders, for example, use mechanoreceptors in their legs to detect vibrations of their webs. A spider can sense whether the vibrations are from small fluttering objects that might be good to eat, like flies or moths; from larger, possibly predatory, animals; or from potential mates. Fish have mechanoreceptors in organs called the lateral lines, which detect movement and vibration of the water around them. Some, such as blind cave fish that spend their lives in total

▶ FIGURE 39-3 **Receptors in the human skin** The diversity of receptors in the skin allows us to perceive mechanical stimuli such as touch, pressure, and vibration, as well as other sensations such as pain, heat, and cold. (See Chapter 31 for a description of the other structures of the skin that are illustrated here.)

hair

Meissner's corpuscle (light touch, slow vibrations)

layers of skin

Pacinian corpuscle (rapid vibration, rapid pressure changes)

free nerve ending (hair movement)

free nerve ending (touch, heat, cold, pain)

Ruffini corpuscle (steady pressure)

darkness, use their lateral lines to navigate through their surroundings without bumping into things.

CHECK YOUR LEARNING

Can you describe the types of stimuli that are detected by mechanoreceptors? Can you give some examples of mechanoreceptors in your body and their functions?

39.3 HOW IS SOUND DETECTED?

The mammalian ear performs two very different functions: perceiving sounds and determining the direction of gravity and the orientation and movement of the head, information that is used to control balance and body posture. In this section, we will describe the role of the ear in sound perception.

The Ear Converts Sound Waves into Electrical Signals

Sound is produced by vibrating objects—drums, vocal cords, or the speakers in your cell phone. Our ears convert the resulting sound waves into electrical signals that our brains interpret as the direction, pitch, and loudness of sound.

The ears of humans and other mammals consist of three parts: the outer, middle, and inner ear (**Fig. 39-4a**).

The **outer ear** consists of the pinna and the auditory canal. The **pinna,** a flap of skin-covered cartilage attached to the surface of the head, collects sound waves. Humans and other large animals determine the direction a sound comes from by differences in when the sound arrives at the two ears and in how loud the sound is in each ear. The shape of the pinna and, in many animals, the ability to swivel it around, further help to determine the location of sounds.

The **auditory canal** conducts sound waves from the pinna to the **middle ear,** which consists of the **tympanic membrane,** or eardrum; three tiny bones called the **hammer** (malleus), **anvil** (incus), and **stirrup** (stapes); and the **auditory tube** (Eustachian tube). The auditory tube in turn connects the middle ear to the pharynx and equalizes air pressure across the eardrum, between the middle ear and the outside atmosphere. The auditory tube may become swollen shut if you have a cold. If this happens, air pressure changes (such as those experienced during takeoff and landing in an airplane) cause the pressure on one side of the eardrum to be higher than on the other side. The resulting pressure difference pushes painfully on the eardrum, and can even cause it to break.

Sound waves traveling through the auditory canal vibrate the tympanic membrane, which in turn moves the hammer, anvil, and stirrup. These small bones then transmit

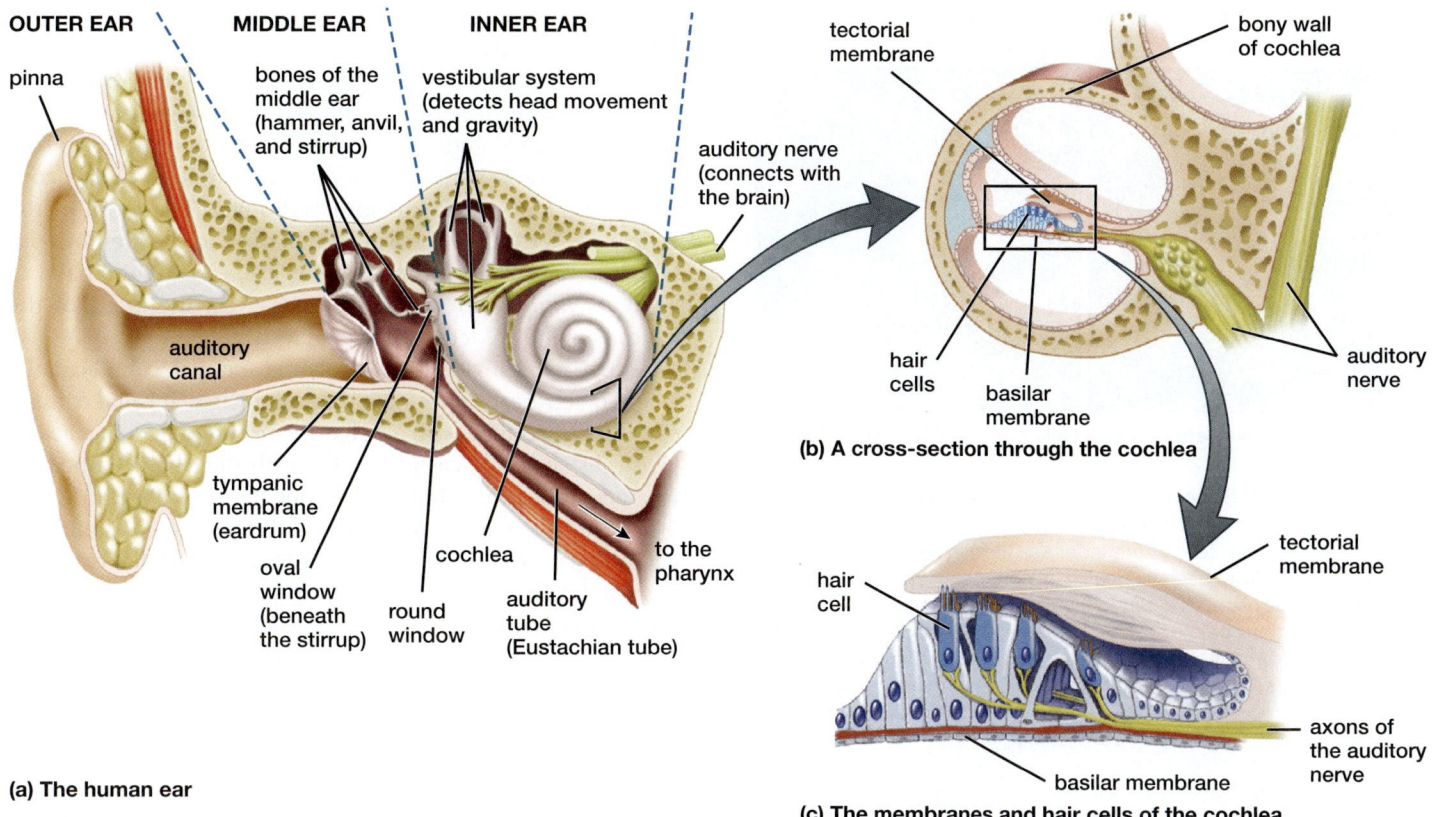

OUTER EAR   **MIDDLE EAR**   **INNER EAR**

pinna

bones of the middle ear (hammer, anvil, and stirrup)

vestibular system (detects head movement and gravity)

auditory nerve (connects with the brain)

auditory canal

tympanic membrane (eardrum)

oval window (beneath the stirrup)

round window

cochlea

auditory tube (Eustachian tube)

to the pharynx

tectorial membrane

bony wall of cochlea

hair cells

basilar membrane

auditory nerve

(b) A cross-section through the cochlea

hair cell

tectorial membrane

axons of the auditory nerve

basilar membrane

(c) The membranes and hair cells of the cochlea

(a) The human ear

▲ FIGURE 39-4 **The human ear** (a) Overall anatomy of the ear. (b) The cochlea consists of three fluid-filled compartments; hair cells sit atop the basilar membrane in the central compartment. (c) The hairs of hair cells span the gap between the basilar and tectorial membranes. Sound vibrations move the membranes relative to one another, bending the hairs and producing a receptor potential in the hair cells. The hair cells then release neurotransmitters that stimulate action potentials in the axons of the auditory nerve.

vibrations to the **inner ear.** The fluid-filled, hollow bones of the inner ear form a spiral-shaped tube called the **cochlea** ("snail" in Latin). The stirrup bone transmits sound waves to the fluid within the cochlea by vibrating the **oval window,** a flexible membrane covering the opening at the beginning of the cochlea. A second membrane, the **round window,** covers an opening at the far end of the cochlea. Sound waves cause the oval window and round window to move in opposite directions. When the crest of a sound wave pushes the oval window inward, the fluid in the cochlea causes the round window to bulge outward; when the trough of a sound wave pulls the oval window outward, the round window flexes inward. Thus the fluid in the cochlea shifts back and forth as the stirrup bone vibrates the oval window.

Vibrations Are Converted into Electrical Signals in the Cochlea

The cochlea consists of three fluid-filled compartments. The central compartment houses the receptors and the supporting structures that activate the receptors in response to sound vibrations. The floor of this central chamber is the **basilar membrane,** on top of which sit mechanoreceptors called **hair cells.** Hair cells have cell bodies topped by hairlike projections that resemble stiff cilia. Some of these hairs are embedded in a gelatinous structure called the **tectorial membrane (Figs. 39-4b, c).** As the fluid in the cochlea moves

back and forth in synchrony with incoming sound waves, it moves the basilar membrane relative to the tectorial membrane. Movement of the membranes bends the hairs, causing receptor potentials in the hair cells. The receptor potentials cause the hair cells to release neurotransmitters onto neurons whose axons form the **auditory nerve.** These axons produce action potentials that travel to auditory centers in the brain.

How do the structures of the inner ear allow us to perceive loudness (the magnitude of sound vibrations) and pitch (the musical note, or the frequency of sound vibrations)? Remember that the type of stimulus is encoded by which nerve cells fire action potentials, and the intensity of the stimulus is encoded by the rate of action potentials. Soft sounds cause small vibrations of the tympanic membrane, the bones of the middle ear, the oval window, and the basilar membrane. In response, the hairs bend only a little, and the hair cells produce small receptor potentials that trigger the release of a tiny bit of neurotransmitter, resulting in a low rate of action potentials in axons of the auditory nerve. Loud sounds cause large vibrations, which cause greater bending of the hairs and a larger receptor potential, producing a high rate of action potentials in the auditory nerve. Very loud sounds can damage the hair cells (**Fig. 39-5**), resulting in hearing loss, a fate suffered by some rock musicians and their fans. In fact, many sounds in our modern environment, including those coming from jets, trains, lawn mowers, and

hair cells   damaged hairs   scar

Normal hair cells (left) and hair cells damaged by loud sound (right)

▲ FIGURE 39-5 **Loud sounds can damage hair cells** Scanning electron micrographs (SEMs) show the effect of intense sound on the hair cells of the inner ear. (left) In a normal guinea pig, hairs emerge from each hair cell in a V-shaped pattern. (right) After a 24-hour exposure to a sound level approached by loud rock music (2,000 vibrations per second at 120 decibels), many of the hairs are damaged or missing, leaving "scars." Hair cells in humans do not regenerate, so such hearing loss would be permanent. Source: SEMs by Robert S. Preston, courtesy of Professor J. E. Hawkins, Kresge Hearing Research Institute, University of Michigan Medical School.

MP3 players, have the potential to damage hearing, especially if they are prolonged.

The perception of pitch is a little more complex. The basilar membrane is narrow and stiff at the end near the oval window, wider and more flexible near the tip of the cochlea. This progressive change in width and stiffness causes each portion of the membrane to vibrate most strongly when stimulated by a particular frequency of sound: high notes near the oval window and low notes near the tip of the cochlea. The brain interprets signals originating in hair cells near the oval window as high-pitched sound; signals from hair cells located progressively closer to the tip of the cochlea are interpreted as progressively lower in pitch. Young people with undamaged cochleas can hear sounds from about 20 vibrations per second (very low bass) to about 20,000 vibrations per second (very, very high treble). People gradually suffer hearing loss with age, especially for high-pitched sounds.

CHECK YOUR LEARNING

Can you describe the parts of the human ear and explain how sound waves move through the ear? Can you describe the structures and mechanisms by which sound waves are converted to electrical activity in the inner ear? Are you able to explain how pitch and loudness are encoded?

39.4 HOW ARE GRAVITY AND MOVEMENT DETECTED?

The inner ear not only detects sound, it also contains structures, collectively called the **vestibular apparatus,** that detect gravity and the orientation and movement of the head. The vestibular apparatus is a fluid-filled tube embedded in the bones of the skull, consisting of the vestibule (a small

chamber at the entrance to the apparatus) and the semicircular canals (**Fig. 39-6;** see also Fig. 39-4a).

The vestibule contains the **utricle** and **saccule,** which detect the direction of gravity and the orientation of the head. The utricle and saccule each consist of a cluster of hair cells. Their hairs are embedded in a gelatinous matrix containing tiny stones of calcium carbonate. The hairs in the utricle are vertical, whereas those in the saccule are horizontal. Gravity pulls the stones downward, causing the hairs to bend in various directions depending on the angle of the head. People can detect about a half-degree tilt.

Beyond the vestibule are three **semicircular canals,** which detect head movement. Each semicircular canal consists of a fluid-filled tube with a bulge at one end, called an *ampulla*. Hair cells sit inside each ampulla, with their hairs embedded in a gelatinous capsule (but without the stones found in the utricle and saccule). Acceleration of the head—for example, if you nod your head "yes" or lurch sideways in a roller coaster—pushes the fluid against the capsule, bending the hairs. The three semicircular canals are arranged perpendicularly to each other, similar to the two walls and the floor in the corner of a room, allowing you to detect head movement in any direction.

As in the cochlea, axons from the auditory nerve innervate the hair cells of the vestibular apparatus. When the hairs bend, the hair cells release neurotransmitter onto the endings of these axons, triggering action potentials that travel in the axons to balance and motion centers in the brain.

CHECK YOUR LEARNING

Can you describe the structures that are used to detect gravity and the orientation and movement of the head? Can you explain how a single type of sensory receptor, the hair cell, can respond to sound, gravity, or movement?

39.5 HOW IS LIGHT PERCEIVED?

The vast majority of animals can detect light. Some, such as flatworms and jellyfish, can distinguish light from dark, but their eyespots, as they are often called, cannot form an image. Most arthropods, some mollusks such as octopus and squid, and almost all vertebrates have eyes that do form images of the world around them.

In all animals, regardless of the sophistication of their eyes, vision begins with cells called **photoreceptors.** These cells contain **photopigments,** receptor molecules that change shape when they absorb light. This shape change sets off chemical reactions inside the photoreceptors that result in receptor potentials.

The Compound Eyes of Arthropods Produce a Mosaic Image

Many arthropods, including insects and crustaceans, have **compound eyes,** which consist of a mosaic of many individual light-sensitive subunits called **ommatidia** (singular, ommatidium; **Fig. 39-7**). Each ommatidium functions as an individual light detector, like the pixels in a digital camera. However, the best arthropod eyes—in dragonflies, for example—contain

calcium carbonate stones

gelatinous matrix

hairs

hair cells

axons from the
auditory nerve

structure of the utricle

semicircular
canals

utricle

saccule

vestibule

auditory
nerve

cochlea

ampullae

ampulla

gelatinous material

hairs

hair cells

axons from the
auditory nerve

structure of an ampulla

◀ FIGURE 39-6 **The vestibular apparatus detects gravity and the orientation and movement of the head** (top) The hairs of hair cells in the utricle and saccule bend under the weight of calcium carbonate stones, providing information about the direction of gravity and the orientation of the head. (bottom) The hairs of hair cells in the ampullae of the semicircular canals bend when head movement causes the fluid in the canals to slosh around.

QUESTION The vestibular apparatus is located completely inside the head. If you close your eyes, how can you tell if the rest of your body is tilted with respect to gravity?

▼ FIGURE 39-7 **Compound eyes** (a) A scanning electron micrograph of the head of a mosquito, showing a compound eye on each side of the head. (b) Each eye is made up of numerous ommatidia. Within each ommatidium are several receptor cells, capped by a lens. Pigmented cells surrounding each ommatidium prevent the passage of light to adjacent receptors.

(a) Compound eyes

lenses

pigmented
cells

receptor
cells

(b) Ommatidia

Single ommatidium

▶ **FIGURE 39-8 A bee's view of the world** At a distance of 2 or 3 feet, a bee sees **(a)** an orchid flower as **(b)** a mosaic of individual blobs of color.

(a) A tropical orchid

(b) The same orchid as seen by a bee

only about 30,000 ommatidia. A human eye, in contrast, contains more than 100 million photoreceptors. By human standards, therefore, the image formed by a compound eye is very pixilated, like an extremely low-resolution digital image that has been enlarged too much (**Fig. 39-8**). Compound eyes are excellent at detecting movement, as light and shadow flicker across adjacent ommatidia, which is an advantage in avoiding predators and in hunting.

The Mammalian Eye Collects and Focuses Light and Converts Light into Electrical Signals

A mammalian eye is built something like a camera (**Fig. 39-9**). The eye consists of two major modules: (1) accessory structures that hold the eye in a fairly fixed shape, control the amount of light that enters, and focus the light rays (comparable to the body and lens of a camera); and (2) the retina, which contains photoreceptors that respond to incoming light (comparable to the image sensors in a digital camera).

The eyeball is surrounded by the **sclera,** a tough connective tissue layer that is visible as the white of the eye and is continuous with the transparent **cornea** at the front. Light enters the eye through the cornea. The light then travels through a chamber filled with a watery fluid called **aqueous humor,** which provides nourishment for the cells of both the lens and the cornea. The light continues through the **pupil,** a circular opening in the center of the colored **iris.** Contraction and relaxation of the muscles of the iris regulate the size of the pupil and, hence, the amount of light entering the rest of the eye. Light next encounters the **lens,** a structure composed of transparent proteins and shaped like a flattened sphere. The lens is suspended behind the pupil by a ring of smooth muscle. Behind the lens is a large chamber filled with **vitreous humor,** a clear jelly-like substance that helps maintain the shape of the eyeball.

After passing through the vitreous humor, light hits the **retina.** Here, light energy is converted into action potentials that are conducted to the brain. Behind the retina is the **choroid.** The choroid's rich blood supply helps nourish the cells

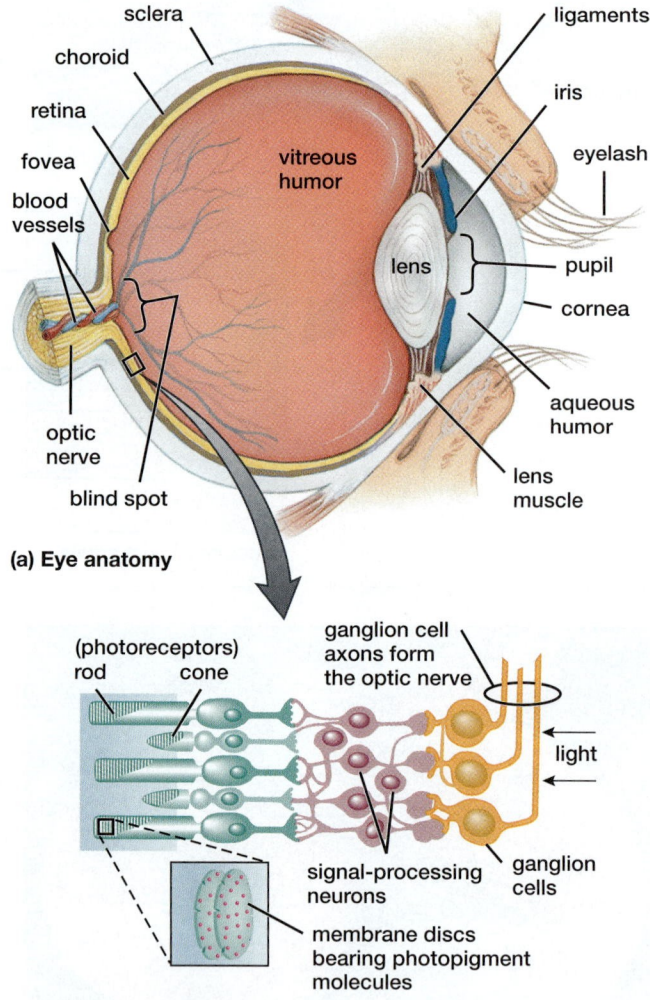

(a) Eye anatomy

(b) Cells of the retina

▲ **FIGURE 39-9 The human eye (a)** The anatomy of the human eye. **(b)** The human retina contains photoreceptors (rods and cones), signal-processing neurons, and ganglion cells. Each rod and cone bears an extension packed with membranes in which light-sensitive molecules (photopigments) are embedded.

► FIGURE 39-10 **Focusing in the human eye** (a) The lens changes shape to focus on objects at different distances. (b) Nearsightedness is corrected by eyeglasses with concave lenses. (c) Farsightedness is corrected by eyeglasses with convex lenses.

QUESTION Today, many nearsighted and farsighted people choose to correct their vision problems with laser surgery on their corneas rather than with corrective lenses. For nearsightedness and farsightedness, respectively, how should corneas be reshaped to correct the problem?

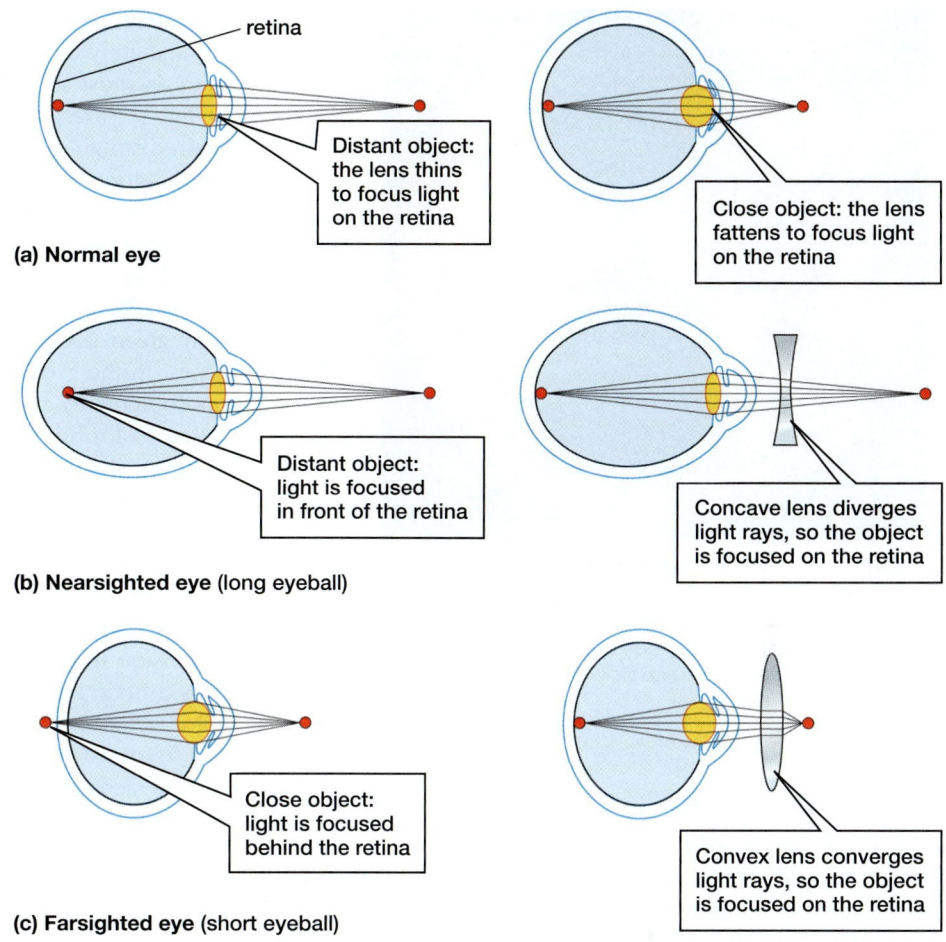

retina

Distant object: the lens thins to focus light on the retina

(a) Normal eye

Close object: the lens fattens to focus light on the retina

Distant object: light is focused in front of the retina

(b) Nearsighted eye (long eyeball)

Concave lens diverges light rays, so the object is focused on the retina

Close object: light is focused behind the retina

(c) Farsighted eye (short eyeball)

Convex lens converges light rays, so the object is focused on the retina

of the retina. In people, the choroid is darkly pigmented. It absorbs stray light, preventing the light from bouncing around inside the eyeball and interfering with sharp vision. In nocturnal animals, the choroid is often reflective rather than dark. By reflecting light back through the retina, a mirror-like choroid gives the photoreceptors a second chance to capture scarce photons of light that they may have missed the first time through. Although light reflected back and forth inside the eyeball degrades the image, at night, it's better to have blurry vision than no vision at all.

The Lens Focuses Light on the Retina

Light coming from an object is focused most sharply on a small area of the retina called the **fovea.** Although focusing begins at the cornea, whose rounded contour bends light rays, the lens is responsible for final, sharp focusing. The shape of the lens is adjusted by its encircling muscle. When viewed from the side, the lens is either rounded, to focus on nearby objects, or flattened, to focus on distant objects (**Fig. 39-10a**).

If your eyeball is too long or your cornea is too rounded, you will be **nearsighted**—light from distant objects will focus in front of the retina, so you will not see them clearly. If your eyeball is too short or your cornea is too flat, you will be **farsighted**—light from nearby objects will focus behind the retina. These conditions can be corrected by either contact lenses or eyeglasses with lenses of the appropriate shape

(**Figs. 39-10b, c**). Both nearsightedness and farsightedness can also be corrected with laser surgery that reshapes the cornea. As people age, the lens stiffens. As a result, the lens can no longer round up enough to focus on nearby objects. By their mid-forties, most people require reading glasses for close work.

The Retina Detects Light and Produces Action Potentials in the Optic Nerve

The photoreceptors, called **rods** and **cones** because of their shapes, are located at the rear of the retina (see Fig. 39-9b). Both rods and cones contain membranes packed with photopigment molecules. Photoreception begins when the photopigments absorb light, which triggers chemical reactions that produce receptor potentials in the photoreceptor cells.

Between the photoreceptors and the incoming light lie several layers of neurons that process signals from the photoreceptors. These neurons enhance our ability to detect edges, movement, and changes in light intensity. The cells nearest the vitreous humor (at the front of the retina) are the **ganglion cells,** whose axons make up the **optic nerve.** Ganglion cells convert signals from the photoreceptors and the intervening neurons into action potentials. To reach the brain, ganglion cell axons pass through the retina at a location called the **blind spot** (**Fig. 39-11;** see also Fig. 39-9a). This area lacks photoreceptors, so images focused there cannot be seen.

▲ FIGURE 39-11 **The human retina** A portion of the human retina, photographed through the cornea and lens of a living person. The blind spot and fovea are visible. Blood vessels supply oxygen and nutrients; note that the vessels are dense over the blind spot (where they won't interfere with vision) and scarce near the fovea.

QUESTION Despite the presence of the blind spot, you do not consciously experience a "hole" in your vision. Why not?

Rods and Cones Differ in Distribution and Light Sensitivity

Although cones are located throughout the retina, they are concentrated in the fovea, where the lens focuses images most sharply (see Figs. 39-9a and 39-11). The fovea looks like a dent near the center of the retina, because the ganglion cells and signal-processing neurons are spread apart, although they still retain their connections with one another and with the photoreceptors. This arrangement allows light to reach the cones of the fovea without having to pass through these other cells.

Human eyes have three varieties of cones, each containing a slightly different photopigment. Each type of photopigment is most strongly stimulated by a particular wavelength of light, corresponding roughly to red, green, or blue. The brain distinguishes color according to the relative intensity of stimulation of different cones. For example, the sensation of blue is produced by wavelengths of light that mostly stimulate blue cones; yellow is produced by fairly equal stimulation of red and green cones, with much less stimulation of blue cones. About 7% of men have difficulty distinguishing red from green, because they possess a defective copy of one of the genes on the X chromosome that code for the red and green photopigments, respectively (see Chapter 10). Although often called "color-blind," these men are more accurately described as "color-deficient." True color blindness, in which a person perceives the world only in shades of gray, is extremely rare.

Rods are most abundant outside the fovea, in the periphery of the retina. Rods are longer than cones and contain far more photopigment, so they are much more sensitive to light than cones are. Therefore, rods are largely responsible for our vision in dim light, such as moonlight. Unlike cones, all rods contain identical photopigments, so rods cannot provide color vision. In dim light, the world appears in shades of gray.

Not all vertebrates have both rods and cones. Those that are active almost entirely during the day (certain lizards, for example) may have all-cone retinas, whereas many nocturnal animals (such as rats and ferrets) and those dwelling in dimly lit habitats (such as deep-sea fishes) have retinas that contain mostly or only rods.

Binocular Vision Allows Depth Perception

Among mammals, the placement of the eyes on the head varies with the lifestyle of the animal. Predators and omnivores usually have both eyes facing forward (**Fig. 39-12a**), producing slightly different but extensively overlapping

(a) Binocular vision

(b) Almost 360° vision

▲ FIGURE 39-12 **Eye position differs in predators and prey** (a) Most predatory mammals, such as this lynx, have eyes in front of the head; both eyes can be focused on a target, providing binocular vision. (b) Most prey animals, such as rabbits, have eyes at the sides, which allows them to scan for predators.

QUESTION Why do some herbivorous or fruit-eating animals, such as monkeys and fruit bats, have both eyes in front?

visual fields. This **binocular vision** allows depth perception, the accurate judgment of the distance of an object. This ability is obviously important to a cat about to pounce on a mouse.

In contrast, most herbivores have one eye on each side of the head (**Fig. 39-12b**), with little overlap in their visual fields. Some depth perception is sacrificed in favor of a nearly 360-degree field of view, allowing prey animals to spot a predator approaching from any direction.

CHECK YOUR LEARNING

Can you describe the structures of the human eye and explain the pathway taken by light from outside the eye to the photoreceptors? Can you explain how color is encoded? Are you able to describe the differences between normal, nearsighted, and farsighted eyes and explain how these examples of defective focusing can be corrected by artificial lenses?

39.6 HOW ARE CHEMICALS SENSED?

Chemoreceptors respond to chemicals in either the internal or external environments. Internal chemoreceptors in some large blood vessels and in the hypothalamus of the brain monitor levels of crucial molecules such as sugar, water, oxygen, and carbon dioxide in the blood. Terrestrial vertebrates have two senses that respond to chemicals from outside the body: Olfaction, the sense of smell, allows animals to detect airborne molecules. Gustation, the sense of taste, allows animals to detect chemicals dissolved in water or saliva in the mouth.

Olfactory Receptors Detect Airborne Chemicals

Receptor cells for olfaction are neurons located in a patch of mucus-covered tissue in the upper portion of the nasal cavity (**Fig. 39-13**). Olfactory receptor neurons bear long dendrites

that protrude into the nasal cavity and are embedded in the mucus. Odorous molecules in the air, such as those generated by coffee, diffuse through the mucus and bind to receptor proteins on the dendrites. Olfactory receptor neurons send axons directly to the olfactory bulb of the brain.

Humans produce 350 to 400 different olfactory receptor proteins. However, any given olfactory receptor neuron bears only one kind of receptor protein. Each receptor protein is specialized to bind a particular type of odor molecule and cause the olfactory neuron to produce a receptor potential. If the receptor potential is large enough, it exceeds threshold, producing action potentials that travel along the neuron's axon to the brain. Many odors are complex mixtures of molecules that stimulate several different receptor proteins, so our perception of odors arises as a result of the brain interpreting signals from many different types of olfactory receptor neurons.

As you know, many other animals detect odors better than we do. These animals have more olfactory receptor neurons, more receptor proteins, or both. Dogs, for example, have 20 to 40 times more olfactory receptor neurons than people do, and twice as many different types of receptor proteins.

Taste Receptors Detect Chemicals Dissolved in Liquids

The human tongue bears about 5,000 **taste buds,** embedded in small bumps, called *papillae* (**Fig. 39-14a**). A few taste buds are also found in the back of the mouth. Each taste bud is a small pit in a papilla, opening into the oral cavity through a taste pore. A taste bud contains a cluster of 50 to 150 cells of several types: supporting cells, stem cells, and taste receptor cells (**Fig. 39-14b**). Supporting cells regulate the composition of the interstitial fluid and help the receptor cells to function properly. The stem cells divide to produce replacement receptor and support cells, following

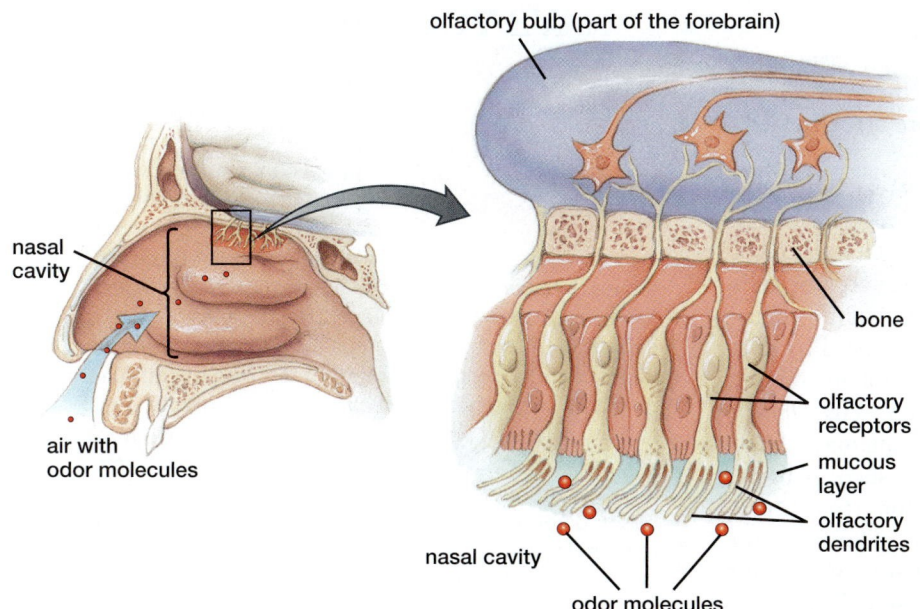

olfactory bulb (part of the forebrain)

nasal cavity

air with odor molecules

nasal cavity

bone

olfactory receptors

mucous layer

olfactory dendrites

odor molecules

◀ **FIGURE 39-13 Human olfactory receptors** The receptors for olfaction in humans are neurons bearing microscopic hairlike projections that protrude into the nasal cavity. The projections are embedded in a layer of mucus in which odor molecules dissolve before contacting the receptors.

(a) The human tongue

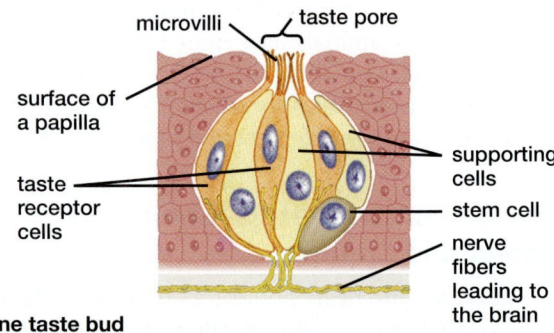

(b) One taste bud

▲ **FIGURE 39-14 Human taste receptors (a)** The human tongue is covered with papillae in which taste buds are embedded. **(b)** Each taste bud consists of taste receptor cells, supporting cells, and stem cells. Microvilli of taste receptor cells protrude into the taste pore. The microvilli bear protein receptors that bind tasty molecules, producing a receptor potential.

normal wear and tear or a close encounter with scalding-hot coffee. The taste receptor cells bear microvilli (thin projections of the plasma membrane) that protrude into the taste pore. Dissolved chemicals enter the pore and contact these microvilli. Although it was once thought that taste buds for specific tastes are concentrated on specific areas of the tongue, the different types of taste buds are actually fairly evenly distributed.

There are five known tastes: sour, salty, sweet, bitter, and *umami* (a Japanese word loosely translated as "delicious"). Sour and salty sensations are caused by hydrogen ions or sodium ions, respectively, entering certain taste receptor cells through ion channels in the plasma membranes of the microvilli. Sweet, bitter, and *umami* sensations are caused by specific organic molecules binding to receptor proteins on the surface of the microvilli of other taste receptor cells. *Umami* receptor cells respond to the amino acid glutamate. High concentrations of glutamate found in foods such as meat, fish, and cheese stimulate the *umami* receptor, producing a sensation sometimes described as "savory." Monosodium glutamate (MSG) has long been used as a seasoning, because it enhances the flavor of meat, fish, and vegetable dishes.

We perceive a great variety of tastes in two ways. First, a food may stimulate two or more receptor types, making the substance taste "sweet and sour," for example. Second, and more important, foods usually release molecules into the air inside the mouth. These odorous molecules bind to olfactory receptors, which contribute an odor component to the basic flavor.

To prove that what we call taste is really mostly olfaction, try holding your nose (and closing your eyes) while you eat different flavors of jelly beans. The flavors—from cherry and pear to buttered popcorn—will be indistinguishably sweet. Likewise, when you have a bad cold, otherwise tasty foods seem bland.

CHECK YOUR LEARNING

Can you define the term *chemoreceptor* and explain the difference between olfaction and taste? Can you explain how different, complex odors and tastes are perceived?

39.7 HOW IS PAIN PERCEIVED?

If you burn, cut, crush, or spill acid on yourself, you will experience the sensation of pain. Pain is a subjective feeling arising in the brain, produced by the stimulation of **pain receptors** (also called *nociceptors*). Pain receptors are found in most parts of the body. Pain perception is crucially important to well-being and survival, teaching humans and other animals to avoid behaviors and objects that may damage the body. A few people are born without the ability to perceive pain. They are highly accident prone, suffering many bruises, cuts, and even broken bones.

There are just about as many types of pain receptors as there are ways to harm your body. For example, some pain receptors respond to high temperatures, with a typical threshold of about 109°F (43°C). Others respond to low temperatures, below about 59°F (15°C). These temperatures may not seem

? HAVE **YOU** EVER WONDERED . . .

Why Chili Peppers Taste Hot?

The "hot" ingredient in chili peppers is a chemical called capsaicin. The heat of chili peppers depends on how much capsaicin they contain, ranging from very mild pimentos to jalapenos (about 20 times hotter) to habaneros (50 times hotter than jalapenos) to the excruciatingly painful Naga (5 to 10 times hotter than habaneros). Capsaicin activates receptor proteins on many pain receptors, including some in the mouth. These receptors are also activated by high temperatures, by acid, and by molecules produced during inflammation (which is why infected injuries often hurt). Recall that what type of stimulus we perceive depends on which sensory cells fire action potentials. Because individual pain receptors respond to both damaging heat and capsaicin, the brain interprets both of these stimuli as burning pain.

to be very hot or very cold, but remember, these are not air temperatures but rather the temperature of the skin where the pain receptors are located. For example, the cold-sensitive pain receptors can experience a temperature as low as 59°F only if the external temperature is so cold that it overwhelms the warming effect of blood flowing just below the surface of the skin.

Some pain receptors are mechanoreceptors that respond to excessive stretching, such as might be caused by bloating in the intestine (gas pains). If you are cut or bruised, damaged cells release their contents, including potassium ions (K^+) and various enzymes. Potassium ions activate still other pain receptors. Some of the enzymes convert proteins in the blood to bradykinin, a chemical that activates some pain receptors. Some individual pain receptors can respond to several different damaging stimuli, including excessive heat, acid, or certain chemicals.

CHECK YOUR LEARNING

Can you name the types of stimuli that are perceived as painful? Do you understand why it is useful to have pain receptors?

CASE STUDY revisited

Bionic Ears

A cochlear implant contains a set of 16 to 22 platinum electrodes threaded through the cochlea (Figure 39-15) and wired to a stimulator. The electrodes are carefully arranged, from the beginning of the cochlea (near the oval window) to its tip. Each electrode can be activated independently of the others. The sound-receiving unit, worn on the outside of the head, contains a tiny microphone and a microprocessor. The microphone picks up sounds and the microprocessor enhances some sounds, such as those in speech, and filters out noise. Then it broadcasts appropriate signals to the stimulator, which sends tiny electric currents to the electrodes in the cochlea. The currents stimulate action potentials in the axons of the auditory nerve.

The microprocessor sends electric currents to the electrodes with a strength that is proportional to the loudness of the sound (although the currents are always tiny): the stronger the current, the faster the auditory nerve axons fire. For low notes, the microprocessor sends electrical current to the electrodes positioned near the tip of the cochlea. For high notes, the microprocessor sends current to the electrodes located near the oval window. These electrical currents stimulate action potentials in roughly the same axons that would have been stimulated by hairs cells in these locations in a functioning cochlea.

As amazing as cochlear implants are, they are still very unsophisticated compared to a functioning biological cochlea. A cochlea contains about 18,000 to 24,000 hair cells, about 3,500 of which are primarily involved in detecting sound and stimulating axons in the auditory nerve. Sixteen to 22 electrodes in an implant might seem hopelessly crude for listening to music, but the brain can do wondrous things with very little information. When Samantha's implant was first turned on, "All I could hear were beeps for every syllable in a word." Over time, however, the beeps resolved into useful sounds. As Samantha puts it, "It takes time to train your brain."

There may always be limitations to what cochlear implants can do. For example, the physics of electricity dictates that the electrodes in the cochlea must be relatively far apart, which sets an upper limit on how many electrodes there can

stimulating electrodes

▲ FIGURE 39-15 **A cochlear implant** Each electrode of the implant can be activated by electric current independently of the others, thereby stimulating only a small portion of the auditory nerve axons that innervate the cochlea. Uncoiled, the implant would be about an inch (25 millimeters) long.

be, which in turn restricts how well a user can hear music and perhaps detect emotional overtones in speech. But in Samantha's words, her cochlear implant ushered her "into a New World."

Consider This

BioEthics Samantha was 16 years old when she received her implant—a decision that she didn't make lightly. However, some very young children are now being given cochlear implants, long before they are old enough to understand their condition and make an informed choice. Is this a decision that parents should make for young children, like whether to have vaccinations? Or should the use of cochlear implants wait until children are old enough to decide for themselves?

CHAPTER REVIEW

Summary of Key Concepts

39.1 How Do Animals Sense Their Environment?

Sensory receptors convert a stimulus from the internal or external environment to an electrical signal called a receptor potential. Either directly or indirectly, receptor potentials result in action potentials in specific axons that connect to appropriate brain regions. The type of stimulus perceived is determined by which neurons in the brain are activated. The intensity of a stimulus is encoded by the frequency of action potentials reaching the brain. Sensory receptors are categorized according to the stimulus to which they respond. Many sensory receptors are contained within sense organs that help the receptors to respond to a specific stimulus.

39.2 How Are Mechanical Stimuli Detected?

Mechanoreceptors detect stimuli such as touch, vibration, pressure, stretch, or sound. Some mechanoreceptors, including those for touch and the sensation of itching, are free nerve endings. Other mechanoreceptors are surrounded by accessory structures that regulate which stimulus, such as pressure, sound, or gravity, is detected.

39.3 How Is Sound Detected?

In the vertebrate ear, air vibrates the tympanic membrane, which transmits vibrations to the bones of the middle ear and then to the oval window of the fluid-filled cochlea in the inner ear. Within the cochlea, vibrations bend the hairs of hair cells, which are receptors located between the basilar and tectorial membranes. This bending produces receptor potentials in the hair cells that release neurotransmitters onto the endings of axons of the auditory nerve, causing action potentials that travel to auditory centers in the brain. The pitch (musical note) of sound is encoded by which hair cells are stimulated by a given frequency of sound vibrations. The intensity (loudness) of sound is encoded by the frequency of action potentials in the axons of the auditory nerve.

39.4 How Are Gravity and Movement Detected?

The vestibular apparatus of the inner ear consists of the utricle and saccule in the vestibule, which detect the direction of gravity and the orientation of the head, and the semicircular canals, which detect movement of the head.

39.5 How Is Light Perceived?

In the vertebrate eye, light enters the cornea and passes through the pupil to the lens, which focuses an image on the retina. The retina contains two types of photoreceptors, rods and cones, which produce receptor potentials in response to light. These signals are processed through several layers of neurons in the retina and are translated into action potentials in the ganglion cells, whose axons make up the optic nerve leading to the brain. Rods are more abundant and more light-sensitive than cones, providing black-and-white vision in dim light. Cones, which are concentrated in the fovea, provide color vision.

39.6 How Are Chemicals Sensed?

Terrestrial vertebrates detect chemicals in the external environment either by smell (olfaction) or by taste (gustation). Each olfactory or taste receptor cell type responds to only one or a few specific types of molecules, allowing discrimination among tastes and odors. The olfactory neurons of vertebrates are located in the nasal cavity. Taste receptors are primarily located in taste buds on the tongue.

39.7 How Is Pain Perceived?

Pain receptors respond to damaging stimuli such as cuts, burns, acid, or extremely high or low temperatures. Some pain receptors also respond to chemicals that are produced by the body during tissue damage. Some of these same pain receptors respond to temperature extremes.

Key Terms

anvil   760
aqueous humor   764
auditory canal   760
auditory nerve   761
auditory tube   760
basilar membrane   761
binocular vision   767
blind spot   765
chemoreceptor   767
choroid   764
cochlea   761
compound eye   762
cone   765
cornea   764
farsighted   765
fovea   765
ganglion cell   765
hair cell   761
hammer   760
inner ear   761
intensity   758
iris   764
lens   764
mechanoreceptor   759
middle ear   760
nearsighted   765
ommatidium (plural, ommatidia)   762

optic nerve   765
outer ear   760
oval window   761
pain receptor   768
photopigment   762
photoreceptor   762
pinna   760
pupil   764
receptor   757
receptor potential   758
retina   764
rod   765
round window   761
saccule   762
sclera   764
semicircular canal   762
sense organ   757
sensory receptor   757
stirrup   760
taste bud   767
tectorial membrane   761
tympanic membrane   760
utricle   762
vestibular apparatus   762
vitreous humor   764

Learning Outcomes

In this chapter, you have learned to . . .

LO1 Describe the important characteristics of sensory receptors, and summarize how nervous systems perceive environmental stimuli.

LO2 Define the five major types of sensory receptors and provide examples of each.

LO3 Describe the structures and functions of the mammalian ear, including how the ear converts sound waves into electrical activity, how it signals the intensity and pitch of sound, and how it signals the direction of gravity and movement of the head.

LO4 Describe the structures and functions of the mammalian eye, including how light is focused, how photoreceptors convert light into electrical activity, and how the retina signals the color and intensity of light.

LO5 Compare and contrast the senses of taste and olfaction.

LO6 Explain how pain is perceived.

Thinking Through the Concepts

Fill-in-the-Blank

1. Sensory receptors respond to an appropriate stimulus with an electrical signal called a(n) _____. Larger stimuli cause these signals to be larger than the signals from small stimuli. Ultimately, the intensity of a stimulus is conveyed to the brain encoded by the _____ of action potentials in axons connected to specific sensory areas of the brain. **LO1**

2. Sensory receptors are finely tuned to respond to specific stimuli. _____ respond to physical deformation, such as touch or pressure. _____ respond to light. The receptors for taste and olfaction are examples of _____. **LO2 LO5**

3. The mammalian ear consists of three parts: the outer, middle, and inner ear. The flap of skin-covered cartilage on the outside of the head is the _____, which collects sound waves and funnels them down the auditory canal to a flexible membrane called the _____. This connects to three small bones, the _____, _____, and _____ (list in the correct order). The final bone vibrates the _____, the beginning of the cochlea. Within the cochlea, the vibrations move the cilia of specialized mechanoreceptors called _____. **LO3**

4. Light enters the human eye through the _____ and _____, and then passes through the pupil, which is a hole in the _____. Light then passes through the _____ and _____ before finally striking the retina. Both the _____ and _____ are involved in focusing light. **LO4**

5. The retina of the human eye contains two types of photoreceptors, called _____ and _____. _____ are larger, are more sensitive to light, and provide black-and-white vision. _____ are mostly located in the central region of the retina, called the _____. These photoreceptors are smaller and less sensitive to light, but provide color vision. **LO4**

6. In humans, the five principal kinds of taste sensations are _____, _____, _____, _____, and _____. Molecules that leave the food and enter the air inside the mouth are detected by the sense of _____, which plays a major role in the brain's perception of taste. **LO5**

Review Questions

1. How do the senses encode the intensity of a stimulus? The type of stimulus? **LO1**

2. What are the names of the specific receptors used for taste, vision, hearing, smell, and touch? **LO2**

3. Why are we apparently able to distinguish hundreds of different flavors if we have only five types of taste receptors? How are we able to distinguish so many different odors? **LO5**

4. Describe the structure and function of the various parts of the human ear by tracing the route of a sound wave from the air outside the ear to action potentials in the auditory nerve. **LO3**

5. How does the structure of the inner ear allow for the perception of pitch? Of loudness? **LO3**

6. Diagram the overall structure of the human eye. Label the cornea, iris, lens, sclera, retina, and choroid. Describe the function of each labeled structure. **LO4**

7. How does the eye's lens change shape to allow focusing of distant objects? What defects make focusing on distant objects impossible, and what is this condition called? What type of lens can be used to correct it, and how does the lens do so? **LO4**

8. Compare and contrast rods and cones. **LO4**

9. List the types of stimuli that activate pain receptors. What properties do these stimuli have in common? **LO6**

Applying the Concepts

1. It has been said that sensory perceptions are purely a creation of your brain. Discuss this statement. Do you agree or disagree? What are the implications for communicating with other humans, with other animals, and with intelligent life elsewhere in the universe?

2. We don't merely identify odors. We also label them good or bad, fragrant or disgusting. What do you think might be the evolutionary advantage of emotional responses to odors? Do you think that all animals have the same emotional responses to odors that humans do?

3. Many people like spicy foods, but most other mammals don't. Historically, the general trend was that the hotter the climate, the spicier the foods a society traditionally favored. Using these facts as a starting point, what might be the selective advantage for plants that can manufacture spicy chemicals? Why might it be useful for humans to tolerate, and even enjoy, spicy food?

Answers to Figure Caption questions and Fill-in-the-Blank questions can be found in the Answers section at the back of the book.

MB *Go to MasteringBiology for practice quizzes, activities, eText, videos, current events, and more.*

CHAPTER 40

Action and Support: The Muscles and Skeleton

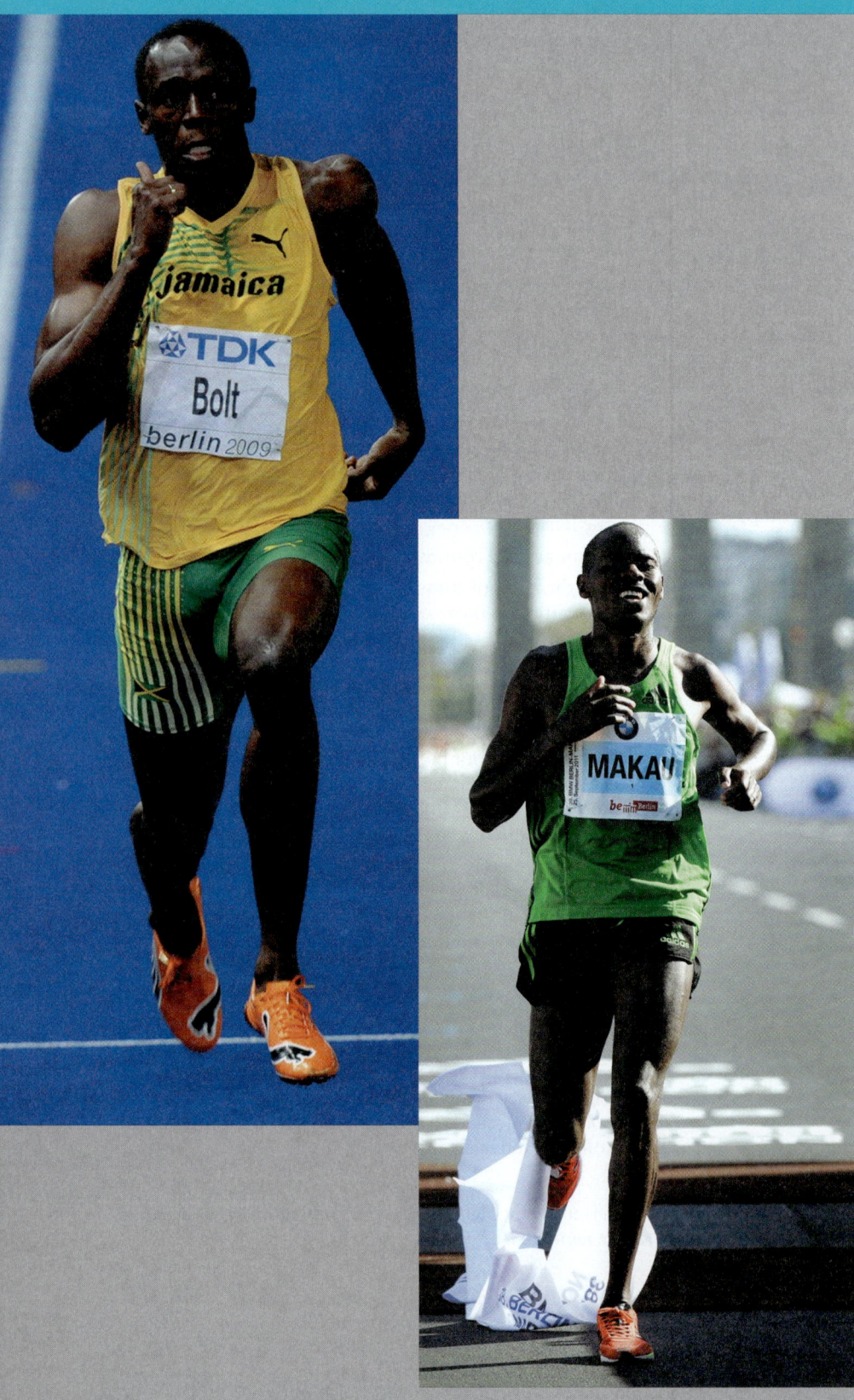

CASE STUDY
Muscles of Gold

WHEN USAIN BOLT (left) blasted out of the starting blocks at the 2008 Beijing Olympics, his leg muscles propelled his body faster than any human had ever gone before—100 meters in only 9.69 seconds. He could have run faster but was so far ahead that he just cruised the last 20 meters to the finish line. The following year, at the World Championships in Berlin, he indeed ran faster—a new world record of 9.58 seconds. Bolt also holds the world record for 200 meters, and won the 100 meter and 200 meter sprints in both the 2008 and 2012 Olympics. Truly, the world's fastest human.

Or is he? In 2011, Patrick Makau Musyoki (right) set a new world record in the marathon—more than 26 miles in just 2 hours, 3 minutes, and 38 seconds. True, Musyoki's average speed was only about 17.6 seconds per 100 meters, but he kept this pace up for over 42,000 meters.

Could Bolt beat Musyoki in a marathon? No way. Nor could Musyoki even qualify for the Olympics in the 100-meter dash. Why not? For the obvious answer, compare their bodies. Bolt weighs over 200 pounds, and Musyoki, just about 130. Bolt has both a larger skeleton and bigger muscles than Musyoki. Bolt's weight would be a serious problem during a marathon, but his large leg muscles are essential for powering off the blocks and driving to the finish of the 100 meters.

Bolt and Musyoki are both incredible athletes, but their bodies are very different. And what about weightlifters or offensive linemen in the National Football League? Are such athletes "born" to their events? Could Bolt have become a marathoner? Could Musyoki bulk up and join the NFL? As you read this chapter, consider the muscles of Usain Bolt, Patrick Musyoki, and your own legs. How do muscles contract? What supplies them with the energy they need? Could you build muscles of gold, like Bolt or Musyoki?

Perfect coordination of muscles and skeletons: Usain Bolt (left) and Patrick Musyoki (right) both qualify as the world's fastest human, each in his own way.

AT A GLANCE

40.1 How Do Muscles Contract?

40.2 How Do Cardiac and Smooth Muscles Differ from Skeletal Muscle?

40.3 How Do Muscles and Skeletons Work Together to Provide Movement?

LEARNING GOALS

LG1 Describe the structure and function of vertebrate skeletal muscles and the mechanisms that cause muscle cells to contract.

LG2 Explain how the nervous system controls the contraction of skeletal muscles, including how the nervous system determines which muscles contract and how forcefully they contract.

LG3 Compare and contrast skeletal, cardiac, and smooth muscle.

LG4 Describe hydrostatic skeletons, exoskeletons, and endoskeletons, and how antagonistic muscles act on these skeletons to move an animal's body.

LG5 Describe the vertebrate skeleton and its principal functions.

LG6 List the three types of connective tissue in the vertebrate skeleton, and describe the composition and function of each.

LG7 Explain the processes by which bones are remodeled and repaired.

40.1 HOW DO MUSCLES CONTRACT?

Almost all animals have **muscle,** a tissue composed of cells that are capable of contracting and thereby moving the parts of the body. Even sponges and sea jellies have cells that can contract. What's more, their cells contract using the same types of proteins, interacting in basically the same way, as in human muscle cells. In this chapter, we will focus on vertebrate muscles, but remember that the basic principles are similar throughout most of the animal kingdom—the ability to move and the fundamental cellular mechanisms that produce movement are extremely ancient.

The three types of vertebrate muscle—skeletal, cardiac, and smooth—differ somewhat in function, appearance, and control. We will begin by describing the structure and function of skeletal muscle. We will describe cardiac and smooth muscle in Section 40.2.

Vertebrate Skeletal Muscles Have Highly Organized, Repeating Structures

Skeletal muscle, so named because it moves the skeleton, appears striated (meaning "striped") when viewed through a microscope. Nearly all skeletal muscle is under voluntary, or conscious, control by the nervous system. Skeletal muscles can produce contractions ranging from quick twitches (blinking your eye) to powerful, sustained tension (carrying an armload of textbooks). An individual skeletal muscle consists of a series of nested, repeating parts, a little like a Russian nesting matryoshka doll (**Fig. 40-1**). Let's start at the outside of a muscle and work our way in.

Skeletal muscles are encased in connective tissue sheaths, which merge to form tough, fibrous **tendons** that attach the ends of muscles to bones. Within a muscle's outer sheath, individual muscle cells, called **muscle fibers,** are

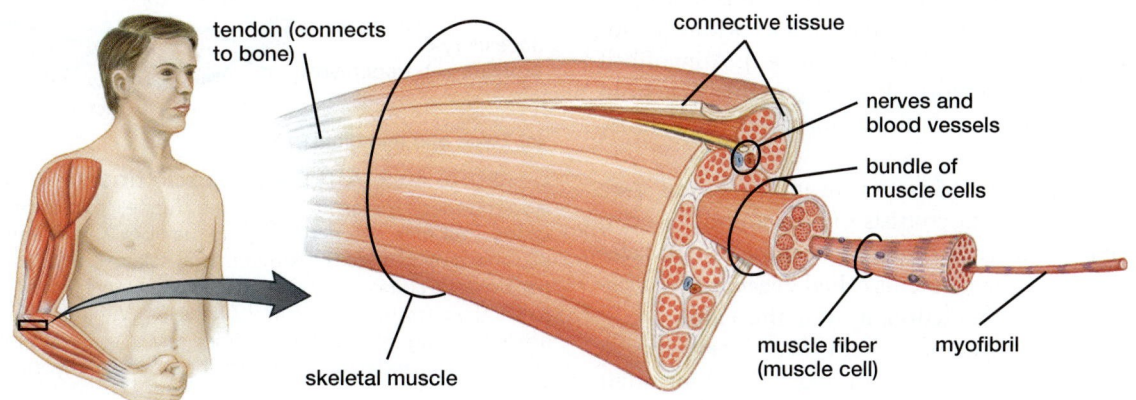

▲ FIGURE 40-1 **Skeletal muscle structure** A muscle is surrounded by connective tissue and is attached to bones by tendons. Muscle cells, called muscle fibers, are packaged into bundles within the muscle. Individual muscle fibers and bundles of fibers are also encased in connective tissue. Each fiber is packed with cylindrical subunits called myofibrils.

grouped into bundles by further coverings of connective tissue. Blood vessels and nerves pass through the muscle in the spaces between the bundles. Each individual muscle fiber also has its own thin connective tissue wrapping. These multiple connective tissue coverings, each joined to the others, provide the strength that keeps a muscle from bursting apart during contraction.

Muscle fibers are among the largest cells in the human body. They range from 10 to 100 micrometers in diameter, and some extend the entire length of a muscle, so some single muscle fibers are over a foot (30 centimeters) long. Each skeletal muscle fiber contains many nuclei, located just beneath the cell's plasma membrane; the largest fibers have several thousand nuclei.

Individual muscle fibers contain many parallel cylindrical structures called **myofibrils** (Fig. 40-2; see also Fig. 40-1). Each myofibril is surrounded by a specialized type of endoplasmic reticulum called the **sarcoplasmic reticulum (SR),** which consists of flattened, membrane-enclosed compartments filled with fluid containing a high concentration of calcium ions (Fig. 40-2a). As we will see, the calcium ions in the SR play a crucial role in muscle contraction. The plasma membrane that surrounds each muscle fiber forms tubes, called **T tubules,** that tunnel deep into the inside of the cell at regular intervals. T tubules encircle the myofibrils, running between, and closely attached to, segments of the SR (see Fig. 40-2a).

Each myofibril, in turn, consists of repeating subunits called **sarcomeres,** aligned end to end along the length of the myofibril (Fig. 40-2b) and connected to one another by protein discs called **Z lines**—as if you had taken thousands of miniature soup cans and glued them together end to end. Within each sarcomere lies a precise arrangement of thin and thick protein filaments. Each **thin filament** is anchored to a Z line at one end. Suspended between the thin filaments are **thick filaments.** The regular arrangement of thin and thick filaments within each myofibril gives a skeletal muscle fiber its striped appearance.

The filaments of myofibrils are composed primarily of two proteins, **actin** and **myosin,** that interact with one another to contract the muscle fiber (Fig. 40-2c). A myofibril also contains smaller amounts of other proteins that hold the fibril together, attach the thin filaments to the Z lines, and regulate contraction.

Individual actin proteins are nearly spherical (see Fig. 40-2c). A thin filament consists of two strands of actin proteins, wound about each other like two pearl necklaces twisted together. Accessory proteins called *troponin* and *tropomyosin*, which regulate contraction, lie atop the actin.

An individual myosin protein is shaped like a hockey stick—a head attached at an angle to a long shaft (see Fig. 40-2c). Unlike the blade of a hockey stick, however, the **myosin head** is hinged to the shaft, and can move back and forth. A thick filament consists of a bundle of myosin proteins, with the shafts in the middle of the bundle and the heads protruding out. The heads on the two ends of a thick filament are oriented in opposite directions (see Fig. 40-2b).

(a) Cross-section of a muscle fiber

(b) A sarcomere

(c) Thick and thin filaments

▲ FIGURE 40-2 A skeletal muscle fiber (a) Each muscle fiber is surrounded by plasma membrane. Tunnels of plasma membrane, called T tubules, extend down into the fiber. Sarcoplasmic reticulum surrounds each myofibril within the muscle cell. **(b)** Each myofibril consists of a series of subunits called sarcomeres, attached end to end by protein bands called Z lines. **(c)** Within each sarcomere are alternating thin filaments, composed of actin, troponin, and tropomyosin; and thick filaments, composed of myosin.

We will begin by describing the movements of the thin and thick filaments that cause an individual muscle fiber to shorten, continue with a description of the control of individual muscle fibers by the nervous system, and conclude by explaining how the nervous system controls the strength and duration of contraction of entire muscles.

Muscle Fibers Contract Through Interactions Between Thin and Thick Filaments

The molecular architecture of thin and thick filaments allows them both to grip and to slide past one another, shortening the sarcomeres and producing muscle contraction using a process called the *sliding filament* mechanism (Fig. 40-3).

Each spherical actin protein has a site that can bind to a myosin head. In a relaxed muscle cell, however, these binding sites on actin are covered by tropomyosin, preventing the myosin heads from attaching (Fig. 40-3 ❶). When a muscle contracts, tropomyosin moves aside, uncovering the binding sites on the actin proteins. Myosin heads bind to the exposed sites, temporarily linking the thick and thin filaments (Fig. 40-3 ❷). Then the myosin heads flex, pulling on the thin filaments and causing them to slide a tiny distance along the thick filament, toward the middle of the sarcomere (Fig. 40-3 ❸). Because the thin filaments are attached to the Z lines at the ends of the sarcomere, this movement shortens the sarcomere (Fig. 40-4). All of the sarcomeres of a muscle

fiber shorten simultaneously, so the whole fiber contracts a little. The myosin heads then release the thin filament, extend, reattach farther along the thin filament (Fig. 40-3 ❹), and flex again, shortening the muscle fiber a little more, much like a sailor hauling in a long anchor line hand over hand, a little at a time. The cycle repeats as long as the muscle is contracting.

Muscle Contraction Uses the Energy of ATP

ATP provides the energy for muscle contraction. You might think that the energy would be used to flex the myosin head and pull the thin filament along. However, the energy of ATP is used not to flex the myosin head, but to extend it (see Fig. 40-3 ❹). This isn't as strange as it might sound. Imagine shooting a pebble with an old-fashioned slingshot made of a Y-shaped stick and a rubber band. It takes energy to stretch out the rubber band, but once you've stretched it, all you need to do is to let go of the band, and it snaps back, shooting out the pebble. Similarly, ATP energy is used to extend the myosin head, storing the energy in this "stretched" position. When the head binds to actin, it's like letting go of

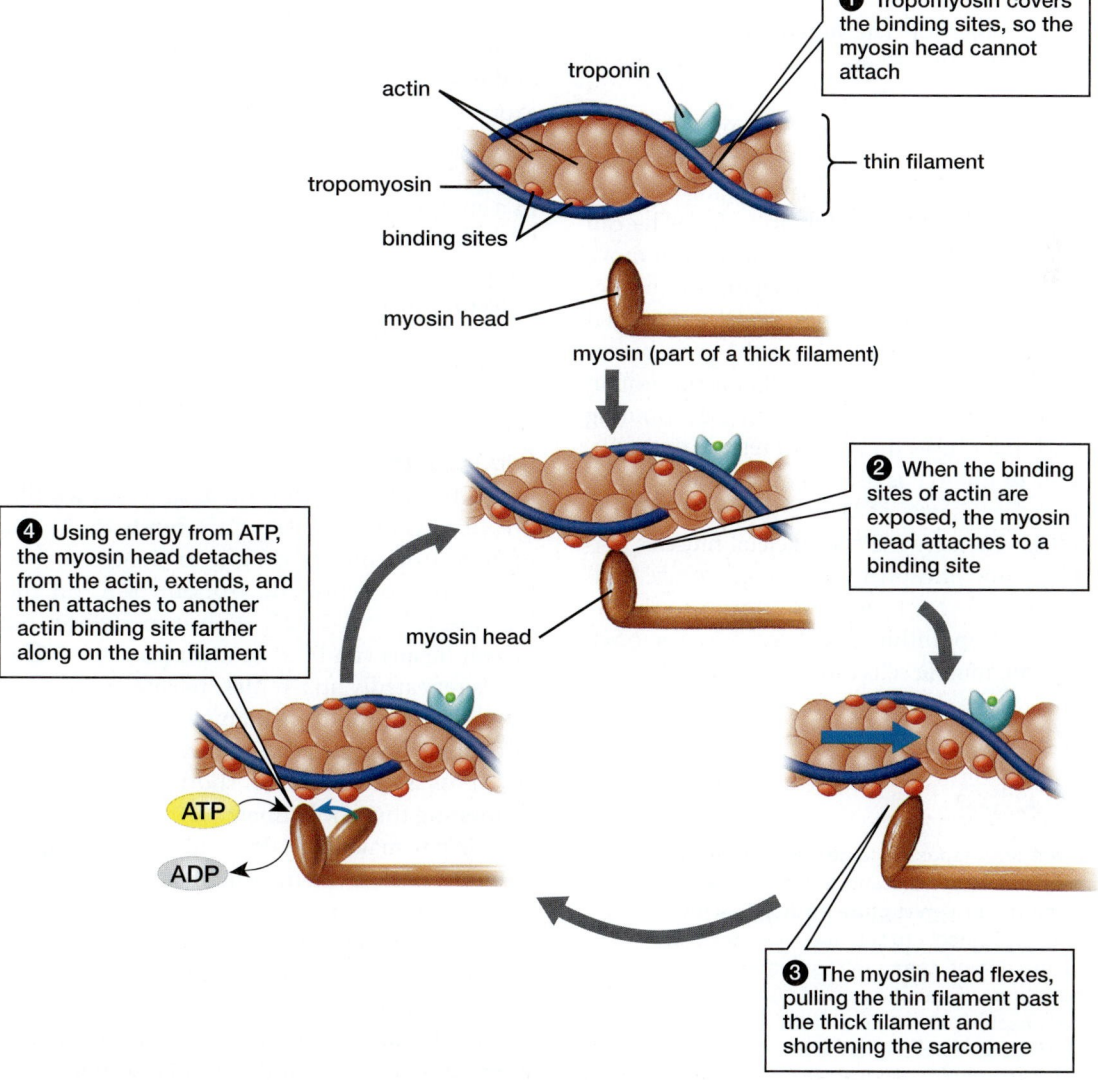

▲ FIGURE 40-3 The sliding filament mechanism of muscle contraction

▲ FIGURE 40-4 Filament sliding shortens sarcomeres
Flexing of the myosin heads pulls the thin filaments toward the center of each sarcomere, shortening the sarcomere.

QUESTION Muscles typically generate maximum force when they start from a relaxed state (neither contracted nor stretched). Why would a highly stretched muscle generate less force when it starts contracting?

the rubber band in a slingshot: The stored energy flexes the myosin head and pulls the thin filament toward the center of the sarcomere.

There is another crucial role for ATP in muscle contraction: When ATP binds to a myosin head, it causes the head to release actin. Once again, picture a sailor hauling in an anchor line. When he has pulled the line in as far as he can with one arm, he must let go of the rope before he can move this arm farther down and grasp the rope again for another pull. Similarly, when a myosin head has flexed and pulled on the thin filament, the head must release the actin before the head can extend and bind again to another actin protein located a little farther along the thin filament. ATP binding to the myosin head first causes the head to release the actin, and then the energy of the ATP is used to extend the head.

A skeletal muscle's reserves of ATP are used up after only a few seconds of high-intensity exercise. Skeletal muscles also stock a supply of creatine phosphate, an energy-storage molecule that can be used to regenerate ATP. However, creatine phosphate is also used up within a few seconds. For brief, high-intensity exertion, muscle cells can generate a little more

CASE STUDY continued

Muscles of Gold

Muscles can contract with amazing force—sometimes more than bones and tendons can withstand. A sprinter coming out of the blocks may injure or even rupture an Achilles tendon (which connects the calf muscle to the heel). Occasionally, weightlifters dislocate a joint, tear a tendon off its attachment to a bone, or break a bone. Anabolic steroids (see Chapter 37) appear to increase muscle strength more than skeletal strength, and as a consequence, make such injuries more likely.

ATP using glycolysis, which does not require oxygen, but is also not very efficient (see Chapter 8). For prolonged or low-intensity exercise, muscle cells produce ATP using cellular respiration, which requires a continuous supply of oxygen, delivered to the muscles by the cardiovascular system.

The Nervous System Controls the Contraction of Skeletal Muscles

Skeletal muscle contraction is voluntary, controlled by the nervous system. The cycle of myosin head movements that cause skeletal muscle fibers to contract begins by moving tropomyosin away from the binding sites on actin (see Fig. 40-3). Therefore, activity in the nervous system must control the position of tropomyosin. How does this work?

Muscle fibers can fire action potentials, much like neurons can (see Chapter 38). As we shall see shortly, action potentials in muscle fibers set off a series of events that move tropomyosin away from the actin binding sites and cause the fibers to contract. The role of the nervous system is to trigger action potentials in muscle fibers.

Motor Neurons Excite Skeletal Muscle Fibers at Neuromuscular Junctions

Skeletal muscle fibers are activated by motor neurons. The cell bodies of most motor neurons are in the spinal cord; their axons exit the cord in spinal nerves and contact muscle fibers at specialized synapses called **neuromuscular junctions** (**Fig. 40-5**). At a neuromuscular junction, an action potential in a motor neuron's axon releases the neurotransmitter acetylcholine onto the muscle fiber. The acetylcholine produces a huge excitatory postsynaptic potential in the muscle fiber, triggering an action potential in the fiber (**Fig. 40-5 ❶**).

Recall that the plasma membrane of a muscle fiber sends T tubules into the fiber alongside the sarcoplasmic reticulum surrounding each myofibril. The muscle fiber's action potential moves down the T tubules to the SR (**Fig. 40-5 ❷**), where it causes calcium ions (Ca^{2+}) to be released from the SR onto the thin and thick filaments of the myofibril (**Fig. 40-5 ❸**). Ca^{2+} binds to the small accessory protein, troponin, causing it to change shape and pull the larger accessory protein, tropomyosin, off the actin binding sites (**Fig. 40-5 ❹**). With tropomyosin out of the way, myosin heads can bind to actin (**Fig. 40-5 ❺**). The myosin heads repeatedly attach, flex, release, extend, and reattach to actin, pulling the thin filaments toward the center of each sarcomere. A single action potential in a muscle fiber causes all of its sarcomeres to shorten simultaneously, slightly shortening the whole fiber.

What makes the fiber stop contracting? When the action potential in the muscle fiber is over (in just a few thousandths of a second), the SR stops releasing Ca^{2+}. Active transport proteins in the membrane of the SR pump Ca^{2+} back into the SR. Ca^{2+} leaves the accessory proteins, which then slide back over the actin binding sites. Therefore, the myosin heads can no longer attach to actin. Contraction stops within a few hundredths of a second.

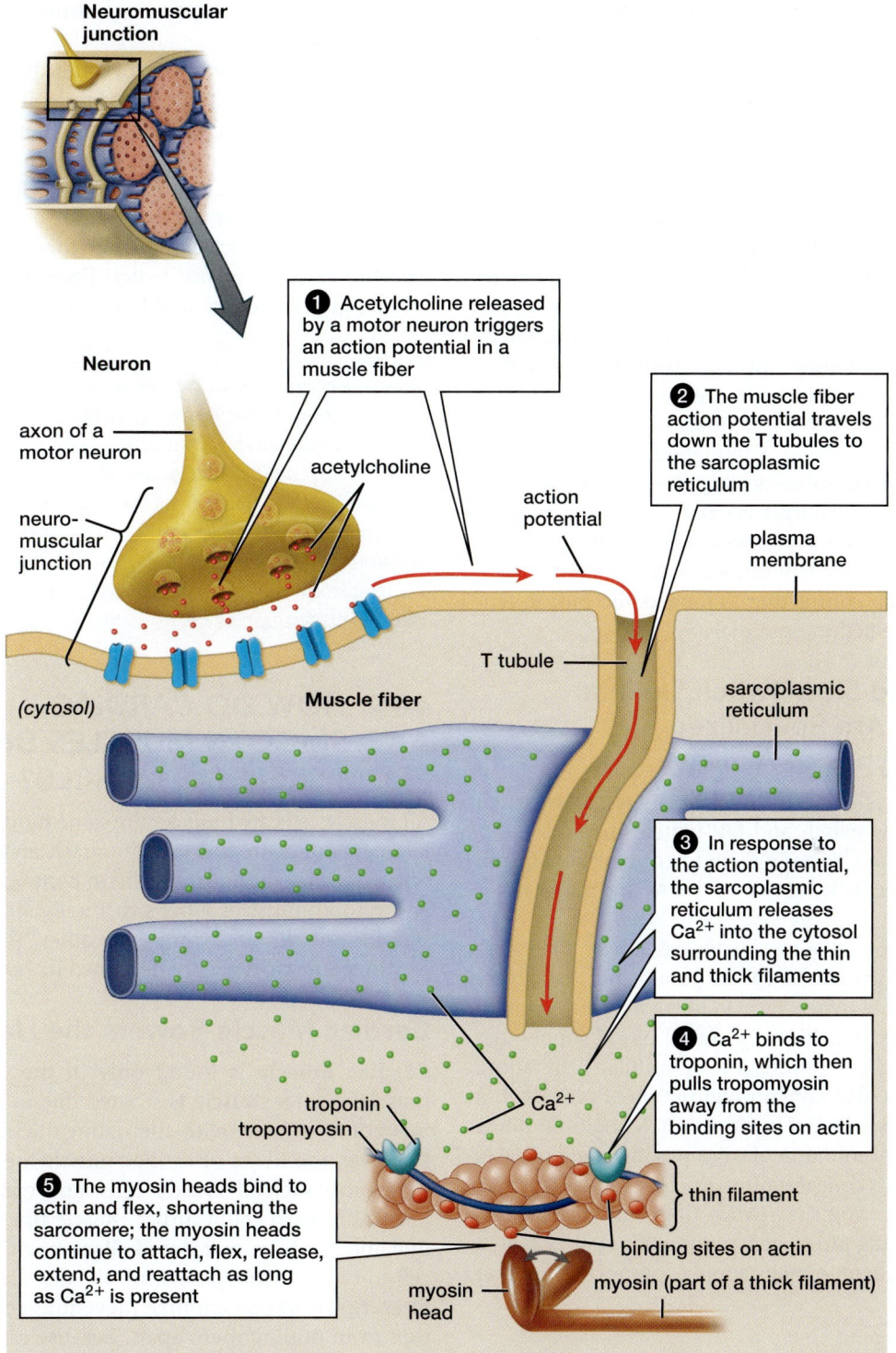

▲ FIGURE 40-5 Activity in a motor neuron stimulates contraction of a skeletal muscle fiber

The Size of Motor Units and the Frequency of Action Potentials Determine the Force of Muscle Contraction

Controlling the force, distance, and duration of muscle contraction requires controlling how many fibers in a muscle contract, how much they contract, and how long they contract.

A single motor neuron typically synapses with several muscle fibers in a single muscle. A motor neuron and all the muscle fibers that it stimulates are called a **motor unit.** Motor units vary in size. In muscles used for fine control, such as those that move the eyes or fingers, motor units are small: A single motor neuron may synapse on just a few muscle fibers. In muscles used for large-scale movements, such as those of the thigh and buttocks, motor units are large: A single motor neuron may synapse on dozens or even hundreds of muscle fibers. To see the difference in control, try tying a piece of chalk to your knee and using it to write your name on a blackboard!

The nervous system controls the force of muscle contraction by varying both the number of muscle fibers stimulated and the rate of action potentials in each fiber. Because motor neurons synapse on multiple muscle fibers in a given muscle, and because the muscle fibers are attached to one another and to the muscle's tendons, a single action potential in a single motor neuron will cause some contraction of the entire muscle. The contractions caused by a single motor neuron firing multiple action potentials in rapid succession add up to a larger contraction. Simultaneously firing several motor neurons that stimulate multiple fibers in the same muscle will also cause a larger contraction of the muscle. Rapid firing of all of the motor neurons that innervate all of the fibers in the muscle will cause a maximal contraction. The muscle will keep contracting as long as the motor neurons continue to fire action potentials.

Fast-Twitch and Slow-Twitch Skeletal Muscle Fibers Are Specialized for Different Types of Activity

Skeletal muscle fibers come in two basic types, slow-twitch and fast-twitch. Slow-twitch and fast-twitch fibers have different forms of myosin, causing them to contract slowly and more rapidly, respectively. However, there are many other differences as well.

Slow-twitch fibers contract with less power than fast-twitch fibers, but they can keep on contracting for a very long time. How? Slow-twitch muscle fibers are thin, contain lots of mitochondria, and have a plentiful blood supply that provides oxygen for ATP production by cellular respiration. Slow-twitch muscles also contain high levels of *myoglobin*, a red-colored protein similar to hemoglobin. Myoglobin binds oxygen when the concentration of oxygen is high, and releases oxygen again when its concentration drops. Myoglobin acts as an oxygen store, allowing slow-twitch fibers to continue cellular respiration during prolonged exertion. Thus, slow-twitch fibers have an excellent oxygen supply, produce abundant

CASE STUDY continued

Muscles of Gold

The legs of champion sprinters like Usain Bolt have about 80% fast-twitch fibers, capable of the rapid, explosive contractions that are so essential to coming off the blocks. The legs of world-class marathoners like Patrick Musyoki, on the other hand, have about 80% slow-twitch fibers, which are less powerful but can rapidly contract again and again, each leg stepping about 11,000 times to complete a fast marathon. Both athletes probably have about the same number of muscle fibers in their legs, but Bolt has bigger muscles, because Bolt's fast-twitch fibers are thick, while Musyoki's slow-twitch fibers are thin.

ATP, and have relatively few myofibrils to use up ATP, so they resist fatigue.

Fast-twitch fibers, on the other hand, have a large diameter, and hence many myofibrils. Therefore, fast-twitch fibers contract with greater force than slow-twitch fibers do. However, fast-twitch fibers have a smaller blood supply, much less myoglobin, and fewer mitochondria, so they cannot keep up cellular respiration for as long as slow-twitch fibers can. Fast-twitch fibers, therefore, fatigue more rapidly. Extreme versions of fast-twitch fibers use mostly glycolysis for energy production, and fatigue very quickly.

CHECK YOUR LEARNING

Are you able to describe the structure of vertebrate skeletal muscle, from a whole muscle through individual muscle cells and subcellular structures down to the proteins involved in muscle contraction? Can you explain the sliding filament mechanism of muscle contraction? Are you able to explain how the nervous system causes the contraction of skeletal muscles and how it determines the strength of contraction? Can you distinguish between fast-twitch and slow-twitch muscles?

40.2 HOW DO CARDIAC AND SMOOTH MUSCLES DIFFER FROM SKELETAL MUSCLE?

All muscle cells are built on the same fundamental principles: Filaments of actin and myosin attach and slide past one another, causing the muscle cells to contract. However, cardiac and smooth muscles differ structurally and functionally from skeletal muscles and from each other. The characteristics of the three muscle types are described in **Table 40-1**.

Cardiac Muscle Powers the Heart

Cardiac muscle is found only in the heart. Like skeletal muscle, cardiac muscle is striated due to its regular arrangement of sarcomeres with alternating thick and thin filaments. However, the fibers of cardiac muscle are branched, smaller than most skeletal muscle cells, and contain only a single nucleus in each cell. Cardiac muscle fibers are connected to one another by **intercalated discs** (see Chapter 32). Strong cell-to-cell attachments within the discs hold cardiac muscle fibers firmly to one another, preventing the forces of contraction from pulling them apart. Because cardiac muscles must contract about 70 times each minute, and sometimes much faster, for your whole life, cardiac muscle fibers require a large, uninterrupted supply of ATP. Therefore, cardiac muscle fibers are surrounded by an extensive network of blood vessels and contain enormous numbers of mitochondria, which may make up 30% to 40% of the mass of cardiac muscle.

Unlike skeletal muscle fibers, cardiac muscle fibers can contract spontaneously, without input from the nervous system, although both the rate and force of contraction can be influenced by the nervous system and by hormones. The ability to contract spontaneously is particularly well developed in the specialized cardiac muscle fibers of the heart's natural pacemaker

TABLE 40-1 Properties of the Three Muscle Types

	Type of Muscle			
Property	**Smooth**		**Cardiac**	**Skeletal**
Muscle appearance	Nonstriated		Striated	Striated
Cell shape	Tapered at both ends		Branched	Tapered at both ends
Number of nuclei	One per cell		One per cell	Many per cell
Speed of contraction	Slow		Intermediate	Slow to rapid
Contraction stimuli	Spontaneous, stretch, nervous system, hormones		Spontaneous	Nervous system
Function	Controls movement of substances through hollow organs and tubes		Pumps blood	Moves the skeleton
Under voluntary control?	No		No	Yes

Skeletal muscle
muscle fiber
nuclei

Cardiac muscle
muscle fiber
intercalated discs with gap junctions link adjacent cells
nuclei

Smooth muscle
muscle fiber
nucleus

(see Chapter 32). Action potentials from the pacemaker spread rapidly through gap junctions in the intercalated discs that interconnect cardiac muscle fibers, so the entire heart contracts as a coordinated unit. Although cardiac muscle is usually not under voluntary control, most people can be trained to regulate their heart rate to a limited extent.

Smooth Muscle Produces Slow, Involuntary Contractions

Smooth muscle surrounds blood vessels and most hollow organs, including the uterus, bladder, and digestive tract.

The cells of smooth muscle are not striated, like those of skeletal and cardiac muscles, because in smooth muscle cells the thin and thick filaments are scattered (see Table 40-1). Like cardiac muscle fibers, smooth muscle fibers each contain a single nucleus. Smooth muscle fibers are directly connected to one another by gap junctions, allowing the cells to contract in synchrony, either slow and sustained (such as the constriction of arteries that elevates blood pressure during times of stress; see Chapter 32) or slow and wavelike (such as the waves of contraction that move food through the digestive tract; see Chapter 34). Contraction of smooth muscles can be stimulated by stretch, hormones, signals from the autonomic nervous system (see Chapter 38), or by combinations of these stimuli. Smooth muscle contraction is generally not under voluntary control.

CHECK YOUR LEARNING

Can you describe the similarities and differences between skeletal, cardiac, and smooth muscles, including their appearance, where they are found, what stimulates them to contract, and what kinds of movements they usually cause?

40.3 HOW DO MUSCLES AND SKELETONS WORK TOGETHER TO PROVIDE MOVEMENT?

Despite enormous differences in body form and structure, nearly every animal—whether earthworm, crab, horse, or human—moves using the same fundamental mechanism: Contracting muscles exert forces on a structure that supports the body, called a **skeleton,** causing the body to change shape.

The Actions of Antagonistic Muscles on Skeletons Move Animal Bodies

Across the animal kingdom, there are three different types of skeletons: hydrostatic skeletons, exoskeletons, and endoskeletons. Coordinated movement of an animal's body is produced by alternating contractions of **antagonistic muscles**—pairs of muscles with opposing actions that squeeze or pull on its skeleton (Fig. 40-6).

Worms, cnidarians (sea jellies, anemones, and their relatives), and many mollusks (slugs, octopuses, and their relatives) have a **hydrostatic skeleton,** which is basically a sac or tube filled with liquid (Fig. 40-6a). "Hydrostatic" means roughly "to stand with water," and that is just what hydrostatic skeletons do for the bodies they support. Think of a water-filled balloon: The balloon "stands up" because it contains water. Puncture the balloon and it collapses. Further, the volume of the balloon is fixed (because liquids cannot be compressed), but you can change its shape by squeezing it.

An animal with a hydrostatic skeleton controls the overall shape of its body by squeezing its watery skeleton with two sets of antagonistic muscles in its body wall—one circular, the other longitudinal. Consider a tubular animal such as an earthworm (see Fig. 40-6a). If it contracts its longitudinal muscles, its body becomes shorter and fatter. If it contracts its circular muscles, its body becomes longer and thinner. How does an earthworm use these muscles to crawl through its burrow? Picture a worm that has anchored its head end by pushing the bristles in its skin into the wall of its burrow. The worm then contracts its longitudinal muscles, pulling the rest of its body up toward its head. Next, it anchors its tail with another set of bristles and releases the bristles at its head. Finally, the worm contracts its circular muscles, making its body long and thin, which pushes its head forward away from its tail. By

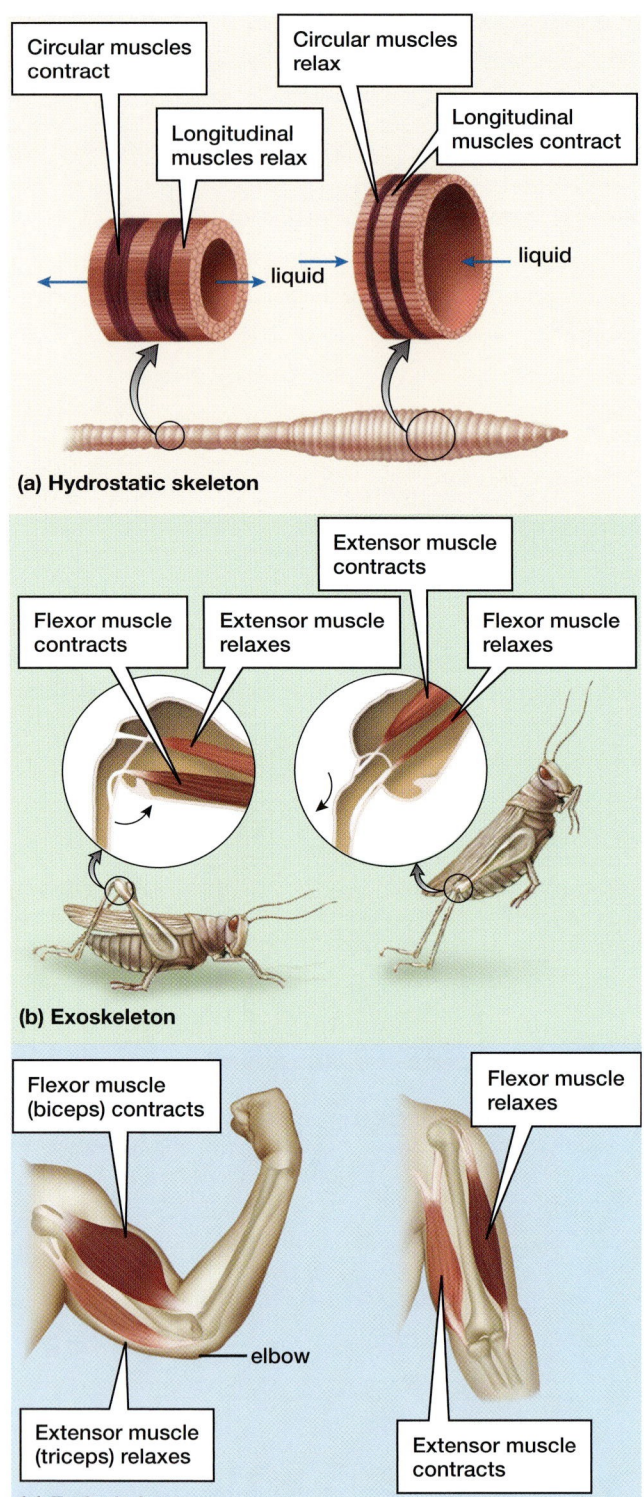

(a) Hydrostatic skeleton

(b) Exoskeleton

(c) Endoskeleton

▲ FIGURE 40-6 Antagonistic muscles move hydrostatic skeletons, exoskeletons, and endoskeletons (a) A hydrostatic skeleton, as found in earthworms, is essentially a liquid-filled tube enclosed within walls containing antagonistic circular and longitudinal muscles. Contraction of the circular muscles makes the tube long and thin (left); contraction of the longitudinal muscles make the tube short and thick (right). (b) Antagonistic flexor and extensor muscles attach to the inner surfaces of an exoskeleton on opposite sides of a flexible joint. Alternating contractions of the antagonistic muscles bend and straighten the joint. (c) Antagonistic flexor and extensor muscles attach to an endoskeleton on opposite sides of the outer surfaces of joints.

repeating the entire process over and over—anchoring the head, pulling up the tail, anchoring the tail, and pushing the head forward—the worm crawls ahead. A real earthworm alternately contracts the longitudinal and circular muscles in fairly short sections of its body in a wavelike pattern, but the principle is the same. Your intestines transport food using a similar pattern of muscular contractions, called peristalsis (see Chapter 34).

The bodies of arthropods (such as spiders, crustaceans, and insects) are encased by rigid **exoskeletons** (literally, "outside skeletons"; **Fig. 40-6b**). Movement of an exoskeleton typically occurs only at **joints** in the legs, mouthparts, antennae, bases of the wings, and body segments, where thin, flexible tissue joins stiff sections of exoskeleton. Antagonistic muscles attach to opposite sides of the exoskeleton, across the inside of a joint. Contraction of a **flexor** muscle bends a joint; contraction of an **extensor** muscle straightens a joint. Alternating contraction of the antagonistic muscles moves joints back and forth, allowing the animal to walk, fly, or eat. Although an exoskeleton provides an effective support system for the muscles that move the body, it comes with a major problem: It cannot significantly expand to accommodate growth. Therefore, an arthropod must periodically molt its exoskeleton so that it can grow (**Fig. 40-7**).

▲ **FIGURE 40-7 A crab molts its exoskeleton** Arthropods, such as this blue crab, must molt their exoskeletons in order to grow. Here a dark, newly molted crab has just crawled out of its old exoskeleton.

QUESTION Why are thick, armor-like exoskeletons found mostly in water-dwelling animals, whereas land-dwelling insects and spiders tend to have thinner exoskeletons?

Endoskeletons ("internal skeletons") are rigid structures found inside the bodies of echinoderms (sea stars and their relatives) and chordates (vertebrates and their relatives). In vertebrates, movement occurs primarily at joints, where two parts of the skeleton are firmly but flexibly attached to one another. Antagonistic muscles, such as the biceps (a flexor) and the triceps (an extensor), attach on opposite sides of the outside of a joint (in this case, the elbow; **Fig. 40-6c**). Antagonistic muscles move joints back and forth, or rotate them in one direction or the other. We will examine the movement of vertebrate endoskeletons in more detail in the next section.

You may wonder why all animals use pairs of antagonistic muscles to move their skeletons. The answer is simple: As we described in Section 40.1, muscles produce force by contracting. No animal has evolved a mechanism to produce force by lengthening a muscle. When a muscle stops contracting, it will lengthen only if it is pulled on by other forces. For example, suppose you contract your biceps muscle, bending your elbow and bringing your hand up to your shoulder. How do you straighten out your arm again? You can let gravity pull your forearm down and stretch out the biceps, or you can contract your triceps muscle and forcefully straighten your arm, which also stretches out the biceps, but you cannot push your arm straight with your biceps. The only way to flex a joint and then forcefully extend it again is to have pairs of antagonistic muscles alternately contracting on opposite sides of the joint.

The Vertebrate Endoskeleton Serves Multiple Functions

The endoskeleton of humans and other vertebrates provides a wide range of functions:

- The skeleton provides a rigid framework that supports the body and protects its internal organs. The brain and spinal cord are almost completely enclosed within the skull and vertebral column, the rib cage protects the lungs and the heart, and the pelvic girdle supports and partially protects abdominal organs.
- The skeleton allows locomotion. Different types of vertebrates have evolved skeletons that permit them to crawl, walk, run, jump, swim, fly, or perform various combinations of these actions.
- In mammals, the bones of the middle ear are essential to hearing; these bones transmit sound vibrations between the eardrum and the cochlea (see Chapter 39).
- Red bone marrow produces red blood cells, white blood cells, and platelets (see Chapters 32 and 36).
- Bones store calcium and phosphorus, absorbing and releasing these minerals as needed.

The vertebrate skeleton consists of two parts (**Fig. 40-8**). The **axial skeleton** includes the bones of the head, vertebral column, and rib cage. The **appendicular skeleton** includes the pectoral and pelvic girdles, and the appendages attached to them: the forelimbs (in humans, the arms and hands) and hind limbs (in humans, the legs and feet). The pectoral

frontalis

trapezius

deltoid

pectoralis
major

biceps

triceps

external
oblique

rectus
abdominis

quadriceps

gastrocnemius

tibialis
anterior

skull

mandible

clavicle

sternum

humerus

rib

intervertebral
discs

vertebrae

pelvis

ulna

radius

coccyx
(tail bone)

carpals

metacarpals

phalanges

femur

patella

tibia

fibula

tarsals

metatarsals

phalanges

◄ **FIGURE 40-8 The human muscular and skeletal systems** The muscular and skeletal systems work together to move the body. Here, some of the major muscles and bones in the human body are illustrated. The human skeleton consists of 206 bones grouped into the axial skeleton (blue) and the appendicular skeleton (beige).

girdle, which consists of the clavicle and scapula in humans, links the arms to the axial skeleton and provides attachment sites for muscles of the trunk and arms. Hip bones form the pelvic girdle, which links the legs to the axial skeleton, helps protect the abdominal organs, and forms attachment sites for muscles of the trunk and legs.

The Vertebrate Skeleton Is Composed of Cartilage, Ligaments, and Bone

Three types of connective tissue—cartilage, ligaments, and bone—make up the skeleton (see Chapter 31 for more information about connective tissue). All three consist of living cells embedded in a matrix of a protein called *collagen*, with various amounts of other substances included in the matrix. **Cartilage** contains large amounts of glycoproteins (glucosamine, a popular dietary supplement, is a precursor for these glycoproteins) and often includes elastic fibers. **Ligaments** consist principally of slightly wavy collagen fibers, arranged parallel to one another. Ligaments do not have significant amounts of glycoproteins or elastic fibers. In addition to collagen, **bone** contains large amounts of minerals, mostly calcium and phosphate, which make it hard and rigid.

Cartilage Provides Flexible Support and Connections

Cartilage plays many roles in the vertebrate skeleton. In some fishes, such as sharks and rays, the entire skeleton is composed of cartilage. During the embryonic development of other vertebrates, the skeleton, except for the skull and collarbone, is first formed from cartilage, which is later replaced by bone (**Fig. 40-9**). Cartilage also covers the ends of bones at joints (**Fig. 40-10**), supports the flexible portions of the nose and external ears, and provides the framework for the larynx, trachea, and bronchi of the respiratory system. In addition,

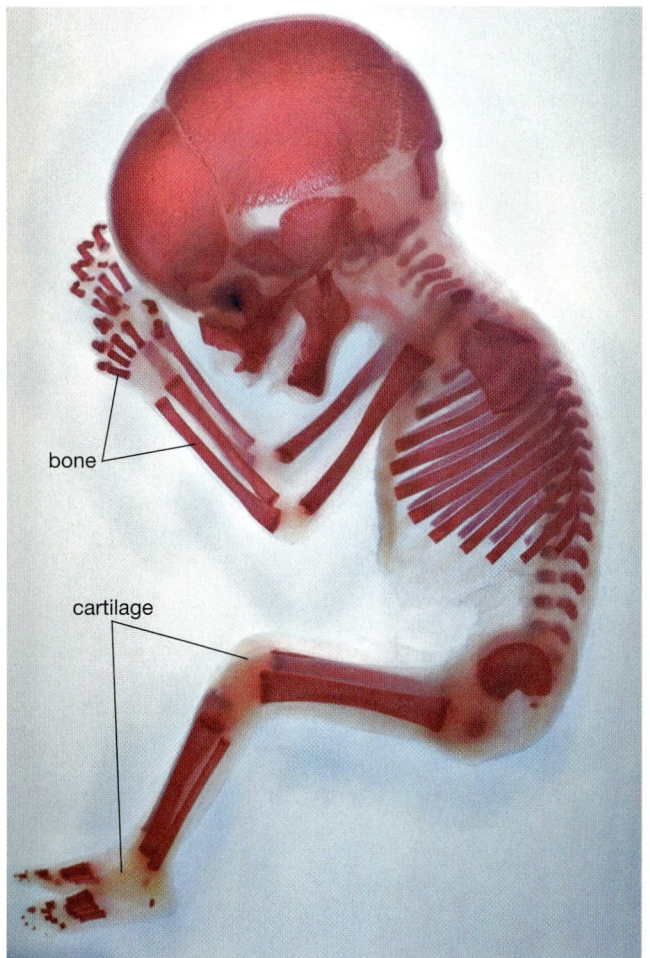

bone

cartilage

▲ **FIGURE 40-9 Bone replaces cartilage during development** In this 16-week-old human fetus, bone is stained magenta. The clear areas at the wrists, knees, ankles, elbows, and breastbone show cartilage that will later be replaced by bone.

cartilage forms the tough, shock-absorbing intervertebral discs between the vertebrae of the backbone (see Fig. 40-8).

The living cells of cartilage are called **chondrocytes.** These cells secrete the glycoproteins and collagen that make up most of the matrix of cartilage (**Fig. 40-10a**). No blood vessels penetrate cartilage. To exchange wastes and nutrients, chondrocytes rely on diffusion of materials through the collagen matrix. As you might predict, cartilage cells have a very low metabolic rate, so damaged cartilage repairs itself very slowly, if at all.

Ligaments Connect Bone to Bone in Joints

The bones of most movable joints are attached to one another by strong, flexible ligaments. The parallel orientation of collagen in ligaments gives them tremendous strength, but only in line with the collagen fibers. Twisting forces or impacts perpendicular to the collagen fibers, as sometimes happen to a knee in sports such as football, basketball, or skiing, can easily rupture a ligament.

Bone Provides a Strong, Rigid Framework for the Body

A bone consists of a hard outer shell of **compact bone** enclosing **spongy bone** in its interior (see Fig. 40-10, middle). Compact bone is dense and strong and provides an attachment site for muscle. Compact bone consists of small tubes called *osteons*, in which collagen and calcium phosphate surround a central canal containing blood vessels (**Fig. 40-10b**). Spongy bone consists of an open network of bony fibers. It is porous, lightweight, and rich in blood vessels. Bone marrow, where blood cells form, is found in the cavities of spongy bone.

There are three types of bone cells: **osteoclasts** (bone-dissolving cells), **osteoblasts** (bone-forming cells), and **osteocytes** (mature bone cells). Early in development, when bone replaces cartilage in the skeleton, osteoclasts invade and dissolve the cartilage. Osteoblasts then secrete a hard matrix of bone and become trapped within it. As bones mature, the trapped osteoblasts mature into osteocytes. Although they are generally not capable of enlarging a bone, osteocytes are essential to bone health, because they constantly rework the calcium phosphate deposits, preventing excessive crystallization that would make the bone brittle and likely to break.

Bone Remodeling Allows Skeletal Repair

Each year, 5% to 10% of all the bone in your body breaks down and is replaced through the coordinated activities of osteoclasts, which secrete acid that dissolves small amounts of bone, and osteoblasts, which secrete new bone. This process, called *bone remodeling*, allows the skeleton to alter its shape in response to the demands placed on it.

Bone remodeling varies with age. Early in life, the activity of osteoblasts outpaces that of osteoclasts, so bones become larger and thicker as a child grows. In the aging body, however, the balance of power shifts to favor osteoclasts, and bones become more fragile. Although both sexes lose bone mass with age, the loss is typically greater in women, as we explore in "Health Watch: Osteoporosis—When Bones Become Brittle."

cartilage

chondrocytes

collagen matrix

(a) Cartilage

spongy bone (contains marrow)

compact bone

blood vessels

central canal

osteon

osteocytes

(b) Compact bone

▲ **FIGURE 40-10 Cartilage and bone** A bone consists of an outer layer of compact bone with spongy bone inside. For simplicity, blood vessels are not shown. **(a)** In cartilage, the chondrocytes, or cartilage cells, secrete an extracellular matrix of the protein collagen, which surrounds them. **(b)** Compact bone is composed of osteons, each including a central canal containing blood vessels. Mature bone cells, called osteocytes, are embedded in the osteons.

QUESTION Compact bone is stronger than spongy bone. Why isn't the entire skeleton made of compact bone?

The ultimate bone remodeling occurs after a fracture. Typically, a physician moves the ends of a broken bone back into their proper alignment and immobilizes the break with a cast or splint. The fracture will have ruptured the thin layer of connective tissue, rich in capillaries and osteoblasts, that surrounds the bone. Healing begins when a large blood clot surrounds the break (**Fig. 40-11 ❶**). Phagocytic cells from the blood and osteoclasts from the damaged bone ingest cellular debris and dissolve bone fragments. Osteoblasts, in conjunction with cartilage-forming cells, secrete a *callus*, a porous mass of bone and cartilage that surrounds the break (**Fig. 40-11 ❷**). The callus replaces the original blood

❶ Blood from ruptured blood vessels forms a clot surrounding the ends of the broken bone

❷ Healing begins when a callus of cartilage replaces the clot

❸ Bone gradually replaces the cartilage in the callus

❹ When mature bone completely replaces the callus and the original shape of the bone has been mostly restored, the fracture is healed

large blood clot

compact bone

spongy bone

▲ **FIGURE 40-11 Bone repair**

Health | Watch

Osteoporosis—When Bones Become Brittle

Bone density in people peaks between ages 25 and 35. In middle age, as the activity of osteoclasts starts to exceed that of osteoblasts, bone density begins a slow decline. In the 12 million Americans with **osteoporosis** (literally, "porous bones"), the loss is severe enough to make their bones vulnerable to fractures and deformities (Fig. E40-1a). In many cases, the vertebrae of individuals with osteoporosis compress, causing a hunchbacked appearance (Fig. E40-1b). In extreme cases, just lifting a shopping bag, opening a window, or even sneezing can break a bone. People with osteoporosis are very likely to break a bone during a fall.

Women are about four times more likely than men to suffer from osteoporosis. Typically their bones are smaller and less dense than those of men to begin with. Further, although the high estrogen levels of premenopausal women stimulate osteoblasts and help maintain bone density, estrogen production drops dramatically after menopause, and bone density therefore decreases rapidly. In men, testosterone stimulates osteoblasts. Although testosterone levels decline with age, they usually do not drop as fast or as far as women's estrogen levels do after menopause. In both sexes, alcoholism and smoking also contribute to bone loss.

People with osteoporosis, in consultation with their physicians, may choose medical therapy to help maintain bone density. Usually, drugs that inhibit osteoclast activity, such as Fosamax®, Actonel®, and Boniva®, are the frontline medications in treating osteoporosis. The hormone calcitonin also inhibits osteoclast activity, although some studies suggest that it is not as effective as these drugs for treating osteoporosis. Although not without other risks, estrogen replacement therapy (usually with very low amounts of estrogen) can slow bone loss in postmenopausal women, as can a drug, raloxifene, that mimics estrogen's effects on bone.

Can osteoporosis be prevented? Calcium, of course, is a major component of bone. In addition, bones thrive on moderate stress. Therefore, regular exercise throughout life, combined with a diet containing adequate calcium and vitamin D (which is required for calcium absorption from the intestine), helps to ensure that bone mass is as high as possible before age-related losses begin, and slows the rate of loss with aging. Older people tend to be less active, resulting in loss of bone minerals, but low-impact, weight-bearing exercise such as walking or dancing can reduce bone loss and sometimes even increase bone mass.

(a) Normal bone   Osteoporotic bone   (b) A victim of osteoporosis

▲ FIGURE E40-1 **Osteoporosis** **(a)** A cross-section of (left) a normal bone compared with (right) a bone from a woman with osteoporosis. **(b)** The devastating effects of osteoporosis extend beyond a hunchbacked appearance; its victims are also at high risk for bone fractures.

---

clot and temporarily holds the ends of the break together. Osteoclasts, osteoblasts, and capillaries invade the callus. Osteoclasts break down cartilage while osteoblasts add new bone (Fig. 40-11 ❸). Finally, osteoclasts remove excess bone, mostly restoring the bone's original shape, although often leaving a slight thickening (Fig. 40-11 ❹). Completely healing a fracture usually takes about 6 weeks.

Antagonistic Muscles Move Joints in the Vertebrate Skeleton

As we noted earlier, almost all animals move by the action of pairs of antagonistic muscles working on a skeleton. Here, we will focus on the arrangement and movement of muscles around the movable joints of the human skeleton

(Fig. 40-12). Not all joints are movable; for example, immobile joints called sutures join the bones of the skull. In movable joints, the portion of each bone that forms the joint is coated with a layer of cartilage. The smooth, resilient cartilage allows the bone surfaces to slide past one another with little friction (see Fig. 40-12). Joints are held together by ligaments, which are strong and flexible, but usually not very elastic. Tendons attach muscles to the bones.

The tendon at one end of a muscle, called the *origin*, is fixed to a bone that remains stationary, while the tendon at the other end of the muscle, the *insertion*, is attached to the movable bone on the far side of the joint. When one of a pair of antagonistic muscles contracts, it moves the bone around its joint, and simultaneously stretches the relaxed opposing muscle. Antagonistic muscles allow for a remarkable

▲ FIGURE 40-12 **The human knee** The human knee, showing antagonistic muscles (the biceps femoris and the quadriceps of the thigh), tendons, and ligaments. The complexity of this joint, coupled with the extreme stresses placed on it during activities such as jumping, running, or skiing, makes it very susceptible to injury.

? HAVE YOU EVER WONDERED ...

| If Exercise Affects Your Skeleton? | Everyone knows that exercise, especially high-intensity exercise like weightlifting, increases muscle size and strength. But how does exercise affect tendons, ligaments, and bones? Weight-bearing exercise increases the activity of |

tendon and ligament-building cells, causing them to produce more of the collagen proteins that make up the bulk of tendons and ligaments. Weight-bearing exercise also stimulates bone-building osteoblasts. As a result, the bone remodels, very much as it does when a broken bone heals. Both tendons and bones become thicker and stronger, especially at the very places where the exercise stressed them. This effect is so pronounced that tennis players have thicker bones in their racket arms. Archaeologists can even tell which skeletons found in medieval battlefields are the remains of English longbowmen—they typically have very broad shoulders and thick left arms, the result of endless hours of shooting bows with pull strengths well over 100 pounds.

range of motions, depending on the configuration of a joint, including moving bones back and forth, moving them side to side, or rotating them.

Hinge joints are located in the elbows, knees, and fingers. Like a hinged door, these joints move in only two dimensions (Fig. 40-13a). When the flexor muscle contracts, it bends the joint; when the extensor muscle contracts, it straightens the

(a) A hinge joint

(b) A ball-and-socket joint

▲ FIGURE 40-13 **Hinge and ball-and-socket joints** (a) The human elbow is a hinge joint. (b) The human hip can rotate because it has a ball-and-socket joint. This consists of the rounded end of the femur (the ball) that fits into a cuplike depression (the socket) in the pelvic bone.

joint. In Figure 40-12, for example, contraction of the biceps femoris (the flexor) bends the leg at the knee, while contraction of the quadriceps (the extensor) straightens it. Alternating contractions of flexor and extensor muscles cause the lower leg bones to swing back and forth at the knee joint.

In **ball-and-socket joints,** such as those of the hip and shoulder, the round end of one bone fits into a hollow depression in another (Fig. 40-13b). Ball-and-socket joints allow movement in several directions—compare the wide range of movement of your shoulder with the limited bending of your knee. The range of motion in ball-and-socket joints is made possible by at least two pairs of antagonistic muscles, oriented at angles to each other, which move the joint in three dimensions.

CHECK YOUR LEARNING

Can you describe hydrostatic skeletons, exoskeletons, and endoskeletons, and explain how antagonistic muscles act on these types of skeletons to cause movement of an animal's body? Are you able to explain the functions of vertebrate skeletons? Can you list and describe the functions of the different types of connective tissue that make up a vertebrate skeleton? Are you able to explain how bones are developed from an embryo to adult? Can you describe the process of bone remodeling and how a bone fracture is repaired?

Muscles of Gold

Muscle fibers differ in their composition and capabilities. Fast-twitch fibers are thick and contract rapidly, but fatigue readily. Slow-twitch fibers are thin and contract slowly, but resist fatigue. These facts raise a few interesting questions.

First, are differences in fiber types what make Usain Bolt the Olympic champion and world record holder in the 100- and 200-meter sprints and Patrick Musyoki the fastest marathoner? Yes and no. Without an overwhelming majority of fast-twitch muscle fibers, Bolt certainly could not win gold in the sprints. However, all the competitors in the sprint finals probably have just as high a proportion of fast-twitch fibers as Bolt does. Likewise, all top marathoners are superficially very similar to Musyoki—extremely thin, with mostly slow-twitch muscle fibers in their legs. Therefore, the right types of muscle fibers are essential, but aren't enough, on their own, to set records.

Second, are sprinters and marathoners born or made? Probably both. The proportions of fast-twitch and slow-twitch fibers vary greatly within the human population. Current research suggests that about half of the variability in fiber type is genetic and half is environmental. People with a high proportion of slow-twitch fibers are likely to excel in, and therefore favor, endurance sports, whereas those with more fast-twitch fibers will probably prefer sports requiring bursts of speed or strength. Intensive training generally does not change the number of muscle fibers a person has, but can convert some slow-twitch fibers into fast-twitch fibers, and vice versa. Bolt and Musyoki train very hard, and in very specialized ways, for their events. Therefore, a world-class marathoner was probably both born with a tendency to develop slow-twitch muscles, and converted many fast-twitch

fibers to slow-twitch through prolonged distance training. The opposite would be true for sprinters.

In addition, it's not just a matter of slow-twitch versus fast-twitch fibers, training regimes, or running through the pain that makes someone an elite athlete. At least 221 chromosomal genes and 18 mitochondrial genes contribute to athletic performance. No one knows which alleles of all of these genes are carried by any elite athlete, but Bolt and Musyoki undoubtedly have a superb set of fitness alleles.

Finally, how do genetics, muscle fiber types, and training apply to regular people? Well, no one can reach the Olympics—let alone bring home the gold—without first winning the genetic lottery, so most people are out of luck in this regard. Further, unlike in the early days of the Olympics, it now takes many years of training both for fitness and technique just to have a chance at competing—the days of the "born athlete" who wakes up one day and decides to win the Olympics are long gone.

Consider This

BioEthics Many parents try to give their children a boost in life—a very understandable urge. For a fee, a couple of companies will test little kids to see which allele they carry for a gene that encodes an important protein in muscle cells. One allele is associated with fast-twitch muscles and ability in sports that require sprinting or strength. Presumably, kids homozygous for this allele should be steered into sprinting or football rather than distance running or soccer. However, about 18% of the population is homozygous for this allele, and hardly any of these people are superathletes. Do you think that such genetic testing is worthwhile, particularly for the kids? What if you could test for, say, 20 fitness genes?

CHAPTER REVIEW

Summary of Key Concepts

40.1 How Do Muscles Contract?

Skeletal muscles consist of cells called muscle fibers, surrounded by connective tissue and attached to bones by tendons. Skeletal muscle fibers consist of myofibril subunits surrounded by sarcoplasmic reticulum. Myofibrils are composed of repeating subunits called sarcomeres, attached end to end. Each sarcomere consists of alternating thick filaments made of myosin, and thin filaments made of actin and two accessory proteins (troponin and tropomyosin). Thin filaments are attached to proteins called Z lines that form the ends of sarcomeres. The thick and thin filaments are arranged in a regular pattern in a skeletal muscle, giving it a striped, or striated, appearance.

A motor neuron innervates each muscle fiber at a synapse called a neuromuscular junction. An action potential in the motor neuron causes it to release acetylcholine onto the muscle fiber, stimulating an excitatory postsynaptic potential that triggers an action potential in the fiber. This action potential stimulates the release of Ca^{2+} from the sarcoplasmic reticulum. Ca^{2+} causes the accessory proteins of the thin filament to move away from binding sites on actin to which myosin heads then

attach. The myosin heads flex, causing the thin and thick filaments to slide past each other, shortening the sarcomere. Shortening of sarcomeres shortens the entire muscle fiber. When the action potential is over, calcium is actively transported back into the sarcoplasmic reticulum, ending contraction. The energy for muscle contraction comes from ATP, which may be produced by cellular respiration, or, for bursts of strenuous activity, by glycolysis, which does not use oxygen.

The degree of muscle contraction is determined by the number of muscle fibers stimulated by a motor neuron and the frequency of action potentials in each fiber. Rapid firing in all of the fibers of a muscle causes maximum contraction.

40.2 How Do Cardiac and Smooth Muscles Differ from Skeletal Muscle?

Cardiac muscle is striated due to its regular arrangement of thick and thin filaments. Its cells contract rhythmically and spontaneously; these contractions are synchronized by action potentials produced by specialized muscle fibers in the heart's pacemaker. Cardiac muscle fibers are interconnected by gap junctions. These conduct action potentials between cells, producing coordinated contraction.

Smooth muscles lack organized sarcomeres, and so they are not striated. Smooth muscle cells are connected to one another by gap junctions. Smooth muscle surrounds hollow organs such as the uterus, digestive tract, bladder, arteries, and veins, producing involuntary, slow, and sustained or rhythmic contractions.

40.3 How Do Muscles and Skeletons Work Together to Provide Movement?

Animals have hydrostatic skeletons, exoskeletons, or endoskeletons. In all cases, pairs of antagonistic muscles act on the skeleton to move the body.

The vertebrate skeleton provides support for the body, attachment sites for muscles, and protection for internal organs. Red blood cells, white blood cells, and platelets are produced in bone marrow. Bone acts as a storage site for calcium and phosphorus. The axial skeleton includes the skull, vertebral column, and rib cage. The appendicular skeleton consists of the pectoral and pelvic girdles and the bones of the arms, legs, hands, and feet.

The vertebrate skeleton consists of cartilage, ligaments, and bone. All consist of cells that secrete collagen and various other substances into an extracellular matrix. Cartilage is located at the ends of bones, and forms pads in joints and the intervertebral discs. It also supports the nose, ears, and respiratory passages. During embryological development, cartilage is the precursor of bone. Ligaments connect bones at movable joints. Bone is formed by osteoblasts, which secrete a collagen matrix that becomes hardened by calcium phosphate. A typical bone consists of an outer shell of compact bone, to which muscles are attached, and inner spongy bone, which may contain bone marrow. Bone remodeling occurs through the coordinated action of osteoclasts, which dissolve bone, and osteoblasts, which create new bone.

In the vertebrate skeleton, movement occurs around joints. Tendons attach pairs of antagonistic muscles to the bones on either side of a joint. The contraction of one muscle moves the joint and stretches out its antagonistic muscle. In hinge joints, contraction of the flexor muscle bends the joint; contraction of its antagonistic extensor straightens it. At ball-and-socket joints, antagonistic muscles rotate one bone around another bone.

Key Terms

actin   774
antagonistic muscle   780
appendicular skeleton   781
axial skeleton   781
ball-and-socket joint   786
bone   783
cardiac muscle   778
cartilage   783
chondrocyte   783
compact bone   783
endoskeleton   781
exoskeleton   781
extensor   781
flexor   781
hinge joint   786
hydrostatic skeleton   780
intercalated disc   778
joint   781
ligament   783
motor unit   777
muscle   773

muscle fiber   773
myofibril   774
myosin   774
myosin head   774
neuromuscular junction   776
osteoblast   783
osteoclast   783
osteocyte   783
osteoporosis   785
sarcomere   774
sarcoplasmic reticulum
   (SR)   774
skeletal muscle   773
skeleton   780
smooth muscle   779
spongy bone   783
T tubule   774
tendon   773
thick filament   774
thin filament   774
Z line   774

Learning Outcomes

In this chapter, you have learned to . . .

LO1 Describe the structure and function of vertebrate skeletal muscles and the mechanisms that cause muscle cells to contract.

LO2 Explain how the nervous system controls the contraction of skeletal muscles, including how the nervous system determines which muscles contract and how forcefully they contract.

LO3 Compare and contrast skeletal, cardiac, and smooth muscle.

LO4 Describe hydrostatic skeletons, exoskeletons, and

endoskeletons, and how antagonistic muscles act on these skeletons to move an animal's body.

LO5 Describe the vertebrate skeleton and its principal functions.

LO6 List the three types of connective tissue in the vertebrate skeleton, and describe the composition and function of each.

LO7 Explain the processes by which bones are remodeled and repaired.

Thinking Through the Concepts

Fill-in-the-Blank

1. The three types of skeletal systems found in animals are _____, _____, and _____. All three move the skeleton using pairs of _____ muscles. **LO4**

2. _____ and _____ muscles are striped because they have a regular alignment of sarcomeres. _____ and _____ muscles are usually not under conscious control. _____ and _____ muscles are interconnected by gap junctions. **LO3**

3. A skeletal muscle cell is called a(n) _____. It consists of bundles of cylindrical structures called _____, which in turn consist of much smaller cylinders, the _____, attached end to end at their _____. **LO1**

4. Muscle contraction results from interactions between thin filaments, composed mainly of the protein _____, and thick filaments, composed mainly of the protein _____. The "heads" of the thick filament protein grab on to the thin filament protein and flex. The energy of _____ is used to extend the head, which stores that energy to power the flexing movement of the head, pulling the thin filament past the thick filament. **LO1**

5. Motor neurons innervate muscle fibers at specialized synapses called _____. The release of acetylcholine at these synapses ultimately triggers an action potential in the muscle fiber, which invades the interior of the fiber through _____, and stimulates the release of calcium ions from the _____. **LO2**

6. Skeletons consist of three types of connective tissue: _____, _____, and _____. All consist of cells embedded in a matrix of _____ protein and other extracellular components, such as glycoproteins or calcium phosphate. **LO5**

7. A(n) _____ joint moves in two dimensions, while a(n) _____ joint has significant motion in three dimensions. At a two-dimensional joint, the _____ muscle bends the joint and the _____ muscle straightens the joint. **LO4 LO5**

Review Questions

1. Sketch a relaxed muscle fiber containing a myofibril, sarcomeres, and thick and thin filaments. How would a contracted muscle fiber look by comparison? **LO1**

2. Describe the process of skeletal muscle contraction, beginning with an action potential in a motor neuron and ending with the relaxation of the muscle. Your answer should include the following words: *neuromuscular junction, T tubule, sarcoplasmic reticulum, calcium, thin filaments, binding sites, thick filaments, sarcomere, Z line,* and *active transport.* **LO1 LO2**

3. Explain the following two statements: Muscles can only actively contract; muscle fibers lengthen passively. **LO1 LO4**

4. What are the three types of skeletons found in animals? For one of them, describe how the muscles are arranged around the skeleton and how contractions of the muscles result in movement of the skeleton. **LO4**

5. Compare the structures of the following pairs: spongy and compact bone, smooth and skeletal muscle, and cartilage and bone. **LO3 LO6**

6. Explain the functions of osteoblasts and osteoclasts. **LO7**

7. Describe a hinge joint and how it is moved by antagonistic muscles. **LO4 LO5**

Applying the Concepts

1. Discuss some of the problems that would result if the human heart were made of skeletal muscle instead of cardiac muscle.

2. Myasthenia gravis is caused by the abnormal production of antibodies that bind to acetylcholine receptors on muscle cells and eventually destroy the receptors. The disease causes muscles to become flaccid, weak, or paralyzed. Drugs such as neostigmine, which inhibits the enzyme that breaks down acetylcholine, are used to treat myasthenia gravis. How does neostigmine restore muscle activity? Do you think that neostigmine could restore normal muscle function for an entire lifetime?

3. Would taking erythropoietin, a hormone that increases the number of red blood cells, primarily benefit sprinters or endurance athletes? Why? Explain your answer in terms of the characteristics of fast-twitch and slow-twitch muscle fibers.

4. In chickens, breast meat is light and leg meat is dark. Explain this difference in terms of muscle fiber types and the normal amounts of activity performed by the breast and leg muscles. Would you expect a wild duck to have the same meat colors in the same locations as a domestic chicken does?

Answers to Figure Caption questions and Fill-in-the-Blank questions can be found in the Answers section at the back of the book.

MB *Go to MasteringBiology for practice quizzes, activities, eText, videos, current events, and more.*

Lulu, a white rhino cow in the Budapest Zoo, sniffs her calf, Layla. Lulu was artificially inseminated with sperm from Easyboy, a male rhino with whom she had lived for years without ever becoming pregnant.

CASE STUDY
To Breed a Rhino

RHINOS ARE MASSIVE, aggressive, and seemingly indestructible. Nevertheless, three of the five species are critically endangered. How can such animals be in danger of extinction? Their name—rhinoceros, or "nose horn"—tells you why: They are being killed for their horns.

Some cultures place enormous value on rhino horn. In Yemen, for example, young men are traditionally given a ceremonial dagger, the jambiya, often with a handle made of rhino horn. In China and some other East Asian cultures, powdered rhino horn is believed to reduce fevers and cure rheumatism, gout, and many other disorders. (Contrary to popular belief, few people use rhino

horn as an aphrodisiac.) Because rhinos are rare, rhino horn is very expensive. In recent years, the black market price soared to about $30,000 a pound, significantly more than the price of gold.

To help stave off extinction of endangered rhinos, scientists at several zoos are breeding rhinos in captivity. By mating carefully selected rhinos, they hope to preserve the rhinos' genetic diversity, which is crucial to a species' survival. Captive breeding sounds simple enough—put a male and female into the same enclosure and let nature take its course. Unfortunately, this often doesn't work with large, aggressive loners like rhinos, who sometimes seem to prefer combat to copulation. Even if a male and female don't fight, they may not mate—Lulu and

Easyboy lived together for 20 years in the Budapest Zoo and never mated. Artificial insemination would avoid both the dangers of natural mating and a possible lack of sexual interest. Mating or artificial insemination will result in pregnancy only if the female has an egg ready to fertilize. Therefore, scientists need to know when the female ovulates naturally or, alternatively, they must be able to induce her to ovulate at the right time.

How do male and female animals produce gametes? Why are the reproductive systems of males and females so different? How do scientists assisting rhinos to reproduce—or human physicians collecting eggs for *in vitro* fertilization—induce ovulation or determine when ovulation occurs naturally?

AT A GLANCE

41.1 How Do Animals Reproduce?

41.2 What Are the Structures and Functions of Human Reproductive Systems?

41.3 How Can People Prevent Pregnancy?

LEARNING GOALS

LG1 Compare and contrast asexual and sexual reproduction, and describe some advantages and disadvantages of each.

LG2 Define *budding, fission,* and *parthenogenesis,* and provide some examples of animals that use each of these mechanisms of reproduction.

LG3 Define *internal* and *external fertilization,* and provide examples of animals that use each of these methods of fertilization.

LG4 Describe the human male reproductive tract, including the gonads, the tubes that conduct sperm, and the glands that provide the fluid portion of semen.

LG5 Describe the human female reproductive tract, including the gonads and the accessory structures that transport sperm and eggs and in which an embryo develops.

LG6 Explain how hormones from the hypothalamus, anterior pituitary gland, and gonads control reproduction in men and women.

LG7 Describe spermatogenesis and oogenesis.

LG8 Describe the structures of sperm and eggs and how fertilization occurs.

LG9 Explain how the principal methods of contraception work.

41.1 HOW DO ANIMALS REPRODUCE?

Animals reproduce either sexually or asexually. In **asexual reproduction,** a single animal produces offspring, usually through repeated mitotic cell divisions in some part of its body. The offspring are therefore genetically identical to the parent. In **sexual reproduction,** organs called **gonads** produce haploid sperm or eggs through meiotic cell division. A sperm and egg—usually from separate parents—fuse to produce a diploid fertilized egg, or **zygote,** which then undergoes repeated mitotic cell divisions to produce an offspring. Because the offspring receives genes from both of its parents, it is not genetically identical to either.

In Asexual Reproduction, an Organism Reproduces Without Mating

Asexual reproduction is efficient in effort (no need to search for mates, court members of the opposite sex, or battle rivals) and materials (no wasted gametes). There are several common methods of asexual reproduction in the animal kingdom.

Budding Produces a Miniature Version of the Adult

Many sponges and cnidarians (such as corals, anemones, and *Hydra*) reproduce by **budding.** A miniature version of the adult—a bud—grows directly on the body of the adult (**Fig. 41-1**). When the bud has grown large enough, it breaks off and becomes independent, eventually growing into an adult in its turn.

Fission Followed by Regeneration Can Produce a New Individual

Many animals are capable of **regeneration,** the ability to regrow lost body parts. Regeneration is part of reproduction in species that reproduce by **fission.** Some flatworms, corals, sea jellies, and brittle stars reproduce by dividing into two or

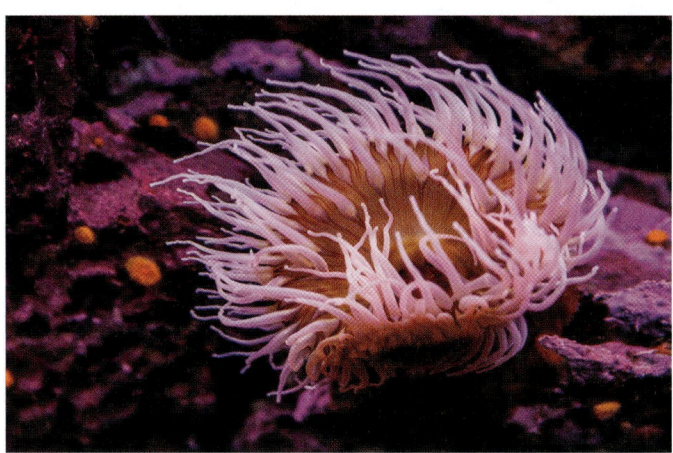

▲ **FIGURE 41-1 Budding** The offspring of some cnidarians, such as this anemone, grow as buds sprouting from the body of the parent. When sufficiently developed, the buds may separate from the parent and live independently.

QUESTION Which type of cell division gives rise to the cells of the bud's body?

more pieces, each of which regenerates the missing parts of a complete body.

During Parthenogenesis, Eggs Develop Without Fertilization

The females of some animal species can reproduce by a process known as **parthenogenesis,** in which egg cells develop into offspring without being fertilized. In some species, parthenogenetically produced offspring are haploid. For example, male honeybees develop from unfertilized haploid eggs, and remain haploid; their diploid sisters develop from fertilized eggs. Some parthenogenetic fish, amphibians, and reptiles produce diploid offspring by doubling the number of chromosomes in the eggs, either before or after

▲ **FIGURE 41-2 A desert grassland whiptail lizard**
This entire species consists of parthenogenetically reproducing females—and no males at all.

meiosis. In most species, all the resulting diploid offspring are female. In fact, some species of fish and lizards, such as the whiptail lizard of the southwestern United States and Mexico (**Fig. 41-2**), consist entirely of parthenogenetically reproducing females. Still other animals, such as some aphids, can reproduce either sexually or parthenogenetically, depending on environmental factors such as the season of the year and the availability of food (**Fig. 41-3**).

In Sexual Reproduction, an Organism Reproduces via the Union of Sperm and Egg

No one is certain why sexual reproduction evolved. However, the most likely explanation is that the genetic variability resulting from sexual reproduction is advantageous, providing offspring with new traits that may enable them to

exploit new habitats or employ new methods of acquiring food, avoiding predation, and foiling parasites.

In most animal species, an individual is either male or female, defined by the type of gamete that it produces. The female gonad, called the **ovary,** produces **eggs,** which are large, haploid cells containing food reserves that provide nourishment for the **embryo,** an offspring in its early stages of development before birth or hatching. Eggs have no independent means of moving about. The male gonad, called the **testis** (plural, testes), produces small haploid **sperm,** which have almost no cytoplasm and hence no food reserves. Sperm can swim by thrashing their tails. **Fertilization,** the union of sperm and egg, produces a diploid zygote.

In some animals, such as earthworms and many snails, single individuals produce both sperm and eggs. Such individuals are called **hermaphrodites.** Most hermaphrodites still engage in sex, with two individuals exchanging sperm (**Fig. 41-4**). Some hermaphrodites, however, can fertilize their own eggs. Many hermaphroditic animals, including tapeworms and some types of snails, are relatively immobile and may find themselves isolated from other members of their species, so self-fertilization may be the only way they can reproduce.

For species with two separate sexes and hermaphrodites that cannot self-fertilize, successful reproduction requires that sperm and eggs from different individuals be brought together for fertilization.

External Fertilization Occurs Outside the Parents' Bodies

In **external fertilization,** sperm and egg unite outside the bodies of the parents. Sperm and eggs are typically released into water, a process called **spawning,** and the sperm swim to reach the eggs. Because sperm and eggs are generally short lived, spawning animals must synchronize their reproductive behaviors, both temporally (male and female spawn at the same time) and spatially (male and female spawn in the same place). Such synchronization may be achieved through environmental cues, chemical signals, courtship behaviors, or some combination of these factors.

offspring being born   adult female aphid

▲ **FIGURE 41-3 A female aphid gives birth** In spring and early summer, when food is abundant, aphid females reproduce parthenogenetically; in fact, the females are born pregnant! In fall, they reproduce sexually, as the females mate with males.

QUESTION Why do aphids switch to sexual reproduction in the fall?

▲ **FIGURE 41-4 Hermaphroditic earthworms exchange sperm**

▲ FIGURE 41-5 **Environmental cues may synchronize spawning** In the Great Barrier Reef of Australia, thousands of corals spawn simultaneously, creating this "blizzard" effect. Spawning in these corals is linked to the phase of the moon.

Most spawning animals rely on environmental cues to some extent. Seasonal changes in day length often stimulate the physiological changes required for reproduction, which restricts breeding to certain times of year. However, the actual release of sperm and egg must be more precisely synchronized. For example, many corals of Australia's Great Barrier Reef synchronize spawning by the phase of the moon, so

▲ FIGURE 41-6 **Courtship rituals synchronize spawning** During spawning, male and female Siamese fighting fish perform a courtship dance that culminates with both partners releasing gametes simultaneously. The male will collect the fertilized eggs in his mouth and spit them into the nest of bubbles visible above the courting fish. The male guards the eggs and the newly hatched babies in the nest until they are able to swim and fend for themselves.

QUESTION In addition to ensuring synchronized release of gametes, what other advantages do courtship rituals provide?

▲ FIGURE 41-7 **Splendid tree frogs mating** The smaller male clutches the female and stimulates her to release eggs. Although they spend most of their lives high up in the trees of Central American rain forests, they descend to small pools of water to breed.

that males and females simultaneously release a blizzard of sperm and eggs into the water (**Fig. 41-5**).

Many animals rely on mating behaviors to synchronize spawning. Most fish, for example, have courtship rituals, ensuring that they release their gametes in the same place and at the same time (**Fig. 41-6**). Frogs and toads mate in shallow water near the edges of ponds and lakes. The male mounts the female and prods the sides of her abdomen (**Fig. 41-7**). This stimulates her to extrude her eggs, while he deposits sperm onto them.

Internal Fertilization Occurs Within the Female's Body

During **internal fertilization,** sperm are placed within the female's moist reproductive tract, where her eggs are fertilized. Internal fertilization is an important adaptation to terrestrial life because sperm quickly die if they dry out. Even in aquatic environments, internal fertilization may increase the likelihood of success, because the sperm and eggs are confined in a small space rather than dispersed in a large volume of water.

Internal fertilization usually occurs by **copulation,** in which the male deposits sperm directly into the female's reproductive tract (**Fig. 41-8**). In a variation of internal fertilization, males of some species, including some salamanders, scorpions, and grasshoppers, package their sperm in a container called a *spermatophore* (Greek for "sperm carrier"). Males and females usually engage in a mating display, after which the male may insert his spermatophore into the female's reproductive tract or drop the spermatophore on the ground for the female to pick up if she chooses. Once inside the female, the spermatophore releases sperm.

In most animals, neither sperm nor eggs live very long. Therefore, **ovulation,** the release of a mature egg cell from the ovary of the female, usually must occur about the same time that sperm are deposited in the female's reproductive

▲ FIGURE 41-8 **Internal fertilization allows reproduction on land** Ladybugs mate on a blade of grass.

CASE STUDY c o n t i n u e d

To Breed a Rhino

To attract a mate, Sumatran rhinos urinate, raise their tails, vocalize, and bump each other on the head and genitals with their snouts. Sometimes this courtship gets violent enough to injure or even kill one or both animals, particularly if the female resists the male's advances and can't escape because the animals are confined in a small area.

tract. Most mammals, for example, use courtship displays to synchronize mating with ovulation. Many copulate only when the female signals readiness to mate, which usually occurs about the same time as ovulation. In some animals, such as rabbits, the act of mating stimulates ovulation, so new, healthy sperm and eggs are almost guaranteed to meet.

CHECK YOUR LEARNING

Can you explain the difference between sexual and asexual reproduction and describe some advantages of each? Can you define *budding, fission,* and *parthenogenesis* and provide some examples of animals that use each of these mechanisms of reproduction? Can you describe internal and external fertilization and provide examples of animals that use each of these methods of fertilization?

41.2 WHAT ARE THE STRUCTURES AND FUNCTIONS OF HUMAN REPRODUCTIVE SYSTEMS?

People and other mammals have separate sexes, copulate, and fertilize their eggs internally. Although most mammals reproduce only during certain seasons of the year, men produce sperm more or less continuously, and women ovulate about once a month.

The Ability to Reproduce Begins at Puberty

Sexual maturation in humans occurs at **puberty,** a stage of development also characterized by rapid growth and the appearance of secondary sexual characteristics. The age of puberty has been dropping for at least 150 years. For example, the average age at which girls have their first menstrual period has dropped from about 17 years in the mid-1800s to about 12 years today, although there is considerable variability: Most girls undergo puberty between the ages of 9 and 13, with a few starting as early as 6 or as late as 15. Boys generally become sexually mature a year or two later than girls do. Improved nutrition and reduced disease are thought to be the major factors causing the lower age of puberty. Some researchers suspect that exposure to environmental chemicals that mimic the effects of estrogen or testosterone may also play a role.

In both sexes, brain maturation causes the hypothalamus to release **gonadotropin-releasing hormone (GnRH),** which stimulates the anterior pituitary to produce **luteinizing hormone (LH)** and **follicle-stimulating hormone (FSH).** Although LH and FSH derive their names from their functions in females, they are equally essential in males. LH and FSH stimulate the testes to produce the male sex hormone **testosterone** and the ovaries to produce the female sex hormone **estrogen.** In response to increases in testosterone, males develop secondary sexual characteristics: The penis and the testes enlarge; pubic, underarm, and facial hair appears; the larynx enlarges (deepening the voice); and muscular development increases. In response to increased estrogen and other hormones that surge at puberty, females develop enlarged breasts, wider hips, pubic and underarm hair, and begin menstruating. Courtship behaviors also start to appear in both sexes during puberty.

The Male Reproductive System Includes the Testes and Accessory Structures

The male reproductive system includes the testes, which produce testosterone and sperm; glands that secrete substances that activate and nourish sperm; and tubes that store sperm and conduct them out of the body (**Fig. 41-9** and **Table 41-1**). Testosterone and sperm production are nearly continuous, beginning at puberty and lasting until death.

Sperm Are Produced in the Testes

The testes are located in the **scrotum,** a pouch that hangs outside the main body cavity. This location keeps the testes about 1° to 6°F (0.5° to 3°C) cooler than the core of the body, depending on whether the man is standing (cooler) or sitting (warmer) and what type of clothing he is wearing. Cooler temperatures promote sperm development. Each testis is nearly filled with coiled, hollow **seminiferous tubules,** in which sperm are produced (**Fig. 41-10a**). **Interstitial cells,** which synthesize testosterone, are located in the spaces between the tubules (**Fig. 41-10b**).

Just inside the wall of each seminiferous tubule lie cells called **spermatogonia** (singular, spermatogonium), which give rise to sperm, and much larger **Sertoli cells,** which

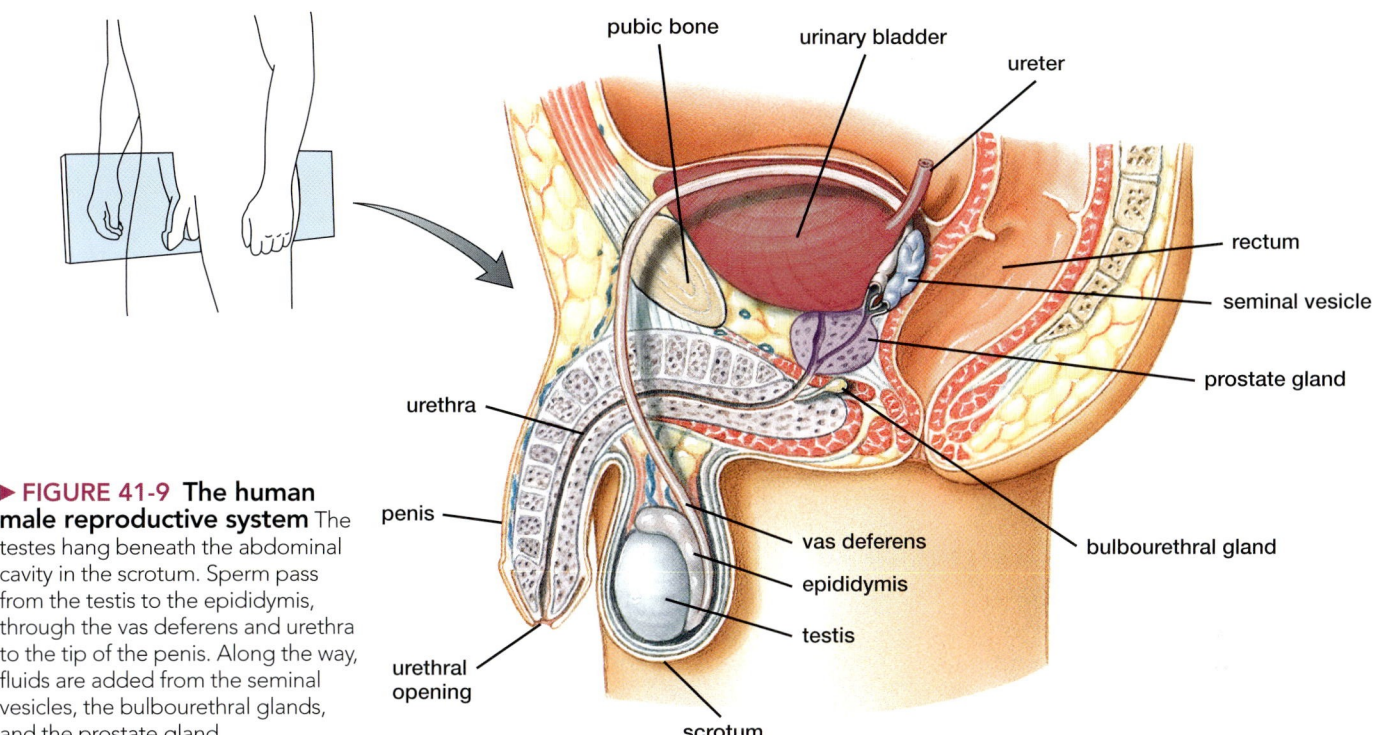

► **FIGURE 41-9 The human male reproductive system** The testes hang beneath the abdominal cavity in the scrotum. Sperm pass from the testis to the epididymis, through the vas deferens and urethra to the tip of the penis. Along the way, fluids are added from the seminal vesicles, the bulbourethral glands, and the prostate gland.

nourish the developing sperm and regulate their growth. Spermatogonia are diploid cells that undergo mitotic cell division, forming two types of daughter cells (**Fig. 41-11**). One daughter cell remains a spermatogonium, ensuring a steady supply throughout a male's life. The other daughter cell becomes committed to **spermatogenesis,** the processes that produce haploid sperm.

TABLE 41-1 The Human Male Reproductive Tract

Structure	Function
Testis (male gonad)	Produces sperm and testosterone
Epididymis and vas deferens (ducts)	Store sperm; conduct sperm from the testes to the urethra
Urethra (duct)	Conducts semen from the vas deferens and urine from the urinary bladder to the tip of the penis
Penis	Deposits sperm in the female reproductive tract
Seminal vesicles (glands)	Secrete fluid into the semen
Prostate gland	Secretes fluids into the semen
Bulbourethral glands	Secrete fluid into the semen

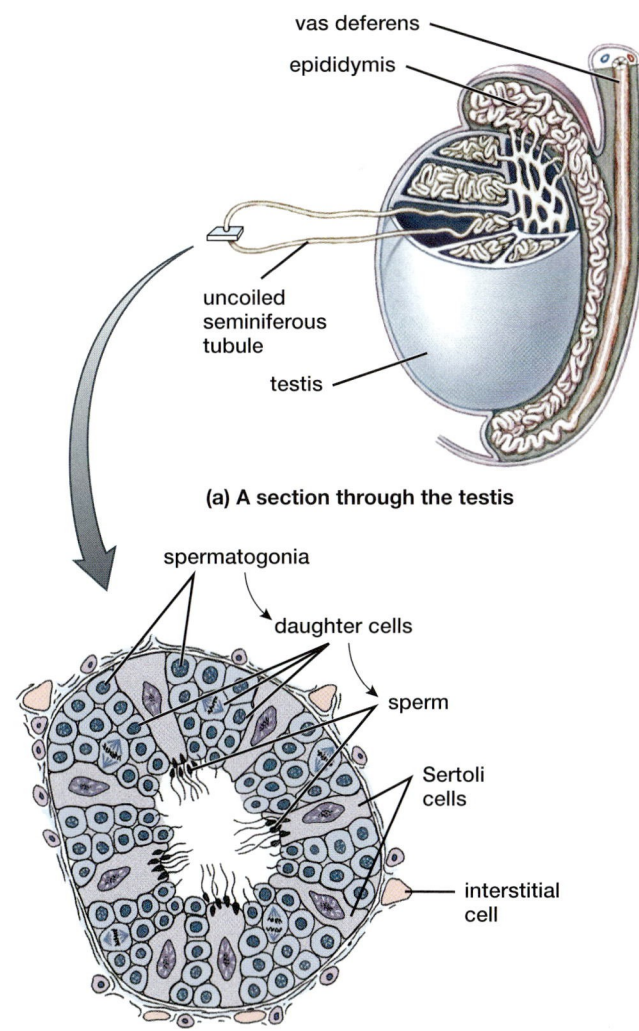

(a) A section through the testis

(b) Cross-section of a seminiferous tubule

► **FIGURE 41-10 The anatomy of the testes (a)** A section of the testis, showing the seminiferous tubules, epididymis, and vas deferens. **(b)** The walls of the seminiferous tubules are lined with Sertoli cells and spermatogonia. As the spermatogonia divide, the daughter cells move inward. Mature sperm are freed into the central cavity, which is continuous with the inside of the epididymis. Testosterone is produced by the interstitial cells.

▶ **FIGURE 41-11 Spermatogenesis**
A spermatogonium divides by mitotic cell division; one of its two daughter cells remains a spermatogonium (curved arrow), while the other enlarges and becomes a primary spermatocyte (straight arrow), which will undergo meiotic cell division followed by differentiation, producing haploid sperm. For clarity, only two pairs of chromosomes are shown; however, human cells have 23 pairs of chromosomes.

Spermatogenesis begins as the committed daughter cell differentiates into a **primary spermatocyte,** a large diploid cell that will undergo meiotic cell division (see Chapter 9). At the end of meiosis I, each primary spermatocyte gives rise to two haploid **secondary spermatocytes.** Each secondary spermatocyte undergoes meiosis II, producing two **spermatids.** Thus, each primary spermatocyte generates a total of four spermatids. Spermatids differentiate into sperm without further cell division. Spermatogonia, spermatocytes, and spermatids are enfolded in the Sertoli cells. As spermatogenesis proceeds, the developing sperm migrate to the central cavity of the seminiferous tubule into which the mature sperm are released (see Fig. 41-10b).

A human sperm (**Fig. 41-12**) is unlike any other cell of the body. There is little cytoplasm, so the haploid nucleus nearly fills the sperm's head. Atop the nucleus lies a sac of enzymes called an **acrosome.** As we will see, the enzymes dissolve protective layers that surround the egg, enabling the sperm to penetrate through them. Behind the head is the midpiece, packed with mitochondria. The mitochondria provide the energy needed to move the tail, which is actually

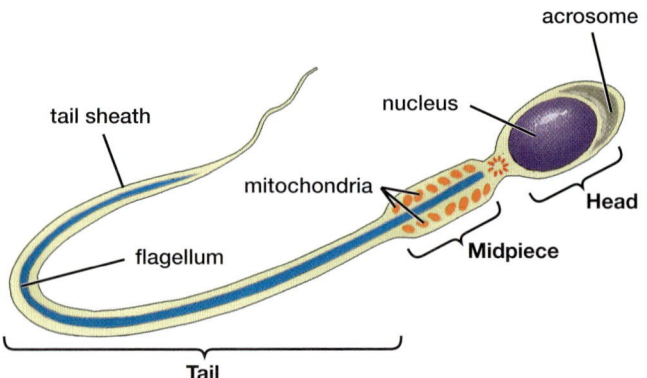

▲ **FIGURE 41-12 A human sperm cell** A mature sperm contains little more than a haploid nucleus, an acrosome (containing enzymes that digest the barriers surrounding the egg), mitochondria for energy production, and a tail (a long flagellum) for locomotion.

a long flagellum. Whiplike movements of the tail propel the sperm through the female reproductive tract.

Hormones from the Anterior Pituitary and Testes Regulate Spermatogenesis

Spermatogenesis is stimulated by GnRH release from the hypothalamus, which causes the anterior pituitary to release LH and FSH. LH stimulates the interstitial cells of the testes to produce testosterone (**Fig. 41-13**). The combination of testosterone and FSH stimulates the Sertoli cells and promotes spermatogenesis.

Testicular function is regulated by negative feedback. Testosterone inhibits the release of GnRH, LH, and FSH, which limits further testosterone production and sperm development. The Sertoli cells, when stimulated by FSH and testosterone, secrete a hormone called inhibin, which inhibits FSH production by the anterior pituitary. This complex feedback process maintains relatively constant levels of testosterone and sperm production.

Accessory Structures Contribute to Semen and Conduct the Sperm Outside the Body

The seminiferous tubules merge to form the **epididymis,** a long folded tube (see Figs. 41-9 and 41-10a). The epididymis leads into the **vas deferens,** a tube that carries sperm out of the scrotum. Most of the roughly hundred million sperm produced each day are stored in the vas deferens and epididymis. The vas deferens joins the **urethra,** which runs to the tip of the **penis.** The urethra conducts urine out of the body during urination and sperm out of the body during ejaculation.

Men ejaculate a fluid, called **semen,** that consists of about 5% sperm mixed with secretions from three types of glands that empty into the vas deferens or the urethra: the seminal vesicles, the prostate gland, and the bulbourethral glands (see Fig. 41-9 and Table 41-1). Paired **seminal vesicles** secrete about 60% of the semen. Liquid from the seminal vesicles is rich in fructose, a sugar that provides energy for the sperm. The liquid's slightly alkaline pH protects the sperm

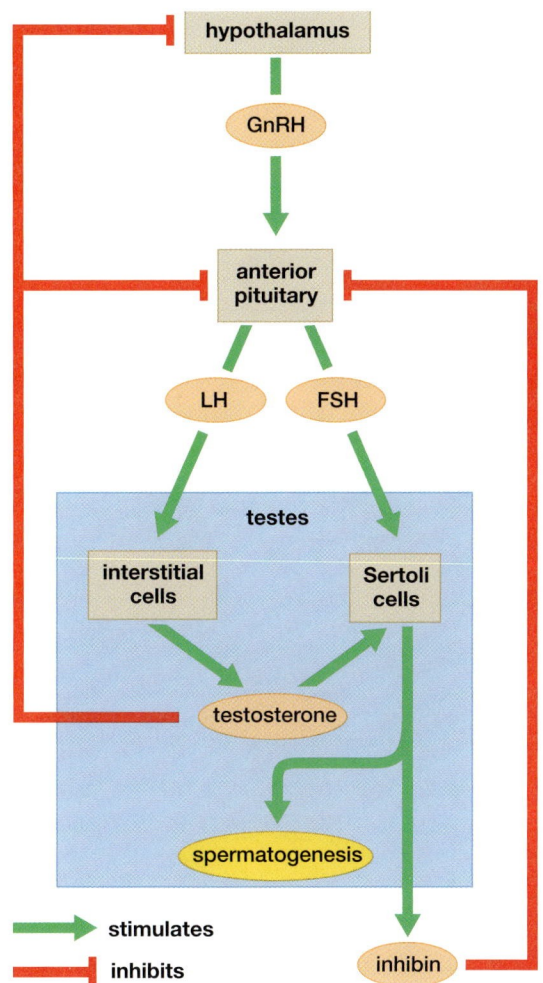

▲ FIGURE 41-13 Hormonal control of spermatogenesis
Negative feedback interactions between the hypothalamus, anterior pituitary, and testes keep the concentration of testosterone and the production of sperm at nearly constant levels.

QUESTION Why would injections of testosterone suppress sperm production?

TABLE 41-2 The Human Female Reproductive Tract	
Structure	Function
Ovary (female gonad)	Produces eggs, estrogen, and progesterone
Fimbriae (at the opening of the uterine tube)	Bear cilia that sweep the egg into the oviduct
Uterine tube	Conducts the egg to the uterus; site of fertilization .
Uterus	Muscular chamber where the embryo develops
Cervix	Nearly closes off the outer end of the uterus; supports the developing embryo during pregnancy
Vagina	Receptacle for semen; birth canal

from the acidity of urine remaining in the man's urethra and from acidic secretions in a woman's vagina. Seminal vesicle fluid also contains prostaglandins (see Chapter 37), which stimulate uterine contractions that help to transport the sperm up the female reproductive tract. The **prostate gland** produces an alkaline, nutrient-rich secretion that makes up about 30% of the semen. The prostate fluid includes enzymes that increase the fluidity of the semen after it is released into the vagina, allowing the sperm to swim more freely. Paired **bulbourethral glands** secrete a small amount of alkaline mucus into the urethra, neutralizing remaining traces of acidic urine.

The Female Reproductive System Includes the Ovaries and Accessory Structures

The female reproductive system consists of the ovaries and structures that accept sperm, conduct the sperm to the egg, and nourish the developing embryo (**Fig. 41-14** and **Table 41-2**).

Egg Development in the Ovaries Begins Before Birth

Oogenesis, the formation of egg cells, begins in the developing ovaries of a female embryo (**Fig. 41-15**). Oogenesis starts with the formation of diploid cells called **oogonia** (singular, oogonium) as early as the 6th week of embryonic development. From about the 9th through the 20th weeks, the oogonia enlarge and differentiate, becoming **primary oocytes.** By about the 20th week, all of the primary oocytes have begun meiotic cell division, but this stops during prophase of meiosis I. None of the primary oocytes will resume meiotic cell division until puberty.

Therefore, a woman is born with her lifetime's supply of primary oocytes—about 1 to 2 million—and no new oocytes are generated later in life. Many of these die each day, but about 400,000 remain at puberty. This is plenty, because only a few oocytes resume meiotic cell division during each month of a woman's reproductive span, from puberty to menopause at about age 50.

The Ovary Produces Eggs, Estrogen, and Progesterone During the Menstrual Cycle

In the mature ovary, each oocyte is surrounded by a layer of smaller cells. Together, the oocyte and these accessory cells make up a **follicle** (**Fig. 41-16 ❶**). Follicle development and ovulation are governed by interactions among hormones produced by the hypothalamus, anterior pituitary, and ovary, in a roughly 28-day cycle called the **menstrual cycle.**

Each month, the hypothalamus secretes GnRH, which stimulates the anterior pituitary to release LH and FSH. FSH stimulates about a dozen follicles to begin developing (**Fig. 41-16 ❷** and **❸**). The small follicle cells multiply, providing nourishment for the developing oocyte. The follicle cells also release estrogen into the bloodstream. Usually, only one follicle completely matures during each menstrual cycle.

The maturing follicle secretes increasing amounts of estrogen, which stimulates a surge of LH that causes the primary oocyte to complete meiosis I, dividing into a single

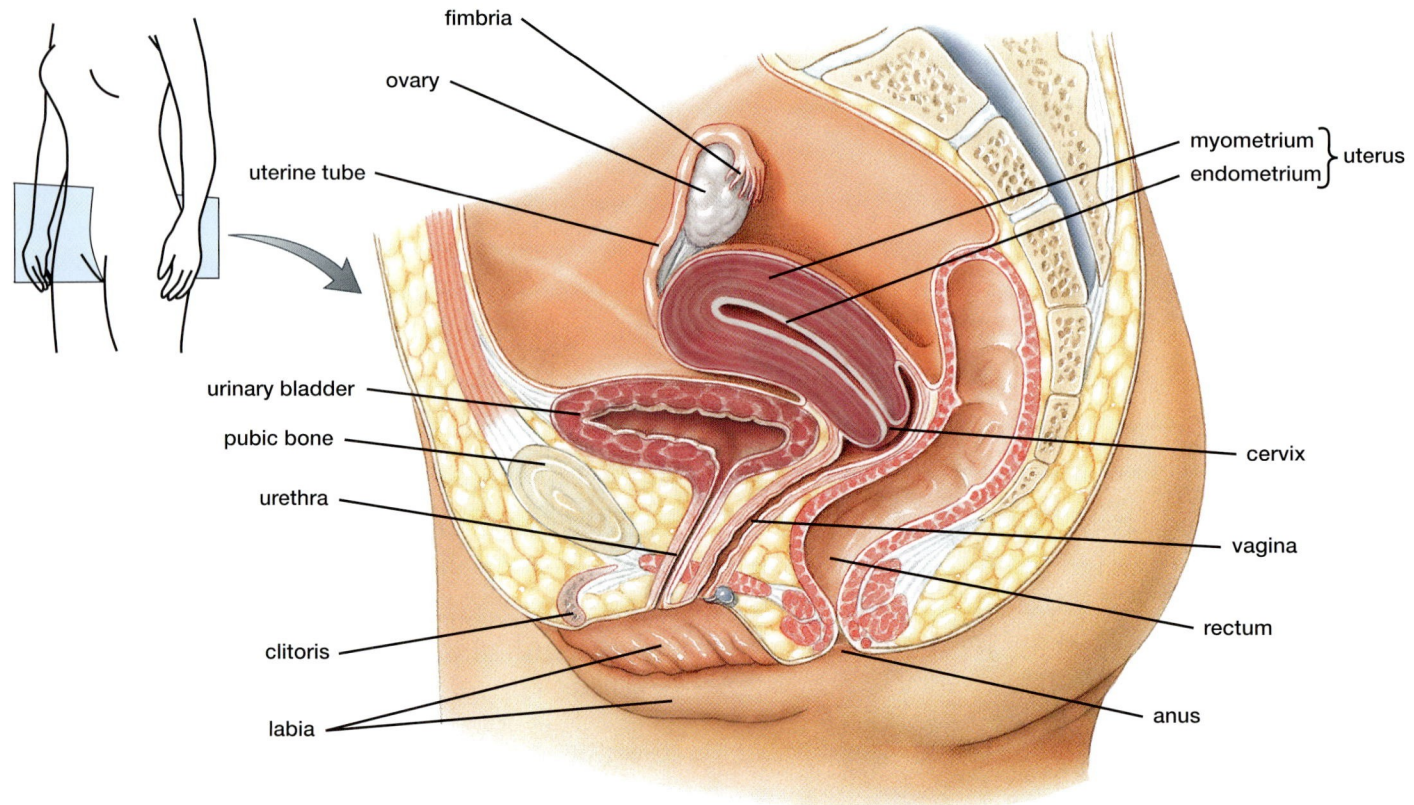

▲ FIGURE 41-14 The human female reproductive system Eggs are produced in the ovaries and enter the uterine tube. Sperm and egg usually meet in the uterine tube, where fertilization and very early development occur. The early embryo implants in the lining of the uterus, where development continues. The vagina receives sperm and serves as the birth canal.

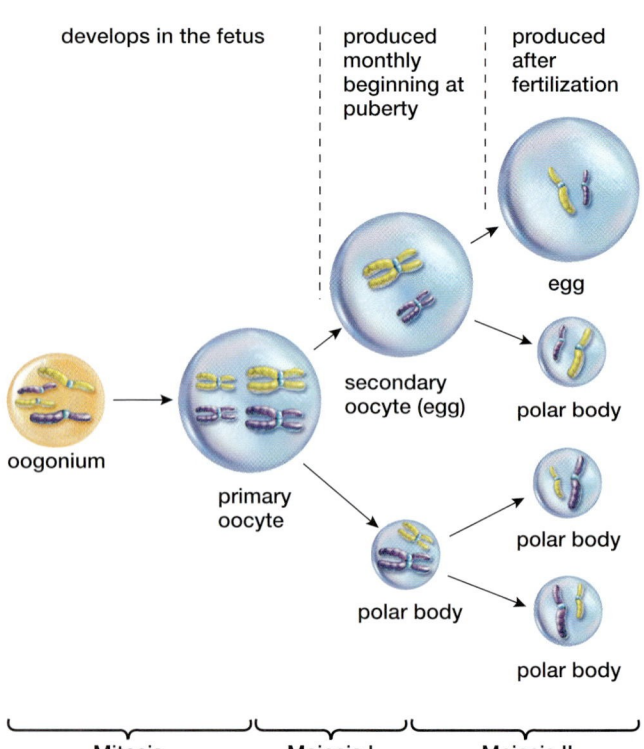

secondary oocyte and the first **polar body** (see Fig. 41-15). The polar body is a small cell, with very little cytoplasm; it cannot be fertilized by sperm. The secondary oocyte will not undergo meiosis II unless it is fertilized. The surge of LH also causes ovulation, as the follicle erupts through the surface of the ovary, releasing its secondary oocyte (**Fig. 41-16 ❹**). For convenience, we will refer to the ovulated secondary oocyte as the egg.

Some of the follicle cells leave the ovary with the egg, but most remain in the ovary (**Fig. 41-16 ❺**), where they enlarge, forming a temporary gland called the **corpus luteum** (**Fig. 41-16 ❻**). The corpus luteum secretes both estrogen and a second hormone called **progesterone.** The combination of estrogen and progesterone inhibits further release of GnRH, LH, and FSH, thereby preventing the development of any more follicles. If the egg is not fertilized, the corpus luteum will degenerate within a few days (**Fig. 41-16 ❼**).

◀ FIGURE 41-15 Oogenesis An oogonium enlarges to form a primary oocyte. At meiosis I, almost all the cytoplasm is included in one daughter cell, the secondary oocyte. The other daughter cell is a small polar body that contains chromosomes but little cytoplasm. At meiosis II, almost all the cytoplasm of the secondary oocyte is included in the egg, and a second small polar body discards the remaining "extra" chromosomes. The first polar body may also undergo the second meiotic division. The polar bodies eventually degenerate. In humans, meiosis II does not occur unless a sperm penetrates the egg.

▲ **FIGURE 41-16 Follicle development** For convenience, this diagram shows all of the stages of follicle development throughout a complete menstrual cycle, going clockwise from lower right. In a real ovary, all the stages would never be present at the same time, and the follicle would not move around the ovary as it develops.

"In Greater Depth: Hormonal Control of the Menstrual Cycle" explains how hormonal interactions among the hypothalamus, anterior pituitary, and ovary drive oogenesis, follicle development, and ovulation.

Accessory Structures Include the Uterine Tubes, Uterus, and Vagina

Each ovary nestles within the open end of a **uterine tube** (also called the oviduct or Fallopian tube), which leads from the ovary to the uterus. The opening to the uterine tube is fringed with ciliated "fingers," called *fimbriae*, that nearly surround the ovary (see Fig. 41-14). The cilia create a current that sweeps the ovulated egg into the uterine tube. There, the egg may encounter sperm and be fertilized. Cilia lining the uterine tube transport the fertilized egg down the tube into the **uterus.**

The wall of the uterus has two layers that correspond to its dual functions of nourishing the developing embryo and delivering a child. The inner lining, or **endometrium,** will form the mother's contribution to the **placenta,** the structure that transfers oxygen, carbon dioxide, nutrients, and wastes between mother and embryo (see Chapter 42). The outer muscular wall of the uterus, called the **myometrium,** contracts during childbirth, expelling the infant out of the uterus.

The lower end of the uterus is nearly closed off by the **cervix,** a ring of connective tissue that encircles a tiny opening. The cervix holds the developing baby in the uterus and then expands during labor, permitting passage of the child. Beyond the cervix is the **vagina,** which opens to the outside. The vagina serves both as the receptacle for the penis and sperm during intercourse and as the birth canal. The vaginal lining is acidic, which reduces the likelihood of infections.

Developing follicles secrete estrogen, which stimulates the endometrium to become thicker and grow an extensive network of blood vessels and glands that secrete carbohydrates, lipids, and proteins. After ovulation, estrogen and progesterone released by the corpus luteum further stimulate the development of the endometrium. Thus, if an egg is fertilized, when the developing embryo reaches the uterus, it encounters a rich

CASE STUDY continued

To Breed a Rhino

Mammalian sperm don't live very long after ejaculation. Therefore, for a rhino cow to become pregnant, the sperm must be placed in her vagina or uterus, whether by mating or by artificial insemination, at about the same time that ovulation occurs. The simplest way to ensure correct timing is to induce ovulation and have mating or artificial insemination occur on that same day. What hormone(s) do you think could be used to induce ovulation? Check the "Case Study Revisited" at the end of the chapter to find out what rhino specialists actually use.

IN GREATER DEPTH Hormonal Control of the Menstrual Cycle

The average length of the menstrual cycle is about 28 days, although many women have cycles as short as 21 days or as long as 35 days. We will use a 28-day cycle for the descriptions that follow. The beginning of menstruation from the previous cycle is designated as day 1 of the new cycle, because menstruation is easily observed. However, the hormonal events that drive the new cycle begin a day or two earlier. The menstrual cycle actually consists of two cycles: the ovarian cycle and the uterine cycle.

The Ovarian Cycle

In the ovarian cycle, interactions of hormones produced by the hypothalamus, the anterior pituitary gland, and the ovaries drive the development of follicles, maturation of oocytes, and production of the corpus luteum. In the description below, the numbered steps correspond to the numbers in black circles in **Figure E41-1a.**

The ovarian cycle begins as the hypothalamus spontaneously releases GnRH (top panel, step ❶). GnRH stimulates the release of FSH (blue line) and LH (red line) by the anterior pituitary (second panel, step ❷). FSH stimulates the development of several follicles within each ovary (third panel, step ❸). The small follicle cells surrounding the oocyte secrete estrogen (purple line, fourth panel, step ❹). Under the combined influences of FSH, LH, and estrogen, the follicles continue to grow. The maturing follicles secrete increasing amounts of estrogen. Estrogen promotes the continued development of one of the follicles and of the primary oocyte within it. Estrogen also stimulates the

hypothalamus to release more GnRH (top panel, step ❺).

Increased GnRH stimulates a surge of LH and FSH at about the 13th or 14th day of the cycle (second panel, step ❻). The resulting high concentration of LH has three important consequences. First, it triggers the resumption of meiosis I in the oocyte, producing the secondary oocyte and the first polar body. Second, the LH surge causes ovulation. Third, it transforms the remnants of the follicle into the corpus luteum (both illustrated in the third panel, step ❼). The corpus luteum secretes both estrogen (purple line) and progesterone (green line; fourth panel, step ❽). The combination of estrogen and progesterone inhibit GnRH production, reducing the release of FSH and LH (first and second panels, step ❾), and thereby preventing the development of more follicles. If the egg is not fertilized, the corpus luteum starts to disintegrate about 10 days after ovulation (third panel, step ❿). This occurs because the corpus luteum cannot survive without stimulation by LH. Estrogen and progesterone secreted by the corpus luteum shut down LH production by the anterior pituitary, so the corpus luteum actually causes its own destruction.

When the corpus luteum degenerates, estrogen and progesterone levels plummet (fourth panel, step ⓫). The reduced levels of estrogen and progesterone no longer inhibit the hypothalamus, so the spontaneous release of GnRH resumes (first panel, back to step ❶). GnRH stimulates the release of FSH and LH, initiating the development of a new set of follicles.

The Uterine Cycle

In the uterine cycle, estrogen and progesterone stimulate the development

of the endometrium of the uterus. In the description below, the numbered steps correspond to the numbers in black circles in **Figure E41-1b.**

On day 1 of a menstrual cycle, the uterus starts to shed the endometrium that had developed during the previous cycle (step ❶). A few days later, estrogen, released by the developing follicle, stimulates renewed development of the endometrium (step ❷). After ovulation, the corpus luteum develops and releases estrogen and progesterone, which further stimulate the growth of the endometrium (step ❸). The degeneration of the corpus luteum at the end of the ovarian cycle causes a sharp drop in the levels of estrogen and progesterone, and the endometrium begins to disintegrate (back to step ❶).

Hormonal Control of the Menstrual Cycle Includes Both Positive and Negative Feedback

The hormones of the menstrual cycle are regulated by both positive and negative feedback. The first half of the cycle is dominated by positive feedback. FSH and LH stimulate estrogen production by the follicles. High levels of estrogen then stimulate the midcycle surge of FSH and LH. This positive feedback causes hormone concentrations to reach high levels. Negative feedback dominates the second half of the cycle. Estrogen and progesterone from the corpus luteum inhibit the release of GnRH, FSH, and LH. Without LH to keep it alive, the corpus luteum dies, shutting down progesterone production and greatly reducing estrogen production.

---

environment for growth. If fertilization does not occur, however, the corpus luteum disintegrates, estrogen and progesterone levels fall, and the enlarged endometrium disintegrates as well. The uterus then contracts (sometimes causing menstrual cramps) and expels the excess endometrial tissue, a process called **menstruation** (from the Latin "mensis," meaning "month").

The Embryo Sustains Its Own Pregnancy

If fertilization occurs, the embryo prevents the degeneration of the corpus luteum that would otherwise end the current menstrual cycle and start up a new one. Shortly after it implants in the uterus, the embryo secretes an LH-like hormone called **chorionic gonadotropin (CG).** This hormone travels in the bloodstream to the ovary, where it functions to keep

the corpus luteum alive. The corpus luteum continues to secrete estrogen and progesterone for a few months during pregnancy. These hormones continue to stimulate the development of the endometrium, nourishing the embryo and sustaining the pregnancy. Some CG is excreted in the mother's urine, where it can be detected to confirm pregnancy.

During Copulation, Sperm Are Deposited in the Vagina

The male role in copulation begins with erection of the penis. Before erection, the penis is relaxed (flaccid), because smooth muscles surrounding the arterioles that supply it with blood are contracted, allowing little blood flow into the penis

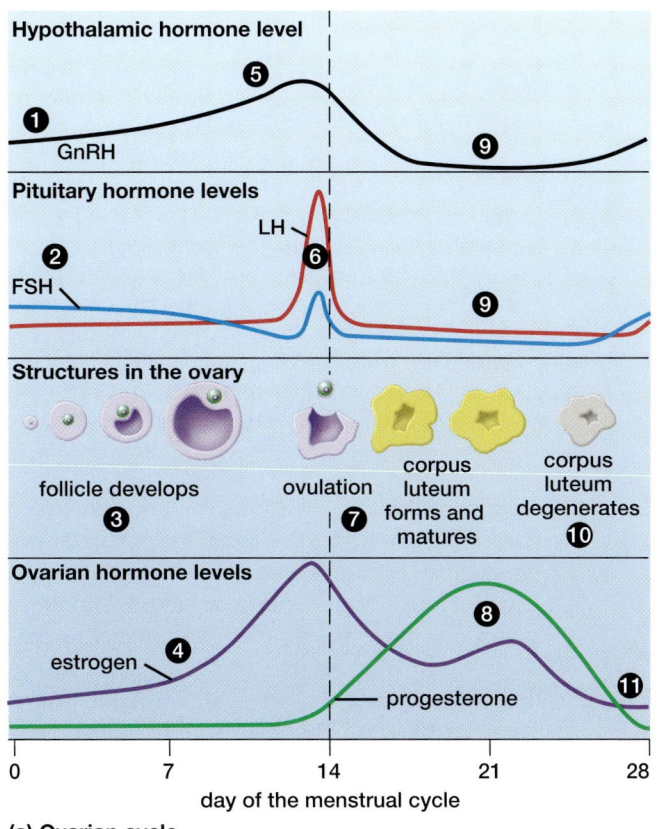

(a) Ovarian cycle

◀ **FIGURE E41-1 Hormonal control of the menstrual cycle** The menstrual cycle consists of two related cycles: **(a)** the ovarian cycle and **(b)** the uterine cycle. The circled numbers refer to the events described in the text.

QUESTION Most birth control pills contain estrogen and progesterone. Why would these hormones prevent pregnancy?

(b) Uterine cycle

(Fig. 41-17a). Under psychological and physical stimulation, signals from the nervous system cause these smooth muscles to relax. The arterioles dilate and blood flows into tissue spaces within the penis. As these tissues swell, they squeeze off the veins that drain the penis (Fig. 41-17b). Blood pressure in the penis increases, causing an erection. After the penis is inserted into the vagina, movements further stimulate touch receptors on the penis, triggering ejaculation. Muscles encircling the epididymis, vas deferens, and urethra contract, forcing semen out through the penis and into the vagina. A typical ejaculation consists of about 2 to 5 milliliters of semen containing about 100 million to 400 million sperm.

In the female, sexual arousal causes increased blood flow to the vagina, to the paired folds of tissue called the **labia** (singular, labium), and to the **clitoris,** a small structure just above the vagina (see Fig. 41-14). Stimulation of the clitoris may result in orgasm. Female orgasm is not necessary for fertilization, but contractions of the vagina and uterus during orgasm probably help to move sperm up toward the uterine tubes.

Intimate contact during copulation provides a favorable environment for transmitting disease organisms, as we describe in "Health Watch: Sexually Transmitted Diseases" on pp. 804–805.

During Fertilization, the Sperm and Egg Nuclei Unite

Sperm, under ideal conditions, may live for 2 to (rarely) 4 days inside the female reproductive tract, and an unfertilized egg

(a) Relaxed

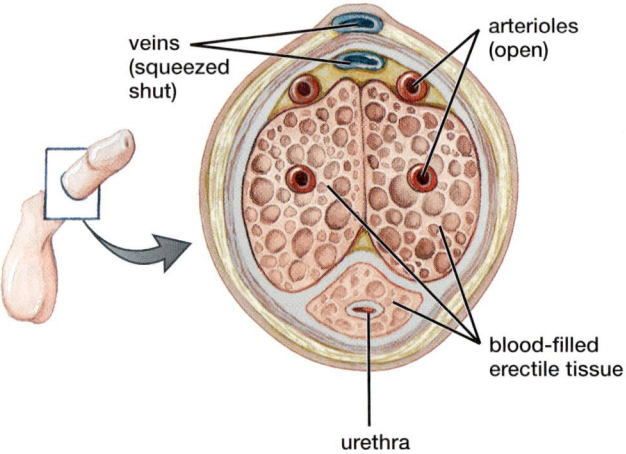

(b) Erect

▲ FIGURE 41-17 **Changes in blood flow in the penis cause erection** (a) Smooth muscles encircling the arterioles leading into the penis are usually contracted, limiting blood flow. (b) During sexual stimulation, these muscles relax, and blood flows into spaces within the penis. The swelling penis squeezes off the veins through which blood would leave the penis, increasing its blood pressure and causing it to become elongated and firm.

(a) An ovulated secondary oocyte

(b) Sperm surrounding an oocyte

▲ FIGURE 41-18 **The secondary oocyte and fertilization** (a) A human secondary oocyte (egg) shortly after ovulation. Sperm must digest their way through the corona radiata and the zona pellucida to reach the oocyte. (b) Sperm surround the oocyte, attacking the corona radiata and zona pellucida.

? HAVE **YOU** EVER WONDERED . . .

How Do Porcupines Mate?

A female porcupine is fertile for only 8 to 12 hours a year. She signals her fertility by scent marking and squeaking. Males converge on her location and sometimes fight for the opportunity to mate. The winner then performs a courtship ritual that includes soaking his potential mate with a high-velocity spray of urine! If the female isn't interested, she shakes off the urine and walks away. If she is ready to mate, she arches her tail over her back and flattens the quills around her rump, so the male doesn't get stabbed. The pair copulates many times over the next few hours. Porcupine mating is highly successful—almost all females become pregnant every fall.

remains viable for a day or so. During intercourse, the penis releases sperm into the vagina. The sperm move through the cervix, into the uterus, and finally, they enter the uterine tubes. If copulation occurs within a day or two of ovulation, the sperm may meet an egg in one of the uterine tubes.

When it leaves the ovary, the egg is surrounded by follicle cells. These cells, now called the **corona radiata,** form a barrier between the sperm and the egg (**Fig. 41-18a**). A second barrier, the jelly-like **zona pellucida** ("clear area"), lies between the corona radiata and the egg. In the uterine tube, hundreds of sperm surround the corona radiata (**Fig. 41-18b**). Each sperm releases enzymes from its acrosome. These enzymes weaken both the corona radiata and the zona pellucida, allowing the sperm to wriggle through to the egg. If there aren't enough sperm, not enough enzymes are released, and none of the sperm will reach the egg. This may be the reason that natural selection has led to the ejaculation of so many sperm. Perhaps 1 in 100,000 reaches the uterine tube, and 1 in 20 of those encounters the egg, so only a few hundred of the several hundred million sperm that were ejaculated join in attacking the barriers around the egg.

When the first sperm contacts the egg's surface, the plasma membranes of egg and sperm fuse, and the sperm's head enters the egg cytoplasm. Sperm entry triggers two critical changes in the egg: First, vesicles near the surface of the egg release chemicals into the zona pellucida that reinforce it and prevent additional sperm from entering. Second, the egg undergoes the second division of meiosis (see Fig. 41-15). Fertilization occurs as the haploid nuclei of sperm and egg fuse, forming the diploid nucleus of the zygote.

Defects in the male or female reproductive system can prevent fertilization. For example, a blocked uterine tube can prevent sperm from reaching the egg. A man with a low sperm count (fewer than 20 million sperm per milliliter of semen) may be unable to impregnate a woman through sexual intercourse because too few sperm reach the egg. Today, many couples who have difficulty conceiving a child seek help through artificial insemination or *in vitro* fertilization (see "Health Watch: High-Tech Reproduction" on p. 807).

CHECK YOUR LEARNING

Can you describe the human male and female reproductive tracts, including the gonads and accessory structures, and the functions of each structure? Can you explain how hormonal interactions control reproduction in men and women? Can you describe spermatogenesis and oogenesis, the locations and timing of these two processes, the important features of sperm and eggs, and how fertilization occurs?

41.3 HOW CAN PEOPLE PREVENT PREGNANCY?

During most of human evolution, child mortality was high, and natural selection favored people who produced enough children to offset this high mortality rate. Today, however, most people do not need to have many children to ensure that a few will survive to adulthood. We nevertheless retain reproductive drives that are perhaps better suited to a more precarious existence.

Historically, limiting fertility has not been easy. In the past, women in some cultures have tried such inventive techniques as swallowing froth from the mouth of a camel or placing crocodile dung in the vagina. Since the 1970s, several effective techniques have been developed for **contraception** (the prevention of pregnancy). All forms of birth control have possible drawbacks. The choice of contraceptive method should always be made in consultation with a health professional.

Sterilization Provides Permanent Contraception

The most foolproof method of contraception is **sterilization,** in which the pathways through which sperm or eggs must travel are blocked or cut (Fig. 41-19). In men, the vas deferens leading from each testis can be severed and the ends tied, clamped, or sealed in an operation called a vasectomy. The surgery is performed under a local anesthetic, in some cases requires no stitches, and has no known effects on health or sexual performance. Sperm are still produced, but they cannot leave the epididymis, where they die. Phagocytic white blood cells and the cells that make up the lining of the epididymis remove the debris.

The somewhat more complex operation known as a tubal ligation renders a woman infertile by clamping or cutting her uterine tubes, and tying or sealing off the cut ends. Ovulation still occurs, but sperm cannot travel to the egg, nor can the egg reach the uterus. An alternative procedure is a tubal implant. Tiny springs are guided through the vagina, cervix, and uterus, and inserted into each uterine tube. Tubal implants require no incisions and only local anesthesia. The springs cause the uterine tubes to form scar tissue that blocks passage of both sperm and eggs.

The vas deferens is severed and its ends are sealed

testis

scrotum

(a) Vasectomy

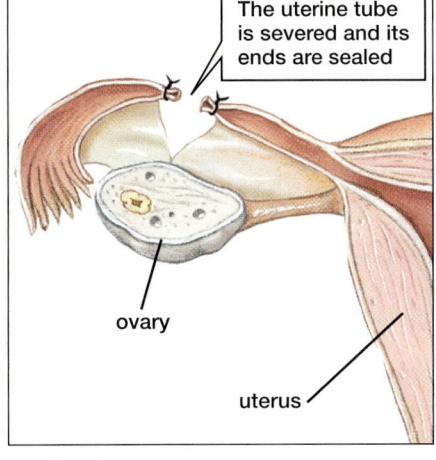

The uterine tube is severed and its ends are sealed

ovary

uterus

(b) Tubal ligation

◄ FIGURE 41-19 Sterilization (a) In a vasectomy, a surgeon cuts the vas deferens and ties, clamps, or seals the cut ends, preventing the passage of sperm. (b) In a tubal ligation, a surgeon clamps or cuts the uterine tube and ties or seals the cut ends, preventing sperm from reaching the oocyte and the oocyte from reaching the uterus.

Health Watch

Sexually Transmitted Diseases

Sexually transmitted diseases (STDs) are caused by bacteria, viruses, protists, or arthropods that infect the sexual organs and reproductive tract. As their name implies, they are transmitted primarily through sexual contact.

Bacterial Infections

The U.S. Centers for Disease Control and Prevention (CDC) estimate that there are about 700,000 new cases of **gonorrhea** each year in the United States. Gonorrhea bacteria can penetrate membranes lining the urethra, anus, cervix, uterus, uterine tubes, and throat. Infected males may experience painful urination and discharge of pus from the penis. Female symptoms are often mild and include a vaginal discharge or painful urination. Gonorrhea can be treated with antibiotics. However, because many infected individuals experience few or no symptoms, they may not realize they have the disease and seek treatment. Therefore, they may continue to infect their sexual partners. Gonorrhea infections may cause a buildup of scar tissue in the uterine tubes, resulting in infertility. The bacteria may also attack the eyes of infants born to infected mothers. Gonorrhea was once a major cause of blindness in infants, but today, most newborns are immediately given antibiotic eyedrops to kill gonorrhea bacteria.

About 45,000 to 50,000 cases of **syphilis** are reported to the CDC each year. Syphilis bacteria can penetrate the mucous membranes of the genitals, lips, anus, or breasts. Syphilis begins with a sore at the site of infection, sometimes followed by a rash, often on a totally different site on the body. If untreated, the bacteria spread, damaging many organs including the skin, kidneys, heart, and brain, sometimes with fatal results. Syphilis is easily cured with antibiotics, but many people in the early stages of syphilis infection have mild symptoms and do not seek treatment. Syphilis can be transmitted to the embryo during pregnancy. Some infected infants are stillborn or die shortly after birth; others suffer damage to the skin, teeth, bones, liver, and central nervous system.

Chlamydia is the most common bacterial STD: The CDC estimates that there are about 2.8 million new infections in the United States each year. Chlamydia causes inflammation of the urethra in males and of the urethra and cervix in females. In many cases, there are no obvious symptoms, so the infection goes untreated and the bacteria may be passed to sexual partners. A chlamydia infection may block the uterine tubes, resulting in sterility. Chlamydia can cause eye inflammation in infants born to infected mothers and is a major cause of blindness in developing countries.

Viral Infections

Acquired immune deficiency syndrome, or **AIDS,** is caused by the human immunodeficiency virus (HIV; see Chapter 36). It is spread primarily by sexual activity, contaminated blood and needles, and from mother to newborn. Syphilis increases the likelihood of HIV infection by two- to fivefold, because syphilis causes sores through which the HIV virus can enter the body. There is no cure for AIDS, but drug combinations can keep the disease under control for many years.

Genital herpes is extremely common; the CDC estimates that 16% of American adults—tens of millions of people—have genital herpes. Genital herpes may cause painful blisters on the genitals and surrounding skin and is transmitted primarily when blisters are present. Even when the blisters have healed, herpes virus remains in the body, emerging unpredictably, possibly in response to stress. Antiviral drugs can reduce the severity of outbreaks. A pregnant woman with an active case of genital herpes can transmit the virus to her developing embryo, in very rare cases causing mental or physical disability or stillbirth. Herpes can also be transmitted to babies during childbirth.

Human papillomavirus (HPV) (**Fig. E41-2a**) infects an estimated 50% of sexually active people at some time in their lives. Most experience no symptoms and never know that they have been infected. However, the virus may cause warts on the labia, vagina, cervix, or anus in women, and on the penis, scrotum, groin, or thighs in men. Certain strains of HPV cause most, and possibly all, cases of cervical cancer, which kills nearly 4,000 women each year in the United States.

If a sterilized woman or man wishes to reverse the operation, a surgeon can attempt to reconnect the uterine tubes or vas deferens. About 70% to 90% of young women are able to become pregnant after their uterine tubes have been reconnected by a skilled, experienced surgeon. Pregnancy rates vary, however, according to the age of the woman (older women have a lower success rate) and the method of the original tubal ligation (higher success rates occur if the uterine tubes were clamped rather than cut or cauterized). Tubal implants are not readily reversible.

To reverse a vasectomy, each vas deferens is surgically reconnected. Sperm reappear in the ejaculate in 70% to 98% of cases, depending mostly on the skill of the surgeon. However, the pregnancy rate of the men's sexual partners is only 30% to 75%. The longer the interval between vasectomy and reconnection, the lower the pregnancy rate. In many cases, a vasectomized man gradually develops an immune response against his sperm, so even if the vasectomy is successfully reversed, he may ejaculate damaged sperm that cannot reach or fertilize an egg.

Temporary Birth Control Methods Are Readily Reversible

Temporary methods of birth control prevent pregnancy in the immediate future, while leaving open the option of later pregnancies. Absolute abstinence, of course, provides complete protection against pregnancy and sexually transmitted diseases (STDs). Other temporary birth control methods fall into three categories: (1) preventing ovulation, (2) preventing sperm and egg from meeting, and (3) preventing implantation in the uterus (Table 41-3).

(a) The human papillomavirus

(b) Pubic lice

▲ **FIGURE E41-2 Agents of sexually transmitted diseases** **(a)** The DNA of some strains of the human papillomavirus becomes inserted into chromosomes of cells of the cervix. Proteins synthesized from the instructions in the viral DNA may promote uncontrolled cell division, causing cancer. **(b)** Pubic lice are called "crabs" because of their shape and the tiny claws on the ends of their legs, which they use to cling to pubic hair.

Two vaccines are now available that protect against the most common cancer-causing forms of HPV. The CDC recommends HPV vaccination for males as well as females. The vaccine can prevent new HPV infections, but cannot cure existing infections.

Protist and Arthropod Infections

Trichomoniasis is caused by a protist that colonizes the mucous membranes lining the urinary tract and genitals of both males and females. The CDC estimates that about 3.7 million Americans have trichomoniasis. Symptoms may include itching or burning sensations and, sometimes, a discharge from the penis or vagina. However, most people don't have any symptoms, so they aren't treated, and may pass the disease to their sexual partners. Prolonged infections can cause sterility.

Crab lice, also called pubic lice (**Fig. E41-2b**), are tiny insects that live and lay their eggs in pubic hair. Their mouthparts are adapted for penetrating skin and sucking blood and body fluids, a process that causes severe itching. "Crabs" can also spread infectious diseases.

Note that birth control methods provide no protection against STDs unless they prevent physical contact between the penis and vagina.

Synthetic Hormones Prevent Ovulation

Birth control pills typically contain synthetic versions of estrogen and progesterone; minipills contain only progesterone. As you learned earlier in this chapter, follicle development is stimulated by FSH, and ovulation is triggered by a midcycle surge of LH. The estrogen in birth control pills prevents FSH release, so follicles do not develop. Even if one were to develop, the progesterone in the pill would suppress the surge of LH needed for ovulation. Progesterone also thickens the cervical mucus, making it more difficult for sperm to move from the vagina into the uterus, and slows down the movement of sperm and egg in the uterine tubes.

Contraceptive patches, rings, injections, or implants containing estrogen and progesterone are also available, as described in Table 41-3. These devices are usually effective for a few weeks to as long as a few years.

Fertilization Cannot Occur If Sperm Cannot Contact the Egg

There are several ways to prevent the encounter of sperm and egg, including what are called barrier methods. Some barrier devices, such as the diaphragm and cervical cap, cover the cervix, preventing the entry of sperm into the uterus. A female condom completely lines the vagina. A male condom prevents sperm from being deposited in the vagina. Both types of condoms offer some protection against the spread of sexually transmitted diseases. Barrier devices are more effective when combined with a spermicide (sperm-killing

TABLE 41-3 Methods of Temporary Contraception

Method	Description and Mechanism	Pregnancy Rate per Year[1]	Protection Against STDs
None	Regular intercourse with no contraception	About 95% for women under 25 years old, declining to about 45% by age 40	None
Hormonal methods that prevent ovulation			
Birth control pill	Pill containing either synthetic estrogen and synthetic progesterone (combination pill) or progesterone only (minipill); taken daily	0.3% to 8%	None
Vaginal ring	Flexible plastic ring containing synthetic estrogen and progesterone, inserted into the vagina around the cervix; replaced every 4 weeks	0.3% to 8%	None
Contraceptive patch	Skin patch containing synthetic estrogen and progesterone; replaced weekly	0.1% to 8%[2]	None
Birth control injection	Injection of synthetic progesterone that blocks ovulation; repeated at 3-month intervals	0.1% to 3%	None
Contraceptive implant	Small plastic rod containing synthetic progesterone that blocks ovulation; replaced every 3 years	<0.1%	None
Methods that prevent sperm and egg from meeting			
Abstinence	No sexual activity	0%	Excellent
Condom (male)	Thin latex or polyurethane sheath placed over the penis just before intercourse, preventing sperm from entering the vagina; more effective when used in conjunction with spermicide	2% to 15%	Good
Condom (female)	Lubricated polyurethane pouch inserted into the vagina just before intercourse, preventing sperm from entering the cervix; more effective when used in conjunction with spermicide	5% to 21%	Good (probably about the same as a male condom)
Sponge	Domed disposable sponge containing spermicide, inserted in the vagina up to 24 hours before intercourse	9% to 20% (failure rates double after giving birth)	Poor
Diaphragm or cervical cap	Reusable, flexible, domed rubber-like barriers; spermicide is placed within the dome, and the diaphragm (larger) or cap (smaller) is fitted over the cervix just before intercourse	6% to 20%	Poor
Spermicide	Sperm-killing foam is placed in the vagina just before intercourse, forming a chemical barrier to sperm	16% to 29%	Probably none
Rhythm	Measuring the time since the last menstruation, or changes in body temperature and cervical mucus, to estimate the time of ovulation so that intercourse can be avoided during the fertile period	4% to 25% (rarely performed correctly)	None
Methods with multiple actions			
IUD (intrauterine device)	Small plastic device treated with hormones or copper and inserted through the cervix into the uterus	0.1% to 1%	None
"Morning after" pill (emergency contraception)	Concentrated dose of the hormones in birth control pills, taken within 72 hours after intercourse	10% to 40% (less effective the later the pill is taken after intercourse)	None

[1]The percentage of women becoming pregnant per year. The low numbers represent the pregnancy rate with consistent, correct contraceptive use; the higher numbers represent the pregnancy rate with more typical use that is not always consistent or correct. It is likely that many women do not report incorrect usage, so the actual failure rates with correct use may be lower.
[2]The patch and the pill are about equally effective; however, the patch is more likely to be used properly. The patch may be less effective in women weighing more than 200 pounds.

substance), in case the barriers are breached and allow a few sperm to enter the vagina or uterus.

There are also ways of preventing sperm from reaching the egg that do not employ impenetrable barriers, but they are generally less effective than barrier methods. For example, the use of spermicides alone or a spermicide-soaked "sponge" inserted into the vagina is less effective than condoms, diaphragms, or cervical caps used together with spermicides. The rhythm method (abstaining from intercourse during the ovulatory period of the menstrual cycle) is also less reliable, because the menstrual cycle may vary from month to month, making ovulation difficult to predict exactly. In addition, sperm

Health|Watch

High-Tech Reproduction

A surprisingly large number of couples have difficulty conceiving a child, but the ability to manipulate biological processes has led to major advances in assisted reproduction. For women who do not ovulate regularly, injection of fertility drugs (which cause the release of extra FSH and LH) stimulates ovulation. However, fertility drugs often cause several eggs to be released simultaneously, with the result that the rate of multiple births in the United States has nearly doubled since 1980. Multiple births are much riskier than single births, for both the mother and her children.

Worldwide, well over 3 million people are now alive who were conceived by *in vitro* fertilization (IVF; literally, "fertilization in glass"). IVF usually starts with giving the woman drugs that stimulate follicle development. When the follicles are about to ovulate, a surgeon inserts a long needle into each ripe follicle and sucks out its oocyte. The oocytes are placed in a dish with freshly collected sperm. The fertilized eggs divide. At the two- to eight-cell stage (**Fig. E41-3**), usually two embryos (in young women) to four embryos (in women over age 40) are gently sucked into a tube and expelled into the uterus. Transplanting more than one embryo increases the success rate, but also increases the probability of multiple births. (Nadya Suleman, the famous "Octomom," had 12 embryos implanted at once, 8 of which developed.) Embryos that are not immediately implanted can be frozen for later use. IVF costs about $10,000 to $15,000 per attempt. With an average success rate of about 30%, a typical conception via IVF costs about $30,000 to $50,000. The success rate is much lower for women over 35 years of age.

Even men whose sperm are incapable of swimming or normal fertilization may be able to father children

▲ **FIGURE E41-4 Intracytoplasmic sperm injection** An egg is held in place on the tip of a smooth glass pipette. A much smaller, sharp pipette injects a single sperm cell directly into the egg's cytoplasm.

through intracytoplasmic sperm injection (ICSI). In ICSI, immature sperm cells are extracted from the testes and injected through a tiny, sharp pipette directly into an egg's cytoplasm (**Fig. E41-4**).

Using sperm-sorting technology, parents can even change the odds of having a male or female child. This may be medically important if the parents are carriers of sex-linked disorders, but some parents are just seeking to balance the genders of their children. A sperm-sorting methodology, now in clinical trials in the United States, provides 80% to 90% accuracy in separating X-bearing sperm from Y-bearing sperm. The sorted sperm could then be used either for artificial insemination or *in vitro* fertilization. For absolute gender certainty during IVF, the gender of an embryo can be determined before implantation with 100% accuracy, by carefully removing one of the cells and analyzing its chromosomes.

As these examples show, biomedical technology has the potential to alter how humans reproduce (wealthy humans, at least). An article in the popular science magazine *New Scientist* half-jokingly suggested that in the not-too-distant future, people would have sex for fun, but for reproduction, they would use IVF combined with preimplantation diagnosis of hereditary diseases and genetic engineering to fix any disorders that are found (see Chapter 13). Assuming that such a future becomes possible, do you think that it's a good idea?

▲ **FIGURE E41-3 A human embryo ready for implantation**

can survive for a few days in the female reproductive tract, so intercourse must be avoided for several days before ovulation.

Highly unreliable methods include withdrawal (removing the penis from the vagina before ejaculation) and douching (attempting to wash sperm out of the vagina before they have entered the uterus).

Some Birth Control Methods Work Through Multiple Mechanisms, Including Preventing Implantation

Some birth control methods may have several modes of action, including one or more of preventing ovulation, hindering survival or motility of sperm, or preventing implantation of the embryo in the uterus. The intrauterine device (IUD)—a small copper or plastic loop, squiggle, or shield inserted through the cervix and into the uterus—interferes with sperm motility or survival and also alters the uterine lining, lessening the likelihood of implantation. "Morning after" pills, which contain high doses of the same hormones that are used in birth control pills, act primarily by delaying or preventing ovulation, but they may also interfere with the development of the corpus luteum and prevent implantation.

Male Birth Control Methods Are Under Development

You may have noticed that most birth control techniques are designed for use by women. There are probably three major reasons for this. First, the woman, not the man, becomes pregnant, bears the health risks of the pregnancy and childbirth, and usually plays a larger role in child care than men do. Second, it was relatively simple to design "use it and forget about it" birth control methods for women, including birth control pills and intrauterine devices, that do not interfere with sexual desire or performance. It has proven much more difficult to design similar methods for men. Third, opinion surveys find that a significant number of men, often more than 25%, claim they would never use hormonal birth control methods, even if they are assured that the hormones would not interfere with their sexual performance or other gender-related characteristics.

Nevertheless, researchers are working on male contraception. One of the options under investigation is to inject reversible plugs into the vas deferens; the plugs block sperm from moving out of the testes to the urethra. Other options include drugs, usually administered by injection, that block the action of GnRH, FSH, or LH, thus preventing sperm from being produced. The drugs currently being tested would need to be supplemented with testosterone injections, because testosterone production by the testes would be inhibited (see Fig. 41-13). Human trials of hormonal methods generally give only 80% to 90% effectiveness, substantially lower than hormonal birth control in women (see Table 41-3).

Thus, the development and marketing of male contraceptives are hindered by lack of male acceptance, lower effectiveness than many female birth control methods, and the fact that many methods require injection or minor surgery. It appears that the widespread availability of safe, effective, and convenient male contraception—except for condoms, which are helpful in preventing both STDs and unwanted pregnancies—is still years away.

CHECK YOUR LEARNING

Can you describe the principal methods of contraception?

CASE STUDY revisited

To Breed a Rhino

Whether using artificial insemination or nature's way, rhino breeders are most successful when they can time sperm transfer to coincide with ovulation. The surest way to achieve this synchrony is to induce the female to ovulate when a male rhino or sperm sample is ready. As you have learned, ovulation in mammals is stimulated by a surge of LH and FSH, which in turn is stimulated by a surge in GnRH. Ovulation in rhino cows can be induced by injecting the females with a synthetic version of GnRH.

The next step depends on how the cow will be fertilized. For normal mating, the biologists wait until the female shows signs of mating interest. Depending on the species, this may include scent marking, whistling, squealing, and swinging her tail. The male and female are then enclosed together and watched carefully for signs of mating or aggression. For artificial insemination, the biologists use ultrasound to check for the presence of ripe follicles. Semen is collected from a bull rhino and inserted into the cow's uterus. For IVF, the biologists suck ripe eggs out of the cow's ovary, again using ultrasound to guide their efforts, and then mix the eggs with sperm.

These procedures have all proven successful. White, black, Indian, and Sumatran rhinos have been successfully bred using one or more of these techniques. With highly endangered species like the Sumatran, Javan, and black rhinos, preserving genetic diversity is extremely important, which means that every male and female should be given the chance to pass on their genes. Artificial insemination using frozen sperm would be much safer and less expensive than shipping rhinos around the world. To test this technique, Lulu, the white rhino cow at the Budapest Zoo, was artificially inseminated with sperm from Simba, a rhino bull living at a zoo in Great Britain. Simba's sperm had been stored for 3 years, frozen in liquid nitrogen. The resulting calf, Lisimba, was the 38-year-old bull's first offspring.

Similar techniques have also been used with other endangered species. For example, at the Smithsonian National Zoo in Washington, D.C., black-footed ferrets were born whose fathers had died almost a decade earlier. Several facilities, including the Audubon Nature Institute and the San Diego Zoo, maintain Frozen Zoos—liquid nitrogen storage tanks containing frozen sperm, tissue samples, and even embryos from endangered species.

Consider This

BioEthics Are exotic methods of preserving endangered species worth it? Wouldn't it be better to preserve habitat and eliminate poaching? Terri Roth, a rhino expert at the Cincinnati Zoo, says, "We do not ever promote technology as a way to save species instead of saving habitat, because that's ridiculous." But Roth and her colleagues feel strongly that any technique is worth it to preserve species until "the wild is safe again and we can put them back out there." Other people believe that endangered species may need to be triaged, and some declared to be hopeless, so that the funds spent to keep a tiny population alive in zoos or liquid nitrogen could be freed to buy habitat or enforce hunting laws and protect species that have a better chance of avoiding extinction. How would you decide?

CHAPTER REVIEW

Summary of Key Concepts

41.1 How Do Animals Reproduce?

Animals reproduce either sexually or asexually. Asexual reproduction—by budding, fission, or parthenogenesis—produces offspring that are usually genetically identical to the parent. In sexual reproduction, haploid gametes, usually from two separate parents, unite and produce offspring that are genetically different from either parent.

During sexual reproduction, a male gamete (a small, motile sperm) fertilizes a female gamete (a large, nonmotile egg). Some species are hermaphroditic, producing both sperm and eggs, but most have separate sexes. Fertilization can occur outside the bodies of the animals (external fertilization) or inside the body of the female (internal fertilization). External fertilization must occur in water so that the sperm can swim to meet the egg. Internal fertilization generally occurs by copulation, in which the male deposits sperm directly into the female's reproductive tract.

41.2 What Are the Structures and Functions of Human Reproductive Systems?

Puberty is triggered when the hypothalamus releases GnRH, which in turn causes release of FSH and LH from the anterior pituitary. LH and FSH stimulate the gonads to produce testosterone (in males) or estrogen (in females). Testosterone and estrogen induce secondary sexual characteristics and the development of sperm or eggs.

The human male reproductive system consists of paired testes, which produce sperm and testosterone; accessory structures that conduct sperm out of the man's body into the female's reproductive system; and three sets of glands. The seminal vesicles, prostate gland, and bulbourethral glands secrete fluids that provide energy for the sperm, activate the sperm to swim, and provide the proper pH for sperm survival. Spermatogenesis and testosterone production are stimulated by FSH and LH, secreted by the anterior pituitary. Spermatogenesis and testosterone production are nearly continuous, beginning at puberty and lasting until death.

The human female reproductive tract consists of paired ovaries, which produce eggs as well as the hormones estrogen and progesterone, and accessory structures that conduct sperm to the egg and receive and nourish the embryo during prenatal development. Oogenesis, hormone production, and development of the lining of the uterus repeat in a monthly menstrual cycle. The cycle is controlled by hormones from the hypothalamus (GnRH), anterior pituitary (FSH and LH), and ovaries (estrogen and progesterone). The production of estrogen and mature eggs begins at puberty and lasts until menopause.

During copulation, the male ejaculates semen into the female's vagina. The sperm swim through the vagina and uterus into the uterine tube, where fertilization usually takes place. The unfertilized egg is surrounded by two barriers, the corona radiata and the zona pellucida. Enzymes released from the acrosomes in the heads of sperm digest these layers, permitting sperm to reach the egg. Only one sperm enters the egg and fertilizes it.

41.3 How Can People Prevent Pregnancy?

Permanent contraception can be achieved by sterilization, usually by severing the vas deferens in males (vasectomy) or the uterine tubes in females (tubal ligation). Temporary contraception

techniques include those that prevent ovulation by delivering estrogen and progesterone—for example, birth control pills, vaginal rings, contraceptive patches and implants, and hormone injections. Sperm may also be prevented from reaching an egg if ovulation occurs by methods such as the diaphragm, cervical cap, sponge, and condom, accompanied by spermicide. Spermicide alone is less effective. Withdrawal and douching are unreliable. The rhythm method, which has a high failure rate, requires abstinence around the time of ovulation. Intrauterine devices may block sperm movement and prevent implantation of the early embryo. The "morning after" pill may prevent ovulation, the development of the corpus luteum, or implantation.

Key Terms

acquired immune deficiency syndrome (AIDS)   804
acrosome   796
asexual reproduction   791
budding   791
bulbourethral gland   797
cervix   799
chlamydia   804
chorionic gonadotropin (CG)   800
clitoris   801
contraception   803
copulation   793
corona radiata   802
corpus luteum   798
crab lice   805
egg   792
embryo   792
endometrium   799
epididymis   796
estrogen   794
external fertilization   792
fertilization   792
fission   791
follicle   797
follicle-stimulating hormone (FSH)   794
genital herpes   804
gonad   791
gonadotropin-releasing hormone (GnRH)   794
gonorrhea   804
hermaphrodite   792
human papillomavirus (HPV)   804
internal fertilization   793
interstitial cell   794
labium (plural, labia)   801
luteinizing hormone (LH)   794
menstrual cycle   797
menstruation   800
myometrium   799

oogenesis   797
oogonium (plural, oogonia)   797
ovary   792
ovulation   793
parthenogenesis   791
penis   796
placenta   799
polar body   798
primary oocyte   797
primary spermatocyte   796
progesterone   798
prostate gland   797
puberty   794
regeneration   791
scrotum   794
secondary oocyte   798
secondary spermatocyte   796
semen   796
seminal vesicle   796
seminiferous tubule   794
Sertoli cell   794
sexual reproduction   791
sexually transmitted disease (STD)   804
spawning   792
sperm   792
spermatid   796
spermatogenesis   795
spermatogonium (plural, spermatogonia)   794
sterilization   803
syphilis   804
testis (plural, testes)   792
testosterone   794
trichomoniasis   805
urethra   796
uterine tube   799
uterus   799
vagina   799
vas deferens   796
zona pellucida   802
zygote   791

Learning Outcomes

In this chapter, you have learned to . . .

LO1 Compare and contrast asexual and sexual reproduction, and describe some advantages and disadvantages of each.

LO2 Define *budding*, *fission*, and *parthenogenesis*, and provide some examples of animals that use each of these mechanisms of reproduction.

LO3 Define *internal* and *external fertilization*, and provide examples of animals that use each of these methods of fertilization.

LO4 Describe the human male reproductive tract, including the gonads, the tubes that conduct sperm, and the glands that provide the fluid portion of semen.

LO5 Describe the human female reproductive tract, including the gonads and the accessory structures that transport sperm and eggs and in which an embryo develops.

LO6 Explain how hormones from the hypothalamus, anterior pituitary gland, and gonads control reproduction in men and women.

LO7 Describe spermatogenesis and oogenesis.

LO8 Describe the structures of sperm and eggs and how fertilization occurs.

LO9 Explain how the principal methods of contraception work.

Thinking Through the Concepts

Fill-in-the-Blank

1. Reproduction by a single animal, without the need for sperm fertilizing an egg, is called _____ reproduction. A(n) _____ is a new individual that grows on the body of the adult and eventually breaks off to become independent. During _____, an adult animal splits in two, with the two halves regenerating the rest of a complete organism. **LO1 LO2**

2. In mammals, the male gonad is called the _____. It produces both sperm and the sex hormone _____. Within the male gonad, spermatogenesis occurs within the hollow, coiled structures called _____. **LO4**

3. A sperm consists of three regions, the head, midpiece, and tail. The head contains very little cytoplasm, and consists mostly of the _____, in which the chromosomes are found, and a sac of enzymes, the _____. Organelles in the midpiece, the _____, provide energy for movement of the tail. **LO7**

4. Sperm are stored in the _____ and _____ until ejaculation, when the sperm, mixed with fluids from three glands, the _____, _____, and _____, flow through the _____ to the tip of the penis. **LO4**

5. In mammals, the female gonad is the _____. It produces eggs and two hormones, _____ and _____. Although usually referred to as an "egg," female mammals actually ovulate a cell called a _____, which has completed only meiosis I. Meiosis I also produces a much smaller cell, the _____, that serves mainly as a way to discard chromosomes. The egg is swept up by ciliated structures, called fimbriae, that form the entrance to the _____. Fertilization usually occurs in

this structure. The fertilized egg then implants in the _____, where it will develop until birth. **LO5**

6. Oocytes develop in a multicellular structure, the follicle. After ovulation, most of the follicle cells remain in the ovary, forming a temporary endocrine gland, the _____. This gland disintegrates about 10 days after ovulation unless stimulated by a hormone, _____, secreted by the developing embryo. **LO5 LO6**

Review Questions

1. Describe the advantages and disadvantages of asexual reproduction and sexual reproduction and note an example of an animal that uses each type. For animals that reproduce sexually, describe the processes of internal and external fertilization, and provide an example of an animal that engages in each type of fertilization **LO1 LO3**

2. Compare the structures of the egg and sperm. What structural modifications do sperm have that facilitate movement, energy use, and gaining access to the egg? **LO8**

3. What is the role of the corpus luteum in a menstrual cycle? In early pregnancy? What determines its survival after ovulation? **LO5 LO6**

4. List the structures, in order, through which a sperm passes, starting with the seminiferous tubules of the testis and ending in the uterine tube of the female. **LO4 LO5**

5. Name the three accessory glands of the male reproductive tract. What are the functions of the secretions they produce? **LO4**

6. Describe the interactions among hormones secreted by the pituitary gland and ovaries that produce the menstrual cycle. **LO6**

7. Describe the principal methods of contraception. Why do current hormonal contraceptive methods mostly focus on women rather than on men? **LO8**

Applying the Concepts

1. Suggest a birth control method for each of the following couples, and explain why it would be appropriate: Couple A, who have intercourse three times a week but never want to have children; Couple B, who have intercourse once a month and may want to have children someday; and Couple C, who have intercourse three times a week and want to have children someday.

2. Would a hypothetical contraceptive drug that blocked receptors for FSH and LH be useful in males? How would it work? What side effects might it have?

3. **BioEthics** Think of all the ways a couple can produce a child, including *in vitro* fertilization using the couple's eggs and sperm, *in vitro* fertilization using a donor's sperm or egg, and insemination of a surrogate mother with sperm from the couple's husband. Imagine some more. Do these options present ethical issues for you? What legal issues might possibly arise? What medical issues?

Answers to Figure Caption questions and Fill-in-the-Blank questions can be found in the Answers section at the back of the book.

MB *Go to MasteringBiology for practice quizzes, activities, eText, videos, current events, and more.*

CASE STUDY

Rerunning the Program of Development

FLATWORMS DO IT. Sea stars do it. To a more limited, but still very impressive, extent, so do some insects, crabs, crayfish, salamanders, juvenile alligators, and lizards.

What can these animals do? They can regrow lost body parts. Flatworms and sea stars can even regenerate most of a lost body, starting with a fairly small part. Cut a flatworm in half, and the tail portion can grow a new head. Cut off the arms of some species of sea stars and each arm can regenerate a whole animal. Insects, crabs, and crayfish can regrow lost legs and antennae. Some salamanders can regenerate their legs and tails (including the tail section of the spinal cord). Juvenile alligators and a few species of lizards can regrow their tails.

Regeneration is a lot more complicated than healing a wound. In wound healing, typically a few cell types proliferate and essentially stitch the edges of the wound together. In adult mammals, the job usually isn't done very well, and a scar remains as a permanent reminder of the injury. For some wounds, such as damage to the spinal cord, healing binds the broken parts together and helps reestablish the blood supply but leaves a scar that prevents the recovery of any significant function. In contrast, regeneration is like a rerun of the developmental program for the lost body part. When a salamander regenerates a lost leg, it must grow bone, tendons, blood vessels, nerves, muscle, and skin, all in the right places and all integrated properly with one another to restore leg function.

Human regeneration has been a dream of physicians for decades. If a human embryo can grow a leg during development, why can't an adult regrow a leg to replace one lost to amputation? To find out if regeneration in humans might be possible, biomedical researchers first need to understand how normal development is controlled.

Salamanders, such as this axolotl, can regrow lost legs. (Inset) Rerunning the program of development to regenerate lost body parts isn't simple; sometimes, the program malfunctions, as in this two-headed flatworm.

AT A GLANCE

LEARNING GOALS

LG1 Define *development* and describe how differentiation of cells is essential to development.

LG2 Compare and contrast direct and indirect development and provide examples of animals that undergo each type of development.

LG3 Describe early development in amphibians, including cleavage, the formation of the blastula, gastrulation, and organogenesis.

LG4 Name the four extraembryonic membranes and describe their functions in reptiles and mammals.

LG5 Explain the roles of transcription factors, gene-regulating substances in the egg cytoplasm, induction between cells, and homeobox genes in development.

LG6 Describe human embryonic development from fertilization to birth.

LG7 Describe the structure and function of the placenta.

LG8 Name the hormones involved in milk production and secretion and describe their roles.

LG9 Define *aging* from a biological perspective and describe the mechanisms thought to cause aging.

42.1 WHAT ARE THE PRINCIPLES OF ANIMAL DEVELOPMENT?

Development is the process by which a multicellular organism grows and increases in organization and complexity. Development is usually considered to begin with a fertilized egg and end with a sexually mature adult. Three principal processes contribute to development. First, individual cells multiply. Second, some of their daughter cells **differentiate,** or specialize in structure and function, for example, as nerve or muscle cells. Third, as they differentiate, groups of cells move to precise locations in the body and become organized into multicellular structures, such as the brain or a biceps muscle.

All of the cells of an individual animal's body (except gametes) are genetically identical to one another and to the fertilized egg from which they came. How can genetically identical cells differentiate into remarkably different structures? As we will see, the solution is to use specific sets of genes in different places in an animal's body, at specific times during an animal's life.

In this chapter, we will focus on early developmental stages, from fertilized egg through birth. We begin with a brief survey of the diverse ways in which different animals develop.

CHECK YOUR LEARNING

Can you define *development* and describe the processes underlying animal development?

42.2 HOW DO INDIRECT AND DIRECT DEVELOPMENT DIFFER?

Chicks, puppies, and human babies develop as they grow up. However, in most respects, baby birds and mammals are miniature versions of the adults of their species, and they undergo a process called **direct development.** Many animal species, however, undergo **indirect development,** in which the newborn has a very different body structure than the adult.

During Indirect Development, Animals Undergo a Radical Change in Body Form

Indirect development occurs in amphibians, such as frogs and toads, and in most invertebrates. In animals that undergo indirect development, females typically produce large numbers of eggs, each containing a small amount of food reserve called **yolk.** The yolk nourishes the developing embryo until it hatches into a small, sexually immature feeding stage called a **larva** (plural, larvae; Fig. 42-1). Because the parents usually provide their offspring with neither food nor protection from predators, most die in their larval stage. After feeding for a few weeks to several years, the handful of survivors undergo a revolution in body form, known as **metamorphosis,** and become sexually mature adults.

Most larvae not only look very different from the adults, but they also play different roles in their ecosystems. For instance, most adult butterflies sip nectar from flowers, whereas their caterpillar larvae eat leaves (see Fig. 42-1). Most toads spend the majority of their adult life on land, eating insects, worms, and snails; their tadpole larvae are aquatic and usually feed on algae.

Newborn Animals That Undergo Direct Development Resemble Small Adults

Other animals, including some snails and fish, and all mammals and reptiles (including birds) undergo direct development, in which a newborn animal resembles an adult (Fig. 42-2). As a young animal matures, it grows much bigger but does not fundamentally change its body form.

For animals of a given adult size, the newborns of directly developing species are typically larger than the newly hatched larvae of indirectly developing species. Therefore, they need more nourishment before emerging into the world. Two strategies have evolved to meet this food requirement. Birds and most other reptiles, and many fish, produce eggs that contain relatively large amounts of yolk, which nourishes the embryo before it hatches. Mammals, some snakes, and a few fish have relatively little yolk in their eggs; instead, the embryos are nourished within the mother's body. Despite

(a) Seahorses

(a) Caterpillar (larva)

(b) Snails

(b) Butterfly (adult)

▲ **FIGURE 42-1 Indirect development** The larvae of animals with indirect development are very different from the adult form, in structure, behaviors, and ecological niches. **(a)** Caterpillars of butterflies, such as the blue morpho caterpillar shown here, feed on leaves, usually of a limited number of host plant species. **(b)** The principal food of most adult butterflies is flower nectar. Although blue morphos will drink nectar, they prefer fluids from fermenting fruits and even decomposing animal carcasses.

(c) Polar bears

▲ **FIGURE 42-2 Direct development** The offspring of animals with direct development closely resemble their parents from the moment of hatching or birth. **(a)** A male seahorse gives birth. The female seahorse deposits her eggs in his pouch, where he releases sperm to fertilize them. The fertilized eggs develop in the pouch for a few weeks, when vigorous muscular contractions of the pouch squirt out as many as 200 young. **(b)** Many land and freshwater snails hatch from small, yolk-rich eggs. **(c)** Mammalian mothers nourish their developing young within their bodies before birth, and with milk from their mammary glands after birth.

often extensive development in the egg or in the mother's body, the young of many directly developing animals, such as those of birds and mammals, require additional care and feeding after birth. As a result of these demands both before and after birth, relatively few offspring are produced. However, the costly parental investment helps a fairly high proportion of the offspring survive to adulthood.

CHECK YOUR LEARNING

Can you describe direct and indirect development and name some animal groups that have each type of development?

42.3 HOW DOES ANIMAL DEVELOPMENT PROCEED?

Most of the mechanisms of development are similar in all animals. Here, we will focus on vertebrate development. We will begin by describing development in amphibians such as frogs, newts, and salamanders. Amphibians have long been favored subjects for the study of development because they can be induced to breed at any time of year; they produce numerous large eggs and embryos; and the embryos develop in water, where they can be easily observed, rather than inside their mothers. Most aspects of amphibian development are comparable to the development of other vertebrates.

Cleavage of the Zygote Begins Development

Development begins when a fertilized egg, or **zygote,** undergoes a series of mitotic cell divisions collectively called **cleavage** (Fig. 42-3a). The zygote is a very large cell; a frog zygote, for example, may be a million times larger than an average cell in an adult frog. During cleavage, there is little or no cell growth between cell divisions, so as cleavage progresses, the available cytoplasm is split up into ever smaller cells, whose size gradually approaches that of cells in an adult. After a few cell divisions, a solid ball of cells, the **morula** (Fig. 42-3b), is formed. As cleavage continues, a cavity opens within the morula, and the cells become the outer covering of a hollow structure called the **blastula** (Fig. 42-3c).

The details of cleavage differ among species and are partly determined by the amount of yolk, because yolk hinders the division of the cytoplasm. In frogs, cells in the yolky portion of the egg (the bottom of the egg in Fig. 42-3a) divide more slowly than cells in the almost yolk-free portion (the top in Fig. 42-3a), so the morula has larger cells on the bottom than on the top. Eggs with extremely large yolks, such as a hen's egg, cannot divide all the way through; in such cases, cleavage produces a flat cluster of cells atop the yolk. Nevertheless, a hollow blastula (or its equivalent; see Section 42.5) is always produced, although in birds it resembles a flattened bag rather than a ball.

Gastrulation Forms Three Tissue Layers

In amphibians and many other animals, the location of cells on the surface of the blastula forecasts their ultimate developmental fate in the adult. In Figure 42-3c, we have colored the cells of the blastula blue, yellow, or pink. These colors indicate the parts of the adult body that the cells are destined to produce during development (see p. 815). Most of the cells are on the surface of the blastula, but most of the structures they will form will be on the inside of the animal. The cells move to their proper destinations during the next stage of development, called **gastrulation** (literally, "producing the stomach").

Gastrulation begins when a dimple called the **blastopore** forms on one side of the blastula (see Fig. 42-3c). The dimple enlarges, going deeper and deeper into the blastula and forming a cavity that will eventually become the digestive tract (Fig. 42-3d). The migrating cells form three tissue layers in the embryo, which is now called a **gastrula** (Fig. 42-3e). The cells that move through the blastopore to line the future digestive tract (yellow) are called **endoderm** (meaning "inner skin"). Endoderm also forms the liver, pancreas, and the lining of the respiratory tract. The cells remaining on the outside (blue) of the developing gastrula are called **ectoderm** ("outer skin"). These cells mostly form surface structures such as skin, hair, and nails. Ectoderm also forms the nervous system. Cells that migrate between the endoderm and ectoderm form the third layer (pink), called **mesoderm** ("middle skin"). Mesoderm forms structures that are generally located between the skin and lining of the digestive tract, including muscles, the skeleton, and the circulatory system. Table 42-1 lists the major structures produced from each layer of cells.

The Major Body Parts Develop During Organogenesis

Organogenesis is the development of the body's organs from the three embryonic layers. Organogenesis proceeds by two major processes. First, a series of "master" genes turn on and off in specific cells. Somewhat like the ignition switch in your car, which turns on the engine, headlights, seat-belt alarm, power steering, and so on, each master gene controls the activity of many individual genes involved in producing, say, an arm or a backbone. We will return to these master genes in Section 42.4.

Second, organogenesis prunes away superfluous cells, much as Michelangelo chiseled away "extra" marble, revealing his statue of David. In development, the sculpting of body parts often requires the death of excess cells. For example, amphibians, reptiles, and mammals pass through embryonic stages in which they have tails and webbed fingers and toes. In humans, these stages appear during the fourth to seventh weeks of development. As the cells of the tail and webbing die, the tail disappears and the fingers and toes become separate.

Development in Reptiles and Mammals Depends on Extraembryonic Membranes

Fish live and reproduce in water. Although many amphibians spend their adult lives on land, they too lay their eggs in water. By being immersed in water, the embryo does not

(a) Zygote

◀FIGURE 42-3 **Gastrulation in the frog** (a) A frog zygote, or fertilized egg, is a huge cell, more than a millimeter in diameter in some species. **(b)** The zygote divides to form a solid ball of cells, the morula. The morula is about the same size as the zygote, although its individual cells are smaller. **(c)** The morula develops into a blastula; in frogs, the blastula is a hollow ball of cells with a lump of yolk-containing cells (yellow) at one end. The blastula is still about the same size as the zygote. The positions of the cells in the frog blastula foretell their future developmental fate (see the color-coded legend). **(d)** During gastrulation, surface cells migrate into the blastula to form the endoderm and mesoderm layers of the gastrula. The cells remaining on the surface will form ectoderm. **(e)** The resulting three-layered embryo is now called a gastrula.

(b) Cleavage of the zygote forms a morula

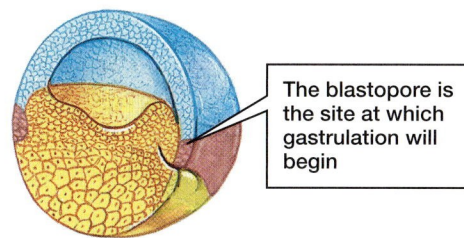

The blastopore is the site at which gastrulation will begin

(c) The blastula just before gastrulation

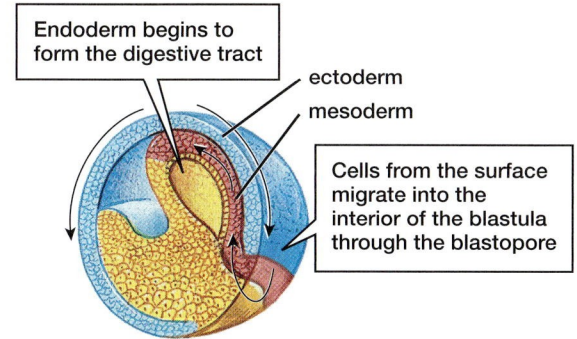

Endoderm begins to form the digestive tract

ectoderm

mesoderm

Cells from the surface migrate into the interior of the blastula through the blastopore

(d) Cells migrate at the start of gastrulation

future digestive tract

(e) A three-layered gastrula has formed

ectoderm   mesoderm   endoderm

TABLE 42-1 Derivation of Adult Tissues from Embryonic Cell Layers	
Embryonic Layer	**Adult Tissue**
Ectoderm	Epidermis of the skin; hair; lining of the mouth and nose; glands of the skin; nervous system
Mesoderm	Dermis of the skin; muscle, skeleton; circulatory system; gonads; kidneys; outer layers of the digestive and respiratory tracts
Endoderm	Lining of the digestive and respiratory tracts; liver; pancreas

dehydrate. In addition, the water surrounding the embryo supplies it with oxygen and carries away its wastes, including carbon dioxide.

Fully terrestrial vertebrate life was not possible until the evolution of the shelled **amniotic egg.** This innovation first arose in reptiles and persists today in that group (including birds) and its descendants, the mammals. Inside the amniotic egg, the embryos of these animals develop in their own "private pond," even if the egg is laid on land. The amniotic egg includes four **extraembryonic membranes:** the chorion, amnion, allantois, and yolk sac (**Table 42-2**). In reptiles, the **chorion** lines the shell and allows for exchange of oxygen and carbon dioxide between the embryo and the air. The **amnion** encloses the embryo in a watery environment. The **allantois** surrounds and isolates wastes. The **yolk sac** contains the yolk.

In mammals (except for platypuses and echidnas, which lay eggs), the embryo develops within the mother's body until birth. Nevertheless, all four extraembryonic membranes still persist, and in fact these membranes are essential for development. Table 42-2 compares the structures and functions of these extraembryonic membranes in reptiles and placental mammals. (Marsupials, such as kangaroos and opossums, have a small, short-lived placenta that differs substantially from the more robust placenta of most mammals.)

CHECK YOUR LEARNING

Can you describe early development in amphibians, including the processes of cleavage, blastula formation, and organogenesis? Can you name the extraembryonic membranes and describe their functions in reptiles and mammals?

TABLE 42-2 Vertebrate Embryonic Membranes

Reptile   Mammal

Membrane	Structure	Function	Structure	Function
Chorion	Membrane lining the inside of the shell	Acts as a respiratory surface; regulates the exchange of gases and water between the embryo and the air	Fetal contribution to the placenta	Provides for the exchange of gases, nutrients, and wastes between the embryo and the mother
Amnion	Sac surrounding the embryo	Encloses the embryo in fluid	Sac surrounding the embryo	Encloses the embryo in fluid
Allantois	Sac connected to the embryonic urinary tract; a capillary-rich membrane lining the inside of the chorion	Stores wastes (especially urine); acts as a respiratory surface	Membranous sac arising from the gut; varies in size	May store metabolic wastes; contributes to the umbilical cord blood vessels
Yolk sac	Membrane surrounding the yolk	Contains yolk as food; digests yolk and transfers its nutrients to the embryo	Small, membranous, fluid-filled sac	Helps absorb nutrients from the mother; forms blood cells; contributes to the umbilical cord

42.4 HOW IS DEVELOPMENT CONTROLLED?

A zygote contains all of the genes needed to produce an entire animal. Nearly every cell of the adult body also contains all of these genes. In any given cell, however, some genes are used, or expressed, while others are not. The differentiation of cells during development arises because of differences in gene expression.

There are several methods of controlling gene expression (see Chapter 12). One important method is to control which genes are copied, or transcribed, into messenger RNA (mRNA), which subsequently directs the synthesis of the proteins encoded by the genes. Every cell contains proteins, called *transcription factors*, that bind to specific genes and turn their transcription on or off. Which genes are transcribed then determines the structure and function of the cell.

Using the techniques of molecular biology, developmental biologists have discovered many of the genes and proteins involved in development, and some of the interactions among them. However, the fundamental principles of development were discovered about a century ago, using tools no

more sophisticated than forceps, tiny knives, and hair from a baby's head, as we explore in "Science in Action: Unraveling the Mechanisms of Animal Development" on p. 818.

Molecules Positioned in the Egg and Produced by Nearby Cells Control Gene Expression During Embryonic Development

How can the daughter cells of a fertilized egg inherit different transcription factors, transcribe different genes, and become structurally and functionally distinct from one another? In animal embryos, the differentiation of individual cells and the development of entire structures are driven by one or both of two processes: (1) the actions of transcription factors and other gene-regulating substances inherited from the mother in her egg, and (2) chemical communication between the cells of the embryo.

Maternal Molecules in the Egg May Direct Early Embryonic Differentiation

Virtually all the cytoplasm in a zygote is already present in an egg before it is fertilized; the sperm contributes little more than a nucleus (see Chapter 41). In most invertebrates and some vertebrates, specific protein molecules become localized in different places in the egg's cytoplasm. Some of these proteins are transcription factors that regulate which genes are turned on and off, in some cases turning on master genes that control the activity of many other genes.

During the first few cleavage divisions, the zygote and its daughter cells divide at specific places and in specific orientations. As a result, these early embryonic cells receive different maternal transcription factors. Therefore, different cells transcribe different genes, start differentiating into distinct cell types, and in many cases, ultimately give rise to specific adult structures. The position of maternal molecules in some eggs so strongly controls development that the egg can be mapped according to the major structures that will be produced by daughter cells inheriting each section of cytoplasm. For example, in the sea squirt egg diagrammed in **Figure 42-4,** cells that receive cytoplasm colored blue will form the skin, cells that receive cytoplasm colored green will form the nervous system, and so on.

Early development in mammals is difficult to study, because the eggs are very small—1/1000th the volume of a frog egg—and are produced in small numbers (one to a dozen or so per ovulation). Current evidence suggests that the cells formed during cleavage in mammals are probably not completely identical, but, unlike the cells of early amphibian and sea squirt embryos, they are not rigidly determined as to which part of the embryo they can form.

Chemical Communication Between Cells Regulates Most Embryonic Development

During later development, the fate of each cell is determined by chemical interactions between cells, in a process called **induction.** During induction, certain cells release chemical messengers that alter the development of other, usually nearby, cells. Specific groups of genes are selectively activated in response to these messengers, inducing the cells to differentiate in specific ways.

In amphibian embryos, it has been known since the early 1900s that a cluster of cells located near the blastopore (see Fig. 42-3c), called the *organizer*, determines whether

nearby cells will become ectoderm or mesoderm, and even where the head and nervous system will form (see "Science in Action: Unraveling the Mechanisms of Animal Development"). How does the organizer accomplish this? The cells of the organizer release proteins that interact with other messenger molecules to stimulate or inhibit the expression of master genes in nearby cells. These master genes often encode transcription factors that alter the transcription of many other genes. Which groups of genes are expressed determines the structures and functions of the cells. As these cells differentiate, they in turn release other chemicals that alter the fate of still other cells, in a cascade that culminates in the development of the tissues and organs of the adult body.

Homeobox Genes Regulate the Development of Entire Segments of the Body

Homeobox genes are found in animals as diverse as fruit flies, frogs, and humans. Although their functions differ somewhat in different animals, homeobox genes are master genes coding for transcription factors that regulate the transcription of many other genes. Each homeobox gene affects the development of a particular region of the body. Homeobox genes were discovered in fruit flies, where specific mutations cause entire parts of the body to be duplicated, replaced, or omitted. For example, one mutant homeobox gene causes the development of an extra body segment, complete with an extra set of wings.

Homeobox genes are arranged on the chromosomes in a head-to-tail order, and are transcribed in cells in the specific locations in the body whose development they control. For example, "head" homeobox genes are transcribed in the head of the embryo, and "tail" homeobox genes are transcribed in the tail (**Fig. 42-5**).

▲ **FIGURE 42-5 Homeobox genes regulate development of body segments** The sequence of homeobox genes on the chromosome corresponds to their role in the development of different body segments. In fruit flies, each homeobox gene is active in the body segment shown in the same color, in a head-to-tail order.

QUESTION Snakes have ribs all the way from just behind the head almost to the tip of the tail. Snakes also lack legs. Propose a possible genetic mechanism for this body structure, based on homeobox genes.

---

egg   **larva**

▲ **FIGURE 42-4 A "fate map" of the sea squirt egg** Gene-regulating substances in the cytoplasm of the egg of a sea squirt control early development. The coloring of the egg and larva shows which parts of the egg will give rise to which parts of the larva.

QUESTION If human development were as thoroughly determined as this, do you think that identical twins would be more or less common?

SCIENCE IN ACTION Unraveling the Mechanisms of Animal Development

Embryologists have long used amphibian eggs to study development. These eggs are very large (often more than a millimeter in diameter) and easy to obtain in large numbers. They can also tolerate a lot of experimental manipulation and still develop fairly normally.

Molecules in the Egg Cytoplasm Regulate Early Development

In the 1920s, the German embryologist Hans Spemann wondered how a fertilized egg, and later the genetically identical cells of an embryo, could produce structures as different as a brain, a leg, or a stomach. Might there be something in the egg cytoplasm that determines how genetic information is used during early development?

Some fertilized amphibian eggs have three distinctively pigmented regions: dark on top, light on the bottom, and a crescent-shaped band of gray in between, on one side of the egg (Fig. E42-1). Usually, the first cleavage division splits this gray crescent, so that both daughter cells receive some of it. Sometimes, however, the first cleavage plane misses the crescent: One daughter cell gets the entire crescent and the other doesn't get any.

Using fine forceps, Spemann lassoed two-celled salamander embryos with strands of hair from his baby daughter. He tightened a noose between each pair of cells to separate them. If both cells contained crescent material, both developed into normal embryos (Fig. E42-1a). But if only one cell contained crescent material, that cell would develop normally, whereas the cell without crescent material would form a blob that Spemann called a "belly piece" (Fig. E42-1b). Spemann concluded that something, not necessarily the gray crescent itself, but in the same location in the cell, regulated the use of the genetic information that the two daughter cells received. In modern terms, gene-regulating molecules, probably transcription factors, are positioned in specific locations in the egg cytoplasm. These transcription factors control gene expression and determine the developmental fates of the daughter cells. If a daughter cell receives the necessary transcription factors,

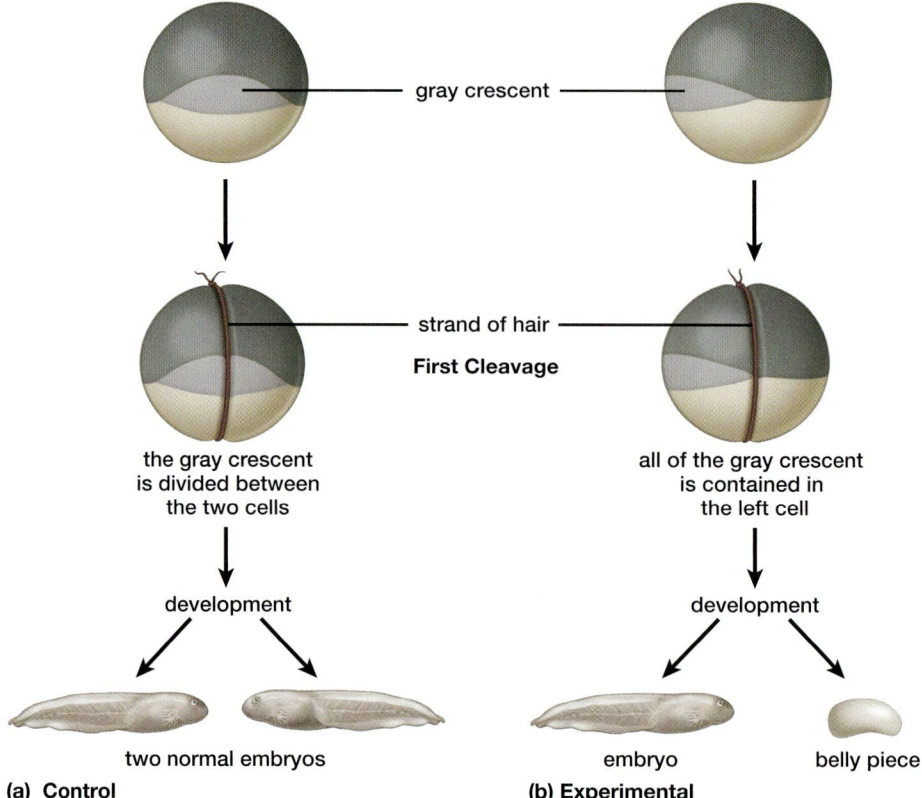

gray crescent

First Cleavage

strand of hair

the gray crescent is divided between the two cells

all of the gray crescent is contained in the left cell

development

development

two normal embryos

embryo   belly piece

(a) Control

(b) Experimental

▲ FIGURE E42-1 **The egg cytoplasm controls early development in amphibians (a)** In salamanders and other amphibians, the first cleavage division usually bisects the gray crescent, so both daughter cells receive some of the crescent material. Separating the two cells produces two normal embryos. **(b)** Sometimes, the first division encloses the entire crescent in the cytoplasm of one daughter cell. If the two cells are separated, only the cell with crescent material can form an embryo.

it develops into an embryo. If not, it still divides, but can form only a blob.

Certain Cells Influence the Development of Their Neighbors

What about later development? Surely the egg can't have infinitely precise positioning of myriads of molecules, parceled out to daughter cells with such accuracy that one particle of egg cytoplasm generates a pinky, another a ring finger, another a thumb, and so on? Indeed not. Spemann and other embryologists had already done experiments in which bits of embryos, usually at the blastula stage, were transplanted to various locations

in other embryos of the same species. They found that the fate of the transplanted cells was, in general, not predetermined, as if the cells' destiny were controlled by the specific molecules they contained. Instead, the transplants assumed the developmental fate of the area into which they were transplanted (Fig. E42-2a).

However, Spemann found that one small patch of cells, called the dorsal lip of the blastopore, followed its own developmental path, no matter where it was transplanted into a recipient embryo. In fact, in one experiment, transplanted dorsal lip cells caused the formation of a second head in the recipient embryo, producing a two-headed tadpole.

1 Cells from a region that normally becomes skin are removed from a light-colored blastula

2 These cells are transplanted to a region of a dark-colored blastula that normally becomes nerve tissue

3 Embryonic development proceeds

4 The light-colored, transplanted cells differentiate into nerve tissue

(a) A typical transplantation experiment

1 The dorsal lip of the blastopore is removed from a light-colored blastula

2 The dorsal lip is transplanted inside a dark-colored blastula

3 Two head regions start to develop; one (left) consists of a mixture of transplanted cells (pale) and host cells (dark); the other (right) consists of host cells only

4 The transplanted dorsal lip of the blastopore causes the formation of a second embryo (left) attached to the host embryo (right); the second embryo consists mostly of host cells (dark)

(b) Transplantation of the dorsal lip of the blastopore

▲ **FIGURE E42-2 Induction** During most of development, the fate of any given cell is strongly influenced by the cells that surround it. **(a)** When part of a blastula is transplanted into a second blastula, the surrounding cells usually induce the transplant to assume the characteristics of the region into which the transplant was placed. **(b)** A transplanted dorsal lip of the blastopore, however, induces the surrounding cells to change their fates and develop into most of the structures of a secondary embryo.

In the 1920s, Spemann and a graduate student, Hilde Mangold, performed a brilliant experiment showing that transplanted dorsal lip cells actually changed the developmental fate of cells of the recipient embryo. Mangold cut out pieces of embryos from pale-colored newts and transplanted them into embryos of dark-colored newts, and vice versa. In this way, she could determine whether a structure that developed in the host embryo consisted of cells from the donor or the host. A transplant from the dorsal lip of the blastopore caused nearby cells of the host embryo to form parts of a second head and, occasionally, to become almost a complete secondary embryo (Fig. E42-2b). Spemann and Mangold concluded that the dorsal lip—now called Spemann's organizer—"organized" surrounding host cells to form the head-to-tail axis of the embryo.

In 1935, Spemann was awarded the Nobel Prize for Physiology or Medicine. Hilde Mangold had died tragically in a fire in 1924. Because Nobel Prizes are not awarded posthumously, she could not share the prize. To acknowledge Mangold's contributions, Spemann always referred to "the experiments of Hilde Mangold" when describing their results.

CHECK YOUR LEARNING

Can you explain the roles of gene-regulating substances in the egg cytoplasm, transcription factors, and induction between cells in development? Can you describe the role of homeobox genes?

CASE STUDY c o n t i n u e d

Rerunning the Program of Development

In salamander development, the sequential activation of specific homeobox genes is required to produce a foreleg with a humerus (the bone found in the upper arm of a person), radius and ulna (lower arm), carpals (wrist), metacarpals (palm), and phalanges (fingers). The same homeobox genes are turned on, in a very similar sequence, during leg regeneration.

42.5 HOW DO HUMANS DEVELOP?

Human development is controlled by the same mechanisms that control the development of other animals. In fact, our development strongly reflects our evolutionary heritage.

Differentiation and Embryonic Growth Are Rapid During the First Two Months

Fertilization of a human egg usually takes place in the uterine tube. The resulting zygote undergoes a few cleavage divisions in the uterine tube, becoming a morula on its way to the uterus (Fig. 42-6). By about the fifth day after fertilization, the zygote has developed into a hollow ball of cells, known as a **blastocyst** (the mammalian version of a blastula; Fig. 42-7a). A mammalian blastocyst consists of an outer layer of cells surrounding a cluster of cells called the **inner cell mass.** The outer cell layer attaches to, and then

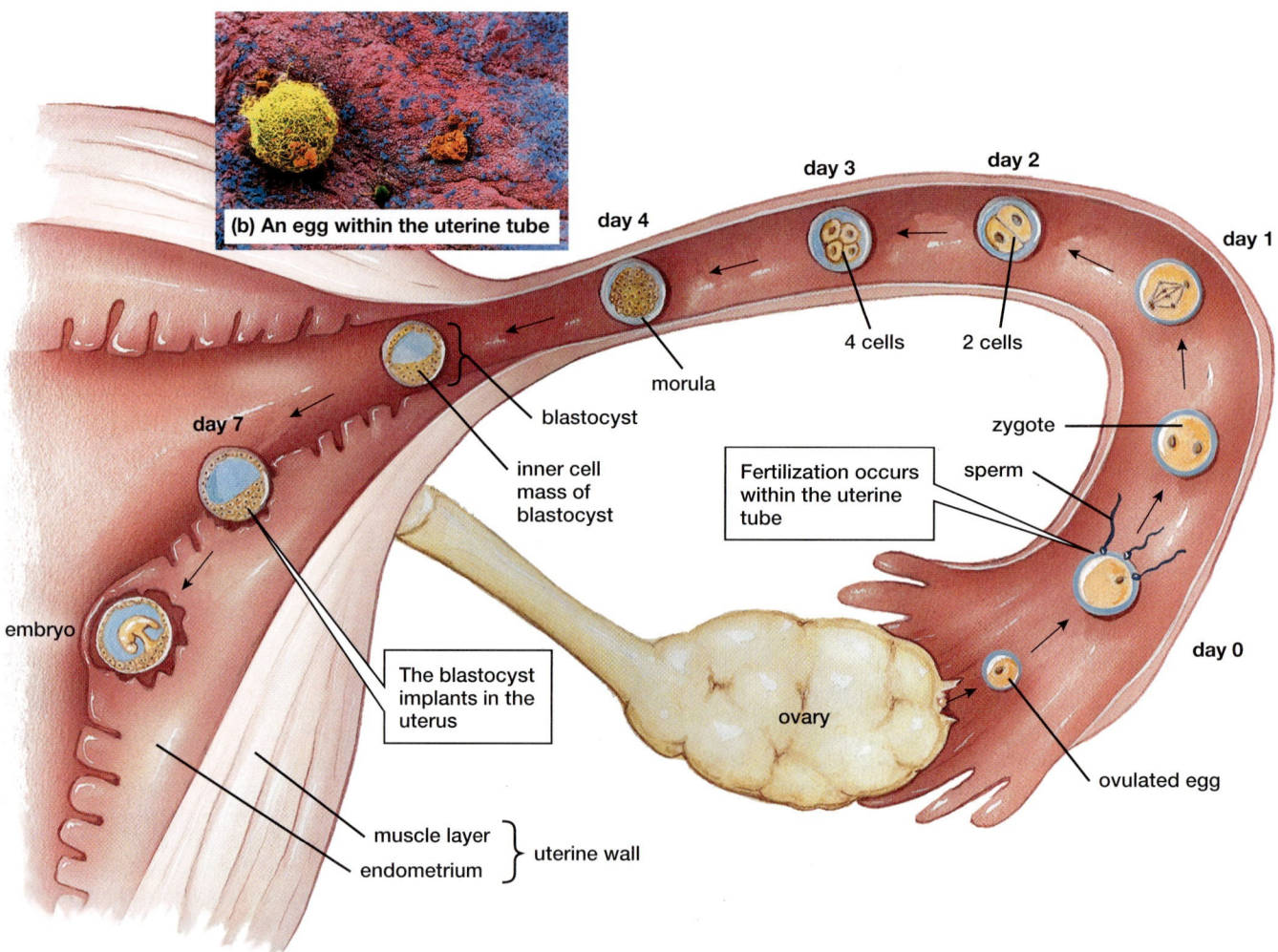

(b) An egg within the uterine tube

(a) The first week of development

▲ FIGURE 42-6 **The journey of the egg** (a) A human egg is fertilized in the uterine tube and slowly travels down to the uterus. Along the way, the zygote divides a few times, becoming first a morula and then a blastocyst. When the blastocyst reaches the uterine endometrium, it burrows in. (b) An egg, surrounded by sperm (yellow in this colored micrograph), is cradled in the uterine tube.

QUESTION The cells of the corona radiata surround a fertilized egg. As the egg develops into a blastocyst, what must happen before the blastocyst can implant in the uterus?

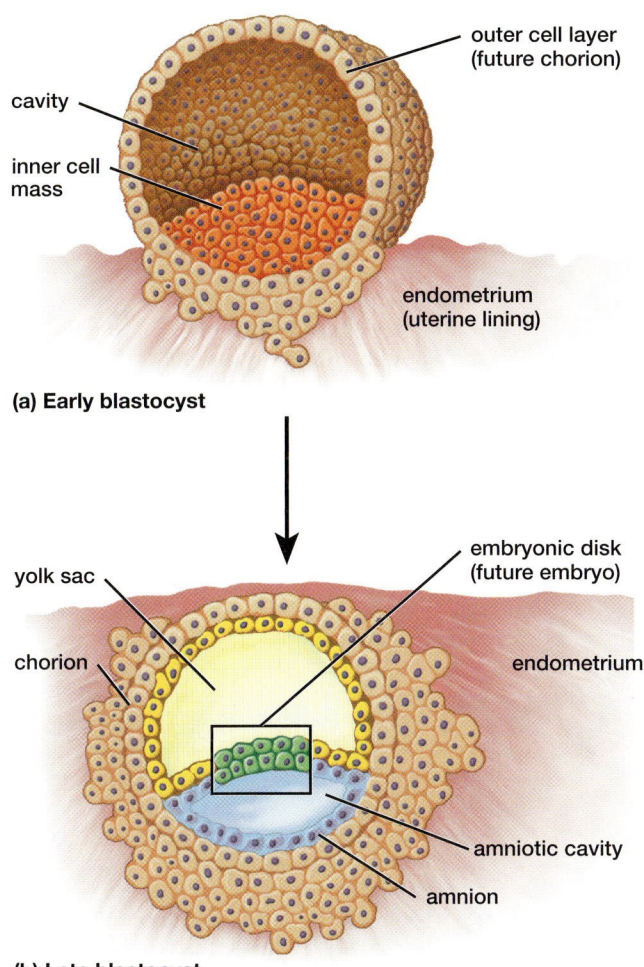

(a) Early blastocyst

outer cell layer (future chorion)

cavity

inner cell mass

endometrium (uterine lining)

yolk sac

embryonic disk (future embryo)

chorion

endometrium

amniotic cavity

amnion

(b) Late blastocyst

▲ **FIGURE 42-7 A blastocyst implants** **(a)** As it burrows into the uterine lining, the outer cell layer of the blastocyst forms the chorion, the embryonic contribution to the future placenta. **(b)** A few days later, the blastocyst has completely submerged beneath the uterine lining. The inner cell mass begins to develop into the yolk sac, the amnion, and the embryonic disk.

burrows into, the lining of the uterus (the endometrium), a process called **implantation** (**Figs. 42-7a, b**). The outer cell layer of the blastocyst will become the chorion. The complex intermingling of the chorion and the endometrium forms the placenta, which we will describe shortly.

All the cells of the inner cell mass have the potential to develop into any type of tissue. This remarkable flexibility allows the inner cell mass to produce both the entire embryo and the three remaining extraembryonic membranes. The inner cell mass is also the source of human embryonic stem cells, which researchers hope may one day be used to replace damaged adult tissues, as described in "Health Watch: The Promise of Stem Cells."

After Implantation, Two Cavities Form and Gastrulation Occurs

During the second week of development, the inner cell mass grows and splits, forming two fluid-filled sacs that are separated by a double layer of cells called the **embryonic disk**

(see Fig. 42-7b). One layer of cells is continuous with the yolk sac, although in placental mammals there is no yolk. The second layer of cells is continuous with the amnion.

Gastrulation begins near the end of the second week. Cells migrate through a slit in the amnion side of the embryonic disk. This slit is the disk's equivalent of the amphibian blastopore. Once inside the disk, the migrating cells form mesoderm, endoderm, and the fourth extraembryonic membrane, the allantois. The cells remaining on the surface become ectoderm.

Organogenesis Begins During Weeks Three to Eight

During the third week of development, the embryo begins to form the spinal cord and brain. The heart starts to beat about the beginning of the fourth week. At this time, the embryo, bathed in fluid contained within the amnion, bulges into the uterine cavity (**Fig. 42-8**). Meanwhile, the umbilical cord

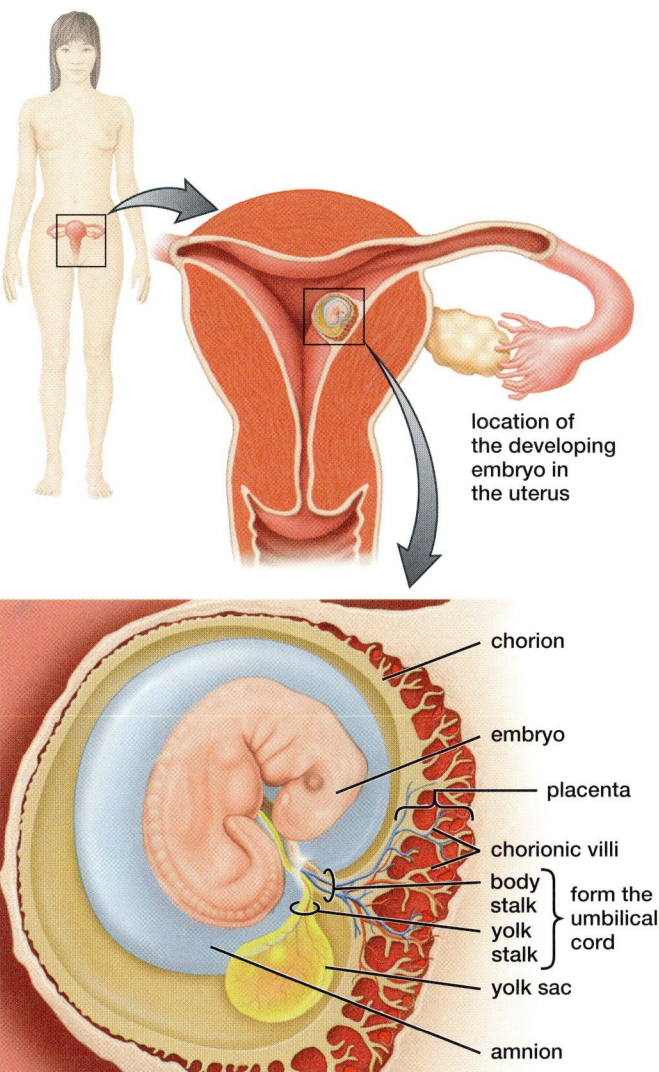

location of the developing embryo in the uterus

chorion

embryo

placenta

chorionic villi

body stalk yolk stalk } form the umbilical cord

yolk sac

amnion

▲ **FIGURE 42-8 Human development during the fourth week** The embryo bulges into the uterus, and the placenta is restricted to one side. The body stalk and yolk stalk combine to form the umbilical cord, which exchanges wastes and nutrients between embryo and mother.

Health|Watch
The Promise of Stem Cells

Most cells in adult animals have become committed to a specific developmental pathway and cannot differentiate into other cell types. In contrast, a **stem cell** has not differentiated, and can give rise to more than one cell type. The medical potential of stem cells is vast—victims of heart attacks, strokes, spinal cord injuries, and degenerative diseases from arthritis to Parkinson's disease would benefit if their damaged tissues could be regenerated.

There are three types of stem cells. The first type, **embryonic stem cells (ESCs),** are derived from the inner cell mass of a blastocyst (see Fig. 42-7). In an intact embryo, the stem cells of the inner cell mass produce all the cell types of the entire body. ESCs grown in cell culture can also generate any cell type in the body, if they are exposed to the correct mixture of proteins and other molecules that nudge them into one differentiation pathway or another (**Fig. E42-3**). The second type of stem cell is present in small numbers in most parts of the body—including muscle, skin, bone marrow, fat, brain, and heart. These are usually called **adult stem cells (ASCs),** although they are also present in children. ASCs can differentiate into only a few cell types, generally the cell types that normally make up the tissue from which they were taken. Stem cells from bone marrow, which normally produce both red and white blood cells, have been used for decades in transplants to treat diseases such as leukemia.

The third type of stem cell can be produced by inserting a handful of genes into adult (non–stem) cells, transforming them into **induced pluripotent stem cells (iPSCs).** IPSCs appear to be able to differentiate into all cell types. Until 2009, generating iPSCs required either inserting at least one cancer-causing gene or using a potentially dangerous virus that inserts its DNA into the chromosomes of the host cell.

Both of these techniques would probably make the resulting iPSCs unsafe to use in people. However, in 2009, researchers discovered how to transform fetal human stem cells into iPSCs using a single, noncarcinogenic gene. Other researchers have developed procedures for transforming the non–stem cells of adults into the equivalent of ASCs, capable of differentiating into several different, but not all, cell types of the body.

There are still significant obstacles to surmount before stem cells can become effective therapies for people. First, embryos must be destroyed to obtain ESCs (at least with today's technology). In most cases, the embryos are "extras" that were created for in vitro fertilization and would eventually be destroyed. Nevertheless, many people object to destroying human embryos to obtain ESCs.

Second, the immune system would reject stem cells that are not genetically identical, or at least a very close match, to the recipient. Rejection is generally not a problem with ASCs, which can often be obtained from the patient, but it seems likely that ESCs would be rejected as foreign. Surprisingly, in 2011 researchers found that the changes that transform normal cells into iPSCs might alter them enough to cause rejection, even when the original cells were taken from the patient's own body.

Third, researchers do not yet know if ESCs and iPSCs are safe to use in people. In early 2009, the media reported that an Israeli boy had been taken to Russia and injected with ESCs three times in an attempt to cure a rare brain disease. However, some of the ESCs grew into a tumor (benign, fortunately). At that time, human safety studies with ESCs had not yet been performed, so many researchers and physicians were appalled at the risks that had been taken. Clinical studies of ESCs that were under way in 2012 should help to answer questions about ESC safety within the next few years.

---

forms from the fusion of the yolk stalk and body stalk. The yolk stalk connects the yolk sac to the embryonic digestive tract. The body stalk contains the allantois, which contributes the blood vessels that will become the umbilical arteries and vein. The umbilical cord now connects the embryo to the placenta, which has formed from the merger of the chorion of the embryo and the lining of the uterus (see page 825).

During the fourth and fifth weeks, the embryo develops a prominent tail and pharyngeal (gill) grooves (indentations behind the head that are homologous to a fish embryo's developing gills; Fig. 42-9). These structures are reminders that we share evolutionary ancestry with other vertebrates that retain their tails and gills in adulthood. In humans, however, the tail and grooves disappear as development continues. By the seventh week, the embryo has rudimentary eyes and a rapidly developing brain, and the webbing between its fingers and toes is disappearing.

After Two Months, the Embryo Is Recognizably Human

As the second month draws to an end, nearly all of the major organs have begun to develop. The gonads appear and develop into testes or ovaries. Sex hormones are secreted—either

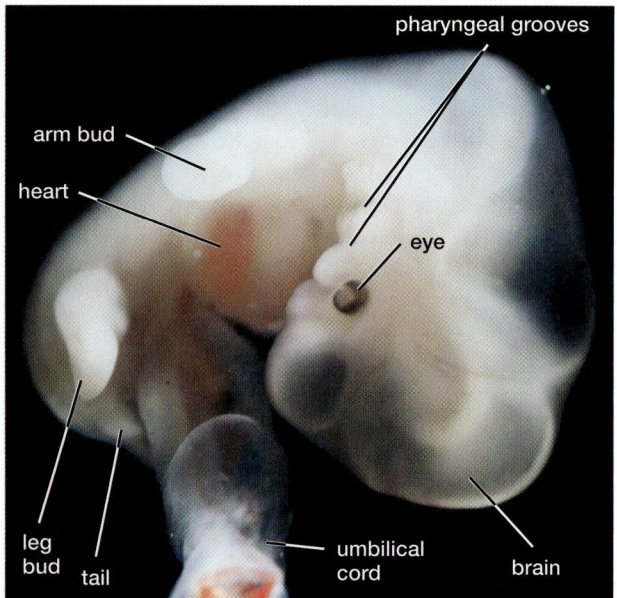

▲ **FIGURE 42-9 A 5-week-old human embryo** At the end of the fifth week, the human embryo is almost half head. A tail and pharyngeal (gill) grooves are clearly visible, evidence of our evolutionary relationship to other vertebrates.

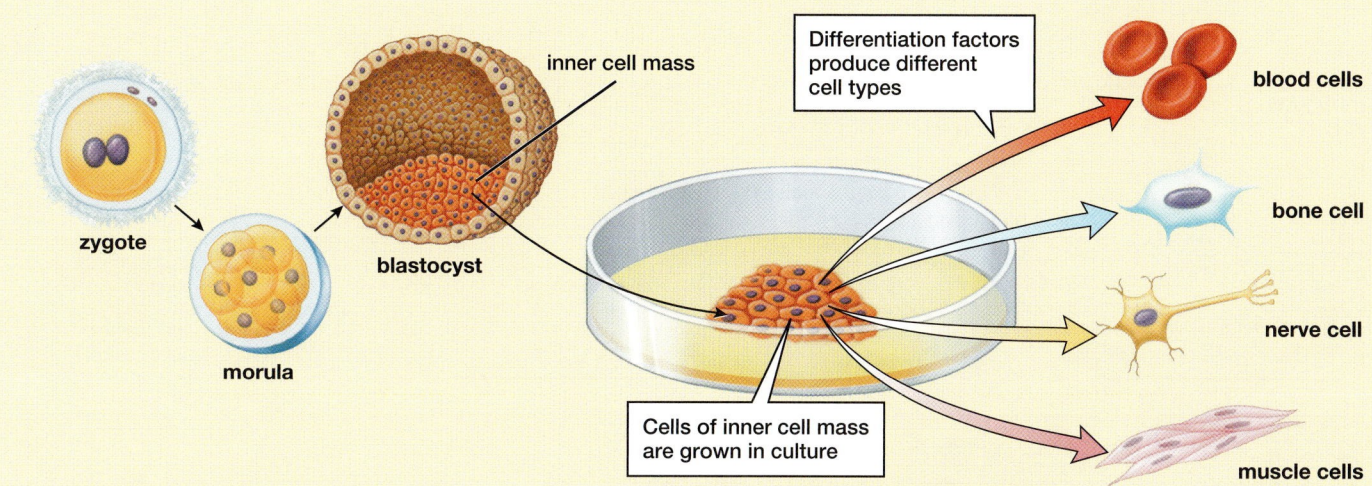

inner cell mass

Differentiation factors
produce different
cell types

blood cells

bone cell

nerve cell

muscle cells

zygote

morula

blastocyst

Cells of inner cell mass
are grown in culture

▲ FIGURE E42-3 **Culturing embryonic stem cells from the inner cell mass of a blastocyst**

ASCs offer the most immediate hope for successful medical treatments. ASCs have been used for almost a decade to treat horses and dogs with bone, ligament, or tendon injuries. Controlled studies show that stem cells injected directly into the injury site help horses with tendon and ligament damage to recover faster and heal more completely. A few people have already been injected with ASCs to treat tendon or ligament injuries. For example, professional baseball pitcher Bartolo Colon received ASCs for a shoulder injury that never fully healed after rotator cuff surgery. A year later, he was pitching for the New York Yankees. Of course, no one really knows if the ASCs were crucial to Colon's recovery or if he would have healed anyway.

ASCs have also been used for tracheal transplants in two patients whose own tracheas had been destroyed by disease. For the first patient, a trachea from an organ donor was stripped of its cells, leaving only a cartilage framework behind. The cartilage was then seeded with the patient's own stem cells. For the second patient, an artificial tracheal skeleton was made from plastic and then seeded with his stem cells. In both cases, the stem cells attached and grew a covering for the new trachea. Because their own cells were used, the patients' bodies didn't reject the transplants.

testosterone from the testes or estrogen from the ovaries. These hormones will affect the development of many structures, including the reproductive organs, brain, and urinary tract. At the end of the second month, the embryo has taken on a generally human appearance, and is now called a **fetus** (Fig. 42-10).

The first two months of pregnancy are a time of extremely rapid differentiation and growth for the embryo, and a time of considerable danger. Although vulnerable throughout development, the rapidly developing organs are especially sensitive to toxic substances during this time.

Growth and Development Continue During the Last Seven Months

Pregnancy usually continues for another 7 months. As the brain and spinal cord grow, they begin to generate behaviors. As early as the third month of pregnancy, the fetus can move, respond to stimuli, and even suck its thumb. The lungs, stomach, intestine, and kidneys enlarge and become functional. Fetal urine, in fact, makes up most of the amniotic fluid during the last 6 months of pregnancy. Although a full-term pregnancy lasts about 38 weeks, nearly all fetuses 32 weeks or older can survive outside the womb with medical

amniotic sac

umbilical cord

placenta

▲ FIGURE 42-10 **An 8-week-old human embryo** At the end of the eighth week, the embryo is clearly human in appearance and is now called a fetus. Most of the major organs of the adult body have begun to develop.

week 1	week 2	week 3	week 4	week 5	week 6
zygote to late blastocyst		embryo			
		0.06–0.1 inch (1.5–2.5 mm)	0.12–0.20 inch (3–5 mm)	0.28–0.35 inch (7–9 mm)	0.32–0.43 inch (8–11 mm)
Cleavage of zygote forms the morula and then the blastocyst, which implants in the uterus.	The blastocyst burrows into the endometrium; forms the yolk sac, amnion, and embryonic disk.	Gastrulation occurs; the notochord and beginning of the neural tube form.	The neural tube closes; arm buds, tail, and pharyngeal (gill) grooves form; the heart beats.	The eyes begin to form; leg buds form; the brain enlarges.	External ears and webbed fingers form; the pharyngeal (gill) grooves and tail disappear.

week 7	week 8	week 10	week 12	week 16
embryo		fetus		
0.67–0.79 inch (1.7–2.0 cm)	0.90–1.10 inches (2.3–2.8 cm)	1.25–1.75 inches (3.2–4.4 cm)	2–3 inches (5–7.6 cm)	4–5 inches (10.2–12.7 cm)
Webbed toes form; bones begin to stiffen; the back straightens; the eyelids begin to form.	All the major organs begin to form; the arms can bend; fingers and toes are distinct. Facial features and outer ears take shape.	After 8 weeks, the embryo is called a fetus. Red blood cells form; toes separate; eyelids have developed; major brain parts are present; the hands can form fists.	The neck is well defined; all organs are present; distinctly male or female external genitals are present; arms and legs move; teeth begin to form; a heartbeat can be detected.	Sucking and swallowing movements occur; the liver and pancreas begin functioning. The body has grown relative to the head; major organs continue developing. The mother may feel movement; weight is about 5 oz.

week 20	week 24	week 30	week 36
fetus			
6–7 inches (15.2–17.8 cm)	8–9 inches (20.3–22.9 cm)	15–16 inches (38.1–40.6 cm)	16–19 inches (40.6–48.3 cm)
The fetus may suck its thumb; arms and legs can punch and kick; the body can change position. Fingernails are formed; fat is deposited under the skin; eyebrows and eyelashes appear.	Brain development continues, hearing develops, and the eyes can move. The fetus can hiccup, squint, smile, and frown. The fetus may have hair on its head. Unique foot- and fingerprints appear. Weight is about 1–1.5 lbs.	Brain development continues; the eyes open and close and see light; the fetus kicks and stretches. Breathing movements occur but the lungs are not mature. Bones are present but flexible. The baby may survive if born.	Eyes open and close corresponding to wake and sleep cycles; body fat increases; lungs and other organs are functional. The fetus can grasp and orient toward light. Weight is about 5–6 lbs, and the child is no longer considered premature if born. Full term is 38 weeks.

▲ FIGURE 42-11 A calendar of development from zygote to birth assistance. Most infants born as early as 26 weeks survive if they are given intensive care, but they may suffer from lifelong disabilities, particularly neurological disorders. Younger infants have a much lower chance of survival.

Figure 42-11 summarizes human embryonic development, from fertilization to the conclusion of pregnancy.

The Placenta Exchanges Materials Between Mother and Embryo

During the first few days after implantation, the embryo obtains nutrients directly from the endometrium of the uterus. During the following week or so, the **placenta** begins to develop from interlocking structures produced by the embryo and the endometrium (Fig. 42-12). The outer layer of the blastocyst forms the chorion, which grows finger-like **chorionic villi** that extend into the endometrium. Blood vessels of the umbilical cord connect the embryo's circulatory system with a dense network of capillaries in the villi. Meanwhile, some of the blood vessels of the endometrium erode away, producing pools of maternal blood that bathe the chorionic villi.

The embryo's blood and the mother's blood remain separated by the walls of the villi and their capillaries, so the two blood supplies do not actually mix to any great extent. However, many small molecules move across the walls of the villi and capillaries between the mother's blood and the embryo's blood. Oxygen diffuses from the mother to the embryo. Nutrients, many aided by active transport, also travel from the mother to the embryo. Carbon dioxide and wastes, such as urea, diffuse from the embryo to the mother.

The walls of the capillaries and chorionic villi act as barriers to some substances, including large proteins and most cells. However, some disease-causing organisms and many harmful chemicals can penetrate the placenta, as described in "Health Watch: The Placenta—Barrier or Open Door?"

Pregnancy Culminates in Labor and Delivery

During the last months of pregnancy, the fetus usually becomes positioned head downward in the uterus, with the crown of the skull resting against, and held up by, the cervix. Childbirth (Fig. 42-13) generally occurs around the end of the ninth month. Birth results from a complex interplay between uterine stretching caused by the growing fetus and maternal and fetal hormones that finally cause **labor,** the contractions of the uterus that result in delivery of the child.

Unlike skeletal muscles, the smooth muscles of the uterus can contract spontaneously, and stretching enhances their tendency to contract. As the baby grows, it stretches the uterine muscles, which occasionally contract weeks before delivery (these are called Braxton Hicks or false labor contractions). No one really knows what triggers labor in humans, but chemical signals from both the placenta and the fetus may be involved. Whatever the initial stimulus, the placenta responds by releasing prostaglandins, which make the uterine muscles more likely to contract. As the uterus

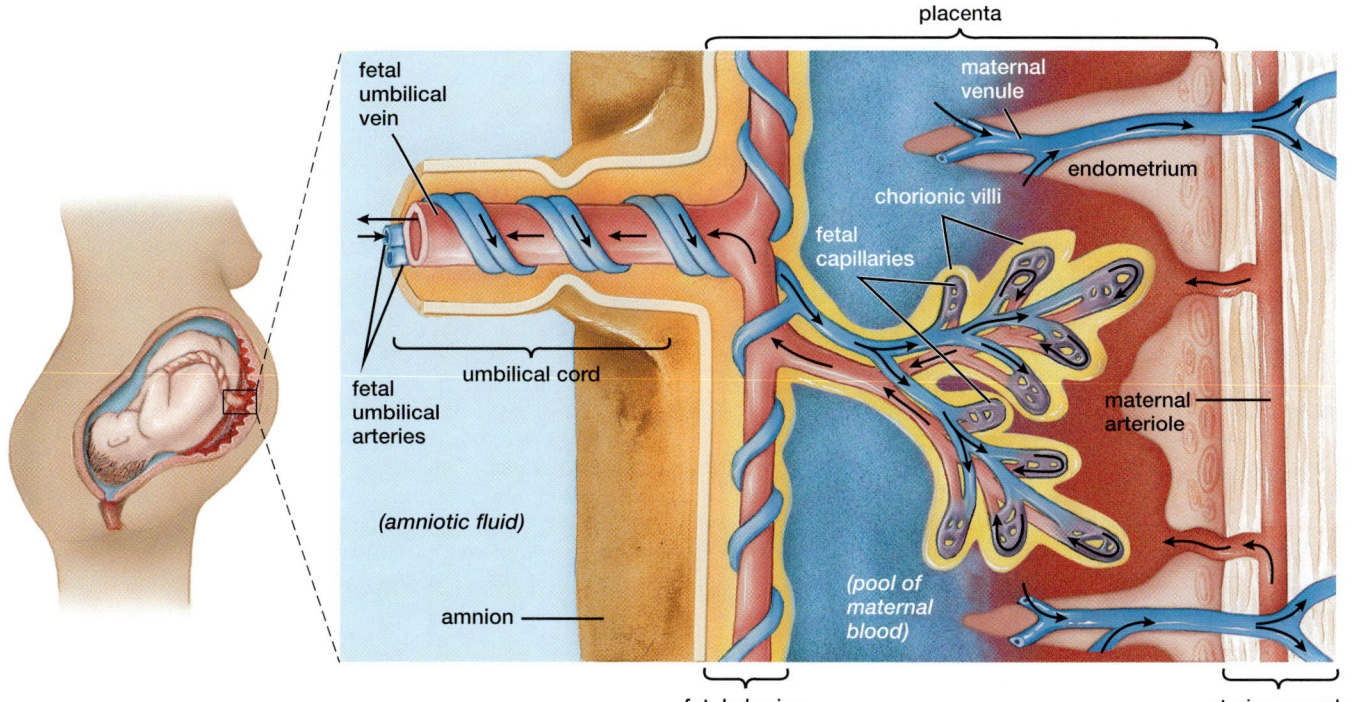

▲ FIGURE 42-12 **The placenta** The placenta allows exchange of wastes and nutrients between fetal capillaries and maternal blood pools, while keeping the fetal and maternal blood supplies separate. The umbilical arteries carry deoxygenated blood from the fetus to the placenta, and the umbilical vein carries oxygenated blood back to the fetus.

QUESTION Marsupial mammals, such as kangaroos, have a much more rudimentary placenta. What would you predict about the degree of development of newborn marsupials?

Health Watch

The Placenta—Barrier or Open Door?

The fetal and maternal blood supplies are separated by several membranes, including the walls of the villi and the fetal capillaries (see Fig. 42-12). However, many molecules, and even some infectious microbes, can pass through these barriers, moving from mother to fetus.

Microbes

Bacterial invaders include syphilis, some types of *Streptococcus* that cause blood infections, and *Listeria*, which is often found in contaminated food. Viral invaders include those that cause genital herpes, German measles (rubella), hepatitis B, and HIV. Many of these microbes cause severe disorders in the fetus. For example, 30% to 40% of fetuses infected with syphilis are stillborn. *Listeria* can cause reduced fetal growth, premature delivery, and meningitis (a sometimes fatal infection of the membranes surrounding the brain). Infections with any of several types of viruses may cause premature birth, mental retardation, eye defects, and many other disorders.

Drugs

Many medicinal and recreational drugs are lipid soluble and can easily move across plasma membranes (see Chapter 5), including those of the placenta. Some lipid-soluble drugs can harm the fetus. Thalidomide, for example, was commonly prescribed in Europe in the late 1950s and early 1960s as a sedative and antinausea medication to combat morning sickness during pregnancy. Its devastating effects on embryos were discovered only when many babies were born with missing or extremely abnormal limbs. In the late 1980s, pregnant women who took isotretinoin, a potent antiacne medication, were found to have a high risk of miscarriage, giving birth prematurely, and bearing babies with significant deformities. Many recreational drugs, such as cocaine and methamphetamine, increase the risk of miscarriage, premature birth, and delivering infants with low birth weight.

Toxic Substances in Cigarette Smoke

Many women who smoke continue the habit during pregnancy, thus exposing their developing children to nicotine, carbon monoxide, and a host of carcinogens. These mothers have a higher frequency of miscarriages and tend to give birth to underweight infants. Children born to smokers have a higher than normal incidence of behavioral and intellectual problems.

Alcohol

Alcohol is soluble both in water and lipids. It dissolves in the blood and then easily passes through the placenta: When a pregnant woman drinks, the blood of her unborn child has the same concentration of alcohol as her own blood does. Worse, the fetus cannot metabolize alcohol

severe, fatal
fetal alcohol syndrome   normal

▲ **FIGURE E42-4 Alcohol impairs brain development**
Comparing the brain of a 6-week-old child with severe fetal alcohol syndrome (left) and the brain of a normal child of the same age (right) shows the devastating effects of alcohol on the developing brain.

as rapidly as an adult, so alcohol's effects in the fetus are greater and longer lasting.

About 12% of U.S. women drink during pregnancy. Each year, they give birth to almost 40,000 infants with fetal alcohol spectrum disorders, a range of disabilities caused by prenatal alcohol exposure. Women who drink during pregnancy tend to have smaller babies that are more likely to grow up to be anxious, depressed, or aggressive. If a pregnant woman drinks heavily on a regular basis (four or more drinks per day), or goes on alcoholic binges (five or more drinks at a time), her child has a significant risk of developing full-blown **fetal alcohol syndrome (FAS),** including below-average intelligence, hyperactivity, irritability, and poor impulse control. Children with FAS may also have reduced growth and defects of the heart and other organs. In extreme cases, children afflicted with fetal alcohol syndrome may have small, improperly developed brains (Fig. E42-4). The damage is irreversible and can be fatal.

Fetal alcohol syndrome is the most common cause of mental retardation in the United States. The U.S. Surgeon General advises pregnant women and those who are likely to become pregnant to avoid alcohol completely.

The Placenta Cannot Be Trusted to Protect the Embryo

A pregnant woman should assume that any drugs she takes will cross the placenta to her developing infant. Crucial stages of embryonic development can occur before a woman even realizes that she is pregnant, so any woman trying to become pregnant, or having unprotected sex, should take the same precautions as if she were already pregnant.

---

contracts, it pushes the fetus' head against the cervix, stretching it. Stretching the cervix sends nervous signals to the mother's brain, causing the release of the hormone oxytocin (see Chapter 37). Oxytocin stimulates contractions of the uterine muscles, pushing the baby harder against the cervix, which stretches further, causing still more oxytocin to be

released. This positive feedback cycle continues until the cervix expands far enough for the baby to emerge.

After a brief rest following childbirth, the uterus resumes its contractions and shrinks remarkably. During these contractions, the placenta is sheared from the uterine wall and expelled through the vagina. The umbilical cord now

❶ The baby orients head downward, facing the mother's side; the cervix begins to thin and expand in diameter (dilate)

❷ The cervix dilates completely to 10 centimeters (almost 4 inches wide), and the baby's head enters the vagina, or birth canal; the baby rotates to face the mother's back

❸ The baby's head emerges

❹ The baby rotates to the side once again as the shoulders emerge

▲ FIGURE 42-13 **Childbirth**

releases prostaglandins that cause the muscles surrounding fetal blood vessels in the umbilical cord to contract and shut off blood flow. Although tying off the umbilical cord is standard practice, it is not usually necessary; if it were, other mammals would not survive birth.

Milk Secretion Is Stimulated by the Hormones of Pregnancy

During pregnancy, large quantities of estrogen and progesterone are secreted by the placenta. Estrogen and progesterone, acting together with several other hormones, stimulate the milk-producing **mammary glands** in the breasts to swell and develop the capacity to secrete milk. The mammary glands are arranged in a circle around the nipple, each with a milk duct leading to the nipple (**Fig. 42-14**).

Prolactin, a hormone secreted by the anterior pituitary gland, promotes both mammary gland enlargement and the actual secretion of milk, a process called **lactation.** Prolactin

release is stimulated by the high levels of estrogen produced by the placenta. Therefore, you might think that milk secretion would begin even before the child is born. However, lactation is inhibited by progesterone; when the placenta is ejected from the uterus right after childbirth, progesterone levels plummet, allowing prolactin to cause lactation.

Milk is released when the infant's suckling stimulates nerve endings in the nipples, which signal the hypothalamus to cause the pituitary gland to release an extra surge of prolactin and oxytocin. Oxytocin causes muscles surrounding the mammary glands to contract, ejecting the milk into the ducts that lead to the nipples (see Chapter 37). The prolactin surge stimulates milk production for the next feeding.

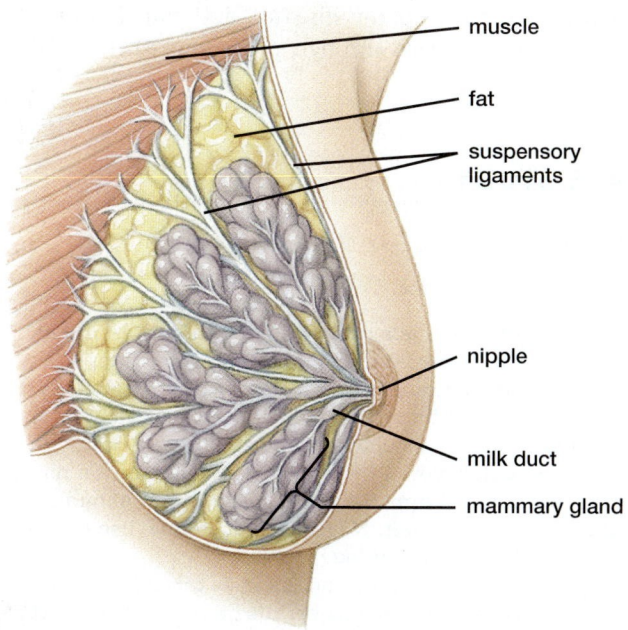

▲ FIGURE 42-14 **The structure of the human breast**
During pregnancy, fatty tissue, milk-secreting glands, and milk ducts all increase in size.

❓ HAVE **YOU** EVER WONDERED . . .

Why Childbirth Is So Difficult?

Compared to the birth of other mammals, childbirth is a lengthy and painful ordeal. Why? Probably for two reasons. First, we walk upright, requiring that the human pelvis be shaped differently than the pelvis of a four-footed mammal. As a result, a woman's pelvic opening is small (for her body size). Second, human infants have enormous heads, which actually must be squashed a bit to fit through the pelvic opening. Monkeys, with relatively small infant heads and large pelvic openings, give birth quickly and almost effortlessly. In contrast, human deliveries usually last several hours, cause considerable pain to the mother and probably the infant, and often require assistance. They don't call it "labor" for nothing!

During the first few days after birth, the mammary glands secrete a yellowish fluid called **colostrum.** Colostrum is high in protein and contains antibodies from the mother that help protect the infant against disease while its immune system develops. Colostrum is gradually replaced by mature milk, which is higher in fat and milk sugar (lactose) and lower in protein.

CHECK YOUR LEARNING

Can you describe human embryonic development from fertilization to birth? Do you understand the structure and function of the placenta? Can you name the hormones involved in milk production and secretion and describe their roles in these processes?

▲ FIGURE 42-15 **Youth meets age**

42.6 IS AGING THE FINAL STAGE OF HUMAN DEVELOPMENT?

Aging (Fig. 42-15) is the gradual accumulation of damage to essential biological molecules, particularly DNA, resulting in defects in cell functioning, declining health, and ultimately death. This damage results from natural errors in DNA replication, radiation from the sun or from the rocks beneath our feet, and chemicals in food, cigarettes, and industrial products. Young bodies can often repair damage or compensate for it. As an animal ages, however, its repair abilities diminish; eventually, the body's tolerance for damage is exceeded.

As an animal ages, muscle and bone mass are lost, skin elasticity decreases, reaction time slows, and senses such as vision and hearing become less acute. A less-robust immune response renders the aging individual more vulnerable to disease. Eventually, no longer able to fight off natural assaults, the animal dies.

For thousands of years, people have attempted to delay aging and extend the human life span. Modern medical care can prevent or cure many diseases, and can fix or replace some damaged organs through procedures such as bypass surgery and transplants. Some dietary changes, particularly not eating very much food (euphemistically called "caloric restriction"), can prolong life, at least in animal experiments. However, what seems to be the maximum possible human life span, about 130 years, has not changed.

Several evolutionary hypotheses suggest that aging is unavoidable. For example, natural selection favors organisms that leave the largest number of healthy, successful offspring. Even a hypothetically immortal animal that wouldn't die from "inside," so to speak, would eventually succumb to predation, accident, or disease. Therefore, perhaps natural selection favors devoting more of the body's resources to reproduction than to the continuous bodily repair required for immortality. The fact that humans can live so long after they stop reproducing is probably evidence of the selective advantage conferred by the care and teaching given to the young by their elders.

CHECK YOUR LEARNING

Can you define *aging* and describe the mechanisms thought to cause it?

CASE STUDY revisited

Rerunning the Program of Development

If so many other animals can regenerate important body parts, why can't we? Well, humans *can* regenerate, a little. Young children, up to the age of 10 or so, can often regenerate a fingertip, nail and all, that has been accidentally cut off. Similar results have been reported in a few adults with severed fingertips. But regeneration never occurs, even in children, if a finger is cut off far enough that no knuckles are left. No human, of course, has ever regenerated a lost arm or leg.

However, things aren't as hopeless as this might indicate. Researchers are trying to uncover the genetic programs that run during the development of specific body parts, and find out how to turn on those programs in adults. A tantalizing hint comes from mice. A mutant strain of mice called MRL can regrow damaged ears, including cartilage, skin, and hair. MRL mice can also partially regenerate amputated toes. Mice are a lot more similar to people than sea stars, salamanders, or lizards are, so these findings offer a real ray of hope that someday, we may discover how to regenerate damaged or lost human body parts.

Consider This

"Knocking out" a single gene, called p21, allows otherwise normal mice to regenerate their ears. The p21 protein is involved in the control of cell division, suppressing the division of cells with damaged DNA. If someone devised a way of selectively inactivating p21 in humans, what do you think the effects might be, both for regeneration and for health?

CHAPTER REVIEW

Summary of Key Concepts

42.1 What Are the Principles of Animal Development?

Development is the process by which an organism grows and increases in organization and complexity, using three main mechanisms: (1) Individual cells multiply; (2) daughter cells differentiate; and (3) groups of cells move about and become organized into multicellular structures.

42.2 How Do Indirect and Direct Development Differ?

Animals undergo either indirect or direct development. In indirect development, eggs (usually with relatively little yolk) hatch into larvae, which undergo metamorphosis to become adults with notably different body forms. In direct development, the newborn animal is sexually immature but otherwise resembles a small adult. Animals with direct development usually either produce large, yolk-filled eggs or nourish the developing embryo within the mother's body.

42.3 How Does Animal Development Proceed?

Animal development occurs in several stages. *Cleavage and blastula formation:* A fertilized egg undergoes cell divisions with little intervening growth. Cleavage results in the formation of the morula, a solid ball of cells. A cavity then opens up within the morula, forming a hollow ball of cells called a blastula. *Gastrulation:* A dimple forms in the blastula, and cells migrate from the surface into the interior of the ball, eventually forming a three-layered gastrula. These three cell layers—ectoderm, mesoderm, and endoderm—give rise to all the adult tissues. *Organogenesis:* The cell layers of the gastrula form organs characteristic of the animal species (see Table 42-1).

In mammals and reptiles (including birds), extraembryonic membranes (the chorion, amnion, allantois, and yolk sac) encase the embryo in a fluid-filled space and regulate the exchange of nutrients and wastes between the embryo and its environment.

42.4 How Is Development Controlled?

All the cells of an animal body contain a full set of genetic information, yet different cells are specialized for particular functions. Cells differentiate by stimulating and repressing the transcription of specific genes, which is controlled by (1) differences in gene-regulating substances inherited from the mother in her egg and/or (2) chemical communication between the cells of the embryo, a process called induction. A particularly important group of genes, called homeobox genes, regulates the differentiation of body segments and their associated structures, such as wings or legs.

42.5 How Do Humans Develop?

A human egg is fertilized in the uterine tube. The resulting zygote develops into a blastocyst and implants in the endometrium. The outer wall of the blastocyst will become the chorion and form the embryonic contribution to the placenta; the inner cell mass develops into the embryo and the other three ex-traembryonic membranes. During the third and fourth weeks of gestation, the digestive tract begins to form, the heart begins to beat, and the rudiments of a nervous system appear. By the end of the second month, the major organs have begun to form, and the embryo—now called a fetus—appears human. In the next 7 months before birth, the fetus continues to grow; the lungs, stomach, intestine, kidneys, and nervous system enlarge, develop, and become functional (see Figure 42-11).

During pregnancy, mammary glands in the mother's breasts enlarge under the influence of estrogen, progesterone, and other hormones, such as prolactin. After about 9 months, uterine contractions are triggered by a complex interplay of uterine stretch and the release of prostaglandin and oxytocin. As a result, the uterus expels the baby and then the placenta. After its birth, the infant begins suckling the breast and activates the release of prolactin, which stimulates milk production, and oxytocin, which in turn triggers milk secretion.

42.6 Is Aging the Final Stage of Human Development?

Aging is the gradual accumulation of cellular damage (particularly to DNA) over time that leads to loss of functionality and, eventually, to death. Some biologists hypothesize that aging results from natural selection favoring a sexually mature animal putting more resources into reproduction than into bodily repair.

Key Terms

adult stem cell (ASC) *822*	fetus *823*
aging *828*	gastrula *814*
allantois *815*	gastrulation *814*
amnion *815*	homeobox gene *817*
amniotic egg *815*	implantation *821*
blastocyst *820*	indirect development *812*
blastopore *814*	induced pluripotent stem cell
blastula *814*	(iPSC) *822*
chorion *815*	induction *817*
chorionic villus (plural,	inner cell mass *820*
chorionic villi) *825*	labor *825*
cleavage *814*	lactation *827*
colostrum *828*	larva (plural, larvae) *812*
development *812*	mammary gland *827*
differentiate *812*	mesoderm *814*
direct development *812*	metamorphosis *812*
ectoderm *814*	morula *814*
embryonic disk *821*	organogenesis *814*
embryonic stem cell	placenta *825*
(ESC) *822*	stem cell *822*
endoderm *814*	yolk *812*
extraembryonic	yolk sac *815*
membrane *815*	zygote *814*
fetal alcohol syndrome	
(FAS) *826*	

Learning Outcomes

In this chapter, you have learned to . . .

LO1 Define *development* and describe how differentiation of cells is essential to development.

LO2 Compare and contrast direct and indirect development and provide examples of animals that undergo each type of development.

LO3 Describe early development in amphibians, including cleavage, the formation of the blastula, gastrulation, and organogenesis.

LO4 Name the four extraembryonic membranes and describe their functions in reptiles and mammals.

LO5 Explain the roles of transcription factors, gene-regulating substances in the egg cytoplasm, induction between cells, and homeobox genes in development.

LO6 Describe human embryonic development from fertilization to birth.

LO7 Describe the structure and function of the placenta.

LO8 Name the hormones involved in milk production and secretion and describe their roles.

LO9 Define *aging* from a biological perspective and describe the mechanisms thought to cause aging.

Thinking Through the Concepts

Fill-in-the-Blank

1. A fertilized egg, also called a _____, is a very large cell. One of the first events of animal development is the division of this cell, a process called _____, to produce a solid ball of much smaller cells, the _____. Soon, a cavity opens in this ball of cells, producing a hollow ball called the _____. Movements of cells in this hollow ball during gastrulation produce the three embryonic tissue layers, the _____, _____, and _____. **LO3**

2. The amniotic egg contains four extraembryonic membranes. In both reptiles and mammals, the _____ encloses the embryo in a watery environment. In mammals, the _____ forms the embryo's part of the placenta. The allantois contributes the blood vessels of the _____, which connects the embryo to the placenta. Finally, the _____ contains stored food in reptile eggs, but in mammals, this is "empty" and principally contributes to the lining of the digestive tract. **LO4**

3. Undifferentiated cells that can multiply and produce daughter cells of many different types are called _____. Embryonic cells of this type can differentiate into any cell type of the body; they are derived from the _____. **LO1 LO6**

4. Genes that control the development of entire body segments and their accompanying limbs (if any) are called _____. They usually code for proteins called _____ that regulate the transcription of many other genes. **LO5**

5. Milk production in the mammary glands is stimulated by a hormone from the pituitary gland called _____. Another hormone, _____, stimulates contraction of both the uterine muscles during childbirth and the muscles around the mammary glands that cause milk to be ejected from the glands. **LO8**

Review Questions

1. Define *development*. Describe the three major processes that drive development in animals. **LO1**

2. Distinguish between indirect and direct development and give examples of each. **LO2**

3. Describe the structure and function of the four extraembryonic membranes found in reptiles, including birds. Are these four present in placental mammals? In what ways are their roles similar in reptiles and mammals? How do they differ? **LO4 LO9**

4. What is gastrulation? Describe gastrulation in amphibians. **LO3**

5. Name two structures derived from each of the three embryonic tissue layers: endoderm, ectoderm, and mesoderm. **LO3**

6. Describe induction. **LO5**

7. In humans, where does fertilization occur, and what stages of development occur before the fertilized egg reaches the uterus? **LO6**

8. Describe how the human blastocyst gives rise to the embryo and its extraembryonic membranes. **LO6**

9. Explain how the structure of the placenta prevents mixing of fetal and maternal blood while allowing for the exchange of substances between the mother and the fetus. **LO7**

10. How do changes in the breast prepare a mother to nurse her newborn? How do hormones influence these changes and stimulate milk production? **LO8**

11. Define *aging* and describe the cellular mechanisms that are thought to contribute to aging. **LO10**

Applying the Concepts

1. A researcher obtains two frog embryos at the gastrula stage of development. She carefully removes a cluster of cells from a location that she knows would normally become neural tube tissue and transplants it into the second gastrula in a location that would normally become skin. Does the second gastrula develop two neural tubes? Explain your answer.

2. In almost all animals that produce eggs with very little yolk, the blastula stage is roughly spherical. Mammalian eggs, however, contain virtually no yolk but have flat embryonic disks, which undergo gastrulation much like spherical blastulas do. Propose an explanation for this difference in shape.

3. **BioEthics** Some people feel that the production of induced pluripotent stem cells means that research on embryonic stem cells is unnecessary. Take a position on this issue and defend it.

Answers to Figure Caption questions and Fill-in-the-Blank questions can be found in the Answers section at the back of the book.

MB *Go to MasteringBiology for practice quizzes, activities, eText, videos, current events, and more.*

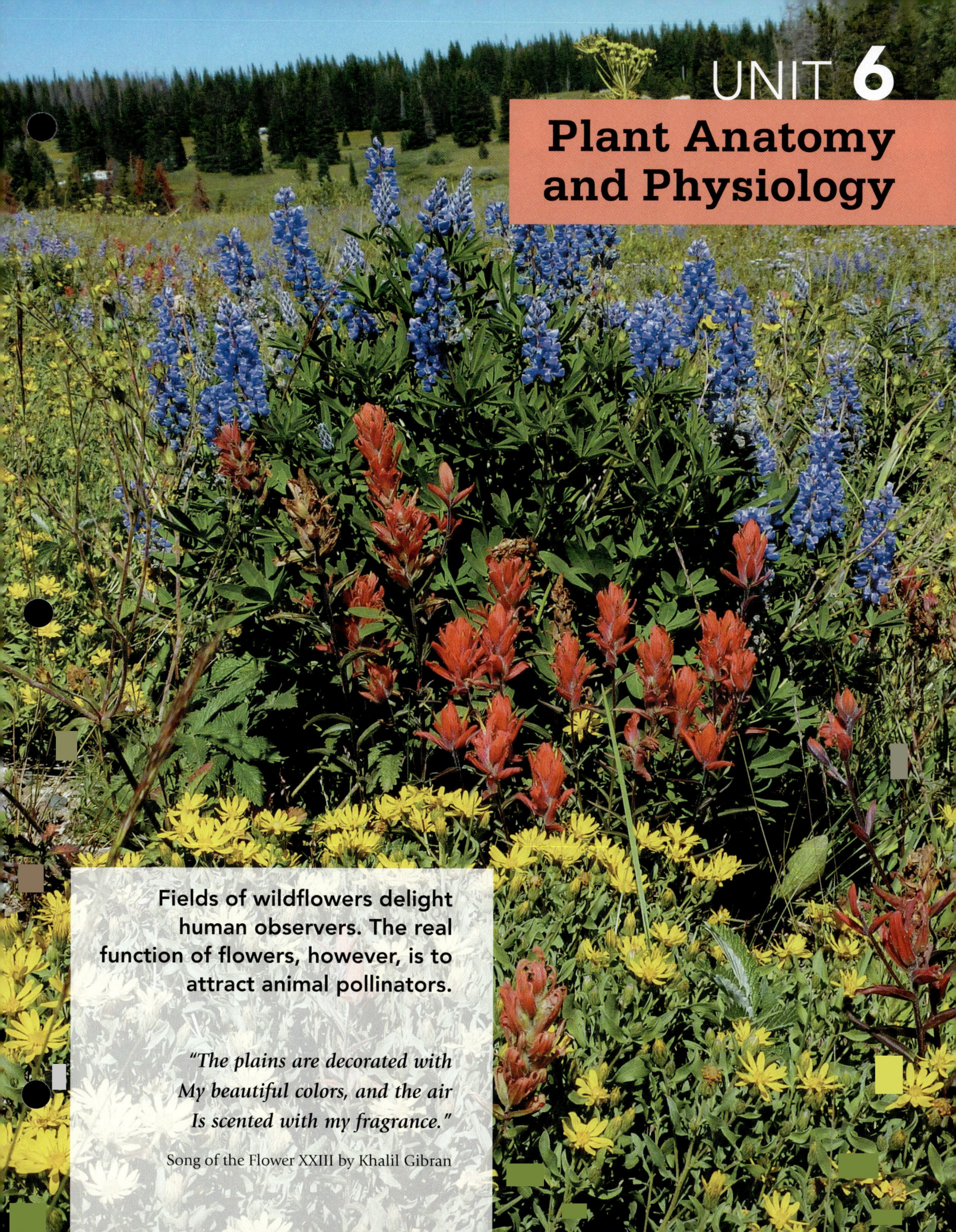

Plant Anatomy and Physiology

Fields of wildflowers delight human observers. The real function of flowers, however, is to attract animal pollinators.

*"The plains are decorated with
My beautiful colors, and the air
Is scented with my fragrance."*

Song of the Flower XXIII by Khalil Gibran

CHAPTER 43 Plant Anatomy and Nutrient Transport

The flaming red colors of maple leaves in autumn are a delight for tourists, but how do the colors benefit the trees?

CASE STUDY

Autumn in Vermont

EACH AUTUMN, more than 3.5 million "leaf peepers" tour Vermont to admire the dazzling red, yellow, and orange trees. New England, upstate New York, and southeastern Canada all boast stunning displays of fall color. Small towns nestled in the valleys of the northern Appalachian Mountains reap an abundant harvest of tourist dollars from the fall foliage as well. But what's in it for the trees? The colors didn't evolve to entertain visitors. The changing colors presumably help the trees in some way—but how?

During most of the growing season, of course, leaves are green, colored by the chlorophyll molecules that are essential for photosynthesis. Other pigments are responsible for fall colors in the leaves of deciduous trees and shrubs. Yellow and orange are the colors of carotenoids, which are responsible for the brilliant hues of daffodils, carrots, oranges, and bananas. Green leaves also contain carotenoids, which assist photosynthesis by absorbing some colors of light that chlorophyll cannot. And if the sunlight is too bright in midsummer, carotenoids siphon off some of the energy, protecting the chlorophyll molecules from heat damage. The red colors come from molecules called anthocyanins. Very few leaves contain anthocyanins during the summer; these pigments are synthesized only in the fall.

Autumn colors prompt a host of questions about plants and their leaves. Why are the leaves of deciduous trees large and flat, and why are they shed in the autumn? Why do they turn color before they fall? Why are carotenoids visible only in the autumn? Why would natural selection favor plants that spend energy synthesizing anthocyanin in leaves that are about to drop off? As Nobel Prize–winning biologist Barbara McClintock says, "Plants are extraordinary." The development of fall colors is just one example of the remarkable abilities of plants that we will explore in this chapter.

AT A GLANCE

LEARNING GOALS

LG1 List the major functions of root and shoot systems in flowering plants.

LG2 Explain the difference between meristem cells and differentiated cells, and describe how each contributes to plant growth.

LG3 Name the three tissue systems in flowering plants, and describe the principal cell types in each system.

LG4 Describe the structure and function of xylem and phloem.

LG5 Describe the structure of a typical dicot leaf, and explain the function of each part.

LG6 Describe the surface structures of a stem.

LG7 Compare and contrast primary and secondary growth in stems, and discuss the principal internal structures formed during each type of growth.

LG8 Describe the structure of a root formed by primary growth.

LG9 Explain how water and minerals are absorbed by roots.

LG10 Explain how fungi and bacteria may help plants to obtain scarce nutrients.

LG11 Explain how water and minerals move in xylem.

LG12 Describe how guard cells control the opening and closing of stomata.

LG13 Explain how water and sugars move in phloem.

43.1 HOW ARE PLANT BODIES ORGANIZED?

Most people admire fields of wildflowers and groves of giant sequoias, but seldom stop to think about the adaptations that allow these plants to thrive. Plants cannot move to seek food or water, escape predators, avoid winter, or find a mate. Yet dwarf willows survive in the harsh tundra of Alaska, mangroves perch in salt water along coastlines throughout the Tropics, and cacti endure the searing heat of the Mojave Desert. Evolution has produced a wide variety of distinctly different types of plants (see Chapter 21). In this chapter, we will focus principally on the flowering plants, or angiosperms, that are the most widespread and diverse group of plants.

The bodies of flowering plants consist of two major parts: the root system and the shoot system (**Fig. 43-1**). The **root system** consists of all the roots of a plant. **Roots** are branched portions of a plant body, usually embedded in the soil, that typically carry out six major functions: (1) Anchor the plant in the ground; (2) absorb water and minerals from the soil; (3) transport water, minerals, sugars, and hormones to and from the shoot; (4) store surplus sugars and starches; (5) produce hormones; and (6) interact with soil fungi and bacteria that help the plant to acquire nutrients.

The **shoot system** is usually located aboveground. The shoot system performs five major functions: (1) Capture sunlight energy and synthesize sugars during photosynthesis; (2) transport materials to and from various parts of the plant; (3) store surplus sugars and starches; (4) reproduce; and (5) produce hormones. In flowering plants, the shoot system consists of one or more **stems,** the principal support structures of a shoot, which bear buds, leaves, and (in season) flowers

and fruits (see Fig. 43-1). A **bud** is an embryonic shoot. Different types and locations of buds may produce branches, flowers, or additional growth at the top of an existing stem. Leaves are the principal sites of photosynthesis in most plants. Flowers are the plant's reproductive organs. Flowers later produce seeds enclosed within fruits, which protect the seeds during their development and aid in seed dispersal (see Chapter 44).

Earth supports an amazing diversity of flowering plants, from tender marigolds to prickly rose bushes, from short grasses to towering oaks. Overall, however, all flowering plants can be placed into one of two large groups, called monocots and dicots (**Fig. 43-2**). **Monocots** include lilies, daffodils, tulips, palm trees, and a wide variety of grasses (not only the familiar lawn grasses but also wheat, rice, corn, oats, and bamboo). **Dicots** include virtually all "broadleafed" plants, such as deciduous trees and bushes, most vegetables, and many flowers in fields and gardens.

There are several differences between monocots and dicots in both root and shoot systems. For example, monocots typically have long, thin leaves and bushy fibrous roots (think of grasses), while dicots have broader leaves and thick tap roots (as in dandelions). However, the characteristic that gives the groups their names is the number of *cotyledons*. Monocots have a single cotyledon ("mono" means "one"), and dicots have two ("di" means "two"). A cotyledon is the part of a plant embryo that absorbs and often stores food reserves in the seed and then transfers the food to the rest of the embryo when the seed sprouts. When you eat a peanut, each half of the "nut" consists almost entirely of a cotyledon, which has stored starches and fats that power the growth of a peanut seedling. Cotyledons are often called "seed leaves," because they sometimes become the first green, photosynthetic leaves

▶ **FIGURE 43-1 The structures and functions of a typical flowering plant** It will be helpful to refer back to this figure as you read the rest of the chapter.

functions

structures

growth and development of plant structures — leaf primordia
— apical meristem

— terminal bud

— lateral bud
— node
— flower

reproduction

body support; transport of water and nutrients — stem

reproduction — fruit

energy acquisition by photosynthesis; gas exchange — blade
— petiole

leaf

— branch

shoot system

— root

acquisition of water and minerals

— branch roots

root system

— root hairs
— root cap

during seed sprouting. (We will examine the structure and function of cotyledons in more detail in Chapter 44.)

CHECK YOUR LEARNING

Can you list the principal functions of shoots and roots?
Can you distinguish between monocots and dicots?

43.2 HOW DO PLANTS GROW?

Animals and plants develop in dramatically different ways. In animals as diverse as earthworms, octopuses, snails, mammals, and birds, newborns closely resemble adults, except,

of course, in size. All parts of a newborn's body become larger as the newborn grows, until it reaches its adult size. In contrast, most flowering plants grow throughout their lives, never reaching a stable adult body size or, in many cases, a stable body shape. Moreover, most plants grow longer or taller only at the tips of their branches and roots.

Plants are composed of two fundamentally different types of cells: meristem cells and differentiated cells. **Meristem cells,** like the stem cells of animals, are not specialized and are capable of mitotic cell division (see Chapter 9). Some of their daughter cells lose the ability to divide and transform into **differentiated cells** with specialized structures and functions.

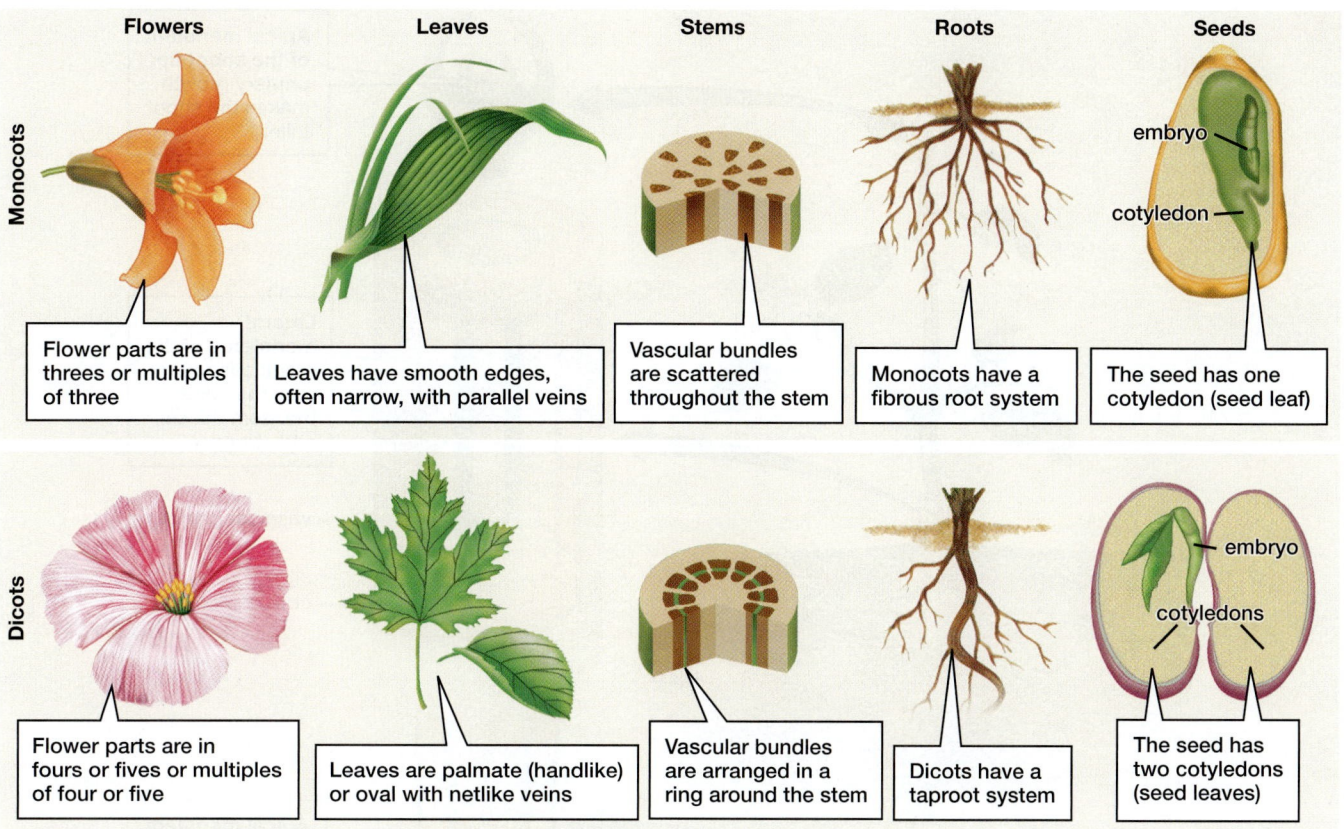

▲ FIGURE 43-2 Characteristics of monocots and dicots

Plants grow as a result of the division of meristem cells found in two general locations in the plant body (**Fig. 43-3**). **Apical meristems** ("tip meristems") are located at the tips of roots, shoots, and their branches. Growth produced by apical meristem cells is called **primary growth:** an increase in the height or length of a shoot or root and the development of specialized parts of the plant, such as leaves and buds. In both roots and shoots, primary growth also produces a complex arrangement of internal structures that we will explore later in the chapter. Primary growth from apical meristems explains why a swing tied to a tree branch doesn't get any higher off the ground over the years—a tree grows taller only at the tips of its branches.

Lateral meristems ("side meristems") are concentric cylinders of meristem cells. There are two types of lateral meristems, *cork cambium* and *vascular cambium* (see Fig. 43-3). Cell division in lateral meristems and the differentiation of the resulting daughter cells produce **secondary growth,** typically an increase in the diameter and strength of roots and shoots. Secondary growth occurs in woody plants, including deciduous trees, many shrubs, and most conifers such as pines and spruces. Some woody plants become very tall and thick and may live for hundreds, or even thousands, of years. Secondary growth is why the branch holding up a swing, and the trunk of the tree bearing the branch, both became thicker and stronger over the years.

Many plants, such as strawberries, grasses, and annual garden flowers, do not undergo secondary growth. As you might predict, most plants that lack secondary growth are soft bodied, with short, flexible stems.

CHECK YOUR LEARNING

Can you explain the difference between meristem cells and differentiated cells? Can you describe primary growth and secondary growth?

? HAVE **YOU** EVER WONDERED . . .

How Trees Can Live So Long?

Some trees, such as giant sequoias and bristlecone pines, can live more than 1,000 years; the oldest known bristlecone pine, called Methuselah, is more than 4,800 years old! Most truly ancient trees have at least three of the following four characteristics. (1) Trees don't have localized, absolutely essential parts, like a brain or a heart. If a branch or part of a root dies, the rest of the tree lives on. (2) Many trees don't really grow old. Because their meristems never stop producing new cells, they always have some young roots and branches. (3) Many trees produce resins or other substances that prevent attack from bacteria or fungi. (4) Ancient trees often live in cold or dry environments, which limit the growth of bacteria and fungi and reduce undergrowth that might catch fire and burn down the trees. Barring accidents such as a lightning strike or major climate change, a bristlecone pine, which has all of these advantages, could potentially live for tens of thousands of years.

Apical meristem of the shoot tip: primary growth makes the shoot taller

Lateral meristems of the stem: secondary growth makes the stem thicker and stronger

vascular cambium

cork cambium

Apical meristem of the root tip: primary growth makes the root longer

▲ **FIGURE 43-3 Meristems** A dicot contains apical meristems at the tips of shoots, shoot branches, roots, and root branches. Lateral meristems occur as cylinders of meristem cells in older stems and roots.

43.3 WHAT ARE THE TISSUES AND CELL TYPES OF PLANTS?

As meristem cells differentiate, they produce a wide variety of cell types. When one or more specialized types of cells work together to perform a specific function, such as conducting water and minerals, they form a **tissue.** Functional groups of more than one tissue are called **tissue systems.** The plant body is composed of three tissue systems (**Table 43-1**). The **dermal tissue system** covers the outer surface of the plant body. The **ground tissue system** makes up most of the body of young plants. Its functions include photosynthesis, storage, and support. The **vascular tissue system** transports fluids throughout the plant body.

The Dermal Tissue System Covers the Plant Body

The **epidermis** is the dermal tissue system formed during primary growth. The epidermis is the outermost cell layer covering the leaves, stems, and roots of young plants and new growth in older plants. Epidermis also covers flowers, seeds, and fruit.

The epidermis of the aboveground parts of a plant is generally composed of tightly packed, thin-walled cells, covered with a waterproof, waxy **cuticle** secreted by the epidermal cells. Cuticle reduces the evaporation of water from the plant and helps protect it from the invasion of disease microorganisms. As we will see later, adjustable pores regulate the movement of water vapor, O_2, and CO_2 across the epidermis. Root epidermis is not covered with cuticle, which would prevent the roots from absorbing water and minerals. In woody plants, **periderm** is the dermal tissue that replaces epidermis on roots and stems as they undergo secondary growth (see Section 43.5).

The Ground Tissue System Makes Up Most of the Young Plant Body

The ground tissue system consists of all of the tissues of the plant body except dermal and vascular tissues. The three types of ground tissues are parenchyma, collenchyma, and sclerenchyma.

Parenchyma tissue, which makes up most of the body of a young plant, consists of cells with thin cell walls (**Fig. 43-4a**). Parenchyma cells carry out most of the plant's

TABLE 43-1   Tissue Systems of Plants

Type	Tissues Within the Tissue System	Functions	Locations of the Tissue Systems
Dermal tissue system	Epidermis	Protects the plant body	
	Periderm (secondary growth)	Regulates the movement of O_2, CO_2, and water vapor between the air and the plant	
Ground tissue system	Parenchyma	Photosynthesizes; principally in leaves and young stems; secretes hormones	
	Collenchyma	Stores nutrients; principally in stems and roots	
	Sclerenchyma	Supports the plant body, as strengthening fibers in both xylem and phloem	
Vascular tissue system	Xylem	Transports water and dissolved minerals from root to shoot	
	Phloem	Transports sugars and other organic molecules, such as amino acids, proteins, and hormones throughout the plant body	

leaf

stem

root

■ dermal tissue
■ ground tissue
■ vascular tissue

metabolic activities, including photosynthesis (most of the cells of a leaf are parenchyma cells), secretion of hormones, and food storage. Potatoes, seeds, fruits, and storage roots such as carrots are packed with parenchyma cells that store various types of sugars and starches. Parenchyma cells, plumped up with water, also help to support the bodies of many non-woody plants, which is why your houseplants wilt and sag if you forget to water them. Some parenchyma cells can divide. Parenchyma cells are also found in periderm and vascular tissues.

Collenchyma tissue is composed of cells that are typically elongated, with thickened, but still flexible, cell walls (**Fig. 43-4b**). Collenchyma cells generally cannot divide.

Collenchyma cells help to support the bodies of young and non-woody plants. Collenchyma also stiffens the leaf stalks, or *petioles*, of all plants. Celery stalks, which are extremely thick petioles, are supported by "strings" composed mostly of collenchyma cells.

Sclerenchyma tissue is composed of cells with thick, hardened walls (**Fig. 43-4c**). Like collenchyma, sclerenchyma cells support and strengthen the plant body; however, unlike collenchyma and parenchyma cells, they die after they differentiate. Their thick cell walls then remain as a source of support. Sclerenchyma cells form nut shells and the outer covering of peach pits. Scattered throughout the parenchyma

(a) **Parenchyma cells in a white potato**   (b) **Collenchyma cells in a celery stalk**   (c) **Sclerenchyma cells in a pear**

▲ **FIGURE 43-4 Ground tissue** (a) Parenchyma cells are living, with thin cell walls. They perform many functions, including photosynthesis, hormone secretion, and storage. (b) Collenchyma cells are living and have thickened, but somewhat flexible, cell walls. They help support the plant body. (c) Sclerenchyma cells have thick, hard walls and die after they differentiate.

cells in the flesh of a pear, sclerenchyma cells give pears their gritty texture. Sclerenchyma cells also form an important component of wood.

The Vascular Tissue System Transports Water and Nutrients

The vascular tissue system of plants conducts water and dissolved substances throughout the plant body. The vascular tissue system consists of two conducting tissues, xylem and phloem, together with sclerenchyma and parenchyma cells.

Xylem Transports Water and Dissolved Minerals from the Roots to the Rest of the Plant

Xylem transports water and dissolved minerals from the roots to all parts of the shoot system. Xylem contains two specialized conducting cell types: tracheids and vessel elements (**Fig. 43-5**). Both tracheids and vessel elements develop thick cell walls and then die as their final stage of differentiation, leaving behind hollow tubes of nonliving cell wall.

   Tracheids are thin, elongated cells, with tapered, overlapping ends. The ends and sides of tracheids contain **pits,** which are porous dimples in the walls that separate adjacent cells. Because the cell wall in a pit is both thin and porous, water and minerals can pass freely from one tracheid to another, or between a tracheid and an adjacent vessel element.

   Vessel elements, which are larger in diameter than tracheids, form pipelines called **vessels.** Vessel elements are stacked end to end. Their adjoining end walls may be perforated with large holes, or the end walls may disintegrate entirely, leaving an open tube.

Phloem Transports Sugars and Other Organic Molecules Throughout the Plant Body

Phloem transports a solution containing a variety of organic molecules, including sugars, amino acids, and hormones, from the structures that synthesize them to other structures that need them. Unlike xylem, in which fluids take a one-way trip from root to shoot, phloem transports fluids up or down

▲ **FIGURE 43-5 Xylem** Xylem contains two types of conducting cells: tracheids and vessel elements. Tracheids are thin with tapered ends; vessel elements are much larger in diameter and usually have blunt ends. Both the ends and sides of adjacent tracheids are connected by pits. Pits in their side walls also interconnect tracheids and vessel elements.

the plant, to or from the leaves and roots, depending on the metabolic state of various parts of the plant at any given time.

   Phloem contains two cell types that cooperate in conducting sugar solutions in a plant: sieve-tube elements and companion cells (**Fig. 43-6**). **Sieve-tube elements** are joined end to end to form pipes called sieve tubes. As sieve-tube elements mature, they lose their nuclei and most other organelles, leaving behind only a thin layer of cytoplasm lining the plasma membrane. The junction between two sieve-tube elements is called a **sieve plate.** Here, membrane-lined

▶ **FIGURE 43-6 Phloem** Phloem includes sieve-tube elements and companion cells. Sieve-tube elements, stacked end to end, form the conducting system of phloem. Where they join at sieve plates, large membrane-lined pores allow fluid to move between them. Each sieve-tube element has a companion cell that nourishes it and regulates its function. In the light micrograph on the right, note that the companion cell has cytoplasm and organelles that are typically found in a living cell, whereas the sieve-tube element is almost empty.

pores connect the insides of the sieve-tube elements, allowing fluid to move from one cell to the next.

How can sieve-tube elements maintain and repair their plasma membranes when they lack nuclei and most other organelles? Life support for sieve-tube elements is provided by adjacent **companion cells,** which are connected to sieve-tube elements by pores called **plasmodesmata** (singular, plasmodesma; see Chapter 5). Companion cells supply sieve-tube elements with proteins and high-energy compounds such as ATP.

CHECK YOUR LEARNING

Can you name the three tissue systems in flowering plants and describe their locations in the plant body? Can you describe the cell types in each tissue system and explain their functions? Are you able to explain the difference between xylem and phloem, both in structure and function?

43.4 WHAT ARE THE STRUCTURES AND FUNCTIONS OF LEAVES?

Leaves are the principal photosynthetic structures of most plants. Photosynthesis uses the energy of sunlight (captured by chlorophyll) to convert water (H_2O) and carbon dioxide (CO_2) into sugar, releasing oxygen (O_2) as a by-product (see Chapter 7). CO_2 enters leaves from the air. Water is absorbed from the soil by a plant's roots and transported to the leaves. You might think that maximum photosynthesis would occur in a porous leaf (which would allow CO_2 to diffuse easily from the air into the leaf) with a large surface area (which would intercept the most sunlight). However, on a hot, dry day, a plant with large, porous leaves might lose more water through evaporation than it could replace from the soil, threatening the entire plant with death by dehydration. The leaves of most flowering plants are an elegant compromise among the conflicting demands of acquiring sunlight and CO_2 while conserving water (**Fig. 43-7**): Most leaves have a large, mostly waterproof surface, with pores that can open and close to admit CO_2 or restrict water evaporation, as needed.

A typical angiosperm leaf consists of a large, flat portion, the **blade,** connected to the stem by a stalk called the **petiole.** Inside the blade and petiole, xylem and phloem connect the leaf to the vascular system of the rest of the plant.

The Epidermis Regulates the Movement of Gases into and out of a Leaf

The leaf epidermis consists of a layer of nonphotosynthetic, transparent cells that secrete a waxy cuticle on their outer surfaces. The cuticle is nearly waterproof and reduces the evaporation of water out of the leaf. The epidermis and cuticle are pierced by openings called **stomata** (singular, stoma) that regulate the diffusion of CO_2, O_2, and water vapor into and out of the leaf. A stoma consists of two sausage-shaped **guard cells,** which enclose, and adjust the size of, the pore between them. Unlike other epidermal cells, guard cells contain chloroplasts and carry out photosynthesis. Later, we will see how guard cells change shape and open or close the central pore, depending on environmental conditions.

◀FIGURE 43-7 **A typical dicot leaf** The cells of the epidermis lack chloroplasts and are transparent, allowing sunlight to penetrate to the chloroplast-containing mesophyll cells beneath. A waterproof cuticle covers the outside surfaces of the epidermal cells. The stomata that pierce the epidermis and the loose, open arrangement of the mesophyll cells ensure that CO_2 can diffuse into the leaf from the air and reach all of the photosynthetic cells.

Photosynthesis Occurs in Mesophyll Cells

The transparent epidermal cells allow sunlight to reach the **mesophyll** ("middle of the leaf"), which consists of loosely packed cells containing chloroplasts. Many leaves contain two types of mesophyll cells—an upper layer of columnar *palisade cells* and a lower layer of irregularly shaped *spongy cells* (see Fig. 43-7). Mesophyll cells carry out most of the photosynthesis that occurs in a leaf. Air spaces between mesophyll cells allow CO_2 from the atmosphere to diffuse to each cell and O_2 produced during photosynthesis to diffuse away.

Veins Transport Water and Nutrients Throughout the Leaf

Vascular bundles (in leaves, these are also called **veins**) containing xylem and phloem conduct materials between the leaf and the rest of the plant body. Xylem delivers water and minerals to the mesophyll cells of the leaf, and phloem carries away the sugar they produce during photosynthesis.

Many Plants Produce Specialized Leaves

Temperature and the availability of water and light have exerted strong selection pressures on leaves. Plants growing

CASE STUDY continued

Autumn in Vermont

All leaves contain carotenoids. Summer leaves are green rather than yellow or orange because they contain about 50 times more chlorophyll than carotenoids. During the fall, chlorophyll breaks down before carotenoids do. These carotenoids, revealed when the more abundant chlorophyll disappears, color many autumn leaves yellow or orange.

in the dim light that reaches the floor of a tropical rain forest often have very large leaves, an adaptation demanded by the low light level and permitted by the abundant water of the rain forest (**Fig. 43-8a**). The leaves of desert-dwelling cacti, in contrast, have been reduced to spines that provide almost no surface area for evaporation (**Fig. 43-8b**). Spines also protect the juicy cactus stem from being eaten by herbivores. The plump leaves of other drought-resistant succulents store water in their cells and are covered with a thick cuticle that greatly reduces water evaporation (**Fig. 43-8c**).

▶ FIGURE 43-8 Specialized leaves **(a)** Some plants that live on the rain-forest floor (such as this elephant ear) have enormous leaves that capture the limited light filtering through the taller trees. **(b)** Spines of desert cacti are nonphotosynthetic leaves whose minuscule surface area reduces evaporation. Sharp spines also protect the plant from browsing animals. **(c)** Succulent desert plants have fleshy leaves that store water from the infrequent rains. **(d)** An onion consists of a short central stem surrounded by thickened leaves that store water and food.

QUESTION Onion plants are biennials, which live for only 2 years. As a gardener, should you harvest onions at the end of their second year so they would be as large as possible?

(a) Elephant ear leaves

(b) Cactus spines

(c) Succulent leaves

leaf bases

stem

(d) Leaves of an onion

Some plants have evolved leaves with surprising structures and functions. Onions and daffodil bulbs consist of extremely short underground stems enveloped by thick, overlapping leaves, which store water, sugars, and other nutrients during the summer to support new growth the following spring (Fig. 43-8d). Carnivorous plants, such as Venus flytraps and sundews, have leaves that are modified into snares that can trap and digest unwary insects (see the opening photo for Chapter 45). Pea plants produce thin, modified leaves called tendrils that coil around other plants, or the trellis in your garden, helping the plant to reach the sunlight.

CHECK YOUR LEARNING

Can you diagram an angiosperm leaf and describe the structure and function of each part? Can you describe some specialized leaf adaptations and explain the function of each?

43.5 WHAT ARE THE STRUCTURES AND FUNCTIONS OF STEMS?

The stems of a plant support its leaves, lifting them up to the sunlight. Stems also transport water and dissolved minerals from the roots to the leaves, and transport sugars produced in the leaves to the roots and other parts of the shoot, such as buds, flowers, and fruits. In most dicots, stems undergo primary growth during their first year and at the tips of their branches throughout life. In perennial dicots, older stems and branches undergo secondary growth.

Primary Growth Produces the Structures of a Young Stem

At the tip of a newly developing shoot sits a **terminal bud** consisting of apical meristem cells surrounded by a wrapping of developing leaves, or **leaf primordia** (singular, primordium) (Fig. 43-9). During primary growth, the meristem cells divide. Most of their daughter cells differentiate into the specialized cell types of leaves, buds, and the structures of the stem. The remaining daughter cells stay meristematic, providing for future growth of the shoot.

Below the terminal bud, the surface of a stem consists of two regions: nodes and internodes. A **node** is the site of attachment of the petiole of a leaf to the stem. When nodes first form from the apical meristem, they are close together, but the stem between the nodes elongates, forming naked **internodes** (literally, "between nodes"). As most dicot shoots grow, small clusters of meristem cells, the **lateral buds,** are left behind at the nodes, usually in the angle between the leaf petiole and the surface of the stem. Under appropriate hormonal conditions, lateral buds sprout and grow into branches (see Chapter 45). As a branch grows, it duplicates the development of the stem, including new leaves and both terminal and lateral buds. During the reproductive season, usually spring or summer, meristem cells form **flower buds,** generally in the same locations in which a terminal or lateral bud would otherwise develop.

The apical meristem of a shoot also produces the internal structures of the stem, typically grouped into four tissues: epidermis, cortex, pith, and vascular tissues. Monocots and dicots differ somewhat in their arrangement of vascular tissues; we will discuss only dicot stems.

The Epidermis of the Stem Reduces Water Loss While Allowing Carbon Dioxide to Enter

The epidermis of a young stem, like that of leaves, secretes a waxy cuticle that reduces water loss. As in leaves, the stem epidermis is perforated by stomata that regulate the movement of carbon dioxide, oxygen, and water vapor into and out of the stem.

The Cortex and Pith Support the Stem, Store Food, and May Photosynthesize

Cortex and pith consist of parenchyma cells (ground tissue) that fill most of the young stem. In dicots, the **cortex** lies between the epidermis and the vascular tissues, and the **pith** fills the central part of the stem, surrounded by the vascular tissues. Cells of the cortex and pith perform three major functions:

- **Support** In very young stems, water filling cortex and pith cells pushes the cytoplasm up against the cell wall, causing turgor pressure (see Chapter 5). Turgor pressure stiffens the cells, much as air inflates a tire, and helps to hold the stem erect.
- **Storage** Parenchyma cells in both cortex and pith convert sugar into starch and store the starch as a food reserve.
- **Photosynthesis** In many young dicot stems, the outer cells of the cortex contain chloroplasts and carry out photosynthesis. In plants such as cacti, in which the leaves are reduced to spines, the cortex of the stem may be the only green, photosynthetic part of the plant.

Vascular Tissues in the Stem Transport Water, Dissolved Nutrients, and Hormones

Most young dicot stems contain bundles of vascular tissue. Each bundle can be thought of as a rod of vascular tissue, often triangular in cross-section, extending the full height of the stem. A dozen or more bundles are arranged in a ring within the stem (Fig. 43-9b, top). The pointed inner edge of each bundle consists of *primary xylem*, and the rounded outside edge consists of *primary phloem* (Fig. 43-9c, top). (These structures are called primary xylem and primary phloem because they are produced from an apical meristem during primary growth.) Xylem carries water and dissolved minerals from the roots up through the stem. Phloem carries sugar from photosynthetic cells in the leaves and stem cortex to all the other parts of the plant.

Secondary Growth Produces Thicker, Stronger Stems

The stems of most conifers and perennial dicots undergo secondary growth, resulting from cell division in the lateral meristems of the vascular cambium and cork cambium (see Fig. 43-9). Such stems may survive for decades, centuries,

(a) Primary and secondary growth in a dicot stem   (b) Stem cross-sections   (c) Vascular tissues

▲ FIGURE 43-9 **Primary and secondary growth in a dicot shoot** A dicot shoot after 2 years of growth, showing primary growth in the upper part of the shoot and secondary growth farther down. **(a)** A vertical slice of the stem, showing its internal structures. Cross-sections of **(b)** the entire stem and **(c)** vascular tissue after primary growth (top) and secondary growth (bottom). During secondary growth, the vascular cambium forms a complete ring around the stem, so the resulting secondary xylem and phloem also form complete rings, rather than the separate vascular bundles of primary xylem and phloem that form during primary growth.

and, in rare cases, even millennia, becoming thicker and stronger each year.

Vascular Cambium Produces Secondary Xylem and Secondary Phloem

The **vascular cambium** forms a cylinder of meristem cells that lies between the primary xylem and primary phloem. Daughter cells of the vascular cambium produced toward the inside of the stem differentiate into secondary xylem; those produced toward the outside of the stem differentiate into secondary phloem (**Fig. 43-9c,** bottom).

The cells of secondary xylem, with their thick cell walls, form the wood in trees and woody shrubs. Young secondary xylem, called **sapwood,** transports water and minerals and is located just inside the vascular cambium (**Fig. 43-10**). Older secondary xylem, the **heartwood,** fills the central portion of older stems. Heartwood no longer carries water and minerals, but continues to provide support and strength to the stem.

Phloem cells are much weaker than xylem cells. Sieve-tube elements and companion cells are split apart as the stem enlarges, and are also crushed between the hard xylem on the inside of the stem and the tough outer covering, called cork (see below). Only a thin strip of recently formed phloem remains alive and functioning.

Cell division in the vascular cambium ceases during the cold of winter. In spring and summer, the cambium cells divide, forming new secondary xylem and phloem. In spring and early summer, when water is plentiful, xylem cells are large, with thin cell walls; in late summer, when water becomes scarce, xylem cells are smaller in diameter, with thicker cell walls. As a result, tree trunks in cross-section often show a pattern of **annual rings** of alternating pale regions (large cells with thin walls) and dark regions (small cells with thick walls), as shown in Figure 43-10. The tree's age can be determined by counting the annual rings. The widths of the rings also provide information about past climate, because wet years produce larger cells and wider rings.

Cork Cambium Replaces the Epidermis with Cork

Epidermal cells are mature, differentiated cells that usually no longer divide. As new secondary xylem and phloem are added each year, increasing the diameter of a stem, the epidermis splits off. Some cells in the epidermis, cortex, or phloem become capable of cell division and form a layer of lateral meristem cells, the **cork cambium** (see Figs. 43-9 and 43-10). These cells divide, forming daughter cells called **cork cells,** with tough, waterproof cell walls that protect the trunk both from drying out and from physical damage. Cork cambium and its daughter cells form the periderm.

(a) Cross-section of a tree trunk

- heartwood (xylem)
- sapwood (xylem)
- vascular cambium
- bark { phloem
- cork cambium
- cork }

late xylem

early xylem

(b) An annual ring

◀ **FIGURE 43-10 Annual rings (a)** Most trees in temperate climates form annual rings of secondary xylem. Early wood, formed during spring and early summer, consists of large cells with thin cell walls, and is pale. Late wood, usually formed in late summer, consists of smaller cells with thicker walls, and is dark. **(b)** The junction between spring and summer wood in a single annual ring.

QUESTION Would you expect a tree from a climate that was uniformly warm and wet all year to have growth rings? Explain your answer.

▶ **FIGURE 43-11 Cork (a)** The cork of a giant sequoia in the Sierra Nevada of California may be up to 3 feet thick (almost a meter). This massive, fire-resistant cork layer contributes to the sequoia's great longevity—sequoias in the Sierra Nevada commonly live to be more than 2,000 years old, and some are more than 3,000 years old. Blackened areas are from past fires. **(b)** A layer of cork is stripped from a cork oak. The cork will regrow and can be harvested again in about a decade.

(a) Cork protects this giant sequoia tree

(b) Harvesting cork from a cork oak

Cork cells die as they mature, forming a protective layer that can be more than 2 feet thick in some tree species, such as the giant sequoia (**Fig. 43-11a**). As the trunk expands from year to year, the outermost layers of cork split apart or peel off, accommodating the growth. Corks used to plug wine bottles are made from the outermost layer of cork from cork oaks, carefully peeled off by harvesters so as not to harm the underlying cork cambium (**Fig. 43-11b**).

The common term **bark** includes all of the tissues outside the vascular cambium: phloem, cork cambium, and cork cells (see Fig. 43-10). Removing a strip of bark all the way around a tree, called girdling, kills the tree because it severs the phloem. Without phloem, sugars synthesized in the leaves cannot reach the roots. Deprived of energy, the roots can no longer take up minerals, and the tree dies.

Many Plants Produce Specialized Stems or Branches

Flowering plants have evolved a variety of stem specializations. For example, many stems are adapted for storage. The fat, bulging trunk of the baobab tree contains water-storing parenchyma cells, which allow it to thrive in climates with sporadic rainfall (**Fig. 43-12a**). Cactus stems both photosynthesize and store water. The common white potato is actually an underground stem, packed with parenchyma specialized to store starch. Each "eye" of a potato is a lateral bud, ready to sprout a branch when conditions become favorable. Strawberries send out horizontal stems, called runners, that spread along the soil, sprouting new strawberry plants from buds on the runners (**Fig. 43-12b**). Certain branches of grapes and Boston ivy form grasping tendrils that coil around trees and trellises or adhere to buildings, giving the plant better access to sunlight (**Fig. 43-12c**). Thorns are branch adaptations that protect certain plants, such as hawthorns and honeylocust trees, from browsing by large herbivores (**Fig. 43-12d**).

CHECK YOUR LEARNING

Can you diagram the internal and external structures of a dicot stem after primary growth and after secondary growth? Can you explain the function of each major structure of the stem? Can you describe some specializations of stems and the functions of each?

43.6 WHAT ARE THE STRUCTURES AND FUNCTIONS OF ROOTS?

Most dicots, such as carrots and dandelions, develop a **taproot system,** consisting of a central root with many smaller roots branching out from its sides (**Fig. 43-13a**). Most monocots, such as grasses and daffodils, produce a **fibrous root system** in which many roots of roughly equal size emerge from the base of the stem (**Fig. 43-13b**).

In young roots of both taproot and fibrous root systems, divisions of the apical meristem and differentiation of the resulting daughter cells give rise to four distinct regions (**Fig. 43-14**). Daughter cells produced on the lower side of the apical meristem differentiate into the root cap. Daughter cells produced on the upper side of the apical meristem differentiate into cells of the epidermis, cortex, and vascular cylinder.

The Root Cap Shields the Apical Meristem

The apical meristem is the source of primary growth in a root. The **root cap,** located at the very tip of the root, protects the apical meristem from being scraped off as the root pushes down between the rocky particles of the soil. Root-cap cells have thick cell walls and secrete a slimy lubricant that helps ease the root between soil particles. Nevertheless, root-cap cells wear away and must be continuously replaced by new cells produced by the apical meristem.

The Epidermis of the Root Is Permeable to Water and Minerals

Young roots absorb water and minerals from the soil. The root's outermost covering of cells is the epidermis, which is in contact with the soil and the water trapped among the soil particles. Unlike stem epidermis, which is covered by a waxy cuticle that reduces evaporation, root epidermis lacks a cuticle. Consequently, the walls of root epidermal cells are highly permeable to water and minerals. Further, many epidermal cells grow **root hairs** into the surrounding soil (**Fig. 43-15;** see also Fig. 43-14). By expanding the root's surface area, root hairs increase its ability to absorb water and minerals. Root hairs may add dozens of square yards of surface area to the roots of even small plants.

(a) Baobab tree

(b) Strawberry plants

(c) Grape vine

(d) Honeylocust tree

▲ **FIGURE 43-12 Specialized stems and branches** (a) The enormously expanded water-storing trunk of the baobab tree allows it to thrive in a dry climate. **(b)** Strawberry plants can reproduce using runners, which are horizontal stems. Where a node of a runner touches the soil, its lateral bud may sprout into a complete plant. **(c)** Tendrils are specialized branches that allow grape vines to cling to trees or trellises. **(d)** Honeylocusts protect themselves with branches modified into strong, sharp, often forked thorns.

The Cortex Stores Food and Controls the Absorption of Water and Minerals into the Root

Cortex occupies most of the inside of a young root, between the epidermis and the vascular cylinder (see Fig. 43-14). Most of the cortex consists of large, loosely packed parenchyma cells with porous cell walls. Sugars produced in the shoot by photosynthesis are transported down to cortex parenchyma cells, where they are converted to starch and stored. Roots of perennial plants store starches through the cold winter

(a) A taproot system   **(b) A fibrous root system**

◀ **FIGURE 43-13 Taproots and fibrous roots** (a) Dicots typically have a taproot system, consisting of a long central root with many smaller, secondary roots branching from it. **(b)** Monocots usually have a fibrous root system, with many roots of equal size.

epidermis

root hair

cortex

endodermis
of cortex

pericycle

xylem

phloem

vascular
cylinder

apical
meristem

root
cap

▲ **FIGURE 43-14 Primary growth in roots** Primary growth in roots results from mitotic cell division in the apical meristem near the tip. The root is composed of the root cap, epidermis, cortex, and vascular cylinder.

months, releasing them the following spring to power new growth of the shoot system. The cortex is especially large in roots specialized for carbohydrate storage, such as those of sweet potatoes, beets, carrots, and radishes (**Fig. 43-16**).

The innermost layer of the cortex is very different from the rest of the cortex. This layer, called the **endodermis,** consists of a ring of small, closely packed cells surrounding the vascular cylinder (see Fig. 43-14). The cell wall of each endodermal cell contains a band of waterproof material called the **Casparian strip.** The Casparian strip is found between the cells (top, bottom, and both sides) but not on the inner or outer faces. Water and dissolved minerals can flow

root
hairs

▲ **FIGURE 43-15 Root hairs** Root hairs, shown here on a sprouting radish, greatly increase a root's surface area, enhancing the absorption of water and minerals from the soil.

▲ **FIGURE 43-16 Specialized roots** Dicot taproots modified for nutrient storage include (left to right) beets, carrots, and radishes.

▲ **FIGURE 43-17 Branch roots** Branch roots emerge from the pericycle of a root. The center of this branch root is already differentiating into vascular tissue.

freely around epidermal and cortex cells by moving through their porous cell walls. However, the Casparian strip blocks water and minerals from traveling between endodermal cells (see Section 43.7).

The Vascular Cylinder Contains Conducting Tissues and Forms Branch Roots

The **vascular cylinder** contains the conducting tissues of xylem and phloem. The outermost layer of the root vascular cylinder is the **pericycle,** located just inside the endodermis of the cortex and outside the xylem and phloem (see Fig. 43-14). Pericycle cells work together with endodermal cells to regulate the movement of minerals and water into the xylem of the vascular cylinder.

The pericycle is also the source of branching in roots. Under the influence of hormones, pericycle cells become meristematic and divide, forming the apical meristem of a **branch root** (**Fig. 43-17**). Branch root development is similar to that of primary roots except that the branch must break out through the cortex and epidermis of the primary root. It does so both by crushing the cells that lie in its path and by secreting enzymes that digest them. The vascular tissues of the branch root connect with the vascular tissues of the primary root.

Roots May Undergo Secondary Growth

The roots of woody plants, including conifers and deciduous trees and shrubs, become thicker and stronger through secondary growth. Although there are some differences between secondary growth in stems and roots, the essentials are similar: Vascular cambium produces secondary xylem and phloem in the interior of the root, and cork cambium produces protective layers of cork cells on the outside of the root.

CHECK YOUR LEARNING

Can you diagram the structure of a dicot root after primary growth and explain the function of each major structure? Do you understand how branch roots form?

43.7 HOW DO PLANTS ACQUIRE NUTRIENTS?

Nutrients are substances obtained from the environment that are required for the growth and survival of an organism. Plants need only inorganic nutrients, because, unlike animals, plants can synthesize all of their own organic molecules. Plants require some nutrients, called macronutrients, in large quantities; collectively these make up more than 99% of the dry weight of the plant body. Others, called micronutrients, are needed only in trace amounts (**Table 43-2**).

Plants obtain carbon from carbon dioxide in the air, oxygen from the air or dissolved in water, and hydrogen from water. These three elements make up more than 95% of the mass of most plants. The other nutrients are obtained when plant roots absorb **minerals,** such as potassium (K^+), calcium (Ca^{2+}), nitrate (NO_3^-), or phosphate (PO_4^{3-}), from the soil.

Finally, a large part of the mass of a living plant is water. Water is also used to dissolve and transport minerals, sugars, hormones, and other organic molecules throughout the plant body. For most plants, the primary source of water is the soil.

Roots Transport Minerals from the Soil into the Xylem of the Vascular Cylinder

Roots absorb minerals from the soil and transport them to the shoot. Soil consists of rock particles, air, water, and organic matter. Although the rock particles and organic matter contain minerals, only minerals dissolved in the soil water can be taken up by roots. Minerals are transported from root to shoot in xylem; therefore, a root must move minerals from the soil water to the xylem in the root's vascular cylinder (**Fig. 43-18**).

A young root is made up of (1) living cells; (2) extracellular space, mostly filled by the walls of these cells; and (3) the tracheids and vessel elements of xylem, which are dead and consist solely of cell walls (see Fig. 43-5). Most plant cell walls, including the walls of the living cells in a root and the tracheids and vessel elements of xylem, are very porous. Therefore, soil water and its dissolved minerals can easily move from the soil through the extracellular space between the cells of the epidermis and cortex (**Fig. 43-18a** ❶, blue pathway), until the water and minerals reach the Casparian strip of the endodermis. Like waterproof mortar in a brick wall, the Casparian strip is a continuous, water-impermeable band in the cell walls of the top, bottom, and sides of the endodermal cells, but not on their inner faces (bordering the vascular cylinder) or their outer faces (bordering the rest of the cortex; **Fig. 43-18b**). The Casparian strip blocks the movement of water and minerals through extracellular space from the cortex into the xylem. How, then do minerals move from the soil water into a root's vascular cylinder?

Minerals absorbed by a root must take a three-step journey (see Fig. 43-18a):

1. **From soil water into a living cell outside the Casparian strip.** Minerals in soil water, including the water filling the extracellular space between the cells of the epidermis and cortex, contact the root hairs, epidermal cells, cortex cells, and outer faces of

TABLE 43-2 Essential Nutrients Required by Plants

Element*	Major Source	Function
Macronutrients		
Carbon	CO_2 in air	Component of all organic molecules
Oxygen	O_2 in air and dissolved in soil water	Component of all organic molecules
Hydrogen	Water in soil	Component of all organic molecules
Nitrogen	Dissolved in soil water (as nitrate and ammonia)	Component of proteins, nucleotides, and chlorophyll
Potassium	Dissolved in soil water	Helps control osmotic pressure, and regulates stomata opening and closing
Calcium	Dissolved in soil water	Component of cell walls; is involved in enzyme activation and the control of responses to environmental stimuli
Phosphorus	Dissolved in soil water (as phosphate)	Component of ATP, nucleic acids, and phospholipids
Magnesium	Dissolved in soil water	Component of chlorophyll; activates many enzymes
Sulfur	Dissolved in soil water (as sulfate)	Component of some amino acids and proteins; component of coenzyme A
Micronutrients		
Iron	Dissolved in soil water	Component of some enzymes; activates some enzymes; is required for chlorophyll synthesis
Chlorine	Dissolved in soil water	Helps maintain ionic balance across membranes; participates in splitting water during photosynthesis
Copper	Dissolved in soil water	Component of some enzymes; activates some enzymes
Manganese	Dissolved in soil water	Activates some enzymes; participates in splitting water during photosynthesis
Zinc	Dissolved in soil water	Component of some enzymes; activates some enzymes
Boron	Dissolved in soil water	Found in cell walls
Molybdenum	Dissolved in soil water	Component of some enzymes involved in nitrogen utilization

*Listed in approximate order of abundance in the plant body.

endodermal cells. Some minerals are taken up across the plasma membranes into the cytoplasm of these cells (**Fig. 43-18a ❷**, black arrows). The concentration of minerals in soil water is usually far lower than their concentration within plant cells, so most minerals enter the cells by active transport (see Chapter 5). Because epidermal cells and their root hairs have the largest surface area, they absorb the most minerals.

2. **Through plasmodesmata interconnecting living cells.** The insides of adjacent living plant cells are interconnected by tiny pores called plasmodesmata, which allow small molecules to pass from the cytoplasm of one cell to the cytoplasm of neighboring cells (see Chapter 5). Active transport of minerals into epidermal cells and their root hairs produces a high concentration of minerals in their cytoplasm, so minerals diffuse through plasmodesmata from the epidermal cells, into and through cortex cells, and into endodermal and pericycle cells (**Fig. 43-18a ❸**, red pathway). Minerals that are absorbed directly into a cortex cell or an endodermal cell take a shorter journey across fewer layers of cells, but in all cases, this intracellular pathway bypasses the Casparian strip by moving minerals through the cytoplasm of endodermal cells.

3. **Into the extracellular space within the vascular cylinder.** Minerals exit across the plasma membrane of an endodermal cell or pericycle cell, either by diffusion

or by active transport, into the extracellular space of the vascular cylinder (**Fig. 43-18a ❹**, black arrows). Remember that the tracheids and vessel elements consist only of porous cell walls, with no plasma membranes and no cytoplasm. Therefore, minerals diffuse freely from the extracellular space into the tracheids and vessel elements of the xylem (**Fig. 43-18a ❺**, blue arrows).

The Casparian Strip Prevents the Loss of Minerals from the Vascular Cylinder

The Casparian strip may appear to be a major obstacle to mineral absorption, but in reality, the Casparian strip plays the essential role of leak-proofing the vascular cylinder. The soil water that has soaked into the cortex of a root has a low concentration of minerals. As described above, the vascular cylinder acquires a much higher concentration of minerals. If minerals could move *between* endodermal cells, minerals would diffuse back out of the vascular cylinder into the soil water down this concentration gradient. However, the Casparian strip prevents minerals from moving between the endodermal cells. Therefore, once inside the vascular cylinder, minerals cannot leak back out again.

Roots Take Up Water from the Soil by Osmosis

Water moves passively across plasma membranes by osmosis, diffusing from areas of high free water concentration

(a) Pathways of mineral and water uptake

(b) Endodermal cells, showing the Casparian strip

▲ FIGURE 43-18 **Mineral and water uptake by roots** (a) Mineral and water movement through a young root, from the soil water to the conducting cells of xylem. Minerals and water flowing through porous cell walls and extracellular space are shown as blue lines and arrows. Minerals and water crossing a plasma membrane are shown as black arrows. Minerals and water flowing from cell to cell through plasmodesmata are shown as red lines and arrows. (b) The location of the Casparian strip in the cell walls of endodermal cells.

(with a low concentration of solutes) to areas of low free water concentration (with a high concentration of solutes; see Chapter 5). Soil water, with a low mineral concentration, has a high free water concentration. Minerals are actively transported into the outer cell layers of the root, move into the extracellular space of the vascular cylinder, and then into the tracheids and vessel elements of the xylem. Therefore, the cytoplasm of the living cells and the extracellular space of the vascular cylinder, including the insides of the tracheids and vessel elements, all have high mineral concentrations and low free water concentrations. Plasma membranes separate the high free water concentration in the soil water from the low free water concentration in the living cells and xylem.

Although no one is certain how much water moves across which plasma membranes, the general principle is clear: Water moves by osmosis from the soil water, across plasma membranes (see Fig. 43-18a ❷, ❹), and into the tracheids and vessel elements (see Fig. 43-18a ❺). In some plants, this osmotic entry of water following mineral uptake is so powerful that it creates **root pressure**: Water entering the vascular cylinder actually pushes the solution of minerals up the root into the shoot. Sometimes, the effect of root pressure is visible as droplets of water, forced out of the tips of leaves (**Fig. 43-19**).

However, in most plants, under most conditions, osmosis following mineral uptake is not the major force that causes water to enter into roots. Instead, water moving up the xylem, powered by the evaporation of water from the leaves, rapidly removes water from the tracheids and vessel elements in the

▲ FIGURE 43-19 **Root pressure** Sometimes, in the early morning, if there is a combination of high soil water, relatively low temperatures, and high humidity, root pressure may force water out of the tips of leaves.

roots (see Section 43.8). Water movement out of the root xylem and up into the stem xylem provides a constant osmotic gradient that moves water molecules from the soil water, across the plasma membranes of root cells, and into the vascular cylinder.

Symbiotic Relationships Help Plants Acquire Nutrients

Many minerals, although abundant in rock particles, are too scarce in soil water to support plant growth. However, most plants have evolved mutually beneficial relationships with fungi that help them acquire these minerals. In addition, nitrogen is almost always in short supply in the soil. Plants called **legumes,** such as peas, soybeans, alfalfa, and clover, house bacteria in their roots that can convert nitrogen in the air into ammonia or nitrate that the plants can use in synthesizing their nitrogen-containing organic molecules.

Fungal Mycorrhizae Help Most Plants to Acquire Minerals

The roots of most land plants form symbiotic relationships with fungi. The resulting complexes, called **mycorrhizae** ("fungus root" in Greek; singular, mycorrhiza), help the plant obtain scarce minerals from the soil (**Fig. 43-20**). Microscopic fungal strands intertwine between the root cells and extend out into the soil. The web of fungal filaments greatly increases the volume of soil from which minerals can be absorbed, compared to the volume in contact with the plant root alone. In addition, the fungus can extract some minerals, particularly phosphate, that are often bound in the rock particles of the soil and cannot be taken up by roots. Minerals absorbed by the fungus are then transferred to the root. The fungus, in return, receives sugars and other organic molecules from the plant. In this way, both the fungus and the plant can grow in places where neither could survive alone, including deserts and high-altitude, rocky soils.

Nitrogen-Fixing Bacteria Help Legumes to Acquire Nitrogen

Amino acids, nucleic acids, and chlorophyll all contain nitrogen, so plants need large amounts of it. Although nitrogen

◀ **FIGURE 43-20 A mesh of fungal strands surrounds and penetrates into a root.**

root

mycorrhizae

gas (N_2) makes up about 78% of the atmosphere and readily diffuses into the air spaces in the soil, plants can use nitrogen only in the form of ammonia (NH_3) or nitrate (NO_3^-). Some **nitrogen-fixing bacteria** in the soil combine atmospheric N_2 with hydrogen to make NH_3, a process called **nitrogen fixation.** The bacteria then use NH_3 to synthesize amino acids and nucleic acids. However, nitrogen fixation requires a lot of energy, using eight ATPs to make a single molecule of NH_3. Consequently, nitrogen-fixing bacteria don't manufacture a lot of extra NH_3.

Legumes and a few other plants enter into a mutually beneficial relationship with certain species of nitrogen-fixing bacteria. The bacteria enter the legume's root hairs and make their way into cortex cells. Both the bacteria and cortex cells multiply, forming a swelling, or **nodule,** composed of cortex cells full of bacteria (**Fig. 43-21**). The bacteria live off the root's food reserves, obtaining so much energy that they produce more NH_3 than they need. The surplus NH_3 diffuses into the host cortex cells, providing the plant with usable nitrogen.

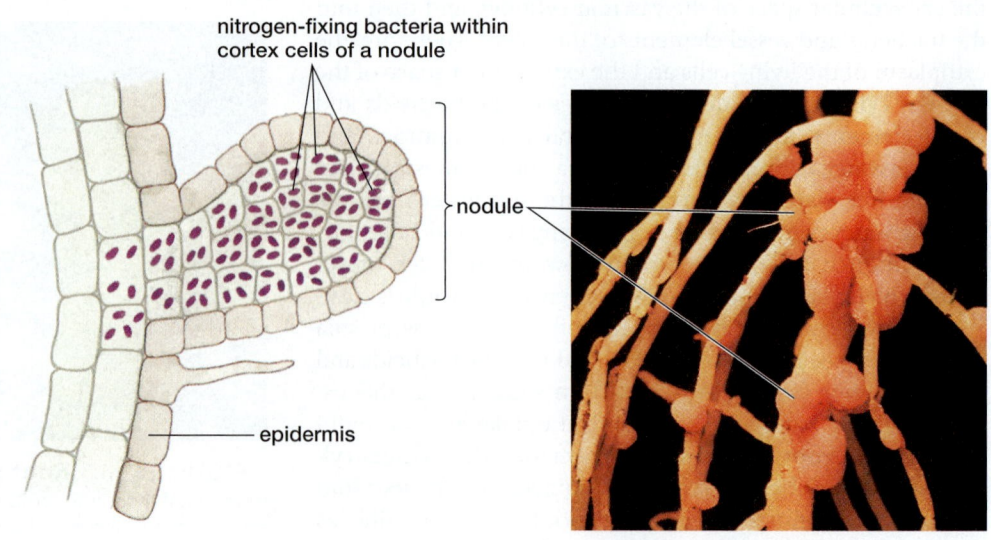

nitrogen-fixing bacteria within cortex cells of a nodule

nodule

epidermis

▶ **FIGURE 43-21 Root nodules house nitrogen-fixing bacteria**

43.8 HOW DO PLANTS MOVE WATER AND MINERALS FROM ROOTS TO LEAVES?

After entering the root xylem, water and minerals must be moved to the topmost parts of a plant—in the tallest redwood tree, a distance of about 380 feet (115 meters). Because minerals are dissolved in the water, they are passively carried along as the water moves upward. But how do plants overcome the force of gravity and make water flow upward? In most plants, at least 90% of the water absorbed by the roots evaporates through the stomata of leaves. This evaporation, called **transpiration,** provides the force that moves water upward through the plant body.

The Cohesion–Tension Mechanism Explains Water Movement in Xylem

According to the **cohesion–tension mechanism,** water is pulled up the xylem by transpiration from the leaves (**Fig. 43-22**). As its name suggests, this mechanism involves two essential processes:

- **Cohesion** Attraction among water molecules holds water together, forming a chain-like column within the xylem tubes.
- **Tension** The chain of water is pulled up the xylem by tension produced by water evaporating from the leaves.

flow of water

1 Water evaporates through the stomata of leaves

2 Cohesion of water molecules to one another by hydrogen bonds creates a "water chain." Tension caused by evaporation from the leaves pulls the chain of water molecules up the xylem

water molecules

3 Water enters the vascular cylinder of the root

◀ FIGURE 43-22 The cohesion–tension mechanism of water flow from root to leaf in xylem

Hydrogen Bonds Between Water Molecules Produce Cohesion

In a molecule of water (H_2O), the oxygen bears a slightly negative charge, while the two hydrogens bear slightly positive charges. As a result, the oxygen of one water molecule is electrically attracted to the hydrogens of nearby water molecules, forming hydrogen bonds (see Chapter 2). Just as individually weak cotton threads together create the strong fabric of your jeans, the network of individually weak hydrogen bonds in water produces a strong **cohesion,** the ability of a substance to resist being pulled apart. The hydrogen bonds create a chain of water in the xylem, extending the entire height of the plant, that is at least as strong as a steel wire of the same diameter.

Transpiration Produces the Tension That Pulls Water Upward

Transpiration provides the force for water movement—the "tension" part of the mechanism. Water evaporates from mesophyll cells into the air spaces within the leaf and out through the stomata into the atmosphere (**Fig. 43-22 ❶**). As water leaves mesophyll cells, their free water concentration decreases. This causes water to move by osmosis from nearby xylem into the mesophyll cells.

Cohesion and Tension Work Together to Move Water Up the Xylem

Water molecules leaving the xylem are linked to other water molecules in the same xylem tube by hydrogen bonds— cohesion. Therefore, as water molecules leave the xylem to replace the water evaporating from mesophyll cells, they pull more water up the xylem—tension (**Fig. 43-22 ❷**). The tension in xylem is strong enough to lift water more than 500 feet (150 meters), much taller than any living tree. Tension pulling water up the xylem continues all the way down to the roots, where water in the extracellular space of the vascular cylinder is pulled in through the porous pits in the walls of the vessel elements and tracheids of the xylem, replenishing the bottom of the chain of water molecules (**Fig. 43-22 ❸**).

CASE STUDY c o n t i n u e d

Autumn in Vermont

The brilliant colors of autumn last just a few weeks, and then the leaves fall from the trees. Why do deciduous trees drop their leaves? In temperate climates, the soil is frozen for long periods of time during the winter. Therefore, the trees cannot absorb water from the soil. However, water slowly evaporates from ice, a process called sublimation—that's why old ice cubes in your freezer get smaller over time. Large leaves, even if frozen for much of the winter, would evaporate a lot of water. By dropping their leaves, deciduous trees reduce winter water loss when they would be unable to absorb replacement water from the frozen soil. A few trees, particularly some oaks, retain their leaves all winter, but these leaves are completely dead and do not draw water from the rest of the tree.

The movement of water into the root xylem causes water to enter the root from the soil. Only the aboveground parts of a plant, usually the shoot, can transpire, so tension always pulls water up the xylem, never down. Therefore, water flow in xylem is unidirectional, from root to shoot.

A large maple tree may transpire about 250 gallons of water a day—almost a ton of water lifted up more than 50 feet, every day. What supplies the energy to transport so much water? The sun. Sunlight warms both the leaves and the air, powering the evaporation of water from the leaves. Now imagine a whole forest, with each tree releasing hundreds of gallons of water into the air through transpiration each day. This massive transpiration can have a major effect on the local climate, as we describe in "Earth Watch: The Surprising Impacts of Forests on Climate and the Growth of Their Own Trees."

Minerals Move Up the Xylem Dissolved in Water

The minerals absorbed by plant roots are charged ions, such as K^+, Ca^{2+}, PO_4^{3-}, or Cl^-. Ions dissolve in water because either the negative or the positive parts of water molecules are electrically attracted to the ions (see Chapter 2). In xylem, electrical attraction loosely holds the minerals to the water. Therefore, the minerals are carried along with the water from root to shoot.

Stomata Control the Rate of Transpiration

Transpiration powers the transport of water and minerals, but transpiration is also a great threat to plants, because it is by far the largest source of water loss—a loss that can be fatal in hot, dry weather. Water loss can be reduced by closing the stomata through which the water evaporates. Closing the stomata, however, also prevents CO_2 from entering the leaf—and CO_2 is the source of the carbon atoms in the sugars produced by photosynthesis. Therefore, a plant must regulate its stomata to achieve a balance between acquiring carbon dioxide and losing water. In most plants, stomata open during the day, when sunlight can power photosynthesis, and close during the night, conserving water. However, if excessive evaporation threatens the plant with dehydration, the stomata will close regardless of the time of day.

Guard Cells Regulate Opening and Closing of Stomata

A stoma consists of a central pore surrounded by two guard cells (**Fig. 43-23**; see also Fig. 43-7). How do plants open and close their stomata? This is actually a two-part question: (1) Mechanically, how is the size of the opening changed; and (2) physiologically, how do guard cells respond to stimuli, such as sunlight or dehydration, and adjust the size of the opening?

Guard cells adjust the size of the opening of a stoma by changing their volume and shape, which they accomplish by taking up or losing water. Guard cells have both an unusual shape and a specific arrangement of cellulose fibers in their cell walls (**Fig. 43-24a**). The two guard cells of a stoma are curved slightly outward, like two sausage links tied together at their two ends. Cellulose fibers in the guard cell

(a) Light micrograph of a stoma

(b) SEM of a stoma

▲ **FIGURE 43-23 Stomata** Stomata as seen through **(a)** a light microscope and **(b)** a transmission electron microscope. The light micrograph shows that guard cells contain chloroplasts (green ovals inside the cells), unlike the other cells of the leaf epidermis, which do not.

(a) Closed stoma

(b) Opening a stoma

❶ K^+ enters the guard cells (red arrows)

❷ Water follows by osmosis (blue arrows)

❸ The guard cells lengthen and bend outward

❹ The pore opens

K^+ ions

cellulose "belts"

pore

guard cells

◀ **FIGURE 43-24 How guard cells open a stoma** Guard cells open and close the pore of a stoma as a result of changes in K^+ concentration, water content, and, hence, cell volume. **(a)** The structure of a stoma, showing the central pore closed. The guard cells have a relatively low K^+ concentration (about the same as other cells in the epidermis), a low water content, and a low volume. **(b)** When K^+ is transported into the guard cells, the resulting high K^+ concentration causes water to enter the guard cells by osmosis, which increases their volume and opens the pore.

QUESTION When the stomata close, how is photosynthesis affected? How is the movement of water into the roots affected?

Earth|Watch

The Surprising Impacts of Forests on Climate and the Growth of Their Own Trees

Tropical rain forests grow in the Amazon River basin because of the year-round warm temperatures and abundant rainfall (**Fig. E43-1**). Where does the rain come from? A lot of it comes from moist air masses that develop near the equator (see Chapter 29). However, tropical rain forests are a significant source of their own high humidity and plentiful rainfall.

In the Amazon rain forest, an acre of soil supports hundreds of towering trees, each bearing hundreds of thousands of leaves. The combined surface area of the leaves dwarfs the surface area of the soil. Therefore, most of the water evaporating from the forest comes from the trees—as little as 2 or 3 gallons a day for small understory trees, but as much as 200 gallons a day for large trees of the canopy. This evaporation produces perpetually high humidity. Some of the transpired water falls again as rain, right there in the rain forest. In fact, up to half of the rainfall in the Amazon region is water transpired by the trees themselves.

When rain forests are clear-cut, the local climate changes dramatically. In the intact forest, water evaporating from the leaves cools the air. As trees are removed and transpiration is reduced, the region becomes not only drier, but hotter as well.

▲ **FIGURE E43-1 The Amazon rain forest** The rain-forest community helps produce and maintain its own environment.

In uncut adjacent forests, rain-forest trees do not thrive under conditions of decreased rainfall and warmer temperatures, so many of these trees begin to die. These uncut but damaged forests also transpire less water, making the local climate even drier and hotter, and the vicious cycle continues.

Rain-forest transpiration affects distant forests as well. For example, the Monteverde Cloud Forest Reserve in Costa Rica depends for its survival on an almost constant shroud of fog. Transpiration from Costa Rica's lowland rain forests pumps moisture into the winds flowing up the mountain slopes. As the air cools, the moisture condenses into fog. However, a century of logging has eliminated about 80% of Costa Rica's rain forests. As a result, the air contains less moisture. Drier air requires colder temperatures for water to condense, which means that the fog forms higher up the mountains than in years past. As a result, the forest in the lower part of the mountains is changing, with different species of trees, and even different species of birds and lizards, colonizing the slopes.

The impacts of trees on their own growth are not limited to tropical rain forests. On Tenerife, one of the Canary Islands off the northwest coast of Africa, rainfall averages less than 8 inches (20 cm) a year, with average high temperatures between 68° and 88°F (20° to 31°C) year-round. Nevertheless, some of Tenerife's mountain slopes support pine forests. How is this possible? Of course, it's a little cooler, with a little more rain, on the mountainsides than on the coast. The major factor, however, is the forest itself. The trees intercept fog drifting across the island from the Atlantic Ocean. Water from the fog condenses on the trees and trickles down to the soil. More than half of the water in Tenerife's streams comes from condensing fog.

The nearby island of Lanzarote has a similar climate, and its mountains used to be covered with fog-catching forest. However, the forests were almost completely cut down long ago to make room for crops, and have not regrown. Modern efforts to restore the native trees have failed repeatedly, because the soil is now too dry. A project is under way to capture the fog with plastic mesh nets and drain the water into storage containers. The trapped water will be used to grow tree seedlings. Researchers hope that, when the trees become large enough, they can capture their own moisture and become part of a self-sustaining forest.

---

walls encircle the cells, as if dozens of tiny belts were buckled firmly around each sausage.

A stoma opens when its guard cells take up water, increasing the volume of the cells. The cellulose belts that surround the guard cells prevent the cells from getting fatter, and so they get longer instead. The curved shape of the guard cells and their attachment at the ends mean that they can get longer only by curving outward even more, which opens the central pore. A stoma closes when its guard cells lose water and decrease in volume. The cells become shorter and less curved, closing the pore.

How does a guard cell change its water content? Water always moves across plasma membranes by osmosis, following a concentration gradient of dissolved solutes. Guard cells create osmotic gradients across their plasma membranes by adjusting the potassium ion (K^+) concentration in their cytoplasm in response to stimuli such as light and CO_2 (see below). Moving K^+ into the cell causes water to enter by osmosis, increasing the volume of the cells and opening the pore (**Fig. 43-24b**). Moving K^+ back out again causes water to leave by osmosis, shrinking the cells and closing the pore.

Three important stimuli control K^+ movement into and out of guard cells:

- **Light** Light is absorbed by pigments in the guard cells, triggering a series of reactions that causes K^+ to enter the cells. Water follows by osmosis, and the guard cells swell, opening the pore. In the dark, the reactions promoting K^+ entry stop, allowing K^+ to diffuse back out of the guard cells. Water follows out by osmosis, closing the pore.

- **Carbon dioxide** Recall that guard cells contain chloroplasts and can carry out photosynthesis (see Fig. 43-23). Carbon fixation during photosynthesis uses up CO_2, reducing its concentration inside the guard cells. Low CO_2 concentrations stimulate K^+ entry into the guard cells. The increased K^+ concentration in the guard cells causes water to enter by osmosis, opening the stoma. At night, photosynthesis stops but cellular respiration continues, so CO_2 levels rise. K^+ transport into the guard cells stops. K^+ diffuses out, water follows out by osmosis, and the guard cells close the pore.

- **Water** When a plant begins to dry out, it synthesizes a hormone called abscisic acid (see Chapter 45). Abscisic acid binds to receptors on the guard cells and shuts down K^+ entry. As K^+ diffuses out of the guard cells, water follows out by osmosis, and the guard cells close the stomata. Dangerous dehydration is most likely to occur on sunny days. As you might expect, the effects of abscisic acid are powerful, overcoming the effects of light and CO_2, and causing the stomata to close.

CHECK YOUR LEARNING

Can you describe how water moves in xylem? Do you understand how minerals are transported in xylem? Can you describe how potassium and water movements open and close stomata and describe the stimuli that trigger these movements?

43.9 HOW DO PLANTS TRANSPORT SUGARS?

Sugars synthesized in the leaves must be moved to other parts of the plant, where they nourish nonphotosynthetic structures such as roots or flowers, or are stored in cortex cells of roots and stems. Sugar transport is the function of phloem.

Botanists studying phloem sometimes employ unlikely lab assistants: aphids. An aphid is an insect that feeds by sucking fluids through a sharp hollow tube called a stylet, similar to the more familiar stylet used by a mosquito to suck blood from its victims. An aphid, however, feeds on the fluid in phloem. An aphid inserts its stylet through the epidermis and cortex of a young stem into a sieve tube (Fig. 43-25). As we will see, phloem fluid is under pressure, so the plant actually drives the fluid into the aphid's digestive tract; sometimes, the aphid inflates like a balloon. After an aphid has penetrated a sieve tube, botanists can anesthetize the aphid, remove most of its body, and leave the stylet in place. Phloem fluid flows out of the stylet for several hours. Chemical analysis shows that the phloem fluid consists of water containing about 10% to 20% dissolved sugar (mostly sucrose), with lower amounts of other substances such as amino acids, proteins, and hormones. What drives the movement of this sugary solution in phloem?

The Pressure-Flow Mechanism Explains Sugar Movement in Phloem

The most widely accepted explanation for fluid transport in phloem is the **pressure-flow mechanism,** in which differences in water pressure drive the flow of fluid through the

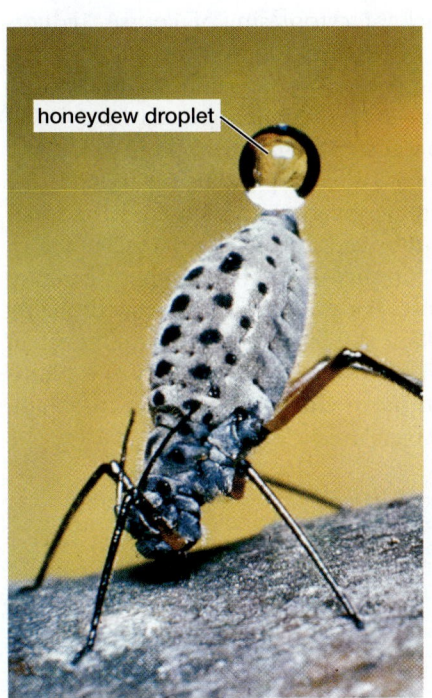

honeydew droplet

(a) An aphid sucks sap

stylet of aphid

(b) A stylet penetrates into phloem

◀FIGURE 43-25 **An aphid feeds on the sugary fluid in phloem sieve tubes** **(a)** When an aphid pierces a sieve tube, pressure in the tube forces the fluid out of the phloem and into the aphid's digestive tract. The aphid excretes excess fluid from its anus, as a sugar-rich "honeydew." This fluid is collected by certain species of ants that, in turn, defend the aphids from predators. **(b)** The flexible stylet of an aphid passes through many layers of cells to penetrate a sieve-tube element.

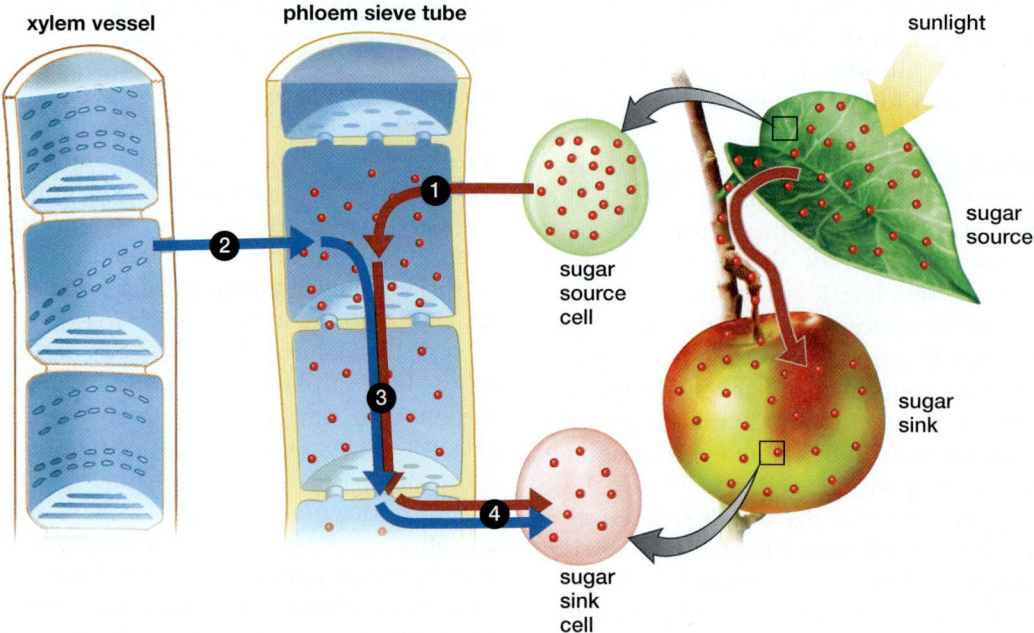

▲ **FIGURE 43-26 The pressure-flow mechanism of sugar transport in phloem**
A photosynthesizing leaf is a sugar source, whereas a developing fruit is a sugar sink. Differences in water pressure drive phloem fluid from the leaf to the fruit. Numbers correspond to the explanations in the text. Red arrows show the movement of sugar; blue arrows show the movement of water. The gradient of blue color in the phloem sieve tube represents water pressure, which is higher in the source end of the tube and lower in the sink end of the same tube.

sieve tubes. These pressure differences are created indirectly by the production and use of sugar in different parts of the plant. Any part of a plant that synthesizes more sugar than it uses, such as a photosynthesizing leaf, is a sugar **source.** Any structure that uses more sugar than it produces is a sugar **sink.** Sinks include meristems and developing flowers and fruits. Sources and sinks can change with the seasons. For example, the roots of deciduous trees are sugar sinks during the summer, when they convert sucrose to starch for storage. The following spring, the roots become sugar sources, as they convert the starch back to sucrose, supplying energy for the development of new leaves.

According to the pressure-flow mechanism, phloem sieve tubes carry fluid away from sugar sources and toward sugar sinks, as illustrated by the numbered steps in **Figure 43-26.** ❶ Sugar produced by a source cell (for example, in a photosynthesizing leaf) is actively transported into a phloem sieve tube. This raises the sugar concentration in the phloem fluid in that portion of the sieve tube. ❷ Water from nearby xylem follows the sugar into the sieve tube by osmosis. Because their rigid cell walls prevent sieve-tube cells from expanding, water entering the sieve tube increases the pressure of the fluid. ❸ The resulting water pressure at a source region drives the fluid through the phloem sieve tubes to regions of lower pressure. ❹ How is lower pressure created? Cells of a sugar sink (such as a fruit)

actively transport sugar out of the phloem. Water follows by osmosis, producing lower water pressure in this part of the sieve tube. The phloem fluid thus moves from a source, where water pressure is high, to a sink, where water pressure is lower, carrying the dissolved sugar with it.

The pressure-flow mechanism explains why sieve-tube elements must have intact plasma membranes but very little cytoplasm. The lack of cytoplasm, of course, reduces resistance to the flow of fluid in a sieve tube. Both the accumulation of sugar and the resulting entry of water into the sieve-tube cells can only occur if the cells have intact plasma membranes, because plasma membranes are generally impermeable to sugar (thus keeping the sugar inside the sieve-tube cells) but are permeable to water, allowing water to move into the sieve-tube cells by osmosis (see Chapter 5).

Note that sinks of sucrose may be either above or below sources of sucrose. For example, a photosynthesizing leaf near the top of a plant will be below the apical meristem but above the roots. The pressure-flow mechanism explains how phloem fluid can move either up or down the plant, from sugar sources to sugar sinks.

CHECK YOUR LEARNING

Can you explain how sugar solutions move in phloem and why they always move from sources to sinks of sugars?

Autumn in Vermont

Of all the fall colors, the striking reds are the most intriguing. In most trees, yellow and orange carotenoids play essential roles in photosynthesis throughout the summer, but the red anthocyanins are synthesized only during the autumn, just a few weeks before the leaves are shed. What good are anthocyanins? They act as sunscreen for leaves.

At the onset of autumn, as temperatures cool but days remain fairly long and the sun is still very bright, a leaf's photosynthetic pathways become less efficient and the leaf is no longer able to use all of the light it absorbs. The excess light energy can harm the chloroplasts and reduce photosynthesis even more. Anthocyanins block ultraviolet and blue light, and act as antioxidants, scavenging harmful free radicals that might be formed by UV light or by inefficient photosynthesis. Therefore, red, anthocyanin-containing leaves are better protected against intense light than are leaves without anthocyanin and, hence, can carry out photosynthesis more effectively. In some plants, red leaves dominate the outer parts of the plant, which receive the most sunlight, whereas shaded inner leaves become orange and yellow.

Why spend energy to protect leaves that are just about to fall off? To reclaim nutrients, especially nitrogen and phosphate, that a plant cannot afford to lose. In the autumn, complex organic molecules in leaves are broken down, and the resulting simpler molecules are transported to storage cells in stems and roots, where they remain throughout the winter. Photosynthesis provides the energy needed for this breakdown and transport. Mutant plants that are unable to synthesize anthocyanin retrieve far less nitrogen from their leaves under conditions of intense light and low temperatures than normal, anthocyanin-producing plants can, indicating that anthocyanin indeed helps a plant to reclaim nutrients from its leaves.

Consider This

Not all trees turn red in the fall. The Rocky Mountains, for example, are famous for their autumn display of bright yellow aspens, despite sunny days and cold nights. How can aspens thrive in the Rockies, subjected to sunlight far more intense than is ever seen in New England? What other hypotheses might explain the yellow leaves of aspens in autumn? How might you test your hypotheses?

CHAPTER REVIEW

Summary of Key Concepts

43.1 How Are Plant Bodies Organized?

The body of a land plant consists of root and shoot systems. Roots are usually underground. Their functions include anchoring the plant in the soil; absorbing water and minerals; storing surplus sugar and starch; transporting water, minerals, sugars, and hormones; producing some hormones; and interacting with soil fungi and microorganisms that provide nutrients. Shoots are generally located aboveground and consist of stems, buds, leaves, and (in season) flowers and fruit. Shoot functions include photosynthesis, transport of materials, storage of sugar and starch, reproduction, and hormone synthesis. The two principal groups of flowering plants are the monocots and dicots. Their major characteristics are shown in Fig. 43-2.

43.2 How Do Plants Grow?

Plant bodies are composed of meristem cells and differentiated cells. Meristem cells are undifferentiated and retain the capacity for mitotic cell division. Differentiated cells arise from divisions of meristem cells, become specialized for particular functions, and usually do not divide. Most meristem cells are located in apical meristems at the tips of roots and shoots and in lateral meristems in the shafts of roots, stems, and branches. Primary growth (growth in length or height and the differentiation of parts) results from the division and differentiation of cells derived from apical meristems. Secondary growth (growth in diameter and strength) results from the division and differentiation of cells derived from lateral meristems.

43.3 What Are the Tissues and Cell Types of Plants?

Plant bodies consist of three tissue systems: dermal, ground, and vascular systems (see Table 43-1). The dermal tissue system forms the outer covering of the plant body. The dermal tissue system of leaves and of primary roots and stems is usually a single cell layer of epidermis. Dermal tissue after secondary growth, called periderm, consists of cork cells and the cork cambium that produces them.

The ground tissue system consists of several cell types, including parenchyma, collenchyma, and sclerenchyma cells. Most are involved in photosynthesis, support, or storage. Ground tissue makes up most of the body of a young plant during primary growth.

The vascular tissue system consists of xylem, which transports water and minerals from the roots to the shoots, and phloem, which transports water, sugars, and other organic molecules throughout the plant body.

43.4 What Are the Structures and Functions of Leaves?

A leaf is composed of a petiole and a blade. The petiole attaches the blade to a stem or branch. The blade consists of an outer epidermis covered by a waterproof cuticle; mesophyll cells, which have chloroplasts and carry out photosynthesis; and vascular bundles of xylem and phloem, which carry water, minerals, and photosynthetic products to and from the leaf. The epidermis is perforated by adjustable pores called stomata that regulate the exchange of gases and water.

43.5 What Are the Structures and Functions of Stems?

Primary growth in dicot stems results in a structure consisting of an outer epidermis, usually covered by a waterproof cuticle; supporting, storage, and sometimes photosynthetic cells of cortex beneath the epidermis; vascular tissues of xylem and phloem; and supporting and storage cells of pith at the center.

Leaves and lateral buds are found at nodes along the surface of the stem. Under the proper hormonal conditions, lateral buds sprout into branches. Secondary growth in stems results from cell divisions in the vascular cambium and cork cambium. Vascular cambium produces secondary xylem and secondary phloem, increasing the stem's diameter. Cork cambium produces waterproof cork cells that cover the outside of the stem.

43.6 What Are the Structures and Functions of Roots?

Primary growth in roots results in a structure consisting of an outer epidermis and an inner vascular cylinder of xylem and phloem, with cortex between the two. The apical meristem near the tip of a root is protected by a root cap. Cells of the root epidermis absorb water and minerals from the soil. Root hairs are projections of epidermal cells that increase the surface area for absorption. Most cortex cells store surplus sugars (usually in the form of starch) produced by photosynthesis. The innermost layer of cortex cells is the endodermis, which controls the movement of water and minerals from the soil into the vascular cylinder. The outer layer of the vascular cylinder is the pericycle, which helps to regulate mineral transport into the vascular cylinder and also produces branch roots. Secondary growth in roots is similar to secondary growth in stems.

43.7 How Do Plants Acquire Nutrients?

Most minerals are taken up from the soil water by active transport into the root hairs. These minerals diffuse from cell to cell into the root through plasmodesmata, ultimately to the endodermis and pericycle. There, they are actively transported into the extracellular space of the vascular cylinder. The minerals diffuse from this extracellular space into the tracheids and vessel elements of xylem.

Water follows the high mineral concentration into the vascular cylinder, moving by osmosis across the plasma membranes of epidermal, cortex, and endodermal cells. Water is rapidly pulled up the xylem of the vascular cylinder, which also promotes water movement into the vascular cylinder by osmosis. Water transport in xylem is usually the most important factor driving water uptake in roots.

Many plants form root–fungus complexes called mycorrhizae that help absorb some soil nutrients. Nitrogen can be absorbed only as ammonium or nitrate, both of which are scarce in most soils. Legumes have evolved a cooperative relationship with nitrogen-fixing bacteria that invade legume roots. The plant provides the bacteria with sugars, and the bacteria use some of the energy in those sugars to convert atmospheric nitrogen to ammonia, which the plant then absorbs.

43.8 How Do Plants Move Water and Minerals from Roots to Leaves?

The cohesion–tension mechanism explains xylem function. Water molecules are electrically attracted to one another, forming hydrogen bonds between molecules. The resulting cohesion holds together the water within xylem tubes. As water molecules evaporate from the leaves during transpiration, the hydrogen bonds pull other water molecules up the xylem to replace them. This movement is transmitted down the xylem to the root, where water loss from the vascular cylinder promotes water movement by osmosis from the soil water across the endodermis and into the vascular cylinder. Minerals move up the xylem dissolved in the water.

Stomata in leaves and young epidermis control the evaporation of water (transpiration). The opening of a stoma is regulated by the shape and volume of the guard cells that form the pore. Open stomata allow more rapid transpiration. In most plants, stomata open during the day (admitting carbon dioxide needed for photosynthesis); the stimuli that trigger opening of the stomata are light and a low carbon dioxide concentration in the guard cells. If the plant is losing too much water, a hormone called abscisic acid is released, causing the guard cells to close the stomata.

43.9 How Do Plants Transport Sugars?

The pressure-flow mechanism explains sugar transport in phloem. Parts of the plant that synthesize sugar (for example, leaves) export sugar into the sieve tube. High sugar concentrations cause water to enter the sieve tube by osmosis, increasing the local water pressure in the phloem. Other parts of the plant (for example, fruits) consume sugar, reducing the sugar concentration in the sieve tube and causing water to leave the tube by osmosis, which reduces pressure. Water and dissolved sugar move in the sieve tubes from high pressure to low pressure.

Key Terms

annual ring   843
apical meristem   835
bark   844
blade   839
branch root   847
bud   833
Casparian strip   846
cohesion   852
cohesion–tension mechanism   851
collenchyma   837
companion cell   839
cork cambium   843
cork cell   843
cortex   841
cuticle   836
dermal tissue system   836
dicot   833
differentiated cell   834
endodermis   846
epidermis   836
fibrous root system   844
flower bud   841
ground tissue system   836
guard cell   839
heartwood   843
internode   841
lateral bud   841
lateral meristem   835
leaf   839
leaf primordium (plural, primordia)   841
legume   850
meristem cell   834
mesophyll   840
mineral   847
monocot   833

mycorrhiza (pl., mycorrhizae)   850
nitrogen fixation   850
nitrogen-fixing bacterium   850
node   841
nodule   850
nutrient   847
parenchyma   836
pericycle   847
periderm   836
petiole   839
phloem   838
pit   838
pith   841
plasmodesma (plural, plasmodesmata)   839
pressure-flow mechanism   855
primary growth   835
root   833
root cap   844
root hair   844
root pressure   849
root system   833
sapwood   843
sclerenchyma   837
secondary growth   835
shoot system   833
sieve plate   838
sieve-tube element   838
sink   856
source   856
stem   833
stoma (plural, stomata)   839
taproot system   844
terminal bud   841

Learning Outcomes

In this chapter, you have learned to . . .

LO1 List the major functions of root and shoot systems in flowering plants.

LO2 Explain the difference between meristem cells and differentiated cells, and describe how each contributes to plant growth.

LO3 Name the three tissue systems in flowering plants, and describe the principal cell types in each system.

LO4 Describe the structure and function of xylem and phloem.

LO5 Describe the structure of a typical dicot leaf, and explain the function of each part.

LO6 Describe the surface structures of a stem.

LO7 Compare and contrast primary and secondary growth in stems, and discuss the principal internal structures formed during each type of growth.

LO8 Describe the structure of a root formed by primary growth.

LO9 Explain how water and minerals are absorbed by roots.

LO10 Explain how fungi and bacteria may help plants to obtain scarce nutrients.

LO11 Explain how water and minerals move in xylem.

LO12 Describe how guard cells control the opening and closing of stomata.

LO13 Explain how water and sugars move in phloem.

Thinking Through the Concepts

Fill-in-the-Blank

1. Plants grow through division of _____ cells and differentiation of the resulting daughter cells. These dividing cells reside in two locations: at the tip of a shoot or root, called the _____, and in cylinders along the sides of roots and stems, called _____. Which is responsible for primary growth? _____ Which is responsible for secondary growth? _____ **LO2 LO7**

2. The three tissue systems of a plant body are _____, _____, and _____. Which covers the outside of the plant body? _____ Which conducts water, minerals, and sugars within the plant body? _____ Which stores starches? _____ **LO3**

3. Water travels upward through plant roots and shoots within hollow tubes of _____, which contains two types of conducting cells, _____ and _____. Water molecules within these tubes are interconnected by forces called _____, which allow a chain of water molecules to be pulled up the plant, driven by the evaporation of water from the leaves, a process called _____. The _____

in the epidermis of a leaf control the rate of evaporation from the leaf. **LO5 LO11 LO12**

4. _____ transports sugars and other organic molecules from sources of sugar to sinks of sugar. The conducting cell type in this tissue is the _____, which is supported by adjacent cells called _____. **LO13**

5. The very tip of a young root is protected by the cells of the _____. The surface area of a young root is increased by projections from epidermal cells called _____, which move minerals from the soil water into their cytoplasm by the process of _____. Many young roots have a mutually beneficial relationship with fungi to help absorb some minerals; these root–fungus complexes are called _____. **LO8 LO10**

Review Questions

1. What are the main functions of roots, stems, and leaves? **LO1**

2. Describe the locations and functions of the three tissue systems in land plants. **LO3**

3. Distinguish between primary growth and secondary growth, and describe the cell types involved in each. **LO2 LO7**

4. Distinguish between meristem cells and differentiated cells. **LO2**

5. Diagram the internal structure of a dicot root after primary growth, labeling and describing the function of epidermis, cortex, endodermis, pericycle, xylem, and phloem. What tissues are located in the vascular cylinder? **LO8**

6. Diagram the internal structure of a dicot stem after primary growth, labeling and describing the function of the epidermis, cortex, pith, xylem, and phloem. How does the structure change after secondary growth? **LO7**

7. Describe xylem and phloem, including their cell types and the solutions they carry. **LO4 LO11 LO13**

8. What types of cells form root hairs? What is the function of root hairs? **LO8 LO9**

9. Diagram the structure of a leaf. What structures regulate water loss and CO_2 absorption by a leaf? **LO5**

10. Describe the daily cycle of opening and closing of stomata. **LO12**

11. Describe how water and minerals are absorbed by a root. **LO9**

12. Describe how water and minerals move in xylem. Why is water and mineral movement in xylem unidirectional? **LO11**

13. Describe how sugars are moved in phloem and explain why phloem fluid may move up or down the plant. **LO13**

14. Describe the surface structures of a stem after primary growth. Where are buds located, and what are their functions? **LO6**

Applying the Concepts

1. An important goal of molecular botanists is to insert the genes for nitrogen fixation into crop plants such as corn or wheat. Why would the insertion of such genes be useful? What changes in farming practices would this technique allow?

2. When a hurricane sweeps over low-lying coastal areas, it leaves behind a lot of salt water. Many land plants die soon thereafter. Why would salt water kill plants?

3. Grasses (monocots) form their primary meristem near the ground surface rather than at the tips of branches the way dicots do. How does this feature allow you to grow a lawn and mow it every week in the summer? What would happen if you had a dicot lawn and mowed it?

Answers to Figure Caption questions and Fill-in-the-Blank questions can be found in the Answers section at the back of the book.

MB *Go to MasteringBiology for practice quizzes, activities, eText, videos, current events, and more.*

CASE STUDY

Some Like It Hot— and Stinky!

MOST FLOWERS ARE PRETTY, delicate, fragrant, and fairly small. And flowers usually remain at the same temperature as their environment. But not all flowers obey these rules: Behold *Amorphophallus titanum*, known in its native Sumatra as bunga bangkai, the corpse flower. (People seldom translate its scientific name in print, but you can probably figure it out.)

Five to 10 feet tall, the corpse flower gets its common name from its odor, variously described as resembling decomposing fish, decaying pumpkin, or just plain rotting carrion. The corpse flower also gets hot. It generates pulses of warm water vapor, at temperatures as high as 97°F (36°C). In the Sumatran forest, the huge flower acts like a chimney, blasting its smelly steam high into the air and dispersing it throughout the vicinity.

Flesh flies and carrion beetles are attracted by the smell of rotting flesh. They swarm around animal carcasses and lay their eggs in them. When the eggs hatch, the larvae eat the decaying meat, then eventually pupate and hatch into another generation of flies and beetles. When a corpse flower emits "parfum de carcass," it attracts these scavengers.

Several other flowers are warm, or stinky, or both. The related dead-horse arum smells up Corsica and other islands in the northern Mediterranean. The stinking corpse lily (see Chapter 21) shares Sumatran forests with the corpse flower. In South Africa, the starfish flower attracts flies with its five-armed, putrid flowers—and believe it or not, some people grow them as houseplants! Several relatives of the corpse flower, including philodendrons and skunk cabbages, bear heat-generating flowers that are not particularly offensive (although the leaves of skunk cabbages do smell bad if they are damaged).

Why do plants produce flowers? Specifically, why would a plant produce an enormous, hot, putrid flower? And how does a plant benefit by deceiving beetles and flies into mistaking its flower for a mass of rotting flesh? We'll explore these questions in this chapter.

Amorphophallus titanum is truly breathtaking—in more ways than one! The flower is often more than 5 feet tall, and it smells like a rotting carcass.

AT A GLANCE

LEARNING GOALS

LG1 Describe the sexual life cycle of plants, and compare and contrast the cycles of ferns and flowering plants.

LG2 Describe the structure of a complete flower and explain the function of each part.

LG3 Describe the development of the male and female gametophytes of flowering plants.

LG4 Explain the processes of pollination and double fertilization in flowering plants.

LG5 Explain how the structures of a fruit and seed develop from a flower.

LG6 Explain why many seeds undergo a period of dormancy before they germinate, and describe the environmental conditions that may break dormancy and allow germination to occur.

LG7 Describe seed germination and the early growth of seedlings in monocots and dicots.

LG8 Describe the adaptations of flowers and their animal pollinators that facilitate efficient pollination.

LG9 Explain how fruits aid in seed dispersal.

44.1 HOW DO PLANTS REPRODUCE?

Many plants can reproduce either sexually or asexually. During asexual reproduction, mitotic cell division in an existing plant produces an offspring that is genetically identical to the parent. For example, aspen trees can sprout new shoots from their roots (see Fig. 9-2). When the arching branches of blackberries or the spreading horizontal runners of strawberries (see Fig. 43-12b) touch the ground, they take root and produce new plants. The bulbs of tulips, daffodils, and amaryllis can reproduce by growing new, smaller bulbs.

Biologists disagree on what factors, or combination of factors, favored the evolution of sexual reproduction. However, the advantages of sexual reproduction must be significant, because most eukaryotes, including all of the plants mentioned above, reproduce sexually at least some of the time. This chapter will focus on sexual reproduction in plants.

The Plant Sexual Life Cycle Alternates Between Diploid and Haploid Stages

The sexual life cycle of plants is described as **alternation of generations,** because it alternates between multicellular diploid and multicellular haploid generations that give rise to each other in a continuing cycle (**Fig. 44-1**). Let's begin at the multicellular diploid stage. Specialized reproductive cells in the diploid stage undergo meiotic cell division, producing haploid cells called **spores** (**Fig. 44-1 ❶**). For this reason, the diploid stage is named the **sporophyte,** or "spore-bearing plant." Why are these cells spores and not gametes? Gametes (sperm or eggs) never divide, but, if the opportunity arises, two gametes fuse to form a diploid fertilized egg, or **zygote.** In contrast, spores never fuse to form a diploid cell; instead, plant spores undergo mitotic cell division to produce a multicellular

haploid stage called the **gametophyte** ("gamete-bearing plant"; **Fig. 44-1 ❷**). Specialized reproductive cells in the gametophyte differentiate into haploid gametes (**Fig. 44-1 ❸**). Two gametes then fuse to form a zygote (**Fig. 44-1 ❹**). The zygote grows by mitotic cell division, becoming the next multicellular diploid generation (**Fig. 44-1 ❺**).

Although alternation of generations is the sexual life cycle of all plants, the relative size, complexity, and life span of the sporophyte and gametophyte stages vary considerably among different types of plants. In mosses, for example, the gametophyte stage is an independent plant that dominates the life cycle (see Chapter 21). A sperm fertilizes an egg that is retained in the gametophyte. The resulting zygote develops into a sporophyte that grows directly on the gametophyte and relies on the gametophyte for nourishment. The sporophyte is never an independent plant.

In ferns, both the haploid and diploid stages are free-living, independent plants (see Chapter 21). Reproductive cells of the sporophyte—the fern commonly seen in moist, shady woods—undergo meiotic cell division, producing clusters of haploid spores, typically on the undersides of the fronds (**Fig. 44-2**). If a spore lands on moist soil, it germinates into a haploid gametophyte. Cells in separate male and female structures on the gametophyte differentiate into sperm and eggs. A sperm fertilizes an egg inside the female reproductive structure, creating a diploid zygote. Cell divisions of the zygote produce an embryonic sporophyte that begins growing atop the gametophyte, but soon the sporophyte develops its own roots and leaves and becomes the dominant stage of the life cycle.

In gymnosperms (conifers and their relatives) and angiosperms (flowering plants), the diploid sporophyte stage is dominant; the haploid gametophyte stage is never a free-living, independent plant.

① Meiotic cell division produces haploid spores

MEIOTIC CELL DIVISION

sporophyte (2n)

n spore

② Mitotic cell division of the spore produces a multicellular gametophyte

⑤ Mitotic cell division of the zygote produces a multicellular sporophyte

diploid generation

haploid generation

zygote 2n

gametophyte (n)

④ Fusion of gametes produces a diploid zygote

FERTILIZATION

n

③ The gametophyte produces gametes

n

gametes (eggs and sperm)

haploid (*n*)
diploid (*2n*)

▲ **FIGURE 44-1 Alternation of generations**

▲ **FIGURE 44-2 Spore production in ferns** In most ferns, clusters of reproductive cells on the undersides of the fronds of the sporophyte stage (left) produce spores that burst out of the capsules in which they are produced (right) and are dispersed by the wind.

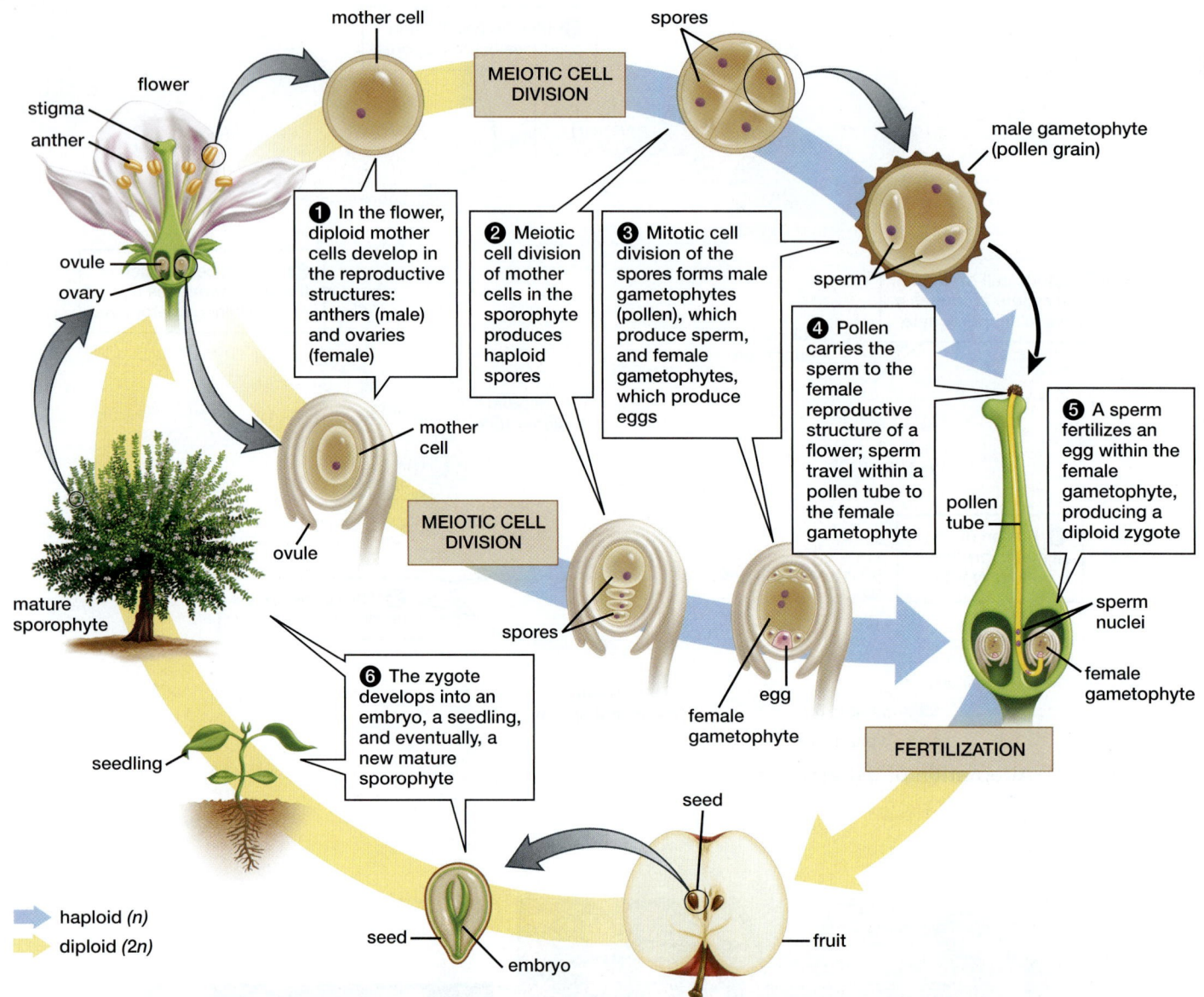

▲ FIGURE 44-3 The sexual life cycle of a flowering plant

Let's look at the sexual life cycle of flowering plants in more detail (**Fig. 44-3**). The multicellular sporophyte stage is the plant you see in gardens, orchards, forests, and fields. At the appropriate time of year, it produces flowers. Male and female reproductive structures in the flower produce specialized *mother cells* (**Fig. 44-3 ❶**) that undergo meiotic cell division to form haploid male and female spores (**Fig. 44-3 ❷**). The spores undergo mitotic cell division to produce haploid male and female gametophytes (**Fig. 44-3 ❸**). The male gametophyte is the **pollen grain.** It is multicellular, but just barely—it consists of only three cells, two of which are sperm. The female gametophyte usually consists of seven cells, one of which is an egg. The female gametophyte remains in the flower. Pollen grains are carried by wind or animals to the female reproductive structure of the flower (**Fig. 44-3 ❹**). One of the pollen's sperm fertilizes an egg, producing a diploid zygote (**Fig. 44-3 ❺**). The zygote undergoes repeated

mitotic cell divisions to form an embryo and eventually a new adult sporophyte plant (**Fig. 44-3 ❻**).

CHECK YOUR LEARNING

Can you describe the plant sexual life cycle? Can you diagram the life cycles of ferns and flowering plants and explain the reproductive function of each stage?

44.2 WHAT ARE THE FUNCTIONS AND STRUCTURES OF FLOWERS?

Angiosperms have many features that distinguish them from gymnosperms, but flowers are one of the most conspicuous differences. Although both gymnosperms and flowering plants package their sperm inside pollen grains, gymnosperm pollen is carried by the wind from one plant to another (see Chapter 21). Relying on the wind to transport

Health | Watch

Are You Allergic to Pollen?

Wind pollination can succeed only if plants release huge quantities of pollen into the air. Unfortunately for allergy sufferers, people often inhale these microscopic male gametophytes. Proteins on the surfaces of pollen grains activate the immune systems of sensitive individuals, causing itchy eyes, runny noses, coughing, and sneezing. If you are among these unlucky people, your own immune system is creating these symptoms in an attempt to rid your body of harmless pollen, which it mistakes for dangerous parasites that should be expelled as quickly as possible (see Chapter 36).

People who suffer from "hay fever" are usually sensitive only to specific types of pollen. In temperate climates, springtime sufferers may be allergic to tree pollen, whereas summertime allergies are often caused by grasses. In North America, however, the most common cause of hay fever is not hay (grass), but ragweed, which releases pollen during late summer and fall (Fig. E44-1). A single ragweed plant can release a billion pollen grains during its lifetime; collectively, ragweed plants release about a million tons of pollen in North America each year. Ragweed pollen has been collected 400 miles out to sea and 2 miles up in the atmosphere, so it is nearly impossible to avoid ragweed pollen completely.

Plants pollinated by bees and other animals rarely cause allergies, because their pollen is sticky and produced in small amounts. Goldenrod, for example, which blooms during the ragweed season and is conspicuously yellow, is often blamed for allergies that are actually caused by ragweed. Goldenrod's yellow blooms attract bee and butterfly pollinators, and most people can enjoy goldenrod in perfect comfort.

▲ FIGURE E44-1 **Inconspicuous ragweed flowers and their pollen**

pollen is a successful reproductive strategy—if it weren't, conifers would not be so abundant—but it is also inefficient, because the vast majority of pollen grains do not reach their targets. Many evolutionary biologists believe that the selective advantage of the first flowers was to entice animals, particularly insects, to carry pollen from one plant to another. Most flowering plants trade a few pollen grains or a sip of nectar as lunch for an insect in exchange for pollen transport. This mutually beneficial relationship led to the evolution of colorful, scented flowers that help the insects to locate them.

The ancestors of modern flowering plants probably depended on insects for pollen transport. However, about 10% of today's flowering plants, including many deciduous trees such as oaks, maples, birches, and cottonwoods, have evolved greatly reduced flowers that release their pollen into the wind. Grasses, sagebrush, and ragweed also produce prodigious amounts of wind-borne pollen. Some of the pollen reaches another flower of the same species. Unfortunately, some pollen ends up inside the noses of allergy sufferers, as we explore in "Health Watch: Are You Allergic to Pollen?"

Flowers Are the Reproductive Structures of Angiosperms

Flowers are the sexual reproductive structures of angiosperms, produced by the diploid sporophyte. A **complete flower** (Fig. 44-4), such as that of a petunia, rose, or lily, consists of four sets of modified leaves: the sepals, petals, stamens, and carpels. The **sepals** are located at the base of the flower. There are two major groups of flowering plants, the dicots and monocots (see Chapter 43). In dicots such as roses, strawberries, or apples, sepals are usually green and leaflike (Fig. 44-4a); in monocots such as daffodils, tulips, or amaryllis (Fig. 44-4b), sepals typically resemble the petals. In both dicots and monocots, sepals surround and protect the flower bud as the remaining three flower structures develop.

Just above the sepals are the **petals,** which are often brightly colored and fragrant, advertising the location of the flower to potential pollinators. The male reproductive structures, the **stamens,** are attached just above the petals. Each stamen usually consists of a slender **filament** bearing an **anther** that produces pollen. In the center of the flower are one or more female reproductive structures, called **carpels.**

(a) A representative dicot flower

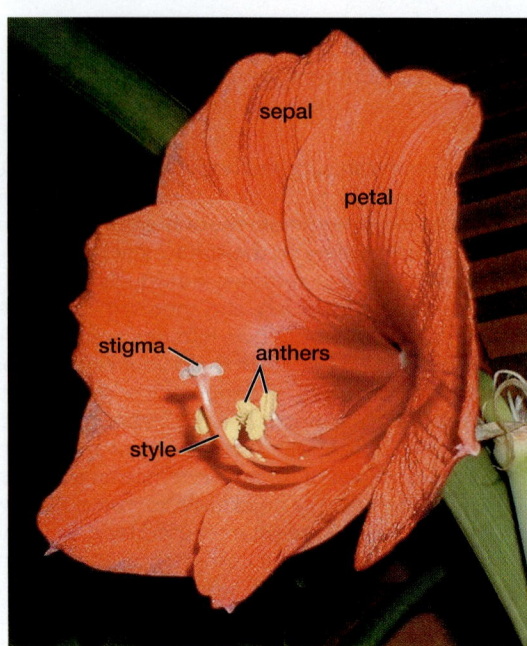

(b) An amaryllis (monocot) flower

▲ FIGURE 44-4 **A complete flower** **(a)** A complete flower has four parts: sepals, petals, stamens (the male reproductive structures), and at least one carpel (the female reproductive structure). This illustration shows a complete dicot flower. **(b)** The amaryllis is a complete monocot flower, with three sepals (virtually identical to the petals), three petals, six stamens, and three carpels (fused into a single structure). The anthers are well below the stigma, making self-pollination unlikely.

A typical carpel is vase shaped, with a sticky **stigma** mounted atop an elongated **style. Pollination** occurs when pollen from an anther of a stamen lands on the stigma of a carpel. The style connects the stigma with the **ovary** at the base of the carpel. Inside the ovary are one or more **ovules;** a female gametophyte develops inside each ovule. After fertilization, each ovule will become a **seed,** consisting of a small, embryonic plant and stored food for the embryo. The ovary will develop into a **fruit** enclosing the seeds.

Incomplete flowers lack one or more of the four floral parts. For example, grass flowers (see Fig. 44-8) lack both petals and sepals. If an incomplete flower lacks either stamens or carpels, it is called an **imperfect flower.**

CASE STUDY continued

Some Like It Hot—and Stinky!

A corpse flower is actually a mass of separate, tiny male and female flowers. In principle, sperm from the male flowers could fertilize eggs in the female flowers of the same plant. However, mating among close relatives can lead to genetic defects if offspring become homozygous for faulty recessive alleles of important genes (see Chapters 10 and 15). Self-fertilization, of course, is the ultimate in mating with a relative. The corpse flower avoids self-fertilization by opening its female flowers a day earlier than its male flowers. The female flowers usually wilt before the male flowers mature. Therefore, pollinators seldom transfer pollen from an anther to a stigma on the same plant.

Plant species with imperfect flowers produce separate male and female flowers, sometimes on a single plant, as in zucchini (**Fig. 44-5**) and the corpse flower. Other species with imperfect flowers produce male and female flowers on separate plants. Only female American holly trees, for

▲ FIGURE 44-5 **Male and female imperfect flowers** Plants of the squash family, such as zucchinis, bear separate female (left) and male (right) flowers. Note the small zucchini (actually a fruit) forming from the ovary at the base of the female flower.

QUESTION In species with separate male and female flowers on the same plant, why would natural selection favor individuals whose male and female flowers bloom at different times?

◄FIGURE 44-6 **Male gametophyte development**

anther

pollen sacs

microspore mother cell

sporophyte

❶ Microspore mother cells develop within the pollen sacs of the anther of a flower

MEIOTIC CELL DIVISION

❷ Meiotic cell division produces four haploid microspores

microspores

tube cell nucleus

mature pollen grain

sperm cells

stigma

immature pollen grain

tube cell cytoplasm

tube cell nucleus

generative cell

❹ The generative cell produces two sperm cells by mitotic cell division; the male gametophyte is now mature

❸ Each microspore produces an immature male gametophyte (a pollen grain) by mitotic cell division

➡ haploid *(n)*

➡ diploid *(2n)*

example, produce decorative red fruit, so female trees are often favored as ornamentals (although a few males must be nearby to pollinate the females).

The Pollen Grain Is the Male Gametophyte

Male gametophytes develop within the anthers of a flower (**Fig. 44-6**). Each anther consists of four chambers called pollen sacs. Within each pollen sac, hundreds to thousands of diploid **microspore mother cells** develop (**Fig. 44-6 ❶**). Each microspore mother cell undergoes meiotic cell division (see Chapter 9) to produce four haploid **microspores** (**Fig. 44-6 ❷**). Each microspore then undergoes one mitotic cell division to produce an immature pollen grain—an immature male gametophyte—consisting of two cells: a large **tube cell** occupying most of the volume of the pollen grain and a smaller **generative cell** that resides within the cytoplasm of the tube cell (**Fig. 44-6 ❸**). Mitotic cell division of the generative cell produces two haploid sperm cells (**Fig. 44-6 ❹**). The mature pollen grain, consisting of the two sperm cells inside the cytoplasm of the tube cell, becomes surrounded by a tough, waterproof coat, often sculpted with an elaborate pattern of pits and protrusions characteristic

of the plant species (**Fig. 44-7**). The coat protects the sperm during their journey to a sometimes-distant female carpel.

When the pollen has matured, the pollen sacs of the anther split open. In wind-pollinated flowers, such as those

▲ **FIGURE 44-7 Pollen grains** The tough outer coats of many pollen grains are elaborately sculptured in species-specific shapes and patterns, as shown in this colorized scanning electron micrograph.

▲ **FIGURE 44-8 Wind-pollinated flowers** The flowers of grasses and many deciduous trees are wind pollinated, with anthers (yellow structures) exposed to the wind.

QUESTION Would you expect wind-pollinated flowers to have colorful petals and sweet scents? Why or why not?

of grasses (Fig. 44-8) and oaks, the anthers usually protrude from small, often inconspicuous flowers. The slightest breeze carries off the pollen grains. Most are lost, but a few reach and pollinate other flowers of the same species. In animal-pollinated flowers, the pollen adheres weakly to the anther until a pollinator comes along and brushes or picks it off.

The Female Gametophyte Forms Within the Ovule

Within the ovary of a carpel, clusters of cells differentiate into ovules (Fig. 44-9). Depending on the plant species, a carpel may have as few as one ovule or as many as several dozen. Each young ovule consists of protective outer layers called **integuments,** which surround a single, diploid **megaspore mother cell** (Fig. 44-9 ❶). The megaspore mother cell undergoes meiotic cell division, producing four haploid **megaspores** (Fig. 44-9 ❷). Only one megaspore survives; the other three degenerate. The nucleus of the surviving megaspore undergoes three rounds of mitosis, producing eight haploid nuclei (Fig. 44-9 ❸). Plasma membranes and cell walls then form, dividing the cytoplasm into the seven (not eight) cells that make up the female gametophyte (Fig. 44-9 ❹). There are three small cells at each end, each with a single nucleus, and one large central cell with two nuclei. The egg is one of the three cells at the lower end, located near an opening in the integuments of the ovule.

Pollination of the Flower Leads to Fertilization

Pollination is necessary for fertilization, but these are two distinct events, just as copulation and fertilization are separate events in mammals. Pollination occurs when a pollen grain lands on the stigma of a flower of the same plant species (Fig. 44-10 ❶). The pollen grain absorbs water from the stigma and swells, splitting open the pollen coat. The tube cell, which makes up most of the pollen grain (see Fig. 44-6), elongates through the opening in the pollen coat and burrows through the style, producing a tube that will conduct sperm down the style to an ovule within the ovary (Fig. 44-10 ❷).

❶ A megaspore mother cell develops within each ovule of the ovaries of a flower

ovule

megaspore mother cell

ovary

integuments

MEIOTIC CELL DIVISION

❷ Meiotic cell division produces four haploid megaspores; three degenerate

megaspores

❸ The single remaining megaspore forms eight nuclei by mitosis

❹ Cytoplasmic division produces the seven cells of the mature female gametophyte

central cell with two nuclei

female gametophyte

egg cell

▲ **FIGURE 44-9 Female gametophyte development**

➡ haploid (n)

➡ diploid (2n)

▲ FIGURE 44-10 **Pollination and fertilization of a flower**

When the pollen tube reaches an opening in the integuments of an ovule and enters the female gametophyte, the tip of the tube ruptures, releasing the two sperm. In a process unique to flowering plants, called **double fertilization,** both sperm fuse with cells of the female gametophyte (Fig. 44-10 ❸). One sperm fertilizes the egg, producing a diploid zygote that will develop into an embryo and eventually into a new sporophyte. The second sperm enters the large central cell, where its nucleus fuses with the two nuclei already present, forming a triploid nucleus, containing three sets of chromosomes. Through repeated mitotic cell divisions, the central cell will develop into the triploid **endosperm,** a food-storage tissue within the seed. The other five cells of the female gametophyte degenerate.

CHECK YOUR LEARNING

Can you diagram the structure of a complete flower and explain the function of each part? Can you describe the development of male and female gametophytes in flowering plants? Are you able to explain the processes of pollination and double fertilization?

44.3 HOW DO FRUITS AND SEEDS DEVELOP?

After double fertilization, the female gametophyte and the surrounding integuments of the ovule develop into a seed. The seed is surrounded by the ovary, which will form a fruit. The petals and stamens shrivel and fall away as the fruit enlarges.

The Fruit Develops from the Ovary

When you eat a fruit, you are consuming the plant's ripened ovary, sometimes accompanied by other flower parts (Fig. 44-11). In a bell pepper, for example, the edible flesh develops from the wall of the ovary. Each of the seeds develops from an individual ovule within the ovary. As you will learn in Section 44.6, fruits are not always edible; some are hard, feathery, winged, spiked, adhesive, or even explosive. These

? HAVE **YOU** EVER WONDERED . . .

When Is a Fruit a Vegetable?

To a biologist, a fruit is the ripened, seed-containing ovary of an angiosperm, often including some other parts of the flower. A vegetable is a nonreproductive part of a plant, such as a leaf or root. The culinary and legal worlds, however, beg to differ. Many foods that a biologist would call fruits are deemed vegetables by chefs and cookbooks: tomatoes, zucchinis, cucumbers, bell and chili peppers, and eggplants, to name just a few. The U.S. Supreme Court ruled on the status of the tomato in 1893, calling it a vegetable, reasoning that vegetables are served as part of a main course and fruits are served for dessert. Arkansas decisively straddles the fence: In 1987, the South Arkansas Vine Ripe Pink Tomato was named both the state fruit and the state vegetable!

▲ FIGURE 44-11 Development of fruit and seeds in a bell pepper Fruits and seeds develop from flower parts. The ovary wall ripens into the fruit flesh. Each bell pepper ovary houses many ovules, which develop into seeds.

various shapes, colors, and textures all serve the same function: They help disperse seeds away from the parent plant, in many cases taking advantage of the mobility of animals.

The Seed Develops from the Ovule

Three distinct developmental processes transform an ovule into a seed (Fig. 44-12a). First, the integuments of the ovule

thicken, harden, and become the **seed coat** that surrounds and protects the seed. Second, the triploid central cell divides rapidly. The resulting daughter cells absorb nutrients from the parent plant, forming a food-filled endosperm. Third, the zygote develops into an embryo.

As the seed matures, the embryo begins to differentiate into shoot and root (Figs. 44-12b, c). The shoot portion consists of a short stem, often with one or two developing

▶ FIGURE 44-12 Seed development (a) Seed formation begins after one sperm fuses with the egg, forming a diploid zygote, and the second sperm fuses with the two nuclei of the central cell. The embryo develops from the zygote. The endosperm develops from the triploid central cell, which undergoes many mitotic cell divisions as it absorbs nutrients from the parent plant. The integuments of the ovule develop into the seed coat. **(b)** Monocot seeds, such as corn, have a single cotyledon and usually retain most of their endosperm until germination. **(c)** Dicot seeds, such as beans, have two cotyledons that usually absorb most of the endosperm before germination. Dicot seeds, therefore, typically consist mostly of the two large cotyledons.

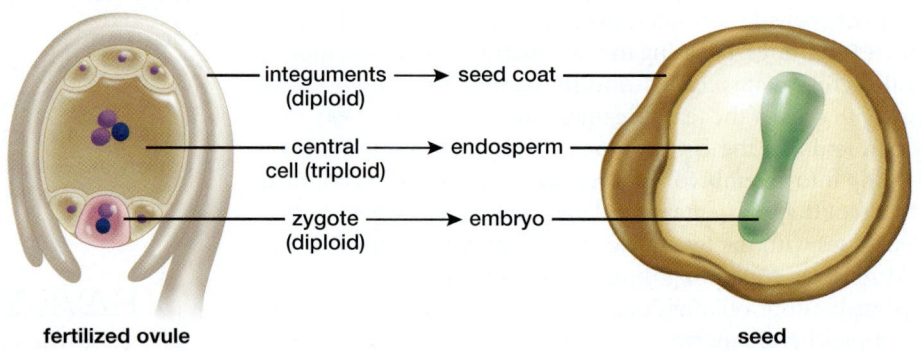

(a) Early development of the seed

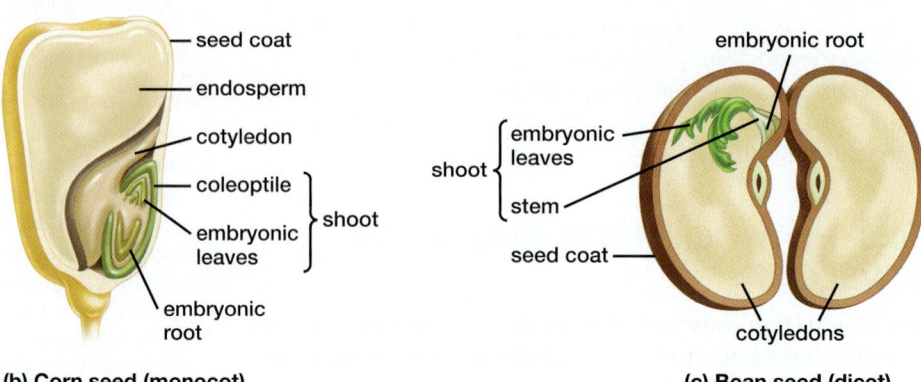

(b) Corn seed (monocot)

(c) Bean seed (dicot)

leaves, and either one or two **cotyledons,** or "seed leaves," that absorb food molecules from the endosperm and transfer them to other parts of the embryo. The seeds of monocots, as their name implies, have a single cotyledon ("mono" means "one"; see Fig. 44-12b). In most monocots, including grasses, rice, corn, and wheat, the cotyledon absorbs some endosperm during development, but most of the endosperm remains in the mature seed until the seed sprouts. Wheat flour is ground-up endosperm; wheat germ is made from the embryo. A monocot embryo is enclosed in a pair of sheaths, one surrounding the developing root and a second, called the **coleoptile,** surrounding the developing shoot tip.

Dicot seeds have two cotyledons ("di" means "two"; see Fig. 44-12c). In the seeds of most dicots, including peas, beans, peanuts, and walnuts, the cotyledons absorb most of the endosperm during seed development, so the mature seed is virtually filled by the cotyledons. If you strip the thin seed coat from a bean or peanut, you will find that the inside splits easily into two halves; each is a cotyledon. The tiny white nub adhering to one of the cotyledons is the rest of the embryo.

CHECK YOUR LEARNING

Are you able to explain how the parts of a flower develop into the parts of a fruit and its seeds? Can you describe the differences between monocot and dicot seeds?

44.4 HOW DO SEEDS GERMINATE AND GROW?

Germination, often called sprouting, occurs when the embryonic plant within a seed grows, breaks out of the seed, and forms a seedling. Seeds need warmth and moisture to germinate, but even under ideal conditions, newly matured seeds may not germinate immediately. Instead, they often enter a period of **dormancy** during which they will not sprout. Dormant seeds are typically able to resist adverse environmental conditions such as freezing and drying.

Seed Dormancy Helps Ensure Germination at an Appropriate Time

Seed dormancy solves two problems. First, it prevents seeds from germinating within a moist fruit, where the sprout might be eaten by a fruit-eating animal or destroyed by mold growing in rotting fruit. Even if they survived, multiple seedlings germinating within a single fruit would grow in a dense cluster, competing with one another for nutrients and light. Second, environmental conditions that are suitable for germination (such as warmth and moisture) may sometimes be followed by conditions that would not allow the seedling to survive and mature. For example, seeds in temperate climates are usually produced in late summer and have a harsh winter ahead of them. Most do not germinate even during mild autumn weather. Instead, the seeds remain dormant through both autumn and winter, so that tender new sprouts do not freeze. Germination usually occurs the following spring. In warm, moist, tropical regions, where environmental conditions are suitable for germination throughout the year, seed dormancy is much less common.

In addition to warmth and water, many plant species have further requirements for seed germination. These requirements are finely tuned to the plant's native environment and dispersal mechanisms. The three most common requirements to break seed dormancy are drying, exposure to cold, and disruption of the seed coat.

- **Drying** Seeds that require drying often are dispersed by fruit-eating animals that cannot digest the seeds. The animals excrete the seeds in their feces. Exposed to the air, the seeds dry out. Later, when temperature and moisture levels are favorable, they germinate.
- **Cold** Seeds of many temperate and arctic plants will not germinate unless they are exposed to prolonged subfreezing temperatures, followed by sufficient warmth and moisture. Requiring a substantial cold spell keeps them from sprouting until the following spring.
- **Seed coat disruption** The seed coat may need to be weathered or partially digested before germination can occur. Some coats contain chemicals that inhibit germination. In deserts, for example, years may go by without enough water for a plant to complete its life cycle. The seed coats of many desert plants have water-soluble chemicals that inhibit germination, and only a hard rainfall can wash away enough of the inhibitors to allow sprouting.

During Germination, the Root Emerges First, Followed by the Shoot

During germination, the embryo absorbs water, which makes it swell and burst its seed coat. The root usually emerges first and grows rapidly, absorbing water and minerals from the soil (Fig. 44-13). Much of the water is transported to the shoot, whose cells elongate and push upward through the soil toward the light.

The energy for germination ultimately comes from the endosperm of the seed. Recall that monocot seeds retain most of their endosperm until germination. During germination, the cotyledon digests the endosperm, absorbing its nutrients and transferring them to the growing embryo. In dicot seeds, the cotyledons have already absorbed most of the endosperm long before germination, so the cotyledons merely transfer these nutrients to the embryo as germination occurs.

Most seeds are buried in soil to some extent, so the embryo, especially its apical meristems, must be protected from being damaged by sharp soil particles during germination. The apical meristem of a root tip is protected by a root cap throughout the life of the plant (see Chapter 43). Shoots, however, are underground for only a brief time during germination, and so their apical meristems need only temporary protection. In monocots, the coleoptile encloses the shoot tip like a glove around a finger (Fig. 44-13a), and pushes aside soil particles as the tip grows. Once out in the air, the coleoptile degenerates, allowing the shoot to emerge. The cotyledon stays belowground in the remnants of the seed.

▶ **FIGURE 44-13 Seed germination** First, the root grows rapidly, absorbing water and minerals. Using these resources, the shoot pushes upward through the soil. **(a)** In monocots such as corn, the shoot tip is protected within a tough coleoptile. **(b)** In some dicots, such as the bean shown here, the hypocotyl bends, forming a hook that pushes through the soil first, protecting the downward-pointing shoot tip. In other dicots, such as the pea, the bend forms in the epicotyl.

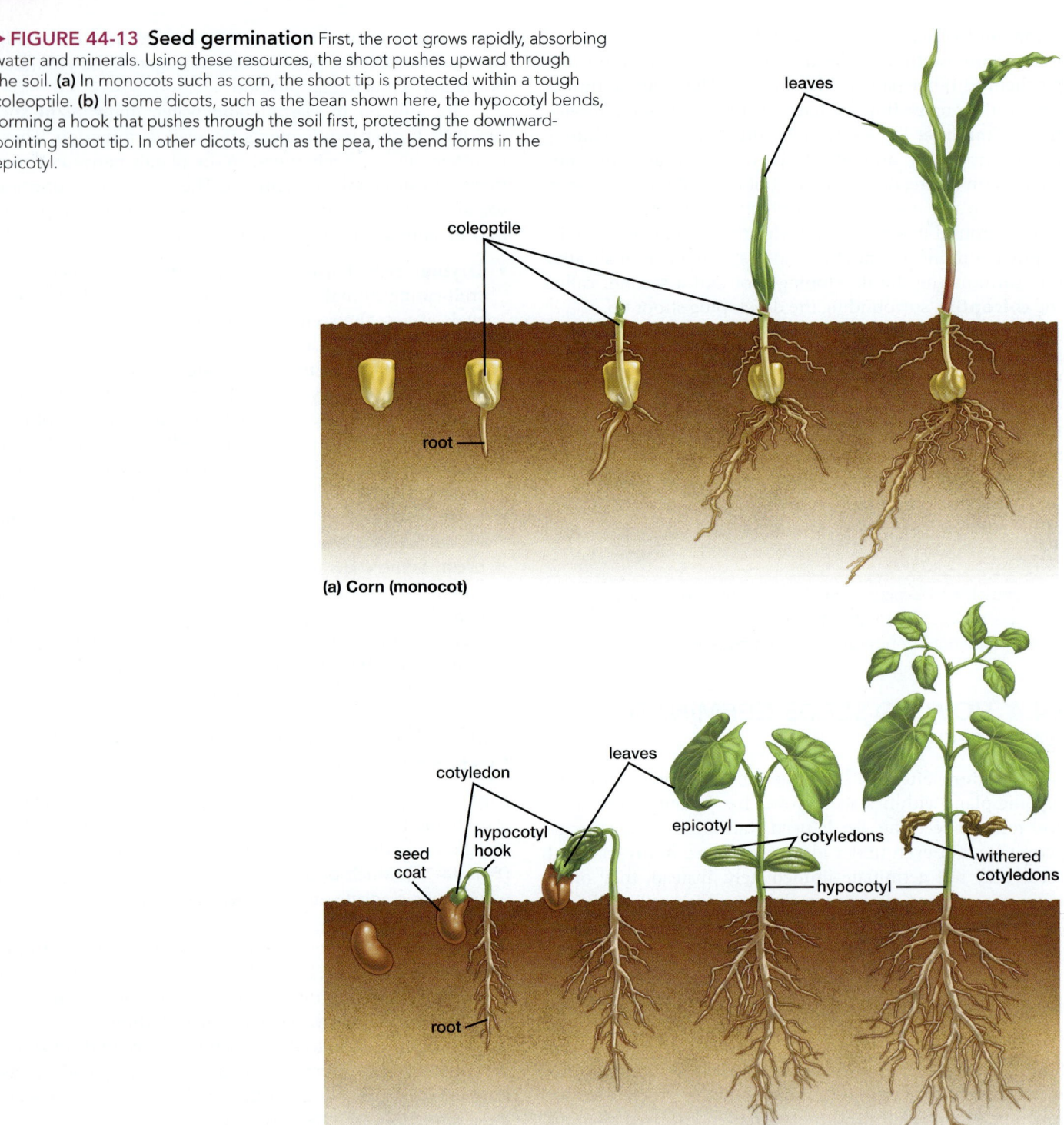

(a) Corn (monocot)

(b) Bean (dicot)

Dicots do not have coleoptiles. Instead, they form a hook in the embryonic shoot, either in the section below the point where the cotyledons attach to the stem, called the **hypocotyl** ("hypo" is Greek, meaning "below"), or in the section above the point where the cotyledons attach to the stem, called the **epicotyl** ("epi" means "above"; Fig. 44-13b). The hook, encased in thick-walled epidermal cells, forces its way up through the soil, clearing the path for the downward-pointing apical meristem and its delicate new leaves. In dicots with hy-

pocotyl hooks, such as zucchini and beans (see Fig. 44-13b), the elongating shoot carries the cotyledons out of the soil into the air. These aboveground cotyledons typically become green and photosynthetic (hence their name of "seed leaves"), and transfer both previously stored food and newly synthesized sugars to the shoot (Fig. 44-14). Eventually, the cotyledons wither away. In dicots with epicotyl hooks, the cotyledons remain below the ground, shriveling up as the embryo absorbs their stored food. In all dicots, the shoot straightens after it

▲ FIGURE 44-14 **Cotyledons nourish the developing plant** In some dicots, such as the zucchini shown here, the cotyledons emerge from the soil, expand, and photosynthesize. The crinkled leaves develop a bit later. Eventually, the cotyledons shrivel up.

emerges, orienting its leaves toward the sunlight and undergoing primary growth (see Chapter 43).

CHECK YOUR LEARNING

Do you understand why many seeds undergo dormancy before germinating? Can you describe the process of germination in monocots and dicots?

44.5 HOW DO PLANTS AND THEIR POLLINATORS INTERACT?

Plants and their pollinators have coevolved; that is, each has acted as an agent of natural selection on the other. Animal-pollinated flowers evolved traits that attract useful pollinators and frustrate undesirable visitors that might eat nectar or pollen without pollinating the flower in return. Pollinators evolved senses and behaviors that help them to locate and identify nutritious flowers and extract the nectar or pollen. Animal-pollinated flowers can be loosely grouped into three categories, on the basis of the benefits, real or perceived, that they offer to potential pollinators: food, sex, or a nursery.

Some Flowers Provide Food for Pollinators

Many flowers provide food for foraging animals such as bees, moths, butterflies, hummingbirds, and even a few mammals, such as bats and lemurs, as we explore later in this chapter in "Earth Watch: Pollinators, Seed Dispersers, and Ecosystem Tinkering" on p. 875. In return for food, the animals unwittingly distribute pollen from flower to flower. Most flying pollinators locate flowers from a distance because the flower colors contrast with the mass of green leaves surrounding them. For example, bees have good color vision, but see a different range of colors than humans do (**Fig. 44-15**). Typically, bees do not perceive red as a distinct color, but they can see ultraviolet light. Therefore, bee-pollinated flowers are usually white, blue, yellow, or orange but not red. Many have markings that reflect UV light, including central spots or lines pointing toward the center, almost like bull's-eyes. We can also thank the bees for most of the sweet-smelling flowers because "flowery" odors attract these pollinators.

(a) A comparison of color vision in humans and bees

human vision   bee vision

(b) Flower color patterns seen by humans and bees

▲ FIGURE 44-15 **Ultraviolet patterns guide bees to nectar** (a) The spectra of color vision for humans and bees overlap considerably, but are not identical. Humans (top) are sensitive to red, which bees (bottom) do not perceive; bees can see near-UV light, which is invisible to the human eye. (b) Many flowers photographed (left) under ordinary daylight and (right) under UV light show striking differences in color patterns. Bees can see the UV patterns that presumably direct them to the nectar- and pollen-containing centers of the flowers.

Bee-pollinated flowers have structural adaptations that help ensure pollen transfer. In the Scotch broom flower, for example, nectar forms in a crevice between enclosing petals. In newly opened flowers, pollen-laden stamens lurk within the crevice. When a bee visits a young flower, the stamens emerge, brushing pollen onto its back as the bee's weight deflects the petals downward (**Fig. 44-16**). In older flowers, the carpel's style elongates, pushing the sticky stigma out through the crevice; therefore, when a pollen-coated bee delves for nectar, she leaves pollen behind on the stigma.

Many flowers adapted for moth and butterfly pollinators have nectar-containing tubes that accommodate the long tongues of these insects. Flowers pollinated by night-flying moths open only in the evening. Most are white,

CASE STUDY continued

Some Like It Hot—and Stinky!

How do corpse flowers get hot? They have special mechanisms for disconnecting cellular respiration from ATP synthesis. In most cells, cellular respiration uses about 40% of the energy in glucose to synthesize ATP, with the rest given off as heat (see Chapter 8). Hot flowers, on the other hand, synthesize very little ATP; instead, almost all of the energy in glucose is released as heat, causing the flower to warm up.

▶ **FIGURE 44-16 Pollinating a pollinator** **(a)** In Scotch broom flowers, a bee finds nectar near the junction of the top and bottom petals. **(b)** The bee's weight bends the bottom petals downward, causing pollen-laden stamens to pop up and cover the bee's hairy back with pollen. The bee will carry the pollen to other Scotch broom flowers, leaving some on their stigmas.

(a) A bee on a Scotch broom flower

stamens

(b) The flower deposits pollen on the bee

which makes them more visible in the dark. Some also give off strong, musky odors that attract moths. Bat-pollinated flowers are also usually white and open at night. Beetles and flies often feed on animal wastes or carrion, so flowers pollinated by these insects frequently smell like dung or rotting flesh. These flowers deceive their pollinators by smelling like a nutritious meal but offering no food. Some flowers, including the corpse flower, also heat up. The warmth attracts pollinators and helps broadcast the flower's foul scents.

Hummingbirds are one of the few important vertebrate pollinators (**Fig. 44-17**), although some mammals also pollinate flowers (see "Earth Watch: Pollinators, Seed Dispersers, and Ecosystem Tinkering"). Because hummingbirds have a poor sense of smell, hummingbird-pollinated flowers seldom synthesize fragrant chemicals. However, they often produce more nectar than other flowers, because hummingbirds need more energy than insects do and will favor flowers that provide it. Hummingbird-pollinated flowers may have a deep, tubular shape that accommodates the birds' long bills and tongues while preventing most insects from reaching the nectar. In addition, these flowers are often red, which is attractive to hummingbirds but is not seen well by bees.

Some Flowers Are Mating Decoys

A few plants, most notably some orchids, take advantage of the mating drives of male wasps, bees, and flies. These orchid flowers mimic female wasps, bees, or flies both in scent (the orchids release a sexual attractant similar to that produced by the female insect) and shape (**Fig. 44-18**). When a male insect lands atop these "faux females" and attempts to copulate, a packet of pollen often becomes attached to the insect. When the amorous insects repeat their mating attempts on other orchids of the same species, they transfer pollen packets from one flower to another.

▲ **FIGURE 44-17 Hummingbirds are effective pollinators** A hummingbird feeds at a hibiscus flower. The flower's anthers are positioned so that they deposit pollen on the bird's head.

QUESTION Why have many plants that are pollinated by hummingbirds evolved flowers shaped like long, narrow tubes?

▲ **FIGURE 44-18 Sexual deception promotes pollination** This male wasp is trying to copulate with an orchid flower. The result is successful reproduction—by the orchid, but not the wasp!

Earth|Watch

Pollinators, Seed Dispersers, and Ecosystem Tinkering

Flowering plants dominate most terrestrial ecosystems largely because of mutually beneficial relationships with animals that pollinate their flowers and disperse their seeds. If pollinators or seed dispersers are eliminated, intentionally or accidentally, entire ecosystems may be endangered.

Sound alarmist? Consider Madagascar. On this island off the coast of Africa, primates called lemurs are the primary pollinators and seed dispersers (Fig. E44-2a). In some cases, no other animal will suffice. For example, the traveler's palm has large, tough flowers that must be ripped apart for pollen to escape or enter. The black-and-white ruffed lemur—except for people, the world's largest pollinator—is probably the only animal on Madagascar with enough strength and dexterity to pollinate traveler's palms. Unfortunately, black-and-white ruffed lemurs are critically endangered.

Other species of lemurs are important seed dispersers for at least 40 species of trees in Madagascar. When lemurs eat the trees' fruit, they swallow the seeds. A few hours later, the seeds are excreted in their feces, usually a few hundred yards away from the parent tree. What's more, passing through a lemur's digestive tract makes the seeds more likely to sprout. Large lemurs are the only animals on Madagascar that are big enough to swallow the seeds of many of the rain-forest trees, so they may be the only animals that disperse the seeds. About a quarter of Madagascar's lemur species are endangered, and if the lemurs disappear, many of the trees may disappear, too. Because 90% of Madagascar's plant species are found nowhere else in the world, the loss of lemurs may mean the extinction of many species of trees.

In many other tropical forests, monkeys and fruit-eating bats are important agents of seed dispersal (Fig. E44-2b). In some Amazon River rain forests, for example, howler monkeys disperse seeds of more than 100 species of plants. In Mexican rain forests, spider monkeys and howler monkeys disperse the seeds of dozens of tree species. As with seeds dispersed by lemurs, a brief trip through the digestive tracts of monkeys or bats helps seeds to germinate.

Most domesticated "ecosystems"—large-scale farms—rely on humans for seed dispersal. However, they are often vulnerable to the loss of pollinators. Some crops, such as corn and wheat, are pollinated by the wind, but most fruits and nuts, and many vegetables, depend on pollination by introduced European honeybees. The U.S. Department of Agriculture estimates that bee pollination provides about $15 billion in crop value, with honeybees responsible for most of that.

Honeybees have been in trouble for some time, plagued by mites (tiny relatives of spiders), parasitic flies, viruses, fungi, and pesticides. However, in the fall of 2006, many whole colonies disappeared, leaving only the queen and larvae behind, a phenomenon named colony collapse disorder. Some beekeepers lost up to 90% of their colonies that winter. No one really knows why colony collapse disorder occurs. A recent hypothesis is that pesticide exposure weakens the bees' immune systems, making them more susceptible to mites or viruses.

There are many native bee species—about 2,000 in California alone—many of which pollinate flowers. If colony collapse disorder continues, can native bees take up the slack? Probably not—at least not right away. Populations of many native bees have been severely reduced by competition from honeybees. Further, the hedgerows, meadows, and forest edges on which native bees depend for food and shelter have often been replaced by enormous, intensively managed farms. Native bees face other threats, too. For example, populations of several species of bumblebees in North America have declined by more than 90%; scientists suspect that fungal infections and certain pesticides are the major culprits.

Flowering plants, their pollinators, and their seed dispersers often form an intricate, interconnected web, with each supporting the others. As ecologist Aldo Leopold wrote in *A Sand County Almanac*, "To keep every cog and wheel is the first precaution of intelligent tinkering." When humans tinker with Earth's ecosystems, we must be careful not to lose some of their critical cogs and wheels.

◀ FIGURE E44-2 Animal seed dispersers are crucial to some ecosystems (a) Endangered black-and-white ruffed lemurs are the only pollinators for traveler's palm trees in Madagascar. (b) A bat eats a ripe fig in Kenya. Without bats and other seed-dispersing animals, some tropical forest communities would be seriously disrupted.

(a) Black-and-white ruffed lemur

(b) Fruit-eating bat

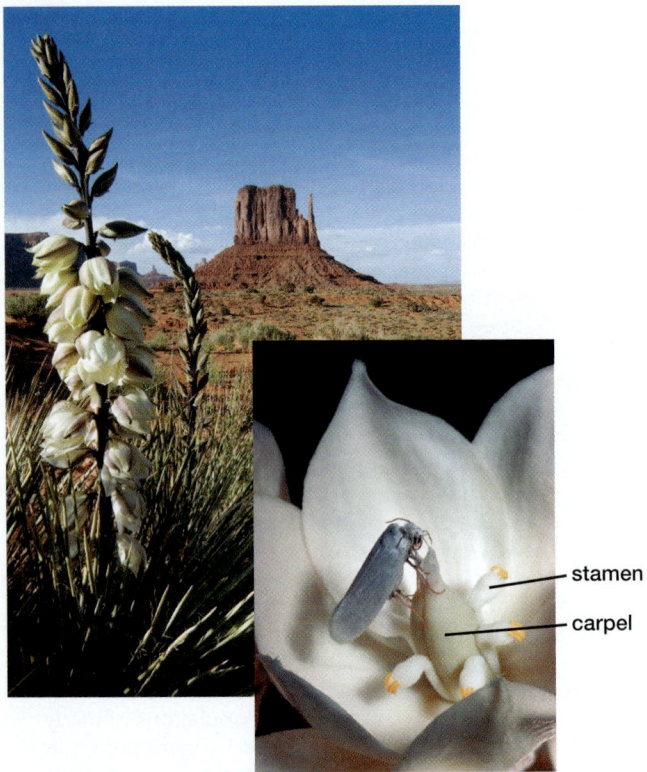

▲ **FIGURE 44-19 A mutually beneficial relationship** Yuccas bloom in Monument Valley, Arizona. (Inset) A yucca moth places pollen on the stigma of the carpel of a yucca flower.

Some Flowers Provide Nurseries for Pollinators

Perhaps the most elaborate relationships between plants and pollinators occur in a few cases in which insects fertilize a flower and then lay their eggs in the flower's ovary. This arrangement occurs between milkweeds and milkweed bugs, figs and fig wasps, and yuccas and yucca moths (Fig. 44-19). For example, when a female moth visits a yucca flower, she collects pollen and rolls it into a compact ball. She carries the pollen ball to another yucca flower, drills a hole in the ovary wall, and lays her eggs inside the ovary. Then she smears the pollen ball on the stigma of the flower. By pollinating the yucca, the moth ensures that the plant will provide a supply of developing seeds for its caterpillar offspring. Because the caterpillars eat only a fraction of the seeds, the yucca also reproduces successfully. The mutual adaptation of yucca and yucca moth is so complete that neither can reproduce without the other.

CHECK YOUR LEARNING

Can you describe how the structures of flowers and their specific animal pollinators facilitate efficient pollination?

44.6 HOW DO FRUITS HELP TO DISPERSE SEEDS?

A plant benefits if its seeds are dispersed far enough away so that its offspring don't compete with it for light and nutrients. In addition, plant predators, from insects to mammals, are often abundant nearby, or are attracted to, the parent plants. Although the parent may be able to withstand some browsing, a deer or a handful of voracious caterpillars might kill a seedling. Finally, plant species will be more successful and widespread if they at least occasionally disperse their seeds to distant habitats. The fruits of flowering plants use a fascinating variety of mechanisms to disperse seeds.

Explosive Fruits Shoot Out Seeds

A few plants develop explosive fruits that eject their seeds far away from the parent plant. Dwarf mistletoes, common parasites of trees, produce fruits that shoot sticky seeds more than 60 feet (about 20 meters). If a seed strikes a nearby tree, it sticks to the bark and germinates, sending rootlike fibers into the vascular tissues of its host, from which it draws its nourishment. Because the proper germination site for a mistletoe seed is not the ground but a tree limb, shooting the seeds, rather than dropping them, is clearly useful. Other plants with explosive fruits include the touch-me-not, witch hazel, and squirting cucumber.

Lightweight Fruits May Be Carried by the Wind

Dandelions, milkweeds, elms, and maples produce lightweight fruits with large wind-catching surfaces. Each hairy tuft on a dandelion puff is a separate fruit bearing a single small seed that it can carry for miles if the winds cooperate (Fig. 44-20a). The single wing of the maple fruit, in contrast, causes its seed to spin like a propeller as it falls, usually taking it only a few yards from its parent tree (Fig. 44-20b).

Floating Fruits Allow Water Dispersal

Many fruits can float on water for a time and may be carried in streams or rivers, although this is usually not their principal method of dispersal. The coconut fruit, however, is a champion floater. Round, buoyant, and waterproof, the

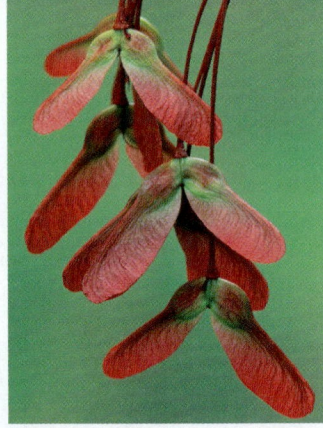

(a) Dandelion fruits   **(b) Maple fruits**

▲ FIGURE 44-20 **Wind-dispersed fruits (a)** Dandelion fruits have filamentous tufts that catch the breezes. **(b)** Maple fruits resemble miniature glider–helicopters, whirling away from the tree as they fall.

▲ FIGURE 44-21 **Water-dispersed fruit** This coconut may have been washed ashore after a long journey at sea. Coconut "meat" and coconut "milk" are two different types of endosperm. The large size and massive food reserves of coconuts are probably adaptations for successful germination and growth on barren, sandy beaches.

QUESTION Although many fruits can float, why is water seldom their primary method of dispersal?

coconut drops from its parent palm, often on a sandy shore. It may germinate there or it may roll into the sea and float for weeks or months until it washes ashore on some distant isle (**Fig. 44-21**). There, it may germinate, possibly establishing a new coconut colony where none previously existed.

Clingy or Edible Fruits Are Dispersed by Animals

Clingy fruits, such as burdocks, burr clover, foxtails, and stick-tights, grasp animal fur (or human clothing) with prongs, hooks, spines, or adhesive hairs (**Fig. 44-22**). The parent plant holds its ripe fruit very loosely, so that even slight contact with fur pulls the fruit off the plant, leaving it stuck to the animal. Some of these fruits later fall off as the animal brushes against objects, grooms itself, or sheds its fur.

Unlike hitchhiker fruits, edible fruits benefit both the plant and the animal disperser. The plant stores sugars, starches, and appealing flavors in a fleshy fruit that surrounds the seeds, enticing hungry animals (**Fig. 44-23**). Some edible fruits, including peaches, plums, and avocados, contain large, hard seeds that animals usually do not eat. Other fruits, such as blackberries, raspberries, strawberries, and tomatoes, have small seeds that animals swallow. These seeds are eventually excreted unharmed. Some have seed coats that must be scraped or weakened by passing through an animal's digestive tract before they will germinate. Besides being transported away from its parent plant, a seed that is swallowed and excreted benefits in another way: It ends up with its own supply of fertilizer!

It is often important for the right type of animal to eat the fruit and seeds. For example, when a mammal eats the seeds of chili peppers, it experiences a painful burning sensation, which discourages mammals from eating the fruit. If mammals do swallow chili seeds, their digestive tracts destroy

▲ FIGURE 44-22 **The cocklebur fruit uses hooked spines to hitch a ride on furry animals** This dog—or its owner—will eventually dislodge the cockleburs, often far away from the parent plant.

the seeds. Birds, on the other hand, are apparently insensitive to the burning chemicals in chili seeds. Further, seeds that have passed through a bird's digestive tract germinate at three times the rate of those that just fall to the ground. Therefore, being eaten by a bird, but not a mammal, ensures the most successful reproduction for chili peppers.

CHECK YOUR LEARNING

Do you understand why it is useful for seeds to disperse away from their parent plant? Can you describe how fruit structures aid in seed dispersal?

▲ FIGURE 44-23 **The colors of ripe fruits attract animals** In Pennsylvania, bright red serviceberries attract cedar waxwings in summer and autumn. Only ripe fruits containing mature seeds are sweet and brightly colored, appealing to animals that feed on them and disperse their seeds.

CASE STUDY revisited

Some Like It Hot—and Stinky!

The corpse flower, starfish flower, stinking corpse lily, and dead-horse arum are pollinated by carrion-loving beetles and flies. Chemical analysis shows that the dead-horse arum produces some of the same putrid chemicals that rotting carcasses do. Decaying carcasses also warm up, a by-product of the metabolism of bacteria decomposing the flesh. Heat production in these flowers is probably another aspect of corpse mimicry; researchers have found that flies prefer warm, smelly flowers to cool, equally smelly ones.

The dead-horse arum has a particularly sophisticated bloom that ensures not just pollination but cross-pollination. Its bloom consists of numerous, separate, female and male flowers contained in a chamber at the base of the flower. The female flowers blossom on the first day that the plant blooms. This is also when the plant emits its putrid odor and warms up. Lured by the smell and the warmth, flies enter the chamber. Inward-pointing spines at the entrance keep them from leaving again. As they blunder about in the chamber, the flies shower the female flowers with any pollen that they may have picked up from an earlier visit to another arum. By the next morning, the female flowers have wilted, the spines have collapsed, the bloom no longer smells, and the male flowers have matured. As the flies leave the chamber, the male flowers dust them with pollen. Later, some of these same flies will be deceived by, and pollinate, other dead-horse arums.

The corpse flower and dead-horse arum seem to offer no reward in exchange for pollination. This is not always the case with warm flowers, however. Many species of philodendron, plants found in tropical rain forests and in many people's houses, produce hot flowers (up to 114°F for one species) with mildly pleasant aromas. The flowers of some species serve as warm, cozy orgy rooms for beetles, who crawl into the bloom in the early evening and spend the night mating (**Fig. 44-24**). The beetles use only half as much energy keeping warm in a philodendron bloom as they would in the cool night air outside. The beetles also feed on pollen and other parts of the flower. Philodendrons, therefore, give the beetles real rewards in exchange for their services as pollinators.

(a) A newly opened philodendron flower

(b) Beetles swarm on a philodendron flower

▲ **FIGURE 44-24 A beetle orgy (a)** This philodendron flower produces scent on the first day of blooming. **(b)** Then it heats up, attracting beetles that congregate, feed, mate, and conserve energy on the warm blossom.

Consider This

Heat-producing flowers are rare, and many are members of evolutionarily ancient groups. Some botanists hypothesize that warmth was an early adaptation that attracted beetle pollinators. Today, most plants bear "fast-food" flowers: They supply their pollinators with a sip of nectar, and then send them on their way with a dusting of pollen. Compare the pollination strategies used by dead-horse arums and philodendrons with the more common fast-food flowers. Why do fast-food flowers predominate today?

CHAPTER REVIEW

Summary of Key Concepts

44.1 How Do Plants Reproduce?

The sexual life cycle of plants, called alternation of generations, includes both a multicellular diploid stage (the sporophyte generation) and a multicellular haploid stage (the gametophyte generation). Meiotic cell division in cells of the diploid sporophyte produces haploid spores. The spores undergo mitotic cell division to produce the haploid gametophyte generation. Reproductive cells of the gametophyte differentiate into sperm and eggs, which fuse to produce a diploid zygote. Mitotic cell division of the zygote gives rise to another sporophyte generation. In mosses, the gametophyte is the dominant stage; the sporophyte develops on the gametophyte and never lives independently. In ferns, the sporophyte is the dominant stage, although both the gametophyte and sporophyte are independent structures. In gymnosperms and angiosperms, the sporophyte is the dominant stage; the gametophytes are very small and never live independently.

44.2 What Are the Functions and Structures of Flowers?

Complete flowers consist of four parts: sepals, petals, stamens (male reproductive structures), and carpels (female reproductive structures). The sepals form the outer covering of the flower bud. Most petals (and in most monocots, the sepals) are brightly colored and attract pollinators to the flower. The stamen consists of a filament that bears an anther, in which pollen develops. The carpel consists of the ovary, in which one or more female gametophytes develop, and a style that bears a sticky stigma to which pollen adheres during pollination. Most flowers are pollinated by animals and have colorful, often scented, petals. Some flowers are wind pollinated; these are usually small, with prominent anthers but often lacking petals.

The male gametophyte of flowering plants is the pollen grain. The diploid microspore mother cell undergoes meiotic cell division to produce four haploid microspores. Each undergoes mitotic cell division to form the pollen grain. The immature pollen grain consists of a tube cell and a generative cell that will later divide to produce two sperm cells.

The female gametophyte develops within the ovules of the ovary. A diploid megaspore mother cell undergoes meiotic cell division to form four haploid megaspores. Three megaspores degenerate; the fourth undergoes three rounds of mitosis, producing the eight nuclei of the female gametophyte. The nuclei become enclosed in seven cells. One of these becomes the egg and another becomes a large central cell with two nuclei.

When a pollen grain lands on a stigma, its tube cell grows through the style to the female gametophyte. The two sperm cells travel down the style within the tube and enter the female gametophyte. One sperm fuses with the egg to form a diploid zygote. The other sperm fuses with the two nuclei of the central cell, producing a triploid cell that will give rise to the endosperm.

44.3 How Do Fruits and Seeds Develop?
The function of fruit is to disperse seeds. A fruit is a ripened ovary, often with contributions from other parts of the flower. Seeds develop from the ovules. The integuments of the ovules form the seed coat. Within the seed coat, a seed contains an embryo and variable amounts of food-storing endosperm.

44.4 How Do Seeds Germinate and Grow?
Seed germination requires warmth and moisture. Energy for germination comes from food stored in the endosperm, which is transferred to the embryo by the cotyledons. Seeds may remain dormant for some time after fruit ripening. To break dormancy and germinate, some seeds require drying, exposure to cold, or disruption of the seed coat. The root emerges first from the germinating seed, absorbing water and nutrients that are transported to the shoot. Monocot shoots are protected by a coleoptile that covers the shoot tip during germination. Dicot shoots form epicotyl or hypocotyl hooks that protect the delicate shoot tip.

44.5 How Do Plants and Their Pollinators Interact?
Plants and their animal pollinators act as agents of natural selection on one another. Flowers attract animals with scent, food such as nectar, and appropriate colors and shapes that make them visible and accessible to their pollinators. Some flowers deceive pollinators, attracting insects with food scents or the shape of a mate. Some plants and their pollinators, such as the yucca plant and yucca moth, are completely dependent on one another.

44.6 How Do Fruits Help to Disperse Seeds?
Fruits disperse seeds in many ways. Explosive fruits shoot seeds away from the parent plant. Lightweight fruits are carried by the wind. Floating fruits are dispersed by water. Clinging fruits adhere to animals. Edible fruits are eaten by animals, with little or no harm to the seeds.

Key Terms

alternation of generations 862	integument 868
anther 865	megaspore 868
carpel 865	megaspore mother cell 868
coleoptile 871	microspore 867
complete flower 865	microspore mother cell 867
cotyledon 871	ovary 866
dormancy 871	ovule 866
double fertilization 869	petal 865
endosperm 869	pollen grain 864
epicotyl 872	pollination 866
filament 865	seed 866
flower 865	seed coat 870
fruit 866	sepal 865
gametophyte 862	spore 862
generative cell 867	sporophyte 862
germination 871	stamen 865
hypocotyl 872	stigma 866
imperfect flower 866	style 866
incomplete flower 866	tube cell 867
	zygote 862

Learning Outcomes
In this chapter, you have learned to . . .

LO1 Describe the sexual life cycle of plants, and compare and contrast the cycles of ferns and flowering plants.

LO2 Describe the structure of a complete flower and explain the function of each part.

LO3 Describe the development of the male and female gametophytes of flowering plants.

LO4 Explain the processes of pollination and double fertilization in flowering plants.

LO5 Explain how the structures of a fruit and seed develop from a flower.

LO6 Explain why many seeds undergo a period of dormancy before they germinate, and describe the environmental conditions that may break dormancy and allow germination to occur.

LO7 Describe seed germination and the early growth of seedlings in monocots and dicots.

LO8 Describe the adaptations of flowers and their animal pollinators that facilitate efficient pollination.

LO9 Explain how fruits aid in seed dispersal.

Thinking Through the Concepts
Fill-in-the-Blank
1. The plant sexual life cycle is called _____. The diploid generation is called the _____. Reproductive cells in this stage of the cycle produce spores through _____ (type of cell division). The spores germinate to produce the haploid generation, called the _____. **LO1**

2. In a flowering plant, the male gametophyte is the _____. It is formed in the _____ of a flower. Pollination occurs when pollen lands on the _____ of a flower of the same plant

species. The pollen grain grows a tube through the _____ of the carpel to the ovary at the base of the carpel. The tube enters an ovule through an opening in the _____ of the ovule. **LO1 LO2 LO3**

3. Double fertilization in flowering plants occurs when one sperm cell fuses with an egg to form a diploid cell, the _____, and a second sperm fuses with the two nuclei of the central cell of the female gametophyte. This triploid cell will divide mitotically to form a food-storage tissue called the _____. This food is transferred to the developing plant embryo by structures called _____. **LO3 LO4 LO5 LO7**

4. The fruit of a flowering plant is formed from the _____ of a flower, possibly with additional contributions from other flower parts. The seed forms from the _____. The seed coat develops from the outer covering, or _____, of this structure. **LO5**

5. _____ is the emergence of the embryo from a seed. Usually, the _____ (structure of the embryo) emerges first. In monocots, the shoot is protected by a sheath called the _____. In dicots, a(n) _____ in the hypocotyl or epicotyl protects the apical meristem and developing leaves. **LO7**

Review Questions

1. Diagram the plant sexual life cycle. Which stages are haploid and which are diploid? At which stage are gametes formed? **LO1**

2. Diagram a complete flower. Where are the male and female gametophytes formed? **LO2**

3. Describe the development of the female gametophyte in a flowering plant. How does double fertilization occur? **LO3 LO4**

4. What is a pollen grain and how is it formed? **LO3**

5. What are the parts of a seed and how does each part contribute to the development of a seedling? **LO5**

6. Describe the characteristics you would expect to find in flowers that are pollinated by the wind, beetles, bees, and hummingbirds, respectively. **LO8**

7. What is the endosperm? From which cell of the female gametophyte is it derived? Is endosperm usually more abundant in the mature seed of a dicot or of a monocot? **LO4 LO5**

8. Describe three mechanisms whereby seed dormancy is broken in different types of seeds. How are these mechanisms related to the typical environment of the plant? **LO6**

9. How do monocot and dicot seedlings protect the delicate shoot tip during seed germination? **LO7**

10. Describe three types of fruits and the mechanisms whereby these fruit structures help disperse their seeds. **LO9**

Applying the Concepts

1. A friend gives you some seeds to grow in your yard. When you plant some, nothing happens. What might you try to get the seeds to germinate?

2. Charles Darwin once described a flower that produced nectar at the bottom of a tube nearly a foot deep (30 centimeters). He predicted that there must be a moth or other animal with a 30-centimeter-long "tongue" to match. He was right; it's a moth. Such specialization almost certainly means that this particular flower could be pollinated only by that specific moth. What are the advantages and disadvantages of such specialization?

3. Many plants that we call weeds were brought from another continent either accidentally or purposefully. In their new environment, they have few competitors or animal predators, so they tend to grow in such large numbers that they displace native plants. Think of several ways in which humans become involved in plant dispersal. To what degree do you think humans have changed the distributions of plants? In what ways is this change helpful to humans? In what ways is it a disadvantage?

Answers to Figure Caption questions and Fill-in-the-Blank questions can be found in the Answers section at the back of the book.

(MB) *Go to MasteringBiology for practice quizzes, activities, eText, videos, current events, and more.*

CHAPTER 45

Plant Responses to the Environment

A fly enters a Venus flytrap.

CASE STUDY
Predatory Plants

IN A MARSH, plants are hungry. Not for sugars and starches, which they can synthesize just as other plants do, but for nitrogen. Plants need nitrogen for the same reasons that we do—to synthesize ATP, nucleic acids, and a huge variety of proteins. But available nitrogen can be scarce in marsh soils, which tend to be acidic. Acidic conditions discourage the growth of nitrogen-fixing bacteria, which capture atmospheric nitrogen and transform it into a form that plants can absorb through their roots. The lack of nitrogen in bogs has selected for an effective way for plants to acquire this nutrient—by ingesting and digesting animals, whose bodies provide abundant nitrogen-containing protein. As a result,

small-scale adventure abounds in a marsh in midsummer.

As a fly lands on the clamshell-like leaves of a seemingly harmless Venus flytrap plant, the leaves suddenly snap together and their spiked edges mesh, trapping the hapless insect. During the next week or so, enzymes digest the protein from the fly's body, and the leaf absorbs the nitrogen-containing molecules before the trap reopens to attract its next victim. Nearby, a green lacewing lands on stalks bearing sweet droplets that resemble dew glistening in the sunlight. Immediately, the insect finds itself struggling helplessly in a sticky mass secreted by a sundew leaf (see Fig. 45-14). The lacewing's struggle stimulates the sundew to secrete a cocktail of digestive enzymes into the

goo, where they rapidly break down the insect's body. The sundew's leaves then absorb the liberated nitrogen compounds. Beneath the surface of the marsh, still another drama unfolds. A bladderwort plant dangles hundreds of pear-shaped chambers into the water (see Fig. 45-15). Each is sealed by a watertight trapdoor whose lower edge is fringed with bristles. A tiny water flea (related to shrimp, but barely visible to us) swims by, brushing the hairs. Within one-sixtieth of a second, the animal is sucked into the chamber and digested.

How do predatory plants detect their prey and then move quickly enough to trap them? How do "ordinary" plants sense environmental stimuli and respond to them? Read on to learn more about the amazing abilities of plants.

AT A GLANCE

45.1 What Are Some Major Plant Hormones?

45.2 How Do Hormones Regulate Plant Life Cycles?

45.3 How Do Plants Communicate, Defend Themselves, and Capture Prey?

LEARNING GOALS

LG1 Name the major plant hormones, and list the functions of each.

LG2 Explain how abscisic acid helps plants withstand unfavorable conditions and how gibberellin stimulates plant growth and germination.

LG3 Define *tropism*, describe three tropisms, and explain what is known about their mechanisms of action and the roles they play in the life of plants.

LG4 Compare the effects of auxin and cytokinin in root and shoot systems, and explain how these effects promote a balance between the two systems.

LG5 Explain how phytochromes in some plants control shoot elongation and flowering in response to day length.

LG6 Describe where florigen is produced and where and how it acts.

LG7 Explain how auxin, gibberellin, and ethylene contribute to the development of seeds and to the development and ripening of fruit.

LG8 Describe senescence and dormancy and how ethylene and abscisic acid influence these states.

LG9 Explain how some plants defend themselves and warn neighboring plants against predators.

LG10 Explain the mechanisms underlying the rapid movements exhibited by sensitive and predatory plants.

45.1 WHAT ARE SOME MAJOR PLANT HORMONES?

Plant hormones are chemicals secreted by specific cells and transported—often via xylem or phloem—to other parts of the plant body. Different plant hormones influence every aspect of the plant life cycle. Each plant hormone can elicit a variety of responses depending on the type of cell it stimulates, the stage of the plant life cycle, the concentration of the hormone, the presence and concentrations of other hormones, and the species of plant. In the sections that follow, we focus on six important types of plant hormones: auxins, gibberellins, cytokinins, ethylene, abscisic acid, and florigens (Table 45-1).

Auxins are a chemically related group of hormones that exert a wide variety of effects. Auxins promote or inhibit

TABLE 45-1 Some Major Plant Hormones and Their Functions

Hormone	Some Major Effects	Major Sites of Synthesis
Auxins	Promote cell elongation in shoots Inhibit growth of lateral buds (apical dominance) Promote root branching Control phototropism and gravitropism in shoots and roots Stimulate vascular tissue development Stimulate fruit development Delay senescence of leaves and fruit	Shoot apical meristem
Gibberellins	Stimulate stem elongation by promoting cell division and cell elongation Stimulate fruit development, and seed germination	Shoot apical meristem Plant embryos Young leaves
Cytokinins	Stimulate cell division throughout the plant Stimulate lateral bud sprouting Inhibit formation of branch roots Delay senescence of leaves and flowers	Root apical meristem
Ethylene	Promotes growth of shorter, thicker stems in response to mechanical disturbance Stimulates ripening in some fruits Promotes senescence in leaves Promotes leaf and fruit drop	Throughout the plant, particularly during stress and aging
Abscisic acid	Causes stomata to close Inhibits stem growth and stimulates root growth in response to drought Maintains dormancy in buds and seeds	Throughout the plant
Florigens	Stimulate flowering in response to day length	Mature leaves

elongation in different target cells of a germinating seed, causing the shoot to bend toward light and the root toward the pull of gravity. Auxin inhibits the formation of stems by lateral buds, but stimulates lateral root branching. Auxin also stimulates the differentiation of vascular tissues (xylem and phloem) in newly formed plant parts. In the mature plant, auxin stimulates fruit development, and it prevents both fruits and leaves from falling prematurely. The primary site of auxin synthesis is the shoot apical meristem (the growing shoot tip). Some auxin is also synthesized in young leaves, as well as in developing seeds and plant embryos, where it influences development and germination.

A variety of synthetic auxins have been developed that allow growers to manipulate plants. An auxin called 2,4-D is widely used to kill dicot plants by disrupting the normal balance among auxin and other plant hormones. Other synthetic auxins are used commercially to promote root formation in plant cuttings, to stimulate fruit development, and to delay fruit fall. Auxin was the first plant hormone to be recognized, as described in "Science in Action: How Were Plant Hormones Discovered?"

Gibberellins form a large group of related hormones with the important role of promoting stem elongation by increasing both cell division and cell elongation. Gibberellins also stimulate fruit production and development, and seed germination. Gibberellins are produced in the shoot apical meristem, in young leaves, and in plant embryos.

Gibberellins are named after the fungus *Gibberella fujikuroi*, which causes "foolish seedling" disease in rice. This disease stimulates rice plants to grow exceptionally tall. In the 1930s, scientists isolated substances produced by the fungus that stimulated plant growth. These and similar molecules are now called gibberellins.

Cytokinins promote cell division, which is required for the growth of all plant tissues. The major site of cytokinin synthesis is in the root apical meristem, although some is also generated in the shoot. In roots, cytokinins inhibit the formation of branch roots; in shoots, cytokinins promote the formation of branches by stimulating cell division in lateral bud meristems. Cytokinins cause nutrients to be transported into plant leaves, stimulating chlorophyll production and delaying aging. Commercially, cytokinin is sprayed on cut flowers to keep them fresh.

Ethylene is an unusual plant hormone because it is a gas. Produced in most plant tissues, it is released in response to a wide range of environmental stimuli and has diverse effects on its target cells. Ethylene serves as a plant "stress hormone" synthesized by plant tissues to stimulate adaptive changes in response to wounding, flooding, drought, and temperature extremes. Ethylene also stimulates the formation of weak-celled abscission layers, which allow leaves, flower petals, and fruit to drop off at the appropriate times of year. Ethylene is best known, and most commercially valuable, for its ability to cause certain fruits to ripen.

Abscisic acid, synthesized in tissues throughout the plant body, helps plants to withstand unfavorable environmental conditions. Abscisic acid causes stomata to close when water is scarce (see Chapter 43). It also promotes root growth and inhibits stem growth under dry conditions. Abscisic acid helps to maintain dormancy in seeds at times when germination would lead to death. The name "abscisic acid" was based on the mistaken hypothesis that it caused leaf abscission, now known to be a function of ethylene.

Florigens are hormones synthesized in leaves that stimulate flowering in response to environmental cues, particularly light. Florigens are a family of very similar proteins that travel in the phloem from leaves to the apical meristem, which produces the flower.

CHECK YOUR LEARNING

Can you list six important types of plant hormones? Can you provide the major site of synthesis for each hormone and examples of the hormone's major actions?

45.2 HOW DO HORMONES REGULATE PLANT LIFE CYCLES?

The life cycle of each plant species results from a complex interplay between its genes and its environment. Hormones are released in response to environmental stimuli and then usually alter gene activity. In the sections that follow, we summarize the effects of the best-studied hormones on plant life cycles.

Each Plant Life Cycle Begins with a Seed

The timing of seed germination is crucial to ensure that new seedlings will have adequate water and appropriate temperatures to thrive. Two hormones play major roles in seed germination: abscisic acid and gibberellin.

Abscisic Acid Maintains Seed Dormancy

Seed dormancy is maintained by abscisic acid. The seeds of many temperate plants require a period of cold temperatures to allow germination; this ensures that seedlings do not emerge until winter is over. In these plants prolonged cold reduces the amount of abscisic acid within the seed, preparing the plant to germinate during the warming days of spring.

In deserts, where lack of water is the greatest threat to seedling survival, some plants produce seeds with high levels of abscisic acid in their coats. It requires a hard rain to wash the hormone away, allowing germination to occur only when enough moisture is available for the plant to develop and complete its life cycle. Abundant winter rains may trigger germination in desert seeds that have remained dormant for years, briefly carpeting some North American deserts with wildflowers (**Fig. 45-1**). In contrast, some plants—particularly those of grasslands, chaparral, and some types of forests—require fire to germinate, as described in "Earth Watch: Where There's Smoke, There's Germination."

Earth | Watch

Where There's Smoke, There's Germination

Wildfires are necessary to maintain ecosystems such as tallgrass prairie and coastal chaparral (see Chapter 29), and to regenerate aging climax forests. Without human intervention, lightning-set fires naturally sweep through temperate forests every few decades, keeping them relatively open and providing a diversity of habitats. But people suppressed fires in the United States for most of the 20th century. As a result, many forests are now overly dense, with less healthy trees competing for space, water, nutrients, and light. When fire comes, it burns these crowded forests with an unnatural intensity, killing far more of the trees, and destroying seeds and root systems that would otherwise survive. Far from being restored, the forest is devastated.

Land plants have been evolving in the presence of fire throughout their evolutionary history. As a result, many plants are adapted to take advantage of the open space, light, and nutrients that become newly available in the aftermath of a fire. Many species of fire-adapted plants are stimulated to germinate by chemicals in smoke; for example, nitrogen dioxide gas in smoke triggers germination in a variety of chaparral species (**Fig. E45-1**). Australian researchers recently discovered a group of compounds (called "karrikins," from the Aboriginal word for "smoke") that are released by burning cellulose and trigger germination in several smoke-responsive species. Scientists are beginning to discover how smoke influences the plant hormones involved in germination. For example, wild tobacco plants, which rapidly colonize burned areas, show a dramatic decrease in abscisic acid levels shortly after smoke exposure, coupled

▲ **FIGURE E45-1 Fires help maintain chaparral ecosystems**

with an increase in sensitivity to gibberellin. In the mustard *Arabidopsis*, karrikins activate genes that cause an increase in gibberellin synthesis.

Seed germination in response to chemicals in smoke ensures that fire-blackened landscapes will speedily regain their carpet of green. Recognizing the value of fire, forest managers now often allow forest fires to continue burning if they don't endanger human lives and property.

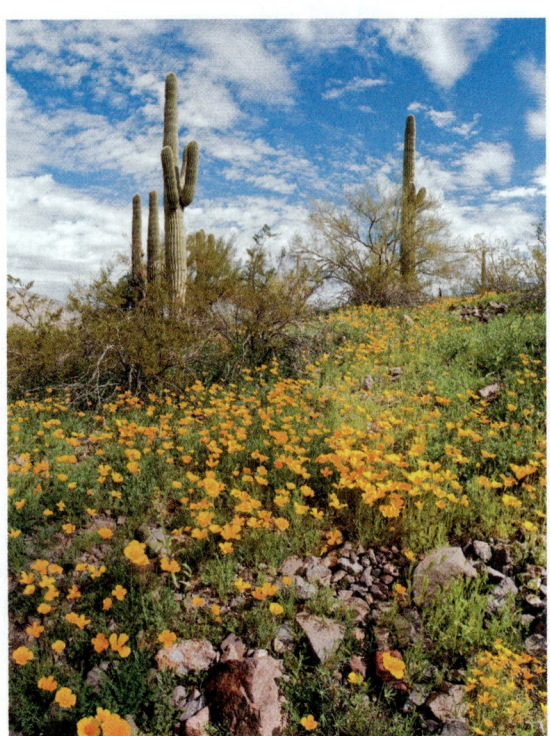

Gibberellin Stimulates Seed Germination

Falling levels of abscisic acid and rising levels of gibberellin produced by the embryo allow germination to occur. Gibberellin activates genes that code for enzymes that break down the stored starch reserves of the endosperm. This releases sugar that is used by the developing embryo to provide both energy and carbon atoms used to synthesize biological molecules.

Auxin Controls the Orientation of the Sprouting Seedling

Nourished by the endosperm, the growing embryo breaks out of its seed coat, and the new seedling must orient its root downward and its shoot upward. Plants use **tropisms,** growth toward or away from stimuli, to respond to directional cues such as light and gravity.

◀ **FIGURE 45-1 A desert in bloom** After abundant rainfall, annual desert plants often germinate in large numbers, grow rapidly, and carpet the desert floor with blossoms, as shown here in Picacho Peak State Park, Arizona.

◄ **FIGURE 45-2 Auxin causes gravitropism**

shoot tip

Auxin, produced in the shoot tip, is distributed evenly across the shoot and root as it travels downward

seed

Auxin from the shoot tip travels down and collects in the root tip, causing it to grow straight down

root tip

Auxin is transported to the lower side of the shoot, where it stimulates cell elongation and causes the stem to bend upward

Auxin is transported to the lower side of the root, where it inhibits cell elongation and causes the root to bend downward

(a) The shoot and root are oriented vertically

(b) The shoot and root are oriented horizontally

(c) A young sunflower plant displays negative gravitropism after growing horizontally in darkness for three days

(d) A germinating corn shoot displays negative gravitropism and its root displays positive gravitropism

Auxin Stimulates Shoot Elongation Away from Gravity and Toward Light

As seeds germinate, auxin causes **gravitropism** (growth toward or away from gravity) and **phototropism** (growth toward or away from light) in both shoots and roots. Auxin is synthesized in the shoot apical meristem and stimulates cells of the stem to elongate. If the stem is vertical, auxin is distributed evenly among cells below the meristem (**Fig. 45-2a**). If the stem is horizontal, auxin is transported to cells in the lower side of the stem (**Fig. 45-2b, d**). In response, the lower cells elongate rapidly, bending the stem upward and away from the pull of gravity (*negative gravitropism*). When the shoot tip becomes vertical, auxin becomes evenly distributed and the stem grows straight up out of the soil. Plants temporarily placed in darkness in a horizontal position will exhibit negative gravitropism (**Fig. 45-2c**).

Auxin also causes phototropism, a directional response to light. Auxin accumulates in the side of the shoot away from the light, causing this side to elongate and bending the shoot toward the light. When the shoot points directly toward the light, auxin becomes evenly distributed, and bending ceases. Seedlings on a windowsill clearly demonstrate *positive phototropism*, which dominates negative gravitropism in the presence of abundant sunlight (**Fig. 45-3**).

Auxin Stimulates Root Elongation Toward Gravity

Roots use positive gravitropism to orient downward into the soil (**Fig. 45-2b, d**). Auxin is transported from the shoot tip down into the root tip. If the root is horizontal, cells in the root tip sense the direction of gravity and transport auxin

▲ **FIGURE 45-3 Radish seedlings show positive phototropism**

SCIENCE IN ACTION How Were Plant Hormones Discovered?

Anyone who keeps houseplants knows that the plants bend toward sunlight streaming through a window. What causes this phototropism?

Charles and Francis Darwin Determined the Source of the Signal

In England, around 1880, Charles Darwin and his son Francis investigated phototropism using grass coleoptiles (protective sheaths that surround monocot seedlings). Upon illuminating the sprout from various angles, they observed that a region just below the tip would bend, pointing the tip toward the light source. When they covered the tip of the coleoptile with an opaque cap, the coleoptile didn't bend (**Fig. E45-2**). A clear cap allowed the stem to bend, even when the bending region was covered by an opaque sleeve (**Fig. E45-3**). The Darwins concluded that the tip of the coleoptile perceives the direction of light, although bending occurs farther down. This suggested that the tip transmits information about light direction down to the bending region.

▲ **FIGURE E45-3** The shoot tip sends a signal to the bending region

Peter Boysen-Jensen Demonstrated That the Signal Is a Chemical

In 1913, Peter Boysen-Jensen of Denmark cut the tips off coleoptiles and found that the remaining stump neither elongated nor bent toward the light. If he replaced the tip and placed the patched-together coleoptile in the dark, it elongated straight up. In the light, it showed normal phototropism. When he inserted a thin layer of gelatin that prevented direct contact but permitted substances to diffuse between the severed tip and the stump, he observed the same elongation and bending. In contrast, an impenetrable barrier eliminated these responses (**Fig. E45-4**). Boysen-Jensen concluded that the factor is a diffusing chemical.

Frits Went Collected the Chemical and Named It Auxin

In 1926, Frits Went, working in the Netherlands, placed the tips of oat coleoptiles on agar, allowing the unidentified chemical to migrate into this gelatinous material (**Fig. E45-5**). He then cut up the agar and placed small pieces on the tops of coleoptile

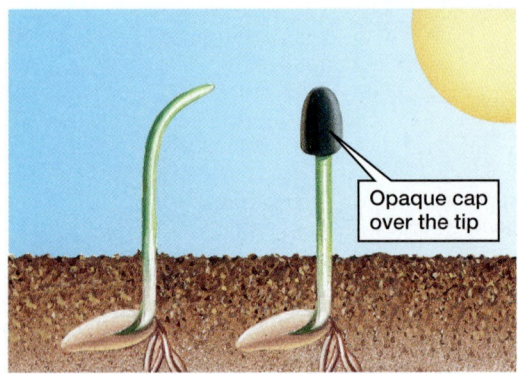

▲ **FIGURE E45-2** The shoot tip senses light

to the lower side of the root. There, auxin inhibits root cell elongation, causing the root to exhibit positive gravitropism, bending downward in the direction of gravity. When the root tip points down, the auxin becomes equally distributed, causing the root to grow straight downward.

How Do Plants Sense Gravity?

Specialized starch-filled plastids called *statoliths* are involved in gravitropism. Statoliths, found in certain cells of the shoot and the root cap, are denser than the surrounding cytosol, so they settle into the lower part of a cell (**Fig. 45-4**). Auxin accumulates on the side of a cell with the greatest density of statoliths. However, some mutant plants that lack statoliths still exhibit gravitropism, suggesting that other mechanisms may also contribute to this response.

The Growing Plant Emerges and Reaches Upward

As a seed germinates, pressure of the root and shoot against the surrounding soil induces ethylene production. Ethylene

causes both the root and shoot to slow their elongation and become thicker, stronger, and better able to force their way through the soil. In dicots, ethylene also causes the emerging

▲ **FIGURE 45-4** Statoliths respond to gravity
Statoliths in root caps fall to the lowest part of the root cells. In a vertically oriented root tip (left), statoliths stimulate elongation straight down. Orienting the root horizontally (right) causes the statoliths to fall to the lower side of the cell and the root tip to bend down.

▲ **FIGURE E45-4 A chemical diffuses down the coleoptile**

Porous gelatin is placed between the tip and shoot

An impenetrable barrier is placed between the tip and shoot

▲ **FIGURE E45-5 The chemical accumulates in agar**

The tips are placed on agar

▲ **FIGURE E45-6 The chemical stimulates growth**

Agar is centered on the stump

Agar is off-center on the stump

stump, the stump bent away from the side with agar (**Fig. E45-6**). Confirming the hypothesis that a diffusing chemical stimulated elongation, he named it "auxin," from a Greek word meaning "to increase."

Kenneth Thimann Determined the Chemical Structure of Auxin

Working at Caltech in the early 1930s, Kenneth Thimann purified auxin and determined its chemical structure. He and other researchers then determined which features of the molecule were important for its action in plants. This knowledge led to the production of synthetic auxins, such as the herbicide 2,4-D, developed in 1946.

The discovery and characterization of auxin spanned two continents over a 50-year period. It relied on clear communication among scientists, and illustrates how each new scientific discovery leads to more questions. Research on auxin continues throughout the world today, steadily opening new lines of inquiry. For example, researchers have recently identified 11 different genes involved in auxin synthesis. This understanding may lead to new ways of enhancing the growth of crop plants.

stumps in darkness. A piece of agar placed squarely atop the stump caused it to elongate straight up; all of the stump cells received equal amounts of the chemical and elongated at the same rate. When Went placed agar on only one side of the cut

shoot to form a hook, so that the tender new leaves point downward. The bend of the hook protects the leaves as the plant forces its way up through the soil (see Fig. 44-13b).

Gibberellins, which are active during many phases of the plant life cycle, stimulate stem elongation and help determine the ultimate height of the plant. Shorter varieties of plants such as wheat and rice have mutations that cause them to produce less gibberellin or be less sensitive to it. Selective breeding of such plants, which are sturdy and resist damage by wind and rain, has allowed large increases in agricultural production.

Plants called *vines* have flexible stems that rely on nearby objects for support as they grow upward toward sunlight. Vines display **thigmotropism,** a directional movement or change in growth in response to touch. Most thigmotropic plants wrap specialized *tendrils* (modified leaves or stems) or the entire stem around supporting structures (**Fig. 45-5**). Twining occurs when cell elongation is inhibited on the side of the stem in contact with the object, while elongation continues on the opposite side. Electrical signals in response to touch seem to be involved in this plant behavior.

Auxin and Cytokinin Control Stem and Root Branching

The size of the root and shoot systems of plants must be balanced so that as the shoot grows, the roots also grow enough to provide adequate anchorage, water, and nutrients to support the needs of the stem. Auxin produced by the shoot apical meristem and cytokinin produced by the root apical meristem promote this balance.

Branching in Stems Is Inhibited by Auxin and Stimulated by Cytokinin

Auxin produced in the shoot apical meristem travels down the stem and inhibits lateral buds from developing into branches, a phenomenon called **apical dominance.** Auxin levels decrease with distance from the shoot tip, so lower lateral buds are more likely to become branches (**Fig. 45-6,** left). As gardeners know, pinching off the tip of a growing plant eliminates apical dominance, allowing many lateral buds to sprout and causing the plant to become bushier (**Fig. 45-6,** right).

▲ FIGURE 45-5 **Thigmotropism** (Left) Red tendrils of this grape plant coil in response to contact. (Right) The entire stem of this morning glory twines around a post. In each case, cells on the side of the stem opposite an object elongate more rapidly than cells in contact with the object.

Cytokinin, produced in the root apical meristem, stimulates the development of lateral stem buds into stem branches. Because the lower buds receive more cytokinin and less auxin, they are more likely to form branches than are buds near the apical meristem.

▲ FIGURE 45-6 **Apical dominance** The bean plant on the left has produced a single stem that is exhibiting apical dominance by suppressing side branches. The bean plant on the right has had its apical meristem cut off, allowing lateral buds to develop into branches.

Branching in Roots Is Stimulated by Auxin and Inhibited by Cytokinin

Auxin, transported down from the shoot apical meristem, stimulates root pericycle cells to divide and form branch roots (see Fig. 43-17). Commercial auxin powder allows gardeners to produce a new plant by dipping the cut end of a stem into the auxin, which stimulates it to develop roots. In contrast, the cytokinin produced in the root apical meristem inhibits root branching. Lower roots receive more cytokinin and less auxin, so their branching is suppressed. Root branching is stimulated closer to the shoot, and inhibited near the root tip. The gradients and counteracting effects of auxin and cytokinin help to keep the size of the root and shoot systems in balance with one another (**Fig. 45-7**).

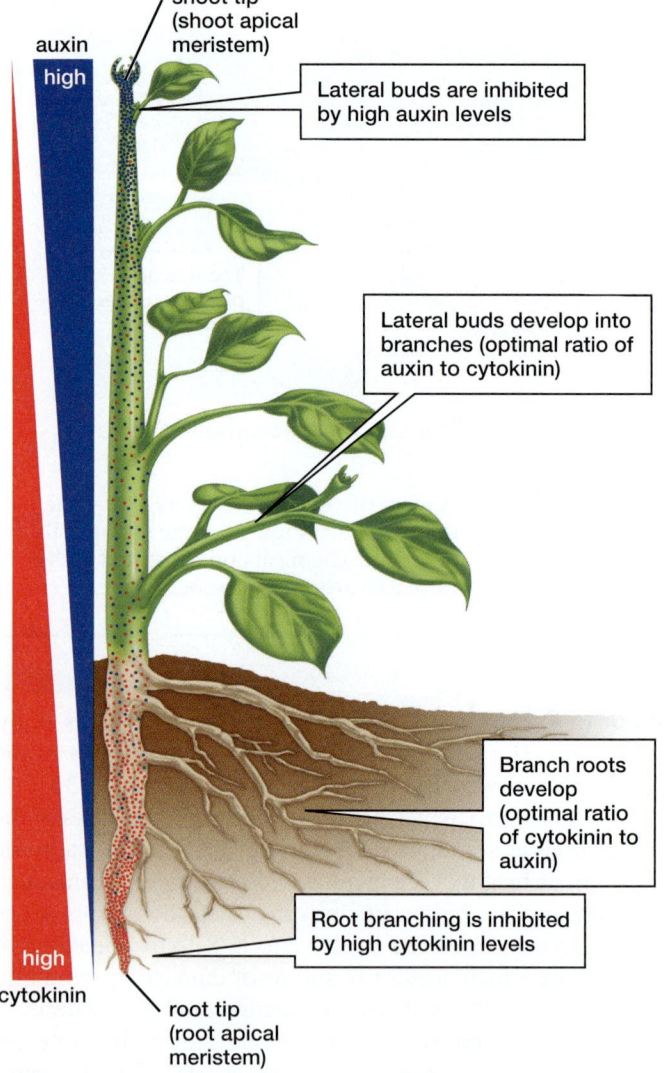

▲ FIGURE 45-7 **The role of auxin and cytokinin in lateral bud sprouting** Auxin (blue) and cytokinin (red) control the sprouting of lateral buds and the development and branching of lateral roots. Auxin from the shoot apical meristem moves downward; cytokinin from the root apical meristem moves upward.

QUESTION If you removed a plant's shoot apical meristem and applied auxin to the cut surface, how would you expect the lateral buds to respond?

Plants Sense and Respond to Light and Darkness

The timing of flowering and seed production must be finely tuned to the environment. In temperate climates, plants must flower early enough so that their seeds mature before the killing frosts of late autumn. Depending on how quickly the seeds develop, flowering may occur throughout the growing season from spring through fall.

How do plants sense the changing seasons? Many cues, such as temperature or water availability, vary quite unpredictably in temperate climates; the only consistent and reliable cue is day length. Days that progressively increase in length herald spring and summer, while shortening days signal the coming fall and winter.

Some plants, called **day-neutral plants** (such as roses, tomatoes, cucumbers, and corn), flower independently of day length under favorable conditions (**Fig. 45-8**). The flowering of other plants is related to day length. Experiments have demonstrated that such plants actually measure the duration of uninterrupted darkness. **Long-day plants** (short-night plants), including iris, lettuce, spinach, and hollyhocks, flower only when uninterrupted darkness is shorter than a specific length of time that varies with the species of plant (see Fig. 45-8). **Short-day plants** (long-night plants), such as cockleburs, chrysanthemums, asters, potatoes, and goldenrod, flower when uninterrupted darkness exceeds a species-specific duration (see Fig. 45-8). For example, a cocklebur, a long-night plant, flowers only if uninterrupted darkness lasts more than 8.5 hours, whereas spinach, a long-day plant, flowers only if uninterrupted darkness lasts 10 hours or less; both will flower with 9-hour nights. How do these plants sense the duration of uninterrupted darkness?

Photopigments Called Phytochromes Detect Light

Organisms detect light using molecules called photopigments, which absorb specific wavelengths of light. As a photopigment absorbs light energy, it alters its chemical configuration, initiating a series of biochemical changes that cause a response in the organism. The eyes of animals contain photopigments that allow vision, whereas the bodies of plants contain photopigments called **phytochromes** (literally, "plant colors") that allow plants to respond to light.

Plants measure the duration of darkness using an internal **biological clock** that relies on complex biochemical reactions. If a plant in darkness is exposed to light of a certain wavelength, its phytochrome molecules will "reset" the clock, something like stopping and resetting a stopwatch. Thus, if an 8-hour night is interrupted after 4 hours with a few minutes of white light, the plant will perceive only the uninterrupted 4 hours of darkness that occur after the interval of light.

◄ FIGURE 45-8 The effects of the length of uninterrupted darkness on flowering

How does this work? A phytochrome occurs in two similar forms—P_r and P_{fr}—that are converted from one to the other by specific wavelengths of light. The P_r form absorbs red light (r; a wavelength of about 660 nm) and converts to P_{fr}, whereas P_{fr} absorbs far-red light (fr; 730 nm) and converts to P_r. The P_r is inactive and remains stable unless it is exposed to red light. The P_{fr} is the active form of phytochrome that triggers responses to light, such as flowering. It is, however, unstable, so that over a period of hours in darkness, P_{fr} spontaneously reverts back to P_r (**Table 45-2**).

Different types of plants respond in different ways to P_{fr}. For example, P_{fr} stimulates flowering in long-day plants, but inhibits flowering in short-day plants. In the short-day cocklebur, interrupting a 9-hour dark period with just a minute or two of red or white light (which includes red wavelengths) prevents flowering by converting P_r to P_{fr}. As you might predict, cocklebur flowering is not affected by interrupting the dark period with far-red light, which is not absorbed by P_r and so does not produce P_{fr}.

Phytochromes Influence Many Plant Responses

Light is life for plants, which use photosynthesis to trap and store light energy as chemical energy (see Chapter 7). Because light is such an important stimulus for plants, it is not surprising that the light-responsive phytochromes influence many aspects of plant development. For example, P_{fr} inhibits shoot elongation. Because P_{fr} spontaneously reverts to P_r in the dark, a shoot that breaks out of its seed in the darkness of the soil will contain no inhibitory P_{fr} and so will elongate rapidly and emerge from the soil. As it encounters the red wavelengths in sunlight, P_{fr} is formed in the shoot. This slows elongation, preventing the seedling from growing too rapidly and becoming weak and spindly. As a dicot shoot emerges from the soil, light triggers P_{fr} formation, which causes the hook in its stem to straighten, orienting the new leaves upward toward the light. P_{fr} also stimulates the leaves to grow and produce chlorophyll.

A seedling growing beneath other plants will be exposed primarily to far-red light, because the green chlorophyll in the leaves above it absorbs most red light but allows far-red light to pass through. In a sun-loving plant, as far-red light converts the inhibitory P_{fr} into inactive P_r, the stem of the shaded seedling will elongate rapidly, which makes it more likely to emerge from the shade into sunlight. There, the red wavelengths present in sunlight convert P_r back to P_{fr}, which inhibits stem elongation and allows the stem to grow thicker and stronger.

Florigen Stimulates Flowering in Response to Light Cues

In experiments that began in the 1930s, plant physiologists found that flowering in long-day or short-day plants can be induced by exposing just a single leaf to the appropriate schedule of light and darkness, even if the apical meristem which produces the flower receives no light. Such experiments supported the hypothesis that an unknown factor must be produced in leaves and transported to the apical meristem, where it promotes flowering. This hypothetical flowering hormone was dubbed "florigen" (from the Latin, "flor" meaning "flower," and "gen," "to produce"). But the florigen molecule itself was not identified until 2007, when researchers found a gene called *FT*, which is active in leaves but not in the apical meristem. FT codes for a protein that is transported from the leaves through the phloem to the apical meristem. The pathway was confirmed in several different plant species, including both monocots and dicots, and the *FT* protein was finally confirmed as the elusive florigen.

Hormones Coordinate the Development and Ripening of Fruits and Seeds

Fruit and seed development and ripening are influenced by auxin, gibberellin, and ethylene.

Auxin promotes the growth of the plant ovary, causing it to store food materials and to develop into a mature fruit. Synthetic auxin may be sprayed on fruits to cause them to grow larger and sweeter than they would under natural conditions or to prevent the fruit from dropping prematurely. Gibberellin likewise contributes to fruit growth. It is commercially applied to grapes, which grow larger and form looser clusters as a result (**Fig. 45-9**).

▲ **FIGURE 45-9 Commercial uses for plant hormones** The grapes on the left were sprayed with gibberellin, producing looser clusters of larger grapes. Those on the right have developed naturally, forming smaller grapes in tighter clusters.

QUESTION Agricultural biotechnologists have developed genetically modified tomato plants in which ethylene production is blocked. Why might such a plant be valuable to tomato growers?

TABLE 45-2 Light and Phytochrome Activity

Light	Resulting Phytochrome	Activity
Red	P_{fr}	Active
Far-red	P_r	Inactive
White (sunlight)	$P_r + P_{fr}$	Active
Darkness	P_r	Inactive

In nature, seed maturation and fruit ripening are closely coordinated, because most seeds are dispersed in fruits by animals. Unripe fruit in general is often green, hard, bitter, and sometimes poisonous. As the seeds mature, however, the fruit ripens and becomes softer, as enzymes weaken its cell walls; sweeter, as different enzymes break down starch into sugar; and more brightly and conspicuously colored, as green chlorophyll is broken down, and yellow, orange, and red pigments are revealed or synthesized. These features attract animals that will disperse the seeds (Fig. 45-10), just as they appeal to people perusing the produce section of the supermarket.

In fruits such as bananas, apples, pears, tomatoes, and avocados, the changes that accompany ripening are stimulated by ethylene. As these fruits ripen, they themselves release ethylene, which hastens the ripening of other fruits nearby. The discovery of the role of ethylene in ripening has revolutionized modern fruit marketing. Bananas grown in Central America can be picked green and tough, then shipped to North American markets, where they are ripened with ethylene. Green tomatoes, which withstand transport better than ripe ones, may also be ripened at central warehouses using ethylene. But after purchasing ripe fruits, we often want the ripening to stop, so our bananas don't become black and our tomatoes mushy as we store them. Fortunately, special produce bags are now available whose plastic contains a substance that absorbs ethylene and slows ripening. So now people can manipulate the ripening process in many fruits from start to finish. Not all fruits ripen when treated with ethylene; those that don't include strawberries, grapes, and cherries.

Senescence and Dormancy Prepare the Plant for Winter

In autumn, under the influence of shortening days and falling temperatures, plants undergo **senescence,** a genetically programmed series of events that prepares the plant for winter. Ethylene production increases, and the production of auxin and cytokinin (which delay senescence) declines.

During senescence, starch and chlorophyll in the leaf are broken down into simpler molecules that are transported to the stem and roots for winter storage.

Although many hormones influence senescence, ethylene is particularly important. Ethylene promotes the breakdown of chlorophyll and triggers the production of enzymes that weaken a layer of cells, called the **abscission layer,** located where fruit or leaf stalks join the stem (Fig. 45-11). This allows the aging leaf or ripe fruit to drop off. Leaves may die and drop suddenly when stress causes a rapid increase in ethylene production. This happens if the leaves are infected by microorganisms or exposed to temperature extremes, or drought—as you may observe if you forget to water your houseplants.

In the autumn, new buds become tightly wrapped and dormant rather than developing into leaves or branches. Dormancy in buds, as in seeds, is maintained by abscisic acid. The plant's metabolism slows and it enters its long winter sleep, awaiting signals of warmth, moisture, and longer spring days before awakening once again.

CHECK YOUR LEARNING

Can you explain the roles of plant hormones in seed dormancy, seed germination, and directional growth of the emerging root and shoot? Can you explain thigmotropism? Do you understand how plants balance root and shoot growth? Do you know how some plants sense and respond to light and darkness? Can you explain how hormones coordinate the maturation of seeds and fruit and control fruit ripening? Do you understand how hormones control senescence?

▲ FIGURE 45-10 **Ripe fruit becomes attractive to animal seed dispersers** A desert tortoise is attracted to the ripe and tasty fruit of a prickly pear cactus.

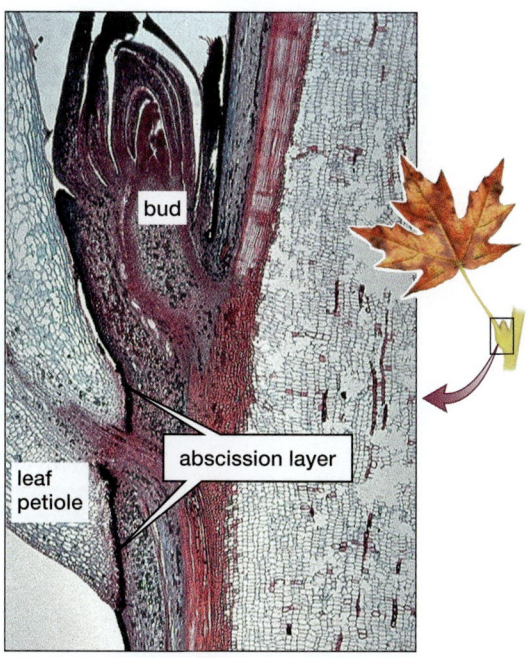

▲ FIGURE 45-11 **The abscission layer** This light micrograph of the base of a maple leaf clearly shows the abscission layer. A lateral bud is visible above the senescing leaf petiole.

45.3 HOW DO PLANTS COMMUNICATE, DEFEND THEMSELVES, AND CAPTURE PREY?

Flowering plants have been coevolving with animals and microorganisms for well over 100 million years. In addition to communicating with their insect pollinators (see Chapter 44), plants have evolved surprisingly sophisticated behaviors under selection pressures imposed by parasitism, predation, and limited nutrients.

Plants May Summon Insect "Bodyguards" When Attacked

When chewed on by hungry insects, many plants release volatile chemicals into the air, producing what could be considered a chemical cry for help. One example is provided by corn leaves attacked by caterpillars (Fig. 45-12). The corn alarm signal is stimulated by a compound called volicitin in the saliva of the caterpillar; damage from other causes (such as hail) will not stimulate this response. Parasitic female wasps are attracted to the corn's alarm call and lay their eggs in the caterpillar's body. Upon hatching, the wasp larvae consume the caterpillar from the inside, then form cocoons from which adult wasps emerge. Similarly, lima bean plants that are attacked by spider mites release a chemical that attracts a carnivorous mite that preys on the spider mite. Wild tobacco plants munched on by hornworms (hawk moth caterpillars) respond by releasing chemicals that differ depending on

CASE STUDY continued

Predatory Plants

Whereas some plants summon insects for help, other plants summon insects for food, luring them with sugary secretions and bright colors. The inner edges of Venus flytrap leaves are studded with nectar glands that produce a sweet-smelling sugar lure, and the inner surfaces of the plant's leaves are often red, giving them a flower-like appearance. Enzymes secreted by the leaves of the Venus flytrap digest the trapped insect, and the plant feasts on the nitrogen-rich nutrients the insect's body provides. Sundews attract their insect prey using an appealing scent, red hairs, and sticky droplets that resemble nectar (see Fig. 45-14).

the time of attack. During the day, when parasitic wasps are active and seeking hornworm caterpillars, the tobacco plants produce wasp-attracting chemicals. At night, when adult hawk moths are active, the tobacco plants release chemicals that deter the moths from laying eggs on them.

Attacked Plants Defend Themselves

Leaf damage from insects causes many plants to produce a signaling molecule, which then travels through the plant body, where it stimulates responses that make the plant more distasteful, difficult to eat, or toxic. For example, when tobacco leaves are damaged, they produce more

❷ Volicitin and leaf damage cause the plant to synthesize and release volatile chemicals

❸ Chemicals attract female parasitic wasps that lay eggs in the caterpillar

❶ A caterpillar chews on a corn leaf, leaving traces of saliva that contains volicitin

❹ Wasps' eggs hatch in the caterpillar; larvae consume the caterpillar and form cocoons from which wasps emerge

▶ FIGURE 45-12 A chemical cry for help

nicotine (a poison used commercially as an insecticide). Radish plants attacked by caterpillars produce a bitter-tasting chemical and grow more spiny hairs on their leaves. Even the offspring of these damaged plants are less attractive to predators than seedlings from uninjured parent plants. Thus, plants not only defend themselves, but also pass on a chemical signal within their seeds that causes specific genes to be activated, triggering the development of more defenses in their offspring.

Plants have a type of immune response that helps defend them against infectious microorganisms. In response to infection, many plants produce salicylic acid, which stimulates production of a variety of proteins that help them fight their current infection and future infections as well.

Wounded Plants Warn Their Neighbors

If plants under siege can summon help, might neighboring plants also get the message? Evidence is accumulating that some healthy plants sense chemicals released by nearby members of their species that are infected by microbes such as viruses. Tobacco plants infected with a virus produce large amounts of salicylic acid, which helps boost their immune responses. Some of the salicylic acid produced by the infected plants is converted to methyl salicylate (wintergreen oil), a highly volatile compound. When healthy plants are exposed to air that had circulated over the infected plants, the healthy plants absorb the methyl salicylate and convert it back to salicylic acid, which boosts their immune defenses, making them better able to resist the virus. Communication among plants may also cross species lines. One study showed that wild tobacco plants exposed to damaged sagebrush leaves produce stronger and more rapid defenses when attacked by the tobacco hornworm than do unexposed tobacco plants.

Sensitive Plants React to Touch

If you touch a sensitive plant such as a *Mimosa* (**Fig. 45-13**), the rows of leaflets along each side of the petiole immediately fold together, and the petiole droops. This rapid and

dramatic movement may surprise and discourage leaf-eating insects. It is stimulated by electrical signals in *motor cells* at the base of each leaflet and where the petiole joins the stem. The signal allows potassium ions, normally maintained at a high internal concentration, to diffuse rapidly out of the motor cells. As water follows by osmosis, the motor cells shrink rapidly, pulling the leaflets together and causing the petioles to droop.

Carnivorous Sundews and Bladderworts Respond Rapidly to Prey

Some plants have evolved ingenious prey-catching behavior. Sundew leaves attract and then trap insects with hairs whose tips secrete sweet glistening gluey globules (**Fig. 45-14**). The vibrations from the insect's struggle to escape produce an electric current in the hairs. The current triggers a

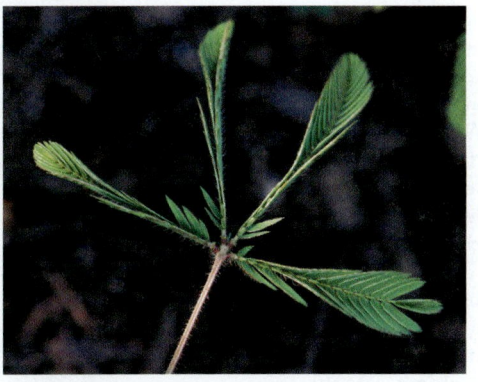

(a) Before the leaves are touched

(b) After the leaves are touched

◄ **FIGURE 45-13 A rapid response to touch (a)** A leaf of the sensitive plant (*Mimosa*) consists of an array of leaflets emerging from a central stalk, which is attached to the main stem by a short petiole. **(b)** Touching causes the leaflets to fold together.

▲ FIGURE 45-14 **A sundew with its insect prey** The leaf of this sundew uses sweet, colorful, and sticky hairs to entice and capture a lacewing insect.

thigmotropic response, causing the hairs to curl around the insect, smothering and then digesting it with their secretions.

The world's speediest plant is the predatory aquatic bladderwort (**Fig. 45-15**). The trapdoor covering each hollow bladder is hinged at the top, opens inward, and is sealed shut by sticky secretions. Cells lining the bladder actively transport ions out of the water in the bladder, causing water to leave the bladder by osmosis. This pulls the bladder walls inward under tension, setting the trap. If a tiny aquatic organism, such as a water flea, bumps into the bristles that surround the trapdoor, the bristles push the door inward and break the seal around it. In a split second, the bladder walls spring outward to their resting position, suddenly increasing the volume of the bladder and sucking in the water flea to its death and eventual digestion.

▲ FIGURE 45-15 **The bladderwort snares tiny aquatic organisms** Bladderworts are studded with prey-catching bladders. (Inset) A water flea has triggered this bladder to expand rapidly. If the animal were smaller, it would have been sucked inside and digested. This particular bladder, however, has bitten off more than it can chew, and the water flea is wedged in its trapdoor.

CHECK YOUR LEARNING

Can you explain how plants that are attacked by predators and parasites defend themselves and warn neighboring plants? Do you understand why and how sensitive plants, sundews, bladderworts, and Venus flytraps produce their rapid movements?

CASE STUDY revisited

Predatory Plants

How do the leaves of a *Venus flytrap* close quickly enough to trap a fly? This plant physics and physiology problem caught the attention of a Harvard researcher, L Mahadevan. His team painted dots of fluorescent paint on the flytrap leaves, and then tracked them with a high speed video camera as the leaves closed. Using these data in a mathematical model, the researchers concluded that a flytrap leaf is maintained in its open state by a difference in the water content between the inner and outer layers of the leaf. When an insect bumps into the hairs more than once in a short time, water moves from the inner layers into the outer layers. This changes the size of the cells, causing the leaf to switch its shape from a slightly convex shape to a strongly concave shape that closes around the insect (**Fig. 45-16**).

If nothing is trapped, reopening the leaves takes a few days, and consumes large quantities of ATP. But the plant has evolved a "fail-safe" mechanism that helps prevent it from closing on an inanimate object. For the leaves to clamp shut, either two hairs must be touched simultaneously, or one hair must be touched twice within about 20 seconds. Hairs bumped by a nectar-seeking

▲ FIGURE 45-16 **Success!**

insect open channels that allow ions to flow, generating an electrical signal that stimulates the trap to clamp shut in less than half a second.

The flytrap still holds mysteries. How do the sensory hairs "know" which ones have been touched and at what interval? How does the electrical signal stimulate the trap to close? Science is a never-ending quest for deeper understanding, because the answer to one question inevitably raises more.

Consider This

Many wetlands in the United States are threatened by runoff water from nearby farms, which may be heavily fertilized or rich in animal wastes. Carnivorous plants thrive in nitrogen-poor marshes partly because other species, which can't trap nitrogen-rich food, cannot compete with them. Explain why runoff from farms poses a threat to carnivorous plants in nearby wetlands.

CHAPTER REVIEW

Summary of Key Concepts

45.1 What Are Some Major Plant Hormones?

Plants have evolved the ability to sense and react to environmental stimuli including touch, gravity, moisture, light, and day length. Plant hormones, secreted in response to changes in the environment, influence every stage of the plant life cycle. Six major types of plant hormones that are important in the flowering plant life cycle are auxins, gibberellins, cytokinins, ethylene, abscisic acid, and florigens (see Table 45-1).

45.2 How Do Hormones Regulate Plant Life Cycles?

Hormones regulate plant growth and development in response to environmental stimuli. Dormancy in seeds is enforced by abscisic acid. Falling levels of abscisic acid and rising levels of gibberellin trigger germination. As the seedling grows, auxin causes the shoot to grow toward light and away from the pull of gravity, and it causes the roots to grow toward gravity. Gibberellin stimulates the stem to elongate. Statoliths help plant shoot and root tip cells detect gravity and influence auxin accumulation toward the pull of gravity. Ethylene slows elongation and promotes thickening of both roots and shoots as the plant pushes through soil. Gibberellin influences plant height by causing stem elongation.

Branching in stems and roots is controlled by auxin and cytokinin, which keep these two systems in balance. Auxin, synthesized in the shoot apical meristem, inhibits stem branching but stimulates root branching. Cytokinin, synthesized in the root apical meristem, inhibits root branching, but stimulates stem branching.

Plants detect light and darkness using photopigments called phytochromes. The stable, inactive form of a phytochrome (P_r) absorbs red light and is converted into the unstable active form (P_{fr}), which absorbs far-red light and converts back to P_r. The P_{fr} phytochrome causes the hook of the emerging shoot to straighten and the seedling to elongate. It also triggers the expansion of new leaves and the synthesis of chlorophyll. Leaf phytochrome measures uninterrupted darkness and helps control flowering in long-day and short-day plants. In response to the appropriate day length, the flowering hormone florigen is produced in plant leaves and travels to the flower bud in the apical meristem where it initiates flowering.

Seed and fruit growth and development are influenced by auxin, cytokinin, and gibberellin. In some fruits, ethylene triggers ripening. During senescence, fruit and leaves fall in response to ethylene, which weakens their abscission layers. Buds become dormant as a result of high concentrations of abscisic acid.

45.3 How Do Plants Communicate, Defend Themselves, and Capture Prey?

Some plants under attack by insects release volatile chemicals that attract other insects that prey on or parasitize the plant predators. Chemicals released by injured or infected plants may also stimulate neighboring plants to increase their defenses. Damaged plants use internal chemical signals to strengthen their own defenses and, sometimes, to produce better-defended offspring.

The aquatic bladderwort generates tension within its bladders. The tension is released when a bladder is touched by prey, expanding the bladder and sucking in water and prey. Sensitive plant leaves fold together when touch sensors in the leaves generate electrical signals that cause specialized cells to rapidly lose water.

Key Terms

abscisic acid *883*	gravitropism *885*
abscission layer *891*	long-day plant *889*
apical dominance *887*	phototropism *885*
auxin *882*	phytochrome *889*
biological clock *889*	plant hormone *882*
cytokinin *883*	senescence *891*
day-neutral plant *889*	short-day plant *889*
ethylene *883*	thigmotropism *887*
florigen *883*	tropism *884*
gibberellin *883*	

Learning Outcomes

In this chapter you have learned to . . .

LO1 Name the major plant hormones, and list the functions of each.

LO2 Explain how abscisic acid helps plants withstand unfavorable conditions and how gibberellin stimulates plant growth and germination.

LO3 Define *tropism*, describe three tropisms, and explain what is known about their mechanisms of action and the roles they play in the life of plants.

LO4 Compare the effects of auxin and cytokinin in root and shoot systems, and explain how these effects promote a balance between the two systems.

LO5 Explain how phytochromes in some plants control shoot elongation and flowering in response to day length.

LO6 Describe where florigen is produced and where and how it acts.

LO7 Explain how auxin, gibberellin, and ethylene contribute to the development of seeds and to the development and ripening of fruit.

LO8 Describe senescence and dormancy and how ethylene and abscisic acid influence these states.

LO9 Explain how some plants defend themselves and warn neighboring plants against predators.

LO10 Explain the mechanisms underlying the rapid movements exhibited by sensitive and predatory plants.

Thinking Through the Concepts

Fill-in-the-Blank

1. Auxin produced by the apical meristem of the shoot _____ sprouting of lateral branches; this phenomenon is called _____. In roots, auxin _____ development of root side branches. The discovery of auxin by the Darwins was based on its ability to stimulate _____. **LO1 LO4**

2. The hormone _____ was discovered by researchers investigating "foolish seedling" disease in rice. During stem elongation, this hormone stimulates both cell _____ and cell _____. **LO1 LO2**

3. The hormone _____ is a gas. During senescence, this hormone stimulates the formation of the _____, which allows dead leaves and ripe fruits to fall. Ethylene hastens the _____ of fruits such as _____ and _____. **LO1 LO7 LO8**

4. The major site of cytokinin synthesis is the _____. This hormone_____ the sprouting of lateral roots, but _____ the sprouting of lateral _____. **LO1 LO4**

5. The hormone _____ causes stomata to close when water is scarce. It maintains _____ in seeds during unfavorable environmental conditions. **LO1 LO2 LO8**

Review Questions

1. How does auxin cause positive phototropism and negative gravitropism in shoots? **LO3**

2. What is apical dominance? How do auxin and cytokinin interact in determining the growth of lateral buds? **LO4**

3. Why are short-day plants more accurately called "long-night" plants? At what season of the year would you expect them to flower? Explain why. **LO5**

4. What is phytochrome? How do the two forms of phytochrome help control the plant life cycle? **LO5**

5. Which hormones cause fruit development? Which hormone causes ripening in bananas? **LO7**

6. What is senescence? Describe some changes that accompany leaf senescence in the fall. Which hormone causes abscission? **LO8**

7. What is a major agricultural impact of gibberellin? Of ethylene? **LO2 LO7**

8. Describe one example of a chemical defense mechanism in plants. **LO9**

9. Describe how a sensitive plant closes its leaves. Why might this behavior have evolved? **LO10**

Applying the Concepts

1. Suppose you got a job in a greenhouse in which the owner was trying to force chrysanthemums (long-night plants) to flower for Mother's Day (in mid-May). You accidentally turned on the light in the middle of the night. Would you be likely to lose your job? Why or why not? What would happen if you turned on the lights during the day?

2. A student reporting on a project said that one of her seeds did not grow properly because it was planted upside down, so that it got confused and tried to grow down. Should the teacher accept this explanation? Why or why not?

3. Bean sprouts for salads must be grown in the dark to form their long stems. Why do seedlings grow long and spindly in the dark? What advantages would growing this way in darkness have in nature?

4. Suppose that on August 4, you discover that both a short-day plant and a long-day plant have bloomed in your garden. Discuss how it is possible for both plants to bloom.

5. Would you expect a predominance of long-day, short-day, or day-neutral plants in the tropics near the equator? Explain.

Answers to Figure Caption questions and Fill-in-the-Blank questions can be found in the Answers section at the back of the book.

MB *Go to MasteringBiology for practice quizzes, activities, eText, videos, current events, and more.*

Biological Vocabulary: Common Roots, Prefixes, and Suffixes

Biology has an extensive vocabulary often based on Greek or Latin rather than English words. Rather than memorizing every word as if it were part of a new, foreign language, you can figure out the meaning of many new terms if you learn a much smaller number of word roots, prefixes, and suffixes. We have provided common meanings in biology rather than literal translations from Greek or Latin. For each item in the list, the following information is given: meaning; part of word (prefix, suffix, or root); example from biology.

a–, an–: without, lack of (prefix); *abiotic*, without life

acro–: top, highest (prefix); *acrosome*, vesicle of enzymes at the tip of a sperm

ad–: to (prefix); *adhesion*, property of sticking to something else

allo–: other (prefix); *allopatric* (literally, "different fatherland"), restricted to different regions

amphi–: both, double, two (prefix); *amphibian*, a class of vertebrates that usually has two life stages (aquatic and terrestrial; e.g., a tadpole and an adult frog)

andro: man, male (root); *androgen*, a male hormone such as testosterone

ant–, anti–: against (prefix); *antibiotic* (literally "against life"), a substance that kills bacteria

antero–: front (prefix or root); *anterior*, toward the front of

aqu–, aqua–: water (prefix or root); *aquifer*, an underground water source, usually rock saturated with water

apic–: top, highest (prefix); *apical meristem*, the cluster of dividing cells at the tip of a plant shoot or root

arthr–: joint (prefix); *arthropod*, animals such as spiders, crabs, and insects, with exoskeletons that include jointed legs

–ase: enzyme (suffix); *protease*, an enzyme that digests protein

auto–: self (prefix); *autotrophic*, self-feeder (e.g., photosynthetic)

bi–: two (prefix); *bipedal*, having two legs

bio–: life (prefix or root); *biology*, the study of life

blast: bud, precursor (root); *blastula*, embryonic stage of development, a hollow ball of cells

bronch: windpipe (root); *bronchus*, a branch of the trachea (windpipe) leading to a lung

carcin, –o: cancer (root); *carcinogenesis*, the process of producing a cancer

cardi, –a–, –o–: heart (root); *cardiac*, referring to the heart

carn–, –i–, –o–: flesh (prefix or root); *carnivore*, an animal that eats other animals

centi–: one hundredth (prefix); *centimeter*, a unit of length, 1 one-hundredth of a meter

cephal–, –i–, –o–: head (prefix or root); *cephalization*, the tendency for the nervous system to be located principally in the head

chloro–: green (prefix or root); *chlorophyll*, the green, light-absorbing pigment in plants

chondr–: cartilage (prefix); *Chondrichthyes*, class of vertebrates including sharks and rays, with a skeleton made of cartilage

chrom–: color (prefix or root); *chromosome*, a threadlike strand of DNA and protein in the nucleus of a cell (*Chromosome* literally means "colored body," because chromosomes absorb some of the colored dyes commonly used in microscopy.)

–cide: killer (suffix); *pesticide*, a chemical that kills "pests" (usually insects)

–clast: break down, broken (root or suffix); *osteoclast*, a cell that breaks down bone

co–: with or together with (prefix); *cohesion*, property of sticking together

coel–: hollow (prefix or root); *coelom*, the body cavity that separates the internal organs from the body wall

contra–: against (prefix); *contraception*, acting to prevent conception (pregnancy)

cortex: bark, outer layer (root); *cortex*, outer layer of kidney

crani–: skull (prefix or root); *cranium*, the skull

cuti: skin (root); *cuticle*, the outermost covering of a leaf

–cyte, cyto–: cell (root or prefix); *cytokinin*, a plant hormone that promotes cell division

de–: from, out of, remove (prefix); *decomposer*, an organism that breaks down organic matter

dendr: treelike, branching (root); *dendrite*, highly branched input structures of nerve cells

derm: skin, layer (root); *ectoderm*, the outer embryonic germ layer of cells

deutero–: second (prefix); *deuterostome* (literally, "second opening"), an animal in which the coelom is derived from the gut

di–: two (prefix); *dicot*, an angiosperm with two cotyledons in the seed

diplo–: both, double, two (prefix or root); *diploid*, having paired homologous chromosomes

dys–: difficult, painful (prefix); *dysfunction*, an inability to function properly

eco–: house, household (prefix); *ecology*, the study of the relationships between organisms and their environment

ecto–: outside (prefix); *ectoderm*, the outermost tissue layer of animal embryos

–elle: little, small (suffix); *organelle* (literally, "little organ"), a subcellular structure that performs a specific function

end–, endo–, ento–: inside, inner (prefix); *endocrine*, pertaining to a gland that secretes hormones inside the body

epi–: outside, outer (prefix); *epidermis*, outermost layer of skin

equi–: equal (prefix); *equidistant*, the same distance

erythro–: red (prefix); *erythrocyte*, red blood cell

eu–: true, good (prefix); *eukaryotic*, pertaining to a cell with a true nucleus

ex–, exo–: out of (prefix); *exocrine*, pertaining to a gland that secretes a substance (e.g., sweat) outside of the body

extra–: outside of (prefix); *extracellular*, outside of a cell

–fer: to bear, to carry (suffix); *conifer*, a tree that bears cones

gastr–: stomach (prefix or root); *gastric*, pertaining to the stomach

–gen–: to produce (prefix, root, or suffix); *antigen*, a substance that causes the body to produce antibodies

glyc–, glyco–: sweet (prefix); *glycogen*, a starch-like molecule composed of many glucose molecules bonded together

gyn, –o: female (prefix or root); *gynecology*, the study of the female reproductive tract

haplo–: single (prefix); *haploid*, having a single copy of each type of chromosome

hem–, hemato–: blood (prefix or root); *hemoglobin*, the molecule in red blood cells that carries oxygen

hemi–: half (prefix); *hemisphere*, one of the halves of the cerebrum

herb–, herbi–: grass (prefix or root); *herbivore*, an animal that eats plants

hetero–: other (prefix); *heterotrophic*, an organism that feeds on other organisms

hom–, homo–, homeo–: same (prefix); *homeostasis*, to maintain constant internal conditions in the face of changing external conditions

hydro–: water (usually prefix); *hydrophilic*, being attracted to water

hyper–: above, greater than (prefix); *hyperosmotic*, having a greater osmotic strength (usually higher solute concentration)

hypo–: below, less than (prefix); *hypodermic*, below the skin

inter–: between (prefix); *interneuron*, a neuron that receives input from one (or more) neuron(s) and sends output to another neuron (or many neurons)

intra–: within (prefix); *intracellular*, pertaining to an event or substance that occurs within a cell

iso–: equal (prefix); *isotonic*, pertaining to a solution that has the same osmotic strength as another solution

–itis: inflammation (suffix); *hepatitis*, an inflammation (or infection) of the liver

kin–, kinet–: moving (prefix or root); *cytokinesis*, the movements of a cell that divide the cell in half during cell division

lac–, lact–: milk (prefix or root); *lactose*, the principal sugar in mammalian milk

leuc–, leuco–, leuk–, leuko–: white (prefix); *leukocyte*, a white blood cell

lip–: fat (prefix or root); *lipid*, the chemical category to which fats, oils, and steroids belong

–logy: study of (suffix); *biology*, the study of life

lyso–, –lysis: loosening, split apart (prefix, root, or suffix); *lysis*, to break open a cell

macro–: large (prefix); *macrophage*, a large white blood cell that destroys invading foreign cells

medulla: marrow, middle substance (root); *medulla*, inner layer of kidney

mega–: large (prefix); *megaspore*, a large, haploid (female) spore formed by meiotic cell division in plants

–mere: segment, body section (suffix); *sarcomere*, the functional unit of a vertebrate skeletal muscle cell

meso–: middle (prefix); *mesophyll*, middle layers of cells in a leaf

meta–: change, after (prefix); *metamorphosis*, to change body form (e.g., developing from a larva to an adult)

micro–: small (prefix); *microscope*, a device that allows one to see small objects

milli–: one-thousandth (prefix); *millimeter*, a unit of measurement of length; 1 one-thousandth of a meter

mito–: thread (prefix); *mitosis*, cell division (in which chromosomes appear as threadlike bodies)

mono–: single (prefix); *monocot*, a type of angiosperm with one cotyledon in the seed

morph–: shape, form (prefix or root); *polymorphic*, having multiple forms

multi–: many (prefix); *multicellular*, pertaining to a body composed of more than one cell

myo–: muscle (prefix); *myofibril*, protein strands in muscle cells

neo–: new (prefix); *neonatal*, relating to or affecting a newborn child

neph–: kidney (prefix or root); *nephron*, functional unit of mammalian kidney

neur–, neuro–: nerve (prefix or root); *neuron*, a nerve cell

neutr–: of neither gender or type (usually root); *neutron*, an uncharged subatomic particle found in the nucleus of an atom

non–: not (prefix); *nondisjunction*, the failure of chromosomes to distribute themselves properly during cell division

oligo–: few (prefix); *oligomer*, a molecule made up of a few subunits (see also *poly–*)

omni–: all (prefix); *omnivore*, an animal that eats both plants and animals

oo–, ov–, ovo–: egg (prefix); *oocyte*, one of the stages of egg development

opsi–: sight (prefix or root); *opsin*, protein part of light-absorbing pigment in eye

opso–: tasty food (prefix or root); *opsonization*, process whereby antibodies and/or complement render bacteria easier for white blood cells to engulf

–osis: a condition, disease (suffix); *atherosclerosis*, a disease in which the artery walls become thickened and hardened

oss–, osteo–: bone (prefix or root); *osteoporosis*; a disease in which the bones become spongy and weak

para–: alongside (prefix); *parathyroid*, referring to a gland located next to the thyroid gland

pater, patr–: father (usually root); *paternal*, from or relating to a father

path–, –i–, –o–: disease (prefix or root); *pathology*, the study of disease and diseased tissue

–pathy: disease (suffix); *neuropathy*, a disease of the nervous system

peri–: around (prefix); *pericycle*, the outermost layer of cells in the vascular cylinder of a plant root

phago–: eat (prefix or root); *phagocyte*, a cell (e.g., some types of white blood cell) that eats other cells

–phil, philo–: to love (prefix or suffix); *hydrophilic* (literally, "water loving"), pertaining to a water-soluble molecule

–phob, phobo–: to fear (prefix or suffix); *hydrophobic* (literally, "water fearing"), pertaining to a water-insoluble molecule

photo–: light (prefix); *photosynthesis*, the manufacture of organic molecules using the energy of sunlight

–phyll: leaf (root or suffix); *chlorophyll*, the green, light-absorbing pigment in a leaf

–phyte: plant (root or suffix); *gametophyte* (literally, "gamete plant"), the gamete-producing stage of a plant's life cycle

plasmo, –plasm: formed substance (prefix, root, or suffix); *cytoplasm*, the material inside a cell

ploid: chromosomes (root); *diploid*, having paired chromosomes

pneumo–: lung (root); *pneumonia*, a disease of the lungs

–pod: foot (root or suffix); *gastropod* (literally, "stomach-foot"), a class of mollusks, principally snails, that crawl on their ventral surfaces

poly–: many (prefix); *polysaccharide*, a carbohydrate polymer composed of many sugar subunits (see also *oligo–*)

post–, postero–: behind (prefix); *posterior*, pertaining to the hind part

pre–, pro–: before, in front of (prefix); *premating isolating mechanism*, a mechanism that prevents gene flow between species, acting to prevent mating (e.g., having different courtship rituals or different mating seasons)

prim–: first (prefix); *primary cell wall*, the first cell wall laid down between plant cells during cell division

pro–: before (prefix); *prokaryotic*, pertaining to a cell without (that evolved before the evolution of) a nucleus

proto–: first (prefix); *protocell*, a hypothetical evolutionary ancestor to the first cell

pseudo–: false (prefix); *pseudopod* (literally, "false foot"), the extension of the plasma membrane by which some cells, such as *Amoeba*, move and capture prey

quad–, quat–: four (prefix); *quaternary structure*, the "fourth level" of protein structure, in which multiple peptide chains form a complex three-dimensional structure

ren: kidney (root); *adrenal*, gland attached to the mammalian kidney

retro–: backward (prefix); *retrovirus*, a virus that uses RNA as its genetic material; this RNA must be copied "backward" to DNA during infection of a cell by the virus

sarco–: muscle (prefix); *sarcoplasmic reticulum*, a calcium-storing, modified endoplasmic reticulum found in muscle cells

scler–: hard, tough (prefix); *sclerenchyma*, a type of plant cell with a very thick, hard cell wall

semi–: one-half (prefix); *semiconservative replication*, the mechanism of DNA replication, in which one strand of the original DNA double helix becomes incorporated into the new DNA double helix

–some, soma–, somato–: body (prefix or suffix); *somatic nervous system*, part of the peripheral nervous system that controls the skeletal muscles that move the body

sperm, sperma–, spermato–: seed (usually root); *gymnosperm*, a type of plant producing a seed not enclosed within a fruit

–stasis, stat–: stationary, standing still (suffix or prefix); *homeostasis*, the physiological process of maintaining constant internal conditions despite a changing external environment

stoma, –to–: mouth, opening (prefix or root); *stoma*, the adjustable pore in the surface of a leaf that allows carbon dioxide to enter the leaf

sub–: under, below (prefix); *subcutaneous*, beneath the skin

sym–: same (prefix); *sympatric* (literally "same father"), found in the same region

tel–, telo–: end (prefix); *telophase*, the last stage of mitosis and meiosis

test–: witness (prefix or root); *testis*, male reproductive organ (derived from the custom in ancient Rome that only males had standing in the eyes of the law; *testimony* has the same derivation)

therm–: heat (prefix or root); *thermoregulation*, the process of regulating body temperature

trans–: across (prefix); *transgenic*, having genes from another organism (usually another species); the genes have been moved "across" species

tri–: three (prefix); *triploid*, having three copies of each homologous chromosome

–trop–, tropic: change, turn, move (suffix); *phototropism*, the process by which plants orient toward the light

troph: food, nourishment (root); *autotrophic*, self-feeder (e.g., photosynthetic)

ultra–: beyond (prefix); *ultraviolet*, light of wavelengths beyond the violet

uni–: one (prefix); *unicellular*, referring to an organism composed of a single cell

vita: life (root); *vitamin*, a molecule required in the diet to sustain life

–vor: eat (usually root); *herbivore*, an animal that eats plants

zoo–, zoa–: animal (usually root); *zoology*, the study of animals

Periodic Table of the Elements

1 **H** 1.008																	2 **He** 4.003
3 **Li** 6.941	4 **Be** 9.012											5 **B** 10.81	6 **C** 12.01	7 **N** 14.01	8 **O** 16.00	9 **F** 19.00	10 **Ne** 20.18
11 **Na** 22.99	12 **Mg** 24.31											13 **Al** 26.98	14 **Si** 28.09	15 **P** 30.97	16 **S** 32.07	17 **Cl** 35.45	18 **Ar** 39.95
19 **K** 39.10	20 **Ca** 40.08	21 **Sc** 44.96	22 **Ti** 47.87	23 **V** 50.94	24 **Cr** 52.00	25 **Mn** 54.94	26 **Fe** 55.85	27 **Co** 58.93	28 **Ni** 58.69	29 **Cu** 63.55	30 **Zn** 65.39	31 **Ga** 69.72	32 **Ge** 72.61	33 **As** 74.92	34 **Se** 78.96	35 **Br** 79.90	36 **Kr** 83.80
37 **Rb** 85.47	38 **Sr** 87.62	39 **Y** 88.91	40 **Zr** 91.22	41 **Nb** 92.91	42 **Mo** 95.94	43 **Tc** (98)	44 **Ru** 101.1	45 **Rh** 102.9	46 **Pd** 106.4	47 **Ag** 107.9	48 **Cd** 112.4	49 **In** 114.8	50 **Sn** 118.7	51 **Sb** 121.8	52 **Te** 127.6	53 **I** 126.9	54 **Xe** 131.3
55 **Cs** 132.9	56 **Ba** 137.3	57 ***La** 138.9	72 **Hf** 178.5	73 **Ta** 180.9	74 **W** 183.8	75 **Re** 186.2	76 **Os** 190.2	77 **Ir** 192.2	78 **Pt** 195.1	79 **Au** 197.0	80 **Hg** 200.6	81 **Tl** 204.4	82 **Pb** 207.2	83 **Bi** 209.0	84 **Po** (209)	85 **At** (210)	86 **Rn** (222)
87 **Fr** (223)	88 **Ra** (226)	89 **†Ac** (227)	104 **Rf** (261)	105 **Db** (262)	106 **Sg** (263)	107 **Bh** (264)	108 **Hs** (265)	109 **Mt** (268)	110 **Ds** (281)	111 **Rg** (280)	112 ****** (277)	113 ******	114 ****** (285)	115 ******			

atomic number (number of protons)

element (chemical symbol)

atomic mass (total mass of protons + neutrons + electrons)

	58 **Ce** 140.1	59 **Pr** 140.9	60 **Nd** 144.2	61 **Pm** (145)	62 **Sm** 150.4	63 **Eu** 152.0	64 **Gd** 157.3	65 **Tb** 158.9	66 **Dy** 162.5	67 **Ho** 164.9	68 **Er** 167.3	69 **Tm** 168.9	70 **Yb** 173.0	71 **Lu** 175.0
*Lanthanide series														
†Actinide series	90 **Th** 232.0	91 **Pa** 231	92 **U** 238.0	93 **Np** (237)	94 **Pu** (244)	95 **Am** (243)	96 **Cm** (247)	97 **Bk** (247)	98 **Cf** (251)	99 **Es** (252)	100 **Fm** (257)	101 **Md** (258)	102 **No** (259)	103 **Lr** (262)

The periodic table of the elements was first devised by Russian chemist Dmitri Mendeleev. The atomic numbers of the elements (the numbers of protons in the nucleus) increase in normal reading order: left to right, top to bottom. The table is "periodic" because all the elements in a column possess similar chemical properties, and such similar elements therefore recur "periodically" in each row. For example, elements that usually form ions with a single positive charge—including H, Li, Na, K, and so on—occur as the first element in each row.

The gaps in the table are a consequence of the maximum numbers of electrons in the most reactive, usually outermost, electron shells of the atoms. It takes only two electrons to completely fill the first shell, so the first row of the table contains only two elements, H and He. It takes eight electrons to fill the second and third shells, so there are eight elements in the second and third rows. It takes 18 electrons to fill the fourth and fifth shells, so there are 18 elements in these rows.

The lanthanide series and actinide series of elements are usually placed below the main body of the table for convenience—it takes 32 electrons to completely fill the sixth and seventh shells, so the table would become extremely wide if all of these elements were included in the sixth and seventh rows.

The important elements found in living things are highlighted in color. The elements in pale red are the six most abundant elements in living things. The elements that form the five most abundant ions in living things are in purple. The trace elements important to life are shown as dark blue (more common) and lighter blue (less common). For radioactive elements, the atomic masses are given in parentheses, and represent the most common or the most stable isotope. Elements indicated as double asterisks have not yet been named.

Metric System Conversions

To Convert Metric Units:	Multiply by:	To Get English Equivalent:
Length		
Centimeters (cm)	0.394	Inches (in)
Meters (m)	3.281	Feet (ft)
Meters (m)	1.094	Yards (yd)
Kilometers (km)	0.621	Miles (mi)
Area		
Square centimeters (cm^2)	0.155	Square inches (in^2)
Square meters (m^2)	10.764	Square feet (ft^2)
Square meters (m^2)	1.196	Square yards (yd^2)
Square kilometers (km^2)	0.386	Square miles (mi^2)
Hectare (ha) (10,000 m^2)	2.471	Acres (a)
Volume		
Cubic centimeters (cm^3)	0.0610	Cubic inches (in^3)
Cubic meters (m^3)	35.315	Cubic feet (ft^3)
Cubic meters (m^3)	1.308	Cubic yards (yd^3)
Cubic kilometers (km^3)	0.240	Cubic miles (mi^3)
Liters (L)	1.057	Quarts (qt), U.S.
Liters (L)	0.264	Gallons (gal), U.S.
Mass		
Grams (g)	0.0353	Ounces (oz)
Kilograms (kg)	2.205	Pounds (lb)
Metric ton (tonne) (t)	1.102	Ton (tn), U.S.
Speed		
Meters/second (mps)	2.237	Miles/hour (mph)
Kilometers/hour (kmph)	0.621	Miles/hour (mph)

To Convert English Units:	Multiply by:	To Get Metric Equivalent:
Length		
Inches (in)	2.540	Centimeters (cm)
Feet (ft)	0.305	Meters (m)
Yards (yd)	0.914	Meters (m)
Miles (mi)	1.609	Kilometers (km)
Area		
Square inches (in^2)	6.452	Square centimeters (cm^2)
Square feet (ft^2)	0.0929	Square meters (m^2)
Square yards (yd^2)	0.836	Square meters (m^2)
Square miles (mi^2)	2.590	Square kilometers (km^2)
Acres (a)	0.405	Hectare (ha) (10,000 m^2)
Volume		
Cubic inches (in^3)	16.387	Cubic centimeters (cm^3)
Cubic feet (ft^3)	0.0283	Cubic meters (m^3)
Cubic yards (yd^3)	0.765	Cubic meters (m^3)
Cubic miles (mi^3)	4.168	Cubic kilometers (km^3)
Quarts (qt), U.S.	0.946	Liters (L)
Gallons (gal), U.S.	3.785	Liters (L)
Mass		
Ounces (oz)	28.350	Grams (g)
Pounds (lb)	0.454	Kilograms (kg)
Ton (tn), U.S.	0.907	Metric ton (tonne) (t)
Speed		
Miles/hour (mph)	0.447	Meters/second (mps)
Miles/hour (mph)	1.609	Kilometers/hour (kmph)

Metric Prefixes

Prefix			Meaning
giga-	G	$10^9 =$	1,000,000,000
mega-	M	$10^6 =$	1,000,000
kilo-	k	$10^3 =$	1,000
hecto-	h	$10^2 =$	100
deka-	da	$10^1 =$	10
		$10^0 =$	1
deci-	d	$10^{-1} =$	0.1
centi-	c	$10^{-2} =$	0.01
milli-	m	$10^{-3} =$	0.001
micro-	µ	$10^{-6} =$	0.000001

water boils (100° C / 212° F)

water freezes (0° C / 32° F)

$$°C = \frac{°F - 32}{1.8}$$

$$°F = (1.8 \times °C) + 32$$

Classification of Major Groups of Eukaryotic Organisms*

Kingdom†	Phylum or Class	Common Name
Excavata	Parabasalia	parabasalids
	Diplomonadida	diplomonads
Euglenozoa	Euglenida	euglenids
	Kinetoplastida	kinetoplastids
Stramenopila	Oomycota	water molds
	Phaeophyta	brown algae
	Bacillariophyta	diatoms
Alveolata	Apicomplexa	sporozoans
	Pyrrophyta	dinoflagellates
	Ciliophora	ciliates
Rhizaria	Foraminifera	foraminiferans
	Radiolaria	radiolarians
Amoebozoa	Tubulinea	amoebas
	Myxomycota	acellular slime molds
	Acrasiomycota	cellular slime molds
Rhodophyta		red algae
Chlorophyta		green algae
Plantae	Marchantiophyta	liverworts
	Anthocerotophyta	hornworts
	Bryophyta	mosses
	Lycopodiopsida	club mosses
	Polypodiopsida	ferns, horsetails
	Gymnospermae	cycads, ginkgos, gnetophytes, conifers
	Anthophyta	flowering plants
Fungi	Chytridiomycota	chytrids
	Neocallimastigomycota	rumen fungi
	Blastocladiomycota	blastoclades
	Glomeromycota	glomeromycetes
	Ascomycota	sac fungi
	Basidiomycota	club fungi
Animalia	Porifera	sponges
	Cnidaria	hydras, sea anemones, sea jellies, corals
	Ctenophora	comb jellies
	Platyhelminthes	flatworms
	Annelida	segmented worms
	Oligochaeta	earthworms
	Polychaeta	tube worms
	Hirudinea	leeches
	Mollusca	mollusks
	Gastropoda	snails
	Pelecypoda	mussels, clams
	Cephalopoda	squid, octopuses
	Nematoda	roundworms
	Arthropoda	arthropods
	Insecta	insects
	Arachnida	spiders, ticks
	Crustacea	crabs, lobsters
	Myriapoda	millipedes, centipedes
	Chordata	chordates
	Urochordata	tunicates
	Cephalochordata	lancelets
	Myxini	hagfishes
	Petromyzontiformes	lampreys
	Chondrichthyes	sharks, rays
	Actinopterygii	ray-finned fishes
	Actinistia	coelacanths
	Dipnoi	lungfishes
	Amphibia	amphibians (frogs, salamanders)
	Reptilia	reptiles (turtles, crocodiles, birds, snakes, lizards)
	Mammalia	mammals

*This table lists only those taxonomic categories described in the textbook.

†Although the major protist groups are not generally called "kingdoms," they are approximately the same taxonomic rank as the kingdoms Plantae, Fungi, and Animalia.

Glossary

abiotic (ā-bī-ah′-tik): nonliving; the abiotic portion of an ecosystem includes soil, rock, water, and the atmosphere.

abscisic acid (ab-sis′-ik): a plant hormone that generally inhibits the action of other hormones, enforcing dormancy in seeds and buds and causing the closing of stomata.

abscission layer: a layer of thin-walled cells, located at the base of the petiole of a leaf, that produces an enzyme that digests the cell walls holding the leaf to the stem, allowing the leaf to fall off.

absorption: the process by which nutrients enter the body through the cells lining the digestive tract.

accessory pigment: a colored molecule, other than chlorophyll *a*, that absorbs light energy and passes it to chlorophyll *a*.

acellular slime mold: a type of organism that forms a multinucleate structure that crawls in amoeboid fashion and ingests decaying organic matter; also called *plasmodial slime mold*. Acellular slime molds are members of the protist clade Amoebozoa.

acid: a substance that releases hydrogen ions (H^+) into solution; a solution with a pH less than 7.

acid deposition: the deposition of nitric or sulfuric acid, either in rain (acid rain) or in the form of dry particles, as a result of the production of nitrogen oxides or sulfur dioxide through burning, primarily of fossil fuels.

acidic: referring to a solution with an H^+ concentration exceeding that of OH^-; referring to a substance that releases H^+.

acquired immune deficiency syndrome (AIDS): an infectious disease caused by the human immunodeficiency virus (HIV); attacks and destroys T cells, thus weakening the immune system.

acrosome (ak′-rō-sōm): a vesicle, located at the tip of the head of an animal sperm, that contains enzymes needed to dissolve protective layers around the egg.

actin (ak′-tin): a major muscle protein whose interactions with myosin produce contraction; found in the thin filaments of the muscle fiber; see also *myosin*.

action potential: a rapid change from a negative to a positive electrical potential in a nerve cell. An action potential travels along an axon without a change in amplitude.

activation energy: in a chemical reaction, the energy needed to force the electron shells of reactants together, prior to the formation of products.

active site: the region of an enzyme molecule that binds substrates and performs the catalytic function of the enzyme.

active transport: the movement of materials across a membrane through the use of cellular energy, normally against a concentration gradient.

adaptation: a trait that increases the ability of an individual to survive and reproduce compared to individuals without the trait.

adaptive immune response: a response to invading toxins or microbes in which immune cells are activated by a specific invader, selectively destroy that invader, and then "remember" the invader, allowing a faster response if that type of invader reappears in the future; see also *innate immune response*.

adaptive immune system: a widely distributed system of organs (including the thymus, bone marrow, and lymph nodes), cells (including macrophages, dendritic cells, B cells, and T cells), and molecules (including cytokines and antibodies) that work together to combat microbial invasion of the body; the adaptive immune system responds to and destroys specific invading toxins or microbes; see also *innate immune response*.

adaptive radiation: the rise of many new species in a relatively short time; may occur when a single species invades different habitats and evolves in response to different environmental conditions in those habitats.

adenine (A): a nitrogenous base found in both DNA and RNA; abbreviated as A.

adenosine diphosphate (a-den′-ō-sēn dī-fos′-fāt; ADP): a molecule composed of the sugar ribose, the base adenine, and two phosphate groups; a component of ATP.

adenosine triphosphate (a-den′-ō-sēn trī-fos′-fāt; ATP): a molecule composed of the sugar ribose, the base adenine, and three phosphate groups; the major energy carrier in cells. The last two phosphate groups are attached by "high-energy" bonds.

adhesion: the tendency of polar molecules (such as water) to adhere to polar surfaces (such as glass).

adipose tissue (a′-di-pōs): tissue composed of fat cells.

adrenal cortex: the outer part of the adrenal gland, which secretes steroid hormones that regulate metabolism and salt balance.

adrenal gland: a mammalian endocrine gland, adjacent to the kidney; secretes hormones that function in water and salt regulation, and in the stress response.

adrenal medulla: the inner part of the adrenal gland, which secretes epinephrine (adrenaline) and norepinephrine (noradrenaline) in the stress response.

adrenocorticotropic hormone (a-drēn-ō-kor-tik-ō-trō′-pik; ACTH): a hormone, secreted by the anterior pituitary, that stimulates the release of hormones by the adrenal cortex, especially in response to stress.

adult stem cell (ASC): any stem cell not found in an early embryo; can divide and differentiate into any of several cell types, but usually not all of the cell types of the body.

aerobic: using oxygen.

age structure diagram: a graph showing the distribution of males and females in a population according to age groups.

aggression: antagonistic behavior, normally among members of the same species, that often results from competition for resources.

aggressive mimicry (mim′-ik-rē): the evolution of a predatory organism to resemble a harmless animal or a part of the environment, thus gaining access to prey.

aging: the gradual accumulation of damage to essential biological molecules, particularly DNA in both the nucleus and mitochondria, resulting in defects in cell functioning, declining health, and ultimately death.

albinism: a recessive hereditary condition caused by defective alleles of the genes that encode the enzymes required for the synthesis of melanin, the principal pigment in mammalian skin and hair; albinism results in white hair and pink skin.

alcoholic fermentation: a type of fermentation in which pyruvate is converted to ethanol (a type of alcohol) and carbon dioxide, using hydrogen ions and electrons from NADH; the primary function of alcoholic fermentation is to regenerate NAD^+ so that glycolysis can continue under anaerobic conditions.

aldosterone: a hormone, secreted by the adrenal cortex, that helps regulate ion concentration in the blood by stimulating the reabsorption of sodium by the kidneys and sweat glands.

alga (al′-ga; pl., algae, al′-jē): any photosynthetic protist.

allantois (al-an-tō′-is): one of the embryonic membranes of reptiles (including birds) and mammals; in reptiles, serves as a waste-storage organ; in mammals, forms most of the umbilical cord.

allele (al-ēl′): one of several alternative forms of a particular gene.

allele frequency: for any given gene, the relative proportion of each allele of that gene in a population.

allergy: an inflammatory response produced by the body in response to invasion by foreign materials, such as pollen, that are themselves harmless.

allopatric speciation (al-ō-pat′-rik): the process by which new species arise following physical separation of parts of a population (geographical isolation).

allosteric regulation: the process by which enzyme action is enhanced or inhibited by small organic molecules that act as regulators by binding to the enzyme at a regulatory site distinct from the active site, and altering the shape and/or function of the active site.

alternation of generations: a life cycle, typical of plants, in which a diploid sporophyte (spore-producing) generation alternates with a haploid gametophyte (gamete-producing) generation.

altruism: a behavior that benefits other individuals while reducing the fitness of the individual that performs the behavior.

alveolate (al-vē′-ō-lāt): a member of the Alveolata, a large protist clade. The alveolates, which are characterized by a system of sacs beneath the cell membrane, include ciliates, dinoflagellates, and apicomplexans.

alveolus (al-vē′-ō-lus; pl., alveoli): a tiny air sac within the lungs, surrounded by capillaries, where gas exchange with the blood occurs.

amino acid: the individual subunit of which proteins are made, composed of a central carbon atom bonded to an amino group ($-NH_2$), a carboxyl group ($-COOH$), a hydrogen atom, and a variable group of atoms denoted by the letter R.

amino acid-derived hormone: a hormone composed of one or two modified amino acids. Examples include epinephrine and thyroxine.

ammonia: NH_3; a highly toxic nitrogen-containing waste product of amino acid breakdown. In the mammalian liver, it is converted to urea.

amniocentesis (am-nē-ō-sen-tē′-sis): a procedure for sampling the amniotic fluid surrounding a fetus: A sterile needle is inserted through the abdominal wall, uterus, and amniotic sac of a pregnant woman, and 10 to 20 milliliters of amniotic fluid are withdrawn. Various tests may be performed on the fluid and the fetal cells suspended in it to provide information on the developmental and genetic state of the fetus.

amnion (am′-nē-on): one of the embryonic membranes of reptiles (including birds) and mammals; encloses a fluid-filled cavity that envelops the embryo.

amniotic egg (am-nē-ōt′-ik): the egg of reptiles, including birds; contains a membrane, the amnion, that surrounds the embryo, enclosing it in a watery environment and allowing the egg to be laid on dry land.

amoeba: an amoebozoan protist that uses a characteristic streaming mode of locomotion by extending a cellular projection called a *pseudopod*. Also known as *lobose amoebas.*

amoeboid cell: a protist or animal cell that moves by extending a cellular projection called a pseudopod.

amoebozoan: a member of the Amoebozoa, a protist clade. The amoebozoans, which generally lack shells and move by extending pseudopods, include the lobose amoebas and the slime molds.

amphibian: a member of the chordate clade Amphibia, which includes the frogs, toads, and salamanders, as well as the limbless caecilians.

amygdala (am-ig′-da-la): part of the forebrain of vertebrates that is involved in the production of appropriate behavioral responses to environmental stimuli.

amylase (am′-i-lās): an enzyme, found in saliva and pancreatic secretions, that catalyzes the breakdown of starch.

anaerobe (an-ə-rōb): an organism that can live and grow in the absence of oxygen.

anaerobic: not using oxygen.

analogous structure: structures that have similar functions and superficially similar appearance but very different anatomies, such as the wings of insects and birds. The similarities are the result of similar environmental pressures rather than a common ancestry.

anaphase (an′-a-fāz): in mitosis, the stage in which the sister chromatids of each chromosome separate from one another and are moved to opposite poles of the cell; in meiosis I, the stage in which homologous chromosomes, consisting of two sister chromatids, are separated; in meiosis II, the stage in which the sister chromatids of each chromosome separate from one another and are moved to opposite poles of the cell.

androgen: a male sex hormone.

angina (an-jī′-nuh): chest pain associated with reduced blood flow to the heart muscle; caused by an obstruction of the coronary arteries.

angiosperm (an′-jē-ō-sperm): a flowering vascular plant.

angiotensin (an-jē-ō-ten-sun): a hormone that functions in water regulation in mammals by stimulating physiological changes that increase blood volume and blood pressure.

annual ring: a pattern of alternating light (early) and dark (late) xylem in woody stems and roots; formed as a result of the unequal availability of water in different seasons of the year, normally spring and summer.

antagonistic muscles: a pair of muscles, one of which contracts and in so doing extends the other, relaxed muscle; this arrangement allows movement of the skeleton at joints.

anterior pituitary: a lobe of the pituitary gland that produces prolactin and growth hormone as well as hormones that regulate hormone production in other glands.

anther (an′-ther): the uppermost part of the stamen, in which pollen develops.

antheridium (an-ther-id′-ē-um; pl., antheridia): a structure in which male sex cells are produced; found in nonvascular plants and certain seedless vascular plants.

antibiotic: chemicals that help to combat infection by destroying or slowing down the multiplication of bacteria, fungi, or protists.

antibody: a protein, produced by cells of the immune system, that combines with a specific antigen and normally facilitates the destruction of the antigen.

anticodon: a sequence of three bases in transfer RNA that is complementary to the three bases of a codon of messenger RNA.

antidiuretic hormone (an-tē-dī-ūr-et′-ik; ADH): a hormone produced by the hypothalamus and released into the bloodstream by the posterior pituitary when blood volume is low; increases the permeability of the distal tubule and the collecting duct to water, allowing more water to be reabsorbed into the bloodstream.

antigen: a complex molecule, normally a protein or polysaccharide, that stimulates the production of a specific antibody.

antioxidant: any molecule that reacts with free radicals, neutralizing their ability to damage biological molecules. Vitamins C and E are examples of dietary antioxidants.

anvil: the second of the small bones of the middle ear, linking the tympanic membrane (eardrum) to the oval window of the cochlea; also called the *incus.*

aphotic zone: the region of the ocean below 200 meters where sunlight does not penetrate.

apical dominance: the phenomenon whereby a growing shoot tip inhibits the sprouting of lateral buds.

apical meristem (āp′-i-kul mer′-i-stem): the cluster of meristem cells at the tip of a shoot or root (or one of their branches).

apicomplexan (ā-pē-kom-pleks′-an): a member of the protist clade Apicomplexa, which includes mostly parasitic, single-celled eukaryotes such as *Plasmodium*, which causes malaria in humans. Apicomplexans are part of a larger group known as the alveolates.

appendicular skeleton (ap-pen-dik′-ū-lur): the portion of the skeleton consisting of the bones of the extremities and their attachments to the axial skeleton; the appendicular skeleton therefore consists of the pectoral and pelvic girdles and the arms, legs, hands, and feet.

aquaporin: a channel protein in the plasma membrane of a cell that is selectively permeable to water.

aqueous humor (ā′-kwē-us): the clear, watery fluid between the cornea and lens of the eye; nourishes the cornea and lens.

aquifer (ok′-wifer): an underground deposit of fresh water, often used as a source for irrigation.

archaea: prokaryotes that are members of the domain Archaea, one of the three domains of living organisms; only distantly related to members of the domain Bacteria.

Archaea: one of life's three domains; consists of prokaryotes that are only distantly related to members of the domain Bacteria.

archegonium (ar-ke-gō′-nē-um; pl., archegonia): a structure in which female sex cells are produced; found in nonvascular plants and certain seedless vascular plants.

arteriole (ar-tēr′-ē-ōl): a small artery that empties into capillaries; constriction of arterioles regulates blood flow to various parts of the body.

artery (ar′-tuh-rē): a vessel with muscular, elastic walls that conducts blood away from the heart.

arthropod: a member of the animal phylum Arthropoda, which includes the insects, spiders, ticks, mites, scorpions, crustaceans, millipedes, and centipedes.

artificial selection: a selective breeding procedure in which only those individuals with particular traits are chosen as breeders; used mainly to enhance desirable traits in domesticated plants and animals; may also be used in evolutionary biology experiments.

ascomycete: a member of the fungus clade Ascomycota, whose members form sexual spores in a saclike case known as an ascus.

ascus (as'-kus; pl., asci): a saclike case in which sexual spores are formed by members of the fungus clade Ascomycota.

asexual reproduction: reproduction that does not involve the fusion of haploid gametes.

atherosclerosis (ath'-er-ō-skler-ō'-sis): a disease characterized by the obstruction of arteries by cholesterol deposits and thickening of the arterial walls.

atom: the smallest particle of an element that retains the properties of the element.

atomic mass: the total mass of all the protons, neutrons, and electrons within an atom.

atomic nucleus: the central part of an atom that contains protons and neutrons.

atomic number: the number of protons in the nuclei of all atoms of a particular element.

ATP synthase: a channel protein in the thylakoid membranes of chloroplasts and the inner membrane of mitochondria that uses the energy of H^+ ions moving through the channel down their concentration gradient to produce ATP from ADP and inorganic phosphate.

atrial natriuretic peptide (ANP) (ā'-trē-ul nā-trē-ū-ret'-ik; ANP): a hormone, secreted by cells in the mammalian heart, that reduces blood volume by inhibiting the release of ADH and aldosterone.

atrioventricular (AV) node (ā'-trē-ō-ven-trik'-ū-lar nōd): a specialized mass of muscle at the base of the right atrium through which the electrical activity initiated in the sinoatrial node is transmitted to the ventricles.

atrioventricular valve: a heart valve that separates each atrium from each ventricle, preventing the backflow of blood into the atria during ventricular contraction.

atrium (ā'-trē-um; pl., atria): a chamber of the heart that receives venous blood and passes it to a ventricle.

auditory canal (aw'-di-tor-ē): a relatively large-diameter tube within the outer ear that conducts sound from the pinna to the tympanic membrane.

auditory nerve: the nerve leading from the mammalian cochlea to the brain; it carries information about sound.

auditory tube: a thin tube connecting the middle ear with the pharynx, which allows pressure to equilibrate between the middle ear and the outside air; also called the *Eustachian tube*.

autoimmune disease: a disorder in which the immune system attacks the body's own cells or molecules.

autonomic nervous system: the part of the peripheral nervous system of vertebrates that synapses on glands, internal organs, and smooth muscle and produces largely involuntary responses.

autosome (aw'-tō-sōm): a chromosome that occurs in homologous pairs in both males and females and that does not bear the genes determining sex.

autotroph (aw'-tō-trōf): literally, "self-feeder"; normally, a photosynthetic organism; a producer.

auxin (awk'-sin): a plant hormone that influences many plant functions, including phototropism, gravitropism, apical dominance, and root branching.

axial skeleton: the skeleton forming the body axis, including the skull, vertebral column, and rib cage.

axon: a long extension of a nerve cell, extending from the cell body to synaptic endings on other nerve cells or on muscles.

B cell: a type of lymphocyte that matures in the bone marrow, and that participates in humoral immunity; gives rise to plasma cells, which secrete antibodies into the circulatory system, and to memory cells.

bacteria (sing., bacterium): prokaryotes that are members of the domain Bacteria, one of the three domains of living organisms; only distantly related to members of the domain Archaea.

Bacteria: one of life's three domains; consists of prokaryotes that are only distantly related to members of the domain Archaea.

bacteriophage (bak-tir'-ē-ō-fāj): a virus that specifically infects bacteria.

ball-and-socket joint: a joint in which the rounded end of one bone fits into a hollow depression in another, as in the hip; allows movement in several directions.

bark: the outer layer of a woody stem, consisting of phloem, cork cambium, and cork cells.

Barr body: a condensed, inactivated X chromosome in the cells of female mammals that have two X chromosomes.

basal body: a structure derived from a centriole that produces a cilium or flagellum and anchors this structure within the plasma membrane.

basal ganglion: a cluster of neurons in the interior of the cerebrum, plus the substantial nigra in the midbrain, that functions in the control of movement. Damage to or degeneration of one or more basal ganglia causes disorders such as Parkinson's disease and Huntington's disease.

base: (1) a substance capable of combining with and neutralizing H^+ ions in a solution; a solution with a pH greater than 7; (2) one of the nitrogen-containing, single- or double-ringed structures that distinguishes one nucleotide from another. In DNA, the bases are adenine, guanine, cytosine, and thymine.

basic: referring to a solution with an H^+ concentration less than that of OH^-; referring to a substance that combines with H^+.

basidiomycete: a member of the fungus clade Basidiomycota, which includes species that produce sexual spores in club-shaped cells known as basidia.

basidiospore (ba-sid'-ē-ō-spor): a sexual spore formed by members of the fungus clade Basidiomycota.

basidium (pl., basidia): a diploid cell, typically club-shaped, formed by members of the fungus clade Basidiomycota; produces basidiospores by meiosis.

basilar membrane (bas'-eh-lar): a membrane in the cochlea that bears hair cells that respond to the vibrations produced by sound.

behavior: any observable activity of a living animal.

behavioral isolation: reproductive isolation that arises when species do not interbreed because they have different courtship and mating rituals.

bilateral symmetry: a body plan in which only a single plane through the central axis will divide the body into mirror-image halves.

bile (bīl): a digestive secretion formed by the liver, stored and released from the gallbladder, and used to disperse fats in the small intestine.

binocular vision: the ability to see objects simultaneously through both eyes, providing greater depth perception and more accurate judgment of the size and distance of an object than can be achieved by vision with one eye alone.

binomial system: the method of naming organisms by genus and species, often called the scientific name, usually using Latin words or words derived from Latin.

biocapacity: an estimate of the sustainable resources and waste-absorbing capacity actually available on Earth. Biocapacity calculations are subject to change as new technologies change the way people use resources.

biodegradable: able to be broken down into harmless substances by decomposers.

biodiversity: the diversity of living organisms; measured as the variety of different species, the variety of different alleles in species' gene pools, or the variety of different communities and nonliving environments in an ecosystem or in the entire biosphere.

biofilm: a community of prokaryotes of one or more species, in which the prokaryotes secrete and are embedded in slime that adheres to a surface.

biogeochemical cycle: the pathways of a specific nutrient (such as carbon, nitrogen, phosphorus, or water) through the living and nonliving portions of an ecosystem; also called a *nutrient cycle*.

biological clock: a metabolic timekeeping mechanism found in most organisms, whereby the organism measures the approximate length of a day (24 hours) even without external environmental cues such as light and darkness.

biological magnification: the increasing accumulation of a toxic substance in progressively higher trophic levels.

biology: The study of all aspects of life and living things.

biomass: the total weight of all living material within a defined area.

biome (bī'-ōm): a terrestrial ecosystem that occupies an extensive geographical area and is characterized by a specific type of plant community; for example, deserts.

bioremediation: the use of organisms to remove or detoxify toxic substances in the environment.

biosphere (bī'-ō-sfēr): all life on Earth and the nonliving portions of Earth that support life.

Biosphere Reserve: designated by the United Nations, a Biosphere Reserve is a region intended to maintain biodiversity and evaluate techniques for sustainable human development while maintaining local cultural values.

biotechnology: any industrial or commercial use or alteration of organisms, cells, or biological molecules to achieve specific practical goals.

biotic (bī-ah'-tik): living.

biotic potential: the maximum rate at which a population is able to increase, assuming ideal conditions that allow a maximum birth rate and minimum death rate.

birth rate: the number of births per individual in a specified unit of time, such as a year.

bladder: a hollow muscular organ that stores urine.

blade: the flat part of a leaf.

blastoclade: a member of the fungus clade Blastocladiomycota, whose members have swimming spores with a single flagellum and ribosomes arranged to form a nuclear cap.

blastocyst (blas'-tō-sist): an early stage of human embryonic development, consisting of a hollow ball of cells, enclosing a mass of cells attached to its inner surface, which becomes the embryo.

blastopore: the site at which a blastula indents to form a gastrula.

blastula (blas'-tū-luh): in animals, the embryonic stage attained at the end of cleavage, in which the embryo usually consists of a hollow ball with a wall that is one or several cell layers thick.

blind spot: the area of the retina at which the axons of the ganglion cells merge to form the optic nerve; because there are no photoreceptors in the blind spot, objects focused at the blind spot cannot be seen.

blood: a specialized connective tissue, consisting of a fluid (plasma) in which blood cells are suspended; carried within the circulatory system.

blood–brain barrier: relatively impermeable capillaries of the brain that protect the cells of the brain from potentially damaging chemicals that reach the bloodstream.

blood clotting: a complex process by which platelets, the protein fibrin, and red blood cells block an irregular surface in or on the body, such as a damaged blood vessel, sealing the wound.

blood vessel: any of several types of tubes that carry blood throughout the body.

body mass index (BMI): a number derived from an individual's weight and height that is used to estimate body fat. The formula is weight (in kg) /height^2 (in meters^2).

bone: a hard, mineralized connective tissue that is a major component of the vertebrate endoskeleton; provides support and sites for muscle attachment.

bone marrow: a soft, spongy tissue that fills the cavities of large bones and generates the cell-based components of blood.

boom-and-bust cycle: a population cycle characterized by rapid exponential growth followed by a sudden massive die-off; seen in seasonal species, such as many insects living in temperate climates, and in some populations of small rodents, such as lemmings.

Bowman's capsule: the cup-shaped portion of a nephron in which blood filtrate is collected from the glomerulus.

brain: the part of the central nervous system of vertebrates that is enclosed within the skull.

branch root: a root that arises as a branch of a preexisting root; occurs through divisions of pericycle cells and subsequent differentiation of the daughter cells.

bronchiole (bron'-kē-ōl): a narrow tube, formed by repeated branching of the bronchi, that conducts air into the alveoli.

bronchus (bron'-kus; pl., bronchi): a tube that conducts air from the trachea to each lung.

bud: in plants, an embryonic shoot, often dormant until stimulated by specific combinations of hormones. In asexually reproducing animals, a miniature copy of an animal that develops on some part of the adult animal's body; usually eventually separates from the adult and assumes independent existence.

budding: asexual reproduction by the growth of a miniature copy, or bud, of the adult animal on the body of the parent. The bud breaks off to begin independent existence.

buffer: a compound that minimizes changes in pH by reversibly taking up or releasing H$^+$ ions.

bulbourethral gland (bul-bō-ū-rē'-thrul): in male mammals, a gland that secretes a basic, mucous-containing fluid that forms part of the semen.

bulk flow: the movement of many molecules of a gas or liquid in unison (in bulk, hence the name) from an area of higher pressure to an area of lower pressure.

bundle sheath cells: cells that surround the veins of plants; in C$_4$ (but not in C$_3$) plants, bundle sheath cells contain chloroplasts.

C3 pathway: in photosynthesis, the cyclic series of reactions whereby carbon from carbon dioxide is fixed as phosphoglyceric acid, the simple sugar glyceraldehyde-3-phosphate is generated, and the carbon-capture molecule, RuBP, is regenerated.

C3 plant: a plant that relies on the C$_3$ pathway to fix carbon.

C4 pathway: the series of reactions in certain plants that fixes carbon dioxide into a four-carbon molecule, which is later broken down for use in the Calvin cycle of photosynthesis. This reduces wasteful photorespiration in hot, dry environments.

C4 plant: a plant that relies on the C$_4$ pathway to fix carbon.

calcitonin (kal-si-tōn'-in): a hormone, secreted by the thyroid gland, that inhibits the release of calcium from bone.

calorie (kal'-ō-rē): the amount of energy required to raise the temperature of 1 gram of water by 1 degree Celsius.

Calorie: a unit used to measure the energy content of foods; it is the amount of energy required to raise the temperature of 1 liter of water 1 degree Celsius; also called a *kilocalorie*, equal to 1,000 calories.

Calvin cycle: in photosynthesis, the cyclic series of reactions whereby carbon from carbon dioxide is fixed as phosphoglyceric acid, the simple sugar glyceraldehyde-3-phosphate is generated, and the carbon-capture molecule, RuBP, is regenerated.

cambium (kam'-bē-um; pl., cambia): a lateral meristem, parallel to the long axis of roots and stems, that causes secondary growth of woody plant stems and roots. See *cork cambium; vascular cambium.*

camouflage (cam'-a-flaj): coloration and/or shape that renders an organism inconspicuous in its environment.

cancer: a disease in which some of the body's cells escape from normal regulatory processes and divide without control.

capillary: the smallest type of blood vessel, connecting arterioles with venules; capillary walls, through which the exchange of nutrients and wastes occurs, are only one cell thick.

carbohydrate: a compound composed of carbon, hydrogen, and oxygen, with the approximate chemical formula $(CH_2O)_n$; includes sugars, starches, and cellulose.

carbon cycle: the biogeochemical cycle by which carbon moves from its reservoirs in the atmosphere and oceans through producers and into higher trophic levels, and then back to its reservoirs.

carbon fixation: the process by which carbon derived from carbon dioxide is captured in organic molecules during photosynthesis.

cardiac cycle (kar'-dē-ak): the alternation of contraction and relaxation of the heart chambers.

cardiac muscle (kar'-dē-ak): the specialized muscle of the heart; able to initiate its own contraction, independent of the nervous system.

carnivore (kar'-neh-vor): literally, "meat eater"; a predatory organism that feeds on herbivores or on other carnivores; a secondary (or higher) consumer.

carotenoid (ka-rot'-en-oid): a red, orange, or yellow pigment, found in chloroplasts, that serves as an accessory light-gathering pigment in thylakoid photosystems.

carpel (kar'-pel): the female reproductive structure of a flower, composed of stigma, style, and ovary.

carrier: an individual who is heterozygous for a recessive condition; a carrier displays the dominant phenotype but can pass on the recessive allele to offspring.

carrier protein: a membrane protein that facilitates the diffusion of specific substances across the membrane. The molecule to be transported binds to the outer surface of the carrier protein; the protein then changes shape, allowing the molecule to move across the membrane.

carrying capacity (K): the maximum population size that an ecosystem can support for a long period of time without damaging the ecosystem; determined primarily by the availability of space, nutrients, water, and light.

cartilage (kar'-teh-lij): a form of connective tissue that forms portions of the skeleton; consists consists principally of cartilage cells and their major extracellular secretion, collagen protein.

Casparian strip (kas-par'-ē-un): a waxy, waterproof band, located in the cell walls between endodermal cells in a root, that prevents the movement of water and minerals into and out of the vascular cylinder through the extracellular space.

catalyst (kat'-uh-list): a substance that speeds up a chemical reaction without itself being permanently changed in the process; a catalyst lowers the activation energy of a reaction.

cell: the smallest unit of life, consisting, at a minimum, of an outer membrane that encloses a watery medium containing organic molecules, including genetic material composed of DNA.

cell body: the part of a nerve cell in which most of the common cellular organelles are located; typically a site of integration of inputs to the nerve cell.

cell cycle: the sequence of events in the life of a cell, from one cell division to the next.

cell division: splitting of one cell into two; the process of cellular reproduction.

cell-mediated immunity: an adaptive immune response in which foreign cells or substances are destroyed by contact with T cells.

cell plate: in plant cell division, a series of vesicles that fuse to form the new plasma membranes and cell wall separating the daughter cells.

cell theory: the scientific theory stating that every living organism is made up of one or more cells; cells are the functional units of all organisms; and all cells arise from preexisting cells.

cell wall: A nonliving, protective, and supportive layer secreted outside the plasma membrane of fungi, plants, and most bacteria and protists.

cellular respiration: the oxygen-requiring reactions, occurring in mitochondria, that break down the end products of glycolysis into carbon dioxide and water while capturing large amounts of energy as ATP.

cellular slime mold: a type of organism consisting of individual amoeboid cells that can aggregate to form a slug-like mass, which in turn forms a fruiting body. Cellular slime molds are members of the protist clade Amoebozoa.

cellulose: an insoluble carbohydrate composed of glucose subunits; forms the cell wall of plants.

central nervous system (CNS): in vertebrates, the brain and spinal cord.

central vacuole (vak'-ū-ōl): a large, fluid-filled vacuole occupying most of the volume of many plant cells; performs several functions, including maintaining turgor pressure.

centriole (sen'-trē-ōl): in animal cells, a short, barrel-shaped ring consisting of nine microtubule triplets; a pair of centrioles is found near the nucleus and may play a role in the organization of the spindle; centrioles also give rise to the basal bodies at the base of each cilium and flagellum that give rise to the microtubules of cilia and flagella.

centromere (sen'-trō-mēr): the region of a replicated chromosome at which the sister chromatids are held together until they separate during cell division.

cephalization (sef-ul-ī-zā'-shun): concentration of sensory organs and nervous tissue in the anterior (head) portion of the body.

cerebellum (ser-uh-bel'-um): the part of the hindbrain of vertebrates that is concerned with coordinating movements of the body.

cerebral cortex (ser-ē'-brul kor-tex): a thin layer of neurons on the surface of the vertebrate cerebrum in which most neural processing and coordination of activity occurs.

cerebral hemisphere: one of two nearly symmetrical halves of the cerebrum, connected by a broad band of axons, the corpus callosum.

cerebrum (ser-ē'-brum): the part of the forebrain of vertebrates that is concerned with sensory processing, the direction of motor output, and the coordination of most of the body's activities; consists of two nearly symmetrical halves (the hemispheres) connected by a broad band of axons, the corpus callosum.

cervix (ser'-viks): a ring of connective tissue at the outer end of the uterus that leads into the vagina.

channel protein: a membrane protein that forms a channel or pore completely through the membrane and that is usually permeable to one or to a few water-soluble molecules, especially ions.

chaparral: a biome located in coastal regions, with very low annual rainfall; is characterized by shrubs and small trees.

checkpoint: a mechanism in the eukaryotic cell cycle by which protein complexes in the cell determine whether the cell has successfully completed a specific process that is essential to successful cell division, such as the accurate replication of chromosomes.

chemical bond: an attraction between two atoms or molecules that tends to hold them together. Types of bonds include covalent, ionic, and hydrogen.

chemical digestion: the process by which particles of food within the digestive tract are exposed to enzymes and other digestive fluids that break down large molecules into smaller subunits.

chemical energy: a form of potential energy that is stored in molecules and may be released during chemical reactions.

chemical reaction: a process that forms and breaks chemical bonds that hold atoms together in molecules.

chemiosmosis (ke-mē-oz-mō'-sis): a process of ATP generation in chloroplasts and mitochondria. The movement of electrons down an electron transport system is used to pump hydrogen ions across a membrane, thereby building up a concentration gradient of hydrogen ions; the hydrogen ions diffuse back across the membrane through the pores of ATP-synthesizing enzymes; the energy of their movement down their concentration gradient drives ATP synthesis.

chemoreceptor: a sensory receptor that responds to chemicals in either the internal or external environment.

chemosynthesis (kē-mō-sin-the-sis): the process of oxidizing inorganic molecules, such as hydrogen sulfide, to obtain energy. Producers in hydrothermal vent communities, where light is absent, use chemosynthesis instead of photosynthesis.

chemosynthetic (kēm'-ō-sin-the-tik): capable of oxidizing inorganic molecules to obtain energy.

chiasma (kī-as'-muh; pl., chiasmata): a point at which a chromatid of one chromosome crosses with a chromatid of the homologous chromosome during prophase I of meiosis; the site of exchange of chromosomal material between chromosomes.

chitin (kī'-tin): a compound found in the cell walls of fungi and the exoskeletons of insects and some other arthropods; composed of chains of nitrogen-containing, modified glucose molecules.

chlamydia (kla-mid'-ē-uh): a sexually transmitted disease, caused by a bacterium, that causes inflammation of the urethra in males and of the urethra and cervix in females.

chlorophyll (klor'-ō-fil): a pigment found in chloroplasts that captures light energy during photosynthesis; chlorophyll absorbs violet, blue, and red light but reflects green light.

chlorophyll a (klor'-ō-fil): the most abundant type of chlorophyll molecule in photosynthetic eukaryotic organisms and in cyanobacteria; chlorophyll a is found in the reaction centers of the photosystems.

chloroplast (klor'-ō-plast): the organelle in plants and plantlike protists that is the site of photosynthesis; is surrounded by a double membrane and contains an extensive internal membrane system that bears chlorophyll.

cholecystokinin (kō'-lē-sis-tō-ki'-nin): a digestive hormone produced by the small intestine that stimulates the release of pancreatic enzymes.

chondrocyte (kon'-drō-sīt): a living cell of cartilage. Together with their extracellular secretions of collagen, chondrocytes form cartilage.

chorion (kor'-ē-on): the outermost embryonic membrane in reptiles (including birds) and mammals. In reptiles, the chorion functions mostly in gas exchange; in mammals, it forms the embryonic part of the placenta.

chorionic gonadotropin (kor-ē-on'-ik gō-nādō-trō'-pin; CG): a hormone secreted by the chorion (one of the fetal membranes) that maintains the integrity of the corpus luteum during early pregnancy.

chorionic villus (kor-ē-on-ik; pl., chorionic villi): in mammalian embryos, a finger-like projection of the chorion that penetrates the uterine lining and forms the embryonic portion of the placenta.

chorionic villus sampling (kōr-ē-on'-ik; CVS): a procedure for sampling cells from the chorionic villi produced by a fetus: A tube is inserted into the uterus of a pregnant woman, and a small sample of villi is suctioned off for genetic and biochemical analyses.

choroid (kor'-oid): a darkly pigmented layer of tissue behind the retina; contains blood vessels and also pigment that absorbs stray light.

chromatid (krō'-ma-tid): one of the two identical strands of DNA and protein that forms a duplicated chromosome. The two sister chromatids of a duplicated chromosome are joined at the centromere.

chromatin (krō'-ma-tin): the complex of DNA and proteins that makes up eukaryotic chromosomes.

chromosome (krō'-mō-sōm): a DNA double helix and associated proteins that help to organize and regulate the use of the DNA.

chylomicron: a particle produced by cells of the small intestine; it consists of proteins, triglycerides, and cholesterol and it transports the products of lipid digestion into the lymphatic system and ultimately into the circulatory system.

chyme (kīm): an acidic, souplike mixture of partially digested food, water, and digestive secretions that is released from the stomach into the small intestine.

chytrid: a member of the fungus clade Chytridomycota, which includes species with flagellated swimming spores.

ciliate (sil'-ē-et): a member of a protist group characterized by cilia and a complex unicellular structure. Ciliates are part of a larger group known as the alveolates.

cilium (sil'-ē-um; pl., cilia): a short, hair-like, motile projection from the surface of certain eukaryotic cells that contains microtubules in a 9 + 2 arrangement. The movement of cilia may propel cells through a fluid medium or move fluids over a stationary surface layer of cells.

citric acid cycle: a cyclic series of reactions, occurring in the matrix of mitochondria, in which the acetyl groups from the pyruvic acids produced by glycolysis are broken down to CO_2, accompanied by the formation of ATP and electron carriers; also called the Krebs cycle.

clade: a group that includes all the organisms descended from a common ancestor, but no other organisms; a monophyletic group.

class: in Linnaean classification, the taxonomic rank composed of related orders. Closely related classes form a phylum.

cleavage: the early cell divisions of embryos, in which little or no growth occurs between divisions; reduces the cell size and distributes gene-regulating substances to the newly formed cells.

climate: patterns of weather that prevail for long periods of time (from years to centuries) in a given region.

climate change: a long-lasting change in weather patterns, which may include significant changes in temperature, precipitation, the timing of seasons, and the frequency and severity of extreme weather events.

climax community: a diverse and relatively stable community that forms the endpoint of succession.

clitoris: an external structure of the female reproductive system that is composed of erectile tissue; a sensitive point of stimulation during sexual response.

clonal selection: the mechanism by which the adaptive immune response gains specificity; an invading antigen elicits a response from only a few lymphocytes, which proliferate to form a clone of cells that attack only the specific antigen that stimulated their production.

clone: offspring that are produced by mitosis and are, therefore, genetically identical to each other.

cloning: the process of producing many identical copies of a gene; also the production of many genetically identical copies of an organism.

closed circulatory system: a type of circulatory system, found in certain worms and vertebrates, in which the blood is always confined within the heart and vessels.

closed system: a hypothetical space where neither energy nor matter can enter or leave.

club fungus: a fungus of the clade Basidiomycota, whose members (which include mushrooms, puffballs, and shelf fungi) reproduce by means of basidiospores.

clumped distribution: the distribution characteristic of populations in which individuals are clustered into groups; the groups may be social or based on the need for a localized resource.

cochlea (kahk'-lē-uh): a coiled, bony, fluid-filled tube found in the mammalian inner ear; contains mechanoreceptors (hair cells) that respond to the vibration of sound.

codominance: the relation between two alleles of a gene, such that both alleles are phenotypically expressed in heterozygous individuals.

codon: a sequence of three bases of messenger RNA that specifies a particular amino acid to be incorporated into a protein; certain codons also signal the beginning or end of protein synthesis.

coelom (sē'-lōm): in animals, a space or cavity, lined with tissue derived from mesoderm, that separates the body wall from the inner organs.

coenzyme: an organic molecule that is bound to certain enzymes and is required for the enzymes' proper functioning; typically, a nucleotide bound to a water-soluble vitamin.

coevolution: the evolution of adaptations in two species due to their extensive interactions with one another, such that each species acts as a major force of natural selection on the other.

cohesion: the tendency of the molecules of a substance to stick together.

cohesion–tension mechanism: a mechanism for the transport of water in xylem; water is pulled up the xylem tubes, powered by the force of evaporation of water from the leaves (producing tension) and held together by hydrogen bonds between nearby water molecules (cohesion).

coleoptile (kō-lē-op'-tīl): a sheath surrounding the shoot in monocot seedlings that protects the shoot from abrasion by soil particles during germination.

collecting duct: a tube within the kidney that collects urine from many nephrons and conducts it through the renal medulla into the renal pelvis. Urine may become concentrated in the collecting ducts if antidiuretic hormone (ADH) is present.

collenchyma (kōl-en'-ki-muh): an elongated plant cell type, with thickened, flexible cell walls, that is alive at maturity and that supports the plant body.

colon: the longest part of the large intestine; does not include the rectum.

colostrum (kō-los'-trum): a yellowish fluid, high in protein and containing antibodies, that is produced by the mammary glands before milk secretion begins.

communication: the act of producing a signal that causes a receiver, normally another animal of the same species, to change its behavior in a way that is, on average, beneficial to both signaler and receiver.

community: populations of different species that live in the same area and interact with one another.

compact bone: the hard and strong outer bone; composed of osteons. Compare with *spongy bone*.

companion cell: a cell adjacent to a sieve-tube element in phloem; involved in the control and nutrition of the sieve-tube element.

competition: interaction among individuals who attempt to utilize a resource (for example, food or space) that is limited relative to the demand for that resource.

competitive exclusion principle: the concept that no two species can simultaneously and continuously occupy the same ecological niche.

competitive inhibition: the process by which two or more molecules that are somewhat similar in structure compete for the active site of an enzyme.

complement: a group of blood-borne proteins that participate in the destruction of foreign cells, especially those to which antibodies have bound.

complementary base pair: in nucleic acids, bases that pair by hydrogen bonding. In DNA, adenine is complementary to thymine, and guanine is complementary to cytosine; in RNA, adenine is complementary to uracil, and guanine to cytosine.

complete flower: a flower that has all four floral parts (sepals, petals, stamens, and carpels).

compound eye: a type of eye, found in many arthropods, that is composed of numerous independent subunits called *ommatidia*. Each ommatidium contributes a piece of a mosaic-like image perceived by the animal.

concentration: the number of particles of a dissolved substance (solute) in a given volume of solvent.

concentration gradient: a difference in the concentration of a solute between different regions within a fluid or across a barrier such as a membrane.

conclusion: in the scientific method, a decision about the validity of a hypothesis, made on the basis of experiments or observations.

conducting portion: the portion of the respiratory system in lung-breathing vertebrates that carries air to the alveoli.

cone: a cone-shaped photoreceptor cell in the vertebrate retina; not as sensitive to light as are the rods. The three types of cones are most sensitive to different colors of light and provide color vision; see also *rod*.

conifer (kon-eh-fer): a member of a group of nonflowering vascular plants whose members reproduce by means of seeds formed inside cones; retains its leaves throughout the year.

conjugation: in prokaryotes, the transfer of DNA from one cell to another via a temporary connection; in single-celled eukaryotes, the mutual exchange of genetic material between two temporarily joined cells.

connection protein: a protein in the plasma membrane of a cell that attaches either to the cytoskeleton inside the cell, to other cells, or to the extracellular matrix.

connective tissue: a tissue type consisting of diverse tissues, including bone, cartilage, fat, and blood, that generally contain large amounts of extracellular material.

conservation biology: the application of knowledge from ecology and other areas of biology to understand and conserve biodiversity.

constant-loss population: a population characterized by a relatively constant death rate; constant-loss populations have a roughly linear survivorship curve.

constant region: the part of an antibody molecule that is similar in all antibodies of a given class.

consumer: an organism that eats other organisms; a heterotroph.

contest competition: a mechanism for resolving intraspecific competition by using social or chemical interactions to limit access of some individuals to important limited resources, such as food, mates, or territories.

contraception: the prevention of pregnancy.

contractile vacuole: a fluid-filled vacuole in certain protists that takes up water from the cytoplasm, contracts, and expels the water outside the cell through a pore in the plasma membrane.

control: that portion of an experiment in which all possible variables are held constant; in contrast to the "experimental" portion, in which a particular variable is altered.

convergent evolution: the independent evolution of similar structures among unrelated organisms as a result of similar environmental pressures; see *analogous structures*.

convolution: a folding of the cerebral cortex of the vertebrate brain.

copulation: reproductive behavior in which the penis of the male is inserted into the body of the female, where it releases sperm.

coral reef: an ecosystem created by animals (reef-building corals) and plants in warm tropical waters.

core reserve: a natural area protected from most human uses that encompasses enough space to preserve most of the biodiversity of the ecosystems in that area.

cork cambium: a lateral meristem in woody roots and stems that gives rise to cork cells.

cork cell: a protective cell of the bark of woody stems and roots; at maturity, cork cells are dead, with thick, waterproof cell walls.

cornea (kor'-nē-uh): the clear outer covering of the eye, located in front of the pupil and iris; begins the focusing of light on the retina.

corona radiata (kuh-rō'-nuh rā-dē-a'-tuh): the layer of cells surrounding an egg after ovulation.

corpus callosum (kor'-pus kal-ō'-sum): the band of axons that connects the two cerebral hemispheres of vertebrates.

corpus luteum (kor'-pus loo-tē'-um): in the mammalian ovary, a structure that is derived from the follicle after ovulation and that secretes the hormones estrogen and progesterone.

cortex: the part of a primary root or stem located between the epidermis and the vascular cylinder.

cortisol (kor'-ti-sol): a steroid hormone released into the bloodstream by the adrenal cortex in response to stress. Cortisol helps the body cope with short-term stressors by raising blood glucose levels; it also inhibits the immune response.

cotyledon (kot-ul-ē'-don): a leaflike structure within a seed that absorbs food molecules from the endosperm and transfers them to the growing embryo; also called *seed leaf*.

countercurrent exchange: a mechanism for the transfer of some property, such as heat or a dissolved substance, from one fluid to another, generally without the two fluids actually mixing; in countercurrent exchange, the two fluids flow past one another in opposite directions, and they transfer heat or solute from the fluid with the higher temperature or higher solute concentration to the fluid with the lower temperature or lower solute concentration.

coupled reaction: a pair of reactions, one exergonic and one endergonic, that are linked together such that the energy produced by the exergonic reaction provides the energy needed to drive the endergonic reaction.

covalent bond (kō-vā'-lent): a chemical bond between atoms in which electrons are shared.

crab lice: an arthropod parasite that can infest humans; can be transmitted by sexual contact.

craniate: an animal that has a skull.

crassulacean acid metabolism (CAM): a biochemical pathway used by some plants in hot, dry climates to increase the efficiency of carbon fixation during photosynthesis. Mesophyll cells capture carbon dioxide at night and use it to produce sugar during the day.

critically endangered species: a species that faces an extreme risk of extinction in the wild in the immediate future.

cross-fertilization: the union of sperm and egg from two individuals of the same species.

crossing over: the exchange of corresponding segments of the chromatids of two homologous chromosomes during meiosis I; occurs at chiasmata.

cuticle (kū'-ti-kul): a waxy or fatty coating on the surfaces of the aboveground epidermal cells of many land plants; aids in the retention of water.

cyclic adenosine monophosphate (cyclic AMP): a cyclic nucleotide, formed within many target cells as a result of the reception of amino acid–derived or peptide hormones, that causes metabolic changes in the cell.

cytokine (sī'-tō-kīn): any of several chemical messenger molecules released by cells that facilitate communication with other cells and transfer signals within and among the various systems of the body. Cytokines are important in cellular differentiation and the adaptive immune response.

cytokinesis (sī-tō-ki-nē'-sis): the division of the cytoplasm and organelles into two daughter cells during cell division; normally occurs during telophase of mitotic and meiotic cell division.

cytokinin (sī-tō-ki'-nin): a group of plant hormones that promotes cell division, fruit development, and the sprouting of lateral buds; also delays the senescence of plant parts, especially leaves.

cytoplasm (sī'-tō-plaz-um): all of the material contained within the plasma membrane of a cell, exclusive of the nucleus.

cytosine (C): a nitrogenous base found in both DNA and RNA; abbreviated as C.

cytoskeleton: a network of protein fibers in the cytoplasm that gives shape to a cell, holds and moves organelles, and is typically involved in cell movement.

cytosol: the fluid portion of the cytoplasm; the substance within the plasma membrane exclusive of the nucleus and organelles.

cytotoxic T cell: a type of T cell that, upon contacting foreign cells, directly destroys them.

daughter cell: one of the two cells formed by cell division.

day-neutral plant: a plant in which flowering occurs as soon as the plant has undergone sufficient growth and development, regardless of day length.

death rate: the number of deaths per individual in a specified unit of time, such as a year.

decomposer: an organism, usually a fungus or bacterium, that digests organic material by secreting digestive enzymes into the environment, in the process liberating nutrients into the environment.

deductive reasoning: the process of generating hypotheses about the results of a specific experiment or the nature of a specific observation.

deforestation: the excessive cutting of forests. In recent years, deforestation has occurred primarily in rain forests in the Tropics, to clear space for agriculture.

dehydration synthesis: a chemical reaction in which two molecules are joined by a covalent bond with the simultaneous removal of a hydrogen from one molecule and a hydroxyl group from the other, forming water; the reverse of hydrolysis.

deletion mutation: a mutation in which one or more pairs of nucleotides are removed from a gene.

demographic transition: a change in population dynamic in which a fairly stable population with both high birth rates and high death rates experiences rapid growth as death rates decline, and then returns to a stable (although much larger) population as birth rates decline.

demography: the study of the changes in human numbers over time, grouped by world regions, age, sex, educational levels, and other variables.

denature: to disrupt the secondary and/or tertiary structure of a protein while leaving its amino acid sequence intact. Denatured proteins can no longer perform their biological functions.

denatured: having the secondary and/or tertiary structure of a protein disrupted, while leaving the amino acid sequence unchanged. Denatured proteins can no longer perform their biological functions.

dendrite (den'-drīt): a branched tendril that extends outward from the cell body of a neuron; specialized to respond to signals from the external environment or from other neurons.

dendritic cell (den-drit'-ick): a type of phagocytic leukocyte that presents antigen to T and B cells, thereby stimulating an adaptive immune response to an invading microbe.

denitrifying bacteria (dē-nī'-treh-fī-ing): bacteria that break down nitrates, releasing nitrogen gas to the atmosphere.

density-dependent: referring to any factor, such as predation, that limits population size to an increasing extent as the population density increases.

density-independent: referring to any factor, such as floods or fires, that limits a population's size regardless of its density.

deoxyribonucleic acid (dē-ox-ē-rī-bō-noo-klā'-ik; DNA): a molecule composed of deoxyribose nucleotides; contains the genetic information of all living cells.

dermal tissue system: a plant tissue system that makes up the outer covering of the plant body.

dermis (dur'-mis): the layer of skin beneath the epidermis; composed of connective tissue and containing blood vessels, muscles, nerve endings, and glands.

desert: a biome in which less than 10 inches (25 centimeters) of rain fall each year; characterized by cacti, succulents, and widely spaced, drought-resistant bushes.

desertification: the process by which relatively dry, drought-prone regions are converted to desert as a result of drought and overuse of the land, for example, by overgrazing or cutting of trees.

desmosome (dez'-mō-sōm): a strong cell-to-cell junction that attaches adjacent cells to one another.

detritivore (de-trī'-ti-vor): one of a diverse group of organisms, ranging from worms to vultures, that live off the wastes and dead remains of other organisms.

deuterostome (doo'-ter-ō-stōm): an animal with a mode of embryonic development in which the coelom is derived from outpocketings of the gut; characteristic of echinoderms and chordates.

developed country: a country such as Australia, Japan, and those in Europe and North America whose residents have relatively high average standards of living and good access to education, technology, and medical care.

developing country: a country (including most in Central and South America, Asia, and Africa) that has a relatively low average living standard and limited access to modern technology and medical care.

development: the process by which an organism proceeds from fertilized egg through adulthood to eventual death.

diabetes mellitus (di-uh-bē'-tēs mel-ī'-tus): a disease characterized by defects in the production, release, or reception of insulin; characterized by high blood glucose levels that fluctuate with sugar intake.

diaphragm (dī'-uh-fram): in the respiratory system, a dome-shaped muscle forming the floor of the chest cavity; when the diaphragm contracts, it flattens, enlarging the chest cavity and causing air to be drawn into the lungs.

diastolic pressure (dī'-uh-stal-ik): the blood pressure measured during relaxation of the ventricles; the lower of the two blood pressure readings.

diatom (dī'-uh-tom): a member of a protist group that includes photosynthetic forms with two-part glassy outer coverings; important photosynthetic organisms in fresh water and salt water. Diatoms are part of a larger group known as the stramenopiles.

dicot (dī'-kaht): short for dicotyledon; a type of flowering plant characterized by embryos with two cotyledons, or seed leaves, that are usually modified for food storage.

differentiate: the process whereby a cell becomes specialized in structure and function.

differentiated cell: a mature cell specialized for a specific function; in plants, differentiated cells normally do not divide.

diffusion: the net movement of solute particles from a region of high solute concentration to a region of low solute concentration, driven by a concentration gradient; may occur within a fluid or across a barrier such as a membrane.

digestion: the process by which food is physically and chemically broken down into molecules that can be absorbed by cells.

digestive system: a group of organs responsible for ingesting food, digesting food into simple molecules that can be absorbed into the circulatory system, and expelling undigested wastes from the body.

dinoflagellate (dī-nō-fla'-jel-et): a member of a protist group that includes photosynthetic forms in which two flagella project through armor-like plates; abundant in oceans; can reproduce rapidly, causing "red tides." Dinoflagellates are part of a larger group known as the alveolates.

diploid (dip'-loid): referring to a cell with pairs of homologous chromosomes.

diplomonad: a member of a protist group characterized by two nuclei and multiple flagella. Diplomonads, which include disease-causing parasites such as *Giardia*, are part of a larger group known as the excavates.

direct development: a developmental pathway in which the offspring is born as a miniature version of the adult and does not radically change in body form as it grows and matures.

directional selection: a type of natural selection that favors one extreme of a range of phenotypes.

disaccharide (dī-sak'-uh-rīd): a carbohydrate formed by the covalent bonding of two monosaccharides.

disruptive selection: a type of natural selection that favors both extremes of a range of phenotypes.

dissolve: the process by which solvent molecules completely surround and disperse the individual atoms or molecules of another substance, the solute.

distal tubule: the last section of a mammalian nephron, following the loop of Henle and emptying urine into a collecting duct; most tubular secretion and a small amount of tubular reabsorption occur in the distal tubule.

disturbance: any event that disrupts an ecosystem by altering its community, its abiotic structure, or both; disturbance precedes succession.

disulfide bond: the covalent bond formed between the sulfur atoms of two cysteines in a protein; typically causes the protein to fold by bringing otherwise distant parts of the protein close together.

DNA cloning: any of a variety of technologies that are used to produce multiple copies of a specific segment of DNA (usually a gene).

DNA helicase: an enzyme that helps unwind the DNA double helix during DNA replication.

DNA ligase: an enzyme that bonds the terminal sugar in one DNA strand to the terminal phosphate in a second DNA strand, creating a single strand with a continuous sugar-phosphate backbone.

DNA polymerase: an enzyme that bonds DNA nucleotides together into a continuous strand, using a preexisting DNA strand as a template.

DNA probe: a sequence of nucleotides that is complementary to the nucleotide sequence of a gene or other segment of DNA under study; used to locate the gene or DNA segment during gel electrophoresis or other methods of DNA analysis.

DNA profile: the pattern of short tandem repeats of specific DNA segments; using a standardized set of 13 short tandem repeats, DNA profiles identify individual people with great accuracy.

DNA replication: the copying of the double-stranded DNA molecule, producing two identical DNA double helices.

DNA sequencing: the process of determining the order of nucleotides in a DNA molecule.

domain: the broadest category for classifying organisms; organisms are classified into three domains: Bacteria, Archaea, and Eukarya.

dominance hierarchy: a social structure that arises when the animals in a social group establish individual ranks that determine access to resources; ranks are usually established through aggressive interactions.

dominant: an allele that can determine the phenotype of heterozygotes completely, such that they are indistinguishable from individuals homozygous for the allele; in the heterozygotes, the expression of the other (recessive) allele is completely masked.

dormancy: a state in which an organism does not grow or develop; usually marked by lowered metabolic activity and resistance to adverse environmental conditions.

dorsal root ganglion: a ganglion, located on the dorsal (sensory) branch of each spinal nerve, that contains the cell bodies of sensory neurons.

double circulation: the separation of circulatory routes between the heart and the lungs and between the heart and the rest of the body.

double fertilization: in flowering plants, the fusion of two sperm nuclei with the nuclei of two cells of the female gametophyte. One sperm nucleus fuses with the egg to form the zygote; the second sperm nucleus fuses with the two haploid nuclei of the central cell to form a triploid endosperm cell.

double helix (hē′-liks): the shape of the two-stranded DNA molecule; similar to a ladder twisted lengthwise into a corkscrew shape.

Down syndrome: a genetic disorder caused by the presence of three copies of chromosome 21; common characteristics include learning disabilities, distinctively shaped eyelids, a small mouth, heart defects, and low resistance to infectious diseases; also called *trisomy 21*.

duodenum: the first section of the small intestine, in which most food digestion occurs; receives chyme from the stomach, buffers and digestive enzymes from the pancreas, and bile from the liver and gallbladder.

duplicated chromosome: a eukaryotic chromosome following DNA replication; consists of two sister chromatids joined at the centromeres.

early loss population: a population characterized by a high birth rate, a high death rate among juveniles, and lower death rates among adults; early loss populations have a concave survivorship curve.

ecdysone (ek-dī′-sōn): a steroid hormone that triggers molting in insects and other arthropods.

ecological economics: the branch of economics that attempts to determine the monetary value of ecosystem services and to compare the monetary value of natural ecosystems with the monetary value of human activities that may reduce the services that natural ecosystems provide.

ecological footprint: the area of productive land needed to produce the resources used and absorb the wastes (including carbon dioxide) generated by an individual person, or by an average person of a specific part of the world (for example, an individual country), or of the entire world, using current technologies.

ecological isolation: reproductive isolation that arises when species do not interbreed because they occupy different habitats.

ecological niche (nitch): the role of a particular species within an ecosystem, including all aspects of its interaction with the living and nonliving environments.

ecology (ē-kol′-uh-jē): the study of the interrelationships of organisms with each other and with their nonliving environment.

ecosystem (ē′-kō-sis-tem): all the organisms and their nonliving environment within a defined area.

ecosystem services: the processes through which natural ecosystems and their living communities sustain and fulfill human life. Ecosystem services include purifying air and water, replenishing oxygen, pollinating plants, reducing flooding, providing wildlife habitat, and many more.

ectoderm (ek′-tō-derm): the outermost embryonic tissue layer, which gives rise to structures such as hair, the epidermis of the skin, and the nervous system.

ectotherm: an animal that obtains most of its body warmth from its environment; body temperatures of ectotherms vary with the temperature of their surroundings.

effector (ē-fek′-tor): a part of the body (normally a muscle or gland) that carries out responses as directed by the nervous system.

egg: the haploid female gamete, usually large and nonmotile; contains food reserves for the developing embryo.

electromagnetic spectrum: the range of all possible wavelengths of electromagnetic radiation, from wavelengths longer than radio waves, to microwaves, infrared, visible light, ultraviolet, x-rays, and gamma rays.

electron: a subatomic particle, found in an electron shell outside the nucleus of an atom, that bears a unit of negative charge and very little mass.

electron carrier: a molecule that can reversibly gain or lose electrons. Electron carriers generally accept high-energy electrons produced during an exergonic reaction and donate the electrons to acceptor molecules that use the energy to drive endergonic reactions.

electron shell: a region in an atom within which electrons orbit; each shell corresponds to a fixed energy level at a given distance from the nucleus.

electron transport chain (ETC): a series of electron carrier molecules, found in the thylakoid membranes of chloroplasts and the inner membrane of mitochondria, that extract energy from electrons and generate ATP or other energetic molecules.

element: a substance that cannot be broken down, or converted, to a simpler substance by ordinary chemical means.

elimination: the expulsion of indigestible materials from the digestive tract, through the anus, and outside the body.

embryo (em′-brē-ō): in animals, the stages of development that begin with the fertilization of the egg cell and end with hatching or birth; in mammals, the early stages in which the developing animal does not yet resemble the adult of the species.

embryonic disk: in human embryonic development, the flat, two-layered group of cells that separates the amniotic cavity from the yolk sac; the cells of the embryonic disk produce most of the developing embryo.

embryonic stem cell (ESC): a cell derived from an early embryo that is capable of differentiating into any of the adult cell types.

emerging infectious disease: a previously unknown infectious disease (one caused by a microbe), or a previously known infectious disease whose frequency or severity has significantly increased in the past two decades.

emigration (em-uh-grā′-shun): migration of individuals out of an area.

endangered species: a species that faces a high risk of extinction in the wild in the near future.

endergonic (en-der-gon′-ik): pertaining to a chemical reaction that requires an input of energy to proceed; an "uphill" reaction.

endocrine communication: communication between cells of an animal in which certain cells, usually part of an endocrine gland, release hormones that travel through the bloodstream to other cells (often in distant parts of the body) and alter their activity.

endocrine disrupter: an environmental pollutant that interferes with endocrine function, often by disrupting the action of sex hormones.

endocrine gland: a ductless, hormone-producing gland consisting of cells that release their secretions into the interstitial fluid from which the secretions diffuse into nearby capillaries; most endocrine glands are composed of epithelial cells.

endocrine hormone: a molecule produced by the cells of endocrine glands and released into the circulatory system. An endocrine hormone causes changes in target cells that bear specific receptors for the hormone.

endocrine system: an animal's organ system for cell-to-cell communication; composed of hormones and the cells that secrete them.

endocytosis (en-dō-sī-tō′-sis): the process in which the plasma membrane engulfs extracellular material, forming membrane-bound sacs that enter the cytoplasm and thereby move material into the cell.

endoderm (en′-dō-derm): the innermost embryonic tissue layer, which gives rise to structures such as the lining of the digestive and respiratory tracts.

endodermis (en-dō-der′-mis): the innermost layer of small, close-fitting cells of the cortex of a root that form a ring around the vascular cylinder; see also *Casparian strip*.

endomembrane system: internal membranes that create loosely connected compartments within the eukaryotic cell. It includes the nuclear envelope, the endoplasmic reticulum, vesicles, the Golgi apparatus, and lysosomes.

endometrium (en-dō-mē′-trē-um): the nutritive inner lining of the uterus.

endoplasmic reticulum (en-dō-plaz′-mik re-tik′-ū-lum; ER): a system of membranous tubes and channels in eukaryotic cells; the site of most protein and lipid synthesis.

endoskeleton (en′-dō-skel′-uh-tun): a rigid internal skeleton with flexible joints that allow for movement.

endosperm: a triploid food storage tissue in the seeds of flowering plants that nourishes the developing plant embryo.

endospore: a protective resting structure of some rod-shaped bacteria that withstands unfavorable environmental conditions.

endosymbiont hypothesis: the hypothesis that certain organelles, especially chloroplasts and mitochondria, arose as mutually beneficial associations between the ancestors of eukaryotic cells and captured bacteria that lived within the cytoplasm of the pre-eukaryotic cell.

endotherm: an animal that obtains most of its body heat from metabolic activities; body temperatures of endotherms usually remain relatively constant within a fairly wide range of environmental temperatures.

energy: the capacity to do work.

energy-carrier molecule: high-energy molecules that are synthesized at the site of an exergonic reaction, where they capture one or two energized electrons and hydrogen ions. They include nicotinamide adenine dinucleotide (NADH) and its relative, flavin adenine dinucleotide ($FADH_2$).

energy pyramid: a graphical representation of the energy contained in succeeding trophic levels, with maximum energy at the base (primary producers) and steadily diminishing amounts at higher levels.

energy-requiring transport: the transfer of substances across a cell membrane using cellular energy; includes active transport, endocytosis, and exocytosis.

entropy (en'-trō-pē): a measure of the amount of randomness and disorder in a system.

environmental resistance: any factor that tends to counteract biotic potential, limiting population growth and the resulting population size.

enzyme (en'-zīm): a biological catalyst, usually a protein, that speeds up the rate of specific biological reactions.

epicotyl (ep'-ē-kot-ul): the part of the embryonic shoot located above the attachment point of the cotyledons but below the tip of the shoot.

epidermal tissue: dermal tissue in plants that forms the epidermis, the outermost cell layer that covers leaves, young stems, and young roots.

epidermis (ep-uh-der'-mis): in animals, specialized stratified epithelial tissue that forms the outer layer of the skin; in plants, the outermost layer of cells of a leaf, young root, or young stem.

epididymis (e-pi-di'-di-mus): a series of tubes that connect with and receive sperm from the seminiferous tubules of the testis, and empty into the vas deferens.

epigenetics: the study of the mechanisms by which cells and organisms change gene expression and function without changing the base sequence of their DNA; usually, epigenetic controls over DNA expression involve modification of DNA, modification of chromosomal proteins, or alteration of transcription or translation through the actions of noncoding RNA molecules.

epiglottis (ep-eh-glah'-tis): a flap of cartilage in the lower pharynx that covers the opening to the larynx during swallowing; directs food into the esophagus.

epinephrine (ep-i-nef'-rin): a hormone, secreted by the adrenal medulla, that is released in response to stress and that stimulates a variety of responses, including the release of glucose from the liver and an increase in heart rate.

epithelial tissue (eh-puh-thē'-lē-ul): a tissue type that forms membranes that cover the body surface and line body cavities, and that also gives rise to glands.

equilibrium population: a population in which allele frequencies and the distribution of genotypes do not change from generation to generation.

erythrocyte (eh-rith'-rō-sīt): a red blood cell, which contains the oxygen-binding protein hemoglobin and thus transports oxygen in the circulatory system.

erythropoietin (eh-rith'-rō-pō-ē'-tin): a hormone produced by the kidneys in response to oxygen deficiency; stimulates the production of red blood cells by the bone marrow.

esophagus: a muscular, tubular portion of the mammalian digestive tract located between the pharynx and the stomach; no digestion occurs in the esophagus.

essential amino acid: an amino acid that is a required nutrient; the body is unable to manufacture essential amino acids, so they must be supplied in the diet.

essential fatty acid: a fatty acid that is a required nutrient; the body is unable to manufacture essential fatty acids, so they must be supplied in the diet.

essential nutrient: any nutrient that cannot be synthesized by the body, including certain fatty acids and amino acids, vitamins, minerals, and water.

estrogen: a female sex hormone, produced by follicle cells of the ovary, that stimulates follicle development, oogenesis, the development of secondary sex characteristics, and growth of the uterine lining.

estuary (es'-choo-ār-ē): a wetland formed where a river meets the ocean; the salinity is quite variable, but lower than in seawater and higher than in fresh water.

ethology (ē-thol'-ō-jē): the study of animal behavior.

ethylene: a plant hormone that promotes the ripening of some fruits and the dropping of leaves and fruit; promotes senescence of leaves.

euglenid (ū'-gle-nid): a member of a protist group characterized by one or more whip-like flagella, which are used for locomotion, and by a photoreceptor, which detects light; are photosynthetic. Euglenids are part of a larger group known as euglenozoans.

euglenozoan: a member of the Euglenozoa, a protist clade. The euglenozoans, which are characterized by mitochondrial membranes that appear under the microscope to be shaped like a stack of disks, include the euglenids and the kinetoplastids.

Eukarya (ū-kar'-ē-a): one of life's three domains; consists of all eukaryotes (plants, animals, fungi, and protists).

eukaryote (ū-kar'-ē-ōt): an organism whose cells are eukaryotic; plants, animals, fungi, and protists are eukaryotes.

eukaryotic (ū-kar-ē-ot'-ik): referring to cells of organisms of the domain Eukarya (plants, animals, fungi, and protists). Eukaryotic cells have genetic material enclosed within a membrane-bound nucleus and they contain other membrane-bound organelles.

eutrophic lake: a lake that receives sufficiently large inputs of sediments, organic material, and inorganic nutrients from its surroundings to support dense communities, especially of plants and phytoplankton; contains murky water with poor light penetration.

evolution: (1) the descent of modern organisms, with modification, from preexisting life-forms; (2) the theory that all organisms are related by common ancestry and have changed over time; (3) any change in the genetic makeup (the proportions of different genotypes) of a population from one generation to the next.

excavate: a member of the Excavata, a protist clade. The excavates, which generally lack mitochondria, include the diplomonads and the parabasalids.

excitatory postsynaptic potential (EPSP): an electrical signal produced in a postsynaptic cell that makes the resting potential of the postsynaptic neuron less negative and, hence, makes the neuron more likely to produce an action potential.

excretion: the elimination of waste substances from the body; can occur from the digestive system, skin glands, urinary system, or lungs.

exergonic (ex-er-gon'-ik): pertaining to a chemical reaction that releases energy (either as heat or in the form of increased entropy); a "downhill" reaction.

exhalation: the act of releasing air from the lungs, which results from a relaxation of the respiratory muscles.

exocrine gland: a gland that releases its secretions into ducts that lead to the outside of the body or into a body cavity, such as the digestive or reproductive system; most exocrine glands are composed of epithelial cells.

exocytosis (ex-ō-sī-tō'-sis): the process in which intracellular material is enclosed within a membrane-bound sac that moves to the plasma membrane and fuses with it, releasing the material outside the cell.

exon: a segment of DNA in a eukaryotic gene that codes for amino acids in a protein; see also *intron*.

exoskeleton (ex'-ō-skel'-uh-tun): a rigid external skeleton that supports the body, protects the internal organs, and has flexible joints that allow for movement.

experiment: in the scientific method, the use of carefully controlled observations or manipulations to test the predictions generated by a hypothesis.

exponential growth: a continuously accelerating increase in population size; this type of growth generates a curve shaped like the letter "J."

extensor: a muscle that straightens (increases the angle of) a joint.

external fertilization: the union of sperm and egg outside the body of either parent.

extinction: the death of all members of a species.

extracellular digestion: the physical and chemical breakdown of food that occurs outside a cell, normally in a digestive cavity.

extraembryonic membrane: in the embryonic development of reptiles (including birds) and mammals, one of the following membranes: the chorion (functions in gas exchange), amnion (provision of the watery environment needed for development), allantois (waste storage), or yolk sac (storage of the yolk).

facilitated diffusion: the diffusion of molecules across a membrane, assisted by protein pores or carriers embedded in the membrane.

family: in Linnaean classification, the taxonomic rank composed of related genera. Closely related families make up an order.

farsighted: the inability to focus on nearby objects, caused by the eyeball being slightly too short or the cornea too flat.

fat (molecular): a lipid composed of three saturated fatty acids covalently bonded to glycerol; fats are solid at room temperature.

fatty acid: an organic molecule composed of a long chain of carbon atoms, with a carboxylic acid (–COOH) group at one end; may be saturated (all single bonds between the carbon atoms) or unsaturated (one or more double bonds between the carbon atoms).

feces: semisolid waste material that remains in the intestine after absorption is complete and is voided through the anus. Feces consist principally of indigestible wastes and bacteria.

feedback inhibition: in enzyme-mediated chemical reactions, the condition in which the product of a reaction inhibits one or more of the enzymes involved in synthesizing the product.

fermentation: anaerobic reactions that convert the pyruvic acid produced by glycolysis into lactic acid or alcohol and CO_2, using hydrogen ions and electrons from NADH; the primary function of fermentation is to regenerate NAD+ so that glycolysis can continue under anaerobic conditions.

fertilization: the fusion of male and female haploid gametes, forming a zygote.

fetal alcohol syndrome (FAS): a cluster of symptoms, including mental retardation and physical abnormalities, that occur in infants born to mothers who consumed large amounts of alcoholic beverages during pregnancy.

fetus: the later stages of mammalian embryonic development (after the second month for humans), when the developing animal has come to resemble the adult of the species.

fever: an elevation in body temperature caused by chemicals (pyrogens) that are released by white blood cells in response to infection.

fibrillation (fi-bri-lā'-shun): the rapid, uncoordinated, and ineffective contraction of heart muscle cells.

fibrin (fi'-brin): a clotting protein formed in the blood in response to a wound; binds with other fibrin molecules and provides a matrix around which a blood clot forms.

fibrinogen (fī-brin'-ō-jen): the inactive form of the clotting protein fibrin. Fibrinogen is converted into fibrin by the enzyme thrombin, which is produced in response to injury.

fibrous root system: a root system, commonly found in monocots, that is characterized by many roots of approximately the same size arising from the base of the stem.

filament: in flowers, the stalk of a stamen, which bears an anther at its tip.

filtrate: the fluid produced by filtration; in the kidneys, the fluid produced by the filtration of blood through the glomerular capillaries.

filtration: within Bowman's capsule in each nephron of a kidney, the process by which blood is pumped under pressure through permeable capillaries of the glomerulus, forcing out water and small solutes, including wastes and nutrients.

first law of thermodynamics: the principle of physics that states that within any closed system, energy can be neither created nor destroyed, but can be converted from one form to another; also called the law of conservation of energy.

fission: asexual reproduction by dividing the body into two smaller, complete organisms.

fitness: the reproductive success of an organism, relative to the average reproductive success in the population.

flagellum (fla-jel'-um; pl., flagella): a long, hair-like, motile extension of the plasma membrane; in eukaryotic cells, it contains microtubules arranged in a 9 + 2 pattern. The movement of flagella propels some cells through fluids.

flavin adenine dinucleotide (FAD or FADH$_2$): an electron-carrier molecule produced in the mitochondrial matrix by the Krebs cycle; subsequently donates electrons to the electron transport chain.

flexor: a muscle that flexes (decreases the angle of) a joint.

florigen: one of a group of plant hormones that may stimulate or inhibit flowering in response to day length.

flower: the reproductive structure of an angiosperm plant.

flower bud: a cluster of meristem cells (a bud) that forms a flower.

fluid: any substance whose molecules can freely flow past one another; "fluid" can describe liquids, cell membranes, and gases.

fluid mosaic model: a model of cell membrane structure; according to this model, membranes are composed of a double layer of phospholipids in which various proteins are embedded. The phospholipid bilayer is a somewhat fluid matrix that allows the movement of proteins within it.

follicle: in the ovary of female mammals, the oocyte and its surrounding accessory cells.

follicle-stimulating hormone (FSH): a hormone, produced by the anterior pituitary, that stimulates spermatogenesis in males and the development of the follicle in females.

food chain: a linear feeding relationship in a community, using a single representative from each of the trophic levels.

food vacuole: a membranous sac, within a single cell, in which food is enclosed. Digestive enzymes are released into the vacuole, where intracellular digestion occurs.

food web: a representation of the complex feeding relationships within a community, including many organisms at various trophic levels, with many of the consumers occupying more than one level simultaneously.

foraminiferan (for-am-i-nif'-er-un): a member of a protist group characterized by pseudopods and elaborate calcium carbonate shells. Foraminiferans are generally aquatic (largely marine) and are part of a larger group known as rhizarians.

forebrain: during development, the anterior portion of the brain. In mammals, the forebrain differentiates into the thalamus, the limbic system, and the cerebrum. In humans, the cerebrum contains about half of all the neurons in the brain.

fossil: the remains of a dead organism, normally preserved in rock; may be petrified bones or wood; shells; impressions of body forms, such as feathers, skin, or leaves; or markings made by organisms, such as footprints.

fossil fuel: a fuel, such as coal, oil, and natural gas, derived from the remains of ancient organisms.

founder effect: the result of an event in which an isolated population is founded by a small number of individuals; may result in genetic drift if allele frequencies in the founder population are by chance different from those of the parent population.

fovea (fō'-vē-uh): in the vertebrate retina, the central region on which images are focused; contains closely packed cones.

free nucleotide: a nucleotide that has not been joined with other nucleotides to form a DNA or RNA strand.

free radical: a molecule containing an atom with an unpaired electron, which makes it highly unstable and reactive with nearby molecules. By removing an electron from the molecule it attacks, it creates a new free radical and begins a chain reaction that can lead to the destruction of biological molecules crucial to life.

fruit: in flowering plants, the ripened ovary (plus, in some cases, other parts of the flower), which contains the seeds.

functional group: one of several groups of atoms commonly found in an organic molecule, including hydrogen, hydroxyl, amino, carboxyl, and phosphate groups, that determine the characteristics and chemical reactivity of the molecule.

gallbladder: a small sac located next to the liver that stores and concentrates the bile secreted by the liver. Bile is released from the gallbladder to the small intestine through the bile duct.

gamete (gam'-ēt): a haploid sex cell, usually a sperm or an egg, formed in sexually reproducing organisms.

gametic incompatibility: a postmating reproductive isolating mechanism that arises when sperm from one species cannot fertilize eggs of another species.

gametophyte (ga-mēt'-ō-fīt): the multicellular haploid stage in the life cycle of plants.

ganglion (gang'-lē-un; pl., ganglia): a cluster of neurons.

ganglion cell: in the vertebrate retina, the nerve cells whose axons form the optic nerve.

gap junction: a type of cell-to-cell junction in animals in which channels connect the cytoplasm of adjacent cells.

gas-exchange portion: the portion of the respiratory system in lung-breathing vertebrates where gas is exchanged in the alveoli of the lungs.

gastric gland: one of numerous small glands in the stomach lining; contains cells that secrete mucus, hydrochloric acid, or pepsinogen (the inactive form of the protease pepsin).

gastrin: a hormone produced by the stomach that stimulates acid secretion in response to the presence of food.

gastrovascular cavity: a saclike chamber with digestive functions, found in some invertebrates such as cnidarians (sea jellies, anemones, and related animals); a single opening serves as both mouth and anus.

gastrula (gas'-troo-luh): in animal development, a three-layered embryo with ectoderm, mesoderm, and endoderm cell layers. The endoderm layer usually encloses the primitive gut.

gastrulation (gas-troo-la'-shun): the process whereby a blastula develops into a gastrula, including the formation of endoderm, ectoderm, and mesoderm.

gel electrophoresis: a technique in which molecules (such as DNA fragments) are placed in wells in a thin sheet of gelatinous material and exposed to an electric field; the molecules migrate through the gel at a rate determined by certain characteristics, most commonly size.

gene: the unit of heredity; a segment of DNA located at a particular place on a chromosome that usually encodes the information for the amino acid sequence of a protein and, hence, a particular trait.

gene flow: the movement of alleles from one population to another owing to the migration of individual organisms.

gene linkage: the tendency for genes located on the same chromosome to be inherited together.

gene pool: the total of all alleles of all genes in a population; for a single gene, the total of all the alleles of that gene that occur in a population.

gene therapy: the attempt to cure a disease by inserting, deleting, or altering a patient's genes.

generative cell: in flowering plants, one of the haploid cells of a pollen grain; undergoes mitotic cell division to form two sperm cells.

genetic code: the collection of codons of mRNA, each of which directs the incorporation of a particular amino acid into a protein during protein synthesis or causes protein synthesis to start or stop.

genetic drift: a change in the allele frequencies of a small population purely by chance.

genetic engineering: the modification of the genetic material of an organism, usually using recombinant DNA techniques.

genetic recombination: the generation of new combinations of alleles on homologous chromosomes due to the exchange of DNA during crossing over.

genetically modified organism (GMO): a plant or animal that contains DNA that has been modified or that has been obtained from another species.

genital herpes: a sexually transmitted disease, caused by a virus, that can cause painful blisters on the genitals and surrounding skin.

genomic imprinting: a form of epigenetic control by which a given gene is expressed in an offspring only if the gene has been inherited from a specific parent; the copy of the gene that was inherited from the other parent is usually not expressed.

genotype (jēn'-ō-tip): the genetic composition of an organism; the actual alleles of each gene carried by the organism.

genus (jē'-nus): in Linnaean classification, the taxonomic rank composed of related species. Closely related genera make up a family.

geographic isolation: reproductive isolation that arises when species do not interbreed because a physical barrier separates them.

germination: the growth and development of a seed, spore, or pollen grain.

ghrelin (grel'-in): a peptide hormone secreted by the stomach that acts via the hypothalamus to stimulate hunger.

gibberellin (jib-er-el'-in): one of a group of plant hormones that stimulates seed germination, fruit development, flowering, and cell division and elongation in stems.

gill: in aquatic animals, a branched tissue richly supplied with capillaries around which water is circulated for gas exchange.

gland: a cluster of cells that are specialized to secrete substances such as sweat, mucus, enzymes, or hormones.

glia: cells of the nervous system that provide nutrients for neurons, regulate the composition of the interstitial fluid in the brain and spinal cord, modulate communication between neurons, and insulate axons, thereby speeding up the conduction of action potentials. Also called *glial cells.*

glomeromycete: a member of the fungus clade Glomeromycota, which includes species that form mycorrhizal associations with plant roots and that form bush-shaped branching structures inside plant cells.

glomerulus (glō-mer'-ū-lus): a dense network of thin-walled capillaries, located within the Bowman's capsule of each nephron of the kidney, where blood pressure forces water and small solutes, including wastes and nutrients, through the capillary walls into the nephron.

glucagon (gloo'-ka-gon): a hormone, secreted by the pancreas, that increases blood sugar by stimulating the breakdown of glycogen (to glucose) in the liver.

glucocorticoid (gloo-kō-kor'-tik-oid): a class of hormones, released by the adrenal cortex in response to the presence of ACTH, that makes energy available to the body by stimulating the synthesis of glucose.

glucose: the most common monosaccharide, with the molecular formula $C_6H_{12}O_6$; most polysaccharides, including cellulose, starch, and glycogen, are made of glucose subunits covalently bonded together.

glycerol (glis'-er-ol): a three-carbon alcohol to which fatty acids are covalently bonded to make fats and oils.

glycogen (glī'-kō-jen): a highly branched polymer of glucose that is stored by animals in the muscles and liver and metabolized as a source of energy.

glycolysis (glī-kol'-i-sis): reactions, carried out in the cytoplasm, that break down glucose into two molecules of pyruvic acid, producing two ATP molecules; does not require oxygen but can proceed when oxygen is present.

glycoprotein: a protein to which a carbohydrate is attached.

goiter: a swelling of the thyroid gland, caused by iodine deficiency, that affects the functioning of the thyroid gland and its hormones.

Golgi apparatus (gōl'-jē): a stack of membranous sacs, found in most eukaryotic cells, that is the site of processing and separation of membrane components and secretory materials.

gonad: an organ where reproductive cells are formed; in males, the testes, and in females, the ovaries.

gonadotropin-releasing hormone (gō-na-dō-trō'-pin; GnRH): a hormone produced by the neurosecretory cells of the hypothalamus, which stimulates cells in the anterior pituitary to release follicle-stimulating hormone (FSH) and luteinizing hormone (LH). GnRH is involved in the menstrual cycle and in spermatogenesis.

gonorrhea (gon-uh-rē'-uh): a sexually transmitted bacterial infection of the reproductive organs; if untreated, can result in sterility.

gradient: a difference in concentration, pressure, or electrical charge between two regions.

grassland: a biome, located in the centers of continents, that primarily supports grasses; also called a *prairie.*

gravitropism: growth with respect to the direction of gravity.

gray matter: the outer portion of the brain and inner region of the spinal cord; composed largely of neuron cell bodies, which give this area a gray color in preserved tissue.

greenhouse effect: the process in which certain gases such as carbon dioxide and methane trap sunlight energy in a planet's atmosphere as heat; the glass in a greenhouse causes a similar warming effect. The result, global warming that causes climate change, is being enhanced by the production of these gases by humans.

greenhouse gas: a gas, such as carbon dioxide or methane, that traps sunlight energy in a planet's atmosphere as heat; a gas that participates in the greenhouse effect.

ground tissue system: a plant tissue system, consisting of parenchyma, collenchyma, and sclerenchyma cells, that makes up the bulk of a leaf or young stem, excluding vascular or dermal tissues. Most ground tissue cells function in photosynthesis, support, or carbohydrate storage.

growth factor: small molecules, usually proteins or steroids, that bind to receptors on or in target cells and enhance their rate of cell division or differentiation.

growth hormone: a hormone, released by the anterior pituitary, that stimulates growth, especially of the skeleton.

growth rate: a measure of the change in population size per individual per unit of time.

guanine (G): a nitrogenous base found in both DNA and RNA; abbreviated as G.

guard cell: one of a pair of specialized epidermal cells that surrounds the central opening of a stoma in the epidermis of a leaf or young stem, and regulates the size of the opening.

gymnosperm (jim'-nō-sperm): a nonflowering seed plant, such as a conifer, gnetophyte, cycad, or gingko.

gyre (jīr): a roughly circular pattern of ocean currents, formed because continents interrupt the flow of the current; rotates clockwise in the Northern Hemisphere and counterclockwise in the Southern Hemisphere.

habitat fragmentation: the process by which human development and activities produce patches of wildlife habitat that may not be large enough to sustain minimum viable populations.

habituation (heh-bich-oo-ā'-shun): a type of simple learning characterized by a decline in response to a repeated stimulus.

hair cell: a type of mechanoreceptor cell found in the inner ear that produces an electrical signal when stiff hairlike cilia projecting from the surface of the cell are bent. Hair cells in the cochlea respond to sound vibrations; those in the vestibular system respond to motion and gravity.

hair follicle (fol'-i-kul): a cluster of specialized epithelial cells located in the dermis of mammalian skin, which produces a hair.

hammer: the second of the small bones of the middle ear, linking the tympanic membrane (eardrum) with the oval window of the cochlea; also called the *malleus*.

haploid (hap'-loid): referring to a cell that has only one member of each pair of homologous chromosomes.

Hardy–Weinberg principle: a mathematical model proposing that, under certain conditions, the allele frequencies and genotype frequencies in a sexually reproducing population will remain constant over generations.

heart: a muscular organ responsible for pumping blood within the circulatory system throughout the body.

heart attack: a severe reduction or blockage of blood flow through a coronary artery, depriving some of the heart muscle of its blood supply.

heart rate: the number of cardiac cycles (heartbeats) per minute.

heartwood: older secondary xylem that usually no longer conducts water or minerals, but that contributes to the strength of a tree trunk.

heat of fusion: the energy that must be removed from a compound to transform it from a liquid into a solid at its freezing temperature.

heat of vaporization: the energy that must be supplied to a compound to transform it from a liquid into a gas at its boiling temperature.

Heimlich maneuver: a method of dislodging food or other obstructions that have entered the airway.

helix (hē'-liks): a coiled, spring-like secondary structure of a protein.

helper T cell: a type of T cell that helps other immune cells act against antigens.

hemocoel (hē'-mō-sēl): a cavity within the bodies of certain invertebrates in which a fluid, called hemolymph, bathes tissues directly; part of an open circulatory system.

hemodialysis (hē-mō-dī-al'-luh-sis): a procedure that simulates kidney function in individuals with damaged or ineffective kidneys; blood is diverted from the body, artificially filtered, and returned to the body.

hemoglobin (hē'-mō-glō-bin): the iron-containing protein that gives red blood cells their color; binds to oxygen in the lungs and releases it in the tissues.

hemolymph: in animals with an open circulatory system, the fluid that is located within the hemocoel and that bathes all the body cells, therefore serving as both blood and interstitial fluid.

hemophilia: a recessive, sex-linked disease in which the blood fails to clot normally.

herbivore (erb'-i-vor): literally, "plant-eater"; an organism that feeds directly and exclusively on producers; a primary consumer.

hermaphrodite (her-maf'-ruh-dit'): an organism that possesses both male and female sexual organs. Some hermaphroditic animals can fertilize themselves; others must exchange sex cells with a mate.

hermaphroditic (her-maf'-ruh-dit'-ik): possessing both male and female sexual organs. Some hermaphroditic animals can fertilize themselves; others must exchange sex cells with a mate.

heterotroph (het'-er-ō-trōf'): literally, "other-feeder"; an organism that eats other organisms; a consumer.

heterozygous (het'-er-ō-zī'-gus): carrying two different alleles of a given gene; also called *hybrid*.

hindbrain: the posterior portion of the brain, containing the medulla, pons, and cerebellum.

hinge joint: a joint at which the bones fit together in a way that allows movement in only two dimensions, as at the elbow or knee.

hippocampus (hip-ō-kam'-pus): the part of the forebrain of vertebrates that is important in emotion and especially learning.

histamine (his'-ta-mēn): a substance released by certain cells in response to tissue damage and invasion of the body by foreign substances; promotes the dilation of arterioles and the leakiness of capillaries and triggers some of the events of the inflammatory response.

homeobox gene (hō'-mē-ō-boks): a sequence of DNA coding for a transcription factor protein that activates or inactivates many other genes that control the development of specific, major parts of the body.

homeostasis (hōm-ē-ō-stā'-sis): the maintenance of the relatively constant internal environment that is required for the optimal functioning of cells.

hominin: a human or a prehistoric relative of humans; the oldest known hominin is *Sahelanthropus*, whose fossils are more than 6 million years old.

homologous chromosome: a chromosome that is similar in appearance and genetic information to another chromosome with which it pairs during meiosis; also called *homologue*.

homologous structure: structures that may differ in function but that have similar anatomy, presumably because the organisms that possess them have descended from common ancestors.

homologue (hō'-mō-log): a chromosome that is similar in appearance and genetic information to another chromosome with which it pairs during meiosis; also called *homologous chromosome*.

homozygous (hō-mō-zī'-gus): carrying two copies of the same allele of a given gene; also called *true-breeding*.

hormone: a chemical that is secreted by one group of cells and transported to other cells, whose activity is influenced by reception of the hormone.

host: the prey organism on or in which a parasite lives; the host is harmed by the relationship.

human immunodeficiency virus (HIV): a pathogenic virus that causes acquired immune deficiency syndrome (AIDS) by attacking and destroying the immune system's helper T cells.

human papillomavirus (pap-il-lo'-ma; HPV): a virus that infects the reproductive organs, often causing genital warts; causes most, if not all, cases of cervical cancer.

humoral immunity: an immune response in which foreign substances are inactivated or destroyed by antibodies that circulate in the blood.

Huntington disease: an incurable genetic disorder, caused by a dominant allele, that produces progressive brain deterioration, resulting in the loss of motor coordination, flailing movements, personality disturbances, and eventual death.

hybrid: an organism that is the offspring of parents differing in at least one genetically determined characteristic; also used to refer to the offspring of parents of different species.

hybrid infertility: a postmating reproductive isolating mechanism that arises when hybrid offspring (offspring of parents of two different species) are sterile or have low fertility.

hybrid inviability: a postmating reproductive isolating mechanism that arises when hybrid offspring (offspring of parents of two different species) fail to survive.

hydrogen bond: the weak attraction between a hydrogen atom that bears a partial positive charge (due to polar covalent bonding with another atom) and another atom (oxygen, nitrogen, or fluorine) that bears a partial negative charge; hydrogen bonds may form between atoms of a single molecule or of different molecules.

hydrologic cycle (hī-drō-loj'-ik): the biogeochemical cycle by which water travels from its major reservoir, the oceans, through the atmosphere to reservoirs in freshwater lakes, rivers, and groundwater, and back into the oceans. The hydrologic cycle is driven by solar energy. Nearly all water remains as water throughout the cycle (rather than being used in the synthesis of new molecules).

hydrolysis (hī-drol'-i-sis): the chemical reaction that breaks a covalent bond by means of the addition of hydrogen to the atom on one side of the original bond and a hydroxyl group to the atom on the other side; the reverse of dehydration synthesis.

hydrophilic (hī-drō'-fil'-ik): pertaining to molecules that dissolve readily in water, or to molecules that form hydrogen bonds with water; polar.

hydrophobic (hī-drō-fō'-bik): pertaining to molecules that do not dissolve in water or form hydrogen bonds with water; nonpolar.

hydrophobic interaction: the tendency for hydrophobic molecules to cluster together when immersed in water.

hydrostatic skeleton (hī-drō-stat'-ik): in invertebrate animals, a body structure in which fluid-filled compartments provide support for the body and change shape when acted on by muscles, which alters the animal's body shape and position or causes the animal to move.

hydrothermal vent community: a community of unusual organisms, living in the deep ocean near hydrothermal vents, that depends on the chemosynthetic activities of sulfur bacteria.

hypertension: arterial blood pressure that is chronically elevated above the normal level.

hypertonic (hī-per-ton'-ik): referring to a solution that has a higher concentration of solute (and therefore a lower concentration of free water) than has the cytosol of a cell.

hypha (hī'-fuh; pl., hyphae): in fungi, a threadlike structure that consists of elongated cells, typically with many haploid nuclei; many hyphae make up the fungal body.

hypocotyl (hī'-pō-kot-ul): the part of the embryonic shoot located below the attachment point of the cotyledons but above the root.

hypothalamus: (hī-pō-thal'-a-mus): a region of the brain that controls the secretory activity of the pituitary gland; synthesizes, stores, and releases certain peptide hormones; and directs autonomic nervous system responses.

hypothesis (hī-poth'-eh-sis): a proposed explanation for a phenomenon, often based on limited evidence, that leads to a prediction that can be tested.

hypotonic (hī-pō-ton'-ik): referring to a solution that has a lower concentration of solute (and therefore a higher concentration of free water) than has the cytosol of a cell.

immigration (im-uh-grā'-shun): migration of individuals into an area.

immune system: a system of cells, including macrophages, B cells, and T cells, and molecules, such as antibodies and cytokines, that work together to combat microbial invasion of the body.

imperfect flower: a flower that is missing either stamens or carpels.

implantation: the process whereby the early embryo embeds itself within the lining of the uterus.

imprinting: a type of learning in which an animal acquires a particular type of information during a specific sensitive phase of development.

incomplete dominance: a pattern of inheritance in which the heterozygous phenotype is intermediate between the two homozygous phenotypes.

incomplete flower: a flower that is missing one of the four floral parts (sepals, petals, stamens, or carpels).

indirect development: a developmental pathway in which an offspring goes through radical changes in body form as it matures.

induced pluripotent stem cell (ploo-rē-pō'-tent; iPSC): a type of stem cell produced from nonstem cells by the insertion or forced expression of a specific set of genes that cause the cells to become capable of unlimited cell division and to be able to be differentiated into many different cell types, possibly any cell type of the body.

induction: the process by which a group of cells causes other cells to differentiate into a specific tissue type.

inductive reasoning: the process of creating a generalization based on many specific observations that support the generalization, coupled with an absence of observations that contradict it.

inflammatory response: a nonspecific, local response to injury to the body, characterized by the phagocytosis of foreign substances and tissue debris by white blood cells and by the walling off of the injury site by the clotting of fluids that escape from nearby blood vessels.

ingestion: the movement of food into the digestive tract, usually through the mouth.

inhalation: the act of drawing air into the lungs by enlarging the chest cavity.

inheritance: the genetic transmission of characteristics from parent to offspring.

inhibiting hormone: a hormone, secreted by the neurosecretory cells of the hypothalamus, that inhibits the release of specific hormones from the anterior pituitary.

inhibitory postsynaptic potential (IPSP): an electrical signal produced in a postsynaptic cell that makes the resting potential more negative and, hence, makes the neuron less likely to fire an action potential.

innate (in-āt'): inborn; instinctive; an innate behavior is performed correctly the first time it is attempted.

innate immune response: nonspecific defenses against many different invading microbes, including phagocytic white blood cells, natural killer cells, the inflammatory response, and fever.

inner cell mass: in human embryonic development, the cluster of cells, on the inside of the blastocyst, that will develop into the embryo.

inner ear: the innermost part of the mammalian ear; composed of the bony, fluid-filled tubes of the cochlea and the vestibular apparatus.

inorganic: describing any molecule that does not contain both carbon and hydrogen.

insertion: the site of attachment of a muscle to the relatively movable bone on one side of a joint.

insertion mutation: a mutation in which one or more pairs of nucleotides are inserted into a gene.

insight learning: a type of learning in which a problem is solved by understanding the relationships among the components of the problem rather than through trial and error.

insulin: a hormone, secreted by the pancreas, that lowers blood sugar by stimulating many cells to take up glucose and by stimulating the liver to convert glucose to glycogen.

integration: the process of adding up all of the electrical signals in a neuron, including sensory inputs and postsynaptic potentials, to determine the output of the neuron (action potentials and/or synaptic transmission).

integument (in-teg'-ū-ment): in plants, the outer layers of cells of the ovule that surround the female gametophyte; develops into the seed coat.

intensity: the strength of stimulation or response.

intercalated disc: junctions connecting individual cardiac muscle cells that serve both to attach adjacent cells to one another and to allow electrical signals to pass between cells.

intermediate filament: part of the cytoskeleton of eukaryotic cells that is composed of several types of proteins and probably functions mainly for support.

intermembrane space: the fluid-filled space between the inner and outer membranes of a mitochondrion.

internal fertilization: the union of sperm and egg inside the body of the female.

interneuron: in a neural network, a nerve cell that is postsynaptic to a sensory neuron and presynaptic to a motor neuron. In actual circuits, there may be many interneurons between individual sensory and motor neurons.

internode: the part of a stem between two nodes.

interphase: the stage of the cell cycle between cell divisions in which chromosomes are duplicated and other cell functions occur, such as growth, movement, and acquisition of nutrients.

interspecific competition: competition among individuals of different species.

interstitial cell (in-ter-sti'-shul): in the vertebrate testis, a testosterone-producing cell located between the seminiferous tubules.

interstitial fluid: fluid that bathes the cells of the body; in mammals, interstitial fluid leaks from capillaries and is similar in composition to blood plasma, but lacking the large proteins found in plasma.

intertidal zone: an area of the ocean shore that is alternately covered by water during high tides and exposed to the air during low tides.

intracellular digestion: the chemical breakdown of food within single cells.

intraspecific competition: competition among individuals of the same species.

intron: a segment of DNA in a eukaryotic gene that does not code for amino acids in a protein; see also *exon*.

invasive species: organisms with a high biotic potential that are introduced (deliberately or accidentally) into ecosystems where they did not evolve and where they encounter little environmental resistance and tend to displace native species.

inversion: a mutation that occurs when a piece of DNA is cut out of a chromosome, turned around, and reinserted into the gap.

invertebrate (in-vert'-uh-bret): an animal that lacks a vertebral column.

ion (ī-on): a charged atom or molecule; an atom or molecule that has either an excess of electrons (and, hence, is negatively charged) or has lost electrons (and is positively charged).

ionic bond: a chemical bond formed by the electrical attraction between positively and negatively charged ions.

iris: the pigmented muscular tissue of the vertebrate eye that surrounds and controls the size of the pupil, through which light enters the eye.

islet cell: a cell in the endocrine portion of the pancreas that produces either insulin or glucagon.

isolating mechanism: a morphological, physiological, behavioral, or ecological difference that prevents members of two species from interbreeding.

isotonic (ī-sō-ton'-ik): referring to a solution that has the same concentration of solute (and therefore the same concentration of free water) as has the cytosol of a cell.

isotope (ī'-suh-tōp): one of several forms of a single element, the nuclei of which contain the same number of protons but different numbers of neutrons.

Jacob syndrome: a set of characteristics typical of human males possessing one X and two Y chromosomes (XYY); most XYY males are phenotypically normal, but XYY males tend to be taller than average and to have a slightly increased risk of learning disabilities.

J-curve: the J-shaped growth curve of an exponentially growing population in which increasing numbers of individuals join the population during each succeeding time period.

joint: a flexible region between two rigid units of an exoskeleton or endoskeleton, allowing for movement between the units.

karyotype: a preparation showing the number, sizes, and shapes of all of the chromosomes within a cell.

kelp forest: a diverse ecosystem consisting of stands of tall brown algae and associated marine life. Kelp forests occur in oceans worldwide in nutrient-rich cool coastal waters.

keratin (ker'-uh-tin): a fibrous protein in hair, nails, and the epidermis of skin.

keystone species: a species whose influence on community structure is greater than its abundance would suggest.

kidney: one of a pair of organs of the excretory system that is located on either side of the spinal column; the kidney filters blood, removing wastes and regulating the composition and water content of the blood.

kin selection: a type of natural selection that favors traits that enhance the survival or reproduction of an individual's relatives, even if the traits reduce the fitness of the individuals bearing them.

kinetic energy: the energy of movement; includes light, heat, mechanical movement, and electricity.

kinetochore (ki-net'-ō-kor): a protein structure that forms at the centromere regions of chromosomes; attaches the chromosomes to the spindle.

kinetoplastid: a member of a protist group characterized by distinctively structured mitochondria. Kinetoplastids are mostly flagellated and include parasitic forms such as *Trypanosoma*, which causes sleeping sickness. Kinetoplastids are part of a larger group known as euglenozoans.

kingdom: the second broadest taxonomic category, consisting of related phyla. Related kingdoms make up a domain.

Klinefelter syndrome: a set of characteristics typically found in individuals who have two X chromosomes and one Y chromosome; these individuals are phenotypically males but are sterile and may have several female-like traits, including broad hips and partial breast development.

Krebs cycle: a cyclic series of reactions, occurring in the matrix of mitochondria, in which the acetyl groups from the pyruvic acids produced by glycolysis are broken down to CO_2, accompanied by the formation of ATP and electron carriers; also called the citric acid cycle.

labium (pl., labia): one of a pair of folds of skin of the external structures of the mammalian female reproductive system.

labor: a series of contractions of the uterus that result in birth.

lactation: the secretion of milk from the mammary glands.

lacteal (lak-tēl'): a lymph capillary; found in each villus of the small intestine.

lactic acid fermentation: anaerobic reactions that convert the pyruvic acid produced by glycolysis into lactic acid, using hydrogen ions and electrons from NADH; the primary function of lactic acid fermentation is to regenerate NAD^+ so that glycolysis can continue under anaerobic conditions.

lactose (lak'-tōs): a disaccharide composed of glucose and galactose; found in mammalian milk.

lactose intolerance: the inability to digest lactose (milk sugar) because lactase, the enzyme that digests lactose, is not produced in sufficient amounts; symptoms include bloating, gas pains, and diarrhea.

lactose operon: in prokaryotes, the set of genes that encodes the proteins needed for lactose metabolism, including both the structural genes and a common promoter and operator that control transcription of the structural genes.

large intestine: the final section of the digestive tract; consists of the colon and the rectum, where feces are formed and stored.

larva (lar'-vuh; pl., larvae): an immature form of an animal that subsequently undergoes metamorphosis into its adult form; includes the caterpillars of moths and butterflies, the maggots of flies, and the tadpoles of frogs and toads.

larynx (lar'-inks): the portion of the air passage between the pharynx and the trachea; contains the vocal cords.

late-loss population: a population in which most individuals survive into adulthood; late-loss populations have a convex survivorship curve.

lateral bud: a cluster of meristem cells at the node of a stem; under appropriate conditions, it grows into a branch.

lateral meristem: a meristem tissue that forms cylinders parallel to the long axis of roots and stems; normally located between the primary xylem and primary phloem (vascular cambium) and just outside the phloem (cork cambium); also called *cambium*.

law of conservation of energy: the principle of physics that states that within any closed system, energy can be neither created nor destroyed, but can be converted from one form to another; also called the first law of thermodynamics.

law of independent assortment: the independent inheritance of two or more traits, assuming that each trait is controlled by a single gene with no influence from gene(s) controlling the other trait; states that the alleles of each gene are distributed to the gametes independently of the alleles for other genes; this law is true only for genes located on different chromosomes or very far apart on a single chromosome.

law of segregation: the principle that each gamete receives only one of each parent's pair of alleles of each gene.

laws of thermodynamics: the physical laws that define the basic properties and behavior of energy.

leaf: an outgrowth of a stem, normally flattened and photosynthetic.

leaf primordium (pri-mor'-dē-um; pl., pri-mor-dia): a cluster of dividing cells, surrounding a terminal or lateral bud, that develops into a leaf.

learning: the process by which behavior is modified in response to experience.

legume (leg'-ūm): a member of a family of plants characterized by root swellings in which nitrogen-fixing bacteria are housed; includes peas, soybeans, lupines, alfalfa, and clover.

lens: a clear object that bends light rays; in eyes, a flexible or movable structure used to focus light on the photoreceptor cells of the retina.

leptin: a peptide hormone released by fat cells that helps the body monitor its fat stores and regulate weight.

leukocyte (loo′-kō-sīt): any of the white blood cells circulating in the blood.

lichen (lī′-ken): a symbiotic association between an alga or cyanobacterium and a fungus, resulting in a composite organism.

life table: a data table that groups organisms born at the same time and tracks them throughout their life span, recording how many continue to survive in each succeeding year (or other unit of time). Various parameters such as sex may be used in the groupings. Human life tables may include many other parameters (such as socioeconomic status) used by demographers.

ligament: a tough connective tissue band connecting two bones.

light reactions: the first stage of photosynthesis, in which the energy of light is captured as ATP and NADPH; occurs in thylakoids of chloroplasts.

lignin: a hard material that is embedded in the cell walls of vascular plants and that provides support in terrestrial species; an early and important adaptation to terrestrial life.

limbic system: a diverse group of brain structures, mostly in the lower forebrain, that includes the thalamus, hypothalamus, amygdala, hippocampus, and parts of the cerebrum and is involved in basic emotions, drives, behaviors, and learning.

limnetic zone: the part of a lake in which enough light penetrates to support photosynthesis.

linkage: the inheritance of certain genes as a group because they are parts of the same chromosome. Linked genes do not show independent assortment.

lipase (lī′-pās): an enzyme that catalyzes the breakdown of lipids such as fats into their component fatty acids and glycerol.

lipid (lī′-pid): one of a number of organic molecules containing large nonpolar regions composed solely of carbon and hydrogen, which make lipids hydrophobic and insoluble in water; includes oils, fats, waxes, phospholipids, and steroids.

littoral zone: the part of a lake, usually close to the shore, in which the water is shallow and plants find abundant light, anchorage, and adequate nutrients.

liver: an organ with varied functions, including bile production, glycogen storage, and the detoxification of poisons.

lobefin: fish with fleshy fins that have well-developed bones and muscles. Lobefins include two living clades: the coelacanths and the lungfishes.

local hormone: a general term for messenger molecules produced by most cells and released into the cells′ immediate vicinity. Local hormones, which include prostaglandins and cytokines, influence nearby cells bearing appropriate receptors.

locus (pl., loci): the physical location of a gene on a chromosome.

logistic population growth: population growth characterized by an early exponential growth phase, followed by slower growth as the population approaches its carrying capacity, and finally reaching a stable population at the carrying capacity of the environment; this type of growth generates a curve shaped like a stretched-out letter "S."

long-day plant: a plant that will flower only if the length of uninterrupted darkness is shorter than a species-specific critical period; also called a *short-night plant.*

long-term memory: the second phase of learning; a more-or-less permanent memory formed by a structural change in the brain, brought on by repetition.

loop of Henle (hen′-lē): a specialized portion of the tubule of some nephrons in birds and mammals that creates an osmotic concentration gradient in the fluid immediately surrounding it. This gradient in turn makes possible the production of urine that is more osmotically concentrated than blood plasma.

lung: in terrestrial vertebrates, one of the pair of respiratory organs in which gas exchange occurs; consists of inflatable chambers within the chest cavity.

luteinizing hormone (loo′-tin-īz-ing; LH): a hormone produced by the anterior pituitary that stimulates testosterone production in males and the development of the follicle, ovulation, and the production of the corpus luteum in females.

lymph (limf): a pale fluid found within the lymphatic system; composed primarily of interstitial fluid and white blood cells.

lymph node: a small structure located on a lymph vessel, containing macrophages and lymphocytes (B and T cells). Macrophages filter the lymph by removing microbes; lymphocytes are the principal components of the adaptive immune response to infection.

lymphatic capillary: the smallest vessel of the lymphatic system. Lymphatic capillaries end blindly in interstitial fluid, which they take up and return to the bloodstream.

lymphatic system: a system consisting of lymph vessels, lymph capillaries, lymph nodes, and the thymus, spleen, tonsils, and bone marrow. The lymphatic system helps protect the body against infection, carries fats from the small intestine to blood vessels, and returns excess fluid and small proteins to the blood circulatory system.

lymphocyte (lim′-fō-sit): a type of white blood cell (natural killer cell, B cell, or T cell) that is important in either the innate or adaptive immune response.

lysosome (lī′-sō-sōm): a membrane-bound organelle containing intracellular digestive enzymes.

macronutrient: a nutrient required by an organism in relatively large quantities.

macrophage (mak′-rō-fāj): a type of white blood cell that engulfs microbes and destroys them by phagocytosis; also presents microbial antigens to T cells, helping stimulate the immune response.

major histocompatibility complex (MHC): a group of proteins, normally located on the surfaces of body cells, that identify the cell as "self"; also important in stimulating and regulating the immune response.

Malpighian tubule (mal-pig′-ē-un): the functional unit of the excretory system of insects, consisting of a small tube protruding from the intestine; wastes and nutrients move from the surrounding blood into the tubule, which drains into the intestine, where nutrients are reabsorbed back into the blood, while the wastes are excreted along with feces.

maltose (mal′-tōs): a disaccharide composed of two glucose molecules.

mammal: a member of the chordate clade Mammalia, which includes vertebrates with hair and mammary glands.

mammary gland (mam′-uh-rē): a milk-producing gland used by female mammals to nourish their young.

marsupial (mar-soo′-pē-ul): a member of the clade Marsupialia, which includes mammals whose young are born at an extremely immature stage and undergo further development in a pouch, where they remain attached to a mammary gland; kangaroos, opossums, and koalas are marsupials.

mass extinction: a relatively sudden extinction of many species, belonging to multiple major taxonomic groups, as a result of environmental change. The fossil record reveals five mass extinctions over geologic time.

mass number: The total number of protons and neutrons in the nucleus of an atom.

mast cell: a cell of the immune system that releases histamine and other molecules used in the body's response to trauma and that are a factor in allergic reactions.

matrix: The fluid contained within the inner membrane of the mitochondrion.

mechanical digestion: the process by which food in the digestive tract is physically broken down into smaller pieces.

mechanical incompatibility: a reproductive isolating mechanism that arises when differences in the reproductive structures of two species make the structures incompatible and prevent interbreeding.

mechanoreceptor: a sensory receptor that responds to mechanical stimuli such as the stretching, bending, or dimpling of a part of the body.

medulla (med-ū′-luh): the part of the hindbrain of vertebrates that controls automatic activities such as breathing, swallowing, heart rate, and blood pressure.

megakaryocyte (meg-a-kar′-ē-ō-sīt): a large cell type in the bone marrow, which pinches off pieces of itself that enter the circulation as platelets.

megaspore: a haploid cell formed by meiotic cell division from a diploid megaspore mother cell; through mitotic cell division and differentiation, it develops into the female gametophyte.

megaspore mother cell: a diploid cell, within the ovule of a flowering plant, that undergoes meiotic cell division to produce four haploid megaspores.

meiosis (mī-ō'-sis): in eukaryotic organisms, a type of nuclear division in which a diploid nucleus divides twice to form four haploid nuclei.

meiotic cell division: meiosis followed by cytokinesis.

melatonin (mel-uh-tōn'-in): a hormone, secreted by the pineal gland, that is involved in the regulation of circadian cycles.

memory B cell: a type of white blood cell that is produced by clonal selection as a result of the binding of an antibody on a B cell to an antigen on an invading microorganism. Memory B cells persist in the bloodstream and provide future immunity to invaders bearing that antigen.

memory T cell: a type of white blood cell that is produced by clonal selection as a result of the binding of a receptor on a T cell to an antigen on an invading microorganism. Memory T cells persist in the bloodstream and provide future immunity to invaders bearing that antigen.

menstrual cycle: in human females, a roughly 28-day cycle during which hormonal interactions among the hypothalamus, pituitary gland, and ovary coordinate ovulation and the preparation of the uterus to receive and nourish a fertilized egg. If pregnancy does not occur, the uterine lining is shed during menstruation.

menstruation: in human females, the monthly discharge of uterine tissue and blood from the uterus.

meristem cell (mer'-i-stem): an undifferentiated cell that remains capable of cell division throughout the life of a plant.

mesoderm (mēz'-ō-derm): the middle embryonic tissue layer, lying between the endoderm and ectoderm, and normally the last to develop; gives rise to structures such as muscles, the skeleton, the circulatory system, and the kidneys.

mesophyll (mez'-ō-fil): loosely packed, usually photosynthetic cells located beneath the epidermis of a leaf.

messenger RNA (mRNA): a strand of RNA, complementary to the DNA of a gene, that conveys the genetic information in DNA to the ribosomes to be used during protein synthesis; sequences of three bases (codons) in mRNA that specify particular amino acids to be incorporated into a protein.

metabolic pathway: a sequence of chemical reactions within a cell in which the products of one reaction are the reactants for the next reaction.

metabolic rate: the speed at which cellular reactions that release energy occur.

metabolism: the sum of all chemical reactions that occur within a single cell or within all the cells of a multicellular organism.

metamorphosis (met-a-mor'-fō-sis): in animals with indirect development, a radical change in body form from larva to sexually mature adult, as seen in amphibians (e.g., tadpole to frog) and insects (e.g., caterpillar to butterfly).

metaphase (met'-a-fāz): in mitosis, the stage in which the chromosomes, attached to spindle fibers at kinetochores, are lined up along the equator of the cell; also the approximately comparable stages in meiosis I and meiosis II.

microbe: a microorganism.

microfilament: part of the cytoskeleton of eukaryotic cells that is composed of the proteins actin and (in some cases) myosin; functions in the movement of cell organelles, locomotion by extension of the plasma membrane, and sometimes contraction of entire cells.

micronutrient: a nutrient required by an organism in relatively small quantities.

microRNA: small molecules of RNA that interfere with the translation of specific genes.

microspore: a haploid cell formed by meiotic cell division from a microspore mother cell; through mitotic cell division and differentiation, it develops into the male gametophyte.

microspore mother cell: a diploid cell contained within an anther of a flowering plant; undergoes meiotic cell division to produce four haploid microspores.

microtubule: a hollow, cylindrical strand, found in eukaryotic cells, that is composed of the protein tubulin; part of the cytoskeleton used in the movement of organelles, cell growth, and the construction of cilia and flagella.

microvillus (mī-krō-vi'-lus; pl., microvilli): a microscopic projection of the plasma membrane, which increases the surface area of a cell.

midbrain: during development, the central portion of the brain; contains most of an important relay center, the reticular formation.

middle ear: the part of the mammalian ear composed of the tympanic membrane, the Eustachian tube, and three bones (hammer, anvil, and stirrup) that transmit vibrations from the auditory canal to the oval window.

mimicry (mim'-ik-rē): the situation in which a species has evolved to resemble something else, typically another type of organism.

mineral: an inorganic substance, especially one in rocks or soil, or dissolved in water. In nutrition, minerals such as sodium, calcium, and potassium are essential nutrients that must be obtained from the diet.

mineralocorticoids: steroid hormones produced by the adrenal cortex that regulate salt retention in the kidney, thereby regulating the salt concentration in the blood and interstitial fluid.

minimum viable population (MVP): the smallest isolated population that can persist indefinitely and survive likely natural events such as fires and floods.

mitochondrion (mī-tō-kon'-drē-un; pl., mitochondria): an organelle, bounded by two membranes, that is the site of the reactions of aerobic metabolism.

mitosis (mī-tō'-sis): a type of nuclear division, used by eukaryotic cells, in which one copy of each chromosome (already duplicated during interphase before mitosis) moves into each of two daughter nuclei; the daughter nuclei are therefore genetically identical to each other.

mitotic cell division: mitosis followed by cytokinesis.

molecule (mol'-e-kūl): a particle composed of one or more atoms held together by chemical bonds; the smallest particle of a compound that displays all the properties of that compound.

molt: to shed an external body covering, such as an exoskeleton, skin, feathers, or fur.

monocot: short for monocotyledon; a type of flowering plant characterized by embryos with one seed leaf, or cotyledon.

monomer (mo'-nō-mer): a small organic molecule, several of which may be bonded together to form a chain called a polymer.

monosaccharide (mo-nō-sak'-uh-rīd): the basic molecular unit of all carbohydrates, normally composed of a chain of carbon atoms bonded to hydrogen and hydroxyl groups.

monotreme: a member of the clade Monotremata, which includes mammals that lay eggs; platypuses and spiny anteaters are monotremes.

morula (mor'-ū-luh): in animals, an embryonic stage during cleavage, when the embryo consists of a solid ball of cells.

motor neuron: a neuron that receives instructions from sensory neurons or interneurons and activates effector organs, such as muscles or glands.

motor unit: a single motor neuron and all the muscle fibers on which it forms synapses.

mouth: the opening through which food enters a tubular digestive system.

multicellular: many-celled; most members of the kingdoms Fungi, Plantae, and Animalia are multicellular, with intimate cooperation among cells.

multiple alleles: many alleles of a single gene, perhaps dozens or hundreds, as a result of mutations.

muscle: an animal tissue composed of cells that are capable of contracting and thereby moving specific parts of the body.

muscle fiber: an individual muscle cell.

muscle tissue: tissue composed of one of three types of contractile cells (smooth, skeletal, or cardiac).

muscular dystrophy: an inherited disorder, almost exclusively found in males, in which defective dystrophin proteins cause the skeletal muscles to degenerate.

mutation: a change in the base sequence of DNA in a gene; often used to refer to a genetic change that is significant enough to alter the appearance or function of the organism.

mutualism (mū'-choo-ul-iz-um): a symbiotic relationship in which both participating species benefit.

mycelium (mī-sēl'-ē-um; pl., mycelia): the body of a fungus, consisting of a mass of hyphae.

mycorrhiza (mī-kō-rī′-zuh; pl., mycorrhizae): a symbiotic association between a fungus and the roots of a land plant that facilitates mineral extraction and absorption.

myelin (mī′-uh-lin): a wrapping of insulating membranes of specialized glial cells around the axon of a vertebrate nerve cell; increases the speed of conduction of action potentials.

myofibril (mū-ō-fī′-bril): a cylindrical subunit of a muscle cell, consisting of a series of sarcomeres, surrounded by sarcoplasmic reticulum.

myometrium (mī-ō-mē′-trē-um): the muscular outer layer of the uterus.

myosin (mī′-ō-sin): one of the major proteins of muscle, the interaction of which with the protein actin produces muscle contraction; found in the thick filaments of the muscle fiber; see also *actin*.

myosin head: the part of a myosin protein that binds to the actin subunits of a thin filament; flexion of the myosin head moves the thin filament toward the center of the sarcomere, causing muscle fiber contraction.

natural causality: the scientific principle that natural events occur as a result of preceding natural causes.

natural increase: the difference between births and deaths in a population. This number will be positive if the population is increasing or negative if it is decreasing.

natural killer cell: a type of white blood cell that destroys some virus-infected cells and cancerous cells on contact; part of the innate immune system's nonspecific internal defense against disease.

natural laws: basic principles derived from the study of nature that have never been disproven by scientific inquiry. Natural laws include the laws of gravity, the behavior of light, and the way atoms interact with one another.

natural selection: unequal survival and reproduction of organisms due to heritable differences in their phenotypes, with the result that better adapted phenotypes become more common in the population.

nearshore zone: the region of coastal water that is relatively shallow but constantly submerged, and that can support large plants or seaweeds; includes bays and coastal wetlands.

nearsighted: the inability to focus on distant objects caused by an eyeball that is slightly too long or a cornea that is too curved.

negative feedback: a physiological process in which a change causes responses that tend to counteract the change and restore the original state. Negative feedback in physiological systems maintains homeostasis.

nephridium (nef-rid′-ē-um; pl., nephridia): an excretory organ found in earthworms, mollusks, and certain other invertebrates; somewhat resembles a single vertebrate nephron.

nephron (nef′-ron): the functional unit of the kidney; where blood is filtered and urine is formed.

nerve: a bundle of axons of nerve cells, bound together in a sheath.

nerve cord: a major nervous pathway consisting of a cord of nervous tissue extending lengthwise through the body, paired in many invertebrates and unpaired in chordates.

nerve net: a simple form of nervous system, consisting of a network of neurons that extends throughout the tissues of an organism such as a cnidarian.

nerve tissue: the tissue that makes up the brain, spinal cord, and nerves; consists of neurons and glial cells.

net primary production: the energy stored in the autotrophs of an ecosystem over a given time period.

neuromuscular junction: the synapse formed between a motor neuron and a muscle fiber.

neuron (noor′-on): a single nerve cell.

neurosecretory cell: a specialized nerve cell that synthesizes and releases hormones.

neurotransmitter: a chemical that is released by a nerve cell close to a second nerve cell, a muscle, or a gland cell and that influences the activity of the second cell.

neutral mutation: a mutation that does not detectably change the function of the encoded protein.

neutron: a subatomic particle that is found in the nuclei of atoms, bears no charge, and has a mass approximately equal to that of a proton.

neutrophil (nū′-trō-fil): a type of white blood cell that engulfs invading microbes and contributes to the nonspecific defenses of the body against disease.

nicotinamide adenine dinucleotide (NAD+ or NADH): an electron carrier molecule produced in the cytoplasmic fluid by glycolysis and in the mitochondrial matrix by the Krebs cycle; subsequently donates electrons to the electron transport chain.

nitrogen cycle: the biogeochemical cycle by which nitrogen moves from its primary reservoir of nitrogen gas in the atmosphere via nitrogen-fixing bacteria to reservoirs in soil and water, through producers and into higher trophic levels, and then back to its reservoirs.

nitrogen fixation: the process that combines atmospheric nitrogen with hydrogen to form ammonia (NH_3).

nitrogen-fixing bacterium: a bacterium that possesses the ability to remove nitrogen (N_2) from the atmosphere and combine it with hydrogen to produce ammonia (NH_3).

no-till: a method of growing crops that leaves the remains of harvested crops in place, with the next year's crops being planted directly in the remains of last year's crops without significant disturbance of the soil.

node: in plants, a region of a stem at which the petiole of a leaf is attached; usually, a lateral bud is also found at a node.

nodule: a swelling on the root of a legume or other plant that consists of cortex cells inhabited by nitrogen-fixing bacteria.

noncompetitive inhibition: the process by which an inhibitory molecule binds to a site on an enzyme that is distinct from the active site. As a result, the enzyme's active site is distorted, making it less able to catalyze the reaction involving its normal substrate.

nondisjunction: an error in meiosis in which chromosomes fail to segregate properly into the daughter cells.

nonpolar covalent bond: a covalent bond with equal sharing of electrons.

nonvascular plant: a plant that lacks lignin and well-developed conducting vessels. Nonvascular plants include mosses, hornworts, and liverworts.

norepinephrine (nor-ep-i-nef-rin′): a neurotransmitter, released by neurons of the sympathetic nervous system, that prepares the body to respond to stressful situations; also called *noradrenaline*.

northern coniferous forest: a biome with long, cold winters and only a few months of warm weather; dominated by evergreen coniferous trees; also called *taiga*.

notochord (nōt′-ō-kord): a stiff, but somewhat flexible, supportive rod that extends along the head-to-tail axis and is found in all members of the phylum Chordata at some stage of development.

nuclear envelope: the double-membrane system surrounding the nucleus of eukaryotic cells; the outer membrane is typically continuous with the endoplasmic reticulum.

nuclear pore complex: an array of proteins that line pores in the nuclear membrane and control which substances enter and leave the nucleus.

nucleic acid (noo-klā′-ik): an organic molecule composed of nucleotide subunits; the two common types of nucleic acids are ribonucleic acid (RNA) and deoxyribonucleic acid (DNA).

nucleoid (noo-klē-oid): the location of the genetic material in prokaryotic cells; not membrane enclosed.

nucleolus (noo-klē′-ō-lus; pl., nucleoli): the region of the eukaryotic nucleus that is engaged in ribosome synthesis; consists of the genes encoding ribosomal RNA, newly synthesized ribosomal RNA, and ribosomal proteins.

nucleotide: a subunit of which nucleic acids are composed; a phosphate group bonded to a sugar (deoxyribose in DNA), which is in turn bonded to a nitrogen-containing base (adenine, guanine, cytosine, or thymine in DNA). Nucleotides are linked together, forming a strand of nucleic acid, by bonds between the phosphate of one nucleotide and the sugar of the next nucleotide.

nucleotide substitution: a mutation in which a single base pair in DNA has been changed.

nucleus (atomic): the central region of an atom, consisting of protons and neutrons.

nucleus (cellular): the membrane-bound organelle of eukaryotic cells that contains the cell's genetic material.

nutrient: a substance acquired from the environment and needed for the survival, growth, and development of an organism.

nutrient cycle: the pathways of a specific nutrient (such as carbon, nitrogen, phosphorus, or water) through the living and nonliving portions of an ecosystem; also called a *biogeochemical cycle*.

observation: in the scientific method, the recognition of and a statement about a specific phenomenon, usually leading to the formulation of a question about the phenomenon.

oil: a lipid composed of three fatty acids, some of which are unsaturated, covalently bonded to a molecule of glycerol; oils are liquid at room temperature.

oligotrophic lake: a lake that is very low in nutrients and hence supports little phytoplankton, plant, and algal life; contains clear water with deep light penetration.

ommatidium (ōm-ma-tid′-ē-um; pl., ommatidia): an individual light-sensitive subunit of a compound eye; consists of a lens and several receptor cells.

omnivore: an organism that consumes both plants and animals.

oogenesis (ō-ō-jen′-i-sis): the process by which egg cells are formed.

oogonium (ō-ō-gō′-nē-um; pl., oogonia): in female animals, a diploid cell that gives rise to a primary oocyte.

open circulatory system: a type of circulatory system found in some invertebrates, such as arthropods and most mollusks, that includes an open space (the hemocoel) in which blood directly bathes body tissues.

open ocean: that part of the ocean in which the water is so deep that wave action does not affect the bottom, even during strong storms.

operant conditioning: a laboratory training procedure in which an animal learns to perform a response (such as pressing a lever) through reward or punishment.

operator: a sequence of DNA nucleotides in a prokaryotic operon that binds regulatory proteins that control the ability of RNA polymerase to transcribe the structural genes of the operon.

operon (op′-er-on): in prokaryotes, a set of genes, often encoding the proteins needed for a complete metabolic pathway, including both the structural genes and a common promoter and operator that control transcription of the structural genes.

optic nerve: the nerve leading from the eye to the brain; it carries visual information.

order: in Linnaean classification, the taxonomic rank composed of related families. Related orders make up a class.

organ: a structure (such as the liver, kidney, or skin) composed of two or more distinct tissue types that function together.

organ system: two or more organs that work together to perform a specific function; for example, the digestive system.

organelle (or-guh-nel′): a membrane-enclosed structure found inside a eukaryotic cell that performs a specific function.

organic: describing a molecule that contains both carbon and hydrogen.

organic molecule: a molecule that contains both carbon and hydrogen.

organism (or′-guh-niz-um): an individual living thing.

organogenesis (or-gan-ō-jen′-uh-sis): the process by which the layers of the gastrula (endoderm, ectoderm, mesoderm) rearrange to form organs.

origin: the site of attachment of a muscle to the relatively stationary bone on one side of a joint.

osmolarity: a measure of the total number of dissolved solute particles in a solution.

osmoregulation: homeostatic maintenance of the water and salt content of the body within a limited range.

osmosis (oz-mō′-sis): the diffusion of water across a differentially permeable membrane, normally down a concentration gradient of free water molecules. Water moves into the solution that has a lower concentration of free water from a solution that has a higher concentration of free water.

osteoblast (os′-tē-ō-blast): a cell type that produces bone.

osteoclast (os′-tē-ō-klast): a cell type that dissolves bone.

osteocyte (os′-tē-ō-sīt): a mature bone cell.

osteoporosis (os′-tē-ō-por-ō′-sis): a condition in which bones become porous, weak, and easily fractured; most common in elderly women.

outer ear: the outermost part of the mammalian ear, including the external ear and auditory canal leading to the tympanic membrane.

oval window: the membrane-covered entrance to the cochlea.

ovary: (1) in animals, the gonad of females; (2) in flowering plants, a structure at the base of the carpel that contains one or more ovules and develops into the fruit.

overexploitation: hunting or harvesting natural populations at a rate that exceeds those populations' ability to replenish their numbers.

ovulation: the release of a secondary oocyte, ready to be fertilized, from the ovary.

ovule: a structure within the ovary of a flower, inside which the female gametophyte develops; after fertilization, it develops into the seed.

oxytocin (oks-ē-tō′-sin): a hormone, released by the posterior pituitary, that stimulates the contraction of uterine and mammary gland muscles.

ozone hole: a region of severe ozone loss in the stratosphere caused by ozone-depleting chemicals; maximum ozone loss occurs from September to early October over Antarctica.

ozone layer: the ozone-enriched layer of the upper atmosphere (stratosphere) that filters out much of the sun's ultraviolet radiation.

pacemaker: a cluster of specialized muscle cells in the upper right atrium of the heart that produce spontaneous electrical signals at a regular rate; the sinoatrial node.

pain receptor: a receptor cell that stimulates activity in the brain that is perceived as the sensation of pain; responds to very high or very low temperatures, mechanical damage (such as extreme stretching of tissue), and/or certain chemicals, such as potassium ions or bradykinin, that are produced as a result of tissue damage; also called *nociceptor*.

pancreas (pan′-krē-us): a combined exocrine and endocrine gland located in the abdominal cavity next to the stomach. The endocrine portion secretes the hormones insulin and glucagon, which regulate glucose concentrations in the blood. The exocrine portion secretes pancreatic juice (a mixture of water, enzymes, and sodium bicarbonate) into the small intestine; the enzymes digest fat, carbohydrate, and protein; the bicarbonate neutralizes acidic chyme entering the intestine from the stomach.

pancreatic juice: a mixture of water, sodium bicarbonate, and enzymes released by the pancreas into the small intestine.

parabasalid: a member of a protist group characterized by mutualistic or parasitic relationships with the animal species inside which they live. Parabasalids are part of a larger group known as the excavates.

paracrine communication: communication between cells of a multicellular organism in which certain cells release hormones that diffuse through the interstitial fluid to nearby cells and alter their activity.

parasite (par′-uh-sīt): an organism that lives in or on a larger organism (its host), harming the host but usually not killing it immediately.

parasympathetic division: the division of the autonomic nervous system that produces largely involuntary responses related to the maintenance of normal body functions, such as digestion; often called the *parasympathetic nervous system*.

parathyroid gland: one of four small endocrine glands, embedded in the surface of the thyroid gland, that produces parathyroid hormone, which (with calcitonin from the thyroid gland) regulates calcium ion concentration in the blood.

parathyroid hormone: a hormone released by the parathyroid gland that stimulates the release of calcium from bone.

parenchyma (par-en′-ki-muh): a plant cell type that is alive at maturity, normally with thin cell walls, and carries out most of the metabolism of a plant. Most dividing meristem cells in a plant are parenchyma.

parthenogenesis (par-the-nō-jen′-uh-sis): an asexual specialization of sexual reproduction, in which a haploid egg undergoes development without fertilization.

passive transport: the movement of materials across a membrane down a gradient of concentration, pressure, or electrical charge without using cellular energy.

pathogen: an organism (or a toxin) capable of producing disease.

pathogenic (path′-ō-jen-ik): capable of producing disease; referring to an organism with such a capability (a pathogen).

pedigree: a diagram showing genetic relationships among a set of individuals, normally with respect to a specific genetic trait.

pelagic (puh-la′-jik): free-swimming or floating.

penis: an external structure of the male reproductive and urinary systems; serves to deposit sperm into the female reproductive system and deliver urine to the outside of the body.

peptide (pep′-tīd): a chain composed of two or more amino acids linked together by peptide bonds.

peptide bond: the covalent bond between the nitrogen of the amino group of one amino acid and the carbon of the carboxyl group of a second amino acid, joining the two amino acids together in a peptide or protein.

peptide hormone: a hormone consisting of a chain of amino acids; includes small proteins that function as hormones.

pericycle (per'-i-sī-kul): the outermost layer of cells of the vascular cylinder of a root.

periderm: the outer cell layers of roots and stems that have undergone secondary growth; consists primarily of cork cambium and cork cells.

peripheral nervous system (PNS): in vertebrates, the part of the nervous system that connects the central nervous system to the rest of the body.

peristalsis: rhythmic coordinated contractions of the smooth muscles of the digestive tract that move substances through the digestive tract.

peritubular capillaries: a capillary network surrounding each kidney tubule that allows the exchange of substances between the blood and the tubule contents during tubular reabsorption and tubular secretion.

permafrost: a permanently frozen layer of soil, usually found in tundra of the Arctic or high mountains.

petal: part of a flower, typically brightly colored and fragrant, that attracts potential animal pollinators.

petiole (pet'-ē-ōl): the stalk that connects the blade of a leaf to the stem.

pH scale: a scale, with values from 0 to 14, used for measuring the relative acidity of a solution; at pH 7 a solution is neutral, pH 0 to 7 is acidic, and pH 7 to 14 is basic; each unit on the scale represents a tenfold change in H^+ concentration.

phagocyte (fā'-gō-sīt): a type of immune system cell that destroys invading microbes by using phagocytosis to engulf and digest the microbes. Also called a *phagocytic cell*.

phagocytosis (fa-gō-sī-tō'-sis): a type of endocytosis in which extensions of a plasma membrane engulf extracellular particles, enclose them in a membrane-bound sac, and transport them into the interior of the cell.

pharyngeal gill slit (far-in'-jē-ul): one of a series of openings, located just posterior to the mouth, that connects the throat to the outside environment; present (as some stage of life) in all chordates.

pharynx (far'-inks): in vertebrates, a chamber that is located at the back of the mouth, shared by the digestive and respiratory systems; in some invertebrates, the portion of the digestive tube just posterior to the mouth.

phenotype (fēn'-ō-tīp): the physical characteristics of an organism; can be defined as outward appearance (such as flower color), as behavior, or in molecular terms (such as glycoproteins on red blood cells).

pheromone (fer'-uh-mōn): a chemical produced by an organism that alters the behavior or physiological state of another member of the same species.

phloem (flō-um): a conducting tissue of vascular plants that transports a concentrated solution of sugars (primarily sucrose) and other organic molecules up and down the plant.

phospholipid (fos-fō-li'-pid): a lipid consisting of glycerol bonded to two fatty acids and one phosphate group, which bears another group of atoms, typically charged and containing nitrogen. A double layer of phospholipids is a component of all cellular membranes.

phospholipid bilayer: a double layer of phospholipids that forms the basis of all cellular membranes. The phospholipid heads, which are hydrophilic, face the watery interstitial fluid, a watery external environment, or the cytosol; the tails, which are hydrophobic, are buried in the middle of the bilayer.

phosphorus cycle (fos'-for-us): the biogeochemical cycle by which phosphorus moves from its primary reservoir—phosphate-rich rock—to reservoirs of phosphate in soil and water, through producers and into higher trophic levels, and then back to its reservoirs.

photic zone: the region of an ocean where light is strong enough to support photosynthesis.

photon (fō'-ton): the smallest unit of light energy.

photopigment (fō'-tō-pig-ment): a chemical substance in a photoreceptor cell that, when struck by light, changes shape and produces a response in the cell.

photoreceptor: a receptor cell that responds to light; in vertebrates, rods and cones.

photorespiration: a series of reactions in plants in which O_2 replaces CO_2 during the Calvin cycle, preventing carbon fixation; this wasteful process dominates when C_3 plants are forced to close their stomata to prevent water loss.

photosynthesis: the complete series of chemical reactions in which the energy of light is used to synthesize high-energy organic molecules, usually carbohydrates, from low-energy inorganic molecules, usually carbon dioxide and water.

photosystem: in thylakoid membranes, a cluster of chlorophyll, accessory pigment molecules, proteins, and other molecules that collectively capture light energy, transfer some of the energy to electrons, and transfer the energetic electrons to an adjacent electron transport chain.

phototropism: growth with respect to the direction of light.

phylogeny (fī-lah'-jen-ē): the evolutionary history of a group of species.

phylum (fī-lum): in Linnaean classification, the taxonomic rank composed of related classes. Related phyla make up a kingdom.

phytochrome (fī'-tō-krōm): a light-sensitive plant pigment that mediates many plant responses to light, including flowering, stem elongation, and seed germination.

phytoplankton (fī'-tō-plank-ten): photosynthetic protists that are abundant in marine and freshwater environments.

pigment molecule: a light-absorbing, colored molecule, such as chlorophyll, carotenoid, or melanin molecules.

pilus (pl., pili): a hairlike protein structure that projects from the cell wall of many bacteria. Attachment pili help bacteria adhere to structures. Sex pili assist in the transfer of plasmids.

pineal gland (pī-nē'-al): a small gland within the brain that secretes melatonin; controls the seasonal reproductive cycles of some mammals.

pinna: a flap of skin-covered cartilage on the surface of the head that collects sound waves and funnels them to the auditory canal.

pinocytosis (pi-nō-sī-tō'-sis): the nonselective movement of extracellular fluids and their dissolved substances, enclosed within a vesicle formed from the plasma membrane, into a cell.

pioneer: an organism that is among the first to colonize an unoccupied habitat in the first stages of succession.

pit: an area in the cell walls between two plant cells where the two cells are separated only by a thin, porous cell wall.

pith: cells forming the center of a root or stem.

pituitary gland: an endocrine gland, located at the base of the brain, that produces several hormones, many of which influence the activity of other glands.

placenta (pluh-sen'-tuh): in mammals, a structure formed by a complex interweaving of the uterine lining and the embryonic membranes, especially the chorion; functions in gas, nutrient, and waste exchange between embryonic and maternal circulatory systems, and also secretes the hormones estrogen and progesterone, which are essential to maintaining pregnancy.

placental (pluh-sen'-tul): referring to a mammal possessing a complex placenta (that is, species that are not marsupials or monotremes).

plankton: microscopic organisms that live in marine or freshwater environments; includes phytoplankton and zooplankton.

plant hormone: a chemical produced by specific plant cells that influences the growth, development, or metabolic activity of other cells, typically some distance away in the plant body.

plaque (plak): a deposit of cholesterol and other fatty substances within the wall of an artery.

plasma: the fluid, noncellular portion of the blood.

plasma cell: an antibody-secreting descendant of a B cell.

plasma membrane: the outer membrane of a cell, composed of a bilayer of phospholipids in which proteins are embedded.

plasmid (plaz'-mid): a small, circular piece of DNA located in the cytoplasm of many bacteria; usually does not carry genes required for the normal functioning of the bacterium, but may carry genes, such as those for antibiotic resistance, that assist bacterial survival in certain environments.

plasmodesma (plaz-mō-dez'-muh; pl., plasmodesmata): a cell-to-cell junction in plants that connects the cytosol of adjacent cells.

plasmodium (plaz-mō'-dē-um): a slug-like mass of cytoplasm containing thousands of nuclei that are not confined within individual cells.

plastid (plas'-tid): in plant cells, an organelle bounded by two membranes that may be involved in photosynthesis (chloroplasts), pigment storage, or food storage.

plate tectonics: the theory that Earth's crust is divided into irregular plates that are converging, diverging, or slipping by one another; these motions cause continental drift, the movement of continents over Earth's surface.

platelet (plāt'-let): a cell fragment that is formed from megakaryocytes in bone marrow; platelets lack a nucleus; they circulate in the blood and play a role in blood clotting.

pleated sheet: a form of secondary structure exhibited by certain proteins, such as silk, in which many protein chains lie side by side, with hydrogen bonds holding adjacent chains together.

pleiotropy (ple'-ō-trō-pē): a situation in which a single gene influences more than one phenotypic characteristic.

point mutation: a mutation in which a single base pair in DNA has been changed.

polar body: in oogenesis, a small cell, containing a nucleus but virtually no cytoplasm, produced by both the first meiotic division (of the primary oocyte) and the second meiotic division (of the secondary oocyte).

polar covalent bond: a covalent bond with unequal sharing of electrons, such that one atom is relatively negative and the other is relatively positive.

pollen: the male gametophyte of a seed plant; also called a *pollen grain*.

pollen grain: the male gametophyte of a seed plant.

pollination: in flowering plants, when pollen grains land on the stigma of a flower of the same species; in conifers, when pollen grains land within the pollen chamber of a female cone of the same species.

polygenic inheritance: a pattern of inheritance in which the interactions of two or more functionally similar genes determine phenotype.

polymer (pah'-li-mer): a molecule composed of three or more (perhaps thousands) smaller subunits called monomers, which may be identical (for example, the glucose monomers of starch) or different (for example, the amino acids of a protein).

polymerase chain reaction (PCR): a method of producing virtually unlimited numbers of copies of a specific piece of DNA, starting with as little as one copy of the desired DNA.

polypeptide: a long chain of amino acids linked by peptide bonds. A protein consists of one or more polypeptides.

polyploidy (pahl'-ē-ploid-ē): having more than two sets of homologous chromosomes.

polysaccharide (pahl-ē-sak'-uh-rīd): a large carbohydrate molecule composed of branched or unbranched chains of repeating monosaccharide subunits, normally glucose or modified glucose molecules; includes starches, cellulose, and glycogen.

pons: a portion of the hindbrain, just above the medulla, that contains neurons that influence sleep and the rate and pattern of breathing.

population: all the members of a particular species within a defined area, found in the same time and place and actually or potentially interbreeding.

population bottleneck: the result of an event that causes a population to become extremely small; may cause genetic drift that results in changed allele frequencies and loss of genetic variability.

population cycle: regularly recurring, cyclical changes in population size.

post-anal tail: a tail that extends beyond the anus and contains muscle tissue and the most posterior part of the nerve cord; found in all chordates at some stage of development.

positive feedback: a physiological mechanism in which a change causes responses that tend to amplify the original change.

posterior pituitary: a lobe of the pituitary gland that is an outgrowth of the hypothalamus and that releases antidiuretic hormone and oxytocin.

postmating isolating mechanism: any structure, physiological function, or developmental abnormality that prevents organisms of two different species, once mating has occurred, from producing vigorous, fertile offspring.

postsynaptic neuron: at a synapse, the nerve cell that changes its electrical potential in response to a chemical (the neurotransmitter) released by another (presynaptic) cell.

postsynaptic potential (PSP): an electrical signal produced in a postsynaptic cell by transmission across the synapse; it may be excitatory (EPSP), making the cell more likely to produce an action potential, or inhibitory (IPSP), tending to inhibit an action potential.

potential energy: "stored" energy including chemical energy (stored in molecules), elastic energy (such as stored in a spring), or gravitational energy (stored in the elevated position of an object).

prairie: a biome, located in the centers of continents, that primarily supports grasses; also called *grassland*.

precapillary sphincter (sfink'-ter): a ring of smooth muscle between an arteriole and a capillary that regulates the flow of blood into the capillary bed.

predation (pre-dā'-shun): the act of eating another living organism.

predator: an organism that eats other organisms.

prediction: in the scientific method, a statement describing an expected observation or the expected outcome of an experiment, assuming that a specific hypothesis is true.

premating isolating mechanism: any structure, physiological function, or behavior that prevents organisms of two different species from exchanging gametes.

pressure-flow mechanism: the process by which sugars are transported in phloem; the movement of sugars into a phloem sieve tube causes water to enter the tube by osmosis, while the movement of sugars out of another part of the same sieve tube causes water to leave by osmosis. The resulting pressure gradient causes the bulk movement of water and dissolved sugars from the end of the sieve tube into which sugar is transported (a source) toward the end of the sieve tube from which sugar is removed (a sink).

presynaptic neuron: a nerve cell that releases a chemical (the neurotransmitter) at a synapse, causing changes in the electrical activity or metabolism of another (postsynaptic) cell.

prey: organisms that are eaten, and often killed, by another organism (a predator).

primary consumer: an organism that feeds on producers; an herbivore.

primary electron acceptor: A molecule in the reaction center of each photosystem that accepts an electron from one of the two reaction center chlorophyll *a* molecules and transfers the electron to an adjacent electron transport chain.

primary growth: growth in length and development of the initial structures of plant roots and shoots; results from cell division of apical meristems and differentiation of the daughter cells.

primary oocyte (ō'-ō-sīt): a diploid cell, derived from the oogonium by growth and differentiation, that undergoes meiotic cell division, producing the egg.

primary spermatocyte (sper-ma'-tō-sīt): a diploid cell, derived from the spermatogonium by growth and differentiation, that undergoes meiotic cell division, producing four sperm.

primary structure: the amino acid sequence of a protein.

primary succession: succession that occurs in an environment, such as bare rock, in which no trace of a previous community is present.

primate: a member of the mammalian clade Primates, characterized by the presence of an opposable thumb, forward-facing eyes, and a well-developed cerebral cortex; includes lemurs, monkeys, apes, and humans.

prion (prē'-on): a protein that, in mutated form, acts as an infectious agent that causes certain neurodegenerative diseases, including kuru and scrapie.

producer: a photosynthetic organism; an autotroph.

product: an atom or molecule that is formed from reactants in a chemical reaction.

profundal zone: the part of a lake in which light is insufficient to support photosynthesis.

progesterone (prō-ge'-ster-ōn): a hormone, produced by the corpus luteum in the ovary, that promotes the development of the uterine lining in females.

prokaryote (prō-kar'-ē-ōt): an organism whose cells are prokaryotic (their genetic material is not enclosed in a membrane-bound nucleus and they lack other membrane-bound organelles); bacteria and archaea are prokaryotes.

prokaryotic (prō-kar-ē-ot'-ik): referring to cells of the domains Bacteria or Archaea. Prokaryotic cells have genetic material that is not enclosed in a membrane-bound nucleus; they also lack other membrane-bound organelles.

prokaryotic fission: the process by which a single bacterium divides in half, producing two identical offspring.

prolactin: a hormone, released by the anterior pituitary, that stimulates milk production in human females.

promoter: a specific sequence of DNA at the beginning of a gene, to which RNA polymerase binds and starts gene transcription.

prophase (prō′-fāz): the first stage of mitosis, in which the chromosomes first become visible in the light microscope as thickened, condensed threads and the spindle begins to form; as the spindle is completed, the nuclear envelope breaks apart, and the spindle microtubules invade the nuclear region and attach to the kinetochores of the chromosomes. Also, the first stage of meiosis: In meiosis I, the homologous chromosomes pair up, exchange parts at chiasmata, and attach to spindle microtubules; in meiosis II, the spindle re-forms and chromosomes attach to the microtubules.

prostaglandin (pro-stuh-glan′-din): a family of modified fatty acid hormones manufactured and secreted by many cells of the body.

prostate gland (pros′-tāt): a gland that produces part of the fluid component of semen; prostatic fluid is basic and contains a chemical that activates sperm movement.

protease (prō′-tē-ās): an enzyme that digests proteins.

protein: a polymer composed of amino acids joined by peptide bonds.

protist: a eukaryotic organism that is not a plant, animal, or fungus. The term encompasses a diverse array of organisms and does not represent a monophyletic group.

protocell: the hypothetical evolutionary precursor of living cells, consisting of a mixture of organic molecules within a membrane.

proton: a subatomic particle that is found in the nuclei of atoms; it bears a unit of positive charge and has a relatively large mass, roughly equal to the mass of a neutron.

protonephridium (prō-tō-nef-rid′-ē-um; pl., protonephridia): the functional unit of the excretory system of some invertebrates, such as flatworms; consists of a tubule that has an external opening to the outside of the body, but lacks an internal opening within the body. Fluid is filtered from the body cavity into the tubule by a hollow cell at the end of the tubule, such as a flame cell in flatworms, and released outside the body.

protostome (prō′-tō-stōm): an animal with a mode of embryonic development in which the coelom is derived from splits in the mesoderm; characteristic of arthropods, annelids, and mollusks.

protozoan (prō-tuh-zō′-an; pl., protozoa): a nonphotosynthetic, single-celled protist.

proximal tubule: the initial part of a mammalian nephron, between Bowman's capsule and the loop of Henle; most tubular reabsorption and a small amount of tubular secretion occur in the proximal tubule.

pseudocoelom (soo′-dō-sēl′-ōm): in animals, a "false coelom," that is, a space or cavity, partially but not fully lined with tissue derived from mesoderm, that separates the body wall from the inner organs; found in roundworms.

pseudoplasmodium (soo′-dō-plaz-mō′-dē-um): an aggregation of individual amoeboid cells that form a slug-like mass.

pseudopod (sood′-ō-pod): an extension of the plasma membrane by which certain cells, such as amoebas, locomote and engulf prey.

puberty: a stage of development characterized by sexual maturation, rapid growth, and the appearance of secondary sexual characteristics.

pulmonary circuit: the pathway of blood from the heart to the lungs and back to the heart.

Punnett square method: a method of predicting the genotypes and phenotypes of offspring in genetic crosses.

pupa (pl., pupae): a developmental stage in some insect species in which the organism stops moving and feeding and may be encased in a cocoon; occurs between the larval and the adult phases.

pupil: the adjustable opening in the center of the iris, through which light enters the eye.

quaternary structure (kwat′-er-nuh-rē): the complex three-dimensional structure of a protein consisting of more than one peptide chain.

question: in the scientific method, a statement that identifies a particular aspect of an observation that a scientist wishes to explain.

radial symmetry: a body plan in which any plane along a central axis will divide the body into approximately mirror-image halves. Cnidarians and many adult echinoderms exhibit radial symmetry.

radioactive: pertaining to an atom with an unstable nucleus that spontaneously disintegrates, with the emission of radiation.

radiolarian (rā-dē-ō-lar′-ē-un): a member of a protist group characterized by pseudopods and typically elaborate silica shells. Radiolarians are largely aquatic (mostly marine) and are part of a larger group known as rhizarians.

rain shadow: a local dry area, usually located on the downwind side of a mountain range that blocks the prevailing moisture-bearing winds.

random distribution: the distribution characteristic of populations in which the probability of finding an individual is equal in all parts of an area.

reactant: an atom or molecule that is used up in a chemical reaction to form a product.

reaction center: two chlorophyll *a* molecules and a primary electron acceptor complexed with proteins and located near the center of each photosystem within the thylakoid membrane. Light energy is passed to one of the chlorophylls, which donates an energized electron to the primary electron acceptor, which then passes the electron to an adjacent electron transport chain.

receptor: (1) a protein, located in a membrane or the cytoplasm of a cell, that binds to specific molecules (for example, a hormone or neurotransmitter), triggering a response in the cell, such as endocytosis, changes in metabolic rate, cell division, or electrical changes; (2) a cell that responds to an environmental stimulus (chemicals, sound, light, pH, and so on) by changing its electrical potential.

receptor-mediated endocytosis: the selective uptake of molecules from the interstitial fluid by binding to a receptor located at a coated pit on the plasma membrane and pinching off the coated pit into a vesicle that moves into the cytosol.

receptor potential: an electrical potential change in a receptor cell, produced in response to the reception of an environmental stimulus (chemicals, sound, light, heat, and so on). The size of the receptor potential is proportional to the intensity of the stimulus.

receptor protein: a protein, located in a membrane or the cytosol of a cell, that binds to specific molecules (for example, a hormone or neurotransmitter), triggering a response in the cell, such as endocytosis, changes in metabolic rate, cell division, or electrical changes.

recessive: an allele that is expressed only in homozygotes and is completely masked in heterozygotes.

recognition protein: a protein or glycoprotein protruding from the outside surface of a plasma membrane that identifies a cell as belonging to a particular species, to a specific individual of that species, and in many cases to one specific organ within the individual.

recombinant DNA: DNA that has been altered by the addition of DNA from a different organism, typically from a different species.

recombination: the formation of new combinations of the different alleles of each gene on a chromosome; the result of crossing over.

rectum: the terminal portion of the vertebrate digestive tube where feces are stored until they are eliminated.

reflex: a simple, stereotyped movement of part of the body that occurs automatically in response to a stimulus.

regeneration: the regrowth of a body part after loss or damage; also, asexual reproduction by means of the regrowth of an entire body from a fragment.

regulatory gene: in prokaryotes, a gene encoding a protein that binds to the operator of one or more operons, controlling the ability of RNA polymerase to transcribe the structural genes of the operon.

regulatory T cell: a type of T cell that suppresses the adaptive immune response, especially by self-reactive lymphocytes, and appears to be important in the prevention of autoimmune disorders.

releasing hormone: a hormone, secreted by the hypothalamus, that causes the release of specific hormones by the anterior pituitary.

renal artery: the artery carrying blood to each kidney.

renal cortex: the outer layer of the kidney, in which the largest portion of each nephron is located, including Bowman's capsule and the distal and proximal tubules.

renal medulla: the layer of the kidney just inside the renal cortex, in which loops of Henle produce a highly concentrated interstitial fluid, allowing the production of concentrated urine.

renal pelvis: the inner chamber of the kidney, in which urine from the collecting ducts accumulates before it enters the ureter.

renal vein: the vein carrying blood away from each kidney.

renin: in mammals, an enzyme that is released by the kidneys when blood pressure falls. Renin catalyzes the formation of angiotensin, which causes arterioles to constrict, thereby elevating blood pressure.

replacement level fertility (RLF): the average number of offspring per female that is required to maintain a stable population.

repressor protein: in prokaryotes, a protein encoded by a regulatory gene, which binds to the operator of an operon and prevents RNA polymerase from transcribing the structural genes.

reproductive isolation: the failure of organisms of one population to breed successfully with members of another; may be due to premating or postmating isolating mechanisms.

reptile: a member of the chordate group that includes the snakes, lizards, turtles, alligators, birds, and crocodiles.

reservoir: the major source and storage site of a nutrient in an ecosystem, normally in the abiotic portion.

resource partitioning: the coexistence of two species with similar requirements, each occupying a smaller niche than either would if it were by itself; a means of minimizing the species' competitive interactions.

respiration: in terrestrial vertebrates, the act of moving air into the lungs (inhalation) and out of the lungs (exhalation); during respiration, oxygen diffuses from the air in the lungs into the circulatory system, and carbon dioxide diffuses from the circulatory system into the air in the lungs.

respiratory center: a cluster of neurons, located in the medulla of the brain, that sends rhythmic bursts of nerve impulses to the respiratory muscles, resulting in breathing.

respiratory membrane: within the lungs, the fusion of the epithelial cells of the alveoli and the endothelial cells that form the walls of surrounding capillaries.

respiratory system: a group of organs that work together to facilitate gas exchange—the intake of oxygen and the removal of carbon dioxide—between an animal and its environment.

resting potential: an electrical potential, or voltage, in unstimulated nerve cells; the inside of the cell is always negative with respect to the outside.

restriction enzyme: an enzyme, usually isolated from bacteria, that cuts double-stranded DNA at a specific nucleotide sequence; the nucleotide sequence that is cut differs for different restriction enzymes.

restriction fragment: a piece of DNA that has been isolated by cleaving a larger piece of DNA with restriction enzymes.

restriction fragment length polymorphism (RFLP): a difference in the length of DNA fragments that were produced by cutting samples of DNA from different individuals of the same species with the same set of restriction enzymes; fragment length differences occur because of differences in nucleotide sequences, and hence in the ability of restriction enzymes to cut the DNA, among individuals of the same species.

reticular formation (reh-tik'-ū-lar): a diffuse network of neurons extending from the hindbrain, through the midbrain, and into the lower reaches of the forebrain; involved in filtering sensory input and regulating what information is relayed to conscious brain centers for further attention.

retina (ret'-in-uh): a multilayered sheet of nerve tissue at the rear of camera-type eyes, composed of photoreceptor cells plus associated nerve cells that refine the photoreceptor information and transmit it to the optic nerve.

rhizarian: a member of Rhizaria, a protist clade. Rhizarians, which use thin pseudopods to move and capture prey and which often have hard shells, include the foraminiferans and the radiolarians.

ribonucleic acid (RNA) (rī-bō-noo-klā'-ik; RNA): a molecule composed of ribose nucleotides, each of which consists of a phosphate group, the sugar ribose, and one of the bases adenine, cytosine, guanine, or uracil; involved in converting the information in DNA into protein; also the genetic material of some viruses.

ribosomal RNA (rī-bō-sō'-mul; rRNA): a type of RNA that combines with proteins to form ribosomes.

ribosome (rī'-bō-sōm): a complex consisting of two subunits, each composed of ribosomal RNA and protein, found in the cytoplasm of cells or attached to the endoplasmic reticulum, that is the site of protein synthesis, during which the sequence of bases of messenger RNA is translated into the sequence of amino acids in a protein.

ribozyme: an RNA molecule that can catalyze certain chemical reactions, especially those involved in the synthesis and processing of RNA itself.

RNA polymerase: in RNA synthesis, an enzyme that catalyzes the bonding of free RNA nucleotides into a continuous strand, using RNA nucleotides that are complementary to those of the template strand of DNA.

rod: a rod-shaped photoreceptor cell in the vertebrate retina, sensitive to dim light but not involved in color vision; see also *cone*.

root: the part of a plant body, normally underground, that provides anchorage, absorbs water and dissolved nutrients and transports them to the stem, produces some hormones, and in some plants serves as a storage site for carbohydrates.

root cap: a cluster of cells at the tip of a growing root, derived from the apical meristem; protects the growing tip from damage as it burrows through the soil.

root hair: a fine projection from an epidermal cell of a young root that increases the absorptive surface area of the root.

root pressure: pressure within a root caused by the transport of minerals into the vascular cylinder, accompanied by the entry of water by osmosis.

root system: all of the roots of a plant.

round window: a flexible membrane at the end of the cochlea opposite the oval window that allows the fluid in the cochlea to move in response to movements of the oval window.

rubisco: in the carbon fixation step of the Calvin cycle, the enzyme that catalyzes the reaction between ribulose bisphosphate (RuBP) and carbon dioxide, thereby fixing the carbon of carbon dioxide in an organic molecule; short for ribulose bisphosphate carboxylase.

rumen fungus: a member of the fungus clade Neocallimastigomycota, whose members have swimming spores with multiple flagella. Rumen fungi are anaerobic and most live in the digestive tracts of plant-eating animals.

ruminant (roo'-min-ant): an herbivorous animal with a digestive tract that includes multiple stomach chambers, one of which contains cellulose-digesting bacteria, and that regurgitates the contents ("cud") of the first chamber for additional chewing ("ruminating").

S-curve: the S-shaped growth curve produced by logistic population growth, usually describing a population of organisms introduced into a new area; consists of an initial period of exponential growth, followed by a decreasing growth rate, and finally, relative stability around a growth rate of zero.

sac fungus: a member of the fungus clade Ascomycota, whose members form spores in a saclike case called an ascus.

saccule: a patch of hair cells in the vestibule of the inner ear; bending of the hairs of the hair cells permits detection of the direction of gravity and the degree of tilt of the head.

sapwood: young secondary xylem that transports water and minerals in a tree trunk.

sarcomere (sark'-ō-mēr): the unit of contraction of a muscle fiber; a subunit of the myofibril, consisting of thick and thin filaments, bounded by Z lines.

sarcoplasmic reticulum (sark'-ō-plas'-mik re-tik'-ū-lum; SR): the specialized endoplasmic reticulum in muscle cells; forms interconnected hollow tubes. The sarcoplasmic reticulum stores calcium ions and releases them into the interior of the muscle cell, initiating contraction.

saturated: referring to a fatty acid with as many hydrogen atoms as possible bonded to the carbon backbone (therefore, a saturated fatty acid has no double bonds in its carbon backbone).

savanna: a biome that is dominated by grasses and supports scattered trees; typically has a rainy season during which most of the year's precipitation falls, followed by a dry season during which virtually no precipitation occurs.

science: the organized and systematic inquiry, through observation and experiment, into the origins, structure, and behavior of our living and nonliving surroundings.

scientific method: a rigorous procedure for making observations of specific phenomena and searching for the order underlying those phenomena.

scientific name: the two-part Latin name of a species; consists of the genus name followed by the species name.

scientific theory: an explanation of natural phenomena developed through extensive and reproducible observations; more general and reliable than a hypothesis.

scientific theory of evolution: the theory that modern organisms descended, with modification, from preexisting life-forms.

sclera: a tough, white connective tissue layer that covers the outside of the eyeball and forms the white of the eye.

sclerenchyma (skler-en′-ki-muh): a plant cell type with thick, hardened cell walls; sclerenchyma cells normally die as the last stage of differentiation; may support or protect the plant body.

scramble competition: a free-for-all scramble for limited resources among individuals of the same species.

scrotum (skrō′-tum): in male mammals, the pouch of skin containing the testes.

second law of thermodynamics: the principle of physics that states that any change in a closed system causes the quantity of concentrated, useful energy to decrease and the amount of randomness and disorder (entropy) to increase.

second messenger: an intracellular chemical, such as cyclic AMP, that is synthesized or released within a cell in response to the binding of a hormone or neurotransmitter (the first messenger) to receptors on the cell surface; brings about specific changes in the metabolism of the cell.

secondary consumer: an organism that feeds on primary consumers; a type of carnivore.

secondary growth: growth in the diameter and strength of a stem or root due to cell division in lateral meristems and differentiation of their daughter cells.

secondary oocyte (ō′-ō-sīt): a large haploid cell derived from the diploid primary oocyte by meiosis I.

secondary spermatocyte (sper-ma′-tō-sīt): a large haploid cell derived from the diploid primary spermatocyte by meiosis I.

secondary structure: a repeated, regular structure assumed by a protein chain, held together by hydrogen bonds; for example, a helix.

secondary succession: succession that occurs after an existing community is disturbed—for example, after a forest fire; secondary succession is much more rapid than primary succession.

secretin: a hormone produced by the small intestine that stimulates the production and release of digestive secretions by the pancreas and liver.

seed: the reproductive structure of a seed plant, protected by a seed coat; contains an embryonic plant and a supply of food for it.

seed coat: the thin, tough, and waterproof outermost covering of a seed, formed from the integuments of the ovule.

segmentation (seg-men-tā′-shun): an animal body plan in which the body is divided into repeated, typically similar units.

selectively permeable: the quality of a membrane that allows certain molecules or ions to move through it more readily than others.

self-fertilization: the union of sperm and egg from the same individual.

semen: the sperm-containing fluid produced by the male reproductive tract.

semicircular canal: in the inner ear, one of three fluid-filled tubes, each with a bulge at one end containing a patch of hair cells; movement of the head moves fluid in the canal and consequently bends the hairs of the hair cells.

semiconservative replication: the process of replication of the DNA double helix; the two DNA strands separate, and each is used as a template for the synthesis of a complementary DNA strand. Consequently, each daughter double helix consists of one parental strand and one new strand.

semilunar valve: a valve located between the right ventricle of the heart and the pulmonary artery, or between the left ventricle and the aorta; prevents the backflow of blood into the ventricles when they relax.

seminal vesicle: in male mammals, a gland that produces a basic, fructose-containing fluid that forms part of the semen.

seminiferous tubule (sem-i-ni′-fer-us): in the vertebrate testis, a series of tubes in which sperm are produced.

senescence: in plants, a specific aging process, typically including deterioration and the dropping of leaves and flowers.

sense organ: a multicellular structure that includes both sensory receptor cells and accessory structures that assist in the detection of a specific stimulus; examples include the eye and ear.

sensory neuron: a nerve cell that responds to a stimulus from the internal or external environment.

sensory receptor: a cell (typically, a neuron) specialized to respond to particular internal or external environmental stimuli by producing an electrical potential.

sepal (sē′-pul): one of the group of modified leaves that surrounds and protects a flower bud; in dicots, usually develops into a green, leaflike structure when the flower blooms; in monocots, usually similar to a petal.

septum (pl., septa): a partition that separates the fungal hypha into individual cells; pores in septa allow the transfer of materials between cells.

Sertoli cell: in the seminiferous tubule, a large cell that regulates spermatogenesis and nourishes the developing sperm.

severe combined immune deficiency (SCID): a disorder in which no immune cells, or very few, are formed; the immune system is incapable of responding properly to invading disease organisms, and the individual is very vulnerable to common infections.

sex chromosome: either of the pair of chromosomes that usually determines the sex of an organism; for example, the X and Y chromosomes in mammals.

sex-linked: referring to a pattern of inheritance characteristic of genes located on one type of sex chromosome (for example, X) and not found on the other type (for example, Y); in mammals, in almost all cases, the gene controlling the trait is on the X chromosome, so this pattern is often called X-linked. In X-linked inheritance, females show the dominant trait unless they are homozygous recessive, whereas males express whichever allele, dominant or recessive, is found on their single X chromosome.

sexual reproduction: a form of reproduction in which genetic material from two parent organisms is combined in the offspring; usually, two haploid gametes fuse to form a diploid zygote.

sexual selection: a type of natural selection that acts on traits involved in finding and acquiring mates.

sexually transmitted disease (STD): a disease that is passed from person to person by sexual contact; also known as *sexually transmitted infection (STI)*.

shoot system: all the parts of a vascular plant exclusive of the root; usually aboveground. Consists of stem, leaves, buds, and (in season) flowers and fruits; functions include photosynthesis, transport of materials, reproduction, and hormone synthesis.

short-day plant: a plant that will flower only if the length of uninterrupted darkness exceeds a species-specific critical period; also called a *long-night plant*.

short tandem repeat (STR): a DNA sequence consisting of a short sequence of nucleotides (usually two to five nucleotides in length) repeated multiple times, with all of the repetitions side by side on a chromosome; variations in the number of repeats of a standardized set of 13 STRs produce DNA profiles used to identify people by their DNA.

short-term memory: a memory that lasts only a brief time, usually while a specific neuronal circuit is active or while a change in biochemical activity temporarily strengthens specific synapses.

sickle-cell anemia: a recessive disease caused by a single amino acid substitution in the hemoglobin molecule. Sickle-cell hemoglobin molecules tend to cluster together, distorting the shape of red blood cells and causing them to break and clog capillaries.

sieve plate: in plants, a structure between two adjacent sieve-tube elements in phloem, where holes formed in the cell walls interconnect the cytoplasm of the sieve-tube elements.

sieve-tube element: in phloem, one of the cells of a sieve tube.

simple diffusion: the diffusion of water, dissolved gases, or lipid-soluble molecules through the phospholipid bilayer of a cellular membrane.

simple epithelium: a type of epithelial tissue, one cell layer thick, that lines many hollow organs such as those of the respiratory, digestive, urinary, reproductive, and circulatory systems.

sink: in plants, any structure that uses up sugars or converts sugars to starch, and toward which phloem fluids will flow.

sinoatrial (SA) node (sī'-nō-āt'-rē-ul nōd): a small mass of specialized muscle in the wall of the right atrium; generates electrical signals rhythmically and spontaneously and serves as the heart's pacemaker.

skeletal muscle: the type of muscle that is attached to and moves the skeleton and is under the direct, normally voluntary, control of the nervous system; also called *striated muscle*.

skeleton: a supporting structure for the body, on which muscles act to change the body configuration; may be external or internal.

small intestine: the portion of the digestive tract, located between the stomach and large intestine, in which most digestion and absorption of nutrients occur.

smooth muscle: the type of muscle that surrounds hollow organs, such as the digestive tract, bladder, and blood vessels; not striped in appearance (hence the name "smooth") and normally not under voluntary control.

sodium-potassium (Na⁺-K⁺) pump: an active transport protein that uses the energy of ATP to transport Na^+ out of a cell and K^+ into a cell; produces and maintains the concentration gradients of these ions across the plasma membrane, such that the concentration of Na^+ is higher outside a cell than inside, and the concentration of K^+ is higher inside a cell than outside.

solute: a substance dissolved in a solvent.

solution: a solvent containing one or more dissolved substances (solutes).

solvent: a liquid capable of dissolving (uniformly dispersing) other substances in itself.

somatic nervous system: that portion of the peripheral nervous system that controls voluntary movement by activating skeletal muscles.

source: in plants, any structure that actively synthesizes sugar, and away from which phloem fluid will be transported.

spawning: a method of external fertilization in which male and female parents shed gametes into water, and sperm must swim through the water to reach the eggs.

speciation: the process of species formation, in which a single species splits into two or more species.

species (spē'-sēs): the basic unit of taxonomic classification, consisting of a population or group of populations that evolves independently of other populations. In sexually reproducing organisms, a species can be defined as a population or group of populations whose members interbreed freely with one another under natural conditions but do not interbreed with members of other populations.

specific heat: the amount of energy required to raise the temperature of 1 gram of a substance by 1°C.

sperm: the haploid male gamete, normally small, motile, and containing little cytoplasm.

spermatid: a haploid cell derived from the secondary spermatocyte by meiosis II; differentiates into the mature sperm.

spermatogenesis: the process by which sperm cells form.

spermatogonium (pl., spermatogonia): a diploid cell, lining the walls of the seminiferous tubules, that gives rise to a primary spermatocyte.

spermatophore: a package of sperm formed by the males of some invertebrate animals; the spermatophore can be inserted into the female reproductive tract, where it releases its sperm.

sphincter muscle: a circular ring of muscle surrounding a tubular structure, such as the esophagus, stomach, or intestine; contraction and relaxation of a sphincter muscle controls the movement of materials through the tube.

spinal cord: the part of the central nervous system of vertebrates that extends from the base of the brain to the hips and is protected by the bones of the vertebral column; contains the cell bodies of motor neurons that form synapses with skeletal muscles, the circuitry for some simple reflex behaviors, and axons that communicate with the brain.

spindle microtubule: microtubules organized in a spindle shape that separate chromosomes during mitosis or meiosis.

spiracle (spi'-ruh-kul): an opening in the body wall of insects through which air enters the tracheae.

spleen: the largest organ of the lymphatic system, located in the abdominal cavity; contains macrophages that filter the blood by removing microbes and aged red blood cells, and lymphocytes (B and T cells) that reproduce during times of infection.

spongy bone: porous, lightweight bone tissue in the interior of bones; the location of bone marrow. Compare with *compact bone*.

spontaneous generation: the proposal that living organisms can arise from nonliving matter.

sporangium (spor-an'-jē-um; pl., sporangia): a structure in which spores are produced.

spore: (1) in plants and fungi, a haploid cell capable of developing into an adult without fusing with another cell (without fertilization); (2) in bacteria and some other organisms, a stage of the life cycle that is resistant to extreme environmental conditions.

sporophyte (spor'-ō-fīt): the multicellular diploid stage in the life cycle of a plant; produces haploid, asexual spores through meiosis.

stabilizing selection: a type of natural selection that favors the average phenotype in a population.

stamen (stā'-men): the male reproductive structure of a flower, consisting of a filament and an anther, in which pollen grains develop.

starch: a polysaccharide that is composed of branched or unbranched chains of glucose molecules; used by plants as a carbohydrate-storage molecule.

start codon: the first AUG codon in a messenger RNA molecule.

startle coloration: a form of mimicry in which a color pattern (in many cases resembling large eyes) can be displayed suddenly by a prey organism when approached by a predator.

stem: the portion of the plant body, normally located aboveground, that bears leaves and reproductive structures such as flowers and fruit.

stem cell: an undifferentiated cell that is capable of dividing and giving rise to one or more distinct types of differentiated cell(s).

sterilization: a generally permanent method of contraception in which the pathways through which the sperm (vas deferens) or egg (oviducts) must travel are interrupted; the most effective form of contraception.

steroid: a lipid consisting of four fused carbon rings, with various functional groups attached.

steroid hormone: a class of hormones whose chemical structure (four fused carbon rings with various functional groups) resembles cholesterol; steroids, which are lipids, are secreted by the ovaries and placenta, the testes, and the adrenal cortex.

stigma (stig'-muh): the pollen-capturing tip of a carpel.

stirrup: the third of the small bones of the middle ear, linking the tympanic membrane with the oval window; the stirrup is directly connected to the oval window; also called the *stapes*.

stoma (stō'-muh; pl., stomata): an adjustable opening in the epidermis of a leaf or young stem, surrounded by a pair of guard cells, that regulates the diffusion of carbon dioxide and water into and out of the leaf or stem.

stomach: the muscular sac between the esophagus and small intestine where food is stored and mechanically broken down and in which protein digestion begins.

stop codon: a codon in messenger RNA that stops protein synthesis and causes the completed protein chain to be released from the ribosome.

stramenopile: a member of Stramenopila, a large protist clade. Stramenopiles, which are characterized by hair-like projections on their flagella, include the water molds, the diatoms, and the brown algae.

strand: a single polymer of nucleotides; DNA is composed of two strands wound about each other in a double helix; RNA is usually single stranded.

stratified epithelium: a type of epithelial tissue composed of several cell layers, usually strong and waterproof; mostly found on the surface of the skin.

stroke: an interruption of blood flow to part of the brain caused by the rupture of an artery or the blocking of an artery by a blood clot. Loss of blood supply leads to rapid death of the area of the brain affected.

stroma (strō'-muh): the semifluid material inside chloroplasts in which the thylakoids are located; the site of the reactions of the Calvin cycle.

structural gene: in the prokaryotic operon, the genes that encode enzymes or other cellular proteins.

style: a stalk connecting the stigma of a carpel with the ovary at its base.

subclimax: a community in which succession is stopped before the climax community is reached; it is maintained by regular disturbance—for example, a tallgrass prairie maintained by periodic fires.

substrate: the atoms or molecules that are the reactants for an enzyme-catalyzed chemical reaction.

succession (suk-seh'-shun): a structural change in a community and its nonliving environment over time. During succession, species replace one another in a somewhat predictable manner until a stable, self-sustaining climax community is reached.

sucrose: a disaccharide composed of glucose and fructose.

sugar: a simple carbohydrate molecule, either a monosaccharide or a disaccharide.

sugar-phosphate backbone: a chain of sugars and phosphates in DNA and RNA; the sugar of one nucleotide bonds to the phosphate of the next nucleotide in a DNA or RNA strand. The bases in DNA or RNA are attached to the sugars of the backbone.

surface tension: the property of a liquid to resist penetration by objects at its interface with the air, due to cohesion between molecules of the liquid.

survivorship curve: the curve that results when the number of individuals of each age in a population is graphed against their age, usually expressed as a percentage of their maximum life span.

survivorship table: a data table that groups organisms born at the same time and tracks them throughout their life span, recording how many continue to survive in each succeeding year (or other unit of time). Various parameters such as gender may be used in the groupings. Human life tables may include many other parameters (such as socioeconomic status) used by demographers.

sustainable development: human activities that meet current needs for a reasonable quality of life without exceeding nature's limits and without compromising the ability of future generations to meet their needs.

sympathetic division: the division of the autonomic nervous system that produces largely involuntary responses that prepare the body for stressful or highly energetic situations; often called the *sympathetic nervous system.*

sympatric speciation (sim-pat'-rik): the process by which new species arise in populations that are not physically divided; the genetic isolation required for sympatric speciation may be due to ecological isolation or chromosomal aberrations (such as polyploidy).

synapse (sin'-aps): the site of communication between nerve cells. At a synapse, one cell (presynaptic) releases a chemical (the neurotransmitter) that changes the electrical potential or metabolism of the second (postsynaptic) cell.

synaptic cleft: in a synapse, a small gap between the presynaptic and postsynaptic neurons.

synaptic communication: communication between cells of an animal in which nerve cells affect the activity of other cells by releasing chemicals called neurotransmitters across a small gap separating the ending of a nerve cell from a target cell (usually another nerve cell, a muscle cell, or an endocrine cell).

synaptic terminal: a swelling at the branched ending of an axon; where the axon forms a synapse.

syphilis (si'-ful-is): a sexually transmitted bacterial infection of the reproductive organs; if untreated, can damage the nervous and circulatory systems.

systematics: the branch of biology concerned with reconstructing phylogenies and with naming clades.

systemic circuit: the pathway of blood from the heart through all the parts of the body except the lungs and back to the heart.

systolic pressure (sis'-tal-ik): the blood pressure measured at the peak of contraction of the ventricles; the higher of the two blood pressure readings.

T cell: a type of lymphocyte that matures in the thymus, and that recognizes and destroys specific foreign cells or substances or that regulates other cells of the immune system.

T-cell receptor: a protein receptor, located on the surface of a T cell, that binds a specific antigen and triggers the immune response of the T cell.

T tubule: a deep infolding of the muscle plasma membrane; conducts the action potential inside a cell.

taiga (tī'-guh): a biome with long, cold winters and only a few months of warm weather; dominated by evergreen coniferous trees; also called *northern coniferous forest.*

taproot system: a root system, commonly found in dicots, that consists of a long, thick main root and many smaller lateral roots that grow from the main root.

target cell: a cell on which a particular hormone exerts its effect.

taste bud: a cluster of taste receptor cells and supporting cells that is located in a small pit beneath the surface of the tongue and that communicates with the mouth through a small pore.

taxonomy (tax-on'-uh-mē): the branch of biology concerned with naming and classifying organisms.

tectorial membrane (tek-tor'-ē-ul): one of the membranes of the cochlea in which the hairs of the hair cells are embedded. In sound reception, movement of the basilar membrane relative to the tectorial membrane bends the hairs.

telomere (tē-le-mēr): the nucleotides at the end of a chromosome that protect the chromosome from damage during condensation, and prevent the end of one chromosome from attaching to the end of another chromosome.

telophase (tēl-ō-fāz): in mitosis and both divisions of meiosis, the final stage, in which the spindle fibers usually disappear, nuclear envelopes re-form, and cytokinesis generally occurs. In mitosis and meiosis II, the chromosomes also relax from their condensed form.

temperate deciduous forest: a biome having cold winters and warm summers, with enough summer rainfall for trees to grow and shade out grasses; characterized by trees that drop their leaves in winter (deciduous trees), an adaptation that minimizes water loss when the soil is frozen.

temperate rain forest: a temperate biome with abundant liquid water year-round, dominated by conifers.

template strand: the strand of the DNA double helix from which RNA is transcribed.

temporal isolation: reproductive isolation that arises when species do not interbreed because they breed at different times.

tendon: a tough connective tissue band connecting a muscle to a bone.

terminal bud: meristem tissue and surrounding leaf primordia that are located at the tip of a plant shoot or a branch.

territoriality: the defense of an area in which important resources are located.

tertiary consumer (ter'-shē-er-ē): a carnivore that feeds on other carnivores (secondary consumers).

tertiary structure (ter'-shē-er-ē): the complex three-dimensional structure of a single peptide chain; held in place by disulfide bonds between cysteines.

test cross: a breeding experiment in which an individual showing the dominant phenotype is mated with an individual that is homozygous recessive for the same gene. The ratio of offspring with dominant versus recessive phenotypes can be used to determine the genotype of the phenotypically dominant individual.

testis (pl., testes): the gonad of male animals.

testosterone: in vertebrates, a hormone produced by the interstitial cells of the testis; stimulates spermatogenesis and the development of male secondary sex characteristics.

tetrapod: an organism descended from the first four-limbed vertebrate. Tetrapods include all extinct and living amphibians, reptiles (including birds), and mammals.

thalamus: the part of the forebrain that relays sensory information to many parts of the brain.

therapeutic cloning: the production of a clone for medical purposes. Typically, the nucleus from one of a patient's own cells would be inserted into an egg whose nucleus had been removed; the resulting cell would divide and produce embryonic stem cells that would be compatible with the patient's tissues and therefore would not be rejected by the patient's immune system.

thick filament: in the sarcomere, a bundle of myosin that interacts with thin filaments, producing muscle contraction.

thigmotropism: growth in response to touch.

thin filament: in the sarcomere, a protein strand that interacts with thick filaments, producing muscle contraction; composed primarily of actin, plus the accessory proteins troponin and tropomyosin.

threatened species: all species classified as critically endangered, endangered, or vulnerable.

threshold: the electrical potential at which an action potential is triggered; the threshold is usually about 10 to 20 mV less negative than the resting potential.

thrombin: an enzyme produced in the blood as a result of injury to a blood vessel; catalyzes the production of fibrin, a protein that assists in blood clot formation.

thylakoid (thī′-luh-koid): a disk-shaped, membranous sac found in chloroplasts, the membranes of which contain the photosystems, electron transport chains, and ATP-synthesizing enzymes used in the light reactions of photosynthesis.

thymine (T): a nitrogenous base found only in DNA; abbreviated as T.

thymosin (thī-mō-sin): a hormone, secreted by the thymus, that stimulates the maturation of T lymphocytes of the immune system.

thymus (thī′-mus): an organ of the lymphatic system that is located in the upper chest in front of the heart and that secretes thymosin, which stimulates maturation of T lymphocytes of the immune system.

thyroid gland: an endocrine gland, located in front of the larynx in the neck, that secretes the hormones thyroxine (affecting metabolic rate) and calcitonin (regulating calcium ion concentration in the blood).

thyroid-stimulating hormone (TSH): a hormone, released by the anterior pituitary, that stimulates the thyroid gland to release hormones.

thyroxine (thī-rox′-in): a hormone, secreted by the thyroid gland, that stimulates and regulates metabolism.

tight junction: a type of cell-to-cell junction in animals that prevents the movement of materials through the spaces between cells.

tissue: a group of (normally similar) cells that together carry out a specific function; a tissue may also include extracellular material produced by its cells.

tissue system: a group of two or more tissues that together perform a specific function.

tonsil: a patch of lymphatic tissue, located at the entrance to the pharynx, that contains macrophages and lymphocytes; destroys many microbes entering the body through the mouth and stimulates an adaptive immune response to them.

trachea (trā-kē-uh): in terrestrial vertebrates, a flexible tube, supported by rings of cartilage, that conducts air between the larynx and the bronchi.

tracheae (tra′-ke-ē): the respiratory organ of insects, consisting of a set of air-filled tubes leading from openings in the body called spiracles and branching extensively throughout the body.

tracheid (trā-kē-id): an elongated cell type in xylem, with tapered ends that overlap the tapered ends of other tracheids, forming tubes that transport water and minerals. Pits in the cell walls of tracheids allow easy movement of water and minerals into and out of the cells, including from one tracheid to the next.

trans fat: a type of fat, produced during the process of hydrogenating oils, that may increase the risk of heart disease. The fatty acids of trans fats include an unusual configuration of double bonds that is not normally found in fats of biological origin.

transcription: the synthesis of an RNA molecule from a DNA template.

transfect: to introduce foreign DNA into a host cell; usually includes mechanisms to regulate the expression of the DNA in the host cell.

transfer RNA (tRNA): a type of RNA that binds to a specific amino acid, carries it to a ribosome, and positions it for incorporation into the growing protein chain during protein synthesis. A set of three bases in tRNA (the anticodon) is complementary to the set of three bases in mRNA (the codon) that codes for that specific amino acid in the genetic code.

transformation: a method of acquiring new genes, whereby DNA from one bacterium (normally released after the death of the bacterium) becomes incorporated into the DNA of another, living bacterium.

transgenic: referring to an animal or a plant that contains DNA derived from another species, usually inserted into the organism through genetic engineering.

translation: the process whereby the sequence of bases of messenger RNA is converted into the sequence of amino acids of a protein.

translocation: a mutation that occurs when a piece of DNA is removed from one chromosome and attached to another chromosome.

transpiration (trans′-per-ā-shun): the evaporation of water through the stomata, chiefly in leaves.

transport protein: a protein that regulates the movement of water-soluble molecules through the plasma membrane.

trial-and-error learning: a type of learning in which behavior is modified in response to the positive or negative consequences of an action.

trichomoniasis (trik-ō-mō-nī′-uh-sis): a sexually transmitted disease, caused by the protist *Trichomonas*, that causes inflammation of the mucous membranes that line the urinary tract and genitals.

triglyceride (trī-glis′-er-īd): a lipid composed of three fatty acid molecules bonded to a single glycerol molecule.

trisomy 21: see *Down syndrome*.

trisomy X: a condition of females who have three X chromosomes instead of the normal two; most such women are phenotypically normal and are fertile.

trophic level: literally, "feeding level"; the categories of organisms in a community, and the position of an organism in a food chain, defined by the organism's source of energy; includes producers, primary consumers, secondary consumers, and so on.

tropical deciduous forest: a biome, warm all year-round, with pronounced wet and dry seasons; characterized by trees that shed their leaves during the dry season (deciduous trees), an adaptation that minimizes water loss.

tropical rain forest: a biome with evenly warm, evenly moist conditions year-round, dominated by broadleaf evergreen trees; the most diverse biome.

tropical scrub forest: a biome, warm all year-round, with pronounced wet and dry seasons (drier conditions than in tropical deciduous forests); characterized by short, deciduous, often thorn-bearing trees with grasses growing beneath them.

tropism: directional growth in response to an environmental stimulus.

true-breeding: pertaining to an individual all of whose offspring produced through self-fertilization are identical to the parental type. True-breeding individuals are homozygous for a given trait.

tube cell: the outermost cell of a pollen grain, containing the sperm. When the pollen grain germinates, the tube cell produces a tube penetrating through the tissues of the carpel, from the stigma, through the style, and to the opening of an ovule in the ovary.

tubular reabsorption: the process by which cells of the tubule of the nephron remove water and nutrients from the filtrate within the tubule and return those substances to the blood.

tubular secretion: the process by which cells of the tubule of the nephron remove wastes from the blood, actively secreting those wastes into the tubule.

tubule (toob′-ūl): the tubular portion of a nephron; includes a proximal portion, the loop of Henle, and a distal portion. Urine is formed from the blood filtrate as it passes through the tubule.

tundra: a biome with severe weather conditions (extreme cold and wind, and little rainfall) that cannot support trees.

turgor pressure: pressure developed within a cell (especially the central vacuole of plant cells) as a result of osmotic water entry.

Turner syndrome: a set of characteristics typical of a woman with only one X chromosome; women with Turner syndrome are sterile, with a tendency to be very short and to lack typical female secondary sexual characteristics.

tympanic membrane (tim-pan′-ik): the eardrum; a membrane that stretches across the opening of the middle ear and transmits vibrations to the bones of the middle ear.

unicellular: single-celled; most members of the domains Bacteria and Archaea and the kingdom Protista are unicellular.

uniform distribution: the distribution characteristic of a population with a relatively regular spacing of individuals, commonly as a result of territorial behavior.

unsaturated: referring to a fatty acid with fewer than the maximum number of hydrogen atoms bonded to its carbon backbone (therefore, an unsaturated fatty acid has one or more double bonds in its carbon backbone).

upwelling: an upward flow that brings cold, nutrient-laden water from the ocean depths to the surface.

urea (ū-rē′-uh): a water-soluble, nitrogen-containing waste product of amino acid breakdown; one of the principal components of mammalian urine.

ureter (ū′-re-ter): a tube that conducts urine from a kidney to the urinary bladder.

urethra (ū-rē′-thruh): the tube leading from the urinary bladder to the outside of the body; in males, the urethra also receives sperm from the vas deferens and conducts both sperm and urine (at different times) to the tip of the penis.

urinary system: the organ system that produces, stores, and eliminates urine. The urinary system is critical for maintaining homeostatic conditions within the bloodstream. In mammals, it includes the kidneys, ureters, bladder, and urethra.

urine: the fluid produced and excreted by the urinary system, containing water and dissolved wastes, such as urea.

uterine tube: the tube leading from the ovary to the uterus, into which the secondary oocyte (egg cell) is released; also called the *oviduct*, or, in humans, the *Fallopian tube*.

uterus: in female mammals, the part of the reproductive tract that houses the embryo during pregnancy.

utricle: a patch of hair cells in the vestibule of the inner ear; bending of the hairs of the hair cells permits detection of the direction of gravity and the degree of tilt of the head.

vaccine: an injection into the body that contains antigens characteristic of a particular disease organism and that stimulates an immune response appropriate to that disease organism.

vagina: the passageway leading from the outside of a female mammal's body to the cervix of the uterus; serves as the receptacle for semen and as the birth canal.

variable: a factor in a scientific experiment that is deliberately manipulated in order to test a hypothesis.

variable region: the part of an antibody molecule that differs among antibodies; the ends of the variable regions of the light and heavy chains form the specific binding site for antigens.

vas deferens (vaz de′-fer-enz): the tube connecting the epididymis of the testis with the urethra.

vascular bundle: a strand of xylem and phloem in leaves and stems; in leaves, commonly called a *vein*.

vascular cambium: a lateral meristem that is located between the xylem and phloem of a woody root or stem and that gives rise to secondary xylem and phloem.

vascular cylinder: the centrally located conducting tissue of a young root; consists of primary xylem and phloem, surrounded by a layer of pericycle cells.

vascular plant (vas′-kū-lar): a plant that has conducting vessels for transporting liquids; also called a tracheophyte.

vascular tissue system: a plant tissue system consisting of xylem (which transports water and minerals from root to shoot) and phloem (which transports water and sugars throughout the plant).

vein: in vertebrates, a large-diameter, thin-walled vessel that carries blood from venules back to the heart; in plants, a vascular bundle in a leaf.

ventricle (ven′-tre-kul): the lower muscular chamber on each side of the heart that pumps blood out through the arteries. The right ventricle sends blood to the lungs; the left ventricle pumps blood to the rest of the body.

venule (ven′-ūl): a narrow vessel with thin walls that carries blood from capillaries to veins.

vertebral column (ver-tē′-brul): a column of serially arranged skeletal units (the vertebrae) that protect the nerve cord in vertebrates; the backbone.

vertebrate: an animal that has a vertebral column.

vesicle (ves′-i-kul): a small, temporary, membrane-bound sac within the cytoplasm.

vessel: in plants, a tube of xylem composed of vertically stacked vessel elements with heavily perforated or missing end walls, leaving a continuous, uninterrupted hollow cylinder.

vessel element: one of the cells of a xylem vessel; elongated, dead at maturity, with thick lateral cell walls for support but with end walls that are either heavily perforated or missing.

vestibular apparatus: part of the inner ear, consisting of the vestibule and the semicircular canals, involved in the detection of gravity, tilt of the head, and movement of the head.

vestigial structure (ves-tij′-ē-ul): a structure that serves no apparent purpose but is homologous to functional structures in related organisms and provides evidence of evolution.

villus (vi′-lus; pl., villi): a finger-like projection of the wall of the small intestine that increases its absorptive surface area.

viroid (vī′-roid): a particle of RNA that is capable of infecting a cell and of directing the production of more viroids; responsible for certain plant diseases.

virus (vī′-rus): a noncellular parasitic particle that consists of a protein coat surrounding genetic material; multiplies only within a cell of a living organism (the host).

vitamin: one of a group of diverse chemicals that must be present in trace amounts in the diet to maintain health; used by the body in conjunction with enzymes in a variety of metabolic reactions.

vitreous humor (vit′-rē-us): a clear, jelly-like substance that fills the large chamber of the eye between the lens and the retina; helps to maintain the shape of the eyeball.

vocal cords: a pair of bands of elastic tissue that extend across the opening of the larynx and produce sound when air is forced between them. Muscles alter the tension on the vocal cords and control the size and shape of the opening, which in turn determines whether sound is produced and what its pitch will be.

vulnerable species: a species that is likely to become endangered unless conditions that threaten its survival improve.

waggle dance: a symbolic form of communication used by honeybee foragers to communicate the location of a food source to their hive mates.

warning coloration: bright coloration that warns predators that the potential prey is distasteful or even poisonous.

water mold: a member of a protist group that includes species with filamentous shapes that give them a superficially fungus-like appearance. Water molds, which include species that cause economically important plant diseases, are part of a larger group known as the stramenopiles.

wax: a lipid composed of fatty acids covalently bonded to long-chain alcohols.

weather: short-term fluctuations in temperature, humidity, cloud cover, wind, and precipitation in a region over periods of hours to days.

wetlands: a region (sometimes called a marsh, swamp, or bog) in which the soil is covered by, or saturated with, water for a significant part of the year.

white matter: the portion of the brain and spinal cord that consists largely of myelin-covered axons and that gives these areas a white appearance.

wildlife corridor: a strip of protected land linking larger areas. Wildlife corridors allow animals to move freely and safely between habitats that would otherwise be isolated by human activities.

work: energy transferred to an object, usually causing the object to move.

working memory: the first phase of learning; short-term memory that is electrical or biochemical in nature.

X chromosome: the female sex chromosome in mammals and some insects.

xylem (zī′-lum): a conducting tissue of vascular plants that transports water and minerals from root to shoot.

Y chromosome: the male sex chromosome in mammals and some insects.

yolk: protein-rich or lipid-rich substances contained in eggs that provide food for the developing embryo.

yolk sac: one of the embryonic membranes of reptiles (including birds) and mammals. In reptiles, the yolk sac is a membrane surrounding the yolk in the egg; in mammals, it forms part of the umbilical cord and the digestive tract but does not contain yolk.

Z line: a fibrous protein structure to which the thin filaments of skeletal muscle are attached; forms the boundary of a sarcomere.

zona pellucida (pel-oo'-si-duh): a clear, noncellular layer between the corona radiata and the egg.

zooplankton: nonphotosynthetic protists that are abundant in marine and freshwater environments.

zygomycete: a fungus species formerly placed in the now-defunct taxonomic group Zygomycota. Zygomycetes, which include the species that cause fruit rot and bread mold, do not constitute a true clade and are now distributed among several other taxonomic groups.

zygosporangium (zī'-gō-spor-an-jee-um): a tough, resistant reproductive structure produced by some fungi, such as bread molds; encloses diploid nuclei that undergo meiosis and give rise to haploid spores.

zygote (zī'-gōt): in sexual reproduction, a diploid cell (the fertilized egg) formed by the fusion of two haploid gametes.

Answers to Figure Caption and Fill-in-the-Blank Questions

(Go to MasterBiology for answers to the Consider This, Review, and Applying the Concepts questions.)

Chapter 1

Figure Caption Questions

Figure 1-10 Global climate change will impact the entire biosphere. The growing human population—whose energy use is driving global climate change and whose demand for space and resources is causing widespread habitat destruction—would also be an appropriate answer.

Figure E1-1 Redi's experiment demonstrated that the maggots were caused by something that was excluded by a gauze cover, but the possibility remained that some agent other than flies produced the maggots. An effective follow-up experiment might involve a series of closed, meat-containing systems that were identical in all respects other than the addition of a single possible causal element. Perhaps one container would have flies added, one roaches, one dust or soot, and so on. And, of course, a control would have nothing added.

Fill-in-the-Blank

1. stimuli; energy, materials; complex, organized; evolve
2. atom; cell; tissues; population; community; ecosystem
3. scientific theory; hypothesis; scientific method
4. evolution; natural selection
5. deoxyribonucleic acid, DNA; genes

Chapter 2

Figure Caption Questions

Figure 2-2 Atoms with outer shells that are not full are unstable. They become stable by filling (or emptying) their outer shells. Except for hydrogen, atoms in biological molecules have outer shells that are stable when filled with eight electrons.

Figure 2-8 No, the salt would not dissolve. You can try this at home. The oil exerts no attraction for the ions in salt, so it remains in solid crystals.

Figure 2-9 Oil forms droplets when submerged in water because oil molecules are nonpolar and hydrophobic. Hydrogen bonds join surrounding water molecules, excluding the oil and causing oil molecules to group together.

Figure 2-10 The sunbathers' skin is reddening because their skin cells are being damaged by ultraviolet rays in sunlight (this causes extra blood flow to the damaged cells). This is dangerous because ultraviolet rays can damage DNA in skin cells, in some cases causing cancer.

Fill-in-the-Blank

1. protons, neutrons; atomic number; electrons, electron shells
2. ion; positive; negative; ionic
3. isotopes; subatomic; energy; elements; radioactive
4. inert; reactive; share
5. polar; hydrogen; cohesion

Chapter 3

Figure Caption Questions

Figure 3-8 During hydrolysis of sucrose, water would be split; a hydrogen atom from water would be added to the oxygen from glucose (that formerly linked the two subunits) and the remaining OH from water would be added to the carbon (formerly bonded to oxygen) of the fructose subunit.

Figure 3-12 The enzyme would hydrolyze the triglyceride into a glycerol and three fatty acids, using up three water molecules.

Figure 3-16 As lipids, steroids are soluble in the lipid-based cell membranes and can cross them to act inside the cell.

Figure 3-21 Heat energy can break hydrogen bonds and disrupt the proteins' three-dimensional structure. This prevents the protein from carrying out its usual function(s).

Fill-in-the-Blank

1. monomer, polymers; polysaccharides; hydrolysis; cellulose, starch, glycogen; sucrose, lactose, maltose
2. oil; wax; fat; steroids; cholesterol; phospholipid
3. dehydration, water; amino acids; primary; helix, pleated sheet; pleated sheet; denatured
4. double bonds (or double covalent bonds); hydrogen bonds; hydrogen bonds and disulfide bonds; hydrogen bonds; peptide bonds
5. ribose sugar, base, phosphate; adenosine triphosphate (ATP); adenine, guanine, cytosine, thymine; deoxyribonucleic acid (DNA), ribonucleic acid (RNA); phosphate

Chapter 4

Figure Caption Questions

Figure 4-7 Fluid would not move upward because the beating of the flagella would direct fluid straight out from the cell membranes. Mucus and trapped particles would accumulate in the trachea.

Figure 4-9 Condensation allows the chromosomes to become organized and separated from one another, so that a complete copy of genetic information can be distributed to each of the daughter cells that results from cell division.

Figure 4-14 The fundamentally similar composition of membranes allows them to merge with one another. This allows molecules in vesicles to be transferred from one membrane-enclosed structure (such as endoplasmic reticulum) to another (such as the Golgi apparatus). The material can be exported from the cell when the vesicle merges with the plasma membrane.

Fill-in-the-Blank

1. phospholipids, proteins; phospholipid; protein
2. cytoskeleton; microfilaments, intermediate filaments, microtubules; microtubules; microtubules; microfilaments; intermediate filaments
3. ribosomes; endoplasmic reticulum; nucleolus; Golgi apparatus; cell wall; messenger RNA
4. rough endoplasmic reticulum; vesicles, Golgi apparatus; carbohydrate; plasma membrane
5. mitochondria; chloroplasts; cell wall; nucleoid; cilia; cytoplasm
6. mitochondria, chloroplasts; double, ATP, size
7. cell walls; nucleoid; plasmids; sex pili

Chapter 5

Figure Caption Questions

Figure 5-7 The sac would be larger in all three cases.

Figure 5-8 The plant cell will not burst because its cell wall will resist the pressure of water entering by osmosis. Animal cells lack a cell wall, and when placed in a highly hypotonic solution will absorb water by osmosis until the cell membrane bursts.

Figure 5-9 The distilled water made the solution hypotonic to the blood cells, causing enough water to flow in to burst their fragile cell membranes.

Figure 5-14 Exocytosis uses cellular energy, whereas diffusion is passive. In addition, while materials move through a membrane during diffusion out of a cell, in exocytosis, materials are expelled without passing directly through the plasma membrane. Exocytosis allows materials that are too large to pass through membranes to be eliminated from the cell.

Fill-in-the-Blank

1. phospholipids; receptor, recognition, enzymes, attachment, transport
2. selectively permeable; diffusion; osmosis; aquaporins; active transport
3. channel, carrier; simple, lipids
4. desmosomes, tight junctions, gap junctions; plasmodesmata
5. simple diffusion, simple diffusion, facilitated diffusion, facilitated diffusion.
6. endocytosis; yes; pinocytosis, phagocytosis; vesicles

Chapter 6

Figure Caption Questions

Figure 6-1   This would be a boring roller coaster because each successive "hill" would need to be much lower than the previous one.

Figure 6-5   Burning glucose is an exergonic reaction.

Figure 6-7   Photosynthesis is an endergonic reaction.

Figure 6-10   No, only exergonic reactions can occur spontaneously after the activation energy is surmounted.

Figure 6-14   Competitive inhibition.

Fill-in-the-Blank

1. created, destroyed; kinetic, potential
2. more, less; organized; entropy
3. exergonic; endergonic; exergonic; endergonic; coupled
4. adenosine triphosphate; energy carrier; adenosine diphosphate, phosphate; energy
5. proteins; catalysts, activation energy; active site; substrates
6. inhibiting; competitive inhibitor; allosteric

Chapter 7

Figure Caption Questions

Figure 7-4   Chlorophyll *a* would appear more green because it absorbs fewer of the green wavelengths than chlorophyll *b*. Chlorophyll *b* would appear more yellowish and less green, because it absorbs more of the green wavelengths.

Figure E7-1   The C_4 pathway is less efficient than the Calvin cycle; C_4 uses one extra ATP per CO_2 molecule (for regenerating PEP). Thus, when CO_2 is abundant and photorespiration is not a problem, C_3 plants produce sugar at lower energy cost, and they outcompete C_4 plants.

Fill-in-the-Blank

1. stomata, oxygen (O_2), carbon dioxide (CO_2); water loss (evaporation); chloroplasts, mesophyll
2. red, blue, violet; green; carotenoids; photosystems, thylakoid
3. reaction center; primary electron acceptor; electron transport chain; hydrogen ions (H^+); chemiosmosis
4. water (H_2O), carbon dioxide (CO_2); Calvin cycle; carbon fixation
5. rubisco, oxygen (O_2); photorespiration; C_4 pathway, CAM (or crassulacean acid metabolism) pathway
6. ATP, NADH, Calvin; RuBP (or ribulose bisphosphate); G3P (glyceraldehyde-3-phosphate), glucose

Chapter 8

Figure Caption Questions

Figure 8-3   Glycolysis results in a net production of two ATP and two NADH for each glucose molecule.

Figure 8-6   Oxygen is the final acceptor in the electron transport chain; if it is not present, electrons cannot proceed through the chain, and production of ATP by chemiosmosis comes to a halt.

Figure 8-8   Glycolysis followed by fermentation results in a net production of two ATP molecules and no NADH.

Figure 8-9   In oxygen-rich environments, both types of bacteria can survive, but aerobic bacteria prevail because their respiration produces more ATP per glucose molecule. In oxygen-poor environments, however, aerobic bacteria are limited by the oxygen shortage, and anaerobes would prevail despite their inefficiency.

Fill-in-the-Blank

1. glycolysis, cellular respiration; cytosol, mitochondria; cellular respiration
2. anaerobic; glycolysis, two; fermentation, NAD^+
3. ethanol (alcohol), carbon dioxide; lactic acid; lactic acid
4. erythropoietin (EPO), red blood, oxygen; cellular respiration
5. matrix, intermembrane space, concentration gradient; chemiosmosis; ATP synthase
6. Krebs; pyruvate; two; NADH, $FADH_2$

Chapter 9

Figure Caption Questions

Figure 9-9   If the sister chromatids of one replicated chromosome failed to separate, then one daughter cell would not receive any copy of that chromosome, while the other daughter cell would receive both copies.

Figure 9-15   If one pair of homologues failed to separate at anaphase I, one of the resulting daughter cells (and the gametes produced from it) would have both homologues and the other daughter cell (and the gametes produced from it) would not have any copies of that homologue.

Fill-in-the-Blank

1. deoxyribonucleic acid (DNA); locus; alleles
2. asexual; sexual; Meiotic
3. prokaryotic fission
4. mitotic; differentiation; Stem
5. homologues OR homologous chromosomes; autosomes; sex chromosomes
6. prophase, metaphase, anaphase, telophase; cytokinesis; telophase
7. kinetochores; polar
8. four; meiosis I; gametes OR sperm and eggs
9. prophase; chiasmata; crossing over
10. shuffling of homologues OR random distribution of homologues; crossing over; fusion of gametes

Chapter 10

Figure Caption Questions

Figure 10-8   Half of the gametes produced by a *Pp* plant will have the *P* allele, and half will have the *p* allele. All of the gametes produced by a *pp* plant will have the *p* allele. Therefore, half of the offspring of a *Pp* × *pp* cross will be *Pp* (purple) and half will be *pp* (white), whereas all of the offspring of a *PP* × *pp* cross will be *Pp* (purple). See Figure 10-9.

Figure 10-11   A plant with wrinkled green seeds has the genotype *ssyy*. A plant with smooth yellow seeds could be *SSYY*, *SsYY*, *SSYy*, or *SsYy*. Set up four Punnett squares to see if the smooth yellow plant's genotype can be revealed by a test cross.

Figure 10-12   Chromosomes, not individual genes, assort independently during meiosis. Therefore, if the genes for seed color and seed shape were on the same chromosome, they would tend to be inherited together, and would not assort independently.

Figure 10-25   One of Victoria and Albert's sons, Leopold, had hemophilia. To be male, Leopold must have inherited Albert's Y chromosome. The X chromosome, not the Y chromosome, bears the gene for blood clotting, so Leopold must have inherited the hemophilia allele from his mother, Victoria.

Fill-in-the-Blank

1. locus; alleles; mutations
2. genotype, phenotype; heterozygous
3. independently; as a group; linked
4. an X and a Y, two X; sperm
5. sex-linked
6. incomplete dominance; codominance; polygenic inheritance

Chapter 11

Figure Caption Questions

Figure 11-5   It takes more energy to break apart a C–G base pair, because these are held together by three hydrogen bonds, compared with the two hydrogen bonds that bind A to T.

Figure E11-7   DNA polymerase always moves in the 3' to 5' direction on a parental strand. Because the two strands of a DNA double helix are oriented in opposite directions, the 5' direction on one strand leads toward the replication fork and the 5' direction on the other strand leads away from the fork. Therefore, DNA polymerase must move in opposite directions on the two strands.

Fill-in-the-Blank

1. nucleotides; sugar (deoxyribose), phosphate, base (order not important)
2. phosphate, sugar (order not important); double helix
3. thymine, cytosine; complementary
4. semiconservative
5. DNA helicase; DNA polymerase; DNA ligase
6. mutations; nucleotide substitution, point mutation

Chapter 12

Figure Caption Questions

Figure 12-3   RNA polymerase always travels in the 3' to 5' direction on the template strand. Because the two DNA strands run in opposite directions, if the other DNA strand were the template strand, then RNA polymerase must travel in the opposite direction (that is, right to left in this illustration).

Figure 12-4   Cells produce far larger amounts of some proteins than others. Obvious examples include cells that produce antibodies or protein hormones, which they secrete into the bloodstream in large quantities, with effects throughout the body. If a cell needs to produce more of certain proteins, it will probably synthesize more mRNA, which will then be translated into that protein.

Figure 12-7   Grouped into codons, the original mRNA sequence visible here is CGA AUC UAG UAA. Changing all G to U would produce the sequence CUA AUC UAU UAA. The two changes are in the first codon (CGA to CUA) and the third codon (UAG to UAU). Refer to the genetic code illustrated in Table 12-3. First, CGA encodes arginine, while CUA encodes leucine, so the first G → U change would substitute leucine for arginine in the protein. Second, UAG is a stop codon, but UAU encodes tyrosine. Therefore, the second G → U change would add tyrosine to the protein instead of stopping translation. The final codon in the illustration, UAA, is a stop codon, so the new protein would end with tyrosine.

Figure E12-1   The mold would be able to grow if any of ornithine, citrulline, or arginine were added to the medium.

Fill-in-the-Blank

1. transcription; translation; ribosome
2. mRNA, tRNA, rRNA (order not important); microRNA
3. three; codon; anticodon
4. RNA polymerase; template; promoter; termination signal
5. start, stop; transfer RNA; peptide
6. point; Insertion; Deletion

Chapter 13

Figure Caption Questions

Figure 13-8   As they do for other genes, each person normally has two copies of each STR gene, one on each of a pair of homologous chromosomes. A person may be homozygous (two copies of the same allele) or heterozygous (one copy of each of two alleles) for each STR. The bands on the gel represent individual alleles of an STR gene. Therefore, a single person can have one band (if homozygous) or two bands (if heterozygous). If a person is homozygous for an STR allele, then he has two copies of the same allele. The DNA from both (identical) alleles will run

in the same place on the gel, and therefore that (single) band will have twice as much DNA as each of the two bands of DNA from a heterozygote. The more DNA, the brighter the band.

Figure 13-9   The genetic material of bacteriophages is DNA. Each bacterial restriction enzyme cuts DNA at a specific nucleotide sequence. A given bacterium is likely to have evolved restriction enzymes that cut DNA at sequences found in bacteriophages but not in their own chromosome.

Fill-in-the-Blank

1. Genetically modified organisms
2. Transformation; plasmids
3. polymerase chain reaction
4. short tandem repeats; length OR size OR number of repeats; DNA profile
5. electrophoresis OR gel electrophoresis; DNA probe; base pairing OR hydrogen bonds

Chapter 14

Figure Caption Questions

Figure 14-6   No. Mutations, the ultimate source of the variation on which natural selection acts, occur in all organisms, including those that reproduce asexually.

Figure 14-7   No. Evolution can include changes in traits that are not revealed in morphology (the physical form of an organism), such as physiological systems and metabolic pathways. More generally, evolution in the sense of changes in a species' genetic make-up is inevitable; genetic evolution is not necessarily reflected in morphological change.

Figure 14-9   Possibilities include:

coccyx (tailbone)—homologous with tail bones of a cat (or any tetrapod)

goose bumps—homologous with erectile hairs of a chimp (or any mammal; used for aggressive displays and insulation)

appendix—homologous with the cecum of a rabbit (and other herbivorous mammals; extension of large intestine used for storage)

wisdom teeth—homologous with grinding rear molars of a leaf-eating monkey (or other mammals)

ear-wiggling muscles (the muscles around the ear that some people can use to make their ears move)—homologous with muscles in a dog (and other mammals) that enable the external ear to orient toward sound

Figure 14-10   Analogous. Bird tails consist of feathers; dog tails do not (they consist of bone/muscle/skin). If the two structures were homologous, they would both consist of bone and muscle, or both consist of feathers

Fill-in-the-Blank

1. wing, arm; analogous, convergent; vestigial
2. common ancestor; amino acids, ATP
3. catastrophism; uniformitarianism; old
4. evolution; mutations, DNA
5. natural selection; artificial selection
6. many traits are inherited; Gregor Mendel

Chapter 15

Figure Caption Questions

Figure 15-3   The surviving colonies would be in different places on each treated plate, and there would be a different number of surviving colonies on each plate (because the antibiotic-caused mutations would arise unpredictably, depending on which bacteria happened to interact with the antibiotic such that mutations were caused). Another possibility would be for all of the colonies to survive (if the antibiotic always caused mutations in every colony).

Figure 15-5   For a locus with two alleles, one dominant and one recessive, there are two possible phenotypes. A mating between a heterozygote and a homozygote-recessive yields offspring with a 50:50 ratio of the two phenotypes.

	B	b
b	Bb (black)	bb (brown)
b	Bb (black)	bb (brown)

Figure 15-6 Allele A should behave roughly as it does in the size 8 population, its frequency drifting to 0 or 1 in almost all cases. But, because the population is a bit larger, the allele should, on average, take a longer time (greater number of generations) to reach a frequency of 0 or 1. The longer period of drift should also allow for more reversals of direction (e.g., frequency drifting down, then up, then down again, etc.) than in the size 8 population.

Figure 15-7 Mutations inevitably and continually add variability to a population and, after the population becomes larger, the counteracting, diversity-reducing effects of drift decrease. The net result is an increase in genetic diversity.

Figure 15-10 Greater for males. A female's reproductive success is limited by her maximum litter size, but a male's potential reproductive success is limited only by the number of available females. When, as in bighorn sheep, males battle for access to females, the most successful males can impregnate many females, while unsuccessful males may not fertilize any females at all. Thus, the difference between the most and least successful male can be very large. In contrast, even the most successful female can have only one litter of offspring per breeding season, which is not that many more offspring than a female who fails to reproduce.

Figure 15-12 There is always a limit to directional selection. As a trait becomes more extreme, eventually the cost of increasing it further outweighs the benefits (for example, the cost of obtaining extra food may outweigh the benefit of larger size), or it may be physically impossible for the trait to become more extreme (for example, a limb's length may be limited by the maximum length that a bone can attain without breaking under its own weight).

Fill-in-the-Blank

1. Hardy–Weinberg principle, equilibrium, allele; no
2. alleles; nucleotides; mutations; homozygous, heterozygous
3. genotype, phenotype; phenotype
4. genetic drift; small; founder effect, population bottleneck; founder effect
5. species; natural selection, coevolution; adaptations
6. reproduction; environment

Chapter 16

Figure Caption Questions

Figure 16-8 Possibilities include continental drift; climate changes (especially glacial advances) that cause habitat fragmentation; formation of islands by volcanic activity or rising sea level; movements of organisms to existing islands (including "islands" of isolated habitats such as lakes, mountaintops, deep-ocean vents); and formation of barriers to movement (e.g., new mountain ranges, deserts, rivers). These processes are indeed sufficiently common and widespread to account for a multitude of speciation events over the history of life.

Figure 16-9 The key question is whether the two populations inhabiting two species of trees (apple and hawthorn) interbreed. Tests might involve careful observation of flies under natural conditions, lab experiments in which captive flies of the two types are provided with opportunities to interbreed, or genetic comparisons to determine the degree of gene flow between the two types of flies.

Figure 16-13 Natural selection cannot look forward and ensure that the only traits that evolve are those that ensure survival of the species as a whole. Instead, natural selection ensures only the preservation of traits that help individuals survive and reproduce more successfully than individuals lacking the trait. So if, in a particular species, highly

specialized individuals survive and reproduce better than less-specialized individuals, the specialized phenotype will eventually dominate, even if it ultimately puts the species at greater risk of extinction.

Fill-in-the-Blank

1. populations, independently; reproductive isolation; asexually
2. behavioral isolation, hybrid inviability, temporal isolation, gametic incompatibility, mechanical incompatibility
3. genetically isolated, diverge; allopatric speciation; genetic drift, natural selection
4. adaptive radiation; habitat (or environment)
5. small, specialized; habitat destruction

Chapter 17

Figure Caption Questions

Figure 17-2 The presence of oxygen would prevent the accumulation of organic compounds by quickly oxidizing them or their precursors. All of the successful abiotic synthesis experiments used oxygen-free "atmospheres."

Figure 17-5 The bacterial sequence would be most similar to that of the plant mitochondrion, because (as the descendant of the immediate ancestor of the mitochondrion) the bacterium shares with the mitochondrion a more recent common ancestor than with the chloroplast or the nucleus.

Figure 17-8 Most likely because of competition with seed plants, which had not yet arisen during the period when ferns and club mosses reached large sizes. After seed plants arose, competition from them eventually eliminated other types of plants from many ecological niches, presumably including those niches that favored evolution of large size.

Figure 17-9 No. The mudskipper merely demonstrates the plausibility of a hypothetical intermediate step in the proposed scenario for the origin of land-dwelling tetrapods. But the existence of a modern example similar in form to the hypothetical intermediate form does not provide information about the actual identity of that intermediate form.

Figure 17-18 The African replacement hypothesis. These fossils are the oldest modern humans found so far, and their presence in Africa suggests that modern humans were present in Africa before they were present anywhere else, which, if true, would mean that they originated in Africa.

Figure E17-1 356.5 million years old. (1:1 ratio means that 1/2 of the original uranium-235 is left, so its half-life has passed.)

Fill-in-the-Blank

1. anaerobic; photosynthesize; poisonous (toxic, harmful); aerobic (cellular); energy
2. RNA (ribonucleic acid); enzymes (catalysts); ribozymes
3. eukaryotic; endosymbiotic; DNA (deoxyribonucleic acid), ribosomes
4. swimming; moist (wet); pollen
5. conifers; wind; flowers, insects; efficient
6. arthropods; exoskeletons; drying
7. reptiles; egg; skin; lungs

Chapter 18

Figure Caption Questions

Figure E18-3 Fungi and animals are monophyletic; protists, great apes, seedless plants, and prokaryotes are not. (Descendants of the most recent common ancestor of protists include all the other eukaryotes; for great apes, include humans; for prokaryotes, include eukaryotes; for seedless plants, include the seed plants.)

Fill-in-the-Blank

1. taxonomy; systematics; clade
2. genus, species; Latin; capitalized, italic
3. domain, kingdom, phylum, class, order, family, genus, species; Archaea, Bacteria, Eukarya
4. anatomy, DNA sequence

5. discover new data on similarities and differences among organisms; reproduce asexually; phylogenetic species concept, phylogeny (evolutionary tree)

6. 1.5 million, 7 million, 10 million, 100 million

Chapter 19

Figure Caption Questions

Figure 19-4 Protective structures like endospores are most likely to evolve in environments in which protection is especially advantageous. Compared to other environments inhabited by bacteria, soils are especially vulnerable to drying out, which can be fatal to unprotected bacteria. Bacteria that could resist long dry periods would gain an evolutionary advantage.

Figure 19-5 Enzymes from bacteria that live in hot environments are active at high temperatures (temperatures that usually denature enzymes in organisms from more temperate environments). This ability to function at high temperatures makes the enzymes useful in test tube reactions (such as the polymerase chain reaction) that are run at high temperatures.

Figure 19-7 The main advantage is efficiency. In prokaryotic fission, every individual produces new individuals. In sexual reproduction, only some individuals (e.g., females) produce offspring. So the average individual of a species produces twice as many offspring by fission as it would by sexual reproduction.

Figure 19-9 The concentration of nitrogen gas would increase, because the major process for removing atmospheric nitrogen would end, while the processes that add nitrogen gas to the atmosphere would continue.

Figure 19-12 Viruses lack ribosomes and the rest of the "machinery" required to manufacture proteins.

Figure 19-13 Viruses replicate by integrating their genetic material into the host cell's genome. Thus, if biotechnologists can insert foreign genetic material into a virus, the virus will naturally tend to transfer the foreign genes to the cells they infect.

Fill-in-the-Blank

1. Bacteria, cell walls, archaea
2. smaller; spherical, rod-shaped, corkscrew-shaped
3. flagella; biofilms; endospores
4. Anaerobic; Photosynthetic
5. prokaryotic fission, conjugation
6. nitrogen-fixing; cellulose
7. tuberculosis, cholera, bubonic plague, syphilis, gonorrhea, pneumonia, Lyme disease, etc.; meat, eggs, produce
8. DNA, RNA, protein; host; bacteriophage

Chapter 20

Figure Caption Questions

Figure 20-2 Sex is the process that combines the genomes of two different individuals. In plants and animals, this mixing of genomes occurs only during reproduction. But in many protists (and prokaryotes), genome mixing may occur through conjugation and other processes that take place independently of reproduction (which in many cases occurs by mitotic cell division).

Figure 20-7 Convergent evolution. Water molds and fungi live in similar environments and acquire nutrients in similar ways. These ecological similarities have fostered the evolution of superficially similar structures, even though the two taxa are only distantly related.

Fill-in-the-Blank

1. pseudopods; decomposers, parasites
2. algae; protozoa
3. secondary endosymbiosis, photosynthetic protist
4. diplomonad (or excavate); apicomplexan (or alveolate); kinetoplastid (or euglenozoan)

5. water mold (or oomycete or stramenopile); amoebozoan
6. dinoflagellates, diatoms; green algae

Chapter 21

Figure Captions Questions

Figure 21-4 Bryophytes lack lignin (which provides stiffness and support) and conducting vessels (which transport materials to distant parts of the body). Vessels and stiff stems seem to be required to achieve heights greater than a few inches.

Figure 21-6 All of the pictured structures are sporophytes. In ferns, horsetails, and club mosses, the gametophyte is small and inconspicuous.

Figure 21-8 The most common adaptations are hard, protective shells and incorporation of toxic and/or distasteful chemicals.

Figure 21-11 Angiosperms

Type of Pollination	Advantages	Disadvantages
Wind	not dependent on presence of animals; no investment in nectar or showy flowers; pollen can disperse over large distances	Larger investment in pollen because most fails to reach an egg; higher chance of failure to fertilize any egg
Animal	Each pollen grain has much greater chance of reaching suitable egg	Depends on presence of animals; must invest in nectar and showy flowers

Both types of pollination persist in angiosperms because the cost–benefit balance, and therefore the most adaptive pollination system, differs depending on the ecological circumstances of a species.

Fill-in-the-Blank

1. green algae; nonvascular plants (bryophytes), vascular plants (tracheophytes); embryos, alternation of generations
2. cuticle, stomata, water; vessels, lignin, water, nutrients
3. swim to the egg; seeds, pollen; gymnosperms, angiosperms; attract pollinators; facilitate seed dispersal.
4. hornworts, liverworts, mosses; club mosses, horsetails, ferns; angiosperms

Chapter 22

Figure Captions Questions

Figure 22-1 Its filamentous shape helps the fungal body to penetrate and extend into its food sources and also maximizes the ratio of surface area to interior volume (which maximizes the area available for absorbing nutrients). The extreme thinness of the filaments ensures that no cell is very far from the surface at which nutrients are absorbed.

Figure 22-8 No. The common source of the two hyphae means that they will both inherit the same mating type. Hyphae must be of different mating types in order to reproduce sexually.

Figure 22-9 Haploid (although there might be some diploid cells in the basidia concealed in the gills of these haploid mushrooms).

Figure 22-20 In nature, bacteria compete with fungi for access to food and living space. The antibiotic chemicals produced by fungi serve as a defense against competition from bacteria.

Fill-in-the-Blank

1. reproduction; spores
2. mycelium, hyphae, septa; chitin
3. mitotic, one; meiotic, zygote, one
4. glomeromycetes; chytrids, rumen fungi, and blastoclades; basidiomycetes
5. Lichens; Mycorrhizae; endophytes
6. wood (cellulose and lignin); yeasts; athlete's foot, jock itch, vaginal infections, ringworm, histoplasmosis, valley fever (many possible answers)

Chapter 23

Figure Caption Questions

Figure 23-6   Sponges are "primitive" only in the sense that their lineage arose early in the evolutionary history of animals and their body plan is comparatively simple. But early origin and simplicity do not determine effectiveness, and the sponge body plan and way of life are clearly suitable for excellent survival and reproduction in many habitats.

Figure 23-8   (a) polyp, (b) medusa, (c) polyp, (d) medusa.

Figure 23-13   Parasitic tapeworms have no gut and absorb nutrients across their body surfaces. Their ribbon-like shape maximizes surface area for absorption and allows the worm body to extend through the greatest possible area of the host's body (to be in contact with as many nutrients as possible).

Figure 23-14   Two openings allow one-way travel of food through the gut, which allows continuous feeding. One-way movement allows more efficient digestion than two-way movement; digestive waste from which all nutrients have been extracted can be excreted quickly without the need for reverse travel back along the gut, and food can be processed more quickly.

Figure 23-15   Water travels easily through the moist epidermis of a leech. When a high concentration of salt is dissolved in the moisture on the outside of a leech's body, water moves rapidly out of the leech's body by osmosis, dehydrating and ultimately killing the animal.

Fill-in-the-Blank

1. consuming other organisms; sexually; cell walls
2. three, endoderm, mesoderm, ectoderm; two, mesoderm
3. cephalized; sensing the environment, ingesting food; cavity, mesoderm
4. protostome; deuterostome, echinoderms, chordates
5. invertebrates, vertebrates; invertebrates; sponges, single-celled organisms; cnidarians; annelids
6. bivalves, gastropods, cephalopods; arthropods; insects, arachnids, crustaceans
7. annelids, arthropods (chordates also correct); closed circulatory; open circulatory, hemocoel
8. mollusks, cnidarians, echinoderms, arthropods

Chapter 24

Figure Caption Questions

Figure 24-7   A freshwater fish's body is immersed in a hypotonic solution, so water tends to continuously enter the body by osmosis. The physiological challenge is to get rid of all this excess water. For a saltwater fish, the challenge is reversed. The surrounding solution is hypertonic, so water tends to leave the body. The physiological challenge is to retain sufficient water.

Figure 24-9   One advantage is that adults and juveniles occupy different habitats and therefore do not compete with one another for resources (the niche occupied by an individual over its lifetime is broadened).

Figure 24-12   Flight is a very expensive trait (consumes a lot of energy, requires many special structures). In circumstances in which the benefits of flight are low, such as in habitats without predators, natural selection may favor individuals that forgo an investment in flight, and flightlessness can arise.

Fill-in-the-Blank

1. hollow, dorsal; anus, notochord
2. tunicates, lancelets, hagfishes (order not important); skull; hagfishes
3. lungfishes; cartilage; ray-finned fishes; jaws
4. mammals, amphibians, reptiles, amphibians
5. monotremes; salamanders; bats

Chapter 25

Figure Caption Questions

Figure 25-1   One possibility is that the variation necessary for selection has never arisen. (If no members of the species by chance gain the ability to discriminate between their own chicks and cuckoo chicks, then selection has no opportunity to favor the novel behavior.) Another possibility is that the cost of the behavior is relatively low. (If parasitism by cuckoos is rare, a parent that feeds any begging chick in its nest is, on average, much more likely to benefit than suffer.)

Figure 25-6   Because the crossbreeding experiment shows that differences in orientation direction between the two populations stem from genetic differences, birds from the western population should orient in a southwesterly direction, regardless of the environment in which they are raised.

Figure 25-19   If females do indeed gain any benefits from their mates, those benefits would necessarily be genetic (as the male provides no material benefits). Male fitness may vary, and to the extent that this fitness can be passed to offspring, females would benefit by choosing the most fit males. If a male's fitness is reflected in his ability to build and decorate a bower, females would benefit by preferring to mate with males that build especially good bowers.

Figure 25-22   Canines forage mainly by smell; apes are mostly visual foragers. Modes of sexual signaling are affected not only by the nature of the information to be encoded, but also by the sensory biases and sensitivities of the species involved. Communication systems may evolve to take advantage of traits that originally evolved for other functions.

Fill-in-the-Blank

1. genes, environment; stimulus; innate
2. energy, predators, injury; practice, skills; adult; brains, learning
3. habituation, repeated; imprinting; sensitive period
4. aggressive; displays, injuring (wounding, damaging); weapons (such as fangs, claws), larger
5. territoriality (territorial behavior); mate, raise young, feed, store food; male; same species
6. For pheromones; any of the following advantages is correct: long lasting, use little energy to produce, are species specific, don't attract predators, may convey messages over long distances. Any of the following disadvantages is correct: cannot convey changing information, convey fewer different types of information. For visual displays, any of the following advantages is correct: instantaneous, can change rapidly, many different messages can be sent, quiet. Any of the following disadvantages is correct: make animals conspicuous to predators, generally ineffective in darkness or dense vegetation, require that recipient be in visual range (close by).
7. species, sex, quality or desirability relative to other males

Chapter 26

Figure Caption Questions

Figure 26-3   Many variables interact in complex ways to produce real population cycles. Weather, for example, affects the lemmings' food supply and thus their ability to survive and reproduce. Predation of lemmings is influenced by both the number of predators and the availability of other prey, which in turn is influenced by multiple environmental variables.

Figure 26-11   Emigration relieves population pressure in an overpopulated area, spreading the migrating animals into new habitats that may have more resources. Human emigration within and between countries is often driven by the desire or need for more resources, although social factors—such as wars and religious or racial persecution—also fuel human emigration. (This subjective question can lead to discussion of the extent to which overpopulation drives human emigration.)

Figure 26-14   Many scenarios are possible, based on different assumptions about increases in technology, changes in birth and death rates, and the resiliency of the ecosystems that sustain us. (This is a subjective question.)

Figure 26-17   When fertility exceeds RLF, there are more children than parents. As the additional children mature and become parents themselves, this more numerous generation produces still more additional children, and so on, in a positive feedback cycle.

Figure 26-18 Projections for 2050 could be inaccurate—in particular, too high for less developed countries—if density-dependent forces of environmental resistance (disease, lack of food or clean water) interacting with density-independent factors (weather extremes, possibly exacerbated by global climate change) increase death rates. Such factors could also stimulate changes in cultural perceptions that might make people eager to limit family size.

Figure 26-19 U.S. population growth is in the rapidly rising "exponential" phase of the S-curve. Stabilization will require some combination of reduction in immigration rates and birth rates. An increase in death rates is less likely, but cannot be ruled out entirely in any future scenario. (Students should be encouraged to speculate about the timing and the reason for various time frames.)

Fill-in-the-Blank

1. Any two of the following: freezing, drought, flood, fire, storms, habitat destruction; predation, competition
2. survivorship curves; early loss; late-loss
3. exponential; no; J-curve; no
4. carrying capacity; logistic, S
5. clumped; uniform; random
6. pyramid (triangle); stable; pyramidal (triangular)
7. births, immigrants, deaths, emigrants; decreases, increases, increases

Chapter 27
Figure Caption Questions

Figure 27-1 An invasive species could occupy a niche nearly identical to the niche of a native species because they did not evolve together. In this case, the invasive species would be likely to eliminate the native species.

Figure 27-3 Examples of coevolution between predators and prey include the keen eyesight of the hawk and the camouflage coloration of a mouse. Grasses have evolved tough silicon embedded in their leaves, and grazing herbivores have teeth that grow continuously so that they are not completely worn down by the abrasive grasses. A moth may detect a bat's echolocation signals and fly erratically, while the bat shifts its echolocation pitch to a pitch that the moth cannot detect. Milkweed plants have evolved a toxin to deter predators, and monarch butterfly caterpillars have evolved resistance to the toxin.

Figure 27-4 The caterpillar's stripes have evolved as warning coloration. This communicates to predators that the caterpillar is poisonous, causing the predators to avoid it.

Figure 27-6 An organism that by chance mutation more closely resembled an object in the environment would be less likely to be seen and eaten by predators, and so it would be more likely to survive and leave offspring. Those offspring that inherited the trait would also be more likely to survive and reproduce, so the trait would gradually become the norm for the species.

Figure 27-15 Forests in which fires are suppressed become unnaturally dense and provide fewer habitats for diverse wildlife. They also accumulate larger quantities of flammable material that burns much hotter when a fire occurs, destroying more of the seeds and root systems needed to initiate secondary succession.

Fill-in-the-Blank

1. natural selection; coevolution; competition, predation, parasitism, mutualism
2. carnivores, herbivores; camouflage; larger, abundant (or numerous)
3. ecological niches; competitive exclusion
4. warning coloration, startle coloration, Batesian mimicry, aggressive mimicry
5. mutualism (or symbiosis), parasitism, predation, parasitism, mutualism; competition
6. succession; primary succession; secondary succession; climax; subclimax

Chapter 28
Figure Caption Questions

Figure 28-1 When energy is converted from one form to another, the amount of useful energy decreases (this is the second law of thermodynamics; see Chapter 6). Much of this energy is lost as heat, which is given off to the environment and cannot be used by living organisms.

Figure 28-2 On land, high productivity is supported by optimal temperatures for plant growth, a long growing season, and plenty of moisture, such as is found in rain forests. Lack of water limits desert productivity. In aquatic ecosystems, high productivity is supported by an abundance of nutrients and adequate light, such as is found in estuaries. Lack of nutrients limits the productivity of the open ocean, even in well-lit surface waters.

Figure 28-4 There are many possibilities, including many combinations that are not linked by arrows in the figure. One example is grass (producer) → grasshopper (primary consumer) → meadowlark (secondary consumer) → hawk (tertiary consumer). Another is grass (producer) → grasshopper (primary consumer) → spider (secondary consumer) → shrew (tertiary consumer).

Figure 28-8 Humanity's need to grow crops to feed our growing population has led to the fixing of nitrogen for fertilizer using industrial processes. Additionally, large-scale livestock feedlots generate enormous amounts of nitrogenous waste. Nitrogen oxides are also generated when fossil fuels are burned in power plants, vehicles, and factories, and when forests are burned. Consequences include the overfertilization of lakes and rivers, and the creation of dead zones in coastal waters that receive excessive nutrient runoff from land. Another important consequence is acid deposition, in which nitrogen oxides formed by combustion produce nitric acid in the atmosphere; this acid is then deposited on land.

Fill-in-the-Blank

1. sunlight, photosynthesis; nutrients, nutrient cycles
2. autotrophs, producers; net primary production
3. trophic levels; food chain; food webs
4. 10
5. heterotrophs, consumers; herbivores, primary consumers; carnivores, secondary consumers; detritivores, decomposers
6. nitrogen-fixing bacteria, denitrifying bacteria; ammonia, nitrate
7. atmosphere, oceans; CO_2 (carbon dioxide); limestone, fossil fuels

Chapter 29
Figure Caption Questions

Figure 29-8 Nutrients are abundant in tropical rain forests, but they are not stored in the soil. The optimal temperature and moisture of tropical climates allow plants to make such efficient use of nutrients that nearly all nutrients are stored in plant bodies, and to a lesser extent, in the bodies of the animals they support. These growing conditions support such a vast array of plants that these, in turn, provide a wealth of habitats and food sources for diverse animals.

Figure 29-12 Both live in arid environments, so their evolution has selected for fleshy, water-storing bodies. In each case, leaf area has been reduced, limiting evaporation. Both of these plants would be attractive to desert animals because they store water; as a result of this predation pressure, both bear defensive spines.

Figure 29-17 Tallgrass prairie is vulnerable to two major encroachments. First, if humans suppress natural wildfires, the adequate rainfall amounts allow forests to take over. Second, these biomes provide the world's most fertile soils and have excellent growing conditions for farming, which has now displaced native vegetation.

Figure 29-29 Coastal ecosystems have an abundance of the two most important limiting factors for life in water: nutrients and light to support photosynthetic organisms. Both upwelling from ocean depths and runoff from the land can provide nutrients, depending on the location of the ecosystem. The shallow water in these areas allows adequate light to penetrate to support rooted plants and/or anchored algae, which in turn provide food and shelter for a wealth of marine life.

Fill-in-the-Blank

1. seasons; ocean currents *or* the presence of the ocean; rain shadow
2. appropriate temperatures, liquid water; light, nutrients, appropriate temperature
3. tropical rain forests; coral reefs
4. tropical rain-forest, soil; temperate deciduous forest; grassland *or* prairie; desert; savanna; tundra
5. littoral zone; phytoplankton, zooplankton; limnetic zone, profundal zone; oligotrophic; eutrophic; wetlands
6. phytoplankton; bioluminescence; chemosynthesis, hydrogen sulfide; high pressure

Chapter 30

Figure Caption Questions

Figure 30-7   Rare species face a greater chance of extinction in a small reserve than do species with larger populations. For example, large carnivores need a large prey population to support them (see energy pyramids in Chapter 28), which in turn need a large area of suitable habitat. Large herbivores are likely to be rare for similar reasons; for example, it takes a very large area of vegetation to support a population of elephants. A chance event (storm, fire, disease) may well kill all of the members of such a population in a small reserve.

Figure 30-10   If a species becomes extinct in one of the small reserves, members of the species in nearby reserves can move through the corridor and recolonize the now-vacant reserve.

Fill-in-the-Blank

1. genetic, species, ecosystem; genetic
2. ecosystem services; many possible answers, including food, wood, medicines, soil formation, erosion and flood control, climate regulation, genetic resources, recreation
3. ecological economics
4. habitat destruction, overexploitation, invasive species, pollution, global climate change; habitat destruction
5. minimum viable population; habitat fragmentation; wildlife corridors
6. sustainable

Chapter 31

Figure Caption Questions

Figure 31-2   If the mammal's heat-sensing nerve endings were rendered nonfunctional, the nervous system would not send a signal to the hypothalamus when the body reached the set-point temperature. Consequently, the hypothalamus would send continuous "turn on" signals to the body's heat-generating and heat-retention mechanisms, which would continue to increase body temperature indefinitely (presumably until stored energy reserves are exhausted or the animal dies).

Figure 31-8   Blood is considered a form of connective tissue because it is composed largely of an extracellular matrix (plasma) in which cells and proteins are suspended.

Fill-in-the-Blank

1. homeostasis; negative feedback
2. cells, tissues, organs, organ systems
3. connective, epithelial, connective, connective, epithelial, muscle, nerve, connective
4. exocrine; endocrine, hormones
5. cardiac, skeletal, cardiac and skeletal, cardiac and smooth, smooth, skeletal

Chapter 32

Figure Caption Questions

Figure 32-1   Insect circulatory systems are not important in gas exchange; in insects this function is assumed by the tracheae.

Figure 32-4   Due to exercise-induced increase in muscle size, the hearts of well-conditioned athletes are larger than those of sedentary people. The heart's growth makes the volume of the ventricles larger, so a greater volume of blood is pumped with each heartbeat. Since the body's resting demand for oxygen remains unchanged, the demand can be met with fewer heartbeats per minute.

Figure 32-8   Iron is a key component of hemoglobin, which is necessary for building red blood cells capable of transporting oxygen. Consumption of iron in the diet is necessary to replace iron that is excreted with body waste because red blood cells are continually broken down after they die, and the iron recycling in the body is not 100% efficient.

Figure 32-10   The hormone erythropoietin stimulates production of additional red blood cells. These extra cells increase the blood's capacity to carry oxygen to muscles, thereby increasing the amount of work that muscles can perform with a given amount of blood.

Figure 32-15   No, because arteries carry oxygen-poor blood (blue) to the lungs, where capillary blood picks up oxygen, and veins carry oxygenated blood (red) away.

Figure 32-16   The principles of diffusion (see Chapter 5) tell us that molecules move down gradients from high to low concentration. Because the tissues through which capillaries pass generally consume oxygen, the concentration of oxygen is higher inside the capillaries than out, and oxygen diffuses along its concentration gradient from the inside of the capillary to the outside. Conversely, cellular respiration generates carbon dioxide, so the concentration of carbon dioxide is higher outside the capillaries than in, so carbon dioxide diffuses along its concentration gradient from the outside of the capillary to the inside.

Figure 32-20   From left to right

Fill-in-the-Blank

1. hemolymph, interstitial; hemocoel
2. left atrium, left ventricle, right atrium, right ventricle, left ventricle
3. sinoatrial node; cardiac muscle cells; atria; atrioventricular node; atrioventricular (AV) bundle; fibrillations
4. capillaries, arteries, veins, venules, arterioles, arteries
5. platelets; thrombin; fibrinogen, fibrin
6. platelets, leukocytes, erythrocytes, leukocytes, erythrocytes, leukocytes, erythrocytes
7. interstitial fluid; capillaries; fats; one-way valves; spleen

Chapter 33

Figure Caption Questions

Figure 33-5   The respiratory systems of amphibians are adapted to moist environments, so amphibians are restricted to habitats that are reasonably moist. Most amphibian larvae develop in water, which limits them to regions where standing water is available.

Figure 33-7   When oxygen binds hemoglobin, the protein changes color, becoming cherry-red. When oxygen leaves hemoglobin, it becomes maroon-red in color, and appears bluish through the skin.

Figure 33-9   In its relaxed state, the diaphragm muscle is expanded and domes upward, reducing the volume of the chest cavity. Contracting it causes it to become smaller and flatten downward, enlarging the chest cavity.

Figure 33-11   It is important for hemoglobin to be able to bind the hydrogen ions released when bicarbonate is formed, because if the hydrogen ions were released into the blood plasma, they would make it acidic (lower pH), which could lead to death.

Fill-in-the-Blank

1. sponges, sea jellies, flatworms; diffusion
2. spiracles; tracheae, circulatory systems
3. the pharynx; the epiglottis; trachea, bronchi, bronchioles, alveoli
4. capillaries; hemoglobin; bicarbonate; cellular respiration
5. alveoli; alveolus, capillary; two; surfactant
6. diaphragm, larger; passive, relax; respiratory center, medulla; carbon dioxide

Chapter 34

Figure Caption Questions

Figure 34-5   Most of the calories in this food come from sugar. Except for providing vitamin C and energy, this food seems to have little nutritional value.

Figure 34-10   Birds use their muscular gizzard, often containing small sharp stones, as a substitute for teeth to grind food.

Figure 34-11   Microorganisms from the rumen are digested along with the food, providing an additional source of nutrients.

Figure 34-16   The required absorptive surface area would have to be provided by some other adaptations, such as a much longer small intestine and a correspondingly larger abdominal cavity to hold it. Or the food might need to spend a much longer time in the intestine, being sloshed back and forth to expose nutrients to the smooth absorptive surface.

Fill-in-the-Blank

1. energy, raw materials; vitamins; minerals; essential
2. intracellular; gastrovascular cavity; tubular; cellulose
3. amylase, starch; protein; pepsinogen, pepsin; small intestine
4. ingestion, mechanical digestion, chemical digestion, absorption, elimination
5. pharynx, digestive, respiratory; epiglottis, swallowing; peristalsis; sphincters
6. bile; liver, gallbladder; epithelial; lacteal, lymphatic
7. gastrin; ghrelin; leptin; secretin, cholecystokinin; gastrin, ghrelin; secretin, cholecystokinin

Chapter 35

Figure Caption Question

Figure 35-7   If you consumed excess water, a drop in blood osmolarity would occur and would be detected by receptors in the hypothalamus. The hypothalamus would signal the posterior pituitary, causing it to reduce its release of ADH. Less ADH would result in fewer aquaporins in the membranes of the distal tubule and collecting duct, making them almost impermeable to water. This would result in production of large amounts of watery urine until the normal blood osmolarity was restored.

Fill-in-the-Blank

1. nephridia; Malpighian tubules; protonephridia; nephridia
2. renal cortex, renal medulla, renal pelvis; ureter, bladder, urethra
3. glomerulus, Bowman's capsule, proximal tubule, loop of Henle, distal tubule
4. proximal tubule, collecting duct, loop of Henle, renal medulla
5. urea, H^+ (hydrogen ions), NaCl (salt), water
6. homeostasis; renin, angiotensin; erythropoietin, red blood cells
7. hypothalamus; posterior pituitary; antidiuretic hormone (ADH); distal tubule, collecting duct, aquaporins, concentrated.

Chapter 36

Figure Caption Questions

Figure 36-4   In most capillaries, there are small gaps between the cells that make up the capillary walls (see Chapter 32). The inflammatory response causes these gaps to become larger, which allows white blood cells to move more easily between cells of the capillary walls.

Figure 36-6   The combination of constant and variable components allows an antibody to accomplish not only functions that are unique to a particular antibody molecule (i.e., recognize and bind to a specific antigen) but also functions that are common to all antibodies in a given class (i.e., carry out the appropriate response to neutralize the invader, such as enhancing phagocytosis). Variable regions provide the unique properties; constant regions provide the universal properties.

Figure 36-13   The same antibodies and bodily reactions that cause allergies also respond to some parasites, such as hookworms that invade the digestive tract. The sneezing, vomiting, and diarrhea that occur as part of an allergic response would help to expel invading parasites or their toxins out of the body.

Fill-in-the-Blank

1. skin; digestive, respiratory, urogenital (order of last three not important)
2. phagocytic, natural killer, inflammatory response, fever
3. antigens; antibodies, T-cell receptors (order of these two not important)
4. light, heavy (order not important); variable, constant (order not important); variable
5. Humoral, plasma cells; Cell-mediated; Cytotoxic; Helper; memory
6. vaccine
7. allergy; immune deficiency; autoimmune disease

Chapter 37

Figure Caption Questions

Figure 37-6   Dwarfism can be treated by administering growth hormone to affected individuals; it is much more difficult to halt the production of excess hormone (especially of a pituitary hormone such as growth hormone). Disruption of the pituitary is likely to disrupt the release of many other hormones as well.

Figure 37-10   There are several possible hypotheses, including (1) a tumor in the anterior pituitary caused multiplication of the TSH-secreting cells; (2) a tumor in the hypothalamus caused multiplication of the cells that secrete TSH-releasing hormone; and (3) an iodine-deficient diet caused the thyroid to produce too little thyroxine, so that the negative feedback system regulating TSH release is shut off.

Figure 37-11   If the receptors could not bind glucagon, then target cells, such as those in the liver, would not be able to respond to glucagon. If blood glucose levels are too low, liver cells and other cells would not convert glycogen to glucose, so blood glucose levels would stay low.

Fill-in-the-Blank

1. endocrine system; target cells, receptors
2. peptide, amino acid-derived, steroid (order not important); Peptide hormones, amino acid-derived hormones (order not important); second messengers; Steroid hormones; changes in gene transcription *or* synthesis of messenger RNA
3. hypothalamus; neurosecretory cells, oxytocin, antidiuretic hormone (order of hormones not important)
4. adrenocorticotropic hormone, follicle-stimulating hormone, luteinizing hormone, growth hormone, prolactin, thyroid-stimulating hormone
5. insulin; diabetes mellitus; Glucagon, glycogen
6. testes, testosterone; ovaries, estrogen, progesterone (order of female hormones not important)
7. glucocorticoids, mineralocorticoids, testosterone (order of hormones not important); norepinephrine, epinephrine (order of hormones not important)

Chapter 38

Figure Caption Questions

Figure 38-4   Based on the evidence, the toxin must act on the postsynaptic part of the synapse, by blocking neurotransmitter receptors. If the toxin acted by blocking release of neurotransmitters from the presynaptic neuron, then adding neurotransmitter to the synapse would have generated a postsynaptic potential. If the toxin acted by preventing sodium channels from opening, then it would not have been possible to stimulate an action potential in the presynaptic neuron.

Figure 38-5   Sensitive areas have a higher density of touch-sensitive sensory neurons. This would allow two types of increased sensitivity: (1) The higher density of receptors would mean that touching any particular place on the sensitive skin area would be more likely to stimulate at least one touch receptor. (2) A higher density of receptors would also mean that a touch of a particular force would stimulate a larger number of neurons. The touch thus would generate a larger number of action potentials in the sensitive area and would be perceived as a stronger stimulus.

Figure 38-10   A severed or damaged spinal cord prevents the sensation of pain from being relayed to the brain but does not disrupt the reflex circuit, which requires only transmission within a small portion of the spinal cord.

Fill-in-the-Blank

1. neuron; dendrite; cell body; axon; synaptic terminal
2. negatively; threshold, positively
3. neurotransmitter; receptors; excitatory postsynaptic potential; inhibitory postsynaptic potential
4. somatic nervous system; autonomic nervous system; sympathetic; parasympathetic
5. cerebellum, pons, medulla (order not important); cerebellum
6. frontal, parietal, temporal, occipital (order not important); occipital; frontal

Chapter 39
Figure Caption Questions

Figure 39-6   The joints and muscles in your neck and the rest of your body have stretch and position receptors. If your body is tilted but you keep your head vertical, position receptors in your neck will be activated. Your brain will compare information from the vestibular apparatus in your head and position receptors in your neck, and calculate the degree of body tilt.

Figure 39-10   For nearsightedness, the cornea is flattened (made less convex), which causes the light rays to converge less before they reach the lens. For farsightedness, the edges of the cornea are reduced, which leaves the cornea with a more rounded, convex shape that causes the light rays to converge more before they reach the lens.

Figure 39-11   Two reasons: (1) The blind spots in your two eyes do not see the same place in your visual field, so one of the eyes always sees the "hole" in the visual field of the other eye. (2) Your eyes constantly move, so no part of the visual field is in the blind spot of either eye for very long. The brain "fills in" the missing visual information with what the eye saw just a fraction of a second before.

Figure 39-12   Distance perception is useful for many behaviors, not just predation. Fruit-eating bats fly around during the daytime, and binocular vision is useful for avoiding obstacles such as branches. Monkeys often jump from branch to branch, and need to know how far it is to the next branch.

Fill-in-the-Blank

1. receptor potential; frequency
2. Mechanoreceptors; Photoreceptors; chemoreceptors
3. pinna; eardrum OR tympanic membrane; hammer (malleus), anvil (incus), stirrup (stapes) (order *is* important); oval window; hair cells
4. cornea, aqueous humor, iris; lens, vitreous humor; cornea, lens
5. rods, cones; Rods; Cones, fovea
6. sweet, salty, sour, bitter, *umami*; olfactory OR smell

Chapter 40
Figure Caption Questions

Figure 40-4   A highly stretched muscle would have the thin filaments pulled so far out that not all of the myosin heads would overlap a thin filament, and hence could not attach to actin when a contraction begins. Fewer myosin heads pulling on the thin filaments would generate less force of contraction than if all of the heads could pull on the thin filaments.

Figure 40-7   Water helps to support some of the weight of a thick exoskeleton. For land-dwelling arthropods, the weight of a thick exoskeleton would require much more energy expenditure just to stand, let alone move.

Figure 40-10   Two reasons: (1) Compact bone is much heavier than spongy bone, so it would require much more energy for an animal to stand or move. (2) Compact bone, being virtually solid, would leave no room for bone marrow, which is where blood cells are produced.

Fill-in-the-Blank

1. hydrostatic skeleton, endoskeleton, exoskeleton (order not important); antagonistic
2. Skeletal, cardiac (order not important); Cardiac, smooth (order not important); Cardiac, smooth (order not important)
3. muscle fiber; myofibrils, sarcomeres, Z lines
4. actin, myosin; ATP
5. neuromuscular junctions; T tubules, sarcoplasmic reticulum
6. cartilage, bone, ligaments (order not important); collagen
7. hinge, ball-and-socket; flexor, extensor

Chapter 41
Figure Caption Questions

Figure 41-1   Mitotic cell division

Figure 41-3   Asexual reproduction produces offspring that are genetically identical to the mother, containing all of her genes. This is an efficient way for her to maximize the number of her genes in the population, but provides no significant opportunity for any of her offspring to have new genotypes that might enhance survival or reproduction. Because the weather is favorable and food is so abundant in the spring and summer, natural selection is not very strong, so this is probably not a major difficulty. When autumn arrives, however, weather and food become more problematic. Sexual reproduction produces genetically variable offspring, some of which may be better adapted to the conditions and be favored by natural selection.

Figure 41-6   Courtship rituals help ensure that animals mate with other individuals of their own species. Courtship rituals are reproductive isolating mechanisms that help prevent potentially disadvantageous hybrid mating. The courtship activities in some species stimulate ovulation.

Figure 41-13   Testosterone inhibits the release of GnRH and FSH. Without adequate FSH, the Sertoli cells are not stimulated, so spermatogenesis slows down or stops.

Figure E41-1   Estrogen and progesterone inhibit the release of GnRH, FSH, and LH. Without enough FSH, follicles will not develop. Even if a follicle somehow does develop, without LH, the follicle will not mature, meiosis I will not be completed, and ovulation will not occur. Without an ovulated secondary oocyte, fertilization and pregnancy cannot occur.

Fill-in-the-Blank

1. asexual; bud; fission
2. testis; testosterone (androgen OK); seminiferous tubules
3. nucleus, acrosome; mitochondria
4. epididymis, vas deferens, seminal vesicles, prostate gland, bulbourethral gland, urethra
5. ovary; estrogen, progesterone; secondary oocyte; polar body; uterine tube OR oviduct OR Fallopian tube; uterus
6. corpus luteum; chorionic gonadotropin (CG)

Chapter 42
Figure Caption Questions

Figure 42-4   Twins would probably never occur, because even the first cell division of cleavage would produce daughter cells with different developmental fates, with the result that each of the daughter cells would be able to give rise to some, but not all, of the structures of the human adult.

Figure 42-5   *Ribs:* Homeobox genes are usually active in a head-to-tail sequence, such that Gene A might be active in the head region, Gene B would be active in the torso, Gene C would be active in the pelvic region, and so on. If, for example, Gene B causes production of ribs, perhaps in snakes, Gene B becomes active all the way from the end of the (very short) neck to the tail segments. *Lack of legs:* A specific homeobox gene would normally be responsible for triggering the production of the entire leg (or at least most of it). A severe mutation in this particular homeobox gene might prevent leg formation.

Figure 42-6   The ovulated egg is surrounded by barriers (the zona pellucida and the corona radiata; see Chapter 41). However, it is the outer layer of the blastocyst that attaches to, and implants in, the uterus. Therefore, the blastocyst must emerge out of these barriers at some point before implantation.

Figure 42-12   Because there is only a very small placenta for exchange of nutrients, wastes, and gases, the embryo is retained in the uterus for a much shorter time. Offspring are born in a much less developed state. Development continues outside the uterus (typically inside an external pouch).

Fill-in-the-Blank

1. zygote; cleavage, morula; blastula; ectoderm, mesoderm, endoderm
2. amnion; chorion; umbilical cord; yolk sac
3. stem cells; inner cell mass
4. homeobox genes; transcription factors
5. prolactin; oxytocin

Chapter 43
Figure Caption Questions

Figure 43-8   The onion stores nutrients for use by the shoot produced during the second year, after which the plant dies. An onion bulb at the end of the second year would be practically nonexistent.

Figure 43-10   In a climate that was truly uniformly warm and wet year-round, the vascular cambium would divide at the same rate year-round, and the daughter secondary xylem cells would grow the same way year-round. All of the secondary xylem cells would have the same size and cell wall thickness, and so they would all be the same color. Thus, there would be no annual rings.

Figure 43-22 CO_2 is required for photosynthesis. When stomata close, photosynthesis in the leaf slows and eventually stops because closed stomata prevent CO_2 in the air from entering the leaf and reaching the mesophyll cells where photosynthesis occurs. Closed stomata also greatly reduce transpiration from the leaf. Because transpiration provides the driving force for the upward movement of water in xylem, which in turn provides the major driving force for water entry into the roots, closing the stomata would slow down water entry into the roots.

Fill-in-the-Blank

1. meristem; apical meristem, lateral meristems; apical meristem; lateral meristem
2. dermal, ground, vascular; dermal; vascular; ground
3. xylem, tracheids, vessel elements; hydrogen bonds, transpiration; stomata
4. Phloem; sieve-tube element, companion cells
5. root cap; root hairs, active transport; mycorrhizae

Chapter 44
Figure Caption Questions

Figure 44-5   Separate bloom times reduce the chances of self-pollination and hence self-fertilization, which would result in inbreeding. Plants that do not inbreed in many cases have more successful offspring. Inbred individuals are more likely to have homozygous recessive alleles that are deleterious.

Figure 44-8   Colorful petals and sweet scents have evolved because they attract pollinators. They would not be useful to a wind-pollinated plant. Further, they require energy to produce, so they would be selected against. Therefore, wind-pollinated plants would not be expected to have colorful petals and sweet scents.

Figure 44-17   When nectar is at the base of a long, tubular flower, it is accessible to hummingbirds (which have long bills and tongues) but not to most insects and other small animals that might like to eat it. A flower whose nectar is inaccessible to insects is more likely to have nectar available for a visiting hummingbird; it is thus more likely that hummingbirds will learn to repeatedly visit this type of flower, ensuring that the flowers will be pollinated and reproduce.

Figure 44-21   Water always flows downhill. Therefore, floating fruits move downhill. A species using water as its primary means of dispersal will move downhill generation after generation, until eventually its most common habitat would be the ocean shore. This, of course, is precisely where coconut palms are found. Other plants, whose habitat is not along the ocean shore, must have means of dispersal other than flowing water.

Fill-in-the-Blank

1. alternation of generations; sporophyte; meiotic cell division; gametophyte
2. pollen grain; anthers; stigma; style; integuments
3. zygote; endosperm; cotyledons
4. ovary; ovule; integuments
5. Germination; root; coleoptile; bend *or* hook

Chapter 45
Figure Caption Questions

Figure 45-7   The lateral buds would be inhibited, just as if the meristem were present to release auxin.

Figure 45-9   With no ethylene present, tomato ripening—and the rotting that follows—would be delayed. Tomatoes would be resistant to bruising during shipping, and shipping and storing could be done at room temperature, saving energy and greatly reducing waste.

Fill-in-the-Blank

1. inhibits; apical dominance; stimulates; phototropism
2. gibberellin; division, elongation
3. ethylene; abscission layer; ripening, (any two of the following are correct, but this listing from the text is not exclusive) bananas, apples, pears, tomatoes, or avocados
4. root apical meristem; inhibits, stimulates, stem buds (or stem branches)
5. abscisic acid; dormancy

Credits

Photo Credits

Sauvanet/Photo Researchers, Inc.; 27-11b: Wildlife GmbH/Alamy Images; 27-11c: Jeff Lepore/Photo Researchers, Inc.; 27-12a: Robert P. Carr/Photoshot Holdings Ltd.; 27-12b: T. Hoffmeister/Bernard Roitberg; E27-2: Stephen P. Yanoviak; 27-13a: Teresa Audesirk; 27-13b: Carol Buchanan/Dreamstime; 27-14a: Tish1/Shutterstock; 27-14b: Jörn Friederich/Alamy; 27-14c: Tom Mangelsen/Nature Picture Library; 27-15a left: Greg Vaughn/Alamy; 27-15a right: Steffen Foerster Photography/Shutterstock; 27-15b left: Lyn Topinka/USGS; 27-15b right: Terry Donnelly/Alamy; 27-15c left: Phil Farnes/Photo Researchers, Inc.; 27-15c right: Jim Zipp/Photo Researchers, Inc.; 27-18a: Teresa Audesirk; 27-19: NASA

Chapter 28 opener: oksana.perkins/Shutterstock; E28-1: Thomas Lazar/Nature Picture Library; 28-10: NASA; 28-11: Monica Schroeder/Photo Researchers, Inc.; 28-12: Peter M. Miller/Photo Researchers, Inc.; 28-15a: USGS; 28-15b: Bruce Molnia/USGS; E28-2a: Arlan Naeg/AFP/Getty Images/Newscom; E28-2b: David Gray/Reuters; 28-17: Accent Alaska.com/Alamy

Chapter 29 opener: Florian Kopp/age fotostock; 29-opener-inset: Tomasz Zajaczlowski/iStockphoto; E29-1 top left: NASA; E29-1 bottom left: NASA; 29-8 top left: Ricardo Arnaldo/Alamy; 29-8 bottom left: Schafer and Hill/Getty Images; 29-8 middle: David Noton Photography/Alamy; 29-8 top right: Willie Davis/Shutterstock; 29-9: Edward Parker/Alamy; 29-10 top left: Erik Kolstad/iStockphoto; 29-10 bottom left: Federico Rizzato/Fotolia; 29-10 middle: EcoPrint/Shutterstock; 29-10 top right: Dirk Freder/iStockphoto; 29-11a: Vladimir Wrangel/Shutterstock; 29-11b: Louise A Heusinkveld/Alamy; 29-12a: Teresa and Gerald Audesirk; 29-12b: Tom McHugh/Photo Researchers, Inc.; 29-13: George H.H. Huey/Alamy; 29-14a: Tom McHugh/Photo Researchers, Inc.; 29-14b: Craig K. Lorenz/Photo Researchers, Inc.; 29-15: Mike Goldwater/Christian Aid/Getty Images; 29-19: Photoshot Archive Creative; 29-17: Clint Farlinger/Alamy; 29-18 top left: Alan Scheer/Shutterstock; 29-18 bottom left: Bill Brooks/Alamy; 29-18 middle: Eric Gevaert/Shutterstock; 29-18 top right: Olivierl/Dreamstime; 29-16: Teresa and Gerald Audesirk; 29-20 top left: Aleksander Bolbot/Shutterstock; 29-20 bottom left: Paul Tessier/iStockphoto; 29-20 middle: Francis Bosse/iStockphoto; 29-20 top right: Greg Okimi/iStockphoto; 29-21 top left: Ro-Ma Stock/Corbis RF/Alamy; 29-21 bottom left: Jon Meier/iStockphoto; 29-21 middle: Mark Wallace/Alamy; 29-21 top right: Photos.com; 29-22 top left: Top-Pics TBK/Alamy; 29-22 bottom left: Tom & Pat Leeson/Photo Researchers, Inc.; 29-22 middle: Wm. Bacon III/National Audubon Society/Photo Researchers, Inc.; E29-2: Nastenok/Shutterstock; 29-23: JK Enright/Alamy; 29-24 top left: Lisa Dearing/Alamy; 29-24 bottom left: Dmitry Deshevykh/iStockphoto; 29-24 middle: Larry Trupp/Alamy; 29-24 top right: Michael Giannechini/Photo Researchers, Inc.; 29-26: Michael P. Gadomski/Photo Researchers, Inc.; 29-29a: Lee Feldstein/iStockphoto; 29-29b: Universal Images Group/SuperStock; 29-29c: NHPA/SuperStock; 29-29d: Vlad61/Shutterstock; 29-30 top left: Aldo Ottaviani/iStockphoto; 29-30 bottom left: Michael Durham/National Geographic Image Collection; 29-30 middle: Michael Patrick O'Neill/Photo Researchers, Inc.; 29-30 top right: Jan Hinsch/Photo Researchers, Inc.; 29-31 top left: Dr. Craig R. Smith; 29-31 bottom left: Peter Batson/Image Quest Marine; 29-31 middle: Greg Rouse; 29-31 top right: Kelvin Aitken/Image Quest Marine; 29-32 top left: Verena Tunnicliffe/Newscom; 29-33-inset: American Association for the Advancement of Science

Chapter 30 opener: Gregory G. Dimijian/Photo Researchers, Inc.; 30-2a: Martin Strmko/Shutterstock; 30-2b: Richard Nowitz/National Geographic Stock; 30-2c: Michael S Nolan/age fotostock America, Inc.; 30-3: Yan Man/Chine Nouvelle/Sipa/Newscom; 30-4: seung jae kim/iStockphoto; E30-1a: South Florida Water Management District; E30-1b: South Florida Water Management; E30-3: Doug Perrine/Nature Picture Library; 30-6a: José Enrique Molina/age fotostock; 30-6b: Earth Sciences & Image Analysis Lab/NASA/Johnson Space Center; 30-7: Bruno Locatelli; E30-4a: Morales/age fotostock/SuperStock; E30-4b: Doug Perrine/Nature Picture Library; 30-8a: Mark Smit/Photo Researchers, Inc.; 30-8b: Franz Pagot/Alamy; 30-9a: Ed Reschke/Getty Images; 30-9b: Jen Chase/Associated Press; 30-9c: Reinhard Dirscherl/Alamy; 30-11: California State Parks; 30-13a: Dr. George Naderman, Former Extension Soil Specialist (retired), College of Agriculture and Life Sciences, NC State University, Raleigh, NC; 30-13b: Dr. George Naderman, Former Extension Soil Specialist (retired), College of Agriculture and Life Sciences, NC State University, Raleigh, NC

Chapter 31 opener: David Stluka/Associated Press; 31-1a: Tom McHugh/Photo Researchers, Inc.; 31-1b: kwan tse/Shutterstock; 31-1c: Ivan Cholakov Gostock-dot-ne/Shutterstock; E31-1 top: NASA; E31-1 bottom: NASA; E31-2: Russell Millner/Alamy; 31-5a: Nina Zanetti, Pearson Education; 31-5b: Ed Reschke/Getty Images; 31-5c: Nina Zanetti, Pearson Education; 31-6: Ed Reschke/Getty Images; 31-7: Steve Gschmeissner/Photo Researchers, Inc.; 31-8: Dr. Yorgos Nikas/Photo Researchers, Inc.; 31-9a: Innerspace Imaging/Photo Researchers, Inc.; 31-9b: Ed Reschke/Getty Images; 31-9c: G W Willis/Getty Images; 31-10a: Dr. Yorgos Nikas/Photo Researchers, Inc.; 31-10b: Riccardo Cassiani-Ingoni/Photo Researchers, Inc.

Chapter 32 opener: Rudy Archuleta/Redux Pictures; 32-opener-inset: Kevin Curtis/Photo Researchers, Inc.; 32-8a: Susumu Nishinaga/Photo Researchers, Inc.; 32-8b: Michael Ross/Photo Researchers, Inc.; 32-8c: Kenneth Kaushansky, M.D., University of California, San Diego Department of Medicine; 32-11: SPL/Photo

Researchers, Inc.; 32-13: CNRI/Photo Researchers, Inc.; 32-16: Science Source/Photo Researchers, Inc.; 32-20: Eric V. Grave/Photo Researchers, Inc.; 32-21: World Health Organization/Image Library, Special Programme for Research and Training in Tropical Diseases/Photo by Chandran

Chapter 33 opener: Torleif Svensson/age fotostock; 33-opener-inset: Caters News/ZUMA Press/Newscom; 33-1a: Leslie Newman & Andrew Flowers/Photo Researchers, Inc.; 33-1b: Terry Audesirk; 33-1c: Andrew J. Martinez/Photo Researchers, Inc.; 33-3: Mary Jonilonis/age fotostock; 33-4b: Molecular Expressions; 33-5a: Kim Taylor and Jane Burton/Dorling Kindersley; 33-5b: Design Pics/Superstock; 33-5c: Mark Kostich/iStockphoto; 33-6c: Hans-Rainer Duncker, University of Giessen, Germany; E33-3a: Matt Meadows/Getty Images; E33-3b: Southern Illinois University/Photo Researchers, Inc.; E33-3c: National Cancer Institute

Chapter 34 opener: Paul Hackett/Reuters; 34-opener-inset: Giulio Di Mauro/EPA/Newscom; 34-1: T. Walker/Photri Images/Alamy; 34-2: Dai Kurokawa/EPA/Newscom; 34-3: Dr. M.A. Ansary/Photo Researchers, Inc.; 34-4: Biophoto Associates/Photo Researchers, Inc.; 34-5: Teresa and Gerald Audesirk; 34-6a: Dennis Sabo/Shutterstock; 34-7a: Roland Birke/Tierbild Okapia/Photo Researchers, Inc.; E34-1: Dr. E. Walker/Science Photo Library/Photo Researchers, Inc.; 34-17: LYDIE/SIPA/Newscom

Chapter 35 opener: New York Daily News/Susan Watts; E35-1: AJPhoto/Photo Researchers, Inc.; 35-08: Susan Watts/Shutterstock; 35-10: New York Daily News/Susan Watts

Chapter 36 opener: Ron Wolfson/Getty Images; 36-1: Dr. Gary Gaugle/Photo Researchers, Inc.; 36-2: Juergen Berger/Photo Researchers, Inc.; 36-3a: NIBSC/Science Photo Library/Photo Researchers, Inc.; 36-3b: S.H.E. Kaufmann & J.R. Golecki/Science Photo Library/Photo Researchers, Inc.; 36-10: Dr. Andrejs Liepins/Photo Researchers, Inc.; E36-2: James Cavallini/Photo Researchers, Inc.; E36-3: Sheila Terry/Photo Researchers, Inc.; 36-14: NIBSC/Science Photo Library/Photo Researchers, Inc.

Chapter 37 opener: Papilio/Alamy; E37-1: UPI Photos/Newscom; 37-6: Feld Entertainment, Inc.; 37-9: Alison Wright/Photo Researchers, Inc.; E37-2: Alan Vajda; 37-13: The Jackson Laboratory, Bar Harbor, Maine.; 37-14: Haruhiko Fujiwara

Chapter 38 opener: United Archives GmbH/Alamy Images; 38-CO-inset: Michael R. Jeffords/Teresa Audesirk; 38-3: Steve Gschmeissner/Photo Researchers, Inc.; E38-3: Alain Dex/Photo Researchers, Inc.; 38-15: BSIP/Photo Researchers, Inc.

Chapter 39 opener: Advanced Bionics LLC; 39-CO-inset: Samantha Brilling; 39-5a: Scanning electron micrographs by Dr. Robert S. Preston and Prof. Joseph E. Hawkins, Kresge Hearing Research Institute, University of Michigan Medical School.; 39-7a: Steve GSteve Gschmeissner/Photo Researchers, Inc.; 39-8a: Teresa and Gerald Audesirk; 39-8b: Teresa and Gerald Audesirk; 39-11: Susan Eichorst/Audesirk; 39-12a: gillmar/Shutterstock; 39-12b: Sekernas/Dreamstime; 39-15: Advanced Bionics LLC

Chapter 40 opener: Thomas Lohnes/Getty Images; 40-CO-inset: Odd Andersonen/AFP/Getty Images/Newscom; 40-7: Tony Florio/Photo Researchers, Inc.; 40-9: Photo Researchers, Inc.; 40-10a: Wayne Schaefer; 40-10b: Michael Abbey/Photo Researchers, Inc.; E40-1a: Michael Klein/Getty Images; E40-1b: Nordic Photos/Superstock

Chapter 41 opener: wAlanb/Getty Images; 41-1: © Richard T. Nowitz/Corbis; 41-2: Suzanne L. Collins/Photo Researchers, Inc.; 41-3: Peter J. Bryant/Biological Photo Service; 41-4: Robin Chittende/Frank Lane Picture Agency; 41-5: Jurgen Freund/Nature Picture Library; 41-6: ZooFari/Creative Commons Attribution-Share Alike 3.0 Unported, http://creativecommons.org/licenses/by-sa/3.0/deed.en; 41-7: Nature's Images/Photo Researchers, Inc.; 41-8: isgaby/iStockphoto; 41-16: C. Edelmann/La Villette/Photo Researchers, Inc.; 41-18a: Lennart Nilsson/Albert Bonniers Forlag/Scanpix Sweden AB; 41-18b: David M. Phillips/Photo Researchers, Inc.; E41-2a: Dr. Linda Stannard, UCT/Photo Researchers, Inc.; E41-2b: Meckes/Ottawa/Photo Researchers, Inc.; E41-3: Ekem; E41-4: CC Studio/Photo Researchers, Inc.

Chapter 42 opener: Jane Burton/Nature Picture Library; 42-CO-inset: Dr. Michael Levin and Dr. Taisaku Nogi; 42-1a: Survivalphotos/Alamy; 42-1b: Tomas Kvidera/Shutterstock; 42-2a: Dr. Paul A. Zahl/Photo Researchers, Inc.; 42-2b: Kim Taylor/Nature Picture Library; 42-2c: Outdoorsman/Dreamstime; 42-6b: Dr. Yorgos Nikas/Photo Researchers, Inc.; 42-9: Dopamine/Photo Researchers, Inc.; 42-10: CRNI/Photo Researchers, Inc.; E42-4: Sterling K. Clarren, M.D., Children's Hospital & Regional Medical Center, University of Washington School of Medicine; 42-15: PhotoLink/Getty Images

Chapter 43 opener: Ojo Images Ltd/Alamy; 43-6b: Biophoto Associates/Photo Researchers, Inc.; 43-8a: Emmanuel Lattes/Alamy; 43-8b: almagami/Shutterstock; 43-8c: Photos.com; 43-8d: Gusto Productions/Photo Researchers, Inc.; 43-9: Ed Reschke/Getty Images; 43-10a: Corbis Premium RF/Alamy; 43-10b: Biology Pics/

Photo Researchers, Inc.; 43-11a: Neale Clark/Robert Harding/Alamy; 43-11b: M. Newton/Alamy; 43-12a: Bernard Gagnon; 43-12b: Kenneth W. Fink/Photo Researchers, Inc.; 43-12c: Charles D. Winters/Photo Researchers, Inc.; 43-12d: KentZilla; 43-13a: Lynwood M. Chace/Photo Researchers, Inc.; 43-13b: Stefan Diller/Photo Researchers, Inc.; 43-14b: M. I. Walker/Photo Researchers, Inc.; 43-14 top: Ed Reschke/Getty Images; 43-15: Teresa and Gerald Audesirk; 43-16: Teresa and Gerald Audesirk; 43-17: J. Robert Waaland, University of Washington/Biological Photo Service; 43-19: Noah Elhardt; 43-20: Biophoto Associates/Science Source/Photo Researchers, Inc.; 43-21: Hugh Spencer/National Audubon Society/Photo Researchers, Inc.; 43-23a: Ray Simon/Photo Researchers, Inc.; 43-23b: Dr. Jeremy Burgess/Science Photo Library/Photo Researchers, Inc.; E43-1: Jordi Bas Casas/Photoshot Archive Creative; 43-25a: Martin H. Zimmermann, Harvard Forest, Harvard University; 43-25b: Martin H. Zimmermann/Harvard Forest

Chapter 44 opener: Lucy Nicholson/AFP/Getty Images; 44-2 left: Tom Joslyn/age fotostock; 44-2 right: Andrew Syred/Photo Researchers, Inc.; E44-1: John Kaprielian/Photo Researchers, Inc.; E44-1-inset: David Scharf/Photo Researchers, Inc.; 44-4: Heather Audesirk; 44-5: Teresa and Gerald Audesirk; 44-7: © Medical-on-Line/Alamy; 44-8: Teresa and Gerald Audesirk; 44-11L: David Rhodes/Purdue University/Dept of Horticulture and Landscape Architecture; 44-11R: Teresa and Gerald Audesirk; 44-14: Teresa and Gerald Audesirk; 44-15L: Leonard Lessin/Photo Researchers, Inc.; 44-15R: Leonard Lessin/Photo Researchers, Inc.; 44-16a: Nuridsany et Perennou/The National Audubon Society Collection/Photo Researchers, Inc.; 44-16b: Nuridsany et Perennou/The National Audubon Society Collection/Photo Researchers, Inc.; 44-17: Anthony Mercieca/Photo Researchers, Inc.; 44-18: Veronica Carter/Alamy; E44-2a: John Warburton Lee/SuperStock; E44-2b: Merlin D. Tuttle/Bat Conservation International/Photo Researchers, Inc.; 44-19: Philippe Clement Nature/Picture Library; 44-19-inset: M. P. L. Fodgen/Photoshot; 44-20a: Teresa and Gerald Audesirk; 44-20b: Stephen Maka/age fotostock; 44-21: Teresa and Gerald Audesirk; 44-22: Scott Camazine/Photo Researchers, Inc.; 44-23: "Doris Dumrauf / Alamy"; 44-24a: Marc Gibernau (CNRS - France). From: "Science News". Vol. 164, No. 24, December 13, 2003, p. 379;44-24b: Marc Gibernau (CNRS - France). From: "Science News". Vol. 164, No. 24, December 13, 2003, p. 379.

Chapter 45 opener: Terry Audesirk; E45-1: David McNew/Getty Images; 45-1: Anton Foltin/Shutterstock; 45-2c: Terry Audesirk; 45-2d: Martin Shields/Alamy; 45-3: Terry Audesirk; 45-5: Terry Audesirk; 45-6: Nigel Cattlin/Alamy; 45-9: The Regents of the University of California, Agriculture and Natural Resources; 45-10: John Cancalosi/Getty Images; 45-11: Biophoto Associates/Photo Researchers, Inc.; 45-13a: Ed Reschke/Getty Images; 45-13b: Ed Reschke/Getty Images; 45-14: Kim Taylor/Photoshot; 45-15: John Clegg/Photo Researchers, Inc.; 45-15-inset: Kim Taylor/Photoshot; 45-16: Terry Audesirk

Text Credits

p. 7: Theodosius Dobzhansky (1900–1975); p. 18: Transactions of The Connecticut Academy of Arts and Sciences, vol. 38, Dec. 1949; p. 27: Loren Eiseley, The Immense Journey, p.15 (Vintage Books, 1959); p. 39: Data from U.S. Food and Drug Administration; p. 50: The Centers for Disease Control and Prevention; p. 56: Robert Hooke, (1635–1703); p. 234: Richard Adams of Texas A&M University, quoted in "Quietly, animal cloning speeds onward", The Christian Science Monitor, 10-15-2001; p. 257: The Origin of Species by Charles Darwin, published on 24 November 1859; p. 266: The Principle Of Population (1798 1st edition) with A Summary View (1830), and Introduction by Professor Antony Flew. Penguin Classics. ISBN 0-14-043206-X; p. 266: On the Origin of Species by Charles Darwin Published: November 24, 1859; p. 272: Data from Monsanto Company; p. 272: Jonathan B. Losos, Kenneth I. Warheitt, & Thomas W. Schoener, Adaptive differentiation following experimental colonization in Anolis lizards, Nature, 387 (1997), pp. 70–73; p. 287: John van Wyhe, editor. 2002-. The Complete Work of Charles Darwin Online. (http://darwin-online.org.uk/); p. 288: The Last Extinction 2nd Edition, edited by Les Kaufman and Kenneth Mallory, MIT Press Classic Series, 1993; Figure 17-11: This Dynamic Earth: The Story of Plate Tectonics by Kious, W. Jacquelyne and Tilling, Robert I., U.S. Geological Survey General Information Product; p. 445: © Copyright 1998 The Washington Post Company; p. 445: J.L.B. Smith, Old Fourlegs: The Story of the Coelacanth, Longmans, Green and Company, 1956; Figure 25-5: Image based on N. Tinbergen, The Wilson Bulletin, March 1948, Vol. 60 No.1 pp. 6–52; Figure 25-17: Based on Tinbergen, N. 1952. The curious behavior of the stickleback. Scientific American 187:22–26. p. 490: On the Origin of Species by Charles Darwin November 24, 1859; Figure 26-4: Data from the U.S. Fish and Wildlife Service. 2008. International recovery plan for the whooping crane. Canadian Wildlife Service and U.S. Fish and Wildlife Service. 2007. International recovery plan for the whooping crane. Ottawa: Recovery of Nationally Endangered Wildlife (RENEW), and U.S. Fish and Wildlife Service, Albuquerque, New Mexico. 162 pp; Figure 26-7: Data from Connell, J. H. 1961. Effects of competition, predation by Thais lapillus and other factors on natural populations of the barnacle Balanus balanoides. Ecological Monographs 31:61–104. © Copyright

by the Ecological Society of America; Figure 26-9: Data from Utida, S. 1957. Cyclic fluctuations of population density intrinsic to the host-parasite system. Ecology 38:442–49. © Copyright by the Ecological Society of America; Figure 26-15: Arias E. United States life tables, 2004. National vital statistics reports; vol 56 no 9. Hyattsville, MD: National Center for Health Statistics. 2007; p. 501: Sep Djekoule, Chicago Tribune, 1999; Figure 26-19: Data from the U. S. Census Bureau, Population Division; Figure 27-1: Data from The Struggle for Existence, G. F. Gause, 1934, Dover Phoenix Editions Dover Books on Biology, Edition 1, G.F. Gause @2003 pp.102; Figure 27-2: Data from MacArthur, R. H. 1958. Population ecology of some warblers of Northeastern coniferous forest. Ecology 39:599-619; Figure 28-14: Dr. Pieter Tans, NOAA/ESRL (www.esrl.noaa.gov/gmd/ccgg/trends/) and Dr. Ralph Keeling, Scripps Institution of Oceanography (scrippsco2.ucsd.edu/); Figure 28-16: Data from IPCC, 2007, Fourth assessment report: Summary for policymakers, Figure SPM.5; Figure 29-4: Korner, Christian, Spehn, Eva M., Global Mountain Biodiversity Assessment. 2002 DIVERSITAS (UNESCO and UNEP); Figure 30-1: Millennium Ecosystem Assessment, 2005. Ecosystems and Human Well-being: Opportunities and Challenges for Business and Industry. World Resources Institute, Washington, DC; Figure 30-5: World Wildlife Fund; Figure 30-5: © 2010 WWF (panda.org). Some rights reserved; Figure 30-5: "The living planet report," in The World Wildlife Fund, the Zoological Society of London, and the Global Footprint Network, 2008, http://www.footprintnetwork.org; p. 587: Duncan Graham-Rowe, Rise in threatened species is accelerating, New Scientist, 11:11, 11-17-2004, Reed Business Information Ltd; p. 592: Caring for the Earth: A Strategy for Sustainable Living. 1991. Gland, Switzerland: Published in partnership by IUCN, UNEP, and WWF; p. 595: Who Will Care for the Earth?, Annual Report 1993: IUCN, International Union for Conservation of Nature & Natural Resources, 1994; p. 633: John Appezzato, Marathoner running with a lot of heart, The Star Ledger, 2008; p. 636: Jeanette Johnson, The Other Side Of The Mountain, MSNBC, 2008; p. 652: Carre Otis, Living With Anorexia: Carré Otis, WebMD, 2005; p. 652: Carre Otis, We're a Nation Under A Spell, Newsweek, 2005; p. 673: Reference to the American T.V. Show "60 Minutes"; p. 676: On the Structure and Use of the Malpighian Bodies of the Kidney, with Observations on the Circulation through That Gland. by William Bowman, Published by: The Royal Society, 1842; p. 678: National Kidney Foundation, 2012; p. 678: Emily Leach, 8-17-2011, The Wall Street Journal; p. 678: 2011 USRDS Annual Data Report, Volume two; Precis: An Introduction to End-Stage renal disease in the United States. http://www.usrds.org/2011/pdf/v2_00_precis_11.pdf, pg. 151; p. 685: JAMA. 2010; 303(10):959–966. doi:10.1001/jama.2010.237 Perioperative Mortality and Long-term Survival Following Live Kidney Donation; p. 686: Kevin Sack, 60 Lives, 30 Kidneys, All Linked, February 18, 2012, New York Times; p. 686: Anthony DeGiulio, Good Samaritan's offer leads to 4-way kidney swap, 8-4-8, New York Daily News; p. 688: The Rime of the Ancient Mariner, Samuel Taylor Coleridge; p. 731: Romeo and Juliet by William Shakespeare; p. 769: Samantha Brilling, Sam's Hearing Life: My First Day into the New World, 2005, samshearinglife.blogspot.com/.../my-first-day-into-new-world.html; p. 808: Terri Roth, How to Breed a 2,000-pound Rhino, Discover Magazine, 2001; p. 831: Song of the Flower XXIII by Khalil Gibran, 1914; p. 869: U.S. Supreme Court, NIX v. HEDDEN, 149 U.S. 304 (1893) 149 U.S. 304 NIX et al. v. HEDDEN, Collector. No. 137. May 10, 1893; p. 869: Arkansas Code, Title 1, Chapter 4, Section 1-4-115. Title 1. General Provisions. Chapter 4. State Symbols, Motto, Etc. Acts 1987, No. 255, § 1; p. 875: Conservation International (2007). "Madagascar and the Indian Ocean Islands". Biodiversity Hotspots. Conservation International; p. 875: Questions and Answers: Colony Collapse Disorder, News & Events, Agricultural Research Service, United States Department of Agriculture; p. 875: Aldo Leopold, A Sand County Almanac, Oxford University Press, New York 1968; p. 875: Lemurs: Ecology and Adaptation edited by Lisa Gould, M.L. Sauther: refs for lemurs as primary pollinators and seed dispersers in Madagascar. African Trees May Be Tied to Lemurs' Fate, by John Roach for National Geographic News, July 26, 2004: lemurs as primary seed dispersers on Madagascar. Garbutt, Nick (2007). Mammals of Madagascar, A Complete Guide. A&C Black Publishers: Black-and-white ruffed lemurs as major pollinator of travelers palm on Madagascar. U.S. Forest Service: http://www.fs.fed.us/wildflowers/pollinators/animals/unusual.shtml. Black-and-white ruffed lemurs as major pollinator of travelers palm on Madagascar. Biotropica 34(2): 261–272 (2002) Primary Seed Dispersal by Red Howler Monkeys and the Effect of Defecation Patterns on the Fate of Dispersed Seeds, by Ellen Andresen: over 100 tree species in Amazon rainforest have seeds dispersed by howler monkeys. International Union for the Conservation of Nature (IUCN) Red List: endangered lemur species; p. 880: Charles Darwin, Fertilisation of Orchids, 1862; p. 883: Darwin, C. R. (1880). The Power of Movement in Plants. London: Murray; p. 886: Growth hormones in plants. Authorized English translation of Die wuchsstofftheorie und ihre bedeutung für die analyse des wachstums und der wachstumsbewegungen der pflanzen. Translated and revised by George S. Avery, Jr., and Paul R. Burkholder with the collaboration of Harriet B. Creighton and Beatrice A. Schee, McGraw Hill, 1936; p. 893: Ilya Raskin, Professor, Rutgers University; p. 894: L Mahadevan, Yoël Forterre, Jan M Skotheim, Jacques Dumais, © 2005, How the Venus Flytrap snaps.

Index

Citations followed by *b* refer to material in boxes; citations followed by *f* refer to material in figures or illustrations; and citations followed by *t* refer to material in tables.